Y0-ARF-135

MOLECULAR MEDICAL MICROBIOLOGY

MOLECULAR MEDICAL MICROBIOLOGY

Edited by

Max Sussman

Department of Microbiology
University of Newcastle upon Tyne

A Harcourt Science and Technology Company

SAN DIEGO • SAN FRANCISCO • NEW YORK • BOSTON
LONDON • SYDNEY • TOKYO

This book is printed on acid-free paper.

Academic Press
A Harcourt Science and Technology Company
Harcourt Place, 32 Jamestown Road, London NW1 7BY, UK
http://www.academicpress.com

Academic Press
A Harcourt Science and Technology Company
525 B Street, Suite 1900, San Diego, California 92101-4495, USA
http://www.academicpress.com

ISBN 0-12-677530-3

Library of Congress Catalog Number: 2001089056

A catalogue record for this book is available from the British Library

Typeset by Newgen Imaging Systems (P) Ltd, Chennai, India
Printed and bound in Spain by Grafos SA Arte Sobre Papel, Barcelona
01 02 03 04 05 06 07 GF 9 8 7 6 5 4 3 2 1

Dedication

ונתתי לה בביתי ובחומתי יד וש.....
Isaiah 56:5

Siegmund Hirsch
Hamburg 14.1.1905 - Auschwitz 30.9.1942

Josepha Alexandra Hirsch-Levy
Hamburg 9.9.1905 - Auschwitz 17.7.1942

Ursula Selma Hirsch
Hamburg 11.8.1931 - Auschwitz 17.7.1942

Shulamith Hirsch
Hamburg 28.11.1933 - Auschwitz 17.7.1942

Uncle, Aunt and Cousins

הי״ד

Editorial Board

Preface

The developments of the last almost fifty years suggested that the time might be ripe for a book dedicated to medical microbiology as seen from a molecular vantage point. Such a project faces the problem that the subject is immense and growing at an intimidating rate even while this book was being edited. Nevertheless, I hope that it is at least a faithful representation of the subject.

Molecular medical microbiology cannot, of course, stand in isolation from the 'whole organism' approach of the pre-molecular age that is so well described in many standard texts. It was felt, therefore, that to make the size of the book manageable, contributors could safely assume that readers in search of information about the more classical phenotypic and descriptive aspects of the subject would profitably turn elsewhere.

This book is in a way a *prolegomenon*. Molecular medical microbiology is not a 'settled subject' and authors were given only limited guidance about the content and structure of their contributions. Each chapter, therefore, reflects the particular insights and priorities of its authors.

My Advisory Editors were most helpful, especially during the initial planning stages of the project and I am grateful to them for their advice, which in some cases was sought at very inconvenient times. I wish also to thank all the contributors and particularly those who submitted early and, later in the gestation, readily agreed to revise their chapters.

My successor Professor Carlos Hormaeche has shown his friendship to me in many ways. I am most fortunate, some twelve years after my retirement, to be able to continue to work in the Department of Microbiology and Immunology of the Newcastle Medical School.

Finally, the journey from manuscript to printed page is long, often tortuous and sometimes mysterious and a multitude of hands is involved. The eagle and expert eye of my copy editors has saved me from many a lapse. I should like to express my thanks to the many at Academic Press who assisted me at various stages. Particularly deserving of gratitude amongst these are, Dr Tessa Picknett for her assistance and understanding, Dr Lilian Leung and Soma Mitra for their help with keeping a semblance of order and Sutapas Bhattacharya for his reassuring efficiency and limitless knowledge about book production.

Max Sussman
Newcastle Upon Tyne
July 2001

List of Contributors

Soman N. Abraham
Depts of Pathology and Microbiology, Duke University Medical Center, Durham NC 27710, USA

Mark Achtman
Max-Planck Institut für Infektionsbiologie and Deutsches Rheuma-Forschungszentrum, Schumannstr 21/22, 10117 Berlin, Germany

Shin-Ichi Aizawa
Dept of Bioscience, Teikyo University 1-1 Toyosatodai, Utsunomiya, 320-8551, Japan

David G. Allison
School of Pharmacy and Pharmaceutical Sciences, University of Manchester, Manchester M13 9PL, UK

Valerie Asche
Menzies School of Health Research, POB 4794, Darwin, NT 0801, Australia

Larry M. Baddour
Dept of Medicine Section of Infectious Diseases, University of Tennessee Medical Centre, Knoxville, TN 37920-6999, USA

Camella C. Bailey
Center for Vaccine Development, University of Maryland School of Medicine, 685 W Baltimore Street, Baltimore, MD 21201, USA

Michael R. Barer
Dept of Microbiology and Immunology, The Medical School, University of Newcastle Upon Tyne, Newcastle Upon Tyne, NE2 4HH, UK
Present address: Dept of Microbiology and Immunology, University of Leicester, P.O. Box 138, Medical Sciences Building, University Road, Leicester LE1 9HN, UK

Arnold S. Bayer
Harbor-UCLA Medical Center, Division of Infectious Diseases, 1000 W Carson St, Box 466, Torrance CA 90509, USA

Blaine L. Beaman
Depts of Medical Microbiology and Immunology, University of California School of Medicine, Davis, CA 95616, USA

Douglas J. Beecher
HMRU, FBI Academy, Building 12, Rm 201, Quantico, VA 22135, USA

Peter M. Bennett
Depts of Pathology and Microbiology, School of Medical Sciences, University of Bristol, University Walk, Bristol BS8 1TD, UK

Sucharit Bhakdi
Insitut für Medizinische Mikrobiologie und Hygiene der Universität, Hochhaus am Augustusplatz, 55101 Mainz, Germany

Svend Birkelund
Dept of Medical Microbiology and Immunology, The Bartholin Building, University of Aarhus, DK-8000 Aarhus C, Denmark

Michael H. Block
AstraZeneca, Alderley Park, Macclesfield SK10 4TG, UK

R.A. Bojar
The Skin Research Centre, Dept of Microbiology, University of Leeds, Leeds LS2 9JT, UK

Stephen J. Bourke
Dept of Respiratory Medicine, Royal Victoria Infirmary, Newcastle Upon Tyne NE1 4LP, UK

C. Josephine Brooke
Dept of Microbiology, University of Western Australia, 35 Stirling Highway, Crawley WA 6009, Australia

Robert R. Brubaker
Dept of Microbiology, 57A Giltner Hall, Michigan State University, East Lansing, MI 48824-1101, USA

Thierry Calandra
Division of Infectious Diseases, BH-19.111, Centre Hospitalier Universitaire Vaudois, CH-1011 Lausanne, Switzerland

John A. Chaddock
CAMR, Porton Down, Salisbury, Wilts SP4 0JG, UK

Babara J. Chang
Dept of Microbiology, University of Western Australia, 35 Stirling Highway, Crawley WA 6009, Australia

Matthew R. Chapman
Depts of Molecular Microbiology and Microbial Pathogenesis, PO Box 8230, Washington University School of Medicine, 600 S Euclid Avenue St Louis, MO 63110-1093, USA

Henrik Chart
Division of Enteric Diseases, PHLS Central Public Health Laboratory, 61 Colindale Avenue, London NW9 5HT, UK

Gunna Christiansen
Dept of Medical Microbiology and Immunology, The Bartholin Building, University of Aarhus, DK-8000 Aarhus C, Denmark

Steven Clegg
Dept of Microbiology, College of Medicine, University of Iowa, 3403 Bowen science Building, Iowa City IA 52242-1109, USA

Frank M. Collins
Laboratory of Mycobacterial Diseases, Bldg 29 Room 505, CBER/FDA, 1401 Rockville Pike, Rockville MD 20857, USA

M.J. Colston
Leprosy and Mycobacterial Research, National Institute for Medical Research, The Ridgeway, London NW7 1AA, UK

Howard N. Cooper
Dept of Medical Microbiology, Imperial College School of Medicine, St Mary's Campus, Norfolk Place, London W2 1PG, UK

Pascal F. Cossart
Unité des Interactions Bactéries-Cellules, Institut Pasteur, 28 rue de Docteur Roux, Paris 75015, France

Alan S. Cross
Dept of Medicine, Division of Infectious Diseases, University of Maryland School of Medicine, 22 S. Greene Street, Baltimore, MD 21201, USA

Sally J. Cutler
Dept of Infectious Diseases, Imperial College School of Medicine, Hammersmith Hospital Campus, Ducane Road, London W12 ONN, UK

Simon M. Cutting
School of Biological Sciences, Royal Holloway University of London, Egham, Surrey TW20 0EX, UK

Jeremy W. Dale
Molecular Microbiology Group, School of Biological Sciences, University of Surrey, Guildford, Surrey GU2 5XH, UK

Antoine Danchin
Hong Kong University Research Centre, Dexter Man Building, 8 Sassoon Road, Pokfulam, Hong Kong
Also at: Regulation of Gene Expression, URA 2171 CNRS, Institut Pasteur, 28, Rue du Dr Roux, 75724 Paris cedex 15, France

Rina Das
Dept Molecular Pathology, Walter Reed Army Institute of Research, Washington DC 20307-5100, USA

Gregory A. Dasch
Viral and Rickettsial Diseases Division, Naval Medical Research Center, Bethesda MD 20889-5607
Present address: Viral and Rickettsial Zoonoses Branch, Division of Viral and Rickettsial Diseases, Centers for Disease Control and Prevention, Atlanta, GA 30333, USA

Charles J. Dorman
Dept of Microbiology, Moyne Institute of Preventitive Medicine, University of Dublin, Trinity College, Dublin 2, Republic of Ireland

Ron J. Doyle
Dept of Microbiology, University of Louisville School of Medicine, Louisville KY 40292, USA

Mette Drasbek
Loke Diagnostics, Science Park, Aarhus, Denmark

Roman Dziarski
Northwest Center for Medical Education, Indiana University School of Medicine, Gary, IN 46408, USA

Jennifer Eardley
Center for Vaccine Development, University of Maryland School of Medicine, 685 W Baltimore Street, Baltimore, MD 21201, USA

Karen L. Elkins
Center for Biologics Evaluation and Research, FDA, Building 29 Room 428, 1401 Rockville Pike, Bethesda MD 20892, USA

Marina E Eremeeva
Dept of Microbiology and Immunology, University of Maryland at Baltimore, School of Medicine, 655 West Baltimore Street, Baltimore MD 21201-1559, USA
Present address: Viral and Rickettsial Zoonoses Branch, Division of Viral and Rickettsial Diseases, Centers for Disease Control and Prevention, Atlanta, GA 30333, USA

Paul Everest
Dept of Veterinary Pathology, Glasgow University, Bearsden, Glasgow G61 1QH, UK

Edward J. Feil
Dept of Biology and Biochemistry, University of Bath, Bath, BA2 7AY, UK

Sydney M. Finegold
Veterans Affairs Medical Center and UCLA School of Medicine, Los Angeles, California 90073, USA

Brett B. Finlay
Biotechnology Laboratory, University of British Columbia, Room 237 Wesbrook Building, 6174 University Boulevard, Vancouver, BC V6T 1Z3 BC, Canada

James Flexman
Dept of Microbiology and Infectious Diseases, Royal Perth Hospital, Wellington Street, Perth, Western Australia, 6000

John W. Foster
Depts of Microbiology and Immunology, University of South Alabama College of Medicine, Mobile AL 36688-0002, USA

Timothy J. Foster
Dept of Microbiology, Moyne Institute of Preventive Medicine, Trinity College, Dublin 2, Ireland

Laura S. Frost
Dept of Biological Sciences, University of Alberta, Edmonton, Alberta T6G 2E9, Canada

J.S.H. Gaston
Rheumatology Unit, Dept of Medicine, Adenbrook's Hopsital, University of Cambridge, Cambridge CB2 2QQ, UK

Curtis G. Gemmell
University Dept of Bacteriology, Glasgow Royal Infirmary, 84–86 Castle Street, Glasgow G4 0SF, UK

Peter Gilbert
School of Pharmacy and Pharmaceutical Sciences, University of Manchester, Manchester M13 9PL, UK

Michel Pierre Glauser
Division of Infectious Diseases, BH-19 111, Centre Hospitalier Universitaire Vaudois, CH-1011 Lausanne, Switzerland

Tana Green
Dept of Genitourinary Medicine, Royal Hallamshire Hospital, Glossop Road, Sheffield S10 2JF, UK

Elwyn Griffiths
Chief, Biologicals, World Health Organisation, CH-1211 Geneva 27, Switzerland

Antje Haase
Menzies School of Health Research, POB 4794, Darwin NT 0801, Australia

Jorg H. Hacker
Instutut für Molekulare Infektions Biologie, Universität Würzburg, Röntgenring 11, D-97070 Würzburg, Germany

Ian C. Hancock
Dept of Microbiology and Immunology, The Medical School, University of Newcastle Upon Tyne, Newcastle Upon Tyne NE2 4HH, UK

Margaret M. Hannan
Department of Medical Microbiology, Imperial College School of Medicine, St Mary's Campus, Norfolk Place, London W2 1PG, UK *and* National Center for HIV, STD and TB Prevention, Centers for Disease Control and Prevention, Atlanta GA 30333, USA

Colin R. Harwood
Dept of Microbiology, The Medical School, University of Newcastle Upon Tyne, Framlington Place, Newcastle Upon Tyne NE2 4HH, UK

R.J. Hay
Guys, Kings and St Thomas School of Medicine, St John's Institute of Dermatology, Block 7, Basement, KCL, St Thomas's Hospital, London SE1 7EH, UK

Robert A. Heinzen
Dept of Molecular Biology, University of Wyoming, P.O. Box 3944, Laramie WY 82071-3944, USA

I. Henderson
DynPort Vaccine Company, LLC, 60 Thomas Johnson Drive, Frederick, MD 21702, USA

Didier Heumann
Division of Infectious Diseases, BH-19.111, Centre Hospitalier Universitaire Vaudois, CH-1011 Lausanne, Switzerland

Nicola J. High
School of Biological Sciences, Stopford Building, University of Manchester, Oxford Road, Manchester M13 9PT, UK

Jan A. Hobot
Medical Microscopy Sciences, University of Wales College of Medicine, Heath Park, Cardiff CF4 4XN, UK

Keith T. Holland
The Skin Research Centre, Dept of Microbiology, University of Leeds, Leeds LS2 9JT, UK

Lan Hu
Laboratory of Enteric and STDs, FDA Center for Biologic Evaluation and Research, Building 29, Room 420, NIH Campus, Bethesda MD 20892, USA

Scott J. Hultgren
Depts of Molecular Microbiology and Microbial Pathogenesis, PO Box 8230, Washington University School of Medicine, 600 S Euclid Avenue, St Louis, MO 63110-1093, USA

Boris Ionin
Dept of Molecular Pathology, Walter Reed Army Institute of Research, Washington DC 20307-5100, USA

John M. Janda
Microbial Diseases Laboratory, Division of Communicable Disease Control, California Dept of Health Services, 2151 Berkeley Way – Room 330, Berkeley CA 94704-1011, USA

Klaus Jann
Max-Planck-Institut für Immunbiologie, Stübweg 51, Postfach 1169, D-79108 Freiburg-Zähringen, Germany
Present address: Tannenweg 47, D-79183 Waldkirch, Germany

Barbara Jann
Max-Planck-Institut für Immunbiologie, Stübweg 51, Postfach 1169, D-79108 Freiburg-Zähringen, Germany
Present address: Tannenweg 47, D-79183 Waldkirch, Germany

Angela M. Jansen
Division of Infectious Diseases, School of Medicine, University of Maryland, 10 South Pine Street, Baltimore, MD 21201-1192, USA

Marti Jett
Dept Molecular Pathology, Walter Reed Army Institute of Research, 503 Robert Grant Road, Silver Spring, MD 20910, USA

James B. Kaper
Center for Vaccine Development, University of Maryland School of Medicine, 685 W Baltimore Street, Baltimore, MD 21201, USA

Alison R. Kerr
Division of Infection and Immunity, Institute of Biomedical Sciences, Joseph Black Building, University of Glasgow, Glasgow G12 8QQ, UK

Julian M. Ketley
Dept of Genetics, Adrian Building, University of Leicester, University Road, Leicester LE1 7RH, UK

Gerald T. Keusch
Associate Director for International Research, Fogarty International Center, NIH Building 31, Room B2 C02, Bethesda MA 20892 USA

G.R. Kinghorn
Dept of Genitourinary Medicine, Royal Hallamshire Hospital, Glossop Road, Sheffield S10 2JF, UK

Dennis J. Kopecko
Laboratory of Enteric and STDs, FDA Center for Biologic Evaluation and Research, Building29, Room 420, NIH Campus, Bethesda, MD 20892, USA

Catherine S. Lachenauer
Channing Laboratory, Harvard Medical School, 181 Longwood Avenue, Boston MA 02115-5899, USA

Teresa Lagergård
Dept of Medical Microbiology and Immunology, University of Göteborg, Guldhedsgatan 10, S-413 46 Göteborg, Sweden

Peter A. Lambert
Depts of Pharmaceutical and Biological Sciences, University of Aston, Birmingham B4 7ET, UK

Marc Lecuit
Unité des Interactions Bactéries-Cellules, Institut Pasteur, 28 rue de Docteur Roux, Paris 75015, France

Xin Li
Division of infectious Diseases, School of Medicine, University of Maryland, 10 South Pine Street, Baltimore, MD 21201-1192, USA

Margaret A. Liu
Bill and Melinda Gates Foundation, PO BOX 23350, Seattle, WA 98102, USA

Reggie Y. C. Lo
Dept of Microbiology, College of Agricultural Science, University of Guelph, Guelph, Ontario, Canada

William Lynn
Dept of Infectious Diseases, Level 8, Ealing Hospital, Uxbridge Road, Southall UB1 3HW, UK

Alastair P. MacMillan
FAO/WHO Collaborating Centre for Brucella Reference and Research, Dept of Bacterial Diseases, Veterinary Laboratories Agency, Weybridge, Surrey KT15 3 NB, UK

Angela C. Martin
Department of Biochemistry, University of Oxford, South Parks Road, Oxford OX1 3QU, UK

Millicent Masters
Institute of Cell and Molecular Biology, University of Edinburgh, King's Buildings, Mayfield Road, Edinburgh EH9 3JR, UK

Bruce A. McClane
Dept of Molecular Genetics and Biochemistry, University of Pittsburgh School of Medicine, Pittsburgh PA 15261, USA

Kathleen A. McDonough
Wadsworth Center, New York State Dept of Health and Dept of Biological Sciences, University of Albany, SUNY PO 22002, 120 New Scotland Ave, Albany, NY 12201-2002, USA

Karen F. McGregor
Dept of Microbiology, University of Western Australia, 35 Stirling Highway, Crawley WA 6009, Australia

Brian J. Mee
Dept of Microbiology, University of Western Australia, 35 Stirling Highway, Crawley WA 6009, Australia

Jack Melling
Karl Landsteiner Institute, Rennweg 95B, A1030 Vienna, Austria

Colin Michie
Consultant Paediatrician, Ealing Hospital, Uxbridge Road, Southall UB1 3HW, UK

Michael F. Minnick
Division of Biological Sciences, University of Montana, Missoula MT 59812-1002, USA

Tim J. Mitchell
Division of Infection and Immunity, Institute of Biomedical Sciences, Joseph Black Building, University of Glasgow, Glasgow G12 8QQ, UK

Albert G. Moat
Home: 778 Roslyn Avenue, Glenside, PA 19038-3805, USA
Institutional Affiliation: Emeritus Professor, Dept of Microbiology, Immunology and Molecular Genetics, Marshall University School of Medicine, Huntington, WV, USA

H. L. Mobley
Division of Infectious Diseases School of Medicine, University of Maryland, 10 South Pine St, Baltimore MD 21201-1192, USA

Cesare Montecucco
Centro CNR Biomembrane and, Dipartimento di Scienze Biomediche, Università di Padova, Via G Colombo 3, 35121 Padova, Italy

Sheldon L. Morris
Laboratory of Mycobacterial Diseases, CBER/FDA, Bethesda MD 20852, USA

J. Gareth Morris FRS
Institute of Biological Sciences, University of Wales, Aberystwyth, SY23 3DA, UK

Donald Morrison
Scottish MRSA Reference Laboratory, Dept of Microbiology, Stobhill Hospital, Balornock Road, Glasgow G21 3UW, UK

Inge Muhldorfer
Byk Gulden Pharmaceuticals, Byk-Gulden Straße 2, D-978467 Konstanz, Germany

Matthew A. Mulvey
Depts of Molecular Microbiology and Microbial Pathogenesis, PO Box 8230, Washington University School of Medicine, 600 S Euclid Avenue, St Louis, MO 63110-1093, USA

Francis E. Nano
Department of Biochemistry and Microbiology, University of Victoria, Victoria, British Columbia, Canada

James P. Nataro
Center for Vaccine Development, Depts of Paediatrics and Microbiology and Immunology, University of Maryland School of Medicine, 685 W Baltimore Street, Baltimore MD 21201, USA

Roger Neill
Dept of Molecular Pathology, Walter Reed Army Institute of Research, Washington, DC 20307-5100, USA

Wright W. Nichols
AstraZeneca, Alderley Park, Macclesfield SK10 4TG, UK
Present address: Director of Microbiology, AstraZeneca R&D Boston, 35 Gatehouse Drive, Waltham, MA 02451, USA

David O'Callaghan
INSERM U-431, Faculte de Medicine, Avenue Kennedy, 30900 Nmes, France

Itzhak Ofek
Dept of Human Microbiology, Sackler Faculty of Medicine, University of Tel Aviv, Israel

Iruka N. Okeke
Center for Vaccine Development, University of Maryland School of Medicine, 685 W Baltimore Street, Baltimore, MD 21201, USA

Steven M. Opal
Dept of Medicine, Division of Infectious Diseases, Brown University School of Medicine, Providence, Rhode Island, USA

Roger Parton
Division of Infection and Immunity, Institute of Biomedical and Life Sciences, Joseph Black

Building, University of Glasgow, Glasgow G12 8QQ, UK

Sheila Patrick
Depts of Microbiology and Immunobiology, School of Medicine, Queen's University of Belfast, Grosvenor Road, Belfast BT12 6BN, UK

John H. Pearce
School of Biological Sciences, University of Birmingham, Birmingham, Birmingham B15 2TT, UK
Present address: 126 Oxford Road, Moseley, Birmingham, B13 9SH, UK

Charles W. Penn
School of Biological Sciences, University of Birmingham, Birmingham B15 2TT, UK

T. Hugh Pennington
Dept of Medical Microbiology, Medical School, University of Aberdeen, Aberdeen AB25 2ZD, UK

Kenneth M. Peterson
Dept of Microbiology, Louisiana State University Health Sciences Center, 1501 Kings Highway, PO Box 33932, Shreveport LA 71130-3932, USA

Wolfgang K. Piepersberg
Bergische Universität GH, Faculty of Chemistry, Dept of Chemical Microbiology, Gauss-Strasse 20, D-42097 Wuppertal, Germany

Carrie A. Poore
Division of Infectious Diseases, School of Medicine, University of Maryland, 10 South Pine Street, Baltimore, MD 21201-1192, USA

Rohit S. Prajapati
School of Biological Sciences, Royal Holloway University of London, Egham, Surrey TW20 0EX, UK

Peter Pujic
Dept of Medical Microbiology and Immunology, University of California School of Medicine, Davis, CA 95616, USA

C. P. Quinn
Centre for Applied Microbiology and Research, Porton Down, Salisbury, UK
Present address: Microbial Pathogenesis and Immune Response Lab, MSPB, Mailstop D11, Centres for Disease Control, 1600 Clifton Road NE, Atlanta, GA 30333, USA

Rino Rappuoli
IRIS Chiron SpA, Via Florentina 1, Sienna 53100, Italy

A. Agneta Richter-Dahlfors
Karolinska Institutet, MTC, Box 280, S-171 77 Stockholm, Sweden

Thomas V. Riley
Dept of Microbiology, University of Western Australia and The Western Australian Centre for Pathology and Medical Research, Queen Elizabeth II Medical Centre, Nedlands 6009, Western Australia

Mark Roberts
Dept of Veterinary Pathology, Glasgow University Veterinary School, Bearsden Road, Glasgow G61 1QH, UK

Roy M. Robins-Browne
Microbiological Research Unit, Royal Children's Hospital, Flemington Road, Parkville, Victoria 3052, Australia

Julian I. Rood
Bacterial Pathogenesis Research Group, Dept of Microbiology, Monash University, Clayton 3800, Victoria, Australia

Ornella Rossetto
Centro CNR Biomembrane and, Dipartimento di Scienze Biomediche, Università di Padova, Via G Colombo 3, 35121 Padova, Italy

R. R. B. Russell
Dept of Oral Biology, Dental School, Newcastle Upon Tyne NE2 4BW, UK

James E. Samuel
Dept of Medical Microbiology and Immunology, College of Medicine, Texas A and M University, College Station, TX 77843-1114, USA

Chihiro Sasakawa
Institute of Medical Science, University of Tokyo, 4-6-1 Shirokanedai Minato Ku, Tokyo 108-8639, Japan

Giampietro Schiavo
Imperial Cancer Research Fund, 61 Lincoln Inns Fields, London WC2A 3PX, UK

Carl A. Schnaitman
1902 E Medlock Drive, Phoenix, Arizona 85016-4127, USA

Ira Schwartz
Dept of Biochemistry and Molecular Biology, New York Medical College, Valhalla NY 10595, USA

Tricia Ann Scurtz Sebghati
Dept of Microbiology, College of Medicine, University of Iowa, 3403 Bowen science Building, Iowa City IA 52242-1109, USA

Nathan Sharon
Dept of Membrane Research and Biophysics, Weizmann Institute of Science , Rehovot, Israel

Richard W.P. Smith
Institut für Molekulare Biotechnologie e.V., Beutenbergstrasse 11, D-07745 Jena, Germany

Noel H. Smith
School of Biological Sciences, University of Sussex, Brighton, BN1 9QG, UK

Rebecca J. Smith
Dept of Microbiology and Immunology, Leicester University, P.O. Box 138, Medical Sciences Building University Road, Leicester LE1 9HN, UK

Valerie A. Snewin
Scientific Programme Officer The Wellcome Trust, The Wellcome Building, 183 Euston Road, London NW1 2BE, UK

Brian G. Spratt
Dept of Infectious Disease Epidemiology, Imperial College School of Medicine, St Mary's Hospital, Norfolk Place, London, W2 1PG, UK

Lola V. Stamm
Program in Infectious Diseases, Dept of Epidemiology, 242 Rosenau Hall, University of North Carolina at Chapel Hill, Chapel Hill NC 27599-7400, USA

Paul M. Sullam
Dept of Infectious Disease, University of California at San Francisco, California, USA

Johnathan D. Sussman
Greater Manchester Neuroscience Centre, Hope Hospital, Eccles Road, Salford, M6 8HD, UK

Max Sussman
Dept of Microbiology, The Medical School, University of Newcastle Upon Tyne, Newcastle Upon Tyne NE2 4HH, UK

Richard W. Titball
Defence Evaluation and Research Agency, CBD Porton Down, Salisbury, Wiltshire SP4 0JG, UK

Kevin J. Towner
Dept of Microbiology and PHLS Laboratory, University Hospital, Queen's Medical Centre, Nottingham NG7 2UH, UK

Peter C. B. Turnbull
Arjemptur Technology Ltd , Porton Down, Science Park, Salisbury SP4 0JQ UK, Contact address: 86 St Mark's Avenue Salisbury, Wiltshire SP1 3DW, UK

Qinning Wang
Dept of Microbiology, University of Western Australia, 35 Stirling Highway, Crawley WA 6009, Australia

Guiqing Wang
Dept of Biochemistry and Molecular Biology, New York Medical College, Valhalla NY 10595, USA

Michael R. Wessels
Channing Laboratory, 181 Longwood Avenue, Boston MA 02115-5899, USA

Brian M. Wilkins
Dept of Genetics, University of Leicester, Leicester LE1 7HN, UK

Anil Wipat
Dept of Microbiology, The Medical School University of Newcastle Upon Tyne, Framlington Place, Newcastle Upon Tyne NE2 4HH, UK

Martin J. Woodward
Dept of Bacterial Diseases, Veterinary Laboratories Agency (Weybridge), Addlestone, Surrey KT15 3NB, UK

Gary P. Wormser
Chief Division of Infectious Disease, New York Medical College, Valhalla NY 10595, USA

Brendan W. Wren
Dept of Infectious and Tropical Diseases, London School of Hygiene and Tropical Medicine, Keppell Street, London WC1E 7HT, UK

Kwok Yung Yuen
Dept of Microbiology, Queen Mary's Hospital, Faculty of Medicine, Hong Kong University, Pokfulam, Hong Kong

Wilma Ziebuhr
Instutut für Molekulare Infektions Biologie, Universität Würzburg, Röntgenring 11 D-97070, Würzburg, Germany

Amy B. Zuppardo
Bureau of Laboratories, Jacksonville, FL, USA
Present address: Dept of Health and Hospitals Office of Public Health, Room 709, 325 Loyola Avenue, New Orleans LA 70112, USA

Contents

Colour plate sections are located as follows

Colour Plates

Plate 1 (right) General structure of the flagellum in Gram-negative bacteria. The components of the basal body are shown inserted into the various layers of the cell envelope. The flagellar filament is attached to the basal body by the hook. Since the flagellar filament is a helical structure, it has a central channel. HAP, hook-associated protein; OM, outer membrane; PL, peptidoglycan layer; CM, cell membrane.

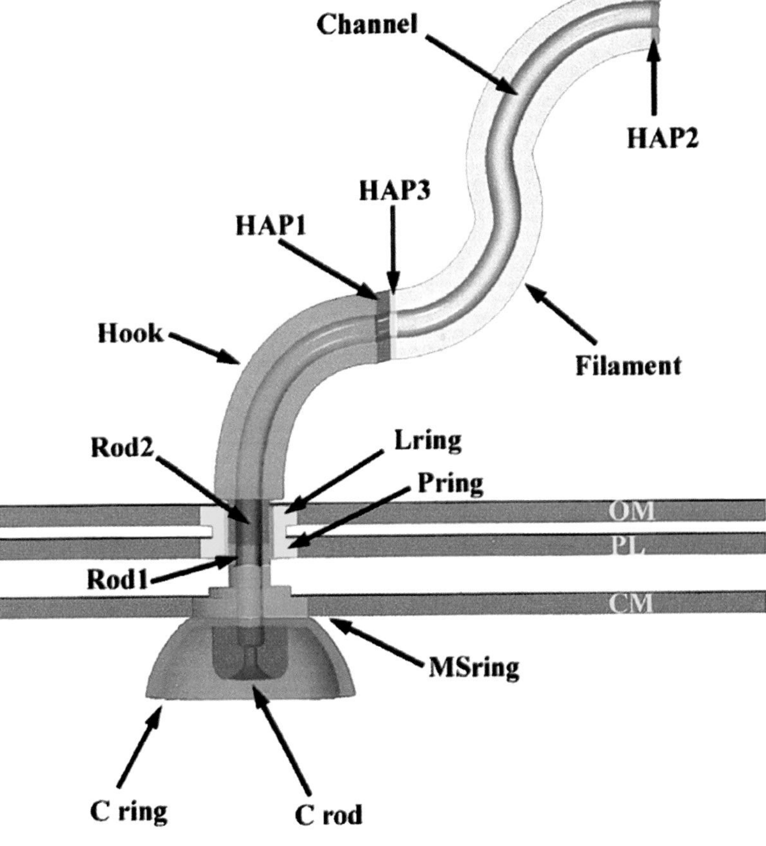

Plate 2 (below) Schematic model of the flagellar motor. The hook and filament are omitted. The basal body, the major structural portion of the flagellar motor, is composed of several structural complexes: the LP ring complex, rod, MS ring complex, C ring complex and C rod.

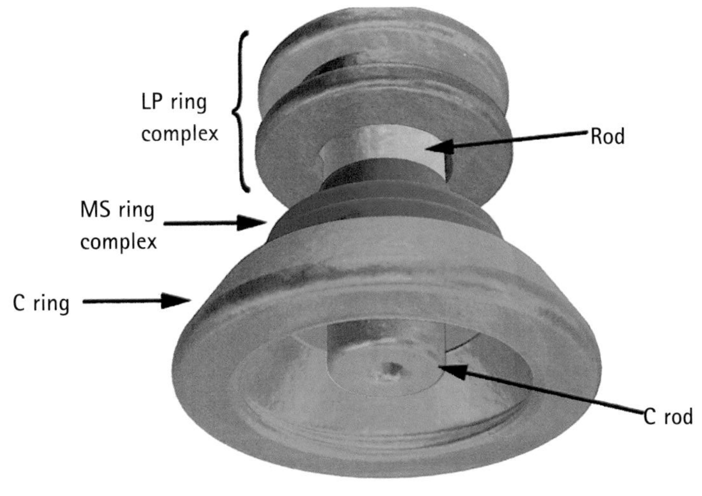

Plate 3 (below) Morphological pathway of flagellar assembly. The MS ring complex is the first structure of flagellar assembly that can be recognized by electron microscopy. After the C ring and C rod have created an export apparatus on the cytoplasmic side of the M ring, flagellar proteins are exported to construct flagella. The Mot complex is formed at the last stage of assembly at the same time as filament growth.

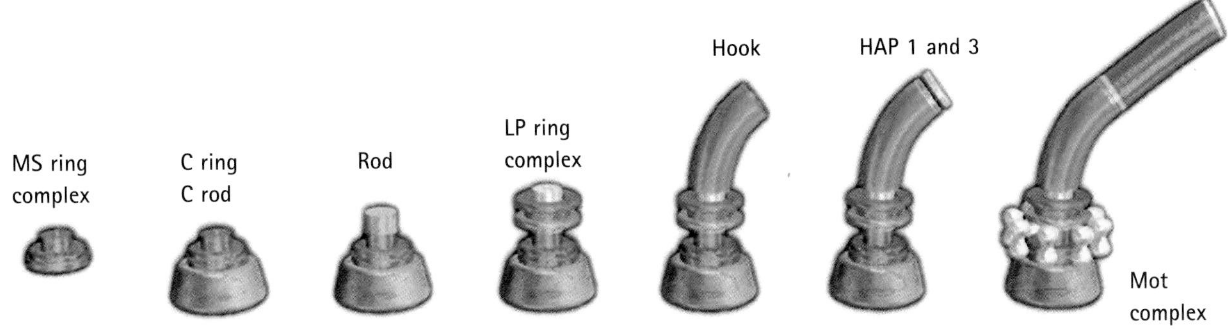

Plate 4 (right) Molecular structure of the flagellum of *S. typhimurium* (cf. Plate 1). The components of the flagellum are shown with the gene that encodes them indicated in brackets. HAP, hook-associated protein; OM, outer membrane; PL, peptidoglycan layer; CM, cell membrane.

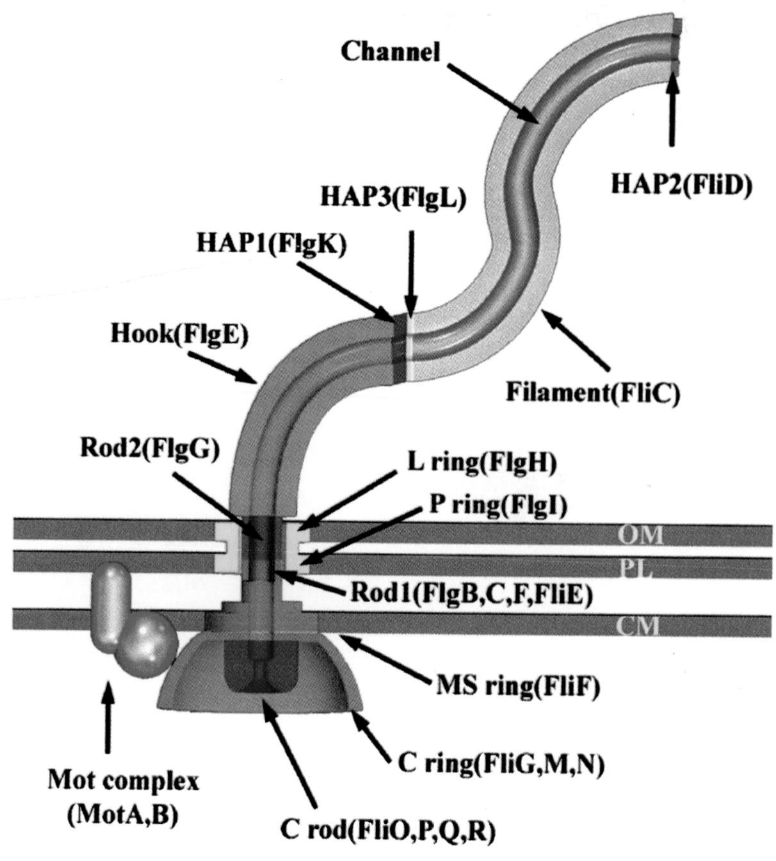

Plate 5 The genetic code.

First position	Second position: U		Second position: C		Second position: A		Second position: G		Third position
	U		C		A		G		
U	UUU	Phe	UCU	Ser	UAU	Tyr	UGU	Cys	U
	UUC	Phe	UCC	Ser	UAC	Tyr	UGC	Cys	C
	UUA	Leu	UCA	Ser	UAA	Stop	UGA	Stop	A
	UUG	Leu	UCG	Ser	UAG	Stop	UGG	Trp	G
C	CUU	Leu	CCU	Pro	CAU	His	CGU	Arg	U
	CUC	Leu	CCC	Pro	CAC	His	CGC	Arg	C
	CUA	Leu	CCA	Pro	CAA	Gln	CGA	Arg	A
	CUG	Leu	CCG	Pro	CAG	Gln	CGG	Arg	G
A	AUU	Ile	ACU	Thr	AAU	Asn	AGU	Ser	U
	AUC	Ile	ACC	Thr	AAC	Asn	AGC	Ser	C
	AUA	Met	ACA	Thr	AAA	Lys	AGA	Arg	A
	AUG	Met	ACG	Thr	AAG	Lys	AGG	Arg	G
G	GUU	Val	GCU	Ala	GAU	Asp	GGU	Gly	U
	GUC	Val	GCC	Ala	GAC	Asp	GGC	Gly	C
	GUA	Val	GCA	Ala	GAA	Glu	GGA	Gly	A
	GUG	Val	GCG	Ala	GAG	Glu	GGG	Gly	G

basic non-polar acidic uncharged

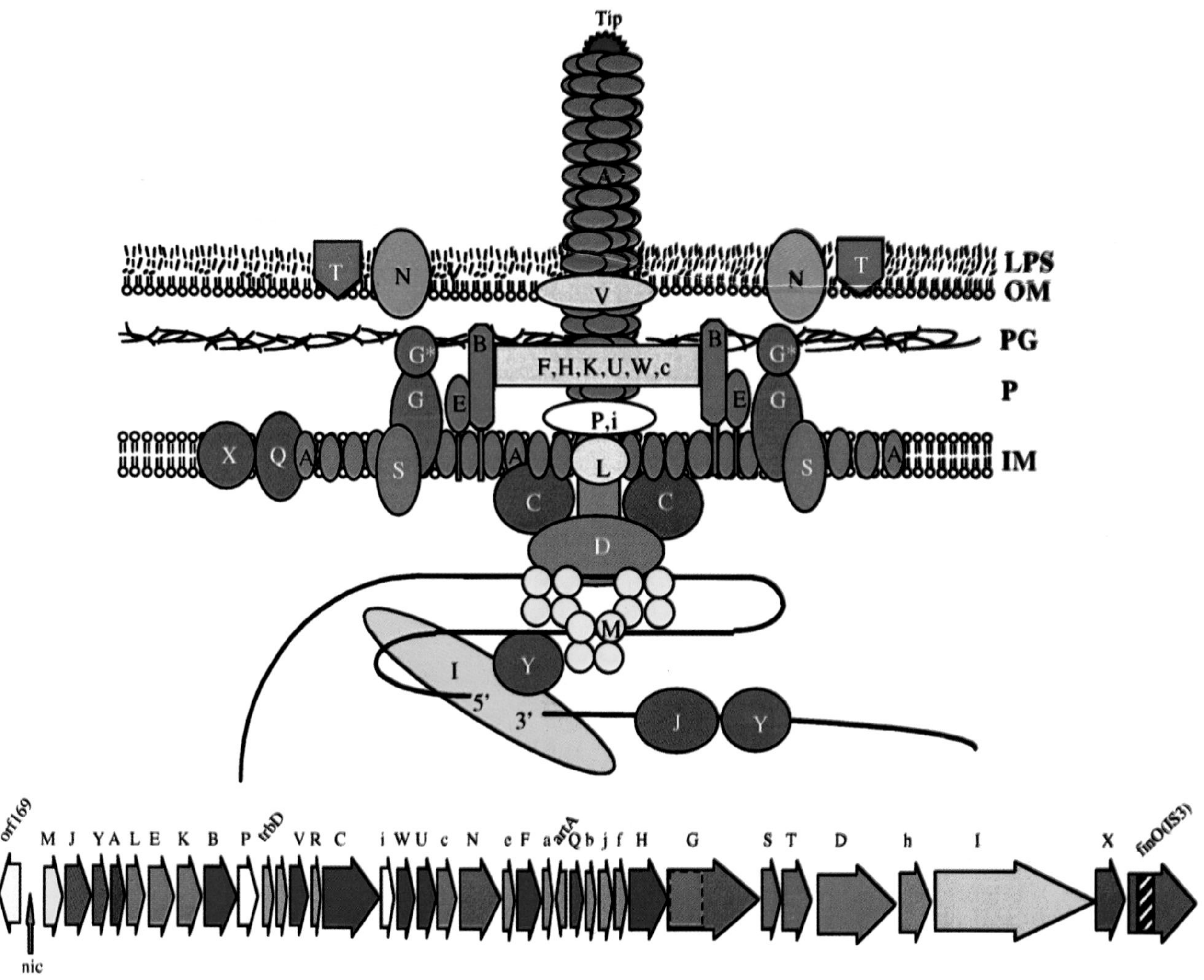

Plate 6 A model of the F transfer apparatus and map of the F transfer region. The gene products for pilus synthesis (blue), mating-pair stabilisation (green), DNA metabolism (yellow to brown) and regulation (purple) are indicated at their approximate location within the cell envelope. The nature of the pilus tip is unknown but is thought to contact a receptor on the surface of the recipient cell. TraV, a lipoprotein in the outer membrane, might give lateral stability to the extended structure. TraF, H, K, U, W and TrbC in the periplasm and TraP and TrbI (white) in the inner membrane also support the pilus and are involved in pilus elongation and retraction. Mutants in *traP* have fragile pili which break off the cell easily, suggesting that TraP anchors the pilus in the inner membrane. Pili of *trbI* mutants are particularly long, implying that TrbI could be involved in retraction. TraB, a TonB/TolA analogue could extend into the periplasmic space and provide stability or mediate an unknown signal from the outer membrane to the inner membrane. TraE is an inner membrane protein of unknown function. TraL is a TraA (pilin) homologue that is expressed at very low levels and might be involved in pilus placement. Pilin is inserted in the inner membrane by the action of TraQ and is acetylated by TraX. TraC is involved in pilus assembly in which the α helices of the pilins are realigned from a vertical position in the inner membrane to a more horizontal position in the assembled pilus. TraN, which contacts OmpA in the recipient cell, might form a transenvelope complex with TraG* and TraG to stabilise the mating pair after contact is initiated. TraS and TraT (red) are inner and outer membrane proteins involved in entry exclusion and surface exclusion, respectively. TraI is an asymmetrical protein which cleaves at *nic* in a site- and strand-specific manner. This cleavage reaction is mediated by TraY and TraM which forms a nucleosome-like complex at *oriT* and interacts with TraD the proposed nucleic acid pump. TraD is positioned at the base of the pilus in an unknown manner. The LPS (lipopolysaccharide), OM (outer membrane), PG (peptidoglycan), P (periplasm) and IM (inner membrane) are indicated on the right. The transferable DNA strand cleaved at *nic* with the 5' end, which precedes the leading region, is indicated. Proteins of unknown function are grey and the lytic transglycosylase, Orf169, is indicated in pink.

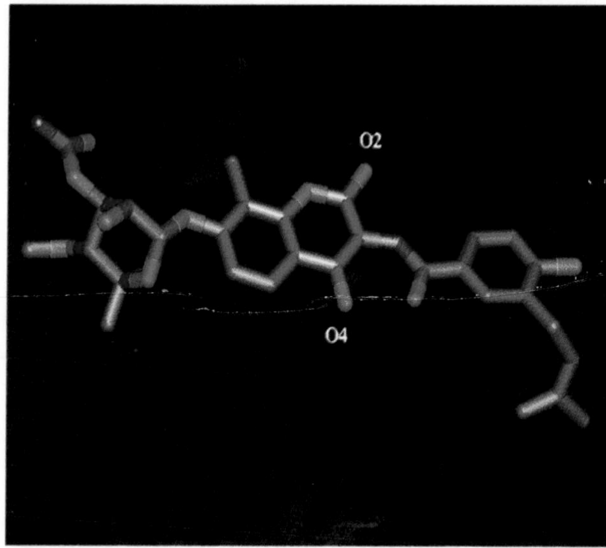

Plate 7 The crystal structure of novobiocin. The colour coding is as follows: green, carbon; blue, nitrogen; red, oxygen. Hydrogen atoms are not shown explicitly: they are to be inferred. The positions 02 and 04 refer to the ring numbering as shown on the hydroxycoumarin ring of structure 2.

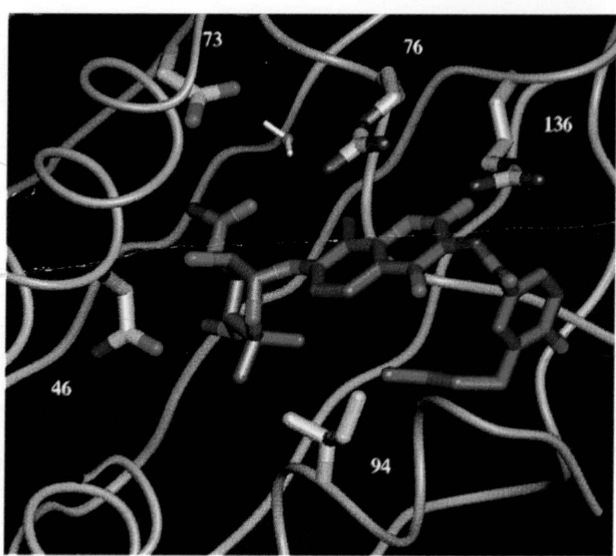

Plate 8 The crystal structure of the complex of novobiocin and the 24-kDa fragment of the B-subunit of DNA gyrase from *E. coli*. The colour coding of novobiocin is the same as in Fig. 2. Again hydrogen atoms are not shown on that molecule. The pink line represents the shape of the peptide backbone. The numbers refer to amino acid positions in the primary sequence of intact GyrB. The side-chains of the numbered amino acids are shown (Asn46, Asp73, Arg76, Ile94 and Arg136), with colour coding as for novobiocin, except that carbon atoms are pale green. Between Asp73 and Arg76 there is a bound water molecule, whose hydrogen atoms are shown in white.

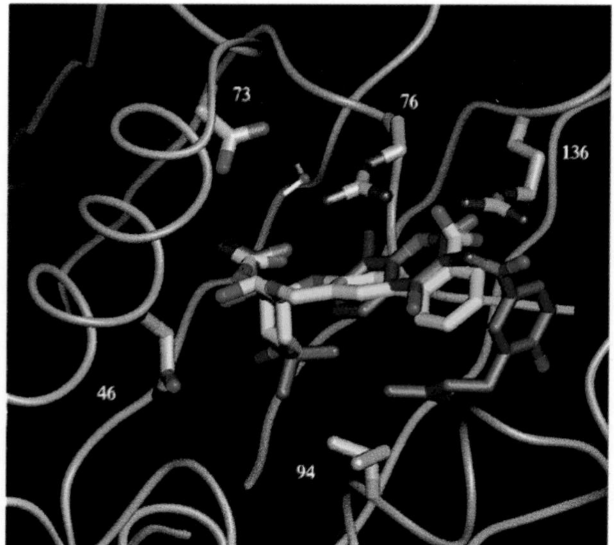

Plate 9 The crystal structure of the complex of compound **7** and the 24-kDa fragment of the B subunit of DNA gyrase from *E. coli*, compared with novobiocin when it occupies the same site. The conventions are as in Fig. 3, except that the carbon atoms of compound **7** are coloured yellow.

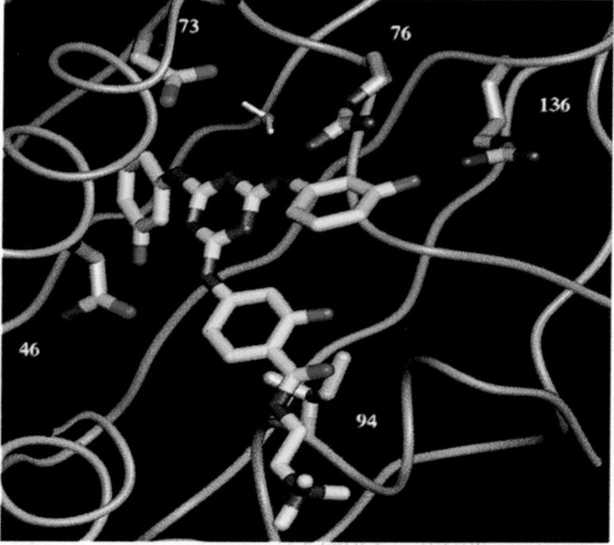

Plate 10 The crystal structure of the complex of the soluble triazine **9** and the 24-kDa fragment of the B subunit of DNA gyrase from *E. coli*. The conventions are as in Fig. 3, except that the carbon atoms of compound **9** are coloured yellow and the fluorine atoms green.

1

Molecular Medical Microbiology: The Concept

Max Sussman

Medical School, Newcastle upon Tyne, UK

Medical microbiology has long had a molecular dimension. From the earliest days of microbiology the phenotypic descriptions of micro-organisms included biochemical and physiological details that were in fact molecular. The term molecular microbiology as it has recently come into common use describes an approach to the study of micro-organisms based principally on their genetic characteristics. The origins of this trend, as of much else in modern biology, can be seen in the seminal description of the structure of DNA by Watson and Crick (1953a) and their immediate recognition that implicit in this structure is the mechanism by which genetic material is to be copied. Most recently this knowledge has led to the sequencing of complete bacterial genomes and the promise that genomics in turn will lead to comprehensive information about individual bacteria.

The work of the half-century since the structure of DNA was elucidated has yielded information that has changed views about bacteria. The development of medical bacteriology can broadly be divided into pre-molecular and molecular phases. An interesting account of the state of medical bacteriology in the late pre-molecular age up to the 1930s is detailed in *A System of Bacteriology in Relation to Medicine* (Fildes and Ledingham, 1930–1931) and the more general history of medical bacteriology up to about the mid-1930s has been presented by Bulloch (1938). It is interesting to recall that at that time virology was still an integral part of bacteriology, as was the then embryonic science of immunology. The intensive study of the bacteriophages, mainly those of *Escherichia coli*, which began in the 1940s, would ultimately lead to some of the most important tools and concepts of molecular biology.

The recent molecular approach to the study of microbiology has had a profound effect on the advance of knowledge not only about bacteria in general but also about pathogenic bacteria in particular; the rate of progress during the last 25 years has been particularly striking. The complexity of intact organisms was the great difficulty and weakness inherent in the analytical study of individual phenotypic characteristics. This contrasts with the power of methods by which genes can be cloned and their phenotype studied in isolation. It now seems remarkable how much was achieved in the past by methods and approaches that are now regarded as slow and laborious.

In the case of pathogens, the most striking advances have been in the understanding of the mechanisms of bacterial pathogenesis (see, for example, Groisman, 2001). It is particularly fascinating that phenotypically quite different organisms may have very similar pathogenetic mechanisms. Indeed, in some cases, the molecular genetic evidence suggests that these systems passed between bacteria by gene transfer at some time during microbial evolution.

The importance of the molecular approach to medical microbiology goes far further than merely providing information about mechanisms of pathogenicity. The central objective of medical science is the prevention of disease. The first great preventive

 doi:10.1006/bkmm.2001.0001

insights, based on epidemiological observations, were recognition by the social pioneers of the 19th century of the importance of clean water and safe sewage disposal. Of almost equal importance was the discovery of toxoids at the end of the 19th century. This, together with later developments in immunology and the manipulation of genes and immunogens, is giving rise to highly successful vaccines. A striking recent example is the vaccine against *Haemophilus influenzae* type b, which promises to eliminate the risk of childhood meningitis due to this virulent pathogen. As will be apparent from the chapters that follow, molecular biology is yielding far-reaching insights into every branch of medical microbiology.

A Brief History

A detailed account of the origins of molecular bacteriology would lead us too far afield, but a brief and necessarily superficial sketch of its development may be helpful. A classical account of the subject is that by Cairns *et al.* (1966).

The rapid development of the genetics of higher plants and animals in the early twentieth century at first had little effect on bacteriology and the mechanisms of inheritance in bacteria were thought to be of quite a different kind. What was then called bacterial 'variation' was recognised but its genetic basis was elusive, though it was suspected by some early bacteriologists (Arkwright, 1930). Contemporary views began to emerge in the 1940s, when biochemical mutants of *Neurospora crassa* were first isolated (Beadle and Tatum, 1941) and the unity of genetics in the biological world was recognised. The foundations of bacterial genetics were laid a little later with the demonstration that bacteriophage-resistant mutants appear spontaneously in bacterial cultures (Luria and Delbrück, 1943).

An earlier step had been the classical landmark discovery by Griffith (1928) of the transformation of *Streptococcus pneumoniae* in mice *in vivo*. This was a prelude to the demonstration by Avery *et al.* (1944) that Griffith's transforming factor was, in fact, DNA and it was also the first demonstration of genetic transfer in bacteria. The solution of the structure of DNA as a double helix (Watson and Crick, 1953a) and the recognition that this structure has genetic implications (Watson and Crick, 1953b), led to the deciphering of the genetic code that determines protein structure. The sequence of the transcription of the DNA message to messenger RNA and its translation into the amino acid sequence of proteins was recognised as sufficiently fundamental that it became known as the 'dogma', though like all dogma it did not long survive unmodified.

The discovery of the phenomenon of host-controlled restriction-modification in bacteriophage led to the isolation of restriction endonucleases, which later became universal tools for the specific manipulation of DNA. Bertani and Weigle (1953) found that most, but not all, λ bacteriophage particles grown on *E. coli* K12 were unable to multiply in a strain of *E. coli* that carries the prophage P1 (*E. coli* K12{P1}). The rare λ phage particles that did succeed to multiply were subsequently able to propagate at full efficiency in *E. coli* K12(P1). This phenomenon was explained by two functions of the prophage P1: first, that it restricted the propagation of λ phage in *E. coli* K12(P1) and, second, that it modified the few survivors and so allowed them to propagate in this strain. Restriction is a general phenomenon due to enzymes (restriction endonucleases) that destroy foreign DNA by cutting it up. The cutting occurs at sites that depend on the specificity of the endonuclease present in particular cells and restriction may be seen as a kind of defence mechanism against extraneous genetic material. Modification is due to the methylation of DNA by methylases to protect it from the action of the endonucleases. The subject has been reviewed by Wilson and Murray (1991).

The restriction endonucleases of different organisms vary in their specificity for the base sequences at which they cut DNA. Since the sequence specificity of endonucleases can be determined, they are potential reagents for cutting DNA at specific sites, so yielding manageable lengths of DNA for study. Similarly, DNA ligase of *E. coli* is a repair enzyme designed to rejoin the nicks that may occur in DNA from time to time. Methods of DNA analysis and techniques for gene transfer became the foundation tools of genetic manipulation and molecular genetics. For details the reader is referred to standard texts on molecular biology, bacterial genetics and genetic manipulation (see also Chapter 22).

Though determination of the base sequence of DNA has been practised for some time, more recently the entire base sequences of the genomes of many pathogenic bacteria have been determined, resulting in the development of the new science of *genomics*. The complete genome sequences so determined often give important insights into the biology of the cell; they are frequently referred to throughout this book (see Chapter 16).

The base sequence of structural genes provides direct evidence of the amino acid sequence of the proteins for which they code. The young science of *proteomics* goes further in that it is concerned with the total protein content in a cell at a particular time. This only in part reflects genomic information, since the population of proteins in a cell at any time depends not only on the informational content of the genome but also on controls on transcription and translation, the effects of protein modification and environmental effects.

It will be evident that a large body of molecular information is progressively becoming accessible to explain many hitherto mysterious characteristics of pathogenic bacteria. It should also be clear that deductions from molecular biology alone are not sufficient to explain many features of microbial behaviour; a complete understanding will ultimately depend on a re-integration of the information obtained from molecular biology into 'whole organism' biology.

Molecular Medical Microbiology

Much of the past practice and thinking in medical and particularly clinical bacteriology was based on the classical views of bacteria and their behaviour. These views are changing rapidly under the influence of accumulating fundamental molecular knowledge. These changes, already evident in the better understanding of bacterial pathogenicity and pathogenesis, are extending into every branch of medical and clinical microbiology, from bacterial morphology at one extreme to ecology, epidemiology and prevention at the other. At the same time molecular techniques are finding an increasing use in the diagnostic laboratory. In addition, the molecular approach has far reaching implications for the understanding of the interaction between host and microbe (Lederberg, 1998a, 1998b).

This book is an attempt to collect a wide range of information about all aspects of bacterial pathogens. Its general structure is, therefore, not unlike that of the classical textbooks; the more fundamental aspects of microbiology are considered before individual pathogens are discussed in detail. It differs in that it has generally been assumed that information of a more traditional and classical kind can be found in many well-known and excellent texts. This has made it possible to concentrate on the more recent molecular information that is increasingly rapidly becoming available.

Medical microbiology, whether molecular or classical, is now frequently practised and researched by scientists rather than by medical practitioners. A number of chapters entitled 'a clinical overview' have, therefore, been included in the belief that medical microbiology cannot be seen in its full perspective without reference to the effect of infection on the host.

This new molecular knowledge can be expected increasingly to influence medical bacteriology from its diagnostic beginning to its therapeutic aim by increasing the speed of the former and the range and specificity of the latter. It will also play an increasing part in reaching epidemiological and preventive objectives. It is interesting to contemplate the possibility that in future the targets of therapy may move from structural components of the bacterial cell to the pathogenetic principles which are the ultimate causes of the disease syndromes that are observed clinically. A first step in this direction, at the beginning of the twentieth century, was the discovery of the toxins of Gram-positive organisms and their neutralising antitoxins. If, as a result of molecular insights, the same were possible for the toxins of Gram-negative bacteria or their secretory delivery systems, could some of the problems of antibiotic resistance be circumvented? Such possibilities would not only cause excitement in the playground of the research laboratory, but they would also dramatically change practices and protocols of the diagnostic laboratory. What seems certain is that the molecular approach will increasingly change and enhance the understanding and practice of medical microbiology.

References

Arkwright JA (1930) Variation. In: Fildes P, Ledingham JCG (eds) *A System of Bacteriology in Relation to Medicine*, London: HMSO, Vol I, pp. 311–374.

Avery OT, MacLeod CM, McCarty M (1944) Studies on the chemical nature of the substance inducing transformation of pneumococcal types. Induction of transformation by a desoxyribonucleic acid fraction isolated from pneumococcus-type III. *J. Exp. Med.* 79: 137–157.

Beadle GW, Tatum EL (1941) Genetic control of biochemical reactions in *Neurospora*. *Proc. Natl Acad. Sci. USA* 27: 499–506.

Bertani G, Weigle JJ (1953) Host controlled variations in bacterial viruses. *J. Bacteriol.* 65: 113–121.

Bulloch W (1938) *The History of Bacteriology*. London: Oxford University Press.

Cairns J, Stent GS, Watson JD (eds) (1966) *Phage and the Origins of Molecular Biology*. New York: Cold Spring Harbour Laboratory of Quantitative Biology.

Fildes P, Ledingham JCG (eds) (1930–1931) *A System of Bacteriology in Relation to Medicine*, London: HMSO, Vols I–IX.

Griffith F (1928) The significance of pneumococcal types. *J. Hyg. Camb.* 27: 113–159.

Groisman EA (ed.) (2001) *Principles of Bacterial Pathogenesis*. San Diego, CA: Academic Press.

Lederberg J (1998a) Infectious agents, hosts in constant flux. *ASM News* 64: 18–22.

Lederberg J (1998b) Parasites face a perpetual dilemma. *ASM News* 64: 77–80.

Luria SE, Delbrück M (1943) Mutations of bacteria from virus sensitivity to virus resistance. *Genetics* 28: 491.

Watson JD, Crick FH (1953a) A structure for deoxyribose-nucleic acid. *Nature* 171: 737–738.

Watson JD, Crick FH (1953b) Genetical implications of the structure of deoxyribonucleic acid. *Nature* 171: 964–967.

Wilson GG, Murray NE (1991) Restriction and modification systems. *Annu Rev Genet.* 25: 585–627.

PART 1

BACTERIAL STRUCTURE

2 Bacterial Ultrastructure

Jan A. Hobot

University of Wales College of Medicine, Cardiff, UK

Bacteria were first studied under the light microscope, using dyes and stains to investigate their structural features. Such methods showed that bacteria were of varying shapes and sizes. There was a cell wall, from which flagella could be observed to originate, a capsule surrounded the bacterial cell, a nuclear area was visible within the cytoplasm, and inclusion bodies were identified on the basis of their colour reaction with certain stains. Later, the use of phase-contrast light microscopy allowed living bacteria to be observed, but even under the best conditions it was not possible to study the *ultra*structure, i.e. the fine structure, of bacteria. The advent of the electron microscope with its high resolving power changed all this and revolutionised the study of bacterial ultrastructure. The use of electron microscopes permitted new insights into bacterial organisation, and these could now be linked to cellular functions.

In recent years, there have been many exciting developments not only in instrumentation (Surman *et al.*, 1996) but also in the procedures used to prepare specimens for microscopy (Hobot, 1990). These have improved and greatly added to our knowledge of bacterial ultrastructure. It has emerged that the bacterial cell is a dynamic entity rather than a static one, and it is the aim of this chapter to highlight these new aspects of bacterial structural studies, and point the way for future investigations at a molecular level.

General Morphology

Bacteria are small, prokaryotic cells, generally of the size of mitochondria. Their length varies from about 0.2 to more than 10 µm, and their width from about 0.2 to 1.5 µm. A variety of bacterial shapes can be observed under the light microscope, including cocci, rods, spiral, and even cubes! These shapes can also be observed by scanning electron microscopy (SEM), which provides both a three-dimensional view of cellular structures and information about their surface topography (Fig. 1). Apart from recording cell shape, SEM investigations can, for example, be coupled with studies on the effects of various drugs on overall morphology (Fig. 2a,b).

External and internal bacterial structures can be investigated at high resolution by transmission electron microscopy (TEM). In a procedure known as negative staining, whole bacteria or isolated structures, such as pili, fimbriae, etc., are attached to metallic support grids covered with a very thin plastic film of collodion or formvar. Then a heavy metal stain, such as uranyl acetate or phosphotungstic acid, is dried down over the preparation and details of surface structural features are revealed. Alternatively, a metal, usually platinum, can be vaporised within a vacuum coater, to cast a shadow over the specimen attached to a support grid, to highlight structural surface features. This is a

 doi:10.1006/bkmm.2001.0002

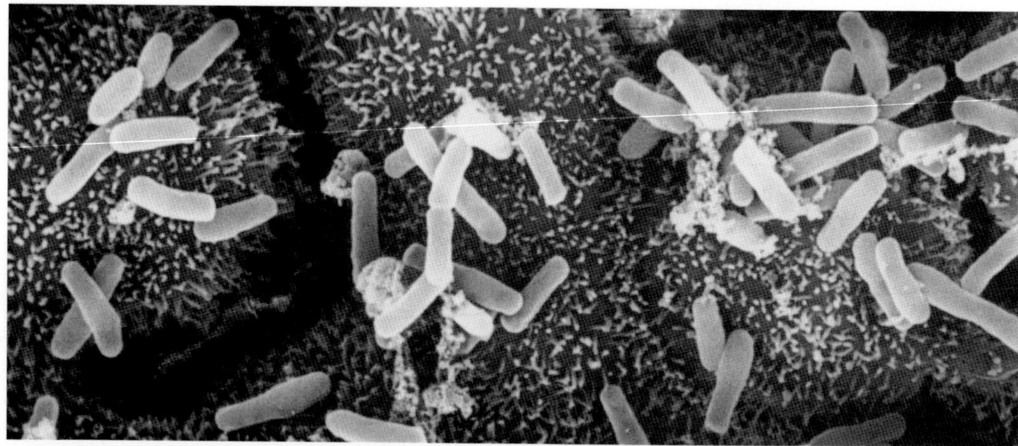

Fig. 1 Scanning electron micrograph of rod-shaped cells of *Clostridium difficile* adhering to the villi of human gut epithelium (Alsaif *et al.*, 1997). Bar marker 2 μm.

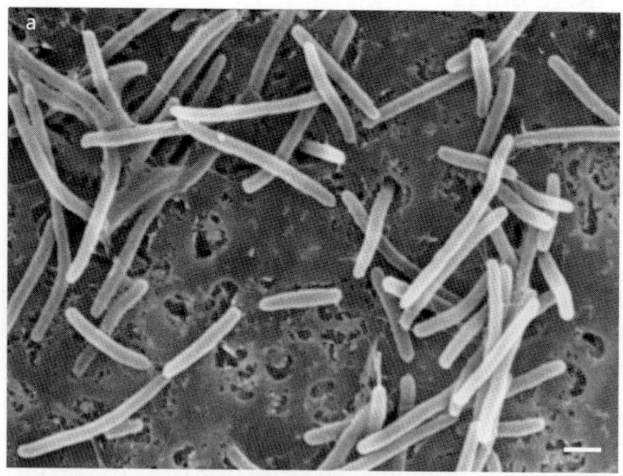

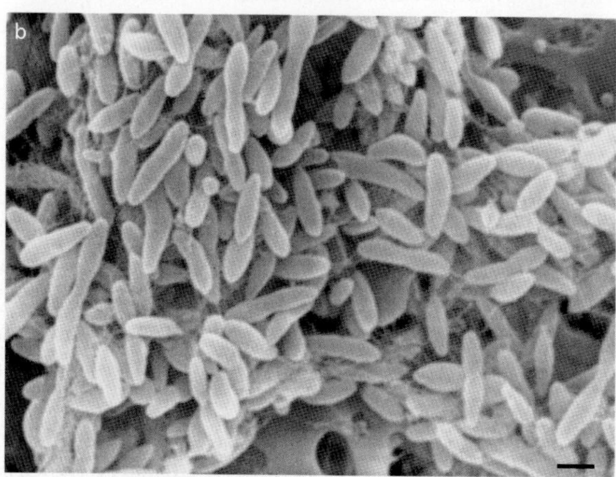

Fig. 2 Scanning electron micrograph of *Escherichia coli*. (a) Untreated control showing typically long, rod-shaped cells. (b) After treatment with meropenem cells are shorter with a more ovoid appearance. Bar marker 2 μm.

procedure known as metal shadowing. In another popular procedure, bacterial preparations are embedded into a resin which, after hardening can be cut into very thin sections or slices with an ultramicrotome. These ultrathin sections, less than 0.1 μm in thickness, can reveal the internal structure and organisation of bacteria (Robinson *et al.*, 1987).

By these means electron microscopy reveals that bacteria have a variety of surface appendages such as flagella, pili, fimbriae, and slime or capsular layers that extend from the cell wall. The inner surface of the cell wall is in contact with a cytoplasmic membrane within which there is the cytoplasm. This contains a nuclear region of DNA, ribosomes and various inclusion bodies (**Fig. 3**). There are no unit membrane-bound structures, as are found in eukaryotic cells, but inclusion bodies such as carboxysomes, lipid bodies, and gas vacuoles, and the complex cell membrane systems of photosynthetic bacteria are surrounded by a non-unit protein layer.

Surface Appendages

Various surface appendages are present on the outer surface of the bacterial cell wall. They are concerned with the relationship between the bacterium and its external environment.

Flagella

Flagella are key structures concerned with bacterial motility. Nevertheless, bacteria that lack flagella may still be motile. A type of gliding motility can be

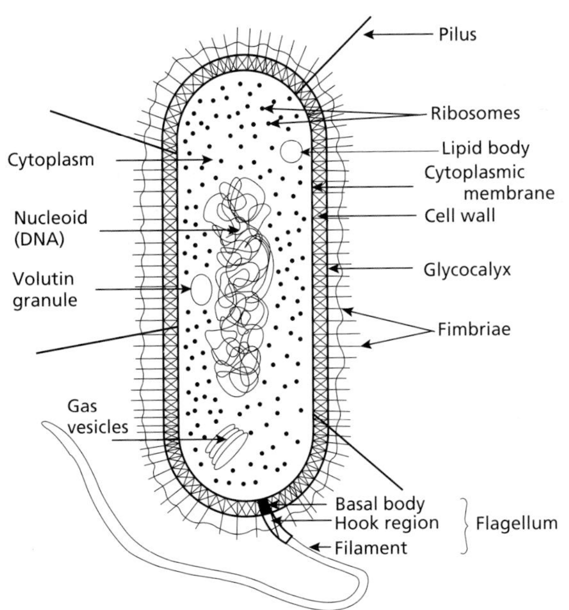

Fig. 3 Generalised diagram of the main features of a bacterial cell.

achieved by the flexible movement of the whole cell. This is mainly observed on solid media, whereas flagellar motion is common to liquid environments. Flagella can be located singly at one cell pole (monotrichous flagella), at both poles (amphitrichous flagella), in large numbers along the length of the cell (peritrichous flagella), or as a tuft of flagella at a polar end (lophotrichous flagella). They usually have a wavy or undulating appearance, and are about 10–15 nm in diameter and 2–15 μm in length (Fig. 4). Flagella are composed of subunits of a low molecular weight protein, flagellin (20–40 kDa) arranged in a helical manner. The filamentous part of the flagellum extends outwards from the bacterial surface, and is anchored to the bacterium by its basal body. This consists of a series of protein ring-like structures that hold the flagellum into the bacterial cell wall. Although it has been suggested that the bottom of this basal body structure penetrates through the cytoplasmic membrane into the cytoplasm, the exact structural relationship between the basal body and cell wall peptidoglycan

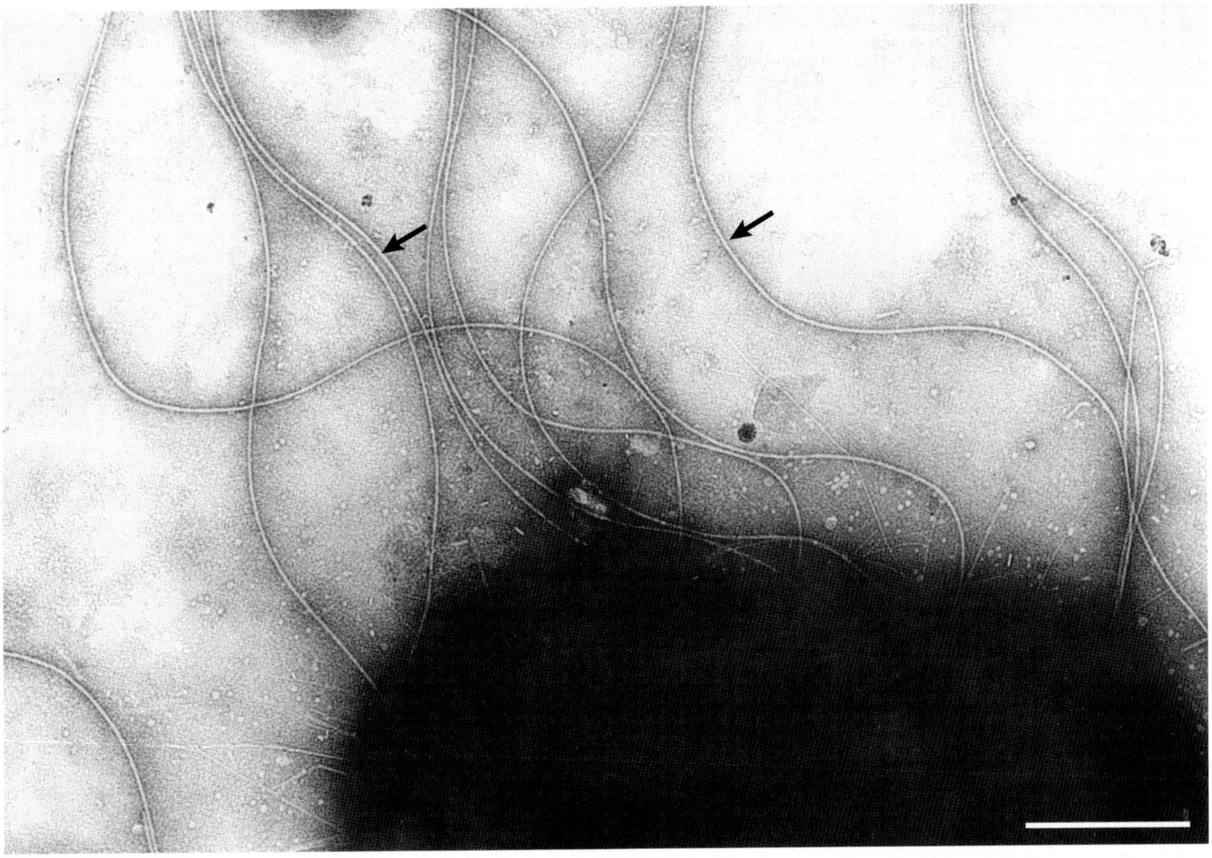

Fig. 4 Peritrichous flagella (arrows) of *Escherichia coli* prepared by negative staining of whole bacteria. Bar marker 0.5 μm. Micrograph courtesy of Dr Nabeel Alsaif.

remains unclear (Dijkstra and Keck, 1996). A short hook region acts as a connection between the long outer filament and basal body, the latter being responsible for flagellar movement. Flagella are discussed in greater detail in Chapter 8.

Pili and Fimbriae

Many bacteria have surface appendages of various sizes and appearances called pili and fimbriae. These terms have been used interchangeably. The term 'pilus' (plural pili), however, should be reserved for appendages involved in conjugation, that is the transfer of genetic material (see Bacterial Conjugation, p. 25), and the term 'fimbria' (plural fimbriae) should be reserved for structures concerned with the adhesion of bacteria to various surfaces, including cell surfaces, and to each other in coaggregation, as for example in dental plaque. Fimbriae and pili are considered in greater detail in Chapter 9.

On average, one to ten conjugative pili and up to more than 400 fimbriae may be present on the

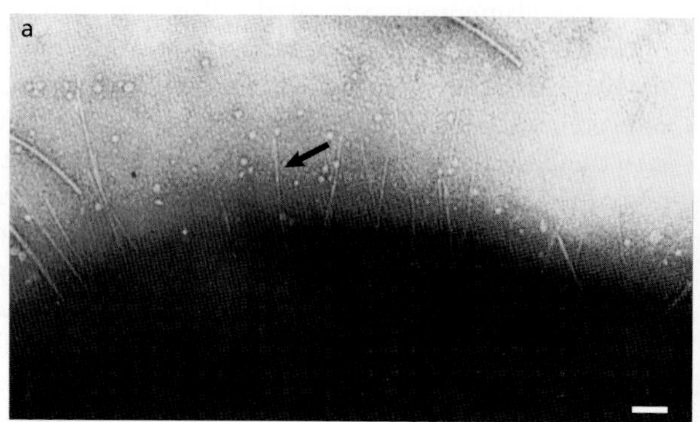

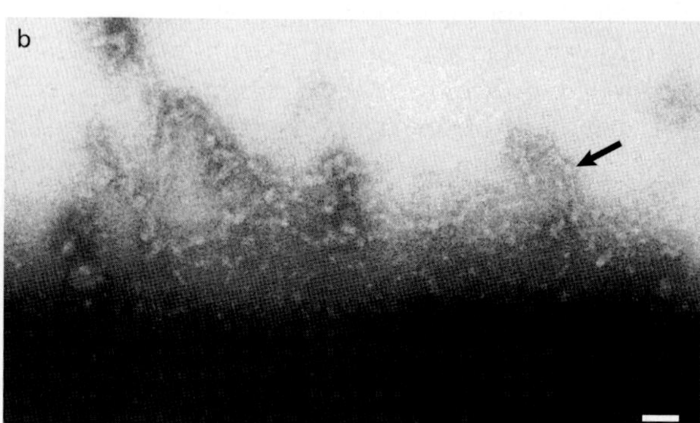

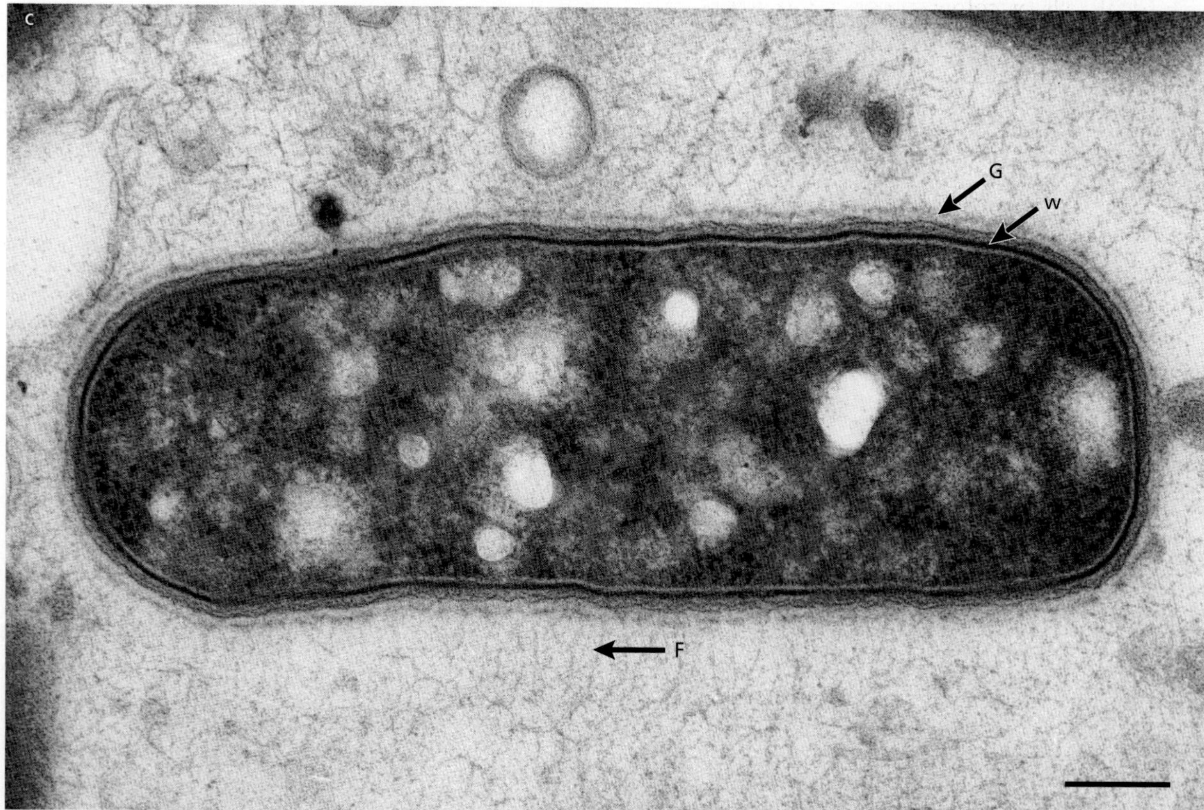

Fig. 5 Negative stain preparations of (a) relatively straight fimbriae (arrows) from *Escherichia coli* (by courtesy of Dr. Nabeel Alsaif); (b) wiry, fibrillar type fimbriae (arrows) from *Prevotella intermedia* (bar marker 0.1 μm); (c) ultrathin section of *P. intermedia* showing the fimbriae (F) from (b) radiating outwards from the cell wall (W)/glycocalyx (G). Bar marker 0.2 μm.

surface of a bacterial cell. Fimbriae may be of one or more types (Levine *et al.*, 1984; Leung *et al.*, 1989). Their diameters vary from 1 to 12 nm, and they may be 0.1–2 μm in length. Some appear as relatively straight rods, while others have a flexible or wiry fibrillar morphology (**Fig. 5**). Conjugative pili are longer than fimbriae and are composed mainly of a protein (pilin, 17–25 kDa) organised into a helical, tube-like structure. During conjugation genetic material passes along the hydrophilic centre or lumen of the pilus (Silverman, 1997). Fimbriae have a similar structure but, because they are involved in cellular adhesion, the presence of strain-specific protein subunits confers a variety of agglutination properties. Recently, fimbriae have been observed to be associated with membrane blebbing in *Prevotella intermedia*, a bacterium found in dental infections (**Fig. 6**). When bacteria are grown under conditions of low haemin concentration, there is a tendency for them to haemagglutinate (Stubbs *et al.*, 1999).

Glycocalyx

The glycocalyx is polysaccharide and protein film that surrounds bacterial cells. It may be found surrounding individual cells or colonies, and it forms part of the complex organisation of bacterial biofilms. The glycocalyx can be divided into two main types: capsules and S-layers.

Capsules

Capsules are well-defined outer layers of uniform diameter on bacterial surfaces (**Fig. 7**). This is in contrast to slime layers, which extend outwards from bacterial surfaces in a loosely organised network of material. Capsular layers are mainly composed of a material made up of an arrangement of polysaccharide chains that form a hydrated 'polysaccharide gel' on the cell surface. The thickness of this material depends upon genetic and environmental factors. The capsule allows for the diffusion of small molecules, but can exclude bacteriophages, protect against phagocytosis or binding of host-mediated antibodies, and in some strains encapsulation may be associated with virulence. Capsules are considered in detail in Chapter 4.

S-layers

S-layers are found in many types of bacteria, including Gram-positive and Gram-negative bacteria, cyanobacteria, and archaebacteria. They are composed of crystalline arrays of protein or glycoprotein subunits (**Fig. 8**; Sleytr *et al.*, 1996). In Gram-positive organisms the S-layer lies on the outer surface of the peptidoglycan layer, while in Gram-negative bacteria it is present on the outer surface of the outer membrane. It is probable that the main functions of the S-layer are to protect the bacterial cell from hostile external environments, and to act as a permeability barrier to macromolecules. To achieve the selective permeability of the S-layer, pores are present within its protein subunits to facilitate the movement of various molecules.

The presence of pores in S-layers can be demonstrated by means of atomic force microscopy (AFM) which, unlike electron microscopy, does not operate

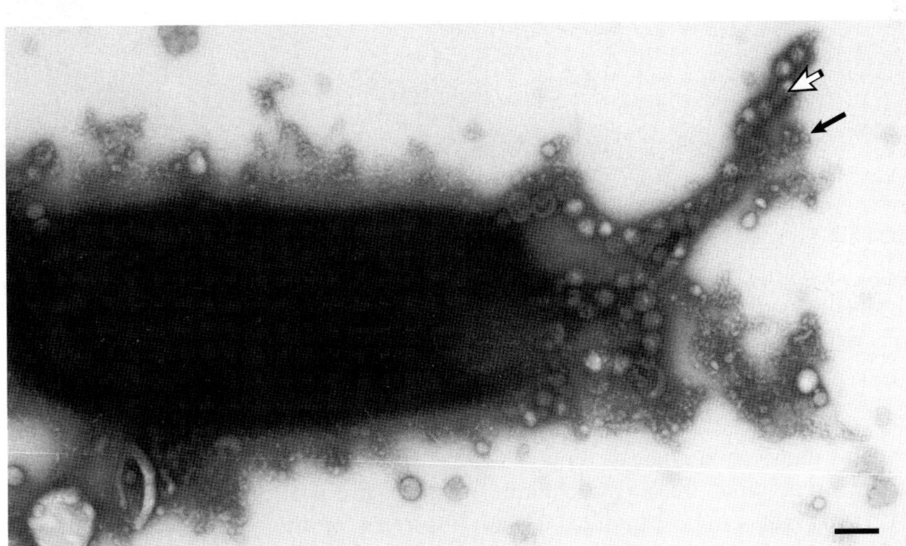

Fig. 6 Negatively stained preparation of *Prevotella intermedia* showing a chain of membrane blebs or small vesicles (open arrow) associated with fimbriae (arrow). Bar marker 0.2 μm.

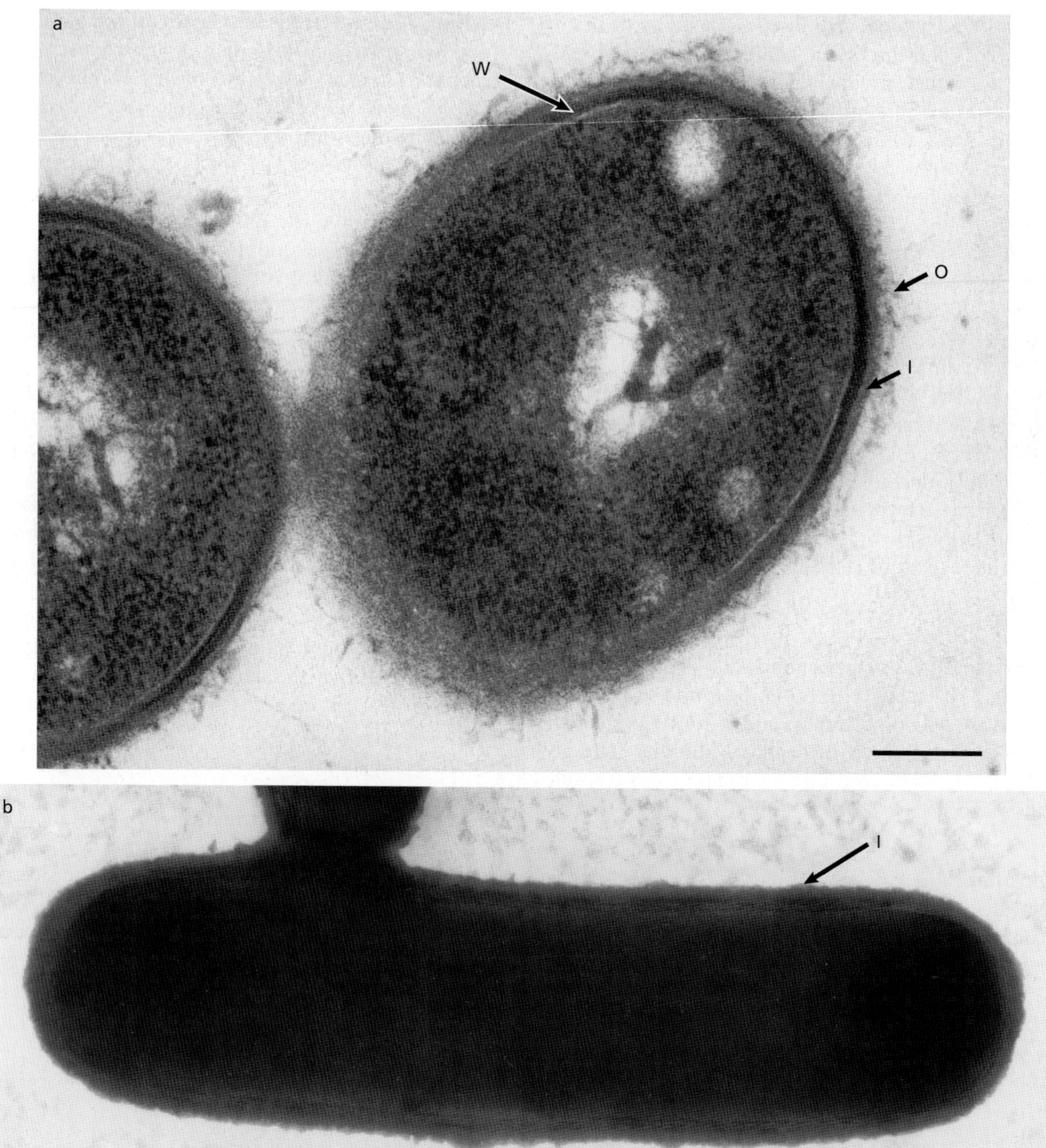

Fig. 7 Capsular layers of *Clostridium difficile*. (a) An ultrathin section showing that the cell wall (W) has a two-layered surface capsule, composed of a compact, uniform inner layer (I) and a loose, fibrillar outer layer (O). Bar marker 0.2 μm. (b) Negative staining of whole bacteria only details the inner capsular layer (I). Bar marker 0.5 μm.

under a vacuum. This allows physiological events to be monitored and by using rapid freezing techniques (Echlin, 1992) certain events can be 'captured'. Using AFM the surfaces of biological structures can be viewed directly, not under a vacuum, in physiologically buffered solutions. The resolution achievable with AFM is very high; for example it can reveal the pattern of protein subunits of the S-layer on the surface of the radiotolerant bacterium *Deinococcus radiodurans* (Fig. 9; Müller *et al.*, 1996). By observing the inner surface of the S-layer of *D. radiodurans* with AFM, Müller *et al.* (1996) were able to demonstrate

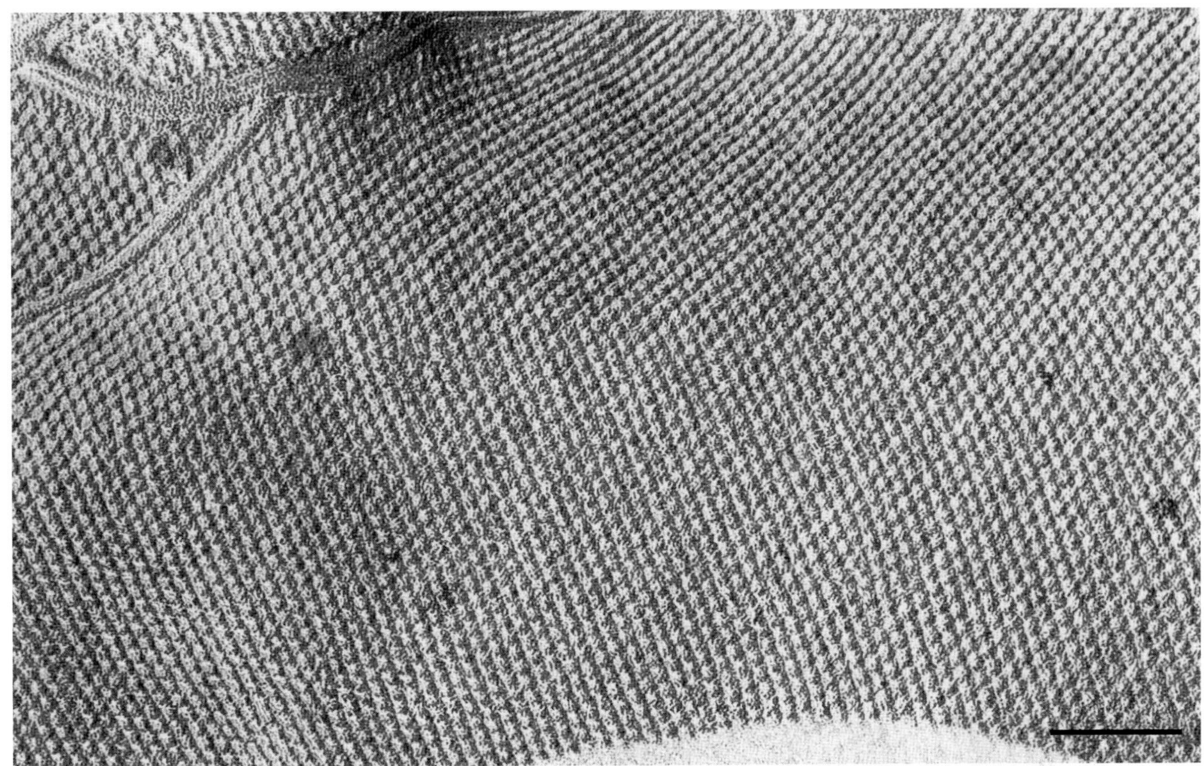

Fig. 8 Freeze-etched preparation of intact cells of *Thermoanaerobacter thermohydrosulfuricus* L110-69 showing the hexagonal S-layer lattice (Sleytr *et al.*, 1996). By courtesy of Dr Paul Messner. Bar marker 0.1 μm. In freeze-fracture, cells are rapidly frozen, fractured by a cold blade at −196°C, the exposed fractured surfaces metal shadowed and, finally, a carbon replica prepared. These replicas reveal fine structural details on wall and membrane surfaces.

the presence of pores within the protein subunits (Figs 10 and 11). These pores could be seen in 'open' and 'closed' conformations, suggesting a role in regulating the passage of macromolecules into the bacterial cell.

Cell Wall

Beneath the S-layer is the cell wall proper. It is approximately 20–40 nm in thickness, and is important for maintaining the bacterial shape. It also has to withstand an internal osmotic pressure of 5–20 atmospheres. This is made possible by the presence of a strong, flexible peptidoglycan layer (see Chapter 7). The cell wall allows for the passage of macromolecules from the external environment to the cytoplasmic membrane, and thence into the cytoplasm. If the cell wall is ruptured, the cytoplasmic contents may be lost and the bacterial cell undergoes lysis.

By light microscopy, based on the application of the Gram stain, bacteria can be divided into two groups: Gram-negative and Gram-positive. The basis of the different staining reaction of the two groups has been established by means of electron microscopy (Beveridge and Davies, 1983). In the case of Gram-positive cells the cell wall peptidoglycan is not removed by the decolorisation step of the Gram-staining procedure, so the cytoplasm retains the crystal violet stain. In the case of Gram-negative organisms the outer membrane is lost and the cytoplasmic membrane becomes permeable, allowing the loss of stain from the cytoplasm. Counter-staining with carbol fuchsin is necessary to observe such Gram-negative bacteria under the light microscope (Beveridge and Davies, 1983).

By the application of new preparative methods for the electron microscope (Hobot *et al.*, 1984; Hobot, 1990), the ultrastructural differences between Gram-negative and Gram-positive cell walls can be studied (Fig. 12). With an understanding of the chemical structure of the cell wall (see below), a more dynamic representation of the organisation of wall polymers within both types can be reconstructed (Fig. 13).

Conventional preparation methods for electron microscopy that involve complete specimen dehydration, before embedding in an epoxy resin, are carried

Fig. 9 High-resolution atomic force microscope scan of the outer surface of the S-layer (hexagonally packed layer, HPI) of *Deinococcus radiodurans*. (a, c) The scan of the HPI layer shows surface topography. Bar marker 10 nm = 0.01 μm. (b, d) Computer-enhanced images of the HPI layer showing that the protein subunits are composed of a central core surrounded by a pattern of 6-fold symmetry. Bar marker 5 nm = 0.005 μm. From Müller *et al.* (1996) by courtesy of the American Society for Microbiology.

out at room temperature and may introduce structural artefacts. Improvements in specimen preparation techniques for electron microscopy overcome these problems. These have emphasised procedures that employ partial dehydration at room temperature or low-temperature processing (below −30°C) into acrylic resins, so that ultrathin sections of bacterial cells can be obtained (Hobot and Newman, 1990; Hobot, 1991). These methods make it necessary to stabilise the biological structure by an initial chemical fixation step, usually with glutaraldehyde. Where chemical fixation may damage delicate cellular structures, such as the nucleoid and its DNA (see below), this can be avoided by using rapid freezing (a physical immobilisation of the structure in ice) (Hobot, 1991; Hobot *et al.*, 1985). The ice is then replaced by acetone at low temperatures (−80°C), followed by processing into an acrylic resin. The procedure is known as 'cryosubstitution'.

Gram-negative Cell Walls

The Gram-negative cell wall consists of an outer membrane overlying an inner peptidoglycan layer (Figs 12a,c and 13).

Outer Membrane

The outer membrane has several important functions. It acts as a permeability barrier to solutes, it hinders the entry of some antibiotics and protects the cell wall peptidoglycan from lysozyme, which can degrade peptidoglycan, leading to cell lysis. The outer membrane is a highly organised structure with an asymmetrical lipid bilayer. One of the major components of the outer leaflet is lipopolysaccharide (LPS) together with a glycolipid. The latter, termed lipid A, anchors the LPS to the outer membrane. The LPS consists of polysaccharide and oligosaccharide chains that are antigenic and characteristic for particular bacterial

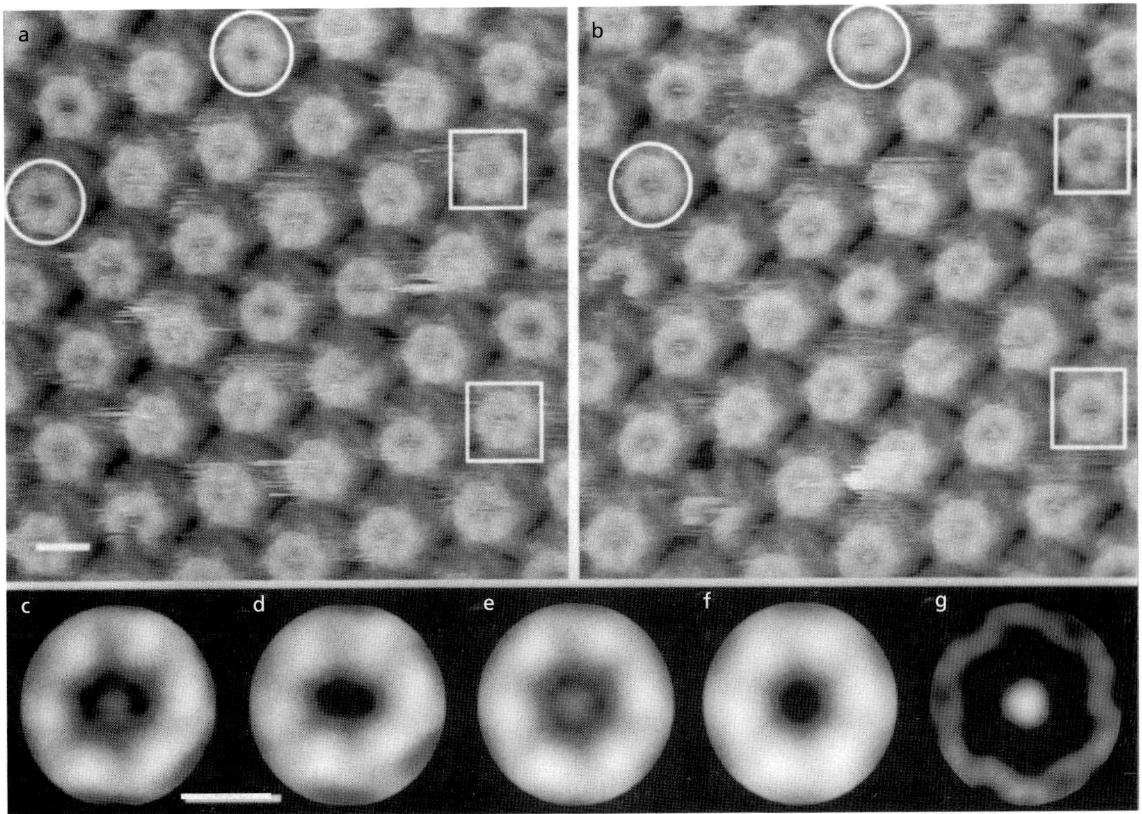

Fig. 10 High-resolution atomic force microscope scan of the inner surface of the S-layer (hexagonally packed layer, HPI) of *Deinococcus radiodurans* showing conformational changes to the HPI layer. (a) Some pores are in the 'open' conformation (circles), others are 'closed' (squares). When scanned by AFM 5 minutes later (b), previously 'open' pores are now 'closed' (circles) and those once 'closed' are now 'open' (squares). Bar marker 10 nm = 0.01 μm. Details of the position of the patterned 6-fold symmetry of the protein subunits are shown in (c–g). Bar marker 6 nm = 0.006 μm. From Müller *et al.* (1996) by courtesy of the American Society for Microbiology.

Fig. 11 (a) High-resolution atomic force microscope scan of the inner surface of the S-layer (hexagonally packed layer, HPI) of *Deinococcus radiodurans*. Bar marker 50 nm = 0.05 μm. (b) Computer-enhanced image of the inner surface of the HPI layer showing pores in an 'open' (central core appears empty) or 'closed' position (filled central core). Bar marker 10 nm = 0.01 μm. From Müller *et al.* (1996) by courtesy of the American Society for Microbiology.

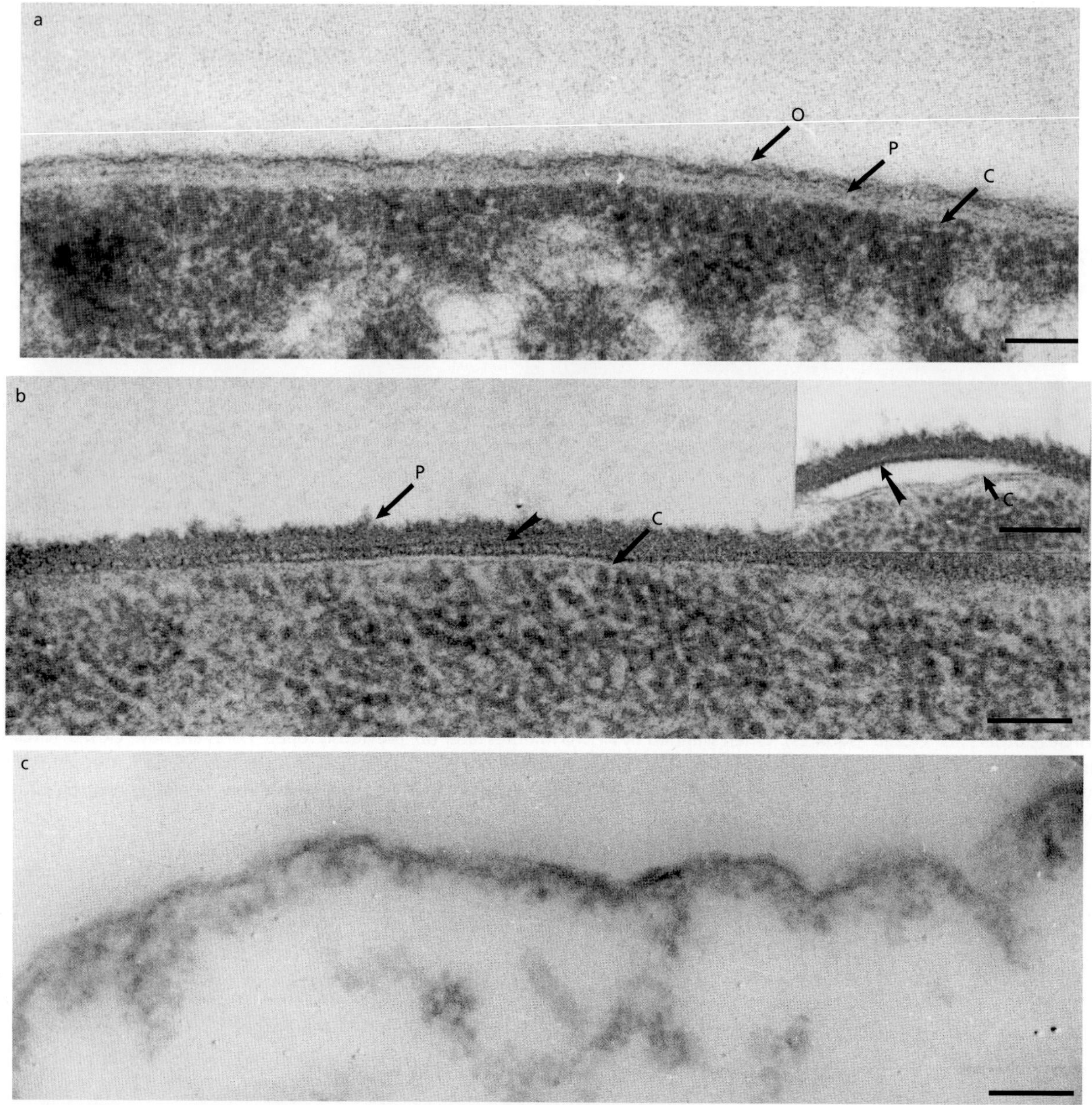

Fig. 12 Ultrathin sections of bacterial cell walls. (a) Gram-negative cell wall of *Escherichia coli*. Between the outer membrane (O) and cytoplasmic membrane (C) lies the periplasm or periplasmic gel (P) of uniform thickness which contains the peptidoglycan and periplasmic proteins. Bar marker 0.1 μm. (b) Gram-positive cell wall of *Bacillus subtilis*. There is no outer membrane but only an external cell wall layer of peptidoglycan and anionic polymers (P) that overly the cytoplasmic membrane (C). There is a thin, electron-dense layer (arrow) with a definite structural pattern immediately adjacent to the outer leaflet of the cytoplasmic membrane, but is part of the inner surface of the cell wall (inset – arrow), possibly the anchorage site of lipoteichoic acid to the cytoplasmic membrane. Bar marker 0.1 μm. (c) Isolated peptidoglycan gel (i.e. the sacculus) from *E. coli* observed not as a thin, single-layered entity, but as a diffuse, hydrated structure. Bar marker 0.1 μm.

species. They are recognised by immunological tests. Lipid A is an endotoxin that may give rise to fever and serious illness in humans and animals.

Other important components of the outer leaflet of the outer membrane are a group of proteins referred to as 'porins' (Jap and Walian, 1996). These span the outer membrane from outer leaflet to inner leaflet. They have a hydrophobic region, which allows integration of the porin as part of the hydrophobic region of the membrane. Porins also possess an internal hydrophilic channel or pore that permits the passage of small macromolecules, such as glucose or maltose

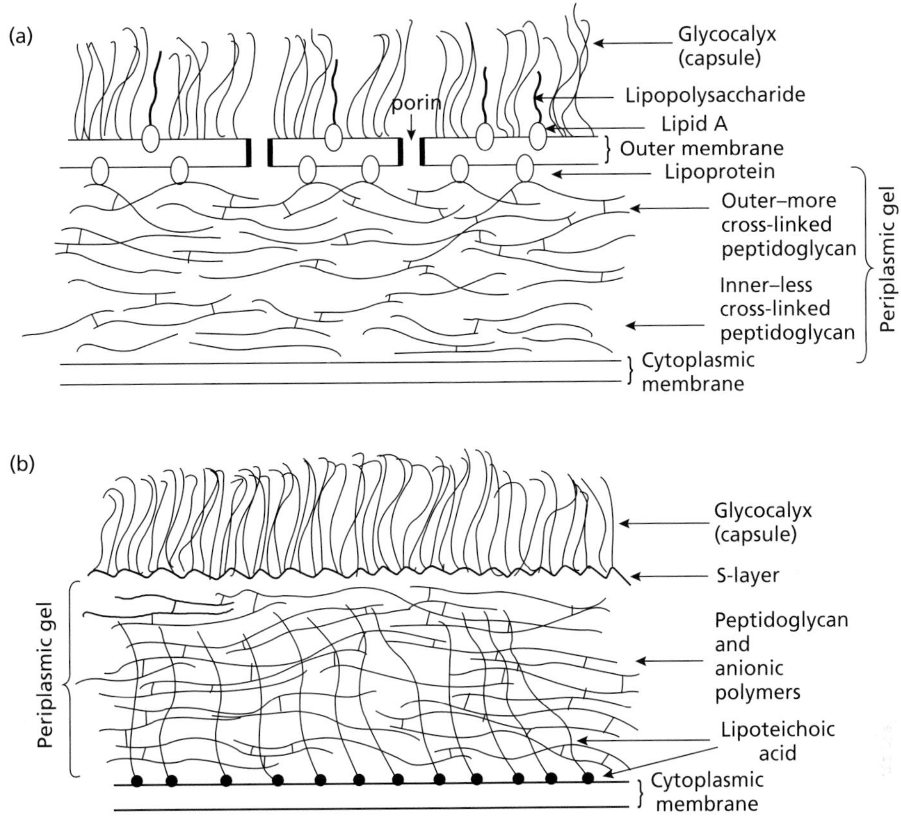

Fig. 13 Diagrammatic representation of (a) Gram-negative and (b) Gram-positive bacterial cell walls.

amongst others, from the external environment to the cell interior. These porins can open or close their pores under cellular control. They account for the permeability properties of the outer membrane (cf. the hexagonally packed layer (HPI) of *Deinococcus radiodurans*, Figs 9–11). The inner leaflet of the outer membrane is made up mainly of phospholipids similar in composition to the lipids found in most cellular cytoplasmic membranes. Lipoproteins are located here and play a role in the attachment of peptidoglycan to the outer membrane (Braun, 1975).

Periplasm and Peptidoglycan Gel

The region between the outer and inner membranes is commonly referred to as the periplasm (Figs 12a and 13), and is a distinct area that contains proteins and enzymes, implying both functionality and dynamism (Graham *et al.*, 1991). The application of improved low-temperature microscope preparation techniques has dramatically altered the structural concept of the periplasm and initiated new studies into its nature (Hobot *et al.*, 1984; Dijkstra and Keck, 1996). The periplasm contains the peptidoglycan responsible for maintaining the cell shape. In its isolated form, this peptidoglycan has been termed the 'sacculus' (Weidel *et al.*, 1960). It is composed of a sugar backbone of repeating units of *N*-acetylglucosamine and *N*-acetylmuramic acid. The sugar backbone is joined by peptide cross-links in which two short chains of amino acids (alanine, glutamic acid, lysine) are cross-linked by a diaminopimelic acid residue. The length of these peptide side-chains may vary depending on the degree of hydration of the peptidoglycan (Braun *et al.*, 1973; Oldmixon *et al.*, 1974), such that the peptidoglycan can fill the space between the outer and inner membranes. The peptidoglycan therefore forms a hydrated gel within the periplasm (Fig. 12c; Hobot *et al.*, 1984). The degree of cross-linking between sugar chains confers structural integrity to the peptidoglycan and can affect its permeability properties by varying the pore size of the gel, thus affecting the size of molecules permitted to diffuse across the cell wall. Specific cell wall-spanning systems are probably also involved in the permeability properties ascribed to peptidoglycan (Dijkstra and Keck, 1996). These may be associated with the porins of the outer membrane.

In order to prevent the osmotic pressure of the cytoplasm from bursting the cell, in addition to

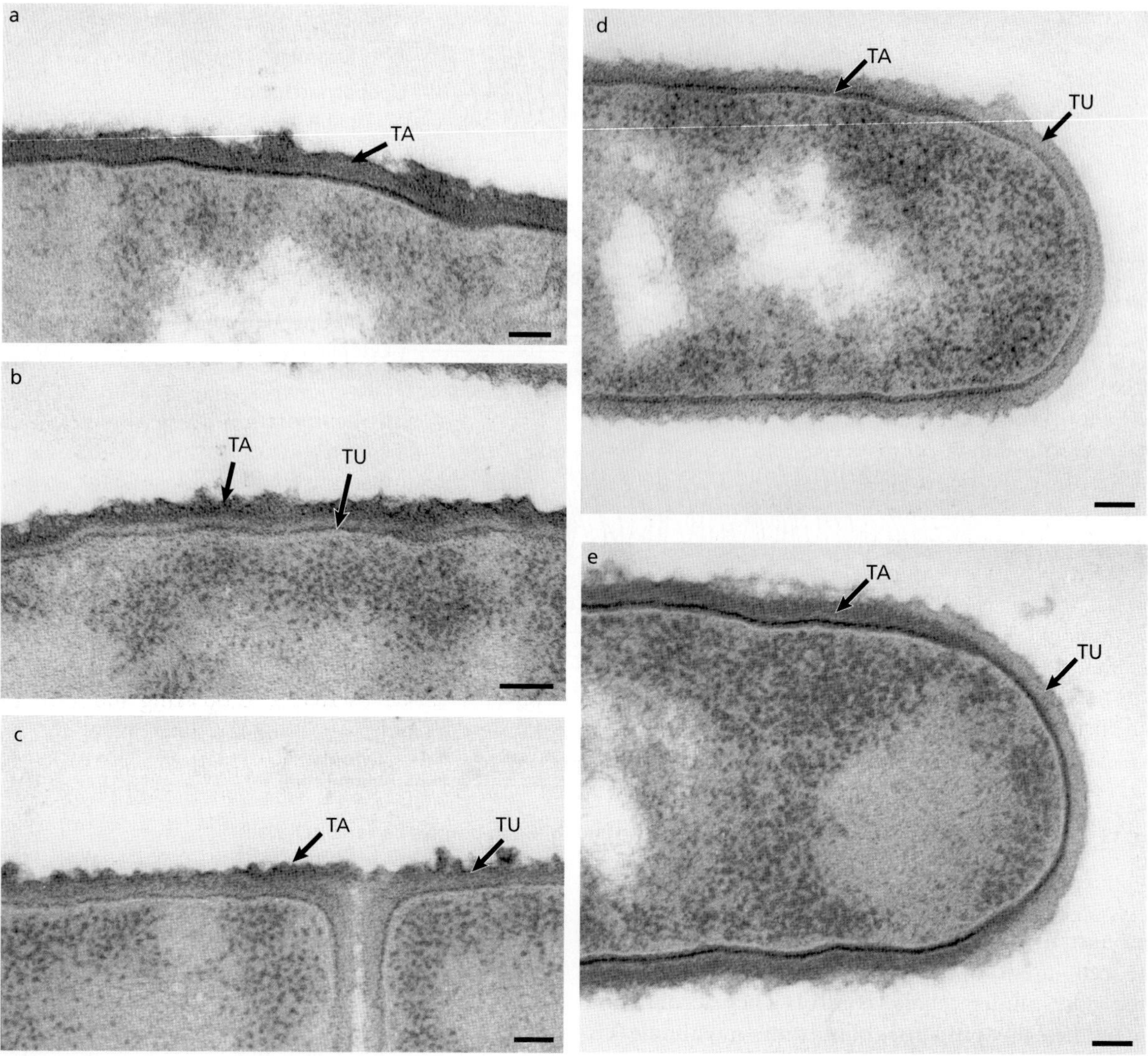

Fig. 14 Ultrathin sections that show incorporation of newly synthesised wall polymers into the cell wall of *Bacillus subtilis*. Bacteria grown in magnesium-limiting cultures have a darkly stained cell wall containing teichoic acid (a – TA, arrow). As growth conditions gradually change to phosphate-limiting concentrations, new wall material that contains teichuronic acid is incorporated along the length of the cylindrical wall as a lightly stained band of material (b – TU, arrow), and eventually replaces the old teichoic acid wall (c). During growth in transition from phosphate-limiting cultures to magnesium-limiting conditions, teichuronic acid is conserved in the polar region, while teichoic acid has been incorporated fully in the cylindrical portion of the cell wall (d and e). Bar marker 0.2 μm.

the peptidoglycan, the cell wall contains a class of macromolecules termed 'membrane-derived oligosaccharides' (MDOs) (Schulman and Kennedy, 1979).

Together, these various components of the periplasm form a layer between outer and inner membrane termed the 'periplasmic gel' (**Fig. 13**; Hobot *et al.*, 1984). The outer part of the periplasmic gel contains a more cross-linked peptidoglycan attached by lipoprotein to the outer membrane. This gives rise to the mechanical integrity of the cell wall (i.e. the sacculus), with an inner, less cross-linked peptidoglycan, which provides a continuous structural link between the outer and inner membranes in the context of a hydrated periplasmic gel (**Fig. 13**).

Gram-positive Cell Walls

The Gram-positive cell wall, unlike its Gram-negative equivalent, lacks an outer membrane but has a larger peptidoglycan layer (**Figs 12b** and **13**).

Peptidoglycan Layer and Periplasmic Gel

The peptidoglycan layer of Gram-positive bacteria is more cross-linked than that of Gram-negative bacteria, but it has a high water content similar to that of Gram-negative peptidoglycan (Gerhardt and Judge, 1964; Humpries *et al.*, 1981). This accounts for the molecular sieving properties attributed to the cell wall layer and its role as a possible permeability barrier for the bacterial cell (Hancock, 1991). It may, therefore, be suggested that the peptidoglycan in Gram-positive cell walls has the organisation of an expanded gel network (Humpries *et al.*, 1981), i.e. that it is a hydrated peptidoglycan gel similar to that found in Gram-negative bacteria.

Teichoic acid, an anionic polysaccharide, is a major water-soluble polymer that accounts for up to 50% of the Gram-positive cell wall. Although it is most widespread among Gram-positive bacteria, in some bacterial species it is replaced by teichuronic acid. It is linked by phosphodiester bridges to the *N*-acetyl-muramic acid residue of the peptidoglycan sugar backbone. Another form of teichoic acid, lipoteichoic acid, is covalently linked to the glycolipid of the cytoplasmic membrane, providing a possible anchor for the cell wall polymers to the bacterial cell membrane (**Figs 12b** and **13**; Hobot, 1990). The teichoic acids are considered in greater detail in Chapter 5.

The cell wall of Gram-negative bacteria is often called the periplasm, because in addition to its cell wall polymers it contains a group of proteins and enzymes (see above). Recently, however, it has been postulated that the Gram-positive cell wall may also function as a periplasm, as it too contains a definite protein fraction that possesses properties similar to those of a periplasm (Merchante *et al.*, 1995). Therefore, since cell walls of Gram-positive and Gram-negative bacteria contain hydrophilic proteins and enzymes together with hydrated peptidoglycan and polysaccharide components, their cell walls may be conceived of as being organised like a periplasmic gel.

Cell Wall Growth

The incorporation of new cell wall material can be observed directly by electron microscopy. Teichoic acid polymers are rich in phosphate residues that readily react with uranyl acetate, a heavy metal stain that provides contrast to biological material in the electron microscope. Teichuronic acid polymers, on the other hand, contain few phosphate residues and consequently react weakly with uranyl acetate. When growing *Bacillus subtilis* cells under continuous culture conditions in a fermenter it is possible, by altering the growth conditions, to influence the proportion of teichoic acid or teichuronic acid polymers produced. By means of modern electron microscopy preparative methods, including improved methods for embedding biological specimens into acrylic resins at room temperature, the incorporation of newly synthesised teichoic or teichuronic acid-containing polymers into the existing cell wall can be monitored directly by staining the cells at various time intervals with uranyl acetate (Hobot and Newman, 1990). In this way Merad *et al.*, 1989 demonstrated uniform 'inside-to-outside' incorporation of wall polymers along the length of the cylindrical bacterial cell wall (**Fig. 14a–c**). Growth of the cell wall at the poles was slower, however, than that along the sides of the bacterial cylinder (**Fig. 14d–e**). This probably reflects the effect of growth rate on the rate of new wall incorporation along the cylindrical sides of the bacteria, but not on growth in polar or possibly septal regions (Clarke-Sturman *et al.*, 1989).

The incorporation of new cell wall material into existing wall requires wall polymer synthesis. This is initiated in the cytoplasm and is associated with the cytoplasmic membrane, across which wall material is transferred into the periplasmic gel. Linkages in the indigenous wall polymers must be broken so that new polymers can be inserted. The breaking and insertion of new wall polymers is under the control of lytic enzymes such as amidases that are present in the cell wall (**Fig. 15**). Their localisation can be demonstrated by electron microscopy coupled with immunocytochemical techniques, so-called immunoelectron microscopy (**Fig. 16**; Hobot and Newman, 1996; Newman and Hobot, 2001). For example, antibodies are raised against particular proteins or enzymes, or even nucleic acids, and these can subsequently be made to react with the complementary antigen present on the surface of resin sections. Similarly, lectins can be used to localise carbohydrate moieties (Roth, 1983). The precise location of cell surface structures can be visualised under the electron microscope with the relevant gold-tagged antibodies or lectins (**Fig. 16**). Immunocytochemistry allows microscopy not only to reveal biological structure, but also to relate structure to functional aspects of molecular and cellular biochemistry, physiology, and genetics.

Cytoplasmic Membrane

The cell wall is separated from the underlying cytoplasm by a cytoplasmic membrane that is 7–8 nm thick. It is structurally organised as a lipid bilayer (unit membrane) and acts as a permeability barrier for the bacterial cell (**Fig. 17**). The cytoplasmic membrane allows the passive diffusion of water and small solutes, but it is also involved in active molecular transport. Active transport systems in the cytoplasmic

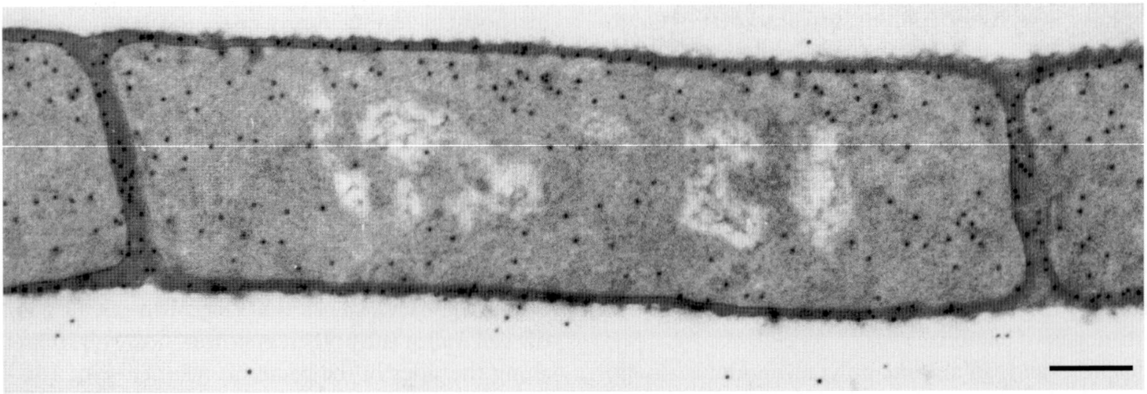

Fig. 15 Immunocytochemical localisation of *N*-acetylmuramyl-L-alanine amidase in *Bacillus subtilis*. (For details of the immunocytochemical method see Fig. 16). The ultrathin resin sections of *B. subtilis* were incubated with a polyclonal rabbit antibody raised against the amidase and its location visualised by 10 nm colloidal gold particles tagged with goat anti-rabbit antibody. The autolytic enzyme amidase is present within the periplasmic gel of the cell wall and septa. Sites of enzyme synthesis in the cytoplasm are also seen (Hobot and Rogers, 1991). Bar marker 0.2 μm.

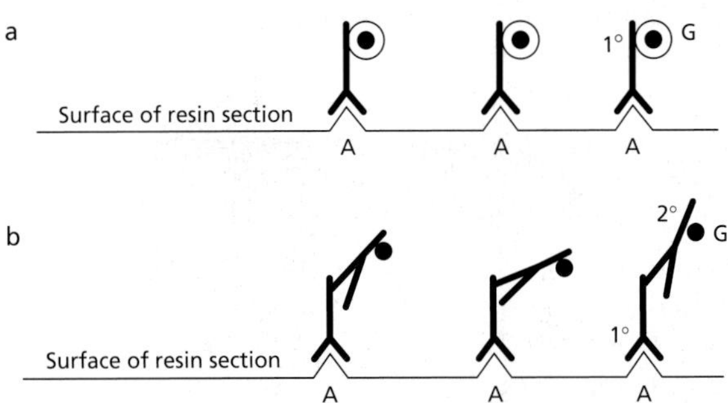

Fig. 16 Diagrammatic representation of the immunocytochemical localisation of specific antibodies on the surfaces of resin sections by means of immunocolloidal gold particles (Newman and Hobot, 2001). (a) Primary antibody (1°) recognises its complementary antigen (A; protein/enzyme) on the surface of the resin section. A colloidal gold (G), coated with protein-A from *Staphylococcus aureus*, which specifically binds to the Fc portion of the antibody, is added to visualise the antibody by electron microscopy. (b) Primary antibody (1°) is visualised by the subsequent attachment of a species-specific secondary antibody (2°) that is already tagged with a colloidal gold particle.

membrane, such as protein pumps, maintain and regulate the internal ionic concentration and osmotic pressure of the bacterial cell. Active uptake of nutrients can be regulated by transport systems located within the cytoplasmic membrane that are made up of several enzymes termed permeases. These systems also provide for the removal of waste material from the bacterial cell and the release of extracellular enzymes and other substances. The proteins and enzymes of the respiratory system can sometimes be located on the cytoplasmic membrane, which corresponds to the membrane invaginations of the mitochondria (cristae). In photosynthetic bacteria invaginations of the cytoplasmic membrane give rise to complex membrane systems.

Another type of membrane invagination that has been observed in bacteria is the mesosome. Using rapid freezing techniques and cryosubstitution to prepare ultrathin sections or by viewing the frozen sections of bacteria directly, it has been demonstrated conclusively that mesosomes are artefacts of conventional preparative methods. As such they are not associated with cytoplasmic membrane functions nor with DNA in cell division (Ebersold *et al.*, 1981; Dubochet *et al.*, 1983; Hobot *et al.*, 1985).

When the medium in which bacteria are suspended has a higher osmotic pressure than that within the cell, the cytoplasmic membrane and the cytoplasm shrink away from the cell envelope, a phenomenon known as plasmolysis. If rapid freezing techniques are applied to such plasmolysed bacterial cells it can be seen that there are no regions of direct contact, such as adhesion sites or bridges, between the outer membrane and the cytoplasmic membrane (**Fig. 18**; Hobot *et al.*, 1984). This indicates that the periplasmic gel is responsible for the functional co-ordination between the outer and inner (cytoplasmic) membranes (Kellenberger, 1990; Hobot, 1991).

Nucleoid

The site in the bacterial cytoplasm that contains the nuclear material or prokaryotic DNA is termed the nucleoid. This is not a nucleus because there is no nuclear membrane, as is found in eukaryotes. Preservation of prokaryotic DNA can be achieved by rapid

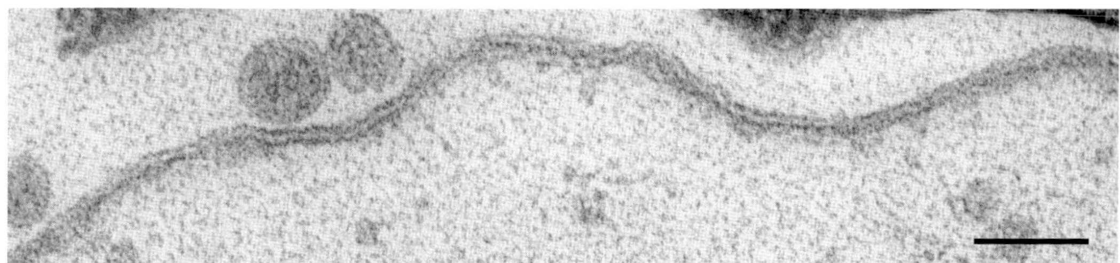

Fig. 17 Ultrathin section of cytoplasmic membrane from *Bacillus subtilis* showing a double-track ('tramline') appearance. This is caused by the hydrophilic reagents and stains used in electron microscopy that react with the hydrophilic groups of the outer and inner leaflet and endow them with electron contrast; they do not react with the hydrophobic centre of the membrane. Bar marker 0.1 μm.

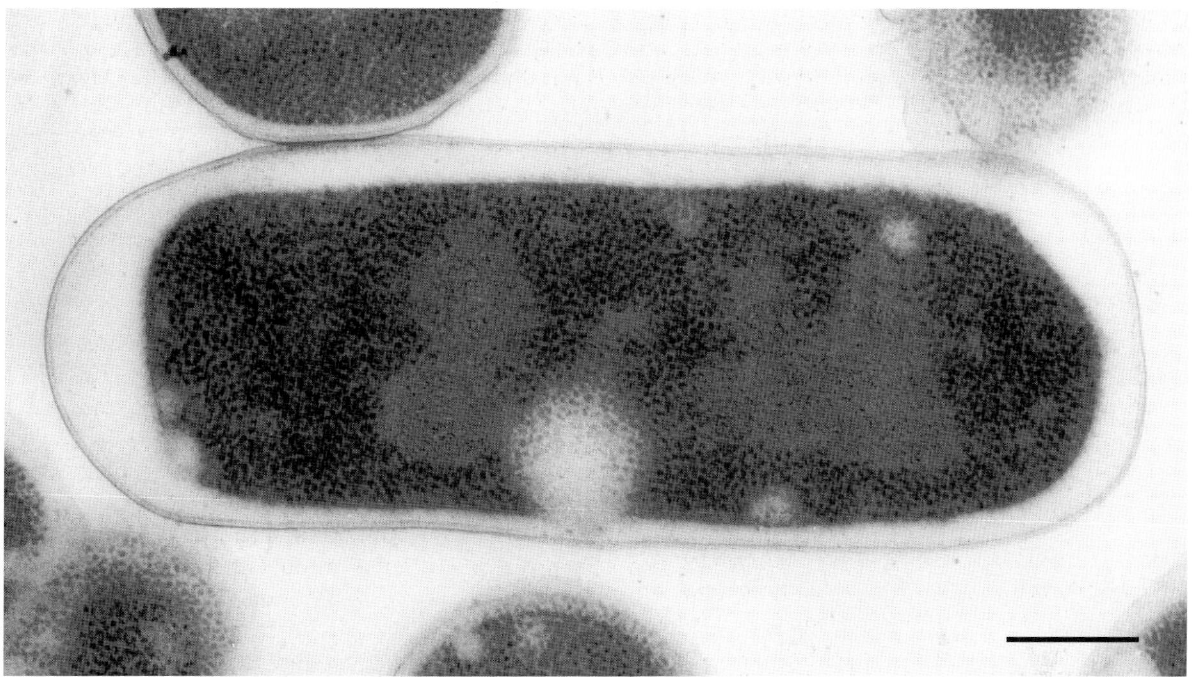

Fig. 18 A plasmolysed cell of *Escherichia coli*, prepared by rapid freezing techniques, cryosubstitution and ultrathin sectioning (for methods see Hobot, 1991). The clear separation between cell wall and cytoplasmic membrane/cytoplasm is shown. Plasmolysis causes the cell wall to separate from the cytoplasmic membrane along the periplasmic gel between outer, more cross-linked peptidoglycan and inner, less cross-linked peptidoglycan (cf. Fig. 13). No adhesion sites/bridges are present. Bar marker 0.2 μm.

freezing and cryosubstitution. This shows the nucleoid as an area in the cytoplasm free of ribosomes, with numerous small clefts that protrude into the cytoplasmic regions (**Fig. 19a**; Hobot *et al.*, 1985). These give the nucleoid a coralline appearance in three-dimensional reconstructions (Bohrmann *et al.*, 1991). The nucleoid contains the single, closed circular bacterial chromosome, seen as very fine strands of DNA-plasm (**Fig. 19b**). The DNA in a typical cell, which contains about 3 million base pairs, is about 1 mm in length. To package this amount of DNA into the cytoplasm, the chromosome is organised into 40–50 supercoiled loops or domains of DNA. The bulk of this is resting double-stranded DNA (**Fig. 20b**), with the active uncoiled, single-stranded DNA (ssDNA) present only at the periphery of the nucleoid (**Fig. 20a**; Hobot *et al.*, 1987).

By means of immunocytochemical electron microscopy techniques, which are more precise and give far higher resolution of localisation than fluorescent labels and light or confocal laser scanning microscopy, a histone-like protein (HU), RNA polymerase, and DNA topoisomerase I have been located to the peripheral region of single-stranded DNA, where topoisomerase I is responsible for relaxing supercoiled DNA (Bjornsti *et al.*, 1986; Dürrenberger *et al.*, 1988). At

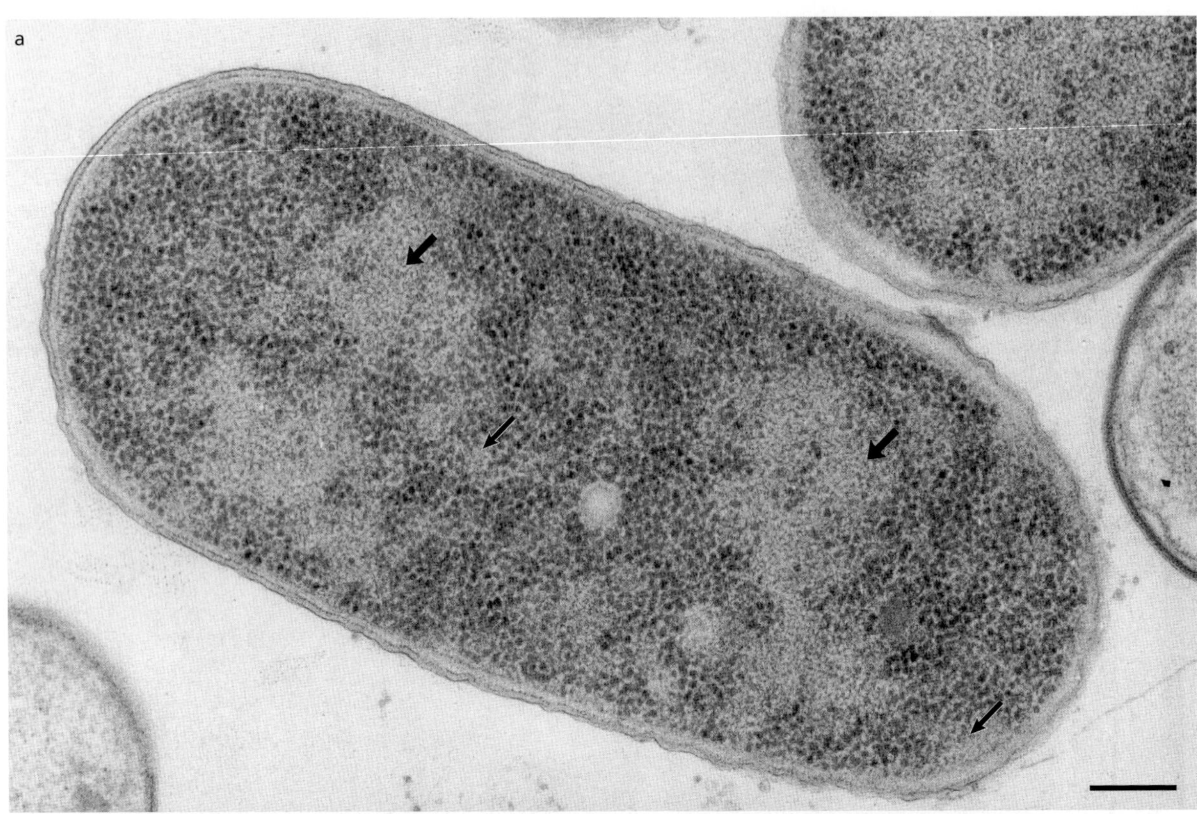

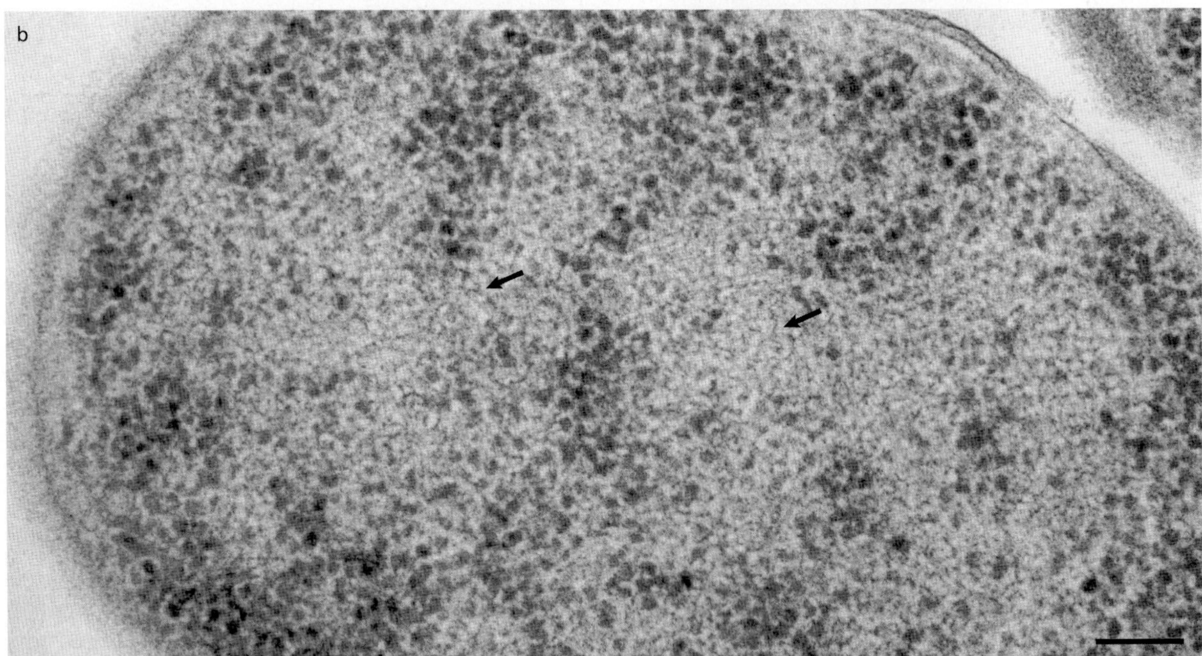

Fig. 19 (a) The nucleoid of *Escherichia coli*, prepared by rapid freezing techniques, cryosubstitution and ultrathin sectioning, is seen as a ribosome-free area within the cytoplasm (large arrows) with cleft protrusions into the cytoplasmic regions (small arrows). Bar marker 0.2 μm. (b) At higher magnifications the DNA-plasma consists of very fine strands (arrows) without the presence of large structures that resemble eukaryotic nucleosomes. Bar marker 0.1 μm.

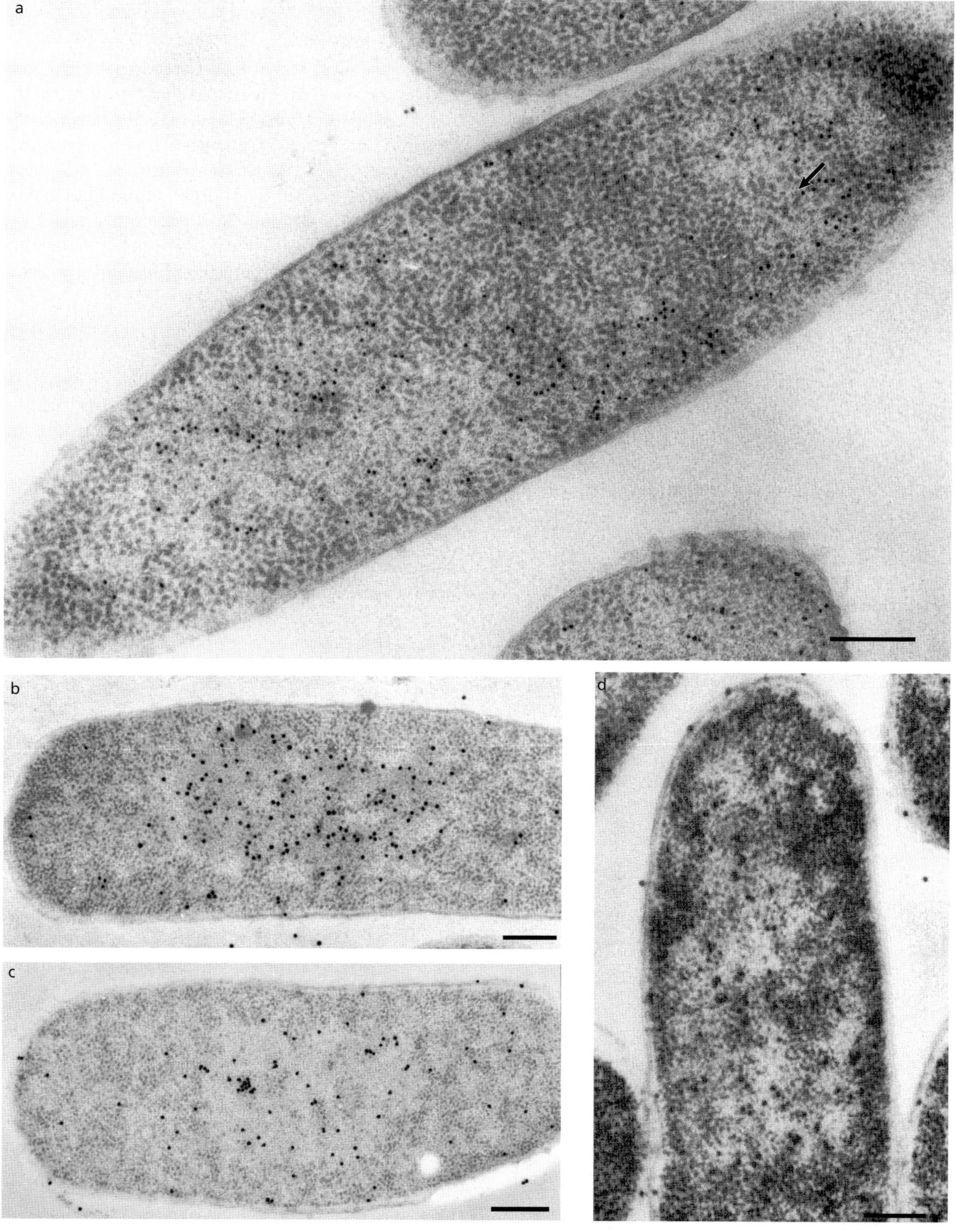

Fig. 20 Immunocytochemical localisation of DNA and associated proteins/enzymes in ultrathin sections of *Escherichia coli* by immunocolloidal gold technology. (a) Single-stranded DNA (ssDNA). Mouse monoclonal antibodies against ssDNA are localised to the periphery of the nucleoid (ribosome-free areas) by 10 nm colloidal gold particles (arrow) tagged with goat anti-mouse antibodies. (b) Gold label for double-stranded DNA is visualised throughout the nucleoid. (c) Gold label for RNA polymerase is localised to the peripheral regions of the nucleoid, as is gold label for HU, a histone-like protein (d). Bar marker 0.2 μm.

the periphery of the nucleoid, single-stranded DNA is associated with histone-like proteins (**Fig. 20d**), and probably forms a stable transcription–translation complex with RNA polymerase and ribosomes (**Fig. 20c**).

Many bacteria carry one or more pieces of DNA called plasmids. Although not essential for the viability, growth, and development of bacteria, plasmids can influence the response of the bacterium to its external environment by allowing the expression of useful genes carried by the plasmid. These may be genes for antibiotic resistance, synthesis of hydrolytic enzymes, and factors that affect virulence. Exchange of plasmids by genetic transfer may occur during bacterial conjugation (see below).

The cytoplasm also contains another nucleic acid, RNA. Ribosomal RNA (rRNA) exists free in the bacterial cytoplasm and, as noted above, it interacts closely with messenger mRNA and ssDNA during protein synthesis. The ribosomes of bacterial cells are smaller (70S) than their counterparts in the eukaryotic cells (80S).

Cytoplasmic Inclusions

Under certain growth conditions bacteria may produce storage products that are seen as inclusion bodies in the cytoplasm. Under the electron microscope they are generally observed as electron-lucent (clear) areas (**Fig. 21**). Several types of inclusion bodies have been identified, some of which can also be seen by light microscopy after staining. Inclusion bodies are not delimited by unit membranes, but some have a non-unit membrane protein layer.

Non-membrane Inclusions

Volutin Granules

Volutin granules, sometimes termed metachromatic granules because of their colour reaction with the dyes used in light microscopy, contain polymerised inorganic phosphate, an energy-rich compound that acts as a reserve store of energy and of phosphate. In some cases, nucleic acid, protein, and lipid are associated with these granules. Under the light microscope volutin granules appear red-violet after staining with toluidine blue or methylene blue, and the cytoplasm appears blue.

Polysaccharide Granules

Polysaccharide granules contain either glycogen or starch as an energy or carbon reserve. After staining with iodine, glycogen granules appear reddish-brown, while starch granules appear blue.

Non-unit Membrane Inclusions

Carboxysomes

Carboxysomes are polyhedral inclusion bodies that contain the enzyme ribulose 1,5-diphosphate carboxylase. This is responsible for carbon dioxide fixation in cyanobacteria.

Lipid Bodies

Lipid is stored as an energy or carbon reserve in the form of poly-β-hydroxybutyric acid. Lipid bodies are highly refractile inclusions in the cytoplasm and can be stained with Sudan black, in contrast to other bacterial contents.

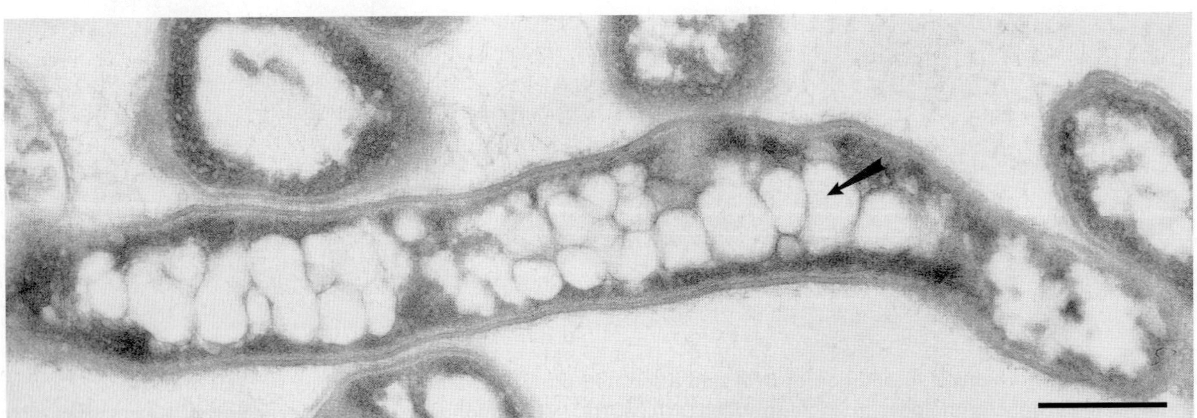

Fig. 21 Ultrathin sections reveal numerous cytoplasmic inclusion bodies (arrow) in the cytoplasm of *Streptomyces lividans* when grown under phosphate-limiting conditions. Bar marker 0.5 μm.

Gas Vesicles

Gas vesicles or vacuoles are found primarily in aquatic bacteria. They are composed of a series of hollow cylinders in the bacterial cytoplasm. An outer protein layer excludes water and ions, but allows gases to enter to provide cell buoyancy.

Photosynthetic Systems

Purple and green bacteria and cyanobacteria are able to carry out photosynthetic functions. Their photosynthetic pigments are located within stacks of membranes termed thylakoids, which are found mainly in the cyanobacteria, or in numerous small vesicles called chlorobium vesicles. While photosynthesis in cyanobacteria produces oxygen as a by-product, this is not so in the purple and green bacteria (anoxygenic photosynthesis), where sulphur plays an important role as electron donor rather than water in the cyanobacteria.

Cell Division

Bacterial cells multiply by simple division into two daughter cells, which are exact copies of the parent. Cell division occurs when the cytoplasmic membrane together with the cell wall invaginates into the cytoplasm and begins to divide the bacterial cell into two (**Fig. 22a**). The invagination elongates to divide the cell (**Fig. 22b,c**) until eventually a cross-wall or septum divides the parent cell in two (**Fig. 22d**). New cell wall material is synthesised within the developing septum, where the septal peptidoglycan may differ from that found in the cylindrical side-wall (Signoretto *et al.*, 1996). The two new daughter cells then segregate (**Fig. 22e**) and the cell cycle proceeds in each until the next stage of division is reached. Cell division is thus a stage in the cell cycle that follows development and elongation, in association with DNA replication. However, the kinetics of cell division are such that under conditions of rapid growth, the formation of dividing septa lags behind DNA replication, which may have occurred more than once, and as many as four distinct nucleoids may be present in one long cell.

Bacterial Conjugation

Transfer of genetic material occurs during the process of bacterial conjugation. During this process, DNA plasmid is transferred from one bacterium (the donor) of a mating pair into another (the recipient) via a pilus. The transfer can take about 90–150 minutes, and can be observed directly by video-enhanced microscopy (**Fig. 23**; Dürrenberger *et al.*, 1991). As one bacterium of a mating pair approaches, pilus binding takes place (**Fig. 23a**), followed by pilus retraction and stabilisation of wall-to-wall contact (**Fig. 23b**). DNA transfer follows (**Fig. 23c**) and then the cells separate (**Fig. 23d**). During wall-to-wall contact of the mating bacteria, DNA transfer takes place. In order to observe the delicate structure of the contact point by electron microscopy, Dürrenberger *et al.* (1991) used rapid freezing and cryosubstitution. Serial sections through the point of contact, termed the conjugational junction, did not reveal specific substructures, such as plasma bridges with fusion of membranes or cell wall to

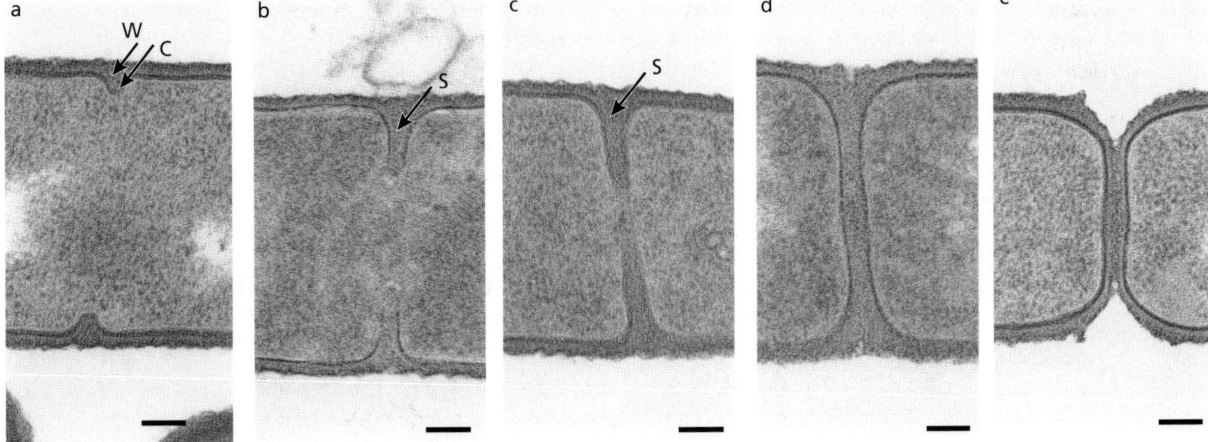

Fig. 22 Cell division in ultrathin sections of *Bacillus subtilis*. (a) The cytoplasmic membrane (C) and cell wall (W) invaginate into the cytoplasm to begin the division process. The developing septum (S) elongates (b, c) until it divides the cell in two (d). Eventually the two daughter cells begin to separate (e). Bar marker 0.1 μm.

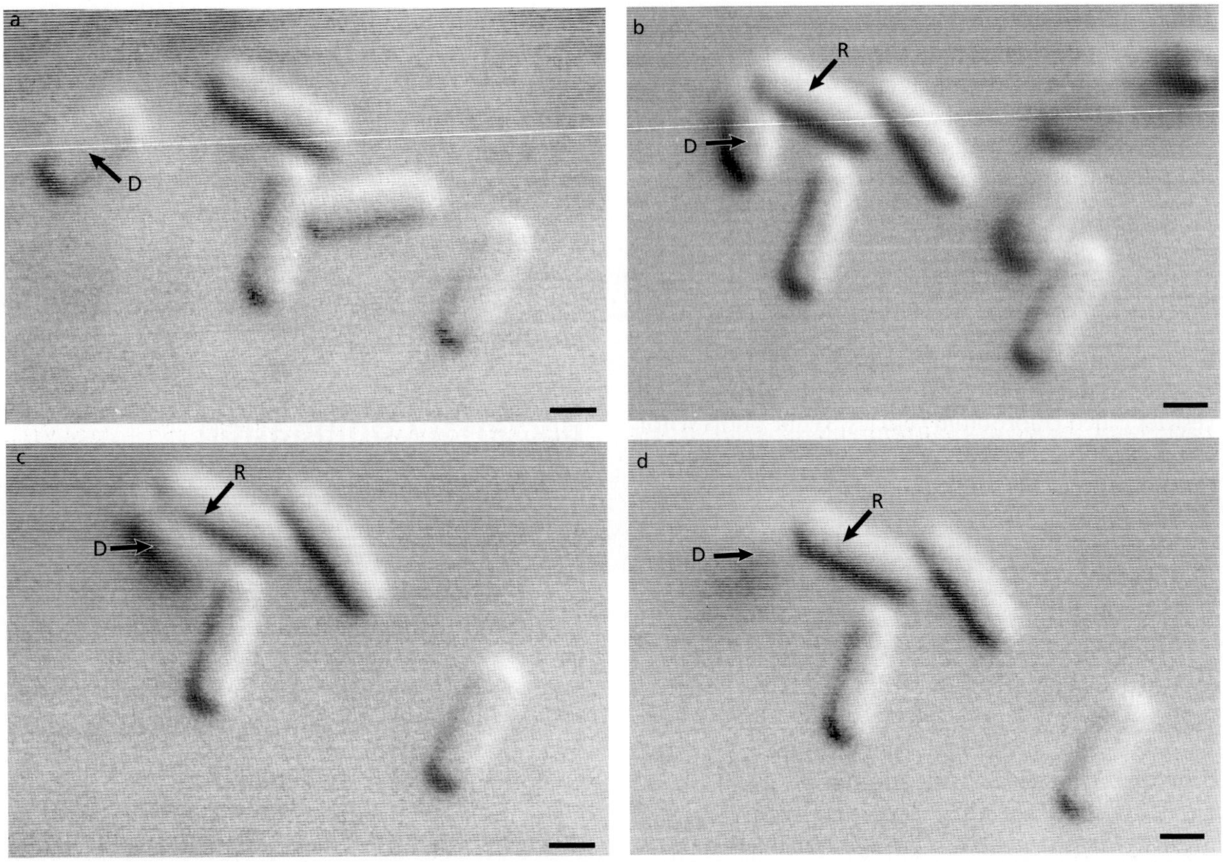

Fig. 23 Video-enhanced microscopy of bacterial conjugation in *Escherichia coli*. (a) Pilus binding occurs as the donor cell (D) approaches. (b) The wall-to-wall contact between donor and recipient (R) cells is stabilised. (c) DNA transfer takes place and (d) disaggregation occurs. Bar marker 1 μm. From Dürrenberger *et al.* (1991) by courtesy of Academic Press Inc.

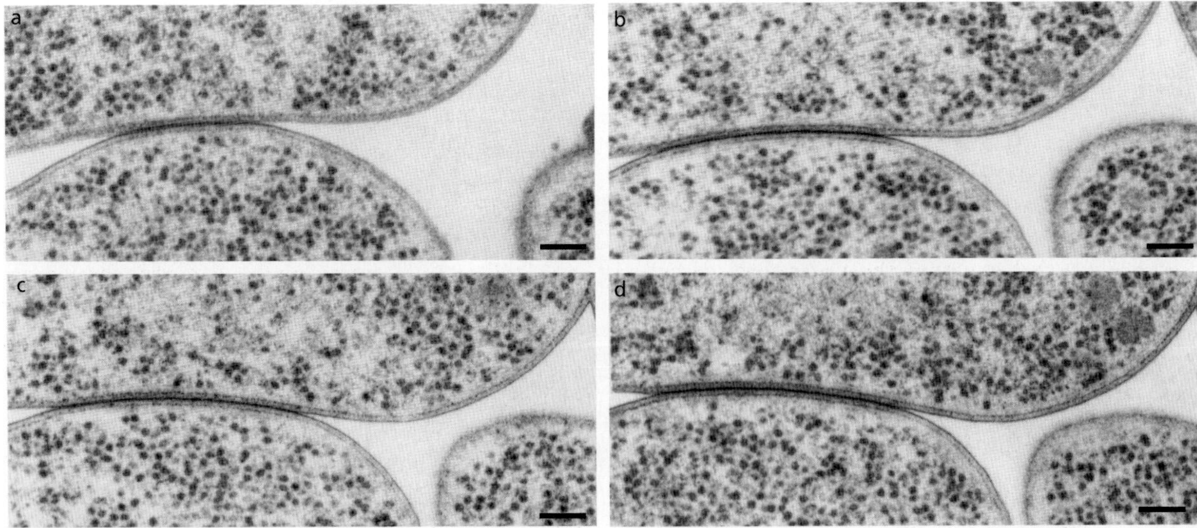

Fig. 24 Serial sections through the conjugational junction (point of wall-to-wall contact) in *Escherichia coli*. The conjugational junction appears as a continuous electron-dark line between the two mating cells. There is no evidence for membrane, cell wall or cytoplasmic fusion taking place throughout the length or depth of the conjugational junction (a–d). Bar marker 0.2 μm. From Dürrenberger *et al.* (1991) by courtesy of Academic Press Inc.

mediate DNA transfer (**Fig. 24**). The conjugational junction resembled the morphology of tight junctions in eukaryotic cells. It is likely, therefore, that the pilus initially triggers a signal for the mechanism of DNA transfer to be initiated, possibly by the formation of a continuous protein channel across the periplasmic gel of the recipient cell wall (Dürrenberger *et al.*, 1991). Subsequently, DNA is transferred to the recipient cell from the donor through the channel.

Bacterial Sporulation

Bacteria of the genera *Bacillus* and *Clostridium* can develop a dormant or resting phase, the endospore. Endospores are highly resistant, specialised structures that can withstand heat, dehydration, freezing, toxic chemicals, enzymes, and ultraviolet radiation. They can survive for very long periods.

Endospores are generally formed when there is a reduction of available nutrients (carbon, nitrogen sources) in the environment, but suitable ionic conditions, moisture, pH, and temperature are usually necessary for sporulation to proceed to completion. Spores are found only among the Gram-positive bacteria, in which a thick cell wall is present. Actively growing (vegetative) cells give rise to one spore per cell. Endospores do not show metabolic activity until conditions for germination (spore outgrowth) prevail. Sporulation consists of several distinct morphological stages (I to IX), which can be followed in the electron microscope (**Fig. 25**).

Bacterial sporulation is initiated when the cytoplasmic membrane, together with the cell wall, invaginates into the cytoplasm to start the formation of a cross-wall (**Fig. 25a** – stage I). This is generally located towards one end of the vegetative cell, and eventually cuts off a small part of the cell cytoplasm from the rest (**Fig. 25b,c** – stages II–III). The new small cell contains a single DNA chromosome, and is engulfed by the whole vegetative cell to form the forespore which lies within the vegetative cell cytoplasm (**Fig. 25d** – stage IV). The forespore is surrounded by two membranes, the outer and inner forespore membranes, and between them is a thin layer of vegetative cell wall (peptidoglycan and anionic wall polymers; **Fig. 25d** – stage IV). With the development of the forespore, the vegetative cell is committed to endospore formation. The thin cell wall layer develops in such a way that a two-layered peptidoglycan is formed. First, an inner thin layer, the germ cell wall is formed, which is destined to become the cell wall or peptidoglycan template for new vegetative cell wall synthesis and germination. Next, there is the formation of an outer layer of peptidoglycan (cortical peptidoglycan) that has been altered chemically to enable the endospore to maintain its properties of resistance (Atrih *et al.*, 1996). This layer is also known as the cortex (**Fig. 25e** – early stage V). During stage V the cortex develops and becomes a much thicker layer (**Fig. 25f** – late stage V). Concurrently, synthesis begins of the outer protein spore coat layers responsible for the resistance properties of the endospore (**Fig. 25e,f** – stage V). The outer spore coat gradually increases in thickness, coupled with further development of the cortex (**Fig. 25g** – stage VI), until eventually a complete endospore is formed within the cytoplasm of the vegetative cell (**Fig. 25h** – stage VII). The central spore cytoplasm also undergoes changes, so that it has a very low water content with only a few enzymes present and no mRNA. At stage VIII degeneration of the vegetative cell takes place with release of the mature endospore, ready to begin its dormant phase (**Fig. 25i** – stage IX). Sporulation is considered in greater detail in Chapter 10.

Bacterial Germination

Bacterial germination is the change of an endospore from its resting stage to an actively growing vegetative cell. This takes place when the external environment is favourable for growth by the provision of essential nutrients and water. Germination is divided into three stages: activation, germination, and outgrowth.

Physical effects, such as heat shock, or chemical agents, such as availability of a carbon source, ions, or minerals, may activate germination. The impermeable barriers of the spore coat of dormant spores (**Fig. 26a**) are attacked by lytic enzymes that disrupt their structure and that of the cortex and the spore cytoplasm begins to rehydrate (**Fig. 26b**). At the beginning of germination, the spore begins to swell and the germ cell wall separates from the cortex (**Fig. 26c**). The germ cell wall then thickens as new vegetative cell wall is laid down, and the spore coat layers and cortex begin to break down (**Fig. 26d**). At first mRNA is not present in the dormant or germinating spore, and it has been shown by immunoelectron microscopy that an indigenous store of cell wall lytic enzyme (*N*-acetylmuramyl-L-alanine amidase) is involved in the initial synthesis and incorporation of new cell wall polymers before new mRNA and protein synthesis takes over to complete cell outgrowth (Hobot and Newman, 1996). The cell then elongates and begins to take on a cylindrical shape. Eventually the spore coat and cortex layers rupture to allow outgrowth of the new cell to begin (**Fig. 26e**). At this

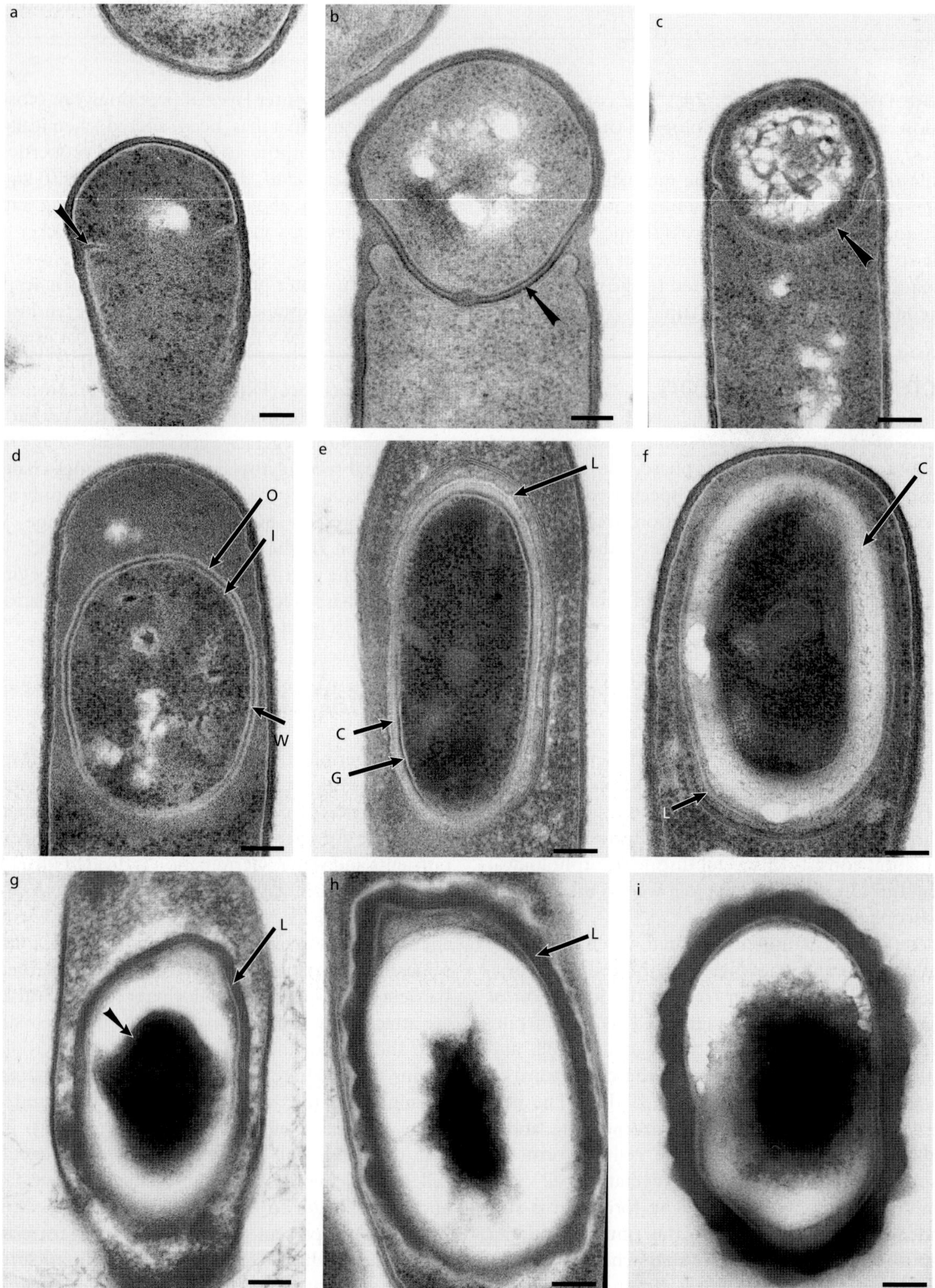

Fig. 25 Bacterial sporulation in *Bacillus subtilis* (a–e) and *Clostridium difficile* (f–i) as shown by ultrathin sections. At stage I (a) the cytoplasmic membrane and cell wall invaginate (arrow) to form a complete cross-wall (arrow) to produce a new small cell (b, c – stages II–III). At stage IV (d) the new small cell is engulfed by the vegetative cell to form the forespore with outer (O) and inner (I) forespore membranes and a thin cell wall layer in-between (W). At early stage V (e) the cell wall develops to form a thin, inner germ cell wall (G) and an outer cortex (C). New spore coat layers (L) are formed and development of the cortex continues (f – late stage V). At stage VI (g) the spore coat layers thicken and the spore cytoplasm (arrow) becomes dehydrated and smaller in volume. By stage VII (h) the mature endospore is almost formed, after which it is released to begin its period of dormancy (i – stage IX). Bar marker 0.2 μm.

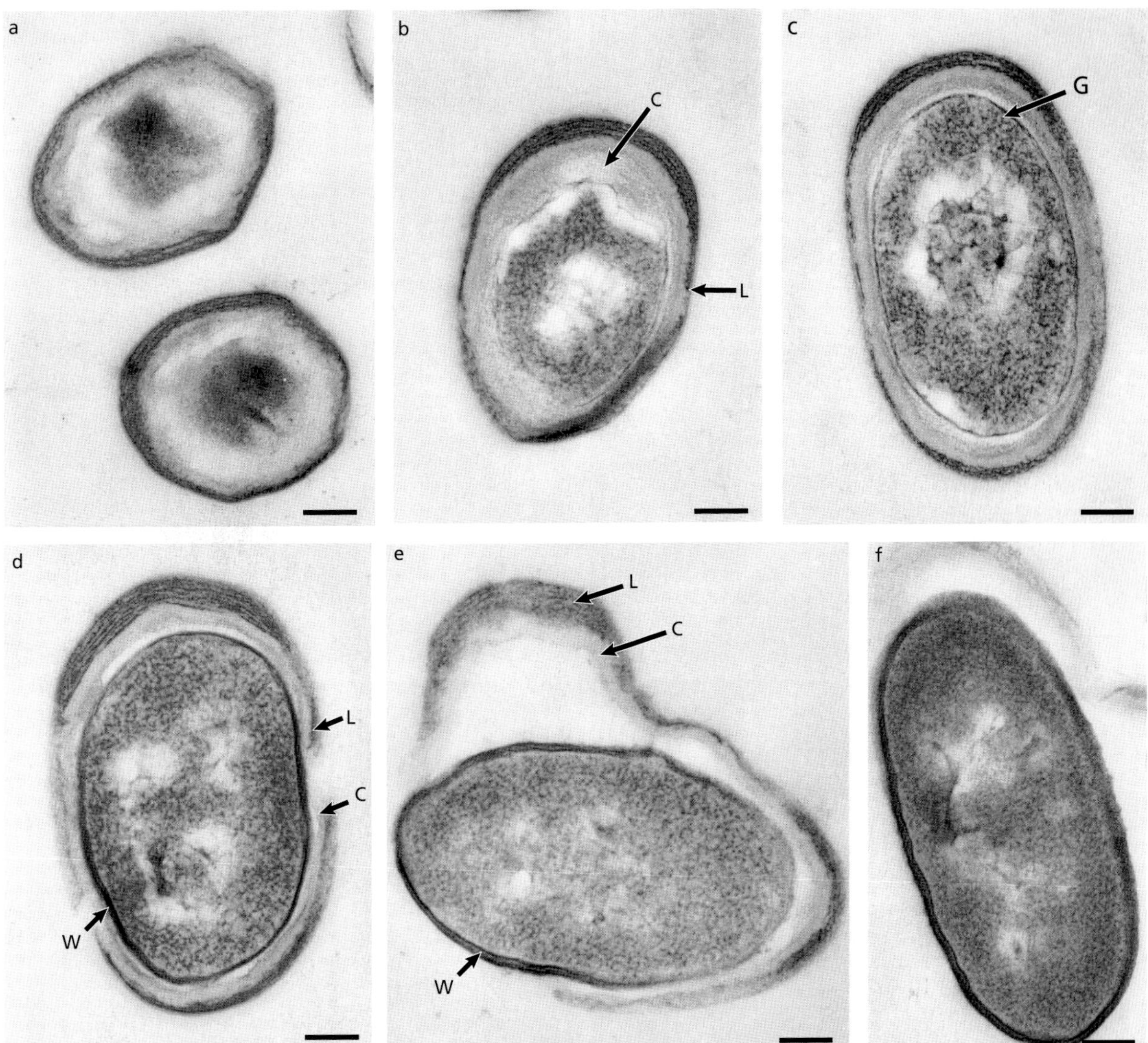

Fig. 26 Spore germination in *Bacillus subtilis* as shown by ultrathin sectioning. (a) Dormant spores. Bar marker 0.2 μm. (b) On activation the spore cytoplasm begins to rehydrate and disruption of the spore wall layers (L) and cortex (C) begins. Bar marker 0.2 μm. (c) The spore swells and the germ cell wall (G) separates from the cortex. Bar marker 0.2 μm. (d) New vegetative cell wall is synthesised (W), the germinating spore elongates and there is a breakdown of the spore coat layers (L) and cortex (C). Bar marker 0.2 μm. (e) The spore outgrowth occurs as the spore coat layers and cortex are shed. Bar marker 0.2 μm. (f) A viable vegetative cell is formed. Bar marker 0.2 μm.

stage the cell resumes active growth, DNA synthesis restarts, and a little later the new vegetative cell undergoes its first division (Fig. 26f).

Microscopy and Microbiology

This chapter has touched on the fascinating insights that microscopy has given to our understanding of the microbial world. The adventure is by no means over. New developments in technology and improved methods for specimen preparation will continue to provide information about bacterial structure to complement advances in other scientific disciplines that will investigate microbiological phenomena in the new millennium (Newman and Hobot, 1999; Fig. 27).

Acknowledgements

The author wishes to thank the following for kindly providing the bacterial cultures used for microscope preparations: Nabeel Alsaif for Figs 1 and 7, Lisa

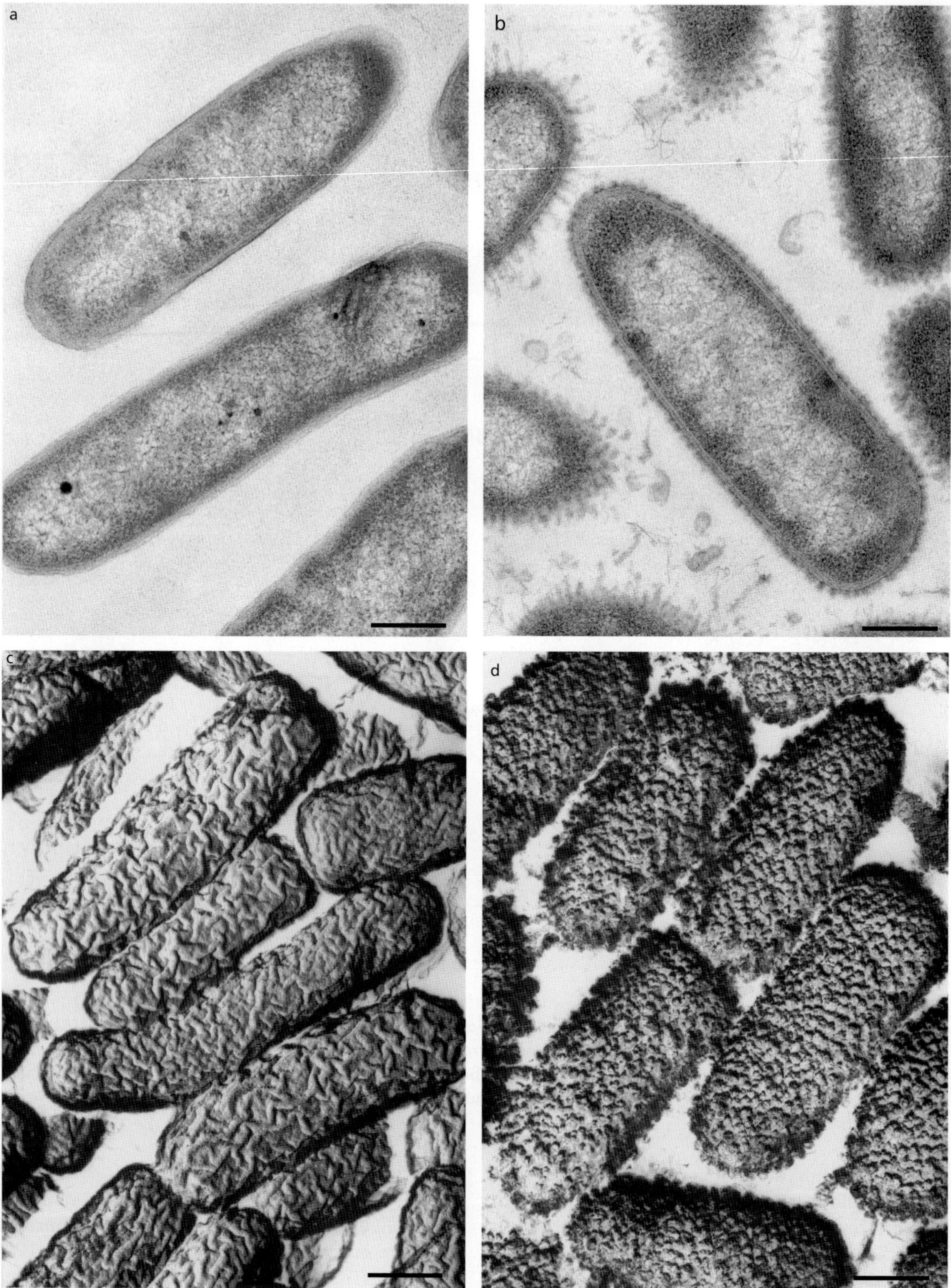

Fig. 27 An example of what can be achieved by using low-temperature procedures and ultrathin sectioning, together with a new approach to preparing surface replicas of bacteria that avoids rapid freezing techniques. *Pseudomonas aeruginosa* cultures were treated with polymixin B. Ultrathin sections show normal cell wall morphology in control cells (a) (cf. Fig. 12a); in (b) antibiotic-treated cells show disruption only of the outer leaflet of the outer membrane, resulting in a regular patchwork appearance on the cell surfaces. Surface replicas, prepared after chemical fixation, dehydration in ethanol, critical point drying, metal shadowing and replica preparation, show a smooth undulating surface in control cultures (c), but a rough appearance of the cell surface is apparent covered with small bumps/protrusions in antibiotic-treated cultures (d). Bar markers 0.5 μm.

Harris for **Fig. 2**, Simon Stubbs for **Figs 5b** and **6**, Colin Harwood for **Fig. 21** and Susu Zughaier for **Fig. 27**.

References

Alsaif NM, Hobot JA, Brazier JS, Duerden BI (1997) Adhesion of *Clostridium difficile* to Caco-2 cell line, a colonization factor. *Rev. Med. Microbiol.* 18: S62–S63.

Atrih A, Zöllner P, Allmaier G, Foster SJ (1996) Structural analysis of *Bacillus subtilis* 168 endospore peptidoglycan and its role during differentiation. *J. Bacteriol.* 178: 6173–6183.

Beveridge TJ, Davies JA (1983) Cellular responses of *Bacillus subtilis* and *Escherichia coli* to the Gram stain. *J. Bacteriol.* 156: 846–858.

Bjornsti MA, Hobot JA, Kelus AS, Villiger W, Kellenberger E (1986) New electron microscopic data on the structure of the nucleoid and their functional consequences. In: Gualerzi CO, Pon CL (eds) *Bacterial Chromatin*. Berlin: Springer-Verlag, pp. 64–81.

Bohrmann B, Villiger W, Johansen R, Kellenberger E (1991) Coralline shape of the bacterial nucleoid after cryofixation. *J. Bacteriol.* 173: 3149–3158.

Braun V (1975) Covalent lipoprotein from the outer membrane of *Escherichia coli*. *Biochim Biophys Acta* 415: 335–377.

Braun V, Gnirke H, Henning U, Rehn K (1973) Model for the structure of the shape maintaining layer of the *Escherichia coli* cell envelope. *J. Bacteriol.* 114: 1264–1270.

Clarke-Sturman AJ, Archibald AR, Hancock IC, Harwood CR, Merad T, Hobot JA (1989) Cell wall assembly in *Bacillus subtilis*: partial conversion of polar wall material and the effect of growth conditions on the pattern of incorporation of new material at the polar caps. *J. Gen. Microbiol.* 135: 657–665.

Dijkstra AJ, Keck W (1996) Peptidoglycan as a barrier to transenvelope transport. *J. Bacteriol.* 178: 5555–5562.

Dubochet J, McDowal AW, Menge B, Schmid EN, Lickfeld KG (1983) Electron microscopy of frozen-hydrated bacteria. *J. Bacteriol.* 155: 381–390.

Dürrenberger M, Bjornsti MA, Uetz T, Hobot JA, Kellenberger E (1988) Intracellular location of the histone-like protein HU in *Escherichia coli*. *J. Bacteriol.* 170: 4757–4768.

Dürrenberger M, Villiger W, Bächi T (1991) Conjugational junctions: morphology of specific contacts in conjugating *Escherichia coli* bacteria. *J. Struct. Biol.* 107: 146–156.

Ebersold HR, Cordier JL, Lüthy P (1981) Bacterial mesosomes: method dependent artefacts. *Arch. Microbiol.* 130: 19–22.

Echlin P (1992) *Low-Temperature Microscopy and Analysis*. New York: Plenum Press.

Gerhardt P, Judge JA (1964) Porosity of isolated cell walls of *Saccharomyces cerevisiae* and *Bacillus megaterium*. *J. Bacteriol.* 87: 945–951.

Graham LL, Beveridge TJ, Nanninga N (1991) Periplasmic space and the concept of the periplasm. *Trends Biochem. Sci.* 16: 328–329.

Hancock IC (1991) Microbial cell surface architecture. In: Mozes N, Handley PS, Busscher HJ, Rouxhet PG (eds) *Microbial Cell Surface Analysis. Structural and Physiochemical Methods*. New York: VCH Publishers, pp. 22–59.

Hobot JA (1990) New aspects of bacterial ultrastructure as revealed by modern acrylics for electron microscopy. *J. Struct. Biol.* 104: 169–177.

Hobot JA (1991) Low-temperature embedding techniques for studying microbial cell surfaces. In: Mozes N, Handley PS, Busscher HJ, Rouxhet PG (eds) *Microbial Cell Surface Analysis. Structural and Physiochemical Methods*. New York: VCH Publishers, pp. 127–150.

Hobot JA, Newman GR (1990) Electron microscopy for bacterial cells. In: Harwood CR, Cutting SM (eds) *Molecular Biological Methods for Bacillus*. Chichester: John Wiley and Sons, pp. 352–362.

Hobot JA, Newman GR (1996) Immunomicroscopy: resin techniques and on-section labelling with immunocolloidal gold or immunoperoxidase – planning a protocol. *Scanning Microsc.* 10: 121–145.

Hobot JA, Rogers HJ (1991) Intracellular location of the autolytic *N*-acetyl-L-alanine amidase in *Bacillus subtilis* 168 and in an autolysis-deficient mutant by immunoelectron microscopy. *J. Bacteriol.* 173: 961–967.

Hobot JA, Carlemalm E, Villiger W, Kellenberger E (1984) Periplasmic gel: new concept resulting from the reinvestigation of bacterial cell envelope ultrastructure by new methods. *J. Bacteriol.* 160: 143–152.

Hobot JA, Villiger W, Escaig J, Maeder M, Ryter A, Kellenberger E (1985) Shape and fine structure of nucleoids observed on sections of ultrarapidly frozen and cryosubstituted bacteria. *J. Bacteriol.* 162: 960–971.

Hobot JA, Bjornsti MA, Kellenberger E (1987) Use of on-section immunolabeling and cryosubstitution for studies of bacterial DNA distribution. *J. Bacteriol.* 169: 2055–2062.

Humpries M, Wilkinson AE, Edwards B, Thompson JS (1981) The densities of bacterial cell walls. *Biochem. Soc. Trans.* 9: 436.

Jap BK, Walian PJ (1996) Structure and functional mechanisms of porins. *Physiol. Rev.* 76: 1073–1088.

Kellenberger E (1990) The 'Bayer bridges' confronted with results from improved microscopy methods. *Mol. Microbiol.* 4: 697–705.

Leung KP, Fukushima H, Sagawa H, Walker CB, Clark WB (1989) Surface appendages, hemagglutination, and adherence to human epithelial cells of *Bacteriodes intermedius*. *Oral Microbiol. Immunol.* 4: 204–210.

Levine MM, Ristaino P, Marley G *et al.* (1984) Coli surface antigens 1 and 3 of colonization factor antigen II-positive enterotoxigenic *Escherichia coli*: morphology, purification, and immune responses in humans. *Infect. Immun.* 44: 409–420.

Merad T, Archibald AR, Hancock IC, Harwood CR, Hobot JA (1989) Cell wall assembly in *Bacillus subtilis*: visualisation of old and new wall material by electron microscopic examination of samples stained selectively for teichoic acid and teichuronic acid. *J. Gen. Microbiol.* 135: 645–655.

Merchante R, Pooley HM, Karamata D (1995) A periplasm in *Bacillus subtilis*. *J. Bacteriol.* 177: 6176–6183.

Müller DJ, Baumeister W, Engel A (1996) Conformational change of the hexagonally packed layer of *Deinococcus radiodurans* monitored by Atomic Force Microscopy. *J. Bacteriol.* 178: 3025–3030.

Newman GR, Hobot JA (1999) Resins for combined light and electron microscopy: a half century of development. *Histochem. J.* 31: 495–505.

Newman GR, Hobot JA (2001) *Resin Microscopy and On-Section Immunocytochemistry*. Heidelberg: Springer-Verlag (2nd. edition).

Oldmixon EH, Glauser S, Higgins ML (1974) Two proposed configurations for bacterial cell wall peptidoglycans shown by space-filling models. *Biopolymers* 13: 2037–2060.

Robinson DG, Ehlers U, Herken R, Herrmann B, Mayer F, Schürmann FW (1987) *Methods of Preparation for Electron Microscopy*. Berlin: Springer-Verlag.

Roth J (1983) Application of lectin gold complexes for electron microscopic localisation of glycoconjugates on thin sections. *J. Histochem. Cytochem.* 31: 987–999.

Schulman H, Kennedy EP (1979) Localization of membrane-derived oligosaccharides in the outer envelope of *Escherichia coli* and their occurrence in other gram-negative bacteria. *J. Bacteriol.* 137: 686–688.

Signoretto C, Di Stefano F, Canepari P (1996) Modified peptidoglycan chemical composition in shape-altered *Escherichia coli*. *Microbiology* 142: 1919–1926.

Silverman PM (1997) Towards a structural biology of bacterial conjugation. *Mol. Microbiol.* 23: 423–429.

Sleytr UB, Messner P, Pum D, Sára M (1996) *Crystalline Bacterial Cell Surface Proteins*. Austin, Texas: RG Landes/Academic Press.

Stubbs S, Hobot JA, Waddington RJ *et al.* (1999) Effect of environmental haemin upon the physiology of *Prevotella intermedia* R78. *Lett. Appl. Microbiol.* 29: 31–36.

Surman SB, Walker JT, Goddard DT *et al.* (1996) Comparison of microscope techniques for the examination of biofilms. *J. Microbiol. Meth.* 25: 57–70.

Weidel WM, Frank H, Martin HH (1960) The rigid layer of the cell wall of *Escherichia coli* strain B. *J. Gen. Microbiol.* 22: 158–166.

3

Bacterial Cell Walls: An Overview

Ian C. Hancock

University of Newcastle upon Tyne, UK

The first interactions between a host and a potentially pathogenic bacterium are critical for the establishment of infection and depend crucially on the properties of the bacterial cell surface. Indeed, a useful working definition of the bacterial 'surface' is the array of molecules involved in this initial interaction. While these molecules are not necessarily structural components of the cell wall, many are associated with the wall and some are covalently attached to it. The surfaces presented by bacteria to their environment vary, not only between strains and species, but also phenotypically within a single strain. Their physicochemical interactions depend on a mosaic of ionic, polar and non-polar domains contributed by cell wall components and a variety of non-covalently linked ions and macromolecules that associate with the cell envelope. Wall components may give rise to diffuse surface properties such as anionic surfaces of Gram-positive bacteria or the hydrophobic surfaces of mycobacteria. On the other hand, they may be responsible for very specific interactions with receptors on other cells or non-biological surfaces.

The dynamic nature of cell wall structure, due to autolysin-catalysed turnover of peptidoglycan, the concomitant release of wall-associated molecules, and the activities of wall-bound proteases, polysaccharases and lipases, ensures the fluidity of surface interactions, and contributes wall macromolecules to the immediate extracellular environment. It is important to recognise the contribution this turnover process makes to the rate at which phenotypic changes can be expressed, particularly in Gram-positive bacteria and in mycobacteria.

The aim of this chapter is to provide an outline of bacterial cell wall architectures and dynamics in the context of medical microbiology. The structures and properties of medically important wall components are dealt with elsewhere this book.

Peptidoglycan – the Structural Framework of the Cell Wall

With the exception of the group known as mollicutes (mycoplasmas), all bacteria have a cell wall whose structural 'scaffolding' consists of peptidoglycan, a macro-polymer of polysaccharide chains, cross-linked by short oligopeptides that have the potential to interlink three polysaccharide chains in three dimensions. This cross-linking creates a single, porous, macromolecular net of peptidoglycan around the whole cell. Due to improved techniques of enzymatic

 doi:10.1006/bkmm.2001.0003

and chemical analysis, knowledge of the structural architecture of this material has escalated in the last few years. Although the main constructional principles are the same in peptidoglycan from all genera of bacteria, the design of the polymer allows a substantial degree of variation in architecture. This variation results from differences in polysaccharide chain length, differences in the degree of cross-linking of polysaccharides through their peptide side-chains and differences in the relative extent to which cross-linked peptide side-chains are attached to one or two further peptide side-chains on other polysaccharide strands. These variations can result in 'single sheet' and 'multi-layered' architectures, with the potential for local or general variations in thickness and in the mesh size of the net. In general, Gram-negative bacteria are characterised by the single sheet form, while Gram-positive bacteria, mycobacteria and actinomycetes possess multi-layered peptidoglycan.

The peptidoglycan is the skeleton on which the cell wall is constructed. In Gram-positive bacteria it supports a variety of covalently and non-covalently attached 'accessory polymers' throughout its thickness. These may be polysaccharides and related macromolecules, and proteins. In mycobacteria the covalently linked accessory polysaccharide in turn serves as the support for the lipid components of the cell surface. In Gram-negative bacteria, a lipoprotein covalently attached to peptidoglycan is the anchor for the outer membrane. Additionally, the thick, porous structure of the peptidoglycan layer in Gram-positive bacteria and mycobacteria acts as an extracellular compartment, functionally similar in many ways to the periplasm of Gram-negative bacteria, in which many other polysaccharides and proteins are temporarily trapped.

These architectural 'styles' and their defining features are described below. It is not surprising, given the key role of peptidoglycan in bacterial cell structure and integrity, and its unique structural features, that its biosynthesis is one of the most selective and effective targets for therapeutic antimicrobial agents. Clinically the most widely used classes are the penicillins, cephalosporins and glycopeptides (vancomycin, teichoplanin), but other known antibiotics (such as D-cycloserine, bacitracin, moenomycin, tunicamycin) target other stages in peptidoglycan synthesis, pointing to additional potential targets for more effective agents.

Gram-negative Bacteria

The cell envelopes of Gram-negative bacteria are characterised by the outer membrane, which gives rise to a compartment of the cell, the periplasm (Oliver, 1996), delimited by the cytoplasmic ('inner') membrane and the outer membrane and containing the peptidoglycan (or 'murein') layer of the wall (**Fig. 1**). The outer membrane is attached to the peptidoglycan by molecules of a specific 7.2-kDa lipoprotein. This has a C-terminal peptide link to the diaminopimelic acid residue of a murein peptide chain and an N-terminal cysteine residue that carries a thioether-linked diacylglycerol residue and is further *N*-acylated. The N-terminal lipid integrates into the inner leaflet of the outer membrane bilayer, completing the connection to the murein. As many as 10% of the murein peptides may carry a covalently linked lipoprotein and additional lipoprotein occurs free in the outer membrane, making it the predominant outer membrane protein (OMP), constituting up to 70% of the total OMPs.

The Outer Membrane Bilayer

The bilayer of the outer membrane is a biologically unique structure (Nikaido, 1996). While the inner leaflet is composed of phospholipids of the same types as those found in the inner, cytoplasmic membrane, the outer leaflet contains little or no phospholipid (at least, in enterobacteria) but instead consists of lipopolysaccharide (LPS) or lipo-oligosaccharide (LOS) molecules (see Chapter 6), orientated with the long acyl chains of their lipid A moieties in the hydrophobic core of the membrane and their oligosaccharide or polysaccharide components creating a highly hydrophilic external surface. This layer also exhibits a substantial negative charge on its outer surface, due to LPS. Carboxylate groups in 2-keto-3-deoxyoctulosonic acid (KDO) and phosphate groups in lipid A

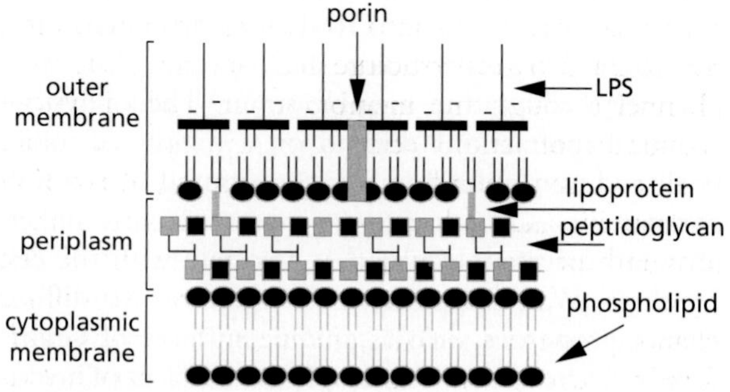

Fig. 1 The structure of the Gram-negative cell wall. The diagram shows the phospholipid bilayer of the inner (cytoplasmic) membrane, and the outer membrane with its outer leaflet of lipopolysaccharide (LPS). The thin layer of peptidoglycan, and the lipoprotein that links it to the outer membrane, occupy the periplasm compartment between the two membranes.

both contribute to this. In the LPS, phosphoethanolamine and aminoarabinose substituents contribute positive charges to the lipid A and core regions (Rietschel *et al.*, 1994) and this can have a profound effect on the physico-chemical properties of the outer membrane. For example, *Escherichia coli* acquires greater resistance to the cationic antibacterial agents polymyxin and protamine as a result of increased substitution in the lipid A and core regions of its LPS with these two cationic groups (Nummila *et al.*, 1995). In those species that contain a true lipopolysaccharide, the polysaccharide chains constitute the important, often serotype-specific O-antigens. The toxic properties of LPS endotoxins depend critically on the pattern of acylation of the lipid A moiety (see Chapter 6).

Outer Membrane Proteins

In addition to the structural lipoprotein mentioned above, the outer membrane contains a variety of OMPs, which confer specific functional properties including enzymatic activities (notably phospholipases), diffusion pores, specific receptor activities and export machinery for extracellular proteins and polysaccharides. As well as being crucial for proper cellular physiology, many of these have attracted interest as potential vaccine targets.

A group of 'porin' proteins that create diffusion pores in the outer membrane constitutes the major OMP class, representing about 10% of OMPs. Although these are integral membrane proteins, with many transmembrane domains, they do not possess any amino acid sequences typical of hydrophobic transmembrane helices. This apparent anomaly has been explained by the X-ray crystallographic solution of the *Rhodobacter capsulatus* porin by Weiss *et al.* (1991). This revealed that the trans-membrane segments are organised into 16 antiparallel β strands, forming a 'β-barrel' structure that constitutes a tubular channel through the membrane. Instead of having sequences of amino acids with hydrophobic side-chains, these β strands have approximately alternating hydrophilic and hydrophobic amino acids. The consequence of the β arrangement is that nearly all the hydrophilic amino acids face into the hydrophilic channel and the alternating hydrophobic amino acids form a hydrophobic outer surface to the barrel. This contributes to the aggregation of many porins into trimers, and to their interaction with the hydrophobic core of the membrane.

Although the primary sequences of classical porins from different species can differ remarkably from one another, the architecture of the pores they form is very well conserved (Cowan *et al.*, 1992). The highly polar nature and the dimensions of the interior of the channels formed by the porins confer some general properties on all porin channels – they exclude lipophilic solutes and large molecules. However, different porins form pores of a variety of sizes and some, such as OprF, the major general diffusion porin of *Pseudomonas aeruginosa*, apparently exist as monomers in the membrane, rather than as a trimer. Although OprF forms quite a large pore, it has a low permeability compared with the general porins OmpC and OmpF of *E. coli*. OmpA, another major outer membrane protein of *E. coli*, has substantial sequence homology with OprF and also forms a transmembrane conductance channel. Its permeability is so low, however, that it does not contribute significantly to the overall permeability of the outer membrane and instead seems to play a role in the structural stability of the outer membrane (see Nikaido, 1996, for a review). In addition to the major general diffusion porins, increasing numbers of minor OMPs are being identified that exhibit more specific diffusion properties, because they have high-affinity receptor sites for specific molecules. In *E. coli*, for example, *lamB*, *scrY* and *tsx* encode trimer porins with selectivity, respectively, for maltose, sucrose and nucleosides. In many cases porin expression is regulated by environmental factors, such as limitation of a specific nutrient or osmolarity.

Gram-positive Bacteria

The distinctive feature of the cell walls of Gram-positive bacteria, and the property that enables them to take up and retain the Gram stain, is the absence of an outer membrane and hence the absence of a membrane-bound periplasm. Instead, they possess multilayered peptidoglycan covalently substituted with the anionic polymers teichoic acid or teichuronic acid (see Chapter 5) (**Fig. 2**). In most species these 'accessory' polymers constitute a very significant proportion of the weight of the cell wall, for example about 50% in *Staphylococcus aureus*. In spite of the absence of true periplasm, Gram-positive walls have a large water capacity and the ability to retain large amounts of proteins, lipoglycans and cations, because of this thick, hydrophilic, porous structure. Moreover, there is ample evidence that the thickness and high degree of cross-linking of the peptidoglycan confers molecular sieving properties on the Gram-positive cell wall. This filtration property, together with the large fixed negative charge conferred by the covalently linked anionic polymers, places significant limits on the types of molecule that can diffuse freely through the cell wall in either direction (Hancock, 1991).

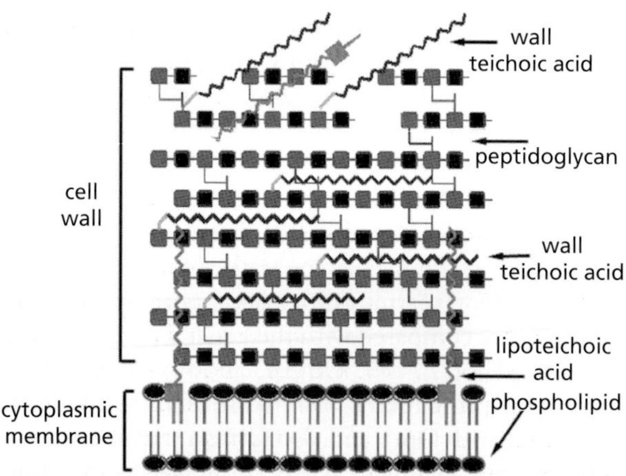

Fig. 2 The structure of a Gram-positive cell wall. The diagram shows the phospholipid bilayer of the cytoplasmic membrane and the thick, multi-layered peptidoglycan. Wall teichoic acid is covalently linked to the peptidoglycan throughout the depth of the wall. The figure also shows lipoteichoic acid, both attached to the outer surface of the cytoplasmic membrane and also being secreted, in a partially deacylated form, through the cell wall.

Despite the presence of anionic covalently linked accessory polymers, and the polar nature of peptidoglycan itself, most Gram-positive bacteria exhibit significant surface hydrophobicity. A complete physicochemical model of this phenomenon is not yet available, but the main wall components contributing to it appear to be wall-bound proteins, which are occasionally covalently linked to peptidoglycan, and lipoteichoic acids (see Chapter 5) and other lipoglycans. The latter probably start their lives as components of the outer surface of the cytoplasmic membrane, but are continuously released into the cell wall and eventually reach its outer surface, perhaps by cell wall turnover (see below), where they can be detected immunochemically. Some of these, like LPS, exhibit endotoxic and immunostimulatory properties that can be important in pathogenicity.

Much interest has focused recently on wall-bound proteins in Gram-positive bacteria. While most of these constitute a small proportion of the weight of the wall, their roles as adhesins and specific receptors involved in host invasion and the establishment of infection are of great importance. Immunological, biochemical and genetic analysis of these proteins has revealed several distinct families that vary in their mode of interaction with the cell wall and how they are displayed at the cell surface.

Covalently linked proteins are typified by the IgG-binding protein A of *Staph. aureus* and have been investigated in depth by Navarre and Schneewind (1999). They showed that their attachment to peptidoglycan is signalled by a specific, conserved amino acid sequence towards the C-terminus of the protein that serves as the target for a 'sortase' enzyme. This catalyses a transpeptidation reaction in which the C-terminus is removed with the concomitant peptide linkage of the rest of the protein to a peptide side-chain of the peptidoglycan. Mechanistically, this closely resembles the attachment of covently linked outer membrane lipoprotein in Gram-negative bacteria. Other conserved regions of these proteins appear to confer appropriate interactions with the peptidoglycan with which they are intercalated, while towards their N-terminus they possess specific binding domains for host proteins. Targets include immunoglobulins, fibrinogen, fibronectin and collagen. Another member of this group, the cell surface 'internalin' protein of *Listeria monocytogenes*, plays a vital role in host cell invasion and colonisation. Strains that lack a functional wall targeting system, and hence secrete the protein, are selected for in populations under some environmental conditions, with potential consequences for their interactions in biofilms. Moreover, surface proteins are prime candidates for regulation by global quorum-sensing systems, as exemplified by the *agr* system of staphylococci (Ji *et al.*, 1995), and are released from their wall-bound state into the environment by cell wall turnover (see below).

Other proteins displayed at the cell surface are bound strongly but non-covalently within the cell wall. Of these, the best understood belong to a family of lipoproteins characterised by a distinctive N-terminal signal peptidase cleavage site containing a cysteine residue. This is the target for acylation and lipid substitution by processes precisely analogous to those involved in outer membrane lipoprotein synthesis in Gram-negative bacteria. A few of these lipoproteins, notably the wall-bound penicillinase of *Bacillus licheniformis*, are well known (Sarvas, 1986). Many others, mainly of unknown functions, are being recognised by sequence homologies in genome sequences. The lipid group of lipoproteins anchors the N-terminus of the polypeptide to the outer surface of the cell membrane, while the rest of the protein is functionally within the cell wall, but there is no evidence for specific interaction with wall components. On the other hand, Lupas *et al.* (1994) have identified a class of surface proteins ('SLH') that contain two helical domains close to the N-terminus and appear to be responsible for direct interaction with peptidoglycan. The SLH domain was first recognised in S-layer proteins (see below). Subsequently it was also found in the protein components that attach cellulosomes to the cell wall in *Clostridium thermocellum*, where it was convincingly shown to confer binding to peptidoglycan (Lemaire *et al.*, 1995).

As mentioned above, some strains of bacteria exhibit considerable hydrophobicity, as observed by an ability to stick to hydrophobic surfaces or to collect at a hydrocarbon–water interface. In many cases this property is due to the presence of species-specific or even strain-specific surface proteins, which in some cases may have evolved as adhesins but in other cases have other receptor functions and only 'accidentally' stabilise adhesion. The increasing recognition of proteins at the surface of Gram-positive bacteria that are specifically targeted to the cell wall, makes the exploration of this phenomenon in relation to protein 'hydrophobins' important, because it imposes particular constraints on the dynamics of the proteins in growing and non-growing bacteria (Navarre and Schneewind, 1999). The ability of some of these hydrophobic proteins, such as the CshA protein of *Streptococcus gordonii*, to form cell surface fibrils extends their sphere of influence around the cell, and thus may increase the chance of hydrophobic interactions with other cells and surfaces by allowing these to occur at a greater distance from the negatively charged surface of the cell wall (McNab *et al.*, 1999).

Protein S-layers

S-layer proteins that form semi-crystalline arrays of globular polypeptide units over the whole surface of some Gram-positive and Gram-negative strains are a very distinctive group of wall proteins (Beveridge *et al.*, 1997). They self-aggregate from aqueous solution into sheet-like structures, but it has been demonstrated in some Gram-positive bacteria, such as *Lactobacillus* and *Bacillus*, that their association with the cell wall is mediated in addition by specific domains interacting with structural wall polymers such as peptidoglycan (via the SLH domain described above) and teichoic acid (Sara *et al.*, 1998). The distribution of S-layer proteins amongst bacterial species is still incompletely investigated and there are few reports of these in bacterial pathogens. Their roles *in vivo* are poorly understood, but it has been demonstrated that they form effective permeability barriers to large molecules. The extent to which they mask other surface molecules and functions is not known in any detail.

Mycobacteria

Nearly 60 species of *Mycobacterium* have been recognised and despite their diverse habits and ecology, which range from soil saprophytes to major human pathogens, they are all characterised by the same unusual wall structure (Brennan and Draper, 1994; Brennan and Nikaido, 1995). This resembles the walls of Gram-positive bacteria in possessing multi-layered peptidoglycan and a covalently linked accessory polymer, arabinogalactan. Its peptidoglycan is, however, distinctive in containing *N*-glycolylmuramic acid instead of *N*-acetylmuramic acid, a property shared only with *Rhodococcus*, *Corynebacterium* and some nocardioform bacteria. But the most important feature is the substantial proportion of its total mass (up to 60%) that is made up of lipid components, particularly the very long chain (C60–C90) mycolic acids. Functionally, this lipid material forms an extremely hydrophobic outer layer in which the cell wall hydrocarbon chains are arrayed perpendicular to the cell surface and tightly packed in a similar way to acyl chains in the hydrophobic core of a conventional lipid bilayer membrane (Nikaido *et al.*, 1993).

The current model of this structure (Minnikin, 1982, 1991) (**Fig. 3**) envisages a lipid bilayer outside the peptidoglycan layer of the wall. The inner leaflet of this layer consists of mycolic acids esterified to the non-reducing termini of arabinogalactan chains that are covalently anchored at their reducing termini to the peptidoglycan. Absorbed to this inner leaflet of mycolic acids is an outer leaflet of non-covalently linked ('extractable') lipids, mainly glycolipids, which contain shorter-chain fatty acids. It is predicted that the inner leaflet of this structure would have very low fluidity like, and for similar reasons, the lipopolysaccharide leaflet of the Gram-negative outer membrane. A consequence of such a structure is low permeability to both hydrophilic and lipophilic solutes, and this property is widely regarded as a major reason for the resistance of mycobacteria to most antibiotics, though other factors undoubtedly contribute to this (Jarlier and Nikaido, 1994).

Mycolylarabinogalactan

The major accessory polymer, and a major antigen, of the mycobacterial cell wall is a neutral polysaccharide, arabinogalactan, which is covalently attached to C6 of a muramic acid residue of peptidoglycan through a reducing-terminal rhamnosyl-*N*-acetylglucosamine-1-phosphate, a similar arrangement to the *N*-acetyl-mannosaminyl-*N*-acetylglucosamine-1-phosphate linkage unit that attaches teichoic acids to peptidoglycan in Gram-positive bacteria (see Chapter 5). The arabinogalactan is unusual amongst bacterial exopolysaccharides in not possessing a subunit structure. Instead, it consists of a linear polymer of

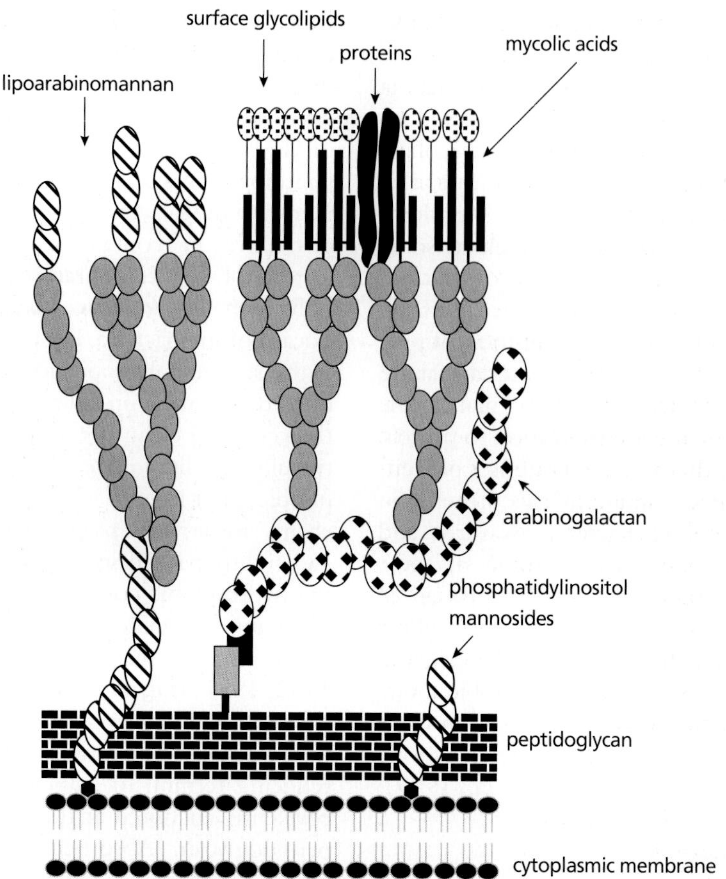

Fig. 3 The structure of a mycobacterial cell wall. The diagram shows the cytoplasmic membrane and the cell wall. The polysaccharides (arabinogalactan and lipoarabinomannan) are magnified to illustrate their structural complexity. Mycolic acids attached to the outer arabinose residues of the arabinogalactan form the inner leaflet of a membrane-like hydrophobic outer layer, the outer leaflet of which is made up of glycolipids. The precise way in which the polysaccharides interact with the peptidoglycan layer is unknown (after Kremer *et al.*, 2000).

galactofuranose residues substituted, towards its non-reducing end, with about two branched polymers of arabinose. The non-reducing ends of about two-thirds of these branches are acylated with four mycolic acids per branch.

The detailed structure of this complex polymer has been elucidated (Besra and Chatterjee, 1994) **(Fig. 4)** and is highly conserved within the genus *Mycobacterium*, though slow-growing, pathogenic species are modified by substitution with probably a single galactosamine residue on each arabinogalactan molecule (Draper *et al.*, 1997). Greater differences are found in other mycolic acid-containing genera. *Nocardia* and *Rhodococcus* species exhibit differences in glycosidic linkage within both the galactan and arabinan segments. In addition, the galactan moiety of the *Nocardia* polymer carries β-glucosyl substituents at C6 of some of the galactose residues. In all cases, however, the overall organisation of the molecule is the same and the linkage to peptidoglycan is fully conserved (Daffé *et al.*, 1993).

Mycolylarabinogalactan (mAG) makes up about 35% of the weight of the mycobacterial cell wall and on average about 10% of peptidoglycan disaccharide repeating units carry an mAG.

The Hydrophobic Wall Layer

As mentioned above, all the current evidence supports a model for the wall in which the distal mycolic acid chains of mAG pack tightly at the outer surface of the wall to form the inner leaflet of a lipid bilayer, the outer leaflet of which is composed of a variety of glycolipids (Besra and Chatterjee, 1994) some with shorter mycolic acids and others with normal fatty acyl chains similar to those in the cytoplasmic membrane phospholipids. Such a structure would be expected to be impermeable to water and hydrophilic solutes, and indeed measurements of the permeation rates of antibiotics and nutrients have indicated that they are about three orders of magnitude lower than those for the *E. coli* outer membrane (Jarlier and Nikaido, 1994). Whatever permeability this hydrophobic layer does display appears to be mediated, at least for hydrophilic solutes, by channel-forming proteins similar in organisation to the porins of Gram-negative outer membranes (Trias *et al.*, 1992; Jarlier and Nikaido, 1994). The relatively low rate of solute diffusion through the hydrophobic wall layer appears to be due to the small number of these porins in the wall, compared with *E. coli* outer membrane, rather than lower permeability of the individual channels. It should be emphasised, however, that the susceptibilities of different mycobacterial species to antibiotics vary widely, and a variety of structural factors in addition to porins are likely to be involved (Barry and Mdluli, 1996). The large variety of other lipids, both amphipathic ones, such as phosphatidylinositol mannosides (PIMs) and trehalose mycolates, and hydrophobic ones, such as phthiocerol dimycocerosates (PDIMs), that are found in the cell wall, provide the potential for great variation of physical properties, though their roles are not yet well understood.

A consequence of the relatively impermeable hydrophobic layer at the outer surface of the wall is that the cells possess a periplasm, containing the hydrophilic polysaccharide and peptidoglycan components of the wall, outside the cytoplasmic membrane. The composition of this region remains poorly understood, but it is undoubtedly the site of the cryptic β-lactamases (Kasik, 1979).

Amphiphilic Wall Polymers

Although mycobacteria do not contain lipoteichoic acids, their envelopes possess large amounts of another serologically active amphiphilic polymer, lipoarabinomannan (LAM), and related 'lipoamphiphiles' are found in other mycolic acid-containing genera (Sutcliffe, 1998). LAM consists of a complex glycan with a phosphatidylinositol anchor. It has a domain organisation reminiscent of that of arabinogalactan, consisting of a mannan attached through its reducing terminus to phosphatidylinositol mannoside and through its non-reducing subterminal mannose to a branched arabinan, whose non-reducing termini are capped with one or more mannose residues (**Fig. 4**) (Besra and Chatterjee, 1994). A related polymer, lipomannan (LM), which lacks the arabinan portion of LAM but is otherwise rather similar in structure, also occurs in the wall.

Relationship Between the Cell Wall and the Cytoplasmic Membrane

Membrane Structure and Composition

The cytoplasmic membranes of bacteria are typical biological membranes based on a fluid phospholipid

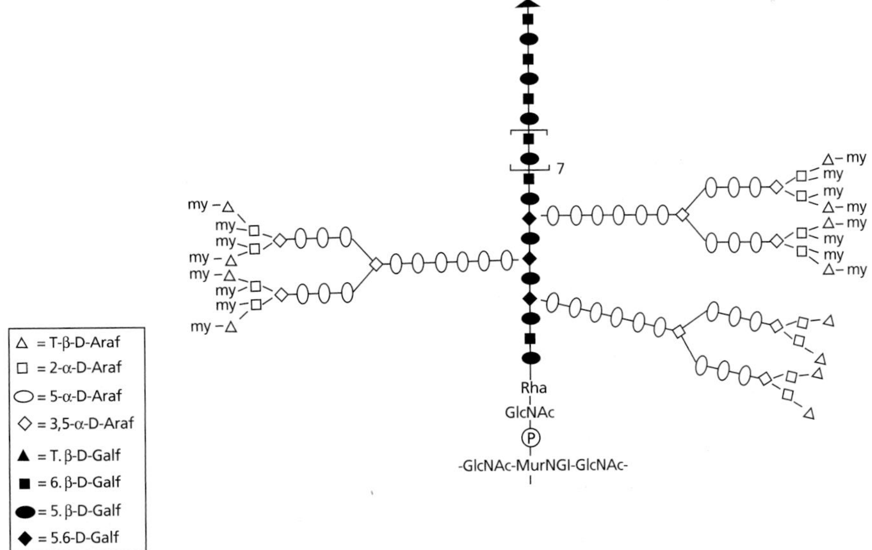

Fig. 4 The structure of mycolylarabinogalactan (mAG). The diagram shows the mAG component of the mycobacterial cell wall and its attachment to *N*-glycolylmuramic acid (MurNGl) component of peptidoglycan. It illustrates the subunit structure of the polysaccharide made up of the sugars arabinose (Ara), galactose (Gal), rhamnose (Rha) and *N*-acetylglucosamine (GlcNAc), and long-chain mycolic acids (my).

bilayer containing a mosaic of integral, membrane-spanning proteins and peripheral proteins. Thorough reviews of bacterial membrane composition and properties are available for Gram-negative bacteria (Kadner, 1996) and Gram-positive bacteria (de Mendoza *et al.*, 1993). Apart from the identification of the biosynthetic pathways and the genes encoding them, bacterial membrane phospholipids have received relatively little attention in recent years. There remain many fundamental questions to be answered about the organisation and roles of bacterial membrane lipids.

Qualitatively, the basic structure of bacterial cytoplasmic membranes differs from that of a eukaryotic cell membrane in several respects: the absence of sterols, the rarity of glycoproteins, and the presence of plant-like glycolipids and phospholipids containing branched-chain fatty acids in many Gram-positive genera. Many of the classes of protein found in the cytoplasmic membrane are involved in communication with the external environment and are common to all bacteria. These include transport proteins, sensor proteins involved in chemotaxis, the sensor components of two-component sensor-regulator systems that regulate gene expression in response to environmental signals, and the enzymes responsible for phospholipid synthesis and the polymerisation of extracellular polysaccharides and peptidoglycan. The presence of other major classes of membrane proteins, such as respiratory electron carriers and ATP synthases, and the membrane anchors of protein structures such as pili and flagella, depends on the physiological status of the species and its growth conditions.

Like other biological membranes, bacterial cytoplasmic membranes are distinctly asymmetrical. This is partly due to the vectorial nature of many membrane processes, such as transport and signalling, and the consequent distribution of proteins, but also to the asymmetrical distribution of particular phospholipid species between the inner and outer leaflets of the lipid bilayer. The latter is a well-known and important feature of eukaryotic cell membranes but has received very little attention in bacteria. Asymmetrical distribution of phosphatidylethanolamine between the inner and outer leaflets of the *Bacillus megaterium* membrane was demonstrated by Rothman and Kennedy (1977a, 1977b) and in *B. subtilis* by Bishop *et al.* (1977), but because of the very different amounts of this phospholipid in the two types of membrane, no generally applicable conclusions were drawn. Equivalent information is not available for Gram-negative bacteria (Kadner, 1996).

Substantial rates of phospholipid turnover have been measured in bacterial membranes. Turnover appears to be a consequence of the roles of some membrane lipids in the biosynthesis of other cell envelope components. Thus, in *E. coli* phosphatidylethanolamine (PE) donates phosphoethanolamine groups for the synthesis of outer membrane LPSs (see above) and periplasmic membrane-derived oligosaccharides (MDOs; see Periplasm, below) and this leads to turnover of about 5% of the PE per generation in growing bacteria (Miller and Kennedy, 1987). Similarly, up to 40% of the other major phospholipid, phosphatidylglycerol (PG), turns over per generation as the result of its role in the synthesis of a third membrane lipid, cardiolipin, the donation of glycerol phosphate groups to MDO and the donation of diacylglycerol groups for lipoprotein synthesis (see below). In Gram-positive bacteria a similar turnover of PG is seen because of its role as the donor of glycerol phosphate groups for lipoteichoic acid synthesis (de Mendoza *et al.*, 1993) (see Chapter 5).

Since molecular techniques for investigating the orientation of membrane-spanning proteins became available, it has been possible to predict, from protein sequences, membrane-spanning sections of polypeptide and hence to identify functional domains located at either surface of the membrane. Studies of this kind have been crucial in studying the mechanisms of action of transport, receptor and signalling proteins. Some proteins of this class have a long N- or C-terminal sequence protruding at the outer surface of the membrane (Kadner, 1996).

In addition to integral, membrane-spanning proteins, membrane-bound lipoproteins are being identified in increasing numbers. These possess an N-terminal cysteine residue with a thioether-linked diacylthioglycerol residue and an *N*-linked long chain fatty acid. The terminal lipid anchors the protein into the lipid bilayer of the membrane. These proteins are synthesised with an N-terminal signal sequence that is cleaved by a prolipoprotein-specific signal peptidase, after translocation to the outer surface of the membrane, at the cysteine residue that subsequently becomes lipid substituted. Thus, they are mainly outer-surface membrane-anchored proteins and have physiological roles in which this orientation is important. This kind of behaviour is typified by lipoprotein β-lactamases of a variety of Gram-positive bacteria and the Prt proteases of *Lactococcus* and *Lactobacillus* (Kehoe, 1994). In Gram-negative bacteria the outer membrane is an alternative point of anchorage for lipoproteins, among which is the structurally important lipoprotein that covalently links the outer membrane to the peptidoglycan layer (see Gram-negative Bacteria, above).

In Gram-positive bacteria, 'lipoamphiphiles', including lipoteichoic acid and other glycolipid-linked carbohydrate polymers (see Chapter 6) are presumed to be anchored to the outer leaflet of the cytoplasmic membrane by their lipid moiety, although no direct proof of this is available. These polymers are polyanionic and their presence on the outer surface of the membrane thus specifically modifies the polyelectrolyte environment at the outer surface. It has been proposed that this plays an important role in regulating the activity of autolytic enzymes and hence in controlling cell wall turnover and cell separation after division (Doyle *et al.*, 1988) (see below).

The Periplasm

As described above, the presence of an outer membrane in Gram-negative bacteria generates an additional cell compartment, the periplasm, between it and the cytoplasmic membrane. The composition and properties of this compartment are receiving growing attention as it becomes evident that it is the site of a range of vital biochemical processes (Oliver, 1996). The physico-chemical properties of the periplasmic contents are of considerable importance because of the need for secreted proteins and polysaccharides to cross the periplasm on their way to the outer membrane and the external environment. These properties have been thoroughly discussed by Koch (1998).

At least 60 gene products in *E. coli* have been described as 'periplasmic proteins' on the basis of their release from the bacteria by osmotic shock under conditions that do not cause leakage of proteins through the cytoplasmic membrane (Oliver, 1996). Many of these are the substrate-binding components of ATP-dependent transport systems and might more properly be regarded as peripheral membrane proteins associated with the outside of the cytoplasmic membrane (Doige and Ames, 1993). Other periplasmic proteins are involved in protein translocation (Driessen *et al.*, 1998). A unique property of the periplasm is its provision of an appropriate redox potential and enzymes for the catalysis of disulphide bond formation in proteins. Thus *E. coli* alkaline phosphatase only becomes active after its export to the periplasm, because of the stabilisation of the catalytically active conformation of the protein by disulphide cross-links (Derman and Beckwith, 1991). The periplasmic proteins DsbA, DsbB and DsbC catalyse disulphide formation (Bardwell *et al.*, 1993). As well as its importance in regulating the topology of enzyme activity in the cell, this phenomenon has also provided a valuable molecular biological tool for determining the location of membrane protein domains by the use of site-specific fusions with the alkaline phosphatase gene.

The periplasm contains a high concentration of macromolecules and is very viscous, thereby severely limiting the rate of lateral protein diffusion (Brass *et al.*, 1986). As well as the proteins mentioned above, it contains peptidoglycan and acidic glucose oligosaccharides known as membrane-derived oligosaccharides (MDOs). Electron microscopy in low-temperature fixed and stained sections reveals electron-dense material between the clear peptidoglycan layer and the cytoplasmic membrane surface. This has been interpreted as being caused by the presence of additional peptidoglycan of low cross-linkage (Hobot *et al.*, 1984), but this has not been detected in detailed studies of peptidoglycan architecture in *E. coli* (see Chapter 7). MDOs, β1–2 glucose oligosaccharides with β1–6 branches and glycerophosphate, ethanolamine phosphate and succinyl ester substituents are major components of the periplasm. They are sufficiently large molecules to be trapped in the periplasm by the outer membrane and are involved in the osmoregulation of the periplasm, and increase in amount on osmotic downshift (Rumley *et al.*, 1992).

Interaction of the Membrane and Cell Wall

The outer surface of the cytoplasmic membrane is surrounded by the cell wall in Gram-positive bacteria and the periplasm, containing the peptidoglycan layer, in Gram-negative bacteria. It is therefore likely that macromolecular components of the outer membrane surface penetrate into, and interact with, these other envelope regions. This is clearly true of membrane-bound structures such as pili, fimbriae and flagella that penetrate the wall and project at the outer surface of the bacterium. It is also the case with some membrane-anchored lipoproteins of high molecular weight, such as the non-fimbrial adhesins SarA and SsaB of oral streptococci (Kehoe, 1994). Similarly, amphiphilic polymers such as lipoteichoic acid that are anchored through a lipid to the membrane surface in Gram-positive bacteria, and MDOs that are distributed throughout the periplasm in Gram-negative bacteria, are likely to penetrate into the hydrophilic mesh-like structure of the peptidoglycan. The nature of the interactions between the peptidoglycan-bound components of the cell wall and these periplasmic and membrane-attached molecules is very poorly understood. Some wall-linked proteins in Gram-positive bacteria, such as staphylococcal protein A, possess a proline-rich region close to their C-terminal peptidoglycan anchor and it has been suggested that this has evolved to provide the flexibility required for the protein to weave its way through the peptidoglycan mesh (Kehoe, 1994). Some hydrolytic enzymes that are periplasmic in Gram-negative bacteria, such as phosphatases, are retained in the cell wall of Gram-positive bacteria, apparently by the possession of a high positive surface charge that attaches them ionically to the fixed negative charges of the teichoic acid–peptidoglycan complex. However, the ways in which less cationic proteins, and anionic polysaccharides interact with the anionic wall, and the effect this might have on their conformation and functional properties, are less predictable, though potentially very important.

The Bacterial Cell Wall as a Dynamic Structure

Chaloupka *et al.* (1962) first demonstrated that the cell wall peptidoglycan of some bacteria continuously turns over during growth, leading to the release of fragments of peptidoglycan and undegraded accessory wall polymers into the culture fluid. The phenomenon is now known to be displayed by many, or possibly all, bacteria (Doyle *et al.*, 1988). Turnover can be as high as 50% of the wall per cell generation and can readily be visualised by electron microscopy, giving a characteristic crenelated appearance to the cell surface. Thus, new cell wall material has to be synthesised much faster than is accounted for by cell growth alone (Merad *et al.*, 1989). The wall turnover rates of Gram-negative bacteria were believed for many years to be much lower than those of Gram-positives. However, the discovery of a highly organised, coupled degradation and uptake system for the re-utilisation of peptidoglycan turnover products in *E. coli* and other Gram-negative bacteria has provided an alternative explanation for the failure to detect peptidoglycan turnover products in the culture fluid of these bacteria (Höltje, 1995). Calculations of turnover rates that take re-utilisation of the products into account indicate similar rates to those of Gram-positive bacteria.

Turnover is catalysed by several classes of enzyme known collectively as autolysins and most species contain several different classes of these (Smith *et al.*, 2000) (Chapter 7). It is evident that the activities of these enzymes must be tightly controlled to avoid cell suicide by wall lysis. Indeed, autolysin-induced cell disruption is a major reason for the bactericidal action of some antibiotics, such as the β-lactams (Goodell *et al.*, 1976; Kitano *et al.*, 1980). The solubilisation of cell wall antigens resulting from antibiotic-induced autolysis has possible implications for pathogenicity (Periti and Mazzei, 1998; van Langevelde *et al.*, 1998). How autolysin action is controlled and localised during normal growth is unknown, but recent evidence indicates mechanisms for the specific localisation of autolysins in *Staph. aureus* (Baba and Schneewind, 1998).

Cell wall turnover has some important implications for pathogenicity. First, and particularly significantly for Gram-positive bacteria with a thick layer of peptidoglycan-based cell wall, shedding an opsonised cell surface by wall turnover may contribute to evasion of the immune response. Secondly, turnover allows a bacterium to adapt by changing the composition of its cell wall much faster than would be possible if new cell wall was only synthesised at the rate of cell growth. This may play a role in such processes as antigen switching and the shedding of pili and fimbriae. Thirdly, wall turnover results in the solubilisation and dissemination of cell wall antigens in an infected host, independent of host defence processes such as phagocytosis or complement-mediated killing (Hancock, 1989; Hancock and Cox, 1991). Although some bacteria effectively recover much of their peptidoglycan turnover products, wall turnover inevitably leads also to the release of other wall components that were attached, covalently or otherwise, to the peptidoglycan (Wiedemann *et al.*, 1998). Very little information is available on the significance of this process.

A combination of controlled wall turnover and different patterns of wall growth (**Fig. 5**) imposed by different cell growth and division strategies has the potential to produce a cell with localised differences in surface chemistry. In *Bacillus*, this process generates cell poles containing wall material that is older than, and hence potentially different from the cylindrical part of the cell surface. It can thus be experimentally manipulated by controlled changes in growth conditions (Clarke-Sturman *et al.*, 1989). In the branched, filamentous actinomycete bacteria the cell wall grows mainly, or solely, at the apical tips of branches, resulting in apical regions of immature cell surface that differ from the rest of the filament in the exposure of peptidoglycan functional groups (Schwedock *et al.*, 1997).

Particularly striking examples of asymmetrical wall composition have been found in the streptococci. Here, the mode of binary fission involves wall growth principally or solely as a septum that develops into hemispheres of the walls of the two daughter cells. The other hemisphere of wall of each daughter cell is the partially conserved wall of the parent cell, and is thus older than the newly formed wall hemisphere. In conjunction with temporal regulation of gene expression, therefore, streptococcal replication can give rise to markedly zonal cell surface composition (**Fig. 6**). This is evident in the assymetrical distribution of protein

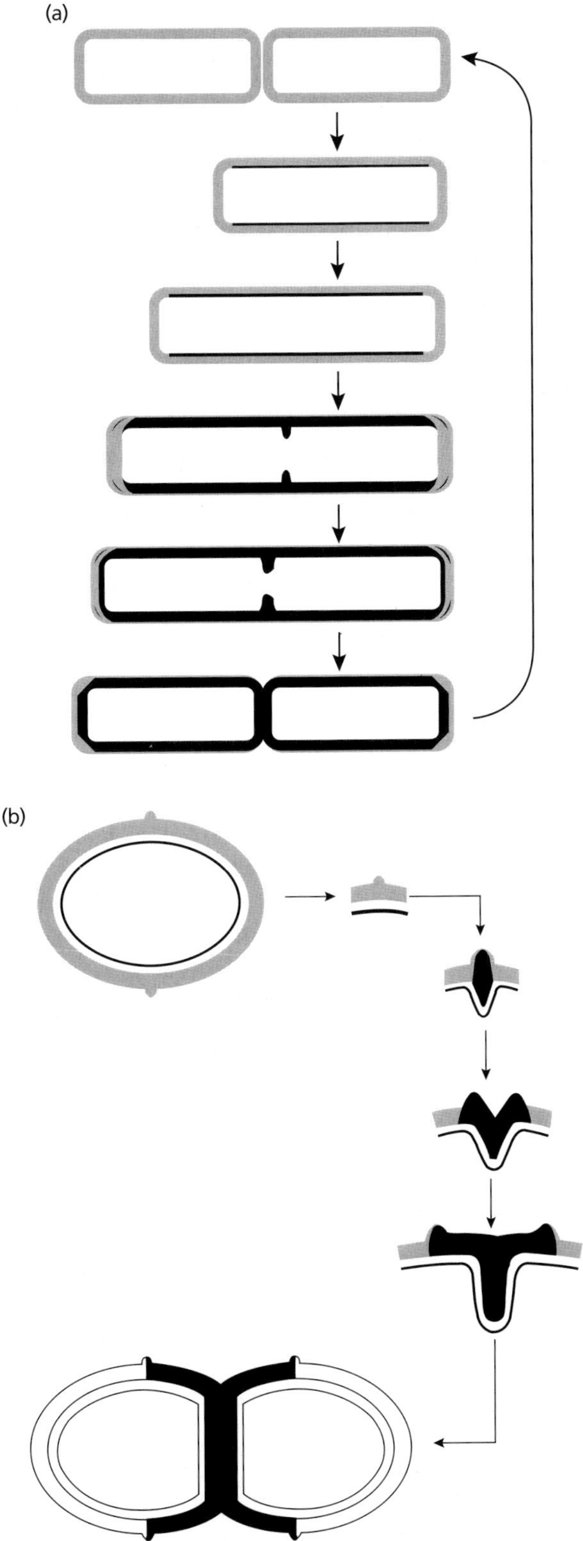

Fig. 5 Comparison of the modes of cell wall growth of bacilli and cocci during division by binary fission. (a) Inside-to-outside growth of the cell wall of a Gram-positive bacillus during one cell division, showing the insertion of new cell wall material (black) into the entire undersurface of the cylindrical wall during elongation and subsequent septation. Old cell wall (grey) is lost continuously by turnover during this process to maintain the wall at constant thickness. Based on Archibald *et al*., 1993. (b) Growth of the cell wall of a coccus (*Enterococcus hirae* illustrated) during one cell division, showing that completely new hemispheres of cell wall (black) develop by septal growth. It should be noted that turnover of the old (grey) cell wall may also occur during this process but is not illustrated. Based on Harold, 1990.

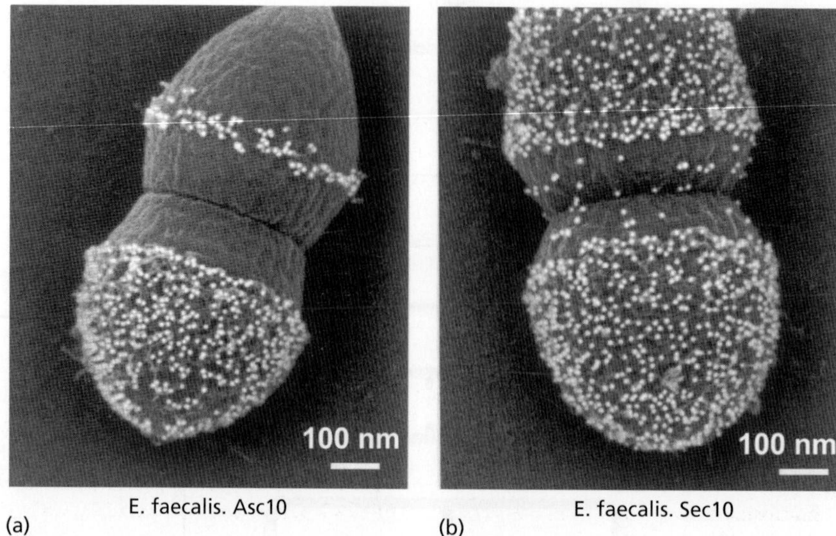

(a) E. faecalis. Asc10 (b) E. faecalis. Sec10

Fig. 6 Asymmetrical distribution of *Enterococcus* mating proteins on the cell wall. The figure shows replicating *Enterococcus faecalis* cells immuno-gold labelled for surface proteins (a) Asc10 or (b) Sec10 that play roles in regulating pheromone-inducible gene transfer involving the conjugative plasmid pINY1801. Comparison with Figure 5b reveals the association of surface expression of the proteins with cell wall formed at particular periods in the cell cycle. Reproduced with permission from Olmsted *et al.* (1993) *Copyright* © 1993 American Society for Microbiology.

structures such as fibrils and fimbria, and polysaccharides on the surface of streptococci (Olmsted *et al.*, 1993; Yamashita *et al.*, 1998). The same phenomenon is theoretically possible, for example, in staphylococci, but has not been described.

References

Archibald AR, Hancock IC, Harwood CR (1993) Cell walls. In: Sonenshein AL, Hoch JA, Losick R (eds) *Bacillus subtilis and Other Gram-Positive Bacteria*. Washington, DC: ASM Press, Ch. 27.

Baba T, Schneewind O (1998) Targeting of muralytic enzymes to the cell division site of Gram-positive bacteria: repeat domains direct autolysin to the equatorial surface ring of *Staphylococcus aureus*. *EMBO J.* 17: 4639–4646.

Bardwell JCA, Lee J-O, Jander G, Martin N, Berlin D, Beckwith J (1993) A pathway for disulfide bond formation *in vivo*. *Proc. Natl Acad. Sci. USA* 90: 1038–1042.

Barry CE III, Mdluli K (1996) Drug sensitivity and environmental adaptation of mycobacterial cell wall components. *Trends Microbiol.* 4: 275–281.

Besra GS, Chatterjee D (1994) Lipids and carbohydrates of *Mycobacterium tuberculosis*. In: Bloom BR (ed.) *Tuberculosis: Pathogenesis, Protection and Control*. Washington, DC: American Society for Microbiology, Ch. 20.

Beveridge TJ, Pouwels PH, Sara M *et al.* (1997) Functions of S-layers. *FEMS Microbiol. Rev.* 20: 99–149.

Bishop DG, Op den Kamp JAG, van Deenan LLM (1977) The distribution of lipids in the protoplast membranes of *Bacillus subtilis*. *Eur. J. Biochem.* 80: 381–391.

Brass JM, Higgins CF, Foley M, Rugman FA, Birmingham J, Garland PB (1986) Lateral diffusion of proteins in the periplasm of *Escherichia coli*. *J. Bacteriol.* 165: 787–794.

Brennan PJ, Draper P (1994) Ultrastructure of *Mycobacterium tuberculosis*. In: Bloom BR (ed.) *Tuberculosis: Pathogenesis, Protection and Control*. Washington, DC: American Society for Microbiology, Ch. 19.

Brennan PJ, Nikaido H (1995) The envelope of mycobacteria. *Annu. Rev. Biochem.* 64: 29–63.

Chaloupka J, Kreckova P, Rihova L (1962) Peptidoglycan turnover in *Bacillus megaterium*. *Experientia* 18: 62–64.

Clarke-Sturman AJ, Archibald AR, Hancock IC, Harwood CR, Merad T, Hobot J (1989) Cell wall assembly in *Bacillus subtilis*: partial conservation of polar material. *J. Gen. Microbiol.* 135: 657–665.

Cowan SW, Schirmer T, Rummel G *et al.* (1992) Crystal structures explain functional properties of two *E. coli* proteins. *Nature* 358: 727–733.

Daffé M, McNeil M, Brennan PJ (1993) Major structural features of the cell wall arabinogalactans of *Mycobacterium*, *Rhodococcus* and *Nocardia* spp. *Carbohydr. Res.* 249: 383–398.

de Mendoza D, Grau R, Cronan JE (1993) Biosynthesis and function of membrane lipids. In: Sonenshein AL, Hoch JA, Losick R (eds) *Bacillus subtilis and Other Gram-Positive Bacteria*. Washington, DC: ASM Press, Ch. 28.

Derman AI, Beckwith J (1991) *Escherichia coli* alkaline phosphatase fails to acquire disulfide bonds when retained in the cytoplasm. *J. Bacteriol.* 173: 7719–7722.

Doige CA, Ames F-LG (1993) ATP-dependent transport systems in bacteria and humans: relevance to cystic fibrosis and multidrug resistance. *Annu. Rev. Microbiol.* 47: 291–319.

Doyle RJ, Chaloupka J, Vinter V (1988) Turnover of cell walls. *Microbiol. Rev.* 52: 554–567.

Draper P, Khoo KH, Chatterjee D, Dell A, Morris HR (1997) Galactosamine in walls of slow-growing mycobacteria. *Biochem. J.* 327: 519–525.

Driessen AJM, Fekkes P, van der Wolk JPW (1998) The Sec system. *Curr. Opin. Microbiol.* 1: 216–222.

Goodell EW, Lopez R, Tomasz A (1976) Suppression of lytic effect of β-lactams in *Escherichia coli* and other bacteria. *Proc. Natl Acad. Sci. USA* 73: 3293–3297.

Hancock IC (1989) Encapsulation of coagulase-negative staphylococci. *Z. Bakteriol.* 272: 11–18.

Hancock IC (1991) Microbial cell surface architecture. In: Mozes N, Rouxhet PG, Busscher HJ, Handley P (eds) *Analysis of Microbial Surfaces.* New York: VCH Publishers, Ch. 2.

Hancock IC, Cox CM (1991) Turnover of cell surface-bound capsular polysaccharide in *Staphylococcus aureus. FEMS Microbiol. Lett.* 77: 25–30.

Harold FM (1990) To shape a cell: an enquiry into the causes of morphogenesis of micro-organisms. *Microbiol. Rev.* 54: 381–431.

Hobot JD, Carlemalm E, Villiger W, Kellenberger E (1984) Periplasmic gel: new concept resulting from the reinvestigation of bacterial cell envelope ultrastructure by new methods. *J. Bacteriol.* 160: 143–152.

Höltje JV (1995) From growth to autolysis – the murein hydrolases in *Escherichia coli. Arch. Microbiol.* 164: 243–254.

Jarlier V, Nikaido H (1994) Mycobacterial cell wall: structure and role in natural resistance to antibiotics. *FEMS Microbiol. Lett.* 123: 11–18.

Ji G, Beavis RC, Novick RP (1995) Cell density control of staphylococcal virulence mediated by an octapeptide pheromone. *Proc. Natl Acad. Sci. USA* 92: 12055–12059.

Kadner RJ (1996) The cytoplasmic membrane. In: Neidhardt FC (ed.) *Escherichia coli, Salmonella typhimurium, Cellular and Molecular Biology*, 2nd edn. Washington, DC: ASM Press, Ch. 7.

Kasik JE (1979) Mycobacterial β-lactamases. In: Hamilton-Miller JMT, Smith JT (eds) *Beta-lactamases.* London: Academic Press, pp. 339–350.

Kehoe MA (1994) Cell-wall-associated proteins in Gram-positive bacteria. In: Ghuysen J-M, Hackenbeck R (eds) *Bacterial Cell Wall.* Amsterdam: Elsevier Science, Ch. 11.

Kitano K, Williamson R, Tomasz A (1980) Murein hydrolase defect in the β-lactam tolerant mutants of *Escherichia coli. FEMS Microbiol. Lett.* 7: 133–136.

Kremer L, Baulard AR, Besra GS (2000) Genetics of mycolic acid biosynthesis. In: Hatfull GF, Jacob WR (eds) *Molecular Genetics of Mycobacteria.* Washington, DC: ASM Press, pp. 173–190.

Koch AL (1998) The biophysics of the Gram-negative periplasmic space. *Crit. Rev. Microbiol.* 24: 23–59.

Lemaire M, Ohayon H, Gounon P, Fujino T, Béguin P (1995) OlpB, a new outer layer protein of *Clostridium thermocellum* and binding of its S-layer-like domains to components of the cell wall. *J. Bacteriol.* 177: 2451–2459.

Lupas A, Engelhardt H, Peters J, Santarius U, Volker S, Baumeister W (1994). Domain structure of the *Acetogenium kivui* surface layer revealed by electron crystallography and sequence analysis. *J. Bacteriol.* 176: 1224–1233.

McNab R, Forbes H, Handley PS, Loach DM, Tannock GW, Jenkinson HF (1999) Cell wall-anchored CshA polypeptide (259 kilodaltons) in *Streptococcus gordonii* forms surface fibrils that confer hydrophobic and adhesive properties. *J. Bacteriol.* 181: 3087–3095.

Miller KJ, Kennedy EP (1987) Transfer of phosphoethanolamine residues from phosphatidyl ethanolamine to the membrane-derived oligosaccharides of *Escherichia coli. J. Bacteriol.* 169: 682–686.

Merad T, Archibald AR, Hancock IC, Harwood CR, Hobot J (1989) Cell wall assembly in *Bacillus subtilis. J. Gen. Microbiol.* 135: 645–655.

Minnikin D.E (1982) Lipids complex lipids, their chemistry, biosynthesis and roles. In: Ratledge C, Stanford J (eds) *The Biology of the Mycobacteria*, Vol. 1. London: Academic Press, pp. 95–184.

Minnikin DE (1991) Chemical principles in the organisation of lipid components in the mycobacterial cell envelope. *Res. Microbiol.* 142: 423–427.

Navarre WW, Schneewind O (1999) Surface proteins of Gram-positive bacteria and mechanisms of targeting to the cell envelope. *Microbiol. Mol. Biol. Rev.* 63: 174–232.

Nikaido H (1996) Outer membrane. In: Neidhardt FC (ed.) *Escherichia coli, Salmonella typhimurium, Cellular and Molecular Biology*, 2nd edn. Washington, DC: ASM Press, pp. 29–47.

Nikaido H, Kim S-H, Rosenberg EY (1993) Physical organisation of lipids in the cell wall of *Mycobacterium chelonae. Mol. Microbiol.* 8: 1025–1030.

Nummila K, Kilpeläinen I, Zahringer U, Vaara M, Helander IM (1995) Lipopolysaccharides of polymyxin B-resistant mutants of *Escherichia coli* are extensively substituted by 2-aminoethylpyrophosphate and contain aminoarabinose in lipid A. *Mol. Microbiol.* 16: 271–278.

Oliver DB (1996) Periplasm. In: Neidhardt FC (ed.) *Escherichia coli, Salmonella typhimurium, Cellular and Molecular Biology*, 2nd edn. Washington, DC: ASM Press, pp. 88–103.

Olmsted SB, Erlandsen SL, Dunny GM, Wells C (1993) High-resolution visualisation by field emission scanning electron microscopy of *Enterococcus faecalis* surface proteins encoded by the pheromone-inducible conjugative plasmid pCF10. *J. Bacteriol.* 175: 6229–6237.

Periti P, Mazzei T (1998) Antibiotic-induced release of bacterial cell wall components in the pathogenesis of sepsis and septic shock: a review. *J. Chemother.* 10: 427–448.

Rietschel ET, Kirikae T, Schade FU *et al.* (1994) Bacterial endotoxin: molecular relationships of structure to activity and function. *FASEB J.* 8: 217–225.

Rothman JE, Kennedy EP (1977a) Rapid transmembrane movement of newly synthesised phospholipids during membrane assembly. *Proc. Natl Acad. Sci. USA* 74: 1821–1825.

Rothman JE, Kennedy EP (1977b) Asymmetrical distribution of phospholipids in the membrane of *Bacillus megaterium*. *J. Mol. Biol.* 110: 603–618.

Rumley MK, Therisod M, Weissborn AC, Kennedy EP (1992) Mechanisms of regulation of membrane-derived oligosaccharides in *Escherichia coli*. *J. Biol. Chem.* 267: 11806–11810.

Sara M, Egelseer EM, Dekitsch C, Sleytr UB (1998) Identification of two binding domains, one for peptidoglycan and another for a secondary cell wall polymer, on the N-terminal part of the S-layer protein SbsB from *Bacillus stearothermophilus* PV72/p2. *J. Bacteriol.* 180: 6780–6783.

Sarvas M (1986) Protein secretion in bacilli. *Curr. Top. Microbiol. Immunol.* 125: 103–125.

Schwedock J, McCormick JR, Angert ER, Nodwell JR, Losick R (1997) Assembly of the cell-division protein FtsZ into ladder-like structures in the aerial hyphae of *Streptomyces coelicolor*. *Mol. Microbiol.* 25: 847–858.

Smith TJ, Blackman SA, Foster SJ (2000) Autolysins of *Bacillus subtilis*: multiple enzymes with multiple functions. *Microbiology* 146: 249–262.

Sutcliffe IC (1998) Cell envelope composition and organisation in the genus *Rhodococcus*. *Antonie van Leewenhoek Int. J. Gen. Mol. Microbiol.* 74: 49–58.

Trias J, Jarlier V, Benz R (1992) Porins in the cell wall of mycobacteria. *Science* 258: 1479–1481.

van Langevelde P, vanDissel JT, Ravensbergen E, Appelmelk BJ, Schrijver IA, Groeneveld PHP (1998) Antibiotic-induced release of lipoteichoic acid and peptidoglycan from *Staphylococcus aureus*: Quantitative measurements and biological reactivities. *Antimicrob. Agents Chemother.* 42: 3073–3078.

Weiss MS, Abele U, Weckesser J, Welte W, Schiltz E, Schultz GE (1991) Molecular architecture and electrostatic properties of a bacterial porin. *Science* 254: 1627–1629.

Wiedemann B, Pfeifle D, Wiegand I, Janas E (1998) β-Lactamase induction and cell wall recycling in gram-negative bacteria. *Drug Resistance Updates* 1: 223–226.

Yamashita Y, Tsukioka Y, Tomihisa K, Nakano Y, Koga T (1998) Genes involved in cell wall localization and side chain formation of rhamnose-glucose polysaccharide in *Streptococcus mutans*. *J. Bacteriol.* 180: 5803–5807.

4

Capsular Polysaccharides

Klaus Jann and Barbara Jann

Max-Planck-Institut für Immunbiologie, Freiburg, Germany

The cell surface is of great importance in the interactions of bacteria with their surroundings – culture media, soil, plant or animal systems. It is the site where receptors for viruses, cells or for soluble ligands, such as complement factors or antibodies, are located. In Gram-negative bacteria the cell is surrounded by the cytoplasmic membrane, on which the peptidoglycan is layered. This is overlaid with a second membrane, the outer membrane, which contains protein and lipopolysaccharide (LPS, also termed O-antigen) as essential surface antigens. The space between the inner and outer membranes, including the peptidoglycan, is termed the periplasm. It contains a number of enzymes, transport proteins and osmotic stabilisers. In contrast, Gram-positive bacteria do not have an outer membrane, and surface components, such as proteins and teichoic acids, are directly associated with the peptidoglycan.

Both Gram-positive and Gram-negative bacteria may produce an additional extracellular compartment, the capsule, which covers the bacterial cell and often prevents the reaction of underlying cell surface antigens with their homologous antibodies. In such cases, capsules are inhibitors of specific serological reactions (Ørskov *et al.*, 1977). The inhibition of LPS-mediated O-agglutination by the overlaying capsule was the basis for the original definition of capsules in *Escherichia coli*. On the basis of their antigenic specificity they were designated as K-antigens (from the German term 'Kapsel') (Ørskov *et al.*, 1977).

Capsules are found in several bacteria of medical importance, especially in extraintestinal and invasive strains (Table 1). With the exception of the poly-γ-glutamate capsule of *Bacillus anthracis*, bacterial capsules are polysaccharides, and in most cases capsular polysaccharides (CPS) are negatively charged (acidic). They form a gelatinous mass around the cell, and contain up to about 90% water. Great care has to be taken, therefore, in the visualisation of capsules and their ultrastructural analysis (Bayer, 1990). Light-microscopy techniques include the use of India ink as a contrasting background, against which the capsule is seen as an unstained halo around the cells, and the use of the capsule 'quellung' reaction, in which antibodies stabilise the capsule and increase its volume. Fluorescent anticapsular antibodies can be used in light microscopy. In electron microscopy (Bayer, 1990), for the analysis of capsular ultrastructure capsules are stabilised with a basic protein, such as ferritin or cationised (esterified) proteins. In immunoelectron microscopy, a more specific variant of electron microscopy, antibodies are used for the stabilisation and demonstration of capsules. Bacteria are either demonstrated as such (whole cell mount) or as ultrathin sections after embedding in a resin, such as Epon or Lowicryl. In immunoelectron microscopy (Bayer, 1990) antibodies are used either alone or in combination with a second antibody labelled with gold spheres of defined sizes (immunogold technique). Electron micrographs of encapsulated *E. coli* using two different techniques are shown in Fig. 1.

The term 'capsule' commonly signifies an extracellular polysaccharide physically associated with the cell surface. In contrast to a polysaccharide capsule,

 doi:10.1006/bkmm.2001.0004

Table 1 Encapsulation of some medically important bacteria

	Capsular antigens		References on Structure
	Number of Serotypes	Chemical Nature	
Gram-negative			
Klebsiella pneumoniae	90	PS	Kenne and Lindberg, 1983
Escherichia coli[a]	70	PS	Jann and Jann, 1990, 1997
Neisseria meningitidis	13	PS	Jennings, 1990
Haemophilus influenzae	6	PS	Kenne and Lindberg, 1983
Salmonella	1	PS	Heyns and Kiessling, 1967; Daniels *et al.*, 1989
Pseudomonas aeruginosa	1	PS	Skjak-Braek *et al.*, 1986
Gram-positive			
Streptococcus pneumoniae[b]	80	PS	Kenne and Lindberg, 1983
Streptococcus group B[c]	5	PS	Jennings, 1990
Streptococcus pyogenes	1	PS	Hök *et al.*, 1984
Staphylococcus aureus	11	PS	Karakawa and Vann, 1982; Sau and Lee, 1996
Bacillus anthracis	1	PG	Troy, 1980

[a] The M polysaccharide (colanic acid) is not included in this list; it forms a capsule but is not a type specific capsular antigen.

[b] Most strains of *Streptococcus pneumoniae* produce a teichoic acid-like polysaccharide (C-PS) in addition to the type specific capsular antigen (Kenne and Lindberg, 1983).

[c] Group B *Streptococcus* also produces a group antigen (C-substance), a multi-antennary polysaccharide that consists of rhamnose and phosphate (Michon *et al.*, 1988).

PS, polysaccharide; PG, polyglutamate.

which is retained by the bacterial cell, slime polysaccharides are released into the medium. Both capsular and slime polysaccharides are negatively charged and the differentiation between them is not strict. Thus, the mucus polysaccharide (colanic acid, M-antigen) produced by many bacteria at low temperature is usually considered to be a capsule. The extracellular polysaccharide (alginate) of *Pseudomonas aeruginosa*, which is an important pathogenic factor in chronic pulmonary infection of cystic fibrosis patients (May *et al.*, 1991; Wang *et al.*, 1996), may also be regarded as a capsule.

Chemistry of Capsular Polysaccharides

Primary Structures

The capsular polysaccharides (CPS) are acidic polysaccharides made up of repeating oligosaccharide units (Karakawa and Vann, 1982; Kenne and Lindberg, 1983; Jann and Jann, 1990, 1997; Jennings, 1990), which may be either linear or branched. The primary structures can be formulated by the respective repeating unit and the joining linkages. The negative charges are due to acidic sugar components or substitution with charged non-carbohydrate moieties. The most prominent acidic components, together with other frequent constituents, are listed in Table 2 and the structures of some CPSs discussed in this chapter are shown in Table 3.

M-antigen (colanic acid) is structurally related to the K30 CPS, which is reflected in the serological cross-reactivity of these polysaccharides (Grant *et al.*, 1969; Garegg *et al.*, 1971).

The CPSs are sometimes substituted with *O*-formyl- or *O*-acetyl-groups (in ester linkage) or with pyruvate (in ketosidic linkage). Amino acids are found linked either to the hydroxyl groups of sugar rings (esters) or to the carboxyl group of hexuronic acids (amides) (see Table 3).

Secondary Structures

Little information is available about the secondary structures of CPSs. The polySia (K1 antigen of *E. coli* and CPS of *Neisseria meningitidis* group B – see Tables 2 and 3) have been reported to be helical, either throughout the polysaccharide chain or as helical regions of the polysaccharide (Jennings *et al.*, 1981, 1993; Brisson *et al.*, 1992). X-ray spectroscopy of the *E. coli* K4 CPS (Atkins, personal communication) and NMR studies on the *E. coli* K5 CPS (Torrii, personal

(a)

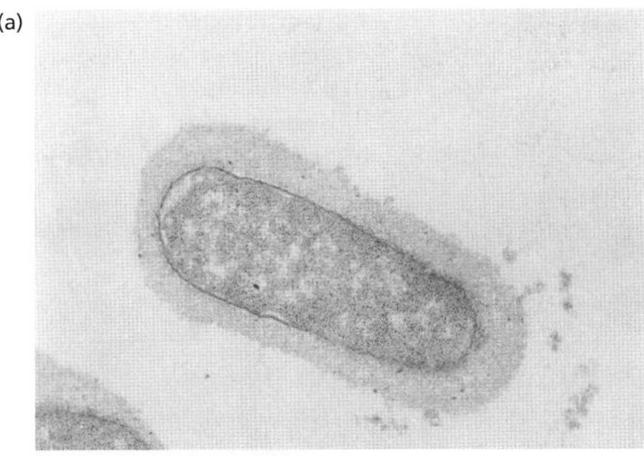

(b)

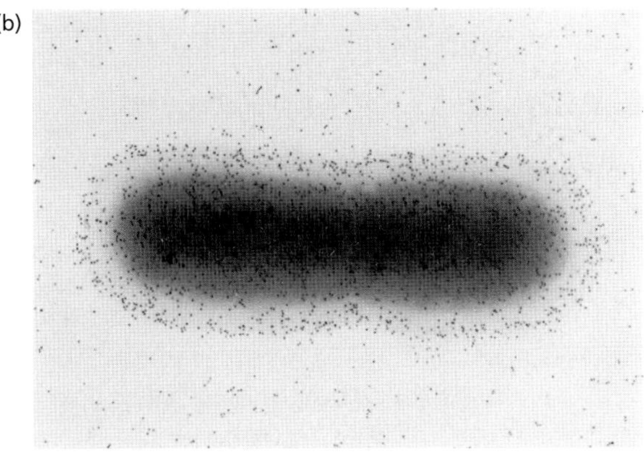

Fig. 1 Electron micrographs of encapsulated *E. coli* O18:K5:H$^-$. (a) Ultrathin section of an Epon-embedded bacterium after treatment with a murine monoclonal anti-K5 antibody. (b) Whole mount of *E. coli* O18:K5:H$^-$ treated first with a murine monoclonal antibody and then with an anti-mouse (second) antibody, conjugated to 10-nm gold beads.

communication) also indicate helical superstructures for these polysaccharides.

In the serotype III CPS of *Streptococcus* group B (Table 3), hydrogen bonding between the side-chain NeuNAc and the main-chain GlcNAc residues keeps the side-chain Gal in an orientation that determines the major epitope of the CPS (Jennings *et al.*, 1981; Wessels *et al.*, 1989).

Lipid Substitution of CPSs and their Attachment to the Cell Surface

In many CPSs of Gram-negative bacteria the reducing end is capped by a lipid substituent. In the group 2 CPSs of *E. coli* (see below), *N. meningitidis* group B and *Haemophilus influenzae* type b, the reducing sugars are substituted with a phosphatidic acid in a phosphodiester linkage (Gotschlich *et al.*, 1981; Schmidt and Jann, 1982; Kuo *et al.*, 1985). In *E. coli K5*, the substituent is phosphatidyl-Kdo (Schmidt and Jann, 1982). The K30 and K40 CPSs of capsule groups 1 and 4 of *E. coli* (see below) and presumably also of

Table 2 Components of CPS

Carbohydrate	Non-carbohydrate
Acidic	
GlcA, GalA, ManA	Serine, threonine
ManNAcA, GalNAcA	Phosphate
Qui4NMal	Succinate
Kdo, NeuNAc	Pyruvate
Neutral	
Glc, Gal, Man	Formate, acetate
Rha, Fuc, Fru	Ribitol
Rib	
GlcNAc, GalNAc,	
ManNAc, QuiNAc	
FucNAc	

Glc, glucose; Gal, galactose; Man, mannose; Rha, rhamnose; Fuc, fucose; Fru, fructose; Rib, ribose; GlcNAc, *N*-acetylglucosamine; GalNAc, *N*-acetylgalactosamine; ManNAc, *N*-acetylmannosamine, QuiNAc; *N*-acetylquinovosamine (2,6-dideoxy-2-*N*-acetylglucose); FucNAc, *N*-acetylfucosamine; GlcA, glucuronic acid; GalA, galacturonic acid; ManA, mannuronic acid; ManNAcA, *N*-acetylmannosuronic acid; GalNAc, *N*-acetylgalacturonic acid; Qui4NMal, 4,6-dideoxy-4-malonyl aminoglucose (Sieberth *et al.*, 1993); Kdo,2-keto-3-deoxy-*manno* octonic acid (Unger, 1981); NeuNAc, *N*-acetylneuraminic acid (also sialic acid, Sia); (Reuter and Gabius, 1996; Schauer, 1991).

Klebsiella have core–lipid A at their reducing end (Jann *et al.*, 1992; Whitfield and Valvano, 1993). Since core–lipid A is a typical part of enterobacterial LPSs, these CPSs appear to be LPSs. The lipid substitutions are shown schematically in Fig. 2.

Only a small portion of the K30 CPS is substituted with core–lipid A, the greater part is either not substituted at all or, more probably, with phosphatidic acid. The presence of core–lipid A and phosphatidic acid on different populations of the same polysaccharide has also been described for the enterobacterial common antigen (ECA), which surrounds the bacterial cell as a (micro)capsule (Kuhn *et al.*, 1988).

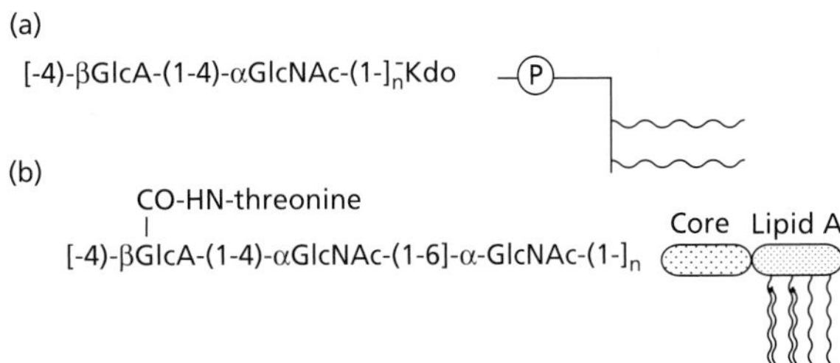

Fig. 2 (a) Substitution of the *E. coli* K5 capsular polysaccharide (group 2) with phosphatidyl-Kdo (Bronner *et al.*, 1993; Schmidt and Jann, 1982). (b) Substitution of the *E. coli* K40 CPS (group 1) with core–lipid A (Jann *et al.*, 1992). The wriggly lines represent fatty acid substituents. In (a) these are predominantly C16 and C18, and in (b) they are C14(βOH) and C12(βOH) (Rietschel *et al.*, 1984).

Table 3 Structures of some capsular polysaccharides (CPS)

CPS of	Structure	References
Escherichia coli K1[a]	→8)–αNeuNAc–(2→	Ørskov *et al.*, 1979
Neisseria meningitidis B		Bhattacharjee *et al.*, 1975
Moraxella nonliquefaciens		Devi *et al.*, 1991
Pasteurella haemolytica A2		Adlam *et al.*, 1987
N. meningitidis C	→9)-αNeuNAc-2-(→	Bhattacharjee *et al.*, 1975
E. coli K92	→8)-αNeuNAc-(2→9)-α-NeuNAc-(2→	Glode *et al.*, 1977
E. coli K4	→4)-βGlcA-(1→3)-βGalNAc-(1→ 2 ↑ 2 βFru	Rodriguez *et al.*, 1988
E. coli K5	→4)-βGlcA-(1→4)-αGlcNAc-(1→	Vann *et al.*, 1981
E. coli K10	→3)-α-LRha-(1→3)-βDQui4NMal-(1→ 2 | OAc	Sieberth *et al.*, 1993
E. coli K54	→3)-βGlcA-(1→3)-α-Rha-(1→ | CO-NH Threonine (Serine)	Hofmann *et al.*, 1985
E. coli K30	→2)-αMan-(1→3)-βGal-(1→ 3 ↑ 1 GlcA-(1→3)-αGal	Chakraborty *et al.*, 1980
E. coli K40	CO-NH-Threonine | →4)-βGlcA-(1→4)-αGlcNAc(1→6)-αGlcNAc-(1→	Dengler *et al.*, 1986
Haemophilus influenzae B	→3)-βRIb-(1→1)-Ribitol-6- O-PO(OH)-O-	Branefors-Hellander *et al.*, 1976
Aerobacter aerogenes DD45 (*Klebsiella*)	→3)-αGal-(1→3)-αMan-(1→3)-αGal-(1→ 2 ↑ 1 βGlcA	Yurewitz *et al.*, 1971
Salmonella typhi (Vi-ag)	→4)-αGalNAcA-(1→ 2 | O-Ac	Heyns and Kiessling, 1967
Pseudomonas aeruginosa[b]	→4)-βManA-(1→ 2 | (O-Ac)	Skjak-Braek *et al.*, 1986
Staphylococcus aureus type 5	→4)-βManNAcA-(1→4)-αLFucNAc-(1→3)-βDFucNAc-(1→ 3 | O-Ac	Moreau *et al.*, 1990
Staphylococcus aureus type 8	→3)-βManNAcA-(1→3)-αLFucNAc-(1→3)-βDFucNAc-(1→ 4 | O-Ac	Fournier *et al.*, 1984
Streptococcus pyogenes	→4)-βGlcA-(1→3)-βGlcNAc-(1→	Hök *et al.*, 1984
Streptococcus B type III	→4)-βGlc-(1→6)-βGlcNAc-(1→3)-βGal-(1→ 4 ↑ 1 αNeuNAc-(2→3)-βGal	Wessels *et al.*, 1987

Table 3 *(Cont.)*

CPS of	Structure	References
Streptococcus pneumoniae type 3	→4-βGlc-(1→3)-βGlcA-(1→	Reeves and Goebel, 1941
Streptococcus pneumoniae type 19F	→4)-βManNAc-(1→4)-αGlc-(1→2)-αRha-(1→OPO$_3^-$	Jennings *et al.*, 1980
Streptococcus pneumoniae type 23F[d]	αRha 1 ↓ 2 →4)-βGlc-(1→4)-βGal-(1→4)-βRha-(1→ 3 | Gro-2-PO$_3^-$O	Richards and Perry, 1988
Bacillus fragilis CPS-A	4,6,pyruvate ⇓ →3)-αSug-(1→4)-αGalNAc-(1→3)-βGal-(1→ 3 ↑ 1 βGal	Baumann *et al.*, 1992
Bacillus fragilis CPS-B	→3)-βQuiNAc-(1→4)-βGalNAc-(1→4)-αQuiNAc-(1→ 3 ↑ 1 α-LFuc-(1→2)-βGlcNAc 4 | O-PO$_3^{++}$-CH$_2$-$\overset{+}{N}$(CH$_3$)$_3$	Baumann *et al.*, 1992

[a] *O*-Acetylation of the K1 CPs at C7 or C8 results in the form variant K1 (OAc$^+$) (Ørskov *et al.*, 1979).

[b] In contrast to the alginates from other sources, that from *P. aeruginosa* does not contain β-L-guluronic acid. Only some of the mannuronic acid (ManA) residues are *O*-acetylated.

QuiNMal, 4,6-dideoxy-4-malonylglucose (*N*-malonylquinosamine); Gro-2-PO$_3^-$, *sn*-glycerol-2-phosphoryl; Sug, 2,4-dideoxy-4-FucNAc.

The lipid substituent of CPS is generally considered to be a cell anchor, which provides an association with the cell surface by hydrophobic interaction with the outer membrane.

In Gram-positive bacteria, which lack an outer membrane, CPSs are attached to the cell wall peptidoglycan (De Cueninck *et al.*, 1982; Sørensen *et al.*, 1988; Sørensen *et al.*, 1990; Daniels *et al.*, 1989). There are indications for a covalent linkage, possibly through a phosphodiester bond to muramic acid.

General Considerations

In general, CPSs are distinctly longer, up to about 100 kDa per chain, than the polysaccharide moieties (O-PS) of the cell wall (outer membrane) O-antigens of Gram-negative bacteria (LPS, up to about 10–20 kDa). In some Gram-negative bacteria the LPS seems to occur in two physically distinct forms with the same primary structure. Whereas one O-PS has the chain length found in LPS (up to about 20 kDa), the other has the larger size usually reported for CPS (around 100 kDa). In *E. coli* O111, which has a neutral O-PS, the long (L)PS apparently extends so far out from the bacterial surface that it exerts biological functions typical of CPS (see below) (Joiner, 1985; Krallmann-Wenzel and Schmidt, 1994). In this case the term LPS capsule seems appropriate. Similarly, *Proteus vulgaris* O19 and O57 form LPS with a long-chain O-PS that is considered to be a capsule (Beynon *et al.*, 1992; Perry and McLean, 1994; Uhrin *et al.*, 1994).

As far as is known, the capsules of most Gram-negative (*Neisseria*, *Haemophilus*, *Klebsiella*) and Gram-positive bacteria (*Streptococcus*, *Staphylococcus*) appear as a more or less homogeneous group of acidic polysaccharides with comparable chemistry, biosynthesis and genetics.

On the basis of biochemical and genetic parameters, the capsules of *E. coli* have been divided into groups, originally termed IA, IB, II and III (sometimes also termed II/III) (Finke *et al.*, 1990; Jann and Jann,

Table 4 Classification of *Escherichia coli* capsules

Characteristic	Groups			
	1	2	3	4
Former K-antigen group	IA	II	I/II or III	IB
Expressed in O groups	O8, O9, O20(O101)	Many	Many	O8, O9
Terminal lipid moiety	Lipid–A-core in K_{LPS} unknown for CPS	Phosphatidyl-Kdo	Phosphatidyl-Kdo?	Lipid–A-core in K_{LPS} unknown for CPS
Chain growth at	Reducing end	Non-reducing end	Non-reducing end	Reducing end
Mode of polymerization	Block	Single chain	Single chain?	Block
Translocation[a] proteins	Wzx (PST[b])	ABC-2 system[c]	ABC-2 system?	Wzx (PST)
Transport[d] proteins	Wza, Wzc	KpsD, KpsE	(KpsE, KpsD)?	Unknown
Elevated CKS	No	Yes	No	No
Gene locus/close to	*cps,/his*	*kps/serA*	*kps/serA*	*rfb/his*
Expression below 20°C	Yes	No	Yes	Yes
Model CPS system	K30	K1, K5	K10, K54	K40
Similar to	*Klebsiella*	*Neisseria, Haemophilus*	*Neisseria, Haemophilus*	*Neisseria*, many genera

[a] Across the cytoplasmic membrane.
[b] Translocation system according to Paulsen *et al.* (1997).
[c] Consisting of KpsM/KpsT (Wzm and Wzt according to Reeves *et al.* (1996)).
[d] Across the outer membrane.
CKS, CMP-Kdo synthetase.
For details see text.
After Whitfield and Roberts (1999).

1990, 1992, 1997). More recent genetic findings have led to a modified classification of *E. coli* capsules including groups 1 to 4 as shown in Table 4 (Whitfield and Roberts, 1999).

CPSs of groups 1 and 4 occur only in a limited number of O groups, notably O8 and O9; CPSs of groups 2 and 3 occur in many O groups. Interestingly, CPSs of groups 1 and 4 are co-expressed on the cell surface with short-chain versions of the same structure, bound to core–lipid A (Jann *et al.*, 1992; Whitfield and Valvano, 1993; Drummelsmith and Whitfield, 1999). The latter have been designated as K_{LPS} (Whitfield and Valvano, 1993). It is not known whether the long-chain CPSs are lipid-substituted at the reducing end. Group 1 CPSs do not contain amino sugars, which are regular constituents of group 4 CPSs.

Group 2 capsular polysaccharides are characteristically substituted at the reducing end with phosphatidyl-Kdo (Schmidt and Jann, 1982). Since the activated form of Kdo is cytidine monophosphate (CMP)-Kdo, which also participates in LPS biosynthesis, strains with group 2 capsules have a high CMP-Kdo synthetase activity. The capsule-specific enzyme, which is closely related to the one that is LPS-specific, has been isolated and characterised (Rosenow *et al.*, 1995; Jelakovic *et al.*, 1996). The expression of group 2 capsules is thermoregulated (expressed at 37°C but not at 20°C) (Ørskov *et al.*, 1984), whereas those of groups 1, 3 and 4 are not thermoregulated. It should be noted that the thermoregulation (expression below 37°C) of the M-antigen (colanic acid) is opposite to that of the group 2 CPS (see below). The genetic *cps* locus of group 1 CPS and the *rfb* locus of group 4 CPS are linked to *his*, the cluster that directs the biosynthesis of histidine. The genetic *kps* locus of group 2 and 3 capsules is close to the *serA* (the gene for phosphoglycerate dehydrogenase in the biosynthesis of serine) (Ørskov and Nyman, 1974; Vimr *et al.*, 1989; Vimr, 1990; Boulnois *et al.*, 1992).

It is noteworthy that some group 4 CPSs that co-exist with an O-antigen (O_{LPS}), such as the O8 antigen, may occur as the sole LPS of another strain, and are thus an O-antigen. For example, the K87-CPS of *E. coli* O8:K87 and the O32-LPS of *E. coli* O32 have the same structure (Jann *et al.*, 1971). The structures of some of the polysaccharides that can occur as O-antigens and as K-antigens are given in Table 5.

Independently of their immunochemical classification as O- and/or K-antigens, LPSs and CPSs may intercalate and interact with each other on the bacterial surface. The formation of LPS–CPS complexes has been demonstrated *in vitro* (Ørskov and Ørskov,

Table 5 Structures of some surface polysaccharides which occur as cell wall LPS and/or CPS

Structure	Antigen	Bacterium	Reference for structures
αCol 1 ↓ 3 →4)-αGlc-(1→4)-αGal-(1→3)-βGlcNAc-(1→ 6 ↑ 1 αCol	O111 K_{LPS}	*Escherichia coli*	Eklund *et al.*, 1979
→4)-βGlcA-(1→3)-LFuc-(1→3)-GlcNAc-(1→6)-Gal-(1→ 4 ↑ 1 βGlc-2-OAc	O32 K87	*E. coli*	Jann *et al.*, 1971
→3)-βGal-(1→3)-βGalNAc-(1→4)-αGal-(1→4)-αNeuNAc-(2→ 9 | O-Ac	O104 K9	*E. coli*	Kogan *et al.*, 1992
→3)-αGlcNAc-(1→2)-αGal-(1→4)-βGal-(1→ || 4,6 Pyruvate	O19 K_{LPS}	*P. vulgaris* O19	Perry and McLean, 1994
αCol-(1→2)-βGal=4,6P(O)OH 1 ↓ 3 →6)-βGlcNAc-(1→4)-αGal-(1→3)-βQuiNAc-(1→ 4 ↑ 1 αCol	O139 K_{LPS}	*Vibrio cholerae* O139 (Bengal)	Knirel *et al.*, 1995

Col, colitose (3,6-dideoxy-L-galactose); QuiNAc, *N*-acetylquinovose (2-acetamido-2,6-dideoxy-D-glucose). If a polysaccharide occurs as O- and K-antigen on the same bacterium, the short version is the O-antigen and the long version is the capsular antigen (termed K_{LPS}) with the same serological specificity (compare Whitfield and Valvano, 1993).

1991). Interaction of CPS with LPS on the bacterial cell surface may explain how the serum resistance of invasive Gram-negative bacteria is apparently affected by the chemical nature of both O- and K-specific polysaccharides (see below).

Genetics of Capsular Polysaccharides

Capsule Genes

In general, the capsule genes cluster together at a specific site on the bacterial chromosome. These clusters comprise genes for the biosynthesis and the surface expression of CPSs. The names given to some capsule gene clusters, together with the literature references, are listed in **Table 6**.

Gram-negative Bacteria

The capsule genes of *E. coli* are divided into three regions: regions 1, 2 and 3. The central region (region 2) contains genes that direct the biosynthesis of the respective CPS. It comprises genes for the synthesis/activation of capsule-specific constituents (e.g. GlcA, NeuNAc or Fru) and genes for their transfer to the polymer. This region is type-specific and its size depends on the complexity of the respective CPS (Boulnois *et al.*, 1987; Roberts *et al.*, 1988; Boulnois and Jann, 1989). The G + C content of the region 2

Table 6 Terminology of capsule gene clusters

	Designation	Reference
Escherichia Coli	*kps*	Boulnois and Roberts, 1990
	cps	Rhan *et al*., 1999
	rfb	Amor and Whitfield, 1997
Neisseria meningitidis	*cps*	Frosch and Edwards, 1993
Haemophilus influenzae	*cap*	Kroll *et al*., 1990
Klebsiella pneumoniae	*kps*	Laakso *et al*., 1988; Arakawa *et al*., 1991
Streptococcus pneumoniae	*cap*	Garcia and Lopez, 1997
	cps	Dillard and Yother, 1994
Streptococcus pyogenes	*has*	Dougherty and van de Rijn, 1993, 1994
Staphylococcus aureus	*cap*	Sau and Lee, 1996

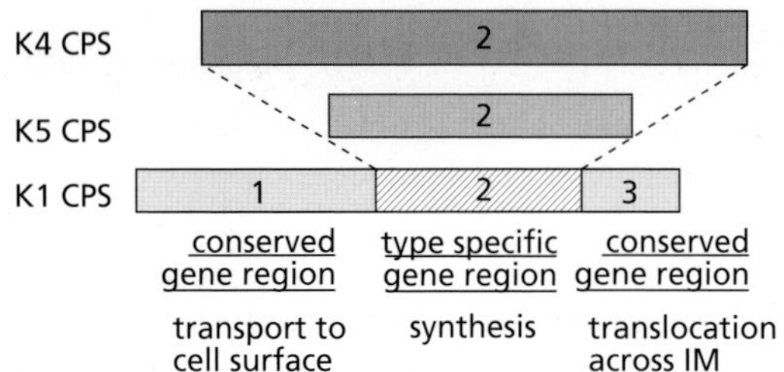

Fig. 3 Organisation of the group 2 capsule genes and the function of the three gene regions. The approximate relative sizes of region 2 from the K1, K4 and K5 antigen gene clusters are indicated. For the structures of the respective CPS see Table 3.

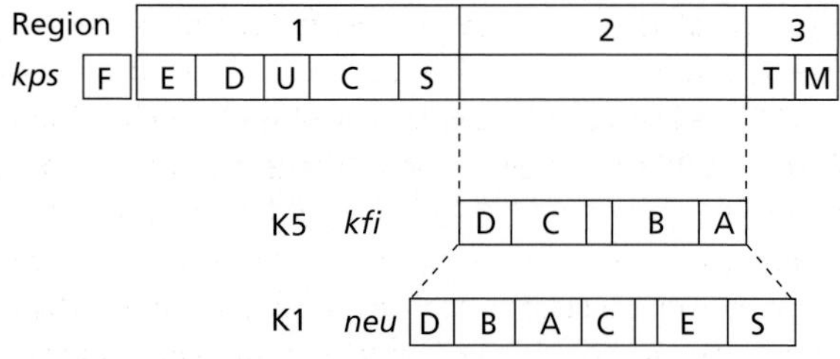

Fig. 4 The capsular gene clusters from *E. coli* K1 and K5 with their common (*kps*) and specific (*kfi, neu*) genes. The specific regions are not drawn to scale.

genes of *E. coli* K1 and K5 is distinctly lower (about 30–35%) than that of the *E. coli* chromosome (about 50%) (Pazzani *et al.*, 1993). This indicates the possibility that they were transferred from non-*coli* source(s).

The synthesis-directing region 2 is flanked by regions 1 and 3 which harbour genes for the common functions of translocation of the CPS across the cytoplasmic membrane and its transport from the cytoplasmic membrane to the cell surface (Jann and Jann, 1990, 1997; Smith *et al.*, 1990; Vimr, 1990; Pavelka *et al.*, 1991; Pazzani *et al.*, 1993; Whitfield and Valvano, 1993; Roberts, 1996; Whitfield and Roberts, 1999). A schematic representation of the capsule genes for the K1, K4 and K5 CPSs (of group 2) with different sizes of repeating units (see **Table 3**) is given in **Fig. 3**.

In *E. coli* with group 2 CPS, the genes of the common regions 1 and 3 are termed the *kps* genes (*FEDUCS* for region 1, and *MT* for region 3) (Vimr, 1990; Pavelka *et al.*, 1991; Pazzani *et al.*, 1993; Silver *et al.*, 1993). The type-specific region 2 genes of *E. coli* K5 are termed *kfi* (K five) (Petit *et al.*, 1995), those of *E. coli* K1 are termed *neu* (neuraminic acid) (Troy, 1992). The complete gene cluster of *E. coli* K1 and K5 is shown in **Fig. 4**.

The gene for UDP glucose dehydrogenase (Ugd) in *E. coli* K5 (*KfiD*) has homology to the *ugd* gene in *E. coli* K30 (group 1) (Drummelsmith *et al.*, 1997).

The distribution of individual expression genes in regions 1 and 3 differs in strains with group 2 CPS and group 3 CPS (Russo *et al.*, 1998; Rhan *et al.*, 1999; Whitfield and Roberts, 1999). This is shown schematically in **Fig. 5**.

The *kpsM* and *kpsT* genes in the group 2 and 3 capsule clusters and the *wzx* gene in the group 1 cluster direct the translocation of the finished polysaccharide (groups 2 and 3) and the lipid A-linked repeating unit (group 1) across the cytoplasmic membrane. The *kpsC* and *kpsS* genes of groups 2 and 3 are involved in this process. For this reason, both genes are marked correspondingly in **Fig. 5**. The *kpsE* and *kpsD* genes of group 2 and 3, and the *wza* and *wzb* genes of group 1 form transport systems that expose CPSs at the cell surface (see below). The gene organisation of the M antigen (colanic acid) in *E. coli* has been described by Stevenson *et al.* (1996).

In *N. meningitidis* group B the capsule genes are principally arranged as in *E. coli* capsule group 3, but

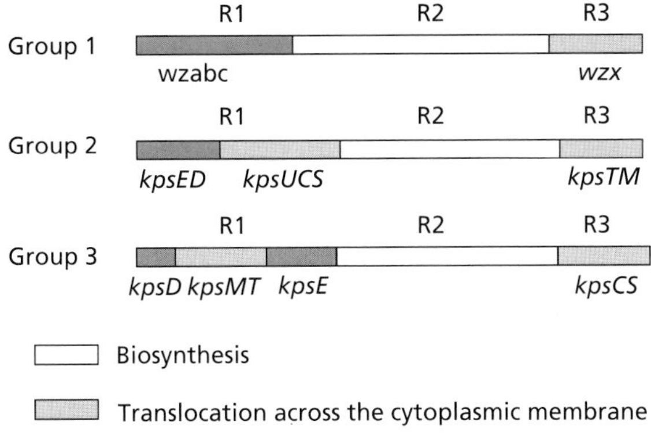

Fig. 5 Distribution of the expression genes in regions (R)1 and 3 of the capsule gene clusters from *E. coli* capsules of groups 1, 2 and 3. The gene regions and their genes are not drawn to scale. For details see text.

are combined with some genes for the lipooligosaccharide (LOS) synthesis (Frosch *et al.*, 1991). The gene cassette for the CPS of *N. meningitidis* group A (poly-(1,6)-α-D-*N*-acetylmannosamine-1-phosphate) has been described (Swartley *et al.*, 1998). Genetic analysis of virulent outbreak strains (Swartley *et al.*, 1997) has indicated a horizontal exchange between strains of groups B and C by the exchange of sialyltransferases of the respective synthesis gene regions. These clinically important strains can, therefore, escape vaccine-induced or natural protective immunity by capsule switching.

Interestingly, although *N. gonorrheae* does not produce a capsule, it contains capsule genes (Petering *et al.*, 1996).

The capsule genes of *H. influenzae* group b are arranged in a tandem sequence of regions 1–3 (Hoiseth *et al.*, 1986; Kroll *et al.*, 1991).

In *Ps. aeruginosa*, the gene cluster for the alginate capsule, at 34 min on the chromosome, contains the genes *algA* for the synthesis of GDPMan, which encodes phosphomannose isomerase and GDPMan pyrophosphorylase; *algC*, which encodes phosphomannomutase; *algD*, which encodes GDP mannose dehydrogenase; and the genes, *alg8*, *alg60*, *alg44*, for polymer synthesis and transport of the polymer, and the modification genes *algF* (acetylase) and *algG* (GlcA-5-epimerase). *AlgE*, which is assumed to encode an outer membrane protein responsible for transport through the outer membrane, is also located to this site. Regulatory genes ($algR_1$, $algR_2$, $algR_3$) are located at 10 min and 13 min, respectively. The *algR* gene, detected between *alg44* and *algE*, was thought to be engaged in the translocation of alginate across the cytoplasmic membrane (Aarons *et al.*, 1997), but this is not so, and recently it has been shown to be involved in its polymerisation (Jain and Ohman, 1998). It has also been shown that *algR* mutants secrete uronic acids, probably mannosuronic acid (Jain and Ohman, 1998). The genetics and synthesis of alginate in *Pseudomonas* have been reviewed (May *et al.*, 1991; May and Chakrabarty, 1994; Gacesa, 1998).

The Vi-antigen (see **Table 3**) of *Salmonella typhi* and *Citrobacter freundii* is determined by two separate gene loci: *viaB* (synthesis and surface expression) and *viaA* (regulation) (Johnson *et al.*, 1965; Hashimoto *et al.*, 1991). The *viaB* locus of *S. typhi* contains the synthesis genes *vipB*, *vipA* and *vipC*, which, it has been proposed, determine UDPGlcNAc-4 epimerase, UDPGalNAc dehydrogenase and GalNAcA polymerase, respectively. *ViaB* also contains the export genes *vexA*, *vexB*, *vexC* and *vexD* (Hashimoto *et al.*, 1991, 1993; Kolyva *et al.*, 1992; Huong *et al.*, 1992). Based on DNA sequencing, *vexB* and *vexC* are respectively homologous to the *E. coli* transport genes *kpsM* and *kpsT*. Their products are regarded as forming an ABC transporter system, which passes the Vi polysaccharide across the cytoplasmic membrane (Hashimoto *et al.*, 1993) (see below). The functions of *vexA* and *vexD* are unknown.

Gram-positive Bacteria

After early general studies on the expression of the capsules of the *Strep. pneumoniae* (Griffiths, 1928), progress was slow until pneumococcal capsule genes could be cloned and analysed in *E. coli* (Dillard and Yother, 1991). To avoid confusion before entering into a discussion of the capsule genes of *Strep. pneumoniae*, it should be mentioned that there are two parallel nomenclatures for the designation of these genes, i.e. *cps* and *cap* (see Table 6). The type 3 specific capsule genes *cps3D*, which encodes UDPGlc dehydrogenase, and *cps3S*, which encodes the type 3 polymerase, are linked in an operon (Garcia *et al.*, 1993; Arrecubieta *et al.*, 1996; Garcia and Lopez, 1997). The genes for phosphoglucomutase and UDPglucose pyrophosphorylase (*cps3M* and *cpsU*, respectively) are also in the *cps3* gene cluster (Rubens *et al.*, 1993; Arrecubieta *et al.*, 1994; Dillard and Yother, 1994; Dillard *et al.*, 1995). Homologous capsule clusters were also found in types 14, 19F, 23F, and 33F (Kolkman *et al.*, 1996, 1997; Llull *et al.*, 1998; Morona *et al.*, 1997a, b, 1999a; Ramirez and Tomasz, 1998). *Strep. pneumoniae* type 19 comprises four subtypes, 19F, 19A, 19B, and 19C (see Table 3 and Lee and Frazer, 1980; Katzenellenbogen and

Jennings, 1983; Beynon *et al.*, 1991). The gene clusters that determine the corresponding CPS structures have recently been cloned and characterised (Morona *et al.*, 1999c). While analysing the *cps* genes of types 19F and 19A, these authors identified two classes of *Strep. pneumoniae* that express the *cpsC,D,E* genes, located at the 5′ end of all *Strep. pneumoniae* capsule clusters, and homologous to either those of type 19F (class I) or to those of type 19A (class II) (Morona *et al.*, 1999b). It should be mentioned that *cpsC* and *cpsD*, together with the adjacent *cpsB* gene, which are present in all pneumococci, appear to represent a regulation system for the extent of CPS expression.

Type 37 is an interesting exception. The polymerisation of its polysaccharide, which consists of β-1, 3 linked units of the disaccharide βGlc-(13)-βGlc, is determined by a single polymerase gene, *tts*, distant from the *cps* site of all the other *Strep. pneumoniae* (Llull *et al.*, 1998). In addition, type 37 strains carry a silent *cps33f* gene locus. *Strep. pneumoniae* type 37 is thus genetically a natural binary strain. Encapsulated pneumococcus strains could be transformed with *tts* into strains that display the binary type of encapsulation (Llull *et al.*, 1998) Binary capsule expression was obtained by transformation of a type 3 strain, which has a defective UDP-dehydrogenase gene, with type 1 capsule genes in which this gene was intact. The resulting pneumococcus strain expressed both type 1 and type 3 capsules (Munoz *et al.*, 1997).

The fact that the capsule clusters are flanked on both sides by common DNA sequences may explain capsule switches in *Strep. pneumoniae* by transformation with DNA. Previous observations on the switching of capsular types (Bernheimer *et al.*, 1967) were later analysed *in vitro* and *in vivo* in a number of serotypes (Coffey *et al.*, 1991, 1998; Kelly *et al.*, 1994; Barnes *et al.*, 1995; Kolkman *et al.*, 1998; Nessin *et al.*, 1998). The changes *in vivo* were predominantly observed in (multi-)drug resistant strains. This plasticity of the pneumococcal genome may be a means of adaptation of the bacteria to their environment (Claverys *et al.*, 2000). It should be mentioned that, in a series of chemical reactions, the type 14 CPS has recently been transformed into oligosaccharide fragments of type 3 (Zou *et al.*, 1998).

In *Strep. pyogenes* (group A), there are three genes (*hasA*, *hasB*, *hasC*) with one promotor, concerned with the expression of the hyaluronate capsule. They determine proteins for substrate synthesis (HasB and HasC for UDPGlcA synthesis) and the hyaluronate synthetase HasA (a transferase with specificity for GlcA and GlcNAc) (DeAngelis and Weigel, 1993, 1994; Dougherty and van de Rijn, 1993, 1994). The *hasB* gene, which determines UDPGlc dehydrogenase, shows homology to the *kfiD* gene of *E. coli* K5 (group 2 CPS) and to the *ugd* gene of *E. coli* K30 (group 1 CPS) (see above). Transcriptional regulation of capsule synthesis has been studied in a highly encapsulated M type 18 strain and a poorly encapsulated M type 3 strain (Alberti *et al.*, 1998). The fine structure of the promotor, especially in the −35 spacer region and in the +2 to +8 positions relative to the start of the *hasA* transcription, proved to be important.

Staph. aureus types 1 and 2 produce copious capsules but are rarely found amongst clinical isolates. In contrast, types 5 and 8, which are frequent clinical isolates, produce only microcapsules. This is because the promotors of types 1 and 2 are about 10 times stronger than those of types 5 and 8. All four types, which produce CPSs of very similar structure, have extended CPS clusters. In *Staph. aureus* type 1 (strain M), 10 capsule genes (*cap1A–cap1D*, *cap1G–cap1M*) have been found (Lee, 1992, 1995; Lin *et al.*, 1994; Ouyang and Lee, 1997). Homology of the gene products Cap1L with VipA from *S. typhimurium* (see above) and of Cap1I with VipB implicate these *cap* genes in the synthesis of GalNAcA, which is related to the Vi-CPS constituent GalNAcA (Lin *et al.*, 1994). The weak homology of Cap1M with VipC points to the *cap1M* gene directing GalNAcA transferase (Lin *et al.*, 1994). *Staph. aureus* types 5 and 8 have been analysed with analogous results (Lee *et al.*, 1994; Sau and Lee, 1996; Sau *et al.*, 1997a, 1997b). The CPSs of types 5 and 8 differ, *inter alia*, in the linkage position of their *O*-acetyl substituent (**Table 3**). The genes for the respective transacetylases have been described, and seem to be CPS-specific (Sau *et al.*, 1997a; Bhasin *et al.*, 1998).

Since Gram-positive bacteria do not have an outer membrane, genes that correspond to region 1 of *E. coli* and related Gram-negative bacteria do not exist (see **Table 4**).

Nomenclature of Polysaccharide Genes

The number of genes that direct identical or similar reactions in different bacteria is very large. Different laboratories have used different terminologies for related genes, which often encode the same biochemical reaction. Gene nomenclature has further been complicated by the different naming by different research groups of the same gene(s) in the same microorganism. To simplify and co-ordinate the nomenclature of genes in bacterial polysaccharide expression, a new nomenclature (w^{***}) for these genes has been proposed (Reeves *et al.*, 1996).

Biosynthesis of Capsular Polysaccharides

Site of Polymerisation

Electron microscopy and the use of oriented membrane vesicles in incubations *in vitro* have revealed that in most cases the assembly of CPSs takes place on the inner side of the cytoplasmic membrane (Bayer, 1990; Kröncke *et al.*, 1990a, 1990b; Bronner *et al.*, 1993; Whitfield and Troy, 1984). For subsequent surface expression, this requires the passage of the CPS through two membranes in Gram-negative bacteria and through one membrane in Gram-positive bacteria.

The Starting Reaction

The starting reaction of CPS biosynthesis, which is distinct from the subsequent polymerisation reactions, is reversible and represents a biochemical switch for capsule expression. CPS biosynthesis in *Aerobacter aerogenes* strain DD45 (*Klebsiella*) starts with the reversible transfer of Gal-1-P to the C55-lipid undecaprenylphosphate (C55-P) (Troy *et al.*, 1971):

$$\text{C55-P} + \text{UDPGal} \rightleftharpoons \text{C55-P-P-Gal} + \text{UMP}$$

This type of initiation reaction was originally demonstrated in the biosynthesis of LPS and peptidoglycan, and is operative in the syntheses of many polymers (Whitfield and Valvano, 1993).

For most CPSs the nature of the starting reaction is unknown. In *E. coli* K1 it was proposed that the *neuS* gene is involved in polySia synthesis (Troy, 1992). The amino acid sequence of the putative NeuS protein, as deduced from the *neuS* DNA sequence, contains a dolichol-binding motif (Troy, 1992). This would also be consistent with the participation of C55-P in the synthesis of polySia. A dolichol-binding motif was not found for any of the region 2 sequences of the *E. coli* K5 capsule gene cluster.

Genes for the synthesis of glycosyl-diphosphoundecaprenol have been detected in the *his*-proximal *cps* genes for the K30 CPS (group 1), and *rfb* genes for the K40 CPS (group 4) of *E. coli* (Whitfield and Valvano, 1993; Drummelsmith and Whitfield, 1999; Whitfield and Roberts, 1999). The respective repeating oligosaccharides on these lipid-linked sugars are assembled from nucleotide diphosphate-activated sugar constituents.

(a)

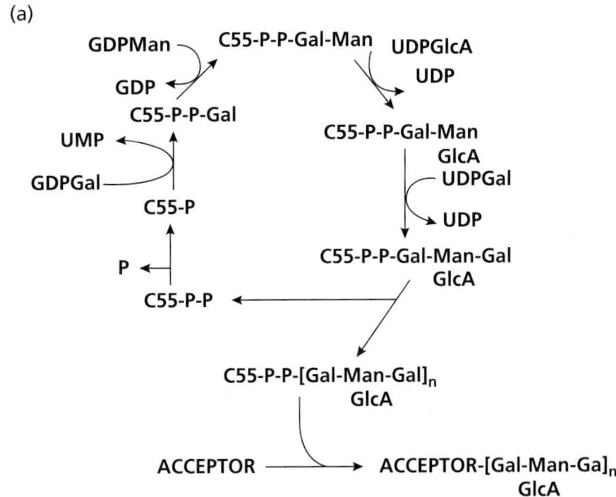

(b)

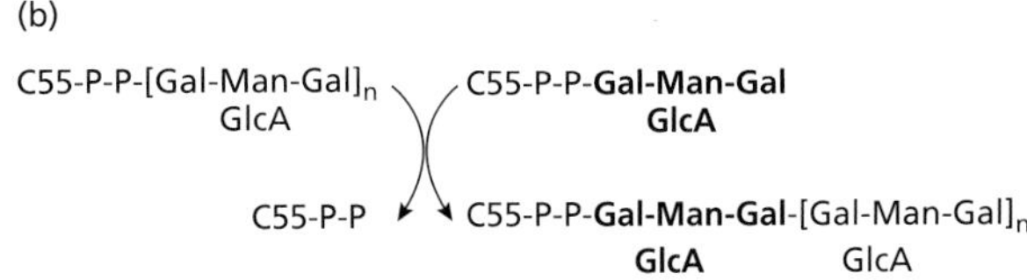

Fig. 6 (a) Biosynthetic pathway of the CPS from *Aerobacter aerogenes* DD-45 (*Klebsiella*). The nature of the acceptor is not known, but is possibly core–lipid A or phosphatidic acid C55-undecaprenol. Drawn with information from Troy *et al.* (1971). (b) Mechanism of chain elongation with the block polymerisation (transfer of the growing chain at the reducing end).

Polymerisation

Block Polymerisation

In block polymerisation, oligosaccharide repeating units (RU) are formed on undecaprenylphosphate (C55-P) with the formation of C55-P-P-RU. The oligosaccharides are assembled by a specific polymerase with the liberation of C55-P-P. The latter is then dephosphorylated and the C55-P so formed is recycled. During the polymerisation, the C55-PP-linked polysaccharide is transferred to newly formed C55-P-P-RU with growth of the chain at its reducing end. The synthesis of the CPS of *K. aerogenes* is shown as an example in Fig. 6.

This mechanism had been reported for the biosynthesis of *Salmonella* LPS (Wright *et al.*, 1967).

The K30 and K40 CPS of *E. coli* capsule groups 1 and 4, respectively, are polymerised by assembly from their undecaprenoldiphosphate-linked oligosaccharides. In contrast to groups 2 and 3, *E. coli* CPS of groups 1 and 4 is polymerised at the outer face of the cytoplasmic membrane (see below) (Whitfield and Roberts, 1999; Rhan *et al.*, 1999). The polymerisation of the K40 CPS depends critically on the previous substitution of the carboxyl group of GlcA with L-serine (Table 3). The seryltransferase gene has been cloned and characterised, and its role in the polymerisation of the K40 repeating units to the K40 CPS has been elucidated (Amor *et al.*, 1999). The role of seryl substitution is noteworthy. As defined by the sequence of sugar transfers, and in distinction from the chemical formulation of the repeating unit shown in Table 3, GlcNAc–GlcA(serine)–GlcNAc is the biological repeating unit. Thus, the signal for polymerisation is at the middle sugar unit and not at the site where the polymerase joins the lipid-linked repeating units (–GlcNAc–GlcNAc–).

Single-chain Polymerisation at the Non-reducing End

In this polymerisation, which has been reported for group 2, and possibly also for group 3 CPS (Jann and Jann, 1997; Whitfield and Roberts, 1999), the polysaccharide chain grows on a hitherto unknown acceptor, possibly undecaprenylphosphate, by sequential transfer of the sugar units from their nucleotide diphosphate-activated forms to the non-reducing end of the growing chain. As an example, polymerisation of the K5 CPS of *E. coli* (Finke *et al.*, 1991; Bronner *et al.*, 1993) is shown in Fig. 7.

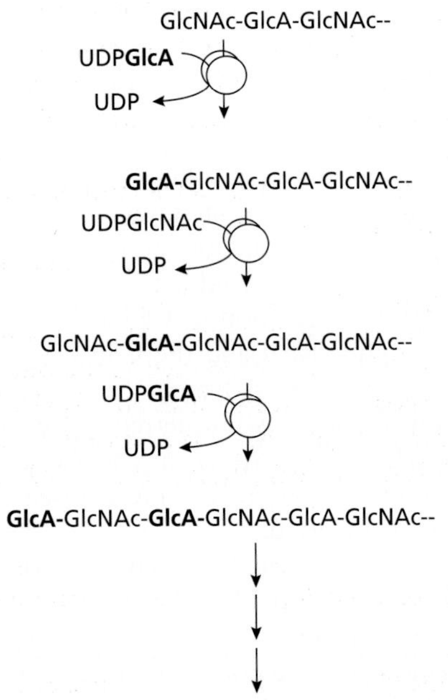

GlcNAc-GlcA-GlcNAc--
UDP**GlcA** → UDP
GlcA-GlcNAc-GlcA-GlcNAc--
UDPGlcNAc → UDP
GlcNAc-**GlcA**-GlcNAc-GlcA-GlcNAc--
UDP**GlcA** → UDP
GlcA-GlcNAc-**GlcA**-GlcNAc-GlcA-GlcNAc--
↓
↓
↓

Fig. 7 Biosynthesis of the K5 CPS by the single-chain mechanism. The sugar components are transferred from their UDP-activated forms directly to the non-reducing end of the polysaccharide chain. The double circles in the reaction arrows indicate the homo-dimer of the GlcA-GlcNAc double transferase.

GlcA and GlcNAc are transferred by one protein, K5 polysaccharide synthetase, which is encoded by *kfiC*. This enzyme is genetically co-regulated with the UDPGlc dehydrogenase, encoded by *kfiD* (Sieberth *et al.*, 1995). The combined functions of a protein with two transferase activities and a nucleotide diphosphate sugar dehydrogenase has also been observed in the biosynthesis of the hyaluronate capsule in *Strep. pyogenes* (Dougherty and van de Rijn, 1993, 1994), and of the type 3 CPS of *Strep. pneumoniae* (Dillard and Yother, 1994; Dillard *et al.*, 1995). A double transferase that corresponds to that in *E. coli* K5 has been described in an eukaryotic system for the synthesis of *N*-acetylheparosan (identical with the K5 CPS) (Kusche *et al.*, 1991; Lidholt *et al.*, 1992, 1994; Lind *et al.*, 1993).

The polymerisation reaction in *E. coli* K1 (Whitfield and Troy, 1984; Troy *et al.*, 1991; Troy, 1992) and also in *N. meningitidis* group B (Masson and Holbein, 1985) has the same overall mechanism as that in *E. coli* K5. The only difference seems to be the apparent participation of C55-P in the polymerisation, at the non-reducing end, of the K1 CPS.

The chondroitin backbone of the *E. coli* K4 CPS is synthesised by the same mechanism as the K5 CPS, also without the apparent participation of C55-P (Hannesson *et al.*, 1996; Lidholt and Fjelstad, 1997).

Single-chain Polymerisation at the Reducing End

This mechanism has been proposed for the synthesis of the CPS of *Strep. pyogenes* (Fig. 8). Here, GlcA and GlcNAc are apparently sequentially and directly transferred from their UDP-activated forms to the reducing end of the hyaluronate chain (Prehm, 1986). In this process, the enzyme-associated polysaccharide chain is transferred to the UDP-sugar that newly enters

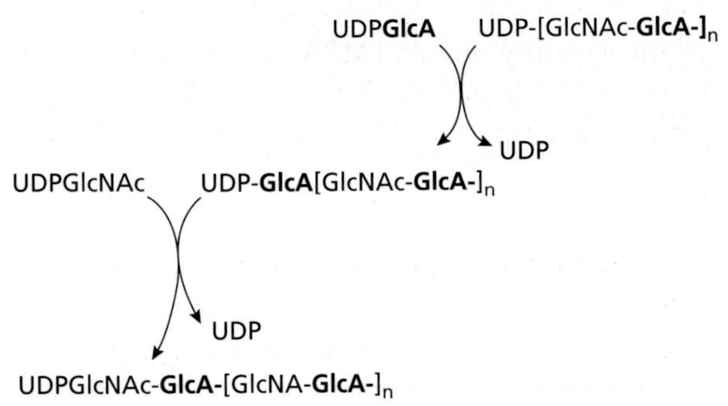

UDP**GlcA** UDP-[GlcNAc-**GlcA-**]$_n$
UDP
UDPGlcNAc UDP-**GlcA**[GlcNAc-**GlcA-**]$_n$
UDP
UDPGlcNAc-**GlcA-**[GlcNA-**GlcA-**]$_n$

Fig. 8 Biosynthesis of the *Streptococcus pyogenes* hyaluronate by the single-chain mechanism. The sugar components are transferred from their UDP-activated forms directly to the reducing end of the polysaccharide chain. Drawn with the information from Prehm (1986).

the active site of the polymerase. A similar mechanism has been proposed for the synthesis of the alginate capsule of *Ps. aeruginosa* (May and Chakrabarty, 1994) and for the CPS of *Strep. pneumoniae* type 3 (Arrecubieta *et al.*, 1999; Llull *et al.*, 1999).

Surface Expression

Gram-negative Bacteria

Surface expression of capsules in Gram-negative bacteria encompasses the translocation of the finished CPS across the cytoplasmic membrane and its subsequent transport to the cell surface.

The mechanism of CPS surface expression in *E. coli* appears to depend on the mechanism of polymerisation. Group 1 and 4 CPSs, which are polymerised in a block mechanism from undecaprenoldiphosphate-linked oligosaccharide-repeating units, are polymerised at the outer surface of the cytoplasmic membrane (see above). In these groups, therefore, only the lipid-linked oligosaccharides, but not the complete CPSs, are translocated. This process is directed by the *wzx* gene, the product of which (Wzx) is a translocase, reminiscent of that operative in LPS expression (Reeves *et al.*, 1996; Drummelsmith and Whitfield, 1999; Whitfield and Roberts, 1999). It is only after this translocation that the polymerase, determined by *wzy*, assembles the CPS. The translocating system was termed the PST protein (polysaccharide-specific transport system), with 12 transmembrane regions (12-spanner) (Paulsen *et al.*, 1997).

The polymerised group 1 and group 4 CPSs are subsequently transported through the periplasm and the outer membrane by a second transport system that consists of two proteins, termed accessory proteins. One of these, Wzc, the cytoplasmic membrane periplasm accessory protein (MPA) is integrated into the cytoplasmic membrane. The other, Wza, the outer membrane-associated accessory protein (OMA), is periplasmic and associated with the outer membrane, and presumably also with the cytoplasmic membrane. Although the expression proteins have been characterised (Paulsen *et al.*, 1997; Drummelsmith and Whitfield, 1999; Whitfield and Roberts, 1999), the molecular details of the processes involved are not known.

CPSs of group 2 have a different mode of expression. Here, the polysaccharide chain is assembled at the inner face of the cytoplasmic membranes by polymerase enzymes. This is followed by the transfer of the polysaccharide chain from a hitherto undefined initial acceptor (possibly undecaprenolphosphate) to phosphatidyl-Kdo. Coupled with this transfer, the complete CPS in its lipid substituted form is translocated across the cytoplasmic membrane by the action of a translocator that consists of two copies of the KpsT and KpsM proteins each, directed by the *kpsT* and *kpsM* genes (Smith *et al.*, 1990; Pavelka *et al.*, 1991; Pazzani *et al.*, 1993). Both are members of the ABC-2 (ATP-binding cassette) transporter system (Reizer *et al.*, 1992; Paulsen *et al.*, 1997; Whitfield and Roberts, 1999), which is a member of the general ABC superfamily (Saier, 1994).

The KpsM protein is a six-spanner membrane protein, presumably forming a specialised channel, and the KpsT protein has an ATP-binding site and provides the energy of translocation by ATP hydrolysis (Smith *et al.*, 1990; Bliss and Silver, 1996, 1997; Roberts, 1996). For *E. coli* K1 the KpsT protein binds reversibly to the polySia (K1 CPS) that is to be translocated (Bliss and Silver, 1997). The translocation of the group 2 CPS not only utilises ATP, but also depends on the transmembrane potential, as demonstrated with *E. coli* K1 and K5 (Kröncke *et al.*, 1990a; Troy *et al.*, 1991). The nomenclature for these translocating proteins in various genera is given in Table 7.

After translocation across the cytoplasmic membrane, the CPSs are transported to the cell surface by two transporter proteins, KpsE and KpsD. These are representatives of the more recently established OMA/MPA 2 system (Paulsen *et al.*, 1997).

Table 7 Genes for the ABC transporter system in *Escherichia coli*, *Neisseria meningitidis* group B, and *Haemophilus influenzae* type B

Strain	Transporter	Energizer	Reference
E. coli K1 and K5	*kpsM*	*kpsT*	Pavelka *et al.*, 1991; Smith *et al.*, 1990
N. meningitidis B	*ctrC*	*ctrD*	Frosch *et al.*, 1991
H. influenzae B	*bexB*	*bexA*	Kroll *et al.*, 1990

kps, K polysaccharide; *ctr*, capsule transport; *bex*, export of group B polysaccharide.

KpsE, which is determined by the *kpsE* gene, is bound to the cytoplasmic membrane by one transmembrane region at the amino-terminal end and by hydrophobic interaction at the carboxy-terminal end. The bulk of the protein is exposed to the periplasm (Rosenow *et al.*, 1995). KpsD is a periplasmic protein that can associate with the cytoplasmic membrane and with KpsE (Silver *et al.*, 1987; Wunder *et al.*, 1994). The genes for the surface expression of group 1, 2 or 3 CPSs on *E. coli* and the processes directed by their products are compared in Table 8.

E. coli with group 4 CPSs presumably have the same expression machinery as those with group 1 CPSs. Since this has not been proved unambiguously, group 4 CPSs are not included in Table 8.

The expression proteins described above may form or be part of the membrane adhesion sites, which

Table 8 Genes for the surface expression of *Escherichia coli* CPS and the processes directed

CPS group	Genes[a]	Process directed	Transport system[b]
1	*wzx*	Translocation of lipid A-linked repeating units across the cytoplasmic membrane	PST-1
	wza, wzb	Transport of the lipid A-linked repeating units from the cytoplasmic membrane to the cell surface	OMA/MPA 1
2, 3[c]	*kpsM, kpsT*	Translocation of the finished phosphatidyl-Kdo linked CPS across the cytoplasmic membrane	ABC-2
	kpsE, kpsD	Transport of the phosphatidyl-Kdo linked CPS to the outer membrane	OMA/MPA 2

[a] The *wz* nomenclature is according to Reeves *et al.* (1996).

[b] The transport systems follows the classification of Paulsen et *al.* (1997).

[c] In these groups, the genes *kpsCS* are proposed to direct the substitution of the CPS with phosphatidyl-Kdo. They are thus involved in the translocation (Table 9 and Fig. 9).

Table 9 Characterisation of the K5 and K1 gene products of *Escherichia coli*

A. Common expression gene products (regions 1 and 3)

Region	*Kps*	Size (kDa)	Location	Putative function
1	*E*	44	im/pp	Transport protein
1	*D*	60.5	pp	Transport protein
1	*U*	27	cp	CMP-Kdo synthetase
1	*C*	76	cp im?	Synthesis of phosphatidyl-Kdo and its ligation with the CPS
3	*T*	25	cp	Translocation energizer
3	*M*	29	im	Translocator (channel)

B. Specific region 2 genes

E. coli	Region 2 gene product	Size (kDa)	Putative function
K5	KfiD	44	UDP glucose dehydrogenase
	KfiC	60	Transferase for GlcA and GlcNAc
	KfiB	66	Unknown
	KfiA	27	Unknown
K1	neuA	50	CMP NeuNAc synthetase
	neuB	39	NeuNAc synthetase
	neuC	44	ManNAc synthetase
	neuD	23	Unknown
	neuE	28	NeuNAc transferase?
	neuS	47	Polymerase

have been shown to be import and export sites for macromolecules (Bayer, 1990).

The polymerisation and surface expression of the (group 2) K1 and K5 CPSs of *E. coli* have been studied in greater detail, and models for the topography of polymerisation and surface expression have been presented. The proteins operative in the expression of the K1 and K5 CPS are listed in **Table 9**, and a model for the assembly of these proteins in *E. coli* K5 (Whitfield and Roberts, 1999) is shown in **Fig. 9**.

A similar model has been presented for the K1 CPS (Bliss and Silver, 1996). Because of their temperature regulation (Ørskov *et al.*, 1984; Jann and Jann, 1990, 1992, 1997), the expression of *E. coli* group 2 capsules, as well as capsule enzyme activities, could be studies in temperature shift experiments (from 18–20°C to 37°C) (Whitfield and Troy, 1984; Kröncke *et al.*, 1990a, 1990b).

Gram-positive Bacteria

The lack of an outer membrane, and hence lack of a periplasmic space, simplifies the surface presentation of CPS. In *Staph. aureus* it has been postulated that the

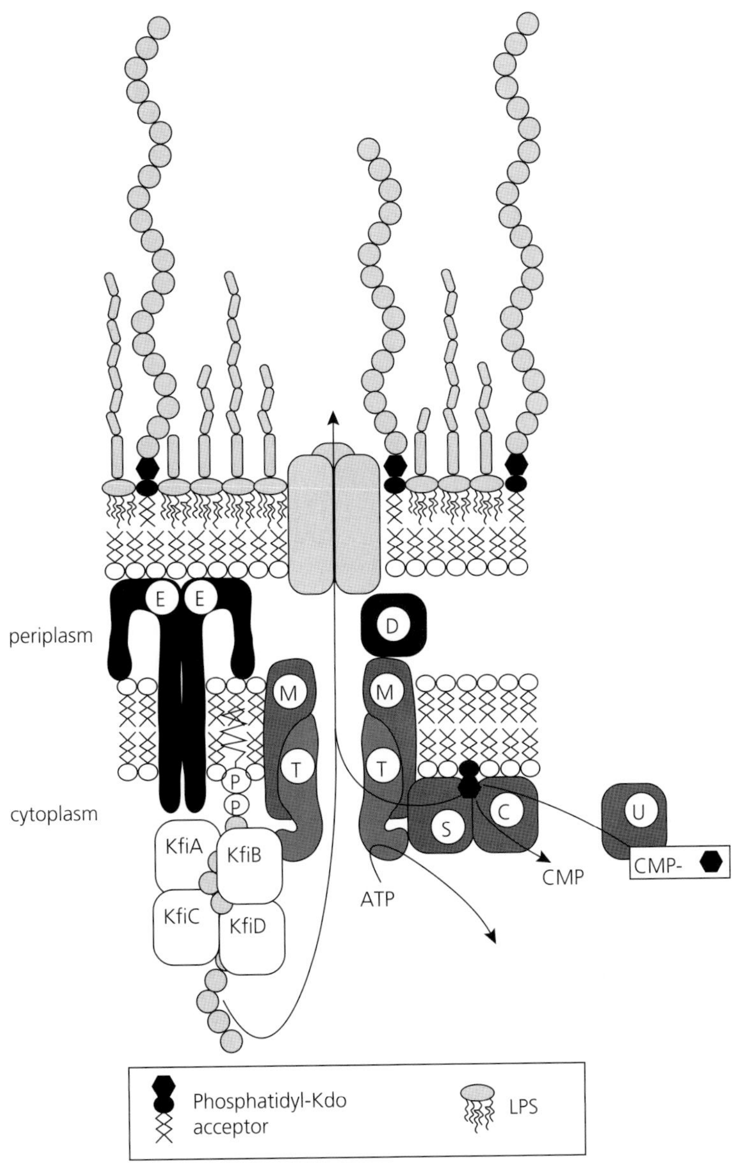

Fig. 9 A model for the assembly and surface expression of the K5 CPS. Proteins involved in CPS synthesis (KfiABCD) interact with a scaffold comprising KpsCSMTED to form a biosynthesis–export–translocation complex. The CPS is shown as growing on an undecaprenyl pyrophosphate carrier (zig-zag-PP), but this remains conjectural. From Whitfield and Roberts (1999), with permission. For the components of the export machinery see Table 9.

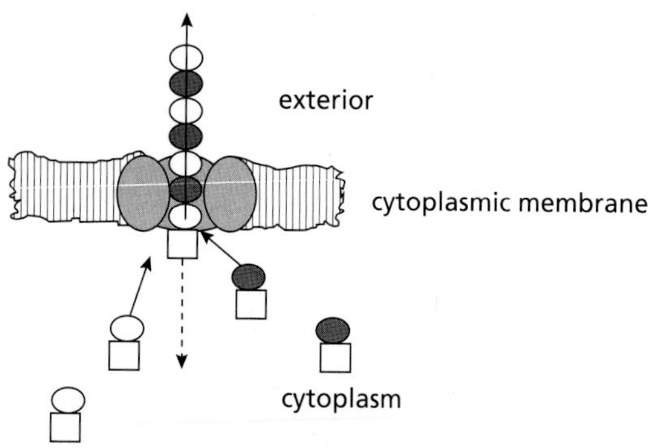

Fig. 10 Model for the export of the CPS (hyaluronate) of *Streptococcus pyogenes*. The hyaluronate polymerase assembles the polysaccharide by alternate addition of glucuronic acid (black circles) and *N*-acetylglucosamine (open circles) from their activating UDP moieties (open boxes). In this process, the polysaccharide stays in contact with the polymerase (double transferase), oscillating between the active sites occupied by nucleotide sugars newly entering the active site(s). The polysaccharide chain is pushed out of the cell, non-reducing end first. Drawn with information from Prehm, 1986.

polysaccharide chain is expelled through a pore formed by the polymerase complex (Prehm, 1986) (see Fig. 10).

Whereas ABC-2 translocators have been postulated for drug export in a drug-resistance system, PST translocators for CPS export have been detected in *Staph. aureus*, *Strep. pneumoniae* and *Strep. agalactiae* (Guidolin *et al.*, 1994; Paulsen *et al.*, 1997). In contrast to Gram-negative bacteria, the energy is provided by a separate cytoplasmic ATP-hydrolysing protein (protein C), and not by the ATPase activity of one of the translocating proteins, such as KpsT.

Regulation and Modification of CPS Expression

Encapsulation may not always be advantageous for virulent bacteria, and so capsules may be present or absent depending on different environments. The expression of CPSs must, therefore, be subject to environmental signals, such as temperature, as in the group 2 and 3 CPSs and the M-polysaccharide of *E. coli* (Gottesman and Stout, 1991; Whitfield and Valvano, 1993), or concentration of certain ions or oxygen tension, as in the alginate capsule of *Ps. aeruginosa* (Bayer, 1990). Although a direct link between the signal and response systems has not been established, a number of regulatory systems have been described.

Two-component Regulation Systems

The temperature dependence of CPS expression in *E. coli K-12* (M-antigen, colanic acid) and the role of the Lon protease was first analysed by Markowitz (1977). Later, Gottesman and Stout, 1991 described a system for the regulation of capsule synthesis (Rcs), which is a member of the family of two-component signal transduction systems (Hoch and Silhavy, 1995). Membrane protein RcsC acts as sensor of an external signal (see Fig. 11). It interacts with the response regulator protein RcsB, which in turn forms a complex with the DNA-binding protein RcsA. The latter activates the promotor of CPS genes.

The level of RcsA is controlled by the Lon protease and also by a balanced interaction between the *rcsA*-silencing gene *hns* (Atlung and Ingmer, 1997), and a small antisilencer RNA, DsrA, which is directed from a downstream region of *rcsA* (Sledjenski and Gottesman, 1995). The response of the Rcs system to the signal was originally thought to be autophosphorylation of RcsC and subsequent phosphorylation of RcsB by the phosphorylated form of RcsC. It is not now clear whether this is the actual course of events, or whether RcsF is a phosphorylase and RcsC is a dephosphorylase.

The Rcs system has also been described for the Vi-antigen of *S. typhi* (Huong *et al.*, 1992) and the K2 and K20 CPS of *Klebsiella* (McCallum and Whitfield, 1991; Wacharotayankun *et al.*, 1992).

Although group 1 CPSs of *E. coli* are not temperature regulated, they are Rcs regulated, whereas the temperature-regulated group 2 CPS are not Rcs regulated (Jayaratne *et al.*, 1993; Keenleyside *et al.*, 1993). This underlines a dissociation between the temperature and Rcs-regulatory pathways (Gottesman and Stout, 1991). The environmental signal for the Rcs circuit is not known.

The *algB* gene and the *algR* gene(s) in *Ps. aeruginosa* regulate the transcription of the *algD* gene (Goldberg and Dahnke, 1992), which encodes the enzyme GDP mannose dehydrogenase (Deretic *et al.*, 1986). The alginate regulation system, however, seems to be more complex than the Rcs system of *E. coli*, since it also comprises components of the heat-shock response system (Goldberg and Ohman, 1987; Flynn and Ohman, 1988; Deretic *et al.*, 1989; Schurr and Deretic, 1997). In addition, the AlgB/AlgR response proteins do not seem to depend for their activity on (auto)phosphorylation (Ma *et al.*, 1998). Exogenous oxygen tension is one of the signals processed, and *alg* genes are up-regulated at higher tension. The left and right cardiac circuits, respectively, have a pO_2 of 80 mmHg and 40 mmHg, with the

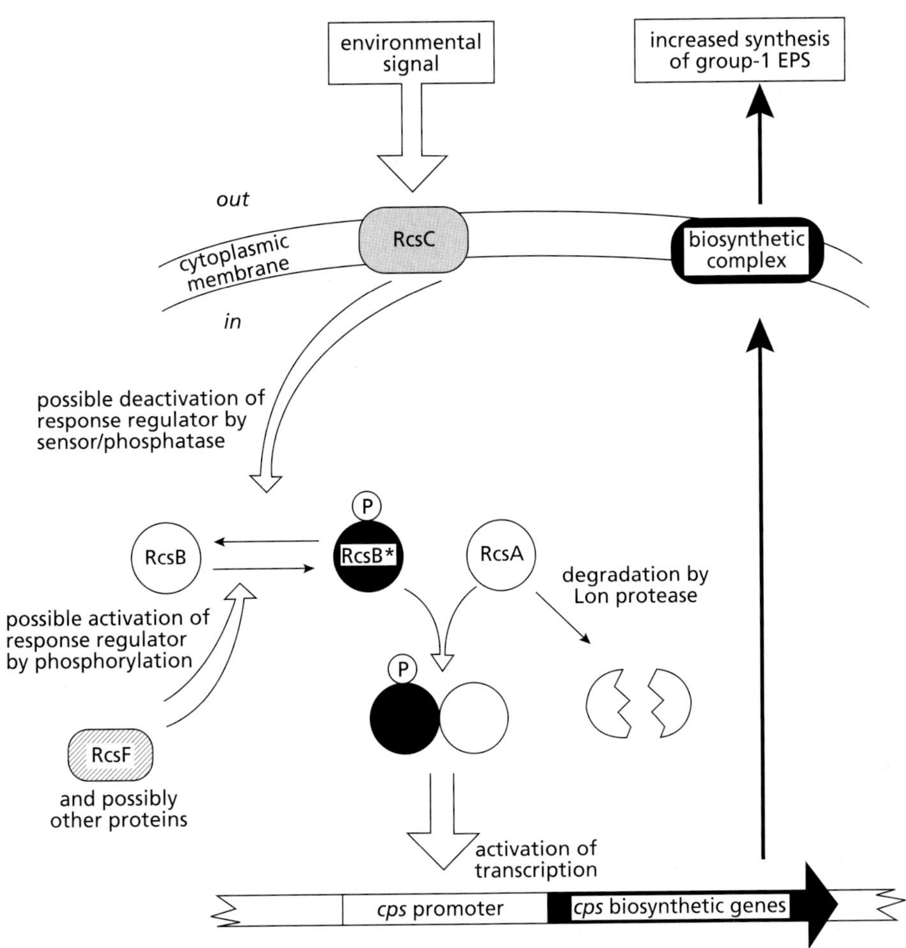

Fig. 11 Proposed mode of action of the Rcs system in the regulation of the group 1-like CPS of *E. coli* K-12. In this model, RcsC is the transmembrane sensor, RcsB is the response regulator and RcsA is an essential accessory protein. The active response regulator is probably an RcsB–RcsA complex. From Whitfield and Valvano (1993), with permission.

result that expression of alginate as virulence factor in bacterial endocarditis is more pronounced in the left side of the heart (aortic and mitral valves) (Bayer, 1990). Unlike *Ps. aeruginosa*, *algR* is not required in *Ps. syringae* (Fakhr *et al.*, 1999).

The RfaH/JUMPstart (*ops*) Control System

A non-coding sequence of 39 base pairs just upstream of several gene clusters, the JUMP sequence, directs the production of various polysaccharides in enteric bacteria (Hobbs and Reeves, 1994). These include the genes for the outer core (*rfa*) and O-antigen (*rfb*) of LPS, the Vi-antigen (*viaB*) of *S. typhi*, and group 2 CPSs (*kps*) of *E. coli* (Fig. 12).

Bailey *et al.*, 1997 later found that the short *ops* (operator polarity suppressor) element is an essential part of the JUMPstart regulon. The expression of group 2 CPS depends on the *rfaH* gene (Stevens *et al.*, 1994), which had originally been described as encoding an antiterminator for LPS expression (Pradel and Schnaitman, 1991). It has been proposed that in group 2 CPS expression, the RfaH protein interacts with the *ops* element contained by the JUMPstart sequence to ensure efficient transcription of the *kps* genes (Stevens *et al.*, 1997). This indicates that the JUMPstart/RfaH is a control system for CPS expression. Since the RfaH/JUMPstart circuit is apparently operative in many gene systems, such as *tra*, *kps*, *viaB*, *rfa*, *rfb*, *hly* (Hobbs and Reeves, 1994; Stevens *et al.*, 1994, 1997), the RfaH protein can be considered to be a global regulator.

Modification of CPS Expression by Gene Deletion and Amplification

Haemophilus influenzae type b can modify the extent of CPS expression by reduction/amplification of its

JUMPstart	GGTAGCTG...AGCCA.GGGCGGTAGCGT..	
K5 *kps*	GGTAGCTGTTAAGCCAAGGGCGGTAGCGTAC	28bp kpsM
K1 *kps*	CGTAGCTGTTAAGCCAAGGGCGGTAGCGTAC	28bp *KPSM*
fra	CGTAGCTGTTGAGCCTGGGGCGGTAGCGTGC	83bp *rfaQ*
hly	GGTAGTTTCCGG . . . AAGGGCGGTAGCATGC	1995bp *hlyC*
tra	G .TGCGTGAAAAGGGATGGGCGGTAGCGTGT	25bp *traR*
	GGTag agcca GGGCGGTAGC gT c	

Fig. 12 Alignment of sequences 5′ of RfaH-regulated operons. The consensus JUMPstart sequence (Hobbs and Reeves, 1994) is shown for comparison. The numbers on the right-hand side refer to the distance, in base pairs, between the last residue shown and the site of translation initiation. Sequences were extracted from GenBank/EMBL entries with accession numbers X35819, U05251, M80599, X07565 and U01159. Gaps (. . .) have been introduced to improve the alignment. The consensus sequence derived from the alignment of the JUMPstart and sequences 5′ to *kps*, *rfa*, *hly* and *tra* genes (Hobbs and Reeves, 1994) is shown in bold. Conserved residues are in upper case, and residues in lower case represent conservation in four out of five aligned sequences. From Stevens *et al.* (1994), with permission.

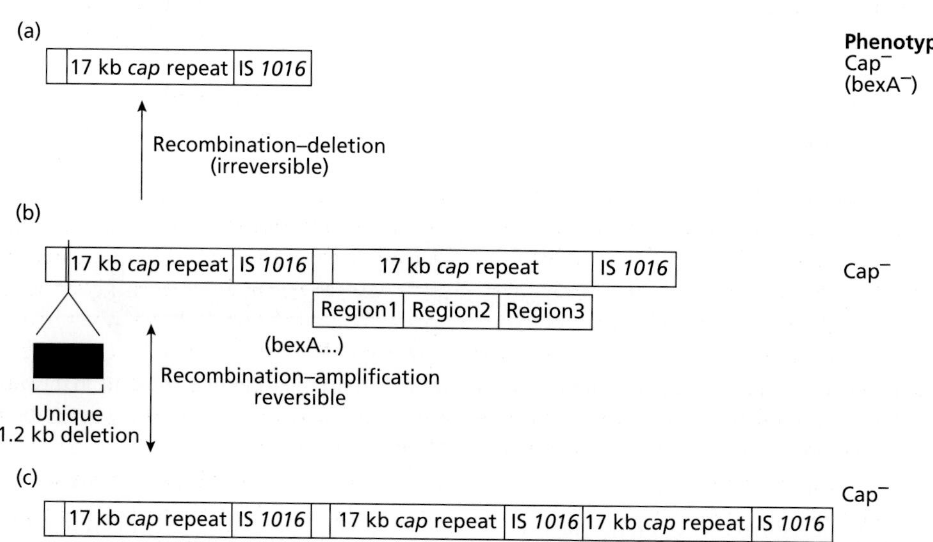

Fig. 13 Modification of CPS expression in *Haemophilus influenzae* type b, by *cap* gene amplification and deletion. (a) A single *cap* locus. (b) The duplicated *cap* locus, with the duplication bounded by the insertion sequence IS*1016*. A unique 1.2-kb region, which consists of the left-hand end of *cap* region 1 and the right-hand side of the insertion element, has been lost from the left-hand end of the duplicated locus. (c) The presence of the insertion element also facilitates amplification of the copy number of *cap*. Not to scale. From Roche and Moxon (1995), with permission.

bex gene cluster (Roche and Moxon, 1995). Normally, this cluster, which consists of regions 1, 2 and 3, is present twice in tandem on the chromosome (Kroll *et al.*, 1990). The duplication is bounded by an insertion element (IS*1016*), Recombination deletion, which includes the loss of the left-hand end of region 1 and the right-hand end of the insertion element, results in irreversible loss of encapsulation (Fig. 13). Amplification of the complete *bex* cluster (including IS*1016*) results in a reversible increase of encapsulation. It is suggested that in this way *H. influenzae* type b can adapt capsule expression, and thus its pathogenicity, to the environment (Roche and Moxon, 1995).

Modification by Reversible Insertion

Genetic analysis of adhesive uncapsulated *N. meningitidis* group B (Hammerschmidt *et al.*, 1996a, 1996b; Hilse *et al.*, 1996) has revealed a reversible inactivation of essential *sia* genes (of region 2). The 844-bp insertion element IS*1303*, which originated from the meningococcal chromosome, inserted specifically into the *siaA* gene, to direct the

C-2-epimerisation reaction: UDPGlcNAc ⇒ ManNAc + UDP. This interrupts the synthesis of CPMNeuNAc in its first specific reaction. Whereas the integration frequency is not known, excision occurs with a frequency of about 10^{-4}. Similarly, the expression of the Vi-antigen, cloned from *C. freundii* into *E. coli* K-12, is abrogated by the integration of an IS*1*-like element into the specific *viaB* gene region (see above) (Ou *et al.*, 1988). In this case, however, the insertion element originates from the *E. coli* K-12 chromosome and not from the *Citrobacter* chromosome.

The meningococcal *siaD* gene, which encodes the sialyltransferase, is inactivated in a different way (Hammerschmidt *et al.*, 1996b). Here, insertion/deletion of one or several cytidine residues into/from a $(dC)_7$ box at the 5′-end of the *siaD* gene, which encodes the polysialyltransferase, results in a slipped gene mismatch and inactivates the gene. This regulation has also been detected for several virulence factors of *Neisseria* (Meyer *et al.*, 1990).

Modulation of CPS Expression by Reversible Tyrosine Phosphorylation

In a recent report (Morona *et al.*, 2000), regulation/modulation of CPS expression by the concerted action of a tyrosine kinase and a specific tyrosine phosphatase was described in *Strep. pneumoniae* 19F and generally discussed for *Strep. pneumoniae* (Morona *et al.*, 1999a). The *cpsD, cpsB*, and *cpsC* genes, located at the 5′ end of the capsule focus (see above) regulate the extent of encapsulation of pneumococci. *cps*D encodes the CpsD protein, which is an autokinase that phosphorylates itself at multiple tyrosine sites. It is a substrate for the tyrosine-specific phosphatase CpsB, the *cpsB* gene product. Morona *et al.* (1999a, 2000) postulated that in encapsulated pneumococci, CpsD in its unphosphorylated state, forms a complex with CpsC which positively regulates the polymerisation of the CPS. On auto-phosphorylation of CpsD, the complex falls apart and capsule formation is then negatively regulated. Re-activation occurs through de-phosphorylation of CpsD-phosphate by CpsB. Presumably, CpsC can alternate between a polymerisation-inducing conformation (in complex with CpsD) and in an inactive, uncomplexed conformation. External signals that induce phosphorylation/de-phosphorylation are not known. It should be pointed out that CpsC is not a polymerase. Rather, it seems to be a polymerisation-regulator. In this context, it is interesting to note that the *cpsC* gene has pronounced homology to the *rol* and *rcl* genes (Batchelor *et al.*, 1992; Bastin *et al.*, 1993), which control the chain length of O polysaccharides in lipopolysaccharide biosynthesis.

Tyrosine autokinases and tyrosine-specific phosphatases, which used to be taken as characteristic of eukaryotic cells, have also been described in *E. coli* K-12 (Wzc and Wzb, respectively) (Vincent *et al.*, 1999) and in *Acinetobacter johnsonii* (Ptk and Ptp, respectively) (Duclos *et al.*, 1996; Grangeasse *et al.*, 1997, 1998). A specific role for both phosphorylating/de-phosphorylating systems was not described.

The polymerisation inducer and the regulator kinase may not only occur as two distinct proteins, like CpsC and CpsD for capsule production in pneumococci (Morona *et al.*, 2000) or EpsC and EpsD for the production of exopolysaccharide in *Strep. thermophilli* (Stingele *et al.* 1996). They may also be combined into one protein. This is the case with the Exo protein of *Rhizobium meliloti*, which regulates the size of the exopolysaccharide that is essential for symbiosis of rhizobia with their host plants (Becker *et al.*, 1995; Gonzales *et al.*, 1998). The amino end of the Exo protein, which is encoded by the 5′ end of the *exoP* gene, represents the polymerisation modulator that corresponds to CpsC of the pneumococci. Its carboxyl end, which is encoded by the 3′ end of the *exoP*, represents the regulating the autokinase that corresponds to CpsD of pneumococci. Down-regulation of the polymerisation modulator by autophosphorylation at the amino end of the protein results in the loss of the *R. meliloti* exopolysaccharide and the appearance of dimers and trimers of its repeating unit. This could be considered to be due to the loss of chain length/polymerisation regulation. A similar situation has been described for the Ams protein in the synthesis of the exopolysaccharide of *Erwinia amylovora* and, more recently, for the Wzc protein in the polymerisation of the K30 CPS of *E. coli* 09a:K30 (Drummelsmith and Whitfield, 1999).

Biological Significance

Role of Capsules as Virulence Factors

In the course of an infection, invasive bacteria encounter various host surroundings, such as the outside of mucosal cells, cytoplasm, the circulation, intracellular matrix amongst others. Bacterial surface components that play a role in an infection are not necessary in all phases of infection. For instance, adhesins and invasins are essential during the contact of the bacteria with their target cells and for invasion. Later, when the bacteria enter the circulation, capsules are required to counteract complement activity. With *Neisseria* and *Haemophilus*, however, capsules cover the cell surface adhesins and so interfere with bacterial adhesion (Stephens and Farley, 1991; Stephens *et al.*, 1993; Virji *et al.*, 1993). Consequently, adhesive bacteria are predominantly uncapsulated, which keeps their cell surface adhesins/invasins exposed, and it is the bacterial population exposed to the circulation that is encapsulated (Virji *et al.*, 1993). The mechanisms described above for the regulation of CPS expression seem important for the adaptation of invasive bacteria to a given host-induced situation, and encapsulation protects the bacteria against the host defence with complement and phagocytes and/or against their recognition by the immune system of the host.

Interference of CPSs with the Non-specific Host Defence

In the immunologically innocent host who has not developed a specific immune defence against an infectious micro-organism, the action of complement provides an important mechanism for the abatement of infection. The complement system may be activated either with the aid of specific antibodies (classical pathway) or in the absence of specific antibodies, by triggering of the properdin system (alternative pathway). Both types of activation result in bacteriolysis and/or phagocytosis, processes that are effected by the terminal membrane attack complex (MAC) and the activated complement component C3 (e.g. C3b), respectively (Loos, 1985; Cross, 1990).

In the absence of specific anticapsular antibodies, capsules prevent the action of complement (Edwards *et al.*, 1980; Kasper, 1986; Kim *et al.*, 1986; Jann and Jann, 1992; Marques *et al.*, 1992). This may be due to interference of CPS with the activation of C3, either by inhibiting the classical C3 convertase, C2bC4b, which is under the control of the C4-binding protein C4bp (Nussenzweig and Melton, 1981), or to interference with the cell association of an activated form of C3 (Pangborn and Müller-Eberhard, 1978). Some CPSs appear to bind, and thus inactivate C4bp. The activation of C3 to C3b is also controlled by factors B and H, which regulate the affinity of C3b for the bacterial surface. Certain polysaccharides, notably those that contain sialic acid, such as the K1 and K92 CPSs of *E. coli*, the CPSs of *N. meningitidis* groups B and C, the CPSs of group B *Streptococcus*, interfere with the association of factor B and C3b and interrupt or impair the subsequent complement cascade (Van Dijk *et al.*, 1979; Edwards *et al.*, 1982; Jarvis and Vedros, 1987).

It has also been postulated that bacterial capsules may just be too thick for activated complement to reach the bacterial cell wall and exert its destructive activity there. This was borne out by studies with the LPS capsule of *E. coli* O111 ($K_{LPS\ O111}$) (Joiner, 1985).

The basis of the resistance of encapsulated *Staph. aureus* and of *Strep. pneumoniae* is probably different. Even if the capsules interact with C3, a sub- or intra-capsular localisation of the C3b formed prevents its physical interaction with complement receptors on phagocytes (Brown *et al.*, 1983; Cross, 1990). The immunogenicity and immunochemistry of *Strep. pneumoniae* has been reviewed (Van Dam *et al.*, 1990).

The role of the capsule in the virulence of *Strep. pneumoniae* was demonstrated by genetic switching of the capsular type in isogenic strains. Virulence was determined by the presence of a CPS from a virulent strain (Kelly *et al.*, 1994). For a discussion see Claverys *et al.* (2000).

In Gram-negative bacteria, the CPS alone does not always appear to protect against complement action. For instance, the K1 CPS of *E. coli* protects against complement lysis, whereas the K5 CPS does not. Also, the combination of a given CPS with the cell wall LPS seems to determine the outcome of bacterium–host interactions (Cross *et al.*, 1986; Jann and Jann, 1992). The above-mentioned complex formation of CPS and LPS (Ørskov and Ørskov, 1991) may be of relevance for this phenomenon. The fact that only a limited number of O–K (LPS–CPS) combinations are encountered in invasive strains of *E. coli* also tends to support this notion.

Immune Responses to CPSs and their Potential as Vaccine Candidates

In general, bacterial CPS induces an immune response that results in the production of protective anticapsular antibodies in the exposed host. With whole bacteria or when coupled to a carrier protein as immunogen, CPS induces a T-cell-dependent immune response that results in the production of IgG antibodies. These can be boosted by subsequent injections of the immunogen. In contrast, pure soluble CPS induces a T-cell-independent response that results in IgM antibodies that cannot be boosted. The immunogenicity of free CPSs seems to depend, *inter alia*, on their molecular size, very large polysaccharides inducing a peripheral tolerance (the 'treadmill effect', discussed by Jann and Westphal, 1984).

The early finding that isolated capsular polysaccharides of pneumococci are immunogenic in mice (Schiemann and Casper, 1927) and in humans (Francis and Tillett, 1930), paved the way to the protection against infection with virulent encapsulated bacteria. The pneumococcal CPSs were the first purified polysaccharides to be used as vaccines (Austrian, 1976,

1985; Robbins, 1978; Jennings, 1983; Van Dam *et al.*, 1990; Keitel *et al.*, 1994). For greater efficiency, combinations of purified capsular polysaccharides were used as polyvalent vaccines, not only against *Strep. pneumoniae* (23 types) (Lee *et al.*, 1995), but also against other encapsulated bacteria such as *N. meningitidis* (4 types) (Cadoz *et al.*, 1985), *Klebsiella* (24 types) and *Pseudomonas* (8 types) (Edelman *et al.*, 1994).

The success of polysaccharide vaccines is hampered by their low immunogenicity in infants (Austrian, 1976; Gotschlich *et al.*, 1977; Peltola *et al.*, 1977). This drawback could, at least in part, be overcome by the use of polysaccharides chemically coupled to carrier proteins. It turned out that such conjugate vaccines generally improved immunogenicity and rendered the response T-cell-dependent (Anderson *et al.*, 1986; Jennings, 1990; Paoletti *et al.*, 1992; Edelman *et al.*, 1994; Kantor *et al.*, 1997). As with pure polysaccharides, the chain length of the conjugated polysaccharides proved to be important, with a mean value of about 20 kDa being optimal (Jann and Westphal, 1984; Jennings, 1990; Paoletti *et al.*, 1992; Kantor *et al.*, 1997; Laferriere *et al.*, 1997). The use of a bacterial toxoid, such as the diphtheria toxoid or the tetanus toxoid as carrier protein in a vaccine (Anderson, 1983; Paoletti *et al.*, 1992; Redhead *et al.*, 1994; Begg *et al.*, 1995) has the additional potential of protecting against the relevant toxin.

The extremely low immunogenicity of the CPSs from some pathogenic bacteria is due to structural relatedness to host components. Antibodies to these CPSs may, in fact, be auto-antibodies, and the corresponding immune cells were probably eliminated during development.

The K1 CPS of *E. coli* and the CPS of *N. meningitidis* group B, *Moraxella nonliquefaciens* and *Pasteurella haemolytica* A2, which all have the same structure (Table 3), are identical to the terminal polySia region of the embryonic form of the neural cell adhesion molecule NCAM (Finne, 1982; Troy, 1992). This phenomenon, known as molecular mimicry, can be overcome by chemical alteration of the CPS such as exchange of an *N*-formyl group for the *N*-acetyl group present in the native CPS (Jennings *et al.*, 1987, 1993). It should be mentioned that a diphtheria toxoid conjugate of the CPS of *N. meningitidis* group C, which differs from the meningococcal group D CPS only in linkage position (Table 3), is immunogenic not only in adults but also in infants (Richmond *et al.*, 1999).

The K5 CPS of *E. coli* (Table 3) has the same structure as *N*-acetylheparosan (Vann *et al.*, 1981), the first polymeric intermediate in the biogenesis of heparin (Lindahl, 1972). NCAM presents its 'bacterial' structure only in its embryonic form, and loses it later on transformation to its adult form, whereas the heparin intermediate *N*-acetylheparosan, a mammalian structural correlate to the K5 CPS, is permanently expressed in the host. Circumvention of the molecular mimicry by the K5 CPS is so far unknown.

Inability and/or impairment of the production of anticapsular antibodies renders the host vulnerable to infection by the respective encapsulated bacteria, for example *E. coli* K1 or K5. In essence, host defence is locked in the pre-immune phase of infection. Together with anticomplementary activity, as discussed above, such bacterial infections are severe.

Non-immune Activities of CPSs

CPSs may exert activities that are not primarily involved in immunological processes. The CPS of *Bacteroides fragilis*, which forms a strong complex of two distinct polysaccharides (Table 3), induces intestinal abscesses (Tzianabos *et al.*, 1993). This is due to the presence of both positive (amino groups) and negative (carboxyl and phosphate groups) charges in the CPS. Removal of either charge abrogates or greatly reduces abscess-forming capacity. Other CPSs that have only negative charges, such as the Vi-antigen of *S. typhi* (Table 3), can be rendered abscess-inducing by introducing a positive charge (de-*N*-acetylation).

The CPSs of *Staph. aureus* types 5 and 8 bind specifically to human epithelial cells (KB), endothelial cells and monocytes. Binding occurs in the absence of normal human serum and is therefore a direct interaction. It induces the formation of cytokines: interleukin-8 (IL-8) with KB cells; IL-6 and IL-8 with endothelial cells; IL-1β, IL-6, IL-8 and tumour necrosis factor (TNFα) with monocytes (Soell *et al.*, 1995). *Staph. aureus* causes bacteraemia in hospitalised and immuno-compromised patients and is also a causative agent of endocarditis, septic arthritis and osteomyelitis. The production of pro-inflammatory cytokines is an interesting aspect in the complications of *Staph. aureus* infections.

The ion-binding capacity of charged CPSs may be important for non-specific processes during bacterial infections. The long-chain LPS of *Proteus vulgaris* O19, which forms an LPS capsule (see above) binds cations, such as Mg^{2+}. This apparently induces the formation of stones during *Proteus* urinary tract infections (McLean *et al.*, 1985; Dumanski *et al.*, 1994). The same has also been found with *Proteus mirabilis* O6 and O57.

References

Aarons SJ, Sutherland IW, Chakrabarty AM, Gallagher MP (1997) A novel gene, *algK*, from the alginate biosynthetic cluster of *Pseudomonas aeruginosa*. *Microbiology* 143: 641–652.

Adlam C, Knight JM, Mugridge A, Williams JM, Lindon JC (1987) Production of colominic acid by *Pasteurella haemolytica* serotype A2. *FEMS Microbiol. Lett.* 42: 23–25.

Alberti S, Ashbaugh CD, Wessels MR (1998) Structure of the has operon promoter and regulation of hyaluronic acid capsule expression in group A streptococus. *Mol. Microbiol.* 28: 343–353.

Amor PA, Whitfield C (1997) Molecular and functional analysis of genes required for expression of group IB K antigens in *Escherichia coli*: characterization of the his-region containing gene cluster for multiple cell-surface polysaccharides. *Mol. Microbiol.* 26: 145–161.

Amor PA, Yethon JA, Monteiro MA, Whitfield C (1999). Assembly of the K40 antigen of *Escherichia coli*: Identification of a novel enzyme responsible for addition of L-serine residues to the glycan backbone and its requirement for K40 polymerization. *J. Bacteriol.* 181: 772–780.

Anderson PW (1983) Antibody responses to *Haemophilus influenzae* type b and diphtheria toxin induced by conjugates of oligosaccharides of the type b capsule with the non toxic protein CRM/197. *Infect. Immun.* 39: 233–238.

Anderson PW, Pichichero ME, Insel RA, Betts R, Eby R, Smith DH (1986) Vaccines consisting of periodate-cleaved oligosaccharides from the capsule of *Haemophilus influenzae* type b coupled to a protein carrier: structural and temporal requirements for priming in the human infant. *J. Immunol.* 137: 1181–1186.

Arakawa Y, Ohta M, Wacharotayankun R *et al.* (1991) Biosynthesis of Klebsiella capsular polysaccharide in *Escherichia coli* HB101 requires the functions of rmpA and the chromosomal cps gene cluster of the virulent strain *Klebsiella pneumoniae* Chedid (O1:K2). *Infect. Immun.* 59: 2043–2050.

Arrecubieta C, Lopez R, Garcia E (1994) Molecular characterization of cap3A, a gene from the operon required for the synthesis of the capsule of *Streptococcus pneumoniae* Type 3: Sequencing of mutations responsible for the unencapsulated phenotype and localization of the capsular cluster on the pneumococcal chromosome. *J. Bacteriol.* 176: 6375–6383.

Arrecubieta C, Lopez R, Garcia E (1996) Type 3-specific synthase of *Streptococcus pneumoniae* (Cap3B) directs type 3 polysaccharide biosynthesis in *Escherichia coli* and in pneumococcal strains of different serotypes. *J. Exp. Med.* 184: 449–455.

Atlung T, Ingmer H (1997) H-NS: a modulator of environmentally regulated gene expression. *Mol. Microbiol.* 24: 7–17.

Austrian R (1976) Vaccines of pneumococcal capsular polysaccharides and the prevention of pneumococcal pneumonia. In: Beers RF, Basset E (eds) *The Role of Immunological Factors in Infections, Allergic and Autoimmune Processes.* New York: Raven Press, pp. 79–89.

Austrian R (1985) Polysaccharide vaccines. *Ann. Inst. Pasteur Microbiol.* 136B: 295–307.

Bailey MJA, Hughes C, Koronakis V (1997) RfaH and the ops element, components of a novel system controlling bacterial transcription elongation. *Mol. Microbiol.* 26: 845–851.

Barnes DM, Whittier S, Gilligan PH, Soares S, Tomasz A, Henderson FW (1995) Transmission of multidrug-resistant serotype 23F *Streptococcus pneumoniae* in group day care: evidence suggesting capsular transformation of the resistant strain *in vivo*. *J. Infect. Dis.* 171: 890–896.

Bastin DA, Stevenson G, Brown PK, Haase A, Reeves PR (1993) Repeat units polysaccharides of bacteria: a model for polymerization resembling that of ribosome and fatty acid synthetase, with a novel mechanism determining chain length. *Mol. Microbiol.* 7: 725–734.

Batchelor RA, Alifano P, Biffali E, Hull SI, Hull RA (1992) Nucleotide sequence of the genes regulating O-polysaccharide antigen chain length (*rol*) from *Escherichia coli* and *Salmonella typhimurium*: protein homology and functional complementation. *J. Bacteriol.* 174: 5228–5236.

Baumann H, Tzianabos AO, Brisson J-R, Kasper DL, Jennings HJ (1992) Structural elucidation of two capsular polysaccharides from one strain of *Bacteroides fragilis* using high-resolution NMR spectroscopy. *Biochemistry* 31: 4081–4089.

Bayer AS, Eftekhar F, Tu J, Nast CC, Speert EP (1990) Oxygen-dependent up-regulation of mucoid exopolysaccharide (alginate) production in *Pseudomonas aeruginosa*. *Infect. Immun.* 58: 1344–1349.

Bayer ME (1990) Visualization of the bacterial polysaccharide capsule. *Curr. Top. Microbiol. Immunol.* 150: 129–157.

Becker A, Niehaus K, Pühler A (1995) Low-molecular weight succinoglycan is predominantly produced by *Rhizobium meliloti* strains carrying a mutated exoP protein characterized by a periplasmic N-terminal domain and a missing C-terminal domain. *Mol. Microbiol.* 16: 191–203.

Begg NT, Miller E, Fairley CK *et al.* (1995) Antibody responses and symptoms after DTP and either tetanus or diphtheria *Haemophilus influenzae* type B conjugate vaccines given for primary immunisation by separate or mixed injection. *Vaccine* 13: 1547–1550.

Bernheimer HP, Wardmundsen IE, Austrian R (1967) Qualitative differences in the behavior of pneumococcal desoxyribonucleic acids transforming to the same capsular type. *J. Bacteriol.* 93: 320–333.

Beynon LM, Richards JC, Perry MB, Kniskern PJ (1991) Antigenic and structural relationships within group 19 *Streptococcus pneumoniae*: chemical characterization of the specific capsular polysaccharides type 19B and 19C. *Can. J. Chem.* 70: 131–137.

Bhasin N, Albus A, Michon F, Livolsi PJ, Park JS, Lee JC (1998) Identification of a gene essential for *O*-acetylation of the *Staphylococcus aureus* type 5 capsular polysacharide. *Mol. Microbiol.* 27: 9–21.

Bhattacharjee AK JHJ, Kenny CP, Martin A, Smith ICP (1975) Structural determination of the sialic acid polysaccharide antigens of *Neisseria meningitidis* serogroup B and C with carbon-13 nuclear magnetic resonance. *J. Biol. Chem.* 250: 1926–1932.

Bliss JM, Silver RP (1996) Coating the surface: a model for expressionn of capsular polysialic acid in *Escherichia coli* K1. *Mol. Microbiol.* 21: 221–231.

Bliss JM, Silver RP (1997) Evidence that KpsT, the ATP-binding component of an ATP-binding cassette, is exposed to the periplasm and associates with polymer during translocation of the polysialic acid capsule of *Escherichia coli* K1. *J. Bacteriol.* 179: 1400–1403.

Boulnois GJ, Jann K (1989) Bacterial polysaccharide capsule synthesis, export and evolution of structural diversity. *Mol. Microbiol.* 3: 1819–1823.

Boulnois GJ, Roberts IS (1990) Genetics of capsular polysaccharide production in bacteria. *Curr. Top. Microbiol. Immunol.* 150: 1–18.

Boulnois GJ, Roberts IS, Hodge R, Hardy K, Jann K, Timmis KM (1987) Analysis of the K1 capsule biosynthesis genes of *Escherichia coli*: definition of three functional regions for capsule production. *Mol. Gen. Genet.* 208: 242–246.

Boulnois GJ, Drake R, Pearce R, Roberts I (1992) Genome diversity at the serA-linked capsule locus in *Escherichia coli. FEMS Microbiol. Lett.* 100: 121–124.

Branefors-Hellander P, Erbing C, Kenne L, Lindberg B (1976) Structural studies of the capsular antigens from *Haemophilus influenzae* type b. *Acta Chem. Scand.* 30: 276–277.

Brisson J-R, Baumann H, Imberty A, Perez S, Jennings HJ (1992) Helical epitope of the group B meningococcal α(2–8)-linked sialic acid polysaccharide. *Biochemistry* 31: 4996–5004.

Bronner D, Sieberth V, Pazzani C *et al.* (1992) Expression of the capsular K5 polysaccharide of *Escherichia coli*: biochemical and electron microscopic analyses of mutants with defects in region 1 of the K5 gene cluster. *J. Bacteriol.* 175: 5984–5992.

Brown EJ, Joiner KA, Cole RM, Berger M (1983) Localization of complement component 3 on *Streptococcus pneumoniae*: anticapsular antibody causes complement deposition on the pneumococcal capsule. *Infect. Immun.* 39: 403–409.

Bugert P, Geider K (1995) Molecular analysis of *ams* operon required for exopolysaccharide synthesis of *Erwinia amylovora*. *Mol. Microbiol.* 15: 917–933.

Cadoz M, Armand J, Arminjon F, Gire R, Lafaix C (1985) Tetravalent (A,C,Y,W135) meningococcal vaccines in children: immunogenicity and safety. *Vaccine* 3: 340–342.

Chakraborty AK, Friebolin H, Stirm S (1980) Primary structure of the *Escherichia coli* serotype 30 capsular polysaccharide. *J. Bacteriol.* 141: 971–972.

Coffey TJ, Dowson CG, Daniels M *et al.* (1991) Horizontal transfer of multiple penicillin-binding protein genes, and capsular biosynthetic genes, in natural populations of *Streptococcus pneumoniae*. *Mol. Microbiol.* 5: 2255–2260.

Coffey TJ, Enright MC, Daniels M *et al.* (1998) Recombinational exchanges at the capsular polysaccharide biosynthetic locus lead to frequent serotype changes among natural isolates of *Streptococcus pneumoniae*. *Mol. Microbiol.* 27: 73–83.

Cross AS (1990) The biological significance of bacterial encapsulation. *Curr. Top. Microbiol.Immunol.* 150: 87–95.

Cross A, Kim KS, Wright DC, Sadoff JC, Gemski P (1986) Role of lipopolysaccharide and capsule in the serum resistance of bacteremic strains of *Escherichia coli*. *J. Infect. Dis.* 154: 497–503.

Daniels EM, Schneerson R, Egan WM, Szu SC, Robbins JB (1989) Characterization of the *Salmonella paratyphi* C Vi polysaccharide. *Infect. Immun.* 57: 3159–3164.

De Cueninck BJ, Shockman GD, Swenson RM (1982) Group B, type III streptococcal cell wall: composition and structural aspects revealed through endo-*N*-acetylmuraminidase-catalyzed hydrolysis. *Infect. Immun.* 35: 572–582.

DeAngelis PL, Weigel PH (1993) Isolation of a *Streptococcus pyogenes* gene locus that directs hyaluronan biosynthesis in acapsular mutants and in heterologous bacteria. *J. Bacteriol.* 268: 14568–14571.

DeAngelis PL, Weigel PH (1994) Immunochemical confirmation of the primary structure of streptococcal hyaluronan synthase and synthesis of high molecular weight product by the recombinant enzyme. *Biochemistry* 33: 9033–9039.

Dengler T, Jann B, Jann K (1986) Structure of the serine-containing capsular polysaccharide K40 antigen from *Escherichia coli* O8:K40:H9. *Carbohydr. Res.* 150: 233–240.

Deretic V, Tomasek T, Darzins A, Chakrabarty AM (1986) Gene amplification induces mucoid phenotype in rec-2 *Pseudomonas aeruginosa* exposed to Kanamycin. *J. Bacteriol.* 165: 510–516.

Deretic V, Dikshit R, Konyecsni WM, Chakrabarty AM, Misra TK (1989) The algR gene, which regulates mucoidy in *Pseudomonas aeruginosa*, belongs to a class of environmentally responsive genes. *J. Bacteriol.* 171: 1278–1283.

Devi SJN, Schneerson R, Egan W, Vann WF, Robbins JB, Shiloach J (1991) Identity between polysaccharide antigens of *Moraxella nonliquefaciens*, group B *Neisseria meningitidis*, and *Escherichia coli* K1 (non-O acetylated). *Infect. Immun.* 59: 732–736.

Dillard JP, Yother J (1991) *Streptococcus pneumoniae* sequences cloned in *Escherichia coli*: effect of promotor strength and transcription terminators. *J. Bacteriol.* 173: 5101–5109.

Dillard JP, Yother J (1994) Genetic and molecular characterization of capsular polysaccharide biosynthesis in *Streptococcus pneumoniae*. *Mol. Microbiol.* 12: 959–972.

Dillard JP, Vandersea MW, Yother J (1995) Characterization of the cassette containing genes for Type 3 capsular polysaccharide biosynthesis in *Streptococcus pneumonioae*. *J. Exp. Med.* 181: 973–983.

Dougherty BA, van de Rijn I (1993) Molecular characterization of hasB from an operon required for hyaluronic acid synthesis in group A streptococci. Demonstration of UDP-glucose dehydrogenase activity. *J. Biol. Chem.* 268: 7118–7124.

Dougherty BA, van de Rijn I (1994) Molecular characterization of hasA from an operon required for hyaluronic acid synthesis in group A streptococci. *J. Biol. Chem.* 269: 169–175.

Drummelsmith J, Whitfield C (1999) Gene products required for surface expression of the capsular form of the group 1 K antigen in *Escherichia coli* (O9a:K30). *Mol. Microbiol.* 31: 1321–1332.

Drummelsmith J, Amor PA, Whitfield C (1997) Polymorphism, duplication, and IS1-mediated rearrangement in the chromosomal his-rfb-gnd region of *Escherichia coli* strains with group 1A capsular antigens. *J. Bacteriol.* 179: 3232–3238.

Duclos B, Grangeasse C, Vagany E, Riberty M, Cozzone AJ (1996) Autophosphorylation of a bacterial protein at tyrosine. *J. Mol. Biol.* 259: 891–895.

Dumanski AJ, Hedelin H, Edin-Liljegren A, Beauchemin D, McLean RJC (1994) Unique ability of the *Proteus mirabilis* capsule to enhance mineral growth in infectious urinary calculi. *Infect. Immun.* 62: 2998–3003.

Edelman R, Taylor DN, Wasserman SS *et al.* (1994) Phase 1 trial of a 24-valent *Klebsiella* capsular polysaccharide vaccine and an eight-valent *Pseudomonas* O-polysaccharide conjugate vaccine administered simultaneously. *Vaccine* 12: 1288–1294.

Edwards MS, Nicholson-Weller A, Baker CJ, Kasper DL (1980) The role of specific antibody in alternative complement pathway-mediated opsonophagocytosis of type III group B *Streptococcus. J. Exp. Med.* 151: 1275–1287.

Edwards MS, Kasper DL, Jennings HJ, Baker CJ, Nicholson-Weller A (1982) Capsular sialic acid prevents activation of the alternative complement pathway by type III, group B streptococci. *J. Immunol.* 128: 1278–1283.

Eklund K, Garegg PJ, Kenne L, Lindberg AA, Lindberg B (1979) Structural studies on the *Escherichia coli* 0111 lipopolysaccharide. In: Abstracts of the IXth International Symposium on Carbohydrates, London, 1978, No. 493. Oxford: Pergamon Press.

Fakhr MK, Penaloza-Vasquez , Chakrabarty AM,Bender CL (1999) Regulation of alginate biosynthesis in *Pseudomonas syringae* pv. *syringae*. *J. Bacteriol.* 181: 3478–3485.

Finke A, Jann B, Jann K (1990) CMP-Kdo synthetase activity in *Escherichia coli* expressing capsular antigens. *FEMS Microbiol.Lett.* 69: 129–134.

Finke A, Bronner D, Nikolaev AV, Jann B, Jann K (1991) Biosynthesis of the *Escherichia coli* K5 polysaccharide, a representative of group II polysaccharides: polymerization *in vitro* and characterization of the product. *J. Bacteriol.* 173: 4088–4094.

Finne J (1982) Occurrence of unique polysialosyl carbohydrate units in glycoproteins of developing brain. *J. Biol. Chem.* 257: 11966–11970.

Flynn JL, Ohman DE (1988) Cloning of genes from mucoid *Psaeudomonas aeruginosa* which control spontaneous conversion to the alginate production phenotype. *J. Bacteriol.* 170: 1452–1460.

Fournier JM, Vann WF, Karakawa WW (1984) Purification and characterization of *Staphylococcus aureus* type 8 capsular polysaccharide. *Infect. Immun.* 45: 87–93.

Francis T Jr, Tillett WS (1930) Cutaneous reactions in pneumonia. The development of antibodies following intradermal injection of type specific polysaccharides. *J. Exp. Med.* 52: 573–585.

Frosch M, Edwards U (1993) Molecular mechanisms of capsule expression in *Neisseria meningitidis* serogroup B. In: Roth J, Rutishauser U, Troy FA (eds) *Polysialic Acid from Microbe to Man*. Basel: Birkhäuser Verlag, pp. 49–57.

Frosch M, Edwards U, Bousset K, Krausse B, Weisgerber C (1991) Evidence for a common molecular origin of the capsule gene loci in Gram-negative bacteria expressing group II capsular polysaccharides. *Mol. Microbiol.* 5: 1251–1263.

Gacesa P (1998) Bacterial alginate biosynthesis – recent progress and future prospects. *Microbiology* 144: 1133–1143.

Garcia E, Lopez R (1997) Molecular biology of *Streptococcus pneumoniae*. *FEMS Microbiol. Lett.* 149: 1–10.

Garcia E, Garcia P, Lopez R (1993) Cloning and sequencing of a gene involved in the synthesis of the capsular polysaccharide of *Streptococcus pneumoniae* type 3. *Mol. Gen. Genet.* 239: 188–195.

Garegg PJ, Lindberg B, Onn T, Sutherland IW (1971) Comparative structural studies on the M-antigen from *Salmonella typhimurium*, *Escherichia coli*, and *Aerobacter cloacae*. *Acta Chem. Scand.* 25: 2103–2108.

Glode MP, Robbins JB, Liu TY, Gotschlich EC, Ørskov I, Ørskov F (1977) Cross antigenicity and immunogenicity between capsular polysaccharides of group C and of *E.coli* K92. *J. Infect. Dis.* 135: 94–102.

Goldberg JB, Dahnke T (1992) *Pseudomonas aeruginosa* AlgB, which modulates the expression of alginate, is a member of the NitrC subclass of procaryotic regulators. *Mol. Microbiol.* 6: 59–66.

Goldberg JB, Ohman DE (1987) Construction and characterization of *Pseudomonas aeruginosa* algB mutants: role of algB in high-level production of alginate. *J. Bacteriol.* 169: 1593–1602.

Gonzalez JE, Semino CE, Wang L-X, Castellano-Torres LE, Walker GC (1998). Biosynthetic control of molecular weight in the polymerization of the octasaccharide subunits of succinoglycan, a symbiotically important exopolysaccharide of Rhizobium meliloti. *Proc. Natl Acad. Sci. USA* 95: 13477–13482.

Gotschlich EC, Goldschneider I, Lepow ML, Gold R (1977) The immune response to bacterial polysaccharides in man. In: Haber E, Krause RM (eds) *Antibody in Human Diagnosis and Therapy*. New York: Raven Press, pp. 391–402.

Gotschlich EC, Frazer BA, Mishimura O, Robbins JB, Liu TY (1981) Lipid on capsular polysaccharides of Gram-negative bacteria. *J. Biol. Chem.* 256: 8915–8921.

Gottesman S, Stout V (1991) Regulation of capsular polysaccharide synthesis in *Escherichia coli* K-12. *Mol. Microbiol.* 62: 1599–1606.

Grangeasse C, Vagany C, Vincent G, Deléage G, Duclos B, Cozonne AJ (1997) Characterization of a bacterial gene encoding an autophosphorylating protein tyrosine kinase. *Gene* 204: 259–265.

Grangeasse C, Doublet P, Vincent C *et al.* (1998) Functional characterization of a low-molecular-mass phosphotyrosine-protein phosphatase of *Acinetobacter johnsonii*. *J. Mol. Biol.* 278: 399–347.

Grant WD, Sutherland IW, Wilkinson JF (1969) Exopolysaccharide colanic acid and its occurrence in the Enterobacteriaceae. *J. Bacteriol.* 100: 1187–1193.

Griffith F (1928) The significance of pneumococal types. *J.Hyg.* 27: 113–159.

Guidolin A, Morona JK, Morona R, Hansman D, Paton JC (1994) Nucleotide sequence analysis of genes essential for capsular polysaccharide biosynthesis in *Streptococcus pneumoniae* type 19F. *Infect. Immun.* 62: 5384–5396.

Hammerschmidt S, Hilse R, van Putten JPM, Gerardy-Schahn R, Unkmeier A, Frosch M (1996a) Modulation of cell surface sialic acid expression in *Neisseria meningitidis* via a transposable element. *EMBO J.* 15: 192–198.

Hammerschmidt S, Müller A, Sillmannn H *et al.* (1996b) Capsule phase variation in *Neisseria meningitiidis* serogroup B by slipped-strand mispairing in the polysialyltransferase gene (siaD): correlation with bacterial invasion and the outbreak of meningococcal disease. *Mol. Microbiol.* 20: 1211–1220.

Hannesson HH, Hagner-McWhirter A, Tiedemann K, Lindahl U, Malmström A (1996) Biosynthesis of dermatan sulphate. Defructosylated *Escherichia coli* K4 capsular polysaccharide as a substrate for the D-glucuronyl-C5-epimerase, and an indication of a two-base reaction mechanism. *Biochemistry* 313: 589–586.

Hashimoto Y, Ezaki T, Li N, Yamamoto H (1991) Molecular cloning of the ViaB region of *Salmonella typhi*. *FEMS Microbiol. Lett.* 90: 53–56.

Hashimoto Y, Li N, Yokoyama H, Ezaki T (1993) Complete nucleotide sequence and molecular characterization of ViaB region encoding Vi antigen in *Salmonella typhi*. *J. Bacteriol.* 175: 4456–4465.

Heyns K, Kiessling G (1967) Strukturaufklärung des Vi-Antigens aus *Citrobacter freundii* (*E.coli*) 5396/38. *Carbohydr. Res.* 3: 340–352.

Hilse R, Hammerschmidt S, Bautsch W, Frosch M (1996) Site-specific insertion of IS1301 and distribution in *Neisseria meningitidis* strains. *J. Bacteriol.* 178: 2527–2532.

Hobbs M, Reeves P (1994) The JUMPstart sequence: a 39 bp element common to several polysaccharide gene clusters. *Mol. Microbiol.* 12: 855–856.

Hoch JA, Silhavy TS (eds) (1995) *Two-component Signal Transduction*. Washington, DC: ASM Press.

Hofmann P, Jann B, Jann K (1985) Structure of the amino acid-containing capsular polysaccharide (K54 antigen) from *Escherichia coli* O6:K54:H10. *Carbohydr. Res.* 139: 261–271.

Hoiseth SK, Moxon ER, Silver RP (1986) Genes involved in *Haemophilus influenzae* type b capsule expression are part of a 18-kilobase tandem duplication. *Proc. Natl Acad. Sci. USA* 83: 1106–1110.

Hök M, Kjellen L, Johanson S (1984) Cell surface glycosaminoglycans. *Annu. Rev. Biochem.* 53: 847–869.

Huong H-SH, Noon KF, Ou JT, Baron LS (1992) Expression of Vi antigen in *Escherichia coli* K-12: characterization of ViaB from *Citrobacter freundii* and identity of ViaA with RcsB. *J. Bacteriol.* 174: 5910–5915.

Jain S, Ohman DE (1998) Deletion of algU in mucoid *Pseudomonas aeruginosa* blocks alginate polymer formation and results in uronic acid secretion. *J. Bacteriol.* 180: 634–641.

Jann B, Jann K (1990) Structure and biosynthesis of the capsular antigens of *Escherichia coli*. *Curr. Top. Microbiol. Immunol.* 150: 19–43.

Jann B, Jann K, Schmidt G, Ørskov F, Ørskov I (1971) Comparative immunochemical studies of the surface antigens of *Escherichia coli* strains O8:K87(B?):H19 and (O32):K87(B?):H45. *Eur. J. Biochem.* 23: 515–522.

Jann K, Jann B (1992) Capsules of *Escherichia coli*, expression and biological significance. *Can. Microbiol.* 38: 705–710.

Jann K, Jann B (1997) Capsules of *Escherichia coli*. In: Sussman M (ed.) *Escherichia coli Mechanisms of Virulence*. Cambridge: Cambridge University Press, pp. 113–143.

Jann K, Westphal O (1984) *Microbial Polysaccharides*. In: Sela M (ed.) *The Antigens*, Vol. 3. New York: Academic Press, pp. 1–125.

Jann K, Dengler T, Jann B (1992) Core-lipid A on the K40 polysaccharide of *Escherichia coli* O8:K40:H9, a representative of group I capsular polysaccharides. *Zbl. Bakt.* 276: 196–204.

Jarvis GA, Vedros NA (1987) Sialic acid of group B *Neisseria meningitidis* regulates alternative complement pathway activation. *Infect. Immun.* 55: 174–180.

Jayaratne P, Keenleyside WJ, MacLachlan PR, Dodgson C, Whitfield C (1993) Characterization of rcsB and rcsC from *Escherichia coli* O9:K30.:H12 and examination of the role of the rcs regulatory system in expression of group I capsular polysaccharides. *J. Bacteriol.* 175: 5384–5394.

Jelakovic S, Jann K, Schulz G (1996) The three-dimensional structure of capsule-specific CMP:2-keto-3-deoxy-manno-octonic acid synthetase from *Escherichia coli*. *FEBS Lett.* 391: 157–161.

Jennings HJ (1983) Capsular polysaccharides as human vaccines. *Adv. Carbohydr. Chem. Biochem.* 41: 155–208.

Jennings HJ (1990) Capsular polysaccharides as vaccine candidates. *Curr. Top. Microbiol. Immunol.* 150: 97–127.

Jennings HJ, Rosell KG, Carlo DJ (1980) Structural determination of the capsular polysaccharide of *Streptococcus pneumoniae* type 19 (19F). *Can. J. Chem.* 58: 1069–1074.

Jennings HJ, Lugowski C, Kasper DL (1981) Conformational aspects critical to the immunospecificity of the type III group B streptococcal polysaccharide. *Biochemistry* 20: 4511–4518.

Jennings HJ, Gamian A, Ashton FE (1987) N-propionylated group B meningococcal polysaccharide mimicks a unique epitope on group B *Neisseria meningitidis. J. Exp. Med.* 165: 1207–1211.

Jennings HJ, Brisson J-R, Kulakowska M, Michon F (1993) Polysialic acid vaccines against meningitis caused by *Neisseria meningitidis* and *Escherichia coli.* In: Roth J, Rutishauser U, Troy FA (eds) *Polysialic acid from Microbes to man.* Basel: Birkhäuser Verlag, pp. 25–38.

Johnson EM, Krauskopf B, Baron LS (1965) Genetic mapping of Vi and somatic antigenic determinants in *Salmonella. J. Bacteriol.* 90: 302–308.

Joiner K (1985) Studies on the mechanism of bacterial resistance to complement mediated killing and on the mechanism of action of bacteriocidal antibody. *Curr. Top. Microbiol. Immunol.* 121: 99–133.

Kantor E, Luxenberg JS, Lucas AH, Granoff DM (1997) Phase I study of the immunogenicity and safety of conjugated *Haemophilus influenzae* type b vaccines in the elderly. *Vaccine* 15: 129–132.

Karakawa WW, Vann WF (1982) Capsular polysaccharides of *S. aureus. Semin. Infect. Dis.* 4: 285–293.

Kasper DL (1986) Bacterial capsules – old dogmas and new tricks. *J. Infect. Dis.* 153: 407–415.

Katzenellenbogen E, Jennings HJ (1983) Structural determination of the capsular polysaccharide of *Streptococcus pneumoniae* type 19A (57). *Carbohydr. Res.* 124: 235–245.

Keenleyside WJ, Bronner D, Jann K, Jann B, Whitfield C (1993) Coexpression of colanic acid and serotype specific capsular polysaccharides in *Escherichia coli* strains with group II K antigens. *J. Bacteriol.* 175: 6725–6730.

Keitel WA, Bond NL, Zahradnik JM, Cramton TA, Robbins JB (1994) Clinical and serological responses following primary and booster immunization with *Salmonella typhi* Vi capsular polysaccharide vaccines. *Vaccine* 12: 185–199.

Kelly T, Dillard JP, Yother J (1994) Effect of genetic switching of capsular type on virulence of *Streptococcus pneumoniae. Infect. Immun.* 62: 1813–1819.

Kenne L, Lindberg B (1983) Bacterial polysaccharides. In: Aspinall GO (ed.) *The Polysaccharides,* Vol. 2. New York: Academic Press, pp. 287–363.

Kim KS, Kang JH, Cross AS (1986) The role of capsular antigens in serum resistance and *in vivo* virulence of *Escherichia coli. FEMS Microbiol. Lett.* 35: 275–278.

Knirel YA, Paredes L, Jansson P-E, Weintraub A, Widmalm G, Albert J (1995) Structure of the capsular polysaccharide of *Vibrio cholerae* O139 synonym Bengal containing D-galactose 4,6-cyclophosphate. *Eur. J. Biochem.* 232: 391–396.

Kogan G, Jann B, Jann K (1992) Structure of the *Escherichia coli* O104 polysaccharide and its identity with the capsular K9 polysaccharide. *FEMS Microbiol. Lett.* 91: 135–140.

Kolkman MAB, Morrison DA, van der Zeist BAM, Nuiten PJM (1996) The capsule polysaccharide synthesis locus of *Streptococcus pneumoniae* serotype 14: identification of the glycosyl transferase gene cpsE. *J. Bacteriol.* 178: 3736–3741.

Kolkman MAB, van der Zeijst BAM, Nuijten PJM (1997) Functional analysis of glycosyl transferases enconded by the capsular polysaccharide locus of *Streptococcus pneumoniae* serotype 14. *J. Biol. Chem.* 272: 19502–19508.

Kolkman MC, van den Zeijst BAM, Nuijten PJM (1998). Diversity of capsular polysaccharide synthesis gene clusters in *Streptococcus pneumoniae. J. Biochem.* 123: 937–945.

Kolyva S, Waxin H, Popoff MY (1992) The Vi antigen of *Salmonella typhi*: molecular analysis of the ViaB locus. *J. Microbiol.* 138: 297–304.

Krallmann-Wenzel U, Schmidt G (1994) A simple procedure to demonstrate the presence of the O-antigen capsule in enteropathogenic *Escherichia coli. Zbl. Bakt.* 281: 140–145.

Kroll JS, Loynds B, Brophy LN, Moxon ER (1990) The bex locus in encapsulated *Haemophilus influenzae*: a chromosomal region involved in capsule polysaccharide export. *Mol. Microbiol.* 4: 1853–1962.

Kroll JS, Loynds BM, Moxon ER (1991) The *Haemophilus influenzae* capsulation gene cluster: a compound transposon. *Mol. Microbiol.* 5: 1459–1460.

Kröncke KD, Boulnois G, Roberts I *et al.* (1990a) Expression of the *Escherichia coli* K5 capsular antigen: Immunonelectron microscopic and biochemical studies with recombinant *E.coli. J. Bacteriol.* 172: 1085–1091.

Kröncke KD, Golecki JR, Jann K (1990b) Further electron microscopic studies on the expression of *Escherichia coli* group II capsules. *J. Bacteriol.* 172: 3469–3472.

Kuhn H-M, Meier-Dieter U, Mayer H (1988) ECA, the enterobacterial common antigen. *FEMS Microbiol. Rev.* 54: 195–222.

Kuo JSC, Doelling VW, Graveline JF, McCoy DW (1985) Evidence for covalent attachment of phospholipid to the capsular polysaccharide of *Haemophilus influenzae* type b. *J. Bacteriol.* 163: 769–773.

Kusche M, Hannesson HH, Lindahl U (1991) Biosynthesis of heparin. Use of *Escherichia Coli* K5 capsular polysaccharide as a model substrate in enzymic polymer-modification reactions. *Biochemistry* 275: 151–158.

Laakso DH, Homonylo MK, Wilmot SJ, Whitfield C (1988) Transfer and expression of the genetic determinants for O and K antigen synthesis *Escherichia coli* O9:K(A)30 and *Klebsiella* spp. O1:K20. *Can. J. Microbiol.* 34: 987–992.

Laferriere CA, Sood RK, de Muys J-M, Michon F, Jennings HH (1997) The synthesis of the *Streptococcus pneumoniae* polysaccharide-tetanus toxoid conjugates and the effect of chain length on immunogenicity. *Vaccine* 15: 179–186.

Lee CJ, Fraser BA (1980) The structures of the cross-reactive types 19 (19F) and (19A) pneumococcal polysaccharides. *J. Biol. Chem.* 255: 6847–6853.

Lee CY (1992) Cloning of genes affecting capsule expression in *Staphylococcus aureus* strain M. *Mol. Microbiol.* 6: 1515–1522.

Lee CY (1995) Association of staphylococcal type-1 capsule-encoding genes with a discrete genetic element. *Gene* 167: 115–119.

Lee H-J, Kang J-H, Henrichsen J *et al.* (1995) Immunogenicity and safety of a 23-valent pneumococcal polysaccharide vaccine in healthy children and in children at increased risk of pneumococcal infection. *Vaccine* 13: 1533–1538.

Lee JC, Xu S, Albus A, Livolsi PJ (1994) Genetic analysis of type 5 capsular polysaccharide expression by *Staphylococus aureus. J. Bacteriol.* 176: 4883–4889.

Lidholt K, Fjelstad M (1997) Biosynthesis of the *Escherichia coli* K4 capsule polysaccharide. A parallel system for studies of glycosyltransferases in chondroitin formation. *J. Biol. Chem.* 272: 2682–2787.

Lidholt K, Weinke JL, Kieser CS *et al.* (1992) A single mutation affects both *N*-acetylglucosaminyl transferase and glucuronyltransferase activities in a chinese hamster ovary cell mutant defective in heparan sulfate biosynthesis. *Proc. Natl Acad. Sci. USA* 89: 2267–2271.

Lidholt K, Fjelstad M, Jann K, Lindahl U (1994) Substrate specificities of glycosyl transferases involved in the formation of heparin precursor and *E.coli* K5 capsular polysaccharides. *Carbohydr. Res.* 255: 87–101.

Lin WS, Cunneen T, Lee CY (1994) Sequence analysis and molecular characterization of genes required for the biosynthesis of type 1 caspular polysaccharide in *Staphylococcus aureus. J. Bacteriol.* 176: 7005–7016.

Lind T, Lindahl U, Lidholt K (1993) Biosynthesis of heparin/heparan sulfate. Identification of a 70 kDa protein catalyzing both the D-glucuronosyl and the *N*-acetyl-D-glucosaminyl transferase reactions. *J. Biol. Chem.* 268: 20705–20708.

Lindahl U (1972) Enzymes involved in the formation of the carbohydrate structure of heparin. *Methods Enzymol.* 28: 676–684.

Llull D, Lopez R, Garcia E, Munoz R (1998) Molecular structure of the gene cluster responsible for the synthesis of the polysaccharide capsule of *Streptococcus pneumoniae* type 33F. *Biochim. Biophys. Acta* 1443: 217–224.

Llull D, Munoz R, Lopez R, Garcia E (1999) A single gene (*tts*) located outside the *cap* locus directs the formation of *Streptococcus pneumoniae* type 37 capsular polysaccharide: type 37 pneumococci are natural, genetically binary strains. *J. Exp. Med.* 190: 241–251.

Loos M (1985) The complement system: activation and control. *Curr. Top. Microbiol. Immunol.* 121: 7–18.

Ma S, Selvaraj U, Ohman DE, Quarlese R, Hassett DJ, Wozniak DJ (1998) Phosphorylation independent activity of the response regulators AlgB and AlgR in promoting alginate biosynthesis in mucoid *Pseudomonas aeruginosa. J. Bacteriol.* 180: 956–968.

McCallum KL, Whitfield C (1991) The rcsA gene of *Klebsiella pneumoniae* O1:K20 is involved in expression of the serotype-specific K (capsular) antigen. *Infect. Immun.* 59: 494–502.

McLean RJC, Nickel JC, Noakes VC, Costerton JW (1985) An in vitro ultrastructural study of infectious kidney stone growth in the urinary tract. *Infect. Immun.* 49: 805–811.

Markowitz A (1977) Genetics and regulation of bacterial capsular polysaccharide synthesis and radiation sensitivity. In: *Surface Carbohydrates of the Procaryotic Cell.* Sutherland IW (ed.) London: Academic press, pp. 415–462.

Marques MB, Kasper DL, Pangburn MK, Wessels MR (1992) Prevention of C3 deposition by capsular polysaccharide is a virulence mechanism of type III group B streptococci. *Infect. Immun.* 60: 3986–3993.

Masson L, Holbein BE (1985) Role of lipid intermediate(s) in the synthesis of serogroup b *Neisseria meningitidis* capsular polysaccharide. *J. Bacteriol.* 161: 861–867.

May TB, Chakrabarty AM (1994) *Pseudomonas aeruginosa*: genes and enzymes of alginate synthesis. *Trends Microbiol.* 2: 151–157.

May TB, Shinabarger DM, Maharaj R *et al.* (1991) Alginate synthesis by *Pseudomonas aeruginosa*: a key pathogenic factor in chronic pulmonary infections of cystic fibrosis patients. *Clin. Microbiol. Rev.* 4: 191–206.

Meyer TF, Gibbs CP, Haas R (1990) Variation and control of protein expression in *Neisseria. Annu. Rev. Microbiol.* 44: 451–477.

Michon F, Brisson J-R, Dell A, Kasper DL, Jennings HJ (1988). Multiantennary group-specific polysaccharide of group B Streptococcus. *Biochemistry* 27: 5341–5351.

Moreau M, Richards JC, Fournier JM, Byrd RA, Karakawa WW, Vann WF (1990) Structure of the type-5 capsular polysaccharide of *Staphylococcus aureus. Carbohydr. Res.* 201: 285–297.

Morona JK, Morona R, Paton JC (1997a) Characterization of the locus encoding the *Streptococcus pneumoniae* type 19F capsular polysaccharide. *Mol. Microbiol.* 23: 751–763.

Morona JK, Morona R, Paton JC (1997b) Molecular and genetic characterization of the capsule biosynthesis locus of *Streptococcus pneumoniae* type 19B. *J. Bacteriol.* 179: 4953–4958.

Morona JK, Miller DC, Coffey TJ *et al.* (1999a) Molecular and genetic characterization of the capsule biosynthesis locus of *Streptococus pneumoniae* type 23F. *Microbiology* 145: 781–789.

Morona JK, Morona R, Paton JM (1999b) Analysis of the 5′ portion of the type 19A capsule locus identifies two classes of *cpsC*, *cpsD*, and *cpsE* genes in *Streptococcus pneumoniae. J. Bacteriol.* 181: 3599–3605.

Morona JK, Morona R, Paton JM (1999c) Comparative genetics of capsular polysaccharide biosynthesis in *Streptococcus pneumoniae* types belonging to the serogroup 19. *J. Bacteriol.* 101: 5355–5364.

Morona JK, Paton JC, Miller DC, Morona R (2000) Tyrosine phosphorylation of CpsD negatively regulates capsular polysaccharide biosynthesis in *Streptococcus pneumoniae*. *Mol. Microbiol.* 35: 1431–1442.

Nesin M, Ramirez M, Tomasz A (1998) Capsular transformation of a multidrug-resistant *Streptococcus pneumoniae in vivo*. *J. Infect. Dis.* 177: 707–713.

Nussenzweig V, Melton R (1981) Human C4 binding protein (C4bp). *Methods Enzymol.* 80: 124–133.

Ou JT, Baron LS, Rubin FA, Kopecko DJ (1988) Specific insertion sequence 1-like DNA element causes the reversible expression of the virulence capsular antigen Vi of *Citrobacter freundii* in *Escherichia coli*. *Proc. Natl Acad. Sci. USA* 85: 4402–4405.

Ouyang S, Lee CY (1997) Transcriptional analysis of type 1 capsule genes in *Staphylococcus aureus*. *Mol. Microbiol.* 23: 473–482.

Ørskov I, Nyman K (1974) Genetic mapping of the antigenic determinants of two polysaccharide antigens K10 and K54 in *Escherichia coli*. *J. Bacteriol.* 120: 43–51.

Ørskov F, Ørskov I (1991) Complex formation between *Escherichia coli* lipopolysaccharide O antigen and capsular K antigen as detected by immunoelectrophoresis. *APMIS* 99: 615–619.

Ørskov I, Ørskov F, Jann B, Jann K (1977) Serology, chemistry and genetics of O and K antigens of *Escherichia coli*. *Bacteriol. Rev.* 41: 667–710.

Ørskov F, Ørskov I, Sutton A *et al.* (1979) Form variation in *Escherichia coli* K1: determined by O-acetylation of the capsular polysaccharide. *J. Exp.Med.* 149: 669–685.

Ørskov F, Sharma V, Ørskov I (1984) Influence of growth temperature on the development of *Escherichia coli* polysaccharide K antigens. *J. Gen. Microbiol.* 130: 2681–2684.

Pangborn MU, Müller-Eberhard HJ (1978) Complement C3 convertase: cell surface restriction of beta1H control and generation of restriction on neuraminidase treated cells. *Proc. Natl Acad. Sci. USA* 75: 2416–2420.

Paoletti LC, Kasper DL, Michon F *et al.* (1992) Effects of chain length on the immunogenicity in rabbits of group B Streptococcus type III oligosaccharide – tetanus toxoid conjugates. *J. Clin. Invest.* 89: 203–209.

Paulsen JT, Beness AM, Saier MH Jr., (1997) Computer-based analyses of the protein constituents of transport systems catalysing export of complex carbohydrates in bacteria. *Microbiology* 143: 2685–2699.

Pavelka MS, Wright LF, Silver RP (1991) Identification of two genes, kpsM and kpsT, in the region 3 of the polysialic acid gene cluster of *Escherichia coli* K1. *J. Bacteriol.* 173: 4603–4610.

Pazzani C, Rosenow C, Boulnois GJ, Bronner D, Jann K, Roberts IS (1993) Molecular analysis of region 1 of the *Escherichia coli* K5 antigen gene cluster: a region encoding proteins involved in cell surface expression of capsular polysaccharides. *J. Bacteriol.* 175: 5978–5983.

Peltola H, Käythy H, Sivonen A, Mäkelä PH (1977) *Haemophilus influenzae* type b capsular polysaccharide vaccine in children : a double blind field study of 100,000 vaccines 3 months to 5 years of age in Finland. *Pediatrics* 60: 730–737.

Perry MB, McLean LL (1994) The structure of the polysaccharide produred by *Proteus vulgaris* (ATCCC 49990). *Carbohydr. Res.* 253: 257–263.

Petering H, Hammerschmidt S, Frosch M, van Putten JPM, Ison CA, Robertson BD (1996) Genes associated with the meningococcal capsule complex are also found in *Neisseria gonorrhoeae*. *J. Bacteriol.* 178: 3342–3345.

Petit C, Rigg GP, Pazzani C *et al.* (1995) Analysis of region 2 of the *Escherichia coli* K5 gene cluster: a region encoding proteins for the biosynthesis of the K5 polysaccharide. *Mol. Microbiol.* 17: 611–620.

Pradel E, Schnaitman C (1991) Effect of rfaH (sfrB) and temperature on expression of rfa genes of *Escherichia coli* K-12. *J. Bacteriol.* 173: 6428–6431.

Prehm P (1986) Mechanism, localization, and inhibition of hyaluronate synthesis. In: Kuettner K (ed.) *Articulate Cartilage Biochemistry.* New York: Raven Press, pp. 81–91.

Rahn A, Drummelsmith J, Whitfield C (1999) Conserved organization in the cps gene clusters for expression of *Escherichia coli* group 1 K-antigens: relationship to the colanic acid biosynthesis locus and the cps genes from *Klebsiella pneumoniae*. *J. Bacteriol.* 181: 2307–2313.

Ramirez M, Thomas A (1998) Molecular characterization of the complete 23F capsular polysaccharide locus of *Streptococcus pneumoniae*. *J. Bacteriol.* 180: 3273–3278.

Redhead K, Sesardic D, Yost SE *et al.* (1994) Interaction of *Haemophilus influenzae* type b conjugate vaccines with diphtheria-tetanus-pertussis vaccine in control tests. *Vaccine* 12: 1460–1466.

Reeves RE, Goebel WF (1941) Chemoimmunological studies on the soluble specific substrate of pneumococcus. V. The structure of the type III polysaccharide. *J.Biol.-Chem.* 139: 511–519.

Reeves P, Hobbs M, Valvano MA *et al.* (1996) Bacterial polysaccharide synthesis and gene nomenclature. *Trends Microbiol.* 4: 495–503.

Reizer J, Reizer A, Saier MH Jr (1992) A new subfamily of bacterial ABC type transport systems catalyzing export of drugs and carbohydrates. *Protein Sci.* 1: 1326–1332.

Reuter G, Gabius H-J (1996) Sialic acids: structure–analysis–metabolism–occurrence–recognition. *Biol. Chem. Hoppe-Seiler* 377: 325–342.

Rhan A, Drummelsmith J, Whitfield C (1999) Conserved organization in the cps gene clusters for expression of *Escherichia coli* group K1 antigens: relationship to the colanic acid biosynthesis locus and the cps genes from *Klebsiella pneumoniae*. *J. Bacteriol.* 181: 2307–2313.

Richards JC, Perry MB (1988) Structure of the specific polysaccharide of *Streptococcus pneumoniae* type 23F. *Biochem. Cell. Biol.* 66: 758–771.

Richmond P, Borrow R, Miller E *et al.* (1999) Meningococcal serogroup C conjugate vaccine is immunogenic in infancy and primes for memory. *J. Infect. Dis.* 179: 1569–1572.

Rietschel ET, Wollenweber H-W, Brade H *et al.* (1984) Structure and conformation of the lipid A component of lipopolysaccharides. In: Proctor RA (ed.) *Handbook of Endotoxin*, Vol. 1: Rietschel ER (ed.) *Chemistry of Endotoxin* Amsterdam: Elsevier pp. 187–220.

Robbins JB (1978) Vaccines for the prevention of encapsulated bacterial diseases: current status, problems, and prospects for the future. *Immunochemistry* 15: 839–854.

Roberts IS (1996) The biochemistry and genetics of capsular polysaccharide production in bacteria. *Annu. Rev. Microbiol.* 50: 285–315.

Roberts IS, Mountford R, Hodge R, Jann KB, Boulnois GJ (1988) Common organization of gene clusters for production of different capsular polysaccharides (K antigens) in *E.coli*. *J. Bacteriol.* 170: 1305–1310.

Roche JR, Moxon RE (1995) Phenotypic variation of carbohydrate surface antigens and the pathogenesis of *Haemophilus influenzae* infections. *Trends Microbiol.* 3: 304–309.

Rodriguez ML, Jann B, Jann K (1988) Structure and serological characteristic of the capsular K4 antigen of *Escherichia coli* O5:K4:H4, a fructose containing polysaccharide with a chondroitin backbone. *Eur. J. Biochem.* 177: 117–124.

Rosenow C, Esumeh F, Roberts IS, Jann K (1995) Characterization and localization of the KpsE protein of *Escherichia coli*, which is involved in polysaccharide export. *J. Bacteriol.* 177: 1137–1143.

Rubens CE, Heggen LM, Haft RF, Wessels MR (1993) Identification of cpsD, a gene essential for type III capsule expresssion in group B streptococci. *Mol. Microbiol.* 8: 843–855.

Russo TA, Wenderoth S, Carlino U, Merck JM, Lesse AJ (1998) Identification, genomic organization, and analysis of the group III capsular polysaccharide genes kpsD, kpsM, kpsT, and kpsE from an extraintestinal isolate of *Escherichia coli* (CP9, O4/K54/H5). *J. Bacteriol.* 180: 338–349.

Saier MHJ (1994) Computer-aided analysis of transport protein sequences gleening evidence concerning function, structure, biogenesis and evolution. *Microbiol. Rev.* 58: 71–93.

Sau S, Lee CY (1996) Cloning of type 8 capsule genes and analysis of gene clusters for the production of different capsular polysaccharides in *Staphylococcus aureus*. *J. Bacteriol.* 178: 2118–2126.

Sau S, Sun J, Lee CY (1997a) Molecular characterization and transcriptional analysis of type 8 capsule genes in *Staphylococcus aureus*. *J. Bacteriol.* 179: 1614–1621.

Sau S, Bhasin N, Wann ER, Lee JC, Foster TJ, Lee CY (1997b) The *Staphylococcus aureus* allelic genetic loci for serotype 5 and 8 capsule expression contain the type-specific genes flanked by common genes. *Microbiology* 143: 2395–2405.

Schauer R (1991) Biosynthesis and function of *N*- and *O*-substituted sialic acids. *Glycobiology* 1: 449–.

Schiemann O, Casper W (1927) Sind die spezifischen präcipitablen Substanzen der 3 Pneumokokkentypen Haptene? *Z. Hyg. Infektionskr.* 108: 220–257.

Schmidt MA, Jann K (1982) Phospholipid substitution of capsular (K) polysaccharide antigens from *Escherichia coli* causing extraintestinal infection. *FEMS Microbiol. Lett.* 14: 74–79.

Schurr MJ, Deretic V (1997) Microbial pathogenesis in cystic fibrosis: co-ordinate regulation of heat shock response and conversion to mucoidy in *Pseudomonas aeruginosa*. *Mol. Microbiol.* 24: 411–420.

Sieberth V, Jann B, Jann K (1993) Structure of the K10 capsular antigen from *Escherichia coli* O11:K10:H10, a polysaccharide containing 4,6-dideoxy-4-malonylamino-D-glucose. *Carbohydr. Res.* 246: 219–228.

Sieberth V, Rigg GP, Roberts I, Jann K (1995) Expression and characterization of UDPGlc dehydrogenase (KfiD), which is encoded in the type specific region of the *E.coli* K5 capsule genes. *J. Bacteriol.* 177: 4562–4565.

Silver RP, Aaronson W, Vann WF (1987) Translocation of capsular polysaccharides in pathogenic strains of *Escherichia coli* requires a 60-kilodalton periplasmic protein. *J. Bacteriol.* 169: 5489–5495.

Silver RP, Annunciato P, Pavelka MS, Pidgeon RP, Wright LF, Wunder DE (1993) Genetic and molecular analyses of the polysialic acid gene cluster of *Escherichia coli*. In: Roth J, Rutishauser U, Troy FA (eds) *Polysialic Acid* Basel: Birkhäuser Verlag, pp. 59–71.

Skjak-Braek G, Grasdalen H, Larsen B (1986) Monomer sequence and acetylation pattern in some bacterial alginates. *Carbohydr. Res.* 154: 239–250.

Sledjenski D, Gottesman S (1995) A small RNA acts as an antisilencer of the H-NS silencec rcsA gene of *Escherichia coli*. *Proc. Natl Acad. Sci. USA* 92: 2003–2007.

Smith A, Boulnois GJ, Roberts IS (1990) Molecular analysis of the *Escherichia coli* K5 kps locus: identification and characterization of an inner membrane capsular polysaccharide transport system. *Mol. Microbiol.* 4: 1863–1869.

Soell M, Diab M, Haan-Archipoff G, Beretz A, Herbelin C, Poutrel B, Klein J-P (1995) Capsular polysaccharide types 5 and 8 of *Staphylococcus aureus* bind specifically to human epithelial (KB) cells, endothelial cells, and monocytes and induce release of cytokines. *Infect. Immun.* 63: 1380–1386.

Sørensen UBS, Blom J, Birch-Andersen A, Henrichsen J (1988) Ultrastructural localization of capsules, cell wall polysaccharide, cell wall proteins, and F antigen in pneumococci. *Infect. Immun.* 56: 1890–1896.

Sørensen UBS, Henrichsen J, Chen H-C, Szu SC (1990). Covalent linkage between the capsular polysaccharide and the cell wall peptidoglycan of *Streptococcus pneumoniae* revealed by immunochemical methods. *Microb. Pathog.* 8: 325–334.

Stephens DS, Farley MM (1991) Pathogenic events during infection of the human nasopharynx with *Neisseria menigitidis*, and *Haemophilus influenzae*. *Rev Infect. Dis.* 13: 22–33.

Stephens DS, Spellman PA, Swartley JS (1993) Effect of the (α2–8)-linked polysialic acid capsule on adherence of *Neisseria meningitidis* to human mucosal cells. *J. Infect. Dis.* 167: 475–479.

Stevens MP, Hänfling P, Jann B, Jann K, Roberts IS (1994) Regulation of *Escherichia coli* K5 capsular polysaccharide expression: evidence for involvement of RfaH in the expression of group II capsules. *FEMS Microbiol. Lett.* 124: 93–98.

Stevens MP, Clarke BR, Roberts IS (1997) Regulation of the *Escherichia coli* capsule gene cluster by transcription antitermination. *Mol. Microbiol.* 24: 1001–1012.

Stevenson G, Andrianopoulos K, Hobbs M, Reeves P (1996) Organization of the *Escherichia coli* K-12 gene cluster responsible for production of the extracellular polysaccharide colanic acid. *J. Bacteriol.* 118: 4885–4893.

Stingele F, Neeser J-R, Mollet B (1996) Identification and characterization of the *eps* (exopolysaccharide) gene cluster from *Streptococcus thermophilis* Sfi6. *J. Bacteriol.* 178: 1680–1690.

Swartley JS, Marfin AA, Edupuganti S *et al.* (1997) Capsule switching of *Neisseria meningitidis*. *Proc. Natl Acad. Sci. USA* 94: 271–276.

Swartley JS, Liu LJ, Miller YK, Martin LE, Edupuganti S, Stephens DS (1998) Characterization of the gene cassette required for biosynthesis of the (alpha-1,6) linked *N*-acetyl-D-mannosamine-1-phosphate capsule of serogroup A *Neisseria meningitidis*. *J. Bacteriol.* 180: 1533–1539.

Troy FA (1980) Chemistry and biosynthesis of the poly (γ-D-glutamyl) capsule in *Bacillus licheniformis*. In: Kleinkauf H, v. Dören H (eds) *Peptide Antibiotics – Biosynthesis and Functions*. Berlin: Walter de Gruyter, pp. 49–83.

Troy FA (1992) Polysialylation: from bacteria to brains. *Glycobiology* 2: 5–23.

Troy FA, Frerman FE, Heath EC (1971) The biosynthesis of capsular polysaccharide in *Aerobacter aerogenes*. *J. Biol. Chem.* 246: 118–133.

Troy FA, Janas T, Janas T, Merker RI (1991) Vectorial translocation of polysialic acid chains across the inner membrane of neuroinvasive *Escherichia coli* K1. *FASEB J.* 5: A1548.

Tzianabos AO, Onderdonk AB, Rosner B, Cisneros RL, Kaspar DL (1993) Strucural features of polysaccharides that induce intra-abdominal abcesses. *Science* 262: 416–419.

Unger FM (1931) The chemistry and biological significance of 3-deoxy-D-manno-2-octulosonic acid (KDO). *Adv. Carbohydr. Chem.* 38: 323–388.

Van Dam JEG, Fleer A, Snippe H (1990) Immunogenicity and immunochemistry of *Streptococcus pneumoniae* capsular polysaccharides. *Antonie von Leeuwenhoek* 58: 1–47.

Van Dijk WC, Verbruch HA, Tol ME, Peters R, Verhoef J (1979) Role of *Escherichia coli* Kl capsular antigens during complement activation. C3 fixation and opsonization. *Infect. Immun.* 25: 603–609.

Vann WF, Schmidt MA, Jann B, Jann K (1981) The structure of the capsular polysaccharide (K5 antigen) of urinary tract infective *Escherichia coli* O10:K5:H4, a polymer similar to desulfoheparin. *Eur. J. Biochem.* 116: 359–364.

Vimr E (1990) Map position and genomic organization of the kps cluster for polysialic acid synthesis in *Escherichia coli* K1. *J. Bacteriol.* 173: 1335–1338.

Vimr ER, Aaronson W, Silver RP (1989) Genetic analysis of chromosomal mutations in the polysialic acid gene cluster of *Escherichia coli* K1. *J. Bacteriol.* 171: 1106–1117.

Vincent C, Doublet P, Grangeasse C, Vagany E, Cozzone AJ, Duclos B (1999) Cells of *Escherichia coli* contain a protein-tyrosine kinase, Wzc, and a phosphotyrosine-protein phosphatase. *J. Bacteriol.* 181: 3472–3477.

Virji M, Makepeace K, Ferguson DJP, Achtman M, Moxon ER (1993) Meningococcal Opa and Opc proteins: their role in colonization and invasion of human epithelial and endothelial cells. *Mol. Microbiol.* 10: 499–510.

Wacharotayankun R, Arakawa Y, Ohta M *et al.* (1992) Involvement of rcsB in *Klebsiella* K2 capsule synthesis in *Echerichia coli* K-12. *J. Bacteriol.* 174: 1063–1067.

Wang J, Lory S, Ramphal R, Jin S (1996) Isolation and characterization of *Pseudomonas aeruginosa* genes inducible by respiratory mucus derived from cystic fibrosis patients. *Mol. Microbiol.* 22: 1005–1012.

Wessels MR, Poszgay V, Kasper DL, Jennings HJ (1987) Structure and immunochemistry of an oligosaccharide repeating unit of the capsular polysaccharide of type III group B *Streptococcus*. *J. Biol Chem.* 262: 8262–8267.

Wessels MR, Craig ER, Vicente-Javier B, Kasper DL (1989) Definition of a bacterial virulence factor: sialylation of the group B streptococcal capsule. *Proc. Natl Acad. Sci. USA* 86: 8983–8987.

Whitfield C, Roberts LS (1999) Structure, assembly and regulation of expression of capsules in *Escherichia coli*. *Mol. Microbiol.* 31: 1307–1319.

Whitfield C, Troy FA (1984) Biosynthesis and assembly of the polysialic acid capsule in *Echerichia coli*. *J. Biol. Chem.* 259: 12769–12775.

Whitfield C, Valvano MA (1993) Biosynthesis and expression of cell-surface polysaccharides in Gram-negative bacteria. *Adv. Microb. Physiol.* 35: 135–246.

Wright A, Dankert M, Fennesay R, Robbins PW (1967) Characterization of a polyisoprenoid compound functional in O antigen biosynthesis. *Proc. Natl Acad. Sci. USA* 57: 1798–1803.

Wunder DE, Aaronson W, Hayes SF, Bliss JM, Silver RP (1994). Nucleotide sequence and mutational analysis of the gene encoding KpsD, a periplasmic protein involved in transport of polysialic acid in *Escherichia coli*. *J. Bacteriol.* 176: 4025–4033.

Yurewitz EC, Ghalambor MA, Heath EC (1971) The structure of the *Aeromonas aerogenes* capsular polysaccharide. *J. Biol. Chem.* 246: 5596–5606.

Zou W, Laferriere CA, Jennings HJ (1998). Oligosaccharide fragments of the type III group B streptococcal polysaccharide derived from *S. pneumoniae* type 14 capsular polysaccharide by a chemoenzymatic method. *Carbohydr. Res.* 309: 297–301.

5

Teichoic Acids of Gram-positive Bacteria

Ian C. Hancock

University of Newcastle upon Tyne, UK

Zeta potential measurements for bacteria reveal that, irrespective of their hydrophobicity, the cells bear a net negative surface charge. In Gram-positive bacteria this charge is provided by wall teichoic acids (WTA) and acidic polysaccharides commonly described as teichuronic acids, polyanionic macromolecules that are covalently linked components of the cell wall. A second class of anionic cell envelope macromolecules, lipoteichoic acids (LTA), possess terminal membrane anchor groups, and although they are not covalently linked to the cell wall, they, too, penetrate the wall network and contribute to the net charge. While all these polymers occur together in the cell walls of a few bacteria, such as *Bacillus licheniformis*, the predominant anionic wall polymers in aerobic and micro-aerophilic species are teichoic acids, in which the negative charges are provided by a phosphodiester group in the regular repeating units of the polymer chain. This group is strongly acidic, with a pK_a of about 1.5, and therefore contributes a full negative charge at all physiological pH values. Growing evidence indicates that a wall-bound anionic polymer and LTA are both essential for viability. In a few Gram-positive genera, notably *Micrococcus*, not known to contain teichoic acids, teichuronic acids (TU), which contribute negative charge by virtue of carboxylate groups in their repeating structure, are present instead. Wall teichoic acids have not been reported in obligately anaerobic Gram-positive species, though an ill-defined detergent-extractable polysaccharide that contains phosphate is present in the envelope of *Clostridium botulinum* (Takumi *et al.*, 1991). Teichoic acids are also absent from mycolic acid-containing actinomycete genera (*Mycobacterium*, *Corynebacterium*, *Rhodococcus*, *Nocardia*) but they are present in a very wide range of amycolate actinomycetes (Naumova and Shashkov, 1997), again with the exception of anaerobic species. Lipoteichoic acids have not been reported in actinomycetes.

The non-covalently attached anionic cell envelope polymer lipoteichoic acid (LTA) is associated with the outer face of the cytoplasmic membranes of many Gram-positive bacteria (Aasjord and Grov, 1980; Fischer, 1988) and also, often in a partly deacylated form (Kessler *et al.*, 1979), in loose association with the cell wall. Although this macromolecule bears a superficial resemblance to a wall teichoic acid, the two differ in their biosynthetic origins, their stereochemistry and their chain termination – attachment to peptidoglycan in WTA, and attachment to a

 doi:10.1006/bkmm.2001.0005

glycolipid in LTA. The amphiphilic nature of LTA is crucial for its biological properties, which more closely resemble those of lipopolysaccharide than WTAs.

Wall teichoic acids and lipoteichoic acids confer a variety of surface properties of great immunological and chemical significance for infection and pathogenesis by Gram-positive bacteria.

Structure of Wall Teichoic Acids and their Attachment to the Cell Wall

Teichoic acids are fairly short linear polymers, usually with a chain length of 30–40 repeating units. They exhibit a diversity of regular repeating polymeric structures in which the common feature is the linkage of the repeating units through phosphodiesters (Fig. 1). The simplest types are those in which the repeating unit is a simple alditol phosphate: glycerol phosphate (Fig. 1a), erythritol phosphate, ribitol phosphate (Fig. 1b), or mannitol phosphate or a sugar-1-phosphate (usually an *N*-acetylhexosamine-1-phosphate) (Fig. 1c). Of these, the glycerol phosphate and ribitol phosphate types ('glycerol teichoic acids' and 'ribitol teichoic acids') are the most widely distributed amongst bacterial species. They frequently bear glycosyl substituents on one of the hydroxyls of the alditol. Other teichoic acids contain more complex repeating units in which a sugar, oligosaccharide or sugar-1-phosphate, linked to an alditol phosphate, constitutes part of the main polymer chain. Enormous structural diversity is thus possible, through variations in the sugar or alditol composition of the repeating unit, and through positional variations of the linkage between components of the repeating unit and of the linkage between the repeating units in the complete polymer chain. Moreover, D-alanine, ester linked through its carboxyl group to a hydroxyl of the repeating unit alditol, is a common feature of a wide range of alditol-containing teichoic acids. In these polymers, the degree of alanylation varies, but is always less than one alanyl residue per repeating unit. The free amino group of the ester-linked alanine contributes a positive charge at physiological pH values and thus results in a reduction in the overall negative charge of the polymer. Other substituents, including lysyl, acetyl and succinyl groups, have been reported, though more rarely (Naumova and Shashkov, 1997).

In contrast to the diversity of polymer chain structure, the mode of covalent attachment of the teichoic acids to the peptidoglycan of the cell wall is highly conserved in Gram-positive bacteria. Invariably, a small oligomeric 'linkage unit' intervenes between the phosphate terminus of the polymer chain and the primary hydroxyl at C6 of a muramic acid residue

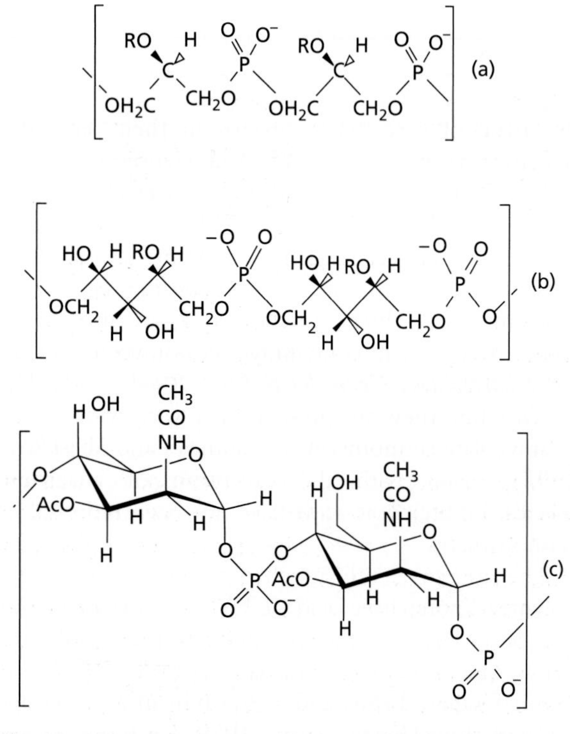

Fig. 1 Structures of common types of wall teichoic acid. Each diagram illustrates the structure of a dimer of the repeating unit of the linear polymer chain. (a) Poly(glycerol phosphate) teichoic acid; (b) poly(ribitol phosphate) teichoic acid; (c) poly-(*N*-acetylmannosamine-1-phosphate) teichoic acid. *R*, glycosyl or D-alanyl substituent; Ac, *O*-acetyl substituent.

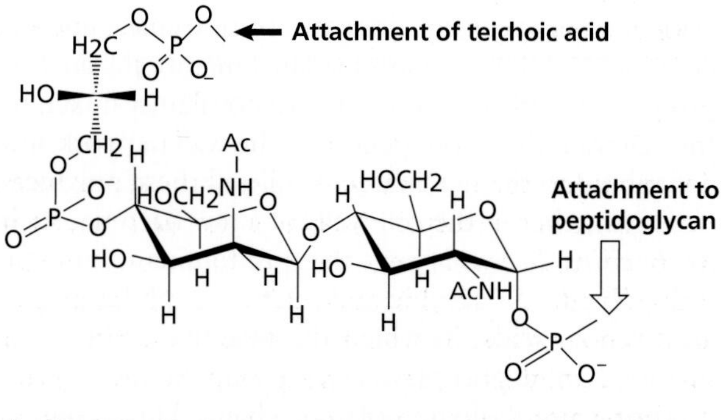

Fig. 2 Structure of the linkage unit that attaches wall teichoic acid to peptidoglycan. The diagram shows the *N*-acetylmannosaminyl-*N*-acetylglucosamine-1-phosphate that is linked through its terminal phosphate to C6 of a *N*-acetylmuramic acid residue in peptidoglycan. One glycerol phosphate residue is shown intervening between the *N*-acetylmannosamine residue and the teichoic acid main-chain, but up to three occur in some species (see text for details).

Table 1 Anionic wall polymers of clinically significant Gram-positive bacteria

Bacterial species	Wall teichoic acid	Reference
Actinomadura madurae, A. pelletieri, A. viridis	Poly(glycosylglycerol phosphate), glycosyl-poly(ribitol phosphate)	Shashkov *et al.*, 1999
Bacillus cereus	Glc-poly(glycerol phosphate), also teichuronic acids	Sasaki *et al.*, 1980; White, 1977
Listeria monocytogenes	GlcNAc-poly(ribitol phosphate), poly(GlcNAc-ribitol phosphate)	Uchikawa *et al.*, 1986; Kamisango *et al.*, 1983
Nocardiopsis dassonvilliae	Poly(GalNAc-glycerol-phospho-glycerol-phosphate)	Tul'skaya *et al.*, 1993
Staphylococcus aureus, Staph. capitis, Staph. haemolyticus, Staph. hominis, Staph. intermedius	GlcNAc-poly(ribitol phosphate)	Endl *et al.*, 1983; Osland *et al.*, 1980
Staph. epidermidis, Staph. warneri, S. cohnii	GlcNAc-poly(glycerol phosphate), Glc-poly(glycerol phosphate)	Endl *et al.*, 1983
Staphylococcus hyicus, Staph. sciuri	Poly(GlcNAc-1-phosphoglycerol phosphate)	Endl *et al.*, 1983
Staph. xylosus, Staph. saprophyticus	GlcNAc-poly(ribitol phosphate), GlcNAc-poly(glycerol phosphate)	Endl *et al.*, 1983
Streptococcus bovis	Teichuronic acid	Pazur *et al.*, 1976
Strep. oralis	Poly[Gal*f*-Gal*p*-Gal*f*-GalNAc-Gal*p*-ribitol phosphate]	Abeygunawardana *et al.*, 1991
Strep. pneumoniae	Poly[Glc-AATGal-GalNAc(-6-choline-P)-GalNAc(-6-choline-P)-ribitol phosphate]	Fischer *et al.*, 1993

AATGal, 2-acetamido-4-amino-2,4,6-trideoxy-D-galactose.

Table 2 Capsular and O-antigenic teichoic acids in Gram-negative bacteria

Bacterial species	Role of teichoic acid	Reference
Actinobacillus pleuropneumoniae	Capsular polymer	Beynon *et al.*, 1991
Campylobacter coli	LPS-associated polymer	Aspinall *et al.*, 1993
Escherichia coli	Capsular polymers O-antigens	Jann *et al.*, 1985
Fusobacterium necrophorum	O-antigen	Hermansson *et al.*, 1993
Hafnia alvei	O-antigen	Katzenellenbogen *et al.*, 1992
Haemophilus influenzae	Capsular polymers	Moxon and Kroll, 1990
Neisseria meningitidis	Capsular polymers	Bundle *et al.*, 1974

in peptidoglycan. This unit consists of between one and three glycerol phosphate residues linked in turn to the *N*-acetylmannosamine residue of the disaccharide *N*-acetylmannosamine-β1,4 *N*-acetylglucosamine-1-phosphate. The 1-phosphate group on the *N*-acetylglucosamine forms a phosphoester linkage to the primary hydroxyl at C6 of the *N*-acetylmuramic acid of the peptidoglycan (Fig. 2). Within the true Gram-positive bacteria, only two isolated variations of this structure have been observed, involving replacement of *N*-acetylmannosamine by glucose (Kaya *et al.*, 1985) or of *N*-acetylglucosamine by glucosamine (Kojima *et al.*, 1985). The reason for evolutionary conservation of this structure is unknown, but its similarity to the mode of attachment of cell wall arabinogalactan in *Mycobacterium* (Brennan and Nikaido, 1995) and of *N*-linked oligosaccharides in glycoproteins of eukaryotes is intriguing.

Although Shashkov *et al.* (1999) have presented good evidence for species-specificity of teichoic acid structure among actinomycete bacteria, there also appears to be strain-dependent variation within species of many Gram-positive bacteria. Within some species, such as *Staphylococcus aureus*, strain variation is limited to alterations in the anomeric configuration of the substituent sugar (*N*-acetylglucosamine); in others, such as *Staph. epidermidis*, the nature of the glycosyl substituent can vary (*N*-acetylglucosamine or glucose); still other species exhibit a diversity of polymer backbone (for example, ribitol and glycerol teichoic acids in different strains of *Staph. saprophyticus*). Thus, teichoic acids do not constitute species-specific markers of currently recognised species of medical interest. The teichoic acids of some representative bacteria of medical importance are shown in Table 1.

Cell wall teichoic acids are absent from Gram-negative bacteria. However, there are many examples of teichoic acids among capsular polymers, notably in *Escherichia coli*, *Haemophilus influenzae* and *Neisseria meningitidis* (Table 2). In these cases the polymer chains are much longer (typically 100–150 repeating units), are not substituted with D-alanine and are synthesised with a terminal phospholipid group that may serve to anchor them to the outer surface of the outer membrane during assembly of the capsule. Several examples of immunological cross-reaction between these capsular 'polysaccharides' and cell wall teichoic acids from Gram-positive bacteria have been reported (Robbins *et al.*, 1972; Myerowitz *et al.*, 1973; Vann *et al.*, 1976).

Structure and Properties of Lipoteichoic Acids

The structure and occurrence of lipoteichoic acids is much more highly conserved than that of wall teichoic acids. LTA makes up between 1 and 2% of the cell dry weight and with only a very few exceptions, such as the LTA-like Forssman antigen of *Streptococcus pneumoniae* (Klein *et al.*, 1994), LTA is based on a poly(glycerol phosphate) backbone assembled from the glycerol phosphate moiety of the membrane phospholipid phosphatidylglycerol. These units have the opposite stereochemistry to those in WTAs, which are derived from the nucleotide CDP-glycerol. Consequently, the glycerol phosphate termini of WTAs and LTAs differ in configuration (Koch and Fischer, 1978). Like WTAs, however, these LTAs frequently bear glycosyl substituents and also almost invariably exhibit some degree of D-alanylation. In *Staph. aureus*, substitution appears to be uniform along the chain (Fischer, 1988). A typical example is shown in Fig. 3.

Although the lipid moiety of the LTA generally appears to be based on a monoglycosyl or diglycosyl diacylglycerol group, more heavily acylated LTAs, in which a phosphatidyl group is attached through a phosphodiester to one of the sugar residues of the glycolipid, have been reported in *Enterococcus* (Tsutsui *et al.*, 1991). In *Bacillus coagulans* and *B. megaterium*, the terminal lipid is diacylglycerol rather than a glycolipid (Iwasaki *et al.*, 1986).

Little is known about the physico-chemical properties of the membrane- and wall-bound forms of LTAs. Immunochemical labelling indicates exposure of the poly(glycerol phosphate) domain of LTA on the outer surface of *Lactobacillus* (Wicken and Knox, 1980). However, the conclusion that the anionic chains must therefore adopt an extended conformation, perpendicular to the membrane surface, allowing them to penetrate the cell wall while the terminal lipid is intercalated with the outer leaflet of the cytoplasmic membrane bilayer, is now recognised to be an oversimplification. LTA makes a major contribution to the hydrophobicity of the bacterial surface (Miörner *et al.*, 1983; Op den Camp *et al.*, 1985), implying exposure of the lipid domains of some of the LTA molecules. Moreover, turnover of LTA through the cell wall into the culture fluid has been clearly demonstrated, and LTA is present in isolated, membrane-free cell walls (Kessler *et al.*, 1979).

In aqueous solution LTAs form micelles with critical micellar concentrations in the range 0.28–5 μM (Wicken *et al.*, 1986; Courtney *et al.*, 1986). Recently the solution properties of LTAs from *Staph. aureus* and from *Strep. pneumoniae* have been studied by X-ray scattering analysis (Labischinski *et al.*, 1991; Fischer *et al.*, 1997). In spite of their very different chemical structures, both LTAs form micelles of about 150 molecules, arranged spherically with the polyanionic chains extending outwards from a lipid interior. The micelles are 22–23 nm in diameter, with an outer hydrophilic shell 8.5 nm thick. Comparison of the thickness of this shell with the calculated fully extended lengths of the polyanionic chains suggests a highly coiled or possibly helical conformation within the chains. These dimensions are remarkably independent of the density of substitution with positively charged alanine or choline, leading to the suggestion (Fischer *et al.*, 1997) that the multi-molecular structure is dominated by the size and packing density of the common diacylglycerol lipid moiety of the LTAs. This implies similar supramolecular structures, in solution, for all lipoteichoic acids and lipoglycans of Gram-positive bacteria. The relevance of this information to the properties of the polymers *in vivo* is indicated by the observation that some biological activities of LTAs are dependent on the ability to form micelles in solution (Courtney *et al.*, 1986). However, the extent to which their solution properties reflect their behaviour *in vivo* in the membrane-attached, non-micellar form, and in the poorly understood wall-associated form, remains to be determined.

Several species of Gram-positive bacteria appear not to contain a lipoteichoic acid, but instead possess lipid-substituted polysaccharides for which the name

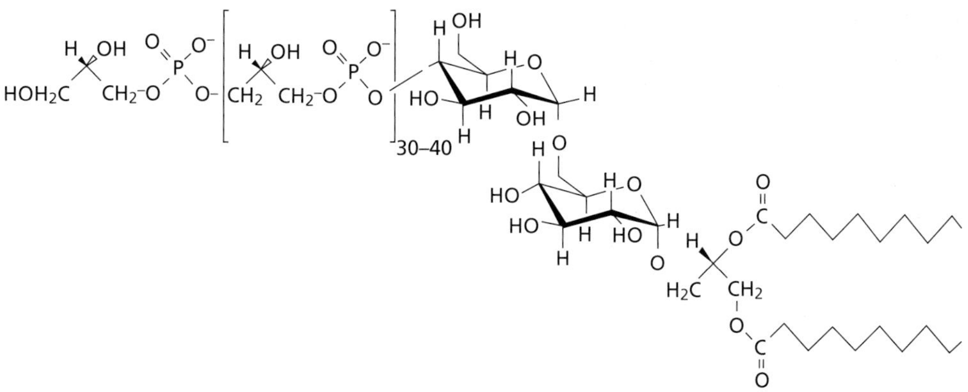

Fig. 3 Structure of the lipoteichoic acid of *Staphylococcus aureus*. The fatty acid composition is variable, and the sugar components of the glycolipid moiety are species-specific. The poly(glycerol phosphate) chain may carry glycosyl and alanyl substituents as in wall teichoic acids (see text).

macroamphiphile has been coined (Sutcliffe and Shaw, 1991). Though they do not contain phosphate, these polymers may be anionic by virtue of acidic substituents such as ester-linked succinate or acetal-linked pyruvate, which provide negatively charged carboxylate groups. Analogous polymers such as lipoarabinomannan (LAM) occur in mycolate actinomycetes (Hunter and Brennan, 1986).

The Cellular Roles of Teichoic Acids

Teichoic acids are quantitatively important constituents of Gram-positive cell walls. In many bacteria they constitute up to 50% of the weight of the cell wall and contain as much as 40% of total cell phosphorus (cf. 50% in nucleic acids). In such cases approximately 10% of all muramic acid residues throughout the peptidoglycan network of the cell wall bear a teichoic acid chain. Their synthesis thus demands a major commitment of metabolic resources. In *Bacillus subtilis*, teichoic acid is not synthesised constitutively; during phosphate-limited growth its synthesis is rapidly inactivated. Under these conditions it is replaced, however, in the cell wall by a phosphate-free teichuronic acid, thus maintaining a substantial surface negative charge. This negative charge, in *Bacillus* (Mauël *et al.*, 1989) and possibly in *Staphylococcus* (Fischer, 1994), is essential for viability. When teichuronic acid cannot substitute for teichoic acid, conditional mutations in teichoic acid synthesis are lethal under non-permissive conditions. Similarly, teichuronic acid mutants die during phosphate limitation (Müller *et al.*, 1997). There is evidence for less dramatic modulation of teichoic acid levels in the cell walls of other species, such as *Staph. aureus*, in response to changes in growth conditions, but there, teichoic acid synthesis appears to be constitutive (Dobson and Archibald, 1978). Lipoteichoic acid-free mutants have never been isolated, an observation that implies the essentiality also of this envelope component.

Despite the evidence for their essential nature, the primary cellular role of teichoic acids remains unknown. However, most of the functions ascribed to them depend on their polyanionic properties, and their contribution to the overall polyelectrolyte properties of the thick, porous, elastic Gram-positive cell wall (Hancock, 1999). The hypothesis that their negatively charged phosphodiester groups, by interacting with mobile cations in the cell wall and the region of contact between wall and membrane, serve to stabilise cation gradients and to buffer the cell membrane against environmental cation fluxes is particularly persuasive. It is supported by a variety of evidence. Magnesium-dependent enzymes in the cytoplasmic membrane of *Bacillus licheniformis* are insensitive to external changes in magnesium ion concentration in the presence of cell wall-bound or LTA-bound magnesium ions, in comparison to the same enzymes in naked membranes, indicating a buffering effect of the wall-bound cations (Hughes *et al.*, 1973). Rose and Hogg (1995) demonstrated the potential for the LTAs of oral streptococci to act as a calcium buffer and to reduce free calcium levels at caries-prone sites on teeth.

Evidence has been presented that cell wall autolysin activity is regulated by a proton gradient in the cell wall, generated by respiration-driven proton efflux from the cell (Joliffe *et al.*, 1980). Autolysin activity in this model is suppressed by a high proton concentration. The high density of negative charge in the wall, provided by the teichoic acids, could stabilise such a gradient, thus accounting for the observation that during normal growth only the outermost region of the cell wall undergoes autolysis (Pooley, 1976; Merad *et al.*, 1989). The extruded protons also compete with metal cations for anionic binding sites in the wall, presumably leading to mobilisation of the metal ions closest to the membrane surface (Mera *et al.*, 1992).

A more direct interaction between teichoic acids and cell wall autolysins was reported by Herbold and Glaser (1975), who found that the kinetics of association of purified *B. subtilis N*-acetylmuramyl-L-alanine amidase with purified cell walls were modified by the presence of wall-bound teichoic acid. The association was also strain-dependent, the autolysins from strains differing in their wall teichoic acids bound more efficiently to homologous cell walls than to those from strains with different teichoic acids. However, the nature of this interaction has never been characterised. Robson and Baddiley (1977) similarly demonstrated defective interaction of autolysins with TU-deficient cell walls of *B. licheniformis*, leading to the growth of cells in chains because of the failure of autolysin-dependent septal cleavage after cell division. Detailed studies of the interaction of the autolytic *N*-acetylmuramyl-L-alanine amidase (LytA) of *Strep. pneumoniae* (Holtje and Tomasz, 1975; Markiewicz and Tomasz, 1990) with teichoic acid have shown that productive binding of the enzyme to the cell wall depends on the choline substituents on the teichoic acid. Since the choline phosphate-substituted LTA of pneumococci also activates the autolysin, it is possible that this is the activator *in vivo* (Briese and Hakenbeck, 1985). Substitution of ethanolamine for choline allows growth to continue, but the autolysin functions abnormally, leading to aberrant morphology. A similar effect has been demonstrated in *Strep. oralis*, which also possesses choline-substituted WTA and LTA (Horne and Tomasz, 1993). Sanz *et al.* (1992) identified a choline-binding domain in the C-terminal region of the *Strep. pneumoniae* LytA amidase and CPL1 lysozyme. The enzymes were more susceptible to proteolytic degradation in the absence of choline, and although engineered enzymes devoid of the binding domain retained catalytic activity, they had lower specific activities.

A variety of anionic amphiphilic molecules, including lipoteichoic acids, inhibit autolysins in aqueous solution, and a regulatory function for LTA at the membrane–wall interface has been proposed (Cleveland *et al.*, 1976). However, as discussed above, it is not clear to what extent the micellar form of LTA in aqueous solution resembles membrane-bound or wall-entrapped LTA in its physico-chemical and biological properties, and the ability of the membrane-bound polymer to regulate autolytic activity remains in doubt. Wecke *et al.* (1996) have found that mutants of *Bacillus subtilis* defective in alanylation of WTA and LTA, but otherwise isogenic with the wild-type strain, exhibit reduced endogenous and β-lactam-induced autolysis. They have speculated that functional autolysin binding to the cell wall involves electrostatic interactions between the cationic enzymes and the negative charge of the wall-membrane region. By reducing this net negative charge, esterified alanine may reduce interaction with the enzyme. This is consistent with the observation (O'Brien *et al.*, 1995) that methicillin-resistant isolates of *Staph. aureus* contained reduced levels of ester-linked D-alanine and that mutants defective in the incorporation of D-alanine into their walls exhibited significantly reduced rates of endogenous autolysis and increased resistance to methicillin. Removal of alanine from LTA *in vitro* increased its inhibitory activity against autolysins from *Staph. aureus* (Fischer *et al.*, 1981).

In addition to the interaction of autolysin with the choline-containing teichoic acids of *Strep. pneumoniae* described above, these teichoic acids have now been demonstrated to act as receptors for at least 12 other surface proteins of this bacterium, recognised by a characteristic C-terminal repeat domain (Masure, unpublished, cited in Masure, 1998). There is evidence that these proteins are involved in virulence.

The Biological Activities of Wall Teichoic Acids

Cell wall teichoic acids are important surface antigens in several pathogenic species. The ribitol teichoic acid (RTA) of *Staph. aureus* is the major immunological determinant of that species and antibodies generally exhibit specificity for the anomeric configuration of the *N*-acetylglucosaminyl substituents on the polymer (Osland *et al.*, 1980; Umeda *et al.*, 1992). Similarly, glycosyl substituents constitute the immunodominant epitopes of *Listeria* wall teichoic acids. Interestingly, in *L. monocytogenes* serotype 1a, whose teichoic acid has both *N*-acetylglucosaminyl and L-rhamnosyl substituents, only the latter group is a significant immunodeterminant (Kamisango *et al.*, 1983). However, IgG antibodies to *Staph. aureus* wall teichoic acid, including antibodies specific for the polyribitolphosphate backbone, have been detected in the serum of a large proportion of healthy adults in many studies (Prokesova *et al.*, 1991) and with lower frequencies in children (Kontiainen *et al.*, 1994). They occur at greatly elevated levels during staphylococcal infections at a variety of sites. In a rabbit study, anti-RTA IgG was detected in the vitreous compartment of the eye during staphylococcal endophthalmitis (Engstrom *et al.*, 1991) and was accompanied by secretory IgA antibodies in tears. In a detailed study in the rat, however, an IgM response predominated and its peak titre correlated with elimination of bacteria from the vitreous compartment (Ravindranath *et al.*, 1995). Vaccination with phage lysates of *Staph. aureus* enhanced the anti-RTA antibody response to this organism but did not afford useful protection against infection (Giese *et al.*, 1996). IgG responses have also been reported for post-operative eye infections caused by *Staph. epidermidis* strains possessing a glycerol wall teichoic acid (Pleyer *et al.*, 1992).

Immunoblotting of electrophoretically separated carbohydrate antigens from the slime of *Staph. epidermidis* with the serum of patients has demonstrated elevated IgG antibody levels in patients with staphylococcal and streptococcal infections, including *Staph. epidermidis* infections of prosthetic joints (Lambert *et al.*, 1996), but with a low level of specificity.

Like many other acidic carbohydrate bacterial polymers, teichoic acids exert a wide range of cellular effects in animals and humans. Purified cell walls, and soluble peptidoglycan–teichoic acid complexes prepared from them, induce a plasma-stimulated production of cytokines such as tumour necrosis factor-α (TNF-α) and interleukin-6 (IL-6) by human monocytes (Heumann *et al.*, 1994). In a study of peptidoglycan and teichoic acid purified from *Staph. epidermidis*, concentration-dependent stimulation of monocyte production of TNF-α, IL-6 and IL-1 was observed with both wall polymers, though teichoic acid had relatively little effect on IL-1 expression (Mattsson *et al.*, 1993). The stimulatory effect of teichoic acids is several orders of magnitude lower than that of lipopolysaccharides. Nevertheless, it has been suggested that teichoic acid, peptidoglycan or covalent complexes of the two, solubilised from staphylococci by autolysis following treatment with β-lactam antibiotics could result in significant cytokine release and contribute to septic shock (Mattsson *et al.*, 1996). Teichoic acid and peptidoglycan cannot, however, be the only staphylococcal envelope components that elicit a cytokine response, since envelope fractions from a wall-less L-form of *Staph. aureus* also stimulated TNF-α production by macrophages (Kuwano *et al.*, 1993). The lethality of peptidoglycan and teichoic acid knockout mutations has so far made unambiguous testing of the roles of the individual surface components impossible.

An interaction between wall-bound pneumococcal teichoic acid and platelet-activating factor (PAF) has been shown to play a role in the induction of inflammation by *Strep. pneumoniae*, but teichoic acid on its own had no PAF-like activity (Cabellos *et al.*, 1992). A role for pneumococcal teichoic acid in activation of dNF-κB, controlling expression of genes of the acute-phase response, was proposed by Cundell *et al.* (1995). The absence of choline from the teichoic acid did not affect the activation process, leading Spellberg *et al.* (1996) to suggest that peptidoglycan rather than WTA is the principal effector.

Like lipopolysaccharides of Gram-negative bacteria, the teichoic acids of Gram-positive bacteria are capable of activating a bradykinin-generating cascade through Hagemann factor and the kallikrein pathway (Maeda and Yamamoto, 1996).

The potential role of wall anionic polymers as adhesins, for both biological and inert materials, has been emphasised in several studies. The ribitol wall teichoic acid of *Staph. aureus* is an adhesin for epithelial cells (Ward, 1992). A *Staph. aureus* adhesin responsible for attachment of the bacteria to HeLa cells also appears to be identical to cell wall teichoic acid. Purified teichoic acid inhibited adhesion, and bound to HeLa cells in a dose-dependent way (Matsuura *et al.*, 1996).

The unusual oligosaccharide phosphate teichoic acids of the oral streptococci *Strep. mitis* and *Strep. oralis* are responsible for the lectin-mediated co-aggregation with *Actinomyces naeslundii* and *A. viscosus* that plays an important part in the microbial colonisation of teeth. In some cases, specificity can be attributed to the specific structure of the teichoic acid. The oligosaccharide units are involved in specific lactose-sensitive interactions with fimbrial surface lectins of *A. naeslundii* involving, depending on the polysaccharide, a β-*N*-acetylgalactosaminyl-galactose or a β-galactosyl-*N*-acetylgalactosamine motif (Kolenbrander and Ganeshkumar, 1993).

Biological Activities of Lipoteichoic Acids

LTA shares many biological properties with lipopolysaccharides (LPS) of Gram-negative bacteria and, as in the case of LPS, its effects are inhibited by polymyxin in a dose-dependent way. Moreover, LTA is also weakly reactive in the *Limulus* amoebocyte lysate assay (Renzi and Lee, 1995). All these similarities are presumably due to the similar amphiphilic nature of the two types of macromolecule (Ohshima *et al.*, 1990). However, LTAs are generally several orders of magnitude less potent, and are correspondingly much less pyrogenic than LPS (Himanen *et al.*, 1993; Takada *et al.*, 1995). Though the types of activities exhibited by LTAs are those expected for macroamphiphiles (Sutcliffe and Shaw, 1991), some depend on more specific structural features of the molecule. In contrast to the fully tetra-acylated LTA-2 component of *Enterobacter hirae*, the less hydrophobic, diacyl, LTA-1 fraction of the lipoteichoic acid was found to be almost devoid of cytokine-inducing activity (Tsutsui *et al.*, 1991; Takada *et al.*, 1995). The net charge on the molecule is also important. Increased alanylation, and hence reduced net negative charge, of LTA was found to be associated with the induction of *vanB*-dependent vancomycin resistance and concomitant increase in tolerance to β-lactam antibiotics in *Enterococcus faecium* (Gutmann *et al.*, 1996), though this contrasts with a report of increased methicillin tolerance associated with a decreased level of alanylation in *Staph. aureus* (O'Brien *et al.*, 1995).

Adhesion

LTAs appear to be important in cell–substrate and cell–cell adhesion by virtue of both their hydrophobic (Op den Camp *et al.*, 1985) and polyanionic properties. Their hydrophobicity has been implicated, for example, in adhesion of oral streptococci to teeth (Hogg and Lightfoot, 1989), of group A streptococci to mucosal surfaces (Courtney *et al.*, 1992) and of *Staph. epidermidis* to pharyngeal epithelial cells (Chugh *et al.*, 1990a) and to fibrin–platelet clots (Chugh *et al.*, 1990b). LTAs are effective ligands for the type 1 macrophage scavenger receptor (MSR), which has broad specificity for polyanionic molecules and has been implicated in a variety of host-defence mechanisms (Dunne *et al.*, 1994). Studies of the competitive effects of different LTAs on the interaction of soluble receptor proteins with a variety of Gram-positive bacteria (Dunne *et al.*, 1994) or purified MSR ligands (Greenberg *et al.*, 1996) have shown that the hydrophobicity and the negative charge density, modulated by substitution with D-alanine and glycerol phosphate, or content of amino sugars with free amino groups, have a profound effect on LTA's potency as a ligand. It has recently been demonstrated that inter-species co-aggregation of the oral *Strep. gordonii* depends on alanylation of its LTA (Clemans *et al.*, 1999), though it is unclear whether the effect of the absence of alanyl groups is due to elimination of a specific interaction or to a more general alteration of surface physico-chemistry.

Cytokine Induction

In mice, staphylococcal LTA stimulates TNF-α induction in monocytes and IL-2 receptor formation in lymphocytes and can thus be regarded as an immunomodifier (Ohshima *et al.*, 1991). Since this work, extensive evidence for an LTA role in cytokine stimulation has appeared. Stimulation by staphylococcal LTA of transcription of the IL-8 gene and its expression by peripheral blood monocytes and by phagocytic monocytes at sites of inflammation in staphylococcal endocarditis has been demonstrated directly (Standiford *et al.*, 1994). The process depends on aggregation of the LTA receptors on the monocyte surface (Mancuso *et al.*, 1994). LTA also induces production of TNF-α mRNA, and of the protein, in murine macrophages (English *et al.*, 1996). Thus, convincing evidence is accumulating that LTA is an important modulator of the host inflammatory response to infection with Gram-positive bacteria.

LTA is the specific cell envelope component responsible for IL-12 induction by Gram-positive bacteria in human monocytes (Cleveland *et al.*, 1996). A competitive LPS antagonist, *Rhodobacter spheroides* LPS, competitively blocked IL-12 induction by both LPS and LTA by a CD14-mediated pathway that was inhibited in both cases by anti-CD14 monoclonal antibodies.

Nitric Oxide Induction

In septic shock similar inhibition of vasoconstriction is observed in the presence of LPS or LPS-free Gram-positive bacterial products. In toxaemia caused by Gram-positive bacteria this has been attributed to nitric oxide production (Lonchampt *et al.*, 1992; Dekimpe *et al.*, 1995; English *et al.*, 1996; Tsuneyoshi *et al.*, 1996), due to induction of nitric oxide synthase by LTA. *Staph. aureus* LTA also causes an elevated level of cGMP in cultured vascular smooth muscle cells, and induction of nitric oxide synthase was blocked by inhibitors of guanyl cyclase (Lonchampt *et al.*, 1992).

A detailed study of the pathway of induction of nitric oxide synthesis by *Staph. aureus* LTA in cultured murine macrophages has been carried out by Kengatharan *et al.* (1996). They concluded that NFκB is the transcriptional activator of nitric oxide synthase, and that LTA mediates its activation by a phospholipase C–tyrosine kinase signal transduction pathway.

Like LTA induction of IL-12, activation of nitric oxide synthase in macrophages by LTA seems to be mediated by CD14. Anti-CD14 monoclonal antibodies depressed the response to *Staph. aureus* LTA, while a mutant cell line defective in CD14 expression exhibited a much reduced response to the LTA (Hattor *et al.*, 1997).

Soluble Teichoic Acids *In Vivo*

It is important to note that nearly all studies of the biological activities of teichoic acids and lipoteichoic acids have been made with purified, water-soluble preparations that have potentially very different physico-chemical properties from the native forms of the polymers. Interpretation of the data must, therefore, take into account the likely form of the polymers during their interactions with the host *in vivo*.

Although anionic wall polymers become covalently linked components of the cell wall at the time they are synthesised, they have a limited life in the wall and are ultimately released in a soluble form by the process of cell wall turnover (see Chapter 3). The residence time of a component in the cell wall is best measured in terms of cell generations, since turnover appears to depend on growth rate. In *Bacillus* the residence time may be less than one cell generation. Thus during balanced growth, within one cell generation an amount of wall polymer equivalent to the complete content of the cell wall is released in soluble form into the culture fluid. The turnover products have only been thoroughly characterised in a few cases, but they appear to consist of intact anionic polymer chains covalently linked to fragments of peptidoglycan resulting from degradation of the peptidoglycan by wall autolysins, notably *N*-acetylmuramyl-L-alanine amidases and the murein hydrolase, endo-*N*-acetylglucosaminidase. Treatment with β-lactam antibiotics can enhance autolysis, but may also cause the release of linear, uncross-linked peptidoglycan and unattached WTA (Waxman *et al.*, 1980; Fischer and Tomasz, 1984). The extent to which these processes result in accumulation and dissemination of soluble antigens during human infection and antibiotic therapy, and mechanisms by which the soluble anionic polymers may subsequently be degraded by bacterial or human enzymes, or by free radical degradation in cells of the immune response, have not been intensively investigated. Immunochemical techniques have been used to demonstrate the presence of microbial turnover products in blood (Gutierrez and Liebana, 1993). Since wall teichoic acids released by autolysin degradation of peptidoglycan remain attached to fragments of peptidoglycan, it is important to assess the activities of these complexes as well as the isolated WTA (see, for example, Himanen *et al.*, 1993), and the properties of insoluble higher complexes (see Martinez *et al.*, 1993).

Lipoteichoic acids may similarly exert their effects both in soluble form and as components of cell envelopes or fragments. In this case the soluble, micellar form may be very different in its properties from the membrane-bound or wall-bound molecules (see Structure of Wall Teichoic Acids and Their Attachment to the Cell Wall above).

Wall turnover products become important components of the slime that surrounds biofilms at sites of bacterial colonisation (Hancock, 1989). Hussain *et al.* (1993) showed that about 80% of the slime accumulated by *Staph. epidermidis*, growing under laboratory conditions on a chemically defined medium, was teichoic acid.

Genetics and Regulation of Anionic Polymer Synthesis

Karamata and colleagues have made a thorough study of the genetics of synthesis of the glycerol teichoic acid of *Bacillus subtilis* 168 (*tag* locus) and the ribitol teichoic acid of *B. subtilis* W23 (*tar* locus). In each case, genes unique for teichoic acid synthesis are clustered at a single chromosomal locus, in two divergently transcribed operons, *tagAB* and *tagDEF* (Mauël *et al.*, 1991). Transcription of both operons is sigma A(σ^A)-dependent. A third σ^A-dependent *tag* operon, *tagGH*, appears to encode a two-component ABC transporter (Lazarevic and Karamata, 1995) of the kind involved in the export of O-antigens and capsular polymers, including the *Haemophilus influenzae* type a and b capsular teichoic acid (Kroll *et al.*, 1990).

In *Bacillus*, teichoic acid synthesis is part of the *Pho* regulon, and is suppressed at growth-limiting exogenous inorganic orthophosphate concentrations through the two-component sensor-regulator system PhoP/PhoR (Qi and Hulett, 1998). Neither the target nor the mechanism of this regulation is fully understood. However, biochemical studies have shown that *in vivo* the regulatory step in teichoic acid synthesis during downshift caused by phosphate limitation is specifically the synthesis of the lipid-bound linkage unit that ultimately attaches teichoic acid to peptidoglycan (Hancock, 1983). This comprises reactions catalysed by the products of genes in both *tag* operons and by an *N*-acetylglucosamine-1-phosphate-undecaprenol phosphate phosphotransferase. This appears to be due to both a direct repression of the *tagAB* and *tagDEF* operons by PhoP~P, and to PhoP~P up-regulation of the teichuronic acid synthesis pathway (Liu and Hulett, 1998; Soldo *et al.*, 1999), which changes in activity reciprocally with teichoic acid synthesis. Moreover, teichoic acid synthesis is regulated at the biochemical level, as well as genetically. Ribitol teichoic acid synthesis, though not linkage unit synthesis, can be reactivated in phosphate-limited bacteria independently of protein synthesis, by addition of phosphate to the culture medium (Hancock, 1983). In *B. subtilis*, wall teichoic acid synthesis (Hancock, 1981; Mauël *et al.*, 1994) and lipoteichoic acid synthesis (Johnstone *et al.*, 1982) are inactivated at the end of exponential growth by unknown mechanisms.

It is important to note that these regulatory processes have only been demonstrated in *B. subtilis*. Although species of more clinical significance, such as *Staph. aureus* employ the same pathway as that in *B. subtilis* for the biosynthesis of wall teichoic acids, and possess similar Pho systems, there is no evidence for Pho regulation of wall synthesis (Fox *et al.*, 1998).

The genes that constitute an operon responsible for the alanylation of lipoteichoic acid in *Lactobacillus casei* (*dltA-E*) have been identified (Heaton and Neuhaus, 1992; reviewed Neuhaus *et al.*, 1996). Related genes have been identified in *Strep. gordonii* (Clemans *et al.*, 1999), *Staph. aureus* (Peschel *et al.*, 1999) and in *B. subtilis* (Perego *et al.*, 1995). The properties of the predicted protein products of this operon suggest that they are involved in trans-membrane translocation of D-alanyl groups and subsequent esterification to teichoic acid at the outer surface of the cytoplasmic membrane (Perego *et al.*, 1995). The identification of the *dlt* operon and the consequent ability to produce knockout mutants offers a new approach to studying the biological significance of teichoic acids in infection and pathogenesis. For example, it has been observed that alanine-deficient mutants of *Staph. aureus* are more resistant to autolysis and to methicillin (O'Brien *et al.*, 1995) but more sensitive to defensins, protegrins, and other antimicrobial peptides (Peschel *et al.*, 1999). Interpretation of these results in molecular terms, however, is limited by the fact that although the immediate substrate for the alanylation reaction appears to be LTA or its biosynthetic precursors (Heaton and Neuhaus, 1992), inactivation of *dlt* genes also results in failure to analylate WTA (Perego *et al.*, 1995). It is unclear whether the same enzymes are directly responsible for alanylation of WTA, or whether, as previously suggested by Haas *et al.* (1984), alanine is transferred to WTA from LTA in a transesterification reaction. Unlike the *tag* operons, the *dlt* operon is transcribed from a σ^D-dependent promoter. Like WTA synthesis, alanylation in *B. subtilis* is turned off at the end of exponential growth, and there is evidence that this is dependent on the AbrB and SpoOA sporulation initiation proteins (Perego *et al.*, 1995). LTA synthesis is inactivated at the onset of sporulation, but is subsequently reactivated (Johnstone *et al.*, 1982) and appears in the spore membrane. Alanylation, however, is not reactivated, resulting in the presence of alanine-free LTA in the spore (Perego *et al.*, 1995). Information about the roles of these molecular mechanisms in bacterial pathogenesis is not yet available.

References

Aasjord P, Grov A (1980) Immunoperoxidase and electron microscopy studies of staphylococcal lipoteichoic acid. *Acta Pathol. Microsc. Scand. B* 88: 47–52

Abeygunawardana C, Bush CA, Cisar JO (1991) Complete structure of the cell-surface polysaccharide of *Streptococcus oralis* C104 – a 600 MHz NMR study. *Biochemistry* 30: 8568–8577.

Aspinall GO, McDonald AG, Pang H, Kurjanczyk LA, Penner JL (1993) An antigenic polysaccharide from *Campylobacter coli* serotype O/30 – structure of a teichoic acid-like antigenic polysaccharide associated with the lipopolysaccharide. *J. Biol. Chem.* 268: 18321–18329.

Beynon LM, Perry MB, Richards JC (1991) Structure of the capsular polysaccharide from *Actinobacillus pleuropneumoniae* serotype 7. *Carbohydr. Res.* 209: 211–223.

Brennan PJ, Nikaido H (1995) The envelope of mycobacteria. *Annu. Rev. Biochem.* 64: 29–63.

Briese T, Hakenbeck R (1985) Interaction of pneumococcal amidase with lipoteichoic acid and choline. *Eur. J. Biochem.* 146: 417–427.

Bundle DR, Smith ICP, Jennings HJ (1974) Determination of the structure and conformation of bacterial polysaccharides by carbon 13 nuclear magnetic resonance. Studies on the group-specific antigens of *Neisseria meningitidis* serogroups A and X. *J. Biol. Chem.* 249: 2275–2281.

Cabellos C, Macintyre DE, Forrest M, Burroughs M, Prasad S, Tuomanen E (1992) Differing roles for platelet activating factor during inflammation of the lung and subarachnoid space – the special case of *Streptococcus pneumoniae*. *J. Clin. Invest.* 90: 612–618.

Chugh TD, Bahr GM, Essa SA, Burns GJ (1990a) Adherence of *Staphylococcus epidermidis* to pharyngeal epithelial cells mediated by lipoteichoic acid. *Curr. Microbiol.* 20: 343–347.

Chugh TD, Burns GJ, Shuhaiber HJ, Bahr GM (1990b) Adherence of *Staphylococcus epidermidis* to fibrin-platelet clots *in vitro* mediated by lipoteichoic acid. *Infect. Immun.* 58: 315–319.

Clemans DL, Kolenbrander PE, Debabov DV *et al.* (1999) Insertional inactivation of genes responsible for the D-alanylation of lipoteichoic acid in *Streptococcus gordonii* DL1 (Challis) affects intrageneric coaggregations. *Infect. Immun.* 67: 2464–2474.

Cleveland RF, Wicken AJ, Daneo-Moore L, Shockman GD (1976) Inhibition of wall autolysins in *Streptococcus faecalis* by lipoteichoic acids and lipids. *J. Bacteriol.* 126: 192–197.

Cleveland MG, Gorham JD, Murphy TL, Tuomanen E, Murphy KM (1996) Lipoteichoic acid preparations of Gram-positive bacteria induce interleukin-12 through a CD14-dependent pathway. *Infect. Immun.* 64: 1906–1912.

Courtney Simpson WA, Beachey EH (1986) Relationship of critical micellar concentration of bacterial lipoteichoic acids to biological activities. *Infect. Immun.* 51: 414–418.

Courtney HS, Vonhunolstein C, Dale JB, Bronze MS, Beachey EH, Hasty DL (1992) Lipoteichoic acid and M-protein – dual adhesins of Group A streptococci. *Microb. Pathogen.* 12: 199–208.

Cundell D, Masure HR, Tuomanen EI (1995) The molecular basis of pneumococcal infection – a hypothesis. *Clin. Infect. Dis.* 21: S204–S212.

Dekimpe SJ, Kengatharan M, Thiemermann C, Vane JR (1995) The cell-wall components peptidoglycan and lipoteichoic acid from *Staphylococcus aureus* act in synergy to cause shock and multiple organ failure. *Proc. Natl Acad. Sci. USA* 92: 10359–10363.

Dobson BC, Archibald AR (1978) Effects of growth conditions on the composition of the cell wall of *Staphylococcus aureus. Arch. Microbiol.* 119: 295–301.

Dunne DW, Resnick D, Greenberg J, Krieger M, Joiner KA (1994) The type-I macrophage scavenger receptor binds to Gram-positive bacteria and recognizes lipoteichoic acid. *Proc. Natl Acad. Sci. USA* 91: 1863–1867.

Endl J, Seidle HP, Fiedler F, Schleifer KH (1983) Chemical composition and structure of cell wall teichoic acids of staphylococci. *Arch. Microbiol.* 135: 215–223.

English BK, Patrick CC, Orlicek SL, McCordic R, Shenep JL (1996) Lipoteichoic acid from viridans streptococci induces the production of tumor-necrosis-factor and nitric-oxide by murine macrophages. *J. Infect. Dis.* 174: 1348–1351.

Engstrom RE, Mondino BJ, Glasgow BJ, Pitchekianhalabi H, Adamu SA (1991) Immune response to *Staphylococcus aureus* endophthalmitis in a rabbit model. *Invest. Ophthalmol. Visual Sci.* 32: 1523–1533.

Fischer H, Tomasz A (1984) Production and release of peptidoglycan and wall teichoic acid polymers by pneumococci treated with β-lactam antibiotics *J. Bacteriol.* 157: 507–513.

Fischer W (1988) Physiology of lipoteichoic acids in bacteria. *Adv. Microb. Physiol.* 29: 233–303.

Fischer W (1994) Lipoteichoic acid and lipids in the membrane of *Staphylococcus aureus. Med. Microbiol. Immunol.* 183: 61–76.

Fischer W, Rosel P, Koch HU (1981) Effect of alanine ester substitution and other structural features of lipoteichoic acids on their inhibitory activity against autolysins of *Staphylococcus aureus. J. Bacteriol.* 146: 467–475.

Fischer W, Behr T, Hartmann R, Peterkatalinic J, Egge H (1993) Teichoic acid and lipoteichoic acid of *Streptococcus pneumoniae* possess identical chain structures – a reinvestigation of teichoic acid (C-polysaccharide). *Eur. J. Biochem.* 215: 851–857.

Fischer W, Markwitz S, Labischinski H (1997) Small-angle X-ray scattering analysis of pneumococcal lipoteichoic acid phase structure. *Eur. J. Biochem.* 244: 913–917.

Fox KF, Stewart GC, Fox A (1998) Synthesis of microcapsule by *Staphylococcus aureus* is not responsive to environmental phosphate concentration. *Infect. Immun.* 66: 4004–4007.

Giese MJ, Adamu SA, Pitchekianhalabi H, Ravindranath RMH (1996) The effect of *Staphylococcus aureus* phage lysate vaccine on a rabbit model of staphylococcal blepharitis, phlyctenulosis and catarrhal infiltrates. *Am. J. Ophthalmol.* 122: 245–254.

Greenberg JW, Fischer W, Joiner KA (1996) Influence of lipoteichoic acid structure on recognition by the macrophage scavenger receptor. *Infect. Immun.* 64: 3318–3325.

Gutierrez J, Liebana J (1993) Methods for the detection of structural components and metabolites of bacteria and fungi. *Ann. Biol. Clin.* 51: 83–90.

Gutmann L, Alobeid S, Billotklein D, Ebnet E, Fischer W (1996) Penicillin tolerance and modification of lipoteichoic acid associated with expression of vancomycin resistance in vanb-type *Enterococcus faecium* D366. *Antimicrob. Agents Chemother.* 40: 257–259.

Haas R, Koch H-U, Fischer W (1984) Alanyl turnover from lipoteichoic acid to teichoic acid in *Staphylococcus aureus. FEMS Microbiol. Lett.* 21: 27–31.

Hancock IC (1981) The biosynthesis of ribitol teichoic acid by toluenised cells of *Bacillus subtilis* W23. *Eur. J. Biochem.* 119: 85–90.

Hancock IC (1983) Activation and inactivation of the synthesis of secondary wall polymers in *Bacillus subtilis* W23. *Arch. Microbiol.* 134: 222–226.

Hancock IC (1989) Encapsulation of coagulase-negative staphylococci. *Z. Bakteriol.* 272: 11–18.

Hancock IC (1999) Bacterial surface composition and properties. In: Newman HN, Wilson M (eds) *Dental Plaque Revisited – Oral Biofilms in Health and Disease.* Cardiff: BioLine, pp. 37–48.

Hattor Y, Kasai K, Akimoto K, Thiemermann C (1997) Induction of NO synthesis by lipoteichoic acid from *Staphylococcus aureus* in J774 macrophages: involvement of a CD14-dependent pathway. *Biochem. Biophys. Res. Commun.* 233: 375–379.

Heaton MP, Neuhaus FC (1992) Biosynthesis of D-alanyl-lipoteichoic acid – cloning nucleotide sequence and expression of the *Lactobacillus casei* gene for the D-alanine-activating enzyme. *J. Bacteriol.* 174: 4707–4717.

Herbold DR, Glaser L (1975) Interaction of *N*-acetylmuramic acid L-alanine amidase with cell wall polymers. *J. Biol. Chem.* 250: 7231–7238.

Hermansson K, Perry MB, Altman E, Brisson JR, Garcia MM (1993). Structural studies of the O-antigenic polysaccharide of *Fusobacterium necrophilum. Eur. J. Biochem.* 212: 801–809.

Heumann D, Barras C, Severin A, Glauser MP, Tomasz A (1994) Gram-positive cell walls stimulate synthesis of tumor necrosis factor alpha and interleukin 6 by human monocytes. *Infect. Immun.* 62: 2715–2721.

Himanen JP, Pyhala L, Olander RM *et al.* (1993) Biological activities of lipoteichoic acid and peptidoglycan teichoic-acid of *Bacillus subtilis* 168 (Marburg). *J. Gen. Microbiol.* 139: 2659–2665.

Hogg DD, Lightfoot I (1989) Interaction of streptococcal lipoteichoic acid with artificial tooth pellicle. *Arch. Oral Biol.* 34: 615–620.

Holtje JV, Tomasz A (1975) Specific recognition of choline residues in cell wall teichoic acid by the *N*-acetylmuramyl-L-alanine amidase of pneumococcus. *J. Bacteriol.* 124: 1023–1027.

Horne DS, Tomasz A (1993) Possible role of a choline-containing teichoic acid in the maintenance of normal cell shape and physiology in *Streptococcus oralis. J. Bacteriol.* 175: 1717–1722.

Hughes AH, Hancock IC, Baddiley J (1973) The function of teichoic acids in cation control in bacterial membranes. *Biochem. J.* 132: 83–93.

Hunter SW, Brennan PJ (1990) Evidence for the presence of a phosphatidylinositol anchor on the lipoarabinomannan and lipomannan of *Mycobacterium tuberculosis. J. Biol. Chem.* 265: 9272–9279.

Hussain M, Hastings JGM, White PJ (1992) Comparison of cell wall teichoic acid with high molecular weight extracellular slime material from *Staphylococcus epidermidis. J. Med. Microbiol.* 37: 368–375.

Iwasaki H, Shimada A, Ito E (1986) Comparative studies of lipoteichoic acids from several *Bacillus* strains. *J. Bacteriol.* 167: 508–516.

Jann B, Dengler T, Jann K (1985). The capsular (K51) antigen of *Escherichia coli* O1 : K51 : H-, an O-acetylated poly-*N*-acetylglucosamine phosphate. *FEMS Microbiol. Lett.* 29: 257–261.

Johnstone K, Simion FA, Ellar DJ (1982) Teichoic acid and lipid metabolism during sporulation of *Bacillus megaterium* KM. *Biochem. J.* 202: 459–467.

Joliffe HK, Doyle RJ, Streips UN (1981) The energised membrane and cellular autolysis in *Bacillus subtilis. Cell* 25: 753–763.

Kamisango K, Fujii M, Okumura H *et al.* (1983) Structural and immunochemical studies of teichoic acids in *Listeria monocytogenes. J. Biochem.* 93: 1401–1409.

Katzenellenbogen E, Romanowska E, Kocharova NA, Knirel YA, Shashkov AS, Kochetkov NK (1992) The structure of a glycerol teichoic acid-like O-specific polysaccharide of *Hafnia alvei* 1205. *Carbohydr. Res.* 231: 249–260.

Kaya S, Araki Y, Ito E (1985) Structural studies on the linkage unit between polygalactosylglycerol phosphate and peptidoglycan in the walls of *Bacillus coagulans. Eur. J. Biochem.* 147: 41–46.

Kengatharan M, Dekimpe SJ, Thiemermann C (1996) Analysis of the signal-transduction in the induction of nitric oxide synthase by lipoteichoic acid in macrophages. *Br. J. Pharmacol.* 117: 1163–1170.

Kessler RE, Van de Rijn I, McCarty M (1979) Characterisation and localisation of the enzymatic deacylation of lipoteichoic acid in group A streptococci. *J. Exp. Med.* 150: 1498–1509.

Klein RA, Hartmann R, Egge H, Behr T, Fischer W (1994) The aqueous-solution structure of the tetrasaccharide-ribitol repeat-unit from the lipoteichoic acid of *Streptococcus pneumoniae* strain R6 determined using a combination of nmr spectroscopy and computer calculations. *Carbohydr. Res.* 256: 189–222.

Koch HU, Fischer W (1978) Acyldiglucosyldiacylglycerol-containing lipoteichoic acid with a poly(3-O-galabiosyl-2-O-galactosyl-sn-glycero-1-phosphate) chain from *Streptococcus lactis* Kiel 42172. *Biochemistry* 17: 5275–5281.

Kojima N, Araki Y, Ito E (1985) Structural studies on the linkage unit of ribitol teichoic acid of *Lactobacillus plantarum. Eur. J. Biochem.* 148: 29–34.

Kolenbrander PE, Ganeshkumar N (1993) Coaggregation-specific adherence among human oral plaque bacteria. *FASEB J.* 7: 406–413.

Kontiainen S, Seppala I, Miettinen A, Kosunen TU, Verkasalo M, Maenpaa J (1994) Antibodies against some bacterial antigens in children. *Acta Paediatr.* 83: 1137–1142.

Kroll JS, Loynds BM, Brophy LN, Moxon ER (1990) The bex locus in encapsulated *Haemophilus influenzae*: a chromosomal region involved in capsular polysaccharide export. *Mol. Microbiol.* 4: 1853–1862.

Kuwano K, Akashi A, Matsuura I, Nishimoto M, Arai S (1993) Induction of macrophage-mediated production of Tumor Necrosis Factor Alpha by an L-form derived from *Staphylococcus aureus. Infect. Immun.* 61: 1700–1706.

Labischinski H, Naumann D, Fischer W (1991) Small and medium-angle X-ray analysis of bacterial lipoteichoic acid phase structure. *Eur. J. Biochem.* 202: 1269–1274.

Lambert PA, VanMaurik A, Parvatham S, Akhtar Z, Fraise AP, Krikler SJ (1996) Potential of exocellular carbohydrate antigens of *Staphylococcus epidermidis* in the serodiagnosis of orthopedic prosthetic infection. *J. Med. Microbiol.* 44: 355–361.

Lazarevic V, Karamata D (1995) The tagGH operon of *Bacillus subtilis* 168 encodes a 2-component ABC transporter involved in the metabolism of 2 wall teichoic acids. *Mol. Microbiol.* 16: 345–355.

Liu W, Hulett FM (1998) Comparison of PhoP binding to the tuaA promoter with PhoP binding to other Pho regulon promoters establishes a *Bacillus subtilis* Pho core binding site. *Microbiology* 144: 1443–1450.

Lonchampt MO, Auguet M, Delaflotte S, Goulinschulz J, Chabrier PE, Braquet P (1992) Lipoteichoic acid – a new inducer of nitric-oxide synthase. *J. Cardiovasc. Pharmacol.* 20: S145–S147.

Maeda H, Yamamoto T (1996) Pathogenic mechanisms induced by microbial proteases in microbial infection. *Biol. Chem. Hoppe-Seyler* 377: 217–226.

Mancuso G, Tomasello F, Ofek I, Teti G (1994) Anti-lipoteichoic acid antibodies enhance release of cytokines by monocytes sensitized with lipoteichoic acid. *Infect. Immun.* 62: 1470–1473.

Markiewicz Z, Tomasz A (1990) Protein-bound choline is released from the pneumococcal autolytic enzyme during adsorption of the enzyme to cell wall particles. *J. Bacteriol.* 172: 2241–2244.

Martinez L, Timmerman CP, Fleer A, Verhoef J (1993) Chemiluminescence of human polymorphonuclear leukocytes after stimulation with whole cells and cell wall components of *Staphylococcus epidermidis*. *J. Med. Microbiol.* 39: 196–203.

Masure HR (1998) 'Comment'. *Trends Microbiol.* 6: 299–300.

Matsuura T, Miyake Y, Nakashima S, Komatsuzawa H, Akagawa Y, Suginaka H (1996) Isolation and characterisation of teichoic acid-like substance as an adhesin of *Staphylococcus aureus* to HeLa cells. *Microbiol. Immunol.* 40: 247–254.

Mattsson E, Verhage L, Rollof J, Fleer A, Verhoef J, VanDijk H (1993) Peptidoglycan and teichoic acid from *Staphylococcus epidermidis* stimulate human monocytes to release tumor necrosis factor alpha, interleukin-1-beta and interleukin-6. *FEMS Immunol. Med. Microbiol.* 7: 281–287.

Mattsson E, VanDijk H, Verhoef J, Norrby R, Rollof J (1996) Supernatants from *Staphylococcus epidermidis* grown in the presence of different antibiotics induce differential release of tumor necrosis factor alpha from human monocytes. *Infect. Immun.* 64: 4351–4355.

Mauël C, Young M, Margot P, Karamata D (1989) The essential nature of teichoic acids in *Bacillus subtilis* as revealed by insertional mutagenesis. *Mol. Gen. Genet.* 215: 388–394.

Mauël C, Young M, Margot P, Karamata D (1991) The essential nature of teichoic acids in *Bacillus subtilis* as revealed by insertional mutagenesis. *Mol. Gen. Genet.* 215: 388–394.

Mauël C, Young M, Monsutti-Grececscu A, Marriott SA, Karamata D (1994) Analysis of *Bacillus subtilis* tag gene expression using transcriptional fusion. *Microbiology* 140: 2279–2288.

Mera MU, Kemper M, Doyle R, Beveridge TJ (1992) The membrane-induced proton motive force influences the metal-binding ability of *Bacillus subtilis* cell walls. *Appl. Environ. Microbiol.* 58: 3837–3844.

Merad T, Archibald AR, Hancock IC, Harwood CR, Hobot JA (1989) Cell wall assembly in *Bacillus subtilis*. *J. Gen. Microbiol.* 135: 645–655.

Miörner H, Johansson G, Fronvall G (1983) Lipoteichoic acid is the major cell wall component responsible for surface hydrophobicity in group A streptococci. *Infect. Immun.* 39: 336–343.

Moxon ER, Kroll JS (1990) The role of bacterial polysaccharide capsules as virulence factors. *Curr. Top. Microbiol. Immunol.* 150: 65–86.

Müller JP, An Z, Merad T, Hancock IC, Harwood CR (1997) Influence of *Bacillus subtilis phoR* on cell wall anionic polymers. *Microbiology* 143: 947–956.

Myerowitz RL, Gordon RE, Robbins JB (1973). Polysaccharides of the genus *Bacillus* cross-reactive with the capsular polysaccharides of *Diplococcus pneumoniae* type III *Haemophilus influenzae* type b and Neisseria meningitidis group A. *Infect. Immun.* 8: 896–900.

Naumova IB, Shashkov AS (1997) Anionic polymers in the cell walls of Gram-positive bacteria. *Biochemistry (Moscow)* 62: 809–941.

Neuhaus FC, Heaton MP, Debabov DV, Zhang QY (1996) The dlt operon in the biosynthesis of D-alanyl-lipoteichoic acid in *Lactobacillus casei*. *Microb. Drug Resist. Mechan. Epidemiol. Dis.* 2: 77–84.

O'Brien MJ, Kuhl SA, Starzyk MJ (1995) Correlation of teichoic acid D-alanyl esterification with the expression of methicillin resistance in *Staphylococcus aureus*. *Microbios* 83: 119–137.

Ohshima Y, Ko HL, Beuth J, Roszkowski K, Roszkowski W (1990) Biological properties of staphylococcal lipoteichoic acid and related macromolecules. *Zentralbl. Bakteriol.* 274: 359–365.

Ohshima Y, Ko HL, Beuth J, Burrichter H, Oette K, Pulverer G (1991) Activation of mononuclear immune cells in response to staphylococcal lipoteichoic acid. *Zentralbl. Bakteriol. Int. J. Med. Microbiol. Virol. Parasitol. Infect. Dis.* 275: 374–381.

Op den Camp HJM, Oosterhof A, Veerkamp JH (1985) Cell surface hydrophobicity of *Bifidobacterium bifidum* subsp. *pennsylvanicum*. *Antonie van Leeuwenhoek J. Microbiol.* 51: 303–312.

Osland M, Grov A, Oeding P (1980) Immunochemical analysis of the teichoic acid from *Staphylococcus aureus*. *Acta Pathol. Microbiol. Scand. B* 88: 121–123.

Pazur JH, Dropkin DJ, Dreher KL, Forsberg LS, Lowman CS (1976) Glycans from streptococcal cell walls. *Arch. Biochem. Biophys.* 176: 257–266.

Perego M, Glaser P, Minutello A, Strauch MA, Leopold K, Fischer W (1995) Incorporation of D-alanine into lipoteichoic acid and wall teichoic acid in *Bacillus subtilis* – identification of genes and regulation. *J. Biol. Chem.* 270: 15598–15606.

Peschel A, Otto M, Jack RW, Kalbacher H, Jung G, Gotz F (1999) Inactivation of the *dlt* operon in *Staphylococcus aureus* confers sensitivity to defensins protegrins and other antimicrobial peptides. *J. Biol. Chem.* 274: 8405–8410.

Pleyer U, Mondino BJ, Adamu SA, Pitchekianhalabi H, Engstrom RE, Glasgow BJ (1992) Immune response to *Staphylococcus epidermidis*-induced endophthalmitis in a rabbit model. *Invest. Ophthalmol. Vis. Sci.* 33: 2650–2663.

Pooley HM (1976) Turnover and spreading of old wall during surface growth of *Bacillus subtilis*. *J. Bacteriol.* 125: 1127–1138.

Prokesova L, Dung DH, Jilek M *et al.* (1991) Class IgG, IgM and IgA antibodies against *Staphylococcus aureus* antigens in human serum and saliva. *Folia Microbiol.* 36: 502–506.

Qi Y, Hulett FM (1998) Role of PhoP in transcriptional regulation of genes involved in cell wall anionic polymer biosynthesis in *Bacillus subtilis*. *J. Bacteriol.* 180: 4007–4010.

Ravindranath RMH, Mondino BJ, Adamu SA, Pitchekianhalabi H, Hasan SA, Glasgow BJ (1995) Immunopathological features of *Staphylococcus aureus* endopthalmitis in the rat. *Invest. Ophthalmol. Vis. Sci.* 36: 2482–2491.

Renzi PM, Lee CH (1995) A comparative study of biological activities of lipoteichoic acid and lipopolysaccharide. *J. Endotoxin Res.* 2: 431–441.

Robbins JB, Myerowitz RL, Whisnant JK *et al.* (1972). Enteric bacteria cross-reactive with *Neisseria meningitidis* groups A and C and *Diplococcus pneumoniae* types I and III. *Infect. Immun.* 6: 651–656.

Robson RL, Baddiley J (1977) Role of teichuronic acid in *Bacillus licheniformis*: defective autolysis due to deficiency of teichuronic acid in a novobiocin-resistant mutant. *J. Bacteriol.* 129: 1051–1058.

Rose RK, Hogg SD (1995) Competitive binding of calcium and magnesium to streptococcal lipoteichoic acid. *Biochim. Biophys. Acta* 1245: 94–98.

Sanz JM, Diaz E, Garcia JL (1992) Studies on the structure and function of the N-terminal domain of the pneumococcal murein hydrolases. *Mol. Microbiol.* 6: 921–931.

Sasaki Y, Araki Y, Ito E (1980) Structure of the linkage region between glycerol teichoic acid and peptidoglycan in *Bacillus cereus* AHU1030 cell walls. *Biochem. Biophys. Res. Commun.* 96: 529–534.

Shashkov AS, Potekhina NV, Naumova IB, Evtushenko LI, Widmalm G (1999) Cell wall teichoic acids of *Actinomadura viridis* VKM AC-1315T. *Eur. J. Biochem.* 262: 688–695.

Soldo B, Lazarevic V, Pagni M, Karamata D (1999) The teichuronic acid operon of *Bacillus subtilis* 168. *Mol. Microbiol.* 31: 795–805.

Standiford TJ, Arenberg DA, Danforth JM, Kunkel SL, Vanotteren GM, Strieter RM (1994) Lipoteichoic acid induces secretion of interleukin-8 from human blood monocytes – a cellular and molecular analysis. *Infect. Immun.* 62: 119–125.

Sutcliffe IC, Shaw N (1991) Atypical lipoteichoic acids of Gram-positive bacteria. *J. Bacteriol.* 173: 7065–7069.

Takada H, Kawabata Y, Arakaki R *et al.* (1995) Molecular and structural requirements of a lipoteichoic acid from *Enterococcus hirae* ATCC 9790 for cytokine inducing antitumor and antigenic activities. *Infect. Immun.* 63: 57–65.

Tsuneyoshi I, Kanmura Y, Yoshimura N (1996) Lipoteichoic acid from *Staphylococcus aureus* depresses contractile function of human arteries *in vitro* due to the induction of nitric oxide synthase. *Anesth. Analg.* 82: 948–953.

Tsutsui O, Kokeguchi S, Matsumura T, Kato K (1991) Relationship of the chemical structure and immunobiological activities of lipoteichoic acid from *Streptococcus faecalis* (*Enterococcus hirae*) ATCC9790. *FEMS Microbiol. Immunol.* 76: 211–218.

Tul'skaya EM, Streshinskaya GM, Naumova IB, Shashkov AS, Terekhova LP (1993) A new structural type of teichoic acid and some chemotaxonomic criteria of two species *Nocardiopsis dasonvillei* and *N. antarcticus*. *Arch. Microbiol.* 160: 299–305.

Uchikawa K, Sekikawa I, Azuma I (1986) Structural studies on teichoic acids in the cell walls of several serotypes of *Listeria monocytogenes*. *J. Biochem.* 99: 315–327.

Umeda A, Yokoyama S, Arizona T (1992) Location of peptidoglycan and teichoic acid on the cell wall surface of *Staphylococcus aureus* as determined by immunoelectron microscopy. *J. Electron Microsc.* 41: 46–52.

Vann WF, Liu T-Y, Robbins JB (1976) Bacillus pumilus polysaccharide cross-reactive with meningococcal group A polysaccharide. *Infect. Immun.* 13: 1654–1662.

Ward TT (1992) Comparison of *in vitro* adherence of methicillin-sensitive and methicillin-resistant *Staphylococcus aureus* to human nasal epithelial cells. *J. Infect. Dis.* 166: 400–404.

Waxman DJ, Yu W, Strominger JL (1980) Linear uncrosslinked peptidoglycan secreted by penicillin-treated *Bacillus subtilis*. *J. Biol. Chem.* 255: 11577–11587.

Wecke J, Perego M, Fischer W (1996) D-Alanine deprivation of *Bacillus subtilis* teichoic acids is without effect on cell growth and morphology but affects the autolytic activity. *Microb. Drug Resist. Mech. Epidemiol. Dis.* 2: 123–129.

White PJ (1977) A survey for the presence of teichuronic acid in walls of *Bacillus megaterium* and *Bacillus cereus*. *J. Gen. Microbiol.* 102: 435–439.

Wicken AJ, Knox KW (1980) Bacterial cell surface amphiphiles. *Biochim. Biophys. Acta* 604: 1–26.

Wicken AJ, Evans JD, Knox KW (1986) Critical micellar concentrations of lipoteichoic acids. *J. Bacteriol.* 166: 72–77.

6

The Genetics and Biosynthesis of Lipopolysaccharides

Carl A. Schnaitman

Retired, formerly of University of Virginia Medical School, Charlottesville, Virginia, and Arizona State University, Tempe, Arizona, USA

History and Genetic Nomenclature

The story of lipopolysaccharide (LPS) genetics began during the early decades of the nineteenth century, as pathologists and bacteriologists attempted to sort out the apparently bewildering variety of enteric pathogens and normal enteric flora. Biochemical tests, such as sugar fermentation and hydrogen sulphide production could differentiate these into broad species, but an underlying complexity was revealed by serological analysis. Three general types of antigens were identified: H antigens, which were related to motility and later shown to be flagellar antigens, K or Vi antigens, which were present in a surface layer which could be removed by extraction with mild solvents, such as saline or hot water, and an underlying layer referred to as somatic or O antigens, which remained tightly bound to the cell surface. Interestingly, it is now known that outer membrane proteins, especially the porins and OmpA protein, are by weight among the most abundant components of the outer membrane, but these very stable proteins are poor antigens and not easily detected by classical serology. The O antigens are immunodominant, almost certainly because of their activity as endotoxins. The term endotoxin refers to certain common features of almost all LPS molecules which allow them to bind to specific receptors and elicit a broad range of host defences, including activation of various components of the immune systems of host animals.

Like other amphipathic membrane components, LPS presents severe solubility problems. Unlike smaller amphipathic molecules such as phospholipids, LPS molecules are very large, highly cross-linked molecules which are insoluble in the mixed organic solvents generally used for thin-layer chromatography. They are acidic and resemble nucleic acids in chemical composition. However, they also contain a lipid which is a powerful membrane anchor, and this prevents them from dissolving in salt solutions like nucleic acids. Instead, in aqueous solution they form large, insoluble micelles which readily bind other cell components. The earliest attempts to purify LPS for use as vaccines or for chemical

 doi:10.1006/bkmm.2001.0006

studies resulted in complex mixtures which were heavily contaminated by proteins and nucleic acids.

After the early serological period, during which the antigenic structure of LPS was worked out in remarkable detail, there were two bursts of research activity. Each began with the introduction of new technology. The first occurred after World War II, when Otto Westphal and Otto Lüderitz discovered that LPS could be extracted and purified by employing mixtures of hot, water-saturated phenol with various organic solvents. After mild acid hydrolysis, the purified material could be separated into a lipid fraction and several water-soluble carbohydrate fractions. Working with others, they found that almost all of the serological specificity resided in the carbohydrates. Moreover, specific serological determinants were found to be associated with specific sugars and linkages. Early attention was focused on *Salmonella enterica*, which had been divided into a large number of species on the basis of serological reactivity. Early on, chemists who studied LPS were not so much interested in the structure of the cell surface as in defining the chemical nature of antigenic determinants, now termed epitopes.

By 1980, the overall chemistry of *S. enterica* LPS was fairly well understood. *S. enterica* sv. Typhimurium has an excellent traditional genetic system, and this was used to identify many of the genes involved in LPS synthesis and to map their genetic loci. The work to this stage is summarised in an excellent review by Mäkelä and Stocker (1984).

The second burst of activity began in the mid to late 1980s, after two highly significant technical advances. The first was the application to the analysis of LPS of sodium dodecyl sulphate (SDS) gel electrophoresis, a technique which was well-established for resolving proteins. The most significant development was the realisation that it was not necessary to extract the LPS. Instead, LPS could be released from whole cells or purified membrane fractions by simple protease digestion in the presence of SDS (Hitchcock and Brown, 1983). The SDS binds to the lipid portion of LPS, and this allows the LPS to be resolved into discrete size classes by electrophoresis in an SDS-containing polyacrylamide gel.

The LPS gel bands are commonly visualised by periodate oxidation followed by silver staining, or by immunoblotting. Since most LPS contains galactose (Gal), it can be specifically radiolabelled by means of polar *galE* mutants, which lack UDP-Gal epimerase but retain low levels of Gal kinase (Schnaitman and Austin, 1990). When a high-affinity Gal transport system is induced by pre-growth of cultures with the gratuitous induced D-fucose, this can be as effective as silver staining for visualizing LPS. When high-resolution SDS gel electrophoresis is used to analyse both the purified cytoplasmic membrane fraction, in which precursor forms of LPS can be visualised, and the purified outer membrane fraction, which contains completed species of LPS, more than 30 different bands can be resolved in LPS from a typical strain. By observing the effect of defined mutations on the presence or distribution of various size classes, the biochemical phenotype of the mutations can be inferred or approximated.

An even greater technical advance has been the ability to clone and sequence DNA. This has allowed not only the precise analysis of gene clusters and isolated genes, but also the construction of defined mutations *in vitro*. These can then be introduced into the chromosome of suitable strains so that their phenotype can be studied *in vivo*. At present, more than 100 unique genes involved in the synthesis of LPS and related capsular polysaccharides have been identified and sequenced in enteric bacteria, and functions are known or proposed for the vast majority of these. Parallel advances have occurred in non-enteric Gram-negative species. It is now appreciated that these polysaccharides are the products of one of the most extensive systems of genetic polymorphism in the world of living organisms.

Significant advances have also been made in the technology of the chemical analysis of lipids and carbohydrates. Foremost among these have been advances in high-resolution nuclear magnetic resonance (NMR) spectrometry, and in the development of new forms of mass spectrometry.

The end result of these recent technological advances has been an exponential increase in knowledge of LPS and related capsular polysaccharides. One problem this has created is that of genetic nomenclature. When the field of LPS genetics was small and limited primarily to *S. enterica* sv. Typhimurium, the nomenclature of LPS genes was fairly simple. The parent was known as a smooth organism, meaning that it produced long LPS molecules and exhibited glistening, moist colonies. Virtually all mutants in the LPS pathway produced truncated LPS molecules, which gave the colonies a flatter, drier appearance. These became known as rough mutants, and the mnemonic 'rf' was used to described the loci of mutations giving rise to this 'rough' phenotype. It was soon appreciated that these genes mapped to several distinct chromosomal sites, some containing large gene clusters. These were assigned serial mnemonics based on their order of discovery, such as *rfa*, *rfb*, *rfc*, etc. This nomenclature is limited in its ability to accommodate all of the unique genes and complex loci which have been identified. A new nomenclature, based on mnemonics beginning with 'w' to indicate a function in cell wall synthesis, has been proposed (Reeves *et al.*,

1996). At present, the names of the genes involved in the synthesis of surface polysaccharides are in flux. Some authors have adopted the new system completely, others partially, and some rarely or not at all. To complicate matters further, those who apply names to genes or organisms generally fall into two classes: the 'lumpers', who apply the same name to related functions or organisms, and the 'splitters', who prefer to apply different names to reflect small but significant differences. The latter seem to be in ascendance, perhaps because of the wealth of precise information deduced from DNA sequences, and many new names are being applied to genes formerly grouped under a single mnemonic.

To establish a firm bridge between the present and the past, I have used the following nomenclature convention. For *Escherichia coli* K-12 and *S. enterica* sv. Typhimurium, for which detailed linkage maps exist, I have used the older (rf) names for all genes for which the linkage is known. I have supplied the newer (w) names in brackets at various points. In related organisms, I have used the rf designation for genes that are clearly homologues of genes listed in the *E. coli* K-12 and *S. enterica* sv. Typhimurium maps and have included the w names supplied by the authors. For genes that are not clearly homologues of these genes, I have used the w designation of the authors. If they are analogous although not strongly homologous to K-12 and Typhimurium genes, I have given the w name, but included the rf designation, followed by a question mark, in brackets. The older nomenclature of *E. coli* K-12 LPS genes will remain the same at least through the 11th edition of the linkage map, which is currently in preparation (M. Berlyn, personal communication). I hope the system I have employed will not prove too confusing.

The Structure of Lipopolysaccharides and their Role in the Outer Membrane

LPS can be divided into three regions from a functional and a biosynthetic standpoint. The innermost region consists of lipid A. This comprises a disaccharide of glucosamine (GlcN) to which are attached 6–7 fatty acids of several types, and 2–3 phosphate groups. Linked to this, via a very acid-labile ketosidic linkage, is a branched oligosaccharide of about 12 sugars. This is termed the core oligosaccharide. It also contains about four phosphate groups and two ethanolamine residues. Some molecules end with the core, others are substituted with a repeating polysaccharide containing, on average, about 80–100 individual sugars. This repeating polysaccharide is termed the O antigen, or O-specific chain.

The chemical structure of a typical LPS molecule from *S. enterica* sv. Typhimurium is shown in **Fig. 1**. The various parts of the molecules described above are shown alongside the structural drawing. The core oligosaccharide has been further subdivided into an inner core region, containing the eight-carbon sugar 2-keto-3-deoxyoctulosonic acid (KDO) and the seven-carbon sugar L-*glycero*-D-*manno*-heptose (Hep), and an outer core region containing only the hexoses glucose (Glc), galactose (Gal) and *N*-acetylglucosamine (GlcNAc). The molecule shown in **Fig. 1** also includes a monomeric structure, called an O-unit or O-repeating unit, which is repeated about 20 times to form the typical O antigen. The particular O-unit of this strain is composed of five hexoses: two molecules of Gal and one molecule each of rhamnose (Rha), abequose (Abe) and mannose (Man).

Computer-generated models that illustrate two views of an energetically favourable conformation of LPS are shown in **Fig. 2**. The LPS molecule is the same as in **Fig. 1**, except that it has an O antigen composed of five O-units, in other words 25 sugar molecules. Note that even the relatively short O antigen is already quite flexible. Extending it to its normal length of about 20 O-units does not change its properties, and it remains a bulky but flexible structure. Many molecules such as this on the surface of a cell give it the appearance of the shag-pile carpets popular in the 1960s – flexible strings of yarn extending from a dense matrix. In contrast, the lipid A region and the core oligosaccharide are brick-shaped and packed together to form a dense matrix on the surface of the outer membrane. This matrix is stabilised by the various phosphate and amino groups on the ends of the lipid A disaccharide and along the inner core. These groups interact, either with each other, or by being bridged by divalent cations, to lend stability to this matrix.

Much of the story of LPS is the story of mutations in the genes which determine its structure. This whole complex molecule is assembled serially, from the disaccharide of lipid A to the tip of the O antigen. Loss of any one of the enzymes involved in this construction results in an altered or truncated molecule. Loss of the O antigen has no effect on the growth of bacteria in the laboratory, but may be devastating in the organism's natural habitat, where that 'shag carpet' prevents molecules such as activated complement components from reaching their outer membrane targets. As the truncation goes deeper, however, it begins to have serious effects, even on growth in the laboratory. The first effect is a loss of detergent resistance, which occurs when the truncation extends into the inner core.

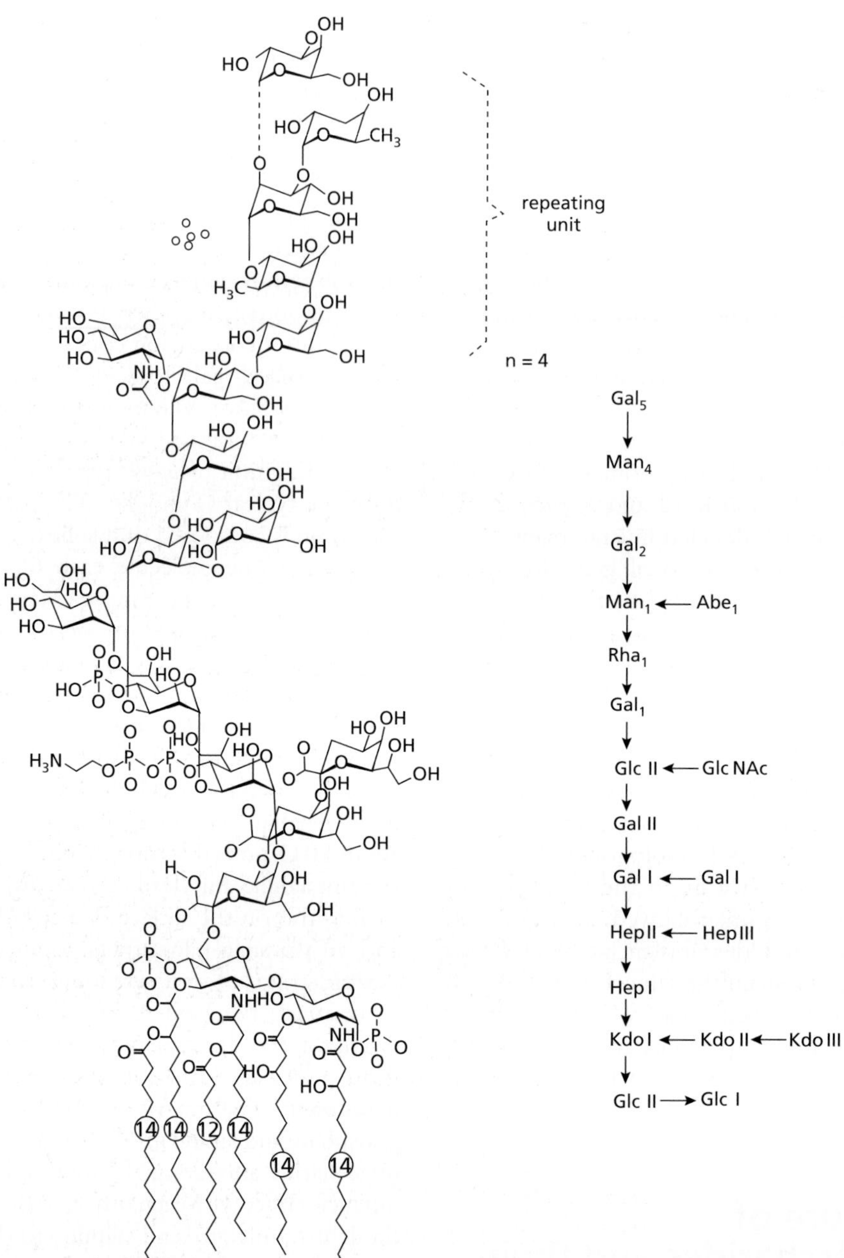

Fig. 1 Structural representation of smooth LPS from *Salmonella enterica* sv. Typhimurium. This structure, with four O antigen repeating units, is the basis for the molecular modelling shown in Fig. 2. The sugars are identified in the diagram to the right of the structure. From Kastowsky *et al.* (1992).

Enteric bacteria normally quite comfortable about growing in the upper reaches of the mammalian intestine, an environment rich in bile salts, lose this resistance. Eventually, growth can take place only at mild temperatures, and finally, not at all. The shortest LPS molecule that will support any growth consists of lipid A substituted with two KDO residues, and strains which produce only this do not thrive. This tells us that the structure of the core oligosaccharide is important to the function of LPS in the outer membrane.

This review will deal almost entirely with enteric bacteria, since these are the organisms in which the genetics and biochemistry of LPS have been studied in most detail. They provide the model by which other Gram-negative bacteria are understood. With the exception of obligate parasites such as *Neisseria* and *Haemophilus*, in which LPS has a different structure and plays a somewhat different role, they provide a good model for other organisms of clinical interest.

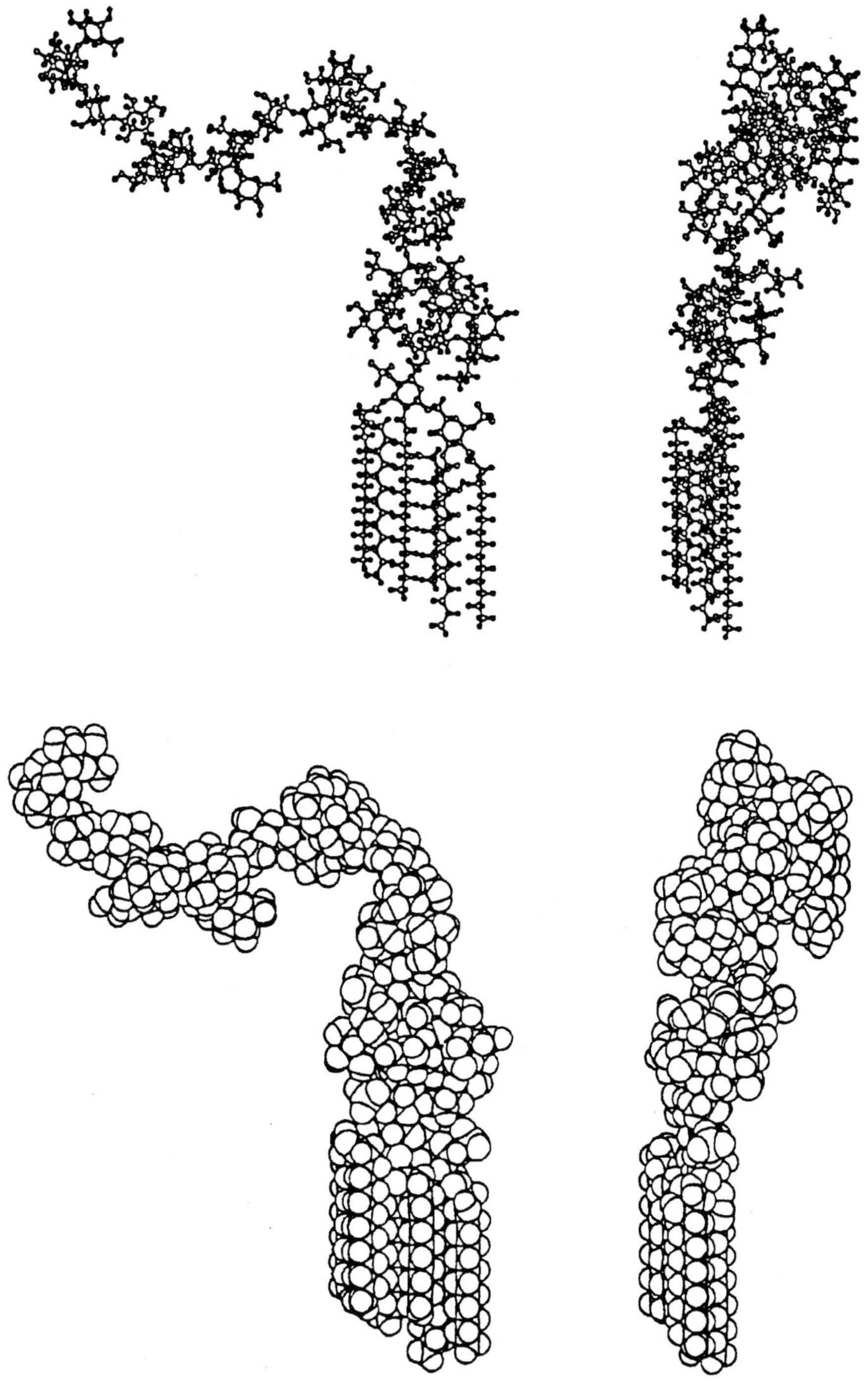

Fig. 2 Side views (90° rotation) of one of the most favourable computer-generated conformations of the truncated smooth LPS molecule (four O repeats) described in Fig. 1. The acyl groups of lipid A are depicted as extending straight down at the bottom of the molecule. The molecule is shown in both stick (top) and space-filling models. From Kastowsky *et al.* (1992).

The Structure and Synthesis of Lipid A

Genetics and Biology of Lipid A

As noted above, the synthesis and export of KDO-substituted lipid A is essential for all growth and macromolecular synthesis. As may be expected for such an important process, most of the genes involved in the synthesis of the lipid A disaccharide and its acylation, initial phosphorylation and export are located in growth-related gene clusters and macromolecular synthesis operons. This is an arrangement which presumably allows them to share growth-related regulatory signals. The basic lipid A pathway and the genes which determine it are strongly conserved across strain and species lines, and at this level of LPS synthesis there is almost no biochemical or genetic polymorphism. In contrast, the genes involved in some of the late steps of lipid A modification are located in special environmentally regulated gene clusters termed virulence clusters. These two types of gene clusters are shown in Fig. 3a.

The gene clusters involved in synthesis and modification of lipid A are both quite distinct from the arrangement of genes for the polysaccharide portion of the lipopolysaccharide molecule (Fig. 3b). In the enteric bacteria and related free-living and facultative parasitic organisms, the polysaccharide biosynthesis genes are arranged in relatively large clusters of 6–25 genes. Although there is some overlap between these clusters, each is primarily devoted to the synthesis of a single kind of oligosaccharide or polysaccharide. This is probably much better suited to the polymorphism of these oligosaccharides and polysaccharides, which appear to have evolved by genetic exchange between multiple ancestral organisms.

In order to understand the differences between the genes that determine the basic lipid A pathway, the virulence-associated genes which modify the basic lipid A structure in enteric bacteria, and the highly polymorphic nature of the gene clusters which encode the synthesis of the oligosaccharides and polysaccharides attached to lipid A, it is necessary to understand the biological and evolutionary relationships that underlie virulence and host-parasite interactions. The basic lipid A structure, or a structure derived from it by subsequent modification, is found in all Gram-negative eubacteria, with the possible exception of some obligate intracellular parasites. This indicates that this structure must have originated very early in the evolution of Gram-negative eubacteria, long before the evolution of the higher plants and animals which were later to serve as hosts or, to put it more bluntly, as a culture medium for these bacteria.

The basic structure of lipid A is more strongly conserved than that of capsular polysaccharides, and of any known outer membrane proteins, including porins, pili, flagella, specialised transport proteins, and receptors for host macromolecules. Thus, as

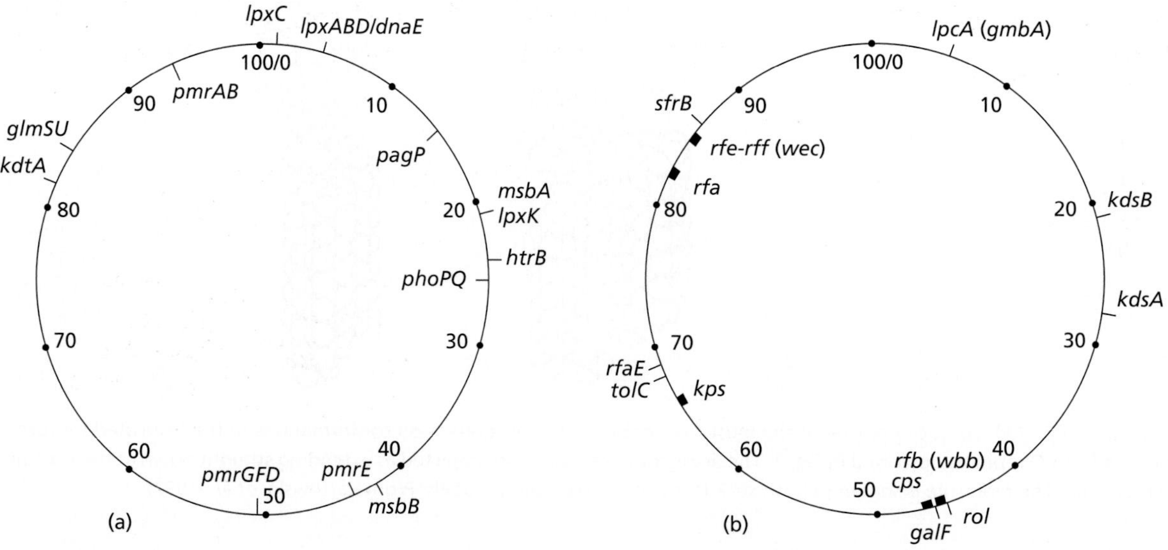

Fig. 3 Locations of genes involved in LPS biosynthesis. (a) Genes involved in synthesis and modification of lipid A. The outer circle shows genes involved in the synthesis of lipid A, and the inner circle shows genes involved in lipid A modifications which increase resistance to natural compounds, such as polymyxin and cationic antimicrobial peptides. (b) Major gene clusters involved in the synthesis of the polysaccharide portions of LPS and polysaccharide (K) capsules (inner circle) and individual genes involved in the synthesis and modification of these polymers (outer circle).

multi-cellular eukaryotic organisms evolved, lipid A served as the primary cell surface target against which host defences were directed. In the simplest of early multi-cellular organisms, this probably consisted of host cell surface receptors directed against lipid A, which served to immobilise, and later to engulf, invading bacteria. As higher organisms have evolved, the host systems, the lipid A receptors, and the responses they trigger, have become more sophisticated. Invertebrates, such as molluscs, lobsters and horseshoe crabs (*Limulus*) have specialised blood cells, sometimes called haemocytes or amaebocytes, which recognise lipid A and agglutinate or become active as phagocytes in the presence of very low levels of LPS.

The history of the study of the biological activity of LPS dates to around the beginning of the twentieth century, when the terms 'endotoxin' and 'pyrogen' were applied, respectively, by Richard Pfeiffer and Eugenio Centanni. Both terms referred to biological activities elicited by crude extracts of Gram-negative bacteria. When LPS purified by hot phenol extraction became available in 1952, it was shown that the collection of phenomena collectively known as endotoxicity or endotoxic shock, and the febrile response known as pyrogenicity, best observed by injecting tiny amounts of material into rabbits, were properties of this material. By using mild acid hydrolysis to cleave the ketosidic linkage between KDO and lipid A, it was possible to separate purified LPS into a soluble oligosaccharide/polysaccharide component and an insoluble lipid component. It was shown that the oligosaccharide/polysaccharide component had neither endotoxic nor pyrogenic activity. Unfortunately, the insolubility of the lipid A, shorn of the polysaccharide which allowed it to be dispersed in aqueous suspension, prevented the effective demonstration that the biological activity resided in the lipid component. It was assumed that it was the lipid A that held the biological activity. Attempts to define the toxic principle by fractionation or degradation of lipid A preparations derived from LPS have not been particularly successful.

Initially, the lipid itself was thought to be toxic. From 1969 to 1984, this view was reinforced by the publication and acceptance of an incorrect structure for lipid A, which showed the oligosaccharide attached to the 3′ hydroxyl of the glucosamine disaccharide backbone. Models of this structure revealed a very awkward, strained lipid molecule which might well be capable of directly damaging the membranes of host cells. The correct structure of lipid A became apparent when the structures of the monosccharide precursors

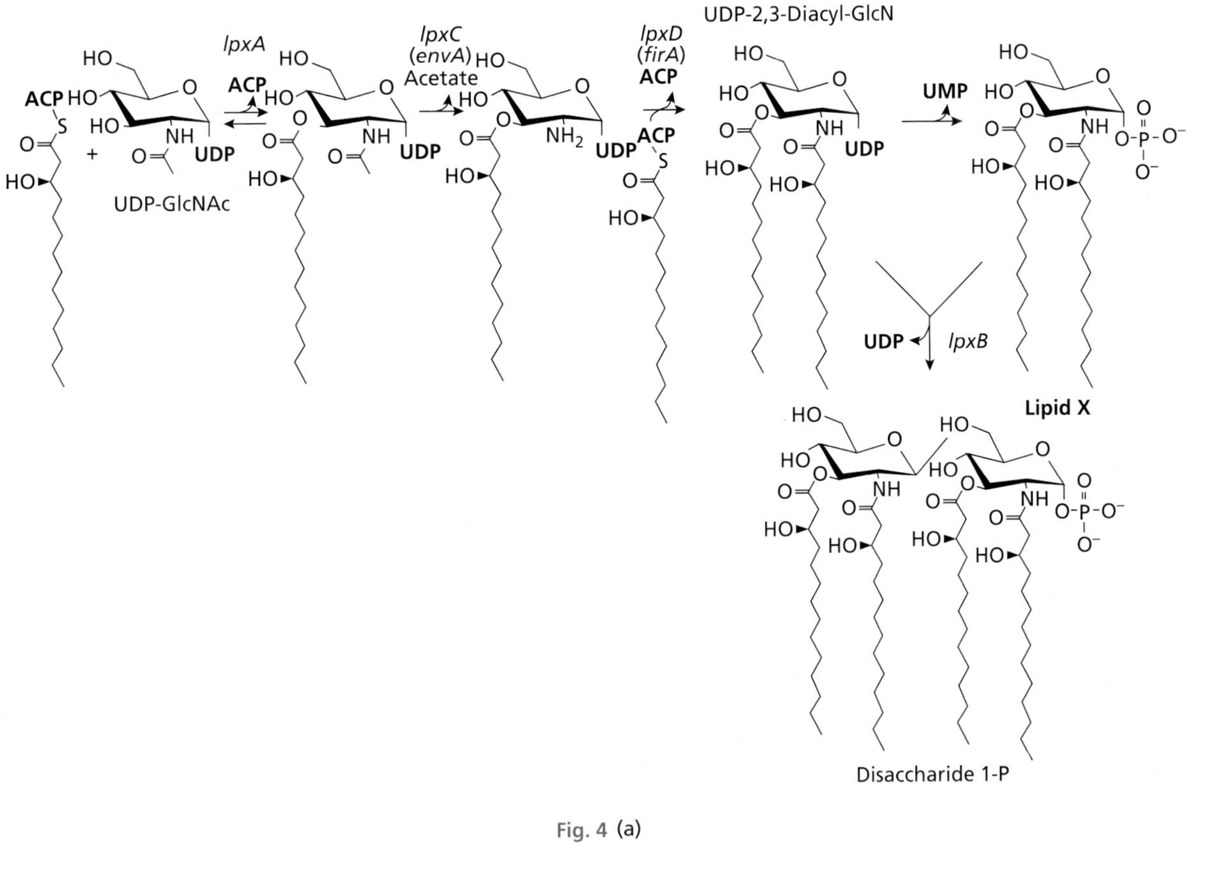

Fig. 4 (a)

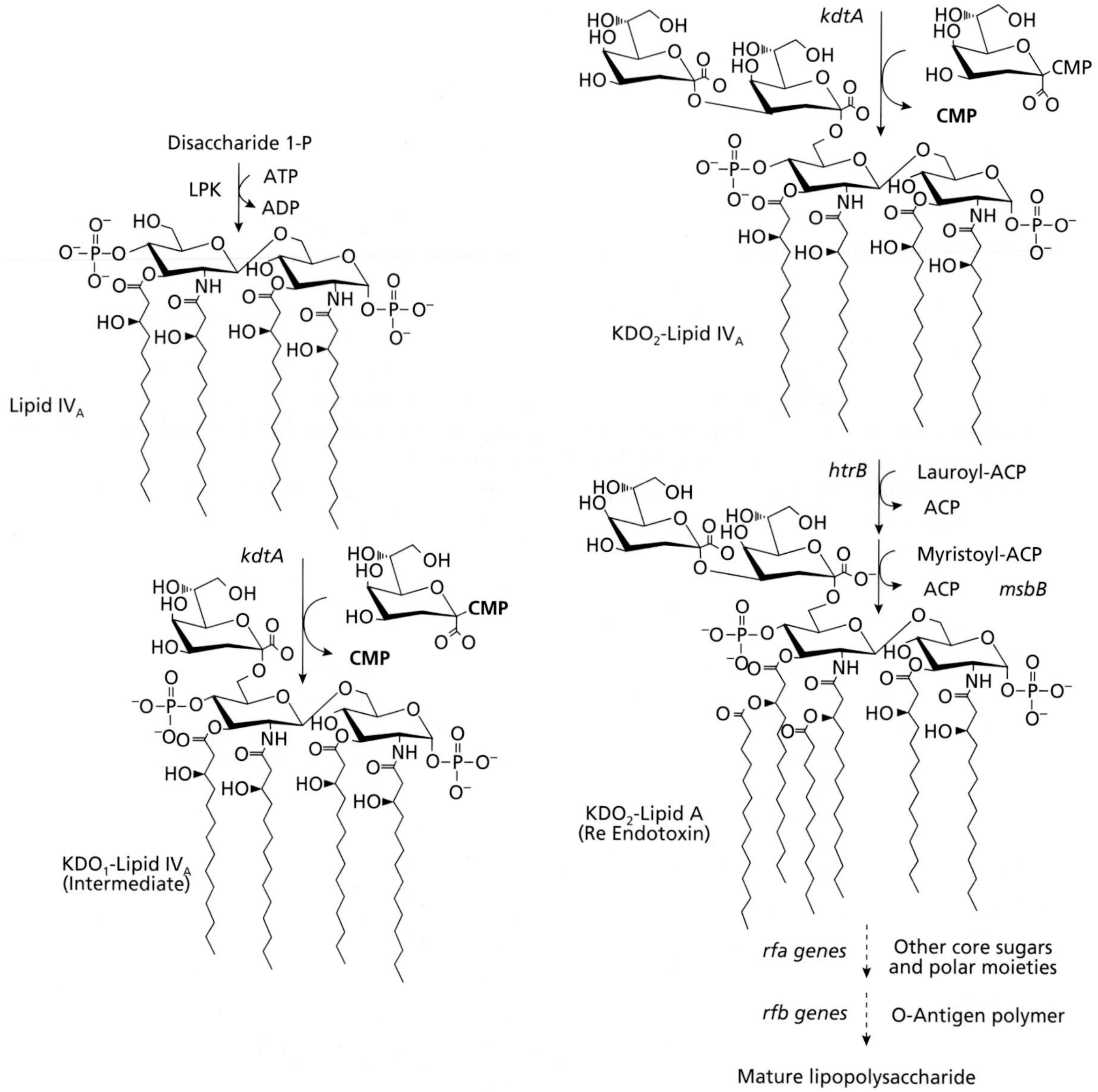

Fig. 4 (b)

were determined (Raetz, 1984). Instead of having to rely on the degradation of LPS to determine the minimally effective structure, it has been possible to build up structures to test this based on the pathways shown in Fig. 4a,b.

At about the same time as the correct structure was being determined, it was becoming apparent that the many of the systemic effects of LPS were due to the release of cytokines such as tumour necrosis factor α (TNFα) and interleukin-1 (IL-1) from macrophages and other cells of the immune system. The implication of these observations was that LPS must bind to specific receptors to elicit the synthesis and release of these agents.

Two major LPS receptor molecules have been identified and studied (reviewed by Raetz, 1996). The first is an acute-phase serum LPS-binding protein (LBP) and the second is a surface receptor called CD14. LPB shares homology with a cholesterol ester transfer protein, and is thought to deliver LPS to receptors such as CD14 on the surface of macrophages and other cells. Binding of LPS or an LPS–LBP complex to CD14 on the cell surface activates a regulatory cascade which ultimately results in the transcription of genes that encode cytokines and LPS-induced enzymes such as nitric oxide synthase, which aid in establishing an inflammatory response. The regulatory cascade appears to involve two isoforms

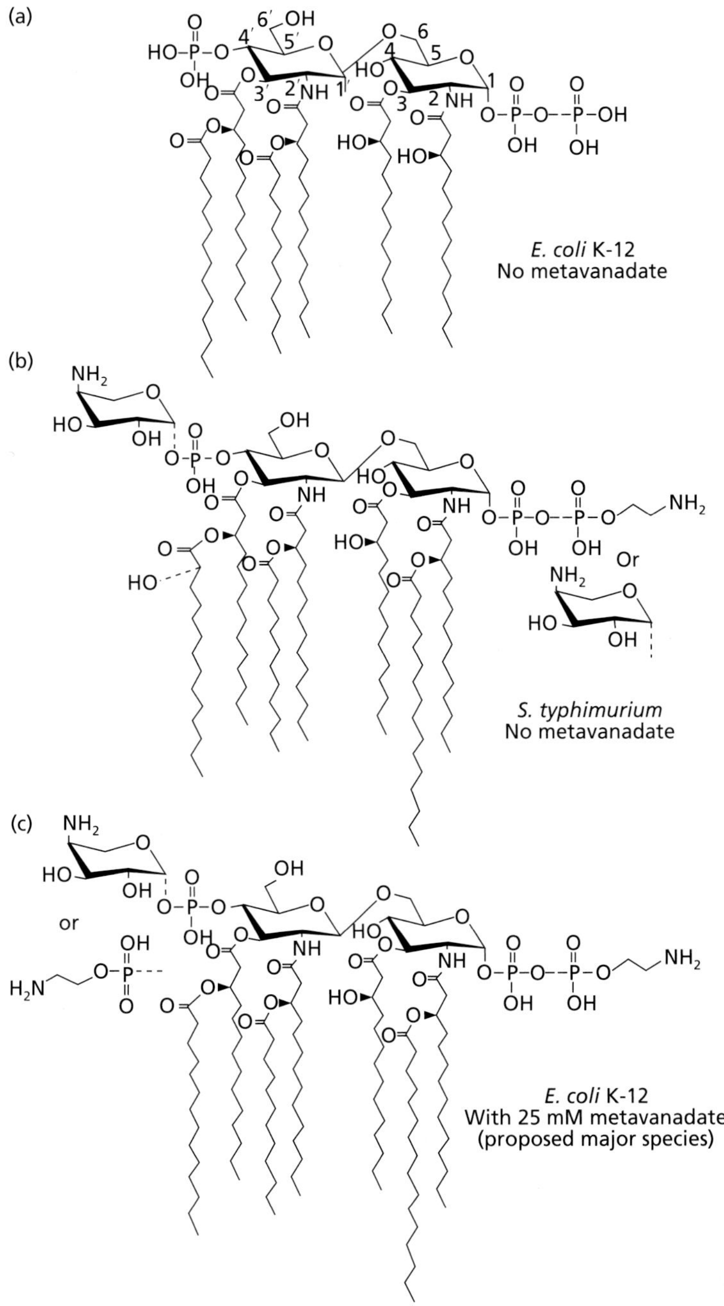

Fig. 4 (c)

Fig. 4 Biosynthesis and modification of lipid A. (a) Synthesis in *E. coli* K-12 of disaccharide-1-P from β-hydroxy acyl-ACP (acyl carrier protein) and UDP-GlcNAc. (b) Synthesis of fully acylated KDO_2-lipid A from disaccharide-1-P. (c) Comparison of the proposed structures of modified forms of *E. coli* K-12 and *S. enterica* sv. Typhimurium lipid A. The top figure shows the structure of lipid A from *E. coli* K-12 grown under ordinary laboratory conditions. The middle figure shows the structure of lipid A from *S. typhimurium* grown under ordinary laboratory conditions. The bottom panel shows the structure of *E. coli* K-12 lipid A from a culture grown with 25 mM ammonium metavanadate, a phosphatase inhibitor, which is thought to activate the PmrAB and/or PhoPQ virulence regulatory systems. From Raetz (1996), and Zhou *et al.* (1999).

of mitogen-activated protein kinases (MAP kinases), which are activated shortly after LPS binding, and tyrosine phosphorylation of a MAP kinase-like intermediate molecule called p38 (Raetz, 1996).

Knowledge of the role of receptors, such as CD14, has made it possible to demonstrate that LPS itself is not toxic. Cells that lack CD14, such as Chinese hamster ovary cell lines, show no effect on exposure to very high concentrations of LPS, and other cell lines which show low levels of activation by LPS show enhanced activation when transfected with the gene for CD14.

The bottom line in all of these studies is that mammalian hosts, and presumably also other higher plants and animals, have evolved specific receptors that recognise the uniquely bacterial and nearly invariant lipid A structures shown in Fig. 4a,b, and that binding of LPS by such receptors elicits a variety of specific biological responses. Support for this receptor-mediated hypothesis comes from observing the effects of the biosynthetic intermediates shown in Fig. 4a, and the effect of less-toxic forms of LPS produced by some bacterial species.

Monosaccharide precursors display no endotoxin activity, but show slight endotoxin-antagonising activity. The under-acylated intermediate lipid IV_A is particularly interesting in that it shows endotoxic activity when tested against human cells, but endotoxin antagonist activity in murine cells. Likewise, some naturally occurring under-acylated lipid A molecules, such as those from *Rhodospirillum sphaeroides*, also function as lipid A antagonists, although these appear to possess similar activity against both murine and human cells. These demonstrations of differential activity in different systems clearly support a receptor-mediated hypothesis for endotoxin activity (Raetz, 1996). Understanding the pathway for lipid A synthesis and the structure of the intermediates has made this a fertile field for drug development, both for the development of lipid A antagonists, which can potentially reduce the severity of endotoxin shock, and for antimicrobial agents directed against the synthesis of lipid A.

Synthesis of the Basic Lipid A Molecule

The early steps in lipid A synthesis, as described by Raetz (1996), is summarised in Fig. 4a, and Table 1 lists the genes involved and some of their alternate names. Since all of the genes in this portion of the pathway are essential, most mutants were selected for a phenotype not directly related to the synthesis of lipid A, such as a temperature-sensitive growth defect or a defect in phospholipid metabolism, and these phenotypes are reflected in their alternative names. The success in assigning functions to these genes is largely due to the availability of *in vitro* assays that use radioactive precursors and thin-layer chromatography to separate and visualise the various intermediates shown in Fig. 4a.

The synthesis of lipid A begins not as a disaccharide, but as unique acylated monosaccharides. Indeed, it was the accumulation of one of these, termed lipid X, in a mutant with a defect in growth and lipid metabolism, which led to the discovery of the pathway. As shown in Fig. 4a, the pathway begins with the sugar nucleotide UDP-GlcNac. This sugar nucleotide plays multiple unique roles in the synthesis of cell surface components. Besides being the precursor of peptidoglycan and lipid A, it is the precursor for the GlcNAc added to the polyisoprenoid lipid carrier undecaprenol phosphate by the *rfe* (*wecA*) gene product, which serves to initiate the synthesis of many O antigens and capsular polysaccharides. In *Salmonella*, UDP-GlcNAc is the donor for the GlcNAc which is added to the terminal glucose of the core to mark it for the addition of an O antigen cap. In a theme repeated throughout lipid A biosynthesis, the genes for the synthesis of this important intermediate are located where the action is. In *E. coli* two of the three unique genes, *glmS* and *glmU*, are located at 84 min near *oriC*, the origin of chromosome replication, and the third, *glmM*, shares a small operon with a tRNA gene and a gene involved in protein secretion.

The other early intermediate in lipid A synthesis is *R*-3-hydroxymyristoyl-acyl carrier protein (ACP). The first reaction that is unique to the lipid A pathway is the transfer of the hydroxymyristoyl group from ACP to the 3-hydroxy group of UDP-GlcNAc. This reaction is catalysed by the product of the *lpxA* gene, which, along with the *lpxB* and *D* genes, lies in a region at 4.4 min, which also includes genes for fatty acid and macromolecular synthesis. *LpxD* protein catalyses the aminoacylation of acyl-GlcN, and *lpxB* encodes the disaccharide synthetase, which forms the tetraacylated lipid A disaccharide.

After the initial acylation of UDP-GlcNac, the next two steps involve removal of the *N*-acetyl group, and its replacement with a second, *N*-linked, hydroxyacyl group. This leads to a UDP-2,3-diacyl-GlcNAc. Half of the diacyl-GlcNAc molecules have their UDP removed to create 2,3-diacyl-GlcNAc-1-phosphate, also known as lipid X. It was the accumulation of lipid X in *lpxB* mutants which provided one of the first clues to the nature of the pathway (Raetz, 1984). Lipid X combines with the other molecule of UDP-2,3-diacyl-GlcNAc to form the tetraacylated disacccharide-1-phosphate. Phosphorylation of the disaccharide on the 4′ position completes the synthesis of lipid IV_A. Up to this point in the biosynthetic pathway, all the genes are essential. This includes the *lpxK* gene, which encodes the kinase which adds the 4′ P. A conditional knockout *lpxK* mutation results in a complete cessation of growth within a few hours after a shift to non-permissive conditions, and this is accompanied by an accumulation of the tetraacylated disaccharide-1 phosphate.

Table 1 Genes involved in *E. coli K-12* lipid A synthesis

Gene name(s)	Map location (min)	Function
lpxA	4.4	UDP-GlcNAc *O*-acyltransferase (1st acylase)
lpxB (*pgsB*)	4.4	Lipid A disaccharide synthase
lpxC (*envA*)	2.3	*O*-Acyl UDP-GlcNAc deacetylase
lpxD (*firA*)	4.3	UDP-GlcNAc *N*-acyltransferase (2nd acylase)
lpxK	20.9	Lipid A 4′ kinase
kdtA (*waaA*)	82.0	Bifunctional KDO I and KDO I transferase
htrB	24.0	3′ Lauroyl oxyacyltransferase (5th acylase)
msbA	20.8	ABC transporter, transports R-LPS molecules across cytoplasmic membrane
msbB	41.8	2′ Myristoyl oxyacyltransferase (6th acylase)
kdsA	27.3	KDO-8-P synthetase
kdsB	20.9	CMP-KDO cytidyltransferase

Attachment of the First Core Sugars and the Completion of Lipid A

The final steps of the lipid A pathway can be thought of as overlapping the first steps of the core oligosaccharide pathway, both in the nature of the reactions and in the location of the genes in one of the large gene clusters devoted to the synthesis of complex carbohydrates. One can think of this as a transition from an essential lipid pathway to a more complex and less clearly essential carbohydrate pathway.

The attachment of two KDO residues to the 6′ end of the lipid A results in a product, known as KDO_2-lipid IV_A (Fig. 4b), which is the smallest lipopolysaccharide molecule that can be exported to the outer membrane to support viability. Both KDO I and KDO II are added by a single enzyme, in what appears to be a concerted reaction. The phosphorylation of the 4′ end of the disaccharide by the *lpxK* kinase is a necessary condition for KDO addition.

This bifunctional enzyme is the product of the *kdtA* gene, which is located in the *rfa* gene cluster. The *kdtA* gene is adjacent to a long operon, which contains genes for biosynthesis of the lipopolysaccharide core, and is the first gene in a two-gene operon. Repeating a theme common to other genes that encode enzymes that act early in the lipopolysaccharide pathway, namely that genes that encode early steps in lipid and LPS synthesis are clustered together, the second gene in the *kdtA* operon is the *coaD* gene. The *coaD* gene encodes the rate-limiting enzyme in coenzyme A biosynthesis (Geerlof *et al.*, 1999). The *kdtA* operon is transcribed divergently from a 10-gene operon, beginning with *rfaQ* (Fig. 5), which encodes enzymes involved in synthesis and modification of the lipopolysaccharide core oligosaccharide. A short noncoding region between *rfaQ* and *kdtA* is almost identical between *S. typhimurium* and *E. coli* K-12, and presumably contains transcription information for both operons.

The final steps in acylation involve attaching the two 'piggyback' (acyloxyacyl) fatty acids (laurate and myristate) to the hydroxyl groups of the two *R*-3-hydroxymyristoyl residues on the 2′ and 3′ positions of lipid A. This generates a hexa-acylated structure that is highly conserved and unique, and which, along with the disaccharide backbone, serves to distinguish lipid A from other biological lipids. Both acyloxyacylation reactions follow the addition of the two KDO residues and require the prior addition of KDO. The genes that encode both acylation enzymes were discovered in a search for genes necessary to survive heat shock. The gene for the transfer of laurate to the *R*-3-hydroxymyristoyl residue on the 4′ position of GlcN is known as *htrB*, and was found in a search for such genes, and the gene for the transfer of myristate to the amide-linked *R*-3-hyroxymyristate residue is known as *msbB* and was found as a cloned multi-copy suppressor of *htrB* (Clementz *et al.*, 1997).

Thus the 'normal' lipid A structure consists of a disaccharide of GlcN, with a polysaccharide beginning with KDO attached to the 6′ position. The 2, 3, 2′ and 3′ positions are acylated with *R*-3-hydroxymyristoyl groups and the 1 and 4′ positions are phosphorylated. Two-thirds of the lipid A from typical *E. coli* K-12 strains has P at the 1 and 4′ positions, while the remainder has P at the 4′ end and pyrophosphate (PP) at the 1 position. As noted above, the 2′ and

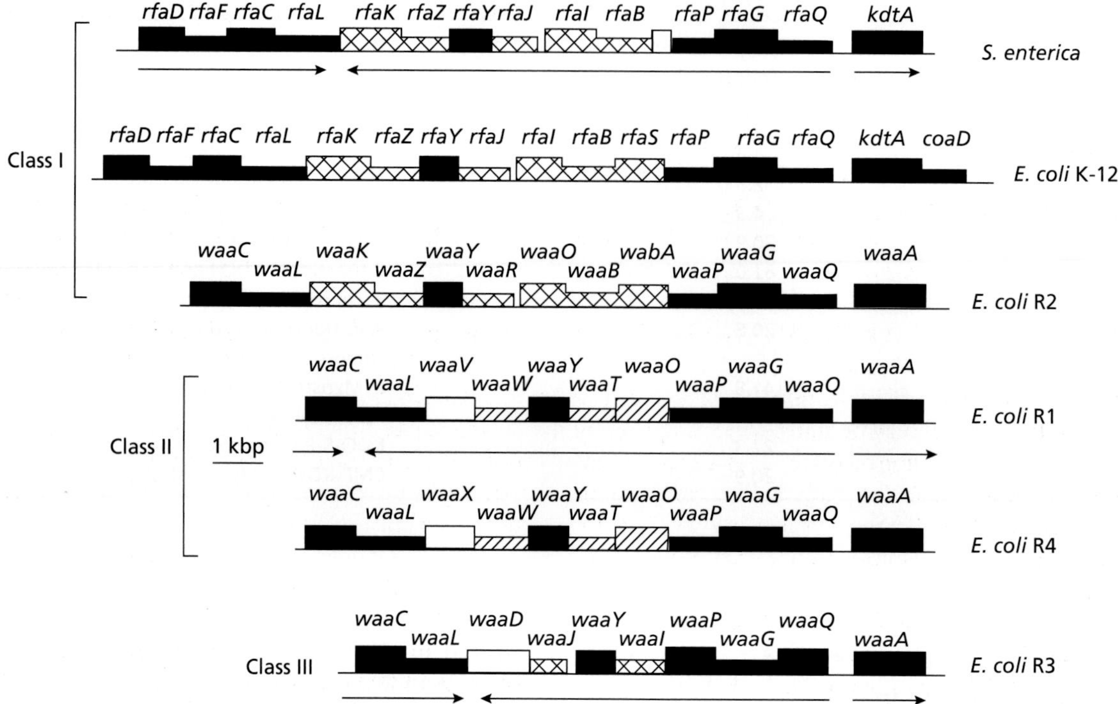

Fig. 5 A comparison of the *rfa* clusters of *S. enterica* sv. Typhimurium and the five known core types *E. coli*, re-drawn from Heinrichs *et al.* (1998c). The dark blocks indicate genes which are conserved among all six core groups, the diamond-shaded blocks indicate genes shared among more than two core groups, the diagonal-shaded blocks indicated genes shared by only two groups, and the unshaded blocks indicate unique genes. The bars on the left indicate the three classes into which these groups have been arranged, as noted in the text. The arrows show the directions of transcription for each of these three classes. The names of the *S. enterica* and *E. coli* K-12 genes are from the genetic linkage maps of these organisms (Sanderson *et al.*, 1996; Berlyn, 1998; M. Berlyn, personal communication), while the names of the genes of *E. coli* core groups R1, R2, R3 and R4 are those used by Heinrichs *et al.* (1998c).

3′ *R*-3-hydroxymyristoyl groups are acylated with laurate and myristate. With the exception of the PP at the 1 position, the entire biosynthetic pathway and all of the genes involved in its synthesis are known (Zhou *et al.*, 1999).

In addition to being the structure that binds to specific receptors and to specific protein components of the mammalian immune system, this relatively invariant structure is the target of antimicrobial agents such as polymyxin, polylysine, protamine, and a variety of cytolytic cationic antimicrobial peptides (CAMPs) which occur in phagocytic cells. In response to these host defences, enteric bacteria capable of surviving phagocytosis and penetrating host tissues have evolved the ability to add a series of 'decorations' or additional modifications to the basic lipid A molecule to make it less of a target for these agents (Guo *et al.*, 1998; Gunn *et al.*, 1998; Zhou *et al.*, 1999). These modifications include the acylation of the *R*-3-hydroxymyristoyl group at position 2 with palmitate, and the addition of ethanolamine or phosphoethanolamine (to generate a PPEtN residue) and L-4-aminoarabinose (Ara4N) to either (or both of) the 1 and 4′ phosphates. The modified products are readily visualised by thin-layer chromatography of lipid A, released from LPS by hydrolysis with pH 4.5 sodium acetate buffer (Zhou *et al.*, 1999). The exact sites of the pyrophosphorylethanolamine (PPEtN) and Ara4N substituents are, however, still a matter of discussion, and the enzymes directly involved in adding these substituents have not been identified.

In *S. enterica* sv. Typhimurium, two different virulence-associated two-component regulatory systems are involved in adding these modifications. The first is the PhoP/PhoQ system encoded by genes located in the 25-min region of the Typhimurium map (Sanderson *et al.*, 1996), which encodes resistance to a number of CAMPs (Miller *et al.*, 1989). The second is the PmrA/PmrB system (Roland *et al.*, 1993) encoded by genes in the 93.5-min region, which is involved in resistance to polymyxin and adds phosphoethanolamine (PEtN) and Ara4N to lipid A. PmrA/PmrB is activated by PhoP/PhoQ, but it is also activated independently by growth in mildly acidic conditions (Soncini and Groisman, 1996). Polymyxin-resistant mutants of *E. coli* K-12 also show elevated

PEtN and are substituted with Ara4N (Nummila *et al.*, 1995).

Two unlinked PmrA/PmrB-regulated loci have recently been identified in *Salmonella* (Gunn *et al.*, 1998). One is a single gene (*pmrE*) while the other (*pmrF*) appears to be part of a PmrA/PmrB-regulated operon containing seven potential genes. Analysis of the predicted gene products suggests that both the single gene and the operon are part of a carbohydrate biosynthetic pathway, which presumably includes the synthesis and attachment of Ara4N. The *pmrF* gene product shows some similarity to a dolichol-phosphomannose synthase of yeast, raising the interesting possibility that Ara4N may be transferred to lipid A by a bactoprenol-linked intermediate. This suggests that Ara4N may be added in the periplasm, rather than on the cyoplasmic face of the cytoplasmic membrane where the synthesis of lipid A takes place (Zhou *et al.*, 1999).

An unlinked gene, *pagP*, has also been found in the PhoP/PhoQ system (Guo *et al.*, 1998. It is regulated by the PhoP/PhoQ system, and appears to be important to the inducible resistance to CAMPs and to the increased acylation of lipid A. This supports the hypothesis that increased acylation with palmitate is a significant part of the CAMP-resistance mechanism.

Ara4N is always present in *S. enterica* LPS, but it is not normally present in wild-type *E. coli* K-12. However, Zho *et al.* (1999) found that growth in the presence of ammonium vanadate induces not only substitution of lipid A with Ara4N, but also with pyrophosphorylethanolamine (PPEtn) and palmitate. These products can be detected by thin-layer chromatography, and this should make it possible to identify the genes involved in their synthesis. It should be possible in future to determine the exact localisation of these modification reactions in the cell, and to determine the connection, if any, between modification of the 1 and 4′ phosphate residues and export of LPS to the outer membrane (Zhou *et al.*, 1998). More recently, Zhou *et al.* (2000) used high-resolution NMR spectroscopy to provide more definitive information on the exact locations of PPEtn and Ara4N. These experiments led to the surprising conclusion that, in fragments purified from closely related species, these substituents are present at opposite ends of the lipid A disaccharide. This leads to an unexpected level of micro-heterogeneity, which may be important in the adaptation of the bacterial surface to different host environments. One possibility is that one, or both, of these substituents may be added to both ends of a given molecules during export, and that they are subseque-ntly removed from one end or the other or replaced dur-ing maturation in response to environmental signals.

The Genetics and Biosynthesis of the LPS Core

The *rfa* Gene Cluster

In contrast to the highly conserved structure of the lipid A region, all portions of the LPS molecule distal to the lipid A disaccharide exhibit genetic polymorphism. In the case of the inner and outer core oligosaccharides, this polymorphism is seen between strains and between species and genera. This is seen very clearly in *Salmonella enterica* and *E. coli*, the best-studied species.

Unlike the lipid A and O antigen regions of the molecule, the sugar backbone of the inner and outer core regions is invariably synthesised processively, by the sequential addition of sugars to the non-reducing end of the growing molecule. Branch sugars and other substituents are added, and in some cases removed or modified, at various points in the growth and maturation of the backbone oligosaccharide. In the enteric bacteria the other substituents include phosphoethanolamine (PEtn), pyrophosphorylethanolamine (PPEtn) and free phosphate (P). There is evidence that the P associated with the core oligosaccharide is formed by hydrolysis of PPEtn during or after export of the LPS to the outer membrane. Since this hydrolysis is often partial, either PPEtn or P may be found at the same site. The relative amounts of these substituents appear to respond to growth conditions in a way which is still not well understood.

In the enteric bacteria, most of the genes for the assembly of the core oligosaccharide are found in a complex cluster of about 16–20 genes. This is termed the *rfa* cluster, and it is located in the 82-min region of the chromosome. The gene products encoded by the *rfa* clusters of *S. enterica* and *E. coli* consist primarily of sugar transferases, including those that add both KDO residues to the lipid A backbone, several Hep transferases, and the hexosyltransferases which assemble the outer core backbone and add the branch sugars. In addition, the *rfa* cluster encodes an isomerase involved in the synthesis of ADP-Hep (the substrate for the Hep transferases), as well as containing genes involved in the addition of phosphate-containing substituents and one gene involved in the first step of O antigen attachment to the core. A complete summary of the genes found in the enteric *rfa* clusters that have been sequenced is shown in Table 2. This table also includes a listing of the known genes involved in core synthesis that lie outside of the *rfa* cluster. The arrangement of genes that lie within the *rfa* gene clusters is shown in Fig. 5, and Figs 6 and 7 provide schematic diagrams showing the functions of some of these genes.

Table 2 Genes involved in the Synthesis of Enteric LPS core polysaccharides

A. *Rfa* genes found in all core classes

K-12 Gene	Other names	Function
rfaD	*waaD, gmhD*	Heptose synthesis (Hep isomerase)
rfaC	*waaC*	Hep I transferase (to KDO II)
rfaF	*waaF*	Hep II transferase (to Hep I)
rfaL	*waaL*	O antigen ligase (O antigen or O-unit to core)
rfaG	*waaG*	Glc I transferase (to Hep II)
rfaP	*waaP*	Adds P substituent to Hep I
rfaY	*waaY*	Adds P substituent to Hep II
rfaQ	*waaQ*	Adds Hep III branch to Hep II
kdtA	*waaA*	Adds KDO I and KDO II to lipid A

B. *Rfa* genes found in class I (*E. coli* K-12, *S. enterica*, *E. coli* R2) groups

K-12 Gene	Other names	Function
rfaK	*waaK*, $waaU_{K12}$	Adds branch sugar (GlcNAc in *enterica* and R2, Hep IV in K-12) to terminal core Glc, necessary for O-ligation
rfaZ	*waaZ*	Unknown, part of putative LOS pathway
rfaB	*waaB*	Adds branch Gal to Glc I
rfaI	*waaI*, $waaO_{K\text{-}12,\ R2}$	Adds Gal I (*enterica*) or Glc II (K-12, R2) to Glc I. Also in class R3
rfaJ	*waaJ*, $waaR_{K12,\ R2}$	Adds Glc II (*enterica*) or Glc III (K-12, R2) to Gal I (enterica) or Glc II (K-12, R2). Also in class R3
rfaS	$wabA_{R2}$	Absent in *enterica*, unknown in R2. Thought to add Rha to KDO II in K-12 (putative LOS pathway)

C. *Rfa* genes found in class II (*E. coli* R1 and *E. coli* R4) groups and in group R4

Name	Core groups	Function
waaW	R1 and R4	Adds terminal Gal II to Gal I
waaV	R1	Adds β-Glc branch to Glc II
waaX	R4	Adds β-Gal branch to Glc II
WaaD	R3	Adds GlcNAc branch to Gal I (?)

D. *E. coli* K-12 genes located outside of *rfa*

Name	Map location (min)	Function
lpcA (gmhA)	5.4	Sedoheptulose-*p*-phosphoheptose isomerase
rfaE	68.8	Bifunctional Hep mutase-ADP transferase
sfrB (rfaH)	86.7	Transcription antiterminator, polysaccharide gene clusters

There are common and highly conserved features which, so far as is known, appear to be present in all of the *rfa* gene clusters of *E. coli* and *S. enterica* (Heinrichs *et al.*, 1998c). The clusters consist of three transcriptional units, a long central operon, transcribed from right to left as shown in Fig. 5, and two shorter flanking transcripts. The left flanking transcript is transcribed convergently with the long operon, and consists of two genes that encode the Hep transferases which add the two backbone heptoses, and the gene for the Hep isomerase. It will be referred to in this review as the *rfa* Hep operon. In *E. coli* K-12 the Hep operon is thought to be transcribed from a promoter located within an adjacent flanking gene *gmhD*, which may be involved in the synthesis of a precursor to glycero-manno-heptose. It was previously known as *htrL*, since it is required for growth at high temperature.

The right flanking transcript is transcribed divergently from the long operon, and includes *kdtA*, the gene for the bifunctional transferase which adds KDO I to the 6′ position of the lipid A backbone disaccharide and KDO II to the 4 position of KDO I. As previously noted, it also contains a smaller gene designated *coaD*, which encodes an essential function in the synthesis of coenzyme A. The two genes will be referred to as the *rfa* KDO operon.

```
          rfaJ       rfaI      rfaG
     Glc II¹→²Gal I¹→³Glc I¹→³Hep II→
rfaK↑1,2            rfaB↑1.6
     GlcNAc              Gal II

     S. enterica sv. Typhimurium
```

```
               waaJ      waaI     waaG
         Glc II¹→²Gal I¹→³Glc I¹→³Hep II→
waaD?↑1,2  waaD?↑1,3
         Glc III     GlcNac

         E. coli R3
```

```
                waaT     waaO   waaG
          Gal I¹→²Glc II¹→³Glc I¹→³Hep II→
waaW↑1,2   waaV↑1,3
          Gal II      βGlc

          E. coli R1
```

```
           rfaJ       rfaI      rfaG
     Glc III¹→²Glc II¹→³Glc I¹→³Hep II→
rfaK↑1,6              rfaB↑1.6
     Hep IV                Gal

     E. coli K-12
```

```
                waaR     waaO   waaG
         Glc III¹→²Glc II¹→³Glc I¹→³Hep II→
waaK↑1,2              waaB↑1,6
         GlcNac             Gal

         E. coli R2
```

```
                waaT     waaO   waaG
          Gal I¹→²Glc II¹→³Glc I¹→³Hep II→
waaW↑1,2   waaX↑1,4
          Gal II      βGal

          E. coli R4
```

```
                        βGlc II⁶→¹βGalA II
                             ↓1,4
Glc I¹→⁴GalA I¹→³Hep II¹→³Hep I→
                   ↑1,7
                   Hep III
        Klebsiella pneumoniae O8
```

Fig. 6 Terminal core polysaccharides from enteric bacteria. The six structures at the top represent the terminal structures (distal to Hep I) resulting from the gene clusters shown in Fig. 5, and the same gene nomenclature has been employed. *K. pneumoniae* O8 is included to show the structure of an enteric core polysaccharide which does not contain phosphate substituents. GalA indicates galacturonic acid. All hexose sugars are D-isomers in the pyranose configuration, and all anomeric linkages are α unless otherwise noted. Redrawn from Heinrichs *et al.* (1998c) and Severn *et al.* (1996).

The long operon is sometimes referred to as the *rfaQ* operon. It includes the genes that encode the hexose transferases, which attach both the core backbone and the branch hexoses. It also includes two genes, *rfaP* and *rfaY*, which attach phosphate-containing substituents, and one or two branch Hep transferases. The long operon is transcribed from a complex promoter including an untranscribed region, which lies between the 5′ end of *kdtA* and the 5′ end of *rfaQ*, the first gene of the long operon. The promoter region probably also includes some of the translated sequence from the 5′ ends of both of these genes. This is based on the observation that the DNA sequence from these coding regions is much more strongly conserved between *E. coli* K-12 and *S. enterica* than is customary for coding regions for homologous genes from these organisms (Schnaitman and Klena, 1993). As DNA sequences of this region of the *rfa* cluster from other strains are acquired, it should be possible to establish very precisely the boundary of this conserved region.

Transcription of the *rfa* long operon is regulated by an antitermination factor encoded by the *sfrB* gene of *E. coli* K-12 (Pradel and Schnaitman, 1991), which was first discovered as a regulator of the long *tra* operon of plasmid F (Beutin *et al.*, 1981). It was later shown that *sfrB* is identical to the gene *rfaH* of *S. enterica*, which was incorrectly thought to encode a core galactosyltransferase, and the mnemonics *sfrB* and *rfaH* tend to be used interchangeably in the literature. It regulates a number of long operons involved in the synthesis of extracellular polysaccharides, including some of those of both capsules and O antigens, and it regulates an operon that encodes the synthesis and excretion of haemolysin, which is present in some enteric strains (Nieto *et al.*, 1996). The SfrB protein is a homologue of the *E. coli* transcription antitermination factor known as NusG, and it interacts with utilisation sites on nascent transcripts called operon polarity suppressors (OPS) (Marolda and Valvano, 1998). These are encoded within a larger DNA sequence element called JUMPstart (just upstream from many polysaccharide gene clusters, Hobbs and Reeves, 1994). The properties of the SfrB system resemble those of λQ antitermination, in that the SfrB utilisation sites are close to the transcription initiation site and not far distal as in the case of λN antitermination. Like Q antitermination, SfrB antitermination extends throughout the operon, through both Rho-dependent and Rho-independent termination sites, to an as yet unidentified termination site at the end of the operon. As a result, if antitermination is studied with Lac operon fusions inserted along the operon, the effect of SfrB on distal genes is much greater than the effect on proximal genes (Pradel and Schnaitman, 1991). This is also reflected in the effect of *sfrB* mutations, which have a complex phenotype in which the effect on gene expression increases with distance from the promoter. It also varies with the

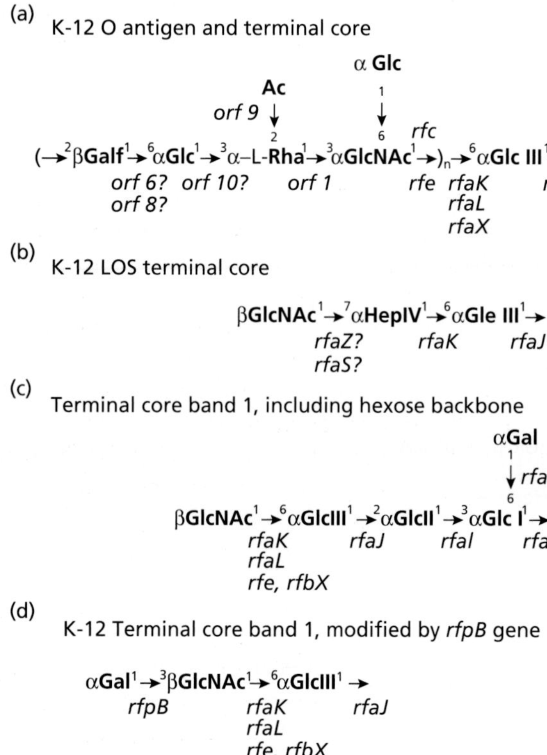

Fig. 7 Products of the split core pathway of *E. coli* K-12. This figure summarises the structure of non-reducing termini of various rough and smooth LPS molecules which are thought to be produced by *E. coli* K-12, and the genes involved in their synthesis. (a) The K-12 O16 antigen (Stevenson *et al.*, 1994) as it is thought to be attached to the terminal core hexose in smooth LPS (Fält *et al.*, 1993; Raetz, 1996). (b) A possible structure for the LOS form of rough LPS (Holst and Brade, 1992). (c) The probable terminal structure of band 1, a defective form of smooth LPS (Klena and Schnaitman, 1993; Fält *et al.*, 1993). (d) The terminal structure of band 1 modified by the product of the *rfpB* gene of *Sh. dysenteriae* 1 (Klena and Schnaitman, 1993; Fält *et al.*, 1993).

leakiness of different *sfrB* alleles. Strains carrying *sfrB* knockout mutations are viable, but not healthy.

The properties of the SfrB antitermination system also present problems for investigators of carbohydrate gene clusters, for whom defined mutations such as insertions and deletions have become the stock-in-trade. SfrB antitermination allows transcription through insertions such as mini-Tn*10*, and eliminates the polarity normally seen with such elements. In addition, a popular strategy has been to clone such clusters into plasmids, and create deletions of genes of interest *in vitro*. In order to cross such constructions back into the chromosome, the deletions are marked by the insertion of drug-resistance cassettes flanked by Rho-independent terminators. The SfrB antitermination system allows transcription to proceed through such constructions into the remainder of the operon. Some of the discrepancies in the literature reflect the failure of investigators to realise that, in antiterminated operons, expression of genes distal to such defined mutations can still occur.

Groups and Classes of Core Structures of *E. coli*, *S. enterica* and *Klebsiella pneumoniae*

All serovars of *S. enterica* were thought to exhibit a common type of core LPS structure, and this was supported by common sensitivity to the core-specific phage FO1 and recognition by core-specific antibodies. A second *S. enterica* core structure has recently been demonstrated for the Arizona O62 serotype (Olsthoorn *et al.*, 1998), but this difference is only a different terminal hexose and the genetics of this modification have yet to be worked out. For the purposes of this review, only the common core type, as exemplified by *S. enterica* sv. Typhimurium, will be considered.

The situation is quite different in *E. coli*, in which five completely different core types are recognised (Heinrichs *et al.*, 1998c). These are designated as the K-12 type and as types R1, R2, R3 and R4. The R1 and R4 types are said to be the most common in clinical isolates (Appelmelk *et al.*, 1994; Amor *et al.*, 2000). A comparison of the *rfa* clusters of *S. enterica* sv. Typhimurium, *E. coli* K-12, and *E. coli* core types R1, R2, R3 and R4 is shown in **Fig. 5**. To place these on an equal footing, schematic outlines of proposed outer core region of smooth molecules – cores of molecules which carry O antigens – from these six groups are shown in **Fig. 6**. It can be seen from the arrangement and nature of the genes, and from the overall structure of the outer core polysaccharides, that these five groups fall into three very distinct classes. For the purposes of this review, these will be termed class I, class II and class III. Class I consists of *S. enterica* sv. Typhimurium, *E. coli* K-12 and *E. coli* R2, while class II includes *E. coli* R1 and *E. coli* R4. Class III contains only *E. coli* R3.

At the core polysaccharide level, class I cores have a terminal hexose backbone that consists of Glc II-Gal II-Glc I or Glc III-Glc II-Glc I and this backbone has an αGal side branch on Glc I, while the class II cores have a terminal hexose backbone that consists of Gal II-Gal I-Glc II-Glc, and Glc II carries a side branch of either βGlc or βGal. In class III, represented by R3, the terminal hexose backbone consists of Glc III-Glc II-Gal I-Glc I and Gal I is substituted by GlcNAc.

For purposes of comparison, the core structure of the less closely related enteric species *K. pneumoniae* (Severn *et al.*, 1996; Süsskind *et al.*, 1998) is also included in **Fig. 6**. The structures of cores from serotypes O1 and O8 have been determined, and are quite similar, suggesting that all *K. pneumoniae* cores may have a common structure. The genetic structure of the *rfa* cluster from this organism has not yet been described.

The Class I *rfa* Gene Cluster Encodes a Split Pathway which Produces Two Distinct Core Structures

Before 1990, almost all knowledge regarding the role of the *rfa* gene cluster in LPS core synthesis was based on the analysis of rough mutants of *S. enterica* sv. Typhimurium. This genetic analysis had been very extensive. Therefore, when the complete DNA sequence of the *E. coli* K-12 *rfa* cluster became available, it was somewhat surprising to find that the long *rfa* operon contained four genes, which had not been predicted from the analysis of rough mutants. These genes were *rfaQ*, *rfaS*, *rfaY* and *rfaZ* (Schnaitman and Klena, 1993). Homologues or analogues of these four genes are also present in the *rfa* cluster for the type R2 core (Heinrichs *et al.*, 1998a). *S. enterica* sv. Typhimurium has three of these genes, but it lacks the *rfaS* gene. The DNA sequence of the *rfaS* region of *S. enterica* sv. Typhimurium reveals a structure termed a 'grey hole', an untranslated gap about one-third the size of *rfaS* (Klena *et al.*, 1993). This gap has a high A + T content and several short reading frames, suggesting that it once contained a functional *rfa* gene. It is likely that this gene is defective in all serovars of *S. enterica*, but this has not been proved directly by DNA sequencing. The implication is that earlier in its history *S. enterica* contained a functional *rfaS* gene.

It has recently been found that *rfaY* is involved in the transfer of a phosphate substituent to Hep II (Yethon *et al.*, 1998). There are two reasons why this gene was not detected earlier. First, it is adjacent to *rfaJ*. Mutations in *rfaJ* are epistatic on *rfaY*, meaning that *rfaJ* mutations would have the same phenotype as *rfaYJ* mutations (Klena *et al.*, 1992b). It would have taken very sophisticated genetic analysis of a number of independent mutants to separate these mutations. Second, it is technically difficult to detect the loss of a phosphate residue from a molecule which has a variable phosphate composition and 4 or 5 other phosphates.

The *rfaQ*, *rfaS* and *rfaZ* genes are not so easily explained. They were studied by constructing defined knockout mutations in each. These were first introduced into a derivative of *E. coli* K-12 strain AB1133, a multiply-marked auxotrophic strain which has been widely used to isolate and map phage- and bacteriocin-resistant mutants. The LPS of this strain can be resolved into three sharp bands, designated bands 1, 2 and 3 in order of increasing mobility. In this strain background an *rfaS* knockout mutation had no detectable phenotype, and an *rfaZ* knockout mutation reduced the amount of band 3 and caused the appearance of very minor bands migrating more rapidly than band 3 (Klena *et al.*, 1992a).

The *rfaQ* mutation had a more interesting phenotype. It increased the mobility of band 3 by an amount consistent with the loss of one sugar, but it did not affect the mobility of bands 1 and 2. The *rfaQ* gene product shares some homology with the heptosyltransferases encoded by *rfaF* and *rfaC*, suggesting that *rfaQ* may also encode a heptosyltransferase (Klena *et al.*, 1992b). It has been shown that the core from an *rfaQ* mutant specifically lacks the Hep III branch substituent previously shown to be present on some core molecules. This indicates that *rfaQ* encodes the Hep III transferase (Yethon *et al.*, 1998).

It had previously been shown by tandem mass spectrometry that about half of the core molecules had the Hep III branch substituent (Parker *et al.*, 1992). The important thing about the shift in band 3 seen in an *rfaQ* mutant was that it showed that the loss of heptose did not occur randomly among all molecules, but was limited to a subset of molecules which could be resolved as a single band by gel electrophoresis. To test this further, advantage was taken of a series of plasmids which contained genes for the synthesis of the *Sh. dysenteriae* 1 (*E. coli* sv. Dysenteriae 1) O antigen and allowed either incomplete or complete O antigens to be produced in *E. coli* K-12. When these plasmids were introduced into an *rfaQ* background, this mutation did not affect the mobility of bands that were identified as molecules whose cores carry incomplete O antigens, nor the bands in the O-ladder that are molecules which carry complete O antigen repeats (Klena *et al.*, 1992a).

The interpretation of these findings is that smooth molecules, or molecules which represent earlier stages of completion in the smooth LPS pathway, do not have Hep III. Since the distribution of this substituent among different molecular species does not appear to be random, this raises the following interesting question. Is HepIII added to all molecules at an early stage of completion, and then trimmed from molecules destined to be smooth LPS, or is it added selectively to molecules which are not part of the smooth pathway? Furthermore, if the second choice is correct, are these molecules marked in some manner? It is known that in *E. coli* K-12 and *S. enterica* two partial substituents, rhamnose (Rha) and phosphoethanolamine (PEtN), can be present on KDO II. Although data are sparse, it is thought that these substituents are mutually exclusive, and are not present on the same molecule.

Because of the chemical reactivity of KDO and the lability of the ketosidic linkage between KDO II and KDO I, the question of which substituents are on which molecules becomes technically very difficult to answer. To make matters worse, the gene (or genes) involved in the addition of PEtN to KDO II have not been identified.

This brings us to the question of the role of the Rha substituent on KDO II, and its relationship to the genes *rfaS* and *rfaZ*. Rha is also present in the K-12 O antigen (O16), and its precursor is the nucleotide sugar dTDP-Rha. This is synthesised by the products of four contiguous genes, *rfbBDAC*, located at the 5′ end of the *E. coli* K-12 *rfb* gene cluster (Stevenson *et al.*, 1994). The same four genes are found in the *rfb* cluster of *S. enterica* sv. Typhimurium, and other enteric bacteria whose O antigens contain Rha.

In contrast to the gel pattern of LPS from strain AB1133, which was described above, the gel pattern of LPS from early prototrophic strains of *E. coli* K-12, such as W1485F and W3110, was very different. Instead of the bands designated 1, 2 and 3, only two bands were visible (see Fig. 4, lanes A and C, in Klena and Schnaitman, 1994), and band 3 was completely absent. A band co-migrating with band 1 of AB1133 was present in W1485F$^-$, but in greatly reduced amount (see lane C in Fig. 4 of Klena and Schnaitman, 1994). Below band 1 was a novel, very heavy band, which migrated slightly more slowly than band 2. When a *rfaQ* mutation was introduced into W1485F$^-$, the migration of this novel band shifted, indicating that it contained Hep III. Thus, it did not appear to represent an earlier stage of completion in the smooth LPS pathway. When this major band was shifted, it was possible to see that a minor band which co-migrated with band 2 had been concealed beneath it.

Experiments with an *E. coli* K-12/Dysenteriae 1 hybrid in an AB1133 strain background indicated that the novel band seen in the LPS from W1485F$^-$ was produced in AB1133 when it carried a plasmid containing DNA from the *rfb* gene cluster of *Sh. dysenteriae* 1. Subcloning of the insert from the plasmid indicated that only one gene from the *rfb* cluster was required, and this gene was identified as *rfbD*. It was known from the *E. coli* K-12 pedigree that most K-12 derivatives carry the *rfbD1* mutation, which arose in strain Y10. Gel analysis of strains, such as W3110 which preceded Y10, indicated that all produced the novel band described above, and that all strains derived from Y10 failed to produce the novel band but instead produced band 3 (Klena and Schnaitman, 1994).

The ability to synthesise dTDP-Rha has a profound effect on the chemical composition of the major core species. Two laboratories independently analysed the core composition of the major LPS species produced by strains derived from Y10, and have proposed the terminal structure shown in Fig. 7c (Fält *et al.*, 1993; Jansson *et al.*, 1981). There is now considerable evidence that this structure is responsible for band 1, and that it is in fact a defective form of smooth LPS. The terminal GlcNAc residue requires four genes from the smooth LPS pathway, *rfaK*, *rfaL*, *rfe* and *rfbX*, for its transfer to Glc III. The most definitive of these are *rfe*, which transfers a GlcNAC residue to bactoprenol, and which appears to initiate the assembly of O antigens in all *E. coli* strains, and *rfbX*, which is thought to encode a 'flippase' that transfers acceptors of O antigens across the cytoplasmic membrane (Feldman *et al.*, 1999; Clarke *et al.*, 1995). Two other laboratories independently analysed the core composition of LPS from W3110, in which the novel band described above is clearly the major species, and they found the structure indicated in Fig. 7b (Holst and Brade, 1992; Feldman *et al.*, 1999). GlcNAc is also present in this molecular species, but it is linked to the 7 position of a heptose molecule (Hep IV), which in turn is linked to the 6 position of Glc III. There is no evidence that this molecule is part of the smooth LPS pathway. In fact, there is circumstantial evidence that it is not, and that the GlcNAc residue is not transferred from bactoprenol. In W1485F$^-$, *rfe* and *rfaL* mutations resulted in the loss of band 1, but did not affect either the amount or the migration of the major band below it.

As noted above, an *rfaS* mutation had no phenotype in AB1133, and *rfaZ* caused only a minor change in band 3. In contrast to this, their phenotypes in strain W1485F$^-$ was dramatic. Both mutations resulted in a complete loss of the novel major band. In the case of *rfaS* the phenotype was identical to the phenotype of the *rfbD*1 mutation, and the mutant strain exhibited an increase in band 1, and the re-appearances of bands 2 and 3. In the *rfaZ* mutation the phenotype was identical with the exception of the alteration in band 3 noted above (Klena and Schnaitman, 1994).

To explain these phenomena, it has been proposed that the core LPS pathway in *E. coli* K-12 is split, in order to produce two different core molecules. One is a core molecule which is designed as an acceptor for O antigen and is present in smooth LPS. It lacks Hep III, and does not have Rha attached to KDO II. The other molecule is destined to remain a rough species. It has a terminal modification that prevents the attachment of O antigen. It contains Hep III and has Rha attached to KDO II. This molecule has been designated a lipo-oligosaccharide (LOS), because of its similarity to the LOS species produced by the genera *Neisseria* and *Haemophilus* (Schnaitman and Klena, 1993). These

LPS molecules end in a complex core oligosaccharide, and they are complete structures which are not designed to accept polymeric O antigens. The LOS is seen as the novel, abundant band produced by strain W3110. A similar abundant rough band, migrating just below the SR band (the semirough band, in which a single O repeat is present), is seen in published gels of smooth LPS from many *E. coli* serotypes. Whether this is an artefact, or a genetically programmed alternative form of LPS, remains a matter of conjecture.

At the genetic level, genes *rfaZ*, *rfaS* and possibly *rfaQ* are part of the LOS pathway. A working model is that *rfaS* encodes the transferase which adds Rha to KDO II, and that *rfaZ* product adds a terminal GlcNAc to the Hep IV residue of molecules marked with Rha. *RfaL* is clearly part of the smooth pathway, since its function in the addition of O antigen is fairly well understood and its loss has no effect on the LOS band. *RfaQ* attaches a Hep III modification, which is present in LOS band and absent in smooth bands, but since some trimming may have gone on, a broader role cannot be ruled out. All of the remaining *rfa* genes appear to be involved in both pathways.

Unfortunately, there is no direct chemical evidence for LOS *per se*. The groups cited above who are still active in the analysis of LPS have, for the most part, chosen to disregard the results of those working in different strain backgrounds. The side-by-side analysis has not been carried out of nearly isogenic strains that differ only in having defined mutations in *rfbD*, *rfaS*, *rfaZ* or *rfe*, nor has an attempt been made to separate out the major core species, which can be done by gel filtration of intact molecules or ion-exchange chromatography of oligosaccharides released by mild hydrolysis. The exact structure of the *E. coli* K-12 core, therefore, remains unknown.

Some support for the LOS hypothesis can be derived from analysis of the other two class I types. The *E. coli* type R2 *rfa* cluster includes an *rfaZ* gene whose product shares some homology with that of *E. coli* K-12. The R2 cluster also includes an intact analogue of *rfaS*, but the product of this gene does not share substantial homology with the product of the K-12 *rfaS* gene (Heinrichs *et al.*, 1998a). One possibility is that the R2 strain has a similar strategy for producing an LOS, but uses a sugar other than Rha to mark the core for additional modification. The gel profile of LPS from an R2 strain shows a significant band below the O-ladder, which might represent LOS. There are two interesting predictions. First, a knockout mutation of the R2 *rfaS* analogue should eliminate or alter the LOS band seen on gels. Second, the R2 strain should produce a rough molecule whose terminal core structure is different from that of *E. coli* K-12, which has a different *rfaK* analogue, or that of *S. enterica* sv. Typhimurium, which has the same *rfaK* analogue but lacks *rfaS* (Heinrichs *et al.*, 1998a).

The gel pattern of smooth *S. enterica* sv. Typhimurium also supports the LOS hypothesis. This organism does not produce a significant rough band at the bottom of the O-ladder, as is seen in smooth strains of K-12 and R2. Instead, it characteristically produces an increased amount of small smooth species (SR, SR + 1, SR + 2, etc.), which are generally absent from gels of LPS from smooth *E. coli* serovars. Since the structure of the *rfaS* region indicates that the loss of a functional *rfaS* gene occurred some time ago, it is possible that this organism has adopted an alternative mechanism to provide areas on the cell surface which have relatively short LPS molecules.

Synthesis of the Class I Inner Core, and the Role of the LPS Core in Outer Membrane Stabilisation

LPS plays an important role in the stabilisation of the outer membrane, and much of this stabilisation involves the generation and manipulation of positive and negative charges. These facilitate charge–charge interactions between LPS molecules and result in extremely stable binding of monovalent, divalent and polyvalent cations. It is an old observation that LPS is extremely difficult to desalt.

One component of this system of charged residues consists of the phosphate residues at the 1 and 4′ positions of the lipid A disaccharide. As mentioned above, these residues are modified further by the addition of phosphate, PEtN and 4-aminoarabinose under the regulation of the PhoP/PhoQ system. In *E. coli* and *S. enterica*, another component of this system of charged residues involves phosphate and PPEtN present in the heptose region of the core. Unfortunately, much less is known about the mechanism by which these substituents are added and possibly modified.

Until very recently, it was difficult to study the synthesis of the heptose region of the core, since the exact identity of the nucleotide precursor and the genes involved in its synthesis were unclear. It is now known that the precursor is ADP-L-*glycero*-D-*manno*-heptose, and this is synthesised in four steps from sedoheptulose-7-phosphate. The first step is catalysed by an isomerase encoded by the *lpcA* (*gmhA*) gene located at 5.4 min in the *E. coli* K-12 chromosome (Brooke and Valvano, 1996a, b). The second and third steps are catalysed by a bifunctional enzyme encoded by the *rfaE* gene located at 68.9 min (Valvano *et al.*, 2000). The final step is an epimerisation carried out by the product of the *rfaD* (*gmhD*) gene, which is the first gene of the Hep operon of the *rfa* gene cluster (Sirisena *et al.*, 1994). The remaining genes in the Hep pathway are *rfaC*, which encodes the transferase that adds Hep I to KDO I, and *rfaF*, which encodes the transferase that adds Hep II to Hep I (Schnaitman and Klena, 1993). Mutations in all of these genes exhibit a deep rough phenotype, as indicated by hypersensitivity to hydrophobic antibiotics such as novobiocin, a reduced content of several outer membrane proteins (especially OmpF protein), an increased outer membrane phospholipid content, sensitivity of outer membrane phospholipids to phospholipase C, and an inability to grow on media containing bile salts or basic dyes as counter-selective agents. It has been shown that the KDO II residue is more heavily substituted with PEtN in a heptose-deficient mutant, which may represent some form of compensation for the loss of phosphate-containing substituents normally present on Hep I and Hep II. Heptose-deficient mutants rapidly acquire a mucoid phenotype, resulting from the induction of synthesis of a colanic acid capsule.

The deep rough phenotype appears not to be a result of the loss of Hep residues *per se*, but to be caused by the loss of the charged groups normally present on Hep I and Hep II. This has been demonstrated very strongly by the successful genetic dissection of *rfaP* (Parker *et al.*, 1992; Yethon *et al.*, 1998). Perhaps more than any other situation in LPS, this illustrates the successful use of defined mutations to study complex pathways. *RfaP* is the third gene in the long operon, and *rfaG*, which encodes the first core hexosyltransferase, is the second. Some of the first mutations used to study *rfaP* were probably polar mutations in *rfaG*, and thus it was unclear whether the deep rough phenotypes of these mutations were due to the loss of Hep I phosphorylation or the absence of the first hexose, Glc I. To further complicate the situation, the *rfaP* function is required for efficient action of the *rfaQ* and *rfaI* transferases, which add, respectively, the branch Hep III to Hep II and the branch Gal to Glc I. This has been studied in K-12 by complementing a long deletion with constructs that carry defined genes, and in an R1 strain, which has an identical inner core, by the use of non-polar insertion mutations to inactivate one gene at a time. The results are in exact agreement. They indicate that *rfaP* is required for the addition of a phosphate-containing substituent to Hep I, and that it is the loss of this substituent which results in the deep rough phenotype.

Studies in strains in which genes distal to *rfaP* are deleted indicate that restoration of *rfaP* function is not sufficient to restore novobiocin-resistance to the wild-type level, even though the outer membrane protein defects and other aspects of the deep rough phenotype have been relieved. In long core deletion mutants, wild-type novobiocin-resistance requires the restoration of genes extending through *rfaY*, the eighth gene of the long operon. *RfaY* is required for the addition of a second phosphate-containing substituent to Hep II (Yethon *et al.*, 1998). This indicates that this second charged site in the inner core also contributes to the structural stability of the outer membrane, although the phenotype of the loss of the Hep II site is much less severe than when the charged sites on both Hep I and Hep II are absent.

At this point, it is appropriate to digress and discuss the core structure of *K. pneumoniae* O8, which is included in Fig. 6. This organism is a free-living opportunistic pathogen, not an obligate parasite-like *E. coli* or *S. enterica*, and thus it appears to be thrifty in its use of phosphate. In the LPS of this organism, the only phosphates present are at the 1 and 4′ ends of the lipid A disaccharides (Severn *et al.*, 1996; Süsskind *et al.*, 1998). Instead of inner core P, the *K. pneumoniae* O8 core has two molecules of galacturonic acid (Gal*p*A), one terminating a branch from Hep I and the other as a distal core backbone sugar. The agreement of the spatial arrangement of these Gal*p*A residues to that of the phosphate-containing substituents on Hep I and Hep II of *E. coli* K-12 and R1 is striking. Gal*p*A has also been found in the LPS cores of members of the genera *Vibrio*, *Rhizobium* and *Rhodocyclus*. *Rhizobium leguminosarium* is even able to carry this phosphate parsimony one step further by re-cycling the phosphate from the ends of lipid A (Raetz, 1996). It produces LPS that is entirely phosphate-free, in which all negative charges are supplied by organic acids such as uronic acids.

Positive charges on LPS do not appear to be as well conserved. This suggests that negative charges, which serve to chelate divalent and polyvalent cations, are the more important contributors to stabilisation of LPS in the outer membrane. Virtually all Gram-negative bacteria are damaged by treatment with EDTA, and chelating agents such as this are rare in the environment. One exception is the water found in swamps and deep forests, which is rich in lignin breakdown products; this type of water is often nearly sterile. On the other hand, free Ca^{2+} is present in most natural environments, and polyamines are present wherever life is abundant.

Very little is known about the mechanism by which phosphate substituents are added to the inner core, and this is further complicated by the heterogeneity of these molecules. As noted previously, three forms of substituents are found in the inner core: free phosphate (P),

PEtN, and PPEtN. It is not at all clear that these are related. There is kinetic evidence that both the phosphate and ethanolamine moieties of the inner core are derived from phosphatidylethanolamine (Hasin and Kennedy, 1982), the most abundant lipid of the inner leaflet of the outer membrane and of the cytoplasmic membrane. It is also known that ATP serves, directly or indirectly, as a phosphate donor under *in vitro* conditions. A logical scenario is that PEtN is transferred directly to the sugar acceptor, while PPEtN is synthesised in two steps: first, the addition of P from ATP, and second, modification of this P by the addition of PEtN. Such a two-step pathway has been proposed for the synthesis of the PPEtN residue on the 1 position of the lipid A disaccharide (Zhou *et al.*, 1999).

An additional difficulty is the heterogeneity of these substituents. In one study, tandem mass spectrometry (MS/MS) was used to study the distribution of these species in the inner core of *E. coli* K-12. In the strain used in this study the long operon genes distal to *rfaP* were deleted, which means that it produced an LPS in which the core was truncated after Glc I and lacked the second phosphate substituent added by the *rfaY* product. Roughly half of the cores had P attached to Hep I and the remainder had PPEtN (Parker *et al.*, 1992). The substituent added by the *rfaY* product has been less thoroughly studied, but it has been reported to be P. The substituent present on the KDO II residue of some molecules has been reported to be only PEtN. Given the number and diversity of phosphate substituents present in LPS, it is very difficult to study this heterogeneity. The most effective technique has been MS/MS, because it can resolve the phosphate composition of individual molecular species, but this technique is expensive and not widely available.

One group of organisms that has been studied by MS/MS includes the genera *Neisseria* and *Haemophilus*. The LOS of these organisms, which is comparable to the enteric LPS core, exhibits much less heterogeneity in terms of phosphate substituents. P is never found, although both PEtN and PPEtN are present. It should be noted that these organisms are very detergent-sensitive. In this context, it should be noted that PEtN and PPEtN are bulky molecules, which are much less effective than P in chelating divalent and polyvalent cations.

Considerable insight into the way these substituents arise and their role in promoting outer membrane stability can be obtained by studying *tolC* mutants (Schnaitman and Klena, 1993). TolC protein is a trimeric outer membrane protein, which forms a channel which is part of a type I export system (Wandersman and Délepelaire, 1990; Koronakis *et al.*, 1997). This export system translocates a specific subset of extracellular proteins, including toxins, proteases and nodulation factors, directly across both the inner and outer membranes. The type I transport system involves at least two other proteins, and it is thought that the three proteins function together as a complex which spans both membranes and the periplasm. It has been best studied as a transport system for a plasmid-encoded haemolysin (Hly). However, in addition to a defective protein export phenotype, *tolC* knockout mutations have a very dramatic LPS phenotype.

TolC knockout mutations are lethal or very deleterious in smooth strains, and Gal labelling experiments in conditionally smooth *galE* strains suggest that a large amount of smooth LPS is shed into the medium. In ordinary rough K-12 strains such mutations are tolerated reasonably well, but the LPS itself has unusual properties. The gel banding pattern is not unusual, but the physical properties of purified LPS are remarkable. LPS isolated from a *tolC* strain by phenol-chloroform–petroleum ether (PCP) extraction followed by lyophilisation is fluffy and easily dispersed in water to make a smooth, transparent, stable suspension, while LPS from an isogenic Tol^+ strain is difficult to disperse in water and almost immediately precipitated as a waxy solid. Preliminary analysis of *tolC* LPS from a strain which produced a core truncated after Glc I, indicated that all of the molecules carried a PPEtN substitution on Hep I (C.A. Schnaitman, unpublished).

TolC knockout mutants also exhibit a very interesting partial deep rough phenotype. They are as sensitive to novobiocin as the most sensitive deep rough mutants, and do not grow on media containing detergents or basic dyes. These mutants have found some value in the biotechnology industry, in which they are used where it is desirable to get hydrophobic molecules into *E. coli*. However, despite their extreme hypersensitivity to hydrophobic compounds, they do not exhibit some of the other properties of deep rough mutants. The outer membrane protein composition is normal, as is the lipid content, and they do not exhibit induction of colanic acid synthesis (Austin *et al.*, 1990).

The *tolC* gene has recently become the subject of detailed genetic analysis, and point mutants exhibiting a number of new phenotypes have been isolated (R. Misra, unpublished). There are mutants that fail to export Hly but do not exhibit novobiocin hypersensitivity, and mutants which are novobiocin-hypersensitive but still export Hly. There are also point mutants that export a defective form of Hly, which is thought to be improperly folded, suggesting that TolC protein may function as a chaperone as well as an export channel. In any case, the export channel function appears to be complex. There are a number of possible explanations for the LPS phenotype of *tolC* mutations but, based on the present state of knowledge of *tolC*, the most promising is that TolC protein is required for the export of a protein that modifies LPS after its export to the outer membrane. The most reasonable modification would be the cleavage of PPEtN in the Hep I–Hep II region to generate P, which would allow LPS to become cross-linked by divalent cations (Schnaitman and Klena, 1993).

Old evidence indicates that LPS undergoes maturation after it reaches the cell surface, and that this maturation allows it to bind divalent cations (Levy and Leive, 1968). These experiments were done with J5, a *galE* mutant of *E. coli* O111, which allowed LPS to be labelled with radioactive Gal. A substantial fraction of the labelled LPS could be released by treatment with EDTA in Tris buffer, a treatment now routinely used to permeabilise cells. However, when the LPS was pulse-labelled with Gal, the newly synthesised LPS could not be released by EDTA. It required a period of time before newly synthesised LPS had 'matured' to an EDTA-releasable form.

A model to explain all of these phenomena is as follows. All LPS is synthesised to contain either PEtN or PPEtN as core substituents. The bulky, somewhat hydrophobic ethanolamine serves as a blocking group on the charged phosphates to allow phosphorylated molecules to be secreted across two membranes. After arrival in the outer leaflet of the outer membrane, the PPEtN substituents are cleaved to release free PEtN and leave P attached to the core where it can bind divalent cations. When this occurs, the LPS becomes segregated into 'islands', where it is cross-linked by divalent cations. The enzyme which carries out this cleavage is translocated to the exterior surface of the cell by the TolC transport system.

Major proteins, such as porins and OmpA protein, also undergo a maturation process (Misra, 1993; de Cock *et al.*, 1996; Kloser *et al.*, 1998). All of these proteins contain a few LPS molecules bound to specific sites, and the binding of LPS to these sites is necessary for correct folding. Presumably these proteins recognise lipid A and the inner core, and the PPEtN substituents or other substituents such as Hep III, which are absent in deep rough mutants may be essential for that recognition. Although there is no evidence for co-translation of newly synthesised proteins and newly synthesised LPS, it seems intuitively likely that LPS would bind to proteins before it becomes modified and cross-linked by divalent cations. Gram-negative cell surface growth takes place in small patches or zones, and it is likely that these contain both types of newly synthesised molecules. It is known that in deep rough mutants, the folding of OmpF is retarded. This results in down-regulation of OmpF synthesis at the translational level. The elevation in phospholipid content is presumably in compensation for the decrease in the surface area occupied by proteins like OmpF. In *tolC* mutants, proteins like OmpF are secreted and folded normally. The hypersensitivity to novobiocin reflects the inability of free LPS, that which is not bound to protein, to be cross-linked into an impermeable structure. Genetic analysis will, it is to be hoped, provide evidence for other actors in this scenario, including the LPS maturation enzyme and the other components of the TolC export pathway.

Evolution of the Class I Core

Evidence from the DNA analysis of *rfa* and *rfb* gene clusters in many strains and species suggests that the majority of these genes evolved in organisms other than the ancestors of present day enteric bacteria. This is based on G + C content. For example, in the span *rfaJ* to *rfaL* the G + C content in *E. coli* K-12 averages about 30–35%, compared with about 50% for the *E. coli* K-12 genome as a whole. The movement of these genes into enteric bacteria from a low-G + C ancestor took place long ago; as long ago as 150 million years ago according to one estimate (A. Wilson, cited in Reeves, 1993). Since then, there has been lateral transfer of these genes among various strains and species to generate what we now see as enteric bacteria. This is well illustrated in a comparison of the *rfa* genes of the three class I strains in which these have been sequenced. A comparison of these DNA sequences and the derived amino acid sequences of the proteins is shown in Fig. 8.

From an overall standpoint, these three gene clusters are very similar, the only striking difference being the loss of the *rfaS* gene in *S. typhimurium*. Otherwise, the arrangement of these genes into operons and the sizes of the individual genes are very similar. From what is known of the functions of these genes, they are all analogous, for example, adding sugars or branch substituents at similar positions in the core, if not strictly identical. What is remarkable is the variation in the degree of nucleotide sequence identity, and more important, the degree of apparent identity or similarity in the deduced protein sequence of the various gene products. For example, if we look at the KdtA (WaaA) protein, which has an identical function in all three organisms, we see that 99.5% of the amino acids are similar or identical (hereafter referred to as the similarity index) between K-12 and R2, and 97.9% are similar between R2 and *S. enterica*. In the case of RfaG

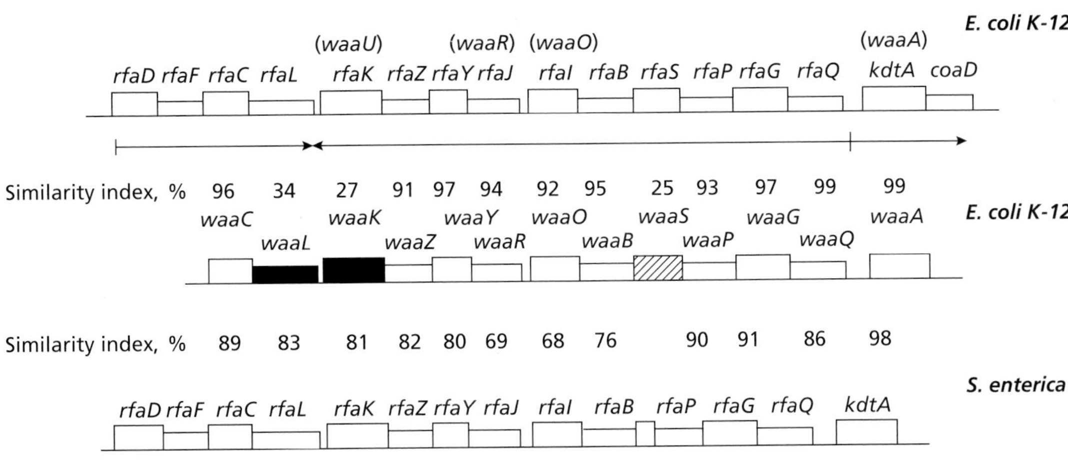

Fig. 8 Evolution of the class I core structure. The structures of the *rfa* clusters from *E. coli* K-12, *E. coli* group R2, and *S. enterica* sv. Typhimurium are shown. In the group R2 cluster, the unshaded blocks indicate genes which share ancestry with *E. coli* K-12, the dark blocks indicate genes which share ancestry with *S. enterica* sv. Typhimurium, and the diagonal shading indicates a gene which does not share obvious ancestry with either *E. coli* K-12 or *S. enterica* sv. Typhimurium. The numbers between the lines indicate the similarity index (% identical amino acids + % similar amino acids) of the encoded proteins. Re-drawn from Heinrichs *et al.* (1998a). The genes names for the R2 genes and the names shown in brackets above the *E. coli* K-12 genes are from that source. Bar indicates 1 kbp.

(WaaG) protein, these numbers are, respectively, 97.1% and 90.9% (Heinrichs *et al.*, 1998a).

The problem begins when *rfaI* and *rfaJ* of *S. enterica* are compared with the analogous pair of genes in the two *E. coli* strains. With reference to the core structures in Fig. 6, it will be seen that there is a difference in core backbone composition between *E. coli* and *S. enterica*. The second backbone heptose, added by *rfaI* and its analogues, is Gal in *Salmonella* and Glc in *E. coli* K-12. If it is assumed that each transferase recognises two substrates, an acceptor molecule and a sugar nucleotide, then each of the *E. coli*/*Salmonella* pairs is half homologous. The sequence homology is interesting. For the two homologous pairs the similarity index is 92.3% and 93.5%, while for the half homologues the numbers are 68.0% and 69.6% (Heinrichs *et al.*, 1998a).

The conundrum of what to name these genes arose about 15 years ago, when a collection of Clarke-Carbon plasmids (random DNA fragments from *E. coli* K-12 ligated into pColE1) were screened for their ability to complement *rfaI* and *rfaJ* mutations in *S. enterica* sv. Typhimurium. Plasmids which complemented both mutations were found, and it was proposed that *E. coli* K-12 has two sets of hexose transferases, one which is expressed to produce *E. coli*-type cores, and one which is normally repressed for the production of *Salmonella* cores. To differentiate these two sets, the names *rfaM* and *rfaN* were proposed for the *E. coli* K-12 genes.

When the *rfa* region was sequenced, this interpretation was found to be incorrect (Pradel *et al.*, 1992; Schnaitman and Klena, 1993). *E. coli* had only a single set of transferases for the second and third backbone hexoses, and had a similarity index of almost 70% to their counterparts in *Salmonella*. As long as both genes from *E. coli* K-12 were added together, they efficiently complemented both *rfaI* and *rfaJ* mutants. The reason is that complementation was measured by the production of smooth LPS, and the enzymes which complete the core and add the O antigen are apparently indifferent to whether the second backbone hexose is Gal or Glc.

At the time the sequence was published, both the *E. coli* K-12 and *Salmonella enterica* sv. Typhimurium Stock Culture Centers concurred that the names of the genes should be the same in both organisms, in spite of the differences in function. This was based on three criteria: first, the rather remarkable degree of homology of the proteins; second, the fact that the proteins performed analogous functions in core synthesis and, third, the longstanding policy that genes with analogous functions should, whenever possible, be given the same names in both organisms. Since then, other names for these genes have been proposed, and they listed in brackets in Fig. 8.

The *rfaK* and *rfaL* genes present a somewhat different picture. *RfaK* is thought to be involved in adding a terminal sugar to the core to mark it for the attachment of an O antigen, and *rfaL* encodes the ligase which attaches O antigens to the marked cores (Schnaitman and Klena, 1993). Previous work had shown that the analogues of these genes from

S. enterica and *E. coli* K-12 show little similarity in amino acid sequence (Klena *et al.*, 1992b). As in the case of *rfaI* and *J*, complementation of *S. enterica* mutants by cloned fragments from *E. coli* K-12 is observed only when both genes are added, suggesting that they work together as a pair. This was supported by the observation that the cloned *rfaK* gene from *E. coli* K-12 inhibited the synthesis of O antigen by wild-type *S. enterica*, which suggests that K-12 gene product adds a modification that blocks the modification added by the *rfaK* product from *S. enterica*.

Thus, if a lateral transfer of these genes were to occur in nature, they would have to be transferred as a pair, and this is exactly what is shown in **Fig. 8**. The *rfaK* and *rfaL* products of *E. coli* R2 and *E. coli* K-12 are not strongly homologous (similarity indices 33% and 26%, respectively). In contrast, the *rfaK* and *rfaL* products of *E. coli* R2 and *S. enterica* show strong homology (similarity indices 81% and 83%). Moreover, the cloned *rfaL* gene of *E. coli* R2 efficiently complements an *S. enterica rfaL* mutant, indicating that the gene products are also functionally homologous. Thus, it appears that the R2 *rfa* gene cluster is a hybrid, resulting from the lateral transfer of genes from an ancestor with an *S. enterica* type *rfa* cluster (Heinrichs *et al.*, 1998a).

A somewhat similar situation is evident for the *rfaS* genes of K-12 and R2, which have a similarity index of only 25.6%. This low an index is of marginal significance in implying an analogous function. As noted in the previous section, it is possible that the R2 *rfaS* gene performs an analogous function in marking the core of a subset of molecules to direct them into an alternate core pathway, but it is likely that it will be a different substituent.

It is tempting to speculate that *E. coli* R2 acquired its unique *rfaS* gene during the same DNA transfer event by which it acquired its unique *rfaK* and *rfaL* genes. Such multiple crossover events are not uncommon in regions such as the central core of the *rfa* cluster, which are flanked by regions of very strong homology. The structure of the putative ancestral *rfaS* gene of *S. enterica* is not known, and the remnants of sequence in that region of the *Salmonella rfa* cluster were not compared with the corresponding sequence of the R2 *rfaS* region, so it is not possible to tell whether the R2 *rfaS* gene was also acquired from a common ancestor of *S. enterica*. It is wise to remember that bacteria are living fossils. As more *rfa* sequences are acquired and analysed, and DNA analysis tools become more sophisticated, it may be possible to reconstruct the ancestral *rfaS* gene of *Salmonella*, just as it has been possible to reconstruct the functional *rfb* gene cluster of *E. coli* K-12.

The three gene clusters shown in **Fig. 8** provide a veritable textbook on the evolution of carbohydrate gene clusters. The flanking *rfa* genes, *rfaD*, *F* and *C* on the left and *rfaG*, *Q* and *kdtA* on the right, exhibit very strong homology at both the DNA and the deduced protein level. This to be expected for genes that carry out essential, strongly conserved functions in all enteric bacteria. If DNA exchange among these bacteria occurs frequently in nature, as all the evidence indicates that it does, the DNA repair coupled to recombination will tend to edit out small sequence changes which result from point mutations.

Ancient evolutionary patterns are seen within these clusters. The *rfaI* and *J* genes exhibit consensus features which define a large family of putative α-glycosyltransferases (Heinrichs *et al.*, 1998a). This family at present includes 16 individual strains, which represent nine Gram-positive and Gram-negative bacterial species, and one higher plant. These genes lie within a block of genes of low G+C content, suggesting an evolutionary origin different from the bulk of the genes in enteric bacteria.

There are also indications of more recent evolution. If *rfaK*, *L* and *S* are excluded, the remaining *rfa* genes have much more similarity between the two *E. coli* strains than between either strain and *S. enterica*. This suggests that the two *E. coli rfa* clusters shared common ancestors more recently than either shared common ancestors with *S. enterica*. The sharp homology discontinuities of the *rfaKL* and *rfaS* blocks indicate that lateral transfer of these genes from other strains has occurred more recently than the overall evolution of the gene clusters. Although the *rfaKL* block of R2 clearly shares ancestry with the corresponding block from *Salmonella*, similar donor homology is not evident for *rfaS*, and it is unknown whether transfer of a novel *rfaS* gene occurred in R2, K-12, or both.

Role of the Core in O-antigen Attachments

Two *rfa* genes, *rfaK* and *rfaL*, participate directly in the process by which O antigens are added to class I cores. This is true whether the O antigen is homopolymeric or heteropolymeric (*rfc* (*wzy*)-dependent or independent) in nature. It is also independent of whether O antigen synthesis is initiated by *rfe*, as seems to be the case in all *E. coli* serovars, or by *rfbP*, as seems to be true of *rfc* (*wzy*)-dependent *S. enterica* serovars.

In *S. enterica* sv. Typhimurium, *rfaK* mutants were first selected on the basis of resistance to the phage Felix O1 (FO1). The cores of such mutants lacked the terminal α1,2-linked GlcNAc normally found on Glc II, and considerable evidence showed that this structure was part of the receptor for FO1 (Mäkelä and

Stocker, 1884). Although these mutants lack readily detectable O antigen, some O antigen was still detected immunologically and the role of *rfaK* in O antigen attachment remained equivocal. Recently, with a hybrid smooth *E. coli* R2 strain expressing *K. pneumoniae* O1 O antigen, it has been shown that a defined *rfaK* knockout mutation completely eliminated the production of O antigen. Thus, the function provided by *rfaK* is essential for attachment of an O antigen.

Whether the GlcNAc residue attached to Glc II by the *rfaK* product in *E. coli* R2 and *S. enterica* remains after the subsequent ligation of an O antigen, or partial O antigen, to the core remains a matter of conjecture. Some early work described GlcNAc as a partial substituent, which is bothersome, considering that almost all of the cores produced by *S. enterica* are substituted with O antigen in smooth strains. And in one study, the ability of *S. enterica* LPS to neutralise phage was far less than predicted, based on the presence of a GlcNAc residue on every molecule (Lindberg and Holm, 1969).

The situation is even more complicated in *E. coli* K-12. In a smooth *E. coli* K-12/Dysenteriae 1 hybrid, *rfaK* was essential for O antigen production (Klena *et al.*, 1992a). In an AB1133 background *rfaK* is required for the production of gel band 1 and the RfpB-modified band, which are known to carry 1-sugar and 2-sugar truncated O antigens, and band 2, which is thought to represent an earlier stage of core completion in the smooth LPS pathway. These results, and the similar structure of the RfaK proteins from the various species, suggests that *E. coli* K-12 *rfaK* performs a similar function with respect to the attachment of O antigen. However, it is clearly not identical. The core Glc III of K-12 does not have an α1,2-linked GlcNAc substituent, but in some strains a β1,6-linked GlcNAc has been reported. K-12 is resistant to phage FO1, but becomes sensitive when transformed by a plasmid carrying *rfaK* from *S. typhimurium* (Klena *et al.*, 1992a). The K-12 RfaK protein shares some homology with known heptosyltransferases, and since the cores of some strains carry an α1,6-linked Hep (Hep IV) attached to Glc III, it has been proposed that the K-12 RfaK protein is the Hep transferase which adds this residue. This is known to be a partial substituent, and it does not appear to be present in some strains. Again, this is difficult to reconcile with an obligate role in O antigen attachment, since the strains, in which it is produced in very low amounts, produce abundant O antigen when they are supplied with a source of functional *rfb* genes.

Studies in three different *E. coli* K-12 systems have indicated that partial O-units assembled on bactoprenol can be efficiently transferred to the core. In all these studies, the first sugar of the partial O-unit was GlcNAc, which had been added to bactoprenol by the Rfe transferase. In one study, the terminal structure of cores to which 1-, 2- and 3-sugar partial O-units were added was determined, and it showed GlcNAc attached directly to Glc III in a β1,6 linkage (Fält *et al.*, 1993). This has not been confirmed. If it is correct, however, and if the K-12 RfaK protein adds Hep IV to the 6 position of Glc III, then the Hep IV must be trimmed off before O antigen can be added. Two other groups have chemically confirmed the addition of partial O-units to the K-12 core, but they did not determine the linkage of these units to the core (Clarke *et al.*, 1995; Feldman *et al.*, 1999).

In all of the core classes, the *rfaL* genes encode hydrophobic proteins with a number of membrane-spanning domains, suggesting that ligation takes place in a membrane environment. The ligation reaction has not been carried out *in vitro*, and little is known about the mechanism of the ligation reaction or whether it takes place on the cytoplasmic or periplasmic face of the membrane. If trimming occurs along with translocation or ligation, there are as yet no candidates for genes that might be involved. Much remains to be done before the final steps of the assembly of LPS and its translocation to the cell surface can be understood.

Lessons from the Class II and Class III Cores

The choice of *E. coli* K-12 as a class teaching strain at Stanford was an extremely fortunate one. This strain carried phage λ and the conjugative plasmid F, which were both critical for the development of bacterial genetics and molecular biology. But it is also an uncommon strain. Two studies have surveyed the distribution of core types among clinical isolates, and the K-12 core was found in only 2–5% of isolates (Appelmelk *et al.*, 1994; Amor *et al.*, 2000). Roughly 75% of the isolates had class II cores, R1 being the most predominant. R3 represents a small percentage of all strains, but it strongly predominates among the verotoxigenic (VTEC) strains, which include serotypes O157, O111 and O26. It also provides the core for the *E. coli* O111 *galE* mutant J5, which has been used in many immunological studies and in studying the synthesis of the outer membrane.

The class II and class III *rfa* clusters (Heinrichs *et al.*, 1998b, c) offer sobering alternatives to the apparent sophistication of core synthesis. With some minor twists, the *rfa* clusters exhibit the same three-operon arrangement, and have the same three conserved genes (*rfaQGP*) at the 5′ end of the long operon. But they have some striking differences. The *rfaL* gene of the class I *rfa* regions is part of the Hep operon, while in class II the *rfaL* gene is transcribed in the opposite direction and is the last gene in the long *rfa* operon. Unless it has its own promoter, this places it under the control of SfrB.

Both the class II and class III long operons are also significantly shorter, by two genes in class II and three genes in class III. The missing genes in both classes are *rfaS* and *rfaZ*, the two genes implicated in the putative LOS pathway, and R3 is also missing a Hex transferase. If these organisms have a mechanism for producing an alternate form of rough LPS, it is not obvious from the structure of their *rfa* gene clusters. Since *rfaQ* mutants played such an important role in pointing out the heterogeneity in the class I core, it would be interesting to study its LPS gel phenotype in class II and class III strains.

It is also not clear whether the class II and class III clusters have genes analogous to *rfaK* of the class I clusters. The best candidate appears to be *rfaV* (*waaV*) of the type R1 cluster. This gene encodes a transferase which adds a branch βGlc to the 3 position of the second backbone hexose, Glc II. *RfaV* (*waaV*) is essential for addition of an O antigen. The chemistry suggests, but does not prove, that the βGlc branch sugar is the acceptor for O antigen, since the results presented are not inconsistent with trimming of the branch sugar and ligation of the O antigen directly to Glc II. As study of the chemistry and genetics of the class I cores continues, it will be interesting to see how much carries over to the other core classes. One thing is already clear, evolution has provided many answers to the question of how to construct core oligosaccharides, many more than might have been anticipated a few decades ago when studies in *S. enterica* made LPS cores seem quite simple.

Properties and Synthesis of O Antigens

The Relationship Between O Antigens and Capsular Polysaccharides

O antigens and capsules both exhibit extensive polymorphism, and there are now 167 O-serogroups and more than 80 distinct polysaccharide K antigens (capsules) in *E. coli* (see Chapter 4). This polymorphism is also seen in other enteric bacteria, and in other genera such as *Vibrio* and *Pseudomonas.*

A new classification scheme for *E. coli* capsules (K antigens) has been described by Whitfield and Roberts (1999). This divides K antigens into four groups, based on details of their synthesis and export. This provides an ideal departure point for describing O antigens, for it is rapidly becoming clear that K and O antigens are alike in more ways than they are different, and in some cases they are essentially the same thing. The latter phenomenon has long been known in *E. coli* O111, which produces a capsule of exactly the same structure as its O antigen, and which differs only in the lipid to which it is attached.

In *E. coli*, both O and K antigens can be roughly divided into two basic types. One type consists of a polysaccharide organised around a backbone consisting of a simple one- or two-sugar repeat, such as a mannan or galactan. Because of their simple repeat structure, these are sometimes referred to as homopolymeric polysaccharides. Since their synthesis does not require the Rfc (Wzy) polymerase, they are also termed Rfc-independent polysaccharides. These are synthesised processively by sugar transferases, which sequentially add single sugars to the non-reducing end of the polymer, essentially a simplified form of the mechanism involved in the synthesis of core oligosaccharides. The polymers are synthesised in the cytoplasm, and with one exception, as noted below, they are exported by a dedicated ABC-2 exporter. O antigens of this type include the O8 and O9 serotypes of *E. coli*, the O1 serotype of *K. pneumoniae*, and *S. enterica* sv. Muenchen. K antigens that fall into this category are group 2 antigens such as K1 and K5, and probably group 3 antigens like K54.

One homopolymeric O antigen, the serotype O54 antigen of *S. enterica* sv. Borreze, does not require an ABC exporter. This polymer is not assembled in the cytoplasm. Instead, after a priming step involving two glycosyltransferases, which add GlcNAc and ManNAc residues to bactoprenol, the bulk of the polysaccharide is polymerised by a single, processive poly-ManNAc synthase that resembles the chitin and cellulose synthases. These are dual-function enzymes which have both export and glycosyltransferase functions, essentially 'squirting' newly synthesised polymer into another cellular compartment during polymerisation (Keenlyside and Whitfield, 1996). It has been proposed that this system be designated a 'synthase-dependent' pathway, to differentiate it from the 'ABC-transporter-dependent' pathway used by other homopolymeric antigens (Whitfield *et al.*, 1997).

The second type of polysaccharide, termed heteropolymeric polysaccharides, are synthesised by assembly of repeat units attached by a pyrophosphate linkage to the lipid carrier bactoprenol. The repeat units, which generally consist of 3–6 sugars, and are often branched. They are assembled into polymers by exchange reactions between bactoprenol molecules, one of which carries only a single repeat unit and another which carries one or more repeat units. For example, a bactoprenol molecule carrying two repeat units will donate these to the non-reducing end of the bactoprenol carrying only a single repeat. The result will be one molecule of bactoprenol-P and a second molecule of bactoprenol carrying three repeats. This can donate its three repeats to another molecule with one repeat, and the result will be a second bactoprenol-P and bactoprenol carrying four repeats. This is continued until the polymer reaches the desired size, of about 20 repeats for most O antigens. The enzyme which carries out these exchange reactions is termed the polymerase, and is the product of a member of the *rfc* (*wzy*) gene family.

For this reason, heteropolymeric antigens are often termed Rfc (Wzy)-dependent polymers. Unlike the Rfc-independent polymers, the site of assembly of the polymer is not known accurately, although it is predicted to be on the periplasmic side of the cytoplasmic membrane. Rfc-dependent antigens do not appear to utilise ABC-2 exporters. Instead they use a PST transport system, sometimes referred to as a 'flippase', encoded by the members of the *rfbX* (*wzx*) gene family. The RfbX system appears to translocate individual or incomplete O-units (Clarke *et al.*, 1995; Feldman *et al.*, 1999), attached to bactoprenol, across the cytoplasmic membrane before polymerisation. Examples of Rfc-dependent O antigens include the sv. Typhimurium O antigen of *S. enterica*, and antigens O7, O16, O111 and O157 of *E. coli*. Rfc-dependent K antigens include the group 1 antigen K30 and related antigens in *Klebsiella* and *Erwinia*, and the group 4 antigen K40 (Whitfield and Roberts, 1999).

There are now a number of examples, in addition to *E. coli* O111 as noted above, in which group one and group four K antigens are actually attached to lipid A instead of to a smaller lipid acceptor. Thus, the distinction between K antigens and O antigens continues to blur, as serology, which originally defined K antigens as surface antigens that masked O antigens, gives way to chemistry and genetics. For K groups 1, 2 and 3, one of the primary distinctions is the location of the gene clusters that encode their synthesis. These are the *cps* gene cluster, which is adjacent to the *rfb* gene cluster near *his*, and the *kps* gene cluster, near *serA* (Whitfield and Roberts, 1999). With some notable exceptions, which will be described in more detail below, the genes for O antigen synthesis are located in the *rfb* cluster, between *cps* and *his*.

O antigens and polysaccharide K antigens share another distinction. Both share a much greater diversity of sugars than is found in other cell polymers. Sugars such as rhamnose, mannose, abequose, paratose and colitose generally occur only in O and K antigens, and occasionally in LPS cores. The total list of these unique sugars runs to more than 30. At the same time, it is very unusual for any of these polymers to have more than two or three of these unique sugars. It would be very inefficient for a cell to have the several hundred genes necessary for the synthesis of all of these unique sugars when it needs fewer than a dozen of these genes to make its own polymers. The solution to this problem is to include unique sugar biosynthetic genes in the gene clusters that encode the synthesis of polymers containing these sugars. For example, if an *rfb* cluster encodes the synthesis of an O antigen containing mannose, rhamnose and tyvelose, the *rfb* cluster is likely to include genes that encode the biosynthetic pathways of these three sugars. This is exactly what is seen in *S. enterica* sv. Typhi (Jiang *et al.*, 1991). The O antigen produced by this strain also includes Gal and Glc, but the nucleotide precursors of these sugars are shared among many pathways. This arrangement also permits mobilisation of these clusters so they can be shared among other strains (Reeves, 1993). One can make a circular argument about whether this arrangement permits transfer out, or exists because it was transferred in.

Since the average strain has one *rfb* cluster and at least two capsule clusters, including the *rfe-rff* (*wec*) cluster, which encodes a generic capsule-like material called enterobacterial common antigen (ECA), one can imagine a situation where an identical sugar biosynthetic pathway will be present in two different gene clusters. This occurs fairly frequently, an example being the mannose pathway in which nearly identical biosynthetic genes are often present in both the *rfb* and *cps* clusters (Stevenson *et al.*, 1996). To a geneticist, the presence of identical genes at different sites in the chromosome is a recipe for mischief. Possible consequences include loss of one of the genes, which will leave one cluster immobilised and dependent on the other, and re-combination between the genes, which can lead to deletion or chromosomal re-arrangement. This possibility is most likely with the *rfa* and *cps* clusters, which are separated by a short segment that contains no essential genes. The similarity between O antigen and capsular polysaccharide gene clusters has contributed to the mix which has led to the present diversity of antigens.

Plasmids and Phage Play a Role in O antigen Diversity

With very few exceptions, the majority of the genes for the production of O antigens are located in the *rfb* cluster, which lies between the *cps* gene cluster and the *his* operon. In *E. coli* K-12 and *S. enterica* sv. Typhimurium, *rfb* is flanked on the 5′ end by the *galF* gene, which encodes a homologue of *galU*. *GalF* cannot replace *galU*, but may improve the production of UDP-Gal, an important precursor of both capsules and O antigens (Stevenson *et al.*, 1996). *GalF* does not appear to be essential for either polymer. The *gnd* and *rol* genes lie at the 3′ end of *rfb*. *Gnd* encodes 6-phosphogluconate dehydrogenase, an enzyme of the pentose phosphate shunt pathway, and has no relationship to *rfb* (Thampapillai *et al.*, 1994). *Rol* (*cld*) encodes a protein involved in determining the length of O antigens and capsules (Batchelor *et al.*, 1991, 1992). The relative order of *gnd* and *rol* with respect to *rfb* and *his* vary among enteric bacteria, but otherwise the structure of this region of the chromosome appears to be reasonably well conserved.

Some organisms, such as various species of *Pseudomonas*, clearly have the ability to produce more than one type of O antigen in response to different environmental conditions. Among the enteric bacteria it is unusual for an organism to produce more than one serotype of LPS at a time, but there are strains which alternate back and forth between two related serotypes. This phenomenon is known as form-variation (Mäkelä and Stocker, 1984), and is presumably similar in mechanism to the phase variation observed for flagellar antigens and some other surface antigens. The point of this is that enteric organisms are normally haploid with respect to *rfb* genes, but they must have existed as diploids or merodiploids at some time to allow lateral gene transfer to occur.

In a few uncommon instances *rfb* genes are found to be on plasmids. Some of these have been studied in considerable detail, and provide some information on how mobile pools of *rfb* genes can exist in the population at large.

One of the best-studied cases is that of *Shigella dysenteriae* 1. A 9-kb high copy number plasmid is invariably associated with clinical isolates of this strain, and loss of the plasmid results in production of rough LPS (reviewed in Schnaitman and Klena, 1993). Two adjacent genes designated as *rfpA* and *rfpB* have been identified on this plasmid (Göhmann *et al.*, 1994). *RfpB* has been identified as the Gal-bactoprenol transferase, which adds the second sugar of the O-unit. The first sugar in the O-unit is GlcNAc, which is added by the GlcNAc-bactoprenol transferase encoded by *rfe*. The *rfpA* gene has no apparent function in *Sh. dysenteriae* 1. Neither of these genes shares homology with known *rfb* genes from *E. coli* or *S. enterica*. The *RfpA* protein shares considerable homology with ExoU of *Rhizobium meliloti*, which is involved in the synthesis of succinoglycan. The RfpB protein shares some homology with the *rfaK* gene product of *Salmonella*, which has an entirely different function, and with a yeast glycosyltransferase. As will be described below, it also shares homology with a protein encoded by the *K. pneumoniae* O1 *rfb* cluster (Clarke *et al.*, 1995).

The *rfb* gene cluster of *Sh. dysenteriae* 1 encodes only two of the four sugar-bactoprenol transferases required for its O-unit (Klena and Schnaitman, 1993). The 3′ end of this *rfb* cluster consists of two short reading frames separated by a short gap, perhaps indicating that the cluster once encoded an additional transferase. Taken at face value, these results suggest that *Sh. dysenteriae* 1 suffered an *rfb* mutation (or mutations) which destroyed its ability to produce smooth LPS, and that it was rescued by a plasmid carrying genes of distant origin, one of which was capable of complementing this mutation. The order in which these events occurred is not evident.

A second anecdote which is perhaps more relevant to the lateral transfer of *rfb* genes concerns *S. enterica* sv. Borreze (Keenleyside and Whitfield, 1995). In this strain, the only functional *rfb* cluster exists on a 6.9-kb plasmid. This makes it the only known *S. enterica* strain in which the *rfb* cluster exists on a natural plasmid. The O-serotype 54 antigen expressed by strains of sv. Borreze, which carry this plasmid, was also expressed when the plasmid was moved into an *E. coli* K-12 background. Other *S. enterica* strains co-express this antigen, and in all cases this expression involves a plasmid. The plasmids found in these other strains vary in size, but all cross-hybridised with the sv. Borreze plasmid.

The plasmid also hybridised with a prototype ColE1 plasmid. Analysis of the plasmid localised the *rfb* gene cluster. In addition, it showed that the plasmid had a ColE1-type *oriV* and all of the features (*mob*, *bom*, RNA I, RNA II, etc.) necessary for maintenance and transfer of a plasmid with the ColE1 type replicon. These features indicate that the plasmid is potentially mobilisable, if suitable conjugative transfer functions are supplied in *trans* by a compatible conjugative plasmid. Thus this plasmid exists in nature as a source of readily transferable *rfb* genes.

Finally, the most detailed analysis has been done on *Shigella sonnei* (*E. coli* sv. Sonnei). Like *S. enterica* sv. Borreze, the *rfb* cluster of *Sh. sonnei* is carried on a plasmid. Both the *rfb* cluster of the plasmid and the defective chromosomal *rfb* region have been partially sequenced (Lai *et al.*, 1998). The Sonnei O antigen does not otherwise occur in *E. coli*, and the *rfb* cluster on the plasmid appears to have been acquired from serotype 17 of *Plesiomonas shigelloides*. The insert in the plasmid hybridises to the chromosomal *rfb* cluster of *P. shigelloides*, and for hundreds of base pairs the sequence of the *P. shigelloides* O17 and Sonnei are identical.

Examination of the restriction map of the *rfb* region of the Sonnei chromosome suggested that a deletion had occurred between an ancestral *cps* gene cluster and the *gnd* gene, which lies between the *rfb* cluster and *his* in most strains, the gene sequence immediately to the left of *gnd* was *manC manB*. These genes encode functions in mannose biosynthesis, and are found in both *cps* and *rfb* clusters of some *E. coli* strains. A recombination event had apparently occurred at a site roughly in the middle of the *manB* gene. The DNA sequence extending from the 3′ end of *manC* to the 5′ end and the middle section of the *manB* gene of Sonnei was 96% identical to the comparable region of the *cps* gene cluster of *E. coli* K-12. The Sonnei DNA sequence extending from the 3′ end of *manC* to the *manC-gnd* intergenic region into the 5′ end of *gnd* was 93% identical to that of *E. coli* O7, in which the gene sequence at the right end of its *rfb* cluster is *manC manB gnd*.

Since this gene sequence may occur in other *E. coli* serovars, PCR primers based on the Sonnei *manB gnd* DNA sequence from this region were used to probe a complete collection of the 166 known *E. coli* serotypes. Fourteen strains had PCR bands identical to that of Sonnei. Of these, the strain with an O62 serotype showed the best DNA sequence agreement, differing from Sonnei by only 1.61%. It is suggested that Sonnei emerged about 10 000 years ago, by loss of the O62 *rfb* cluster and capture of a plasmid carrying the *rfb* cluster of *P. shigelloides* O17 (Lai *et al.*, 1998).

In addition to plasmids, bacteriophage are agents that affect the diversity of O antigens. It is very common among temperate phage that use a major surface component as their receptor for the phage genome to encode a mechanism for modifying or eliminating their receptor once they enter the lysogenic state. This phenomenon is known as lysogenic conversion. Although lysogenic conversion was first studied with respect to LPS (Mäkelä and Stocker, 1984), it can occur with any major surface component, including capsules and major outer membrane proteins. An example of the latter is phage PA-2, which uses OmpC porin protein as its receptor. In the lysogenic state, the phage genome encodes a new porin which replaces OmpC protein in the outer membrane. It is often stated, probably incorrectly, that the function of lysogenic conversion is to prevent superinfection by homologous phage. Since this has almost no consequences, a much more likely reason is the lysogenic conversion prevents the neutralisation of progeny phage by cell debris when a lysogen is induced. It has been shown that after induction of a prophage in which a gene essential for lysogenic conversion has been mutated, the yield of phage is significantly reduced compared with the yield after induction of the wild-type (C. Schnaitman, unpublished). Many temperate phage which use LPS receptors encode an LPS depolymerase, which makes up part of the tail or tail plate assembly and degrades the host LPS to gain access to the outer membrane. The site at which the LPS is modified during the lysogenic state is generally adjacent or identical to the site at which the depolymerase cleaves the host LPS (Lindberg, 1977).

LPS undergoes two forms of lysogenic conversion. The first appears to occur only in the Rfc-dependent O antigens, and involves changing the linkage between the O-units. This can involve changing the site of the linkage, changing the anomeric configuration of the sugar at the reducing end of the O-unit, or both. Since the linkage is determined by the Rfc polymerase, temperate phage that carry out this form of lysogenic conversion must carry a different member of the family of *rfC* genes, and this gene must be expressed in the lysogenic state. Presumably, the linkage between the O-units created by the host-encoded Rfc polymerase is part of the phage receptor. Thus, for lysogenic conversion to be effective in eliminating the phage receptor activity of the host LPS, the lysogen must also have a gene that encodes an inhibitor of the host Rfc polymerase which is expressed in the lysogenic state.

The existence of phage carrying O antigen polymerase genes may explain the wandering nature of the members of the *rfc* gene family, which are found at different sites within the *rfb* cluster as well as being located outside of the *rfb* region, as in *S. enterica* sv. Typhimurium. The relationship between such genes and phage sequences has not been established. The assistance of phage could also be invoked in the generation of the differences observed between *S. enterica* groups B, D1, D2 and D3, which all produce a common type of O-unit backbone but differ primarily in the nature of the linkage between O-units (Curd *et al.*, 1998).

The second form of lysogenic conversion involves the attachment of modifications to backbone or branch sugars. The commonest modifications are glycosylation and acetylation. These types of modifications are widespread, and are observed with both Rfc-dependent and Rfc-independent polymers. Phage involvement has not been demonstrated in the case of all of these modifications, but should be suspected whenever the genes for attachment of modifications or branch Glc units lie outside of the *rfb* gene cluster.

This has been studied in most detail in *Shigella flexneri* (Guan *et al.*, 1999; Allison and Verma, 2000), where as many as 16 different serotypes may have resulted from this phenomenon. A number of the phages involved in this process have now been characterised, and they are morphologically diverse. However, they share common features. Several phages share the attachment site of P22 in the tRNA gene *thrW*, and their *int* and *xis* genes and core *attP* site are similar or identical to those of P22. A related cryptic prophage in *E. coli* K-12 is integrated into the tRNA gene *argW*. Thus they resemble other temperate enteric phages in their use of redundant tRNA genes as attachment sites.

Little is known about the mechanism of acetylation. A single gene appears to be involved in the process and, as expected, this shares homology with a broad family of genes involved in *O*-acetylation from other sources. There may be more than one type of phage-encoded acetylation. In some strains all O-units are uniformly acetylated, suggesting that acetylation takes place in the cytoplasm, while in other strains only the larger units are acetylated, suggesting that acetylation takes place after completion of the molecule.

The situation with respect to glycosylation is different, and more interesting (Guan *et al.*, 1999; Allison and Verma, 2000). The prototype for studying the glycosylation process has been phage SfX, which converts the parental serotype Y, to serotype X. The parental O antigen has a repeating unit consisting of (Rha III-Rha II-Rha I-GlcNAc)$_n$ and the phage conversion adds an α1,6-Glc modification to Rha I. In all the phage glycosylation systems that have been studied, three genes are required. In phage SfX these are termed *gtrA*, *gtrB* and *gtrX*. The *gtrA* and *gtrB* gene products are substantially the same from all phages, but *gtrX* is specific for SfX, and in turn, is thought to be specific for the exact O-unit modification carried out by that phage.

The donor of the Glc is known to be Glc attached to bactoprenol, presumably by a pyrophosphate linkage as in the case of O-units. The *gtrA* protein is a small, hydrophobic protein with four apparent transmembrane segments. It is thought to act like *rfbX* (*wzx*) as a 'flippase' to translocate the Glc attached to bactoprenol across the cytoplasmic membrane, where it can be attached to previously synthesised LPS or O antigen polymers. It is proposed that *gtrB* is involved in the addition of Glc to bactoprenol, and that *gtrX*, which encodes a somewhat larger hydrophobic protein, is involved in the transfer of Glc to the O antigen. This would explain the polymorphism of phage genes analogous to *gtrX*, since each would have to recognise a specific site on a specific polysaccharide.

The study of the SfX system in *Sh. flexneri* and analogous systems, such as the glycosylation mediated by the *oafA* gene of *S. enterica* group B (Mäkelä and Stocker, 1984) is far more important than just the study of a system for generating the serological diversity of O antigens. First, it may be part of a system whereby new pathogenic strains arise by acquiring properties that enable them to colonise different niches in their hosts. Temperate phage have long been known to encode toxins which are expressed in the lysogenic state, and there are recent reports assigning phages other roles in the pathogenic process (Cheetham and Katz, 1995; Karaolis *et al.*, 1999). This may explain in part why novel O-serotypes and pathogenesis seem to travel together. Second, it is the premier system for studying the sidedness and location of various components of the LPS biogenesis system. It is common for authors to claim that, in the biosynthesis of Rfc-dependent O antigens , the *rfbX* (*wzx*) gene product flips O-units attached to bactoprenol across the cytoplasmic membrane (Liu *et al.*, 1996) and that polymerisation of O antigens occurs on the periplasmic face. In fact, none of these has unequivocally been demonstrated.

Three basic methods have been used to study these export processes. These are inhibition by energy poisons, such as uncouplers that collapse the transmembrane potential, localisation of precursors and products by electron microscopy and immunochemistry, and the use of various vesicle preparations and/or spheroplasts. All of these have serious problems. Complete collapse of the transmembrane potential causes serious and often irreversible damage to the integrity of the cytoplasmic membrane. Once such inhibitors are added, cells are essentially dead. Immunochemistry lacks the resolution to differentiate the cytoplasmic and periplasmic surfaces of the cytoplasmic membrane, and when deposits of products or intermediates build up in cell compartments it is generally impossible to determine whether these represent a normal localisation or a default pathway for disposing of material which cannot be translocated or processed normally. Vesicle systems are not available in which sidedness is preserved or can be unequivocally assigned. Spheroplasts represent one of the most misleading systems, since the outer membrane does not detach from the cell, and may only rupture transiently, if at all, when spheroplasts are formed. The cytoplasmic and outer membranes remain closely attached over a significant proportion of the spheroplast surface, and this tight attachment is presumably mediated by exactly the transport systems the investigator wishes to study.

Given the history of the study of LPS, it is unfortunate that dogma continues to be cited as if it had been solidly proven, until someone comes along who has been courageous enough or fortunate enough to come up with a better system. The O antigen glycosylation systems would be an excellent point of departure. Another exciting new system to study localisation reactions is the synthesis of *rfc*-independent O antigens and capsules (Clarke *et al.*, 1995) which is described below.

Synthesis of *rfc*-independent O Antigens

For historical reasons, the synthesis of *rfc*-independent O antigens has not been as intensively studied as that of the synthesis of the *rfc*-dependent clusters. The first of these is that among the well-studied serovars of *S. enterica*, the group C1 serovar Montevideo was the only one that produced an *rfc*-independent O antigen. This serovar was also differentiated from the remainder of commonly studied *S. enterica* serovars by its requirement for the *rfe* gene. The structure of the *S. enterica* sv. Montevideo repeat unit is shown in Fig. 9. The backbone consists of a simple mannan with alternating pairs of α1,2 and β1,2 linkages. Every fourth mannose is glycosylated. The gene for this glycosylation lies outside of the *rfb* cluster, and is possibly one of the phage-encoded modifications described above. A single GlcNAc probably occurs at the reducing end of the molecule, although some references (see Reeves, 1993) show this residue at the end of each 4-Man repeat. Thus, it can be considered a simple homopolymer. The Montevideo *rfb* gene cluster (not shown) is considerably shorter than those of the well-studied *rfc*-dependent *S. enterica* and *E. coli* serotypes (see below and Reeves, 1993). It contains only seven open reading frames, two of which have been identified as encoding genes of the mannose biosynthetic pathway. As expected, this *rfb* cluster lacks an analogue of the *rfc* polymerase gene, and it also lacks an analogue of *rfbX*, the gene that encodes a putative bactoprenol flippase.

The second reason that homopolymeric O antigens have received less study is that it was clear that the widely accepted Robbins model for the polymerisation of O-units assembled on bactoprenol, which resulted in the addition of new units to the reducing end of the growing polymer, could not explain the processive growth of homopolymeric O antigens at the non-reducing end of the polymer.

A new model suitable for *rfc*-independent O antigens has arisen based on studies in *K. pneumoniae* O1 (Clarke *et al.*, 1995). This strain produces a simple galactan whose structure is a repeating disaccharide as shown in Fig. 9. Its *rfa* cluster is composed of six genes organised into a single operon. When a plasmid

$^{3}\beta$Man$^{1}\rightarrow^{2}$Man$^{1}\rightarrow^{2}$Man$^{1}\rightarrow$
E. coli O8

^{3}Man$^{1}\rightarrow^{3}$Man$^{1}\rightarrow^{2}$Man$^{1}\rightarrow^{2}$Man$^{1}\rightarrow^{2}$Man$^{1}\rightarrow$
E. coli O9

$^{4}\beta$ManNAc$^{1}\rightarrow^{3}\beta$ManNAc$^{1}\rightarrow$
S. enterica sv. Borreze

$^{2}\beta$Gal*f*$^{1}\rightarrow^{2}\beta$Gal$^{1}\rightarrow$
Klebsiella pneumoniae O1

^{2}Man$^{1}\rightarrow^{2}$Man$^{1}\rightarrow^{2}\beta$Man$^{1}\rightarrow^{2}\beta$Man$^{1}\rightarrow$
S. enterica sv. Montevideo

Fig. 9 Structures of Rfc-independent (homopolymeric) O antigens. All sugars are D-isomers, and unless indicated, the anomeric linkages are α and they are in the pyranose configuration.

carrying these genes was introduced into *E. coli* K-12 strain AB1133, it had two effects. One was the production of smooth LPS carrying the *K. pneumoniae* O1 O antigen, and the other was the production of a novel core species, which consisted of a single βGal*f*1–3αGal*p* disaccharide added to the K-12 core. Both of these required the participation of *rfe*. Both also required the last gene in the *rfb* operon, which is designated as *rfbF*$_{(\mathrm{KpO1})}$.

A plasmid carrying only the *rfbF*$_{(\mathrm{KpO1})}$ gene downstream from an expression vector promoter also resulted in the addition of a modification to the K-12 core, but this modification consisted of a single Gal*p* residue instead of a Gal*f*-Gal*p* disaccharide. This indicates that the addition of the disaccharide takes place in two steps, and only the first of these involves the *rfbF*$_{(\mathrm{KpO1})}$ gene product. Interestingly, the carboxy-terminal region of the *rfbF*$_{(\mathrm{KpO1})}$ gene product is similar in structure to that of the *rfpB* protein, the Gal*p*-bactoprenol transferase encoded by the small *Sh. dysenteriae* 1 plasmid. It was known that *rfpB* resulted in a similar modification of the K-12 core (Klena *et al.*, 1992a; Klena and Schnaitman, 1993). To complete this comparison, it was shown by Clarke *et al.* (1995) that that a plasmid carrying the *rfbF*$_{(\mathrm{KpO1})}$ gene under the direction of an expression vector promoter was capable of replacing the *rfpB* gene in the production, in K-12, of smooth LPS carrying the *Sh. dysenteriae* O1 O antigen.

A plasmid which carries the last four genes of the *K. pneumoniae* O1 *rfb* cluster downstream from a Tet promoter, when introduced into AB1133, is capable of producing the disaccharide-modified core, but not the complete O1 O antigen. This indicates that the last four genes contain the information for the synthesis of the disaccharide and its addition to the LPS core, while the first two genes provide information necessary for the polymerisation of the O antigen and/or its export. To test this system further, the production of these products was examined in *E. coli* K-12 strain SØ874, which carries a long deletion that removes the entire *E. coli* K-12 *rfb* region. It had previously been shown that this deletion removes the *E. coli* K-12 *rfbX* gene, which is necessary for both the *rfe*-dependent modification of the K-12 core by GlcNAc, and the subsequent addition of Gal by the *rfpB* protein. When the plasmid carrying the full-length *rfb* insert was introduced into strain SØ874, the ladder of smooth LPS was still produced but the modification of the LPS core with the disaccharide was absent. Likewise, the core disaccharide modification was absent when the plasmid carrying the last four genes of the *rfb* cluster was introduced into strain SØ874. This indicates that the modification of the core by the galactan disaccharide is a side reaction resulting from the presence of the *rfbX* gene in the host strain, and not an intermediate step in the production of smooth O1 LPS.

The model that emerged from this work is the following: Synthesis of the O1 antigen is initiated by the addition of GlcNAc to bactoprenol by the Rfe transferase. This takes place on the cytoplasmic surface of the membrane and is followed by the addition of the first Gal*p* residue by the $rfbF_{(\mathrm{KpO1})}$ gene product, and the second Gal*f* residue by the product of one of the other three distal genes of the *rfb* cluster. This serves as a primer for the remaining O1-polysaccharide, which is synthesised by repeating these steps in a processive fashion. These steps also take place on the cytoplasmic surface of the membrane. Ultimately, the full-length polysaccharide is ligated to the LPS core by the RfaL ligase and is translocated to the cell surface by an as yet unknown ABC transport system. This is similar to the mechanism proposed for the synthesis of group 2 capsular polysaccharides (Whitfield and Roberts, 1999).

The order in which these steps take place and the exact site of ligation to the LPS core are not known. When the *K. pneumoniae* O1 *rfb* cluster is expressed in an *E. coli* K-12 background, the presence of *rfbX* in the host *rfb* cluster allows a side reaction to take place in which the Gal disaccharide linked to bactoprenol is captured by the *rfbX* product and is sequestered, or translocated across the cytoplasmic membrane. This allows the Gal disaccharide to be ligated to the LPS core. Although this does not prove that the role of RfbX protein is to translocate bactoprenol-linked molecules across the cytoplasmic membrane, it is consistent with that hypothesis.

Structure of Gene Clusters for *rfc*-dependent O Antigens

More than 20 gene clusters that encode enzymes for the synthesis of *rfc*-dependent O antigens have been sequenced, and common features are beginning to emerge. Some of these are shown in **Fig. 10**. The gene clusters are arranged in single, long operons and, in the cases examined, include a JUMPstart sequence upstream (Wang *et al.*, 1998b; Marolda and Valvano, 1998). This is distinct from the shorter *rfc*-independent gene clusters that have so far as been reported, and they do not appear to be antiterminated by the SfrB system.

In all five of the gene clusters shown in **Fig. 10**, most of the genes appear to be involved in the synthesis O-specific sugars, including galactofuranose, mannose, the methylpentoses rhamnose and fucose, the 6-deoxyamino sugar perosamine and the dideoxy sugars abequose, colitose and *N*-acetylviosamine. The next most abundant group of enzymes are the sugar transferases which assemble the sugars on bactoprenol. Only the *S. enterica* sv. Typhimurium *rfb* cluster contains the transferase for the first sugar added to bactoprenol, the *rfbP* gene which encodes a Gal-bactoprenol transferase. The four *E. coli* strains all use *rfe*, which encodes a GlcNAc-bactoprenol transferase to initiate synthesis of the O-unit.

In four of the five strains, a polymerase of the *rfc* (*wzy*) family is encoded within the *rfb* cluster, *S. enterica* sv. Typhimurium being the exception. The position of the *rfc* gene is not conserved, and the protein sequences of the *rfc* genes do not share strong homology. All five strains encode an *rfbX* (*wzx*) translocase (Liu *et al.*, 1996), and as in the case of the *rfc* genes, the various members of the *rfbX* family do not share strong homology and the position of the gene is not conserved.

Rhamnose is present in the O antigens of three of the five organisms (*E. coli* O7 and *E. coli* K-12, and *S. enterica* sv. Typhimurium), and in all of these a block of four highly conserved genes (*rfbBDAC*) is present at the 5′ end of the gene cluster. Four of the *rfb* clusters include homologues of the *manCB* genes which are involved in the synthesis of mannose and the other O-specific sugars derived from intermediates of the Man pathway. In three of these four clusters, the *manCB* block is located at the 3′ end of the *rfb* cluster. This suggests that these clusters arose by an evolutionary process involving an exchange of blocks of genes. These are interspersed by the more rapidly evolving *rfc* and *rfbX* genes, which encode very hydrophobic proteins in which amino acid sequence is less strongly conserved.

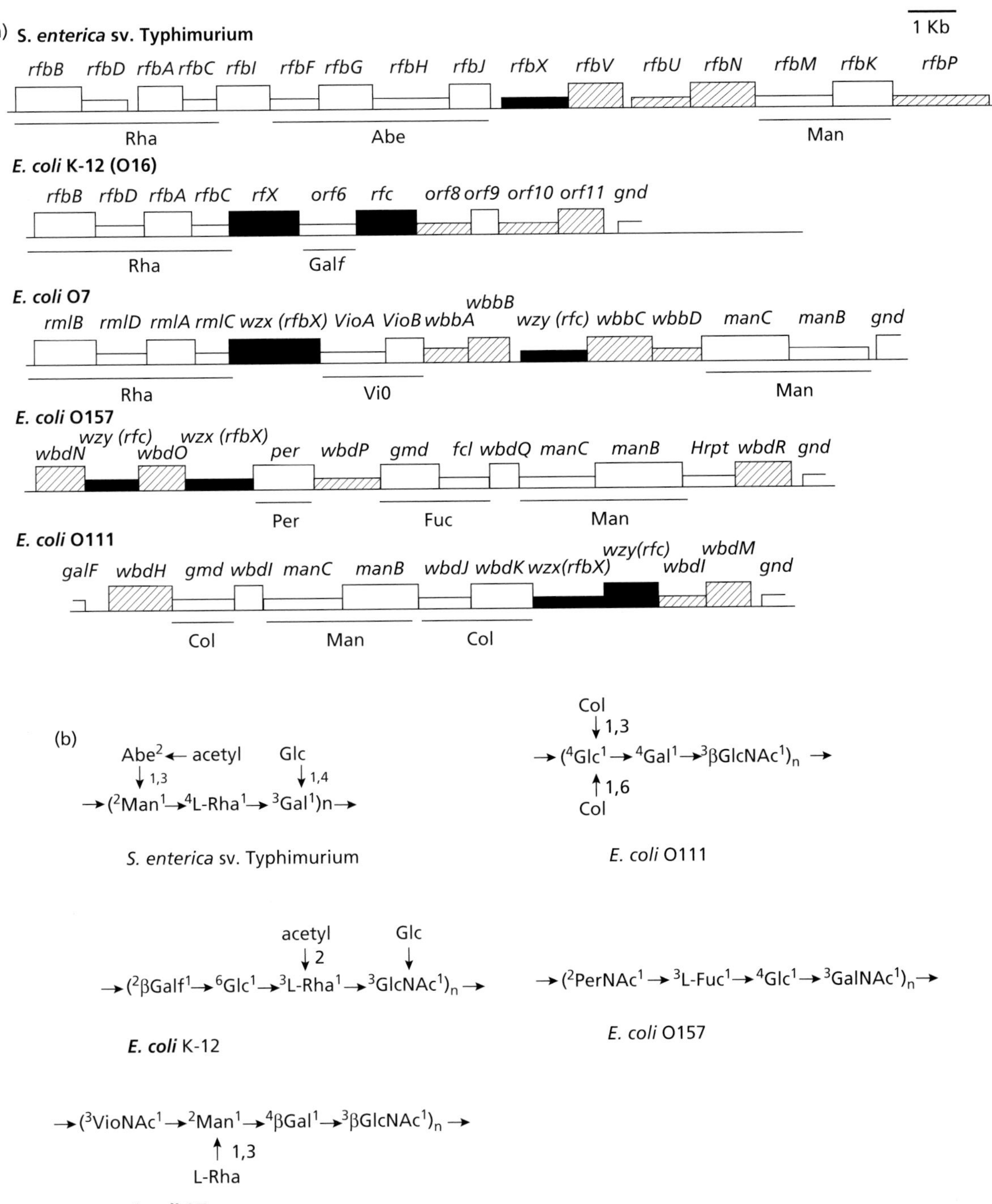

Fig. 10 Structure of the *rfb* clusters and O-units of five representative Rfc-dependent (heteropoymeric) O antigens. (a) The *rfb* clusters. The dark blocks indicate the *rfbX* and *rfc* genes, which encode very hydrophobic transmembrane proteins. The diagonal-shaded blocks indicate genes thought to encode the sugar-bactoprenol transferases which are used to assemble the O-units. Hrpt indicates a partial open reading frame derived from the Hrpt IS element. Genes or gene clusters involved in the biosynthesis of O-unit sugars are underlined and the sugars are indicated by the following abbreviations: Rha, rhamnose; Abe, abequose; Man, mannose; Galf, galactofuranose; Vio, *N*-acetylviosamine; Per, *N*-acetylperosamine; Fuc, L-fucose; Col, colitose. (b) The structure of the O-units. Unless otherwise indicated, all sugars are in the pyranose form and are the D-isomer, and all anomeric linkages are α. References are as follows: *S. enterica* sv. Typhimurium (Jiang *et al.*, 1991; Reeves, 1993), *E. coli* K-12 (Stevenson *et al.*, 1994), *E. coli* O7 (Marolda *et al.*, 1999), *E. coli* O111 (Bastin and Reeves, 1995; Wang *et al.*, 1998a), and *E. coli* O157 (Wang and Reeves, 1998).

The very strong homology among the rhamnose, dideoxyhexose and mannose blocks of the five *rfc*-dependent *rfb* clusters of *S. enterica* compared by Reeves (1993) also roughly matched the alignment of regions of different G+C content observed in the sequence. This suggests that these clusters evolved over a long time by the exchange of blocks of genes contributed by ancestors which had a different G+C content from that of enteric bacteria.

As was described in a previous section on the evolution of the type R2 *rfa* genes, there is strong evidence that the *rfb* gene cluster of group D2 of *S. enterica* arose by lateral gene exchange and recombination between the *rfb* clusters of groups E and D1 (Xiang *et al.*, 1994). The structure of the 5′ end of theD2 *rfb* cluster is identical to that of the 5′ end of the D1 cluster to a point just to the right of gene *rfbV* (Fig. 11). Beginning with the *rfc* gene, the remainder of the sequence extending to the 3′ end of the D2 *rfb* has an identical structure to that of the 3′ end of the E1 cluster. The similarity of structure of the ends of D2 to the corresponding ends of D1 and E1 is not limited to an identical gene order, but is also seen in nearly identical restriction sites and very strong conservation of DNA sequence.

A single open reading frame, which does not appear to be an *rfb* sequence, is located in the D2 sequence between the regions of similarity to D1 and E1. It is derived from the transposable element H-rpt. Xiang *et al.* (1994) present a plausible mechanism by which the hybrid D2 *rfb* cluster could have arisen by a double recombinational event during transposition of the H-rpt element from a plasmid carrying the E1 *rfb* into the chromosomal *rfb* cluster of a group D1 host. It is interesting that a similar H-rpt is found in the *rfb* cluster of *E. coli* O157 (see Fig. 10).

Given the importance of *E. coli* K-12 as a genetic strain, a note should be added about the successful reconstruction of its *rfb* gene cluster. *E. coli* K-12, which was isolated in 1922 in a hospital near Stanford University, was used there for many years as a teaching strain. Stocks were maintained on nutrient agar slants, which selects for rough mutations. The strain was first used for genetic experiments in the 1940s, and its use for this purpose was first published in 1944, when it was already a rough strain. In order to deduce its original *rfb* structure, the *rfb* region from several early strains was cloned and analysed. The majority of these, including strains W3110 (the Kohara phage strain), AB311 and EMG2, had an IS*5* insertion (termed the *rfb-50* mutation) in a gene adjacent to *gnd* at the 3′ end of *rfb*. In screening through a number of early strains with a probe derived from IS*5*, it was found that one strain, WG1, had a significantly different restriction pattern. An early Wisconsin subclone of K-12, WG1, lacked the *rfb-50* mutation. Instead, it carried a significant deletion (termed *rfb-51*) at the 5′ end of the cluster, which removed much of the *rfbB* gene. Since the two mutations were well separated, it was possible to combine DNA from W3110 and WG1 to reconstruct the wild-type K-12 *rfb* region, as shown in Fig. 10 (Liu and Reeves, 1994; Stevenson *et al.*, 1994). Analysis of the LPS product and comparison with *Sh. flexneri* serotypes indicates that it is serotype O16 (Yao and Valvano, 1994).

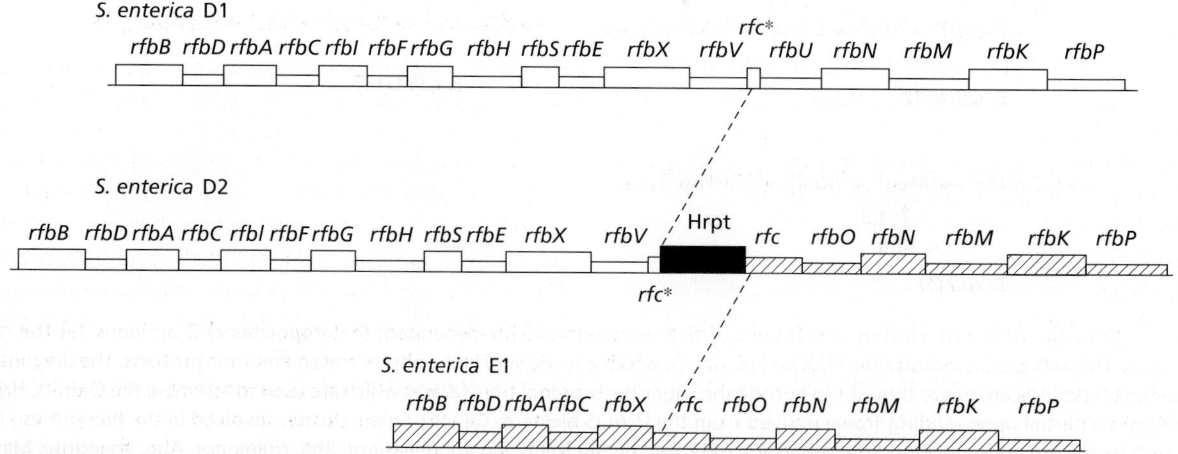

Fig. 11 The origin of the *rfb* cluster of *S. enterica* group D2 by IS-mediated recombination between the *rfb* clusters of groups E1 and D1. The structures of the *rfb* clusters of the three serogroups are shown, and the probable sites of recombination during transposition of the Hrpt IS element are shown by dashed lines. Sequence derived from the Hrpt element is shown in the black box, the diagonally-shaded boxes indicate sequences derived from *S. enterica* E1, and the unshaded boxes indicate sequence derived from *S. enterica* D1. These assignments are based on DNA and derived protein sequences and on conservation of restriction sites. The designation *rfc*∗ indicates a partial Rfc open reading frame. Re-drawn from Xiang *et al.* (1994).

Regulation of O Antigen and Capsular Chain Length

A capsule can be thought of as an athlete's jersey, designed to tear away when tugged upon. But for O antigens, which are essential structures necessary for the integrity of the outer membrane, a tug-of-war is always going on. At one end is the lipid A, with its acyl groups firmly embedded in the outer lipid leaflet of the outer membrane and its highly charged polar region, and a core oligosaccharide beginning with a series of heptose residues which are bound to proteins or cross-linked by divalent cations. At the other end is a long polysaccharide, which binds a great deal of water and exerts a formidable tug away from the cell surface. The delicate balance between these is best illustrated by the lethal phenotype of *tolC* knockout mutations in smooth strains, where a weakening of the inter-molecular bonds between the heptose residues causes the cell to begin to shed its longer LPS molecules (C.A. Schnaitman, unpublished). Thus, it would be deleterious for the O-chains to be infinitely long. At the same time, it is deleterious for the O-chains to be too short, or for their length to be unregulated. Long O-chains are thought to provide complement resistance, presumably by preventing the access of LPS-activated complement components to the outer membrane surface (Goldman and Leive, 1980; Joiner, 1985), and they are essential determinants of virulence (Van Den Bosch *et al.*, 1997).

Since Rfc-dependent O antigens are polymerised by the sequential addition of pre-formed multi-sugar units, LPS molecules carrying these antigens always show a striking, ladder-like array of bands when analysed on SDS gels. A good example of this is shown in Fig. 1 of Whitfield *et al.* (1997). For reasons that are less clear, LPS that carries Rfc-independent O antigens also regularly presents a ladder-like array of bands, in spite of the continuous, processive mode of synthesis of these O antigens. This is thought to be because these molecules often contain regularly alternating anomeric configurations of their backbone sugars, which result from the alternating action of pairs or sets of *rfb*-encoded sugar transferases, and side-branches which are attached to the backbone at regular intervals by the products of genes lying outside of the *rfb* clusters (Whitfield *et al.*, 1997).

When the size distribution of the O antigen-containing bands in both Rfc-independent and Rfc-dependent ladders is examined, the distribution is not that predicted if polymerisation and export occurred at random. The size distribution of the products of random polymerisation and export can be predicted by mathematical modelling (Goldman and Hunt, 1990). This predicts a distribution in which the largest number of molecules would consist of a core with only a single repeat unit (C + 1), and there would be a continuous decrease in the number of molecules as the molecules became larger. In other words, the relationship between size and distribution would be as follows: $(C+1) > (C+2) > (C+3) > (C+4)$ etc. Instead, what is seen with both Rfc-dependent and Rfc-independent polymers is a bimodal distribution. There is a small band of C + 1 molecules, sometimes called semi-rough molecules, followed by bands of larger species that diminish in negative exponential fashion as the molecules increase in size. Further up the gel is a second cluster of dense-staining bands, which represents the bulk of the smooth molecules. This second cluster appears to be organised around a small number of bands that exhibit an optimal size; this might be C + 14, C + 15 and C + 16 in one organism and C + 21, C + 22, C + 23, and C + 24 in another. The size of the median member of this cluster is sometimes referred to as the 'modal size'.

The genetics behind the determination of these size classes is complex. In the Rfc-independent O antigens and capsules it has not been possible to obtain mutants in which the size distribution phenotype has been dissociated from the polymerisation or export of the polysaccharide. It is thought that the regulation of the polysaccharide size may be coupled to the ABC export systems which these molecules employ, but so far it has not been possible to demonstrate this experimentally (Whitfield *et al.*, 1997). In the case of the synthase-dependent O54 antigen of *S. enterica*, chain length does not appear to be regulated, since the size distribution follows the random model.

The situation is different with respect to the Rfc-dependent O antigens and related Rfc-dependent K antigens. A gene lying just outside of the *rfb* clusters for these antigens regulates the modal length of the larger molecules seen in the O-ladders. This gene was first discovered when cloned *rfb* genes were used to complement a deletion extending beyond the *rfb* region. This gene was localised to the region between *rfb* and *his*, and it has been termed *rol* for regulator of length (Batchelor *et al.*, 1991). The null *rol* phenotype is the loss of the bimodal distribution of O antigen size classes, and the production of a unimodal distribution of O antigen size classes, as predicted by the random distribution model of Goldman and Hunt (1995) and observed in *E. coli* and *S. enterica* (Batchelor *et al.*, 1992). This is also true for the K40 capsular antigen (Dodgson *et al.*, 1996), which is a member of the group 4 capsular antigens that use *rfb* genes for their synthesis (Whitfield and Roberts, 1999). Rol also regulates chain length in *Sh. flexneri*, and this, in turn, is required for virulence (Van Den Bosch *et al.*, 1997).

The *rol* gene (also known as *cld*, for chain length distribution, or more recently as *wzz*) plays a direct role in the determination of the chain length. This is most clearly shown by a series of experiments in which *rol* genes from several different serotypes were cloned and introduced into a common *E. coli* O111 strain background. Each different *rol* gene produced a different modal value of O-length, which resembled that of the serotype from which it had been cloned. Moreover, amino acid substitutions introduced into Rol proteins by mutagenesis *in vitro*, resulted in different modal values (Franco *et al.*, 1998). The allele specificity demonstrated in each of these sets of experiments argues for a direct role of the Rol protein, but does not shed any light on a potential mechanism.

The Rol protein is a transmembrane protein which spans the cytoplasmic membrane twice and has a large domain on the periplasmic side. Rol analogues have been found in a number of enteric species, two species of *Yersinia*, and *Pseudomonas aeruginosa* (Whitfield *et al.*, 1997). A *rol* analogue has been described in the *rfe-rff* (*wec*) gene cluster, where it determines the length of the enterobacterial common antigen (Barr *et al.*, 1999). Otherwise, *rol* genes are not generally associated with regulation of capsule length, with the general exception of Rfc-dependent capsules attached to the LPS core. Proteins that share limited homology with Rol have been reported in polysaccharide gene clusters of a wide variety of species, but these have not been shown to have Rol activity, and in some cases they have other functions. Models for Rol mechanisms have been proposed (Bastin *et al.*, 1993; Morona *et al.*, 1995) but have not yet received experimental support. Perhaps the biggest problem with Rol is that it may well function by interacting with the *rfaL* (ligase) and *rfc* (polymerase) proteins. The Rol, RfaL and Rfc proteins are transmembrane proteins whose general functions are well understood, but whose exact functions, substrates and mechanisms remain completely obscure. As in the case of so many other systems involving polysaccharides at the juncture of synthesis, assembly and export, the facts will be known only when the reactions have been re-constituted *in vitro* with defined components.

Synthesis and Assembly of *rfc*-dependent O Antigens: Bringing it All Together

As is true throughout the LPS field of study, the best or earliest studied systems are often the least well understood. The reason is that these systems were studied before such innovations as polyacrylamide gel electrophoresis, NMR spectroscopy, and DNA cloning and sequencing could be brought to bear on them. This is especially true of the reactions involved in the assembly of Rfc (Wzy)-dependent O antigens synthesised on bactoprenol, and their transfer to the LPS core.

There are five main actors in the Rfc-dependent smooth LPS pathway. The first is the MsbA protein, an ABC family transport system which plays an essential role in translocating lipid A and its attached core polysaccharide out of the cytoplasm (Zhou *et al.*, 1998). The second is the RfaL (WaaL) protein, which plays a critical role in ligating complete and incomplete O antigens to the core (Schnaitman and Klena, 1993). The third is the RfbX (Wzx) protein, which is thought to play a role parallel to that of MsbA in translocating O-sugars attached to bactoprenol out of the cytoplasm (Liu *et al.*, 1996; Feldman *et al.*, 1999). The fourth is the Rfc (Wzy) protein, which is involved in assembling complete O-units carried on bactoprenol into long polymers, and possibly also in ligating these long O-polymers to single O-units attached to the LPS core (Schnaitman and Klena, 1993; Raetz, 1996). The fifth is the Rol protein, which was described above (Whitfield *et al.*, 1997). All of these proteins have membrane-spanning domains, and are thought to be anchored in the cytoplasmic membrane. The important consideration for each of these main actors is its specificity. These are summarised in Table 3.

The best way to illustrate the interations between these and other critical proteins is to enclose the steps in four distinct boxes, as shown in Fig. 12. Each of these boxes has a definite point of entry and a definite point of egress, and each is thought to have a defined location in the cell.

Box 1 encloses the lipid A pathway. The entry to this pathway is through LpxA, which transfers the first 3-hydroxymyristoyl group from acyl carrier protein to the 3 position of UDP-GlcNAc (Raetz, 1996). All of the reactions that occur in box 1 can be thought of as parts of an unusual membrane lipid pathway, with the polar ends of the β-hydroxyacyl moieties and the GlcNAc disaccharide functioning as the polar head group, analogous to that of a conventional glycolipid. All of the intermediates are lipid-soluble, and all are thought to be arranged on the cytoplasmic face of the cytoplasmic membrane. The point of egress from box 1 is the LpxK kinase, which adds the 4′ phosphate to the tetra-acylated disaccharide-1-P to create lipid IV_A (Garrett *et al.*, 1998).

Box 2 contains the core biosynthetic pathway. The entry to this pathway is through KdtA, which adds the two KDO residues to form KDO_2-lipid IV_A, the

Table 3 Specificity of hydrophobic membrane proteins

Protein/function	Minimum structure recognised	Structures not recognised
MsbA/ABC transporter	Tetraacetyldisaccharide-P with two KDO residues	Fifth and sixth acyl residues, Hep residues or distal core sugars, inner core P or PPEtN
RfaI/O-antigen ligase	Single sugar attached to bactoprenol,[a] or single O-unit attached to bactoprenol	Additional O-unit sugars or multiple O-units attached to bactoprenol[b]
RfbX/O-unit transporter	Same as RfaI, above	Additional O-unit sugars attached to bactoprenol
Rfc/O-unit polymerase	Complete O-units attached to bactoprenol, or single O-units attached to LPS core[b]	Glucosyl or acetyl branch substituents on O-units
Rol/regulator of ligation	Unknown	Rfc-independent O antigens

[a] This has been tested in *E. coli*, in which the first O-unit sugar is GlcNAc added by the *rfe* transferase, but not in *S. enterica*, in which the first O-unit is Gal added by the *rfbP* transferase.

[b] The substrates for Rfc and RfaL depend on which model for ligation is correct. In the traditional Robbins–Wright–Uchida model, ligase transfers a single O-unit to the terminal core hexose. In the two-step model, ligase transfers a single O-unit to the terminal core hexose, and polymerised O antigens are added to this single O-unit by the Rfc polymerase. These two ligation models are discussed by Schnaitman and Klena (1993).

smallest molecule that can function as LPS (Clementz and Raetz, 1991). This is also the minimal structure that can be recognised and translocated across the cytoplasmic membrane by the MsbA transport complex; it is assumed that MsbA, like other ABC transporters, functions as part of a membrane complex. All of the intermediate molecules in box 2 are bound to the inner face of the cytoplasmic membrane.

The early reactions that take place in box 2 are the addition of the two final acyloxyacyl moieties by the HtrB and MsbB acyltransferases (Clementz *et al.*, 1997), and the assembly and phosphorylation of the Hep inner core region. The core is completed by the backbone and branch hexosyltransferases, the branch Hep transferases and a final step of phosphorylation.

Given the remarkable lack of specificity of the MsbA transport system for the reducing ends of the LPS molecules it translocates, one can ask what kind of core completion step initiates the translocation. The best answer probably is that no one step initiates MsbA translocation, so long as the molecule has the basic structure of KDO_2-lipid IV_A at its core. A better way of

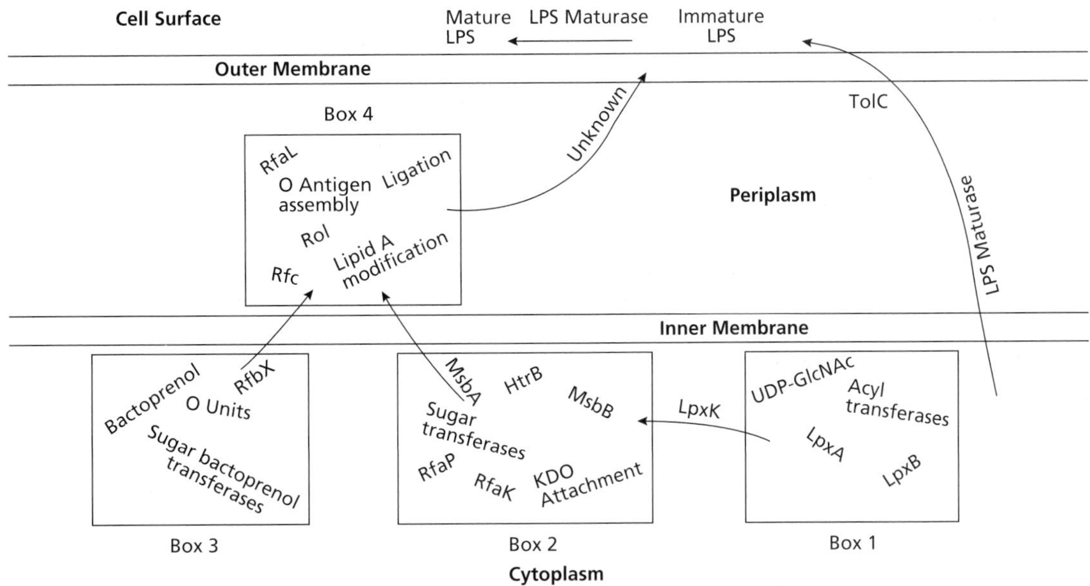

Fig. 12 A series of boxes, as described in the text, indicating the way in which the reactions of smooth LPS biosynthesis are compartmented and integrated. Typical reactions and gene products are shown in each box to indicate the kinds of processes found in each. The arrows show the flow of LPS intermediates through the pathway.

thinking about the translocation process is that the various transferases and kinases that assemble the core remain bound to the growing LPS molecule until they have completed their assembly steps. They thereby retard translocation until these assembly complexes have dissociated. The old 'ducks on a pond' model for sugar transferases, in which transferases floated around on the membrane as individual species until they encountered their incomplete LPS substrates, was popular in the 1970s but is now somewhat out of vogue (Schnaitman and Klena, 1993). In its place ternary or higher-order complexes are envisaged. For example, the complex for construction of the Hex II–Hex III region of the outer core may consist of a complex of RfaB, RfaI and RfaJ proteins together with the appropriate sugar nucleotides and a growing LPS molecule. Normally, this complex would not dissociate until all steps are complete, when it might pass the LPS molecule along to another complex that contains the RfaY kinase protein, ATP and a phosphoethanolamine transferase, and so on.

This sequence could be interrupted at any point by the loss of an essential component. For example, if RfaB protein were to be eliminated by an in-frame deletion of *rfaB*, the complex would not form and further modification of the LPS intermediate would come to a halt after the addition of Glc I. Without bound complexes to retard translocation, MsbA would translocate the truncated molecule across the cytoplasmic membrane and start it on its way to the outer membrane. This is exactly how the system appears to work when defined mutations are introduced at various steps in the core biosynthetic pathway. An accumulation of molecules at a stage of completion just before the stage that is blocked is seen in the outer membrane.

In the wild-type cell, the normal steps for the egress from box 2 involve the actions of the RfaK or RfaZ proteins, which complete the final steps in the smooth LPS and LOS pathways. RfaK marks the translocated molecules for box 4 and the ligation of an O antigen, while RfaZ marks the molecules of the LOS pathway to be sent along to the outer membrane without additional modification.

Box 3 represents the pathway for the synthesis of O-units. The entry into box 3 is through the Rfe (WecA) protein, or its analogue, the RfbP protein of *S. enterica*. These proteins are, respectively, GlcNAc- or Gal-bactoprenol transferases. In their respective systems, these enzymes transfer the first O-sugar from the appropriate sugar nucleotide to bactoprenol, resulting in what is thought to be a pyrophosphate linkage (Raetz, 1996). The remainder of the synthesis of the O-unit is similar to the biosynthesis of the LPS core. Individual sugar transferases (usually encoded by the *rfb* cluster) sequentially add each remaining backbone sugar, and in some cases branch sugars, to the reducing end of the previous sugar on the bactoprenol carrier. The reactions in box 3 are thought to occur on the cytoplasmic face of the cytoplasmic membrane.

At completion, translocation of the O-unit bound to bactoprenol out of the cytoplasm is thought to involve the function of the RfbX transporter. This appears to function in a manner analogous to that of MsbA, but RfbX protein is a member of the PST(2) subfamily of complex carbohydrate transporters (Paulsen *et al.*, 1997). The two transport systems also differ in that MsbA is thought to transport glycophospholipids as well as, or possibly along with LPS, while no such additional lipid translocation activity has been reported for RfbX.

Like MsbA, RfbX shows no apparent specificity for the length or state of completion of the O-unit. Several studies have now demonstrated the ability of cells to ligate one-, two- or three-sugar partial O antigens to the LPS core by an RfbX-dependent mechanism (Klena and Schnaitman, 1993; Clarke *et al.*, 1995; Feldman *et al.*, 1999). Exactly the same model can be proposed for RfbX as was described above for MsbA. So long as a complex containing sugar transferases and their substrates is bound to the non-reducing end of the growing O-unit, it is blocked from translocation. In the wild-type situation, egress of the bactoprenol-bound O-unit from box 3 is initiated via RfbX as soon as the last available sugar transferase has completed its work.

Box 4 contains the reactions thought to be located on the periplasmic face of the cytoplasmic membrane. Bactoprenol-bound O-units and LPS molecules with a completed core of appropriate composition to serve as an attachment site for O antigens enter the box 4 pathway by the RfbX and MsbA transport systems, respectively. Reactions in box 4 include the Rfc-dependent polymerisation of O-units into polymers of a length specified by Rol, the RfaL-dependant ligation of partial O-units, complete O-units and O-polymers to the LPS core, and several possible forms of modification of the completed O antigen, the core oligosaccharide, and lipid A.

O antigen modifications would include the phage-encoded GtrAB-dependent addition of branch Glc units, and some forms of acetylation (Allison and Verma, 2000). Core modifications may include the trimming of branch sugars added by the RfaK and RfaQ transferases, as noted above under core biosynthesis. By analogy to the lipid A modifications noted below, it is also possible that the generation of PPEtN by the addition of phosphoethanolamine residues to the inner core phosphate groups added by the RfaP and RfaY kinases takes place in the periplasmic compartment. Presumably, this would involve phosphatidylethanolamine as the donor of the phosphoethanolamine (Hasin and Kennedy, 1982).

Lipid A modifications would include the PmrAB-dependent addition of 4-aminoarabinose (Zhou *et al.*, 1999), and the addition of phosphoethanolamine to either the 1- or 4′-phosphates along with the PhoPQ-dependent addition of a seventh acyl group (Guo *et al.*, 1998; Gunn *et al.*, 1998). The proposal that these reactions occur on the periplasmic face of the membrane is supported by the evidence that at least one of the required gene products in the 4-aminoarabinose addition pathway is secreted into the periplasm.

Unfortunately, nothing is known about the temporal order of the various processes described above, since they are studied by themselves and not as a set of simultaneous events which share a common space. In many cases, the substrates are not known either. For example, it is not clear whether the *rfaL* ligase adds both single O-units and long polymers to the core, or whether it adds only single O-units and longer polymers are added to the core by the *rfc* polymerase in a manner analogous to the polymerisation of O-units linked to bactoprenol (Schnaitman and Klena, 1993). These questions become very important when attempts are made to re-constitute these processes *in vitro*, where success depends on choosing the correct substrates.

Egress from the box 4 pathways involves translocation of LPS to the outer membrane, a process about which little is known. Presumably, it requires the assembly of a multimeric complex in the outer membrane similar to that visualised in the translocation of a group 1 capsular polysaccharide (Drummelsmith and Whitfield, 2000). Unfortunately, the LPS outer membrane translocation system is likely to be an essential pathway, unlike that of capsular polysaccharides. One cannot always be fortunate enough to pick up essential pathways as multi-copy suppressors, as in the case of *msbA*. Perhaps the most rational strategy would be to create conditional mutations in genes which appear to share homology with other known complex carbohydrate transport functions.

Genetic Polymorphism of LPS as a Tool in Epidemiology and Public Health

The clonal nature of enteric diseases has frequently been shown. In other words, a pathogenic strain of *E. coli* represents a clone that has acquired a specific set of properties, and these allow it to colonise a particular niche in the host and cause a specific disease. These properties tend to be called pathogenic determinants. LPS is present in all Gram-negative bacteria, and while it has some features which may contribute to the disease process, it is seldom singled out as a pathogenic determinant. What it presents instead is a stable and highly polymorphic set of genetic markers which can be used for rapid and very specific identification of pathogenic clones, at least to the serotype level.

Serological analysis was used for this purpose in the past, but it is expensive, slow and difficult to apply to fast-moving diseases of the present day. It may take months to prepare and test a new batch of antiserum. Serum is difficult to store, and storage requires extensive refrigeration which is hard to provide in some less-developed countries. It has a finite shelf life, and stocks must be renewed periodically. The interpretation of serological analysis can be complicated, and requires highly trained laboratory personnel. It also requires a large collection of antisera. Thus, it is only suited for large, central laboratories. In the United States, this means Federal facilities such as the Centers for Disease Control and some of the larger State laboratories.

In contrast, DNA analysis, especially that employing the polymerase chain reaction (PCR), can be provided on a much broader scale at a much lower cost per specimen. The technique is simple, the equipment is inexpensive, it can be performed by laboratory staff with limited training, and there is no need for extensive refrigeration. The information for PCR primers can be stored in a computer file, and primers can be synthesised on demand in almost any well-equipped research centre and easily shipped to field laboratories. Most important, PCR analysis does not require pure cultures. It can be done on primary enrichment cultures or pooled field samples. Under optimum conditions, definitive results can be obtained within 24 hours. It is ideally suited for field analysis, and can be easily set up in remote areas.

Two recent studies (Wang and Reeves, 1998; Wang *et al.*, 1998a) have demonstrated the utility of this approach, by showing how the diversity of the *rfb* gene cluster allows the development of clone-specific PCR probes. The first organism studied was *E. coli* O111, which was found to be associated with a series of pathogenic clones, differentiated by flagellar serotype, isolated from outbreaks of infantile diarrhoea in the United States shortly after World War II. This was the first example in which an *E. coli* serotype was shown to be associated with a particular group of diseases. Sporadic outbreaks of disease associated with this serotype have continued up to the present, and the strains are now classified as enterohaemorrhagic (summarised in Wang *et al.*, 1998a).

The second study deals with *E. coli* O157, which was found in clones specifically associated with an enterohaemorrhagic disease. This is characterised by an exceptionally severe haemolytic-uraemic syndrome, and an unusually low infectious dose. The most prominent outbreak occurred in 1993 in the United States, and was traced to hamburger patties produced by a single, large meat-processing facility. Sporadic outbreaks have continued worldwide.

The structure of the O111 gene cluster is shown in **Fig. 10**. Three genes for specific sugar transferases (*wbdH*, *wbdL* and *wbdM*), the *rfc*(*wzy*) polymerase gene, and the *rfbX* (*wzx*) membrane translocase gene were chosen as potentially unique genes, and from these 13 pairs of PCR primers were constructed. These were tested against 28 pools of DNA, which together represented 166 O-serotypes of *E. coli*, 11 of *Yersinia pseudotuberculosis*, 12 of *Shigella boydii* and 13 of *Salmonella enterica*.

Of the 13 pairs of PCR primers, 12 showed no reaction with the 28 pooled DNA samples, but all reacted with DNA from three different isolates of *E. coli* O111 to produce the predicted fragment sizes. The one PCR primer with sequence from *wbdL* was non-specific, and reacted with the pooled DNA samples. This indicates that it is possible to produce PCR probes highly specific for the *E. coli* O111 *rfb* gene cluster, even though many of the genes present in this cluster are present in other strains.

In the O157 study, a general strategy for the rapid cloning of *rfb* clusters was devised. This consisted of using PCR to amplify long fragments extending from the JUMPstart sequence at the 5′ end of *rfb* to the *gnd* gene which flanks the 3′ end. The O157 *rfb* cluster was obtained on a 14-kb fragment. A set of 16 pairs of primers were produced with a strategy identical to that described above and they were assayed against a large bank of pooled DNA samples as above. Nine of the primer pairs were completely specific in that they showed no reactivity, while the remainder showed bands quite different in size from those obtained with O157 DNA.

All enterohaemorrhagic *E. coli* strains produce Shiga toxin (Stx), and pathogenic clones of serotype O157 are prominent among them. Since not all clones of serotype O157 are serious pathogens, Wang and Reeves (1998) suggest that the search for the desired strains could be narrowed by using a probe for the *stx* gene and an *rfb*-specific probe. I would like to suggest that other carbohydrate-specific genes and gene clusters could be used as targets for clone-specific PCR probes. The *rfa* clusters are not genetically linked to *rfb*, and show considerable polymorphism, as is also true of the *kps* clusters. Another possibility is to use sequences located on integrated prophages, such as those described by Allison and Verma (2000). Integrated prophages, or defective prophages, are very stable genetic markers, and exhibit a level of genetic polymorphism at least as great as that of O antigens (Hendrix *et al.*, 1999).

References

Allison GE, Verma NK (2000) Serotype-converting bacteriophages and O antigen modification in *Shigella flexneri*. *Trends Microbiol.* 8: 17–23.

Amor K, Heinrichs DE, Frirdich E, Ziebell K, Johnson RP, Whitfield C (2000) Distribution of core oligosaccharide types in lipopolysaccharides from *Escherichia coli*. *Infect. Immun.* 68: 1116–1124.

Appelmelk BJ, An Y-Q, Hekker TAM, Thijs LG, MacLaren D, de Graff J (1994) Frequencies of lipopolysaccharide core types in *Escherichia coli* strains from bacteraemic patients. *Microbiology* 140: 1119–1124.

Austin EA, Graves JF, Hite LA, Parker CT, Schnaitman CA (1990) Genetic analysis of lipopolysaccharide core biosynthesis by *Escherichia coli* K-12: insertion mutagenesis of the *rfa* locus. *J. Bacteriol.* 172: 5312–5325.

Barr K, Klena J, Rick PD (1999) The modality of enterobacterial common antigen polysaccharide chain lengths is regulated by o349 of the *wec* gene cluster of *Escherichia coli* K-12. *J. Bacteriol.* 181: 6564–6568.

Bastin DA, Reeves PR (1995) Sequence and analysis of the O antigen gene (*rfb*) cluster of *Escherichia coli O111*. *Gene* 154: 17–23.

Bastin DA, Stevenson G, Brown PK, Haase A, Reeves PR (1993) Repeat unit polysaccharides of bacteria: a model for polymerization resembling that of ribosomes and fatty acid synthetase, with a novel mechanism for determining chain length. *Mol. Microbiol.* 7: 725–734.

Batchelor RA, Haraguchi GE, Hull RA, Hull SI (1991) Regulation by a novel protein of the bimodal distribution of lipopolysaccharide in the outer membrane of *Escherichia coli*. *J. Bacteriol.* 173: 5699–5704.

Batchelor RA, Alifano P, Biffali E, Hull SI, Hull RA (1992) Nucleotide sequence of the genes regulating O polysaccharide antigen chain length (*rol*) from *Escherichia coli* and *Salmonella typhimurium*. Protein homology and functional compartmentation. *J. Bacteriol.* 174: 5228–5236.

Berlyn MKB (1998) Linkage map of *Escherichia coli* K-12, Edition 10: the traditional map. *Microbiol. Mol. Biol. Rev.* 62: 814–984.

Beutin L, Manning PA, Achtman M, Willits N (1981) *SfrA* and *sfrB* of *Escherichia coli* K-12 are transcription control factors. *J. Bacteriol.* 145: 840–844.

Brooke JS, Valvano MA (1996a) Biosynthesis of inner core lipopolysaccharide in enteric bacteria: identification and characterization of a conserved heptose isomerase. *J. Biol. Chem.* 271: 3608–3614.

Brooke JS, Valvano MA (1996b) Molecular cloning of the *Haemophilus influenzae gmhA* (*lpcA*) gene encoding a phosphoheptose isomerase required for lipooligosaccharide biosynthesis. *J. Bacteriol.* 178: 3339–3341.

Cheetham BF, Katz ME (1995) A role for bacteriophages in the evolution and transfer of bacterial virulence determinants. *Mol. Microbiol.* 18: 201–208.

Clarke BR, Bronner D, Keenleyside WJ, Severn WB, Richards JC, Whitfield C (1995) Role of Rfe and RfbF in the initiation of biosynthesis of D-galactan I, the lipopolysaccharide O antigen from *Klebsiella pneumoniae* Serotype O1. *J. Bacteriol.* 177: 5411–5418.

Clementz T, Raetz CRH (1991) A gene coding for 3-deoxy-D-*manno*-octulosonic-acid transferase in *Escherichia coli. J. Biol. Chem.* 266: 9687–9696.

Clementz T, Zhou Z, Raetz CRH (1997) Function of the *Escherichia coli msbB* gene, a multicopy suppressor of *htrB* knockouts, in the acylation of lipid A. *J. Biol. Chem.* 272: 10353–10360.

Curd H, Liu D, Reeves P (1998) Relationships among the O antigen gene clusters of *Salmonella enterica* groups B, D1, D2, and D3. *J. Bacteriol.* 180: 1002–1007.

de Cock H, van Blokland S, Tommassen J (1996) *In vitro* insertion and assembly of outer membrane protein PhoE of *Escherichia coli* K-12 into the outer membrane. *J. Biol. Chem.* 271: 12885–12890.

Dodgson C, Amor P, Whitfield C (1996) Distribution of the *rol* gene encoding the regulator of lipopolysaccharide O-chain length in *Escherichia coli* and its influence on the expression of Group I capsular K antigens. *J. Bacteriol.* 178: 1895–1902.

Drummelsmith J, Whitfield C (2000) Translocation of group 1 capsular polysaccharide to the surface of *Escherichia coli* requires a multimeric complex in the outer membrane. *EMBO J.* 19: 57–66.

Fält IC, Schweda EKH, Weintraub A, Sturm S, Timmis KN, Lindberg AA (1993) Expression of the *Shigella dysenteriae* type-1 lipopolysaccharide repeating unit in *Escherichia coli* K-12/*Shigella dysenteriae* type-1 hybrids. *Eur. J. Biochem.* 213: 573–581.

Feldman MF, Marolda CL, Monteiro MA, Perry MB, Parodi AJ, Valvano MA (1999) The activity of a putative polyisoprenol-linked sugar translocase (Wzx) involved in *Escherichia coli* O antigen assembly is independent of the chemical structure of the repeat. *J. Biol. Chem.* 274: 35129–35138.

Franco AV, Liu D, Reeves PR (1998) The Wzz (Cld) protein in *Escherichia coli*: amino acid sequence variation determines O antigen chain length specificity. *J. Bacteriol.* 180: 2670–2675.

Garrett TA, Que NLS, Raetz CRH (1998) Accumulation of a lipid A precursor lacking the 4′-phosphate following inactivation of the *Escherichia coli lpxK* gene. *J. Biol. Chem.* 273: 12457–12465.

Geerlof A, Lewendon A, Shaw W (1999) Purification and characterization of phosphopantothene adenyltransferase from *Escherichia coli. J. Biol. Chem.* 274: 27105–27111.

Göhmann S, Manning PA, Alpert CA, Walker MJ, Timmis KN (1994) Lipopolysaccharide O antigen biosynthesis in *Shigella dysenteriae* serotype 1: analysis of the plasmid-carried *rfp* determinant. *Microb. Pathog.* 16: 53–64.

Goldman RC, Hunt F (1990) Mechanism of O antigen distribution in lipopolysaccharides. *J. Bacteriol.* 172: 5352–5359.

Goldman RC, Leive L (1980) Heterogeneity of antigenic side-chain length in lipopolysaccharides from *Escherichia coli* O111 and *Salmonella typhimurium. Eur. J. Biochem.* 107: 145–153.

Gunn JS, Lim KB, Krueger J *et al.* (1998) PmrA-PmrB-regulated genes necessary for 4-aminoarabinose lipid A modification and polymyxin resistance. *Mol. Microbiol.* 27: 1171–1182.

Guan S, Bastin DA, Verma NK (1999) Functional analysis of the O antigen glucosylation gene cluster of *Shigella flexneri* bacteriophage SfX. *Microbiology* 145: 1263–1273.

Guo L, Lim KB, Poduje CM *et al.* (1998) Lipid A acylation and bacterial resistance against vertebrate antimicrobial peptides. *Cell* 95: 189–198.

Hasin M, Kennedy EP (1982) Role of phosphatidylethanolamine in the biosynthesis of pyrophosphoethanomamine residues in lipopolysaccharide of *Escherichia coli. J. Biol. Chem.* 257: 12475–12477.

Heinrichs DE, Monteiro MA, Perry MB, Whitfield C (1998a) The assembly system for the lipopolysaccharide R2 core-type of *Escherichia coli* is a hybrid of those found in *Escherichia coli* K-12 and *Salmonella enterica. J. Biol. Chem.* 273: 8849–8859.

Heinrichs DE, Yethon JA, Amor PA, Whitfield C (1998b) The assembly system for the outer core portion of R1- and R4-type lipopolysaccharides of *Escherichia coli. J. Biol. Chem.* 273: 29497–29505.

Heinrichs DE, Yethon JA, Whitfield C (1998c) Molecular basis for structural diversity in the core regions of the lipopolysaccharides of *Escherichia coli* and *Salmonella enterica. Mol. Microbiol.* 30: 221–232.

Hendrix RW, Smith MC, Burns RN, Ford ME, Hatfull GF (1999) Evolutionary relationships among diverse bacteriophages and prophages: all the world's a phage. *Proc. Natl Acad. Sci. USA* 96: 2192–2197.

Hitchcock PJ, Brown TM (1983) Morphological heterogeneity among *Salmonella* lipopolysaccharide chemotypes in silver-stained polyacrylamide gels. *J. Bacteriol.* 154: 269–277.

Hobbs M, Reeves PR (1994) The JUMPstart sequence: a 39 bp element common to several polysaccharide clusters. *Mol. Microbiol.* 12: 855–856.

Holst O, Brade H (1992) Chemical structure of the core region of lipopolysaccharides. In: Morrison DC and Ryan JL (eds) *Bacterial Endotoxic Lipopolysaccharides*, Vol. 1. New York: Elsevier, pp. 134–170.

Jansson P-E, Lindberg AA, Lindberg B, Wollin R (1981) Structural studies on the hexose region of the core in lipopolysaccharides from enterobacteriaceae. *Eur. J. Biochem.* 115: 571–577.

Jiang X-M, Neal B, Santiago F, Lee SJ, Romana LK, Reeves PR (1991) Structure and sequence of the *rfb* (O antigen) gene cluster of *Salmonella* serovar typhimurium (strain LT2). *Mol. Microbiol.* 5: 695–713.

Joiner KA (1985) Studies on the mechanism of bacterial resistance to complement-mediated killing and the mechanism of action of bactericidal antibody. *Curr. Top. Microbiol. Immunol.* 121: 99–133.

Karaolis DKR, Somara S, Maneval DR Jr, Johnson JA, Kaper JB (1999) A bacteriophage encoding a pathogenicity island, a type IV pilus and a phage receptor in cholera bacteria. *Nature* 399: 375–379.

Kastowsky M, Gutberlet T, Bradaczek H (1992) Molecular modeling of the three-dimensional structure and conformational flexibility of bacterial lipopolysaccharide. *J. Bacteriol.* 174: 4798–4806.

Keenleyside WJ, Whitfield C (1995) Lateral transfer of *rfb* genes: a mobilizable ColE1-type plasmid carries the $rfb_{O:54}$ (O:54 antigen biosynthesis) gene cluster from *Salmonella enterica* serovar Borreze. *J. Bacteriol.* 177: 5247–5253.

Keenleyside WJ, Whitfield C (1996) A novel pathway for O-polysaccharide biosynthesis in *Salmonella enterica* serovar Borreze. *J. Biol. Chem.* 271: 28581–28592.

Klena JD, Schnaitman CA (1993) Function of the *rfb* gene cluster and the *rfe* gene in the synthesis of O antigen by *Shigella dysenteriae* 1. *Mol. Microbiol.* 9: 393–402.

Klena JD, Schnaitman CA (1994) Genes for TDP-rhamnose synthesis affect the pattern of lipopolysaccharide heterogeneity in *Escherichia coli* K-12. *J. Bacteriol.* 176: 4003–4010.

Klena JD, Ashford RS II, Schnaitman CA (1992a) Role of *Escherichia coli* K-12 genes and the *rfp* gene of *Shigella dysenteriae* 1 in generation of lipopolysaccharide core heterogeneity and attachment of O antigen. *J. Bacteriol.* 174: 7297–7307.

Klena JD, Pradel E, Schnaitman CA (1992b) Comparison of lipopolysaccharide biosynthesis genes *rfaK*, *rfaL*, *rfaY*, and *rfaZ* of *Escherichia coli* K-12 and *Salmonella typhimurium*. *J. Bacteriol.* 174: 4746–4752.

Klena JD, Pradel E, Schnaitman CA (1993) The *rfaS* gene, which is involved in production of a rough form of lipopolysaccharide core in *Escherichia coli* K-12, is not present in the *rfa* cluster of *Salmonella typhimurium* LT2. *J. Bacteriol.* 175: 1524–1527.

Kloser A, Laird M, Deng M, Misra R (1998) Modulations in lipid A and phospholipid biosynthesis pathways influence outer membrane protein assembly in *Escherichia coli* K-12. *Mol. Microbiol.* 27: 1003–1008.

Koronakis V, Li J, Koronakis E, Stauffer K (1997) Structure of TolC, the outer membrane component of the bacterial-type 1 efflux system, derived from two-dimensional crystals. *Mol. Microbiol.* 23: 617–626.

Lai V, Wang L, Reeves PR (1998) *Escherichia coli* clone Sonnei (*Shigella sonnei*) had a chromosomal O antigen gene cluster prior to gaining its current plasmid-born O antigen gene. *J. Bacteriol.* 180: 2983–2986.

Levy SB, Leive L (1968) An equilibrium between two fractions of lipopolysaccharide in *Escherichia coli*. *Proc. Natl Acad. Sci. USA* 61: 1435–1439.

Lindberg AA (1977) Bacterial surface carbohydrates and bacteriophage adsorption. In: Sutherland I (ed.) *Surface Carbohydrates of the Prokaryotic Cell.* New York: Academic Press, pp. 289–356.

Lindberg AA, Holme T (1969). Influence of O side chains on the attachment of the Felix O-1 bacteriophage to *Salmonella* bacteria. *J. Bacteriol.* 99: 513–519.

Liu D, Reeves PR (1994) *Escherichia coli* regains its O antigen. *Microbiology* 140: 49–57.

Liu D, Cole RA, Reeves PR (1996) An O antigen processing function for Wzx (RfbX): a promising candidate for O-unit flippase. *J. Bacteriol.* 178: 2102–2107.

Mäkelä PH, Stocker BAD (1984) Genetics of lipopolysaccharide. In: Rietschel ET (ed.) *Handbook of Endotoxin*, Vol. 1: *Chemistry of Endotoxin*. Amsterdam: Elsevier, pp. 59–137.

Marolda CL, Valvano MA (1998) Promoter region of the *Escherichia coli* O7-specific lipopolysaccharide gene cluster: structural and functional characterization of an upstream untranslated mRNA sequence. *J. Bacteriol.* 180: 3070–3079.

Marolda CL, Feldman MF, Valvano MA (1999). Genetic organization of the O7-specific lipopolysaccharide biosynthesis cluster of *Escherichia coli* VW187 (O7:K1). *Microbiology* 145: 2485–2495.

Miller SI, Kukral AM, Mekalanos JJ (1989). A two-component regulatory system *phoP phoQ* controls *Salmonella typhimurium* virulence. *Proc. Natl Acad. Sci. USA* 86: 5054–5058.

Misra R (1993). OmpF assembly mutants of *Escherichia coli* K-12: Isolation, characterization and suppressor analysis. *J. Bacteriol.* 175: 5049–5056.

Morona R, Van Den Bosch L, Manning PA (1995) Molecular, genetic, and topological characterization of O antigen chain length regulation in *Shigella flexneri*. *J. Bacteriol.* 177: 1059–1068.

Nieto JM, Bailey MJA, Hughes C, Koronakis V (1996) Suppression of transcription polarity in the *Escherichia coli* haemolysin operon by a short upstream element shared by polysaccharide and DNA transfer determinants. *Mol. Microbiol.* 19: 705–713.

Nummila K, Kilpelainen I, Zahringer U, Vaara M, Helander IM (1995) Lipopolysaccharides of polymyxin B-resistant mutants of *Escherichia coli* are extensively substituted by 2-aminoethyl pyrophosphate and contain aminoarabinose in lipid A. *Mol. Microbiol.* 16: 271–278.

Olsthoorn MMA, Petersen BO, Schlecht S, Haverkamp J, Bock K, Thomas-Oates JE, Holst O (1998) Identification of a novel core type in *Salmonella* lipopolysaccharide. *J. Biol. Chem.* 273: 3817–3829.

Parker CT, Kloser AW, Schnaitman CA, Stein M, Gottesman S, Gibson BW (1992) Role of *rfaG* and *rfaP* genes in determining the lipopolysaccharide core structure and cell surface properties of *Escherichia coli* K-12. *J. Bacteriol.* 174: 2525–2538.

Paulsen IT, Beness AM, Saier MH Jr (1997) Computer-based analysis of the protein constituents of transport systems catalysing export of complex carbohydrates in bacteria. *Microbiology* 143: 2685–2699.

Pradel E, Schnaitman CA (1991) Effect of *rfaH* (*sfrB*) and temperature on the expression of *rfa* genes of *Escherichia coli* K-12. *J. Bacteriol.* 173: 6428–6431.

Pradel E, Parker CT, Schnaitman CA (1992) Structures of the *rfaB*, *rfaI*, *rfaJ* and *rfaS* genes of *Escherichia coli* K-12 and their roles in assembly of the lipopolysaccharide core. *J. Bacteriol.* 174: 4736–4745.

Raetz CRH (1984) *Escherichia coli* mutants that allow elucidation of the precursors and biosynthesis of lipid A. In: Rietschel ET (ed.) *Handbook of Endotoxin*, Vol. 1: *Chemistry of Endotoxin*. Amsterdam: Elsevier, pp. 248–268.

Raetz CRH (1996) Bacterial lipopolysaccharides: a remarkable family of bioactive macroamphiphiles. In: Neidhardt FC (ed.) *Escherichia coli and Salmonella typhimurium: Cellular and Molecular Biology*, 2nd edn, Vol. 1. Washington, DC: American Society for Microbiology, pp. 1035–1063.

Reeves P (1993) Evolution of *Salmonella* O antigen variation involved interspecific gene transfer on a large scale. *Trends Genet.* 9: 17–22.

Reeves PR, Hobbs M, Valvano MA *et al.* (1996) Bacterial polysaccharide synthesis and gene nomenclature. *Trends Microbiol.* 4: 495–503.

Roland KL, Martin LE, Esther CR, Spitznagel JK (1993) Spontaneous *pmrA* mutants of *Salmonella typhimurium* LT2 define a new two-component regulatory system with a possible role in virulence. *J. Bacteriol.* 175: 4154–4164.

Sanderson KE, Hessel A, Liu S-L, Rudd KE (1996) The genetic map of *Salmonella typhimurium*, Edition VIII. In: Neidhardt FC (ed.) *Escherichia coli and Salmonella typhimurium: Cellular and Molecular Biology*, 2nd edn, Vol. 2. Washington, DC: American Society for Microbiology, pp. 1903–1999.

Severn WB, Kelly RF, Richards JC, Whitfield C (1996) Structure of the core oligosaccharide in the serotype O8 lipopolysaccharide from *Klebsiella pneumoniae*. *J. Bacteriol.* 178: 1731–1741.

Schnaitman CA, Austin EA (1990) Efficient incorporation of galactose into lipopolysaccharide by *Escherichia coli* K-12 strains with polar *galE* mutations. *J. Bacteriol.* 172: 1256–1287.

Schnaitman CA, Klena JD (1993) Genetics of lipopolysaccharide biosynthesis in enteric bacteria. *Microbiol. Rev.* 57: 655–682.

Sirisena DM, MacLachlan PR, Liu S-L, Hessel A, Sanderson KE (1994) Molecular analysis of the *rfaD* gene, for heptose synthesis, and the *rfaF* gene, for heptose transfer, in lipopolysaccharide synthesis in *Salmonella typhimurium*. *J. Bacteriol.* 176: 2379–2385.

Soncini FC, Groisman EA (1996). Two-component regulatory systems can interact to process multiple environmental signals. *J. Bacteriol.* 178: 6796–6801.

Stevenson G, Neal B, Liu D *et al.* (1994) Structure of the O antigen of *Escherichia coli* K-12 and the sequence of its *rfb* cluster. *J. Bacteriol.* 176: 4144–4156.

Stevenson G, Andrianopoulos K, Hobbs M, Reeves PR (1996) Organization of the *Escherichia coli* K-12 gene cluster responsible for production of the extracellular polysaccharide colanic acid. *J. Bacteriol.* 178: 4885–4893.

Süsskind M, Brade L, Brade H, Holst O (1998) Identification of a novel heptoglycan of $\alpha 1 \rightarrow 2$-linked D-*glycero*-D-*manno*-heptopyranose. *J. Biol. Chem.* 273: 7006–7017.

Thampapillai G, Lan R, Reeves PR (1994) Molecular evolution of the *gnd* locus of *Salmonella enterica*. *Mol. Biol. Evol.* 11: 813–824.

Valvano MA, Marolda CL, Bittner M, Glaskin-Clay M, Simon TL, Klena JD (2000) The *rfaE* gene from *Escherichia coli* encodes a bifunctional protein involved in biosynthesis of the lipopolysaccharide core precursor ADP-L-*glycero*-D-*manno*-heptose. *J. Bacteriol.* 182: 488–497.

Van Den Bosch L, Manning PA, Morona R (1997) Regulation of O-chain length is required for *Shigella flexneri* virulence. *Mol. Microbiol.* 23: 765–775.

Wandersman C, Délepelaire P (1990) TolC, an *Escherichia coli* outer membrane protein required for hemolysin secretion. *Proc. Natl Acad. Sci. USA* 87: 4776–4780.

Wang L, Reeves PR (1998) Organization of *Escherichia coli* O157 O antigen gene cluster and identification of its specific genes. *Infect. Immun.* 66: 3545–3551.

Wang L, Curd H, Qu W, Reeves PR (1998a) Sequencing of *Escherichia coli* O111 O antigen gene cluster and identification of O111-specific genes. *J. Clin. Microbiol.* 36: 3182–3187.

Wang L, Jensen S, Hallman R, Reeves PR (1998b) Expression of the O antigen gene cluster is regulated by RfaH through the JUMPstart sequence. *FEMS Microbiol. Lett.* 165: 201–206.

Whitfield C, Roberts IS (1999) Structure, assembly and regulation of expression of capsules in *Escherichia coli*. *Mol. Microbiol.* 31: 1307–1319.

Whitfield C, Amor PA, Köplin R (1997) Modulation of the surface architecture of Gram-negative bacteria by the action of surface polymer : lipid A-core ligase and by determinants of polymer chain length. *Mol. Microbiol.* 23: 629–638.

Xiang S-H, Hobbs M, Reeves PR (1994) Molecular analysis of the *rfb* gene cluster of a group D2 *Salmonella enterica* strain: evidence for its origin from an insertion sequence-mediated re-combination event between Group E and D1 strains. *J. Bacteriol.* 176: 4357–4365.

Yethon JA, Heinrichs DE, Monteiro MA, Perry MB, Whitfield C (1998) Involvement of *waaY*, *waaQ*, and *waaP* in the modification of *Escherichia coli* lipopolysaccharide and their role in the formation of a stable outer membrane. *J. Biol. Chem.* 273: 26310–26316.

Yao Z, Valvano MA (1994) Genetic analysis of the O-specific lipopolysaccharide biosynthesis region (*rfb*) of *Escherichia coli* K-12 W3110: identification of genes that confer group 6 specificity to *Shigella flexneri* serotypes Y and 4a. *J. Bacteriol.* 176: 4133–4143.

Zhou Z, Lin S, Cotter RJ, Raetz CRH (1999) Lipid A modifications characteristic of *Salmonella typhimurium* are induced by NH_4VO_3 in *Escherichia coli* K-12. *J. Biol. Chem.* 274: 18503–18514.

Zhou Z, White KA, Polissi A, Georgopoulos C (1998) Function of *Escherichia coli* MsbA, an essential ABC family transporter, in lipid A and phospholipid biosynthesis. *J. Biol. Chem.* 273: 12466–12475.

7

The Bacterial Cell: Peptidoglycan

Ron J. Doyle and Roman Dziarski

University of Louisville, Louisville, KY, USA
Indiana University, Gary, IN, USA

Most bacteria, except for *Mycoplasma*, *Ureaplasma* and *Acholeplasma*, possess an external skeleton composed of peptidoglycan. As the name implies, peptidoglycan consists of glycan chains to which short peptide stems are bound. When the stems are cross-linked, the peptidoglycan becomes an insoluble, mat-like organelle that gives the cell its characteristic shape. Numerous cell wall antibiotics, such as penicillin and its analogues, prevent peptidoglycan synthesis and assembly. The peptidoglycan is thus weakened, resulting in either cell lysis or cessation of growth.

Various names have been used for peptidoglycan, including murein, mucopeptide, glycopeptide and basal structure. The word 'murein' is widely used, as it reflects a substance containing muramic acid and peptide bonds. The name murein thus implies a muramic acid-containing derivative of 'protein'. At present, many European journals use murein in preference to peptidoglycan, whereas in the Americas and Asia the reverse is true.

In Gram-positive bacteria the peptidoglycan may comprise up to 50% of the dry weight of the cell. The wall layer of Gram-positive bacteria is 20–40 nm thick, and is usually exposed to the environment. Some Gram-positive bacteria may, however, possess protein surface arrays that overly the peptidoglycan. The wall of Gram-positive bacteria is thus capable of interaction with antibiotics, antibody, complement proteins and enzymes. In Gram-negative bacteria, the peptidoglycan-containing layer is typically covered by an outer membrane. This outer membrane limits access of high molecular weight materials, and usually compounds of $>600-700$ Da have difficulty diffusing through the outer membrane. The peptidoglycan layer of the Gram-negative cell comprises only 1–10% of the cellular dry weight and is just a few nanometres thick. In Gram-positive bacteria, the internal solute concentration is about one molar, whereas the Gram-negative cell may have one-fifth as much internal solute. In terms of physical pressure, the molar concentration is equivalent to >20 atmospheres.

A number of talented scientists have contributed to the knowledge of structure and function of bacterial peptidoglycans. Early pioneers were Rogers and Work (UK), Salton (UK, Australia and USA) and Shockman and Higgins (USA). Biosynthetic reactions were largely described by Gilvarg, Neuhaus, Park and Strominger (USA). The physical properties of peptidoglycan have been studied by Koch (Indiana). Höltje (Tubingen) has contributed contemporary knowledge

 doi:10.1006/bkmm.2001.0007

to peptidoglycan structure and assembly. Several books and reviews are available detailing all aspects of peptidoglycan (Rogers *et al.*, 1980; Ghuysen and Hakenbeck, 1994; Höltje, 1998; DePedro *et al.*, 1998).

Physical Properties of Peptidoglycan

Peptidoglycan is insoluble in non-hydrolysing solvents. It has been referred to as the material which gives the cell wall its rigidity. In fact, peptidoglycan is not rigid, and can expand and contract depending on the pH and ionic environment. The concept of rigid bacterial walls has been a textbook error for decades (Doyle and Marquis, 1994). Since peptidoglycan contains charged groups it is sensitive to changes in solvent (the composition will be discussed in the following section). Peptidoglycan can be purified by bursting bacterial suspensions, centrifuging and treating the insoluble materials with hot detergents such as sodium dodecyl sulphate (SDS). By several cycles of centrifugation and detergent treatment, a residue remains that consists of peptidoglycan and its attendant or accessory polymers. The purified wall residue

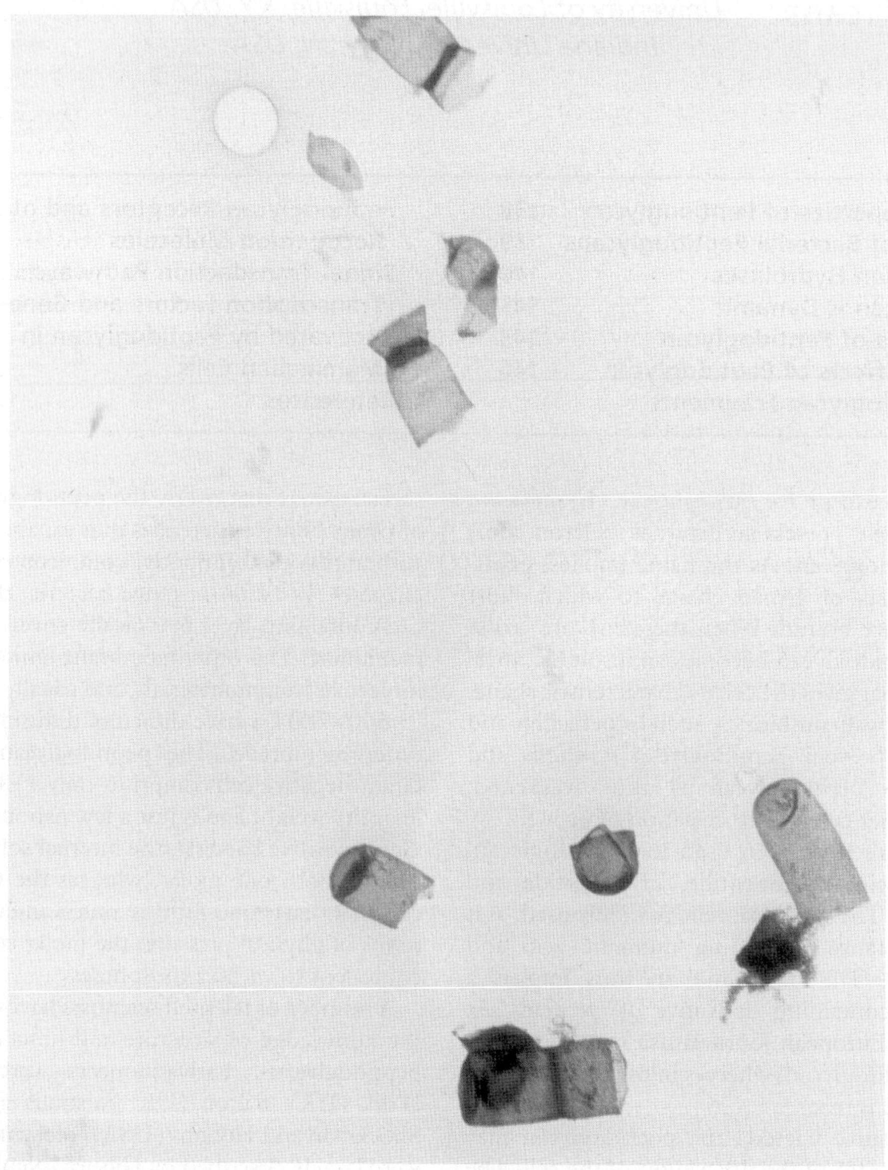

Fig. 1 Electron micrograph of cell walls of *Bacillus subtilis*. A suspension of the bacteria was broken and the insoluble walls isolated. The walls were boiled in detergent, washed and stained. Note that the walls retain the characteristic shapes of side walls, septa and poles. Peptidoglycan is the shape-determining matrix material of bacterial cells.

retains the characteristic shape of the intact cell (Fig. 1). The wall, once formed, is therefore resistant to changes in shape. Purified walls can be used to study absolute composition, cross-linking, peptide sequences and biological properties. Sometimes the peptidoglycan-containing wall structure is referred to as the sacculus. The porosity of peptidoglycan layers is such that proteins of <50 kDa are usually able to diffuse into the growth medium. Most secreted proteins of bacteria are 30–40 kDa molecular weight. On a gram-for-gram basis, peptidoglycan is stronger than steel.

Structures of Bacterial Peptidoglycans

Knowledge of peptidoglycan structure is now in its fifth decade. In the 1950s, researchers purified and acid-hydrolysed various peptidoglycans. The hydrolysates were resolved on the amino acid analyser and by paper chromatography. It soon became clear that most peptidoglycans consist of two amino sugars and just a few (3–7) amino acids. The amino sugars were identified as D-glucosamine and D-muramic acid. The latter is identical to glucosamine except that the muramic acid contains a lactyl group on the number 3 carbon of the sugar. In some cases, acetate was released from the amino sugars during hydrolysis, reflecting acetylation of the ammonium function. These peptidoglycans therefore contained *N*-acetylglucosamine and *N*-acetylmuramic acid. On acid hydrolysis, one mole of *N*-acetylglucosamine yields one mole of glucosamine, detectable on the amino acid analyser. If, however, the walls are first derivatised by fluorodinitrobenzene, any non-acetylated amino sugar would be converted to the *N*-dinitrobenzylate derivative, stable to acid hydrolysis and not detectable in the analyser. In this manner it was possible to determine that some peptidoglycans contained acetylated glucosamine and muramic acid. Similarly, amino acids released by acid hydrolysis could be identified and quantitated on the analyser. By hydrolysis, dinitrobenzylation, treatment with peptidases and carboxypeptidases, followed by paper chromatography and the amino acid analyser, it was possible to determine amino acid composition, peptide sequences, C-termini and whether any of the peptides contained amino acids with unsubstituted amino functions. By wet chemistry and limited instrumentation, sequences could be determined for peptidoglycans. Modern methodology utilises high-performance liquid chromatography (HPLC), mass spectroscopy (MS) and nuclear magnetic resonance (NMR) to characterise bacterial peptidoglycans (Rosenthal and Dziarski, 1994).

The glycan strand is composed of the glucosamine and muramic acid linked by β-1,4 glycosidic bonds. The short peptides ('stem' peptides) are covalently bound to the glycan by an amide linkage between an L-alanine (the most common amino acid, but not the only one) and the lactyl group of a muramic acid. A representative peptidoglycan structure is shown in Fig. 2.

In the glycan strand, which may be from 5–10 disaccharides to over 100 disaccharides in length, the most common modification is acetyl substituted on the ammonium function of the amino sugars. In *Escherichia coli*, *Bacillus subtilis* and various staphylococci, ammonium acetylation is common. In pyogenic streptococci, however, acetylation is uncommon. Members of the genera *Corynebacterium* and *Mycobacterium* may contain *N*-substituted glycolyl groups instead of acetyl groups. The primary hydroxyl of muramic acid may be substituted by acetyl, to create an *O*-ester. Such *O*-acetylations are common in *Staphylococcus aureus* and members of the genera *Proteus* and *Providencia*. In Gram-positive bacteria, it is also common for teichoic acids or teichuronic acids to be bonded covalently to the primary hydroxyl of muramic acid.

The peptide stems are subject to significant variations (Fig. 2). The amide-linked L-alanine may be replaced by a Gly or Ser. The Glu may be amidated on its free carboxylate (α-carboxylate). Gly or $GlyNH_2$ may also replace the Glu. The third amino acid is most often lysine or α,ϵ-diaminopimelic acid (Dap). The latter is a lysine residue with an additional carboxylate added to the ϵ-carbon. It thus has two carboxylates and two amino groups and as such occurs in L, L and L, D (*meso*) forms. Dap is most often found in Gram-negative peptidoglycans. Either Lys or Dap are common in Gram-positive bacteria. L-Ornithine may also be found as the third amino acid of some peptidoglycans. The fourth amino acid is always D-Ala. The common motif for stem peptides is L, D, L, D, and the stem peptides are resistant to animal proteases, probably an evolutionary advantage for bacteria.

The greatest variability in peptidoglycan structure originates in the cross-linking of the stem peptides. These may be directly cross-linked, as in *B. subtilis* and *E. coli*, or intervening amino acids may be present, as in the pentaglycine of *Staph. aureus* (Fig. 3). Penicillin prevents the cross-linking reaction, which involves amino acid 3 and the carboxylate of terminal D-Ala.

Since there are so many variations in peptidoglycan structure, a coherent classification scheme is necessary. The classification method of Schleifer and Kandler, (1972) has been adopted by most taxonomists and cell wall researchers. Table 1 shows a truncated version for

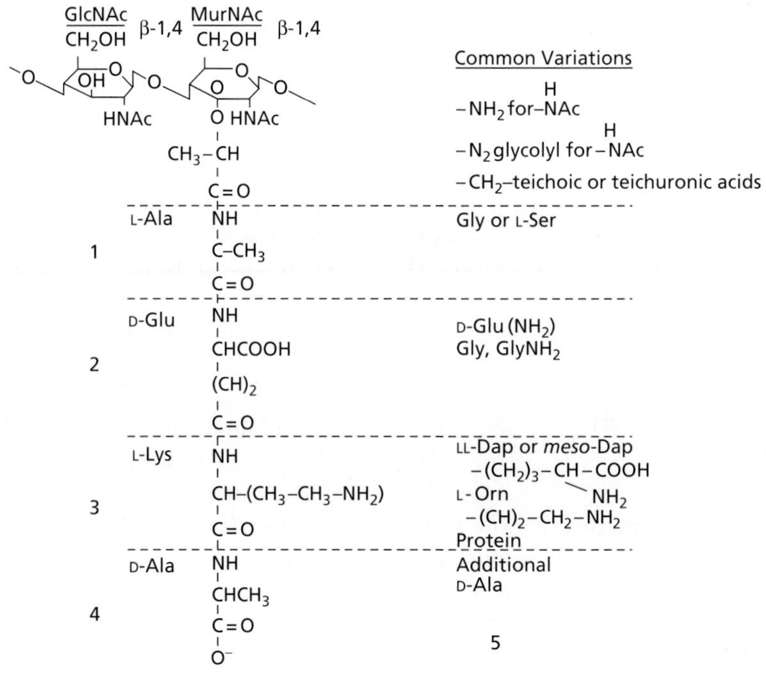

GlcNAc β-1,4 MurNAc β-1,4
CH_2OH CH_2OH
O OH O O O O O O
HNAc O HNAc
CH_3–CH
C=O
L-Ala NH
1 C–CH_3
C=O
D-Glu NH
2 CHCOOH
$(CH)_2$
C=O
L-Lys NH
3 CH–(CH_3–CH_3–NH_2)
C=O
D-Ala NH
4 $CHCH_3$
C=O
O^-

Common Variations
H
–NH_2 for–NAc
H
–N_2 glycolyl for–NAc
–CH_2–teichoic or teichuronic acids
Gly or L-Ser
D-Glu (NH_2)
Gly, GlyNH_2
LL-Dap or *meso*-Dap
–$(CH_2)_3$–CH–COOH
L-Orn NH_2
–$(CH)_2$–CH_2–NH_2
Protein
Additional
D-Ala
5

Fig. 2 A segment of a peptidoglycan. The numbers indicate positions of amino acids of the stem peptides. In the peptides, the most common variations are noted. Cross-linking amino acids must be multiply functional, containing at least two ammonium groups. When protein is found covalently bound to the peptidoglycan of *Enterobacteriaceae*, it is usually from a *meso*-Dap on position 3. Prior to cross-linking, a stem peptide contains five amino acids. Variations may occur in the glycan as well, particularly on the ammonium function and on the primary hydroxyl of Mur. *O*-Acetylation of the primary hydroxyl is common in several bacteria, both Gram-positive and Gram-negative. In Gram-positive bacteria, the 6-C hydroxyl may contain a pyrophosphate linkage to teichoic acid or teichuronic acid.

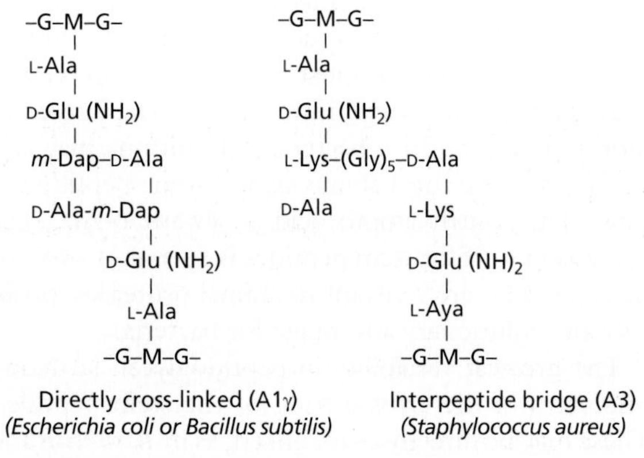

Fig. 3 Examples of direct cross-linking (classified as A1γ) and cross-linking involving a bridge (A3). G, *N*-acetylglucosamine; M, *N*-acetylmuramic acid.

peptidoglycan classification that allows the orderly placement of most pathogens. The scheme depends mainly on stem amino acid 3 and cross-linking. In group A, to which most pathogens belong, amino acid 3 of one stem is cross-linked to amino acid 4 of another stem. In subgroup A1 bridging amino acids are absent. Subgroup A1α requires that amino acid 3 is L-Lys. The variations α, β and γ depend on the presence of L-Lys, L-Orn and *meso*-Dap, respectively. In group B, the cross-linking occurs between amino acid 2 and amino acid 4. Regardless of the placement of a particular peptidoglycan in a group, subgroup or variant, cross-linking is required for the peptidoglycan to withstand cellular turgor pressure. Surprisingly, no classification considers the glycan portion of the structure. The Schleifer–Kandler classification does not take into account modifications that may be found in glucosamine and muramic acid.

Peptidoglycan Hydrolases

One of the first enzymes recognised to degrade bacterial walls was lysozyme, a muramidase capable of cleaving the glycosidic bond on the reducing side of muramic acid (Fig. 4). Lysozyme is one of the most

Table 1 A classification scheme for bacterial peptidoglycans of medical importance

Group A	Cross-linkage occurs between number 3 amino acid of one stem peptide with amino acid number 4 of another stem peptide.
Subgroup A1	Direct cross-link without intervening amino acids.
Variation A1α	L-Lys is number 3 amino acid
Variation A1β	L-Orn is number 3 amino acid
Variation A1γ	*m*-Dap is number 3 amino acid
Subgroup A2	Polymerised stem peptides serve to cross-link
Subgroup A3	Cross-linking amino acids are monocarboxylic L-amino acids or Gly
Variation A3α	L-Lys is number 3 amino acid
Variation A3β	L-Orn is number 3 amino acid
Variation A3γ	*m*-Dap is number 3 amino acid
Subgroup A4	Interpeptide bridge contains a dicarboxylic amino acid
Group B	Cross-linkage occurs between number 2 and number 4 amino acids
Subgroup B1	Interpeptide bridge is an L-diamino acid
Variation B1α	L-Lys is number 3 of stem peptide
Variation B1β	L-Homoserine is number 3 of stem peptide
Variation B1γ	L-Glu is number 3 of stem peptide

evolutionarily conserved enzymes known, and is found in human serum, in saliva, in tears, in plants and even in some bacteriophages. A requirement for the rapid action of lysozyme on peptidoglycan is that the amino sugars are *N*-acetylated and that the primary hydroxyl be unsubstituted. Most human oral bacteria are lysozyme resistant because their peptidoglycans contain free ammonium groups on the amino sugars, or in the case of Gram-negative cells, because the enzyme cannot penetrate the outer membrane. The lysozyme in urine may be a natural defence mechanism, which keeps down bacterial counts in the kidney and bladder. In catheterised patients, biofilms containing lysozyme-resistant bacteria, for example *Proteus vulgaris*, may develop, necessitating removal of the catheter. When lysozyme is added to suspensions of susceptible bacteria, the cells lyse because their peptidoglycans have been degraded. When such cells are subjected to lysozyme in the presence of an osmotic stabilising agent, spherical wall-free forms can be observed (**Fig. 5**). When the cells have been completely freed of wall materials, they are called protoplasts. These do not contain wall antigens, and they burst or lyse when diluted with water. Cells that retain vestiges of wall materials are also susceptible to lysis in hypotonic media but they possess wall antigens and are called spheroplasts. Protoplasts and spheroplasts are both spherical because of the lack of peptidoglycan, which determines bacterial shape. As a rule, protoplasts cannot revert to vegetative bacteria, suggesting the need for peptidoglycan in normal division processes. Spheroplasts, in contrast, may revert to normal vegetative cells if the lysozyme is removed. Spheroplasts may also form on treatment of bacterial suspensions

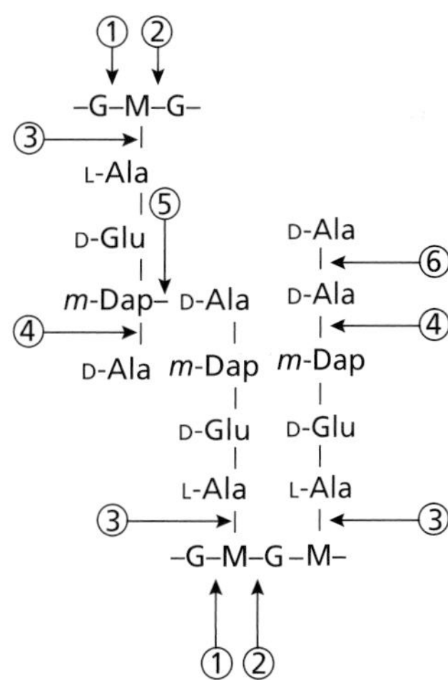

Fig. 4 Sites where peptidoglycan hydrolases may act. 1, *N*-Acetylglucosaminidase; 2, *N*-acetylmuramidase (lysozyme-like) and may be transglycosylase; 3, *N*-acetylmuramoyl-L-alanine amidase; 4, D,L-carboxypeptidase; 5, endopeptidase; 6, D,D-carboxypeptidase.

with cell wall antibiotics, such as penicillin or other β-lactams.

In some cases, the addition of an antibiotic to a bacterial suspension may lead to wall-defective forms. One type of wall-defective bacteria are called L-forms (after the Lister Institute, London). These are thought to occur in several bacteria, especially group A streptococci and *Borrelia burgdorferi* that causes Lyme disease, and are sometimes associated with the prescription of antibiotics. For example, a patient with a streptococcal sore throat may be given a 10-day regimen of antibiotic. After 4–5 days the patient may become symptom-free and stop taking the medication. If the patient suffers a relapse a few days later, the initial course of penicillin may have created L-forms, which revert to vegetative cells on the cessation of the penicillin challenge. L-Forms may or may not require osmotic stabilisation. In the case of L-forms induced in patients by penicillin, osmotic protection does not seem to be necessary. Several researchers have also attempted to ascribe a variety of neurological diseases, such as multiple sclerosis and Parkinson's disease, to *B. burgdorferi* L-forms. These tend to be hydrophobic and may more readily penetrate nerve tissue than wall-sufficient bacteria. Much more research is necessary before a relationship between *B. burgdorferi*, L-forms and neurological diseases can be confirmed.

Some bacteriolytic enzymes are called autolysins and cleave various sites on peptidoglycan (**Fig. 4**). Some of these enzymes hydrolyse glycosidic linkages in a manner similar to lysozyme. Others act on the stem peptides, whereas others may act on bridging peptides. During normal growth, autolysins are well-regulated. Perturbations in growth by antibiotics, loss of protonic potential or diminution in carbon source may result in autolysis. An important area of cell wall research is to develop means to de-regulate autolysins, so that pathogens may be eliminated by autolysis. This may be difficult to achieve, because the autolysins of various bacteria differ.

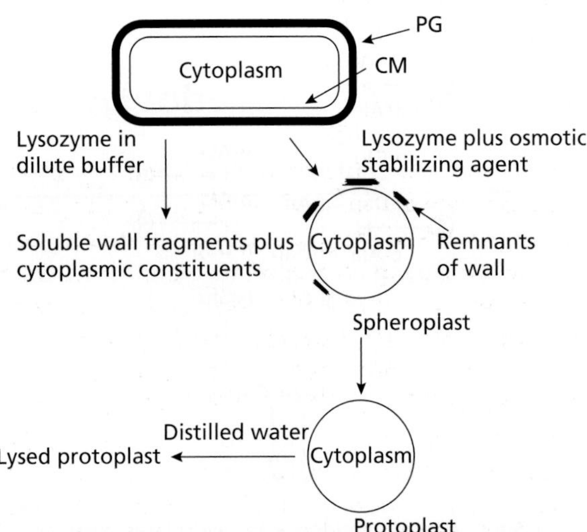

Fig. 5 Hydrolysis of peptidoglycan (PG) in the presence and absence of osmotic protection. CM, cytoplasmic membrane.

Several functions have been ascribed to autolysins, including bacterial competence, sporulation, cell wall turnover, creation of new growth sites to allow for surface expansion, production of bioactive peptidoglycan fragments, cell separation during division processes, regulation of peptidoglycan synthesis by truncation of peptide stem size and as scavenger enzymes capable of solubilising the cell walls of competing bacteria. The fact that mutants completely lacking autolysins have not been found, argues that these enzymes play an essential role during normal growth. Mutants with low autolysin activity tend to form chains of cells. In some cases, these mutants exhibit increased resistance to cell wall antibiotics.

Most bacteria have more than one autolysin. In *E. coli* at least seven enzymes capable of acting on peptidoglycan are known. One of these, lytic transglycosylase, hydrolyses the reducing side of muramic acid, but in doing so, forms a 1,6-anhydromuramic terminus. Transglycosylases capable of adding peptidoglycan precursors to pre-existing wall are not the same as the lytic transglycosylases.

Autolytic activities may be assayed in various ways, including quantitation of the reducing power on isolated walls or detection of free amino groups (**Table 2**). If soluble peptidoglycan fragments can be isolated and fractionated, HPLC and physical methods can be employed to determine their structures, and in this way the specificity of an autolysin can be determined. Since autolysin will clear the gel at the site of migration of the

Table 2 Major enzymes acting on bacterial peptidoglycans

Enzyme	Comments
N-Acetylmuramoyl-2-alanine amidase ('amidase')	Cleaves the amide linkage between the lactyl group of muramic acid and L-alanine. May be monitored by measuring release of L-Ala amino-termini
N-Acetylmuramidase (lysozyme)	Cleaves the β-1,4 glycosidic linkages on the reducing side of muramic acid. Assayed by appearance of reducing power
N-Acetylglucosaminidase	Similar to lysozyme, except the glycosidic linkage on the reducing side of glucosamine is broken
Endopeptidase	Hydrolyses peptide bridges involved in cross-linking stem peptides
D,D-Carboxypeptidase	Releases D-Ala from stem peptides as the peptides become cross-linked
D,L-Carboxypeptidase	Liberates D-Ala · D-Ala from newly-synthesised stem pentapeptide
Lytic transglycosylase	A muramidase, one product of which is 1,6-anhydromuramic acid

enzyme, a convenient way to assess the number of autolysins of a bacterium is electrophoresis of cell extracts in a gel that contains cell wall. This is referred to as a zymogram. The method assumes that, if multiple autolysins are present in the extract, they are each of unique molecular weight. The method further assumes that the autolysins are stable under the conditions of electrophoresis, which may involve the use of detergents and reducing agents.

Peptidoglycan is Dynamic

During growth, some bacterial autolysins excise segments of peptidoglycan, resulting in the solubilisation of muramic acid-containing peptides. In Gram-positive bacteria, such as in *B. subtilis* or *Staph. aureus*, the solubilised peptidoglycans are lost into the environment, or 'turned over' and the solubilised peptidoglycans are known as turnover product. Turnover also occurs in Gram-negative bacteria, such as the Enterobacteriaceae and the Neisseriaceae. Enterobacteriaceae turnover product cannot diffuse through the outer membrane and is returned to the cytoplasm via oligopeptide permeases, to be re-utilised. In both Gram-positive and Gram-negative rods, the turnover is about 50% per generation, and a growing body of literature suggests that cell wall turnover products may mediate inflammatory events. It is known that muramyl peptide, in combination with lipopolysaccharide, provokes a synergistic pyrogenic (fever) response in humans. For this reason, steroids are frequently administered simultaneously with cell wall antibiotics to patients in whom bacterial meningitis is suspected. The responses of mammals to muramyl peptides will be considered below. The purpose of peptidoglycan turnover is not clear. Presumably, turnover or loss of preformed wall makes it possible for new peptidoglycans to be added to the wall, but some bacteria, such as the streptococci, have very little, if any, turnover during a normal cell division.

In *E. coli* peptidoglycan there are no reducing MurNAc groups. All the MurNAc ends appear to be the 1,6-anhydro derivative, probably a product of well-regulated lytic transglycosylases. The predominant turnover products are 1,6-anhydroMurNAc-β-1,4-GlcNAc tetra- and tripeptides. A membrane permease, Amp G, takes the muramyl peptides into the cell. Once within the cell, they are further degraded by an amidase (Amp D) and a β-*N*-acetylglucosaminidase, which releases the disaccharide and peptides. Amp D acts only on anhydromuramyl peptides and has no activity on precursor-containing MurNAc. A periplasmic amidase (Ami A) may also act on the turnover products. When Ami A degrades the muropeptides, the released peptides are transported into the cell by a specific muropeptide transport protein (Mpp) and a general peptide permease (Opp). Growth of the bacteria in the presence of penicillin results in the depression of a β-lactamase (Amp C). A regulatory protein, Amp R, interacts with 1,6-anhydromuramyl peptides, resulting in the induction of the Amp C. By 'sensing' the cytoplasmic concentrations of anhydromuramyl peptides, the cell responds to the penicillin challenge by producing the β-lactamase.

Peptidoglycan turnover may also be responsible for the generation of outer membrane-containing vesicles in Gram-negative bacteria. Vesicles contain outer membrane components, periplasmic constituents and muramyl peptides (Zhou *et al.*, 1998). The turned over muramyl peptides may create turgor on the outer membrane, resulting in blebbing or vesicle formation. Since the vesicles contain both lipopolysaccharide and muramyl peptide, they may contribute to inflammatory reactions.

Biosynthesis of Peptidoglycan

One of the most complicated routes to macromolecular synthesis in the whole of biochemistry is the sequence that leads to peptidoglycan assembly. Bacteria have a strategy for peptidoglycan biosynthesis that utilises reactions in the cytoplasm, the cytoplasmic membrane and the pre-existing wall. Basically, the strategy is to synthesise precursors on a membrane-soluble carrier. The precursors, along with carrier, then diffuse through the cytoplasmic membrane. On reaching the external face of the outer leaflet of the membrane, the precursors are inserted into the wall. Continued additions of new wall to old wall result in bacterial surface expansion. Some antibiotics are capable of inhibiting specific steps in wall synthesis and assembly. Historically, antibiotics that prevent peptidoglycan synthesis have been the most effective and widely used. Indeed, all clinically useful antibiotics that act on the bacterial cell wall inhibit peptidoglycan synthesis. Many of the wall antibiotics are small enough to diffuse through the outer membrane to gain access to sensitive sites. Unfortunately, the widespread usage of antibiotics has selected for resistant bacteria capable of division in the presence of therapeutic concentrations of the antibiotics. The study of peptidoglycan biosynthesis is essential for the development of new kinds of antibiotics and the application of many currently employed antibiotics.

An overview of the synthesis of the peptidoglycan by *Staph. aureus* is given in **Figs 6** and **7**. Early steps

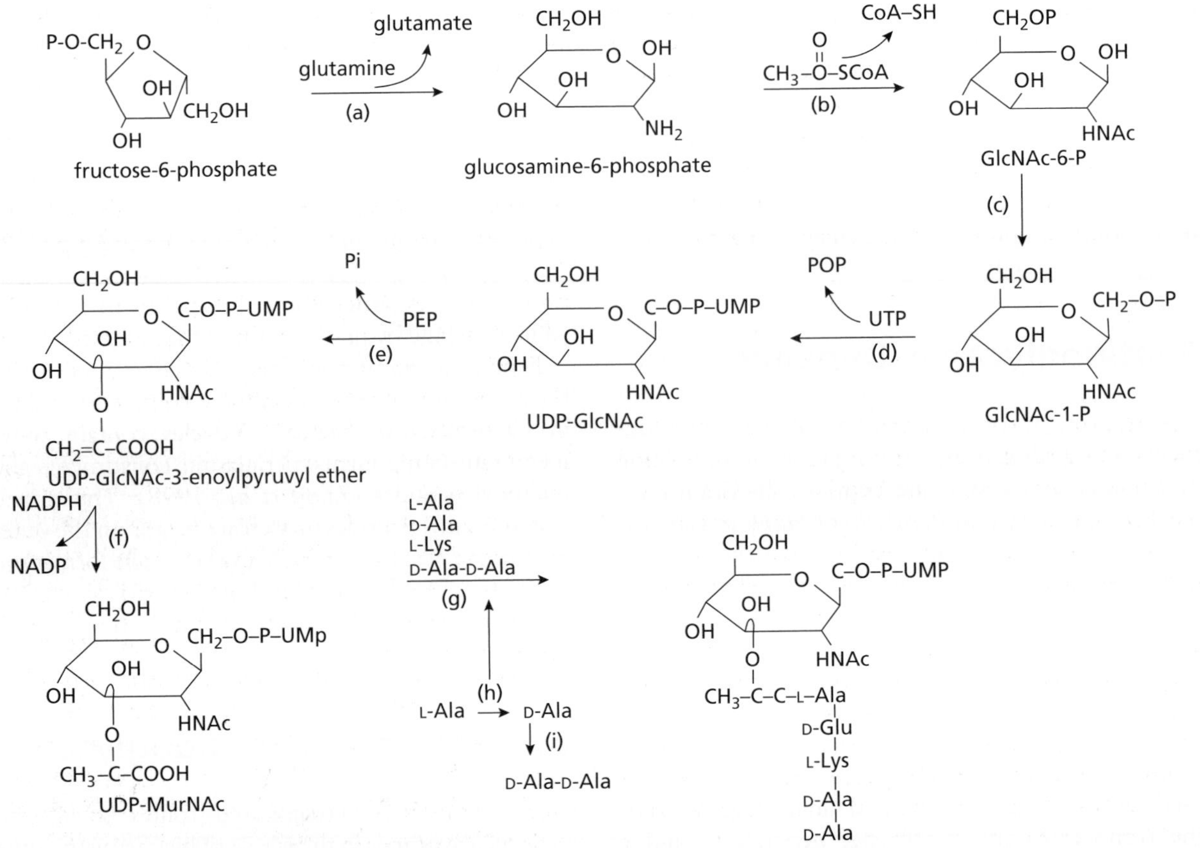

P-O-CH₂ O / OH / CH₂OH / OH — fructose-6-phosphate —(a) glutamine → glutamate→ glucosamine-6-phosphate (CH₂OH, O, OH, OH, OH, NH₂) —(b) CH_3–CO–SCoA → CoA–SH→ GlcNAc-6-P (CH₂OP, O, OH, OH, OH, HNAc)

(c) ↓

GlcNAc-1-P (CH₂OH, O, CH₂–O–P, OH, OH, HNAc) —(d) UTP → POP→ UDP-GlcNAc (CH₂OH, O, C–O–P–UMP, OH, OH, HNAc) —(e) PEP → Pi→ UDP-GlcNAc-3-enoylpyruvyl ether (CH₂OH, O, C–O–P–UMP, OH, OH, HNAc, O, CH₂=C–COOH)

(f) NADPH → NADP ↓

UDP-MurNAc (CH₂OH, O, CH₂–O–P–UMp, OH, OH, HNAc, O, CH₃–C–COOH)

—(g) L-Ala, D-Ala, L-Lys, D-Ala-D-Ala →

L-Ala —(h)→ D-Ala —(i)→ D-Ala-D-Ala

CH₂OH, O, C–O–P–UMP, OH, OH, HNAc, O, CH₃–C–C–L-Ala – D-Glu – L-Lys – D-Ala – D-Ala

Fig. 6 Biosynthesis of a unit peptidoglycan of *Staphylococcus aureus* in the cytoplasm. A stem pentapeptide is linked to MurNAc. Letters refer to enzymes involved in the biosynthetic pathway. a, Glucosamine: fructose-6-phosphate aminotransferase; b, glucosamine-6-phosphate acetylase; c, *N*-acetylglucosamine phosphomutase; d; UDP-*N*-acetylglucosamine pyrophorylase; e, enoylpyruvate transferase; f, UDP-*N*-acetylenolpyruvoylglucosamine reductase; g, represents a collection of enzymes which sequentially add each amino acid to the nascent stem, each addition of which requires an ATP and a specific enzyme; h, alanine racemase; i, D-Ala-D-Ala synthetase, an enzyme requiring an ATP.

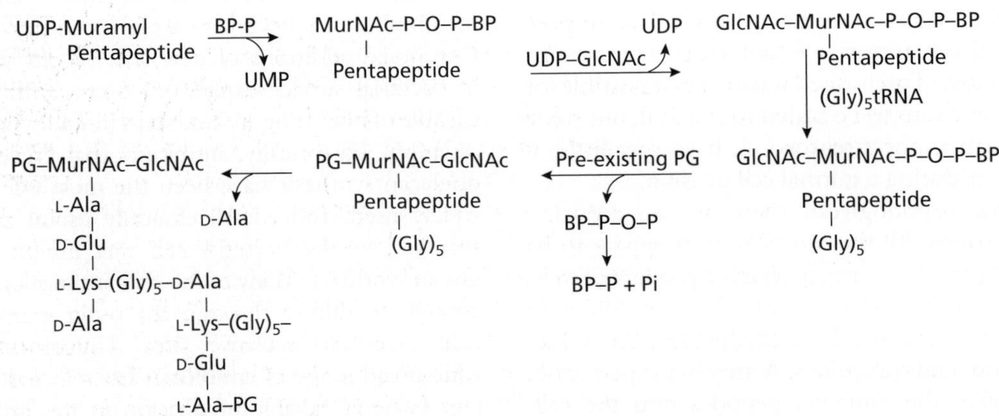

Fig. 7 Route to peptidoglycan from the cytoplasmic precursors. BP-P, bactoprenol phosphate, a membrane-soluble undecaprenol phosphate carrier; P_i, inorganic phosphate; PG, peptidoglycan. The MurNAc-pentapeptide is coupled with BP-P, thereby becoming membrane-soluble. A GlcNAc is transferred to the non-reducing side of MurNAc-pentapeptide, after which a $(Gly)_5$tRNA complex transfers $(Gly)_5$ to the ϵ-amino group of lysine. The growing PG segment is then attached to pre-existing PG at the reducing side of MurNAc, releasing BP-P-O-P which is converted to BP-P plus P_i by a phosphatase. The phosphatase is inhibited by bacitracin, a polypeptide antibiotic frequently used on superficial cuts and abrasions.

in the cytoplasm require the synthesis of GlcNAc, MurNAc and amino acids (Fig. 6). Both GlcNAc and MurNAc are synthesised from fructose-6-phosphate. It should be noted that peptidoglycan synthesis is costly to the cell in terms of energy, and requires high-energy intermediates at numerous steps. The costs to the bacterium, however, are minor when confronted with the alternatives of wall synthesis or cell death.

In the cytoplasm, the general strategy for all peptidoglycan-containing bacteria is to synthesise a disaccharide consisting of MurNAc and GlcNAc. Sequentially amino acids of the stem peptide are added to the MurNAc. The nascent stem peptide is synthesised as a pentapeptide containing a D-Ala · D-Ala terminus. Once the disaccharide–pentapeptide is formed, it is transferred to the membrane-soluble isoprenoid alcohol phosphate carrier, bactoprenol phosphate (BP) or undecaprenyl phosphate. Cross-linking amino acids are then added to the membrane-solubilised disaccharide–pentapeptide. The disaccharide–pentapeptide-containing cross-linking peptide, with its attendant bactoprenol phosphate, then diffuse through the cytoplasmic membrane, where they are added to pre-existing wall. The pre-existing wall is neither a primer nor a template. It simply serves as the site for the surface expansion that accompanies cell division. As new wall is added to old wall, there is no turgor on the newly inserted disaccharide–peptide units. On cross-linking, however, the newly added peptidoglycan assumes its primary function of protecting the protoplast from lysis. Several steps that lead to wall synthesis are sensitive to antibiotics and antimicrobial agents (Table 3). Phosphonomycin, a reactive analogue of phosphoenolpyruvate, has achieved some success as an antibiotic-like drug, especially against pneumococci. Cycloserine, a structural analogue of D-Ala, is a competitive inhibitor of any reaction involving D-Ala. At one time, cycloserine was useful against *Mycobacterium tuberculosis*, but most isolates are now resistant to it.

All bacteria, regardless of morphology, must possess a strategy for laying down new wall (Koch *et al.*, 1982). Is new wall added at only one site, such as a septum, or can wall be added at numerous sites? It is quite easy to distinguish old wall from new if one or other contains a radioactive or morphological marker. In streptococci, a new-born cell ('unit cell') contains a small wall band centred circumferentially on the wall. Cell division processes begin when peptidoglycan of the wall band is degraded by an autolysin. New wall is then added on both sides of the cleaved wall band, ultimately giving rise to two new unit cells, each of which has one-half the wall of the mother cell (Fig. 8). Old wall appears to be slightly thicker in streptococci than recently assembled wall. Although autolysin is essential for normal division processes, there is no turnover of peptidoglycan.

In rod-shaped bacteria, new wall can be added at numerous sites in the cell cylinder or side wall, resulting in cell elongation (Fig. 9). In the relatively thick Gram-positive wall, newly added peptidoglycan is found on the wall facing the cytoplasmic membrane.

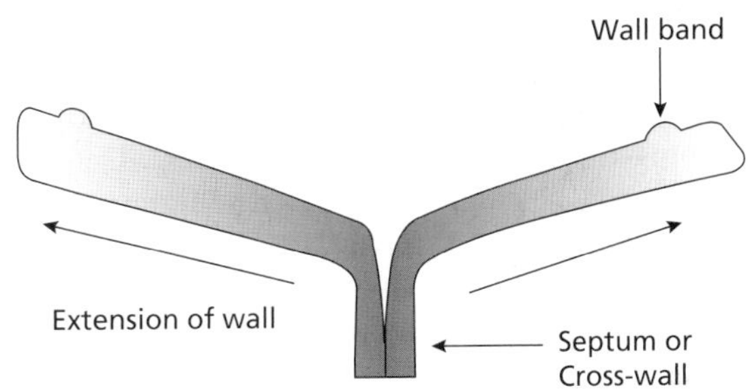

Fig. 8 Insertion of new wall in a streptococcus. As new wall is added at a growth zone, old wall is pushed away. The unit cell (the just born, or youngest cell) thus contains one-half of mother cell wall and one-half new wall. Young wall becomes progressively thicker as it ages. An equatorial cell marker defines the site for a new round of division. Adapted from Higgins and Shockman (1970).

Table 3 Antibiotics affecting peptidoglycan (PG) synthesis and assembly

Antibiotic	Inhibitory site
Phosphonomycin (Fosfomycin®)	Analogue of phosphoenolpyruvate (PEP) which prevents synthesis of MurNAc from GlcNAc
Cycloserine	A structural analogue of D-Ala which serves as a competitive inhibitor
Vancomycin and Ristocetin	Tends to inhibit the ligation of new PG onto pre-existing PG
Bacitracin	This polypeptide inhibits the dephosphorylation of bactoprenol pyrophosphate
Penicillin and other β-lactam antibiotics (cephalosporins, carbapenems and monobactams)	Inhibits both DD-carboxypeptidase and transpeptidase activities, thereby preventing PG cross-linking

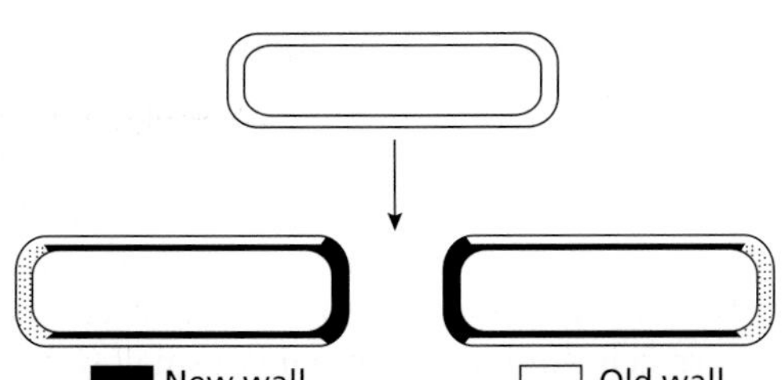

Fig. 9 Addition of peptidoglycan to side walls of rod-shaped bacteria results in surface expansion. In the Gram-positive rod, peptidoglycan is added at a large number of sites in the cell cylinder. Addition of even more peptidoglycan causes stretching of older wall resulting in cell elongation. Very old wall (on the periphery) may be turned over to alleviate stress (Koch *et al.*, 1982). Septa may be assembled in a manner analogous to streptococcal wall growth. In the Gram-negative rod, excision of wall material is accompanied by insertion of new wall. Elongation occurs, not because of stretching of peptidoglycan, but by addition of more wall than is excised by turnover (Höltje, 1998).

When more peptidoglycan is added, older wall is 'pushed' toward the surface and stretches, thus enabling the cell to elongate without the need of a defined growth zone. Autolysins clip the oldest wall, thereby relieving some of the stress caused by the stretching of the peptidoglycan (turnover). In Gram-negative rods, new wall is also added at numerous sites in the cell cylinder. Autolytic activity is required to hydrolyse some sites as new wall is added. The turned over peptidoglycan is processed, however, and re-utilised for growth.

Biological Effects of Peptidoglycan and Peptidoglycan Fragments

Peptidoglycan is indispensable for the structural integrity of virtually all bacteria, and for this reason it was once considered merely as an inert 'corset' surrounding the cell without significance for bacterial virulence and the pathogenesis of infection. During the last two decades, however, it has became apparent that the innate immune system of the host has evolved to recognise peptidoglycan and to eliminate it by hydrolysis, or to respond to it by activation of various cells and other components of the immune system (Table 4). Indeed, peptidoglycan can by itself reproduce most of the major clinical manifestations of bacterial infection, including inflammation, pus formation, fever, leucocytosis, sleepiness, decreased appetite, hypotension and arthritis (Table 5).

Peptidoglycan induces this large number of biological effects indirectly, by triggering the release of various mediators from host cells (Table 4). Other bacterial components, such as lipopolysaccharide (LPS) from the outer membrane of Gram-negative bacteria, or lipoteichoic acids from the cell walls of Gram-positive bacteria, can also induce this release of mediators from the same target cells, and can also reproduce all major clinical manifestations of bacterial infection. Thus, the innate immune system has evolved to recognise several major components of both Gram-positive and Gram-negative bacteria, as well as mycobacteria, spirochetes, *Mycoplasma* and other micro-organisms, and to respond to them with production of similar mediators.

Macrophages are the main target cells in the host that are activated by peptidoglycan. It has been known for several years that peptidoglycan induces the secretion of cytokines by macrophages, primarily IL-1, TNFα, and IL-6 (Dziarski *et al.*, 2000). Recent studies have indicated, however, that peptidoglycan stimulates macrophages to secrete even larger amounts of chemokines, primarily IL-8, MIP-1α, MIP-1β, MIP-2α and MCP-1 (Wang *et al.*, 2000). Peptidoglycan also induces production of nitric oxide (NO), a vasoactive product. Induction of NO production by peptidoglycan, and other microbial stimuli, is only observed in the mouse, however, and not in human macrophages.

Other targets for peptidoglycan include neutrophils, endothelial and epithelial cells, platelets and B lymphocytes (Table 4). Neutrophils express low levels of the peptidoglycan receptor CD14 (see below), and they are, therefore, potential targets for peptidoglycan. However, CD14-mediated effects of peptidoglycan on neutrophils have not been studied.

Endothelial and epithelial cells are targets and the effector cells involved in inflammation because they both can, if appropriately activated, produce mediators of inflammation and they also respond to several mediators of inflammation, such as the cytokines TNFα and IL-1, by producing more mediators and by expressing adhesion molecules. These adhesion molecules are used by leucocytes to attach to endothelial cells during inflammation. Endothelial cells do not express CD14, but they can be activated by complexes of soluble CD14 and LPS (Frey *et al.*, 1992; Pugin *et al.*, 1993). However, even though peptidoglycan forms stable complexes with soluble CD14, these complexes do not activate endothelial and epithelial cells (Jin *et al.*, 1998), and, therefore, these cells are not direct targets for peptidoglycan. Nevertheless, in the presence of blood, endothelial and epithelial cells are very strongly activated by the cytokines

Table 4 Targets for peptidoglycan and lipopolysaccharide and the mediators produced

Target	Mediators	Peptidoglycan	Lipopolysaccharide
Macrophages	Chemokines		
	IL-8	+	+
	MIP-1α	+	+
	MIP-1β	+	+
	MIP-2α	+	+
	MCP-1	+	+
	Cytokines		
	TNFα	+	+
	IL-1	+	+
	IL-6	+	+
	IL-10	?	+
	IL-12	+	+
	CSF	+	+
	Other mediators		
	NO	+	+
	PAF	?	+
	Tissue factor	?	+
Neutrophils	PAF	?	+
	MIP-1	?	+
Endothelial and some epithelial cells (+ sCD14)	IL-6	–	+
	IL-8	–	+
	PAF	?	+
	NO	?	+
	E-selectin, P-selectin	–	+
	ICAM-1	–	+
	VCAM-1	–	+
Platelets	Vasoactive amines	+	+
	PAF	?	+
B lymphocytes	Polyclonal antibodies	+	+
Complement	C3a	+	+
	C5a	+	+
Hageman factor	Kinins (bradykinin)	+	+

CSF, colony-stimulating factor; ICAM, intercellular adhesion molecule; IL, interleukin; MCP, monocyte chemoattractant protein; MIP, macrophage inflammatory protein; NO, nitric oxide; PAF, platelet-activating factor; TNF, tumour necrosis factor; VCAM, vascular cellular adhesion molecule; +, mediator induced; –, mediator not induced; ?, the effect unknown. Modified from Dziarski *et al.* (2000).

TNFα and IL-1 induced by peptidoglycan in blood monocytes (Jin *et al.*, 1998).

Peptidoglycan induces polyclonal activation of B lymphocytes, which results in B-cell proliferation and secretion of polyclonal antibodies (Dziarski, 1980). However, the significance *in vivo* of this polyclonal activation is unclear. Peptidoglycan also causes platelet aggregation, which can result in the release of their vasoactive products.

Peptidoglycan may interact with the humoral components of the host defence system. By a so far unidentified mechanism, it induces activation of both the alternative and classical complement pathways, which results in the generation of the pro-inflammatory anaphylatoxic and chemotactic complement split products, C3a and C5a. Peptidoglycan also induces generation of kinins in plasma, but the mechanism of this induction is not completely clear.

Induction of the above mediators by peptidoglycan or by whole bacteria or other bacterial components *in vivo*, is responsible for all the major clinical manifestations of bacterial infections. The types of clinical manifestations, their severity, and their duration depend on the types, amounts and sites of mediator

Table 5 Summary of biological effects induced by peptidoglycan

Biological effect	Mediators
Inflammation (oedema, erythema, induration)	IL-1, TNFα, IL-6, IL-8, MIP-1, MIP-2, MCP-1, NO, PAF, C3a, C5a, eicosanoids, adhesion molecules, etc.
Pus	IL-8, MIP-1, MIP-2, MCP-1
Arthritis	IL-1, TNFα, kinins, chemokines (?)
Fever	IL-1, TNFα, IL-6
Sleepiness	IL-1, TNFα
Decreased appetite	IL-1, TNFα
Hypotension	PAF, NO, TNFα, IL-1, bradykinin, eicosanoids
Leucocytosis	IL-1, TNFα, CSF
Thrombocytopenia	PAF, clotting factors
Toxicity to tracheal epithelial cells	IL-1, NO
Immune adjuvant	IL-1

For abbreviations see Table 4. Modified from Dziarski *et al.*, 2000.

release (Table 5). Most of the above-mentioned mediators are pro-inflammatory and induce oedema, erythema and induration, which are due to increased vascular permeability and infiltration of the sites of mediator release with inflammatory cells (polymorphs, monocytes and lymphocytes). Very vigorous attraction of phagocytic cells to the sites of infection, primarily caused by large amounts of chemokines induced by bacteria, and also by bacterial chemotactic factors, such as formylated peptides, is responsible for the highly pyogenic nature of many bacterial infections.

Fever is an immediate systemic effect of peptidoglycan-induced release of cytokines (TNFα, IL-1, IL-6) (Heymer *et al.*, 1985), and it is caused by a cytokine-induced increase in prostaglandin synthesis in the thermoregulation centre in the brain. The same cytokines are also responsible for the sleepiness (Krueger, 1990) and decreased appetite (Biberstine and Rosenthal, 1994) associated with bacterial infections. The sleep-promoting factor, first isolated from urine of sleep-deprived humans, turned out to be a naturally released peptidoglycan turnover fragment (GlcNAc-β-1,4-anhydroMurNAc linked to L-Ala-D-isoglutamine-*meso*-Dap-D-Ala), and it is believed to be required by mammals for proper sleep modulation. Since this peptidoglycan fragment cannot be synthesised by the host, it has been compared to a bacterially generated vitamin. Other peptidoglycan fragments, as well as synthetic analogues of peptidoglycan fragments, such as muramyl dipeptide (MDP), also have somnogenic activity and induce slow-wave sleep in experimental animals.

Another effect of a soluble peptidoglycan fragment released from *Bordetella pertussis*, known as tracheal cytotoxin, is toxic to ciliated tracheal epithelial cells. Tracheal cytotoxin is a disaccharide tetrapeptide, and its effect is mediated through a local production of IL-1-induced NO (Heiss *et al.*, 1994).

Leucocytosis is a prominent clinical manifestation of bacterial infections. It can be induced by peptidoglycan through the release of the bone marrow reserve pool of polymorphs, and later by the increased production of cells in the bone marrow, which are both caused by peptidoglycan-generated cytokines and colony-stimulating factors (Heymer *et al.*, 1985). Peptidoglycan can also cause hypotension, but peptidoglycan alone does not induce the circulatory shock often associated with sepsis or endotoxaemia (De Kimpe *et al.*, 1995). This lack of shock-inducing activity of peptidoglycan is most probably due to the relatively low capacity of peptidoglycan to induce TNFα (Wang *et al.*, 2000), the main mediator of septic shock. Peptidoglycan can, however, contribute to the induction of circulatory shock by acting synergistically with other bacterial products, such as lipoteichoic acids (De Kimpe *et al.*, 1995).

Bacteria can cause various forms of chronic relapsing arthritis, such as the arthritis associated with streptococcal rheumatic fever, gastrointestinal infections (*Shigella*, *Yersinia*, etc.), urinary tract infections, gonorrhoea, bacterial overgrowth after intestinal bypass surgery, inflammatory bowel disease and Lyme disease. A unifying feature of these diseases is the absence of local bacterial infection of the joints. The polyarthritis observed in these diseases can be reproduced in an animal model of chronic relapsing arthritis, in which rats are given one systemic injection of bacterial cell walls or cell wall-derived peptidoglycan–polysaccharide complex (Cromartie *et al.*, 1977). One day after this injection, an acute inflammation of the joints develops, it then wanes after 4–5 days, and it is followed after 10–14 days by several bouts of chronic, erosive polyarthritis, which mainly involves peripheral joints and histologically resembles human rheumatoid arthritis and Reiter's syndrome.

The active component of bacterial cell walls that induces this chronic relapsing arthritis is peptidoglycan (Stimpson *et al.*, 1987). The development of arthritis is dependent on the resistance of peptidoglycan to biodegradation, which is due to covalently bound polysaccharide or extensive *O*-acetylation, and on its ability to localise in joints and to persist in macrophages. Relapses are caused by the release of peptidoglycan from Kupffer cells in the liver and from macrophages in other organs, such as spleen and peritoneal cavity (Lichtman *et al.*, 1993).

The acute phase of arthritis is caused by the pro-inflammatory properties of peptidoglycan and can be reproduced by intra-articular or systemic injection of peptidoglycan (Heymer *et al.*, 1985; Stimpson *et al.*, 1987). The joint inflammation is most probably caused by the release of pro-inflammatory cytokines, mainly IL-1, TNFα and IL-6, from macrophages (Schwab *et al.*, 1991). However, relapses are not solely dependent on macrophages, because they also require intact T cells. Similar joint swelling can be induced by an intra-articular injection of other cytokine-inducing agents, such as LPS or the cytokines themselves (Stimpson *et al.*, 1988). Inhibition of cytokine release in the joints is an effective treatment for arthritis (Makarov *et al.*, 1996). These other cytokine-inducing agents, or cytokines themselves, do not induce chronic relapsing arthritis after a systemic injection, because they do not localise in the joints and they do not persist in tissues. However, they can induce relapses of arthritis in peptidoglycan-injured joints (Stimpson *et al.*, 1988).

Peptidoglycan is also arthritogenic in adjuvant arthritis, another model of arthritis, in which monophasic symmetrical polyarthritis follows a single intradermal injection of a water-in-oil emulsion of peptidoglycan or peptidoglycan-containing bacterial fragments (Kohashi *et al.*, 1976). The oil emulsion allows for a slow release and long persistence of peptidoglycan in the tissues.

Peptidoglycan Receptors and other Recognition Molecules

Animals have several cellular and humoral molecules that can recognise peptidoglycan (Table 6). The primary cellular recognition molecules for peptidoglycan are CD14 and Toll-like receptor-2 (TLR2). CD14 is a cell surface glycosylphosphatidylinositol (GPI)-linked 55-kDa glycoprotein expressed predominantly on myelomonocytic cells, including monocytes, macrophages and Langerhans cells, and also at much lower levels on neutrophils. Its structure contains 10 repeats with some similarity to the leucine-rich glycoprotein repeats. CD14 was first identified as the macrophage receptor for LPS on the outer membrane of Gram-negative bacteria (Wright *et al.*, 1990). It soon became apparent, however, that CD14 functions as the receptor for Gram-positive cell walls and their peptidoglycan component (Pugin *et al.*, 1994; Weidemann *et al.*, 1994; Gupta *et al.*, 1996). The function of CD14 as the peptidoglycan receptor is supported by evidence that: (1) activation of CD14-positive cells (monocytes or CD14 transfectants) by peptidoglycan is inhibited by anti-CD14 monoclonal antibodies; (2) peptidoglycan-unresponsive CD14-negative cells become peptidoglycan-responsive after transfection with CD14 and expression of membrane CD14; (3) CD14 binds to peptidoglycan with high affinity ($K_d = 25$ nM), forms stable complexes with CD14 at a molar ratio of approximately 1 : 1 and can completely displace LPS from CD14; and (4) binding of CD14 to peptidoglycan is inhibited by anti-CD14 monoclonal antibodies.

CD14 functions as the cell-activating receptor not only for LPS and peptidoglycan, but also for other

Table 6 Peptidoglycan receptors and recognition molecules

Receptor	Main effect	Other ligands
CD14	Activation of macrophages to secrete mediators	LPS, LAM, LTA, bacterial cell walls, lipoproteins, lipopeptides polyManU, phospholipids
TLR2	Activation of macrophages to secrete mediators	LPS, LTA, LAM, lipoproteins, lipopeptides
PGRP	Inhibition of Gram-positive bacterial growth in PMNs Activation of PO cascade in insects	None
Lysozyme	Hydrolysis of PG	None
Amidase	Hydrolysis of PG	None

CD, cluster of differentiation; LAM, lipoarabinomannan; LTA, lipoteichoic acid; PGRP, peptidoglycan recognition protein; PMN, polymorphonuclear leucocyte; PO, prophenoloxidase; polyManU, poly β-1,4-D-mannuronic acid; TLR, Toll-like receptor.

microbial and synthetic macrophage activators, including lipoteichoic acids from Gram-positive bacteria, lipoarabinomannan from mycobacteria, lipoproteins from spirochetes and mycobacteria, synthetic lipopeptides, poly β-1,4-D-mannuronic acid from Gram-negative bacteria, whole cell walls from Gram-positive bacteria, rhamnose-glucose polymer from streptococcal cell walls, and synthetic poly β-1,4-D-glucuronic acid (Dziarski *et al.*, 2000). For this reason, CD14 is a prototypical pattern recognition receptor that can recognise shared features of a wide variety of micro-organisms. However, CD14 also interacts with non-microbial ligands, such as phospholipids, and mammalian cells undergoing apoptosis.

It is still not precisely understood how a single molecule (CD14) can bind with high affinity to so wide a variety of molecules. However, attempts to map and compare the LPS- and peptidoglycan-binding sites on CD14 suggest that these binding sites are conformational and that they are partially identical and partially different. Therefore, the recognition of different patterns may be encoded in somewhat different regions of CD14 that are partially identical and partially different for various ligands (Gupta *et al.*, 1996; Dziarski *et al.*, 1998).

CD14 of identical sequence as membrane CD14, but without the GPI anchor, is also present in a soluble form in normal serum. Soluble CD14 forms complexes with LPS, and these complexes activate membrane CD14-negative cells, such as vascular endothelial cells, epithelial cells, vascular smooth muscle cells, fibroblasts and astrocytes (Frey *et al.*, 1992; Pugin *et al.*, 1993). By contrast, the complexes of soluble CD14 with peptidoglycan do not activate CD14-negative cells (Jin *et al.*, 1998). Another function of soluble CD14 is to enhance the responses of CD14-positive cells. Recent data indicate that the responses of CD14-positive cells to both LPS and peptidoglycan are enhanced by soluble CD14.

The CD14-mediated responses to LPS are greatly enhanced by LPS-binding protein (LBP), another protein present in normal serum, which transfers single LPS molecules catalytically from LPS aggregates normally present in an aqueous environment to CD14 (Tobias *et al.*, 1995; Dziarski *et al.*, 2000). The CD14-mediated cell activation by peptidoglycan is, however, not enhanced by LBP, and for this reason much higher concentrations of peptidoglycan than of LPS are needed to activate cells.

Although there is no doubt that CD14 functions as the cell-activating receptor for LPS, peptidoglycan and other microbial macrophage activators, it was long suspected that another molecule (or receptor) may be also involved in cell activation by peptidoglycan (and other activators) for the following reasons. First, CD14 by itself cannot transmit the activating signal into the cell, because CD14 is a GPI-linked molecule that does not have transmembrane and cytoplasmic domains. Second, studies with LPS partial structures that act as agonists or antagonists indicate that CD14 by itself cannot always discriminate between agonistic and antagonistic structures (Delude *et al.*, 1995). Third, expression of CD14 only in cells of myeloid origin, but not in cells of non-myeloid origin, makes them responsive to peptidoglycan, indicating that non-myeloid cells do not possess some essential component(s) necessary for the CD14-mediated response (Jin *et al.*, 1998). Fourth, CD14 knockout mice are more than 1000 times less sensitive to LPS than the wild-type mice in several, but not in all, LPS-induced responses (Haziot *et al.*, 1996; 1998), and cells from these CD14 knockout mice are only five to ten times less sensitive to peptidoglycan than the wild-type cells (Haziot *et al.*, 1999). This, again, points to the possibility of CD14-independent responses to peptidoglycan and LPS.

An extensive search for other receptors or signal-transducing molecules has recently resulted in the discovery that Toll-like receptors (TLRs) can fulfil this function. Toll receptors were first discovered in the fruit fly (*Drosophila*), in which they are indispensable for the initiation of dorsal–ventral patterning in the embryos, and for anti-microbial defences in the adult flies. Several mammalian homologues of insect Toll receptors, named TLRs, were then discovered. Their intracellular domains are homologous to the intracellular domain of the IL-1 receptor, but they have different extracellular domains that contain multiple leucine-rich repeats.

TLR2 serves as the cell-activating receptor for peptidoglycan, as well as intact Gram-positive and Gram-negative bacteria, *Mycoplasma* spp., and a number of other microbial cell wall components, such as LPS, lipoteichoic acids, lipoarabinomannan, lipoproteins and lipopeptides (Brightbill *et al.*, 1999; Lien *et al.*, 1999; Schwandner *et al.*, 1999; Takeuchi *et al.*, 1999; Underhill *et al.*, 1999; Yoshimura *et al.*, 1999). Evidence for the function of TLR2 as the receptor for peptidoglycan, Gram-positive bacteria and other microbial activators, is 2-fold. First, expression of TLR2 in TLR2-negative cells makes these cells responsive to these stimulants (Schwandner *et al.*, 1999; Yoshimura *et al.*, 1999), and second, TLR2 knockout mice are unresponsive to these stimulants (Takeuchi *et al.*, 1999). When over-expressed in transfected cells, TLR2 can activate cells in the absence of CD14. The TLR2-mediated responses are, however, usually enhanced by the presence of CD14, and in normal cells that do not over-express TLR2, cell activation may require both CD14 and TLR2 and/or some other TLR. Thus, TLR2 is another pattern recognition receptor that responds to a variety of microbial products, and the range of stimulants that activate cells through TLR2 and CD14 is very similar, if not identical.

TLR4 is the second mammalian Toll receptor with a known function; it is the cell-activating receptor for LPS. Although both TLR2 and TLR4 can function as the LPS receptors in TLR2- or TLR4-transfected cell lines, TLR4 is the primary and dominant LPS receptor *in vivo*. This is supported by studies that show that: (1) TLR4 knockout mice are unresponsive to LPS (Hoshino *et al.*, 1999); (2) a mutation in the TLR4 gene, which occurred naturally in C3H/HeJ and in C57BL/10ScCr mice, makes these mice hyporesponsive to LPS (Poltorak *et al.*, 1998; Qureshi *et al.*, 1999); and (3) TLR2-deficient mice or cells remain responsive to LPS (Takeuchi *et al.*, 1999; Heine *et al.*, 1999). For its function as the LPS receptor, TLR4 requires a helper molecule, MD-2 (Shimazu *et al.*, 1999). TLR4 does not, however, function as the receptor for peptidoglycan, Gram-positive bacteria and a variety of other stimulants that activate cells through TLR2 and CD14. Therefore, TLR4 is a more selective receptor than TLR2.

Another peptidoglycan recognition molecule, named peptidoglycan recognition protein (PGRP), has recently been discovered in insects (Yoshida *et al.*, 1996). Production of this 19-kDa protein in insects is enhanced by a bacterial challenge. PGRP is present in the insect haemolymph and cuticle, it binds peptidoglycan, and it triggers the prophenoloxidase cascade, that is a part of antimicrobial defence system in insects. Recently, insect PGRP as well as mouse and human PGRP homologues have been cloned, thus demonstrating that this protein is highly conserved from insects to mammals (Kang *et al.*, 1998). However, because mammals do not have the prophenoloxidase system, mammalian PGRP must play another role. Like insect PGRP, mammalian PGRP binds with high affinity to peptidoglycan and Gram-positive bacteria. In contrast to CD14, PGRP is highly specific for peptidoglycan and does not bind to Gram-negative bacteria, LPS or other bacterial components (Liu *et al.*, 2000). Recent results indicate that mammalian PGRP is exclusively expressed in polymorphs and that it inhibits growth of Gram-positive bacteria. However, PGRP is not bactericidal and has no lytic activity for peptidoglycan or bacteria. Therefore, in mammals, PGRP may function in neutrophils as an antibacterial protein (Liu *et al.*, 2000).

There are other peptidoglycan recognition molecules in the host. Most notably these are peptidoglycan-lytic enzymes, lysozyme and amidases, present in the serum, body secretions and the lysosomes of phagocytic cells. The function of these enzymes is to lyse peptidoglycan and peptidoglycan-containing bacteria. Other peptidoglycan recognition molecules that, for example, result in peptidoglycan-induced complement activation, platelet aggregation or polyclonal activation of B lymphocytes, have not been so far identified.

Signal Transduction Pathways, Transcription Factors and Genes Activated by Peptidoglycan in Mammalian Cells

Peptidoglycan activates host cells through the CD14 and TLR2 receptors, and this activation induces a signal that is amplified and transmitted into the nucleus, where it causes activation of a number of genes. The signal transduction pathways, transcription factors and genes that are activated by peptidoglycan (and other components of Gram-positive bacteria) are only now beginning to be identified.

The most complete and the best-characterised signal transduction pathway induced by peptidoglycan is the pathway triggered by the TLR2 receptor (**Fig. 10**), which is analogous to the signal transduction pathway induced by the IL-1 receptor (Wang *et al.*, 2001). This similarity of the TLR2- and IL-1 receptor-induced signal transduction pathways stems from the homology of the cytoplasmic domains of these receptors. The homologous pathway is also activated in insects through the Toll receptors. This pathway starts with the recruitment of three adapter molecules, myeloid differentiation protein (MyD88), IL-1 receptor-associated kinase (IRAK) and TNF receptor-associated kinase-6 (TRAF6). These adapter molecules activate an enzyme complex, composed of NF-κB-inducing kinase (NIK) and IκB kinases (IKKα and β), which phosphorylate IκB. IκB is an inhibitor of the ubiquitous transcription factor NF-κB and is bound to it in the cytoplasm. Phosphorylation of IκB results in its degradation and the release of NF-κB, which translocates to the nucleus and activates chemokine (IL-8) and cytokine (TNFα) genes (Wang *et al.*, 2001). Although NF-κB is required for the activation of IL-8 or TNFα genes, other transcription factors may also participate in or be required for the activation of these chemokine and cytokine genes. For example, it appears that EGR-1 and CREB/ATF1

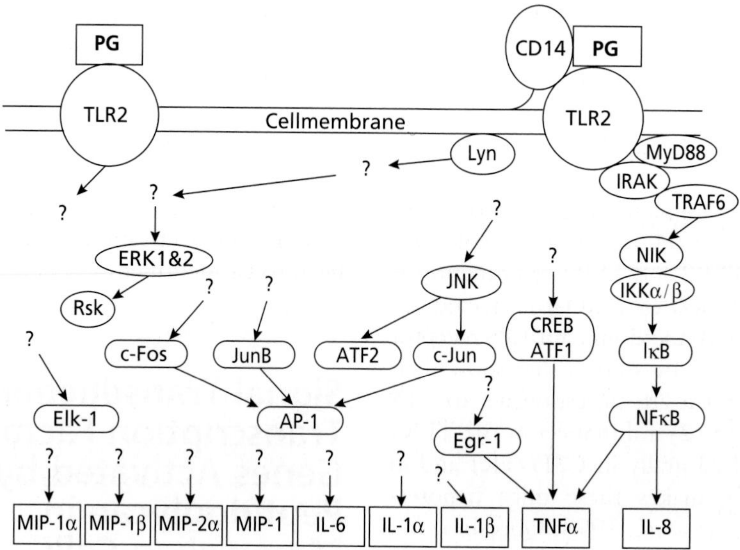

Fig. 10 CD14- and TLR2-mediated signal transduction pathways, transcription factors and main genes activated by peptidoglycan.

transcription factors also participate in activation of the TNFα gene.

Peptidoglycan induces activation or synthesis of a number of other transcription factors, such as c-Fos, Jun B, ATF 2, cJun and Elk-1 (Gupta *et al.*, 1999), but it is still not clear which signal transduction pathways result in the activation of these transcription factors, or expression of which genes is regulated by these transcription factors in peptidoglycan-activated cells. Similarly, peptidoglycan induces tyrosine phosphorylation of a receptor-associated tyrosine kinase, Lyn, and also activation and tyrosine phosphorylation of two families of mitogen-induced protein kinases, ERK and JNK (Gupta *et al.*, 1996; Dziarski *et al.*, 1996). However, the pathways leading to the activation of these kinases in peptidoglycan-activated cells, as well as the downstream pathways activated by these kinases are still not known.

Peptidoglycan induces expression of more than 120 genes in human monocytes (Wang *et al.*, 2001). The most-induced genes are the chemokine genes, and the second most-induced genes are the cytokine genes. However, the regulation of the expression of this vast number of genes activated by peptidoglycan remains unknown.

References

Biberstine KJ, Rosenthal RS (1994) Peptidoglycan fragments decrease food intake and body weight gain in rats. *Infect. Immun.* 62: 3276–3281.

Brightbill HD, Libraty DH, Krutzik SR *et al.* (1999) Host defense mechanisms triggered by microbial lipoproteins through toll-like receptors. *Science* 285: 732–736.

Cromartie WJ, Craddock JG, Schwab JH, Anderle S, Yang C (1977) Arthritis in rats after systemic injection of streptococcal cell walls. *J. Exp. Med.* 146: 1585–1602.

De Kimpe SJ, Kengatharan M, Thiemermann C, Vane JR (1995) The cell wall components peptidoglycan and lipoteichoic acid from *Staphylococcus aureus* act in synergy to cause shock and multiple organ failure. *Proc. Natl Acad. Sci. USA* 92: 10359–10363.

Delude RL, Savedra R, Zhao H, Thieringer R, Yamamoto S, Fenton MJ, Golenbock DT (1995) CD14 enhances cellular responses to endotoxin without imparting ligand-specific recognition. *Proc. Natl Acad. Sci. USA* 92: 9288–9292.

DePedro MA, Höltje J-V, Löffelhardt W (eds) (1998) *Bacterial Growth and Lysis.* New York: Plenum Press.

Doyle RJ, Marquis RE (1994) Elastic, flexible peptidoglycan and bacterial cell wall properties. *Trends Microbiol.* 2: 57–60.

Dziarski R (1980) Polyclonal activation of immunoglobulin secretion in B lymphocytes induced by staphylococcal peptidoglycan. *J. Immunol.* 125: 2478–2483.

Dziarski R, Jin Y, Gupta D (1996) Differential activation of extracellular signal-regulated kinase (ERK) 1, ERK2, p38, and c-Jun NH_2-terminal kinase mitogen-activated protein kinases by bacterial peptidoglycan. *J. Infect. Dis.* 174: 777–785.

Dziarski R, Tapping RI, Tobias P (1998) Binding of bacterial peptidoglycan to CD14. *J. Biol. Chem.* 273: 8680–8690.

Dziarski R, Ulmer AJ, Gupta D (2000) Interactions of bacterial lipopolysaccharide and peptidoglycan with mammalian CD14. In: Doyle RJ (ed.) *Glycomicrobiology.* New York: Kluwer Academic/Plenum Publishers, pp. 145–186.

Frey EA, Miller DS, Jahr TG *et al.* (1992) Soluble CD14 participates in the response of cells to lipopolysaccharide. *J. Exp. Med.* 176: 1665–1671.

Ghuysen J-M, Hakenbeck R (eds) (1994) *Bacterial Cell Wall*. Amsterdam: Elsevier.

Gupta D, Kirkland TN, Viriyakosol S, Dziarski R (1996) CD14 is a cell-activating receptor for bacterial peptidoglycan. *J. Biol. Chem.* 271: 23310–23316.

Gupta D, Wang Q, Vinson C, Dziarski R (1999) Bacterial peptidoglycan induces CD14-dependent activation of transcription factors CREB/ATF and AP-1. *J. Biol. Chem.* 274: 14012–14020.

Haziot A, Ferrero E, Kontgen F *et al.* (1996) Resistance to endotoxin shock and reduced dissemination of Gram-negative bacteria in CD14-deficient mice. *Immunity* 4: 407–414.

Haziot A, Lin XY, Zhang F, Goyert SM (1998) The induction of acute phase proteins by lipopolysaccharide uses a novel pathway that is CD14-independent. *J. Immunol.* 160: 2570–2572.

Haziot A, Hijiya N, Schultz K, Zhang F, Gangloff SC, Goyert SM (1999) CD14 plays no major role in shock induced by *Staphylococcus aureus* but down-regulates TNFα production. *J. Immunol.* 162: 4801–4805.

Heine H, Kirschning CJ, Lien E, Monks BG, Rothe M, Golenbock DT (1999) Cells that carry a null allele for Toll-like receptor 2 are capable of responding to endotoxin. *J. Immunol.* 162: 6971–6975.

Heiss LN, Lancaster JR, Corbett JA, Goldman WE (1994) Epithelial autotoxicity of nitric oxide: role in the respiratory cytopathology of pertussis. *Proc. Natl Acad. Sci. USA* 91: 267–270.

Heymer B, Seidl PH, Schleifer KH (1985) Immunochemistry and biological activity of peptidoglycan. In: Stewart-Tull DES, Davis M (eds) *Immunology of the Bacterial Cell Envelope*. New York: J. Wiley & Sons, pp. 11–46.

Higgins ML, Shockman GD (1970) Model for cell wall growth of *Streptococcus faecalis*. *J. Bacteriol.* 101: 643–648.

Höltje J-V (1998) Growth of the stress-bearing and shape-maintaining murein sacculus of *Escherichia coli*. *Microbiol. Mol. Biol. Rev.* 62: 181–203.

Hoshino K, Takeuchi O, Kawai T *et al.* (1999) Toll-like receptor 4 (TLR4)-deficient mice are hyporesponsive to lipopolysaccharide: evidence for TLR4 as the *lps* gene product. *J. Immunol.* 162: 3749–3752.

Jin Y, Gupta D, Dziarski R (1998) Endothelial and epithelial cells do not respond to complexes of peptidoglycan with soluble CD14, but are activated indirectly by peptidoglycan-induced tumor necrosis factor-α and interleukin-1 from monocytes. *J. Infect. Dis.* 177: 1629–1638.

Kang D, Liu G, Lundstrom A, Gelius E, Steiner H (1998) A peptidoglycan recognition protein in innate immunity conserved from insects to humans. *Proc. Natl Acad. Sci. USA* 95: 10078–10082.

Koch AL, Higgins ML, Doyle RJ (1982) The role of surface stress in the morphology of microbes. *J. Gen. Microbiol.* 128: 927–945.

Kohashi O, Pearson CM, Watanabe Y, Kotani S, Koga T (1976) Structural requirements for arthritogenicity of peptidoglycans from *Staphylococcus aureus* and *Lactobacillus plantarum* and analogous synthetic peptides. *J. Immunol.* 116: 1635–1639.

Krueger JM (1990) Somnogenic activity of immune response modifiers. *Trends Pharmacol. Sci.* 11: 122–126.

Lichtman SN, Bachmann S, Munoz SR *et al.* (1993) Bacterial cell wall polymers (peptidoglycan-polysaccharide) cause reactivation of arthritis. *Infect. Immun.* 61: 4645–4653.

Lien E, Sellati TJ, Yoshimura A, Flo TH *et al.* (1999) Toll-like receptor 2 functions as a pattern recognition receptor for diverse bacterial products. *J. Biol. Chem.* 274: 33419–33425.

Liu C, Gelius E, Liu G, Steiner H, Dziarski R (2000) Mammalian peptidoglycan recognition protein binds peptidoglycan with high affinity, is expressed in neutrophils, and inhibits bacterial growth. *J. Biol. Chem.* 275: 24490–24499.

Makarov SS, Olsen JC, Johnston WN *et al.* (1996) Suppression of experimental arthritis by gene transfer of interleukin 1 receptor antagonist cDNA. *Proc. Natl Acad. Sci. USA* 93: 402–406.

Poltorak A, He X, Smirnova I *et al.* (1998) Defective LPS signaling in C3H/HeJ and C57BL/10ScCr mice: mutations in *Tlr4* gene. *Science* 282: 2085–2088.

Pugin J, Schurer-Maly C-C, Leturq D, Moriarty A, Ulevitch RJ, Tobias PS (1993) Lipopolysaccharide activation of human endothelial and epithelial cells is mediated by lipopolysaccharide-binding protein and soluble CD14. *Proc. Natl Acad. Sci. USA* 90: 2744–2748.

Pugin J, Heumann D, Tomasz A *et al.* (1994) CD14 is a pattern recognition receptor. *Immunity* 1: 509–516.

Qureshi ST, Lariviere L, Leveque G *et al.* (1999) Endotoxin-tolerant mice have mutations in Toll-like receptor 4 (*Tlr4*). *J. Exp. Med.* 189: 615–625.

Rogers HJ, Perkins HR and Ward JB (1980) *Microbial Cell Wall and Membranes*. London: Chapman and Hall.

Rosenthal RS, Dziarski R (1994) Isolation of peptidoglycan and soluble peptidoglycan fragments. *Methods Enzymol.* 235: 253–285.

Schleifer KH, Kandler O (1972) Peptidoglycan types of bacterial cell walls and their taxonomic implications. *Bacteriol. Rev.* 36: 407–477.

Schwab JH, Anderle SA, Brown RR, Dalldorf FG, Thompson RC (1991) Pro- and anti-inflammatory roles of interleukin-1 in recurrence of bacterial cell wall-induced arthritis in rats. *Infect. Immun.* 59: 4436–4442.

Schwandner R, Dziarski R, Wesche H, Rothe M, Kirschning CJ (1999) Peptidoglycan- and lipoteichoic acid-induced cell activation is mediated by Toll-like receptor 2. *J. Biol. Chem.* 274: 17406–17409.

Shimazu R, Akashi S, Ogata H *et al.* (1999) MD-2, a molecule that confers lipopolysaccharide responsiveness on Toll-like receptor 4. *J. Exp. Med.* 189: 1777–1782.

Stimpson SA, Lerch RA, Cleland DR *et al.* (1987) Effect of acetylation on arthropathic activity of group A streptococcal peptidoglycan-polysaccharide fragments. *Infect. Immun.* 55: 16–23.

Stimpson SA, Dalldorf FG, Otterness IG, Schwab JH (1988) Exacerbation of arthritis by IL-1 in rat joints previously injured by peptidoglycan-polysaccharide. *J. Immunol.* 140: 2964–2969.

Takeuchi O, Hoshino K, Kawai T *et al.* (1999) Differential roles of TLR2 and TLR4 in recognition of Gram-negative and Gram-positive bacterial cell wall components. *Immunity* 11: 443–451.

Tobias PS, Soldau K, Gegner JA, Mintz D, Ulevitch RJ (1995) Lipopolysaccharide binding protein-mediated complexation of lipopolysaccharide with soluble CD14. *J. Biol. Chem.* 270: 10482–10488.

Underhill DM, Ozinsky A, Smith KD, Aderem A (1999) Toll-like receptor-2 mediates mycobacteria-induced proinflammatory signaling in macrophages. *Proc. Natl Acad. Sci. USA* 96: 14459–14463.

Wang Q, Dziarski R, Kirschning CJ, Muzio M, Gupta D (2001) Micrococci and peptidoglycan activate TLR2/MyD88/IRAK/TRAF/NIK/IKK/NF-κB signal transduction pathway that induces transcription of IL-8. *Infect. Immun.* 69: 2270–2276.

Wang Z-M, Liu C, Dziarski R (2000) Chemokines are the main pro-inflammatory mediators in human monocytes activated by *Staphylococcus aureus*, peptidoglycan, and endotoxin. *J. Biol. Chem.* 275: 20260–20267.

Weidemann B, Brade H, Rietschel ET *et al.* (1994) Soluble peptidoglycan-induced monokine production can be blocked by anti-CD14 monoclonal antibodies and by lipid A partial structures. *Infect. Immun.* 62: 4709–4715.

Wright SD, Ramos RA, Tobias PS, Ulevitch RJ, Mathison JC (1990) CD14, a receptor for complexes of lipopolysaccharide (LPS) and LPS binding protein. *Science* 249: 1431–1433.

Yoshida H, Kinoshita K, Ashida M (1996) Purification of a peptidoglycan recognition protein from hemolymph of the silkworm, *Bombyx mori*. *J. Biol. Chem.* 271: 13854–13860.

Yoshimura A, Lien E, Ingalls RR, Tuomanen E, Dziarski R, Golenbock D (1999) Recognition of Gram-positive bacteria by the innate immune system occurs via Toll-like receptor 2. *J. Immunol.* 163: 1–5.

Zhou L, Srisatjaluk R, Justus DE, Doyle RJ (1998) On the origin of membrane vesicles in Gram-negative bacteria. *FEMS Lett.* 163: 223–228.

8
Flagella

Shin-Ichi Aizawa

Teikyo University, Utsunomiya, Japan

Flagella are protein structures with a strictly defined morphology and are unique organs of motility present in many bacterial species. They endow bacteria with the ability to swim actively at up to 25 μm/second in liquid environments and to respond to a variety of stimuli by altering their swimming behaviour. Flagella may be externally located, with their proximal end built into the cell envelope, as in *Escherichia coli*, *Salmonella* spp. and other Gram-negative organisms. In *Treponema pallidum* (Chapter 85) they lie within the periplasmic space and endow the organisms with the ability to move through viscous environments. Motility is not, however, restricted to bacteria that possess flagella. 'Twitching and social gliding motility', which is comparatively slow (0.1 μm/second) depends on type IV fimbria that attach to surfaces and, by retracting forcibly they allow the bacteria to crawl on surfaces much as millipedes do, and also to form adherent colonies. This type of motility can be observed in *Neisseria gonnorhoeae* (Merz *et al.*, 2000) and *Pseudomonas aeruginosa* (Mattick *et al.*, 1996). A mechanism for this type of motility has been proposed (Sun *et al.*, 2000). Swarming motility is most famously observed in *Proteus* spp. but it also occurs in *P. aeruginosa* (Rashid and Kornberg, 2000). Only the common bacterial flagella are considered in this chapter.

Eukaryotes also have flagella, but prokaryotic and eukaryotic flagella differ in a number of respects. Bacterial flagella derive their energy from a proton motive force (PMF) and rotate, but the eukaryotic structures are far more complicated, flexible, beat like a whip and obtain their energy from ATP. Each bacterial flagellum is driven by a rotary device termed the flagellar motor, which is located at its base. As would be expected from the structural simplicity of bacterial flagella, they are no longer able to move when removed from the bacterial cell.

Motile bacteria are able to sense temperature (Imae *et al.*, 1984) and gradients of chemical substances (Macnab and Koshland, 1972) in their environment, and to move towards favourable stimuli and away from stimuli unfavourable for survival. Such movement, known as taxis, depends on the control of flagellar activity by sensing mechanisms, which determine that organisms swim actively towards, for example, potentially nutrient sugars and amino acids (positive chemotaxis). To achieve such movement, cells show zig-zag movements; favourable directions of movement are achieved by swimming for longer in a favourable direction than in other directions, with the net result that a favourable direction results after a series of 'trials'. Exposure to potentially toxic substances leads to the cessation of directional swimming and 'tumbling' occurs as a means for changing direction; after tumbling, the cell moves directionally away from the unfavourable stimulus. Bacterial chemotaxis has been reviewed by Blair (1995) and Stock and Surette (1996).

The machinery for chemotaxis is determined by about 50 genes that code for chemoreceptors, enzymes in the signal transduction system and flagella. Some 40 of these genes are required for the construction of flagella, particularly for the flagellar

 doi:10.1006/bkmm.2001.0008

motor, which represents the major part of the flagellum. The genetic mechanisms that control the expression of these many genes are sophisticated and fascinating.

The details of the flagellar genetics are not described in this chapter, but a brief outline of the gene arrangement is necessary. The genes for flagella, motility and chemotaxis in *E. coli* are arranged on the bacterial chromosome in three regions or levels, I, II and III; the latter consists of two sub-regions, IIIA and IIIB. In *Salmonella typhimurium* there is an additional region IV. Genes of level I are required to stimulate the transcription of the other genes. The gene *fliA* of level II is a sigma factor which augments the transcription of other genes in level II and is necessary for the transcription of the genes in level III. The gene *flgM* codes for an anti-sigma factor that inhibits *fliA* and can be pumped out of the cell by the flagellar export system when construction of basal body–hook structure is complete. Flagellar structure and function, and its control and genetics have been reviewed by Macnab (1996).

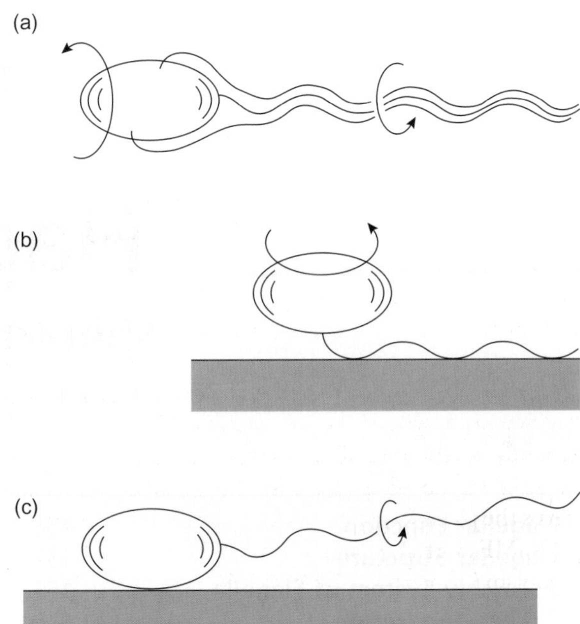

Fig. 1 Three methods for measuring the speed of flagellar rotation: (a) In free-swimming cells the cell body and the flagellar bundle rotate in the opposite directions. The rotational speed of the bundle is ~80 Hz. (b) When a cell is tethered to a surface by a flagellum, only the cell body rotates. The rotational speed of the cell body is ~20 Hz. (c) When a cell body is tethered to a surface, its flagellum rotates without a load. The rotational speed of the flagellum is ~200 Hz.

Flagellar Function

In terms of its physical properties and the regulation of the movement, the rotational motion of flagella is different from the sliding movement found in other motility systems. The rotational force (torque) generated by the flagellar motor cannot be measured directly, but it can be estimated from the swimming speed of cells or the rotational speed of the flagella. Methods of measuring the rotational speeds of a single flagellum or a bundle of flagella are illustrated in **Fig. 1**.

Rate of Rotation

For free-swimming *E. coli* cells in suspension, the average rotational frequency of flagellar bundles, as measured by light scattering and Fourier analysis, is about 80 Hz (revolutions per second) (Lowe *et al.*, 1987). Cells tethered to the surface of a glass slide by a single flagellum rotate at 20 Hz or less, suggesting that the cell is too heavy for a single motor to move. When cells are attached to a glass surface by their cell body, each flagellum rotates freely at its maximum frequency of 200 Hz as measured by laser dark-field microscopy (Kudo *et al.*, 1990). A maximum frequency of 1700 Hz (102 000 revolutions/minute) has been recorded for a polar flagellum on a swimming cell of *Vibrio alginolyticus.* Though this appears an extraordinarily high frequency, the swimming speed is not as rapid as would be expected from the measured value, which indicates that the force is mostly dissipated in the slippage between the flagellum and water molecules (Magariyama *et al.*, 1994).

Tumbling

The zig-zag motion of a swimming cell is a consequence of two successive events, smooth swimming and tumbling to a stop. In the case of a cell with a single polar flagellum, the flagellum can either 'push' or 'pull' the cell according to the rotational direction of the helical filament. The push and pull modes both cause smooth swimming. The pause between the two modes is very brief and results in a continuous to-and-fro motion.

In the push mode of a cell with many flagella, these form a bundle behind the cell body. The reverse rotation of some flagella in the bundle, however, causes a brief tangling of flagella followed by an unbundling of the flagella in reverse rotation. During this complicated sequence of events, the cell appears to tumble and stop until it once again resumes smooth swimming in a new direction.

Energy Source

The flagellar motor is powered by PMF created by a flow of protons from the outside to the inside of the cell membrane. The mechanism by which PMF is converted into the torque of the flagellar motor is one of the great unsolved mysteries in biology (Manson *et al.*, 1977). PMF is also used by many other systems in the membrane, such as amino acid and sugar export. Sodium ion motive force (NaMF) can be also employed for some export systems. None of the systems has yet been solved at the atomic level, so how these different systems can function in a similar way remains a puzzle. The choice of PMF or NaMF depends on the individual mechanism but not on the bacterial species. For example, the flagellar motor that drives the polar flagellum of *V. alginolyticus* is powered by NaMF, but that of the lateral flagella of the same organism is energised by PMF (Atsumi *et al.*, 1992).

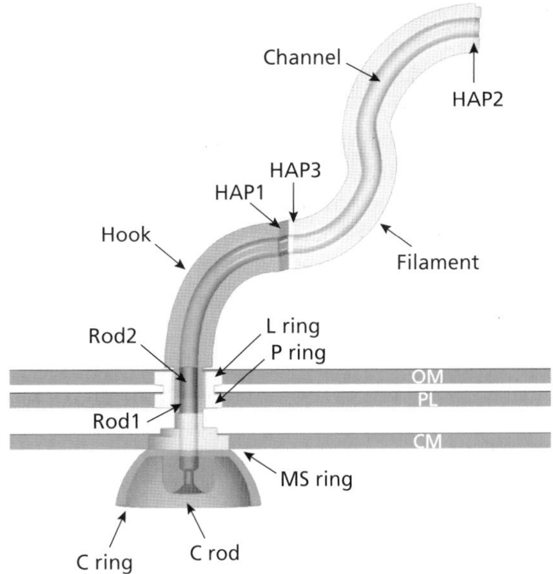

Fig. 2 General structure of the flagellum in Gram-negative bacteria. The components of the basal body are shown inserted into the various layers of the cell envelope. The flagellar filament is attached to the basal body by the hook. Since the flagellar filament is a helical structure, it has a central channel. HAP, hook-associated protein; OM, outer membrane; PL, peptidoglycan layer; CM, cell membrane (See also Colour Plate 1).

Flagellar Structure

The term flagellum, in its conventional and historical sense, suggests a helical filament extending from the cell surface. The helix in most species of enteric bacteria is left-handed, but in some, such as *Rhodobacter*, it is right-handed; how helix handedness was selected in evolution is unknown. It may have depended on environmental conditions, such as viscosity, pH and salt concentration.

The flagellar filament differs considerably in terms of function and structure from the other cell surface appendages pili and fimbriae (Chapter 9), which are organs of adhesion. The flagellar filament is helical, whereas the filaments of the other two structures are straight. Their interactions with the bacterial membrane are also very different; flagella are firmly anchored in both the outer and inner bacterial membranes by a structure called the basal body, whereas pili and fimbriae have their origin in the outer membrane.

Briefly, a flagellum consists of three structural parts: the filament, the hook and the basal body. The general structure of a typical flagellum is shown in Fig. 2. The junction between the flagellar filament and the basal body, as observed by electron microscopy, is a sharply curved structure termed the hook, which has particularly interesting physical features. The unique features of each of these parts of the flagellum are described below, but first it is helpful to consider the function of the flagellum as a molecular machine.

The apparent physical features of the flagellar motor are not very different from those of the electric motors in everyday use. Functionally, the flagellar filament is like the propeller of a motor boat, the basal body is the motor and the hook is the universal joint between the two. The screw must be rigid in order to exert the necessary thrust to push the bulk of the boat, whereas the connector must be flexible to join the two parts and to convey torque efficiently at different angles. The same is true of the flagellum; the filament is rigid, whereas the hook is flexible. The flagellar motor rotates at high speed, just like a man-made motor (e.g. 20 000 revolutions/minute). The flagellar motor can also rotate in both directions, either to push the cell forward or to pull it backwards.

Filament

Flagellin

Flagellin is the building block of the flagellar filament. The gene that encodes flagellin was previously termed *hag* (from H-antigen), but is now called *fliC*, according to the unified terminology for the flagellar genes of *E. coli* and *Salmonella* (Iino *et al.*, 1988). In some bacteria the filaments consist of a single kind of flagellin; in others they consist of several kinds of flagellin which are closely related isomers (Table 1). In general, one of the flagellin isomers is the major component of a filament, but the role of the other isomers has not been clarified, because deletion of any of many isomer genes does not usually affect filament formation and motility.

The bacterial kingdom can be divided into two groups, eubacteria and archaea, according to the molecular architecture of their flagellin, which reflects the unique features of the molecule. The archaea flagellins tend to be small, with N-terminal signal sequences that can be cleaved by a secretion system, whereas those of enteric bacteria are large, have no signal sequence and are transported by the flagella-specific export system (**Table 1**) (see below).

It has been claimed (Jarrel *et al.*, 1996) that archaeal flagella originate from the outer membrane in the same manner as fimbriae and pili, suggesting that the rotation of this type of flagellum is not generated by a flagellar motor. If this is correct, it is possible that another type of rotary system exists. The account that follows deals only with the flagella of eubacteria, on the basis of information derived from *S. typhimurium*, since *Salmonella* flagella have been most extensively studied and in many respects they are representative of eubacterial flagella.

The flagellin peptide can be divided into three regions according to its structural features: two terminal regions and the central region (**Fig. 3**). The sequences of the terminal regions are well conserved, but the central region is variable. The conserved sequences of the terminal regions have been used to calculate a distance matrix and to create a phylogenetic tree, much as in the case of the prokaryotic phylogeny based on 16S rRNA sequences (DeShazer *et al.*, 1997).

The terminal regions of the flagellin peptide are indispensable for filament formation, and the predicted

Table 1 Flagellar filament and flagellins from various bacterial species

Bacterial group	Bacterial species	Flagellation pattern	Helical filament type	Flagellin (kDa)
Proteobacteria, γ-subdivision	*Escherichia coli*	Peritrichous	Normal	
	Salmonella typhimurium	Peritrichous	Normal	58
	Vibrio parahaemolyticus	Polar-monotrichous (sheathed) Lateral-peritrichous	Curly	40
	Pseudomonas aeruginosa	Peritrichous	Normal	a: 45, 52; b: 53
Proteobacteria, α-subdivision	*Rhizobium meliloti*[a]	Peritrichous (complex)	Curly	A: 40.459, B: 41.044
	Rhodobacter sphaeroides	Lateral-monotrichous	Curly $\leftrightarrow$ Coiled	
Proteobacteria, δ- and ϵ-subdivisions	*Helicobacter pylori*	Polar	Curly?	57, 56
	Campylobacter jejuni	Polar		59.538, 59.909
Low (G + C) gram-positive bacteria	*Bacillus subtilis*	Peritrichous	Normal	
	Listeria monocytogenes			29
	Clostridium difficile			39
Spirochaetes	*Spirochaeta aurantia*	Periplasmic		31, 32, 33, 34, 36
	Borrelia burgdorferi	Periplasmic	Normal?	37, 41
Euryarchaeota	*Methanococcus thermolithotrophicus*			26, 44, 62
	Methanospirillum hungatei			24, 25
	Halobacterium halobium	Polar-tuft		A1: 30, A2: 26, B1: 26, B2: 36, B3: 26
	Treponema pallidum			

[a] Pleier and Schmitt (1989).

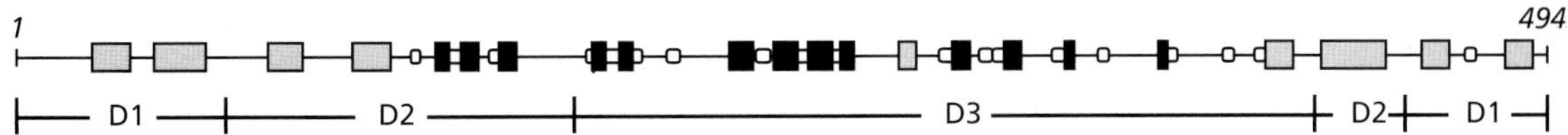

Fig. 3 Domain structures of the flagellin molecule. The *Salmonella* i-type flagellin molecule (494 amino acids) is drawn as a straight line. Predicted secondary structures are shown as grey boxes (α helix), black boxes (β sheets), and small open boxes (turns). The molecule is divided into three domains: D1 in each terminal region, D3 in the central region, and between these the D2 region. D1 is indispensable for filament polymerisation. D3 is a super-variable region that contains H antigenic epitopes.

secondary structures of these regions are β helices. Circular dichroism (CD) measurements of flagellin, however, show a typical α helix profile, and the peaks of the profile are enhanced in the filament form, indicating that the α helix content in flagellin increases when it is polymerised into filament (Vonderviszt *et al.*, 1989); this is also supported by nuclear magnetic resonance measurements. About 60 amino acids in the terminal regions of flagellin are as mobile as free amino acids, but in filaments the same regions take on a stable α-helical structure (Aizawa *et al.*, 1990). In filaments, these α-helical rods interact circularly with one another to form the inner core of the filament (Mimori *et al.*, 1995). The significance of the dynamic change of the secondary structure is considered below.

The sequence of the central region of flagellin is variable not only between different bacterial species but also between *Salmonella* serovars, and this is responsible for H antigenic specificity (Smith and Selander, 1990). The central region of flagellin occupies the outermost part of the filament and is exposed to the environment, and its epitopes are recognised as H antigens by the immune system of humans and animals. The flagellins of environmental bacteria, on the other hand, do not have such regions, presumably because they are not subject to the effects of antibodies. Consequently, a part of the region can be deleted without affecting filament formation, which tends to confirm that H antigens are an extra portion of the filament. It is, therefore, to be expected that peptides with an appropriate sequence can be inserted into the deleted part of the central variable region. However, attempts to overproduce peptides as large as the central region have been unsuccessful, and only small peptides, such as for example an epitope of cholera toxin can be expressed at as high a level as that of ordinary flagellin (Newton *et al.*, 1989).

Phase Variation

Salmonella species usually have two flagellin genes, *fliC* and *fljB*, at separate locations on the chromosome. These can be expressed as the major flagellins, but not at the same time in any given cell. The two flagellins, H1 and H2, have significantly different antigenic specificity, resulting in two types of cells with completely different flagellar antigens. This alternative expression of two different flagellins with different antigenic specificity, a phenomenon known as phase variation, allows the cells to avoid attack by antibodies *in vivo*. The frequency of alternate expression of the two genes is about 10^{-3}–10^{-5} per cell per generation, and switching between the two is controlled by the *fljB* operon. The mechanism of the switch has been reviewed by Macnab (1987).

The *fljB* operon The *fljB* operon includes the genes *fljA* and *fljB*. The latter encodes H2 flagellin and the *fljA* gene is the repressor of *fliC* transcription. An upstream region of the operon, the H region, is flanked by two 26 bp sequences, *hixL* (repeat L) and *hixR* (repeat R), each of which consists of two imperfect 13 bp repeats and thus can serve as sites of recombination. The promoter for *fljB–fljA* is located within the H region, and in one orientation can transcribe this operon. When the *fljB–fljA* operon is transcribed, the FljA repressor prevents transcription of the *fliC* gene and prevents the formation of the H1 flagellin (Kutsukake and Iino, 1980).

The *hin* gene encodes the site-specific recombinase that can flip the H region so that the promoter for *fljB–fljA* is no longer oriented in the proper direction to transcribe the *fljB–fljA* genes. Inversion by the Hin recombinase is stimulated by a second protein, Fis, which binds to enhancer sequences in the H region. The Hin–Fis complex, termed an invertasome, gives rise to H-region inversion by a looping mechanism similar to that used to regulate expression of alternative genes in many other organisms (Silverman and Simon, 1980).

Filament Structure

The flagellar filament is a rigid helical tube composed of chemically identical subunits of a single protein, flagellin, arranged as a helical tube like a piano wire, as indicated by measurements of quasi-elastic light-scattering in solution (Fujime *et al.*, 1972). The flagellar filament is the largest part of a flagellum and it has been studied most extensively. The physical properties, lattice structures and the primary structure of flagellin have been solved.

Direct observation of flagella in suspension by dark-field microscopy provides empirical evidence for the rigidity of the flagellar filament. The filaments are not subject to structural thermal motion like actin–meromyosin filaments, which are very flexible and constantly changing in shape (Nagashima and Asakura, 1980). Since the rigidity of the flagellin filament appears not to be inherent in flagellin itself, it must be derived from the way in which the molecules are arranged in the filament.

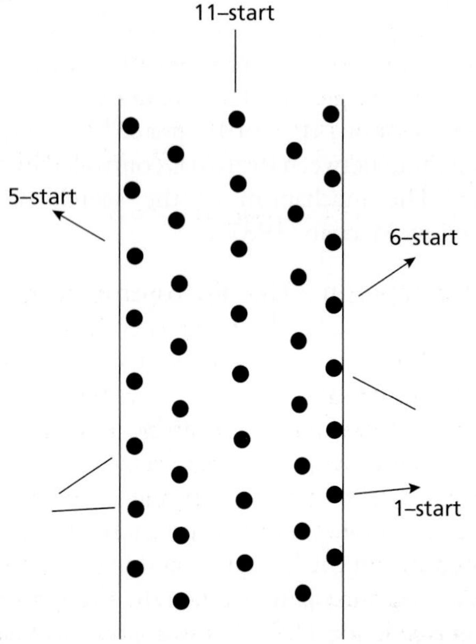

Fig. 4 Lattice structure of a tubular polymer. Dots represent the positions of identical subunits. There are several groups of lines that cover all the dots. The basic (or 1-start) helix can trace all dots with a single line, and 5 (or 6)-start helices can cover all dots with a group of 5 (or 6) parallel lines. 11-start helices are also called proto-filaments, though this is imaginary (see text).

For technical reasons only straight forms, not helical forms, can be analysed by optical diffraction to determine the packing pattern of subunits in the flagellar filament. Optical diffraction patterns of electron micrograph images of straight filaments consist of several pairs of layer lines, indicating that the packing of flagellin in a filament is helical. If the filament were to be cut on one side and opened out to lie flat, a set of lines running on arrays of subunits would constitute a lattice. A single line that passes through all the subunits is called the basic helix (**Fig. 4**).

Similarly, lines that cover all subunits with a set of 5 lines are called 5-start helices. A set of lines closest to the filament axis are 11-start helices. Although these lines are often called 'proto-filaments', this is merely imaginary; there are no such filaments. These sets of n-start helices correspond to the layer lines of diffraction patterns. The equation that governs the relationship between these layer lines is $l = tn + sm$. It is called the 'selection rule' and it defines a helical lattice. The variables l, m, and n correspond to co-ordinates of spots in diffraction (Fourier) space and they describe the situation of a given subunit in real space. The parameters t and s (which must be integers) uniquely define a helix; a helix has s subunits in t turns of the basic helix. For example, an R-type straight filament of *Salmonella* is described as $l = 17n + 93m$; that is, 93 subunits are equally placed on a line which covers a certain length of a tube in 17 turns (**Fig. 5**).

The structure of the R-type Straight filament has been solved by electron cryomicroscopy at a resolution of 9 Å. Filaments were frozen-hydrated at the temperature of liquid helium and observed with a stable field emission source. The structure so revealed

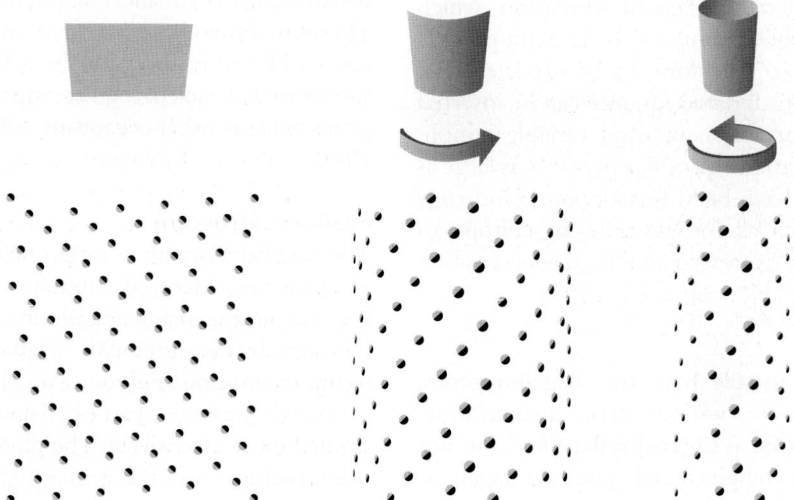

Fig. 5 Lattice structure of R-type straight filament. The selection rule of this lattice is $l = 17n + 93m$. Dots on a flat sheet (left) are placed equally according to the selection rule. By turning the sheet into a tube (right), the surface structure of a filament is seen.

showed a densely packed core surrounded by sparsely packed outer arms (Mimori *et al.*, 1995). Flagellin in the filament was divided into four domains (D0–D3) (Fig. 6). The domains were assigned to the regions of a stretched molecule; the conserved terminal regions belong to D0 and the hypervariable central region belongs to D3.

Filament Helicity

The handedness of filament helicity can be determined from the slanting patterns of the top and bottom surfaces of a helix, as observed by dark-field microscopy. Mutants that produce filaments of different shapes have been isolated and the helicity parameters of the filaments have been determined (Table 2).

The left-handed helix is termed 'Normal', because at an early stage of this research the left-handed helix was most frequently found in nature. (Note that the names of filament shapes are indicated by adjectives starting with capital letters.) A typical filament shape of *S. typhimurium* is Normal, and this organism has been used in many studies on filaments. It should, however, be noted that right-handed helices (Curly filaments) also occur naturally in many other bacteria such as *Caulobacter cresentus* and *Rhizobium melilotti*, but Curly has been regarded as a variant of Normal (Leifson, 1960). In *Rhodobacter sphaeroides* the stable shape of filaments isolated from cells is left-handed, Coiled, but it is converted into Curly when rotating

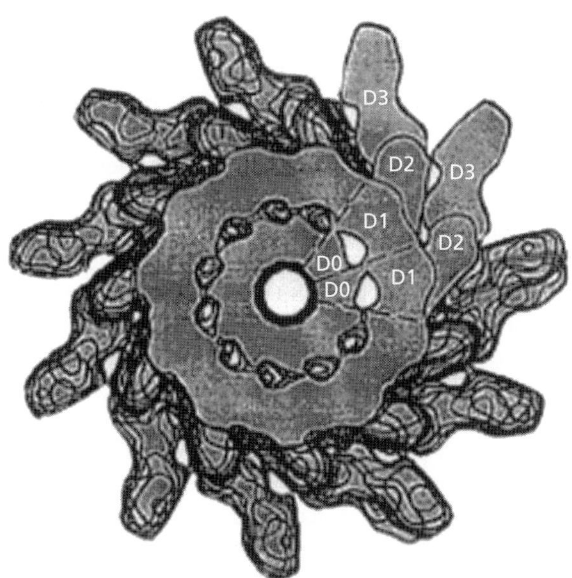

Fig. 6 Domain structures of the flagellar filament. The cross-section of the filament is drawn schematically. In this figure, the original D1 of Fig. 2 is divided into two domains: D0 and D1. A continuous area covered by D0–D3 corresponds to a flagellin molecule.

Table 2 Helical parameters of flagellar filaments

Name	Handedness	Pitch (μm)	Diameter (μm)
Straight[a] (L-type)	–	∞	0
Normal	Left	2.55	0.6
Coiled	Left	0[b]	1.0
Semi-coiled	Right	1.29	0.5
Curly I	Right	1.20	0.2
Curly II	Right	1.00	0.15
Straight[a] (R-type)	–	∞	0

[a] Two types of Straight filaments exist. They do not have helical handedness as macroscopic filaments, but they have internal helicity (L- and R-type) derived from packing patterns of flagellin subunits.
[b] The pitch of Coiled is theoretically close to zero.

on a cell. This conversion of helical handedness of a filament is called polymorphism, which is described in detail below.

By definition, there is no bending stress in a straight filament, but twisting stress can be applied along the axis. Theoretically three types of twisting modes, right (R)-handed, left (L)-handed, and zero-handed are possible but only the first two exist in nature. The differences between the two are so slight that only analysis of diffraction patterns can distinguish them. From the position of the first layer line in the diffraction pattern, tilt angles of 11-start helices from the axis have been calculated as $+7°$ for the R-type and $-2.5°$ for the L-type. These filaments do not function as a screw and consequently give rise to non-motile cells (Mimori *et al.*, 1995).

Small changes in either of the 494-amino acid terminal regions of the flagellin molecule causes dramatic changes of filament shape, but they usually occur in the C-terminal region. For example, point mutations that lead to 449Ala → Val for an R-type flagellin and 426Gly → Ala for an L-type flagellin give rise to straight filaments (Kanto *et al.*, 1991). Similarly, other single amino acid substitutions at other positions give rise to Curly, Coiled and other forms of filament.

Filament structure is physically rigid but heat-labile and acid-labile, and flagellin monomer obtained by disassembly of the helix at 60°C or treatment at pH 4 self-assembles into filaments *in vitro*. When the flagellin thus obtained, and small fragments of filaments prepared by sonication, are incubated overnight at room temperature (~26°C), long helical filaments are obtained, which have exactly the same shape as the filament from which the flagellin was obtained. Thus, flagellin obtained from Normal filaments self-assembles into Normal filaments, and flagellin from Curly into Curly filaments (Asakura *et al.*, 1964).

The flagellin molecules maintain filament helicity. Thus, if a mixture of two different flagellins is allowed to polymerise into filaments, the reconstituted filaments take on various helical forms, depending upon the ratio of the two flagellins. Striking results are obtained if flagellins from two types of straight filaments are co-polymerised. Thus, R-flagellin and L-flagellin each by itself self-assembles into R- or L-straight filaments respectively. In a mixture that contains 10–25% of R-flagellin, Normal filaments are formed, whereas a mixture that contains 50% of R-flagellin reconstitutes into Curly filaments. The rule is that the larger the proportion of a given flagellin, the closer the reconstituted filaments are to the original form of the filament from which the flagellin was isolated. It should be noted that the helices that appear in such experiments are discrete and discontinuous, irrespective of the ratios of two flagellins; helices of intermediate shape do not occur (Kamiya *et al.*, 1982).

The self-assembly of flagellar filaments is an example of fundamental rules that determine the construction of supra-molecular structures. This self-assembly mechanism is not, however, sufficient to produce flagella on growing cells *in vivo*, and an elaborate mechanism is used to make their construction more efficient. A helper protein, HAP2, is necessary for filament formation; without it flagellin does not polymerise into filaments and monomers are secreted (see below).

Polymorphism

The number of filament shapes allowed in nature is limited (**Table 2**), but although the filament is rigid, under certain conditions it can change its shape. This characteristic of filaments is termed polymorphism. The most frequent conversion so far observed is between Normal and Curly. In this conversion, there are no other stable filament forms and this indicates that Normal and Curly are the stable forms under different conditions (Kamiya *et al.*, 1982).

Polymorphism can be induced by various agents, including acid and alkaline pH, salt, temperature, alcohol, and mechanical forces. Detailed observations on polymorphism over a wide range of pH and potassium chloride concentrations can be summarised in a phase diagram (Kamiya and Asakura, 1977) (**Fig. 7**).

At the pH values between 6 and 8, Normal filaments remain unchanged irrespective of potassium chloride concentration. In both acidic and alkaline solutions, Coiled filaments become dominant, but at extreme values of pH (above pH 12 and below pH 4), Curly filaments appear. Eventually Straight filaments briefly appear just before melting begins. A similar sequence of polymorphic transitions – Normal → Coiled → Curly → Straight – occurs when other agents are employed.

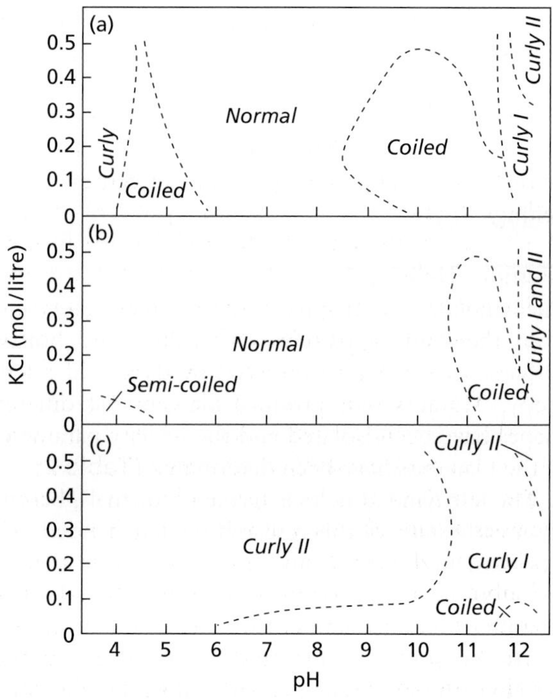

Fig. 7 Phase diagram of polymorphic flagella. Three antigen-types of flagella are shown: (a) i-type; (b) n-type; and (c) c-type flagella. In boundary areas (dotted lines), two different shapes of filaments co-exist.

In practice, polymorphism is important for flagellar function, because it is essential for the survival of cells in adverse environments. When a cell rotates a number of flagella in the pushing mode, the filaments come together into a bundle to establish harmonious rotation. Filaments never become entangled with one another, because the cross-over points between any two filaments move towards the far free ends where crossings disappear. If some of the filaments rotate in the contrary direction, however, the bundles of filaments end up in a disastrous state; filaments push in opposite directions and become entangled with one another. If all the filaments rotate in the pulling mode, cross-overs between filaments travel upwards towards the proximal ends where they are attached to the cell, and here they become 'stuck'. Fortunately, if the cell is sufficiently active, this situation does not last long. Twisting stresses that accumulate in filaments bring about conformational changes in the flagellin subunits of the filaments and consequently convert left-handed Normal filaments into the right-handed Curly filaments. Waves of Curly filaments in the pulling mode then travel down towards the distal ends of the filaments and tangled cross-overs disappear. When filaments become free of a bundle, Curly returns to

Normal because stresses are released (Macnab and Ornston, 1977).

Calladine Model

Professor C.R. Calladine, a structures specialist in the Department of Engineering at Cambridge University, took a different view to explain the polymorphism of flagellar filaments. He regarded the flagellar filament as being built from flagellin-protein 'bricks'. It is straightforward to design a brick that will self-assemble to construct a straight tube having the known surface lattice with its 11 near-longitudinal 'strands' of subunits. In such a straight structure, each unit is in an identical environment with respect to its neighbours. But the helical flagellar filament is actually *curved*, and so the strands must have slightly different lengths as one goes around the circumference.

The crucial problem is how can this happen if the subunits are all identical. Calladine showed that a single, mechanically bi-stable 'switch' feature of the building-brick, such as alternate neighbouring bonding-points at inter-unit connections, could explain all of the known polymorphic features of flagellar filaments. According to this model, the geometry of the subunit is such that they will not connect into a tube if all of the ambivalent connections are either of 'type A', or of 'type B'; but a tube can be built if some are of one type and some are of the other. The different connecting arrangements make strands of different lengths, and the elastic energy of the assembly turns out to be least when the two types of connection 'cluster' into two zones around the circumference. The packing pattern propagates along the 11-start lattice lines. Flagellins from different strains and mutants have very slightly different geometry as a result of discrete amino-acid substitutions and so the ratio of the two types of connection varies, thereby producing different helical forms for different mutants. Figure 8 shows schematically the 12 possible packing patterns with different numbers (ranging from 0 to 11) of strands of type B on the left, and the corresponding helical forms (at a much smaller scale) on the right. In the centre is a plot of the curvature of the filaments against twist. Twist changes steadily as the number of strands of type B increases, while curvature is zero at the two extremes, and maximum in mid-range. The detailed dimensions of the 'bricks' were chosen so that some of the 12 theoretical states corresponded to the then experimentally observed polymorphic helical forms, some left-handed and some right-handed; since 1978 several more of the theoretical states have been found to exist.

This mechanical model explains polymorphism on account of small amino acid changes in the flagellin, changes of pH and temperature and co-polymerisation of flagellin from two species and mechanical torque. For details of the model see Calladine (1978).

Hook

The problems presented by the hook are as interesting as those of the flagellar filament. The hook is the short connector between the filament and the basal body and is said to function as a universal joint, which transmits torque from one axis to the other when they are fixed at an angle with respect to each other. In machines, a device to achieve perfect transmission requires a complex combination of rotary gears. Such a sophisticated mechanism cannot be expected in the hook, since it is made of a single kind of protein. A much simpler mechanism is therefore necessary, and the problems this presents will now be considered.

The hook is flexible to bending but rigid to twisting. This characteristic of the hook may be seen by analogy to the behaviour of rubber. The hook is, like rubber tubing, at one and the same time flexible to bending but rigid to twisting. Though a sheet of rubber is extendable to any direction, rubber tubing if sufficiently thick is resistant to twisting, but not to bending. In a similar way, the interactions between the hook subunits are only extendable.

The wild-type hook is of a relatively well-controlled average length of 55 nm with a standard deviation of 6 nm (Hirano *et al.*, 1994), but the length of filaments varies over a wide range. This raises the problem of how such an accuracy is achieved, the molecular 'scales' by which the length is determined and whether the defined length of the hook is related to its function.

If the principal function of the hook is that of a universal joint, it would be unnecessary in the flagella of certain organisms. For example, a single polar flagellum does not bend when rotating, and the cell would be able to go forwards or backwards even without a hook. Nevertheless, all flagella so far studied have a hook and this raises the question as to whether hooks have any other roles.

Hook Protein

Shearing forces applied to bacterial cells often lead to the co-isolation of hooks and filaments, which suggests that structurally the hook belongs to the filament. The *flgE* gene that encodes the hook protein is, however, located in the *flgB* operon of nine genes (*flgB–J*) that encode the component proteins of the basal body. The operon belongs to level II in the transcriptional hierarchy under control of the FliA–FlgM regulatory system (Kutsukake and Iino, 1994). In genetic terms, therefore, the hook belongs to the basal body group of structures.

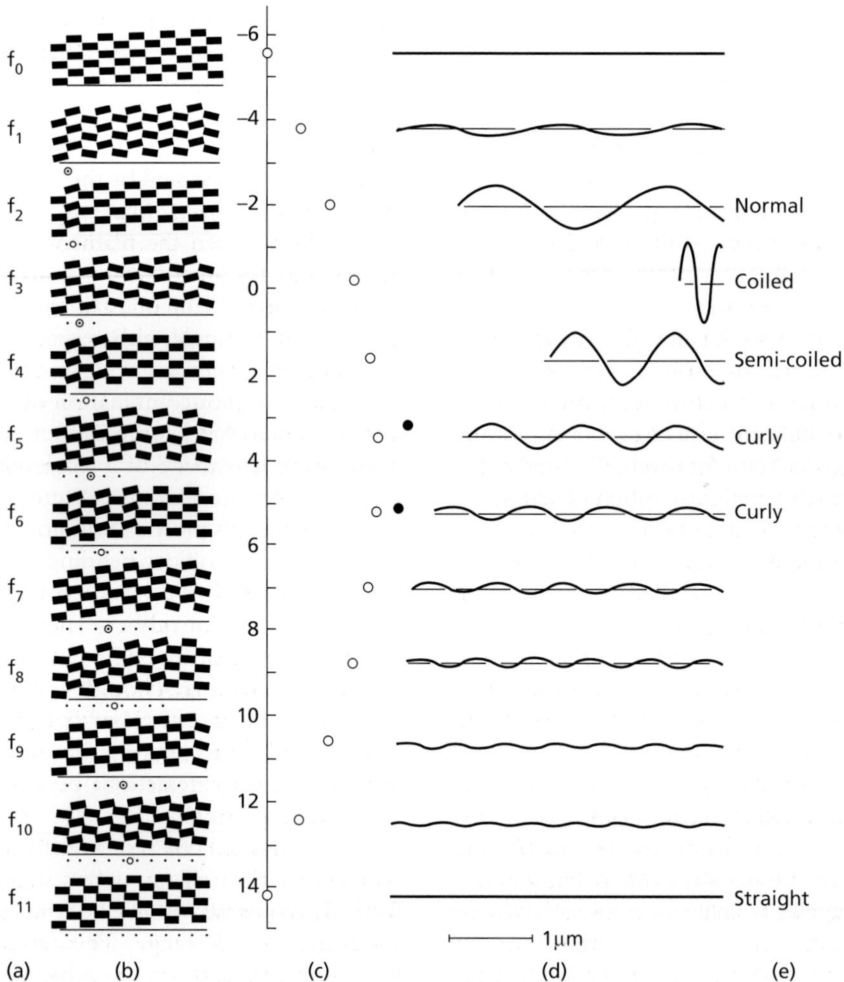

Fig. 8 (a) and (b): 12 distinct surface packing patterns adopted by A subunits. The packing patterns are designated f_i, where i is the number of longitudinal strands with alternative bonding. Strands marked (•) are slightly shorter than the others and impart curvature (κ) to the filament and cluster as shown to minimise elastic strain-energy. The centre of the cluster is marked (○). Twist (τ) is imparted to the filament by tilting longitudinal columns. Left-handed tilt (e.g. f_1) gives left-handed twist and a left-handed helix; the converse is true for right-handed tilt (e.g. f_{10}). (c) Plot of κ against τ for the filaments built by the 12 packings with equal scales for κ and τ. The ideal waveforms (○) are equally spaced on a half-sine wave and register with the patterns in (b). (d) The 12 ideal waveforms in register with (b) and (c). (e) The conventional descriptions of the waveforms. Some of the theoretical waveforms have not been observed. (After Calladine (1978), with permission).

The molecular size of *FlgE* in various species is very diverse. In *S. typhimurium* it consists of 402 amino acids, and it ranges from 26 kDa in *B. subtilis* to 70 kDa in *Helicobacter pylori*. Like flagellin, the hook molecule can be divided into two terminal regions and the central region. The amino acid sequences of the two terminal regions are well conserved, but those of the central region are to a great extent diverse. The size of the conserved sequence in the N-terminal region is not large, but it contains the sequence ANNLAN, which is commonly found in flagellin, hook-associated proteins (HAPs) and rod proteins, the flagellar axial proteins.

The secondary structure of *FlgE* is mostly β sheet, unlike that of flagellin which, according to CD measurements, contains a high content of α helices. As we have seen, not only the molecular architectures of the filament and hook but also the lattice structures of their polymer forms look very similar (Wagenknecht *et al.*, 1982). The distinctive differences between the physical properties and functions of these two types of tubular polymers may be derived from the differences between their secondary structures.

Polyhook

A group of mutants produce elongated hooks that have been termed polyhooks, although this term is misleading, because a polyhook is not a block-polymer of hook-units but a continuous polymer of hook protein.

The mutations responsible for polyhook mutants are located in *fliK* and/or *flhB* but not in *flgE*. This means that polyhook protein is identical to the ordinary hook protein FlgE. Since polyhooks can be much longer than wild-type hooks, they have often been used for biochemical analysis of hook protein. The *fliK*-polyhook mutants do not usually produce filaments, so the cells are non-motile, whereas *flhB*-polyhook mutants not only produce polyhook but also have filament attached and are motile. When observed by the critical dry method polyhook is a right-handed helix, whereas by negative staining electron microscopy, in which the helix is pressed down on the surface of the specimen grid, wavy filaments are seen. Ordinary hook is not, therefore, just a bending tube but also twisted tube with a quarter the wavelength of polyhook.

Under appropriate conditions *in vitro*, hook proteins obtained from polyhooks self-assemble into polyhooks of indefinite lengths (Kato *et al.*, 1982). However, in the same way that HAP2 (*FliD*) is necessary for filament formation, FlgD is necessary for hook formation. Without FlgD, hook proteins are secreted into media as monomers (Homma *et al.*, 1984), and polyhooks therefore self-assemble *in vitro* but not *in vivo*.

Since FlgD is always found at the tip of hooks in various mutants that lack filaments (Ohnishi *et al.*, 1994), it may be one of the HAP (see below) and act as a scaffolding protein for the hook. But FlgD is not found in mature flagella, which suggests that it may be replaced by other proteins. The most plausible candidate is HAP1, which in mature flagella locates at the distal end of the hook.

Hook-Associated-Proteins

Three kinds of minor proteins are present in the hook-filament region. These proteins were originally found at the tip of the hook in certain Fla-mutants and were, therefore, termed the hook-associated proteins (HAPs). Two of these, HAP1 and HAP3, are always found between the filament and hook and the other, HAP2, is found either at the tip of the filament or the hook in the absence of filaments. The copy numbers are estimated to be, respectively, HAP1 10–15, HAP3 10–30 and HAP2 6–12 (Ikeda *et al.*, 1987).

HAP1 polymerises on the hook when it reaches its mature length. The filament does not grow directly from the hook even though the lattices of these two structures are very similar. Since HAP1 mutants retain hooks of the same length as those of wild-type, HAP1 is not necessary to stop hook growth. It may be concluded, therefore, that HAP1 is not involved in the control of hook length, but its exact function is not known. Since HAP2 is found at the tip of the filament, it is termed the filament cap protein. Hexamers of HAP2 form a plug-like structure and seems to bind to the distal end of the filament. Without HAP2, flagellin is secreted into media as monomer, which suggests that HAP2 helps flagellin to polymerise into filaments (Ikeda *et al.*, 1996).

HAP3 functions as a filament stabiliser. A mutation in HAP3 allows flagellin to grow into apparently normal filaments, but the mutant cells swim in a strange way. Dark-field microscopy shows that flagella of these mutants cannot hold the Normal shape for long, but they frequently change their helical shape while swimming, indicating that polymorphic transitions due to twisting stresses are occurring all the time (Fahner *et al.*, 1994). How filament structural stability is maintained in the wild-type remains a mystery.

Basal Body

The basal body is the structure that represents the flagellar motor. Though this suggests that the structure and the motor are one and the same, this cannot be stated with confidence, since some structural components necessary for the full function of the motor are missing from the current model of the basal body. Consequently, each time a new component has been found, the definition of the basal body has been refined. The correspondence between the structure and function of the basal body is summarised in Table 3. The two types of terminology used in Table 3 often lead to confusion because they are often used without definition. In what follows, the structural terminology will be used as far as possible.

The basal body was originally described as a granular particle at the base of the flagellum. It was isolated by DePamphilis and Adler (1973) and defined as a structure that consists of rings with a rod passing

Table 3 Relationship between function and structure of the flagellum

Function	Structure	Protein
Flagellar motor	*Basal body*	
Bushing	LP ring complex	FlgH, FlgI
Shaft	Rod	FlgB, FlgC, FlgF, FLgG
Rotor[a]	MS ring complex	FliF
	C ring	FliG, FliM, FliN
Stator[a]	Mot complex	MotA, MotB
Export Apparatus	C rod	FliO, FliP, FliQ, FliR
	Chaperones	FliS, FliT, FliJ, FlgN

[a] The stator and rotor have not been identified. This assignment is tentative.

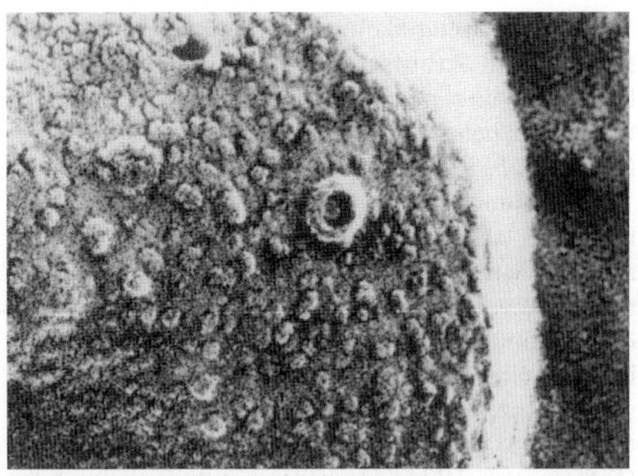

Fig. 9 Electron micrograph of the C ring shown by the quick-freeze deep-etching replica method. The view is from the cytoplasmic surface of a *Salmonella* cell.

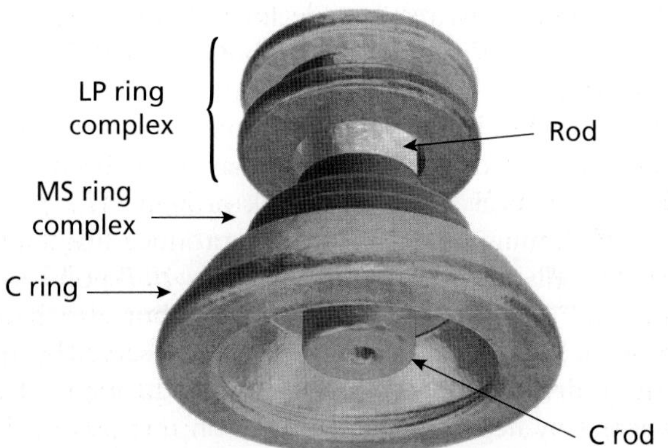

Fig. 10 Schematic model of the flagellar motor. The hook and filament are omitted. The basal body, the major structural portion of the flagellar motor, is composed of several structural complexes: the LP ring complex, rod, MS ring complex, C ring complex and C rod (See also Colour Plate 2).

through them. In *E. coli* there are four rings, but in *B. subtilis* there are only two. The difference in the number of rings in these two organisms relates to the difference in the cell envelope structures between Gram-negative and Gram-positive bacteria. Some organisms may have more than four rings, but the significance of the extra ring(s) is not clear.

Under isolation conditions milder than those originally used to purify the basal body (Khan *et al.*, 1991), a cup-shaped structure, called the C ring, was found to be attached to the cytoplasmic side of the M ring. The C ring can be visualised by quick-freeze deep-etching replica electron microscopy (**Fig. 9**) (Katayama *et al.*, 1996). When *in situ*, C rings look similar to those isolated from cells by biochemical means, indicating that the structure is sufficiently stable to tolerate treatment with detergent. At the centre of the C ring is a rod-like structure designated the C rod, which may be the export apparatus that has been sought.

In summary, therefore, the current model of the basal body of Gram-negative bacteria consists of five sub-structures, the LP ring complex, the MS ring complex, a periplasmic rod, the C ring and the C rod (**Fig. 10**).

The LP Ring Complex

The LP ring complex functions as a bushing through the outer membrane of Gram-negative bacteria; corresponding structures do not exist in Gram-positive bacteria. The complex is extraordinary rigid and it is stable over a range of pH from 2 to 12 and for a few minutes at boiling point (Aizawa *et al.*, 1985). The diameter of the complex ($\sim$50 nm) is sufficiently uniform that a two-dimensional sheet of the complexes can extend up to a few micrometres without cracking (Akiba *et al.*, 1991). The inner surface of the complex is very smooth, because it sustains friction caused by rotation of the rod at speeds as high as 1700 Hz. The super-lubrication mechanism of the LP ring complex is unknown.

The LP ring complex consists of two proteins, FlgH (L ring) and FlgI (P ring). The secretion of these two proteins is exceptional for flagellar proteins. They have signal peptides that can be cleaved by the primary secretion system, whereas the other flagellar proteins are exported by the flagella-specific export apparatus without cleavage (Jones *et al.*, 1989). FlgH is a lipoprotein and the lipoyl modification is necessary for its proper function (Schoenhals and Macnab, 1996). FlgI is said to bind to the peptidoglycan layer, but little is known about this interaction.

The LP ring complex is constructed around the rod as a bushing. Since the secretion of FlgI and FlgH is different from that of other flagellar component proteins, they may first be released in the periplasm and then gather diffusively at the rod. FlgI self-assembles directly on to the rod to form a collar-like P ring. At the same time the P ring becomes connected to the peptidoglycan layer. Then, FlgH self-assembles directly on to the rod, at a position a little above the P ring. When the L ring is formed and the C-terminal ends of each FlgH subunit interact with the P ring, the two rings undergo major conformational changes to form an independent complex that releases itself from the rod.

Rod

The rod is a tubular polymer. There is no direct evidence for this, but the structure must be a tube to allow flagellin to pass through during filament elongation.

The tube is not uniform, but electron microscopy shows that it consists of two parts, a distal and a proximal part.

The rod consists of four component proteins: FlgB, FlgC, FlgF and FlgG. The proximal half, which belongs to the MS ring complex, consists of the M ring, the S ring and the proximal rod. It was long thought that the M and S rings play an important role in the function of the motor, and that torque is generated between the two (Berg, 1974) but in fact the two rings are actually constructed of a single protein, FliF (Ueno *et al.*, 1992). In both its terminal regions FliF has sequences that span the membrane, and about 36 FliF subunits self-assemble to form an MS ring complex in the cytoplasmic membrane. The lower half of the M ring is buried in the membrane, and the other parts are exposed to the periplasmic space. If the 36 FliF subunits are arranged in a circle, the inner diameter of the ring will be about 10 nm if the distance between the membrane-spanning helices is assumed to be 1 nm. The MS ring complex is the first structure to be assembled during flagellar morphogenesis and it is the 'construction base' on which all other sub-structures are built (see below).

The distal part of the flagellum is constructed of FlgG. The locations of the minor components, FlgB, FlgC and FlgF, is unknown, but only about six copies of each molecule are found in mature flagella (Jones *et al.*, 1990). Flagella sheared from cells retain the distal rod attached to the hook-filament complex, indicating that the joint between the two parts of the rod is mechanically weak (Okino *et al.*, 1989).

The C ring

The C (cytoplasmic) ring is a cup-shaped structure attached to the cytoplasmic surface of the M ring (**Fig. 10**) (Katayama *et al.*, 1996). Under mild conditions it can be isolated from the basal body fraction as an intact structure, but it is fragile under conditions of high salt concentrations and high pH (Khan *et al.*, 1991).

The C ring is the switch complex. It contains all three switch components, FliG, FliM and FliN, that have been observed by immunoelectron microscopy (Francis *et al.*, 1994) and the genes that encode these proteins have been identified as responsible for switching the rotational directions of the flagellar motor. In mutants of any one of these genes, the rotational direction of flagella is in a fixed direction, clockwise or counter-clockwise. Genetic analysis of these genes has revealed that they are responsible not only for switching but also for flagellum formation and torque generation (Yamaguchi *et al.*, 1986). The detailed roles of the switch complex in torque generation and flagella formation have, however, not been clarified. The physical interactions between these three proteins has been directly demonstrated *in vitro* with the purified proteins (Oosawa *et al.*, 1994).

The genes *fliF* and *fliG* are located next each other with an 8-bp overlap. A mutant with a 7-bp in-frame deletion in this overlapping region results in a fusion of the proteins FliF and FliG (Francis *et al.*, 1992). Since this mutant is still able to swim, the FliF–FliG must rotate together, which is contrary to the hypothesis that the M ring is the rotor or the torque generator. The C-terminus of FliF is predicted to be located on the outside of the M ring, which suggests that FliG assembles at the periphery of the M ring. FliG binds to the cytoplasmic surface of the M ring.

Of the switch proteins that have been purified, FliM interacts only with phosphorylated CheY, a signal protein from the sensory transduction system (Welch *et al.*, 1993). Conformational changes in FliM induced by binding of CheY seem to make the flagellar motor go into reverse rotation. The copy numbers of FliM and FliG are of the same order, 30–40, whereas that of FliN is estimated to be in hundreds. FliN subunits assemble at the periphery of the C ring to form a cup-shaped structure, and a mutant in which FliM and FliN are fused can swarm, indicating that FliN is not the torque generator.

C rod

The C rod is a candidate for the flagella-specific export apparatus, because of its position in the central space of the C ring.

The major part of the C rod consists of at least four membrane proteins: FliO, FliP, FliQ and FliR (Ohnishi *et al.*, 1997). These proteins self-assemble into the C rod in a small space on the cytoplasmic side of the M ring that surrounds the C ring. Whether there is a membrane constituent in this space or the C rod directly binds to the M ring is not known. The C rod has never been observed by conventional methods other than quick-freeze techniques (Katayama *et al.*, 1996). It has been elusive, probably because it is very fragile and, therefore, missing from C ring preparations. FlhA and FlhB are expected to be components of the C rod, but their location relative to the other components is not known.

Chaperones

Efficient export of flagellar component proteins requires chaperones. Three soluble chaperones are known: FliS for flagellin, FliT for HAP2 and FlgN for HAP1/HAP3 (Fraser *et al.*, 1999). Another soluble component, FliI, has ATPase activity, suggesting that it has a role in the gate of export (see below).

The Mot Proteins

The Mot complex, which is composed of the proteins MotA and MotB, has been regarded as the stator of the flagellar motor. This remains hypothetical, however, because attempts to visualise the structure have been unsuccessful. On theoretical grounds, a motor should have a stator. The requirements are that it should be fixed on a rigid wall of the motor and that it must convert energy in the interaction with the rotor. If this is correct, the Mot complex is the only candidate for the stator.

MotA and MotB are both membrane proteins. MotA has four membrane-spanning regions that form a proton channel (Blair and Berg, 1990). MotB has one membrane-spanning region in the N-terminal region, and a consensus sequence that is found in many other proteins that interact with the peptidoglycan layer (Chun and Parkinson, 1988). The Mot complex functions as the stator and as the torque-generating unit.

Some bacterial species have more than two Mot proteins. Thus, *V. alginolyticus* has PomA and PomB, and also MotX and MotY. PomA and PomB are the sodium conducting channel components of the sodium driven motor, while MotX and MotY are the proton-conducting channel of the proton-driven motor. *Rhizobium meliloti* has MotA, MotB, MotC and MotD. MotA and MotB correspond to the proteins in *E. coli*. The roles of MotC and MotD are probably minor (Platzer *et al.*, 1997).

Mot proteins are not necessary for flagella formation: Mot mutants have intact but paralysed flagella. Expression of Mot proteins from a plasmid leads to recovery from paralysis in a corresponding Mot mutant; at first the paralysed cells swim slowly but eventually they reach full speed (Block *et al.*, 1984). This so-called 'resurrection' experiment indicates that the motor does not work until the final construction steps in flagellum formation are completed.

Assembly System of Flagella

As in the case of bacteriophages, self-assembly has been regarded as the mechanism of flagellum assembly, because these two supra-molecular structures are of similar size and composed of many components with rotational symmetry. Recent studies on the assembly of flagella have, however, revealed that flagellum formation is more complicated. The differences between flagella and bacteriophages are mostly derived from their sites of construction; bacteriophages are reproduced within the cell, whereas flagella grow outside the cell.

Morphogenesis

Fla$^-$ Mutants

Flagellar mutants are conventionally selected in semi-solid agar in which wild-type cells swarm but mutants do not. Three types of non-motile mutants are known, most of which do not have any filaments (Fla$^-$ mutants), but in some mutants filaments are present but paralysed (Mot$^-$ mutants). In some other mutants, filaments rotate but the switching of rotational directions is abnormal (Che$^-$ mutants).

When observed by electron microscopy, some Fla$^-$ mutants retain partial structures of the basal body (Suzuki *et al.*, 1978; Kubori *et al.*, 1992). If the partial structures of the basal body of Fla$^-$ mutants are aligned in order of size, the morphological pathway of the basal body is revealed (**Fig. 11**).

Distal Growth

The flagellum grows from its distal end, unlike pili and fimbriae which grow at their proximal end. Distal growth is typically observed in experiments with filaments both *in vitro* and *in vivo*. If filaments are

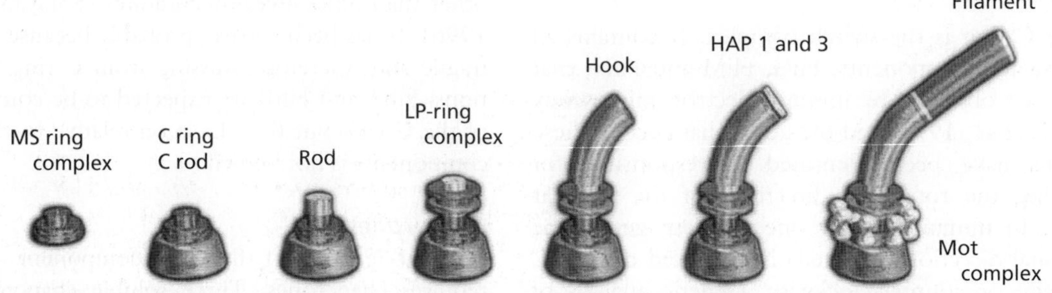

Fig. 11 Morphological pathway of flagellar assembly. The MS ring complex is the first structure of flagellar assembly that can be recognized by electron microscopy. After the C ring and C rod have created an export apparatus on the cytoplasmic side of the M ring, flagellar proteins are exported to construct flagella. The Mot complex is formed at the last stage of assembly at the same time as filament growth (See also Colour Plate 3).

reconstituted *in vitro*, they grow from only one end of the filament-fragments (Asakura *et al.*, 1964). As observed by electron microscopy, the fragments have asymmetrical ends but the proximal end is pointed like an arrow-head, whereas the distal end has a 'fish tail' appearance. Under *in vivo* conditions, Curly filaments are produced if *p*-fluorophenylalanine is added to the growth medium. These are found attached to the distal end of Normal filaments, indicating that new filaments growth occurs distally (Iino, 1969).

Non-Self Assembly

As we have seen above, filaments self-assemble *in vitro*, but not *in vivo*. The same is true of hooks. Both structures require helper proteins to elongate *in vivo*. Though flagellin has an intrinsic ability to polymerise into a filament spontaneously, the optimal conditions for polymerisation *in vitro* and *in vivo* are different.

In vitro, filaments grow most rapidly at 26°C, whereas *in vivo* filaments grow at 37°C. The growth rate *in vitro* is 10-fold slower than that *in vivo* (Aizawa and Kubori, 1998). Furthermore, since under optimal conditions a cell divides every 20 minutes, filament elongation must proceed as fast as cell division, but it takes overnight for the filaments to grow even a few microns *in vitro*. *In vivo*, the specific export apparatus is composed of the C ring and the C rod is necessary for the efficient assembly of flagella.

Export Apparatus

Type III Secretion System

In *Salmonella*, the invasion system secretes pathogenicity factor proteins into the growth medium without their being processed. This group of proteins is known as the type III secretion system.

Consensus Sequence

In the N-terminal regions of the flagellar axial proteins, rod proteins, hook protein, HAPs and flagellin, there is a consensus sequence (ANNLAN). Although there are no signal sequences in flagellar proteins, the consensus sequence is recognised by the export apparatus. In common with these flagellar proteins, certain proteins related to pathogenicity do not have the signal sequences.

FliI ATPase Activity

FliI has sequence similarity to the catalytic β subunit of the F_0F_1-ATPase (Vogler *et al.*, 1991) (see also below). It has even greater similarity to the Spa47 protein of *Shigella flexneri* and the InvC protein of *S. typhimurium*. These proteins are believed to play a role in the export of virulence proteins. Furthermore, an antibody directed against the F1 β subunit of *E. coli* cross-reacts with FliI, which confirms that these proteins are structurally related (Dreyfus *et al.*, 1993; Fan and Macnab, 1996).

Control of Hook Length

The hook is an extracellular polymer, and it is a typical example where the export apparatus may play a central role in the control of hook length. The *fliK* gene has been regarded as controlling hook length, because the mutational sites of polyhook mutants locate in *fliK* but not in *flgE*. The most popular mechanism for control of length at molecular level is by the use of a molecular scale, as seen in tobacco mosaic virus and bacteriophage tail. Even the null mutant of *fliK* gives rise to polyhook, however, implying that *FliK* is not the molecular ruler for the hook (Williams *et al.*, 1996). Since hook length appears not to be regulated by a molecular scale, there is presumably another mechanism, such as control of length by the amount of proteins exported (Makishima *et al.*, 2001).

Although *flhB* mutants do not produce polyhooks, some *flhB* mutants produce polyhook-filaments that swim clumsily. The null mutant of *flhB* has no flagella, suggesting that *FlhB* is involved in the export apparatus.

Morphological Pathway

So far, the flagellum has been dissected and analysed from the filament to the basal body. Now, to see the whole story, flagellar growth is described in the sequence that occurs naturally *in vivo* (Aizawa, 1996). The structure of a flagellum and its component proteins is illustrated in Fig. 12.

In the Cytoplasm

The MS ring complex is the first structure in flagellar construction. Of the proteins expressed at the very beginning of flagellum formation, FliF is the first to self-assemble in the cytoplasmic membrane into a supra-molecular structure, the MS complex. Then, the switch proteins, FliG, FliM and FliN, assemble to form the C ring at the periphery of the cytoplasmic surface of the M ring part of the complex. Independently, at the central part, proteins FliO, FliP, FliQ and FliR assemble into the C rod, which functions as an export

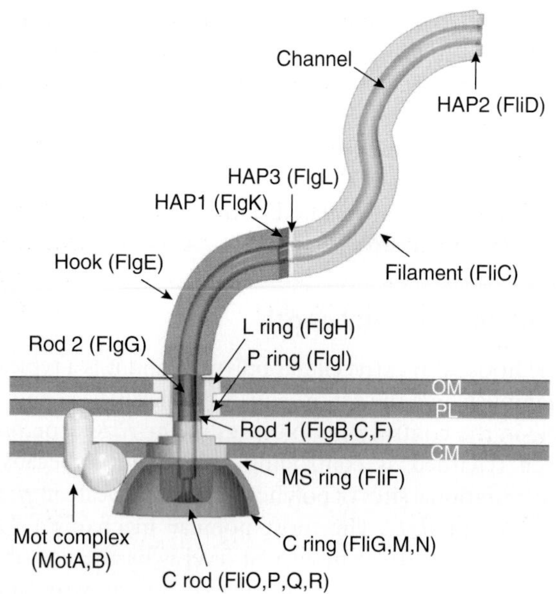

Fig. 12 Molecular structure of the flagellum of *S. typhimurium* (cf. Fig. 2). The components of the flagellum are shown with the gene that encodes them indicated in brackets. HAP, hook-associated protein; OM, outer membrane; PL, peptidoglycan layer; CM, cell membrane (See also Colour Plate 4).

apparatus specific for flagellar proteins. The roles of the C ring at this stage are unclear, though it is known to be essential at the beginning of flagellum formation.

In the Periplasmic Space

The locations of FlgB, FlgC, FlgG and FlgF, the four proteins that comprise the rod, are not known. To form a rod, all must be present at the same time; the lack of any one of them causes a failure of rod construction (Kubori *et al.*, 1992). FlgF forms the proximal half of the rod and FlgG the distal half, leaving the other components to connect the two parts.

When the rod becomes mature, the hook begins to grow at its tip, but growth is soon arrested, presumably by physical obstruction of the outer membrane. To overcome this physical barrier, the outer P and L rings must be formed in order to form a hole through the outer membrane. Unlike other flagellar proteins, these two proteins are secreted in a conventional signal-peptide-dependent manner. A precursor of the P ring assembles near the mid-point of the rod to form a structure resembling a neck collar, which seems to be firmly fixed on the rod.

The assembly of the L ring proteins on to the P ring seems to relieve the P ring from the constraint. Then the two rings join by an unknown mechanism to form a rigid complex, which is now extremely stable to acid, heat and enzymatic action. Once the outer ring complex is formed, the interaction between the rod and the complex becomes negligible, allowing them to rotate with respect to each other and the ring complex to slip off the rod when the hook is absent.

Outside the Cell

The hook grows continuously with the help of FlgD once the physical barrier is removed. Hook growth stops when the hook length has reached a certain value, but the mechanism of length control is unknown. When the hook reaches its mature length, FlgD is replaced by HAP1. FlgD is therefore a scaffolding protein. HAP1 stays at the distal end of the hook to allow HAP3 to assemble on to it. The HAPs assemble in the order HAP1–HAP3–HAP2, but whether the monomers assemble in a proximal-to-distal fashion within each zone, as in the case of the hook, is not known. The inability to form extended HAP1 or HAP3 polymers *in vitro*, even with very high monomer concentrations, suggests that HAP1 and HAP3 self-regulate the lengths of their respective zones.

Filament formation requires HAP2, which functions much as FlgD does in hook growth by adding new flagellin subunits beneath a HAP2 cap. Unlike FlgD, HAP2 stays at the tip of filaments indefinitely unless the filaments are mechanically broken. Filament length is also regulated, but this regulation is not as strict as it is in hook growth. The length distribution suggests that filaments do not grow at a constant rate but, as in the case of the hook, they grow very rapidly at first and then slow down exponentially.

As long as their structure is not mechanically broken, flagella seem to be used indefinitely. If the filaments are broken, they may re-grow by sending a new set of HAP2 and new flagellins to the tip. If the flagellum is broken at the basal body, however, as often happens, it will not re-grow, and new flagella appear (Aizawa and Kubori, 1998).

Origin of Flagella

Type III Secretion System

The structural complexity of the flagellar motor and the involvement of more than 40 genes in the flagellar system naturally invites the question of their origin. A line of evidence suggests that the flagellum shares a common ancestor with other systems, such as the virulence factor secretion systems of pathogenic bacteria (Lee, 1997). Virulence factors are secreted without signal sequences by the type III secretion systems of *Salmonella*, *Shigella* and *Yersinia*, and many genes are involved (Lee, 1997). The regulation of these virulence factors has been reviewed by Lucas and Lee (2000). It is also striking that the components of the type III secretion system and the basal body components are homologous (Hueck, 1988; Aizawa, 2001) (**Table 4**).

Needle-like structures protruding from the cell have been observed in *Salmonella* and they have been purified (Kubori *et al.*, 1998) (**Fig. 13**). When purified to homogeneity these needle-like structures show structural similarity to the flagellar basal body, except that the hook portion looks very straight and is therefore called a needle (**Fig. 14**).

The major protein components of the needle structure, PrgK, PrgH and InvG, have been identified as virulence gene products (Kubori *et al.*, 1998). As far as their structure and function are concerned, the export apparatus of the flagellar motor and the needle structure may have evolved from a common ancestor.

Table 4 Homologues of the type III secretion apparatus needle complex of various bacterial species and the flagellar basal body

Type III export apparatus core components	*Salmonella*		*Yersinia*	*Shigella*	EPEC[b]	*Pseudomonas*	*Ralstonia solanacearum*	Flagella
	SPI-1[a]	SPI-2						
1	SpaP	SsaR	YscR	Spa24	EscR	PscR	HrcR	FliP
2	SpaQ	SsaS	YscS	Spa9	EscS	PscS	HrcS	FliQ
3	SpaR	SsaT	YscT	Spa29	EscT	PscT	HrcT	FliR
4	SpaS	SsaU	YscU	Spa40	EscU	PscU	HrcU	FlhB
5	InvA	SsaV	LcrD	MxiA	EscV	PcrD	HrcV	FlhA
6	InvC	SsaN	YscN	Spa47	EscN	PscN	HrcN	FliI

[a] SPI-1 and SPI-2 are the two *Salmonella* pathogenicity islands.
[b] Enteropathogenic *Escherichia coli*.

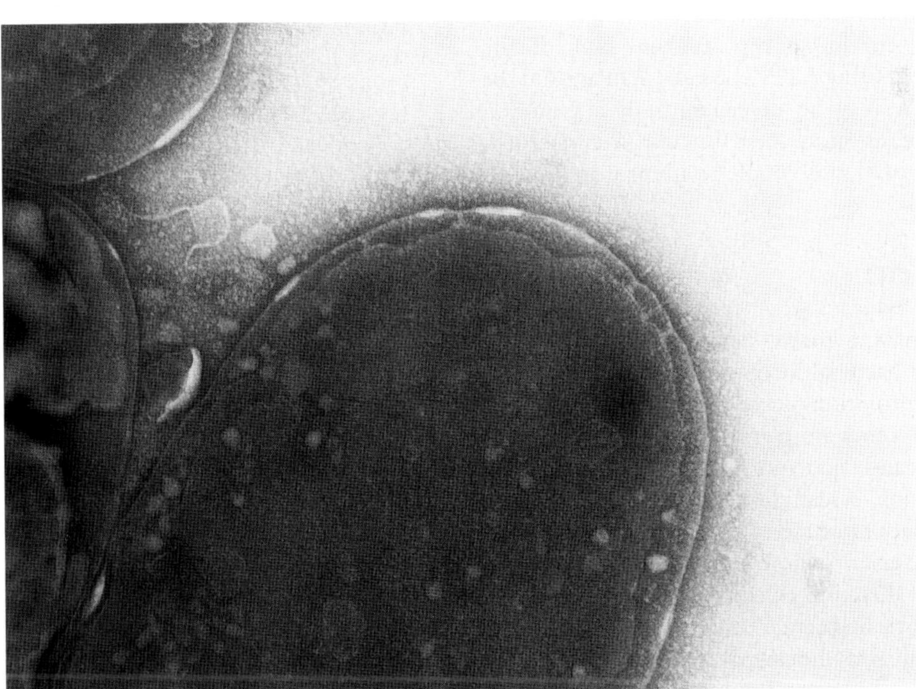

Fig. 13 Electron micrograph of the needle complexes of *S. typhimurium* protruding from the cell surface. The needle complexes appear to originate from the inner membrane.

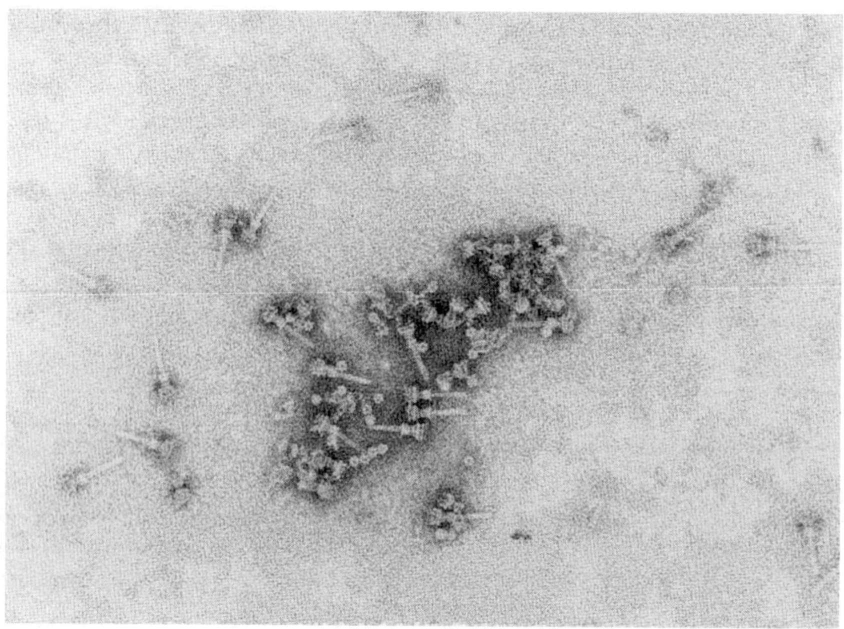

Fig. 14 Electron micrograph of the purified needle complexes of *S. typhimurium*.

F_0F_1-ATPase

Since its discovery in 1974, the flagellar motor has been regarded as 'the smallest rotary device in nature'. This striking description is now applied to the F_0F_1-ATPase, and Noji *et al.* (1997) showed that it is able to rotate. It is an ATP synthetase that is energised by PMF and is also found in mitochondria. In the reverse reaction, the enzyme pumps protons out of the membrane by utilising ATP. The F_0F_1-ATPase and the flagellar motor may be descendants of a common ancestor, since, as far as their function is concerned, in both cases rotation is coupled with proton flow.

Conclusion

The flagellum is a rotary device that has evolved exclusively for bacterial locomotion. It rotates at several hundred revolutions per second and uses the PMF. More than 40 genes are involved in the construction of flagellum, and an export apparatus specific for flagellar proteins, a scaffolding protein and capping proteins are elaborated for efficient construction. The number of flagella, helical handedness of filament and rotational direction of the flagellar motor determine the behavioural range of each bacterial species. The structural and functional similarities between the flagellum, the type III export apparatus and F_0F_1-ATPase, are suggestive of their evolutionary history.

References

Aizawa S-I (1996) Flagella assembly in *Salmonella typhimurium*. *Mol. Microbiol.* 19: 1–5.

Aizawa S-I (2001) Bacterial flagella and type III secretion system. *FEMS Microbiol. Lett.* (in press).

Aizawa S-I, Kubori T (1998) Bacterial flagellation and cell division. *Genes to Cells* 3: 1–10.

Aizawa S-I, Dean GE, Jones CJ, Macnab RM, Yamaguchi S (1985) Purification and characterization of the flagellar hook-basal body complex of *Salmonella typhimurium*. *J. Bacteriol.* 161: 836–849.

Aizawa S-I, Vonderviszt F, Ishima R, Akasaka K (1990) Termini of *Salmonella* flagellin are disordered and become organized upon polymerization into flagellar filament. *J. Mol. Biol.* 211: 673–677.

Akiba T, Yoshimura H, Namba K (1991) Monolayer crystallization of flagellar L-P rings by sequential addition and depletion of lipid. *Science* 252: 1544–1546.

Asai Y, Kojima S, Kato H *et al.* (1997) Putative channel components for the fast-rotating sodium-driven flagellar motor of a marine bacterium. *J. Bacteriol.* 179: 5104–5110.

Asakura S, Eguchi G, Iino T (1964) Reconstitution of bacterial flagella *in vitro*. *J. Mol. Biol.* 10: 42–56.

Atsumi T, McCarter L, Imae Y (1992) Polar and lateral flagellar motors of marine *Vibrio* are driven by different ion-motive forces. *Nature* 355: 182–184.

Berg HC (1974) Dynamic properties of bacterial flagellar motors. *Nature (London)* 249: 77–79.

Blair DF (1995) How bacteria sense and swim. *Annu. Rev. Microbiol.* 49: 489–522.

Blair DF, Berg HC (1990) The MotA protein of *Escherichia coli* is a proton-conducting component of the flagellar motor. *Cell* 60: 439–449.

Block SM, Blair DF, Berg HC (1984) Successive incorporation of force-generating units in the bacterial rotary motor. *Nature* 309: 470–472.

Calladine CR (1978) Change of waveform in bacterial flagella: The role of mechanics at the molecular level. *J. Mol. Biol.* 118: 457–479.

Chun SY, Parkinson JS (1988) Bacterial motility: membrane topology of the *Escherichia coli* MotB protein. *Science* 239: 276–278.

DePamphilis ML, Adler J (1971) Fine structure and isolation of the hook-basal body complex of flagella from *Escherichia coli* and *Bacillus subtilis*. *J. Bacteriol.* 105: 384–395.

DeShazer D, Brett PJ, Carlyon R, Woods DE (1997) Mutagenesis of *Burkholderia pseudomallei* with Tn5-OT182: Isolation of motility mutants and molecular characterization of the flagellin structural gene. *J. Bacteriol.* 179: 2116–2125.

Dreyfus G, Williams AW, Kawagishi I, Macnab RM (1993) Genetic and biochemical analysis of *Salmonella typhimurium* FliI, a flagellar protein related of the catalytic subunit of the F_0F_1 ATPase and to virulence proteins of mammalian and plant pathogens. *J. Bacteriol.* 175: 3131–3138.

Fahner KA, Block SM, Krishnaswamy S, Parkinson JP, Berg HC (1994) A mutant hook-associated protein (HAP3) facilitates torsionally induced transformations of the flagellar filament of *Escherichia coli*. *J. Mol. Biol.* 238: 173–186.

Fan F, Macnab RM (1996) Enzymatic characterization of FliI-An ATPase involved in flagellar assembly in *Salmonella typhimurium*. *J. Biol. Chem.* 271: 31981–31988.

Francis NR, Irikura VM, Yamaguchi S, DeRosier DJ, Macnab RM (1992) Localization of the *Salmonella typhimurium* flagellar switch protein FliG to the cytoplasmic M-ring face of the basal body. *Proc. Natl Acad. Sci. USA* 89: 6304–6308.

Francis NR, Sosinsky GE, Thomas D, DeRosier DJ (1994) Isolation, characterization and structure of bacterial flagellar motors containing the switch complex. *J. Mol. Biol.* 235: 1261–1270.

Fraser GM, Bennett JCQ, Hughes C (1999) Substrate-specific binding of hook-associated proteins by FlgN and FliT, putative chaperones for flagellum assembly. *Mol. Microbiol.* 32: 569–580.

Fujime S, Maruyama M, Asakura S (1972) Flexural rigidity of bacterial flagella studied by quasielastic scattering of laser light. *J. Mol. Biol.* 68: 511–522.

Hirano T, Yamaguchi S, Oosawa K, Aizawa S-I (1994) Roles of FliK and FlhB in determination of flagellar hook length in *Salmonella typhimurium*. *J. Bacteriol.* 176: 5439–5449.

Homma M, Fujita H, Yamaguchi S, Iino T (1984) Excretion of unassembled flagellin by *Salmonella typhimurium* mutants deficient in hook-associated proteins. *J. Bacteriol.* 159: 1056–1059.

Iino T (1969) Polarity of flagellar growth in *Salmonella*. *J. Gen. Microbiol.* 56: 227–239.

Iino T, Komeda Y, Kutsukake K *et al.* (1988) New unified nomenclature for the flagellar genes of *Escherichia coli* and *Salmonella typhimurium*. *Microbiol. Rev.* 52: 533–535.

Ikeda T, Homma M, Iino T, Asakura S, Kamiya R (1987) Localization and Stoichiometry of Hook-associated proteins within *Salmonella typhimurium* flagella. *J. Bacteriol.* 169: 1168–1173.

Ikeda T, Oosawa K, Hotani H (1996) Self-assembly of the filament capping protein, FliD, of bacterial flagella into an annular structure. *J. Mol. Biol.* 259: 679–686.

Imae Y, Mizuno T, Maeda K (1984) Chemosensory and thermosensory excitation in adaptation-deficient mutants of *Escherichia coli*. *J. Bacteriol.* 159: 368–374.

Jarrel KF, Bayley DP, Kostyukova AS (1996) The archaeal flagellum: a unique motility structure. *J. Bacteriol.* 178: 5057–5064.

Jones CJ, Homma M, Macnab RM (1989) L-, P-, and M-ring proteins of the flagellar basal body of *Salmonella typhymurium*: gene sequences and deduced protein sequences. *J. Bacteriol.* 171: 3890–3900.

Jones CJ, Macnab RM, Okino H, Aizawa S-I (1990) Stoichiometric analysis of the flagellar hook-(basal-body) complex of *Salmonella typhimurium*. *J. Mol. Biol.* 212: 377–387.

Joys TM, Stocker BAD (1963) Mutation and recombination of flagella antigen I of *Salmonella typhimurium*. *Nature (London)* 197: 413–414.

Kamiya R, Asakura S (1977) Flagellar transformations at alkaline pH. *J. Mol. Biol.* 108: 513–518.

Kamiya R, Hotani H, Asakura S (1982) Polymorphic transition in bacterial flagella. In: Amos WB, Duckett JD (eds) *Prokaryotic and Eukaryotic Flagella*. Cambridge: Cambridge University Press, pp. 53–76.

Kanto S, Okino H, Aizawa S-I, Yamaguchi S (1991) Amino acids responsible for flagellar shape are distributed in terminal regions of flagellin. *J. Mol. Biol.* 219: 471–480.

Katayama E, Shiraishi T, Oosawa K, Baba N, Aizawa S-I (1996) Geometry of the flagellar motor in the cytoplasmic membrane of *Salmonella typhimurium* as determined by stereo-photogrammetry of quick-freeze deep-etch replica images. *J. Mol. Biol.* 255: 458–475.

Kato S, Aizawa S-I, Asakura S (1982) Reconstitution *in vitro* of the flagellar polyhook from *Salmonella*. *J. Mol. Biol.* 161: 551–560.

Khan S, Khan IH, Reese TS (1991) New structural features of the flagellar base in *Salmonella typhimurium* revealed by rapid-freeze electron microscopy. *J. Bacteriol.* 173: 2888–2896.

Kubori T, Shimamoto N, Yamaguchi S, Namba K, Aizawa S-I (1992) Morphological pathway of flagellar assembly in *Salmonella typhimurium*. *J. Mol. Biol.* 226: 433–446.

Kubori T, Matsushima Y, Nakamura D *et al.* (1998) Supramolecular structure of the *Salmonella typhimurium* type III protein secretion system. *Science* 280: 602–605.

Kudo S, Magariyama Y, Aizawa S-I (1990) Abrupt changes in flagellar rotation observed by laser dark-field microscopy. *Nature* 346: 677–680.

Kutsukake K, Iino T (1980) A trans-acting factor mediates inversion of a specific DNA segment in flagellar phase variation of *Salmonella*. *Nature* 284: 479–481.

Kutsukake K, Iino T (1994) Role of the FliA-FlgM regulatory system on the transcriptional control of the flagellar regulon and flagellar formation in *Salmonella typhimurium*. *J. Bacteriol.* 176: 3598–3605.

Lee CA (1997) Type III secretion systems: machines to deliver bacterial proteins into eukaryotic cell? *Trends Microbiol.* 5: 148–156.

Leifson E (1960) *Atlas of Bacterial Flagellation*. New York: Academic Press, 171 pp.

Lowe G, Meister M, Berg HC (1987) Rapid rotation of flagellar bundles in swimming bacteria. *Nature* 325: 637–640.

Lucas RL, Lee CA (2000) Unravelling the mysteries of virulence gene regulation in *Salmonella typhimurium*. *Mol. Microbiol.* 36: 1024–1033.

Macnab RM (1987) Flagella. In: Neidhardt FC (ed.) Escherichia coli *and* Salmonella typhimurium: *Cellular and Molecular Biology*. Washington, DC: American Society for Microbiology, pp. 70–83.

Macnab RM (1996) Flagella and motility. In: Neidhardt FC (ed.) Escherichia coli *and* Salmonella typhimurium: *Cellular and Molecular Biology*, 2nd edn. Washington, DC: American Society for Microbiology, pp. 123–145.

Macnab RM, Koshland DE Jr (1972) The gradient-sensing mechanism in bacterial chemotaxis. *Proc. Natl Acad. Sci. USA* 69: 2509–2512.

Macnab RM, Ornston MK (1977) Normal-to-curly flagellar transitions and their role in bacterial tumbling: Stabilization of an alternative quaternary structure by mechanical force. *J. Mol. Biol.* 112: 1–30.

Magariyama Y, Sugiyama S, Muramoto K *et al.* (1994) Very fast flagellar rotation. *Nature* 371: 752.

Makishima S, Komoriya K, Yamaguchi S, Aizawa S-I (2001) Length of the flagellar hook and the capacity of the Type III export apparatus. *Science* 291: 2411–2413.

Manson MD, Tedesco PM, Berg HC, Harold FM, van der Drift C (1977) A proton motive force drives bacterial flagella. *Proc. Natl Acad. Sci. USA* 74: 3060–3064.

Mattick JS, Whitchurch CB, Aim RA (1996) The molecular genetics of type 4 fimbriae in *Pseudomonas aeruginosa* – a review. *Gene* 179: 147–155.

Merz AJ, So M, Sheetz MP (2000) Pilus retraction powers bacterial twitching motility. *Nature* 407: 98–102.

Mimori Y, Yamashita I, Murata K *et al.* (1995) The structure of the R-type straight flagellar filament of *Salmonella* at 9 Å resolution by electron cryomicroscopy. *J. Mol. Biol.* 249: 69–87.

Nagashima H, Asakura S (1980) Dark-field light microscopy study of the flexibility of F-actin complexes. *J. Mol. Biol.* 136: 169–182.

Newton SMC, Jakob CO, Stocker BAD (1989) Immune response to cholera toxin epitope inserted in *Salmonella* flagella. *Science* 244: 70–73.

Noji H, Yasuda R, Yoshida M, Kinoshita K (1997) Direct observation of the rotation of F1-ATPase. *Nature* 386: 299–302.

Ohnishi K, Ohta Y, Aizawa S-I, Macnab RM, Iino T (1994) FlgD is a scaffolding protein needed for flagellar hook assembly in *Salmonella typhimurium*. *J. Bacteriol.* 176: 2272–2281.

Ohnishi K, Fan F, Schoenhals GJ, Kihara M, Macnab RM (1997) The FliO, FliP, FliQ, and FliR proteins of *Salmonella typhimurium*: putative components for flagellar assembly. *J. Bacteriol.* 179: 6092–6099.

Okino H, Isomura M, Yamaguchi S *et al.* (1989) Release of flagellar filament-hook-rod complex by a *Salmonella typhimurium* mutant defective in the M ring of the basal body. *J. Bacteriol.* 171: 2075–2082.

Oosawa K, Ueno U, Aizawa S-I (1994) Overproduction of the bacterial flagellar switch proteins and their interactions with the MS ring complex *in vitro*. *J. Bacteriol.* 176: 3683–3691.

Pugsley AP (1993) The complete general secretory pathway in Gram-negative bacteria. *Microbiol. Rev.* 57: 50–108.

Rashid MH, Kornberg A (2000) Inorganic phosphate is needed for swimming, swarming and twitching motilities of *Pseudomonas aeruginosa*. *Proc. Natl Acad. Sci. USA* 97: 4885–4890.

Schoenhals GL, Macnab RM (1996) Physiological and biochemical analysis of FlgH, a lipoprotein forming the outer membrane L ring of the flagellar basal body of *Salmonella typhimurium*. *J. Bacteriol.* 178: 4200–4207.

Silverman M, Simon MI (1974) Flagellar rotation and the mechanism of bacterial motility. *Nature (London)* 249: 73–74.

Silverman M, Simon M (1980) Phase variation: genetic analysis of switching mutants. *Cell* 19: 845–854.

Smith NH, Selander RK (1990) Sequence invariance of the antigen-coding central regions of the phase I flagellar filament gene (fliC) among strains of *Salmonella typhimurium*. *J. Bacteriol.* 172: 603–609.

Stock JB, Surette MG (1996) Chemotaxis. In: Escherichia coli *and* Salmonella typhimurium: *Cellular and Molecular Biology*. Washington, DC: American Society for Microbiology, pp. 1103–1129.

Sun H, Zusman DR, Shi W (2000) Type Iv pilus of *Myxococcus xanthus* is a motility apparatus controlled by the *frz* chemosensory system. *Curr. Biol.* 10: 1143–1146.

Suzuki T, Iino T, Horiguchi T, Yamaguchi S (1978) Incomplete flagellar structures in nonflagellate mutants of *Salmonella typhimurium*. *J. Bacteriol.* 133: 904–915.

Ueno T, Oosawa K, Aizawa S-I (1992) The M ring, S ring and proximal rod of the flagellar basal body of *Salmonella typhimurium* are composed of subunits of a single protein, FliF. *J. Mol. Biol.* 227: 672–677.

Vogler A, Homma M, Irikura VM, Macnab RM (1991) *Salmonella typhimurium* mutants defective in flagellar filament regrowth and sequence similarity of FliI to F0F1, vacuolar, and archaebacterial ATPase subunits. *J. Bacteriol.* 173: 3564–3572.

Vonderviszt F, Kanto S, Aizawa S-I, Namba K (1989) Terminal regions of flagellin are disordered in solution. *J. Mol. Biol.* 209: 127–133.

Wagenknecht T, DeRosier DJ, Aizawa S-I, Macnab RM (1982) Flagellar hook structures of *Caulobacter* and *Salmonella* and their relationship to filament structure. *J. Mol. Biol.* 162: 69–87.

Welch M, Oosawa K, Aizawa S-I, Eisenbach M (1993) Phosphorylation-dependent binding of a signal molecule to the flagellar switch of bacteria. *Proc. Natl Acad. Sci. USA* 90: 8787–8791.

Williams AW, Yamaguchi S, Togashi F *et al.* (1996) Mutations in fliK and flhB affecting flagellar hook and filament assembly in *Salmonella typhimurium*. *J. Bacteriol.* 178: 2960–2970.

Yamaguchi S, Fujita H, Ishihara A, Aizawa S-I, Macnab RM (1986) Subdivision of flagellar genes of *Salmonella typhimurium* into regions responsible for assembly, rotation, and switching. *J. Bacteriol.* 166: 187–193.

9

Fimbriae and Pili

Matthew R. Chapman, Matthew A. Mulvey and Scott J. Hultgren

Washington University School of Medicine, St. Louis, MO, USA

Pili or fimbriae are proteinaceous, filamentous, polymeric organelles expressed on the surface of bacteria. Pili were first noted in early electron microscopic investigations as non-flagellar, filamentous appendages of bacteria. In 1955 Duguid designated these appendages 'fimbriae' (plural, from Latin for thread or fibre) and correlated their presence with the ability of *Escherichia coli* to agglutinate red blood cells (Duguid *et al.*, 1955). Ten years later Brinton introduced the term 'pilus' (singular, from Latin for hair) to describe the fibrous structures (the F pilus) associated with the conjugative transfer of genetic material between bacteria (Brinton, 1965). Since then, 'pilus' has become a generic term used to describe all types of non-flagellar filamentous appendages, and it is used interchangeably with the term 'fimbriae'.

The numerous types of pili assembled by both Gram-negative and Gram-positive organisms also aid in the adaptation, survival and spread of commensal bacteria. Pili can act as receptors for bacteriophages (Bradley and Williams, 1982; Montag *et al.*, 1987; Bradley and Whelan, 1989; Blond-Elguindi *et al.*, 1993; Frost, 1993; Karaolis *et al.*, 1999), facilitate DNA uptake and transfer (conjugation) (Paranchych and Frost, 1988; Frost, 1993; Baron and Zambryski, 1996; Firth *et al.*, 1996), and at least one type of pilus, type 4, can function in cellular motility (Whitchurch *et al.*, 1991; Mattick *et al.*, 1996). The primary function of most pili, however, is to act as scaffolding for the presentation of specific adhesive moieties. Adhesive pilus subunits (adhesins) are often incorporated as minor components into the tips of pili, but major structural subunits can also function as adhesins. Adhesins can mediate the interaction of bacteria with each other, with inanimate surfaces, and with tissues and cells in susceptible host organisms.

The expression of pili can have substantial impact on the establishment and persistence of pathogenic bacteria within their hosts. For many bacterial pathogens, adhesive pili play a key role in the colonisation of host tissues. Type 1 pili-mediated attachment to bladder epithelial cells is critical in the establishment of urinary tract infections by uropathogenic *E. coli* (Minion *et al.*, 1989; Krogfelt *et al.*, 1990; Langermann *et al.*, 1997). Similarly, binding of P pili to Galα(1–4)Gal sugar moieties in the kidney are necessary for development of pyelonephritis (Roberts *et al.*, 1994). Enteric pathogens produce a wide variety of adhesive pili that facilitate bacterial colonisation of the intestinal tract. These include the K88 (Mooi *et al.*, 1983, 1986; Jacobs *et al.*, 1987), K99 (Smit *et al.*, 1984), and 987P pili (Schifferli *et al.*, 1987, 1991) made by enterotoxigenic

 doi:10.1006/bkmm.2001.0009

Escherichia coli (ETEC) strains (Hacker, 1992), the long polar fimbriae (LPF) (Bäumler and Heffron, 1995) and plasmid-encoded fimbriae (PEF) of *Salmonella typhimurium* (Friedrich *et al.*, 1993), and the aggregative adherence fimbriae (AAF) of enteroaggregative *E. coli* (Mooi and de Graaf, 1985; de Graaf and Mooi, 1986; Hacker, 1992; de Graaf and Gaastra, 1994). In the small intestine, toxin co-regulated pili (TCP) are essential for the attachment of *Vibrio cholerae* to gut epithelial cells (Sun *et al.*, 1990). These pili also act as receptors for the cholera toxin phage (CTX\Φ), a lysogenic phage that encodes the two subunits of the cholera toxin (Karaolis *et al.*, 1999). This phage, with its encoded toxin, is transferred between *V. cholera* strains via interactions with TCP within the small intestine. Other pili also function in the acquisition of virulence factors. The uptake of DNA facilitated by type 4 pili and DNA transfer directed by conjugative pili can provide pathogens with accessory genes, enabling them to synthesise a wider repertoire of virulence factors and giving them resistance to a greater number of antibiotics. Biofilm formation, which in some cases appears to require pili such as type 1, type 4 or curli, can also increase the resistance of bacteria to antibiotic treatments and may aid bacterial colonisation of tissues and medical implants (Austin *et al.*, 1998; Kolter and Losick, 1998; O'Toole and Kolter, 1998; Vidal *et al.*, 1998).

Pili are not necessarily static organelles and dynamic alterations of pilus structures during the infection process may influence the pathogenicity of piliated bacteria. For example, electron microscopic studies of mouse bladders infected with type 1-piliated uropathogenic *E. coli* showed that the pili mediating bacterial adherence to the bladder epithelial cells were 10–20 times shorter than typical type 1 pili (Mulvey *et al.*, 1998). It is possible that the shorter type 1 pili observed are the result of pilus retraction, or breakage, during the infection process. The shortening of type 1 pili may provide a means for reeling bacteria in towards their target host cells, allowing the bacteria to make intimate contact with the bladder epithelium after initial attachment at a distance. Within the gut, type 4B pili (BFP) promote the autoaggregation of enteropathogenic *E. coli* (EPEC) strains, a phenomenon that probably facilitates initial adherence of EPEC to the intestinal epithelium. Recent work by Bieber and co-workers suggests that, after initial attachment, an energy-dependent conformational change in the quaternary structure of BFP is needed for the further dispersal of EPEC over human intestinal cells and for the full virulence of this pathogen (Bieber *et al.*, 1998).

During the infection process, adhesive pili are often situated at the interface between host and pathogen where they can potentially mediate cross-talk between the two organisms. In recent years, a few examples of pilus attachment inducing signal transduction pathways in host eukaryotic cells have been reported. Binding of the type 4A pili of *Neisseria* to host receptors (probably CD46) on target epithelial cells has been shown to stimulate the release of intracellular Ca^{2+} stores, a signal known to control a multitude of eukaryotic cellular responses (Kallstrom *et al.*, 1998; Muller *et al.*, 1999). Likewise, the attachment of P pili to Galα(1–4)Gal-containing host receptors on target uroepithelial cells can trigger the intracellular release of ceramides, important second messenger molecules that are capable of activating a variety of protein kinases and phosphatases involved in signal transduction processes (Svanborg Eden *et al.*, 1983; O'Hanley *et al.*, 1985; Svanborg *et al.*, 1996). The signals induced within uroepithelial cells upon binding of P-piliated bacteria eventually result in the secretion of several immunoregulatory cytokines. Binding of type 1-piliated bacteria to mannosylated receptors on uroepithelial cells can similarly induce the release of cytokines, although apparently through signalling pathways different to those stimulated by P pilus binding (Connell *et al.*, 1996). Pili can also transduce signals into bacterial cells. This was demonstrated by Zhang and Normark who, in 1996, showed that the binding of P pili to host receptors stimulated the activation of iron-acquisition machinery in uropathogenic *E. coli* (Zhang and Normark, 1996). This probably increases the ability of uropathogens to obtain iron and to survive in the iron-poor environment of the urinary tract. Understanding how pili can transmit signals into bacterial cells, and the consequences of such signalling, awaits future studies.

Historically, pili have been named and grouped based on phenotypic traits such as adhesive and antigenic properties, distribution among bacterial strains, and microscopic characterisations. In the pioneering work of Duguid and co-workers, pili expressed by different *E. coli* strains were distinguished on the basis of their ability to bind to and agglutinate red blood cells (haemagglutination) in a mannose-sensitive (MS) versus a mannose-resistant (MR) fashion (Duguid *et al.*, 1979; Duguid and Old, 1980). In this article we have classified pili expressed by Gram-negative bacteria into six different groups, according to the mechanisms by which they are assembled. This classification scheme is not all-inclusive, but provides a convenient means for discussing the diverse types of pili, their functions, structures and assembly. Representatives of various pilus types assembled by the different pathways discussed in the following sections are listed in **Table 1**.

Table 1 Pilus assembly pathways

Structure	Assembly gene products	Organism	Disease(s) associated with pilus expression
1. Chaperone/usher pathway			
Thick, rigid pili	*FGS chaperone/usher*		
P pili	PapD/PapC	*E. coli*	Pyelonephritis/cystitis
Prs pili	PrsD/PrsC	*E. coli*	Cystitis
Type 1 pili	FimC/FimD	*E. coli* *Salmonella* spp. *K. pneumoniae*	Cystitis
S pili	SfaE/SfaF	*E. coli*	UTI, newborn meningitis
F1C pili	FocC/FocD	*E. coli*	Cystitis
H. influenzae fimbriae	HifB/HifC	*H. influenzae*	Otitis media, meningitis
H. influenzae biogroup aegyptius fimbriae	HafB/HafE	*H. influenzae*	Brazilian purpuric fever
Type 2 and 3 pili	FimB/FimC	*B. pertussis*	Whooping cough
MR/P pili	MrpD/MrpC	*P. mirabilis*	Nosocomial UTI
PMF pili	PmfC/PmfD	*P. mirabilis*	Nosocomial UTI
Long polar fimbriae	LpfB/LpfC	*S. typhimurium*	Gastroenteritis
Pef pili	PefD/PefC	*S. typhimurium*	Gastroenteritis
Ambient-temperature fimbriae	AftB/AftC	*P. mirabilis*	UTI
987P fimbriae	FasB/FasD	*E. coli*	Diarrhoea in piglets
REPEC fimbriae	RalE/RalD	*E. coli*	Diarrhoea in rabbits
Thin, flexible pili			
K99 pili	FaeE/FaeD	*E. coli*	Neonatal diarrhoea in calves, lambs, piglets
K88 pili	FanE/FanD	*E. coli*	Neonatal diarrhoea in piglets
F17 pili	F17D/F17papC	*E. coli*	Diarrhoea
MR/K pili	MrkB/MrkC	*K. pneumoniae*	Pneumonia
Atypical structures	*FGL chaperone/usher*		
CS31A capsule-like protein	ClpE/ClpD	*E. coli*	Diarrhoea
Antigen CS6	CssC/CssD	*E. coli*	Diarrhoea
Myf fimbriae	MyfB/MyfC	*Y. enterolitica*	Enterocolitis
pH 6 antigen	PsaB/PsaC	*Y. pestis*	Plague
CS3 pili	CS3-1/CS3-2	*E. coli*	Diarrhoea
Envelope antigen F1	Caf1M/Caf1A	*Y. pestis*	Plague
Non-fimbrial adhesins I	NfaE/NfaC	*E. coli*	UTI, newborn meningitis
SEF14 fimbriae	SefB/SefC	*S. enteritidis*	Gastroenteritis
Aggregative adherence fimbriae I	AggD/AggC	*E. coli*	Diarrhoea
AFA-III	AfaB/AfaC	*E. coli*	Pyelonephritis
II. Alternate chaperone pathway			
CS1 pili	CooB/CooC	*E. coli*	Diarrhoea
CS2 pili	CotB/CotC	*E. coli*	Diarrhoea
CS4 pili		*E. coli*	Diarrhoea
CS14 pili		*E. coli*	Diarrhoea
CS17 pili		*E. coli*	Diarrhoea
CS19 pili		*E. coli*	Diarrhoea
CFA/I pili	CfaA/CfaC	*E. coli*	Diarrhoea
Cable type II pili		*B. cepacia*	Opportunistic in cystic fibrosis patients

Table 1 (*Cont.*)

Structure	Assembly gene products	Organism	Disease(s) associated with pilus expression
III. Type II secretion pathway			
Type 4A pili	General secretion apparatus (main terminal branch) 14 to > 20 proteins	*Neisseria* spp.	Gonorrhoea
		P. aeruginosa	Meningitis
		Moraxella sp.	Opportunistic pathogen
		D. nodosus	Conjuntivitis
		E. corrodens	Respiratory infections
		Azoarcus sp.	Ovine footrot
Type 4B pili:			
Bundle-forming pili	General secretion apparatus (main terminal branch) 14 to > 20 proteins	*E. coli*	Diarrhoea
Longus		*E. coli*	Diarrhoea
CFA/III		*E. coli*	Diarrhoea
R64 pili		*E. coli*	
Toxin co-regulated pili		*V. cholerae*	Cholera
IV. Type III secretion pathway			
Hrp pili	Type III secretion apparatus, ~20 proteins	*P. syringae*	Hypersensitive response (in resistant plants)
EspA pilus-like structures		*E. coli*	Diarrhoea
V. Extracellular nucleation/precipitation pathway			
Curli	CsgG/CsgE/Csg F	*E. coli*	Sepsis
		S. enteritidis	
VI. Conjugative pilus assembly pathway (type IV secretion pathway)			
F pili (IncF1)	Type IV export apparatus, 12–16 proteins	*E. coli*	Antibiotic resistance
IncN, IncP, IncW-encoded pili		*E. coli*	
T pili		*A. tumefaciens*	Crown gall disease

UTI, urinary tract infection.

Chaperone/usher and Alternate Chaperone/usher Pathways

All pilins destined for assembly on the surface of Gram-negative bacteria must be translocated across the inner membrane, through the periplasm, and across the outer membrane. To accomplish these steps, various adhesive organelles in many different bacteria require two specialised assembly proteins: a periplasmic chaperone and an outer membrane usher (Fig. 1a). Chaperone/usher assembly pathways are involved in the biogenesis of over 30 different structures, including composite pili, thin fibrillae and non-fimbrial adhesins (Hung and Hultgren, 1998). The prototypical members of this class are P and type 1 pili, which are known to be important determinants of disease. Uropathogenic *E. coli* require type 1 pili to attach to conserved, mannose-containing host receptors expressed by the bladder epithelium and help to prevent the bacteria from being washed from the body with the flow of urine. P pili may serve a similar function in the kidneys, inhibiting the clearance of pyelonephritic *E. coli* from the upper urinary tract. Here, we will focus on the structure and assembly mechanisms of the P and type 1 pilus chaperone/usher systems.

P and type 1 pili are both composite structures consisting of a thin fibrillar tip joined end to end to a right-handed helical rod (Gong and Makowski, 1992; Bullitt and Makowski, 1995) (Fig. 1d,e). Chromosomally located gene clusters that are organisationally and functionally homologous encode P and type 1 pili (Fig. 1b,c). The P pilus tip is a 2-nm-wide structure composed of a distally located adhesin (PapG), a tip pilin (PapE) and adapter pilins (PapF and PapK) (Kuehn *et al.*, 1992; Jacob-Dubuisson *et al.*, 1993). The PapG adhesin binds to Galα(1–4)Gal moieties present in the globoseries of glycolipids found on the surface of erythrocytes and kidney cells (Leffler and Svanborg-Eden, 1980; Striker *et al.*, 1995), and consistent with this binding specificity, P pili are major virulence factors associated with pyelonephritis caused by uropathogenic *E. coli* (Kallenius *et al.*, 1981).

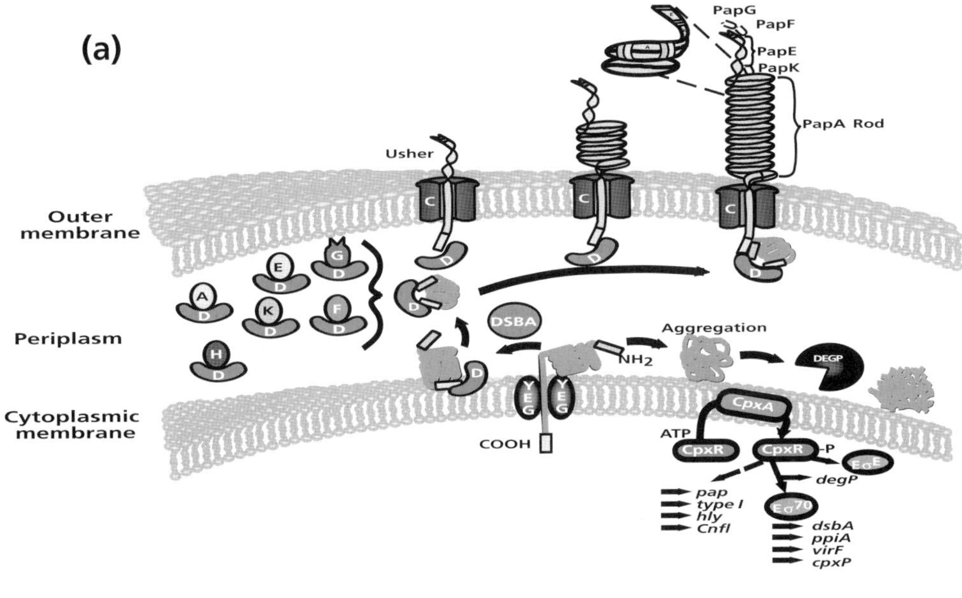

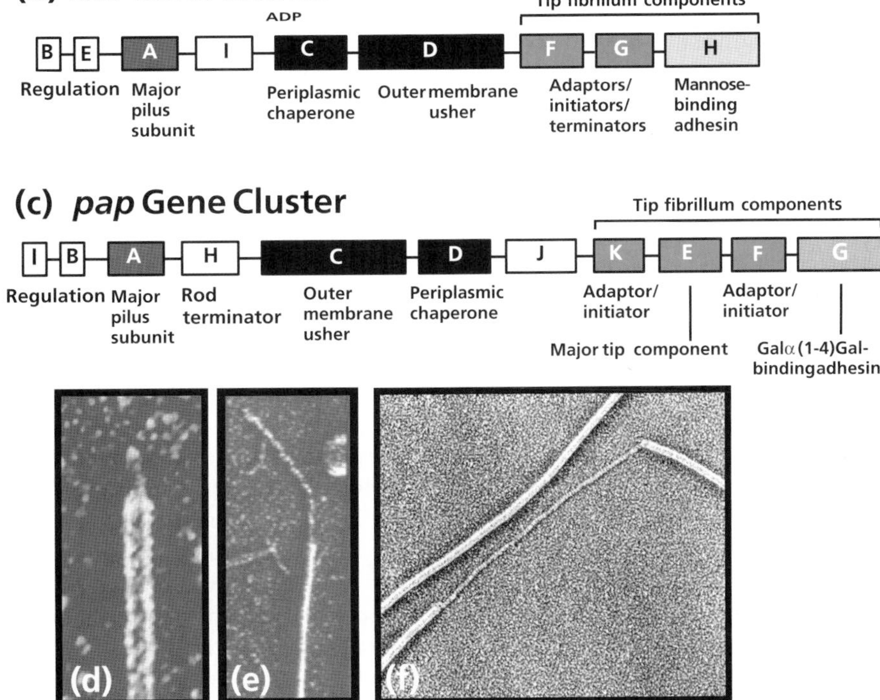

Fig. 1 (a) Model of P pilus assembly by the chaperone/usher pathway. Structural subunits for the pilus tip (PapG, PapF, PapE and PapK) and for the pilus rod (PapA) are translocated across the inner membrane of *Escherichia coli* via the *sec* system. On the periplasmic side of the inner membrane they interact with the chaperone PapD, which facilitates the folding and release of the subunits from the inner membrane. DsbA is required for the proper folding of both the subunits and the PapD chaperone. In the absence of the chaperone, the subunits aggregate and are degraded by the periplasmic protease DegP. The build-up of unchaperoned subunits results in the activation of the Cpx system, increasing the production of periplasmic protein folding and degradation proteins. Subunit–chaperone complexes are targeted to the outer membrane usher, PapC, where the chaperone is released and subunit–subunit interactions occur. The PapC usher forms a 2-nm-wide pore through which the assembled pilus structure is extruded as a linear fibre across the outer membrane. Once on the exterior of the cell, the linear PapA polymer forms a thick, right-handed helical rod that is unable to slip back through the usher pore. The formation of the coiled PapA rod is thought to help drive the outward growth of the pilus. (b,c) The type 1 (*fim*) and P (*pap*) gene clusters are depicted in (b) and (c), respectively. These gene clusters share organisational as well as functional homologies. High-resolution electron micrographs showing the pilus rod and tip fibrillum structures of (d) type 1 and (e) P pili. Unravelling of a portion of a P pilus rod into a linear fibre is shown in (f). The images shown in (d), (e) and (f) are at different magnifications.

The minor pilin PapF is thought to join the PapG adhesin to the tip fibrillum, the bulk of which is made up of a polymer of PapE subunits. PapK is thought to terminate the growth of the PapE polymer and to join the tip structure to the rod (Kuehn *et al.*, 1992; Jacob-Dubuisson *et al.*, 1993). The pilus rod is composed of multiple PapA subunits joined end to end and then coiled into a right-handed, 6.8-nm-thick helical rod, having a pitch distance of 24.9 Å and 3.28 subunits per turn. The rod is terminated by a minor subunit, PapH, which may serve to anchor the pilus in the membrane (Baga *et al.*, 1987).

Similar to the P pilus structure, the type 1 pilus has a short, 3-nm-wide fibrillar tip made up of the mannose-binding adhesin FimH and two additional pilins, FimG and FimF (Klemm and Christiansen, 1987; Maurer and Orndorff, 1987; Russell and Orndorff, 1992). The FimH adhesin mediates attachment to mannosylated receptors expressed on a wide variety of cell types and has been shown to be a significant virulence determinant for the development of cystitis (Ofek *et al.*, 1977; Salit and Gotschich, 1977; Tewari *et al.*, 1993; Mulvey *et al.*, 1998). The type 1 tip fibrillum is joined to a rod comprised predominantly of FimA subunits arranged in a 6–7-nm-diameter helix with a pitch distance of 23.1 Å and 3.125 subunits per turn. Both type 1 and P pilus rods have central axial holes with diameters of 2–2.5 Å and 1.5 Å, respectively.

Despite architectural similarities, type 1 pili appear to be more rigid and prone to breaking than P pili (Ponniah *et al.*, 1991). Some reports have argued that, unlike the P pilus system in which the tip subunits are thought to be located only within the tip, some of the type 1 tip subunits may also be occasionally intercalated within the rod structure (Ponniah *et al.*, 1991).

The major pilin subunits in both P and type I pili are organised in a head-to-tail manner. Additional quaternary interactions between subunits in adjacent turns of the helical rod appear to stabilise the structure and may help drive the outward growth of the organelle during pilus assembly (see below). Disruption of these latter interactions by mechanical stress or by incubation in 50% glycerol can cause the pilus rod to reversibly unwind into a 2-nm-thick linear fibre similar in appearance to the tip fibrillum (Abraham *et al.*, 1992; Bullitt and Makowski, 1995) (**Fig. 1f**). Bullitt and Makowski (1995) have proposed that the ability of the pilus rods to unwind allows them to support tension over a larger range of lengths (Bullitt and Makowski, 1995). This may help P and type 1 pili to withstand stress, such as shearing forces from the bulk flow of fluid through the urinary tract, without breaking.

The three-dimensional crystal structure of the PapD chaperone has been solved, revealing two immunoglobulin-like (Ig) domains oriented into a boomerang shape such that a subunit-binding cleft is created between the two domains (Holmgren and Brändén, 1989). Using genetics, biochemistry and crystallography, PapD was found to bind in part to a highly conserved motif present at the C-terminus of all subunits assembled by PapD-like chaperones (Soto *et al.*, 1998).

Recently, the PapD and FimC chaperones have been crystallised in complex with subunit proteins PapK and FimH, respectively (Choudhury *et al.*, 1999; Sauer *et al.*, 1999). The chaperone-subunit structures of PapD–PapK and FimC–FimH have suggested a likely mechanism for pili assembly. Like PapD, the FimC chaperone consists of two Ig domains oriented in an L-shape (Choudhury *et al.*, 1999). The FimH adhesin also consists of two domains: a receptor-binding domain and a pilin domain. The receptor-binding domain, which does not interact significantly with the chaperone, is an elongated 11-stranded β-barrel with a putative mannose-binding pocket at its distal end (Choudhury *et al.*, 1999; Sauer *et al.*, 1999). The pilin domain of FimH and the single domain of the PapK subunit have Ig folds that lack the seventh (C-terminal) β-strand present in canonical Ig folds (**Fig. 2**). The absence of this strand produces a deep groove along the surface of the pilin domain and exposes its hydrophobic core – hence the instability of pilus subunits in the absence of the chaperone. In the chaperone–subunit complex, the chaperone contributes its G1 β-strand to complete the Ig fold of the subunit by occupying the groove, an interaction termed donor strand complementation (**Fig. 2**) (Choudhury *et al.*, 1999; Sauer *et al.*, 1999).

The groove of subunits has also been implicated in subunit–subunit interactions. Thus, in addition to stabilising the subunit by shielding its hydrophobic core, the donor strand complementation interaction also caps one of the subunit interactive surfaces and prevents premature pilus formation in the periplasm. Indeed, the folding, stabilisation and capping of the subunit may be part of a single dynamic process that ensures that interactive surfaces of subunits are not accessible until pilus assembly occurs. By donating the missing strand to the Ig fold of the subunit, the chaperone may provide a context for the proper folding of the subunit and thus ensure that the fully folded interactive groove only forms on the chaperone, where it would remain until the subunit is incorporated into the pilus (Sauer *et al.*, 1999).

Pilus assembly is thought to proceed by a donor strand exchange mechanism (Sauer *et al.*, 1999). Subunits have an N-terminal extension (residues 1–13 in

(a)

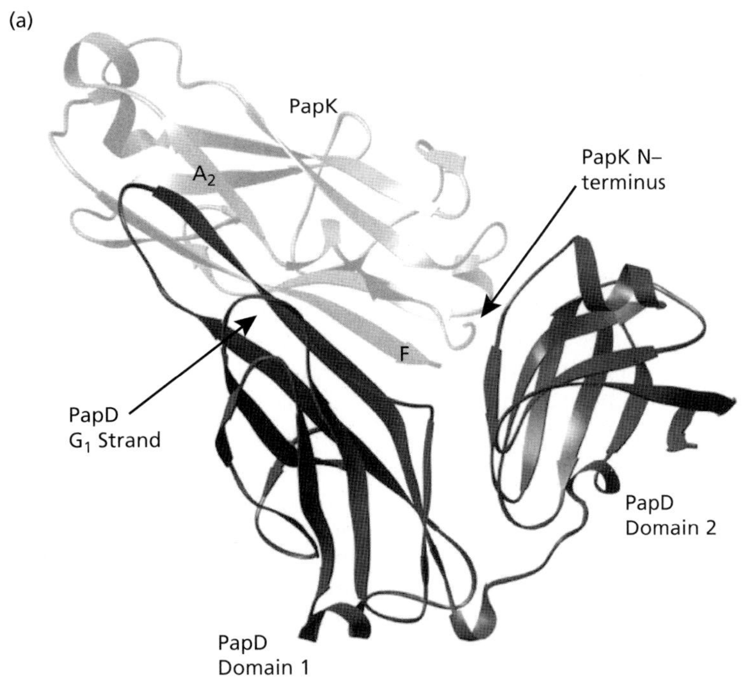

(b)

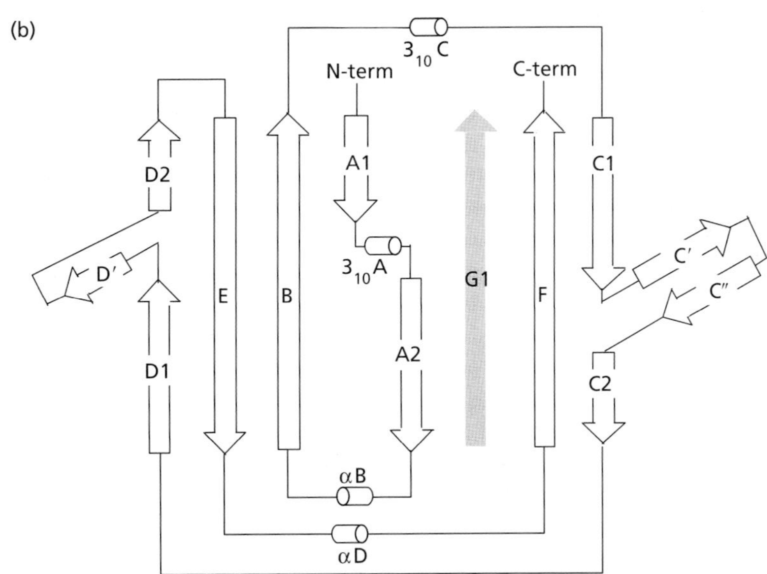

Fig. 2 (a) Ribbon diagram of the PapD (black) and PapK (grey) derived from the crystal structure of the PapD–K complex. PapD is black and PapK is grey. The G1 β-strand of PapD is shown as it sets in the groove created by the A2 and F strands of PapK, thus completing the immunoglobulin-like fold of PapK. The length of the loop region connecting the G1 and F1 β-strands of PapD is one of the defining characteristics marking PapD as a member of the FGS subfamily of periplasmic chaperones. (b) Topology of PapK. The β-strands are indicated with arrows and the helices are shown as cylinders.

PapK, for example) with a highly conserved alternating hydrophobic motif that has been shown to participate in subunit–subunit interactions. This motif is similar to an alternating hydrophobic motif present in the G1 β-strand of the chaperone that participates in donor strand complementation. The N-terminal extension does not contribute to the Ig fold of the subunit but rather projects away from the rest of the pilin domain where it would be free to interact with another subunit. During pilus biogenesis, the N-terminal extension of one subunit is thought to displace the chaperone G1 strand and insert into the groove of the neighbouring subunit. The mature pilus would thus consist of a linear array of canonical Ig domains, in which each fold is completed by a strand from the neighbouring subunit.

Once formed in the periplasm, chaperone–subunit complexes are targeted to the outer membrane usher where the chaperone is released, exposing interactive surfaces on the subunits that facilitate their assembly into the pilus. Studies in the P and type 1 pilus systems have demonstrated that the adhesin–chaperone complexes, PapDG or FimCH, bind tightest and fastest to the usher and that the adhesins are the first subunits assembled into the pilus (Saulino *et al.*, 1998). Binding of the chaperone–adhesin complex induces a conformational change in the usher, possibly priming it for pilus assembly. Additional subunits are incorporated into the pilus depending, in part, upon the kinetics with which they are partitioned to the usher in complex with the chaperone (Saulino *et al.*, 1998). Conserved N-terminal regions, in addition to the conserved C-terminal motif of the pilus subunits, mediate subunit–subunit interactions within the mature pilus. Differences in the complementary surfaces in these conserved regions from one subunit to another may help dictate which of the subunits can be joined to one another during pilus assembly. Thus, the order of the subunits within the final pilus structure is determined by the specific contacts made between the different pilus subunits and also by the differential affinities of the various chaperone–subunit complexes for the usher.

Besides acting as an assembly platform for the growing pilus, the usher protein appears to have additional roles in pilus biogenesis. High-resolution electron microscopy revealed that the PapC usher is assembled into a 15-nm-diameter ring-shaped complex with a 2-nm-wide central pore (Thanassi *et al.*, 1998b). PapC and other usher family members are thought to have a predominantly β-sheet secondary structure, typical of outer membrane pore-forming proteins, and they are predicted to present large regions to the periplasm for interaction with chaperone–subunit complexes (Thanassi *et al.*, 1998a). After dissociating from the chaperone at the usher, subunits are incorporated into a growing pilus structure that is predicted to be extruded as a one-subunit-thick linear fibre through the central pore of the usher complex. Packaging of the linear pilus fibre into a thicker, helical rod on the outside surface of the bacterium may provide a driving force for the translocation of the pilus across the outer membrane, possibly acting as a sort of ratcheting mechanism to force the pilus to grow outwards. Combined with the targeting affinities of the chaperone–subunit complexes for the usher and the binding specificities of the subunits for each other, this may provide all the energy and specificity needed for the ordered assembly and translocation of pili across the outer membrane.

In addition to composite structures exemplified by P and type 1 pili, chaperone–usher pathways also mediate assembly of thin fibrillae such as K88 and K99 pili and non-fimbrial adhesins. K88 and K99 pili are 2- to 4-nm-thick fibres that mediate adherence to receptors on intestinal cells. They are significant virulence factors expressed by enterotoxigenic *E. coli* (ETEC) strains that cause diarrhoeal diseases in livestock. These pili were given the 'K' designation after being mistakenly identified as K-antigens in *E. coli*. In contrast to P and type 1 pili, the adhesive properties of K88 and K99 pili are associated with the major pilus subunits. The receptor-binding epitopes on the individual major pilus subunits are exposed on the pilus surface and available for multiple interactions with host tissue. In general, pili with adhesive major subunits, such as K88 and K99 pili, are thin, flexible fibrillar structures. In comparison, pili with specialised adhesive tip structures, such as P and type 1 pili, are relatively rigid and rod-like.

A variation of the chaperone/usher pilus assembly pathway has been identified in strains of enterotoxigenic *E. coli* (ETEC). These bacteria are major pathogens associated with diarrhoeal diseases of travellers, infants and young children. ETEC strains produce several types of uniquely assembled adhesive pili that are considered to be important mediators of bacterial colonisation of the intestine. The best studied of these pili is CS1, which appears to be composed predominantly of a major subunit, CooA, with a distally located minor component, CooD (Sakellaris *et al.*, 1996). Several CS1-like pili have been identified and include CS2, CS4, CS14, CS17, CS19, and CFA/I pili expressed by various ETEC strains and the cable type II pili of *Burkholderia cepacia*, an opportunistic pathogen of cystic fibrosis patients (Goldstein *et al.*, 1995; Sajjan *et al.*, 1995). Four linked genes, *cooA*, *cooB*, *cooC* and *cooD*, are the only specific genes required for the synthesis of functional CS1 pili (Jordi *et al.*, 1992; Froehlich *et al.*, 1994, 1995).

Homologous genes required for the production of CS2 and CFA/I pili have also been cloned and sequenced (Froehlich *et al.*, 1995). Electron microscopic examination reveals that the CS1-like pili are architecturally similar to P and type 1 pili assembled by the chaperone/usher pathway (Fig. 3b), although none of the proteins involved in the biogenesis of CS1-like pili have any significant sequence homologues to

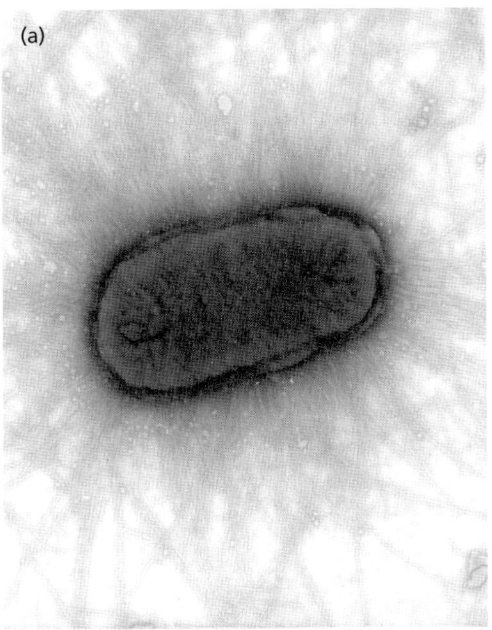

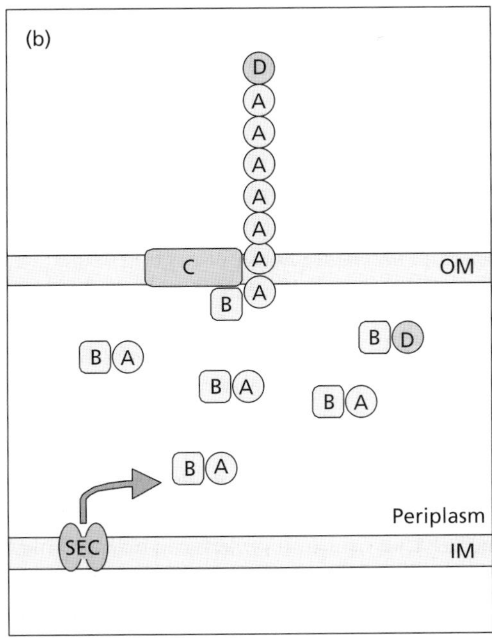

Fig. 3 (a) CS2 pili (~2-nm-thick pili) assembled by the alternate chaperone pathway in *Escherichia coli*. Photo courtesy of Harry Sakellaris and June R. Scott. (b) Assembly of CS1 pili from *E. coli* via the alternate chaperone pathway. The inner and outer membranes are designated IM and OM, respectively. The CooB chaperone forms periplasmic complexes with the main components of the pilus, CooA and CooD. It also appears to bind and stabilise the outer membrane protein CooC in the absence of subunits. CooC appears to function as an outer membrane channel for passage of the pilin fibre.

those of any other pilus system (Froehlich *et al.*, 1995; Sakellaris *et al.*, 1996).

The assembly of CS1-like pili depends upon a specialised set of periplasmic chaperones that are distinct from those of the chaperone/usher pathway described above. Therefore, we refer to this mode of pilus assembly as the alternate chaperone pathway. In the case of CS1 pili, the chaperone CooB binds to and stabilises the major and minor pilin subunits, CooA and CooD, which enter into the periplasm in a *sec*-dependent fashion (Sakellaris and Scott, 1998) (Fig. 3b). Both CooA and CooD share a conserved sequence motif near their C-termini that may function as a chaperone recognition motif (Sakellaris and Scott, 1998). One of the functions of CooB appears to be the delivery of the pilin subunits to an outer membrane protein, CooC, which may function as a channel, or usher, for the assembly of pilus fibres. In addition to the pilin subunits, CooB also binds to and stabilises CooC in the outer membrane. In the absence of the CooB chaperone, CooC and the pilin subunits are degraded. Although currently less well defined, the assembly of CS1 and related structures appears similar in many respects to the assembly of pili by the classic chaperone/usher pathway. Since CS1-like pili do not appear to be related to those assembled by the chaperone/usher pathway, it has been suggested that these two pilus assembly systems arose independently through convergent evolution (Sakellaris and Scott, 1998).

Type III Secretion Pathway

The various pilus assembly pathways described in the previous sections all rely upon components of the *sec* machinery for the translocation of their respective pilus subunits across the inner membrane. Recently, two new types of pili that are assembled by a *sec*-independent pathway known as type III secretion have been identified (Roine *et al.*, 1997; Knutton *et al.*, 1998; Neves *et al.*, 1998). The type III secretion system is encoded by numerous Gram-negative pathogens and enables these bacteria to secrete and inject pathogenic effector molecules into the cytosol of host eukaryotic cells. About 20 gene products, most of which are inner membrane proteins, comprise the type III secretion system. Components mediating type III secretion are conserved in pathogens as diverse as *Yersinia* and *Erwinia*, but the secreted effector proteins vary significantly between species (Hueck, 1998). The type III secretion apparatus, which appears to span the periplasmic space, resembles the basal body of a flagellum connected to a straight rod that extends across the outer membrane (Cornelis, 1998). Interestingly, all type III secretion systems encode some components with homologies to proteins involved in flagellar assembly (Hueck, 1998). The secretion of proteins by the type III system is an ATP-dependent process that involves no distinct periplasmic intermediates.

EPEC encodes four proteins, EspA, EspB, EspD and Tir, which are secreted by a type III pathway (Collington *et al.*, 1998; Knutton *et al.*, 1998). These proteins facilitate intimate contact between the pathogen and host intestinal cells and are required for the formation of specific (attaching and effacing) lesions. One of these proteins, EspA, can assemble into 7- to 8-nm-thick peritrichously expressed pilus-like fibres that are organised into ~50-nm-wide bundles. These extend up to 2 μm from the bacterial surface and appear to be made up of only EspA molecules. Interestingly, EspA shares substantial sequence identity with a flagellin from *Yersinia enterolitica*. During the infection process, the EspA fibres appear to mediate contact between EPEC and the host cell surface prior to the establishment of more intimate bacterial attachment (Neves *et al.*, 1998). The EspA fibres seem to assist the translocation of EspB effector molecules into host cells where they can subvert host signal transduction pathways (Kenny *et al.*, 1996).

In *Pseudomonas syringae* and other plant pathogens, hypersensitive response and pathogenicity (*hrp*) genes control the ability of these bacteria to cause disease in susceptible plants and to elicit the hypersensitive response in resistant plants. The hypersensitive response is a phenomenon characterised by rapid localised host cell death at the site of infection that appears to limit the spread of a pathogen in an infected plant. A subset of the *hrp* genes, recently renamed *hrc* genes, encode components of a type III secretion system. In 1997, Roine and co-workers showed that one of the proteins, HrpA, secreted by the Hrp type III secretion system is assembled into 6- to 8-nm-wide, peritrichously expressed pili (Roine *et al.*, 1997). It was proposed that these pili, known as Hrp pili, are involved in mediating bacteria–plant interactions in the intercellular spaces of the host plant. In addition, Hrp pili may assist the delivery of effector proteins into host plant cells. The exact nature and functions of the Hrp pili of *P. syringae* and the EspA-containing pili of EPEC remain to be elucidated.

Extracellular Nucleation/Precipitation Pathway

Fibres that coil into a fuzzy adhesive mass on the bacterial surface are referred to as thin aggregative pili or, as is now the accepted nomenclature, curli (Romling *et al.*, 1998a). Curli are a class of thin (~2–4 nm diameter), irregular and highly aggregated surface fibres expressed by *Escherichia coli* and *Salmonella* spp. (Olsen *et al.*, 1989; Collinson *et al.*, 1991) (Fig. 4a). These distinct organelles mediate binding to the dye Congo red and a variety of host proteins, including plasminogen (Sjobring *et al.*, 1994), laminin (Arnqvist *et al.*, 1992), fibronectin (Olsen *et al.*, 1989) and human contact phase proteins (Ben Nasr *et al.*, 1996). Curli are also involved in bacterial colonisation of inert surfaces and have been implicated in biofilm formation (Romling *et al.*, 1998b; Vidal *et al.*, 1998). Curli are highly stable structures that can only be depolymerised by means of extreme chemical treatment.

The formation of curli represents a departure from the other modes of pilus assembly discussed in the previous sections. The genes for curli biogenesis are located on two divergently transcribed operons, expressed maximally at 26°C and during stationary-phase growth (Olsen *et al.*, 1993b; Hammar *et al.*, 1995). The *csgBA* operon encodes two homologous proteins, CsgA and CsgB, both of which can be transported across the outer membrane as soluble proteins (Fig. 4b). The major component of *E. coli* curli is a 15.3-kDa CsgA protein, which shares over 86% primary sequence similarity to its counterpart in *Salmonella enteritidis*, AgfA (Romling *et al.*, 1998a). CsgB, a minor component of the curli structure, nucleates CsgA subunits in a process that is poorly understood (Bian and Normark, 1997). Bacteria that express the entire *csg* operon may nucleate curli as subunits which are secreted from the outer membrane, such that curli subunits remain in close proximity with each other and the bacterial cell surface. However, it has been observed that a $CsgA^{+}CsgB^{-}$ 'donor' strain can secrete CsgA subunits which can be assembled into curli on the surface of a $CsgA^{-}CsgB^{+}$ 'recipient' strain in a process termed 'interbacterial complementation'. This suggests that bacteria can secrete subunits in a nucleation competent form into the cellular milieu, and after secretion these subunits are assembled on the bacterial surface. Presumably, interbacterial curli assembly is contingent on outer membrane-bound CsgB in the $CsgA^{-}CsgB^{+}$ recipient strain as a $CsgA^{+}CsgB^{-}$ strain cannot act as a recipient (Hammar *et al.*, 1996).

Curli produced by interbacterial complementation are morphologically different from wild-type curli, appearing as thicker and less branched structures (Hammar *et al.*, 1996). Morphologically similar structures were observed by anchoring the CsgB protein to the outer membrane as a transmembrane fusion

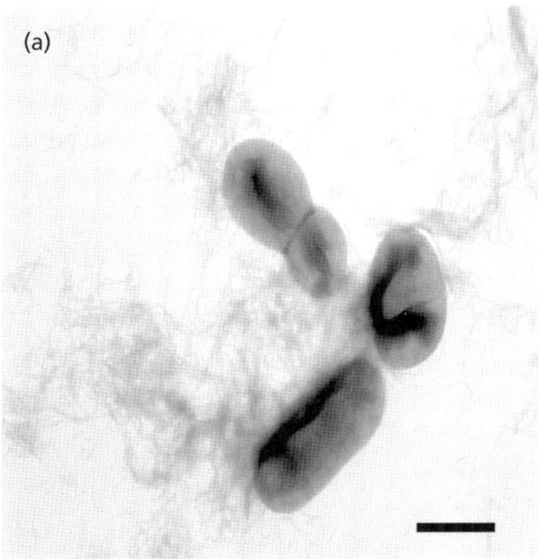

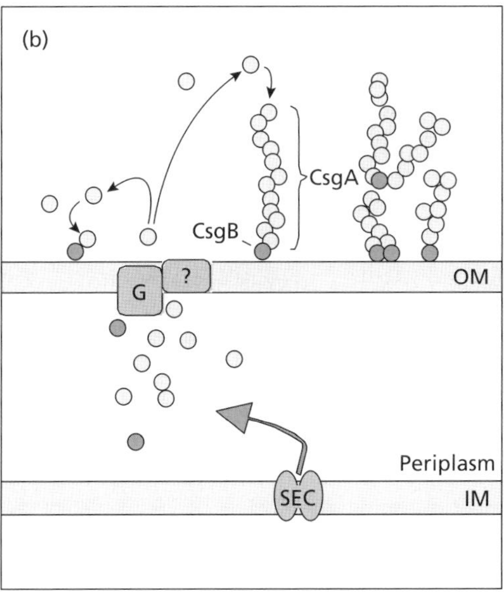

Fig. 4 (a) *Escherichia coli* expressing curli, <2-nm-wide structures assembled by the extracellular nucleation/precipitation pathway. (b) Model of curli assembly by the extracellular nucleation/precipitation pathway. CsgA, the main component of curli from *E. coli*, is secreted across the outer membrane. Surface localised CsgB serves to nucleate CsgA assembly. CsgB is also found distributed along the curli fibre, where it may serve to initiate branching of the fibre. G (CsgG) is an outer membrane-localised lipoprotein that is required for the stabilisation and secretion of CsgA and CsgB, although its exact function is not yet clear.

protein with the periplasmic maltose-binding protein. This suggests that CsgB interspersed along the length of the curli fibre might initiate branching. In support of this, CsgB-specific antibodies appear to localise at the branch points of curli using immunoelectron microscopy (Bian and Normark, 1997). Curli fibres consisting of CsgA polymers with no intercalated CsgB bind fibronectin and Congo red poorly (Bian and Normark, 1997), suggesting that CsgB, or a combination of CsgA and CsgB contribute to adhesion.

The *csgDEFG* operon encodes CsgD, a transcriptional activator of curli synthesis, and three genes encoding putative assembly factors. One of these factors, CsgG, has recently been shown to be a lipoprotein that is localised to the periplasmic side of the outer membrane (Loferer *et al.*, 1997) (Fig. 4b). In the absence of CsgG, curli assembly does not take place and CsgA and CsgB are rapidly degraded. The precise role of CsgG is not known at this time, although Normark and colleagues have suggested that CsgG might act as a chaperone-like protein that protects CsgA and CsgB from degradation within the periplasm, thereby facilitating their secretion (Loferer *et al.*, 1997). It is also possible that CsgG assembles into multimers that could function as a Csg-specific channel within the outer membrane. The roles of the CsgE and CsgF gene products are not yet known.

Escherichia coli and *Salmonella* spp., the only bacteria that are known to express curli, are clinically important pathogens able to cause severe bacterial diseases including pneumonia, meningitis, diarrhoeal diseases, pyelonephritis and cystitis. In addition, these bacteria must adapt to an environment outside the mammalian host. The role of curli during the life cycle of *E. coli* and *Salmonella* spp. is unclear, but the observation that curli are maximally expressed at 26°C (Olsen *et al.*, 1993a, 1993b) is consistent with the idea that curli are required for growth and survival of these bacteria outside the host. However, curli expression in *Salmonella typhimurium* can be induced at 37°C by iron depletion, and thus it is possible that curli-proficient *E. coli* may express the organelle *in vivo* at 37°C (Romling *et al.*, 1998b).

Type II Secretion Pathway for Type 4 Pilus Assembly

Type 4 pili are multi-functional structures expressed by a wide diversity of bacterial pathogens, including *Pseudomonas aeruginosa*, *Neisseria gonorrhoeae* and *N. meningitidis*, *Moraxella* spp., *Azoarcus* spp., *Dichelobactor nodus*, and many other species classified in these and other genera (Tennent and Mattick, 1994). Type 4 pili are significant colonisation factors and have been shown to mediate bacterial interactions with animal, plant and fungal cells (Dorr *et al.*, 1998; Strom and Lory, 1993). In addition, these pili can modulate target cell specificity (Jonsson *et al.*, 1994) and biofilm formation (O'Toole and Kolter, 1998), and can act as receptors for bacteriophage (Bradley, 1974; Bradley and Pitt, 1974). Type 4 pili are also associated with a flagella-independent form of bacterial locomotion called twitching motility that allows for the lateral spread of bacteria across a surface (Darzins and Russell, 1997).

Type 4 pili are 6-nm-wide structures typically assembled at one pole of the bacterium. They can extend up to several micrometres in length and are composed primarily, if not completely, of a small subunit usually in the range of 145–160 amino acids (Strom and Lory, 1993). These subunits have distinctive features, including a short (6–7 amino acids), positively charged leader sequence that is cleaved during assembly, *N*-methylphenylalanine as the first residue of the mature subunit, and a highly conserved, hydrophobic N-terminal domain (**Fig. 5a**). The adhesive properties of type 4 pili are, in general, determined by the major pilus subunit. Additional minor components, however, may associate with these pili and alter their binding specificities. In the case of *Neisseria*, a tip-localised adhesin, PilC1, appears to mediate bacterial adherence to epithelial cells (Rudel *et al.*, 1995a, b).

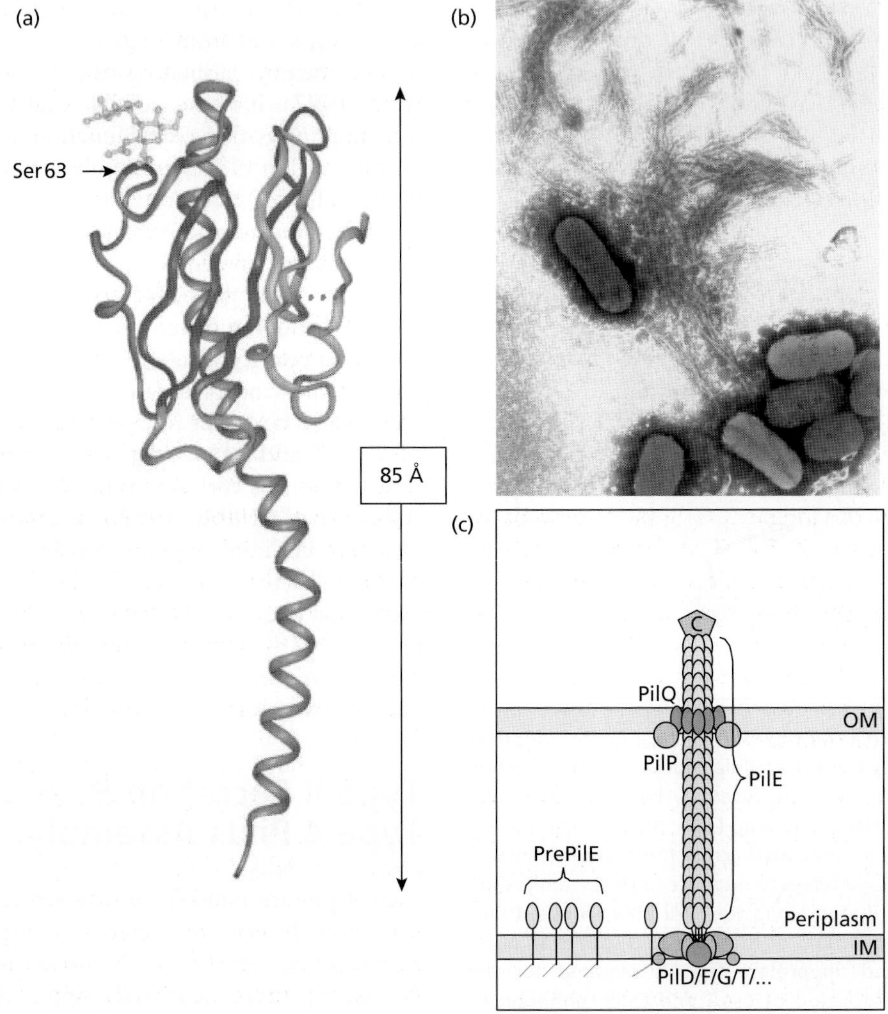

Fig. 5 (a) Ribbon model of the type 4A pilin, PilE, from *Neisseria gonorrhoeae*. Secondary structural elements include a hydrophobic, conserved N-terminal α-helical spine connected to a variable domain containing an extended sugar loop with an *O*-linked disaccharide at position Ser63 and a disulphide-containing C-terminal region. The disulphide bridge is signified by a dotted line. (b) The polar type 4B bundle forming pili (~6-nm-wide pili) of EPEC. Photo courtesy of Dave Bieber and Gary Schoolnik. (c) Model of type 4 pilus assembly by *N. gonorrhoeae*. The PilE prepilin is translocated into the periplasm aided by the *sec* machinery. PilE is processed by the PilD signal peptidase, which cleaves the positively charged leader sequence from the N-terminus of the pilin subunit. An inner membrane assembly complex then assembles the mature PilE subunit into a pilus fibre. PilQ mediates translocation of the pilus through the outer membrane, possibly with the assistance of other factors such as PilP. The PilC adhesin, which appears to be incorporated at the tip of the growing pilus fibre, also seems to be required for translocation of the pilus through the outer membrane.

Recently, a second class of type 4 pili, referred to as class B or type 4B, has been defined. Type 4B pili were initially characterised in enteric pathogens and include the toxin co-regulated pilus (TCP) of *Vibrio cholerae* (Faast *et al.*, 1989), the bundle-forming pilus (BFP) of enteropathogenic *E. coli* (EPEC) (Giron *et al.*, 1991), and the longus and CFA/III pili of ETEC (Giron *et al.*, 1994; Taniguchi *et al.*, 1995). In comparison with the typical type 4 pilins (referred to as class A or type 4A) described above, the known type 4B pilins are somewhat larger and have a longer (13–30 amino acids) leader sequence (Fig. 5b). Also, in place of *N*-methylphenylalanine as the first amino acid in the mature pilus subunit, type 4B subunits have other methylated residues such as *N*-methylmethionine for TCP and *N*-methylleucine for BFP. TCP, BFP and longus pili form large polar bundles over 15 μm in length. In contrast, CFA/III pili are 1–10 μm long and are peritrichously expressed. The number of pili classified as type 4B is increasing and now includes the R64 thin pilus, an organelle involved in bacterial conjugation (Kim and Komano, 1997).

Parge and co-workers solved the crystal structure of the type 4A pilin subunit (PilE) from *N. gonorrhoeae*, greatly advancing the understanding of the structure, function and biogenesis of type 4 pili (Parge *et al.*, 1995) (Fig. 5a). PilE contains 158 amino acids and was determined to have an over-all ladle-shape, comprising a β-roll with a long hydrophobic N-terminal β-helical spine (residues 2–54). The C-terminal domain of type 4 pilins possesses hypervariable regions that affect the binding specificities and antigenicity of type 4 pili (Strom and Lory, 1993). In PilE, these hypervariable regions include (1) a sugar loop (residues 55–77) with an *O*-linked disaccharide at position Ser63 and (2) a disulphide-containing region (residues 121–158) that, despite having a hypervariable nature, adopts a regular β-hairpin structure (β_5–β_6) followed by an extended C-terminal tail (Parge *et al.*, 1995). A disulphide-containing C-terminal hypervariable region is common among the type 4 pilins. The remainder of PilE was shown to consist of two β-hairpins forming a four-stranded anti-parallel β-sheet (residues 78–93 and 103–122) with a connecting β_2–β_3 loop region (residues 94–102) (Parge *et al.*, 1995).

Through systematic modelling, Parge and co-workers showed that PilE was probably assembled into pili as monomers arranged in a helix with about five PilE subunits per turn and a pitch distance of approximately 41 Å (Parge *et al.*, 1995). PilE subunits are predicted to be packed into pili as a three-layered assemblage consisting of an inner core of coiled conserved hydrophobic α-helices surrounded by β-sheets and an outermost layer composed of the disaccharide and hypervariable regions. Hydrophobic packing of the inner core of α-helices along with the flexibility of these helices may permit type 4 pili to bend and adopt twisted, bundled conformations as seen in Fig. 5b. Hydrogen bonds throughout the middle layer of β-sheets may provide much of the mechanical stability for the pilus. The hypervariable outermost layer is not an integral part of the pilus structure and associates with the middle layer of β-sheets through only a few conserved interactions. Thus, the outermost region can be structurally pliant and can accommodate extreme amino acid changes that may lead to antigenic variation and altered binding specificities, without disrupting the assembly of the pilus.

The antigenic characteristics of type 4 pili synthesised by *N. gonorrhoeae* can be modified extensively by a remarkable mechanism (Haas and Meyer, 1987; Seifert *et al.*, 1988). This pathogen encodes more than 15 different silent pilin genes termed PilS that lack the invariant N-terminal domain present in PilE. By recombination of silent PilS genes with the PilE locus a single neisserial strain can theoretically express over 10 million PilE variants.

The biogenesis of type 4 pili is substantially more complicated than pilus assembly by the chaperone/usher or alternate chaperone pathways. Type 4 pilus assembly requires the expression of a myriad of different genes that are usually located in various unlinked regions on the chromosome. Exceptions include TCP (Faast *et al.*, 1989; Gupta, 1990; Shaw and Taylor, 1990), BFP (Nataro *et al.*, 1987; Giron *et al.*, 1991; Donnenberg *et al.*, 1992), and the R64 thin pilus (Kim *et al.*, 1993; Kim and Komano, 1997), which are currently the only type 4 pili for which the majority of the genes required for pilus biogenesis are located within a single genetic locus. While chromosomally located genes encode most type 4A pili, all known type 4B pili, with the exception of TCP, are encoded by plasmids. The number of genes essential for type 4 pilus biogenesis and function ranges from 14, for pili such as BFP, to over 20, for structures such as the type 4A pili of *N. gonorrhoeae*. In *P. aeruginosa* it is estimated that about 0.5% of the bacterium's genome is involved in the synthesis and function of type 4 pili (Alm and Mattick, 1997). Among the various different bacterial species expressing type 4 pili, the genes encoding the type 4 pilus structural components are similar while the regulatory components surrounding them are typically less conserved.

Several gene products are currently known to be central to the assembly of type 4 pili. These include (1) a prepilin peptidase that cleaves off the leader peptide from nascent pilin subunits (Nunn and Lory, 1991; Strom *et al.*, 1991, 1993); (2) a polytopic inner membrane protein that may act as a platform for pilus assembly (Kaufman *et al.*, 1993; Sohel *et al.*, 1993; Tonjum *et al.*, 1995); (3) a hydrophilic nucleotide-binding protein located in the cytoplasm or associated with the cytoplasmic face of the inner membrane that may provide energy for pilus assembly and/or twitching motility (Nunn *et al.*, 1990; Whitchurch *et al.*, 1991; Whitchurch and Mattick, 1994); (4) an outer membrane protein complex that forms a pore for passage of the pilus to the exterior of the bacterium (Hobbs and Mattick, 1993; Mattick *et al.*, 1996; Pugsley, 1993). Many of the components involved in type 4 pilus assembly share homology with proteins that are part of DNA uptake and protein secretion systems collectively known as the main terminal branch of the general secretory (*sec*-dependent) pathway, or type II secretion. These secretion pathways encode proteins with type 4 pilin-like characteristics and other proteins with homology to type 4 prepilin peptidases and outer membrane pore-forming proteins. Whether type II secretion systems can assemble pili or a pilus-like structure is not known.

Recently, Meyer and colleagues have described a model for type 4 pilus assembly in *N. gonorrhoeae* (Fussenegger *et al.*, 1997) (**Fig. 5c**). The PilE propilin subunits are transported into the periplasm by the *sec* translocation machinery. Following translocation, the propilin subunits remain anchored in the inner membrane by their hydrophobic N-terminal α-helical domains, with their hydrophilic C-terminal heads oriented towards the periplasm. Removal of the positively charged PilE propilin leader sequence by the PilD signal peptidase drives the hydrophobic stems of the PilE subunits to associate to form a pilus. An inner membrane assembly complex made up of several proteins including PilD, PilF, PilG and PilT aids this process. The assembled pilus penetrates the outer membrane through a gated pore formed by the multimeric complex Omc. The PilC adhesin associated with the tips of type 4 pili produced by *N. gonorrhoeae* may facilitate passage of the nascent pili through the Omc pore. One implication of this assembly model is that the N-terminal region of PilE resides in a continuous hydrophobic environment during both inner membrane transport and pilus assembly. This may allow polymerisation and, interestingly, depolymerisation of the pilus to proceed with only minimal energy requirements. Continued extension and retraction of type 4 pili by rounds of polymerisation and depolymerisation reactions are proposed to be the basis of twitching motility, one of the functions of type 4 pili. This process could be controlled by nucleotide-binding proteins such as PilT, PilB and PilU associated with the inner membrane assembly complex. The capacity of type 4 pili to depolymerise may also provide a means for transforming DNA, which could potentially interact with the type 4 pilus fibre, to enter into the bacterial cell.

Type IV Secretion and Conjugative Pilus Assembly

Type IV secretion systems (not to be confused with the secretion of autotransporters, such as IgA proteases, which is also known as type IV secretion) represent a major pathway for the transfer of both nucleic acids and proteins between cells (Zupan *et al.*, 1998). Conjugative pili, arguably the most well-studied structure assembled by type IV secretion systems, facilitate the interbacterial transfer of DNA by allowing donor and recipient bacteria to make specific and stable intercellular contacts. In some cases, conjugative pili may also form the conduits for intercellular DNA transfer. Horizontal gene transfer, or conjugation, mediated by conjugative pili is inextricably associated with the spread of antibiotic resistance among bacterial pathogens. Conjugative pili are generally encoded by self-transmissible plasmids that are capable of passing a copy of their genes to a recipient bacterium. Closely related plasmids, with similar replication control systems, are unable to co-exist within the same cell. This property has been termed 'incompatibility' and provides the primary basis for cataloguing conjugal plasmids and the pili that they encode (Datta, 1975). Thus far, in *E. coli* alone, over 25 incompatibility groups comprising well over 100 different plasmids have been defined (Bradley, 1980). Plasmids within a particular incompatibility group usually encode conjugative pili with similar antigenic properties, sensitivities to pilus-specific phages and morphologies.

Among the multitude of known incompatibility groups, three morphologically and functionally distinct types of conjugative pili have been defined: (1) rigid (Bradley *et al.*, 1982); (2) thick, flexible (Bradley, 1984) and (3) thin, flexible pili (Bradley, 1980; Frost, 1993). Rigid conjugative pili are 8- to 11-nm-wide structures that are usually specified by conjugal DNA transfer systems that function well only on solid surfaces, while thick, flexible pili typically promote conjugation on solid surfaces and in liquid media equally well. Conjugal DNA transfer promoted by rigid or thick, flexible pili can be enhanced, in some cases, by the presence of thin, flexible pili. These are similar in appearance to type 4 pili and at least one member of the thin, flexible pilus group (the R64 thin pilus) has been identified at the molecular level as a type 4B pilus (as described above). Thin, flexible pili appear to function primarily in the stabilisation of bacterial mating-pairs, increasing the rate of DNA transfer (Bradley, 1984). Conjugation does not occur in the presence of thin, flexible pili alone or in the absence of rigid or thick, flexible pili (Bradley, 1984).

Currently, the most thoroughly studied conjugative pilus is the F pilus encoded by the self-transmissible, broad host range F (fertility) plasmid, a member of the F1 incompatibility (IncF1) group of plasmids borne by *E. coli*. The F pilus system is prototypical for numerous other conjugation systems and F pilus biogenesis is distinct from type 4 and other pilus assembly pathways. F pili are 8-nm-thick, flexible helical filaments composed primarily, if not completely, of repeating 7.2-kDa (70-amino-acid) TraA pilin subunits (Frost *et al.*, 1984, 1985). Donor (F^+) cells typically express one to three F pili that are usually 1–2 μm long. Each F pilus possesses a 2-nm-wide central channel lined by basic, hydrophilic residues, which could potentially interact with negatively charged DNA or RNA molecules during conjugation (Paiva *et al.*, 1992).

TraA is organised into pentameric, doughnut-shaped discs that are stacked within the pilus such that successive discs are translated 1.28 nm along the pilus axis and rotated 28.8° with respect to the lower disc (Paiva *et al.*, 1992). The TraA pilin has two hydrophobic domains located towards the centre and at the C-terminus of the pilin. The hydrophobic domains are thought to extend as anti-parallel α-helices from the central axis to the periphery of the pilus shaft. These domains are separated by a short, basic region that appears to form the hydrophilic wall of the central channel of the pilus. The N-terminal domain of TraA is predicted to face the exterior of the pilus. However, this domain is antigenically masked when the N-terminal residue of TraA is acetylated during maturation of the pilin (see below) (Frost *et al.*, 1984; Finlay *et al.*, 1985). This modification is common among all known F-like pilins and appears to cause the N-terminal domain to be tucked back into or along the pilus shaft (Finlay *et al.*, 1985; Worobec *et al.*, 1985, 1986; Frost *et al.*, 1986).

Acetylation is not essential for F pilus assembly or function, but does help to prevent aggregation of F-like pili and affects the phage-binding characteristics of these organelles (Grossman and Silverman, 1989; Grossman *et al.*, 1990; Moore *et al.*, 1993). Although masked within the pilus shaft, the acetylated N-terminal domain of TraA appears to be exposed in unassembled pilin subunits and at the distal tips of pili (Frost *et al.*, 1986; Paiva *et al.*, 1992). The F pilus tip is believed to initiate contact between donor and recipient cells during conjugation. Alterations in the N-terminal sequence of TraA provide the primary basis for the antigenic diversity observed among various F-like pili (Frost *et al.*, 1985).

At least 16 gene products encoded by the F plasmid are involved in F pilus assembly and an additional 20 or more are needed for conjugation. Two gene products, TraQ and TraX, mediate the processing of the TraA pilin to its mature form (Moore *et al.*, 1993). TraA is synthesised as a 12.8-kDa (121-amino-acid) cytoplasmic propilin that is translocated across the inner membrane where it is proteolytically processed by host signal peptidase I to yield the 7.2-kDa pilin form (Frost *et al.*, 1984). TraQ, an inner membrane protein, facilitates the translocation process and may help to position the TraA propilin for processing into mature pilin (Moore *et al.*, 1982). In the absence of TraQ, the translocation of TraA is disrupted and most of the pilin is degraded. After processing, the N-terminal residue (alanine) of TraA is acetylated by TraX, a polytopic inner membrane protein (Moore *et al.*, 1993). While TraQ and TraX are involved in the maturation of the TraA pilin, 13 additional gene products (TraL, TraB, TraE, TraK, TraV, TraC, TraW, TraU, TraF, TraH, TraG, TrbC and TrbI) affect the assembly of TraA into the pilus filament. Most of these proteins appear to associate with either the inner or outer bacterial membrane and they may constitute a pilus assembly complex that spans the periplasmic space (Firth *et al.*, 1996).

The exact mechanism by which TraA is assembled into pili is not yet defined. The mature TraA pilin accumulates in the inner membrane with its N-terminus facing the periplasm (Paiva *et al.*, 1992). Both hydrophobic domains of TraA span the inner membrane, with the hydrophilic region of TraA connecting them on the cytoplasmic side (Paiva *et al.*, 1992). Small clusters of TraA also accumulate in the outer membrane and these may function as intermediates in F pilus assembly and disassembly (Sowa *et al.*, 1983). Large portions of the TraA sequence have the propensity to assume both β-sheet and α-helical structures, although the α-helical conformation is known to predominate in assembled pili (Armstrong *et al.*, 1980; Date *et al.*, 1977). Frost and co-workers have suggested that a shift between β-sheet and α-helical conformations may drive pilus assembly and disassembly (Frost *et al.*, 1984). F pilus assembly is energy dependent and the depletion of ATP levels by respiratory poisons such as cyanide results in F pilus depolymerisation and retraction (O'Callaghan *et al.*, 1973; Novotny and Fives-Taylor, 1974). It has been postulated that TraA is normally cycled between pili and periplasmic and inner membrane pools by rounds of pilus outgrowth and subsequent retraction. During conjugation, F pilus retraction is thought to serve a stabilising function by shortening the distance between bacterial mating pairs and allowing for more intimate contact (Harrington and Rogerson, 1990; Frost, 1993).

Several components of the F pilus assembly machinery share significant homology with proteins encoded by other conjugative systems. These include proteins specified by broad-host-range plasmids in other incompatibility groups (such as IncN, IncP and IncW) and many of the proteins encoded by the Ti plasmid-specific *vir* genes of the plant pathogen *Agrobacterium tumefaciens* (Baron and Zambryski, 1996). These bacteria elaborate 10-nm-wide promiscuous conjugative pili, called T pili, which direct the interkingdom transfer of a specific genetic element, known as T-DNA, into plant and yeast cells. The introduction of T-DNA into plant cells induces plant tumour formation. T pilus assembly by *A. tumefaciens* requires the expression of at least 12 *vir* gene products encoded by the Ti plasmid. VirB2 is the major, and possibly only, component of the T pilus and it is predicted to be structurally homologous to the F pilus subunit TraA (Pohlman *et al.*, 1994). Recently, the structures of the VirB2 and TrbC (IncP conjugative pili) have been solved using mass spectrometry (Eisenbrandt *et al.*, 1999). These pilins are cyclised via an intramolecular covalent bond after final processing is complete. This circular structure is thought to represent the primary configuration of the pilins that are assembled into the conjugative apparatus.

Besides stabilising donor–recipient interactions, it is not yet clear how F and T pili or any pilus structures function in conjugative DNA transfer processes. However, substantial evidence exists, at least in the case of F pili, suggesting that pilus components or the pilus itself can serve as a specialised channel for the transmission of DNA and any accompanying pilot proteins across the donor and possibly the recipient cell membranes (Firth *et al.*, 1996). In light of this possibility, it is interesting to note that many components of the conjugative pilus systems have homologues to other type IV secretion systems responsible for the secretion of proteins across the bacterial membrane (Covacci *et al.*, 1999). These secretion systems often constitute major virulence determinants of pathogenic bacteria. For example, the Ptl proteins responsible for the export of the multiple subunit toxin of *Bordetella pertussis* are homologous to proteins in the F pili. Furthermore, these secretion systems are distantly related to transport systems used by *Legionella pneumophila* and *Helicobacter pylori* to inject virulence factors into host eukaryotic cells (Covacci *et al.*, 1999). Conjugative pilus systems such as those encoding F and T pili thus appear to be representative of a larger family of macromolecular transport systems. Understanding how conjugative pili help to mediate the intercellular transfer of macromolecules remains a significant challenge.

References

Abraham SN, Land M, Ponniah S, Endres R, Hasty DL, Babu JP (1992) Glycerol-induced unraveling of the tight helical conformation of *Escherichia coli* type 1 fimbriae. *J. Bacteriol.* 174: 5145–5148.

Alm RA, Mattick JS. (1997) Genes involved in the biogenesis and function of type-4 fimbriae in *Pseudomonas aeruginosa*. *Gene* 192: 89–98.

Armstrong GD, Frost LS, Sastry PA, Paranchych W (1980) Comparative biochemical studies on F and EDP208 conjugative pili. *J. Bacteriol.* 141: 333–341.

Arnqvist A, Olsen A, Pfeifer J, Russell DG, Normark S (1992) The Crl protein activates cryptic genes for curli formation and fibronectin binding in *Escherichia coli* HB101. *Mol. Microbiol.* 6: 2443–2452.

Austin JW, Sanders G, Kay WW, Collinson SK (1998) Thin aggregative fimbriae enhance *Salmonella enteritidis* biofilm formation. *FEMS Microbiol. Lett.* 162: 295–301.

Baga M, Norgren M, Normark S (1987) Biogenesis of *E. coli* Pap pili: PapH, a minor pilin subunit involved in cell anchoring and length modulation. *Cell* 49: 241–251.

Baron C, Zambryski PC (1996) Plant transformation: a pilus in *Agrobacterium* T-DNA transfer. *Curr. Biol.* 6: 1567–9.

Bäumler AJ, Heffron F (1995) Identification and sequence analysis of lpfABCDE, a putative fimbrial operon of *Salmonella typhimurium*. *J. Bacteriol.* 177: 2087–2097.

Ben Nasr A, Olsen A, Sjobring U, Muller-Esterl W, Bjorck L (1996) Assembly of human contact phase proteins and release of bradykinin at the surface of curli-expressing *Escherichia coli*. *Mol. Microbiol.* 20: 927–935.

Bian Z, Normark S (1997) Nucleator function of CsgB for the assembly of adhesive surface organelles in *Escherichia coli*. *EMBO J.* 16: 5827–5836.

Bieber D, Ramer SW, Wu CY *et al.* (1998) Type IV pili, transient bacterial aggregates, and virulence of enteropathogenic *Escherichia coli*. *Science* 280: 2114–2118.

Blond-Elguindi S, Cwirla SE, Dower WJ *et al.* (1993) Affinity panning of a library of peptides displayed of bacteriophages reveals the binding specificty of BiP. *Cell* 75: 717–728.

Bradley DE (1974) Adsorption of bacteriophages specific for *Pseudomonas aeruginosa* R factors RP1 and R1822. *Biochem. Biophys. Res. Commun.* 57: 893–900.

Bradley DE (1980) Morphological and serological relationships of conjugative pili. *Plasmid* 4: 155–169.

Bradley DE (1984) Characteristics and function of thick and thin conjugative pili determined by transfer-derepressed plasmids of incompatibility groups I1: I2: I5 B, K and Z. *J. Gen. Microbiol.* 130: 1489–1502.

Bradley DE, Pitt TL (1974) Pilus-dependence of four *Pseudomonas aeruginosa* bacteriophages with non-contractile tails. *J. Gen. Virol.* 24: 1–15.

Bradley DE, Whelan J (1989) *Escherichia coli* tolQ mutants are resistant to filamentous bacteriophages that adsorb to the tips, not the shafts, of conjugative pili. *J. Gen. Microbiol.* 135: 7.

Bradley DE, Williams PA (1982) The TOL plasmid is naturally derepressed for transfer. *J. Gen. Microbiol.* 128: 3019–3024.

Bradley DE, Sirgel FA, Coetzee JN, Hedges RW, Coetzee WF (1982) Phages C-2 and J: IncC and IncJ plasmid-dependent phages, respectively. *J. Gen. Microbiol.* 128: 2485–2498.

Brinton Jr CC (1965) The structure, function, synthesis, and genetic control of bacterial pili and a model for DNA and RNA transport in gram negative bacteria. *Trans. N.Y. Acad. Sci.* 27: 1003–1165.

Bullitt E, Makowski L (1995) Structural polymorphism of bacterial adhesion pili. *Nature* 373: 164–167.

Choudhury D, Thompson A, Stojanoff V *et al.* (1999) X-ray structure of the FimC-FimH chaperone-adhesin complex from uropathogenic *Escherichia coli. Science* 285: 1061–1066.

Collington GK, Booth IW, Donnenberg MS, Kaper JB, Knutton S (1998) Enteropathogenic *Escherichia coli* virulence genes encoding secreted signalling proteins are essential for modulation of Caco-2 cell electrolyte transport. *Infect. Immun.* 66: 6049–6053.

Collinson SK, Emody L, Muller K-H, Trust T, Kay WM (1991) Purification and characterization of thin, aggregative fimbriae from *Salmonella enteritidis. J. Bacteriol.* 173: 4773–4781.

Connell H, Agace W, Hedlund M, Klemm P, Shembri M, Svanborg C (1996) Fimbriae-mediated adherance induces mucosal inflammation and bacterial clearance. In: Ofek KA (ed.) *Toward Anti-adhesion Therapy for Microbial Diseases.* New York: Plenum Press.

Cornelis GR (1998) The *Yersinia* deadly kiss. *J. Bacteriol.* 180: 5495–5504.

Covacci A, Telford JL, Del Giudice G, Parsonnet J, Rappuoli R (1999) Helicobacter pylori virulence and genetic geography. *Science* 284: 1328–1333.

Darzins A, Russell MA (1997) Molecular genetic analysis of type-4 pilus biogenesis and twitching motility using *Pseudomonas aeruginosa* as a model system – a review. *Gene* 192: 109–115.

Date T, Inuzuka M, Tomoeda M (1977) Purification and characterization of F pili from *Escherichia coli. Biochemistry* 16: 5579–5585.

Datta N (1975) Microbiology-1974. In: Schlessinger D (ed.) *Microbiology-1974.* Washington, DC: American Society for Microbiology, pp. 9–15.

de Graaf FK, Gaastra W (1994) Fimbriae of enterotoxigenic *Escherichia coli.* In: Klemm P (ed.) *Fimbriae, Adhesion, Genetics, Biogenesis and Vaccines.* Ann Arbor MI: CRC Press, pp. 53–70.

de Graaf FK, Mooi FR (1986) The fimbrial adhesins of *Escherichia coli. Adv. Microb. Physiol.* 28: 65–143.

Donnenberg M, Giron J, Nataro J, Kaper J (1992) A plasmid encoded type IV fimbrial gene of enteropathogenic *Escherichia coli* associated with localized adherence. *Mol. Microbiol.* 6: 3427.

Dorr J, Hurek T, Reinhold-Hurek B (1998) Type IV pili are involved in plant-microbe and fungus-microbe interactions. *Mol. Microbiol.* 30: 7–17.

Duguid JP, Old DC (1980) Adhesive properties of Enterobacteriacae. In: Beachy EH (ed.) *Bacterial Adherence Receptors and Recognition.* London: Chapman and Hall, pp. 186–217.

Duguid JP, Smith IW, Dempster G, Edmunds PN (1955) Non-flagellar filamentous appendages ('fimbriae') and hemagglutinating activity in bacterium coli. *J. Pathol. Bacteriol.* 70: 335–348.

Duguid JP, Clegg S, Wilson MI (1979) The fimbrial and non-fimbrial haemagglutinins of *Escherichia coli. J. Med. Microbiol.* 12: 213–227.

Eisenbrandt R, Kalkum M, Lai EM, Lurz R, Kado CI, Lanka E (1999) Conjugative pili of IncP plasmids, and the Ti plasmid T pilus are composed of cyclic subunits. *J. Biol. Chem.* 274: 22548–22555.

Faast R, Ogierman A, Stroeher UH, Manning PA (1989) Nucleotide sequence of the structural gene, tcpA, for a major pilin subunit of *Vibrio cholerae. Gene* 85: 227–231.

Finlay BB, Frost LS, Paranchych W, Parker JM, Hodges RS (1985) Major antigenic determinants of F and ColB2 pili. *J. Bacteriol.* 163: 331–335.

Firth N, Ippen-Ihler K, Skurray RA (1996) Structure and function of the F factor and mechanism of conjugation. In: *Escherichia coli and Salmonella typhimurium: Cellular and Molecular Biology.* Neidhardt FC, Curtiss R III, Ingraham JL *et al.* (eds). Washington, DC: American Society for Microbiology, pp. 2377–2401.

Friedrich MJ, Kinsey NE, Vila J, Kadner RJ (1993) Nucleotide sequence of a 13.9 kb segment of the 90 kb virulence plasmid of *Salmonella typhimurium*: the presence of fimbrial biosynthetic genes. *Mol. Microbiol.* 8: 543–558.

Froehlich BJ, Karkashian A, Melsen LR, Wakefield JC, Scott JR (1994) CooC and CooD are required for assembly of CS1 pili. *Mol. Microbiol.* 12: 387–401.

Froehlich BJ, Karakashian A, Sakellaris H, Scott JR (1995) Genes for CS2 pili of enterotoxigenic *Escherichia coli* and their interchangeability with those for CS1 pili. *Infect. Immun.* 63: 4849–4856.

Frost L (1993) Conjugative pili and pilus-specific phages. In: Clewell DB (ed.) *Bacterial Conjugation.* New York: Plenum Press, pp. 189–221.

Frost LS, Paranchych W, Willetts NS (1984) DNA sequence of the F traALE region that includes the gene for F pilin. *J. Bacteriol.* 160: 395–401.

Frost LS, Finlay BB, Opgenorth A, Paranchych W, Lee JS (1985) Characterization and sequence analysis of pilin from F-like plasmids. *J. Bacteriol.* 164: 1238–1247.

Frost LS, Lee JS, Scraba DG, Paranchych W (1986) Two monoclonal antibodies specific for different epitopes within the amino-terminal region of F pilin. *J. Bacteriol.* 168: 192–198.

Fussenegger M, Rudel T, Barten R, Ryll R, Meyer TF (1997) Transformation competence and type-4 pilus biogenesis in *Neisseria gonorrhoeae* – a review. *Gene* 192: 125–134.

Giron JA, Ho ASY, Schoolnik GK (1991) An inducible bundle-forming pilus of Enteropathogenic *Escherichia coli*. *Science* 254: 710–713.

Giron JA, Levine MM, Kaper JB (1994) Longus: a long pilus ultrastructure produced by human enterotoxigenic *Escherichia coli*. *Mol. Microbiol.* 12: 71–82.

Goldstein R, Sun L, Jiang RZ, Sajjan U, Forstner JF, Campanelli C (1995) Structurally variant classes of pilus appendage fibers coexpressed from *Burkholderia (Pseudomonas) cepacia*. *J. Bacteriol.* 177: 1039–1052.

Gong M, Makowski L (1992) Helical structure of pap adhesion pili from *Escherichia coli*. *J. Mol. Biol.* 228: 735–742.

Grossman TH, Silverman PM (1989) Structure and function of conjugative pili: inducible synthesis of functional F pili by *Escherichia coli* K-12 containing a lac-tra operon fusion. *J. Bacteriol.* 171: 650–656.

Grossman TH, Frost LS, Silverman PM (1990) Structure and function of conjugative pili: monoclonal antibodies as probes for structural variants of F pili. *J. Bacteriol.* 172: 1174–1179.

Gupta RS (1990) Sequence and structural homology between a mouse T-complex protein TCP-1 and the 'chaperonin' family of bacterial (GroEL, 60–65 kDa heat shock antigen) and eukaryotic proteins. *Biochem. Int.* 20: 833–841.

Haas R, Meyer TF (1987) Molecular principles of antigenic variation in *Neisseria gonorrhoeae*. *Antonie Van Leeuwenhoek* 53: 431–434.

Hacker J (1992) Role of fimbrial adhesins in the pathogenesis of *Escherichia coli* infections. *Can. J. Microbiol.* 38: 720–727.

Hammar M, Arnqvist A, Bian Z, Olsen A, Normark S (1995) Expression of two csg operons is required for production of fibronectin- and congo red-binding curli polymers in *Escherichia coli* K-12. *Mol. Microbiol.* 18: 661–670.

Hammar M, Bian Z, Normark S (1996) Nucleator-dependent intercellular assembly of adhesive curli organelles in *Escherichia coli*. *Proc. Natl Acad. Sci. USA* 93: 6562–6566.

Harrington LC, Rogerson AC (1990) The F pilus of *Escherichia coli* appears to support stable DNA transfer in the absence of wall-to-wall contact between cells. *J. Bacteriol.* 172: 7263–7264.

Hobbs M, Mattick JS (1993) Common components in the assembly of type 4 fimbriae, DNA transfer systems, filamentous phage and protein-secretion apparatus: a general system for the formation of surface-associated protein complexes. *Mol. Microbiol.* 10: 233–243.

Holmgren A, Brändén C (1989) Crystal structure of chaperone protein PapD reveals an immunoglobulin fold. *Nature* 342: 248–251.

Hueck CJ (1998) Type III protein secretion systems in bacterial pathogens of animals and plants. *Microbiol. Mol. Biol. Rev.* 62: 379–433.

Hung DL, Hultgren SJ (1998) Pilus biogenesis via the chaperone/usher pathway: an integration of structure and function. *J. Struct. Biol.* 124: 201–220.

Jacob-Dubuisson F, Heuser J, Dodson K, Normark S, Hultgren S (1993) Initiation of assembly and association of the structural elements of a bacterial pilus depend on two specialized tip proteins. *EMBO J.* 12: 837–847.

Jacobs AAC, Roosendaal B, van Breemen JFL, de Graaf FK (1987) Role of phenylalanine 150 in the receptor binding domain of the K88 fibrillar subunit. *J. Bacteriol.* 169: 4907–4911.

Jonsson AB, Ilver D, Falk P, Pepose J, Normark S (1994) Sequence changes in the pilus subunit lead to tropism variation of *Neisseria gonorrhoeae* to human tissue. *Mol. Microbiol.* 13: 403–416.

Jordi BJ, Willshaw GA, van der Zeijst BA, Gaastra W (1992) The complete nucleotide sequence of region 1 of the CFA/I fimbrial operon of human enterotoxigenic *Escherichia coli*. *DNA Sequence* 2: 257–263.

Kallenius G, Svenson SB, Hultberg H *et al.* (1981) Occurrence of P fimbriated *Escherichia coli* in urinary tract infection. *Lancet* ii, 1369–1372.

Kallstrom H, Islam MS, Berggren PO, Jonsson AB (1998) Cell signaling by the type IV pili of pathogenic *Neisseria*. *J. Biol. Chem.* 273: 21777–21782.

Karaolis DK, Somara S, Maneval DR, Jr Johnson JA, Kaper JB (1999) A bacteriophage encoding a pathogenicity island, a type-IV pilus and a phage receptor in cholera bacteria. *Nature* 399: 375–379.

Kaufman MR, Shaw CE, Jones ID, Taylor RK (1993) Biogenesis and regulation of the *Vibrio cholerae* toxin-coregulated pilus: analogies to other virulence factor secretory systems. *Gene* 126: 43–49.

Kenny B, Lai LC, Finlay BB, Donnenberg MS (1996) EspA, a protein secreted by enteropathogenic *Escherichia coli*, is required to induce signals in epithelial cells. *Mol. Microbiol.* 20: 313–323.

Kim SR, Komano T (1997) The plasmid R64 thin pilus identified as a type IV pilus. *J. Bacteriol.* 179: 3594–3603.

Kim SR, Funayama N, Komano T (1993) Nucleotide sequence and characterization of the traABCD region of IncI1 plasmid R64. *J. Bacteriol.* 175: 5035–5042.

Klemm P, Christiansen G (1987) Three fim genes required for the regulation of length and mediation of adhesion of *Escherichia coli* type 1 fimbriae. *Mol. Gen. Genet.* 208: 439–445.

Knutton S, Rosenshine I, Pallen MJ *et al.* (1998) A novel EspA-associated surface organelle of enteropathogenic *Escherichia coli* involved in protein translocation into epithelial cells. *EMBO J.* 17: 2166–2176.

Kolter R, Losick R (1998) One for all and all for one. *Science* 280: 226–227.

Krogfelt KA, Bergmans H, Klemm P (1990) Direct evidence that the FimH protein is the mannose specific adhesin of *Escherichia coli* type 1 fimbriae. *Infect. Immun.* 58: 1995–1999.

Kuehn MJ, Heuser J, Normark S, Hultgren SJ (1992) P pili in uropathogenic *E. coli* are composite fibres with distinct fibrillar adhesive tips. *Nature* 356: 252–255.

Langermann S, Palaszynski S, Barnhart M *et al.* (1997) Prevention of mucosal *Escherichia coli* infection by FimH-adhesin-based systemic vaccination. *Science* 276: 607–611.

Leffler H, Svanborg-Eden C (1980) Chemical identification of a glycosphingolipid receptor for *Escherichia coli* attaching to human urinary tract epithelial cells and agglutinating human erythrocytes. *FEMS Microbiol. Lett.* 8: 127–134.

Loferer H, Hammar M, Normark S (1997) Availability of the fibre subunit CsgA and the nucleator protein CsgB during assembly of fibronectin-binding curli is limited by the intracellular concentration of the novel lipoprotein CsgG. *Mol. Microbiol.* 26: 11–23.

Mattick JS, Whitchurch CB, Alm RA (1996) The molecular genetics of type-4 fimbriae in *Pseudomonas aeruginosa* – a review. *Gene* 179: 147–155.

Maurer L, Orndorff PE (1987) Identification and characterization of genes determining receptor binding and pilus length of *Escherichia coli* type 1 pili. *J. Bacteriol.* 169: 640–645.

Minion FC, Abraham SN, Beachey EH, Goguen JD (1989) The genetic determinant of adhesive function in type 1 fimbriae of *Escherichia coli* is distinct from the gene encoding the fimbrial subunit. *J. Bacteriol.* 165: 1033–1036.

Montag D, Riede I, Eschbach ML, Degen M, Henning U (1987) Receptor-recognizing proteins of T-even type bacteriophages. Constant and hypervariable regions and an unusual case of evolution. *J. Mol. Biol.* 196: 165–174.

Mooi FR, de Graaf FK (1985) Molecular biology of fimbriae of enterotoxigenic *Escherichia coli*. *Curr. Top. Microbiol. Immunol.* 118: 119–138.

Mooi FR, Wijfies A, de Graaf FK (1983) Identification of precursors in the biosynthesis of the K88ab fimbriae. *J. Bacteriol.* 154: 41–49.

Mooi FR, Claassen I, Bakker D, Kuipers H, de Graaf FK (1986) Regulation and structure of an *Escherichia coli* gene coding for an outer membrane protein involved in export of K88ab fimbrial subunits. *Nucleic Acids Res.* 14: 2443–2457.

Moore D, Sowa BA, Ippen-Ihler K (1982) A new activity in the Ftra operon which is required for F-pilin synthesis. *Mol. Gen. Genet.* 188: 459–464.

Moore D, Hamilton CM, Maneewannakul K, Mintz Y, Frost LS, Ippen-Ihler K (1993) The *Escherichia coli* K-12 F plasmid gene traX is required for acetylation of F pilin. *J. Bacteriol.* 175: 1375–1383.

Muller A, Gunther D, Dux F, Naumann M, Meyer TF, Rudel T (1999) Neisserial porin (PorB) causes rapid calcium influx in target cells and induces apoptosis by the activation of cysteine proteases. *EMBO J.* 18: 339–352.

Mulvey MA, Lopez-Boado YS, Wilson CL *et al.* (1998) Induction and evasion of host defenses by type 1-piliated uropathogenic *Escherichia coli*. *Science* 282: 1494–1497.

Nataro JP, Maher KO, Mackie P, Kaper JB (1987) Characterization of plasmids encoding the adherence factor of enteropathogenic *Escherichia coli*. *Infect. Immun.* 55: 2370–2377.

Neves BC, Knutton S, Trabulsi LR *et al.* (1998) Molecular and ultrastructural characterisation of EspA from different enteropathogenic *Escherichia coli* serotypes. *FEMS Microbiol. Lett.* 169: 73–80.

Novotny CP, Fives-Taylor P (1974) Retraction of F pili. *J. Bacteriol.* 117: 1306–1311.

Nunn DN, Lory S (1991) Product of the *Pseudomonas aeroginosa* gene pilD is a prepilin peptidase. *Proc. Natl Acad. Sci. USA* 88: 3281.

Nunn D, Bergman S, Lory S (1990) Products of three accessory genes, pilB, pilC, and pilD, are required for biogenesis of *Pseudomonas aeruginosa* pili. *J. Bacteriol.* 172: 2911–2919.

O'Callaghan RJ, Bundy L, Bradley R, Paranchych W (1973) Unusual arsenate poisoning of the F pili of *Escherichia coli*. *J. Bacteriol.* 115: 76–81.

O'Hanley P, Lark D, Falkow S, Schoolnik G (1985) Molecular basis of *Escherichia coli* colonization of the upper urinary tract in BALB/c mice. Gal-Gal pili immunization prevents *Escherichia coli* pyelonephritis in the BALB/c mouse model of human pyelonephritis. *J. Clin. Invest.* 75: 347–360.

O'Toole GA, Kolter R (1998) Flagellar and twitching motility are necessary for *Pseudomonas aeruginosa* biofilm development. *Mol. Microbiol.* 30: 295–304.

Ofek I, Mirelman D, Sharon N (1977) Adherence of *Escherichia coli* to human mucosal cells mediated by mannose receptors. *Nature* 265: 623–625.

Olsen A, Jonsson A, Normark S (1989) Fibronectin binding mediated by a novel class of surface organelles on *Escherichia coli*. *Nature* 338: 652–655.

Olsen A, Arnqvist A, Hammar M, Normark S (1993a) Environmental regulation of curli production in *Escherichia coli*. *Infect. Agents Dis.* 2: 272–274.

Olsen A, Arnqvist A, Hammar M, Sukupolvi S, Normark S (1993b) The RpoS sigma factor relieves H-NS-mediated transcriptional repression of csgA, the subunit gene of fibronectin-binding curli in *Escherichia coli*. *Mol. Microbiol.* 7: 523–536.

Paiva WD, Grossman T, Silverman PM (1992) Characterization of F-pilin as an inner membrane component of *Escherichia coli* K12. *J. Biol. Chem.* 267: 26191–26197.

Paranchych W, Frost LS (1988) The physiology and biochemistry of pili. *Adv. Microb. Physiol.* 29: 53–114.

Parge HE, Forest KT, Hickey MJ, Christensen DA, Getzoff ED, Tainer JA (1995) Structure of the fibre-forming protein pilin at 2.6 Å resolution. *Nature* 378: 32–38.

Pohlman RF, Genetti HD, Winans SC (1994) Common ancestry between IncN conjugal transfer genes and macromolecular export systems of plant and animal pathogens. *Mol. Microbiol.* 14: 655–668.

Ponniah S, Endres RO, Hasty DL, Abraham SN (1991) Fragmentation of *Escherichia coli* type 1 fimbriae exposes crytic D-mannose-binding sites. *J. Bacteriol.* 173: 4195–4202.

Pugsley A (1993) The complete general secretory pathway in Gram negative bacteria. *Microbiol. Rev.* 57: 50–108.

Roberts JA, Marklund B-I, Ilver D *et al.* (1994) The Gal α(1–4) Gal-specific tip adhesin of *Escherichia coli* P-fimbriae is needed for pyelonephritis to occur in the normal urinary tract. *Proc. Natl Acad. Sci. USA* 91: 11889–11893.

Roine E, Wei W, Yuan J *et al.* (1997) Hrp pilus: an hrp-dependent bacterial surface appendage produced by Pseudomonas syringae pv. tomato DC3000. *Proc. Natl Acad. Sci. USA* 94: 3459–3464.

Romling U, Bian Z, Hammar M, Sierralta WD, Normark S (1998a) Curli fibers are highly conserved between *Salmonella typhimurium* and *Escherichia coli* with respect to operon structure and regulation. *J. Bacteriol.* 180: 722–731.

Romling U, Sierralta WD, Eriksson K, Normark S (1998b) Multicellular and aggregative behaviour of *Salmonella typhimurium* strains is controlled by mutations in the agfD promoter. *Mol. Microbiol.* 28: 249–264.

Rudel T, Boxberger HJ, Meyer TF (1995a) Pilus biogenesis and epithelial cell adherence of *Neisseria gonorrhoeae* pilC double knock-out mutants. *Mol. Microbiol.* 17: 1057–1071.

Rudel T, Scheuerpflug I, Meyer TF (1995b) *Neisseria* PilC protein identified as type 4 pilus tip-located adhesin. *Nature* 373: 357–359.

Russell PW, Orndorff PE (1992) Lesions in two *Escherichia coli* type 1 pilus genes alter pilus number and length without affecting receptor binding. *J. Bacteriol.* 174: 5923–5935.

Sajjan US, Sun L, Goldstein R, Forstner JF (1995) Cable (cbl) type II pili of cystic fibrosis-associated *Burkholderia (Pseudomonas) cepacia*: nucleotide sequence of the cblA major subunit pilin gene and novel morphology of the assembled appendage fibers [published erratum appears in *J. Bacteriol.* 1995 June; 177(11): 3558]. *J. Bacteriol.* 177: 1030–1038.

Sakellaris H, Scott JR (1998) New tools in an old trade: CS1 pilus morphogenesis. *Mol. Microbiol.* 30: 681–687.

Sakellaris H, Balding DP, Scott JR (1996) Assembly proteins of CS1 pili of enterotoxigenic *Escherichia coli. Mol. Microbiol.* 21: 529–541.

Salit IE, Gotschich EC (1977) Type 1 *Escherichia coli* pili: characterization of binding to monkey kidney cells. *J. Exp. Med.* 146: 1182–1194.

Sauer FG, Futterer K, Pinkner JS, Dodson KW, Hultgren SJ, Waksman G (1999) Structural basis of chaperone function and pilus biogenesis. *Science* 285: 1058–1061.

Saulino ET, Thanassi DG, Pinkner JS, Hultgren SJ (1998) Ramifications of kinetic partitioning on usher-mediated pilus biogenesis. *EMBO J.* 17: 2177–2185.

Schifferli DM, Abraham SN, Beachey EH (1987) Use of monoclonal antibodies to probe subunit- and polymer-specific epitopes of 987P fimbriae of *Escherichia coli. Infect. Immun.* 55: 923–930.

Schifferli DM, Beachy EH, Taylor RK (1991) Genetic analysis of 987P adhesion and fimbriation of *Escherichia coli*: the fas genes link both phenotypes. *J. Bacteriol.* 173: 1230–1240.

Seifert HS, Ajioka RS, Marchal C, Sparling PF, So M (1988) DNA transformation leads to pilin antigenic variation in *Neisseria gonorrhoeae. Nature* 336: 392–395.

Shaw CE, Taylor RK (1990) *Vibrio cholerae* O395 tcpA pilin gene sequence and comparison of predicted protein structural features to those of type 4 pilins. *Infect. Immun.* 58: 3042–3049.

Sjobring U, Pohl G, Olsen A (1994) Plasminogen, absorbed by *Escherichia coli* expressing curli or by *Salmonella enteritidis* expressing thin aggregative fimbriae, can be activated by simultaneously captured tissue-type plasminogen activator (t-PA). *Mol. Microbiol.* 14: 443–452.

Smit H, Gaastra W, Kamerling JP, Vliegenthart JFG, de Graaf FK (1984) Isolation and structural characterization of the equine erythrocyte receptor for enterotoxigenic *Escherichia coli* K99 fimbrial adhesin. *Infect. Immun.* 46: 578.

Sohel I, Puente JL, Murray WJ, Vuopio-Varkila J, Schnoolnik GK (1993) Cloning and characterization of the bundle-forming pilin gene of enteropathogenic *Escherichia coli* and its distribution in *Salmonella* serotypes. *Mol. Microbiol.* 7: 563.

Soto GE, Dodson KW, Ogg D *et al.* (1998) Periplasmic chaperone recognition motif of subunits mediates quaternary interactions in the pilus. *EMBO J.* 17: 6155–6167.

Sowa BA, Moore D, Ippen-Ihler K (1983) Physiology of F-pilin synthesis and utilization. *J. Bacteriol.* 153: 962–968.

Striker R, Nilsson U, Stonecipher A, Magnusson G, Hultgren SJ (1995) Structural requirements for the glycolipid receptor of human uropathogenic *Escherichia coli. Mol. Microbiol.* 16: 1021–1029.

Strom MS, Lory S (1993) Structure–function and biogenesis of type IV pili. *Annu. Rev. Microbiol.* 47: 565–596.

Strom MS, Nunn D, Lory S (1991) Multiple roles of the pilus biogenesis protein, PilD: involvement of PilD inexcretion of enzymes from *Pseudomonas aeruginosa. J. Bacteriol.* 173: 1175.

Strom MS, Nunn DN, Lory S (1993) A single bifunctional enzyme, PilD, catalyzes cleavage and N-methylation of proteins belonging to the type IV pilin family. *Proc. Natl Acad. Sci. USA* 90: 2404.

Sun DX, Mekalanos JJ, Taylor RK (1990) Antibodies directed against the toxin-coregulated pilus isolated from *Vibrio cholerae* provide protection in the infant mouse experimental cholera model. *J. Infect. Dis.* 161: 1231–1236.

Svanborg C, Hedlund M, Connell H *et al.* (1996) Bacterial adherence and mucosal cytokine responses. Receptors and transmembrane signaling. *Ann. NY Acad. Sci.* 797: 177–190.

Svanborg Eden C, Andersson B *et al.* (1983) Receptor analogues and anti-pili antibodies as inhibitors of bacterial attachment *in vivo* and *in vitro*. *Ann. NY Acad. Sci.* 409: 580–592.

Taniguchi T, Fujino Y, Yamamoto K, Miwatani T, Honda T (1995) Sequencing of the gene encoding the major pilin of pilus colonization factor antigen III (CFA/III) of human enterotoxigenic *Escherichia coli* and evidence that CFA/III is related to type IV pili. *Infect. Immun.* 63: 724–728.

Tennent JM, Mattick JS (1994) Type 4 pili. In: Klemm P (ed.) *Fimbriae: Adhesion, Genetics, Biogenesis and Vaccines.* London: CRC Press, pp. 127–147.

Tewari R, MacGregor JI, Ikeda T, Little JR, Hultgren SJ, Abraham SN (1993) Neutrophil activation by nascent FimH subunits of type 1 fimbriae purified from the periplasm of *Escherichia coli*. *J. Biol. Chem.* 268: 3009–3015.

Thanassi DG, Saulino ET, Hultgren SJ (1998a) The chaperone/usher pathway: a major terminal branch of the general secretory pathway. *Curr. Opin. Microbiol.* 1: 223–231.

Thanassi DG, Saulino ET, Lombardo M-J, Roth R, Heuser J, Hultgren SJ (1998b) The PapC usher forms an oligomeric channel: implications for pilus biogenesis across the outer membrane. *Proc. Natl Acad. Sci. USA* 95: 3146–3151.

Tonjum T, Freitag NE, Namork E, Koomey M (1995) Identification and characterization of pilG, a highly conserved pilus-assembly gene in pathogenic *Neisseria*. *Mol. Microbiol.* 16: 451–464.

Vidal O, Longin R, Prigent-Combaret C, Dorel C, Hooreman M, Lejeune P (1998) Isolation of an *Escherichia coli* K-12 mutant strain able to form biofilms on inert surfaces: involvement of a new ompR allele that increases curli expression. *J. Bacteriol.* 180: 2442–2449.

Whitchurch CB, Mattick JS (1994) *Escherichia coli* contains a set of genes homologous to those involved in protein secretion, DNA uptake and the assembly of type-4 fimbriae in other bacteria. *Gene.* 101: 9–15.

Whitchurch CB, Hobbs M, Livingston SP, Krishnapillai V, Mattick JS (1991) Charachterization of a *Pseudomonas aeruginosa* twitching motility gene and evidence for a specialized protein export system widespread in eubacteria. *Gene.* 101: 33–44.

Worobec EA, Paranchych W, Parker JM, Taneja AK, Hodges RS (1985) Antigen-antibody interaction. The immunodominant region of EDP208 pili. *J. Biol. Chem.* 260: 938–943.

Worobec EA, Frost LS, Pieroni P *et al.* (1986) Location of the antigenic determinants of conjugative F-like pili. *J. Bacteriol.* 167: 660–665.

Zhang JP, Normark S (1996) Induction of gene expression in *Escherichia coli* after pilus-mediated adherence. *Science* 273: 1234–1236.

Zupan JR, Ward D, Zambryski P (1998) Assembly of the VirB transport complex for DNA transfer from *Agrobacterium tumefaciens* to plant cells. *Curr. Opin. Microbiol.* 1: 649–655.

10

Spores, Sporulation and Germination

Rohit S. Prajapati and Simon M. Cutting

Royal Holloway, University of London, Egham, Surrey, UK

Spores are produced by bacterial cells as a response to profound environmental changes, most commonly nutrient deprivation. Spores are robust, dormant life forms, designed to outlast adverse conditions, and to provide a potential advantage in a competitive environment. The best-known and most studied spore formers are members of the Gram-positive genera *Bacillus* and *Clostridium*, but many other bacteria also form spores (Table 1). The bacteria listed in Table 1 form 'true spores', which are termed endospores because the spore is formed *within* the cell. Endospores are eventually released from the sporangium ('mother cell') and so represent the germ line. The spores are dehydrated, making them particularly resistant to extremes of heat, desiccation, ultraviolet radiation and toxic chemicals; the primary purpose of endospores appears to be survival under environmental extremes.

Other bacteria produce dormant life forms with the primary purpose of facilitating dispersion, and in many respects they are functionally analagous to fungal spores. The micro-organisms that produce these dormant life forms include the Gram-negative myxobacteria, including the genera *Myxococcus* and *Stigmatella*, and the Gram-positive filamentous actinomycetes, including the genera *Streptomyces* and *Actinomyces*. The spores formed by these organisms are referred to as exospores. They are formed by differentiation of the entire bacterial cell into a spore, and although they are produced as a response to nutrient deprivation, they are not as robust as endospores and are less able to survive environmental challenges. Since dormant exospores are not regarded as 'true' spores, they will not be discussed further in this chapter.

How long a spore can survive is a controversial and interesting question (Gest and Mandelstam, 1987). Recently, spores of a new species related to *Bacillus sphaericus* were recovered and cultured from the abdominal contents of extinct bees preserved in Dominican amber. Remarkably, the amber was dated as 25–40 million years old and, if the results can be accepted, they illustrate the extraordinary longevity of these life forms (Cano and Borucki, 1995).

Spores are of considerable scientific and medical interest for a number of reasons.

1. *Pathogenicity*: A number of spore formers are pathogens, and some can cause life-threatening illnesses, such as tetanus and anthrax that are transmitted by contact with the spores of *Clostridium tetani* and *Bacillus anthracis*. In these diseases the spore plays an important role in transmission, because the micro-organism remains viable for long periods under adverse conditions until the host becomes infected. Similarly, the sheer robustness of spores presents numerous problems in their eradication, for example by sterilisation and disinfectants.
2. *As models for cell differentiation*: Spores are produced by a process of cell differentiation in which the bacterium undergoes an irreversible programme of development that leads to the

 doi:10.1006/bkmm.2001.0010

Table 1 Endospore-forming bacteria

Genus	Characteristics
Gram-positive	
Bacillus	Aerobic/facultative anaerobes, rods
Sporolactobacillus	Microaerophilic, rods
Clostridium	Anaerobic, rods, major human and animal pathogens
Desulfotomaculum	Sulfate-reducing, rods
Thermoanerobacter	Thermophilic (65–70°C), anaerobic, rods
Thermoactinomyces	Thermophilic, rods, filamentous
Oscillospira	Anaerobic, rods
Sporohalobacter	Halophile, anaerobic, rod
Anaerobacter	Anaerobic, rod, fixes N_2, produces up to five spores per cell
Alicyclobacillus	Anaerobic, acidophilic, rods
Amphibacillus	Anaerobic, alkaliphilic, rods
Heliobacterium	Anaerobic, phototrophic, rods
Heliophilum	Anaerobic, phototrophic, rods
Syntrophospora	Anaerobic, degrades fatty acids, rods
Desulfitobacterium	Anaerobic, dechlorinates chlorophenols, rods
Sporosarcina	Aerobic, cocci, arranged in tetrads
Gram-negative	
Sporomusa	Anaerobic, rod

creation of an alternate cell type. This programme of cell differentiation has for many decades provided a model for the study of unicellular development where morphological changes can be correlated with an underlying programme of gene expression and regulation providing useful, and often seminal, insights into the development of higher organisms.

3. *Clinical use*: Spores have long been used as edible probiotics, that is, bacterial products designed to enhance the natural intestinal microflora. Consumption of significant quantities of spores has been claimed to prevent mild diarrhoea following antibiotic usage.

Spore Structure

Although spore-specific stains exist, endospores can easily be seen by phase-contrast microscopy as white, highly refractile bodies. They can be seen either as enclosed within the sporangium, in which case they are still undergoing development, or as free bodies that have been released from the mother cell. Typically, spores are about 0.5–1.5 μm in length and will have either an ellipsoidal (ovoid) or spherical shape. During maturation within the sporangium the spore may exist centrally, at one end of the cell (terminally) or at a central and lateral position. Finally, some spores cause a distinct bulge or swelling of the sporangium. These characteristics differ from species to species and provide a simple taxonomic reference. For example, *B. subtilis* spores are ellipsoidal, centrally placed and do not swell the sporangium.

Electron microscopic analysis of spores has revealed a complex structure (**Fig. 1**). This differs somewhat between species, but within all spores is a chamber known as the core or spore protoplast. Here bacterial DNA is found in a highly condensed form. It is thought that at least some proteins referred to as small acid-soluble proteins (SASP) are able to bind to the DNA and protect it from ultraviolet radiation, as well as aiding its folding. Although this has not been proved, it offers an obvious analogy to the eukaryotic histone proteins. The core also contains a high concentration of dipicolinic acid (DPA), much of which is chelated with Ca^{2+} ions. This complex is unique to spores and may be involved in protection against heat damage. The core is encased in a layer of phospholipid membrane. Over this membrane there is a thin layer of peptidoglycan, referred to as the primordial germ cell wall (PGCW). On spore germination this peptidoglycan layer will form the new wall of the outgrowing cell. Above the PGCW there is a thicker layer of peptidoglycan, termed the cortex. This cortical layer is markedly different from the PGCW peptidoglycan and is mostly devoid of pentapeptide cross-linking, so rendering it 'looser'. The cortex is immediately apparent in electron micrographs as a white layer between the core and spore coats. Lying above the cortex is the spore coat. The coat can comprise one or more distinct layers of protein (Driks, 1999). The primary function of the coat appears to be in resistance to solvents and toxic chemicals. It must be partially permeable because germination (see below) is thought to arise from catalytic reactions occurring in the PGCW and cortex. Considering the function of the spore coat, it is remarkable that in some species, such as *B. subtilis*, more than 18 distinct spore coat polypeptides have been identified, while in *B. megaterium* only one polypeptide appears to be present. In *B. subtilis* at least two major layers can be identified by microscopy, the inner coat, which has a lamellated appearance, and an outer coat that is electron dense and appears dark grey in electron micrographs. With some spore formers, the outer coat layer shows a distinctive symmetrical and crystalline surface similar to the S-layers of Archaea bacteria (Aronson and Fitz-James, 1976). This may represent an ordered array of one unique cross-linked coat protein.

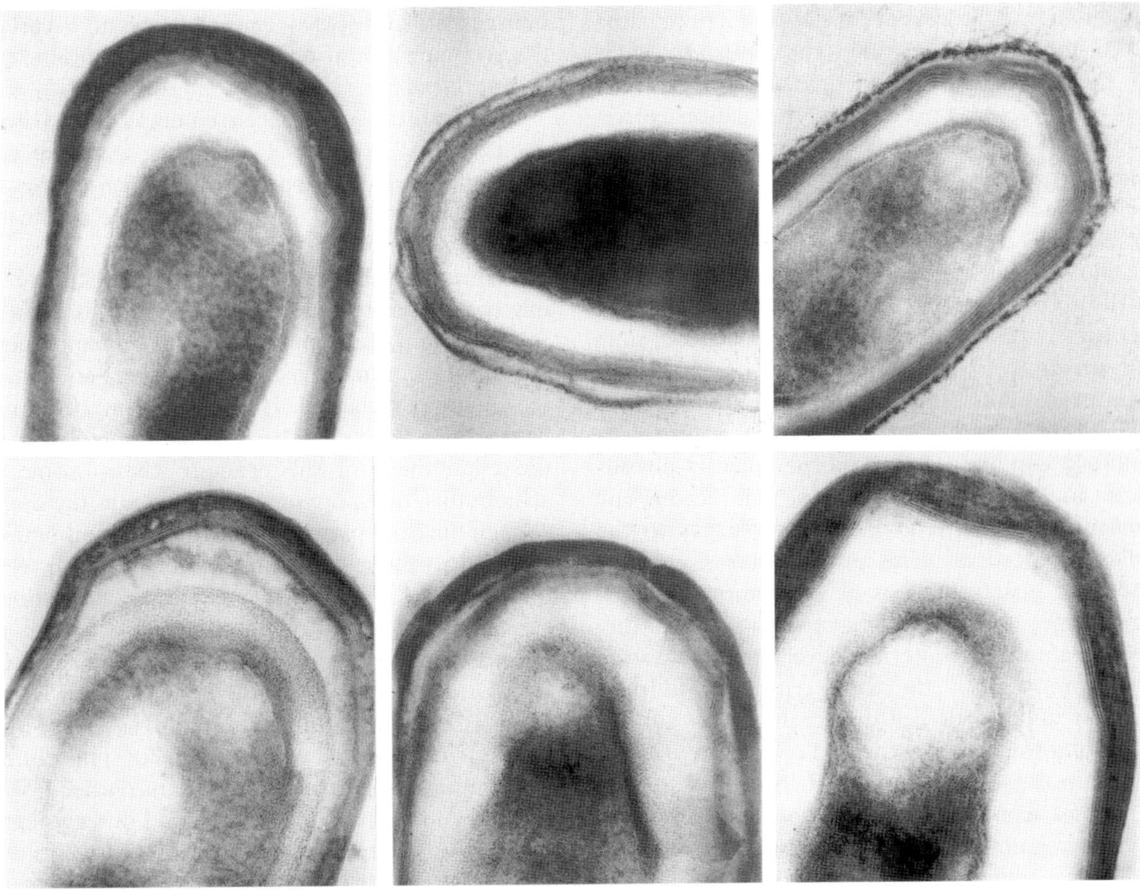

Fig. 1 Mature endospores. The figure shows electron micrographs of mature endospores from six *Bacillus* spp. Clockwise from top left: *B. subtilis* strain 168; *B. pumilus var. biosubtyl*; *B. clausii*; *B. subtilis* strain W23; *B. licheniformis*; *B. subtilis var. niger*. In each figure the basic structure is similar: an inner core, an electron transparent cortex, a lamellated inner coat and an electron-dense outer coat. However, differences can be detected, e.g. the exosporial layers lying above the outer coat in *B. pumilus var. biosubtyl* and *B. clausii* and the 'sausage-like' outer coat of *B. licheniformis*.

Finally, some spores may contain an additional layer termed the exosporium, which is loosely attached to the outer coat as a thin layer that resembles a phospholipid membrane. The composition of the exosporium is unclear, and its function, if any, is equally intriguing since in some 'laboratory' strains this structure is often lost, presumably as a result of repeated passaging.

Spore Formation

The morphological, physiological and biochemical changes that accompany bacterial endospore formation have been well characterised in *B. subtilis* and to a lesser extent in other *Bacillus* and *Clostridium* species (for a review see Piggot and Coote, 1976). Originally it was thought that spore formation was simply a response to starvation, particularly of sources of carbon, nitrogen and phosphorous. It is now known that the primary trigger that induces cell differentiation is a fall in the intracellular concentration of guanosine 5′-triphosphate (GTP). In addition to nutritional signals, however, the population density of a starving culture appears to be important for efficient spore formation. Specifically, cells will sporulate with greater efficiency at a higher population density. This is because vegetative cells grown to a high density produce an oligopeptide pheromone that is imported by neighbouring cells by way of an oligopeptide transport system, which in effect tells them that other cells are starving or about to starve. Extracellular signalling of this kind plays an important role in development of myxobacteria, and makes communication between cells possible (Kroos, 1994). Finally, it has been shown conclusively that the decision to sporulate is directly linked to the starving cell reaching a precise juncture in its cell cycle and having two chromsomes present in the cell (Ireton and Grossman, 1994).

Together, the three inputs of starvation, population density and cell cycle provide tight control for entry into the sporulation programme. This control is important because, once begun, sporulation is irreversible and the cell is said to be 'committed' to entering dormancy. While sporulation may, at first, appear to be a logical response to adverse conditions, it may also be argued that in a highly competitive environment, when appropriate conditions do return and the spore germinates, it may face a plethora of other bacterial competitiors and be placed at a distinct disadvantage.

The process of sporulation from commitment to release of the mature spore lasts about 8 hours under laboratory conditions. The morphological, physiological and biochemical changes that accompany sporulation have been used to assign stages to the development process. This provides a basis for the subsequent genetic analysis of sporulation (**Fig. 2**). Stage 0 represents the undifferentiated bacterial cell. As we have noted, to enter the developmental cascade the cell must satisfy the three criteria of nutrient depletion, appropriate cell density and possession of two chromosomes. First, the two genomes condense and then polarise, each to one pole of the cell, and at this point a septum is synthesised at one pole of the cell. Synthesis of the septum is similar to the creation of a vegetative septum, with invagination of the cytoplasmic membrane. However, the major difference between the spore septum and the vegetative septum is the asymmetric positioning of the spore septum at one end of the cell.

At stage II the spore septum is complete, and divides the cell into two distinct chambers, a smaller compartment termed the prespore and a larger chamber known as the mother cell (note that the designation 'stage I' is no longer used). Each chamber contains its own chromosome and at this stage the cell is irreversibly committed to spore formation. At this point the peptidoglycan lying between the two membranes of the spore septum is partially digested, so that the septum loses its rigidity. This is important, because the next stage involves synthesis of more membrane and the engulfment of the prespore chromosome. To achieve this, the spore septum bulges into the mother cell and the two points where the septal membranes are attached to the cytoplasmic membrane move to one pole of the cell. Finally, when the two septal membrane attachment points meet, the membranes fuse to produce a compartment enclosed within the sporangial cell or mother cell, so defining the end of stage III. The engulfed chamber is often referred to as the forespore and is surrounded by two layers of phospholipid membrane. These membranes are of reverse polarity (**Fig. 2**) and are termed the inner and outer forespore membranes, respectively (IFM and OFM).

In stage IV, two distinct events occur. First, the primordial germ cell wall is deposited between the

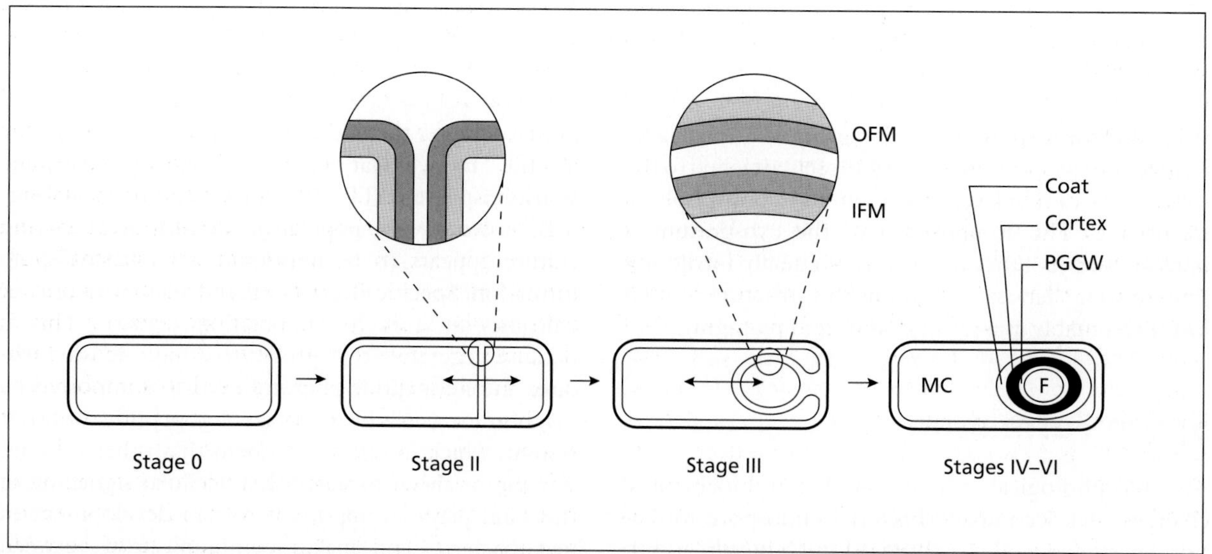

Fig. 2 The principal stages of spore formation. The entire process takes approximately 8 hours under laboratory conditions. The sporulation process is described in the text. The enlargement shows the two phospholipid membrane bilayers. The shading indicates the two monolayers and shows how they are of reversed polarity. The release of the mature spore after stage VI is not shown here. IFM, inner forespore membrane; OFM, outer forespore membrane; PGCW, primordial germ cell wall; F, forespore; MC, mother cell.

IFM and OFM, and this peptidoglycan layer remains closely attached to the IFM. Next, the cortex is formed between the primordial germ cell wall and the OFM. At this point, the forespore becomes apparent by phase-contrast microscopy as a phase grey body within the sporangial cell, and Ca^{2+}-DPA begins to accumulate in the core, which begins to dehydrate. During stage V the coat proteins are laid down onto the OFM. There is now compelling evidence that assembly of the coats is complex and requires, (1) self-assembly of some proteins, (2) proteolytic processing of larger coat polyprotein precursors, (3) deposition of coat proteins as distinct layers and (4) covalent cross-linking of some coat polypeptides (Driks, 1999). At this point the spore starts to become resistant to solvents, heat and, finally, to lysozyme. At the same time spores become more apparent under phase-contrast microscopy as refractile bodies and are termed phase white or phase bright.

During stage VI the spores mature further and the exosporium is formed. Also at this time the mother cell chromosome is broken down by endonucleases. Finally, at stage VII the mature spore is released from the sporangium by lysis of the cell envelope.

Control of Spore Formation

Bacillus subtilis has been studied extensively at the genetic level as an example of unicellular differentiation to understand how the cell orchestrates a programme of morphogenetic changes (Errington, 1993; Losick *et al.*, 1986). *B. subtilis* has provided a model system with which to study spore development in other spore formers. In part, this is a consequence of the excellent genetics of this organism, and the fact that it can naturally become competent for take up exogenous DNA (DNA-mediated transformation). In turn, this has allowed the development of a large number of genetic techniques, such as mapping mutations, gene cloning, gene expression, etc. Recently, the entire *B. subtilis* genome has been sequenced (Kunst *et al.*, 1997), and the European Union has sponsored ambitious projects aimed at identifying the function of all the genetic loci on the *B. subtilis* genome. As a consequence, enormous progress has been made and we now have an excellent understanding at the genetic level of how the spore is formed. The systematic dissection of sporulation has followed a logical chronological order: first, definition of the morphological and biochemical changes that a developing cell undergoes; second, identification and mapping of developmental loci by mutation analysis; third, cloning and characterisation of developmental genes; and finally, functional analysis of the gene products.

Most genetic loci involved in sporulation have been identified by classical genetic methods (mutagenesis) and are termed *spo* loci. About 50 of these exist and they are characterised according to the stage in which their encoded gene products are involved. For example, *spoII* and *spoIII* loci are involved in stages II and III of sporulation, respectively, and mutations in these loci prevent sporulation proceeding beyond these stages. Some developmental genes have no obvious phenotype when disrupted; these have been identified by 'reverse genetics', using the protein sequence to identify the gene. These genes include many that encode structural components of the spore. Since many of these gene products are redundant, they cannot be identified by classical mutation analysis. In this group are the coat genes (*cot*) involved in spore coat biosynthesis and the *ssp* genes that encode the SASP proteins, which are involved in protecting the spore genome and chromosome condensation. In addition to these loci, genes involved in germination (*ger*) have been identified, as well as many other genes which were originally identified as vegetative loci involved in normal cell growth and division, and which play a role in spore formation.

Almost all sporulation genes in *B. subtilis* have now been extensively characterised. They have been cloned and sequenced and gene expression measured by examining either sporulation-specific mRNA or, more commonly, β-galactosidase synthesis originating from a *gene-lacZ* reporter construct. The outcome of these studies is an elegant programme of gene expression (**Fig. 3**), in which sporulation genes are switched on and off at distinct times during development (temporal gene expression) and, in many cases, in only one chamber of the developing cell (spatial gene expression). This makes sense, because gene products may only be needed at precise times during sporulation and then only in one chamber. For example, the small acid-soluble proteins (SASP) that bind to and condense with the forespore chromosome are required only in the forespore chamber (Setlow, 1992, 1993). Likewise, the coat proteins are expected to assemble onto the outer membrane of the forespore and thus are synthesised in the mother cell chamber.

Temporal and spatial expression of sporulation genes is controlled by a change in the transcription machinery of the cell. During vegetative cell growth, RNA polymerase holoenzyme consists of two basic components, core enzyme associated with the sigma A (σ^A) subunit. Core is a heteroligomeric complex which is responsible for synthesising mRNA, while σ^A is responsible for recognising the promoter sequence of

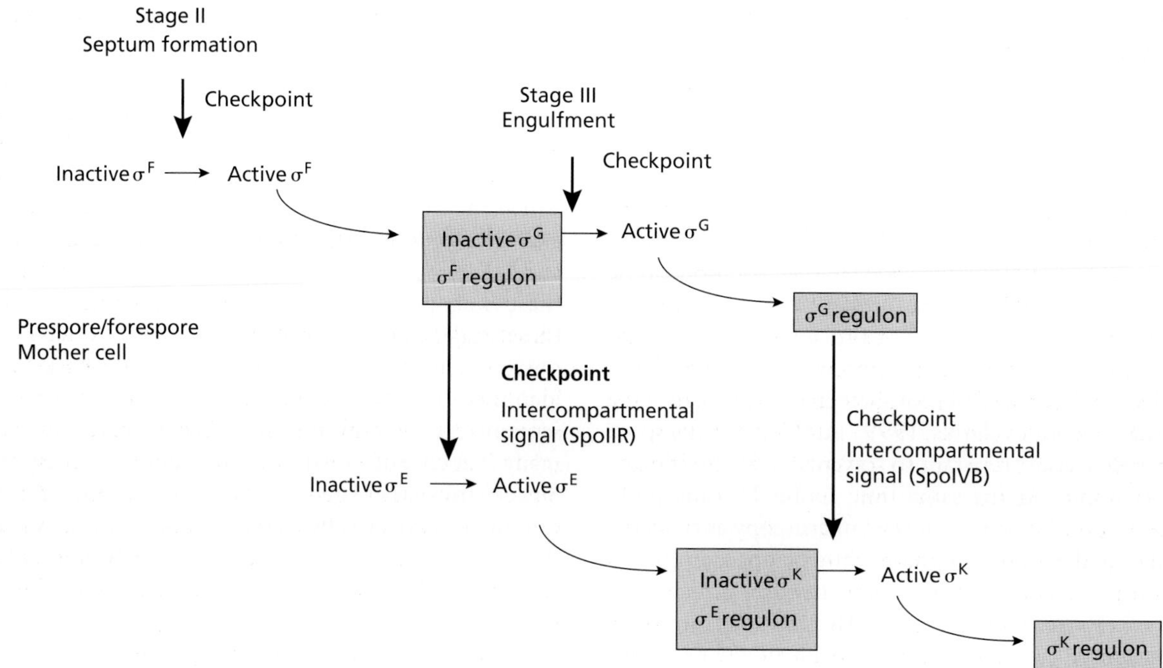

Fig. 3 Regulated gene expression during *Bacillus subtilis* development, showing the regulatory events that occur in the sporulating cell. Developmental gene expression is controlled by the action of five sigma factors (σ) which direct the expression of specific subsets of genes (regulons). Four of these sigma factors act in either the prespore/forespore or mother cell chambers, but not both. σ^H acts in the pre-septational cell, and for clarity, is not shown here. Each sigma factor is made in an inactive form, either by being encoded as a functionally inactive protein (σ^E and σ^K) or by binding to an inhibitor (σ^F). Activation of sigma is brought about by a specific developmental cue, formation of the spore septum, engulfment of the pre-spore chamber to form the forespore or inter-compartmental signalling from the opposed chamber. These controls are termed 'checkpoints' and ensure appropriate fine tuning of the developmental process and an increased fidelity to the morphological process.

genes to be expressed during normal cell growth, and on promoter recognition it is discarded to enable core enzyme to produce mRNA (Haldenwang, 1995). On the initiation of spore formation, new sigma factors are produced which replace σ^A and enable the specific expression of subsets of genes (regulons) that contain unique promoter sequences. This has been referred to as a 'cascade of sigma factors' and demonstrates a basic framework in which the control of gene expression is used to drive a complex developmental process depending on scores of genes (Stragier and Losick, 1990; Kroos *et al.*, 1999).

In *B. subtilis* five major sigma factors are employed during spore formation, σ^H, σ^F, σ^E, σ^G and σ^K. σ^H is used first and appears to be active in the pre-septational cell. σ^F is active only in the pre-spore chamber at stage II. It is responsible for transcription, at stage III, of the structural gene *spoIIIG*, which encodes σ^G exclusively in the forespore. Similarly, in the mother cell compartment, σ^E is active at stage II and enables the transcription, at stage III–IV, of the structural gene for σ^K, the final acting sigma factor in the mother cell. Remarkably, sigma factor activity is controlled not only at the level of expression of the sigma factor gene, but also by post-translational events. These are referred to as 'checkpoint controls' and ensure that gene expression is tightly coupled to development. It is now clear that σ^F and σ^G are made in inactive forms and are activated by physiological cues (Stragier, 1992). σ^F is actually present in the pre-septational cell, but is activated only in the smaller pre-spore chamber (Lewis *et al.*, 1996). σ^F is normally inhibited by the SpoIIAB protein, which binds to the sigma factor. At septation a further protein, SpoIIAA, is dephosphorylated by the SpoIIE protein, which is embedded in the pre-spore septum. As a consequence of this, SpoIIAA can bind to SpoIIAB as well as ADP, forming a SpoIIAA–SpoIIAB–ADP complex and releasing σ^F as an active molecule (Min *et al.*, 1993). Since SpoIIE is present in the membrane, completion of the spore septum presents a cue for σ^F activation. This process is, of course, far more complex than stated here, but it illustrates the complexity of a seemingly simple regulatory step.

How σ^G is activated is not yet known, but the morphological cue appears to be completion of engulfment at stage III and σ^G may be activated using a process similar to that used to activate σ^F. However, in the case of σ^G it is known that activation depends upon events that occur in the mother cell, because mutations in some mother cell-expressed genes prevent σ^G activation.

σ^E and σ^K are made as inactive protein precursors (Lu *et al.*, 1990; Stragier *et al.*, 1988). Activation is brought about by proteolytic processing of the N-terminal leader sequence to render the sigma factor active. In the case σ^E and σ^K, proteolytic cleavage of the sigma factor is controlled by proteolytic enzymes that are embedded in the spore septum (SpoIIGA protease for σ^E) and forespore membranes (SpoIVFB for σ^K). These proteases are responsive to a signal that originates in the opposed chamber of the compartment (Cutting *et al.*, 1990; Londono-Vallejo and Stragier, 1995). In the case of σ^E, SpoIIR is the signal protein which is expressed in the pre-spore and is the product of a σ^F-transcribed gene (Karow *et al.*, 1995; Londono-Vallejo and Stragier, 1995). With σ^K, the signal is the SpoIVB protein, encoded by the forespore-expressed gene *spoIVB* transcribed by RNA polymerase associated with σ^G (Cutting *et al.*, 1991).

Such intercompartmental signalling imposes tight control on a developmental process (Losick and Stragier, 1992; Losick and Shapiro, 1993). Since expression of a new sigma factor elicits expression of a downstream regulon, it is imperative that the developing cell is at the appropriate state of readiness. In other words, mis-expression of sporulation genes could lead to deleterious effects on spore development. Indeed, inadvertent expression of the σ^K regulon leads to the formation of spores of aberrant form that are unable to germinate properly, and this is presumably due to the mis-assembly of the spore coats (Cutting *et al.*, 1990).

The outcome of these genetic studies has been an essentially complete understanding of a unicellular morphological process.

Germination

Spores must germinate to enable them to re-enter the normal cycle of growth and cell division. As mentioned above, the decision to sporulate is under tight control and the developmental programme itself is regulated by numerous regulatory controls to ensure that, if a spore is to be made, it must be made without error. Such checks and balances ensure the viability of the germ line. It is not surprising that the decision to germinate is also subject to specific requirements.

Spore germination can be divided into three basic steps: activation, germination and outgrowth (Moir and Smith, 1990). Activation is usually a heat shock and, for example, with *B. subtilis* spores, a shock of 75°C for 45 minutes is sufficient. It should be stressed that activation is very much a laboratory artefact and that spores will germinate without a heat-activation step if a germinant (i.e. a molecule that triggers germination) is present, but at a greatly reduced frequency. Germination is the breakage of the spore coat and cortical layers. At this stage the coats are simply 'shattered' and a presumably *de novo* reaction must take place to facilitate cleavage of the spore coats and cortex. Germination can be viewed by phase-contrast microscopy as the change from a phase-bright spore to a phase-dark spore. The entire process is rapid, taking from 90 seconds in *B. subtilis* to up to 6 minutes in *B. megaterium*.

Germination requires presence of a germinant, that is, one or more molecules that trigger the ensuing changes in spore structure. Common germinants are L-alanine and L-alanine analogues, or AGFK (a mixture of asparagine, glucose, fructose and KCl), while other germinants tend to be more species-specific such as riboside and inosine. It is likely that specific receptors responsive to these germinants exist. These are thought to lie in the inner forespore membranes of the spore, which implies that the spore coats and cortical layers are permeable to these molecules. The decision to germinate is rapid, and only brief exposure is required. As the spore coats break, the spore core rapidly rehydrates and DPA is lost from the spore. At this time the SASP proteins contained within the core are rapidly degraded to provide a source of amino acids for the resumption of protein synthesis.

The final step is outgrowth. Here, RNA and protein synthesis resumes followed by DNA replication and cell division. At this point the core, now termed the protoplast, emerges through the broken coat layers, elongates and finally resumes normal vegetative cell growth.

Spore-borne Diseases

The soil is the natural habitat of most spore formers, but some (e.g. *Clostridium* spp.) are also found in the intestinal tract of humans and animals. Under appropriate conditions some of these organisms can cause serious disease and in most cases the spore is the primary agent in disease transmission

(Table 2). The spore-forming pathogens are considered in detail in other chapters.

It is worth noting that some *Bacillus* species are considered non-pathogenic but evidence is accumulating that they can, under appropriate conditions, cause potentially fatal diseases including pneumonia, bacteraemia and meningitis (Farrar, 1963; Ihde and Armstrong, 1973). The infections caused by these apparently non-pathogenic species fall into three groups:

1. *local infections* at previously damaged sites, e.g. *B. subtilis* infections of the cornea (Bijsterveld and Richards, 1965),
2. *mixed infections*, in which the organism is present together with another known pathogen, and
3. *disseminated infections*, where the organism is recovered in pure culture from the blood or spinal fluid of infected patients.

It should be stressed, however, that bacteraemia is extremely rare and probably represents an opportunistic infection in immunocompromised patients or infection with significant inocula of *Bacillus* spp. by contamination.

Finally, it should be mentioned that occupational asthma in laundry workers has been attributed to *B. subtilis* (Dijkman *et al.*, 1973; Pepys *et al.*, 1973). Many laundry products contain enzymes mixed with the detergent which are used at the 'pre-soak' stage. The 'biological' component of these washing powders is routinely derived from *B. subtilis* cultures and contains a number of contaminating cellular and spore components. Inhalation of large quantities of these bacterial products has been shown to produce allergic reactions and in a number of cases there have been fatalities due to anaphylatic shock.

Table 2 Diseases caused by spore-forming bacteria[a]

Organism	Disease
Bacillus anthracis	Anthrax
B. cereus	Food poisoning
B. subtilis[b]	Asthma
	Local and opportunistic infections
Clostridium tetani	Tetanus
C. botulinum	Botulism
C. perfringens	Gas gangrene, skin ulcers bacteraemia, food poisoning
C. difficile	Food poisoning

[a] This list is not exhaustive.

[b] *B. subtilis* is not generally considered a pathogen (see text).

Probiotics – Spores for Eating

Bacillus spores are currently being used as a probiotic – 'a promoter of the normal enteric microbiotic flora' – for oral bacteriotherapy and bacterioprophylaxis of gastrointestinal disorders, many of which lead to diarrhoea in humans (Tannock, 1999). Ingestion of significant quantities of spores has been claimed to restore the normal microbial flora after illness or antibiotic usage (for review see Mazza, 1994). These products are very similar to the more commonly used *Lactobacillus* probiotics marketed in the form of various preparations in Europe (see Table 3). Although sold for human consumption, the bacterial species contained in these preparations are poorly characterised and in some cases do not correspond to the species cited (Green *et al.*, 1999; Hamilton-Miller *et al.*, 1999).

Little is understood of how *Bacillus* spores may exert their 'therapeutic' effect. It is thought, but not proven, that the robustness of spores enables them to pass the gastric acid barrier, and allows a proportion of spores to germinate in the small intestine and populate the intestinal tract, albeit only briefly (Mazza, 1994). Although the intestine is essentially anaerobic, low oxygen levels may exist, especially in the duodenum, and these may allow low levels of spore germination and subsequent growth.

Other well-known bacterial probiotics, such as the lactobacilli, are normally found in the gut (Fuller, 1989, 1991). This makes the mode of action of *Bacillus* spore probiotics particularly intriguing. Three

Table 3 Commercial spore probiotics for human consumption

Product	Country[a]	Organism[b]
Enterogermina®	Italy	*Bacillus clausii*[c] (Green *et al.*, 1999)
Lactipan Plus®	Italy	*Sporolactobacillus* spp.[d]
Domuvar®	Italy	*Bacillus subtilis*
Bactisubtil®	France	*Bacillus cereus*
Subtyl®	Vietnam	*Bacillus subtilis*
Biosubtyl®	Vietnam/ Indonesia	*Bacillus pumilus* (Green *et al.*, 1999)

[a] Country of manufacture.

[b] Where the bacterial species has been identified by analysis of 16S rRNA the reference is listed.

[c] This was identified by 16S rRNA analysis to be related to *B. alcalophilus* but further work has shown it to be *B. clausii* (M. Oggioni; personal communication).

[d] This product is listed as *Lactobacillus sporogenes*.

modes of action have been proposed: (1) competition with potential intestinal pathogens for adhesion sites, (2) immunostimulation, which could inhibit potential pathogens, and (3) secretion of unique antimicrobials. Understanding the fate of orally administered spores is clearly the first step in understanding their therapeutic action.

References

Aronson AI, Fitz-James P (1976) Structure and morphogenesis of the bacterial spore coat. *Bacteriol. Rev.* 40: 360–402.

Bijsterveld OP, Richards RD (1965) *Bacillus* infections of the cornea. *Arch. Ophthalmol.* 74: 91–95.

Cano RJ, Borucki M (1995) Revival and identification of bacterial spores in 25- to 40-million-year-old Dominican amber. *Science* 268: 1060–1064.

Cutting S, Driks A, Schmidt R, Kunkel B, Losick R (1991) Forespore-specific transcription of a gene in the signal transduction pathway that governs pro-σ^K processing in *Bacillus subtilis. Genes Dev.* 5: 456–466.

Cutting S, Oke V, Driks A, Losick R, Lu S, Kroos L (1990) A forespore checkpoint for mother cell gene expression during development in *B. subtilis. Cell* 62: 239–250.

Dijkman JH, Borghans JGA, Savelberg PJ, Arkenbout PM (1973) Allergic bronchial reactions to inhalation of enzymes of *Bacillus subtilis. Am. Rev. Respir. Dis.* 107: 387–394.

Driks A (1999) *Bacillus subtilis* spore coat. *Microbiol. Mol. Biol. Rev.* 63: 1–20.

Errington J (1993) *Bacillus subtilis* sporulation: regulation of gene expression and control of morphogenesis. *Microbiol. Rev.* 57: 1–33.

Farrar WE (1963) Serious infections due to 'non-pathogenic' organisms of the genus *Bacillus. Am. J. Med.* 34: 134–141.

Fuller R (1989) Probiotics in man and animals. *J. Appl. Bacteriol.* 66: 365–378.

Fuller R (1991) Probiotics in human medicine. *Gut* 32: 439–442.

Gest H, Mandelstam J (1987) Longevity of micro-organisms in natural environments. *Microbiol. Sci.* 4: 69–71.

Green DH, Wakeley PR, Page A, Barnes A, Baccigalupi L, Ricca E, Cutting SM (1999) Characterization of two *Bacillus* probiotics. *Appl. Environ. Microbiol.* 65: 4288–4291.

Haldenwang WG (1995) The sigma factors of *Bacillus subtilis. Microbiol. Rev.* 59: 1–30.

Hamilton-Miller JMT, Shah S, Winkler JT (1999) Public health issues arising from microbiological and labelling quality of foods and supplements containing probiotic microorganisms. *Public Health Nutr.* 2: 223–229.

Ihde DC, Armstrong D (1973) Clinical spectrum of infection due to *Bacillus* species. *Am. J. Med.* 55: 839–845.

Ireton K, Grossman AD (1994) A developmental checkpoint couples the initiation of sporulation to DNA replication in *Bacillus subtilis. EMBO J.* 13: 1566–1573.

Karow ML, Glaser P, Piggot PJ (1995) Identification of a gene, *spoIIR*, which links activation of σ^E to the transcriptional activity of σ^F during sporulation in *Bacillus subtilis. Proc. Natl Acad. Sci. USA* 92: 2012–2016.

Kroos L, Zhang B, Ichikawa H, Nicco Yu Y-T (1999) Control of s factor activity during *Bacillus subtilis. Mol. Microbiol.* 31: 1285–1294.

Kroos LCS (1994) Intercellular and intracompartmental communication during *Bacillus subtilis* sporulation. In: Piggot P, Moran CP, Youngman P (eds). *Regulation of Bacterial Differentiation.* Washington, DC: American Society for Microbiology, pp. 155–180.

Kunst F, Ogasawara N, Moszer I (1997) The complete genome sequence of the Gram-positive bacterium *Bacillus subtilis. Nature* 390: 249–256.

Lewis PJ, Magnin T, Errington J (1996) Compartmentalised distribution of the proteins controlling the prespore-specific transcription factor σ^F of *Bacillus subtilis. Genes Cells* 1: 881–894.

Londono-Vallejo J-A, Stragier P (1995) Cell–cell signaling pathway activating a developmental transcription factor in *Bacillus subtilis. Genes Dev.* 9: 503–508.

Losick R, Shapiro L (1993) Checkpoints that couple gene expression to morphogenesis. *Science* 262: 1227–1228.

Losick R, Stragier P (1992) Crisscross regulation of cell-type-specific gene expression during development in *B. subtilis. Nature* 355: 601–604.

Losick R, Youngman P, Piggot PJ (1986) Genetics of endospore formation in *Bacillus subtilis. Annu. Rev. Genet.* 20: 625–669.

Lu S, Halberg R, Kroos L (1990) Processing of the mother-cell sigma factor, σ^K, may depend on events occurring in the forespore during *Bacillus subtilis* development. *Proc. Natl Acad. Sci. USA* 87: 9722–9726.

Mazza P (1994) The use of *Bacillus subtilis* as an antidiarrhoeal microorganism. *Boll. Chim. Farmaceutico* 133: 3–18.

Min K-T, Hilditch CM, Diederich B, Errington J, Yudkin MD (1993) σ^F, the first compartment-specific transcription factor of *B. subtilis*, is also regulated by an anti-s factor that is also a protein kinase. *Cell* 74: 735–742.

Moir A, Smith DA (1990) The genetics of bacterial spore germination. *Annu. Rev. Microbiol.* 44: 531–553.

Pepys J, Wells ID, D'Souza MF, Greenberg M (1973) Clinical and immunological responses to enzymes of *Bacillus subtilis* in factory workers and consumers. *Clin. Allergy* 3: 143–160.

Piggot PJ, Coote JG (1976) Genetic aspects of bacterial endospore formation. *Bacteriol. Rev.* 40: 908–962.

Setlow P (1992) I will survive: protecting and repairing spore DNA. *J. Bacteriol.* 174: 2737–2741.

Setlow P (1993) Spore structural proteins. In: Sonenshein AL, Hoch JA, Losick R (eds) *Bacillus subtilis and Other Gram-positive Bacteria.* Washington, DC: American Society for Microbiology, pp. 801–809.

Stragier P (1992) Establishment of forespore-specific gene expression during sporulation of *Bacillus subtilis*. In: Mohan S, Dow C, Cole JA (eds) *Prokaryotic Structure and Function: a New Perspective. Society for General Microbiology Symposium*, pp. 297–310. Cambridge, UK: Cambridge University Press.

Stragier P, Losick R (1990) Cascades of sigma factors revisited. *Mol. Microbiol.* 4: 1801–1806.

Stragier P, Bonamy C, Karmazyn-Campbelli C (1988) Processing of a sporulation sigma factor in *Bacillus subtilis*: how morphological structure could control gene expression. *Cell* 52: 697–704.

Tannock GW (1999) *Probiotics: A Critical Review*. Norfolk: Horizon Scientific Press.

PART 2

BACTERIAL CELL FUNCTION

11 Bacterial Growth, Culturability and Viability

M.R. Barer

Medical School, Newcastle upon Tyne, UK

During the process of infection, growth of microbes in a host leads directly or indirectly to tissue damage or disordered physiology in that host. From this it follows that growth of microbes is critical to the establishment and development of infection. It also seems likely that failure to control growth of the infecting microbe will lead to more severe disease and reduce the likelihood of a favourable outcome for the host. Those microbes that are not obligate human parasites will also need to grow outside the human host and this capacity may be important for maintaining transmission cycles. (Note that in this personal rubric, specific infections are defined by the genetic complements of the microbes and the host, by the preceding immunological experience and physiological state of the host and by the route of exposure to the infective agent. Infections may be asymptomatic or symptomatic and the latter group comprise those individuals recognised as suffering from an infective disease.)

From a practical point of view, growth of microbes *in vitro* remains central to the fulfilment of the Henle-Koch postulates, the diagnosis of many infections and the determination of microbial viability. The development of molecular methods in microbiology to determine the molecular basis of growth and its regulation, and to develop such methods to detect growth and viability presents a challenge. Ideally, such methods should be applicable to the microbial populations in the human host. In this chapter I consider, from the bacteriological perspective, progress towards these ideals and some of the key problems that need to be resolved.

Bacterial Growth

Growth involves the accumulation of biomass and may include genomic replication, cell division and an increase in the number of propagules of the organism concerned. For most bacteria it is generally held that, after division, a newly formed cell placed in an environment favourable to growth will double its mass then divide to form two equal sized progeny by binary fission. This process has been subjected to intensive analysis and will be discussed here from a highly selective viewpoint. For more comprehensive discussions, the reader is referred to recent reviews (Neidhardt *et al.*, 1990; Gerhardt, 1994; Russell and Cook, 1995; Koch, 1997; Kovarova-Kovar and Egli, 1998; Madigan *et al.*, 2000).

Patterns of Growth and Sources of Information

Most information about bacterial growth has been obtained by the study of selected organisms in broth cultures. Liquid cultures are convenient; most variables can be precisely controlled and the scale can be

 doi:10.1006/bkmm.2001.0011

adjusted to provide sufficient biomass for almost any form of analysis. In achieving reproducible results between laboratories the development of chemically defined media, consistent inocula and the recognition of growth states that can be detected by sequential optical density or turbidity measurements has provided a platform for further development. The widely accepted terminology of lag, exponential (or log) and stationary phases of growth in batch culture provide essential physiological points of reference and these are often applied, with scant justification, to bacterial cells and populations outside the highly defined laboratory environments indicated.

A detailed analysis of the energetics and stoichiometry of bacterial growth has been made possible by analysing bacterial populations growing at constant rates in chemostat or turbidostat cultures (Pirt, 1975; Gerhardt, 1994; Kovarova-Kovar and Egli, 1998). These systems provide a relatively reproducible gold standard in which a state referred to as 'balanced exponential growth' can be achieved for extended times. Similar levels of reproducibility can be achieved with mid-exponential batch cultures and inocula prepared by dilution from the last of multiple successive rounds of exponential phase culture. The end result in each case is generally believed to be a uniform population of cells, all of which are growing at similar rates. Thus, it is considered legitimate that analyses of cells in balanced exponential growth can be divided equally amongst all the cells present in the sample to yield estimates of content or activity per cell.

An important alternative approach has been to start by considering the bacterial cell cycle, which starts with the birth of a cell by binary fission of a parental cell and ends with the division of the new cell. This kind of work draws substantially on an understanding of the eukaryotic cell cycle where the biochemical and physiological events have been separated into distinct phases (G_1, S, G_2, M with or without G_0) and has been pursued with techniques that provide large populations of cells which are all at the same stage of the cycle. While some controversy continues, it is generally thought that events considered critical for progression through the cell cycle in eukaryotes, such as initiation and termination of DNA synthesis, are not similarly regulated in bacteria. Rather, the short-term fate of a cell is determined by the rate at which it accumulates biomass, and at particular size:growth rate ratios division is initiated. Interestingly, Cooper (1991, 2000), who has championed this view, has recently argued that even the eukaryotic cell cycle does not really have distinct phases.

For obvious reasons, most knowledge obtained with these methods refers to large cell populations ($>10^7$) of readily culturable bacteria in exponential growth phase. Here, we are primarily concerned with the behaviour of pathogens during infections and as they survive between causing infections. It is clear that not only will these organisms rarely be in a simple suspension phase, but also it is unlikely that the environment will be conducive to unimpeded exponential growth. Evidently, the degree to which most knowledge of bacterial growth is applicable to the environments that primarily concern us must be limited.

Laboratory studies on bacterial growth have also provided limited information regarding growth in colonies on or in solidified laboratory media (Wimpenny, 1988; McKay *et al.*, 1997) and in biofilms (Gottenbos *et al.*, 1999; McLean *et al.*, 1999). While information on the growth of bacteria in colonies and in broth may be valuable in the designing of isolation and culture media for medically important bacteria (Gerhardt, 1994), growth in biofilms is probably the principal mode of bacterial propagation in natural communities. In a medical context, this is most relevant to the colonisation of medical devices, particularly intravascular cannulae.

The challenge for medical microbiology is, therefore, to study the growth of bacterial populations *in situ* in the range of complex and highly specialised environments provided by humans, and in environments such as food and water, which provide sources and reservoirs for infection. Only then will the relevance of understanding based on detailed physiological studies of defined broth cultures to infection become clear.

Molecular Information Related to Bacterial Growth

In spite of the problems discussed above, studies on carefully defined broth cultures remain the principal reliable source of information about the molecular basis of bacterial growth. As key genes involved in growth and its regulation have been identified through recent pre- and post-genomic studies, the possibility of determining the importance of these genes to infection through deletion, over-expression and reporter studies has been extensively exploited. A somewhat arbitrary selection of genes whose expression has been related to growth in various ways is reviewed below. Ultimately, it should be possible to recognise all the genes required for growth in specific environments. It seems likely that these will fall into two categories: those required in all circumstances and those required only for special environments. The challenge for medicine is to recognise genes that may represent targets for diagnosis, therapy or prevention.

Ribosomal RNA

A single *Escherichia coli* chromosome generally carries seven copies of the genes that encode ribosomal RNA. In contrast, the *Mycobacterium tuberculosis* chromosome encodes only one copy. Given the greater than 10-fold difference in minimum doubling times between these organisms (0.3 hours versus 6 hours), it seems likely that this is not an accident. The 16S, 23S and 5S genes, and some transfer RNA genes, are located in tandem and are initially transcribed into RNA as a single molecule, which therefore includes the so-called intergenic transcribed sequences (ITS). The transcript is then processed into the recognised subunit components and these combine with ribosomal proteins to form functional ribosomes. Apart from the central role now occupied by the 16S molecule and the ITSs in the classification of Bacteria and Archaea, the rate at which these genes are transcribed and the 16S rRNA content of bacterial cells has been directly correlated with bacterial growth rates *in vitro* (Bremer and Dennis, 1987; Delong *et al.*, 1989). Analysis of these genes and their products in samples therefore presents opportunities both to identify and make some inferences about the protein synthetic capacity and growth rate(s) of the organisms present.

Chromosome Replication

Chromosome replication requires more time to complete than that available between cell divisions during rapid growth of *E. coli*. The organism circumvents the potential problem of producing cells with less than a single complete genome by initiating rounds of chromosome replication at intervals compatible with the cell replication rate. Initiation always starts at the same locus (*oriC*) and proceeds bi-directionally to the terminus region (Messer and Weigel, 1996). One consequence of this is that cells in rapidly growing populations contain more than one chromosome replication fork in progress and the largest cells present (i.e. those close to fission), have a chromosomal DNA content in excess of two copies of the complete genome.

The mechanism that determines the interval between initiating rounds of chromosomal replication is not understood, but several gene products are known to be essential. Amongst these the DnaA protein, a DNA-, ATP- and ADP-binding protein, has been most extensively studied and appears to play a central role in assembly of the initiation complex (Messer and Weigel, 1996). A further consequence of the pattern of replication is suppression of transcription of specific genes as the replication fork passes through. This leads to apparent cell cycle-related gene regulation in synchronised cultures (Zhou *et al.*, 1997).

Cell Division

Our understanding of the molecular basis of bacterial cell division has advanced dramatically in recent years. Progress has been fuelled by the development of immuno-cytochemical techniques for bacteriology and by the use of translational reporter fusions with Gfp the 'green fluorescent protein'. The *gfp* gene encodes a jellyfish protein and can be fused to many other proteins without unduly disrupting their functions. The Gfp protein has intrinsic fluorescence properties, so that the location of the fusion protein can be followed by fluorescence microscopy.

These developments have made possible the localisation of key molecules that determine the site and process of cell division. Amongst these the tubulin-like molecule FtsZ has been extensively studied. Around 10 000 molecules of this key protein are present in each *E. coli* cell and, like its eukaryotic counterpart, it is present in soluble and polymerised forms. Location of FtsZ polymers into ring structures indicates the site of prospective septum formation and, with *ftsZ::gfp* translational fusions, it has been possible to observe, in real time, the formation and subsequent contraction of the FtsZ ring in parallel with septum formation and cell division (Margolin, 1998; Sun and Margolin, 1998) (**Fig. 1**). Although FtsZ possess GTPase activity, it is not known whether it provides the physical force required for septation and fission. Inhibition of FtsZ polymerisation by a protein produced as part of the SOS response (SulA) in growing cells leads to filamentation, thereby illustrating the key role of FtsZ in fission. All bacteria so far studied possess FtsZ homologues and the relative abundance of the molecule makes it an attractive target for study in clinical samples. The presence of FtsZ rings indicates active cell division and, in *Bacillus subtilis*, asymmetric positioning of the ring indicates the onset of sporulation (Levin and Grossman, 1998).

Global Regulatory Proteins

Global regulatory proteins direct differential gene expression, either by binding to DNA or to components of the transcription/translation apparatus. Their own levels of expression and activity are modulated by a variety of internal and external stimuli. It would be impractical to discuss even a small minority of these molecules here, but the levels and/or activities of some prominent examples in relation to bacterial growth are outlined in **Table 1**.

The complex regulatory hierarchy and network that is emerging from the study of these proteins and their

Table 1 Regulatory proteins associated with different aspects of growth in bacteria

Category	Protein	Gene(s)	Some relationships to growth	Function
Regulation via DNA topology[a]	H-NS	*osmZ*	Levels in constant ratio to DNA content during growth. Depressed in stationary phase	Histone-like DNA-binding protein that represses transcription of multiple genes
	LRP	*lrp*	Repressed by growth in rich medium	Selective repression and activation of genes appropriate to available nutrient sources
	IHF	*ihfA (himA)* *ihfB (himD)*	Induction on entry into stationary phase. Expression dependent on ppGpp	Interaction with DNA induces 180° bend enabling long-range interactions
SOS response[b]	RecA LexA	*recA* *lexA*	Activation of RecA by DNA damage induces cleavage of LexA and de-repression of SOS genes and arrest of cell division	RecA controlled genes effect DNA repair and maintain λ-like phage lysogeny
Alternate σ factors[c] (required for transcription)	σ^{70}	*rpoD*	σ^S levels increase on entry into stationary phase and on sudden growth arrest. Alternate σ factors appear to compete with σ^{70} for binding to a limited amount of core RNA polymerase. Promoter specificity is modulated by alternate σ factor binding in combination with many other factors (e.g. H-NS, LRP, IHF and ppGpp). In *B. subtilis* a succession of alternate σ factors directs the programme of gene expression in sporulation	Main RNA polymerase σ subunit
	σ^{38} (σ^S)	*rpoS*		Stationary phase and stress-induced
	σ^{32}	*rpoH*		Heat shock-induced
	σ^{24} (σ^E)	*rpoE*		Induced by extreme heat shock and regulates extracellular proteins
	σ^{28}	*fliA*		Flagellar gene regulation
	(σ^F) σ^{54}	*rpoN(glnF)*		Control of nitrogen metabolism
Universal stress response proteins[d]	UspA	*uspA*	Induced in late exponential phase and by all known stress responses (σ^{70}- and ppGpp-dependent)	Regulation via phosphorylation of target proteins?
	UspB	*uspB*	Induced during transition phase (σ^{38}- and ppGpp-dependent)	Unknown
General metabolic regulators[e]	CRP	*crp*	Low glucose (e.g. on entry into stationary phase) leads to increased cAMP levels	CRP–cAMP complexes activate or repress specific genes
	FNR	*fnr (nirA)*	FNR senses oxygen (anaerobic/aerobic growth)	Reduced FNR activates or represses specific genes

Information presented predominantly relates to studies on *E. coli*; IHF, integration host factor; USP, universal stress protein; CRP, cAMP receptor protein; General sources: Neihardt *et al.* (1990); Moat and Foster (1995); Neihardt (1996); Salgado *et al.* (2000); [a]Aviv *et al.* (1994); Bouvier *et al.* (1998); Dorman *et al.* (1999); Marshall *et al.* (1999); [b]Walker (1996); [c]Errington (1993); Farewell *et al.* (1998a, 1998b); Levin and Grossman (1998) HenggeAronis (1999); Raivio and Silhavy (1999); Severinov (2000); [d]Nystrom and Neidhardt, 1994; Freestone *et al.* (1997); Farewell *et al.* (1998); [e]Botsford and Harman (1992); Kiley and Beinert (1998); Saier (1998); Ferenci (1999); Sawers (1999).

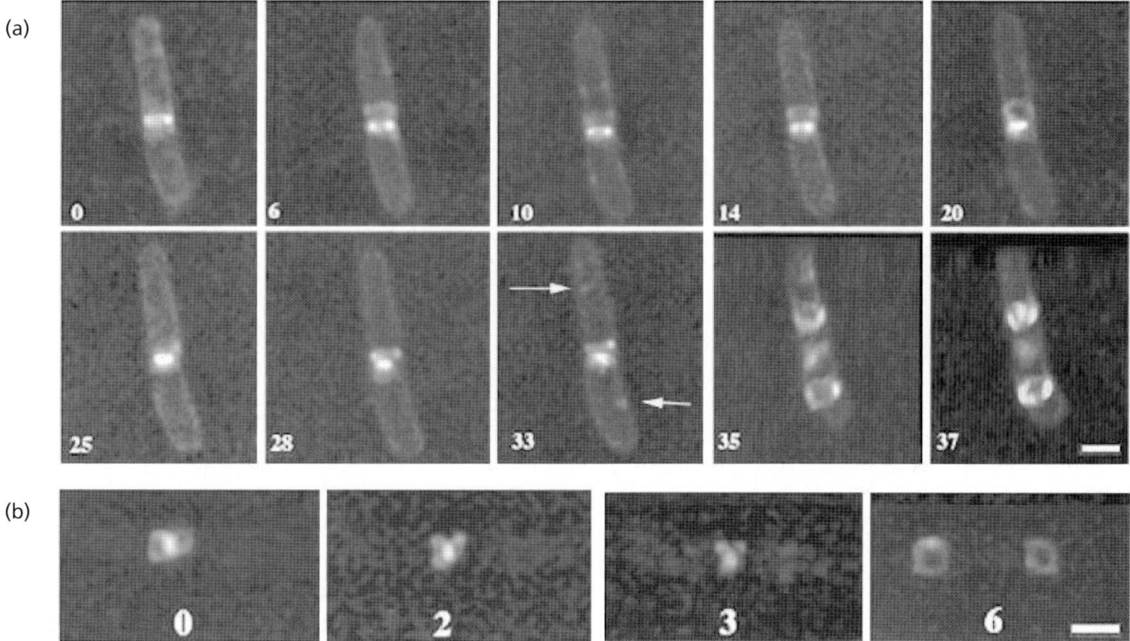

Fig. 1 Dynamics of FtsZ ring formation in *Escherichia coli*. (a) Successive three-dimensional reconstructions based on digital fluorescence images of a dividing cell expressing an *ftsZ::gfp* reporter fusion. Bright fluorescence indicates the sites at which the FtsZ rings form, and a duller secondary fluorescent membrane label indicates the cell outline. Numbers indicate the time in minutes after the first observation. Scale bars represent 1.3 μm. Note that nucleation sites form in daughter cells before fission. The images at 35 and 37 minutes are shown at a tilted angle and demonstrate the ring structure. (b) This shows more detail at the time of division. Reproduced with permission from Sun and Margolin (1998).

cognate regulons is beyond the scope of this chapter. The painstaking process of analysing their respective roles is really only in its early stages and many regulators are undoubtedly yet to be recognised and more functions to be defined. The relationships of these molecules to growth, suggested in Table 1, emerges essentially from studies on samples from populations in specific growth phases during growth in defined media. With a few notable exceptions, information on the expression of regulatory proteins in contexts relevant to infection is very limited (Marshall *et al.*, 2000).

It should not be forgotten that many other classes of molecule regulate bacterial phenotype. The underlying point here is that phenotype and growth state cannot necessarily be inferred from the detection of selective mRNA profiles. At the macromolecular level selective proteases (Gottesman, 1999) and antisense RNAs (Wassarman *et al.*, 1999) have received much attention. Small molecules such as cyclic AMP (cAMP) (Botsford and Harman, 1992) and guanosine tetraphosphate (ppGpp) (Cashel *et al.*, 1996; JoseleauPetit *et al.*, 1999) are also recognised to have important regulatory roles. Many gene products affect their intracellular levels and have pleiotropic allosteric effects on their respective binding proteins. The role of ppGpp, the key product of the stringent response, deserves special mention here since, by binding to the B subunit of DNA-dependent RNA polymerase, it provides another means of directing selective gene expression. The stringent response is stimulated by amino acid starvation and is generally associated with growth arrest (Cashel *et al.*, 1996).

In this selective survey the obviously important areas of energy metabolism, cell envelope biosynthesis and assembly and the so-called housekeeping genes have largely been ignored. However, the process of relating the expression of genes to bacterial growth could be extended to cover the entire genome, and this can serve little purpose until an adequate interpretive framework is available. The global approaches offered by proteomics and arrays provide realistic prospects that this will be achieved. The present is clearly an intense information-gathering phase, which should eventually lead to a new level of understanding.

Growth and Stasis

The growth phases of bacteria in batch culture have been reviewed extensively elsewhere (Pirt, 1975;

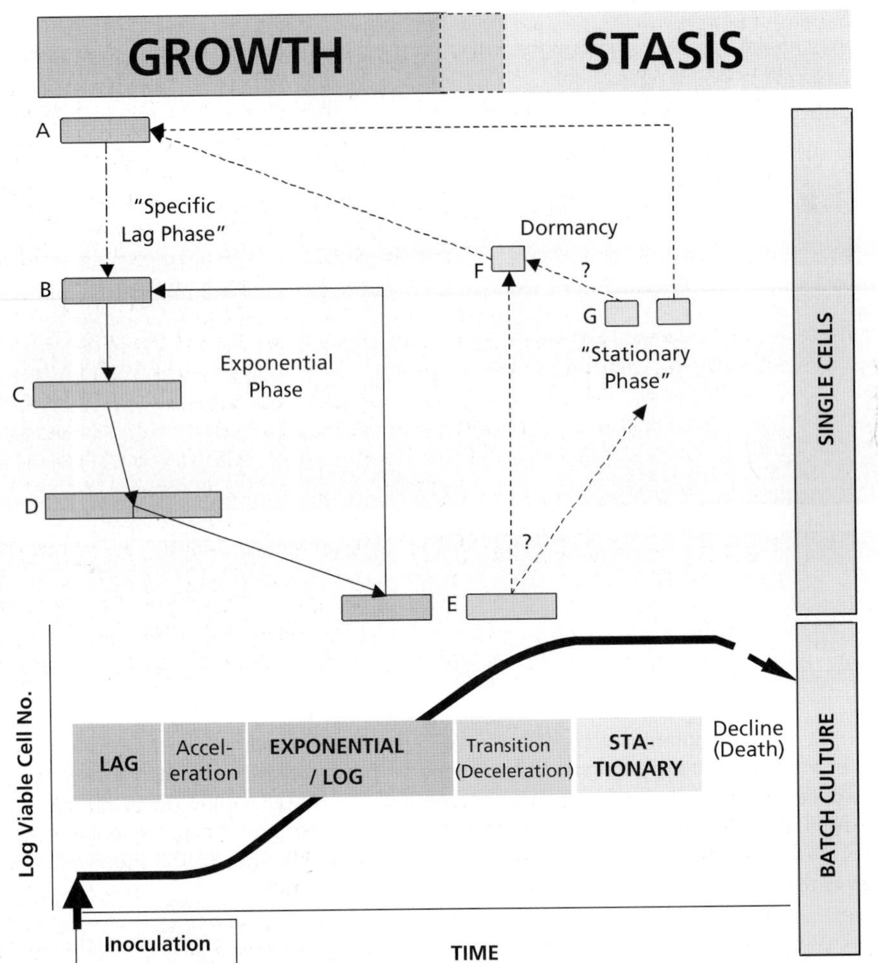

Fig. 2 Diagram comparing recognised states of growth and stasis in single cells and populations of bacteria. Note that while the upper (single cell) and lower (population) sections of the diagram can be compared, this is not so for all elements (see text). Lettering of cells represents different stages discussed in the text. '?' indicates a possible process for which there is little information.

Neidhardt *et al.*, 1990; Gehardt, 1994; Moat and Foster, 1995). Here the focus will be on individual cells and their populations. The aim is to introduce a framework within which cells in physiological states of particular significance to medicine can be recognised and to cross-reference this to the classical growth phases.

The various physiological states that can be recognised in relation to the growth of bacteria are outlined diagrammatically in Fig. 2. Laboratory cultures can be observed at the population or cellular level and a comparison between these is attempted in Fig. 2. The possible medical significance of some of the states identified is outlined in Table 2. A central dichotomy is suggested between cells growing or committed to grow and those in some form of non-growing state (stasis). All the states possess potential for further growth and replication and should therefore be considered viable.

Exponential Phase

State A in Fig. 2 envisages a hypothetical cell committed to growth, with appropriate resources available but not yet growing detectably. This cell may be adapting to a new environment or recovering from injury. Eventually the cell achieves state B in which its phenotype is adapted to commence growth in its current environment and is seen as equivalent to the product of division in a growing culture. This cell grows as indicated through state C, in which septum formation is initiated, to state D in which the septum is completed and fission is in progress. The separation of the progeny (E) into growing and static cells is

Table 2 Some implications of different growth phases of bacteria for medicine

Growth phase	Significance for laboratory studies	Medical significance
Lag	Heterogeneous, may contain many dead cells, affects timing of experiments	A consideration in determining safety margins when performing risk assessments on inevitably or potentially contaminated materials (e.g. food, drinks or pharmaceuticals)
Exponential (log)	When carefully prepared, contains the most defined and least heterogeneous population of cells. Demonstrably reproducible between labs	Most of our knowledge of metabolism, physiology and antibiotic susceptibility based on these populations. Regulation of growth rate well understood *in vitro* but little known *in vivo*. Rapidly growing cells are probably responsible for acute disease
Stationary	Requires a defined preceding exponential phase to obtain populations with reproducible properties. Extensive population heterogeneity and may include dormant cells	Distinctive properties in terms of virulence factor production and susceptibility to stress. Susceptibility to antibiotics is often reduced. Spores are highly stress resistant and are not susceptible to antibiotics and this may also be true of other, less distinctive dormant cell types. Dormant cells may be responsible for latent disease states

arbitrary and serves only to illustrate alternate pathways. If conditions have been conducive for continued growth, both progeny would be expected to continue in exponential growth.

The period between A and B is tentatively referred to here as the specific lag phase. In operational terms, the lag phase is measured as the interval between inoculation and the onset of detectable growth, and can include an initial period of cell death and growth below the limit of detection. This period may include the times indicated between G and A or F and A, i.e. the time taken for non-growing cells to adapt and become committed to growth (see below). Sometimes a more specific estimate of the time between inoculation and onset of net growth is made by extrapolating back from the log-linear exponential growth phase to intersect with the inoculum level (Fig. 2). The lag phase is particularly relevant to food microbiology where, in combination with estimated growth rates, it may form a component of the time taken before various foods are considered unsafe for consumption.

An enormous amount of knowledge has been gained about populations dominated by cells in the exponential phase of the growth cycle indicated in Fig. 2. The rapidity of biomass accumulation is often breathtaking, with doubling times of less than 30 minutes readily achievable by many pathogens that cause acute infection. It seems likely that such growth rates could underpin the rapid development of some infective conditions. The gradient of the exponential phase depends on the environmental conditions and the organism.

Unrestricted (exponential) growth *in vitro* in chemostat cultures is amenable to quite sophisticated mathematical analysis (e.g. Pirt, 1975; Gerhardt, 1994; Kovarova-Kovar and Egli, 1998). It has been possible to measure the growth yield on defined substrates and the effects of increasing and decreasing specific nutrients and modifying other culture conditions. The overall result has been a fairly comprehensive and quantitative view of how bacterial metabolism is integrated during exponential growth. The achievements in this field from the 1950s onwards have given rise to confidence that the growth of at least some bacteria can be predicted with a high degree of accuracy. However, when the predictive models that emerged from this work were tried out in the context of industrial-scale production and in the food industry it was found that changes in conditions required for 'real life' settings were outside the parameters modelled. Thus, while in relation to current knowledge of bacterial metabolism a fairly good estimate can be made of the input and output from chemostat cultures, the predictions begin to unravel even in the relatively controlled environment of industrial fermentations and the food industry.

During infections, bacteria encounter an, as yet, almost completely undefined set of nutritional environments. In view of this and the modest predictive power of the available models outside the controlled environments in which they were developed, it seems unlikely that knowledge based on the study of unrestricted growth *in vitro* can in the near future have much to say about growth *in vivo.*

Exponential Phase Inocula: a Key Resource in Bacteriology

The importance of studying bacteria in exponential growth should not be underestimated, principally because it provides a standardised approach to providing cell populations in a defined physiological state that can be reproduced in different laboratories. This is as true for fundamental physiological and biochemical studies as it is for antibiotic susceptibility and pathogenicity studies.

Stress Responses

An important aspect of bacterial physiology that has mainly been studied with exponential phase cells is the capacity of organisms to respond to environmental changes. Where these changes are potentially lethal, the responses are referred to as stress responses. The genetic basis for the phenotypic changes elicited by environmental change have been studied extensively, initially by mutational and reporter analysis and at the proteomic level (Ishihama, 1997; Booth and Louis, 1999; Foster, 1999; HenggeAronis, 1999; Storz and Imlay, 1999; Yura and Nakahigashi, 1999). Latterly, genomic and subgenomic arrays have afforded an attractive approach to studying these adaptive responses at a global transcriptional level (Tao *et al.*, 1999). Depending on the nature of the environmental change or stress, the changes in gene expression elicited may involve between tens and hundreds of different genes. Where the change is not stressful (as defined above) growth is substantially slowed down or arrested and resumes after the adaptation is complete. In some cases, notably where nutrient depletion precludes further growth, changes in the gene expression pattern are not confined to a single shift but a sequential programme of change is initiated (Kjelleberg *et al.*, 1993). Where this results in a defined morphological adaptive change, such as in sporulation, it is referred to as differentiation (Errington, 1996). In contrast, if the change is stressful, such as a substantial pH change or temperature increase, a proportion of the population is killed. The adaptive response in the survivors makes them at least temporarily more resistant (a higher proportion of survivors) to further similar stress.

Stress responsive genes may be activated by one or more stimuli, and in some cases multiple stimuli. Three examples of such multiply responsive genes, *rpoS*, *uspA* and *uspB*, are cited in Table 1. It is conspicuous that all three are upregulated in stationary phase. This lends weight to the view that the arrest of net growth in batch culture referred to as stationary phase is itself a form of stress response related either to nutrient depletion or accumulation of toxic metabolites. Several stress responses confer cross-protection against other stresses (e.g. pH and heat) and stationary-phase populations are generally more resistant to stress than exponential-phase populations (Antelmann *et al.*, 1996; HenggeAronis, 1999).

Stationary Phase

Although the classical view of the stationary phase is that it reflects growth arrest associated with nutrient exhaustion, recent work has raised the possibility that cessation of growth may sometimes be more ‘elective’. Indeed, there is evidence that when they enter stationary phase bacterial populations have not exhausted all the nutrients potentially available for growth (Barrow *et al.*, 1996), and that they may produce specific autocrine signals that tell cells not to grow (Lazazzera, 2000). The possible roles for cell-to-cell communication in regulating growth are further discussed below.

Returning to Fig. 2, some of the reasons are identified why cells may enter the stasis section of the cycle. Apart from the ability to reproduce the phenomenon of the stationary phase in batch cultures and to alter the kinetics of its onset by the composition of the growth medium, little is known about how the transition from exponential to stationary phase is regulated. A summary of the recognised influences is given in Fig. 3. It is certainly clear that genes such as *uspA*, *uspB* and *rpoS* are up-regulated before or at the beginning of the cessation of net detectable growth (cf. Table 1), and it seems likely that they play significant regulatory roles. It must be emphasised, however, that even *in vitro* it is very

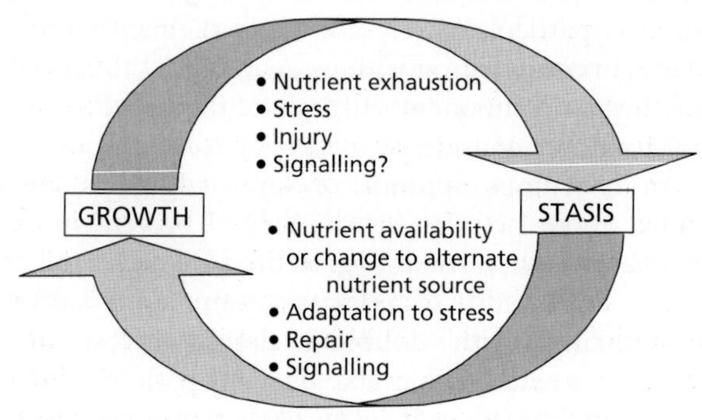

Fig. 3 Diagrammatic illustration of factors that affect growth and stasis at the population level.

difficult to achieve uniform bacterial populations and that the standard methods of analysis reflect only gross changes in biomass and dominant biochemical properties. It is, therefore, entirely possible that in batch culture, cell populations characterised as in exponential or stationary phase, in fact comprise mixtures of cells in all of the states identified in Fig. 2 but in different proportions. For example, the ratio of growing cells : static cells in exponential phase may be 100 : 1 or it may be 1 : 100. Only in the chemostat or tubidostat or after multiple rounds of growth and dilution before stationary phase will the static population be kept to a minimum, though still not eliminated.

An important consequence of the last point is that it is very difficult to attribute specific patterns of gene expression to specific cell populations when the measurements have been performed at the population level. Thus, for example, when changes are observed in the patterns of protein expression on two-dimensional gels in response to a stress, it is not possible to be certain that individual changes are occurring in all the cells sampled. Indeed, it is quite possible that multiple subpopulations are represented. Moreover, the relationships between gene expression and phenotype are rarely determined in such experiments. So, where the stress is lethal to a portion of the population, one cannot even be certain that the changes are taking place in cells that are going to survive or those that are going to die. These problems are not insurmountable, but they do show some limitations to the global analytical approach.

From the above it should be apparent that while the stationary phase can be recognised as a phenomenon and characterised at the molecular level in batch cultures (Matin *et al.*, 1989; Kolter *et al.*, 1993; Spector, 1998), the notion of a 'stationary phase cell' is by no means precise. It is well known that smaller cells with lower ribosomal content dominate stationary phase cultures, but similar statements could be made about chemostat cultures at very low dilution rates. In the case of stationary phase resulting from various nutrient limitations, the smaller cells appear to result from reductive cell divisions (fission without cell growth) (Kjelleberg *et al.*, 1993) but it is not certain that this is always the case. One important feature of stationary phase cultures is that, in terms of maintaining colony-forming unit (CFU) levels, they are generally more resistant to multiple stresses (e.g. removal of C, N or P from the medium, heat shock or antibiotic treatment). This may well relate to up-regulation of genes like *rpoS*, the expression of which is associated with several different stress responses. Teleologically this makes sense, since at least notionally the stationary phase implies a relative lack of resources for the bacteria, and it would seem prudent for them to be protected against multiple noxious influences when the capacity to respond is reduced.

Cells that persist in stasis can be considered to be ageing. When stasis is associated with nutrient exhaustion, the capacity for turnover and repair in cells is limited. Recent studies have provided evidence that accumulation of oxidative damage to proteins is a critical aspect of survival under these conditions (Dukan and Nystrom, 1998, 1999). These and related studies have been drawn together into a framework for understanding metabolic adaptations to stasis in bacteria that may have far reaching implications about how the control of non-replicating bacterial populations of concern to medicine and public health should be approached (Nystrom, 1998, 1999).

Finally, stationary phase cultures are by no means inert. Stress responses can be detected and, at least in the case of the case of acid stress, a response specific to the stationary phase can be demonstrated (Foster, 1999). More significantly, at least in terms of Fig. 2, stationary phase cultures are not exclusively composed of non-growing cells. A phenomenon referred to as 'GASP' (growth adapted to stationary phase) has been described, in which long-term stationary phase cultures contain successive growing subpopulations. These populations apparently arise from successive mutations in genes such as *rpoS*. The GASP phenotype can be recognised by inoculating an early stationary phase culture with a small number of cells of an antibiotic-resistant, but otherwise isogenic strain taken from a late stationary phase culture. The mutant then grows to dominate the culture into which it has been inoculated (Zambrano and Kolter, 1996). These experiments illustrate the point that, while the total colony count may remain the same, dynamic events occurring within the study population may be concealed.

Dormancy and Sporulation

In contrast to the notional stationary phase cell shown in Fig. 1, there are at least some defined examples of dormant cells. Here, the term dormant is used to denote cells in which there has been a reversible shutdown of metabolic activity (Kaprelyants *et al.*, 1993). The bacterial spore provides the clearest example of a dormant bacterial cell. Sporulation is a differentiation pathway involving sequential activation of genes, initially in the mother cell and then selectively in the mother and developing spore cell (Errington, 1993; Stragier and Losick, 1996). The process provides a genetic paradigm for differentiation and dormancy in bacteria and has been studied most extensively in *B. subtilis.* In particular, recognition of the importance of switching between alternate sigma factors has provided a useful framework for the study of adaptation and differentiation in bacteria. Sporulation leads to the production of highly stress-resistant cells and can be viewed as an extreme form of the adaptations that occur in the stationary phase. Indeed, sporulation is initiated in the transition phase and by the factors identified in Fig. 3. It should be noted that several peptide signalling factors have been defined in this context (Lazazzera and Grossman, 1998). Exactly what determines whether a cell enters stationary phase or dormancy (G or F, Fig. 2) is not defined, but it is certainly the case that spores and vegetative cells are both present in stationary phase cultures. This heterogeneous response reinforces the points made above concerning the multiple populations that may be present in bacterial monocultures. Other heterogeneous responses including competency and motility also occur during transition phase in *B. subtilis.*

The dormant cells of *Micrococcus luteus* described by Kaprelyants *et al.*, 1994 are somewhat less well defined. These cells develop slowly after maintenance of stationary phase cultures for several months. In classical terms, the decline phase is well established in these cultures since colony counts have generally fallen by several orders of magnitude by the time the dormant cells can be demonstrated. Although some morphological changes are recognised in the populations that contain dormant *M. luteus* cells, Kaprelyants and colleagues explicitly recognise dormant cells by two properties: first, by their substantially reduced capacity to take up the membrane energisation-sensitive fluorescent probe rhodamine 123, when compared with exponential or early stationary phase populations, and second, by the capacity to be cultured by colony formation or in broth. Through painstakingly careful experiments, these workers were able to demonstrate that dormant cells that could not be cultured by conventional means could, none the less, be resuscitated by exposure to cell-free supernatants from growing *M. luteus* cultures. Subsequently, these supernatants were shown to contain a protein (17 kDa) which they termed resuscitation promoting factor (Rpf) and the structural gene (*rpf*) was cloned and sequenced (Mukamolova *et al.*, 1998a).

An important distinction must be made between dormancy, as exemplified by sporulation, and that of the *M. luteus* model. Sporulation clearly results from a programme of gene expression that can legitimately be described as differentiation. The *M. luteus* cells meet an operational definition of dormancy (see below), but there is no evidence that this results from a specific genetic programme or that it confers survival advantage in the way that spores clearly do. Indeed, there is evidence that the dormant cells are in fact 'injured' or 'moribund', since their permeability properties are demonstrably 'repaired' during the resuscitation process (Kaprelyants and Kell, 1993).

Nucleotide sequence data are now available from many of the genes involved in sporulation and germination and from *rpf* in *M. luteus.* Thus the opportunity to determine whether other bacteria encode homologues of these genes arises. Sporulation gene homologues have been found in bacteria that have not been demonstrated to produce spores (DeMaio *et al.*, 1996). Caution must be exercised, however, before concluding that these organisms have dormant forms that have not been recognised. Several genes first recognised in the context of sporulation are now known to have important functions in vegetative cells.

In contrast, whether or not dormant *M. luteus* cells are comparable to spores, studies on purified Rpf have shown that *M. luteus* has distinctive growth-enhancing properties, notably a shortening of the lag phase, that have led its discoverers to describe it as the first 'bacterial cytokine'. *Rpf* homologues appear to be confined to Gram-positive organisms with a high G + C content, and it is particularly noteworthy that the genomes of *Mycobacterium tuberculosis* and *M. leprae* both encode multiple homologues (Mukamolova *et al.*, 1998a).

It will be noted that a discussion of the decline or death phase of batch cultures has so far been avoided. This is a highly variable phenomenon depending on the organism, strain and medium used. Classically the total cell number in the culture is maintained while the colony count declines. It should be apparent from the foregoing that the stationary phase is only stationary in respect of total cell counts and net biomass. Moreover, the properties of the culture population become less well defined with increasing time after the end of the exponential phase and little of substance can be said about the molecular events that occur after the first few days of this period. However, the recent trend to question whether cells that do not produce colonies on the standard culture medium for the organism concerned, such as those that develop during the decline phase, may none the less be considered 'viable', makes it possible to defer this discussion to the final section of this chapter.

Exit from Dormant or Stationary Cellular States and Re-entry into Growth

An important feature of Fig. 2 is the implication that static cells do not re-enter the growth cycle simply by reversing the process by which they entered into stasis. This is known to be the case with spores, where germination is clearly not the opposite of sporulation; a comparable process is suggested for other forms of dormancy and stasis. Germination has been shown to depend on one or more specific germination signals, but it is not known whether this is so for other forms of static cell (Moir and Smith, 1990). None the less, it does seem clear that requirements for growth initiation are somewhat distinct from those necessary for its maintenance.

Defining the nutrients and signals and other conditions necessary for initiating growth is of considerable medical importance. Not only are there several diseases, notably tuberculosis, that have 'latent' phases in which it is thought that the pathogen may itself be dormant (Grange, 1992; Barer, 1997), but also the reliable determination of the presence of organisms in clinical and environmental samples by culture remains absolutely central to patient management and public health monitoring (Barer *et al.*, 1993). Where dormancy does operate, a complete knowledge of the factors that activate and de-activate growth could make it possible to recognise why latent disease re-activates. Moreover, it might be possible specifically to activate dormant cells to render them susceptible to standard chemotherapy. With respect to cultivation, it is only when a comprehensive understanding is available of the factors necessary for organisms to initiate growth, that one can be confident that culture-based detection is at its most sensitive.

Apart from germination, very little is known about the process of transition from stasis to growth. The discovery of Rpf provides exciting opportunities to study one example of this process, particularly because a key measurable effect of this molecule is a reduction in the lag phase. Kaprelyants and Kell (1996) have reviewed the evidence relating to other molecules that have effects comparable to Rpf. Growth-stimulatory molecules that affect both Gram-positive and Gram-negative organisms have been identified with varying degrees of certainty. The recognition that *N*-acyl homoserine lactones may be stimulatory in some species is particularly significant in view of the well-recognised and diverse signalling functions attributable to this group of molecules.

It seems likely that signalling molecules play a significant role in growth regulation in at least some species in some environments. It has even been suggested that growth of some species may be completely dependent on molecules like Rpf (Kaprelyants and Kell, 1996). While this is an interesting proposal, demonstration of Rpf-dependency requires very specific environmental conditions and the natural physiological roles for the molecule and its homologues are far from understood. Whatever these may be, the significance of signalling in bacterial growth cannot be ignored; in particular the possibility that growth is a 'social', 'communal' or 'quorum-dependent' process in bacteria is now firmly on the agenda.

Culturability and Viability

Isolation of single bacterial strains in pure culture has provided the foundation stones of medical microbiology. Strategies for enrichment and selective isolation perhaps represent the best-developed aspects of the subject throughout the twentieth century (Gerhardt, 1994; Mackie *et al.*, 1996) and they will not be reviewed here, particularly because the molecular requirements of isolation procedures are not defined. It is self-evident that isolation and culture have provided abundant and relevant information and given us many insights into infection. Here, two key questions that are brought into sharp relief by the application of molecular techniques are addressed:

1. How well does culture represent the bacteria present in a particular sample?
2. What are the limitations of molecular techniques in representing the bacteria present in a particular sample?

These two questions provide starting points from which culturability and viability may be considered. Discussion of the problems that beset terminology in this area will not be repeated here (Barer *et al.*, 1993; Kaprelyants *et al.*, 1993; Barer, 1997; Kell *et al.*, 1998; Barer and Harwood, 1999). In order that the reader should at least recognise the framework within which the problematic terms are used here, a table of definitions is provided (Table 3). Clear distinctions are drawn between use of terms in a conceptual (theoretical) and operational (practical) contexts.

Culturability

From the early days of medical microbiology it was recognised that there were organisms that defied attempts at propagation in pure cultures *in vitro*. *Mycobacterium leprae* and *Treponema pallidum* still stand as monuments to this fact a century after they were first described. These are, however, exceptions against a backdrop of numerous successful studies that have led to the isolation of many medically significant organisms. Over the last 30 years the examples of

Table 3 Definitions of key terms relating to growth, culturability and viability of bacteria

Term	Conceptual definition	Operational definition
Viable	Retaining the capacity for replication over a stated or generally accepted time frame	Explicit demonstration of replication in a validated laboratory system
Replication	Genomic replication and segregation into a new self-propagating unit (propagule)	Observed cell fission or increase in number of propagules
Culturable	Capable of detectable replication in a realisable laboratory system	Detected replication in a validated laboratory system
Growth	Accumulation of biomass	Demonstrated accumulation of biomass
Dormancy	A reversible state of low metabolic activity in a unit that maintains viability	A demonstrated reversible low state of metabolic activity demonstrated by a specific technique or set of techniques in an operationally viable unit
Resuscitation	Transition from a temporary state in which the specified unit had lost the capacity for self-replication to regain that capacity	A demonstrated transition from a temporary state of non-culturability in a defined system to culturability in that system. The procedure must exclude regrowth as a possible explanation
Regrowth	Growth and/or replication within a population from below to above to levels of propagules that can be detected by a defined culture test.	Growth and/or replication demonstrated by one method that led to detection of the organism by another, less sensitive, method. If the latter were applied alone, resuscitation would appear to have occurred
Cryptic growth	Growth and/or replication of a sub-population within the study population that cannot be detected by the methods applied	Growth and/or replication demonstrated by one method that was not detected by another, less sensitive, method
Suicide	An irreversible process by which a viable unit determines the loss of its own viability by a specific mechanism	A non-reversible response in which a population of propagules can be shown to have lost demonstrable viability as a result of a process intrinsic to that response
Activity	A metabolic or behavioural process occurring within the unit under consideration. The unit may be viable or non-viable	A demonstrated behavioural, biochemical or physiological process occurring within the unit under consideration
Survival	Maintenance of viability	Maintenance of operational viability

Campylobacter, *Legionella* and *Helicobacter* serve as reminders that persistent attempts at isolation have borne fruit when appropriate conditions are eventually achieved. Conversion of laborious research isolation methods to routinely applicable detection procedures has led to recognition of these agents as major pathogens whose roles in the diseases they cause were previously unsuspected.

Two conclusions may be drawn from these observations. First, efforts to isolate 'novel' organisms in association with human disease frequently bear fruit and there seems to be little reason to suspect that such approaches will not continue to be fruitful. Second, failure to obtain isolates of organisms known or suspected to be present in samples often relates to a failure in preparing a culture medium with sufficient selective or growth supportive properties for the organism in question.

These observations should be set against a background of dramatic developments in the field of microbial ecology. It had been recognised for many years that when comparisons are made between the total number of bacterial cells present in almost any environmental sample and the total number of colony-forming units obtained by non-selective agar culture, the former exceed the latter by anything from one to three orders of magnitude. This excess of cells that fail to form colonies has been referred to as the 'great plate count anomaly' (Staley and Konopka, 1985) and it has provided a new generation of microbial ecologists with the incentive to study this phenomenon with molecular techniques.

'As Yet Uncultured' Bacteria

The findings have been striking. It has been possible specifically to recognise and classify many bacteria that have not been recovered or characterised in laboratory cultures (**Fig. 4**) by methods based on direct recovery of 16S rRNA subunit genes from samples (Giovannoni *et al.*, 1990) and in-situ hybridisation to ribosomes at the single cell level (Amann *et al.*, 1995). Of the rDNA sequences recovered in studies of this sort, the majority are apparently derived from organisms that have not previously been characterised and this has fuelled the view that culture methods do not adequately represent the range or numbers of bacteria present in most samples that contain a diverse microbial flora.

To what extent are these observations applicable to medical microbiology? Has there been a failure to recover the majority of human pathogens in culture and is there ignorance of the most abundant components of the normal flora? The available data do not allow clear answers to these questions, but early studies in which the methods of molecular microbial ecology have been applied have borne impressive fruit and the case for their further application is overwhelming; some examples are discussed below.

Ribosomal RNA gene recovery studies in medical microbiology have been particularly successful when applied to samples from well-defined clinical syndromes. The ground-breaking studies of Whipple's disease and the description of *Tropheryma whippelii* on the basis of 16S rRNA sequence data have clearly demonstrated the potential of molecular approaches to the resolution of some long-standing problems (Wilson *et al.*, 1991; Relman *et al.*, 1992; Relman, 1999). Taken together with the studies on bacillary angiomatosis and the description of *Mycobacterium genavense* (Relman *et al.*, 1990; Bottger *et al.*, 1992), the *T. whippelii* story would seem to endorse the view that molecular approaches could make it possible to work with organisms that permanently defy attempts at cultivation. It should be noted, however, that both *M. genavense* and *T. whippelii* have now been propagated *in vitro*, though in the case of the latter, not in axenic culture (Coyle *et al.*, 1992; Schoedon *et al.*, 1997).

It seems, therefore, that the message from *Campylobacter*, *Legionella* and *Helicobacter* – that intensive effort towards culture of a specific target organism regularly leads to its isolation – is applicable also to organisms recognised by molecular methods.

It is against the background of these examples that the phenomenon of 'as yet uncultured' (AYU) bacteria and the meaning and use of the term 'non-culturable' should be considered. As discussed above, many studies provide evidence for a substantial excess of AYU bacteria and the observations that are encapsulated in the phrase 'the great plate count anomaly' have been repeated on numerous occasions. In both cases, bacteria that are operationally non-culturable are detected. There is, however, little evidence that the non-culturable populations in the two cases are one and the same. In any given study the cells that are not

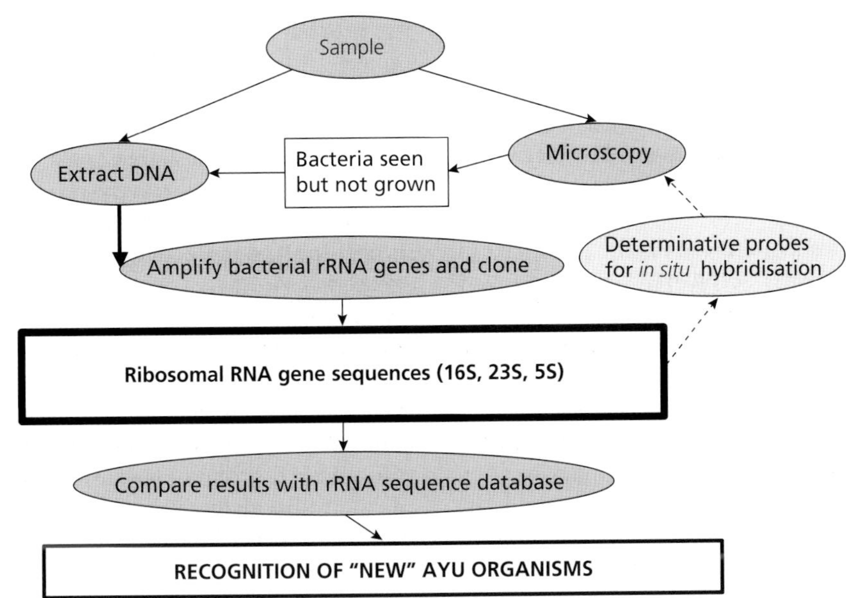

Fig. 4 Diagrammatic outline of the recognition and characterisation of 'new' organisms by microscopy and rRNA gene recovery.

recovered in culture potentially comprise two populations: those belonging to species for which no method of culture has yet been devised and cells of organisms that can normally be cultured by the method applied but which cannot for one reason or another be recovered. These latter 'non-recoverable' cells may include dead, moribund, injured and other 'temporarily non-culturable' cells, and they will be discussed more extensively below. Here, it should be noted that cells of many readily culturable organisms become difficult to recognise and isolate when they are part of complex microbial communities. It should also be noted that it would be impractical to apply all the possible, relevant and available cultural methods to such complex communities. The range of nutritional requirements imposed by the diversity of taxa and physiological states present is simply too great.

Cells of Culturable Organisms not Recovered in Culture

As indicated above, some bacterial cells can be detected by microscopy or molecular methods that are operationally non-culturable, yet they do not belong to the AYU group. These cells belong to groups of organisms that can be cultured, but isolation and or culture fail for a variety of reasons. These reasons include:

1. *Competition with other organisms*: Growth of the organism is suppressed or obscured by other organisms present in the same environment.
2. *Sublethal injury*: The organism is in the process of recovering from some form of damage, such as the effect of an antibiotic. During the recovery period the organisms may not be recoverable on conventional isolation media but, if they are provided with appropriate conditions, they can regain 'culturability'.
3. *Lethally injured, moribund and 'dead' cells*: Such cells are non-viable but they may still be detected by molecular or microscopic methods (see below).

Cells in categories 2 and 3 are readily produced in laboratory experiments. However, their explicit recognition and significance in natural samples remains problematic. In classical operational terms, sublethally injured cells fail to grow under certain selective isolation conditions, but can still grow under non-selective conditions. The framework for these effects is well developed in food microbiology (Mackey, 2000). Overall, the effect of injury detected this way is to render the injured cells more fastidious in their growth requirements. It also appears that injured cells incubated under inappropriate conditions do not simply fail to grow, they actually lose viability.

While this approach provides a practical means of recognising injury, it is clear that patterns of damage and repair can occur that do not fit into this pattern. For example, DNA damage that elicits the SOS response causes an arrest of DNA replication and cell division, while SOS-mediated repair continues (Walker, 1996). This process is not detected by simple colony counts, because growth is only delayed, not prevented. Nonetheless, this process provides a second paradigm for injury and repair that is, perhaps, more familiar from an anthropomorphic perspective. Cells affected by some inimical stimulus, such as an antibiotic or an immune effector system, acquire damaged components that must be replaced by a process of repair. During the repair process, conditions should be optimised for turnover and replacement of damaged components rather than for growth. This conceptual framework has led to the development of 'recovery' media specifically to fulfil this role.

No satisfactory generalisations can be made about the recovery of injured bacteria. What matters from a medical perspective is whether they retain pathogenic potential and, if so, is this is altered as compared with 'healthy' cells, and how the injured cells may be detected. At the time of writing the available evidence suggests that the pathogenic potential of sublethally injured cells is reduced (LeChevallier *et al.*, 1985; Singh *et al.*, 1986; Caro *et al.*, 1999). Optimal recovery conditions appear to be specific to the nature of the injury. In one study, the culture conditions giving the highest yield from starved cultures were shown to vary almost day by day (Mukamolova *et al.*, 1998b).

The third group of non-culturable cells identified above have clearly lost all ability to initiate new infections. They may, however, retain some metabolic capacity, which may, in special circumstances, include toxin production and release (Rahman *et al.*, 1996; Kimmitt *et al.*, 1999). They may, of course, also have associated endotoxic activity.

A general problem with the classification of cells into these three groups is that assignment can only be made in retrospect, after recovery by cultivation *in vitro* has succeeded or failed (see operational definitions in Table 3 and discussions in Kell *et al.* (1998) and Barer and Harwood (1999). Moreover, the assignments can only be crude estimates, since the technical difficulties involved in determining the correlation between cells counted by microscopy or molecular methods and colony counts are very substantial. In consequence, one can only speculate about the contributions made by such cells to the problems faced by medical microbiologists. None the less, it is clear that currently available molecular detection methods make none of the distinctions indicated. Moreover, with some notable exceptions (Sheridan *et al.*, 1998; Desjardin *et al.*, 1999), such methods do not currently distinguish between viable and non-viable organisms.

Viability

Determinations of viability have played a central role in bacteriology (Postgate, 1967, 1969; Barer and Harwood, 1999). In medical microbiology one is concerned primarily with measuring the effects of antimicrobial agents, sterilisation and disinfection regimens, and with immune effector mechanisms. In addition, only viable bacteria are considered capable of spreading infection. For many years it has been the accepted practice to equate viability with culture-based estimations such as colony or most probable number (MPN) counts. The MPN count is a well-standardised limiting dilution method (Department of Health and Social Security, 1969; Button *et al.*, 1993; Mackie *et al.*, 1996). However, partly fuelled by some of the problems raised in the previous section, a number of workers have pointed out that culture may not be adequate to the task it is intended to perform (Roszak and Colwell, 1987; Oliver, 1995; McDougald *et al.*, 1998). Moreover, it has been suggested that some methods other than culture, sometimes termed indirect methods (Postgate, 1967), may better reflect the distribution and activity of bacteria (Oliver, 1993). The resulting discussion has often focused on the suggestion that many bacteria may enter a state referred to as 'viable but non-culturable' (VBNC) (Xu *et al.*, 1982; Barer *et al.*, 1993; Kaprelyants *et al.*, 1993; Barer, 1997; Kell *et al.*, 1998).

The discussion must start with a definition of viability (Table 3). Established precedent indicates that a viable entity has the capacity to persist into the future and to carry out the functions with which it is normally associated (Postgate, 1967; Kell *et al.*, 1998; Barer and Harwood, 1999). Since the lifespan of individual bacteria is essentially the time between cell divisions, and this is a variable that depends on growth state, it has generally been accepted that the capacity for replication is implicit in bacterial viability. This capacity is unambiguously demonstrated in colony counts and estimates of MPN. However, dependence on culture tests for viability determinations leaves us without practical means of assessing viability in AYU bacteria and imposes time delays on analytical work. These problems have provided incentives to develop indirect methods further.

Previous discussions of indirect techniques have concentrated on divisions between different technical approaches. Here, a classification based on the different aspects of cellular and community function is presented (Table 4). The problems relating to these assays have been reviewed elsewhere (Kell *et al.*, 1998; Barer and Harwood, 1999). To summarise, it is possible both in theory and in practice to recognise circumstances where any of these tests fail to identify viable or non-viable cells as defined above. None the less, very good correlations can be obtained in specific settings between indirect test results and culture tests (see for example, Kaprelyants and Kell, 1992; Jepras *et al.*, 1995). Problems arise when the context in which the test has been applied is altered. In particular, when the assay is applied to different organisms or with different inimical stimuli for which the mechanism of cell death may be different, it cannot be assumed that correlations will be as good; re-validation is desirable with respect to every change in conditions.

In spite of these problems, indirect tests can be very useful. In particular, the recognition of trends in test results can be informative. Moreover, many problems fall away if one ignores the potential relationships of test results to viability and simply report them as tests of function with all the usual range of considerations regarding confounding factors and the need for appropriate controls (McFeters *et al.*, 1995). That such test results have some relationship to viability is not denied, but the relationship is variable, and justifies no more than rather vague statements about the probability that a given cell is viable or that a given culture is more or less viable than a comparable culture. Of course, the statements can be more precise when the conditions are closer to those in which the test was validated. In one well-studied *in-vitro* model, however, the best that could be achieved was a likelihood that a positive test of cytoplasmic membrane energisation indicated a 19% probability that a given cell could be recovered in culture (Kaprelyants *et al.*, 1996).

A central conclusion from the foregoing is that culturability currently remains the only unambiguous operational means of demonstrating viability. This clearly presents a problem for AYU bacteria, and the only rigorously defendable view is that, for the most part, the viability of cells and populations belonging to AYU taxa must be considered indeterminate. There are two riders to this view. First, the viability of *M. leprae* can clearly be recognised by propagation of the organism in a suitable animal host. Second, it is clear that when cells of an organism are detected in a sample, they must have got there from somewhere and that they must have been the progeny of a viable cell. Depending on the dynamics of the system sampled, the necessary proximity of the parental cell can be inferred. In many systems, including the human body, the likelihood of engulfment and destruction by a eukaryotic cell is high. Thus, the presence of a cell in a

Table 4 A functional classification of indirect tests that have been applied to assess the viability of bacteria

Function assayed	General method	Examples
Cell 'integrity'	Quantitation of cells	Optical density, nephelometry, flow cytometry, Coulter counting
	Microscopy	Morphology of cell by light and electron microscopy. Change in phase-contrast appearance
Cell permeability	Dye exclusion	Propidium iodide exclusion
	Retention of intracellular components	Assay of released DNA or enzyme (e.g. lactate dehydrogenase)
Cell nucleic acid content	Cell associated DNA and RNA	DAPI staining of DNA. rRNA in-situ hybridisation
Cell enzyme content	Modification of chromogenic or fluorogenic substrate	Fluorescein diacetate hydrolysis, tetrazolium reduction
Energy status	Presence of property that requires constant energy input	ATP content, ATP/ADP ratio. Rhodamine 123 accumulation (membrane energisation)
Evidence of integrated function	Assay of property that requires function of multiple cell systems	Reporter gene expression (e.g. *luxAB*). Inducible reporter gene expression (e.g. *lacZ*). Labelled precursor incorporation into specific cell product via multiple processing steps (amino acid incorporation, CO_2 production, $[^3H]$thymidine incorporation)
Transcription	Detection of mRNA	Uracil incorporation. Specific transcript detection by hybridisation or RT PCR
Evidence of growth potential	Detection of properties that always occur in growing cells	Presence of septa or multiple nucleoids. Cell elongation (Kogure test)

Specific references to these methods may be found in Kell *et al.* (1998); Barer and Harwood, 1999.
RT PCR, reverse-transcription polymerase chain reaction.

high turnover system implies that either the cell is viable, that viable 'relatives' are close by or that the cell is resistant to destruction.

Ultimately, the aim must be to produce validated tests that explicitly demonstrate bacterial replication in natural environments. To some extent, it has been possible to do this with culturable bacteria by recognising the dilution of replication of non-replicating markers (Maw and Meynell, 1968; Smith, 2000), but the technical challenge of achieving this with AYU bacteria remains substantial.

The 'Viable but Non-culturable' Controversy

Against the background outlined above, a substantial body of published work has accumulated over the last 20 years relating to bacterial cells that are described as 'viable but non-culturable' (VBNC) or said to be in a VBNC state. Some authors have applied this term to AYU bacteria. Although semantically, VBNC bacteria that belong to AYU taxa must exist, as noted above, the viable cells cannot explicitly be identified. Moreover, the area of controversy is concentrated on operationally non-culturable cells of culturable taxa, and adherents to the VBNC terminology predominantly apply the phrase in this context. The key issues have been reviewed extensively elsewhere and a brief summary is provided below (Roszak and Colwell, 1987; Barer *et al.*, 1993; Oliver, 1993; Kell *et al.*, 1998; McDougald *et al.*, 1998; Barer and Harwood, 1999).

The term VBNC was first applied by Colwell and associates to cultures of *Vibrio cholerae* that retained some form of demonstrable cellular activity, but which failed to grow in standard culture tests (Xu *et al.*, 1982). The work had been stimulated by attempts to define the environmental reservoir for this organism.

Indeed, the possibility that pathogens persist in environmental reservoirs in non-culturable forms remains an attractive hypothesis for any bacterial infection in which the distribution of cases is not concordant with the potential sources and reservoirs of the causal organism as identified by culture (Barer *et al.*, 1993). Similarly, it has been tempting to suggest that related phenomena may be responsible for periods when infections such as tuberculosis are non-patent, i.e. not detectable by culture (Barer, 1997).

While the VBNC hypothesis has proved highly stimulating in relation to expanding views of bacterial physiology and the transmission of infection, the degree to which putative VBNC cells can be recognised as separate from any of the categories identified in the previous section and the general value of the terminology are contentious.

Non-culturable bacterial cells have been described as VBNC on the basis of retained cellular integrity or activity, and by demonstration of their return to culturability. The latter has been achieved by special recovery methods or by animal passage (Nilsson *et al.*, 1991; Oliver and Bockian, 1995; Rigsbee *et al.*, 1997; Whitesides and Oliver, 1997). At one level it is clear that some of the reported phenomena can be attributed to injured cells that were temporarily non-culturable (Bogosian *et al.*, 2000). Furthermore, if cells are identified as VBNC by ultimately returning them to culture they were never non-culturable in the first place, but only temporarily so. Finally, it should also be noted that it can be very difficult to recognise the transition from an authentic temporarily non-culturable state to a culturable state, and that the experimental protocols of several published studies can be criticised on this basis (Kell *et al.*, 1998). Thus, the validity of the VBNC terminology can be questioned on the basis of established terminology, semantics and experimental technique.

The key issue for medical microbiology remains the mechanism by which cells can become non-culturable. The VBNC debate has raised the possibility that some bacteria can effectively give rise to non-culturable forms by differentiation rather than by injury or death. If this happens one is faced with the possibility that culture tests do not accurately reflect the destructive effects of antibiotics and immune effector mechanisms. While in this area many intriguing phenomena have been described, the available evidence does not justify such concern. Moreover, no genetic programme has been specifically identified that gives rise to temporarily non-culturable cells with a long-term potential for survival.

The author's view is that phenomena that have been considered to reflect VBNC cells are transient and result from cell injury and degeneration rather than adaptation or differentiation.

Conclusions

Progress towards the ideals of recognising bacterial growth and viability in medically relevant samples has been limited and, while there are important leads, there are at present no established molecular approaches explicitly to detect these phenomena on anything approaching a comprehensive scale. At present, knowledge is heavily biased towards laboratory studies on *E. coli* and *B. subtilis*, and the degree to which this information is applicable to all medically important bacteria *in vivo* remains to be established (**Fig. 5**).

Recent studies have detected many as yet uncultured or highly fastidious bacteria that associate with humans (Wilson and Blitchington, 1996; Millar *et al.*, 1996; Kroes *et al.*, 1999; Suau *et al.*, 1999; Beswich *et al.*, 1999). These studies have also extended the range of physiological states recognised in bacteria and they have identified large cell populations for which the viability is either indeterminate or marginal. The implications of these populations for medical microbiology have yet to be established. Indeed, as has already been the case with the molecular analysis of virulence determinants, it seems certain that the Henle-Koch postulates will have to be further modified to encompass the causal role in human disease of organisms that are detected only by molecular methods. The very considerable challenges to be faced in making progress with these problems seem an appropriate test of strength for the era of molecular medical microbiology.

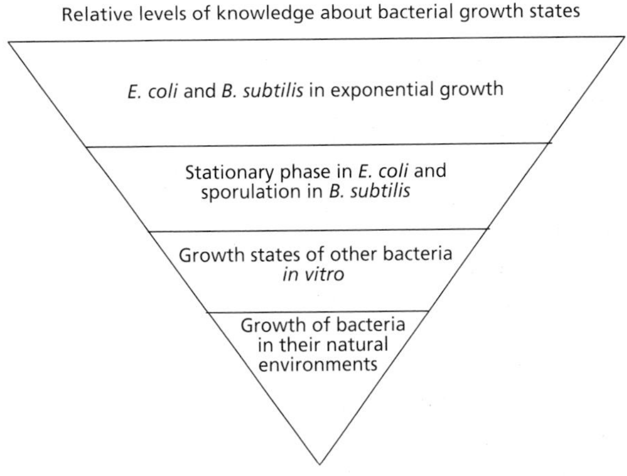

Fig. 5 Inverted pyramid of knowledge about the growth of bacteria.

References

Amann RI, Ludwig W, Schleifer KH (1995) Phylogenetic identification and in situ detection of individual microbial cells without cultivation. *Microbiol. Rev.* 59: 143–169.

Antelmann H, Engelmann S, Schmid R, Hecker M (1996) General and oxidative stress responses in *Bacillus subtilis*: Cloning, expression, and mutation of the alkyl hydroperoxide reductase operon. *J. Bacteriol.* 178: 6571–6578.

Aviv M, Giladi H, Schreiber G, Oppenheim AB, Glaser G (1994) Expression of the genes coding for the *Escherichia coli* integration host factor are controlled by growth phase, rpoS, ppGpp and by autoregulation. *Mol. Microbiol.* 14: 1021–1031.

Barer MR (1997) Viable but non-culturable and dormant bacteria: time to resolve an oxymoron and a misnomer? *J Med. Microbiol.* 46: 629–631.

Barer MR, Harwood CR (1999) Bacterial viability and culturability. *Adv. Microb. Physiol.* 40.

Barer MR, Gribbon LT, Harwood CR, Nwoguh CE (1993) The viable but non-culturable hypothesis and medical microbiology. *Rev. Med. Microbiol.* 4: 183–191.

Barrow PA, Lovell MA, Barber LZ (1996) Growth suppression in early-stationary-phase nutrient broth cultures of *Salmonella typhimurium* and *Escherichia coli* is genus specific and not regulated by sigma S. *J. Bacteriol.* 178: 3072–3076.

Beswich AJ, Lawley B, Fraise AP, Pahor AL, Brown NL (1999) Detection of *Alloiococcus otitis* in mixed bacterial populations from middle-ear effusions of patients with otitis media. *Lancet* 354: 386–389.

Bogosian G, Aardema ND, Bourneuf EV, Morris PJL, O'Neil JP (2000) Recovery of hydrogen peroxide-sensitive culturable cells of *Vibrio vulnificus* gives the appearance of resuscitation from a viable but non-culturable state. *J. Bact.* 182: 5070–5075.

Booth IR, Louis P (1999) Managing hypoosmotic stress: aquaporins and mechanosensitive channels in *Escherichia coli*. *Curr. Opin. Microbiol.* 2: 166–169.

Botsford JL, Harman JG (1992) Cyclic AMP in prokaryotes. *Microbiol. Rev.* 56: 100–122.

Bottger EC, Teske A, Kirschner P, Bost S, Chang HR, Beer V, Hirschel B (1992) Disseminated *Mycobacterium genavense* infection in patients with AIDS. *Lancet* 340: 76–80.

Bouvier J, Gordia S, Kampmann G, Lange R, HenggeAronis R, Gutierrez C (1998) Interplay between global regulators of *Escherichia coli*: effect of RpoS, Lrp and H-NS on transcription of the gene osmC. *Mol. Microbiol.* 28: 971–980.

Bremer H, Dennis PP (1987) Modulation of chemical composition and other parameters of the cell by growth rate. In: Neidhardt FC, Ingraham JL, Low KB *et al.* (eds) *Escherichia coli and Salmonella typhimurium: Cellular and Molecular Biology*, Vol.2. Washington, DC: ASM Press, pp. 1527–1542.

Button DK, Schut F, Quang P, Martin R, Robertson BR (1993) Viability and isolation of marine bacteria by dilution culture theory, procedures, and initial results. *Appl. Environ. Microbiol.* 59: 881–891.

Caro A, Got P, Lesne J, Binard S, Baleux B (1999) Viability and virulence of experimentally stressed nonculturable *Salmonella typhimurium*. *Appl. Environ. Microbiol.* 65: 3229–3232.

Cashel M, Gentry DM, Hernandez VJ, Vinella D (1996) The stringent response. In: Neidhardt FC (ed.) *Escherichia coli and Salmonella: Cellular and Molecular Biology*, Vol. 1. Washington, DC: ASM Press, pp. 1458–1496.

Cooper S (1991) *Bacterial Growth and Division: Biochemistry and Regulation of Prokaryotic and Eukaryotic Division Cycles.* San Diego: Academic Press.

Cooper S (2000) Toward a standard system for the mammalian cell cycle. *ASM News* 66: 71–75.

Coyle MB, Carlson LC, Wallis CK *et al.* (1992) Laboratory aspects of *Mycobacterium genavense*, a proposed species isolated from AIDS patients. *J. Clin. Microbiol.* 30: 3206–3212.

Delong EF, Wickham GS, Pace NR (1989) Phylogenetic stains – ribosomal RNA-based probes for the identification of single cells. *Science* 243: 1360–1363.

DeMaio J, Zhang Y, Ko C, Young DB, Bishai WR (1996) A stationary-phase stress-response sigma factor from *Mycobacterium tuberculosis*. *Proc. Natl Acad. Sci. USA* 93: 2790–2794.

Department of Health and Social Security, Ministry of Housing and Local Government (1969) *The Bacteriological Examination of Drinking Water Supplies.* London: HMSO.

Desjardin LE, Perkins MD, Wolski K *et al.* (1999) Measurement of sputum *Mycobacterium tuberculosis* messenger RNA as a surrogate for response to chemotherapy. *Am. J. Respir. Crit. Care Med.* 160: 203–210.

Dorman CJ, Hinton JC D, Free A (1999) Domain organization and oligomerization among H-NS-like nucleoid-associated proteins in bacteria. *Trend. Microbiol.* 7: 124–128.

Dukan S, Nystrom T (1998) Bacterial senescence: stasis results in increased and differential oxidation of cytoplasmic proteins leading to developmental induction of the heat shock regulon. *Genes Dev.* 12: 3431–3441.

Dukan S, Nystrom T (1999) Oxidative stress defense and deterioration of growth-arrested *Escherichia coli* cells. *J. Biol. Chem.* 274: 26027–26032.

Errington J (1993) *Bacillus subtilis* sporulation: regulation of gene expression and control of morphogenesis. *Microbiol. Rev.* 57: 1–33.

Errington J (1996) Determination of cell fate in *Bacillus subtilis*. *Trends Genet.* 12: 31–34.

Farewell A, Kvint K, Nystrom T (1998) Negative regulation by RpoS: a case of sigma factor competition. *Mol. Microbiol.* 29: 1039–1051.

Farewell A, Kvint K, Nystrom T (1998) uspB, a new sigma S-regulated gene in Escherichia coli which is required for stationary-phase resistance to ethanol. *J. Bacteriol.* 180: 6140–6147.

Ferenci T (1999) Regulation by nutrient limitation. *Curr. Opin. Microbiol.* 2: 208–213.

Foster JW (1999) When protons attack: microbial strategies of acid adaptation. *Curr. Opin. Microbiol.* 2: 170–174.

Freestone P, Nystrom T, Trinei M, Norris V (1997) The universal stress protein, UspA, of *Escherichia coli* is phosphorylated in response to stasis. *J. Mol. Biol.* 274: 318–324.

Gerhardt P (1994) Section II: Growth. In: Gerhardt P, Murray RGE, Wood WA, Krieg NR (eds) *Methods for General and Molecular Bacteriology.* Washington, DC: American Society for Microbiology, pp. 135–294.

Giovannoni SJ, Britschgi TB, Moyer CL, Field KG (1990) Genetic diversity in Sargasso Sea bacterioplankton. *Nature* 345: 60–63.

Gottenbos B, vanderMei HC, Busscher HJ (1999) Models for studying initial adhesion and surface growth in biofilm formation on surfaces. *Methods Enzymol.* 310: 523–534.

Gottesman S (1999) Regulation by proteolysis: developmental switches. *Curr. Opin. Microbiol.* 2: 142–147.

Grange JM (1992) The mystery of the mycobacterial 'persistor'. *Tubercle Lung Dis.* 73: 249–251.

HenggeAronis R (1999) Interplay of global regulators and cell physiology in the general stress response of *Escherichia coli. Curr. Opin. Microbiol.* 2: 148–152.

Ishihama A (1997) Adaptation of gene expression in stationary phase bacteria. *Curr. Opin. Genet. Dev.* 7: 582–588.

Jepras RI, Carter J, Pearson SC, Paul FE, Wilkinson MJ (1995) Development of a robust flow cytometric assay for determining numbers of viable bacteria. *Appl. Environ. Microbiol.* 61: 2696–2701.

JoseleauPetit D, Vinella D, Dari R (1999) Metabolic alarms and cell division in *Escherichia coli. J. Bacteriol.* 181: 9–14.

Kaprelyants AS, Kell DB (1992) Rapid assessment of bacterial viability and vitality by rhodamine 123 and flow cytometry. *J. Appl. Bacteriol.* 72: 410–422.

Kaprelyants AS, Kell DB (1993) Dormancy in stationary-phase cultures of *Micrococcus luteus* – flow cytometric analysis of starvation and resuscitation. *Appl. Environ. Microbiol.* 59: 3187–3196.

Kaprelyants AS, Kell DB (1996) Do bacteria need to communicate with each other for growth. *Trends Microbiol.* 4: 237–242.

Kaprelyants AS, Gottschal JC, Kell DB (1993) Dormancy in non sporulating bacteria. *FEMS Microbiol. Rev.* 104: 271–286.

Kaprelyants AS, Mukamolova GV, Kell DB (1994) Estimation of dormant *Micrococcus luteus* cells by penicillin lysis and by resuscitation in cell free spent culture medium at high dilution. *FEMS Microbiol. Lett.* 115: 347–352.

Kaprelyants AS, Mukamolova GV, Davey HM, Kell DB (1996) Quantitative analysis of the physiological heterogeneity within starved cultures of *Micrococcus luteus* by flow cytometry and cell sorting. *Appl. Environ. Microbiol.* 62: 1311–1316.

Kell DB, Kaprelyants AS, Weichart DH, Harwood CR, Barer MR (1998) Viability and activity in readily culturable bacteria: a review and discussion of the practical issues. *Antonie Van Leeuwenhoek* 73: 169–187.

Kiley PJ, Beinert H (1998) Oxygen sensing by the global regulator, FNR: the role of the iron-sulfur cluster. *FEMS Microbiol. Rev.* 22: 341–352.

Kimmitt PT, Harwood CR, Barer MR (1999) Induction of type 2 shiga toxin synthesis in *Escherichia coli* O157 by 4-quinolones. *Lancet* 353: 1588–1589.

Kjelleberg S, Albertson N, Flardh K *et al.* (1993) How do non-differentiating bacteria adapt to starvation? *Antonie van Leeuwenhoek* 63: 333–341.

Koch AL (1997) Microbial physiology and ecology of slow growth. *Microbiol. Mol. Biol. Rev.* 61: 305.

Kolter R, Siegele DA, Tormo A (1993) The stationary phase of the bacterial life cycle. *Annu. Rev. Microbiol.* 47: 855–874.

Kovarova-Kovar K, Egli T (1998) Growth kinetics of suspended microbial cells: From single-substrate-controlled growth to mixed-substrate kinetics. *Microbiol. Mol. Biol. Rev.* 62: 646–666.

Kroes I, Lepp PW, Relman DA (1999) Bacterial diversity within the human subgingival crevice. *Proc. Natl Acad. Sci. USA* 96: 14547–14552.

Lazazzera BA (2000) Quorum sensing and starvation: signals for entry into stationary phase. *Curr. Opin. Microbiol.* 3: 177–182.

Lazazzera BA, Grossman AD (1998) The ins and outs of peptide signaling. *Trends Microbiol.* 6: 288–294.

LeChevallier MW, Singh A, Schiemann DA, McFeters GA (1985) Changes in virulence of waterborne enteropathogens with chlorine injury. *Appl. Environ. Microbiol.* 50: 412–419.

Levin PA, Grossman AD (1998) Cell cycle and sporulation in *Bacillus subtilis. Curr. Opin. Microbiol.* 1: 630–635.

McDougald D, Rice SA, Weichart D, Kjelleberg S (1998) Nonculturability: adaptation or debilitation? *FEMS Microbiol. Ecol.* 25: 1–9.

McFeters GA, Yu FP P, Pyle BH, Stewart PS (1995) Physiological assessment of bacteria using fluorochromes. *J. Microbiol. Methods* 21: 1–13.

McKay AL, Peters AC, Wimpenny JWT (1997) Determining specific growth rates in different regions of *Salmonella typhimurium* colonies. *Lett. Appl. Microbiol.* 24: 74–76.

Mackey BM (2000) Injured bacteria. In: Lund BM, Baird-Parker A, Gould GM (eds) *Food Microbiology.* Maryland: Aspen Publishers.

McLean RJ C, Whiteley M, Hoskins BC, Majors PD, Sharma MM (1999) Laboratory techniques for studying biofilm growth, physiology, and gene expression in flowing systems and porous media. *Methods Enzymol.* 310: 248–264.

Mackie TJ, McCartney JE, Collee JG (1996) *Mackie & McCartney Practical Medical Microbiology.* London: Churchill Livingstone.

Madigan MT, Martinko JM, Parker J (2000) Microbial Growth. In: *Brock Biology of Microorganisms.* Upper Saddle River, NJ: Prentice-Hall, pp. 740–772.

Margolin W (1998) A green light for the bacterial cytoskeleton. *Trends Microbiol.* 6: 233–238.

Marshall DG, Sheehan BJ, Dorman CJ (1999) A role for the leucine-responsive regulatory protein and integration host factor in the regulation of the Salmonella plasmid virulence (spv) locus in *Salmonella typhimurium*. *Mol. Microbiol.* 34: 134–145.

Marshall DG, Bowe F, Hale C, Dougan G, Dorman CJ (2000) DNA topology and adaptation of *Salmonella typhimurium* to an intracellular environment. *Phil. Trans. R. Soc. Lond. B Biol. Sci.* 355: 565–574.

Matin A, Auger EA, Blum PH, Schultz JE (1989) Genetic basis of starvation survival in nondifferentiating bacteria. *Annu. Rev. Microbiol.* 43: 293–316.

Maw J, Meynell GG (1968) The true division and death rates of *Salmonella typhimurium* in the mouse spleen determined with superinfecting phage. *Br. J. Exp. Pathol.* 49: 597–613.

Messer W, Weigel C (1996) Initiation of chromosome replication. In: Neidhardt FC (ed.) *Escherichia coli and Salmonella: Cellular and Molecular Biology*, Vol. 2. Washington, DC: ASM Press, pp. 1579–1601.

Millar MR, Linton CJ, Cade A, Glancy D, Hall M, Jalal H (1996) Application of 16S rRNA gene PCR to study bowel flora of preterm infants with and without necrotizing enterocolitis. *J. Clin. Microbiol.* 34: 2506–2510.

Moat AG, Foster JW (1995) *Microbial Physiology*, 3rd edn. New York: J. Wiley.

Moir A, Smith DA (1990) The genetics of bacterial spore germination. *Annu. Rev. Microbiol.* 44: 531–553.

Mukamolova GV, Kaprelyants AS, Young DI, Young M, Kell DB (1998a) A bacterial cytokine. *Proc. Natl Acad. Sci. USA* 95: 8916–8921.

Mukamolova GV, Yanopolskaya ND, Kell DB, Kaprelyants AS (1998b) On resuscitation from the dormant state of *Micrococcus luteus*. *Antonie Van Leeuwenhoek* 73: 237–243.

Neidhardt FC (1996) *Escherichia coli and Salmonella: Cellular and Molecular Biology*. Washington, DC: ASM Press.

Neidhardt FC, Ingraham JL, Schaechter M (1990) *Physiology of the Bacterial Cell: A Molecular Approach.* Sunderland, MA: Sinauer Associates.

Nilsson L, Oliver JD, Kjelleberg S (1991) Resuscitation of *Vibrio vulnificus* from the viable but nonculturable state. *J. Bacteriol.* 173: 5054–5059.

Nystrom T (1998) To be or not to be: the ultimate decision of the growth-arrested bacterial cell. *FEMS Microbiol. Rev.* 21: 283–290.

Nystrom T (1999) Starvation, cessation of growth and bacterial ageing. *Curr. Opin. Microbiol.* 2: 214–219.

Nystrom T, Neidhardt FC (1994) Expression and role of the universal stress protein, Uspa, of *Escherichia coli* during growth arrest. *Mol. Microbiol.* 11: 537–544.

Oliver JD (1993) Formation of viable but nonculturable cells. In: Kjelleberg S (ed.) *Starvation in Bacteria.* NewYork, Plenum Press, pp. 239–272.

Oliver JD (1995) The viable but non-culturable state in the human pathogen *Vibrio vulnificus. FEMS Microbiol. Lett.* 133: 203–208.

Oliver JD, Bockian R (1995) In vivo resuscitation, and virulence towards mice, of viable but nonculturable cells of *Vibrio vulnificus. Appl. Environ. Microbiol.* 61: 2620–2623.

Pirt SJ (1975) *Principles of Microbe and Cell Cultivation.* Oxford: Blackwell Scientific.

Postgate J (1967) Viability measurements and the survival of microbes under minimum stress. *Adv. Microb. Physiol.* 1: 1–21.

Postgate JR (1969) Viable counts and viability. *Methods Microbiol.* 1: 611–628.

Rahman I, Shahamat M, Chowdhury MA, Colwell RR (1996) Potential virulence of viable but nonculturable *Shigella dysenteriae* type 1. *Appl. Environ. Microbiol.* 62: 115–120.

Raivio TL, Silhavy TJ (1999) The sigma(E) and Cpx regulatory pathways: overlapping but distinct envelope stress responses. *Curr. Opin. Microbiol.* 2: 159–165.

Relman DA (1999) The search for unrecognized pathogens. *Science* 284: 1308–1310.

Relman DA, Loutit JS, Schmidt TM, Falkow S, Tompkins LS (1990) The agent of bacillary angiomatosis – an approach to the identification of uncultured pathogens. *N. Engl. J. Med.* 323: 1573–1580.

Relman DA, Schmidt TM, Macdermott RP, Falkow S (1992) Identification of the uncultured bacillus of Whipples disease. *N. Engl. J. Med.* 327: 293–301.

Rigsbee W, Simpson LM, Oliver JD (1997) Detection of the viable but nonculturable state in *Escherichia coli* O157:H7. *J. Food Safety* 16: 255–262.

Roszak DB, Colwell RR (1987) Survival strategies of bacteria in the natural environment. *Microbiol. Rev.* 51: 365–379.

Russell JB, Cook GM (1995) Energetics of bacterial-growth – balance of anabolic and catabolic reactions. *Microbiol. Rev.* 59: 48–62.

Saier MH (1998) Multiple mechanisms controlling carbon metabolism in bacteria. *Biotechnol. Bioeng.* 58: 170–174.

Salgado H, SantosZavaleta A, GamaCastro S, MillanZarate D, Blattner FR, ColladoVides J (2000) RegulonDB (version 3.0): transcriptional regulation and operon organization in *Escherichia coli* K-12. *Nucleic Acids Res.* 28: 65–67.

Sawers G (1999) The aerobic/anaerobic interface. *Curr. Opin. Microbiol.* 2: 181–187.

Schoedon G, Goldenberger D, Forrer R *et al.* (1997) Deactivation of macrophages with interleukin-4 is the key to the isolation of *Tropheryma whippelii. J. Infect. Dis.* 176: 672–677.

Severinov K (2000) RNA polymerase structure-function: insights into points of transcriptional regulation. *Curr. Opin. Microbiol.* 3: 118–125.

Sheridan GE, Masters CI, Shallcross JA, MacKey BM (1998) Detection of mRNA by reverse transcription-PCR as an indicator of viability in *Escherichia coli* cells. *Appl. Environ. Microbiol.* 64: 1313–1318.

Singh A, Yeager R, McFeters GA (1986) Assessment of *in vivo* revival, growth, and pathogenicity of *Escherichia coli* strains after copper- and chlorine-induced injury. *Appl. Environ. Microbiol.* 52: 832–837.

Smith H (2000) Questions about the behaviour of bacterial pathogens *in vivo. Phil. Trans. R. Soc. Lond. B Biol. Sci.* 355: 551–564.

Spector MP (1998) The starvation-stress response (SSR) of *Salmonella. Adv. Microb. Physiol.* 40: 233–279.

Staley JT, Konopka A (1985) Measurement of *in situ* activities of nonphotosynthetic microorganisms in aquatic and terrestrial habitats. *Annu. Rev. Microbiol.* 39: 321–346.

Storz G, Imlay JA (1999) Oxidative stress. *Curr. Opin. Microbiol.* 2: 188–194.

Stragier P, Losick R (1996) Molecular genetics of sporulation in *Bacillus subtilis. Annu. Rev. Genet.* 30: 297–341.

Suau A, Bonnet R, Sutren M, Godon JJ, Gibson GR, Collins MD, Dore J (1999) Direct analysis of genes encoding 16S rRNA from complex communities reveals many novel molecular species within the human gut. *Appl. Environ. Microbiol.* 65: 4799–4807.

Sun Q, Margolin W (1998) FtsZ dynamics during the division cycle of live *Escherichia coli* cells. *J. Bacteriol.* 180: 2050–2056.

Tao H, Bausch C, Richmond C, Blattner FR, Conway T (1999) Functional genomics: Expression analysis of *Escherichia coli* growing on minimal and rich media. *J. Bacteriol.* 181: 6425–6440.

Walker GC (1996) The SOS response of *Escherichia coli.* In: Neidhardt FC (ed.) *Escherichia coli and Salmonella: Cellular and Molecular Biology*, Vol. 1. Washington, DC: ASM Press, pp. 1400–1416.

Wassarman KM, Zhang AX, Storz G (1999) Small RNAs in *Escherichia coli. Trends Microbiol.* 7: 37–45.

Whitesides MD, Oliver JD (1997) Resuscitation of *Vibrio vulnificus* from the viable but nonculturable state. *Appl. Environ. Microbiol.* 63: 1002–1005.

Wilson KH, Blitchington RB (1996) Human colonic biota studied by ribosomal DNA sequence analysis. *Appl. Environ. Microbiol.* 62: 2273–2278.

Wilson KH, Blitchington R, Frothingham R, Wilson JAP (1991) Phylogeny of the Whipples-disease-associated bacterium. *Lancet* 338: 474–475.

Wimpenny JW T (1988) The bacterial colony. In: *CRC Handbook of Laboratory Modelling Systems for Microbial Ecosystems*, Vol. II. Boca Raton, FL: CRC Press, pp. 109–139.

Xu HS, Roberts N, Singleton FL, Attwell RW, Grimes DJ, Colwell RR (1982) Survival and viability of nonculturable *Escherichia coli* and *Vibrio cholerae* in the estuarine and marine environment. *Microb. Ecol.* 8: 313–323.

Yura T, Nakahigashi K (1999) Regulation of the heat-shock response. *Curr. Opin. Microbiol.* 2: 153–158.

Zambrano MM, Kolter R (1996) GASPing for life in stationary phase. *Cell* 86: 181–184.

Zhou P, Bogan JA, Welch K *et al.* (1997) Gene transcription and chromosome replication in *Escherichia coli. J. Bacteriol.* 179: 163–169.

12

Vital Energy for Bacterial Growth and Reproduction

J.G. Morris

University of Wales, Aberystwyth, Wales

All micro-organisms necessarily expend free energy to maintain their substance and vitality, and to fuel their growth and reproduction. Only in profoundly dormant states, for example the endospores of certain bacteria whose dehydrated condition precludes any metabolic activity, can their viability be sustained without constant acquisition of free energy from the environment and its transformation into a form capable of performing work on behalf of the living cell.

Ultimately it is solar radiation that maintains life upon this earth, and some organisms (phototrophs) can intercept and directly avail themselves of incident sunlight. Others, including the bacteria with which we shall be especially concerned in this chapter, are chemotrophs that exploit the energy 'locked into' the molecular structures of metabolisable compounds. Most usually, though not exclusively, they release this energy in the course of oxidation–reduction reactions in which these substances, or metabolic derivatives thereof, serve as electron donors and acceptors.

The free energy so 'liberated' is conserved by being transformed into one or both of two types of 'biologically available energy' capable of driving energy-dependent cellular processes. The first is the '(bio)chemical energy' of intracellular metabolites that are characterised by high group transfer potentials. All of these, of which ATP is the most notable example, are produced by 'soluble' enzymes located in the cell cytoplasm. The second is the 'biologically utilisable' free energy captured by energy-transducing cell membranes in the form of transmembrane electrochemical gradients of ions (especially protons).

In bacteria, the free energy-conserving membrane is the cytoplasmic membrane (the inner membrane in Gram-negative organisms). The transmembrane electrochemical proton gradient can be generated by the operation of membrane-integrated electron carriers whose serial employment in respiratory (or photosynthetic) processes is associated with differential accumulation of protons at the outer surface of the membrane. This could be the result of vectorial expulsion of protons (proton pumping) and/or due to oxidation of substrates at the outer surface of the membrane releasing protons and electrons, with the latter returning across the membrane to its inner surface and leaving the protons behind (Fig. 1).

A proton motive force, Δp, is therefore created which has an electrical potential component, $\Delta\Psi$ (internal negative) and a thermodynamic activity (concentration) component, ΔpH (usually internal alkaline with respect to the environment). In the absence of a compensating ion flow (anions out, or

 doi:10.1006/bkmm.2001.0012

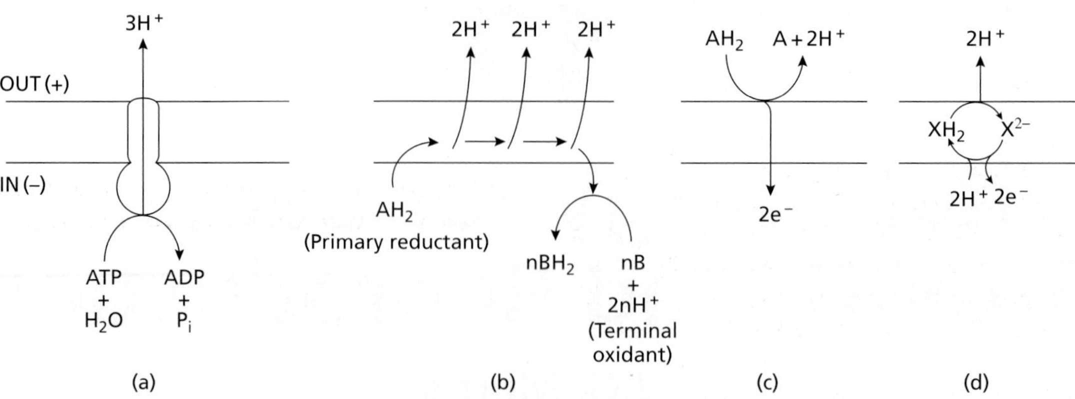

Fig. 1 Diagrammatic representation of different ways of creating a transmembrane electrochemical proton gradient (exterior, electropositive and acid: interior, electronegative and alkaline). (a) proton expulsion driven by ATP hydrolysis at an H^+-F_1F_0 ATPase, as in fermentative anaerobic bacteria; (b) proton pumping at coupling sites in a respiratory electron transport pathway; (c, d) oxidation of reductant at outer surface of the membrane releases H^+, with balancing electrons being transported across the membrane for release at the inner surface. (Bacteria have an average Δp of about 140–200 mV.)

cations in) the protons that are pumped out of the cell will create a $\Delta\Psi$ rather than a ΔpH. However, a ΔpH will be produced when electroneutrality is conserved by balancing movements of other ions. For most bacteria growing at, or near, neutral pH, $\Delta\Psi$ will contribute about 70–80% of the total proton motive force.

A reversible, membrane-embedded, proton-translocating ATPase provides a channel for the passage of protons through the otherwise proton-impermeable and electrically insulating membrane. When operating as an ATP hydrolase (Fig. 1a), the hydrolysis of ATP that yields ADP and inorganic phosphate causes protons to be expelled from the inside to the outside face of the cell membrane. This creates the same type of proton gradient (outside electropositive and acidic compared with the inside) as that generated by respiratory electron transport. When operating in the reverse manner, as an ATP synthase, the same enzyme complex produces ATP from ADP and inorganic phosphate, this synthetic reaction being powered by free energy deriving from the spontaneous inflow of protons (through the proton channel provided by the enzyme). This return flow of protons is in turn impelled by the proton motive force generated by the disparities in electrical charge and ion concentration across the two sides of the intervening membrane. The H^+-ATPase thus serves as a crucial link between the twin currencies of biological energy exchange.

Amongst bacteria we find obligately fermentative organisms whose chief or sole source of free energy capture is via the cytoplasmically localised substrate-level phosphorylation reactions that yield ATP, or their equivalents. Yet these organisms still have the need to generate transmembrane electrochemical ion gradients which are required to sustain a wide range of essential processes, including selective substrate uptake and excretion, cellular homeostasis and even motility. It is the hydrolysis of part of the cytoplasmically generated ATP at the membrane H^+-ATPase that fulfils this need by producing a stoichiometric outflow of protons. Conversely, although in a respiring bacterium the bulk of its free energy will initially be conserved via creation of a transmembrane electrochemical ion gradient, this can be called upon to yield cytoplasmic ATP as required, again as a consequence of the operation of the H^+-ATPase (though this is now acting as an ATP synthase).

The bacterial world also provides the most impressive evidence of evolutionary experimentation with different mechanisms of free energy conservation. An extreme example is the supremely elegant, though evidently not universally appropriate, light-motivated, direct proton pump (based on bacteriorhodopsin) that is a feature of the cell membrane in some halophilic archaea. Also amongst archaea there are the methanogens that utilise carbon dioxide and hydrogen (or acetate) as sole source of energy for growth by employing a highly unusual dual reductive and oxidative energy-conserving metabolic pathway. In the oxidative limb, methyl transfer occurs from N^5-methyltetrahydro-methanopterin to coenzyme M with coupled generation of an electrochemical gradient of Na^+ ions. In the reductive segment of the pathway, exergonic reduction of the heterodisulphide of coenzyme M and coenzyme B produces a transmembrane electrochemical H^+ potential. Such diversity and ingenuity reflect the competitive and evolutionary pressures upon bacteria to exploit every conceivable free energy source present in their environments.

In focusing our attention on bacteria of medical significance, we shall primarily need only to consider how free energy is conserved during fermentative and respiratory processes. Even so, variations in detail abound even in these processes as practised by different bacterial species, reflecting exceptional opportunities or constraints posed by habitats and relations with companion or competing organisms. Being unable in this chapter to consider more than a very few such differences of detail, we shall take *Escherichia coli* as our primary bacterial exemplar. It is a common gut organism that can grow both in the presence and in the absence of oxygen. It can ferment and it can respire, both aerobically and anaerobically. It thus well illustrates the salient features of chemotrophic bacterial energetics, enabling only significant variations to merit special comment. But before considering mechanisms, we must remind ourselves of some fundamental principles of free energy conservation and exploitation.

Background: Thermodynamic Principles

Spontaneity and Exergony

Free energy is that part of the total energy content of a system that is capable of performing work, and a chemical reaction, catalysed or not, can only occur spontaneously if its occurrence is associated with a decrease in the free energy of the system. The magnitude of this loss in an exergonic reaction is expressed as the decrease in Gibbs free energy ($-\Delta G$), expressed as kJ/mol of reactant transformed, and is a measure of the extent to which the participants in the reaction were initially displaced from their equilibrium state. If the ΔG value were positive, then the reaction would be endergonic and would only proceed if there were an input of the requisite free energy from some external source.

One could predict whether the following reaction could occur spontaneously or not from knowledge of the starting and final (at equilibrium) concentrations of the participants,

$$\mathrm{aA + bB = cC + dD}$$

for if initially the participants were present at certain (observed) concentrations, e.g. $[A]_{obs}$, then their initial mass action ratio Γ (capital gamma) would be:

$$\Gamma = \frac{[C]^c_{obs}\,[D]^d_{obs}}{[A]^a_{obs}\,[B]^b_{obs}}\,\mathrm{mol}^{(c+d-a-b)}$$

and this would be different from the equilibrium state, represented by the equilibrium constant K_{eq}:

$$K_{eq} = \frac{[C]^c_{eq}\,[D]^d_{eq}}{[A]^a_{eq}\,[B]^b_{eq}}\,\mathrm{mol}^{(c+d-a-b)}$$

The free energy change associated with the reaction (ΔG) would then be:

$$\Delta G = -2.3RT\log_{10}\mathrm{K}/\Gamma \text{ (in kJ/mol)}$$

where T is the temperature in degrees kelvin, and R has a constant value such that at 25°C the factor $2.3RT$ approximately equals 5.7 kJ/mol.

In the so-called 'standard state' of the system, each of the components is present at its unit activity (ideal 1 M concentration if a solute or 1 atm pressure if a gas). In this standard (though unrealistic) state the mass action constant (Γ) would equal 1 and the 'standard free energy change' would have a unique value (indicated by a superscript as ΔG°), with

$$\Delta G^\circ = -2.3RT\log_{10}K_{eq}$$

In biological systems, the assumption is often made that the pH of the system is 7 (rather than the excessively acidic pH 0 that would represent the standard unit activity of protons). For reactions at pH 7 a modified standard free energy change is therefore reported, distinguished by a prime, i.e. $\Delta G^{\circ\prime}$.

Although any exergonic reaction could proceed spontaneously, whether or not it will do so at a significant rate is *not* predictable from the magnitude of the associated decrease in Gibbs free energy (i.e. $-\Delta G$ value). For example, were an enzyme to catalyse the reaction it would proceed vastly more quickly but with no change in the outcome (terminal equilibrium state) and no difference in the value of the associated $-\Delta G$; the enzyme would merely have taken the reaction via a different path (of lower activation energy).

Many of the reactions that take place in a living cell are endergonic ($+\Delta G$ value) in character; for example, the synthesis of nucleic acids, proteins and other essential cell components. How then can they ever be accomplished? The answer lies in the coupling of any one such endergonic reaction to another of greater exergony (larger negative value of ΔG) so that the net reaction is exergonic overall. As a simple example, one might consider the synthesis of glucose 6-phosphate from glucose and inorganic phosphate. Under 'modified steady state' conditions of 37°C and pH 7 the ΔG value of this reaction (now formally written as $\Delta G^{0\prime}$) would equal +13.8 kJ/mol and the reaction would therefore be deemed endergonic. Under the same conditions of temperature and pH the hydrolysis of ATP that yields ADP and inorganic phosphate would be exergonic ($\Delta G^{0\prime} = -31$ kJ/mol). It follows that under these conditions the phosphorylation of glucose by ATP would be exergonic and hence capable of proceeding spontaneously (and rapidly when catalysed by the enzyme glucokinase):

$$\text{Glucose} + \text{P}_i = \text{Glucose 6-phosphate} \quad \Delta G^{0\prime} = +13.8\,\text{kJ/mol}$$
$$\text{ATP} = \text{ADP} + \text{P}_i \quad \Delta G^{0\prime} = -31.0\,\text{kJ/mol}$$
$$\text{Glucose} + \text{ATP} = \text{Glucose 6-phosphate} + \text{ADP} \quad \Delta G^{0\prime} = -17.2\,\text{kJ/mol}$$

For convenience rather than accuracy, biochemists have tended to think of metabolites such as ATP as 'energy-rich' compounds whose participation as reactants in metabolic processes would tend to render these the more exergonic. Other 'energy-rich' compounds such as fatty acid thioesters of coenzyme A, or aminoacyl-tRNAs may similarly be considered to be 'activated' in comparison to the component-free fatty acids or amino acids.

Although the practice incurred the disdain of physical chemists, the bond(s) cleaved when these 'energy-rich' compounds participated in metabolic reactions, became known as 'energy-rich bond(s)' and were designated as ~. Thus ATP could be represented as AMP~P~P, with the two terminal phosphoanhydride bonds being distinguished from the 'normal' phosphoester bond in AMP whose hydrolysis under the conditions normally prevailing in living cells was accompanied by a much lesser decrease in Gibbs free energy.

It was clear, however, that the free energy liberated was not to be identified with the actual bond strength (which is actually reflected in the enthalpy change ΔH associated with formation or breakage of the bond). Though deeply flawed, the 'high-energy bond' concept served biologists well as a useful form of shorthand that enabled them the better to comprehend how free energy could be conserved by the formation of certain key metabolites and expended when these 'energy carriers' performed their coupling (driving) roles, in the course of which their immediate product(s) were liberated in a form (e.g. ADP or CoA or tRNA) that could be recharged by phosphorylation or adenylation or acylation. On a broader canvas the picture so built up illustrated how free energy could be acquired by living cells from their environment and 'parcelled up' for metabolic expenditure in a host of anabolic processes.

In these terms, bioenergetics could be considered to be primarily concerned with mechanisms of free energy acquisition and expenditure and the efficiencies thereof, with the necessary accountancy being based on a quantum of free energy equivalent to that required under normal cellular conditions to synthesise one mole of ATP from ADP and inorganic phosphate. It must always be borne in mind, however, that the exergony of ATP hydrolysis (or the converse endergony of ATP synthesis) will depend on the mass–action ratio in which the components of the reaction are actually instantaneously present in the biological system under consideration. The more removed this is from the equilibrium condition the greater will be the 'apparent' ΔG of ATP hydrolysis. Thus at pH 7 in the presence of P_i and Mg^{2+} ions both at 10^{-2} concentration, the 'apparent' equilibrium constant of ATP hydrolysis is about 10^5 M. If in this situation the ratio of concentrations [ATP]/[ADP] was 10^{-2}, then the ΔG of hydrolysis of ATP would be −28.5 kJ/mol. If, however, in otherwise the same situation, ATP were to be present at 10 times the concentration of ADP, then the ΔG of its hydrolysis would be much greater (at −45.6 kJ/mol). Thus the 'quantum of free energy' associated with the hydrolysis of ATP to ADP and P_i, which in practice is usually considered to be about −31 kJ/mol, is not constant in all circumstances so that use of the term 'quantum' is actually inappropriate in this context.

Indeed, although for many purposes it is helpful to conceive of ATP as the currency of energy exchange in contemporary living organisms, it is open to debate whether, in the course of evolution, some other compound(s) may have played equivalent roles. Even in some present day micro-organisms we find vestiges of what 'might have been' in this regard. In particular, there are examples of inorganic pyrophosphate acting in like manner. Thus a pyrophosphate-dependent phosphofructokinase (PP_i-PFK) has been isolated from microbes as diverse as the bacterium *Propionibacterium freudenreichii* and the protozoan *Giardia lamblia*. Even in *Escherichia coli*, inorganic pyrophosphate can effect the phosphorylation of certain proteins, whilst addition of pyrophosphate to a glucose minimal medium has been reported to increase the growth yield of *E. coli* and prolong its survival in the stationary phase of batch culture.

Free Energy and Oxidation–Reduction Processes

The majority of fermentations and the totality of respiratory (and phototrophic) modes of free energy conservation are reliant on oxidation reactions as the source of free energy. Any oxidation consists of electron donation by one redox couple to another that displays a greater affinity for electrons. The redox couple, whether it serves as electron donor (reducing agent) or electron acceptor (oxidising agent), is the equivalent of the half cell of an electrical battery and consists of a reductant plus the conjugant oxidant that is formed when the reductant loses electron(s):

$$\text{reductant} = \text{oxidant} + n\text{e}$$

The redox couple operates reversibly, so that whether or not the above reaction favours production of the oxidant or reductant will depend on whether another redox couple is present to which it is able to donate electrons, or from which it can receive electrons. The outcome will evidently depend on the relative potentials of the two redox couples in terms of their electron affinities.

If a reductant when it is oxidised loses one electron per molecule, then one mole of that reductant will lose (donate) the Avogadro number of electrons (6.023×10^{23}). Aggregated together, this number of electrons constitutes an electrical charge of 96 487 coulombs, so that the quantity of electricity generated per mole of reductant that is oxidised (or that is required to reduce one mole of oxidant) is 96 487 C/mol. This is so useful a unit of measurement in electrochemistry that it has been given its own designation as the Faraday constant (F).

If it is now imagined that two redox couples are connected so as to make up an electrical cell in which n electrons are transferred from the reducing to the oxidising half cell (couple) for every molecule of primary reductant that is oxidised, then at maximum efficiency this cell (battery) could in theory perform a quantity of electrical work equal to $-nF\Delta E_h$ C volt/mol ($= -nF\Delta E_h$ J/mol) where ΔE_h, measured in volts is the difference in redox potentials (redox span) between the two couples. Importantly this quantity equals the decrease in Gibbs free energy associated with the (isothermal) oxidation reduction reaction between the two redox couples.

In summary therefore:

$$\Delta G = -nF\Delta E_h \text{ J/mol}$$

The ability to compare the electron affinities of redox couples by measuring the voltage generated when they are coupled to a standard electrochemical half-cell enables any couple to be assigned a value on a redox scale (whose unit is the volt) which has as its reference zero point the 'standard hydrogen half-cell'. The more positive is this scalar value (redox potential) the greater affinity the couple displays for electrons and therefore the more potent its ability as an oxidant.

The actual redox potential of the redox couple in any situation will depend upon the ratio of concentrations in which its oxidised and reduced constituents are present therein, but when this ratio is unity, i.e. when the oxidised and reduced forms are present in equal concentration, then the redox potential of the couple will display its so-called midpoint value E_m. Then, at the same temperature and pH, when its components are present in any other ratio of concentrations, the redox potential of that couple will be given by:

$$E_h = E_m + 2.3\frac{RT}{nF}\log_{10}\frac{[\text{ox}]}{[\text{red}]}$$

The importance of always declaring the pH at which any redox value is determined and of employing in the above equation the value of E_m at that pH, derives from the fact that in some redox couples the gain of electrons by the oxidant is accompanied by a simultaneous acquisition of one or more protons. Thus, for example, NAD^+ undergoes a two-electron reduction with the gain of one proton:

$$NAD^+ + 2e^- + H^+ = NADH$$

On the other hand, ubiquinone (UQ), which can also undergo a two-electron reduction, does so with the simultaneous addition of two protons:

$$UQ + 2e^- + 2H^+ = UQH_2$$

Consideration of the E_m values of the participant redox couples in any electron transfer process can suggest the sequence in which they might interact. However, it should not be forgotten that it is the E_h values of the couples that are important in practice. Thus although at pH 7 the E'm of the NAD^+/NADH couple at −320 mV would suggest that it should accept electrons from the oxidised/reduced ferredoxin couple whose E'_m at pH 7 is less than −400 mV, in several bacterial fermentations hydrogen gas is produced as a consequence of NADH reducing ferredoxin. This apparent reversal of what might have been expected, is explained by the immediate reoxidation of reduced ferredoxin by protons in a reaction producing hydrogen gas. The actual concentrations of the participants (and thence their working ratios) in these circumstances is such that the E_h value of the $NAD^+/NADH$ couple is actually more negative than the E_h value of the ferredoxin couple, so that NADH can supply the electrons for ferredoxin reduction and ultimately hydrogen formation.

The rates of growth and reproduction of aerobically respiring bacterial cultures can often be constrained by limitation of the rate of supply of oxygen, since this gas is only poorly soluble in water. Indeed, the rapid utilisation of dissolved oxygen by such cultures is evidenced by the fact that the culture E_h (measured with a submerged platinum electrode) will be considerably lower than the +820 mV that would be the E'_m value of the O_2/H_2O couple at pH 7. Following inoculation, the E_h of even a well-aerated batch culture of aerobic bacteria will fall progressively from an initial value of somewhere around +350 mV to perhaps −50 mV, before it rises somewhat as its stationary phase is approached. In the case of facultatively anaerobic bacteria, though initially the culture E_h may fall at just about the same rate, after passing through a zone (around −200 mV) wherein the rate of decrease is slower, it accelerates until the E_h may attain a low final value (about −350 mV is not unusual). Obligately anaerobic bacteria may not even be able to initiate growth and respiration until their growth medium has been deoxygenated and brought to a suitably low initial E_h (below −150 mV) by the addition of an appropriate reducing agent. Thereafter as growth proceeds, the culture E_h will quite rapidly fall still further until, as in those instances in which hydrogen gas is generated, it may approach a value more negative than −400 mV. Monitoring the course of E_h change in a growing bacterial culture may therefore provide information regarding the nature of its energy-conserving metabolism.

Bacterial Fermentations

The chief purpose of any fermentation is to produce ATP by the operation of one or more substrate-level phosphorylation (SLP) reactions. The sequence of cytoplasmic reactions that constitutes the fermentation process therefore always contrives to generate one or more SLP substrates, of which there are relatively few despite the great variety of organic compounds that are subject to fermentation by one living microorganism or another. Of these SLP substrates, carbamoyl phosphate and N^{10}-tetrahydrofolate are important intermediates in relatively few fermentations, for example of allantoin or arginine, and of histidine or various purines, respectively. The four most common SLP reactions are in fact those catalysed by ADP-phosphorylating kinases whose phosphate donators are either phosphoacyl anhydrides (such as acetyl phosphate, butyryl phosphate or bisphosphoglycerate) or the phosphoenolester phosphoenolpyruvate.

A minority of fermentations are hydrolytic in character. An example would be the fermentation of arginine via the arginine deiminase (ADI) pathway that is widely distributed in prokaryotes including lactic acid bacteria, *Bacillus licheniformis*, *Pseudomonas aeruginosa*, *Streptococcus faecalis*, clostridia and spirochaetes.

This is essentially a three-step pathway (**Fig. 2**) whereby arginine is first converted to citrulline (by arginine deiminase) and this is then converted (by ornithine transcarbamylase) into ornithine plus carbamoyl phosphate. An ATP-generating SLP reaction then follows, with carbamoyl phosphate donating its phosphate group to ADP (the reaction being catalysed by carbamate kinase). The free energy-conserving formation of ornithine, carbon dioxide and ammonium ions from citrulline ($\Delta G^{\circ\prime} = -19$ kJ/mol) is further 'pushed' by the preceding irreversible hydrolysis of arginine to citrulline and ammonium ions ($\Delta G^{\circ\prime} = -37.7$ kJ/mol). In most of the bacteria that can operate this pathway, synthesis of its component enzymes is repressed if some alternative means of ATP generation is available to the organism. This could be the presence in the growth medium of an alternative fermentable substrate, or the opportunity for aerobic or anaerobic respiration, as in the case of the denitrifying facultative anaerobes *B. licheniformis* and *P. aeruginosa*. Energy deprivation and the presence of arginine in the culture medium are therefore the chief triggers for the co-ordinated synthesis of the constituent enzymes of this pathway.

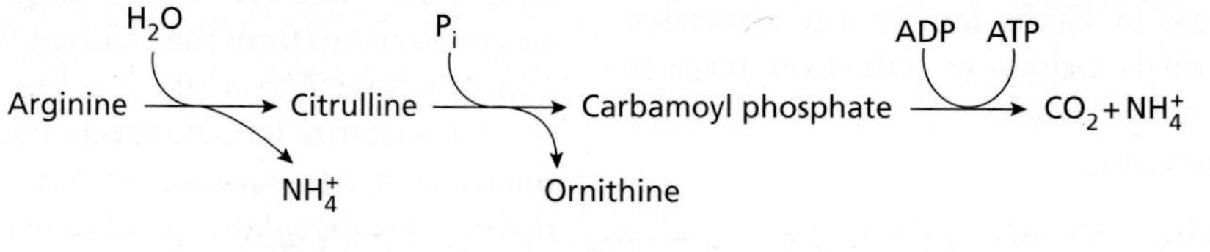

H_2O ; P_i ; ADP → ATP

Arginine → Citrulline (+ NH_4^+) → Carbamoyl phosphate (+ Ornithine) → $CO_2 + NH_4^+$

Fig. 2 Fermentation of L-arginine via the arginine deiminase pathway. Arginine is hydrolysed to form citrulline, which is then converted to ornithine and the SLP substrate carbamoyl phosphate. Phosphotransfer from this intermediate to ADP yields ATP with concurrent production of CO_2 and NH_4^+.

The great majority of microbial fermentations are balanced oxidation-reduction processes, however, with the major organic fermentation products being the reduced compounds formed from those substances (intermediates or added compounds) that serve as the terminal electron acceptors. In the type of amino acid fermentation undertaken by many proteolytic clostridia one amino acid is oxidised at the expense of a second amino acid that is simultaneously reduced (a process named after its discoverer as the Stickland reaction). Stickland himself in the early 1930s, using washed cell suspensions of *Clostridium sporogenes*, noted that when a suitable pair of 'hydrogen-donating' and 'hydrogen-accepting' amino acids was available, the donor amino acid was oxidised and deaminated whilst the acceptor was deaminated but reduced. Thus alanine was oxidised to give rise to acetate, carbon dioxide and ammonia when ornithine was concurrently reduced to give ammonia and γ-aminovaleric acid. By means of a non-amino acid utilising fermentation the anaerobe *Clostridium kluyveri* also can obtain its free energy from a supplied mixture of reductant and oxidant, in this case the C_2 compounds ethanol and acetate, the chief end product of this fermentation is the C_4 fatty acid butyrate, serving as a reminder that not all fermentations are degradative in character.

More frequently though, both reductant and oxidant are produced in the course of the fermentative metabolism of a single substrate that serves as their common precursor. Thus in the homolactic fermentation of glucose, pyruvate is formed by its glycolytic oxidation via the Embden-Meyerhof-Parnas (EMP) pathway. However this oxidation can only be sustained by the balancing reduction of the pyruvate to yield the excreted and accumulated lactic acid (**Fig. 3**).

It is this use of pyruvate as the terminal electron acceptor that regenerates NAD^+ from the NADH produced during oxidation of the primary electron donor (glyceraldehyde 3-phosphate), and it is this 'virtuous cycle' of NAD^+ reduction and NADH re-oxidation that enables this couple to serve as the electron carrier of the EMP pathway. But it is the two SLP reactions of the EMP pathway that are its *raison d'être*, namely, the phosphate transfers to ADP first from 1,3-bisphosphoglycerate and then again from phosphoenolpyruvate. Preliminary 'activation' of the glucose to bring it into the condition (as fructose 1,6-bisphosphate) that it could be cleaved to yield two molecules of triose phosphate consumes two ATP molecules per glucose utilised. With the subsequent SLP reactions yielding four ATP molecules per glucose utilised, the overall (net) yield of ATP via this fermentation route is two ATP molecules per glucose. It is interesting that the EMP pathway has been highly conserved throughout evolution and functions in many organisms, both uni- and multi-cellular. These include not only many fermentative micro-organisms, but also respiring cells including those of higher plants and animals wherein it can serve as the source of pyruvate which in turn yields the acetyl-CoA that is fully oxidised to carbon dioxide and water by the tricarboxylic (citric) acid cycle.

Various routes of fermentation of the one substrate can be employed by different organisms. For example, while the fermentation of glucose to ethanol and carbon dioxide when it is accomplished by the brewing yeast *Saccharomyces cerevisiae* proceeds via the

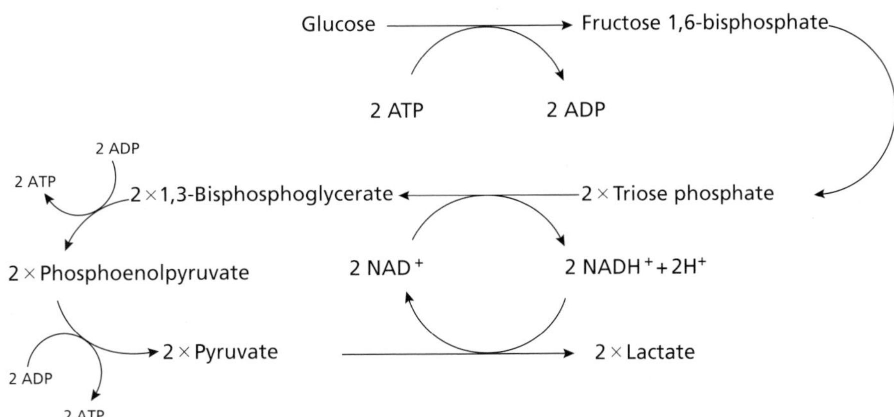

Fig. 3 Homolactic fermentation of glucose via the Embden-Meyerhof-Parnas (EMP) glycolytic pathway in which NAD^+/NADH serves as the 'redox carrier' and there is a net yield of 2 ATP/glucose utilised. Production of ATP from ADP is accomplished in two SLP reactions, in which the fermentation intermediates 1,3-bisphosphoglycerate and phospho*enol*pyruvate serve as the phosphodonors.

EMP pathway and yields 2ATP/glucose, the same fermentation when undertaken by the bacterium *Zymomonas mobilis* proceeds via a different route (the Entner-Doudoroff pathway) which gives rise to only 1ATP/glucose. In heterolactic bacteria, glucose is fermented by a third pathway (the phosphoketolase pathway) to yield some lactate but also acetate and ethanol plus carbon dioxide. In this route, initial phosphorylation of glucose is followed by oxidation to yield 6-phosphogluconate (a dehydrogenation reaction generally employing $NADP^+$ as electron acceptor). Decarboxylation of the 6-phosphogluconate yields ribulose 5-phosphate which is converted to xylulose 5-phosphate which, by the action of a phosphoketolase, is split into glyceraldehyde 3-phosphate and acetyl phosphate. This causes bifurcation of the pathway, for the glyceraldehyde 3-phosphate can proceed via the EMP route to pyruvate and thence to lactate, whilst the acetyl phosphate, after conversion to acetyl-CoA, can be reduced to ethanol. The net yield of ATP in this fermentation would be 1ATP/glucose. However, if an alternative (ancillary) electron acceptor was available, then the necessary redox balance could be sustained without the need to employ all of the acetyl phosphate as an electron sink. The residue could be directly used in an SLP reaction to generate ATP (with acetate now being produced in place of ethanol). To the extent therefore that acetate appears in the products of this heterolactic fermentation, the yield of ATP is greater than the minimal 1ATP/glucose. This example helps to illustrate the greater flexibility of the branched fermentation pathway whose end products may be formed in different proportions depending on, amongst other things, the availability of ancillary electron acceptors. This means that it is not always easy to predict the energy efficiency of such fermentations, though, as a rule of thumb, the more oxidised the end product, the greater the likely specific yield of ATP.

This principle is again evidenced in the fermentations of starch or sugars by anaerobic saccharolytic clostridia such as *Clostridium butyricum* (Fig. 4). The sugar is glycolysed in the usual manner via the EMP pathway and the pyruvate is decarboxylated to yield acetyl-CoA plus hydrogen and carbon dioxide. The acetyl-CoA can either (a) be converted to acetate with concomitant production of 1ATP/acetyl-CoA so utilised, or (b) produce acetoacetyl-CoA which is thereafter reduced in a series of reactions leading to butyrate but generating only 0.5ATP/acetyl-CoA. In practice, the acetyl-CoA is partitioned between the two branch routes such that under normal batch growth conditions 1 mol glucose gives rise to 0.6 mol acetate and 0.7 mol butyrate (together with carbon dioxide and hydrogen) so that about 3.3 mol ATP would

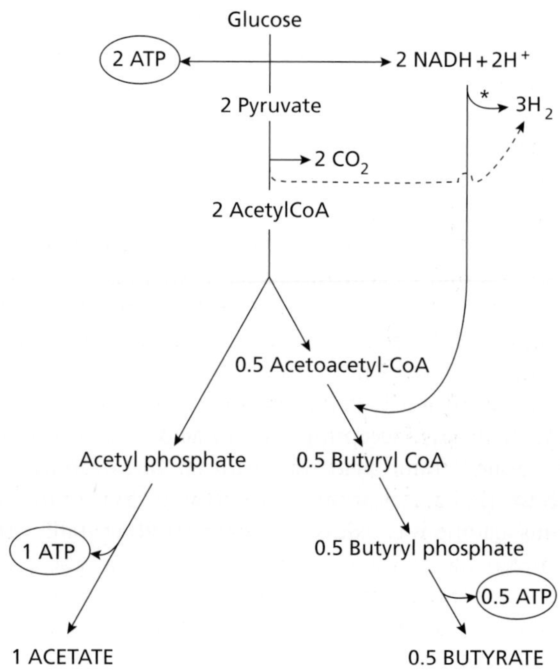

Fig. 4 The branched acidogenic pathway of glucose fermentation by *Clostridium butyricum*. The activity of the NADH-ferredoxin reductase (*) determines what fraction of the available NADH is diverted to the production of hydrogen gas. In turn, this decides both the ratio in which acetate and butyrate are formed and (consequently) the net yield of ATP/glucose consumed.

simultaneously be generated. This regulation of the efficiency of free energy conservation is accomplished by control exercised primarily on the activity of the NADH-ferredoxin oxidoreductase which serves to divert some of the NADH formed in the course of the fermentation to the formation of more gaseous hydrogen than the 2 mol H_2/mol glucose that would result from the ferredoxin-dependent oxidation of pyruvate to acetyl-CoA. Since the conversion of acetyl-CoA to butyrate is dependent upon the availability of NADH, the proportion of this that is diverted to H_2 production will affect the yield of butyrate. Both NADH and acetyl-CoA therefore stand at branch points in this fermentation pathway and their fates are intertwined due to activation by acetyl-CoA of the NADH-ferredoxin oxidoreductase and reinforcing inhibition of this enzyme by free coenzyme A.

Many anaerobic bacteria are indirectly benefited by their ability to produce gaseous H_2 for it serves to sustain a desirably low redox potential in their growth environment and may enable the producer organism (hydrogenogen) to enter into mutually beneficial associations with other H_2-consuming microbes (hydrogenotrophs). The much more direct benefit, however, is that the ability to dispose of excess reducing equivalents as H_2 gas relieves the organism from the necessity to generate (or to obtain from elsewhere) sufficient organic electron acceptor(s) to cope with all of the electron flux emanating from fermentative utilisation of the reduced substrate. This means that more highly oxidised end products can be accumulated with proportionately greater specific yields of ATP. It may also allow the organism to ferment more highly reduced substrates whose utilisation would present problems for organisms which did not possess such means of disposing of the 'excess' reducing equivalents.

An alternative strategy to the controlled release of H_2 gas is evidenced in the once commercially important acetone-butanol fermentation of starch or sugars by the anaerobe *Clostridium acetobutylicum*. During growth in batch culture this bacterium carries out a biphasic fermentation. The first acidogenic phase produces both acetic and butyric acids (with H_2 and CO_2), so causing the pH of the culture medium progressively to fall; as in the similar fermentation undertaken by *C. butyricum* this phase is associated with the production of about 3.3ATP/glucose. However when the pH has fallen sufficiently, a second solventogenic phase of the fermentation ensues, during which some of the previously accumulated acetate and butyrate is utilised alongside freshly metabolised glucose with production of the more reduced neutral solvents acetone plus *n*-butanol. During this solventogenesis there is correspondingly less formation of gaseous H_2 and the net yield of ATP is also smaller. In an even more extreme example of use of a 'first formed' fermentation product as an ancillary electron sink, the so-called homoacetogenic anaerobic bacteria are able to reduce two molecules of carbon dioxide to produce one of acetate.

The point has already been made that some substances that are 'too reduced' to be fermentatively exploited by some bacteria may become available for this purpose if the bacterium is able to employ some other constituent of its growth medium as an electron sink. Thus the growth of many heterolactic bacteria is aided by the presence of acetate which can serve as an electron acceptor (forming ethanol), thus enabling them to ferment even sugar alcohols such as mannitol. A variant sees the fermentation substrate itself taking on the role of electron sink, as in the fermentation of fructose by *Lactobacillus brevis* where part of the fructose is simultaneously reduced to mannitol. In the presence of an alternative and preferred electron acceptor (for example acetoin) a much higher proportion of the fructose is fermented, making for greater efficiency of its utilisation.

In summary therefore, linear (unbranched) fermentation pathways would at first sight seem to be predictable in the efficiency of their free energy conservation. Thus the homolactic, EMP-based fermentation of glucose would be associated with the production by SLP reactions of 2ATP/glucose. Yet it proves to be rather more complicated and hence uncertain than this would indicate. For example, it would be rare to find an organism operating an entirely homolactic fermentation if only because any diversion of NADH to the reduction of biosynthetic intermediates or of 'exogenous' oxidants present in the environment would allow some acetate to be produced in addition to the lactate. Additionally, under certain circumstances more free energy conservation (equivalent to the production of additional ATP) could arise from the excretion of product(s) of the fermentation if this translocation were to generate a transmembrane ion motive force (see p. 250).

The second, perhaps even more unexpected complication, comes from the fact that there are many examples of fermentation processes that generate the same products from the same substrates and hence, it might be assumed, would do so with the same free energy conserving efficiency. Yet this superficial identity may conceal very different pathways of metabolism and quite disparate specific yields of ATP. The different ways in which yeast and *Zymomonas* produce ethanol and carbon dioxide from glucose have already been noted, and similar examples of very different metabolic routes arriving at the same end products are to be found in fermentations of non-carbohydrate substrates also. For example, the anaerobic bacterium *Clostridium tetanomorphum* ferments glutamate via methylaspartate in a vitamin B_{12}-dependent pathway, whilst another anaerobic bacterium, *Fusobacterium fusiformis*, ferments glutamate to the same end products but by a wholly different route in which α-hydroxyglutarate and glutaconyl-CoA are key intermediates. On the other hand, different fermentative bacteria can produce distinctively different end products from the same substrate even though they employ the same main fermentative pathway. The differences then result from the distinctive ways in which they regenerate the intermediary electron carrier. Thus pyruvate formed by the EMP pathway can, by different bacteria, be employed either directly or indirectly as the terminal electron sink, forming, as we have noted, lactate (directly) or acetate and ethanol (after decarboxylation), but also malate (after preliminary carboxylation) or a host of other derivatives (see **Fig. 5**). Which of these products are formed and in what proportions will determine the overall specific ATP yield of the fermentation.

Finally it must be recognised that environmental influences can exert quite dramatic effects on

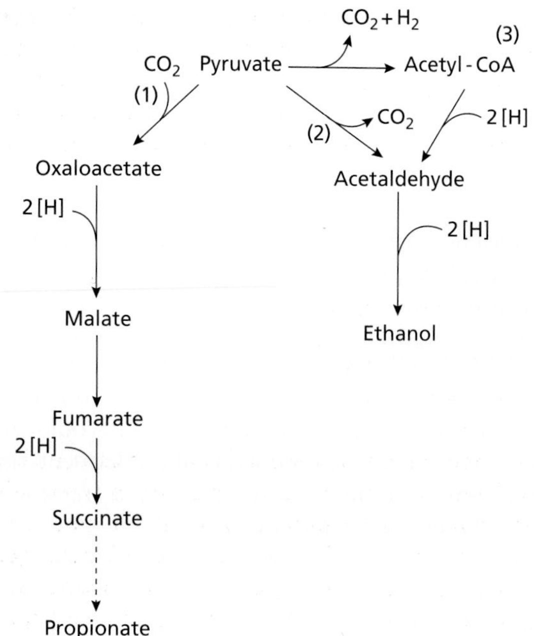

Fig. 5 Some alternatives to the direct homolactic reduction of pyruvate (that consumes 2[H]/pyruvate). (1) Following carboxylation of pyruvate to yield oxaloacetate, reduction of this to produce succinate consumes 4[H]/initial pyruvate. (2) Decarboxylation of pyruvate can yield acetaldehyde whose reduction to ethanol consumes 2[H]/initial pyruvate. (3) Oxidative decarboxylation of pyruvate to acetyl-CoA followed by reduction of this to ethanol consumes 4[H]/initial pyruvate. Other pathways yield other reduced products such as acetoin, butanediol, butanol and isopropanol.

fermentation processes and specific yields of ATP. For example, under conditions of glucose limitation, as the growth rates of cultures of *Streptococcus cremoris* and *Streptococcus lactis* decrease, their fermentations switch from a homolactic to a heterolactic type. Or again, when the pH of its growth medium is changed from a slightly acidic to a slightly alkaline value, lactate formation by *Streptococcus faecalis* subsp. *liquefaciens* is halved, with alternative formation of acetate, ethanol and formate.

The Mixed Acid Fermentation of Glucose by *Escherichia coli* and Other Enterobacteria

The anaerobic fermentation of glucose by *Escherichia coli* yields acetic, formic, lactic and succinic acids together with some ethanol. As much as 85% of the glucose that is fermented is initially converted to pyruvate via the EMP pathway. The pyruvate is then in part degraded by the action of a CoA-dependent pyruvate formate lyase, encoded by the *pfl* gene, whose expression is dependent on the activities of phosphoglucoisomerase and phosphofructokinase. The pyruvate formate lyase is an exceptionally interesting enzyme which cleaves pyruvate as follows:

$$CH_3COCOOH + CoASH \rightarrow CH_3COSCoA + HCOOH$$

The reaction proceeds by a unique homolytic mechanism that involves a glycyl free radical, the requirement for which renders the enzyme susceptible to irreversible damage by exposure to oxygen. To preserve the pyruvate formate lyase from this fate it can be reversibly deactivated by another enzyme that forms part of an Fe^{2+}-requiring polymeric protein complex that is encoded by the *adhe* gene, so designated because the same protein complex is responsible for the aldehyde- and alcohol-dehydrogenase activities of *Escherichia coli*. The task of reactivating the deactivated pyruvate formate lyase falls to yet another enzyme which requires *S*-adenosyl methionine and dihydroflavodoxin as co-substrates.

The pyruvate formate lyase is only synthesised when the bacterium is cultured anaerobically, for transcription of its encoding gene is under the control of the FNR and ArcA proteins that are pleiotropic activators of genes specifically implicated in the anaerobic metabolism of the organism. The flux of organic intermediates through the fermentation pathway (which possibly finds expression in the intracellular concentration of pyruvate) also affects the expression of the pyruvate formate lyase of *E. coli*. The multilayered control of the synthesis and activity of this anaerobic enzyme proclaims its key role and illustrates the fact that rates of fermentative metabolism (and hence of ATP generation) can be subject to a number of controls that ensure that the immediate needs of the organism are being met in an appropriate manner.

In both *Escherichia coli* and *Enterobacter aerogenes* the formate so produced can be catabolised (via two enzymes working in concert and known collectively as the formate hydrogenlyase complex) to give rise to carbon dioxide and hydrogen. *Salmonella typhi* does not possess this ability and hence produces and excretes formate. The acetyl-CoA can either be converted to the SLP substrate acetyl phosphate or it can act as an electron acceptor and be reduced to acetaldehyde and then ethanol. The D-lactic acid of the fermentation product mix arises as the result of direct reduction of some pyruvate by an NADH-linked D-lactate dehydrogenase whose synthesis is induced under anaerobic conditions, especially at low pH values. *E. coli* actually possesses three lactate dehydrogenases but two of these are membrane-bound flavoproteins (one for L- and the other for D-lactate) that enable lactate to be oxidised by the respiratory chain. The fermentative D-lactate dehydrogenase (encoded by the *ldhA* gene) is however located in the cytoplasm and is allosterically activated by its substrate pyruvate.

The enterobacteria clearly illustrate how microbial taxonomists have traditionally exploited differences in fermentation patterns (both in the products formed from the one substrate and in abilities to utilise different substrates) for the purposes of species identification. Thus, unlike *Escherichia coli*, *Enterobacter aerogenes* produces some acetoin during its fermentation of glucose. *Salmonella typhimurium* will grow anaerobically on citrate as sole source of carbon and energy, but *E. coli* can only ferment citrate when an ancillary carbon source (glucose, pyruvate, fumarate) is additionally provided. This is because the citrate-cleaving enzyme (citrate lyase) of *E. coli* though induced in the presence of citrate is, in the absence of a suitable co-substrate, quickly inactivated by deacetylation of its active centre.

An initially surprising finding was that under conditions of phosphate imbalance, when metabolism of glucose was still proceeding through the early (triose phosphate-forming) reactions of the EMP pathway, the later steps in this pathway could be bypassed by the formation of methylglyoxal from dihydroxyacetone phosphate. The methylglyoxal in turn was reduced to D-lactate and this was oxidised to produce pyruvate. This 'methylglyoxal bypass' (involving the enzymes methylglyoxal synthase, glyoxalase and flavin-linked D-lactate dehydrogenase) would effect the conversion of triose phosphate to pyruvate without the concomitant synthesis of ATP (Fig. 6). However, by liberating inorganic phosphate from dihydroxyacetone phosphate it would provide the phosphate required to generate ATP by oxidation of the companion glyceraldehyde 3-phosphate.

Chemiosmosis and Electrophosphorylation in Bacteria

As we have noted, the second major means of biological free energy conservation depends on the establishment and maintenance of unequal concentrations of particular ions across a cell membrane that is impermeable to these ions. The unequal concentrations (activities) of the ion in the two compartments separated by the impermeable membrane is the basis of an ion motive force which would tend to eradicate this inequality.

Two major questions can be asked regarding any such energetic transforming process: (a) how is the transmembrane ion gradient established and sustained and (b) how is the consequential ion motive force exploited to perform useful work?

Early studies of respiratory processes in living cells recognised the key role played by membrane-localised electron transport through sequences of electron carriers of progressively increasing redox potential.

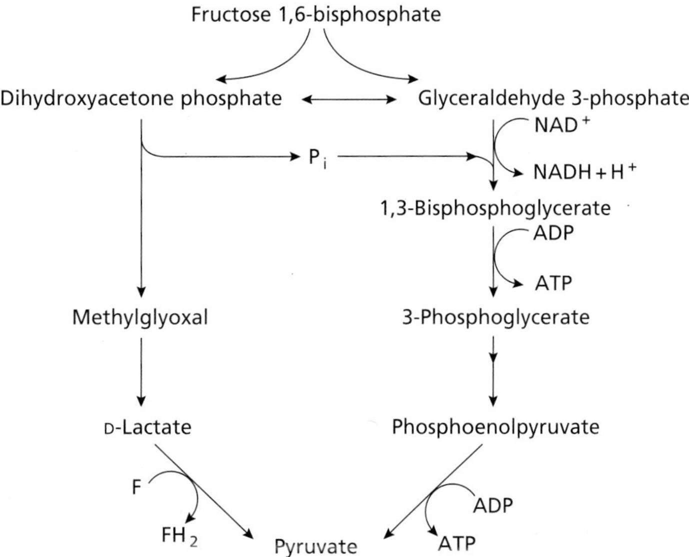

Fig. 6 A variant of EMP glycolysis that involves operation of the 'methylglyoxal bypass'. Note that the D-lactate dehydrogenase is flavin-linked and that the conversion of methylglyoxal to pyruvate is not associated with any (SLP) generation of ATP.

In the case of the oxidation of NADH by molecular oxygen (in aerobic respiration) the primary electron donor (E'_m at pH 7 of NADH/NAD$^+$ couple $= -320$ mV) is ultimately oxidised by O_2 (E'_m at pH 7 of O_2[1atm]/$2H_2O$ [55 M] couple $=$ +820 mV) which theoretically should mean that free energy of up to about 220 kJ/mol could be available to the respiring cell. Initially, with the mechanism of fermentative substrate level phosphorylation very much in mind, investigators instituted a search for one or more 'high-energy compounds' whose formation, if linked to the operation of the membrane-localised electron transport chain, could explain the transformation of part of the free energy of this oxidation process into the 'metabolic free energy' represented by an SLP substrate. However the search was frustratingly futile and it took the remarkable insight of Peter Mitchell to make the connection between respiratory electron transport in the membrane and the concurrent generation of a transmembrane ion motive force in which part of the 'liberated' free energy is conserved. He was led to this view by his previous studies on the mechanisms of 'active transport' of both ions and uncharged compounds across biological membranes in the face of countervailing concentration gradients of the transported molecule. Indirectly, these studies also led to his recognition that a transmembrane electrochemical gradient of one ion (for example, protons) could be exploited to effect co-transport or counter-transport of other ions and molecules. When a proton gradient is created (whether across a bacterial cell membrane, or the equivalent in eukaryotic mitochondria and plastids) this energy-transducing membrane: (1) is virtually proton impermeable; (2) accommodates a sequence of couples of increasing redox potential with passage of electrons along this chain being associated with the vectorial movement of protons from one side of the membrane to the other; and (3) provides a route for the spontaneous return of the accumulated protons in the form of a proton-conducting ATP synthase which is thereby impelled to generate ATP by the coupled phosphorylation of ADP.

The proton electrochemical gradient ($\Delta\mu_{H+}$) measured in kilojoules per mole is the thermodynamic expression of the magnitude of the protonic disequilibrium across the membrane. It is this that is the basis of the proton motive force which seeks to drive the system to equilibrium and so eradicate the difference in proton concentration on both sides of the membrane. It is usual therefore for bioenergetic purposes to express the magnitude of the proton electrochemical gradient ($\Delta\mu_{H+}$) as its equivalent in millivolts and to term this the proton motive force (Δp). So,

$$\Delta p = \Delta\Psi - Z\Delta\text{pH}$$

where $\Delta\Psi$ is the electrical charge difference, ΔpH is the pH difference, and Z is the constant $2.3RT/F$, in which R is the gas constant, T the temperature in degrees kelvin, and F the Faraday constant. At 25°C, $Z\Delta$pH therefore has a numerical value of approximately 60 mV per pH unit. It should also be noted that, because the direction of spontaneous proton flow will be from the acid to the more alkaline compartment (of higher pH), the $Z\Delta$pH component acquires a negative sign in the above equation. Though both components (i.e. $\Delta\Psi$ and $-Z\Delta$pH) contribute to the proton motive force and, in bacteria at least, are interconvertible, they are evidently not identical and the rates at which the respiring membrane can generate these components differ markedly. With the biological insulating membrane having a low electrical capacity, only a few charges have to be vectorially translocated for the generation of a substantial $\Delta\Psi$. However, many more protons would have to be unidirectionally moved to produce from a $-Z\Delta$pH component alone the same millivoltage difference across the membrane. This is particularly the case in a bacterium whose cytoplasm is so buffered that removal of protons will only gradually cause an increase in pH, whilst expulsion of these protons into a large external volume ('Pacific Ocean') may only slightly decrease the external pH. The common situation is that $\Delta\Psi$ is the predominant component of Δp though there may be occasions where Δp exists solely as a ΔpH across the energy-transducing membrane, in which case the inner (cytoplasmic) N-phase may be as much as three pH units more alkaline than the outer P-phase.

The maintenance and effectiveness of the transmembrane proton motive force are dependent on the membrane displaying a high resistance to the diffusion of protons through it. Synthetic compounds that destroy this background impermeability by carrying protons through the membrane (for example, dinitrophenol) are known as protonophores, but also as uncoupling agents because they serve to disengage synthesis of ATP from the initial generation of a proton electrochemical gradient by dissipating the latter. Other organic compounds can act as ionophores by forming lipid-soluble complexes with cations (such as K^+ or Na^+) and rapidly equilibrating these across the cell membrane, thereby short circuiting the membrane potential. Valinomycin acts in this fashion to equalise the concentrations of K^+ ions across a cell membrane and so, by electrogenically carrying K^+ into the cell, it establishes a potential opposite to that of the membrane potential $\Delta\Psi$, and hence nullifies it. Nigericin, on the other hand, accomplishes an electroneutral exchange of K^+ for H^+, so dissipating the ΔpH but not the $\Delta\Psi$. Various types of channel-forming proteins (several bacteriocins) owe their lethality to their ability to discharge the membrane potential of a susceptible bacterium. Thus colicins A, B, E1, Ia, Ib and N all produce voltage-gated ion-conducting channels across the cytoplasmic membrane of *E. coli*, causing its death by the resulting dissipation of the transmembrane Δp.

We can now begin to comprehend why this form of biological free energy conservation has become associated with terms such as 'chemiosmosis', which emphasises the conservation of free energy in the form of unequal ion charge and concentration differences across an insulating impermeable membrane, and 'electrophosphorylation', which directs attention to the secondary capture of some of the free energy of the ionmotive force by the coupling of spontaneous ion flux to the phosphorylation of ADP by the membrane-associated ATP synthase.

Although the primary ion gradient that is established across a biological energy-transducing membrane is generally a proton gradient, in some instances another ion can take on the same role. For example, the thermophilic, amino acid-fermenting anaerobic bacterium *Clostridium fervidus* relies exclusively upon its creation of a transmembrane electrochemical gradient of sodium ions, created by the activity of a membrane Na^+-ATPase.

The Electron Transport Chain in Aerobic Respiration and its Coupling to Transmembrane Proton Translocation

Although there is great diversity in the detail of the electron transport chains in energy-transforming membranes, especially in the case of bacteria, some basic features are common to all. The over-riding principle is that electrons are passed along a sequence of redox couples of increasing electron affinity which intervene between a low potential primary electron donor and a terminal electron sink of higher potential. The purpose is to convert a redox potential difference (ΔE_h) into a proton electrochemical potential difference (Δp), and the employment of several serial electron carriers enables an oxidative process that covers a large redox span to yield up its associated free energy loss in a stepwise manner. Only certain of these steps however are so-called 'coupling sites' associated with the vectorial translocation of protons.

Initial studies of (respiratory) electron transport chains were therefore concerned not only with identifying the nature of their redox components and the sequence in which these interacted, but also with identifying which of the reduction-oxidation steps was proton motive and by what mechanism was this proton movement accomplished. It emerged that however different they might be in detail, respiratory chains were constructed to a basic plan in which some membrane-associated dehydrogenase complex catalyses the transfer of reducing equivalents from a primary electron donor (for example, NADH) to a common, mobile pool of quinone within the membrane. This in turn is re-oxidised by one or more oxidase complexes that reduce the final electron acceptor (dioxygen in aerobic respiration, but some alternative such as nitrate, sulphate or fumarate in the respiration practised by certain anaerobic bacteria).

Now, electrons can be transferred from a reductant to an oxidant either alone (by an e^- carrier) or together with protons (by an H^+ plus e^- carrier); examples of both modes are found in respiratory chains. The question was therefore asked whether concurrent transport of protons was the consequence of the operation of 'proton motive loops' or 'proton pumps'. In the former case (the loop) it was proposed that electron transfer would occur from an H^+ plus e^- carrier to a pure electron (e^-) carrier, these being arranged within the respiratory membrane in such a fashion that the accompanying release of protons (H^+) was directed to the outer (P) side of the membrane. Fresh reduction of the H^+ plus e^- donor would be effected from an electron (e^-) carrier of lesser redox potential, with the required protons being abstracted from the inner (N) side of the membrane.

Proton translocation via a conformational proton pump was, in contrast, envisaged to involve a redox carrier embedded within a more flexible protein which underwent redox-induced changes in acidity (pK_a value). In this way it was conceivable that redox change in the carrier could be co-ordinated with changes in its protonation and conformation which would ensure the ejection of the liberated protons on the outer (P-side) of the membrane. Studies performed with the respiratory mitochondria of plant and animal cells showed that both loop and pump types of proton translocators were indeed implicated and that three of the four major complexes served as coupling sites.

Electron transport in the respiratory bacterial membrane is catalysed by multi-protein enzyme complexes, the electrons being transferred between prosthetic groups bound to certain of these proteins (Fig. 7). Flavoproteins containing FMN or FAD serve as hydrogen carriers, iron/sulphur proteins as electron carriers, quinones as hydrogen carriers and

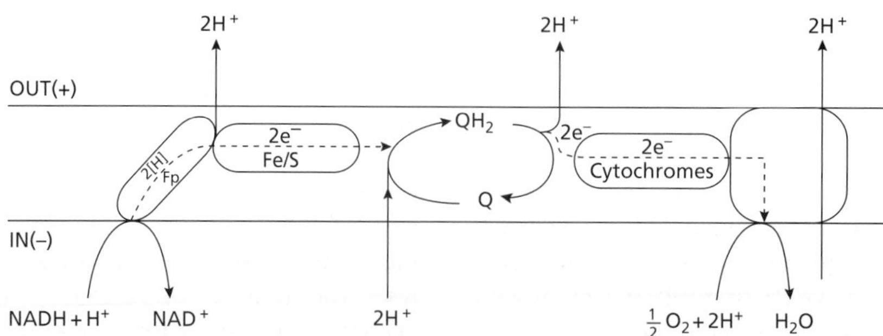

Fig. 7 Outline scheme of a cell membrane-located respiratory pathway effecting the aerobic oxidation of $NADH_2$ (supplied from the cytoplasm), and demonstrating the participation of both 'loop' and 'pump' types of proton translocation. The $NADH + H^+$ donates 2[H] to the flavoprotein dehydrogenase complex (FP) which serves as a coupling site evicting $2H^+$ to the outside of the membrane. The residual $2e^-$ are accepted by an iron–sulphur protein complex, and are then employed, together with $2H^+$ imported from the cytoplasm, to reduce a mobile (diffusible) quinone (Q). The reduced quinone (QH_2) delivers up $2H^+$ at the outer surface of the membrane, with the $2e^-$ being accepted by cytochrome(s). The quinone 'loop' is completed by return of Q to the inner membrane surface. The $2e^-$ are accepted from the cytochrome(s) by a cytochrome oxidase which at the inner membrane surface employs them, with $2H^+$ abstracted from the cytoplasm, to reduce oxygen to water. Simultaneously $2H^+$ are exported by the cytochrome oxidase acting as a proton 'pump'.

cytochromes (haem proteins) as electron carriers. The usual arrangement is that an initiating dehydrogenase complex transfers electrons from a primary electron donor and donates these (along with protons) to a quinone. The reduced quinone hands on the electrons to a terminal oxidase complex via intervening cytochromes. The nature of the terminal respiratory oxidant determines the nature of the final oxidoreductase in the electron transfer chain. This can be an oxidase, as when dioxygen is reduced by a cytochrome oxidase, or a specific membrane-located reductase, as when nitrate or fumarate serve as terminal electron sinks in anaerobic respiration. Complications can also arise at the opposite (start) end of the respiratory chain, for whilst reductants such as $NADH_2$ or $FADH_2$ are synthesised in the cytoplasm and oxidised at the inner surface (N-side) of the cell membrane, there are many instances in Gram-negative bacteria of reductants being oxidised in the periplasm on the P-side of the cell membrane. Such substrates include hydrogen gas, methane, methanol, formate and both ferrous (FeII) and sulphide ions.

The 'mobile pool' of quinone in the respiratory membrane thus plays a most important role, as it can freely move between its partner electron donors and acceptors. In bacteria, these quinones are of two types: Gram-positive and anaerobic Gram-negative bacteria contain only menaquinone (MK, or vitamin K_2) or demethylmenaquinone (DMK), while the majority of obligately aerobic Gram-negative bacteria contain only ubiquinone (UQ). Facultatively anaerobic bacteria contain both types. Thus *E. coli* possesses UQ-8, MK-8 and DMK-8 (where the numeral indicates the number of isoprenoid side-chains on the core quinone). When the bacterium grows aerobically, the intracellular UQ-8 concentration is almost five times greater than the summed concentrations of MK-8 and DMK-8, whilst under anaerobic growth conditions the latter are some 3-fold greater than the UQ-8. Reduction and oxidation of the quinone involves transfer of two electrons with the accompanying addition or release of two protons, so that ubiquinone (UQ) can be reduced to ubiquinol (UQH_2) and this re-oxidised to UQ. However a semiquinone radical ($UQH^{\cdot -}$) can be produced by removal or addition of a single electron and proton and this intrinsically very reactive intermediate can be stabilised by being bound to an appropriate protein, for example the UQH_2 oxidase (cytochrome *bd'*) of *E. coli*. This enables the quinones to link one-electron and two-electron transferring components of a respiratory chain.

Whereas UQ can accept electrons from a broad range of primary electron donors including NADH, succinate, D-lactate, glycerol 3-phosphate, formate and even hydrogen, MK is somewhat more selective; it does not, for example, serve as an oxidant for succinate or D-lactate. On the other hand, in *E. coli*, MK is essential for formate-dependent nitrite reduction and, more generally, functions in anaerobic respiration involving fumarate, dimethylsulphoxide or trimethylamine oxide. Either UQH_2 or MKH_2 (but not $DMKH_2$) can transfer electrons to the nitrate reductase employed during nitrate respiration. In part, the significant difference between the two major quinone redox couples is that the UQ/UQH_2 couple with an E'_m of +113 mV is more oxidising than the MK/MKH_2 couple whose E'_m is −74 mV.

Routes for respiratory electron transfer in the (energised) cell membranes of bacteria can vary greatly in detail, not only between different organisms but also in the one organism depending upon how it is grown and on what substrates. In particular, bacterial respiratory chains may be branched, offering a choice of routes for electron transport which may differ in the number of coupling sites associated with them. Furthermore, their composition can be determined by the rate and manner of growth of the bacterium as well as by the nature of the respiratory reductants and terminal oxidants that it employs. It is evident that quite sophisticated regulatory elements operating at the level of gene expression must control the responses of the organism to the opportunities presented to it by the possible presence in its growth environment of potentially competing respiratory substrates. For example, certain dehydrogenases that function at the start of bacterial respiratory chains may only be synthesised anaerobically, and certain enzymes that may be crucial under aerobic growth conditions may not be synthesised under anaerobic conditions, and vice versa. The more extensive is its repertoire of energy-conserving mechanisms the more important to a bacterium will be the part played by such regulatory mechanisms in effecting their co-ordination and efficient operation.

The Respiratory Chain of *Escherichia coli* – a Facultatively Anaerobic Bacterium

Escherichia coli is highly versatile in the ways that it can procure the free energy that it requires to sustain its vitality, growth and reproduction. In the absence of any alternative way of making ATP, it is capable anaerobically of breaking down many sugars and related compounds in acid-yielding fermentations. Yet if oxygen is present it will preferentially fully oxidise these and other organic compounds by aerobic respiration. Even when oxygen is not available this bacterium can press into service several alternative oxidants to serve as terminal electron sinks in an equivalent anaerobic respiration. Such alternative oxidants include nitrate, fumarate, trimethylamine oxide and dimethylsulphoxide. If two or more of these compounds are available, the organism does not utilise them simultaneously. Instead it consumes them in an order of preference dictated by their relative electron affinities. Thus nitrate is used in preference to fumarate which in turn would be consumed before trimethylamine oxide. The exercise of this serial choice is evidently of service to the bacterium in that it ensures that the free energy yield from any (reduced) respiratory substrate is maximised.

The regulatory mechanism which ensures that oxygen, if present, is always the first choice oxidant and that others are used in order of decreasing redox potential involves controls exerted on the expression of genes specifying key enzymes of respiration (and fermentation). Regulatory DNA-binding proteins (such as Fnr and Arc) in their active states serve as transcriptional control elements responsive to the availability of oxygen in the environment. Other regulatory controls ensure that specific terminal respiratory oxidases are only synthesised when their appropriate electron-accepting substrates are available and when no preferred alternative is present. Thus dioxygen represses the synthesis of fumarate and nitrate reductases, whilst nitrate represses the synthesis of fumarate reductase.

The respiratory chain of aerobically grown *E. coli* consists of membrane-associated dehydrogenases (quinone reductases) that reduce the mobile ubiquinone in the cytoplasmic membrane, and two ubiquinol oxidases (cytochrome oxidases bo′ and bd), that reduce dioxygen to water (Fig. 8).

The existence of two terminal oxidases creates a branch point in the electron transport pathway. When NADH serves as the primary electron donor to the pathway it can be oxidised by either of two membrane-associated NADH dehydrogenases (NDH-1 and NDH-2). Only the first of these (NDH-1, encoded by the *nuo*A–N operon) serves as a coupling site that helps to generate the proton motive force, the stoichiometry of proton expulsion being $4H^+/2e^-$. The second NADH dehydrogenase (NDH-2, encoded by the *ndh* gene) has a much higher *K*m for NADH and could serve to regulate the intracellular NADH concentration independent of free energy conservation. If this were so, then it would be likely to be important

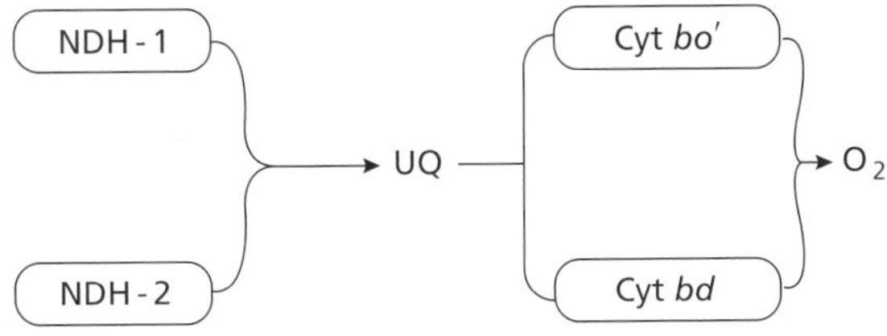

Fig. 8 Diagrammatic representation of the doubly branched, NADH-utilising aerobic respiratory chain of *Escherichia coli*. NADH dehydrogenase 1 (encoded by *nuo*A–N) or NADH dehydrogenase 2 (encoded by *ndh*) effect the reduction of quinone (Q) which is re-oxidised via the terminal cytochrome oxidase complexes, cytochrome bo′ (encoded by *cyo*A–D) or, at lower oxygen tensions, by cytochrome bd (encoded by *cyd*AB). Of the two NADH dehydrogenase complexes, only NDH-1 serves as a coupling site ($4H^+/2e^-$) and is preferentially employed to reduce MQ during fumarate- or dimethylsulphoxide-dependent anaerobic respiration.

when the organism is energy replete. Certainly NDH-2 is the preferred NADH dehydrogenase of aerobic and nitrate respiration, whilst NADH-1 is employed during MK-dependent fumarate or dimethylsulphoxide respiration. With both NDH complexes present together in the *E. coli* cell membrane, the ratio in which they are active will help to determine the efficiency of coupling and hence of net free energy conservation. Doubtless the partitioning of available NADH between them is not arbitrary but is responsive to environmental and growth conditions.

At the other end of the respiratory chain, the cytochrome bo′ complex, which is composed of four polypeptide units with two copper atoms and two b-type haems, predominates at high oxygen tensions. It functions as a proton pump with a stoichiometry of $2H^+/2e^-$, with an additional $2H^+/2e^-$ deriving from the operation of the UQ loop. The cytochrome bd complex, which has a greater affinity for oxygen, consists of two polypeptide chains with two b-type and one d-type haems. It does not act as a proton pump, though $2H^+/2e^-$ are still translocated by the UQ loop. Together with the choice available between NDH-1 and NDH-2, and assuming that ATP synthase operates with a stoichiometry (H^+/ATP) of about three, then the ATP yield associated with the aerobic oxidation of NADH could in *E. coli* vary 4-fold (from 2/3 to 8/3 mol ATP/NADH oxidised. In nitrogen-fixing aerobic bacteria such as *Azotobacter vinelandii* and *Klebsiella pneumoniae* an oxygen-scavenging terminal oxidase unrestrained by coupling to proton translocation serves to protect their oxygen-sensitive nitrogenase. Though in *E. coli* the function of cytochrome oxidase bd is less clear, circumstances can be envisaged wherein it would prove useful to the bacterium to be able to maintain a high rate of oxygen consumption with less associated proton translocation than would result from the alternative use of the proton-pumping cytochrome bo′ complex.

The question of how the components of the respiratory chain can be adjusted to serve the present needs of the organism is tied up with another question: how is the flux of electrons along the chain controlled? The answer would seem to be that this is responsive to thermodynamic disequilibria between the Δp and the redox spans between the proton-translocating regions of the respiratory chain. When respiration continues in the absence of ATP synthesis, its rate is self-adjusted so as just to compensate for any background proton leakage through the cell membrane.

At 25°C and pH 7.5, *E. coli* would sustain a $\Delta\Psi$ of about 140 mV across its cytoplasmic (respiratory) membrane. The internal (cytoplasmic) pH would be no more than 0.5 pH units more alkaline than the external P-phase (periplasm). Together, this would result in a transmembrane proton motive force of about 170 mV, equivalent to a ΔG_p of −40 kJ/mol.

When considering respiratory processes, some thought has necessarily to be given to the source of the primary electron donor which initiates the electron transfer sequence. In the case that this is NADH, then in many organisms the tricarboxylic (citric) acid cycle, wherein acetyl-CoA is fully oxidised, will be a most important source of this reductant. In *E. coli* this cycle operates in the aerobically growing cell but not when the organism is growing anaerobically by fermentation. Even aerobically, if the rate of consumption of excess carbohydrate allows abundant ATP production by glycolysis alone, the cycle is non-functional. This is due to the repression of synthesis of several of its key component enzymes, namely 2-oxoglutarate dehydrogenase, succinyl-CoA dehydrogenase and succinate dehydrogenase. The consequence is a branched ('horseshoe shaped') non-cyclic pathway whereby acetyl-CoA is consumed at a diminished rate (Fig. 9). The oxidative branch forms 2-oxoglutarate for biosynthetic purposes whilst the reductive branch, which converts oxaloacetate first to fumarate (via malate or aspartate) and then to succinate, serves as a balancing electron sink. Under these circumstances, pyruvate formed by glycolysis must be carboxylated and then

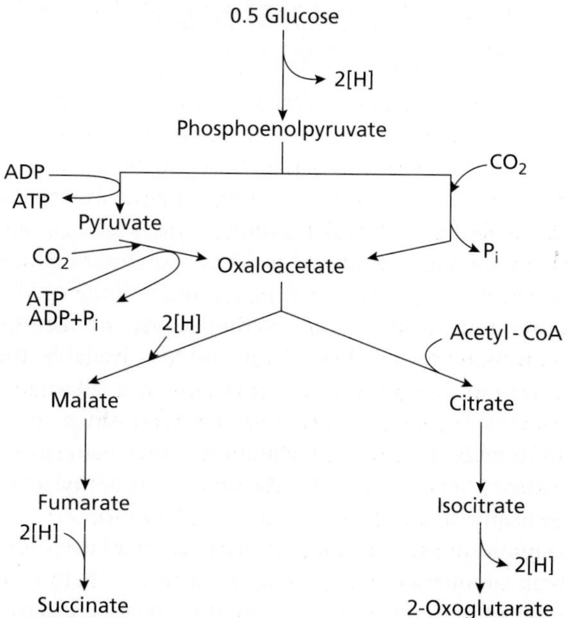

Fig. 9 Non-cyclic derivative of the TCA cycle operating to generate 2-oxoglutarate and succinate for biosynthetic purposes. This pathway may be used when ATP production is already excessive or under anaerobic conditions wherein conversion of 2-oxoglutarate to succinyl-CoA (and thence succinate) does not occur.

yield fumarate and succinate, or by the action of a pyruvate formate lyase enzyme give rise to acetyl-CoA and formate. The further activity of a formate hydrogen lyase complex can convert the formate to CO_2 and H_2, whilst the acetyl-CoA can give rise to acetate or ethanol. Some additional 'reducing equivalents' can be disposed of by the direct reduction of pyruvate to lactate. The outcome is a mixed acid fermentation in which glucose is predominantly converted to equimolar amounts of acetate, ethanol, and two times the molar equivalents of CO_2 and H_2.

Thus when glucose is anaerobically fermented by *E. coli* the anticipated ATP yield would be about 3ATP/mol glucose utilised (2ATP from the glycolytic EMP pathway plus 1ATP from acetate production). Some additional ATP could arise from the Δp generation associated with the net proton export that accompanies lactate excretion at an acid pH, but this would not normally be a major contribution. This relative inefficiency of fermentative ATP generation contrasts markedly with the situation that exists when under aerobic respiratory conditions the citric acid cycle is fully functioning. With the necessity to dispose of the 24 [H]/mol glucose that is generated, the respiratory electron transport pathway could be responsible for the generation of as much as 31–38ATP/mol glucose. When fumarate is provided in excess under anaerobic conditions its ability to serve as an electron sink relieves pyruvate and acetyl-CoA of the need to adopt this role. In consequence, in this anaerobic respiratory mode the pyruvate can be converted wholly to acetyl-CoA and this, in turn, via acetyl phosphate to acetate with an equimolar production of ATP, so that in total 4ATP/mol glucose would be produced as the result of SLP reactions. Additionally the 8[H]/mol glucose would feed into the respiratory chain for which fumarate could serve as the terminal electron acceptor. The operation of this anaerobic respiration would lead to the generation of proportionately additional ATP. When nitrate is employed as a respiratory electron sink, it is reduced to nitrite by a membrane-bound nitrate reductase whose activity sustains proton translocation from the cytoplasm (N-phase) to the periplasm (P-phase) by a loop mechanism associated with UQH_2 oxidation. The nitrite is reduced to ammonium ions by a periplasmic nitrite reductase that contains six c-type haem groups.

F_1F_0 ATP Synthase: a Membrane Transformer

Energy-transducing membranes (respiratory, photosynthetic) that generate a transmembrane proton motive force can directly employ this to impel ion and substrate translocations and, in the case of motile bacteria, to power the flagellar motor. Universally, however, part of the Δp is employed to generate ATP by the action of a membrane-associated protein complex. This consists of an F_0 component that spans the membrane and provides a channel for proton flow and an attached F_1 component that protrudes from the N-side of the membrane as a round head on a slender stalk. It is this latter F_1 component that is the catalytic subunit of the complex and is responsible for the synthesis (or hydrolysis) of ATP. Although the complex is generally termed the proton-translocating ATP synthase it is important to realise that it can also work in reverse as a proton-pumping ATP hydrolase (ATPase) with proton ejection to the P-side of the membrane being sustained by the hydrolysis at F_1 of cytoplasmic ATP. The F_1F_0 ATP synthase serves therefore as a reversible transformer interconverting energy available in the form of an ion gradient into the metabolically convenient form of ATP, and vice versa. When F_1 is detached from F_0 and is therefore decoupled from the Δp, it catalyses only ATP hydrolysis.

The F_1F_0 ATP synthase of *E. coli* actually contains eight different polypeptides. Five of these subunits make up the F_1 domain (α, β, γ, δ and ϵ in the ratio 3:3:1:1:1). Both the α and the β subunits carry nucleotide-binding sites and of these six sites the three that are catalytically active are those associated with the β subunits. The F_0 domain on the other hand is made up of the so-called a, b and c subunits (in the ratio 1:2:9–12), the last being the site of interaction with the specific inhibitor of ATP synthase, dicyclohexylcarbodiimide (DCCD).

How ATP synthesis is impelled by the proton motive force was for many years the subject of much speculation, little of which initially anticipated the ingenuity of the rotational catalysis that appears to explain the actual mechanism. The F_1F_0 ATP synthase in fact operates as a macromolecular rotary motor. The central rotor is formed by an axial 'drive shaft' made up of the γ subunit together with subunits c and ϵ. This runs in a stator composed of a hexagonal ring of α–β subunits (with a, b and δ), and together they constitute what has been termed 'the world's smallest nanoengine' capable of transforming electrochemical, through mechanical, to chemical energy and vice versa.

Although there has been some argument concerning the efficiency of the F_1F_0 ATP synthase transformer, the current view is that it could operate with an H^+/ATP ratio of 3. That is to say, three protons would be transported through F_0 for every ATP molecule synthesised by F_1. It should be noted though that the H^+/ATP stoichiometry of the energy-transducing membrane is likely to be greater than 3. This is not only because of the other uses to which Δp must be put, but also because of 'slippage' and of a certain background leakiness to protons of the energised membrane. Thus, in the complete oxidation of glucose by *E. coli* which aerobically proceeds via glycolysis plus the citric acid cycle, the maximum P/O ratio can be no more than 2.58 and the total yield of ATP per glucose is 31. Yet the effects of passive proton leakage through the energy-transducing membrane will ensure that these maxima are not attained during normal growth of the bacterium.

One might note that prior to the discovery of the mechanism of operation of the F_1F_0 ATP synthase, the title of smallest rotary motor was held by the bacterial flagellar motor that was also driven by the transmembrane proton motive force (in some bacteria such as species of *Vibrio*, by a sodium ion motive force). Although similarly miraculous in its construction, this is much larger in size (40 by 60 nm) and, for swimming purposes, achieves a much higher speed of rotation of the flagellar filaments.

What of bacteria that primarily produce an energised cell membrane by the establishment of a transmembrane gradient of some ion other than H^+? These may employ an ATP synthase whose operation is specifically coupled to the flow of that ion (as in the case of the Na^+-ATP synthase of the obligately anaerobic bacterium *Propiogenium modestum*) or they could employ an H^+-ATP synthase with the required transmembrane electrochemical gradient of protons being secondarily established by the activity of a membrane-located ion/proton antiporter (as in the case of the Na^+/H^+ antiporter of the alkaliphilic bacterium *Bacillus firmus*).

Other Means of Creating a Proton Motive Force

Transport-coupled Membrane Energisation

Certain end products of bacterial fermentations are excreted from the cell by carrier-mediated transmembrane transport with concomitant vectorial ion translocation. If the ion is H^+ and the species to be effluxed is a monovalent anion (A^-), then its symport with $1H^+$ would enhance the ΔpH across the membrane (but not $\Delta\Psi$). Symport of A^- with $> 1H^+$ would increase both ΔpH and $\Delta\Psi$.

Evidence that this mechanism is indeed exploited by some bacteria was initially provided by studies of the excretion of the lactic acid produced during the course of glucose fermentation by *Streptococcus cremoris* (Fig. 10). Under conditions in which the lactate concentration in the medium was very low and the environmental pH was greater than about 6.7 the monovalent lactate$^-$ anion was exported electrogenically in symport with two protons.

The resultant enhancement of free energy conservation (due to the additional contribution to the Δp) was evident in the greater biomass yield achieved by a culture of the bacterium growing under energy-limited conditions. It could be that this behaviour would therefore be particularly advantageous to such a lactic acid producer when growing in mixed culture with companion lactate-consuming microbes.

It has been proposed that *E. coli* might derive similar energetic benefit from the activity of its amino acid decarboxylases if this were coupled to suitable antiporters. For example, in the case of glutamate decarboxylation the glutamate could enter the cell as its

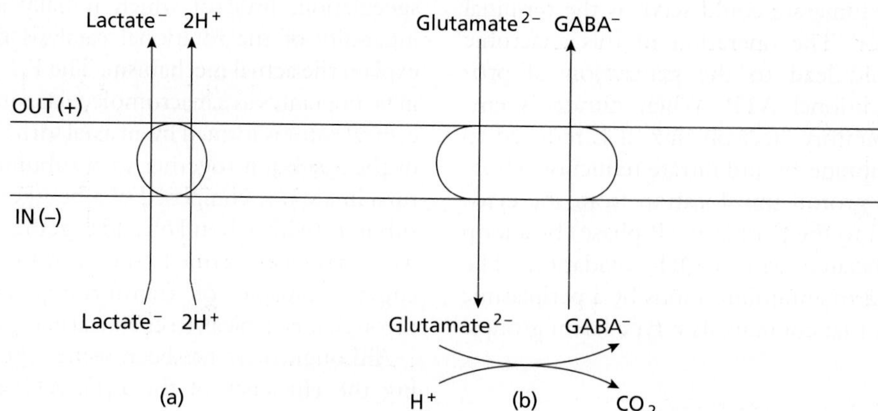

Fig. 10 Examples of transport-coupled energisation of the bacterial cell membrane. (a) Co-transport of lactate in its monoanionic form with a greater than equimolar proportion of H^+ ions. (b) Antiporter activity in which import of a glutamate dianion is coupled to the export of the γ-aminobutyrate (GABA) monoanion. Cytoplasmic decarboxylation of glutamate^{2-} to GABA$^-$ consumes $1H^+$/glutamate^{2-} and so contributes to the transmembrane proton motive force.

dianion (glutamate^{2-}), be decarboxylated in the cytoplasm (with uptake of a proton) to produce the γ-aminobutyrate monoanion (GABA$^-$) which is then expelled from the cell via a glutamate^{2-}/GABA$^-$ antiporter (**Fig. 10**). This electrogenic exchange and the proton consumption associated with the decarboxylation reaction would serve to generate both a $\Delta\Psi$ and a ΔpH. It is relevant to note, however, that the glutamate decarboxylase of *E. coli* is unusual in displaying a pH optimum of 4–4.6. Indeed, the enzyme is not active at pH values above 6, so it is most likely that its major role is to contribute to cytoplasmic pH homeostasis when the bacterium encounters an acid environment.

That such speculation, even in respect of *E. coli*, is not fanciful, is illustrated by the more extreme example provided by the anaerobic bacterium *Oxalobacter formigenes*, which is wholly reliant on the electrogenic antiporter activity that serves as its only means of free energy conservation (**Fig. 11**). This bacterium is an inhabitant of the gut of humans and animals, wherein its role seems to be the detoxification of oxalate. which it decarboxylates to produce formate. The oxalate is taken up as its dianion (oxalate^{2-}) and the formate is simultaneously exported (via the same membrane carrier) as its monoanion (formate$^-$). The exchange is therefore electrogenic, producing a $\Delta\Psi$ (positive outside). Additionally, the consumption of a proton in the decarboxylation reaction, with the CO_2 being liberated in gaseous form and not as HCO_3^-, means that some ΔpH (alkaline inside) will also be developed. The outcome is a sufficiently large transmembrane proton potential to satisfy the energy needs of the organism.

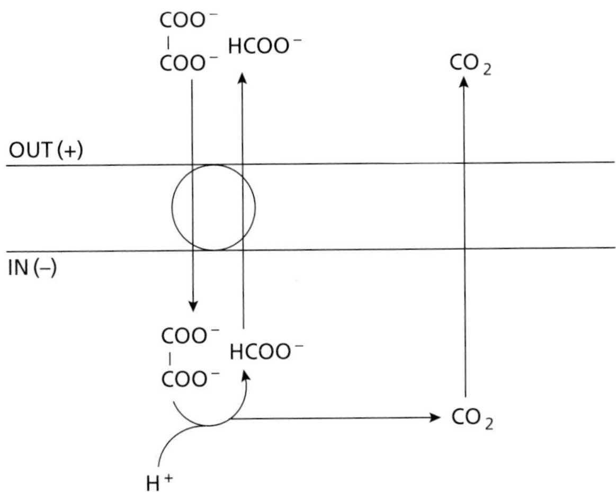

Fig. 11 Transmembrane electrogenic antiporter activity in *Oxalobacter formigenes*. The growth substrate (oxalate^{2-}) is imported as its dianion in exchange for exported formate$^-$ in its monoanionic form. The latter is generated in the cytoplasm by the decarboxylation of oxalate^{2-} that consumes 1H$^+$ and liberates CO_2.

Sodium Ion-pumping Membrane-located Decarboxylases

Several anaerobic bacteria have been found to possess membrane-associated, biotin-dependent decarboxylase enzymes that function as Na^+ pumps. The decarboxylation reactions that they catalyse are highly exergonic and part of the 'liberated' free energy is conserved in the transmembrane electrochemical Na^+ ion gradient.

Veillonella alcalescens is one such anaerobe. Present in the gut and mouth, it cannot ferment carbohydrates but ferments carboxylic acids such as lactate, malate and fumarate to produce propionate, acetate, CO_2 and H_2. Pyruvate and methylmalonyl-CoA are formed as intermediates in these fermentations and the pyruvate is converted to acetate, CO_2 and H_2 in a (SLP) reaction that produces ATP. The methylmalonyl-CoA is decarboxylated by a membrane-located enzyme (**Fig. 12**) that simultaneously translocates Na^+ ions to the outside of the membrane (2Na^+/methylmalonyl-CoA decarboxylated).

The product propionyl-CoA in another ATP-yielding (SLP) reaction is converted into the propionate. The same decarboxylase is active in *Propiogenium modestum*, an anaerobic bacterium present in human saliva. Indeed, when this bacterium ferments succinate it is entirely reliant on the operation of this Na^+-pumping decarboxylase as its only source of ATP (produced by a membrane-integrated Na^+-ATP synthase). A glutaconyl-CoA decarboxylase that generates an electrochemical transmembrane gradient of Na^+ ions is found in several bacteria including

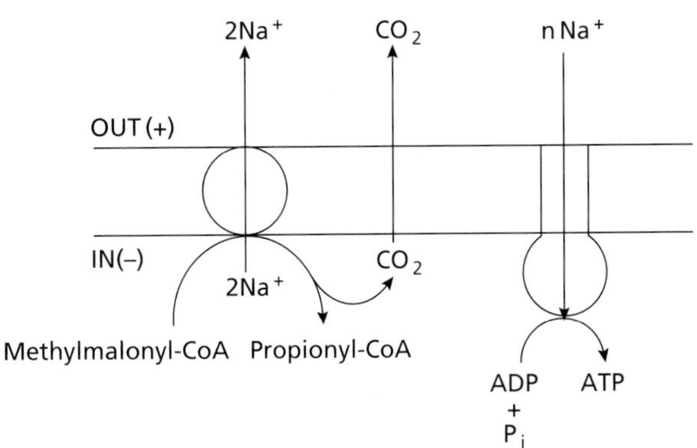

Fig. 12 Coupling of sodium ion-translocating methylmalonyl-CoA decarboxylase activity with ATP synthesis by an Na^+-F_1F_0 ATPase.

Acidaminococcus fermentans, which produces glutaconyl-CoA as an intermediate in its fermentation of glutamate that produces acetate and butyrate. An oxaloacetate decarboxylase that serves as a Na^+ ion pump functions in *Klebsiella pneumoniae*. The bacterium, amongst other uses, exploits the Na^+ ion potential so created to import the citrate that is the substrate of the fermentation.

Exploitation of Free Energy by Bacteria

The free energy that has been 'captured' by the bacterium in the forms of various 'high-energy' metabolites (including ATP) and in the various transmembrane electrochemical gradients, is thereafter available for all the life-sustaining processes that otherwise could not proceed in the absence of a suitably coupled driving force. These include biosynthetic conversions, motility and general cell maintenance.

Active Transport Processes

Particularly significant demands on free energy resources are made to effect selective import and export of ions and uncharged molecules against concentration gradients that would tend to translocate them in the opposite direction. In a free-living bacterium much depends on the effective operation of such active transport mechanisms, for example in maintaining cytoplasmic homeostasis, acquiring nutrients and carbon and/or energy substrates, evicting products of metabolism whose intracellular accumulation would be harmful. It is not surprising that in bacteria such transport processes exploit virtually all of the major ways of harnessing conserved free energy.

ATP Hydrolysis-driven Active Transport

Transport of K^+ ions by *Escherichia coli* is important, since controlled K^+ accumulation (chiefly as potassium glutamate) is the major mechanism employed by the bacterium for turgor regulation, whilst its cytoplasmic K^+ concentration also plays an important role in maintaining the intracellular pH. The organism possesses at least four K^+ ion uptake systems and two or more K^+ efflux systems. Free energy is expended to maintain an intracellular K^+ concentration that is substantially greater than that of K^+ in the growth medium. The high-affinity Kdp uptake system, whose synthesis is induced when the external K^+ ion concentration is very low, is composed of three integral membrane proteins which together form a K^+-translocating ATPase (Fig. 13). Another (Trk) uptake system, whose mechanism is at present still somewhat uncertain, operates at higher external K^+ concentrations.

In Gram-negative bacteria, periplasmic proteins that specifically, and with high affinity, bind required substrates, can play a crucial role in the importation into the cell of these compounds. In *E. coli* such substrates include phosphate, and sulphate ions, ribose, maltose and histidine. The protein undergoes conformational change when it binds its substrate and in this state it carries the molecule to a specific transporter located in the cytoplasmic membrane. This in turn, when

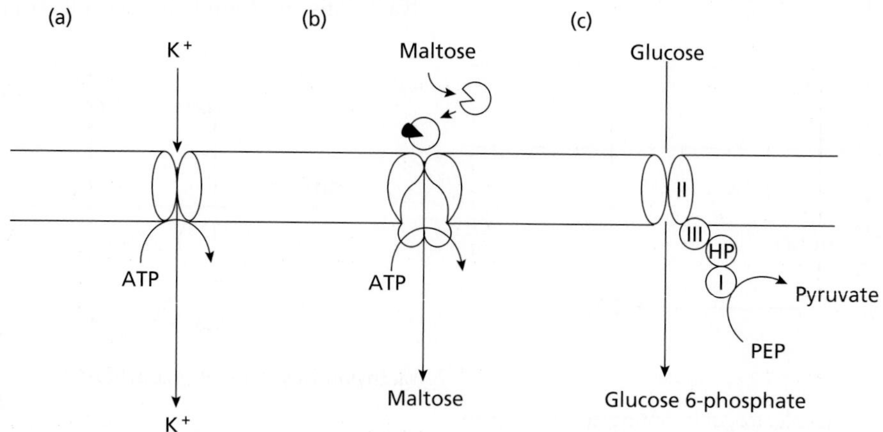

Fig. 13 Schematic representation of three active transport (import) mechanisms in the (inner) cell membrane of *Escherichia coli*. (a) Uptake of K^+ ions by a K^+-translocating ATPase (Kdp-type). (b) ATP-dependent uptake of maltose after initial attachment to a specific periplasmic binding protein. (c) Phosphoenolpyruvate (PEP)-dependent phosphotransferase (PTS) system importing glucose as glucose 6-phosphate.

energised by ATP hydrolysis, translocates the substrate through a gated channel in the membrane that is opened only for this purpose and which closes when the periplasmic protein releases the bound molecule. The transporter then returns to its unstimulated conformation while it awaits the arrival of another substrate-charged periplasmic binding protein molecule.

Thus, for example, in *E. coli* the maltose transport system consists of a binding protein, two channel-forming transmembrane proteins and an energy-transducing protein that hydrolyses ATP (**Fig. 13**). This MalK transporter (which imports 1 maltose/2ATP hydrolysed) is a member of the extensive ABC family of transporters (for ATP-binding cassette). Other members of this ABC family are concerned with export of substrates, some functioning as multi-drug efflux pumps that can confer drug-resistance properties on the bacteria that possess them. In Gram-positive bacteria the equivalents of the periplasmic-binding proteins appear to be anchored to the cytoplasmic membrane.

PEP-driven Group Translocation

In bacteria, especially facultative and obligately anaerobic bacteria that operate an EMP glycolysis pathway, a PEP-sugar phosphotransferase system (PTS) actively imports numerous sugars and sugar alcohols, for example, glucose and mannitol (**Fig. 13**). A protein phosphorylation cascade is initiated by phosphate group transfer from PEP to a cytoplasmic enzyme (E I). This, in turn, hands on the phosphate group to a second soluble protein (H-Pr) which becomes phosphorylated at a key histidine residue. Transfer of the phosphate group from phosphorylated H-Pr to a membrane-integrated E II is catalysed either by E II itself or, in the case of a subset of substrates, by an additional E III. E II subunits specifically bind the sugar or polyol which is thereafter released on the inner side of the cytoplasmic membrane in its phosphorylated form, e.g. glucose 6-phosphate or mannitol 1-phosphate.

It has been calculated that at equal concentrations of pyruvate and PEP the equilibrium internal concentration of the phosphorylated sugar would be as much as one hundred million times greater than the external concentration of the sugar. Yet the PTS system is not only concerned with active transport of sugars; it can also serve a regulatory role. Thus in *E. coli* the activity of the membrane-bound adenylate cyclase that synthesises the signal molecule cyclic AMP is sensitive to whether the E III^{Glucose} is phosphorylated or not. When glucose is not available for transportation this PTS component will be maximally phosphorylated and in this state activation of the adenylate cyclase ensures that cAMP is present in relatively high concentration in the cell. This in turn, via the formation of a cAMP-CAP (catabolite activator protein) complex, can promote the synthesis of alternative carbohydrate-utilising enzymes.

H^+-electrochemical Gradient-driven Transport

Secondary transport processes, working via membrane-located 'porters' (**Fig. 14**), can exploit the free energy available in the primary transmembrane ion gradient to drive the selective inward or outward movement of solutes (whether ionised or electroneutral). By the uniport mechanism an ion is actively transported unidirectionally along its electrochemical gradient. A symport mechanism on the other hand

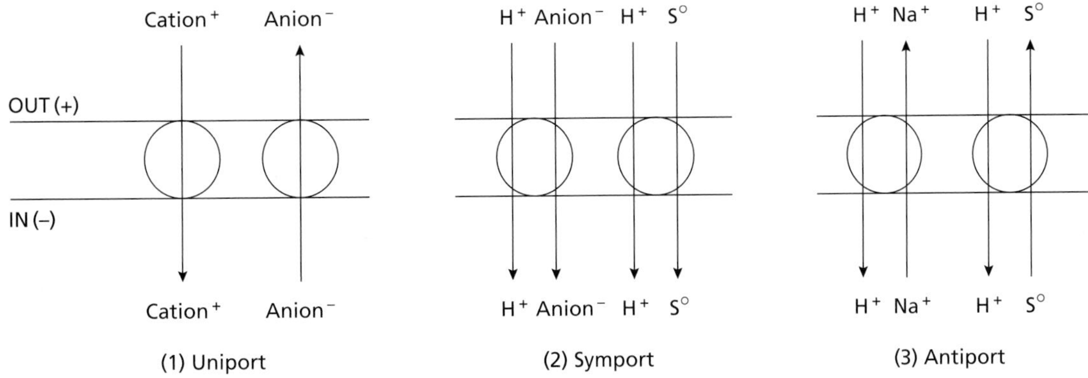

Fig. 14 Transmembrane ion and substrate movements by three types of porter. (1) Uniporters causing electrophoretic movements of ions in response to membrane potential. (2) Symporters: (left) electroneutral co-transport of anion(s) impelled by ΔpH, and (right) electrogenic cotransport of an uncharged substance impelled by both $\Delta\Psi$ and ΔpH. (3) Antiporters: export of cations (electroneutral exchange) or of neutral substances (electrogenic exchange) coupled to the driven importation of protons.

exploits the unidirectional movement of one substrate (ion) along its concentration gradient to effect simultaneous flow of a second substrate in the same direction, transport of both substrates being effected by a common carrier. An antiporter mechanism mediates the active counter flow of two substrates via a single carrier. Chemicals that uncouple electron flow in the membrane from proton translocation, or as in the case of protonophores, that collapse the Δp, also prevent all such forms of portered transport.

Uptake of lactose by *E. coli* is accomplished by a symport mechanism powered by the transmembrane Δp (1 lactose/1 H^+), and at the usual value of Δp an internal/external lactose concentration ratio of 3000 can be developed. The discovery of the LacY lactose transport protein, followed by its cloning and structural analysis, proved to be especially important in increasing knowledge of the modes of action of such transporters. Proton symporters in *Escherichia coli* that are specific for sugars, such as galactose (GalP), arabinose (AraE) and xylose (XylE), form one of the 18 groups of the so-called major facilitator (MF) superfamily.

Antiporters also function in *Escherichia coli*. The KefB and KefC K^+ ion efflux channels operate as antiporters which link K^+ expulsion to H^+ importation, and the need to sustain a relatively low cytoplasmic concentration of Na^+ ions is met by the activities of at least two Na^+/H^+ antiporters. The efflux of tetracycline that contributes to the resistance to this antibiotic displayed by several bacteria is driven by the transmembrane Δp via a specific tetracycline/H^+ antiporter, and various other examples of drug resistances can be explained by similar active exclusion mechanisms.

Antiporters which link uptake of a precursor substrate to excretion of a product of its metabolism are also relatively common. One antiporter in *E. coli* couples uptake of hexose phosphate or glycerol phosphate to the efflux of phosphate ions from the cell. The operation of such antiporters can be of considerable benefit to the cell by ensuring tight linkage of substrate uptake and product efflux so that there is no possibility of harmful intracellular accumulation of excessive concentrations of the product. Furthermore, electroneutral antiportered exchange can be especially helpful to bacteria suffering from energy limitation since no additional free energy is required for their operation. For example, in the lytic fermentation of arginine undertaken by *Enterococcus faecalis*, the arginine is taken into the bacterial cytoplasm and product ornithine is released into the growth medium by the action of an electroneutral antiporter. Since the arginine deiminase fermentation pathway yields only 1ATP net per mol arginine consumed, energy economy in substrate acquisition and product release is evidently helpful. Similarly when *E. coli* respires anaerobically using fumarate as terminal electron acceptor, this is acquired and the succinate that is produced is exported via an electroneutral antiporter.

Na^+ Ion Gradient-linked Transport

In some bacteria, as we have previously noted, the development of a transmembrane electrochemical gradient of Na^+ ions is the primary means of membrane energisation and ATP may be generated by a Na^+-ATP synthase. Yet even when the primary gradient is of H^+ ions, as in *E. coli*, specific transporters can be powered by a Na^+ electrochemical gradient maintained by an electroneutral Na^+/H^+ antiporter that equilibrates the Na^+ and H^+ gradients. Thus in *E. coli* uptake of the sugar melibiose or of the amino acid proline is effected by symport with Na^+.

Energisation of Outer Membrane Transporters in Gram-negative Bacteria

Gram-negative bacteria are surrounded by an outer membrane separated from the (inner) cytoplasmic membrane by a fluid-filled periplasmic space. This outer membrane provides protection against detergents and adverse enzymic activity, and it also excludes many compounds whilst allowing free passage, by diffusion through membrane-spanning pores, of small molecules (<200–600 Da). Since however this outer membrane is itself not energised, the question arises as to how substrate transport through it can be effected when the mechanism requires some input of free energy.

In *E. coli*, the Δp across the inner cytoplasmic membrane provides the free energy required to transport through the outer membrane such large molecules as iron siderophore complexes, or haem, or vitamin B_{12}. This is made possible by the intervention of at least three proteins (TonB, ExbB and ExbD). TonB and ExbD are located in the periplasm and are anchored by their N-termini to the cytoplasmic membrane. ExbB spans the cytoplasmic membrane three times with its N-terminus in the periplasm and most of the remainder of the protein in the cytoplasm. Somehow (the mechanism is as yet unknown) this Ton complex responds to the Δp of the cytoplasmic membrane. It has been proposed that TonB whilst interacting with ExbB and ExbD is also able to bind to specific outer membrane proteins that form energised, gated channels that recognise and internalise specific, actively transported substrates. If so, it could be that

TonB is converted under the influence of the Δp into some energised conformation and in this activated state interacts with the transporter sites on the outer membrane, then returning to the ExbB/D on the inner membrane to become re-energised. If this were the case, then TonB would, in effect, serve as the energy transducer that enables gated porin channels in the outer membrane to open and close during selective active transport (import and export) of certain substances.

The ferric citrate transport system of *E. coli* illustrates the sequence of events (**Fig. 15**). The substrate is first bound by an outer membrane protein (FecA) and, via the intervention of the Ton energy-transfer transducer, effects entry into the periplasm where it is avidly bound by the FecB-binding protein. Active transport through the cytoplasmic membrane then involves membrane proteins FecC and FecD and FecE serving as an ATPase (ABC transporter).

Active export through the outer membrane of a Gram-negative bacterium can be similarly impelled. Thus, to secrete the toxin aerolysin, *Aeromonas hydrophila* relies on two cytoplasmic membrane proteins, ExeA and ExeB. In both amino acid sequence and topology, the latter resembles the TonB of *E. coli*, but ExeA is a novel ATP-binding protein. Together these two proteins co-operate to transduce energy and convey it to an active secretion channel in the outer membrane.

Free Energy Coupling

Studies of bacterial growth and reproduction have shown that these organisms are not always seeking to maximise the efficiency of ATP production and the specific biomass yield but may be more concerned to maximise their growth rates, especially in environments in which there is no shortage of possible sources of free energy. Indeed Harold, (1990) concluded that 'the bacterial solution to energy surplus is in general to waste it'. Coupling of free energy conservation to its useful exploitation is therefore least when growth is severely restricted, as in the case of respiring, washed suspensions of bacteria. Mechanisms of free energy dissipation (besides the potentially harmful generation of heat) include wasteful 'overflow metabolism' and 'futile metabolic cycles' whereby ATP is consumed in the synthesis of some metabolite which is then catabolised to its precursor(s) without recovery of the expended free energy. Other mechanisms rely on non-productive ion leakage through cell membranes or intracellular accumulation of reserve materials (such as glycogen) whose synthesis is endergonic but which, under conditions of continuing energy excess, will not thereafter be consumed during the lifetime of the organism.

The efficiency of free energy conservation too can be modulated via 'slippage' in electron transport-associated phosphorylation and the alternative usage

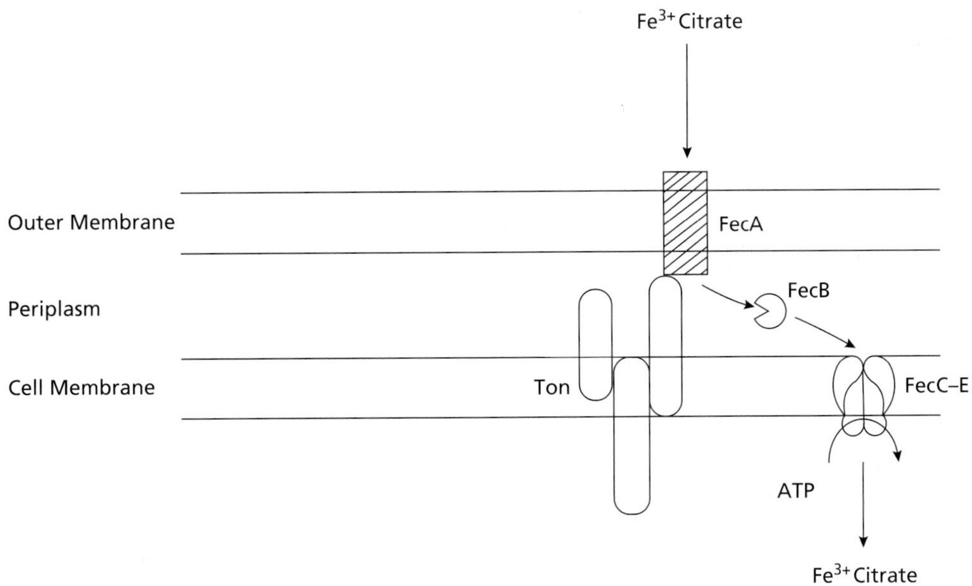

Fig. 15 Schematic representation of the uptake mechanism for iron citrate in the Gram-negative bacterium *Escherichia coli*. The outer membrane protein FecA serves as a gated channel energised by the Ton complex of the cytoplasmic membrane. Once inside the periplasm, the iron citrate is avidly bound by protein FecB which delivers it to the FecC–E complex. This spans the inner cytoplasmic membrane and acts as an ABC-type transporter motivated by ATP hydrolysis.

of branch(es) of the respiratory chain which possess fewer coupling sites.

Taken together, such flexibility and adaptability of both free energy conserving and utilising mechanisms help to ensure the success of free-living bacteria in rapidly changing environments. Again as Harold, (1990) stated,

> When resources are scarce but dependable their efficient use will confer selective advantage, but in other circumstances when resources are ample but only temporarily available competition and reproductive advantage favour organisms that can rapidly acquire a larger share of these resources even if this involves waste. Most bacteria endure a 'feast-or-famine' existence in which the race is to the swift.

Further Reading

Background information on the subject of cellular bioenergetics will be found in any textbook of general biochemistry, but the following more specialist texts should prove especially helpful:

Harris DA (1995) *Bioenergetics at a Glance.* Oxford: Blackwell Science [ISBN 0632023880].

Harold FM (1990) *The Vital Force: A Study of Bioenergetics.* Oxford: W.H. Freeman [ISBN 0716717344].

Nicholls DG, Ferguson SJ (1997) *Bioenergetics 2.* London: Academic Press [ISBN 0125181248].

Neidhardt FC (ed.) (1999) *Escherichia coli and Salmonella: Cellular and Molecular Biology,* 2nd edn. Washington, DC: ASM Press [ISBN 1555811647].

White D (1999) *The Physiology and Biochemistry of Prokaryotes,* 2nd edn. Oxford: Oxford University Press [ISBN 0195125797].

Wrigglesworth JM (1997) *Energy and Life.* New York: Taylor and Francis [ISBN 0748404333].

13 Biosynthesis

Albert G. Moat[1] and John W. Foster[2]

[1] Marshall University School of Medicine, Huntington, WV, USA
[2] University of South Alabama College of Medicine, Mobile, AL, USA

All the biosynthetic pathways in the cell originate from just 12 precursor metabolites (Table 1). The catabolic dissimilation of glucose not only gives rise to these core metabolites but also generates the energy needed for all the work carried out by the cell to accomplish this task. The pathways by which these compounds are distributed into the many amino acids, purines, pyrimidines and other compounds such as the B vitamins is shown in Fig. 1. The fatty acids, which are integral components of lipopolysaccharides, phospholipids and glycolipids, are synthesised from acetyl-CoA. If micro-organisms are grown on so-called poor carbon sources such as malate, succinate, acetate or glycerol, the cell must synthesise hexoses by gluconeogenesis to maintain the production of these important metabolites. This chapter describes each of the biosynthetic pathways in detail, starting with fatty acid biosynthesis.

Table 1 Twelve precursor metabolites that initiate biosynthetic pathways

Glucose-6-phosphate
Oxaloacetate
Fructose-6-phosphate
Succinyl-CoA
Pentose-5-phosphate
α-Ketoglutarate
Erythrose-4-phosphate
Acetyl-CoA
Triose phosphate
Pyruvate
3-Phosphoglycerate
Phosphoenolpyruvate

Biosynthesis of Fatty Acids

The major fatty acid-containing components of microbial cells are the lipopolysaccharides, phospholipids and glycolipids present in microbial membranes. Other lipids play an important role as electron carriers, enzyme co-factors, and light-absorbing pigments. Acetyl-CoA is the starting point for the synthesis of fatty acids and related compounds.

Acetyl-CoA carboxylase, a biotin-containing enzyme that catalyses the ATP-dependent fixation of CO_2 into acetyl-, propionyl- and butyryl-CoA, carries out the first committed reaction in the de-novo synthesis of fatty acids (Fig. 2). The reaction occurs in two stages: the carboxylation of biotin with bicarbonate, catalysed by biotin carboxylase, and transfer of the CO_2 group from carboxy-biotin to acetyl-CoA to form malonyl-CoA, mediated by carboxyltransferase. This initial step is followed by a series of reactions

 doi:10.1006/bkmm.2001.0013

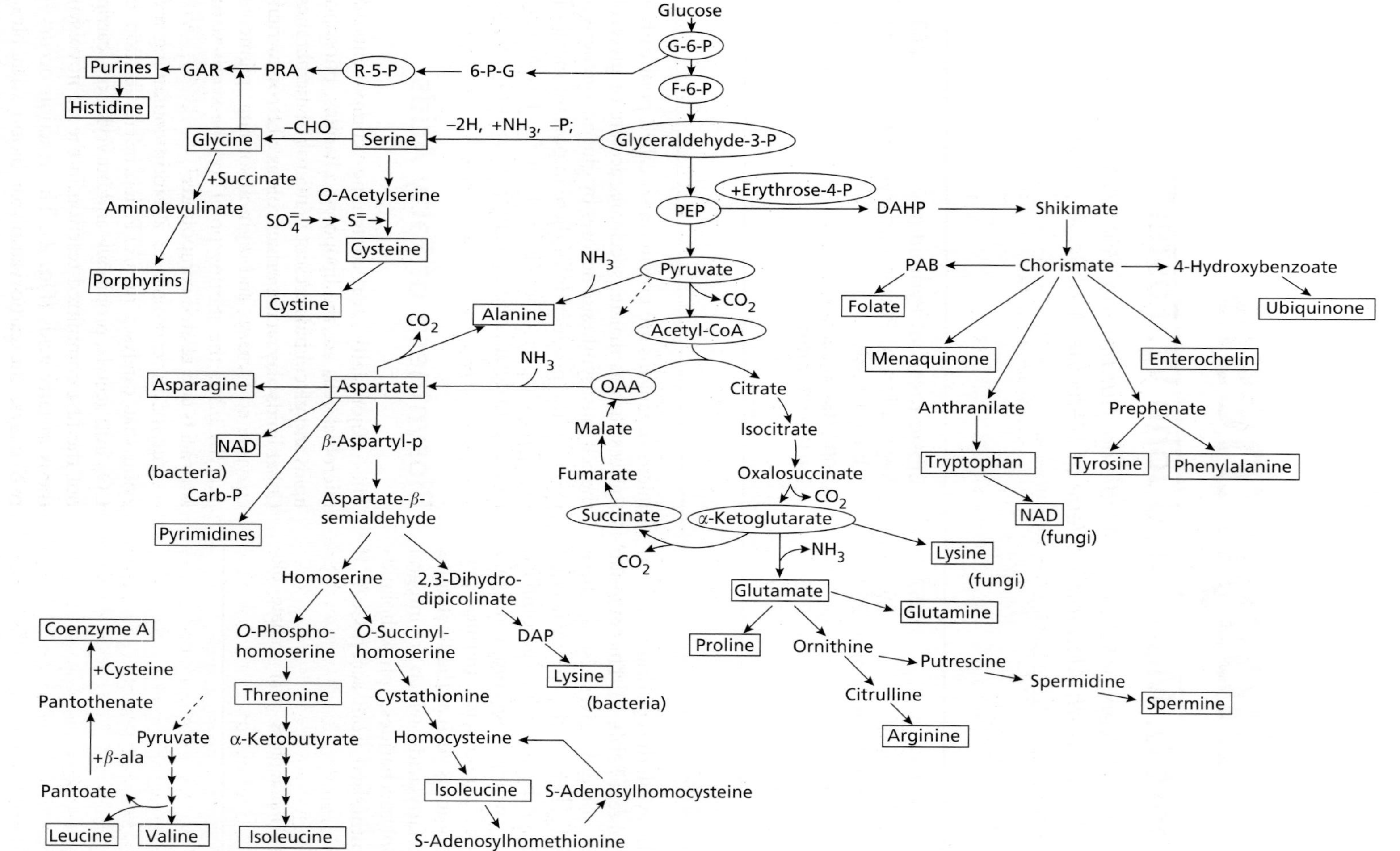

Fig. 1 Biosynthetic pathways leading to major cellular components. The highlighted intermediates are the 12 key compounds that serve as biosynthetic precursors for a variety of essential end products. From Moat AG, Foster JW (1995) *Microbial Physiology*, 3rd edn. New York: John Wiley & Sons.

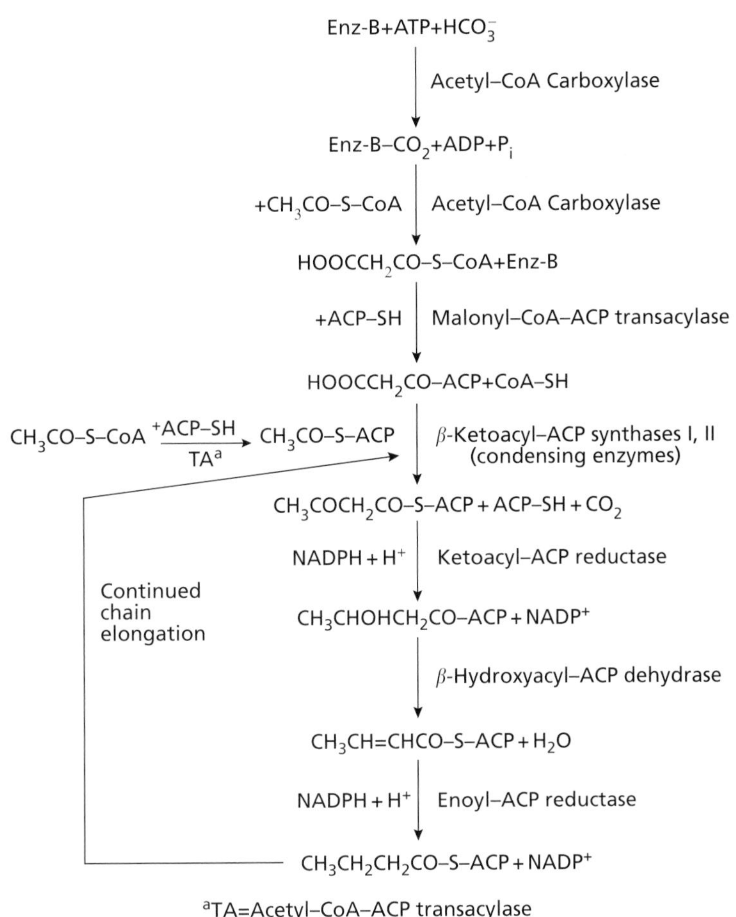

Fig. 2 Reactions involved in the biosynthesis of fatty acids. Enz-B, enzyme biotin; CoA, coenzyme A; ACP, acyl carrier protein; TA, acetyl-CoA-ACP transacylase. The *E. coli* structural genes and enzymes in this pathway are: *fabE*, acetyl-CoA carboxylase (see Fig. 3 for more detail); acetyl-CoA-ACP transacylase; *fabD*, malonyl-CoA-ACP transacylase; *fabB*, β-ketoacyl-ACP synthase I (active with C_2-C_{14} ACP, but inactive with C_{16} ACP, inactive with CoA derivatives); *fabF*, β-ketoacyl-ACP synthase II (active with C_2-C_{14} ACP and $16{:}1^{\Delta 9}$-ACP); β-ketoacyl-ACP reductase; *fabA*, β-hydroxyacyl-ACP dehydrase; enoyl-ACP reductase. From Moat AG, Foster JW (1995) *Microbial Physiology*, 3rd edn. New York: John Wiley & Sons.

that ultimately lead to the synthesis of long-chain fatty acids at the C_{16} (palmitic) and C_{18} level (Fig. 2). The CO_2 that initially enters the system to form malonyl-CoA, is removed in the condensation step. This liberation of CO_2 shifts the equilibrium in the direction of synthesis.

In *Escherichia coli*, the four components of the acetyl-CoA carboxylase system function as an enzyme complex. The genes that encode the biotin carboxyl carrier protein (BCCP), *accB*, and biotin carboxylase, *accC*, are co-transcribed (*accBC*) and map at 72 minutes on the chromosome. The genes that encode the carboxyltransferase α (*accA*) and β (*accD*) subunits, map at different regions of the chromosome. The rates of transcription of the genes encoding all four subunits of acetyl-CoA carboxylase are directly related to the rate of cell growth.

The fatty acid synthetase of *E. coli* exists as a soluble complex (designated type II) from which at least eight individual enzyme components are readily separated and purified. A readily dissociable fatty acid synthetase has been characterised in *Propionibacterium shermanii*. This system requires acetyl-CoA, malonyl-CoA, NADH, NADPH, and a small acyl carrier protein (ACP) similar to that from *E. coli*. Type I fatty acid synthetase, which occurs in mammals and yeast, is a dimer of two giant polypeptides that cannot be dissociated into its individual components.

The type I fatty acid synthetases of yeasts, fungi and mycobacteria exist as multiple copies of apparently identical subunits with molecular weights of approximately 290 000. The complex from *Neurospora crassa* consists of 11–12 equally sized subunits, 6 of which possess the acyl carrier protein function. The lack of a low-molecular-weight acyl carrier protein component is comparable to the system found in yeast.

In *E. coli*, at least six genes (*fabA* to *fabF*) encode the enzymes involved in fatty acid biosynthesis (Fig. 2). The enzyme β-ketoacyl-ACP synthetase I (encoded by *fabB*) is required for the elongation of unsaturated acyl-ACP. Mutants that lack synthetase I activity are unable to synthesise either palmitoleic or *cis*-vaccenic acid, and require unsaturated fatty acids for growth. β-Keto-ACP synthetase II (encoded by *fabF*) is responsible for the formation of *cis*-vaccenic acid. β-Keto-ACP synthetase III catalyses the formation of acetoacetyl-ACP. The *fabA* gene encodes 3-hydroxy-decanoyl-ACP dehydratase. The enzyme malonyl-CoA-ACP transacylase (MCT) is encoded by *fabD* and catalyses the transacylation of ACP with malonate. The *fabD* gene is part of an operon that consists of at least three genes involved in fatty acid biosynthesis. The genes *fabB* and *fabF*, which encode β-ketoacyl-ACP synthetases (KAS), are located immediately upstream and downstream, respectively, of *fabD*, suggesting that *fab*D is part of an operon encoding several components of the fatty acid synthetase system.

In *E. coli* saturated and unsaturated fatty acids are synthesised simultaneously. The divergence point occurs at the C_{10} stage (β-hydroxydecanoate). At this step there is apparently a competition between β-hydroxyacyl-ACP dehydrase, which forms an α, β-*trans* double bond, and β-hydroxydecanolylthioester dehydrogenase, which forms α,β,γ-*cis* double bonds (Fig. 3). Continued elongation leads to the formation of an unsaturated fatty acid. As a result of this mode of unsaturated fatty acid synthesis, only monounsaturated fatty acids are formed, and *cis*-vaccenic acid ($18{:}1^{\Delta 11}$) is the major unsaturated fatty acid formed by organisms that use the anaerobic pathway (see Fig. 3 and Table 2).

In eukaryotes and some highly aerobic bacteria (*Alcaligenes*, *Mycobacterium*, *Corynebacterium* and *Bacillus*) the synthesis of long-chain unsaturated fatty acids involves an oxygen-dependent oxidative desaturation (aerobic pathway) that introduces double bonds after the saturated fatty acid has been

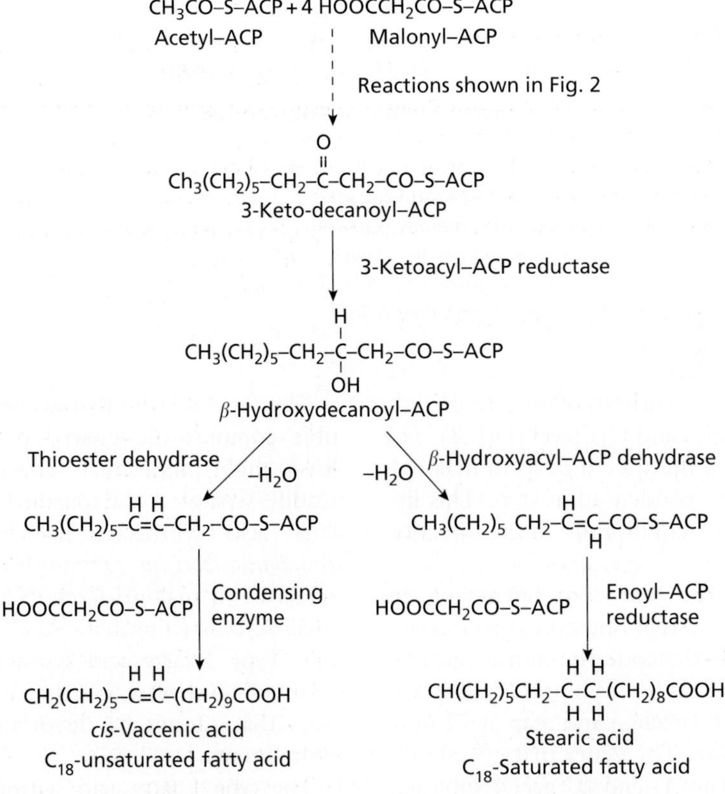

Fig. 3 Divergence of the saturated and unsaturated fatty acid biosynthetic pathways under anaerobic conditions. The pathway to saturated fatty acids shown in Fig. 2 is followed until the C_{10} level (β-hydroxydecanoyl-ACP) at which point there is a competition between the enzymes involved in forming saturated and unsaturated fatty acids. From Moat AG, Foster JW (1995) *Microbial Physiology*, 3rd edn. New York: John Wiley & Sons.

synthesised:

$$\mathrm{CH_3(CH_2)_7{-}\underset{H}{\overset{H}{C}}{-}\underset{H}{\overset{H}{C}}{-}(CH_2)_7CO{-}SCoA + 0.5O_2}$$

Stearoyl–CoA (18:0)

$$\downarrow$$

$$\mathrm{CH_2(CH_2)_7{-}\overset{H}{C}{=}\overset{H}{C}{-}(CH_2)_7CO{-}SCoA + H_2O}$$

Oleoyl–CoA(18:1$^{\Delta 11}$)

Approximately 70% of the fatty acids in membrane lipids of *Saccharomyces cerevisiae* consist of the unsaturated fatty acids palmitoleic acid (16:1) and oleic acid (18:1). The remaining fatty acids are saturated, and consist primarily of palmitic acid (16:0) and lesser amounts of stearic (18:0) and myristic (14.0) acids. Most other fungi contain the di- and trienoic acids linoleic acid (18:2) and α-linolenic acid (18:3). In yeast the Δ-9 desaturase catalyses the formation of the initial double bond between the ninth and tenth carbons of both palmitoyl (16:0) and stearoyl (18:0) CoA substrates, to make 16:1 and 18:1 unsaturated fatty acids. The activity of the Δ-9 desaturase is markedly reduced by the addition of unsaturated fatty acids, presumably because the level of desaturase mRNA is repressed under these conditions.

Table 2 Distribution of pathways for biosynthesis of long-chain monounsaturated fatty acids

Anaerobic pathway	Aerobic pathway
Escherichia coli	*Alcaligenes faecalis*
Salmonella typhimurium	*Corynebacterium diphtheriae*
Serratia marcescens	*Mycobacterium phlei*
Azotobacter agilis	*Bacillus* (several species)
Agrobacterium tumefaciens	*Micrococcus luteus*
Lactobacillus plantarum	*Beggiatoa*
Staphylococcus haemolyticus	Myxobacteria
Clostridium pasteurianum	*Leptospira canicola*
Clostridium butyricum	*Saccharomyces cerevisiae*
Caulobacter crescentus	*Neurospora crassa*
Propionibacterium	*Candida lipolytica*
Cloroflexus auranticus	*Stigmatella aurantica*
Chlorobium limicola	

From Moat AG, Foster JW (1995) *Microbial Physiology*, 3rd edn. New York: John Wiley & Sons.

Biosynthesis of Phospholipids

Phospholipids such as cardiolipin and phosphatidylethanolamine, more correctly referred to as phosphoglycerides, are present in microbial cell membranes. Their biosynthesis starts with the formation of *sn*-glycerol-3-phosphate (G3P). In *E. coli* G3P may be formed from dihydroxyacetone phosphate (DOHAP) derived from hexose cleavage by the action of the biosynthetic G3P dehydrogenase encoded by *gpsA*:

$$\mathrm{DOHAP + NAD(P)H + H^+ \rightleftharpoons G3P + NAD(P)^+}$$

G3P may also be produced from glycerol by the action of glycerol kinase encoded by *glpK*:

$$\mathrm{Glycerol + ATP \rightarrow G3P + ADP}$$

The first committed steps in phosphoglyceride biosynthesis involve the coupling of two molecules of fatty acid to G3P (Fig. 4). This is catalysed by two acyltransferases. Glycerol-3-phosphate acyltransferase, encoded by the *plsB* gene, adds one saturated fatty acid to form 1-acylglycerol-3-phosphate. An unsaturated fatty acid is added to the C2 position by 1-acylglycerol-3-phosphate acyltransferase, encoded by *plsK*, to form phosphatidic acid. In bacteria a fatty acid-ACP is the donor rather than a fatty acid-acyl-CoA. *E. coli* normally contains approximately 70% phosphatidyl ethanolamine, 25% phosphatidylglycerol and 5% cardiolipin. The next step involves formation of cytidine diphosphoglyceride (CDP-diglyceride) by phosphatidate cytidyltransferase as shown in Fig. 4.

At this point the pathway branches. Addition of L-serine by phosphatidylserine synthase (encoded by *pssA*) yields phosphatidylserine. Decarboxylation of phosphatidylserine by phosphatidylserine decarboxylase (encoded by *psd*) yields phosphatidylethanolamine, one of the major phospholipids found in eubacteria. In the other branch of the pathway, addition of a second molecule of G3P to the CDP-diglyceride by phosphatidylglycerophosphate synthase (encoded by *pgsA*) results in the release of CMP and the formation of phosphatidylglycerol phosphate. Removal of P_i by phosphatidylglycerol phosphate phosphatase (encoded by *pgsA*, *pgsB*) yields phosphatidylglycerol. In bacteria, two molecules of phosphatidylglycerol are condensed by cardiolipin synthase (encoded by *cls* in *E. coli*), to release glycerol to form cardiolipin (diphosphatidylglycerol).

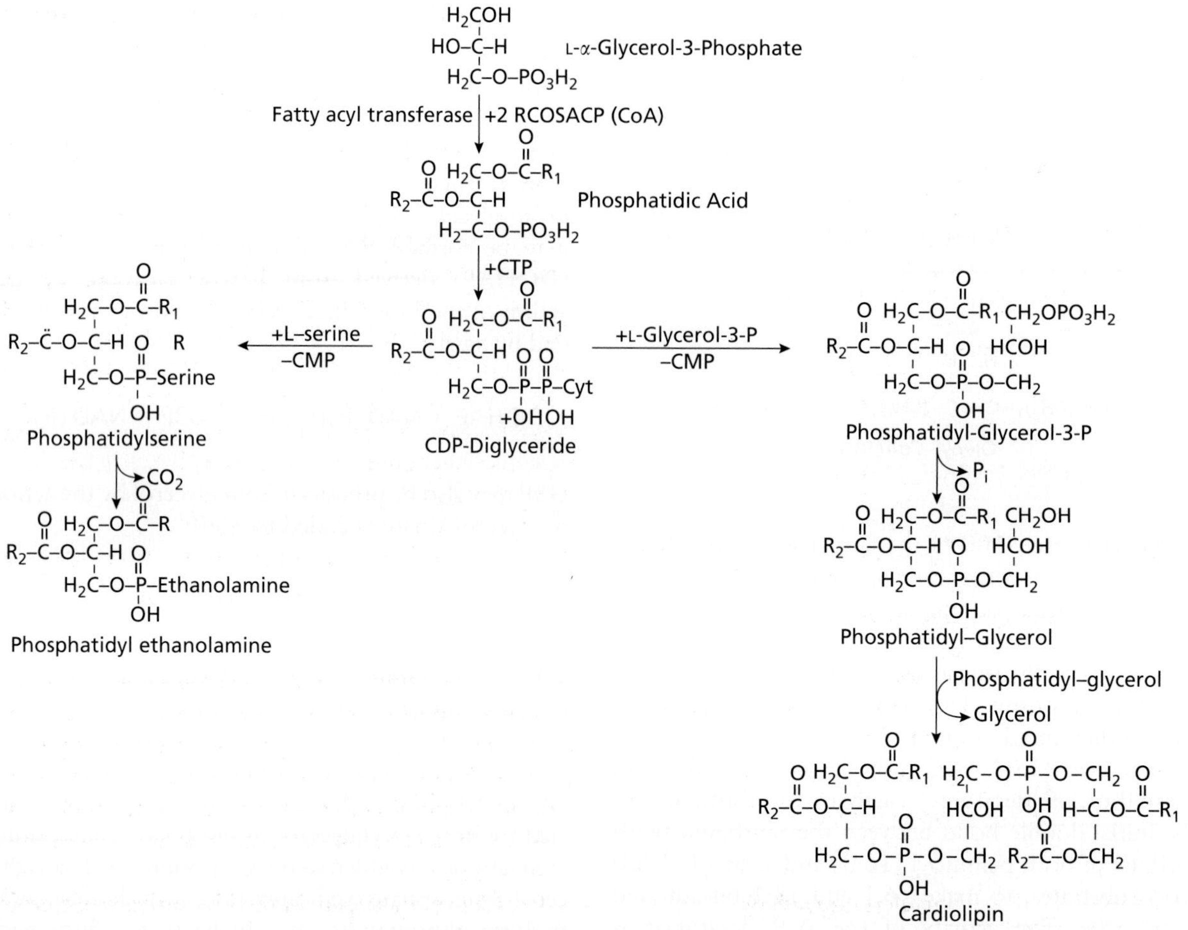

H_2COH
HO–C–H L-α-Glycerol-3-Phosphate
$H_2C–O–PO_3H_2$
Fatty acyl transferase +2 RCOSACP (CoA)
Phosphatidic Acid
+CTP
CDP-Diglyceride
+L-serine −CMP
Phosphatidylserine
CO_2
Phosphatidyl ethanolamine
+L-Glycerol-3-P −CMP
Phosphatidyl-Glycerol-3-P
P_i
Phosphatidyl–Glycerol
Phosphatidyl–glycerol
Glycerol
Cardiolipin

Fig. 4 Biosynthesis of phosphoglycerides (phospholipids). The *E. coli* enzymes and structural genes are: *plsB*, *sn*-glycerol-3-phosphate acyltransferase; *plsK*, 1-acylglycerol-3-phosphate acyltransferase; phosphatidate cytidyltransferase; *pssA*, phosphatidylserine synthase; *psd*, phosphatidylserine decarboxylase; *pgsA*, phosphatidylglycerophosphate synthase; *pgpA*, phosphatidylglycerophosphate phosphatase; *cls*, cardiolipin (diphosphatidylglycerol) synthase. R_1 and R_2 represent fatty acid substitutions. From Moat AG, Foster JW (1995) *Microbial Physiology*, 3rd edn. New York: John Wiley & Sons.

Biosynthesis of Mevalonate, Squalene and Sterols

In fungi, sterols are an important component of the cell membrane. In these organisms, an offshoot of the pathway of fatty acid synthesis leads to the formation of mevalonic acid, the first intermediate in the biosynthesis of sterols. Through a complex series of reactions, mevalonic acid is converted to farnesyl pyrophosphate and then to squalene. Squalene then undergoes several cyclisation reactions, ultimately to give rise to lanosterol, the precursor for the formation of many sterol derivatives. The pathways leading to the formation of the intermediates and to the ultimate formation of sterols are shown in Figs 5 and 6.

The pathway to mevalonic acid (Fig. 5) has been demonstrated in yeast as well as in mammalian liver. The condensing enzyme responsible for the coupling of acetyl-CoA and acetoacetyl-CoA has been isolated and purified from yeast. Hydroxymethylglutaryl-CoA can also arise from malonyl-CoA by reactions that appear to be identical to those responsible for the initiation of the biosynthesis of fatty acids. Free hydroxymethylglutarate is not readily utilised by yeast or mammalian liver as a precursor of steroids, presumably because of the absence of an activating enzyme. *Phycomyces blakesleeanus* can readily incorporate hydroxymethylglutarate, indicating that this organism has the necessary activating enzyme.

The *Mycoplasmataceae* are separated into two groups on the basis of their nutritional requirements for lipid components: *Mycoplasma* species require sterols for growth, while *Acholeplasma* species synthesise carotenoid compounds that assume the functional capacity of sterols. Sterol and carotenol

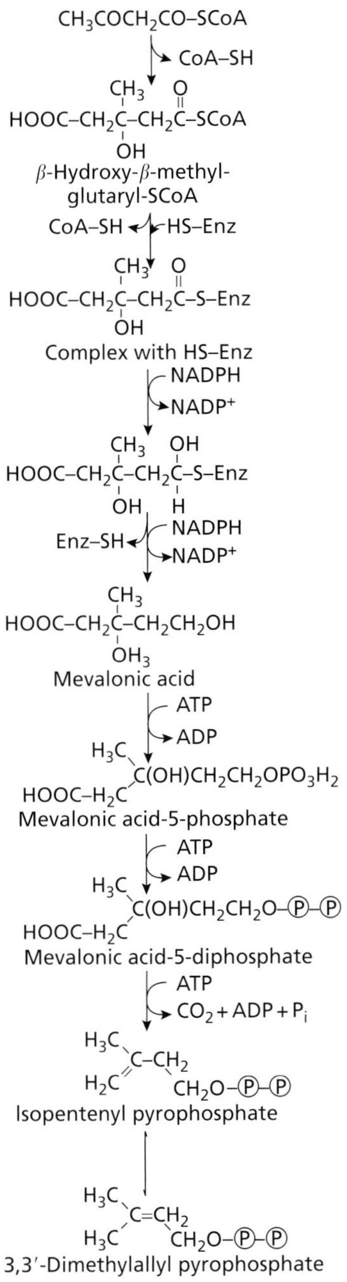

$CH_3COCH_2CO{-}SCoA$

↓ ← CoA–SH

$HOOC{-}CH_2C(CH_3)(OH){-}CH_2C(=O){-}SCoA$

β-Hydroxy-β-methyl-glutaryl-SCoA

↓ CoA–SH ← HS–Enz

$HOOC{-}CH_2C(CH_3)(OH){-}CH_2C(=O){-}S{-}Enz$

Complex with HS–Enz

↓ NADPH → $NADP^+$

$HOOC{-}CH_2C(CH_3)(OH){-}CH_2C(OH)(H){-}S{-}Enz$

↓ Enz–SH ←; NADPH → $NADP^+$

$HOOC{-}CH_2C(CH_3)(OH_3){-}CH_2CH_2OH$

Mevalonic acid

↓ ATP → ADP

$(H_3C)(HOOC{-}H_2C)C(OH)CH_2CH_2OPO_3H_2$

Mevalonic acid-5-phosphate

↓ ATP → ADP

$(H_3C)(HOOC{-}H_2C)C(OH)CH_2CH_2O{-}Ⓟ{-}Ⓟ$

Mevalonic acid-5-diphosphate

↓ ATP → $CO_2 + ADP + P_i$

$(H_3C)(H_2C{=})C{-}CH_2{-}CH_2O{-}Ⓟ{-}Ⓟ$

Isopentenyl pyrophosphate

↓

$(H_3C)(H_3C)C{=}CH_2{-}CH_2O{-}Ⓟ{-}Ⓟ$

3,3′-Dimethylallyl pyrophosphate

Fig. 5 Biosynthesis of mevalonic acid and conversion of mevalonic acid to the isoprenoid structural form. From Moat AG, Foster JW (1995) *Microbial Physiology*, 3rd edn. New York: John Wiley & Sons.

apparently serve in the same functional capacity, since cholesterol added to the culture medium is utilised in place of carotenoids, whose synthesis is repressed under these growth conditions. *Acholeplasma laidlawii* synthesises carotenoids from acetate. The organism contains a specific acetokinase and phospho-transacetylase, both of which are necessary for the synthesis of acetyl-CoA and a β-ketothiolase and CoA transferase required for the synthesis of acetoacetyl-CoA. The presence of β-hydroxy-β-methylglutaryl-CoA condensing enzyme and reductase activities in *Acholeplasma*, together with the ability to incorporate ^{14}C-labelled acetate into mevalonic acid, indicates that these organisms utilise the same pathway as yeast for the synthesis of mevalonic acid. The absence of β-hydroxy-β-methylglutaryl-CoA condensing enzyme and reductase activities in *Mycoplasma hominis* explains its growth requirement for sterol. In other species of *Mycoplasma* (e.g. *M. gallisepticum*), the metabolic block occurs after the mevalonic acid step in the biosynthetic pathway to terpenoids. Mevalonic acid is incorporated by *Lactobacillus casei* predominantly into bactoprenol, a C_{55}-isoprenol concerned with cell wall biosynthesis:

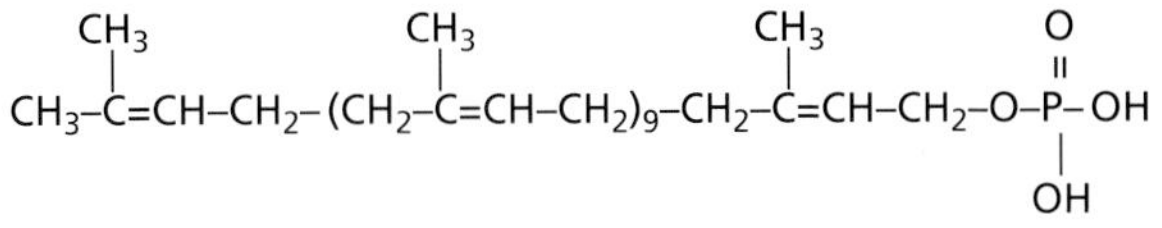

$$CH_3{-}C(CH_3){=}CH{-}CH_2{-}(CH_2{-}C(CH_3){=}CH{-}CH_2)_9{-}CH_2{-}C(CH_3){=}CH{-}CH_2{-}O{-}P(=O)(OH){-}OH$$

The distribution of tritiated bactoprenol in *L. casei* is relatively uniform throughout the cell, except in organisms that are in the process of undergoing cell division. In dividing cells there is a concentration of labelled bactoprenol in the septal region, indicating that the septal membrane is synthesised at the site of cell division.

In order to participate in the biosynthesis of sterols, mevalonic acid must first be converted into the isoprenoid structural form as shown in Fig. 6. This sequence of reactions has been demonstrated in yeast. There is some evidence that 5-diphospho-3-phosphomevalonic acid, a very unstable intermediate, is formed, accounting for the re-arrangement of the double bonds in the formation of the isopentenyl structure. Isopentenyl pyrophosphate and dimethylallyl pyrophosphate are condensed with the elimination of pyrophosphate to form geranyl pyrophosphate. This compound condenses with another molecule of isopentenyl pyrophosphate to yield nerolidol pyrophosphate and, ultimately, *trans*, *trans*-farnesyl pyrophosphate (Fig. 6). The reaction sequence that leads to the formation of nerolidol and farnesyl pyrophosphate has been demonstrated in yeast.

Squalene is formed by the condensation of either two molecules of farnesyl pyrophosphate or, more probably, one molecule of farnesyl pyrophosphate and one molecule of nerolidol pyrophosphate as shown in Fig. 6. The conversion of squalene to ergosterol in yeast or to cholesterol in mammals involves a number of intermediary steps. In yeast, an epoxide intermediate precedes the formation of lanosterol. At least eight additional steps are required

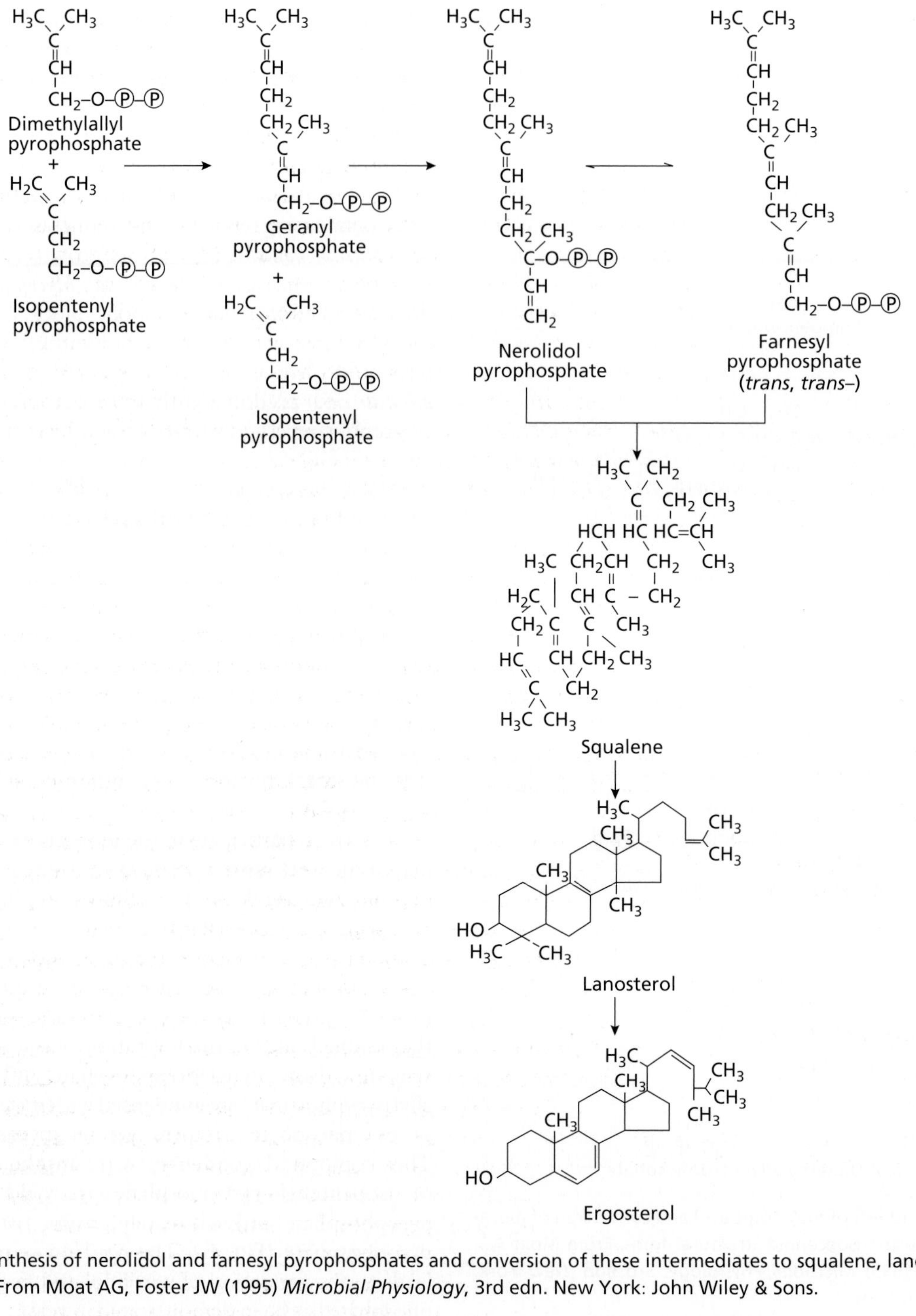

Fig. 6 Biosynthesis of nerolidol and farnesyl pyrophosphates and conversion of these intermediates to squalene, lanosterol and ergosterol. From Moat AG, Foster JW (1995) *Microbial Physiology*, 3rd edn. New York: John Wiley & Sons.

to convert lanosterol to ergosterol. The additional carbon (C_{28}) is donated from *S*-adenosylmethionine at the lanosterol level. In mammalian systems, the conversion of squalene to cholesterol requires at least two sterol carrier proteins (SCP_1 and SCP_2) that bind the substrate and make it reactive to the sterol-synthesising enzymes present in the microsomes. These sterol carrier proteins differ in several respects from the acyl carrier protein that functions in the biosynthesis of fatty acids. However, both of these compounds appear to serve the analogous function of maintaining the solubility and reactivity of the substrates.

Saccharomyces cerevisiae and other yeasts require sterols for growth. Under aerobic conditions, most yeasts are able to synthesise the required level of ergosterol. Under anaerobic conditions, yeast cannot synthesise sterols or unsaturated fatty acids and require their addition to the growth medium. Yeast mutants auxotrophic for sterols and unsaturated fatty acids are altered in the fatty acid composition of their mitochondrial phospholipids. The incorporation of fatty acids into phospholipids varies with the sterol and unsaturated fatty acids supplied. Ergosterol, in the presence of linoleic or linolenic acids or a mixture of palmitoleic and oleic acids, permits excellent growth. Substitution of other sterols such as cholesterol or addition of oleic acid as the sole fatty acid, results in poor growth. The genetic basis for limiting sterol synthesis in *S. cerevisiae* resides in the regulation of hydroxymethylglutaryl-CoA reductase activity. A decrease in the specific activity of this enzyme correlates with accumulation of squalene during supplementation with ergosterol or mevalonolactone. The addition of ergosterol results in feedback inhibition of hydroxymethylglutaryl-CoA reductase.

Amino Acid Biosynthesis

Amino acid biosynthesis is most conveniently discussed on the basis of groups of amino acids that originate from a common precursor:

1. *Glutamate or α-ketoglutarate*: Glutamate, glutamine, glutathione, proline, arginine, putrescine, spermine, spermidine and, in fungi, lysine. δ-Aminolevulinate, a tetrapyrrole (haem) precursor, arises from glutamate in some organisms.
2. *Aspartate*: Aspartate, asparagine, threonine, methionine, isoleucine and, in bacteria, lysine.
3. *Pyruvate*: Alanine, valine, leucine and isoleucine.
4. *Serine-glycine or triose*: Serine, glycine, cysteine, cystine. In fungi, mammals and some bacteria δ-aminolevulinate is formed by condensation of glycine and succinate.
5. *Aromatic amino acids*: Phenylalanine, tyrosine, tryptophan. *p*-Aminobenzoate, enterochelin, ubiquinone, menaquinone and NAD can also originate via branches from the common aromatic amino acid pathway.
6. *Histidine*: Emanates from the purine pathway.

Amino Acids Emanating from Glutamate or α-Ketoglutarate

Glutamine and Glutathione Synthesis

Glutamate and glutamine play a central role in amino acid biosynthesis by the ready transfer of amino or amide groups, respectively, in the synthesis of other amino acids by transamination or transamidation reactions. Glutamine is synthesised from glutamate with the participation of ammonia and ATP. Glutathione, a disulphide-containing amino acid whose functions have only recently begun to be formulated in a precise manner, is synthesised in two steps. The coupling of L-glutamate and L-cysteine in the presence of ATP to form γ-glutamylcysteine is catalysed by a specific synthetase. Glutathione synthetase uses glycine and ATP to form glutathione from γ-glutamylcysteine (**Fig. 7**).

The Proline Pathway

The pathway to proline involves formation of γ-glutamylphosphate from L-glutamate and ATP by γ-glutamyl kinase (**Fig. 7**). In the presence of NADPR, γ-glutamylphosphate is reduced to glutamate γ-semialdehyde. Glutamate γ-semialdehyde can cyclise spontaneously to form l-pyrroline-5-carboxylate. The latter is converted to proline by a specific reductase.

δ-Aminolevulinate Synthesis

The tetrapyrroles include important compounds such as vitamin B_{12}, haem and bacteriochlorophyll. In *Salmonella typhimurium* and *E. coli* δ-aminolevulinic acid (ALA), the first committed precursor to tetrapyrroles, arises from glutamate. This pathway, termed the C_5 pathway, which was originally thought to occur primarily in plants and algae, is now firmly established as a major route to ALA in several bacterial species. In earlier studies it had been considered that the condensation of glycine and succinyl-CoA by the enzyme ALA synthase was the only route of ALA synthesis. This is still the major route of ALA formation in mammals, fungi, and certain bacteria such as *Rhodobacter sphaeroides*, *R. capsulatus* and *Bradyrhizobium japonicum*.

The C_5 pathway to ALA involves activation of glutamyl-tRNA^{Glu} by glutamyl-tRNA synthase, reduction to glutamate γ-semialdehyde by an NADPH-dependent glutamyl-tRNA reductase (encoded by *hemA*), and transamination by glutamate γ-semialdehyde aminomutase (encoded by *hemL*) to form ALA. The details of the pathway of haem biosynthesis will be considered below.

The Arginine Pathway

Bacteria and fungi synthesise ornithine by way of a series of *N*-acetyl derivatives (**Fig. 8**). The function of the *N*-acetyl groups is to prevent the premature cyclisation of l-pyrroline-5-carboxylate to proline. The pathway diverges in different organisms, depending on the manner in which the acetyl group is removed. In the *Enterobacteriaceae* and *Bacillaceae*, *N*-acetylornithine is deacetylated by acetylornithine deacetylase encoded by *argE*. In *Neisseria gonorrhoeae*, the *Pseudomonadaceae*, cyanobacteria, photosynthetic bacteria, and yeasts and moulds, the acetyl group of *N*-acetyl-ornithine is recycled by ornithine acetyltransferase.

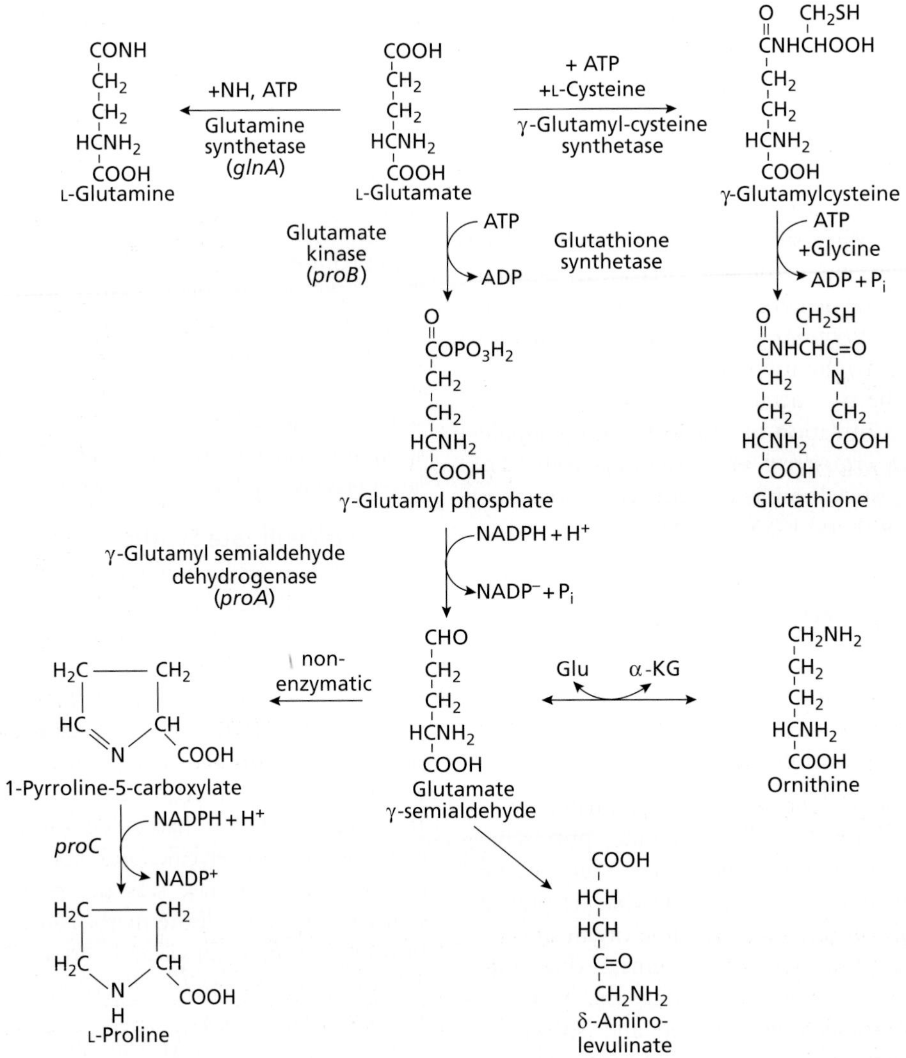

Fig. 7 Pathways to glutamine, glutathione, proline, ornithine and aminolevulinate. The transaminase that inter-converts ornithine and glutamate γ-semialdehyde is reversible. However, it is generally considered that this reaction serves for ornithine degradation. Ornithine is normally synthesised via the pathway shown in Fig. 8. Gene designations are for *E. coli*. *proC*, 1-pyrroline-5-carboxylate reductase. From Moat AG, Foster JW (1995) *Microbial Physiology*, 3rd edn. New York: John Wiley & Sons.

The eight enzymes involved in the arginine pathway (**Fig. 8**) are found in *Neurospora crassa*, *Aspergillus niger*, *S. cerevisiae*, *C. albicans*, *E. coli*, *S. typhimurium*, *B. subtilis*, *B. sphaericus*, various species of *Proteus*, *Streptococcus bovis*, *Neisseria gonorrhoeae*, and several species of *Pseudomonas*.

Carbamoyl phosphate is a common precursor in the biosynthesis of arginine and pyrimidine. *E. coli* and *S. typhimurium* produce a single carbamoyl phosphate synthase that catalyses the reaction:

$$2ATP + HCO_3^- + \text{L-glutamine} + H_2O \xrightarrow[Mg^{2+}]{K^+} NH_2COOPO_3H_2 + 2P_i + 2ADP + \text{glutamate}$$

Ammonia can replace glutamine as a nitrogen donor *in vitro*, but glutamine is the physiologically preferred substrate. The enzyme is composed of two non-identical subunits. The smaller subunit, encoded by *carA*, acts as a glutamine amidotransferase. The larger subunit, encoded by *carB*, carries out the remaining functions. As might be expected, since carbamoyl phosphate is involved in two major biosynthetic pathways, expression of the *carAB* operon is regulated by both arginine and pyrimidines. The enzyme is also subject to allosteric control by intermediates in both pathways. UMP is inhibitory, whereas ornithine stimulates enzyme activity. The enzyme is also activated by inosine monophosphate (IMP) and phosphoribosyl pyrophosphate (PRPP), co-ordinating its activity with purine biosynthesis as well.

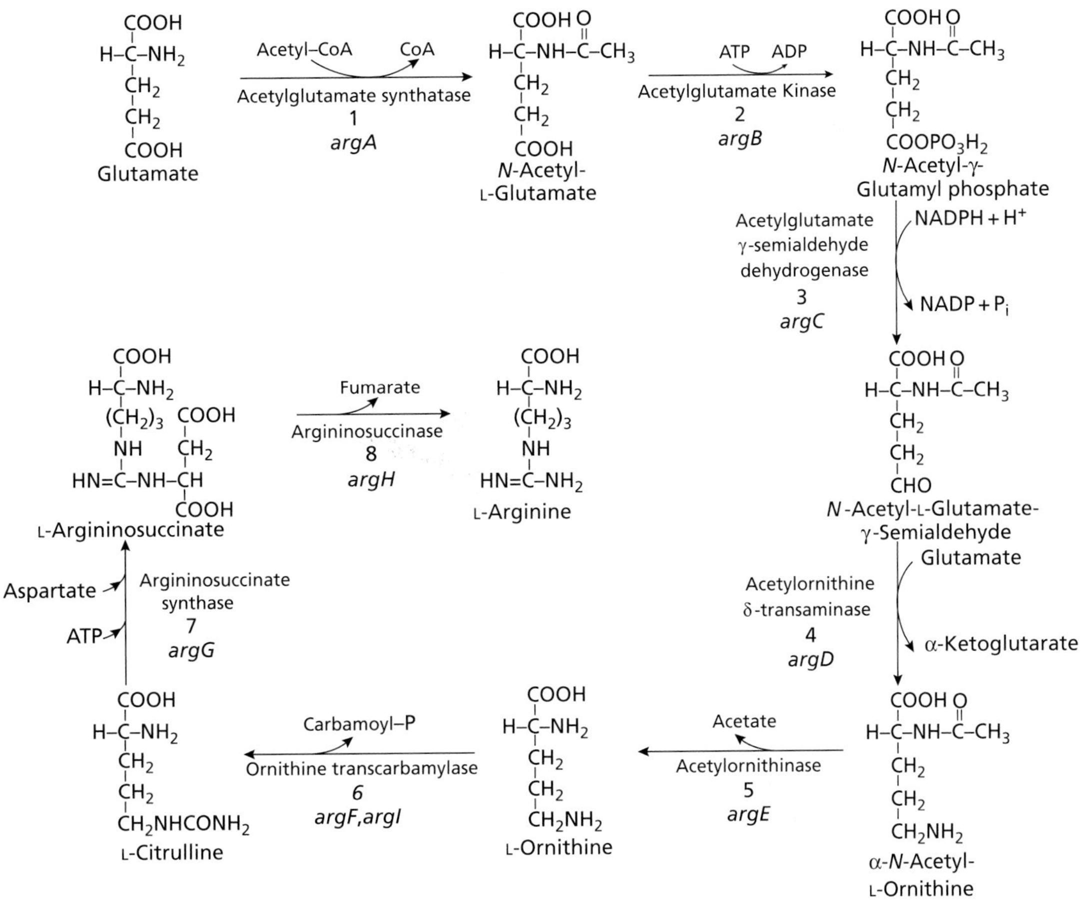

Glutamate → (Acetyl-CoA, CoA; Acetylglutamate synthatase, 1, *argA*) → *N*-Acetyl-L-Glutamate → (ATP, ADP; Acetylglutamate Kinase, 2, *argB*) → *N*-Acetyl-γ-Glutamyl phosphate → ($NADPH + H^+$, $NADP + P_i$; Acetylglutamate γ-semialdehyde dehydrogenase, 3, *argC*) → *N*-Acetyl-L-Glutamate-γ-Semialdehyde → (Glutamate, α-Ketoglutarate; Acetylornithine δ-transaminase, 4, *argD*) → α-*N*-Acetyl-L-Ornithine → (Acetate; Acetylornithinase, 5, *argE*) → L-Ornithine → (Carbamoyl-P; Ornithine transcarbamylase, 6, *argF*, *argI*) → L-Citrulline → (Aspartate, ATP; Argininosuccinate synthase, 7, *argG*) → L-Argininosuccinate → (Fumarate; Argininosuccinase, 8, *argH*) → L-Arginine

Fig. 8 Pathway of arginine biosynthesis in micro-organisms. The structural gene designations for the enzymes are for *E. coli*. In yeasts, moulds, *E. coli*, *Proteus mirabilis*, *Serratia marcescens* and certain other enterobacteria, step 5 occurs as shown. In *Pseudomonas fluorescens*, *Micrococcus glutamicus*, *Anabaena variabilis* and several other bacteria, step 5 involves trans-acetylation between glutamate and acetylornithine. From Moat AG, Foster JW (1995) *Microbial Physiology*, 3rd edn. New York: John Wiley & Sons.

In fungi such as *S. cerevisiae* and *N. crassa* there are two carbamoyl phosphate synthases. One enzyme, carbamoyl phosphate synthase A, is linked to the arginine pathway and is subject to repression by arginine. The other enzyme, carbamoyl phosphate synthase P, is linked to the pyrimidine biosynthesis pathway and is subject to both repression and feedback inhibition by pyrimidines. Localisation of the arginine pathway carbamoyl phosphate synthase in mitochondria of *N. crassa* seems to play a major role in the channelling of this precursor. In yeast there seems to be little channelling of carbamoyl phosphate since both carbamoyl phosphate synthases are in the cytoplasm and contribute to a common pool of carbamoyl phosphate.

Polyamine Biosynthesis

Polyamines are widely distributed in bacteria, yeasts and moulds, as well as in higher forms. Although a number of growth processes are affected by polyamines, their precise role in governing cell growth and differentiation has not been established. The major pathway to polyamines is via the decarboxylation of ornithine to putrescine (Fig. 9). Ornithine decarboxylase is rate-limiting in the pathway to polyamines. The ornithine decarboxylase pathway appears to be common to all cells.

In bacteria and plants an alternate pathway to putrescine proceeds via decarboxylation of arginine to

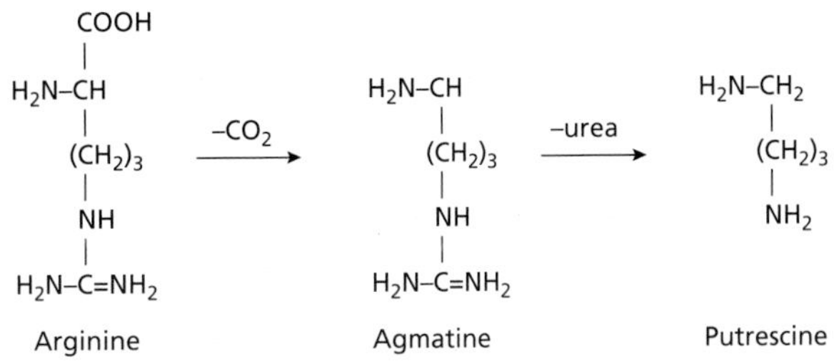

Arginine ($H_2N{-}C{=}NH_2$–NH–$(CH_2)_3$–$H_2N{-}CH$–COOH) → ($-CO_2$) → Agmatine ($H_2N{-}C{=}NH_2$–NH–$(CH_2)_3$–$H_2N{-}CH$) → (−urea) → Putrescine (NH_2–$(CH_2)_3$–$H_2N{-}CH_2$)

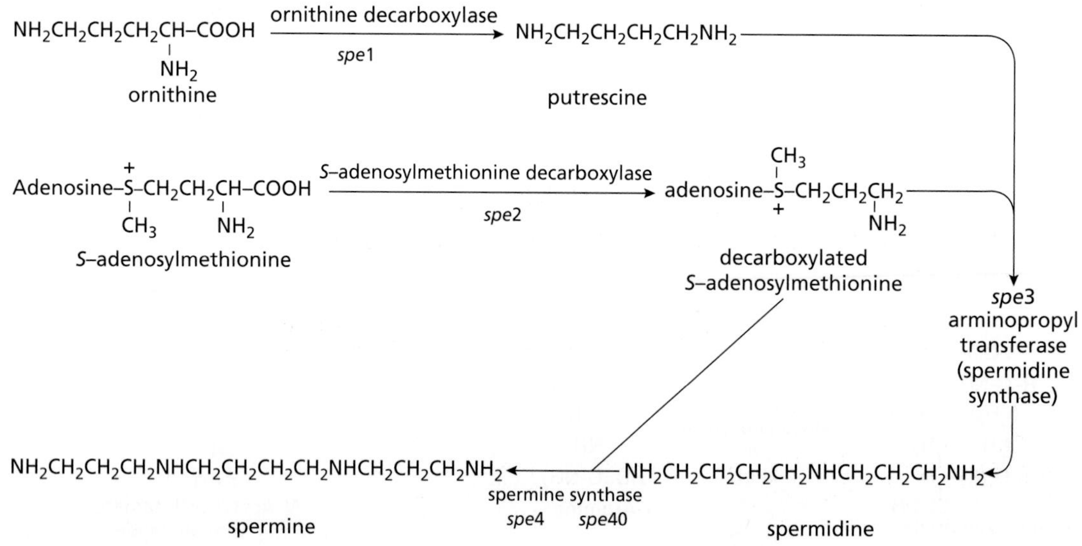

Fig. 9 Biosynthetic steps and genes involved in polyamine biosynthesis in *Saccharomyces cerevisiae*. From Tabor CW, Tabor H, Tyagi AK, Cohn MS (1982) The biochemistry, genetics, and regulation of polyamine biosynthesis in *Saccharomyces cerevisiae*. *Fedn Proc.* 41: 3084–3088.

agmatine by a biosynthetic arginine decarboxylase (encoded by *speA* in *E. coli*). Agmatine ureohydrolase removes urea from agmatine to yield putrescine:

Exogenous arginine acts as a signal for the selective utilisation of this pathway in *E. coli*, which lacks arginase and cannot convert arginine to ornithine. When arginine is added to the growth medium, ornithine levels decline because of the inhibiton of arginine biosynthesis.

Lysine Biosynthesis in Fungi

In yeasts and moulds the biosynthetic pathway to lysine emanates from α-ketoglutarate (**Fig. 10**). The bacterial pathway to lysine, which is initiated by the condensation of pyruvate and aspartate β-semialdehyde, is part of the aspartate family of amino acid biosynthetic routes (**Fig. 11**). These completely divergent routes of lysine biosynthesis represent a major phylogenetic difference between bacteria and fungi. The series of reactions from homocitrate to α-ketoadipate are analogous to the reactions involved in the conversion of citrate to α-ketoglutarate in the citric acid cycle. The formation of homocitrate from an α-ketoglutarate and acetyl-CoA is inhibited by lysine, indicating a feedback control mechanism in the pathway. The initial step in the pathway is also subject to repression by lysine.

The biosynthesis of β-lactam antibiotics (penicillins and cephalosporins) occurs by a branch from the lysine pathway in *Penicillium chrysogenum*, *Cephalosporium acremonium*, *Streptomyces clavuligerus*, and related organisms.

Amino Acids Emanating from Aspartate and Pyruvate

The aspartate and pyruvate families of amino acids are discussed together, because there is a distinct overlap in the enzymes involved in the terminal steps of the biosynthesis of the branched-chain amino acids (**Fig. 12**). The valine and leucine carbon chains are derived from pyruvate. In bacteria, threonine, isoleucine, methionine and lysine all emanate from aspartate. Fungi utilise similar pathways with the exception of lysine, which as we have noted above is synthesised by a completely different pathway.

Asparagine Synthesis

Asparagine is synthesised from aspartate, ammonia (or glutamine) and ATP by asparagine synthase:

$$\text{Aspartate} + NH_3 + \text{ATP} \xrightarrow[\text{AsnA, AsnB}]{} \text{Asparagine} + \text{ADP}$$

Two asparagine synthases are present in *E. coli*. They are encoded by two unlinked structural genes designated *asnA* (84 min) and *asnB* (15 min). Asparaginase A (AsnA) is ammonia-dependent, whereas asparaginase B (AsnB) can use either ammonia or glutamine as the nitrogen donor. Mutant strains that carry only the *aspB* gene can assimilate ammonia into asparagine in the presence of excess ammonia. The glutamine-dependent enzyme is regulated by the

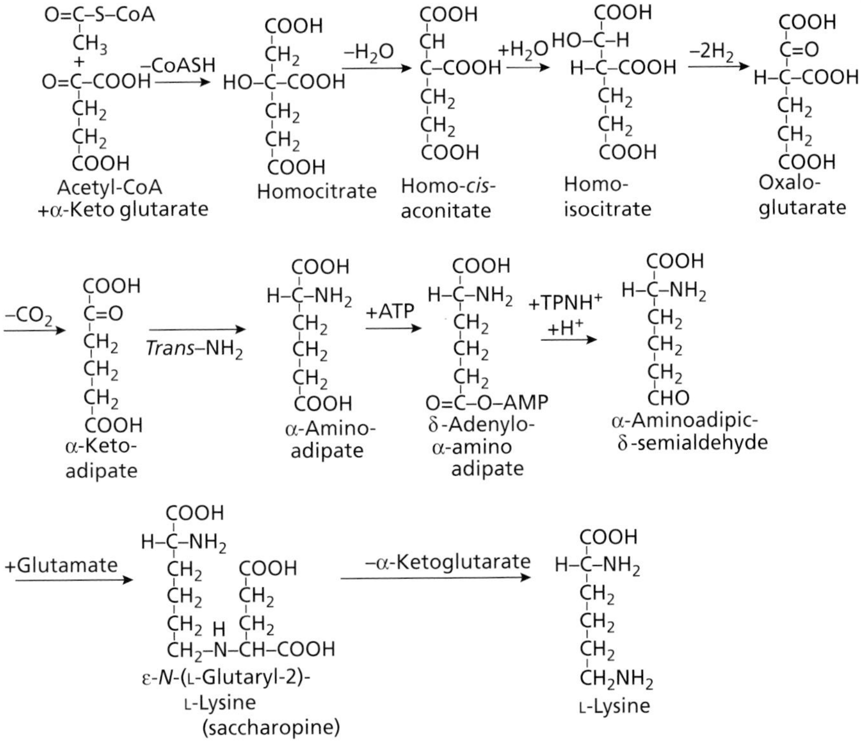

Acetyl-CoA + α-Keto glutarate —(−CoASH)→ Homocitrate —($-H_2O$)→ Homo-*cis*-aconitate —($+H_2O$)→ Homo-isocitrate —($-2H_2$)→ Oxalo-glutarate —($-CO_2$)→ α-Keto-adipate —(*Trans*–NH_2)→ α-Amino-adipate —(+ATP)→ δ-Adenylo-α-amino adipate —(+TPNH$^+$ +H$^+$)→ α-Aminoadipic-δ-semialdehyde —(+Glutamate)→ ε-*N*-(L-Glutaryl-2)-L-Lysine (saccharopine) —(−α-Ketoglutarate)→ L-Lysine

Fig. 10 Biosynthesis of lysine in yeasts and moulds. This pathway is termed the α-aminoadipate pathway. Genetic studies of this pathway have been conducted mainly in *Saccharomyces cerevisiae*, *S. lipolytica* and *Neurospora crassa*. The structural genes for *S. cerevisiae* are: LYS7, homocitrate dehydratase; LYS4, homo-*cis*-aconitate hydratase; LYS12, homo-isocitrate dehydrogenase; LYS2, LYS5, α-aminoadipate reductase; LYS9, saccharopine reductase; LYS1, saccharopine dehydrogenase. From Moat AG, Foster JW (1995) *Microbial Physiology*, 3rd edn. New York: John Wiley & Sons.

Pyruvate + Aspartate-β-semialdehyde —($-H_2O$, *dapA*)→ [−H_2O ⇅ +H_2O: Dihydrodipicolinate] —(NADPH+H$^+$, *dapB*)→ [−H_2O ⇅ +H_2O: Tetrahydrodipicolinate] —(Succinyl–CoA, *dapC*)→ *N*-Succinyl-ϵ-keto-L-α-aminopimelate —(PALP, Glutamate, *dapD*)→ *N*-Succinyl-α,ϵ-diaminopimelate —(−Succinate, *dapE*)→ L-α-Diamino-pimelate —(PALP, Epimerase)→ *Meso*-Diamino-pimelate —($-CO_2$, PALP, *lysA*)→ L-Lysine

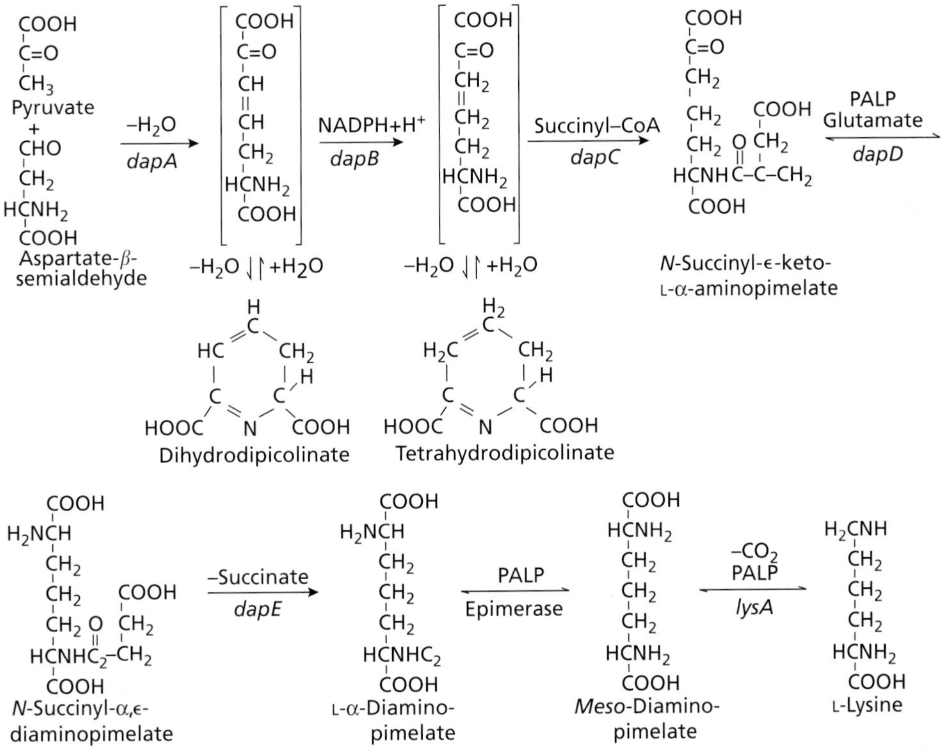

Fig. 11 Biosynthesis of lysine in bacteria. The structural gene designations given are those for *E. coli*. *dapA*, dihydrodipicolinate synthase; *dapB*, dihydrodipicolinate reductase; *dapC*, tetrahydrodipicolinate succinylase; *dapD*, succinyl diaminopimelate aminotransferase; *dapE*, succinyl diaminopimelate desuccinylase; *lysA*, diaminopimelate decarboxylase. PALP, pyridoxal phosphate. From Moat AG, Foster JW (1995) *Microbial Physiology*, 3rd edn. New York: John Wiley & Sons.

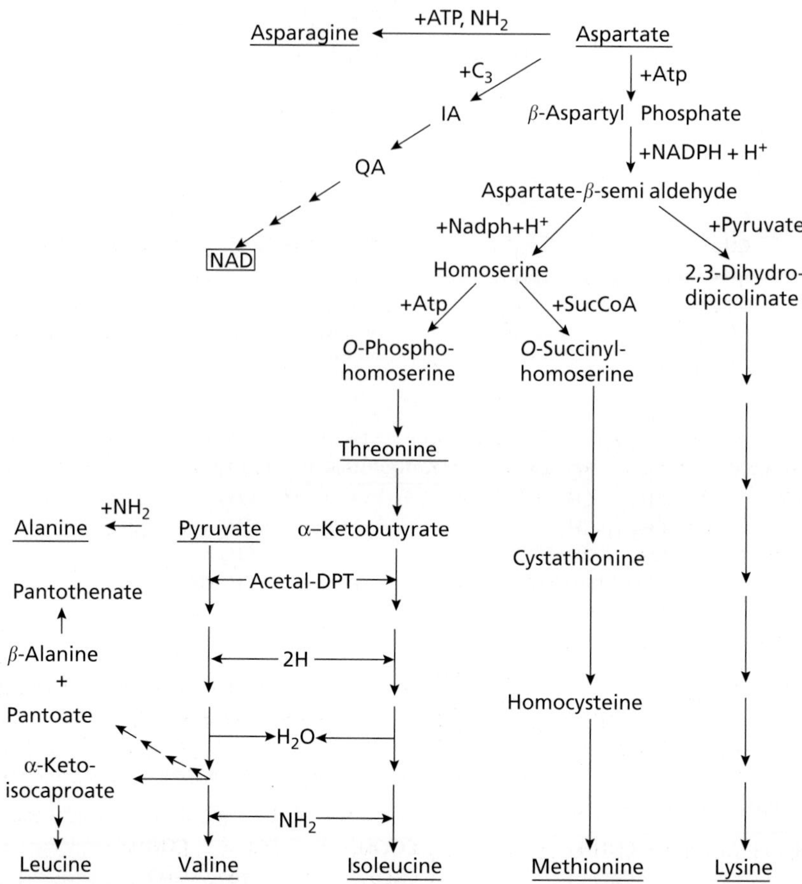

Fig. 12 Outline of the pyruvate and aspartate families of amino acids. From Moat AG, Foster JW (1995) *Microbial Physiology*, 3rd edn. New York: John Wiley & Sons.

asparagine content of the intracellular asparagine pool. There are also multiple asparagine synthases in yeast. Although the yeast enzymes can use ammonia, affinity studies show that glutamine is the preferred substrate.

The Aspartate Pathway

In bacteria and fungi, aspartokinase and aspartate semialdehyde dehydrogenase initiate the aspartate pathway of amino acid biosynthesis (**Fig. 13**). In *E. coli* there are three aspartokinases, each of which is specific to the end product of the pathway involved in its synthesis. Aspartokinase I (ThrA) is specific for the threonine branch. Aspartokinase II (MetL) is specific for the methionine branch and aspartokinase III (LysC) is specific for the lysine branch. Aspartate semialdehyde dehydrogenase (Asd) forms aspartate β-semialdehyde by removal of phosphate from aspartyl phosphate and reduction with NADPH. Aspartate β-semialdehyde stands at the first branch-point in the multi-branched pathway. Reduction of aspartate β-semialdehyde to homoserine is catalysed by homoserine dehydrogenase I (ThrA), which is specific for the threonine branch of the pathway, or homoserine dehydrogenase II (MetL), which is specific for the methionine branch.

The Bacterial Pathway to Lysine

Condensation of aspartate β-semialdehyde with pyruvate yields dihydrodipicolinate, the first intermediate in the bacterial pathway to lysine (**Fig. 11**). This pathway is of special interest because dipicolinic acid is produced during sporulation in *Bacillus* spp. and diaminopimelic acid or lysine is present in the peptidoglycan structures of all prokaryotes that produce a rigid cell wall. This pathway to lysine occurs in *Enterococcus faecalis*, *Staphylococcus aureus*, *E. coli*, *S. typhimurium*, *B. subtilis* and the cyanobacteria, and, by inference, it is regarded as occurring in most other bacteria.

The biosynthetic pathway depicted in **Fig. 11** involves the formation of succinylated derivatives of ϵ-keto-L-α-aminopimelate and α,ϵ-diaminopimelate as intermediates in the pathway. This pathway is

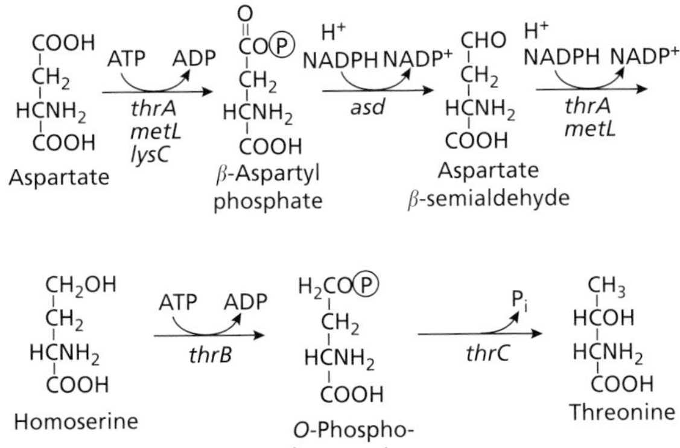

Fig. 13 Initial steps in the aspartate pathway of amino acid biosynthesis. In *E. coli* there are three aspartokinase enzymes. Their designated structural genes are *thrA*, aspartokinase I; *metL*, aspartokinase II; *lysC*, aspartokinase III. The structural gene for aspartate semialdehyde dehydrogenase is *asd*. *E. coli* contains two homoserine dehydrogenases: *thrA*, which is specific for the threonine branch of the pathway and *metL*, specific for the methionine branch. Homoserine is converted to *O*-phosphohomoserine by homoserine kinase (ThrB). Threonine synthase (ThrC) converts *O*-phosphohomoserine to threonine. From Moat AG, Foster JW (1995) *Microbial Physiology*, 3rd edn. New York: John Wiley & Sons.

utilised by *E. coli*, *S. typhimurium* and *Staph. aureus* and a number of other bacteria. Certain *Bacillus* spp. appear to form exclusively acetylated derivatives. A few organisms utilise a pathway in which tetrahydrodipicolinate (piperideine-2, 6-dicarboxylate) is converted in a single step to D,L-diaminopimelate by diaminopimelate dehydrogenase. This pathway is operative in *B. sphaericus* and *Corynebacterium glutamicum*. However, in the latter the pathway that uses succinylated intermediates appears to function along with the direct dehydrogenase pathway. This organism is of interest because of its use for the commercial production of lysine. The dehydrogenase pathway is apparently a pre-requisite for handling increased flow of metabolites to D,L-diaminopimelate and lysine.

Escherichia coli, *Shigella* and *Salmonella* can synthesise an iron-chelating siderophore, aerobactin, from lysine. The iron-regulated aerobactin operon is found on a ColV-K30 plasmid. This operon consists of at least five genes for synthesis (*iuc*, iron uptake chelate) and transport (*iut*, iron uptake transport) of aerobactin. The biosynthetic pathway starts with oxidation of lysine to N^{ϵ}-hydroxylysine by N^{ϵ}-lysine monooxygenase (IucD), acetylation of N^{ϵ}-hydroxylysine to form N^{ϵ}-acetyl-N^{ϵ}-hydroxylysine by the action of an *N*-acetyltransferase enzyme (IucB). The final stage of citrate addition and condensation to form the aerobactin ring is catalysed by aerobactin synthase (IucC).

Threonine, Isoleucine and Methionine Formation

Another branch-point in the aspartate pathway occurs at homoserine. One branch leads to the formation of threonine and, ultimately, isoleucine. Homoserine can also be converted to methionine by several alternate routes. In *E. coli*, *S. typhimurium* and other enteric organisms, conversion of homoserine to homocysteine involves either of two alternative pathways (Fig. 14). Homoserine succinyltransferase (MetA) condenses succinyl-CoA and homoserine to yield α-succinylhomoserine. Cystathionine γ-synthase is apparently capable of catalysing the direct reaction of H_2S or methylsulphide with α-succinylhomoserine to form homocysteine.

The methylation of homocysteine to form methionine may occur via either of two enzymes. One enzyme (MetH) is vitamin B_{12}-dependent and requires NADH, FAD, *S*-adenosylmethionine, and either 5-methyltetrahydrofolate or its triglutamyl derivative. A vitamin B_{12}-independent enzyme (MetE) requires only Mg^{2+} and 5-methyltetrahydropteroyl triglutamate. The 5-methyltetrahydrofolate is formed from serine and tetrahydrofolate (THF) by the action of serine hydroxymethyltransferase (GlyA), which converts THF to 5,10-methylene-THF. The enzyme 5,10-methylenetetrahydrofolate reductase (MetF) uses reduced FAD ($FADH_2$) to convert 5,10-methylene-THF to 5-methyl-THF.

In *Neurospora*, the conversion of homoserine to homocysteine occurs via an analogous sequence of reactions, in which homoserine and acetyl-CoA are condensed to form *O*-acetylhomoserine. The reaction is catalysed by homoserine acetyltransferase. In this sequence, cystathionine γ-synthase exchanges cysteine for the acetyl group in α-acetylhomoserine to release acetate and form cystathionine. This pathway is also utilised by *Aspergillus nidulans*.

An alternate route of homocysteine formation involving the direct sulphhydrylation of α-acetylhomoserine, α-succinylhomoserine or α-phosphohomoserine, is mediated by homocysteine synthase:

$$O\text{-Acetylhomoserine} + H_2S \rightarrow \text{L-homocysteine} + \text{acetate}$$

$$O\text{-succinylhomoserine} + H_2S \rightarrow \text{L-homocysteine} + \text{succinate}$$

$$O\text{-Phosphohomoserine} + H_2S \rightarrow \text{L-homocysteine} + P_i$$

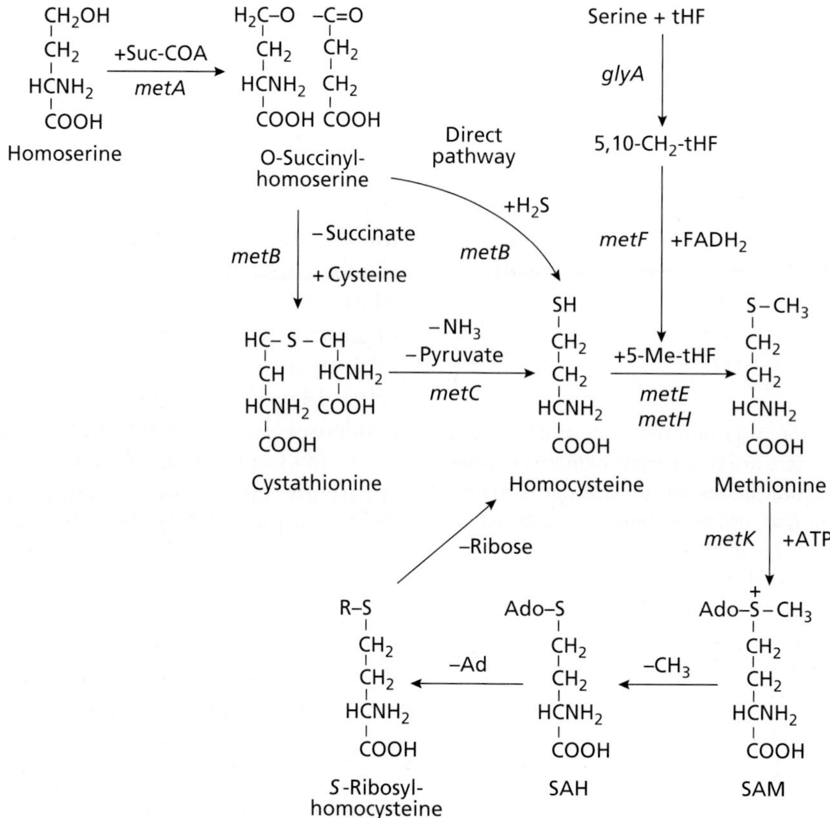

Fig. 14 Alternate pathways from homoserine to methionine. The structural gene designations are for *E. coli*. *metA*, homoserine acyltransferase; *metBI*, cystathionine-γ-synthase; *metC*, cystathionase. 5-Methyltetrahydrofolate (5-Me-THF) is formed from serine and tetrahydrofolate (THF) via serine hydroxymethyltransferase (*glyA*), which yields 5,10-methylenetetrahydrofolate (5,10-CH_2-THF) upon reduction by reduced flavin adenine dinucleotide ($FADH_2$) catalysed by 5,10-methylenetetrahydrofolate reductase (*metF*). The vitamin B_{12}-dependent homocysteinemethylase (*metH*) requires catalytic amounts of *S*-adenosylmethionine and $FADH_2$. *S*-Adenosylmethionine synthase (*metK*) catalyses the synthesis of *S*-adenosylmethionine (SAM) from methionine and ATP. From Moat AG, Foster JW (1995) *Microbial Physiology*, 3rd edn. New York: John Wiley & Sons.

The hydrogen sulphide presumably arises from cysteine desulphhydrase action or the sulphate reduction pathway.

One or both of these activities has been reported in *N. crassa*, *S. cerevisiae* and *E. coli*. The first of these reactions appears to be the main pathway in yeast, but alternative routes for methionine biosynthesis are also available.

Isoleucine, Leucine and Valine Biosynthesis

The isoleucine carbon skeleton is derived, in part, from aspartate via the deamination of threonine. It is convenient to discuss the biosynthesis of all three branched-chain amino acids, isoleucine, leucine and valine, because of the close inter-relationship of the pathways. In the isoleucine–valine pathway, four of the steps in both sequences are catalysed by the same enzymes (**Fig. 15**). The immediate precursor of valine, α-ketoisovalerate, represents another branch-point. Condensation with acetyl-CoA initiates the series of reactions leading to leucine synthesis (**Fig. 15**). Via another series of reactions, α-ketoisovalerate is converted to pantoic acid, which is a precursor of pantothenic acid. The basic series of reactions leading to the formation of the branched-chain amino acids appears to be quite similar in most micro-organisms so far examined. For example, the isopropylmalate pathway to leucine is widespread among diverse organisms capable of leucine biosynthesis.

Regulation of the Aspartate Pathway

Regulation of the biosynthesis of the amino acids of the aspartate pathway is complex because of its multiple branches and, to some extent, the inter-relationships between the aspartate and pyruvate families. In *E. coli* K-12, primary regulation is exerted at two

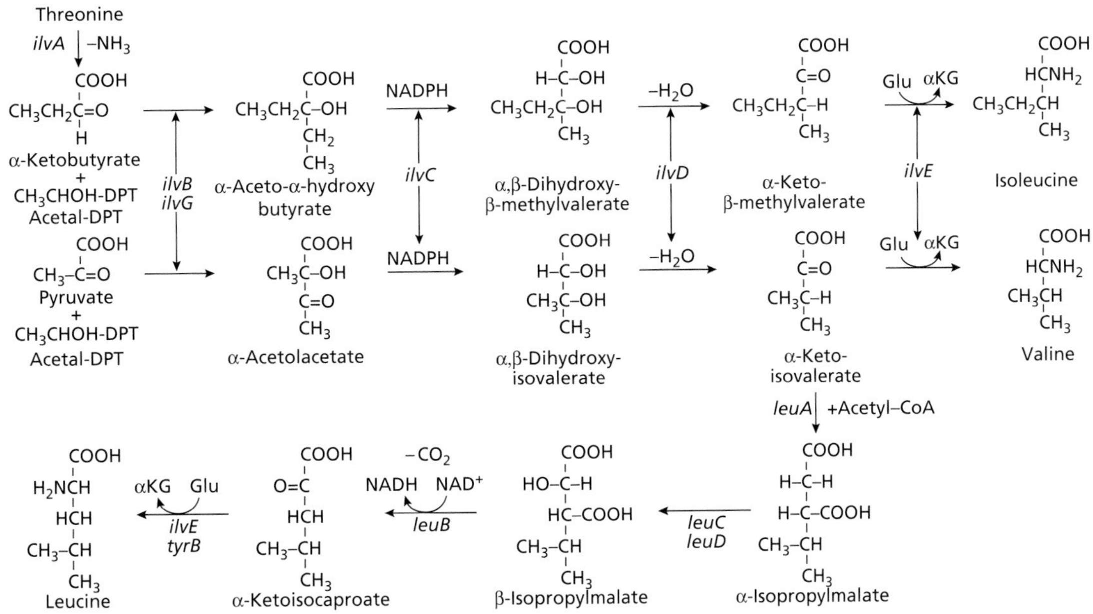

Threonine
ilvA $-NH_3$
COOH
$CH_3CH_2C{=}O$
H
α-Ketobutyrate
+
CH_3CHOH-DPT
Acetal-DPT
ilvB
ilvG
COOH
CH_3CH_2C-OH
CH_2
CH_3
α-Aceto-α-hydroxy butyrate
NADPH
ilvC
COOH
H-C-OH
CH_3CH_2C-OH
CH_3
α,β-Dihydroxy-β-methylvalerate
$-H_2O$
ilvD
COOH
C=O
CH_3CH_2C-H
CH_3
α-Keto-β-methylvalerate
Glu αKG
ilvE
COOH
$HCNH_2$
CH_3CH_2CH
CH_3
Isoleucine

COOH
CH_3-C=O
Pyruvate
+
CH_3CHOH-DPT
Acetal-DPT
COOH
CH_3C-OH
C=O
CH_3
α-Acetolacetate
NADPH
COOH
H-C-OH
CH_3C-OH
CH_3
α,β-Dihydroxy-isovalerate
$-H_2O$
COOH
C=O
CH_3C-H
CH_3
α-Keto-isovalerate
Glu αKG
COOH
$HCNH_2$
CH_3CH
CH_3
Valine

leuA +Acetyl-CoA

COOH
H_2NCH
HCH
CH_3-CH
CH_3
Leucine
αKG Glu
ilvE
tyrB
COOH
O=C
HCH
CH_3-CH
CH_3
α-Ketoisocaproate
$-CO_2$
NADH NAD^+
leuB
COOH
HO-C-H
HC-COOH
CH_3-CH
CH_3
β-Isopropylmalate
leuC
leuD
COOH
H-C-H
H-C-COOH
CH_3-CH
CH_3
α-Isopropylmalate

Fig. 15 Biosynthesis of the branched-chain amino acids, isoleucine, leucine and valine. The structural gene designations and the enzymes for *E. coli* are: *ilvA*, threonine deaminase; *ilvB*, valine-sensitive acetohydroxy acid synthase I; *ilvG*, acetohydroxy acid synthase II; *ilvC*, acetohydroxy acid isomeroreductase; *ilvD*, dihydroxy acid dehydratase; *ilvE*, branched-chain amino acid aminotransferase; *leuA*, α-isopropylmalate synthase; *leuC*, *leuD*, iso-propylmalate dehydratase; *leuB*, β-isopropylmalate dehydrogenase; *tyrB*, aromatic amino acid aminotransferase. From Moat AG, Foster JW (1995) *Microbial Physiology*, 3rd edn. New York: John Wiley & Sons.

points. The major one is the aspartokinase reaction, which regulates the flow of carbon to all of the amino acids involved (Fig. 16). The second site of primary regulation is the conversion of aspartate β-semialdehyde to homoserine, a reaction catalysed by homoserine dehydrogenases I and II. Aspartokinase I and homoserine dehydrogenase I activities are associated with a single multi-functional enzyme, ThrA, which is subject to allosteric inhibition by threonine. Aspartokinase I–homoserine dehydrogenase I synthesis is repressed by a combination of threonine and isoleucine. The structural genes for aspartokinase I–homoserine dehydrogenase I and homoserine kinase (*thrB*) and threonine synthase (*thrC*) lie in close proximity to one another on the *E. coli* map and constitute an operon. Aspartokinase II and homoserine dehydrogenase II activities are also catalysed by a bifunctional protein (MetL). MetL transcription is regulated by methionine. However, neither of the two activities is feedback-inhibited by methionine, threonine, *S*-adenosylmethionine, nor by combinations of these compounds. Aspartate kinase III (*lysC*), is regulated by lysine by both feedback and repression mechanisms. Comparable regulatory systems in the aspartate pathway have been found in other members of the *Enterobacteriaceae*, particularly *Salmonella*, *Enterobacter*, *Edwardsiella*, *Serratia* and *Proteus*.

Biosynthesis of Serine, Glycine and Cysteine

Serine is a precursor of L-cysteine (Fig. 17). Glycine contributes carbon and nitrogen in the biosynthesis of purines, porphyrins and other metabolites. Hydroxymethyltransferase (SHMT), the *glyA* gene product, converts serine to glycine and 5,10-methylenetetrahydrofolate, a major contributor of one-carbon units in the formation of methionine, purines and thymine. Oxidative cleavage of glycine by the glycine cleavage (GCV) enzyme system provides a second source of one-carbon units:

$$CH_2NH_2COOH \rightarrow NH_3 + CO_2 + 5,10\text{-methylenetetrahydrofolate}$$

The GCV system has been described in mammalian liver and in several bacterial species, including *E. coli*, *Peptococcus glycinophilus* and *Arthrobacter globiformis*. It consists of four proteins: the P protein, a pyridoxal phosphate enzyme; the H protein, a lipoate-containing hydrogen carrier; the T protein that transfers

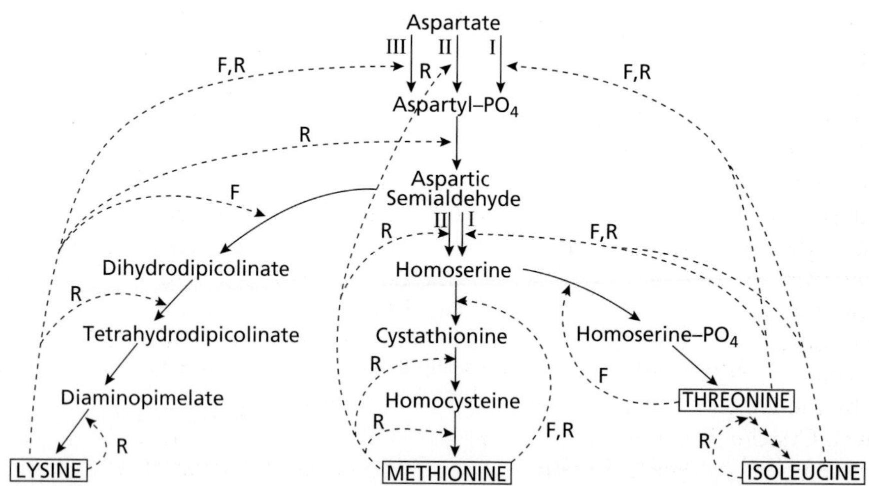

Fig. 16 End product regulation of the aspartate family of amino acids in *Escherichia coli*. F, feedback inhibition; R, repression of enzyme synthesis. Aspartokinase I-homoserine dehydrogenase I is a bifunctional enzyme encoded by *thrA*. Aspartokinase II-homoserine dehydrogenase II is a bifunctional enzyme encoded by *metL*. Aspartokinase III is encoded by *lysC*. From Moat AG, Foster JW (1995) *Microbial Physiology*, 3rd edn. New York: John Wiley & Sons.

$CH_2O(P)$–CH_2OH–COOH (3-PGA) —+NAD, *serA*→ $CH_2O(P)$–C=O–COOH (3-Phosphohydroxypyruvate) —$+NH_2$, *serC*→ $CH_2O(P)$–$CHNH_2$–COOH (3-Phosphoserine) —$-P_i$, *serB*→ CH_2OH–$CHNH_2$–COOH (L-Serine) —–HCO, *glyA*→ CH_2NH_2–COOH (Glycine)

L-Serine —*cysE*, +Acetyl–CoA→ H_2C–O–CO–CH_3–$CHNH_2$–COOH (*O*-Acetylserine)

SO_4^{2-} —*cysA* Sulphate permease→ SO_4^{2-} —+ATP, *cysD*→ APS —+ATP, *cysC*→ PAPS —+NADPH, *cysH*→ SO_3^{2-} —+NADPH, *cysG cysI cysJ*→ S^{2-}

O-Acetylserine + S^{2-} —–Acetate, *cysK*, *cysM*→ H_2C–SH–$CHNH_2$–COOH (L-Cysteine)

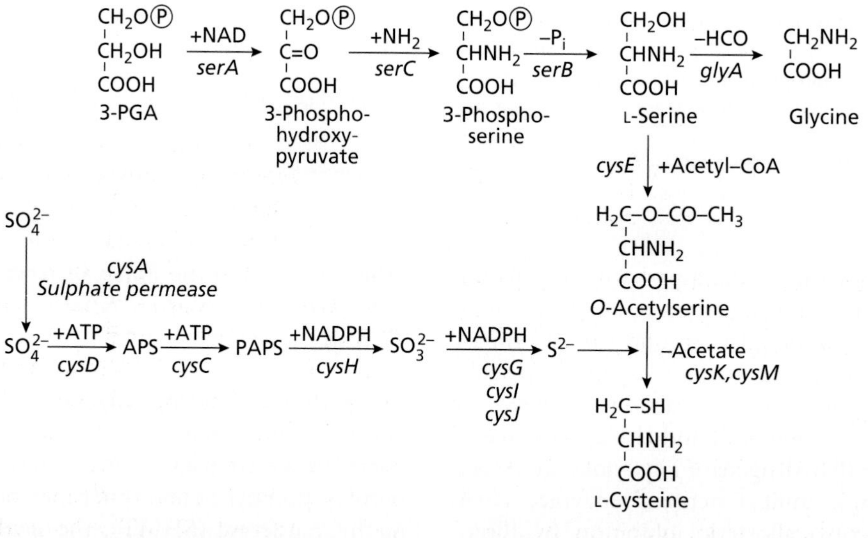

Fig. 17 Biosynthesis of serine, glycine and cysteine. The structural gene designations in *E. coli* are: *serA*, 3-phosphoglycerate (3-PGA) dehydrogenase; *serC*, phosphoserine aminotransferase; *serB*, phosphoserine phosphatase; *glyA*, serine hydroxymethyltransferase; *cysA*, sulphate permease; *cysD*, sulphate adenyltransferase; *cysC*, adenylsulphate (APS) kinase; *cysH*, 3′-phosphoadenyl sulphate (PAPS) reductase; *cysG*, *cysI*, *cysJ*, sulphite reductase; *cysE*, serine acetyltransferase; *cysK*, acetylserine sulphhydrylaseA; *cysM*, acetylserine sulphhydrylase B. *cysB* is a regulator of the *cys* operon. From Moat AG, Foster JW (1995) *Microbial Physiology*, 3rd edn. New York: John Wiley & Sons.

the methylene carbon to tetrahydrofolate; and the L protein, a lipoamide dehydrogenase encoded by *lpd* that is common to the pyruvate and α-ketoglutarate dehydrogenase complexes. The genes that encode these enzymes form an operon that maps at 65.2 minutes on the *E. coli* chromosome. The regulatory proteins Lrp, GcvA and PurR are involved in the regulation of the glycine-inducible GCV system.

Serine, glycine and cysteine are synthesised by a pathway that emanates from 3-phosphoglycerate and proceeds through a series of phosphorylated intermediates (Fig. 17). Serine may be derived from glycine formed by transamination of glyoxylate. Deamination of threonine to α-ketobutyrate with cleavage to acetyl-CoA and glycine can also give rise to serine.

Baker's yeast possesses the phosphorylated pathway to serine from 3-phosphoglycerate as well as the glyoxylate pathway. Regulation of the phosphorylated pathway by serine feedback inhibition of 3-phosphoglycerate dehydrogenase and regulation of the glyoxylate transaminase has been demonstrated in *S. cerevisiae*. In bacteria the phosphorylated pathway appears to be the main route of serine formation. The bacterial pathway is also regulated by feedback inhibition of 3-phosphoglycerate dehydrogenase.

The converging pathways of sulphate reduction and the formation of α-acetylserine shown in **Fig. 17** appear to be the most common route for the synthesis of cysteine. However, cysteine may also arise from serine and H_2S by the action of serine sulphhydrase (cysteine synthase). Cysteine is also formed by transsulphuration between homocysteine and serine:

$$\underset{\text{Homocysteine}}{\begin{array}{c}CH_2SH\\|\\CH_2\\|\\HCNH_2\\|\\COOH\end{array}} + \underset{\text{Serine}}{\begin{array}{c}HOCH_2\\HCNH_2\\COOH\end{array}} \longrightarrow \underset{\text{Cystathionine}}{\begin{array}{cc}CH_2-S-&CH_2\\|&|\\CH_2&HCNH_2\\|&|\\HCNH_2&COOH\\|&\\COOH&\end{array}}$$

$$\longrightarrow \underset{\text{Homoserine}}{\begin{array}{c}CH_2OH\\|\\CH_2\\|\\HCNH_2\\|\\COOH\end{array}} + \underset{\text{Cysteine}}{\begin{array}{c}CH_2SH\\|\\HCNH_2\\|\\COOH\end{array}}$$

Under conditions where methionine serves as the sole source of sulphur, homocysteine is formed by a pathway involving *S*-adenosylmethionine (SAM):

$$\text{Methionine} + \text{ATP} \longrightarrow \text{SAM} \xrightarrow{-CH_3} S\text{-adenosylhomocysteine} \xrightarrow{-\text{ado}} \text{homocysteine}$$

Cysteine biosynthesis is regulated by the genes that code for the enzymes in the pathways of sulphate reduction and the formation of α-acetylserine from serine (**Fig. 17**). In *S. typhimurium* each of the enzymes in the sequence is repressed by cysteine. The first enzyme in the sulphate reduction pathway (sulphate permease, CysA) and serine acetyltransferase (CysE) are also inhibited by cysteine by a feedback mechanism. *O*-Acetylserine induces the transport of sulphate by sulphate permease and also induces the rest of the enzymes in the sulphate reduction sequence. Although the biosynthesis of cysteine and methionine are obviously inter-related, it is the sulphate transport system and the pathway involved in the activation of sulphur that regulate the synthesis of cysteine through positive control mechanisms, whereas methionine biosynthesis is under negative control. It has been proposed, however, that the *O*-acetyl derivatives play a parallel role in the regulation of the synthesis of methionine and cysteine by inducing the initial steps in the pathway. Biochemical and genetic studies indicate that three of the genes that control the pathway of sulphate activation (*cysIJH*) are arranged in a cluster that constitutes an operon.

Aminolevulinate and the Pathway to Tetrapyrroles

δ-Aminolevulinate (ALA) can be synthesised by the C_4 route by the condensation of glycine and succinyl-CoA by ALA synthase, or the C_5 pathway from glutamate. In either case, ALA is the first committed step in the formation of tetrapyrroles (**Fig. 18**). The pathway diverges at uroporphyrinogen III (UroIII), one branch leading to the synthesis of sirohaem and vitamin B_{12} and the other leading to the synthesis of protoporphyrin IX. At this point the pathway branches again, one branch leading to the synthesis of haem and the other to bacteriochlorophyll formation.

The Aromatic Amino Acid Pathway

The Common Aromatic Amino Acid Pathway

In bacteria and fungi chorismate serves as the precursor of eight or more aromatic compounds. The aromatic amino acids phenylalanine, tyrosine and tryptophan and several other related aromatic compounds are produced by a common pathway which begins with the condensation of erythrose-4-phosphate and phosphoenolpyruvate to form 3-deoxy-D-arabino-heptulosonate 7-phosphate (DAHP) (**Fig. 19**). This compound is converted to shikimate and then to chorismate. The name chorismate comes from a Greek word meaning 'to branch'. This compound represents the branch-point leading to the formation of other aromatic amino acids and several other related metabolites. The formation of many additional essential aromatic compounds via branches of this pathway emphasises the need for a complex regulatory system. Furthermore, a mutant blocked in the common pathway before the branch-point may show multiple nutritional requirements for the end products of the branches.

Although there are many similarities in the aromatic amino acid pathway in all bacteria and fungi, there are several differences in detail with regard to the manner in which tryptophan, tyrosine and phenylalanine are formed. Even more apparent are the number of variations in the ways in which other aromatic compounds are formed through branches or extensions from the common pathway.

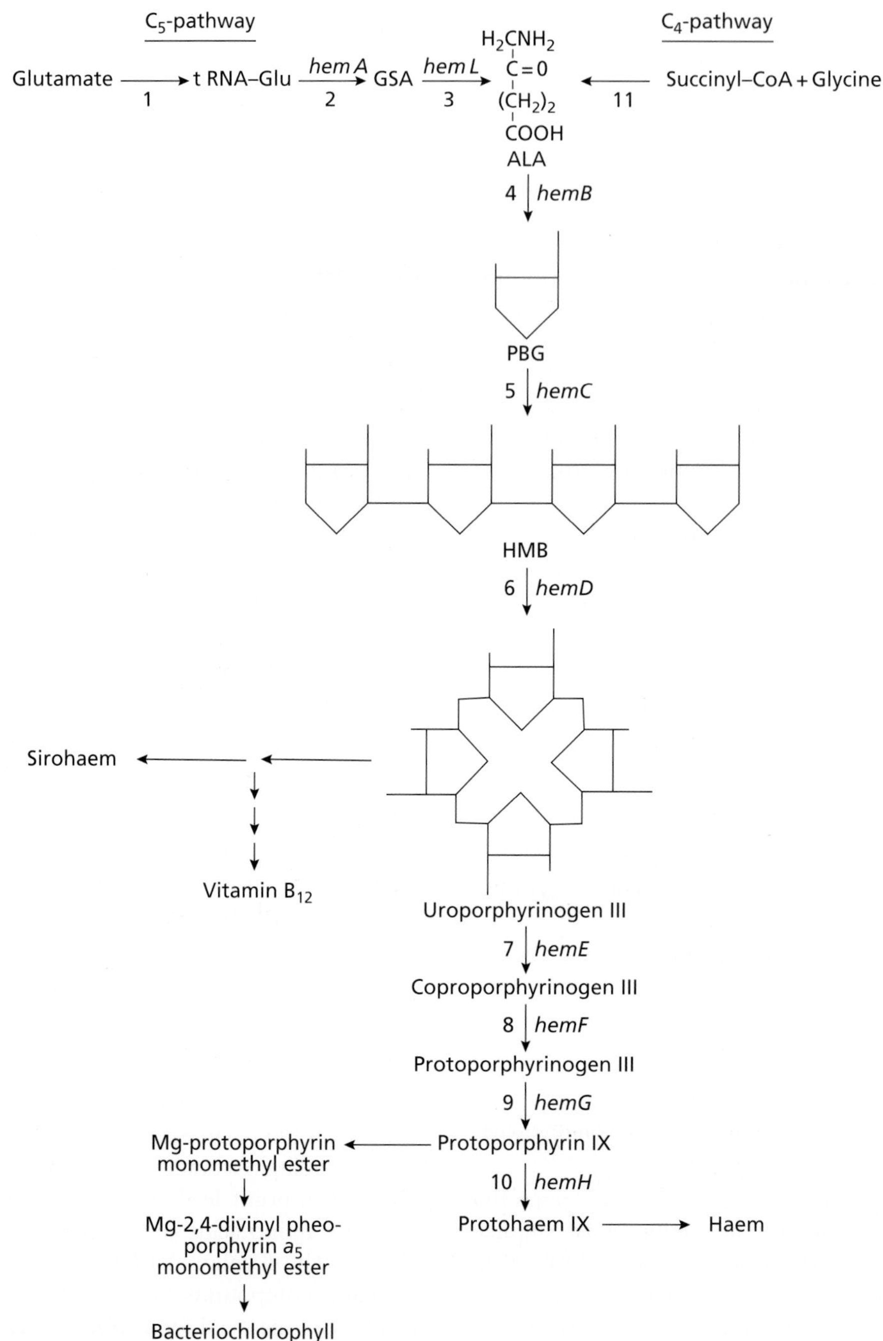

Fig. 18 Tetrapyrrole biosynthetic pathway. The enzymatic steps in the pathway are catalysed by: 1, glutamyl-tRNA synthase; 2, NAD(P)H: glutamyl-tRNA reductase; 3, GSA (glutamate γ-semialdehyde) 2,1-aminotransferase; 4, PBG (porphobilinogen) synthase; 5, HMB (hydroxymethylbilane) synthase; 6, UroIII (uroporphyrinogen) synthase; 7, uroporphyrinogen III decarboxylase; 8, coproporphyrinogen III oxidase; 9, protoporphyrinogen IX oxidase; 10, ferrochetolase; 11, ALA synthase. The structural gene designations are for *E. coli*. From Moat AG, Foster JW (1995) *Microbial Physiology*, 3rd edn. New York: John Wiley & Sons.

In *E. coli* and *S. typhimurium*, regulation of the common aromatic pathway is modulated through three unlinked genes, *aroF*, *aroG* and *aroH*. These genes encode isozymes of the first enzyme, DAHP synthase, which are sensitive to tyrosine, phenylalanine and tryptophan, respectively. Although all three DAHP synthases are regulated transcriptionally, feedback inhibition is quantitatively the major control mechanism *in vivo*. The structural gene for the tryptophan-inhibitable isozyme (*aroF*)

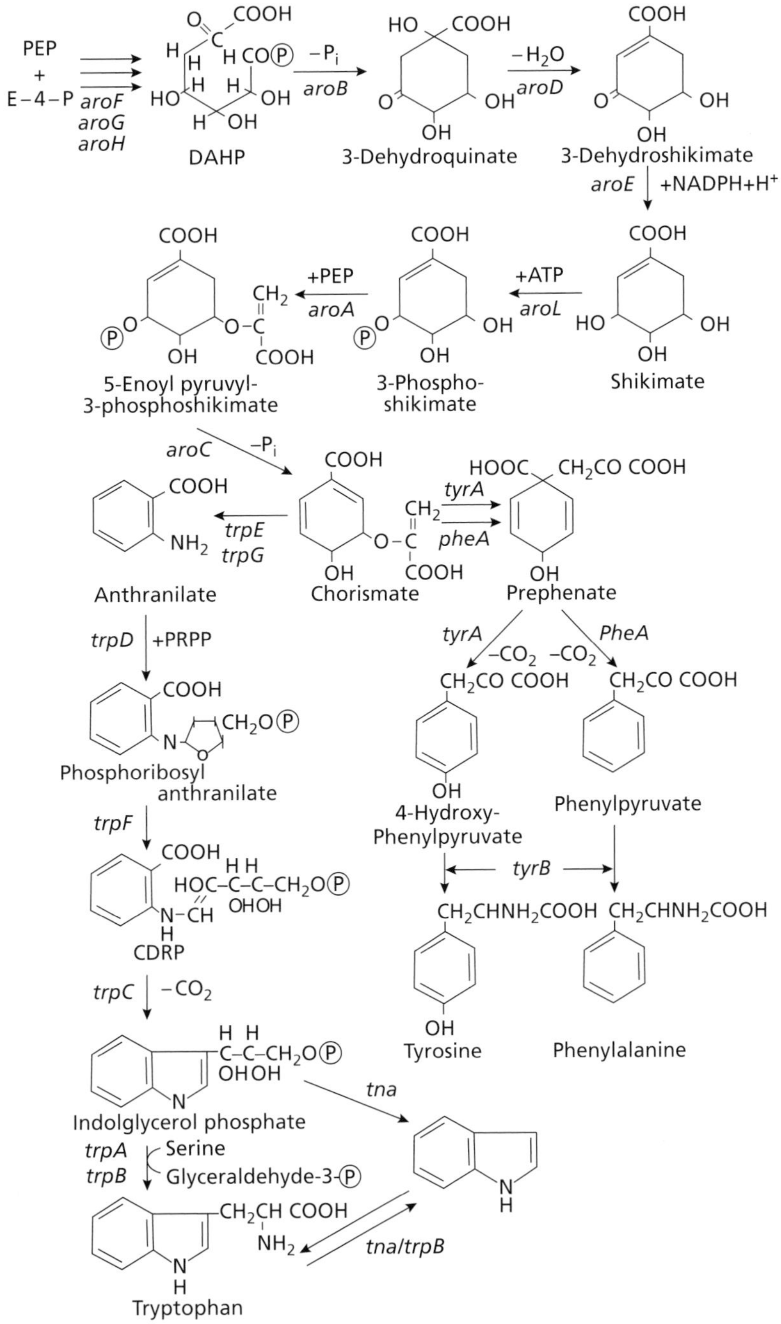

Fig. 19 Biosynthesis of the aromatic amino acids. The structural gene designations for the enzymes in *E. coli* are: *aroF, aroG, aroH*, 3-deoxy-D-arabinoheptulosonate 7-phosphate (DAHP) synthase; *aroB*, 3-dehydroquinate synthase; *aroD*, 3-dehydroquinate dehydratase; *aroE*, shikimate dehydrogenase; *aroL*, shikimate kinase; *aroA*, 5-enolpyruvoylshikimate 3-phosphate synthase; *aroC*, chorismate synthase; *pheA*, chorismate mutase P; *tyrA*, chorismate mutase T; *pheA*, prephenate dehydratase; *tyrA*, prephenate dehydrogenase; *tyrB*, tyrosine aminotransferase; *trpE, trpG*, anthranilate synthase; *trpD*, anthranilate phosphoribosyltransferase; *trpF*, Phosphoribosyl anthranilate isomerase; *trpC*, indolglycerol phosphate synthase; *trpA* + *trpB*, tryptophan synthase; *tna*, tryptophanase. CDRP, 1-(*O*-carboxyphenylamino)-1-deoxyribulose 5-phosphate; PEP, phosphoenolpyruvate; E-4-P, erythrose-4-phosphate. From Moat AG, Foster JW (1995) *Microbial Physiology*, 3rd edn. New York: John Wiley & Sons.

is linked to *aroE*, whereas that for the tyrosine-inhibitable isozyme (*aroG*) is linked to *pheA* and *tyrA*. The *pheA* and *tyrA* gene products are the respective branch-point enzymes that lead to the phenylalanine and tyrosine pathways. However, only the tyrosine-related isozyme (aroG) is completely repressed, and feedback inhibited, by low levels of tyrosine.

A similar pattern of isozymic control of DAHP synthase and inter-relationships of the genes that govern the aromatic pathway is present in *E. coli*. Expression of *aroF* and *aroG* is repressed by the *Tyr* repressor, the *tyrR* gene product, mediated by tyrosine and phenylalanine, respectively. TyrR also regulates the expression of several other genes concerned with aromatic amino acid biosynthesis or transport. These include *aroL*, *aroP*, *tyrB*, *tyrP* and *mtr* (resistance to the tryptophan analogue, 5-methyltryptophan). Expression of *aroH* is controlled by the Trp repressor (encoded by *trpR*), which also regulates the expression of the *trp* operon, the *mtr* gene, as well as expression of *trpR* itself.

In a wide variety of fungi, the genes that code for enzymes 2 to 6 in the pre-chorismate pathway (Fig.19) are arranged in a cluster, designated the *arom* gene cluster. In all of the fungal species so far examined, these enzymes sediment as a complex aggregate on centrifugation in sucrose density gradients. The sedimentation coefficients for these enzymes are very similar in representative examples of different fungi. By comparison, the five enzymes that catalyse the conversion of DAHP to chorismate in bacteria are physically separable in *E. coli*, *S. typhimurium*, *E. aerogenes*, *Anabaena variabilis* and *Chlamydomonas reinhardii*. In *Euglena gracilis* an enzyme aggregate with five activities has been isolated.

In *B. subtilis*, a cluster of contiguous genes for tryptophan biosynthesis represent a polycistronic operon. The *trpEDCFBA* genes form an operon similar to operons found in enteric bacteria. However, in *B. subtilis* the *trp* genes are part of a 'supra-operon' that contains *trpEDCFBA-hisH-tyrA-aroE* (Fig. 20a). It is likely that *aroFBH* may also be present at the 5′-end of this supra-operon. All *aroF* mutants (chorismate synthase) also lack dehydroquinate synthase (AroB) activity. The gene that specifies AroB is closely linked to the gene that codes for the AroF enzyme. Both genes are part of the *aro* gene cluster. Mutants lacking chorismate mutase activity also lack DAHP synthase and shikimate kinase activity, presumably as a result of their aggregation in a multi-enzyme complex. As an indication of the complexity of the 'supra-operon', the *mtrAB* operon of *B. subtilis* encodes GTP cyclohydrolase I (MtrA), an enzyme involved in folate biosynthesis. *mtrB* is a *trans*-acting RNA-binding regulatory protein activated by tryptophan. Transcription termination at the attenuator preceding the *trp* gene cluster presumably occurs as a consequence of the binding of the activated MtrB protein to the nascent transcript (see discussion on Interpathway Regulation, p. 286).

The genes that control tryptophan synthesis in coliform organisms are also arranged in an operon. Arrangement of the *E. coli* operon and the enzymes under its control are shown in Fig. 20b. *E. coli* and *S. typhimurium* each have five structural genes for the enzymes in the operon and they are induced co-ordinately in response to the availability of tryptophan. In *E. coli* the activities of the first two enzymes are catalysed by an aggregate formed from the products of the first two genes in the operon. The activities of the next two steps (3 and 4) are catalysed by a single protein. Tryptophan synthase converts indoleglycerol phosphate to indole and then couples indole to serine to form tryptophan. The enzyme is a complex formed from the products of the last two genes (*trpA* and *trpB*). A regulatory site preceding the structural genes is found to regulate the synthesis of $mRNA^{Trp}$ and the enzymes of the tryptophan pathway, thus reducing operon expression. This attenuator function occurs at the level of transcription by providing a region in which transcription is terminated in the presence of excess charged $tRNA^{Trp}$. However, the degree to which it is used and the manner in which the alternative pathways are regulated differ widely from one species to another.

Pseudomonas aeruginosa and other pseudomonads possess co-existing alternative pathways for the formation of tyrosine and phenylalanine. These account for the unusual resistance of these organisms to analogues of the aromatic amino acids. Under certain circumstances, the aerogenate pathway appears to function as an unregulated overflow route for tyrosine production. In some coryneforms (*Corynebacterium glutamicum*, *Brevibacterium flavum* and *B. ammoniogenes*) the arogenate pathway is an obligatory route for the formation of tyrosine. By comparison, arogenate seems to be a dead-end metabolite in *N. crassa*, the prephenate pathway being used as the major route to tyrosine and phenylalanine.

p-Aminobenzoate and Folate Biosynthesis

Many bacteria and fungi, including *E. coli*, *B. subtilis*, *Streptomyces griseus* and *Pseudomonas acidovorans*, convert chorismate to *p*-aminobenzoate (PAB) via PAB synthase (Fig. 21). In *E. coli* the enzyme complex termed *p*-aminobenzoate synthase consists of two non-identical subunits designated components I and II. Component I is encoded by *pabB* and catalyses the synthesis of 4-amino-4-deoxychorismate from chorismate and ammonia. Component II, encoded by *pabA*, is a glutamine amidotransferase that uses the amide nitrogen of glutamine to convert chorismate to 4-amino-4-deoxychorismate.

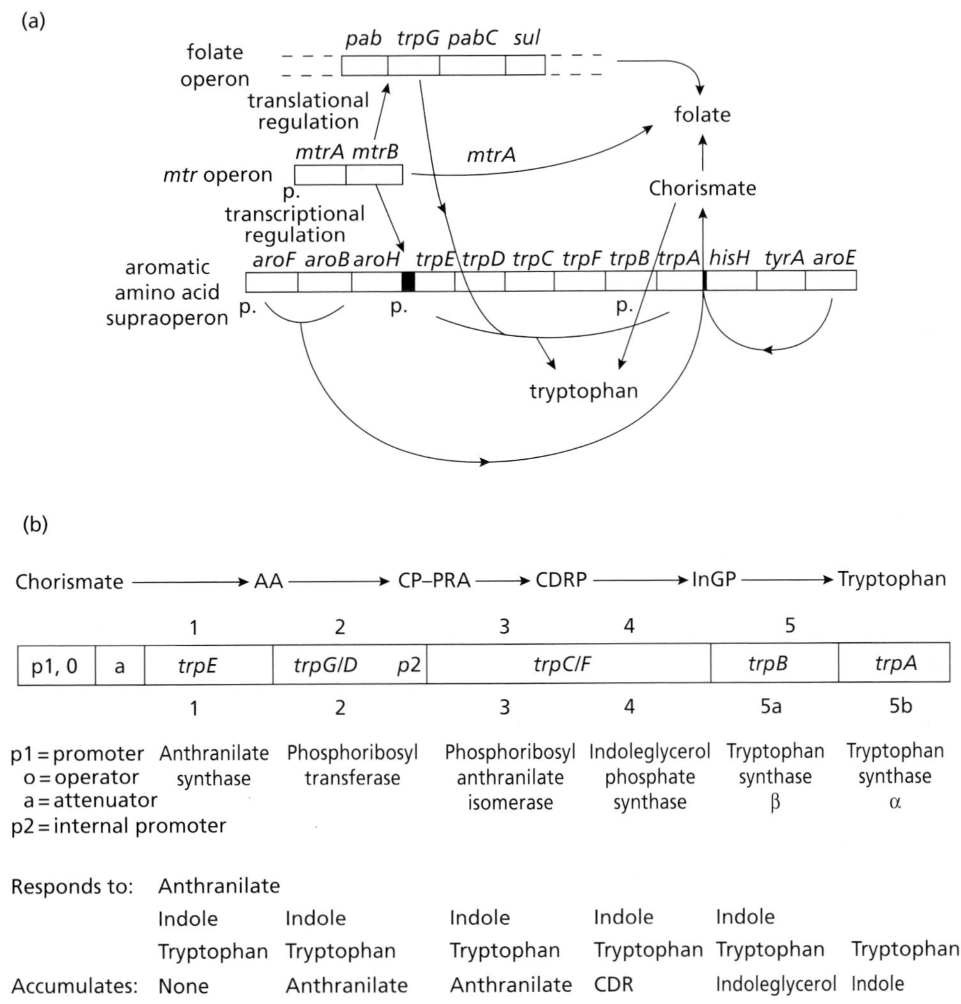

Fig. 20 Organisation of the genes for tryptophan biosynthesis in *Bacillus subtilis* and *Escherichia coli*. (a) The aromatic amino acid supra-operon of *B. subtilis* showing inter-relationships with the genes from the folic acid operon and the *mtr* operon. MtrA (GTP cyclohydrolase I) catalyses formation of the initial pteridine ring (dihydroneopterin triphosphate) of folic acid. AroF, AroB and AroE are required for chorismate synthesis. Pab, TrpG and PabC are involved in *p*-aminobenzoate synthesis from chorismate. Sul catalyses condensation of *p*-aminobenzoate and the pteridine ring. MtrB is considered to bind to a target sequence that overlaps the *trpG* ribosome-binding site, resulting in a translational regulation, and to the *trpE* leader transcript, resulting in transcriptional regulation of the *trp* gene cluster. The *trpEDCFBA* and *trpG* gene products are necessary for tryptophan biosynthesis from chorismate. Note that AroH is involved in phenylalanine and tyrosine biosynthesis from chorismate in some strains of *Bacillus*. p marks the positions of known promoters. Double dashed lines indicate that open reading frames exist upstream of *pab* and downstream of *sul*. From Babitzke P, Gollnick P, Yanofsky C (1992) The *mtrAB* operon of *Bacillus subtilis* encodes GTP cyclohydrolase I (MtrA), an enzyme involved in folic acid biosynthesis, and MtrB, a regulator of tryptophan biosynthesis. *J. Bacteriol.* 174: 2059–2064. (b) Organisation of the genes of the tryptophan operon of *Escherichia coli*. AA, anthranilate; CP-PRA, *N*-(*o*-carboxyphenyl)-phosphoribosylamine; CDRP, 1-(*o*-carboxyphenylamino)-1-deoxyribulose 5′-phosphate; InGP, indoleglycerol phosphate. From Moat AG, Foster JW (1995) *Microbial Physiology*, 3rd edn. New York: John Wiley & Sons.

The latter is converted to PAB by aminodeoxychorismate lyase, encoded by *pabC*. PAB is then condensed with 6-hydroxymethyl-7,8-pterin pyrophosphate by dihydropteroate synthase (FolP). 7,8-Dihydropteroate is then converted to 7,8-dihydrofolate by the additionof glutamate by dihydrofolate synthase (FolC). After 7,8-dihydrofolate is reduced to tetrahydrofolate by dihydrofolate reductase, the synthase enzyme adds additional glutamate units to form folylpolyglutamate.

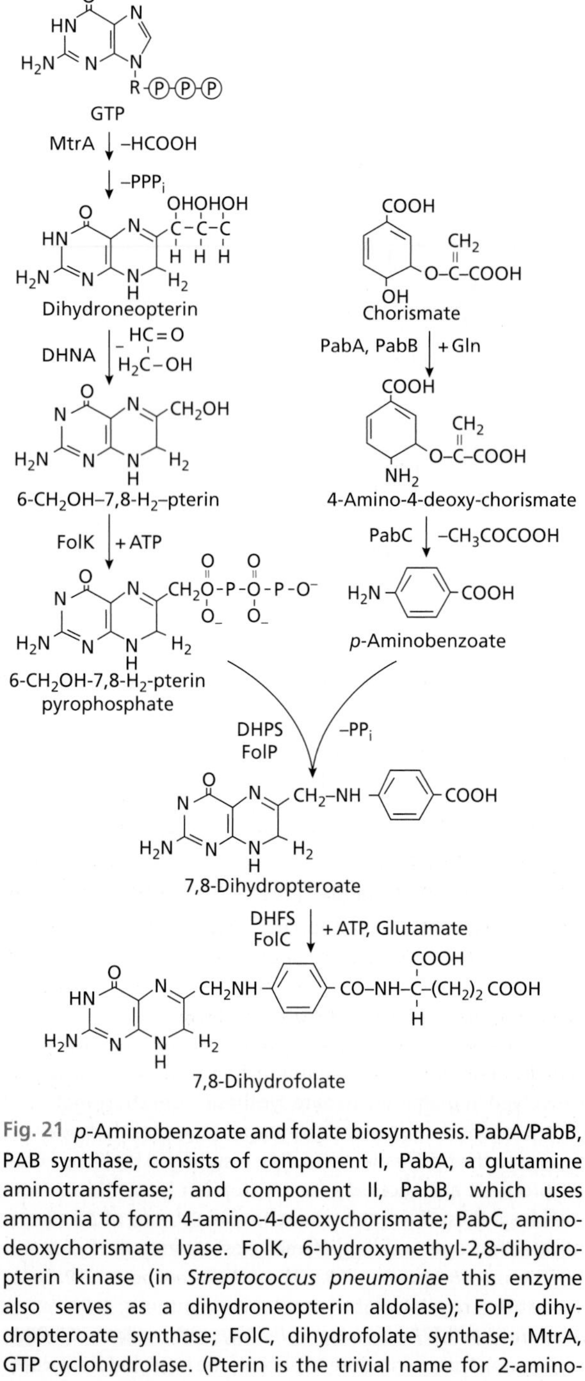

Fig. 21 *p*-Aminobenzoate and folate biosynthesis. PabA/PabB, PAB synthase, consists of component I, PabA, a glutamine aminotransferase; and component II, PabB, which uses ammonia to form 4-amino-4-deoxychorismate; PabC, amino-deoxychorismate lyase. FolK, 6-hydroxymethyl-2,8-dihydropterin kinase (in *Streptococcus pneumoniae* this enzyme also serves as a dihydroneopterin aldolase); FolP, dihydropteroate synthase; FolC, dihydrofolate synthase; MtrA, GTP cyclohydrolase. (Pterin is the trivial name for 2-amino-4-hydroxypteridine; neopterin is the trivial name for 6-[D-erythro-l′,2′,3′-trihydroxypropyl]pterin). DHNA, 1,4-dihydroxy-2-naphthoate. From Moat AG, Foster JW (1995) *Microbial Physiology*, 3rd edn. New York: John Wiley & Sons.

The pteridine component of folate is derived from guanosine triphosphate (GTP). The first step in this sequence is the removal of C8 from GTP as formic acid by GTP cyclohydrolase I to form dihydroneopterin triphosphate. In *B. subtilis* this enzyme is the product of the *mtrA* gene. GTP cyclohydrolase I has been identified in a number of other bacterial species, including *E. coli* and *S. typhimurium*. Dihydroneopterin triphosphate is converted to 7,8-dihydroneopterin by removal of the phosphate residues. 7,8-Dihydropterin is then converted to 6-hydroxymethyl-7,8-dihydropterin by the action of 6-hydroxymethyl-7,8-dihydropterin kinase, the product of the *folK* gene in *E. coli* (*sulD* in *Streptococcus pneumoniae*). In *S. pneumoniae*, *SulD* is a bifunctional enzyme with both 6-hydroxylmethyl-7,8-dihydropterin kinase and 7.8-dihydroneopterin aldolase activities.

Biosynthetic Pathway to Ubiquinone

Ubiquinone (coenzyme Q), the only non-protein component of the electron transport chain, is synthesised from chorismate (**Fig. 22**). The first precursor specific to the ubiquinone pathway, 4-hydroxybenzoate, is formed by the action of chorismate lyase. In some organisms tyrosine can serve as a source of 4-hydroxybenzoate through the intermediary formation of 4-hydroxyphenylpyruvate.

Menaquinone (Vitamin K) Biosynthesis

The biosynthesis of menaquinone originates with the addition of α-ketoglutarate to the ring of chorismate and concomitant removal of pyruvate and CO_2. The details of the pathway are shown in **Fig. 23**. All naturally occurring menaquinones have *trans* configurations in the double bonds of the prenyl side-chain. Menaquinone serves as an important electron carrier during anaerobic growth of *E. coli*. It is essential for electron transfer to fumarate.

Biosynthesis of Nicotinamide Adenine Dinucleotide (NAD)

NAD and NADP, the functional forms of nicotinic acid, are formed from tryptophan in mammals, *N. crassa* and *S. cerevisiae* (**Fig. 24**). *Xanthomonas pruni* also synthesises NAD from tryptophan. However, most bacteria utilise an entirely different pathway to NAD that involves the condensation of aspartate and a three-carbon compound with subsequent conversion of the condensation product into quinolinate (**Fig. 24**). The pathway from quinolinate to NAD appears to be identical in all organisms, regardless of the mode of quinolinate formation. *E. coli*, *S. typhimurium*, *M. tuberculosis*, and other common bacteria

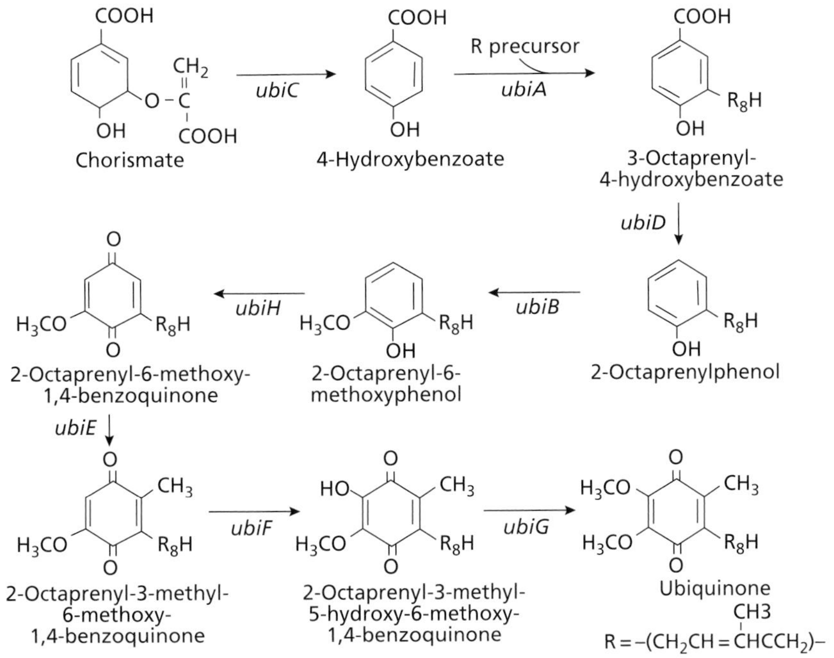

Chorismate → (ubiC) → 4-Hydroxybenzoate → (R precursor, ubiA) → 3-Octaprenyl-4-hydroxybenzoate → (ubiD) → 2-Octaprenylphenol → (ubiB) → 2-Octaprenyl-6-methoxyphenol → (ubiH) → 2-Octaprenyl-6-methoxy-1,4-benzoquinone → (ubiE) → 2-Octaprenyl-3-methyl-6-methoxy-1,4-benzoquinone → (ubiF) → 2-Octaprenyl-3-methyl-5-hydroxy-6-methoxy-1,4-benzoquinone → (ubiG) → Ubiquinone

$R = -(CH_2CH=C(CH_3)CH_2)-$

Fig. 22 Biosynthetic pathway to ubiquinone (coenzyme Q). Gene designations are those for *E. coli*. From Moat AG, Foster JW (1995) *Microbial Physiology*, 3rd edn. New York: John Wiley & Sons.

use the aspartate–dihydroxyacetone phosphate pathway. *Clostridium butylicum* follows a unique pathway by condensation of aspartate and formate into the intermediate *N*-formylaspartate. Further addition of acetate and ring closure yields quinolinate.

Organisms with a specific nutritional requirement for nicotinamide utilise a unique route of NAD biosynthesis that involves conversion of nicotinamide to NAD without previous deamidation:

$$\text{Nicotinamide} + \text{PRPP} + \text{ATP} \longrightarrow \text{nicotinamide mononucleotide (NMN)}$$

$$\text{NMN} + \text{ATP} \longrightarrow \text{NAD}$$

Lactobacillus fructosus and *Haemophilus influenzae* use this route for NAD biosynthesis.

Histidine Biosynthesis

The biosynthesis of histidine occurs by a unique pathway that is more closely linked to the metabolism of pentoses and purines than to any of the other amino acid families. The pathway of histidine biosynthesis (Fig. 25), is initiated by the coupling of phosphoribosyl pyrophosphate (PRPP) with ATP at N_1, of the purine ring followed by opening of the ring. The N_1 and N_2 of the imidazole ring of histidine are thus derived from the N_1 and N_2 of the adenine ring. The aminoimidazole carboxamide ribonucleotide derivatives are normally bound to the enzymes involved in their formation and rearrangement. Cleavage of the open-ringed structures gives rise to imidazoleglycerolphosphate (IGP) and aminoimidazole carboxamide ribonucleotide (AICRP), which is recycled via the purine biosynthetic pathway to reform the purine nucleotide. There do not appear to be branches from this pathway to other end products.

The chemical intermediates and the enzymes involved are virtually identical in a variety of bacteria and fungi, but the arrangement of the genes that control the synthesis and operation of the pathway differ markedly from one organism to another.

In *S. typhimurium* the 10 genes that control the synthesis of the enzymes of the histidine pathway are arranged in a single operon. This is also true for the genes in *Staphylococcus aureus.*

In yeast, the *his4* region specifies three of the enzyme activities of the histidine pathway, namely, the second, third and tenth steps. The enzymes responsible for these reactions are tightly associated in a multi-functional complex. There is no evidence for aggregation of the remainder of the enzymes, nor are the genes that specify the other enzymes contiguous with these or with each other. This contrasts with the close linkage of the 10 *his* genes in *S. typhimurium*, but it is comparable with many other gene–enzyme relationships in fungi. The equivalent genes coding for the second, third and tenth steps in *Neurospora* are also

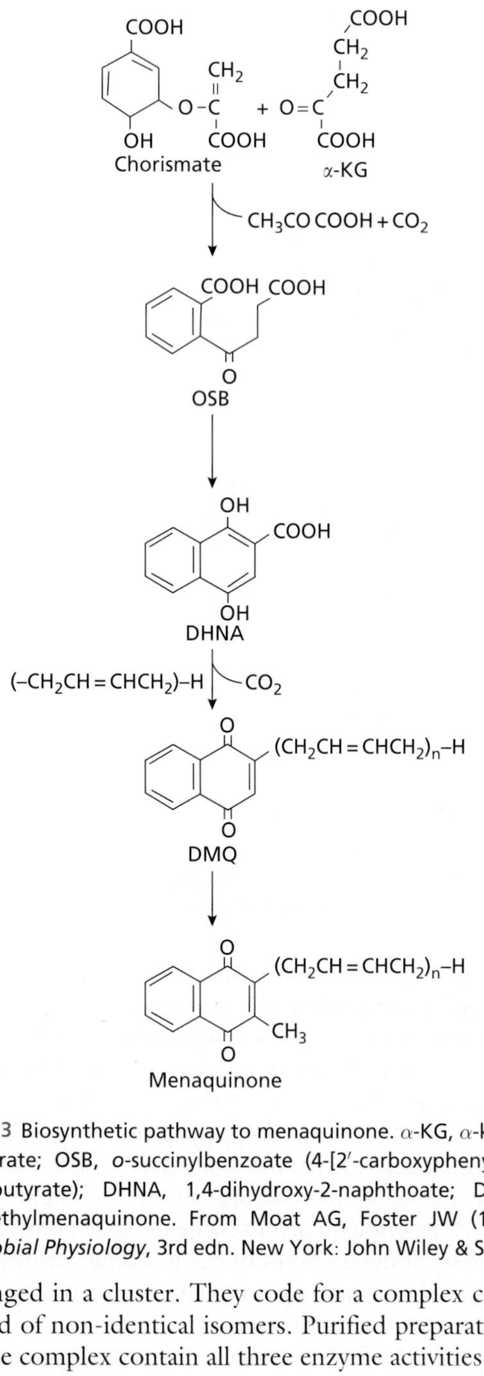

Fig. 23 Biosynthetic pathway to menaquinone. α-KG, α-ketoglutarate; OSB, *o*-succinylbenzoate (4-[2′-carboxyphenyl]-4-oxo-butyrate); DHNA, 1,4-dihydroxy-2-naphthoate; DMQ, demethylmenaquinone. From Moat AG, Foster JW (1995) *Microbial Physiology*, 3rd edn. New York: John Wiley & Sons.

arranged in a cluster. They code for a complex composed of non-identical isomers. Purified preparations of the complex contain all three enzyme activities.

Biosynthesis of Purines

In all organisms so far studied, the biosynthetic pathway to purine nucleotide formation appears to be remarkably similar. Only the mechanisms of regulation seem to vary from one organism to another. The formation of phosphoribosylamine (PRA) from PRPP and glutamine initiates the pathway to purine formation (Fig. 26). It should be noted that the initial steps are concerned with the formation of the ribonucleotides. Glycine carbon and nitrogen are added to form GAR (glycinamide ribonucleotide). Two additional steps result in the formation of the imidazole ring. Sequential addition of carbon and nitrogen atoms and ring condensation complete the formation of IMP, the first purine nucleotide to be formed (Fig. 26).

The enzyme responsible for the conversion of AMP to AMP-S is the same as that which catalyses the conversion of SAICAR to AICAR. The formation of GMP from IMP starts with the dehydrogenation of IMP to form XMP. Then, by way of an amidotransferase reaction, the amido-nitrogen of glutamine is transferred to the 2 position to form GMP.

The amino acid histidine and the essential B vitamins, folate, riboflavin and thiamine, may be regarded as products of the purine biosynthetic pathway, in that a portion of each of their structures is derived from purines. In histidine biosynthesis, as described earlier, ATP is the initial substrate. Coupling of PRPP at N_1 of the purine ring of ATP and ring cleavage results in formation of the histidine precursor, imidazoleglycerolphosphate, and AICAR (Fig. 25). AICAR is recycled through the purine biosynthetic pathway.

GTP is the starting point for the synthesis of the pterin moiety of folate (Fig. 21). GTP cyclohydrolase I initiates the pathway that leads to the formation of a pterin derivative which, when coupled to PAB, forms 7,8-dihydropteroate, the precursor of 7,8-dihydrofolate. Details of folate synthesis were discussed above in relation to the derivation of PAB from chorismate as one of several branches of the common aromatic amino acid biosynthetic pathway.

The pyrimidine ring and the ribityl side-chain of riboflavin are derived from GTP (Fig. 27). GTP cyclohydrolase II, encoded by *ribA*, cleaves the imidazole ring of GTP to form 2,5-diamino-6-ribosylamino-4-(3*H*)-pyrimidinone 5′-phosphate ('phosphoribosyl-aminopyrimidine' or PRP) with the release of formate and PP_i. Ultimately, condensation of 5-amino-6-ribitylamino-2,4-(1*H*,3*H*)-pyrimidinedione with 3,4-dihydroxy-2-butanone 4-phosphate yields 6,7-dimethyl-8-ribityllumazine, the direct precursor of riboflavin (Fig. 27). Ribulose 5-phosphate is the precursor of 3,4-dihydroxy-2-butanone 4-phosphate, which forms the xylene ring of the riboflavin molecule.

The pyrimidine ring of thiamine (vitamin B_1) is synthesised from aminoimidazole ribonucleotide (AIR) as a branch of the purine biosynthetic pathway. Thiamine contains a pyrimidine ring and a thiazole ring. Two precursors (4′-amino-hydroxymethyl-2-methylpyridine pyrophosphate and 4-methyl-5-(β-hydroxyethyl)

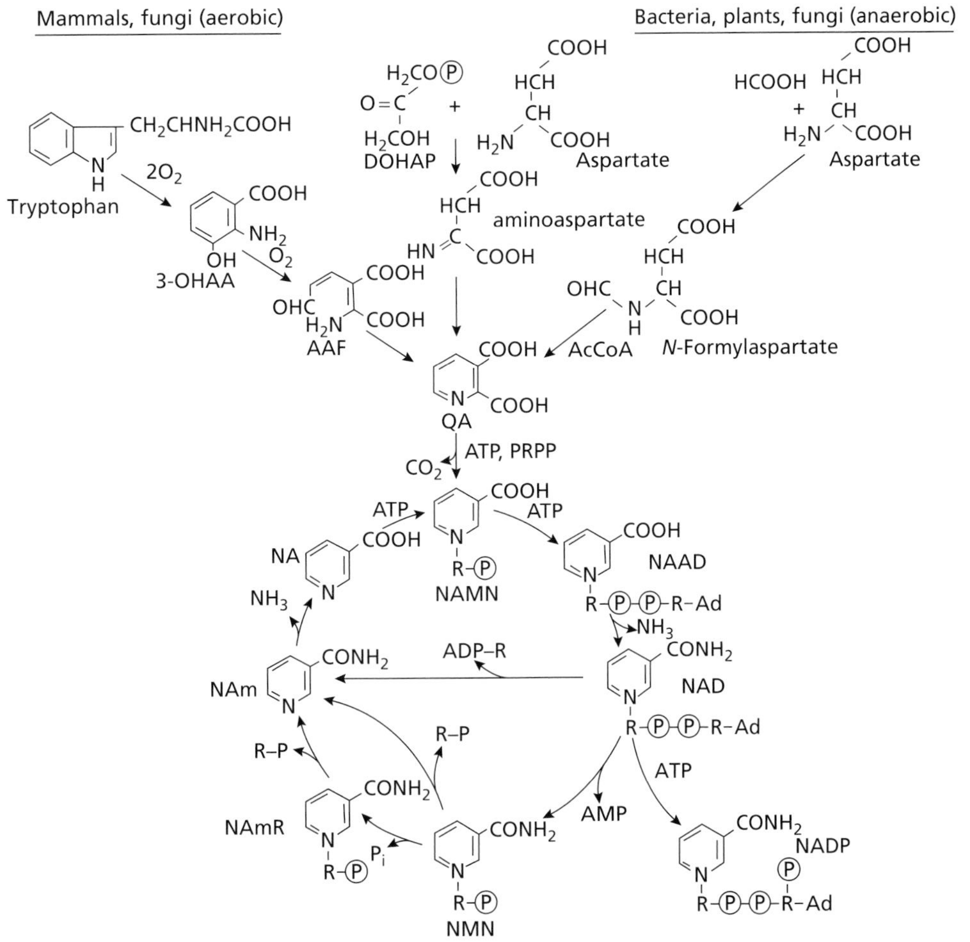

Fig. 24 Alternate pathways for HAD biosynthesis and the pyridine nucleotide cycles. 3-OHAA, 3-hydroxyanthranilate; AAF, 2-acroleyl-3-amino-fumarate; DOHAP, dihydroxyacetone phosphate; AcCoA, acetyl-coA; QA, quinolinate; NAMN, nicotinic acid mononucleotide; NAAD, nicotinic acid adenine dinucleotide; Ad, adenine; NAD, nicotinamide adenine dinucleotide; NADP, nicotinamide adenine dinucleotide phosphate; NMN, nicotinamide mononucleotide; NAmR, nicotinamide riboside; NA, nicotinic acid; NAm, nicotinamide. From Moat AG, Foster JW (1995) *Microbial Physiology*, 3rd edn. New York: John Wiley & Sons.

thiazole monophosphate) are synthesised separately and coupled to form thiamine monophosphate (THI-P). This compound is then phosphorylated to form thiamine pyrophosphate (THI-PP) (Fig. 28). *S. typhimurium* mutants blocked before the AIR step require both purine and thiamine for growth. Isotope labelling studies have confirmed that AIR is the source of the carbon and nitrogen of the pyrimidine ring of thiamine. The genes that code for enzymes in the THI-PP pathway are a tightly linked cluster, *thiCEFGH*, at 90 minutes on the *E. coli* genetic map. The *thiC* gene product is required for the synthesis of the hydroxymethyl-pyrimidine precursor of THI-PP. The *thiEFGH* gene products catalyse the synthesis of the thiazole precursor.

Under anaerobic conditions, *S. typhimurium* can form the pyrimidine moiety of thiamine independently of the *purF* locus. In the absence of oxygen, exogenous pantothenate satisfies the thiamine requirement of *purF* mutants. Only the PurF enzyme is bypassed, however, and PurD, PurG and PurI are still required for the formation of AIR and its conversion to the pyrimidine ring of thiamine.

Biosynthesis of Pyrimidines

Pyrimidine biosynthesis is initiated by the formation of carbamoyl phosphate (Fig. 29). In view of the importance of carbamoyl phosphate in both arginine and pyrimidine synthesis, co-ordinating the channelling of this intermediate into the two pathways is critical. In *Neurospora*, two carbamoyl phosphate gradients are maintained by two separate carbamoyl phosphate synthases.

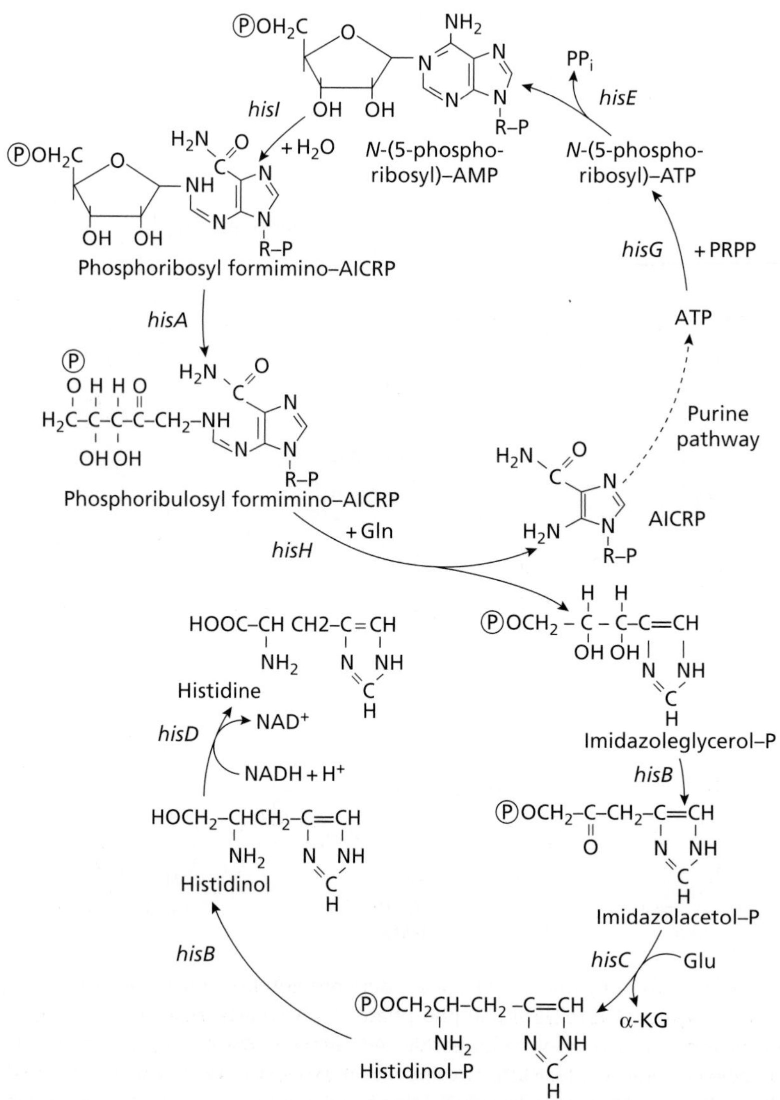

Fig. 25 Biosynthesis of histidine. AICRP, 5-aminoimidazole-4-carboxamide ribonucleotide; PRPP, 5-phosphoribosyl-1-pyrophosphate. The structural gene designations for *E. coli* are: *hisG*, ATP-PRPP transferase; *hisE*, phosphoribosyl-ATP pyrophosphohydrolase; *hisI*, phosphoribosyl-AMP cyclohydrolase; *hisA*, phosphoribosyl formimino-5-amino-1-phosphoribosyl-4-imidazole carboxamide ribotide isomerase; *hisH*, amidocyclase; *hisC*, histidinol phosphate aminotransferase; *hisB*, imidazoleglycerolphosphate dehydrase and histidinolphosphate phosphatase; *hisD*, histidinol dehydrogenase. From Moat AG, Foster JW (1995) *Microbial Physiology*, 3rd edn. New York: John Wiley & Sons.

Carbamoyl phosphate synthase A provides a pool of carbamoyl phosphate for the arginine pathway, as discussed in the section on amino acid biosynthesis, while carbamoyl phosphate synthase P provides a carbamoyl phosphate pool specifically for the pyrimidine pathway. Other organisms, for example, yeast, apparently achieve by regulation what *Neurospora* accomplishes by compartmentation.

Aspartate and carbamoyl phosphate are coupled to form carbamoyl aspartate (ureidosuccinate) which is cyclised by a separate enzyme to form dihydroorotate (Fig. 29). After oxidation to orotate, the nucleotide is formed by phosphoribosyltransferase. It should be noted that, in comparison with purine biosynthesis, the nucleotide stage is established *after* completion of ring formation. Uridine-5′-phosphate (UMP) is converted to UTP, which is then aminated to form cytidine triphosphate (CTP). No known enzyme converts cytosine to CMP. For cytosine to serve as a nutritional source of pyrimidines, it must be deaminated to uracil.

The deoxynucleoside diphosphates are formed from the corresponding ribonucleoside diphosphate (NDP) by the action of thioredoxin, a sulphhydryl-containing protein co-factor, and ribonucleotide diphosphate

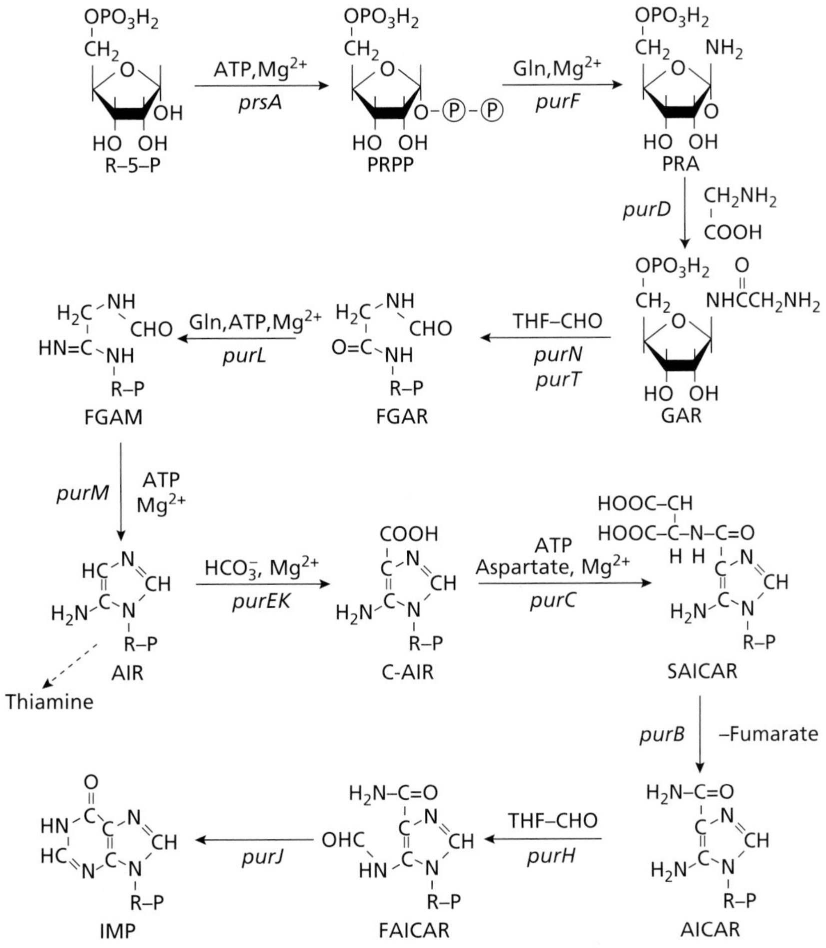

Fig. 26 The pathway of purine biosynthesis. R-5-P, ribose-5-phosphate; PRPP, 5-phosphoribosyl-1-pyrophosphate; PRA, 5-phosphoribosylamine; GAR, glycinamide ribonucleotide; FGAR, 1-*N*-formylglycinamide ribonucleotide; FGAM, α-*N*-formylglycinamidine ribonucleotide; AIR, aminoimidazole ribonucleotide; C-AIR, 5-amino-4-carboxyimidazole ribonucleotide; SAICAR, 5-amino-4-imidazole-(-*N*-succinylo-)-carboxamide ribonucleotide; AICAR, 5-aminoimidazole-4-carboxamide ribonucleotide; FAICAR, 5-formamidoimidazole-4-carboxamide ribonucleotide; IMP, inosine 5′-monophosphate. Structural gene designations for *Escherichia coli* and *Salmonella typhimurium* are: *prs*, ribose phosphate pyrophosphokinase or PRPP synthase; *purF*, aminophosphoribosyltransferase; *purD*, phosphoribosylglycinamide synthase; *purI*, *purL*, phosphoribosylglycinamide formyltransferase; *purG*, phosphoribosylformylglycinamidine synthase; *purI*, *purM*, phosphoribosylaminoimidazole synthase; *purE*, phosphoribosylaminoimidazole carboxylase; *purC*, phosphoribosylaminoimidazole succinocarboxamide synthase; *purE*, adenylosuccinate lyase; *purH*, phosphoribosylaminoimidazole carboxamide formyltransferase; *purJ*, IMP cyclohydrolase. From Moat AG, Foster JW (1995) *Microbial Physiology*, 3rd edn. New York: John Wiley & Sons.

reductase:

$$\text{Thioredoxin}(-\text{SH})_2 + \text{NDP} \rightarrow \text{thioredoxin}(-\text{S}-\text{S}-) + \text{dNDP} + \text{H}_2\text{O}$$

Thymidylate (TMP) is formed from dUMP by transfer of a methyl group from N^5,N^{10}-methylenetetrahydrofolate. In the formation of hydroxymethyldeoxycytidylate, apparently no reduction accompanies the C_1 phosphate transfer from N^5,N^{10}-methylenetetrahydrofolate so that the complete hydroxymethyl group is added. Both the purine and pyrimidine derivatives must be converted to the trinucleotide stage before they can be incorporated into nucleic acid.

In *E. coli* and *S. typhimurium*, dCTP deaminase forms most of the dUTP, the precursor of thymidylate. dUTP is degraded by dUTPase, the product of the *dut* gene, yielding PP_i and dUMP, the substrate for thymidylate synthase. The breakdown of dUTP prevents its incorporation into DNA by DNA polymerase in place of dTTP. Accumulation of dUTP or incorporation of uracil into DNA is probably not

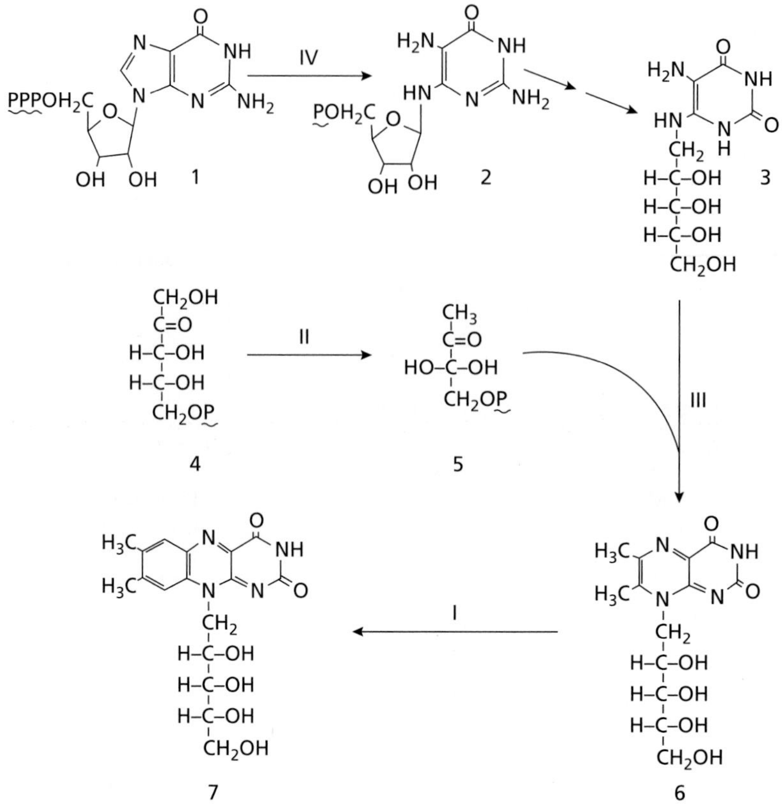

Fig. 27 Biosynthetic pathway to riboflavin. The intermediates and respective enzymes involved are 1, GTP; 2, 5-amino-6-phosphoribosylamino-8-aminopyrimidine; 3, aminoribitylamino-2,4-(*1H,3H*)-pyrimidinedione; 4, ribulose 5-phosphate; 5,3,4-dihydroxy-2-butanone-4-phosphate (DHBP); 6, 6,7-dimethyl-8-ribityllumazine; 7, riboflavin. I, riboflavin synthase; II, DHBP synthase; III, lumazine synthase; IV, GTP cyclohydrolase II (encoded by *ribA*). PO_3 groups are indicated by P. From Lee CY, O'Kane DJ, Meighen EA (1994) Riboflavin synthesis genes are linked with the *lux* operon of *Photobacterium phosphoreum. J. Bacteriol.* 176: 2100–2104.

lethal until at least 10% of DNA thymine is replaced by uracil. Above this level, excess uracil in DNA may result in degradation by repair enzymes causing lethal double-strand breaks. Uracil-containing DNA may not be recognised by DNA-binding enzymes or it may turn off protein synthesis by a regulatory system. A DNA repair system deals with this problem under most circumstances.

Regulation of Purine and Pyrimidine Biosynthesis

Both feedback inhibition and repression–derepression control systems regulate the activity of the purine and pyrimidine biosynthetic pathways. The steps in the purine pathway subject to regulation are the first step, encoded by *purF*, and the enzymes immediately after the branch-point at IMP (encoded by *purA*, *purE*, *quaE*, *quaA*) (Fig. 30). PurF activity is inhibited by either AMP or GMP. Purine-requiring mutants that accumulate intermediates in the culture medium no longer do so when purines or the nucleotides are added. The enzymes at the IMP branch-point are affected by the specific nucleotide being synthesised. IMP dehydrogenase is inhibited by feedback and its synthesis is repressed by GMP. The two enzymes that convert IMP to GMP are coded for by the *quaBA* operon and are co-ordinately controlled. The two genes that encode the enzymes responsible for conversion of IMP to AMP, *purA* and *purE*, are subject to end-product repression. AMP can inhibit phosphoribosylamine (PRA) synthesis by a feedback mechanism. Regulation of the enzymatic steps that lead from IMP to AMP and GMP forms a metabolic 'figure-of-eight' insofar as the end-products of the two branches loop back to inhibit formation of the XMP or AMP-S intermediates (Fig. 30).

The purine pathway genes are distributed throughout the *E. coli* chromosome as single genes and small operons. They are negatively regulated at the transcriptional level by the PurR protein and its co-repressors. *purR* encodes an aporepressor that combines with the purine co-repressors hypoxanthine

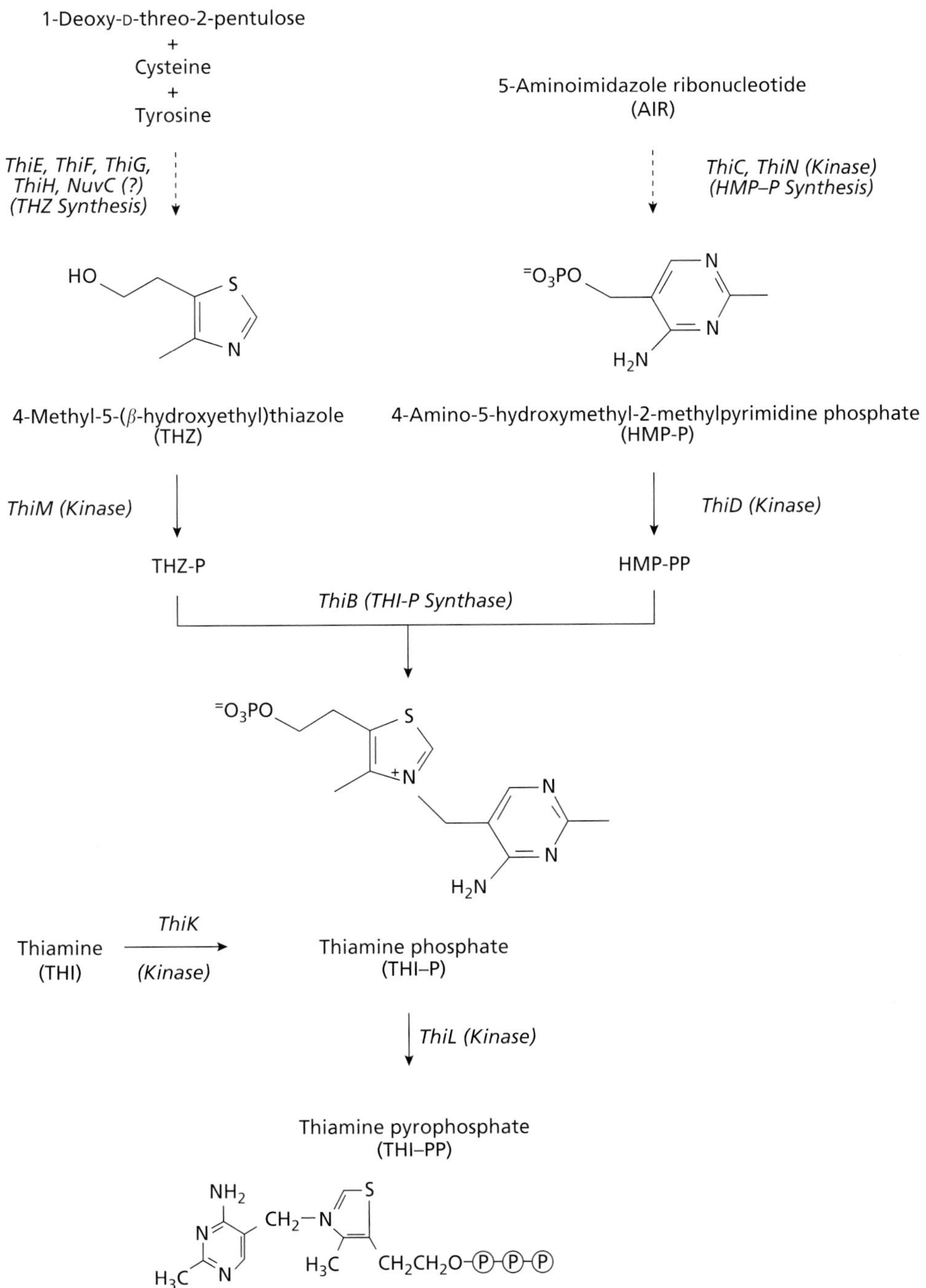

Fig. 28 Pathways of thiamine (vitamin B_1) biosynthesis. The gene designations and enzymes are those in *E. coli* and *S. typhimurium*. Modified from Van der Horn PB, Backstrom AD, Stewart VC and Begley TD (1993) Structural genes for thiamine biosynthesis enzymes (*thi*CEFGH) in *Escherichia coli* K-12. *J. Bacteriol.* 175: 982–992.

and guanine, resulting in an increased affinity for a 16-base pair palindromic operator in each of the eight operons of the purine regulon. Other genes that supply intermediates for purine nucleotide synthesis or are involved in the synthesis or salvage of pyrimidine nucleotides are also part of the *purR*-regulated regulon. This cross-pathway regulation is necessary to ensure the proper supply of precursors for each of the pathways involved. These genes include *glyA* (encoding serine hydroxymethyltransferase, a major contributor of C_1), *codA* (encoding cytosine deaminase, a pyrimidine salvage enzyme), *pyrC* (encoding dihydroorotase), *pyrD* (encoding dihydroorotate dehydrogenase), *prsA* (encoding PRPP synthase),

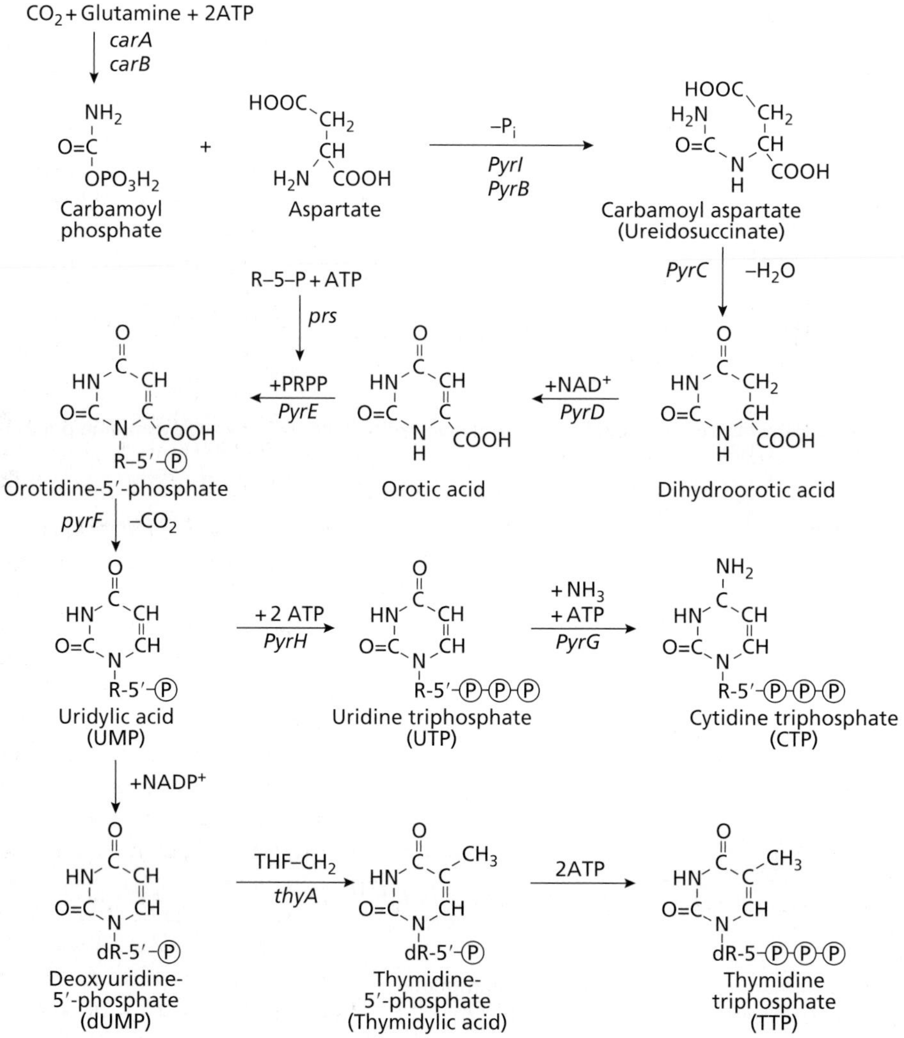

Fig. 29 Pyrimidine biosynthesis. Structural gene designations for *E. coli* are: *carA*, *carB*, carbamoyl phosphate synthases A and B; *pyrI*, *pyrB*, aspartate carbamoyltransferase; *pyrC*, dihydroorotase; *pyrD*, dihydroorotate dehydrogenase; *prs*, PRPP synthase; *pyrE*, orotate phosphoribosyltransferase; *pyrF*, OMP decarboxylase; *pyrH*, pyridine nucleotide phosphorylase; *pyrG*, cytidine triphosphate synthase; *thyA*, thymidylate synthase. From Moat AG, Foster JW (1995) *Microbial Physiology*, 3rd edn. New York: John Wiley & Sons.

speA (encoding arginine decarboxylase) and *glnE* (encoding PII protein in nitrogen regulation). Each of these enzymes contains a site to which PurR binds *in vitro* and they are co-regulated *in vivo* by *purR*. The expression of *E. coli purR* is auto-regulated. Auto-regulation at the level of transcription requires two operator sites, designated $purRo_1$ (O_1) and $purRo_2$ (O_2). Operator O_1 is in the region of DNA between the transcription start site and the site for translation initiation, and O_2 is in the protein-coding region. Operator site O_2 located within the *purR* coding sequence binds the repressor with 6-fold lower affinity than O_1 but still appears to make an important contribution to auto-regulation *in vivo*.

In *B. subtilis* the genes that encode the enzymes for the 10 steps to IMP and the four genes that encode the conversion of IMP to AMP and GMP, form a 12-gene *pur* operon. This is subject to dual control by adenine and guanine compounds. An adenine compound represses transcription initiation and a guanine compound regulates transcription termination-antitermination in an mRNA leader region that precedes the first structural gene.

In considering the regulation of purine biosynthesis, complexities arise as a result of the branch-points that lead to the pyrimidine ring of thiamine, the pterin moiety of folate, the ribityl-amino pyrimidine of riboflavin, and a portion of the ring structure of histidine (**Fig. 31**). PRPP synthase, encoded by *prsA*,

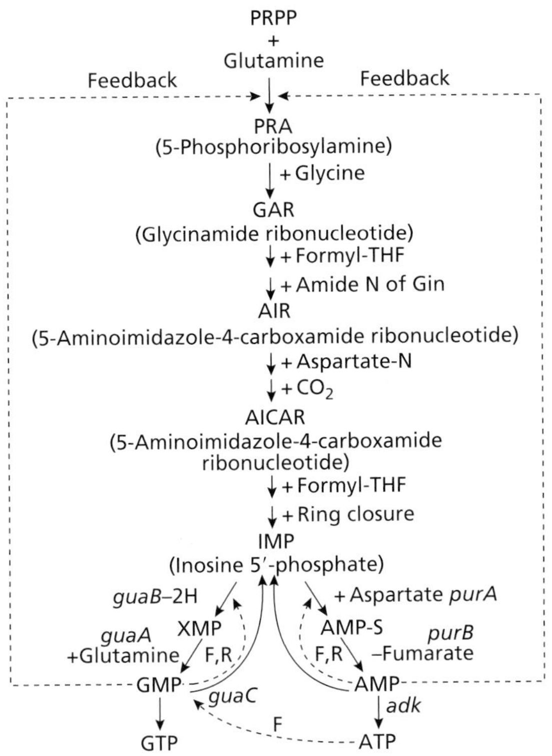

Fig. 30 Regulatory mechanisms in purine biosynthesis. PRPP, phosphoribosyl pyrophosphate; XMP, xanthosine monophosphate; GMP, guanosine monophosphate; GTP, guanosine triphosphate; AMP-S, succinyloadenosine monophosphate. The structural genes for *E. coli* are: *guaA*, GMP synthase; *guaB*, IMP dehydrogenase; *guaC*, GMP aminotransferase; *purA*, adenylosuccinate synthase; *purB*, adenylosuccinate lyase; *adk*, adenylate kinase. Each of the individual enzymatic steps is repressed by AMP. From Moat AG, Foster JW (1995) *Microbial Physiology*, 3rd edn. New York: John Wiley & Sons.

participates in several biosynthetic pathways. It is involved in the biosynthesis of purine and pyrimidine nucleotides, tryptophan, histidine and pyridine nucleotides [NAD(P)]. PRPP synthase is subject to repression by pyrimidine nucleotides and is also under the regulatory control of PurR. In view of the increasing number of genes that are subject to cross-pathway regulation, it seems likely that an even greater number of genes that contain sequences related to PurR-binding sites will be revealed by further research in this area.

Aspartate transcarbamylase (ATCase), the first enzyme specific to the pyrimidine pathway, represents one major site of regulatory control. In *E. coli*, the *pyrBI* operon encodes the catalytic (*pyrB*) and regulatory (*pyrI*) subunits of ATCase. ATCase activity is subject to allosteric regulation by the activator ATP and the inhibitors CTP and UTP, which bind to the regulatory subunit and alter substrate binding at the catalytic site. This dual control by purines and pyrimidines provides the cell with an efficient mechanism for maintaining the proper ratio of these compounds. This system serves as a prototype for allosteric regulatory systems. ATCase activity is also controlled by the level of *pyrBI* expression, which is negatively regulated by pyrimidine availability. This regulation occurs primarily through UTP-sensitive attenuation control, with additional pyrimidine-mediated regulation occurring independently at the level of transcription initiation. Of two promoters, P_1 and P_2 located upstream of *pyrB*, more than 95% of the *pyrBI* transcripts are initiated at promoter P_2 with only a small portion of the level of ATCase influenced by the P_1 promoter.

In *S. typhimurium* and *E. coli*, expression of *pyrC* (encoding dihydroorotase) and *pyrD* (encoding dihydroorotase dehydrogenase) is regulated in response to fluctuations in the intracellular CTP/GTP pool ratio. High CTP/GTP pool ratios repress expression by production of an mRNA initiated with a CTP downstream of the leader region. In the presence of excess pyrimidine an mRNA stem–loop structure is formed (Fig. 32). This will sequester a ribosome-binding site and prevent synthesis of the *pyrC* product.

Purine and Pyrimidine Salvage Pathways

Although it is often assumed that adenine and guanine or their ribonucleotides are freely inter-converted by micro-organisms, this is not actually the case. There is a limited inter-conversion ability between one base and another, and an even more limited ability to inter-convert nucleotides and nucleosides (Figs 33 and 34). Adenine and guanine cannot directly be converted from one to the other by any means (Fig. 33). The free purine bases are converted directly to their nucleotides by pyrophosphorylases (phosphoribosyltransferases) by reaction with PRPP. Two distinct purine-utilising enzymes, one specific for the conversion of adenine to AMP and the other specific for the conversion of hypoxanthine and guanine to IMP and GMP, respectively, are present in a variety of bacteria. These enzymes are subject to feedback inhibition and may play a role in the regulation of purine transport and incorporation into nucleic acids. Nucleotidases and nucleosidases are involved in the sequential degradation of the nucleotides to the free bases and are primarily catabolic enzymes. Glycohydrolases may degrade the nucleotides to the free bases and ribose phosphate.

The inter-conversion of the purine bases may be quite complex and often occurs only through rather devious routes (Fig. 33). As a specific example,

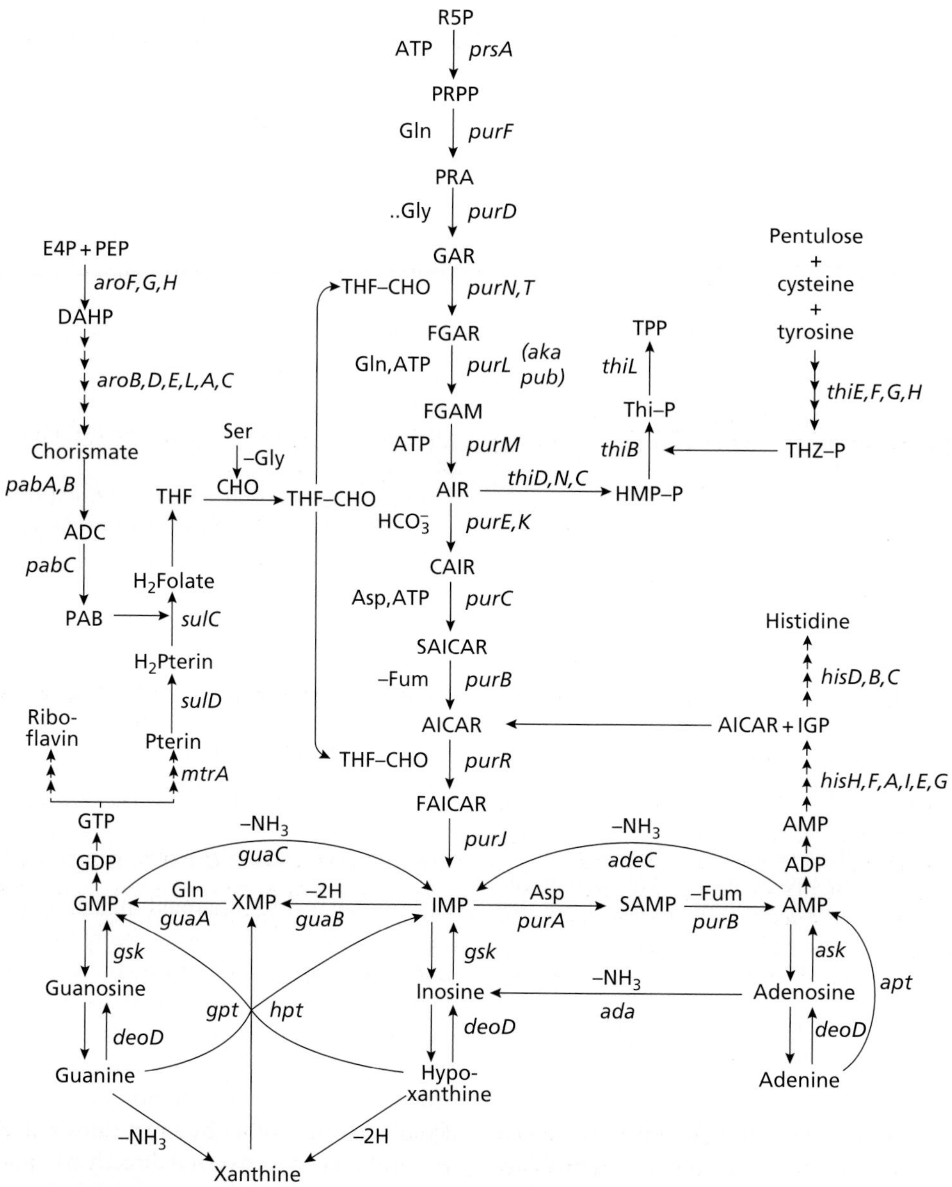

Fig. 31 Biosynthetic pathways leading from the de-novo purine biosynthetic pathway. R5P, ribose 5-phosphate; PRPP, 5-phosphoribosyl pyrophosphate; PRA, 5-phosphoribosylamine; GAR, glycinamide ribonucleotide; FGAR, 1-*N*-formylglycinamide ribonucleotide; FGAM, α-*N*-formylglycinamidine ribonucleotide; AIR, 5-aminoimidazole ribonucleotide; SAICAR, 5-amino-4-imidazole-(-*N*-succinylo-)-carboxamide ribonucleotide; AICAR, 5-aminoimidazole-4-carboxamide ribonucleotide; FAICAR, 5-formamidoimidazole-4-carboxamide ribonucleotide; IMP, inosine 5′-monophosphate; SAMP, *N*-succinylo-AMP; AMP, adenosine 5′-monophosphate; XMP, xanthosine 5′-monophosphate; GMP, guanosine 5′-monophosphate; GTP, guanosine triphosphate; ATP, adenosine triphosphate; RF, riboflavin; H_4folate or THF, tetrahydrofolate; THF-CHO, N_5,N_{10}-tetrahydrofolate; E_4P, erythrose 4-phosphate; PEP, phosphoenolpyruvate; DAHP, 3-hydroxy-D-arabino-heptulosonate 7-phosphate; ADC, 4-amino-4-deoxychorismate; PAB, *p*-aminobenzoate; H_2pterin, 6-CH_2OH-7,8-dihydroxypteroate; H_2folate, 7,8-dihydrofolate; IGP, imidazoleglycerol phosphate; HMP-P, 4-amino-5-hydroxymethylpyrimidine phosphate; THZ-P, 4-methyl-5-(β-hydroxyethyl) thiazole phosphate; Thi-P, thiamine monophosphate; TPP, thiamine pyrophosphate. From Moat AG, Foster JW (1995) *Microbial Physiology*, 3rd edn. New York: John Wiley & Sons.

consider the pathway by which *S. typhimurium* converts exogenous adenine to guanine nucleotide (GMP). Adenine is converted to adenosine, followed by deamination to inosine and subsequent phosphorolysis to hypoxanthine. Hypoxanthine is then converted to inosine monophosphate (IMP) by inosine monophosphate pyrophosphorylase. IMP is then converted to XMP and GMP.

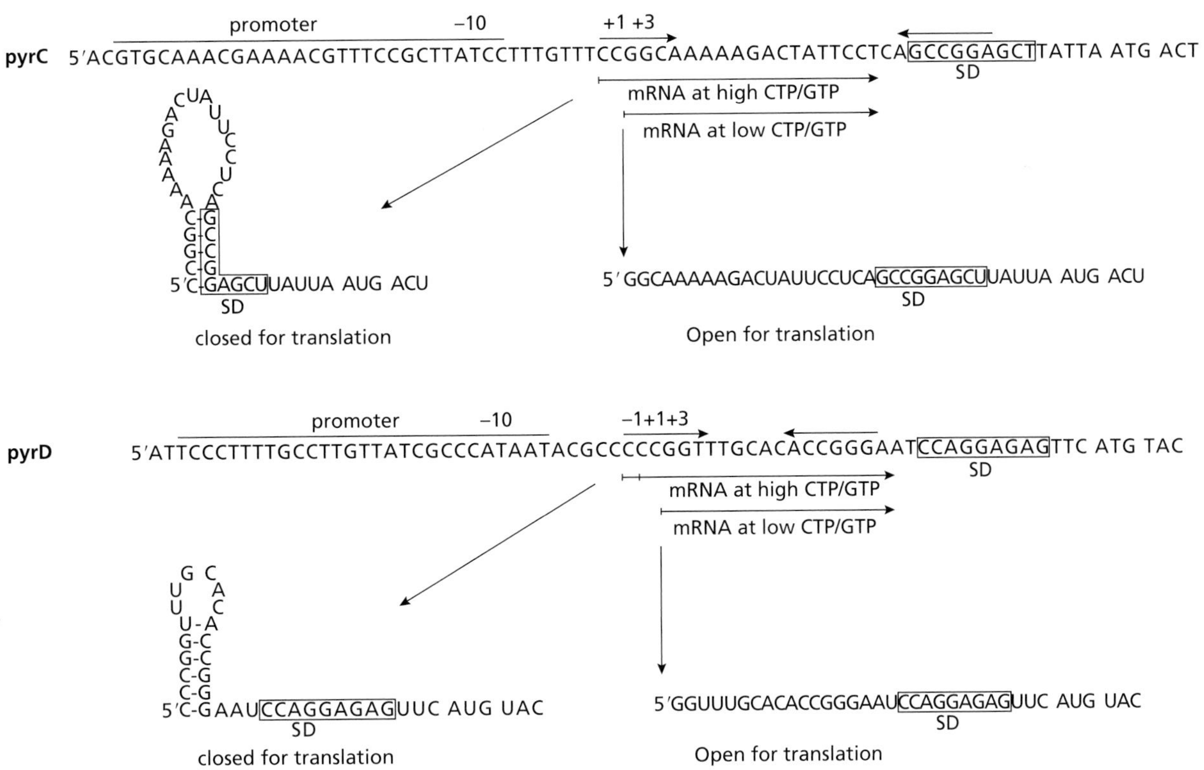

Fig. 32 Structure of the *S. typhimurium pyrC* and *pyrD* promoter-leader regions and the proposed regulatory mechanisms. Nucleotides are numbered relative to the promoter −10 region (overlined) where +1 at the consensus spacing of 7 bp downstream of the −10 element defines the *in-vivo* transcriptional initiation point in repressing conditions. The Shine–Delgarno (SD) regions are boxed, and the regions of hyphenated dyad symmetry are indicated by arrows above the sequence. Arrows below the sequence represent the transcripts arising in conditions of different CTP/GTP pool ratios. The putative secondary structures formed at the 5′-ends of the transcripts are shown. From Sørensen KI, Baker KW, Elln RA, Neuhard J (1993) Nucleotide pool-sensitive selection of transcriptional start site *in vivo* at the *Salmonella typhimurium pyr*C and *pyr* promoters. *J. Bacteriol.* 175: 4137–4144.

At the nutritional level, the ability of a given organism to inter-convert adenine and guanine is reflected in the ability of the two purines to promote the growth of purine-requiring organisms. Adenine serves quite readily as the sole purine nutrient for many purine-requiring organisms, while in this regard guanine shows a limited capability. The capacity of organisms to inter-convert purines and their nucleotides also influences the action of a variety of purine antagonists. Regulation of the *gua* operon, which controls the conversion of IMP to XMP to GMP, is effected by adenine and guanine nucleotides rather than the free bases, and may explain the diverse effects noted for various bases.

Some micro-organisms, such as the lactobacilli, possess nucleoside-*N*-glycosyltransferases, which can catalyse ribosyl and deoxyribosyl transfer reactions from nucleosides to free bases. However, a wide variety of bacteria, including many of the enteric bacteria, appear to inter-convert nucleosides and free bases through the coupling of nucleoside phosphorylases and do not possess nucleoside-*N*-glycosyltransferases. Both purine and pyrimidine bases may be converted to the nucleoside and nucleotide stages by the combined action of nucleoside phosphorylases and nucleoside kinases:

$$\text{Pyrimidine} + \text{R}-5-\text{P} \rightleftharpoons \text{pyrimidine}-\text{ribose} + \text{P}_i$$
$$\text{Pyrimidine}-\text{ribose} + \text{ATP} \rightleftharpoons \text{pyrimidine}-\text{ribose}-\text{P} + \text{ADP}$$

Conclusion

Although specific attention may not have been drawn to them, there are a number of points of relevance between our knowledge of biosynthetic pathways and molecular medical microbiology. For example,

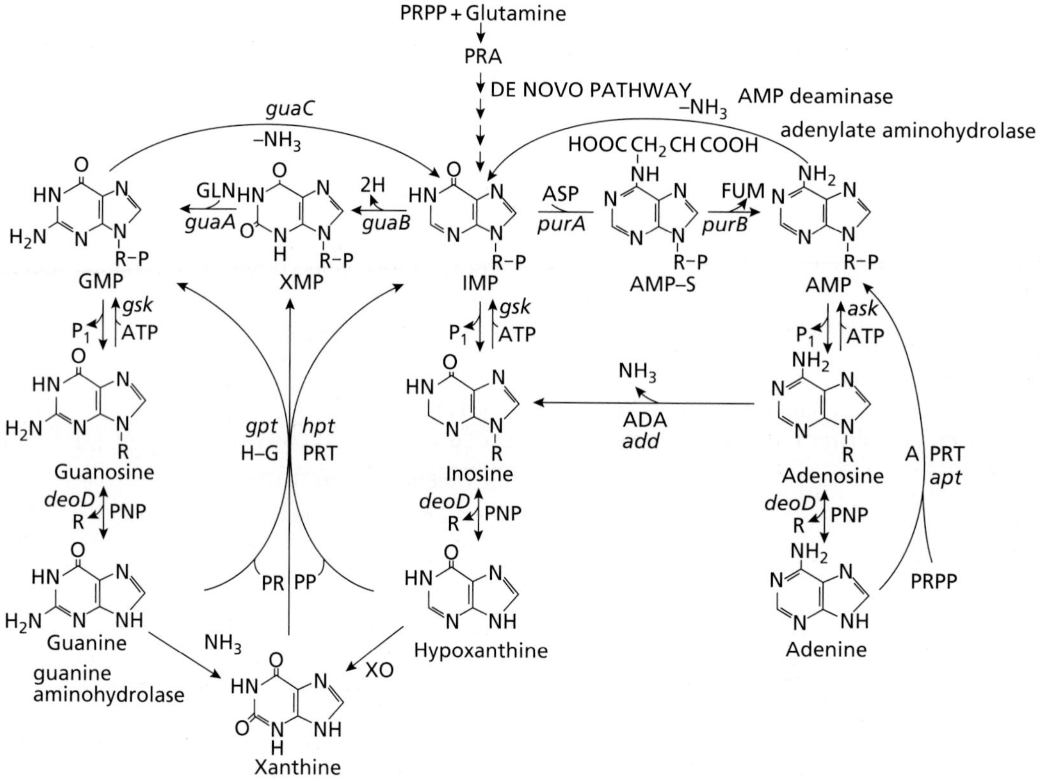

Fig. 33 Inter-conversion of purines and their derivatives. PRPP, 5-phosphoribosyl pyrophosphate; PRA, phosphoribosylamine; GMP, guanosine monophosphate; XMP, xanthine monophosphate; IMP, inosine monophosphate; AMP-S, succinyloadenosine monophosphate; AMP, adenosine monophosphate; *quaB*, IMP dehydrogenase; *quaA*, GMP synthase; *quaC*, GMP reductase; *purA*, adenylosuccinate synthase; *purB*, adenylosuccinatelyase; *hpt*, hypoxanthine phosphoribosyltransferase (H-PRT); *gpt*, guanine/xanthine phosphoribosyltransferase (G-PRT); A-PRT, adenine phosphoribosyltransferase (*apt*); XO, xanthine oxidase; ADA, adenosine deaminase (*add*); GLN, glutamine; ASP, aspartate; FUM, fumarate. PNP purine nucleoside phosphorylase (*deoG*); *ask*, adenosine kinase; *gsk*, quaknosine-inosine kinase. Genes from *Salmonella typhimurium*. From Moat AG, Foster JW (1995) *Microbial Physiology*, 3rd edn. New York: John Wiley & Sons.

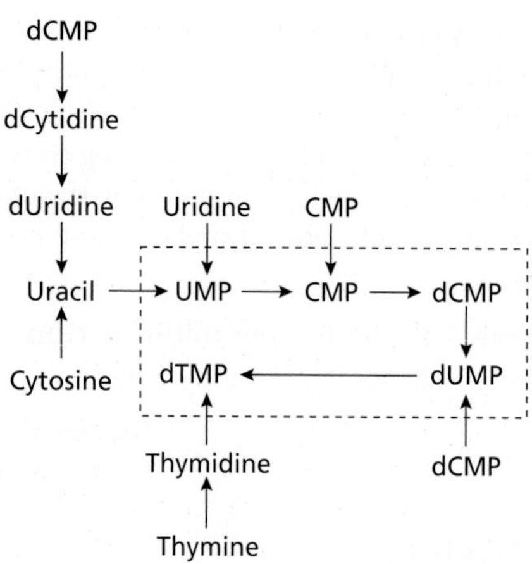

Fig. 34 Inter-conversion of pyrimidines and their derivatives. dCMP, deoxycytosine monophosphate; CMP, cytosine monophosphate; UMP, uridine monophosphate; dTMP, thymidine monophosphate. From Moat AG, Foster JW (1995) *Microbial Physiology*, 3rd edn. New York: John Wiley & Sons.

auxotrophs that cannot supply their nutritional needs from the host will be avirulent. Specific pathways are vulnerable to attack by chemotherapeutic agents. The sulphonamide group of compounds inhibit dihydrofolate synthase. These agents have been used successfully, either alone or in combination with other antimicrobial agents, to treat bacterial infections. Pyrimethamine and trimethoprim inhibit dihydrofolate reductase. These compounds are especially useful when combined with sulphonamides. Many antiviral and anticancer agents interfere with purine and pyrimidine biosynthesis or inter-conversion.

Further Reading

Fatty Acids, Phospholipids and Sterols

Jackowski S, Cronan JE Jr, Rock CO (1991) Lipid metabolism in procaryotes. In: Vance DE, Vance JE (eds)

Biochemistry of Lipids, Lipoproteins, and Membranes. Amsterdam: Elsevier Science, pp. 43–85.

Moat AG, Foster JW (1995) *Microbial Physiology*, 3rd edn. New York: John Wiley & Sons.

Amino Acids – General References

Herrmann KM, Somerville RL (eds) (1983) *Amino Acids. Biosynthesis and Genetic Regulation.* Reading, MA: Addison-Wesley.

Hinnebusch AG (1992) General and pathway-specific regulatory mechanisms controlling the synthesis of amino acid biosynthetic enzymes in *Saccharomyces cerevisiae.* In: Jones EW, Pringle JR, Broach JR (eds) *The Molecular and Cellular Biology of the Yeast Saccharomyces: Gene Expression*, Vol. II, pp. 319–414.

Moat AG, Foster JW (1995) *Microbial Physiology*, 3rd edn. New York: John Wiley & Sons.

Glutamate Family

Cunin R, Glansdorff N, Pierard A, Stalon V (1986) Biosynthesis and metabolism of arginine in bacteria. *Microbiol. Rev.* 50: 314–352.

Davis RH (1986) Compartmental and regulatory mechanisms in the arginine pathways of *Neurospora crassa* and *Saccharomyces cerevisiae. Microbiol. Rev.* 50: 280–313.

Davis RH, Morris DR, Coffino P (1992) Sequestered end products and enzyme regulation: The case of ornithine decarboxylase. *Microbiol. Rev.* 56: 280–290.

Demain AL (1991) Production of beta-lactam antibiotics and its regulation. *Proc. Natl Sci. Council, ROC. Part B: Life Sci.* 15: 252–265.

Mora J (1990) Glutamine metabolism and cycling in *Neurospora crassa. Microbiol. Rev.* 54: 293–304.

Tabor CW, Tabor H (1985) Polyamines in micro-organisms. *Microbiol. Rev.* 49: 81–99.

Aspartate and Pyruvate Families

de Lorenzo V, Bindereif A, Paw BH, Neilands JB (1986) Aerobactin biosynthesis and transport genes of plasmid ColV-K30 in *Escherichia coli* K-12. *J. Bacteriol.* 165: 570–578.

Graves LM, Switzer RL (1990) Aspartokinase III, a new isozyme in *Bacillus subtilis* 168. *J. Bacteriol.* 172: 218–223.

Schrumpf B, Schwarzer A, Kalinowski J, Pühler A, Eggeling L, Sahm H (1992) A functionally split pathway for lysine synthesis in *Corynebacterium glutamicum. J Bacteriol* 173: 4510–4516.

Zhang J-J, Hu FM, Chen NY, Paulus H (1990) Comparison of the three aspartokinase isozymes in *Bacillus subtilis* Marburg and 168. *J. Bacteriol.* 172: 701–708.

Zhang J-J, Paulus H (1990) Desensitization of *Bacillus subtilis* aspartokinase I to allosteric inhibition by meso-diaminopimelate allows aspartokinase I to function in amino acid biosynthesis during exponential growth. *J. Bacteriol.* 172: 4690–4693.

Serine–Glycine Family

Avissar YJ, Beale SI (1989) Identification of the basis for o-aminolevulinic acid auxotrophy in a hemA mutant of *Escherichia coli. J. Bacteriol.* 171: 2919–2924.

Biel AJ (1992) Oxygen-regulated steps in the *Rhodobacter capsulatus* tetrapyrrole biosynthetic pathway. *J. Bacteriol.* 174: 5272–5274.

Elliot T, Avissar YL, Rhie G-E, Beale SJ (1990) Cloning and sequence of the *Salmonella typhimurium* hemL gene and identification of the missing enzyme in *hemL* mutants as glutamate-1-semialdehyde aminotransferase. *J. Bacteriol.* 172: 7071–7084.

Escalante-Semerena BC, Suh S-J, Roth JR (1990) *CobA* function is required for both *de novo* cobalamin biosynthesis and assimilation of exogenous corrinoids in *Salmonella typhimurium J. Bacteriol.* 172: 273–280.

Hansson M, Hederstedt L (1992) Cloning and characterization of the *Bacillus subtilis hemEHY* gene cluster, which encodes protoheme IX biosynthetic enzymes. *J. Bacteriol.* 174: 8081–8093.

Hansson M, Rutberg L, Schroder I, Hederstedt L (1991) The *Bacillus subtilis hemAXCDBL* gene cluster, which encodes enzymes of the biosynthetic pathway from glutamate to uroporphyrinogen III. *J. Bacteriol.* 173: 2590–2599.

Wilson RL, Stauffer LT, Stauffer GV (1993) Roles of the GcvA and PurR proteins in negative regulation of the *Escherichia coli* glycine cleavage enzyme system. *J. Bacteriol.* 175: 5129–5134.

Aromatic Amino Acid Family

Babitzke P, Gollnick P, Yanofsky C (1992) The *mtrAB* operon of *Bacillus subtilis* encodes GTP cyclohydrolase I (MtrA), an enzyme involved in folic acid biosynthesis, and MtrB, a regulator of tryptophan biosynthesis. *J. Bacteriol.* 174: 2059–2064.

Crawford IP (1989) Evolution of a biosynthetic pathway: the tryptophan paradigm. *Annu. Rev. Microbiol.* 43: 567–600.

Green JM, Merkel WK, Nichols BP (1992) Characterization and sequence of *Escherichia coli* pabC, the gene encoding aminodeoxychorismate lyase, a pyridoxal phosphate-containing-enzyme. *J. Bacteriol.* 174: 5317–5323.

Hagervall TG, Jonsson YH, Edmonds CG, McCloskey JM, Bjork GR (1990) Chorismic acid, a key metabolite in modification of tRNA. *J. Bacteriol.* 172: 252–259.

Lopez P, Greenberg B, Lacks SA (1990) DNA sequence of folate biosynthesis gene *sulD*, encoding hydroxymethyldihydropterin pyrophosphokinase in *Streptococcus pneumoniae*, and characterization of the enzyme. *J. Bacteriol.* 172: 4766–4774.

Lopez P, Lacks SA (1993) A bifunctional protein in the folate biosynthetic pathway of *Streptococcus pneumoniae* with dihydroneopterin aldolase and hydroxymethyldihydropterin pyrophosphokinase activities. *J. Bacteriol.* 175: 2214–2220.

Moat AG, Foster JW (1987) Biosynthesis and salvage pathways of pyridine nucleotides. In: Dolphin D, Poulson R, Avramovic O (eds) *Pyridine Nucleotide Coenzymes: Chemical, Biochemical, and Medical Aspects*, Vol. 2B. New York: John Wiley & Sons.

Muday GK, Herrmann KM (1990) Regulation of the *Salmonella typhimurium aroF* gene in *Escherichia coli. J. Bacteriol.* 172: 2259–2266.

Pyne C, Bognar AL (1992) Replacement of the *folC* gene, encoding folylpolyglutamate synthetase-dihydrofolate synthetase in *Escherichia coli*, with genes mutagenized *in vitro. J. Bacteriol.* 174: 1750–1759.

Richter G, Ritz H, Katzenmeier G *et al.* (1993) Biosynthesis of riboflavin: Cloning, sequencing, mapping, and expression of the gene coding for GTP cyclohydrolase II in *Escherichia coli. J. Bacteriol.* 175: 4045–4051.

Sharma V, Suvarna K, Meganathan R, Hudspeth MES (1992) Menaquinone (vitamin K_2) biosynthesis: nucleotide sequence and expression of the *menB* gene from *Escherichia coli. J. Bacteriol.* 174: 5057–5062.

Talarico TL, Ray PH, Dev IK, Merrill BM, Dallas WS (1992) Cloning, sequence analysis, and, overexpression of *Escherichia coli folK*, the gene coding for 7, 8-dihydro-6-hydroxymethylpterin-pyrophosphokinase. *J. Bacteriol.* 174: 5791–5977.

Tran PV, Nichols BP (1991) Expression of *Escherichia coli pabA. J. Bacteriol.* 173: 3680–3687.

Zalkin H, Ebbole DJ (1988) Organization and regulation of genes encoding biosynthetic enzymes in *Bacillus subtilis. J. Biol. Chem.* 263: 1595–1598.

Histidine

Delorme C, Ehrlich SD, Renault P (1992) Histidine biosynthesis genes in *Lactococcus lactis* subsp. *lactis. J. Bacteriol.* 174: 6571–6579.

Purines and Pyrimidines

Bacher A (1990) Biosynthesis of flavins. In: Muller F (ed.) *Chemistry and Biochemistry of Flavoenzymes*, Vol. 1. Boca Raton, FL: CRC Press, pp. 215–259.

Downs DM (1992) Evidence for a new, oxygen–regulated biosynthetic pathway for the pyrimidine moiety of thiamine in *Salmonella typhimurium. J. Bacteriol.* 174: 1515–1521.

He B, Choi KY, Zalkin H (1993) Regulation of *Escherichia coli glnB*, *prsA*, and *speA* by the purine repressor. *J. Bacteriol.* 175: 3598–3606.

Kowasaki Y (1993) Copurification of hydroxyethylthiazole kinase and thiamine pyrophosphorylase of *Saccharomyces cerevisiae*. Characterization of hydroxyethyl thiazole kinase as a bifunctional enzyme in the thiamine biosynthetic pathway. *J. Bacteriol.* 175: 5153–5158.

Reuke B, Korn S, Eishenreich H, Bacher A (1992) Biosynthetic precursors of deazaflavins. *J. Bacteriol.* 174: 4042–4049.

Richter G, Volk R, Krieger C, Lahm H-W, Röthlisberger U, Bacher A (1992) Biosynthesis of riboflavin: Cloning, sequencing, and expression of the gene coding for 3,4-dihydroxy-2–butanone-4-phosphate synthase of *Escherichia coli. J. Bacteriol.* 174: 4050–4056.

Vander Horn PB, Backstrom AD, Stewart V, Begley TP (1993) Structural genes for thiamine biosynthetic enzymes (*thiCEFGH*) in *Escherichia coli* K-12. *J. Bacteriol.* 175: 982–992.

14

Bacterial Chromosomes and Their Replication

Richard W.P. Smith[1] and Millicent Masters[2]

[1]*Institut für Molekulare Biotechnologie e.V., Jena, Germany*
[2]*Institute of Cell and Molecular Biology, University of Edinburgh, Edinburgh, UK*

DNA is the central database that stores and provides the information for making the macromolecules that compose the cell and allow it to function. Because it has a structure that facilitates accurate and straightforward duplication, it is ideally suited for its other role as the hereditary material, which is copied and passed from mother to daughter cell. This chapter is concerned primarily with the replication of DNA. First it is convenient to consider the nature and arrangement of the information stored on the DNA.

Bacterial Genomes and Chromosomes

Most bacteria contain a single circular chromosome, but exceptions to this rule have recently come to light. Since bacterial chromosomes are not paired, bacteria are haploid, and they lack structural proteins. Bacterial chromosomes can be very short; the intracellular pathogen *Mycoplasma genitalium*, which parasitises its host for metabolites, has a chromosome 0.58 Mb long, which encodes only about 470 proteins. The chromosome of *Escherichia coli*, which has been fully sequenced, is 4.6 Mb long, a size close to that of several other free-living bacteria, such as *Bacillus subtilis* (4.2 Mb) and *Mycobacterium tuberculosis* (4.5 Mb). In comparison, the yeast genome is 12 Mb long.

The bacterial genome is frequently augmented by separate DNA structures called plasmids, which vary in length from under 5 kb to more than 100 kb in length. Although the plasmids that inhabit *E. coli* are not normally essential to the cell and can therefore best be understood as forms of 'selfish' DNA, other bacterial species may have multiple plasmids that are not easily displaced, and these may make significant contributions to host survival. Although there are usually only one or two copies of the larger plasmids per cell, there may be many copies of smaller ones. Plasmids control the frequency of their own replication, generally through the action of diffusible substances, proteins or RNA, so that the number of copies per cell of a particular plasmid is maintained. The control of copy number is at the replication initiation stage (see below), and consequently the mechanism by which plasmid replication is initiated often differs in detail from that of the host chromosome. However, the machinery involved in the later stages of replication is borrowed entirely from the host. Plasmids encode a complex machinery that makes possible their transfer between cells (see Chapter 20).

 doi:10.1006/bkmm.2001.0014

The extended *E. coli* chromosome is 1000 times the length of the cell and, in rapidly growing cells, is fairly well dispersed throughout the cell. In non-growing cells it forms a much more compact structure, which can easily be seen as a discrete body by the use of appropriate stains. In order to fit into a cell so much shorter than itself, and in order that all parts of it are immediately accessible to the transcription/translation apparatus, the DNA must be both compacted and organised in space. Part of this compaction is achieved by supercoiling (see below), which reduces the length of DNA. In addition, the DNA appears to have a domain structure, with each domain thought to be about 50 kb in length. With care, chromosomes can be isolated as intact structures termed folded chromosomes, and these appear to retain their domain structure. Such preparations are not viscous as are preparations of unfolded DNA. What maintains the structure of the folded chromosome is, however, still not understood. Although *E. coli* possesses a variety of small, basic, histone-like DNA-binding proteins, none of these is essential for cell survival.

What Does the DNA Sequence Tell Us?

Gene Arrangement

Determination of the *E. coli* chromosome sequence has recently been completed. What does it tell us? Firstly, that the chromosome is densely packed with genes: 88% of the DNA encodes proteins (~4300 genes), ~1% encodes stable RNA and another 1% consists of non-coding repeated sequences. Functions are known or can be inferred with some confidence for about 60% of protein coding genes. Non-coding DNA is mostly located in short sections between genes and is probably concerned with the control of gene expression. The majority of genes are included in multi-gene operons, but a substantial minority, as many as one-third, do not appear to be co-transcribed with another gene.

Since chromosome replication takes up most of the cell cycle, genes replicated early are present in higher concentrations than those replicated late. The faster the cells are growing, the greater the difference in ratio between early and late replicated genes, and this is exploited by the cell. Gene products that are required in higher concentration in rapidly growing cells, such as those involved in translation and transcription, are encoded by genes that are replicated early.

The origin and terminus of replication divide the chromosome into two halves, one of which is replicated in a clockwise direction and the other anticlockwise with respect to the genetic map. Many features of the chromosome are oriented with respect to replication. The seven ribosomal RNA operons are transcribed and replicated in the same direction, but there is only a small tendency for the same to be true for other genes (55% of protein-coding genes are aligned with replication). The leading strands in replication (see below) have base compositions significantly different from the lagging strands, with G being 2% more abundant than C. This is sharply reversed at the origin and terminus. The 8-bp sequence chi, which is important for homologous recombination, is heavily skewed towards the leading strand, suggesting that recombination and replication are likely to be linked in some way.

Other Repeated Sequences

REP is a 30–40 bp palindromic sequence that can form a stable stem–loop structure. In the chromosome there is a total of 581 REP sequences, which occur singly or in groups and account for 0.54% of the genome. The function they may have is not known. Indeed, since they are always transcribed, though never translated, it is not clear whether the role of REP is to modify the structure of DNA, RNA or both. Possible roles include translational termination (in RNA) and interaction with DNA-binding proteins to maintain the domain structure of the chromosome.

Horizontally Transmitted Sequences

Insertion sequences (IS) and transposons are mobile DNA elements that encode the enzymes needed for them to move themselves from one DNA location or molecule to another. IS come in many varieties, all roughly 0.8–1.5 kb in length. In the chromosome of the sequenced *E. coli* strain MG1655, there are 10 different species of IS element, which range in copy number from 1 to 12, but there are no transposons. ISs move or are lost at frequencies well above those at which point mutations occur, and strains vary greatly in the numbers and positions of particular elements. Spontaneous mutations are often the result of the insertion of an IS within a gene. Transposons are complex relatives of ISs, and have acquired 'passenger' sequences, which mostly confer drug resistance or another selective advantage. Both transposons and ISs occur frequently on plasmids and are transferred by them within natural bacterial populations. Transposon transfer by plasmids is responsible for infectious drug resistance (see Chapter 20).

If plasmids are regarded as cytoplasmic DNA parasites, bacteriophages (phages) are chromosomal parasites. The DNA of bacteriophages, such as the well-known λ, can integrate into the continuity of the bacterial chromosome and remain there as a passenger, replicated by the host, until a later event causes it to excise, multiply and destroy the host cell. As the result of mutation, some phage genomes have lost the capacity to excise, and are known as cryptic bacteriophages. In the course of evolution, the genes of cryptic bacteriophages may be co-opted into cellular roles. The *E. coli* chromosome contains eight such cryptic bacteriophage genomes, which account for several per cent of the DNA in the chromosome.

Bacterial DNA Replication

All organisms must transmit their genetic material to the next generation, which requires accurate duplication of their chromosomes before these can be passed on. Bacteria multiply asexually by the division of parental cells to yield daughter cells. Before division, the chromosome must undergo replication so that segregation to the progeny can occur. These events have been the subject of intense investigation for nearly 50 years.

The recognition that the structure of DNA is a double helix, composed of antiparallel complementary strands, suggested the now well-established mechanism of semi-conservative DNA replication. In the course of this replication, one strand acts as the template for the synthesis of the other by polymerisation of single nucleotides to form the daughter strand. This mechanism, or variations of it, operates in all organisms, whether prokaryotic, eukaryotic or viral, and is catalysed by enzymes known as DNA polymerases. With very few exceptions, cellular DNA replication is tightly controlled, so that the entire genetic material is copied exactly once before each cell division. This is achieved in a variety of ways, which generally render the action of polymerases dependent on other events that can be regulated by cell cycle cues.

Analysis of DNA replication in bacteria has largely focused on *E. coli*, because of the extensive genetic knowledge available for this organism and the ease of its manipulation. This has allowed the identification of the genes involved in replication and the study of the physiological effects when mutations occur in them in the living organism, so-called studies *in vivo*. This approach has been complemented by the purification of replication proteins and the successful use of these to reconstitute DNA replication with purified enzymes *in vitro*. This has led to detailed analysis of the replication process. Only relatively recently has replication been studied with any vigour in other bacteria. However, in spite of the overall similarity of the replication process in all organisms, these studies are uncovering details of replication not so far observed in *E. coli*.

DNA replication can be divided into stages: initiation, elongation and termination, followed by segregation of the replicated DNA, so that one chromosome is inherited by each of the daughter cells at the subsequent division. Each of these processes is discussed in some detail below.

Replication begins with initiation. During this process a protein complex is formed at a particular DNA sequence (the origin, or *oriC*), which results in the separation of the strands of the helix to make available single strands that are able to act as templates for daughter DNA synthesis. Helicases, enzymes that bind to one strand of the DNA and move along it to separate the strands to expose further template, then bind at the origin region. A separate helicase molecule binds to each strand, and the two helicases move apart in opposite directions. This creates two replicating forks that will also move in opposite directions around the chromosome to replicate the DNA bidirectionally. The strands of the double helix, once separated, are maintained in the separated state by single-strand DNA-binding proteins, until two complete replicating complexes, poised to move in opposite directions, are established at the origin. The single-strand DNA-binding proteins are eventually displaced by polymerases. This completes the initiation stage. The replicating complexes consist of the multi-subunit DNA polymerase III which, once associated with it, probably remains tethered to the DNA throughout replication, and a primase, whose continuing role is to synthesise RNA primers as required for the discontinuously replicated lagging strand, i.e. the strand made in the $3' \rightarrow 5'$ direction. Primer synthesis is necessary because DNA polymerases cannot initiate a DNA chain *de novo*; they attach bases only to a pre-existing primer, that is an oligonucleotide already hydrogen-bonded to the template. One DNA strand must be replicated discontinuously because polymerisation can occur only in a single direction, $5' \rightarrow 3'$. In contrast to DNA polymerase, primase associates at the fork when required, and disassociates from it as each lagging-strand fragment is completed. The elongation stage consists of the activities of these enzymes at the progressing fork. After primer RNA has been replaced with DNA by a DNA polymerase, the DNA fragments made discontinuously on the lagging strand, termed Okazaki fragments after their discoverer, are joined by the nick-sealing DNA ligase.

Table 1 Enzymes involved in the chromosomal replication of *Escherichia coli*

Enzyme	Gene(s)	Size (kDa)	Function
Initiation			
DnaA	*dnaA*	52	*oriC* binding
DnaC	*dnaC*	6×29	DnaB loading
HU	*hupA, hupB*	$9.5 + 9.5$	DNA bending
IHF	*himA, himD*	$11.2 + 10.6$	DNA bending
FIS	*fis*	2×11.2	DNA bending
SeqA	*seqA*	24	Regulation of initiation
DNA adenine methyltransferase	*dam*	32	Regulation of initiation
Initiation and elongation			
DnaB	*dnaB*	6×50	Helicase
DnaG	*dnaG*	65.6	Primase
DNA gyrase	*gyrA, gyrB*	$2 \times 97 + 2 \times 90$	Supercoiling of DNA
Topoisomerase I	*topA*	100	Relaxes supercoiling
SSB	*ssb*	4×18.9	Binds ssDNA
Elongation			
DNA polymerase III $(\alpha\epsilon\theta)_2(\beta_2)_2\tau_2\gamma_2\delta\delta'\chi\psi$		918.4	Replicative polymerase
α	*dnaE*	129.9	Polymerase
ϵ	*dnaQ*	27.5	Exonuclease
θ	*holE*	8.6	Unknown
τ	*dnaX*	71.1	Dimerises core
γ	*dnaX*	47.5	Clamp loading
δ	*holA*	38.7	Clamp loading
δ'	*holB*	36.9	Clamp loading
χ	*holC*	16.6	Clamp loading
ψ	*holD*	15.2	Clamp loading
β	*dnaN*	40.6	Sliding clamp
DNA polymerase I	*polA*	103	Primer removal
DNA ligase	*lig*	75	Joins Okazaki fragments
Termination			
Tus	*tus*	35.8	Inhibition of helicase
XerC	*xerC*	33.8	Resolution of dimeric
XerD	*xerD*	28.9	Chromosomes
Topoisomerase IV	*parC*	81.2	Decatenation

Eventually, the divergently moving forks meet on the far side of the chromosome. Then termination – the joining of the DNA of the opposing forks – occurs, followed by the topological separation of parental and daughter DNA molecules. Finally, in the segregation step, the fully replicated chromosomes will move apart.

Initiation of Replication in *E. coli*

In addition to the DNA to be replicated and the enzymes that carry out the replication, initiation of DNA replication requires a variety of other proteins that transform DNA into a template for replication. This section begins by describing the structure of DNA and the enzymes that help to maintain that structure. Then the special structure at which replication is initiated, the origin of replication, is described, and the enzymes and proteins involved in the initiation process.

The Template

Supercoiling of the DNA Helix

Double-stranded (ds) DNA is normally found as a B-form, right-handed, double helix, with close to 10 bases per turn of the helix. In most, if not all, organisms the DNA is found in a supercoiled state. Supercoiling (or superhelicity) is the term used to describe the twisting of the axis of the double helix upon itself. Supercoiling is an important consideration in the energetics of DNA unwinding in processes such as transcription, recombination and replication.

When a circular dsDNA molecule lies flat on an imaginary plane surface, it is said to be in a relaxed state; this indicates the absence of supercoiling. If the molecule were to be broken and one end twisted by one or more complete revolutions while the other remained fixed and the two ends were then rejoined, the molecule would adopt a new topological state, depending on the direction and degree of twisting. If the molecule had been twisted so as partially to unwind the double helix, the resulting structure would on average have more bases per turn of the helix. This change in helical pitch may be distributed evenly over the entire molecule or, alternatively, it may be concentrated in certain areas resulting in regional unpaired stretches of DNA. Underwinding may also give rise to a molecule with the same helical pitch as the relaxed form, but with the strain taken up by supercoiling, so that the double helix is wrapped around itself and no longer lies in an imaginary plane (Fig. 1). Bacterial DNA is normally underwound.

Supercoiling can be either negative, as in the above example, or positive, with additional twists in the same direction as the helix, resulting in an overwound structure. Linear and circular DNA may both be supercoiled.

Topoisomerase I

The supercoiled state of the chromosome is maintained by a number of enzymes that have the catalytic function to break and rejoin DNA strands so as to change the topology of the substrate. The first of these enzymes to be identified in *E. coli* was topoisomerase I, a 100-kDa protein that relaxes negatively supercoiled DNA. That is, it inserts turns by passing one strand of the helix through its complementary strand. The mechanism involves cleavage of one strand of DNA followed by the formation of a covalent bond between the free 5′ phosphate of the DNA and a tyrosine residue in the enzyme. By additional non-covalent interactions, the complementary strand is passed through the cleavage site and the broken strand is rejoined. The result is a reduction in negative supercoiling by one turn of the helix.

DNA Gyrase

DNA gyrase, or topoisomerase II, the target of the antimicrobials nalidixic acid and coumermycin, antagonises the effect of topoisomerase I by introducing negative supercoils into DNA (i.e. it removes turns). DNA gyrase binds DNA as a tetramer (A_2B_2) and wraps its substrate around itself into a positive supercoil. In contrast to the action of topoisomerase I, both strands of the helix are cleaved and covalent bonds are formed between both 5′ phosphates and tyrosine residues within the two gyrase A subunits. A section of DNA that lies near the enzyme–DNA complex is then translocated through the break and the cleaved strands are re-sealed. This process results in an increase in negative supercoiling by the equivalent of two turns of the helix. In other words, a positive supercoil can be changed into a negative one.

The gene for DNA gyrase is essential, and mutations of both the quick-stop and slow-stop varieties have been found (for DNA mutant classification see below). These indicate a role for the gyrase in initiation of replication and in elongation, stages that are affected differently depending on the severity of the gyrase defect. In initiation the gyrase maintains a negatively supercoiled template for assembly of the initiation complex and initial helix unwinding. As DNA is unwound during the elongation phase, a compensatory positive supercoiling builds up ahead of the replication fork. Gyrase acts as ‘swivel’ to relieve the resulting topological strain. Without this action, an intolerable accumulation of positive supercoiling would rapidly bring the replication machinery to a halt.

The supercoiling reaction is driven by the binding of ATP, and ATP hydrolysis serves to reset the enzyme for a second round of catalysis. In the absence of

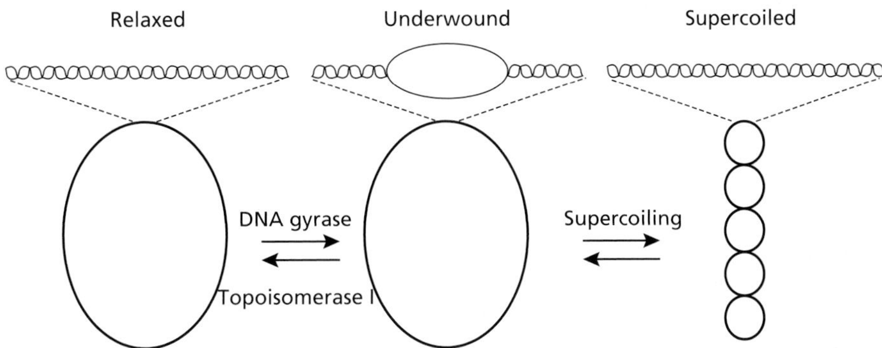

Fig. 1 Schematic illustration of supercoiling. Thick lines represent closed circular dsDNA, which can be unwound by the action of DNA gyrase resulting either in a molecule with a single-stranded region or in the adoption of an alternative, so-called supercoiled conformation with the dsDNA twisted around itself.

ATP gyrase removes negative supercoils from DNA, making the intracellular ATP/ADP concentrations, which are sensitive to the extracellular environment, a determining factor in the level of chromosomal supercoiling.

The Origin of Replication

In all bacteria replication is initiated at a defined site on the chromosome, known as the origin of replication. Although bacteria vary in their lifestyles, and so also in their growth and survival strategies, major features of the origin are remarkably conserved. This implies a common mechanism of initiation that has been preserved for over a billion years. The best-studied example of an origin of replication is that of *E. coli*, and its elements will be discussed and used to highlight similarities and differences with its counterparts in other organisms.

Features of the Origin in *Escherichia coli*

The minimal element in *E. coli* required to initiate bidirectional replication is a 258-bp sequence named *oriC* (Fig. 2a). This sequence is sufficient to promote replication of extrachromosomal DNA, so-called minichromosomes or *oriC* plasmids, and it has therefore been termed an autonomously replicating sequence (ARS). Comparison of *oriC* with ARS elements from other Enterobacteriaceae (*Salmonella typhimurium*, *Klebsiella pneumoniae*, *Enterobacter aerogenes* and *Vibrio harveyi*) has revealed a number of highly conserved 9-bp sequences separated by sequences of variable composition but identical length. These sequences form the binding site for the initiator

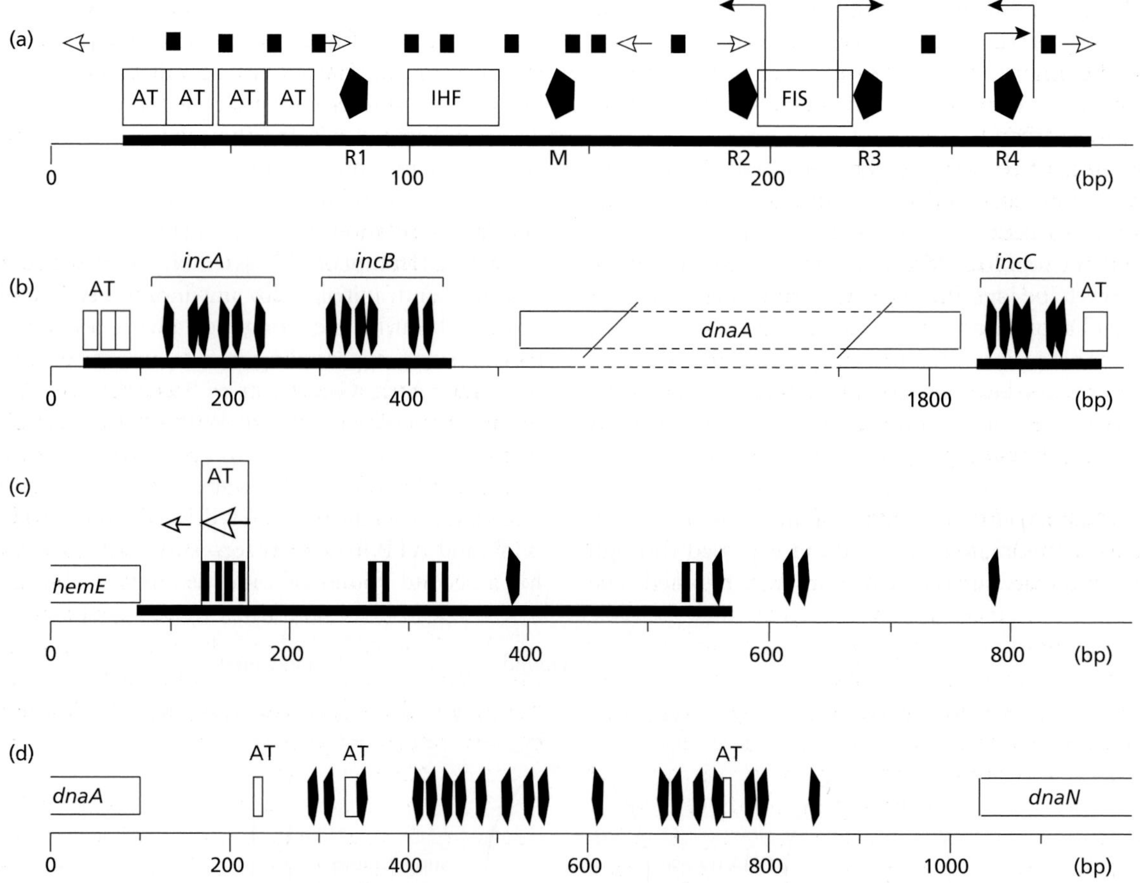

Fig. 2 Origins of replication of various organisms: (a) *E. coli*. Black squares indicate target sites (GATC) for methylation. Binding sites for Fis and IHF are also shown. Open arrowheads represent promoters; black arrows show the positions and direction of the start of DNA synthesis. (b) *B. subtilis*. *incA*, *B* and *C* are clusters of DnaA-boxes, which when present on plasmids, are responsible for incompatibility with the chromosome. (c) *C. crescentus*. Connected black rectangles represent the bipartite recognition sequences for the CtrA protein. (d) *Streptomyces lividans*. Black pentagons denote DnaA-boxes; AT denotes AT-rich regions; thick horizontal lines show the minimal sequences required for autonomous replication. Arrowheads represent promoters, the larger one representing the stronger promoter. Note that different scales are used. See text for details.

protein, DnaA, and are commonly referred to as DnaA-boxes. A second conserved feature of the enterobacterial origins are four AT-rich sequences to the left of the origin which play a role in duplex unwinding. Thirdly, each of the enterobacterial origins contains 9–13 GATC repeats, 8 of which are conserved in position in all of these origins. GATC is the recognition site for DNA adenine methyltransferase (Dam), which methylates the adenine residue within this sequence and plays an important role in control of initiation of replication in *E. coli*.

Other features of importance in the *E. coli* origin are binding sites for two proteins, integration host factor (IHF) and a factor for inversion stimulation (Fis), that introduce bends into DNA. Their ability to do this is thought to assist DNA duplex unwinding during initiation.

Several promoters have been identified within the *E. coli* minimal origin, and these promote transcription of RNAs that terminate after about 150 bp. Additional transcripts originate at the promoters of genes that flank the origin, and they both traverse and terminate within *oriC*. Although the role of these transcripts in initiation of DNA replication is uncertain, a stimulatory effect of transcription away from *oriC* has been shown, especially under sub-optimal conditions for DNA unwinding.

Finally, *oriC* has affinity for certain cell membrane fractions, and specific sites within and around the minimal origin are bound by membrane proteins. Although the exact nature of such membrane complexes has not been determined, compelling evidence suggests that the membrane plays a crucial role in regulating the timing of the initiation event.

Chromosomal Environment of Replication Origins

The molecular events at initiation of replication have mainly been studied in *E. coli*, and to a lesser extent in *Bacillus subtilis*; they remain largely unstudied in other bacteria. Nevertheless, putative origins of replication are being described for an increasing number of organisms. These are identified either by sequence similarity with known origins or by the isolation of ARS elements that are able to promote the replication of genes attached to them. The chromosomal location and DNA sequence of many of these have been determined. A comparison of the genetic organisation of the *oriC* region of various bacteria is shown in **Fig. 3**. Clearly, although the size and sequence of individual origins may vary, the surrounding genes are similar in many cases. Nearly all origins are found in close proximity to the *dnaA* and *dnaN* genes, both of which encode essential replication proteins. A notable exception are the enterobacterial origins, in which these genes are located approximately 42 kb anticlockwise from *oriC* on the genetic map.

In cases where the orientation of transcription of origin proximal genes has been studied, such as in *B. subtilis* and *Pseudomonas putida*, it has been established that genes on either side of *oriC* are transcribed away from the origin so that the direction of transcription is always in concert with that of bidirectional replication, thus avoiding collisions between protein complexes involved in replication and transcription. In *E. coli* the situation is the same, with the exception of a few genes in the immediate proximity of *oriC*. This trend is, however, reversed in the region around the *dnaA* gene. This discrepancy and the separation of *dnaA* and *oriC* in Enterobacteriaceae can be explained by postulating the translocation of a chromosomal fragment during the evolution of this group of bacteria.

Other exceptions to the proximity of the origin and *dnaA* have been found in *Sinorhizobium meliloti*, the cyanobacterium *Synechocystis* sp. PCC6803 and the related *Prochlorococcus marinus*, but as yet little detail is known about the genetic environment of these replication origins.

The DnaA-box

A common feature of all known origins is the presence of sequences identical or similar to the 9-bp DnaA-box first identified in the Enterobacteriaceae. Consensus sequences for the *E. coli* DnaA-box have been suggested and refined on the basis of binding assays. DnaA protein is specifically retained by filter-bound DNA fragments that contain the asymmetric sequence TTAT(C/A)CA(C/A)A, with certain mismatches tolerated at positions 3 and 8. A detailed study of binding constants of the protein to various DnaA-box permutations on short DNA fragments (oligonucleotides) produced the slightly more stringent consensus TT(A/T)TNCACA, where N can be any base. Importantly, both studies showed that the DnaA-box is not the sole determinant for DnaA binding, because flanking sequences significantly influence the protein–DNA interaction.

DnaA-box sequence conservation is remarkable. The distantly related Gram-positive *B. subtilis* has a DnaA-box identical to that of *E. coli*. *Mycobacterium* spp. have been assigned the consensus (T/C)T(G/A)TCC(A/C)CA and *Streptomyces spp.* (T/C)(T/C)(G/A/C)TCCACA. The differences between these and that of *E. coli* may simply reflect adaptation to the higher genomic G + C content of these species.

It is apparent from **Fig. 2** that the arrangement of DnaA-boxes at the origins of different species varies

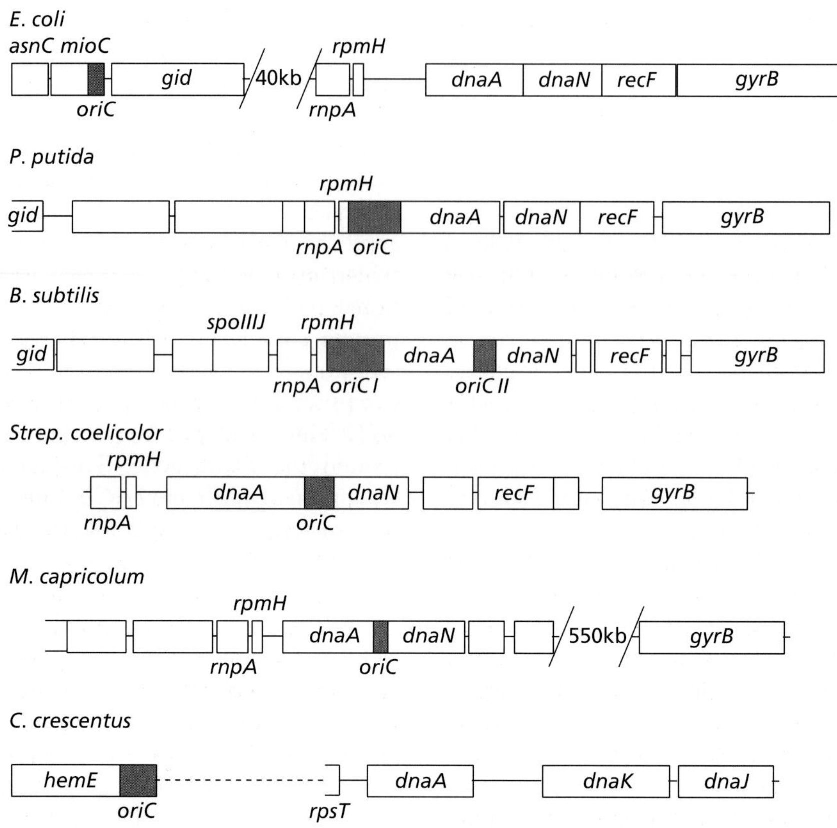

Fig. 3 A comparison of chromosomal regions that harbour the origins of replication in various bacteria. Origins are shaded and conserved genes are indicated. *E. coli* and *P. putida* are Gram-negative bacteria; *B. subtilis* and *S. coelicolor* are Gram-positive. *M. capricolum* is thought to be derived from the Gram-positive group of bacteria. The dashed line in the *C. crescentus* diagram represents non-sequenced DNA.

widely. However, as detailed analysis of replication origins other than that of *E. coli* has not been carried out, little can be said about the significance of these differences. Mutation of individual DnaA-boxes in the origin of *Mycobacterium smegmatis* leads to a drastic reduction in ARS function, indicating that, in contrast to *E. coli*, integrity of all DnaA-boxes is important, probably reflecting differences in assembly of the respective initiation complexes. In most organisms several clusters of DnaA-boxes exist in the origin region and in *B. subtilis* two such clusters, separated by the *dnaA* gene, act together to form the replication origin; neither shows ARS activity when cloned separately.

Proteins Which Participate in Initiation

The genetic analysis of DNA replication in *E. coli* was begun by the decision of several groups of workers to seek temperature-sensitive mutants able to synthesise DNA at low temperature (permissive temperature, 30°C) but unable to do so at high temperature (restrictive temperature, usually 42°C). Such mutants, which have proved extremely useful in the study of essential genes, stem from mutations that render particular proteins thermally unstable. Characterisation of temperature-sensitive mutants resulted in their division into two groups according to their behaviour on a shift to the restrictive temperature. In the so-called quick-stop mutants, DNA replication ceases immediately on raising the temperature, but the slow-stop mutants show some residual DNA synthesis at the restrictive temperature. Mutants belonging to the former group were shown to be deficient in the elongation process of chromosomal replication, whereas the slow-stop mutants were unable to initiate DNA synthesis at the restrictive temperature but could complete rounds of replication already in progress. The isolation of many initiation mutants has been reported, most of which reside in three genes designated *dnaA, dnaB* and *dnaC*. This screening procedure will, of course, identify as initiation genes only genes the products of which are needed only during initiation and not during subsequent elongation.

Proteins needed during initiation and elongation will by this test be grouped with those needed for elongation only. Thus, certain of the initiation products to be described below were not identified as such in the original screens.

The DnaA Protein

Since its discovery in the mid 1960s DnaA has been extensively studied, because it is the protein which both acts first to initiate replication and because it has subsequently proved to have an important role in determining when initiation will occur. Its role is therefore considered in some detail here.

Sequence Conservation

In almost all eubacteria replication requires the ubiquitous DnaA protein. The *dnaA* genes of *E. coli,* and of many other bacteria, have been sequenced. *E. coli dnaA* encodes a basic protein of 467 amino acids. There is no apparent sequence similarity between DnaA and other known DNA-binding proteins active in initiation of DNA replication, such as λ phage O protein, plasmid-encoded Rep proteins and archaebacterial or eukaryotic ORC complex subunits. The homology amongst DnaA proteins from various bacteria is, however, striking. A comparison of the amino-acid sequences of the DnaA proteins of 30 bacteria, some of which diverged over a billion years ago, clearly shows that these proteins are homologous, and vary from 30% to 87% in identity with *E. coli* DnaA. It is also apparent that certain regions of the protein are more highly conserved than others, and on this basis the protein was divided into four domains that differ in degree of similarity between species. Various mutational and biochemical studies with the *E. coli* protein have since assigned distinct functional roles to these domains (Fig. 4).

A short N-terminal sequence (domain I) is conserved among most DnaA proteins; mutations within it render DnaA inactive in replication. This domain mediates interactions between DnaA protein molecules, which are necessary for initiation (see below).

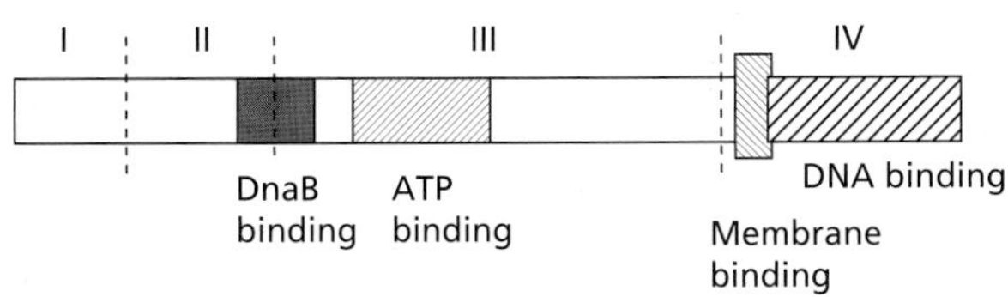

Fig. 4 Domain structure of the DnaA protein. Domains I–IV, determined on the basis of sequence conservation, are indicated. Regions involved in interaction with DnaB, ATP, the membrane and DNA are shown by shading.

Domain II shows little similarity in sequence amongst the various organisms, and has not been functionally characterised, but a potential site of interaction with DnaB helicase has been identified in this part of the *E. coli* protein. Differences in length amongst DnaA homologues are due mainly to variations in the length of domain II. This is especially clear with the *Streptomyces* spp. in which this domain is extended by 120 amino acids, as compared with *E. coli.* Also in domain II the sequence divergence between organisms with genomes of high or low G+C content is most apparent. For instance, *Streptomyces* spp. and *Micrococcus luteus* have acquired a large number of proline and alanine residues which contain G and C in the first two positions of the corresponding codons.

The highest degree of conservation is found in domain III, which contains an amino acid sequence typical of nucleotide-binding proteins. DnaA contains a good match to the consensus of this sequence (GXXGXGKT where X is any amino acid), which is thought to interact with the phosphate moieties of ATP; mutations within this sequence drastically reduce the ATP-binding affinity of the protein.

Domain IV, which is close to the C-terminal part of the protein, is involved in the binding of DNA and is well conserved. The 94 C-terminal amino acids of *E. coli* DnaA, included in domain IV, and the corresponding region of the *B. subtilis* protein, are sufficient for specific binding to DNA. On the basis of secondary structure predictions, a DNA binding motif has been proposed that is unlike any known from other DNA binding proteins. Mutations that map within this putative structure, which it is predicted is present in most known DnaA proteins, abolish DNA binding of the *E. coli* and the *Streptomyces lividans* homologues. A region has also been identified within Domain IV that is involved in interactions between DnaA and the cell membrane.

Binding of DNA

DnaA protein binds dsDNA that carries *oriC,* and binding is enhanced by supercoiling of the DNA. The results of numerous electron-microscopy studies and DNA-binding assays have shown that 20–40 monomers of DnaA, in conjunction with other proteins, form a complex with *oriC.* The kinetics of binding of the DnaA protein to *oriC* suggest that binding is a co-operative process, and that monomers bound to the DNA facilitate the addition of further monomers to the complex through protein interactions. Binding assays with various DNA fragments have shown that interaction of DnaA is not limited to *oriC* but can occur at DnaA-boxes elsewhere on the chromosome, but it is often not clear under what conditions such complexes are formed *in vivo* and what functional significance they may have.

Binding of Nucleotides and Interaction with Phospholipids

DnaA has a high and almost equal affinity for ATP and ADP. Complexes with either nucleotide are present in solution in monomeric or dimeric form, whereas the nucleotide-free protein aggregates. DnaA · ATP and DnaA · ADP both bind oriC DNA with similar kinetics and form complexes that are indistinguishable, though they are clearly distinct from that formed with nucleotide-free DnaA. A clear difference between the two nucleotide-bound forms of DnaA concerns their ability to promote replication at oriC; DnaA · ATP is required for an active initiation complex and DnaA · ADP is inactive. In addition to its capacity to bind nucleotides, DnaA possesses a weak DNA-dependent ATPase activity, resulting in slow hydrolysis of bound ATP to ADP within the initiation complex. ADP remains bound to the protein and renders it inactive for replication.

During purification of DnaA from *E. coli* cells, approximately 20% of the protein is recovered in membrane fractions, and forms a complex with acidic phospholipids, which are components of the membrane. An important property of these phospholipids is their ability to effect a rapid release of bound nucleotides from DnaA. In the presence of ATP acidic phospholipids can in this way catalyse the rejuvenation of an inactive initiation complex by providing a means of regenerating DnaA · ATP from inert DnaA · ADP. The underlying mechanism for this reactivation, which occurs only if the protein is complexed with *oriC* DNA, is unknown. Although first demonstrated with biochemical assays *in vitro*, strong genetic evidence points to an essential role for acidic phospholipids in initiation of replication in the cell.

Nothing is known about the requirement for nucleotides in DnaA-mediated initiation of replication in other species, but the *B. subtilis* DnaA homologue binds both ATP and ADP with high affinity and DNA-dependent ATP hydrolysis occurs.

DnaB Helicase

Progression and initiation of the replication complex requires that the DNA duplex is progressively unwound to allow the access of polymerase and primase to the template. In *E. coli* unwinding is accomplished by the DnaB protein, a helicase that moves in a 5′→3′ direction along the DNA strand to which it is bound. The active form of the protein is a hexamer and shows DNA-dependent nucleoside triphosphatase (NTPase) activity, which fuels the translocation process. DnaB, which is not bound to DNA, is in a stoichiomeric complex with DnaC hexamers, a protein required to load the helicase on to the template. DnaC mutations were the second most common type of temperature-sensitive initiation mutants to be found. Since DnaB is needed throughout replication, it is surprising to find mutant alleles that appear to be defective only in initiation. Presumably these mutant proteins fail to load normally on to the DNA, but once loaded they retain helicase activity.

Duplex unwinding by DnaB is processive, meaning that it remains on the template once bound and is independent of other replication proteins, although it is stimulated by single-strand-binding proteins (SSB; see below), primase and DNA polymerase III holoenzyme. In addition to DnaB, which appears to be the major replicative helicase in *E. coli*, two other helicases, PriA and Rep, have also been implicated in replication but their exact roles are uncertain.

DnaG Primase

DnaG is the *E. coli* primase that acts during initiation of replication to prime leading strand synthesis and, throughout the elongation phase of replication, it primes Okazaki fragment synthesis on the lagging strand. It acts as a monomer but requires the presence of DnaB on the template for activity; it is also through interaction with DnaB that the primase is attracted to the replication fork. It is thought that, once recruited, primase scans the DNA for a 3′ GTC 5′ sequence, which is the signal for formation of a primer 10–12 bases long starting opposite the T. The primer may be composed partly of deoxyribonucleotides, because DnaG can incorporate these as well as ribonucleotides at all but the first two positions. Since primase is required throughout replication and is not processive, all *dnaG* mutants behave as elongation mutants.

Single-strand Binding Protein

SSB stabilises unwound DNA and plays an essential role in replication as well as in other important processes such as recombination and DNA repair. The *E. coli* SSB is a homotetramer of 18.8-kDa subunits that binds co-operatively to single-stranded (ss)DNA in a sequence-non-specific fashion. Its roles in replication are to contribute to the opening and unwinding of the DNA duplex and probably to protect single-stranded regions from nucleolytic attack. SSB stimulates the rate and processivity of DNA polymerase III holoenzyme, most probably by melting regions of secondary structure obstructive to passage of the replication complex, but possibly also by direct interactions with other replication proteins.

Proteins That Bend DNA

Protein HU is a non-specific DNA-binding protein found in association with DNA. It is a heat-stable, basic heterodimer that is very well conserved between species and is capable of binding both ssDNA and dsDNA. Binding of HU protein to dsDNA causes bending and at sufficiently high concentration can condense the DNA. Because of these properties, HU is referred to as a 'histone-like' protein, but its role is certainly not identical to that of the histones that compact chromosomes in the nuclei of eukaryotic cells.

Initiation at *oriC in vitro* is stimulated by low levels of HU protein and *in vivo* mutations in the genes that encode the HU subunits show defects in maintaining mini-chromosomes. The exact role of this protein in initiation is unknown, but it probably assists in duplex unwinding and it also appears to stimulate binding of IHF to its target site in *oriC*.

IHF was discovered through its role in site-specific recombination. It is a dimer with 35% sequence identity to HU, but it binds DNA with sequence specificity. It induces a bend in the DNA at its 13-bp recognition site, of which there is one in *oriC*; mutation of this site inactivates the origin in mini-chromosomes. Although IHF is not essential for cell viability, mutants show asynchrony of initiation, meaning that in contrast to normal cells, not all copies of *oriC* in the cell initiate replication simultaneously.

The homodimeric DNA-binding protein Fis was discovered as a participant in site-specific recombination, but it is structurally unrelated to IHF. Nevertheless, mutations within it show the same asynchronous initiation phenotype as IHF mutants and, furthermore, Fis mutants, in common with HU mutants, do not stably maintain mini-chromosomes. Fis binds the origin of replication at a specific site that is necessary for initiation in mini-chromosomes.

The Initiation Process

The Initial Complex

DnaA protein binds *in vitro* to five DnaA-boxes in *oriC*, R1–R4 and M, but the interaction with R3 and M is significantly weaker than that with the other three DnaA-boxes. The complex can be seen by electron-microscopy and its size depends on the amount of DnaA added, but a particular structure that contains about 20 DnaA monomers per *oriC* template has been shown to be capable of promoting replication. Although complexes at *oriC* can be formed on relaxed as well as on supercoiled DNA, DnaA·ATP and negative superhelicity of the DNA are essential for the complex to advance through its subsequent stages. Studies of growing cell cultures *in vivo* showed that the DnaA-boxes R1, R2 and R4 are protected throughout the cell cycle, whereas R3 is bound only during the initiation phase of replication.

Mutational analysis revealed that the DnaA-boxes are essential for the formation of the initial complex; a single DnaA-box is insufficient. Because of the difficulty of manipulating a structure the size of the *E. coli* chromosome, most studies of origin function have been carried out with mini-chromosomes, plasmids that use *oriC* as replication origin. In contrast to many other plasmids, mini-chromosomes show cell cycle-specific regulation of initiation similar to that of the chromosome, making them attractiv model systems for studies of replication. Extensive mutational analysis of *oriC* on *E. coli* mini-chromosomes revealed that various mutations in individual DnaA-boxes within *oriC* have little or no effect, but that particular combinations of such mutations can drastically reduce origin function; mutation of all four boxes abolishes it. Alteration in the spacing between pairs of DnaA-boxes has a drastic effect on origin function. Most changes in the distances between DnaA-boxes inactivate *oriC* except in the right half of the origin, where alterations by 10–12 bp between R2, R3 and R4 are tolerated. Such a change is equivalent to the deletion or insertion of one turn of the helix, indicating the need for a particular orientation of these DnaA-boxes relative to the helical axis.

It seems likely that proteins other than DnaA, such as the histone-like protein HU and the DNA bending proteins IHF and Fis, form part of the initial complex. A stimulatory effect of the histone-like protein HU has been found and may be involved in the formation or maintenance of the structure of the complex. IHF may be involved in creating a bend in the DNA in such a way as to facilitate an interaction between DnaA protein and the AT-rich region in the formation of an open complex at *oriC*.

The Open Complex

A key step in initiation is strand separation at the origin to allow assembly of the components required to prime and continue DNA synthesis. A common feature of many origins, including those of some plasmids, prokaryotic and eukaryotic viruses as well as the eukaryotic budding yeast *Saccharomyces cerevisiae*, is the presence of one or more AT-rich tracts. These provide a region of helical instability necessary for initial duplex unwinding during initiation of replication, and they are therefore also known as DNA unwinding elements (DUE).

In *E. coli* this function is provided by four AT-rich sequences, three 13-mers and a 12-mer, at the right of the minimal replication origin. The location of initial duplex opening coincides with that of the AT-rich sequences and deletion of either the left or middle 13-mers results in a non-functional origin. Duplex opening still occurs to some extent in such mutants but is abolished if the right 13-mer is deleted. This, and consideration of the energetics involved in the strand separation reaction, led to a model where the 13-mers are opened sequentially starting from the right.

Inherent instability of the AT-rich region is demonstrated by its ability to unwind to some extent *in vitro* without the action of the replication proteins if the DNA is sufficiently supercoiled and in the absence of magnesium ions that stabilise dsDNA. In their presence melting depends on the addition of DnaA protein. Several studies suggest an active role for DnaA in formation of the open complex, which is thought to arise by a direct interaction with one strand of the right and middle 13-mers. This means that the DnaA protein has a dual sequence specificity in binding, first for the DnaA-boxes and second for the 13-mer sequence. There is strong evidence for this, because DnaA interacts specifically with the 13-mers in binding assays. In addition, the integrity of the sequence of the right and middle 13-mers is important, as is their spacing with regard to the DnaA-boxes.

Open complex formation *in vitro* requires a relatively high concentration of hydrolysable ATP, presumably to produce energy for strand opening or subsequent helicase entry. However, only a low level of hydrolysis is observed. It may be that the DnaA monomers in the initial complex are assigned different roles that depend on their exact locations; some will interact only with a DnaA-box and other monomers, whereas others will also interact with the AT-rich 13-mers and yet others may interact with the DnaB · DnaC complex at a later stage. Possibly ATP hydrolysis occurs only in a subset of molecules.

DNA Supercoiling and Transcriptional Activation

The presence of active promoters adjacent to *oriC* increases the efficiency of replication of mini-chromosomes. Because of this, a role for RNA polymerase in stimulating open complex formation, 'transcriptional activation', has been proposed. RNA polymerase unwinds the DNA as it transcribes and thereby generates positive supercoils ahead of it and negative supercoils behind it. For this reason, transcription away from the AT-rich region could stimulate open complex formation by increasing local negative supercoiling. In keeping with such a model a number of viral origins shows a requirement for promoters adjacent to their replication origins.

Although the presence of promoters adjacent to the origin in mini-chromosomes stimulates replication, similar studies with the *E. coli* chromosome do not show such an effect. It appears that mini-chromosomes, although used as models for chromosomal replication, are inherently less efficiently replicated though they carry the same origin sequence as the chromosome. This idea is supported by the fact that the chromosomal origin can tolerate a DnaA-box mutation that inactivates mini-chromosomal replication, and that genetic studies have shown that chromosome replication is less dependent on auxiliary factors such as HU, Fis and IHF.

It is likely that the difference in efficiency between the two systems stems from a difference in supercoiling. Because of its size, the chromosome may be more able to maintain the optimal superhelicity at the origin by dissipating superhelical stress over greater stretches of DNA. Under conditions where superhelicity is drastically altered, which can be brought about by changes in pH, temperature or aeration, it may be that chromosomal replication also becomes dependent on auxiliary factors. Although this has not yet been demonstrated, it has been shown that if the chromosomal origin is rendered inefficient by mutation, it becomes dependent on transcription away from the AT-rich region.

The Pre-priming and Priming Complexes

Once initial strand separation at the origin has been achieved, the DnaB protein is recruited further to unwind the DNA. DnaB is a helicase with affinity for single-stranded DNA, and binding is improved through an allosteric effect of ATP complexed with the protein. DnaB proceeds in the $5' \rightarrow 3'$ direction along the strand of DNA and its action requires a hydrolysable ribonucleoside triphosphate. DnaB is required *in vitro* to direct the replication machinery to *oriC*, because without its presence DNA polymerase III and primase do not show a preference for establishing themselves at that particular site. A complex that contains only DnaA, DnaB and HU can be isolated from an *in vitro* replication reaction and is known as the pre-priming complex. Electron microscopy reveals that it is larger than the open complex, since an additional 50 bp from the left of *oriC* are involved in the structure.

The functioning of DnaB depends entirely on the presence of another protein, DnaC, which appears to have no function other than to deliver DnaB to the initiation complex. DnaB and DnaC form a tight complex in the presence of ATP. A DnaB hexamer binds six DnaC monomers, which in turn each bind one molecule of ATP; nucleotide binding stabilises the complex but ATP hydrolysis is required for the release of DnaC, which is necessary for subsequent events.

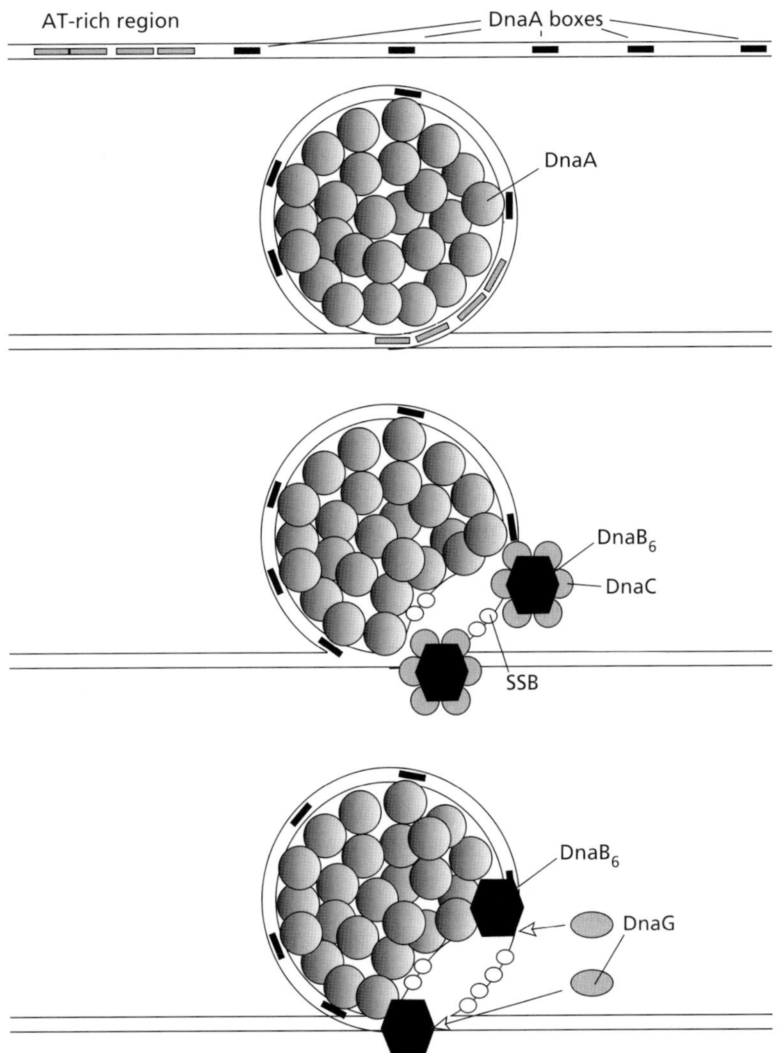

Fig. 5 Simplified scheme of the sequence of events (from top to bottom) during initiation of replication at *oriC* in *E. coli*. Features of *oriC* are shown. The DNA becomes wrapped around a scaffold that consists mainly of DnaA protein, the DNA becomes partially unwound at the AT-rich regions and single-stranded stretches are stabilised by SSB. DnaB helicase is loaded from a complex with DnaC through an interaction with DnaA protein. The DNA becomes further unwound and DnaG primase is recruited. Finally, DNA polymerase III (not shown) is brought to the two divergent replication forks. Proteins are not drawn to scale.

DnaA protein is probably involved in the formation of the pre-priming complex, since an interaction has been demonstrated between DnaA and DnaB, either alone or in complex with DnaC. This interaction may help to relocate DnaB from its presumed site of entry at the 13-mers in the left of the origin, to where the start sites for DNA synthesis have been found at the right of *oriC*. Then unwinding of the template occurs through strand separation by the helicase and the swivelling action of gyrase to remove positive supercoils that would otherwise accumulate. The next step appears to involve a direct interaction of DnaB and DnaG primase to form the priming complex. The events that follow formation of a pre-priming complex no longer depend on the action of DnaA, and *in vitro* the protein can at this stage be recycled to promote initiation on a fresh template. The sequence of events at *oriC*, from open complex formation to priming, is illustrated in Fig. 5.

Timing and Control of the Initiation of Chromosomal Replication

The *E. coli* Cell Cycle and its Parameters

During a single cycle of growth and division, the genetic material present in the cell must be duplicated once and once only. The frequency of chromosome replication is determined at the level of initiation of DNA synthesis and the understanding of the mechanism involved in this has been the aim of much research into the *E. coli* cell cycle.

Observations in the 1950s and 1960s laid the foundation on which the present models of cell cycle control are based. Importantly, this work showed that initiation of replication is a very regular process as each part of the chromosome is replicated at intervals of exactly one generation time. Furthermore, it was revealed that initiation of replication occurs in the cell cycle at a particular time, which changes with growth rate. To describe the timing of key events in the bacterial cell cycle, certain parameters were defined. The most important of these are the time required to replicate the chromosome (C) and the time interval between termination of replication and cell division (D). Both of which are virtually constant at 37°C and generation times of less than 60 minutes; C is approximately 40 minutes and D 20 minutes. At slower growth rates C and D gradually increase, but under usual laboratory experimental conditions, C and D are essentially constant and total roughly 60 minutes. However, *E. coli* can grow with a generation time as short as 20 minutes. To achieve this the organism has developed the capacity to initiate a second round of replication before the previous round has terminated. As many as six replication forks may, therefore, be active on a single chromosome, a process termed dichotomous replication. An understanding of replication control must also take into account the more recent observation that all origins present in a normal rapidly growing cell initiate replication simultaneously.

Over a wide range of growth rates, initiation of replication occurs when the ratio of overall cell mass to the number of origins present in the cell reaches a defined value. In other words, the amount of cellular protein per origin at the time of initiation (initiation mass), is independent of growth rate. The constancy of this parameter led to two basic concepts of the control of the timing of initiation: the 'inhibitor dilution model' and the 'initiator accumulation model'. The former requires synthesis of a stable initiation repressor in a burst immediately after initiation of replication. Cell growth would then dilute the repressor until its concentration drops below a threshold at which initiation is no longer repressed, one generation time later. The initiator accumulation model proposes that a positive factor increases to a critical level in order to trigger replication, and that this factor is then inactivated.

Experimental evidence now suggests that timing of initiation is, indeed at least in part, governed by the rate of synthesis of a positively acting factor, the DnaA protein, which is thought to trigger initiation at all available origins once it reaches a threshold concentration. Further initiations within the same division cycle are then prevented by a number of negatively acting factors.

Positive Control by DnaA

By using a strain in which the *dnaA* gene is expressed from an inducible promoter to allow the manipulation of cellular DnaA content, DnaA was shown to have a positive role in controlling initiation. This strain showed that the initiation mass is increased when DnaA levels are lowered, but reduced at increased levels of *dnaA* gene expression, in other words increasing the intracellular amount of DnaA causes initiation to occur earlier in the cell cycle. Reduction in initiation mass of up to 50% can be achieved before other factors become limiting. At *dnaA* expression levels close to normal the initiation mass is very sensitive to small changes in the rate of DnaA protein synthesis, showing that DnaA, or more accurately DnaA · ATP, normally acts as a rate-limiting substance.

Inactivation of DnaA

Initiation at each origin in the cell is triggered simultaneously, exactly once in each cell cycle, which implies that there must be a mechanism to prevent re-initiation once replication from a particular origin is under way. This could be accomplished either by inactivating the origin itself or by inactivating an essential initiation factor, such as DnaA. In fact, both forms of negative regulation play a role.

When DnaA is exposed to extracts from normal cells it is rapidly inactivated for initiation in a replication system *in vitro*. Inactivation occurs by accelerated hydrolysis of ATP bound to DnaA, which results in an accumulation of the inactive DnaA · ADP complex. Recently it was demonstrated that the main effector of this process is the β subunit of DNA polymerase III, the major replicative polymerase in *E. coli*. As will be described below, the β subunit forms a sliding clamp around the DNA, which tethers the polymerase to its substrate. It is interesting that inactivation of DnaA is efficient only if this subunit is present as part of an actively replicating polymerase. The mechanism of inactivation is still obscure and does not necessarily involve a direct interaction between DnaA and the β subunit, since other factors are known to be required. Nevertheless, the process can be seen as a communication pathway between initiator and the machinery for the elongation stage of replication, and could be a means of preventing re-initiation at a recently triggered origin.

DNA Adenine Methyltransferase

Although not directly involved in DNA replication, the DNA adenine methyltransferase protein (Dam) plays an important part in regulation of the initiation event at *oriC* in *E. coli*, and probably other enteric bacteria, and is also involved in the process of post-replicative mismatch repair. Dam, which is the product of the *dam* gene, catalyses the transfer of a methyl group from *S*-adenosylmethionine to the N^6 position of adenine residues in the sequence GATC. The methylation process allows a temporary distinction to be made between the methylated template strand and the newly formed unmethylated daughter strand, before this too becomes methylated. This distinction directs the process of mismatch repair to the daughter DNA strand. In addition, the hemimethylated state of newly replicated *oriC* DNA makes it a target for a membrane-associated replication inhibition activity.

Negative Control by Membrane Attachment and Methylation

The clearest example of negative control comes from study of the effects of methylation on initiation. By manipulation of *dnaA* mutants, multiple initiations can be made to occur during a single division cycle. Initiations under these circumstances are triggered in waves, where the first and subsequent rounds of initiation occur at intervals. This suggests that it takes some time for the cell to prepare itself for a new burst of initiation, even if all the necessary proteins are present. The spacing between bursts of initiation is reduced on overproduction of Dam. Other evidence for a positive role of the Dam protein in replication came from the observation that mini-chromosomes cannot be introduced into strains deficient in Dam, because they fail to be replicated. It was subsequently discovered, however, that this is true only of hemimethylated or fully methylated mini-chromosomes, and that unmethylated DNA poses no problem for replication. The factor responsible for the inhibition of replication of hemimethylated DNA is associated with the cell membrane. Clues as to the nature of the inhibitor came from the observation that it is possible to isolate complexes of *oriC* and outer membrane fractions. These membrane fractions specifically bind hemimethylated DNA and they can inhibit DNA replication *in vitro* on hemimethylated *oriC* templates.

All DNA is transiently in a state of hemimethylation immediately after passage of the replication fork, before Dam has had the opportunity to methylate its target sites on the newly synthesised strand. A key discovery was that this state of hemimethylation persists for a period longer than normal at two specific chromosomal locations. Methylation at most sites is complete within 5 minutes after replication but *oriC* and the promoter of the *dnaA* gene, but not sites between them, fail to be remethylated for up to 30% of the cell cycle; this period can be reduced by overproduction of Dam protein. This leads to a model in which *oriC*, immediately on being replicated, is rendered refractory to further initiations as a result of sequestration by the membrane, and it can serve as a template for initiation only once released, presumably through the action of Dam. Thus levels of initiation proteins such as DnaA need not be exactly titrated, but can rise beyond the threshold that triggers replication without giving rise to immediate re-initiation. Sequestration of the *dnaA* promoter would ensure that no further DnaA protein is produced until replication-competent levels present in the cytoplasm have been sufficiently reduced, by dilution or inactivation.

The SeqA Protein: a Negative Factor in Initiation

Studies to identify factors involved in the sequestration process revealed a 24-kDa protein, SeqA. Mutants that lack SeqA are viable but initiate replication asynchronously at multiple origins and sometimes use origins twice in a single cell cycle. In a manner consistent with a role in sequestration, *seqA* mutants show a reduced lag in methylating *oriC* and the *dnaA* promoter, and can initiate replication at hemimethylated origins. The SeqA protein has been purified and, although it is absolutely required for binding of the membrane to hemimethylated *oriC* DNA, it does not have the characteristics of a membrane protein nor does it exclusively bind hemimethylated origin DNA. This led to the idea that additional factors are involved and, accordingly an as yet uncharacterised protein, SeqB, which is also necessary for the sequestration process, has been identified.

In addition to its role in preventing premature re-initiation, SeqA must also have an inhibitory effect on initiation at fully methylated origins, since SeqA-deficient cells initiate replication at a lower initiation mass. It is not clear at what stage this negative effect is exerted, but genetic experiments suggest that SeqA is an antagonist of DnaA, because cells that lack SeqA become less sensitive to DnaA mutations, as if they now required less DnaA. It therefore appears likely that SeqA acts at or before formation of the pre-priming complex.

Initiation of Replication in Other Bacteria

Bacillus subtilis

The origin of *B. subtilis* consists of three clusters of DnaA-boxes that flank the *dnaA* gene, *incA, incB* and *incC* (Fig. 2b). All these clusters bind DnaA protein and none of them alone is sufficient for initiation of replication. The nature of the initiation complex *in vitro* suggests that DnaA mediates loop formation between the various binding sites. Open complex formation occurs by DnaA-mediated unwinding at an AT-rich 27-mer downstream of the *dnaA* gene, to the right of *incC*, and not by unwinding of three essential AT-rich 16-mers at the extreme left of the origin, as might be expected by analogy to *E. coli*.

Because there is no *in vitro* replication system, little is known about the precise molecular events that lead up to the establishment of the replication forks, nor have all the participating enzymes been identified. It has, however, long been established that an association of the origin with the cell membrane is essential for initiation to occur. This interaction is mediated by a protein named DnaB for which there is no homologue in *E. coli*, and which should not be confused with the *E. coli* helicase.

Regulation of initiation is mediated by DnaA which, as in *E. coli*, is a limiting positively acting factor. DnaA protein appears to be more precisely titrated in *B. subtilis* than in *E. coli*, because in *B. subtilis* it is impossible to maintain mini-chromosomes, or even plasmids that bear the *incA/B/C* DnaA-box clusters, without inhibiting chromosomal replication. This phenomenon, known as 'origin incompatibility', has been attributed to competition between the various origins present for a limiting amount of DnaA. In contrast, *E. coli* tolerates the presence in the cell of extra copies of the origin. It may be that in *B. subtilis* the overall level of DnaA is more tightly controlled to compensate for its lack of a methylation system and an associated sequestration mechanism to prevent re-replication.

Caulobacter crescentus

C. crescentus is a model organism that has proved very useful in the study of cell cycle and differentiation, because it undergoes asymmetric cell division. A so-called motile swarmer cell is unable to initiate the cell cycle until it differentiates into a non-motile stalked cell. The stalked cell can then initiate replication and it eventually divides asymmetrically to produce a replication-competent stalked cell and a swarmer cell. During cell division the cell must distinguish between its two cell poles and regulate the partitioned daughter chromosomes accordingly.

The origin, *Cori*, contains five potential DnaA-boxes (TGATCCACA) one of which is essential, indicating that DnaA plays a crucial role in initiation. Other features are five repeats of a novel 9-mer motif, GTTAA-N7-TTAA, of which at least three are essential, and a strong promoter (of the *hemE* gene), transcription from which is also necessary for initiation (Fig. 2c). In the swarmer portion of the pre-divisional cell,

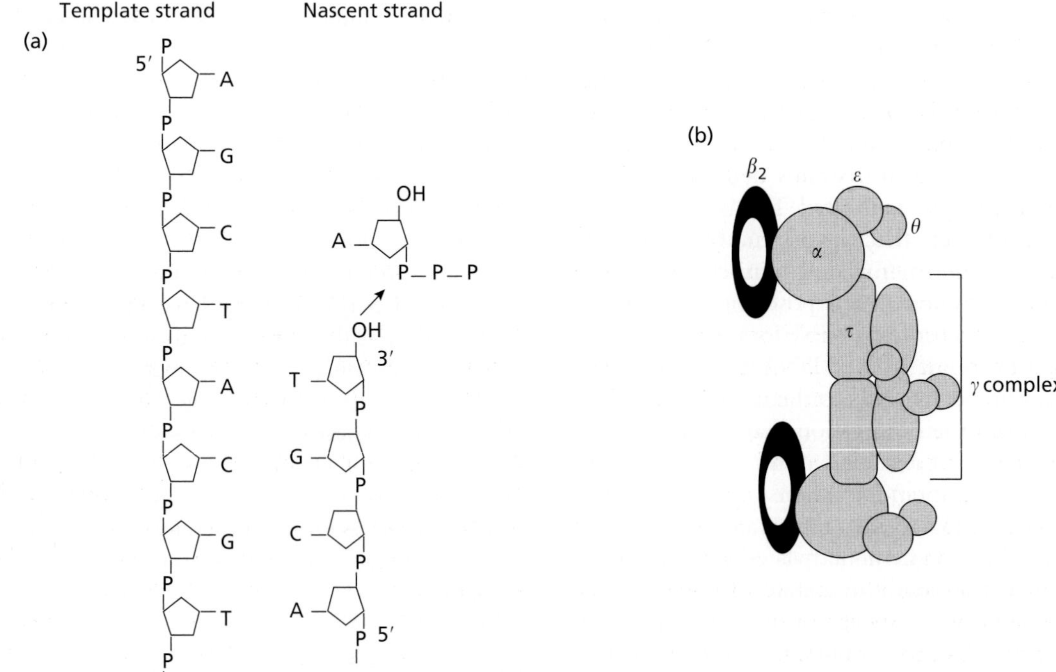

Fig. 6 (a) All DNA polymerases catalyse the reaction shown, between the 3′-hydroxy of the growing chain and the α-phosphate of the incoming nucleoside triphosphate. Thus, chain growth is exclusively in the 5′→3′ direction. (b) Structure of the DNA polymerase III heterodimer. The figure indicates interactions known to occur between the various subunits.

transcription from the *hemE* promoter is selectively repressed by a protein factor, CtrA, by binding to the 9-mer motifs. On differentiation to a stalked cell, CtrA is degraded and replication ensues.

For differentiation and cell cycle progression to occur normally, CtrA activity must be controlled both temporally and spatially. It is not known how this occurs but some features of regulation are coming to light. To be active, CtrA must be phosphorylated on an aspartate residue, and is also a target for proteolysis by specific proteases. Co-ordination of these two activities ultimately determines cell cycle progression, but how CtrA activity is restricted to one half of the pre-divisional cell is still unclear. Phosphorylation and proteolysis are central to eukaryotic cell cycle control and although the prokaryotic and eukaryotic regulatory proteins differ, the above example shows that the general concept is common to both classes of organisms.

Streptomyces Species

Streptomycetes are Gram-positive bacteria with a mycelial life cycle and differ from most other bacteria in possessing a large (6–8 Mb) GC-rich linear chromosome. Linear chromosomes occur in some viruses and in almost all eukaryotes, and various strategies exist for their replication. The simplest is end-to-end replication of both DNA strands without the need for lagging strand synthesis, because both strands are primed by several possible methods at the 5′ end and are replicated continuously to the 3′ end. Variations of this method are used by a number of bacteriophages and viruses. Eukaryotes generally replicate their chromosomes bidirectionally from multiple internal origins, which necessitates co-ordinated synthesis of leading and lagging strands.

Members of the *Streptomyces* family employ a method akin to that of the eukaryotes, but they initiate replication at a single origin located centrally on the chromosome between the *dnaA* and *dnaN* genes (**Fig. 2d**). The origin has been isolated as an ARS sequence and supports bidirectional replication. The sequence is characterised by a large number (19) of DnaA-boxes but, since the extensive AT-rich tracts found in most other origins are lacking, the mechanism of strand separation remains unclear. As in *B. subtilis*, the *Streptomyces* origin shows strong incompatibility towards plasmid-borne copies of the chromosomal origin. A special mechanism is required to replicate the chromosomal ends (see below).

Synechocystis sp. PCC6803

Replication in this freshwater cyanobacterium is interesting because it is light-dependent. Its *dnaA* gene lies in an unusual environment and is transcribed in a light-dependent manner. Surprisingly, mutational inactivation of *dnaA* does not render the cells non-viable. A search for possible origins did not uncover DnaA-box clusters on the chromosome, so it seems likely that this organism uses an unusual mechanism to initiate DNA replication.

Elongation of DNA Chains

Synthesis of DNA is a polymerisation reaction in which deoxynucleoside triphosphates (dNTP) are added to a chain of deoxynucleoside monophosphates (**Fig. 6a**). The terminal hydroxy group (3′-OH) of the nascent chain forms a 3′,5′-phosphodiester bond with the α-phosphate of the dNTP, by nucleophilic attack. This results in the release of inorganic pyrophosphate which is then cleaved by a pyrophosphatase to render the polymerisation reaction irreversible. DNA synthesis is catalysed by a group of enzymes known as DNA polymerases, which are necessarily present in all living organisms and many viruses. Although evolutionary divergence is huge, some important generalisations apply to all known DNA polymerases. First, the reaction catalysed by all of these enzymes is that outlined above; no polymerases have been found that synthesise DNA in the 3′→5′ direction. Second, DNA cannot be synthesised *de novo*, meaning that a 3′-OH group must be present to accept the first incoming dNTP, a process known as ‘priming’. The primer is usually a short stretch of RNA, but some virus replication strategies make use of a DNA or even of a protein primer. DNA polymerases differ in processivity, a term that describes the number of nucleotides added by a polymerase before it dissociates from the template DNA. A highly processive enzyme can synthesise of the order of a megabase after a single association, whereas a non-processive or distributive enzyme may add only a single nucleotide. As described below, some polymerases associate with factors that increase processivity.

Replication from all known bacterial (and eukaryotic) chromosomal origins is bidirectional, meaning that a single origin gives rise to two divergent replication forks, at each of which both strands of DNA are being replicated in the same direction, i.e. away from the origin. Owing to the structure of DNA as an antiparallel double helix and given the fact that DNA polymerases exclusively synthesise DNA in the 5′→3′ direction, only one strand, the leading strand, can be made in a continuous fashion from a single primer at *oriC*. The other, the lagging strand, is synthesised discontinuously by a repetitive process of priming, elongation and ligation. Thus, the lagging strand is composed of a series of short stretches of DNA, each of which is synthesised in a direction opposite to that of leading strand growth, but the overall direction of synthesis is the same as that of the leading strand. The lagging strand fragments of DNA formed between successive priming events are known as Okazaki fragments which, in bacteria, are of the order of 1–2 kb in length. Since the rate of polymerisation in *E. coli* is approximately 1000/s, a new Okazaki fragment must be primed every 1–2 seconds. Consequently, lagging strand synthesis requires the constant removal of old primers and the repair of the resulting gaps between consecutive DNA fragments.

DNA Polymerase III

Although the two strands are replicated by different mechanisms, the leading strand continuously and the lagging strand discontinuously, the protein complex involved in both cases is much the same. *E. coli*, the only bacterium in which replication has been studied in detail, again serves to illustrate the process and its enzymology. The replicative DNA polymerase in *E. coli* is the DNA polymerase III holoenzyme, a complex that consists of 10 different proteins with the following stoichiometry: $(\alpha\epsilon\theta)_2\tau_2\gamma_2\delta\delta'\chi\psi(\beta_2)_2$ (**Table 1**, **Fig. 6b**).

The structure of the holoenzyme is thought to be an asymmetric dimer and consists of various subassemblies, most notably the core ($\alpha\epsilon\theta$) and the γ complex ($\gamma_2\delta\delta'\chi\psi$). The τ subunit is responsible for dimerisation of the enzyme complex. The core contains the polymerase subunit α and a $3'\rightarrow5'$ exonuclease activity, ϵ, which removes mis-incorporated bases and therefore contributes greatly to the fidelity of replication. Recognition and subsequent excision of wrongly incorporated bases is referred to as proofreading, or editing; an editing function is associated with many replicative polymerases of prokaryotic, eukaryotic or viral origin. The α and ϵ subunits enhance the activity of each other when present together in the complex. A role for the third core subunit, θ, has not yet been found. The core itself, although capable of polymerising DNA, has extremely low processivity, and dissociates from the DNA after polymerising only about seven nucleotides. In marked contrast, the holoenzyme shows a processivity of more than 0.5 Mb *in vitro*. This conversion to high processivity is achieved by transfer of the β_2 subunits to the primer template complex by the ATP-dependent action of the γ complex. The β subunit dimerises in a head-to-tail manner and forms a ring-like structure that encloses the DNA duplex and, by an interaction with the α subunit, tethers the polymerase to the template. The task of the γ complex can be thought of as opening and reclosing the β_2 clamp to load it on to the template DNA strand.

In *B. subtilis* a number of genes that encode subunits of its polymerase III have been identified, largely by sequence similarity with the corresponding *E. coli* genes. Surprisingly, *B. subtilis* appears to possess two polymerising subunits, one, like DnaE, without proofreading capacity, and a second, PolC, with polymerising and $3'\rightarrow5'$ exonuclease activities in a single subunit.

Co-ordinated Synthesis of Leading and Lagging Strands

Synthesis of leading and lagging strands is a simultaneous process, with the difference in length of the two nascent strands that of a single Okazaki fragment. The heterodimeric nature of polymerase III suggests that one dimer may be synthesising the leading strand and the other the lagging strand (**Fig. 7**). The opposite orientation of the strands would then require that the lagging strand is looped round for the direction of synthesis of an Okazaki fragment to be aligned with that of the leading strand. Although this is an appealing model, there is no direct evidence for it but each successive Okazaki fragment is made by one and the same polymerase core, which is held at the fork by protein interactions and cycled to each new primer. In addition, there is evidence for a DNA loop at the replication fork of viral replication and, bearing in mind the universal conservation of general features of DNA replication, this is a strong indication that the same may be true of bacterial replication.

Given that both strands are synthesised by polymerase III, the question arises as to what leads to the difference in processivity of the two polymerisation processes, and what signals the synthesis of a new primer. Experimental evidence shows that, unlike the polymerase cores, primase and the polymerase III β subunit act distributively during lagging strand synthesis, which means that for each priming event these proteins are recruited anew from a pool of free protein. It appears that the frequency of Okazaki fragment synthesis is determined by the rate of association and dissociation of DnaG primase at the replication fork. Experiments with a reconstituted replication system *in vitro* show that variation in primase concentration inversely influences the length of Okazaki fragments. Many details of lagging-strand synthesis have been deduced from such *in vitro* experiments, leading to the following model for control of Okazaki-fragment synthesis. During elongation of a nascent Okazaki fragment, DnaG is recruited by

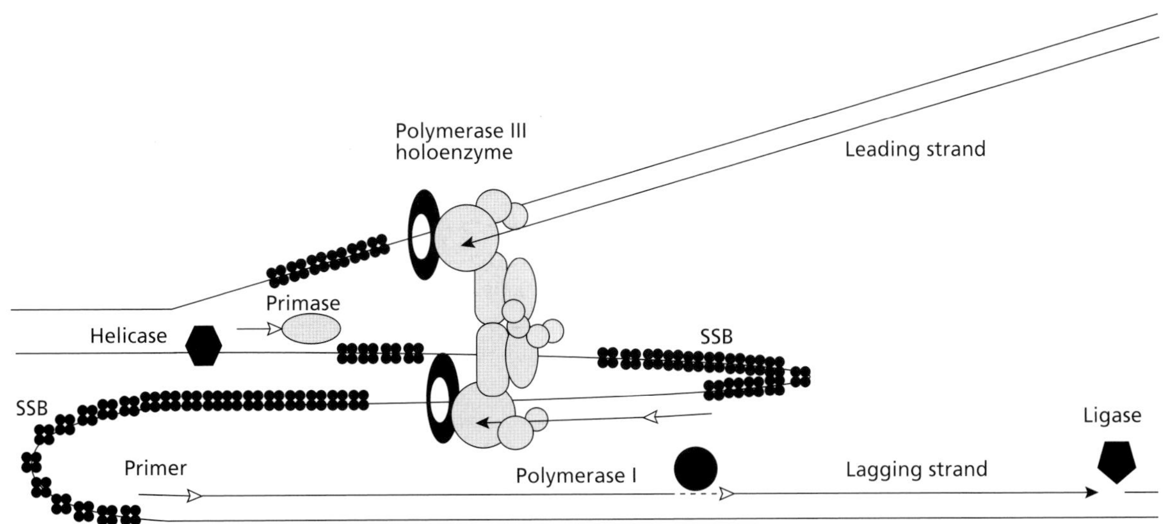

Fig. 7 Schematic representation of a replication fork and associated proteins in *E. coli*. DNA polymerase III simultaneously synthesises the leading and the lagging strands. The DNA template is unwound by the DnaB helicase and synthesis of the lagging strand is primed at intervals by primase (DnaG). Single-stranded regions are stabilised by SSB. Primers are removed by the exonucleolytic action of DNA polymerase I and Okazaki fragments are joined by DNA ligase.

an interaction with the helicase DnaB, and begins primer synthesis. After completion of the primer, DnaG remains bound by interaction with SSB (single-strand binding protein) before it is actively released by the formation of a fresh β_2 sliding clamp at the primer. The replacement of DnaG by a β_2 ring is mediated by the χ subunit of polymerase III which displaces DnaG from SSB. Then, once the synthesis of the nascent Okazaki fragment is completed, the lagging strand polymerase core is cycled from its old β_2 clamp to the newly prepared one at the next primer. Interactions with the rest of the polymerase complex prevent dissociation from the replication fork. As the leading strand is synthesised processively, the β_2 clamp associated with it does not need to be recycled and stays attached to the DNA and the polymerase complex.

To create a continuous daughter strand, the RNA primers must be removed and the DNA fragments connected. Removal of the primers requires the action of DNA polymerase I, which in addition to polymerisation activity also comprises a $3'\rightarrow5'$ exonuclease (proof-reading) function as well as a $5'\rightarrow3'$ exonuclease activity. It appears that this is the sole activity required of polymerase I, because its polymerising activity is not necessary for cell viability. After primer removal the resulting gaps are filled in, but it is not clear by which polymerase(s), and finally a further enzyme, known as DNA ligase, seals the nick by creating a phosphodiester bond between the last nucleotide to be added in the primer gap and the first of the adjacent Okazaki fragment.

It is interesting that in *B. subtilis* the polymerase III complex is localised to specific sites within the cell. It therefore appears that the DNA, which is distributed throughout most of the cell, must be spooled through an immobile polymerase, rather than the enzyme complex tracking freely along the DNA. A correlation exists between the growth rate of the cell and the number of these replication foci, and this is consistent with the fact that rapidly growing cells initiate multiple, overlapping, rounds of replication. Slow-growing cells exhibit one or two foci at the cell centre and at higher growth rates show additional foci at the cell quarters. It will be intriguing to learn what is responsible for anchoring the polymerase to these particular sites.

Fidelity of DNA Replication

Survival of an organism requires that DNA be accurately replicated. This is achieved both by fidelity mechanisms that operate during or immediately after DNA replication, and by a variety of repair mechanisms that recognise and correct damage which occurs because of post-replicative chemical changes to DNA. Fidelity mechanisms operate at three stages during *E.coli* replication. First, the polymerase must select the correct dNTP to be inserted. Second, an incorrectly selected base will be removed by exonucleolysis during the proof-reading or editing process. Third, mispaired bases, which escape both of these mechanisms, can be removed by the mismatch repair system after the replication fork has passed. Together these

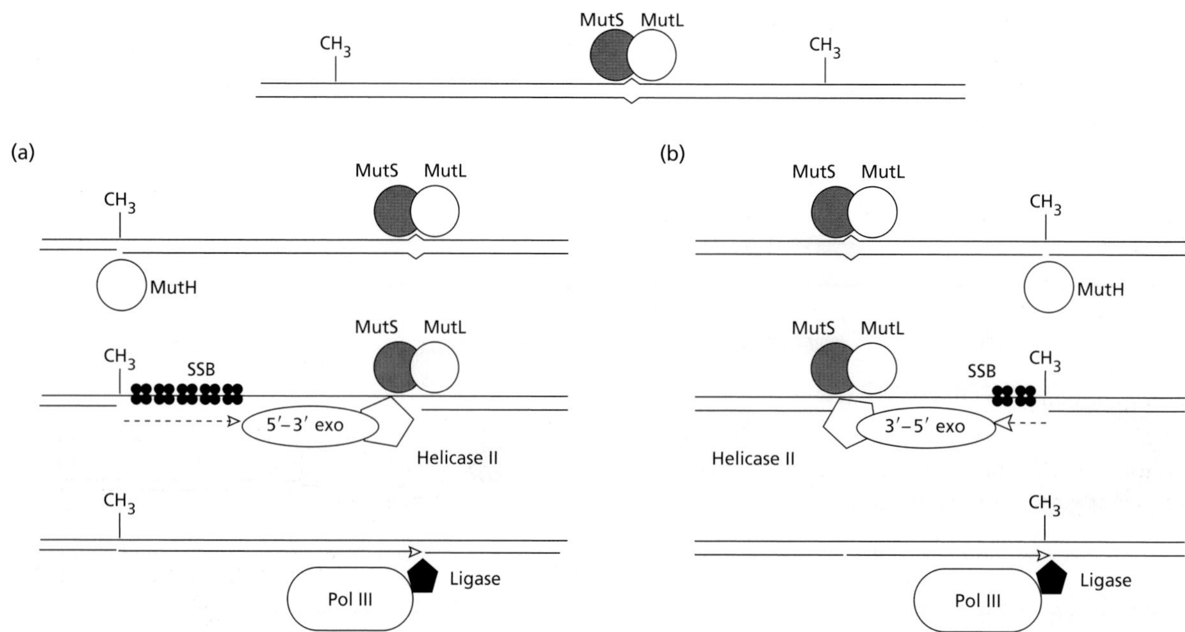

Fig. 8 Mismatch repair in *E. coli*. A mismatch in the DNA is recognised by MutS and MutL. MutH then cleaves the non-methylated strand 5′ to a GATC sequence. (a) (left) upstream of the damage; (b) (right) downstream of the damage. This is stimulated by MutL and may involve a direct interaction between the two proteins (with the intervening DNA looped). The combined action of helicase II (a helicase which acts in repair) and an exonuclease removes the DNA between the methylation site and the mismatch. This gap is then repaired by DNA polymerase III and DNA ligase.

processes ensure an error frequency of only one base in 10^9 or 10^{10}.

It appears that base selection is governed by energetic and geometric principles. The former relies on the fact that hydrogen-bonding and base-stacking interactions between the template base and the incoming base will distinguish a correct (Watson–Crick, AT or GC) base pair from a mismatch, since the former is more stable. The degree to which this is effective depends on the ability of the incoming base to bond with the template, and this will differ depending on the base combination. Thus, for instance, a GT mispair efficiently forms hydrogen bonds and fits well into a standard DNA helix. It is, therefore, unlikely that energy considerations alone are sufficient to account for the high selectivity shown by polymerases. A more important selection principle is likely to be recognition of particular geometric features, such as bond distances and angles, which distinguish correct from incorrect base-pairing. Indeed, errors by DNA polymerase III most often involve base-pairs that are closest to the Watson–Crick geometry.

The proof-reading function relies on competition between the polymerase activity of the α subunit and the exonucleolytic activity of ϵ. This balance is influenced, first, by the fact that a mis-inserted base is an unfavourable substrate for chain elongation which may in turn be due to geometric considerations, and second, because of a preference by the exonuclease for ssDNA over dsDNA. This means that mis-insertion results in a better substrate for the proof-reader and, in addition, a delay in further polymerisation allows more time for it to act. By the same reasoning, correctly paired bases will be protected from removal by favouring rapid elongation by the polymerase over excision by the exonuclease.

If errors are not removed by proof-reading, mismatches become incorporated into duplex DNA and correction requires that the mis-inserted base be distinguished from the template base, which is, by definition, correct. This is done by the mismatch repair mechanism (**Fig. 8**), which utilises the fact that parental DNA is methylated by the action of Dam, which specifically methylates the palindromic GATC sequence. After passage of the replication fork, there is a lag before the newly replicated strand is methylated, and during this time unmethylated daughter and methylated parental strands can be distinguished. A series of enzymes acts together to reverse the polymerase error. MutS recognises the mismatch and together with MutL activates an endonuclease, MutH, to cleave the unmethylated strand at a hemimethylated GATC site. Unidirectional exonucleolytic degradation of the daughter strand past the mismatch then occurs by the action of one of a number of exonucleases, depending on the orientation of the GATC site relative to the mismatch. The resulting gap, which can be as long as several kilobases, is finally repaired by DNA polymerase III holoenzyme, SSB and DNA ligase. Many bacteria do not possess a homologue of the *dam* gene, and it is not known whether these organisms have a form of mismatch repair to supplement proof-reading in order to ensure fidelity of replication.

Although the cell is well equipped to ensure accurate genome duplication, errors still occur. It should be noted, however, that although mutations are usually detrimental, some may prove advantageous by allowing a bacterial population to adapt to environmental challenges. Absolutely accurate duplication is, therefore, in the long term no more desirable than insufficiently accurate replication.

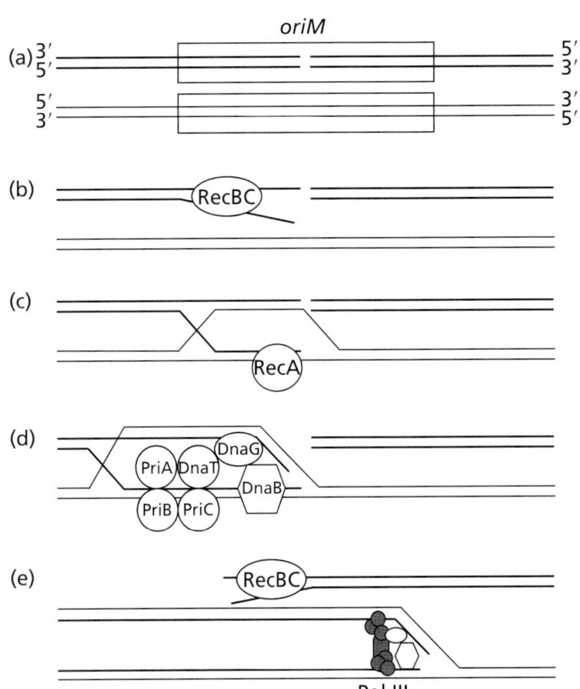

Fig. 9 Initiation of induced stable DNA replication (iSDR) in *E. coli*. Initiation requires two homologous DNA molecules, one of which contains a double strand break (a) At the break the DNA is unwound by the RecBC helicase (b) and the free 3′ end is bound by RecA which catalyses pairing with the complementary strand in the homologous molecule (c). This forms the D-loop that is the site of priming by the primsomal proteins (PriA, PriB, PriC and DnaT), which recruit DnaB helicase to the fork, which in turn recruits DnaG primase (d) DNA polymerase III can then start to synthesise DNA. A replication fork in the opposite direction can be formed in the same way (e).

Inducible Stable DNA Replication

As has been explained above, in all but one species of bacteria so far studied, replication depends absolutely on the DnaA protein. Under certain circumstances, however, DNA replication in *E. coli* can take place independently of both DnaA and *oriC*. This occurs after DNA damage that has induced the cell to express a particular set of genes, the so-called SOS response. This is a transient state during which cell division is delayed while repair proteins mend DNA damage, thus preventing transmission of damaged chromosomes to daughter cells. DnaA, *oriC*-independent replication also differs from normal replication in that it does not require protein synthesis for its initiation and has for this reason been termed inducible stable DNA replication (iSDR).

The mechanism of initiation used during iSDR is shown in Fig. 9. Although iSDR is independent of the DnaA protein, it requires the recombination proteins RecA and the RecBC complex, a set of proteins not normally required for DNA replication. RecBC helicase recognises the end of a linear DNA molecule and unwinds it to produce a single-stranded 3′ tail that is bound by RecA and introduced into an intact duplex DNA molecule with the same sequence. The recipient duplex must be partially unwound in order for base-pairing with the invading strand to take place, which results in a structure called a D-loop. Such a structure is used by iSDR to initiate DNA replication in conjunction with another set of proteins known as the primosome assembly proteins, PriA, PriB, PriC and DnaT, which were originally discovered as participants in bacteriophage replication. These proteins, which are dispensable for normal replication at *oriC*, are capable of forming a complex on DNA and loading the DnaB helicase, thus taking over one task that DnaA fulfils at the normal replication origin. DnaB then interacts with the DnaG primase to complete assembly of the primosome. DnaA is unnecessary during iSDR, since strand separation and helicase recruitment, both functions normally carried out by DnaA, are respectively taken over by D-loop formation and the primosome assembly complex. iSDR is initiated at specific sites on the chromosome, *oriM*s, two of which coincide with, but are not identical to, *oriC* and one of which is located at the chromosome terminus. D-loop formation requires breakage of both DNA strands in one molecule to produce the strand necessary to invade a homologous molecule. It is not known how this process is initiated but a possibility would be that an *oriM* sequence is recognised and cleaved by a specific endonuclease expressed as a result of the SOS response to DNA damage.

Various forms of DNA damage can cause one or both strands of the DNA to become nicked, resulting in a double-strand break on passage of the replication fork. DNA synthesis will cease because the replication machinery requires a continuous template. iSDR could reform a replication complex at such a double-strand break and at the same time repair the damaged chromosome.

iSDR differs in a number of ways from normal replication. First, it lacks the tight control of DnaA-mediated initiation at *oriC* and chromosomes for replication are chosen at random. Second, for unknown reasons iSDR is particularly error prone, although it makes use of DNA polymerase III. Clearly, therefore, iSDR is not the replication mechanism of choice, but it may be the remnant of an evolutionarily less developed system, which offers the advantage that it functions on damaged DNA though at the expense of efficiency and fidelity. It has yet to be determined whether such a mechanism operates in other bacterial species. It is interesting to note that the mitochondria of eukaryotic cells, which are thought to be derived from ancient prokaryotes, normally initiate replication by a D-loop mechanism.

Termination of Replication

Since most bacteria have circular chromosomes, the genome will be completely replicated when the two forks which began at the origin meet and fuse. If both forks travel at the same rate, which is approximately true for bacteria that have been studied, completion of replication will occur at a position diametrically opposite the starting point, which is known as the replication terminus. Organisms with linear chromosomes, such as *Streptomyces* spp. and almost all eukaryotes, face the problem of how to replicate completely the ends of the DNA molecules because of the need to replace the final primer used in lagging-strand synthesis.

Termination and Resolution of Dimeric Chromosomes in *E. coli*

The terminus region of *E. coli* contains several specific sequences, known as *ter* sites, which are recognised by a *trans*-acting factor, the Tus protein. Binding of Tus forms an asymmetric complex that impedes replication forks approaching from one direction but not from the other. This allows replication forks to enter the terminus region from either side, but blocks continuation of replication from the terminus toward the origin. Seven *ter* sites have been found, spread over approximately 1000 kb, but those predominately used are the first to be encountered in the appropriate orientation

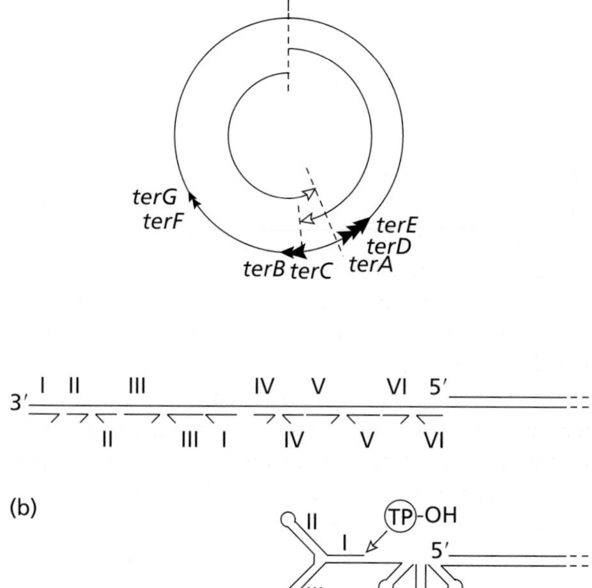

Fig. 10 (a) location of *ter* sites on the *E. coli* chromosome and the location of replication fork arrest, which occurs mainly at the first *ter* site encountered in either direction. (b) Termination of replication in *Streptomyces*. The 3′ end of the leading strand forms a complex hairpin structure that is recognised by the terminal priming protein, TP. This provides the hydroxy group required to prime synthesis of the lagging strand. Complementary sequences are indicated by numbered half arrows in the top part of the figure. Sequences with identical numbers can base-pair with each other.

by either replication fork, i.e. *terC* for a clockwise fork and *terA* for an anticlockwise fork. As a result the region within which both eventually become trapped encompasses about 300 kb (**Fig. 10a**).

ter sites are 23-bp sequences that are also found in a number of plasmids. Tus is a highly basic 35.7-kDa protein with little similarity to other DNA-binding proteins and it binds with high affinity as a monomer to the asymmetric *ter* sequence. This mode of binding probably accounts for the polar effect of the complex, because the replication machinery will be exposed to a different face of the protein, depending on the direction of approach. The mechanism of action of the complex has not yet been completely explained. The available biochemical evidence suggests that the replicative helicase, DnaB, is inhibited by specific protein interactions with Tus, rather than formation of a physical barrier to helicase progression by virtue of the stability of the Tus · *ter* complex.

The role of the replication-fork arrest mechanism is an intriguing question. This is because it has long been known that the entire terminus region of the *E. coli* chromosome can be deleted with little effect on cell viability and, moreover, that the gene that encodes Tus can be disrupted without measurable ill-effect on the cell under a variety of growth conditions. Evidently replication can be completed efficiently in the absence of any specific arrest mechanism. Nevertheless, it is likely that in the minority of cases where one replication fork becomes stalled, there is an advantage in trapping the quicker fork at the terminus ensuring this becomes the last region of the chromosome to be replicated. A possible reason for this is to ensure the resolution of chromosomal dimers that may be formed during replication, as the result of an uneven number of recombination events between parent and daughter molecules. The terminus contains a specific site, known as the *dif* locus, which, in conjunction with the XerC and XerD proteins, permits a site-specific recombination event to resolve chromosome dimers into monomers. As detailed below, chromosomes are physically moved apart during replication, so that if segregation of dimers is to be successful, the *dif* locus must be contained within the last region of the chromosomes to be separated. The proposed role of the Tus-*ter* system is to ensure that this happens. That such a system is beneficial is borne out by the fact that *S. typhimurium*, which diverged from *E. coli* some 160 million years ago, shows remarkable conservation of *ter* sites in an otherwise poorly conserved terminus region and that *E. coli ter* sites are functional in an orientation-dependent manner in *S. typhimurium*. More impressively, *B. subtilis* appears independently to have evolved a functionally similar termination system.

Termination in *B. subtilis*

On the *B. subtilis* chromosome, as in *E. coli*, termination occurs at specific sequences, also termed ter sites. These are unrelated in sequence to their *E. coli* counterparts and they consist of two overlapping 16-bp recognition sequences, A and B, for a small basic protein, RTP (replication termination protein), which shows no homology to the *E. coli* Tus protein. RTP binds as a symmetric dimer to the high affinity B site and is then joined by a further dimer at the lower affinity A site, which further enhances binding of RTP at the B site. As in *E. coli*, the effect of these sites on an advancing replication fork is polar and appears to rely on specific interactions with the replicative helicase to cause fork arrest. Analysis of the structure of RTP and mutational studies show that polarity is probably due to the difference in affinity for the two binding sites. A model has been proposed whereby a helicase, which first encounters the tight RTP · B complex, will be physically stalled and then inactivated by exposure to a particular face of the protein, whereas a helicase coming from the other direction will encounter the weaker RTP · A complex and may not be stalled long enough to suffer inactivation. In addition, the different mode of binding to A and B may not quite expose the same domains to the helicase. The functional similarity between the *E. coli* and *B. subtilis* systems is highlighted by the lack of any homology between Tus and RTP, or between the corresponding binding sites. This appears to constitute an impressive example of convergent evolution, suggesting that confining termination to the terminus region has an importance in nature which is obscured by growth under laboratory conditions.

In addition to its action at the terminus, the RTP protein plays a role in control of replication by causing replication fork stalling or arrest at sites approximately 200 kb either side of *oriC*. This appears to be a mechanism by which replication is inhibited under starvation conditions and also by which over-replication is avoided. In this case RTP utilises binding sites, called *STer*, which are different from the AB motif, but which have sequence similarity to the B sequence. It is as yet unclear why these sites function only under particular conditions where replication must be prematurely stalled or aborted, unlike the *ter* sites which are active during each round of replication.

Termination in Bacterial Species with Linear Chromosomes

Several bacterial species have recently been shown to contain linear chromosomes. The best-studied of these are the streptomycetes, which replicate their genomic DNA bidirectionally from an internal origin towards the two ends, or telomeres, of the chromosomes. Leading-strand synthesis will simply terminate when the replicative polymerase runs out of template, but there is a problem with completion of lagging-strand synthesis; how can the terminal RNA primer be replaced by DNA to replicate the extreme 5′ end of the chromosome? Failure to replace it would result in a double-stranded chromosome with unreplicated 3′ overhangs at both ends.

The way in which chromosomal replication in *Streptomyces* is completed is not yet certain, but studies on linear *Streptomyces* plasmids suggests a possible mechanism. *Streptomyces* chromosomes and plasmids contain highly conserved regions at their telomeres, which contain a number of palindromic sequences. The 3′ terminal 13 bp are complementary to an internal sequence with which it is thought to pair. This pairing, in combination with the pairing of the other terminal palindromes, is thought to form a structure (Fig. 10b) that is recognised by a terminal priming protein, TP, which primes synthesis of the incomplete complementary strand. TP allows the covalent binding of the first nucleoside monophosphate to a hydroxy group on the protein, which obviates the need for an RNA or DNA primer and so permits replication to start at the first base of the template strand. The complex secondary structure of the leading strand 3′ overhang must be melted, through replication of the complementary strand or otherwise, for it to act as the template for the completion of lagging strand synthesis. The resulting molecule is a completely replicated chromosome with a molecule of TP covalently linked to the 5′ terminus of each strand.

Although this method of lagging-strand synthesis resembles that of some viruses, it is dissimilar to the strategy employed by eukaryotes. The telomeres of eukaryotic chromosomes contain long repetitive sequences. These are added on by the enzyme telomerase, which contains a short stretch of RNA that is repetitively used as a template for telomere extension. Since these telomeric sequences are redundant, it is of no consequence if their replication is incomplete.

Spirochaetes of the genus *Borrelia* also have linear chromosomes which, unlike those described above, are joined at the ends to form hairpins. This conformation is also shared by some mitochondrial DNAs. In essence, a *Borrelia* telomere resembles a single-stranded circle of DNA, which is base-paired along its full length. *Borrelia* chromosomes encode DnaA homologues and have potential DnaA binding sites in the middle of the chromosome, and bidirectional replication probably proceeds in the normal way. A mechanism has been proposed for replicating the ends, which would involve replicating around the ends, followed by cutting and religating in a way that would result in conservatively (as opposed to semi-conservatively) replicated ends. A bacterial exception to Watson–Crick rules would be of interest, and further information is awaited.

Segregation of Chromosomes

After the completion of replication three events must take place to ensure that both daughter cells inherit a copy of the chromosome. First, as detailed above, dimeric chromosomes must be resolved into monomers by recombination. Second, interlinked chromosomes must be separated by a topoisomerase and, finally, the chromosomes must be moved towards opposite cell poles before completion of the cell division septum.

Decatenation of Chromosomes

Replication of circular DNA molecules results in two topologically intertwined progeny molecules, or catenanes, and these must be separated before partitioning can be completed. Decatenation is catalysed by a type II topoisomerase, which is distinct from DNA gyrase and in *E. coli* is known as topoisomerase IV. Although its mode of action is similar to that of gyrase, in that it produces a transient double-stranded break in the DNA, it differs in that it favours inter- rather than intra-molecular strand passage. This is attributed to the fact that topoisomerase IV, unlike gyrase, does not wrap the DNA around itself before one strand is translocated through the gapped duplex.

Homologues of topoisomerase IV have been identified in a number of bacteria including *H. influenzae*, *N. gonorrhoeae*, *Staph. aureus* and *Strep. pneumoniae*. They are of interest because they form additional targets for a class of 4-quinolone antibacterial agents, which were originally identified as inhibitors of DNA gyrase.

Chromosome Partitioning

After termination of replication, sister chromosomes are partitioned to daughter cells by a highly accurate process that results in fewer than 1 in 10^4 cells lacking a copy of the chromosome. Little is known about how partitioning is brought about in bacteria. In contrast, detailed knowledge is available about the corresponding process, mitosis, in eukaryotic cells, during which a spindle consisting of microtubules actively separates sister chromosomes by drawing them towards opposite cell poles. Since subcellular structures equivalent to a spindle have not been identified in prokaryotes, it was long assumed that chromosome partitioning is a passive process in which attachment of the chromosome to the cell envelope would effect gradual separation as the cell grew. Such a simple partitioning model is no longer tenable, since it is now known that the cell wall and membrane grow by insertion of material over their entire surface. Indeed, recent experimental evidence points to a more active mechanism.

The recent development of fluorescence techniques permit the determination of the location of particular parts of the chromosome in single cells. This technique has shown that in both *B. subtilis* and in *E. coli* chromosomes are oriented with their origins toward the cell poles. After replication has been initiated, one of the two origins rapidly moves toward the opposite cell pole, which indicates that there is an active apparatus for separating the origin regions of the parent and daughter chromosomes. In *B. subtilis*, a protein known as Spo0J co-localises with the origin region of the chromosome and binds to a 16-bp sequence, *parS*, which occurs several times in the origin proximal part of the chromosome. Genetic studies show that both elements are involved in the partitioning process. Furthermore, studies with *C. crescentus* clearly demonstrate that ParB, a homologue of Spo0J, is involved in chromosome segregation by binding to a sequence near the replication origin. Sequence comparisons show that homologues of these proteins exist in a number of bacteria with the notable exception of *E. coli*.

Although mutation of Spo0J causes observable defects in partitioning, *spo0J* is not an essential gene, which clearly indicates that additional factors are involved. Importantly, this class of protein does not have the features of a motor protein and would therefore not be expected to be responsible for the observed movement of origin regions; its role may only be to maintain the orientation of the chromosome. Possible candidates for motor proteins have been found both in *E. coli* (MukB) and in *B. subtilis* (Smc) and although their involvement in partitioning has been demonstrated, the nature of their activity is as yet unclear. In summary, it appears that chromosome segregation in bacteria is governed by an active system, the details of which remain to be discovered.

Further Reading

The following list of recent monographs and review articles provides more detailed information as well as further references to the topics covered in this chapter.

General

DePamphilis ML (ed.) (1996) *DNA Replication in Eukaryotic Cells*. New York: Cold Spring Harbor Laboratory Press.

Kornberg A, Baker TA (1992) *DNA Replication*. New York: WH Freeman.

DnaA and Initiation

Kaguni JM (1997) *Escherichia coli* DnaA protein: the replication initiator. *Mol. Cells* 7: 145–157.

Marczynski GT, Shapiro L (1993) Bacterial chromosome origins of replication. *Curr. Opin. in Gen. Dev.* 3: 775–782.

Messer W, Weigel C (1996) Initiation of chromosome replication. In: Neidhardt FC *et al.* (eds) *Escherichia coli and Salmonella, Cellular and Molecular Biology*. Washington DC: American Society for Microbiology, pp. 1597–1616.

Moriya S, Imai Y, Hassan AKM, Ogasawara N (1999) Regulation of initiation of *Bacillus subtilis* chromosome replication. *Plasmid* 41: 17–29.

Ogasawara N, Moriya S, Yoshikawa H (1991) Initiation of chromosome replication: structure and function of oriC and DnaA protein in eubacteria. *Res. Microbiol.* 142: 851–859.

Skarstad K, Boye E (1994) The initiator protein DnaA: evolution, properties and function. *Biochem. Biophys. Acta* 1217: 111–130.

Yoshikawa H, Ogasawara N (1991) Structure and function of DnaA and the DnaA-box in eubacteria: evolutionary relationships of bacterial replication origins. *Mol. Microbiol.* 5: 2589–2597.

Regulation

Asai T, Bates DB, Boye E, Kogoma T (1998) Are minichromosomes valid model systems for DNA replication control? Lessons learned from *Escherichia coli*. *Mol. Microbiol.* 29: 671–675.

Autret S, Levine A, Holland IB, Séror SJ (1997) Cell cycle checkpoints in bacteria. *Biochimie* 79: 549–554.

Crooke E (1995) Regulation of chromosomal replication in E. coli: sequestration and beyond. *Cell* 82: 877–880.

Domian IJ, Quon KC, Shapiro L (1996) The control of temporal and spatial organisation during the *Caulobacter* cell cycle. *Curr. Opin. in Gen. Dev.* 6: 538–544.

Levin PA, Grossman AD (1998) Cell cycle: the bacterial approach to coordination. *Curr. Biol.* 8: R28–R31.

Elongation

Kelman Z, O'Donnell M (1995) DNA polymerase III holoenzyme: structure and function of a chromosomal replicating machine. *Ann. Rev. Biochem.* 64: 171–200.

Losick R, Shapiro L (1998) Bringing the mountain to Mohammed. *Science* 282: 1430–1431.

Marians KJ (1996) Replication fork propagation. In: Neidhardt FC *et al.* (eds) *Escherichia coli and Salmonella, Cellular and Molecular Biology*. Washington DC: American Society for Microbiology, pp. 749–763.

Fidelity and Repair

Echols H (1991) Fidelity mechanisms in DNA replication. *Ann. Rev. Biochem.* 60: 477–511.

Friedberg EC, Walker GC, Siede W (1995) *DNA Repair and Mutagenesis*. Washington DC: American Society for Microbiology.

Rupp WD (1996) DNA repair mechanisms. In: Neidhardt FC *et al.* (eds) *Escherichia coli and Salmonella, Cellular and Molecular Biology.* Washington DC: American Society for Microbiology, pp. 2277–2294.

Stable DNA Replication

Kogoma T (1997) Stable DNA replication: interplay between DNA replication, homologous recombination and transcription. *Microbiol. Mol. Biol. Rev.* 61: 212–238.

Termination

Chen CW (1996) Complications and implications of linear chromosomes. *Trends Genet.* 12: 192–196.

Hill TM (1996) Features of the chromosomal terminus region. In: Neidhardt FC *et al.* (eds) *Escherichia coli and Salmonella, Cellular and Molecular Biology.* Washington DC: American Society for Microbiology, pp. 1602–1614.

Yoshikawa H, Wake RG (1993) Initiation and termination of chromosome replication. In: Sonenshein AL, Hoch JA, Losick R. (eds) *Bacillus subtilis and Other Gram-Positive Bacteria: Biochemistry, Physiology and Molecular Genetics.* Washington DC: American Society for Microbiology, pp. 507–528.

Partitioning and Segregation

Schmid MB, von Freiesleben U (1996) Nucleoid segregation. In: Neidhardt FC *et al.* (eds) *Escherichia coli and Salmonella, Cellular and Molecular Biology.* Washington DC: American Society for Microbiology, pp. 1662–1671.

Wheeler RT, Shapiro L (1997) Bacterial chromosome segregation: is there a mitotic apparatus? *Cell* 88: 577–579.

15

Bacterial Protein Synthesis

Colin R. Harwood and Anil Wipat

The Medical School, Newcastle upon Tyne, UK

The unicellular existence of micro-organisms places much greater physiological demands on these organisms with respect to changes in their immediate environment than does the multi-cellular existence of higher eukaryotes. Micro-organisms generally inhabit environments in which the supply of nutrients is discontinuous and their range variable. Consequently, bacteria and unicellular eukaryotic organisms often exhibit a relatively broad metabolic potential when compared with that of multi-cellular organisms. In order to reduce the resulting additional genetic load, the expression of their genes is modulated to match, as far as possible, the metabolic demands of the organism for a given environment. Micro-organisms are also exposed to stresses or agents, such as nutrient deprivation, high osmolarity, heat, antibiotics or irradiation, that are potentially life-threatening. They therefore need to mount appropriate responses which attempt to overcome or at least reduce the worst aspects of these conditions. Finally, pathogenic bacteria have evolved a variety of traits that enable them to infect target host organisms. These traits, usually referred to as virulence factors, are often only induced in the presence of a suitable host or host cell.

In general, bacteria respond to changes in their environment by modulating the synthesis of specific proteins. Three categories of regulation have been identified to achieve this: specific regulation, in which metabolites or catabolites regulate genes encoding enzymes involved in their own metabolism; global regulation, in which the cells respond to general stimuli such as limitations in nutrient sources, or physical or chemical insult (DNA damage, heat shock, osmotic shock, etc.); and temporal regulation, in which regulation is coupled to other events, such as the cell cycle or a differentiation pathway (e.g. sporulation, bacteriophage development). Bacteria deploy an array of strategies for controlling the expression of their genes, often combining several distinct mechanisms to permit different types and degrees of response.

Bacterial Genomes

Chromosomes are the principal repositories of the cell's genetic information, the site of gene expression and the vehicle of inheritance. The term chromosome, meaning dark-staining body, was originally applied to the structures visualised in eukaryotic organisms by light microscopy. This term has now been extended to describe the physical structures that encode the genetic (hereditary) information in all organisms. The term genome is used in the more abstract sense to refer to the sum total of the genetic information of an organism. The genetic material consists of double-stranded (ds)DNA that is decoded, usually in the form of polypeptides. The nucleotides are decoded in sets of three called codons. Since there are four nucleotides (adenosine, cytosine, guanine and thymine), there are 64 (i.e. 4^3) possible codons in the genetic code. Sixty-one codons code for the 20 natural amino acids, the remaining three codons (stop codons) signal the termination of translation. Although each codon specifies a single amino acid, an amino acid can be specified by

 doi:10.1006/bkmm.2001.0015

First position	Second position: U		Second position: C		Second position: A		Second position: G		Third position
U	UUU	Phe	UCU	Ser	UAU	Tyr	UGU	Cys	U
	UUC		UCC		UAC		UGC		C
	UUA	Leu	UCA		UAA	Stop	UGA	Stop	A
	UUG		UCG		UAG	Stop	UGG	Trp	G
C	CUU	Leu	CCU	Pro	CAU	His	CGU	Arg	U
	CUC		CCC		CAC		CGC		C
	CUA		CCA		CAA	Gln	CGA		A
	CUG		CCG		CAG		CGG		G
A	AUU	Ile	ACU	Thr	AAU	Asn	AGU	Ser	U
	AUC		ACC		AAC		AGC		C
	AUA		ACA		AAA	Lys	AGA	Arg	A
	AUG	Met	ACG		AAG		AGG		G
G	GUU	Val	GCU	Ala	GAU	Asp	GGU	Gly	U
	GUC		GCC		GAC		GGC		C
	GUA		GCA		GAA	Glu	GGA		A
	GUG		GCG		GAG		GGG		G

basic   Non-polar   acidic   Uncharged

Fig. 1 The genetic code (See also Colour Plate 5).

Table 1 Comparative properties of viral, bacterial and fungal chromosomes with respect to size, composition and topography

Organism	Type	Number	Size	Nucleic acid	Topology
MS2	Bacteriophage	1	3.6 knt	ssRNA	Circular
ϕX174	Bacteriophage	1	5.4 knt	ssDNA	Linear
Lambda	Bacteriophage	1	48.5 kbp	dsDNA	Linear
T4	Bacteriophage	1	174 kbp	dsDNA	Linear
Mycoplasma genitalium	Eubacterium	1	580 kbp	dsDNA	Circular
Borrelia burgdorferi	Eubacterium	1	1.4 Mbp	dsDNA	Linear
Campylobacter jejuni	Eubacterium	1	1.7 Mbp	dsDNA	Circular
Rhodobacter sphaeroides	Eubacterium	2	3.0 Mbp + 0.9 Mbp	dsDNA	2 × circular
Bacillus subtilis	Eubacterium	1	4.2 Mbp	dsDNA	Circular
Escherichia coli	Eubacterium	1	4.6 Mbp	dsDNA	Circular
Myxococcus xanthus	Eubacterium	1	9.45 Mbp	dsDNA	ND
Methanococcus jannaschii	Archaea	1	1.66 Mbp	dsDNA	Circular
Archaeoglobus fulgidus	Archaea	1	2.8 Mbp	dsDNA	Circular
Schizosaccharomyces pombe	Eukaryote	3	3.5, 4.6 and 5.7 Mbp, total 18.8 Mbp	dsDNA	Linear
Saccharomyces cerevisiae	Eukaryote	16	0.2 to 2.2 Mbp, total 12.43 Mbp	dsDNA	Linear

ND, not determined; dsDNA, double-stranded DNA; ssDNA, single-stranded DNA; bp, nucleotide base pair(s); M, million; nt, nucleotide(s); k, thousand.

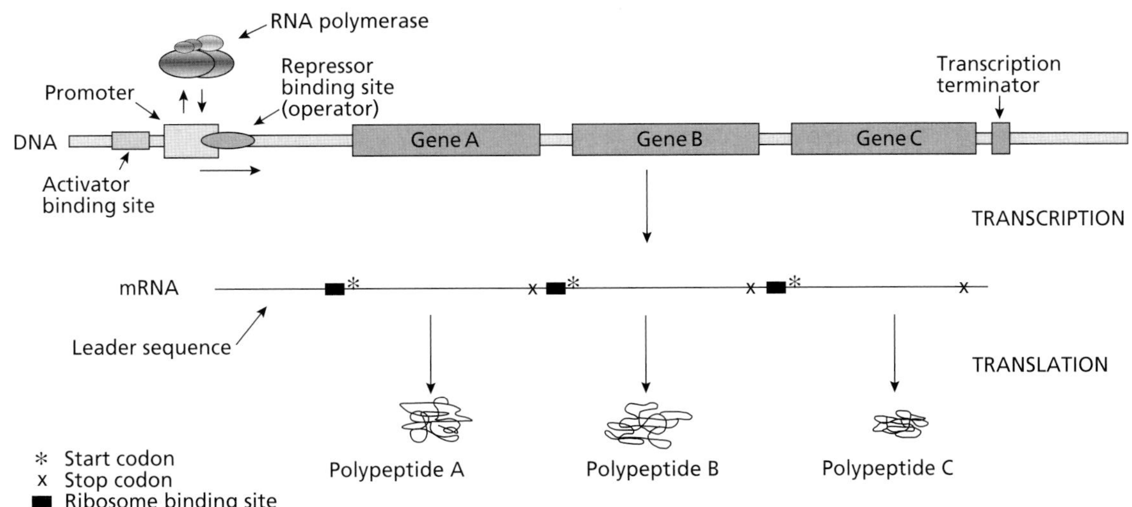

Fig. 2 Diagrammatic representation of the main features of control in the expression of a bacterial operon. The various structural components are not shown to scale.

between one and six codons (Fig. 1). The genetic code is therefore said to be degenerate.

Microbial chromosomes range in size over several orders of magnitude and vary in number, composition and topology (Table 1). Genome sizes tend to reflect the organism's structural complexity and life style. Obligate bacterial parasites such as *Mycoplasma genitalium* (580 kbp) tend to have small genomes, while bacteria with complex life cycles such as *Myxococcus xanthus* (9.45 Mbp) tend to have large genomes.

The chromosome of *Escherichia coli* is typical of many bacteria. It is 1100 μm in length and its 4.6 Mbp of DNA codes for about 4400 polypeptides. The chromosome has a single set of genes, excepting those encoding the ribosomal RNA (rRNA). At least 90% of the DNA encodes polypeptides or stable RNA while the remaining 10% either is used for controlling gene expression or has a purely structural function. Protein-coding sequences can be on either strand of the DNA, although there is a preference for an orientation in the direction in which the DNA is replicated. To enable bacterial cells to regulate gene expression, the DNA is organised into transcriptional units or operons with distinct control sequences and transcriptional and translational start and stop points (Fig. 2). One consequence of this is that genes of related function are often, but not always, clustered together on the chromosome.

Overview of Protein Synthesis

The synthesis of proteins in bacteria is essentially a two-stage process involving transcription, the synthesis of a messenger RNA (mRNA) intermediate using one strand of the duplex DNA as the template, and translation, the decoding of the information in the mRNA into an ordered arrangement of amino acids to form a polypeptide (Fig. 3). The DNA strand that acts as the template for the mRNA (and to which it is complementary) is known as the anticoding or template strand; the DNA strand that bears the same sequence (excepting for the replacement of thymine by uracil) is known as the coding strand.

In eukaryote cells it is necessary to have an mRNA intermediate in the DNA-directed synthesis of proteins because of the spatial separation between the chromosomes, located in the membrane-bound nucleus, and the site of protein synthesis, the ribosome, located in the cytoplasm. In such organisms the role of mRNA is therefore to carry a copy of the genetic information encoding a specific polypeptide from the nucleus to the ribosome. Although the sites of transcription and translation are not physically separated in prokaryotic organisms, there is no less a need for such an intermediate. However, the role of mRNA in prokaryotes is somewhat different. In the first place, without the mRNA intermediate molecule, the process of protein synthesis (translation) would interfere with that of DNA replication. This is because in rapidly growing bacteria, a replication fork passes any particular point on the chromosome approximately every 20 minutes. The presence of multiple, large ribonucleoprotein complexes (ribosomes) would interfere with replication by occupying sites on the DNA. Messenger RNA helps to obviate this problem by separating the processes of replication and polypeptide synthesis.

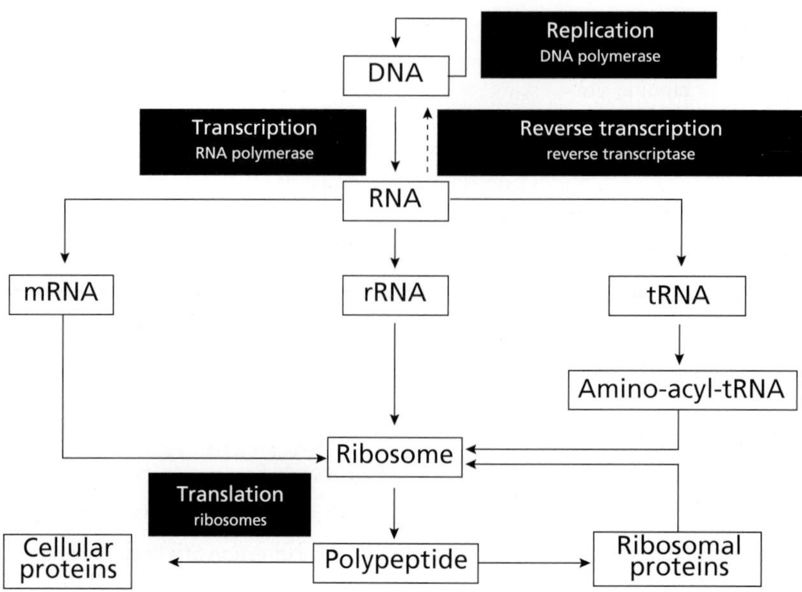

Fig. 3 Overview of protein synthesis.

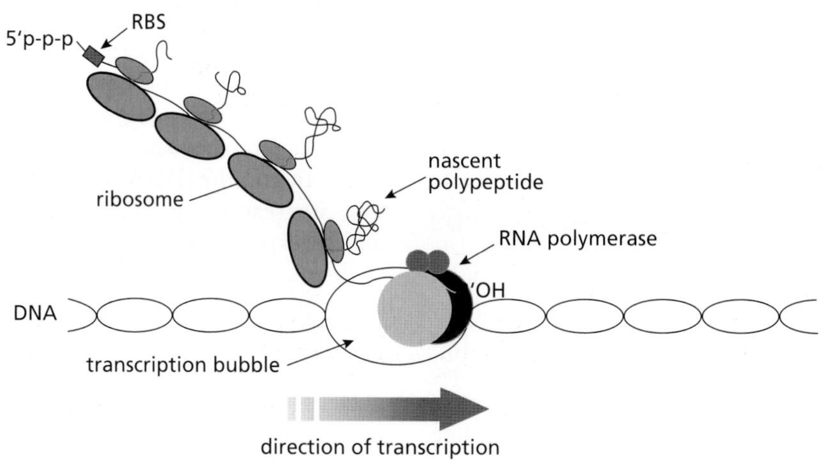

Fig. 4 Coupled transcription and translation.

A second requirement for an intermediate relates to the need for bacteria to control the expression of their genes rapidly in response to changes in their environment. This is achieved by modulating mRNA stability; since the average half-life of mRNA is about 2 minutes, the synthesis of a specific polypeptide will only be maintained if its cognate mRNA is continuously synthesised. If the synthesis of a specific mRNA is reduced or stopped then, after a short delay, so will the synthesis of the corresponding polypeptide(s). In contrast, eukaryotic mRNA is relatively stable, in part a reflection of their reduced need to respond rapidly to environmental changes.

Although the processes by which proteins are synthesised are similar in prokaryotic and eukaryotic organisms, they differ in a number of ways that are of relevance to the control of gene expression. Bacterial mRNA is often polygenic; a single molecule of mRNA often encodes more than one polypeptide. In contrast, eukaryotic mRNA is usually monogenic. In the former case, the synthesis of polygenic mRNA is one of the mechanisms that bacteria use to co-ordinate the synthesis of proteins required for closely related functions, such as the enzymes of a particular metabolic pathway. However, one of the most significant difference is the close coupling of transcription and translation in prokaryotes (Fig. 4). Ribosomes attach to bacterial mRNA molecules and begin translation even before the transcript is fully synthesised. Translation generally keeps pace with transcription; at 37°C the rate of transcription is approximately 42 nucleotides or 14 codons per second, while that of translation is about 15 amino acids per second. The sequence of the mRNA coding a polypeptide will generally be loaded with ribosomes moving 'nose-to-tail' along the molecule from the $5'$ end towards the $3'$ end of the molecule (**Fig. 4**). The ribosomes therefore have protruding polypeptides that increase in length with increasing distance from the $5'$ end of the mRNA molecule. The close coupling of transcription and translation in bacteria ensures the rapid synthesis of proteins and is employed as a mechanism to control the expression of specific genes or operons at the transcriptional and post-transcriptional levels.

Transcription

Transcription is the synthesis of RNA using DNA as a template. The process is carried out by the enzyme RNA polymerase (**Fig. 5**). The same enzyme is responsible for the transcription of all of the genes in a bacterial cell, including mRNA, rRNA and transfer RNA (tRNA). The core RNA polymerase (E) is a multi-protein complex comprising two α, one β and one β' subunit (**Table 2**). RNA polymerase initiates and terminates transcription at specific points in the DNA. Transcription is initiated downstream of specific sequences called promoters, sites that are recognised and bound by RNA polymerase. RNA polymerase is able to recognise several classes of promoter and specificity for these difference classes is determined by an additional subunit, the sigma (σ) factor. When a sigma factor associates with a core enzyme, the resulting complex is referred to as the holoenzyme ($E\sigma$). The sigma factor associated with the transcription of the majority of genes in *E. coli* is referred to as σ^{70}. Other sigma factors are induced under specific growth conditions, often in response to various stresses. These 'alternative' sigma factors therefore lead to the induction of specific subsets of genes.

The process of transcription can be divided into a series of stages: template recognition, initiation, elongation and termination.

Template Recognition and Initiation of Transcription

E. coli varies the frequency of transcription initiation over four orders of magnitude. Some genes are transcribed just once every generation while others are transcribed on average once every second. Template recognition is a multi-stage process (**Fig. 6**) in which RNA polymerase (R) binds initially to non-specific DNA (D) and then to a specific promoter (P) sequence in the form of a closed complex ($RP_{C1/2}$). Finally, open complex formation (RP_o) is induced, in which highly localised separation of the DNA strands takes place. Each of these processes has its own equilibrium or rate constant:

$$R + D + P \rightleftharpoons RD + P \overset{K_B}{\rightleftharpoons} D + RP_{C1} \overset{k_{f1}}{\rightleftharpoons} RP_{C2} \overset{k_{f2}}{\rightleftharpoons} RP_o$$

where K_B is the equilibrium constant for binding and k_f is the rate constant for open complex formation.

Template recognition and initiation are frequently the point at which gene expression is controlled. RNA polymerase core enzyme (E) binds non-specifically to DNA and is not able to initiate transcription on its

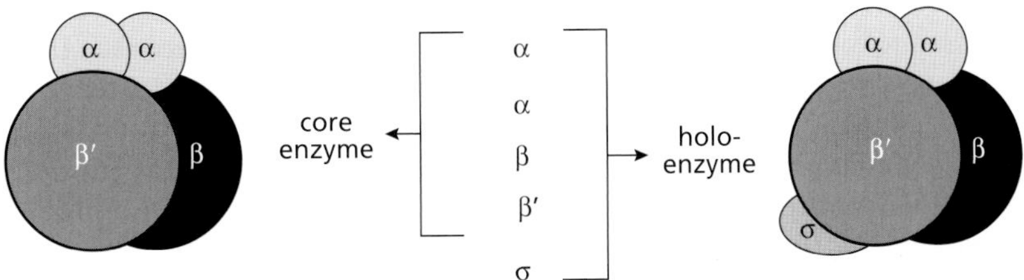

Fig. 5 RNA polymerase.

Table 2 RNA polymerase of *Escherichia coli*

Subunit	Gene	Mass (kDa)	No.	Function
α	*rpoA*	36	2	Enzyme assembly
β	*rpoB*	150	1	Nucleotide binding/polymerase
β'	*rpoC*	155	1	Template binding
σ^{70}	*rpoD*	70	1	Promoter recognition and binding

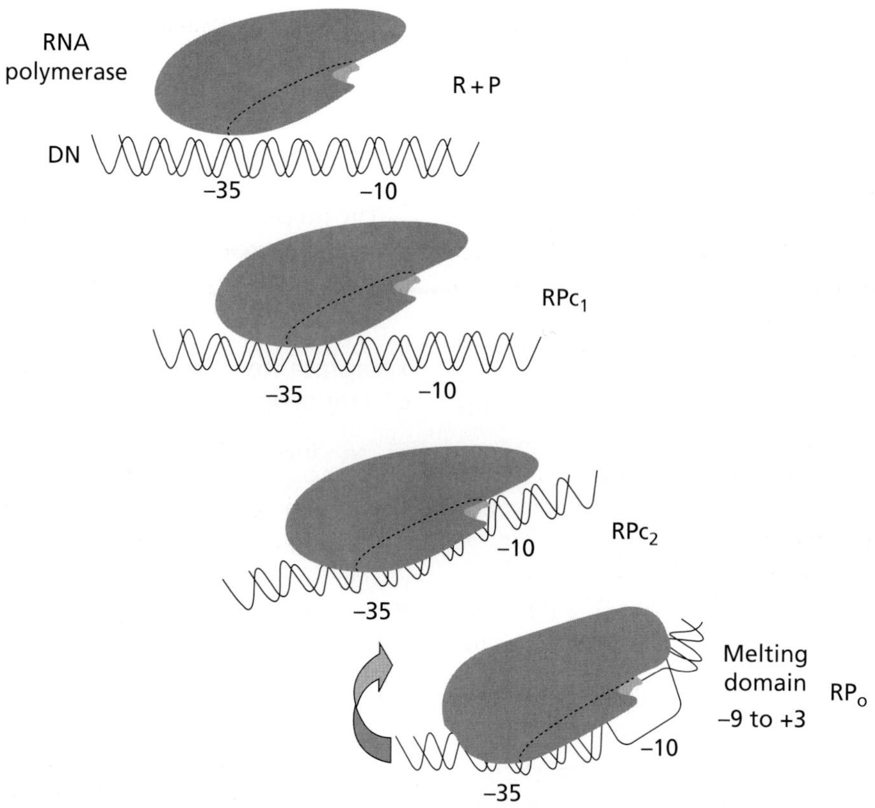

Fig. 6 Initiation of transcription: interactions between RNA polymerase and the promoter.

own. In contrast, the holoenzyme (Eσ) has a reduced affinity for non-specific sites but a greatly increased affinity for specific promoter sequences. The affinity of Eσ for a particular promoter sequence is dependent on the sigma factor associated with the enzyme. The vegetative promoters of bacteria such as *E. coli* and *Bacillus subtilis* are defined primarily by the bases located around −10 and −35 bases upstream (5′) from the transcription initiation site (= +1). Although these sequences cannot be defined precisely, it is possible to define a consensus sequence that appears to represent the optimal binding site for a particular RNA polymerase core enzyme/sigma factor combination. In general, the frequency at which transcription is initiated from a particular promoter (promoter strength) is a function of the affinity of Eσ for the promoter and the rate at which open complex formation occurs.

When *E. coli* is growing exponentially, the vast majority of RNA polymerase holoenzymes contain a sigma factor with a molecular mass of 70 kDa, referred to as σ^{70} (the equivalent sigma factor in *B. subtilis* has a mass of 43 kDa and is referred to as σ^{A}). The consensus sequence recognised by Eσ^{70} (and *B. subtilis* Eσ^{A}) is:

$$5'\text{-}\overset{-35}{\text{TTGaca}}-17\text{ bases }-\overset{-10}{\text{TAtaaT}}-5\text{ to }9\text{ bases}-\text{RNA initi}\overset{+1}{\text{a}}\text{tion site -}3'$$

where the upper case bases are highly conserved and the lower case bases are the main sites of variation between different promoters.

If *E. coli* is subjected to heat shock, it responds by turning down/off the expression of its vegetative genes and instead induces the expression of 17 heat-shock genes. The heat-shock genes have promoters with consensus sequences (5′-CNCTTGAA – 13 to 15 bases – CCCATNT-3′) that are different from those of vegetative genes and are recognised by an RNA polymerase enzyme with an alternative sigma factor, σ^{32}. Other sigma factors have been shown to control responses to nitrogen stress in *E. coli*, sequential gene expression in bacteriophage SPO1 development, differentiation in *Streptomyces* and sporulation in *B. subtilis*.

The binding of the RNA polymerase to the promoter (RP) is followed by extensive conformational changes in both the protein and the DNA (binary complex) (**Fig. 6**). This is reflected in the transition from the closed complex (RP$_c$), in which the two strands of DNA in the promoter region are base-paired, to the open complex (RP$_o$), in which local strand separation has occurred. RNA polymerase has a different footprint on the promoter region in the closed complex as compared with the open complex.

The initial closed complex (RP_{C1}) only involves sequences upstream of +1 and the RNA polymerase only contacts one face of the promoter in the region from −5 to −55. This is followed by a major conformational change (RP_{C2}) in which RNA polymerase-binding involves DNA sequences extending from −55 to +20. The sequence from −55 to −10 is located in a groove on the surface enzyme, while sequences from −10 to +20 are buried in a channel within the structure of the enzyme.

The closed complex is followed by the open complex (RP_o), involving a major conformational change in both the RNA polymerase and DNA sequences within the promoter sequences. The DNA is bent, in part to facilitate localised DNA melting, and the strands of DNA within the duplex, from about −10 to +3, become physically separated.

Transcription is initiated 6–8 bp downstream of the −10 hexamer, usually at a purine (A or G) and only rarely (7%) at a pyrimidine. RNA polymerase catalyses the polymerisation of ribonucleotides in a net 5′ to 3′ direction – new bases are therefore added at the 3′ end of the growing mRNA chain; the 5′ end is characterised by the presence of the original triphosphate moiety of the initiating ribonucleotide substrate. Unlike DNA replication, a primer is not required to initiate transcription. This is because the linkage of the first two bases is catalysed at a specific site within the RNA polymerase molecule, and distinct from the site involved in the polymerisation of the remaining bases in the mRNA chain. This initial dimer, pppXpY, effectively forms the primer for the subsequent polymerisation reactions. This initial step is uniquely sensitive to the antibiotic rifamicin, and this antibiotic has been used extensively as a molecular tool for studying transcription initiation and for determining the stability (half-life) of mRNA molecules. Seven to 12 nucleotides need to be incorporated before the RNA polymerase clears the promoter. Depending on the availability of ribonucleotide substrates, frequent abortive cycling may occur in which transcription is initiated and terminated, with concomitant release of the nascent transcript. After release, transcription is re-initiated without dissociation of the RNA polymerase from the region of the promoter.

Transcription Elongation

Transition to elongation phase occurs when the nascent mRNA chain reaches 11–13 nucleotides in length. It is accompanied by three important changes: a substantial increase in the stability of the nascent transcript, the dissociation of the sigma factor from the RNA polymerase and the movement of RNA polymerase from the region of the promoter (Fig. 7). Nucleotides are added to the 3′ end of the nascent mRNA molecule at a rate of about 43 nucleotides per second. The resulting ternary complex, consisting of DNA, mRNA and RNA polymerase, is highly resistant

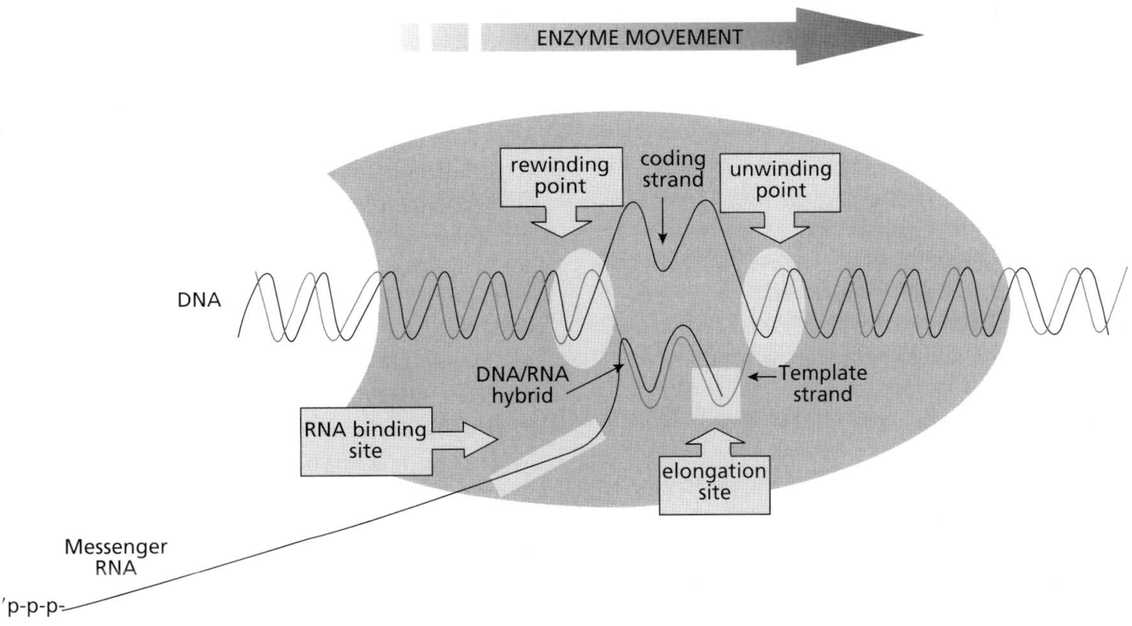

Fig. 7 Transcription elongation complex.

to dissociation and consequently special signals and factors are required for transcription termination. Within the elongation complex, the DNA is unwound for about 18 bp and the DNA/RNA polymerase complex extends to about 40 bp (Fig. 7).

Structural features within the transcribed DNA cause the RNA polymerase to pause, probably to allow translation and transcription to be coupled. NusA and NusG are both elongation factors. NusA binds to the core RNA polymerase, possibly at a location that overlaps with the sigma factor-binding site. NusA acts to prolong pausing at these pause sites and has an inhibitory effect on chain elongation that is only overcome by high concentrations of ribonucleotides. NusG appears to enhance elongation by reducing pausing.

Transcription Termination

Termination results in the release of nascent RNA from the ternary complex. It can either occur spontaneously (factor-independent or intrinsic) at specific termination sequences or is mediated by a termination protein (factor-dependent). Intrinsic termination sequences exhibit a distinct motif in the DNA template which includes a G + C-rich stem of about 20 bp followed by a run of about eight A bases. The G + C stem is highly stable and forms at the expense of the RNA/DNA hybrid within the ternary complex. The run of about eight U bases in the translated mRNA lead to low stability between DNA and RNA to further destabilise the ternary complex. In contrast rho (ρ) factor-dependent termination (Fig. 8) usually causes transcription termination only in regions of RNA that are not being translated. Rho is an RNA-dependent ATPase with RNA–DNA helicase activity and that causes nucleotide strand separation. Rho binds to RNA at G + C-rich *rut* sites, but only if they are not actively being translated. Rho moves in a 5′ to 3′ direction at the expense of ATP hydrolysis, ‘chasing’ the moving RNA polymerase. When Rho catches up with the RNA polymerase it causes it to pause at a potential termination site and then unwinds the RNA–DNA hybrid, causing the RNA polymerase to be released.

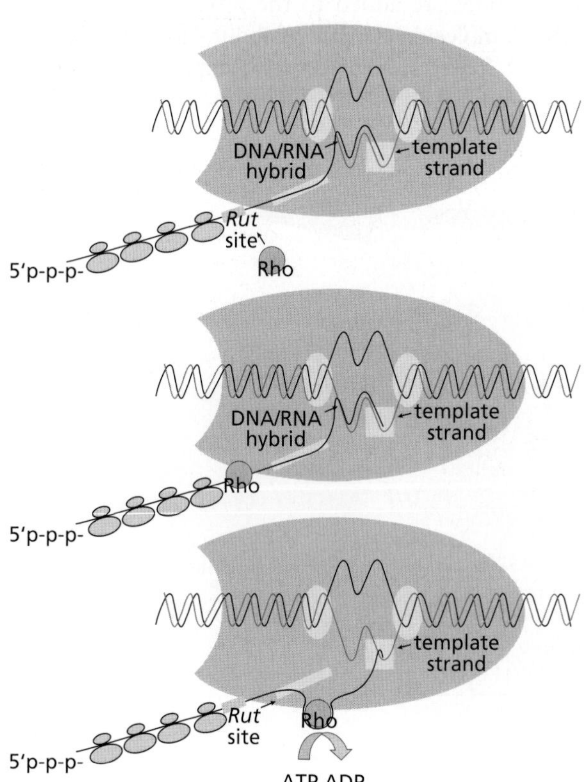

Fig. 8 Rho-dependent transcription termination.

Translation

The linear sequence of nucleotides in mRNA is decoded and translated into a linear sequence of peptide bond-linked amino acids that make up the equivalent protein. The process of translation is carried out on large ribonuclear–protein complexes called ribosomes. Because of their large size, it is possible to isolate ribosomes from the other components of a bacterial cell by ultracentrifugation. This has led to the main components of ribosomes being referred to by their sedimentation characteristics, expressed as Svedberg or S units, within a centrifugal field. Eubacterial ribosomes are 70S in size (cf. 80S in eukaryotes), but readily dissociate under appropriate conditions into subunits of 30S and 50S (Fig. 9), the small and large ribosomal subunits, respectively. The small subunit comprises a single rRNA molecule, the 16S rRNA (1.54 kb), and 21 proteins (small subunit proteins S1–S21). The large subunit comprises two rRNA molecules, the 5S (0.12 kb) and 23S (2.9 kb) rRNAs, and 31 proteins (large subunit proteins L1–L31). Except when associated with mRNA, the individual subunits exist as independent entities within the cell.

The role of the ribosome is to bind to mRNA at the start of a coding region and, by moving in a net 5′ to 3′ direction, to synthesise a polypeptide that is co-linear in sequence to that of the mRNA. The key molecule in this process is tRNA, which provides the link between the encoded genetic information and the appropriate amino acid (Fig. 10). Each tRNA molecule encodes a sequence, the anticodon, that is complementary, and

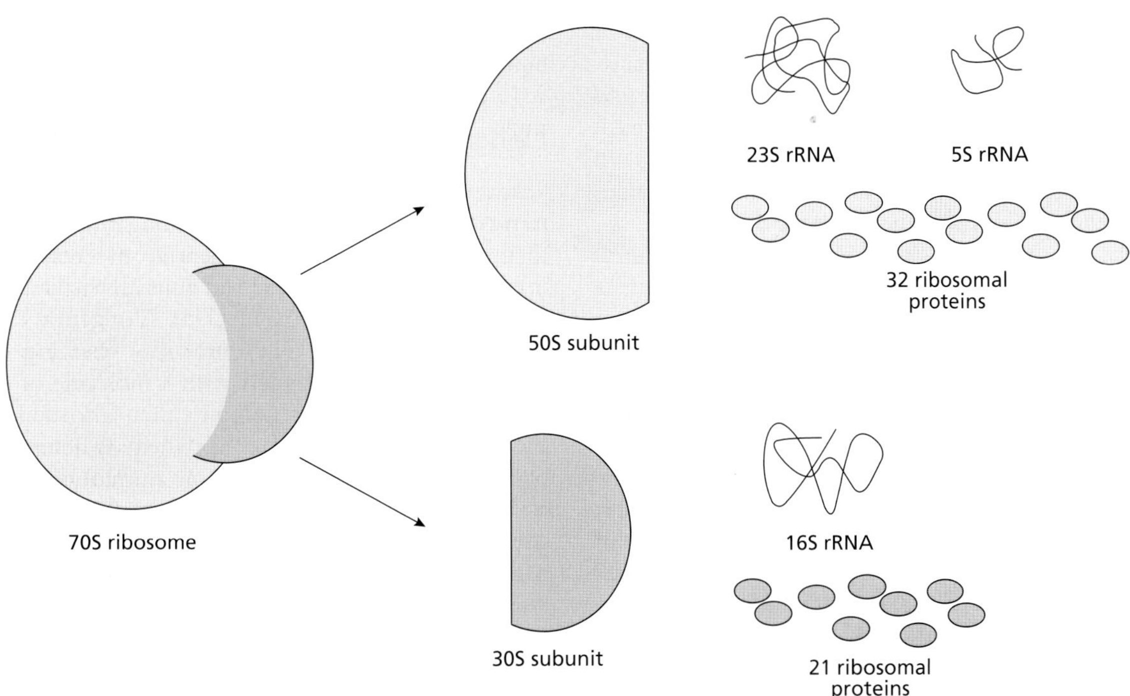

Fig. 9 Composition of bacterial ribosomes.

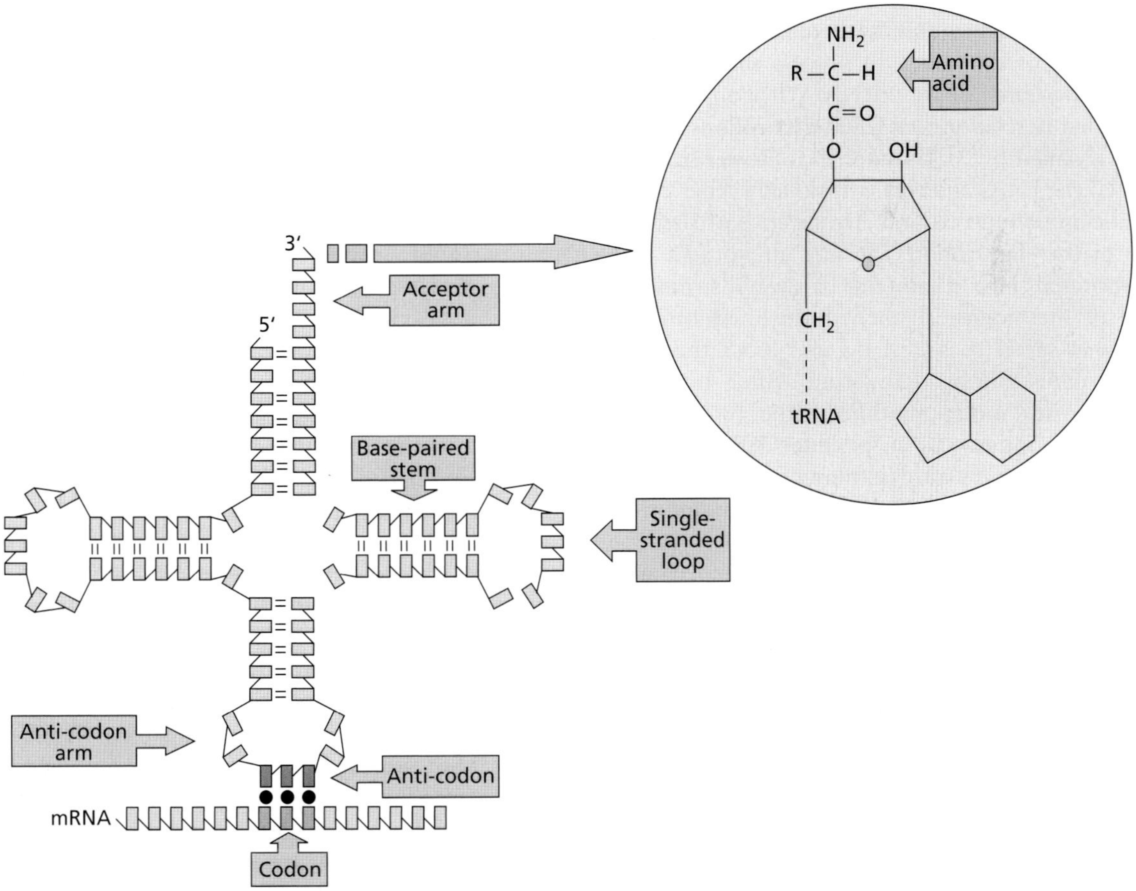

Fig. 10 Structure of transfer RNA.

can therefore bind to, a specific codon. Each type of tRNA molecule is recognised by its cognate aminoacyl-tRNA synthetase that attaches the appropriate amino acid to the 3′-OH end of the molecule. Thus each aminoacyl-tRNA molecule combines elements of transcription, the anticodon sequence, and translation, the amino acid residue attached at the 3′ end.

Like transcription, the process of translation requires various protein factors and, although a dynamic process, can be divided into three distinct phases: initiation, elongation and termination.

Initiation of Translation

Translation initiation involves the recognition, by the 30S ribosome subunit, of a translation initiation region (TIR) in the mRNA immediately upstream of the coding sequence. In polygenic mRNA, each coding region has its own TIR. Although TIRs are difficult to define with precision, they are known to include the site at which the 30S subunit binds (ribosome-binding site (RBS) or Shine–Dalgarno sequence) and the start codon. Because each amino acid is specified by three ribonucleotides (a codon), each strand of nucleic acid can potentially be translated in three different reading frames. The start codon is therefore an important component of the initiation process since it defines the reading frame of the mRNA. When appropriately positioned in relation to a RBS, start codons are normally specified by AUG and GUG. However, in rare cases CUG, UUG and very occasionally other codons can also initiate translation. Irrespective of the sequence of the start codon, the first amino acid to be incorporated into a protein is always a modified form of methionine, namely formylmethionine (f-Met) that serves to both label and block the amino (N)-terminus during translation. After translation this amino acid is often modified, either by cleavage of the formyl moiety by a peptide deformylase, or removed completely with methionine aminopeptidase.

The first stage in the initiation of translation is a specific interaction between the 30S subunit and the TIR, involving base-pairing between sequences at the 3′ end of the 16S rRNA and the RBS (**Fig. 11**). In *E. coli* the RBS consensus, 5′-AGGAGGU-3′, is located approximately 5–8 nucleotides upstream of the start codon. This interaction involves initiation factors 1 (IF1) and 3 (IF3). GTP and initiation factor 2 (IF2) now facilitate the binding of f-Met-tRNA. The initiating aminoacyl-tRNA is unique in binding to the 30S subunit at the peptidyl (P) site rather than the acylation (A) site to which all subsequent aminoacyl-tRNA molecules bind. This reaction stimulates the hydrolysis of GTP to GDP, which is released together with the three initiation factors. This allows the 50S subunit to bind to and complete the initiation complex.

Elongation Reaction

Once the ribosome has been assembled at the TIR it moves into a mode of cyclical elongation reactions that continue until the polypeptide has reached its full length (**Fig. 12**). These reactions are facilitated by elongation factors (EF) and GTP. Elongation factor Tu (EF-Tu) binds GTP and the resulting EF-Tu∼GTP complex interacts with a molecule of aminoacyl-tRNA. It is this complex that is loaded into the A-site on the ribosome. Specificity is determined by interactions between the three nucleotides of the codon at the A-site and the anticodon of the incoming aminoacyl-tRNA. Whilst base-pairing interactions between the first two bases of the codon/anticodon interaction are highly specific, specificity at the third position is more relaxed so that U in the anticodon recognises G in addition to A, and G recognises U in addition to C. This so-called ‘wobble’ reflects the degeneracy of the genetic code in which the nucleotide at the third position is often irrelevant (**Fig. 1**).

Codon–anticodon recognition stimulates a change in the conformation of the tRNA molecule that leads to the hydrolysis of GTP to GDP and the release of the EF-Tu∼GDP complex. EF-Tu is recycled with the aid of EF-Ts: GDP is replaced by GTP and the regenerated EF-Tu∼GTP complex is now ready to interact with another aminoacyl-tRNA molecule.

The release of EF-Tu∼GDP stimulates the elongation of the peptide chain. The amino acid attached to the tRNA at the P-site is linked via a peptide bond to the amino acid residue on the tRNA at the A-site. The necessary peptidyltransferase activity is a property of the 30S ribosome subunit. In the first elongation reaction, it is the initiating f-Met that is peptide-bonded to the amino acid at the A-site. In all subsequent reactions it is the growing peptide chain that is transferred to the amino acid at the A-site.

Peptidyltransferase activity is followed by the release of the tRNA from the P-site and the translocation of the ribosome along the mRNA by three nucleotides. Translocation requires a third elongation factor, EF-G and GTP, and it is the hydrolysis of this nucleotide to GDP that seems to trigger the movement of the ribosome. This results in the remaining tRNA molecule, with attached peptide chain, becoming relocated at the P-site. EF-G and GDP are then released, leaving the A-site clear for the attachment of a new EF-Tu∼GTP–aminoacyl-tRNA complex that represents the start of a new round of elongation. The binding of EF-Tu and EF-G are mutually exclusive.

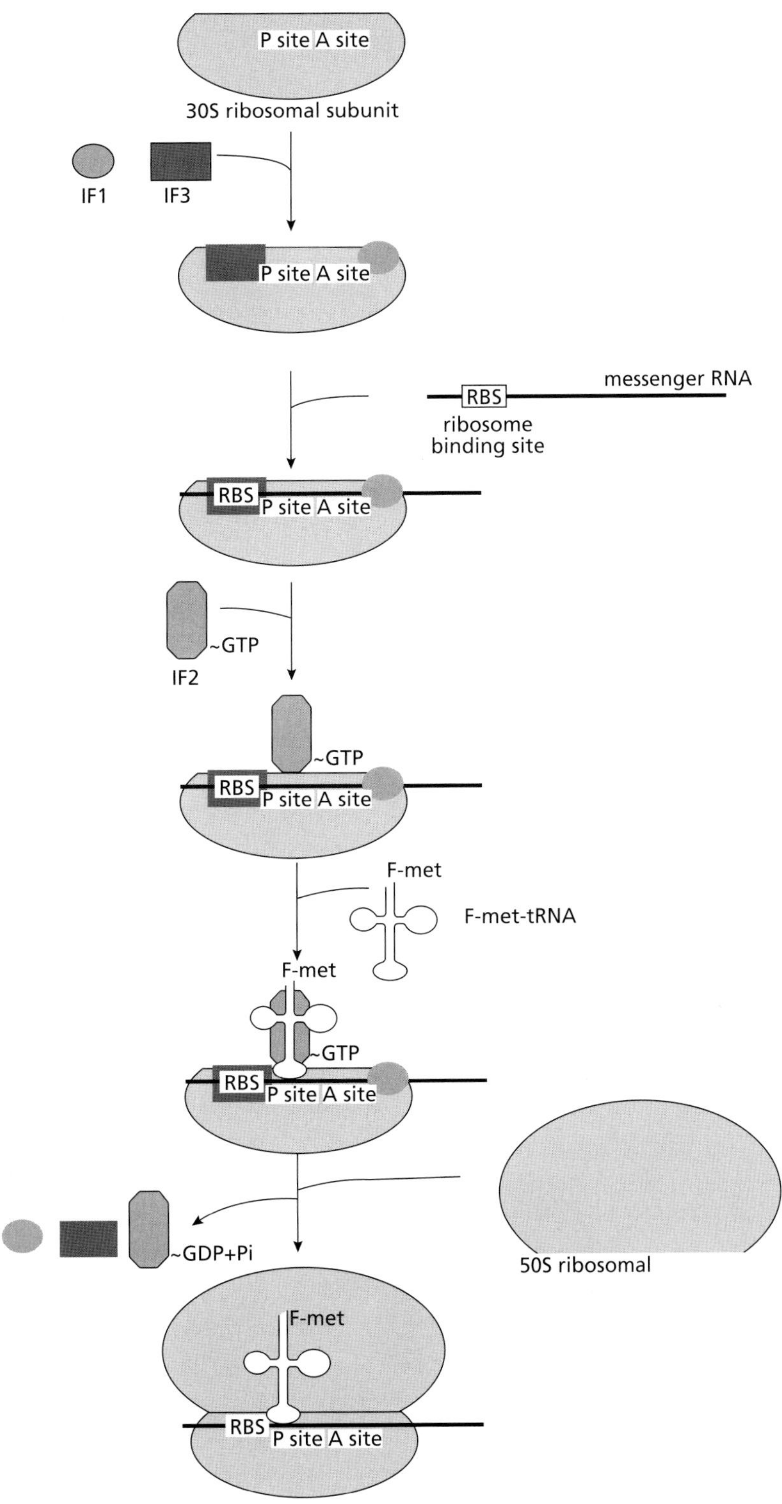

Fig. 11 Initiation of translation.

Termination of Translation

Translation is terminated when one of three stop codons (UAA, UGA or UAG) occupies the A-site on the ribosome and requires the involvement of three release factors, RF-1, RF-2 and RF-3 (**Fig. 13**). Because the termination codons are non-coding with respect to an amino acid, they do not have cognate

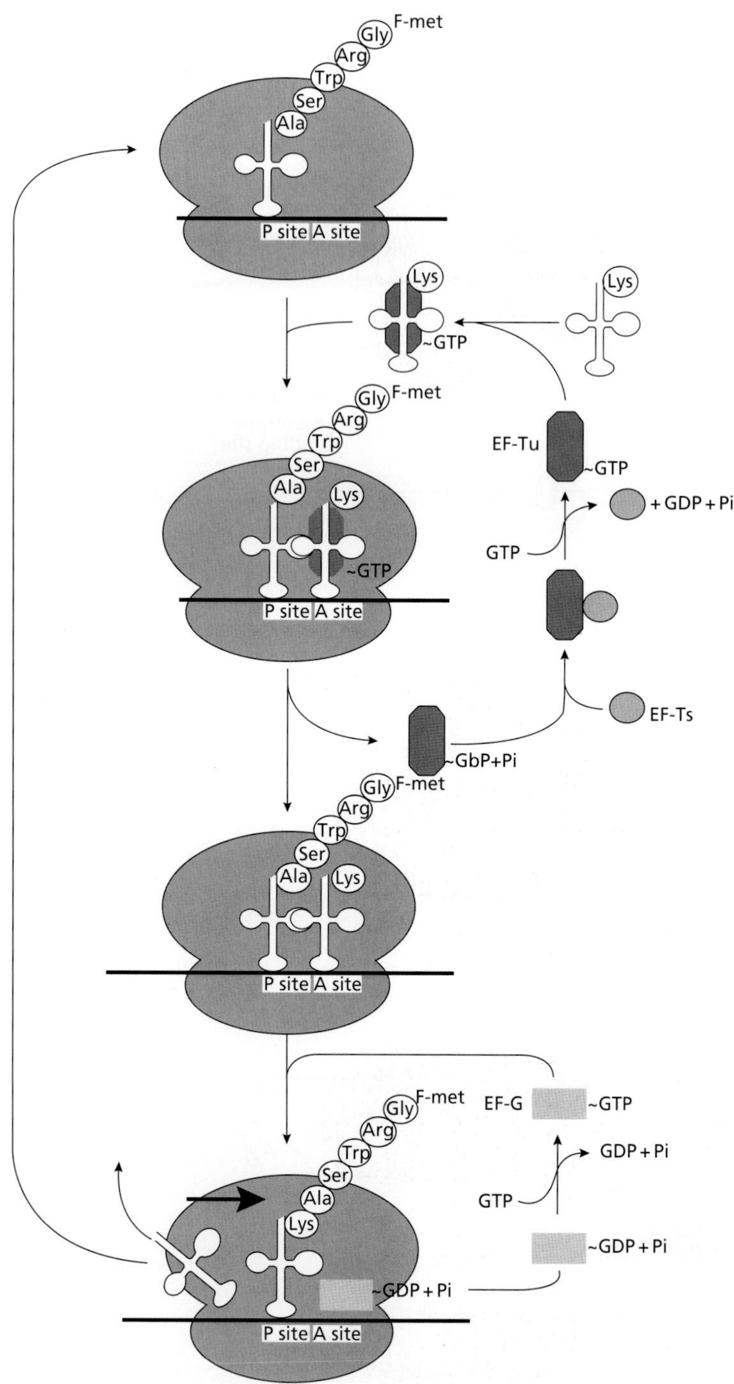

Fig. 12 Translation elongation cycle.

tRNA molecules. RF-1 and RF-2 act at the A-site and are specific for particular codons, RF-1 interacts with either UAA or UAG, while RF-2 recognises UAA or UGA. Either factor stimulates the release of the full-length polypeptide by hydrolysing the bond between the terminal amino acid residue at the C-terminus and the tRNA at the P-site. Release of RF-1 and 2 from the A-site is facilitated by RF-3 which, like EF-Tu and EF-G, is also a GTP-binding protein that appears to mimic tRNA. Finally, the ribosome recycling factor (RRF), EF-G and IF3 act to release the tRNA from the P-site and dissociate the ribosome into 30S and 50S subunits, ready for another round of translation.

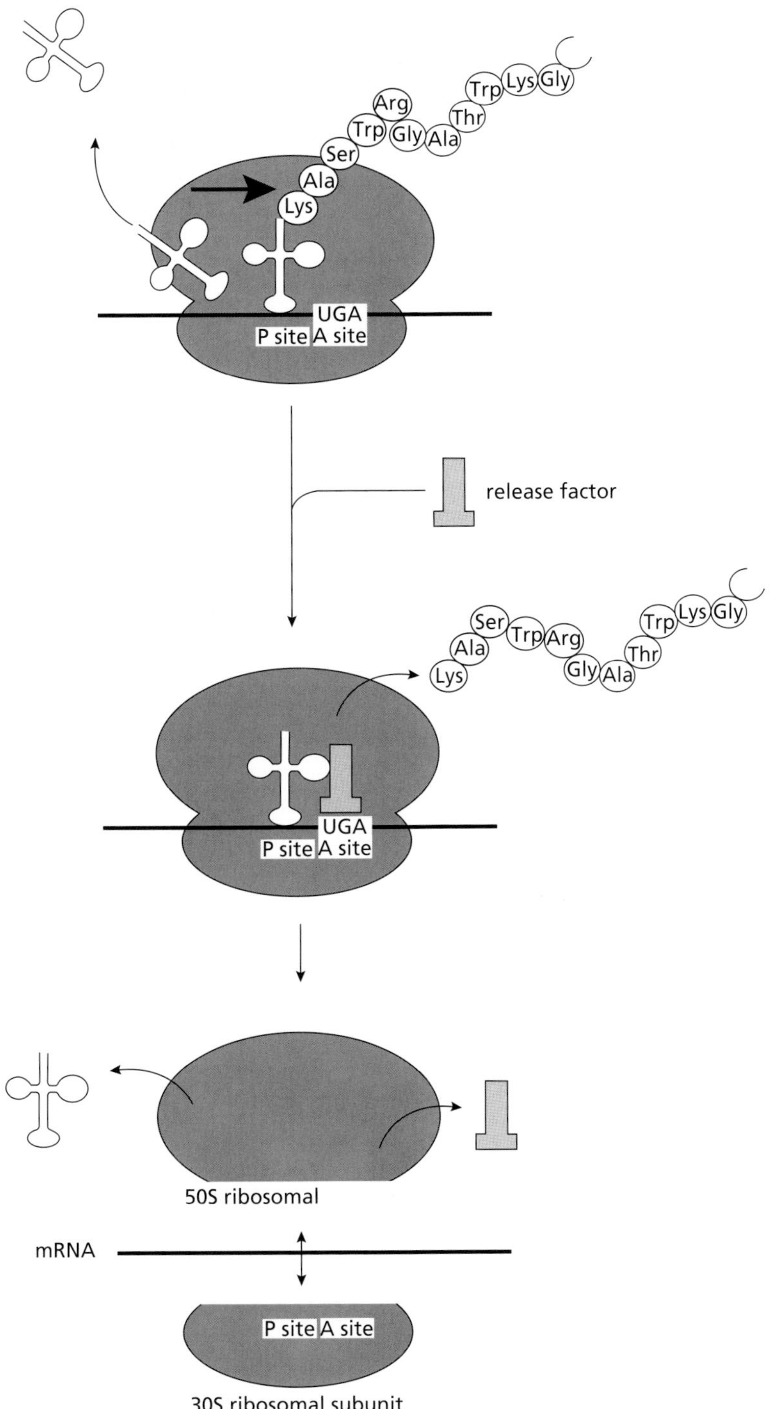

Fig. 13 Translation termination.

Processing Truncated mRNA Molecules

Prematurely truncated mRNA molecules represent a major problem for bacterial cells because the absence of a stop codon means that the normal termination mechanism is unable to operate. Additionally, truncated polypeptides could interfere with cellular functions by forming non-productive interactions. Under conditions where truncated mRNA molecules form (e.g. heat stress) a specific type of tRNA, called transfer-messenger (tm)RNA is inserted at the vacant A-site on the ribosome. tmRNA contains both a tRNA-like structure at its 5′ and 3′ ends, and an internal reading frame encoding a short 'tag' peptide and stop codon. The ribosome translates the tmRNA, facilitating the release of the nascent incomplete polypeptide by providing a stop codon, and endows its C-terminus with a specific peptide tag that directs it to specific proteases for degradation.

Translational Coupling

Occasionally, the translation of adjacent genes on a single mRNA molecule is coupled so that translational initiation of the downstream gene is dependent on the passage of a ribosome through the terminal region of the upstream gene. This happens when the TIR and associated start codon of the downstream gene is sequestrated within a stem–loop structure. Movement of the upstream ribosome as it arrives at its stop codon causes the disruption of the stem–loop, exposing the RBS for ribosome-binding.

Control of Protein Synthesis

The expression of most genes, when examined in detail, is usually found to be controlled at various levels and by a variety of control mechanisms. Some elements of control will be of more apparent significance than others, representing mechanisms of coarse and fine control. Irrespective of the type of control, or the level at which it is exercised, the basic unit at which it is applied is an operon, a gene or group of genes encoded by a single mRNA transcript, a regulon, a group of operons controlled by the same regulatory protein and a stimulon, a group of regulons controlled by the same stimulus. Gene expression can be controlled at various levels in the process of protein synthesis.

Transcriptional Control

Transcription initiation is the first stage of protein synthesis and is one of the main levels at which control is exerted. This makes sense from an energetic point of view since the prevention (or reduction) of synthesis of a particular mRNA species (i.e. that coding for a specific polypeptide or set of polypeptides) saves the very considerable energy required for mRNA synthesis. When control is at subsequent stages, such as transcription termination, mRNA stability or translation, synthesis of mRNA represents a considerable metabolic drain which must be set against the advantages of controlling protein synthesis later in the process.

The affinity of some RNA polymerase holoenzymes for sequences at the promoter is low, leading to a low frequency of transcription initiation. In certain cases the frequency of initiation can be increased by the presence of an activator protein (**Fig. 14**). Activation is

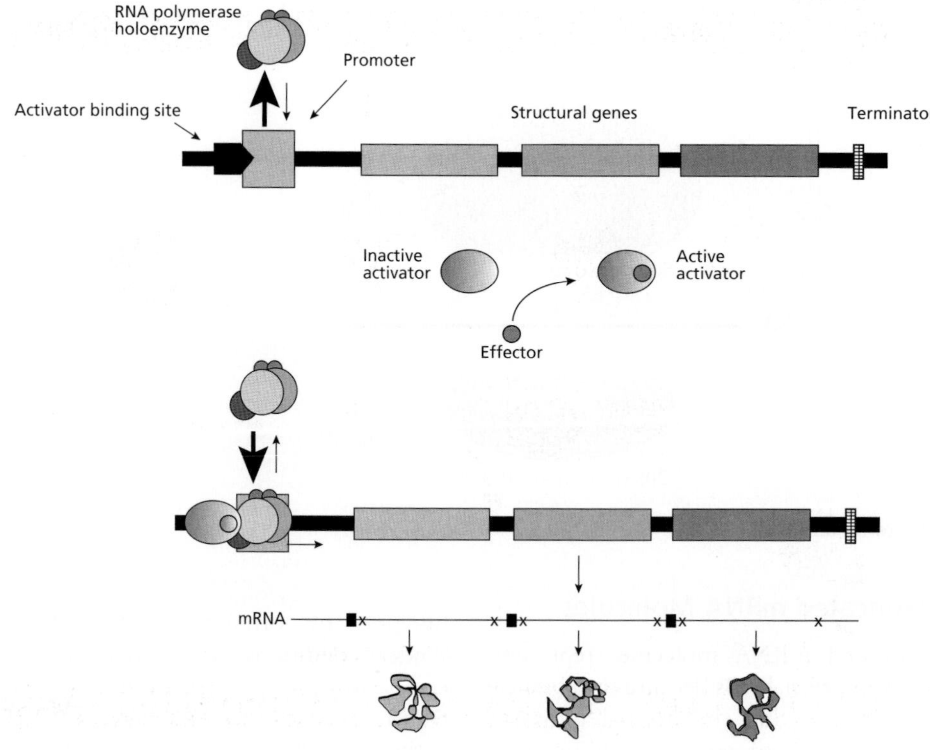

Fig. 14 Activation of transcription.

an example of positive regulation since, in its absence, little or no transcription occurs. Activators bind to an activator-binding site usually located close to the −35 region of the promoter. Activators function by increasing the affinity for RNA polymerase for the promoter and/or enhancing the rate of open complex formation. In some well-studied cases, the activator causes the DNA in this region to bend, possibly initiating localised strand separation.

Repression is a form of negative control in which a repressor protein either excludes RNA polymerase from the promoter or prevents it from moving from the promoter (Fig. 15). In most cases the repressor binds to a symmetrical site, referred to as the operator, that is usually located within or immediately downstream of the promoter. Like many proteins that bind to DNA, repressors often have a helix-turn-helix (HTH) motif that enables them to locate within the major groove of the DNA double helix. The edges of the base pairs within the operator are accessible from the major groove, exposing a characteristic chemical signature that interacts with specific amino acids within one of the α helices. The activity of the repressor is often modulated by a small molecular co-factor (the inducer) that influences its affinity for the operator, and it is the presence of this inducer that determines whether the operon is induced or repressed.

The expression of various genes has been shown to be altered in strains lacking the enzyme Dam methylase. This enzyme methylates the adenine in the sequence GATC, interfering with transcription initiation if this sequence occurs in a promoter. Because methylation occurs subsequent to DNA replication, *dam* methylation sites are transiently hemi-methylated following replication. When this occurs promoters with *dam* methylation sites become transiently derepressed, coupling the resulting gene expression to replication in a cell cycle-dependent manner.

RNA polymerase binding and subsequent open complex formation is sensitive to the level of supercoiling of the DNA in the region of the promoter. Supercoiling, either positive or negative, is the

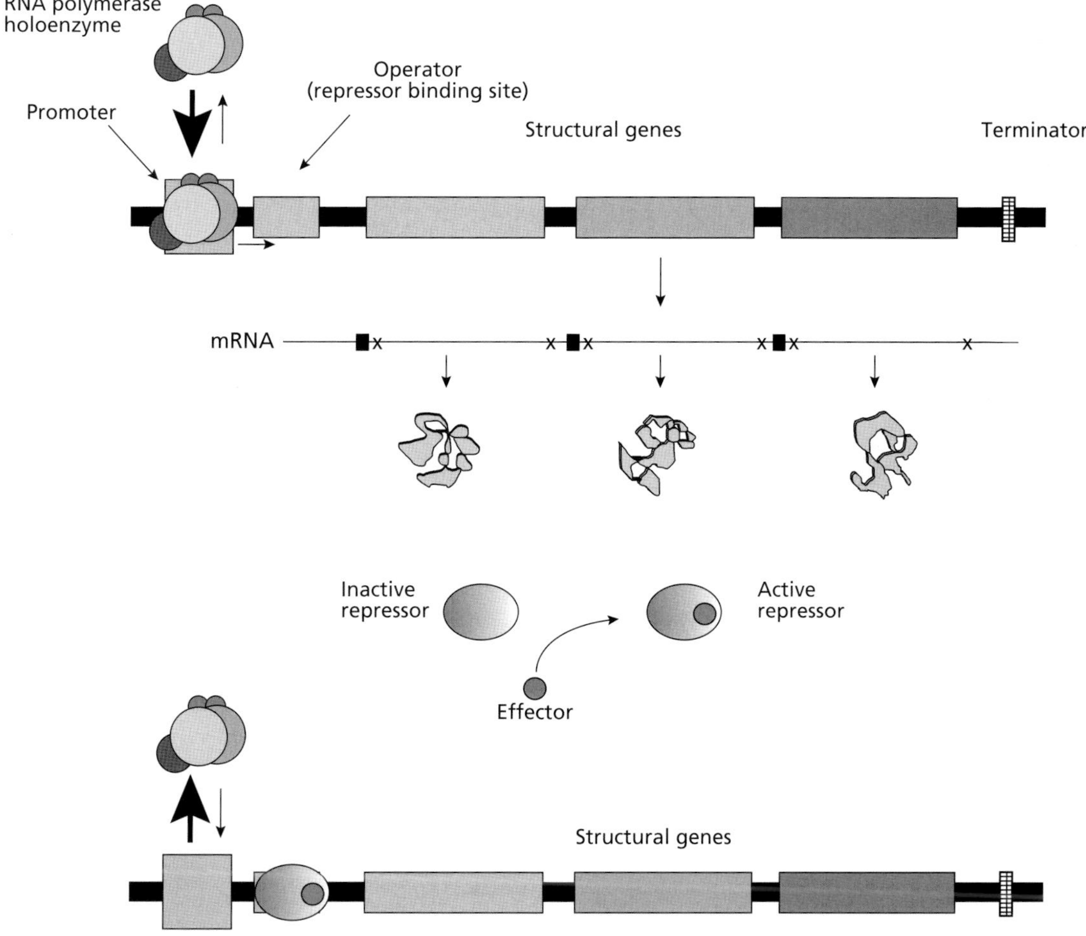

Fig. 15 Repression of transcription.

introduction of twists in the DNA over and above its normal α helicity. Supercoils are introduced into the DNA during replication and transcription, as a result of the activity of DNA gyrases. Regulation of gene expression by alterations in the level of DNA supercoiling, including that of the genes encoding the gyrases themselves, appears to play a role in the cell's response to certain environmental stimuli, including osmolarity, the availability of oxygen and temperature.

While there are no clear examples of control at the level of elongation, transcription termination has been shown to be a relatively common mechanism for controlling the expression of particular groups of genes. One such mechanism is attenuation in which only a proportion of initiated transcripts form full-length transcripts; the remainder being terminated at a site called the attenuator (**Fig. 16**). The attenuator is usually located in the leader sequence between the promoter and the first structural gene. Crucial to the functioning of attenuation is the close coupling of transcription and translation that occurs in prokaryotic cells, but which is absent in eukaryotes. In genes that are controlled by attenuation, the region between the transcription initiation site and the start codon of the first structural gene codes for a small polypeptide.

The attenuator, which is located towards one end of this coding region, contains four complementary regions (labelled 1–4) that allow stem–loop structures to form in the mRNA; 1 : 2 forms a transcriptional stall signal, 2 : 3 an antiterminator and 3 : 4 a rho-independent transcription terminator. The formation of the 2 : 3 loop precludes the formation of the 3 : 4 loop and allows the transcripts to read through the attenuator region. All transcripts initiated at the promoter form the 1 : 2 transcriptional pause signal. With the RNA polymerase paused, a ribosome starts to translate the leader mRNA. As the ribosome approaches the paused RNA polymerase, the latter is released from its pause site and transcription continues; transcription and translation are now intimately coupled. In the presence of a plentiful supply of amino acids the ribosomes keep pace with the synthesis of transcript. The presence of a ribosome towards the end of the leader mRNA prevents the formation of the antiterminator (2 : 3) and instead the termination structure (3 : 4) is formed and transcription is terminated. When the cells are starved for a particular amino acid, its cognate aminoacyl-tRNA is in limited supply and the ribosome stalls at adjacent codons specifying the amino acid in question. This leads to the decoupling of transcription and translation. The antitermination structure (2 : 3) forms, transcription termination within the attenuator is prevented and instead the transcript continues to the terminator at the end of the operon.

Transcription termination can also be prevented by specific antitermination factors. These interact with RNA polymerase, causing it to ignore termination sequences. Antitermination therefore allows conditional expression of contiguous genes from the same promoter, particularly in developmental pathways when genes are required to be expressed sequentially.

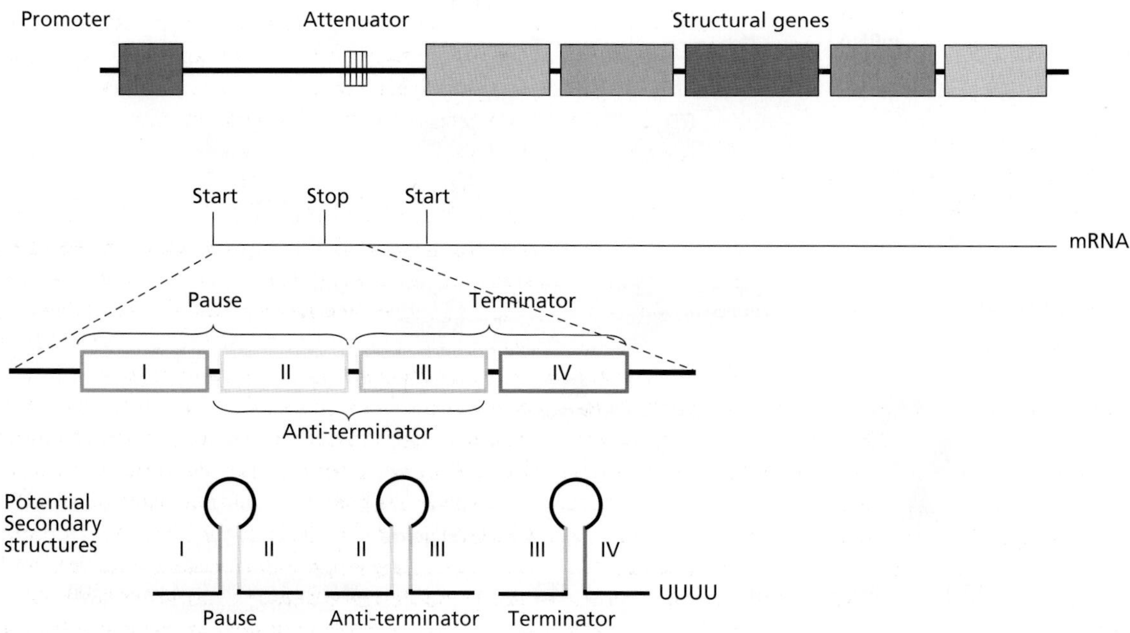

Fig. 16 Attenuation.

Translational Control

In recent years a number of examples of post-transcriptional control have been identified and characterised. Although energetically unfavourable, in that it requires the continuous synthesis of mRNA, post-transcriptional has the advantage of permitting the cells to respond rapidly to environmental changes. This could be particularly advantageous if the change results in chemical insult to the cell, and it is significant that some of the best-characterised examples of translational control are for genes encoding antibiotic resistance.

Not all molecules that block the initiation of translation are proteins, and there are now several examples of small RNA molecules regulating gene expression. Like protein repressors, these molecules are independently synthesised and diffuse to the target site, usually mRNA, to which they anneal by complementary base-pairing. Antisense RNAs interfere with their target site activity by blocking its interaction with another molecule or by preventing it from forming functional secondary structures. Antisense RNA has been shown to control the replication of various plasmids by blocking the synthesis of the enzyme responsible for initiating replication. In other examples, the antisense RNA is complementary to the region of the mRNA containing the TIR and start codon. Binding therefore prevents translation initiation by excluding the 30S ribosome subunit from the RBS.

Post-translational Modification

There are now many examples of post-translational changes to proteins that lead to an alteration in their activity or activation state; these are often the result of specific proteolytic cleavage or modification by, for example, the addition of phosphoryl or methyl residues. In many cases the proteins involved influence the synthesis of other gene products.

Transcription and Translation as Targets for Antimicrobials

Differences in the structure and activities of the prokaryotic and eukaryotic RNA polymerase and ribosomes have led to them being targeted by a variety of antimicrobial agents. Whilst they include many antibiotics that are effective against disease-causing bacteria, elements of similarity still exist and these antibiotics tend to be more toxic to animals than those targeted to structures, such as peptidoglycan, that are unique to bacterial cells.

Inhibitors of Transcription

Rifampicin binds to the β subunit of RNA polymerase where it blocks the initiation of transcription. Once the synthesis of a transcript has been initiated, rifampicin has no inhibitory effect. In contrast, streptolydigin, which also binds to the β subunit of RNA polymerase, inhibits the elongation reaction. Actinomycin D blocks transcription by binding directly to DNA. The toxicity of this antibiotic precludes its use as an antimicrobial agent, but it is used as an antitumour agent.

Inhibitors of Translation

Bacterial ribosomes are the target of a number of antibiotic that inhibit the various active centres of this ribonuclear–protein complex. Streptomycin and aminoglycosides bind to the 30S subunit, possibly directly to the 16S rRNA. Streptomycin distorts the A-site and leads to an increase in the level of misreading of pyrimidine bases in the codon. Puromycin mimics part of the aminoacyl-tRNA in which an aromatic amino acid is linked to a nucleotide. It therefore causes premature termination of translation. Fusidic acid stabilises the ribosome-EF-G~GDP complex, preventing its release and subsequent incorporation of the next aminoacyl-tRNA at the A-site. Chloramphenicol prevents the formation of peptide bonds by inhibiting the peptidyltransferase activity of ribosomes. Tetracycline inhibits the binding of aminoacyl-tRNA at the A-site.

Final Comments

Protein synthesis in prokaryotes resembles that in eukaryotes. However, differences in the mechanics of transcription and translation are sufficient to permit these processes to be used as targets for chemicals that preferentially or exclusively inhibit the growth of bacteria. An important component in the survival and dissemination of free-living bacteria is their ability to match the synthesis of their proteins to the demands of the environment. This is achieved by regulatory networks that control gene expression at a variety of levels. This requires many sensory and regulatory proteins that monitor and respond to physical and chemical changes in the external and intracellular environments. The result is an overlapping network of control that responds exquisitely by appropriate and efficient adjustments to the cell's metabolism and structure.

A typical bacterium with a genome size of 4 million base-pairs encodes about 4000 proteins. By use of a combination of experimental evidence and homology to proteins in other organisms, the functions of a little over half of these proteins have been identified. The identification of the biological role of proteins currently of unknown function and the identification of the complete set of proteins expressed at any point in time and under a particular environment remains a challenge for the immediate future. The term proteome has recently been used to describe the proteins specified by the genome, and the characterisation of the proteome provides a link between the organism's genetics and physiology. It also helps to validate the genome sequence, aids the identification of operons, regulons and stimulons and allows the response of the organism to its environment to be evaluated. The current system of choice for the analysis of bacterial proteomes is a combination of two-dimensional gel electrophoresis combined with polypeptide microsequencing or mass spectrometry techniques.

The genes being transcribed by an organism at any particular point in time are referred to as the transcriptome. The ability to identify which genes are expressed under a particular set of environmental conditions contributes to an understanding of the processes involved in controlling cellular physiology and may contribute to the identification of the activity of genes of unknown function. The availability of complete microbial genome sequences, together with advances in microfabrication technology, has led to the development of technology which facilitates the global analyses of microbial transcriptional patterns. Oligonucleotide hybridisation probes or PCR products for all the ORFs on the genome are bound in ordered arrays to glass or nylon supports. Miniaturisation of the process, using techniques developed for the construction of electronic microchips, means that oligonucleotides can be synthesised *in situ* and arrays in excess of 100 000 probes can be attached to a single 'DNA chip'. Whole-cell mRNA can then be labelled and their presence detected by hybridising to the individual probes that are bound to the chips. Array technology can be used for a number of applications, including simultaneously monitoring the expression of every gene in a genome, the detection of polymorphisms and mutations, speciation and the diagnosis of diseases.

Further Reading

Cotter PA, DiRita VJ (2000) Bacterial virulence gene regulation: an evolutionary perspective. *Annu. Rev. Microbiol.* 54: 519–565.

Lewin B (2000) *Genes VII*. Oxford: Oxford University Press.

Snyder L, Champness W (1997) *Molecular Genetics of Bacteria*. Washington, DC: American Society for Microbiology.

PART 3 GENETICS

16

Bacterial Genomics in the Study of Virulence

Antoine Danchin[1,2] and Kwok Yung Yuen[1,3]

[1]Hong Kong University Research Centre, Pokfulam, Hong Kong and
[2]Institut Pasteur, Paris, France
[3]Queen Mary's Hospital, Hong Kong University, Hong Kong

Steady progress in the technique of DNA sequencing, molecular genetics, and computer sciences and technology has allowed biologists to create a complete inventory of the basic components of cells. The most important part of this inventory is represented in the metaphorical form of a text written with a four-letter alphabet – the complete genome sequence of an organism. The first published genome sequences were eagerly awaited. Some claimed that knowledge of the genome texts could change our vision of the genetic information contained in the bacterial chromosome and, accordingly, would help us to understand the biological functions of living organisms, including the most elusive functions such as virulence and pathogenicity. Although this hope cannot be reconciled with the very nature of biology, which does not proceed from structure – and even less from sequence – to function (Danchin, 1999), it is nevertheless clear that knowledge of genome sequences is a revolution, not only in the way life is understood, but also in the way experiments are conceived and designed to yield further understanding. Dealing with pathogens today is dramatically different from what it was 20 years ago, but this does not mean that we have won the fight against disease. We have to find ways to put together the kind of knowledge obtained from the study of pathogens in their natural environments, i.e. their hosts, and what can be learned from the genome sequence. Fortunately, 6 years after the first complete genome sequence was deciphered (Fleischmann *et al.*, 1995), with the number of fully sequenced microbial genomes approaching 100, we have not been disappointed. New information continues to accumulate. Each new genome, however, contains a riddle; a large proportion of the genetic information is completely unknown. Indeed, even in non-pathogenic model bacteria, such as *Escherichia coli* and *Bacillus subtilis*, which have been subjected to thousands of different experimental studies, more than 30% of genes have no attributed function (Blattner *et al.*, 1997; Kunst *et al.*, 1997), and for genome regions involved in pathogenesis, this figure may be as high as 80%. How can the riddle be solved, and into which categories can we place the corresponding functions? In this chapter we review some of the facts and fictions that tell us how important it is to know and study genomes for an understanding of virulence.

Several dozens of genome sequences are now known, generally from bacteria, but also from eukaryotic cells, such as baker's yeast (Goffeau *et al.*, 1996)

 doi:10.1006/bkmm.2001.0016

and even from multicellular organisms such as the nematode worm *Caenorhabditis elegans* (Wilson, 1999). The genome of *Drosophila melanogaster* has been made public, and after the publication of a draft sequence in 2000 we expect to have deciphered most of the *Homo sapiens* genome by the year 2003. Genome sequences have accumulated rapidly since those of *Haemophilus influenzae* and *Mycoplasma genitalium* were deciphered (Fleischmann *et al.*, 1995; Fraser *et al.*, 1995). It should be remembered, however, that just as the map of the sky is not cosmology, the genome text does not constitute genomics. With the genome text as a reference, the aim of this new sector of biology is nothing less than to understand, with its deepest implications, what life is in all of its manifestations.

In order to understand what follows, we must stress that, perhaps counter-intuitively, biology is extremely abstract; it is the study not of a collection of its elements, but rather of their relationships (Danchin, 1998). Three thousand years ago people used to go to the temple of Delphi, in Greece, to ask its oracle about their future. In her turn, she was in the habit of questioning passers-by. One enigma was put as follows:

> I have a wooden boat. As time passes the planks that make up the boat rot one after another, and are replaced by new ones. After some time, none of the original planks remains: is it the same boat?

Of course, the owner would say it is, and we would probably agree with him. The boat is not limited to the matter of which it is made. It is also something else, in that it has perenniality, which informs the matter of the planks about the relationships established between the planks. In a similar way, in order to understand life, is it possible to consider only the objects that constitute living organisms? If this were so, the identity of an organism would not be permanent, because its constituents are in constant flux. As in the case of the boat, where the ordered architecture of its constituent parts allows it to float, we must consider that the organised relationships between the various objects of a cell give it life. Note that the metaphor of the boat used here is different from the alphabetic metaphor of the genome text. Indeed, it may be helpful, when in this frame of mind, to think of life as if we were reading texts written in Chinese characters – pictograms, rather than alphanumeric symbols. The new knowledge of genome sequences guides us to understand relationships between biological entities (genes, gene products, hyperstructures, cells, etc.), in a way that would have been quite unthinkable 20 years ago. Not only does it allow gene manipulation (in the laboratory and also for industrial purposes), but above all it allows us to understand better how the cell's architecture and dynamics are constituted and related to the abstract nature of the genome text. This abstract property of life must therefore be analysed at a highly conceptual level (Danchin, 1996). In addition, the enormous amount of data provided by the sequencing projects makes manual analysis of sequences unworkable as a standard mode of genome analysis. In recent years, computational biology – biology *in silico* (Danchin *et al.*, 1991) – has therefore been a necessary complement to biology *in vivo*. We next review the general properties of *in silico* genome analysis to see how this new type of investigation can be used to understand pathogens.

Rationale for *In silico* Genome Analysis

DNA specifies a programme for the construction of cells, but there is no reason explicitly to link the abstract nature of this programme with some physical reality of material objects. This explains why our knowledge has been distributed into two entirely separate views of life: that of abstract and symbolic genetics, and that which is concrete and accessible to the experiments of biochemistry or physiology. The very idea of a 'symbol' postulates a separation between a sign and what it represents. In other words, in the case of genomes it states that in general the sequence *alone* cannot give access to the function. This is in conflict with a widespread notion, based on the generally mistaken idea that structure is informative of function; in fact, a structure becomes interesting only when the function it specifies is understood. How, then, is a function created? Evolution proceeds through steps involving association and dissociation of interactions, using only the concrete material objects which are available at a given time. This explains why, when the genome of extant organisms is deciphered, it is necessary to guess the gene functions in much the same way as one would try to guess the functions of artefacts created by a handyman. One might be tempted to extrapolate from other sciences, such as physics, but the concept of function in this sense is not used in physics, because it ascribes a purpose to structures or dynamics. Therefore, before we can understand a function the various relationships linking the relevant biological objects must first be explicitly described. In order to understand gene functions, therefore, it is necessary to create a conceptual and technical approach, in particular in bioinformatics for *in silico* analysis of genomes, which differs widely from computer science approaches such as those used in the management of bank accounts, or for the computation of a three-dimensional structure from a set of X-ray diffraction patterns. In particular, if the aim is to interfere with the growth of pathogenic bacteria, it is essential to identify *processes* that are necessary for the expression of virulence.

There are not many ways of carrying out this task, because a function cannot be deduced from the simple existence of biological entities. As in the case of a medical diagnosis, we must act as Sherlock Holmes does by inferring the nature of a disease from circumstantial evidence. To this end, all kinds of knowledge must be combined, and extrapolated from what is known about them. When considering genes, evolutionary kinships (similarities between genes and gene products) can be uncovered. Similarly, all kinds of gene 'neighbours' in the 'geographical' sense must be explored. These include, for example, genes located next to each other in the sequence, proteins that are part of a multi-component system, and objects located in a common cell compartment. Of course, in the case of pathogenic bacteria, the similarities between the signs of the disease they cause can be identified and compared to those induced by known models. For example, oedema may be the result of several processes, among which is the uncontrolled synthesis of cyclic AMP (cAMP) (Warren *et al.*, 1993), which suggests that the genome may specify gene products that alter host cAMP synthesis or even synthesise toxic adenylate cyclases.

Finally, the spread of knowledge in the scientific community, as expressed in the literature, must be used. For example, knowing that genes function together may lead to the extremely important discovery of a new function (Nitschké *et al.*, 1998). There is no limit to this quest for neighbours, because this is the only way in which to reason inductively and to make discoveries. The existence of pathogenicity islands (PAIs), which is discussed below, is a concrete illustration of the importance of the concept of neighbourhood in the context of the study of bacterial virulence.

A word of caution is necessary. This general approach is very new and difficult to implement in practice. It requires the investigation of an enormous variety of knowledge. The concepts of computer science, which are necessary for this type of work, are used to uncover the symbolic nature of the genome text by relating it, by way of experimental validation *in vivo* and *in vitro*, to the various functions of a living cell. Thus, the study of genome texts involves mathematics, statistics and computer science. Well-integrated with biological knowledge, it must always be validated by crucial experiments *in vitro* and *in vivo*, in particular with reverse genetics, a strategy in which genes are modified and replaced *in situ* by homologous DNA recombination. Finally, all this scattered and diverse knowledge must be organised; this requires considerable work to construct user-friendly specialised databases (Moszer, 1998; Rudd, 2000). In the case of pathogenic bacteria, an efficient way to proceed is to compare the genome of the pathogen with that of well-studied models: this is known as subtractive genomics.

Subtractive Genomics

In practice, the functional exploration of genomes should be undertaken in the following methodological order. The organisms of interest must first be identified, and their 'basic', 'essential' or 'characteristic' behaviour identified: for example, whether the organism is virulent or non-pathogenic, whether it grows in acidic or alkaline media, its utilisation of the different oxidation states of sulphur and iron, light production, and so on. Once the genome sequences have been deciphered, their DNA sequence is analysed *in silico*. More and more programs have been developed for sequence analysis by very diverse methods; for example, programs such as BLAST (Altschul *et al.*, 1997) or FASTA (Pearson, 2000) for similarity searches. At this point, genomes are grouped by similarity. For example, Enterobacteriaceae and Vibrionaceae go together; other groups are the Bacillaceae, the mycobacteria, and the *Mycoplasma* group. In each group one distinguishes between the model organism and the pathogen of interest. For example, one could study the model *Escherichia coli* K12 in parallel with *E. coli* O157:H7, or *B. subtilis* in parallel with *B. anthracis*.

It then becomes interesting to compare what is conserved in the genomes, and what is divergent. It has repeatedly been observed that, in related bacteria, genome sequence similarities vary widely throughout the genome text. Regions that are highly conserved are interspersed with regions that are highly variable in sequence and even in length. This allows the identification of regions that have been maintained as a result of selection pressures common to the organisms being compared. This helps in turn to identify control regions, promoters, coding sequences, etc. (see for example Trotot *et al.*, 1996). In the strategy of subtractive genomics, the subsequent strategy is to modify the regions of interest by *in situ* recombination, often by deletion, insertion or gene fusion with reporters such as the *lacZ* or *cat* gene, in such a way as that we can then propose functions for these signals or genes. This *reverse genetics* step makes it possible to compare normal cell function with that of the mutant, as a follow-up to the study of the expression products of the genes of the wild-type organism in parallel with those of the modified mutant. For example, this can be carried out by two-dimensional electrophoresis, a process of characterising proteins according to their mass and electrical charge – a class of biochemical techniques that has recently been termed 'proteomics' (Wasinger *et al.*, 1995). The identification of the predicted catalytic activities or the characterisation, with specific antibodies, of the cell localisation of a protein complex allows further substantiation of functional hypotheses, as a complement to optical or electron microscopy, as well as other techniques based on physics.

An Illustration: Functional Islands in Genomes

Bacterial pathogens use a number of strategies to cause infection and, occasionally, cause disease in their hosts. They express virulence factors which, together with their regulatory elements, can be divided into a number of common themes based on the conservation of similar mechanisms that are found throughout bacterial genera. Virulence is expressed by a sequence of several processes: escape from the immune system of the host; adhesion; cell invasion; expression of toxins (see for example Cossart and Lecuit, 1998). In spite of a large number of host targets, there are only a few general types of toxins, many of which interfere with cAMP synthesis or degradation. Invasion of host cells by bacteria is a complex process, but there are several common invasion themes that are used by a variety of micro-organisms. Similarly, the theme of cell adhesion is expressed in only a few conserved ways, by the building of bacterial pili and fimbrial adhesins. Once inside a host cell, pathogens seem to have only a limited number of ways to ensure their survival, whether they remain within a host vacuole or escape into the cytoplasm. Avoidance of the host's immune defences is the key to a pathogen's success. Again there are several common themes, including antigenic variation, camouflage by binding host molecules and enzymatic degradation of components of the host defence.

Most virulence factors are found at the surface of bacteria or are secreted into their immediate environment. They operate through any of the known bacterial secretion systems. Some secretion systems, such as type III (Lee and Schneewind, 1999) or RTX ABC-permeases, are often markers of pathogenicity (Coote, 1992). In spite of the fact that bacterial pathogenicity is controlled by complex regulatory circuits, pathogens use only a small number of biochemical families of proteins to express distinct functional factors at the appropriate time during infection. It is remarkable that genes for virulence factors are often maintained on mobile genetic elements. Indeed, it has generally been observed that these genes are clustered in the chromosomes of the virulent strains or carried by bacteriophages (see for example Waldor and Mekalanos, 1996).

The term PAI has been used to refer to large chromosomal regions in pathogenic bacteria which encode clusters of virulence genes. The study of uropathogenic *E. coli* strain 536 (O6:K15:H31) has shown that it carries two unstable DNA regions on its chromosome which encode two haemolysins and adhesion-related fimbriae (see also Chapter 9). Interestingly, both PAIs appear to be incorporated into tRNA-specific loci, one into the tRNA gene for selenocysteine (*selC*), and the other into the leucine-specific tRNA locus *leuX* (Blum *et al.*, 1994). Instability of the PAIs was demonstrated when it was found that a mutant strain had lost both of them, which appeared to have been deleted together with the intact tRNA genes. Although the parent was virulent, the mutant had lost a number of properties, including virulence. Interestingly, the expression of other gene products that affect metabolic properties in addition to virulence was shown to depend on the presence of the intact tRNA loci *selC* and *leuX*. It was shown that the selenocysteine-specific tRNA ($tRNA^{SeCys}$) directly influenced the ability of the bacteria to grow under anaerobic conditions. This was easily accounted for, because selenocysteine is part of formate dehydrogenase (FDH), which is required for mixed acid fermentation. The rare leucine-specific $tRNA_5^{Leu}$ affected a number of properties, including fimbria production, flagella production and motility, iron scavenging and serum resistance. $tRNA^{SeCys}$ was directly involved in the production of FDH, but *leuX*-specific $tRNA_5^{Leu}$ appeared to be necessary for survival *in vivo* (Dobrindt *et al.*, 1998).

PAIs can confer complex virulence phenotypes that often combine several discrete functions. A case in point is the cyclolysin (toxic adenylate cyclase) of *Bordetella pertussis*, which combines the toxic action of a calmodulin-activated adenylate cyclase with the Gram-negative haemolysin RTX secretion–invasion system (Glaser *et al.*, 1988) (see Chapter 75). The origin of PAIs is, therefore, one of the most interesting questions to touch on the driving forces that led to the evolution of bacteria.

Since the first observation of Médigue and co-workers in the non-pathogenic *E. coli* K12, it has repeatedly been found that genes present in the bacterial chromosome could be divided into three classes, according to their bias in using the genetic code

(Médigue *et al.*, 1991; Lawrence and Ochman, 1998). One such class is clearly associated with horizontal gene transfer; in pathogenic strains, for example those that express the haemolysin toxin, the genes for the toxin are members of this class (Boyd and Hartl, 1998). This indicates that many virulence genes can be acquired by bacteria from unrelated organisms, leading to interesting hypotheses about how bacterial pathogens have evolved. It is likely that mechanisms that generate PAIs operate continuously, and this certainly contributes to the emergence of bacterial pathogens with new virulence properties. The combinatorial nature of PAIs ensures that new strains of pathogens constantly evolve. *Helicobacter pylori* provides a good example (see Chapter 64). About half the world's human population carries this Gram-negative spiral bacterium, which colonises the stomach. The link between *H. pylori* and ulceration, as well as its association with the development of both gastric cancer and lymphoma of mucosa-associated lymphoid tissue, is a serious public health problem. Strains of *H. pylori* from patients with peptic ulcer or gastric cancer contain a 40-kb fragment of DNA that is not present in isolates from carriers with asymptomatic infections. The discovery of this chronic active gastritis (CAG) PAI suggests that virulence has evolved in quantum leaps by the acquisition of one or more DNA insertions. Understanding of these common themes in microbial pathogenicity is critical for the understanding of bacterial virulence and is necessary for the development of the new antivirulence agents that are needed to replace current antibiotics (Axon, 1999).

In the last decade of the twentieth century, a new theme emerged in the field of bacterial virulence. In a variety of Gram-negative organisms, complex secretion processes (type II, type III and type IV secretion systems) phylogenetically related to pili formation systems have been found (Lee and Schneewind, 1999; Nunn, 1999; Wall and Kaiser, 1999). These systems, the genes of which are generally located in PAIs, often specialise in the export of virulence factors for delivery directly to host cells. These factors contribute to invasion by subverting normal host cell functions in ways that are beneficial to the invading bacteria. The genes are inserted DNA segments that confer a variety of virulence features on the host bacterium, such as the ability to scavenge iron from the environment (Vasil and Ochsner, 1999) and to adhere to or enter host cells (Galan and Collmer, 1999). Many of these DNA segments appear to have been acquired in a single step by lateral transfer from a foreign source. This remarkable observation provides an interesting explanation for the saltatory rather than continuous nature of bacterial evolution. Since virulence-related genes are located in a single continuous DNA segment, the ability to acquire complex virulence characteristics in a single genetic event, rather than by natural selection over many generations, provides a mechanism for sudden radical changes in bacterium–host interactions. This controversial generalisation may account for many of the concerted changes that appear to confer adaptive properties on living organisms. The modular nature of genes associated with the construction of complex architectures is an efficient means for the transmission of adaptive properties in a single event. Type III secretion systems and PAIs must have played critical roles in the evolution of known pathogens (Hueck, 1998) and are very likely to lead to the emergence of novel infectious diseases in the future.

What is the Origin of Gene Islands?

If genes are grouped together in genomes, and if this is not simply the result of chance, it is important to try to understand the origin of the selection pressures that drive the genes into clusters. In general, legitimate and illegitimate recombination processes, transposition of insertion sequences (ISs), and insertion of prophages all tend to disrupt gene organisation in genomes. Finding genes remaining clustered together in the course of evolution indicates that they are subject to a significant selection pressure for clustering. Cooperation for a function such as virulence probably plays the major role, when horizontal gene transfer occurs at a high rate. Indeed, in this case, any disruption of the cluster will generally result in the lateral transfer of only part of the genes needed to ensure virulence. This will restrict the ecological niche of the organism, and, generally, be deleterious for the corresponding species (but see below). Do we have indications of the corresponding transfer rates and rearrangement rates in nature? The study of base composition in genomes provides us with some estimates. In general, bacteria relevant to medicine display a particular bias in base composition, centred at the origin of replication. The leading strands are G + T-rich, whereas the lagging strands are A + C rich (Rocha *et al.*, 1999c). This is true in the non-pathogen *B. subtilis* in which, apart from occasional insertion of prophages, the stability of the genome appears to be high, but it is also true for *E. coli* where ISs move rapidly and where conjugation results in the shuffling of many sequences. This indicates that there must be a further selection pressure, which tends to maintain the general organisation of the chromosome. And, indeed, Louarn and co-workers have found that the simple inversion of genes on one side of the origin of replication impairs growth, whereas inversion centred on the origin of replication, such as that in one strain of *E. coli* (W3110) as compared with reference strain MG1655, has no apparent effect (Rebollo *et al.*, 1988; Francois *et al.*, 1990).

What may be the nature of this pressure? We have seen that replication maintains a strong difference between the leading and the lagging strands. Transcription is the next important step for the life cycle. Genes are transcribed by an enzyme complex, RNA polymerase, into RNA molecules which are subsequently translated into proteins on ribosomes. Translation depends on the correspondence between the codons of mRNA and the amino acids that constitute the proteins. This correspondence is mediated by tRNA molecules, which carry an amino acid that corresponds to the anticodon of the tRNA (each codon has a complementary anticodon). Of the 64 possible codons, 61 specify the 20 amino acids. The 3 remaining codons, in appropriate contexts, specify a signal that triggers the end of the translation process (Rocha *et al.*, 1999b). On average, therefore, a given amino acid in a protein is defined by 3 codons: each gene may use codons in a specific way. As a consequence, the codon frequency for a given amino acid varies according to the gene; the codon usage is biased. In *E. coli*, for example, the codon AUA, which codes for isoleucine, is much less frequent in some genes than AUC or AUU (which also code for isoleucine). In the same way, glycine is often coded for by GGC and GGU, but rarely by GGG or GGA.

Analysis of the *E. coli* and *B. subtilis* genomes has shown that the codon usage preference is associated with the functional nature of the genes. Several distinct gene classes appear. In particular, Claudine Médigue has demonstrated the existence of a class of genes unique to *E. coli*; about 15% of the genome of the laboratory reference strain corresponds to genes which come from mobile elements that have been transmitted 'horizontally' from one species to another. Interestingly, among these laterally transferred genes, some appear to be implicated in the fidelity of DNA replication, and therefore in anti-mutagenic processes (Médigue *et al.*, 1991). This has certain important consequences for the process of speciation in bacteria, and several authors have proposed the idea that the mutator state of a bacterium is often associated with pathogenicity (Taddei *et al.*, 1997; Field *et al.*, 1999).

Moszer *et al.* (1999) observed that the codon usage bias was linked to the architecture of the cell. This was accounted for as follows. A bacterial cell contains 15 000–30 000 ribosomes/μm^3, and a DNA thread highly coiled on itself. Since this thread is 1 mm in length, it is 1000 times as long as the cell and, although it is flexible, its 'real' length would make it occupy a much larger volume, typically a sphere 10 µm in diameter; this indicates that there are further organising principles. From this DNA thread, mRNA molecules are constantly pulled out by the translation/transcription machinery. The physico-chemical properties of the cytoplasm are those of a dense gel-like network through which the mRNAs pass. The ribosome network dictates the mode of expression of the genes. As a mRNA leader sequence reaches the surface of the RNA polymerase–DNA complex (it may be anchored to the polymerase by its 5′ end), it becomes associated with a ribosome that scans the mRNA until a translation initiation codon is found, which signals the start of the protein coding sequence. The ribosome unfolds the message, even as it is being copied by RNA polymerase from one of the DNA strands. This unfolds DNA locally and results in the movement of the transcribed region to the surface of the chromoid, so facilitating access of RNA polymerase to other genes and facilitating their transcription. Finally, when an appropriate signal reaches the ribosome, the mRNA is degraded, ending its expression process. This scenario indicates that mRNAs with a strong codon preference will bias the concentration of tRNAs corresponding to this preference. As a consequence, a ribosome that translates a highly biased mRNA will behave as an 'attractor' of the cognate tRNA molecules, creating conditions for the optimisation of translation of those other mRNA molecules that fit the bias (Danchin *et al.*, 1999). Thus, for a given codon, the cell will appear locally to be organised into successive layers along a decreasing gradient for the preference of this codon, in parallel with the concentration of the corresponding tRNAs. Indeed, genes with similar codon preference are not distributed randomly along the genome text, but are clustered at specific places. One can even visualise several RNA polymerases transcribing simultaneously, yoked together like draught animals (Rocha *et al.*, 2000a). Moreover, when a particular function requires the participation of several proteins distributed in several cell compartments, the codon preference gradient for the proteins translated for this function appears to follow the architecture of the cell. As a consequence, genes coding for elements that participate in a given structure tend to be clustered together and, conversely, genes that are clustered together correspond to products which are synthesised next to each other. This may trigger possible interactions that will be stabilised by natural selection if they confer an advantage to the organism. In this way bacteria tend to explore and become adapted to a given biotope; pathogenicity is a special case of an ecological niche.

Bacteria and their Biotope

In the wild, living organisms usually interact not only with their kin, but also with many other organisms. In fact, it is rare in nature to find pure bacterial cultures, although *B. subtilis* often produces pure cultures on the surface of leaves (phylloplane) (Arias *et al.*, 1999) and the insect pathogen *Photorhabdus luminescens* makes pure cultures in the insect it infects (Akhurst and Dunphy, 1993). When this happens, one can be sure either that the environment is very unusual and that not many living organisms can thrive there, or that the pure culture cells synthesise a variety of antibiotics which help them to keep the environment to themselves. The general strategy of bacterial interactions is, therefore, that of a species keeping the largest part of a biotope for itself. It is likely that antimicrobials of various types are an early evolutionary weapon designed to achieve this outcome. When bacteria had to cope with multi-cellular organisms, endowed with elaborate defence mechanisms, they evolved virulence, in order to colonise as much as possible of the new niche these organisms provided. As in the case of all predator–prey lethal strategies, however, virulence has a cost (Cressman *et al.*, 1986); the host will die, and this is of no long-term benefit to the progeny of the pathogen. Game theory helps us to understand that the next step is to attenuate virulence, giving rise ultimately to parasitism (Marrow *et al.*, 1996) – a situation that is still deleterious to the host. A further strategy is symbiosis, where the former pathogen is now a symbiont that profits the host in such a way that both live in harmony. Finally, one reaches the situation where most micro-organisms live together, each gaining on the whole from the existence of the other. This is the commensalism that is observed, for example, in the normal gut flora.

It is important to understand these different levels of interaction between a host and the variety of bacteria that live with it from time to time. The study of virulence and pathogenicity benefits from an understanding of the ecological principles that underlie the life cycle of putative pathogens. Indeed, the frontiers between commensalism, symbiosis, parasitism and virulence are imprecise. Opportunistic pathogens explore the borders of virulence in specific hosts, where the biotope is different from that of the normal host or where, for example, there is a poor immune response, unusual cell surface markers, a high blood glucose level or the like. The study of bacterial genomes may help us better to understand the underlying biological functions.

Sequenced Genomes of Pathogenic Bacteria: Early Results

At the time of writing, many bacterial genomes are in the process of being sequenced and 59 complete genome sequences have been published, including those of 23 pathogens. Among the 250 bacterial genomes being sequenced 97 have already been completed; these include near-duplicates of published sequences, which are still under investigation before publication. Only a partial list is given here, limited to the organisms for which genome knowledge has already provided interesting information. A comprehensive database of ongoing genome programs is maintained by Nikos Kyrpides (http://wit.integratedgenomics.com/GOLD).

Enterobacteria and related bacteria

The genome of *E. coli* K12, a non-pathogen, has been known since the late summer of 1997 (Blattner *et al.*, 1997). It has been compared with the genome of several pathogenic *E. coli*, including strain O157:H7. On several public occasions, Frederick Blattner has presented comparisons between different *E. coli* strains. A major outcome has been the observation that a core genome sequence remains substantially undisturbed, while variable islands mark the strain differences. Remarkably, the location of these islands appears to be more or less fixed, one island being exchanged for another at the same location. The fact that functionally related gene islands are often carried by phages is consistent with this observation, since phages often have specific attachment sites. This may even account for the selection pressure for maintaining attachment sites for phages in the chromosome. As a complement to RFLP of lambdoid prophages, knowledge of the genomes allowed Kim and co-workers to devise a new diagnostic method for the O157:H7 strain variability, based on oligonucleotide motif frequency (Kim *et al.*, 1999).

The dangerous pathogens *Salmonella typhi* and *Yersinia pestis* have also been the subject of genome sequencing programs. Although not much has yet been found in the case of *S. typhi*, except in the study of phylogenetic relationships in enterobacteria (McClelland and Wilson, 1998), the study of PAIs in *Y. pestis* has given remarkable information about the origin of the plague pandemics in historical times and more recently (Buchrieser *et al.*, 1999). Briefly, it appears that, contrary to simple expectation, *Y. pestis* evolves by inactivating some virulence-associated genes which are present and active in the less pathogenic strain *Y. pseudotuberculosis*. In addition, we have found that the genome of *Y. pestis* contains three classes of the genes of adenylate cyclases: the enterobacterial class (Trotot *et al.*, 1996), the toxic class (Glaser *et al.*, 1988), and class IV, already identified in *Aeromonas hydrophila* (Sismeiro *et al.*, 1998). It is tempting to speculate that this may be linked to the variety of environments in which the plague agent has to multiply.

The genome of *H. influenzae* was the first to be sequenced. Several pieces of data were used to account for the way in which competence for the uptake of DNA is expressed (Karlin *et al.*, 1996). The genome sequence also revealed interesting information about antigenic variation in this organism (Hood *et al.*, 1996). It is remarkable that in *H. influenzae* the number of DNA repeats longer than 25 bases is low as compared, for instance, with the number found in *E. coli*. This may be a general character of bacteria competent for DNA transformation (Rocha *et al.*, 1999a).

Helicobacter pylori

Two strains of *H. pylori* have been sequenced in their entirety. The genome size of strain 26695 is 1 1697 767 bp and that for strain J99 is 1 143 331 bp, and 89% of the predicted coding sequences are common to the two strains. A region containing approximately 45% of *H. pylori* strain-specific coding sequences, termed the plasticity zone, is present in both chromosomes. Frequent alteration of nucleotides in the third position of the codons and various copies of insertion sequences in the individual chromosomes contribute to distinct polymorphic fingerprints among strains analysed by a variety of typing methods. Except for the data that suggest genome plasticity and generation of antigenic variation, however, it is not yet clear how this polymorphism is associated with virulence (Ge and Taylor, 1999; Taylor, 1999).

The publication of the genome sequences of two strains of *H. pylori* has allowed direct evaluation of the reported previously genetic diversity with respect to gene organisation and nucleotide variability from strain to strain. Cloning and functional characterisation of the genes involved in survival in acidic environments, the production of vacuolating toxin, the CAG PAI, motility, attachment to epithelial cells, competence for DNA transformation and the biosynthesis of lipopolysaccharides have considerably increased our understanding of the molecular basis for the pathogenesis of *H. pylori*. In addition, although it was not possible functionally to identify the origin of replication of the genome, the study of its complete sequence has permitted the placement of the origin in the circular chromosome (Rocha *et al.*, 1999c). The homopolymeric nucleotide tracts and dinucleotide repeats, which potentially regulate the on and off status of the target genes by the strand-slipped mispairing mechanism, are often found in the genes that encode the outer-membrane proteins or the enzymes for lipopolysaccharide synthesis, and within DNA modification/restriction systems. These genes may, therefore, be involved both in antigenic variation and in the *H. pylori*–host interaction (Ge and Taylor, 1999).

Mycobacterium tuberculosis and *Mycobacterium leprae*

One-third of the world population is infected by *M. tuberculosis* (see Chapter 81). The genome of this pathogen consists of 4 411 129 bp with 3324 coding sequences (Cole *et al.*, 1998). About 6% of the predicted proteins have no homology with proteins found in other genomes, and may encode proteins involved in the pathogenesis of tuberculosis. Homology studies have identified several potential virulence factors, including lipases, proteases, esterases, cutinases and a PE-PGRS family of proteins, which may be responsible for the intracellular survival of *M. tuberculosis* in macrophages (Espitia *et al.*, 1999). About 250 enzymes are associated with fatty acid metabolism, and these may represent potential new targets for antimycobacterial therapy. About 10% of the genome encodes two families of acidic glycine-rich proteins with a repetitive structure. These proteins are implicated in antigenic variation, and perhaps with interference in antigen processing during tuberculous infection (Bishai, 1998). Remarkably, the *M. tuberculosis* genome contains genes of the family that synthesises peptide antibiotics, polyketides and fatty acids. Some may be responsible for the immunosuppression induced by the tubercle bacillus. There were two unexpected findings: the apparent absence of PAIs, which suggests that tubercle bacilli are evolutionarily highly adapted to mammalian hosts, and the absence of mismatch repair genes, which may contribute to the genetic variation in the repetitive sequences (Mizrahi and Andersen, 1998). Comparative genomic studies of the virulent *M. tuberculosis* H37Rv and the avirulent *M. bovis* (BCG) revealed the absence in the latter of two novel 12.7-kb fragments (Zumarraga *et al.*, 1999). Large-scale approaches using the genome sequence may soon provide further understanding of the pathogenesis of this very ancient disease (Brosch *et al.*, 1999). In particular, the genome sequence of *M. bovis* and of its BCG derivative is eagerly awaited.

A genome sequencing programme for *M. leprae* was launched in 1995 (Cole, 1998b), and the genome sequence has been published (Cole *et al.*, 2001). Preliminary observations show that the genome is a reduced form of the *M. tuberculosis* genome, where many genes are in the process of being inactivated (Cole, 1998a). This reductive evolution process may be at the core of the way in which *M. leprae* escapes the immune surveillance of the host. It is interesting to note that loss of genes is a common mechanism among the pathways to the development of virulence.

Mycoplasma, Rickettsia and *Chlamydia* genomes

Four mycoplasma genome sequences have been completed, those of *Mycoplasma genitalium*, *Mycoplasma pneumoniae*, *Ureaplasma urealyticum* and *Mycoplasma pulmonis*. An interesting outcome of these genome programmes is that *Mycopl. genitalium* contains the smallest known genome (0.580 Mb) (Razin, 1997). Apart from the identification of repeats that may be important for pathogenicity (Rocha *et al.*, 1999a), little has been found to account for the virulence of these strains. It may be that the process of reductive evolution, which these organisms so clearly exemplify, enables them to cope with hostile environments, while retaining the functions essential to life. A case in point, in *Shigella*, is the inactivation of the lysine decarboxylase gene (Maurelli *et al.*, 1998). Remarkably, the gene complement of *Mycopl. genitalium* is not far from the lowest possible limit for life, counting the number of genes needed for replication, transcription and translation. A further reduction is seen in other genomes, which can no longer sustain life in the absence of pre-existing life and thrive as obligatory endosymbionts (Zomorodipour and Andersson, 1999). It is interesting to see that genome contraction is probably an ongoing process in *Rickettsia* (Andersson and Andersson, 1999). It is likely, therefore, that study of the genome of these organisms will facilitate the development of vaccines and, possibly, of specific metabolic targets for antimicrobial drugs. These cannot be numerous, however, precisely because the degenerate nature of the genome of these organisms makes it necessary for them to exploit the metabolism of the host. In a way they are similar to viruses, which are notoriously difficult to control by pharmaceutical means.

Impact of Microbial Genomics on Infectious Diseases

In addition to the understanding of the biology of microbes, a significant proportion of the knowledge generated by microbial genome programs will translate into value-added deliverables such as novel diagnostic tests, antibiotics, vaccines, strain typing techniques and new concepts of environmental or epidemiological control to prevent the acquisition or transmission of pathogenic microbes. Conventional microbiology has achieved a profound change in the management of infectious diseases, mainly by means of phenotypic approaches. In the last 20 years, advances in molecular biology have allowed the development of rapid diagnostic tests for detection of microbes or identification of resistance. These approaches combine gene hybridisation or amplification, genotyping methods, recombinant serodiagnostics and vaccines. However, the discovery of novel genes relies largely on the screening of expression libraries for a known phenotype or subsequently by using known DNA sequences to design degenerate probes (for hybridisation) or primers (for Race PCR). Information on the organisation, co-operation or overall control of individual genes or gene clusters has so far been either fragmented or meagre. With advances in automated sequencing and computer technology, the recent approach of microbial genomics through sequencing followed by analysis *in vivo*, *in vitro* and *in silico* of the whole genome of microbes is rectifying these deficiencies (see Table 1). In addition to the discovery of genes which were not previously identifiable by traditional methods, the identification of PAIs and other metabolic gene clusters associated with the mechanisms of co-operation between different gene clusters in the genome as a whole will initiate new approaches to research on antibiotics and immunisation. It may soon be possible to design live attenuated vaccines with the controlled deletion of most pathogenic attributes, while retaining their immunogenicity and ability to colonise, or recombinant vaccines with large gene fragments of PAIs. An improved antibiotic that only switches off PAIs, without inhibiting or killing off the normal flora, will decrease the risk of super-infection by more resistant microbes. Already we have novel typing methods that utilise the oligonucleotide-motif frequency in *E. coli* O157:H7. Antigenic or pathogenic shifts and drifts would be monitored by studying gene cluster evolution rather than the study of individual genes or antigens.

Table 1 Clinical impact of advances in microbial genomics

Study approaches	Major techniques	Clinical impact and deliverables
Conventional microbiology (phenotypic)	Mainly phenotypic	Gram and other stains Culture Serodiagnosis Antimicrobial susceptibility tests Strain typing by phenotypic techniques Whole cell/component vaccine
Molecular microbiology (phenotypic to genotypic)	Mainly genotypic	Gene cloning, sequencing and characterisation Probe hybridisation diagnostics PCR diagnostics Recombinant serodiagnostics Genotyping and molecular epidemiology Recombinant vaccines
Microbial genomics (genomotypic to genotypic and phenotypic)	Complete genome sequencing, whole genome DNA microarray expression, functional proteomics (*in silico*, *in vitro* and *in vivo* analysis)	Discovery of new and clinically important genes not achievable with standard molecular microbiological approaches, leading to novel diagnostics tests Recombinant vaccines by means of gene clusters Novel attenuated live vaccines with controlled deletion of gene clusters (pathogenicity islands) Novel antibiotics that selectively switch off pathogenic islands without affecting the normal flora Novel typing methods, such as oligonucleotide-motif frequency Novel epidemiological monitoring of gene cluster evolution on antigenic shift or drift

Conclusion

Two broad conclusions can be drawn from the analysis of the literature and scientific meetings dedicated to the future of genomics. First, it is clear that the sequencing of genomes is just the first step in a new voyage of discovery, and must be accompanied by functional genomics. It is estimated that around 40% of the coding sequences in a fully sequenced organism have no known function at the biochemical level and are unrelated to any known gene, but at the phenotype level the proportion with known properties is much less. Consequently, a shift of emphasis is now taking place, from genome mapping and sequencing to the determination of genome function. This is the area known as functional genomics, which has been described as the development and application of global (genome-wide or system-wide) experimental approaches to assess gene function by making use of the information and reagents provided by genome sequencing and mapping. Functional genomics has recently emerged as a new discipline that employs major innovative technologies for genome-wide analysis, supported by information technology. These activities depend on both experimental and computational methods. High-throughput experimental technologies generate data on gene expression, protein structure and interaction, etc., and powerful information systems and computer-assisted experimental strategies are necessary to manage the resulting data. It is also necessary to integrate this with information from a variety of other sources. The combination of genome-sequencing projects and functional analysis is also producing a flow of data which focuses particularly on the biological relationships between sequences. Hence, the need for models, methods and tools to manipulate sequence networks or clusters is becoming increasingly important.

At present there are only a few major projects on the functional analysis of model genomes; the most notable are those for *B. subtilis* and *S. cerevisiae* (Eurofan). A project for the best-known model bacterium, *E. coli*, is being set up in Japan, where, in parallel, functional analysis of the model of the cyanobacterium *Synechocystis* PCC 6803 is being carried out at the Kazusa Institute. In these projects, each gene of the organism is individually disrupted and fused to a reporter gene, in the hope of discovering phenotypes that may help to identify its function. We do not have heuristics that allow direct access to unknown functions and, apart from two preliminary studies known to the author, no similar *in silico* work is being undertaken anywhere in the world. An excellent illustration of the concept of 'neighbourhood' has, however, been developed with the Entrez software, created by D. Lipman and colleagues at the National Center for Biological Information. We believe that the approach involving the concept of neighbourhood, as described in this chapter, will help to extract valuable information from genome sequences, particularly in terms of medical applications.

In this approach, one must take into account the fact that genes are distributed according to their function, their mode of expression and the location of the protein they specify in the diverse cell compartments. This accounts *a posteriori* for some effects of the gene context that have remained elusive. It is known, for example, that the expression of a gene changes when its location on the chromosome is changed, even when its control elements – its promoter in particular – have been conserved. Beyond this general methodology for exploring 'neighbourhoods', leading to the assignment of functions, an essential fact emerges: the analysis of genome texts can teach us much more than the 'elementary' (or 'primitive') function of the genes. An example might be the mechanisms that underlie the catalytic activity of a specific gene product. Evidence continues to accumulate that an intermediate level of integration is required at the mesoscopic scale – between that of molecules and that of the whole cell – in order to understand the organisation and functions of the bacterial cell (Danchin, 1997, 1999; Norris *et al.*, 1999; Rocha *et al.*, 2000). The objects at this 10–50 nm scale, typical for instance of the dimension of ribosomes or of ATP synthase, may be termed nanosomes, for those that are compact, and reticulosomes (or plectosomes) for the fibrillar organised structures that are spread throughout the cell. An understanding of the cell at this mesoscopic level requires a study of the factors involved both in the formation and in the interactions of these structures. As different objects compete for space in the cytoplasm and at the membrane, *de novo* synthesis of one structure may trigger the destruction or displacement of another. A replication reticulosome may, for example, yield to a nano- or reticulosome associated with division (Norris *et al.*, 1999). Many experimental data indicate that nanosomes or reticulosomes control the electron transport chain, ATP synthesis (Sambongi *et al.*, 1999), DNA partitioning (Jensen and Gerdes, 1997; Yamazoe *et al.*, 1999), ribosome synthesis (Agrawal and Frank, 1999), mRNA degradation (Carpousis *et al.*, 1999) and sulphur transport and metabolism (Rocha *et al.*, 2000b). The exploration of relationships between nanosomes or reticulosomes will prove indispensable to understanding the bacterial cell. In the study of bacterial pathogens, a case in point is the type III secretion system that mounts the virulence response of many invasive bacteria.

References

Agrawal RK, Frank J (1999) Structural studies of the translational apparatus. *Curr. Opin. Struct. Biol.* 9: 215–221.

Akhurst RJ, Dunphy GB (1993) Tripartite interactions between symbiotically associated entomopathogenic bacteria, nematodes, and their insect hosts. In: Beckage NE, Thompson SN, Federici B (eds) *Parasites and Pathogens of Insects.* New York: Academic Press, pp. 1–23.

Altschul SF, Madden TL, Schaffer AA *et al.* (1997) Gapped BLAST and PSI-BLAST: a new generation of protein database search programs. *Nucleic Acids Res.* 25: 3389–3402.

Andersson JO, Andersson SG (1999) Genome degradation is an ongoing process in *Rickettsia*. *Mol. Biol. Evol.* 16: 1178–1191.

Arias R, Sagardoy M, van Vuurde J (1999) Spatio-temporal distribution of naturally occuring *Bacillus* spp. and other bacteria on the phylloplane of soybean under field conditions. *J. Basic Microbiol.* 39: 283–292.

Axon AT (1999) Are all helicobacters equal? Mechanisms of gastroduodenal pathology and their clinical implications. *Gut* 45 Suppl. 1: I1–4.

Bishai W (1998) The *Mycobacterium tuberculosis* genomic sequence: anatomy of a master adaptor. *Trends Microbiol.* 6: 464–465.

Blattner FR, Plunkett G 3rd, Bloch CA *et al.* (1997) The complete genome sequence of *Escherichia coli* K-12. *Science* 277: 1453–1474.

Blum G, Ott M, Lischewski A *et al.* (1994) Excision of large DNA regions termed pathogenicity islands from tRNA-specific loci in the chromosome of an *Escherichia coli* wild-type pathogen. *Infect. Immun.* 62: 606–614.

Boyd EF, Hartl DL (1998) Chromosomal regions specific to pathogenic isolates of *Escherichia coli* have a phylogenetically clustered distribution. *J. Bacteriol.* 180: 1159–1165.

Brosch R, Philipp WJ, Stavropoulos E *et al.* (1999) Genomic analysis reveals variation between *Mycobacterium tuberculosis* H37Rv and the attenuated *M. tuberculosis* H37Ra strain. *Infect. Immun.* 67: 5768–5774.

Buchrieser C, Rusniok C, Frangeul L *et al.* (1999) The 102-kilobase pgm locus of *Yersinia pestis*: sequence analysis and comparison of selected regions among different *Yersinia pestis* and *Yersinia pseudotuberculosis* strains. *Infect. Immun.* 67: 4851–4861.

Carpousis AJ, Vanzo NF, Raynal LC (1999) mRNA degradation. A tale of poly(A) and multiprotein machines. *Trends Genet.* 15: 24–28.

Cole ST (1998a) Comparative mycobacterial genomics. *Curr. Opin. Microbiol.* 1: 567–571.

Cole ST (1998b) The *Mycobacterium leprae* genome project. *Int. J. Lepr. Other Mycobact. Dis.* 66: 589–591.

Cole ST, Brosch R, Parkhill J *et al.* (1998) Deciphering the biology of *Mycobacterium tuberculosis* from the complete genome sequence. *Nature* 393: 537–544.

Cole ST, Eiglmeier K, Parkhill J *et al.* (2001) Massive gene decay in the leprosy bacillus. *Nature* 409: 1007–1011.

Coote JG (1992) Structural and functional relationships among the RTX toxin determinants of gram-negative bacteria. *FEMS Microbiol. Rev.* 8: 137–161.

Cossart P, Lecuit M (1998) Interactions of *Listeria monocytogenes* with mammalian cells during entry and actin-based movement: bacterial factors, cellular ligands and signaling. *EMBO J.* 17: 3797–3806.

Cressman R, Dash AT, Akin E (1986) Evolutionary games and two species population dynamics. *J. Math. Biol.* 23: 221–230.

Danchin A (1996) On genomes and cosmologies. In: Collado-Vides J, Magasanik B, Smith TF (eds) *Integrative Approaches to Molecular Biology*, pp. 91–111. Boston, MA: MIT Press.

Danchin A (1997) Comparison between the *Escherichia coli* and *Bacillus subtilis* genomes suggests that a major function of polynucleotide phosphorylase is to synthesize CDP. *DNA Res.* 4: 9–18.

Danchin A (1998) The Delphic boat or what the genomic texts tell us [editorial]. *Bioinformatics* 14: 383.

Danchin A (1999) From protein sequence to function. *Curr. Opin. Struct. Biol.* 9: 363–367.

Danchin A, Medigue C, Gascuel O, Soldano H, Henaut A (1991) From data banks to data bases. *Res. Microbiol.* 142: 913–916.

Danchin A, Guerdoux-Jamet P, Moszer I, Nitschké P (1999) Mapping the bacterial cell architecture in the chromosome. *Phil. Trans. Royal Soc. B* 354: 1–12.

Dobrindt U., Cohen PS, Utley M., Muhldorfer I., Hacker J (1998) The *leuX*-encoded tRNA5(Leu) but not the pathogenicity islands I and II influence the survival of the uropathogenic *Escherichia coli* strain 536 in CD-1 mouse bladder mucus in the stationary phase. *FEMS Microbiol. Lett.* 162: 135–141.

Espitia C, Laclette JP, Mondragon-Palomino M *et al.* (1999) The PE-PGRS glycine-rich proteins of *Mycobacterium tuberculosis*: a new family of fibronectin-binding proteins? *Microbiology* 145: 3487–3495.

Field D., Magnasco MO, Moxon ER *et al.* (1999) Contingency loci, mutator alleles, and their interactions. Synergistic strategies for microbial evolution and adaptation in pathogenesis. *Ann. NY Acad. Sci.* 870: 378–382.

Fleischmann RD, Adams MD, White O *et al.* (1995) Whole-genome random sequencing and assembly of *Haemophilus influenzae* Rd. *Science* 269: 496–512.

Francois V, Louarn J, Patte J, Rebollo JE, Louarn JM (1990) Constraints in chromosomal inversions in *Escherichia coli* are not explained by replication pausing at inverted terminator-like sequences. *Mol. Microbiol.* 4: 537–542.

Fraser CM, Gocayne JD, White O *et al.* (1995) The minimal gene complement of *Mycoplasma genitalium*. *Science* 270: 397–403.

Galan JE, Collmer A (1999) Type III secretion machines: bacterial devices for protein delivery into host cells. *Science* 284: 1322–1328.

Ge Z, Taylor DE (1999) Contributions of genome sequencing to understanding the biology of *Helicobacter pylori*. *Ann. Rev. Microbiol.* 53: 353–387.

Glaser P, Sakamoto H, Bellalou J, Ullmann A, Danchin A (1988) Secretion of cyclolysin, the calmodulin-sensitive adenylate cyclase-haemolysin bifunctional protein of *Bordetella pertussis*. *EMBO J.* 7: 3997–4004.

Goffeau A, Barrell BG, Bussey H *et al.* (1996) Life with 6000 genes. *Science* 274: 546, 563–567.

Hood DW, Deadman ME, Jennings MP *et al.* (1996) DNA repeats identify novel virulence genes in *Haemophilus influenzae*. *Proc. Natl Acad. Sci. USA* 93: 11121–11125.

Hueck CJ (1998) Type III protein secretion systems in bacterial pathogens of animals and plants. *Microbiol. Mol. Biol. Rev.* 62: 379–433.

Jensen RB, Gerdes K (1997) Partitioning of plasmid R1. The ParM protein exhibits ATPase activity and interacts with the centromere-like ParR-ParC complex. *J. Mol. Biol.* 269: 505–513.

Karlin S, Mrazek J, Campbell AM (1996) Frequent oligonucleotides and peptides of the *Haemophilus influenzae* genome. *Nucleic Ac. Res.* 24: 4263–4272.

Kim J, Nietfeldt J, Benson AK (1999) Octamer-based genome scanning distinguishes a unique subpopulation of *Escherichia coli* O157:H7 strains in cattle. *Proc. Natl Acad. Sci. USA* 96: 13288–13293.

Kunst F, Ogasawara N, Moszer I *et al.* (1997) The complete genome sequence of the gram-positive bacterium *Bacillus subtilis*. *Nature* 390: 249–256.

Lawrence JG, Ochman H (1998) Molecular archaeology of the *Escherichia coli* genome. *Proc. Natl Acad. Sci. USA* 95: 9413–9417.

Lee VT, Schneewind O (1999) Type III secretion machines and the pathogenesis of enteric infections caused by *Yersinia* and *Salmonella* spp. *Immunol. Rev.* 168: 241–255.

Marrow P, Dieckmann U, Law R (1996) Evolutionary dynamics of predator–prey systems: an ecological perspective. *J. Math. Biol.* 34: 556–578.

Maurelli AT, Fernandez RE, Bloch CA, Rode CK, Fasano A (1998) 'Black holes' and bacterial pathogenicity: a large genomic deletion that enhances the virulence of *Shigella* spp. and enteroinvasive *Escherichia coli*. *Proc. Natl Acad. Sci. USA* 95: 3943–3948.

McClelland M, Wilson RK (1998) Comparison of sample sequences of the *Salmonella typhi* genome to the sequence of the complete *Escherichia coli* K-12 genome. *Infect. Immun.* 66: 4305–4312.

Médigue C, Rouxel T, Vigier P, Hénaut A, Danchin A (1991) Evidence for horizontal gene transfer in *Escherichia coli* speciation. *J. Mol. Biol.* 222: 851–856.

Mizrahi V, Andersen SJ (1998) DNA repair in *Mycobacterium tuberculosis*. What have we learnt from the genome sequence? *Mol. Microbiol.* 29: 1331–1339.

Moszer I (1998) The complete genome of *Bacillus subtilis*: from sequence annotation to data management and analysis. *FEBS Lett.* 430: 28–36.

Moszer I, Rocha EP, Danchin A (1999) Codon usage and lateral gene transfer in *Bacillus subtilis*. *Curr. Opin. Microbiol.* 2: 524–528.

Nitschké P, Guerdoux-Jamet P, Chiapello H *et al.* (1998) Indigo: a World-Wide-Web review of genomes and gene functions. *FEMS Microbiol. Rev.* 22: 207–227.

Norris V, Alexandre S, Bouligand Y *et al.* (1999) Hypothesis: hyperstructures regulate bacterial structure and the cell cycle. *Biochimie* 81: 915–920.

Nunn D (1999) Bacterial type II protein export and pilus biogenesis: more than just homologies? *Trends Cell Biol.* 9: 402–408.

Pearson WR (2000) Flexible sequence similarity searching with the FASTA3 program package. *Methods Mol. Biol.* 132: 185–219.

Razin S (1997) Comparative genomics of mycoplasmas. *Wien Klin. Wochenschr.* 109: 551–556.

Rebollo JE, Francois V, Louarn JM (1988) Detection and possible role of two large nondivisible zones on the *Escherichia coli* chromosome. *Proc. Natl Acad. Sci. USA* 85: 9391–9395.

Rocha E, Danchin A, Viari A (1999a) Analysis of long repeats in bacterial genomes reveals alternative evolutionary mechanisms in *Bacillus subtilis* and other competent prokaryotes. *Mol. Biol. Evol.* 16: 1219–1230.

Rocha EP, Danchin A, Viari A (1999b) Translation in *Bacillus subtilis*: roles and trends of initiation and termination, insights from a genome analysis. *Nucleic Acids Res.* 27: 3567–3576.

Rocha EP, Danchin A, Viari A (1999c) Universal replication biases in bacteria. *Mol. Microbiol.* 32: 11–16.

Rocha EP, Guerdoux-Jamet P, Moszer I, Viari A, Danchin A (2000a) Implication of gene distribution in the bacterial chromosome for the bacterial cell factory. *J. Biotechnol.* 78: 209–219.

Rocha EP, Sekowska A, Danchin A (2000b) Sulphur islands in the *Escherichia coli* genome; markers of the cells architecture? *FEBS Lett.* 476: 8–11.

Rudd KE (2000) EcoGene: a genome sequence database for *Escherichia coli* K-12. *Nucleic Acids Res.* 28: 60–64.

Sambongi Y, Iko Y, Tanabe M *et al.* (1999) Mechanical rotation of the c subunit oligomer in ATP synthase (F0F1): direct observation. *Science* 286: 1722–1724.

Sismeiro O, Trotot P, Biville F, Vivares C, Danchin A (1998) *Aeromonas hydrophila* adenylyl cyclase 2: a new class of adenylyl cyclases with thermophilic properties and sequence similarities to proteins from hyperthermophilic archaebacteria. *J. Bacteriol.* 180: 3339–3344.

Taddei F, Matic I, Godelle B, Radman M (1997) To be a mutator, or how pathogenic and commensal bacteria can evolve rapidly. *Trends Microbiol.* 5: 427–428; discussion 428–429.

Taylor DE (1999) *Helicobacter pylori* and its genome: lessons from the treasure map. *Can. J. Gastroenterol.* 13: 218–223.

Trotot P, Sismeiro O, Vivares C, Glaser P, Bresson-Roy A, Danchin A (1996) Comparative analysis of the cya locus in enterobacteria and related gram-negative facultative anaerobes. *Biochimie* 78: 277–287.

Vasil ML, Ochsner UA (1999) The response of *Pseudomonas aeruginosa* to iron: genetics, biochemistry and virulence. *Mol. Microbiol.* 34: 399–413.

Waldor MK, Mekalanos JJ (1996) Lysogenic conversion by a filamentous phage encoding cholera toxin [see comments]. *Science* 272: 1910–1914.

Wall D, Kaiser D (1999) Type IV pili and cell motility. *Mol. Microbiol.* 32: 1–10.

Warren JB, Wilson AJ, Loi RK, Coughlan ML (1993) Opposing roles of cyclic AMP in the vascular control of edema formation. *FASEB J* 7: 1394–1400.

Wasinger VC, Cordwell SJ, Cerpa-Poljak A *et al.* (1995) Progress with gene-product mapping of the Mollicutes: *Mycoplasma genitalium*. *Electrophoresis* 16: 1090–1094.

Wilson RK (1999) How the worm was won. The *C. elegans* genome sequencing project. *Trends Genet.* 15: 51–58.

Yamazoe M, Onogi T, Sunako Y *et al.* (1999) Complex formation of MukB, MukE and MukF proteins involved in chromosome partitioning in *Escherichia coli*. *EMBO J.* 18: 5873–5884.

Zomorodipour A, Andersson SG (1999) Obligate intracellular parasites: *Rickettsia prowazekii* and *Chlamydia trachomatis*. *FEBS Lett.* 452: 11–15.

Zumarraga M, Bigi F, Alito A, Romano MI, Cataldi A (1999) A 12.7 kb fragment of the *Mycobacterium tuberculosis* genome is not present in *Mycobacterium bovis*. *Microbiology* 145: 893–897.

17

Mechanisms of Gene Exchange Between Bacteria

Brian M. Wilkins[1] and Laura S. Frost[2]

[1] *University of Leicester, Leicester, UK*
[2] *University of Alberta, Edmonton, Alberta, Canada*

Gene movement between bacteria is achieved by the three quasi-sexual processes of bacterial conjugation, transduction and transformation. Each involves unidirectional transfer of just part of the genome of a donor bacterium to the recipient cell but is distinct by mechanism and randomness of the genomic fragment transferred. Bacterial conjugation, the dedicated transfer process of plasmids, requires a specific contact between the donor and recipient bacterium and results in the ordered transmission of a unique DNA strand to the recipient cell. In contrast, natural transformation and transduction support transfer of non-specific genomic fragments; the systems differ considerably at the mechanistic level, since transformation involves uptake of naked DNA from the extracellular environment whereas transduction entails transfer of non-viral DNA within a bacteriophage particle.

Molecular characterisation and technological exploitation of bacterial DNA transfer systems was largely achieved with *Escherichia coli*, *Salmonella typhimurium* (official designation *Salmonella enterica* serovar Typhimurium) and *Bacillus subtilis*. As a result, these organisms have become pre-eminent as experimentally accessible models for studying many fundamental aspects of bacterial physiology, biochemistry and genetics. DNA transfer also contributes importantly to bacteria in natural populations by bringing genes of different clonal lineages into the same cell, thereby potentiating rapid evolution of novel adaptive clones. The supernumerary genes carried by conjugative elements and temperate phages comprise a particularly dynamic part of the bacterial genome. These genes, collectively described as the horizontal gene pool, include a plethora of medically, environmentally and commercially important loci. Phenotypes encompass resistances to antibacterial agents, metabolic activities and functions affecting the ecology of the host organism, such as toxin production.

This chapter focuses on the transfer of genes, but gene movement is unproductive without the integration of entrant DNA into the genome of the recipient cell. Integration is achieved by a variety of processes. Homologous recombination allows exchange of identical, or nearly identical, nucleotide sequences by a mechanism involving breakage and re-union of DNA chains. A key structural intermediate is the heteroduplex joint consisting of complementary strands from different DNA molecules aligned by base pairing. There is also a requirement in homologous recombination for a protein of the *E. coli*. RecA type to

 doi:10.1006/bkmm.2001.0017

facilitate the essential strand pairing and exchange reactions (Camerini-Otero and Hsieh, 1995). Genomes can also expand in size by various processes of additive recombination. Mechanisms include transposition, site-specific integration of prophages and plasmids, and replication-maintenance systems enabling stable inheritance of plasmids as extra chromosomal units (Craig, 1996; Helinski *et al.*, 1996; Nash, 1996).

The objective of this chapter is to review molecular aspects of gene movement between bacteria, focusing on medical implications. The first section provides a backdrop by considering the significance of gene transfer and transmissible elements to medically important bacteria. Subsequent sections address molecular understanding of gene transfer by bacterial conjugation, transformation and transduction. The penultimate section reviews barriers that operate, following transfer itself, to inhibit the integration of transferred DNA into the genome of the recipient cell. It is outside the remit of this article to relate in-depth mechanistic understandings of gene transfer to the growing body of information on the ecology of gene transfer in natural environments, but guides to appropriate literature are provided.

Significance of Gene Transfer to Medically Important Bacteria

The genetics of laboratory, wild and pathogenic strains of *E. coli* and *S. typhimurium* have been examined in unparalleled detail, making these closely related organisms particularly appropriate as models for assessing the impact of gene transfer on natural populations of bacteria. One approach to appraising the importance of gene exchange is to use nucleotide sequence data to compare the compositional structure of individual genes with the rest of the genome. Important parameters are percentage content of G + C, or some higher-order motif, and codon adaptation index (CAI). The latter measures the degree to which codon usage is biased towards the pattern observed for very highly expressed genes of the species. Atypical genes are candidates of acquisitions by interspecies transfer. On this basis, some 10% of the sequenced genes of *Salmonella* spp. and about 6% of 500 *E. coli* genes are inferred to have originated from a distantly related organism (Ochman and Lawrence, 1996; Whittam, 1996). The values underestimate the contribution of horizontal transfer to evolution in that genes exchanged between closely related organisms may have the same composition, whereas foreign genes will lose atypical characteristics as they become exposed over long periods of time to the mutational and selection pressures prevailing in the new host.

Another line of evidence for gene transfer comes from comparative sequencing of the same chromosomal region from different natural isolates of the same species. For example, comparisons of ECOR (E. *coli* reference) strains has revealed that genes such as *trp* (tryptophan biosynthesis), *phoA* (alkaline phosphatase) and *gnd* (gluconate dehydrogenase) have mosaic structures, wherein one prominent sequence type is punctuated variably by short sectors of other sequence types. The implication is that a frequent outcome of transfer in natural populations is localised recombination of short segments of homologous DNA in the range of 0.1–1 kb (Maynard Smith, 1995; Milkman, 1996; Whittam, 1996). Mosaic gene structures are particularly apparent in some naturally transformable bacteria, such as *B. subtilis* and *Neisseria* spp., indicating that transforming DNA can be particularly recombinogenic.

Evidence of additive recombination is provided by a comparison of the genetic maps of *E. coli* and *S. typhimurium*. The maps are extensively congruent with the exception of some 15 locations, totalling almost 14% of the genome or 630 kb, where a segment of the genetic map of one species must be looped out to maintain register elsewhere with that of the other species. Loops are generally regarded as insertions of transferred DNA, rather than of duplicated resident sequences, and contain a range of metabolic genes. A tendency of loops to be associated with insertion sequences suggests DNA acquisition by transposition (Riley and Krawiec, 1987).

Additive recombination contributes importantly as a process for rapid evolution of pathogenic bacteria. Cassettes of genes contributing to a particular virulence phenotype and ranging in size from 5 to > 100 kb have been mapped to specific chromosomal locations. Such pathogenicity islands (PAIs) are often distinct from the rest of the genome by compositional structure and presence of flanking repeats or insertion sequences, the hallmarks of additive recombination. Horizontal transfer of a single PAI can convert an avirulent strain into a pathogen, and spontaneous deletion of such a cassette is associated with a reduction of virulence (Groisman and Ochman, 1996; Finlay and Falkow, 1997; Hacker *et al.*, 1997).

One well-characterised PAI named LEE (locus of enterocyte effacement) is found in enteropathogenic *E. coli* as a 36-kb locus determining a unique virulence mechanism known as attaching and effacing (AE), a type III secretion system, secreted virulence proteins, and an adhesin. LEE is inserted 16 bp downstream of the selenocysteine tRNA gene (*selC*), and found as a conserved region in other bacterial species that commonly produce the AE lesion (Elliott *et al.*, 1998). PAI-1, containing genes for α haemolysin (*hly*), is inserted at the same position as LEE but in a uropathogenic strain of *E. coli*. Other PAIs are inserted at yet different tRNA loci. Hence, the regions appear as cassettes that are disseminated by horizontal transfer frequently to insert in specific tRNA loci. The transfer mechanism is unknown but findings that *selC* is the insertion site of the defective retron-containing prophage ΦR7, and a cryptic integrase gene, implicate bacteriophages as vectors.

In addition to PAIs, a variety of phages and plasmids insert at tRNA loci. The advantage of such loci may be their sequence conservation, which would provide a constant integration site in different organisms. Another possibility is that dyad symmetries in such loci, which encode the stems of the 'hairpin' structures in tRNAs, are particularly appropriate recognition sites for the integrase enzymes of phages and plasmids. The 3′ end of the tRNA gene often comprises the attachment site: integrative crossing-over between this region in the phage or plasmid and the chromosomal homologue will maintain a functional copy of the tRNA gene 'upstream' of the insertion, while displacing the 3′ end of the original chromosomal locus 'downstream' of the insertion as an inactive pseudogene (Campbell, 1992; Hopwood and Kieser, 1993).

The contribution of transmissible genes to pathogenicity is demonstrated further by the disproportionately high frequency of virulence genes carried on plasmids and bacteriophages. Examples are the large ~200 kb plasmids of the FII incompatibility (Inc) group found in invasive strains of *Shigella sonnei*, *S. flexneri* and *E. coli* (Galán and Sansonetti, 1996). Plasmids also determine the heat-labile (LT) and heat-stable (ST) toxins produced by enterotoxigenic *E. coli*; some of these plasmids encode yet other pathogenicity factors, such as fimbrial colonisation-factor antigens and antibiotic resistances (O'Brien and Holmes, 1996).

Virulence determinants are also found on temperate bacteriophages. Phage conversion, a process in which the host phenotype is altered by acquisition of a prophage, is responsible for several important toxins, such as diphtheria toxin, streptococcal erythrogenic toxin, botulinum toxin and staphylococcal enterotoxin A. The relevant *tox* gene is often located close to the attachment or integration site in the phage DNA, implying that the gene was picked up from an adjacent chromosomal location when a prophage excised imprecisely, as in specialised transduction (Cheetham and Katz, 1995; see section on Transduction, below). The same class of toxin may be determined by more than one phage type: for example, enterohaemorrhagic *E. coli* phages specifying shiga-like toxins are morphologically diverse with at least one type belonging to the lambdoid family of coliphages (O'Brien and Holmes, 1987, 1996). The cholera toxin-encoding phage CTX is very different in being a small filamentous phage that contains single-stranded (ss) DNA. CTX is related to coliphage M13 but is unusual in integrating at a specific chromosomal site as well as replicating autonomously. Like other filamentous phages, CTX attaches to an extracellular pilus during infection. Intriguingly, the receptor used is the toxin-co-regulated pilus (TCP) of *Vibrio cholerae*, which is a virulence factor required for colonisation of the small intestine (Waldor and Mekalanos, 1996; Mekalanos *et al.*, 1997).

The importance of horizontal gene transfer to medical microbiology is further illustrated by the crisis of transmissible antibiotic resistance where phages, conjugative plasmids and conjugative transposons act as the natural vectors of resistance genes. It is beyond the scope of this chapter to review the extensive literature on the emergence of resistant strains in response to the widespread and often indiscriminate use of antibiotics in the last half century. Reviews include those of DeFlaun and Levy (1989), van Elsas (1992) and Courvalin (1994), and case histories are illustrated in the book by Levy (1992) and the article by Elwell (1994) which traces the penetration of resistance plasmids into the genera *Haemophilus* and *Neisseria*.

Bacterial Conjugation

Bacterial conjugation was discovered in 1946 through the seminal studies of Joshua Lederberg and Edward Tatum on genetic recombination in *E. coli*. The process is not only a cornerstone of bacterial genetics (see Willetts, 1993), but provides a natural route for the horizontal transfer of plasmids and a more recently discovered group of mobile elements called conjugative transposons. Mechanistically the process involves the direct transfer of DNA between cells that have come into contact with one another and requires complex sets of genes on the transmissible element both for its transfer and for its subsequent establishment in the recipient cell. Excellent overviews are provided in books devoted to the subject edited by Clewell (1993c) and Thomas (1999). Conjugation of Gram-negative bacteria is a function of plasmids encoding a conjugative pilus, which is an extracellular filament thought to establish the 'mating' contact between the plasmid-bearing donor cell and the recipient cell. The pilus has also been implicated in forming the channel through which the DNA is transferred, although the evidence for this remains circumstantial. Considerable biochemical and genetic information also exists for the steps involved in the initiation and transport of DNA during pilus-mediated conjugation. Understanding of Gram-positive bacterial conjugation is less complete and focuses on mechanisms for mating-pair formation. Much effort has been expended to understand specialised features of conjugative transposition, which involves specific excision of the element from the donor genome, its transfer to a new cell and integration using enzymes resembling the integrases of many lysogenic phages.

With the advent of computer technology and large databases, sequences from diverse conjugation systems have been compared and aligned with each other and with gene products from a wide variety of seemingly different processes. Similarity at the sequence level has been found for many transfer proteins of Gram-negative conjugation systems for which data are available, as well as with proteins involved in conjugation of Gram-positive bacteria, in transport of the T-DNA sector of the *Agrobacterium tumefaciens* Ti plasmid into the plant nucleus during tumorigenesis, and in partitioning of the bacterial chromosome into the prespore during sporulation in *Bacillus subtilis* (Wu *et al.*, 1995; Firth *et al.*, 1996; Christie, 1997). Interestingly, although the conjugative processes of pilus formation and DNA 'packaging' into the recipient cell have many similarities to bacteriophage replication and the assembly of filamentous phages, little evidence of homology between phage and conjugation systems has been found. We return to this theme in the concluding section.

The grouping of plasmids into incompatibility groups has been a generally useful method of classification (Shapiro, 1977). Incompatibility is manifest by the inability of plasmids to co-exist indefinitely in the same cell line and reflects the sharing of a replication-control system of the same specificity. The conjugation systems of plasmids seem to have co-evolved with the replicons since conjugative plasmids of a given incompatibility (Inc) group usually share similar if not identical transfer systems. In plasmids as diverse as pAD1 of *Enterococcus faecalis* and the IncP elements of Gram-negative bacteria, transfer and vegetative replication are co-regulated which lends credibility to this observation.

Bacterial conjugation can involve: (1) self-transmission of the conjugative plasmid or transposon; (2) mobilisation of a co-resident but physically-independent element, which usually supplies its own genes for DNA transport but relies on the mating system of a self-transmissible plasmid to establish the intercellular bridge; or (3) transfer of a cointegrate formed by the fusion of a conjugative element to another plasmid or to the bacterial chromosome as found in 'high-frequency recombination' (Hfr) donors.

Conjugative Plasmids of Gram-negative Bacteria

Conjugative plasmids found in Gram-negative bacteria can be divided roughly into two groups: those that transfer DNA as a nucleoprotein complex and as such comprise a type IV secretion system, and those that transfer DNA alone. The former plasmids are exemplified by RP4, an IncPα plasmid, and the latter by F of the IncFI group; these plasmids will be used as paradigms in the discussion of the two types of transfer systems since the sequences of the transfer regions are known (Frost *et al.*, 1994; Pansegrau *et al.*, 1994a) and considerable work on various aspects of conjugation has been done. In general, the initial contact between donor and recipient cells in Gram-negative conjugation is mediated either by collision or by the conjugative pilus whose tip recognises a receptor on the recipient cell (**Fig. 1**). Contacts are then stabilised and a conjugation pore is constructed between the two cells. DNA transfer is initiated at a strand-specific cleavage site (*nic*) within the origin of transfer (*oriT*) region. The cleaved strand is then transferred with $5' \rightarrow 3'$ polarity by a rolling circle mechanism associated with synthesis of a replacement strand in the donor cell and a complementary strand in the recipient. After one round of transfer, the transferred strand is circularised by specific religation at *nic*. The mating pair then disaggregates leaving two donor cells.

A remarkable property of many conjugation systems is their ability to support transfer between a broad range of genera, which may exceed the replication-maintenance host range of the element. F-like plasmids are apparently limited by host range to genera of the Enterobacteriaceae (Wilkins, 1995), whereas IncP plasmids are maintained in a very wide range of Gram-negative bacteria (Guiney, 1982) and can transfer effectively and by authentic conjugation from *E. coli* to *Streptomyces* sp. and to the microbial eukaryote yeast (Giebelhaus *et al.*, 1996; Bates *et al.*, 1998).

Transfer frequency reflects the efficiency of mating-pair formation, which in turn is a function of the transfer system encoded by that plasmid. Conjugative systems can be divided into: (1) those that operate equally efficiently in liquid or on semi-solid media as

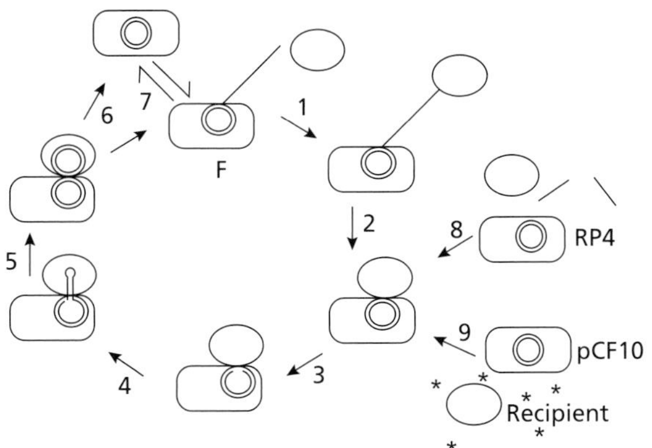

Fig. 1 Summary of the conjugation process for universal (plasmid F) and surface-preferred (plasmid RP4) systems in Gram-negative bacteria and the pCF10 system of Gram-positive bacteria. In universal systems, the pilus attaches to a receptor on the recipient cell surface (1) and retracts to form a stable mating pair or aggregate (2). DNA transfer is initiated (3), causing transport of a single strand in the 5′→3′ direction (4). Transfer is associated with synthesis of a replacement DNA strand in the donor cell and a complementary strand in the recipient (5). The process is terminated by disaggregation of the cells, each carrying a copy of the plasmid (6). The transfer systems of conjugative plasmids in Gram-negative bacteria can be repressed (7) or derepressed (constitutive; 8). Cells carrying RP4 and related plasmids express pili constitutively but the pili are not seen attached to the bacteria. Such cells form mating pairs by collision on a solid surface (8). In Gram-positive bacteria, such as the enterococci, the donor senses the presence of pheromone (*) released by the recipient cell, which triggers mating pair formation and DNA transfer (9). Donor cells are shown as oblongs and recipient cells as ovals. (From Frost, 2000; used by permission of Academic Press).

universal systems, and (2) those that function preferentially on the surface of a semi-solid medium which, experimentally, is usually a filter placed on top of an agar surface. In Gram-negative bacteria, these groups correlate respectively with: (1) thick flexible pili found attached to most donor cells and exemplified by the pili of F and H complex plasmids, and (2) rigid pili usually found free in the medium and very rarely visualised attached to a donor cell, as specified by IncP, N and W plasmids (Bradley *et al.*, 1980). The latter group presumably requires the pilus for stabilisation of mating pairs and/or DNA transport but not as a 'grappling hook' to attach to free-swimming recipient cells in liquid media. Some rigid-pilus systems, typified by IncI complex plasmids, are associated with a thin flexible pilus required not for DNA transfer itself but to increase the efficiency of conjugation in liquid media, presumably by promoting the formation of stable mating pairs (Bradley, 1984).

Physiology of Conjugation

Several physiological parameters are known to affect conjugation including cell number, temperature, oxygen levels and stationary phase. A donor/recipient ratio of 1 : 10 at a density of approximately 10^8 cells/mL is usually employed experimentally to ensure good cell-to-cell contact in liquid or on solid media. Oxygen levels can have a considerable effect, depending on the medium. Stallions and Curtiss (1972) reported that donors of plasmid F in a broth medium mated very efficiently – and slightly better under anaerobic rather than aerobic conditions – whereas donors in a synthetic medium conjugated relatively poorly especially under anaerobiosis. Biebricher and Düker (1984) observed that under anaerobic conditions, broth-grown cells had 3–4 very long F pili which persisted into stationary phase, whereas cells grown in a synthetic medium produced very few pili. Cells grown with aeration in both media had maximum piliation in exponential phase but lost visible piliation at later stages. The latter observation correlates with findings that aerated stationary-phase cultures of F-containing cells behave as fertile recipients termed F^- phenocopies. These cells apparently maintain the transferosome (pilus assembly/conjugative pore) in the cell envelope while the cytoplasmic components of the relaxosome (a nucleoprotein complex at *oriT*) decrease to low levels. The addition of fresh media re-activates the potential for transfer in minutes (Frost and Manchak, 1998). Oxygen levels also appear to affect donors of IncP plasmids since the low transfer frequency of RP4 in liquid media can be improved dramatically by vigorous shaking of the mating culture (Frost and Simon, 1993).

Transfer of F-like plasmids as well as RP4 has a temperature optimum of 37°C and a pH optimum of 7.5 which reflect the ability of the host cell to express the conjugative pilus. The F pilus disappears at temperatures below 25°C and above 55°C (Novotny and Fives-Taylor, 1974), although quick chilling on ice maintains the pilus structure. The IncH mating systems can be extremely thermosensitive with a temperature optimum of 20°C which correlates with pilus production (Maher *et al.*, 1993). The *virB* region of the Ti plasmid expresses a pilus at 19°C, which appears to be important for T-DNA transfer to the infected plant nucleus (Fullner *et al.*, 1996).

Conjugative pili are often expressed on the cell surface at low levels of 10^{-2}–10^{-3} per bacterium, suggesting that a regulatory system exists to limit pilus production. Other systems appear to be expressed constitutively giving one to five pili per cell. Overexpression of conjugative pili can increase the number of pili per cell to about 20, but the most noticeable effect is the increase in the number of pili accumulating in the medium (Grossman and Silverman, 1989; Maneewannakul *et al.*, 1992a; Haase *et al.*, 1995). The pED208 plasmid, which expresses the F transfer operon from a promoter within an IS*2* element, is truly multi-piliated with up to 50 pili per cell (Armstrong *et al.*, 1980).

F Pilus Structure and Assembly

The conjugative pilus is a branch on a complex evolutionary tree that encompasses many aspects of specialised protein secretion, pilus-mediated cell movement and DNA transport. The F-pilus is an extracellular filament 1–5 μm in length, which is composed of identical 7.2-kDa pilin subunits arranged in a helical array of five subunits per turn with a rise per turn of 1.28 nm. The pilus is 8.0 nm in diameter with an internal lumen of 2.0 nm, a space that could accommodate a single strand of transferring DNA. A basal knob is characteristic of many types of pili. Other features used to characterise pili are aggregation and rigidity, as well as the presence of a pointed tip (Paranchych and Frost, 1988).

Studies from Karin Ippen-Ihler's laboratory have shown that F propilin (121 amino acids) has a 51 amino acid leader sequence which is cleaved by LepB (leader peptidase B) of the host. The nascent propilin polypeptide is inserted in the inner membrane in a Sec-independent manner which requires TraQ, a chaperone-like inner-membrane protein of 10.9 kDa (Majdalani and Ippen-Ihler, 1996). Tn*phoA* fusion protein analysis has determined the orientation of the pilin subunit in the membrane before assembly, with the N- and C-termini exposed in the periplasm (Paiva *et al.*, 1992). The length of the leader sequence is not important for propilin processing, since it can function in a TraQ-independent manner to transport β lactamase to the periplasm (Majdalani *et al.*, 1996). After cleavage, the pilin subunit undergoes acetylation at the N-terminus by TraX, which forms the major epitope on the pilin protein. Oddly, this epitope is not detectable on the surface of the pilus; furthermore, acetylation is unimportant in phage or recipient-cell- binding and *traX* mutants mate with normal efficiency (Moore *et al.*, 1993).

The first step in pilus maturation is the formation of a specialised structure at the cell surface which appears to be the unextended pilus tip. The formation of this tip is affected by mutations in *traL, E, K, C, G* as measured by filamentous phage infection, which requires only this tip structure for infection (Schandel *et al.*, 1987). The presence of this tip structure is not sufficient for mating, even on solid surfaces, suggesting that pilus outgrowth is required for DNA transfer. Pilus extension requires TraB, F, H, W, V, and TraP has been associated with pilus stability (Anthony *et al.*, 1999). The extended pilus contains pilin subunits organised in such a way that the C-terminus is exposed laterally and the N-terminus is buried within the structure. The C-terminus has been used as a site for fusing foreign epitopes or peptides which support epitope-specific phage infection (Borrebaeck *et al.*, 1997; Rondot *et al.*, 1998). Two clusters of positively charged amino acids, which separate two α-helical segments, are oriented towards the interior and exterior of the pilus (Silverman, 1997). The first cluster is involved in phage attachment whereas mutations in the second cluster, intriguingly, affect DNA transfer (J. Manchak and L.S. Frost, unpublished results). The phenotype of these latter mutants resembles that of *traU* mutations which affect DNA transfer rather than pilus assembly (Moore *et al.*, 1990).

In the IncH system, pilus outgrowth is a relatively slow process and RNase-treated phage particles attached to nascent pili emerging from the cell surface remain attached to the distal end of the pilus as it increases in length, supporting the model for pilus outgrowth from its base (Maher *et al.*, 1993).

The identity of pilin proteins in other transfer systems has been less studied. The transfer system of RP4 is encoded by two transfer regions, Tra1 and Tra2, containing genes called *tra* and *trb*, respectively (Lessl *et al.*, 1992). Pilus assembly in this system requires all but *trbK* of the Tra2 core region (i.e. *trbB, C, D, E, F, G, H, I, J, K* and *L*) plus *traF* of Tra1 (Lessl *et al.*, 1993). The remaining Tra1 core genes are involved in DNA processing during transfer (Pansegrau *et al.*, 1994a). There appear to be 20 pilus-assembly sites per *E. coli* cell, as monitored by the pilus-specific phage PRD1. The meaning of this observation in light of the relatively low copy number of RP4 is unclear since presumably only one site is needed for plasmid transfer (Kotilainen *et al.*, 1993). The pilin subunit of the RP4 transfer system, TrbC, undergoes three cleavage events before its assembly into the RP4 pilus, which are carried out by the host leader peptidase as well as TraF, a leader peptidase homologue (Haase and Lanka, 1997). A surprising discovery is the circularisation of the pilin polypeptide in a head-to-tail fashion for RP4 pilin and T-pilin of the Ti plasmid (Eisenbrandt *et al.*, 1999). The packing of these subunits into the pilus will be of considerable interest.

Another extensively studied system is that of the thin flexible pilus of IncI1 plasmids such as ColIb-P9 and R64. These plasmids encode thin flexible and thick rigid pili as products of separate transfer regions. The thin I1 pilus is not required for DNA transport itself but promotes the formation or stabilisation of mating pairs in liquid media. As judged by the sequence of the major pilin subunit (PilS, 19 kDa) and assembly components, the thin I1 pilus is related to type IV host-encoded pili (Yoshida *et al.*, 1998, see Concluding remarks). Thin I1 pili also have a minor protein, PilV, whose sequence is changed by re-arrangements of a ~2-kb multiple DNA inversion region called the shufflon. This locus also encodes a site-specific recombinase (Rci) of the phage λ integrase family, which mediates these re-arrangements. Site-specific recombinations in the shufflon select one of several C-terminal segments for the PilV protein (PilVA, VA′, VB etc.). The various forms of PilV affect the conjugation efficiency of different enterobacteria in liquid, probably by determining recognition of different LPS receptors on recipients. For example, PilVA′ promotes recognition of *E. coli* whereas PilVB′ interacts with *S. typhimurium* LT2 (Gyohda *et al.*, 1997). The possibility that PilV is located at the pilus tip as an adhesin is an intriguing speculation.

Pilus Retraction

Binding of the F pilus tip to the recipient cell is thought to trigger pilus retraction, although there is little evidence for this event and no data on the route of the pilus subunits during the process. Pilus attachment to a recipient cell might trigger retraction by briefly altering the membrane potential of the donor cell, or the process might depend on the constant outgrowth and retraction of the pilus. Retraction of the F pilus has been observed at room temperature, but not with quick chilling on ice, at high temperature or in the presence of metabolic poisons such as cyanide or arsenate (Paranchych and Frost, 1988). Mutants altered in F *trbI* give extremely long pili as if they have a retraction defect but remain transfer proficient (Maneewannakul *et al.*, 1992b); this is reminiscent of the non-retractile pili of the F-like plasmid ColB2 (Frost *et al.*, 1985). Pulse-chase experiments from the Ippen-Ihler laboratory have demonstrated that turnover of membrane F-pilin is slow but its loss can be accelerated by blending the cells, suggesting that pilus release into the medium is normally negligible (Sowa *et al.*, 1983). Although mating-pair formation and transfer efficiency are greatly affected by the addition of sodium dodecyl sulphate (SDS) to F^+ cultures, the number of transconjugants rises with time suggesting that collisions between cells are sufficient for transfer to occur and that pilus retraction promotes mating pair formation whereas pilus extension is important for DNA transport (Frost and Simon, 1993).

Other pili, such as that of RP4, are usually found detached from cells but are as essential for transfer as the F pilus (Fig. 1). Since the RP4 pilus is required for both intrageneric and intergeneric transfer, the pilus system must be important in forming the intercellular transport apparatus, or transferosome. Systems such as that of RP4, which preferentially cause mating of cells on surfaces, do not have the property of pilus retraction and establish cell-to-cell contact by collision.

The conjugative pilus is the receptor for a wide range of bacteriophages, including the small spherical RNA-containing phages, the filamentous ssDNA-containing phages and a number of double-stranded (ds) DNA phages which may or may not have associated lipid (Fig. 2; Frost, 1993). The RNA phages usually attach to the side of the pilus but the filamentous ssDNA phages attach either to the tip, as is the case for f1, fd and M13 adsorption to the F pilus, or to the side, as found for Pf3 attachment to the RP4 pilus. The dsDNA phages usually attach near the tip of the pilus, often to the conical point at the tip which is clearly visible on a number of the rigid pilus types. Pilus retraction has been shown to be important for Ff (F-specific filamentous) phage infection (Jacobson, 1972). However, the need for retraction in the infection process for other phage has not been established. There is some information on the route of entry of the nucleic acid of the F-specific phages. The RNA phage R17 and its relatives – but not Qβ, a closely related phage of a separate group – have been shown to require the TraD protein, one of the essential proteins for conjugative transfer of F. Entry of Ff phage into the cell requires a second set of receptors consisting of the bacterial TolQRA protein complex based in the inner membrane (Webster, 1996). The phage attach to the tip of the pilus via the D1–D2 domains of the g3p attachment protein. On contact with the cell surface, the D1 domain, which is now exposed, interacts with the TolA protein extending into the periplasm and the phage is imported via the TolQRA uptake system (Click and Webster, 1997; Riechmann and Holliger, 1997). Recently, the crystal structure of the TolA–g3p complex has been solved (Lubkowski *et al.*, 1999).

Mating-pair Formation

After F-pilus retraction, cells come into close contact to form ‘mating junctions’ which are clearly visible in the thin-section electron micrographs of Dürrenberger *et al.* (1991). A junction is characterised by an electron-dense region, probably of protein, between the cells; there is no obvious fusion of the cell envelopes or pore structure (Fig. 3a). One donor cell can form junctions with several surrounding recipient cells, presumably giving complexes equivalent to the mating aggregates described by Achtman *et al.* (1978). These tight aggregates render the cells resistant to the addition of SDS and to the shear forces generated

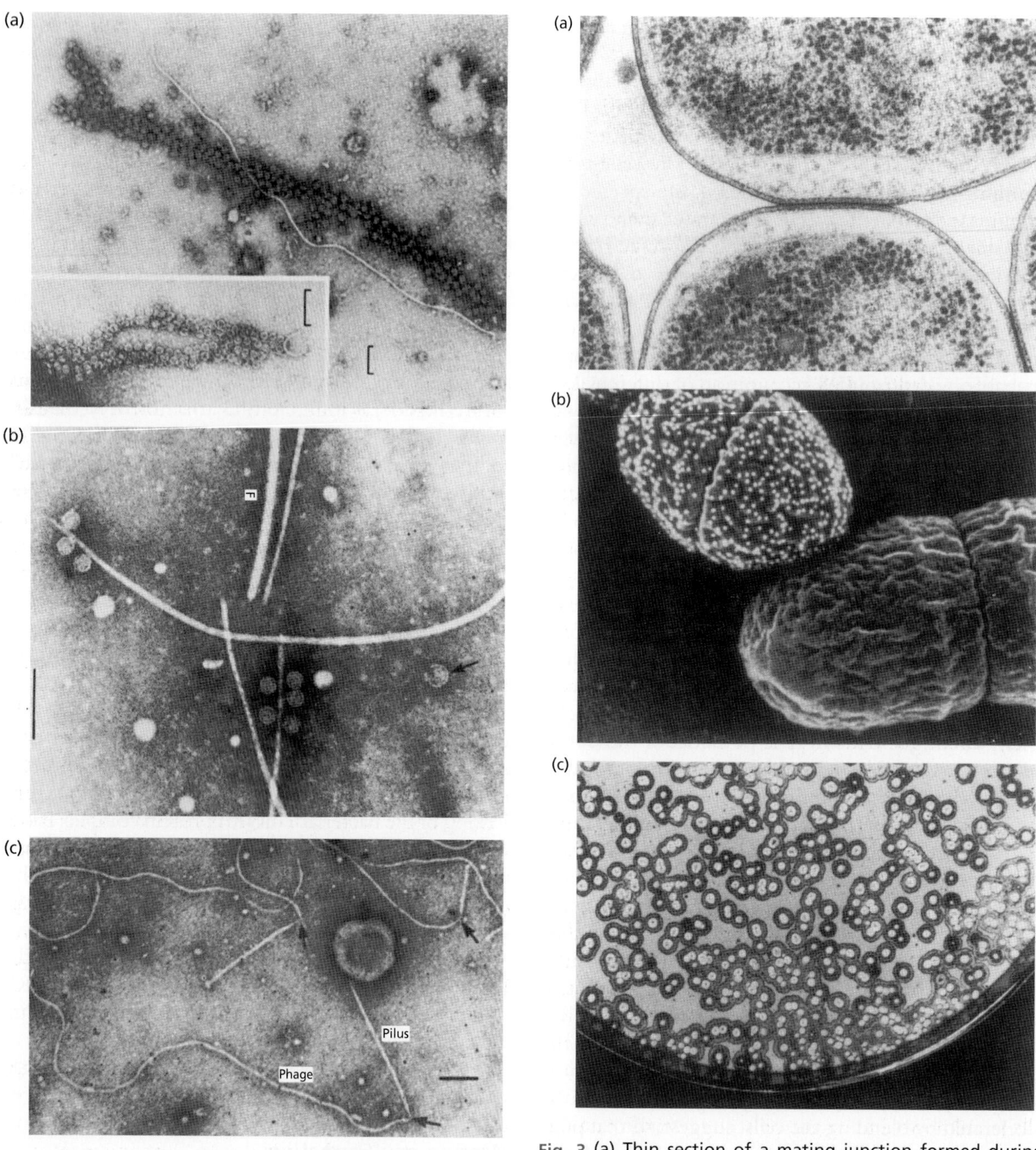

Fig. 2 Diversity of conjugative pili and their phages: (a) Thick flexible pilus of plasmid F (IncFI) coated with MS2 RNA phage lying next to an uncoated F0*lac* (IncFIV) pilus; inset shows pili expressed by R711b (IncX) coated with MS2 virions treated with formalin to allow attachment to otherwise resistant pili (Bradley and Meynell, 1978; used by permission of *Microbiology*). (b) Thick flexible T pili expressed by plasmid pIN25 with tRNA phages attached to their distal tapered points; F indicates a flagellum (Bradley *et al.*, 1981a; used by permission of *Microbiology*). (c) Rigid N pili with filamentous X virions attached to the tips; the pili have a nail-like appearance (Bradley *et al.*, 1981b; used by permission of *Microbiology*).

Fig. 3 (a) Thin section of a mating junction formed during plasmid F-mediated conjugation (Dürrenberger *et al.*, 1991; used by permission of *Journal of Structural Biology*). (b) An *Enterococcus faecalis* mating pair determined by low-voltage scanning electron microscopy using a back-scatter detector. The donor is labelled with monoclonal antibodies against 'aggregation substance' (Asc10) conjugated to 12 nm gold particles (Ruhfel *et al.*, 1997; From *Bacteria as Multicellular Organisms*, edited by J.A. Shapiro and M. Dworkin. Copyright 1997 by Oxford University Press, Inc. Used by permission of Oxford University Press, Inc.). (c) Pocks formed by the presence of occasional spores of a *Streptomyces* donor of plasmid SCP2* on a dense lawn of a plasmid-free recipient strain of *S. coelicolor* (photograph courtesy of M. Bibb, John Innes Centre, Norwich, UK).

by pipetting. Hfr donors form very durable mating aggregates which must be disrupted by mechanical agitation while aggregates involving F^+ donors actively dissociate soon after transfer. Fully piliated donor bacteria clump to form large cell aggregates that are visible with the naked eye, but these aggregates are easily dispersed by vortexing.

The junction between mating cells requires that the outer membranes of the two cells come into close apposition. Presence of O-side chain in the LPS of the recipient cell can dramatically reduce the mating efficiency for some plasmids (e.g. F but not RP4), suggesting the LPS physically hinders mating-pair formation (Bradley, 1989). Stabilisation of F-mediated mating pairs requires two transfer gene products, TraN and TraG, in the donor cell and the LPS inner core and OmpA protein in the recipient cell (Anthony *et al.*, 1994). TraN (60 kDa) is essential for transfer and recognises different moieties on the recipient cell; for example F TraN interacts with OmpA (Klimke and Frost, 1998). TraG is a 103-kDa inner membrane protein with a large periplasmic domain which is thought to undergo post-translational processing to release a 53-kDa fragment called TraG* into the periplasm. The N-terminal domain is implicated in F-pilus synthesis, and the C-terminal domain, possibly TraG*, is responsible for mating-pair stabilisation, although there is considerable overlap between the two functions. Plasmid specificity is exhibited by F and R100-1 TraG for their respective entry exclusion protein, TraS, when the latter is expressed in the recipient cell. Possible mechanisms include translocating TraG alone, or exposed at the pilus tip, into the inner membrane of the recipient cell (Anthony *et al.*, 1999). Firth and Skurray (1992) have suggested that TraN and TraG could form a protein bridge that spans the entire cell envelope from inner to outer membrane of the donor cell with the proteins for tip formation and pilus assembly associated with this complex (Fig. 4).

The mechanism of active disaggregation of mating cells is unknown but might be related to the finding that transfer of IncI1 plasmids is terminated by a process requiring *de novo* protein synthesis in the newly infected cell (Boulnois and Wilkins, 1978). Achtman (reviewed in Achtman and Skurray, 1977) inferred that active disaggregation involved a gene product from the distal portion of the *tra* operon, making TraT, the surface exclusion protein in the outer membrane, a candidate. Riede and Eschbach (1986) proposed that TraT functions to block access to the OmpA protein by the transfer apparatus of a second donor cell, resulting in reduced mating of plasmid-containing cells. TraT might have a second function in interfering with TraN–OmpA interactions, thereby leading to dissolution of the mating junction. Such a process would require rapid expression of TraT in the transconjugant cell.

Entry or Surface Exclusion

Most conjugative elements encode an exclusion (Exc) system that reduces redundant transfer between cells harbouring the same or a closely related plasmid. The function may be limited to systems allowing very efficient conjugation, since conjugative transposons – which transfer infrequently – appear to lack an Exc system. Plasmid F has two exclusion genes, *traT* and *traS*. TraT, a 26-kDa outer-membrane lipoprotein, reduces stabilisation of mating pairs in a process known as surface exclusion (Sfx; Perumal and Minkley, 1984). TraS, a 16.9-kDa inner-membrane protein, is thought to block DNA import from a potential donor cell (entry exclusion; Eex) by preventing the triggering of DNA transfer-replication in the donor (Manning *et al.*, 1981). Thus the signal that initiates transfer appears to be derived from the inner membrane of the recipient cell. This is an attractive hypothesis since the energised inner membrane of a living cell is one feature common to all conjugative recipient cells from bacteria to microbial eukaryotes and plants.

RP4 has a single entry exclusion gene, *trbK*, which encodes a small 6-kDa lipoprotein situated in the inner membrane. TrbK acts after mating-pair formation and is believed to block steps in the formation of the conjugation pore within the recipient cell envelope (Haase *et al.*, 1996). The exclusion functions of IncI1 plasmids and pheromone-responding enterococcal plasmids likewise function after the formation of mating pairs (Hartskeerl and Hoekstra, 1984; Dunny *et al.*, 1985), as does F TraS. Interestingly, F TraT has also been associated with serum resistance (Sukupolvi and O'Connor, 1990); possibly the protein evolved as a virulence determinant.

Exclusion functions are also involved in protecting cells from 'lethal zygosis', a process in which recipient cells are killed in the presence of high numbers of Hfr donor cells apparently due to membrane damage caused by prolonged mating contacts. Transconjugants acquire immunity to lethal zygosis (Ilz) in matings with F^+ donors through receipt of the plasmid *ilz* loci (Skurray and Reeves, 1974), subsequently found to encompass *traS* and *traT* (Ou, 1980). Thus entry exclusion not only reduces redundant transfer but is seen through the Ilz phenotype to protect plasmid-containing cells at high densities from potentially lethal matings with sibling donor cells. The Ilz phenotype may be important in conferring rapid protection on newly formed transconjugant cells when a plasmid sweeps through a colony by epidemic

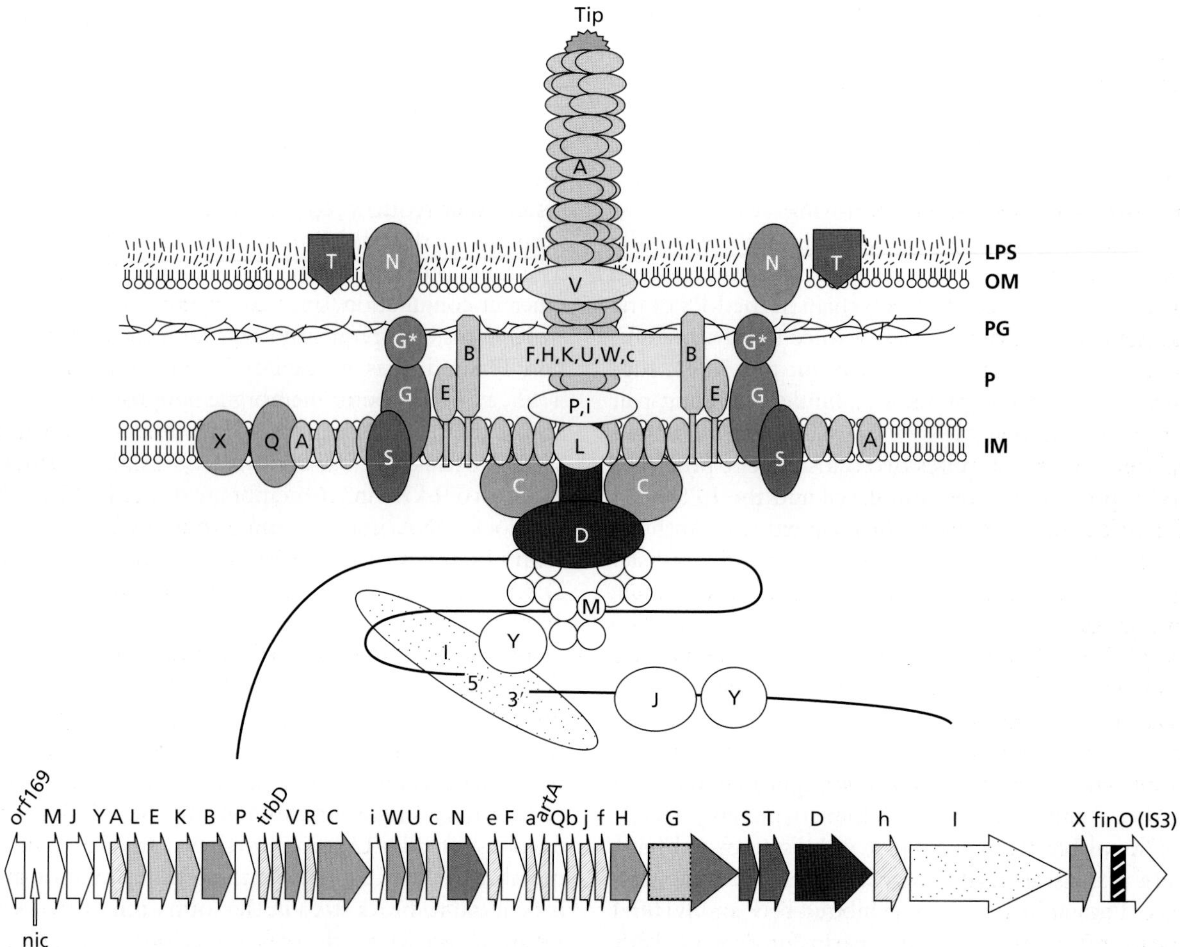

Fig. 4 A model of the F transfer apparatus and map of the F transfer region. The gene products for pilus synthesis (blue), mating-pair stabilisation (green), DNA metabolism (yellow to brown) and regulation (purple) are indicated at their approximate location within the cell envelope. The nature of the pilus tip is unknown but is thought to contact a receptor on the surface of the recipient cell. TraV, a lipoprotein in the outer membrane, might give lateral stability to the extended structure. TraF, H, K, U, W and TrbC in the periplasm and TraP and TrbI (white) in the inner membrane also support the pilus and are involved in pilus elongation and retraction. Mutants in *traP* have fragile pili which break off the cell easily, suggesting that TraP anchors the pilus in the inner membrane. Pili of *trbI* mutants are particularly long, implying that TrbI could be involved in retraction. TraB, a TonB/TolA analogue could extend into the periplasmic space and provide stability or mediate an unknown signal from the outer membrane to the inner membrane. TraE is an inner membrane protein of unknown function. TraL is a TraA (pilin) homologue that is expressed at very low levels and might be involved in pilus placement. Pilin is inserted in the inner membrane by the action of TraQ and is acetylated by TraX. TraC is involved in pilus assembly in which the α helices of the pilins are re-aligned from a vertical position in the inner membrane to a more horizontal position in the assembled pilus. TraN, which contacts OmpA in the recipient cell, might form a transenvelope complex with TraG* and TraG to stabilise the mating pair after contact is initiated. TraS and TraT (red) are inner and outer membrane proteins involved in entry exclusion and surface exclusion, respectively. TraI is an asymmetrical protein which cleaves at *nic* in a site- and strand-specific manner. This cleavage reaction is mediated by TraY and TraM which forms a nucleosome-like complex at *oriT* and interacts with TraD the proposed nucleic acid pump. TraD is positioned at the base of the pilus in an unknown manner. The LPS (lipopolysaccharide), OM (outer membrane), PG (peptidoglycan), P (periplasm) and IM (inner membrane) are indicated on the right. The transferable DNA strand cleaved at *nic* with the 5′ end, which precedes the leading region, is indicated. Proteins of unknown function are grey and the lytic transglycosylase, Orf169, is indicated in pink. (See also Colour Plate 06).

spread. Fertility inhibition sets in about six generations later to repress the transfer system to a very low constitutive level until a new source of recipient cells is encountered (Simonsen, 1990).

Relaxosomes and DNA Transport

Formation of a functional mating junction in Gram-negative bacterial conjugation prompts the rapid initiation of DNA transfer into the recipient cell without prior transcription and translation. This process proceeds at about 45 kb per minute at 37°C; thus F transfer occurs in just over 2 minutes. The mechanism of transfer initiation is fairly well understood for both self-transmissible and mobilisable plasmids (Lanka and Wilkins, 1995; Zechner *et al.*, 1999). Initiation requires formation of the relaxosome, a nucleoprotein complex consisting of negatively supercoiled *oriT* DNA and at least one transfer-region protein called relaxase. This enzyme mediates a strand- and site-specific cleavage-joining reaction at *nic* within *oriT*. The nicking reaction results in covalent attachment of the relaxase to the 5′-terminal nucleotide of the cleaved strand; the 3′ end can act as a primer for replacement strand synthesis by a DNA polymerase via a rolling circle mechanism (Pansegrau *et al.*, 1990). The nucleotide sequence around the *nic* site of a number of plasmids can be aligned to give five groups typified by F, RP4 and three mobilisable plasmids, namely RSF1010, ColE1 and the streptococcal plasmid pMV158 (Lanka and Wilkins, 1995; Guzmán and Espinosa, 1997). A single group may contain conjugative and mobilisable plasmids from both Gram-negative and Gram-positive bacteria, attesting to the evolutionary relatedness of transfer systems operating in diverse bacterial taxa.

The *oriT* region of a plasmid can range from 38 to ~500 bp in size and usually contains sequences that dictate intrinsic bends as well as direct or inverted repeats which act as binding sites for proteins that alter the structure of the DNA and presumably bring the *nic* site into register with the relaxase and transport proteins. Binding sites for bacterial protein integration host factor (IHF) and TraY have been located in F in the region 3′ of *nic*, which is theoretically the last portion of the plasmid to enter the recipient cell (Gao *et al.*, 1994). Either the dimer or monomer form of TraY (15.2 kDa), which resembles the Arc and Mnt phage repressors, binds high-affinity sites near *oriT* and the P*traY* promoter region at the head of the major 33 kb transfer operon (Nelson and Matson, 1996). Matson and co-workers have shown a stepwise assembly in which TraY in conjunction with IHF binds at *oriT* and correctly integrates TraI, the relaxase, to allow specific cleavage at the nick site (Howard *et al.*, 1995).

TraY is also required for optimal expression of *traM* (Penfold *et al.*, 1996) and the long *traY-traI* operon (Silverman and Sholl, 1996). TraM (14.5 kDa) is an essential transfer protein which forms a tetramer and binds to three sites near *oriT*, two of which coincide with promoters for *traM*, resulting in autoregulation of its expression (Penfold *et al.*, 1996). On the basis of its ability to alter the superhelical density of supercoiled DNA, TraM is thought to aid formation of a nucleosome-like structure which might have a role in strand separation prior to unwinding by the TraI relaxase-helicase (Zechner *et al.*, 1999). Using affinity chromatography, TraM has been shown to interact with TraD (81.7 kDa), an inner membrane protein implicated in the transport of DNA during conjugation (Disqué-Kochem and Dreiseikelmann, 1997). TraM may direct the relaxosome to the base of the transferosome, which contains TraD and possibly the pilus assembly/retraction apparatus. TraM is specific for its cognate DNA-binding sites but is promiscuous in its ability to interact with other plasmids' transfer machinery including TraD which does not share the property of plasmid specificity (Kupfelweiser *et al.*, 1998; Sastre *et al.*, 1998). This result has been confirmed using chimeras of the F and R100-1 *oriT* regions, which show that TraM must bind to the plasmid destined for transfer (Klimke *et al.*, 1997).

The observations of Gordon *et al.* (1997) further complicate the connection between the transferosome and the relaxosome. Using a multi-copy *lacO*/LacI-GFP reporter system, they analysed the position of the F plasmid throughout the cell cycle. Their findings suggest that F replicates at the midpoint of the cell and abruptly moves to the 1/4 and 3/4 positions just before cell division. These results suggest that potential transferosomes are not randomly arranged over the cell surface but occur at discrete positions that coincide with the position of the plasmid. Furthermore, conjugation might only occur at certain times in the cell cycle and would be co-ordinately regulated with vegetative replication as well as the phase of cell growth. Although no connection between F vegetative replication and conjugation has been noted, such a connection has been demonstrated for other systems such as RP4. Thus the relaxosome might be free of the transferosome except in response to the establishment of a stable mating pair. For F, TraM and TraI are located in the cytoplasm and preliminary experiments suggest that TraI, but not TraM, is present in the membrane in response to the addition of recipient cells (R. Fekete and L.S. Frost, unpublished results) suggesting that the relaxosome associates with the transferosome only during mating.

Although there are five prototype *nic* sequences, there are four classes of relaxase (F, RP4, RSF1010 and pMV158; Zechner *et al.*, 1999) grouped on the basis of homology and sequence specificity for *nic*. F TraI, a large (192-kDa) cytoplasmic protein, has at least two functional domains. The N-terminal domain contains the relaxase activity and the large C-terminal domain has DNA helicase or unwinding activity. The protein cleaves at *nic* and enters into a phosphodiester bond with the 5′ phosphoryl group, through one of the two pairs of tyrosines in the N-terminal domain. TraI is also involved in the termination reaction: here the protein is thought to use its coupled cleavage-joining activity to process the transfer intermediate and generate a monomeric single-stranded circle of DNA that is released into the recipient cell. Termination requires an inverted repeat in the trailing portion of a transferring strand, which is located to the 5′ side of *nic* (Gao *et al.*, 1994). Genetic evidence suggests that F DNA travels through the conjugation pore at a rate of ~750 nucleotides/s at 37°C, which approximates to the rate of TraI-mediated DNA unwinding *in vitro*. Conjugative DNA synthesis is not required for transfer; an attractive model suggests that the helicase activity of TraI drives transport through a pore formed by TraD and pilus assembly proteins.

Events occurring at *oriT* of the RP4 plasmid have been worked out in detail by Lanka and associates (Lanka and Wilkins, 1995; Zechner *et al.*, 1999). Relaxosome assembly in this system does not depend on any host protein such as IHF, which perhaps contributes to the broader transfer range that IncP plasmids enjoy. Four proteins are required, TraH, TraI, TraJ and TraK, which are encoded by the relaxase operon of the Tra1 core region. RP4 relaxase is the 82-kDa TraI protein, coincidentally given the same name as the F counterpart. The enzyme carries out the phosphotransfer reaction becoming covalently linked via tyrosine-22 to the 5′ end at the specific cleavage site in *oriT*. Histidine-116, located in a motif conserved in rolling-circle replication-initiation proteins, is also required for relaxase function and probably activates the hydroxyl group of Tyr22 within the active site of the enzyme. Relaxosome formation involves cascade-like events in which TraJ binds to a specific site in *oriT* to give a complex allowing binding of TraI. TraH stabilises this complex through interactions with TraJ/TraI (Pansegrau *et al.*, 1994b); TraK is similar to F TraM in that multiple molecules bind to an intrinsically bent region of *oriT* to cover some 180 bp to form a nucleosome-like structure observable by electron microscopy (Ziegelin *et al.*, 1992).

RP4 TraG belongs to an important family of inner membrane proteins that are necessary for DNA transfer in diverse Gram-negative and -positive systems and include F TraD, pTi VirD4 and TrwB of R388 (IncW). These homologues contain at least one consensus nucleotide-binding motif essential for function and are thought to use the energy of NTP hydrolysis to couple the activated relaxosome to the transport machinery (Lanka and Wilkins, 1995). RP4 TraG is essential for conjugative DNA export but not for pilus production or donor-specific phage propagation (Lessl *et al.*, 1993).

One interesting aspect of the transfer systems of IncP and IncI1 plasmids, among others, is the presence of a DNA primase gene (*traC* in RP4, *sog* in ColIb-P9; Lanka and Wilkins, 1995). The genes encode two large in-frame translational products which are highly anisometric. The proteins are not essential for conjugation, but increase its efficiency in some recipient strains. The larger TraC polypeptide and both Sog proteins are transported in large amounts per DNA strand into the cytoplasm of the recipient cell where they facilitate synthesis of the complementary DNA strand, presumably by generating RNA primers. They are the only proteins clearly observed to be transported from the donor to recipient cell in bacterial conjugation systems (Rees and Wilkins, 1990). The situation is reminiscent of T-DNA transport by the Vir system, where a nucleoprotein complex consisting of VirD2 bound to the 5′ end of the T-DNA, as well as the single-stranded DNA-binding VirE2 protein, is transported into the plant cell. An important question is the mechanism of transport of these plasmid primase proteins: are they bound to the transferring DNA strand, giving a complex of considerable mass for transport through the pilus, or is the protein transported separately to bind the DNA in the recipient cell? The latter model parallels the independent transfer of VirE2 and T-DNA into plant cells, which might be the normal mode of transfer during tumourigenesis (Sundberg *et al.*, 1996). Plasmid primases characterised to date have been associated with rigid pilus, surface-preferred, mating systems suggesting that there may be some fundamental differences in the nature of the conjugation pore between these and universal systems mediated by flexible pili.

Structure of the Conjugative Pore

Comparison of the genes for transport of the *Bordetella pertussis* toxin (*ptl*), the T-DNA complex (*virB*), nucleoprotein transport by the RP4 (*trb*) and DNA transport by F (*tra*) reveals some interesting clues to function (Table 1). The simplest of these systems, *ptl*, consists of nine genes whose products comprise a pore for release of the assembled toxin from the periplasm to the exterior of the cell. Homologues for pilin and pilus tip assembly are present but no extended pilus is expressed. The next level of complexity is the *vir* region of the Ti plasmid of *A. tumefaciens*, which promotes transfer over a considerable phylogenetic distance from a bacterium to a plant cell. The transport apparatus for the T-DNA nucleoprotein complex requires the expression of 11 genes within the *virB*

Table 1 Increasing levels of complexity of type IV secretion and conjugative transfer systems

Characteristics	Pore[a]	+Pilus	+Sensing	+Retraction
System	Ptl	Vir	P (Trb)	F (Tra)
'Lysozyme'[b]		B1	N	Orf169
Pilin	A	B2	C	A
Nucleator	B	B3	D	L
Tip/secretin (NTP)[b]	C	B4	E	C
Tip/secretin		B5	F	E
Anchor	D	B6	L	
OM lipoprotein	I	B7	H	V
Structure	E	B8		
Structure	F	B9		
Structure	G	B10	I	B
Transport (NTP)	H	B11	B	
Structure			G	K
Structure			J	
Retraction				U,W,F,H,G^N,TrbC,I
Entry exclusion			K	S
Mating-pair stabilisation				G^C,N
Surface exclusion				T
DNA transport		D4	TraG	D
Relaxase (NTP)		D2	TraI	I
Helicase				I
DNA primase			TraC (Pri)	
ssDNA-binding protein		E2	SSB (RP4)	SSB (F)

[a] See text for details.
[b] An ATP/NTP binding motif has been predicted/demonstrated.
G^N and G^C refer to the N- and C-termini of TraG.

operon for T-pilus expression in response to a complex signalling pathway between a wounded plant and the bacterium. The signalling pathway induces transcription and translation of *vir* genes rather than immediate conjugative DNA synthesis and transfer via a pre-existing transport system. The RP4 pilus, expressed by the *trb* genes, is slightly more complicated; here pilus outgrowth is constitutive and contact with a recipient cell is sensed to trigger immediate DNA transfer. RP4 pilus retraction has not been demonstrated, which matches the impaired ability of RP4 donor cells to mate in liquid media. Lastly the F-like systems support efficient mating in liquid media; compared to RP4, these systems have extra genes which could permit pilus retraction in addition to the ability to detect effective mating.

Although there is some homology between the F and RP4 transfer systems, a case can be made that the two systems are quite distinct. The RP4 system can transport protein–DNA complexes and is a true type IV secretion system, whereas the F system appears to transfer DNA alone. Both require pilus outgrowth for DNA transfer, but RP4 does not have an F TraG homologue, which is essential to F transfer and might be translocated to the recipient cell perhaps by a mechanism akin to the syringe-like phage tails of T-even phages. The energy for DNA transfer appears to be different in the RP4 and F systems. Although both have homologues associated with pilus assembly (TrbE and TraC, respectively) and DNA transport is envisioned to be supplied by TraG or TraD, F lacks a homologue of TrbB, a protein whose relatives belong to type IV secretion systems. On the other hand, RP4 apparently lacks a helicase activity such as that supplied by the C-terminal domain of F TraI, the relaxase. One solution to this puzzle could involve an assembly including TrbB, which rotates and drives the RP4 DNA into the recipient cell, similar to that proposed for phage-head packaging (Hendrix, 1998) while unwinding the DNA by the helicase supplies the force needed to drive the DNA through the pore during F transfer.

On the basis of genetic data, sequence comparisions and limited knowledge of protein–protein interactions, a model for the F transfer apparatus can be constructed (Fig. 4). The outer membrane contains TraN, a possible secretin, which is thought to interact with TraG to form a bridge spanning the donor cell envelope. Since TraG has a role in pilus synthesis, other proteins required for pilus biogenesis (TraA, L, E, K, B, V, C, W, U, F, Q, H, X and TrbC) should be associated with this structure. Contact with the recipient cell could allow TraN to translocate TraG to the recipient cell with the pilus transport apparatus embedded in this structure. TraC is involved in pilus elongation after the pilus tip structure, composed of a separate protein or an alternate assembly of TraA pilin subunits, is constructed and exposed at the cell surface. This model predicts that TraD should interact with one or more of the proteins involved in pilus assembly, as well as with TraM which anchors the relaxosome to TraD. TraV is a lipoprotein inserted in the outer membrane to face the periplasmic space (Doran *et al.*, 1994); it might be tethered to the peptidoglycan and give lateral stability to the pilus structure as it extends from the cell. TraB is a periplasmic protein with a cytoplasmic membrane anchor; it is predicted to have an extended structure based on its overall similarity to the TolA and TonB proteins, which also extend into the periplasmic space, and its sequence similarity to VirB10 (Christie, 1997). An attractive hypothesis is that TraB might help sense cell-to-cell contact and transmit this signal to the transfer apparatus. Keeping in mind that the structure of phage tails, flagella and type III secretion needles are composed of interlocking cylindrical assemblies of proteins, it is expected that a similar arrangement will be found for transfer systems.

The nature of the conjugation pore has been under debate for many years, because the ultimate destination of the DNA during conjugation is the cytoplasm of the recipient cell. Mutations in recipient cells that limit DNA receipt to the periplasm have not been found. If the conjugation pore extends to the cytoplasmic membrane of the recipient cell, it would be interesting to know whether there is a pilot protein that leads the DNA into the recipient cell, perhaps to establish a new transferosome at the point of contact with the cytoplasmic membrane. Such a protein has yet to be detected but a candidate is the relaxase, tethered to the 5′ end of the DNA.

Establishment of Plasmids in the Recipient Cell

The immigrant DNA must overcome the hostile environment often present in the new host as the prelude to establishing a site for replication and partitioning during cell division, as well as for further transfer. The leading region, which is the first portion of a plasmid to enter the recipient cell, encodes a number of interesting proteins for plasmid survival (Bates *et al.*, 1999; Manwaring *et al.*, 1999). In F-like plasmids, the first gene (F *orf169;* R1 *gene19*) to enter the recipient cell encodes an inner-membrane protein having a motif indicative of a family of bacterial transglycosylases involved in peptidoglycan chemistry (Bayer *et al.*, 1995). This protein is interesting in view of the need for a mechanism to allow the pilus and/or conjugation tube to penetrate the peptidoglycan layer of the cell envelope, as well as to construct a new pilus assembly site in the transconjugant cell. Mutations in this gene do not greatly reduce conjugation efficiency and have a modest effect on RNA phage infection in the laboratory. The same motif is found in proteins of other transfer systems, which include TrbN (IncP), TraL (IncN), PilT (IncI1) and VirB1 (pTi). The F leading region also carries the *flm* plasmid maintenance system, which is similar to the *hok/sok* addiction system of plasmid R1 (Loh *et al.*, 1988), as well as *ssiE,* one of the several ssDNA replication-initiation sites located on the plasmid and utilised by host enzymes during DNA replication (Nomura *et al.*, 1991). Presumably some such site is necessary in systems lacking a plasmid primase to allow initiation of complementary strand synthesis. The *ssiE* site also functions as a novel type of promoter (designated F*rpo*) that specifically initiates transcription of single stranded DNA (Masai and Arai, 1997).

The leading regions of many plasmids of the F and I complexes carry *ssb* and *psiB* genes that are strikingly conserved yet inessential for basic plasmid functions (Golub *et al.*, 1988; Chilley and Wilkins, 1995). The *ssb* gene encodes a ssDNA-binding protein of the type essential for bacterial DNA replication, and the *psiB* locus specifies a protein that prevents induction of the bacterial SOS stress as a function of its intracellular concentration. Expression of these *ssb* and *psiB* genes is transiently induced in the newly infected cell to give a burst of expression that presumably prevents SSB starvation and SOS induction by the entrant DNA (Bagdasarian *et al.*, 1992; Jones *et al.*, 1992). IncP plasmids carry a *ssb*-like gene outside the leading region. Although the gene is located next to *trfA* in the replication module, it is strongly expressed following entry into a new cell giving transient over-production of the SSB protein (see section on regulation, below).

In addition to *ssb* and *psiB,* the leading regions of several types of plasmid (IncB, FV, I1, K and N, but not FI or P) determine an ArdA antirestriction system that protects the immigrant plasmid from type I restriction enzymes (see section on barriers to the horizontal transmission of genes). Again *ardA* is an 'early' gene that is rapidly expressed to give high levels of transcription within a minute or so of entry into the recipient cell. This rapid response appears to involve transcription of the incoming DNA strand (Althorpe *et al.*, 1999). Sequence data supports such a mechanism since, with the exception of *orf169*, all ORFs in the leading regions of F and ColIb are orientated such that the transcribed strand is the same as the conjugatively transferred strand. Moreover, the F leading region contains a second presumptive single strand promoter as a homologue of F*rpo* (Manwaring *et al.*, 1999), and there are three F*rpo* homologues dispersed in the ColIb leading region (Bates *et al.*, 1999).

Regulation of F Transfer Genes

Although temperature, oxygen levels and stationary phase affect transfer ability, the underlying regulatory mechanisms are unknown. Under ideal growth conditions, the transfer genes of F-like plasmids are usually repressed by a process termed fertility inhibition to a low level such that only a fraction of cells are competent donors. In the presence of recipient cells, fertile donors transfer immediately; there is no evidence that this is an induced response requiring signalling molecules released by recipients to trigger transfer gene expression. Newly formed transconjugants are fertile for approximately six generations before repression is established and this high frequency of transfer (HFT) state potentiates rapid spread of the plasmid throughout a recipient population.

In the F plasmid there are three primary promoters for the Tra operon, P_{traM}, P_{traJ} and P_{traY}. P_{traY} initiates transcription of the complete *tra* operon from *traY* to *traI* and possibly beyond (**Fig. 5**). There are secondary, less well-characterised promoters for the FinP antisense RNA which controls TraJ expression, as well as for *artA, trbF, traS, traT and traD*. P_{traY} is positively regulated by TraJ, TraY, SfrA and IHF whereby TraJ interacts with a protein–DNA complex and alleviates repression of the P_{traY} promoter (Silverman *et al.*, 1991b; Ptashne and Gann, 1997; Taki *et al.*,

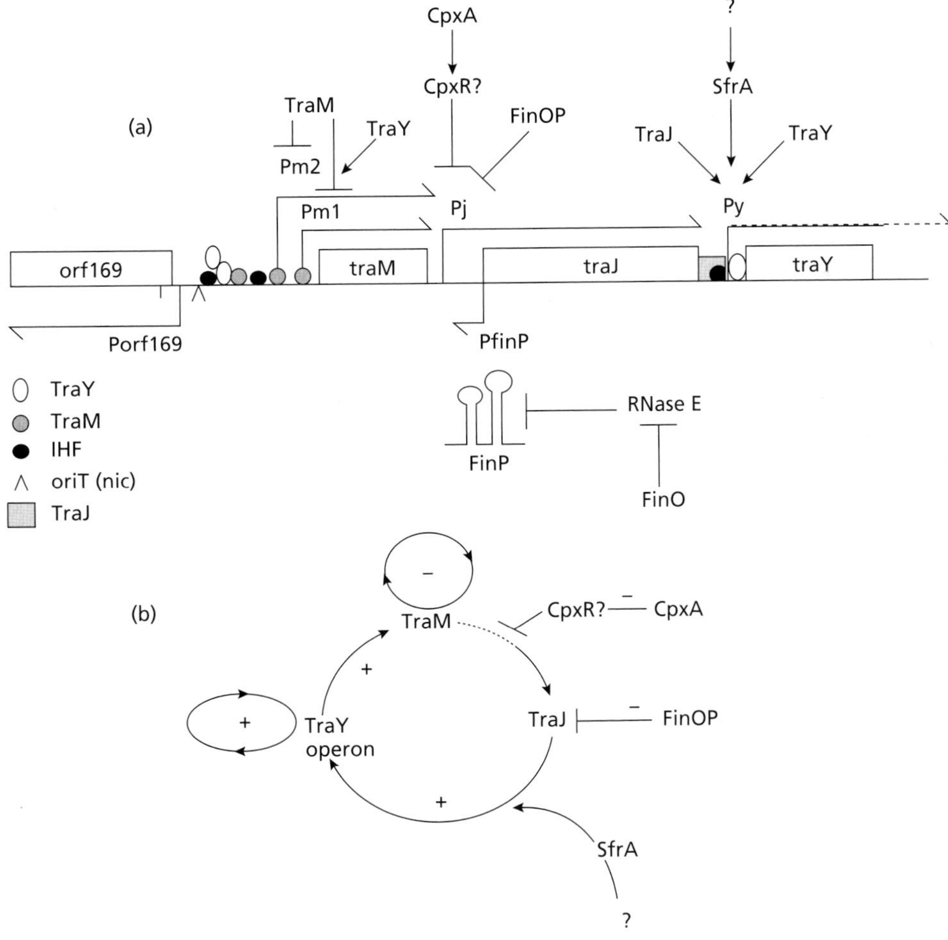

Fig. 5 Regulation of the F transfer region: (a) Control of the promoters at the head of the transfer region. (b) Interconnection between TraY operon expression and *traM* expression (adapted from Pölzleitner *et al.*, 1997). See text for details.

1998). The requirement for TraJ can be bypassed by substituting a foreign promoter for P_{traY} (Maneewannakul *et al.*, 1996).

Host factors affecting F transfer include ArcA (or SfrA), which has a binding site upstream of the P_{traY} promoter. This is required for maximal P_{traY} expression and is a response regulator belonging to the two-component signal transduction family (Strohmaier *et al.*, 1998). ArcA is also the regulator of aerobic respiration, but different domains of the protein seem to function in fertility and aerobic respiration control, answering to different sensor proteins (Silverman *et al.*, 1991a). A second two-component regulatory pair has been implicated in regulating *traJ* expression; the sensor, CpxA (conjugative pilus expression; Silverman *et al.*, 1993) and CpxR, a regulator that responds to CpxA and senses environmental stress outside the cytoplasmic compartment, including protein denaturation (Danese and Silhavy, 1997; Pogliano *et al.*, 1997) and phospholipid composition (Mileykovskaya and Dowhan, 1997). TraJ expression appears not to require CpxA under normal circumstances; instead, TraJ is down-regulated as part of the stress response. The SfrB protein, also known as RfaH, is a NusG homologue (Bailey *et al.*, 1996) which has anti-termination activity and ensures efficient transcription of the long (33 kb) *tra* operon.

The fertility inhibition system of F-like plasmids requires *finO* and *finP*, which comprise an antisense system for controlling *traJ*. The *finP* promoter is very weak compared to the opposing *traJ* promoter and the two promoters are sensitive to Dam methylation which partially relieves F repression (Torreblanca *et al.*, 1999). FinP is a ~78-nucleotide antisense RNA molecule which is complementary to the untranslated portion of the *traJ* transcript and has two stem–loop structures that interact with complementary loops in the TraJ mRNA. These interactions trigger duplex formation and occlusion of the ribosome binding site for TraJ translation (van Biesen and Frost, 1994). Alleles of *finP* show plasmid specificity, reflecting the need for pairing of complementary RNA molecules. The *finO* gene is located at the distal end of the transfer region. FinO protein binds to the FinP countertranscript and alters its rate of decay by blocking the action of the host endoribonuclease, RNase E (Jerome and Frost, 1999; Jerome *et al.*, 1999); the increased intracellular concentration of FinP efficiently blocks TraJ translation. In F an IS*3* element is inserted within *finO* (Cheah and Skurray, 1986), leading to constitutive expression of *traJ* and hence the transfer operon. However, FinO can be supplied *in trans* by a number of IncF plasmids. A regulatory circuit has been established for F whereby FinOP control of TraJ affects TraY production, which in turn co-ordinates expression of the promoters for *traY-I* and *traM*, an essential transfer gene (Penfold *et al.*, 1996). The system for the normally repressed plasmid R1 differs in that TraY-dependent transcription of *traM* positively affects *traJ* transcription with further stimulation of P_{traY} (Koraimann *et al.*, 1996; Pölzleitner *et al.*, 1997).

A number of fertility inhibition systems have been identified among unrelated plasmids, which define a hierarchy for transfer. For instance, F is repressed by *finO*$^+$ F-like plasmids, and it in turn represses RP4 via the PifC protein (Santini and Stanisich, 1998) which is part of a locus that interferes with phage T7 infection (Garcia and Molineux, 1995). RP4 in turn blocks the transfer of IncW plasmids via the *fiw* locus (Pansegrau *et al.*, 1994a). A number of plasmids interfere with F transfer using very different mechanisms (Firth *et al.*, 1996). Because of heterogeneity in these transfer inhibition mechanisms, research in this area has been slow and the importance of fertility inhibition to gene flow in the environment has probably been underrated.

Regulation of the RP4 Transfer Regions

There is no self-imposed fertility inhibition system operating on IncPα plasmids, in contrast to the F-like plasmids. Further differences are that the IncP system appears to be remarkably independent of host control and that transfer gene expression is co-ordinated with plasmid replication. RP4 transfer is not positively regulated but rather undergoes derepression via a complex control circuit (Motallebi-Veshareh *et al.*, 1992; Pansegrau *et al.*, 1994a; Jagura-Burdzy and Thomas, 1997). Transcription of the Tra1 core region involves divergent promoters (P_{traJ} and P_{traK}) arranged in back-to-back orientation in the *oriT* region, as well as the P_{traG} promoter. The Tra2 region is transcribed from P_{trbA} and the adjacent downstream signal P_{trbB}. The P_{trbA} promoter overlaps a divergent strong promoter (P_{trfA}) responsible for transcription of the replication region consisting of the *ssb* gene and the *trfA* locus encoding replication-activation proteins. Expression from the P_{trfA} and P_{trbA} promoters is integrated by the overlapping control region, thereby co-ordinating vegetative replication and transfer of the plasmid (Fig. 6).

There has been considerable interest in unravelling the control mechanism at the P_{trfA}/P_{trbA} junction (Jagura-Burdzy and Thomas, 1997). KorA protein, which is one of the global regulators of the plasmid encoded by the central control region, acts as a switch. At low concentrations of KorA, as in a new transconjugant, the extremely strong P_{trfA} promoter dominates over P_{trbA}, favouring plasmid replication. However, the reasonably strong P_{trbB} promoter is activated in

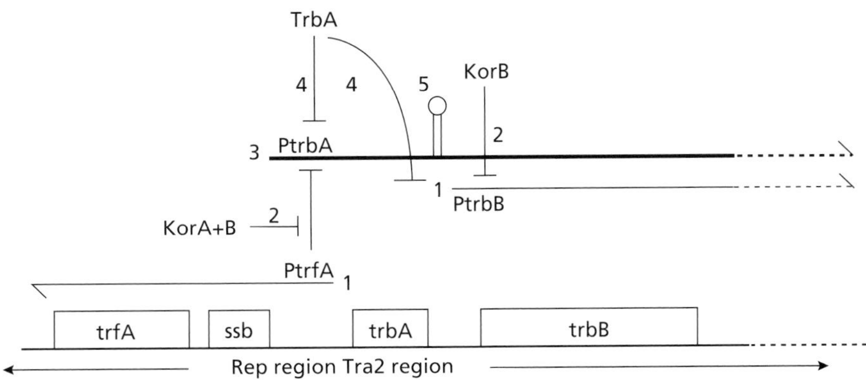

Fig. 6 Regulation of the RP4 Tra2 region and its coupling to control of the TrfA replication operon. 1. On entering a recipient cell, *PtrfA* dominates over the divergent and overlapping *PtrbA* promoter while high-level transcription of the Tra2 region occurs from *PtrbB*. 2. As levels of KorA and KorB global regulators rise, transcription from *PtrfA* and *PtrbB* becomes blocked but transcription of Tra2 occurs from *PtrbA* (3). 4. TrbA, as a third global regulator, controls *PtrbA* and *PtrbB* along with KorA,B to achieve steady-state levels of transfer proteins in the cell. 5. Expression of TrbB is limited by a hairpin which affects *trbB* translation.

the transconjugant cell allowing establishment of the mating apparatus. At higher concentrations, KorA plus a second regulator, KorB, interact and repress transcription from P_{trfA} and stimulate transcription of the Tra2 region from P_{trbA} (Kostelidou *et al.*, 1999). TrbA protein, the product of the first gene in the Tra2 core, acts as another global regulator to repress transcription from P_{trbB} and the three promoters in the Tra1 core. Thus, the P_{trbA} promoter determines steady-state expression of the transfer system, presumably maintaining a balance between the number of transferosomes and relaxosomes in the cell. A further complexity in the regulation of TrbB expression is the presence of a large hairpin in mRNA originating from P_{trbA}, which blocks *trbB* translation.

In addition to global regulation by TrbA, the Tra1 core is subject to local regulation that modulates the concentration of relaxosomal proteins. Specific binding of TraK to the *oriT* region leads to down-regulation of P_{traJ} and P_{traK} by excluding RNA polymerase, while TraJ binding reduces the activity of P_{traJ}, presumably by blocking the elongation of transcripts initiated at the promoter (Zatyka *et al.*, 1994).

Extracellular Signalling

The IncF and IncP systems involve pre-existing conjugation proteins and appear to trigger DNA transfer in direct response to mating-pair formation, but other systems require transcription and translation of the conjugation genes in response to extracellular inducer molecules. Interestingly, the use of an extracellular signal to trigger conjugation is not a widespread phenomenon. There is tantalising evidence that secreted peptide pheromones are used in a number of Gram-positive systems (see below), but the only known parallels in Gram-negative bacteria are the Ti-like plasmids of *A. tumefaciens*. These exploit non-peptide signalling molecules to initiate both Vir-mediated transfer of T-DNA to plant cells and Tra-mediated transfer of the entire plasmid between agrobacteria. T-DNA transfer from the bacterium to the plant is controlled by a two-component regulatory system, VirA and VirG, which senses phenolic compounds released by wounded plant tissue to trigger transcription of the *vir* operons (Winans *et al.*, 1994). Successful tumour formation in the plant results in the release of opines which constitute a food source for the bacteria. Detection of these opines induces expression of the positive regulator of the conjugation genes, TraR. This is a member of the LuxR family of regulators that responds to extracellular accumulation of diffusible compounds of the *N*-acyl homoserine lactone group. TraR responds to *N*-(3-oxooctanoyl)-L-homoserine lactone, also known as *Agrobacterium* autoinducer (AAI), which is synthesised by the plasmid TraI protein. At a critical level of accumulation, AAI interacts with TraR to stimulate transcription of the conjugative transfer genes, thereby conferring fertility on agrobacterial donor cells within the rhizosphere (Hwang *et al.*, 1994; More *et al.*, 1996; Wirth *et al.*, 1996; Kleerebezem *et al.*, 1997).

Conjugation of Gram-positive Bacteria

Plasmid-mediated conjugation of Gram-positive bacteria differs from the paradigms discussed above in that there is no evidence for a conjugative pilus. Secondly, a number of systems require induction, resulting in a time lag, amounting to about 30 minutes for the pheromone-inducible systems of *Enterococcus faecalis*, before transfer is detectable. Conjugative plasmids have been isolated from strains of *Enterococcus*,

Lactococcus, *Staphylococcus*, *Bacillus*, *Clostridium*, *Streptomyces* and *Nocardia* among others (Clewell, 1993c). In general, known Gram-positive plasmids can be characterised into two classes. One contains relatively small conjugative plasmids (< 20 kb) that transfer at modest rates (10^{-3} or less per donor) by surface-preferred mechanisms. This group is typified by the so-called broad host-range plasmids isolated from MLS (macrolides, lincosamides and streptogramin)-resistant streptococci and from staphylococci (Macrina and Archer, 1993). The second class includes relatively narrow host-range, large (> 60 kb) plasmids found mostly in the enterococci and typified by pAD1 and pCF10 from *E. faecalis*. Such elements transfer between cells in liquid media with high efficiencies of about 10^{-1}–10^{-2} per donor and the phenomenon of cell aggregation or clumping is observed routinely. While induction of the conjugation process and the mechanism of mating-pair formation have been studied in detail, the molecular basis of DNA transfer is almost completely unknown but is presumed to be similar to that for Gram-negative systems.

Conjugation of *Enterococcus* spp.

Conjugation of Gram-positive bacteria was first demonstrated for strains of *Enterococcus* carrying a haemolysin/bacteriocin determinant later identified to be a plasmid (Clewell, 1993b). When donor and recipient cells were added together in liquid culture, the donor cells would develop a 'fuzzy' appearance and after about 30 minutes form mating aggregates visible with the naked eye. The donor culture could be induced to form the fuzzy coat and mate almost immediately if pre-treated for several hours with sterile spent medium from the recipient culture. The component in the spent broth was called clumping-inducing-agent (CIA), later identified as a small hydrophobic peptide pheromone encoded by the chromosome (**Fig. 1**). Aside from attracting donor cells, the function of these peptides remains elusive although they appear to be derived from the signal sequences of membrane proteins.

Enterococci have a complex programme of pheromone release (Dunny *et al.*, 1995). Each strain is estimated to release half a dozen different pheromones at about 10^{-8}–10^{-9} mol/L which are active at minute concentrations (10^{-11}–10^{-12} mol/L). Usually only closely related plasmids within the same incompatibility group recognise the same pheromone. Once a cell has acquired a plasmid, release of the cognate pheromone is shut down and a small increase in exogeneous pheromone concentration is sufficient to trigger the programme of gene expression that prepares the cell for conjugation. Questions that have interested researchers in this field have centred on the nature of pheromone recognition and its ability to initiate the conjugation process.

Each plasmid studied in detail – pAD1, pPD1 and pCF10 – recognises an unmodified hydrophobic peptide of seven to eight amino acids usually containing a single hydroxyamino acid. The specificity of the pheromone appears to reside at the N-terminal end of the peptide. Each pheromone is named to reflect the plasmid which recognises it (i.e. cAD1, cPD1 and cCF10). In the pAD1 system, pheromone cAD1 is expressed more strongly by recipient cells under anaerobic conditions, although the ecological advantage of the oxygen effect is unclear (Weaver and Clewell, 1991). A small inhibitor peptide (iAD1, iPD1 or iCF10, respectively) is secreted by the donor at levels that neutralise any pheromone that has escaped from the cell, thereby blocking accidental triggering of the transfer system. However, an increase of only five molecules of pheromone per cell in the surrounding medium can trigger the mating response in these systems. Other systems exist to prevent a plasmid-containing cell from responding to its own pheromone. For example, the membrane-bound PrgY (pheromone responsive gene) protein of pCF10 prevents pheromone released from the precursor from being taken up by the same cell (Dunny and Leonard, 1997), while production of pheromone by pAD1 donors is controlled by the TraB 'shutdown' protein of the plasmid (An and Clewell, 1994).

The pCF10 conjugation region (~25–30 kb) is divided into sectors for regulation, surface exclusion, aggregation and plasmid transfer, consisting of several operons (Bensing and Dunny, 1997; Bensing *et al.*, 1997). A typical pCF10 donor cell expresses the inhibitor iCF10, consisting of 7 amino acids processed from a 23-amino acid precursor. This peptide is encoded by the 5′ end of a 400-nucleotide *prgQ* gene transcript called QS. Exogenous cCF10 pheromone (LVTLVFV) released from recipients is thought to be imported by a process involving its binding to the plasmid PrgZ protein on the cell surface and uptake via the chromosome-encoded oligopeptide permease system (**Fig. 7**; Leonard *et al.*, 1996).

Pheromone import activates production of transfer region proteins by a complex programme of gene control involving antitermination and activated translation of transcripts emanating from the *prgQ* promoter (Dunny and Leonard, 1997). The proposed antitermination mechanism allows extension of transcripts to give long-range transcription of the conjugation genes. An intriguing feature of the positive control system is the production of a longer 530-nucleotide form of the *prgQ* transcript, called QL,

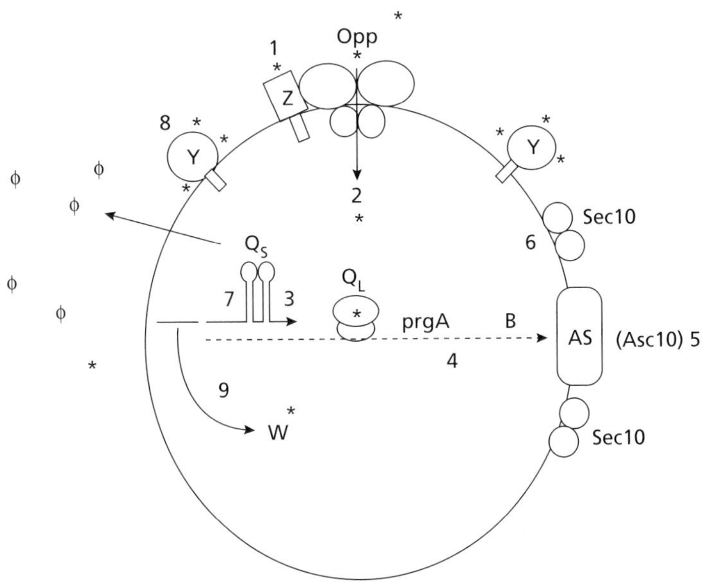

Fig. 7 Regulation of pCF10 conjugative functions. 1. PrgZ detects minute amounts of exogenous pheromone peptide (*) and promotes its uptake by the host Opp system (2). Pheromone is detected and the Q_S transcript is elongated via an antitermination mechanism (3) to give Q_L. 4. Q_L and pheromone modify ribosomes which preferentially translate the mRNA for *prgA* (specifying Sec10) and *prgB* (specifying AS or Asc10), which are deposited on the cell surface (5) as proteins involved in entry exclusion and cell aggregation, respectively. After establishment of the plasmid in the recipient cell, Q_S is expressed (7) giving the inhibitor peptide, iCF10 (ϕ). PrgY prevents uptake of released pheromone (8). PrgW requires pheromone to initiate plasmid replication (9).

which modifies the ribosomes and allows efficient translation of downstream mRNA. In addition, pheromone associates with ribosomal proteins to cause a second translation-enhancing modification (Bensing and Dunny, 1997; Bensing *et al.*, 1997).

The model for the control of pAD1 transfer genes differs from that described above in involving negative regulation by TraA. This is the only protein in the pAD1 conjugation system known to bind pheromone. TraA is thought to bind pAD1 sequences cause local termination of transcription at structures upstream of the positive control gene *traE1*. Import of the pheromone causes the removal of TraA from the DNA, resulting in read-through transcription of *traE1*. TraE1 protein then activates its own transcription and promoters of the conjugation genes, including the locus for aggregation substance and DNA transfer itself (Bastos *et al.*, 1997; Fujimoto and Clewell, 1998).

A further intriguing complexity is a connection between the control of conjugation and plasmid replication. Hedberg *et al.* (1996) described a 6.9-kb region in pCF10 which is required for negative control of mating functions and yet contains the basic plasmid replicon. This region specifies PrgW, the putative replication-initiation protein which can bind pheromone, suggesting that endogenous pheromone may be required for replication. Interestingly, the *oriT* for pAD1 has been mapped within *repA*, which is thought to encode the replication-initiation protein of the plasmid, implying that RepA may have a role in transfer as well as plasmid maintenance (An and Clewell, 1997).

The conjugation proteins of these plasmids include the 137-kDa aggregation substance (AS) known as Asc10 or PrgB of pCF10, and Asa1 of pAD1. This protein forms a fuzzy coat which is deposited asymmetrically on the surface of the bacteria (Fig. 3b). A second surface protein of ~100 kDa (Sec10 or PrgA of pCF10, and Sea1 of pAD1) acts in entry exclusion to block plasmid transport between donor cells at a stage after cell clumping (Clewell, 1993b). The exclusion protein is produced at low levels constitutively and at elevated levels in the pheromone-activated donor cell, and deposited in a similar way to AS on the cell surface (Olmsted *et al.*, 1993). Cell aggregation occurs via an interaction between AS on the donor cell and binding substance (BS) on the recipient which is thought to be lipoteichoic acid (LTA). The orientation of the AS–BS interaction is unimportant; for instance, LTA-deficient recipient cells expressing AS will bind and conjugate efficiently with AS-deficient donors. Close contact is mediated by divalent cations and the negative surface charge of the bacteria (Clewell, 1993b; Dunny *et al.*, 1995).

AS and BS proteins have been implicated in increased adhesion and invasiveness of *E. faecalis*, as well as enhanced virulence in a model system for endocarditis (Schlievert *et al.*, 1998). Enterococci show increased virulence when grown in serum which contains pheromone-like molecules while subinhibitory concentrations of antibiotics cause a clumping response in the absence of pheromone (Wu *et al.*, 1999). Asa1 has two RGD motifs suggesting that it might be involved in attachment to the integrin-like proteins on eukaryotic cells. Binding to pig kidney tubular cells has been demonstrated to be AS-dependent and this binding can be reversed by the addition of RGD-containing peptides (Wirth *et al.*, 1996). In addition, pheromones have been involved in chemotaxis of neutrophils where the iAD1 inhibitor but not the cAD1 pheromone acts to simulate neutrophil chemotaxis (Clewell, 1993a).

Conjugation of Other Non-filamentous Gram-positive Bacteria

Transfer functions in related organisms (*Lactococcus*, *Streptococcus*, *Staphylococcus*) have not been studied as extensively as for the enterococci. Interest in the conjugative systems of these genera reflects the increased incidence of strains carrying multiple antibiotic resistance and virulence determinants, and the rise of untreatable nosocomial infections in particular. The increased frequency of these 'superbugs' appears to be the result of gene transfer mediated by conjugative plasmids (Macrina, 1993; Macrina and Archer, 1993) and conjugative transposons (Clewell and Flannagan, 1993).

Conjugative plasmids studied in some detail include the streptococcal plasmids pAMβ1 and pIP501 (Wang and Macrina, 1995a, b), and the staphylococcal plasmids pGO1 (Morton *et al.*, 1993; Climo *et al.*, 1996) and pSK41 (Berg *et al.*, 1998). The nucleotide sequence of 14.4 kb conjugation-associated regions have been determined for pGO1 and pSK41, which, in pGO1, is located at a distance from a 2.2-kb module containing the *oriT* and putative relaxase gene. In general, these plasmids transfer with moderate frequencies (10^{-3}–10^{-6} per donor) on solid surfaces. All of the conjugation systems seem to cause bacterial aggregation but none appears to involve pheromone induction. These bacteria might, however, supply pheromones that trigger intergeneric transfer from other organisms. For example, pSK41 specifies a secreted pheromone-like peptide derived from the signal peptide of a lipoprotein (TraH) encoded by the conjugation region. The peptide can act as a pheromone in matings between *Staphylococcus* and *Enterococcus* but there is no evidence it functions likewise in matings of staphylococci (Firth *et al.*, 1994).

The *Lactococcus lactis* 712 sex factor and pRS01 are similar conjugative elements that can mobilise the bacterial chromosome and non-conjugative plasmids after co-integrate formation. One specific co-integration gives a re-arrangement allowing very high frequencies of transfer associated with a cell-aggregation phenotype. The predicted 135-kDa clumping-factor protein has the general structural features of a Gram-positive surface protein and strongly resembles AS determined by enterococcal plasmids (Godon *et al.*, 1994).

Intriguingly, analyses of the transfer region of these lactococcal elements has revealed the presence of a group II intron in the putative relaxase gene, which is removed by functional splicing to provide an active transfer machine (Mills *et al.*, 1996; Shearman *et al.*, 1996). Group II introns possess ribozymic activity and encode a polyprotein related to reverse transcriptase, which promotes movement to other genomic locations and 'homing' on intron-less alleles (Cousineau *et al.*, 1998). Introns of this type are commonly found in the mitochondria and chloroplasts of fungal and plant cells but have only recently been detected in bacteria, where they are associated with potentially mobile DNA (Michel and Ferat, 1995). Thus class II introns appear to be colonists that spread between cells as passengers in transmissible DNA. This interpretation is supported by the detection of another intron of this type in a transposition gene of a conjugative transposon of *Clostridium difficile* (Mullany *et al.*, 1996).

There is scant information about self-transmissible elements in other Gram-positive bacteria. Conjugative plasmids have been described for *Clostridium* spp. which are spore-forming anaerobes of medical and commercial importance (Rood and Cole, 1991; Macrina, 1993). Interestingly, one such plasmid, pIP401, carries a mobilisable resistance transposon, Tn*4451* (6.3 kb). The transposon contains a presumptive relaxase gene and upstream *oriT* region, identified by sequence similarities to the mobilisation determinants of staphylococcal plasmids, and can excise precisely to give a circular molecule as the presumptive transfer intermediate (Bannam *et al.*, 1995). Another commercially important bacterium is *Bacillus thuringiensis* which expresses a variety of insecticidal toxins. A large (~200 kb) toxin-encoding plasmid of this species has been described which transfers with high efficiency in liquid cultures; the conjugation system involves a plasmid-encoded aggregation protein exposed at the cell surface but there is no evidence of an inducing pheromone (Jensen *et al.*, 1995).

Conjugation of *Streptomyces*

The genus *Streptomyces* is medically important because of its ability to synthesise a complex array of pigmented and non-pigmented antibiotics comprising about 60% of the available antibiotics now used therapeutically. *Streptomyces* also encode antibiotic resistance mechanisms for their own survival and have been singled out as a potential source of resistance genes which can be transferred to other bacteria. The organisms have a complex life cycle involving the development of aerial and substrate mycelia on solid media in response to nutrient limitation at the onset of stationary phase. The mycelia are multi-nucleated filamentous structures which eventually septate and form uninucleate spores.

Conjugation for this genus was first noticed through the fertility of auxotrophic strains in producing recombinants via chromosome mobilisation. Since then, a rich diversity of conjugative plasmids has been discovered, as reviewed by Hopwood and Kieser (1993). The plasmids vary considerably in size and topology, ranging from small circular plasmids such as pIJ101 (8.8 kb) to the 'giant' linear plasmid SCP1 (350 kb). All transfer at high frequency on solid surfaces. One class of integrating plasmid, typified by SLP1 (17 kb), exists as a chromosomal element that can excise in the donor, transfer and integrate precisely into the genome of the recipient cell. The attachment site for such integrative and excisive recombination contains the 3′ region of an essential tyrosinyl tRNA gene of *S. lividans* (Vögtli and Cohen, 1992; see section on significance of gene transfer, above). Although conjugative plasmids in *Streptomyces* have the propensity for chromosome mobilisation, circular plasmids are less efficient than linear ones at transferring virtually the whole chromosome between bacteria.

Conjugation of *Streptomyces* can involve intramycelial transfer in addition to first-step transfer between mycelia. In the first step, a tight junction is formed between growing mycelia before the development of the aerial form or the induction of secondary metabolites, which is followed by plasmid transfer into a single recipient compartment (Pettis and Cohen, 1996). The second step is peculiar to the actinomycetes and involves the dissemination or spreading of the plasmid within the hyphae of the recipient strain, ensuring that each compartment receives a copy of the plasmid. This step is characterised by the formation of 'pocks' (**Fig. 3C**) which represent the transient repression of cell growth in mycelia containing the newly acquired plasmid. Pocks may also be characterised by premature production of secondary metabolites such as pigmented antibiotics (Hopwood and Kieser, 1993).

Intermycelial transfer and fertility are remarkable in requiring only one essential gene, exemplified by *tra* of pIJ101 (Kendall and Cohen, 1988), *traB* of the 11 kb pSN22 plasmid (Kosono *et al.*, 1996) and *traA* of SCP2*, a plasmid with exceptionally high chromosome-mobilising ability (Brolle *et al.*, 1993). Intramycelial transfer and pock formation by pIJ101 additionally requires at least two 'spread' genes, *spdA* and *spdB* (Kendall and Cohen, 1988). Pock formation by pSN22 has similar genetic requirements (Kataoka *et al.*, 1991). The products of the essential *tra* gene in these systems vary in size and sequence but have both A- and B-type Walker NTP-binding motifs, as do transport proteins belonging to other conjugation systems (F TraD, RP4 TraG and pSK41 TraK; see Firth *et al.*, 1993). The *Streptomyces* Tra proteins show significant sequence similarities to SpoIIIE of *B. subtilis*, which is a protein required for translocation of the bacterial chromosome through the septum that partitions the pre-spore compartment of the sporulating cell (Wu *et al.*, 1995). These findings, together with the apparent simplicity of actinomycete conjugation and the lack of any functional or sequence indications that the plasmids specify a relaxase, raise the intriguing possibility that conjugation of *Streptomyces* differs fundamentally from other paradigms in that dsDNA is the substrate for transfer.

Chromosome Mobilisation and Hfr Donors

Many conjugative plasmids have 'chromosome-mobilisation ability' (Cma), defined as the ability to cause transfer of the bacterial chromosome with frequencies in the range of 10^{-3}–10^{-8} per donor cell (Holloway, 1979). Cma is important not only in allowing horizontal transfer of chromosomal genes but as a historic landmark leading to the discovery of plasmids as self-transmissible elements. Much higher transfer frequencies are achieved with Hfr donors, where the plasmid is integrated stably into the bacterial chromosome as by the Campbell model (see section on transduction). Plasmid F can integrate at some 20 chromosomal sites which are non-random but dispersed around the chromosome. Integration generally involves recombination of homologous sequences, such as the multiple copies of insertion sequences located on plasmids and the bacterial chromosome. Transconjugant cells rarely, if ever, receive a complete copy of the Hfr chromosome, because transfer terminates randomly during a cycle of transfer presumably at a spontaneous ssDNA break (Low, 1996).

Most chromosome mobilisation from an F^+ strain is independent of any sub-population of Hfr cells. Such transfers may involve transient covalent linkage of the plasmid to the chromosome or recognition of a sequence that fortuitously resembles the plasmid's transfer origin. It is notable that few wild-type plasmids apart from F can form stable Hfr donors, even when regions of the bacterial chromosome are spliced into the plasmid to facilitate homologous recombination. The difficulty can be bypassed using genetic strategies that inactivate initiation of replication in one of the co-integrated replicons, suggesting that fusion of a plasmid to the chromosome is generally a deleterious event (Reimmann and Haas, 1993; Wilkins, 1995). One exception is the IncJ incompatibility group of plasmids, which preferentially integrate into the chromosome and appear to be related to conjugative transposons (Murphy and Pembroke, 1999).

Integrated plasmids can excise from the chromosome. Excisive recombination at sites other than those involved in integration generates a 'prime' plasmid, formally similar to a specialised transducing phage (see section on transduction). F′ plasmids vary considerably in size and may include as much as 30% ($\sim$1400 kb) of the *E. coli* chromosome, as well as a full complement of plasmid genes (Holloway and Low, 1996). Prime plasmids contribute to the dynamic nature of bacterial genomes by amplifying the copy number of sectors of the bacterial chromosome, potentiating their evolution through provision of a duplicate that is mutable without perturbing cellular functions, and enabling productive exchange of chromosomal genes between cells without the need for homologous recombination in the recipient bacterium.

Mobilisable Plasmids

Many small plasmids are mobilisable by other self-transmissible plasmids. In general, the mobilisable (Mob) plasmid encodes the proteins required for recognising and cleaving its own *nic* site and directing the complex to the transferosome determined by the conjugative element. A Mob$^+$ element may be mobilisable by several apparently unrelated conjugation systems but with different efficiencies. For instance, ColE1 is mobilised efficiently by IncFI, IncI1 and IncPβ plasmids but less effectively and with different requirements by an IncW plasmid (Warren *et al.*, 1979). ColE1, which was the subject of pioneering studies on relaxosomes (see Lovett and Helinski, 1975), encodes four genes for mobilisation (*mbeA, B, C* and *D*) with the first three encoding components of the relaxosome (Chan *et al.*, 1985; Boyd *et al.*, 1989). MbeA is the relaxase; MbeB and MbeC may function in the assembly of the relaxosome or might direct it to the transferosome. MbeD protein functions in entry exclusion as well as in transfer (Yamada *et al.*, 1995). It is noteworthy that *mob* genes are deleted from pBR322, pBR325 and pBR328, which are vectors derived from ColE1-type plasmids, and all but pBR322 lack *oriT*. The latter plasmid can be mobilised by RP4. ColE1 requires TraD and the pilus assembly and mating pair-stabilisation gene products of plasmid F for successful mobilisation. CloDF13, a related plasmid, does not require F TraD and apparently supplies an equivalent protein (Willetts, 1980).

One of the most remarkable mobilisable plasmids is RSF1010 ($\sim$8.6 kb; Fig. 8) and its close relatives R1162 and R300B (sequence presented in Scholz *et al.*, 1989; Sakai and Komano, 1996). These plasmids, classified into the IncQ group, have an extremely broad host range and are mobilised efficiently by plasmids of the IncP group. They are also mobilised into plant cells and between agrobacteria by the Vir, but not Tra, system of the Ti plasmid (Buchanan-Wollaston *et al.*, 1987; Beijersbergen *et al.*, 1992). IncQ plasmids have a 38 bp *oriT* region, which is related by sequence to the *oriT*s of Gram-positive conjugative plasmids pIP501 and pGO1, and has been shown genetically to transfer by a replicative

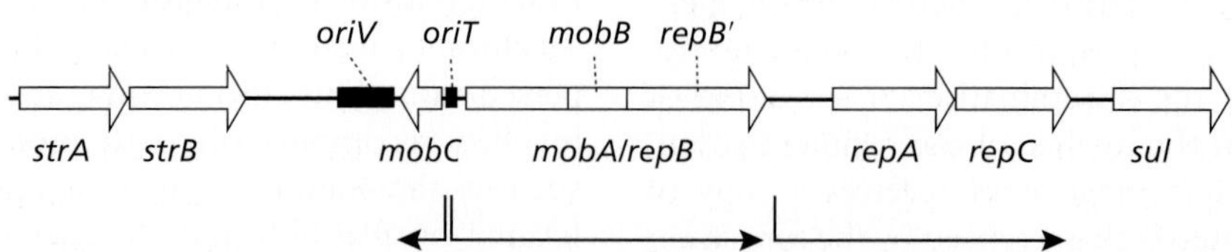

Fig. 8 Plasmid RSF1010 (8684 bp) shown as a linear structure. The simplicity of this plasmid belies its remarkably broad host range and ability to be mobilised at various levels of efficiency by conjugative plasmids from many incompatibility groups. The origins of vegetative replication (*oriV*) and transfer (*oriT*) are indicated as black boxes. Boxes indicate replication (Rep) and mobilisation (Mob) genes. *repB′* is expressed as the 3′-region of *mobA* which is identical to *repB*. The 5′-region of *mobA* encodes relaxase. *mobB* is translated in a different frame from *mobA*. *sul* and *str* confer resistance to sulfonamide and streptomycin. Arrows indicate the direction and length of transcripts in the Mob/Rep regions (after Scholz *et al.*, 1989).

rolling-circle mechanism. The plasmids encode three mobilisation proteins with MobA being the relaxase (see Lanka and Wilkins, 1995). MobA also contains a primase distal to the relaxase which is essential for conjugation (Henderson and Meyer, 1999), MobC promotes strand separation near *nic* (Zhang and Meyer, 1997) and MobB is required to stabilise the relaxosome (Perwez and Meyer, 1999). Mobilisation of RSF1010 by RP4 requires *traG* in the Tra1 region in addition to the normal complement of mating-pair formation genes (Lessl *et al.*, 1993). Cabezon *et al.* (1994, 1997) have suggested that mobilisable plasmids such as RSF1010 and ColE1 interact with proteins of the TraG (RP4), TraD (F) and TrwC (R388) family with different affinities, which could be the basis for the differing efficiencies of mobilisation by the conjugative plasmids.

Study of mobilisation plasmids in Gram-positive bacteria is less well developed than in Gram-negative organisms (Macrina and Archer, 1993). One of the best-described families consists of small plasmids that replicate vegetatively by a rolling-circle mechanism. An example is the pMV158 streptococcal plasmid which is mobilisable by pIP501 and pAMβ1 (Priebe and Lacks, 1989). This plasmid replicates by a rolling-circle mechanism which initiates at a precise location. Since conjugation is an equivalent process, the co-ordination of these two processes is of interest. pMV158 encodes a relaxase termed MobM that cleaves at a unique *oriT* site in a manner similar to other relaxases. The *oriT* sequence contains multiple inverted repeats and is conserved in several Gram-positive mobilisable plasmids, providing another structural paradigm of an origin of transfer (Guzmán and Espinosa, 1997; Grohmann *et al.*, 1999).

Conjugative Transposons

Conjugative transposons comprise an important group of self-transmissible elements carried as insertions in the bacterial chromosome or a plasmid (reviewed by Clewell and Flannagan, 1993; Salyers *et al.*, 1995a, b; Scott and Churchward, 1995). They are disseminated widely and carry genes conferring multiple drug resistance on clinically important strains as well as other traits, such as nisin production and sucrose metabolism in lactococci. The transposons appear to be particularly prevalent in Gram-positive cocci as well as in one genus of Gram-negative bacteria – the anaerobic *Bacteroides.* A wider distribution is indicated by discovery of CTn*scr94* as an enterobacterial conjugative transposon encoding a sucrose metabolic pathway (Hochhut *et al.*, 1997). A particularly well-characterised family contains Tn*916*, a 18-kb element originally identified in *E. faecalis*, plus Tn*1545*, discovered in *Strep. pneumoniae*; members of the family carry the *tetM* tetracycline resistance locus and have been found in or introduced into some 25 genera of Gram-positive and Gram-negative bacteria (Clewell *et al.*, 1995). Interest in this group of elements has surpassed that in conjugative plasmids of Gram-positive bacteria because of their link to antibiotic resistance, ability to transfer productively over a broad host range, and their genetic value in insertional mutagenesis.

Conjugative transposition is thought to involve excision of the element to give a 'transposon circle', which is incapable of vegetative replication, followed by transfer of a single strand to the recipient cell by a replicative process and integration of the resulting duplex at one of a preferred number of sites apparently consisting of bent DNA. The elements can also transpose to new sites in the donor cell. Transfer frequencies are low – in the range of 10^{-4}–10^{-9} per donor – but transconjugants frequently contain more than one copy of the element. The *oriT* region has been identified in the Tn*916* sequence (Flannagan *et al.*, 1994; Jaworski and Clewell, 1995). There is no evidence of any entry exclusion function. Conjugative transposons superficially resemble other transposons in their ability to integrate and excise, but closer examination has revealed important differences. The first clue was the lack of a short signature repeat of host sequence at the borders of the element, which is a characteristic of classical transposons. Secondly, conjugative transposons move by a phage λ-like process involving integration and excision genes. The integrase protein has been purified and found to function by a mechanism resembling that of the integrase family of site-specific recombinases (Taylor and Churchward, 1997). In conjugation, the enzyme is required only in the donor suggesting that it may be co-transferred with the DNA (Bringel *et al.*, 1992).

A complex excision reaction involving integrase and excisionase (Scott and Churchward, 1995; Marra and Scott, 1999) results in staggered cleavage within non-identical border sequences or 'coupling sequences' flanking the element to produce short single-stranded ends. These interact to form a mispaired heteroduplex of 6 bp as a prelude to covalent closure of the excised element. One of the donor coupling sequences is inherited by the transconjugant, suggesting that it is selected on the basis of orientation in the donor; only one strand of Tn*916* is transferred to the recipient. However, the finding that circular Tn*916* molecules excised in *E. faecalis* contain a homoduplex joint usually corresponding to the left coupling sequence indicates that one strand of the heteroduplex may undergo preferential repair before transfer (Manganelli *et al.*, 1997).

Another well studied class includes the large conjugative transposons found in Gram-negative anaerobes, particularly *Bacteroides* spp., and typified by Tc^r Em^r DOT (Salyers *et al.*, 1995a, b). This 70-kb element carries *tetQ* and *ermF*, encoding tetracycline resistance and erythromycin resistance, respectively. An intriguing feature of these elements is the transcriptional induction of excision and transfer in response to the presence of low concentrations of tetracycline, which is thought to have contributed to the near universal resistance of *Bacteroides* to this drug (Stevens *et al.*, 1993). Tc^r Em^r DOT transposons are also involved in triggering the excision of NBUs (non-replicating *Bacteroides* units) and supporting their mobilisation. The units are 10–12 kb in size and are judged by nucleotide sequence to be discrete from the conjugative transposons. NBUs encode a single transfer protein, Mob, which binds and nicks at *oriT* and directs the complex to the conjugative machinery of the transposon. Excision and integration are phage-like processes involving a circular intermediate with an integration site of the same sequence as the chromosomal site and an enzyme of the λ integrase family. Unlike conjugative transposons of *Bacteroides*, NBUs have been reported to transfer from *Bacteroides* to *E. coli* and to integrate therein. Furthermore, they are mobilisable by IncP conjugative plasmids. The ability of these various self-transmissible transposons and mobilisable units to transfer widely and be inherited as passengers in replicons native to the host cell make them formidable adversaries in the battle to control bacterial infections through the use of antibiotics.

Conjugation in Natural Environments

As illustrated in the section on the significance of gene transfer (above), there is compelling evidence that bacterial conjugation is an ecologically and evolutionarily important gene transfer system. Conjugation has been shown to occur in diverse environments including the intestinal tracts of humans and other homeothermic animals (see for example Freter, 1986; Levy, 1986; Doucet-Populaire *et al.*, 1992), surfaces of animals and plants, soils and sediments, and aquatic and wastewater systems. Further details can be found in the monographs edited by Levy and Miller (1989) and Wellington and van Elsas (1992), and the review by van Elsas *et al.* (1999). There remain significant questions concerning transfer frequencies *in situ*, and the identification and quantification of environmental factors and population densities affecting transfer rates in different ecosystems. There is also a need to determine the hosts that are naturally available to plasmids; use of laboratory strains of *E. coli* K-12, which are atypical of natural isolates of *E. coli*, in conjunction with IncP or IncF plasmids, chosen for their promiscuity or historical precedent respectively, is seen to be inappropriate for the general modelling of gene transfer *in situ*. Understanding of conjugation in the ecological context will require more information on the range of hosts that have supported the evolution of specific plasmid types, as well as the contemporary hosts providing the reservoirs for their current maintenance.

Natural Bacterial Transformation Systems

Bacterial transformation was the first genetic transfer process to be described for bacteria. The historic landmark was the discovery by Frederick Griffith in 1928 that mutant *Streptococcus pneumoniae* lacking a capsule acquired the ability to produce a capsule when mixed with heat-killed capsulated cells and injected into the mouse peritoneal cavity. Subsequent work reported by Oswald Avery and colleagues in 1944 led to DNA of the donor cell being identified as the substance responsible for transformation. Mechanistic studies of natural transformation have focused on *Strep. pneumoniae* and *B. subtilis* as representative Gram-positive organisms, and on *H. influenzae* and *N. gonorrhoeae* as Gram-negative paradigms. Useful reviews include those by Dreiseikelmann (1994), Lacks (1988, 1999) and Dubnau (1991, 1997). Information on other natural transformation systems is provided by Stewart (1992) and Lorenz and Wackernagel (1994). Artificial transformation systems applied to organisms which are not naturally transformable, such as *E. coli*, have been reviewed by Hanahan and Bloom (1996).

Natural transformation systems differ from conjugation in that the numerous proteins involved are specified by the recipient rather than the donor cell and the recipient needs to acquire a physiological state called competence in order to take up exogenous DNA. A further difference is that although entrant DNA is single-stranded in both transfer processes, it is generated by a degradative mechanism in transformation in contrast to the rolling-circle mode of strand displacement in the general model of conjugation. The transferring strand enters the recipient with a leading 3′ terminus in at least some transformation systems, which is opposite to the polarity of strand transfer in conjugation.

Regulation of Competence Development

In most systems, competence is a transient condition associated with induced synthesis of a small number of specific proteins. In *Strep. pneumoniae*, the condition develops during exponential growth in response to a secreted 17-residue pheromone described as competence-stimulating polypeptide (CSP). This factor contributes to a cell-density-sensing mechanism in which basal levels of CSP derived by transcription from a constitutive promoter trigger transcription from a second promoter that is autoinducible in being stimulated by elevated levels of its product. This arrangement leads to a cascade of CSP production at high cell densities, which induces competence development in nearly every bacterium in a mid-exponential phase culture. The extracellular concentration of the CSP pheromone is sensed by a two-component signal transduction system encoded by the operon containing the gene for the CSP precursor. CSP ultimately leads to production of a competence-specific sigma factor for expression of genes for DNA uptake and processing (Pestova *et al.*, 1996; Kleerebezem *et al.*, 1997). Involvement of pheromones and cell-density sensing mechanisms provides a mechanistic link to certain conjugation systems (see section on conjugation, above).

Competence development by *B. subtilis* differs in being a post-exponential response that occurs in only a small fraction (~ 10%) of cells and involves a complex interplay of signals related to growth phase, nutritional conditions and metabolic state. Competence optimally occurs some 2 hours after a transition from exponential growth into a glucose-supplemented minimal medium and is associated with a cell type that is relatively dormant in macromolecular synthesis. Some 40 genes have been identified for the development of competence. About half belong to the 'late' *comC*, *comE*, *comF* and *comG* loci required for biosynthesis of the DNA binding and uptake apparatus. Other *com* genes function earlier in an elaborate complex of regulatory networks that converge to control expression and activation of the competence transcription factor, ComK (Fig. 9). This protein stimulates expression of its own gene (*comK*) by autoregulation and is singularly required to enhance transcription of the late *com* loci and cause competence-related expression of *comK* and other recombination loci. Transcription of *comK* is influenced by other DNA-binding proteins (AbrB, CodY, DegU and SinR) acting as positive and negative regulators to integrate different signals and information on nutrient availability (Grossman, 1995; Hahn *et al.*, 1996).

ComK is the key regulator (van Sinderen *et al.*, 1995). It forms an inhibitory complex with MecA and ClpC proteins but is released from the complex during competence development via convergent pathways initiated by two cell density-sensing pheromones (Turgay *et al.*, 1997). One such pheromone is a modified 9–10 amino acid peptide called ComX. Extracellular ComX is sensed by a two-component regulatory system (ComP–ComA) causing phosphorylation of the ComA response regulator. Phosphorylated ComA then acts as a transcription factor to activate the *comS* gene. In turn, ComS protein releases ComK from the inhibitory complex with MecA and ClpC. The second pheromone is called competence stimulating factor (CSF). This is a 5 amino acid peptide which is transported into the cell by the Spo0K oligopeptide permease transporter. Intracellular CSF modulates the level of phosphorylated ComA by inhibiting RapC as a phosphatase that converts ComA into the unphosphorylated form (Magnuson *et al.*,

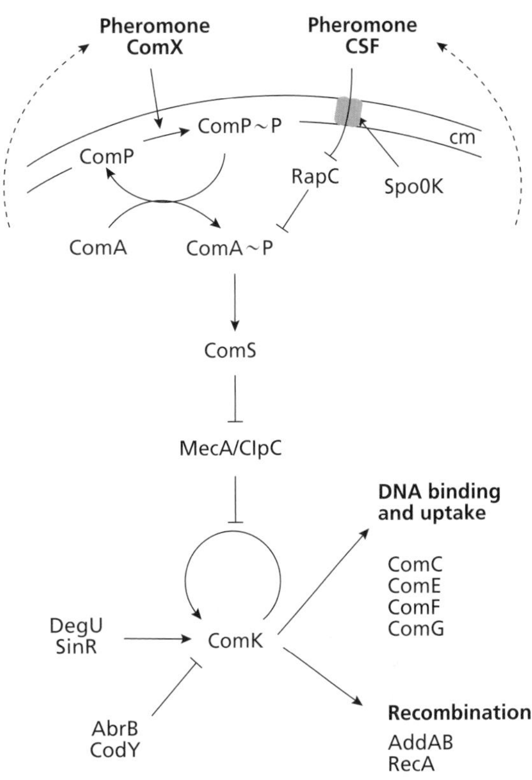

Fig. 9 Model for pheromone-mediated regulation of competence development in *B. subtilis*. Pheromones ComX and CSF are cell density-sensing peptide pheromones that release ComK from inhibitory interactions with MecA and ClpC. ComK is an autoregulatory transcription factor required for expression of the *comC, E, F* and *G* loci encoding the DNA binding and uptake system, plus competence-enhanced transcription of recombination loci. Blocked lines indicate negative action. Details are in the text (from Wilkins and Meacock, 2000; used by permission of Academic Press).

1994; Solomon *et al.*, 1996; Kleerebezem *et al.*, 1997).

Regulation of competence in *H. influenzae* is quite different in that almost every cell can develop the state as a stable internally regulated condition. The same applies to some other Gram-negative bacteria (Stewart, 1992). The response in *H. influenzae* is triggered by a transition into unbalanced growth following a shift-down from a rich to a nutrient-limited medium and involves accumulation of the cyclic AMP-catabolite activator protein complex as a key regulator. This triggers expression of the *tfoX/sxy* gene which activates late competence genes involved in DNA uptake and processing (Karadapurum and Barcak, 1997). Other Gram-negative bacteria, such as *N. gonorrhoeae*, are competent throughout the growth cycle (Lorenz and Wackernagel, 1994).

DNA Uptake by *B. subtilis* and *Strep. pneumoniae*

Considerable diversity exists between systems with regard to the DNA transport apparatus and its specificity. *B. subtilis* and *Strep. pneumoniae* take up DNA without specificity for nucleotide sequence. Typically uptake is initiated by the binding of dsDNA to about 50 competence-specific receptor sites on the surface of the cell via a process with specific energy and ionic requirements. Bound DNA is exposed to the extracellular medium, as indicated by its sensitivity to shear and exogenous DNases. The DNA then accumulates random double strand breaks, which may be caused by successive nicking of opposing DNA strands. The endonucleases responsible are thought to be located at DNA uptake sites. DNA uptake commences 1–2 minutes after binding and is thought to occur through an aqueous pore in a surface-exposed complex of proteins. Uptake is associated with the intracellular appearance of ssDNA fragments in the size range of about 5–15 kb and release into the medium of oligonucleotides as degradation products of the second strand (see Dubnau, 1991, 1997; Lacks, 1999).

DNA uptake by *Strep. pneumoniae* is potentiated by a constitutively produced 30-kDa endonuclease, EndA, located in the cell membrane. The enzyme acts *in vitro* to produce oligonucleotide fragments from single- and double-stranded DNA (Rosenthal and Lacks, 1980). The transforming DNA strand enters the cell with a leading $3'$ terminus at a rate of $\sim$100 nucleotides per second at 31°C, while the complementary strand is degraded with similar kinetics in the $5' \rightarrow 3'$ direction (Mejean and Claverys, 1993).

The bioenergetics of DNA transport have been reviewed by Palmen and Hellingwerf (1997). The classic proposal that DNA uptake is powered by hydrolysis of the second strand remains unproven, since the reaction might be required to generate the substrate for the transporter. Several models have been suggested, some of which may operate co-operatively. These invoke one or both components of the proton-motive force; an anion exchange reaction dependent on proton cycling; involvement of a poly-β-hydroxybutyrate : polyphosphate : Ca^{2+} complex as DNA transporter, as proposed for DNA uptake in artificially competent *E. coli* (see Hanahan and Bloom, 1996); and an ATP-dependent machine acting on ssDNA.

The genetics of DNA transport has been studied extensively by the Dubnau laboratory using *B. subtilis* (see Dubnau, 1997). The DNA translocation apparatus (**Fig. 10**) is encoded by the dispersed *comC, comE, comF* and *comG* loci which contain some 15 open reading frames (ORFs). Some of these gene products show intriguing similarities at the sequence level to proteins involved in the biosynthesis and function of type 4 bacterial pili, and secretion of exoenzymes (see concluding section). For example, the N-terminal moieties of ComGC, ComGD, ComGE and ComGG have prepilin-like signal sequences and potential processing sites. These proteins, together with ComGF, are membrane-associated and essential for DNA binding, and may be assembled on the outer face of the cell membrane to provide some type of structure for DNA passage across the cell wall. Processing of the proteins is mediated by ComC which resembles the prepilin protease family. ComGA and ComGB are similar to proteins required for type IV pilus biogenesis, with the former containing a conserved Walker-box motif characteristic of nucleotide-binding proteins. Dubnau (1997) has raised the interesting speculation that, by analogy with the role of type IV pili in twitching motility, the ComG apparatus may cycle between an assembled and dissembled form to present transforming DNA to receptors on the outer surface of the cell membrane. The presumptive ATPase activity of ComGA might power this dynamic process.

ComEA and ComEC are integral membrane proteins. Mutational studies suggest that ComEC contributes to the DNA uptake channel and that ComEA functions additionally to facilitate DNA-binding. ComFA resembles the DEAD family of DNA helicases, in particular PriA which is an *E. coli* replication protein that can translocate along ssDNA in an ATP-dependent manner. Thus ComFA is envisaged to use ATP hydrolysis to translocate ssDNA through an uptake channel consisting of ComEC and containing ComEA at the entry port (**Fig. 10**).

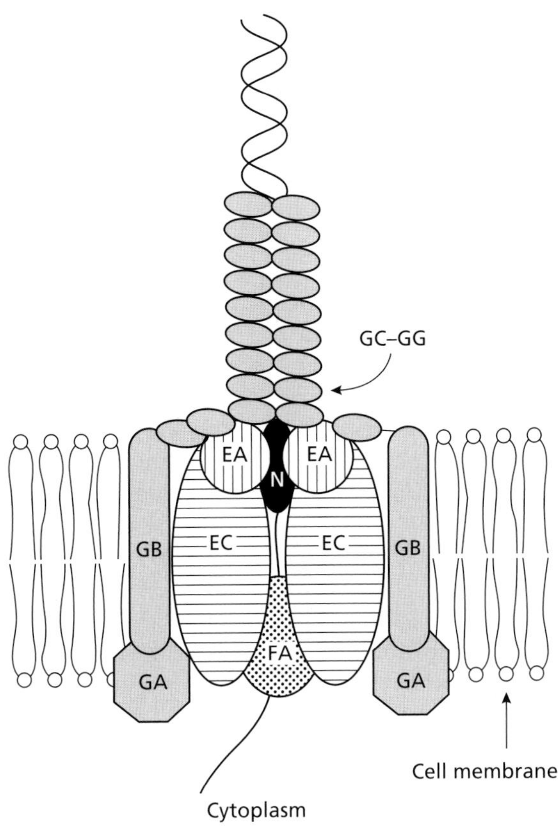

Fig. 10 Schematic representation of the transformation apparatus of *B. subtilis*. ComGC, ComGD, ComGE and ComGG proteins form a complex on the outside of the cell to aid passage of DNA across the cell wall. Other proteins and their putative roles are: ComGA, GB, assembly/disassembly of the GC-GG structure; ComEA, DNA binding and uptake; ComEC, uptake channel; ComFA, DNA translocation (from Wilkins and Meacock, 2000; used by permission of Academic Press).

DNA Uptake by *H. influenzae* and *N. gonorrhoeae*

DNA uptake by the *H. influenzae* system differs from the *B. subtilis-Strep. pneumoniae* paradigm in showing preference for homologous DNA. Specificity is determined by a 9-nucleotide uptake sequence (US), 5′-AAGTGCGGT-3′, present in about 1500 copies in the 1.8×10^3 kb *H. influenzae* genome. The USs are widely distributed but tend to be localised in inverted repeats downstream of genes. Here they might contribute to transcription termination (Smith *et al.*, 1995). *H. influenzae* transformation additionally requires a specific membrane structure called a transformasome. There are several of these structures per competent cell: each has a diameter of ~20 nm and consists of a membrane vesicle that emanates from a zone of adhesion of inner and outer membranes to protrude at the cell surface. dsDNA is inferred to bind to a protein receptor on the outside of the transformasome and to enter the structure where it is protected from exogenous DNase. ssDNA is released from the transformasome into the cytoplasm in a polar manner with a leading 3′ terminus (Kahn and Smith, 1984).

The major transformation system of *N. gonorrhoeae* also shows specificity of DNA uptake determined by the uptake sequence, 5′-GCCGTCTGAA-3′, which is also found in transcriptional terminators (Goodman and Scocca, 1988). An intriguing finding is that DNA uptake across the outer cell membrane is dependent on the structural components of the bacterial type IV pilus, including PilC which is associated with the cell surface and pilus tip. Possibly the pilus system is required to present the US receptor at the cell surface. Alternatively, depolymerisation of the pilus fibre, as in twitching motility, or an activity of its transport apparatus may function to translocate the DNA across the outer cell membrane. Passage of transforming DNA across the remainder of the cell envelope involves competence proteins apparently unassociated with the pilus. Of these, ComL and Tpc are thought to cause localised re-structuring of the polysaccharide–peptide polymers of the cell wall to allow DNA import. An integral membrane protein (ComA) resembling at the sequence level the ComEC DNA uptake protein of *B. subtilis* aids transport across the inner membrane (see Fussenegger *et al.*, 1997).

Another transfer system described for *N. gonorrhoeae* involves membrane vesicles called blebs. These structures, consisting of inner or outer membrane material, have been found to mediate nuclease-resistant transfer of antibiotic-resistance plasmids between *N. gonorrhoeae* cells, as well as to other Gram-negative bacteria. Bleb-mediated transfer might be a common phenomenon since DNA is packaged within membrane vesicles of a range of Gram-negative bacteria (Dorward and Garon, 1990).

Integration of Transforming DNA

Chromosomal ssDNA entering the cell is integrated into the recipient genome by homologous recombination to give a region of heteroduplex DNA, ranging in size from about 40 bp to 10 kb or more. Integration is efficient in that >70% of the ssDNA mass is incorporated. *B. subtilis* and *Strep. pneumoniae* are known to specify a competence-induced ssDNA-binding protein ('eclipse' protein) homologous to *E. coli* SSB protein, which coats the entrant DNA possibly to protect the transforming strand from intracellular nucleases and facilitate formation of recombination intermediates. Competence-dependent induction of *recA* genes occurs in a number of systems, presumably enhancing the efficiency of integration (Dubnau, 1991; Palmen and Hellingwerk, 1997).

Heteroduplex DNA formed from genetically different strands derived from closely related organisms will contain one or more mispaired regions. In pneumococcal transformation, these regions can be repaired by the *hex* system. The efficiency of correction depends both on the mispaired bases and on the proximity of adjacent repairable mismatches. The donor contribution is corrected preferentially, thereby reducing the transformation efficiency of some markers by as much as 20-fold. The repair system might be targeted to the donor strand by transient gaps at the ends of the heteroduplex where the newly inserted strand abuts the resident strand (Lacks, 1988; Humbert *et al.*, 1995). The classic model proposes that mismatch repair involves excision of the mispaired region followed by infilling by DNA synthesis, as for the correction of replication errors. Mismatch repair enzymes are considered further in the section on barriers to the horizontal transmission of genes, below.

Transformation by Plasmids

As is the case for chromosomal DNA, plasmids are taken up as linear single-stranded molecules which are subject to extensive fragmentation. Plasmid DNA gives rise to very few transformants in natural systems, presumably due to the difficulty of forming circular unit-length molecules from ssDNA fragments. The frequency of transformation by plasmid monomers is dependent on the square of the DNA concentration, suggesting that intact plasmids are regenerated in the transformed cell from sections of more than one monomer. The mechanism might involve annealing of complementary portions of different fragments of (+) and (−) strands, followed by infilling by DNA synthesis to give an intact duplex (Saunders and Guild, 1981; de Vos *et al.*, 1981).

In contrast, transformation frequencies of plasmid multimers show a linear dependency on DNA concentration, implying that a single multimer can give rise to a transformant. One plausible explanation is that the larger transforming fragments include a complete genome and that DNA polymerase switches templates at the terminal overlap in the entrant strand to generate a nascent strand of unit size (Saunders and Guild, 1981; Lacks, 1988). Alternatively, fragments of (+) and (−) polarity derived from a single multimeric molecule might anneal to give a monomeric circle (de Vos *et al.*, 1981).

Ecological Implications of Transformation

Many genera other than the paradigms cited here acquire natural competence. The list spans Gram-positive and Gram-negative (eu)bacteria, including cyanobacteria, and archaea. The phylogenetic diversity of these organisms, coupled with the mechanistic complexity of DNA uptake systems, indicates that transformation has an important role in natural populations. DNA of microbial origin is plentiful in seawater, freshwater and sediments, and transformation has been demonstrated in such environments. DNA can acquire some protection from ubiquitous extracellular nucleases through its capacity to bind via hydrogen bonds and electrostatic interactions to the surfaces of mineral particles in soils and sediments in a form that confers protection yet allows interactions with cell surfaces (Stewart, 1992; Lorenz and Wackernagel, 1994; Yin and Stotzky, 1997).

Natural transformation may also be facilitated by active participation of donor cells, allowing temporal coupling of the release of DNA to its uptake by recipients. DNA release might result from the activity of autolytic enzymes or reflect an active export system compatible with cellular survival. Coupling of DNA release and uptake occurs in cell contact transformation: this process, defined as bidirectional, DNase I-sensitive gene transfer within mats of cells lacking detectable plasmids or phages, has been observed for several Gram-positive and Gram-negative organisms (Stewart *et al.*, 1983; Stewart, 1992; Lorenz and Wackernagel, 1994).

Although natural transformation is viewed to be an evolutionary strategy that potentiates genetic diversity, the process may have other roles. One hypothesis is that transformation evolved as a mechanism for recombinational repair of recipient genomes damaged by environmental agents. A prediction of this hypothesis is that competence should be induced by DNA-damaging treatments which activate DNA repair genes of the inducible SOS regulon. However, no such relationship is observed in *B. subtilis*, although the converse holds in that classic SOS functions, including strong derepression of *recA*-type genes, occurs spontaneously during competence development (Dubnau, 1991). Furthermore, DNA uptake was not enhanced when *B. subtilis* was irradiated with ultraviolet (UV) light; rates of homologous recombination were increased but this response might be a direct consequence of UV-generated recombinogenic lesions in the recipient genome (see Michod and Wojciechowski, 1994). Another notion is that competence is an adaptation to starvation conditions, whereby DNA uptake potentiates alternate metabolic processes. According to this hypothesis, nucleosides and bases salvaged from the DNA uptake process and from breakdown of non-recombined entrant DNA are channelled into the nucleotide pools or catabolised further to provide alternative sources of carbon, nitrogen and energy.

Transduction

Bacterial transduction involves the packaging of a fragment of non-viral DNA into a bacteriophage particle during lytic growth of the phage in the donor cell and its transfer in the encapsidated state to the recipient cell. Thus the process typically involves transport of duplex DNA rather than ssDNA as in the general models of conjugation and transformation. The process characteristically lacks a requirement for cell-to-cell contact and is resistant to nucleases added to cell-free culture supernatants of donor bacteria. Transduction was discovered in 1952 by Norton Zinder and Joshua Lederberg in a search for gene transfer and recombination in *S. typhimurium*; the agent responsible is a temperate bacteriophage known as P22. Other well-characterised transducing viruses are coliphages P1 and λ. These phages are temperate, meaning that on cellular infection they either enter the lytic or establish a quiescent relationship with the host bacterium described as lysogeny (Campbell, 1996a, b; Yarmolinksy and Sternberg, 1988). In the lysogenic state, expression of most phage genes is repressed but the phage DNA – called prophage – can be induced into lytic growth by certain physiological conditions such as damage to cellular DNA.

Phage λ is described as a specialised transducing phage, since it transduces a particular region of the bacterial chromosome (Weisberg, 1996). In contrast, phages P1 and P22 have the capacity to transfer many different chromosomal regions and, hence, are described as generalised transducing phages (Masters, 1996). As discussed below, an important molecular difference between phages capable of specialised and generalised transduction is the mechanism used for packaging DNA into phage particles.

Specialised Transduction

Specialised transduction is mediated by lysogenic phages that integrate as prophages at a specific site in the bacterial chromosome. Integration is described by the classic Campbell model wherein a circular molecule of phage DNA recombines by reciprocal exchanges with a site in the bacterial chromosome to produce a single circular structure called a cointegrate. Phage λ integrates by recombination of a complex 'attachment' (*att*P or POP′) site in the phage and a structurally simpler partner site (*att*λ or BOB′) located in an intergenic region between the *gal* and *bio* operons in the bacterial chromosome. The two sites are not identical in that POP′ has a minimal size of 238 bp, whereas the chromosomal BOB′ site is 21 bp. Recombination occurs within a 15-bp identity sequence common to both sites and is mediated by the phage-encoded integrase (Int) enzyme and a host protein, integration host factor. Excision is formally the reverse process but the recombination involves hybrid sites (BOP′ and POB′) at the prophage–chromosome junctions and requires the phage-encoded excisionase (Xis) protein in addition to integrase. Synthesis of Int and Xis proteins is tightly controlled in a lysogenic cell by the λ *c*I repressor and only occurs when the lytic programme is induced (Campbell, 1992, 1996a, b; Nash, 1996).

Occasionally, with a frequency of $\sim 10^{-5}$, excisive recombination occurs aberrantly at sites other than *att*. One type of excision involves a site within the prophage and a second site distal to the bacterial *gal* or *bio* genes located in the flanking segments of chromosomal DNA. Excised molecules can be encapsidated into infective particles, provided they contain the packaging-initiation locus (*cos*) of the phage and are of packageable size. Appropriate sizes range from 76 to 105% of the 49-kb λ genome. One consequence of the size constraint in packaging is that transducing DNA molecules frequently lack a complete viral genome, since inclusion of bacterial DNA is compensated by a deficiency of phage DNA. A further consequence is that individual phage particles can mediate transduction of either the *gal* or *bio* regions but not both (**Fig. 11**).

After infection of a recipient cell, transducing DNA circularises at the *cos* sites and recombines with the bacterial chromosome. Integration by the Int system is impeded by the fact that transducing DNA contains BOP′ or POB′, rather than the POP′ substrate present in the normal λ genome. Thus integration is commonly mediated by homologous recombination of the transduced portion of bacterial DNA and the matching region in the recipient chromosome. The resulting cointegrate contains a small duplication of the *gal* or *bio* region.

Generalised Transduction

Headful Packaging of DNA

Generalised transduction involves transfer of a non-specific fragment of the bacterial chromosome without associated phage DNA. Inclusion of bacterial rather than phage DNA in the viral particle results from errors

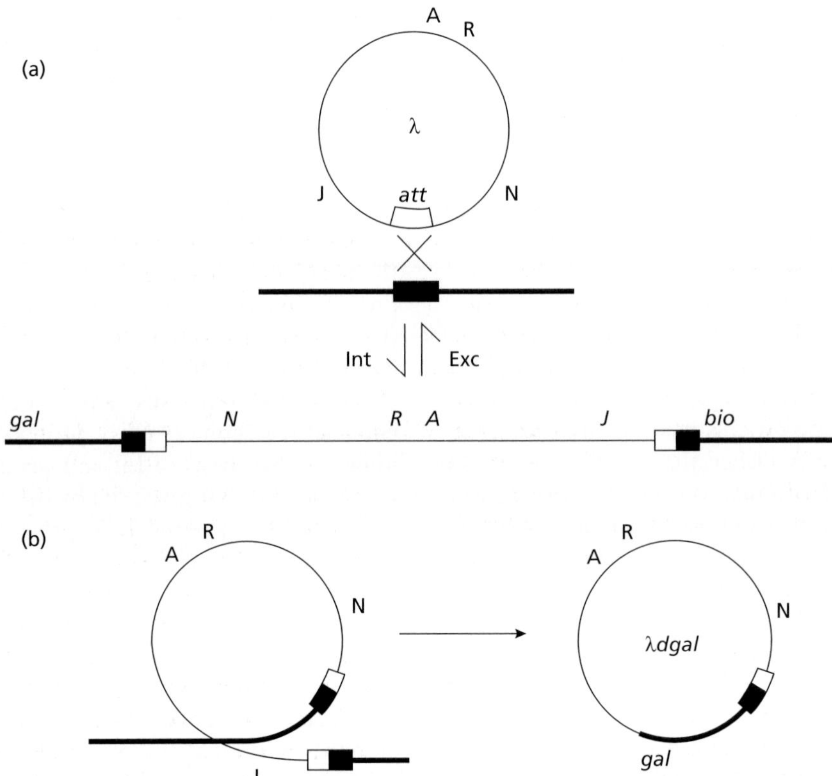

Fig. 11 The Campbell model for the production of specialised λ transducing phages: (a) Normal integration and excision of the prophage is mediated by site-specific recombination of *att* sites (shown as boxes). (b) An abnormal recombination outside the *att* sites gives λ*gal* transducing DNA deleted of essential phage genes in the A–J region but carrying flanking bacterial *gal* genes. λ*dbio* as well as λ*bio* plaque-forming phages can likewise be generated by abnormal excisive recombination.

in the substrate selected for packaging by the 'headful' mechanism. The substrate for headful packaging is a concatemer of phage DNA, defined as a covalently continuous structure containing more than one genomic length of DNA. Such structures are generated by recombination of unit-length molecules or by rolling-circle replication, which is the predominant mode of replication at late stages of the lytic cycle. Linear phage DNA molecules are packaged from the concatemer by a mechanism that measures the amount of DNA that can be packaged into the head of a phage particle. Typically the packaged DNA is slightly larger than one genomic length. As a consequence, each viral DNA molecule has terminally repeated sequences but the genetic region comprising the repeat varies in different phage particles.

The packaging mechanisms of phages P1 and P22 encapsidate 100 kb and 44 kb of DNA per particle, respectively. Packaging is initiated at a unique site called *pac*. P1 *pac* is within a 161 bp segment containing several copies of a hexanucleotide motif thought to promote recognition and binding of the cleavage protein (Casjens and Hayden, 1988). The initial cleavage potentiates unidirectional packaging of a headful of DNA which is terminated by a second less precise cleavage described as the 'headful' cut. The cycle may be repeated to allow two or three more rounds of packaging from the same concatemer. Such a processive encapsidation mechanism generates a population of viral chromosomes with terminal repeats that vary relative to the genetic map of the phage (Campbell, 1996a).

Generalised Transducing Particles

The density- and radioactivity-labelling experiments of Ikeda and Tomizawa (1965) were seminal in establishing that P1 transducing phages comprise ~0.3% of the particles in a lysate, contain cellular DNA synthesised prior to infection but no phage DNA, and include DNA molecules that are covalently associated with protein. Single-burst experiments have revealed that packaging can be initiated at more than one site per cell and that chromosomal DNA may be packaged processively for several headfuls (Harriman, 1972). P1 transduction frequencies can vary by ~20-fold but the variation is attributed primarily to recombination effects occurring in the recipient cell rather than selectivity at packaging (Masters *et al.*,

1984). In contrast, transduction frequencies by phage P22 can vary by 10^3-fold, evidently because the P22 packaging system is more fastidious than the P1 system in requiring *pac*-like sequences in the chromosome. Indeed, bias can be introduced into the P1 system by inserting a P1 *pac* site into the bacterial chromosome.

Inheritance of Transduced DNA

Analysis of the fate of labelled DNA has shown that only about 2% of P1-transduced DNA is integrated into the bacterial chromosome. RecA-dependent homologous recombination is responsible for either splicing a sector of dsDNA into the recipient molecule or inserting a length of ssDNA as heteroduplex DNA. At least half of the recombinational replacements are predicted to extend over ~23 kb, which corresponds to the spacing between markers that are cotransduced with a frequency of 50% (Masters, 1996). This size estimate is consistent with molecular analyses using restriction-fragment-length polymorphisms, which indicated that single incorporations of >25 kb are commonly generated by P1 transduction of DNA between homogenic strains. Much shorter segments were incorporated in inter-strain transductions, possibly due to cleavage of entrant DNA by restriction systems native to the recipient cell (McKane and Milkman, 1995).

Approximately 75% of P1-transduced DNA remains in the extrachromosomal state (Sandri and Berger, 1980), which correlates with the genetic observation that as many as 90% of the fragments transduced by phages P1 and P22 give rise to abortive rather than complete transductants. Abortive transductants appear as microcolonies and are thought to contain a non-replicating, transcriptionally-active DNA fragment that is transmitted unilinearly to one daughter bacterium at cell division. The high frequency and persistence of abortive transductants raises the question of how the DNA fragment is protected from nucleases. DNA abortively transduced by P1 has the electrophoretic and sedimentation properties of a circular molecule and is converted into a linear form by protein-denaturing conditions, as if the ends of a linear fragment are clamped together by one or more proteins. Three phage proteins are implicated in the closure of P22-abortively transduced DNA. The proteins (gp7, 16 and 20) are thought to be injected in a complex with the viral DNA as a normal part of the infection process to aid translocation of the DNA across the cell envelope and its subsequent circularisation. An abortive transductant is viewed as a trapped intermediate of such a circle-closure reaction (Benson and Roth, 1997).

Host Range of Transducing Vectors

The ability of a phage to disseminate genes by transduction is limited by transfer host range, defined as its ability to adsorb to a cell and inject DNA. Specificity of cell-surface receptors is a critical factor. These structures have been identified by isolating and characterising phage-resistant mutants, and found to include diverse components of the outer cell membrane (Nikaido, 1996). Examples are repeat units of the O antigen polysaccharide portion of the lipopolysaccharide (recognised by phage P22), a terminal glucose of the LPS core (phage P1), classical porin proteins (phage T2 and the lambdoid 434 phage), specific channel-forming proteins (phages T6 and λ) and high-affinity receptor proteins (phages T1, T5).

Some infections require specific components of both cell membranes. For example, phage λ requires LamB protein, which forms a channel in the outer membrane for high-affinity uptake of maltose, as well as a component of the mannose phosphotransferase (PTS) transport system in the inner membrane. It is unknown whether the cellular transport systems provide incipient pores for DNA translocation across the two membranes or act as receptors to target a phage tail through the cell envelope as a DNA-delivery tube.

Compared to many conjugative plasmids, phages have a narrow host range. However, phages can infect a broader range of hosts than those capable of supporting the lytic response. Hence, host range defined by plaque-forming ability may underestimate the capacity of a phage to disseminate transposable elements and transduced DNA among different species. Phage P1 illustrates the point, since it can infect without progeny production a range of organisms outside the family Enterobacteriaceae, including members of the genera *Agrobacterium*, *Alcaligenes*, *Flavobacterium*, *Myxococcus*, *Pseudomonas* and *Yersinia* (see Yarmolinsky and Sternberg, 1988).

Transduction in Natural Environments

Reviews in this area include those of Kokjohn (1989), Miller and Sayler (1992) and Yin and Stotzky (1997). Transduction has been observed in laboratory conditions for more than 50 Gram-positive and -negative organisms, suggesting that the process is a universal phenomenon in natural environments. This conclusion is supported by the recognition that phage concentrations in aquatic environments are considerably higher than previously believed and may exceed bacterial abundancies. Although free phages may have a finite 'lifetime', their persistence in terrestrial and aquatic habitats is prolonged by adsorption to clay minerals and other particles. Levels of free particles for the many temperate transducing bacteriophages can be maintained by spontaneous prophage induction in lysogenic bacteria. However, transduction is not a process restricted to temperate phages, since virulent phages employing headful packaging can participate in generalised transduction provided that lysis of incipient transductants by superinfecting phages is prevented. The necessary protection occurs at low multiplicities of infection.

Despite persuasive evidence of the occurrence of transduction in natural environments, the impact of the process as a means for disseminating genes is limited by the specificity of phage adsorption to bacterial cell surfaces. Transduction may have biological roles additional to potentiating genetic diversity. It seems reasonable to assume that, since prophages are induced into lytic growth by agents that damage DNA, the frequency of transducing phages will be elevated in bacterial communities suffering environmental stress. Under such conditions, transduction may function as a gene-transfer process that promotes recombinational repair of DNA.

Barriers to the Horizontal Transmission of Genes – Genetic Isolation

The previous sections focus on mechanisms enabling gene transfer between bacteria. Here we address the antithesis: processes that prevent the integration of transferred DNA into the genome of the recipient cell. One such process is the activity of host-restriction systems, viewed as surveillance processes that destroy incoming DNA identified as foreign by its lack of cognate methylation. The destructive effects of restriction are contrasted with phage and plasmid-encoded systems that have evolved to allow these transmissible elements to evade restriction. Genetic isolation is also promoted by a second barrier that operates at the level of homologous recombination to limit exchanges between sequence-diverged DNAs. Important components of this second barrier are nucleotide sequence divergence *per se* and mismatch repair enzymes functioning to prevent recombinational heteroduplexes with mispaired bases. These processes maintain the genetic isolation of bacterial species (Matic *et al.*, 1995).

DNA Restriction

Role and Diversity of Restriction–Modification (R–M) Systems

Classical R–M systems consist of a restriction endonuclease and a matching modification enzyme or DNA methyltransferase which recognise the same nucleotide sequence. Methylation of one or more specific bases within the recognition sequence protects DNA from cleavage by the restriction endonuclease. Modification enzymes can methylate DNA *de novo,* but their primary role is to maintain the methylation of DNA by converting hemimethylated sequences generated by replication into the fully methylated state.

R–M systems can be classified into three types on the basis of their enzyme composition, cofactor requirements, symmetry of recognition sequence and DNA cleavage position (Bickle and Krüger, 1993; King and Murray, 1994; Redaschi and Bickle, 1996). Type I systems are the most complex in consisting of three different Hsd subunits – HsdR, HsdM and HsdS. The HsdS polypetide determines the specificity of the nucleotide sequence recognised. The enzyme complex catalyses either modification or restriction depending on the methylation status of the substrate DNA. DNA cleavage occurs at variable and often considerable distances from the specificity sites. Type II R–M systems consist of separate methylases and endonucleases and are familiar as the enzymes widely used in recombinant DNA technology. The enzymes are ubiquitous and diverse, being detected in more than a quarter of all bacterial isolates and comprising more than 200 different specificities (Roberts and Macelis, 1994). Few type III systems have been identified to date. The main enzyme is a heteromultimer of Res and Mod subunits possessing modification and restriction activities. DNA cleavage occurs $\sim$25 bp distant from the recognition sequence.

One widely cited view is that R–M systems evolve as cellular defence mechanisms against foreign DNA, particularly that of virulent phages. This role may be important in aiding a bacterial strain to colonise a new environment containing previously unencountered phages. However, any advantage may be transient because a restriction system is not an absolute barrier to infecting DNA and a phage DNA molecule escaping destruction can acquire modification (Korona *et al.*, 1993).

R–M systems may have other ecological roles. One hypothesis is that restriction potentiates homologous recombination in natural populations of bacteria by cleaving large entrant DNA molecules into recombinogenic fragments (King and Murray, 1994). Such fragmentation of DNA may underlie the mosaic pattern of localised DNA replacements that punctuate the clonal structure of homologous genes in closely related natural isolates (McKane and Milkman, 1995). Another hypothesis is that type II R–M systems evolve as genomic parasites that force their maintenance on bacteria by killing segregant cells. This notion would explain why R–M genes are frequently carried on plasmids (Naito *et al.*, 1995). Type I R–M genes do not behave as selfish units and their evolution must be driven by some alternative selective pressure (O'Neill *et al.*, 1997).

Effect of Restriction on the Transmission of Bacterial Genes

The prevalence and diversity of R–M systems imply that restriction is a commonly encountered barrier to productive gene exchange. Patterns of inheritance of transferred chromosomal DNA are disrupted severely in a restricting recipient cell. Yields of recombinants in conjugative and transductional crosses are depressed and genetic linkages break down, presumably reflecting fragmentation of entrant DNA and digestion of fragments by exonucleases such as RecBCD (exonuclease V) enzyme (Boyer, 1971; Iida *et al.*, 1987). In contrast, restriction has little effect on the inheritance of chromosomal DNA transferred by natural transformation systems; the entrant DNA is presumably refractory to cleavage because it is single-stranded and protected in the heteroduplex recombinant structures by cognate methylation of the recipient DNA strand (Trautner *et al.*, 1974). Consistent with this explanation, plasmids transferred as duplex DNA by artificial transformation procedures are sensitive to restriction.

Restriction Avoidance by Bacteriophages and Plasmids

Restriction systems are not absolute barriers to productive transfer. The escaping fraction in a phage infection is in the range of 10^{-1}–10^{-5}, depending on factors such as the number of specificity sites per DNA molecule and the physiological state of the bacteria. Not surprisingly, phages have evolved strategies for avoiding restriction (Bickle and Krüger, 1993; King and Murray, 1994). The evasion systems are diverse and include proteins which are produced rapidly after infection to combat type I enzymes by abolishing their activities (e.g. Ocr or gene *0.3* protein of phage T7) and to enhance *de novo* modification of DNA generated by replication of infecting DNA escaping restriction (e.g. λ Ral function). Other avoidance processes operate through the donor cell to confer unusual chemical modifications on the infecting DNA, as found for T-even phages, or to provide protective proteins that are co-infected with DNA (e.g. P1 Dar proteins).

Plasmids also specify restriction-avoidance mechanisms that operate during conjugation. The classic example is the ArdA system encoded by I1 plasmids and a few other Inc groups (B, FV, K and N) of enterobacterial plasmid (Chilley and Wilkins, 1995). The system selectively inhibits the restriction activity of type I enzymes by a process requiring *ardA* expression in the conjugatively infected cell. Such a mode of action raises the question of how ArdA protein accumulates in time to prevent restriction of the immigrant plasmid. The explanation includes early transfer of *ardA* as part of the plasmid leading region, coupled with its rapid expression by transcription of the incoming DNA strand (see section on bacterial conjugation, above). This arrangement allows accumulation of ArdA protein before the entrant plasmid strand is converted into duplex DNA, the substrate of restriction enzymes (Bates *et al.*, 1999).

In addition, phage and plasmid genomes are often found to be deficient in sequences recognised by the restriction systems of putative hosts. Such findings are interpreted as the outcome of a passive defence strategy whereby selection leads to progressive accumulation of spontaneous mutants with a reduced number of restriction targets (Sharp, 1986; Korona *et al.*, 1993; Wilkins *et al.*, 1996). Counter-selection of sites occurs when the phage or plasmid shuttles between non-restricting and restricting bacterial strains, which repeatedly exposes DNA in the unmodified state to the cognate restriction system. R–M systems are sufficiently ubiquitous for such selections to occur in transfers between closely related strains, rather than in a history of promiscuous transfers between phylogenetically distant organisms.

Although phages and plasmids become adapted to specific restriction systems by loss of cognate target sites, bacteria can acquire the defence of a new restriction system via transfer of R–M genes. Efficient acquisition of novel R–M genes requires delayed production of the restriction function to allow the bacterial genome to acquire the protection of modification. Transfer of genes for a type I R–M system was found to be associated with prompt expression of the modification phenotype in the naive recipient cell but a delay in the onset of restriction by as many as 15 generations (Prakesh-Cheng *et al.*, 1993). Such efficient transmission of novel *hsd* genes re-quires a two-component protease encoded by the bacterial chromosome. The protease complex consists of ClpX, active as a molecular chaperone that attracts and prepares the polypeptide substrate for hydrolysis, and ClpP as a serine protease. The ClpXP complex is thought to target the HsdR subunit for proteolysis after the restriction endonuclease has bound to unmodified chromosomal target sites but before the DNA cleavage response (Makovets *et al.*, 1999).

Type II R–M genes can also be transferred to naive recipient cells by conjugation, transduction and transformation without extensive cell killing. Regulation by preferential expression of the methyltransferase gene at the transcriptional and post-transcriptional levels has been implicated, as has a passive mechanism in which M polypeptides are promptly active as monomers whereas restriction polypeptides must accumulate and dimerise to provide the functional form. There is also evidence that transient low levels of DNA cleavage can be tolerated (Redaschi and Bickle, 1996).

Recombinational Barriers to Interspecies Gene Exchange

Recombination barriers have been studied extensively in crosses of *E. coli* and *S. typhimurium*. The genomes of these closely related organisms have diverged by about 15% at the nucleotide sequence level, primarily by accumulation of synonymous codon changes that have no effect on polypeptide function. In such interspecies crosses, chromosomal re-combination frequencies were found to be deficient by about five orders of magnitude in both Hfr-mediated conjugation and transduction (Rayssiguier *et al.*, 1989). Similar effects have been observed in transformation, as shown by a ~1000-fold reduction in *B. subtilis* transformation frequency when the donor and recipient DNAs are divergent by 15% at the sequence level (Majewski and Cohan, 1998).

RecA protein can mediate, albeit with reduced frequencies, *in vitro* formation of joints between extensively diverged DNA molecules. However, joint formation *in vivo* is less tolerant of mismatches. A single mismatch in a 31-bp region of homology, which is close to the minimum threshold size for recombination, can impede initiation of exchanges, and numerous mismatches in longer segments disrupt homologous recombination via a process involving enzymes of the mismatch repair system (Shen and Huang, 1989).

Mismatch repair was originally defined as a system that maintains the fidelity of the bacterial genome by removing DNA replication errors. The repair process involves removal and resynthesis of long patches of nucleotides by a mechanism requiring the *mut* gene products. Of these, MutS and MutL participate in mismatch recognition. The barrier to recombination in interspecies conjugation is relaxed in *mutS* and *mutL* mutants of *E. coli*, suggesting that the Mut proteins prevent the formation of mispaired heteroduplexes in recombination or subsequently eliminate them (Rayssiguier *et al.*, 1989; Shen and Huang, 1989; Matic *et al.*, 1995). The former possibility is supported by findings that purified MutS and MutL proteins inhibit RecA-mediated exchanges *in vitro* between sequence-diverged DNAs (Worth *et al.*, 1994), and that artificial transformation of *E. coli* by plasmids containing a mispaired heteroduplex was only marginally impeded in mut^+ cells relative to *mut* mutants (Westmoreland *et al.*, 1997).

The pneumococcal Hex system of mismatch repair was described in the section on natural bacterial transformation as a mechanism that removes mismatches in heteroduplex intermediates formed in transformation. The Hex system has limited efficacy since it becomes saturated by multiple mismatches in a single heteroduplex region (Humbert *et al.*, 1995). Furthermore, mutational studies indicate that mismatch repair in transformation makes only a minor contribution to the recombinational isolation of different *Bacillus* species (Majewski and Cohan, 1998). The same may apply to conjugative crosses of different species, since suppression of mismatch repair in matings of *E. coli* Hfr donors and *S. typhimurium* recipients enhanced recombinant production but not to the level observed in homospecific controls (Rayssiguier *et al.*, 1989). Evidently the major barrier to the genetic exchange of sequence-diverged chromosomal genes lies in the strand exchange reactions fundamental to the recombination process.

Concluding Remarks – Relationships Between DNA Transfer and Protein Secretion Systems

The nature of the channels for DNA translocation across cell envelopes remains a major mystery. Molecular details are scarce not only for bacterial conjugation and transformation, but also for transduction where the DNA injection systems of many phage vectors remain ill defined. Despite obvious functional diversity between transfer processes, rapidly expanding genetic databases are providing compelling evidence of evolutionary relationships between different DNA transport systems and, more intriguingly, links to protein secretion pathways (**Fig. 12**).

Protein secretion systems have been grouped into different pathways according to shared characteristics. Type IV systems, as originally proposed by Salmond (1994), describe contact-mediated transport of nucleoprotein complexes in conjugation processes. Examples are transfer of the *A. tumefaciens* T-complex by the Vir system and of the bulky plasmid primase–DNA complexes by the I1 and P conjugation systems. Sequence alignments further indicate evolutionary relationships between proteins involved in the biosynthesis of the VirB complex and several types of

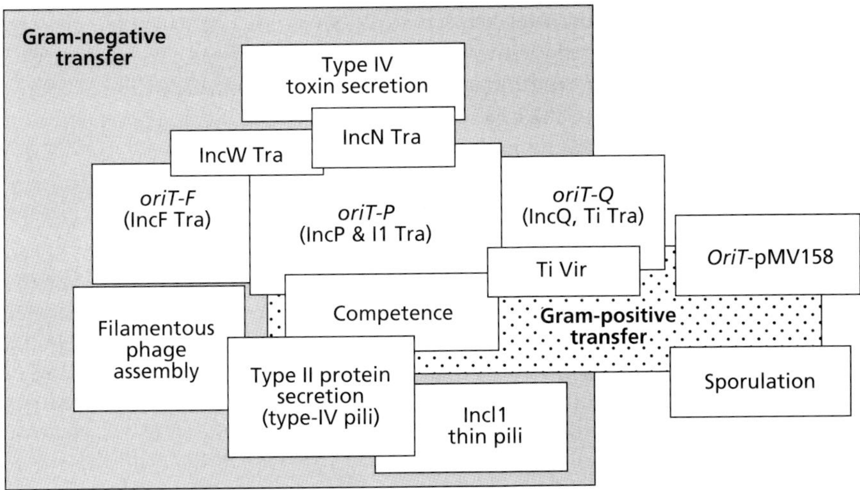

Fig. 12 Relationships between systems for DNA translocation across bacterial cell envelopes and protein secretion. Gram-negative and -positive conjugation systems overlap at the level of their DNA transport apparatus; *oriT* sequences are divided here into four groups named *oriT-F* (IncF, N and W plasmids); *-P* (IncP and I1 systems plus Ti Vir); *-Q* (IncQ, Ti Tra and some Gram-positive conjugative plasmids); and -pMV158 (Gram-positive mobilisable plasmids). Macromolecular transport systems are grouped on the basis of similarities in amino acid sequence. Thus, the type IV protein secretion pathway for release of pertussis toxin is part of a superfamily containing the transporters for the T-complex and for plasmids of the IncN, P and W groups. Further relationships exist between type II protein secretion pathways and the biosynthetic machinery for the thin I1 conjugative pilus, the DNA uptake system for *B. subtilis* transformation, and the extrusion of filamentous phages. Type IV pili function directly in the transformation system of *N. gonorrhoeae*. The essential Tra protein of *Streptomyces* plasmids is similar to the *B. subtilis* SpoIIIE protein involved in partitioning the chromosome to the prespore.

enterobacterial conjugative pilus, viewed as DNA transporters, and the protein-transport systems of a number of pathogens. Hence, some toxin transporters are classified as type IV secretion systems (Salmond, 1994; Baron *et al.*, 1997; Christie, 1997).

For example, the Ptl system of *Bordetella pertussis*, which liberates the pertussis (whooping cough) toxin across the outer cell membrane, is classified as type IV because the nine genes of the *ptl* operon are homologous and co-linear with various genes of the Ti Vir and the F, N, P, and W plasmid transport machineries (Christie, 1997). Moreover, certain of these transport genes are related to six genes in the *cag* pathogenicity island of pathogenic *H. pylori* strains associated with peptic ulcers. Thus *cag* is thought to specify a type IV secretion module for contact-dependent signalling of virulent *H. pylori* to epithelial cells, possibly inducing remodelling of the cell surfaces and pedestal formation (Covacci *et al.*, 1999). A third example that truly blends pathogenicity with conjugation is a virulence system determined by pathogenic strains of *L. pneumophila* that infect intracellular vacuoles. The *dot* and *icm* genes of such strains contribute to a system believed to secrete macromolecules to the vacuolar membrane or cytoplasm of host cells. Certain *dot/icm* genes additionally cause transfer of mobilisable plasmids between bacteria, providing a functional connection between conjugation and a virulence secretory system (Segal *et al.*, 1998; Vogel *et al.*, 1998).

The type III secretion pathway shares with conjugation systems the important trait of contact-mediated transport of proteins into a second cell. However, no evidence of homology has been found between the two kinds of system. Well-studied type III systems are those for secretion of *Yersinia* outer proteins (Yops) into the cytosol of the host cell and translocation of the intimin receptor of enteropathogenic *E. coli* to epithelial cell membranes (Galán and Collmer, 1999). Electron micrographs of type III secretion apparatuses have been obtained for both the EPEC system (Knutton *et al.*, 1998) and a similar system in *S. typhimurium* which is evolutionarily related to the flagellum-synthesis apparatus (Kubori *et al.*, 1998). The appearance of these ‘needle-like’ supramolecular structures is striking and very reminiscent of phage tails, which again suggests common factors in the genesis of these transport systems.

A range of relationships exists between DNA transport systems and the type II secretion pathway which is the main terminal branch of the general secretory pathway (GSP; see Hobbs and Mattick, 1993). Type II secretion involves an *E. coli* Sec-like system for export of pre-secretory proteins across the inner cell membrane and a complex of some 14 or more integral and peripheral membrane proteins for translocation of exoproteins to the exterior medium (Filloux, 1999). Some Gsp proteins resemble components involved in type 4 piliation and the pilin subunit itself. Type 4 pili mediate attachment of bacteria to host cells and cell movement across surfaces by 'twitching' motility (Darzins, 1994).

One link between DNA transfer and the type II secretory pathway is provided by sequence similarities of GspD, which forms a pore-like 'secretin' in the outer membrane, and gene IV protein of phage M13. The latter is thought to comprise a gated pore for the extrusion of ssDNA within the protein filament of the maturing phage particle (Russel, 1995; Linderoth *et al.*, 1997). Another link exists to the competence machinery of natural transformation systems, as evidenced by striking similarities between the N-terminal moieties of the ComGC, ComGD, ComGE and ComGG proteins of *B. subtilis* and type 4 prepilin-like proteins. Furthermore, ComGA, ComGB and ComC resemble respectively the GspE, GspF and GspO homologues involved in processing and assembly/ translocation of the type 4 pilin subunit (Dubnau, 1997). Direct involvement of a type 4 pilus in DNA transfer processes is seen through the requirement of the organelle for uptake of transforming DNA by *N. gonorrhoeae* (see section on natural bacterial transformation), and sequence evidence indicating that the thin I1 conjugative pilus is a type 4 pilus (Yoshida *et al.*, 1998; see section on bacterial confugation).

Taken together (**Fig. 12**), the evidence indicates that common ancestral genes have been recruited on an evolutionary timescale into diverse import–export pathways, giving superfamilies of systems for traffic of naked DNA, proteins and DNA–protein complexes into and out of bacterial cells.

Acknowledgements

BMW was supported by MRC grant G9321196MB and LSF by MRC of Canada MT 11249 and NSERC 139684. We thank Mervyn Bibb, David Bradley, Keith Chater, David Dubnau, Markus Dürrenberger, Gary Dunny and Leendert Hamoen for invaluable help in supplying material and ideas for figures.

References

Achtman M, Skurray R (1977) A redefinition of the mating phenomenon in bacteria. In: Reissig JL (ed.) *Microbial Interactions: Receptors and Recognition.* London: Chapman & Hall, Vol. 3, pp. 233–279.

Achtman M, Kennedy N, Skurray R (1977) Cell–cell interactions in conjugating *Escherichia coli*: role of *traT* protein in surface exclusion. *Proc Natl Acad. Sci. USA* 74: 5104–5108.

Achtman M, Morelli G, Schwuchow S (1978) Cell–cell interactions in conjugating *Escherichia coli*: role of F pili and fate of mating aggregates. *J. Bacteriol.* 135: 1053–1061.

Althorpe NJ, Chilley PM, Thomas AT, Brammar WJ, Wilkins BM (1999) Transient transcriptional activation of the IncI1 plasmid anti-restriction (*ardA*) and SOS inhibition gene (*psiB*) early in conjugating recipient bacteria. *Mol. Microbiol.* 31: 133–142.

An FY, Clewell DB (1994) Characterization of the determinant (*traB*) encoding sex pheromone shutdown by the hemolysin/bacteriocin plasmid pAD1 in Enterococcus faecalis. *Plasmid* 31: 215–221.

An FY, Clewell DB (1997) The origin of transfer (*oriT*) of the enterococcal, pheromone-responding, cytolysin plasmid pAD1 is located within the *repA* determinant. *Plasmid* 37: 87–94.

Anthony KG, Sherburne C, Sherburne R, Frost LS (1994) The role of the pilus in recipient cell recognition during bacterial conjugation mediated by F-like plasmids. *Mol. Microbiol.* 13: 939–953.

Anthony KG, Klimke WA, Manchak J, Frost LS (1999) Comparison of proteins involved in pilus synthesis and mating pair stabilization from the related plasmids F and R100-1: insights into the mechanism of conjugation. *J. Bacteriol.* 181: 5149–5159.

Armstrong GD, Frost LS, Sastry PA, Paranchych W (1980) Comparative biochemical studies on F and EDP208 conjugative pili. *J. Bacteriol.* 141: 333–341.

Bagdasarian M, Bailone A, Angulo JF, Scholz P, Bagdasarian M, Devoret R (1992) PsiB, an anti-SOS protein, is transiently expressed by the F sex factor during its transmission to an *Escherichia coli.* K-12 recipient. *Mol. Microbiol.* 6: 885–893.

Bailey MJA, Hughes C, Koronakis V (1996) Increased distal gene transcription by the elongation factor RfaH, a specialized homologue of NusG. *Mol. Microbiol.* 22: 729–737.

Bannam TL, Crellin PK, Rood JI (1995) Molecular genetics of the chloramphenicol-resistance transposon Tn*4451* from *Clostridium perfringens*: the TnpX site-specific recombinase excises a circular transposon molecule. *Mol. Microbiol.* 16: 535–551.

Baron C, Llosa M, Zhou S, Zambryski PC (1997) VirB1, a component of the T-complex transfer machinery of Agrobacterium tumefaciens, is processed to a C-terminal secreted product, VirB1. *J. Bacteriol.* 179: 1203–1210.

Bastos M-C-F, Tanimoto K, Clewell DB (1997) Regulation of transfer of the *Enterococcus faecalis* pheromone-responding plasmid pAD1: temperature-sensitive transfer mutants and identification of a new regulatory determinant, *traD*. *J. Bacteriol.* 179: 3250–3259.

Bates S, Cashmore AM, Wilkins BM (1998) IncP plasmids are unusually effective in mediating conjugation of

Escherichia coli and *Saccharomyces cerevisiae*: involvement of the Tra2 mating system. *J. Bacteriol.* 180: 6538–6543.

Bates S, Roscoe RA, Althorpe NJ, Brammar WJ, Wilkins BM (1999) Expression of leading region genes on IncI1 plasmid ColIb-P9; genetic evidence for single-stranded DNA transcription. *Microbiol.* 145: 2655–2662.

Bayer M, Eferl R, Zellnig G *et al.* (1995) Gene 19 of plasmid R1 is required for both efficient conjugative DNA transfer and bacteriophage R17 infection. *J. Bacteriol.* 177: 4279–4288.

Beijersbergen A, Dulk-Ras AD, Schilperoort RA, Hooykaas PJJ (1992) Conjugative transfer by the virulence system of *Agrobacterium tumefaciens*. *Science* 256: 1324–1327.

Bensing BA, Dunny GM (1997) Pheromone-inducible expression of an aggregation protein in *Enterococcus faecalis* requires interaction of a plasmid-encoded RNA with components of the ribosome. *Mol. Microbiol.* 24: 295–308.

Bensing BA, Manias DA, Dunny GM (1997) Pheromone cCF10 and plasmid pCF10-encoded regulatory molecules act post-transcriptionally to activate expression of downstream conjugation functions. *Mol. Microbiol.* 24: 285–294.

Benson NR, Roth J (1997) A *Salmonella* phage-P22 mutant defective in abortive transduction. *Genetics* 145: 17–27.

Berg T, Firth N, Apisiridej S, Hettiaratchi A, Leelaporn A, Skurray R (1998) Complete nucleotide sequence of pSK41: evolution of staphylococcal conjugative multiresistance plasmids. *J. Bacteriol.* 180: 4350–4359.

Bickle TA, Krüger DH (1993) Biology of DNA restriction. *Microbiol. Rev.* 57: 434–450.

Biebricher CK and Düker E-M (1984) F and type I piliation of *Escherichia coli*. *J. Gen. Microbiol.* 130: 951–957.

Borrebaeck CAK, Soderlind E, Frost LS, Malmborg A-C (1997) Combining bacterial and phage display: tapping the potential of molecular libraries for gene identification. In: Guttry P (ed.) *Display Technologies: Novel Targets and Strategies*. Southborough, MA: IBC, pp. 197–208.

Boulnois GJ, Wilkins BM (1978) A ColI-specified product, synthesized in newly infected recipients, limits the amount of DNA transferred during conjugation of *Escherichia coli* K-12. *J. Bacteriol.* 133: 1–9.

Boyd AC, Archer JA, Sherratt DJ (1989) Characterization of the ColE1 mobilization region and its protein products. *Mol. Gen. Genet.* 217: 488–498.

Boyer HW (1971) DNA restriction and modification mechanisms in bacteria. *Ann. Rev. Microbiol.* 25: 153–176.

Bradley DE (1984) Characteristics and function of thick and thin conjugative pili determined by transfer-derepressed plasmids of incompatibility groups I1, I2, I5, B, K and Z. *J. Gen. Microbiol.* 130: 1489–1502.

Bradley DE (1989) Interaction of drug resistance plasmids and bacteriophages with diarrheagenic strains of *Escherichia coli*. *Infect. Immun.* 57: 2331–2338.

Bradley DE, Meynell E (1978) Serological characteristics of pili determined by the plasmids R711b and F_0lac. *J. Gen. Microbiol.* 108: 141–149.

Bradley DE, Taylor DE, Cohen DR (1980) Specification of surface mating systems among conjugative drug resistance plasmids in *Escherichia coli* K-12. *J. Bacteriol.* 143: 1466–1470.

Bradley DE, Coetzee JN, Bothma T, Hedges RW (1981a) Phage t: a group T plasmid-dependent bacteriophage. *J. Gen. Microbiol.* 126: 397–403.

Bradley DE, Coetzee JN, Bothma T, Hedges RW (1981b) Phage X: a plasmid-dependent, broad host range, filamentous bacterial virus. *J. Gen. Microbiol.* 126: 389–396.

Bringel F, van Alstine GL, Scott JR (1992) Conjugative transposition of Tn*916*: the transposon *int* gene is required only in the donor. *J. Bacteriol.* 174: 4036–4041.

Brolle D-F, Pape H, Hopwood DA, Kieser T (1993) Analysis of the transfer region of the *Streptomyces* plasmid SCP2*. *Mol. Microbiol.* 10: 157–170.

Buchanan-Wollaston V, Passiatore JE, Cannon F (1987) The *mob* and *oriT* mobilization functions of a bacterial plasmid promote its transfer to plants. *Nature* 328: 172–174.

Cabezon E, Lanka E, de la Cruz F (1994) Requirements for mobilization of plasmids RSF1010 and ColE1 by the IncW plasmid R388: *trwB* and RP4 *traG* are interchangeable. *J. Bacteriol.* 176: 4455–4458.

Cabezon E, Sastre JI, de la Cruz F (1997) Genetic evidence of a coupling role for the TraG protein family in bacterial conjugation. *Mol. Gen Genet.* 254: 400–406.

Camerini-Otero RD, Hsieh P (1995) Homologous re-combination proteins in prokaryotes and eukaryotes. *Ann. Rev. Genet.* 29: 509–552.

Campbell AM (1992) Chromosomal insertion sites for phages and plasmids. *J. Bacteriol.* 174: 7495–7499.

Campbell AM (1996a) Bacteriophages. In: Fields BN, Knipe DM, Howley PM *et al.* (eds) *Fields' Virology*, 3rd edn. Philadelphia: Lippincott-Raven, pp. 587–607.

Campbell A.M. (1996b) Bacteriophages. In: Neidhardt FC (ed.) *Escherichia coli and Salmonella: Cellular and Molecular Biology*, 2nd edition. Washington, DC: American Society for Microbiology, pp. 2325–2338.

Casjens S, Hayden M (1988) Analysis *in vivo* of the bacteriophage P22 headful nuclease. *J. Mol. Biol.* 199: 467–474.

Chan PT, Ohmori H, Tomizawa JI, Lebowitz J (1985) Nucleotide sequence and gene organization of ColE1 DNA. *J. Biol. Chem.* 260: 8925–8935.

Cheah K-C, Skurray RA (1986) The F plasmid carries an IS*3* insertion within *finO*. *J. Gen. Microbiol.* 132: 3269–3275.

Cheetham BF, Katz ME (1995) A role for bacteriophages in the evolution and transfer of bacterial virulence determinants. *Mol. Microbiol.* 18: 201–208.

Chilley PM, Wilkins BM (1995) Distribution of the *ardA* family of antirestriction genes on conjugative plasmids. *Microbiology* 141: 2157–2164.

Chilley PM, Wilkins BM (1997) Transcriptional induction of leading region genes during bacterial conjugation: measurement by competitive reverse-transcription PCR. *Plasmid* 37: 219.

Christie PJ (1997) *Agrobacterium tumefaciens* T-complex transport apparatus: a paradigm for a new family of multifunctional transporters in eubacteria. *J. Bacteriol.* 179: 3085–3094.

Clewell DB (1993a) Bacterial sex pheromone-induced plasmid transfer. *Cell* 73: 9–12.

Clewell DB (1993b) Sex pheromones and the plasmid-encoded mating response in *Enterococcus faecalis*. In: Clewell DB (ed.) *Bacterial Conjugation*. New York: Plenum Press, pp. 349–367.

Clewell DB (1993c) *Bacterial Conjugation*. New York: Plenum Press.

Clewell DB, Flannagan SE (1993) The conjugative transposons of gram-positive bacteria. In: Clewell DB (ed.) *Bacterial Conjugation*. New York: Plenum Press, pp. 369–393.

Clewell DB, Flannagan SE, Jaworski DD (1995) Unconstrained bacterial promiscuity: the Tn*916*-Tn*1545* family of conjugative transposons. *Trends Microbiol.* 3: 229–236.

Click EM, Webster RE (1997) Filamentous phage infection: required interactions with the TolA protein. *J. Bacteriol.* 179: 6464–6471.

Climo MW, Sharma VK, Archer GL (1996) Identification and characterization of the origin of conjugative transfer (*oriT*) and a gene (*nes*) encoding a single-stranded endonuclease on the staphylococcal plasmid pGO1. *J. Bacteriol.* 178: 4975–4983.

Courvalin P (1994) Transfer of antibiotic resistance genes between Gram-positive and Gram-negative bacteria. *Antimicrob. Agents Chemother.* 38: 1447–1451.

Cousineau B, Smith D, Lawrence-Cavanagh S *et al.* (1998) Retrohoming of a bacterial group II intron: mobility via complete reverse splicing, independent of homologous DNA recombination. *Cell* 94: 451–462.

Covacci A, Telford JL, Del Giudice G, Parsonnet J, Rappuoli R (1999) *Helicobacter pylori* virulence and genetic geography. *Science* 284: 1328–1333.

Craig NL (1996) Transposition. In: Neidhardt FC (ed.) Escherichia coli *and* Salmonella: *Cellular and Molecular Biology*, 2nd edn. Washington DC: American Society for Microbiology, pp. 2339–2362.

Danese PN, Silhavy TJ (1997) The sigma E and the Cpx signal transduction systems control the synthesis of periplasmic protein-folding enzymes in *Escherichia coli*. *Genes Dev.* 11: 1183–1193.

Darzins A (1994) Characterization of a *Pseudomonas aeruginosa* gene cluster involved in pilus biosynthesis and twitching motility: sequence similarity to the chemotaxis proteins of enterics and the gliding bacterium *Myxococcus xanthus*. *Mol. Microbiol.* 11: 137–153.

DeFlaun MF, Levy SB (1989) Genes and their varied hosts. In: Levy SB, Miller RV (eds) *Gene Transfer in the Environment*. New York: McGraw-Hill, pp. 1–32.

de Vos WM, Venema G, Canosi U, Trautner TA (1981) Plasmid transformation in *Bacillus subtilis*: fate of plasmid DNA. *Mol. Gen Genet.* 181: 424–433.

Disqué-Kochem C, Dreiseikelmann B (1997). The cytoplasmic DNA-binding protein TraM binds to the inner membrane protein TraD *in vitro*. *J. Bacteriol.* 179: 6133–6137.

Doran TJ, Loh SM, Firth N, Skurray RA (1994). Molecular analysis of the F plasmid traVR region: traV encodes a lipoprotein. *J. Bacteriol.* 176: 4182–4186.

Dorward DW, Garon CF (1990) DNA is packaged within membrane-derived vesicles of gram-negative but not gram-positive bacteria. *Appl. Environ. Microbiol.* 56: 1960–1962.

Doucet-Populaire F, Trieu-Cuot P, Andremont A, Courvalin P (1992) Conjugal transfer of plasmid DNA from *Enterococcus faecalis* to *Escherichia coli* in digestive tracts of gnotobiotic mice. *Antimicrob. Agents Chemother.* 36: 502–504.

Dreiseikelmann B (1994) Translocation of DNA across bacterial membranes. *Microbiol. Rev.* 58: 293–316.

Dubnau D (1991) Genetic competence in *Bacillus subtilis*. *Microbiol. Rev.* 55: 395–424.

Dubnau D (1997) Binding and transport of transforming DNA by *Bacillus subtilis*: the role of type IV pilin-like proteins – a review. *Gene* 192: 191–198.

Dunny GM, Leonard BAB (1997) Cell–cell communication in gram-positive bacteria. *Ann. Rev. Microbiol.* 51: 527–564.

Dunny GM, Zimmerman DL, Tortorello ML (1985) Induction of surface exclusion (entry exclusion) by *Streptococcus faecalis* sex pheromones: use of monoclonal antibodies to identify an inducible surface antigen involved in the exclusion process. *Proc. Natl Acad. Sci. USA* 82: 8592–8586.

Dunny GM, Leonard BAB, Hedberg PJ (1995) Pheromone-inducible conjugation in *Enterococcus faecalis*: interbacterial and host-parasite chemical communication. *J. Bacteriol.* 177: 871–876.

Dürrenberger MB, Villiger W, Bachi T (1991) Conjugational junctions: morphology of specific contacts in conjugating *Escherichia coli* bacteria. *J. Struct. Biol.* 107: 146–156.

Eisenbrandt R, Kalkum M, Lai EM, Lurz R, Kado CI, Lanka E (1999) Conjugative pili of IncP plasmids, and the Ti plasmid T pilus are composed of cyclic subunits. *J. Biol. Chem.* 274: 22 548–22 555.

Elliott SJ, Wainwright LA, McDaniel TK *et al.* (1998) The complete sequence of the locus of enterocyte effacement (LEE) from enteropathogenic *Escherichia coli* E2348/69. *Mol. Microbiol.* 28: 1–4.

Elwell LP (1994) R plasmids and antibiotic resistances. In: Miller VL, Kaper JB, Portnoy DA, Isberg RR (eds) *Molecular Genetics of Bacterial Pathogenesis*. Washington DC: American Society for Microbiology, pp. 17–41.

Filloux A (1999) Type II protein secretion: the main terminal branch of the general secretory pathway. In: Broome-Smith JK, Baumberg S, Stirling CJ, Ward FB (eds) *Transport of Molecules across Microbial Membranes*. Cambridge: Cambridge University Press, pp. 80–109.

Finlay BB, Falkow S (1997) Common themes in microbial pathogenicity revisited. *Microbiol. Mol. Biol. Rev.* 61: 136–169.

Finlay BB, Frost LS, Paranchych W, Willetts NS (1986) Nucleotide sequences of five IncF plasmid *finP* alleles. *J. Bacteriol.* 167: 754–757.

Firth N, Skurray R (1992) Characterization of the F plasmid bifunctional conjugation gene, *traG*. *Mol. Gen Genet.* 232: 145–153.

Firth N, Ridgway KP, Byrne ME *et al.* (1993) Analysis of a transfer region from the staphylococcal conjugative plasmid pSK41. *Gene* 136: 13–25.

Firth N, Fink PD, Johnson L, Skurray RA (1994) A lipoprotein signal peptide encoded by the staphylococcal conjugative plasmid pSK41 exhibits an activity resembling that of *Enterococcus faecalis* pheromone cAD1. *J. Bacteriol.* 176: 5871–5873.

Firth N, Ippen-Ihler K, Skurray RA (1996) Structure and function of the F factor and mechanism of conjugation. In: Neidhardt FC (ed.) *Escherichia coli and Salmonella: Cellular and Molecular Biology*, 2nd edn. Washington DC: American Society for Microbiology, pp. 2377–2401.

Flannagan SE, Zitzow LA, Su YA, Clewell DB (1994) Nucleotide sequence of the 18-kb conjugative transposon Tn*916* from *Enterococcus faecalis*. *Plasmid* 32: 350–354.

Freter R (1986) The need for mathematical models in understanding colonization and plasmid transfers in the mammalian intestine. In: Levy SB, Novick RP (eds) *Banbury Report 24: Antibiotic Resistance Genes: Ecology, Transfer, and Expression*. New York: Cold Spring Harbor Laboratory, pp. 81–92.

Frost LS (1993) Conjugative pili and pilus-specific phages. In: Clewell DB (ed.) *Bacterial Conjugation*. New York: Plenum Press, pp. 189–221.

Frost LS (2000) Conjugation, bacterial. In: Lederberg J (ed.) *Encyclopedia of Microbiology*, 2nd edn. San Diego: Academic Press, pp. 847–862.

Frost LS, Simon J (1993) Studies on the pili of the promiscuous plasmid RP4. In: Kado CI, Crosa JH (eds) *Molecular Mechanisms of Bacterial Virulence*. Amsterdam: Kluwer, pp. 47–65.

Frost LS, Manchak J (1998) F^- phenocopies: Characterization of expression of the F transfer region in stationary phase. *Microbiology* 144: 2579–2587.

Frost LS, Finlay BB, Opgenorth A, Paranchych W, Lee JS (1985) Characterization and sequence analysis of pilin from F-like plasmids. *J. Bacteriology* 164: 1238–1247.

Frost LS, Ippen-Ihler K, Skurray RA (1994) Analysis of the sequence and gene products of the transfer region of the F sex factor. *Microbiol. Rev.* 58: 162–210.

Fujimoto S, Clewell DB (1998) Regulation of the pAD1 sex pheromone response of *Enterococcus faecalis* by direct interaction between the cAD1 peptide mating signal and the negatively regulating, DNA-binding TraA protein. *Proc. Natl. Acad. Sci USA* 95: 6430–6435.

Fullner KJ, Lara JC, Nester EW (1996) Pilus assembly by *Agrobacterium* T-DNA transfer genes. *Science* 273: 1107–1109.

Fussenegger M, Rudel T, Barten R, Ryll R, Meyer TF (1997) Transformation competence and type 4 pilus biosynthesis in *Neisseria gonorrhoeae* – a review. *Gene* 192: 125–134.

Galán JE, Collmer A (1999) Type III secretion machines: bacterial devices for protein delivery into host cells. *Science* 284: 1322–1328.

Galán JE, Sansonetti PJ (1996) Molecular and cellular bases of *Salmonella* and *Shigella* interactions with host cells. In: Neidhardt FC (ed.) *Escherichia coli and Salmonella: Cellular and Molecular Biology*, 2nd edn, Washington DC: American Society for Microbiology, pp. 2757–2773.

Gao Q, Luo Y, Deonier RC (1994) Initiation and termination of DNA transfer at F plasmid *oriT*. *Mol. Microbiol.* 11: 449–458.

Garcia LR, Molineux IJ (1995) Incomplete entry of bacteriophage T7 DNA into F plasmid-containing *Escherichia coli*. *J. Bacteriol.* 177: 4077–4083.

Giebelhaus LA, Frost L, Lanka E, Gormley EP, Davies JE, Leskiw B (1996) The Tra2 core of the IncPα plasmid RP4 is required for intergeneric mating between *Escherichia coli* and *Streptomyces lividans*. *J. Bacteriol.* 178: 6378–6381.

Godon J-J, Jury K, Shearman CA, Gasson MJ (1994) The *Lactococcus lactis* sex-factor aggregation gene *cluA*. *Mol. Microbiol.* 12: 655–663.

Golub EI, Bailone A, Devoret R (1988) A gene encoding an SOS inhibitor is present in different conjugative plasmids. *J. Bacteriol.* 170: 4392–4394.

Goodman SD, Scocca JJ (1988) Identification and arrangement of the DNA sequence recognized in specific transformation of *Neisseria gonorrhoeae*. *Proc. Natl Acad. Sci. USA* 85: 6982–6986.

Gordon GS, Sitnikov D, Webb CD *et al.* (1997) Chromosome and low copy number plasmid segregation in *E. coli*: visual evidence for direct mechanisms. *Cell* 90: 1113–1121.

Grohmann E, Guzman LM, Espinosa M (1999) Mobilization of the streptococcal plasmid pMV158: interaction of MobM protein with its cognate *oriT* region. *Mol. Gen Genet.* 261: 707–715.

Groisman EA, Ochman H (1996) Pathogenicity islands: bacterial evolution in quantum leaps. *Cell* 87: 791–794.

Grossman AD (1995) Genetic networks controlling the initiation of sporulation and the development of genetic competence in *Bacillus subtilis*. *Ann. Rev. Genet.* 29: 477–508.

Grossman TH, Silverman PM (1989) Structure and function of conjugative pili: inducible synthesis of functional F pili by *Escherichia coli* K-12 containing a lac-tra operon fusion. *J. Bacteriol.* 171: 650–656.

Guiney DG (1982) Host range of conjugation and replication functions of the *Escherichia coli* sex plasmid F'*lac*: comparison with the broad host range plasmid RK2. *J. Mol. Biol.* 162: 699–703.

Guzmán LM, Espinosa M (1997) The mobilization protein, MobM, of the streptococcal plasmid pMV158 specifically cleaves supercoiled DNA at the plasmid *oriT*. *J. Mol. Biol.* 266: 688–702.

Gyohda A, Funayama N, Komano T (1997) Analysis of DNA inversions in the shufflon of plasmid R64. *J. Bacteriol.* 179: 1867–1871.

Haase J, Kalkum M, Lanka E (1996). TrbK, a small cytoplasmic membrane lipoprotein, functions in entry exclusion of the IncPα plasmid RP4. *J. Bacteriol.* 178: 6720–6729.

Haase J, Lanka E (1997) A specific protease encoded by the conjugative DNA transfer systems of IncP and Ti plasmids is essential for pilus synthesis. *J. Bacteriol.* 179: 5728–5735.

Haase J, Lurz R, Grahn AM, Bamford DH, Lanka E (1995) Bacterial conjugation mediated by plasmid RP4: RSF1010 mobilization, donor-specific phage propagation, and pilus production require the same Tra2 core components of a proposed DNA transport complex. *J. Bacteriol.* 177: 4779–4791.

Hacker J, Blum-Oehler G, Mühldorfer I, Tschäpe H (1997) Pathogenicity islands of virulent bacteria: structure, function and impact on microbial evolution. *Mol. Microbiol.* 23: 1089–1097.

Hahn J, Luttinger A, Dubnau D (1996) Regulatory inputs for the synthesis of ComK, the competence transcription factor of *Bacillus subtilis*. *Mol. Microbiol.* 21: 763–775.

Hanahan D, Bloom FR (1996) Mechanisms of DNA transformation. In: Neidhardt FC (ed.) *Escherichia coli and Salmonella: Cellular and Molecular Biology*, 2nd edn. Washington DC: American Society for Microbiology, pp. 2449–2459.

Hartskeerl RA, Hoekstra WPM (1984) Exclusion in IncI-type *Escherichia coli* conjugations: the stage of conjugation at which exclusion operates. *Antonie van Leeuwenhoek* 50: 113–124.

Harriman PD (1972) A single-burst analysis of the production of P1 infectious and transducing particles. *Virology* 48: 595–600.

Hedberg PJ, Leonard BAB, Ruhfel RE, Dunny GM (1996) Identification and characterization of the genes of *Enterococcus faecalis* plasmid pCF10 involved in replication and in negative control of pheromone-inducible conjugation. *Plasmid* 35: 46–57.

Helinski DR, Toukdarian AE and Novick RP (1996) Replication control and other stable maintenance mechanisms of plasmids. In: Neidhardt FC (ed.) *Escherichia coli and Salmonella: Cellular and Molecular Biology*, 2nd edn. Washington DC: American Society for Microbiology, pp. 2295–2324

Henderson D, Meyer R (1999) The MobA-linked primase is the only replication protein of R1162 required for conjugal mobilization. *J. Bacteriol.* 181: 2973–2978.

Hendrix RW (1998) Bacteriophage DNA packaging: RNA gears in a DNA transport machine. *Cell* 94: 147–150.

Hobbs M, Mattick JS (1993) Common components in the assembly of type 4 fimbriae, DNA transfer systems, filamentous phage and protein-secretion apparatus: a general system for the formation of surface-associated protein complexes. *Mol. Microbiol.* 10: 233–243.

Hochhut B, Jahreis K., Lengeler JW, Schmid J (1997) CTn*scr94*, a conjugative transposon found in enterobacteria. *J. Bacteriol.* 179: 2097–2102.

Holloway BW (1979) Plasmids that mobilize bacterial chromosome. *Plasmid* 2: 1–19.

Holloway BW, Low KB (1996) F-prime and R-prime factors. In: Neidhardt FC (ed.) *Escherichia coli and Salmonella: Cellular and Molecular Biology*, 2nd edn. Washington DC: American Society for Microbiology, pp. 2413–2420.

Hopwood DA, Kieser T (1993) Conjugative plasmids of *Streptomyces*. In: Clewell DB (ed.) *Bacterial Conjugation*. New York: Plenum, pp. 293–311.

Howard MT, Nelson WC, Matson SW (1995) Stepwise assembly of a relaxosome at the F plasmid origin of transfer. *J. Biol. Chem.* 270: 28381–28386.

Humbert O, Prudhomme M, Hakenbeck G, Dowson CG, Claverys J-P (1995) Homologous recombination and mismatch repair during transformation in *Streptococcus pneumoniae*: Saturation of the Hex mismatch repair system. *Proc Natl Acad. Sci. USA* 92: 9052–9056.

Hwang I, Li P-L, Zhang L *et al.* (1994) TraI, a LuxI homologue, is responsible for production of conjugation factor, the Ti plasmid *N*-acylhomoserine lactone autoinducer. *Proc Natl Acad. Sci. USA* 91: 4639–4643.

Iida S, Streiff MB, Bickle TA, Arber W (1987) Two DNA antirestriction systems of bacteriophage P1, *darA,* and *darB*: characterization of *darA*$^-$ phages. *Virology* 157: 156–166.

Ikeda H, Tomizawa J (1965) Transducing fragments in generalized transduction by phage P1. I. Molecular origin of the fragments. *J. Mol. Biol.* 14: 85–109.

Jacobson A (1972) Role of F pili in the penetration of bacteriophage fl. *J. Bacteriol.* 10: 835–843.

Jagura-Burdzy G, Thomas CM (1997) Dissection of the switch between genes for replication and transfer of promiscuous plasmid RK2: basis of the dominance of *trfAp* over *trbAp* and specificity for KorA in controlling the switch. *J. Mol. Biol.* 265: 507–518.

Jaworski DD, Clewell DB (1995) A functional origin of transfer (*oriT*) on the conjugative transposon Tn916. *J. Bacteriol.* 177: 6644–6651.

Jensen GB, Wilcks A, Petersen SS, Damgaard J, Baum JA, Andrup L (1995) The genetic basis of the aggregation system in *Bacillus thuringiensis* subsp. israelensis is located on the large conjugative plasmid pXO16. *J. Bacteriol.* 177: 2914–2917.

Jerome LJ, Frost LS (1999) *In vitro* analysis of the interaction between the FinO protein and FinP antisense RNA of F-like conjugative plasmids. *J. Biol. Chem.* 274: 10356–10362.

Jerome LJ, van Biesen T, Frost LS (1999) Degradation of FinP antisense RNA from F-like plasmids: The RNA-binding protein, FinO, protects FinP from ribonuclease E. *J. Mol. Biol.* 285: 1457–1473.

Jones AL, Barth PT, Wilkins BM (1992) Zygotic induction of plasmid ssb and *psiB* genes following conjugative transfer of IncI1 plasmid ColIb-P9. *Mol. Microbiol.* 6: 605–613.

Kado CI (1994) Promiscuous DNA transfer system of *Agrobacterium tumefaciens*: role of the *virB* operon in sex pilus assembly and synthesis. *Mol. Microbiol.* 12: 17–22.

Kahn ME, Smith HO (1984) Transformation in *Haemophilus*: a problem in membrane biology. *J. Membr. Biol.* 81: 89–103.

Karudapuram S, Barcak GJ (1997) The *Haemophilus influenzae dprABC* genes constitute a competence-inducible operon that requires the product of the *tfoX* (*sxy*) gene for transcriptional activation. *J. Bacteriol.* 179: 4815–4820.

Kataoka M, Seki T, Yoshida T (1991) Regulation and function of the *Streptomyces* plasmid pSN22 genes involved in pock formation and inviability. *J. Bacteriol.* 173: 7975–7981.

Kendall KJ, Cohen SN (1988) Complete nucleotide sequence of the *Streptomyces lividans* plasmid pIJ101 and correlation of the sequence with genetic properties. *J. Bacteriol.* 170: 4634–4651.

King G, Murray NE (1994) Restriction enzymes in cells, not eppendorfs. *Trends Microbiol.* 2: 465–469.

Kleerebezem M, Quadri LE, Kuipers OP, de Vos WM (1997) Quorum sensing by peptide pheromones and two-component signal-transduction systems in Gram-positive bacteria. *Mol. Microbiol.* 24: 895–904.

Klimke WA, Frost LS (1998) Genetic analysis of the role of the transfer gene *traN* of the F and R100-1 plasmids in mating pair stabilization during conjugation. *J. Bacteriol.* 180: 4036–4043.

Klimke WA, Fekete R, Manchak J, Anthony K, Frost LS (1997) Plasmid specificity and interaction: The similarities and differences between the two transfer regions of two compatible plasmids F and R100–1. In: Syvanen M, Kado CI (eds) *Horizontal Gene Transfer*. London: Chapman & Hall, pp. 25–39.

Knutton S, Rosenshine I, Pallen MJ *et al.* (1998) A novel EspA-associated surface organelle of enteropathogenic *Escherichia coli* involved in protein translocation into epithelial cells. *EMBO J.* 17: 2166–2176.

Kokjohn TA (1989) Transduction: mechanism and potential for gene transfer in the environment. In: Levy SB, Miller RV (eds) *Gene Transfer in the Environment*. New York: McGraw-Hill, pp. 73–97.

Koraimann G, Teferle K, Markolin G, Woger W, Hogenauer G (1996) The FinOP repressor system of plasmid R1: analysis of the antisense RNA control of traJ expression and conjugative DNA transfer. *Mol. Microbiol.* 21: 811–821.

Korona R, Korona B, Levin BR (1993) Sensitivity of naturally occurring coliphages to type I and type II restriction and modification. *J. Gen. Microbiol.* 139: 1283–1290.

Kosono S, Kataoka M, Seki T, Yoshida T (1996) The TraB protein, which mediates the intermycelial transfer of the *Streptomyces* plasmid pSN22, has functional NTP-binding motifs and is localized to the cytoplasmic membrane. *Mol. Microbiol.* 19: 397–405.

Kostelidou K, Jones AC, Thomas CM (1999) Conserved C-terminal region of global repressor KorA of Broad-host-range plasmid RK2 is required for co-operativity between KorA and a second RK2 global regulator, KorB. *J. Mol. Biol.* 289: 211–221.

Kotilainen MM, Grahn AM, Bamford JK, Bamford DH (1993) Binding of an *Escherichia coli* double-stranded DNA virus PRD1 to a receptor coded by an IncP-type plasmid. *J. Bacteriol.* 175: 3089–3095.

Kubori T, Matsushima Y, Nakamura D *et al.* (1998) Supramolecular structure of the *Salmonella typhimurium* type III protein secretion system. *Science* 280: 602–605.

Kupelweiser G, Schwab M, Högenauer G, Koraimann G, Zechner EL (1998) Transfer protein TraM stimulates TraI-catalyzed cleavage of the transfer origin of plasmid R1 *in vivo*. *J. Mol. Biol.* 275: 81–94.

Lacks SA (1988) Mechanisms of genetic recombination in gram-positive bacteria. In: Kucherlapati R, Smith GR (eds) *Genetic Recombination*. Washington, DC: American Society for Microbiology, pp. 43–86.

Lacks SA (1999) DNA uptake by transformable bacteria. In: Broome-Smith JK, Baumberg S, Stirling CJ, Ward FB (eds) *Transport of Molecules across Microbial Membranes*. Cambridge: Cambridge University Press, pp. 138–168.

Lanka E, Wilkins BM (1995) DNA processing reactions in bacterial conjugation. *Ann. Rev. Biochem.* 64: 141–169.

Leonard BA, Podbielski A, Hedberg PJ, Dunny GM (1996) *Enterococcus faecalis* pheromone binding protein, PrgZ, recruits a chromosomal oligopeptide permease system to import sex pheromone cCF10 for induction of conjugation. *Proc. Natl Acad. Sci. USA* 93: 260–264.

Lessl M, Lanka E (1994) Common mechanisms in bacterial conjugation and Ti-mediated T-DNA transfer to plant cells. *Cell* 77: 321–324.

Lessl M, Balzer D, Lurz R, Waters VL, Guiney DG, Lanka E (1992) Dissection of IncP conjugative plasmid transfer: definition of the transfer region Tra2 by mobilization of the Tra1 region in trans. *J. Bacteriol.* 174: 2493–2500.

Lessl M, Balzer D, Weyrauch K, Lanka E (1993) The mating pair formation system of plasmid RP4 defined by RSF1010 mobilization and donor-specific phage propagation. *J. Bacteriol.* 175: 6415–6425.

Levy SB (1986) Ecology of antibiotic resistance determinants. In: Levy SB, Novick RP (eds) *Banbury Report 24: Antibiotic Resistance Genes: Ecol., Transfer and Expression*. New York: Cold Spring Harbor Laboratory, pp. 17–30.

Levy SB (1992) *The Antibiotic Paradox: How Miracle Drugs Are Destroying the Miracle*. New York: Plenum Press.

Levy SB, Miller RV (eds) (1989) *Gene Transfer in the Environment*. New York: McGraw-Hill.

Linderoth NA, Simon MN, Russel M (1997) The filamentous phage pIV multimer visualized by scanning transmission electron microscopy. *Science* 278: 1635–1638.

Loh SM, Cram DS, Skurray RA (1988) Nucleotide sequence and transcriptional analysis of a third function (Flm) involved in F-plasmid maintenance. *Gene* 66: 259–268.

Lorenz MG, Wackernagel W (1994) Bacterial gene transfer by natural genetic transformation in the environment. *Microbiol. Rev.* 58: 563–602.

Lovett MA, Helinski DR (1975) Relaxation complexes of plasmid DNA and protein. II. Characterization of the proteins associated with the unrelaxed and relaxed complexes of plasmid ColE1. *J. Biol. Chem.* 250: 8790–8795.

Low KB (1996) Hfr strains of *Escherichia coli* K-12. In: Neidhardt FC (ed.) *Escherichia coli and Salmonella: Cellular and Molecular Biology*, 2nd edn. Washington DC: American Society for Microbiology, pp. 2402–2405.

Lubkowski J, Hennecke F, Pluckthun A, Wlodawer A (1999) Filamentous phage infection: crystal structure of g3p in complex with its coreceptor, the C-terminal domain of TolA. *Struct. Fold Design* 7: 711–722.

Macrina FL (1993) Conjugal transfer in anaerobic bacteria. In: Clewell DB (ed.) *Bacterial Conjugation*. New York: Plenum Press, pp. 331–348.

Macrina FL, Archer GL (1993) Conjugation and broad host range plasmids in Streptococci and Staphylococci. In: Clewell DB (ed.) *Bacterial Conjugation*. New York: Plenum Press, pp. 313–329.

Magnuson R, Solomon J, Grossman AD (1994) Biochemical and genetic characterization of a competence pheromone from *B. subtilis*. *Cell* 77: 207–216.

Maher D, Sherburne R, Taylor DE (1993) H-pilus assembly kinetics determined by electron microscopy. *J. Bacteriol.* 175: 2175–2183.

Majdalani N, Ippen-Ihler K (1996) Membrane insertion of the F-pilin subunit is Sec independent but requires leader peptidase B and the proton motive force. *J. Bacteriol.* 178: 3742–3747.

Majdalani N, Moore D, Maneewannakul S, Ippen-Ihler K (1996) Role of the propilin leader peptide in the maturation of F pilin. *J. Bacteriol.* 178: 3748–3754.

Majewski J, Cohan FM (1998) The effect of mismatch repair and heteroduplex formation on sexual isolation in Bacillus. *Genetics* 148: 13–18.

Makovets S, Doronina VA, Murray NE (1999) Regulation of endonuclease activity by proteolysis prevents breakage of unmodified bacterial chromosomes by type I restriction enzymes. *Proc Natl Acad. Sci. USA* 96: 9757–9762.

Maneewannakul K, Maneewannakul S, Ippen-Ihler K (1992a) Sequence alterations affecting F plasmid transfer gene expression: a conjugation system dependent on transcription by the RNA polymerase of phage T7. *Mol. Microbiol.* 6: 2961–2973.

Maneewannakul S, Maneewannakul K, Ippen-Ihler K (1992b) Characterization, localization, and sequence of F transfer region products: the pilus assembly gene product TraW and a new product, TrbI. *J. Bacteriol.* 174: 567–574.

Maneewannakul K, Kathir P, Endley S *et al.* (1996) Construction of derivatives of the F plasmid pOX-*tra715*: characterization of *traY* and *traD* mutants that can be complemented *in trans*. *Mol. Microbiol.* 22: 197–205.

Manganelli R, Ricci S, Pozzi G (1997) The joint of Tn*916* circular intermediates is a homoduplex in Enterococcus faecalis. *Plasmid* 38: 71–78.

Manning PA, Morelli G, Achtman M (1981) TraG protein of the F sex factor of *Escherichia coli* K-12 and its role in conjugation. *Proc Natl Acad. Sci. USA* 78: 7487–7491.

Manwaring NP, Skurray RA, Firth N (1999) Nucleotide sequence of the F plasmid leading region. *Plasmid* 41: 219–225.

Marra D, Scott JR (1999) Regulation of excision of the conjugative transposon Tn*916*. *Mol. Microbiol.* 31: 609–621.

Masai H, Arai K (1997) F*rpo*: a novel single-stranded DNA promoter for transcription and for primer RNA synthesis of DNA replication. *Cell* 89: 897–907.

Masters M (1996) Generalized transduction. In: Neidhardt FC (ed.) *Escherichia coli and Salmonella: Cellular and Molecular Biology*, 2nd edn. Washington DC: American Society for Microbiology, pp. 2421–2441.

Masters M, Newman BJ, Henry CM (1984) Reduction of marker discrimination in transductional recombination. *Mol. Gen Genet.* 196: 85–90.

Matic I, Rayssiguier C, Radman M (1995) Interspecies gene exchange in bacteria: the role of SOS and mismatch repair systems in evolution of species. *Cell* 80: 507–515.

Maynard Smith J (1995) Do bacteria have population genetics? In: Baumberg JPW, Young EMH, Wellington S, Saunders JR (eds) *Population Genetics of Bacteria*. Cambridge: Cambridge University Press, pp. 1–12.

McKane M, Milkman R (1995) Transduction, restriction and recombination patterns in *Escherichia coli*. *Genetics* 139: 35–43.

Mejean V, Claverys J-P (1993) DNA processing during entry in transformation of *Streptococcus pneumoniae*. *J. Biol. Chem.* 268: 5594–5599.

Mekalanos JJ, Rubin EJ, Waldor MK (1997) Cholera: molecular basis for emergence and pathogenesis. *FEMS Immunol. Med. Microbiol.* 18: 241–248.

Michel F, Ferat J-L (1995) Structure and activity of group II introns. *Ann. Rev. Biochem.* 64: 435–461.

Michod RE, Wojciechowski MF (1994) DNA repair and the evolution of transformation IV. DNA damage increases transformation. *J. Evol. Biol.* 7: 147–175.

Mileykovskaya E, Dowhan W (1997) The Cpx two-component signal transduction pathway is activated in *Escherichia coli* mutant strains lacking phosphatidylethanolamine. *J. Bacteriol.* 179: 1029–1034.

Milkman R (1996) Recombinational exchange among clonal populations. In: Neidhardt FC (ed.) Escherichia coli *and* Salmonella: *Cellular and Molecular Biology*, 2nd edn. Washington DC: American Society for Microbiology, pp. 2663–2684.

Miller RV, Sayler GS (1992) Bacteriophage-host interactions in aquatic systems. In: Wellington EMH, van Elsas JD (eds) *Genetic Interactions among Microorganisms in the Natural Environment*. Oxford: Pergamon Press, pp. 176–193.

Mills DA, McKay LL, Dunny GM (1996) Splicing of a group II intron involved in the conjugative transfer of pRS01 in lactococci. *J. Bacteriol.* 178: 3531–3538.

Moore D, Hamilton CM, Maneewannakul K, Mintz Y, Frost LS, Ippen-Ihler K (1993) The *Escherichia coli* K-12 F plasmid gene *traX* is required for acetylation of F-pilin. *J. Bacteriol.* 175: 1375–1383.

Moore D, Maneewannakul K, Maneewannakul S, Wu JH, Ippen-Ihler K, Bradley DE (1990) Characterization of the F-plasmid conjugative transfer gene *traU*. *J. Bacteriol.* 172: 4263–4270.

More MI, Finger LD, Stryker JL, Fuqua C, Eberhard A, Winans SC (1996) Enzymatic synthesis of a quorum-sensing autoinducer through use of defined substrates. *Science* 272: 1655–1658.

Morton TM, Eaton DM, Johnston JL, Archer GL (1993) DNA sequence and units of transcription of the conjugative transfer gene complex (*trs*) of *Staphylococcus aureus* plasmid pGO1. *J. Bacteriol.* 175: 4436–4447.

Motallebi-Veshareh M, Balzer D, Lanka E, Jagura-Burdzy G, Thomas CM (1992) Conjugative transfer functions of broad-host-range plasmid RK2 are coregulated with vegetative replication. *Mol. Microbiol.* 6: 907–920.

Mullany P, Pallen M, Wilks M, Stephen JR, Tabaqchali S (1996) A group II intron in a conjugative transposon from the Gram-positive bacterium, *Clostridium difficile*. *Gene* 174: 145–150.

Murphy DB, Pembroke JT (1999) Monitoring of chromosomal insertions of the IncJ elements R391 and R997 in *Escherichia coli* K-12. *FEMS Microbiol. Lett.* 174: 355–361.

Naito T, Kusano K, Kobayashi I (1995) Selfish behaviour of restriction-modification systems. *Science* 267: 897–899.

Nash HA (1996) Site-specific recombination: integration, excision, resolution, and inversion of defined DNA segments. In: Neidhardt FC (ed.) Escherichia coli *and* Salmonella: *Cellular and Molecular Biology*, 2nd edn. Washington DC: American Society for Microbiology, pp. 2363–2376.

Nelson WC, Matson SW (1996) The F plasmid *traY* gene product binds DNA as a monomer or a dimer: structural and functional implications. *Mol. Microbiol.* 20: 1179–1187.

Nikaido H (1996) Outer membrane. In: Neidhardt FC (ed.) Escherichia coli *and* Salmonella: *Cellular and Molecular Biology*, 2nd edn. Washington DC: American Society for Microbiology, pp. 29–47.

Nomura N, Masai H, Inuzuka M *et al.* (1991) Identification of eleven single-strand initiation sequences (*ssi*) for priming of DNA replication in the F, R6K, R100 and ColE2 plasmids. *Gene* 108: 15–22.

Novotny CP, Fives-Taylor P (1978) Effects of high temperature on *Escherichia coli* F pili. *J. Bacteriol.* 133: 459–464.

O'Brien AD, Holmes RK (1987) Shiga and Shiga-like toxins. *Microbiol. Rev.* 51: 206–220.

O'Brien AD, Holmes RK (1996) Protein toxins of *Escherichia coli* and *Salmonella*. In: Neidhardt FC (ed.) Escherichia coli *and* Salmonella: *Cellular and Molecular Biology*, 2nd edn. Washington DC: American Society for Microbiology, pp. 2788–2802.

O'Neill M, Chen A, Murray NE (1997) The restriction-modification genes of *Escherichia coli* K-12 may not be selfish: They do not resist loss and are readily replaced by alleles conferring different specificities. *Proc Natl Acad. Sci. USA* 94: 14596–14601.

Ochman H, Lawrence JG (1996) Phylogenetics and the amelioration of bacterial genomes. In: Neidhardt FC (ed.) Escherichia coli *and* Salmonella: *Cellular and Molecular Biology*, 2nd edn. Washington DC: American Society for Microbiology, pp. 2627–2637.

Olmsted SB, Erlandsen SL, Dunny GM, Wells CL (1993) High-resolution visualization by field emission scanning electron microscopy of *Enterococcus faecalis* surface proteins encoded by the pheromone-inducible conjugative plasmid pCF10. *J. Bacteriol.* 175: 6229–6237.

Ou JT (1980) Role of surface exclusion genes in lethal zygosis in *Escherichia coli* K12 mating. *Mol. Gen Genet.* 178: 573–581.

Paiva WD, Grossman T, Silverman PM (1992) Characterization of F-pilin as an inner membrane component of *Escherichia coli* K12. *J. Biol. Chem.* 267: 26191–26197.

Palmen R, Hellingwerf KJ (1997) Uptake and processing of DNA by *Acinetobacter calcoaceticus* – a review. *Gene* 192: 179–190.

Pansegrau W, Ziegelin G, Lanka E (1990) Covalent association of the gene product of plasmid RP4 with the 5′-terminal nucleotide at the relaxation nick site. *J. Biol. Chem.* 265: 10637–10644.

Pansegrau W, Lanka E, Barth PT *et al.* (1994a) Complete nucleotide sequence of Birmingham IncPα plasmids. Compilation and comparative analysis of sequence data. *J. Mol. Biol.* 239: 623–663.

Pansegrau W, Schroder W, Lanka E (1994b) Concerted action of three distinct domains in the DNA cleaving-joining reaction catalyzed by relaxase (TraI) of conjugative plasmid RP4. *J. Biol. Chem.* 269: 2782–2789.

Paranchych W, Frost LS (1988) The physiology and biochemistry of pili. *Adv. Microb. Physiol.* 29: 53–114.

Penfold SS, Simon J, Frost LS (1996) Regulation of the expression of the *traM* gene of the F sex factor of *Escherichia coli*. *Mol. Microbiol.* 20: 549–558.

Perumal N, Minkley EG Jr (1984) The product of the F sex factor *traT* surface exclusion gene is a lipoprotein. *J. Biol. Chem.* 259: 5357–5360.

Perwez T, Meyer R (1999) Stabilization of the relaxosome and stimulation of conjugal transfer are genetically distinct functions of the R1162 protein MobB. *J. Bacteriol.* 181: 2124–2131.

Pestova EV, Håvarstein LS, Morrison DA (1996) Regulation of competence for genetic transformation in *Streptococcus pneumoniae* by an auto-induced peptide pheromone and a two component regulatory system. *Mol. Microbiol.* 21: 853–862.

Pettis GS, Cohen SN (1996) Plasmid transfer and expression of the transfer (*tra*) gene product of plasmid pIJ101 are temporally regulated during the *Streptomyces lividans* life cycle. *Mol. Microbiol.* 19: 1127–1135.

Pohlman RF, Genetti HD, Winans SC (1994) Common ancestry between IncN conjugal transfer genes and macromolecular export systems of plant and animal pathogens. *Mol. Microbiol.* 14: 655–668.

Pogliano J, Lynch AS, Belin D, Lin ECC, Beckwith J (1997) Regulation of *Escherichia coli* cell envelope proteins involved in protein folding and degradation by the Cpx two-component system. *Genes Dev.* 11: 1169–1182.

Pölzleitner E, Zechner E, Renner W *et al.* (1997) TraM of plasmid R1 controls transfer gene expression as an integrated control element in a complex regulatory network. *Mol. Microbiol.* 25: 495–507.

Prakash-Cheng A, Chung SS, Ryu J (1993) The expression and regulation of hsd_K genes after conjugative transfer. *Mol. Gen Genet.* 241: 491–496.

Priebe SD, Lacks SA (1989) Region of streptococcal plasmid pMV158 required for conjugative mobilization. *J. Bacteriol.* 171: 4778–4784.

Ptashne M, Gann A (1997) Transcriptional activation by recruitment. *Nature* 386: 569–577.

Pugsley AP (1993) The complete general secretory pathway in gram-negative bacteria. *Microbiol. Rev.* 57: 50–108.

Rayssiguier C, Thaler DS, Radman M (1989) The barrier to recombination between *Escherichia coli* and *Salmonella typhimurium* is disrupted in mismatch-repair mutants. *Nature* 342: 396–401.

Redaschi N, Bickle TA (1996) DNA restriction and modification systems. In: Neidhardt FC (ed.) Escherichia coli *and* Salmonella: *Cellular and Molecular Biology*, 2nd edn. Washington DC: American Society for Microbiology, pp. 773–781.

Rees CE, Wilkins BM (1990) Protein transfer into the recipient cell during bacterial conjugation: studies with F and RP4. *Mol. Microbiol.* 4: 1199–1205.

Reimmann C, Haas D (1993) Mobilization of chromosomes and nonconjugative plasmids by cointegrative mechanisms. In: Clewell DB (ed.) *Bacterial Conjugation*. New York: Plenum Press, pp. 137–188.

Riechmann L, Holliger P (1997) The C-terminal domain of TolA is the coreceptor for filamentous phage infection of *E. coli*. *Cell* 90: 351–360.

Riede I, Eschbach M-L (1986) Evidence that TraT interacts with OmpA of *Escherichia coli*. *FEBS Lett.* 205: 241–245.

Riley M, Krawiec S (1987) Genome organization. In Neidhardt FC (ed.) Escherichia coli *and* Salmonella: *Cellular and Molecular Biology*, 2nd edn. Washington DC: American Society for Microbiology, pp. 967–981.

Roberts RJ, Macelis D (1994) REBASE-restriction enzymes and methylases. *Nucleic Acids Res.* 22: 3628–3639.

Rondot S, Anthony KG, Dubel S *et al.* (1998) Epitopes fused to F-pilin are incorporated into functional recombinant pili. *J. Mol. Biol.* 279: 509–603.

Rosenthal AL, Lacks SA (1980) Complex structure of the membrane nuclease of *Streptococcus pneumoniae* revealed by two-dimensional electrophoresis. *J. Mol. Biol.* 141: 133–146.

Rood JI, Cole ST (1991) Molecular genetics and pathogenesis of *Clostridium perfringens*. *Microbiol. Rev.* 55: 621–648.

Ruhfel RE, Leonard BAB, Dunny GM (1997). Pheromone-inducible conjugation in *Enterococcus faecalis*. In: Shapiro J, Dworkin M (eds) *Bacteria as Multicellular Organisms*. New York: Oxford University Press, pp. 53–68.

Russel M (1995) Moving through the membrane with filamentous phages. *Trends Microbiol.* 3: 223–228.

Sakai H, Komano T (1996) DNA replication of IncQ broad-host-range plasmids in gram-negative bacteria. *Biosci., Biotechnol. Biochem.* 60: 377–382.

Salmond GPC (1994) Secretion of extracellular virulence factors by plant pathogenic bacteria. *Ann. Rev. Phytopathol.* 32: 181–200.

Salyers AA, Shoemaker NB, Li LY (1995a) In the driver's seat: the Bacteroides conjugative transposons and the elements they mobilize. *J. Bacteriol.* 177: 5727–5731.

Salyers AA, Shoemaker NB, Stevens AM, Li LY (1995b) Conjugative transposons: an unusual and diverse set of integrated gene transfer elements. *Microbiol. Rev.* 59: 579–590.

Sandri RM, Berger H (1980) Bacteriophage P1-mediated generalized transduction in *Escherichia coli*: structure of abortively transduced DNA. *Virology* 106: 30–40.

Santini JM, Stanisich VA (1998) Both the *fipA* gene of pKM101 and the *pifC* gene of F inhibit conjugal transfer of RP1 by an effect on traG. *J. Bacteriol.* 180: 4093–4101.

Sastre JI, Cabezon E, de la Cruz F (1998) The carboxyl terminus of protein TraD adds specificity and efficiency to F-plasmid conjugative transfer. *J. Bacteriol.* 180: 6039–6042.

Saunders CW, Guild WR (1981) Monomer plasmid DNA transforms *Steptococcus pneumoniae*. *Mol. Gen Genet.* 181: 57–62.

Schandel KA, Maneewannakul S, Ippen-Ihler K, Webster RE (1987) A *traC* mutant that retains sensitivity to f1 bacteriophage but lacks F pili. *J. Bacteriol.* 169: 3151–3159.

Schlievert PM, Gahr PJ, Assimacopoulos AP *et al.* (1998) Aggregation and binding substances enhance pathogenicity in rabbit models of *Enterococcus faecalis* endocarditis. *Infect. Immun.* 66: 218–223.

Scholz P, Haring V, Wittmann-Liebold K, Bagdasarian M, Scherzinger E (1989) Complete nucleotide sequence and gene organization of the broad-host-range plasmid RSF1010. *Gene* 75: 271–288.

Scott JR, Churchward GG (1995) Conjugative transposition. *Ann. Rev. Microbiol.* 49: 367–397.

Segal G, Purcell M, Shuman HA (1998) Host cell killing and bacterial conjugation require overlapping sets of genes within a 22-kb region of the *Legionella pneumophila* genome. *Proc. Natl Acad. Sci. USA* 95: 1669–1674.

Shapiro JA (1977) Bacterial plasmids. In: Bukhari AI, Shapiro JA, Adhya SL (eds) *DNA Insertion Elements, Plasmids, and Episomes*. Cold Spring Harbor: Cold Spring Harbor Press, pp. 601–670.

Sharp PM (1986) Molecular evolution of bacteriophages: evidence of selection against recognition sites of host restriction enzymes. *Mol. Biol. Evol.* 3: 75–83.

Shearman C, Godon J-J, Gasson M (1996) Splicing of a group II intron in a functional transfer gene of *Lactococcus lactis*. *Mol. Microbiol.* 21: 45–53.

Shen P, Huang HV (1989) Effect of base pair mismatches on recombination via the RecBCD pathway. *Mol. Gen Genet.* 218: 358–360.

Silverman PM (1997) Towards a structural biology of bacterial conjugation. *Mol. Microbiol.* 23: 423–429.

Silverman PM, Sholl A (1996) Effect of *traY* amber mutations on F-plasmid *traY* promoter activity *in vivo*. *J. Bacteriol.* 178: 5787–5789.

Silverman PM, Rother S, Gaudin H (1991a) Arc and Sfr functions of the *Escherichia coli* K-12 *arcA* are genetically and physiologically separable. *J. Bacteriol.* 173: 5648–5652.

Silverman PM, Wickersham E, Rainwater S, Harris R (1991b) Regulation of the F plasmid traY promoter in *Escherichia coli* K-12 as a function of sequence context. *J. Mol. Biol.* 220: 271–279.

Silverman PM, Tran L, Harris R, Gaudin HM (1993) Accumulation of the F plasmid TraJ protein in *cpx* mutants of *Escherichia coli*. *J. Bacteriol.* 175: 921–925.

Simonsen L (1990) Dynamics of plasmid transfer on surfaces. *J. Gen. Microbiol.* 136: 1001–1007.

Skurray RA, Reeves P (1974) F factor-mediated immunity to lethal zygosis in *Escherichia coli* K-12. *J. Bacteriol.* 117: 100–106.

Smith HO, Tomb J-F, Dougherty BA, Fleischmann RD, Venter JC (1995) Frequency and distribution of DNA uptake signal sequences in the *Haemophilus influenzae* Rd genome. *Science* 269: 538–540.

Solomon JM, Lazazzera BA, Grossman AD (1996) Purification and characterization of an extracellular peptide factor that affects two different developmental pathways in *Bacillus subtilis*. *Genes Dev.* 10: 2014–2024.

Sowa BA, Moore D, Ippen-Ihler K (1983) Physiology of F-pilin synthesis and utilization. *J. Bacteriol.* 153: 962–968.

Stallions DR, Curtiss R III (1972) Bacterial conjugation under anaerobic conditions. *J. Bacteriol.* 111: 294–295.

Stevens AM, Shoemaker NB, Li LY, Salyers AA (1993) Tetracycline regulation of genes on Bacteroides conjugative transposons. *J. Bacteriol.* 175: 6134–6141.

Stewart GJ (1992) Transformation in natural environments. In: Wellington EMH, van Elsas JD (eds) *Genetic Interactions among Microorganisms in the Natural Environment*. Oxford: Pergamon Press, pp. 216–234.

Stewart GJ, Carlson CA, Ingraham JL (1983) Evidence for an active role of donor cells in natural transformation of *Pseudomonas stutzeri*. *J. Bacteriol.* 156: 30–35.

Strohmaier H, Noiges R, Kotschan S *et al.* (1998) Signal transduction and bacterial conjugation: characterization of the role of ArcA in regulating conjugative transfer of the resistance plasmid R1. *J. Mol. Biol.* 277: 309–316.

Sukupolvi S, O'Connor CD (1990) TraT lipoprotein, a plasmid-specified mediator of interactions between gram-negative bacteria and their environment. *Microbiol. Rev.* 54: 331–341.

Sundberg C, Meek L, Carroll K, Das A, Ream W (1996) VirE1 protein mediates export of the single-stranded DNA-binding protein VirE2 from *Agrobacterium tumefaciens* into plant cells. *J. Bacteriol.* 178: 1207–1212.

Taki K, Abo T, Ohtsubo E (1998) Regulatory mechanisms in the expression of the *traY-I* operon of sex factor plasmid R100: involvement of *traJ* and *traY* gene products. *Genes Cells* 3: 331–345.

Taylor KL, Churchward G (1997) Specific DNA cleavage mediated by the integrase of conjugative transposon Tn*916*. *J. Bacteriol.* 179: 1117–1125.

Thomas CM (1999) *The Horizontal Gene Pool: Bacterial Plasmids and Gene Spread*. Amsterdam: Harwood.

Torreblanca J, Marques S, Casadesus J (1999) Synthesis of FinP RNA by plasmids F and pSLT is regulated by DNA adenine methylation. *Genetics* 151: 31–45.

Trautner TA, Pawlek B, Bron S, Anagnostopoulos C (1974) Restriction and modification in *B. subtilis*. *Mol. Gen Genet.* 131: 181–191.

Turgay K, Hamoen LW, Venema G, Dubnau D (1997) Biochemical characterization of a molecular switch involving the heat shock protein ClpC, which controls the activity of ComK, the competence transcription factor of *Bacillus subtilis*. *Genes Dev.* 11: 119–128.

van Biesen T, Frost LS (1994) The FinO protein of IncF plasmids binds FinP antisense RNA and its target, *traJ* mRNA, and promotes duplex formation. *Mol. Microbiol.* 14: 427–436.

van Elsas JD (1992) Antibiotic resistance gene transfer in the environment: an overview. In: Wellington EMH, van Elsas JD (eds) *Genetic Interactions among Microorganisms in the Natural Environment*. Oxford: Pergamon Press, pp. 17–39.

van Elsas JD, Fry J, Hirsch P, Molin S (2000) Ecology of plasmid transfer and spread. In: Thomas CM (ed.) *The Horizontal Gene Pool: Bacterial Plasmids and Gene Spread*, Chapter 4. Amsterdam: Harwood, pp. 175–206.

van Sinderen D, Luttinger A, Kong L, Dubnau D, Venema G, Hamoen L (1995) *comK* encodes the competence transcription factor, the key regulatory protein for competence development in *Bacillus subtilis*. *Mol. Microbiol.* 15: 455–462.

Vogel JP, Andrews HL, Wong SK, Isberg RR (1998) Conjugative transfer by the virulence system of *Legionella pneumophila*. *Science* 279: 873–876.

Vögtli M, Cohen SN (1992). The chromosomal integration site for the *Streptomyces* plasmid SLP1 is a functional $tRNA^{Tyr}$ gene essential for cell viability. *Mol. Microbiol.* 6: 3041–3050.

Waldor MK, Mekalanos JJ (1996) Lysogenic conversion by a filamentous phage encoding cholera toxin. *Science* 272: 1910–1914.

Wang A, Macrina FL (1995a) Characterization of six linked open reading frames necessary for pIP501-mediated conjugation. *Plasmid* 34: 206–210.

Wang A, Macrina FL (1995b) Streptococcal plasmid pIP501 has a functional *oriT* site. *J. Bacteriol.* 177: 4199–4206.

Warren GJ, Saul MW, Sherratt DJ (1979) ColE1 plasmid mobility: essential and conditional functions. *Mol. Gen Genet.* 170: 103–107.

Weaver KE, Clewell DB (1991) Control of *Enterococcus faecalis* sex pheromone cAD1 elaboration: effects of culture aeration and pAD1 plasmid-encoded determinants. *Plasmid* 25: 177–189.

Webster RE (1996) Biology of the filamentous bacteriophage. In: Kay BK *et al.* (eds) *Phage Display of Peptides and Proteins.* San Diego, CA: Academic Press. pp. 1–20.

Weisberg RA (1996) Specialized transduction. In: Neidhardt FC (ed.) *Escherichia coli and Salmonella: Cellular and Molecular Biology*, 2nd edn. Washington DC: American Society for Microbiology, pp. 2442–2448.

Wellington EMH, van Elsas J (1992) *Genetic Interactions among Microorganisms in the Natural Environment.* Oxford: Pergamon Press.

Westmoreland J, Porter G, Radman M, Resnick MA (1997) Highly mismatched molecules resembling recombination intermediates efficiently transform mismatch repair proficient *Escherichia coli. Genetics* 145: 29–38.

Whittam TS (1996) Genetic variation and evolutionary processes in natural populations of *Escherichia coli.* In: Neidhardt FC (ed.) Escherichia coli *and* Salmonella: *Cellular and Molecular Biology*, 2nd edn. Washington DC: American Society for Microbiology, pp. 2708–2720.

Wilkins BM (1995) Gene transfer by bacterial conjugation: diversity of systems and functional specializations. In: Baumberg S, Young JPW, Wellington EMH, Saunders JR (eds) *Population Genetics of Bacteria.* Cambridge: Cambridge University Press, pp. 59–88.

Wilkins BM, Lanka E (1993) DNA processing and replication during plasmid transfer between gram-negative bacteria. In: Clewell DB (ed.) *Bacterial Conjugation.* New York: Plenum Press, pp. 105–136.

Wilkins BM, Meacock PA (2000) Transformation, genetic. In: Lederberg J (ed.) *Encyclopedia of Microbiology*, 2nd edn. San Diego CA: Academic Press, pp. 651–665.

Wilkins BM, Chilley PM, Thomas AT, Pocklington MJ (1996) Distribution of restriction enzyme recognition sequences on broad host range plasmid RP4: molecular and evolutionary implications. *J. Mol. Biol.* 258: 447–456.

Willetts N (1980) Interactions between the F conjugal transfer system and CloDF13::Tna plasmids. *Mol. Gen Genet.* 180: 213–217.

Willetts N (1993) Bacterial conjugation, a historical perspective. In: Clewell DB (ed.) *Bacterial Conjugation.* New York: Plenum Press, pp. 1–22.

Winans SC, Mantis NJ, Chen CY, Chang CH, Han DC (1994) Host recognition by the VirA, VirG two-component regulatory proteins of *Agrobacterium tumefaciens. Res. Microbiol.* 145: 461–473.

Wirth R, Muscholl A, Wanner G (1996) The role of pheromones in bacterial interactions. *Trends Microbiol.* 4: 96–103.

Worth L Jr., Clark S, Radman M, Modrich P (1994) Mismatch repair proteins MutS and MutL inhibit RecA-catalyzed strand transfer between diverged DNAs. *Proc Natl Acad. Sci. USA* 91: 3238–3241.

Wu K, An FY, Clewell DB (1999) *Enterococcus faecalis* pheromone-responding plasmid pAD1 gives rise to an aggregation (clumping) response when cells are exposed to subinhibitory concentrations of chloramphenicol, erythromycin, or tetracycline. *Plasmid* 41: 82–88.

Wu LJ, Lewis PJ, Allmansberger R, Hauser PM, Errington J (1995) A conjugation-like mechanism for prespore chromosome partitioning during sporulation in *Bacillus subtilis. Genes Dev.* 9: 1316–1326.

Yamada Y, Yamada M, Nakazawa A (1995) A ColE1-encoded gene directs entry exclusion of the plasmid. *J. Bacteriol.* 177: 6064–6068.

Yarmolinksy MB, and Sternberg N (1988) Bacteriophage P1. In: Calendar R (ed.) *The Bacteriophages*, Vol. 1. New York: Plenum Press, pp. 291–438.

Yin X, Stotzky G (1997) Gene transfer among bacteria in natural environments. *Adv. Applied Microbiol.* 45: 153–212.

Yoshida T, Feruya N, Ishikura M *et al.* (1998) Purification and characterization of thin pili of IncI1 plasmids ColIb-P9 and R64: formation of PilV-specific cell aggregates by type IV pili. *J. Bacteriol.* 180: 2842–2848.

Zatyka M, Jagura-Burdzy G, Thomas CM (1994) Regulation of transfer genes of promiscuous IncPλ plasmid RK2: repression of Tra1 region transcription both by relaxosome proteins and by the Tra2 regulator TrbA. *Microbiology* 140: 2981–2990.

Zechner EL, de la Cruz F, Eisenbrandt R *et al.* (2000) Conjugative DNA transfer processes. In: Thomas CM (ed.) *The Horizontal Gene Pool: Bacterial Plasmids and Gene Spread*, Chapter 3 Amsterdam: Harwood, pp. 87–174

Zhang S, Meyer R (1997) The relaxosome protein MobC promotes conjugal plasmid mobilization by extending DNA strand separation to the nick site at the origin of transfer. *Mol. Microbiol.* 25: 509–516.

Ziegelin G, Pansegrau W, Lurz R, Lanka E (1992) TraK protein of conjugative plasmid RP4 forms a specialized nucleoprotein complex with the transfer origin. *J. Biol. Chem.* 267: 17279–17286.

18

The Mobility of Genetic Material

Peter M. Bennett

School of Medical Sciences, University of Bristol, Bristol, UK

Bacteria are noted for their adaptability, both in the long term, as witnessed by their colonisation of virtually all possible habitats on earth, irrespective of the harshness of the environment (Madigan *et al.*, 1997), and in the short term, as exemplified by the development of antibiotic resistance in most, if not all, clinically important species during little more than 60 years. The basis of this adaptability is fast generation times, relatively high mutation rates, and the ability to acquire DNA from diverse bacterial sources and incorporate the foreign sequences into the bacterial genome. Indeed, such is the potential of the various known bacterial systems of DNA transfer and recombination that, in principle, essentially any gene in one bacterial cell might be acquired by another bacterial cell should the need arise. It is self-evident that in practice there are constraints, in that frequent, random gene transfer between bacteria of different species would rapidly erode species differences, which is clearly not the case. None the less, a considerable amount of gene flow occurs between bacteria. This is particularly well documented for bacteria of clinical importance, e.g. the movement of antibiotic-resistance genes, but there is no reason to believe that dissemination of resistance genes is unique. Rather, research into bacterial gene transfer has tended to concentrate on resistance genes and associated systems because of their obvious importance and interest to the community at large and because each confers on the host a readily detected phenotype, i.e. resistance to a particular antibiotics(s), so these genes are easily tracked in laboratory experiments That genes conferring other characteristics, e.g. virulence determinants (Hacker and Kaper, 2000) or additional nutritional possibilities (Van der Meer *et al.*, 1992; Davison, 1999), are also subject to horizontal gene transfer is well documented (see also Syvanen and Kado, 1998; Thomas, 2000).

Genes can be transferred between bacterial cells by one of three processes: transformation, transduction and conjugation. Transferred genes may travel between bacterial cells as DNA fragments, either as free DNA or within a bacteriophage capsid, as part of a bacteriophage genome or, more commonly, as part of a bacterial plasmid or conjugative transposon. In addition to moving between bacterial cells, genes can also translocate from one site on a DNA molecule to another on the same or on a different DNA molecule. Such genetic mobility usually involves discrete DNA elements, principally transposable elements and gene cassettes. This chapter describes the various bacterial gene transfer systems and mobile genetic elements and attempts to illustrate each with clinically relevant examples.

The bacterial genome can be defined as the set of genes carried by a bacterial cell, irrespective of their location. Most will be intrinsic to the microbe, i.e. all representatives of the species will have them, but in particular strains a few genes may be temporary acquisitions; kept, in working order, so long as there is

 doi:10.1006/bkmm.2001.0018

an appropriate selective pressure. Antibiotic-resistance genes are good examples of this itinerant genetic material. In bacteria, genes may be carried on a single, large DNA molecule (usually circular), the bacterial chromosome, or may be distributed among two or more DNA molecules, of which one is usually considerably larger than the rest and is referred to as the chromosome; the others are secondary chromosomes or plasmids (Cole and Saint-Girons, 1999). The distinction is that plasmids encode functions that are not essential to the cell's survival (except in particular circumstances), so a plasmid is dispensable, whereas one or more genes on a secondary chromosome is essential and can be considered part of the core gene complement of the cell, i.e. the minimum housekeeping gene pool consistent with continued reproduction, loss of which would be lethal in the organism's normal ecological niche(s). The genes on all these structures comprise the bacterial gene pool and many are inherently genetically mobile in that they can, and do, move naturally from one bacterial cell to another or from one DNA molecule to another. This movement can have significant clinical repercussions in terms of bacterial identification, bacterial pathogenicity and antibiotic resistance.

Transformation

That genes can transfer from one bacterial cell to another has been known for many years: the phenomenon was first demonstrated unwittingly in the transformation experiments of Griffith in the late 1920s (Griffith, 1928), when he showed that some genetic factor released by a heat-killed virulent strain of pneumococcus (*Streptococcus pneumoniae*) could be acquired by an avirulent strain, restoring virulence to the latter, as judged by its recovered ability to lethally infect mice. Almost two decades later Avery and his colleagues (1944) showed that the transforming factor was DNA, demonstrating unequivocally that a cell's genes are contained in its DNA.

The term transformation originally defined the change in phenotype that a cell underwent as the consequence of acquisition and expression of a new DNA sequence(s) by uptake of DNA released into the environment by another cell, e.g. an avirulent strain of *Strep. pneumoniae* was transformed to a virulent strain in the experiment performed by Griffith (1928). The term transformation is now used also to describe the process of gene transfer responsible for the change in phenotype, particularly when artificial transformation systems are involved (see e.g. Maniatis *et al.*, 1982).

Strep. pneumoniae is one of a small number of clinically important bacterial species that has been shown to be able to take up DNA from its environment naturally (Lorenz and Wackernagel, 1994) (**Table 1**) and to rescue some of the captured sequence by integrating it into its own genome. Cells that can do this are said to be competent, and natural competence is a genetically determined physiological state that facilitates the uptake of DNA (Dubnau, 1999). By and large, the DNA captured comes from bacteria of the same or a closely related species, probably released as a consequence of death and lysis, because the newly acquired sequence normally shows significant homology to a sequence previously present at the same location on the genome of the recipient cell. Sequences that have little or no homology with a genomic sequence in the recipient cell are rarely, if ever, rescued, because the new sequence cannot be incorporated into the genome of the recipient cell by homologous recombination. This form of recombination requires precise alignment of the incoming DNA with the related genome locus (Kowalczykowski *et al.*, 1994), which happens only when the two sequences are extremely similar, e.g. originate from related species. The result of the recombination is sequence replacement where like is replaced by 'almost like'.

There are several variations of natural transformation. All depend on homologous recombination to rescue and preserve the acquired sequence as a heritable trait. Differences are most apparent in the initial steps, when the DNA is bound to the cell surface and is processed for internalisation. These processes are not yet fully understood, although considerable detail is known (Dubnau, 1999).

Strep. pneumoniae is not permanently competent; rather, it becomes competent at a particular point in the growth cycle. When competent, double-stranded

Table 1 Naturally transformable clinical bacteria

Clinical isolates	References
Campylobacter jejuni	Wang and Taylor (1990)
Campylobacter coli	Wang and Taylor (1990)
Haemophilus influenzae	Barany and Kahn (1985)
Haemophilus parainfluenzae	Gromkova and Goodgal (1979)
Helicobacter pylori	Smeets *et al.* (2000)
Moraxella spp.	Bootsma *et al.* (2000)
Neisseria gonorrhoeae	Wolfgang *et al.* (1999)
Neisseria meningitidis	Swartley *et al.* (1997)
Staphylococcus aureus	Rudin *et al.* (1974)
Streptococcus pneumoniae	Avery *et al.* (1944)
Streptococcus sanguis	Behnke (1981)
Streptococcus mutans	Shah and Caufield (1993)

(ds)DNA fragments can be bound to its outer surface; then one strand of the bound fragment is internalised where, if it is sufficiently similar to a sequence already present with which recombination may occur, it is used to re-fashion that part of the genome. Binding and movement of DNA across the cell envelope require special competence factors that are produced in a growth-dependent fashion. *Strep. pneumoniae* can take up heterologous DNA, i.e. from unrelated organisms, as well as homologous DNA from the same or closely related species (Lorenz and Wackernagel, 1994), but is only able to rescue sequences that show significant homology to its own genome. None the less, this can result in genes being extensively remodelled, to an extent that the hybrid sequence encodes a product with new and useful properties. Such re-modelling may involve several sections of the original gene and imported sequences from more than one donor, creating what is known as a gene mosaic (**Fig. 1**). The penicillin-binding protein (PBP) genes of penicillin-resistant strains of *Strep. pneumoniae* are examples of gene mosaics (Dowson *et al.*, 1994). PBPs, so called because they bind penicillin covalently, are enzymes that catalyse the final steps in the synthesis of peptidoglycan (Spratt, 1983), the bacterial cell envelope component that confers its mechanical strength. The PBPs of penicillin-sensitive strains of *Strep. pneumoniae* are exquisitely sensitive to penicillin, and are readily acylated by it. The PBPs of penicillin-resistant *Strep. pneumoniae* are not (Dowson *et al.*, 1994). Comparisons of the PBP gene sequences of penicillin-sensitive and penicillin-resistant strains revealed that the latter have undergone extensive rearrangements, consistent with parts of each PBP gene being replaced with similar but different sequences from unknown sources (Dowson *et al.*, 1994). Some of the PBP genes of penicillin-resistant *Strep. pneumoniae* have clearly been reconstructed from small DNA segments from several different, but related, bacterial species.

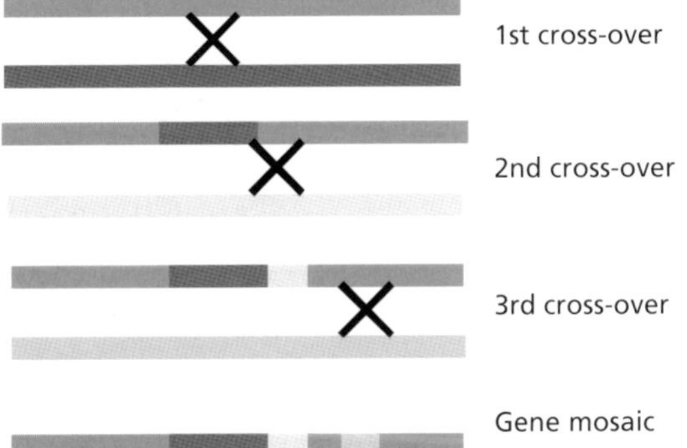

Fig. 1 Generation of gene mosaics. The different shaded sections indicate DNA sequences from different sources.

Some naturally transformable bacteria, such as *Haemophilus* and *Neisseria* species, can only take up DNA from the same or a closely related species. Heterologous DNA is not recovered. This specificity is determined by short nucleotide signatures that are found frequently on the genomes of the species in question. *Neisseria gonorrhoea* and *Haemophilus influenzae* are two such. In the former case the signature is a 10-bp sequence, whereas for *H. influenzae* the signature is a 9-bp sequence contained within a 29-bp domain (Lorenz and Wackernagel, 1994; Kroll *et al.*, 1998).

For all transformations, dsDNA fragments bind to the surface of the bacterial cell, There, in naturally competent systems, the DNA is processed in a species-specific manner. In the case of *Strep. pneumoniae*, one strand of the DNA fragment is internalised where, if possible, the sequence is saved by homologous recombination. The complementary strand is degraded. With *H. influenzae* and *N. gonorrhoeae*, the process may occur in two stages: (1) uptake of dsDNA into the periplasm, followed by transport of a single strand across the inner membrane, coupled to degradation of the complementary strand, as for *Strep. pneumoniae* (Stewart and Carlson, 1986; Dubnau, 1999). Again, sequences are rescued by homologous recombination. Like all genome changes, transformation events are random; changes beneficial to individual cells are consolidated in the population by subsequent selection.

Laboratory transformation systems are used in genetic manipulation of bacteria. This work often employs bacterial strains that show little or no natural competence, e.g. the intensively studied and used *E. coli* K12 and its derivatives. With these systems a state of competence is created by manipulating the cationic composition of the cell envelope and the transforming DNA is bound to the cell surface at low temperature (0°C). Transport of the DNA into the cell is achieved by a sudden sharp increase in temperature (to 37–42°C) (Maniatis *et al.*, 1982). Alternatively DNA can be introduced into bacterial cells by applying a high electrical voltage (2.5 kV) to a mixture of cells and DNA, a process called electroporation (electrotransformation) (Ausubel *et al.*, 1990). In these cases dsDNA is taken up by the cell and transformation with intact plasmid DNA, i.e. a self-replicating DNA molecule (one that does not need to undergo recombination to be rescued), tends to give much higher levels of transformation than fragments of DNA. These laboratory transformation systems are artificial and the mechanisms differ from natural transformation systems. What they have in common is the participation of free DNA.

Mechanisms of natural transformation differ principally in the type of DNA that will bind to the cell surface, i.e. homologous or heterologous DNA, and the method of DNA uptake into the cell (Stewart and Carlson, 1986). Incorporation of the recovered DNA sequence into the genome by homologous recombination is achieved in essentially the same way, regardless of means of entry. The process of bacterial transformation can be thought of as involving the following steps:

1. Release of DNA by the donating cell. This is highly likely to be a consequence of cell death and lysis, although it has been reported that bacterial cells can release DNA into the environment and remain viable (Leduc *et al.*, 1982), but the mechanism is unknown.
2. Development of competence in the recipient cell. This is genetically controlled and may be restricted to a particular part of the culture cycle or occur as the result of a detrimental change in nutritional status of the growth medium. Only one bacterial species, *N. gonorrhoeae*, is known to be constitutively competent, i.e. competent throughout the culture cycle (Lorenz and Wackernagel, 1994). Achieving natural competence requires the production of specific competence factors (Dubnau, 1991, 1999).
3. Attachment of free dsDNA to the surface of the cell. With some bacterial species, e.g. *Strep. pneumoniae*, this process is non-specific, but with others, e.g. *H. influenzae* and *N. gonorrhoeae*, the DNA to be bound must carry an appropriate signature/binding sequence (Lorenz and Wackernagel, 1994; Kroll *et al.*, 1998).
4. Transfer of the bound DNA across the bacterial cell envelope. In some cases this probably involves a form of transport pore, e.g. *Strep. pneumoniae*; in others the DNA is first internalised in a membrane vesicle termed a transformasome, e.g. *H. influenzae* (Kahn *et al.*, 1983) for transport to the periplasm, and from there is transported to the cytoplasm (Dubnau, 1999).
4. Integration of the captured DNA sequence into the genome of the recipient cell by homologous recombination.

The evolution of penicillin-resistant pneumococci, through the generation of PBP gene mosaics, as examples of transformation-recombination in operation in natural systems has already been described. Transfer of plasmids encoding β-lactamase among *H. influenzae* and *N. gonorrhoeae* is thought to have involved transformation as well as conjugation (Saunders *et al.*, 1986). Recently, it has been reported that virulence genes have been transferred between *Haemophilus* and *Neisseria* spp. by natural transformation (Kroll *et al.*, 1998).

Conjugation

The second mechanism of gene transfer is conjugation. With this mechanism DNA is passed directly from one cell to another. Conjugation was discovered by Joshua Lederberg and his associates in the mid 1940s (Lederberg and Tatum, 1946; Lederberg, 1947), when it was shown that genes can be transferred from one strain of *E. coli* to another by a mechanism that does not involve free DNA, nor a bacteriophage (see later), but does require physical contact between donor and recipient cells (Table 2). Elegant genetic experiments demonstrated that gene transfer is essentially unidirectional and defined the element responsible as the fertility or F-factor. Some years later it was shown that the F-factor is a bacterial plasmid (see section on plasmids, below). Since then numerous studies have reported and described a multitude of plasmids in a great number of bacterial species, many of which, like the F-factor, can transfer between bacterial cells. Such elements are called conjugative plasmids and mediate their own transfer between bacterial cells (Wilkins, 1995). Many other plasmids are also able to transfer with the assistance of a conjugative plasmid, a phenomenon called plasmid mobilisation (Cabezón *et al.*, 1994, 1997; Albiger *et al.*, 2000).

Table 2 Comparisons of requirements of natural bacterial DNA transfer systems

	Transformation	Transduction	Conjugation
Specific requirement	Species-related free DNA	Transducing bacteriophage	Conjugative plasmid or transposon
Donor–recipient contact	No	No	Yes
Re-combination for DNA recovery	Yes	Yes	No
Transfer strongly species-dependent	Yes	Yes	No

Conjugation Between Gram-negative Bacteria

The best-studied conjugative plasmid is the F-factor, or simply F. F is a relatively large plasmid of 90 kb, of which approximately 30 kb, termed the *tra* region, is devoted to plasmid transfer functions. This region encodes approximately two dozen genes, the products of which are necessary for plasmid transfer. Most are required to produce a proteinaceous filament, called a sex pilus, that projects from the cell surface and acts to couple the plasmid donor cell to a potential recipient. The sex pilus is essentially a thin tube constructed of helically arranged protein (pilin) subunits. Once contact has been made, via the tip, the pilin subunits depolymerise into the membrane (Frost *et al.*, 1994), drawing the donor and recipient cells together. At this point it is believed the envelopes of the two cells fuse locally to create a DNA transport pore through which the plasmid DNA can pass (**Fig. 2**) (Wilkins, 1995). With Gram-negative cells the pore must pass through four biological membranes. How this is achieved is not yet known. It should be noted that cell coupling often involves several donors and recipients in a multi-cell mating aggregate (Achtman, 1975), rather than being restricted to a single donor and recipient.

Conjugation is a combined DNA transfer and replication mechanism. Transfer involves only one strand of the DNA duplex and is coupled to synthesis of a replacement strand in the donor cell and a complementary strand in the recipient (**Fig. 2**) (Wilkins, 1995).

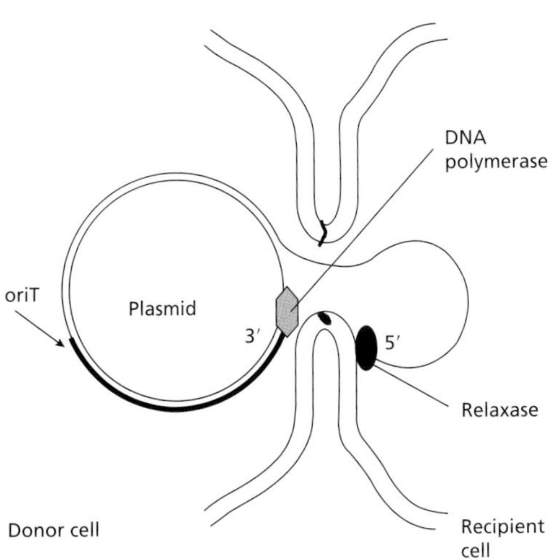

Fig. 2 Conjugative DNA transfer. Original plasmid DNA is depicted as thin strands; newly synthesised DNA is depicted as a thick strand.

Plasmids that transfer by conjugation, be they conjugative or mobilisable, each possess a short, specific nucleotide sequence, *oriT*, that serves as the origin of transfer (hence the designation) and as the origin of replication for synthesis of the replacement strand (**Fig. 2**) (Zechner *et al.*, 2000). At *oriT* one strand of the plasmid DNA is nicked, i.e. a single phosphodiester bond is cleaved, by an enzyme called a relaxase (Wilkins, 1995; Zechner *et al.*, 2000). The enzyme derives its name from the fact that cleavage of the single phosphodiester bond relaxes the supercoiling of the target DNA. In the process the relaxase is covalently joined to the 5′-phosphate group generated by the cleavage. This protein–DNA complex anchors one end of what is formally a linear DNA strand at the DNA transfer pore site. Whether the relaxase–DNA complex passes into the recipient cell to anchor the 5′ end of the transferring DNA strand on the far side of the pore (**Fig. 2**) is not known. If this is the case, then the internal diameter of the pore need only be wide enough to accommodate a single DNA strand; in contrast, if the relaxase complex remains in the donor cell at the entrance to the pore, then the pore would have to be wide enough to accommodate two non-complementary strands of DNA, one of which would be essentially static, anchored by the relaxase, while the other would move past it. The former model (**Fig. 2**) seems more likely.

The phosphodiester bond cleavage also generates an –OH group at the 3′ end of the linearised strand from which DNA synthesis can proceed, in a continuous manner, using the intact circular strand of the plasmid as the template (**Fig. 2**). As the new DNA strand is synthesised, the old copy is displaced and transferred to the recipient. Note that the newly synthesised DNA is a head-to-tail extension of the DNA strand transferred to the recipient cell. This form of DNA replication is called rolling circle (RC) replication and can generate multiple tandem repeats joined head to tail, as we shall see. Transfer is believed to be driven by RC replication, which also requires a topoisomerase to unwind the DNA prior to replication. There is no evidence that any additional energy is required for DNA transfer. Once the transferred strand is in the recipient cell, a complementary strand is synthesised, discontinuously, to generate a dsDNA molecule. Because synthesis of the complementary strand of DNA in the recipient cell is discontinuous, RNA primers are required. These may synthesised by a plasmid-encoded primase that is produced in the donor and transported to the recipient, presumably through the DNA transport pore (Wilkins and Lanka, 1993; Zechner *et al.*, 2000), or by an appropriate host-cell enzyme. When single-strand replication of the plasmid in the donor is complete and a second copy of *oriT* enters the recipient, it is thought that relaxase cleaves the DNA again at *oriT* to release the transferred sequence from the newly synthesised one, and joins the ends of the transferred DNA together to generate the circular form of the plasmid. Hence, after conjugation, both the donor and recipient cells have copies of the plasmid. Plasmid establishment in the recipient also requires single-strand (ss) DNA-binding protein to protect the transferred strand as complementary strand is synthesised and DNA ligase to seal the circle.

It should be noted that conjugative plasmids mostly transfer intact from donor to recipient, i.e. the transfer is precise. If the donor cell carries two or more conjugative plasmids, a single cell coupling event may result in transfer of two (or more) different plasmids.

Conjugation between Gram-positive Bacteria

For many years, conjugation was thought to be restricted to Gram-negative bacteria, but Clewell and his colleagues (Dunny *et al.*, 1978) demonstrated that plasmid transfer could occur between enterococcal strains by a conjugation mechanism. As with Gram-negative systems, DNA transfer requires cell–cell contact; however, enterococci use a different strategy to that of Gram-negative bacteria to achieve cell–cell contact. Plasmid-free enterococci excrete several small hydrophobic peptides in tiny amounts ($2-5 \times 10^{-11}$ mol/L). These function as sex pheromones, each of which is specific for a different plasmid type (Dunny *et al.*, 1978; Clewell, 1993). Each pheromone triggers the expression of transfer functions of a particular plasmid type in the donors, including synthesis of cell surface aggregating substances that cover the donor cells and promote clumping of donor and recipient cells. Once the plasmid has been transferred, production of the peptide pheromone-specific to the plasmid is repressed in the transconjugant, while production of alternative pheromones-specific for other plasmid types continues (Clewell, 1993). In the absence of pheromone production, expression of the majority of transfer genes is repressed. In addition, the cells also produce another small peptide which specifically inhibits the ability of the cell to respond to the particular pheromone. Hence the transconjugant remains receptive to other plasmids but not to ones closely related to the one acquired, minimising the number of potentially redundant transfer events.

As with Gram-negative bacteria, mating appears to involve aggregates of donor and recipient cells, rather than simple mating couples. As with Gram-negative plasmids, transfer of enterococcal plasmids is believed to proceed by ssDNA transfer driven by RC replication, since they also encode relaxases and carry *oriT* sequences (Guzman and Espinosa, 1997). Hence, the transfer mechanisms appear to differ mainly in the ways in which the donor and recipient cells make contact with each other to establish the conditions for DNA transfer.

Given the variety of plasmids in bacterial cells, conjugation is a powerful mechanism for horizontal gene transfer, particularly as many plasmids are highly promiscuous, i.e. they can be shown in the laboratory to transfer to and from bacteria of many different species and even across genera (see Thomas, 1989). Some have been shown to be capable of crossing the Gram-negative Gram-positive divide and some will even mediate DNA transfer to eukaryotic cells (Courvalin *et al.*, 1998; Salyers *et al.*, 1998). In the medical setting plasmid spread is responsible for much of the antibiotic resistance found in clinical bacterial isolates (Talbot *et al.*, 1980; O'Brien *et al.*, 1986; Mayer, 1988).

Transduction

The third mechanism of DNA transfer between bacterial cells is transduction, which can be defined as a genotypic modification giving rise to a phenotypic change as the result of horizontal DNA transfer mediated by a bacteriophage, i.e. a bacterial virus. Some, but not all, bacteriophages, are capable of acting as transducing agents. The phenomenon is a consequence of rare mistakes made in phage reproduction. Two types of transduction have been described, generalised and specialised (Masters, 1996; Weisberg, 1996; Bennett and Howe, 1998), exemplified by transduction mediated by the coliphages (phages that grow in *E. coli*) P1 and λ(lambda), respectively. Generalised transduction means that all genes in a cell can be transferred by the phage, and specialised transduction describes the situation where the phage only transfers a few specific genes, from host to recipient.

Generalised Transduction

P1 infects *E. coli* and closely related Enterobacteriaceae. When it does so the linear, dsDNA genome carried in the phage particle is injected into the host cell where it is circularised, replicated (ultimately by RC replication to produced long linear concatenated copies – multiple copies of the phage genome joined head to tail) and expressed (Sternberg and Hoess, 1983). The concatamers are subsequently cut into genome-sized fragments and incorporated into new phage particles. Transducing particles arise from errors in phage particle assembly, when host DNA is accidentally incorporated into the head of the new particle in place of a copy of the phage genome. Any part of the host genome can, in principle, be incorporated. Since the P1 genome is almost 100 kb in size, the phage head can accommodate approximately 100 kb or 2% of host (*E. coli*) DNA, equivalent to about 100 average-size (1 kb) genes. The transducing particles that result contain no phage DNA, but are otherwise functional. They can adhere to and inject their DNA into a new host cell. Transduction occurs when one or more of the genes on the injected DNA is rescued by incorporation into the genome of the infected cell, a process that, like transformation, requires homologous recombination. Since phages tend to have narrow host ranges, horizontal gene transfer by this mechanism also involves only closely related species. This form of transduction can also mediate transfer of whole bacterial plasmids between bacteria, providing the plasmid DNA can be accommodated in the head of the phage. In this case, recombinational rescue of the DNA is unnecessary to establish the DNA in the transductant because the DNA has its own origin of replication.

Perhaps the best example of horizontal gene transfer by bacteriophages in the clinical context is the transfer of resistance genes between cells of *Staphylococcus aureus*, as demonstrated by (Lacey, 1973, 1984). Many antibiotic-resistance genes in *Staph. aureus* are found singly on small plasmids, transfer of which is mediated, at least in part, by transducing phages. Miller (1998) has also demonstrated the capacity for transduction of antibiotic-resistance genes among Gram-negative bacteria in fresh lake water, but the contribution to dissemination of these genes in clinical and non-clinical environments is unknown.

Specialised Transduction

Like P1, λ bacteriophage also infects *E. coli* and closely related bacteria. Again a linear, dsDNA genome is injected into the new host cell and circularised. One of two things then happen. The genome can be committed to the phage lytic cycle, when it is replicated and expressed to produce a new batch of identical phage particles. For λ this process is accurate, and production of generalised transducing phage particles does not occur. Alternatively, λ can lysogenise the host, when it is integrated into the host genome at one particular site in one orientation by a site-specific recombination system, to become a prophage, i.e. a state in which most phage genes are silenced, no phage particles are made, and the λ genome is replicated passively as part of the bacterial chromosome. The prophage state is aborted when the cell's DNA is damaged. DNA damage triggers a repair cascade, the SOS response (Walker, 1996). One consequence of this is that the integration step is reversed, releasing the phage genome as a circular DNA species, a process that is mediated by essentially the same site-specific mechanism that integrated the phage into the chromosome in the first place (Campbell, 1996). Occasionally, a mistake is made; instead of the phage genome bring excised precisely, a small section of the chromosomal DNA attached to one end of the phage genome is also removed. Since the length of DNA that can be packaged into the phage head is more or less constant, acquisition of additional sequences at one end is matched by a compensating loss of phage DNA at the other end of the phage genome. Phage particles that contain such DNA are fully competent to deliver the hybrid DNA into a susceptible cell but phage development may be impaired due to loss of essential phage functions (Weisberg, 1996). When the DNA in such particles is injected into a new host, the chromosomal markers on the hybrid DNA molecule may be rescued by lysogenisation. The modified phage genome can be inserted into the host cell chromosome, by phage-encoded integrase, as if it was a normal phage genome. Alternatively, non-phage DNA sequences may be rescued by recombination with homologous sequences on the host genome, using the cell's normal recombination system. Such transduction is specialised because the only host genes that can be captured and transferred are those that flank the phage insertion site: in the case of *E. coli* and λ, the genes of the *gal* operon (encodes enzymes for galactose catabolism) on one side and those of the *bio* operon (biotin synthesis) on the other.

The modified genomes of specialised transducing phages may have lost genes necessary for phage development and so cannot by themselves produce infectious particles. In these cases phage development depends on a helper phage, e.g. the wild-type phage which supplies the missing functions. To propagate the mutant phage, cells are co-infected and both normal phages and transducing phage particles are produced and released when the cell lyses. The resulting phage preparation effects high-frequency transduction of the markers carried (Weisberg, 1996).

In *E. coli* there is a natural insertion site for λ, att^{λ}, on the chromosome between the *gal* and *bio* gene clusters. If this is deleted, then the phage may insert, at lower frequency, into one of several secondary sites that resemble att^{λ}. Transducing particles generated from such lysogens carry *E. coli* chromosomal genes flanking the secondary site of insertion. Before the advent of gene cloning using restriction enzymes and transformation, specialised λ transduction phages isolated following insertion at secondary sites were used to capture and transfer a variety of *E. coli* genes, initially for complementation analyses and later for gene-sequencing studies, notably for analysis of genes encoding ribosome and RNA polymerase components (Nomura and Morgan, 1977).

Mobile DNA Elements

The term mobile DNA refers to a DNA sequence that can move between bacterial cells or between DNA molecules, i.e. different replicons. A replicon is defined as a DNA molecule that can be replicated independently of other DNA molecules. Replicons in bacteria may be chromosomes, plasmids or bacteriophage genomes.

When discussing mobile DNA, one class is generally ignored or dealt with superficially, namely, bacteriophage DNA. None the less, these 'DNA elements' obviously transfer from cell to cell and, in cases where there is a potential to lysogenise the host, there is clearly an opportunity to influence the host phenotype. Indeed important clinical phenotypes are determined by phage-carried genes; witness production of diphtheria toxin (Freeman, 1951; Murphy *et al.*, 1974) and cholera toxin (Waldor and Mekalanos, 1996), which in both cases is due to lysogenic conversion (Bennett and Howe, 1998).

Excepting bacteriophage genomes, there are two classes of DNA elements that transfer from one bacterial cell to another: plasmids and conjugative transposons.

Plasmids

A plasmid is a dsDNA molecule, often circular, that can be replicated independently of the host cell's chromosome (see Grinsted and Bennett, 1988). Plasmid size can differ enormously, from a small molecule with just sufficient genetic capacity to code for one average-sized gene to DNA molecules that encode hundreds of genes and that may in some cases be better considered as bacterial mini-chromosomes (Cole and Saint-Girons, 1999). Plasmids are categorised in a number of ways that reflect the interests shown in them and the manner in which they influence the host cell, e.g. catabolic plasmids confer degradative abilities, resistance (R) plasmids confer resistance to an antibiotic(s), virulence plasmids confer pathogenic traits, etc. (Top *et al.*, 2000). In terms of mobility, plasmids fall into one of three classes: (1) conjugative plasmids, which encode a fully functional transfer system for their own mobilisation; (2) mobilisable plasmids, which lack the genes that specify the machinery for establishing cell-to-cell contact but do encode functions for their own cell-to-cell transfer, the minimum requirement being an origin of transfer, *oriT*; and (3) non-mobilisable plasmids, i.e. those that lack even an *oriT*. Even in these cases plasmids may be transferred by a process termed conduction, which involves recombination between the non-mobilisable plasmid and one that can be conjugally transferred to generate a plasmid fusion. Such recombination is often mediated by transposable elements, particularly IS elements (see later). The recombination creates a single, hybrid structure that is transferred by the transfer machinery of the conjugal component. Once in the recipient cell, the components of the hybrid structure may disengage essentially to regenerate both the participating plasmids.

In a conjugation system involving a Gram-negative donor, cell–cell contact is established by a sex pilus. The majority of transfer genes carried by conjugative plasmids are involved in sex pilus construction; relative few are needed for DNA transfer. On the sex factor F, approximately a third of its genetic capacity, about 30 kb (sufficient for about two dozen genes), is devoted to DNA transfer functions (Frost *et al.*, 1994; Zechner *et al.*, 2000). These genes are arranged as a single cluster that is found not only on F but also on several R plasmids, e.g. R1 and R100. Other conjugative plasmids encode unrelated but similar transfer systems, e.g. the Tra system encoded by R plasmid RP4 (Zechner *et al.*, 2000). Several others have been described (Zechner *et al.*, 2000). All employ multiple genes accounting for 20 kb of DNA or more. Accordingly, conjugative plasmids of Gram-negative bacteria are never small, with a minimum size of about 25 kb. Many, like F, are much larger.

In contrast, many mobilisable Gram-negative plasmids are quite small, such as the *E. coli* colicin plasmid ColE1. The mobilisation genes on such an element account for less than 3 kb of the coding sequence, the entire plasmid being only 7.8 kb (Boyd *et al.*, 1989). ColE1 and related plasmids have assumed a degree of eminence because they have been used as the basis of many of the plasmid cloning vectors used in genetic engineering. Since it is desirable that recombinant plasmids constructed with these vectors be contained within the bacterial hosts for which they have been designed, part of vector construction involves the removal of mobilisation genes and the *oriT* sequence of the plasmid.

A number of very small natural Gram-negative plasmids appear to be mobilisable, but require the assistance not only of a conjugative plasmid but also another mobilisable plasmid, because although they possess an *oriT* site, they lack one or more of the functions needed for DNA processing and strand transfer. These missing functions are supplied by the second mobilisable plasmid, while the functions needed to promote mating pair formation are provided by the conjugative plasmid.

In the case of Gram-positive conjugative plasmids, many also are quite large, ~50–60 kb (Zechner *et al.*, 2000), although the transfer regions are smaller than those of Gram-negative systems, probably because the cell-cell coupling mechanism is less complicated and so requires fewer functions, i.e. there is no pilus to assemble and then depolymerise.

Conjugative Transposons

Conjugative transposons, like conjugative plasmids, mediate their own transfer from one cell to another, i.e. they encode their own set of transfer functions. They differ from plasmids in that they do not replicate independently of other replicons, i.e. they lack an origin of vegetative replication, *oriV*; rather they are replicated passively as a component of a larger replicon (Salyers *et al.*, 1995; Rice, 1998). Thus these elements shuttle in and out of other replicons, like other transposable elements. To transfer from one cell to another, a conjugative transposon first disengages from its carrier replicon to create a free circular double-stranded form of the element. Transfer to a new host then follows by a mechanism thought to be essentially the same as that used for plasmid transfer, i.e. single-strand transfer from an *oriT*, coupled to replacement synthesis by RC replication. Circularisation of the transferred strand and synthesis of the complementary strand completes the transfer process and recreates the free circular form of the element, which is then efficiently inserted into one of the replicons of the new host (Salyers *et al.*, 1998).

Transposable Elements

As plasmids and their like transfer between bacterial cells, so transposable elements transfer between DNA replicons. Accordingly, transposition is a form of recombination. Unlike homologous recombination, however, there is generally no need for sequence homology between the old and new sites of insertion, nor between the site of insertion and the transposable element. A transposable element is a structure defined by its terminal sequences that are recombination 'hot spots'. These sequences are usually short perfect or near-perfect inverted repeats (IRs) and serve as binding sites for the element-specific transposase which mediates the excision–integration reactions needed to accomplish transposition of the element. Transpositions are necessarily recombinations since each event involves breaking and reforming phosphodiester bonds at the ends of the element; the consequence is DNA sequence re-assortment.

There are two basic mechanistic themes in transposition (Bennett, 2000; Merlin *et al.*, 2000): (1) translocation by some form of cut-and-paste mechanism whereby the element, in the form of dsDNA, is moved to a new genetic site; and (2) translocation involving replication, where the element is duplicated and one copy remains at the original site and one is inserted somewhere else on the same or on a different replicon (**Fig. 3**). Transposable elements are responsible for much of the macromolecular DNA rearrangements detected in bacteria (insertions, deletions, inversions, chromosomal translocations) and much of the dissemination of antibiotic-resistance genes (**Tables 3** and **4**) (Bennett, 2000; Merlin *et al.*, 2000).

Transposable elements can be grouped in four large classes: (1) insertion sequences (IS elements) (Mahillon and Chandler, 1998); (2) transposons (see Berg and Howe, 1989); (3) conjugative transposons (Salyers *et al.*, 1995; Rice, 1998) and (4) transposing bacteriophages (Pato, 1989). The last group comprise a set of DNA bacteriophages that replicate their genomes by transposition. The type element of this group is the *E. coli* temperate phage Mu. Although transposing bacteriophages can mediate transposition of host replicon DNA, i.e. non-phage sequences, to date they have not been implicated in DNA rearrangements of clinical importance; accordingly, they will be considered no further here. For further information see Toussaint and Résibois (1983) and Pato (1989).

IS Elements

IS elements are the smallest transposable elements, with sizes ranging from approximately 800 bp to 2000 bp, with an average size of about 1400 bp (Mahillon and Chandler, 1998). Most have short, terminal IRs that serve as binding sites for the cognate transposase. These flank one, or at most two, genes that encode the element's transposase and in some cases a transposition regulatory function. The transposase mediates the breakage and formation of the phosphodiester bonds at the extremities of the element

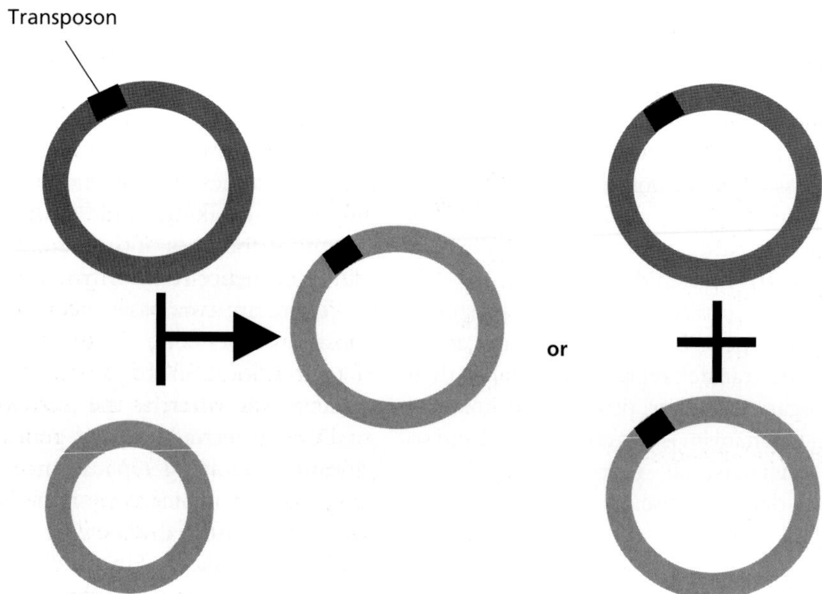

Fig. 3 Transposition of antibiotic-resistance genes.

Table 3 Some prokaryotic composite transposons

Transposon	Size (kb)	Terminal elements	Target duplication (bp)	Marker(s)[a]
Elements from Gram-negative bacteria				
Tn*5*	5.7	IS*50* [IR][b]	9	KmBlSm
Tn*9*	2.5	IS*1* [DR]	9	Cm
Tn*10*	9.3	IS*10* [IR]	9	Tc
Tn*903*	3.1	IS*903* [IR]	9	Km
Tn*1525*	4.4	IS*15* [DR]	8	Km
Tn*1681*	4.7	IS*1* [IR]	9	HST
Tn*2350*	10.4	IS*1* [DR]	9	Km
Tn*2680*	5.0	IS*26*[c] [DR]	8	Km
Elements from Gram-positive bacteria				
Tn*3851*	5.2	NR	NR	GmTbKm
Tn*4001*	4.7	IS*256* [IR]	8	GmTbKm
Tn*4003*	3.6	IS*257* [DR]	8	Tm

[a]Resistance genes: Bl, bleomycin; Cm, chloramphenicol; Gm, gentamicin; Km, kanamycin; Sm, streptomycin; Tb, tobramycin; Tc, tetracycline. HST, heat-stable enterotoxin; NR, not recorded.
[b]Dr, direct repeat; IR, inverted repeat.
[c]IS*26* and IS*6* are synonyms.

to disengage the element from one site and insert it into a new one. The regulatory function, if one is encoded, serves to damp down the frequency of transposition. This is desirable. Since new sites of insertion are selected essentially at random, transposition is mutagenic and potentially fatal to the host and a high frequency of transposition would inevitably result in lethal genetic damage to the host cell. IS elements themselves carry no genes of clinical importance. Indeed, they are too small to accommodate any genes other than those needed for transposition. However, IS elements can, and do, form composite transposons, many of which carry genes for antibiotic resistance (Table 3) (see next section).

Table 4 Some Tn*3*-like transposons

Transposon	Size (kb)	Terminal IRs (bp)	Target (bp)	Marker (s)[a]
Elements from Gram-negative bacteria				
Tn*1*	5	38/38	5	Ap
Tn*3*	4.957	38/38	5	Ap
Tn*21*	19.6	35/38	5	HgSmSu
Tn*501*	8.2	35/38	5	Hg
Tn*1000*	5.8	36/37	5	None
Tn*1721*[b]	11.4	35/38	5	Tc
Tn*1722*[b]	5.6	35/38	5	None
Tn*2501*	6.3	45/48	5	None
Tn*3926*	7.8	36/38	5	Hg
Tn*4651*	56	32/38	5	xyl
Elements from Gram-positive bacteria				
Tn*551*	5.3	35	5	Ery
Tn*917*	5.3	38	5	Ery
Tn*4430*	4.2	38	5	None
Tn*4451*	6.2	12	NR	Cm
Tn*4556*	6.8	38	5	None

[a]Ap, ampicillin; Cm, chloramphenicol; Ery, erythromycin; Hg, mercuric ions; Sm, streptomycin; Su, sulphonamide; Tc, tetracycline; xyl, xylose catabolism.
[b]Tn*1721* is a composite structure that utilises Tn*1722* as its basis for transposition.

Transposons

A transposon is a transposable element that, in addition to encoding the functions needed for its own transposition, also carries at least one gene that encodes a product with a function unconnected with transposition. These non-transposition functions alter the host cell phenotype in a predictable manner, e.g. render the cell resistant to a particular antibiotic(s) or confer a new catabolic activity, unlike IS elements that can alter cell phenotype by gene insertional inactivation, but do so randomly. Many of the transposons described carry drug-resistance genes (Tables 3 and 4).

Composite Transposons

Structurally, transposons fall into one of two distinct classes, composite or non-composite (complex). Composite elements have a distinctive modular structure with long direct or inverted terminal repeats formed from a pair of identical or near identical IS elements. These flank a central unique core sequence that may encode any one of a number of diverse functions (Table 3). Many such elements from both Gram-negative and Gram-positive bacteria have been described, and are responsible for much of the spread of antibiotic-resistance genes among bacteria of clinical importance. One or both of the terminal IS elements may be transpositionally active and each may transpose independently of the whole ensemble. The composite structure transposes because it is necessarily delineated by the same short IRs that delineate each of the participating IS elements. Since a pair of IRs is the only structural feature needed for transposition, then any sequence bounded by such sequences can, in principle, transpose, given provision of the appropriate transposase; witness the construction of artificial transposons as research tools (Berg *et al.*, 1989; Herrero *et al.*, 1990; de Lorenzo *et al.*, 1990). There are limits, however; as the distance between the IRs increases the frequency of transposition drops off (Chandler *et al.*, 1981). Hence, in practice, most composite transposons are relatively small, with sizes in the range 4–10 kb (Table 3).

Many composite transposons transpose by a cut-and-paste mechanism, in which the entire structure is completely disengaged from the donor replicon by breakage of the four phosphodiester bonds that anchor it in the host replicon. The released structure may be circularised before it is re-inserted at a new site, or the ends may simply be synapsed by the transposase to form a pseudo-circular DNA molecule, e.g. Tn*10* (Bender and Kleckner, 1986; Kleckner *et al.*, 1996; Kennedy *et al.*, 1998). Such cut-and-paste mechanisms move the element, as a dsDNA structure, from one genetic site to another. The gap left at the vacated site may or may not be sealed. If it is sealed, a deletion results; if not, the double strand break inhibits replication and that copy of the replicon is lost. Elements that use the latter variation can truly be described as 'selfish DNA', since each transposition is at the expense of the previous host replicon.

Non-composite (Complex) Transposons

Complex, or non-composite, transposons, do not have modular structures. They do possess short terminal IRs, usually of the order of 20–40 bp although longer ones are known, and they encode one or more functions necessary for their own transposition. In addition, like the composite transposons, each encodes at least one non-transposition function, e.g. resistance to an antibiotic (Table 4). When transposition occurs, the entire ensemble moves; the sequence encoding the transposition functions is incapable of independent transposition, unlike the terminal elements of composite transposons. In crude terms, in relation to transposition, non-composite transposons are more akin to the insertion sequences that form the terminal elements of composite transposons than to the composite elements themselves. Indeed, pairs of non-composite transposons, such as Tn*3* (Table 4), can also mediate transposition of the intervening DNA sequence. This has been demonstrated in the laboratory (Bennett *et al.*, 1986), and observed in nature; witness Tn*1721*, a composite transposon encoding tetracycline resistance derived from the non-composite element Tn*1722* (Grinsted *et al.*, 1990).

Although it is convenient to refer to non-composite transposons as a group because they have structural similarities, it should not be thought that members of the set are necessarily related phylogenetically. The elements represent several independent lines of evolutionary descent. This is clearly reflected in the finding that different elements transpose by different mechanisms. Just as the term composite transposon denotes no more than that the element has a modular structure, as described above, so the terms non-composite or complex transposon indicate simply that the element does not have a modular form and is not a transposing bacteriophage.

The typical non-composite transposon is Tn*3* (Table 4), first found on the R plasmid R1 (Heffron *et al.*, 1977). Essentially the same element, Tn*1*, had previously been identified on RP4 (Hedges and Jacob, 1974). Both encode a TEM β lactamase, TEM-1 on Tn*3* and TEM-2 on Tn*1*. These enzymes differ by a single amino acid, a Gln to Lys substitution at position 37 in TEM-2, a change that has little effect on enzyme activity. The TEM β lactamase, TEM-1, was first reported in a clinical isolate of *E. coli* (Datta and Kontomichalou, 1965). Since then it has been found worldwide in numerous bacterial species, including clinically important members of the Enterobacteriaceae, *Acinetobacter* spp., *Pseudomonas aeruginosa*, *Stenotrophomonas maltophilia*, *H. influenzae* and *N. gonorrhoeae*. Undoubtedly its widespread distribution reflects not only spread of the microbes but also its transposon-plasmid carriage. This worldwide spread has left an unwanted legacy; a huge gene reservoir for evolutionary development of the β lactamase. The last 10 years has seen the emergence of approximately 60 gene variants, encoding so-called extended-spectrum β lactamases derived from both TEM-1 and TEM-2 by acquisition of simple point mutations (Bush *et al.*, 1995; Du Bois *et al.*, 1995). The mutations subtly alter the configuration of the active site of the enzyme so that it productively accommodates β lactams such as the third-generation cephalosporins, ceftazidime and cefotaxime, against which TEM-1 and TEM-2 show no activity. The mutations, of course, occurred at random, but with the genes so widespread, especially in clinical settings, it was inevitable that, at some time and place, mutation and selection would coincide to amplify the new resistance genes.

In the last few years the evolution of these genes has proceeded one stage further. To counter the growing problem of β-lactamase-mediated resistance to β-lactam antibiotics, β-lactamase inhibitors, such as the β-lactam clavulanic acid, were developed. When these are co-administered with a therapeutic β lactam, antibiotic efficacy can be restored. Inhibitor-resistant forms of TEM have now evolved, again by simple point mutation (Amyes, 1997). The speed of evolution of the TEM gene to generate a plethora of extended-spectrum and inhibitor-resistant forms is staggering, but it reflects the large gene pool that exists, thanks largely to horizontal spread of the genes by transposon-plasmid vectors.

Tn*3* is the type element of an extended family of phylogenetically related transposons, with members in both Gram-negative and Gram-positive bacteria (Table 4). Most, but not all, of those characterised encode resistance to one or more antibiotics. One subset of note, the Tn*21*-like transposons (Grinsted *et al.*, 1990), are responsible for much of the spread of several antibiotic-resistance genes, especially among the Enterobacteriaceae. Tn*21* is a moderately large transposon (approximately 20 kb) that confers resistance to streptomycin, spectinomycin, sulphonamides and mercury (de la Cruz and Grinsted, 1982). The reason for the success of the Tn*21*-like elements, with respect to dissemination of antibiotic-resistance genes is that, in addition to them being transposons and largely carried on plasmids, each encodes its own *in vivo* genetic engineering kit termed an integron (Bennett, 1999). That on Tn*21*, In1, is a class 1 integron.

Integrons and Gene Cassettes

An integron comprises a gene, *int*, for a site-specific re-combination enzyme called integrase and an adjacent site, *attI* (functionally equivalent to the λ integration site), at which resistance (and other) genes can be inserted by site-specific re-combination mediated by the integrase (Fig. 4). An integron may or may not be associated with other genes. If not, then the integration site *attI* is vacant. Genes are inserted into *attI*, or deleted from the array by site-specific re-combination (Fig. 4). The insertion/deletion units are called gene cassettes because, in general, each carries a single gene, together with a short sequence called a 59-base element, which is essential for site-specific

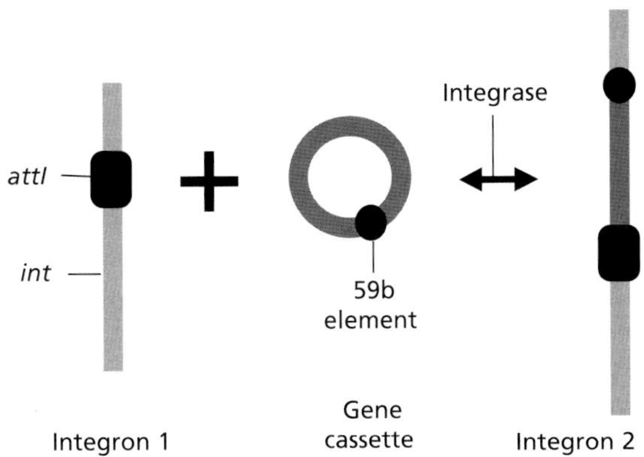

Fig. 4 Bacterial integrons.

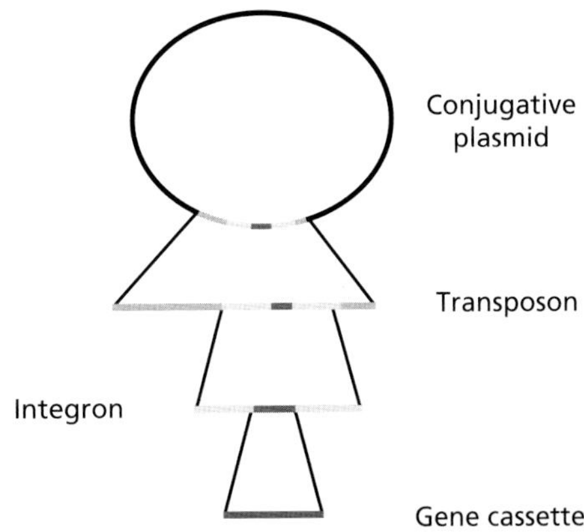

Fig. 5 Anatomy of a resistance plasmid.

recombination. Accordingly, gene cassettes rarely exceed 1 kb in size. In linear array on Tn*21* the integron In1 has the form

int attI aadA 59b sul

where *aadA* encodes an aminoglycoside adenylyltransferase conferring resistance to streptomycin and spectinomycin, *59b* is a 59-base element and *sul* encodes a sulphonamide-resistant dihydropteroate synthetase which was once encoded on a gene cassette that has subsequently degenerated so that it is no longer mobile. Class 1 integrons all have the *sul* gene, but may or may not have *aadA*, which is part of a functional gene cassette that can, therefore, be excised by integrase. When this happens integrase mediates a site-specific recombination between *attI* on one side of *aadA* and the *59b* on the other, releasing the gene cassette as a circular dsDNA structure carrying only *aadA* and a *59b* element (**Fig. 4**). The cassette is then free to reinsert into the vacated *attI* or into another *attI* site on another integron. Insertion is not only site-specific but also orientation-specific. This is necessary because genes on gene cassettes generally lack promoters. Expression of these is driven from a promoter(s), at the promoter-distal end of *int*, oriented so as to drive transcription through *attI* into genes that have been acquired by the integron. Resistance gene arrays can be assembled and dismantled by insertion and deletion of different gene cassettes, respectively. Arrays that persist are those that confer a useful set of phenotypic properties and are selected by environmental conditions. Arrays that confer no advantage will not rise to prominence.

Most gene cassettes accommodate a single gene, although a few have clearly arisen from fusions of two cassettes as a consequence of deletion of the intervening *59b* element. Approximately 40 different gene cassettes have been isolated and sequenced (Recchia and Hall, 1995). The majority carry antibiotic-resistance genes, conferring resistance to aminoglycosides, β lactams, chloramphenicol or trimethoprim, but a few confer resistance to disinfectants used in hospitals and one or two are cryptic, i.e. the function of the genes on the cassettes are unknown. With the variety of gene cassettes known, and undoubtedly more to be discovered, the integron system is very flexible. With its transposon-plasmid locations it is well suited to disseminate antibiotic-resistance genes. Although the great majority of cassettes described so far encode either antibiotic or disinfectant resistance, there is no compelling reason to believe that cassette genes are restricted to these types. Rather, like transposons and plasmids, where resistance types also predominate, the bias is likely to reflect our intense interest in matters that affect our health. Most, if not all, of the integrons described to date have been found on transposons and plasmids taken from bacteria obtained from clinical samples.

Between them, gene cassettes, integrons, transposons and plasmids provide a formidable set of genetic tools for re-fashioning DNA sequences (**Fig. 5**), whereby genes can be moved about from one replicon to another and redesigned replicons can be moved from one bacterial cell to another.

Genomic Islands

In the last few years the genomes of a number of bacterial species have been fully sequenced. Analyses of these sequences indicate that they are not homogeneous. Rather, each consists of a homogeneous core sequence, reflected in a homogeneous G+C content. The core specifically accommodates housekeeping genes, i.e. those essential for growth and reproduction, and these display common codon usage, indicative of co-evolution. Punctuating the core sequence are sequences with different G+C contents, indicating different evolutionary descents, which have probably been acquired by horizontal gene transfer. Among the genes in human pathogens that have been acquired in this way are those essential for virulence. Such sequences have been called pathogenicity islands (PAIs) and they are a subset of what can generally be called genomic islands (Hacker and Kaper, 2000). The origins of some PAIs are more evident than others, in that they are clearly lysogenic bacteriophages (which may or may not still be active), such as those that encode the shiga toxin of enterobacteria and the neurotoxins of *Clostridium botulinum* (Cheetham and Katz, 1995), or transposable elements.

The existence of PAIs in bacterial genomes was first recorded for human pathogens among the Enterobacteriaceae, but examples in animal and plant pathogens have now also been identified (Hacker and Kaper, 2000). PAIs often themselves appear to be sequence mosaics that have been assembled in multiple steps, as judged by heterogeneity of G + C content, and carry genes encoding one or more virulence factors, such as adherence factors, invasion factors, iron uptake systems, protein secretion systems and toxins (Hacker and Kaper, 2000). They tend to occupy relatively large genomic regions of 10–200 kb, are often flanked by short direct sequence repeats, indicative of acquisition by transposition or site-specific recombination mechanisms, and are frequently found next to tRNA genes. PAIs also carry genes encoding functions related to transposases, integrases and IS elements. In the majority of cases it is not known if these genes are still active, but it is presumed the products were in some way instrumental to DNA transfer. A few PAIs have been shown to move from one integration site to another and some may be horizontally transmitted by bacteriophages, e.g. those of *Staph. aureus* (Lindsay *et al.*, 1998) and *Vibrio cholera* (Karaolis *et al.*, 1999). Possession of particular PAIs is what distinguishes uropathogenic (UPEC), enteropathogenic (EPEC) and enterohaemorrhagic (EHEC) *E. coli* from non-virulent strains of the same microbe (Hacker and Kaper, 2000).

PAIs are sub-types of genomic islands, distinguished by carriage of one or more genes encoding pathogenicity factors. More generally, genomic islands have been found to encode a diversity of functions, including xenobiotic catabolism, nitrogen fixation and antibiotic resistance (Hacker and Kaper, 2000). One important example of the last mentioned is methicillin resistance in methicillin-resistant *Staph. aureus* (MRSA) strains that are endemic in many hospitals throughout the world. These strains possess genomic islands of 50–60 kb that include *mecA*, a gene encoding a PBP with extremely low affinity for most β lactams, including methicillin. Such strains are resistant to most, if not all, β lactams of clinical importance because the new PBP, PBP2′, can assume the essential functions of the normal PBP complement of *Staph. aureus*. Where the *mecA* genomic island originates and how it was acquired by *Staph. aureus* is not known (Archer and Niemeyer, 1994). Initially the spread of MRSA was essentially clonal, but the genomic island containing *mecA* has now been disseminated among a number of strains of *Staph. aureus* and even other *Staphylococcus* species.

References

Achtman M (1975) Mating aggregates in *Escherichia coli* conjugation. *J. Bacteriol.* 123: 505–515.

Albiger B, Comanducci A, Dodd HM *et al.* (2000) PUB2380: Characterization of a ColD-like resistance plasmid. *Plasmid* 44: 54–65.

Amyes SG (1997) Antibiotic resistance. Resistance mediated by inhibitor-resistant and extended-spectrum TEM and SHV beta-lactamases. *J. Med. Microbiol.* 46: 454–457.

Archer GL, Niemeyer DM (1994) Origin and evolution of DNA associated with resistance to methicillin in staphylococci. *Trends Microbiol.* 2: 343–347.

Ausubel FM, Brent R, Kingston RE *et al.* (eds) (1990) *Current Protocols in Molecular Biology*, Vol. 1. New York: John Wiley.

Avery OT, MacLeod CM, McCarty M (1944) Studies on the chemical nature of the substance inducing transformation of pneumococcal types. Induction of transformation by a deoxyribonucleic acid fraction isolated from *Pneumococcus* Type III. *J. Exp. Med.* 79: 137–158.

Barany F, Kahn ME (1985) Comparison of transformation mechanisms of *Haemophilus parainfluenzae* and *Haemophilus influenzae*. *J. Bacteriol.* 161: 72–79.

Behnke D (1981) Plasmid transformation of *Streptococcus sanguis* (Challis) occurs by circular and linear molecules. *Mol. Gen. Genet.* 181: 490–497.

Bender J, Kleckner N (1986) Genetic evidence that Tn10 transposes by a nonreplicative mechanism. *Cell* 45: 801–815.

Bennett PM (1999) Integrons and gene cassettes: a genetic construction kit for bacteria. *J. Antimicrob. Chemother.* 43: 1–4.

Bennett PM (2000) Transposable elements. In: Lederberg J (ed.) *Encyclopaedia of Microbiology*, 2nd edn. San Diego, CA: Academic Press,Vol. 4. pp. 704–724.

Bennett PM, Howe TGB (1998) Bacterial and bacteriophage genetics. In: Collier L, Balows A, Sussman M (eds) *Topley & Wilson's Microbiology and Microbial Infections*, 9th edn. *Systematic Bacteriology*. London: Arnold, Vol. 2. pp. 231–294.

Bennett P, Heritage J, Perry A, Harvey J, Zagaglia C (1986) Transposition and plasmid evolution: variations on a theme. In: Levy SB, Novick RP (eds) *Banbury Report 24: Antibiotic Resistance Genes: Ecology, Transfer and Expression*. Cold Spring Harbor, NY: Cold Spring Harbor Laboratory, pp. 355–366.

Berg CM, Berg DE, Groisman EA (1989) Transposable elements and the genetic engineering of bacteria. In: Berg DE, Howe MM (eds) *Mobile DNA*. Washington, DC: ASM Press, pp. 879–925.

Berg DE, Howe MM (1989) *Mobile DNA*. Washington, DC: ASM Press.

Bootsma HJ, van Dijk H, Vauterin P, Verhoef J, Mooi FR (2000) Genesis of BRO beta-lactamase-producing *Moraxella catarrhalis*: evidence for transformation-mediated horizontal transfer. *Mol. Microbiol.* 36: 93–104.

Boyd AC, Archer JA, Sherratt DJ (1989) Characterization of the ColE1 mobilization region and its protein products. *Mol. Gen. Genet.* 217: 488–498.

Bush K, Jacoby GA, Medeiros AA (1995) A functional classification scheme for β-lactamases and its correlation with molecular structure. *Antimicrob. Agents Chemother.* 39: 1211–1233.

Cabezón E, Lanka E, de la Cruz F (1994) Requirements for mobilization of plasmids RSF1010 and ColE1 by the IncW plasmid R388: trwB and RP4 traG are interchangeable. *J. Bacteriol.* 176: 4455–4458.

Cabezón E, Sastre J, de la Cruz F (1997) Genetic evidence of a coupling role for the TraG protein family in bacterial conjugation. *Mol. Gen. Genet.* 254: 400–406.

Campbell A (1996) Bacteriophages. In: Neidhardt FC (ed.) *Escherichia coli and Salmonella: Cellular and Molecular Biology*, 2nd edn. Washington, DC: ASM Press, pp. 2325–2338.

Chandler M, Clerget M, Caro L (1981) IS1 promoted events associated with drug resistance plasmids. *Cold Spring Harbor Symp. Quant. Biol.* 45: 157–165.

Cheetham BF, Katz ME (1995) A role for bacteriophages in the evolution and transfer of bacterial virulence determinants. *Mol. Microbiol.* 18: 201–208.

Clewell DB (1993) Bacterial sex pheromone-induced plasmid transfer. *Cell* 73: 9–12.

Cole S, Saint-Girons I (1999) Bacterial genomes – all shapes and sizes. In: Charlebois RL (ed.) *Organization of the Prokaryotic Genome*. Washington, DC: ASM Press, pp. 35–62.

Courvalin P, Goussard S, Grillot-Courvalin C (1998) Gene transfer from bacteria to mammalian cells. In: Syvanen M, Kado CI (eds) *Horizontal Gene Transfer*. London: Chapman & Hall, pp. 107–117.

Datta N, Kontomichalou P (1965) Penicillinase synthesis controlled by infectious R factors in enterobacteriaceae. *Nature* 208: 239–242.

Davison J (1999) Genetic exchange between bacteria in the environment. *Plasmid* 42: 73–91.

de la Cruz F, Grinsted J (1982) Genetic and molecular characterization of Tn21, a multiple resistance transposon from R100.1. *J. Bacteriol.* 151: 222–228.

Dowson CG, Coffey TJ, Spratt BG (1994) Origin and molecular epidemiology of penicillin-binding-protein-mediated resistance to β-lactam antibiotics. *Trends Microbiol.* 2: 361–366.

Du Bois SK, Marriott MS, Amyes SG (1995) TEM- and SHV-derived extende-spectrum beta-lactamases: relationship between selection, structure and function. *J. Antimicrob. Chemother.* 35: 7–22.

Dubnau D (1991) The regulation of genetic competence in *Bacillus subtilis*. *Mol. Microbiol.* 5: 11–18.

Dubnau D (1999) DNA uptake in bacteria. *Annu. Rev. Microbiol.* 53: 217–244.

Dunny GM, Brown BL, Clewell DB (1978) Induced cell aggregation and mating in *Streptococcus faecalis*: evidence for a bacterial sex pheromone. *Proc. Natl Acad. Sci. USA* 75: 3479–3483.

Freeman VJ (1951) Studies on the virulence of bacteriophage-infected strains of *Corynebacterium diphtheria*. *J. Bacteriol.* 61: 675–678.

Frost LS, Ippen-Ihler K, Skurray RA (1994) Analysis of the sequence and gene products of the transfer region of the F sex factor. *Microbiol. Rev.* 58: 162–210.

Griffith F (1928) Significance of pneumococcal types. *J. Hygiene* 27: 113–159.

Grinsted J, Bennett PM (1988) *Plasmid Technology*, 2nd edn. London: Academic Press.

Grinsted J, de la Cruz F, Schmitt R (1990) The Tn21 subgroup of bacterial transposable elements. *Plasmid* 24: 163–189.

Gromkova R, Goodgal S (1979) Transformation by plasmid and chromosome DNAs in *Haemophilus parainfluenzae*. *Biochem. Biophys. Res. Commun.* 88: 1428–1434.

Guzman LM, Espinosa M (1997) The mobilization protein, MobM, of the streptococcal plasmid pMV158 specifically cleaves supercoiled DNA at the plasmid *oriT*. *J. Mol. Biol.* 266: 688–702.

Hacker J, Kaper JB (2000) Pathogenicity islands and the evolution of microbes. *Annu. Rev. Microbiol.* 54: 641–679.

Hedges RW, Jacob A (1974) Transposition of ampicillin resistance from RP4 to other replicons. *Mol. Gen. Genet.* 132: 31–40.

Heffron F, Bedinger P, Champoux JJ, Falkow S (1977) Deletions affecting the transposition of an antibiotic resistance gene. *Proc. Natl Acad. Sci. USA* 74: 702–706.

Herrero M, de Lorenzo V, Timmis KN (1990) Transposon vectors containing non-antibiotic resistance selection markers for cloning and stable chromosome insertion of foreign genes in Gram-negative bacteria. *J. Bacteriol.* 172: 6557–6567.

Kahn ME, Barany F, Smith HO (1983) Transformasomes: specialised membraneous structures which protect DNA during *Haemophilus* transformation. *Proc. Natl Acad. Sci. USA* 80: 6927–6931.

Karaolis DKR, Somara S, Maneval DR Jr, Johnson JA, Kaper JB (1999) A bacteriophage encoding a pathogenicity island, a type-IV pilus and a phage receptor in cholera bacteria. *Nature* 399: 375–379.

Kennedy AK, Guhathakurta A, Kleckner N, Haniford DB (1998) Tn10 transposition via a DNA hairpin intermediate. *Cell* 95: 125–134.

Kleckner N, Chalmers RM, Kwon D, Sakai J, Bolland S (1996) Tn10 and IS10 transposition and chromosome rearrangements: mechanism and regulation *in vivo and in vitro*. *Curr. Top. Microbiol. Immunol.* 204: 49–82.

Kowalczykowski SC, Dixon DA, Eggleston AK, Lauder SD, Rehrauer WM (1994) Biochemistry of homologous recombination in *Escherichia coli*. *Microbiol. Rev.* 58: 401–465.

Kroll JS, Wilks KE, Farrant JL, Langford PR (1998) Natural genetic exchange between *Haemophilus* and *Neisseria*: intergeneric transfer of chromosomal genes between major human pathogens. *Proc. Natl Acad. Sci. USA* 95: 12381–12385.

Lacey RW (1973) Genetic basis, epidemiology, and future significance of antibiotic resistance in *Staphylococcus aureus*: a review. *J. Clin. Pathol.* 26: 899–913.

Lacey RW (1984) Antibiotic resistance in *Staphylococcus aureus* and streptococci. *Br. Med. Bull.* 40: 77–83.

Lederberg J (1947) Gene recombination and linked segregations in *Escherichia coli*. *Genetics* 32: 505–525.

Lederberg J, Tatum EL (1946) Novel genotypes in mixed cultures of biochemical mutants of bacteria.. *Cold Spring Harbor Symposia on Quantitative Biology* 11: 113–114.

Leduc M, Kasra R, van Heijenoort J (1982) Induction and control of the autolytic system of *Escherichia coli*. *J. Bacteriol.* 152: 26–34.

Lindsay JA, Ruzin A, Ross HF, Kurepina N, Novick RP (1998) The gene for toxic shock toxin is carried by a family of mobile pathogenicity islands in *Staphylococcus aureus*. *Mol. Microbiol.* 29: 527–543.

Lorenz MG, Wackernagel W (1994) Bacterial gene transfer by natural genetic transformation in the environment. *Microbiol. Rev.* 58: 563–602.

Lorenzo V de, Herrero M, Jakubzik U, Timmis KN (1990) Mini-Tn5 transposon derivatives for insertion mutagenesis, promoter probing, and chromosomal insertion of cloned DNA in Gram-negative bacteria. *J. Bacteriol.* 172: 6568–6572.

Madigan MT, Martinko JM, Parker J (1997) *Brock: Biology of Microorganism*, 8th edn. Englewood Cliffs, NJ: Prentice Hall.

Mahillon J, Chandler M (1998) Insertion sequences. *Microbiol. Mol. Biol. Rev.* 62: 725–774.

Maniatis T, Fritsch EF, Sambrook J (1982) *Molecular Cloning. A Laboratory Manual.* Cold Spring Harbor, New York: Cold Spring Harbor Laboratory.

Masters M (1996) Generalised transduction. In: Neidhardt FC (ed.) *Escherichia coli and Salmonella: Cellular and Molecular Biology*, 2nd edn. Washington, DC: ASM Press, pp. 2421–2441.

Mayer LW (1988) Use of plasmid profiles in epidemiologic surveillance of disease outbreaks and in tracing the transmission of antibiotic resistance. *Clin. Microbiol. Rev.* 1: 228–243.

Merlin C, Mahillon J, Nešvera J, Toussaint A (2000) Gene recruiters and transporters: the modular structure of bacterial mobile elements. In: Thomas CM (ed.), *The Horizontal Gene Pool: Bacterial Plasmids and Gene Spread.* Amsterdam: Harwood Academic Publishers, pp. 363–409.

Miller RV (1998) Bacterial gene swapping in nature. *Sci. Am.* 278: 66–71.

Murphy JR, Pappenheimer AM Jr, Borms ST de (1974) Synthesis of diphtheria *tox*-gene products in *Escherichia coli* extracts. *Proc. Natl Acad. Sci. USA* 71: 11–15.

Nomura M, Morgan EA (1977) Genetics of bacterial ribosomes. *Annu. Rev. Genet.* 11: 297–347.

O'Brien TF, Mayer KH, Hopkins JD (1986) Global surveillance of the deployment of antibiotic resistance genes and plasmids. In: Levy SB, Novick RP (eds) Banbury Report 24: *Antibiotic Resistance Genes: Ecology, Transfer, and Expression.* Cold Spring Harbor, NY: Cold Spring Harbor Laboratory, pp. 3–15.

Pato ML (1989) Bacteriophage Mu. In: Berg DE, Howe MM (eds) *Mobile DNA.* Washington, DC: ASM Press, pp. 23–52.

Recchia GD, Hall RM (1995) Gene cassettes: a new class of mobile element. *Microbiology* 141: 3015–3027.

Rice LB (1998) Tn916 family conjugative transposons and dissemination of antimicrobial determinants. *Antimicrob. Agents Chemother.* 42: 1871–1877.

Rudin L, Sjöström JE, Lindberg M, Philipson L (1974) Factors affecting competence for transformation in *Staphylococcus aureus*. *J. Bacteriol.* 118: 155–164.

Salyers AA, Shoemaker NB, Stevens AM, Li L-Y (1995) Conjugative transposons: an unusual and diverse set of integrated gene transfer elements. *Microbiol. Rev.* 59: 579–590.

Salyers AA, Cooper AJ, Shoemaker NB (1998) Lateral broad host range gene transfer in nature: how and how much? In: Syvanen M, Kado CI (eds) *Horizontal Gene Transfer.* London: Chapman & Hall, pp. 40–52.

Saunders JR, Hart CA, Saunders VA (1986) Plasmid-mediated resistance to β-lactam antibiotics in Gram-negative bacteria: the role of *in vivo* recyclization reactions in plasmid evolution. *J. Antimicrob. Chemother.* 18(Suppl C): 57–66.

Shah GR, Caufield PW (1993) Enhanced transformation of *Streptococcus mutans* by modifications in culture conditions. *Ann. Biochem.* 214: 343–346.

Smeets LC, Bijlsma JJ, Boomkens SY, Vandenbroucke-Grauls CM, Kusters JG (2000) *comH*, a novel gene essential for natural transformation of *Helicobacter pylori*. *J. Bacteriol.* 182: 3948–3954.

Spratt BG (1983) Penicillin-binding proteins and the future of beta-lactam antibiotics. *J. Gen. Microbiol.* 129: 1247–1260.

Sternberg N, Hoess K (1983) The molecular genetics of bacteriophage P1. *Annu. Rev. Genet.* 17: 123–154.

Stewart GJ, Carlson CA (1986) The biology of natural transformation. *Annu. Rev. Microbiol.* 40: 211–235.

Swartley JS, Marfin AA, Edupuganti S *et al.* (1997) Capsule switching of *Neisseria meningitidis. Proc. Natl Acad. Sci. USA* 94: 271–276.

Syvanen M, Kado CI (eds) (1998) *Horizontal Gene Transfer.* London: Chapman & Hall.

Talbot HW, Yamamoto DK, Smith MW, Seidler RJ (1980) Antibiotic resistance and its transfer among clinical and non-clinical *Klebsiella* strains in botanical environments. *Appl. Environ. Microbiol.* 39: 97–104.

Thomas CM (ed.) (1989) *Promiscuous Plasmids of Gram-negative Bacteria.* New York, Academic Press.

Thomas CM (ed.) (2000) *The Horizontal Gene Pool: Bacterial Plasmids and Gene Spread.* Amsterdam: Harwood Academic Publishers.

Top EM, Moënne-Loccoz Y, Pembroke T, Thomas CM (2000) Phenotypic traits conferred by plasmids. In: Thomas CM (ed.) *The Horizontal Gene Pool: Bacterial Plasmids and Gene Spread.* Amsterdam: Harwood Academic Publishers, pp. 249–285.

Toussaint A, Résibois A (1983) Phage Mu: transposition as a life-style. In: Shapiro JA (ed.) *Mobile Genetic Elements.* New York: Academic Press, pp. 103–158.

Van der Meer JR, de Vos WM, Harayama S, Zehnder AJB (1992) Molecular mechanisms of genetic adaptation to xenobiotic compounds. *Microbiol. Revs* 56: 677–694.

Waldor MK, Mekalanos JJ (1996) Lysogenic conversion by a filamentous phage encoding cholera toxin. *Science* 272: 1910–1914.

Walker GC (1996) The SOS response of *Escherichia coli.* In: Neidhardt FC (ed.) *Escherichia coli and Salmonella: Cellular and Molecular Biology,* 2nd edn. Washington, DC: ASM Press, pp. 1400–1416.

Wang Y, Taylor DE (1990) Natural transformation in *Campylobacter* species. *J. Bacteriol.* 172: 949–955.

Weisberg RA (1996) Specialized transduction. In: Neidhardt FC (ed.) *Escherichia coli and Salmonella: Cellular and Molecular Biology,* 2nd edn. Washington, DC: ASM Press, pp. 2442–2448.

Wilkins BM (1995) Gene transfer by bacterial conjugation: diversity of systems and functional specializations. In: Baumberg S, Young JPW, Wellington EMH, Saunders JR (eds) *Society for General Microbiology Symposium 52, Population Genetics of Bacteria.* Cambridge: Cambridge University Press, pp. 59–88.

Wilkins BM, Lanka E (1993) DNA processing and replication during plasmid transfer between Gram-negative bacteria. In: Clewell D (ed.) *Bacterial Conjugation.* New York: Plenum Press, pp. 105–136.

Wolfgang M, van Putten JP, Hayes SF, Koomey M (1999) The *comP* locus of *Neisseria gonorrhoeae* encodes a type IV prepilin that is dispensible for pilus biogenesis but essential for natural transformation. *Mol. Microbiol.* 31: 1345–1357.

Zechner EL, de la Cruz F, Eisenbrandt R *et al.* (2000) Conjugative-DNA transfer processes. In: Thomas CM (ed.) *The Horizontal Gene Pool: Bacterial Plasmids and Gene Spread.* Amsterdam: Harwood Academic Publishers, pp. 87–174.

19

Co-ordination of Virulence

Charles J. Dorman

Moyne Institute of Preventive Medicine, University of Dublin, Trinity College, Dublin, Republic of Ireland

The composition of a bacterium at any point in space or time is determined by the expression of the genetic material within it. In adapting to a novel set of environmental circumstances, the bacterium initiates, terminates or simply modulates the expression of a subset of its genes. The information required to ensure an appropriate resetting of the gene expression programme within the bacterium is supplied by the environment and the organism acquires this information through a wide variety of sensing mechanisms. Changes in gene expression involve variations in gene transcription, mRNA translation and post-translational regulatory events. Studies in which the total protein profiles of cells grown under different conditions were compared by two-dimensional gel electrophoresis, or where reporter-gene fusion methods were used to measure gene expression, have demonstrated that very large numbers of genes are affected simultaneously when external conditions change. It has come to be appreciated that such wholesale changes in expression are co-ordinated in the cell by a combination of local and global gene regulatory mechanisms (Dorman, 1994). Much of the pioneering work in this field has been carried out with *Escherichia coli* K-12 or with non-pathogenic strains of *Salmonella typhimurium* (that is *S. enterica* serovar Typhimurium). Subsequently, it was recognised that virulence genes in pathogenic bacteria are also regulated and that their regulatory mechanisms are frequently similar, and sometimes identical, to those in use in these two intensely studied bacteria (Miller *et al.*, 1989a; Mekalanos, 1992). Virulence genes are rarely expressed constitutively. Instead, the expression of a virulence gene is reserved for sets of environmental circumstances which indicate to the bacterium that it is probably in contact with its host and is also likely to be at the site in or on the host where the product of this particular virulence gene is most required. Many pathogens express not just one but an array of virulence genes and this expression is co-ordinated. The products of the genes may be required at successive stages of the infection and subsets of the genes may have to be expressed simultaneously, and the products may be needed in different quantities. The bacteria have evolved sophisticated methods of ensuring that the expression programme operates correctly and to their benefit. This chapter will review current understanding of how this is achieved, with well-studied examples of virulence gene regulation to illustrate the key principles.

 doi:10.1006/bkmm.2001.0019

Bordetella pertussis and the BvgA/BvgS Regulatory Proteins

Bordetella pertussis, the pathogen responsible for whooping cough, expresses several virulence factors (see Chapter 75 and has been an important subject for study by those interested in how virulence gene expression is co-ordinated in response to environmental signals (Scarlato *et al.*, 1993a). During respiratory tract infection, *B. pertussis* attaches to ciliated epithelia and this adherence is associated with expression of fimbriae, filamentous haemagglutinin and an outer membrane protein, pertactin (Charles *et al.*, 1989; Relman *et al.*, 1989; Willems *et al.*, 1990). Expression of pertussis toxin and adenylate cyclase by the bacterium results in elevated levels of cyclic AMP (cAMP) in host cells, and this effectively inhibits the ability of macrophages and neutrophils to deal with the infecting organism. The ciliated epithelia are themselves killed by the bacterially encoded tracheal cytotoxin (Weiss and Hewlett, 1986). Experiments *in vitro* have shown that expression of some virulence genes is inhibited when the bacteria are grown at low temperature or in the presence of magnesium sulphate or nicotinic acid (Lacey, 1960; Weiss and Hewlett, 1986). This phenomenon is termed phenotypic variation. Virulent strains are also capable of giving rise to avirulent variants, a process which is reversible and is called phase variation. Genetic experiments have revealed that phenotypic and phase variation both act through the same genetic locus, *bvg*. This regulates some genes positively, while repressing the expression of others (Knapp and Mekalanos, 1988; Miller *et al.*, 1989a).

The *bvg* locus has been cloned and sequenced and the nature of the products it encodes has been deduced. The locus is composed of two genes, *bvgAS*, and it is arranged as an operon and is located upstream of and transcribed in the opposite direction to the *fhaB* gene, which encodes filamentous haemagglutinin. The BvgA and BvgS proteins are members of a large family of environmental signal transducers known as 'two-component systems' (Stock *et al.*, 1989; Uhl and Miller, 1995). These protein partnerships typically consist of a sensor polypeptide that receives one or more environmental signals and then transmits this information to a response regulator protein, which in turn elicits a biological response. The communication between the two proteins involves a phosphotransfer reaction (Boucher *et al.*, 1994; Uhl and Miller, 1994).

BvgS is a membrane-associated protein of 1238 amino acids with two transmembrane domains in its N-terminal domain. These flank a periplasmic domain of 514 amino acids. The remainder of the protein lies in the cytoplasm, and contains an ATP-binding domain and a histidine residue (number 729) on which the protein becomes phosphorylated. BvgS also possesses a domain with sequence similarity to the N-terminus of BvgA. This so-called receiver domain may function to modulate the strength of the signal transmitted by the activated form of BvgS to its partner (Stibitz and Yang, 1991) (**Fig. 1**). BvgA is a 209-amino-acid protein divided into two domains. The N-terminus receives the signal from BvgS through phosphorylation on an aspartate residue (number 54) while the C-terminus binds DNA through a helix–turn–helix motif (Aricò *et al.*, 1989; Stibitz and Yang, 1991). This DNA binding occurs at the operator sites (consensus sequence: 5′-TTTCCTA-3′) found upstream of BvgA-regulated genes and is potentiated by the phosphorylation of BvgA (Roy and Falkow, 1991; Scarlato *et al.*, 1990).

Transcription of *bvgAS* is under the positive control of BvgA. A weak promoter known as P2 maintains a basal level of *bvg* transcription, and a stronger promoter (P1) is responsible for high-level transcription. These promoters never operate simultaneously, indicating a complex interplay between them. A third promoter, P3, has been identified as contributing to *bvg* transcription and an antisense RNA transcribed from promoter P4 may inhibit *bvg* gene expression by interaction with the mRNA (Roy *et al.*, 1990; Scarlato *et al.*, 1990). By these means, the cell maintains a low level of the *bvgAS* gene products until environmental signals induce amplified expression.

Other promoters under the positive control of BvgAS can be distinguished by the rapidity with which they respond to environmental signals. In addition to the *bvg* promoters, those of *fha* (filamentous haemagglutinin) and *fim* (fimbriae) are activated within 10 minutes. On the other hand, the promoters of the *cya* (adenylate cyclase) and *ptx* (pertussis toxin) genes require more than 5 hours to become fully activated (Scarlato *et al.*, 1991). It is possible that factors additional to BvgA may be required to activate the slower promoters, or that very high intracellular levels of BvgA are required and that these take time to accumulate (Huh and Weiss, 1991; Scarlato *et al.*, 1993a, 1993b). This two-stage activation can be rationalised by a model which regards the products of the *fha* and *fim* genes as required at the outset of the interaction of *B. pertussis* with its host, and the products of the *cya* and *ptx* genes as required at later stages (Scarlato *et al.*, 1991).

BvgAS-repressed genes have been detected in *B. pertussis* and in the closely related species

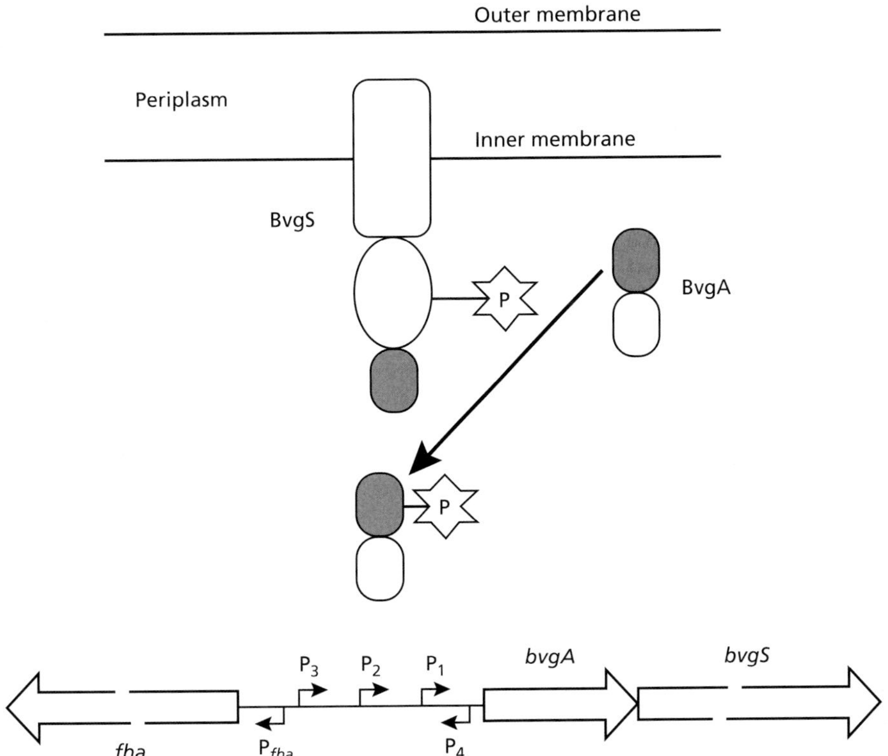

Fig. 1 The BvgS/BvgA two-component regulatory system of *B. pertussis*. The BvgS sensor protein is shown located in the inner membrane. It is phosphorylated in the cytoplasmic domain and the phosphate group is represented by the star labelled 'P'. The shaded area in the cytoplasmic domain is related in its amino acid sequence to the N-terminal domain of the partner protein, BvgA, which is also shown shaded. This is the site of phosphorylation within BvgA. Once activated by phosphotransfer, BvgA can activate the promoters of the Bvg regulon. This includes the divergently transcribed *bvgAS* and *fha* genes shown here. The three promoters involved in *bvg* gene expression are numbered P_1, P_2 and P_3. Promoter P_4 is believed to drive expression of an antisense RNA; it is oriented in the same direction as the promoter of the *fha* gene, P_{fha}.

B. bronchiseptica (the cause of atrophic rhinitis in pigs and kennel cough in dogs); the genes of *B. bronchiseptica* are the better characterised. Here siderophore production has been found to be regulated negatively by the *bvg* locus (Uhl and Miller, 1995) and transcription of two motility loci, *flaA* and *frlAB*, is repressed by BvgAS (Akerley and Miller, 1993).

Expression of *fim* gene-encoded fimbriae in *B. pertussis* is regulated positively by BvgAS but is also subject to individual regulation by a phase variation mechanism. This control mechanism operates through small insertions or deletions within a 15-bp tract of C residues located within the promoter region (Willems *et al.*, 1990). Thus, the *fim* genes of *B. pertussis* belong to the Bvg regulon but also experience local control through a 'private' regulatory mechanism.

Salmonella typhimurium

The enteric pathogen *S. typhimurium* has been the subject of intensive investigation and the source of many important insights into the co-ordination of virulence gene expression. One reason is that it is very amenable to genetic analysis and another that many methodologies have been devised to allow it to be studied while engaged in infection processes, either *in vitro* or *in vivo*. Another benefit is that the organism has been studied as a laboratory subject for decades by investigators with no immediate interest in its contributions to disease but whose discoveries have, nevertheless, informed medical and veterinary problems.

Variations in gene expression can be studied in *S. typhimurium* by analysing total protein expression profiles by two-dimensional gel electrophoresis, by creating fusions at random to easily assayable reporters of changes in transcription, or by comparing the mRNA profiles of cultures grown under different conditions. Genetic requirements for survival in the host can be tested by mutagenesis. The mutations can be made in genes chosen because their products are thought likely to be needed in the host, or methods can be employed to create mutations at random and then to discover which had the most severe effect on the survival of the bacteria. The signature-tagged transposon mutagenesis method uses transposons to inactivate at random *S. typhimurium* genes, or genes in other pathogens (Hensel *et al.*, 1995). Each transposon is distinguishable from every other by having a short, unique, DNA sequence (its signature) incorporated within it. Bacteria that fail to survive in the host can be detected by comparing the population of organisms before and after infection. A signature present before infection that vanishes once the bacteria enter the host is likely to be associated with a transposon insertion in a gene required for survival. In preliminary trials, this method showed great promise by re-detecting genes already known to be required for survival *in vivo*, as well as detecting many novel genes (Hensel *et al.*, 1995).

In-vivo expression technology (IVET) uses a different approach in an attempt to detect genes required for survival in the host. Here, promoters are cloned at random in front of an essential gene in a mutant deficient in expression of that gene. If the promoter is activated in the host, the essential gene is expressed and these bacteria are enriched in the population. This method at first showed great promise, but the genes emerging from the screens have tended to be housekeeping genes involved in central metabolic pathways or genes whose products at best play an ancillary role in bacterial pathogenicity (Mahan *et al.*, 1993).

Knowledge of the environmental stresses likely to be encountered by *S. typhimurium* during infection has helped in the design of screens for genes which would assist in bacterial adaptation to those conditions. Mutants deficient in responses to oxidative stress, acid stress, heat shock and many other environmental insults have been accumulated over the years by bacterial geneticists working with laboratory strains of *S. typhimurium*, and the mutations have subsequently been tested for effects on virulence in animal and tissue culture studies. These studies have identified both structural and regulatory genes. In each case, the discovery of genes of one class has permitted predictions to be made about the likely involvement of genes of the other. The outcomes of such studies are reviewed briefly below.

PhoP/PhoQ

The PhoQ protein is a histidine protein kinase and the PhoP protein is its response regulator partner. The proteins communicate by phosphotransfer and constitute a two-component regulatory system comparable to the BvgA/BvgS system of *B. pertussis* described above. Expression of approximately 40 genes is controlled by PhoP/PhoQ. The environmental signals transduced through this system include pH and starvation for phosphate, carbon, nitrogen and sulphur. PhoQ is located in the cytoplasmic membrane and is thought to detect the environmental signals here before transmitting them to PhoP. PhoP is a DNA-binding protein that can regulate some genes positively and others negatively (**Fig. 2**).

Null mutations in *phoP* result in a loss of virulence in *S. typhimurium*. Such mutants are compromised in their ability to survive in macrophages, and they display enhanced susceptibility to defensins and low pH. Defensins are peptides of low molecular mass that breach the membranes of their bacterial victims. The simplest interpretation of these findings is that PhoP is required to activate expression of genes that code for products involved in resistance to these environmental threats or to repress expression of genes involved in making the bacterium more susceptible.

PhoP and PhoQ are not concerned exclusively with the control of gene expression during infection. These genes are conserved in a wide variety of enteric bacteria, including non-pathogens, and probably play a role in helping bacteria to adapt to changes in the environment which have no direct relevance to infection. This can be illustrated by considering the case of the *phoN* gene, a member of the PhoP regulon which was discovered early on and which encodes a periplasmically located non-specific acid phosphatase (Kier *et al.*, 1977). Mutations in the *phoN* gene have no affect on virulence (Fields *et al.*, 1989), although the gene requires PhoP for induction is response to starvation for phosphate, carbon, nitrogen or sulphur (Groisman and Hefferon, 1995). The *phoN* gene is even more interesting for having been acquired by horizontal transfer, probably plasmid-mediated conjugation, from a bacterium other than *Salmonella* (Groisman *et al.*, 1992). Thus, this bacterium finds it advantageous to co-regulate virulence genes with housekeeping genes, even housekeeping genes acquired relatively recently in genome evolution.

Several genes have been described as regulated by the PhoP/PhoQ two-component system and as contributing to bacterial virulence. An early example was the *pagC* gene, which codes for an outer membrane protein required for survival of *Salmonella* in macrophages (Miller *et al.*, 1989b). This protein resembles others that contribute to bacterial virulence, such as the Ail protein of *Yersinia enterocolitica*, which confers serum complement resistance (Pulkkinen and Miller, 1991). The *pag* genes are so-called because they are PhoP-activated. PhoP-repressed genes are known by the designation *prg*. An example of this class of PhoP-regulated genes is *prgH*, which contributes to invasion of epithelial cells. In general, its regulation is the reciprocal of the *pagC* gene, although aspects of the growth conditions, such as anaerobiosis, can conspire to upset this neat reciprocal relationship (reviewed by Groisman and Hefferon, 1995). Nevertheless, there is a clear parallel here with the situation in *Bordetella* described above in which the Bvg two-component regulatory system can activate transcription of some genes while repressing that of others.

PhoP/PhoQ is also noteworthy for regulating yet another two-component system, which provides an interesting example of two such systems organised in a regulatory cascade. This second two-component system is PmrA/PmrB and is concerned with regulating resistance to cationic peptides and, indirectly, with the modification of lipopolysaccharide. The *pmrApmrB* genes are co-transcribed with *pagB*, a PhoP-activated gene coding for a membrane protein which may interact with PmrB in the cytoplasmic membrane to confer peptide resistance (Gunn and Miller, 1996) (**Fig. 2**). The environmental signal detected by the

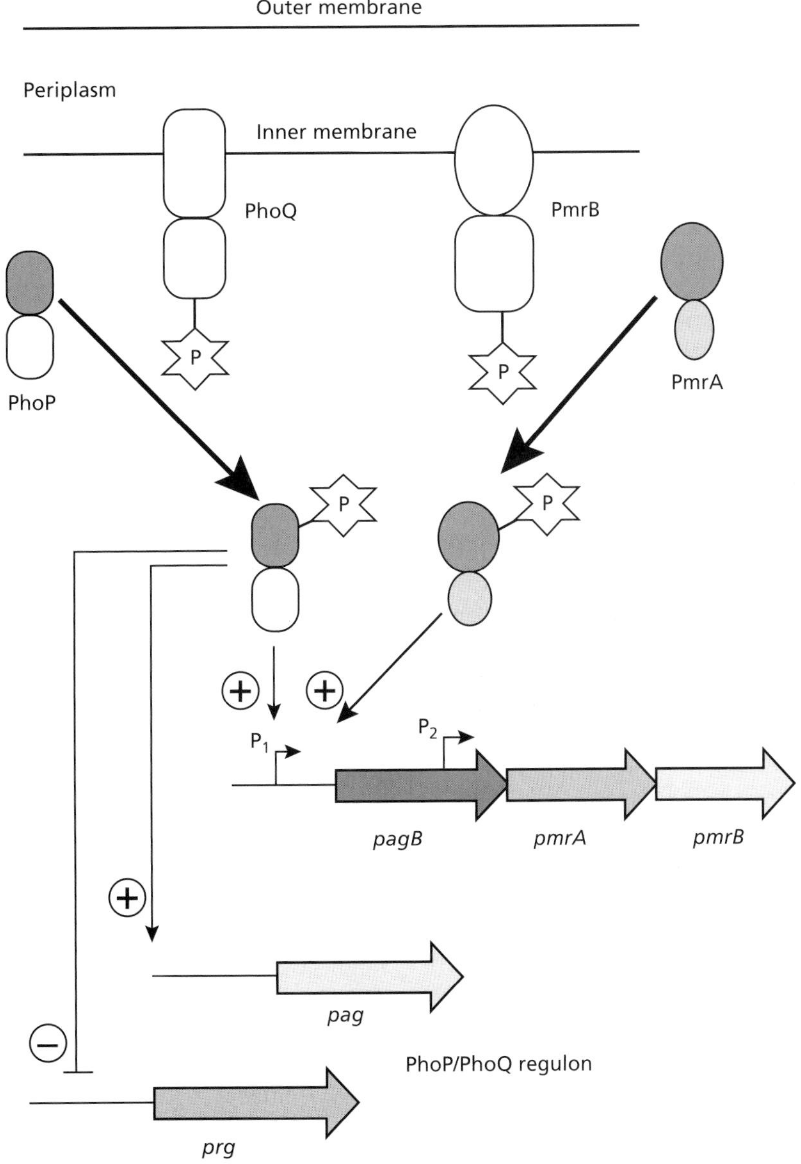

Fig. 2 The complex PhoP/PhoQ regulon of *S. typhimurium*. The PhoQ sensor protein is located in the inner membrane and transfers a phosphate group to the N-terminal domain of its partner protein, PhoP. Activated PhoP regulates negatively transcription of *prg* genes but regulates positively transcription of *pag* genes, one of which, *pagB*, forms an operon with genes coding for another two-component system, PmrA/PrmB. The sensor component, PrmB, is believed to be located in the inner membrane and the activated form of its partner protein, PrmA, positively autoregulates *prmAB* gene expression. The star symbols labelled 'P' represent phosphate groups; + and − indicate positive and negative regulation, respectively.

membrane-located PmrB sensor has not been fully characterised, but includes low pH. Up to seven genes previously identified as PhoP-regulated are now known to be regulated by PmrA, suggesting that this complex cascade may extend the range of environmental signals to which the structural genes respond (Soncini and Groisman, 1996).

OmpR/EnvZ

The OmpR and EnvZ proteins form yet another two-component regulatory system in *S. typhimurium*. Like PhoP/PhoQ, this system is found throughout the enteric bacteria and it regulates genes that code for products required for housekeeping functions, as well as contributing to virulence. The OmpR/EnvZ system is interesting in that it has been implicated in the control of virulence in *Shigella flexneri* and in *S. typhimurium*.

EnvZ is a cytoplasmic membrane-associated protein with a periplasmic domain (composed of most of the N-terminus of the protein), two transmembrane helices and a cytoplasmic domain made up of the C-terminus. This is a histidine protein kinase, which becomes phosphorylated when the bacteria are subjected to osmotic stress. Current models envisage a role for the N-terminus in sensing osmotic pressure in the periplasm and a role for the transmembrane helices in transmitting that signal to the catalytic domain in the cytoplasm. The phosphate group is passed from EnvZ to the N-terminal domain of the partner response regulator protein OmpR, which thus becomes proficient for DNA-binding. The DNA-binding domain of the OmpR protein is in the C-terminus. Like PhoP/PhoQ, the EnvZ/OmpR two-component system controls expression of a regulon of genes and, once again, reciprocal control is a feature (**Fig. 3**).

The best-studied members of the regulon are the *ompC* and *ompF* genes, which code for two of the major outer membrane porins of *Salmonella*, and other enteric bacteria such as *E. coli*. Under conditions of high osmolarity, *ompC* is transcribed and *ompF* is repressed; at low osmolarity the opposite is the case. Reciprocal control is partly a function of the intracellular concentration of the phosphorylated form of OmpR (OmpR-P) and partly a function of the number, quality and location of binding sites for OmpR-P upstream of the *ompC* and *ompF* genes. Two types of sites have been described, one with high and the other with low affinity for OmpR binding. The high-affinity sites are effective in binding the small quantities of OmpR-P found in cells grown at low osmolarity. These sites occur upstream of *ompF* and serve to activate transcription of that gene. Low-affinity sites are found upstream of the *ompC* gene and these must be occupied if *ompC* is to be transcribed. Since this is only likely to occur when OmpR-P is plentiful, *ompC* transcription is confined to growth conditions in which the cell experiences osmotic stress. Repression of *ompF* at high osmolarity appears to involve a low-affinity site, which is only occupied by OmpR-P under those growth conditions. When this site is occupied, a loop forms in the DNA which represses the *ompF* promoter.

The need for OmpC and OmpF porins has been rationalised on the basis that the bacterium needs to admit low-molecular-mass hydrophilic solutes for nutrition, but simultaneously to exclude harmful small molecules, such as bile salts. Since the latter are found in the mammalian gut, which is a high-osmolarity environment, expression of the OmpC protein with a pore diameter of 1.08 nm, rather than OmpF with a diameter of 1.16 nm, permits solute uptake while excluding harmful molecules. In low-osmolarity environments, which by definition have low concentrations of nutrients, it is more advantageous for the bacterium to express the wider-diameter pore as an aid to scavenging. Expression of the correct porins in each environment encountered by the bacterium may be regarded as a crucial component in adaptation. It is, therefore, reasonable to propose that a pathogen which has lost this ability might be at a disadvantage during infection.

This was tested in the case of *S. typhimurium* by comparing a wild-type strain and a derivative with a null mutation in *ompR* for virulence in a mouse model system. Here, the ompR mutant was found to be attenuated strongly, and failed to kill mice when delivered by the oral route (Dorman *et al.*, 1989). Mutants deficient in either the *ompC* or the *ompF* gene were not attenuated, whereas *ompCompF* double mutants showed attenuation in the mouse model infection system, albeit to a lesser extent than the *ompR* null mutant (Chatfield *et al.*, 1991). This suggests that OmpR-dependent genes additional to *ompC* and *ompF* are involved in virulence. What these genes might be remains unknown, but a role has been described for the OmpR/EnvZ system in regulating expression of the *S. typhi* Vi polysaccharide, a capsular antigen thought to be involved in virulence (Pickard *et al.*, 1994).

The Virulence Plasmid-encoded *spv* Operon

The non-typhoid serovars of *Salmonella* possess a high-molecular-mass plasmid that encodes functions required by the bacteria to establish a systemic infection in the host. The structure of this plasmid varies between serovars, but all have a copy of the *spv* operon

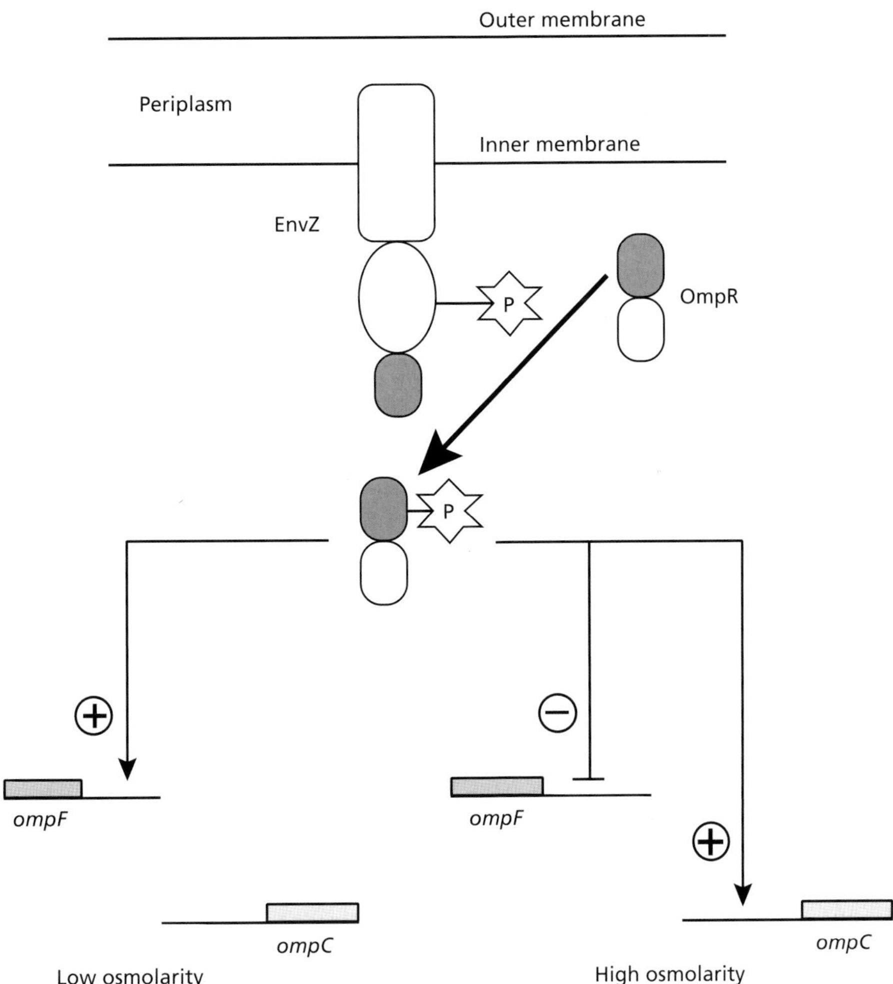

Fig. 3 The OmpR/EnvZ two-component system of *S. typhimurium*. The sensor protein EnvZ is located in the inner membrane where it transfers a phosphate group to the N-terminal domain of the DNA-binding protein OmpR. Phosphorylated OmpR is abundant under high-osmolarity growth conditions where it activates transcription of the *ompC* gene and contributes to the repression of *ompF*. Under low-osmolarity growth conditions, the phosphorylated form of OmpR is in low concentration and only high-affinity binding sites at the *ompF* promoter are likely to be bound; *ompC*, with its low affinity OmpR-binding sites, remains unexpressed. The star symbols labelled 'P' represent phosphate groups; + and − indicate positive and negative regulation, respectively.

in common (reviewed by Guiney *et al.*, 1995). This operon consists of four co-transcribed structural genes, *spvABCD* and a closely linked but independently transcribed gene, *spvR*,which codes for a positive regulator related to the LysR family of DNA binding proteins. The *spv* genes are largely responsible for the virulence phenotype associated with possession of the plasmid, although other plasmid-encoded functions are also important. One of these, the Rck protein, is related to the chromosomally specified PagC protein referred to above and confers high-level serum complement-resistance when cloned into plasmid vectors (Heffernan *et al.*, 1992).

The precise contribution made to virulence by the *spv* gene products remains obscure. Nevertheless, the regulation of their expression has attracted considerable interest. These genes are expressed during the stationary phase of growth in bacteria grown in batch culture. Transcriptional activation of *spvABCD* requires the SpvR protein, and this protein also regulates positively its own gene, *spvR* (Pullinger *et al.*, 1989; Caldwell and Gulig, 1991; Krause *et al.*, 1991, 1992). It has been reported that the *spvR* gene is subject to functional polymorphisms which can affect its potency as a transcription inducer (Taira *et al.*, 1995). This may reflect adaptation by different serovars of *Salmonella* to different hosts. The interaction of SpvR protein with DNA has been studied *in vitro*, and binding has been detected in the promoter regions of *spvR* and *spvA* (Krause *et al.*, 1995; Grob and Guiney, 1996). This is consistent with results from genetic studies with *spvR* mutants in conjunction with reporter gene fusions to *spv* genes or Northern blot analysis of *spv* gene transcripts.

In addition to regulation by the plasmid-encoded SpvR protein, *spv* gene expression is also affected by several factors encoded by genes located on the bacterial chromosome. A central role has been established for the sigma factor RpoS in the positive control of *spv* gene expression (Fang *et al.*, 1992; Norel *et al.*, 1992; Kowarz *et al.*, 1994; Chen *et al.*, 1995). This sigma factor is required at both the *spvR* and the *spvA* promoters, where it helps to activate transcription in association with SpvR (Chen *et al.*, 1995). RpoS exerts wide-ranging effects on the transcriptional profile of the cell and is itself subject to complex regulation at both the transcriptional and post-transcriptional levels (Loewen and Hengge-Aronis, 1994). This subject is developed in more detail below. Other regulators with wide-ranging effects on gene expression, which influence *spv* expression, include the nucleoid-associated protein H-NS, and the cAMP-CRP complex (O'Byrne and Dorman, 1994a; 1994b). Both H-NS and cAMP-CRP negatively affect *spv* expression, but since both contribute to the complex control of *rpoS* gene expression, it is not known whether their effects on *spv* are direct or indirect. Like RpoS, each has also been implicated in controlling virulence in *S. typhimurium* (see below).

RpoS and MviA

The complexities of *rpoS* gene regulation have already been referred to. Expression of this gene is induced on entry into the stationary phase of growth, when starvation stimulates expression at both the transcriptional and post-transcriptional level; it is also induced by osmotic stress through a post-transcriptional mechanism (Lange and Hengge-Aronis, 1994). Mutations in *rpoS* attenuate the virulence properties of *S. typhimurium*, and the mutant is 1000 times less virulent than its parent (Fang *et al.*, 1992). Since loss of the virulence plasmid, with its associated *spv* genes, results in only a 100-fold reduction in virulence, RpoS is thought to be involved in the control of other genes that contribute to *Salmonella* virulence (Fang *et al.*, 1992).

An unlinked gene, *mviA*, influences *rpoS* expression and plays a role in the mouse virulence of *S. typhimurium* (Bearson *et al.*, 1996). The MviA protein is required for *S. typhimurium* to infect susceptible strains (so-called *Ity*^{s} strains) of inbred mice (Benjamin *et al.*, 1991). The *mviA* gene is allelic with the *sprE* and the *rssB* genes of *E. coli* (Muffler *et al.*, 1996; Pratt and Silhavy, 1996), and the product of *mviA* (*sprE*, *rssB*) is a member of the response regulator protein family. At the time of writing, its partner kinase has still to be detected. Mutations in this response regulator result in increased stability of the RpoS protein. Thus, it contributes to proteolytic turnover of the sigma factor and this turnover depends on the ClpX/ClpP protease (Muffler *et al.*, 1996; Pratt and Silhavy, 1996). In relation to mouse virulence, the *mviA* gene modulates the induction of RpoS expression in response to acid shock, an environmental stress which is likely to be highly significant during infection of the host. The acid-shock results in an MviA-dependent stabilisation of RpoS and the increased intracellular concentration of the sigma factor permits the activation of genes coding for acid stress proteins (Bearson *et al.*, 1996).

Additional regulatory inputs have been described. The nucleoid-associated protein H-NS affects expression of RpoS by a post-transcriptional mechanism and guanosine-3′,5′-bispyrophosphate is required for normal function of the *rpoS* promoter (Barth *et al.*, 1995; Lange *et al.*, 1995). Thus, the *rpoS* gene is under complex and multi-factorial control and this complicates analysis of other genes that depend on RpoS for normal expression.

H-NS

At 20 000 copies per cell, the H-NS protein is one of the most abundant nucleoid-associated proteins found in enteric bacteria (Ussery *et al.*, 1994). It is a DNA-binding protein that seems to perform both a structural function, organising prokaryotic chromatin, and a gene-regulatory function in the cell. In most cases where detailed information is available, H-NS seems to control gene expression negatively. H-NS also negatively regulates its own expression, with *hns* transcription active during periods in which DNA replication occurs, and repressed when replication ceases (Free and Dorman, 1995). The other genes whose expression it influences have little in common apart from contributing to the ability of bacteria to adapt to environmental stress. Several of these genes contribute directly to bacterial virulence, including the *spv* genes of *S. typhimurium* described above.

Other evidence indicates that *hns* may play a more general role in controlling expression of the virulence phenotype in *S. typhimurium*. A Tn*phoA* mutation isolated in a mouse-virulent strain of *S. typhimurium* attenuated virulence in a mouse infection model, to cause the bacteria to express phosphatase (the *phoA* gene codes for alkaline phosphatase, so this was not unexpected) and to involve insertional inactivation of *hns* (Harrison *et al.*, 1994). Since H-NS is expressed in the cytoplasm and alkaline phosphatase must be secreted to become active, the generation of an H-NS-PhoA fusion could not be invoked to explain the expression of phosphatase. Instead, the *hns* mutation was found to have resulted in derepression of the *phoN* gene (discussed above as a component of the PhoP/PhoQ regulon) which encodes acid phosphatase. This was confirmed by inactivating *phoN* and thus removing the ability of the bacteria to express phosphatase activity. Furthermore, the *hns* lesion could be transduced to wild-type in some strains without restoring full virulence. This showed that the *hns*::Tn*phoA* mutation has an 'enabling' effect in the bacteria, permitting secondary mutations to occur which were (presumably) directly responsible for attenuation of the virulence phenotype.

Adenylate Cyclase and cAMP-CRP

CRP is a DNA-binding protein of enteric bacteria which binds to a specific sequence upstream of a large number of genes. DNA binding requires CRP to dimerise and to bind cAMP. Intracellular levels of cAMP reflect the activity of adenylate cyclase, the product of the *cya* gene. cAMP-CRP binding usually results in activation of transcription, but, in some cases, repression has been reported. CRP contacts RNA polymerase to effect isomerisation of a closed to an open transcription initiation complex; in some cases bending of the DNA at the site of CRP binding makes a significant contribution to this process (Botsford and Harman, 1992).

S. typhimurium possesses copies of *cya* and of *crp*, the gene encoding CRP. Mutations in these genes result in attenuation of mouse virulence, and the resulting attenuated strains can be used as live vaccines to deliver heterologous antigens (Curtiss and Kelly, 1987; Curtiss *et al.*, 1988). Deletion mutations in *cya* and *crp* are likely to have extremely pleiotropic effects (Botsford and Harman, 1992), and it is very difficult to establish the likely direct cause of the loss of virulence.

CRP is now recognized as belonging to a family of related DNA-binding proteins. Another member of this family which has been studied in detail is FNR, a regulator of some genes expressed under anaerobic growth conditions. Interestingly, *fnr* null mutants are not affected in their mouse virulence characteristics (C.J. Dorman and G. Dougan, unpublished data). In contrast, the PrfA protein of *Listeria monocytogenes* is also related to CRP and plays a key role specifically in controlling virulence gene expression (see below).

Motility Gene Regulators FlgM and FliA

Bacterial motility represents a complex phenotype and involves regulation at the transcriptional and the post-transcriptional levels. The *S. typhimurium flgM* gene encodes a negative regulator of flagellin biosynthesis (Gillen and Hughes, 1991). The FlgM protein acts as an anti-sigma factor to inhibit the activity of a positive regulator, FliA, a mechanism that involves direct binding of FlgM protein to the FliA sigma factor (Ohnishi *et al.*, 1992). Mutations in *flgM* attenuate mouse virulence in *S. typhimurium* and reduce its ability to survive in macrophages; these effects are repressed by mutations in the *fliA* gene (Schmitt *et al.*, 1994). These findings suggest that the influence of the FlgM–FliA regulatory partnership extends beyond flagellin biosynthesis to include genes making a direct contribution to bacterial virulence.

Shigella flexneri and Enteroinvasive *Escherichia coli*

The facultative intracellular pathogens *Shigella flexnei* and enteroinvasive *E. coli* (EIEC) share a dependence on products encoded by a high-molecular-mass plasmid of the FIIA incompatibility group in order to express an invasive phenotype in the host (Hale, 1991). *Sh. flexneri* is the causative agent of bacillary dysentery and has been studied much more intensively than EIEC from the point of view of virulence gene regulation. Nevertheless, it would appear that both organisms share the same virulence gene control mechanisms.

Expression of the invasive phenotype in *Sh. flexneri* is regulated in response to several environmental signals, which are presumed to inform the bacterium that it has arrived in the appropriate host location for expression of this phenotype. The key signals are a temperature of 37°C, a level of osmolarity equivalent to that of physiological saline and a neutral pH. The invasion genes are located on the high-molecular-mass plasmid; their regulators are also found on this plasmid and at different sites on the chromosome. Some regulators appear to be dedicated exclusively to controlling invasion gene expression, whereas others also contribute to the housekeeping functions of the cell. In this way, virulence gene expression is co-ordinated with that of the housekeeping genes, possibly as a means of ensuring the overall fitness of the bacterium for its host-associated environment.

Virulence Gene Regulatory Cascade

Sh. flexneri and EIEC invasion gene expression is regulated at the level of transcription by a cascade of control elements (Fig. 4). The primary regulator is the

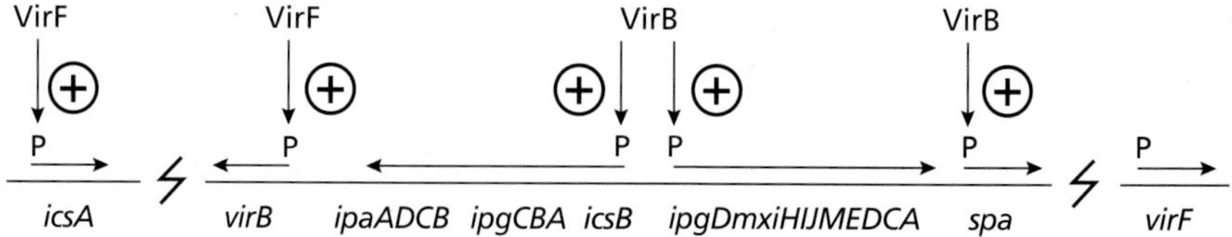

Fig. 4 The plasmid-linked virulence genes of *Sh. flexneri*. The plasmid-encoded regulatory proteins VirF and VirB regulate positively transcription of the virulence genes. The direction and extent of the transcription units are represented by horizontal arrows and the promoters by P. The + show that the regulatory inputs are all positive. Gaps between *icsA* and *virB* and between *spa* and *virF* show that long distances separate these genes on the plasmid.

plasmid-encoded VirF protein, which belongs to the large family of AraC-like DNA-binding proteins (Sakai *et al.*, 1986; Dorman, 1992). The VirF protein activates transcription of the intermediate regulatory gene *virB*, and binding of this protein to the promoter region of the *virB* gene has been demonstrated *in vitro*. The footprint, obtained with a MalE-VirF fusion protein, extends from −17 to −117 of the promoter, numbered with reference to the transcription start site at position +1. In *Sh. sonnei*, the *virB* gene is known as *invE* (Watanabe *et al.*, 1990). The VirB protein is required for expression of the genes that code for the invasion proteins, and the genes that encode the proteins which secrete the invasins to the bacterial cell surface (Adler *et al.*, 1989; Tobe *et al.*, 1991, 1993). Activation is antagonised by the H-NS nucleoid-associated protein, which has been shown to bind to the *virB* promoter region *in vitro* (Tobe *et al.*, 1993). H-NS binds between positions −20 to +20 at the promoter, a pattern that obscures the pribnow box and suggests that H-NS simply excludes RNA polymerase from the promoter by steric hindrance.

Transcriptional activation of the cascade is modulated by temperature and osmolarity, and both regulatory features are perturbed severely in *hns* mutants (Hromockyj *et al.*, 1992; Porter and Dorman, 1994). Each of these environmental factors alters the supercoiling of bacterial DNA, and DNA supercoiling has been implicated in the control of this virulence-gene regulatory cascade (Dorman *et al.*, 1990; Ní Bhriain and Dorman, 1993), with the *virB* promoter showing sensitivity to variations in supercoiling (Tobe *et al.*, 1993, 1995). Activation of the *virB* promoter depends absolutely on the VirF protein and it is thought that the role of temperature is to supercoil the *virB* promoter negatively to a point where the VirF protein is able to activate it. In support of this hypothesis, introduction of negative supercoils at the virB promoter by coupling it to a divergently transcribed inducible promoter allows VirF to activate *virB* even at the normally non-permissive temperature of 30°C (Tobe *et al.*, 1995).

Although *virF* regulation occurs over a short range *in vivo*, it seems highly likely that VirF protein is available in the cell under all growth conditions and will activate the *virB* promoter once its DNA has acquired an appropriate topology. The promoters which VirB regulates display a number of individual quirks, and they are not induced uniformly by the environmental signals which up-regulate *virB* expression. This suggests that opportunities exist for fine-tuning the promoters after activation. The range of regulation displayed by each promoter correlates with its location in the cascade. At the top, *virF* is regulated over a 2-fold range, the vassal regulator *virB* is controlled over a 10-fold range and the structural genes, which code for the invasins and their type III secretory system, are controlled over a 100-fold range (M.E. Porter and C.J. Dorman, unpublished results). The looseness of transcriptional control at the *virF* promoter has already been referred to. Regulation seems to become tighter as the cascade is descended. This may be because inappropriate expression of the primary regulatory gene *virF* imposes a modest cost on the cell, whereas inappropriate expression of the invasins and their secretory system would be much more deleterious. Therefore, the primary regulator is loosely controlled so that it is available and primed to respond to environmental signals immediately whereas the structural genes are only activated under conditions truly likely to be encountered at the site within the host where virulence gene expression is most likely to benefit the bacterium.

OmpR/EnvZ in *Sh. flexneri*

Mutations in the chromosomal *ompB* locus of *Sh. flexneri*, which consists of *envZ* and *ompR*, have been shown severely to impair virulence and to alter expression of the plasmid-encoded virulence genes (Bernardini *et al.*, 1990). It is unlikely that *ompB* regulates virulence gene expression directly; for example, no evidence has been presented for OmpR binding to DNA sequences associated with the plasmid-encoded genes. It is assumed that, as in *E. coli* and

S. typhimurium, *ompB* primarily regulates porin gene expression in *Sh. flexneri*. As in *S. typhimurium*, porin gene expression is required for full virulence in *Sh. flexneri*. Here, mutations which abolish expression of the *ompC* gene attenuate the virulence of this pathogen, although mutations which abolish *ompF* expression alone do not (Bernardini *et al.*, 1993). Thus, *Sh. flexneri* and *S. typhimurium* share a need for normal expression of porin genes for full virulence, although the details of the contributions made by these genes differ in each species.

CpxA/CpxR

Another chromosomally encoded two-component regulatory system which controls *Shigella* virulence gene expression is CpxA/CpxR. In *Shigella sonnei*, expression of *virF* is modulated by pH at the level of transcription, and expression is repressed at pH 6.0 and active at pH 7.4 (Nakayama and Watanabe, 1995). pH also modulates *virF* expression in *Sh. flexneri* (M.E. Porter and C.J. Dorman, unpublished results). A mutation in the *cpxA* gene de-represses *virF* transcription at the lower pH while reducing it slightly at pH 7.4. The CpxA protein is the sensor partner of a two-component system identified originally as playing a role in F plasmid conjugation (McEwen and Silverman, 1980). A mutation in the gene that codes for the response regulator partner protein CpxR abolishes *virF* expression (Nakayama and Watanabe, 1995). These findings reveal a role for pH in modulating virulence gene expression and for the CpxA/CpxR two-component regulatory system in the control of *virF*. Whether or not the regulatory link between CpxA/CpxR and *virF* is direct or is not at present known.

Virulence Plasmid Integration at *metB*

The large plasmids of *Sh. flexneri* and EIEC are able to integrate into the chromosome at a specific site, the *metB* gene. Integration is reversible and can be either precise or imprecise. These processes require an active *recA* gene, indicating that homologous recombination is involved (Zagaglia *et al.*, 1991). Integration renders the bacterium auxotrophic for methionine and suppresses expression of the plasmid-encoded virulence functions. It has been proposed that this may give the bacterium a survival advantage when living away from its host. Under these conditions, the virulence genes may be regarded collectively as being in an archived state. In EIEC strains with an integrated plasmid the *virF* gene continues to be expressed normally, although expression of *virB* is strongly reduced. It has been proposed that the topological state of the integrated plasmid favours transcriptional repression of *virB*, probably by H-NS (Colonna *et al.*, 1995). This type of mechanism is an excellent example of co-ordinated regulation of virulence gene expression achieved through a single molecular event, that is, RecA-dependent recombination between the chromosome and the virulence plasmid.

RpoS

Sh. flexneri possesses a homologue of the *E. coli rpoS* gene which encodes the stationary-phase sigma factor RpoS. Mutations in the *Sh. flexneri rpoS* locus do not strongly affect the expression of the plasmid-linked virulence genes (M.E. Porter and C.J. Dorman, unpublished data). The *Sh. flexneri rpoS* gene is, however, required for acid and base tolerance (Small *et al.*, 1994). Presumably, this points to the existence in *Sh. flexneri* of a regulon of RpoS-dependent genes involved in adaptation to pH-imposed stress. Such a regulon of genes is likely to be very important during infection, but is also likely to assist the cell when growing away from its host.

Vibrio cholerae

The intestinal pathogen *Vibrio cholerae* is the causative agent of cholera in humans. The virulence gene regulon of *V. cholerae* represents a regulatory paradigm, illustrating mechanisms for the detection and transduction of environmental signals to the genome and the operation of a sophisticated regulatory cascade (Fig. 5). At the heart of the virulence system is the *ctxAB* operon which encodes the A and B subunits of cholera toxin. Historically, it was attempts to understand more completely how expression of this toxin is expressed that led to the elucidation of the regulatory cascade which is now recognised as also controlling many other virulence genes in this pathogen.

The *ctxAB* Operon

Post-transcriptional regulation plays a critical role in determining the relative number of cholera toxin A and B subunits in the cell. Cholera toxin is 84 kDa in molecular mass and is composed of the A and B subunits in a ratio of 1 : 5. The *ctxA* and *ctxB* genes overlap so that the first two nucleotides of the translation termination codon of *ctxA* are also the last two nucleotides of the translation start signal of *ctxB*. Each gene has its own ribosome-binding site, with that for *ctxB* being within the 3′-end of the *ctxA* open reading frame. Both genes sharing the same promoter and differential translational efficiency is thought to contribute to the different levels of CtxA and CtxB

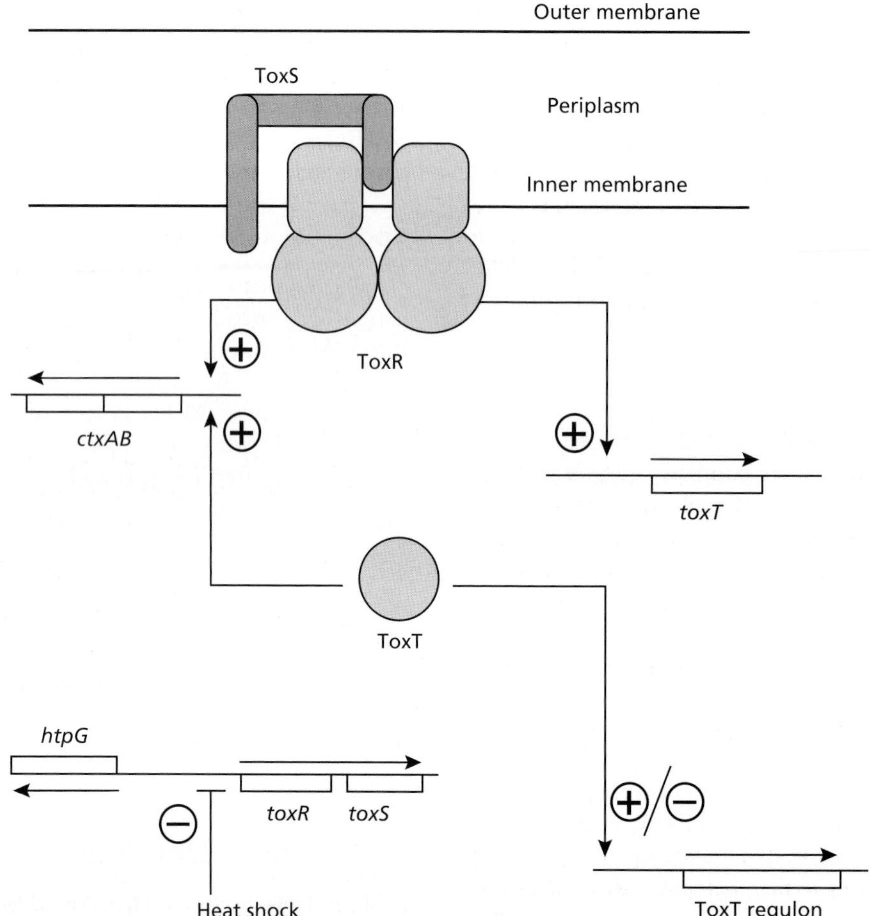

Fig. 5 The ToxR/ToxS/ToxT regulon of *V. cholerae*. The ToxS protein is shown assisting dimerization of ToxR in the inner membrane. ToxR regulates *ctxAB* and *toxT* transcription directly, with ToxT regulating *ctxAB* and the other genes of the regulon. Activation of the heat-shock gene *htpG* has a negative influence on transcription of the *toxRS* genes. + and – indicate positive and negative regulation, respectively.

protein expressed in the cell; the ribosome-binding site for *ctxB* is about nine times more efficient than that of *ctxA* (Mekalanos *et al.*, 1983).

The *ctxAB* operon is also subject to transcriptional regulation. Upstream of the promoter is a regulatory region consisting of multiple copies of the sequence 5′-TTTTGAT-3′. Deletion of these repeated sequences results in a loss of normal *ctxAB* expression. These constitute a *cis*-acting regulatory element and they are bound by a positively acting transcriptional regulator called ToxR (Miller *et al.*, 1987). In addition to regulating *ctxAB* transcription, ToxR is involved in the control of several other *V. cholerae* genes.

ToxR, the Primary Regulator of the *ctxAB* Operon

Approximately one-third of the 32-kDa ToxR protein is located in the periplasm and two-thirds in the cytoplasm (Miller *et al.*, 1987), which is an unusual situation for a DNA-binding protein with a role in transcription regulation (Fig. 5). The putative DNA-binding domain of ToxR shows identity with the corresponding domain of OmpR (Miller *et al.*, 1987; Ottemann *et al.*, 1992), with the difference that DNA-binding in ToxR occurs in its N-terminus, whereas in OmpR this is a feature of the C-terminus of the protein. The finding that ToxR is required to transduce a number of environmental signals to *ctxAB* helps explain its unusual cellular location. These include osmolarity, temperature, pH, and the presence or absence of the amino acids asparagine, arginine, glutamate, and serine (Miller and Mekalanos, 1988). The C-terminal portion of ToxR binds to the 5′-TTTTGAT-3′ repeats of the *ctxAB* promoter and this binding is modulated by the periplasmic domain. The simplest interpretation of these findings is that the periplasmic domain receives information about the external environment and transmits this to the *ctxAB* promoter region via the DNA-binding activity of the N-terminus of ToxR. This model and the relationship the N-terminal domain of ToxR to the DNA-binding domain of OmpR led to ToxR being regarded briefly as a one-component version of the more familiar two-component regulator systems. The discovery that other regulatory proteins must collaborate with ToxR for normal function eventually, however, ruled out this intriguing possibility (DiRita, 1992; see below).

The Secondary Regulator ToxT

Several lines of evidence point to a role for ToxR in controlling transcription of a group of genes in *V. cholerae*. The hypothesis that ToxR governs expression of a regulon of genes of which the *ctxAB* operon was just one member suffered a setback, however, when it proved difficult to demonstrate a direct interaction between ToxR and any other promoter. Other candidates for membership of the ToxR regulon include the genes that code for the outer membrane proteins, OmpT and OmpU, as well as the adhesins known as the toxin co-regulated pilus (TCP) and the accessory colonisation factor (ACF) (Miller and Mekalanos, 1988). ToxR also contributes to control of lipoprotein production and the expression of serum resistance (Parsot *et al.*, 1991). It was eventually discovered that ToxR operates via an intermediate regulatory called ToxT or TcpN (DiRita *et al.*, 1991; Higgins *et al.*, 1992; Ogierman and Manning, 1992). This 32-kDa protein is a member of the AraC family of transcription regulators and the gene that encodes it, *toxT*, is regulated positively by ToxR (**Fig. 5**). In spite of this fact, *toxT* shows little DNA sequence similarity to *ctxAB* at its regulatory region. The explanation is likely to involve the recognition by ToxR of a structural motif, rather than a primary DNA sequence, which is common to *toxT* and *ctxAB* (Higgins and DiRita, 1996). In addition to activating transcription of the *tcp* and *acf* genes, ToxT also contributes to the activation of *ctxAB* (DiRita *et al.*, 1991; Higgins and DiRita, 1994).

ToxS, a Membrane-located Partner for ToxR

Sequencing of the DNA downstream of the *toxR* gene on the *V. cholerae* chromosome revealed the presence of *toxS*, a gene that codes for a 19-kDa product required for the normal function of ToxR when the latter is encoded from a single-copy gene. ToxS interacts with ToxR in the periplasm and assists in the oligomerisation of ToxR proteins in the membrane (DiRita and Mekalanos, 1991; Miller *et al.*, 1991) (**Fig. 5**). Thus, the Tox regulatory system appears to be a three-component system, with ToxS assisting in correct oligomerisation of ToxR in the membrane, and ToxT serving to relay signals passing through ToxS/ToxR to various structural genes coding for virulence factors, including the *ctxAB* operon which codes for cholera toxin. In addition, ToxR can activate *ctxAB* directly, raising the possibility that the involvement of both ToxR and ToxT at the *ctxAB* promoter reflects a need to amplify the inducing signal under some circumstances.

This regulatory system may exert effects that extend beyond the virulence genes. The *aldA* gene, which encodes the cytoplasmically located aldehyde dehydrogenase, is controlled by ToxR and this enzyme is clearly a component of metabolism rather than a dedicated virulence factor (Parsot and Mekalanos, 1991). The significance of *aldA* control by ToxR is unclear, but it raises the possibility that expression of other housekeeping genes may be co-ordinated with that of the virulence genes.

The Heat-shock Gene, *htpG*

Upstream from *toxR* and transcribed from the opposite DNA strand is the heat-shock gene *htpG*. This codes for a member of the Hsp90 family of heat-shock proteins and it is induced at the level of transcription by shifting the culture from a temperature of 22°C to 37°C (Parsot and Mekalanos, 1990). As expression of *htpG* increases, expression of *toxR* decreases in proportion. Similar results are obtained with these genes cloned in *E. coli*, where expression of *htpG* depends on the heat-shock sigma factor RpoH. It has been proposed that the equivalent heat-shock sigma factor acts in *V. cholerae* to regulate negatively *toxR* transcription via an indirect mechanism that involves the thermal activation of the upstream *htpG* gene (Parsot and Mekalanos, 1990). In this way, thermal signals can modulate *toxR* transcription without the need for it to evolve a complicated regulatory element specific for its own promoter. A problem with these findings is that growth at 37°C would appear to be repressive for *toxR* transcription and this is counter-intuitive given the temperature of the host intestine where ToxR-dependent gene expression is required. Perhaps temperature inputs are themselves modulated by additional signals, which are uniquely associated with those microenvironments where the ToxR regulon must be expressed.

DNA Topology and *acf* Gene Expression

The possibility that a role exists for DNA topology in controlling virulence gene expression in *V. cholerae* has not been rigorously explored. Evidence has, however, been presented that DNA supercoiling affects expression of the divergently transcribed accessory colonisation factor genes *acfA* and *acfD* (Parsot and Mekalanos, 1992). Expression of these ToxT-dependent genes is aberrant when they are cloned in plasmid vectors, indicating a sensitivity to the DNA context. This is supported by data that show a dependence on DNA supercoiling, suggesting that in addition to the presence of the AraC-like protein ToxT, the promoters of these genes also require a particular DNA topology in order to be expressed normally. This is reminiscent of the situation at the *virB* promoter of *Sh. flexneri*, a gene regulated positively by another AraC-like protein (VirF) and with a strong dependency on DNA topology (Tobe *et al.*, 1993, 1995).

Listeria monocytogenes and the PrfA Regulon

The Gram-positive bacterium *Listeria monocytogenes* is a facultative intracellular pathogen (see Chapter 69) capable of invading and multiplying in different types of phagocytic and non-phagocytic eukaryotic cells. Like *Sh. flexneri*, it uses an actin-based propulsion system to move rapidly from one host cell to another. This enables the bacterium to spread through host tissues while evading the immune response. *L. monocytogenes* encodes several enzymes and other proteins that promote invasion of host cells, cause damage to host tissues allowing the bacteria to escape from intracellular vacuoles, and which recruit and polymerise host actin for use in the cell-to-cell spreading mechanism (reviewed by Sheehan *et al.*, 1994).

The genes that code for the *L. monocytogenes* virulence factors are located on the chromosome and their expression is regulated co-ordinately at the level of transcription in response to growth phase, temperature and cellobiose (Mengaud *et al.*, 1991; Leichmeister-Wächter *et al.*, 1992; Park and Kroll, 1993) (**Fig. 6**). A master control gene has been identified and designated *prfA* (Leichmeister-Wächter *et al.*, 1990). The PrfA protein is related to the Crp/Fnr family of gene regulatory proteins (Lampidis *et al.*, 1994). In common with the other members of this family, PrfA is usually a positively acting transcription factor, although it may act as a repressor in some cases. Its efficacy as a transcription factor is not equivalent for

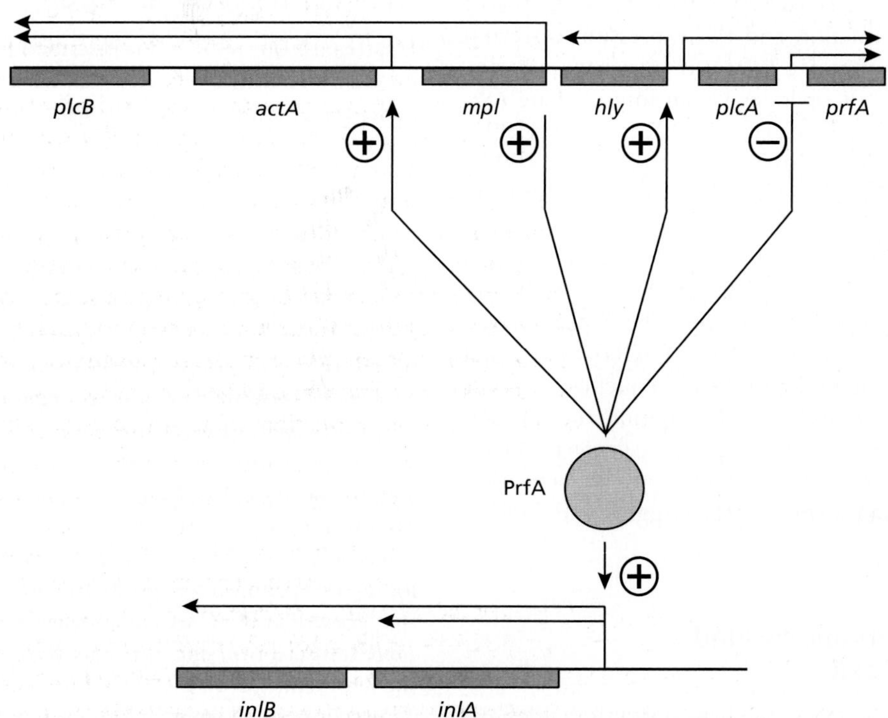

Fig. 6 The PrfA regulon of *L. monocytogenes*. Two chromosomal loci are involved. The *prfA* gene is located at the larger locus and codes for a protein involved in positive control of transcription of the virulence gene promoters at both loci. There may be a role for PrfA in repressing transcription of its own gene. The horizontal arrows show the direction and extent of the transcription units; + and − indicate positive and negative regulation, respectively.

all promoters under its control. The divergently transcribed *hly* and *plcA* genes, which code for a pore-forming cytolysin and a phosphatidylinositol-specific phospholipase, respectively, are activated most efficiently by PrfA. Much weaker is activation of the *actA* (actin assembly) and *mpl* (possibly a metalloprotease) genes. Also poorly activated by PrfA is *inlA*, the gene coding for internalin, a protein required for *L. monocytogenes* entry into cultured epithelial cells. This differential sensitivity of promoters to PrfA may be physiologically relevant in the sense that it helps meet the requirement of the pathogen to express different virulence determinants at different stages of infection (Sheehan *et al.*, 1995).

The molecular basis of this differential regulation may lie in the sequences to which PrfA is thought to bind at individual virulence gene promoters. In the case of the *hly* and *plcA* genes, the putative binding sites each have a twofold axis of symmetry and they are identical to one another. At the less strongly activated genes, there are mismatches in the regions of symmetrical sequence which may attenuate the efficiency with which PrfA can interact with the DNA and (possibly) with RNA polymerase (Sheehan *et al.*, 1995).

At present little is known about regulatory events extending beyond the PrfA regulon, because gene regulatory studies in *L. monocytogenes* have tended to focus on the expression of its many virulence factors. Consequently, much less is known about the control of housekeeping genes in this bacterium. This contrasts with the situation in Gram-negative pathogens, such as *S. typhimurium*, where a rich deposit of knowledge about genetic composition and regulation allows virulence gene control to be considered in a wider context. Obtaining information about genetic events beyond the PrfA-controlled genes is likely to help in obtaining a more complete appreciation of how *L. monocytogenes* adapts to its host and causes disease.

Co-ordination within Populations

It is now recognised that regulatory mechanisms operate in bacteria which can co-ordinate gene expression not just in individual cells but in the population as a whole. These mechanisms may even extend across species barriers in the case of some mixed populations. They involve the production of chemical signals that are released by bacteria, taken up by their neighbours and produce changes in the gene expression profile of these cells. This phenomenon has been called quorum sensing, to illustrate its co-operative aspect and the need for a minimum number of participating cells to be present for the mechanism to operate. It has also been given the less eloquent name of density-dependent regulation.

At the heart of this regulatory mechanism lies a family of low-molecular-mass diffusible compounds known as auto-inducers. These molecules interact with transcription regulatory proteins both in the bacterium which synthesized them and in neighbouring bacteria. The interaction occurs above a critical concentration threshold and results in the activation of gene expression. The early work in this field was carried out with bioluminescent marine bacteria, some of which have symbiotic, rather than pathogenic, relationships with fish, whereas others are free-living. Examples of these are *V. fischeri* (symbiont) and *V. harveyi* (free-living). These bacteria synthesise homoserine lactone derivatives which perform the signalling function. Only when the density of the bacterial culture reaches a critical level will the local concentration of homoserine lactone autoinducer be high enough to induce transcription of the genes involved in expressing bioluminescence (Bassler and Silverman, 1995).

Subsequently it was recognised that this form of regulation is not restricted to marine organisms. Other bacteria synthesise related auto-inducer compounds and have proteins closely related to those in the marine vibrios which interact with those compounds. They contribute to the regulation of diverse functions, such as antibiotic biosynthesis in *Erwinia carotovora* (Bainton *et al.*, 1992), conjugative transfer of a plant virulence plasmid in *Agrobacterium tumefaciens* (Piper *et al.*, 1993; Zhang *et al.*, 1993) and the regulation of virulence factor expression in the pathogens *Erwinia carotovora* and *Pseudomonas aeruginosa* (Jones *et al.*, 1993; Passador *et al.*, 1993).

In the case of *Ps. aeruginosa*, an opportunistic pathogen of humans, the virulence factor which is under auto-inducer control is elastase. This is an extracellular enzyme which degrades or inactivates components of host tissue and of the immune system. It is encoded by the *lasB* gene and transcriptional activation depends on the LasR protein working in concert with the *Pseudomonas* auto-inducer, PAI. The autoinducer is expressed by the *lasI* gene and is regulated positively by both PAI and LasR (Seed *et al.*, 1995). Additional regulatory factors are also involved, including a second auto-inducer, *N*-butyrylhomoserine lactone (Pearson *et al.*, 1995). It is likely that many more examples of virulence factor control by auto-inducers remain to be discovered.

Phase Variation

The examples of gene regulation considered so far have involved mechanisms of control that will apply in every cell in a bacterial population in response to a shared shift in environmental circumstances. This type of regulation is sometimes called 'stereotypic', since it involves a uniform response by the population. Simultaneously, other events occur in the population which involve just a few cells. These events seem to occur at random in the sense that one cannot predict which cell will display the behaviour, in sharp contrast to stereotypic control where every cell will behave in a predictable way. Phase variation represents an example of such a random process. Here, expression of a cellular component is switched on or off in a temporally and spatially unpredictable way in a few members of the population, effectively creating a mixed population of cells. Some components of the bacterial cell surface display phase-variable expression and this pattern of behaviour can aid in pathogenesis. The surface component may assist the bacterium to colonise the host; adhesins are important examples of virulence factors which are frequently subject to phase variable control.

A disadvantage of expressing such surface components is that they may be immunogenic and their expression may compromise the attempts of the bacterium to circumvent host defences. By placing the surface component under phase-variable expression, the population is likely at any time and place to contain some members which express and some which do not express this cellular component. The environment then selects that sub-population best adapted to the prevailing conditions.

It is becoming apparent that phase-variable regulation frequently involves a reversible change in the structure of the DNA in the regulatory sequences, or the coding sequences (or both) of the gene which expresses the surface component. This change in structure can involve an extensive rearrangement or a change to just one residue (Robertson and Meyer, 1992). It is also becoming apparent that phase variable systems do not in a regulatory sense operate in isolation. Control elements known to contribute to stereotypic gene regulatory circuits are now recognised as modulators of phase variable gene expression. This raises the interesting possibility that the random aspect of phase variation may be influenced under certain circumstances, perhaps to favour one outcome over another.

Pap Pili

Pyelonephritis-associated pili (Pap) are expressed phase variably in uropathogenic strains of *E. coli* (see Chapter 72). A cluster of nine genes specifies the synthesis and assembly of the pilus at the bacterial cell surface. Expression of these genes is controlled at the level of transcription (Blyn *et al.*, 1989) and the site of regulation lies in the region between the divergently transcribed *papI* and *papB* genes (Kaltenbach *et al.*, 1995) (**Fig. 7**).

Phase variation involves roles for DNA methylation through the Dam methylase and for the leucine responsive regulatory protein, Lrp (Blyn *et al.*, 1990; Braaten *et al.*, 1992). The Lrp protein binds to six sites in the *papI-B* intergenic region, two of which are also sites for Dam methylation. These sites of methylation are designated GATC-I (distal from the *papB* promoter) and GATC-II (proximal to the *papB* promoter). In phase-'off' bacteria, GATC-I is fully methylated and GATC-II is unmethylated and bound by Lrp. The same is true of the Lrp-binding sites to either side of GATC-II. In this position, Lrp overlaps the *papB* promoter and inhibits transcription. Transition to the 'on' phase involves a role for the PapI protein. This can bind to the DNA-bound form of Lrp, but not the free form, and alters its affinity for its binding site such that the PapI–Lrp complexes move from GATC-II to a hemimethylated form of GATC-I which is generated by DNA replication. A further round of DNA replication leaves GATC-I fully unmethylated. Meanwhile, the Dam methylase has methylated Lrp-free GATC-II, producing the phase 'on' state in which *papB* is transcribed (Kaltenbach *et al.*, 1995) (**Fig. 7**).

The involvement of DNA methylation links *pap* expression to the replication of the genetic material in the cell. The involvement of Lrp has the potential to link *pap* expression to the large number of other genes whose expression is modulated by Lrp (Calvo and Matthews, 1994). This 18.8-kDa DNA-binding protein can activate or repress transcription. It bends DNA and can also serve as an architectural element, organising the structures of nucleoprotein complexes. The amino acid leucine can enhance, antagonise or have no effect on the input of Lrp, depending on the system under investigation. In the case of *pap*, leucine has almost no effect on the ability of Lrp to influence the system (Braaten *et al.*, 1992).

In addition to regulation by the global effector, Lrp, *pap* gene expression is also modulated by another global regulator, H-NS. This protein participates in the thermal control of *pap* gene expression. In wild-type cells, *pap* genes are repressed at 26°C but transcribed at 37°C. If the *hns* gene is mutated, the *papI* and *papBA* genes are expressed at the lower temperature (Göransson *et al.*, 1990). Thus, the *pap* genes belong not only to the Lrp regulon, but also to

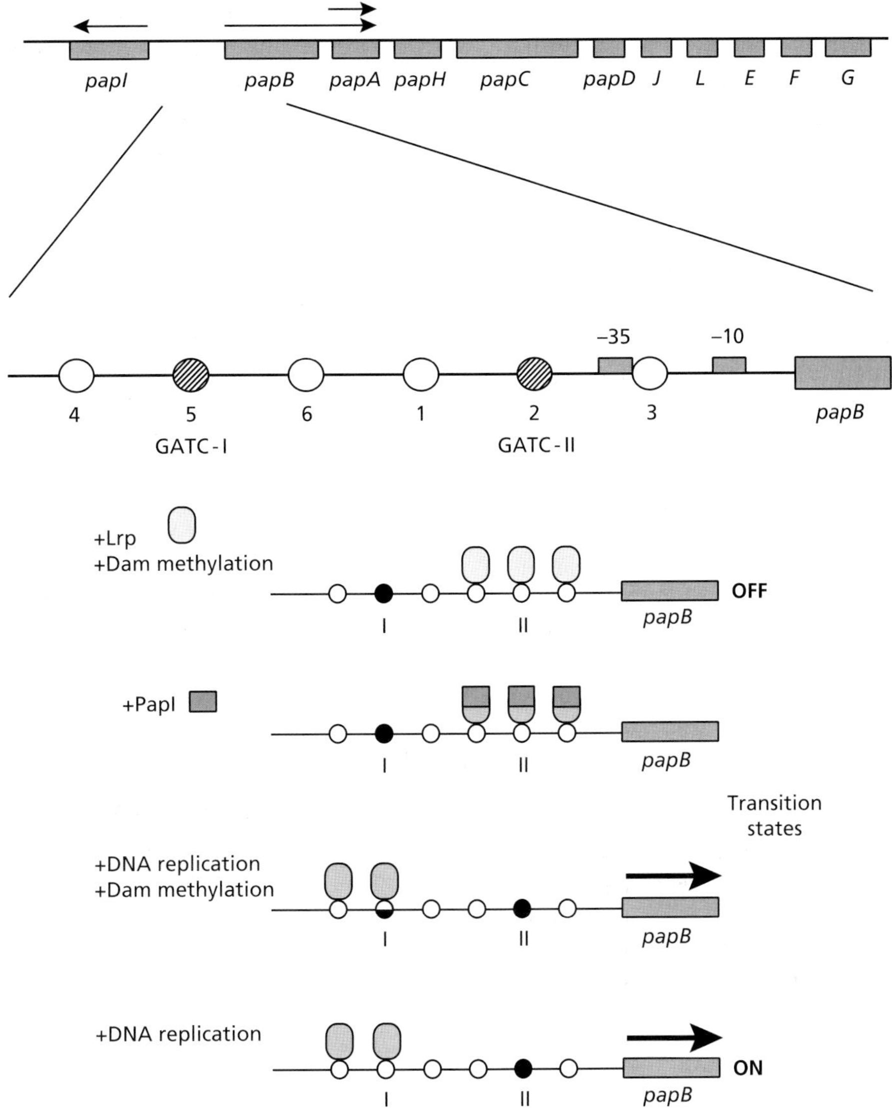

Fig. 7 The *pap* locus of uropathogenic *E. coli*. The entire *pap* operon is shown at the top of the figure. The regulatory region between the divergently-transcribed *papI* and *papB* genes is shown in expanded form below. Six binding sites for Lrp are located here, shown using the numbering system of Kaltenbach *et al.* (1995). Dam methylation occurs at GATC-I and GATC-II within sites 5 and 2, respectively. Site 2 is closely adjacent to the promoter of the *papB* gene and site 3 is located within that promoter. In the OFF phase, sites 1, 2 and 3 are bound by Lrp and RNA polymerase is excluded from the *papB* promoter; GATC-I is fully methylated. The PapI protein modifies Lrp and this relocates to sites 4 and 5, allowing GATC-II to be methylated by Dam and RNA polymerase to gain access to the *papB* promoter, resulting in a transition to the ON phase. DNA replication is the mechanism through which unmethylated DNA strands are produced.

the H-NS regulon. This situation is complicated further by the observation that the pleiotropic regulator cAMP-Crp positively regulates *pap* transcription (Forsman *et al.*, 1989), making these pilus genes members of yet a third broadly based regulon. As with other cAMP-Crp-regulated genes, the *pap* genes show sensitivity to changes in the carbon source used for growth (Blyn *et al.*, 1989).

Type 1 Fimbriae in *Escherichia coli*

Like the Pap pili, type 1 fimbriae in *E. coli* are also subject to phase-variable regulation. Unlike the *pap* genes, the *fim* genes are controlled by a DNA rearrangement. This involves inversion of a 314-bp DNA element that contains the promoter required for transcription of the type 1 fimbrial subunit gene, *fimA*

(Abraham *et al.*, 1985). This DNA element, known as the *fim* switch, is flanked by 9-bp inverted repeats, which form part of the binding sites for the FimB and FimE regulatory proteins (Dove and Dorman, 1996; Gally *et al.*, 1996). These are related to the site-specific recombinase Int, which is specified by bacteriophage λ and which drives the integration and excision of λ from the *E. coli* chromosome by site specific recombination (Dorman and Higgins, 1987; Eisenstein *et al.*, 1987). When the promoter within the *fim* switch is directed towards *fimA*, this gene is transcribed; when the promoter is directed away from *fimA*, this gene is silent and the bacterium does not produce type 1 fimbriae on its surface. Protein FimB switches the promoter in either direction with almost equal efficiency, whereas the FimE protein has a preference for switching the promoter to the 'off' orientation (Blomfield *et al.*, 1991) (**Fig. 8**).

The site-specific recombinases do not act alone to drive inversion of the *fim* switch. They require the accessory proteins Lrp and the integration host factor (IHF) (Dorman and Higgins, 1987; Eisenstein *et al.*, 1987; Blomfield *et al.*, 1993; Gally *et al.*, 1994). Lrp has been described in the previous section on Pap pili. In *E. coli*, IHF is a heterodimeric DNA-binding protein which binds to DNA at specific sequences and introduces there a bend in the DNA helix of up to 140° (Kosturko *et al.*, 1989). This protein shares with Lrp an ability to function as an organiser of genome structure and as a regulator of transcription. It is required for the normal function of the *fimA* promoter, as well as for inversion of the *fim* switch (Dorman and Higgins, 1987; Eisenstein *et al.*, 1987). Two binding sites for Lrp have been characterised within the switch, whereas one IHF-binding site is located within the switch just upstream of the *fimA* promoter and another just outside the switch at the left-hand inverted repeat (**Fig. 8**). When occupied, these presumably contribute to the formation of a nucleoprotein complex which promotes *fim* inversion in a manner analogous to the promotion of λ integration into the chromosome by the formation of the intasome structure.

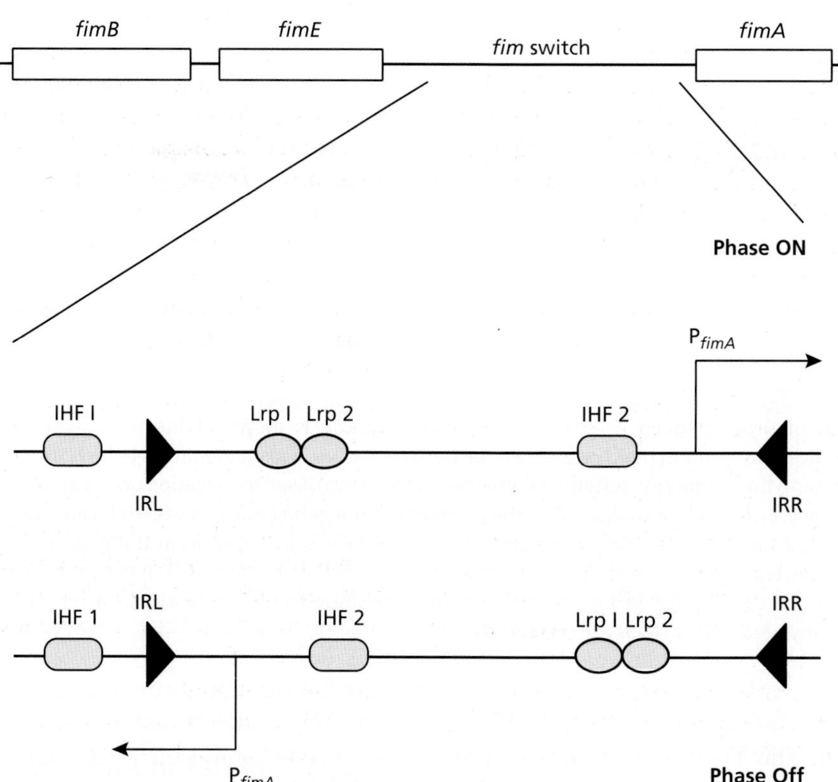

Fig. 8 The *fim* switch of *E. coli*. The genes coding for the site-specific recombinases FimB and FimE and the gene encoding the type-1 fimbrial subunit protein, FimA, are shown at the top of the figure. The region between *fimE* and *fimA* contains the *fim* switch and this is shown in an expanded form below. The invertible region is bounded by 9 bp inverted repeats (filled triangles) and carries the promoter for *fimA* transcription (P_{fimA}). In phase ON, this promoter is directed towards *fimA* and in phase OFF it is directed away. Binding sites for IHF are located just outside (IHF1) and inside (IHF2) the *fim* switch. There are also two binding sites for Lrp within the invertible element.

A role for protein H-NS has been discovered in which it inhibits inversion of the *fim* switch. In mutants incapable of expressing H-NS, inversion rates are greatly accelerated (Kawula and Orndorff, 1991). This protein also has an inhibitory effect on the function of the *fimA* promoter (Dorman and Ní Bhriain, 1992). Moreover, it represses the promoters of the *fimB* and *fimE* genes, which code for the site-specific recombinases (Olsen and Klemm, 1994). The *fimB* gene has a promoter with the unusual feature of being activated by DNA relaxation. Expression of topoisomerase I and DNA gyrase is required for normal rates of inversion of the *fim* switch. In mutants deficient in topoisomerase I, inversion virtually ceases; in strains in which gyrase activity is inhibited, inversion becomes directional and biased towards the 'on' phase (Dove and Dorman, 1994). These observations show that the level of DNA supercoiling in the vicinity of the *fim* switch modulates *fim* gene expression substantially.

Environmental influences play a large role in controlling *fim* gene expression. The composition of the growth medium is critical to determining the rate of *fim* switch inversion. In part this reflects the ability of the amino acid leucine to accentuate the positive effect of the Lrp protein on the inversion mechanism. Solid or liquid media influence *fim* gene expression differently. Some strains of *E. coli* display differential expression of the regulatory genes *fimB* and *fimE* under solid phase or liquid growth conditions, with important consequences for the directionality of the *fim* inversion event (Schwan *et al.*, 1992). Temperature modulates the ability of the recombinases to invert the *fim* switch. At low temperatures, the FimE protein is more effective at switching the system to the 'off' phase, whereas at 37°C FimB activity is dominant (Gally *et al.*, 1993). The *fimA* promoter is more active at 37°C than at 30°C (Dorman and Ní Bhriain, 1992.) Growth phase is another important determinant of *fim* gene expression. The stationary phase sigma factor RpoS is required for normal regulation of *fimA* and *fimB* and it influences the rate of inversion of the *fim* switch negatively (probably indirectly) (Dove *et al.*, 1997). The supply of the rare $tRNA_5^{Leu}$, encoded by the *leuX* gene, is critical. The *fimB* mRNA contains five codons for this tRNA and so the rate of its translation is limited by the supply of $tRNA_5^{Leu}$ (Klemm, 1986; Ritter *et al.*, 1995).

Clearly, *fim* gene expression is under complex, multi-factorial control. The regulatory mechanism consists of the randomly inverting *fim* switch and the modulating influences of a large number of more conventional regulatory elements such as Lrp, IHF, and H-NS, which link the control of *fim* gene expression to that of many other genes in the cell. In concert with the equally essential but less-well-defined input of DNA supercoiling, these factors serve to bias the inversion towards particular outcomes under certain environmental conditions. Since these events may be assumed to occur in all of the cells in the population simultaneously, they override the truly random nature of the phase variation in the population and cause it to mimic a stereotypic response. This suggests that mechanisms may have evolved in *E. coli* which suspend the stochastic element in *fim* gene expression when it is in the interest of the population as a whole to do so. Similar strategies may be applied to other phase-variable systems in this or in other bacteria.

Perspective

The gene-regulatory mechanisms reviewed in this chapter have come from just a small number of organisms. They have been selected because they have been the subject of intensive study over many years and are now understood in some detail at the molecular level. In spite of the variety of diseases caused by these bacteria, they have much in common at the level of virulence gene regulation. Shared themes which have emerged in this brief survey include the possession by regulons of virulence genes of a private regulator which governs their expression and has little to contribute to the control of other genes in the cell. Examples include the VirF and VirB proteins of *Shigella* and the BvgA and BvgS proteins of *B. pertussis.* Expression of these same regulons may in turn be modulated by regulatory proteins which make very widespread contributions to gene regulation throughout the genome. For instance, the H-NS protein is involved in controlling expression of the *pap* genes of uropathogenic *E. coli*, but it is also known to affect transcription of many other genes coding for products which aid in bacterial adaptation to environmental change (Ussery *et al.*, 1994). The private regulators exert a co-ordinating influence within the select group of the virulence regulon but proteins such as H-NS link these otherwise self-contained gene groups to very many others in the cell. The degree of stringency with which these very busy proteins can modulate co-ordinately gene expression is probably relatively low, and so they must rank below the specific control elements in a hierarchy of regulatory factors.

Lower still in the regulatory hierarchy is the co-ordinating influence of DNA topology. This varies in the cell in response to disturbances in the homeostatically determined levels of the countervailing activities of DNA topoisomerase I and DNA gyrase. If the DNA becomes too relaxed, more gyrase is expressed and if the DNA becomes too supercoiled more topoisomerase I is expressed (Drlica, 1992). This reflects the differential responses of the topoisomerase I and gyrase gene promoters to changes in the level of DNA supercoiling. Thermodynamic regulation of supercoiling links gyrase activity to general metabolism. Here, the phosphorylation potential of the cell, as measured by the ratio of ATP to ADP, has a strong influence on the activity of this ATP-dependent DNA topoisomerease (van Workum *et al.*, 1996). Finally, DNA transactions such as transcription create local domains of positively or negatively supercoiled DNA and these modulate promoter function (Liu and Wang, 1987; Chen *et al.*, 1992). It is likely that the role of supercoiling in co-ordination of gene regulation is a crude one, setting a background level of free energy in DNA which is available to enhance transcription from some promoters while inhibiting others. From the point of view of virulence gene expression, its significance lies in the fact that several such genes possess supercoiling-sensitive promoters and the thermodynamic regulation of supercoiling is influenced strongly by the types of environmental stresses which bacteria encounter during infection. These include changes in pH, osmolarity, temperature and oxygen availability (Dorman, 1995). Hence, the capacity of supercoiling to form the basement level of a regulatory hierarchy which can co-ordinate the expression of virulence genes with one another and with the other genes of the bacterial cell. The result is less a collection of independently regulated genetic units and more a co-ordinated and integrated entity capable of optimising the fitness of the bacterium for any environment it encounters.

Acknowledgements

Research in the author's laboratory is supported by the Wellcome Trust (UK), the European Union (DG XII), the Health Research Board of Ireland and Enterprise Ireland.

References

Abraham JM, Freitag CS, Clements JR, Eisenstein BI (1985) An invertible element of DNA controls phase variation of type 1 fimbriae of *Escherichia coli*. *Proc. Natl Acad. Sci. USA* 82: 5724–5727.

Adler B, Sasakawa C, Tobe T *et al.* (1989) A dual transcriptional activation system for the 230 kb plasmid genes coding for the virulence-associated antigen of *Shigella flexneri*. *Mol. Microbiol.* 3: 627–635.

Akerley BJ, Miller JF (1993) Flagellin transcription in *Bordetella bronchiseptica* is regulated by the BvgAS virulence control system. *J. Bacteriol.* 175: 3468–3479.

Aricò B, Miller JJ, Roy C *et al.* (1989) Sequences required for expression of *Bordetella pertussis* virulence factors share homology with prokaryotic signal transduction proteins. *Proc. Natl Acad. Sci. USA* 86: 6671–6675.

Bainton NJ, Bycroft BW, Chhabra SR *et al.* (1992) A general role for the *lux* autoinducer in bacterial cell signalling: control of antibiotic biosynthesis in *Erwinia*. *Gene* 116: 87–91.

Barth M, Marschall C, Muffler A, Fischer D, Hengge-Aronis R (1995) Role of the histone-like protein H-NS in growth phase-dependent and osmotic regulation of σs and many σs-dependent genes in *Escherichia coli*. *J. Bacteriol.* 177: 3455–3464.

Bassler BL, Silverman MR (1995) Intercellular communication on marine *Vibrio* species: density-dependent regulation of the expression of bioluminescence. In: Hoch JA Silhavy TJ (eds) *Two-Component Signal Transduction*. Washington, DC: American Society for Microbiology Press, pp. 431–445.

Bearson SM, Benjamin Jr DWH, Swords WE, Foster JW (1996) Acid shock induction of RpoS is mediated by the mouse virulence gene *mviA* of *Salmonella typhimurium*. *J. Bacteriol.* 178: 2572–2579.

Benjamin Jr WH, Yother J, Hall P, Briles DE (1991) The *Salmonella typhimurium* locus *mviA* regulates virulence in *Ity*^{s} but not *Ity*^{r} mice: functional *mviA* results in avirulence; mutant (nonfunctional) *mviA* results in virulence. *J. Exp. Med.* 174: 1073–1083.

Bernardini ML, Fontaine A, Sansonetti PJ (1990) The two-component regulatory system OmpR-EnvZ controls the virulence of *Shigella flexneri*. *J. Bacteriol.* 172: 6274–6281.

Bernardini ML, Sanna MG, Fontaine A, Sansonetti PJ (1993) OmpC is involved in invasion of epithelial cells by *Shigella flexneri*. *Infect. Immun.* 61: 3625–3635.

Blomfield IC, McClain MS, Princ JA, Calie PJ, Eisenstein BI (1991) Type 1 fimbriation and *fimE* mutants of *Escherichia coli* K-12. *J. Bacteriol.* 173: 5298–5307.

Blomfield IC, Calie PJ, Eberhardt KJ, McClain MS, Eisenstein BI (1993) Lrp stimulates phase variation of type 1 fimbriation in *Escherichia coli* K-12. *J. Bacteriol.* 175: 27–36.

Blyn LB, Braaten BA, Low DA (1990) Regulation of *pap* pilin phase variation by a mechanism involving fifferential *dam* methylation states. *EMBO J.* 9: 4045–4054.

Blyn LB, Braaten BA, White-Ziegler CA, Rolfson DH, Low DA (1989) Phase-variation of pyelonephritis-associated pili in *Escherichia coli*: evidence for transcriptional regulation. *EMBO J.* 8: 613–620.

Botsford JL, Harman JG (1992) Cyclic AMP in prokaryotes. *Microbiol. Rev.* 56: 100–122.

Boucher PE, Menozzi FD, Locht C (1994) The modular architecture of bacterial response regulators: insights into the activation mechanism of the BvgA transactivator of *Bordetella pertussis*. *J. Mol. Biol.* 241: 363–377.

Braaten BA, Platko JV, van der Woude MW *et al.* (1992) Leucine-responsive regulatory protein controls the expression of both the *pap* and *fan* pili operons in *Escherichia coli*. *Proc. Natl Acad. Sci. USA* 89: 4250–4254.

Caldwell AL, Gulig PA (1991) The *Salmonella typhimurium* virulence plasmid encodes a positive regulator of a plasmid-encoded virulence gene. *J. Bacteriol.* 173: 7176–7185.

Calvo JM, Matthews RG (1994) The leucine-responsive regulatory protein, a global regulator of metabolism in *Escherichia coli*. *Microbiol. Rev.* 58: 466–490.

Charles IG, Dougan G, Pickard D *et al.* (1989) Molecular cloning and characterization of protective outer membrane protein P.69 from *Bordetella pertussis*. *Proc. Natl Acad. Sci. USA* 86: 3554–3558.

Chatfield SN, Dorman CJ, Hayward C, Dougan G (1991) Role of *ompR*-dependent genes in *Salmonella typhimurium* virulence: mutants deficient in both *ompC* and *ompF* are attenuated *in vivo*. *Infect. Immun.* 59: 449–452.

Chen C-Y, Buchmeier NA, Libby S, Fang FC, Krause M, Guiney DG (1995) Central regulatory role for the RpoS sigma factor in expression of *Salmonella dublin* plasmid virulence genes. *J. Bacteriol.* 177: 5303–5309.

Chen D, Bowater R, Dorman CJ, Lilley DMJ (1992) Activity of a plasmid-borne *leu500* promoter depends on the transcription and translation of an adjacent gene. *Proc. Natl Acad. Sci. USA* 89: 8784–8788.

Colonna B, Casalino M, Fradiani PA *et al.* (1995) H-NS regulation of virulence gene expression in enteroinvasive *Escherichia coli* harboring the virulence plasmid integrated into the host chromosome. *J. Bacteriol.* 177: 4703–4712.

Curtiss IIIR, Kelly SM (1987) *Salmonella typhimurium* deletion mutants lacking adenylate cyclase and cyclic AMP receptor protein are avirulent and immunogenic. *Infect. Immun.* 55: 3035–3043.

Curtiss IIIR, Goldschmidt RM, Fletchall NB, Kelley SM (1988) Avirulent *Salmonella typhimurium cya crp* oral vaccine strains expressing a streptococcal colonization and virulence antigen. *Vaccine* 6: 155–160.

DiRita VJ, (1992) Coordinate expression of virulence genes by ToxR in *Vibrio cholerae*. *Mol. Microbiol.* 6: 451–458.

DiRita VJ, Mekalanos JJ (1991) Periplasmic interaction between two membrane regulatory proteins, ToxR and ToxS, results in signal transduction and transcriptional activation. *Cell* 64: 29–37.

DiRita VJ, Parsot C, Jander G, Mekalanos JJ (1991) Regulatory cascade controls virulence in *Vibrio cholerae*. *Proc. Natl Acad. Sci. USA* 88: 5403–5407.

Dorman CJ (1992) The VirF protein from *Shigella flexneri* is a member of the AraC transcription factor superfamily and is highly homologous to Rns, a positive regulator of virulence genes in enterotoxigenic *Escherichia coli*. *Mol. Microbiol.* 6: 1575.

Dorman CJ (1994) *Genetics of Bacterial Virulence*. Oxford: Blackwell Science.

Dorman CJ (1995) DNA topology and the global control of bacterial gene expression: implications for the regulation of virulence gene expression. *Microbiology* 141: 1271–1280.

Dorman CJ, Higgins CF (1987) Fimbrial phase variation in *Escherichia coli*: dependence on integration host factor and homologies with other site-specific recombinases. *J. Bacteriol.* 169: 3840–3843.

Dorman CJ, Ní Bhriain N (1992) Thermal regulation of *fimA*, the *Escherichia coli* gene coding for the type 1 fimbrial subunit protein. *FEMS Microbiol. Lett.* 99: 125–130.

Dorman CJ, Chatfield S, Higgins CF, Hayward C, Dougan G (1989) Characterization of porin and *ompR* mutants of a virulent strain of *Salmonella typhimurium*: *ompR* mutants are attenuated *in vivo*. *Infect. Immun.* 57: 2136–2140.

Dorman CJ, Ní Bhriain N, Higgins CF (1990) DNA supercoiling and environmental regulation of virulence gene expression in *Shigella flexneri*. *Nature* 344: 789–792.

Dove SL, Dorman CJ (1994) The site-specific recombination system regulating expression of the type 1 fimbrial subunit gene of *Escherichia coli* is sensitive to changes in DNA supercoiling. *Mol. Microbiol.* 14: 975–988.

Dove SL, Dorman CJ (1996) Multicopy *fimB* gene expression in *Escherichia coli*: binding to inverted repeats *in vivo*, effect on *fimA* gene transcription and DNA inversion. *Mol. Microbiol.* 21: 1161–1173.

Dove SL, Smith SGJ, Dorman CJ (1997) Control of *Escherichia coli* type 1 fimbrial gene expression in stationary phase: a negative role for RpoS. *Mol. Gen. Genet.* 254: 13–20.

Drlica K (1992) Control of bacterial DNA supercoiling. *Mol. Microbiol.* 6: 425–433.

Eisenstein BI, Sweet D, Vaughn V, Friedman DI (1987) Integration host factor is required for the DNA inversion event that controls phase variation in *Escherichia coli*. *Proc. Natl Acad. Sci. USA* 84: 6506–6510.

Fang FC, Libby SJ, Buchmeier NA *et al.* (1992) The alternative σ factor KatF (RpoS) regulates *Salmonella* virulence. *Proc. Natl Acad. Sci. USA* 89: 11978–11982.

Fields PI, Groisman EA, Heffron F (1989) A *Salmonella* locus that controls resistance to microbicidal proteins from phagocytic cells. *Science* 243: 1059–1062.

Forsman K, Göransson M, Uhlin BE (1989) Autoregulation and multiple DNA interactions by a transcriptional regulatory protein in *E. coli* pili biogenesis. *EMBO J.* 8: 1271–1277.

Free A, Dorman CJ (1995) Coupling of *Escherichia coli hns* mRNA levels to DNA synthesis by autoregulation: implications for growth phase control. *Mol. Microbiol.* 18: 101–113.

Gally DL, Rucker TJ, Blomfield IC (1994) The leucine-responsive regulatory protein binds to the *fim* switch to control phase variation of type 1 fimbrial expression in *Escherichia coli* K-12. *J. Bacteriol.* 176: 5665–5672.

Gally DL, Leathart J, Blomfield IC (1996) Interaction of FimB and FimE with the *fim* switch that controls the phase variation of type 1 fimbriae in *Escherichia coli* K-12. *Mol. Microbiol.* 21: 725–738.

Gillen KL, Hughes KT (1991) Molecular characterization of *flgM*, a gene encoding a negative regulator of flagellin biosynthesis in *Salmonella typhimurium*. *J. Bacteriol.* 173: 6453–6495.

Göransson M, Sondén B, Nilsson P *et al.* (1990) Transcriptional silencing and thermoregulation of gene expression in *Escherichia coli*. *Nature* 344: 682–685.

Grob P, Guiney DG (1996) *In vitro* binding of the *Salmonella dublin* virulence plasmid regulatory protein SpvR to the promoter regions of *spvA* and *spvR*. *J. Bacteriol.* 178: 1813–1820.

Groisman EA, Hefferon F (1995) Regulation of *Salmonella* virulence by two-component regulatory systems. In Hoch JA, Silhavy TJ (eds) *Two-Component Signal Transduction*. Washington, DC: American Society for Microbiology Press, pp. 319–332.

Groisman EA, Saier Jr MH, Ochman H (1992) Horizontal transfer of a phosphatase gene as evidence for mosaic structure of the *Salmonella* geneome. *EMBO J.* 11: 1309–1316.

Guiney DG, Libby S, Fang FC, Krause M, Fierer J (1995) Growth-phase regulation of plasmid virulence genes in *Salmonella*. *Trends Microbiol.* 3: 275–279.

Gunn JS, Miller SI (1996) PhoP-PhoQ activates transcription of *pmrAB*, encoding a two-component regulatory system involved in *Salmonella typhimurium* antimicrobial peptide resistance. *J. Bacteriol.* 178: 6857–6864.

Hale TL (1991) Genetic basis of virulence in *Shigella flexneri*. *Microbiol. Rev.* 55: 206–224.

Harrison JA, Pickard D, Higgins CF *et al.* (1994) Role of *hns* in the virulence phenotype of pathogenic salmonellae. *Mol. Microbiol.* 13: 133–140.

Heffernan EJ, Harwood J, Fierer J, Guiney DG (1992) The *Salmonella typhimurium* virulence plasmid complement resistance gene *rck* is homologous to a family of virulence-related outer membrane protein genes including *pagC* and *ail*. *J. Bacteriol.* 174: 84–91.

Hensel M, Shea JE, Gleeson C, Jones MD, Dalton E, Holden DW (1995) Simultaneous identification of bacterial virulence genes by negative selection. *Science* 269: 400–403.

Higgins DE, DiRita VJ (1994) Transcriptional control of *toxT*, a regulatory gene in the ToxR regulon of *Vibrio cholerae*. *Mol. Microbiol.* 14: 17–29.

Higgins DE, DiRita VJ (1996) Genetic analysis of the interaction between *Vibrio cholerae* transcription activator ToxR and *toxT* promoter DNA. *J. Bacteriol.* 178: 1080–1087.

Higgins DE, Nazareno E, DiRita VJ (1992) The virulence gene activator ToxT from *Vibrio cholerae* is a member of the AraC family of transcriptional activators. *J. Bacteriol.* 174: 6974–6980.

Hromockyj AE, Tucker SC, Maurelli AT (1992) Temperature regulation of *Shigella* virulence: identification of the repressor gene *virR*, an analogue of *hns*, and partial complementation by tyrosyl transfer RNA (tRNA1tyr). *Mol. Microbiol.* 6: 2113–2124.

Huh YJ, Weiss AA (1991) A 23-kilodalton protein, distinct from BvgA, expressed by virulent *Bordetella pertussis* binds to the promoter region of *vir*-regulated genes. *Infect. Immun.* 59: 2389–2395.

Jones S, Yu B, Bainton NJ *et al.* (1993) The *lux* autoinducer regulates the production of exoenzyme virulence determinants in *Erwinia carotovora* and *Pseudomonas aeruginosa*. *EMBO J.* 12: 2477–2482.

Kaltenbach LS, Braaten BA, Low DA (1995) Speciifc binding of PapI to Lrp-*pap* DNA complexes. *J. Bacteriol.* 177: 6449–6455.

Kawula TH, Orndorff PE (1991) Rapid site-specific DNA inversion in *Escherichia coli* mutants lacking the histone-like protein H-NS. *J. Bacteriol.* 173: 4116–4123.

Kier LD, Weppleman R, Ames BN (1977) Regulation of two phosphatases and a cyclic phosphodiesterase of *Salmonella typhimurium*. *J. Bacteriol.* 130: 420–428.

Klemm P (1986) Two regulatory *fim* genes, *fimB* and *fimE*, control the phase variation of type 1 fimbriae in *Escherichia coli*. *EMBO J.* 5: 1389–1393.

Knapp S, Mekalanos JJ (1988) Two *trans*-acting regulatory genes (*vir* and *mod*) control antigenic modulation in *Bordetella pertussis*. *J. Bacteriol.* 170: 5059–5066.

Kosturko LD, Daub E, Murialdo H (1989) The interaction of *E. coli* integration host factor and λ *cos* DNA: multiple complex formation and protein-induced bending. *Nucleic Acids Res.* 17: 317–334.

Kowarz L, Coynault C, Robbe-Saule V, Norel F (1994) The *Salmonella typhimurium katF* (*rpoS*) gene: cloning, nucleotide sequence, and regulation of *spvR* and *spvABCD* virulence plasmid genes. *J. Bacteriol.* 176: 6852–6860.

Krause M, Roudier C, Fierer J, Harwood J, Guiney DG (1991) Molecular analysis of the virulence locus of the *Salmonella dublin* plasmid pSDL2. *Mol. Microbiol.* 5: 307–316.

Krause M, Fang FC, and Guiney DG (1992) Regulation of plasmid virulence gene expression in *Salmonella dublin* involves an unusual operon structure. *J. Bacteriol.* 174: 4482–4489.

Krause M, Fang FC, El-Gedaily A, Libby S, and Guiney DG (1995) Mutational analysis of SpvR binding to DNA in the regulation of the *Salmonella* plasmid virulence operon. *Plasmid* 34: 37–47.

Lacey BW (1960) Antigenic modulation of *Bordetella pertussis*. *J. Hygiene* 58: 57–91.

Lampidis R, Gross R, Sokolovic Z, Goebel W, Kraft J (1994) The virulence regulator protein of *Listeria ivanovii* is highly homologous to PrfA from *Listeria monocytogenes* and both belong to the Crp-Fnr family of transcription regulators. *Mol. Microbiol.* 13: 141–151.

Lange R, Hengge-Aronis R (1994) The cellular concentration of the σs subunit of RNA polymerase in *Escherichia coli* is controlled at the levels of transcription, translation and protein stability. *Genes Dev.* 8: 1600–1612.

Lange R, Fischer D, Hengge-Aronis R (1995) Identification of transcriptional start sites and the role of ppGpp in the expression of *rpoS*, the structural gene for the σs subunit of RNA polymerase in *Escherichia coli*. *J. Bacteriol.* 177: 4676–4680.

Leichmeister-Wächter M, Haffner C, Domann E, Goebel W, Chakraborty T (1990) Identification of a gene that positively regulates expression of lysteriolysin, the major virulence factor of *Listeria monocytogenes*. *Proc. Natl Acad. Sci. USA* 87: 8336–8340.

Leichmeister-Wächter M, Domann E, Chakraborty T (1992) The expression of virulence genes in *L. monocytogenes* is thermoregulated. *J. Bacteriol.* 174: 947–952.

Liu LF, Wang JC (1987) Supercoiling of the DNA template during transcription. *Proc. Natl Acad. Sci. USA* 84: 7024–7027.

Loewen PC, Hengge-Aronis R (1994) The role of the sigma factor σs (KatF) in bacterial global regulation. *Ann. Rev. Microbiol.* 48: 53–80.

Mahan MJ, Slauch JM, Mekalanos JJ (1993) Selection of bacterial virulence genes that are specifically induced in host tissues. *Science* 259: 686–688.

McEwen J, Silverman P (1980) Genetic analysis of *Escherichia coli* K-12 chromosomal mutants defective in expression of F-plasmid functions: identification of genes *cpxA* and *cpxB*. *J. Bacteriol.* 144: 60–67.

Mekalanos JJ (1992) Environmental signals controlling expression of virulence determinants in bacteria. *J. Bacteriol.* 174: 1–7.

Mekalanos JJ, Swartz DL, Pearson GDN, Harford N, Groyne F, de Wilde M (1983) Cholera toxin genes: nucleotide sequence, deletion analysis and vaccine development. *Nature* 306: 551–557.

Mengaud J, Dramsi S, Gouin E, Vazquez-Boland JA, Milon G, Cossart P (1991) Pleiotropic control of *Listeria monocytogenes* virulence factors by a gene which is autoregulated. *Mol. Microbiol.* 5: 2273–2283.

Miller JF, Mekalanos JJ, Falkow S (1989a) Coordinate regulation and sensory transduction in the control of bacterial virulence. *Science* 243: 916–922.

Miller SI, Kukral AM, Mekalanos JJ (1989b). A two-component regulatory system (*phoP phoQ*) controls *Salmonella typhimurium* virulence. *Proc. Natl Acad. Sci. USA* 86: 5054–5058.

Miller VL, Mekalanos JJ (1988) A novel suicide vector and its use in construction of insertion mutations: osmoregulation of outer membrane proteins and virulence determinants in *Vibrio cholerae* requires *toxR*. *J. Bacteriol.* 170: 2575–2683.

Miller VL, Taylor RK, Mekalanos JJ (1987) Cholera toxin transcriptional activator ToxR is a transmembrane DNA binding protein. *Cell* 48: 271–279.

Miller VL, DiRita VJ, Mekalanos JJ (1991) Identification of *toxS*, a regulatory gene whose product enhances ToxR-mediated activation of the cholera toxin promoter. *J. Bacteriol.* 171: 1288–1293.

Muffler A, Fischer D, Altuvia S, Storz G, Hengge-Aronis R (1996) The response regulator RssB controls stability of the σs subunit of RNA polymerase in *Escherichia coli*. *EMBO J.* 15: 1333–1339.

Nakayama S-I, Watanabe H (1995) Involvement of *cpxA*, a sensor of a two-component regulatory system, in the pH-dependent regulation of expression of *Shigella sonnei virF* gene. *J. Bacteriol.* 177: 5062–5069.

Ní Bhriain N, Dorman CJ (1993) Isolation and characterization of a *topA* mutant of *Shigella flexneri*. *Mol. Microbiol.* 7: 351–358.

Norel F, Robbe-Saule V, Popoff MY, Coynault C (1992) The putative sigma factor KatF (RpoS) is required for the transcription of the *Salmonella typhimurium* virulence gene *spvB* in *Escherichia coli*. *FEMS Microbiol. Lett.* 99: 271–276.

O'Byrne CP, Dorman CJ (1994a). Transcription of the *Salmonella typhimurium spv* virulence locus is regulated negatively by the nucleoid-associated protein H-NS. *FEMS Microbiol. Lett.* 121: 99–106.

O'Byrne CP, Dorman CJ (1994b). The *spv* operon of *Salmonella typhimurium* LT-2 is regulated by the cyclic AMP (cAMP)-cAMP receptor protein system. *J. Bacteriol.* 176: 905–912.

Ogierman MA, Manning PA (1992) Homology of TcpN, a putative regulatory protein of *Vibrio cholerae*, to the AraC family of transcriptinal activators. *Gene* 116: 93–97.

Ohnishi K, Kutsukake K, Suzuki H, Iiono T (1992) A novel transcription regulation mechanism in the flagellar regulon of *Salmonella typhimurium*: an anti-sigma factor, σf. *Mol. Microbiol.* 6: 3149–3157.

Olsen PB, Klemm P (1994) Localization of promoters in the *fim* gene cluster and the effect of H-NS on the transcription of *fimB* and *fimE*. *FEMS Microbiol. Lett.* 116: 95–100.

Ottemann KM, DiRita VJ, Mekalanos JJ (1992) ToxR proteins with substitutions in residues conserved with OmpR fail to activate transcription from the cholera toxin promoter. *J. Bacteriol.* 174: 6807–6814.

Park SF, Kroll RG (1993) Expression of listeriolysin and phosphatidylinositol-specific phospholipase C is repressed by the plant-derived molecule cellobiose in *Listeria monocytogenes*. *Mol. Microbiol.* 8: 653–661.

Parsot C, Mekalanos JJ (1990) Expression of ToxR, the transcriptional activator of the virulence factors in *Vibrio cholerae*, is modulated by the heat shock response. *Proc. Natl Acad. Sci. USA* 87: 9898–9902.

Parsot C, Mekalanos JJ (1991) Expression of the *Vibrio cholerae* gene encoding aldehyde dehydrogenase is under the control of ToxR, the cholera toxin transcriptional activator. *J. Bacteriol.* 173: 2842–2851.

Parsot C, Mekalanos JJ (1992) Structural analysis of the *acfA* and *acfD* genes of *Vibrio cholerae*: effects of DNA topology and transcriptional activators on expression. *J. Bacteriol.* 174: 5211–5218.

Parsot C, Taxman E, Mekalanos JJ (1991) ToxR regulates the production of lipoproteins and the expression of serum resistance in *Vibrio cholerae*. *Proc. Natl Acad. Sci. USA* 88: 1641–1645.

Passador L, Cook JM, Gambello MJ, Rust L, Iglewski BH (1993) Expression of *Pseudomonas aeruginosa* virulence genes requires cell-to-cell communication. *Science* 260: 1127–1130.

Pearson JP, Passador L, Iglewski BH, Greenberg EP (1995) A second *N*-acylhomoserine lactone signal produced by *Pseudomonas aeruginosa*. *Proc. Natl Acad. Sci. USA* 92: 1490–1494.

Pickard D, Li J, Roberts M *et al.* (1994) Characterization of defined *ompR* mutants of *Salmonella typhi*: *ompR* is involved in the regulation of Vi polysaccharide expression. *Infect. Immun.* 62: 3984–3993.

Piper KR, von Bodman SB, Farrand SK (1993) Conjugation factor of *Agrobacterium tumefaciens* regulates Ti plasmid transfer by autoinduction. *Nature* 362: 448–450.

Porter ME, Dorman CJ (1994) A role for H-NS in the thermo-osmotic regulation of virulence gene expression in *Shigella flexneri*. *J. Bacteriol.* 176: 4187–4191.

Pratt LA, Silhavy TJ (1996) The response regulator SprE controls the stability of RpoS. *Proc. Natl Acad. Sci. USA* 93: 2488–2492.

Pullinger GD, Baird GD, Williamson GM, Lax AJ (1989) Nucleotide sequence of a plasmid gene involved in the virulence of salmonellas. *Nucleic Acids Res.* 17: 7983.

Pulkkinen WS, Miller SI (1991) A *Salmonella typhimurium* virulence protein is similar to a *Yersinia enterocolitica* invasion protein and a bacteriophage lambda outer membrane protein. *J. Bacteriol.* 173: 86–93.

Relman DA, Domenighini M, Tuomanen E, Rappouli R, Falkow S (1989) Filamentous hemagglutinin of *Bordetella pertussis*: nucleotide sequence and crucial role in adherence. *Proc. Natl Acad. Sci. USA* 86: 2637–2641.

Ritter A, Blum G, Embödy L *et al.* (1995) tRNA genes and pathogenicity islands: influence on virulence and metabolic properties of uropathogenic *Escherichia coli*. *Mol. Microbiol.* 17: 109–121.

Robertson BD, Meyer TF. (1992) Genetic variation in pathogenic bacteria. *Trends Genet.* 8: 422–427.

Roy CR, Falkow S (1991) Identification of *Bordetella pertussis* regulatory sequences required for transcriptional activation of the *fhaB* gene and autoregulation of the *bvgAS* operon. *J. Bacteriol.* 173: 2385–2392.

Roy CR, Miller JF, Falkow S (1990) Autogenous regulation of the *Bordetella pertussis bvgABC* operon. *Proc. Natl Acad. Sci. USA* 87: 3763–3767.

Sakai T, Sasakawa C, Makino S, Kamata K, Yoshikawa M (1986) Molecular cloning of a genetic determinant for Congo red binding ability which is essential for the virulence of *Shigella flexneri*. *Infect. Immun.* 51: 476–482.

Scarlato V, Prugnola A, Aricò B, Rappuoli R (1990) Positive transcriptional feedback at the *bvg* locus controls expression of virulence factors in *Bordetella pertussis*. *Proc. Natl Acad. Sci. USA* 87: 6753–6757.

Scarlato V, Aricò B, Prugnola A, Rappuoli R (1991) Sequential activation and environmental regulation of virulence genes in *Bordetella pertussis*. *EMBO J.* 10: 3971–3975.

Scarlato V, Aricò B, Rappuoli R (1993a) DNA topology affects transcriptional regulation of the pertussis toxin gene of *Bordetella pertussis* in *Escherichia coli* and *in vitro*. *J. Bacteriol.* 175: 4764–4771.

Scarlato V, Aricò B, Domenighini M, Rappuoli R, (1993b) Environmental regulation of virulence factors in *Bordetella* species. *BioEssays* 15: 99–104.

Schmitt CK, Darnell SC, Tesh VL, Stocker BAD, O'Brien AD (1994) Mutation of *flgM* attenuates virulence of *Salmonella typhimurium*, and mutation of *fliA* represses the attenuated phenotype. *J. Bacteriol.* 176: 368–377.

Schwan WR, Seifert HS, Duncan JL (1992) Growth conditions mediate differential transcription of *fim* genes involved in phase variation of type 1 pili. *J. Bacteriol.* 174: 2367–2375.

Seed PC, Passador L, Iglewski BH (1995) Activation of the *Pseudomonas aeruginosa lasI* gene by LasR and the *Pseudomonas* autoinducer PAI: an autoinduction regulatory hierarchy. *J. Bacteriol.* 177: 654–659.

Sheehan B, Kocks C, Dramsi S *et al.* (1994) Molecular and genetic determinants of the *Listeria monocytogenes* infectious process. *Curr. Top. Microbiol. Immunol.* 192: 187–216.

Sheehan B, Klarsfeld A, Msadek T, Cossart P (1995) Differential activation of virulence gene expression by PrfA, the *Listeria monocytogenes* virulence regulator. *J. Bacteriol.* 177: 6469–6476.

Small P, Blankenhorn D, Welty D, Zinser E, Slonczewski JL (1994) Acid and base resistance in *Escherichia coli* and *Shigella flexneri*: role of *rpoS* and growth pH. *J. Bacteriol.* 176: 1729–1737.

Soncini FC, Groisman EA (1996) Two-component regulatory systems can interact to process multiple environmental signals. *J. Bacteriol.* 178: 6796–6801.

Stibitz S, Yang MS (1991) Subcellular localization and immunological detection of proteins encoded by the *vir* locus of *Bordetella pertussis*. *J. Bacteriol.* 173: 4288–4296.

Stock J, Ninfa AJ, Stock AM (1989) Protein phosphorylation and regulation of adaptive responses in bacteria. *Microbiol. Rev.* 53: 450–490.

Taira S, Heiskanen P, Hurme R, Heikkilä H, Riikonen P, Rhen M (1995) Evidence for functional polymorphism of the *spvR* gene regulating virulence gene expression in *Salmonella*. *Mol. Gen. Genet.* 246: 437–444.

Tobe T, Sasakawa C, Okada N, Honma Y, Yoshikawa M (1991) Temperature-regulated expression of invasion genes in *Shigella flexneri* is controlled through the transcriptional activation of the *virB* gene on the large plasmid. *Mol. Microbiol.* 5: 887–893.

Tobe T, Yoshikawa M, Mizuno T, Sasakawa C (1993) Transcriptional control of the invasion regulatory gene *virB* of *Shigella flexneri*: activation by VirF and repression by H-NS. *J. Bacteriol.* 175: 6142–6149.

Tobe T, Yoshikawa M, Sasakawa C (1995) Thermoregulation of *virB* transcription in *Shigella flexneri* by sensing of changes in local DNA superhelicity. *J. Bacteriol.* 177: 1094–1097.

Uhl MA, Miller JF (1994) Autophosphorylation and phosphotransfer in the *Bordetella pertussis* BvgAS signal transduction cascade. *Proc. Natl Acad. Sci. USA* 91: 1163–1167.

Uhl MA, Miller JF. (1995) *Bordetella pertussis* BvgAS virulence control system. In: Hoch JA, Silhavy TJ (eds) *Two-Component Signal Transduction*. Washington, DC: American Society for Microbiology Press, pp. 333–349.

Ussery DW, Hinton JCD, Jordi BJAM *et al.* (1994) The chromatin-associated protein H-NS. *Biochimie* 76: 968–980.

van Workum M, van Dooren SJM, Oldenburg N *et al.* (1996) DNA supercoiling depends on the phosphorylation potential in *Escherichia coli*. *Mol. Microbiol.* 20: 351–360.

Watanabe H, Arakawa E, Ito K-I, Kato J-I, Nakamura A (1990) Genetic analysis of an invasion region by use of a Tn*3-lac* transposon and identification of a second positive regulator gene, *invE*, for cell invasion of *Shigella sonnei*: significant homology of InvE with ParB of plasmid P1. *J. Bacteriol.* 172: 619–629.

Weiss AA, Hewlett EL (1986) Virulence factors of *Bordetella pertussis*. *Annu. Rev. Microbiol.* 40: 661–686.

Willems R, Paul A, van der Heide HGJ, ter Avest AR, Mooi FR (1990) Fimbrial phase variation in *Bordetella pertussis*: a novel mechanism for transcriptional regulation. *EMBO J.* 9: 2803–2809.

Zagaglia C, Casalino M, Colonna B, Conti C, Calconi A, Nicoletti M (1991) Virulence plasmids of enteroinvasive *Escherichia coli* and *Shigella flexneri* integrate into a specific site on the host chromosome: integration greatly reduces expression of plasmid-carried virulence genes. *Infect. Immun.* 59: 792–799.

Zhang L, Murphy PJ, Kerr A, Tate ME (1993) *Agrobacterium* conjugation and gene regulation by *N*-acyl-L-homoserine lactone. *Nature* 362: 446–448.

20

Population Genetics of Bacterial Pathogens

Brian G. Spratt, Edward J. Feil and Noel H. Smith

Oxford University, Oxford, UK

Population genetics is concerned with describing the patterns of genetic variation that occur within populations, and attempting to explain how they have arisen in terms of the known evolutionary processes of mutation, recombination, selection, drift, etc. This may seem of little concern to medical microbiologists, but out of this apparently unpromising material have come important insights into the origins and evolution of bacterial pathogens. Studies of the population genetics of bacteria started in the 1970s when Milkman (1973) analysed genetic variation within *Escherichia coli* in an attempt to test Kimura's neutral theory of evolution. The impact of population genetics on medical microbiology is, however, mainly due to the pioneering work of Selander and his colleagues who, in the 1980s, analysed the population structures of a number of bacterial pathogens (Selander and Levin, 1980; Ochman *et al.*, 1983; Whittam *et al.*, 1983a, b; Selander *et al.*, 1985; Caugant *et al.*, 1986; Musser *et al.*, 1986). These studies established that bacterial populations were essentially clonal, implying that rates of recombination are low in nature, and showed that for many bacterial pathogens a large number of distinct lineages, or clones, can be identified, but only a few clones, or clusters of closely related clones, cause most serious disease (Selander and Musser, 1990). In some cases, particular clones may be associated with the different diseases caused by a single species. The great variation in pathogenic potential among isolates of many bacterial species is of immediate practical importance, because it allows a search for the factors that determine the increased virulence of particular clones.

The population-genetic approach has also been important in showing what type of variation should be used, and what variation should not be used, to assess the genetic relatedness of isolates of a bacterial pathogen. For example, it has stressed the inadequacy of serological methods for typing isolates of bacterial pathogens and the need to use genetic variation that is likely to be selectively neutral (Selander and Musser, 1990).

Most subsequent studies have confirmed, or have been interpreted as confirming, that bacterial

 doi:10.1006/bkmm.2001.0020

populations are clonal, but in recent years the generality of this view has been challenged (Maynard Smith *et al.*, 1993; Guttman, 1997; Spratt and Maiden, 1999), and it is now accepted that the population structures of bacteria range from the highly clonal to the non-clonal. In this chapter we review the properties of clonal and non-clonal populations, the methods by which the population structures of bacteria can be determined, the relevance of population structure to the evolution of bacterial pathogens, and the way in which molecular typing and molecular epidemiology should be approached. It is not possible in the available space to cover all of the bacterial pathogens whose population or evolutionary biology have been addressed, and we will focus predominantly on studies of *Salmonella enterica* and *E. coli*, as examples of clonal species, and of *Neisseria gonorrhoeae* and *N. meningitidis*, as examples of basically non-clonal species. Several useful general summaries of bacterial population genetics have been published in recent years (Young, 1989; Selander and Musser, 1990; Maynard Smith, 1996; Musser, 1996; Haubold and Rainey, 1997; Spratt and Maiden, 1999). A number of relevant articles are also included in a symposium volume (Baumberg *et al.*, 1995). The population genetics of bacteria, fungi, and protozoan parasites, have many similarities and these have been reviewed by Tibayrenc (1996).

Species and Populations

One of the difficulties of bacterial population genetics is the lack of any satisfactory concept of species (Maynard Smith, 1996). In eukaryotes, species are defined in terms of exchange of genes, which occurs between members of a species, but not (or very rarely) between members of different species. Attempts have been made to define bacterial species in terms of gene flow. For example, Dykhuizen and Green (1991) suggested that a group of bacterial isolates could be placed in the same species if analysis of the sequences of their housekeeping genes showed evidence of recombination (non-congruence of gene trees; see below), and in different species if the sequences provided no evidence for recombination between the isolates (congruent gene trees). This approach works reasonably well with enterobacteria (Dykhuizen and Green, 1991), and can be used to justify the species status of *N. gonorrhoeae* (Vázquez *et al.*, 1993), but it is not generally applicable, as recombination may be very rare within some bacteria (Dykhuizen *et al.*, 1993), whereas in others recombination occurs between the housekeeping genes of isolates that differ by 20% in sequence, and which mani-festly should be called different species (Zhou *et al.*, 1997).

Consideration of gene flow within and between populations cannot provide general rules that can help microbiologists to define species. We have to take a pragmatic attitude and define bacterial species as clusters of isolates that occupy a defined sequence space, and which are clearly separable from related clusters of isolates. Population genetics provides the quantitative methods that allow the genetic relatedness between isolates of bacterial species to be determined, and which can be used to decide whether named bacterial species fulfil the above criteria. Studies of this kind have shown that some named bacterial species contain genetically distinct subpopulations (cryptic species) which may deserve to be given species status (e.g. Selander *et al.*, 1985; Quentin *et al.*, 1993). Conversely, it has shown that some named species, or subspecies, are clearly inappropriate groupings of isolates. For example, population genetics has shown that isolates of *Shigella* 'species' are not genetically distinct from *E. coli* isolates (Ochman *et al.*, 1983), and that *Haemophilus influenzae biogroup aegyptius* consists of several lineages within *H. influenzae* that are not particularly closely related to each other (Musser and Selander, 1990).

It would be convenient to be able to define species as groups of isolates that share a certain level of sequence similarity, but this approach is too rigid, since some clearly delineated species may occupy a very narrow sequence space (e.g. *N. gonorrhoeae*), whereas others may occupy a much wider sequence space (e.g. *Helicobacter pylori*). In the case of several pathogens, for understandable reasons, species are defined by the diseases they cause, resulting in some cases in a single human-adapted clone being raised to species status (e.g. *Mycobacterium tuberculosis*). Our concept of bacterial species is therefore pragmatic, and sometimes leads to assemblages of isolates that appall population biologists, who prefer alternative assemblages that would in turn appall medical microbiologists.

Clonal and Non-clonal Population Structures

The population structures of bacteria depend on a number of factors. Crucially they depend on the frequency with which chromosomal genes in the population recombine. Recombination in bacteria is very different from that in higher organisms. Firstly, it is unlinked to reproduction, and alleles in the population are therefore not inevitably shuffled in each generation as they are in most higher organisms, and the extent of recombination in bacteria could therefore either be very low (or absent), or very high. Secondly, recombination in bacteria is thought to be highly localised, usually involving the unidirectional replacement of a small part of the chromosome of a recipient bacterium (typically a few kilobases) with the corresponding region from the genome of another isolate of the same species (Maynard Smith *et al.*, 1991).

If recombination is totally absent, the population will exist as an assemblage of independent lineages, or clones. These lineages will be stable over long periods of time as evolutionary change occurs predominantly by mutation, although diversity may also arise by other mechanisms, such as the gain and loss of plasmids, or the movement of insertion sequences. Differences in the frequencies of particular lineages in the population will occur over time as some are lost, or increase in frequency, by chance. Additionally, mutations that result in increased fitness will arise at intervals and the lineages within which these mutations occur will increase in frequency within the population as a consequence of natural selection. In time these successful lineages may spread worldwide, and the identification of the same strain in different continents, in different decades, is a typical feature of clonal populations. The continuous emergence and expansion of fitter clones inevitably results in the loss of other lineages and acts to purge the genetic variation within clonal populations (periodic selection; Atwood *et al.*, 1951; Levin, 1981).

The other extreme population structure arises when exchange of chromosomal segments in the population is very frequent, and evolutionary change occurs much more commonly by recombination than by mutation. Frequent recombinational exchanges lead to a constantly varying collection of genotypes within these populations and stable clones are not present. As a consequence, populations in which recombinational exchanges are very frequent will be non-clonal. In these populations, any fitter variants that arise will begin to increase in frequency within the population as they do in clonal populations. Frequent localised recombinational exchanges will, however, rapidly 'pock' the initially uniform genomes of these emerging lineages, resulting in transient clones whose members may be distinguishable over a number of years, but which eventually become so variable that they can no longer be recognised as the descendants of the ancestral lineage. In evolutionary terms, these transient clones diversify so rapidly that the emergence of fitter lineages does not reduce the variation within the population, since the variation is rescued by recombination. The ineffectiveness of periodic selection at reducing genetic variation, and the re-assortment of alleles by recombination, should lead to a much larger number of genotypes in a non-clonal population compared to a clonal population of similar antiquity. It is important to stress that the term non-clonal is slightly misleading. As we will discuss later, bacterial populations may be basically non-clonal, but lineages may still persist for lengths of time that are significant in the context of the epidemiology of infectious disease.

The highly clonal and the non-clonal population represent the extremes of possible population structures. Between these extremes there will be a gradient of population structures, depending on the extent to which evolutionary change is brought about by recombination compared to mutation (Maynard Smith *et al.*, 1993). Many bacterial species probably occupy this middle ground, where recombination is significant in the evolution of the population, but is not sufficiently frequent, compared with point mutation, to break down clonal associations.

One of the most important consequences of the variation in the extent of recombination within bacteria is the difference in the rates at which genotypes diversify, leading to populations in which genotypes persist virtually unchanged for hundreds or thousands of years and, at the other extreme, populations where genotypes are ephemeral entities that may persist for only a few months or years (**Fig. 1**).

In highly clonal species, the sequences of housekeeping genes, or electrophoretic variation indexed by multi-locus enzyme electrophoresis (MLEE; see below), can be used to construct dendrograms, which should approximate true phylogenetic histories. The phylogeny can then be used as a framework on to which other characteristics (pathogenicity, serology, host preference, presence of virulence genes, etc.) can be mapped. In this way important insights into the origins of the pathogenicity and host adaptation of the lineages within highly clonal species can be made. This is exemplified by studies with *S. enterica* that will be described below.

In non-clonal species, dendrograms constructed in this way are not phylogenetically informative, and the trees obtained with different types of variation will be non-congruent. Construction of dendrograms can still be useful for weakly clonal species, as isolates that cluster together on trees obtained using different types of variation can identify recently emerged transient clones (as observed with meningococci; see below). In weakly clonal species, however, the long-term evolution of the population is dominated by recombination, and the genetic relationships between the different transient clones, and between these clones and other isolates, cannot be inferred, and the trees provide no deep phylogenetic information (Holmes *et al.*, 1999; Feil *et al.*, 2001).

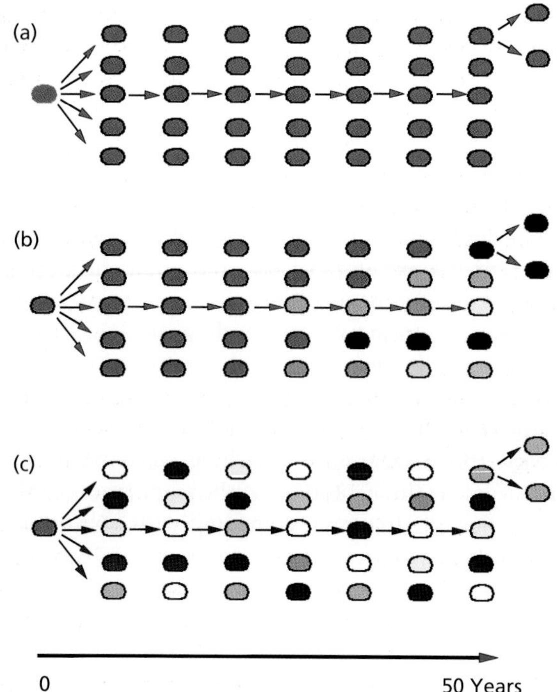

Fig. 1 Diversification of genotypes in clonal, weakly clonal, and non-clonal populations. In each panel an ancestral bacterium is shown at time zero, and the accumulation of electrophoretic variation in housekeeping enzymes (detected by MLEE) in a sample of the progeny over a 50-year period is represented by the changes in colour. (a) In a highly clonal population, electrophoretic variation accumulates slowly by mutation, and after 50 years the isolates are indistinguishable by MLEE from the ancestral bacterium. (b) In a weakly clonal population, variation accumulates more rapidly by localised recombinational exchanges, which occasionally involve one of the loci whose variation is indexed by MLEE, and the isolates after 50 years are therefore not identical, but form a cluster of closely related genotypes. (c) In a non-clonal population, variation accumulates rapidly as recombinational exchanges are very frequent and, after 50 years the isolates have become so diverse that their derivation from the same ancestral bacterium cannot be discerned. On the right of each panel are shown the genotypes of isolates from epidemiologically linked cases of disease. Even in the non-clonal population, the rate of diversification of genotypes is sufficiently slow that these isolates, recovered within a few weeks of each other, are likely to be almost indistinguishable, allowing short term epidemiological studies to be carried out. Longer term epidemiological studies will not be possible in the non-clonal population, since the genotypes diversify so rapidly that, after a few years, the progeny of the ancestral bacterium can no longer be shown to be related to each other.

Sex in bacteria has many of the advantages and disadvantages of sex in higher organisms. Thus, it facilitates the bringing together of advantageous mutations that occur within different lineages, and results in the maintenance of more genetic variation, and more genotypes, than occur in asexual populations. It also allows advantageous mutations, such as chromosomal mutations that provide increased resistance to antibiotics, to spread rapidly into other lineages. In highly clonal populations the spread into other lineages will be much slower, and selection applied by antibiotic usage will lead to an expansion of the lineage within which the antibiotic resistance mutation(s) occurred.

Sex may also have disadvantages, as a combination of alleles at different loci that results in a highly successful lineage will be broken down by recombination. Linkage between all loci is not necessarily broken down even by frequent recombinational exchanges. For example, Gupta *et al.* (1996) have provided a model in which, under certain conditions, immune selection from the host can structure a population into strains that are heterogeneous in overall genotype, but which have non-overlapping combinations of alleles at loci encoding immunoprotective cell-surface antigens.

Mechanisms of Sexual Exchange in Bacteria

Recombination between the chromosomal genes of different lineages within bacterial populations obviously depends on the availability of sexual mechanisms. Three mechanisms that can lead to the exchange of chromosomal genes are well documented – conjugation, transduction and transformation. Mobilisation of chromosomal genes by conjugative transposons has also been reported in *Enterococcus faecalis* (Torres *et al.*, 1991).

Conjugation differs from transduction and transformation since, at least in the laboratory, it can result in the replacement of large segments of bacterial chromosomes with those from a donor strain. However, the horizontal transfer of chromosomal genes by conjugation depends on the abundance in nature of plasmids that can mobilise chromosomal genes. The F factor of *E. coli* is the prototype of this type of plasmid, but it is probably not a good model for most conjugal plasmids (Wilkins, 1995). Firstly, it is derepressed for transfer functions, so that all bacteria within a population have the potential to mate with appropriate recipient bacteria, whereas the transfer genes of other conjugal plasmids are repressed, and only the few bacteria that happen to be expressing their transfer genes are able to initiate mating. Secondly, F is able to integrate into the chromosome to produce transient, or stable, Hfr strains that can efficiently mobilise chromosomal genes into recipient bacteria.

Attempts to produce stable Hfr strains with other conjugal plasmids have usually failed, and transfer of chromosomal genes by most conjugal plasmids is difficult to achieve in the laboratory (Wilkins, 1995). Transfer of chromosomal genes by conjugation is, therefore, expected to be rare in nature.

A more significant mechanism by which chromosomal segments may be exchanged within bacterial species is transduction. Many bacteriophages occasionally package a chromosomal segment rather than their own genome into the phage head, and these defective transducing particles can introduce the chromosomal segment into a recipient bacterium, where parts of the segment may replace the corresponding segment of the recipient by homologous recombination. Transduction has the potential to result in recombinational replacement sizes up to the packaging limit of the transducing phage, which in some cases may be 100 kb or more. In most bacteria, recombinational replacement sizes appear to be very much smaller than the size of the chromosomal segments that would be expected to be introduced by transduction, and it has been suggested that this may be due to differences between the restriction-modification systems of donor and recipient isolates which may result in the cleavage of the incoming DNA by restriction endonucleases (DuBose *et al.*, 1988; McKane and Milkman, 1995).

Several important bacterial pathogens, including *N. gonorrhoeae* and *N. meningitidis*, *H. influenzae*, *He. pylori*, *Campylobacter jejuni* and *Streptococcus pneumoniae*, are naturally transformable and can take up naked DNA present in their environment (Solomon and Grossman, 1996). In some cases, such as *Haemophilus* and *Neisseria*, uptake of DNA appears to be selective, such that species can recognise their own DNA, or that of closely related members of the genus, by the presence of specific uptake sequences that are present many times in their own genome, but which occur very rarely in other genomes (Danner *et al.*, 1980; Goodman and Scocca, 1988). In other species, such as *Strep. pneumoniae*, uptake appears to be indiscriminate and selectivity is presumably only achieved by the requirement for the introduced DNA to be sufficiently similar in sequence to that of the recipient for homologous recombination to be efficient. The size of DNA that can be exchanged by transformation is relatively small (a few kilobases), especially in nature where chromosomal DNA released by autolysis on to mucosal surfaces is likely to be degraded, and where recipient and donor bacteria may differ in restriction-modification systems.

It is impossible to decide which mechanism of genetic exchange has been responsible for any recombinational exchanges observed within bacteria recovered from nature. It is generally assumed that recombinational exchanges in naturally transformable species are due to this process, and that transduction is the likely mechanism in non-transformable species. Bacteriophages are widespread in nature and their potential importance in gene transfer is suggested by the role that they play in the horizontal spread of virulence genes between isolates of bacterial pathogens (Pappenheimer, 1977; O'Brien *et al.*, 1984; Weeks and Ferretti, 1984; Waldor and Mekalanos, 1996; Karaolis *et al.*, 1999).

Sexual mechanisms may be present in all isolates of a species (e.g. all *N. gonorrhoeae* isolates are believed to be naturally transformable), or they may be found in only a minority of isolates (e.g. depending on the presence of plasmids that can promote the conjugal transfer of chromosomal genes), or alternatively may depend on the existence, and abundance, of bacteriophage which can mediate gene exchange by transduction. Mel and Mekalanos (1996) suggested that horizontal gene transfer in nature may be more widespread than expected from laboratory studies, since plasmid transfer functions, competence, or production of phage receptors might be regulated by signals that are only produced during infection.

Species in which sexual mechanisms are absent will inevitably be highly clonal. The reverse is not necessarily true, however, since the frequency of effective genetic exchange in nature, that is between distinct genotypes, depends not only on the availability and efficiency of genetic exchange mechanisms, but also on the extent to which different lineages meet each other. This aspect of the population biology of bacteria may result in non-uniformity of the population structure within a single species, as discussed below for *H. influenzae* and *N. meningitidis*.

Recombinational exchanges in bacterial genes are likely to be predominantly between closely related isolates, typically member of the same species. In most cases, transduction between different species is likely to be limited by the lack of a common phage receptor. In transformation, DNA uptake sequences may reduce the uptake of DNA from other species, although this may be a relatively ineffective barrier since closely related species may share the same uptake sequence, as occurs in *N. gonorrhoeae* and *N. meningitidis*. Even if chromosomal DNA gains entry into a related species, the efficiency of homologous recombination falls off sharply with increasing sequence divergence between donor and recipient DNA (Zawadski *et al.*, 1995), because of reduced pairing between diverged DNA molecules, abortion of mispaired DNA duplexes by mismatch repair systems and, in some cases, differences in restriction-modification systems. These barriers to recombination between species have been reviewed elsewhere (Matic *et al.*, 1996). Inter-species recombinational exchanges, involving sequences that differ by up to about 25% in nucleotide sequence, have been found in some genes, although in most cases these events are believed to be very rare and have been observed because they are strongly selected by the human immune response (Reeves, 1993), or by antibiotic usage (Maynard Smith *et al.*, 1991).

Additive recombination that introduces novel genes into the chromosome has also been important during the evolution of pathogens. In several studies (Groisman and Ochman, 1996; Hacker *et al.*, 1997), chromosomal regions have been identified that are present in only some isolates of a species, or are in one species but are absent from a closely related species, and which contain genes implicated in virulence (pathogenicity islands). In most cases the origins of these regions are unknown, but some appear to be from distantly related species, because the introduced sequences differ significantly in base composition from that of the rest of the chromosome (Lan and Reeves, 1996).

Estimating the Extent of Recombination within Bacterial Populations

Recombination and Levels of Linkage Disequilibrium

How do we decide where a bacterial species fits in the spectrum of possible population structures? The usual approach is to look for linkage disequilibrium between the alleles at different chromosomal loci in a sample of the bacterial population. If recombination is frequent, there will be no association between the alleles at different loci (linkage equilibrium). In contrast, in the absence of recombination, the alleles within the population will be in strong linkage disequilibrium.

To apply this approach we need to be able to distinguish the alleles present at a number of loci within a bacterial population. For historical reasons, allelic variation within genes in bacterial populations is usually detected indirectly, by equating the electrophoretic mobilities of their gene products on starch gels with alleles at the corresponding gene loci (Selander and Musser, 1990). A large number of gene products can be detected on starch gels with specific enzyme stains (Selander *et al.*, 1986) and, typically, variation in about 15–25 enzymes is used (multi-locus enzyme electrophoresis; MLEE). MLEE is a powerful and general technique that can be applied to large numbers of isolates and provides the basic data required for the analysis of levels of linkage disequilibrium (Selander and Musser, 1990). The alleles present at each locus define a multi-locus genotype, or electrophoretic type (ET), and MLEE also provides a powerful technique for quantifying the genetic relatedness between isolates of clonal species by measuring the proportion of alleles shared by different ETs.

Housekeeping enzymes are used for MLEE, as it can be assumed that most of the variation in their electrophoretic mobility is likely to be selectively neutral, or nearly so. MLEE also assumes that electrophoretic variation is due to differences in the amino acid sequences of the enzyme (rather than post-translational modifications), which have arisen by the gradual divergence of the underlying alleles by the process of mutation. In most cases these assumptions appear to be reasonable (Boyd *et al.*, 1994), and isolates of a bacterial species that express the same electrophoretic variant at each of 20 loci are almost certainly very closely related, whereas those that differ at 10/20 loci are likely to be distantly related. These assumptions will, of course, not be valid if recombination is frequent (Feil *et al.*, 1995). This is unimportant for the analysis of linkage disequilibrium, as the statistical tests are specifically designed to address the extent of recombination in the population, but it does invalidate MLEE as a way of quantifying the degree of genetic relatedness of the major lineages in non-clonal populations. This problem is, however, not restricted to MLEE, as the relatedness between isolates of non-clonal species cannot be quantitated by means of any technique, since these populations are not evolving in a bifurcating, tree-like, manner.

A number of tests is available to measure whether the alleles in a bacterial population depart from linkage equilibrium (Sved, 1968; Brown *et al.*, 1980; Whittam *et al.*, 1983a; Hedrich and Thomson, 1986). Some of these look for significant associations between pairs of alleles, but the most sensitive methods are those that assess the linkage between all alleles (Maynard Smith *et al.*, 1993). In a freely recombining population, the distribution of the pairwise differences between strains (the proportion of the enzymes assayed at which two strains express different electrophoretic variants) should approximate the normal distribution. In the absence of recombination, linkage associations are not broken down and, as a result of repeated episodes of periodic selection, a few genotypes rise to high frequency and many genotypes are lost from the population. Consequently, the distribution of allelic mismatches in a clonal population deviates significantly from the normal distribution, with an excess of isolates

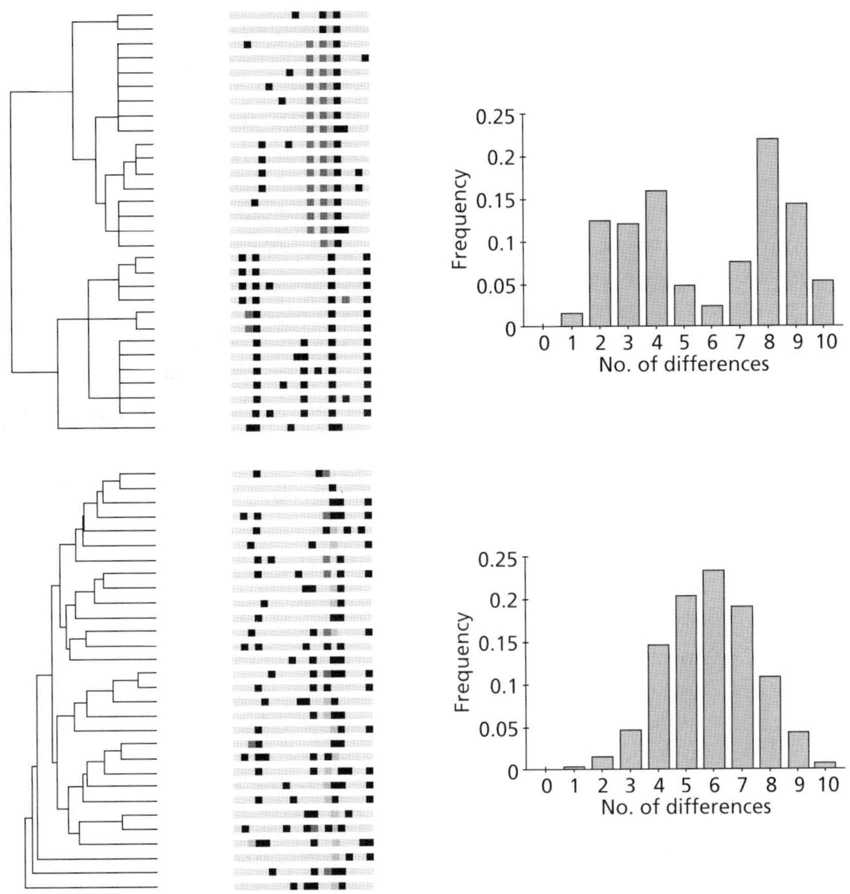

Fig. 2 Linkage associations in clonal and non-clonal populations. The top panel shows the dendrogram constructed from allelic variation at 20 loci within a highly clonal population. The chromosomes of the isolates are shown to the right, with allelic differences at the 20 loci shown by shading. The distribution of allelic mismatches is bimodal as a consequence of linkage disequilibrium, giving rise to an inflated variance. The bottom panel shows the data from a non-clonal population where, as a consequence of frequent recombination, the alleles at different loci are randomised among isolates, and the mismatch distribution is unimodal, with a smaller variance. The index of association (I_A) detects linkage disequilibrium by an increase in the variance of the distribution of allelic mismatches from that expected when alleles are randomly associated. Reproduced with permission from Whittam (1995).

that are very similar to each other and an excess of isolates that are very different (**Fig. 2**).

The *index of association* (I_A) provides a simple statistic that measures whether the observed distribution of allelic mismatches significantly departs from the null hypothesis of free recombination between alleles (Sved, 1968; Brown *et al.*, 1980; Whittam *et al.*, 1983b; Maynard Smith *et al.*, 1993; Haubold *et al.*, 1998). The statistic is set up such that $I_A = 0$ for a population at linkage equilibrium. The significance of any apparent linkage disequilibrium ($I_A > 0$) is assessed by measuring I_A for n random datasets, containing the same number of alleles, and the same frequency of each allele, as in the real dataset. If the value of I_A obtained with the real data is greater than that observed with any of the randomly generated datasets, the linkage disequilibrium is significant at the level of $p < 1/n$. The significance of I_A is estimated in this way, as the previously published method for determining the standard error of I_A (Brown *et al.*, 1980) is now known to be erroneous (J. Maynard Smith, personal communication; see also Haubold *et al.*, 1998).

One of the pitfalls of this approach is that linkage disequilibrium between alleles can be present in populations in which recombination is frequent. For example, a named species may include two quite distinct populations (cryptic species), and recombination may be common within each population, but very rare between populations. Analysis of a sample of this species by MLEE will, therefore, detect significant levels of linkage disequilibrium between alleles, leading to an erroneous conclusion about the extent of recombination and the population structure. Population subdivision can be detected by constructing a dendrogram with the electrophoretic data, and linkage disequilibrium can then be analysed in the separate subpopulations (Maynard Smith *et al.*, 1993). Linkage disequilibrium can also arise if the population structure is dominated by recombination in the long term, but highly successful lineages arise, and increase in frequency, before being disrupted by recombination (transient clones). This problem can be significant in those pathogens where most isolates rarely cause disease, but hypervirulent isolates occasionally emerge, and rise to high frequency within the small subpopulation of the species that cause disease. The over-representation of these transient hypervirulent clones can introduce artefactual linkage disequilibrium if the population that is studied consists of isolates recovered from episodes of disease (Maynard Smith *et al.*, 1993).

Similarly, a sample of a basically non-clonal population that contains many antibiotic-resistant isolates may show linkage disequilibrium, since the latter isolates will have emerged and increased in frequency in the population very recently (on an evolutionary time scale), and are likely to belong to a small number of closely similar ETs. This kind of sampling problem can be detected by showing that linkage disequilibrium disappears, or is greatly reduced, if only one example of each ET (rather than every isolate) is included in the analysis (Maynard Smith *et al.*, 1993).

Linkage equilibrium is less likely to be caused by sampling artefacts and, provided the sample of the population studied is sufficiently large and variable, to provide a statistically robust test of linkage relationships, the absence of any association between the alleles at different loci implies that recombination must be frequent, and that distinct lineages must meet each other in nature.

Other Indicators of Clonal Population Structures

Other features of a population, besides linkage disequilibrium, are often taken as hallmarks of a clonal population: for example, the existence of isolates with an identical genotype in samples of the population recovered at different times from geographically distant locations (Selander and Musser, 1990). Although this may be a good indicator that the population contains stable clones, it should be remembered that the methods used to discriminate between isolates must have sufficient resolving power that apparently identical isolates would not be expected by chance. For example, O'Rourke and Stevens (1993) identified clusters of gonococci that appeared to be identical by MLEE with nine enzymes. This finding does not, however, imply a clonal population structure, since these isolates possessed the most common alleles at each locus, and were present in the population at the frequencies expected from the observed allele frequencies if recombination was frequent (Maynard Smith *et al.*, 1993; O'Rourke and Stevens, 1993).

A good indication of the degree of clonality within bacterial populations can in some cases be obtained from a knowledge of the rate of diversification of lineages (**Fig. 1**). Care must be taken in the type of diversification that is studied, since some variation may accumulate rapidly even in highly clonal species. Variation detected by MLEE is well suited for studying the diversification of lineages. In some species all isolates of a clonal lineage recovered over 50 years have the same ET, whereas in other species considerable variation seems to occur within a matter of years. Of course, the problem with this approach is that it is rarely possible to date the origin of a clone, but in some cases this is possible. For example, clones of multiple antibiotic-resistant pneumococci cannot pre-date the introduction of antibiotics, and are probably only about 20 years old (Coffey *et al.*, 1995a). Isolates of the most well-characterised clone, the multi-resistant Spanish serotype 23F clone (Muñoz *et al.*, 1991), differ by MLEE at up to 4/20 loci, and the rapid accumulation of variation within this very young clone appears to be due to recombinational exchanges (McDougal *et al.*, 1992; Coffey *et al.*, 1996). The rapid diversification of this clone suggests that recombination is frequent and that the population structure of *Strep. pneumoniae* is unlikely to be highly clonal.

MLEE studies of a non-clonal population should detect more alleles per locus than are found within a strictly clonal population with the same level of sequence diversity. Novel electrophoretic variants of a housekeeping enzyme arise in strictly clonal populations only by the rise to prominence of rare non-synonymous mutations that alter electrophoretic mobility without impairing enzyme function. In non-clonal populations, new electrophoretic variants are also produced by intragenic recombination. Localised recombinational exchanges between alleles which encode enzymes with, for example, a −3 net negative charge, and a +3 net positive charge, can lead to a considerable number of new alleles with a variety of net electrophoretic charges.

Non-clonal populations should also contain more ETs than clonal populations with the same level of sequence diversity, since frequent recombinational replacements between different lineages has the effect of shuffling the alleles in non-clonal populations to produce an ever-changing array of ETs. In clonal populations, periodic selection results in the expansion of successful lineages and the loss of other lineages. Consequently, a clonal population generally has relatively few ETs compared with a non-clonal population. Furthermore, MLEE data from a highly clonal population usually generate a dendrogram with more structure (deep branches between major lineages and large inter-nodal distances) than those produced from non-clonal populations, as the homogenising effect of recombination results in the major lineages appearing to be approximately equally related to each other.

Recombination and Nucleotide Sequences

Recombination within bacterial populations can be detected by analysing nucleotide sequences. There are two general approaches. First, if recombination is absent, the evolutionary history of every gene within a bacterium should be the same. The genetic relatedness between isolates of a highly clonal species inferred from the nucleotide sequences of one gene should, therefore, be the same as that inferred from any other gene, or from MLEE data. Non-congruence between gene trees is indicative of a history of recombination. This approach should work well if recombinational exchanges are relatively large, such that whole genes are often replaced (assortative recombination). However, if recombinational exchanges occur, but the replacement sizes are considerably less than the lengths of the sequences used for the construction of gene trees, they may not lead to non-congruence between the trees. Second, frequent intragenic recombinational replacements may result in trees that are poorly supported statistically, since different parts of the genes have different evolutionary histories, making it difficult to show non-congruence. Holmes *et al.* (1999) and Feil *et al.* (2001) provide a good example of the use of non-congruence between trees to infer high rates of recombination in a bacterial species.

Recombinational exchanges can also be inferred from the distribution along a gene of the nucleotide sequence differences between isolates (polymorphic sites). In the absence of recombination, these differences (or, more accurately, those at synonymous sites which do not alter the amino acid sequence) should be approximately randomly distributed along a gene. Intragenic recombination between genetically distinct isolates results in a non-random distribution of these differences (mosaic structure). For example, a gene from two isolates may be identical, except in a 300-bp region where the sequences differ substantially because of the occurrence in one isolate of a recombinational exchange from another isolate. Several methods can be used to detect clustering in the distribution of polymorphic sites and to evaluate its statistical significance (Stephens, 1985; Sawyer, 1989; Maynard Smith, 1992; Sneath, 1995). Additional methods are available that detect putative recombinational events by the presence of regions within a gene (Grassly and Holmes, 1997), or the presence of an excessive number of polymorphic sites within a gene (Maynard Smith and Smith, 1997), that do not support the optimal tree derived from the whole sequence.

The choice of gene for these analyses is crucial. The chance of observing a recombinant allele in a small sample of bacterial isolates depends both on the rate at which recombination actually occurs at the locus and the strength of selection for, or against, the recombinant allele. If positive selection for recombinant alleles is strong, these alleles may be observed even if the frequency of recombination is vanishingly small. Positive selection for variation, or selection for the horizontal spread of alleles through the population, is likely to occur in genes that encode cell surface structures, or some virulence factors. In almost all bacterial species, including those that are believed to have strongly clonal population structures (e.g. *S. enterica*), mosaic structure, and non-congruence between gene trees and trees derived from MLEE data has been observed for this type of gene (e.g. Kroll and Moxon, 1990; Smith *et al.*, 1990; Feavers *et al.*, 1992; Jauris-Heipke *et al.*, 1995; Kapur *et al.*, 1995). Evidence for recombination within cell-surface genes, or virulence genes, or genes encoding the targets of antibiotics (Dowson *et al.*, 1989; Spratt *et al.*, 1989), tells us little about the frequency of recombinational exchanges, compared with point mutation, within the species.

A better estimate of the extent of recombination within bacterial species can be obtained by looking at housekeeping genes, which are unlikely to be subject to positive selection for variation, or any kind of strong directional selection. Recombination within housekeeping genes is likely to result in enzymes that are as well adapted as those of their parents, or perhaps slightly less well. Bacterial populations are vast, and if recombination is very rare, most of these recombinant alleles should be lost from the population, and only a very small proportion will, by chance, drift to an observable frequency. It is, therefore, unlikely that sequencing of a housekeeping gene from 20 isolates of a population in which recombination is extremely rare will uncover many examples of recombinant (mosaic) alleles. The frequent presence of mosaic alleles in housekeeping genes, or non-congruence between the trees constructed for different housekeeping genes, is therefore indicative of a relatively high rate of recombination.

The Ratio of Recombination to Mutation

Measures of linkage disequilibrium, and the detection of recombination in gene sequences, provide little information about the relative rates of recombination in different bacterial species. An approach to this problem was suggested by Guttman and Dykhuizen (1994) who analysed variation along a region of the chromosome in *E. coli* isolates that were known to be very closely related. The ratio of recombinational events to mutational events can be obtained if the sequence differences among isolates that are descended from a very recent common ancestor can be partitioned into mutational events and recombinational events. The ratio of recombinational exchanges to point mutations can then be compared between bacterial species. Recombinational exchanges in two species that differ in sequence diversity will introduce very different numbers of nucleotide differences. Therefore, the relative impact of recombination, compared with point mutation, in the diversification of clones of different bacterial species is more appropriately estimated by measuring the probability that a single nucleotide site in a neutral locus, such as a housekeeping gene, will change by recombination rather than by point mutation (Guttman and Dykhuizen, 1994).

The large amounts of nucleotide sequence information obtained in a new molecular typing procedure (multi-locus sequence typing – MLST; see section below, 'Use of neutral variation for typing') has allowed this per-site recombination/mutation parameter to be measured in different species. In this technique, which is based on the principles of MLEE, the sequences of internal fragments of seven housekeeping genes are determined and, for each locus, each unique sequence is assigned as a distinct allele. Isolates of a bacterial species are, therefore, unambiguously defined by the alleles (sequences) at each of the seven loci. Since the method is highly discriminatory, isolates that are identical at all seven loci (clones), and variants of these clones that differ at only a single locus, can be assumed to be derived from a recent common ancestor. The nucleotide sequence differences between the alleles normally present in a clone, and the allele in each single locus variant of the clone, can be used to distinguish which of these initial events in clonal diversification is likely to be due to a recombinational exchange, and which to a point mutation. As a first approximation, differences at multiple nucleotide sites can be considered to be due to recombinational exchanges and single nucleotide differences the result of a point mutation. However, some recombinational events between similar alleles may introduce only a single nucleotide difference, and may appear to be point mutations. Corrections have to be applied to distinguish which of the alleles that differ at a single site are due to recombination and which due to point mutation (**Fig. 3**).

Using this approach on large MLST datasets, it has been estimated that a single nucleotide site within a *N. meningitidis* housekeeping gene is 80 times more likely to change by recombination than by mutation, and this ratio is about 50 in *Strep. pneumoniae* (Feil *et al.*, 2000). Thus, in both of these naturally transformable species, recombination plays a far greater role in clonal diversification than point mutation. This approach promises to provide meaningful estimates of the relative impact of recombination, compared with point mutation, on clonal diversification in different bacterial pathogens.

Population-genetic Studies of Selected Clonal Pathogens

Salmonella enterica – the Paradigm for Clonal Pathogens

The population genetics of *S. enterica* are discussed here in some detail because this species provides a paradigm for highly clonal bacterial species. *Salmonella* are a large group of intracellular pathogens closely related to *E. coli*. They are found associated with a wide variety of vertebrates, including lizards, birds and mammals (Ewing, 1986). The nomenclature of the *Salmonella* has recently been rationalised, leading to a reduction in the number of recognised species from over 2000 to just 2 – *S. enterica* and *S. bongori* (Le Minor and Popoff, 1987; Reeves *et al.*, 1989). The new nomenclature is cumbersome – *S. typhimurium* becomes *S. enterica subsp. enterica serovar Typhimurium* (Le Minor and Popoff, 1987) – and we will refer to such isolates as serovar Typhimurium, or just Typhimurium (Ewing, 1986).

The group of organisms that comprise the *Salmonella* have been allocated to seven subspecies on the basis of biochemical tests and DNA–DNA hybridisation (Ewing, 1986). This subdivision is fully supported by MLEE studies on representatives of each subspecies (Boyd *et al.*, 1996). Subspecies 1 contains 60% of the serovars and accounts for the vast majority of disease associated with humans. The identification of *Salmonella* strains by serology follows the recommendations of the Kauffmann–White serotyping scheme, which employs the antigenic reactions of the cell surface

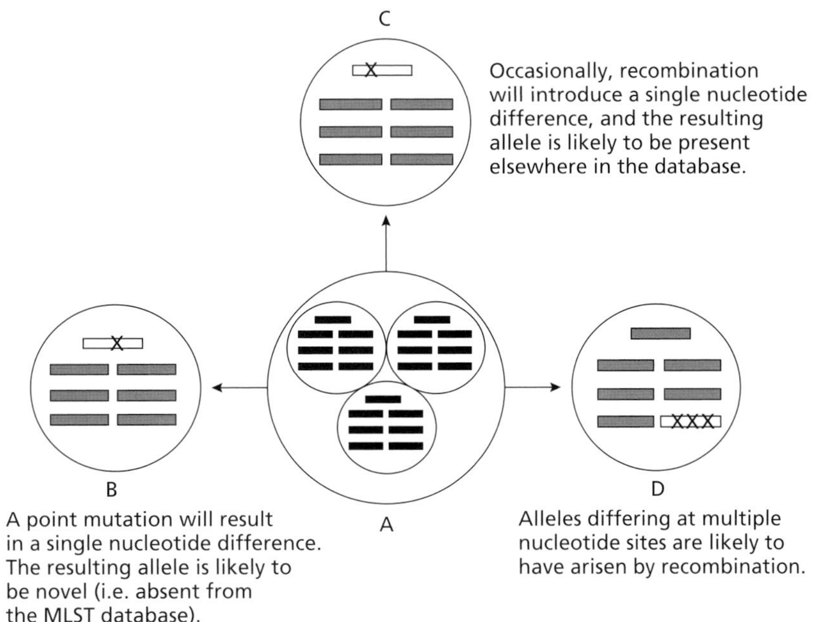

Fig. 3 Estimating the ratio of recombination to mutation. The alleles at the seven loci that are typically found in the members of a clone (A) are given as dark bars, and the variant alleles in the three single locus variants (B–D) are given as light bars. The nucleotide differences between the allele found in the members of the clone and each variant allele are shown as crosses. A point mutation will generate a single nucleotide difference (B), whereas a recombinational exchange is likely to introduce multiple nucleotide differences (D). Single nucleotide differences may be introduced by recombination and most can be identified because the donor allele is likely to be present elsewhere in a large MLST dataset (C). This approach can provide an estimate of the ratio of recombination compared to mutation in generating new alleles and, by counting the numbers of nucleotide differences introduced at each event, the relative likelihood that a single nucleotide site will change by recombination compared with point mutation.

lipopolysaccharide (the O antigen), as well as the phase 1 and phase 2 flagellins (H antigens), to generate an antigenic formula (Selander *et al.*, 1996). The antigenic formula of a strain is used to designate a serotype (=serovar) name, which in the past was assigned species status. For example, a strain bearing a type 11 lipopolysaccharide antigen, in combination with an r antigen on the phase 1 flagellin, and e, n, x antigens on the phase 2 flagellin, is identified as *S. enterica serovar Rubislaw* (Ewing, 1986).

The population structure of the *Salmonella* was first studied by Beltran *et al.* (1988) who used MLEE to analyse variation in 23 enzymes in 1527 isolates of 8 common serovars of subspecies 1. It was shown that some serovars (e.g. Typhimurium and Choleraesuis) are monophyletic, with all isolates belonging to a cluster of closely related ETs (Fig. 4). In other cases, members of the same serovar were only distantly related (polyphyletic serovars). Furthermore, strains with dissimilar serotypes were found to be very similar by MLEE (Fig. 4). This basic pattern, in which most serovars are monophyletic and consist of a few closely related ETs, has been confirmed by numerous MLEE studies involving over 4000 strains of *Salmonella*. However, the observation of polyphyletic serovars invalidates the Kauffmann–White scheme as a natural classification scheme for the salmonellae (Selander *et al.*, 1990b).

The population structure of *S. enterica* has been found to be highly clonal. This conclusion is based on the observation of high levels of linkage disequilibrium, the frequent association of a single antigen profile with a single ET, and the widespread geographical and temporal distribution of predominant ETs (Selander *et al.*, 1996). Extensive nucleotide sequencing studies have been carried out, and show occasional examples of recombinational exchanges between lineages, but the dendrograms derived from the sequences of housekeeping genes show good congruence with each other and with those derived from MLEE data (Nelson *et al.*, 1991; Nelson and Selander, 1992; Boyd *et al.*, 1994). All of these observations are compatible with a highly clonal population in which replacement of an allele by recombination is infrequent relative to mutation.

What is a *Salmonella* Serovar?

It is apparent that the enduring success and utility of the Kauffmann–White serotyping scheme is largely due to the limited recombination detected in

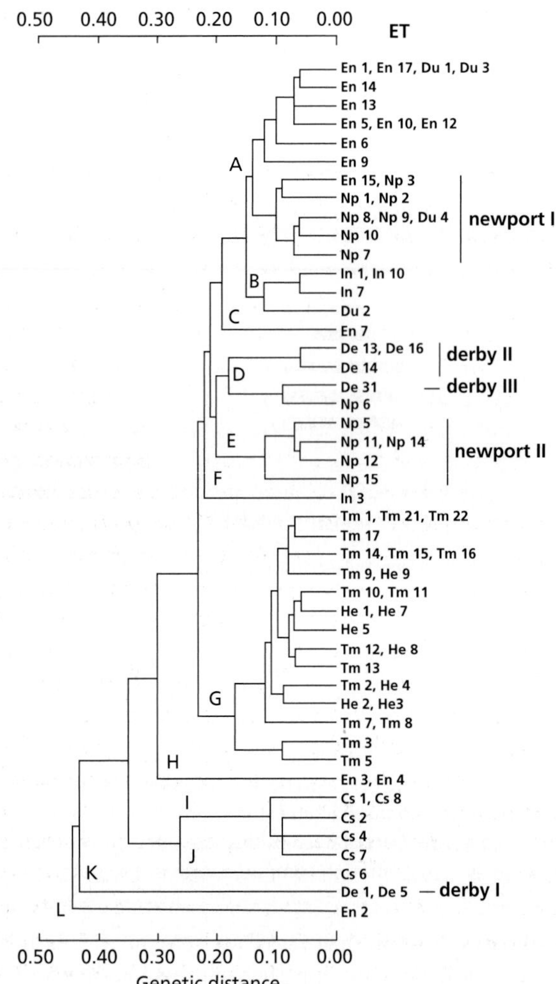

Fig. 4 Genetic relationships among isolates of eight serovars of *S. enterica*. The dendrogram was constructed with MLEE data (23 enzymes) from 1527 isolates of *S. enterica* serovars Enteritidis (En), Dublin (Du), Newport (Np), Infantis (In), Derby (De), Typhimurium (Tm), Heidelberg (He), and Choleraesuis (Cs). The lineages were truncated at a genetic distance of 0.04, corresponding to a difference between ETs at a single enzyme. Isolates at the end of each terminal branch of the tree are therefore either indistinguishable by MLEE or differ at 1/23 enzymes. In this highly clonal species the dendrogram should provide a good indication of the evolutionary relationships between the ETs and is therefore a phylogenetic tree. Reproduced with permission from Beltran *et al.* (1988).

S. enterica. The tight association between flagellin and lipopolysaccharide antigens and pathogenicity, host adaptation, or the phenotype of a strain, has been exploited in the *Salmonella* to generate one of the most effective bacterial serotyping schemes. The efficacy of the Kauffmann–White serotyping scheme is, however, not without its minor drawbacks. In particular, Kauffmann's belief that each serotype represents a single species is no longer accepted, and the majority of the *Salmonella* are now designated as a single species, *S. enterica* (Le Minor and Popoff, 1987). The genetic diversity of *S. enterica* is, however, nearly twice that found within *E. coli* (Selander *et al.*, 1996). Indeed, the genetic diversity within *S. enterica* subspecies 1 alone is similar to that within *E. coli*.

Most of the serovars of *S. enterica* subspecies 1 represent clonal lineages of one or more closely related ETs that bear a distinct combination of flagellin and lipopolysaccharide antigens. For example, 76% of 340 isolates of serovar Typhimurium, collected from worldwide sources, were identical by MLEE using 24 enzymes (Selander *et al.*, 1990b), and the most distantly related strains differed at only 3/24 enzymes (Fig. 4). The genetic similarity of Typhimurium isolates implies that they all had a recent common ancestor which bore the combination of antigens that we now recognise within the Kauffmann-White serotyping scheme as serovar Typhimurium.

Why Does the Kauffmann–White Scheme Work?

It may seem serendipitous that all members of a clonal lineage, such as serovar Typhimurium, should bear a unique combination of antigens that we can use for identification of the clone. This finding is, however, a consequence of the efficacy of periodic selection which, in a highly clonal species, can lead to the entire chromosome in which a favourable allele arises being swept to high frequency in the population. During this process any minor sequence variation at non-selected loci is carried to high frequency along with the selected allele, and will be present in all descendants of the ancestral bacterium in which the favourable allele arose. If periodic selection happens repeatedly, then a clonal lineage may accumulate many fixed, but minor, differences from other lineages. A useful microbial classification scheme will identify these differences, and use them as a marker for the entire chromosome, and for the clonal lineage. The Kauffmann–White scheme works, therefore, because it applies a simple but effective technique for identifying minor differences between strains to a group of organisms with a highly clonal population structure.

The power of serology to detect minor differences is exemplified by the comparative sequencing of the flagellin genes of *S. enterica* serovars Dublin and Enteritidis. These two closely related serovars are distinguished in the Kauffmann–White scheme only by differences in a single flagellin antigen. Comparative sequencing of the flagellin genes from representative samples of these two populations shows that three non-synonymous substitutions are responsible for this antigenic difference (Selander *et al.*, 1992). Thus, on the basis of these three amino-acid differences in the flagellin, a microbiologist is able to distinguish all strains of serovar Dublin, a strongly host-adapted serovar, with a unique virulence plasmid, and a requirement for nicotinic acid, from serovar Enteritidis, a serovar with broad host range and few nutritional requirements.

When Does the Kauffmann–White Scheme Fail?

Most strains of serovar Enteritidis are found within a single clonal lineage, but there are strains with the same antigenic formula, which are also therefore called Enteritidis, that are only very distantly related to the major lineage (**Fig. 4**). Other examples of polyphyletic serovars include Paratyphi B, Derby, Newport and Infantis (Beltran *et al.*, 1988). For strains of Enteritidis, the Kauffmann–White scheme has combined two distantly related lineages. Furthermore, very closely related lineages, such as Typhimurium and Heidelberg, may have different serovar names (Beltran *et al.*, 1991). Thus in some cases the Kauffmann–White scheme fails to describe the true phylogenetic relationships between clones and so obscures some of the evolutionary patterns that population genetics is able to uncover.

The cause of these inconsistencies between serotyping and phylogenies derived from MLEE has been explored in a series of sequencing studies involving both the phase 1 flagellin gene (*fliC*) and the lipopolysaccharide locus (*rfb*). Initially it was found that the nucleotide sequence of the antigenically variable region of the *fliC* gene of Typhimurium was invariant between isolates of this serovar, but differed at 19% of sites from the *fliC* gene of the very closely related serovar Heidelberg (Smith and Selander, 1990). Serovars Heidelberg and Typhimurium are distinguished in the Kauffmann–White scheme solely by antigenic differences in the phase 1 flagellin. It could be suggested that the differences in the nucleotide sequences of these genes merely represents changes that have accumulated during the descent of these two serovars from their recent common ancestor. Sequences virtually identical to the *fliC* genes of Heidelberg and Typhimurium could, however, be recovered from strains that were phylogenetically only distantly related to these serovars (Smith *et al.*, 1990). The only reasonable explanation for these observations is that flagellin genes have been transferred by recombination between serovars. There is also evidence that recombination at the *rfb* locus, which encodes the lipopolysaccharide O antigen, has disrupted the association between serology and phylogenetic relatedness (Reeves, 1993).

It is pertinent to ask why so much recombination has been detected in the genes that determine serotypes, whereas MLEE studies suggest that *S. enterica* is highly clonal. Two factors contribute to this apparent anomaly. First, the examples of recombination in these genes were actively sought out by the experimentalist, and the sampling of genes for nucleotide sequencing is not equivalent to the sampling of strains used in most MLEE studies. Second, the frequency of recombinational exchanges at the flagellin or lipopolysaccharide genes is probably no higher than that at other chromosomal loci, but the products of recombinational exchanges that involve genes which encode surface exposed structures are probably subject to frequency dependent selection imposed by the host immune system. Thus, even very rare recombinational events involving the flagellin or lipopolysaccharide genes may be brought to high frequency in the population (Selander *et al.*, 1996). The observation of recombination at loci encoding cell surface structures is not, therefore, incompatible with the highly clonal population structure of *S. enterica*. Nor is the rate of recombination at flagellin and lipopolysaccharide genes sufficiently frequent totally to invalidate the Kauffmann–White scheme as a method for the identification of clonal lineages. In most other species, recombination is more frequent than in *S. enterica*, and this leads to major inconsistencies between serological classifications based on variation in cell surface molecules and genetic relatedness assessed by population-genetic approaches.

Salmonella Serovars that Cause Enteric Fever

Serovars of *Salmonella* that are primarily human pathogens represent only a tiny proportion of the large number of serovars within subspecies 1. Clinically the most important human-adapted serovar is Typhi, the causative organism of typhoid fever. Other serovars, namely Paratyphi A, B and C, can also cause typhoid-like enteric fevers, although these serovars are not exclusively limited to humans. A population-genetic analysis of 761 isolates of these, and related serovars, revealed no close phylogenetic relationships between any of the major serovars causing enteric fever, implying that clones of human invasive *Salmonella* have independently acquired this pathogenic characteristic (Selander *et al.*, 1990a, b).

All 334 strains of serovar Typhi were shown to be very closely related to each other and two major clones could be distinguished. Strains of ET-Tp1 were recovered from worldwide sources, whereas strains of ET-Tp2 were exclusively from Africa. Atypical strains of Typhi from Indonesia that express abnormal flagellar antigens were found to be identical to Tp1 by MLEE. Neither the presence nor absence of the Vi antigen identified a discrete subset of Typhi strains, suggesting that this characteristic can be rapidly lost or acquired. Overall, *S. enterica serovar typhi* can be considered as a single, globally distributed, clonal lineage within which minor variants can be recognised (Selander *et al.*, 1990b).

In contrast, strains of Paratyphi B form a heterogeneous group and this serovar is one of the small group of polyphyletic *Salmonella* serovars identified by MLEE (Selander *et al.*, 1990a). However, strains of Paratyphi B that cause enteric fever in humans were found to be a single clone, ET-Pb1. Presumably Pb1 has adapted to humans relatively recently, as members of this clone are very similar by MLEE to other clones of Paratyphi B that have a broad host range, and which cause enteritis, rather than enteric fever.

Although strains identified serologically as Paratyphi B are not necessarily closely related, serovars Paratyphi A and C form natural groupings. All isolates of Paratyphi C are very similar, and are allied to serovars Choleraesuis and Typhisuis, suggesting an evolutionary scenario in which a swine-associated serovar, Choleraesuis, gave rise to two host adapted forms – Typhisuis, which causes swine paratyphoid fever, and Paratyphi C, which has adapted to humans. This conclusion is supported by the observation that serovar Choleraesuis can occasionally cause severe enteric fevers in humans (Jegathesan, 1984). Paratyphi A strains form a group closely related to serovar Sendai and are descended from the same lineage as serovar Panama. This group shows a spectrum of human adaptation from the generalist serovar, Panama, to the more human-adapted Paratyphi A and Sendai.

In general, population-genetic studies of the human-adapted serovars of *Salmonella* have successfully identified the phylogenetic relationship between these serovars, and have provided hints of the evolutionary pathway that leads to host adaptation. It is quite apparent that human adaptation is a polyphyletic characteristic, independently acquired in several distinct clonal lineages. Furthermore, with the exception of Typhi, for which no closely related serovar has been identified, all human-adapted salmonellae are derived from lineages that cause invasive disease in other hosts. This observation suggests that the determinants of invasive disease preceded those of host adaptation and that host switching followed by further restriction in host range is common in the *Salmonella*.

The Population Structure of *Escherichia coli*

Studies of *E. coli* populations provided the first evidence that bacterial populations are clonal. Initial indications came from the association of particular *E. coli* serotypes with disease (Ørskov *et al.*, 1976), and were reinforced by the studies of Selander and Levin (1980), which identified isolates that were indistinguishable by MLEE, although they were recovered many years apart. For example, an isolate taken from an infant in Massachusetts in 1979 had the same ET as an *E. coli* K12 laboratory strain that had been isolated 50 years previously from an infant in California. More extensive MLEE studies of both pathogenic *E. coli* isolates, and predominantly commensal isolates, have provided further evidence that *E. coli* is clonal and demonstrated strong linkage disequilibrium between the alleles within *E. coli* populations (reviewed by Whittam, 1995, 1996).

Patterns of nucleotide sequence variation have been studied in numerous *E. coli* housekeeping genes and this has resulted in the emergence of a more complex view of the significance of recombination in the evolution of *E. coli* than originally suggested by MLEE studies. Most of the sequencing studies have used strains from the *E. coli* reference collection (ECOR), which originated from two main sources (Ochman and Selander, 1984). Of the ECOR strains, 40 were from a larger collection of isolates recovered primarily from the faeces of domesticated, wild and zoo animals from all over the world. The other 32 strains were from humans, and the complete set was chosen to represent a diverse range of host preference, geographic distribution and MLEE genotype, although none of the strains was recovered from disease. All of the 72 ECOR strains have been analysed by MLEE, which has provided an inferred phylogeny of these strains.

Sequencing of *phoA*, which encodes alkaline phosphatase, from nine *E. coli* strains, was carried out by DuBose *et al.* (1988). The statistical approaches of Sawyer (1989), and Stephens (1985), both revealed that the distribution of polymorphic sites within *phoA* was significantly non-random. The authors suggested four highly localised recombinational events which could have generated the observed clustering of polymorphisms. Similarly, other studies detected localised recombinational exchanges in a region near the *trp* operon (Stoltzfus *et al.*, 1988), and in the *gnd* gene (Bisercic *et al.*, 1991; Dykhuizen and Green, 1991).

The evidence for recombinational exchanges within these *E. coli* housekeeping genes appeared to conflict with the view that recombination needs to be rare to lead to the clonal population structure inferred by MLEE studies. This conflict is, however, easily resolved. MLEE with 20 enzymes, which have an average molecular mass of 50 000, analyses variation in about 0.7% of the *E. coli* genome. Thus, if recombinational replacements are very small (a few hundred base pairs), they will only rarely involve one of the loci whose variation is indexed by MLEE. Relatively frequent localised recombinational replacements can, therefore, occur without dissipating linkage disequilibrium (DuBose *et al.*, 1988; Maynard Smith *et al.*, 1991).

Similarly, localised recombinational exchanges will not necessarily disrupt the overall congruence between gene trees expected in a clonal species. This is well illustrated by studies on the housekeeping loci *putP* (proline permease), *mdh* (malate dehydrogenase) and *gapA* (glyceraldehyde-3-phosphate dehydrogenase) from the same ECOR strains, all of which provide evidence of some recombinational replacements, although the dendrograms constructed from these gene sequences are largely in accordance with the dendrogram produced from the MLEE dataset (Nelson *et al.*, 1991; Nelson and Selander, 1992; Boyd *et al.*, 1994). Although it is hard to refute the evidence from sequence data that recombination occurs in nature in the *E. coli* population, it has been proposed that recombination does not occur frequently enough, and is generally too localised, to have anything other than a minor effect on the clonal divergence of this species. This view is further supported by the congruence between the dendrogram obtained for the ECOR strains using MLEE and that obtained using molecular typing methods (Desjardins *et al.*, 1995).

Recombination and Hitch-hiking at the *E. coli gnd* Locus

A major exception to the view that recombination is relatively infrequent in *E. coli* housekeeping genes is the *gnd* gene, which encodes 6-phosphogluconate dehydrogenase. Electrophoretic variation at this locus is about three times greater than in other enzymes of similar size, and there is a correspondingly high level of nucleotide sequence variation within *gnd*. Dykhuizen and Green (1991) showed that the dendrogram obtained with *gnd* sequences was completely non-congruent with that obtained with *phoA* or *trp* sequences from the same strains. More extensive studies by Nelson and Selander (1994) confirmed that phylogenetic relationships between *E. coli* strains based on the sequence of *gnd* are grossly non-congruent with those inferred from other housekeeping genes, and from MLEE data. Moreover, Nelson and Selander (1994) provided evidence that several *E. coli* strains harboured replacements at *gnd* which originated from outside the *E. coli* population (e.g. from *Citrobacter*). Inter-species recombinational exchanges presumably explain the unusually high level of sequence variation at this locus.

Several authors have noted that the most plausible explanation for the high level of recombinational replacements at *gnd*, as well as the high level of recombinational promiscuity, is the tight linkage of *gnd* to loci believed to be under diversifying selection from the host immune response (Bisercic *et al.*, 1991; Dykhuizen and Green, 1991). Of particular relevance is the *rfb* region, which lies at a distance of between 200 bp and 2 kb from *gnd* and mediates the biosynthesis of the highly antigenic outer membrane lipopolysaccharide O antigen (Bisercic *et al.*, 1991; Reeves, 1993). Recombinational replacements at the *rfb* locus that lead to antigenically novel lipopolysaccharide epitopes are likely to be strongly selected, as they may allow decreased recognition by the host immune system. As expected, the *rfb* region of *E. coli* is highly variable and this has been attributed to high levels of both intra- and inter-species recombination at this locus (Reeves, 1993). Recombination is blind to gene boundaries, and it is thought that the variation at *gnd* is a consequence of recombinational replacements 'spilling over' from *rfb* – a process known as hitch-hiking. The high frequency of recombinational replacements observed at *gnd* need not be explained in terms of selection acting on *gnd* itself, but on the indirect effect of diversifying selection conferred by neighbouring loci. Although the analysis of *gnd* has provided an important insight into the potential importance of hitch-hiking, and the possibility of inter-species recombination in *E. coli*, the high levels of recombination at this locus must be regarded as a special case.

The Concept of the Clonal Frame

A comprehensive study of sequence variation in *E. coli* has been provided by Milkman and colleagues (reviewed by Milkman, 1996). Analysis of the alleles present at 15 loci in the ECOR collection, using restriction-fragment length polymorphisms (RFLPs), led to the clonal frame model of the *E. coli* population (Milkman and Bridges, 1990; Milkman, 1996). The authors proposed that the ratio of recombination to mutation is sufficiently low that periodic selection is effective in bringing those chromosomes in which favourable alleles arise to high frequency before they became pocked by recombinational replacements. Subsequently, the divergence of these chromosomes occurs by the accumulation of recombinational replacements, and by mutational events. Isolates that have a recent common ancestor therefore share a common clonal frame, which initially includes the whole chromosome but becomes a decreasing fraction of the chromosome as recombinational replacements continue to accumulate as the clone becomes older.

The RFLP analysis suggested that the *E. coli* population could be classified into a hierarchy of clones, some very recent, sharing a common clonal frame, and some much older so that their clonal framework can hardly be discerned, as it has become greatly reduced by recombinational replacements, and obscured by mutational events. In particular, most of the ECOR strains fell into three major groupings which presumably represent the greatly diverged descendants of three ancient periodic selection events within the population (Milkman and Bridges, 1990). This conclusion was confirmed by sequencing the 12.7-kb region around the *trp* operon in 40 ECOR strains and the laboratory K12 strain (Milkman and McKane Bridges, 1993; Milkman, 1996). On the basis of this dataset, 31 of the strains belong to one of three clusters, each showing very high levels of overall sequence similarity (< 1% average pairwise divergence). This grouping is in accordance with evidence from MLEE data, and so is highly likely to represent the true phylogenetic descent of these strains. However, as indicated by the earlier RFLP analysis (Milkman and Bridges, 1990), many of these sequences are pocked by recombinational replacements.

It should be stressed that although these replacements appear common, they are too localised to obscure the true phylogenetic grouping of these strains, and it is the predominance of clonal frame DNA in the 40 ECOR strains that allows the grouping of these strains into three phylogenetically meaningful sequence clusters. Strains belonging to the same cluster based on this dataset share a large proportion of the same clonal frame DNA, with diverged blocks of sequence being due to recombinational replacements. These clusters are examples of what Milkman has termed meroclones, isolates that share the same recent clonal ancestry, and which differ predominantly according to the recombinational replacements they have picked up since the lineage first emerged, rather than through differences arising from mutation. The steady decline of the clonal frame DNA by recombinational replacement will eventually result in the disintegration of the meroclones, and a possible loss of fitness, thus paving the way for the emergence of new fitter lineages that may rise in frequency through periodic selection, and for the cycle to start over again. In essence, this scheme combines the classic notion of independent clonal lineages diverging through mutation with the evidence for recombinational exchanges from gene sequences to engender a sort of 'leaky' clonality (or more properly, meroclonality) brought about by the progressive superimposition of localised replacements on the clonal frame.

How Much Recombination Occurs in *E. coli*?

Guttman and Dykhuizen (1994) have argued that recombination in *E. coli* may be more common than is generally thought. They took a more systematic approach by focusing on four closely linked loci, *sppA*, *gapA*, *pabB* and *zwf*, which span a total distance of about 90 kb. By comparing closely linked loci in this manner, they were able to assess the possibility that single putative recombinational replacements effected more than one locus. Significantly, they suggested that one replacement encompassed three loci, a total distance of over 70 kb. Although it is possible that this replacement is non-contiguous, this suggestion offers a sharp contrast to the very localised replacements reported in other sequencing studies.

The sequences of all 4 genes were determined for 12 ECOR strains, including 5 group A strains known to be very closely related from MLEE data, and from previous sequencing studies. As expected, most of the group A strains shared 100% sequence identity at each of the 4 loci. The data are, however, not entirely consistent with strict clonality, as dendrograms constructed with the sequence data placed at least one of the 5 group A strains outside of the group A cluster at each of the 4 loci examined, and only 2 of the strains clustered together at all 4 loci. These results imply that several presumed recombinational exchanges, but no nucleotide substitutions, have occurred at these loci since the 5 group A strains diverged from their recent common ancestor, which Guttman and Dykhuizen (1994) estimate existed no more than 2400 years ago. The divergence within housekeeping genes at the tips of the *E. coli* phylogenetic tree therefore appears to be due more to recombinational exchanges than to mutation.

Clonal Structure and Ecotypic Structure

If recombinational exchanges occur within a gene more frequently than nucleotide substitutions, how can we reconcile this with the evidence for linkage disequilibrium inferred from MLEE studies? Guttman (1997) claimed that these differences cannot be reconciled, and that it is the presence of strong linkage disequilibrium that must be challenged, as the evidence for recombination from *E. coli* sequence data is direct and unequivocal. The initial analysis of MLEE datasets from *N. meningitidis* (discussed elsewhere in this chapter) was proposed as evidence for clonality in this species, but a re-evaluation of the data revealed that the apparent linkage disequilibrium was in fact an artefact resulting from sampling bias and unrecognised complexities within the meningococcal population. Guttman (1997) has argued that the apparent linkage disequilibrium in *E. coli* may similarly be the result of structuring within the *E. coli* population. Although this proposal needs to be evaluated, it should be remembered that Maynard Smith (1994) estimated that linkage disequilibrium can be present in bacterial populations, in the absence of population substructuring, even when recombination is more significant than mutation in bringing about evolutionary change at a locus.

It is conceivable that structuring within populations may be due to geographic isolation. If the rate of migration within the *E. coli* population was very low, subpopulations could remain isolated geographically, and would therefore appear to diverge independently from one another, leading to the observed linkage disequilibrium, even if recombination occurred at a very high rate within each subpopulation. However, evidence from studies on *E. coli* suggests a high rate of migration (Whittam *et al.*, 1983a), and this effectively rules out any possible restriction to gene flow through geographic barriers.

A second, more favoured, alternative is that population structuring is ecologically based. Rather than treating *E. coli* as a single ecological entity, this argument assumes that single strains of *E. coli* occupy one of numerous discrete subniches exploited by the species (Maynard Smith, 1991; Reeves, 1992; Guttman, 1997). Although the exclusivity of these niches has yet to be demonstrated, it is certainly clear that *E. coli* can be recovered as a bowel commensal, a pathogen in the urinary tract, blood or meninges, or as a free-living bacterium in contaminated soil or water. Additionally, *E. coli* is found in a wide range of animal hosts. It has been proposed that an isolate harbouring a mutation conferring an improved ability to exploit a specific niche will initiate a periodic selection event that will purge the variation within that niche, but not within other niches (Maynard Smith, 1991; Reeves, 1992). This bacterial cell will, therefore, act as a founder for a specifically niche-adapted subpopulation, or ecotype, all the members of which will originate as strict clonemates. The repeated sampling of bacteria carrying specific niche adaptations in this manner could therefore result in the structuring of the population assumed to be responsible for the apparent linkage disequilibrium. An expectation from this model would be the association of ecotypes with specific clonal groups, but this has generally not been demonstrated as yet, except for pathogenic strains of *E. coli* where isolates that cause a specific disease in a specific host are often very closely related (Whittam, 1996).

In conclusion, it is clear from numerous sequencing studies that recombination plays a significant role in the evolution and population structure of *E. coli*, and that the assumption of a highly clonal paradigm based on MLEE data must be refined in order to accommodate this new knowledge. Recombination is also found within *S. enterica* housekeeping genes, but it appears that recombination is more frequent in *E. coli*, and that this species is markedly less strongly clonal than *Salmonella*. It will probably require further studies, with careful sampling of the *E. coli* population, based on ecological considerations, to resolve the controversies surrounding the population structure of this species.

The Population Structure of *Shigella*

The organisms found within the genus *Shigella* are the causative organisms of bacillary dysentery or shigellosis. Members of this genus are placed in four main species by serology, and cause a range of disease syndromes from mild diarrhoea to severe dysentery. The key feature of the pathogenicity of the *Shigella* is the production of a highly potent toxin (shiga toxin). This feature is not, however, unique to *Shigella* spp., since similar toxins are found in some pathogenic strains of *E. coli*.

It has long been recognised that *Shigella* are closely related to *E. coli* (Brenner *et al.*, 1969). An MLEE study of 123 isolates of *Shigella* and 1600 isolates of *E. coli* demonstrated that none of the *Shigella* species is sufficiently different from *E. coli* to justify their species status, let alone their inclusion within a distinct genus (Ochman *et al.*, 1983). In this population-genetic study, the strains of *E. coli* were grouped into three main clusters, and strains of *Shigella* were generally located between two of these clusters (Whittam *et al.*, 1983b). The *Shigella* were further subdivided by MLEE into two main groups, neither of which uniquely corresponded to any of the *Shigella* species. For example, strains identified as *Sh. boydii* were found to be identical by MLEE to strains of *Sh. flexneri*, or *Sh. dysenteriae*. An exception to the general polyphyletic nature of *Shigella* species were strains of *Sh. sonnei*, as 20 strains collected from widely distributed geographic sources were all identical by MLEE using 12 enzymes. *Sh. sonnei*, therefore represents a single clonal lineage of *E. coli*, a conclusion supported by Karaolis *et al.* (1994) who showed that 46 strains of *Sh. sonnei* collected from widely different temporal and geographic sources were virtually identical by a number of molecular techniques. In summary, population-genetic analysis of the *Shigella* has underscored the inability of biochemical or immunological techniques to elucidate phylogenetic relationships between bacteria, or to identify real biological groupings.

Diversity of *E. coli* Isolates from Disease

E. coli isolates are classified serologically by variation in lipopolysaccharide (O antigens) and flagellin (H antigens). Isolates expressing about 12 different O antigens (serogroups) have been associated with outbreaks of infantile diarrhoea (EPEC strains). EPEC isolates of a single serogroup consist of diverse collections of strains, but individual clones, or clusters of related clones, which often express a single flagellar antigen usually predominate (Ørskov *et al.*, 1990). For example, 70 EPEC isolates of serogroup O55, which is classically associated with infantile diarrhoea, were resolved by MLEE using 20 enzymes into 7 ETs, which formed 3 divergent clusters of strains (Rodrigues *et al.*, 1996). However, 74% of the isolates were members of ET-1, which were predominantly serotype O55:H6, or ET-5, which were non-motile O55:H- or O55:H7. Additionally, these two predominant clones were characterised by particular combinations of virulence factors and distinct patterns of adherence to HEp-2 cells. The origins and evolution of EPEC strains have been discussed by Whittam (1995).

Other *E. coli*-associated disease in humans and farm animals appears to be due to a much more diverse collection of strains Whittam (1996). For example, Maslow *et al.* (1995) showed that 187 bloodstream isolates, recovered from 3 locations, were resolved into 101 ETs using 20 enzymes. Among this diversity, a few ETs, or clusters of related ETs, were found repeatedly, and from different locations, and one cluster corresponded to a cluster of isolates previously shown to predominate among isolates from pyelonephritis and neonatal meningitis.

Enterohaemorrhagic *E. coli* Serovar O157:H7

In the early 1980s, *E. coli* strains of serotype O157:H7 were recognised as the cause of bloody diarrhoea (haemorrhagic colitis) and haemolytic uraemic syndrome in humans (Riley *et al.*, 1983). Sporadic and large scale outbreaks of disease associated with this serotype have been reported in Europe, North America and elsewhere. The pathogenesis of this strain of *E. coli* is mediated by several virulence factors including the production of shiga-like cytotoxins.

To determine the genetic relatedness of strains of O157:H7 to other enteropathogenic *E. coli*, a population-genetic study was undertaken by Whittam *et al.* (1993). This study, of 1300 isolates of pathogenic *E. coli*, including over 500 strains bearing the O157 antigen, found that strains of O157:H7 were not closely related to other enteropathogenic shiga-toxin producing, clones of *E. coli*. The O157 characteristic by itself is not phylogenetically useful as strains bearing this antigen could differ at more than half of the enzyme-encoding loci studied by MLEE. However, all strains of *E. coli* O157:H7 were closely related, and 95% were of a single ET. The enterohaemorrhagic O157:H7 strains were most closely related to strains of serotype O55:H7 (ET-5) which are associated with outbreaks of infantile diarrhoea (see previous section). The majority of the remaining strains bearing the O157 antigen were only distantly related to this clone, and formed a clonal group frequently bearing the H43 flagellar antigen.

In addition to a close phylogenetic relationship, strains of O157:H7 and O55:H7 also share similarities in the mechanism for adherence to intestinal cells (localised adherence, discussed by Whittam *et al.*, 1993). The heightened virulence of the enterohaemorrhagic O157:H7 lineage, compared with enteropathogenic O55:H7 strains, is probably mediated by the presence of one or two shiga-like toxins in O157:H7 strains. These observations have led to a preliminary scheme for the emergence of the O157:H7 clonal lineage (Whittam *et al.*, 1993; Feng *et al.*, 1998), which proposes that it is derived from an ancestral O55:H7-like strain that was already capable of producing infantile diarrhoea and exhibited the localised adherence phenotype. The emergence of the O157:H7 clone required the acquisition of a virulence plasmid, and a Shiga-like toxin gene (probably carried on bacteriophage from other enteropathogenic *E. coli*), and a change in the O antigen from O55 to O157 (Whittam, 1995, 1996). Since then, the O157:H7 lineage has diverged further. It is now possible to identify non-motile descendants (Feng *et al.*, 1996), as well as those with a second Shiga-like toxin (Tarr *et al.*, 1989). However, bacterial population genetics has clearly shown that these variants are all descended from a recent common ancestor (Whittam, 1995, Whittam, 1995, 1996).

Population-genetic Studies of Some Other Clonal Pathogens

Streptococcus pyogenes

Group A streptococci (*Strep. pyogenes*) are the cause of a spectrum of diseases of varying clinical severity, including pharyngitis, rheumatic fever, poststreptococcal glomerulonephritis and sepsis. The incidence of severe invasive group A streptococcal disease declined after the introduction of antibiotics, but in the 1980s it increased significantly, with outbreaks of rheumatic fever, and the recognition of a rapidly progressing multi-organ disease, streptococcal toxic-shock-like syndrome (TSLS), with tissue necrosis and erythematous rash (Stevens, 1992). Population-genetic studies have provided insights into the reason for this apparent shift in the severity of disease (Musser, 1996).

Several virulence factors have been identified in strains of *Strep. pyogenes.* The M protein, a surface-exposed cell-membrane-anchored dimeric fibrillar protein, has long been recognised as a major virulence factor. The extensive variation within the M proteins (and M-like proteins) of different isolates provides the basis of the typing system for group A streptococci (Fischetti, 1989). As found with most typing schemes based on cell-surface components, isolates within a single M type may be genetically diverse, and isolates that are genetically indistinguishable may have the same M type. However, certain M types have been long associated with severe disease, and most strains causing TSLS belong to the M-1 or M-3 serotypes (Stevens, 1992).

Musser *et al.* (1991) analysed the clonal relationships between 108 isolates of *Strep. pyogenes* recovered in the USA from patients with TSLS and other invasive disease. Although 33 ETs were resolved, approximately half of the isolates, and more than two-thirds of those from cases of TSLS, were members of two closely related clones, ET1 (serotype M-1) and ET2 (serotype M-3). Serotype M-1 strains have been particularly associated with the recent increase in serious invasive disease in Europe, the USA and elsewhere (Stevens, 1992). Electrophoretic variation in 12 enzymes distinguished 6 ETs among 126 M-1 isolates recovered worldwide, but almost all recent episodes of invasive disease were caused by a single ET (Musser *et al.*, 1995). Further discrimination among ET-1 distinguished a subclone that appears to have been responsible for the recent worldwide resurgence in serious invasive disease (Musser *et al.*, 1995).

Three pyrogenic exotoxins have been identified in *Strep. pyogenes.* Two of the exotoxin genes, *speA* (encoding the scarlet fever toxin) and *speC*, are carried on bacteriophage, and are present in only some strains, whereas *speB* is chromosomal and is present in all strains, although it is not expressed in some (Weeks and Ferretti, 1984; Yu and Ferretti, 1991). In several studies, expression of *speA* appears to be associated with severe invasive disease (Yu and Ferretti, 1989; Musser *et al.*, 1991). A correlation between expression of *speB*, but not of *speC*, and serious invasive disease has also been found in some studies (Talkington *et al.*, 1993). Sequencing of *speA* has shown that one allele (*speA1*) is found in many clonal lineages, and is present in strains of ET1 and ET2 from the 1920s and 1930s, whereas the recent isolates of these ETs that are associated with the resurgence of serious invasive disease have variant alleles (*speA2* and *speA3*), that each differ from *speA1* by a single non-synonymous substitution (Musser *et al.*, 1993). One possibility is that these variant alleles have evolved recently from *speA1* to result in forms of exotoxin A with increased capacity to cause tissue damage. The gene that encodes a streptococcal superantigen has also been found in about 50% of *Strep. pyogenes* isolates. Although the gene was not present in the ET1 serotype M-1 clone, or in older isolates of the ET2 serotype M-3 clone, it was present in recent isolates of this clone from patients with TSLS (Reda *et al.*, 1994).

Both the *speA* and *speC* genes appear to spread horizontally between *Strep. pyogenes* lineages, presumably by bacteriophage-mediated transfer (Kapur *et al.*, 1992). Additionally, genes encoding other known, or putative, virulence factors appear to have been distributed horizontally (Kehoe *et al.*, 1996). The ability of these genes to spread horizontally, combined with substitutions within these genes that increase the virulence potential of their products, may occasionally result in strains that have a particularly high level of virulence. Population-genetic and molecular studies suggest that this type of situation has occurred within *Strep. pyogenes*, to produce the successful clones that are responsible for the recent resurgence of serious invasive disease worldwide.

The overall population structure of *Strep. pyogenes* has not been fully addressed, as most MLEE studies have used isolates from serious invasive disease, which are likely to be a biased sample of the whole population. Recombination is clearly apparent in *Strep. pyogenes*, from the poor congruence between M type and ETs (Whatmore and Kehoe, 1994), the mosaic structure within genes encoding the M proteins and other virulence genes (Whatmore and Kehoe, 1994; Kapur *et al.*, 1995; Kehoe *et al.*, 1996), and the distribution of virulence genes among ETs (Musser *et al.*, 1991, 1993, 1995; Reda *et al.*, 1996). These genes are likely to be under selection, however, and give little indication of the extent of recombination. Additionally, the *speA* and *speC* genes are carried on bacteriophage which is likely to promote their horizontal transfer within the population.

Haemophilus influenzae

Encapsulated (typable) isolates of *H. influenzae* are associated with invasive disease (cellulitis, epiglottitis, meningitis and septicaemia), but the great majority of serious disease, in countries that do not use the new conjugate vaccine, is caused by isolates that express the serotype b capsule. Population genetic studies, by analysis of the electrophoretic variation in 17 enzymes in 2206 isolates, have shown that the classification of *H. influenzae* into 6 capsular types (serotypes a–f) is broadly congruent with the genetic relationships inferred from MLEE (Musser *et al.*, 1988). Thus, each of the 280 ETs contained strains of only a single serotype and, in most cases, isolates of a particular serotype belonged to clusters of closely related ETs. As found with many pathogens, most invasive disease was caused by isolates of a small number of ETs. For example, 70% of the serotype b isolates were represented by only four ETs. In this large study no clear association between ET and the type of invasive disease was found.

The congruence between serotype and ET is not complete, since some isolates of serotype a and b were only distantly related to the major groups of isolates of these serotypes. The expression of the type-b capsule in two distantly related groups of isolates has been studied further by Kroll and Moxon (1990), who showed that the serotype-specific region of the capsular locus in the two groups are extremely similar, suggesting that at some time in the past a recombinational exchange has introduced the serotype-specific genes from a serotype b isolate into a genetically distinct lineage of *H. influenzae*.

The broad congruence between serotype and genetic relatedness inferred by MLEE suggests that recombination is rare in encapsulated *H. influenzae* (Musser *et al.*, 1988), and this view is supported by the presence of strong linkage disequilibrium between alleles, which does not appear to be due to sampling bias (Maynard Smith *et al.*, 1993). The lack of evidence for frequent recombination in encapsulated *H. influenzae* is puzzling, since these isolates are naturally transformable, and the genome contains 1465 copies of the specific sequence that is believed to promote uptake of DNA for transformation, and several genes that are thought to be specifically required for this process (Smith *et al.*, 1995). Recombination may, however, be more significant in non-encapsulated (non-typable) strains of *H. influenzae* associated with otitis media, chronic bronchitis and pneumonia. In a few instances non-encapsulated isolates may be derived from encapsulated strains, but they are mostly a distinct population that cause a different spectrum of disease (Musser *et al.*, 1986). These isolates appear to be more diverse and may be less clonal than the encapsulated isolates. A weakly clonal population structure was first proposed by Porras *et al.* (1986), who found that almost every non-encapsulated isolate was a distinct ET, using MLEE with only 6 enzymes. Additionally, none of the isolates from the USA was identical to any from Sweden, and the levels of linkage disequilibrium were lower than those found in encapsulated *H. influenzae* or in *E. coli* populations. In a further MLEE study, using 15 enzymes, Musser *et al.* (1986) showed that each of 65 non-encapsulated isolates from North America was a unique ET, even though the isolates were from episodes of disease (mostly otitis media). The failure to find identical isolates in this sample is suggestive of a non-clonal population, since non-encapsulated *H. influenzae* are common constituents of the flora of the upper respiratory tract, and the isolates from disease are, therefore, likely to be a highly biased sample of the population, which might be expected to be enriched for any clones that have an increased ability to cause disease. The I_A value obtained using the MLEE data from Musser *et al.* (1986) is significantly greater than zero, however, suggesting that alleles in the population are not randomly associated (B. G. Spratt, unpublished results).

Non-encapsulated *H. influenzae* therefore appear to have a less clonal population structure than encapsulated isolates. A similar difference in population structure is found within *N. meningitidis*, where serogroup A isolates are more clonal than isolates of other serogroups. In both *N. meningitidis* and *H. influenzae* the isolates that have a higher propensity to cause invasive disease are more clonal than those that are frequently carried asymptomatically within the population. In meningococci the higher rate of transmission of serogroup A isolates, and lower rates of carriage, have been proposed to reduce the disruptive effect of recombination on clonal integrity and a similar argument can be applied to *Haemophilus*. In both meningococci and *Haemophilus*, however, there may also be differences in the competence for transformation between isolates expressing different capsular polysaccharides, or which lack capsules.

Haemophilus Isolates from Brazilian Purpuric Fever

Brazilian purpuric fever is a serious invasive disease that was first recognised in 1984 and was caused by an organism identified as *H. influenzae* biogroup aegyptius, which had previously been associated with outbreaks of conjunctivitis. MLEE analysis of isolates from the outbreak, other isolates from this biogroup and encapsulated *H. influenzae* isolates recovered worldwide showed that biogroup aegyptius strains are polyphyletic, clustering into three lineages that were within *H. influenzae* but were not closely related to each other. The outbreak isolates were members of a single clone, distinct from other biogroup aegyptius isolates, which was similar to serotype c isolates of *H. influenzae* (Musser and Selander, 1990).

Population Genetics of Weakly Clonal or Non-clonal Pathogens

The Pathogenic *Neisseria*

Human *Neisseria* consist of a number of commensal species (*N. lactamica*, *N. polysaccharea*, *N. cinerea*, *N. flavescens*, etc.), and two important pathogenic species, *N. meningitidis* (the meningococcus) and *N. gonorrhoeae* (the gonococcus). Meningococci and human commensal *Neisseria* colonise the nasopharynx, although meningococci have the ability occasionally to invade the bloodstream and cerebrospinal fluid to cause septicaemia and meningitis. The gonococcus occupies a distinct ecological niche from all other *Neisseria*, and is predominantly recovered from the genital tract and rectum. Although the diversity of human sexual activity will result in gonococci meeting meningococci and the other human *Neisseria*, the gonococcus is likely to be relatively ecologically isolated.

Neisseria gonorrhoeae

Populations of *N. gonorrhoeae* are basically non-clonal (Spratt *et al.*, 1995). Alleles in gonococcal populations do not depart significantly from linkage equilibrium, suggesting that recombination between distinct gonococcal lineages must be frequent in nature, and that infections with more than one strain are relatively common (Maynard Smith *et al.*, 1993; O'Rourke and Stevens, 1993; Vázquez *et al.*, 1993). This would not be surprising within the highly sexually active core groups who are believed to maintain gonorrhoea within communities (Yorke *et al.*, 1978). The non-clonal population structure of gonococci is supported by the non-congruence between the clustering of isolates obtained using MLEE, auxotyping or serology (de la Fuente and Vázquez, 1994). Unfortunately, the sequences of gonococcal housekeeping genes are too uniform to provide statistically significant evidence for frequent recombination (Spratt *et al.*, 1995).

There are reports of gonococci that appear to be similar in phenotype (usually their combination of auxotype and serovar) and have been suggested to be clones (Sarafian and Knapp, 1989). Many of these clones are probably isolates that happen by chance to have the same auxotype and serovar, since this way of classifying gonococci is not very discriminatory. For example, if 20% of gonococci are proline-requiring (P^-), and 20% are serovar IB-2, frequent recombination would result in 4% of isolates having the P^-/IB-2 auxotype/serovar combination by chance, rather than by common ancestry. To evaluate critically whether there are clones of gonococci, methods must be used that are able to discriminate between the large number of distinct genotypes expected in a recombining population (O'Rourke and Spratt, 1994). Interestingly, there is good evidence that arginine-, hypoxanthine- and uracil-requiring gonococci, which have long been recognised as having a number of distinctive phenotypic properties (Knapp and Holmes, 1975), are a relatively stable clonal lineage, even though the overall population structure appears to be non-clonal (Gutjahr *et al.*, 1997). How a clonal lineage is maintained in the gonococcal population is not clear, but possible reasons have been discussed elsewhere (Gutjahr *et al.*, 1997).

Neisseria meningitidis

The population structure of *N. meningitidis* is complex (Maiden, 1993; Maynard Smith *et al.*, 1993; Maiden and Feavers, 1995; Spratt *et al.*, 1995). First, there are serious sampling problems that are not present in the case of *N. gonorrhoeae*, since approximately 10% of the population may carry meningococci in their nasopharynx, and yet the incidence of invasive meningococcal disease in the developed world is typically only about 2 cases per 100 000 of population. Thus, invasive isolates represent only about 0.0002% of the total meningococcal population, and the organism is basically a commensal that has the ability to cause devastating disease occasionally. From the perspective of population genetics, a sample of 200 isolates chosen to be representative of the total meningococcal population would contain no isolates from invasive disease. However, most samples of meningococci that have been used for population-genetic studies consist of isolates that are almost all from invasive disease (Caugant *et al.*, 1986, 1987) . Similar sampling problems are also likely to occur in studies of other pathogens that are frequently carried but rarely cause disease (e.g. *Strep. pneumoniae* or *Staph. aureus*). This type of sampling problem is important, since it may make bacterial species appear to be much more clonal than they really are.

A second complication is that serogroup A meningococci have a very different epidemiology from isolates of the other important serogroups (predominantly serogroup B and C meningococci). Whereas serogroup B and C isolates cause endemic and hyperendemic disease, and small outbreaks of disease, serogroup A isolates cause epidemic and pandemic meningitis with very high attack rates that in recent decades have been restricted mainly to the developing world – particularly sub-Saharan Africa (Achtman, 1994).

Studies by MLEE have shown that populations of serogroup B and C meningococci from invasive disease

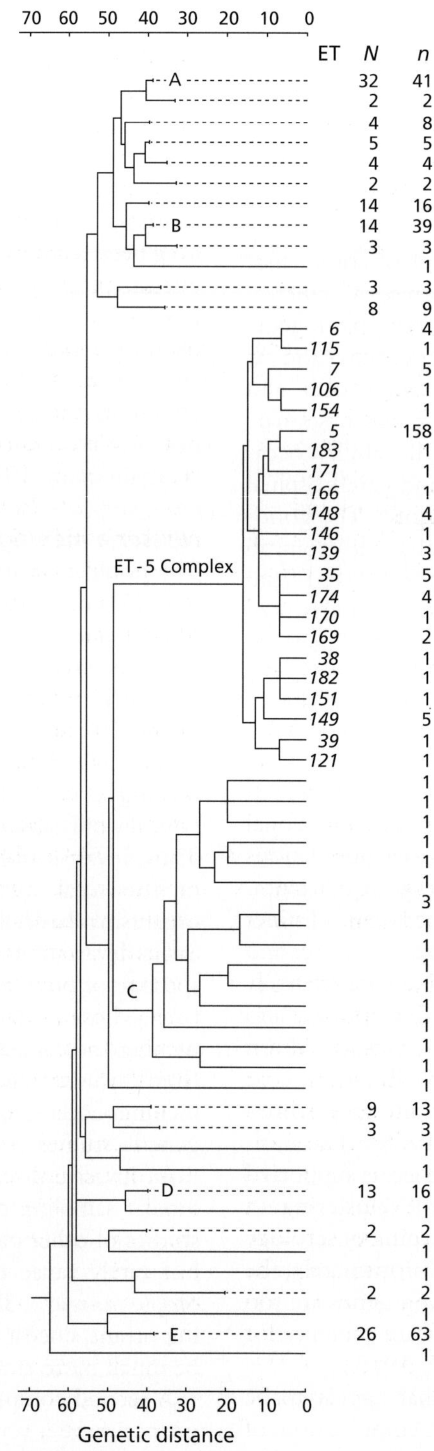

Fig. 5 Dendrogram based on MLEE data from *N. meningitidis* isolates. The dendrogram was constructed with MLEE data (15 enzymes) from 460 serogroup B and C *N. meningitidis* isolates recovered from invasive disease. In some cases terminal branches of the tree contain more than one ET, and these clusters of closely related ETs have been truncated at the positions marked by the dashed lines. The number of ETs in the truncated branches are shown (*N*). The number of isolates in single ETs, or in truncated clusters of ETs, are indicated (*n*). In this basically non-clonal species, the dendrogram does not correspond to a phylogenetic tree, and the relationships between the major lineages, but not the clustering of the identical or very similar isolates, would be different if the dendrogram had been constructed with variation in other housekeeping enzymes, or some other type of neutral variation. The dendrogram is, however, useful for identifying the large cluster of closely related ETs, which represents the isolates of the unstable hyperendemic ET-5 clonal complex which was responsible for much of the invasive meningococcal disease in Europe at the time this population of meningococcal strains was assembled. Reproduced with permission from Caugant *et al.* (1986).

are highly diverse, but contain large clusters of isolates that have very similar genotypes (Caugant *et al.*, 1986, 1987). The first of these hyperendemic lineages to be recognised was the ET-5 clonal complex (a cluster of closely related ETs), which was responsible for increased rates of meningococcal disease in many countries during the 1980s (Caugant *et al.*, 1986; Fig. 5). More recently, other hyperendemic lineages of serogroup B and C meningococci have been documented, e.g. the ET-37, A4 and lineage 3 clonal complexes (Scholten *et al.*, 1994; Maiden and Feavers, 1995). The high incidence of these successful hyperendemic lineages in the highly biased samples of the meningococcal population taken from invasive disease inevitably results in significant linkage disequilibrium in the MLEE data ($I_A > 0$). Maynard Smith *et al.* (1993) showed that the linkage disequilibrium detected by the I_A statistic is greatly reduced if only one example of each meningococcal ET is included in the analysis, and it essentially disappears if only one example of each major lineage is included. The rationale behind this approach is to try and move back in the tree derived from the MLEE data to a time when the hyperendemic clones had not emerged. This approach does not reduce the I_A value to zero in truly clonal species (e.g. *S. enterica*), but care must be taken, as the use of only some of the data may decrease the statistical robustness of the approach (Maynard Smith *et al.*, 1993).

These results have been interpreted to mean that the serogroup B and C meningococcal population is basically non-clonal, and that the long-term evolution of the population is dominated by recombination (Maynard Smith *et al.*, 1993). Occasionally strains arise that are more virulent, or more transmissible, than other meningococci and these increase in frequency to become common in the highly biased sample of isolates recovered from invasive disease. The over-representation of these hyperendemic strains in meningococcal populations taken from invasive disease makes it look as though meningococci are clonal. However, if steps are taken to remove this sampling bias, the population appears to be basically non-clonal.

If recombination is common in serogroup B and C meningococci, the hyperendemic clones should be unstable, since they will become increasingly diverse as their originally uniform genomes accumulate recombinational exchanges from genetically distinct isolates. This appears to be the case since meningococcal clones are much less stable than the clones seen in highly clonal species, which are believed to persist with little change for thousands of years. In most cases it is difficult to know the age of the hyperendemic clones, but members of the ET-5 hyperendemic clonal complex have not been identified before the 1970s. At the time that ET-5 was identified as a major cause of meningococcal disease (Caugant *et al.*, 1986), there were already 22 distinct ETs within the clonal complex (Fig. 5), and since then further diversification has occurred. Variation at the glucose-6-phosphate dehydrogenase locus in members of the ET-5 complex has been analysed and, as expected, is due to recombinational exchanges, rather than mutation (J. Zhou and B.G. Spratt, unpublished data). Similarly, isolates of the hyperendemic lineage III clonal complex were first recognised in 1980, and the early isolates were indistinguishable by MLEE, but since 1984 lineage III has diversified markedly and now includes a number of ETs (Scholten *et al.*, 1994). This very rapid rate of diversification should be contrasted with the stability of clones in a highly clonal species such as *S. enterica*.

Also, unlike in *Salmonella*, there is a poor congruence between the clustering of meningococcal isolates obtained with MLEE and serology (Caugant *et al.*, 1987). Thus, isolates of the ET-5 clonal complex are usually serogroup B, and those of ET-37 are usually serogroup C, but serogroup C isolates of ET-5, and serogroup B isolates of ET-37 are not uncommon, presumably as a result of recombinational exchanges within the capsular gene locus (Caugant *et al.*, 1987). Similarly, meningococcal isolates of the same hyperendemic clone often differ in serotype and serosubtype. For example, isolates of lineage III from 1980 were of a single serogroup, serotype and serosubtype (B:4:P1.4) but, since 1984, isolates that are serotypes 14 and 15, and subtypes P1.2, P1.7, and P1.12 have been identified.

As mentioned above, non-clonal populations should consist of a huge number of distinct genotypes, since the variation present within the population is not purged by periodic selection, and is constantly shuffled by recombinational exchanges. A good illustration of the diversity in meningococci is provided by a recent study of serogroup B meningococci (Seiler *et al.*, 1996). Even though the 232 meningococci analysed were from invasive disease, and consequently included multiple examples of the currently circulating hyperendemic clones, MLEE using only 13 enzymes identified 197 distinct ETs (Fig. 6). If the hyperendemic isolates are removed almost every isolate in this basically non-clonal population is distinct.

A high rate of recombination in meningococci is also supported by studies of housekeeping genes, which typically show highly significant mosaic structure, and a gross lack of congruence between different gene trees (Zhou and Spratt, 1992; Feil *et al.*, 1995; Zhou *et al.*,

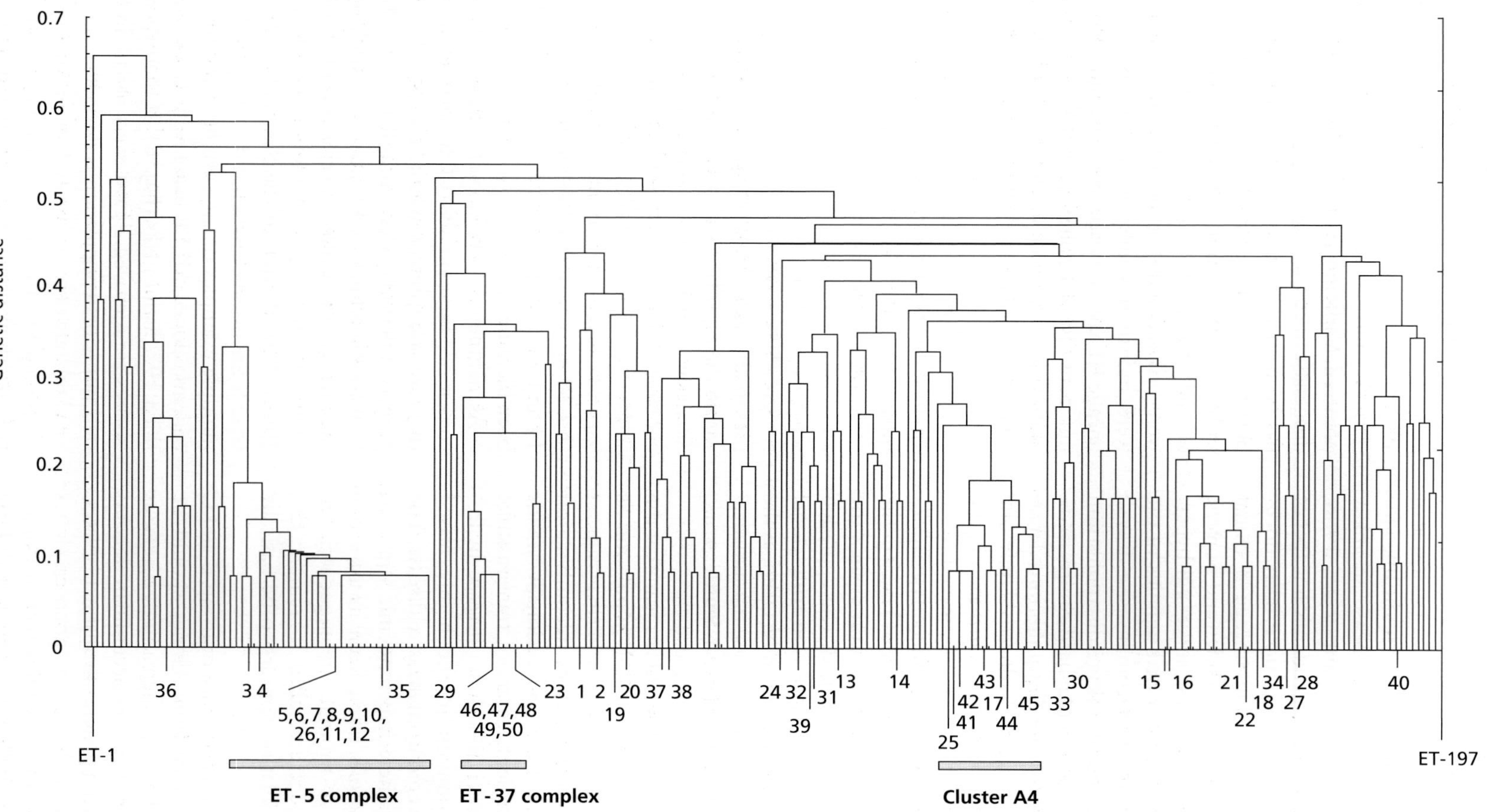

Fig. 6 Diversity of serogroup B meningococci. The dendrogram was constructed from variation in housekeeping enzymes within a sample of serogroup B meningococci recovered from invasive disease. The clusters of related ETs that define the hyperendemic lineages (ET-5, ET-37 and A4) that occasionally arise within the meningococcal population are shown. In this basically non-clonal population, recombination is believed to be frequent, such that the population should consist of a large number of distinct genotypes. In this case, indexing variation at only 13 loci, distinguishes 192 different ETs among the 232 isolates. Reproduced with permission from Seiler *et al.* (1996).

1997; Holmes *et al.*, 1999). Feil *et al.* (1999) used the approach outlined above (see above, 'The ratio of recombination to mutation') to show that during clonal diversification a nucleotide site within a meningococcal housekeeping gene is about 80 times more likely to change by recombination than by point mutation.

Interestingly, the studies of meningococcal housekeeping genes have also shown a surprisingly high rate of recombination between meningococci and the closely related human commensal *Neisseria* spp. Inter-species recombinational replacements from *N. flavescens* and *N. cinerea* were first detected in the penicillin-binding protein 2 genes of relatively penicillin-resistant meningococci (Spratt *et al.*, 1989, 1992; Bowler *et al.*, 1994). Similar events have subsequently been shown to occur in a number of housekeeping genes (Fig. 7), and it has been estimated that about 1 in 20 housekeeping genes from a typical meningococcus contains small recombinational replacements from commensal *Neisseria* spp. (Feil *et al.*, 1995; Zhou *et al.*, 1997). A history of inter-species recombination in some *Neisseria* housekeeping genes appears to have led to major distortions in the inferred phylogenetic relationships between the human *Neisseria* species (Feil *et al.*, 1996; Zhou *et al.*, 1997; Smith *et al.*, 1999).

Inter-species recombinational exchanges have not been found in gonococcal housekeeping genes, although they are found in genes under strong selection (Spratt, 1988; Halter *et al.*, 1989; Vázquez *et al.*, 1995), presumably because these events occur much more rarely in gonococci than meningococci, since the former are relatively ecologically isolated in the genital tract and rectum, and meet other *Neisseria* species much less often than do the latter (Vázquez *et al.*, 1993).

Serogroup A meningococci may possibly be more strongly clonal than serogroup B and C meningococci, as a direct result of the increased rate of transmission of these strains, and the low levels of nasopharyngeal carriage of meningococci in sub-Saharan Africa (Olyhoek *et al.*, 1987; Achtman, 1994; Maiden and Feavers, 1995; Spratt *et al.*, 1995). Both these factors decrease the opportunity for serogroup A meningococcal clones to be broken up by recombinational exchanges with other meningococci. Serogroup A isolates show much more congruence between MLEE data and serology than other meningococci, as expected if recombination is less frequent in this population (Suker *et al.*, 1994). Serogroup A meningococci are genetically rather uniform and have probably emerged relatively recently (Morelli *et al.*, 1997).

Genetic Isolation Between *N. gonorrhoeae* and *N. meningitidis*

N. gonorrhoeae and *N. meningitidis* are very closely related, differing at only about 2% of nucleotides sites within typical housekeeping genes, and might be combined as a single species were it not for the important differences in the diseases they cause. In fact, the separation into distinct species is entirely justified on population-genetic grounds, since both MLEE data and sequence data show that gonococci are clearly distinct from meningococci (Vázquez *et al.*, 1993). The clear distinction between such similar populations is at first sight surprising, given the evidence for frequent recombination within both gonococci and meningococci, but is probably brought about by the relative ecological isolation between these populations, resulting in a lower rate of recombination between populations than within populations. Gonococcal housekeeping genes are very uniform and it is likely that they emerged relatively recently as a lineage of *N. meningitidis* that can successfully colonise the genital mucosa (Vázquez *et al.*, 1993; Feavers and Maiden, 1998).

Helicobacter pylori

He. pylori is the causative agent of acute gastritis and peptic ulcer disease (Blaser, 1992), and has been associated with gastric malignancies (Bayerdorffer *et al.*, 1995). *He. pylori* isolates are naturally transformable (Wang *et al.*, 1993b), and population-genetic studies have provided evidence for a non-clonal population structure. The evidence for frequent recombination, presumably mediated by transformation, is mainly derived from a MLEE study of a sample of *He. pylori* recovered from 74 patients in the Houston area (Go *et al.*, 1996). Even though these isolates were recovered in one hospital, they were extremely diverse, and using only 6 enzymes, 73 different ETs were distinguished. The number of alleles per locus, and the genetic diversity within the sample, was higher than that found in MLEE studies of any other bacterial species. In addition to the diversity in housekeeping genes, diversity appears to be introduced by chromosomal re-arrangements, as the gene order differs between *He. pylori* isolates (Jiang *et al.*, 1996). The large number of genotypes in the population, and the lack of any linkage disequilibrium between alleles, suggests that *He. pylori* is non-clonal. Recent studies applying the homoplasy test of Maynard Smith and Smith (1998) to nucleotide sequence data from *He. pylori* housekeeping genes have also suggested a very high rate of recombination in this species (Suerbaum *et al.*, 1998). As expected from the inferred

```
         111111111111222222222222222222222223333333333333333334444444444444444555555555555555555555555555555556666666666666666
       180145566688890000444445556777788899990000122223555667791123455566778999001223333444445555566677778899990000123334555556
       345040656846951456045694597045623646745796146731476902014054139291435681431812460478924678147036928147802398113552478903
NM-1   GTCGTGTCGAGGCCCGGTTCATTGGGCTGGCTCACGGCGACAGGCGAGTCGTTATGGCCCAGCACGTGGGGTCCGTTCATCGAAAGGAAAGGGGAGGACCCGCCACCGACGGCGGCAGAC
NM-2   ---A------A---------G-C-----------------------C---CCCC--------T-TACAC-TC------------G-C------A---G------------C---------
NM-3   A---------A--ATCAC--G-C-----------------------C---------------T-TACAC-TC------------G-C------A---G------------C------A--
NM-14  -CG--------A--------G-C-A---------------------T-----------T---T-TAC-TATC-------------------------G----------------------
NM-17  ----------A--ATCAC--G-C-----------------------C---------------T-TACAC-TC------------G-C------A--------------------------
NM-18  -CG--AGTCTA---T-----G-C-A----------------G-----------G-C-T----T-TACAC---T-----------------------TC------------T------A-T
NM-19  ----------A---------G-C-------T---------------T------------T------------------------------------------------------------
NM-22  AC--CAG-----T-T---CA-C-AAATCAA-CATGAAAACAGCAGCCAGG--CCC--G----T-TACAC-TC------------G-C---------TC---A------------------
NM-23  A---------A--ATCAC--G-C-----------------------C------------T------------------------G-C------A---G----A-------C------A--
NM-27  ----------A---------G-C-------T---------------T------------T------------------------------------TC------------C---------
NM-29  ----------------------------------------------C---------------T-TACAC-TC------------G-C---------TC------------C---------
NM-32  ------------------------------------------------------------------------------------G-C------A---G------------C---------
NFLAV  AC--CAG-----T-T---CA-C-AAATCAA-CATGAAAACAGCAGCCAGG--CCC-CG-TCCTG--CAC-AC-GACCAGATTGCCCCCGCAACCGC-GTGTA-GTTATGA-AGACGC-CT
```

Fig. 7 Intra- and inter-species recombination in a meningococcal housekeeping gene. The sequence of the *mtgA* gene, which encodes a transglycosylase, was determined for 50 serogroup B meningococci, selected from the strains used to construct the dendrogram depicted in Fig. 5. The polymorphic (variable) sites in the 12 different *mtgA* alleles found in these strains (NM-1, etc.), and in the type strain of *N. flavescens* (NFLAV), are shown. The nucleotide present at each polymorphic site is shown for strain NM-1. For the other strains, only the nucleotides that differ from those of strain NM-1 are shown. The position of each polymorphic site in the sequence of *mtgA* is shown, in vertical format, above each site. The sequence from strain NM-22 contains a very clear example of an inter-species recombinational replacement (nucleotides 1–413) from *N. flavescens.* The mosaic structure resulting from the inter-species recombinational exchange in the *mtgA* allele of strain NM-22 is highly significant ($p < 0.0001$) by the statistical test of Maynard Smith (1992). The statistical test of Sawyer (1979) detected highly significant evidence for intra-species recombinational exchanges within the other 11 meningococcal sequences ($p < 0.0001$). (E.J. Feil, J. Zhou and B.G. Spratt, unpublished results).

high rate of recombination between lineages within this species, mixed infections with genetically distinct *He. pylori* isolates are relatively common (Hurtado and Owen, 1994; Weel *et al.*, 1996).

The severity of gastric disease is believed to depend on differences between *He. pylori* isolates in the distribution, or expression, of a number of virulence factors (*cagA*, *vacA*, etc.; Blaser, 1992). Frequent recombination might be expected to distribute these genes within the population, and this may be the case, as no association was found between the severity of gastric disease and ET, or clusters of related ETs, although the sample size was probably too small to detect weak associations (Go *et al.*, 1996).

Other Weakly Clonal or Non-clonal Bacteria

There are indications that recombination may be frequent in another naturally transformable species, *Strep. pneumoniae*, as low levels of linkage disequilibrium between alleles have been reported (Lomholt, 1995; Hall *et al.*, 1996; Takala *et al.*, 1996). A further indication of relatively rapid rates of recombination is provided by the high rate of diversification of antibiotic-resistant clones of *Strep. pneumoniae* (see above). As discussed in the next section, serotype changes also appear to occur frequently in nature (Coffey *et al.*, 1991), by recombinational exchanges at the capsular locus (Coffey *et al.*, 1998b). Direct evidence has been provided for a high rate of recombination by showing that a single nucleotide site in a pneumococcal housekeeping gene is about 50 times more likely to change by recombination than by point mutation (Feil *et al.*, 2000).

A non-clonal population structure has also been proposed for *Pseudomonas aeruginosa* (Selander and Musser, 1990). This suggestion is supported by the lack of correlation, except in isolates of one serotype, between ribotyping and electrophoretic variants at esterase loci (Denamur *et al.*, 1993). Similarly, non-encapsulated *H. influenzae* may have a weakly clonal population structure, as discussed above.

Many of the species in which non-clonal, or weakly clonal, population structures have been inferred are naturally transformable. This feature extends to non-pathogenic species, since recombination also appears to be frequent in some populations of naturally transformable *Bacillus* spp. (Duncan *et al.*, 1994). There are, however, several exceptions, as *Ps. aeruginosa* is not known to be naturally transformable, and recombination appears to be frequent in some populations of *Rhizobium* (Souza *et al.*, 1992), *Ps. fluorescens* (Haubold and Rainey, 1996) and *Burkholderia cepacia* (Wise *et al.*, 1995), where natural transformation has also not been documented.

Also, the availability of natural transformation as a mechanism for genetic exchange does not inevitably lead to rates of recombination that are sufficient to disrupt clonal structure, since encapsulated *H. influenzae* populations appear to be highly clonal, although non-encapsulated isolates are less clonal.

Population Genetics and Antibiotic Resistance

Antibiotic resistance in both nosocomial and community-acquired bacterial pathogens has become an increasing problem in recent years (Neu, 1992). In many cases resistance genes are carried on plasmids that can be transferred readily to other isolates, or other species (Davies, 1994). Resistance to some antibiotics, in some species, is due to alterations within chromosomal genes (Spratt, 1994), and the horizontal spread of resistance is likely to be less frequent. Even in clonal species, however, the horizontal spread of chromosomally mediated resistance is to be expected, since very rare recombinational events will be significant as they will be strongly selected by antibiotic usage. Studies of antibiotic resistance in bacterial pathogens need, therefore, to consider both the clonal spread of resistant isolates, and the creation of new resistant isolates by the horizontal spread of the resistance determinants into new lineages. Population genetic studies have in several cases provided insights into the origins and spread of antibiotic-resistant isolates.

A good example is provided by penicillin (and cephalosporin) resistance in pneumococci (Coffey *et al.*, 1995a). In this species, resistance is due entirely to the development of altered forms of penicillin-binding proteins (PBPs) that have decreased affinity for β-lactam antibiotics (Hakenbeck *et al.*, 1980). High-level resistance to penicillin requires the development of altered forms of PBPs 1A, 2B and 2X (Barcus *et al.*, 1995, whereas high-level resistance to extended-spectrum cephalosporins requires alterations of only PBP 1A and 2X (Coffey *et al.*, 1995b). Interestingly, the genes that encode low-affinity PBPs have emerged by inter-species recombinational exchanges, presumably mediated by transformation, which replace parts of the PBP genes from susceptible pneumococci with the corresponding regions from the homologous PBP genes of closely related commensal streptococcal species (Dowson *et al.*, 1989; Coffey *et al.*, 1995a).

The altered PBPs of penicillin-resistant pneumococci are very variable and have clearly arisen on numerous occasions (Coffey *et al.*, 1995a). By indexing the relatedness of the penicillin-resistant strains (e.g. using MLEE), and their altered PBP 1A, 2B and 2X genes, it has been possible to identify successful penicillin-resistant and multiple antibiotic-resistant clones, which in some cases have spread worldwide (Muñoz *et al.*, 1991; McDougal *et al.*, 1992; Sibold *et al.*, 1992; Soares *et al.*, 1993). In Spain, a country with a high incidence of antibiotic resistance, there are currently about six major clones of penicillin-resistant or multiple resistant pneumococci (Coffey *et al.*, 1996). The best characterised is the multi-resistant Spanish serotype 23F clone which probably arose in Spain in the late 1970s or early 1980s, and has subsequently spread worldwide (Muñoz *et al.*, 1991; Coffey *et al.*, 1995a). By analysing the relatedness of the resistant strains and their altered PBP genes, it has been possible to demonstrate the significance of recombinational events, involving both PBP genes and capsular genes, in the emergence of new resistant clones, or clonal variants (Coffey *et al.*, 1991). Thus, strains have been identified that are indistinguishable from the multi-resistant Spanish serotype 23F clone by MLEE, antibiotic-resistance profile, and the sequences of their altered PBP genes, but which are serotype 19F (Coffey *et al.*, 1991), 19A (Coffey *et al.*, 1998a), or 14 (Barnes *et al.*, 1995). Recent studies have established that these serotype variants have arisen by recombinational exchanges at the capsular gene locus (Coffey *et al.*, 1998b). Furthermore, the horizontal spread of altered PBP genes into unrelated pneumococcal lineages has been inferred from the recovery of penicillin-resistant serotype 9V pneumococci in Spain that possess the same altered PBP 1A, 2B and 2X genes as the multi-resistant Spanish serotype 23F clone, but are not closely related by MLEE, and differ in serotype and antibiotic-resistance profile (Coffey *et al.*, 1991).

Studies of methicillin-resistant *Staph. aureus* (MRSA) have similarly shown that horizontal spread of chromosomal resistance determinants is important. MRSA isolates possess a gene (*mecA*) that encodes a very low-affinity PBP which is not present in penicillin-susceptible isolates and may have been introduced by an illegitimate recombinational event from a staphylococcal isolate related to *Staph. sciuri* (Wu *et al.*, 1996). The *mecA* gene in diverse MRSA strains is identical, arguing that it was introduced into *Staph. aureus* on a single occasion. Population-genetic studies have shown that the *mecA* gene is now carried by many phylogenetically diverse *Staph. aureus* lineages, implying that the *mecA* gene has been disseminated by horizontal transfer and recombination (Musser and Kapur, 1992).

The Age of Bacterial Species

Attempts have been made to determine the antiquity of some bacterial pathogens from the amount of divergence between gene sequences. This approach assumes that substitutions in homologous genes, such as the 16S ribosomal RNA (rRNA) gene, occur at essentially the same rate in different lineages of a phylogenetic tree (the molecular clock hypothesis), and the main problem is to relate DNA sequence divergence to evolutionary time. In higher organisms, the first appearance of species in the fossil record provides an approximate time scale for the dendrogram based on rRNA gene sequences, but this approach is clearly not possible with bacteria. One approach has been to study *Buchnera* species that live as obligate maternally transmitted bacterial endosymbionts, and have co-speciated with their aphid hosts for at least 100 million years. The phylogenetic trees derived from the small subunit rRNA genes of this bacterial endosymbiont species and its aphid hosts are congruent, and the dates of appearance of aphid species in the fossil record has allowed the rate of substitution at the 16S rRNA locus of *Buchnera* to be estimated at 2–4% per 100 million years (Moran and Baumann, 1994). This estimate is clearly subject to considerable uncertainty, and is probably too high, since genes in the endosymbiont *Buchnera* species appear to evolve about twice as rapidly as those of free-living *Buchnera* species (Moran *et al.*, 1995). However, a similar estimate for the rate of substitution within rRNA genes (2% per 100 million years) has been suggested by Ochman and Wilson (1987), who assumed that the divergence between *E. coli* and *S. enterica* occurred at around the time at which mammals appeared (140 million years ago).

Using this substitution rate, the sequences of rRNA genes can provide estimates of the dates at which different genera diverged, but rRNA genes evolve slowly, and usually provide too little resolution to give reliable information about recent evolutionary events. Housekeeping genes evolve more rapidly and the levels of sequence diversity at these loci within populations may give some information on the age of bacterial species. In many cases the extensive sequence diversity within the housekeeping genes of bacterial pathogens show that they are ancient, but there are some examples, particularly among species that infect only humans, where the levels of diversity are extremely low, and in these cases it is possible that they have emerged relatively recently.

A good example is *M. tuberculosis*, where extensive sequencing has shown an almost total lack of sequence diversity among isolates recovered over a wide geographic range, and almost no divergence between *M. tuberculosis* and *M. bovis*. There are considerable uncertainties in using clock rates to estimate the age of bacterial species that have almost no sequence variation, but it has been suggested (Kapur *et al.*, 1994; Sreevatsan *et al.*, 1997) that *M. tuberculosis* may be a clone of *M. bovis* that became established in humans around the time that cattle were domesticated (ca. 10 000–15 000 years ago). Some caution is required in making these assertions, since a lack of sequence diversity in a species simply means that the species has been through a severe population bottleneck, which could have coincided with speciation, or alternatively, in highly clonal species, could be a recent global episode of periodic selection. Housekeeping genes in gonococci are also very uniform and this species may be relatively young (Vázquez *et al.*, 1993). Since gonococci are non-clonal, purging of variation by periodic selection is not expected to be effective, and is unlikely to provide a reason for the low levels of sequence diversity within the species.

Relevance of Population Genetics to Molecular Typing and Epidemiology

Molecular typing is concerned with the nature of isolates that cause disease, and their relationship to similar disease-causing isolates recovered in a community, a country or worldwide. Typing methods are used to ask two very different kinds of question. First, are the isolates recovered from a localised outbreak of disease the same, or different (short-term epidemiology)? Second, how do strains that cause disease in one geographic area relate to strains that cause disease globally (long-term or global epidemiology)? To answer these questions is relatively straightforward with highly clonal pathogens, since clones are stable over decades, and with truly non-clonal pathogens, such as *He. pylori* or *N. gonorrhoeae*, only the first question can be posed (see below).

Typing of weakly clonal pathogens, such as *N. meningitidis*, is more problematic, since clones diversify relatively rapidly by the accumulation of recombinational exchanges, and the ease with which typing can be performed, and the techniques that need to be used, depend on the amount of time since the isolates being studied had a common ancestor. The identity of isolates recovered from a localised outbreak caused by the same strain can be established with any number of sufficiently discriminatory methods, since the common ancestor of these isolates occurred so recently that there has been no time for any significant variation to accumulate (**Fig. 1**). It becomes more difficult to show whether an outbreak strain belongs to a transient clone that has been associated with disease worldwide by comparing it to reference isolates from other countries, as substantial variation in pulsed-field gel patterns, MLEE profile and serology will have occurred since the outbreak strain and the other members of the transient clone had a common ancestor. In this case, a more thorough evaluation of the relationship of the outbreak strain to the members of any known transient clones, and the overall population, with quantitative methods (e.g. MLEE) may be needed. This problem becomes more acute as the time from the origin of the transient clones found within weakly clonal populations increases, as these clones diversify progressively until their members can no longer be distinguished from the background population (**Fig. 1**).

Use of Neutral Variation for Typing

Discrimination between different isolates of bacterial species depends on the identification of variable phenotypic or genotypic characters. Traditionally, medical microbiologists have taken advantage of variation in cell-surface components to develop typing schemes, such as serogroups, serotypes or phage types. Serotypes and serogroups are important to medical microbiologists, since in many cases virulence is associated with particular serotypes/groups. Furthermore, vaccines are commonly based on cell-surface components, and a knowledge of the distribution of serogroups or types may be important in the context of vaccine development. However, one of the important contributions of bacterial population genetics has been to demonstrate the inappropriateness of using this sort of variation for typing (Selander and Musser, 1990), although this message, voiced on numerous occasions, is often still ignored.

The cell-surface components of pathogens are in most cases believed to be variable as a consequence of selection for variation by the host immune system. Mutations in cell-surface genes that alter amino acids in immunodominant epitopes, or recombinational events that introduce novel sequences, will often be strongly selected if they allow the organism to escape detection by the host immune system. Similarly, the different polysaccharide capsules recognised by some serological reagents may vary in the extent to which they are recognised as foreign by the immune system (e.g. because of molecular mimicry by some capsular polysaccharides), or in their ability to protect against the non-specific arms of the host immune system (e.g. complement-mediated phagocytosis). Recombinational exchanges may be selected that result in the replacement of a highly immunogenic, or weakly antiphagocytic capsule, with a less immunogenic, or more antiphagocytic capsule.

Variation in cell-surface proteins, lipopolysaccharides, capsular carbohydrates, bacteriophage receptors or virulence factors cannot be used to provide a reliable indication of the genetic relationships between isolates of bacterial species, since there is strong selection for mutational and recombinational events at these loci. Even in species where horizontal gene transfer is believed to be very rare, the strong selection for recombinational exchanges at loci that encode cell surface structures is sufficient to lead to a lack of congruence between serology, phage type etc., and the genetic relatedness between isolates.

In some cases, typing based on variation in cell-surface structures works quite well. For example, it may be appropriate to answer very short-term epidemiological questions since, in most cases, isolates recovered over a few weeks from an outbreak of disease caused by a single strain are unlikely to differ in serology or phage type. Even for this purpose other techniques are likely to be more reliable and more discriminatory. Typing based on cell-surface variation may also work well for global epidemiology in some species where recombination is very rare (highly clonal species), and in which selection for variation may be relatively weak. For example, the relative stability of serological markers in *S. enterica* is the basis of the success of the Kauffmann–White scheme for the identification of serovars (see above). In this case the scheme works fairly well, since recombination at the loci that define serovars is rare, and the 'hit and run' nature of human salmonelloses may impose less strong immune selection for mutations that alter lipopolysaccharide and flagellin epitopes than occurs in pathogens that are exposed for longer periods to the immune systems of their hosts. Thus, in the short term, serology can define *S. enterica* clones. In the longer term, however, recombination at the loci that define serovars has been highly significant, and serology provides a very poor indication of the phylogenetic relationships between different *S. enterica* clones (**Fig. 4**; see above). Similarly, in other species, isolates from outbreaks of disease, or members of recently emerged clones, may be uniform in serology, but in the longer term the relationship between serology and genetic relatedness breaks down.

Unless the reliability of the method has been carefully evaluated, it is inappropriate to use variation in cell-surface components for global typing schemes since, for the reasons given above, isolates that appear to be the same by serology may be distantly related, and those that appear to be very different by serology may be very closely related. Similarly, molecular typing methods based on variation within genes encoding cell-surface structures should not be used for global epidemiology.

Typing schemes should be highly discriminatory, and should use variation that is likely to be selectively neutral and, for clonal species, allows the phylogenetic relationships between isolates to be inferred. MLEE provides a good example of such a typing method and is highly discriminatory if variation in an appropriate number of enzymes is measured. The number of enzymes that need to be used depends on the number of alleles per locus, and the frequencies of the different alleles in the population. Indexing of variation in 15 enzymes that have an average of 4 distinguishable alleles per locus, present in approximately equal frequency, allows the discrimination of $> 10^9$ distinct ETs. In practice, alleles are not present in equal frequencies and this considerably reduces the degree of discrimination. However, provided a sufficient number of suitably polymorphic enzymes are used, isolates that are indistinguishable by MLEE can be assumed to be members of a single clone, and to be very closely related in overall genotype. This level of resolving power is useful for the characterisation of non-clonal species, since essentially every epidemiologically unrelated isolate of such a species should be genetically distinct. If the variation being indexed by a typing method is very limited, and is capable of resolving only a small number of genotypes, isolates of non-clonal species will inevitably appear to be indistinguishable and may be considered to be clones. Surprisingly, papers are often published that use typing methods which can resolve only a very limited number of genotypes, and which may result in very misleading inferences about the relatedness of isolates and the structure of the population.

Molecular typing methods based on uncharacterised genomic variation, such as PFGE, or PCR with arbitrary primers, can also be highly discriminatory, and probably detect variation that is largely neutral, although the quantitation of the genetic relationships between isolates can be problematic. These methods often distinguish between isolates that are indistinguishable by MLEE, presumably because variation within a clonal lineage accumulates more rapidly by deletions, insertions or inversions, which probably cause much of the variation detected by the above techniques, than does variation in the electrophoretic mobility of housekeeping enzymes (Wang *et al.*, 1993a; Desjardins *et al.*, 1995; Navarro *et al.*, 1996).

Typing methods can be based on variation that accumulates too fast, as well as on variation that accumulates too slowly. Analysis of variation in the sequence of rRNA is very useful for plumbing the depths of the phylogenetic tree of life, but it evolves slowly, and it would clearly be inappropriate for typing isolates of a bacterial species, since all isolates are likely to have the same 16S rRNA sequence. At the other extreme, typing using variation that accumulates too fast may make almost every isolate distinct even in clonal species. Thus, PFGE of isolates from typhoid fever might lead to the conclusion that these isolates were a heterogeneous collection of unrelated strains. To see that Typhi isolates are closely related, the variation uncovered by PFGE needs to be put in the context of the variation in pulsed-field gel patterns within the whole of *S. enterica*. Variation in the electrophoretic mobility of housekeeping enzymes accumulates relatively slowly, and MLEE has been shown in numerous studies to be highly suitable for the identification of clones within bacterial populations.

Although MLEE has been invaluable for providing the population-genetic framework of bacterial populations, it is indirect and old fashioned. In MLEE, the alleles at housekeeping loci are implied from the electrophoretic mobilities of their gene products. Clearly, when the enzymes have different mobilities their alleles will be distinct, but the opposite is not the case. Identical mobilities may be due to identity, or similarity, of the DNA sequence, but it is also possible for enzymes with indistinguishable mobilities to be encoded by very different DNA sequences. This is particularly likely to be a problem when comparing distantly related isolates of a species, and this can result in a loss of resolution, and reliability, in the deeper branches of phylogenies inferred from MLEE data.

The indexing of allelic variation at multiple housekeeping loci remains the most appropriate approach to bacterial population genetics, as it provides information that is ideally suited for molecular typing, and the determination of population structure and phylogenetic relationships. The problems inherent in the indirect assignment of alleles by MLEE can now be overcome by the direct assignment of alleles, by sequencing 450-bp internal fragments from multiple housekeeping genes, in hundreds of isolates. Assignment of alleles by DNA sequencing is unambiguous and has the advantage that more alleles are distinguished per locus, allowing high discrimination between isolates to be achieved using less loci than are required for MLEE. For example, sequence analysis of 7 housekeeping gene fragments, each with 20 alleles per locus, would allow the resolution of >1 billion genotypes. Direct sequencing of 450-bp PCR products can be carried out rapidly with automated DNA sequencers, and the complete assignment of alleles at 7 loci (sequencing on both DNA strands) can be accomplished using only 14 lanes or capillaries of an automated sequencer.

The development of a method that uses the sequences of seven housekeeping gene fragments for strain characterisation (MLST) has been described by Maiden *et al.* (1998) and has proved to be excellent for identifying the major virulent meningococcal lineages (**Fig. 8**). MLST schemes have also been described for *Strep. pneumoniae* (Enright and Spratt, 1998), and have been developed for *Staph. aureus* and *Strep. pyogenes* (Enright *et al.*, 2001).

MLST provides a method that, like MLEE, quantifies the genetic relatedness between isolates of clonal species but, in contrast to MLEE, is ideally suited to the comparison of strains being studied in different laboratories via the internet. This allows microbiologists from any laboratory that has the ability to sequence DNA to relate their local strains of a pathogen to those found globally, by submitting the 450-bp sequences from the housekeeping genes to a central web site containing the MLST database for that species (http://www.mlst.net). The data obtained for MLST can also be used to address aspects of bacterial population and evolutionary biology. For example, it is providing estimates of the relative ratio of recombination to mutation in different bacterial species during clonal diversification (see section on 'The ratio of recombination to mutation', above; Feil *et al.*, 1999, 2000). For clonal species, the phylogenetic relationships between isolates can also be inferred from the consensus tree constructed from the gene sequences.

Although MLST may be suitable for addressing some short-term epidemiological questions (e.g. whether cases of disease from a localised outbreak are due to the same strain), which can readily be answered by a number of other molecular techniques, it should be ideal for determining the relatedness of an outbreak strain to those isolated in other outbreaks, within the same country or worldwide. MLST may not be suitable for all species of pathogen (e.g. *M. tuberculosis* which has too little sequence variation in housekeeping genes), but acceptance of the technique as the gold standard for typing appropriate bacterial pathogens would resolve the highly unsatisfactory situation where different typing methods are often used for the same pathogens in different laboratories and, even when the same methods are used, the data that are produced are difficult to compare from one laboratory to another,

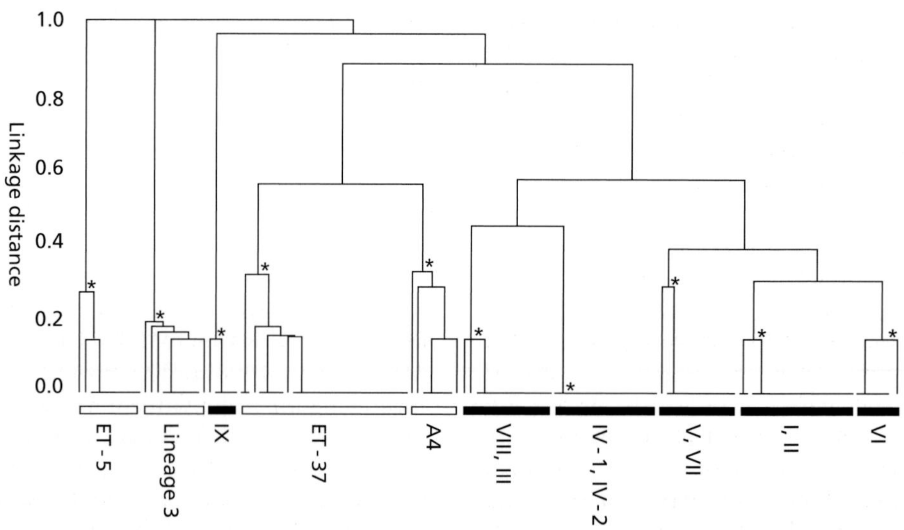

Fig. 8 Use of MLST to identify the clonal complexes associated with invasive meningococcal disease. The different sequences at each of the 7 housekeeping genes used in MLST are assigned as distinct alleles and, for each isolate, the alleles at the 7 loci define its allelic profile. The dendrogram was constructed from the matrix of percentage pairwise differences between the allelic profiles of 126 isolates of *N. meningitidis.* The horizontal lines at the tips of the dendrogram indicate multiple isolates that have identical allelic profiles. The rectangles show the clonal complexes and their designations; those in black are serogroup A clonal complexes and those in white are serogroup B and C. The asterisks show the ancestral node for each clonal complex. From Feil *et al.* (1999).

and are of little use for evolutionary, phylogenetic, or population-genetic studies.

Population Structure and Molecular Typing

A knowledge of the extent of recombination in bacterial pathogens is important since low levels of recombination result in highly clonal populations, where lineages persist with little variation over hundreds or thousands of years, whereas, at the other extreme, high rates of recombination lead to non-clonal populations in which lineages diversify so rapidly that the isolates recovered in one decade may be completely different from those recovered in the next (**Fig. 1**). Clearly, the way in which these two populations evolve, and the methods that should be used to study their molecular epidemiology will differ, as will the types of epidemiological questions that we can sensibly ask. The population structures of bacterial pathogens should therefore be considered when devising methods to type isolates of bacterial species.

For highly clonal pathogens such as *S. enterica* the stability of clones is very high and MLEE provides little variation that can be used to study the global or local epidemiology of the infections caused by the clones associated with human disease. For example, isolates from typhoid fever are indistinguishable by MLEE, or differ at only a single locus (Selander *et al.*, 1990b). Detailed epidemiological studies of typhoid fever need to make use of the micro-variation that inevitably accumulates even within a highly clonal lineage. In this case, ribotyping and PFGE have been shown to discriminate between epidemiologically unrelated isolates (Altwegg *et al.*, 1989; Navarro *et al.*, 1996).

In weakly clonal pathogens, such as serogroup B and C isolates of *N. meningitidis*, the diversification of clones is rapid compared with those of *S. enterica*, but is sufficiently slow that isolates of the ET-5, ET-37 or A4 clonal complexes can still be recognised with appropriate typing methods, even though these lineages may be more than 20 years old (Maiden and Feavers, 1995; Spratt *et al.*, 1995). In this case, variation that accumulates relatively slowly (e.g. MLEE or MLST) may be needed to show whether a strain is a member of one of the recognised clonal complexes, whereas more discriminatory methods, such as PFGE, may be needed for asking more detailed epidemiological questions.

For non-clonal species, the population consists of a large and ever-changing collection of unique genotypes and, in the absence of stable clones, long-term epidemiological questions – for example, tracking the global spread of particular strains – will not be possible. Fortunately, no pathogens are likely to diversify so rapidly that short–term epidemiological questions cannot be addressed. In the case of gonococci, it is possible to obtain information about the transmission of gonorrhoea among sexual partners, or within core groups (O'Rourke *et al.*, 1995). Similarly, with *He. pylori* it is possible to ask questions about the transmission of the organism within families, or whether recurrence of colonisation after antibiotic treatment is due to re-colonisation by a different strain or treatment failure (Shortridge *et al.*, 1997). Whether longer term epidemiological questions can be addressed with these non-clonal species requires a better understanding of the rate of diversification of lineages.

As well as considering the rate of diversification of lineages when choosing a typing method, the different levels of sequence diversity found among bacterial species must be considered. Both *He. pylori* and *N. gonorrhoeae* are non-clonal, and typing methods are required that can distinguish the large number of genotypes expected within these populations. This is not a problem with *He. pylori*, as the species is extremely diverse and many different typing methods provide sufficient variation for typing (Go *et al.*, 1996). There is little diversity in housekeeping genes of gonococci and it is more difficult to obtain a highly discriminatory typing method based on neutral variation (Spratt *et al.*, 1995). An extreme example of species uniformity is provided by *M. tuberculosis* which is so uniform (essentially a single clone) that the usual molecular typing methods are not discriminatory, and typing has relied on variation between isolates in the numbers and sites of integration of the insertion sequence IS6110 (van Soolingen *et al.*, 1994).

Concluding Remarks

The emergence of new pathogens, or strains of old pathogens with increased potential to cause disease, or increased resistance to antibiotics, and the resurgence of diseases that were believed to be under control, has given rise to much concern in recent years. A huge number of studies have been carried out on the characterisation and molecular epidemiology of bacterial strains causing disease, and on the development of new molecular typing methods. Relatively few of these studies have used methods that allow molecular typing and epidemiology to be placed within the framework of population biology. This is unfortunate, as the population-genetic approach has provided important insights into the origins and evolution of virulence in populations of bacterial pathogens, and the role of mutation, recombination and gene acquisition in the continuing evolution of these populations. These studies are of obvious relevance to medical microbiology and to the long-term success of new conjugate polysaccharide vaccines which target only a fraction of the isolates that cause disease.

In recent years the role of recombination in the evolution of bacterial pathogens has become more apparent. Recombination has certainly been important in the evolutionary history of most, if not all, bacterial species since, as far as we are aware, evidence for recombinational exchanges within genes under strong selection has been detected in almost all bacterial pathogens. These events have been crucial in allowing bacterial species to avoid the immune response, in distributing among the population genes that increase virulence or transmission between hosts, or in providing increased resistance to antibiotics. In some cases, the acquisition of blocks of genes (pathogenicity islands) by illegitimate recombinational events has clearly been of paramount importance in launching an isolate of a species into a new and more virulent form. For many species, however, recombinational exchanges appear to be sufficiently rare, in comparison with mutation, that the populations are dominated by the presence of relatively stable clones. In a few species, recombination is clearly much more frequent and dominates the evolutionary biology of these populations. In many cases, we do not have a clear view of the relative importance of mutation and recombination in the evolution of bacterial pathogens, and further studies, taking into account the differences in ecology of different bacterial populations, are required.

Acknowledgements

Work from our laboratory was supported by the Wellcome Trust. B.G.S. is a Wellcome Trust Principal Research Fellow.

References

Achtman M (1994) Clonal spread of serogroup A meningococci – a paradigm for the analysis of microevolution in bacteria. *Mol. Microbiol.* 11: 15–22.

Altwegg M, Hickman-Brenner FW, Farmer JJ (1989) Ribosomal RNA gene restriction patterns provide increased sensitivity for typing *Salmonella typhi* strains. *J. Infect. Dis.* 160: 145–149.

Atwood KC, Schneider LK, Ryan FJ (1951) Periodic selection in *Escherichia coli. Proc Natl Acad. Sci. USA* 37: 146–155.

Barcus VA, Ghanekar K, Yeo M, Coffey TJ, Dowson CG (1995) Genetics of high level penicillin resistance in clinical isolates of *Streptococcus pneumoniae. FEMS Microbiol. Lett.* 126: 299–304.

Barnes DM, Whittler S, Gilligan PH, Soares S, Tomasz A, Henderson FW (1995) Transmission of multidrug-resistant serotype 23F *Streptococcus pneumoniae* in group day care – evidence suggesting capsular transformation of the resistant strain *in vivo*. *J. Infect. Dis.* 171: 890–896.

Baumberg S, Young JPW, Wellington EMH, Saunders JR (eds) (1995) *Population Genetics of Bacteria*. Cambridge: Cambridge University Press.

Bayerdorffer EA, Neubauer A, Rudolph B *et al.* for MALT Lymphoma Study Group (1995) Regression of primary gastric lymphoma of mucosa-associated lymphoid tissue type after cure of *Helicobacter pylori* infection. *Lancet* 345: 17771–17777.

Beltran P, Musser JM, Helmuth R *et al.* (1988) Toward a population genetic analysis of *Salmonella* – genetic diversity and relationships among strains of serotypes *S. choleraesuis*, *S. derby*, *S. Dublin*, *S. enteritidis*, *S. Heidelberg*, *S. infantis*, *S. newport*, and *S. typhimurium*. *Proc Natl Acad. Sci. USA* 85: 7753–7757.

Beltran P, Plock SA, Smith NH, Whittam TS, Old DC, Selander RK (1991) Reference collection of strains of the *Salmonella typhimurium* complex from natural populations. *J. Gen. Microbiol.* 137: 601–606.

Bisercic M, Feutrier JY, Reeves PR (1991) Nucleotide sequences of the *gnd* genes from nine natural isolates of *Escherichia coli*: evidence of intragenic recombination as a contributing factor in the evolution of the polymorphic *gnd* locus. *J. Bacteriol.* 173: 3894–3900.

Blaser MJ (1992) *Helicobacter pylori*: its role in disease. *Clin. Infect. Dis.* 15: 386–393.

Bowler LD, Zhang Q-Y, Riou J-Y, Spratt BG (1994) Interspecies recombination between the *penA* genes of *Neisseria meningitidis* and commensal *Neisseria* species during the emergence of penicillin resistance in *N. meningitidis*: natural events and laboratory simulation. *J. Bacteriol.* 176: 333–337.

Boyd EF, Nelson K, Wang F-S, Whittam TS, Selander RK (1994) Molecular genetic basis of allelic polymorphism in malate dehydrogenase (MDH) in natural populations of *Escherichia coli* and *Salmonella enterica*. *Proc Natl Acad. Sci. USA* 91: 1280–1284.

Boyd EF, Wang FS, Whittam TS, Selander RK (1996) Molecular genetic relationships of the Salmonellae. *Appl. Env. Microbiol.* 62: 804–808.

Brenner DJ, Fanning GR, Johnson KE, Citarella RV, Falkow S (1969) Polynucleotide sequence relationships among members of Enterobacteriaceae. *J. Bacteriol.* 98: 637–650.

Brown AHD, Feldman MW, Nevo E (1980) Multilocus structure of natural populations of *Hordeum spontaneum*. *Genetics* 96: 523–536.

Caugant DA, Froholm LO, Bovre K *et al.* (1986) Intercontinental spread of a genetically distinctive complex of clones of *Neisseria meningitidis* causing epidemic disease. *Proc Natl Acad. Sci. USA* 83: 4927–4931.

Caugant DA, Mocca LF, Frasch CE, Froholm LO, Zollinger WD, Selander RK (1987) Genetic structure of *Neisseria meningitidis* populations in relation to serogroup, serotype, and outer membrane protein pattern. *J. Bacteriol.* 169: 2781–2792.

Coffey TJ, Dowson CG, Daniels M *et al.* (1991) Horizontal transfer of multiple penicillin-binding protein genes, and capsular biosynthetic genes, in natural populations of *Streptococcus pneumoniae*. *Mol. Microbiol.* 5: 2255–2260.

Coffey TJ, Dowson CG, Daniels M, Spratt BG (1995a) Genetics and molecular biology of β-lactam-resistant pneumococci. *Microb. Drug Res.* 1: 25–30.

Coffey TJ, Daniels M, McDougal LK, Dowson CG, Tenover FC, Spratt BG (1995b) Genetic analysis of clinical isolates of *Streptococcus pneumoniae* with high-level resistance to third generation cephalosporins. *Antimicrob. Agents Chemother.* 39: 1306–1313.

Coffey TJ, Berrón S, Daniels M *et al.* (1996) Multiply-antibiotic-resistant *Streptococcus pneumoniae* recovered from Spanish hospitals (1988–1994): novel major clones of serotypes 14, 19F and 15F. *Microbiology* 142: 2747–2757.

Coffey TJ, Enright MC, Daniels M *et al.* (1998a) Serotype 19A variants of the Spanish serotype 23F multiresistant clone of *Streptococcus pneumoniae*. *Microb. Drug Res.* 4: 51–55.

Coffey TJ, Enright MC, Daniels M *et al.* (1998b) Recombinational exchanges at the capsular polysaccharide biosynthetic locus lead to frequent serotype changes among natural isolates of *Streptococcus pneumoniae*. *Mol. Microbiol.* 27: 73–83.

Danner DB, Deich RA, Sisco KL, Smith HO (1980) An 11-base-pair sequence determines the specificity of DNA uptake in *Hemophilus* transformation. *Gene* 11: 311–318.

Davies JE (1994) Inactivation of antibiotics and the dissemination of resistance genes. *Science* 264: 375–382.

de la Fuente L, Vázquez JA (1994) Genetic structures of non-penicillinase-producing *Neisseria gonorrhoeae* strains in relation to auxotype and serovar class. *J. Infect. Dis.* 170: 696–700.

Denamur E, Picard B, Decoux G, Denis JB, Elion J (1993) The absence of correlation between allozyme and *rrn* RFLP analysis indicates a high gene flow rate within human clinical *Pseudomonas aeruginosa* isolates. *FEMS Microbiol. Lett.* 110: 275–280.

Desjardins P, Picard B, Kaltenbock B, Elion J, Denamur E (1995) Sex in *Escherichia coli* does not disrupt the clonal structure of the population – evidence from random amplified polymorphic DNA and restriction fragment length polymorphism. *J. Mol. Evol.* 41: 440–448.

Dowson CG, Hutchison A, Brannigan JA *et al.* (1989) Horizontal transfer of penicillin-binding protein genes in penicillin-resistant clinical isolates of *Streptococcus pneumoniae*. *Proc Natl Acad. Sci. USA* 86: 8842–8846.

DuBose RF, Dykhuizen DE, Hartl DL (1988) Genetic exchange among natural isolates of bacteria: recombination within the *phoA* gene of *Escherichia coli*. *Proc Natl Acad. Sci. USA* 85: 7036–7040.

Duncan KE, Ferguson N, Kimura K, Zhou X, Istock CA (1994) Fine scale genetic and phenotypic structure in natural populations of *Bacillus subtilis* and *Bacillus licheniformis* – implications for bacterial evolution and speciation. *Evolution* 48: 2002–2025.

Dykhuizen DE, Green L (1991) Recombination in *Escherichia coli* and the definition of biological species. *J. Bacteriol.* 173: 7257–7268.

Dykhuizen DE, Polin DS, Dunn JJ *et al.* (1993) *Borrelia burgdorferi* is clonal: implications for taxonomy and vaccine development. *Proc. Natl Acad. Sci. USA* 90: 10163–10167.

Edwards PR (1986) *Edward's and Ewing's Identification of the Enterobacteriaceae*, 4th edn. New York: Elsevier.

Enright MC, Spratt BG (1998) A multilocus sequence typing scheme for *Streptococcus pneumoniae*: identification of clones associated with serious invasive disease. *Microbiology* 144: 3049–3060.

Enright MC, Day NPJ, Davies CE, Peacock SJ, Spratt BG (2000) Multilocus sequence typing for the characterization of methicillin-resistant (MRSA) and methicillin-susceptible (MSSA) clones of *Staphylococcus aureus*. *J. Clin. Microbiol.* 38: 1008–1015.

Enright MC, Spratt BG, Kalia A, Cross JH, Bessen DE (2001) Multilocus sequence typing of *Streptococcus pyogenes* and the relationships between emm-type and clone. *Infect. Immun.* 69: 2416–2427.

Feavers IM, Heath AB, Bygraves JA, Maiden MCJ (1992) Role of horizontal genetic exchange in the antigenic variation of the class-1 outer-membrane protein of *Neisseria meningitidis*. *Mol. Microbiol.* 6: 489–495.

Feavers IM, Maiden MCJ (1998) A gonococcal *porA* pseudogene: implications for understanding the evolution and pathogenicity of *Neisseria gonorrhoeae*. *Mol. Microbiol.* 30: 647–656.

Feil E, Carpenter G, Spratt BG (1995) Electrophoretic variation in adenylate kinase of *Neisseria meningitidis* is due to inter- and intra-species recombination. *Proc Natl Acad. Sci. USA* 92: 10535–10539.

Feil E, Zhou J, Maynard Smith J, Spratt BG (1996) A comparison of the nucleotide sequences of the *adk* and *recA* genes of pathogenic and commensal *Neisseria* species: evidence for extensive inter-species recombination within *adk*. *J. Mol. Evol.* 43: 631–640.

Feil E, Maiden MCJ, Achtman M, Spratt BG (1999) The relative contributions of recombination and mutation to the divergence of clones of *Neisseria meningitidis*. *Mol. Biol. Evol.* 16: 1496–1502.

Feil EJ, Maynard Smith J, Enright MC, Spratt BG (2000) Estimating recombinational parameters in *Streptococcus pneumoniae* from multilocus sequence typing data. *Genetics* 154: 1439–1450.

Feil EJ, Holmes EC, Bessen DE *et al.* (2001) Recombination within natural populations of pathogenic bacteria: Short-term empirical estimates and long-term phylogenetic consequences. *Proc. Natl. Acad. Sci. USA* 98: 182–187.

Feng P, Fields PI, Swiminathan B, Whittam TS (1996) Characterization of nonmotile variants of *Escherichia coli* O157 and other serotypes using an antiflagellin monoclonal antibody. *J. Clin. Microbiol.* 34: 2856–2859.

Feng P, Lampel KA, Karch H, Whittam TS (1998) Genotypic and phenotypic changes in the emergence of *Escherichia coli* O157:H7. *J. Infect. Dis.* 177: 1750–1753.

Fischetti VA (1989) Streptococcal M protein: molecular design and biological behavior. *Clin. Microbiol. Rev.* 2: 285–314.

Go MF, Kapur V, Graham DY, Musser JM (1996) Population genetic analysis of *Helicobacter pylori* by multilocus enzyme electrophoresis – extensive allelic diversity and recombinational population structure. *J. Bacteriol.* 178: 3934–3938.

Goodman SD, Scocca JJ (1988) Identification and arrangement of the DNA sequence recognized in specific transformation of *Neisseria gonorrhoeae*. *Proc Natl Acad. Sci. USA* 85: 6982–6986.

Grassly NC, Holmes EC (1997) A likelihood method for the detection of selection and recombination using nucleotide sequences. *Mol. Biol. Evol.* 14: 239–247.

Groisman EA, Ochman H (1996) Pathogenicity islands – bacterial evolution in quantum leaps. *Cell* 87: 791–794.

Gupta S, Maiden MCJ, Feavers IM, Nee S, May RM, Anderson RM (1996) The maintenance of strain structure in populations of recombining infectious agents. *Nature Medicine* 2: 437–442.

Gutjahr T, O'Rourke M, Ison CA, Spratt BG (1997) Arginine, hypoxanthine, uracil-requiring isolates of *Neisseria gonorrhoeae* are a clonal lineage within a non-clonal population. *Microbiology* 143: 633–640.

Guttman DS (1997) Recombination and clonality in natural populations of *Escherichia coli*. *Trends Ecol. Evol.* 12: 16–22.

Guttman DS, Dykhuizen DE (1994) Clonal divergence in *Escherichia coli* as a result of recombination. *Science* 266: 1380–1383.

Hacker J, Blum-Oehler G, Mühldorfer I, Tschäpe H (1997) Pathogenicity islands of virulent bacteria: structure, function and impact on microbial evolution. *Mol. Microbiol.* 23: 1089–1097.

Hakenbeck R, Tarpay M, Tomasz A (1980) Multiple changes of pencillin-binding proteins in penicillin-resistant clinical isolates of *Streptococcus pneumoniae*. *Antimicrob. Agents Chemother.* 17: 364–371.

Hall LMC, Whiley RA, Duke B, George RC, Efstratiou A (1996) Genetic relatedness within and between serotypes of *Streptococcus pneumoniae* from the United Kingdom – analysis of multilocus enzyme electrophoresis, pulsed-field gel electrophoresis, and antimicrobial resistance patterns. *J. Clin. Microbiol.* 34: 853–859.

Halter R, Pohlner J, Meyer TF (1989) Mosaic-like organization of IgA protease genes in *Neisseria gonorrhoeae* generated by horizontal genetic exchange *in vivo*. *EMBO J.* 8: 2737–2744.

Haubold B, Rainey PB (1996) Genetic and ecotypic structure of a fluorescent *Pseudomonas* population. *Mol. Ecol.* 5: 747–761.

Haubold B, Rainey PB (1997) Towards an understanding of the population genetics of plant-colonzing bacteria. *Adv. Bot. Res.* 24: 335–351.

Haubold B, Travisano M, Rainey PB, Hudson RR (1998) Detecting linkage disequilibrium in bacterial populations. *Genetics* 150: 1341–1348.

Hedrich PW, Thomson G (1986) A two-locus neutrality test: applications to humans *E. coli*, and lodgepole pine. *Genetics* 112: 135–156.

Holmes EC, Urwin R, Maiden MCJ (1999) The influence of recombination on the population structure and evolution of the human pathogen *Neisseria meningitidis*. *Mol. Biol. Evol.* 16: 741–749.

Hurtado A, Owen RJ (1994) Identification of mixed genotypes in *Helicobacter pylori* from gastric biopsy tissue by analysing urease gene polymorphism. *Med. Microbiol.* 8: 307–313.

Jauris-Heipke S, Liegl G, Preac-Mursic V, Soutschek E, Will G, Wilske B (1995) Molecular analysis of genes encoding outer surface protein C (*ospC*) of *Borrelia burgdorferi* Sensu Lato: relationship to *ospA* genotype and evidence for lateral gene exchange of *ospC*. *J. Clin. Microbiol.* 33: 1860–1866.

Jegathesan M (1984) *Salmonella* serotypes isolated from man in Malaysia over the 10-year period 1973–1982. *J. Hygiene* 92: 395–399.

Jiang Q, Hiratsuka K, Taylor DE (1996) Variability of gene order in different *Helicobacter pylori* strains contributes to genome diversity. *Mol. Microbiol.* 20: 833–842.

Kapur V, Nelson K, Schlievert PM, Selander RK, Musser JM (1992) Molecular population genetic evidence of horizontal spread of two alleles of the pyrogenic exotoxin C gene (*speC*) among pathogenic clones of *Streptococcus pyogenes*. *Infect. Immun.* 60: 3513–3517.

Kapur V, Whittam TS, Musser JM (1994) Is *Mycobacterium tuberculosis* 15,000 years old? *J. Infect. Dis.* 170: 1348–1349.

Kapur V, Kanjilal S, Hamrick MR *et al.* (1995) Molecular population genetic analysis of the streptokinase gene of *Streptococcus pyogenes*: mosaic alleles generated by recombination. *Mol. Microbiol.* 16: 509–519.

Karaolis DK, Lan R, Reeves PR (1994) Sequence variation in *Shigella sonnei* (Sonnei), a pathogenic clone of *Escherichia coli*, over four continents and 41 years. *J. Clin. Microbiol.* 32: 796–802.

Karaolis DR, Somara S, Maneval DR, Johnson JA, Kaper JB (1999) A bacteriophage encoding a pathogenicity island, a type IV pilus and a phage receptor in cholera bacteria. *Nature* 399: 375–379.

Kehoe MA, Kapur V, Whatmore AM, Musser JM (1996) Horizontal gene transfer among group A streptococci – implications for pathogenesis and epidemiology. *Trends Microbiol.* 4: 436–443.

Knapp JS, Holmes KK (1975) Disseminated gonococcal infections caused by *Neisseria gonorrhoeae* with unique growth requirements. *J. Infect. Dis.* 132: 204–208.

Kroll JS, Moxon ER (1990) Capsulation in distantly related strains of *Haemophilus influenzae* type b – genetic drift and gene transfer at the capsulation locus. *J. Bacteriol.* 172: 1374–1379.

Lan R, Reeves PR (1996) Gene transfer is a major factor in bacterial evolution. *Mol. Biol. Evol.* 13: 47–55.

Le Minor L, Popoff MY (1987) Designation of *Salmonella enterica* sp. nov., nom. rev., as the type and only species of the genus *Salmonella*. *Int. J. Syst. Bacteriol.* 37: 465–468.

Levin BR (1981) Periodic selection, infectious gene exchange, and the genetic structure of *E. coli* populations. *Genetics* 99: 1–23.

Lomholt H (1995) Evidence of recombination and an antigenically diverse immunoglobulin A1 protease among strains of *Streptococcus pneumoniae*. *Infect. Immun.* 63: 4238–4243.

Maiden MCJ (1993) Population-genetics of a transformable bacterium – the influence of horizontal genetic exchange on the biology of *Neisseria meningitidis*. *FEMS Microbiol. Lett.* 112: 243–250.

Maiden MCJ, Feavers IM (1995) Population genetics of *Neisseria meningitidis*. In Baumberg S, Young JPW, Wellington EMH, Saunders JR (eds). *The Population Genetics of Bacteria*. Cambridge: Cambridge University Press, pp. 269–293.

Maiden MCJ, Bygraves JA, Feil E *et al.* (1998) Multilocus sequence typing: a portable approach to the identification of clones within populations of pathogenic microorganisms. *Proc Natl Acad. Sci. USA* 95: 3140–3145.

Maslow JN, Whittam TS, Gilks CF, Wilson RA, Mulligan ME, Adams KS (1995) Clonal relationships among bloodstream isolates of *Escherichia coli*. *Infect. Immun.* 63: 2409–2417.

Matic I, Taddei F, Radman M (1996) Genetic barriers among bacteria. *Trends Microbiol.* 4: 69–73.

Maynard Smith J (1991) The population genetics of bacteria. *Proc. Roy. Soc. Lond. B* 245: 37–41.

Maynard Smith J (1992) Analysing the mosaic structure of genes. *J. Mol. Evol.* 34: 126–129.

Maynard Smith J (1994) Estimating the minimum rate of genetic transformation in bacteria. *J. Evol. Biol.* 7: 525–534.

Maynard Smith J (1996) Population genetics: an introduction. In Neidhardt FC *et al.* (ed.) *Escherichia coli and Salmonella*. Washington DC: ASM Press, pp. 2685–2690.

Maynard Smith J, Smith NH (1998) Detecting recombination from gene trees. *Mol. Biol. Evol.* 15: 590–599.

Maynard Smith J, Dowson CG, Spratt BG (1991) Localised sex in bacteria. *Nature* 349: 29–31.

Maynard Smith J, Smith NH, O'Rourke M, Spratt BG (1993) How clonal are bacteria? *Proc Natl Acad. Sci. USA* 90: 4384–4388.

McDougal LK, Facklam R, Reeves M *et al.* (1992) Analysis of multiply antimicrobial-resistant isolates of *Streptococcus pneumoniae* from the United States. *Antimicrob. Agents Chemother.* 36: 2176–2184.

McKane M, Milkman R (1995) Transduction, restriction and recombination patterns in *Escherichia coli*. *Genetics* 139: 35–43.

Mel SF, Mekalanos JJ (1996) Modulation of horizontal gene transfer in pathogenic bacteria by *in vivo* signals. *Cell* 87: 795–798.

Milkman R (1973) Electrophoretic variation in *Escherichia coli* from natural sources. *Science* 182: 1024–1026.

Milkman R, Bridges MM (1990) Molecular evolution of the *Escherichia coli* chromosome. III. Clonal frames. *Genetics* 126: 505–517.

Milkman R, McKane Bridges M (1993) Molecular evolution of the *Escherichia coli* chromosome. IV. Sequence comparisons. *Genetics* 133: 455–468.

Milkman R (1996) Recombinational exchange among clonal populations. In Neidhardt FC *et al.* (ed.) *Escherichia coli and Salmonella*. Washington DC, pp. 2663–2684.

Moran NA, Baumann P (1994) Phylogenetics of cytoplasmically inherited microorganisms of arthropods. *Trends Ecol. Evol.* 9: 15–20.

Moran NA, von Dohlen CD, Baumann P (1995) Faster evolutionary rates in endosymbiont bacteria than in cospeciating insect hosts. *J. Mol. Evol.* 41: 727–731.

Muñoz R, Coffey TJ, Daniels M *et al.* (1991) Intercontinental spread of a multiresistant clone of serotype 23F *Streptococcus pneumoniae*. *J. Infect. Dis.* 164: 302–306.

Morelli G, Malorny B, Muller K *et al.* (1997) Clonal descent and microevolution of *Neisseria meningitidis* during 30 years of epidemic spread. *Mol. Microbiol.* 25: 1047–1064.

Musser JM, Barenkamp SJ, Granoff DM, Selander RK (1986) Genetic relationships of serologically nontypable and serotype b strains of *Haemophilus influenzae*. *Infect. Immun.* 55: 183–191.

Musser JM, Kroll JS, Moxon ER, Selander RK (1988) Clonal population structure of encapsulated *Haemophilus influenzae*. *Infect. Immun.* 56: 1837–1845.

Musser JM, Selander RK (1990) Brazilian purpuric fever: evolutionary genetic relationships of the case clone of *Haemophilus influenzae* biogroup aegypticus to encapsulated strains of *Haemophilus influenzae*. *J. Infect. Dis.* 161: 130–133.

Musser JM, Hauser AR, Kim MH, Schlievert PM, Nelson K, Selander RK (1991) *Streptococcus pyogenes* causing toxic-shock-like syndrome and other invasive diseases: clonal diversity and pyrogenic exotoxin expression. *Proc Natl Acad. Sci. USA* 88: 2668–2672.

Musser JM, Kapur V (1992) Clonal analysis of methicillin-resistant *Staphylococcus aureus* strains from intercontinental sources: association of the *mec* gene with divergent phylogenetic lineages implies dissemination by horizontal gene transfer and recombination. *J. Clin. Microbiol.* 30: 2058–2063.

Musser JM, Kapur V, Kanjilal S *et al.* (1993) Geographic and temporal distribution and molecular characterization of two highly pathogenic clones of *Streptococcus pyogenes* expressing allelic variants of pyrogenic exotoxin A (scarlet fever toxin). *J. Infect. Dis.* 167: 337–346.

Musser JM, Kapur V, Szeto J, Pan X, Swanson DS, Martin DR (1995) Genetic diversity and relationships among *Streptococcus pyogenes* strains expressing serotype M1 protein: recent intercontinental spread of a subclone causing episodes of invasive disease. *Infect. Immun.* 63: 994–1003.

Musser JM (1996) Molecular population genetic analysis of emerged bacterial pathogens – selected insights. *Emerg. Infect. Dis.* 2: 1–17.

Navarro F, Llovet T, Echeita MA *et al.* (1996) Molecular typing of *Salmonella enterica* serovar Typhi. *J. Clin. Microbiol.* 34: 2831–2834.

Nelson K, Whittam TS, Selander RK (1991) Nucleotide polymorphism and evolution in the glyceraldehyde-3-phosphate dehydrogenase gene (*gapA*) in natural populations of *Salmonella* and *Escherichia coli*. *Proc Natl Acad. Sci. USA* 88: 6667–6671.

Nelson K, Selander RK (1992) Evolutionary genetics of the proline permease gene (*putP*) and the control region of the proline utilization operon in populations of *Salmonella* and *Escherichia coli*. *J. Bacteriol.* 174: 6886–6895.

Nelson K, Selander RK (1994) Intergeneric transfer and recombination of the 6-phosphogluconate dehydrogenase gene (*gnd*) in enteric bacteria. *Proc Natl Acad. Sci. USA* 91: 10227–10231.

Neu HC (1992) The crisis in antibiotic resistance. *Science* 257: 1064–1073.

O'Brien AD, Newland J, Miller SF, Holmes RK, Smith HW, Formal SF (1984) Shiga-like toxin converting phages from *Escherichia coli* strains that cause hemorrhagic colitis or infantile diarrhea. *Science* 226: 694–696.

Ochman H and Selander RK (1984) Standard reference strains of *E. coli* from natural populations. *J. Bacteriol.* 157: 690–693.

Ochman H and Wilson AC (1987) Evolution in bacteria – evidence for a universal substitution rate in cellular genomes. *J. Mol. Evol.* 26: 74–86.

Ochman H, Whittam TS, Caugant DA, Selander RK (1983) Enzyme polymorphism and genetic population structure in *Escherichia coli* and *Shigella*. *J. Gen. Microbiol.* 129: 2715–2726.

Olyhoek T, Crowe BA, Achtman M (1987) Clonal population structure of *Neisseria meningitidis* serogroup A isolated from epidemics and pandemics between 1915 and (1983) *Rev. Infect. Dis.* 9: 665–692.

O'Rourke M, Spratt BG (1994) Further evidence for the non-clonal population structure of *Neisseria gonorrhoeae*: extensive genetic diversity within isolates of the same electrophoretic type. *Microbiology* 140: 1285–1290.

O'Rourke M, Stevens E (1993) Genetic structure of *Neisseria gonorrhoeae* populations: a non-clonal pathogen. *J. Gen. Microbiol.* 139: 2603–2611.

O'Rourke M, Ison CA, Renton AM, Spratt BG (1995) Opa-typing – a high resolution tool for studying the epidemiology of gonorrhoea. *Mol. Microbiol.* 17: 865–875.

Ørskov F, Ørskov I, Evans DJ, Sack RB, Sack DA, Wadstrom T (1976) Special *Escherichia coli* serotypes among enterotoxigenic strains from diarrhoea in adults and children. *Med. Microbiol. Immunol.* 162: 73–80.

Ørskov F, Whittam TS, Cravioto A and Ørskov I (1990) Clonal relationships among classic enteropathogenic *Escherichia coli* (EPEC) belonging to different O groups. *J. Infect. Dis.* 162: 76–81.

Pappenheimer AM (1977) Diptheria toxin. *Ann. Rev. Microbiol.* 46: 69–94.

Porras O, Caugant DA, Gray B, Lagergård T, Levin BR, Svanborg-Edén C (1986) Difference in structure between type b and non-typable *Haemophilus influenzae* populations. *Infect. Immun.* 53: 79–89.

Quentin R, Martin C, Musser JM, Pasquier-Picard N, Goudeau A (1993) Genetic characterization of a cryptic genospecies of *Haemophilus* causing urogenital and neonatal infections. *J. Clin. Microbiol.* 31: 1111–1116.

Reda KB, Kapur V, Mollick JA, Lamphear JG, Musser JM (1994) Molecular characterization and phylogenetic distribution of the streptococcal superantigen gene (*ssa*) from *Streptococcus pyogenes. Infect. Immun.* 62: 1867–1874.

Reda KB, Kapur V, Goela D, Lamphear JG, Musser JM (1996) Phylogenetic distribution of streptococcal superantigen *ssa* allelic variants provides evidence for horizontal transfer of *ssa* within *Streptococcus pyogenes. Infect. Immun.* 64: 1161–1165.

Reeves PR (1992) Variation in O-antigens, niche-specific selection and bacterial populations. *FEMS Microbiol. Lett.* 100: 509–516.

Reeves PR (1993) Evolution of *Salmonella* O antigen variation by interspecific gene transfer on a large scale. *Trends Genet.* 9: 17–22.

Reeves MW, Evins GM, Heiba AA, Plikaytis BD, Farmer JJ (1989) Clonal nature of *Salmonella typhi* and its genetic relatedness to other salmonellae as shown by multilocus enzyme electrophoresis, and proposal of *Salmonella bongori* comb. nov. *J. Clin. Microbiol.* 27: 313–320.

Riley LW, Remis RS, Helgerson SD *et al.* (1983) Hemorrhagic colitis associated with a rare *Escherichia coli* serotype. *New Engl. J. Med.* 308: 681–685.

Rodrigues J, Scaletsky ICA, Campos LC, Gomes TAT, Whittam TS, Trabulsi LR (1996) Clonal structure and virulence factors in strains of *Escherichia coli* of the classic serogroup O55. *Infect. Immun.* 64: 2680–2686.

Sarafian SK, Knapp JS (1989) Molecular epidemiology of gonorrhea. *Clin. Microbiol. Rev.* 2: S49–S55.

Sawyer SA (1989) Statistical tests for detecting gene conversion. *Mol. Biol. Evol.* 2: 539–556.

Scholten RJPM, Poolman JT, Valkenburg HA, Bijlmer HA, Dankert J, Caugant DA (1994) Phenotypic and genotypic changes in a new clone complex of *Neisseria meningitidis* causing disease in The Netherlands, 1958–1990. *J. Infect. Dis.* 169: 673–676.

Seiler A, Reinhardt R, Sarkari J, Caugant DA, Achtman M (1996) Allelic polymorphism and site-specific recombination in the *opc* locus of *Neisseria meningitidis. Mol. Microbiol.* 19: 841–856.

Selander RK, Levin BR (1980) Genetic diversity and structure in *Escherichia coli* populations. *Science* 210: 545–547.

Selander RK, Musser JM (1990) Population genetics of bacterial pathogenesis. In Iglewski BH, Clark VL (eds) *Molecular Basis of Bacterial Infections.* San Diego: Academic Press, pp. 11–36.

Selander RK, McKinney RM, Whittam TS *et al.* (1985) Genetic structure of populations of *Legionella pneumophila. J. Bacteriol.* 163: 1021–1037.

Selander RK, Caugant DA, Ochman H, Musser JM, Gilmour MN, Whittam TS (1986) Methods of multilocus enzyme electrophoresis for bacterial population genetics and systematics. *Appl. Env. Microbiol.* 51: 873–884.

Selander RK, Beltran P, Smith NH *et al.* (1990a) Genetic population structure, clonal phylogeny, and pathogenicity of *Salmonella paratyphi* B. *Infect. Immun.* 58: 1891–1901.

Selander RK, Beltran P, Smith NH *et al.* (1990b) Evolutionary genetic relationships of clones of *Salmonella* serovars that cause human typhoid and other enteric fevers. *Infect. Immun.* 58: 2262–2275.

Selander RK, Smith NH, Li J *et al.* (1992) Molecular evolutionary genetics of the cattle-adapted serovar *Salmonella dublin. J. Bacteriol.* 174: 3587–3589.

Selander RK, Li J, Nelson K (1996) Evolutionary genetics of *Salmonella enterica.* In Neidhardt FC *et al.* (ed.) *Escherichia coli and Salmonella.* Washington DC: ASM Press, pp. 2691–2707.

Shortridge VD, Stone GG, Flamm RK *et al.* (1997) Molecular typing of *Helicobacter pylori* isolates from a multicenter U.S. clinical trial by *ureC* restriction fragment length polymorphism. *J. Clin. Microbiol.* 35: 471–473.

Sibold C, Wang J, Henrichsen J, Hakenbeck R (1992) Genetic relationships of penicillin-susceptible and -resistant *Streptococcus pneumoniae* strains isolated on different continents. *Infect. Immun.* 60: 4119–4126.

Smith HO, Tomb JF, Dougherty BA, Fleischmann RD, Venter JC (1995) Frequency and distribution of DNA uptake signal sequences in the *Haemophilus influenzae* Rd genome. *Science* 269: 538–540.

Smith NH, Selander RK (1990) Sequence invariance of the antigen-coding central region of the phase 1 flagellar filament gene (*fliC*) among strains of *Salmonella typhimurium. J. Bacteriol.* 172: 603–609.

Smith NH, Beltran P, Selander RK (1990) Recombination of *Salmonella* phase I flagellin genes generates new serovars. *J. Bacteriol.* 172: 2209–2216.

Smith NH, Holmes EC, Donovan GM, Carpenter GA, Spratt BG (1999) Networks and groups within the genus *Neisseria*: analysis of *argF*, *recA*, *rho* and 16S rRNA sequences from human *Neisseria* species. *Mol. Biol. Evol.* 16: 773–783.

Sneath PHA (1995) The distribution of the random division of a molecular sequence. *Binary Comput. Microbiol.* 7: 148–152.

Soares S, Kristinsson KG, Musser JM, Tomasz A (1993) Evidence for the introduction of a multiresistant clone of serotype 6B *Streptococcus pneumoniae* from Spain to Iceland in the late 1980s. *J. Infect. Dis.* 168: 158–163.

Solomon JM and Grossman AD (1996) Who's competent and when – regulation of natural genetic competence in bacteria. *Trends Genet.* 12: 150–155.

Souza V, Nguyen TT, Hudson RR, Piñero D, Lenski RE (1992) Hierarchical analysis of linkage disequilibrium in

Rhizobium populations: evidence for sex? *Proc Natl Acad. Sci. USA* 89: 8389–8393.

Spratt BG (1988) Hybrid penicillin-binding proteins in penicillin-resistant gonococci. *Nature* 332: 173–176.

Spratt BG (1994) Resistance to antibiotics mediated by target alterations. *Science* 264: 388–393.

Spratt BG, Maiden MCJ (1999) Bacterial population genetics, evolution and epidemiology. *Phil. Trans. Roy. Soc. Lond. B* 354: 701–710.

Spratt BG, Zhang Q-Y, Hutchison A, Jones DM, Brannigan JA, Dowson CG (1989) Recruitment of a penicillin-binding protein gene from *Neisseria flavescens* during the emergence of penicillin resistance in *Neisseria meningitidis. Proc Natl Acad. Sci. USA* 86: 8988–8992.

Spratt BG, Bowler LD, Zhang Q-Y, Zhou J, Maynard Smith J (1992) Role of inter-species transfer of chromosomal genes in the evolution of penicillin resistance in pathogenic and commensal *Neisseria* species. *J. Mol. Evol.* 34: 115–125.

Spratt BG, Smith NH, Zhou J, O'Rourke M, Feil E (1995) The population genetics of the pathogenic *Neisseria*. In Baumberg S, Young JPW, Wellington EMH, Saunders JR (eds) *The Population Genetics of Bacteria*, SGM Symposium No. 52, Cambridge: Cambridge University Press, pp. 143–160.

Sreevatsan S, Pan X, Stockbauer KE *et al.* (1997) Restricted structural gene polymorphism in the *Mycobacterium tuberculosis* complex indicates evolutionarily recent global dissemination. *Proc Natl Acad. Sci. USA* 94: 9869–9874.

Stevens DL (1992) Invasive group A streptococcal infections. *Clin. Infect. Dis.* 14: 2–13.

Stephens JC (1985) Statistical methods of DNA sequence analysis – detection of intragenic recombination or gene conversion. *Mol. Biol. Evol.* 2: 539–556.

Stoltzfus A, Leslie JF, Milkman R (1988) Molecular evolution of the *Escherichia coli* chromosome. II. Analysis of structure and natural variation in a previously uncharacterized region between *trp* and *tonB*. *Genetics* 120: 345–358.

Suerbaum S, Smith JM, Bapumia K *et al.* (1998) Free recombination within *Helicobacter pylori*. *Proc Natl Acad Sci USA* 95(21): 12619–12624.

Suker J, Feavers IM, Achtman M, Morelli G, Wang JF, Maiden MCJ (1994) The *porA* gene in serogroupA meningococci – evolutionary stability and mechanism of genetic variation. *Mol. Microbiol.* 12: 253–265.

Sved JA (1968) The stability of linked systems of loci with a small population size. *Genetics* 59: 543–563.

Takala AK, Vuopiovarkila J, Tarkka E, Leinonen M, Musser JM (1996) Subtyping of common pediatric pneumococcal serotypes from invasive disease and pharyngeal carriage in Finland. *J. Infect. Dis.* 173: 128–135.

Talkington DF, Schwartz B, Black CM *et al.* (1993) Association of phenotypic and genotypic characteristics of invasive *Streptococcus pyogenes* isolates with clinical components of streptococcal toxic shock syndrome. *Infect. Immun.* 61: 3369–3374.

Tarr PI, Neill MA, Clausen CR, Newland JW, Neil RJ, Mosely SL (1989) Genotypic variation in pathogenic *Escherichia coli* O157:H7 isolated from patients in Washington, 1984–1987. *J. Infect. Dis.* 159: 1019–1024.

Tibayrenc M (1996) Towards a unified evolutionary genetics of microorganisms. *Ann. Rev. Microbiol.* 50: 401–429.

Torres OR, Korman RZ, Zahler SA, Dunny GM (1991) The conjugative transposon Tn925 – enhancement of conjugal transfer by tetracycline in *Enterococcus faecalis* and mobilization of chromosomal genes in *Bacillus subtilis* and *E. faecalis. Mol. Gen. Genet.* 225: 395–400.

van Soolingen D, DeHaas PEW, Hermans PWM, van Embden JDA (1994) DNA fingerprinting of *Mycobacterium tuberculosis. Meth. Enzymol.* 235: 196–205.

Vázquez JA, de la Fuente L, Berrón S *et al.* (1993) Ecological separation and genetic isolation of *Neisseria gonorrhoeae* and *Neisseria meningitidis. Curr. Biol.* 3: 567–572.

Vázquez JA, Berrón S, O'Rourke M *et al.* (1995) Interspecies recombination in nature: a meningococcus that has acquired a gonococcal PIB porin. *Mol. Microbiol.* 15: 1001–1007.

Waldor MK, Mekalanos JJ (1996) Lysogenic conversion by a filamentous phage encoding cholera toxin. *Science* 272: 1910–1914.

Wang G, Whittam TS, Berg CM, Berg DE (1993a) RAPD (arbitrary primer) PCR is more sensitive than multilocus enzyme electrophoresis for distinguishing related bacterial strains. *Nucleic Acids Res.* 21: 5930–5933.

Wang Y, Roos KP, Taylor DE (1993b) Transformation of *Helicobacter pylori* by chromosomal metronidazole resistance and by a plasmid with a selectable chloramphenicol resistance marker. *J. Gen. Microbiol.* 139: 2485–2493.

Weeks CR, Ferretti JJ (1984) The gene for type A streptococcal exotoxin (erythrogenic toxin) is located in bacteriophage T12. *Infect. Immun.* 46: 531–536.

Weel JFL, van der Hulst RWM, Gerrits Y *et al.* (1996) The interrelationship between cytotoxin-associated gene A, vacuolating cytotoxin, and *Helicobacter pylori*-related diseases. *Infect. Immun.* 173: 1171–1175.

Whatmore AM, Kehoe MA (1994) Horizontal gene transfer in the evolution of group A streptococcal *emm*-like genes – gene mosaics and variation in *vir* regulons. *Mol. Microbiol.* 11: 363–374.

Whatmore AM, Kapur V, Sullivan DJ, Musser JM, Kehoe MA (1994) Non-congruent relationships between variation in *emm* gene sequences and the population genetic structure of group A streptococci. *Mol. Microbiol.* 14: 619–631.

Whittam TS, Ochman H, Selander RK (1983a) Geographic components of linkage disequilibrium in natural populations of *Escherichia coli. Mol. Biol. Evol.* 1: 67–83.

Whittam TS, Ochman H, Selander RK (1983b) Multilocus genetic structure in natural populations of *Escherichia coli. Proc Natl Acad. Sci. USA* 80: 1751–1755.

Whittam TS, Wolfe ML, Wachsmuth IK, Ørskov F, Ørskov I, Wilson RA (1993) Clonal relationships among

Escherichia coli strains that cause hemorrhagic colitis and infantile diarrhea. *Infect. Immun.* 61: 1619–1629.

Whittam TS (1995) Genetic population structure and pathogenicity in enteric bacteria. In Baumberg S, Young JPW, Wellington EMH, Saunders JR (eds) *The Population Genetics of Bacteria*, SGM Symposium No. 52, Cambridge: Cambridge University Press, pp. 217–245.

Whittam TS (1996) Genetic variation and evolutionary processes in natural populations of *Escherichia coli*. In Neidhardt FC *et al.* (ed.) *Escherichia coli and Salmonella*. Washington DC: ASM Press, pp. 2708–2720.

Wilkins BM (1995) Gene transfer by bacterial conjugation: diversity of systems and functional specializations. In Baumberg S, Young JPW, Wellington EMH, Saunders JR (eds) The Population Genetics of Bacteria, SGM Symposium No. 52, Cambridge: Cambridge University Press, pp. 59–88.

Wise MG, Shimkets LJ, McArthur JV (1995) Genetic structure of a lotic population of *Burkholderia (Pseudomonas) cepacia*. *Appl. Env. Microbiol.* 61: 1791–1798.

Wu SW, Piscitelli C, deLencastre H, Tomasz A (1996) Tracking the evolutionary origin of the methicillin resistance gene: Cloning and sequencing of a homologue of *mecA* from a methicillin susceptible strain of *Staphylococcus sciuri*. *Microb. Drug Res.* 2: 435–441.

Yorke JA, Hethcote WHJ, Nold A (1978) Dynamics and control of the transmission of gonorrhea. *Sex. Transm. Dis.* 5: 51–56.

Young JPW (1989) The population genetics of bacteria. In Hopwood DA, Chater KF (eds) *Genetics of Bacterial Diversity*. New York: Academic Press, pp. 417–438.

Yu CE, Ferretti JJ (1989) Molecular epidemiological analysis of the type A streptococcal exotoxin (erythrogenic toxin) gene (*speA*) in clinical *Streptococcus pyogenes* strains. *Infect. Immun.* 57: 3715–3719.

Yu CE, Ferretti JJ (1991) Frequency of the erythrogenic toxin B and C genes (*speB* and *speC*) among clinical isolates of group A streptococci. *Infect. Immun.* 59: 211–215.

Zawadski P, Roberts MS, Cohan FM (1995) The log-linear relationship between sexual isolation and sequence divergence in *Bacillus* transformation is robust. *Genetics* 140: 917–932.

Zhou J, Spratt BG (1992) Sequence diversity within the *argF*, *fbp* and *recA* genes of natural isolates of *Neisseria meningitidis*: interspecies recombination within the *argF* gene. *Mol. Microbiol.* 6: 2135–2146.

Zhou J, Bowler LD, Spratt BG (1997) Interspecies recombination, and phylogenetic distortions, within the glutamine synthetase and shikimate dehydrogenase genes of *Neisseria meningitidis* and commensal *Neisseria* species. *Mol. Microbiol.* 23: 799–812.

21

A Phylogenetic Perspective on Molecular Epidemiology

Mark Achtman

Max-Planck Institut für Infektionsbiologie and Deutsches Rheuma-Forschungszentrum Berlin, Germany

Bacterial epidemiology is a descriptive discipline. It characterises pathogenic bacteria by using markers that can subdivide bacterial species. Epidemiology has been practised for over 100 years and, in addition to its intrinsic scientific interest, has aided disease prevention and contributed to improved understanding of parasite–host interactions. Epidemiological investigations are tedious, often involving repetitive testing of numerous isolates, and are particularly difficult in areas with poor medical infrastructure or when dealing with fastidious organisms that are difficult to cultivate. Yet, over the decades, most of our knowledge of bacterial disease patterns has been obtained by exactly such analyses and reliable and interpretable information is continuously supplied by classical methods, based on empirically derived serological tools.

Molecular epidemiology attempts bacterial classification on the basis of natural genetic variation. During the last two decades, molecular epidemiology has been used as an adjunct and increasingly as a replacement for the classical methods. The most important new insight has probably been the recognition that numerous diseases are caused by a limited number of clonal groupings. Classical epidemiology, molecular biology, population genetics and phylogenetics have all influenced contemporary molecular epidemiology. The result is a highly diverse field, unified only by the use of DNA-based methods. DNA-based methods are also the only available choice for non-cultivatable organisms. High throughput can be achieved at reasonable cost with these methods, and they are amenable to automation.

Potentially, molecular epidemiology has the power to elucidate the mechanisms leading to bacterial outbreaks and epidemics and to supply a durable global framework for bacterial classification based on genetic relationships. Sequences from complete genomes will allow rational typing schemes to be developed, using loci that reflect phylogenetic history. Innumerable publications possibly classifiable as molecular epidemiology, however, do not reflect this potential. Novel typing methods of questionable value are continuously being described (Achtman, 1996) and the existence of considerable natural variation among related organisms is often ignored in the interpretation of molecular data (Tenover *et al.*, 1995). The suitability of an approach for recognising single-source outbreaks has become one of the primary goals of method development in this field and methods are often evaluated solely in terms of reproducibility and discrimination rather than their potential for global long-term epidemiology. Increasingly, attempts are being made to institutionalise international networks for strain identification based on typing methodology that is not necessarily suitable for phylogenetic classification.

This chapter, written during the summer of 1997 and updated during the autumn of 2000, presents a highly personal view of molecular bacterial

 doi:10.1006/bkmm.2001.0021

epidemiology. It summarises approaches that are informative from the phylogenetic viewpoint, and describes results from selected pathogenic species where the available information seems to be somewhat more complete. It is doubtful whether many currently available, DNA-based methods will be of more than historical interest within a decade and therefore I largely ignore novel methods that have not yet been investigated intensively. Another chapter in this volume (see Chapter 20) presents a superb review of the population biology of bacterial pathogens. Population-biological approaches are therefore not be considered here, except where my interpretations and conclusions differ from those in that chapter.

Classification and Typing

The term *classification* is used here for top-down methodologies that differentiate within bacterial species, and *typing* for bottom-up approaches that distinguish between isolates (**Fig. 1**). Ideally, classification should be based on properties that possess the potential to detect relationships at the species level whereas typing is often dedicated to distinguishing the maximum number of isolates. Long-term, global epidemiology requires classification whereas short-term, local epidemiology can often be performed by either classification or typing.

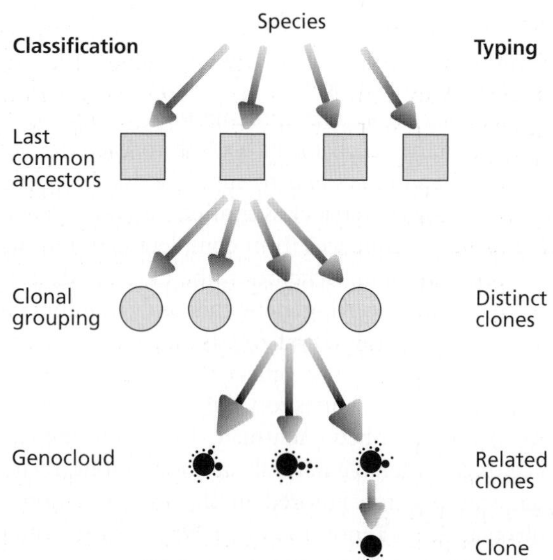

Fig. 1 Classification versus typing. Classification (left) is a top-down strategy which subdivides species into clonal groupings, each potentially representing the descendants of last common ancestors. In any one country at a particular time, the isolates of a particular clonal grouping form a genocloud. A genocloud consists of the frequent genotype (larger circle) plus rare variants (dots). Different genoclouds from one clonal grouping may exist in different countries or at different time points. Typing (right) is a bottom-up strategy which can assign very closely related variants to so-called 'clones' and can potentially recognise that genoclouds within a clonal grouping are related. Typing can only rarely elucidate the phylogenetic relationships between different clonal groupings.

Traditional typing of bacterial pathogens by serological or other methods (antibiotic resistance, phage typing, etc.) may be relevant to disease treatment and to vaccine development but is increasingly unsuitable for modern epidemiology because it is based on variable phenotypes that are not necessarily related to genetic descent. Horizontal genetic exchange leads to the acquisition of genes encoding foreign lipopolysaccharides (LPS) (Reeves, 1993; Comstock *et al.*, 1996; Davies *et al.*, 1997; Mooi and Bik, 1997; Shepherd *et al.*, 2000) or capsular polysaccharides (Kroll and Moxon, 1990; Swartley *et al.*, 1997, 1998; Morona *et al.*, 1999). Antibiotic resistance is also spread by horizontal genetic exchange (Dowson *et al.*, 1994; Reichmann *et al.*, 1997; Bootsma *et al.*, 2000) as well as by mobile genetic elements (Hall and Collis, 1995). Selectively neutral markers that are not under selection are more likely to index the genetic relatedness between isolates and less likely to be imported by gene transfer than are serological markers. A classification scheme that is based on selectively neutral markers will also reflect the population biology of the species under investigation and can be used for phylogenetic analyses.

The population biology of bacterial pathogens varies dramatically. The terms clonal, panmictic and epidemic have been used to describe three extreme types of population structures typified by *Salmonella enterica*, *Neisseria gonorrhoeae* and *Neisseria meningitidis*, respectively (Maynard Smith *et al.*, 1993). In a truly clonal species, all genetic changes would result from sequential accumulation of mutations after descent from a common ancestor. All classification schemes of sufficient resolution should yield comparable results for a clonal species, and typing could be performed with any method that is easy, quick and reliable. Very few bacterial species are truly clonal, however, and occasional horizontal genetic exchange has occurred even in prototypic clonal species such as *S. enterica* (Ochman and Lawrence, 1996) and *Escherichia coli* (Milkman and Bridges, 1993; Guttman and Dykhuizen, 1994; Guttman, 1997). Classification should therefore be based on multiple neutral loci scattered around the chromosome in order to ensure that strains are reliably grouped on the basis of genetic relationships.

Some panmictic species, such as *Helicobacter pylori* (Go *et al.*, 1996; Suerbaum *et al.*, 1998) or

N. gonorrhoeae (O'Rourke and Spratt, 1994), are so variable that identical strains are only found among isolates from direct contacts. Recombination within such panmictic species is so frequent that apparently identical multilocus enzyme electrophoresis (MLEE) electrophoretic types (ETs) may reflect convergence through horizontal genetic exchange rather than clonal descent (Maynard Smith *et al.*, 1993). For *H. pylori*, the frequency of homoplasies is as high as would be expected from free recombination (Suerbaum *et al.*, 1998). Owing to the extensive recombination it may be impossible to devise a phylogenetic classification of such species, but typing schemes based on phylogenetic principles can still yield valuable information. Furthermore, even panmictic species can contain clonal groupings (Gutjahr *et al.*, 1997) or show geographical specialisation (Achtman *et al.*, 1999a). A good typing scheme should be able both to reliably differentiate between unrelated bacteria and to allow recognition of the clonal groupings.

In panmictic species, clonal groupings should be continuously disrupted by frequent recombination. The designation 'epidemic' was therefore evoked to account for apparently clonal groupings that do exist in some panmictic species (Maynard Smith *et al.*, 1993). According to this concept, the apparent clonal groupings are 'short-lived'; they are so fit that they have multiplied extensively and are frequently isolated even if they will eventually be destroyed by recombination. Note that 'short-lived' refers to an evolutionary perspective and such clonal groupings can be long-lived from the viewpoint of molecular epidemiology. For example, it has been estimated that the last common ancestor of certain 'short-lived' clones in *N. meningitidis* existed approximately a century ago (Morelli *et al.*, 1997). Classification schemes are thus appropriate for epidemic clones in panmictic species although typing would be preferable for other isolates.

Clearly, different approaches are needed for different species due to differing population structures. Durable classification and typing schemes can therefore only be developed if they can cope with species-specific peculiarities.

Methods that Index Genetic Variation

Analysis of the sequence variation of multiple genes encoding housekeeping enzymes is probably the most reliable method for the classification of clonal bacteria into genetically related groupings (Boyd *et al.*, 1996; Maiden *et al.*, 1998). Sequence variation in housekeeping genes is usually neutral, i.e. not subject to natural selection, and therefore the differences between alleles of housekeeping genes are more likely to reflect the time since phylogenetic descent than are those of genes whose products are under selection (antibiotic resistance, pathogenesis). Most of the sequence variation between closely related bacteria results from recombination rather than mutation in *E. coli* (Guttman and Dykhuizen, 1994), *N. meningitidis* (Feil *et al.*, 1999) and *Streptococcus pneumoniae* (Feil *et al.*, 2000). Because bacterial species are so prone to recombination, multiple genes need to be sequenced, even for highly clonal bacteria, in order to ensure that horizontal genetic exchange has not obscured genetic relationships among strains.

The housekeeping genes of some clonal bacterial species are too uniform to differentiate between different isolates because there has been insufficient time to accumulate variation. The last common ancestor of *Mycobacterium tuberculosis* existed as recently as 15 000 years ago (Kapur *et al.*, 1994) and *Yersinia pestis* is a clone of *Yersinia pseudotuberculosis* that evolved 1500–20 000 years ago (Achtman *et al.*, 1999b). Similarly, essentially no sequence variation seems to exist in *Bacillus anthracis* (Keim *et al.*, 1997). For such species, strain typing for epidemiological purposes can be very difficult. Typing of *M. tuberculosis* has been based on DNA fingerprints of the variable chromosomal locations of transposable insertion elements (Van Embden *et al.*, 1993) (restriction-fragment length polymorphism, RLFP) or other repetitive elements (Kremer *et al.*, 1999), or the sequence variability of genes involved in antibiotic resistance (Bifani *et al.*, 1997). The validity of this approach remains to be demonstrated on a global scale, although it has been very useful for recognising individual clones (Bifani *et al.*, 1997; Kiers *et al.*, 1997; Kremer *et al.*, 1999). RFLP of IS elements also correlates well with biotype for *Y. pestis* (Achtman *et al.*, 1999b).

Bacterial classification by the sequencing of multiple housekeeping gene fragments from numerous strains (multilocus sequence typing, MLST) has only recently been introduced (Maiden *et al.*, 1998). Each unique sequence (allele) is assigned a distinct numerical designation. For each isolate tested, seven loci scattered around the genome are sequenced and the isolate is assigned to a 'sequence type' (ST), consisting of the allele numbers at all seven loci. The number of nucleotides differing between alleles is ignored and therefore point mutations and horizontal genetic exchange are weighted equally. The data are stored in a database that is globally accessible via the internet (http://www.mlst.net). MLST classification has already become the gold standard for *N. meningitidis* because it uses the same proven principles as MLEE (see below) but can be used by many laboratories around the globe. MLST has also been established for *Strep. pneumoniae* (Enright and Spratt, 1998; Enright *et al.*, 1999) and is in progress for *Staphylococcus aureus*, *Strep. pyogenes*, *Campylobacter jejuni*, *Haemophilus influenzae*, *Strep. agalactiae* and *Moraxella catarrhalis*. MLST is still expensive and labour-intensive, however, and requires a modicum of technical competence. Furthermore, the method is so new that only few publications based on it have appeared; former epidemiological studies have been based on simpler and cheaper methods summarised below. None is as reliable or as simple to interpret as is MLST and I am convinced that they will be progressively replaced by sequence-based methods once the price and difficulty of multiple sequencing has been reduced by technological developments.

Multilocus Enzyme Electrophoresis

MLEE is a fairly simple method of detecting a proportion of the non-synonymous sequence variation in housekeeping genes by the electrophoresis of cell extracts on starch gels or cellulose acetate sheets followed by staining for specific enzymes (Selander *et al.*, 1986). This method has been used extensively for the analyses of bacterial species since the mid-1980s and has allowed the reliable recognition of genetically related bacteria within clonal species such as *E. coli* (Ochman and Selander, 1984) and *S. enterica* (Boyd *et al.*, 1996). A statistical test, the I_A test (see http://soft.ice.mpg.de/lian/ for details), tests whether the MLEE data indicate a clonal or panmictic species (Maynard Smith *et al.*, 1993; Haubold *et al.*, 1998).

As with MLST, MLEE data are recorded by assigning unique allele designations to each distinguishable electrophoretic variant (Fig. 2). Each unique combination of protein variants is referred to as an electrophoretic-type (ET) and phylogenetic analyses, such as UPGMA cluster analysis (Sneath and Sokal, 1978), are performed on the numbers of pairwise differences. Significant clusters containing ETs that are distinct from other clusters are considered to represent clonal groupings. Usually at least 10 enzymes are used for MLEE analyses, in order to provide a representative sample of genes scattered around the genome.

MLEE has been used to screen thousands of isolates and poses minimal requirements for data storage and analysis. As the number of known protein variants increases, however, it becomes increasingly difficult to combine data from different studies because each new variant must be compared with all formerly identified variants to ensure that it is distinct. For the same reason, laboratories would need to exchange all reference strains expressing each known protein variant in order to maintain a uniform nomenclature. Furthermore, the method is not very portable: minor differences in electrophoretic conditions can result in partial non-reproducibility between independent laboratories. Defined sets of reference strains that represent the diversity revealed by MLEE classification have been defined for *E. coli* (Ochman and Selander, 1984),

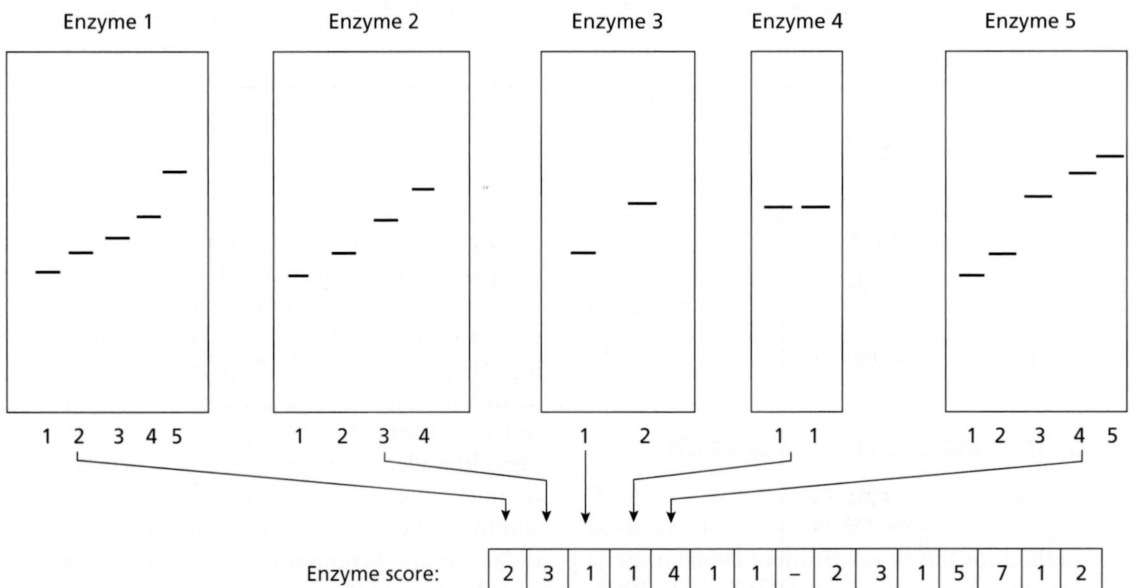

Fig. 2 Scoring of MLEE data. Top: fictitious electrophoretic variants seen within a fictitious species for five enzymes. Enzyme 4 is monomorphic and all strains express variant 1; the other enzymes are polymorphic and strains express one of a number of variants (5 for enzyme 1, 4 for enzyme 2, etc.). Bottom: The variant expressed by one strain for all enzymes tested (15 in this fictitious example) are noted as a 15-digit enzyme score. Enzyme 8 was not expressed by this strain and is recorded as missing data.

S. enterica (Selander *et al.*, 1996) and epidemic clones of *N. meningitidis* (Seiler *et al.*, 1996). Standardised MLEE classifications and defined reference strains are invaluable for screening the degree of genetic variability of individual genes within a species (Ochman and Groisman, 1996; Seiler *et al.*, 1996; Boyd *et al.*, 1997).

Among the cheap and simple techniques currently available, MLEE correlates most closely with clonal descent and phylogeny and MLEE data have been used extensively for the population biology of bacteria. Dendrograms generated from MLEE data are most significant at the level of clonal groupings, however, and the distances between the deep branches are less likely to reflect the true phylogeny of a species than are sequence data based on numerous loci (Boyd *et al.*, 1996). Note that since the speciation of *E. coli* 100 million years ago, a large proportion of the genome is thought to have been imported from other species and another large proportion that was present in the last common ancestor is thought to have been lost (Lawrence and Ochman, 1997). Thus, even with sequence-based data, it may not be possible to reconstruct the true phylogeny of any species that has existed for a reasonable length of time.

DNA Fingerprint Techniques

A plethora of techniques based on variation in fingerprint patterns (Fig. 3) have been described, some of which show fairly good correlation with indices of clonal descent. Fingerprint methods can be very useful for typing but are probably unsuitable for classification because of the lack of reproducibility between different laboratories. The primary role of DNA fingerprinting in the future will probably be for the rapid identification of single source outbreaks; attempts to use such data for phylogenetic analyses will be replaced by sequence-based methods.

Many flavours of DNA fingerprinting have been described, all of which depend on polymorphic restriction sites. Depending on the particular implementation, these sites are within single-copy housekeeping genes (Guibourdenche *et al.*, 1997), flank rRNA operons (ribotyping) (Popovic *et al.*, 1993; Karaolis *et al.*, 1994a; Bennekov *et al.*, 1996; Louie

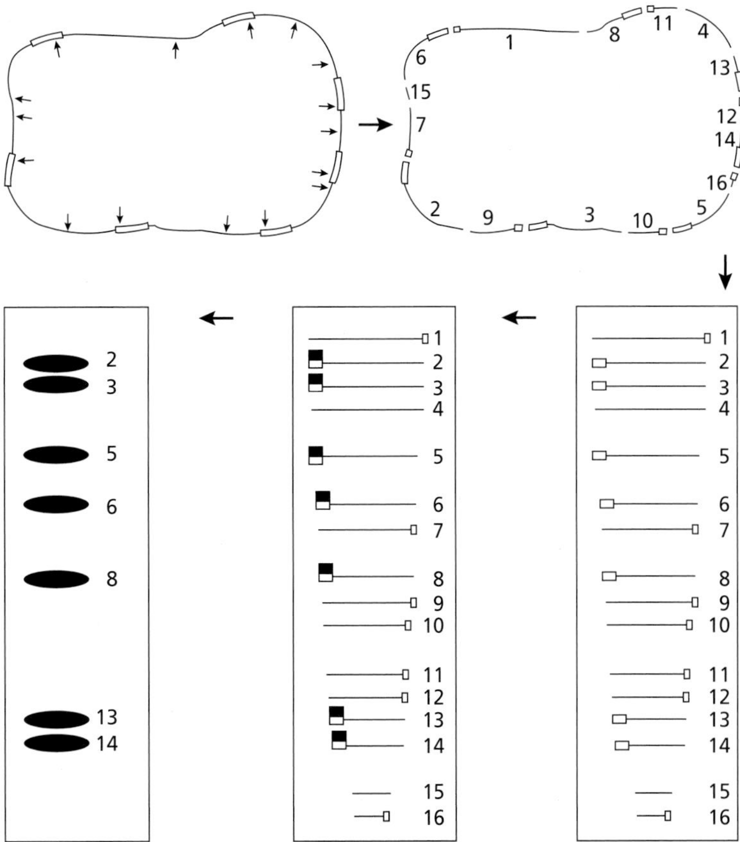

Fig. 3 DNA fingerprinting using hybridisation to an insertion element present in multiple copies on the genome. Top, left to right: Positions of the insertion element and generation of restriction fragments with a rare-cutting enzyme. Bottom, right to left: separation of fragments by PFGE, hybridisation with a specific labelled probe and detection. From Braden (1997) with permission.

et al., 1996; Swaminathan *et al.*, 1996; Smith-Vaughan *et al.*, 1997), within rRNA operons (Smith-Vaughan *et al.*, 1995; Sriprakash and Gardiner, 1997), scattered around the chromosome (pulsed-field gel electrophoresis, PFGE) (Beall *et al.*, 1995; Tenover *et al.*, 1995; Hall *et al.*, 1996; Louie *et al.*, 1996; Swaminathan *et al.*, 1996; Mahenthiralingam *et al.*, 1996a; Van Belkum *et al.*, 1997) or flank repetitive DNA, including transposable elements (Van Embden *et al.*, 1993; Van Soolingen *et al.*, 1994; Kamerbeek *et al.*, 1997).

DNA segments are not uniform in their variability. The most conserved segments are those encoding rRNA because the multiple copies of the 16S and 23S rRNA genes scattered around bacterial chromosomes are often identical and rRNA evolution is fairly slow. rRNA sequences are particularly useful for taxonomic studies but do not discriminate well between closely related species or within species. Single-copy genes encoding enzymes are more variable but are also usually conserved due to functional constraints on the encoded proteins. Mutations in such DNA segments tend to be synonymous (Milkman and McKane, 1995); most non-synonymous mutations change enzymatic activity and are eliminated by natural selection. Virulence genes and genes encoding immunogenic antigens are still more variable, probably because of selection for variants which can evade the host immune response or with altered specificities (Smith *et al.*, 1995; Selander *et al.*, 1996). Still more variable are intergenic DNA segments where not only point mutations, but also deletions and insertions, are quite common (Milkman and McKane, 1995; Seiler *et al.*, 1996; Klee *et al.*, 2000).

Microevolution can be defined as any combination of minor genetic changes that occurs in a relatively short period of time (Achtman, 1994). In addition to variation introduced by horizontal genetic exchange, mutations can accumulate at amazingly high frequencies in short periods of time (Duim *et al.*, 1996, 1997; Bifani *et al.*, 1997; Morelli *et al.*, 1997; Zhu *et al.*, 2001). Microevolution results in differences between related isolates and can obscure their relationships, particularly for any fingerprinting technique that is based on a single gene. Other fingerprinting techniques use restriction sites that are scattered around the chromosome and have been devised to detect a maximum of variability. Such variability is most likely to be concentrated in exactly those regions which are most variable as a result of microevolution and least related to genetic descent. For example, multiple novel restriction sites were introduced by horizontal genetic exchange into variable genes during spread of *N. meningitidis* (Morelli *et al.*, 1997; Zhu *et al.*, 2001) and the PFGE pattern changed during long-term carriage of *E. coli* O157:H7 (Karch *et al.*, 1995). Bacterial chromosomes are mobile; they can change in size and organisation (Bergthorsson and Ochman, 1995; Dempsey *et al.*, 1995; Liu and Sanderson, 1995; Gibbs and Meyer, 1996; Jiang *et al.*, 1996; Frøholm *et al.*, 2000). Pathogenicity islands of up to 200 kb and integrated bacteriophages add to the genomic plasticity (Cheetham and Katz, 1995; Datz *et al.*, 1996; Hacker *et al.*, 1997; Klee *et al.*, 2000) as does the movement of transposable elements (Boyd and Hartl, 1997; Achtman *et al.*, 1999b). All these sources of variability will combine to reduce the correlation between fingerprinting results and phylogenetic descent.

PCR-Based Techniques

Numerous PCR-based typing methods have been described, many of which depend on restriction-site polymorphism (e.g. Smith-Vaughan *et al.*, 1995; Guibourdenche *et al.*, 1997) and have the same basic properties as other fingerprinting methods. In contrast, arbitrary primer PCR, the random amplified polymorphic DNA (RAPD) method (Welsh and McClelland, 1990) (**Fig. 4**), is based on homologies to short oligonucleotide primers. RAPD is an inexpensive, efficient and sensitive method which has now been applied to numerous pathogenic organisms (Akopyanz *et al.*, 1992; Versalovic *et al.*, 1993; Linton *et al.*, 1994; Makino *et al.*, 1994; Woods *et al.*, 1994; Coelho *et al.*, 1995b; Lin *et al.*, 1996; Mahenthiralingam *et al.*, 1996a,b; Bart *et al.*, 1998; Achtman *et al.*, 2001). It usually consists of PCR using a single short primer which amplifies segments scattered around the chromosome where pairs of sites located within a few kilobases happen to hybridise with the primer used. After PCR amplification, the products are resolved by gel electrophoresis. Chromosomal sequences with identity to the primer will generate PCR products more efficiently than sequences with only partial homology. Bands of differing intensity are produced and the reliability of this method depends on using standardised conditions. Even the most reproducible RAPD methods differ to some extent with the thermocycler used and with other conditions that are often impossible to standardise between laboratories.

The primary use of RAPD has been for rapid strain typing. In a laboratory where the method has been standardised, it can readily be applied to diverse species and differentiates between unrelated strains. As with fingerprinting methods, consistent assignments of molecular weights are essential if strains tested at different times are to be compared, and such comparisons suffer because of the lack of reproducibility of

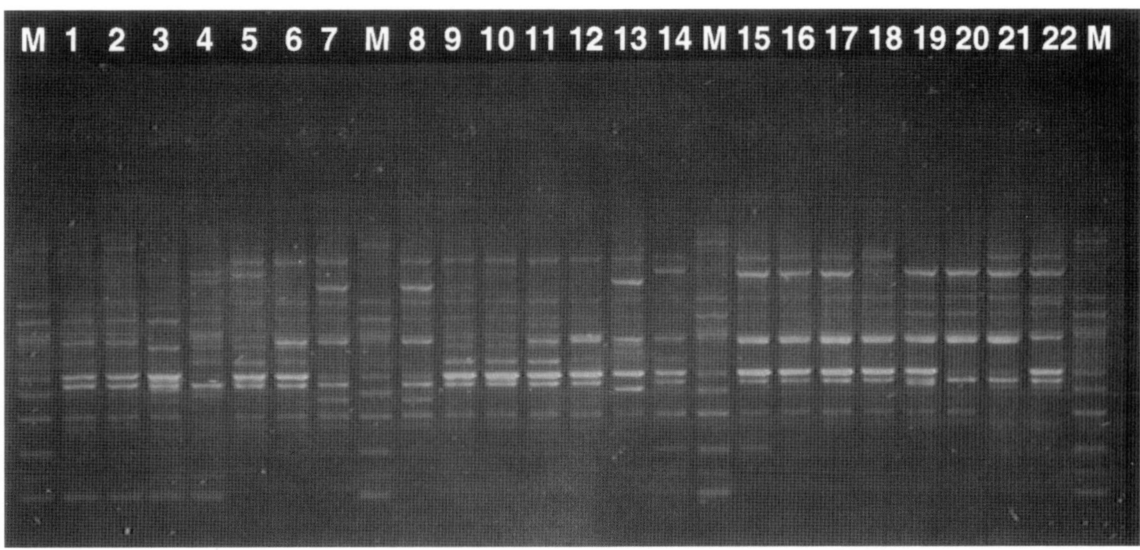

Fig. 4 RAPD gel electrophoresis. PCR with a single short primer was performed with chromosomal DNA from serogroup A meningococci (tracks 1–22) and a serogroup B strain as a consistency marker (M) and separated by agarose gel electrophoresis (Bart *et al.*, 1997). With permission from Arie van der Ende.

gel electrophoresis. In theory, RAPD should not be very suitable for strain classification because it is based on sequences that are not necessarily well conserved. A comparison of the results from RAPD, MLST and RFLP of PCR products has shown that RAPD is useful for rapid distinction between unrelated groupings but is less reliable for the grouping of closely related isolates (Achtman *et al.*, 2001).

Amplified-fragment Length Polymorphism

Amplified-fragment length polymorphism (AFLP) is a newly described method (Vos *et al.*, 1995) that potentially allows the genotyping of numerous bacterial species (Savelkoul *et al.*, 1999). In its most common application, chromosomal DNA is digested with a mixture of two enzymes that cut at different frequencies, such as *Eco*RI plus *Taq*I. Double-stranded oligonucleotide adapters that are specific for each of the two restriction ends and whose ligation products do not regenerate the original restriction sites are ligated to the digested genomic DNA. Digestion and ligation can be performed concurrently. PCR products are then generated using primers that are complementary to the adapters, coupled to a fluorescent dye. The number of PCR products can be reduced to a more manageable number by using primers that extend beyond the adapter sequences at the 3′ end by one to three nucleotides. The PCR products are then analysed by gel electrophoresis and cluster analysis (Fig. 5).

AFLP can be used for any genomic DNA. Band intensities vary much less than for other PCR-based methods such as RAPD, and the sizes of the PCR products can be quantified on sequencing gels using internal size markers labelled with a different fluorescent dye. AFLP of *E. coli* K-12 yielded all 48 bands that were predicted from the complete genome sequences and the measured band sizes were within 1–2 bp of the known sizes (Arnold *et al.*, 1999). AFLP distinguishes between *C. coli* and *C. jejuni* (de Boer *et al.*, 2000b) (Fig. 5) and yields a similar phylogenetic tree to MLEE for the EcoR collection of *E. coli* (Arnold *et al.*, 1999). Comparison of AFLP with other molecular methods for a variety of different species has shown that they yield similar groupings but that AFLP is the most discriminatory (Doignon-Bourcier *et al.*, 2000; Jonas *et al.*, 2000; Nair *et al.*, 2000; Poussier *et al.*, 2000; Thyssen *et al.*, 2000). AFLP may therefore become the method of choice for identifying major groupings by rapid analysis of limited numbers of isolates. As with other genomic fingerprinting methods, however, the fine discrimination between individual isolates by AFLP depends to a considerable extent on the most variable sequences within the genome and can contain so much random noise that informative, phylogenetic signals are obscured (see for example Achtman *et al.*, 2001).

Nomenclature, Strain Variants and Clonal Groupings

A clinical laboratory may need to determine whether a few strains represent a clonal outbreak. Such a laboratory could well use any of the methods described above for a rapid comparison of strains collected locally

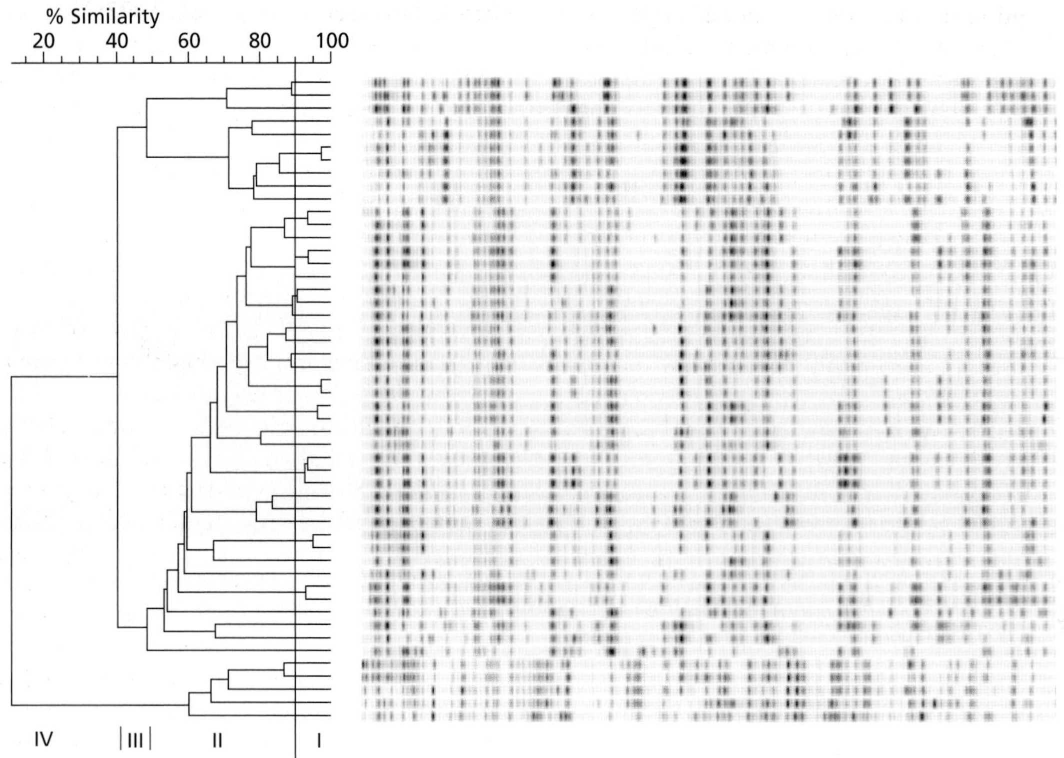

Fig. 5 AFLP of *C. jejuni* and *C. coli* (bottom five lanes). Right, photograph of AFLP patterns of isolates from humans and chickens. Left, UPGMA tree of similarities between the fingerprints and indicating four windows of similarity (I, >90%; II, >50%; III, >40%; IV, <40%). The data summarised in this figure are presented in greater detail by Duim *et al.* (2000). With permission from Birgitta Duim.

with control isolates defined as endemic strains according to epidemiological criteria. The unstated assumption, which is not necessarily justified (Salih *et al.*, 1990b), is that endemic disease is always caused by diverse bacteria. A second assumption that may not always be justified is that an outbreak is caused by the introduction of a particular (virulent) strain and represents the recent clonal expansion from a single cell. In fact, outbreaks can also reflect environmental changes and multiple, distantly related strains can sometimes be isolated even from apparent single-source outbreaks (Feavers *et al.*, 1999).

Indistinguishable or highly similar bacteria isolated from an outbreak are often referred to as a 'clone'. Originally, the term 'clone' was introduced to designate genetically related natural isolates that are descended from a common ancestor (Selander and Levin, 1980; Caugant *et al.*, 1981; Achtman *et al.*, 1983). The usage of that term may have been unfortunate because it is also used in molecular genetics to describe the progeny of a single cell. All the cells from a single colony need not be identical, because mutations can arise during the development of the colony. Similarly, all the related isolates from nature need not be identical, because minor genetic changes (micro-evolution) accumulate with time. However, the time scale since the last common ancestor is certainly much greater among the natural isolates than among the sister cells from the colony, and the natural isolates are also exposed to other events, including horizontal genetic exchange, which can increase genetic diversity. Thus, natural clones are not expected to be as uniform as sister cells from a single colony. The degree of variation between seemingly identical natural isolates awaits complete genome sequencing of apparently related bacteria. However, even the limited sequence data currently available demonstrate variation among related organisms, regardless of the species examined. I therefore now recommend the term 'clonal grouping', corresponding to a monophyletic clade in eukaryotic population biology. A clonal grouping consists of a group of natural isolates which are closely related by methods that index genetic variation, and which is distinct from other clonal groupings (**Fig. 1**). Alternative terms that have been used are clonal complex or lineage. Clonal groupings can contain even more closely related subgroupings that have been termed 'genoclouds' (Zhu *et al.*, 2001) (**Fig. 1**).

A description of a few strains from a single clonal outbreak is difficult to relate to data from other outbreaks in the absence of a global scheme of classification and nomenclature. And nomenclature is most useful when it reflects hierarchical genetic relationships and is sufficiently dynamic that new strain variants and groupings can readily be incorporated. Rigid schemes which arbitrarily define the limits of variation between related bacteria (Tenover *et al.*, 1995) are counterproductive and ignore phylogenetic descent and species-dependent differences in diversity.

Global classifications and hierarchical nomenclatures require the analysis of large, representative strain collections and consensus between different laboratories. Different geographical regions may harbour different clonal groupings (Achtman and Pluschke, 1986; Musser *et al.*, 1990a; Souza *et al.*, 1992; Coelho *et al.*, 1995a; Achtman *et al.*, 1999a) and different clonal groupings may well predominate at different periods of time (Wang *et al.*, 1992; Whittam *et al.*, 1993; Karaolis *et al.*, 1995; Achtman *et al.*, 2001). My experience has been that such global collections can be initiated or expanded by requesting strains from national reference laboratories after extensive literature searches (Achtman *et al.*, 1983; Olyhoek *et al.*, 1987). Alternatively, different national laboratories can pool their local strain collections (Van Embden *et al.*, 1993). Classification methods that allow entries via the internet (Maiden *et al.*, 1998) make it possible for a global community of scientists to pool their resources. In contrast, methods that rely on continued comparison with reference strains require the continued exchange of isolates between laboratories and do not have the same potential for continued expansion.

Data are best stored in a computerised database. It must be possible to export the data in a non-proprietary format to enable analysis with a variety of programs working under different computer operating systems. With time, the data from many thousands of strains will be included, necessitating considerable organisation and long-term planning (Woelffer *et al.*, 1996). Numerical data are rarely a major problem in the short term; for example, for MLEE or MLST analysis, individual electrophoretic variants are assigned unique numbers and each strain is characterised by its combination of numbers (Fig. 2). In contrast, fingerprint data depend on the storage of image files, which consume large amounts of storage space. Electrophoretic data from fingerprints or RAPD tests can be reduced to numerical data (molecular weights) by commercially available programs, such as GelCompar or its successor Bionumerics (2000), which also allow computerised searches through historical data sets (Bifani *et al.*, 1997; Van Soolingen *et al.*, 1999). Molecular weight assignments are inherently inaccurate, however, and it is difficult to maintain consistent numerical assignments for collections containing thousands of isolates. Strain similarities from independent analyses based on fingerprint data require independent confirmation, i.e. re-testing strains from which data has been entered in pairwise fashion or by using additional tests to confirm relatedness, before the similarities can be believed.

Closely related isolates can be detected by generating dendrograms from unweighted pair group mean average (UPGMA) or neighbour-joining (NJ) cluster analysis by using statistical software, phylogenetic programs or dedicated programs such as GelCompar. For MLEE or MLST, the data used to generate dendrograms consist of the number of identical alleles between pairs of isolates, whereas for fingerprint data the number or proportion of matching bands is used. Dendrograms always yield trees with branches, even if there is no true phylogenetic tree-like structure within the data. Alternative tests such as Splits graph decomposition analysis (Bandelt and Dress, 1992; Huson, 1998) or congruence analysis (Feil *et al.*, 2001) can and should be used to test whether the data are tree-like or represent recombinational networks. Tree topologies will change with additional data. It may therefore be advisable to wait until considerable data are available before assigning names to particular clonal groupings. In an ideal situation, the existence of a clonal grouping should be confirmed by various methods and should fit with epidemiological observations.

Nomenclature is useful only for stable clonal groupings. Probably only spurious hierarchical classifications are possible for panmictic bacteria such as most strains of *H. pylori*, *N. gonorrhoeae* or *N. meningitidis*. In such cases, the classification scheme could be usefully restricted to clonal subsets of these species, such as geographically distinct groups of *H. pylori*, HAU gonococci or serogroup A meningococci.

Epidemiological Patterns

Epidemic Disease Caused by Serogroup A Meningococci

Not all meningococci are highly pathogenic. In some European countries, the nasopharynx of 10% or more of the healthy population is colonised by meningococci; many of them are non-encapsulated or express capsular polysaccharides of serogroups 29E, W135, X,

Y or Z (Cartwright *et al.*, 1987; Caugant *et al.*, 1994). These bacteria might be considered to represent commensals since they rarely cause disease. Unlike the majority of isolates from healthy carriers, most of the isolates from endemic meningococcal disease and localised outbreaks belong to hypervirulent clonal groupings and normally express the serogroup B or C polysaccharides (Caugant, 1998). For example, serogroup B ET-5 complex meningococci have been responsible for hyperendemic disease levels in Norway and elsewhere (Caugant, 1998) and ET-37 complex serogroup C meningococci have caused major outbreaks in Czechoslovakia (Kriz *et al.*, 1999) and Canada (Ashton *et al.*, 1991). These generalities are not universal, and the association between capsular serogroup and clonal grouping is not immutable. A serious outbreak associated with the Haj pilgrimage to Saudi Arabia in 2000 was caused by ET-37 complex meningococci expressing the W-135 capsular polysaccharide (Popovic *et al.*, 2000; Taha *et al.*, 2000) even though most ET-37 complex meningococci normally express the serogroup C polysaccharide.

In the past, most large meningitis epidemics were associated with serogroup A meningococci. Epidemic meningococcal disease has occurred repeatedly since its initial description at the beginning of the nineteenth century (Vieusseux, 1806). Epidemics were frequent in numerous countries until World War II (Peltola, 1983) but large epidemics are now rare except for certain areas of the developing world, in particular the Sahel region of Africa called the Meningitis Belt (Lapeyssonnie, 1963; Achtman, 1990, 1995; Tikhomirov *et al.*, 1997). Epidemic waves affecting several countries have recurred every 10–20 years in the Meningitis Belt since the beginning of the twentieth century (Waddy, 1957; Mpairwe and Matovu, 1971; Greenwood, 1987; Achtman, 1990; Zhu *et al.*, 2001). Epidemic waves also occurred in China every 10 years between the 1950s and the 1980s (Wang *et al.*, 1992). These epidemics have affected large numbers of individuals. Ten million cases of meningococcal meningitis occurred in China between 1950 and 1989 (Wang *et al.*, 1992). In the Meningitis Belt, more than 500 000 deaths occurred between the 1930s and the 1980s (Lapeyssonnie, 1963; Greenwood, 1987) and 500 000 cases were documented in the dry seasons of 1996–1997 (Tikhomirov *et al.*, 1997).

Classification

The population structure of most serogroup A meningococci is clonal (Bart *et al.*, 2001). MLEE analysis of 500 isolates from a global strain collection using 15 cytoplasmic allozymes plus 4 outer membrane proteins (Olyhoek *et al.*, 1987; Wang *et al.*, 1992) defined 9 related clonal groupings called subgroups I–III, IV-1, IV-2, and V–VIII. A tenth related clonal grouping, subgroup X, was revealed by RAPD plus MLST analyses of isolates from Moscow (Achtman *et al.*, 2001). In addition, the data from all three methods show that additional, unrelated serogroup A bacteria exist (subgroup IX plus single isolates of a variety of MLST types) (Bart *et al.*, 1998; Achtman *et al.*, 2001) (**Fig. 6**). The three classification methods yield the same general tree structure (Achtman *et al.*, 2001; Bart *et al.*, 2001) except that subgroups I/II, III/VIII and IV-1/IV-2 are distinguished only by MLEE. Possibly subgroups II, VIII and IV-2 represent particular genoclouds of subgroups I, III and IV-1, respectively, but since only few isolates are available, almost all of which were isolated before 1970, further analyses of these bacteria may be primarily of historical interest.

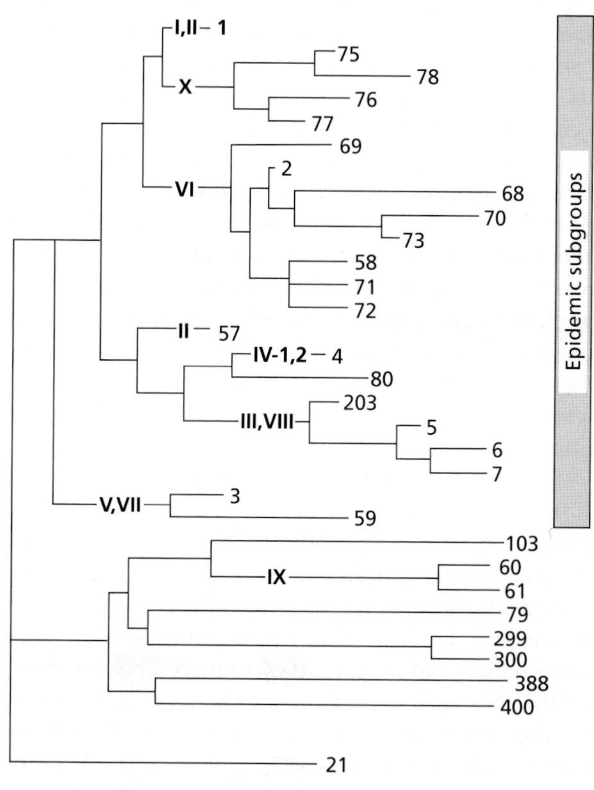

Fig. 6 Genetic structure and recent differentiation of serogroup A meningococci. Neighbour-joining tree of the numbers of different alleles for seven housekeeping gene fragments between different STs. ST designations are at the tips of each twig and subgroup designations are written over the appropriate branches. Epidemic subgroups are indicated at the right. Modified from Achtman *et al.* (2001).

Evolutionary Descent

The tree in Fig. 6 suggests that the epidemic serogroup A subgroups are closely related and share a common ancestor. Of the seven housekeeping gene fragments used to generate this tree, six were identical in many of the isolates even though numerous other alleles are present in diverse meningococci (Achtman *et al.*, 2001). Similarly, the 1240 bp segment containing *opcA* (Zhu *et al.*, 1999) plus flanking DNA was very uniform within the epidemic subgroups (Seiler *et al.*, 1996). Most of the strains contained *opcA1* but subgroups V and VII contain *opcA2* and subgroup IV-2 contains *opcA3*. These three alleles differ by only 1 to 2 nucleotides whereas other *opcA* alleles from other meningococci were much more diverse (Seiler *et al.*, 1996).

Even more sequence data support the recent descent from a common ancestor of an internal branch of the tree containing subgroups III, IV-1 and IV-2 (Morelli *et al.*, 1997). Older strains of subgroups III (1966), IV-1 (1966) and IV-2 (1917) possess identical sequences at four chromosomal loci (*opaA*, *opaB*, *opaD* and *opaJ*) encoding Opa (opacity) proteins (total of 3 kb), the *iga* gene encoding IgA1 protease (3 kb) and a 4 kb stretch downstream of *opaB*, containing three housekeeping genes and an intergenic region. The only exceptions were a deletion of the *opaJ* locus in subgroup IV-1 and a single nucleotide polymorphism in an intergenic region in the subgroup IV-2 strain (Morelli *et al.*, 1997). Sequence variants of all of these regions were frequent among more recent isolates of the three subgroups (see below), indicating that the three subgroups had existed for only a few decades before the earliest dates of isolation. Therefore, the common ancestor of subgroups III, IV-1 and IV-2 probably existed as recently as the nineteenth century (Morelli *et al.*, 1997). By extrapolation, the epidemic serogroup A subgroups may have evolved shortly before 1805 when the first meningococcal epidemics were documented (Achtman *et al.*, 2001).

Microevolution

Microevolution has occurred repeatedly within the epidemic subgroups. In order to measure the frequency of such events, fragments of four genes that encode two immunogenic outer membrane proteins and two housekeeping genes were sequenced from subgroup IV-1 isolates from a single African epidemic (Linz *et al.*, 2000). These four genes were chosen from a 25 kb region in order to allow an analysis of the history of recombination events. Of the approximately 100 subgroup IV-1 isolates tested, half were re-combinants that had imported a median length of 5 kb (range 1–10 kb) from related commensal neisseria and other meningococci. Together with other data, these results have led to the conclusion that a global gene pool exists within the neisseria (Fig. 7), resulting in mosaic genes as well as complete replacements (Maiden *et al.*, 1996). The frequency of import was apparently particularly high in the subgroup IV-1 analysis (Linz *et al.*, 2000) owing to immune selection for variants of the immunogenic TbpB protein, part of

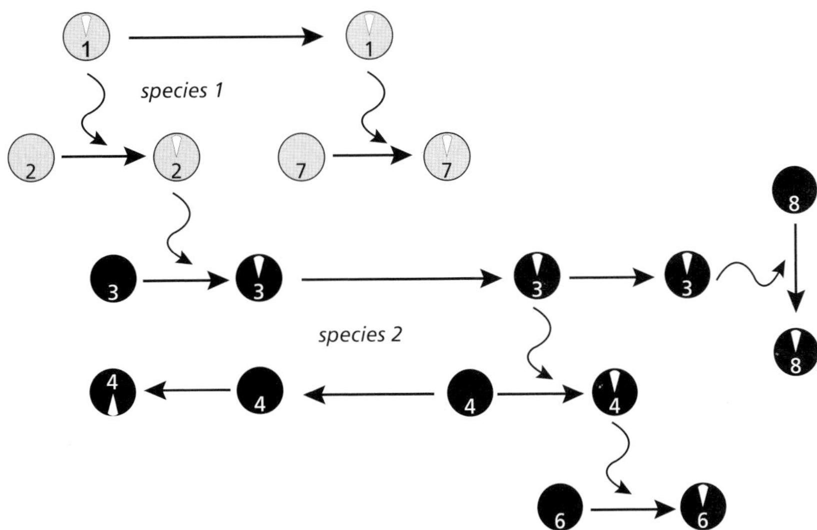

Fig. 7 A global gene pool in *Neisseria*. A particular allele arises in one strain (1) and is transmitted to other strains of the same species (2, 7) and of another species that shares the same global gene pool (3), where it continues to spread (4, 6, 8). Finding the same allele in two strains (e.g. 6 and 8) does not indicate that either is the donor and it is generally impossible to determine the donor of any allele imported by horizontal genetic exchange. Modified from Maiden *et al.* (1996).

the transferrin-binding complex. All the recombinants had imported a foreign *tbpB* allele, mostly from a protein family that differed from the one originally present in ancestral subgroup IV-1 strains. In a subsequent analysis (Zhu *et al.*, 2001), sequence variation was investigated in 500 subgroup III bacteria isolated between 1966 and 2000 from global sources during three pandemic waves. In addition to *tbpB*, *opaB*, *opaD* and *iga*, an IS*1106*A insertion element that maps near *opcA* and the *pgm* housekeeping gene encoding phosphoglucomutase were also investigated. All the genes were polymorphic and a total of 57 genotypes were identified. The results indicated that sequence variants arise continuously due to import, translocation (*opa* genes) and occasional mutations, resulting in a genocloud structure consisting of a frequent genotype and its descendent variants (Fig. 8). Sequence variants are often less fit than their parents and are lost due to competition and bottlenecks associated with epidemic spread (Morelli *et al.*, 1997; Zhu *et al.*,

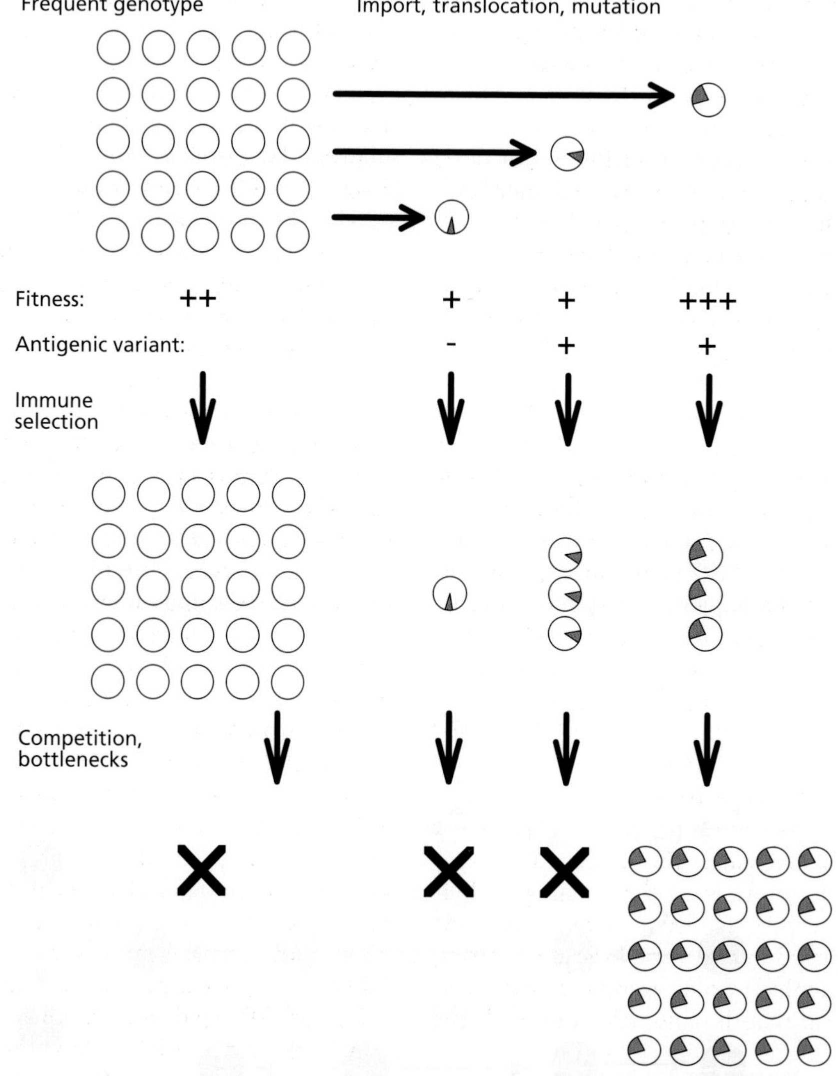

Fig. 8 Genocloud structure and replacement. Rare variants with different levels of fitness evolve by microevolution from frequent genotypes. The frequent genotype plus its variants constitutes a genocloud. Immune selection can result in amplification of rare antigenic variants, even if they are less fit than the frequent genotype. Competition for transmission and bottlenecks during transmission between different geographical areas purify populations and remove all less fit variants, including those that were temporarily selected by immune pressure. Thus, spread from country to country results in the frequent genotype being isolated repeatedly whereas the rare genotypes tend to be specific to particular countries. However, occasionally fitter variants arise that may also possess antigenic differences. These can out-compete the original frequent genotype, and competition and bottlenecks can then result in replacement of the original frequent genotype by a new frequent genotype. It then also gives rise to new variants, resulting in a new genocloud.

2001). Fitter variants result in new genoclouds and a total of nine genoclouds were found during the three decades of pandemic spread. The genoclouds differed from the parental genotype at up to four of the six loci and differed from each other by 1–6 loci. Each genocloud only seems to persist for a limited period of time and the oldest genoclouds have not been isolated in the last decade.

The results summarised above indicate that the apparent clonality (Bart *et al.*, 2001) of the epidemic subgroups reflects the narrow window of observation available to microbiologists. The epidemic subgroups are continuously evolving and genoclouds apparently exist for a maximum of only one or two decades before being replaced by fitter genoclouds from the same or a different subgroup. Sequence variants are most frequent for genes that are under diversifying selection, such as those encoding immunogenic outer membrane proteins. Neutral housekeeping genes also evolve, but more slowly. Because of the genocloud structure, most isolates from any one epidemic or pandemic are uniform. These purification mechanisms also ensure that genetic variation is either rapidly eliminated or fixed, leading to a progressive accumulation of random differences between genoclouds and between subgroups. In agreement, earlier analyses showed that most members of a subgroup and even closely related subgroups were uniform for epitopes on the PorA and PorB outer membrane porins and on pilin, with certain exceptions (Wang *et al.*, 1992). As might be predicted, however, certain of the epidemic subgroups possess distinctive *porA*, *opcA* and *iga* genes as well as distinctive pili and lipopolysaccharide (Suker *et al.*, 1994; Morelli *et al.*, 1997) and distinctive alleles at housekeeping genes (Fig. 6).

Epidemics in African Countries

The Berlin strain collection contains serogroup A meningococci isolated since the 1960s from epidemics in various countries of the Meningitis Belt (Olyhoek *et al.*, 1987). It also includes numerous meningococci isolated in two west African countries, Mali and the Gambia, from healthy carriers and sick patients during both epidemic and endemic conditions.

Most of the serogroup A meningococci isolated from diseased patients during an epidemic wave belong to the same genocloud. Members of a genocloud are indistinguishable by MLEE, MLST or PFGE (Olyhoek *et al.*, 1987; Crowe *et al.*, 1989; Guibourdenche *et al.*, 1996; Morelli *et al.*, 1997; Nicolas *et al.*, 2001) with only few exceptional isolates that are closely related to the frequent genotype. Temporally distinct epidemic waves were often due to different subgroups, however. Thus, serogroup A meningococci isolated from African epidemics in the 1960s and 1970s were subgroup I, epidemics in the 1980s were associated with subgroup IV-1 and epidemics since the late 1980s with subgroup III. The subgroup I epidemics were part of a pandemic wave in which these bacteria spread from the Mediterranean region to the rest of the globe (Olyhoek *et al.*, 1987). The subgroup III epidemics were part of two pandemic waves that have spread from east Asia (Zhu *et al.*, 2001).

Subgroup IV-1 not only caused epidemic disease in the 1980s but has also been routinely isolated from endemic disease in Africa. Subgroup IV-1 bacteria were also isolated from an epidemic otherwise associated with subgroup I (Niger, early 1960s) and from endemic disease where most of the isolated were serogroup C meningococci of the ET-37 complex (Mali, 1990–1992) (Wang *et al.*, 1993; Delvig *et al.*, 1995). Thus, these bacteria have been resident in Africa since at least the 1960s. Outside Africa, subgroup IV-1 has been isolated only in India (Wang *et al.*, 1992). These results indicate that epidemic waves are not only associated with the import of novel meningococci from other geographical regions but can also be caused by the multiplication of resident bacteria.

Although most cases of disease in any epidemic were caused by one subgroup, other, unrelated, meningococci were isolated from the same geographical areas. Swabbing exercises in 1983–1984, during and up to 1 year after an epidemic in the Gambia, showed that 15% of healthy individuals carried meningococci in their nasopharynx (Hassan-King *et al.*, 1988). The carriage rate dropped to about 1% thereafter and remained low for over a decade. Swabbing exercises in Mali between 1990 and 1993 during an endemic period when most disease was caused by serogroup C, ET-37 or, more rarely, serogroup A, subgroup IV-1 meningococci (unpublished data) also showed that only 1% of the familial contacts of patients carried meningococci in the nasopharynx (unpublished data). Similar low carriage rates were observed in 1999–2000 (S. Gagneux, personal communication) after a subgroup III epidemic in Ghana in 1997–1998 (Gagneux *et al.*, 2000). Thus, the carriage rate for meningococci seems to be very low in West Africa during endemic periods. Some of the meningococci isolated from healthy carriers are indistinguishable from those organisms that cause disease but multiple other unrelated clonal groupings have also been isolated. Half of the carrier isolates from the Gambia belonged to a few clonal groupings of serogroups W135, 29E and Y (Crowe *et al.*, 1989). Similarly, serogroup 29E, Y, and X meningococci of a limited number of clonal groupings were isolated in Mali (1990–1993) (unpublished) and X meningococci were isolated from Ghana (S. Gagneux, personal communication). The existence of these diverse clonal groupings among healthy carriers indicates that sufficient genetic diversity exists in Africa for occasional horizontal genetic exchange during co-colonisation in the healthy population. During epidemic periods the colonisation rate increases, as do opportunities for horizontal genetic exchange.

Serogroup C and W135 isolates of the ET-37 complex have been isolated from endemic disease (Wang *et al.*, 1993; Kwara *et al.*, 1998) and during serogroup A epidemics (Guibourdenche *et al.*, 1996). A major outbreak was caused by W135 ET-37 complex meningococci during the annual Haj pilgrimage to Saudi Arabia of 2000 (Popovic *et al.*, 2000; Taha *et al.*, 2000). Even the rare serogroups such as X have now also caused major outbreaks in West Africa (Etienne *et al.*, 1990; Campagne *et al.*, 1999). Thus, the former truism that outbreaks and epidemic disease in Africa are exclusively associated with serogroup A meningococci is no longer tenable.

Unresolved Questions

Very little is known about what causes epidemic disease. The majority of individuals encountering meningococci during an epidemic become healthy carriers; disease is a rare event. Co-infection with microorganisms that modulate the immune response has been implicated as a causal factor for epidemic disease (Griffiss, 1982; Cartwright *et al.*, 1991) but the importance of their role remains uncertain. It is equally unclear why most developed countries have been spared from epidemic meningococcal meningitis since World War II, especially since epidemic microorganisms have been imported into many of those countries.

Why does epidemic disease recur in waves in China and the Meningitis Belt? The simplest explanation would be the combination of the waning of herd immunity due to new births plus antibody decay combined with a lack of immunity to the novel antigens expressed by different clonal groupings (Moore, 1992). Similar explanations might also explain replacement of one genocloud by a second, such as has been recently documented for subgroup III in Africa (Zhu *et al.*, 2001). However, bactericidal tests with human sera did not distinguish between subgroups III and IV-1 (Achtman *et al.*, 1992) and the age distribution of epidemic disease in Africa is less biased toward small children than that of endemic disease in developed countries (Greenwood *et al.*, 1987; Salih *et al.*, 1990a).

Epidemic Disease Caused by *Vibrio cholerae*

Many of the patterns found for serogroup A meningococci may also apply to *Vibrio cholerae*. Unlike meningococci which exclusively colonise humans, *V. cholerae* have a large environmental reservoir and can be isolated from water as well as from crustaceans. Most attention has focused on epidemic strains, which until recently were thought always to possess the O1 LPS serogroup, and relatively little is known about other serotypes, usually from environmental sources, which were thought to lack epidemic potential. An epidemic caused by *V. cholerae* O139 (Nair *et al.*, 1996) and recent molecular genetic analyses of horizontal genetic exchange have stimulated a re-evaluation of genetic relationships among pathogenic *V. cholerae*.

Pandemics caused by *V. cholerae* have been subdivided into seven periods (1817–1823, 1829–1837, 1852–1860, 1863–1875, 1881–1896, 1899–1923, 1961–present). Recent pandemics have been caused by biotypes called 'classical' (fifth and sixth pandemics) and 'El Tor' (seventh pandemic) strains. The O139 bacteria from recent epidemics in India arose from El Tor strains through horizontal genetic exchange by substitution of a 35 kb region encoding O139 LPS for a 22 kb region encoding O1 LPS (Comstock *et al.*, 1996; Sozhamannan *et al.*, 1999). MLEE, ribotyping, RAPD and sequencing of housekeeping genes have shown that El Tor, Classical *V. cholerae* strains and *V. cholerae* O1 strains isolated in the Gulf states of the USA are closely related clonal groupings that differ at only a few loci (Karaolis *et al.*, 1994a; Karaolis *et al.*, 1995; Byun *et al.*, 1999). An additional clonal group has been found in the Amazon basin (Salles and Momen, 1991; Coelho *et al.*, 1995a) but these isolates have apparently not yet been investigated by sequence analysis. The differences between the clonal groupings seem to reflect evolution by recombination rather than by mutation (Byun *et al.*, 1999).

The extent of microevolution in *V. cholerae* has not yet been extensively investigated, although numerous hints exist that it occurs frequently (Mooi and Bik, 1997; Reidl and Mekalanos, 1995). The CTXϕ phage containing the *ctxA* and *ctxB* genes that encode cholera toxin subunits has apparently been introduced from separate sources into El Tor and classical strains (Boyd *et al.*, 2000). The classification of *V. cholerae* according to phylogenetic relationships has not yet become widely established and nomenclature remains associated with serological and other phenotypic properties. The basis for pandemic spread and geographical isolation remain as unexplained as for meningococci.

Endemic Disease Caused by *Mycobacterium tuberculosis*

Tuberculosis has always been one of the major life-threatening diseases in developing countries and it has been estimated that one-third of the global population has been infected by these bacteria (Bloom and Murray, 1992). AIDS and other immunocompromised patients are particularly susceptible to clinical tuberculosis. Together with a recent upsurge of antibiotic resistance among tubercle bacilli, the increased number of susceptible individuals has resulted in a resurgence of tuberculosis morbidity and mortality rates in developed countries, after decades of decreasing morbidity rates and effective antibiotic treatment (McKinney *et al.*, 1998). A mathematical model describing the intrinsic dynamics of tuberculosis (Blower *et al.*, 1995) indicates that waves of disease lasting over a century can result from increased morbidity rates, suggesting that tuberculosis will remain a major problem for decades to come.

M. tuberculosis is highly uniform (Sreevatsan *et al.*, 1997), as if a last common ancestor existed as recently as 15 000 years ago (Kapur *et al.*, 1994; Kremer *et al.*, 1999). In the absence of sufficient conserved diversity, it has not been possible to devise sturdy classification schemes and instead typing methods have been implemented for epidemiological analyses. Most analyses depend on DNA fingerprinting of the restriction sites flanking a variable number of IS6110 insertion elements (Van Embden *et al.*, 1993) (**Fig. 3**). The resolution of this method is low for strains possessing five or fewer IS6110 elements, and additional fingerprinting of other repetitive elements is used for finer classification (Braden, 1997; Burman *et al.*, 1997; Kamerbeek *et al.*, 1997). A number of different laboratories have data sets containing thousands of IS6110 fingerprints and are collaborating to improve recognition of individual strains which have spread across international barriers (Kremer *et al.*, 1999). It will probably not be possible to develop long-term epidemiological classification systems with these techniques, however, because the repetitive DNAs are unstable and changes accumulate relatively rapidly (Yeh *et al.*, 1998; de Boer *et al.*, 1999, 2000a). Also, gel-based technology is not inherently suitable for permanent classification systems.

Many or most isolates from any one location are distinct according to DNA fingerprinting (Yang *et al.*, 1995; Van Soolingen *et al.*, 1999) but occasional strains are isolated from several patients, particularly in some developing countries (Hermans *et al.*, 1995; Bifani *et al.*, 1997; Kiers *et al.*, 1997; Sola *et al.*, 1997; Kremer *et al.*, 1999). The speed of spread of such clonal outbreaks in immunocompromised patients and healthcare personnel can be very dramatic (Daley *et al.*, 1992). One large clonal outbreak in New York City, including variants which had acquired point mutations resulting in antibiotic resistance, has been described in greater detail (Bifani *et al.*, 1997). Fingerprinting has also helped retrospective epidemiological tracing of a cluster of cases in the Netherlands (Kiers *et al.*, 1997; Van Soolingen *et al.*, 1999). Three clonal groupings that have spread internationally have also been described (Kremer *et al.*, 1999).

Sequence analyses of regions involved in antibiotic resistance revealed two polymorphic sites in *katG* and *gyrA* which may characterise clones that have evolved in the last 15 000 years since the descent of *M. tuberculosis sensu stricto* from the last common ancestor of the *M. tuberculosis* complex (Sreevatsan *et al.*, 1997). It is interesting to note that 26% of the *M. tuberculosis* strains most closely related to that last common ancestor seemed to possess the same IS6110 at one particular chromosomal locus, suggesting that some IS6110 insertions can be stable over long periods of time.

The relevance to phylogenetic analysis of the available data is unclear because of the limited sequence variability within this species and the limits of the typing methods used. I wonder whether *M. tuberculosis sensu stricto* should not be considered to represent a clonal grouping rather than a species, similarly to other clonal groupings with species designations, such as *Shigella sonnei* (Karaolis *et al.*, 1994b) and *Y. pestis* (Achtman *et al.*, 1999b).

Molecular methods have now become very tightly integrated with traditional epidemiological tracing of tuberculosis and provide a paradigm for such integration at the local and international levels. In contrast, the different epidemiologies associated with acute disease and recurrent disease do not seem to have been subjected to detailed molecular analyses and the information on global patterns of spread is still scanty.

Endemic Disease Caused by Panmictic Bacteria

Molecular techniques are gaining increasing acceptance for tracing epidemiologically linked contacts. Spread between sexual partners of *N. gonorrhoeae* can be traced by fingerprinting of *opa* alleles (O'Rourke *et al.*, 1995). RAPD has been used to distinguish infection with a new strain of *H. pylori* from recrudescence after antibiotic therapy (van der Hulst *et al.*, 1997). In other analyses, RAPD was used to show that individuals can be colonised concurrently with multiple strains of *H. pylori* (Berg *et al.*, 1997; Kersulyte *et al.*, 1999). The frequency of clonal groupings in panmictic species (Gutjahr *et al.*, 1997) will become clearer as more strains are investigated, but it seems likely that classification methods will remain irrelevant to these bacteria because of their high frequency of horizontal genetic exchange. However, recent data have shown that despite free recombination in panmictic species such as *H. pylori* (Suerbaum *et al.*, 1998), populations from different geographical areas can be distinguished by multilocus sequences of housekeeping genes (Achtman *et al.*, 1999a). Apparently, recombination between different continents has not been sufficiently frequent in this species to eradicate phylogenetic signals due to former geographical isolation.

Macroevolution

A number of reviews have dealt with the molecular aspects of the evolution of virulence and antibiotic-resistance (Dowson *et al.*, 1994; Moxon *et al.*, 1994; Arthur *et al.*, 1996; Versalovic *et al.*, 1996; Virji, 1996; Groisman and Ochman, 1996, 1997; Hacker *et al.*, 1997; Musser and Krause, 1997). There is also an extensive older literature on transmissible plasmids and bacteriophages encoding antibiotic resistance and occasionally virulence factors (Bennett, 1995; Cheetham and Katz, 1995; Clewell *et al.*, 1995; Wilkins, 1995).

Clearly, acquisition of a large lysogenic bacteriophage, plasmid or pathogenicity island (Pai) (Hacker *et al.*, 1997) warrants use of the term macroevolution, rather than microevolution, even if it occurred in one step, and the expression of proteins encoded by the novel DNA might increase the efficiency of colonisation, spread or survival of the host bacteria. Macroevolution will also result after the import of novel chromosomal genes and operons by horizontal genetic exchange from unrelated species (Lawrence and Ochman, 1997; Ochman *et al.*, 2000). The relative frequency of integration of chromosomal DNA versus acquisition of plasmids and bacteriophages probably reflects the relative frequencies of transformation, conjugation and transduction within different species.

Certain bacterial diseases have first been recognised in recent decades (menstrual toxic shock syndrome, haemolytic uraemic syndrome, legionellosis) or have increased dramatically in frequency (streptococcal diseases, antibiotic resistance of staphylococci, enterococci, or tubercle bacilli), leading to considerable interest in 'emerging and re-emerging diseases'. Are these phenomena related to *recent* macroevolution of pathogenicity and spread of virulence determinants?

Pathogenicity Islands

The 40 kb Pais, SPI-1 and SPI-2, were apparently imported into *S. enterica* 100 million years ago, soon after speciation (Ochman and Groisman, 1996) and have since undergone considerable amounts of progressive evolution (Boyd *et al.*, 1997; Groisman and Ochman, 1997). Similarly, the *cag* PAI has been present in *H. pylori* for sufficient time to allow considerable microevolution (Censini *et al.*, 1996). The 35 kb LEE Pai is present in unrelated strains of *E. coli*, *Citrobacter freundii* and *Hafnia alvei* (McDaniel *et al.*, 1995), all isolated from gastrointestinal disease involving attaching and effacing lesions where the actin cytoskeleton of infected cells is re-modelled. Although it has been concluded that the LEE element was acquired by different clonal groupings in the recent past (McDaniel *et al.*, 1995), that acquisition was so long ago, at least thousands of years, that it pre-dated numerous subsequent events of microevolution (Reid *et al.*, 2000). Furthermore, the LEE element is integrated at different chromosomal sites in unrelated clonal groupings (Wieler *et al.*, 1997; Reid *et al.*, 2000), indicating that it was imported independently into those bacteria. Of the other Pais that are currently known, none seems to have been responsible for the emergence of major pathogenic clones in recent decades and it remains to be seen whether any emerging or re-emerging disease can be attributed to acquisition of a Pai.

Macroevolution of *Escherichia coli* O157:H7

Enteropathogenic strains of particular O:H (LPS:flagellar) serogroups were the first bacteria for which a clonal structure was recognised (Ørskov *et al.*, 1976; Ørskov and Ørskov, 1983). These serogroup designations continue to be used for *E. coli* isolated from enteric disease supplemented by acronyms which have attained a confusing level of diversity for the non-expert: enteropathogenic *E. coli* (EPEC), enteroaggregative *E. coli* (EAggEC), enteroinvasive *E. coli* (EIEC) and enterohaemorrhagic *E. coli* (EHEC) (see Chapter 70). The acronyms reflect the disease potential due to the expression of products encoded by the LEE element, by a plasmid (pili and EHEC haemolysin) and by a bacteriophage (shiga toxins 1 and 2 called Stx1, Stx2, formerly SLT or VT; see Calderwood *et al.* (1996). Shiga-toxin-producing *E. coli* (STEC) are variously referred to as EHEC, VTEC (verotoxin-producing *E. coli*), and formerly SLTEC (shiga-like-toxin-producing *E. coli*).

MLEE has been used to classify thousands of enteropathogenic strains, resulting in still more acronyms (DEC1–DEC15) (Whittam *et al.*, 1993). These various pathogenic bacteria can be readily accommodated within the MLEE scheme defined by the EcoR collection of non-pathogenic *E. coli* (Pupo *et al.*, 1997) but a species-wide classification scheme has not gained general acceptance. Only limited data from multilocus sequence typing are so far available (Pupo *et al.*, 2000; Reid *et al.*, 2000).

The use of acronyms which focus on disease source and on possession of variable DNA which can be readily transferred by horizontal genetic exchange obscures genetic relationships and makes it very difficult to obtain a global picture of the epidemiology of diarrhoeal disease. The lack of a uniform classification based on the diversity of *E. coli* further complicates issues. Current efforts also seem to concentrate on typing schemes based on PFGE in order to determine whether individual outbreaks have a common source (Izumiya *et al.*, 1997). Useful as such analyses are for public health officials concerned with tracing and stopping contaminated food sources, they will not contribute markedly to our understanding of the phylogenetics and long-term epidemiology of these bacteria.

Of the multitude of *E. coli* causing enteric disease, recent macroevolution has only been claimed for O157:H7 EHEC bacteria. *E. coli* O157:H7 was first identified as the cause of haemorrhagic colitis in 1983, after two outbreaks associated with fast-food beef in the US (Riley *et al.*, 1983). Soon thereafter, *E. coli* expressing a toxin similar to shiga toxin were implicated as a cause of haemolytic uraemic syndrome (Karmali *et al.*, 1985) and a large number of outbreaks caused by these bacteria have since been reported from diverse countries (for outbreaks up to 1993 see citations in USDA:APHIS (1994)).

The oldest known *E. coli* O157:H7 were isolated from bloody diarrhoea in humans in 1975 (Riley *et al.*, 1983) and from cattle in 1977 (Ørskov *et al.*, 1997). The first outbreaks attributed to these organisms occurred in 1982 (Riley *et al.*, 1983), and haemorrhagic colitis and haemolytic uraemic syndrome are considered to represent an emerging disease. Although other pathogenic *E. coli* isolated from enteric disease are genetically diverse, most O157:H7 strains belong to only two ETs (DEC3 and DEC4), which are almost identical to an ET including *E. coli* O55:H7 (Whittam *et al.*, 1993). *E. coli* O157:H7 and O55:H7 contain an *eae* gene encoding intimin and presumably all possess the LEE element (McDaniel *et al.*, 1995; Rodrigues *et al.*, 1996). Whittam has suggested that *E. coli* O157:H7 arose recently from *E. coli* O55:H7 by acquisition of a phage encoding Stx2 and a plasmid encoding adhesins, followed by subsequent evolution to yield other strains which cannot ferment sorbitol and which express Stx1 plus Stx2 (Whittam, 1995).

It seems conceivable that the macroevolution of *E. coli* O157:H7 occurred in the 1970s, shortly before the first outbreaks associated with those bacteria, and that recent evolution is responsible for the emergence of haemorrhagic colitis and haemolytic uraemic syndrome. The correlation is probably an artefact, however. O serogroups were assigned only to *E. coli* that were commonly isolated from disease, and only common isolates were investigated serologically (F. Ørskov, personal communication). The fact that QJ;*E. coli* O157 was first defined in 1972–1973 therefore indicates only that these bacteria rarely caused human disease before that period, not that they did not exist in healthy animals or people. Before the 1970s, in the absence of serotyping reagents, rare disease caused by *E. coli* O157:H7 would not have been noticed, and the date of evolution of O157:H7 bacteria might have been several thousand years ago. MLST data indicate that *E. coli* O157:H7 separated from *E. coli* K-12 about 4 million years ago (Reid *et al.*, 2000) and that virulence genes were acquired subsequently. Given the amount of sequence variation since acquisition of virulence genes, it seems highly likely that these events occurred thousands of years earlier than the earliest recorded outbreaks.

Dissociating the date of evolution from the earliest recorded outbreaks has precedents in other diseases. For example, menstrual toxic shock syndrome first reached epidemic proportions about 1980 and was strongly associated with one clonal grouping of *Staph. aureus* (Musser *et al.*, 1990b). The disease was, however, apparently induced by the use of particular tampons (Musser and Krause, 1997), rather than the evolution of a novel pathogen. Similarly, streptococcal scarlet fever and other diseases declined in frequency between 1910 and the 1980s in the USA and Europe before re-emerging in the 1980s (Musser and Krause, 1997). Recent streptococcal disease has often been associated with the so-called M1T1 clone. In contrast, the first historically documented outbreaks of plague in 541 CE may well reflect the evolution of *Y. pestis* from *Y. pseudotuberculosis* (Achtman *et al.*, 1999b) and the first epidemics of meningococcal meningitis in 1805 may reflect the evolution of *N. meningitidis* serogroup A (Achtman *et al.*, 2001).

Concluding Remarks

Leprosy, diphtheria, whooping cough, borelliosis, brucellosis and numerous other bacterial diseases with fascinating epidemiological features have not been mentioned here because I am not aware of extensive molecular epidemiological analyses that are relevant to a phylogenetic perspective. Similarly, only selected references illustrating a phylogenetic perspective have been cited, even for those few organisms that were selected for description. Evolutionary analyses have been included only where they illuminated aspects of molecular epidemiology, and the interested reader is referred to the extensive treatment of these topics by Spratt *et al.* (Chapter 20). Finally, the data on serogroup A meningococci have been described most extensively, partly because I feel most competent in this area. In spite of these shortcomings, it is hoped that this chapter offers useful insights for the practising and future molecular epidemiologist.

Molecular epidemiology needs to combine two partially contradictory approaches, classification and typing, in order to yield the maximal information about bacterial disease. Classification should be based on conservative properties that index the phylogenetic descent of the organisms, whereas typing should use the most sensitive methods available, including variable virulence determinants, to enable fine discrimination of individual isolates. In the absence of classification schemes, it will not be possible to accumulate a global over-view of changes in disease patterns; but lacking typing methods, we would never be able to discern the details of person-to-person transmission. The two concepts should be considered complementary, rather than alternative.

Future classification will certainly be based on extensive sequencing and chip-based methods that reflect sequence variation and typing will come to depend on methods that are more reliable and easier to interpret than current fingerprinting technologies. The conversion to these future technologies will depend on curators maintaining large, well-characterised strain collections and collaborative international analyses.

A large part of the literature deals with bacteria isolated in the USA and Europe although most bacterial disease occurs in the developing world. Obtaining representative material from the developing world continues to pose logistic problems and suffers from lack of adequate funding and communication. A considerable portion of the global diversity of bacterial pathogens is quite likely to be present in developing countries that have not yet been sampled extensively. The epidemiology of bacterial diseases in the developed world is not representative of the whole globe, and increasingly, international communications will result in spread and re-import of disease. Cholera and tuberculosis provide examples of global collaboration on international problems. Global interactions need to be extended to the whole gamut of bacterial disease, and local reference laboratories need increased interaction with and access to centres of international excellence where global strain collections and special expertise are available.

Progress in molecular epidemiology has been rapid in its short history of only two decades. Future progress will hopefully continue to accelerate such that this field can fulfil its promise in the current century.

Acknowledgements

I wish to thank Michael W. Heuzenroeder, Sebastian Suerbaum, Howard Ochman, Peter M. Small, Lothar Beutin and Giovanna Morelli for taking the time to read the first version of this manuscript and provide suggestions for its improvement. Several figures were kindly gifted: thanks to Christopher R. Braden for Fig.3, Arie van der Ende for Fig. 4 and Birgitta Duim for Fig. 5. I also gratefully acknowledge the patience of Max Sussman in waiting for this excessively tardy review.

References

Achtman M, Mercer A, Kusecek B *et al.* (1983) Six widespread bacterial clones among *Escherichia coli* K1 isolates. *Infect. Immun.* 39: 315–335.

Achtman M, Pluschke G (1986) Clonal analysis of descent and virulence among selected *Escherichia coli*. *Annu. Rev. Microbiol.* 40: 185–210.

Achtman M (1990) Molecular epidemiology of epidemic bacterial meningitis. *Rev. Med. Microbiol.* 1: 29–38.

Achtman M, Kusecek B, Morelli G *et al.* (1992) A comparison of the variable antigens expressed by clone IV-1 and subgroup III of *Neisseria meningitidis* serogroup A. *J. Infect. Dis.* 165: 53–68.

Achtman M (1994) Clonal spread of serogroup A meningococci. A paradigm for the analysis of microevolution in bacteria. *Mol. Microbiol.* 11: 15–22.

Achtman M (1995) Global epidemiology of meningococcal disease. In: Cartwright K (ed.) *Meningococcal Disease.* Chichester, UK: Wiley, pp. 159–175.

Achtman M (1996) A surfeit of YATMs? *J. Clin. Microbiol.* 34: 1870.

Achtman M, Azuma T, Berg DE *et al.* (1999a) Recombination and clonal groupings within *Helicobacter pylori* from different geographical regions. *Mol. Microbiol.* 32: 459–470.

Achtman M, Zurth K, Morelli G *et al.* (1999b) *Yersinia pestis*, the cause of plague, is a recently emerged clone of *Yersinia pseudotuberculosis*. *Proc. Natl Acad. Sci. USA* 96: 14043–14048.

Achtman M, Ende A van der, Zhu P *et al.* (2001) Molecular epidemiology of four successive waves of serogroup A meningitis in Moscow, Russia between 1969 and 1997. *Emerg. Infect. Dis.* 7: 420–427.

Akopyanz N, Bukanov NO, Westblom TU, Kresovich S, Berg DE (1992) DNA diversity among clinical isolates of *Helicobacter pylori* detected by PCR-based RAPD fingerprinting. *Nucleic Acids Res.* 20: 5137–5142.

Arnold C, Metherell L, Willshaw G, Maggs A, Stanley J (1999) Predictive fluorescent amplified-fragment length polymorphism analysis of *Escherichia coli*: high-resolution typing method with phylogenetic significance. *J. Clin. Microbiol.* 37: 1274–1279.

Arthur M, Reynolds P, Courvalin P (1996) Glycopeptide resistance in enterococci. *Trends Microbiol.* 4: 401–407.

Ashton FE, Ryan JA, Borczyk A *et al.* (1991) Emergence of a virulent clone of *Neisseria meningitidis* serotype 2a that is associated with meningococcal group C disease in Canada. *J. Clin. Microbiol.* 29: 2489–2493.

Bandelt H-J, Dress AWM (1992) Split decomposition: A new and useful approach to phylogenetic analysis of distance data. *Mol. Phylogenet. Evol.* 1: 242–252.

Bart A, Schuurman IGA, Achtman M *et al.* (1998) Random amplified polymorphic DNA (RAPD) genotyping of serogroup A meningococci yields similar results to multilocus enzyme electrophoresis (MEE) and reveals new genotypes. *J. Clin. Microbiol.* 36: 1746–1749.

Bart A, Barnabé C, Achtman M *et al.* (2001) Strong linkage disequilibrium between different genetic markers challenges the epidemic clonality model in Neisseria meningitidis serogroup A isolates. *Infection, Genetics and Evolution* in press.

Beall B, Cassiday PK, Sanden GN (1995) Analysis of *Bordetella pertussis* isolates from an epidemic by pulsed-field gel electrophoresis. *J. Clin. Microbiol.* 33: 3083–3086.

Bennekov T, Colding H, Ojeniyi B, Bentzon MW, Hoiby N (1996) Comparison of ribotyping and genome fingerprinting of *Pseudomonas aeruginosa* isolates from cystic fibrosis patients. *J. Clin. Microbiol.* 34: 202–204.

Bennett PM (1995) The spread of drug resistance. In: Baumberg S *et al.* (ed.) *Population genetics of bacteria.* Cambridge, Cambridge University Press, pp. 317–344.

Berg DE, Gilman RH, Lelwala-Guruge J *et al.* (1997) *Helicobacter pylori* populations in Peruvian patients. *Clin. Infect. Dis.* 25: 996–1002.

Bergthorsson U, Ochman H (1995) Heterogenity of genome sizes among natural isolates of *Escherichia coli. J. Bacteriol.* 177: 5784–5789.

Bifani PB, Plikaytis BB, Kapur V *et al.* (1997) Origin and interstate spread of a New York City multidrug-resistant *Mycobacterium tuberculosis* clone family. *JAMA* 275: 452–457.

Bionumerics (2000) Ver. 2.0. Kortrijk, Belgium: Applied Maths BVBA.

Bloom BR, Murray CJ (1992) Tuberculosis: commentary on a reemergent killer. *Science* 257: 1055–1064.

Blower SM, McLean AR, Porco TC *et al.* (1995) The intrinsic transmission dynamics of tuberculosis epidemics. *Nat. Med.* 1: 815–821.

Bootsma HJ, Dijk H Van, Vauterin P, Verhoef J, Mooi FR (2000) Genesis of BRO beta-lactamase-producing *Moraxella catarrhalis*: evidence for transformation-mediated horizontal transfer. *Mol. Microbiol.* 36: 93–104.

Boyd EF, Wang F-S, Whittam TS, Selander RK (1996) Molecular genetic relationships of the Salmonellae. *Appl. Environ. Microbiol.* 62: 804–808.

Boyd EF, Hartl DL (1997) Nonrandom location of IS1 elements in the genomes of natural isolates of *Escherichia coli. Mol. Biol. Evol.* 14: 725–732.

Boyd EF, Li J, Ochman H, Selander RK (1997) Comparative genetics of the *inv-spa* invasion gene complex of *Salmonella enterica. J. Bacteriol.* 179: 1985–1991.

Boyd EF, Heilpern AJ, Waldor MK (2000) Molecular analyses of a putative CTX precursor and evidence for independent acquisition of distinct CTXs by toxigenic *Vibrio cholerae. J. Bacteriol.* 182: 5530–5538.

Braden CR (1997) Current concepts in *Mycobacterium tuberculosis* DNA fingerprinting. *Infect. Dis. Clin. Pract.* 6: 89–95.

Burman WJ, Reves RR, Hawkes AP *et al.* (1997) DNA fingerprinting with two probes decreases clustering of *Mycobacterium tuberculosis. Am. J. Resp. Crit. Care Med.* 155: 1140–1146.

Byun R, Elbourne LD, Lan R, Reeves PR (1999) Evolutionary relationships of pathogenic clones of *Vibrio cholerae* by sequence analysis of four housekeeping genes. *Infect. Immun.* 67: 1116–1124.

Calderwood SB, Acheson DWK, Keusch GT *et al.* (1996) Proposed new nomeclature for SLT (VT) family. *ASM News* 62: 118–119.

Campagne G, Schuchat A, Djibo S *et al.* (1999) Epidemiology of bacterial meningitis in Niamey, Niger, 1981–96. *Bull. WHO* 77: 499–508.

Cartwright KAV, Stuart JM, Jones DM, Noah ND (1987) The Stonehouse survey: nasopharyngeal carriage of meningococci and *Neisseria lactamica. Epidemiol. Infect.* 99: 591–601.

Cartwright KAV, Jones DM, Smith AJ *et al.* (1991) Influenza A and meningococcal disease. *Lancet* 338: 554–557.

Caugant DA, Levin BR, Selander RK (1981) Genetic diversity and temporal variation in the *E. coli* population of a human host. *Genetics* 98: 467–490.

Caugant DA, Høiby EA, Magnus P *et al.* (1994) Asymptomatic carriage of *Neisseria meningitidis* in a randomly sampled population. *J. Clin. Microbiol.* 32: 323–330.

Caugant DA (1998) Population genetics and molecular epidemiology of *Neisseria meningitidis. Acta Pathol. Microbiol. Immunol. Scand. B* 106: 505–525.

Censini S, Lange C, Xiang Z *et al.* (1996) *cag*, a pathogenicity island of *Helicobacter pylori*, encodes type-I specific and disease-associated virulence factors. *Proc. Natl Acad. Sci. USA* 93: 14648–14653.

Cheetham BF, Katz ME (1995) A role for bacteriophages in the evolution and transfer of bacterial virulence determinants. *Mol. Microbiol.* 18: 201–208.

Clewell DB, Flannagan SE, Jaworski DD (1995) Unconstrained bacterial promiscuity: the Tn916-Tn1545 family of conjugative transposons. *Trends Microbiol.* 3: 229–236.

Coelho A, Andrade JRC, Vicente ACP, Salles CA (1995a) New variant of *Vibrio cholerae* O1 from clinical isolates in Amazonia. *J. Clin. Microbiol.* 33: 114–118.

Coelho A, Vicente ACP, Baptista MAS *et al.* (1995b) The distinction of pathogenic *Vibrio cholerae* groups using arbitrarily primed PCR fingerprints. *Res. Microbiol.* 146: 671–683.

Comstock LE, Johnson JA, Michalski JM, Morris JG Jr, Kaper JB (1996) Cloning and sequence of a region encoding a surface polysaccharide of *Vibrio cholerae* O139 and characterization of the insertion site in the chromosome of *Vibrio cholerae* O1. *Mol. Microbiol.* 19: 815–826.

Crowe BA, Wall RA, Kusecek B *et al.* (1989) Clonal and variable properties of *Neisseria meningitidis* isolated from cases and carriers during and after an epidemic in the Gambia, West Africa. *J. Infect. Dis.* 159: 686–700.

Daley CL, Small PM, Schecter GF *et al.* (1992) An outbreak of tuberculosis with accelerated progression among persons infected with the human immunodeficiency virus. An analysis using restriction-fragment-length polymorphisms. *N. Engl. J. Med.* 326: 231–235.

Datz M, Janetzki-Mittmann C, Franke S *et al.* (1996) Analysis of the enterohemorrhagic *Escherichia coli* O157 DNA region containing lambdoid phage gene p and Shiga-like toxin structural genes. *Appl. Environ. Microbiol.* 62: 791–797.

Davies RL, Arkinsaw S, Selander RK (1997) Evolutionary genetics of *Pasteurella haemolytica* isolates recovered from cattle and sheep. *Infect. Immun.* 65: 3585–3593.

de Boer AS, Borgdorff MW, Haas PE de *et al.* (1999) Analysis of rate of change of IS6110 RFLP patterns of *Mycobacterium tuberculosis* based on serial patient isolates. *J. Infect. Dis.* 180: 1238–1244.

de Boer AS, Kremer K, Borgdorff MW *et al.* (2000a) Genetic heterogeneity in *Mycobacterium tuberculosis* isolates reflected in IS6110 restriction fragment length polymorphism patterns as low-intensity bands. *J. Clin. Microbiol.* 38: 4478–4484.

de Boer P, Duim B, Rigter A, van Der PJ, Jacobs-Reitsma WF, Wagenaar JA (2000b) Computer-assisted analysis and epidemiological value of genotyping methods for *Campylobacter jejuni* and *Campylobacter coli*. *J. Clin. Microbiol.* 38: 1940–1946.

Delvig AA, Koumaré B, Glaser RW *et al.* (1995) Comparison of three human-murine heteromyeloma cell lines for formation of human hybridomas after electrofusion with human peripheral blood lymphocytes from meningococcal cases and carriers. *Hum. Antibod. Hybridom.* 6: 42–46.

Dempsey JAF, Wallace AB, Cannon JG (1995) The physical map of the chromosome of a serogroup A strain of *Neisseria meningitidis* shows complex rearrangements relative to the chromosomes of two mapped strains of the closely related species *N. gonorrhoeae*. *J. Bacteriol.* 177: 6390–6400.

Doignon-Bourcier F, Willems A, Coopman R *et al.* (2000) Genotypic characterization of *Bradyrhizobium* strains nodulating small Senegalese legumes by 16S–23S rRNA intergenic gene spacers and amplified fragment length polymorphism fingerprint analyses. *Appl. Environ. Microbiol.* 66: 3987–3997.

Dowson CG, Coffey TJ, Spratt BG (1994) Origin and molecular epidemiology of penicillin-binding-protein-mediated resistance to β-lactam antibiotics. *Trends Microbiol.* 2: 361–366.

Duim B, Vogel L, Puijk W *et al.* (1996) Fine mapping of outer membrane protein P2 antigenic sites which vary during persistent infection by *Haemophilus influenzae*. *Infect. Immun.* 64: 4673–4679.

Duim B, Bowler LD, Eijk PP *et al.* (1997) Molecular variation in the major outer membrane protein P5 gene of nonencapsulated *Haemophilus influenzae* during chronic infections. *Infect. Immun.* 65: 1351–1356.

Duim B, Ang CW, Van Belkum A *et al.* (2000) Amplified fragment length polymorphism analysis of *Campylobacter jejuni* strains isolated from chickens and from patients with gastroenteritis or Guillain–Barre or Miller Fisher syndrome. *Appl. Environ. Microbiol.* 66: 3917–3923.

Enright MC, Spratt BG (1998) A multilocus sequence typing scheme for *Streptococcus pneumoniae*: identification of clones associated with serious invasive disease. *Microbiology* 144: 3049–3060.

Enright MC, Fenoll A, Griffiths D, Spratt BG (1999) The three major Spanish clones of penicillin-resistant *Streptococcus pneumoniae* are the most common clones recovered in recent cases of meningitis in Spain. *J. Clin. Microbiol.* 37: 3210–3216.

Etienne J, Sperber G, Adamou A, Picq JJ (1990) Epidemiological notes: meningococcal meningitis of serogroup X in Niamey (Niger). *Med. Trop.l (Marseilles)* 50: 227–229.

Feavers IM, Gray SJ, Urwin R *et al.* (1999) Multilocus sequence typing and antigen gene sequencing in the investigation of a meningococcal disease outbreak. *J. Clin. Microbiol.* 37: 3883–3887.

Feil EJ, Maiden MC, Achtman M, Spratt BG (1999) The relative contribution of recombination and mutation to the divergence of clones of *Neisseria meningitidis*. *Mol. Biol. Evol.* 16: 1496–1502.

Feil EJ, Holmes EC, Bessen DE *et al.* (2001) Recombination within natural populations of pathogenic bacteria: short-term empirical estimates and long-term phylogenetic comparisons. *Proc. Natl Acad Sci. USA*: 98: 182–187.

Feil EJ, Maynard Smith J, Enright MC, Spratt BG (2000) Estimating recombinational parameters in *Streptococcus pneumoniae* from multilocus sequence typing data. *Genetics* 154: 1439–1450.

Frøholm LO, Kolstøh AB, Berner JM, Caugant DA (2000) Genomic rearrangements in *Neisseria meningitidis* strains of the ET-5 complex. *Curr. Microbiol.* 40: 372–379.

Gagneux S, Hodgson A, Erhard I *et al.* (2000) Microheterogeneity of serogroup A (subgroup III) *Neisseria meningitidis* during an outbreak in northern Ghana. *Trop. Med. Internat. Health* 5: 280–287.

Gibbs CP, Meyer TF (1996) Genome plasticity in *Neisseria gonorrhoeae*. *FEMS Microbiol. Lett.* 145: 173–179.

Go MF, Kapur V, Graham DY, Musser JM (1996) Population genetic analysis of *Helicobacter pylori* by multilocus enzyme electrophoresis: extensive allelic diversity and recombinational population structure. *J. Bacteriol.* 178: 3934–3938.

Greenwood BM (1987) The epidemiology of acute bacterial meningitis in tropical Africa. In: Williams JD, Burnie J (eds) *Bacterial Meningitis*. London : Academic Press, pp. 61–91.

Greenwood BM, Greenwood AM, Bradley AK *et al.* (1987) Factors influencing susceptibility to meningococcal disease during an epidemic in the Gambia, West Africa. *J. Infect.* 14: 167–184.

Griffiss JM (1982) Epidemic meningococcal disease: Synthesis of a hypothetical immunoepidemiologic model. *Rev. Infect. Dis.* 4: 159–172.

Groisman EA, Ochman H (1996) Pathogenicity islands: bacterial evolution in quantum leaps. *Cell* 87: 791–794.

Groisman EA, Ochman H (1997) How *Salmonella* became a pathogen. *Trends Microbiol.* 5: 343–349.

Guibourdenche M, Høiby EA, Riou J-Y *et al.* (1996) Epidemics of serogroup A *Neisseria meningitidis* of subgroup III in Africa, 1989–1994. *Epidemiol. Infect.* 116: 115–120.

Guibourdenche M, Giorgini D, Guèye A *et al.* (1997) Genetic analysis of meningococcal population using polymorphism of *pilA/pilB* locus: a molecular approach for meningococcal epidemiology. *J. Clin. Microbiol.* 35: 745–750.

Gutjahr TS, O'Rourke M, Ison CA, Spratt BG (1997) Arginine-, hypoxanthine-, uracil-requiring isolates of *Neisseria gonorrhoeae* are a clonal lineage within a non-clonal population. *Microbiology* 143: 633–640.

Guttman DS (1997) Recombination and clonality in natural populations of *Escherichia coli*. *Tree* 12: 16–22.

Guttman DS, Dykhuizen DE (1994) Clonal divergence in *Escherichia coli* as a result of recombination, not mutation. *Science* 266: 1380–1383.

Hacker J, Blum-Oehler G, Mühldorfer I, Tschäpe H (1997) Pathogenicity islands of virulent bacteria: Structure, function and impact on microbial evolution. *Mol. Microbiol.* 23: 1089–1097.

Hall LMC, Whiley RA, Duke B, George RC, Efstratiou A (1996) Genetic relatedness within and between serotypes of *Streptococcus pneumoniae* from the United Kingdom: Analysis of multilocus enzyme electrophoresis, pulsed-field gel electrophoresis, and antimicrobial resistance patterns. *J. Clin. Microbiol.* 34: 853–859.

Hall RM, Collis CM (1995) Mobile gene cassettes and integrons: Capture and spread of genes by site-specific recombination. *Mol. Microbiol.* 15: 593–600.

Hassan-King MKA, Wall RA, Greenwood BM (1988) Meningococcal carriage, meningococcal disease and vaccination. *J. Infect.* 16: 55–59.

Haubold B, Travisano M, Rainey PB, Hudson RR (1998) Detecting linkage disequilibrium in bacterial populations. *Genetics* 150: 1341–1348.

Hermans PWM, Messadi F, Guebrexabher H *et al.* (1995) Analysis of the population structure of *Mycobacterium tuberculosis* in Ethiopia, Tunisia and the Netherlands: Usefulness of DNA typing for global tuberculosis epidemiology. *J. Infect. Dis.* 171: 1504–1513.

Huson DH (1998) SplitsTree: Analyzing and visualizing evolutionary data. *Bioinformatics* 14: 68–73.

Izumiya H, Terajima J, Wada A *et al.* (1997) Molecular typing of enterhemorrhagic *Escherichia coli* O157:H7 isolated in Japan by using pulsed-field gel electrophoresis. *J. Clin. Microbiol.* 35: 1675–1680.

Jiang Q, Hiratsuka K, Taylor DE (1996) Variability of gene order in different *Helicobacter pylori* strains contributes to genome diversity. *Mol. Microbiol.* 20: 833–842.

Jonas D, Meyer HG, Matthes P *et al.* (2000) Comparative evaluation of three different genotyping methods for investigation of nosocomial outbreaks of Legionnaires' disease in hospitals. *J. Clin. Microbiol.* 38: 2284–2291.

Kamerbeek J, Schouls L, Kolk A *et al.* (1997) Simultaneous detection and strain differentiation of *Mycobacterium tuberculosis* for diagnosis and epidemiology. *J. Clin. Microbiol.* 35: 907–914.

Kapur V, Whittam TS, Musser JM (1994) Is *Mycobacterium tuberculosis* 15,000 years old. *J. Infect. Dis.* 170: 1348–1349.

Karaolis DK, Lan R, Reeves PR (1994a) Molecular evolution of the seventh-pandemic clone of *Vibrio cholerae* and its relationship to other pandemic and epidemic *V. cholerae* isolates. *J. Bacteriol.* 176: 6199–6206.

Karaolis DKR, Lan R, Reeves PR (1994b) Sequence variation in *Shigella sonnei* (Sonnei), a pathogenic clone of *Escherichia coli*, over four continents and 41 years. *J. Clin. Microbiol.* 32: 796–802.

Karaolis DK, Lan R, Reeves PR (1995) The sixth and seventh cholera pandemics are due to independent clones separately derived from environmental, nontoxigenic, non-O1 Vibrio cholerae. *J. Bacteriol.* 177: 3191–3198.

Karch H, Russmann H, Schmidt H, Schwarzkopf A, Heesemann J (1995) Long-term shedding and clonal turnover of enterohemorrhagic *Escherichia coli* O157 in diarrheal diseases. *J. Clin. Microbiol.* 33: 1602–1605.

Karmali MA, Petric M, Lim C *et al.* (1985) The association between idiopathic hemolytic uremic syndrome and infection by verotoxin-producing *Escherichia coli*. *J. Infect. Dis.* 151: 775–782.

Keim P, Kalif A, Schupp J *et al.* (1997) Molecular evolution and diversity in *Bacillus anthracis* as detected by amplified fragment length polymorphism markers. *J. Bacteriol.* 179: 818–824.

Kersulyte D, Chalkauskas H, Berg DE (1999) Emergence of recombinant strains of *Helicobacter pylori* during human infection. *Mol. Microbiol.* 31: 31–43.

Kiers A, Drost AP, Van Soolingen D, Veen J (1997) Sherlock Holmes and DNA fingerprinting. Use of DNA fingerprinting in international source case finding during a large outbreak of tuberculosis in the Netherlands. *Internat. J. Lung Dis.* 1: 239–245.

Klee SR, Nassif X, Kusecek B *et al.* (2000) Molecular and biological analysis of eight genetic islands that distinguish *Neisseria meningitidis* from the closely related pathogen *Neisseria gonorrhoeae*. *Infect. Immun.* 68: 2082–2095.

Kremer K, Soolingen D Van, Frothingham R *et al.* (1999) Comparison of methods based on different molecular epidemiological markers for typing of *Mycobacterium tuberculosis* complex strains: Interlaboratory study of discriminatory power and reproducibility. *J. Clin. Microbiol.* 37: 2607–2618.

Kriz P, Giorgini D, Musilek M, Larribe M, Taha MK (1999) Microevolution through DNA exchange among strains of *Neisseria meningitidis* isolated during an outbreak in the Czech Republic. *Res. Microbiol.* 150: 273–280.

Kroll JS, Moxon ER (1990) Capsulation in distantly related strains of *Haemophilus influenzae* tybe b: Genetic drift and gene transfer at the capsulation locus. *J. Bacteriol.* 172: 1374–1379.

Kwara A, Adegbola RA, Corrah PT *et al.* (1998) Meningitis caused by a serogroup W135 clone of the ET-37 complex of *Neisseria meningitidis* in West Africa. *Trop. Med. Internat. Health* 3: 742–746.

Lapeyssonnie L (1963) La méningite cérébrospinale en Afrique. *Bull. WHO* 28 (Suppl): 53–114.

Lawrence JG, Ochman H (1997) Amelioration of bacterial genomes: rates of change and exchange. *J. Mol. Evol.* 44: 383–397.

Lin AW, Usera MA, Barrett TJ, Goldsby RA (1996) Application of random amplified polymorphic DNA analysis to differentiate strains of *Salmonella enteritidis*. *J. Clin. Microbiol.* 34: 870–876.

Linton CJ, Jalal H, Leeming JP, Millar MR (1994) Rapid discrimination of *Mycobacterium tuberculosis* strains by random amplified polymorphic DNA analysis. *J. Clin. Microbiol.* 32: 2169–2174.

Linz B, Schenker M, Zhu P, Achtman M (2000) Frequent interspecific genetic exchange between commensal neisseriae and *Neisseria meningitidis*. *Mol. Microbiol.* 36: 1049–1058.

Liu S-L, Sanderson KE (1995) Rearrangements in the genome of the bacterium *Salmonella typhi*. *Proc. Natl Acad. Sci. USA* 92: 1018–1022.

Louie M, Jayaratne P, Luchsinger I *et al.* (1996) Comparison of ribotyping, arbitrarily primed PCR, and pulsed-field gel electrophoresis for molecular typing of *Listeria monocytogenes*. *J. Clin. Microbiol.* 34: 15–19.

Mahenthiralingam E, Campbell ME, Foster J, Lam JS, Speert DP (1996a) Random amplified polymorphic DNA typing of *Pseudomonas aeruginosa* isolates recovered from patients with cystic fibrosis. *J. Clin. Microbiol.* 34: 1129–1135.

Mahenthiralingam E, Campbell ME, Henry DA, Speert DP (1996b) Epidemiology of *Burkholderia cepacia* infection in patients with cystic fibrosis: Analysis by randomly amplified polymorphic DNA fingerprinting. *J. Clin. Microbiol.* 34: 2914–2920.

Maiden MCJ, Malorny B, Achtman M (1996) A global gene pool in the neisseriae. *Mol. Microbiol.* 21:1297–1298.

Maiden MCJ, Bygraves JA, Feil E *et al.* (1998) Multilocus sequence typing: a portable approach to the identification of clones within populations of pathogenic micro-organisms. *Proc. Natl Acad. Sci. USA* 95: 3140–3145.

Makino S-I, Okada Y, Maruyama T, Kaneko S, Sasakawa C (1994) PCR-based random amplified polymorphic DNA fingerprinting of *Yersinia pseudotuberculosis* and its practical applications. *J. Clin. Microbiol.* 32:65–69.

Maynard Smith J, Smith NH, O'Rourke M, Spratt BG (1993) How clonal are bacteria? *Proc. Natl Acad. Sci. USA* 90: 4384–4388.

McDaniel TK, Jarvis KG, Donnenberg MS, Kaper JB (1995) A genetic locus of enterocyte effacement conserved among diverse enterobacterial pathogens. *Proc. Natl Acad. Sci. USA* 92: 1664–1668.

McKinney JD, Jacobs WR Jr, Bloom BR (1998) Persisting problems in tuberculosis. In: Krause RM (ed.) *Emerging Infections.* New York: Academic Press, pp. 51–146.

Milkman R, Bridges MM (1993) Molecular evolution of the *Escherichia coli* chromosome. IV. Sequence comparisons. *Genetics* 133: 455–468.

Milkman R, McKane M (1995) DNA sequence variation and recombination in *E. coli*. In: Baumberg S *et al.* (ed.) *Population Genetics of Bacteria.* Cambridge: Cambridge University press, pp. 127–142.

Mooi FR, Bik EM (1997) The evolution of epidemic *Vibrio cholerae* strains. *Trends Microbiol.* 5: 161–165.

Moore PS (1992) Meningococcal meningitis in sub-Saharan Africa: A model for the epidemic process. *Clin. Infect. Dis.* 14: 515–525.

Morelli G, Malorny B, Müller K *et al.* (1997) Clonal descent and microevolution of *Neisseria meningitidis* during 30 years of epidemic spread. *Mol. Microbiol.* 25: 1047–1064.

Morona JK, Miller DC, Coffey TJ *et al.* (1999) Molecular and genetic characterization of the capsule biosynthesis locus of *Streptococcus pneumoniae* type 23F. *Microbiology* 145: 781–789.

Moxon ER, Rainey PB, Nowak MA, Lenski RE (1994) Adaptive evolution of highly mutable loci in pathogenic bacteria. *Curr Biol.* 4: 24–33.

Mpairwe Y, Matovu HL (1971) Cerebrospinal meningitis in East Africa 1911–1965. *Trans. Roy. Soc. Trop. Med. Hyg.* 65: 70–75.

Musser JM, Krause RM (1997) The revival of group A streptococcal diseases, with a commentary on staphylococcal toxic shock syndrome. In: Krause RM (ed) *Emerging Infections*. San Diego, CA: Academic Press, pp. 185–218.

Musser JM, Kroll JS, Granoff DM *et al.* (1990a) Global genetic structure and molecular epidemiology of encapsulated *Haemophilus influenzae*. *Rev. Infect. Dis.* 12: 75–111.

Musser JM, Schlievert PM, Chow AW *et al.* (1990b) A single clone of *Staphylococcus aureus* causes the majority of cases of toxic shock syndrome. *Proc. Natl Acad. Sci. USA* 87: 225–229.

Nair CC, Albert MJ, Shimada T, Takeda Y (1996) *Vibrio cholerae* O139 Bengal: the new serogroup causing cholera. *Rev. Med. Microbiol.* 7: 43–51.

Nair S, Schreiber E, Thong KL, Pang T, Altwegg M (2000) Genotypic characterization of *Salmonella typhi* by amplified fragment length polymorphism fingerprinting provides increased discrimination as compared to pulsed-field gel electrophoresis and ribotyping. *J. Microbiol. Meth.* 41: 35–43.

Nicolas P, Décousset L, Riglet V *et al.* (2001) Multilocus sequence typing of serogroup A meningococci isolated in Africa between 1988 and 1999, shows the clonal expansion of ST5 and emergence of ST7.*Emerg. Infect. Dis.* in press.

O'Rourke M, Spratt BG (1994) Further evidence for the non-clonal population structure of *Neisseria gonorrhoeae*: Extensive genetic diversity within isolates of the same electrophoretic-type. *J. Gen. Microbiol.* 140: 1285–1290.

O'Rourke M, Ison CA, Renton AM, Spratt BG (1995) Opa-typing: a high-resolution tool for studying the epidemiology of gonorrhoea. *Mol. Microbiol.* 17: 865–875.

Ochman H, Selander RK (1984) Standard reference strains of *Escherichia coli* from natural populations. *J. Bacteriol.* 157: 690–693.

Ochman H, Groisman EA (1996) Distribution of pathogenicity islands in *Salmonella* spp. *Infect. Immun.* 64: 5410–5412.

Ochman H, Lawrence JG (1996) Phylogenetics and the amelioration of bacterial genomes. In: Neidhardt FC *et al.* (ed.) *Escherichia coli* and *Salmonella typhimurium*: *Cellular* and *Molecular Biology*, 2nd edn. Washington, DC: American Society for Microbiology, pp. 2627–2637.

Ochman H, Lawrence JG, Groisman EA (2000) Lateral gene transfer and the nature of bacterial innovation. *Nature (London)* 405: 299–304.

Olyhoek T, Crowe BA, Achtman M (1987) Clonal population structure of *Neisseria meningitidis* serogroup A isolated from epidemics and pandemics between 1915 and 1983. *Rev. Infect. Dis.* 9: 665–692.

Ørskov F, Ørskov I, Evans DJ Jr, Sack RB, Sack DA, Wadström T (1976) Special *Escherichia coli* serotypes among enterotoxigenic strains from diarrhoea in adults and children. *Med. Microbiol. Immunol.* 162: 73–80.

Ørskov F, Ørskov I (1983) Summary of a workshop on the clone concept in the epidemiology, evolution of the Enterobacteriaceae and other bacteria. *J. Infect. Dis.* 148: 346–357.

Ørskov F, Wilson RA (1993) Clonal relationships among *Escherichia coli* strains that cause hemorrhagic colitis and infantile diarrhea. *Infect. Immun.* 61: 1619–1629.

Ørskov F, Ørskov I, Villar JA (1997) Cattle as reservoir of verotoxin-producing *Escherichia coli* O157:H7. *Lancet* ii: 276.

Peltola H (1983) Meningococcal disease: Still with us. *Rev. Infect. Dis.* 5: 71–91.

Popovic T, Bopp C, Olsvik O, Wachsmuth K (1993) Epidemiologic application of a standardized ribotype scheme for *Vibrio cholerae* O1. *J. Clin. Microbiol.* 31: 2474–2482.

Popovic T, Sacchi CT, Reeves MW *et al.* (2000) *Neisseria meningitidis* serogroup W135 isolates associated with the ET- 37 complex. *Emerg. Infect. Dis.* 6: 428–429.

Poussier S, Trigalet-Demery D, Vandewalle P *et al.* (2000) Genetic diversity of *Ralstonia solanacearum* as assessed by PCR-RFLP of the *hrp* gene region, AFLP and 16S rRNA sequence analysis, and identification of an African subdivision. *Microbiology* 146(7): 1679–1692.

Pupo GM, Karaolis DKR, Lan RT, Reeves PR (1997) Evolutionary relationships among pathogenic and nonpathogenic *Escherichia coli* strains inferred from multilocus enzyme electrophoresis and *mdh* sequence studies. *Infect. Immun.* 65: 2685–2692.

Pupo GM, Lan R, Reeves PR (2000) Multiple independent origins of Shigella clones of *Escherichia coli* and convergent evolution of many of their characteristics. *Proc. Natl Acad. Sci. USA* 97: 10567–10572.

Reeves P (1993) Evolution of *Salmonella* O antigen variation by interspecific gene transfer on a large scale. *Trends Genet.* 9: 17–22.

Reichmann P, König A, Liñares J *et al.* (1997) A global gene pool for high-level cephalosporin resistance in commensal *Streptococcus* species and *Streptococcus pneumoniae*. *J. Infect. Dis.* 176: 1001–1012.

Reid SD, Herbelin CJ, Bumbaugh AC, Selander RK, Whittam TS (2000) Parallel evolution of virulence in pathogenic *Escherichia coli*. *Nature (London)* 406: 64–67.

Reidl J, Mekalanos JJ (1995) Characterization of *Vibrio cholerae* bacteriophage K139 and use of a novel mini-transposon to identify a phage-encoded virulence factor. *Mol. Microbiol.* 18: 685–701.

Riley LW, Remis RS, Helgerson SD *et al.* (1983) Hemorrhagic colitis associated with a rare *Escherichia coli* serotype. *N. Engl. J. Med.* 308: 681–685.

Rodrigues J, Scaletsky ICA, Campos LC *et al.* (1996) Clonal structure and virulence factors in strains of *Escherichia coli* of the classic serogroup O55. *Infect. Immun.* 64: 2680–2686.

Salih MAM, Ahmed HS, Karrar ZA *et al.* (1990a) Features of a large epidemic of group A meningococcal meningitis in Khartoum, Sudan in 1988. *Scand J. Infect. Dis.* 22: 161–170.

Salih MAM, Danielsson D, Bäckman A *et al.* (1990b) Characterization of epidemic and non-epidemic *Neisseria meningitidis* serogroup A strains from Sudan and Sweden. *J. Clin. Microbiol.* 28: 1711–1719.

Salles CA, Momen H (1991) Identification of *Vibrio cholerae* by enzyme electrophoresis. *Trans. Roy. Soc. Trop. Med. Hyg.* 85: 544–547.

Savelkoul PHM, Aarts HJM, Haas J de *et al.* (1999) Amplified-fragment length polymorphism analysis: the state of an art. *J. Clin. Microbiol.* 37: 3083–3091.

Seiler A, Reinhardt R, Sarkari J, Caugant DA, Achtman M (1996) Allelic polymorphism and site-specific recombination in the *opc* locus of *Neisseria meningitidis. Mol. Microbiol.* 19: 841–856.

Selander RK, Levin BR (1980) Genetic diversity and structure in *Escherichia coli* populations. *Science* 210: 545–547.

Selander RK, Caugant DA, Ochman H *et al.* (1986) Methods of multilocus enzyme electrophoresis for bacterial population genetics and systematics. *Appl. Environ. Microbiol.* 51: 873–884.

Selander RK, Li J, Nelson K (1996) Evolutionary genetics of *Salmonella enterica.* In: Curtiss R III *et al.* (ed.) *Escherichia coli and Salmonella.* 2nd edn, Washington, DC: ASM Press, vol. 2R, pp. 2691–2707.

Shepherd JG, Wang L, Reeves PR (2000) Comparison of O antigen gene clusters of *Escherichia coli* (*Shigella*) Sonnei and *Plesiomonas shigelloides* O17: Sonnei gained its current plasmid-borne O antigen genes from *P. shigelloides* in a recent event. *Infect. Immun.* 68: 6056–6061.

Smith NH, Maynard Smith J, Spratt BG (1995) Sequence evolution of the *porB* gene of *Neisseria gonorrhoeae* and *Neisseria meningitidis*: Evidence of positive Darwinian selection. *Mol. Biol. Evol.* 12: 363–370.

Smith-Vaughan HC, Sriprakash KS, Mathews JD, Kemp DJ (1995) Long PCR-ribotyping of nontypable *Haemophilus influenzae. J. Clin. Microbiol.* 33: 1192–1195.

Smith-Vaughan HC, Sriprakash IS, Mathews JD, Kemp DJ (1997) Nonencapsulated *Haemophilus influenzae* in aboriginal infants with otitis media: Prolonged carriage of P2 porin variants and evidence for horizontal P2 gene transfer. *Infect. Immun.* 65: 1468–1474.

Sneath PHA, Sokal RR (1978) *Numerical Taxonomy.* San Francisco: W.H. Freeman.

Sola C, Horgen L, Goh KS, Rastogi N (1997) Molecular fingerprinting of *Mycobacterium tuberculosis* on a Caribbean island with IS6110 and DRr probes. *J. Clin. Microbiol.* 35: 843–846.

Souza V, Nguyen TT, Hudson RR, Pinero D, Lenski RE (1992) Hierarchical analysis of linkage disequilibrium in *Rhizobium* populations: evidence for sex? *Proc. Natl Acad. Sci. USA* 89: 8389–8393.

Sozhamannan S, Deng YK, Li M *et al.* (1999) Cloning and sequencing of the genes downstream of the *wbf* gene cluster of *Vibrio cholerae* serogroup O139 and analysis of the junction genes in other serogroups. *Infect. Immun* . 67: 5033–5040.

Sreevatsan S, Pan X, Stockbauer K *et al.* (1997) Restricted structural gene polymorphism in the *Mycobacterium tuberculosis* complex indicates evolutionarily recent global dissemination. *Proc. Natl Acad. Sci. USA* 94: 9869–9874.

Sriprakash KS, Gardiner DL (1997) Lack of polymorphism within the rRNA operons of group A streptococci. *Mol. Gen. Genet.* 255: 125–130.

Suerbaum S, Maynard Smith J, Bapumia K *et al.* (1998) Free recombination within *Helicobacter pylori. Proc. Natl Acad. Sci. USA* 95: 12619–12624.

Suker J, Feavers IM, Achtman M *et al.* (1994) The *porA* gene in serogroup A meningococci: evolutionary stability and mechanism of genetic variation. *Mol. Microbiol.* 12: 253–265.

Swaminathan B, Matar GM, Reeves MW *et al.* (1996) Molecular subtyping of *Neisseria meningitidis* serogroup B: Comparison of five methods. *J. Clin. Microbiol.* 34: 1468–1473.

Swartley JS, Marfin AA, Edupuganti S *et al.* (1997) Capsule switching of *Neisseria meningitidis. Proc. Natl Acad. Sci. USA* 94: 271–276.

Swartley JS, Liu LJ, Miller YK *et al.* (1998) Characterization of the gene cassette required for biosynthesis of the (1 → 6)-linked *N*-acetyl-D-mannosamine-1-phosphate capsule of serogroup A *Neisseria meningitidis. J. Bacteriol.* 180: 1533–1539.

Taha M-K, Achtman M, Greenwood B *et al.* (2000) Serogroup W135 meningococcal disease in Hajj pilgrims. Implications for surveillance and vaccination. *Lancet* 356: 2159.

Tenover FC, Arbeit RD, Goering RV *et al.* (1995) Interpreting chromosomal DNA restriction patterns produced by pulsed-field gel electrophoresis: criteria for bacterial strain typing. *J. Clin. Microbiol.* 33: 2233–2239.

Thyssen A, Eygen S Van, Hauben L *et al.* (2000) Application of AFLP for taxonomic and epidemiological studies of *Photobacterium damselae* subsp. *piscicida. Internat. J. Syst. Evol. Microbiol.* 50(3): 1013–1019.

Tikhomirov E, Santamaria M, Esteves K (1997) Meningococcal disease: public health burden and control. *World Health Statistics Quarterly* 50: 170–176.

USDA: APHIS (1994) *Escherichia coli O157: H7.* Fort Collins, CO: Centers for Epidemiology and Animal Health.

Van Belkum A, Leeuwen W Van, Verkooyen R *et al.* (1997) Dissemination of a single clone of methicillin-resistant *Staphylococcus aureus* among Turkish hospitals. *J. Clin. Microbiol.* 35: 978–981.

van der Hulst RWM, Rauws EAJ, Köycü B *et al.* (1997) *Helicobacter pylori* reinfection is virtually absent after successful eradication. *J. Infect. Dis.* 176: 196–200.

Van Embden JDA, Cave MD, Crawford JT *et al.* (1993) Strain identification of *Mycobacterium tuberculosis* by DNA fingerprinting: recommendation for a standardized methodology. *J. Clin. Microbiol.* 31: 406–409.

Van Soolingen D, Haas PEW De, Haagsma J *et al.* (1994) Use of various genetic markers in differentiation of *Mycobacterium bovis* strains from animals and humans and for studying epidemiology of bovine tuberculosis. *J. Clin. Microbiol.* 32: 2425–2433.

Van Soolingen D, Borgdorff MW, Haas de PE *et al.* (1999) Molecular epidemiology of tuberculosis in the Netherlands: a nationwide study from 1993 through 1997. *J. Infect. Dis.* 180: 726–736.

Versalovic J, Kapur V, Mason EO Jr *et al.* (1993) Penicillin-resistant *Streptococcus pneumoniae* strains recovered in Houston: Identification and molecular characterization of multiple clones. *J. Infect. Dis.* 167: 850–856.

Versalovic J, Swanson DS, Musser JM (1996) Nucleic acid sequencing studies of microbial pathogens: insights into epidemiology, virulence, drug resistance and diversity. In: Persing DH (ed.) *PCR Protocols for Emerging Infectious Diseases.* Washington, DC: ASM Press, pp. 59–88.

Vieusseux M (1806) Mémoire sur le maladie qui a régné à Genève au printemps de 1805. *J. Med., Chirurg Pharm.* 11: 163–182.

Virji M (1996) Meningococcal disease: epidemiology and pathogenesis. *Trends Microbiol.* 4: 466–469.

Vos P, Hogers R, Bleeker M, van de Reijans M *et al.* (1995) AFLP: a new technique for DNA fingerprinting. *Nucleic Acids Res.* 23: 4407–4414.

Waddy BB (1957) African epidemic cerebrospinal meningitis. Part II. Cerebrospinal meningitis in other related African territories. *Trop. Med. Hyg.* 60: 218–222.

Wang J-F, Caugant DA, Li X *et al.* (1992) Clonal and antigenic analysis of serogroup A *Neisseria meningitidis* with particular reference to epidemiological features of epidemic meningitis in China. *Infect. Immun.* 60: 5267–5282.

Wang J-F, Caugant DA, Morelli G, Koumaré B, Achtman M (1993) Antigenic and epidemiological properties of the ET-37 complex of *Neisseria meningitidis. J. Infect. Dis.* 167: 1320–1329.

Welsh J, McClelland M (1990) Fingerprinting genomes using PCR with arbitrary primers. *Nucleic Acids Res.* 18: 7213–7218.

Whittam TS, Wolfe ML, Wachsmuth IK *et al.* (1993) Clonal relationships among *Escherichia coli* strains that cause hemorrhagic colitis and infantile diarrhea. *Infect. Immun.* 61: 1619–1629.

Whittam TS (1995) Genetic population structure and pathogenicity in enteric bacteria. In: Baumberg S *et al.* (eds) *Population Genetics of Bacteria.* Cambridge: Cambridge University Press, pp. 217–245.

Wieler LH, McDaniel TK, Whittam TS, Kaper JB (1997) Insertion site of the locus of enterocyte effacement in enteropathogenic and enterohemorrhagic *Escherichia coli* differs in relation to the clonal phylogeny of the strains, *FEMS Microbiol. Lett.* 156: 49–53.

Wilkins BM (1995) Gene transfer by bacterial conjugation: Diversity of systems and functional specializations. In: Baumberg S *et al.* (eds) *Population genetics of bacteria.* Cambridge: Cambridge University Press, pp. 59–88.

Woelffer GB, Bradford WZ, Paz A, Small PM (1996) A computer-assisted molecular epidemiologic approach to confronting the reemergence of tuberculosis. *Am. J. Med. Sci.* 311: 17–22.

Woods JP, Kersulyte D, Tolan RW Jr, Berg CM, Berg DE (1994) Use of arbitrarily primed polymerase chain reaction analysis to type disease and carrier strains of *Neisseria meningitidis* isolated during a university outbreak. *J. Infect. Dis.* 169: 1384–1389.

Yang ZH, De Haas PEW, Wachmann CH *et al.* (1995) Molecular epidemiology of tuberculosis in Denmark in 1992. *J. Clin. Microbiol.* 33: 2077–2081.

Yeh RW, Ponce DL, Agasino CB *et al.* (1998) Stability of *Mycobacterium tuberculosis* DNA genotypes. *J. Infect. Dis.* 177: 1107–1111.

Zhu P, Morelli G, Achtman M (1999) The *opcA* and *opcB* regions in *Neisseria*: genes, pseudogenes, deletions, insertion elements and DNA islands. *Mol. Microbiol.* 33: 635–650.

Zhu P, Ende A van der, Brieske N *et al.* (2001) *Fit* genotypes and escape variants of subgroup III *Neisseria meningitidis* during three pandemics of epidemic meningitis. *Proc. Natl. Acad. Sci. USA* 98: 5234–5239.

22 Genetic Manipulation

Jeremy W. Dale

University of Surrey, Guildford, UK

The aim of this chapter is to explain the principles underlying the molecular genetic techniques used to elucidate the key characteristics of bacterial pathogens and their products, specifically antigenic structure, antibiotic resistance mechanisms, and virulence determinants. There are three strands to these techniques:

- cloning and expression of genes in other hosts
- genetic manipulation of the pathogen itself
- analysis of gene expression – especially those genes expressed during infection rather than in laboratory culture.

In the final section, I examine some of the genetic approaches that have been used for the identification of virulence determinants. In order to achieve adequate focus on these objectives, I assume that the reader has some understanding of the general basis of genetic manipulation and the associated aspects of molecular biology. Supplementary information will be found in Old and Primrose (1994), Brown (1995), and Dale (1998).

Cloning and Characterization of Bacterial Genes

Gene cloning involves the ligation of fragments of DNA to a vector molecule that is capable of replication in the chosen host organisms (usually *Escherichia coli*), introducing the recombinant molecules into the host cells (by transformation or electroporation, or, in the case of bacteriophage λ vectors, by *in vitro* packaging), and identifying the clones that carry the required insert. When attempting to clone an unknown gene, it is not necessary to purify the required DNA fragment for cloning; instead, a mixture of fragments from the entire genome can be used, generating a large number of clones each of which carries a different fragment. This is known as a gene library. (More specifically, this is a genomic library, to distinguish it from a cDNA library, as described later; however, cDNA libraries are less commonly used for bacterial sources.)

The gene library therefore represents the entire genetic composition of the target organism, and the separation of any one gene from all the others depends on the ability to screen large numbers of clones. This is commonly done by hybridisation with a labelled gene probe; for some purposes antibody screening is also valuable.

When the DNA sequence is known, or can be inferred, the polymerase chain reaction (PCR) provides a more direct way of cloning a specific gene. The basis of this procedure is also described below.

Construction and Screening of Gene Libraries

Choice of Vector

In order to have a reasonable chance of isolating the required gene, we need to know how many independent clones are required to constitute a representative gene library. This is not as straightforward as it might seem; as the number of clones increases, the likelihood of duplicating information also increases. The number

 doi:10.1006/bkmm.2001.0022

of clones required can be determined by the formula

$$N = \frac{\ln(1-P)}{\ln(1-f)}$$

where N is the number of clones required, P is the probability (e.g. $P = 0.99$ means there is a 99% probability of your gene being included) and f is the average fraction of the genome represented by each clone, i.e. if the genome size is 4 Mb and the average insert size is 4 kb, then $f = 10^{-3}$.

Using the values indicated shows that nearly 5000 clones are needed to achieve the desired probability of obtaining the required clone. This refers to the number of independent clones, each of which arises from a single cell acquiring a unique recombinant DNA molecule; the library will be amplified before use, by subculturing the mixture of bacteria or by propagating the phage, but this (although increasing the number of bacteria or phage) does not increase the effective size of the library. Note also that the calculation assumes an equal probability of cloning any fragment of DNA. If the required fragment is damaging to the host cell, or is unstable for other reasons, it may be much more difficult to obtain.

The choice of vector to be used is influenced by this calculation, as well as by the method to be used in screening the library. A conventional plasmid vector (e.g. pUC18) is capable of carrying 4 kb inserts, and screening 5000 clones is far from impossible. However, it is usually preferable to reduce the size of the library required, and for genomes larger than those in bacteria it becomes essential. There is no strict limit on the insert size that such a plasmid can carry, but the efficiency of the process becomes poor and there is an increasing risk of instability of the plasmid constructs. Genomic libraries are therefore more commonly constructed using vectors that can readily accommodate larger DNA fragments.

λ Replacement Vectors One class of vectors, based on bacteriophage λ, that are suitable for this purpose are known as replacement vectors (**Fig. 1**). Unlike plasmid vectors, which are cut at a single site by the chosen restriction enzyme, λ replacement vectors are cut at two positions. Since λ DNA is linear rather than circular, this results in three fragments: the left and right arms, which carry all the genes necessary for lytic growth of the phage, plus a third fragment known as the stuffer fragment. The purpose of this fragment is related to the biology of the phage itself. λ phage particles require a certain amount of DNA within them for mechanical stability. The left and right arms of the vector are too small to provide this, so the addition of the stuffer fragment is necessary for phage viability (although it contains no genes that are required for growth of the phage).

After digestion, the left and right arms are separated by gel electrophoresis from the stuffer fragment, which is discarded. The arms are too small to produce viable particles, unless they are ligated with an insert fragment above a minimum size. This strategy has two advantages: (1) the removal of large portions of λ DNA that are not needed for lytic growth (such as those concerned with lysogeny) frees up a lot of space for insertion of foreign DNA, and (2) the minimum packaging limit not only ensures that the progeny are recombinant phages, but also provides selection for a decent size of insert. In the example shown, the insert size must be between 8 kb and 23 kb; smaller than 8 kb and the DNA is below the minimum limit; larger than 23 kb and the DNA is too big to fit into the phage head.

After ligation of the arms with the insert fragments, the DNA is combined with a mixture of phage heads and tails which provide all the components necessary for packaging λ DNA into phage particles, which are then used to infect a suitable host strain of *E. coli*. This process, *in vitro* packaging, is highly efficient, and results in much greater yields of recombinant clones than can be obtained by transformation with recombinant plasmids.

A further advantage of using λ vectors to create gene libraries is that the screening processes (see below) are much easier and give better results; very high densities of phage plaques can be screened effectively.

Cosmids Larger fragments can be cloned using another class of vectors known as cosmids, which also utilise some of the features of λ biology. A cosmid is a plasmid vector that carries a λ *cos* site, i.e. the region of λ that is recognised by the packaging system (**Fig. 2**). A cosmid library therefore can be packaged into λ phage particles which can be used to

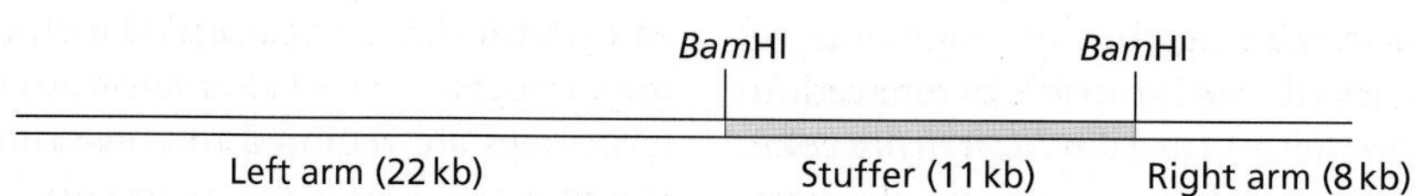

Fig. 1 Structure of EMBL4: a λ replacement vector. The stuffer fragment is removed for cloning, and replaced by the cloned DNA fragment, which must be between 8 kb and 23 kb to yield viable phage particles.

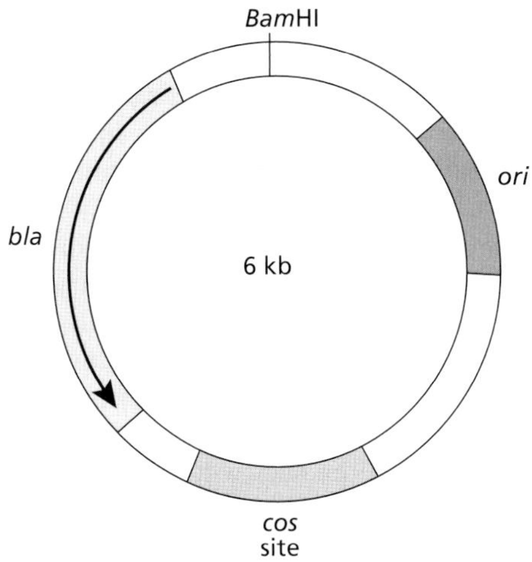

Fig. 2 Structure of a cosmid. The *cos* site enables the plasmid to be packaged into λ phage heads, but only if a suitably large DNA fragment (in this case between 32 kb and 46 kb) has been inserted at the *Bam*H1 site. The *bla* gene codes for β-lactamase production, enabling selection by ampicillin resistance.

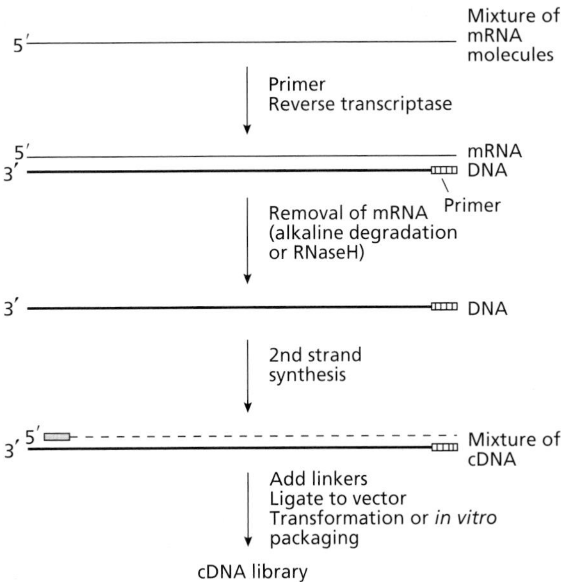

Fig. 3 Construction of a cDNA library. The population of mRNAmolecules from the starting material is copied into a corresponding mixture of DNA molecules. Addition of linkers facilitates cloning using a suitable vector, resulting in a library of clones from which specific genes can be recovered.

'infect' the bacterial host. Since the cosmid contains none of the genes needed for production of phage particles, however, we have to look for bacterial colonies carrying the cosmid, rather than phage plaques.

The use of *in vitro* packaging means that cosmids are subject to the same packaging limits that apply to λ vectors. The cosmid itself is much too small to yield stable phage particles, but insertion of a large DNA fragment (in the example shown, between 32 kb and 46 kb) puts the size of the recombinant molecule into the appropriate range. Cosmid vectors therefore also provide positive selection for large inserts.

Other Vectors Even larger fragments can be cloned by employing other vector systems, notably those based on the bacteriophage P1 (which have a cloning capacity of 95 kb) and the so-called bacterial artificial chromosomes (BAC), which are based on the F plasmid and can accommodate 300 kb. These are not routinely used for gene libraries where the purpose is the isolation of individual genes from a bacterial source, as inserts of this size are rather unwieldy. They are, however, extremely useful for establishing physical maps of the chromosome and as an adjunct to genome sequencing projects.

Construction of cDNA Libraries

The gene libraries described above are referred to as genomic libraries, and represent the complete genetic composition of the target organism. In some circumstances, it is preferable to produce a library representing only those genes that are expressed under defined environmental conditions, and for these purposes a different type of library is used in which the starting material is the mRNA rather than DNA. This requires copying the mRNA into the complementary DNA (cDNA); since this is the reverse of the normal transcription process it is known as reverse transcription.

The basis of the procedure for establishing a cDNA library is shown in Fig. 3. The first step, the production of a single strand of cDNA, is carried out by a reverse transcriptase (formally designated as an RNA-directed DNA polymerase) and results in a RNA : DNA hybrid molecule. Removal of the RNA strand (for example by digestion with RNaseH) is followed by synthesis of the second DNA strand, using a DNA polymerase (or exploiting the DNA-directed DNA polymerase activity of some reverse transcriptases). Each DNA polymerisation step requires a primer; for production of a cDNA library, containing a complete mixture of all cDNA species as possible, short random oligonucleotides can be used. If a specific cDNA sequence is to be obtained, then specific primers can be employed, and the procedure can be coupled with PCR to amplify the required product. This procedure, known as reverse transcript PCR (RT-PCR), and its applications, are described in a subsequent section.

The final stage in the production of a cDNA library is to ligate the mixture of cDNA molecules with an appropriate vector. To increase the efficiency of ligation, short synthetic oligonucleotides, known as linkers, are joined to the blunt ends of the cDNA molecules. The linker molecules are designed to contain a recognition site for a restriction endonuclease so that subsequent digestion with that enzyme will convert the previously blunt-ended fragments (which ligate with very low efficiency) into fragments with cohesive ends, which are much easier to ligate to the vector.

Libraries of cDNA molecules are widely used for cloning mammalian genes, where they have two main advantages over genomic libraries. Firstly, the use of mRNA rather than DNA as the starting material means that introns have been removed; this reduces the size of the 'gene' very considerably, and also makes it easier to express the cloned product in a bacterial host. Secondly, since a cDNA library represents only the expressed genes, the number of clones required to constitute a complete library containing the required gene is much smaller than is the case for a genomic library. Neither of these factors is of much importance with a bacterial gene library. Not only do prokaryotic genes rarely contain introns, but also the smaller size of the genome means that a genomic library is an easily manageable size. Genomic libraries are also much easier to prepare. A further factor is that bacterial mRNA is comparatively short-lived (many bacterial messages have a half-life of only a few minutes), so it may be quite difficult to obtain a representative library of cDNA molecules of anything approaching full length. The combination of these factors results in bacterial cDNA libraries being used only rarely, for specific purposes. One such purpose, covered later in this chapter, is for identification of genes that are differentially expressed under specific environmental conditions, since sequences will only be represented in such a library if the genes concerned are transcribed.

Screening a Library with Gene Probes

Having produced a gene library, which is stored as a mixture of a large number of clones, the next task is to identify the clone carrying the required gene. This is most commonly done by hybridisation with a labelled nucleic acid probe, as illustrated in **Fig. 4**.

The tricky part of this strategy is obtaining a suitable gene probe. Since the purpose of the exercise is to

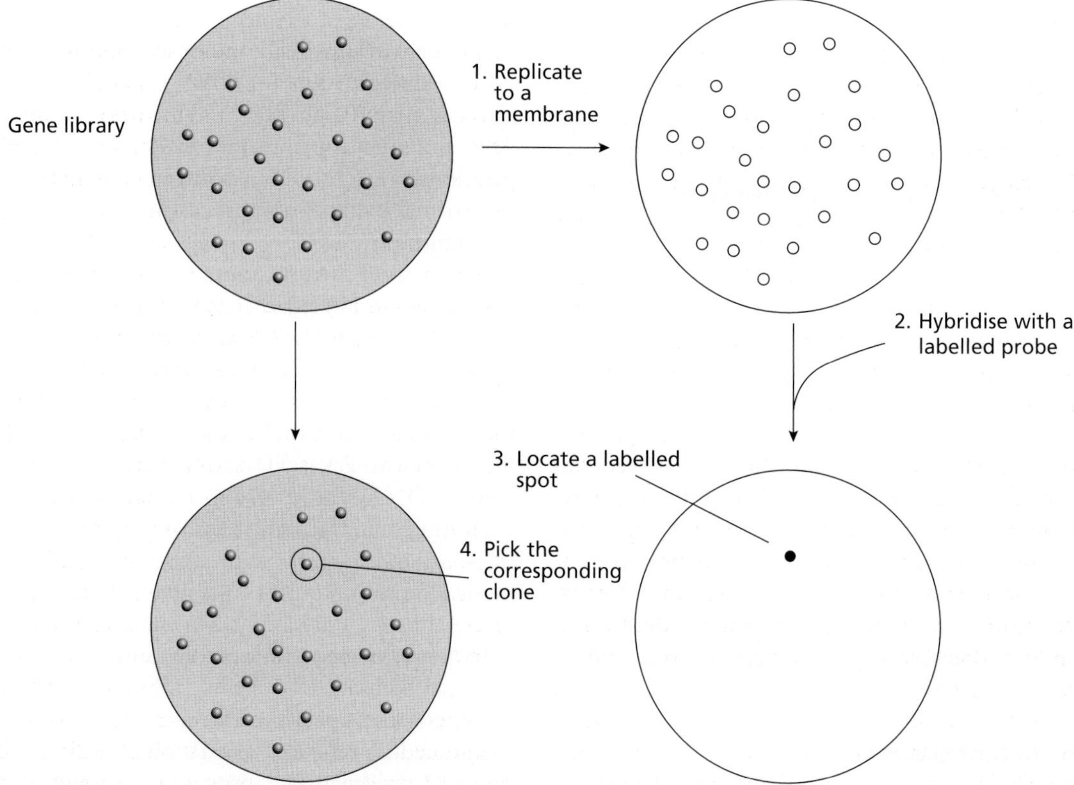

Fig. 4 Screening a gene library. The mixture of clones constituting the gene library is replicated on to a membrane, which is then treated to release the DNA and bind it to the membrane. Hybridisation with a labelled probe will identify the required clone, which can then be recovered from the original plate. A similar procedure can be used with specific antibodies to detect clones that express a recombinant protein antigen.

isolate a gene that has not previously been cloned (at least not from that specific source), there is no homologous DNA available that could serve as a probe. (There are situations in which the object is to re-isolate a known gene, for example in order to study the variation in its structure between different strains, but in that case the procedure would be to sequence the known gene and use PCR to isolate the equivalent gene from other isolates – see below.) The following are some of the most commonly used strategies; none of them are guaranteed to be successful.

Adjacent Sequences In order to isolate a DNA region that is near to a previously characterised gene, that gene can be used as a probe to identify cosmid clones which are likely to carry the region of interest. This technique (or PCR-based derivatives of the same principle) is often used in human genetics to identify genes responsible for inherited disorders when they can be shown to be linked to more easily characterised sequences. It is in general less useful for bacteria, where more direct approaches are available.

Heterologous Probes Genes responsible for the same function are usually remarkably similar, even across quite wide taxonomic boundaries, so it is possible to use a gene from another source to probe a gene library. Since the sequences will probably not be identical, it is necessary to use lower stringency in the hybridisation, i.e. reduce the specificity by reducing the temperature or increasing the ionic strength of the buffers in the washing steps that remove the non-specifically bound DNA probe from the filter. Of course, this also increases the risk of spurious hits, and too low a stringency may result in such a high background as to render the result unreadable. This approach is therefore most likely to be useful if the two organisms are very closely related.

PCR-generated Probes A variant of the above approach is to use the PCR to generate a probe. Comparison of a number of sequences (usually of the encoded protein) of the target function from a wide variety of sources will often show that several short regions (motifs) are very similar or even identical in all cases, i.e. they are highly conserved. It is then a reasonable expectation that the same conserved sequences will be present in the target gene. These conserved regions can then be used to generate primers for a PCR reaction, using DNA from the target organism as the template. If successful, this generates a DNA fragment that can be labelled and used as a probe. Since this is derived from the same source as the gene library, it is a homologous probe, and can be used at high stringency, avoiding the problems encountered with heterologous probes.

This approach is not without its problems. These conserved regions are most likely to be substrate-binding sites, and may also be present in other enzymes with different functions but using the same substrate. For example, many enzymes use ATP as a substrate, and ATP-binding sites are conserved between quite distinct enzymes. When we attempted to use this approach to isolate the dihydrofolate reductase (DHFR) gene from *Mycobacterium tuberculosis*, the sequence of the resultant clone showed very little similarity to DHFR genes from other bacteria.

Reverse Translation of Amino Acid Sequence A totally different approach is to determine part of the amino acid sequence of the protein encoded by the gene concerned. This sequence can then be reverse translated, by working out the DNA sequence of the gene that codes for that sequence of amino acids. From that DNA sequence, a probe can be synthesised.

One problem is that the redundancy of the genetic code means that it may not be possible to assign a codon unambiguously on the basis of the amino acid incorporated. For example, the presence of leucine in the protein could imply any of six codons in the DNA (TTA, TTG, CTT, CTC, CTA, CTG). It is possible to program a DNA synthesiser to incorporate ambiguities into the sequence, but this degree of ambiguity (multiplied over perhaps 10 codons) would result in a probe that was neither sensitive nor specific. Two strategies are used to counter this problem. The first is to select a region of amino acid sequence where there are fewer ambiguities – for example methionine is coded for only by ATG and tryptophan only by TGG; several other amino acids have only two possible codons. Secondly, where there are a number of possible codons, they are not all equally likely. There is usually some degree of codon bias, which applies to some extent to all genes within that organism. Even if only a few genes from your target organism have been sequenced, it is possible to work out a table of codon usage, from which you can infer that certain codons occur so rarely that they can be discounted for the purpose of reverse translation.

This method of generating probes used to be very laborious, as determining even part of a protein sequence required the purification of milligram quantities of the relevant protein to a very high degree of purity. Protein sequencing procedures have improved dramatically, however, so that it is now often possible to obtain sufficient sequence data from a spot isolated from a two-dimensional gel.

Screening Gene Libraries with Antibodies

Gene cloning can be used to identify the major antigenic proteins produced by a bacterium, without requiring purification of those proteins or any information about their function. The key characteristic is that an experimentally infected animal (or indeed a human patient suffering from an infection) produces antibodies to these proteins. These antibodies provide a route for identification of the corresponding clones in a gene library.

The procedure is in principle similar to that using gene probes; the filter is reacted with the primary antibody, and the position of the bound antibody is determined by a second labelled antibody. However, whereas gene probes react directly with the cloned DNA, antibodies detect the products of gene expression. This requires a modification of the cloning strategy to ensure that the cloned fragment is expressed. The vector that has been most commonly adopted for this purpose is the λ vector gt11 (Fig. 5), which has been engineered to contain a β-galactosidase gene and its promoter. Note that λ gt11 is an insertion vector (it contains just one *Eco*RI site and no stuffer fragment) in contrast to the replacement vectors described above; the cloning capacity is less (up to about 7 kb), although this is not a problem as larger insert fragments are not useful with this strategy.

Insertion of a DNA fragment at the *Eco*RI site puts it under the control of the *lac* promoter. If the insert is in the correct orientation and in the right reading frame, the ribosomes translating the *lacZ* message will continue to translate the message derived from the inserted fragment, resulting in the production of a fusion protein consisting of part of the β-galactosidase protein and partly derived from the insert. We do not expect this fusion protein to have the biological activity associated with the cloned fragment in its original host, since we are merely looking for its antigenic properties, which are associated with comparatively short peptide sequences. It is therefore possible, indeed advantageous, to use cloned fragments that are much smaller than those employed for conventional genomic libraries.

Even small parts of a foreign protein may be damaging to the host cells, however. The use of a λ vector is helpful in this respect, since prolonged growth of the bacteria carrying the recombinant construct is not required – the phage infection will kill them anyway. The use of λ gt11 provides an additional safeguard, since the *lac* promoter is inducible; the recombinant phage can therefore be propagated initially under conditions that do not allow expression, and the inducing agent is only added when the library is to be screened.

This strategy was used to good effect in identifying several important antigenic components of *M. tuberculosis* and *M. leprae* (Young *et al.*, 1985a, 1985b), without having to purify any of these antigens. Using whole-cell extracts as immunogens, a panel of monoclonal antibodies had been produced. Screening the appropriate λ gt11 libraries with these monoclonal antibodies resulted in the identification of the relevant clones, enabling subsequent characterisation by sequence determination as well as the production of large quantities of these proteins in pure form from recombinant *E. coli* clones.

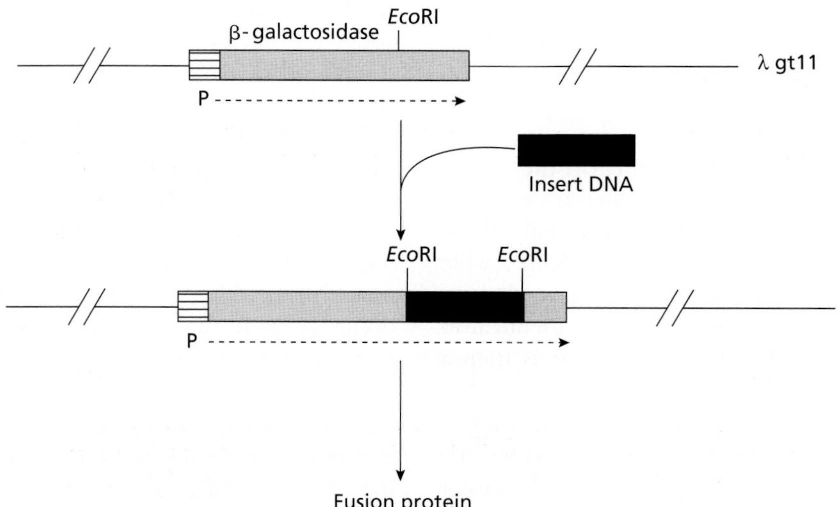

Fig. 5 λ gt11 as a cloning vector. The vector contains a single *Eco*RI site, within a β-galactosidase (*lacZ*) gene. Insertion of a DNA fragment inactivates the *lacZ* gene, so recombinant phage can be readily detected. If the insert is in the correct reading frame, the encoded peptide sequence will be expressed as part of a fusion protein, which can be detected with suitable antibodies.

There are a number of factors that limit the usefulness of this system, especially the fact that it will only detect linear peptide epitopes. Conformational epitopes will not be detected, since the fusion protein is unlikely to adopt anything approaching the natural conformation of the original antigen. Furthermore, antibodies that recognise modified amino acid sequences (e.g. glycosylated epitopes) will not be successful in screening a λ gt11 library. Obviously antibodies recognising non-protein antigens (e.g. carbohydrate antigens) cannot be used. These limitations apply mainly to the identification of specific antigens with monoclonal antibodies. Antisera from convalescent patients, for example, are likely to contain a variety of antibodies recognising antigenic determinants of different types; screening λ gt11 libraries with such antisera provides a valuable means of characterising the antigens recognised in the course of natural infection. Conventional antisera (whether from human subjects or experimental animals), however, commonly contain significant levels of anti-*E. coli* antibodies, which would result in a high background. These have to be absorbed out before the antisera can be used.

Cloning PCR Products

Since complete bacterial genomic libraries contain a manageable number of clones, and can be screened quite rapidly, it is in general not worth attempting to isolate and clone a specific fragment. It is usually just as easy to produce and screen a complete library, with the added advantage that the library can be stored as a resource that can be re-used for other purposes later on. There are two notable exceptions to this general rule. The first of these, the production of cDNA from transcripts that are only produced under defined conditions (e.g. genes that are expressed during infection but not in laboratory growth), is dealt with later in this chapter. The second exception arises where we already have sequence information from a closely related organism. For example, a putative virulence gene may have been cloned and sequenced and we want to characterise the equivalent gene in a related non-pathogenic bacterium. Or we may want to examine the differences in the structure of the relevant (protein) antigen between different serotypes. Since we already know the sequence of one version of this gene, we can use that information to design primers for PCR amplification of the corresponding region of the genome of the variant bacteria. The PCR product can then be directly sequenced (see below), or it can be cloned and characterised.

Cloning of the PCR product can be facilitated by including restriction enzyme cleavage sites within the primer sequence (Fig. 6), since only the 3′ portion of the primers needs to match the target sequence. The amplified product will then carry restriction sites near to the ends; these can be cleaved and ligated to an appropriate vector. It is of course necessary to ensure that this enzyme does not have a recognition site within the target sequence.

This strategy is also invaluable in subsequent manipulation of DNA sequences, for example where it is necessary to attach a promoter to a characterised gene in order to increase expression in the recombinant host. The flexibility and precision of this strategy makes it possible to produce carefully designed constructs.

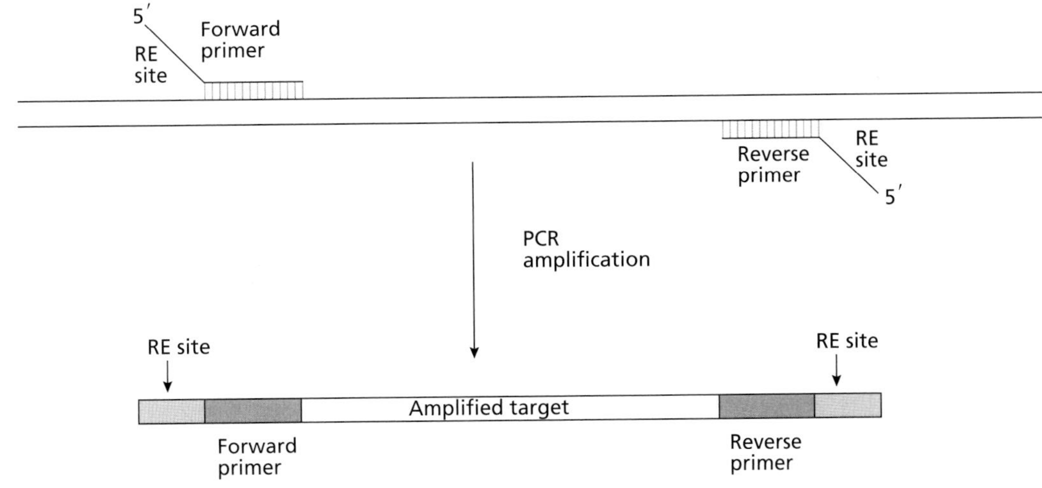

Fig. 6 Adding a restriction endonuclease cleavage site to a PCR product. Incorporation of a restriction site within the 3′ portion of a PCR primer results in a PCR product that is flanked by DNA containing a cleavage site. This facilitates cloning of the amplified product.

Sequence Analysis

Determination of DNA sequence depends on the chain-terminating properties of 2′,3′-dideoxynucleotide triphosphates (ddNTPs, Fig. 7), when used in a DNA synthesis reaction. The absence of the normal 3′-OH group prevents the addition of any further nucleotides. If the reaction is supplied with a mixture of ddATP and the correct substrate dATP (together with the three other dNTPs), then at the first position when dATP should be used a small proportion of molecules will incorporate ddATP instead and will terminate. The remainder will continue until the next T is reached on the template strand, when the same thing will happen. The consequence is a set of molecules of different sizes, each ending with an A residue. These can be separated on a denaturing acrylamide gel, on the basis of their size, thus indicating the position of the A residues on the synthesised strand (or the T residues on the template).

Three other reactions are carried out in parallel, using the same template and primer but using ddCTP, ddGTP or ddTTP respectively. Running the four reactions side by side on the same gel enables the sequence to be read directly, as indicated in Fig. 7.

In manual sequencing protocols, the use of a radioactively labelled substrate enables the bands to be visualised by autoradiography. Most sequencing is now carried out with automated sequencers, with either the primer or the ddNTPs being labelled with a fluorescent dye; the signals from the dye are read as the bands pass a fixed point on the gel.

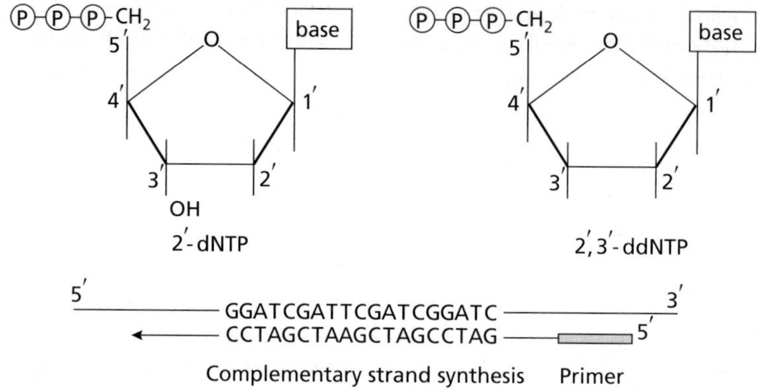

Random termination by ddATP will lead to a mixture of fragments:

AGCTAAGCTAGCCTAG — 5′
AAGCTAGCCTAG — 5′
AGCTAGCCTAG — 5′
AGCCTAG — 5′
AG — 5′

Similar reactions are carried out with ddGTP, ddCTP and ddTTP. The fragments from the four reactions are separated on an acrylamide gel:

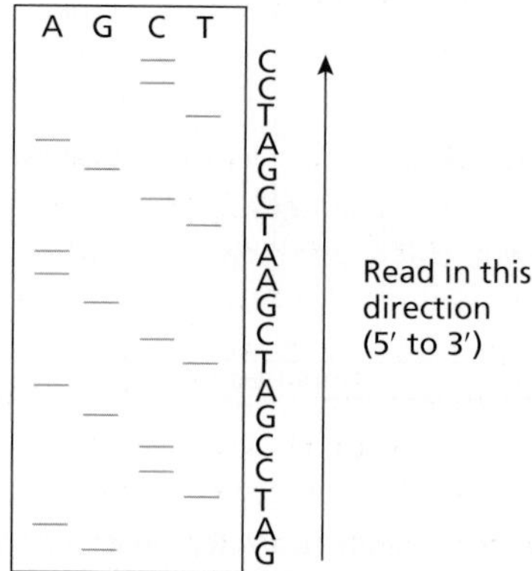

Fig. 7 Determination of DNA sequence.

With either manual or automated sequencing, a sequence of several hundred bases can be read from a single reaction. Longer sequences can be obtained by cloning and sequencing a large number of smaller subfragments, usually at random. These sequences are compared by a computer, and where overlaps are found the overlapping sequences are joined together to form a contig. Eventually, it is hoped that this contig will cover the whole of the sequence to be determined. However, as the sequence determined grows in length it becomes more and more difficult to find new sequences using a random approach. At this stage, more specific strategies can be invoked. For example, specific primers can be synthesised, on the basis of the known sequence, so as to specifically extend the sequence at the end of the contig. Or, where it is believed that two contigs are close together but the overlap has not been obtained, primers from those two contig sequences can be used in a PCR reaction to obtain a DNA fragment that bridges the gap.

Genome Sequencing

Recent advances, both in the capacity of automated DNA sequencers and in the computational ability to handle large amounts of sequence data (most notably in the assembly of large contigs), have enabled sequence analysis to move from the characterisation of individual genes to the determination of the complete DNA sequence of an organism. From the announcement of the first two complete genome sequences (*Haemophilus influenzae* and *M. genitalium*) in 1995, the number of completely sequenced genomes has increased rapidly (see Table 1). The challenge then is to extract biologically useful information from this wealth of data. The function of many genes can be inferred by comparison with known sequences in the databases, which provides answers to a variety of interesting questions – but that still leaves a high proportion of genes with no known function, either because they do not match anything known, or because the matches are to genes that also have no known function.

Table 1 A selection of completed microbial genomes

Species	Size (Mb)
Bacillus subtilis	4.2
Borrelia burgdoferi	1.44
Campylobacter jejuni	1.7
Chlamydia pneumoniae	1.23
Chlamydia trachomatis	1.05
Escherichia coli	4.60
Haemophilus influenzae	1.83
Helicobacter pylori	1.66
Mycobacterium tuberculosis	4.40
Mycoplasma genitalium	0.58
Mycoplasma pneumoniae	0.81
Neisseria meningitidis	2.3
Pseudomonas aeruginosa	5.9
Rickettsia prowazekii	1.1
Streptococcus pneumoniae	2.20
Treponema pallidum	1.14
Ureaplasma urealyticum	0.75
Vibrio cholerae	2.5

Information obtained from The Institute for Genome Research (TIGR): http://www.tigr.org

Sequence Variation

If genome sequencing represents an advance in a vertical direction made possible by improved sequencing technology, the same technology has also made possible significant developments in a horizontal mode. Once we know the sequence of a gene, it is quite easy to amplify a selected region of that gene, using PCR, from many independent isolates of the source bacterium. These PCR products can be directly sequenced, enabling us to build up a very extensive and detailed picture of the evolution of that gene and its relationship to the characteristics of the host strain. This can be applied in many different ways, of which I shall give two examples.

Bacterial Taxonomy The structure of ribosomal RNA genes has been widely used as an aid to the classification of bacteria and other organisms. These genes contain portions that are highly conserved, whatever the species, and other regions that are more variable between species. PCR primers based on the variable regions therefore provide specific diagnostic tests, and the use of conserved regions enables amplification of DNA from a wide range of sources. However, in the latter case, although an amplified product is obtained the sequence of that product will vary from one species to another. Determination of the sequence of the product from a range of isolates will enable their taxonomic relationships to be inferred (Vandamme *et al.*, 1996). Alternatively, if the sequence has already been determined for a set of characterised isolates, a comparison of the sequenced product with that database will enable the identity of the new isolate to be determined. For example, genus-specific primers can be used to amplify a portion of the rRNA gene from any mycobacterial species; since the product will differ, according to species, at a limited number of positions, the species can be identified either by hybridisation with specific oligonucleotide probes or by direct sequence analysis (Böddinghaus *et al.*, 1990; Rogall *et al.*, 1990).

The strength of this procedure is the speed with which identity can be established, using only a small portion of a single colony. It is not even necessary to cultivate the organism under investigation, so the technique can in principle be applied directly to clinical specimens, or for establishing the identity of novel or non-culturable organisms. The application of this procedure for investigating the possible infectious cause of diseases of unknown aetiology has been reviewed by Fredricks and Relman (1996).

A further extension of the principle is to characterise bacteria present in a complex mixture, without requiring isolation of individual organisms. Using primers designed to recognise highly conserved rRNA sequences, virtually any rRNA genes will be amplified. Cloning the mixture of PCR products then enables the determination of the sequence of each individual component, and hence their taxonomic relationships. This strategy has been most widely applied to studies of microbial diversity in the environment, where it has been estimated that as much as 99% of the microbial community is non-culturable (Giovannoni *et al.*, 1990; Ward *et al.*, 1990; Wang and Wang, 1996). The results need to be treated with care, as PCR amplification of a mixture of similar target molecules can yield chimeric products (Wang and Wang, 1996).

Mutational Resistance to Antibiotics Resistance to antibiotics can develop not only through the acquisition of plasmids but also through mutation of chromosomal genes. This is the main, if not the only, mode of resistance to some antibiotics; for example, rifampicin resistance is commonly due to mutation in the *rpoB* gene (encoding the β subunit of RNA polymerase), and resistance to the quinolone antibiotics such as ciprofloxacin is often associated with mutations in the *gyrA* gene (coding for the A subunit of DNA gyrase). The extensive use of sequence analysis for characterising antibiotic resistance mutations in mycobacteria (especially *M. tuberculosis*, which does not have naturally occurring plasmids and all resistance is mutational) has been reviewed by Musser (1995). Sequencing of amplified PCR products from a number of resistant isolates of *M. tuberculosis* has shown that the majority of these mutations occur at a few specific sites, and that these alterations do not occur in sensitive strains.

Expression of Cloned Genes

In addition to the identification and characterisation of the gene responsible for a specific characteristic, one of the main purposes to which gene cloning is applied is to obtain expression of that gene in the recombinant host. This may enable the product to be obtained from a more convenient bacterial host, which is especially useful when working with genes from pathogenic bacteria, as it avoids the hazards associated with large-scale culture of dangerous pathogens. Furthermore some of these bacteria (notably *Treponema pallidum* and *M. leprae*) are difficult, or even impossible, to grow in the laboratory. An additional benefit is that the expression of the cloned gene can be manipulated so as to obtain high levels of product, facilitating studies of the structure, and *in vitro* function of the protein; expression in an alternative host also makes it easier to investigate its antigenicity as a purified product free of any other antigens from the original source.

The expression of bacterial genes is primarily regulated at the transcriptional level, and this is mainly determined by the presence and nature of a promoter site (to which RNA polymerase binds). Some of the more subtle aspects of promoter regulation are considered in a subsequent section; the relevant characteristics at the moment are firstly that promoters vary in their ability to bind RNA polymerase, and hence the level of expression of the attached gene will vary considerably, and secondly that promoters are to some extent species specific, so that moving a gene to an alternative host, even with its own promoter attached, will not necessarily result in expression of that gene.

Expression Vectors

The simplest way of ensuring that a cloned gene will be transcribed at an acceptable level is to use an expression vector, as exemplified by **Fig. 8**. This contains a defined promoter adjacent to the cloning site, so that insertion of a foreign DNA fragment puts it under the control of that promoter. This is known as a transcriptional fusion. It should be noted, however, that this does not inevitably ensure production of a functional protein, which requires both the activity of translational signals (especially the ribosome binding site) from the insert,

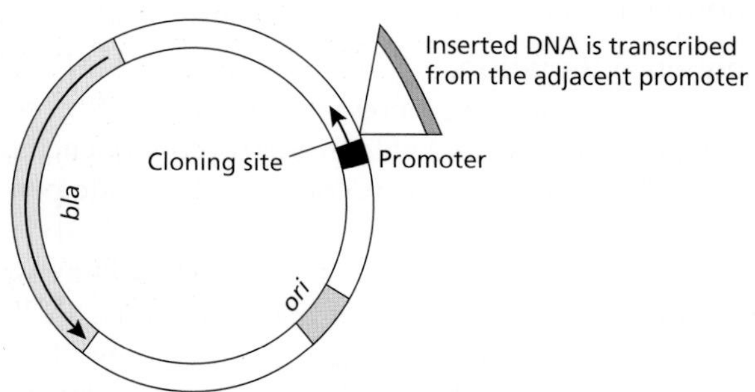

Fig. 8 An expression vector for transcriptional fusions. *bla*, β-lactamase (ampicillin resistance); marker for selecting transformants; *ori*, origin of replication.

and the correct folding (and possibly additional post-translational modification) of the translated polypeptide. The first of these problems can be overcome by the use of a different type of expression vector, in which the vector provides the translation signals as well (**Fig. 9**) resulting in a translational fusion. The cloning strategy has to be designed more carefully in this case, as it is necessary to ensure that the insert is in the same reading frame as the translational start codon.

The second problem, that of folding and post-translational modification, is difficult to predict or to counter. In essence, you can only tell if it works by trying it.

A further problem that is commonly encountered is that expression of a foreign protein, especially at a high level, may be damaging to the host, resulting in reduced growth rate, instability of the recombinant (through selection for loss of the plasmid or the insert) or even death of the recombinant. This problem is often particularly marked with membrane proteins. This can be counteracted by the use of expression vectors that allow control of the promoter activity. For example the *lac* promoter in *E. coli* will be repressed until an inducing agent is added; the recombinant can therefore be grown without induction until a suitable density is achieved, when addition of the inducing agent will allow the activity of the promoter to be de-repressed.

Genetic Manipulation of Bacterial Pathogens

The approaches outlined above are in general satisfactory for the characterisation of individual genes in isolation, and can therefore provide useful information about the structure of a specific protein antigen or the mechanism of antibiotic resistance. In order to investigate the biological function of many such genes, however, or to analyse the genetic basis of more complex characteristics such as virulence, it becomes necessary to examine the behaviour of these genes in their original host. The ability to manipulate the original pathogen, i.e. to be able to introduce new or modified genes, is invaluable for this purpose. With some species, particularly amongst the Enterobacteriaceae, the techniques required are merely an extension of those used with *E. coli*, and can take advantage of the full range of tools available. For other species, the development of suitable procedures for even basic manipulations can be an onerous task.

Cloning Vectors and Transformation

The fundamental requirements for cloning in any host are a cloning vector that will be able to replicate in the chosen host, which must carry a gene that will make

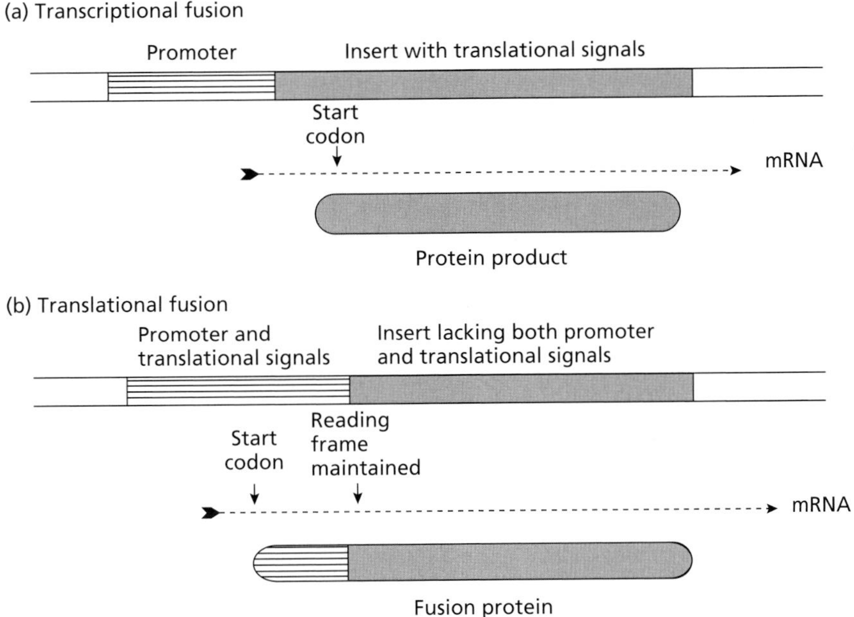

Fig. 9 Expression vectors: transcriptional and translational fusion: (a) In a transcriptional fusion, the inserted DNA provides its own translational signals, leading to production of the native protein. (b) In a translational fusion, translation as well as transcription is initiated from the vector sequences, leading to a fusion protein in which the N-terminal portion of the protein is derived from vector sequences.

the host antibiotic resistant, and an ability to introduce that vector into the host. The most difficult problem is that these requirements – a replication origin, an antibiotic resistance gene, and a transformation system – are inseparable, and it is impossible to develop one without the others. Even the resistance marker is not necessarily a straightforward matter, since many of the interesting pathogens are already resistant to a range of antibiotics, and in addition (for the reasons outlined above) it is impossible to be certain that a resistance gene from another source will function in the target species. A further problem is that mutation to resistance may occur at a frequency that results in an unacceptably high background.

The most likely source of a replication origin for a vector is a plasmid that occurs naturally in the target species, or at least in a closely related organism. For example, in order to develop a cloning system applicable to *M. tuberculosis*, which has no known naturally occurring plasmids, the plasmid pAL5000, which was obtained from *M. fortuitum*, was used. However, this plasmid carried no antibiotic resistance genes, nor were any plasmid-mediated resistance genes known in mycobacteria. Furthermore, mycobacteria have a high degree of intrinsic resistance to many of the more common antibiotics, which limited the choice of suitable markers. Ultimately, it proved possible to construct a suitable vector by fusing pAL5000 with an *E. coli* plasmid carrying a kanamycin resistance gene (Snapper *et al.*, 1988).

Without an extensive development programme, the transformation systems developed for *E. coli* for introducing the vector into the target species are too species specific to carry much likelihood of success. Protoplast transformation systems are rather easier to optimise for a novel host, but these have been largely superseded by electroporation as the method of choice. It is in principle possible to use electroporation for any cells (eukaryotic as well as prokaryotic), although finding the right conditions for preparation of the cells as well as optimising the electroporation parameters can be a problem. However, once you achieve at least a small measure of success, all these problems are soluble. The difficulty is that failure to obtain transformants may be due to (1) failure of the plasmid to replicate, (2) failure of the marker gene to cause sufficient resistance to the antibiotic used, or (3) failure of the electroporation system. A good deal of luck is necessary.

Shuttle Vectors

In order to construct a suitable vector molecule, it is necessary to be able to replicate it. Of course it is not possible to do this in a novel host until a suitable cloning system is available. The way out of this apparent impasse is to have two replication origins on the plasmid: one which will function in the new host, and a second which will function in *E. coli*. Such a plasmid is known as a shuttle vector.

Shuttle vectors remain a valuable tool even after the transformation system has been optimised, as almost inevitably any subsequent modifications, such as introducing additional genes or site-directed mutagenesis, are much easier to carry out in *E. coli*. Then, when the assembly of the new plasmid is complete, it can be transformed into the new host.

In addition to the widely used dual-origin shuttle vectors, it is often possible to use vectors based on broad-host-range plasmids. Although many plasmids are unable to replicate in more than a restricted range of bacteria, some (notably the P group of plasmids, such as RP4) are much less specific, and have been introduced successfully into many different bacterial species, both Gram positive and Gram negative.

Suicide Vectors and Conditional Vectors

For some of the purposes discussed subsequently, including transposon mutagenesis and gene replacement, continued replication of the plasmid is not required or even desired. Instead, the process requires insertion of part of the plasmid into the chromosome. A vector that is used to introduce DNA into the host without subsequent replication is referred to as a suicide plasmid.

A more subtle approach allows replication after transformation, but only under certain conditions. After isolation and purification of the transformants, the conditions are changed so that the plasmid can no longer replicate. Most commonly, these conditionally replicating plasmids are temperature sensitive, so that replication is allowed at a low growth temperature (the permissive temperature), and the plasmid fails to replicate after the culture is shifted to a higher (restrictive) temperature.

Recombinant Vaccines

One application of these techniques that has considerable potential is the expression of antigens from other pathogens in an established vaccine strain, in order to create a live recombinant vaccine that is capable of producing protective immunity against a range of diseases. The carrier that has been most widely used for this purpose is the smallpox vaccine virus, vaccinia, but bacterial vaccine strains (notably attenuated strains of *Salmonella typhi* and the mycobacterial vaccine BCG) are also used as carriers.

One problem that can be encountered, if the foreign genes are incorporated into a shuttle plasmid, is that the plasmid may not be sufficiently stable under conditions of use. This can be overcome by integration of the recombinant DNA into the chromosome of the bacterial host. If efficient recombination systems are available, this can be done by adapting the gene replacement technique (see below). Where this has not been possible, systems have been used that force integration into the chromosome. For example, Stover *et al.* (1991) used a suicide plasmid that could replicate in *E. coli* but not in BCG, and incorporated into the plasmid the attachment site and integrase gene from a mycobacteriophage. After electroporation of BCG, selection for the kanamycin resistance marker from the plasmid yielded colonies in which the plasmid had become integrated into the chromosome through the action of the integrase system.

An alternative approach (Dellagostin *et al.*, 1993) is to construct a composite transposon, in which two copies of an insertion sequence flank a kanamycin resistance gene and the gene for the foreign antigen, again using a suicide vector. In this case, kanamycin-resistant colonies of BCG arise through transposition of the composite transposon into the BCG chromosome.

Transposon Mutagenesis

One way of determining the biological function of a gene is to produce mutants with a defined phenotype and then to analyse the nature of the mutations. Classical genetics relies heavily on the use of point mutations for this purpose, but ascertaining the site of a specific point mutation, for example by crossing pairs of mutant strains, is laborious, even in organisms with well-characterised genetic systems. In the majority of bacterial species, in which genetic systems have not been developed, the task is next to impossible. Fortunately, we can exploit the ability of mobile genetic

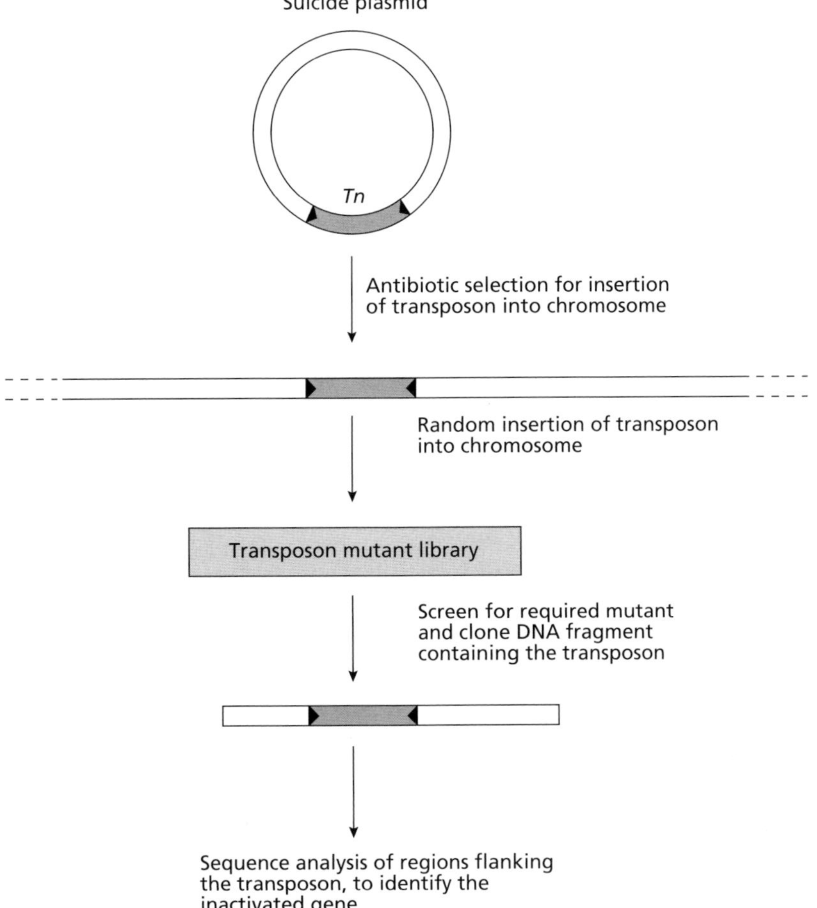

Fig. 10 Transposon mutagenesis. Using a suicide plasmid, antibiotic resistance arises through transposition of the transposon into the chromosome, potentially causing mutation of the gene into which it is inserted. After identification of the required mutant, the inactivated gene can be identified by cloning the transposon together with the flanking DNA.

elements such as transposons to cause mutations by inserting into the DNA. The presence of the transposon at the mutated site then provides an easily detected signal for locating the lesion.

The principle of the procedure, outlined in Fig. 10, involves the use of a suicide vector carrying a transposon such as Tn*5* (which contains a kanamycin resistance gene, *aph*). Since the vector is unable to replicate in the chosen host species, selection for kanamycin-resistant transformants will yield colonies in which the transposon has become inserted into the chromosome. A library of these transposants can then be screened for colonies showing the required mutant phenotype.

Identifying the site of insertion also makes use of the kanamycin resistance gene. Random fragments of DNA from the chosen mutant are ligated with a plasmid vector, followed by transformation of *E. coli*, with selection for recombinant plasmids carrying the *aph* gene. These transformants will carry not only the transposon, but also some of the flanking DNA. These flanking sequences can either be used as a probe to obtain the non-mutated gene from a gene library of the wild-type strain, or they can be sequenced to provide PCR primers for amplification of the wild-type gene. As the number of sequenced genomes increases, this strategy becomes easier, as the determination of the sequence of a small fragment will be sufficient to tell you the nature of the whole gene, by a simple computer search.

One of the limitations of this approach is that the extent of transposition from a suicide vector may be quite limited. Higher yields of transposition products can be obtained if a conditionally replicating vector (as described above) is used, enabling the production of a large population of cells carrying the plasmid before shifting the culture to the restrictive temperature and selecting for transposition. A further limitation is that transposon mutagenesis with Tn*5* is only applicable to bacteria in which the transpositional mobility of this transposon is exhibited. Although a wide range of other transposons exists, and such transposons are found, and are mobile, in many bacterial species, they are not all suitable for mutagenesis. Tn*5* is widely used because it inserts in a highly random fashion, giving rise to a wide range of mutants. In contrast, many other transposons show a higher degree of target site specificity, so that very few different genes can be inactivated.

Gene Replacement

Transposon mutagenesis is the method of choice for producing mutants if the phenotype is readily identifiable but the nature of the gene responsible is unknown, or if there are several genes that could be involved. However, some phenotypes (notably loss of virulence) are not easily selected, which makes transposon mutagenesis a cumbersome approach. An alternative is to attempt to identify a candidate gene, and to test its involvement by specifically inactivating it. The procedure for doing this is known variously as gene replacement, allelic replacement, or more colloquially as gene knockout.

This procedure, outlined in Fig. 11, relies on the tendency of similar sequences to undergo homologous recombination, in which the related sequences are broken, crossed over, and re-joined. The details of this mechanism (or mechanisms, since there are actually several pathways) do not need to be considered here.

A suicide vector is constructed carrying a copy of the target gene which has been disrupted by *in vitro* insertion of an antibiotic resistance gene – again using *aph* as an example. Recombination is often stimulated by using linearised DNA rather than the intact plasmid. Transformation of the host cells is followed by selection for kanamycin resistance; since the plasmid is unable to replicate, kanamycin resistance occurs only if the *aph* gene has been inserted into the chromosome, hopefully by recombination between the homologous sequences flanking the *aph* gene.

Recombination at one site (a single cross-over) will cause insertion of the whole plasmid into the chromosome. This may result in inactivation of the target

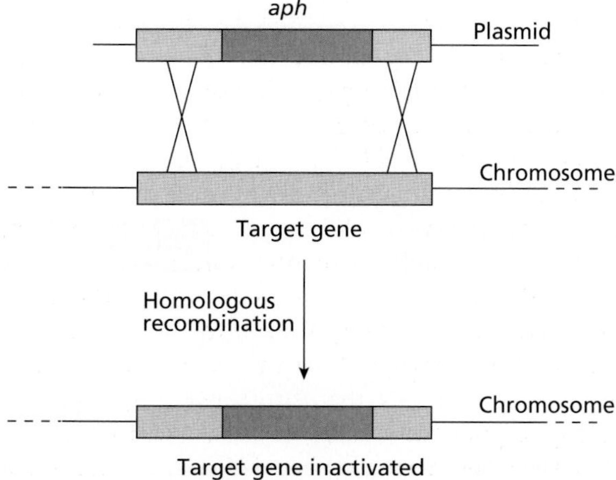

Fig. 11 Gene replacement by homologous recombination. A plasmid is constructed in which an antibiotic resistance marker is flanked by two pieces of the target gene. If the plasmid is unable to replicate in the chosen host, antibiotic resistance arises by homologous recombination, incorporating the antibiotic resistance marker into the chromosome and replacing the original target gene. *aph*, aminoglycosidephosphotransferase (kanamycin resistance).

gene (depending on how the construct was designed), but is likely to be unstable, as further recombination events can excise the plasmid and restore the intact target gene. The desired event is a double cross-over, in which recombination has occurred at either side of the *aph* gene. It may be possible to screen a number of recombinants to identify which are double cross-overs; if these occur rarely, it will then be necessary to incorporate additional markers that can identify those cells in which only the *aph* gene has become incorporated, and not the rest of the plasmid.

This description is highly idealised. Although gene knockouts have been widely used for many species (including laboratory animals), the technique can encounter unexpected difficulties – for example with *M. tuberculosis*, it was initially found that most of the kanamycin-resistant colonies arose through an ill-defined process of non-homologous (or illegitimate) recombination, in which the *aph* gene had become randomly inserted into the chromosome rather than specifically inactivating the selected gene. These difficulties have subsequently been overcome in various ways (Norman *et al.*, 1995; Reyrat *et al.*, 1995; Balasubramanian *et al.*, 1996; Baulard *et al.*, 1996; McFadden, 1996).

Complementation

The desired outcome of either transposon mutagenesis or gene replacement is a mutant strain with a defined alteration in its phenotype, which may be a simple characteristic such as requiring an individual amino acid for growth (auxotrophy), a more complex alteration in its physiology, such as an altered cell wall, a reduction in virulence, or alteration in antigenic composition. Some of these characters may go together, i.e. the strain may show a number of differences (the mutation may be pleiotropic). At the same time, we hope to be able to demonstrate that the strain carries a definable mutation in a single gene, whether it is the insertion of a transposon or the replacement of that gene by a disrupted derivative. It is tempting at this stage to conclude that the change in that gene is directly responsible for the observed alteration in the phenotype, and that the affected gene therefore codes for a key constituent of that process. This temptation must be resisted.

The most obvious problem is that other changes may have occurred. The transposon may have inserted at several sites, not just the one we have characterised, or we may have inadvertently selected a spontaneous mutant and the transposon or gene replacement has nothing to do with the observed changes. The extent of this problem can be reduced most simply by testing whether the same effect occurs in a number of independent mutants, obtained in separate experiments, as well as by tests to verify that only a single copy of the transposon is present.

A further problem is that many mutations, especially those involving transposons, show polar effects, i.e. the mutation affects not only the gene directly altered but also other genes downstream from it. This most commonly occurs when a number of genes are transcribed into a single mRNA (i.e. they form an operon). The mutation may not only inactivate the gene concerned, but may disrupt the transcription of the whole operon. If this is the case, some or all of the observed phenotypic changes may be due to the loss of function of the other genes in the operon, rather than to the gene we have mutated.

The easiest way of testing this is by complementation. If we construct a shuttle plasmid carrying a functional version of the gene we have inactivated, then inserting this plasmid into the mutant should result in restoration of the wild-type phenotype, if the alteration is solely due to loss of that function.

Even if we can show that complementation restores the original phenotype, we still have to be cautious about ascribing functions to the inactivated gene. The absence of a specific gene function may affect many aspects of intermediary metabolism, or the transport of nutrients, which in turn may affect either the activity or the regulation of other genes. In these situations, complementation may still restore the original phenotype, by reinstating the original regulatory pathways, but does not prove that the manipulated gene has a direct role in the phenomena observed. There is no easy or generally applicable way out of this impasse, except that if the conclusions make biological sense then we are probably on safe ground. If we inactivate a gene that is thought to be involved in leucine biosynthesis (by comparison with genes in other bacteria), and the resulting mutant is a leucine auxotroph, then we probably need to look no further for an explanation. On the other hand, if this mutant shows reduced ability to attach to epithelial cells, we may want to be cautious about suggesting that the product of this gene is an adhesin.

Analysis of Gene Expression

A prime target for the application of genetic techniques to bacterial pathogens is the identification of genes that are involved with virulence, since this knowledge will assist with understanding of the mechanisms of pathogenesis, and also, through gene inactivation, the production of specifically attenuated strains for vaccine development.

Unfortunately, the identification of these genes is not always readily amenable to conventional genetic analysis, which usually relies on being able to identify the mutant phenotype in the laboratory. There are some obvious exceptions, notably protein toxins, but in many cases the molecular basis of pathogenicity is poorly understood and is likely to be multi-factorial, so that the effect of any isolated component will be inconclusive. A common strategy in such situations relies on the premise that virulence determinants are likely to be selectively expressed during infection. Therefore, attempts are made to identify genes that are active when the bacterium is multiplying within a host animal (or, for intracellular pathogens, within macrophages or other cells in culture), but are inactive, or expressed at a much lower level, in laboratory culture. In this section, I describe some of the ways in which we can study the expression of genes under different conditions, apart from the direct but insensitive method of analysing the protein profile of the cells.

Transcript Analysis: Hybridisation and RT-PCR

It is often convenient to analyse the expression of specific genes by examining the mRNA produced rather than measuring the protein product. The basic procedure involves extracting RNA from the cells, removing the DNA, and hybridising the purified RNA with a specific nucleic acid probe. The RNA to be hybridised can be applied directly to a filter (dot blot), or transferred to a filter after electrophoresis (northern blot). The northern blot procedure is more specific and will give more reliable results. A semi-quantitative result can be obtained by hybridisation in parallel with a probe that detects a constitutively produced RNA.

The main limitation of these procedures, especially in relation to the analysis of potential virulence determinants, is the amount of mRNA required for a clear result. Although suitable amounts can be obtained from *in vitro* cultures, it is difficult or effectively impossible to obtain enough material from macrophage-grown bacteria, or from tissue samples. In such cases, it is preferable to use the PCR, adapted for use with an RNA template by the addition of a reverse transcription step – hence the technique is known as reverse transcript PCR (RT-PCR). The principle of this procedure is outlined in Fig. 12. The sensitivity of this method enables the detection of very small amounts of message; procedures are also available for obtaining a quantitative result, although the results need to be interpreted with care. It must, however, be remembered that the PCR step will also amplify any remaining traces of DNA, so not only does the purification of the RNA have to be carried out rigorously but also suitable controls must be included to ensure that the signal obtained is indeed the result of amplification of the specific message.

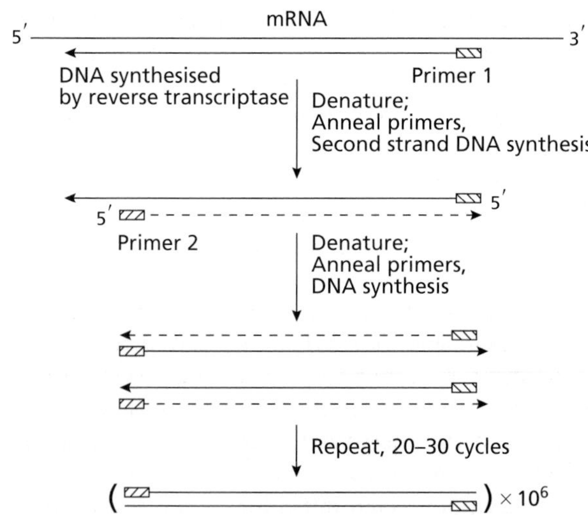

Fig. 12 Reverse transcript PCR (RT-PCR). The PCR technique can be adapted to amplify RNA, by first copying it into DNA using reverse transcriptase.

Although RT-PCR can detect very small amounts of mRNA, the sensitivity will be substantially reduced by the presence of other RNA species, notably ribosomal RNA which will form the bulk of the RNA extracted. Thus even this procedure may not be adequate for the detection of transcripts from poorly expressed genes, although enrichment of mRNA by subtractive hybridisation of RNA with antisense rRNA (Plum and Clark-Curtiss, 1994) can improve the situation.

Quantitative RT-PCR

RT-PCR provides a sensitive method for qualitatively detecting the presence of a specific transcript, but is less suitable for quantitative analysis of the amount of such a transcript present in the cell. It is possible to adapt conventional RT-PCR procedures to obtain quantitative results, for example using competitive assays or dilution end-points, but these are rather cumbersome and it can be difficult to obtain reliable results.

More satisfactory results can be obtained using newer procedures known as real-time PCR. The general characteristic of these procedures is that as the PCR reaction proceeds, a fluorescent signal develops which is dependent on the accumulation of product. The signal can be followed continuously during the reaction; the time at which the signal starts to appear is a function of the amount of template present (assuming equal efficiency of the PCR in each case). One form of real-time PCR uses a dye that fluoresces when it complexes with double-stranded DNA. An alternative is to use a 'molecular beacon', which is essentially a fluorescent energy-transfer-labelled nucleic acid probe. This means that it has a fluorescent label at one end and a quencher at the other end. Initially this probe forms a hairpin so that the fluorophore and quencher are adjacent, and hence no fluorescence is detected. As the reaction proceeds, the probe hybridises to the product, separating the fluorophore and the quencher and allowing a signal to be detected. This allows an extra degree of specificity, but carries the disadvantage that the molecular beacon has to be synthesised specifically for the sequence to be tested. Energy-transfer-labelled primers can be used instead of hybridisation probes, so that the fluorophore becomes separated from the quencher as the primer is incorporated into the product.

The optimal use of these systems for real-time PCR requires special equipment that can read the fluorescence developing while the PCR proceeds. It is possible to carry out quantitative PCR without such equipment, by using an end-point method, i.e. measuring the fluorescence individually after a fixed number of PCR cycles, but standardisation is then less reliable, and the range over which the assay extends is much narrower.

These procedures are well suited for comparison of levels of a specific transcript under different conditions. Comparison of the levels of different transcripts needs to be interpreted with care, as the results will be influenced by the efficiency of the different PCR reactions.

Reporter Genes and Promoter-probe Vectors

A major limitation of RT-PCR for analysing the expression of a specific gene under different circumstances is that it requires the isolation of mRNA from the cells. Following changes over time is restricted by the number of samples that can be processed, which may be a serious limitation if expression occurs over a short time period. Also, analysis of expression within different cells in a population (for example in different parts of a colony) is extremely difficult if not impossible. In such a situation, reporter genes provide a different approach, which enables the behaviour of individual cells to be examined directly.

A reporter is a gene, coding for an easily identified product, from which the expression signals have been removed and replaced by those from the gene under investigation (see **Fig. 13**). In principle, therefore, the expression of the reporter gene should mimic that of the target gene. The *lacZ* gene from *E. coli*, coding for the enzyme β-galactosidase, has been widely used as a reporter gene, since the addition of the chromogenic substrate known as X-gal results in a blue colour as the X-gal is hydrolysed. Fluorogenic substrates are also available, and can be used for monitoring the expression of the *lacZ* reporter gene by bacterial cells within macrophages.

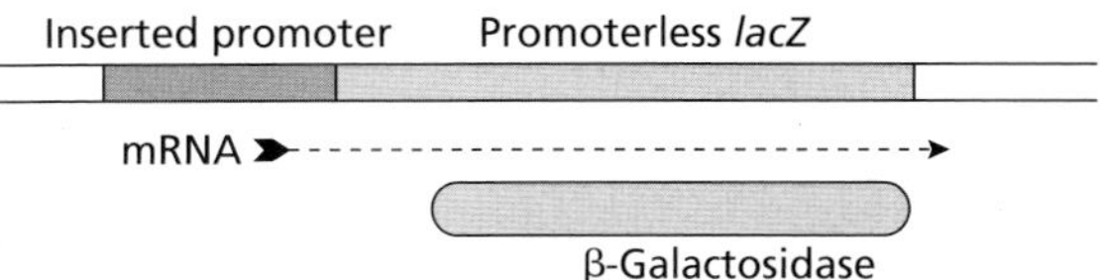

Fig. 13 Expression of *lacZ* as a reporter gene. Attachment of a *lacZ* gene (coding for β-galactosidase) to a specific promoter region enables the activity of that promoter to be monitored *in situ* under a variety of environmental conditions.

An even more versatile reporter gene, which does not require the addition of a substrate, is that coding for the intrinsically fluorescent protein known as green fluorescent protein (GFP) (Chalfie *et al.*, 1994).

The alkaline phosphatase gene *phoA* is also used as a reporter gene, specifically for investigating signals associated with protein secretion, since the activity of the enzyme is normally only manifested after secretion. In this case, translational fusions are used, rather than transcriptional fusions, so as to join the signal peptide from the target gene to the N-terminus of the alkaline phosphatase.

There are many other genes that can be used as reporters, the main requirement being that expression can be directly detected without having to isolate the bacteria and produce cell extracts. Expression can therefore be monitored at different stages of growth of a colony, or using different growth conditions, or even of individual cells *in situ* in varying environmental niches. This envisages that we are investigating the effect of the expression signals from a specific gene when attached to the reporter gene.

The principle can be extended to the identification of individual fragments of DNA from a random mixture of fragments which have the ability to drive expression of the reporter gene under defined conditions. For this purpose, it is more usual to use a reporter gene that provides a selectable phenotype, usually an antibiotic resistance gene. A plasmid is therefore constructed carrying a promoterless resistance gene with a cloning site immediately upstream. Random fragments of DNA inserted at this cloning site will therefore result in the production of antibiotic resistant colonies only if the inserted fragment has promoter activity; for this reason these plasmids are known as promoter-probe vectors. These vectors therefore enable us to search the genome for promoters which have activity in a specified situation (such as anaerobiosis, stationary phase, or intracellular growth) in contrast to reporter gene constructs that allow us to follow the behaviour of a defined promoter under a range of conditions.

The results obtained with promoter-probe vectors need to be treated with caution. The ability of DNA fragments to promote transcription is context-sensitive, so fragments that have promoter activity when inserted at this position on a small, supercoiled multicopy plasmid do not necessarily function as promoters in their original chromosomal location; the reverse also applies, but false-negative results are less likely to be misleading. Furthermore, the choice of antibiotic resistance reporter gene has to be made with care, as low levels of transcription of some markers can result in a resistant phenotype, and these levels can be obtained even after removal of the normal promoter. We have observed this effect with a kanamycin resistance gene (*aph*, aminoglycoside phosphotransferase); other genes which require a higher level of expression to cause resistance (such as chloramphenicol acetyltransferase, *cat*) are a better choice.

Genomes, Transcriptomes and Proteomes

Although (in the time scale of molecular biology) promoter-probe vectors have a long history of use for detecting promoters with specified activity, they yield a selection of such promoters (plus the artefacts referred to above) rather than providing an overview of the total transcriptional activity of the cell.

The availability of genome sequences has provided a way of doing this. Computer analysis of the genome sequence is used to predict all the possible open reading frames (ORFs) that can be translated into proteins. For each ORF a pair of PCR primers is synthesised and used to direct a PCR reaction. This gives rise to a set of several thousand PCR products which can be arrayed on a series of filters (macro-array) or as smaller, closely-spaced spots on a coated glass slide (micro-array). Total mRNA from cells under the required condition is then subjected to RT-PCR, and the product is used to probe the array, giving a pattern of signals corresponding to the transcripts that are present in the original cells. Since the identity of each spot is known, it is then quite easy to determine which genes are transcribed under those conditions.

As the term *genome* is used to describe the total DNA of the cell, the analogous term *transcriptome* is used to refer to the set of transcripts that are made under particular conditions.

This development does not render promoter-probe technology completely obsolete. Some transcripts may be too unstable to be amplified, or they may be present transiently or at very low levels; the computer program may have failed to identify a product (for example if the protein produced is smaller than the limit set, or if the product is RNA rather than protein). In such cases, a promoter-probe may detect activity that is not disclosed by array technology. In addition, promoter probes can be employed with organisms that have not yet been sequenced.

A further development is the study of the overall protein expression capability of a cell under defined conditions (*proteome*). This is most commonly performed by two-dimensional gel electrophoresis; the technology for this and accompanying procedures has now improved to the extent that it is possible to identify specific spots from the two-dimensional gel, using predictions of protein sequence from the genome data.

Identification of Virulence Genes

One of the most important applications of genetic manipulation techniques, as applied to bacterial pathogens, is in the identification of genes that are required for virulence or pathogenicity. This knowledge serves the dual purpose of shedding light on the nature of virulence itself, as well as providing a route for rational attenuation of the pathogen for vaccine construction. Since virulence is not a laboratory phenomenon, standard genetic techniques are rarely applicable, although in some cases a laboratory characteristic such as colony morphology can be used as a marker of virulence. The most notable example of this was the restoration of virulence to rough mutants of *Streptococcus pneumoniae* (Griffith, 1928). In this context, I am excluding consideration of genes coding for bacterial toxins, which are in principle amenable to the standard cloning techniques described above.

Further discussion of the techniques in this section, and additional examples, is to be found in the review by Hensel and Holden (1996).

In vivo Expression Technology (IVET)

The concepts developed in the previous section have been exploited to identify expression signals that are active only during infection. The simplest procedure is to use a promoter-probe vector containing a promoterless marker (such as the *cat* gene) that can be used for selection *in vivo*. When random fragments of DNA

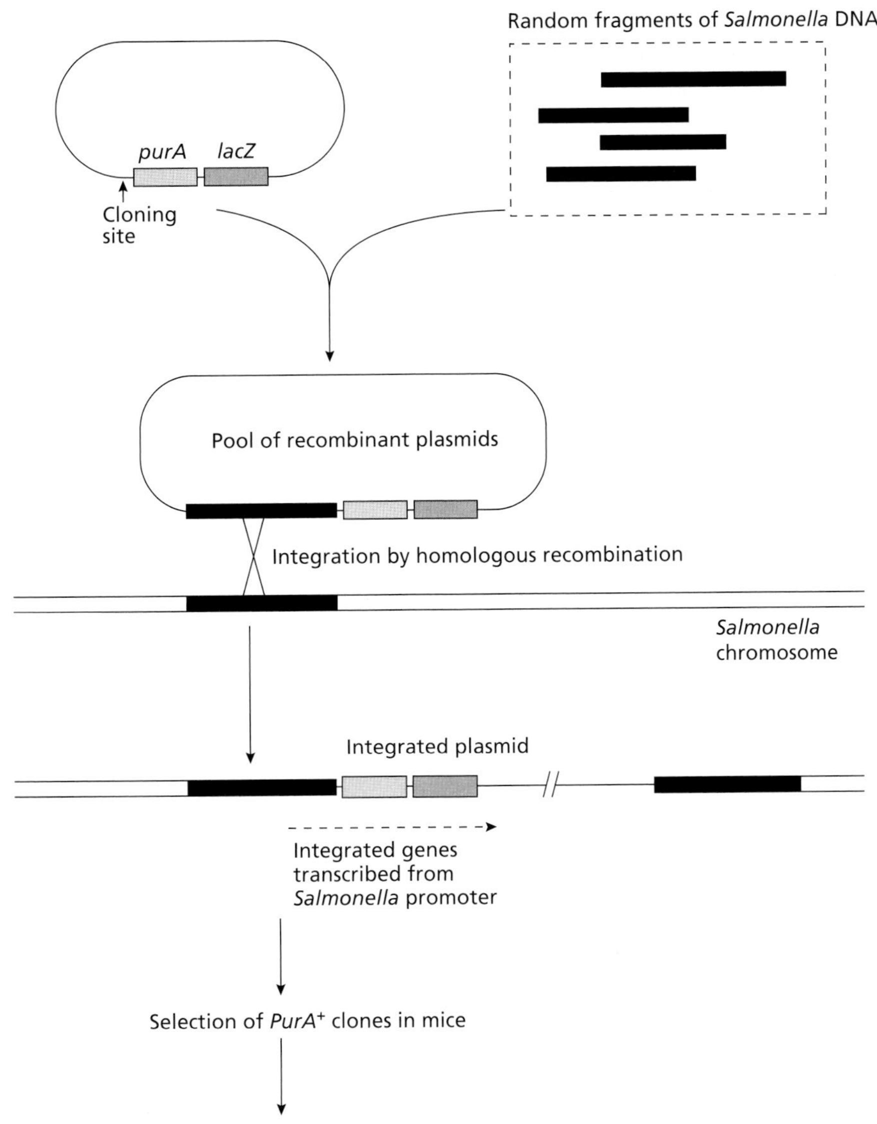

Fig. 14 *In vivo* expression technology (IVET). This procedure enables the identification of promoters that are specifically active during infection rather than in laboratory culture. For explanation see text.

are inserted into such a plasmid, only those with promoter activity will yield chloramphenicol-resistant bacterial clones. In a conventional promoter-probe experiment, this pool of clones would be subjected to chloramphenicol selection *in vitro*; however, it is possible to infect an experimental animal with the pooled bacteria, and apply selection by treating the animal with chloramphenicol. Alternatively, the pooled clones can be used to infect cultured animal cells with selection for expression of the cat gene by the intracellular bacteria; in one example (Staendner *et al.*, 1995), this procedure was used to identify *S. typhi* promoters that were activated by invasion of eukaryotic cells.

This simple technique has a number of limitations, and Mahan *et al.* (1993) therefore used a more sophisticated approach, to which they gave the name *in vivo* expression technology (IVET), the main features of which are outlined in **Fig. 14**. In this case, the selection system involved a *purA* mutant of *S. typhimurium*; complementation of this mutation by expression of an added *purA* gene restores the ability to grow within the experimental host. The artefacts that can arise from the use of plasmids, as referred to above, were avoided by use of a suicide vector, so that maintenance of the selective markers required integration into the bacterial chromosome by homologous recombination.

In order to facilitate measurement of the promoter activity, a second promoterless gene (*lacZ*) was included in the plasmid, downstream from the promoterless *purA*, so that the selected promoter would drive the expression of both genes.

The pooled bacteria containing the integrated *purA–lacZ* fusion (10^6 bacteria) were injected into a mouse; of the bacteria recovered from the spleen, the overwhelming majority were Lac^+ *in vitro*, and therefore were driven by promoters that were active both *in vitro* and (presumably) *in vivo*. About 5% were Lac^- *in vitro*, suggesting that although the promoters were active within the mouse they had little activity on laboratory media; this was confirmed by analysis of β-galactosidase levels in bacteria recovered from the spleens of infected animals. From this information, it is then possible to characterise those genes that are selectively expressed during infection, and subsequently to determine the role of those gene products in the infective process.

It must be recognised, however, that genes that are necessary for infection are not inevitably transcribed only *in vivo*. Apart from general housekeeping genes, and genes involved with fundamental biosynthetic processes, even genes concerned with specific virulence characteristics may have other roles *in vitro* (or may not be completely switched off *in vitro* even though they have no significant role). Furthermore, those factors that are required for initial stages of infection may need to be active to some extent during *in vitro* growth for infection to be successful. The results from IVET analysis or similar use of promoter-probe vectors should not therefore be taken as providing a complete inventory of genes required for virulence.

Differential and Subtractive Hybridisation

The differential expression of genes in specific environments can also be addressed by examination of the mRNA species produced. One form this analysis can take is to use reverse transcriptase to produce cDNA from mRNA templates obtained from bacteria grown under different conditions – for example, broth culture versus macrophage grown bacteria. (It is highly desirable to remove most of the rRNA, e.g. by subtractive hybridisation, to avoid the results being overwhelmed by rRNA signals.) Duplicate filters, of Southern blots of digested chromosomal DNA or of a gene library, can then be hybridised to each of these mixtures of cDNA probes. Bands, or clones, that hybridise to the cDNA from the macrophage-grown culture but not to that from the broth-grown culture, will then represent genes that are selectively expressed within macrophages.

In practice, this differential hybridisation tends to be insensitive and poorly reproducible. Better results have been obtained by subtractive hybridisation, for example as described by Plum and Clark-Curtiss, (1994). In this example (see **Fig. 15**), mRNA was extracted from *M. avium* grown in broth and in macrophages, and (after subtractive hybridisation to remove much of the rRNA) reverse transcribed to cDNA. Adapter sequences were ligated to the cDNA populations to provide PCR primer recognition sites. The amplified cDNA from the broth-grown culture was labelled with biotin, and the two amplified cDNA populations were denatured, mixed and re-annealed. The presence of the biotin label allowed the removal of

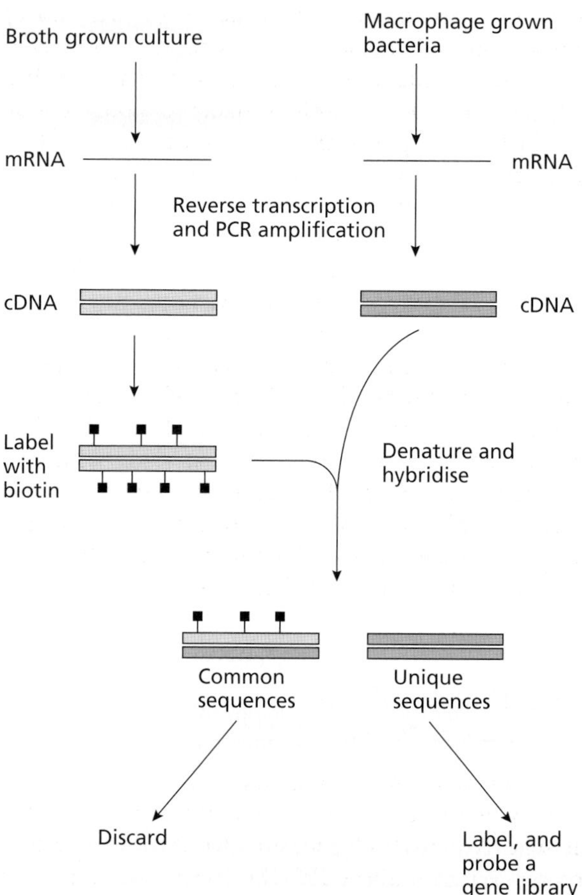

Fig. 15 Subtractive hybridisation. This is a procedure for detection of genes that are selectively expressed during bacterial growth within macrophages. The relevant mRNA will therefore be present only in the macrophage culture; after copying into cDNA and PCR amplification, the two DNA populations are denatured, mixed and allowed to anneal. Sequences present in both cultures will be able to hybridise, and can be removed as shown, while sequences present only in the macrophage-grown culture will not hybridise to the biotin-labelled DNA and will therefore be enriched in the supernatant.

broth-grown cDNA (using streptavidin-coated paramagnetic particles), together with annealed DNA from the macrophage-grown culture. As a result, the DNA in solution was enriched with cDNA produced from mRNA that was present only in the macrophage-grown culture. This step can be repeated as often as required. Finally, the subtracted cDNA can be labelled and used to probe a gene library to identify the genes involved.

Signature-tagged Mutagenesis

A different approach to the identification of virulence determinants, not dependent on assumptions of differential gene expression, is signature-tagged mutagenesis (Hensel *et al.*, 1995), which involves the use of transposon mutagenesis to generate attenuated mutants. Such mutants cannot be selected directly, but the insertion of a highly variable oligonucleotide within the transposon prior to mutagenesis, so that each transposon insertion carries a different sequence tag, enables a distinction to be made between the successful and unsuccessful mutants (Fig. 16).

In this procedure, as applied to *S. typhimurium*, a pool of transposon mutants was used to infect a mouse; bacteria recovered from the spleen represent a mixture of the mutants that were able to multiply successfully. From this mixture of bacteria, the sequence tags were amplified by PCR and labelled, giving rise to a mixture of probes containing the sequence tags from all the successful bacteria. Hybridisation of this mixed probe to an array of each of the original mutants enabled identification of the unsuccessful mutants, as they failed to hybridise to the sequence tag probe.

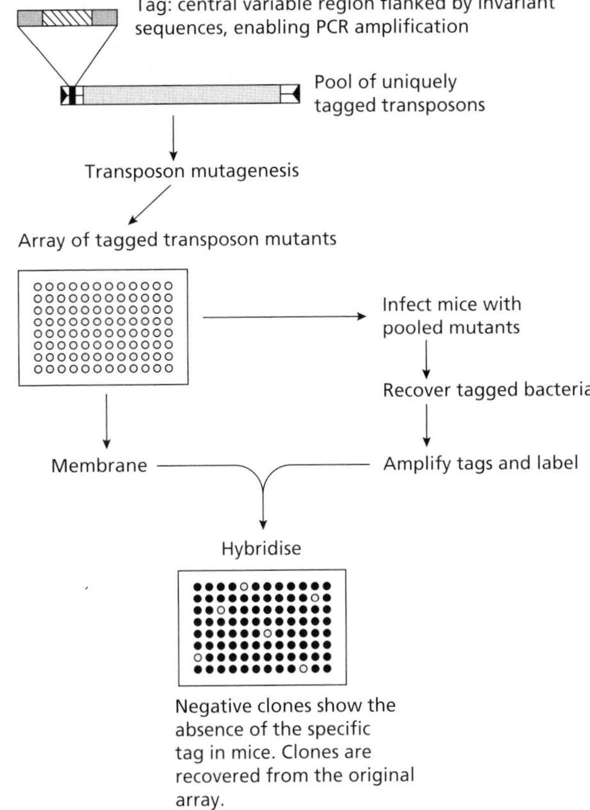

Fig. 16 Signature-tagged mutagenesis. Transposon mutagenesis is performed with a pool of transposons each carrying a unique sequence tag. When the pooled transposon mutants are used to infect mice, any mutants that are defective in essential virulence genes will fail to grow, and will therefore not be recovered in the final mixture. The tags from this mixture are used to probe the original array of mutants, and those mutants that are absent (because of inactivation of a virulence determinant) will be identified by the absence of hybridisation.

Identification of Virulence Genes by Functional Tests

If an attenuated mutant is available, it is possible to use a complementation test to identify the nature of the genetic defect, and hence to identify the supposed virulence determinant. This involves transforming the attenuated strain with a library of shuttle plasmids carrying random fragments of DNA from a virulent strain and infecting mice with pooled transformants. Any transformant that carries a DNA fragment restoring virulence will multiply more effectively in the infected animal, and will therefore be enriched in the population of bacteria recovered from the animal (to a greater or lesser extent, depending on the degree of attenuation of the mutant). This procedure has enabled the identification of genes that restore or enhance the virulence of attenuated strains of *M. tuberculosis* (Pascopella *et al.*, 1994) and *M. bovis* (Collins *et al.*, 1995).

For bacteria whose virulence depends on the ability to invade host cells, a similar procedure can be carried out with cultured animal cells. Thus Isberg and Falkow (1985) were able to identify a gene from *Yersinia pseudotuberculosis* that conferred on *E. coli* the ability to invade animal cells, and Arruda *et al.* (1993) cloned a DNA fragment from *M. tuberculosis* that enabled *E. coli* to invade HeLa cells (although the relationship of this gene to virulence is still unclear).

Conclusion

Genetic manipulation and associated molecular techniques have radically changed the study of medical microbiology, not only in the directions discussed in this chapter (such as the provision of purified protein antigens, the analysis of gene structure and regulation, and the identification of virulence determinants) but also in other respects discussed elsewhere in this book, notably molecular methods for diagnosis and epidemiology. The rapidly growing list of complete genome sequences of bacterial pathogens adds a further dimension to our knowledge. As we find new and better ways of integrating this molecular knowledge with the properties of the bacteria themselves we can expect even more dramatic advances in our understanding of the nature of virulence and host responses, which we hope will lead to equally dramatic advances in the treatment and control of infectious diseases.

References

Arruda S, Bomfim G, Knights R, Huima-Byron T, Riley LW (1993) Cloning of an *M. tuberculosis* DNA fragment associated with entry and survival inside cells. *Science* 261: 1454–1457.

Balasubramanian V, Pavelka Jr MS, Bardarov SS *et al.* (1996) Allelic exchange in *Mycobacterium tuberculosis* with long linear recombination substrates. *J. Bacteriol.* 178: 273–279.

Baulard A, Kremer L, Locht C (1996) Efficient homologous recombination in fast-growing and slow-growing mycobacteria. *J. Bacteriol.* 178: 3091–3098.

Böddinghaus B, Rogall T, Flohr T, Blöcker H, Böttger EC (1990) Detection and identification of mycobacteria by amplification of rRNA. *J. Clin. Microbiol.* 28: 1751–1759.

Brown TA (1995) *Gene Cloning: an Introduction*, 3rd edn. London: Chapman & Hall.

Chalfie M, Tu Y, Euskirchen G, Ward WW, Prasher DC (1994) Green fluorescent protein as a marker for gene expression. *Science* 263: 802–805.

Collins DM, Kawakami RP, De Lisle GW, Pascopella L, Bloom BR, Jacobs Jr WR (1995) Mutation of the principal σ factor causes loss of virulence in a strain of the *Mycobacterium tuberculosis* complex. *Proc. Natl Acad. Sci. USA* 92: 8036–8040.

Dale JW (1998) *Molecular Genetics of Bacteria*, 3rd edn. Chichester: John Wiley.

Dellagostin OA, Wall S, Norman E, O'Shaughnessy T, Dale JW, McFadden J (1993) Construction and use of integrative vectors to express foreign genes in mycobacteria. *Mol. Microbiol.* 10: 983–993.

Fredricks DN, Relman DA (1996) Sequence-based identification of microbial pathogens: A reconsideration of Koch's postulates. *Clin. Microbiol. Rev.* 9: 18–33.

Giovannoni SJ, Britschgi TB, Moyer CL, Field KG (1990) Genetic diversity in Sargasso Sea bacterioplankton. *Nature* 345: 60–63.

Griffith F (1928) The significance of pneumococcal types. *J. Hygiene* 27: 141–144.

Hensel M, Holden DW (1996) Molecular genetic approaches for the study of virulence in both pathogenic bacteria and fungi. *Microbiology* 142: 1049–1058.

Hensel M, Shea JE, Gleeson C, Jones MD, Dalton E, Holden DW (1995) Simultaneous identification of bacterial virulence genes by negative selection. *Science* 269: 400–403.

Isberg RR, Falkow S (1985) A single genetic locus encoded by *Yersinia pseudotuberculosis* permits invasion of cultured animal cells by *Escherichia coli* K-12. *Nature* 317: 262–264.

Mahan MJ, Slauch JM, Mekalanos JJ (1993) Selection of bacterial virulence genes that are specifically induced in host tissues. *Science* 259: 686–688.

McFadden J (1996) Recombination in mycobacteria. *Mol. Microbiol.* 21: 205–211.

Musser JM (1995) Antimicrobial agent resistance in mycobacteria: molecular genetic insights. *Clin. Microbiol. Rev.* 8: 496–514.

Norman E, Dellagostin OA, McFadden J, Dale JW (1995) Gene replacement by homologous recombination in *Mycobacterium bovis* BCG. *Mol. Microbiol.* 16: 755–760.

Old RW, Primrose SB (1994) *Principles of Gene Manipulation*, 5th edn. Oxford: Blackwell Scientific.

Pascopella L, Collins FM, Martin JM *et al.* (1994) Use of *in vivo* complementation in *Mycobacterium tuberculosis* to identify a genomic fragment associated with virulence. *Infect. Immun.* 62: 1313–1319.

Plum G, Clark-Curtiss JE (1994) Induction of *Mycobacterium avium* gene expression following phagocytosis by human macrophages. *Infect. Immun.* 62: 476–483.

Reyrat JM, Berthet FX, Gicquel B (1995) The urease locus of *Mycobacterium tuberculosis* and its utilization for the demonstration of allelic exchange in *Mycobacterium bovis* bacillus Calmette-Guerin. *Proc. Natl Acad. Sci. USA* 92: 8768–8772.

Rogall T, Flohr T, Böttger EC (1990) Differentiation of *Mycobacterium* species by direct sequencing of amplified DNA. *J. Gen. Microbiol.* 136: 1915–1920.

Snapper SB, Lugosi L, Jekkel A *et al.* (1988) Lysogeny and transformation in mycobacteria: stable expression of foreign genes. *Proc. Natl Acad. Sci. USA* 85: 6987–6991.

Staendner LH, Rohde M, Timmis KN, Guzmán CA (1995) Identification of *Salmonella typhi* promoters activated by invasion of eukaryotic cells. *Mol. Microbiol.* 18: 891–902.

Stover CK, de la Cruz VF, Fuerst TR *et al.* (1991) New use of BCG for recombinant vaccines. *Nature* 351: 456–460.

Vandamme P, Pot B, Gillis M, de Vos P, Kersters K, Swings J (1996) Polyphasic taxonomy, a consensus approach to bacterial systematics. *Microbiol. Rev.* 60: 407–438.

Wang GC, Wang Y (1996) The frequency of chimeric molecules as a consequence of PCR co-amplification of 16S rRNA genes from different bacterial species. *Microbiology* 142: 1107–1114.

Ward DM, Weller R, Bateson MM (1990) 16S rRNA sequences reveal numerous uncultured microorganisms in a natural community. *Nature* 345: 63–65.

Young RA, Bloom BR, Grosskinsky CM, Ivanyi J, Thomas D, Davis RW (1985a) Dissection of *Mycobacterium tuberculosis* antigens using recombinant DNA. *Proc. Natl Acad. Sci. USA* 82: 2583–2587.

Young RA, Mehra V, Sweetser D *et al.* (1985b) Genes for the major protein antigens of the leprosy parasite *Mycobacterium leprae*. *Nature* 316: 450–452.

23
Electrophoretic Typing

T. Hugh Pennington
University of Aberdeen, Aberdeen, UK

Historical Background

Why 'electrophoretic' typing? Surely, what is meant is 'molecular', or possibly 'genotypic'? Of course, but 'electrophoretic' is a particularly appropriate term, for it highlights the technique common to most of the enormous variety of typing approaches that fall under these rubrics. It also encourages us to consider their historical development. This is particularly important because such an analysis helps to explain why and when bacteriologists adopted the techniques now used and because it provides a useful starting point for an account of some of the unfinished business that still remains in this field. In any case, one cannot give an account of methods for characterising micro-organisms without making electrophoresis a major player, because of the essential and central role that it plays in molecular biology – in the purification and characterisation of proteins and DNA molecules, in DNA manipulation (genetic engineering), and in DNA sequencing. All these, and genomics and proteomics, would be impossible without it.

Electrophoresis was developed by Arne Tiselius in The Svedberg's department at the University of Uppsala (Kay, 1988). His first apparatus was unveiled in 1937. It was massive – 20 feet long and five feet high (6×1.5 m) – and expensive to construct. It was demanding to operate and very slow, because only a single sample could be analysed during each of its long run times. Nevertheless, it was successful. Thus Tiselius' inaugural paper not only described the apparatus but also for the first time demonstrated α-, β- and γ-globulins. His Nobel Prize for Chemistry followed in 1948. Perhaps the other most influential analysis done on an apparatus of this design was the study of sickle-cell haemoglobin conducted by Itano and others in Pauling's laboratory at Caltech in the late 1940s (Pauling *et al.*, 1949). For the first time it linked the results of classical genetics with molecular abnormalities, and showed sickle-cell anaemia to be a molecular disease. The limitations of the usefulness of the Tiselius apparatus were its high capital cost, low productivity and technical complexity, features that greatly restricted its availability and prevented its becoming a routine tool. More important were the limits imposed by its moving boundary methodology on its ability to resolve complex mixtures, because separation occurred in free solution. Major advances towards solving these problems were provided by the development of zone electrophoresis in the late 1940s and early 1950s (Tiselius and Flodin, 1953). In this approach the reduction of diffusion was achieved by using a porous support – usually paper, starch or agar – which itself was arranged in a small bench-top apparatus that could be easily and cheaply made in hospital and university workshops. Polyacrylamide made its appearance as an electrophoresis medium in 1959.

By the mid-1960s the users of electrophoresis had divided into two camps. One comprised biochemists and molecular biologists. Their electrophoretic work focused on the cataloguing and characterisation of proteins by molecular weight. The demonstration

 doi:10.1006/bkmm.2001.0023

by Shapiro, Viñuela and Maizels in 1967 of a direct relationship between the electrophoretic mobility in polyacrylamide gels (PAGE) and the molecular weight of proteins treated with sodium dodecyl sulphate (SDS) – a powerful charged denaturing detergent that caused proteins to migrate as anions – was particularly influential. By the end of the decade the routine achievement of high resolution and analytical accuracy had been made possible by Ornstein and Davis' (Ornstein, 1964) invention of multiple buffer systems for acrylamide gels that cause proteins to stack at the beginning of an electrophoretic run in concentrated 'discs' (leading to major improvements in resolution) and Laemmli's introduction of SDS into these buffers – in one of the most quoted papers of all time (Laemmli, 1970). It was greatly assisted by the development of slab gel formats, popularised notably by Studier (1973). This greatly increased productivity and has ever since been a central feature of electrophoretic techniques for all kinds of macromolecule. Despite its great power SDS–PAGE was easy to use. Its rapid adoption everywhere lent support to the notion – first promulgated by Erwin Chargaff – that molecular biologists were performing biochemistry without a licence!

Geneticists formed the other camp. They also became enthusiastic users of electrophoresis, but in a different way from biochemists and molecular biologists. The pioneer work in the early 1960s of Harris (1966) on human proteins and Lewontin and Hubby (1966) on *Drosophila* used non-denaturing zone electrophoresis systems and histochemical enzyme stains to show that variation in the electrophoretic mobility of the same housekeeping enzymes from different individuals occurred at frequencies far higher than expected. These revolutionary findings have been confirmed and extended over and over again – for bacteria as well as for eukaryotes. Although this approach – termed multilocus enzyme electrophoresis – has provided geneticists with a simple and rapid method for quantitating the degree of genetic difference between members of a population, its main impact has been to stimulate an enormous amount of theoretical work aimed at explaining why genetic polymorphisms are so common. These studies are still far from complete.

A kind of reunion of the two camps occurred from the early 1970s onwards, when the agarose and low-concentration acrylamide electrophoretic methods, which had been developed for separating small DNA fragments, were enthusiastically exploited by geneticists to analyse the products of restriction enzyme digestion. Like the protein methods of biochemists, these gave size-based rather than charge-based separations. These electrophoretic techniques, and developments from them such as pulsed-field gel electrophoresis (PFGE; see below), are now extensively used in typing systems that detect DNA sequence polymorphisms.

The award of the 1978 Nobel Prize for Medicine or Physiology to Hamilton Smith, Werner Arber and Daniel Nathans marked the seminal contributions of these workers to our understanding of restriction enzymes and their utility as molecular biology reagents. Building on Arber's pioneer work on restriction itself, Smith showed that type II restriction enzymes recognise specific sequences in DNA molecules and then cleave them at that site. Nathans went on to use a *Haemophilus influenzae* enzyme to analyse the genome of simian virus 40 (SV40) (Danna and Nathans, 1971). In this paper he was the first to describe the pattern of bands in a gel as a 'fingerprint'. In 1972, also with Danna, he was the first to type a micro-organism by the restriction enzyme fragment analysis of its genome. Four strains of SV40 were distinguished by band pattern differences, probably caused by small deletions. The analysis of restriction-fragment length polymorphisms (RFLPs) now plays a central role in the DNA typing of bacterial pathogens.

Another Nobel prize-winning technique (Kary B. Mullis, Chemistry, 1993) that has had a revolutionary impact on the typing of bacteria is the polymerase chain reaction (PCR; see Chapter 24; Rebecca Smith). Starting with almost any amount of DNA, it is possible to amplify a particular sequence up to microgram quantities, an amount easily detectable by electrophoresis and simple staining procedures. If some simple conditions are met the product can be sequenced directly.

Typing – General Principles

By the end of the 1970s a number of well-validated, simple, efficient, economical and relatively rapid – but none the less very powerful – electrophoretic methods for the characterisation of macromolecules had become available. How have they been used since then by bacteriologists for typing? To answer this question properly it is first necessary to consider typing itself – its aims, the scientific philosophy that underpins the thinking of those who conduct it and use its results, and the problems that must be taken into account when selecting appropriate methods and interpreting their results.

E.S. Anderson's (1959) account of the aims of typing and the associated methodological requirements that need to be satisfied, has never been bettered:

> Communicable bacterial diseases present epidemiologists with the task of determining the avenues by which particular epidemic strains of bacteria have gained access to patients. Moreover, in a given outbreak of an infectious disease such as typhoid fever, it is important to know whether the organisms isolated from patients all have a common origin, that is, whether the epidemic springs from one or multiple sources. Clearly, in an infection caused by a single bacterial species, what is needed is a method of classification at the intraspecific level. In other words, the species should be divisible into 'types'. In an efficient typing scheme, the organisms isolated from patients, from vehicles of infection such as food or water and from the carrier responsible for the outbreak of the disease under study, all react identically. If an organism is widely disseminated, and especially if it is commonly carried by otherwise healthy persons, epidemiological study is impossible without such a method. In order to be practically useful a typing method must satisfy the following requirements: the types and the typing reagents used for their recognition should be stable; the organism should be subdivisible into an adequate number of types; the technique should be simple and the results reproducible and easy to read. The method should be capable of standardisation in order that, wherever it is employed, the results are comparable; the results of the test should be available quickly; and, before being accepted for general use, the reliability of the method should be established by exhaustive epidemiological trials.

The universal applicability of a method to all isolates is an important criterion not listed by Anderson, probably because of its self-evident nature and his particular interest in phage typing schemes which met it. His omission of cost as a criterion reflects his good fortune in working at a time when a public health laboratory was seen primarily as a protector of the public health rather than, as now, a charge on the public purse.

Although many developers of electrophoretic typing methods have paid full and complete attention to these principles, some have not. Anderson himself provides part of the explanation:

> As far as the pathogenic bacteria are concerned, much of the existing classification has resulted from expediency – the search for methods of distinguishing with confidence the pathogenic from the nonpathogenic flora with which they may be fortuitously associated, and of obtaining them in pure culture for diagnostic study. But the final validity of bacterial species, in the sense in which the term is employed in the classification of the higher organisms, still remains in doubt.

In other words, clinical bacteriologists have been particularly pragmatic in their use of microbial characteristics for classification. They have been aided, indeed driven, to this by the lack of a well-established theoretical – and even evidential – underpinning for bacterial systematics. The power of electrophoretic methods readily to generate large amounts of information has combined with this approach to tempt some to apply them uncritically – in particular, before the development of an appropriate understanding of the genetic basis of these methods.

Are Types Clones?

A pragmatic approach to the use of electrophoretic data has also been encouraged by the view that, because bacteria reproduce asexually, their population structures must be clonal. In the early 1980s, multilocus enzyme electrophoresis studies provided strong evidence that this was true for *E. coli* and for *Salmonella* (Selander *et al.,* 1996). Evidence for clonality in *Salmonella* was the markedly non-random association of enzyme electropherotypes for particular strains. In genetic terms, this is 'strong linkage disequilibrium among alleles at enzyme loci'. The association of particular O and H serotypes with one or only a small number of types defined by enzyme electropherotyping ('multilocus enzyme genotypes'), and the global distribution of some of these genotypes, indicates that these types have long lives – lifetimes long enough to allow them to spread, unchanged, over great distances. The close correlation between genotype, as initially defined by enzyme electrophoresis and more recently and more directly by sequencing studies, and serotype explains the epidemiological success of *Salmonella* serotyping. This was epitomised in the 1934 statement by the *Salmonella* Subcommittee of the Nomenclature Committee of the International Society for Microbiology:

> ... belief in serology as the ultimate criteria in the taxonomy of (*Salmonella*) is the consequence of the experience of the very great practical services serology has rendered, turning a field of uncertainties and pitfalls into one in which identification has become easy, certain and closely correlated with pathology and epidemiology.

It has to be said that the paradigmatic role played by *Salmonella* in typing studies has not always been helpful for bacteriology as a whole. For other pathogens, and for some workers, it has raised the expectation that if only the right markers could be found, a typing scheme with all the excellent properties of the Kauffman–White schema would automatically emerge. But if all bacterial pathogens had as stable a clonal population structure as *Salmonella*, any set of markers – whether generated by serotyping, biotyping or electrophoretic typing – could be used to identify strains (Smith and Smith, 1999). Experience has, however, shown that for many taxa, strain identifications made by more than one of these methods do not map well on to each other. Why is this so? It is due to variations in clonal lifetimes from species to species. For all bacteria the structure of genetic variation is the result of recombination superimposed on the repeated formation and spread of clones (Milkman, 1996). The integrity of clones, as defined by their DNA sequence, will be lost more rapidly if recombination occurs more often. In practical terms this means that the close relationship between different markers, as seen for organisms like *Salmonella*, will also be lost. Clearly, some bacteria are enjoying parasex – transformation, transduction and conjugation – more often than was thought! The essential genetic problem for typers of pathogens has been nicely summarised by Maynard Smith (1996):

> A sample of bacterial strains is analysed electrophoretically or by some other technique, and the resulting data and some standard package are used to construct a phylogenetic tree. This construction is justified only if the objects being classified have arisen by a process of branching without recombination. This is often near enough the truth. But ... it is by no means always true. The population genetics of bacteria is difficult because recombination is neither absent nor very common.

Another problem for analysts is that recombination processes are both 'subtle and multifarious' (Milkman, 1996). Strong evidence has emerged in recent years (Maynard Smith, 1996) from the gene sequences of related bacteria of a mosaic structure, in which the chromosomal DNA of a set of related bacterial strains consists of a basic clonal frame peppered with short inserted sequences. In practice this means that the phylogenetic trees of different genes and different parts of the same genome will not necessarily be identical – in other words, that typing of the same strain done with different markers may give different results.

The main conclusion to be drawn from all this is that because the genomes of different pathogens are evolving in different, and individualistic, ways, it is very important to get as much information as possible about the population genetic structure of a pathogen, and reflect on it, before choosing a typing method and determining the rules for the interpretation of results. Two of Anderson's requirements for a satisfactory typing system – that the types should be stable and that the organism should be subdivisible into an adequate number of types – are, of course, primarily determined by the genetic structure of a bacterial population and its evolution. Clonal lifetimes and the number of clones in a population under investigation are key parameters here. In practice, information about them for pathogens is still very sparse. For the latter, investigators have relied on data generated by the typing scheme which they are studying to assess whether it provides an adequate degree of discrimination between different strains. A good example of the highly empirical approach adopted as a guide to practice is the use of Simpson's index of diversity (D) to compare typing methods and select the most discriminatory one (Hunter and Gaston, 1998). The index is given by the following equation:

$$D - 1 = \frac{1}{N(N-1)} \sum_{j=1}^{s} n_j (n_j - 1)$$

where N is the total number of strains in the sample population, s is the number of types described, and n_j is the number of strains belonging to the jth type. Hunter and Gaston recommend 'that an index of > 0.9 would seem to be desirable if the typing results are to be interpreted with confidence.' This *ex cathedra* statement has been widely accepted as a guide to practice.

The lack of information about the clonal lifetimes of pathogens contributes to a particular problem associated with contemporary electrophoretic typing systems. It is that the quantitative interpretation of results for most of them is significantly hampered by a paucity of information about the frequency of occurrence of types, their geographical distribution and their longevity. One reason for this is that few large long-term studies have so far been done to estimate these parameters. In addition, much of the data that is available comes primarily from the analysis of strains collected during outbreaks, a not entirely representative sample.

How Do the Users of Electrophoresis Define Types?

Anderson's criteria for a practically useful typing system refer not only to the nature and properties of types themselves, but the techniques used to identify them. They should be 'simple and the results reproducible and easy to read ... capable of standardisation ... the results should be available quickly.' In practice this means that typing methods are compromises. Even the most information-rich electrophoretic typing systems sample only a very small proportion of the phenotypic and genotypic information and variation that is there. Nevertheless, the assumption is usually made that electrophoretic typing data accurately indexes genotype. It is assumed, for example, that the nucleotide sequence variation responsible for differences in restriction fragment band patterns is representative of that in the genome as a whole. It may be that misunderstandings about the terms 'genotype' and 'genotyping' contribute to overconfidence here. Strictly speaking, DNA sequence data is, of course, phenotypic – at the molecular level (Lewontin, 1992). Detecting a sequence difference does not necessarily tell us anything about genotype.

Comparative genome sequence data from different strains of pathogens will eventually allow the resolution of questions about the relationship between typing results and genotype and the representative nature of markers. It will allow types to be genetically defined with confidence. Until it is available, pragmatic approaches to the design of typing systems, and hence the definition of types, will continue to be dominant. Typing schemes, and the choice of markers they employ, will continue to be judged in large part by their utility. It is fortunate, therefore, that most medical microbiologists have no hang-ups about the fundamentals of systematics. They follow G.S. Wilson's approach in that they 'subordinate general principles to convenience' (quoted in Cowan, 1962). In a philosophical sense they are followers of William of Ockham, not only as adherents to his razor '*entia non sunt multiplicanda praeter neccesitatem*' – which may be paraphrased as, 'for the job in hand give preference to the simplest system' – but also in being subscribers to one of the two major mediaeval philosophical systems – nominalism – as far as classification goes. This means that at the end of the day they regard bacterial types and species as nothing more than laboratory constructs designed for a particular purpose – the diagnosis and investigation of disease. The other great mediaeval philosophical position, realism, which claimed that entities such as species have a special defining essence, is eschewed. Even those who hanker after this view regarding types are 'schizophrenic' about it in the sense that like a successfully treated patient they may still hallucinate but they don't worry about or act on their pathological thoughts. Certainly, the literature is dominated by the empirical view. Sometimes this lack of concern about the general principles that determine the population genetic structure of a pathogen, and hence its types, together with a desire for simple solutions, gets bacteriologists into trouble. As Francis Crick has pointed out, reliance on Ockham's razor is dangerous in biology because organisms have not been designed, but have evolved. This means that counterintuitive complexity is common.

Nevertheless, types are currently defined as the entities that appear to work best for the task in hand. Two kinds of task present themselves. As soon as an outbreak control team has isolated multiple strains of a pathogen, they want to know whether they are identical. Turn-round times for typing are of the highest priority. If a high degree of discrimination can be made compatible with this, so much the better. Tests with a high Simpson diversity index will be favoured, even if they use very polymorphic and rapidly evolving markers, because it is reasonable to assume that a point-source outbreak will be caused by a single clone which will undergo rather little genetic variation during the life of the outbreak. Different requirements operate for investigations on the patterns of infection across a country and over time. The reproducibility of results in different laboratories at different periods is far more important. The increased likelihood that genetic changes will be occurring during the study period must also be taken into account when making choices about typing methods and rules for the interpretation of results. The differences in these requirements – real-time information needed for the management of infection on the one hand and long-term monitoring data on the other – and a desire to meet them simultaneously explains the popularity of hierarchical typing schemes. These often use a combination of non-electrophoretic methods that have a low Simpson diversity index but a certain reproducibility that comes from using internationally available reagents such as phage typing and an electrophoretic method with a high index. Although attractive to some epidemiologists because results are expressed in clear and traditional nomenclatorial terms that are easy to understand, hierarchical systems that use serology or phage typing must be regarded as undesirable interim compromises because with a few exceptions the types defined by these phenotypic methods cannot be confidently related to genotype in a measurable way.

Protein Electrophoresis

As a routine and generally applicable typing methodology, protein electrophoresis enjoyed a vogue in the late 1980s. Since then it has been almost entirely supplanted by DNA typing methods. Nevertheless, there are two reasons why as a typing approach it still merits consideration here. First, some of the basic scientific considerations that underpin it are general, in that they apply to all electrophoretic methods. Second, protein electrophoretic approaches and developments based on them still play a limited but valuable role for some pathogens.

The use of protein electrophoresis for taxonomic purposes, including the typing of bacteria, was predicted by Sokal and Sneath (1963). It fitted well with their proposal at that time that organisms should be classified by a method that they called 'numerical taxonomy', since termed 'phenetics'. Dissatisfied with what they saw as Procrustean attempts to force classification schemes into patterns derived from phylogenies, which were at best intuitively based on a combination of dubious evidence and guesswork, they proposed that the classification of organisms should be estimated independently of phylogenetic considerations, and be based on a large number of characters, preferably more than 60 and never less than 40, which would all be given equal weight. Taxonomic entities would be constructed by computing coefficients of relationship to generate clusters with high degrees of similarity. They emphasised the strictly empirical nature of their scheme. This feature is not the only similarity between the principles of numerical taxonomy and the way that electropherotypers of bacterial pathogens have worked, and indeed, continue to operate. Thus, the clustering methods used by numerical taxonomists lent themselves to the representation of similarities, and relationships in general, by dendrograms. Likewise, electropherotypers have also found this to be an attractive way of visually presenting large amounts of data.

An early and influential concatenation of phenetics and bacterial protein electrophoresis was the study of Kersters and DeLey (1975) in which a scanner was used to generate data from polyacrylamide gel electrophoresis of solubilised proteins in order to assess taxonomic relationships. This was done using Dice's coefficient of similarity, S_D. It is given by the following equation:

$$S_D = 2nJK / (2nJK + u)$$

where nJK is the number of characters in which both of two types (or species) being compared is positive, and U is the number of characters positive in one and negative in the other. These techniques are now widely used to calculate similarities between DNA restriction enzyme band patterns. Kersters and DeLey were also among the first to use techniques to ameliorate the problem of achieving a high degree of reproducibility from gel to gel – or even from track to track in the same gel. They 'normalised' electrophoretograms using markers of standard size to index positions at the top and bottom of gels. It has to be said that for all gel-electrophoretic methods, and for DNA as well as proteins, problems of variation between, and sometimes within, gels still remain. Standardisation of reagents and obsessive attention to detail have reduced, but not eliminated, them.

During the 1980s many typing systems for pathogens were developed which used sodium dodecyl sulphate polyacrylamide gel electrophoresis and Laemmli's buffers (SDS–PAGE). They relied almost exclusively on the detection of variation in polypeptide molecular weight. Some examined proteins extracted from whole cells by boiling in SDS (Paterson *et al.*, 1987). In others, the characteristic detergent solubility properties of outer membrane proteins were used to concentrate them and discard cytoplasmic ones before electrophoresis (Barenkamp *et al.*, 1981). Some used western blotting to detect protein polymorphisms (Krikler and Pennington, 1986). In this technique, electrophoresed proteins were transferred to nitrocellulose membranes and reacted with antisera – often those contained in pooled human serum or plasma. Most systems used a simple detection system for bands in gels – staining with Coomassie blue – although the radioactive labelling of bacterial cultures with ^{35}S-methionine allowed the development of semi-automated detection systems (Stephenson *et al.*, 1986). Western blot detection systems were introduced because they had greater diversity indexes than Coomassie blue staining of the same bacterial protein extracts. It is reasonable to assume – although never put to formal test – that the explanation for this is that the proteins being detected in blots were particularly immunogenic because of their exposure at the bacterial cell surface. This location exposed them to immune attack and so accounted for their polymorphic nature through strong selection pressure. It has been shown that cell surface proteins are evolving at nearly three times the rate of cytoplasmic enzymes and proteins involved in transport, DNA replication and the regulation of transcription, and 10 times faster than ribosomal proteins (Whittam, 1996). With the exception of outer-membrane-protein typing schemes – which themselves suffer from the limitation of defining types on the basis of a rather small number of extremely polymorphic proteins (Forbes *et al.*, 1992) – a problem for SDS–PAGE typing systems is the difficulty of referring variations in band patterns to allelic variations because of an almost total absence of information about the identity of bands in gels. A typing system that does not suffer from this disadvantage is the multilocus enzyme electropherotyping system much used by geneticists (Selander *et al.*, 1986). Because the electrophoretic mobility of the water soluble cellular enzymes examined in this system is determined by amino acid sequence, protein variants – allozymes – can be interpreted as simple allelic products of a gene. Despite – or maybe because of – this advantage, multilocus enzyme electropherotyping of bacterial pathogens has been used more by geneticists than by medical microbiologists. However, it has been extensively employed in investigations of the population genetic structure of *Neisseria meningitidis* (Yakubu *et al.*, 1999). These have demonstrated that particular electropherotypes – hypervirulent lineages – often dominate as causes of disease. This typing method is particularly suited to long-term and geographically wide-scale epidemiological monitoring. Its diversity index is relatively low. It examines only a small number of markers. Even for them it underestimates genetic variation because not all amino acid replacements lead to changes in electrophoretic mobility and because a proportion of nucleotide sequence changes are silent from a coding point of view. However, none of these limits to resolving power have turned out to be major problems in practice, unlike the technical difficulty of achieving inter-laboratory reproducibility. This has made it difficult to compare results obtained in different centres. The development of multilocus sequence typing (MLST) by Maiden *et al.* (1998) has resolved this problem, because it generates DNA sequence data – which are, of course, electronically transferable. It has also increased the resolving power of the multilocus approach, retaining the advantages of studying known proteins that are not under particularly strong selection pressure, an important feature for long-term studies. Fragments of about 470 bp from the major loci known to mark types associated with disease are amplified by PCR and sequenced. With *N. meningitidis* this identified between 26 and 156 variable bases per fragment, and an average of 17 alleles per locus. Testing six loci identified as many as 24×10^6 MLST types.

DNA Typing Methods

A major feature of the second half of the twentieth century was the replacement of analogue technologies by digital ones. This trend has been paralleled in molecular biology by a similar trend in the development of approaches for DNA analysis. Analogue methods, such as DNA–DNA hybridisation, were developed early. They have played a role – if a controversial one – in working out phylogenies and classifications at the supra-specific level, but have never been particularly useful as typing tools. Their resolution is far too low because they average out genetic differences to give median sequence divergences over whole genomes. The only bacterial DNA molecules which present themselves in the form of units small enough to be analysed directly by gel electrophoresis are plasmids. They can often be characterised easily by this method, and their number and size has been widely used as supplementary typing information (Farrar, 1983).

Genomic DNA differs from plasmids and proteins in that it has to be processed before the information it contains can be used for typing. So the aim of all DNA typing methods has been to generate electrophoresable-sized fragments of genomes by methods which index sequence variation. They have been so successful that for many pathogens they have become not only the methods of choice but the gold standard for other methodologies. Commercial firms have been very active in providing a wide range of reagents and equipment for them. Every bacterial isolate has a unique genome sequence because of recombination, mutation and other genetic processes. The ability of DNA typing to characterise this uniqueness gives it an unassailable advantage over all other methods. SDS–PAGE typing has joined shellac and vinyl records – continuing our analogue-to-digital parallel – in being swept away by the 'perennial gales of creative destruction', to use the economist Joseph Schumpeter's famous description of industrial innovation. Two different approaches used singly or in combination form the basis of most DNA typing methods. Genomic DNA can be cut with restriction enzymes and the resulting fragments analysed directly by electrophoresis. Alternatively, small regions of the genome can be selectively amplified by the PCR using primers and amplification protocols designed to detect particular polymorphisms.

Typing by Restriction Enzyme Digest Fragment Analysis

One of the earliest applications of restriction enzyme analysis to bacterial pathogens was a study of *Leptospira* serovars with *Eco*RI (Marshall *et al.*, 1981). Although serovars *hardjo* and *balcanica* were easily distinguished by the visual inspection of stained gels, the very large number of fragments produced by this enzyme resulted in extremely complex band patterns with a high background. This made quantitation impossible. Since this early work a number of approaches have been developed to circumvent these disadvantages. The majority have focused on ways that operationally reduce the number of bands in a gel and so facilitate pattern analysis. Southern blotting, in which the chromosomal DNA digest is transferred to a membrane and then hybridised to a labelled probe homologous to selected genes, is one way of achieving this. Ribotyping, in which the probes are derived from the 16S and 23S rRNA genes, has been widely used (Grimont and Grimont, 1986). The relatively low degree of polymorphisms in these genes and the added technical complexity of blotting has in general favoured an alternative approach, which is to use restriction enzymes that cut the DNA genomes under analysis relatively rarely and so produce a conveniently small number of fragments. A genetically based approach (Forbes *et al.*, 1991), to select the most appropriate restriction enzyme for the analysis of whole genomes, has been to calculate the frequency of restriction enzyme recognition sites from the frequencies of di- and trinucleotides in sequenced genes by Markov chain analysis. Band patterns are compared by dividing them into sections with DNA size standards, scoring the number of bands in a few of these sections and comparing scores to establish similarities. Although this approach is flexible and rapid, a more empirical approach has been favoured in practice. It is to digest the DNA under study with restriction enzymes that cut rarely because of the large size or unusual sequence configuration of their recognition sites. This approach generates large DNA fragments. It was only made possible by the development of pulsed-field gel electrophoresis (PFGE) by Schwartz and Cantor (1984). Their initial work showed that by using alternately pulsed, perpendicularly oriented non-uniform electric fields it was possible to achieve a molecular-weight-dependent separation of DNA fragments larger than 50 kb – a hitherto impossible task using traditional undirectional uniform fields.

PFGE is currently regarded by many as providing the gold standard for all DNA typing methods. It holds this position for four reasons. First, the original aim of producing a small and analytically optimum number of bands in gels is nearly always easy to achieve in practice; the ability to fine tune by varying electrophoresis conditions, and the availability of a wide range of restriction enzymes, helps here. Second, the reproducibility of PFGE from gel to gel and from laboratory to laboratory is as least as good as for other restriction enzyme typing methods, and is better than some. Third, experience has shown that it can be used to type nearly all human pathogens. Fourth, it has turned out to have an unreasonably high discriminatory power – a high Simpson diversity index. What is meant here by unreasonable is that although PFGE looks only at a very small number of restriction enzyme sites, it reveals nevertheless a seemingly large amount of genetic polymorphism and it does this in organisms as taxonomically diverse as *Escherichia coli*, *Neisseria meningitidis*, *Mycobacterium avium*, and *Staphylococcus aureus* (Olive and Bean, 1999). Insufficient genome sequence data are available at present to account for this as a general phenomenon. However, a genome sequence comparison at two unrelated isolates of *Helicobacter pylori* (Alm *et al.*, 1999) has shown that PFGE may overestimate the true extent of genetic diversity in this organism. Although the overall genomic organisation, gene order and predicted sets of proteins encoded by the genomes of the two isolates were quite similar, with 310 proteins (out of about 1500) having $\geq$98% amino acid conservation, their PFGE profiles were completely different. The differences were mainly due to silent nucleotide variation within genes. One strain contained twice as many sites for the rare-cutter restriction enzyme *Not*I as the other, and silent nucleotide changes in the former were responsible for the absence of six of the seven sites unique to the latter; differences at the seventh site resulted in the alteration of a single amino acid.

These results with *H. pylori* clearly indicate that great care must be exercised before data from restriction enzyme analyses are used to make judgements about *quantitative* differences between strains. General genetic considerations also support this view. Restriction endonuclease data occurs in two forms: either the presence or absence of recognition sequences for particular enzymes as map locations on the genome – restriction site data – or the presence or absence of restriction fragments of a given length – restriction fragment data. At present, virtually all typing data falls into the second category, because appropriate restriction maps are not available for the pathogens under study. Phylogeneticists have made the clear recommendation that this kind of data should not be used for phylogenetic analysis (Swofford and Olsen, 1990). This is because it violates the crucial assumption of character independence. Swofford and Olsen summarise the problem succinctly:

> If a new site evolves between two pre-existing sites, one (longer) fragment disappears and two new (shorter) ones appear. Thus even though two (types) may share two of the three restriction sites, they have no fragments in common, a potentially serious source of error. Some authors have recognised this difficulty and argue that it can be overcome by looking at 'enough' fragment data so that each occurrence of this kind of error will be swamped by other data. We remain unconvinced by this argument, however, because there is no guarantee that if something is done inappropriately enough times, all will work out in the end.

Swofford and Olsen also point out that two other problems attend the use of restriction fragment data for phylogenetic analysis: insertions and deletions are difficult to handle, and the asymmetry in gain and loss of restriction sites (it is easier to lose a particular site than to gain it) needs special analytical attention.

The main conclusion to be drawn from these genetic considerations is that the users of PFGE – or any other electrophoretic typing method that generates patterns of bands which have not been mapped onto genomes or identified genetically in other ways – should focus on epidemiological considerations rather than evolutionary ones. After all, that is what typing is for. In practice it means that whenever band-pattern typing data lacking genetic validation are used to construct a dendrogram it should be remembered that the only purpose is presentational. It is not an evolutionary tree but a convenient way of summarising data in a visually appealing form. When genetic validation is there, an abundance of statistical approaches is available to calculate strain relationships (Swofford *et al.*, 1996).

Tenover *et al.* (1995) have proposed guidelines for the interpretation of PFGE band patterns. They take into account all the problems discussed above, and their guidelines are based not only on the fundamental genetic principles that underpin the method, but also on experience gained from the typing of strains during outbreaks. Rightly, it is emphasised that the guidelines are

> intended for use in analysing discrete sets of isolates obtained during epidemiological studies of potential outbreaks in hospitals or communities spanning relatively short periods (1 to 3 months). The criteria for strain identity are stringent and are not appropriate for studies of large populations of organisms collected over extended periods of 1 year or longer.

Even with stringent criteria for strain identity, pragmatic and qualitative elements remain. Thus the four categories of relatedness of strains that Tenover lists are 'indistinguishable' (zero fragment difference), 'closely related' (2–3 fragment difference, with an epidemiological interpretation of 'probably part of the outbreak'), 'possibly related' (4–6 fragment difference, with an epidemiological interpretation 'possibly part of the outbreak') and 'different', with ≥ 7 fragment difference. This flexibility is right and proper. It echoes and is in the best tradition of the *Origin of Species*. In one of the few comments Darwin made there about how to define a taxonomic entity he said,

> in determining whether a form should be ranked as a species or a variety, the opinion of naturalists having sound judgement and wide experience seems the only guide to follow.

The Polymerase Chain Reaction

The availability of a rapid low-cost method for amplifying small lengths of DNA to easily detectable levels has led to the development of a range of typing methods (Olive and Bean, 1999). The only limit to their variety has been the ingenuity of their inventors. These methods can be divided into two basic categories: those that require prior knowledge of genome sequence data, and those that don't. Two methods are typical of the latter approach. In the random amplified polymorphic DNA method (RAPD) (Welsh and McClelland, 1990; Williams *et al.*, 1990), short arbitrarily chosen sequence primers are used at low annealing temperatures. When two of them anneal to sites a few kilobases apart in the proper orientation, a product results with its length corresponding to the distance between them. The method can be used for typing because the distributions of these random sites vary from strain to strain. Intellectually this method is attractive because in theory it has universal applicability. In practice, the extreme sensitivity of the amplification process to temperature leads to variability in banding patterns which have significantly reduced its utility.

A completely different approach is the amplified fragment length polymorphism assay (Vos *et al.*, 1995). It combines RFLP assays and PCR. Linkers containing sequences homologous to PCR primer binding sites are ligated to restriction fragments. Their sequences have been designed so that only a subset of restriction fragments will be amplified. The PCR primers are labelled so that the products of amplification can be easily detected; fluorescent labels allow the reproducibility and information storage facility of automated DNA sequences to come into play.

The majority of PCR approaches utilise DNA sequence information known for the genomes under study. Most target repetitive sequences (Versalovic *et al.*, 1991). These are ubiquitous in bacterial genomes. These sequences may be DNA common to a species, or may derive from DNA introduced by horizontal transfer – accessory genetic elements (Bachelier *et al.*, 1996; Deonier, 1996). Much of our current knowledge about them comes from work with *E. coli*, although data from genome sequencing projects is rapidly adding to information on them for other species. Large repeats include rRNA genes – used for ribotyping – and insertion sequences (IS). These accessory elements vary enormously from taxon to taxon. Thus, for example, two well-characterised laboratory strains of *Salmonella typhimurium* (LTZ) and *E. coli* (K12), respectively, contain only one type, and between 8 and 12 different types, of IS element. Because IS elements can transpose, their status as genetic loci is less secure than conventional genes, although the position of some is quite strongly conserved. Small repeats are scattered over the chromosome. They have imperfectly palindromic sequence structures. On the basis of variations in these structures they can be classified into several different families. Their number and distribution varies a great deal from strain to strain; no transposition mechanism has been demonstrated for them.

Repeat sequences used for typing include Iss; repetitive extragenic palindromic (REP) elements (38 bp sequences made up of six degenerate positions and a 5 bp pair variable loop between each side of a conserved palindromic stem); enterobacterial repetitive intergenic consensus sequence (ERIC) (126 bp elements located in extragenic regions of the genome and containing a highly conserved central inverted repeat); and the mosaic repetitive elements called BOX, made up of various combinations of three subunit sequences. They are also located in extragenic regions. This is a particularly dynamic field from the developmental point of view. Other repetitive sequences are currently being investigated as typing tools, and new and improved methods for detecting polymorphisms associated with them are also under active development.

PCR has also been used to amplify single polymorphic genes to generate produces which can be characterised for typing by RFLP analysis or sequencing. However, the smaller the proportion of genetic variation in an organism sampled by a typing technique the less likely is the prospect that the results will accurately index variation in the whole genome. A good example of this has been provided by sequence studies on the tandem flagellin genes *fla*A and *fla*B of *Campylobacter jejuni* (Harrington *et al.*, 1997). The results provided clear evidence of intergenomic recombination between *fla*A genes of different strains and intragenomic recombination between *fla*A and *fla*B genes. Although typing by flagellin gene PCR amplification and restriction enzyme digestion has been popular because of its speed and discriminatory power, this demonstration that these genes undergo sequence reassortment at high rates by horizontal gene transfer and recombination – probably reflecting the strong selection pressure exerted by the host immune response – makes the prospect of accurate long-term epidemiological surveillance by this system an unlikely one because sequence variation in them clearly does not accurately reflect the overall genetic structure of the organism.

How to Choose a Typing Method

A large number of powerful electrophoretic typing methods is available. To this must be added the other more traditional approaches such as serotyping and phage typing, because they still have their proponents, despite their frequent inability to index genotype, relatively poor reproducibility, lack of applicability to all strains, and dependence on variable biological reagents that are expensive to produce and have extremely restricted availability. How are we to choose between methods, particularly if no particularly strong advantages or disadvantages emerge after consideration of Anderson's criteria and the population genetics of the organism under consideration? A simple rule of thumb which can be used to indicate that certain typing methods are more likely to be sampling markers representative of genotype is the consilience of inductions. First enunciated by William Whewell more than 150 years ago, it states that 'the maxim by which all systems professing to be natural must be tested is this: '*that the arrangement obtained from one set of characters coincides with the arrangement obtained from another set*' (Pennington, 1994). But where a multiplicity of methods exist, and no clear choice is apparent from a desk analysis, a paradigm to follow is the one that mycobacteriologists have adopted. The importance of disease caused by *Mycobacterium tuberculosis*, the lack of phenotypic markers useful for discriminatory strains, and a certain cohesiveness among the mycobacteriological community combined to stimulate international collaborative work in the 1960s under the auspices of the World Health Organization to facilitate the development and standardisation of typing methods. Support from the WHO Program for Vaccine Development and the European Community Program on Science, Technology and Development continued this collaborated approach for DNA typing. It allowed workers in Holland, the UK, France, Spain, Ethiopia and the United States to develop a standardised technique which exploits variability in the number and genomic position of the insertion sequence IS*6110* to generate strain-specific patterns (van Embden *et al.*, 1993). The use of a standard operating procedure with internal and external molecular size markers permitted inter-laboratory comparison. It has been widely adopted. Inter-laboratory collaborative work has continued, and in a 1999 study all the then currently known typing methods (five RFLP methods and seven PCR assays) were compared for reproducibility, discrimination and specificity using a panel of 90 strains from 38 countries. Strain differentiation by IS*6110* RFLP emerged as a method of choice – and remains so (Kremer *et al.*, 1999).

The value of this kind of structured and rigorous inter-laboratory comparison cannot be overemphasised. Typing is done for practical reasons such as outbreak control and the tracking of the spread of an organism in real time. So those who practice it do not usually want to wait for the normal scientific consensus mechanisms to run their full course – which may take several years – before making decisions about the utility of new developments, particularly if they promise real improvements in the prevention of infection or the management of outbreaks. Inter-laboratory comparisons, properly done, are the best way of resolving this problem. Not only do they do this directly, but they moderate those properties of molecular biology as a subject that have favoured the development of the currently available Babel-like mass of typing methods. The most important of these is its sociological structure – a large number of rather small but intensely competitive groups (Knorr-Cetina, 1999).

Electrophoretic patterns are often described as a bar-code. This is best considered to be aspirational, because the bar-code has properties superior to any currently available electrophoretic typing system. They include a degree of redundancy (accurate decoding does not need certain bars), built in error-detecting features which allows scanners to detect and reject bad patterns, 'wandability' (the ability to read a pattern with a simple hand-held device) omni-directional readability (a design feature which allows patterns to be interpreted whether they are read from left to right or vice-versa) and the almost effortless generation of data that can be transmitted and decoded electronically. Bar-code scanners require no skill from the operator. But even with the best equipment available, running an electrophoresis gel still requires green fingers. For techniques like PFGE, 'tacit knowledge' – the unarticulated craft aspect of science – (Collins, 1974) is important for success. To get this method to work many laboratories have had to obtain the necessary skills by sending staff for training to successful laboratories. At least as important and much more useful heuristically to electropherotypers at the present time is the lesson that can be drawn from the way the bar-code was developed and introduced (Brown, 1997). The checkout reader technology came first, being made possible by the invention of the laser and the hologram and the development of integrated circuits and microprocessors. It was essentially ready by 1970. But it then took 3 years of hard negotiation to develop the US Uniform Grocery Product Code – eventually the Universal Product Code. The enormous difficulty of achieving consensus among aggressively competitive and entrepreneurial businessmen is nicely summarised in Brown's account of an early meeting in 1969 at a Cincinnati motel

> where the menu writers' skill far surpassed the chefs'. At the meeting one group said, 'eleven digits, significant,' the other responded, 'seven digits, non-significant'. After staring at each other for a time, the groups adjourned the meeting and went their separate ways.

Yet because of continuing and vigorous work consensus was achieved and the first bar-code-labelled item, a packet of Wrigley's chewing gum, crossed a scanner in Troy, Ohio in June 1974. Less than 2 years later 75% of the items in a typical US supermarket carried a bar-code. The bar-code has brought enormous benefits to its users, primarily through the electronic handling of data. It is international. Similar advantages will follow similar approaches for electrophoretic typing. For *M. tuberculosis* it is already possible to use the World Wide Web to search a database of RFLP patterns based on IS*6110*-associated DNA polymorphisms and compare fingerprints on an international basis. It is a paradox that the global consensus that made this possible has not yet emerged for other pathogens, even though most of them are easier to grow, easier to break open and extract DNA from, and safer to handle.

The Future

The typing of bacterial pathogens is a dynamic enterprise. Like the subjects of its study, it changes by evolutionary processes. This makes its future development very difficult to predict. It is difficult to know whether emerging techniques such as DNA on a chip or the optical mapping of RFLPs (Schwartz, 1993) will replace current methods. Nevertheless, we can be sure that the importance of genome sequencing, and genomics in general, will continue to increase and that it will make the genetic basis of typing schemes more and more secure. So we can be certain that typers will see themselves less and less as interpreters of empirically generated patterns, and more and more as applied geneticists.

References

Alm RA and 22 other authors (1999) Genomic-sequence comparison of two unrelated isolates of the human gastric pathogen *Helicobacter pylori*. *Nature* 397: 176–180.

Anderson ES (1959) Use of phages in epidemiological studies. In: Adams MH (ed.) *Bacteriophages*. New York: Interscience, pp. 395–420.

Bachellier S, Gilson E, Hofnung M, Hill CW (1996) Repeated sequences. In: Neidhardt FC *et al.* (eds) *Escherichia coli* and *Salmonella*. Washington, DC: ASM Press, pp. 2012–2040.

Barenkamp SJ, Munson RS, Granoff DM (1981) Subtyping isolates of *Haemophilus influenzae* type b by outer-membrane protein profiles. *J. Infect. Dis.* 143: 668–676.

Brown SA (1997) *Revolution at the Checkout Counter*. Cambridge, MA: Harvard University Press.

Collins HM (1974) The TEA set: tacit knowledge and scientific networks. *Sci. Stud.* 4: 165–186.

Cowan ST (1962) The microbial species – a macromyth. In: Ainsworth GC, Sneath PHA (eds) *Microbial Classification*. Cambridge: Cambridge University Press, pp. 433–455.

Danna KJ, Nathans D (1971) Specific cleavage of simian virus 40 DNA by restriction endonuclease of *Haemophilus influenzae*. *Proc. Natl Acad. Sci. USA* 68: 2913–2917.

Deonier RC (1996) Native insertion sequence elements: locations, distributions, and sequence relationships. In: Neidhardt FC *et al.* (eds) *Escherichia coli* and *Salmonella*. Washington, DC: ASM Press, pp. 2000–2011.

Farrar WE (1983) Molecular analysis of plasmids in epidemiological investigations. *J. Infect. Dis.* 148: 1–6.

Forbes KJ, Bruce KD, Jordens JZ, Ball A, Pennington TH (1991) Rapid methods in bacterial DNA fingerprinting. *J. Gen. Microbiol.* 137: 2051–2058.

Forbes KJ, Bruce KD, Ball A, Pennington TH (1992) Variation in length and sequence of porin (*omp* P2) alleles of non-capsulate *Haemophilus influenzae*. *Mol. Microbiol.* 6: 2107–2112.

Grimont F, Grimont PAD (1986) Ribosomal ribonucleic acid gene restriction patterns as potential taxonomic tools. *Ann. Inst. Pasteur Microbiol.* 137B: 165–175.

Harris H (1966) Enzyme polymorphisms in man. *Proc. Roy. Soc. London B* 164: 298–310.

Harrington CS, Thomson-Carter FM, Carter PE (1997) Evidence for recombination in the flagellin locus of *Campylobacter jejuni*: implications for the flagellin gene typing scheme. *J. Clin. Microbiol.* 35: 2386–2392.

Hunter PR, Gaston MA (1988) Numerical index of the discriminatory ability of typing systems: an application of Simpson's index of diversity. *J. Clin. Microbiol.* 26: 2465–2466.

Kay LE (1988) Laboratory technology and biological knowledge. The Tiselius electrophoresis apparatus. *Hist. Philos. Life Sci.* 10: 51–72.

Kersters K, DeLey J (1975) Identification and typing of bacteria by numerical analysis of their electrophoretic protein pattern. *J. Gen. Microbiol.* 87: 333–342.

Knorr-Cetina K (1999) *Epistemic Cultures*. Cambridge, MA: Harvard University Press.

Kremer K, van Soolingen D, Frotthingham R *et al.* (1999) Comparison of methods based on different molecular epidemiological markers for typing of *Mycobacterium tuberculosis* complex strains: interlaboratory study of discriminatory power and reproducibility. *J. Clin. Microbiol.* 37: 2607–2618.

Krikler SJ, Pennington TH, Petrie D (1986) Typing of strains of *Staphylococcus aureus* by Western blot analysis of culture supernates. *J. Med. Microbiol.* 21: 169–171.

Laemmli UK (1970) Cleavage of structural proteins during the assembly of the head of bacteriophage T4. *Nature* 227: 680–685.

Lewontin R (1992) Genotype and phenotype. In: Keller EF, Lloyd EA (eds) *Keywords in Evolutionary Biology*. Cambridge, MA: Harvard University Press, pp. 137–144.

Lewontin RC, Hubby JL (1966) A molecular approach to the study of genic heterozygosity in natural population. II Amount of variation and degree of heterozygosity in natural populations of *Drosophila pseudoobscura*. *Genetics* 54: 595–609.

Maiden MCJ, Bygraves JA, Feil E *et al.* (1998). Multilocus sequence typing: a portable approach to the identification of clones within population of microorganisms. *Proc. Natl Acad. Sci. USA* 95: 3140–3145.

Marshall RB, Wilton BE, Robinson AJ (1981) Identification of *Leptospira* serovars by restriction-endonuclease analysis. *J. Med. Microbiol.* 14: 163–166.

Milkman R (1996) Recombinational change among clonal populations. In: Neidhardt FC *et al.* (eds) *Escherichia coli and Salmonella*. Washington DC: ASM Press, pp. 2663–2684.

Nathans D, Danna KJ (1972) Studies of SV40 DNA. 3 Differences in DNA from various strains of SV40. *J. Mol. Biol.* 64: 515–518.

Olive DM, Bean P (1999) Principles and application of methods for DNA-based typing of microbial organisms *J. Clin. Microbiol.* 37: 1661–1669.

Ornstein L (1964) Disc electrophoresis 1. Background and theory. *Ann. NY Acad. Sci.* 121: 321–349.

Paterson AJ, Macsween KF, Pennington TH (1987) *Haemophilus influenzae* subtyping by SDS – PAGE of whole cell polypeptides. *Epidemiol. Infect.* 99: 179–189.

Pauling LC Itano HC, Singer SJ, Wells IC (1949) Sickle-cell anaemia, a molecular disease. *Science* 110: 543–548.

Pennington TH (1994) Molecular systematics and traditional medical microbiologists – problems and solutions. *J. Med. Microbiol.* 41: 371–373.

Salmonella subcommittee of the Nomenclature Committee of the International Society for Microbiology (1934) The genus *Salmonella* Lignières, 1900. *J. Hygiene* 34: 333–350.

Schwartz DC, Cantor CR (1984) Separation of yeast chromosome-sized DNAs by pulsed field gradient gel electrophoresis. *Cell* 37: 67–75.

Schwartz DC, Li X, Hernandez LI, Ramnarain SP, Huff EJ, Wang Y-K (1993) Ordered restriction maps of *Saccharomyces cerevisiae* chromosomes constructed by optical mapping. *Science* 262: 110–114.

Selander RK, Caugant DA, Ochman H, Musser JM, Gilmour MN, Whittam TS (1986). Methods of multilocus enzyme electrophoresis for bacterial population genetics and systematics. *Appl. Environ. Microbiol.* 51: 873–884.

Selander RK, Li J, Nelson K (1996) Evolutionary genetics of *Salmonella enterica*. In: Neidhardt FC *et al.* (eds) *Escherichia coli* and *Salmonella*. Washington DC: ASM Press, pp. 2691–2707.

Shapiro AL, Viñuela E, Maizels JV (1967) Molecular weight estimation of polypeptide chains by electrophoresis in SDS–polyacrylamide gels. *Biochem. Biophys. Res. Commun.* 28: 815–820.

Smith JM (1996) Population genetics: an introduction. In: Neidhardt FC *et al.* (eds) *Escherichia coli* and *Salmonella*. Washington, DC: ASM Press, pp. 2685–2690.

Smith JM, Smith N (1999) The genetic population structure of pathogenic bacteria. In: Stearns SS (ed.) *Evolution in Health and Disease*. Oxford: Oxford University Press, pp. 183–190.

Sokal RR, Sneath PH (1963) *Principles of Numerical Taxonomy*. Freeman, San Francisco.

Stephenson JR, Crook SJ, Tabaqchali S (1986). New method for typing *Staphylococcus aureus* resistant to methicillin based on sulphur-35 methionine labelled proteins: its application in an outbreak. *BMJ* 293: 581–583.

Studier FW (1973) Analysis of bacteriophage T7 early RNAs and proteins on slab gels. *J. Mol. Biol.* 79: 237–248.

Swofford DL, Olsen GJ (1990) Phylogeny reconstruction. In: Hillis DM, Moritz C (eds) *Molecular Systematics*. Sunderland, MA: Sinauer Associates, pp. 411–501.

Swofford DL, Olsen GJ, Waddell PJ, Hillis DM (1996) Phylogenetic inference. In: Hillis DM, Moritz C (eds) *Molecular Systematics*. Sunderland, MA: Sinauer Associates, pp. 407–514.

Tenover FC, Arbeit RD, Goering RV *et al.* (1995) Interpreting chromosomal DNA restriction patterns produced by pulsed-field gel electrophoresis: criteria for bacterial strain typing. *J. Clin. Microbiol.* 33: 2233–2239.

Tiselius A, Flodin P (1953) Zone electrophoresis. *Adv. Protein Chem.* 8: 461–486.

van Embden JDA, Cave MD, Crawford JT *et al.* (1993) Strain identification of *Mycobacterium tuberculosis* by DNA fingerprinting: recommendations for a standardized methodology. *J. Clin. Microbiol.* 31: 406–409.

Versalovic J, Koenth T, Lupski JR (1991) Distribution of repetitive DNA sequences in eubacteria and application to fingerprinting of bacterial genomes *Nucleic Acids Res.* 19: 6823–6831.

Vos P, Hogers R, Bleeker M *et al.* (1995) AFLP: a new concept for DNA fingerprinting. *Nucleic Acids Res.* 21: 4407–4414.

Welsh J, McClelland M (1990) Fingerprinting genomes using PCR with arbitrary primers. *Nucleic Acids Res.* 18: 7213–7218.

Whittam TS (1996) Genetic variation and evolutionary processes in natural populations of *Escherichia coli*. In: Neidhardt FC *et al.* (eds) *Escherichia coli* and *Salmonella*. Washington DC: ASM Press, pp. 2708–2720.

Williams JG, Kubelik AR, Livak KJ, Ratalsky JA, Tinqey SV (1990) DNA polymorphisms amplified by arbitrary primers are useful genetic markers. *Nucleic Acids Res.* 18: 6531–6535.

Yakubu DE, Abadi FJR, Pennington TH (1999) Molecular typing methods for *Neisseria meningitidis*. *J. Med. Microbiol.* 48: 1–10.

24

Nucleic Acid Probes and The Polymerase Chain Reaction

Rebecca J. Smith

Medical School, Newcastle upon Tyne, UK

Over the last 10 years, diagnostic molecular methods have made a large impact in the diagnosis and management of a variety of infectious diseases. Not only do these techniques provide tools for highly sensitive and specific, culture-independent detection of infectious agents in clinical specimens, but certain genetic characteristics (the presence of virulence, toxin or antimicrobial resistance genes) may also be determined. The most widely used methods include nucleic acid probe hybridisation and polymerase chain reaction (Wolcott, 1992; Morse, 1996). The methodology, applications and limitations of these techniques will be discussed in this chapter.

The Polymerase Chain Reaction

The polymerase chain reaction (PCR) is a rapid procedure for amplification of specific DNA or RNA sequences *in vitro*. It is a highly sensitive and specific technique, with which minute quantities of target DNA, theoretically as little as a single molecule, can be amplified until a sufficient concentration of nucleic acid is obtained, which is greater than the threshold of the detection system being used (Persing, 1991). Since PCR was first described (Mullis *et al.*, 1986; Mullis and Faloona, 1987), this ingeniously simple yet elegant procedure has been refined and modified to allow the improved detection and quantification of specific DNA and RNA sequences. The methodology and clinical applications of PCR and other PCR-based nucleic acid amplification methods such as multiplex PCR, nested PCR, reverse-transcription PCR, broad-range and quantitative PCR. Other nucleic acid amplification techniques such as nucleic acid-based sequence amplification (NABSA), strand displacement amplification and the Qβ replicase system will only be briefly mentioned.

Methodology

An overview of the polymerase chain reaction is shown in Fig. 1. Primer extension that depends on the mode of action of DNA polymerase is the principle that underlies the method. The sequential addition of deoxyribonucleotide triphosphates (dNTPs) to the free 3′ hydroxyl group of a short oligonucleotide primer is initiated by DNA polymerase, one of the primary enzymes involved in DNA replication (see Chapter 14). As a result, the extension of the DNA molecule takes place in a 5′ to 3′ direction and the sequence of bases of the new strand is complementary to the base sequence of the template strand. Primer extension is the basis of many nucleic acid labelling reactions and therefore has particular importance in the generation of probes, which will be discussed below.

 doi:10.1006/bkmm.2001.0024

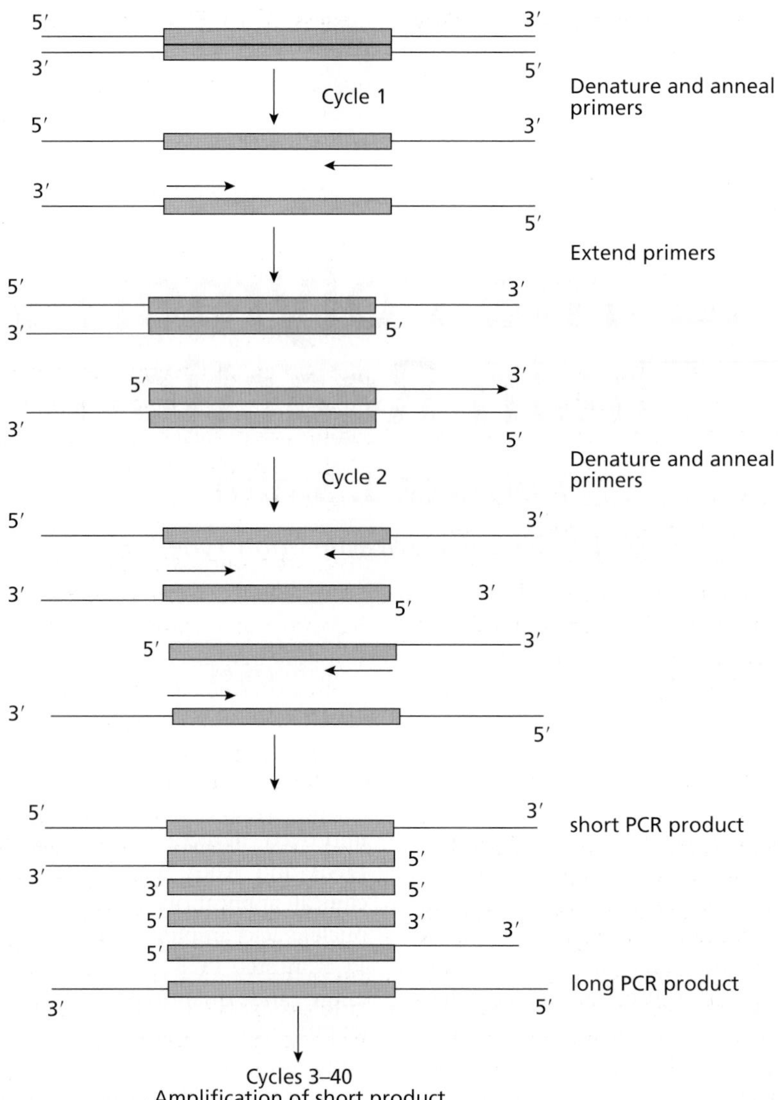

Fig. 1 Schematic diagram of the Polymerase Chain Reaction.

A typical amplification reaction includes the sample of target DNA, a thermostable DNA polymerase, two primers (forward and reverse), dNTPs, reaction buffer and magnesium. The PCR is a three-step process that involves repeated cycles of (1) heating to 90–95°C to denature double-stranded target DNA; (2) cooling (to 30–65°C) to allow primer annealing to complementary sequences on the single stranded DNA, and (3) primer extension which is usually carried out at 72°C. The specificity of the PCR is provided by two oligonucleotide primers whose sequences correspond to short regions of known nucleic acid sequence that flank the gene or non-coding region of the target DNA. After the denaturation step, the primers hybridise to the opposite strands of DNA and are orientated with their 3′ ends facing each other. The subsequent primer extension reaction directs DNA synthesis across the region between the two primers. The extension products serve as a template for the other primer and therefore after each successive cycle, usually of 30–50 cycles, the amount of PCR product is doubled and this results in the exponential accumulation of the specific target fragment.

The length of the PCR product is dictated by the distance between the 5′ termini of the oligonucleotide primers, which become incorporated in to the DNA strand. This results in the production of DNA fragments that are of identical molecular size.

PCR end-products can be detected by several methods. Theoretically, in a matter of a few hours, each PCR cycle doubles the amount of target DNA sequence (amplicon) in the reaction; 10 cycles theoretically multiply the amplicon by a factor of 10^3, 20 cycles by a factor of 10^6 and so on. In practice, however, the efficiency of amplification is less then perfect and, as the number of cycles increases, the amplification efficiency decreases. A number of factors contribute to this decrease in efficiency, including inactivation of the polymerase, accumulation of enzyme inhibitors such as pyrophosphate, inefficiency of primer annealing and reduction in efficiency of denaturation. This loss of exponential amplification during a PCR is termed the plateau effect.

Several methods are used to detect and quantify the amplified PCR product. Usually, the products are stained with fluorescent intercalating dyes, such as ethidium bromide or SYBR green, separated by gel electrophoresis and visualised with short-wave ultraviolet radiation. The luminescence of ethidium bromide can also be used to quantify the amount of PCR product. Electrophoretically separated end-products may be detected with a labelled probe (Southern blot), complementary to all or part of the amplified DNA sequence.

In the original PCR protocols DNA polymerase isolated from *Escherichia coli* was used, but this enzyme is thermolabile and inactivated during the denaturation step and therefore requires the addition of fresh enzyme during each cycle (Erlich *et al.*, 1988). For this reason, PCR was quite limited until heat-stable DNA polymerases and full automation by means of programmable thermal cyclers became widely available and allowed this procedure to reach its full potential. The heat-stable DNA polymerase isolated from *Thermus aquaticus*, a thermophilic organism that lives in hot springs, is known as Taq polymerase (Saiki *et al.*, 1988). It is, however, the recombinant form of this high-temperature polymerase, purified in large quantities from genetically engineered *E. coli*, that is now most widely used for PCR.

Other recombinant DNA polymerases are becoming available for more specialised PCR procedures, including the following: (1) *Tth* DNA polymerase from *Thermus thermophilus*, a thermostable DNA polymerase that eliminates organic pyrophosphate created during the incorporation of dNTPs. Organic pyrophosphate is thought to inhibit DNA polymerase activity and to cause the degradation of DNA under elevated temperature conditions and *Tth* can therefore result in significant improvements in DNA amplification. Under certain conditions *Tth* has a reverse transcriptase activity. (2) Proof-reading thermostable DNA polymerases, such as *Pwo*, from *Pyrococcus woesei*, and *Pfu*, from *Pyrococcus furiosus*, unlike Taq polymerase, possess 3′ to 5′ exonuclease proof-reading activity that enables the polymerase to correct nucleotide mis-incorporation errors, so increasing the fidelity of the PCR.

Technical Problems

The PCR provides a rapid yet highly sensitive and specific method for nucleic acid amplification. Theoretically, it is possible to detect a single target DNA molecule. The threshold of detection is decreased by one order of magnitude when amplifying target molecules from clinical specimens or environmental samples and this is one of the main strengths and weaknesses of the PCR. Contamination of samples by ‘rogue’ nucleic acid molecules is one of the main problems of PCR, particularly if it is being used for diagnostic purposes. The most common source of contamination is from other samples, previous amplification procedures or contaminated reagents, such as *taq* polymerase, buffer, primers and water. Contamination with minute quantities of DNA may lead to false-positive test results and a mis-diagnosis. To combat this problem, especially in clinical laboratories, the different stages of PCR should be carried out in separate rooms, good laboratory practice should be used and ultraviolet light, which inactivates DNA, should be used in sample preparation areas.

Although the PCR itself is relatively simple, the acquisition of target molecules from samples can be time consuming and complicated. Samples are treated to release DNA or RNA and the DNA/RNA is then ‘cleaned up’ to remove potential inhibitors of PCR, which reduce the efficiency of the reaction and are often present in clinical specimens. When the specimen contains large amounts of the target nucleic acid, sample preparation is not as important. For example, a bacterial colony can be suspended in appropriate buffer, boiled and added directly to the PCR tube. The specificity and sensitivity of the PCR may, however, be compromised when the target is present in a low concentration, especially if the sample volume is too large (> 2 mL) or too small (< 20 μL). Processing of such samples may require complex protocols, which increases the risk of contamination.

Modifications of the PCR

Multiplex PCR

The use of multiple primer pairs (multiplex PCR) allows more than one target sequence to be amplified per single reaction and has the potential to produce considerable savings of time and effort in a diagnostic laboratory. Multiplex PCR assays have been developed to detect multiple pathogens in single clinical samples, for example in the detection of common bacterial or viral agents in respiratory tract infections (Echevarrìa *et al.*, 1998; Gröndahl *et al.*, 1999; Osiowy, 1998) and meningitis (Read *et al.*, 1997). The individual genes are detected by probing with labelled species or gene-specific probes, such as Southern blot.

Multiplex PCR has also been applied to the detection of multiple genes in one organism. For example, the identification of methicillin-resistance in methicillin-resistant *Staphylococcus aureus* (MRSA) from clinical samples. These organisms are an important nosocomial pathogen, capable of causing life-threatening infections and hospital outbreaks (see Chapter 39) and a rapid and accurate identification of this pathogen is critical for patient management. Primers are directed towards the *mecA* gene (methicillin resistance) and *femA* (coagulase), and the resulting PCR products can distinguish between methicillin-resistant and -sensitive strains and between coagulase-positive and -negative staphylococci; MRSA is *mecA* +, *femA*+ (Read *et al.*, 1997).

Care should be taken, however, when designing primer sets, since the primers may hybridise to each other rather than to the target molecule, so reducing the efficiency of the reaction. If the sets of primers have dramatically different annealing temperatures, the annealing temperature of the lower primer set would have to be used, thereby reducing the stringency of the primer with the higher annealing temperature, leading to non-specific amplification of the target DNA. As well as detecting target genes of interest, internal controls may also be included to indicate reaction (Jackson *et al.*, 1996; Read *et al.*, 1997). Many of the diagnostic multiplex PCR assays have been developed in-house. To date, commercially available multiplex PCR assays are only available for *Chlamydia trachomatis*, *Neisseria gonorrhoeae* and human herpes virus (Elnifro *et al.*, 2000).

Nested PCR

The nested PCR has been designed to increase sensitivity and specificity, allowing detection of smaller quantities of target DNA. Nested PCR is a two-step process using two sets of primers. One set of primers is used to amplify a region on the target DNA molecule, then the resulting amplification products are subjected to a second round of amplification with another set of primers that lie within the internal sequence amplified by the first primer set. The DNA products from the first round of amplification possess the hybridisation sites for the second primer pair, i.e. they are 'nested' within the sequence amplified by the first primer set. Because of the dual amplification, nested PCR is highly sensitive, and the second primer set confirms the specificity of the first-round product.

The major disadvantage of this method is the risk of contamination that may occur when the first round products are transferred to a separate tube for the second amplification step. This problem has been circumvented either by separating the two primer sets with oil or wax and mixing the two phases after the first amplification step or by designing primers that have substantially different annealing temperatures (Picken *et al.*, 1996). Conventional PCR does not detect latent human herpes virus 6 (HHV-6) DNA in peripheral blood mononuclear cells but with the increased sensitivity of the nested PCR the DNA of latent HHV-6 can be detected in healthy individuals (Cuende *et al.*, 1994). This is also the case for the detection of *Streptococcus pneumoniae* in community-acquired pneumonia (Lorente *et al.*, 2000).

Reverse Transcriptase PCR

The technique of reverse transcription PCR (RT-PCR) extends the ability of PCR to detect specific RNA sequences in samples. In this method RNA targets, mRNA, rRNA and viral genomic RNA, are converted to complementary DNA (cDNA) by reverse transcriptase and then amplified by the PCR. In an actively growing organism, whether bacterial, fungal or viral, the copy number of the target RNA molecule will be much higher than a single DNA copy on the chromosome, thereby increasing the sensitivity of this technique over conventional PCR. This technique also provides an insight into the physiological status of the organism; the detection of cDNA by RT-PCR of mRNA encoded by a pathogen may be evidence of active infection. Conventional PCR cannot distinguish between living and dead organisms, since they both contain DNA. RT-PCR has played an important role in the diagnosis of and the monitoring for viral infections such as HIV-1 (Coombs *et al.*, 1996) and for the detection of viable *Mycobacteria* and the effectiveness of antimicrobial therapy (Jou *et al.*, 1997).

Like DNA polymerase, RNA-dependant DNA polymerase (reverse transcriptase, RT) extends RNA in a 5′ to 3′ direction and requires the presence of an annealed primer to initiate synthesis. Total cellular RNA is isolated, purified and analysed. Prokaryotic mRNA differs from eukaryotic mRNA in that it does not possess a polyadenylated tail and it is also extremely labile and readily becomes degraded. It is, therefore, necessary to determine the quality of isolated RNA by denaturing agarose gel electrophoresis. The presence of the 23S and 16S rRNA subunits indicates the quality of RNA and whether RT-PCR can be carried out. Several ways of priming RNA for cDNA synthesis are

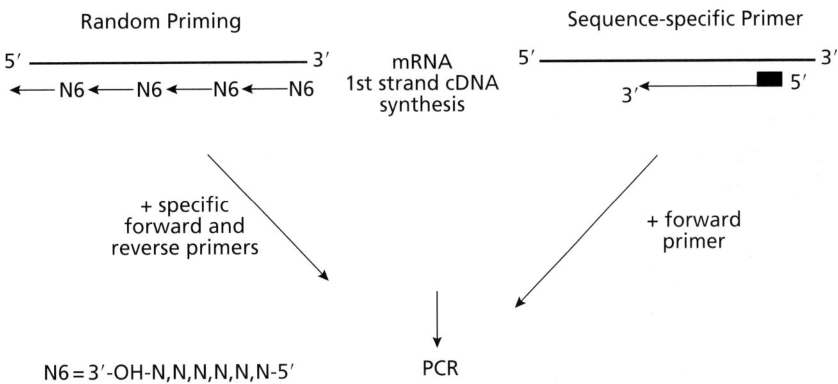

Fig. 2 Reverse transcription PCR (RT-PCR). Figure shows synthesis of cDNA from mRNA molecules using (i) random priming, sequence specific primers for the gene of interest are used in the subsequent PCR. (ii) sequence specific priming using a specific primer for the target nucleic acid which anneals to the 3′ end of the RNA molecule and is extended with reverse transcriptase. This primer along with a suitable 5′ primer can be used for the subsequent PCR step.

available (Fig. 2): (1) Random priming with short oligonucleotides, usually a 6-mer, that possess all possible nucleotides at each position, which anneal to the RNA randomly, priming cDNA synthesis from the free 3′ hydroxyl group on the 6-mer oligonucleotide. Specific primers for the gene of interest are then used in the subsequent PCR. (2) Specific priming with a primer specific for the mRNA of interest, which anneals to the 3′ end of the mRNA molecule in question. DNA synthesis then proceeds in a 5′ to 3′ direction. This 3′ primer, along with a suitable 5′ primer, can be used for the subsequent PCR.

Although this technique is highly sensitive, it is technically demanding. RNA, especially bacterial mRNA, is extremely labile and the presence of RNases must be minimised at all times. Conventional RT-PCRs use thermolabile RTs, such as avian myeloblastosis virus (AMV) or Moloney murine leukaemia virus (M-MLV) RTs, and therefore an initial RT step is required. The resulting cDNA is then added to a normal PCR reaction. Alternatively, some thermostable DNA polymerases, such as *Tth* DNA polymerase, possess reverse transcriptase activity under certain conditions, namely when a cation such as manganese is used instead of magnesium, which allows RT-PCR to be carried out in a single step (Myers and Gelfand, 1991). The development of a one-step RT-PCR has broadened its application in diagnostic microbiology, and commercial kits are available that exploit single-enzyme technology for HIV-1 (e.g. RT-PCR Amplicor Moniter Assay, Roche Diagnostic Systems, California, USA).

Quantitative PCR

PCR and RT-PCR are generally used as a qualitative method for sample analysis, but they can also be used to measure the amount of target DNA or RNA in a sample. Data obtained from quantitative PCR can be used to monitor the microbial load of an infecting pathogen and may yield prognostic information about the progress of a disease. Responses to therapeutic agents can also be assessed and gene expression can be characterised.

Quantitative PCR requires that the measurement is taken during the exponential phase of the reaction, that is before the onset of the plateau effect. This point must be determined for different reactions, since numerous factors can affect amplification efficiency. Most difficulties with quantitative PCR arise because only a small number of cycles in the PCR reaction contain useful information, and fewer amplification cycles are used than with conventional PCR and, therefore, detection of the end-products may be a problem.

In traditional quantitative PCR, the sample is divided into aliquots and placed in multiple reaction tubes. One tube is removed for analysis after each cycle, and the products are analysed by agarose gel electrophoresis. The amount of DNA at each cycle is measured by staining with ethidium bromide or SYBR green and the result is plotted against cycle number. The cycle number at which the sample is first detected (the 'threshold') is compared with the threshold of a known standard.

In some methods, internal standards (endogenous or exogenous) are included to monitor the efficiency of amplification. One approach is to compare the amplification of target DNA with the amplification of DNA that encodes a 'housekeeping' gene, such as glyceraldehyde-3-phosphate dehydrogenase. A housekeeping gene is used on the premise that expression remains constant among the samples being compared (Murphy *et al.*, 1990; Gaudette and Crain, 1991). A product ratio between the amplified target and internal standard is measured; data must be obtained before amplification of either the internal standard or the target DNA reaches a plateau.

The main advantage of this method is that amplification of the housekeeping genes verifies that the target nucleic acid and reaction components were of acceptable quality, but it does not account for differences in amplification efficiencies and primer annealing efficiency between the internal standard and target being quantified. To deal with this problem an exogenous internal standard is used. This technique might also be described as 'competitive PCR'. In this technique, a dilution series of the internal standard is added to a constant amount of target DNA. This template uses the same primer pair as the target DNA but yields a distinguishable molecule (e.g. of a different size). The internal standard and target DNA are compared after amplification, usually by agarose gel electrophoresis, and their yields are measured. The relative amounts of target DNA and internal standard are assumed to be equal in the reaction tube that produces equal amounts of product, and because the initial amount of internal standard in this tube is known, the initial amount of the target DNA can be inferred.

Recently, advances in PCR technology have allowed the monitoring of a PCR in real time (Lightcycler^{TM}, Roche). Two methods of this kind are currently available: DNA detection with SYBR green 1 dye or PCR monitoring with hybridisation probes.

SYBR green 1 dye has a relatively low fluorescence in the unbound state but when bound to dsDNA it fluoresces brightly. During the extension step, dye molecules bind to the newly synthesised DNA, and if the reaction is monitored continuously, an increase in fluorescence is observed in real time. The fluorescence is measured at the end of each extension step to monitor the amplification of the target molecule. It should be noted, however, that SYBR green is a non-specific dye that will bind to a non-specific product, such as primer dimer, thus increasing the background fluorescence and thereby reducing the sensitivity and specificity.

Increased specificity is achieved by monitoring the PCR with fluorescence-labelled sequence-specific oligonucleotide probes, which allows highly specific detection of the amplification product. The increasing amount of fluorescence is proportional to the increasing amount of DNA. Generation of specific fluorescence is only generated by the hybridisation of both oligonucleotides to defined sequences within the amplification product. Non-specific products, such as primer dimers, will not, therefore, be detected and so minute quantities of target DNA or RNA can be detected.

Broad-range PCR

Broad-range PCR is fast becoming the method of choice for identifying aetiological agents of microbial infection. Although conventional methods for characterising and identifying clinical isolates are still used, they require the isolation and cultivation of the causal agent and this may pose problems if this agent cannot be cultured or if it is fastidious. Where these methods have failed to identify the pathogen, broad-range PCR has successfully identified novel, fastidious and non-culturable agents from human tissue or blood (Tang *et al.*, 1997). Certain functional domains within the genome (e.g. 16S ribosomal RNA) that possess phylogenetic information are highly conserved in the prokaryote kingdom. This allows amplification of target DNA from virtually all prokaryote organisms with a single set of primers. Subsequent nucleic acid sequencing of the amplified DNA, and alignment of these sequences against reference sequences (i.e. homologous sequences of 'known' organisms), allows the identification of the pathogen or the possible discovery of a 'new' organism. The aetiology of bacillary angiomatosis (Relman *et al.*, 1990), the unculturable and previously unidentifiable bacillus associated with Whipple's disease, *Tropheryma whippelii* (Relman *et al.*, 1992) and *Ehrlichia* (Anderson *et al.*, 1991; Chen *et al.*, 1994) were identified by amplification of 16S ribosomal DNA. Many other molecular targets have been used to identify and determine the phylogenetic and taxonomic relationships between infectious agents. These include the heat shock protein 65 (*hsp*65) and *groEL* gene to characterise clinical mycobacterial isolates and rickettsia-like agents such as the causal agent of human erlichiosis (Kolbert and Persing, 1999).

Broad-range PCR provides a rapid tool for the identification of pathogenic micro-organisms directly from infected human tissue or blood. Potentially serious drawbacks may, however, lead to misdiagnosis. Background contamination may lead to false-positive results or the amplification of the 'wrong' target. Human tissue and blood may contain inhibitory substances, resulting in false-negative reactions where amplification has been inhibited. Extensive modification of DNA extractions would have to be carried out to circumvent this problem. An obviously major drawback is the cost; broad-range PCR is expensive compared with conventional diagnostic techniques and is, therefore, still not the method of choice for routine identification in clinical laboratories. Commercial systems that may reduce costs, such as The MicroSeq^{TM} (Perkin Elmer and Applied Biosystems, Foster City, California, USA) are, however, being introduced.

Other Nucleic Acid Amplification Techniques

Alternative nucleic acid amplification techniques for use *in vitro* are becoming established to meet various requirements for improved detection of both DNA and RNA, but the different types of amplification are largely equivalent. Established methods include nucleic acid-based sequence amplification (NABSA) (Kwoh *et al.*, 1989), the ligase chain reaction (Wu and Wallace, 1989), strand displacement amplification (Walker *et al.*, 1992) and the Qβ replicase system (Lizardi *et al.*, 1988; Kramer and Lizardi, 1989).

Nucleic Acid Probes

Techniques that require DNA/RNA probes have been in use for almost three decades. A nucleic acid probe is a sequence of single-stranded DNA or RNA that can hybridise specifically with its complementary sequence via base-pairing on the target nucleic acid molecule. Hybridisation analysis is carried out with a labelled probe and target DNA which has been immobilised on a solid phase. Common applications of DNA/RNA probes include Southern (DNA) and northern (RNA) blotting, dot/slot blots (DNA/RNA), colony and plaque hybridisation and *in situ* hybridisation. With the increasing availability of bacterial genome sequences and the development of micro-array systems, it is possible to detect and analyse expression levels of thousands of genes by parallel hybridisation using cDNA probes.

Methodology

Hybridisation was originally performed with long (100–1000 bp), radioactively labelled DNA probes. Today the probe variety is much greater (RNA, oligonucleotide and cDNA) and the choice of probe is largely dependent on the type of application (Southern, northern, micro-array) and the level of sensitivity of detection required (single gene copy detection of amplified products). Probes are generated by the incorporation of a labelled nucleotide (e.g. ^{32}P[dCTP], digoxigenin-11-dUTP) throughout the probe (DNA, RNA, cDNA by random priming or nick translation) or by attaching the label to the ends of the probe (5′ and 3′ end-labelling of oligonucleotides). RNA probes are constructed by cloning the gene of interest and performing *in vitro* transcription; the label is incorporated using a labelled ribonucleotide. cDNA probes are generated by reverse transcribing RNA to DNA, using a labelled deoxyribonucleotide.

Traditionally, probes were labelled with radioisotopes and detected using autoradiography. Recently, greater emphasis has been placed on the development and use of non-radioactive methods because of safety concerns and disposal problems associated with radioisotopes. Non-isotopic labels also have a longer shelf-life than radioisotopes: over 2 years compared with $<$15 days for ^{32}P.

Although a number of non-isotopic labels have been described, biotin and digoxigenin are used most frequently and are commercially available. These labels are detected by a secondary agent such as an antibody (antidigoxigenin) of a ligand (strepavidin in the case of biotin). Enzymes such as alkaline phosphatase and horseradish peroxidase are conjugated to the secondary agent, and in the presence of suitable substrate the label can be detected colorimetrically or by chemiluminescence (more sensitive). DNA can also be rendered fluorescent by the incorporation of nucleotides labelled with a fluorophore, which fluoresces when activated. Oligonucleotides labelled in this way are used in 96-well plate-based amplification methods, flow cytometry and micro-arrays.

Fig. 3 shows an overview of DNA probe hybridisation. The membrane is incubated in blocking buffer (usually denatured salmon sperm DNA) to prevent non-specific binding of the probe to the membrane. It is then incubated with the probe overnight at an optimal temperature for hybridisation. The hybridisation temperature is important, as this will dictate the specificity at which the probe will hybridise to the target DNA. Unbound probe is removed by a series of washes that can be high or low stringency depending on the degree of specificity required. The membrane is then assayed for radioactive or chemiluminescent signals by placing against X-ray film.

Applications

Nucleic acid probes have been applied in diagnostic microbiology, in particular probes are being used for culture conformation as an alternative to conventional laboratory techniques (Daly *et al.*, 1991). DNA probes are also used to detect slow-growing organisms, such as *Mycobacteria* spp. The probes currently available allow culture conformation for several clinically important mycobacteria (*M. avium*, *M. tuberculosis*, *M. kansasii*) (Drake, 1987; Musial *et al.*, 1988) and fastidious organisms, e.g. *Neisseria gonorrhoeae* and *Chlamydia trachomatis* (Limberger *et al.*, 1992). Direct probing from clinical specimens may, however, be problematic, the main stumbling block being the reduced sensitivity of nucleic acid probes. In the absence of amplification procedures, probes may

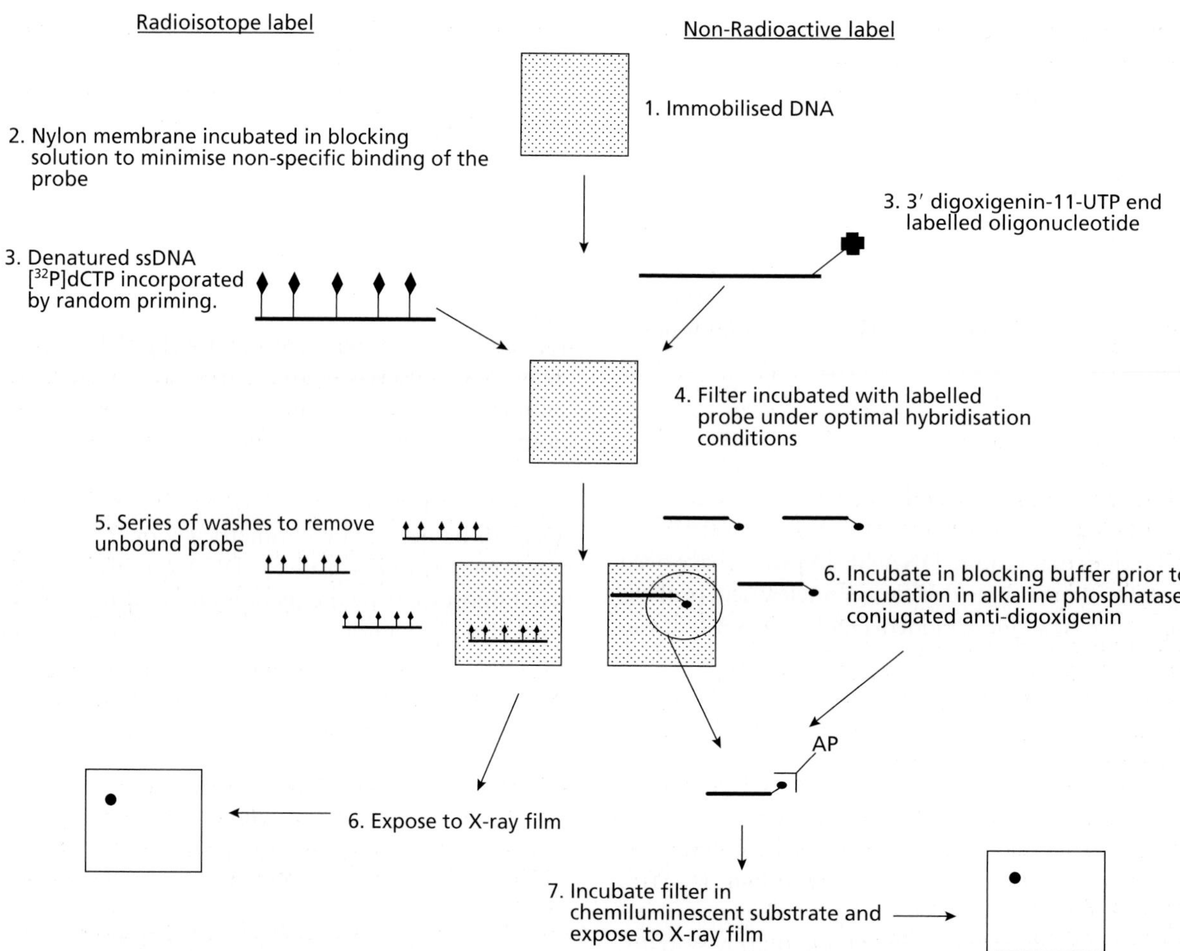

Fig. 3 Schematic diagram of DNA probe hybridisation using radioactive and nonradioactive labelled probes.

detect as few as 10^4 target sequences, while amplification used along with probe detection may detect <10 target sequences. When probes are used to confirm the identity of a cultured organism, the number of target copies will far exceed the threshold of sensitivity, however this may not be the case in clinical specimens.

Probes have also been used for the direct detection of genetically encoded virulence factors in culture isolates and specimens, e.g. detection of Shiga-like toxins in *Escherichia coli* O157 (Newland and Heill, 1988). It must be reiterated that direct probing of clinical specimens will probably require initial amplification of the target sequence.

An important advancement in molecular techniques is that of micro-array technology. Using this technique total gene expression, such as the response of *M. tuberculosis* to chemotherapy, can be monitored (Wilson *et al.*, 1999). The availability of complete bacterial genome sequences has allowed this development to take place. A micro-array contains DNA fragments corresponding to most of the open reading frames present in an organism. These fragments are isolated to a solid phase, such as a poly-L-lysine-coated microscope slide, and immobilised; the co-ordinates of each open reading frame (ORF) on the slide is known. Total RNA can then be isolated and during a reverse transcriptase step the cDNA is labelled with a fluorescent nucleotide. Following hybridisation, the slides are scanned for fluorescence, positive spots indicate the ORF, and the gene being expressed can be determined.

In situ hybridisation allows detection of pathogens in tissues or biopsies. This technique can reveal infectious agents and their hallmarks of infection (i.e. *M. tuberculosis* within a tuberculous granuloma). However this technique is only as good as the physical availability of target nucleic acid to the labelled probe. Also size of the probe may limit its ability to reach target nucleic acid sequences. *In situ* hybridisation has its greatest application in the detection of organisms that are difficult to grow in culture, especially viral agents such as HIV and hepatitis B (Naoumov *et al.*, 1993; An *et al.*, 1999; Patterson *et al.*, 1999).

References

An SF, Groves M, Giometto B, Beckett AAJ, Scaravilli F (1999) Detection and localisation of HIV-1 DNA and RNA in fixed adult AIDS brain by polymerase chain reaction/*in situ* hybridisation technique. *Acta Neuropathol.* 98: 481–487.

Anderson BE, Dawson JE, Jones DC, Wilson KH (1991) *Erlichia chaffeensis*, a new species associated with human erlichiosis. *J. Clin. Microbiol.* 29: 2838–2842.

Chen S, Dumler JS, Bakken JS, Walker DH (1994) Identification of a granulocytotrophic *Erlichia* species as the aetiological agent of human disease. *J. Clin. Microbiol.* 32: 589–595.

Coombs RW, Welles SL, Hooper C *et al.* (1996) Association of plasma human immunodeficiency virus type-1 RNA level with risk of clinical progression in patients with advanced infection. *J. Infect. Dis.* 174: 704–712.

Cuende JI, Ruiz J, Civeira MP, Prieto J (1994) High prevalence of HHV-6 DNA in peripheral-blood mononuclear cells of healthy individuals detected by nested-PCR. *J. Med. Virol.* 43: 115–118.

Daly JA, Clifton NL, Seskin KC, Gooch WMI (1991) Use of rapid nonradioactive DNA probes in culture confirmation tests to detect *Streptococcus agalactiae, Haemophilus influenzae* and *Entrococcus* spp. from pediatric patients with significant infections. *J. Clin. Microbiol.* 29: 80–82.

Drake TA, Hindler JA, Berlin OGW, Bruckner DA (1987) Rapid identification of Mycobacterium-Avium complex in culture using DNA probes. *J. Clin. Microbiol.* 25: 1442–1445.

Echevarrìa JE, Erdmann DD, Swierosz EM, Holloway BP, Anderson LJ (1998) Simultaneous detection and identification of human parainfluenza viruses 1,2, and 3 from clinical samples by multiplex PCR. *J. Clin. Microbiol.* 36: 1388–1391.

Elnifro EM, Ashshi AM, Cooper RJ, Klapper PE (2000) Multiplex PCR: optimization and application in diagnostic virology. *Clin. Microbiol. Rev.* 13: 559–570.

Erlich HA, Gelfand DH, Saiki RK (1988) Specific DNA amplification. *Nature* 331: 461–462.

Gaudette MF, Crain WR (1991) A simple method for quantifying specific messenger-RNAs in small numbers of early mouse embryos. *Nucl. Acids Res.* 19: 1879–1884.

Gröndahl B, Puppe W, Hoppe A, Kühne I, Weigl JAI, Schmitt H-J (1999) Rapid identification of nine microorganisms causing acute respiratory tract infections by single tube multiplex reverse transcription-PCR: feasibility study. *J. Clin. Microbiol.* 37: 1–7.

Jackson R, Morris DJ, Cooper RJ, Bailey AS, Klapper PE, Cleator GM (1996) Multiplex polymerase chain reaction for adenovirus and herpes simplex virus in eye swabs. *J. Virol. Methods* 56: 41–48.

Jou NT, Yoshimori RB, Mason GR, Louie JS, Liebling MR (1997) Single-tube nested, reverse transcriptase PCR for detection of viable *Mycobacterium tuberculosis*. *J. Clin. Microbiol.* 35: 1161–1165.

Kolbert CP, Persing DH (1999) Ribosomal DNA sequencing as a tool for identification of bacterial pathogens. *Curr. Opin. Microbiol.* 2: 299–305.

Kramer FR, Lizardi PM (1989) Replicatable RNA reporters. *Nature* 339: 401–402.

Kwoh DY, Davis GR, Whitfield KM (1989) Transcription based amplification system and detection of amplified human deficiency virus type 1 with a bead-based sandwich hybridisation format. *Proc. Natl Acad. Sci.* 86: 1173–1177.

Limberger RJ, Biega R, Evancoe A, McCarthy L, Slivienski L, Kirkwood M (1992) Evaluation of culture and the Gen-Probe PACE 2 assay for detection of *Neisseria gonorrhoeae* and *Chlamydia trachomatis* in endocervical specimens transported to a state health laboratory. *J. Clin. Microbiol.* 30: 1162–1166.

Lizardi PM, Guerra C, Lomeli H, Tussie-Luna I, Kramer FR (1988) Exponential amplification of recombinant RNA hybridisation probes. *Biotechnology* 6: 1197–1202.

Lorente MLL, Falguera M, Nogues A, Gonzalez AR, Merino MT, Caballero MR (2000) Diagnosis of pneumococcal pneumonia by polymerase chain reaction (PCR) in whole blood: a prospective clinical study. *Thorax* 55: 133–137.

Morse SS (1996) Importance of molecular diagnostics in the identification and control of emerging infections. *Emerg. Infect. Dis.* 1: 201–206.

Mullis KB, Faloona FA (1987) Specific synthesis of DNA *in vitro* via a polymerase-catalysed chain reaction. *Meth. Enzymol.* 155: 335–350.

Mullis K, Faloona F, Scharf S, Saiki R, Horn G, Erlich H (1986) Specific amplification of DNA *in vitro*: the polymerase chain reaction. *Cold Spring Harbor Symp. Quant. Biol.* 51: 263–273.

Murphy LD, Herzog CE, Rudick JB, Fojo AT, Bates SE (1990) Use of the polymerase chain-reaction in the quantitation of Mdr-1 gene-expression. *Biochemistry* 29: 10351–10356.

Musial CE, Tice LS, Stockman L, Roberts GD (1988) Identification of mycobacteria from culture using the Gen-Probe rapid diagnostic system for *Mycobacterium avium* complex and *Mycobacterium tuberculosis* complex. *J. Clin. Microbiol.* 26, 2120–2123.

Myers TW, Gelfand DH (1991) Reverse transcription and DNA amplification by a *Thermus thermophilus* DNA polymerase. *Biochemistry* 30: 7661–7666.

Naoumov NV, Daniels HM, Davison F, Eddleston A, Alexander GJM, Williams R (1993) Identification of hepatitis-B virus-DNA in the liver by *in situ* hybridization using a biotinylated probe – relation to Hbcag expression and histology. *J. Hepatol.* 19: 204–210.

Newland JW, Heill JR (1988) DNA probes for shiga-like toxin-converting bacteriophages. *J. Clin. Microbiol.* 26: 1292–1297.

Osiowy C (1998) Direct detection of respiratory syncitial virus, parainfluenza virus and adenovirus specimens by a multiplex reverse-transcription-PCR assay. *J. Clin. Microbiol.* 36: 3149–3154.

Patterson BK, Czerniewski MA, Pottage J, Agnoli M, Kessler H, Landay A (1999) Monitoring HIV-1 treatment in immune-cell subsets with ultrasensitive fluorescence-*in-situ* hybridisation. *Lancet* 353: 211–212.

Persing DH (1991) Polymerase chain reaction: trenches to benches. *J. Clin. Microbiol.* 29: 1281–1285.

Picken MM, Picken RN, Han D, Cheng Y, Strle F (1996) Single tube nested polymerase chain reaction assay based on flagellin gene sequences for detection of *Borrelia burgdorfi sensu lato*. *Eur. J. Clin. Microb. Infect. Dis.* 15: 489–498.

Read SJ, Jeffery KJM, Bangham CRM (1997) Aseptic meningitis and encephalitis: the role of PCR in the diagnostic laboratory. *J. Clin. Microbiol.* 35: 691–696.

Relman DA, Loutit JS, Schmidt TM, Falkow S, Tompkins LS (1990) The agent of bacillary angiomatosis. *N. Engl. J. Med.* 323: 1573–1580.

Relman DA, Schmidt TM, MacDermott RP, Falkow S (1992) Identification of the uncultured bacillus of Whipple's disease. *N. Engl. J. Med.* 327: 293–301.

Saiki RK, Gelfand DH, Stoffel S *et al.* (1988) Primer-directed enzymatic amplification of DNA with a thermostable DNA-polymerase. *Science* 239: 487–491.

Tang YW, Procop GW, Persing DH (1997) Molecular diagnostics of infectious diseases. *Clin. Chem.* 43: 2021–2038.

Walker GT, Fraiser ML, Schram JL, Little MC, Nadeau JG, Malinoswski DP (1992) Strand displacement amplification – an isothermal, *in vitro* DNA amplification technique. *Nucleic Acids Res.* 20: 1691–1696.

Wilson M, DeRisi J, Kristensen H *et al.* (1999) Exploring drug-induced alterations in gene expression in *Mycobacterium tuberculosis* by microarray hybridisation. *Proc. Natl Acad. Sci. USA* 96: 12833–12838.

Wolcott MJ (1992) Advances in in nucleic acid-based detection methods. *Clin. Microbiol. Rev.* 5: 370–386.

Wu DY, Wallace RB (1989) The ligation amplification reaction (LAR)-amplification of specific DNA sequences using sequential rounds of template-dependent ligation. *Genomics* 4: 560–569.

PART 4

ANTIBACTERIAL AGENTS

25

Endogenous Antimicrobial Molecules: An Ecological Perspective

Wolfgang Piepersberg

Bergische Universität GH, Wuppertal, Germany

The ecology of micro-organisms, including bacteria, archaea, protists, lower algae and fungi, in a given biotope remains a mystery. The reason for this is the experimental inaccessibility of natural micro-communities and their interaction with their immediate environment. The situation is not unlike Heisenberg's uncertainty principle, in that each experimental approach alters the conditions in an unknown way. Moreover, as seen from the vantage point of a living cell, conditions in microhabitats are subject to continuous highly dynamic changes. Also, recent investigations of the composition of natural populations by techniques based on the polymerase chain reaction (PCR) have made it obvious that many more different microbial inhabitants exist in all investigated habitats than was previously thought. Many of these microbial species, even of the higher taxa of the groups mentioned above, are unknown and currently non-cultivatable (Handelsman *et al.*, 1998; Hugenholtz *et al.*, 1998). It is not, however, the purpose of this chapter to address the matter of the ecological impact of the use of antimicrobials in humans and animals, or their influence on the ecology of antibiotic resistance (Guillemot, 1999)

In this situation, where direct information is not available, it is necessary to look for indirect but clear evidence for a particular ecological behaviour and for the meaning of phenotypic traits in the 'molecular imprint' of ecologically relevant factors in individual organisms, i.e. in their genetic information. Particularly in ecosystems with drastically and rapidly changing environmental conditions, such as soils, plant surfaces and animals, it is to be expected that a large proportion of microbial genetic material must be reserved for rapid metabolic responses. These responses would act to protect against adverse physical factors, including temperature and other stresses, chemicals such as toxins, and predators. They can also be used to facilitate genetic adaptation to intermediate-term changes by responding with altered phenotypes that depend on alterations in genotype. This in turn requires a highly responsive gene pool and frequent exchanges with the global and immediately accessible interacting population.

 doi:10.1006/bkmm.2001.0025

Table 1 Examples of natural substances with antimicrobial potential against typical bacterial targets, their producers and resistance mechanisms encountered in the producers and non-producers

Antibiotic (natural product)	Molecular target (mode of action)	Producing micro-organism	Resistance mechanism
Penicillins	D-Ala metabolising enzyme	*Penicillium, Aspergillus*	β lactamase, porin
Cephalosporins	D-Ala metabolising enzyme	*Acremonium*	β lactamase, porin
Carbapenems	D-Ala metabolising enzyme	*Streptomyces* sp.	β lactamase, porin, export
Clavulanic acid	β-lactamase inhibitor	*Streptomyces* sp.	(No antibiotic)
Erythromycin (14-member macrolide)	50S ribosome peptidyltransferase	*Saccharopolyspora* sp.	23S rRNA MT, export, MGT
Tylosin (16-member macrolide)	50S ribosome peptidyltransferase	*Streptomyces* sp.	23S rRNA MT, export, MGT
Daunorubicin (other aromatic PK)	DNA metabolism	*Streptomyces* sp.	Export
Vancomycin	Transglycosylase in murein synthesis	*Streptomyces* sp.	Relief of target site by modified murein
Gentamicin (kanamycin, neomycin, tobramycin)	30S ribosome	*Micromonospora* sp., *Streptomyces* sp.	APH, ANT, AAC, 16S rRNA MT
Spectinomycin	30S ribosome	*Streptomyces* sp.	APH, ANT
Streptomycin	30S ribosome	*Streptomyces* sp.	APH, ANT, export (?)
Kasugamycin	30S ribosome	*Streptomyces* sp.	AAC
Tetracyclin	30S ribosome, EF-G	*Streptomyces* sp.	alternat. EF-G, export
Rifamycin	RNA-polymerase (RNP)	*Amycolatopsis* sp.	n.d. (mutant RNP)
Chloramphenicol	Peptidyltransferase	*Streptomyces* sp.	CAC, export
Lincomycin	Peptidyltransferase	*Streptomyces* sp.	23S rRNA MT, export
Nalidixic acid	Gyrase	*Streptomyces* sp.	Alternative gyrase
Puromycin	A-/P-site, 50S ribosome	*Streptomyces* sp.	PAC, export
Phosphinothricin	Glutamine synthetase	*Streptomyces* sp.	PAT

AAC, aminoglycoside acetyltransferase; APH, aminoglycoside phosphotransferase; ANT, aminoglycoside nucleotidylyltransferase; CAC, chloramphenicol acetyltransferase; EF-G, bacterial translation (elongation) factor G; MGT, macrolide glucosyltransferase; MT, methyltransferase; PAC, puromycin acetyltransferase; PAT, phosphinothricin acetyltransferase; PK, polyketide; n.d. = no (transferable) resistance gene detected.

As we will see below, many of these requirements and genetic and phenotypic factors that are potentially vulnerable to the environment can be identified in the genomes of micro-organisms. This is true in particular for the production of the low-molecular-weight antimicrobial factors we know as 'antibiotics' (for examples see Table 1), and resistance against such factors. It is not reasonable to assume that the production of natural substances with antimicrobial activity and the microbial resistance against them has been changed by the selection pressure of human influence in the recent past. Rather, antibiotics are old-established natural products that have had common, but probably changing and manifold, physiological uses throughout evolutionary time. It may even be that some antibiotics existed and had essential functions in the pre-cellular biological world, for example the 'RNA world', before the emergence of living cells at least 3.5–4 billion years ago (Chadwick *et al.*, 1992).

The Ecology of the Antimicrobial Producers – Possible Ecological Roles of Antimicrobials

In the present day an ecosystem might be investigated purely on the basis of its macroscopically visible and countable organisms, though in such an exercise it would be too difficult to include only higher plants and animals and higher fungi (basidiomycetes). Such an investigation might include a determination of the number of organisms, the development of populations over time, and ecologically relevant behaviour, including, for example, nutritional requirements, participation in the creation of resources and effects on other populations of organisms. Investigations of this kind must be regarded as incomplete if populations of micro-organisms are disregarded, but even the numbers of taxa of these are largely unknown (Pace, 1997; Hugenholtz *et al.*, 1998; Whitman *et al.*, 1998; Stencel, 2000). This is also the case for the lower eukaryotes, and perhaps even for still unknown kingdoms of life forms. The same is true for the physiology and biological relevance of particular large- and small-scale conversions of matter. In other words, we are far from having a safe basis on which to form judgements of the real biodiversity that forms the ecological units or networks in communities.

Only in the few cases of well-studied two-organism interactions, such as those of host–parasite relations in infectious diseases of humans, higher animals and plants, can ecologically relevant data be regarded as known. Even in these cases, however, there are still major gaps in our knowledge, such as: (1) Even when 'Koch's postulates' are fulfilled, do the single strains of micro-organisms thought to be responsible for particular pathological phenomena form the relevant 'ecosystem'? (2) Are there further relevant so-called pathogenicity factors than those that are currently known, and how does the pathogen profit from its survival in the particular infection? (3) Does the pathogenic interaction in the main and real 'ecosystem' occur in the regular lifestyle of the infective micro-organism, or part of that lifestyle when the organism has several different biotopes for the individual stages of the life cycle? If this is not the case, what is the normal habitat and mode of life? Even less is known about free-living micro-organisms in habitats where macro-organisms do not play a major role. This is the case, for example, on the surfaces of macro-organisms and in soils and aqueous habitats, where many different micro-organisms are present.

Another current problem is that we do not really know anything about the dynamics of genetic innovation in the major groups of organisms. For instance, it is generally believed that lateral gene transfer and adaptation to new environments occurs only in prokaryotes (Eubacteria, Archaea) and not in eukaryotes (Lawrence, 1999; Ochman *et al.*, 2000). It is clear that there is a major difference between prokaryotes and eukaryotes in terms of the rates of gene exchange, but there may also be much more in the way of gene flux than has so far been observed. If nothing else, the main mechanisms and carriers of genetic material, such as viruses, transposons and plasmids, are also present in eukaryotes. In addition, the many intracellular bacterial and eukaryotic parasites could give rise to gene exchange between distant groups of organisms. Thus, the universal gene pool may even now be accessible to higher plants and animals for long-term genetic innovations. We may find evidence of much later events than, for example, the major events of endosymbiont uptake that led to mitochondrial and plastid organelles and the concomitant gene donation to the eukaryotic nucleus. Such evidence of putative bacterial gene incorporation may become available from genome sequences, when more of these are available. One of these examples may be that of the amidinotransferase encoding genes (see later). Thus, an intracellular actinomycete parasite of lower vertebrate cells or of their immediate ancestors, similar for instance to *Mycobacterium tuberculosis*, could have introduced the amidinotransferase gene into vertebrates or other species. One could, of course, also argue in such instances that either the gene or enzyme is an invention of the particular eukaryotic branch of organisms and the gene flux occurred in the opposite direction, that is from vertebrate to bacterium, or that independent and convergent evolutionary events have led to the very similar sequences and functions of nucleotides and proteins in very distant groups of organisms. This would, however, principally mean that no single phylogenetic tree can be accepted as giving real information about biological history (Doolittle, 1999; Stencel, 2000).

How does this relate to known microbial natural products and the discovery of new and still unknown drugs, including antibiotic-like substances, secondary metabolites or individualites (Handelsman *et al.*, 1998; Osburne *et al.*, 2000)? Their functions and evolutionary history are still largely unknown (Vining, 1990; Chadwick *et al.*, 1992; Piepersberg, 1993). Antimicrobials and other therapeutically useful natural products have not arisen from human selection, or somehow awaited the arrival of our species. Rather, because of a still largely unstudied ecological role, they have a powerful selective value for their microbial producers that has been disguised by the anthropomorphic term 'antibiotics' (**Table 1**). This hypothesis, though still unproven, must be inferred from several indicative facts which are outlined below in greater detail. Briefly, it must be assumed that these substances are part of an extracellular system of metabolic communication, similar to the system of hormones, cytokines, defensins and the like in human body fluids. Such 'cross-talk' seems always to exist in complex associations of organisms; it is present everywhere in nature and is encountered particularly in soils, especially in mycorrhizas and other interactions between microbes and higher eukaryote. A complementary hypotheses would be that these natural products have multiple purposes. They might be molecular weapons, repellants, attractants, 'nutriophores' (e.g. siderophores), co-catalysts, enzyme or transport inhibitors, etc., shaped by a long evolutionary history for highly specific interaction with their

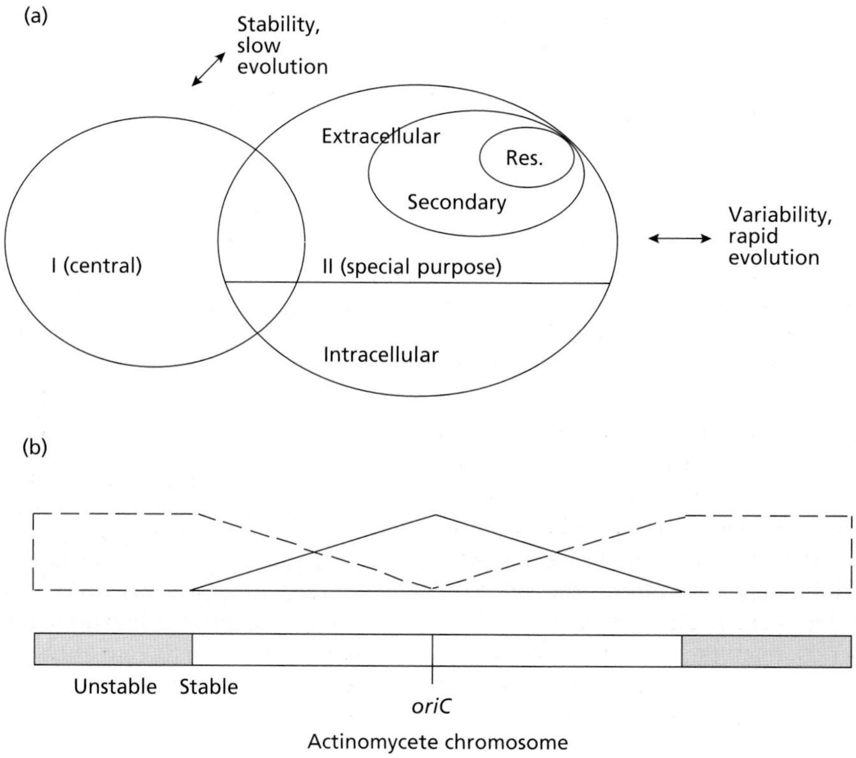

(a)

Stability, slow evolution

Extracellular

Res.

Secondary

Variability, rapid evolution

I (central)

II (special purpose)

Intracellular

(b)

Unstable Stable

oriC

Actinomycete chromosome

Fig. 1 Metabolic compartments and corresponding genetics of a typical producer of antimicrobials (actinomycete, fungus, plant): (a) Hypothetical metabolic compartments are regarded as basically bipartite. A central compartment (I) serves all the essential needs of the macromolecular (DNA, RNA, protein, membrane and cell wall), cell cycle (differentiation, regulation), intermediary, and energy delivering metabolism. A second, highly variable and adaptive compartment (II), called 'special purpose', which may be regarded as subdivided into intra- and extra-cellular targeted metabolic traits, contains the so-called secondary metabolism, defined as the production of low-molecular-weight bioactive molecules secreted into the environment and having their target outside the producing cell; antimicrobial resistance mechanisms are a sub-fraction of this sub-compartment (see text). (b) The genome of an actinomycete seems to exhibit a characteristic correspondence to the above metabolic divisions by having increasingly higher concentrations of genetic material encoding functions in compartment II (gene pool II; broken line indicates relative fraction of gene pool) towards both ends of the single linear chromosome; these ends are also highly variable and contain large segments (more than 1 Mb of DNA) which are non-essential for survival in a laboratory environment. In contrast, the central chromosomal segment grouped around the origin of replication (*oriC*) contains most, or all, of the essential 'housekeeping' genes for compartment I (gene pool I, solid line indicates relative fraction of gene pool). (Res., antimicrobial resistance.)

molecular targets (receptors) and genetically highly regulated.

In order to understand, at least in part, the need for these communication metabolites it is necessary to take a closer look at the metabolic compartments and their underlying genetics, especially of organisms living in complex sessile associations that are often the subject of extreme changes in environmental conditions from which autonomous escape to more favourable niches is not possible. The most typical and abundant habitat hosting such associations is the bacterial–fungal communities living in the rhizosphere of higher plants. Similar associations seem to be characteristic of the environment of relatively sessile marine organisms such as sponges, coelenterates, corals and algae. The participating filamentous or differentiated soil bacteria, fungi, and plants are – not by chance – among the groups that are the most highly creative producers of bioactive natural products such as anti-infectives, cytotoxins, enzyme inhibitors and immune modulators.

In the light of our ignorance, substances such as the antibiotics produced by microbes and the special substances that accumulate in plant cell vacuoles have been called 'idiolites' or 'secondary metabolites'. These are misleading terms that, at best, take cognisance only of the fact that these substances are produced under special conditions (e.g. growth phases) in only a few organisms, which are mostly strain or biovar-specific rather than taxon-specific. However, terms such as 'individualites' (Piepersberg, 1993), which attempt to incorporate the biological, especially the ecological, function of these special-purpose products of communication metabolism, have not as yet been generally accepted.

A generalised view might try to focus on the compartments of metabolism and their underlying genetics by answering some questions: (1) what is the 'secondary' (individual) fraction of metabolism, (2) which properties does the gene pool need for its inheritance and dissemination in the population (Fig. 1) and (3) what are the selective conditions and dynamics of its use by a micro-organism to cope with the environment? Additional, but less direct, information for our attention could stem from the questions (4) how and in which organisms did the complex natural substances originate ('molecular fossil' versus 'playground of evolution' hypotheses; see below) and (5) did they in the distant past have other functions (Zähner *et al.*, 1988; Chadwick *et al.*, 1992; Piepersberg, 1993)? Another problem is that probably not even 1% of living micro-organisms, including higher taxa, are recognised, which may mean that the chemical diversity of individualites is equally far from being explored (Handelsman *et al.*, 1998; Hugenholtz *et al.*, 1998).

Secondary Metabolism as Communication Metabolism

Theoretical treatment of the questions concerning the present or possibly extinct functions of communication (secondary) metabolism has suggested a multiplicity of possible purposes (reviewed by Piepersberg, 1993) (Table 2; Fig. 2).

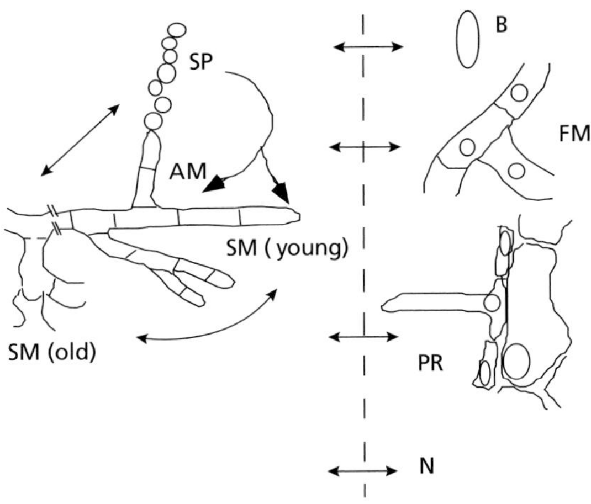

Fig. 2 Possible routes and targets of extracellular communication via special-purpose (secondary) metabolites (SPMs). An example, sketched in the left panel, is the situation inside and outside a producing micro-colony of a streptomycete, e.g. in soil. SM (old), substrate mycelium, the vegetative mycelium outgrowing after spore germination in suitable growth conditions; SM (young), the still active vegetative mycelium at the periphery, after cell differentiation has already started; AM, aerial mycelium, with which cell differentiation starts; SP, chains of arthrospores, which form after segmentation and further differentiation at the tips of AM (Champness and Chater, 1994). The arrows in the left panel indicate routes of intra-species communication (see examples in the text). Possible biological and abiotic targets of special purpose (secondary) metabolites are shown in the right panel. B, bacterial cells; FM, fungal mycelium; PR, plant roots; N, nutrients, the acquisition of which is an example for abiotic chemical interactions with the environment in which typical SPMs play a role.

Table 2 Possible functions and evolution of secondary or individualisation metabolites (individualites)

Function(s)	Selective advantage	Target(s) of product
Early evolution (molecular fossils)		
(1) Ancient catalysts, precursors of proteins, coenzymes, etc.	Extinct (essential)	General ligands (substrate or otherwise); some still existent?
Late evolution (recent inventions)		
(2) 'Selfish metabolism', or products of 'selfish DNA' activity	None	None to anything, specificity by chance
(3) Hormones (autoregulators, modulators of macromolecule synthesis, alarmones or pheromones)	Yes (reduced energy consumption; individualisation), for the producer itself via product	Receptors of signal transduction; high specificity by co-evolution
(4) Repellants or attractants	Yes (exclusion of competition versus symbiosis), for the producer itself via product	Essential complex/ receptors; high specificity by co-evolution
(5) Detoxification, or replacement metabolism	Yes, for the producer itself via metabolic shunt	None
(6) Genetic reserve material, 'playground of evolution'	Yes, but not direct; for the cell lines sharing a gene pool	None to anything under changing environmental conditions

Repellants or Molecular Weapons

One ecologically relevant aspect is that individualites or secondary metabolites may be repellants or molecular weapons – metabolites of extra-organismal warfare directed against competitors and predators in the resident habitat. An 'antibiotic' would best fit into this category.

A special group in this category are the many extracellular enzyme inhibitors of hydrolases, e.g. proteases, glycosidases, esterases, and lipases, which do not have typical antimicrobial effects. These hydrolases might interfere with the nutrient supply of competitors in a highly competitive ecological situation.

Another group generally to be considered in this class are the many specific protein-based toxins and other macromolecules secreted by micro-organisms that may be targeted against near relatives (e.g. bacteriocins) or distant taxa (general toxins), which may have a role in invasive attacks, quite apart from being protective.

Hormones

Individualites may also be hormones (autoregulators, modulators of macromolecular synthesis, alarmones, pheromones, attractants), i.e. substances that transmit specific biochemical signals or information to other cells of the same or another organism and require a specific physiological reaction similar to the variously differentiated cells in a multicellular organism (**Fig. 2**).

The Playground of Evolution

Individualites might also be chance products of a steadily changing biosynthetic potential, based on a rapidly fluctuating genetic 'playground of evolution'. According to this view the selective advantage would lie not in the product or the pathway of its synthesis, but rather in the metabolic adaptability that allows rapid biochemical 'invention' under changing environmental conditions. A larger group of organisms sharing a common exchangeable gene pool would participate in this to their advantage.

Pro-primary Metabolites

Some typical members of the extracellularly targeted substances might be called 'pro-primary metabolites', since in a way they stand at the boundary between an essential central metabolic function and a more individualised special purpose; examples might be the many chemically different siderophores encountered in bacteria.

(a)

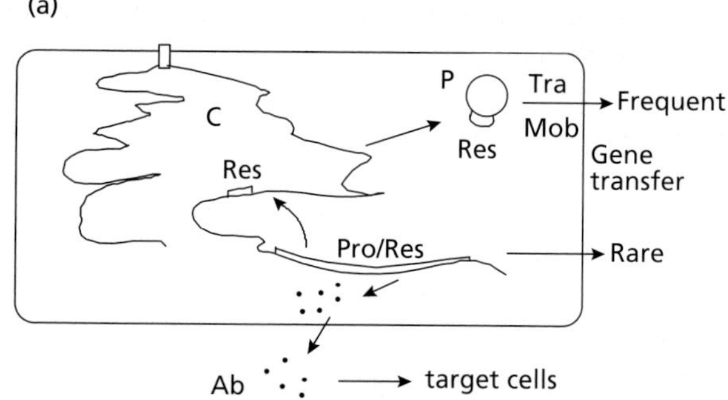

(b)

Fig. 3 An antimicrobial-producing (bacterial) cell and its corresponding genetics: (a) Originally, the genes for production of (Pro) and resistance to (Res) antimicrobials (Ab) seem to reside on the bacterial chromosome (C). Production gene clusters (Pro/Res) and single resistance determinants (Res) can be subjected to horizontal gene transfer. For chromosomal genes this is a relatively rare event, but it may be greatly facilitated by first recombining into a plasmid (P) which is self-transferable (Tra, conjugation) or mobilisable by outside factors (Mob) and thereby may move to other potential producer cells or to pathogens with high frequency. (b) Typical production gene clusters for antibiotics in bacteria are multi-locus assemblies which not only contain the biosynthetic genes (unshaded areas), but mostly also encode the functions for specific regulation (hatched areas), resistance against the toxic end-product (black areas) and active transport (export) out of the cell (light grey areas). Resistance genes may be derived from biosynthetic or transport genes (arrows).

In the author's opinion, in nature a mixture of functions that includes all four of the aspects mentioned above is realised, at least in the microbial individualite producers. A similar, but slightly different situation may exist in the higher plants, where the 'extracellular' world includes the large cellular vacuoles into which most of the diffusible individualite substances are secreted. In animals, individualites in the sense used in this chapter are less abundant, though they exist for instance in form of the low-molecular-weight hormones, immune modulators, and defensins. Their function is largely replaced by protein-based mechanisms, such as the immune system. We return to this subject at the end of the chapter.

Independent Evolution

An attractive hypothesis is the independent evolution in bacteria of their diffusible individualites and the currently recognised secondary metabolic pathways, in parallel with their surface components and their biosynthesis. An indicator for this would be the use of the same gene pool (Piepersberg, 1992). The theory would include all substances that play a role in the build-up of glycan and other modified surface layers, lipids, murein, (glyco-)proteins (e.g. S-layers), polysaccharides, teichoic acids, etc., which could have been precursors of antimicrobials and other extracellulary targeted secondary metabolites, or vice versa.

This theory would predict that all components that can form and variably modify, that is 'individualise', the outer cell surfaces in a bound state, i.e. in polymer or membrane-integrated form, could be also used to produce diffusible low-molecular-weight compounds with potential value for communication. Molecular archaeology, therefore, should make it possible to investigate whether products of similar pathways have been used for both purposes and still can give rise to either of the two bioactive functions.

Two examples are the highly variable components of lipopolysaccharides (LPS) in Gram-negative bacteria, and the polyketide–fatty acid relationship.

Variable Components of LPS

These are mainly sugars composed in a strain-specific, rather than species-specific or higher taxa-specific, fashion. They resemble oligosugars, such as aminoglycosides, in their structure and in the gene pool used (see below). It is, therefore, possible to ask whether oligosugars, which when polymerised make-up LPS side chains (O-antigens) can be used to design new bioactive molecules. The inner core structures, 'endotoxins' (lipid A), are themselves bioactive molecules.

Polyketide–Fatty Acid Relationship

Mycobacteria, particularly the mammal-associated *M. tuberculosis* (see genome evaluation below), have recruited polyketide pathways similar to those for the production of many individualites to produce specific lipid components that are integrated into their rigid and impermeable cell walls, and which seem to be a major prerequisite for their intracellular survival as pathogenic parasites. These polyketide pathways appear to be the only pathways present in these organisms that are closely related to the antimicrobial-producing higher actinomycetes (Cole *et al.*, 1998).

Gene Pools and Gene Fluxes

Bacterial, fungal, and plant genetics currently provide the major sources of insight into the resident tools (genes, clusters, enzymes, pathways, regulators, resistance determinants, etc.) of the ecologically relevant metabolism of extracellular compounds such as the antimicrobials. This means that retrospective inference must be used to interpret molecular data from observations in individual organisms and to draw conclusions indirectly about the ecological interrelationships of free-living micro-organisms.

Information from the first fully sequenced bacterial genomes stresses the dominant role of horizontal gene transfer in the dissemination of such traits (Lawrence, 1999; Mozner, 1999) (**Fig. 3**). Even speciation is no longer connected with these highly variable and adaptive genotypes or phenotypes and the underlying genetic material. Thus, 'species' are easily and rapidly split into cell lineages with very different phenotypic properties in terms of extracellularly targeted metabolism but also, for example, in terms of the cytoplasmic properties, such as consumption of various carbon sources and many other nutrients. In future, therefore, it may be necessary to give up the concept of species as the basic organismal unit in favour of that of 'ecotypes'.

End Regions of Chromosomes

'Chromosome ends' near the sites of replication termination in circular chromosomes, as in *Escherichia coli* and *Bacillus subtilis*, and the protein-bound ends of linear chromosomes, as in streptomycetes, seem to be primary invasion regions with hot-spots of recombinative rearrangements of horizontally transmitted incoming genetic material (**Fig. 3**). A scenario for a kind of 'downstream processing' of this genetic material can be imagined. If it is selected for usefulness, structurally stabilised and collected in a compact gene cluster, and becomes functionally adapted to environmental needs and harmonised with the resident metabolic and regulatory needs, it can become integrated into the pre-existing gene pool in a stable manner.

The highly adaptive extracellularly targeted metabolism of antibiotic producers, as we have sketched it above, requires a corresponding, highly flexible genetic basis. This seems to be the case in typical producers, as exemplified by the case of the streptomycetes and other differentiating mycelial actinomycetes (Hopwood, 1999). The first complete genome of a model typical actinomycete producer of secondary metabolites, *Streptomyces coelicolor* (see below), will be completed in 2001: as of September 2000, about 80% of it is annotated in the EMBL database. However, it can already be concluded that most of the essential genes of central ('primary') metabolism (pool I in **Fig. 1a**) are centred around the genetically much more stable segment that contains the origin of vegetative DNA replication. This is localised in about the middle of the single linear chromosome of ~8 Mb in length, which is about double the size of the typical enterobacterial circular *E. coli* genome (Figs. 1b and 3b).

Gene pool I is ubiquitous and constant, at least in a given taxon, and the expression of most genes is constitutive or dependent on the essential needs of the stages of the cell cycle. Its purposes or targets are replication, transcription, translation, cell-cycle control (differentiation), central intermediary and energy-generating metabolism and synthesis of the essential parts of the cell wall envelope. Gene pool II, in contrast, is strain-specific, highly variable and used only under specific circumstances. Its individual genes or functional gene sets (e.g. production gene clusters) are highly regulated by environmental conditions, and may additionally be dependent on cell differentiation. The situation in a typical bacterial producer cell, such as *Bacillus* sp. or *Streptomyces* sp., is outlined in Fig. 3. The cell contains the genes for production (Pro), normally on the chromosome(s) (C), and mostly in clustered gene assemblies (Pro/Res) that also contain other genes directly coupled to the productive metabolism – often, for example, those for resistance, regulation and transport (Fig. 3b). These gene clusters facilitate horizontal genetic transfer of either the whole or part of the production process, which is reflected by the modular use and the 'mixed' phylogenies of the individual genes when compared with other gene assemblies used for biosynthesis of similar end-products or substructures.

Higher rates of horizontal gene exchange, based on the abundant presence of plasmid transfer (Tra), plasmid mobilisation (Mob), transposons (Tn), insertion sequences (IS) and similar functions or elements, seem to be a rule in these variable gene pools. Otherwise, a location on a plasmid, which could be favourable for transfer, seems to be rare in actinomycetes both for secondary metabolite production gene clusters and for resistance genes (see below). Some gene clusters for secondary metabolites have been found integrated into plasmids (P in Fig. 3). The product (Ab in Fig. 3) is formed in the cytoplasm, often in inactivated form by modification through a resistance mechanism (Res) or biosynthetic step. Generally the active antimicrobial is actively released into the medium where it can reach its target cells by diffusion.

Organisation of the Genome

The genome structure and dynamics of streptomycetes and mycelial actinomycetes reflect the needs of their high flexibility in having a larger genome (~8 Mbp) than other actinomycetes, such as corynebacteria (2.5–3 Mb) and mycobacteria (4 Mb). These higher actinomycetes are able to rearrange, lose or acquire DNA segments of more than 1 Mbp. They have linear chromosomes with highly variable end segments that could represent almost half of the genome (see below). They are also able to exchange large segments of genetic material, at high frequency and by a wide variety of mechanisms, with relatives and even distant organisms such as *E. coli* (Piepersberg, 1993; Hopwood, 1999; Hu *et al.*, 2000). Thus they have a particularly large genetic reservoir for gene mobilisation, and for secondary and other special-purpose metabolism. A similar situation seems to exist in the Gram-negative myxobacteria, which in the differentiating (fruiting-body formation) taxa have genome sizes of around 10 Mb and produce a large variety of individualites (Reichenbach *et al.*, 1988).

Unfortunately, none of these genomes is as yet fully sequenced, but the chromosomal DNA sequence of *S. coelicolor* A3(2), which is expected to be available as a complete version by early 2001, provides some early clues. This strain, together with its close relative *S. lividans* 66, which is frequently used in genetic experiments, taxonomically belongs to the species *S. violaceoruber* (Redenbach *et al.*, 1996). A further wealth of genome data, which also contain a more or less high proportion of genetic or biochemical tools for secondary metabolic activities, comes from other bacterial chromosomes, e.g. those of *M. tuberculosis* HRv (Cole *et al.*, 1998); *B. subtilis* 168, a low-G + C organism (Kunst *et al.*, 1997; Mozner *et al.*, 1999); *Deinococcus radiodurans* R1, a moderately high-G + C bacterium (White *et al.*, 1999); *E. coli* K-12 (Blattner *et al.*, 1997); *Pseudomonas aeruginosa* PA01 (Stover *et al.*, 2000); and *Synechocystis* sp. PCC6803 (Kaneko *et al.*, 1996).

In the preliminarily analysis of *S. coelicolor* A3(2) chromosomal DNA, as of 21 September 2000 (D.A. Hopwood, personal communication) there were 7031 predicted genes in 7 510 386 bp. This shows a typically high bacterial density of gene packing (~1 gene/kb of DNA) and an expected total of about 7500 genes. Almost half of the genes (47%) have a predictable function, of which 6% had already been analysed and 41% are significantly similar to functionally analysed genes from other genomes. Of the rest, 30% of the predicted reading frames encode proteins that are so-called 'conserved hypotheticals' which are also found in other organisms, and 23% are completely new, in the sense that they do not have any relatives from current sequence comparisons. No major differences in the gene arrangement and composition were observed in the assemblies at the unstable ends of the linear chromosome as compared with the central 4–5.5 Mb region. The genes with unknown function are equally distributed between core and end regions of the chromosome, but almost all of the genes involved in macromolecular metabolism (replication and repair, transcription, translation) are located in the more stable central section; there are exceptions, however. For example, as well as the gene considered to be the 'normal' housekeeping gene for ribosomal protein L32, rpmF1 (SC7A1.15), there is a second one, rpmF2 (SCF51A.14), residing in the unstable segment at the left-hand end and not as usual in a ribosomal protein gene cluster. Also the rpmF1 (SC7A1.15) gene lies in an area close to the right end of the chromosome, which could already be less stable. In the genome of *M. tuberculosis* the rpmF gene lies between two copies of the highly repetitive PE_PRGS genes (Cole *et al.*, 1998), which are attributed to phase variation phenomena, in an area where no other translational genes are found. Taken together, these phenomena could mean that some of the actinomycetes are in the process of altering or extending the composition of their translational apparatus, either by differential regulation or by protein exchange, e.g. to be able to form different populations of ribosomes under certain conditions. There are strikingly high proportions of ABC transporter systems (~150; see below) and other transporters, regulators of all kinds (e.g. ~70 two-component response regulator systems), transposon or insertion element-related sequences (more than 49). There are numerous transport systems together with other probably membrane-integrated proteins, which points to a very active communicative metabolism between the cytoplasm and the external environment. The high density of DNA-binding regulators and other cellular control components (such as several protein kinases, etc.) also indicates that there is a need for a very complex control of response of gene expression to changing environmental conditions.

The chromosome of *S. coelicolor* A3(2) was further analysed by two screens. The first was designed to look for significantly similar relatives of the biosynthetic enzyme proteins encoded by 112 representative antibiotic-production genes from the pathways of mainly sugar-containing antibiotics and other bioactive actinomycete molecules (streptomycin, spectinomycin, kasugamycin, lincomycin and acarbose), and 84 antibiotic resistance proteins. The second looked at the context of all reading frames in the still incomplete sequences of the first 1.5–2 Mb segments at both ends. Examples of secondary metabolic traits obtained from this analysis, especially those related to antimicrobial product formation and resistance, are listed in Tables 3 and 4. Further data on functional analysis of secondary metabolic gene clusters (provided by L. Hopwood and G.L. Challis) have been included.

The biosynthetic enzymes and the antibiotic resistance proteins were compared by sequence identity with the most similar protein identified in BLAST searches and classified into those that were (a) highly similar (more than 35% identity), (b) moderately similar (25–35% identity) and (c) insignificantly similar (less than 25% identity). The observations (unpublished) revealed the following. (1) Of the biosynthetic enzymes 12.5% were in group (a), 63.4% in group (b) and 24.1% in group (c), whereas for the antibiotic-resistance proteins the distribution was (a) 49%, (b) 37% and (c) 14%. (2) Most of the hits in categories (a) and (b) were not associated with other related genes in obvious gene clusters. (3) Only a few of the possible resistance genes exerted a resistance phenotype or are known resistance genes (Kieser *et al.*, 2000); *S. coelicolor* A3(2) is sensitive to other antibiotics (e.g. thiostrepton, aminoglycosides) of which category (a) relatives are present. (4) The overall hit density is much higher than for any currently completed bacterial genome sequences (see Table 4). Most of the productive secondary metabolic gene clusters that are also present in a stable state in *S. lividans* 66 are in the more stable core region (cf. Table 3). Once more comparisons for different pathways and more actinomycete and other bacterial genome sequences are available from active producers of natural products, such as myxobacteria, it may be possible to evaluate more precisely the meaning of these results for the biosynthetic capacity. The resistance genes are discussed further below.

Gene Clusters and Isolated Genes

S. coelicolor A3(2) produces four known antibiotically active compounds (Table 3; Kieser *et al.*, 2000). Two of these are the well-known pigments: actinorhodin (ACT, the 'blue pigment'; an aromatic polyketide of type II; *act* gene cluster) and products related to prodigiosin (RED, the 'red pigment'; derived from the amino acids proline, serine, and glycine and some other units, they form a mixture of undecylprodigiosin and other components; also produced by some Gram-negative bacteria; *red* gene cluster) (A. Cerdeno, M.J. Bibb and G.L. Challis, personal communication). The others are the calcium-dependent antibiotic (CDA; a non-ribosomal peptide; *cda* gene cluster), and methylenomycin (cyclopentanone; *mmy* gene cluster). The *mmy* genes are not located on the chromosome, but rather on an integrated linear plasmid, SCP1, which is not present in the wild type and which has moved from another related strain *S. lividans* 66. It is interesting to note that, except for *mmy*, these gene clusters are located in the central region of the

Table 3 Examples of production gene clusters for potentially antimicrobial secondary metabolites and resistance-related genes in the actinomycete genome of *S. coelicolor* A3(2)[a]

A Production gene clusters

Chemical class of secondary metabolite[b]	Product (if known)	Cosmid (code no. in database)[c]	Location in chromosome[d]
Non-ribosomal peptide (NRP)	Calcium-dependent antibiotic (CDA)	SCE63/SCE29	cr
	Coelichelin (siderophore)	SCF34	le
	Unknown	SC1A6	re
	Unknown	SC4C2	re
Cyclopentanone	Methylenomycin	—	SCP1 (le/re)
Modular polyketide (PK type I)	Unknown	SC4A9	re
	Unknown	SCAH10	re
Mixed NRP + PK type I	Prodiginines	SC3F7/SC2E9	le
Aromatic polyketide (type II)	Actinorhodin		cr
	Spore pigment	SC6G9	re
Aromatic polyketide (type II)	Unknown	SC5C11	re
Aromatic polyketide (chalcone type PK)	Unknown	SCG58	le
	Unknown	SC2H12	re
Secondary oligosugar	Unknown	SCF62	le
Triterpene	Unknown (hopanoids)	SC6A5	re
Sesquiterpene	Unknown (pentalenene/geosmin?)	SC7E4	cr
	Unknown (pentalenene/geosmin?)	SC9B1	re
Carotenoids	Unknown	SC6A5	re
	Unknown	SC10G8	re

B Resistance-related genes

Resistance mechanism[e]	Similar to known resistance gene (resistance against)[e]	Gene (code no. in database)[c]	Location in chromosome[d]
Exporter, ABC	*carA, srmB, tlrC* (macrolides)	(StJ9)	le
	carA, srmB, tlrC (macrolides)	SCF12.01c	le
	drrA/drrB (daunorubicin)	SCGD3.25/26	cr
	drrC (daunorubicin)	SC5F2A.02c	re
	lmrC (lincomycin)	SC5F2A.03c	re
	lmrC (lincomycin)	SCF56.20	le
	strV/W (streptomycin)	SC3D11.05c/06c	re
	strV/W (streptomycin)	SCG8A.01/02	le
Exporter, MFS	*cmlV* (chloramphenicol)	SC8G12.02	re
	cml (chloramphenicol)	(St10F4)	re
	cmcT (cephamycin)	SCC30.17c	cr
	lmrA (lincomycin)	SCC30.17c	cr
	otrB (tetracyclin)	SCC8A.02c	cr
β lactamase	*blaB* (β lactams)	SCJ11.17c	le
	ytnB (β lactams)	SCJ21.11	le
	blaB (β lactams)	SC5F8.11c	le
	blaES (β lactams)	SCH24.34	cr
	ampH (β lactams)	SCF43A.20c	le
rRNA methyltransferase	*carB, ermSF, lmrB, tlrD* (macrolides, lincosamides)	SCE66.28c	cr
Antibiotic acetyltransferase	*aacC* (AAC[3]; AGA)	SCC22.09	cr
	cac (capreomycin)	SCC22.03c	cr
	bacA (blasticidin)	SCE66.14	cr
	pat (phosphinothricin)	SCE66.14	cr
Antibiotic phosphotransferase	*aphD* (APH[6]; AGA)	SCD49.05c	cr
Antibiotic glucosyltransferase	*mgtA, mgtB* (macrolides)	SCJ4.21	le
EF-G, alternative target site	*tetO/tetM* (tetracyclin)	(3StF60)	le
Gyrase B, alternative target site	*gyrB2* (novobiocin, nalidixic acid)	SCH18.11c	cr

[a]The genome sequence of the linear chromosome of about 8 Mb length was still incomplete at the time this chapter was written (roughly 80% annotated sequences).
[b]Product of putative metabolism as inferred by analogy to known products in other producers.
[c]The codes of the 339 cosmid-cloned DNA segments to be introduced into the EMBL database are used and of their reading frame (gene) annotations.
[d]le, left end segment (~2 Mb); cr, central region (~4 Mb); re, right end segment (~2 Mb); SCP1, linear plasmid integrated into the chromosome ends of certain strains.
[e] cf. Table 1; AGA, aminoglycoside antibiotics.

Table 4 Comparison of the genetic potential for typical biosyntheses of individualites and antimicrobial resistance traits in some representative bacterial genomes[a]

	S. coelicolor A3(2)	*M. tuberculosis* HRv	*B. subtilis* 168	*D. radiodurans* R1	*E. coli* K-12	*Ps. aeruginosa* PA01	*Synechocystis* sp. PCC6803
A. *Biosynthetic pathway/product*							
PK, type I	≥2	2 (Pps; Mas)	1 (17 pr.)	–	–	–	–
PK, plant-like	≥2	≥2	1 (BcsA)	–	–	–	–
PK, type II	≥2	–	–	–	–	–	–
NRP	≥3	4 (Mbt-cl.)	2 (Pps, Dhb)	–	1 (Ent)	5 (2 clusters)	– (?)
PK/NRP	≥1	1	–	–	–	–	–
RP	– (?)	– (?)	1 (SunT,A)	–	–	–	– (?)
Oligosugar (incl. cell wall)	≥4 (Mur, LPS, etc.)	1 (Mur)	? (Mur, EPS?, Tag/Tua)	1 (Mur)	3 (Mur, LPS, EPS, ECA)	≥2 (Mur, LPS, EPS?)	≥2 (Mur, LPS)
Aminoglycoside	Several genes	–	–	1 (Strl)	–	–	–
Macrolide	Several genes	–	–	1 (Ery)	–	–	–
Anthracyclin	Several genes	–	–	1 (Dnm)	–	–	–
Lincosamide	Several genes	–	–	1 (LmbJ)	–	–	–
Nucleoside	Several genes	–	– (?)	– (?)	–	–	–
Isoprenoid	≥4	–	– (?)	– (?)	–	– (?)	≥5 (Crt)
Siderophore	≥2	2 (mbt,exo)	≥1	≥1	1 (ent,fec,fep)	≥1	2 (fec, ent)
Other	≥2	3 (Lmb)	– (?)	–	–	–	1 (iturin)
B. *Resistance determinant*							
ABC, drug export	≥20	4	7	7	4	8	7
(ABC transp., total)	(≥150)	(34)	(69)	(36)	(67)	(64)	(37)
MFS, drug export	≥14	13	20	≥1 (?)	17	20	≥3
RND, drug export	≥1	–	1	1	4	10	3–4
Other export (Mdr)	≥6	2	4	4	3	3	3
AAC	≥3 [AAC(3/6′)]	1 [AAC(2′)]	1 [AAC(3)]	3 [AAC(3/6′)]	–	–	–
APH (3′/3″)	≥1	2	–	2	–	1	–
APH (6)	≥2	–	–	–	–	2	–
ANT	≥1 [ANT(3″/9)]	–	1 [ANT(6)]	–	–	–	–
β Lactamase	≥4	1 (BlaC)	3 (PenP)	3	1 (AmpC)	3 (AmpC)	–
CAC, SAC	≥2 [Sat/Cat]	–	–	1 (Cat)	–	1 (Cat)	–
Gyrase B	≥1	–	–	–	–	–	–
MLS/TSR, 23S MT	≥2	1 (?)	2 (LmrB)	3-5 (SpoU)	–	–	–
PAC	≥1	1	–	–	–	–	–
PAT	≥1	–	1 (YwnH)	–	–	–	–
STAT	≥	–	1 (YyaR)	–	–	–	–
VPH	≥1 [VPH]	–	1 (YcbJ)	–	–	–	–

Table 4 *Continued*

	***S. coelicolor* A3(A2)**	***M. tuberculosis* HRv**	***B. subtilis* 168**	***D. radiodurans* R1**	***E. coli* K-12**	***Ps. aeruginosa* PA01**	***Synechocystis* sp. PCC6803**
Acetazolamide	?	–	–	–	–	–	1 (Zam)
Bacitracin	?	1	1 (YubB)	1	1 (BacA)	1 (BacA)	–
Bicyclomycin	?	–	1 (YdgK)	–	1	–	–
Fosfomycin	?	–	1 (YndN)	1	–	1	–
Fosmidomycin	?	–	1 (YfnC)	1	1	–	–
Macrotetrolide	≥1	1	–				–
Polymyxin	?	–	–	–	1 (PmrD)	–	–
Pyrazinamide	–	1 (PncA)	–	–	–	–	–
Other, unknown	?	2	11		10		–

[a]The eubacterial genomes have the following genome sizes (Mbp)/numbers of identified protein-encoding genes (stable RNAs): *Streptomyces coelicolor* A3(2) = ca. 8.0/7 to 8,000 (D.A. Hopwood, personal communication); *Mycobacterium tuberculosis* HRv = 4.41/3918 (Cole *et al.*, 1998); *Bacillus subtilis* 168 = 4.21/4100 (Kunst *et al.*, 1997; Mozner *et al.*, 1999); *Pseudomonas aeruginosa* PAO1 = 6.26/5565 (Stover *et al.*, 2000); *Escherichia coli* K-12 = 4.64/4289 (Blattner *et al.*, 1997); *Deinococcus radiodurans* R1 = 2.65/2580 (White *et al.*, 1999); *Synechocystis* sp. PCC6803 = 3.57/3169 (Kaneko *et al.*, 1996); Clusters of genes are counted as one.

AAC = aminoglycoside acetyltransferase; APH = aminoglycoside phosphotransferase; ANT = aminoglycoside nucleotidylyltransferase; CAC = chloramphenicol acetyltransferase; EF-G = bacterial translation (elongation) factor G; MLS = macrolide-lincosamide-streptogramin resistance; MT = methyltransferase; PAC = puromycin acetyltransferase; PAT = phosphinothricind acetyltransferase; PK = polyketide; (N)RP = (non-)ribosomal peptide; STAT = streptothricin acetyltransferase; TSR = thiopeptide resistance; VPH = viomycin phosphotransferase.

chromosome which seems to be the stable region (see above). These individualites are, perhaps, more essential for the producing organism, and their active and stable expression is guaranteed only in the respective genomic environment. In addition, the chromosome is predicted to encode the biosynthesis of at least two siderophores, coelichelin and desferrioxamine (Imbert *et al.*, 1995; Challis and Ravel, 2000).

Several typical secondary metabolic gene clusters were also found, of which neither the producing ability nor the possible product is known (Tables 3 and 4). In addition, many isolated genes for possible antibiotic production or resistance enzymes were found in the flanking parts (chromosome ends) and the central region. Phenotypes are known for some of the resistance determinants. They can be expressed fully (e.g. chloramphenicol resistance; unstable), or inducibly, or they may be silent and expressed only after heterologous gene cloning (Cundliffe, 1992a; Hotta *et al.*, 1995).

There are many putative secondary metabolic genes in the *S. coelicolor* genome of which the encoded proteins have their nearest counterparts in other organisms (non-actinomycete, non-high-G + C DNA) and not in the related enzymes or proteins of the same family. This is most easily explained by the frequent horizontal transfer of such genes between distant groups of organisms.

SCF62 Glycosylation Cluster

The left-end segment cosmid SCF62 in the *S. coelicolor* chromosome contains a set of genes that may make up a partial secondary metabolic glycosylation gene cluster (encoded proteins SCF62.07–27), including a gene encoding SCF62.08. This is related to the uronic-acid-forming UDP-glucose/GDP-mannose 6-dehydrogenase family, and is most closely related to a protein (AlgD; 49.2% identity in 445 amino acids; SW:ALGD_PRESH) from *Pseudomonas syringae* pv. phaseolicola involved in alginate biosynthesis. A similar protein is also encoded in the lincomycin gene cluster of *S. lincolnensis* (LmbL; 30% identity in 259 amino acids; SPTREMBL:Q54365). The SCF62 glycosylation cluster also includes three genes (SCF62.19, SCF62.21, SCF62.26) that encode proteins closely related to three equivalent proteins encoded by the (5′-hydroxy-)streptomycin cluster of *S. glaucescens*. These are StrQ (CDP-glucose synthase; Beyer *et al.*, 1998), StrP (putative NDP-hexose 4-oxidoreductase or epimerase) and StrX (putative NDP-hexose 3,5-epimerase), the genes of which are also in a subcluster. These three genes are absent from the respective *str*-cluster of the streptomycin producer *S. griseus*, and do not even seem to exist in the whole genome of that organism. This led to the speculation that part of the streptomycin pathway follows different biochemical routes in these two streptomycetes (Piepersberg, 1997; Beyer *et al.*, 1998); this gene cluster may be in the process of being adapted to some specific use in *S. coelicolor*.

The SCF62 glycosylation cluster also encodes four glycosyltransferases (GTs) (SCF62.07, SCF62.09, SCF62.17 and SCF62.24) which are related to different families of proteins but are all involved in features of primary and secondary metabolism. SCF62.07 is related to the membrane-spanning GTs nearest to the UDP–galactose–lipid carrier transferase of *Erwinia amylovora* (AmsG; 33% identity in 403 amino acids; SW: AMSG_ERWAM), SCF62.09 is related to the mannosyltransferase B of *E. coli* (23.3% identity in 288 amino acids; TR:Q47594), SCF62.17 is related to an amylovoran biosynthesis protein of *E. amylovora* (AmsK; 23.3% identity in 288 amino acids; SW:AMSK_ERWAM) and SCF62.24 is related to a *Rickettsia prowazekii* glycosyl transferase (LgtD; 29.9% identity in 214 amino acids; GI:3861003).

Although it hosts surprisingly many relatives of known secondary metabolic genes or enzymes from other organisms (see above), the *S. coelicolor* A3(A2) genome does not contain significantly related counterparts to all of these – far from it. An extreme example is the amidinotransferases and their genes; they seem to occur only in organisms where a secondary transfer of the amidino group (R-C(NH)-NH_2), derived from L-arginine, is used for introduction into special metabolites, such as streptomycin and myomycin. These rare enzymes do not seem to occur as general housekeeping proteins that are stably present in any of the larger taxonomic groups known so far, except in vertebrates (see case study below). In bacteria, from which they seem to have originated, these are only resident enzymes present in a non-taxonomic distribution in the producers of certain secondary metabolites (Mehling *et al.*, 1995).

A Case Study

Evolutionary Impact of the Amidinotransferase Enzyme Family

Structure and Function of Amidinotransferases

The amidinotransferases (ADT; EC 2.1.4.–) are unique in structure and catalytic mechanism. They catalyse the transfer of the amidino group from L-arginine to the amino groups of other compounds to create a new guanidino group in the product and do not require a coenzyme. The primary and tertiary structures of these enzymes are well conserved in the mammalian and bacterial (streptomycetes) sources where they have so far been found (Humm *et al.*, 1997; Fritsche *et al.*, 1998). Catalytically the most relevant amino acid residue seems to be a cysteine close to the C-terminus of these enzymes, which seems to interact directly with the carbon atom of the amidino group that binds it by a thioether bond. At least two more distantly related catabolic enzyme families exist in nature which are involved in the metabolism of the guanidino group of L-arginine or of its metabolite derivatives, the arginine deiminases (AIH; arginine iminohydrolase; EC 3.5.3.6) and N^G,N^G-dimethylarginine dimethylaminohydrolases (DDAH; dimethylarginase; EC 3.5.3.18). The functions of the AIHs and DDAHs are, respectively, the hydrolysis of L-arginine and N^G,N^G-dimethyl-L-arginine to L-citrulline and ammonia or dimethylamine (Fig. 4). The AIHs have a

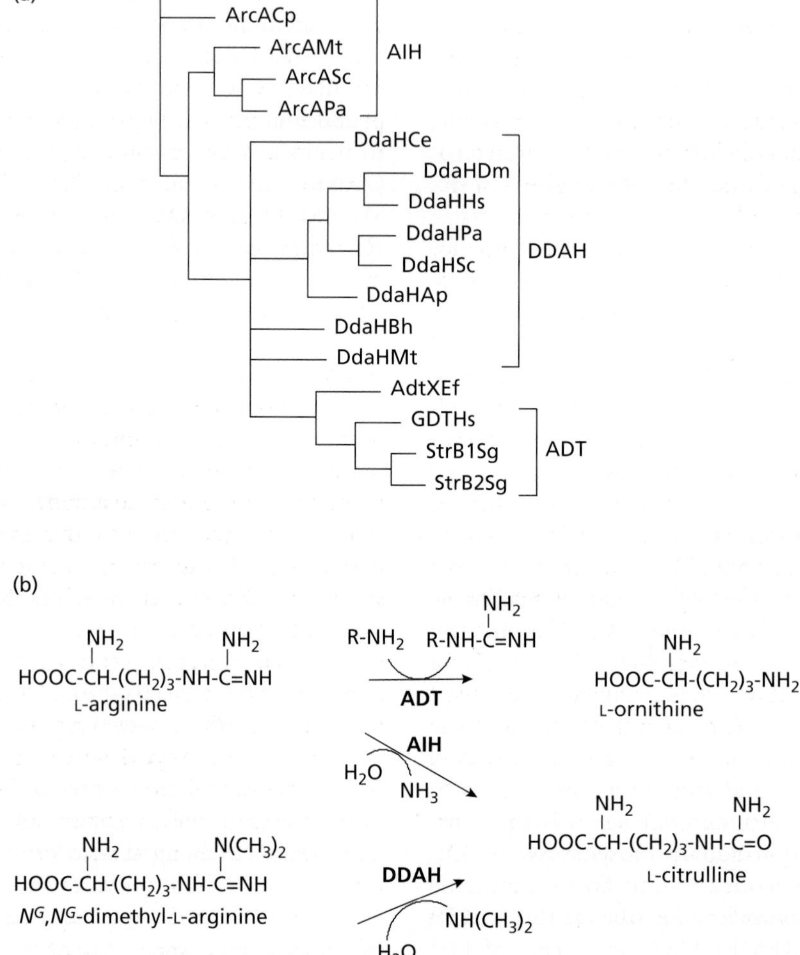

Fig. 4 Phylogenetic tree and catalytic activity of the L-arginine iminase (AIH), dimethylarginase (DDAH), and amidinotransferase (ADT) superfamily of enzymes. The range of sequence identity of pairwise aligned sequences in the subfamilies is: AIH 28–61%, DDAH 20–50%, ADT 39–61% (the AdtXEf is only 13–25% identical to other ADTs). (a) The sequences used in the phylogram are from the following known or postulated proteins (database accession codes in brackets): 1. AIHs: ArcAMa = *Mycoplasma arginini* (X54141); ArcACp = *Clostridium perfringens* (CAA66275); ArcAMt = *M. tuberculosis* (D70602); ArcASc = *S. coelicolor* (SCF55.37); ArcAPa = *Ps. aeruginosa* (X14694) 2. DDAH's: DdaHCe = *C. elegans* (M75933); DdaHDm = *Dr. melanogaster* (AE003492.1); DdaHHs = *H. sapiens* (BAA37117.1); DdaHPa = *Ps. aeruginosa* (CAC09950.1); DdaHSc = *S. coelicolor* (AL049587.2); DdaHAp = *Ae. pernix* (BAA81618.1); DdaHBh = *Bacillus halodurans* (AP001513.1); DdaHMt = *M. tuberculosis* (P71889) 3. ADTs: AdtXEf = *Ent. faecalis* (gnl|TIGR|gef_10493); GDTHs = *H. sapiens* (P50440); StrB1gl = *S. glaucescens* (CAA07380.1); StrB2gl = *S. glaucescens* (CAA07390.1). The figures give the percentage with which the nodes were found in a bootstrapping of 500 repetitions (program PAUP3.1). (b) Schematic representation of the action of the three enzyme families on the guanidino group of L-arginine.

clear function in an anaerobic energy metabolism based on the catabolism of L-arginine in several bacteria (Maghnouj *et al.*, 1998), but the function of the DDAHs in bacteria is not yet clear.

Both the AIH and DDAH enzyme proteins share overall primary structure similarity and particularly a C-terminal domain with the ADTs, which also contain the conserved cysteine residue. Their genes are both present in many but not all bacterial genomes (*E. coli* does not have either), whereas only the DDAH-encoding genes seem to be also stably present in animal genomes (found in *Caenorhabditis elegans*, *Drosophila melanogaster*, and *Homo sapiens*) and in one archaean species, *Aeropyrum pernix*. In cells of mammalian endothelium, and possibly other tissues, the DDAHs function to convert asymmetric N^G,N^G-dimethylarginine, which is an inhibitor of nitric oxide synthase and seems to be a new risk factor in atherosclerosis. This metabolite co-regulates the nitric oxide control of blood pressure in endothelium and other effects, such as cell reactions during mucosal infections, for example by *Ps. aeruginosa* (Dowling *et al.*, 1998; Kimoto *et al.*, 1998). The three enzyme families, which act differently on the guanidino group of arginine, therefore seem to be of common evolutionary origin and may share common features in their substrate-binding and catalytic mechanisms. The three enzyme families are non-taxonomically distributed (**Fig. 4**). Also, there seem to be more enzymes around in some organisms that share at least the catalytic domain of these enzymes. For instance, a non-attributed member of the superfamily from *Enterococcus faecalis*, the only hit found by BLAST similarity searches in microbial genomes with a bacterial ADT sequence, has almost equal similarity to members of all three enzyme families but its function is as yet unknown. In addition, besides a gene encoding an AIH (SCF55.37) and one encoding a DDAH (SC5F2A.01c) in the genome of our reference bacterium for secondary metabolism, *S. coelicolor* A3(A2), there is at least a third gene in an as-yet-unnotated DNA segment that encodes a member of this protein superfamily.

Development of Secondary Metabolic Pathways

What do these puzzling facts tell us about the development of secondary metabolic pathways? In organisms of which the genomes have been sequenced or are currently being sequenced, genes that encode ADTs have so far been found only in mammalian genomes (human chromosome 15) and seem to be present uniformly and are highly conserved (93% sequence identity to rat GDT protein). The amidinotransferases are rare and typically secondary metabolic enzymes in bacteria, and the genes have been found only in a few actinomycete strains (mainly *Streptomyces* sp.; Piepersberg, 1997) that are producers of streptomycin-related aminoglycosides, where they are encoded in the respective production gene clusters. No gene of this family has so far been found in the genome of *S. coelicolor* A3(A2), apart from genes encoding a DDAH (SC) and a AIH (SC). In vertebrates such as mammals these enzymes are essential because of the development of the creatine phosphate pathway as an energy delivery system. Thus, the glycine ADT (GDT) has become an essential housekeeping enzyme in this branch of the animal kingdom. Very similar ADT-encoding genes of the same family are expected to be present in other secondary metabolite producers that synthesise compounds, such as myomycin, which have guanidino groups but are otherwise unrelated to streptomycin (Piepersberg, 1997).

Most of the questions arising from these facts cannot be answered at present, but the following are relevant here: (1) What is the evolutionary origin of ADT-encoding genes and how did these genes pass from bacteria to vertebrates or vice versa? (2) How could ADT enzymes be developed and selected for this particular specificity in absence of a basic requirement for this particular enzyme mechanism, i.e. an ubiquitously present enzyme family? (3) Is there, or has there been in the past, an alternative enzyme family to fulfil this catalytic need in either the bacteria or vertebrates? (4) What is the origin and the function of the second and less conserved ADTs, StrB2 proteins that lack the catalytic cysteine (Fritsche *et al.*, 1998), which are encoded in the streptomycin gene clusters? (5) Was there originally a catabolic role for the ADTs? (6) Was creatine originally a 'secondary metabolite'?

The view that would best fit the current data on the evolution and use of the ADTs is the following. A primordial amidino group able to transfer RNA (ribosyme) or an ineffective protein may have pre-existed the ADTs presently observed and acted catalytically when this was required in the early stages of cellular life, when substances with guanidino groups, such as streptomycin, were 'molecular fossils' (Chadwick *et al.*, 1992). Later, after protein-catalysed metabolism of L-arginine became widely established in early bacterial life, the enzyme structures of AIHs and DDAHs, possibly derived from ADTs, evolved into an enzyme family similar to others involved in the catabolism of more generally distributed substrates. Loss of ancestral ADTs used in earlier pathways that utilised amidinotransfer reactions, or development of new pathways that created a need for a new, 'modern' ADT enzyme, possibly derived from DDAHs or

AIHs in individualite-producing bacteria, would have provided the selective pressure for their development and further evolution.

It seems most likely that the genes were transmitted horizontally into vertebrate ancestors from some secondary metabolic source in an unknown prokaryote, possibly from an actinomycete or a Gram-positive parasite, such as *Ent. faecalis*. This may have occurred at a later time when the creatine pathway became established in animal development, possibly during the establishment of an intracellular parasite or an endosymbiont in specialised eukaryotic cells.

The first and second introduction of a guanidino group into the streptomycin precursor streptidine-6-phosphate, but based on individually encoded and mono-functional enzymes, may be taken as an excellent example similar to domain recruitment in polyketide synthases (PKS) and non-ribosomal peptide synthetases (NRPS). This is so because both require a successive series of five individual enzymatic steps, which starts again only after the first round has been completed (Walker, 1975, 1990; Piepersberg, 1993, 1997). These five catalytic steps, starting with the (amino-)cyclitol precursor, are: dephosphorylation, oxidation by dehydrogenation, transamination, phosphorylation and transamidination from arginine (**Table 2**). Only the two amidinotransferases (StrB1 and StrB2) appear to originate from a gene duplication event that occurred during the process of evolution of the streptidine moiety. An alternative solution to the same chemical problem had obviously been solved in parallel, or because of an early loss of the capacity to synthesise the second guanidino group, on the basis of the common intermediate *N*-amidino-*scyllo*-inosamine by introduction of a carbamoyl group into the cyclitol derivative bluensidine in the streptomycin-related aminoglycoside bluensomycin (Walker, 1990; Piepersberg, 1993, 1997). The other enzymes for the pairs of analogous steps in streptidine biosynthesis, the postulated gene products StrO/(StsD?) (cyclitol-phosphate phosphatases), StrI/StsB (cyclitol dehydrogenases), StsC/StrS or StsA (cyclitol transaminases) and StrN/StsE (cyclitol kinases) are all only very distantly related. They are most probably not derived from recent gene duplication events that could be related to the gene/enzyme development of the streptomycin pathway as it is now seen in streptomycetes.

Secondary Metabolic Aminotransferases and Dehydratases

Many similar enzyme families could be considered in the same way, and several of these have clearly split into subfamilies of 'secondary metabolic enzymes' with highly variable functions. So-called secondary metabolic aminotransferases and dehydratases (SMAT; Piepersberg, 1994, 1997) have been detected that belong to and form a special class (class VI) within the large group of pyridoxalphosphate (PLP)- or pyridoxaminephosphate (PMP)-requiring enzymes. At least two members of this class are encoded by the *E. coli* K-12 genome, many in actinomycete secondary metabolic gene clusters, but no obvious member is encoded by the *S. coelicolor* A3(A2) genome. This large and widely distributed group of PLP/PMP-dependent enzymes, to which the proteins StsA, StsC and StrS also belong, is interesting in this context because they form the only class of the enzyme superfamily that does not have members for intracellularly targeted housekeeping functions. All five other aminotransferases have highly conserved members in nearly all the genomes both of prokaryotes and of eukaryotes. In contrast, all SMATs seem to be involved in the modification of extracellularly targeted molecules, mostly of glycosidic residues. The products biosynthesised with the participation of SMATs are either constituents of cell-wall-attached polysaccharides (e.g. lipopolysaccharides in Gram-negative organisms) and proteins (e.g. the glycosylated pilin of *Neisseria meningitidis*; Power *et al.*, 2000) or are typically diffusible individualites, such as antibiotics.

Aminotransferases for secondary metabolic purposes that have very similar catalytic functions have also been recruited from the other five classes of PLP/PMP-dependent enzymes. For example, class I aminotransferases are involved in the biosynthesis of lincomycin and nikkomycin and class III aminotransferases in the biosynthesis of spectinomycin and acarbose. Why then do SMATs (class VI aminotransferases) exist at all, and which of their properties account for their preferential use in ecological functions?

The reasons for the large differences that seem to govern recruitment of individual genes and enzymes for primary and secondary metabolic purposes may also be quite different, and these too will be discussed for ADTs and SMATs.

The preferred route of evolution for more abundantly used reaction types, such as dehydrogenation, (de)phosphorylation or oxygenation (P450 enzymes), may be not to select from recent gene duplications, because more divergently developed enzymes are more abundantly available than the rarely used ADTs. Analogous substrates among the streptidine precursors for the recently developed ADTs may differ too much to be recognisable by the same enzyme.

Whenever the intermediates converted by the ADTs are significantly toxic to the cell, the need to acquire new and effective enzymes of the appropriate type could have become much greater for selection on horizontal gene transfer. The transamidination steps are very rarely used in nature and no regular reaction of this type is known from most groups of organisms. Since creatine formation involves a transamidination step from arginine to glycine, only vertebrates have developed a basic need for this reaction because of their use of the creatine phosphate system for relaying chemical energy. There is also less little difference between the substrates for the two amidinotransferases involved in the production of the antibiotic streptomycin (Walker, 1975), and it seems that even the single (first-step) enzyme present in the bluensomycin producer may have both activities (Walker, 1990; Fritsche *et al.*, 1998).

All these hypotheses will remain speculative, however, until clear genetic and biochemical proof for the functions of larger sets of these enzymes becomes available.

Recruitment of Genes for Secondary Metabolism

The example of the ADTS shows that actinomycetes appear to be very efficient collectors and users of special purpose (secondary metabolic) gene pools. Further examples are given in **Tables 3** and **4**. In turn, they seem to disseminate their genes to non-related organisms, and even to vertebrates; there are many examples of this (Mazodier and Davies, 1991). This does not, however, necessarily mean that they are also the inventors of these special metabolic traits, such as the production of antibiotics or resistance genes. These secondary biosyntheses or protective defence mechanisms may have arisen in other creative bacterial or eukaryotic groups, pre-existing the higher actinomycetes, or even in ‘molecular fossils’ from the pre-cellular period of the evolution of life, because of long-lasting and uniform selective pressures (Chadwick *et al.*, 1992).

In the current state of the biosphere, however, the actinomycetes, together with other sessile microbial groups similarly active in the continuing needs of extracellular reshaping, may play a major role in collecting the genetic potential for individualite production, recombining it, in part newly inventing it and distributing it by horizontal transfer throughout all natural populations (**Fig. 3**). For this purpose, the genes would be assembled into clusters which, apart from biosynthetic genes, also contain genes for regulation, resistance, transport and, possibly, the extracellular processing of the biosynthetic product. Several lines of indicative evidence exist for either of these speculative scenarios, but a detailed discussion of these is beyond the scope of this chapter.

In view of the highly dynamic nature of the total accessible gene pool, it seems reasonable to assume that environmental selection by other organisms and abiotic factors, as well as self-selection by the interplay between cells of different types in organisms with a complex life cycle, such as the higher actinomycetes, creates new genotypes and phenotypes in rapid succession. This, however, happens to the whole actinomycete group of organisms on the basis of a relatively stable and easily accessible arsenal of metabolic mechanisms and products, selected and adapted during long-term evolution. This would also explain why patterns of secondary metabolites differ significantly between groups of Gram-positive bacteria, such as those of high G + C content (e.g. Actinomycetales) and low G + C content (e.g. Bacillaceae), the myxobacteria, the pseudomonads and the cyanobacteria, or the various groups of filamentous or fruiting-body-forming fungi or higher plants.

At the same time, however, individual products can occur in very different groups, such as the β lactams produced by bacteria and filamentous fungi and the anticancer drug taxol produced both by fungi and by plants. It is therefore reasonable to predict that the arsenal of tools for offence, defence and other types of communicative interplay with different cell types in the environment of actinomycetes and other microorganisms and plants will turn out not to be taxonomically distributed.

Moreover, the modern biosynthetic pathways will not necessarily have co-evolved with their present-day metabolic end-products. They could either have pre-existed or have been ‘invented’ later under a selective pressure based on a strong requirement for the particular compound(s). The retroselective build-up of an efficient and modern pathway would require use of the ‘playground’ offspring of recent chance products of side-reactions of enzymes with other functions as intermediates, and of the ‘molecular fossil’ derivation of long-existing and horizontally acquired enzymes for intermediary steps required for a particular end-product. These assemblies could be directed towards the formation of new or old substances for old or new targets, i.e. ancient versus modern ligand–target site combinations. In particular, the fine-tuning and shaping of the product to fit it better for a given purpose, for example by stepwise restructuring of the basic chemical structure or by addition, reshaping or loss of side groups, could be the result of such a continuous mixing of tools. Compounds produced in this way could eventually then become an essential part of ‘primary’ metabolism in certain groups of organisms.

Examples in bacteria are auto-regulators, siderophores, etc. For instance, the synthesis of β-lactam antibiotics in Gram-negative bacteria, and other secondary metabolic traits, are regulated in their producers by the typical ecologically relevant quorum-sensor pheromones, the *N*-acyl-homoserinelactones, which may be regarded as typical individualites (Eberl, 1999; McGowan *et al.*, 1999).

Antimicrobial Resistance Frequently Co-evolves with the Antimicrobials

Antimicrobial resistance in the susceptible target organism can be by mutation or by acquisition of genetic material, so-called resistance genes. In the clinical situation, both of these general means of acquiring resistance are used by pathogens under the selective conditions of the therapeutic use of a given antibiotic, but spread of resistance among different strains in a given population is almost exclusively associated with the presence of mobile resistance genes which are encoded on plasmids or transposons and often in so-called multi-resistance cassettes.

In most bacterial producers of antibiotic-like compounds the antibiotic products are 'self-toxic', which makes self-protection necessary. In practically all cases so far studied this is achieved by specific resistance genes, in many cases even by the presence of more than one regulatable resistance determinant. Examples of some of the most common resistance mechanisms and distinct resistance genes used by various organisms are listed in Tables 3 and 4.

The genes for resistance are mostly clustered together with the production and regulatory genes (Strohl, 1997). In actinomycetes this means that nearly all the resistance mechanisms are chromosomally encoded. Much evidence has also accumulated that antimicrobial-producing organisms, such as the actinomycetes, are the evolutionary origin of the mobile resistance elements found in hospital facultative pathogens and endemic and pandemic infectious organisms such as *Vibrio cholerae* and other enterobacteria that are non-producers.

Origin and Purpose of Resistance Genes

It has been pointed out that export mechanisms, which are a special case although the most common, were probably developed primarily to take the compounds to their targets outside the cells that produced them. Only later were they also adopted as resistance mechanisms, after their encoding genes had spread to non-producing organisms.

Two major groups of ubiquitous transporters are the major resource of antibiotic exporters (often called 'multi-drug resistance proteins', Mdr); they are used for many different substrates, and for import and export purposes. They are the ATP-dependent primary transporters of the huge ABC transporter superfamily (Fath and Kolter, 1993; for example LmrC of *S. lincolnensis*) and secondary transporters of the pH-gradient-dependent major facilitator superfamily (MFS; Paulsen *et al.*, 1996; Pao *et al.*, 1998; for example LmrA of *S. lincolnensis*). More than 100 different families of other trans-membrane transporters have so far been identified and some of these comprise multi-drug efflux systems. Examples in *E. coli* are the resistance–nodulation–division (RND) Mdr (AcrB), and the small multi-drug resistance family (SMR) (EmrE). The RND family is exceptionally abundant only in *Ps. aeruginosa* PA01 (Stover *et al.*, 2000). None of these is, however, as important or as much used by most other bacteria, particularly the actinomycetes, as the ABC and MFS families (Table 4).

ABC Superfamily

Exporters of the ABC transporter family are especially abundant in the actinomycetes, as mentioned earlier for *S. coelicolor* A3(2) (see above; Table 3), and many of these ATP-dependent trans-membrane complexes are more or less specific antibiotic exporters in producers and in non-producers. Less specific members of this subgroup are called multi-drug resistance (Mdr) determinants. Others are very specific for a particular molecule and mostly encoded by genes coupled to the production gene clusters of the producer strains, for example *LmrC* in the lincomycin-producing *S. lincolnensis.*

At least three subfamilies of relevant ABC exporters for antimicrobials are found. The first subfamily has two ABC domains in one protein chain, but lacks the trans-membrane domains, e.g. those encoded by resistance genes in the gene clusters for lincomycin (*LmrC*), tetracenomycin (*TcmA*), nucleoside antibiotic A201 (*Ard1*), the 14-membered macrolide oleandomycin (*OleB*) and the 16-membered macrolides carbomycin (*CarA*) and tylosin (*TlrC*). Many proteins of this class with significant relationship to antibiotic exporters are encoded on the *S. coelicolor* chromosome (Tables 3 and 4). These cytoplasmic domain exporters are postulated to associate with a suitable, but non-specific, trans-membrane complex, which seems to be present at least in all tested actinomycetes, but they may be ubiquitous.

In the second subfamily a dimer (or tetramer) of a single ABC-domain protein and a specific transmembrane protein are necessary to form the complex. The genes that encode these proteins are mostly linked in transcription units. Examples are found in the producers of the antibiotics daunorubicin (*DrrA*, *DrrB*), mithramycin (*MtrA*, *MtrB*) and oleandomycin (*OleC4*, *OleC5*; note that oleandomycin or its glucoside are exported by two different members of the ABC transporter superfamily in the same producing organism *S. antibioticus*). Again, the *S. coelicolor* genome encodes many significantly related counterparts (Tables 3 and 4).

A third subfamily of ABC exporters seems to be that represented by the StrV and StrW proteins encoded by genes in the streptomycin production clusters of different *Streptomyces* spp. (Piepersberg, 1995, 1997).

Major Facilitator Superfamily

The MFS superfamily of exporters (also called sugar-porters) comprise exporters of, for example, chloramphenicol (*cmlV* in the chloramphenicol producer *S. venezuelae* or *cml* in *S. coelicolor*), lincomycin (*lmrA* in the producer *S. lincolnensis*) and tetracycline (*otrB* in the oxytetracyclin producer *S. rimosus*).

Some of the genes modifying target sites, such as those that encode rRNA methyltransferases, clearly originate from essential rRNA/tRNA methyltransferase genes. Examples are the methyltransferases that confer resistance to thiostreptone and the members of the macrolide-lincosamide-streptogramin (MLS) family of antibiotics, which have close relatives among the conserved essential enzymes of this group present in most or all bacteria. Several example of these conserved genes can be found in the genomes of *S. coelicolor* and *M. tuberculosis*.

Antibiotic Transferases

The *antibiotic transferases*, such as acetyl-, phospho- or glycosyltransferases were among the first resistance determinants detected (Benveniste and Davies, 1973; Umezawa, 1974; Walker, 1975). They may frequently have originated from biosynthetic enzymes by duplication and divergent evolution of their genes. Examples are the closely related biosynthetic phosphotransferase StrN in streptomycin producers, such as *S. griseus*, and the spectinomycin-resistance-conferring phosphotransferase SpcN in the spectinomycin producers *S. spectabilis* and *S. flavopersicus* (Piepersberg, 1997; Piepersberg and Distler, 1997).

In cases where phosphate groups are introduced into the antibiotic precursors during biosynthesis, for example (dihydro-)streptomycin-6-phosphate in the streptomycin producer, these seem to have the side-effect of being protective for the cytoplasmic target sites of the active dephosphorylated compound. Also, they are specifically cleaved off after release of the precursor through the cell wall. In this way a resistance mechanism is born. Another enzyme that introduces a phosphate group into the same position of the active inhibitor is streptomycin 6-phosphotransferase (AphD, StrA, APH(6) in streptomycin-producing *S. griseus*). This is used as a resistance mechanism and a specific extracellular phosphatase (StrK) removes the phosphate to release the active antibiotic into the environment. This is clearly a targeted extracellular behaviour of the producer. This same resistance mechanism, through APH(6), can then be seen as also spreading resistance determinants to genomes of unrelated non-producers (e.g. *D. radiodurans* R1, Table 4) or to plasmids and transposons (e.g. Tn5) of Gram-negative bacteria.

Another similar and intriguing case is that of the antibiotic glycosyltransferases (GTs) which play a crucial role in many biosyntheses, but have also been adopted for resistance (and 'transport tagging') purposes (Cundliffe *et al.*, 1992b; Vilches *et al.*, 1992; Piepersberg, 1994; Solenberg *et al.*, 1997; Trefzer *et al.*, 2000). All these enzymes belong to the same family of proteins, which is typically secondary metabolic. They seem easily to be adapted to new substrate (NDP-sugar) and co-substrate (aglycone) specificities, since in some well-studied cases their encoding genes are likely to have co-evolved by gene duplication and subsequent drift by mutation and selection towards new functions within gene clusters for a particular biosynthesis that involves the attachment of several different sugar side-chains (Solenberg *et al.*, 1997; Trefzer *et al.*, 2000).

The macrolide glucosyltransferases (Mgt; Cundliffe *et al.*, 1992b; Vilches *et al.*, 1992) appear to have been derived from the biosynthetic GTs and seem to have been developed primarily to protect the producer cell as well as to tag the end-product of the pathway for specific export. Outside the cell the glucose residue is removed by hydrolysis by a specific glucosidase that is also encoded in the gene clusters of macrolides (e.g. for erythromycin and oleandomycin; see for example Hopwood *et al.*, 1997; Kieser *et al.*, 2000). The occurence of these enzymes in other non-producing organisms, such as *S. lividans* 66 (MgtA; Cundliffe, 1992b) or *S. coelicolor* A3(2), is the third step observed in the distribution to other organisms, where they may serve as sole resistance mechanisms.

In general, the well-studied resistance-conferring phospho- and acetyltransferases for antimicrobials are clear indicators that these genes are not only ubiquitously disseminated by plasmids, but that they also reside as a reaction reserve in the chromosomes of unrelated organisms, which are not likely to be under continuous selection pressure from all the compounds against which they might be used. Most of these organisms have been isolated from sites, or at times, where human antibiotic use could not have played a role. Also, most of the potential resistance enzymes seem to be inactive in terms of constitutive resistance development (see also below for the ANT mechanisms).

Several examples of transferase genes in the bacterial genomes of non-producers are mentioned in **Table 4**. (1) AAC(2′) enzymes seem to be universally present in the mycobacteria, including *M. tuberculosis* (Ainsa *et al.*, 1997). (2) The enzyme subfamilies of the kanamycin/neomycin resistance-conferring APH(3′) and streptomycin-resistance-conferring APH(6) and APH(3′), respectively, seem to have members (related genes) of various compositions in *S. coelicolor* A3(A2), *M. tuberculosis* HRv, *D. radiodurans* R1, and *Ps. aeruginosa* PA01, but most of these organisms, except for *Ps. aeruginosa* PA01, are sensitive to aminoglycosides (Hachler *et al.*, 1996). (3) AAC(3) enzymes could be encoded by *S. coelicolor* A3(A2), *D. radiodurans* R1, and *B. subtilis* 168. (4) In general, free-living microorganisms seem to have acquired at least as many potential resistance genes as their mammalian-associated counterparts (e.g. *M. tuberculosis* HRv and *E. coli* K-12).

It has been pointed out above that many of the transferases which inactivate antimicrobial compounds that act on central macromolecular (i.e. nucleic acid and protein) metabolism are related to enzymes that modify nucleic acids and proteins (Piepersberg *et al.*, 1988; Piepersberg, 1997). For instance, the many different antibiotic kinases belong to the same kinase family as the ubiquitous and mostly regulatory protein kinases. This is strongly supported by the clear correspondence of three-dimensional structures of the two groups. The antibiotic kinases are phylogenetically much more distant from each other than the protein kinases, which may reflect a longer evolution, such that they are an older or even ancestral group of the more recent protein subfamilies. It is interesting to note that our model streptomycete, *S. coelicolor*, hosts genes for both subgroups, where phenotypes for resident protein kinases are so far known, but not for any of the few possible antibiotic kinases (**Table 2**).

Resistance Bypass Mechanisms

The resistant second target sites that mimic the normal sites and create bypass mechanisms for resistance, as also found in the normally susceptible pathogens, are not in any way phenotypes newly selected by chemotherapeutic use, e.g. of the translational or gyrase inhibitors such as tetracyclines, nalidixic acid, novobiocin and synthetic quinolones. On the contrary, exactly the same resistance mechanisms and genes are used in the producers; resistant counterparts of the essential protein factors involved in macromolecular metabolism (e.g. elongation factor G or gyrase; see **Tables 1**, **2** and **3**).

Another example is the low-level antibiotic resistance which is easily acquired by mechanisms that could also be regarded as bypasses, e.g. enhanced production of a target site. An example is the widespread bacitracin-resistance via enhanced production of the undecaprenol phosphotransferase, BacA (Cain *et al.*, 1993). Many such examples have been also found in clinical situations and might, therefore, also be regarded as ecologically relevant (see e.g. Bryan and Kwan, 1983; Hancock *et al.*, 1991).

An Evolutionary Scenario

As a preliminary conclusion, the following scenario could be imagined for the evolution of widespread resistance genes: (1) Housekeeping genes are recruited for the production of individualites, such as biosynthesis or export mechanisms. (2) In a second phase these genes may become further duplicated and reselected for the resistance-conferring activity of their products, e.g. in case of the antibiotic-modifying enzymes. (3) These genes successively become independent of their gene clusters and their function in the producers after horizontal transfer to other organisms, e.g. by introduction into plasmids, transposons or viruses, or by simple DNA release and uptake ('natural transformation') which seems, for example, to be a common alternative in soils. (4) In other cases genes encoding target sites have been duplicated and mutated to form resistant alternative target complexes, thus bypassing a blocked biochemical pathway when the active inhibitory compound is synthesised inside the cells (e.g. the inhibitors of gyrase, translation factors, etc.). The second target molecules may not function as well, so that they are kept for specific expression only in the production phase when this is required, but apart from the regular, sensitive target site complex. (5) A special example seem to be the case of β lactams and β lactamase, where both the target proteins and the general mechanism of interaction with them is adopted to build up an extracellular resistance mechanism. The

β lactams and the related β lactones all interact with the bacterial cell-wall biosynthetic transpeptidases/ carboxypeptidases by splitting the β-lactam ring and forming a relatively stable covalent protein–inhibitor intermediate. The β lactamases are derivatives of these sensitive peptidases which rapidly release the covalent intermediate by hydrolysis, thereby rapidly inactivating these molecules. This more resembles a hormone–target-cell interaction and may also be used in that way in natural bacterial populations, since both the production of this class of compounds and the presence of chromosomally encoded β lactamases is widespread among bacteria. Even bacteria such as *E. coli* harbour a β-lactamase gene (*ampC*) in their chromosome. The production of β lactams by fungi (e.g. *Penicillium* sp.) must then be regarded as a secondary evolutionary phase after recruitment of the production genes from bacteria, possibly acquired for the warfare against competing bacterial populations.

In some cases, such as the aminoglycoside nucleotidylyltransferases (ANT), which are widespread as transferable resistance determinants in sensitive pathogenic bacteria, equivalents have not as yet been found in antibiotic producers (see Tables 1, 3 and 4). The only occurrence of a genomic ANT enzyme, which encodes a streptomycin-6-adenylyltransferase (ANT(6)) expressed only in genetic variants of the original strain, is known in *B. subtilis* 168 (Kono *et al.*, 1987) (Table 4). It is not known why the actinomycetes that produce aminoglycosides could not also have made use of this modification mechanism. Many essential nucleotidylyltransferases, such as for sugars and proteins, are also known in these organisms. This may mean either that they were not originally used as biosynthetic mechanisms or that another large group of still unknown aminoglycoside producers, which has developed this tool of resistance, may yet exist somewhere in nature. An alternative explanation would be that these resistance enzymes are an example of a production-independent evolution in non-producers. It is necessary to determine the conditions and selective means that created them, since the other aminoglycoside resistance mechanisms (acetylation and phosphorylation) are available and are widespread in the same populations. They have often been found encoded by the same multi-resistance plasmids or transposons in Gram-negative and Gram-positive nosocomial pathogens.

Conclusions

All antimicrobials are probably an essential part of the generally present and ever-changing extracellular communication metabolism among living cells in nature. This implies that (1) all cells of living organisms, even unicellular micro-organisms of the same species, must communicate with each other; (2) this communication process depends on the production of an individualised biochemistry on the cell surfaces and in the secretion of individualites in all organisms; (3) in micro-organisms the soluble and diffusible secretory products called 'antimicrobials' may be part of an inhibitory mechanism. This may be directed as a hormone-like blocker of unwanted metabolism in differently differentiated non-producing cells of the same cell aggregate (e.g. mini-colonies of differentiated cells in the higher actinomycetes or fruiting bodies of the myxobacteria) or used in the biological warfare against other competing or predatory (micro-)organisms in the same micro-habitat (the classical defence hypothesis).

The selective pressure for the invention and conservation of the individual communication substance is exerted by the end-product of a secondary metabolic pathway; the secondary metabolic pathway itself is not a prerequisite, nor is it the initiating fact for the production of a particular product. New products of biosynthetic potential that change by chance developments, or new uses, may create new selective situations for ongoing evolution of new individualites, for shifts towards primary metabolism or for loss of old products that are no longer useful.

Resistance against antimicrobials is based on tools that have co-evolved with the pathways for their production and have spread to become ubiquitous at a low level of incidence in natural microbial populations as a reactive 'genetic reserve material'. This situation changes rapidly, however, under highly selective conditions such as the widespread use of antibiotics in human environments, especially in medicine and in animal breeding.

Since resistance determinants are part of the mobile gene pool they are accessible to almost every micro-organism. Our hope must be that it will in the future be possible to keep valuable therapeutics in a useful balance in relation to resistance phenomena, and to find new agents that can supplement the spectrum of alternatives necessary to cope with important infectious diseases. This will largely depend on an improved understanding of the ecological roles and dynamics of these interesting bioactive substances in their micro-environments.

Acknowledgements

This text is devoted to my teachers in the field of secondary metabolism: A. Böck, J.E. Davies, and

D.A. Hopwood, and to my coworkers, represented by J. Distler, A. Stratmann, and U. Wehmeier. The communication of preliminary data from the *S. coelicolor* A3(A2) genome project by G. Challis and D.A. Hopwood is gratefully acknowledged.

References

Ainsa JA, Perez E, Pelicic V *et al.* (1997) Aminoglycoside 2′-*N*-acetyltransferase genes are universally present in myco bacteria: characterization of the aac(2′)-Ic gene from *Mycobacterium tuberculosis* and the aac(2′)-Id gene from *Mycobacterium smegmatis. Mol. Microbiol.* 24: 431–441.

Beyer S, Mayer G, Piepersberg W (1998) The StrQ protein encoded in the gene cluster for 5′-hydroxystreptomycin of *Streptomyces glaucescens* GLA.O is a α-D-glucose-1-phosphate cytidyltransferase (CDP-glucose synthase). *Eur. J. Biochem.* 258: 1059–1067.

Benveniste R, Davies J (1973) Aminoglycoside antibiotic inactivating enzymes in actinomycetes similar to those present in clinical isolates of antibiotic resistant bacteria. *Proc. Natl Acad. Sci. USA* 70: 2276–2280.

Blattner FR, Plunkett G III, Bloch CA *et al.* (1997) The complete genome sequence of *Escherichia coli* K-12. *Science* 277: 1453–1474.

Bryan LE, Kwan S (1983) Roles of ribosomal binding, membrane potential, and electron transport in bacterial uptake of streptomycin and gentamicin. *Antimicrob. Agents. Chemother.* 23: 835–845.

Cain BD, Norton PJ, Eubanks W, Nick HS, Allen CM (1993) Amplification of the *bacA* gene confers bacitracin resistance to *Escherichia coli. J. Bacteriol.* 175: 3784–3789.

Chadwick D, Whelan J, Widdows K (1992) Secondary metabolites: their function and evolution. In: Chadwick D, Whelan J, Widdows K (ed.) *Secondary Metabolites: Their Function and Evolution,* Ciba Foundation Symp. 171. Chichester: John Wiley.

Challis GL, Ravel J (2000) Coelichelin, a new peptide siderophore encoded by the *Streptomyces coelicolor* genome: structure prediction from the sequence of its non-ribosomal peptide synthetase. *FEMS Microbiol. Lett.* 187: 111–114.

Champness WC, Chater KF (1994) Regulation and interaction of antibiotic production and morphological differentiation in *Streptomyces* spp. In: Piggot P, Moran CP Jr, Youngman P (eds) *Regulation of Bacterial Differentiation.* Washington, DC: American Society for Microbiology, pp. 61–94.

Cole ST, Brosch R, Parkhill J *et al.* (1998) Deciphering the biology of *Mycobacterium tuberculosis* from the complete genome sequence. *Nature* 393: 537–544.

Cundliffe E (1992a) Self-protection mechanisms in antibiotic producers. In: Davies J, Chadwick D, Whelan J, Widdows K (eds) *Secondary Metabolites: Their Function and Evolution,* Ciba Foundation Symp. 171. Chichester: John Wiley, pp. 199–214.

Cundliffe E (1992b) Glycosylation of macrolide antibiotics in extracts of *Streptomyces lividans. Antimicrob. Agents Chemother.* 36: 348–352.

Doolittle WF (1999) Phylogenetic classification and the universal tree. *Science* 284: 2124–2129.

Dowling RB, Newton R, Robichaud A *et al.* (1998) Effect of inhibition of nitric oxide synthase on *Pseudomonas aeruginosa* infection of respiratory mucosa *in vitro. Am. J. Respir. Cell. Mol. Biol.* 19: 950–958.

Eberl L (1999) *N*-acyl homoserinelactone-mediated gene regulation in gram-negative bacteria. *Syst. Appl. Microbiol.* 22: 493–506.

Fath MJ, Kolter R (1993) ABC transporters: bacterial exporters. *Microbiol. Rev.* 57: 995–1017.

Fritsche E, Bergner A, Humm A, Piepersberg W, Huber R (1998) Crystal structure of L-arginine:inosamine-phosphate amidinotransferase StrB1 from *Streptomyces griseus*: an enzyme involved in streptomycin biosynthesis. *Biochemistry* 37: 17664.

Guillemot D (1999) Antibiotic use in humans and bacterial resistance. *Curr. Opin. Microbiol.* 2: 494–498.

Hachler H, Santanam P, Kayser FH (1996) Sequence and characterization of a novel chromosomal aminoglycoside phosphotransferase gene, aph (3′)-IIb, in *Pseudomonas aeruginosa. Antimicrob. Agents Chemother.* 40: 1254–1256.

Hancock REW, Farmer SW, Li Z, Poole K (1991) Interaction of aminoglycosides with outer membranes and purified lipopolysaccharide and OmpF porin of *Escherichia coli. Antimicrob. Agents Chemother.* 35: 1309–1314.

Handelsman J, Rondon MR, Brady SF, Clardy J, Goodman RM (1998) Molecular biological access to the chemistry of unknown soil microbes: a new frontier for natural products. *Chem. Biol.* 5: R245–249.

Hopwood DA (1999) Forty years of genetics with *Streptomyces*: from *in vivo* through *in vitro* to *in silico. Microbiology* 145: 2183–2202.

Hopwood DA, Chater KF, Bibb MJ (1995) Genetics of antibiotic production in *Streptomyces coelicolor* A3(2), a model streptomycete. *Biotechnology* 28: 65–102.

Hopwood DA *et al.* (1997) Special volume with several reviews on the biosynthesis of polyketides and non-ribosomal peptides. *Chem. Rev.* 97: 2465–2705.

Hotta K, Davies J, Yagisawa M (1995) Aminoglycosides and aminocyclitols (other than streptomycin). In: Vining L, Stuttard C (eds) *Biochemistry and Genetics of Antibiotic Biosynthesis.* Stoneham, UK: Butterworth-Heinemann, pp. 571–595.

Hu Z, Hopwood DA, Khosla C (2000) Directed transfer of large DNA fragments between *Streptomyces* species. *Appl. Environ. Microbiol.* 66: 2274–2277.

Hugenholtz P, Goebel BM, Pace NR (1998) Impact of culture-independent studies on the emerging phylogenetic view of bacterial diversity. *J. Bacteriol.* 180: 4765–4774.

Humm H, Fritsche E, Steinbacher S, Huber R (1997) Crystal structure and mechanism of human L-arginine:glycine amidinotransferase: a mitochondrial enzyme involved in creatine biosynthesis. *EMBO J.* 16: 3373–3385.

Imbert M, Bechet M, Blondeau R (1995) Comparison of the main siderophores by some species of *Streptomyces. Curr. Microbiol.* 31: 129–133.

Kaneko T, Sato S, Kotani H *et al.* (1996) Sequence analysis of the genome of the unicellular cyanobacterium *Synechocystis* sp. strain PCC6803. II. Sequence determination of the entire genome and assignment of potential protein-coding regions. *DNA Res.* 3: 109–136.

Kieser T, Bibb MJ, Buttner MJ, Chater KF, Hopwood DA (2000) *Practical Streptomyces genetics.* Norwich, UK: The John Innes Foundation.

Kimoto M, Miyatake S, Sasagawa T *et al.* (1998) Purification, cDNA cloning and expression of human NG, NG-dimethylarginine dimethylaminohydrolase. *Eur. J. Biochem.* 258: 863–868.

Kono M, Ohmiya K, Kanda T, Noguchi N, O'Hara K (1987) Purification and characterization of chromosomal streptomycin adenyltransferase from derivatives of *Bacillus subtilis* Marburg 168. *FEMS Microbiol. Lett.* 40: 223–228.

Kunst F, Ogasawara N, Moszer I *et al.* (1997) The complete genome sequence of the Gram-positive bacterium *Bacillus subtilis. Nature* 390: 249–256.

Lawrence JG (1999) Gene transfer, speciation, and the evolution of bacterial genomes. *Curr. Opin. Microbiol.* 2: 519–523.

Maghnouj A, De Sousa Cabral TF, Stalon V, Vander Wauven C (1998) The *arcABC* gene cluster of *Bacillus licheniformis*, and its activation by the arginine repressor ArgR. *J. Bacteriol.* 180: 6468–6475.

Mazodier P, Davies J (1991) Gene transfer between distantly related bacteria. *Annu. Rev. Genet.* 25: 147–171.

McGowan SJ, Holden MT, Bycroft BW, Salmond GP (1999) Molecular genetics of carbapenem antibiotic biosynthesis. *Antonie Van Leeuwenhoek* 75: 135–141.

Mehling A, Wehmeier UF, Piepersberg W (1995) Nucleotide sequence of *Streptomyces* 16S ribosomal DNA, towards a specific identification system for streptomycetes using PCR. *Microbiology* 141: 2139–2147.

Mozner I, Rocha EPC, Danchin A (1999) Codon usage and lateral gene transfer in *Bacillus subtilis. Curr. Opin. Microbiol.* 2: 524–528.

Ochman H, Lawrence JG, Groisman EA (2000) Lateral gene transfer and the nature of bacterial innovation. *Nature* 405: 299–304.

Osburne MS, Grossman TH, August PR, MacNeil IA (2000) Tapping into microbial diversity for natural products drug discovery. *ASM News* 66: 411–417.

Pace N (1997) A molecular view of microbial diversity. *Science* 276: 734–740.

Pao SS, Paulsen IT, Saier MH (1998) Major facilitator superfamily. *Microbiol. Mol. Biol. Rev.* 62: 1–34.

Paulsen IT, Brown MH, Skurray RA (1996) Proton-dependent multidrug efflux pumps. *Microbiol. Rev.* 60: 575–608.

Piepersberg W (1992) Metabolism and cell individualization. In: Chadwick D, Whelan J, Widdows K (eds) *Secondary Metabolites: Their Function and Evolution.* Ciba Foundation Symposium 171. Chichester: John Wiley, pp. 294–299.

Piepersberg W (1993) Streptomycetes and corynebacteria. In: Rehm H-J, Reed G, Pühler A, Stadler P (eds) *Biotechnology*, 2nd ed, Vol. 1. Weinheim: VCH, pp. 433–468.

Piepersberg W (1994) Pathway engineering in secondary metabolite-producing actinomycetes. *Crit. Rev. Biotechnol.* 14: 251–285.

Piepersberg W (1995) Streptomycin and related aminoglycosides. In: Vining L, Stuttard C (eds) *Biochemistry and Genetics of Antibiotic Biosynthesis.* Oxford: Butterworth-Heinemann, pp. 531–570.

Piepersberg W (1997) Molecular biology, biochemistry, and fermentation of aminoglycoside antibiotics. In: Strohl WR (ed): *Biotechnology of Industrial Antibiotics*, 2nd edn. New York, Marcel Dekker, pp. 81–163.

Piepersberg W, Distler J (1997) Aminoglycosides and sugar components in other secondary metabolites. In: Rehm H-J, Reed G, Pühler A, Stadler P (series eds); Kleinkauf H, v. Döhren H (volume eds) *Biotechnology*, 2nd edn, vol. 7, Products of Secondary Metabolism. Weinheim: VCH, pp. 397–488.

Piepersberg W, Distler J, Heinzel P, Perez-Gonzalez J-A (1988) Antibiotic resistance by modification: Many resistance genes could be derived from cellular control genes in actinomycetes – hypothesis. *Actinomycetology* 2: 83–98.

Power PM, Roddam LF, Dieckelmann M *et al.* (2000) Genetic characterization of pilin glycosylation in *Neisseria meningitidis. Microbiology* 146: 967–979.

Redenbach M, Kieser HM, Denapaite D *et al.* (1996) A set of ordered cosmids and a detailed genetic and physical map for the 8 Mb *Streptomyces coelicolor* A3(2) chromosome. *Mol. Microbiol.* 21: 77–96.

Rehm H-J, Reed G, Pühler A, Stadler P (series eds); Kleinkauf H, v. Döhren H, (volume eds) (1997) *Biotechnology*, 2nd edn, vol. 7, *Products of Secondary Metabolism.* Weinheim: Wiley-VCH, p. 7.

Reichenbach H, Gerth K, Irschik H, Kunze B, Höfle G (1988) Myxobacteria: a source of new antibiotics. *Trends Biotechnol.* 6: 115–121.

Solenberg PJ, Matsushima P, Baltz R (1997) Production of hybrid glycopeptide antibiotics *in vitro* and in *Streptomyces toyocaensis. Chem. Biol.* 4: 195–202.

Stencel C (2000) Microbial diversity: eying the big picture. *ASM News* 66: 142–146.

Stover CK, Pham X-QT, Erwin AL *et al.* (2000) Complete genome sequence of *Pseudomonas aeruginosa* PA01, an opportunistic pathogen. *Nature* 406: 959–964.

Strohl WR (1997) *Biotechnology of Industrial Antibiotics*, 2nd edn. New York: Marcel Dekker.

Trefzer A, Hoffmeister D, Kunzel E *et al.* (2000) Function of glycosyltransferase genes involved in urdamycin A biosynthesis. *Chem. Biol.* 7: 133–142.

Umezawa H (1974) Biochemical mechanisms of resistance to aminoglycoside antibiotics. *Adv. Carbohydr. Chem. Biochem.* 30: 183–225.

Vilches C, Hernandez C, Mendez C, Salas JA (1992) Role of glycosylation and deglycosylation in biosynthesis of and resistance to oleandomycin in the producer organism, Streptomyces antibioticus. *J. Bacteriol.* 174: 161–165.

Vining LC (1990) Functions of secondary metabolites. *Annu. Rev. Microbiol.* 44: 395–427.

Walker JB (1975) ATP:streptomycin 6-phosphotransferase. *Meth. Enzymol.* 43: 628–632.

Walker JB (1990) Possible evolutionary relationships between streptomycin and bluensomycin biosynthetic pathways: Detection of novel inositol kinase and O-carbamoyltransferase activities. *J. Bacteriol.* 172: 5844–5851.

White O, Eisen JA, Heidelberg JF *et al.* (1999) Genome sequence of the radioresistant bacterium *Deinococcus radiodurans* R1. *Science* 286: 1571–1577.

Whitman WB, Coleman DC, Wiebe WJ (1998) Perspective. Prokaryotes: the unseen majority. *Proc. Natl Acad. Sci. USA* 95: 6578–6583.

Zähner I, Drautz H, Fiedler P *et al.* (1988) Ways to new metabolites from actinomycetes. In: Okami Y, Beppu T, Ogawara H (eds) *Biology of Actinomycetes '88.* Tokyo: Japan Scientific Societies Press pp. 171–177.

26

Antimicrobial Action: Basic Principles

D.G. Allison,[1] P. Gilbert[1] and P.A. Lambert[2]

[1] *University of Manchester, Manchester, UK*
[2] *Aston University, Birmingham, UK*

Antimicrobial agents are chemical substances that can either kill or inhibit the growth of micro-organisms. Such substances may be natural products or synthetic chemicals, and they are of undoubted value for the treatment of infection, and in well-defined circumstances they may be used for the prevention of infection. Agents that kill micro-organisms are termed *cidal* (e.g. bactericidal), whereas those that do not kill but merely inhibit growth are referred to as *static* agents (e.g. bacteriostatic). In each case the prefix term indicates the type of organisms affected. In patients with an intact immune system either type of antibiotic is likely to be effective, allowing the host immune system to deal with and eliminate the invading micro-organism. Where immunocompromised patients are concerned, however, bacteriostatic antibiotics are largely ineffective. Classification of antimicrobial agents on this basis is not recommended, since some agents that are capable of killing certain species are only bacteriostatic for others. For example, under identical growth conditions chloramphenicol, an inhibitor of bacterial protein biosynthesis, inhibits the growth of *Escherichia coli* but kills *Haemophilus influenzae*.

Microbial disease is the result of complex interactions between the pathogen and the host it is attempting to colonise (Gemmell, 1996). The complex dynamics of this situation can be further complicated by the use of antibiotics and the interaction may be regarded as an interdependent triangle, in which any effect on one side of the triangle causes effects on the other two sides (Fig. 1).

This and the next two chapters will consider the interactions between antimicrobial agents and micro-organisms at a biochemical and molecular level.

Selective Toxicity and Spectrum of Activity

The essence of antimicrobial chemotherapy is *selective toxicity*, a term introduced by Paul Ehrlich in 1908. The invading micro-organism was regarded as being killed or its growth inhibited without the active agent causing harm to the patient. In order to be effective, antimicrobial agents must, therefore, possess a selective action on micro-organisms without toxicity to the cells of the host. Selective toxicity is most efficient when a similar target does not exist in the host (Greenwood, 2000).

The sensitivity or susceptibility of micro-organisms to antimicrobial agents also varies. Gram-positive bacteria are usually more sensitive to antibiotics than are Gram-negative bacteria. Antibiotics that act on both Gram-positive and Gram-negative bacteria are termed *broad-spectrum antibiotics* and generally have a wider medical usage than *narrow-spectrum*

 doi:10.1006/bkmm.2001.0026

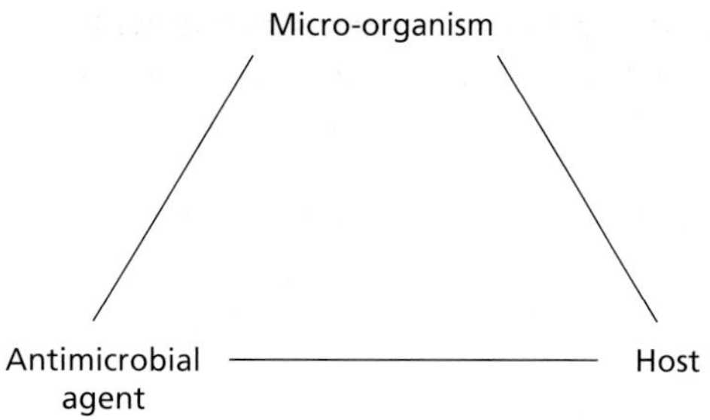

Fig. 1 Interactive triangle between micro-organisms, host tissue and antimicrobial agents.

antibiotics, which act on only a specific group of micro-organisms or indeed only a single microbial species. Narrow-spectrum antibiotics are valuable on account of their ability to control micro-organisms that fail to respond to other, more widely used antibiotics and because they have a limited effect on the normal flora of the host (Solomkin and Miyagawa, 1994; Greenwood, 2000).

Selective Toxicity

Selective toxicity is achieved by exploiting differences between the structure and metabolism of micro-organisms and host cells. Ideally, antimicrobial agents should have target-dependent selectivity, whereby the agent has an inhibitory action that is highly specific and directed towards sites and/or enzymes that are represented only in the target organism or cell. This is more likely to be achieved in prokaryotic micro-organisms than in eukaryotes, because the former are more distinct from host cells (Russell and Chopra, 1996). The best example of this type of compound is the β-lactam group of antibiotics, the selectivity of which is based on the absence from eukaryotic cells of components similar to bacterial peptidoglycans and their associated biosynthetic enzymes. By contrast, viruses are extremely difficult to attack because of their obligate intracellular lifestyle, which depends on host biosynthetic mechanisms. To be successful, an antiviral agent must be able to enter the host cell but inhibit and damage only a virus-specific target.

Targets of antimicrobial agents are diverse and include various cellular functions. In the case of prokaryotic cells five main target sites for antibacterial action are recognised: (1) the cell wall; (2) nucleic acid synthesis; (3) protein synthesis; (4) folic acid synthesis and (5) cell membrane function. These targets differ to a greater or lesser degree from similar targets in the host cell and so allow inhibition of the bacterial cell without concomitant inhibition of the equivalent mammalian cell targets. Each target site encompasses a series of biosynthetic reactions, each of which may be specifically inhibited by an antibacterial agent. Moreover, a range of chemically diverse molecules may inhibit reactions at the same target site, as is demonstrated by inhibitors of protein synthesis. In addition, where the target appears to be common both to the host and to the micro-organism, there may be a selective advantage (Woodkin and Morrison, 1994; Smith and Bagg, 1998). This is the case for inhibitors of protein synthesis, which exploit the differences between prokaryotic (70S) and eukaryotic (80S) ribosomes. Although they generally catalyse a similar series of reactions, these two types of ribosome differ significantly in composition and, as a consequence, they show differences in function. This can be illustrated by streptomycin, which binds to a specific protein on the 30S subunit of the bacterial ribosome that is absent from the corresponding eukaryotic 80S ribosomal subunit.

Similarly, although all cytoplasmic membranes possess much the same general composition, differences exist in the nature of the macromolecules involved. Indeed, a significant difference in physiology between prokaryotic and eukaryotic cytoplasmic membranes is the presence of sterols in the latter. This difference appears to account for the selective action of the polyene antibiotics.

The selective toxicity of antimicrobial agents may, however, often be marginal. For example, the activity of the polymyxins, which disrupt bacterial cell membrane function, is due to a distribution of phospholipids in the bacterial cytoplasmic membrane which is different from that in host cell membranes (McDonnell and Russell, 1999). Similarly, selectivity may be concentration-dependent, when the antibacterial agent, though highly specific, is directed against sites and or enzymes that are represented both in the target bacterial cell and in the host cell. The level of activity against the target is, however, significantly greater. Only rarely do agents have a single mode of action. Rather, the range and type of target varies with concentration. Hence, a critical feature is that the concentration in the host cells should remain unaffected by the antimicrobial agent. Selective target sites can, therefore, be further defined in terms of whether they are unique to the micro-organism or different in structural and metabolic terms from those found in the host (**Table 1**).

Selective toxicity may also occur when agents are excluded from reaching the target site. An example of this type of selectivity is shown by chloramphenicol, which inhibits protein synthesis by 70S ribosomes but not 80S ribosomes – with the exception of the 70S ribosomes present in the mitochondria of eukaryotic

Table 1 The nature of antimicrobial selective target sites

Unique to micro-organism	Different from host
Peptidoglycan	Ribosomes
Folic acid biosynthesis	Nucleic acid
	Cytoplasmic membrane

cells, which have a bacterial origin in evolution. Tetracycline antibiotics also inhibit protein synthesis by preventing aminoacyl-tRNA from entering the ribosomal acceptor sites. This action is not, however, selectively toxic for prokaryotic cells since inhibition of protein synthesis occurs in cell-free protein synthesis systems that contain eukaryotic ribosomes. The selectivity of tetracyclines appears to be based on their mechanism of uptake by bacterial cells, which is much greater than by human cells and is driven by an active transport process.

Both Gram-positive and Gram-negative cells walls can modulate the entry of antimicrobial agents. The peptidoglycan content of the cell wall, together with the outer membrane in Gram-negative bacteria, functions as a molecular sieve for large hydrophilic molecules. Such action depends on the thickness of the peptidoglycan matrix and the degree of cross-linking. In Gram-negative cells, the lipophilic outer membrane is of primary significance in excluding hydrophilic materials, and this probably accounts for the enhanced resistance of Gram-negative cells towards agents such as the β-lactams. It is worth noting that the composition and physico-chemical nature of the bacterial cell envelope is modulated by the growth environment. Thus, although an antibiotic may enter a cell under one set of growth conditions, this does not mean that it will necessarily enter or be as effective against the same organism cultivated under different growth conditions.

In summary, antibacterial agents can be classified in one of three ways: (1) whether they are bactericidal or bacteriostatic; (2) by the level of selective toxicity at the target site and (3) by chemical structure. Classification on this basis of the latter alone is not of practical value, since antibacterial agents include a wide range of different chemical structures. Rather, a combination of target site and chemical structure provides a useful working classification (Cuddy, 1997).

Nature of Injury

The nature of the injury caused by the interaction between antibacterial agents and the target can be identified either as direct or as a secondary or indirect consequence (Woodkin and Morrison, 1994). In the former the agent itself interacts directly with the site of injury, for example surfactant-mediated solubilisation of the cytoplasmic membrane. With indirect injury, such as that caused by cycloserine, the damage is remote from the interactive site. Direct drug–target interactions can be further complicated by virtue of the fact that some drugs bind irreversibly, whereas others are reversibly bound. In the case of reversible binding, permanent injury is not caused at the interactive site. If the agent is removed from the surrounding environment, cell function is automatically restored. Examples of such reversible agents include the sulfonamides and ionophores. By contrast, irreversible binding of the antimicrobial agent to the active site can result either in a permanent injury brought about by a structural change in the target or reversible damage, regardless of whether the agent is removed from the environment. Examples of agents that cause permanent injury include the β-lactams, which disrupt cell wall synthesis (Mascaretti *et al.*, 1999) and agents such as bleomycin which cause DNA strand breakage (Hecht, 2000). In order for cells to recover from injury, the target must be re-made (i.e. de-novo synthesis). Indirect drug–target interactions, on the other hand, lead to the synthesis of defective or incomplete structural cell components. This is perhaps best illustrated by antibiotics that act on the peptidoglycan component of the cell wall (see below). In some instances target sites can be affected directly or indirectly depending upon the class of antibacterial agent (Nagarajan, 1991; Mascaretti *et al.*, 1999).

The mode of action of antibiotics can be further classified according to the injuries experienced by the bacterial cell (Solomkin and Miyagawa, 1994; Smith and Bagg, 1998). These can be considered to be structural or metabolic. Structural effects result in the loss of integrity and function of a key ultrastructural component, such as the cell wall or cytoplasmic membrane, necessary to maintain cell viability. Metabolic interactions, on the other hand, lead to the inhibition or destruction of specific enzymes or enzymatic pathways or transport processes in the target cell. Metabolic injury will probably result in the cessation of growth of the target cell, but not necessarily its death. Thus, for any antibacterial agent, the mode of action can be identified as belonging to one of four different combinations depending on the type of interaction and the injury caused.

Direct Metabolic Injury

Direct metabolic injury occurs when the antimicrobial agent interacts directly with the target site, which is normally a specific enzyme or enzyme pathway unique in some manner to the micro-organism. A classical example of such an interaction is provided by the sulfonamides, which inhibit folic acid biosynthesis. Since mammalian cells do not synthesise folic acid, the sulfonamides do not interfere with host cell metabolism. The sulfonamides are analogues of *p*-aminobenzoic acid, which is itself part of folic acid. They act by the competitive inhibition of dihydropteroate reductase, which is responsible for the catalytic condensation of *p*-aminobenzoic acid with dihydropteridine to form dihydropteroic acid. Use of a sulfonamide, such as sulfanilamide, results in the formation of a 7,8-dihydropterin-sulfonamide adduct. The sulfonamides are broad-spectrum antibacterials, which are relatively slow-acting and predominantly bacteriostatic. Indeed, several generations of bacterial growth are required to deplete the folate pool before inhibition of growth takes place (Woodkin and Morrison, 1994).

Trimethoprim also inhibits folate synthesis and is one of the few antibacterial agents developed according to rational biochemical principles. It has a high affinity for dihydrofolate reductase and it competitively prevents conversion of the metabolically inactive dihydrofolate to the active form of the vitamin tetrahydrofolate. Since sulfonamides and trimethoprim act at different stages in the same metabolic pathway, they are often administered in a synergistic combination. The most commonly used combination is sulfamethoxazole and trimethoprim (co-trimoxazole), which inhibits bacteria at much lower concentrations than does either agent alone (see Chapter 26).

Direct Structural Injury

Direct structural forms of injury occur by the interaction of an active agent with its target site. This can result in a number of different effects to the target, including disruption, deformation or disassembly and defective synthesis. In each instance, the normal function of the target is impaired and leads to a loss in cell viability.

A variety of antibacterial agents are known to modify some component of the cytoplasmic membrane function by direct structural damage. The cytoplasmic membrane is an extremely important bacterial cell structure, since growth and survival of micro-organisms occurs only if bacteria are able to transport nutrients required for growth from the external medium, across the cytoplasmic membrane and into the cytoplasm. Often this occurs against a concentration gradient. Whereas materials of low molecular weight, such as oxygen, carbon dioxide and ammonium ions, enter the cytosol by passive diffusion, the passage of molecules of larger molecular weight is mediated by specific transport proteins. The osmotic barrier function of the cytoplasmic membrane can be attributed to its bilayer arrangement of its phospholipids, in which numerous structural and functional proteins are located. Direct interaction of numerous antibacterial agents with components of the cytoplasmic membrane can result in a non-specific increase in membrane permeability for various solutes or inhibition of membrane-associated enzymes. Agents acting in this way include antibiotics such as the polymyxins, which bind electrostatically in place of divalent cations to negatively charged amphipathic molecules in the membrane, such as cardiolipin. The effect of this interaction is likely to disrupt the packing of membrane lipid, leading to structural disorganisation of the membrane and alteration of its selective permeability (McDonnell and Russell, 1999).

Indirect Metabolic Injury

In many instances the action of antibacterial agents that cause indirect metabolic injury culminates in a direct structural defect in the bacterial cell. The synthesis of a structural target site can be prevented by interaction of the antibiotic with an associated biosynthetic enzyme remote from the site of injury. Inhibition of peptidoglycan pentapeptide chain synthesis provides an excellent example of this type of drug–target interaction.

The synthesis of peptidoglycan is a complex process which begins in the cytoplasm, proceeds across the cytoplasmic membrane and leads to attachment of new wall units to the growing peptidoglycan. Biosynthesis is carried out by an integrated multienzyme system (see Chapters 7 and 27), which provides many different enzyme targets for attack by antibacterial agents. Since the end product is a structure, impairment of its synthesis can lead to an imbalance in the extension process during cell growth and division. This results in a weakened wall, which in turn will cause the cells to lyse because of the inability of the defective peptidoglycan to restrain the high intracellular osmotic potential. An example of this type of interaction occurs with the antibiotic cycloserine, which structurally resembles the peptidoglycan component D-alanine. As a result, cycloserine inhibits two enzymes that catalyse cell-wall synthesis, namely D-alanyl-D-alanine synthetase and alanine racemase. This prevents the incorporation of D-alanine as the two terminal amino acid residues on the entapeptide cell-wall precursor. The synthesis of peptidoglycan also involves the membrane-bound lipid intermediate C-S5-isoprenyl pyrophosphate, which function by transporting the subunits of the polymer across the cytoplasmic membrane. These molecules also provide a potential target for drug action and have been shown to be inhibited by bacitracin.

The action of the quinolone class of antibiotics, examples of which include nalidixic acid, norfloxacin and ciprofloxacin, provide a further example of indirect metabolic interactions that cause structural injury (Corey and Taylor, 1992). These agents specifically inhibit DNA gyrase and block chromosomal replication. DNA gyrase is unique to bacteria and unwinds a supercoiled loop from a cut strand of DNA before replication and transcription. It can also produce negative supercoiling in closed circular DNA (see Chapter 28).

Inhibition of bacterial protein synthesis is also caused by the action of indirect metabolic inhibitors. Tetracyclines inhibit protein synthesis by preventing enzymatic and non-enzymatic binding of aminoacyl-tRNA to the acceptor (A) sites on the 30S ribosome subunit. Chloramphenicol blocks the action of peptidyl transferase, and inhibits the binding of aminoacyl-tRNA to the 50S ribosome A site, thereby preventing peptide bond formation. Aminoglycosides inhibit and kill bacteria by selectively interfering with the S12 ribosomal protein in the 30S subunit, resulting in misreading of mRNA and also preventing the formation of initiation complexes from which protein synthesis proceeds.

Indirect Structural Injury

Perhaps the best example of antibacterial agents causing indirect structural injury to a cell is the action of the β-lactam antibiotics. These act on the terminal stages of peptidoglycan synthesis, by inhibiting the cross-linking of pentapeptide chains. β-Lactam antibiotics bind to enzymes known as penicillin-binding proteins. These are carboxypeptidase and transpeptidase enzymes which are responsible for the final stages of cross-linking of the bacterial cell wall structure. Inhibition of one or more of these essential enzymes results in an accumulation of precursor cell wall units, which causes the autolytic system of the cell to be activated, and results in the weakening of the newly synthesised wall and finally in cell lysis (see Chapter 27).

Therapeutic Index

Antimicrobial agents are undoubtedly valuable for the treatment of infection and, in well-defined circumstances, in its prevention. In addition to microbiological efficacy, however, other equally important factors must be taken into consideration before an antibiotic is adopted for clinical use. The pharmacological properties of an ideal antibiotic are that it should be non-toxic to the host, not interfere with other drugs, possess a long plasma half-life, have good distribution in tissues including the cerebrospinal fluid, have low plasma-protein binding and can be administered orally and parenterally. The costs of development and manufacture are also a consideration (Desnottes, 1996). A measure of the ratio of the harmful concentration to the therapeutic concentration (therapeutic index) is a significant factor that determines the usage of many antibiotics. Often there is a fine balance between the beneficial, therapeutic effects of an antibiotic and possible harmful side effects. Allergy or a toxic reaction may be a serious problem with many antibiotics. Skin rashes and fever are the most common hypersensitivity reactions, but more severe manifestations are anaphylactic reactions, which may occasionally be fatal. In addition to hypersensitivity reactions, severe toxic reactions may occur. For example, the toxic effects of the prolonged use of streptomycin can be damage to hearing. Other examples include the renal toxicity of polymyxins, the irreversible aplastic anaemia caused by chloramphenicol, photosensitivity in some adults taking tetracyclines and irreversible nerve damage associated with aminoglycosides.

References

Corey SV, Taylor GC (1992) Penicillins, cephalosporins, quinolones. *Clin. Pediatr. Med. Surg.* 9: 385–407.

Cuddy PG (1997) Antibiotic classification: implications for drug selection. *Crit. Care Nurs. Q.* 20: 89–102.

Desnottes JF (1996) New targets and strategies for the development of antibacterial agents. *Trends Biotechnol.* 14: 134–140.

Gemmell CG (1996) Antibiotics and the host–parasite relationship (a short review). *Acta Microbiol. Immunol. Hung.* 43: 107–114.

Greenwood D (2000) *Antimicrobial Chemotherapy*, 4th edn. Oxford: Oxford University Press.

Hecht SM (2000) Bleomycin: new perspectives on the mechanism of action. *J. Nat. Prod.* 63: 158–168.

Mascaretti OA, Danelon GO, Laborde M, Mata EG, Setti EL (1999) Recent advances in the chemistry of beta-lactam compounds as selected active-site serine beta-lactamase inhibitors. *Curr. Pharm. Des.* 5: 939–953.

McDonnell G, Russell AD (1999) Antiseptics and disinfectants: activity, action, and resistance. *Clin. Microbiol. Rev.* 12: 147–179.

Nagarajan R (1991) Antibacterial activities and modes of action of vancomycin and related glycopeptides. *Antimicrob. Agents Chemother.* 35: 605–609.

Russell AD, Chopra I (1996) *Understanding Antibacterial Action And Resistance*, 2nd edn. New York: Ellis Horwood.

Smith A, Bagg J (1998) An update on antimicrobial chemotherapy: 1. The mechanisms of action of antibiotics. *Dent. Update* 25: 147–150.

Solomkin JS, Miyagawa CI (1994) Principles of antibiotic therapy. *Surg. Clin. North Am.* 74: 497–517.

Woodkin KA, Morrison SH (1994) Antibiotics: mechanisms of action. *Pediatr. Rev.* 15: 440–447.

Further Reading

Hamilton-Miller JM (1994) Dual-action antibiotic hybrids. *J. Antimicrob. Chemother.* 33: 197–200.

Hwang PM, Vogel HJ (1998) Structure–function relationships of antimicrobial peptides. *Biochem. Cell Biol.* 76: 235–246.

Mazzei T, Mini E, Novelli A, Periti P (1993) Chemistry and mode of action of macrolides. *J. Antimicrob. Chemother.* 31 (Suppl C): 1–9.

Schnappinger D, Hillen W (1996) Tetracyclines: antibiotic action, uptake, and resistance mechanisms. *Arch. Microbiol.* 165: 359–369.

27

Antibiotics that Act on the Cell Wall and Membrane

P.A. Lambert[1], D.G. Allison[2] and P. Gilbert[2]

[1]*Aston University, Birmingham, UK*
[2]*University of Manchester, Manchester, UK*

Inhibition of Peptidoglycan Biosynthesis

Peptidoglycan is a vital component of virtually all bacterial cell walls (see Chapter 7.) It accounts for approximately 50% of the weight of Gram-positive bacterial walls, around 30% of mycobacterial cell walls and 10–20% of the Gram-negative envelope. The peptidoglycan polymer is responsible for the shape, mechanical strength and integrity of bacterial cells. If the synthesis of peptidoglycan is selectively blocked by antibiotic action, bacteria undergo a number of changes in shape and they ultimately die as the result of cell lysis. Mammalian cells do not possess a cell wall or macromolecular structures that resemble peptidoglycan. Consequently, antibiotics that interfere with peptidoglycan synthesis have excellent selective toxicity. A few organisms of medical importance produce cell walls without peptidoglycan, for example *Chlamydia trachomatis* and *Chlamydia pneumoniae*. Some other organisms are capable of surviving in certain environments without their cell wall. For example, the L-forms of *Escherichia coli* and *Proteus* spp. are said to be involved in certain kinds of urinary tract infections. Finally, certain organisms, such as *Mycoplasma pneumoniae*, which causes respiratory tract infections, do not naturally produce a protective cell wall. Bacterial L-forms and the mycoplasmas can survive only in environments where the osmotic pressure is not extreme.

Peptidoglycan contains sugar (glycan) chains, which are cross-linked by short peptide bridges. The glycan chains consist of alternating units of *N*-acetylglucosamine (G) and *N*-acetylmuramic acid (NAMA; M) joined by glycosidic linkages (Fig. 1). The glycan chains vary in length from 20 disaccharides in staphylococci to around 200 in bacilli. Each M unit is substituted by a short peptide chain, some of which are cross-linked either by a direct peptide bond, as in many Gram-negative bacteria, or by additional short bridging stem peptide chains, as in many Gram-positive bacteria. The precise nature of the peptide cross-links varies between organisms but the essential structure of the peptidoglycan macromolecule is the same. The cross-linkage of the peptides between the glycan chains gives the peptidoglycan macromolecule the mechanical strength needed to support the high osmotic pressure that exists in bacterial cells. New peptidoglycan is synthesised and inserted into the cell wall during cell growth and division. The position in the cell wall at which new peptidoglycan is inserted and linked to existing material determines the ultimate shape of the cells and the point of cell division. Clearly, organisms must control this process to maintain balanced growth and morphology. Since peptidoglycan plays a key role in maintaining the shape and structural integrity of the

 doi:10.1006/bkmm.2001.0027

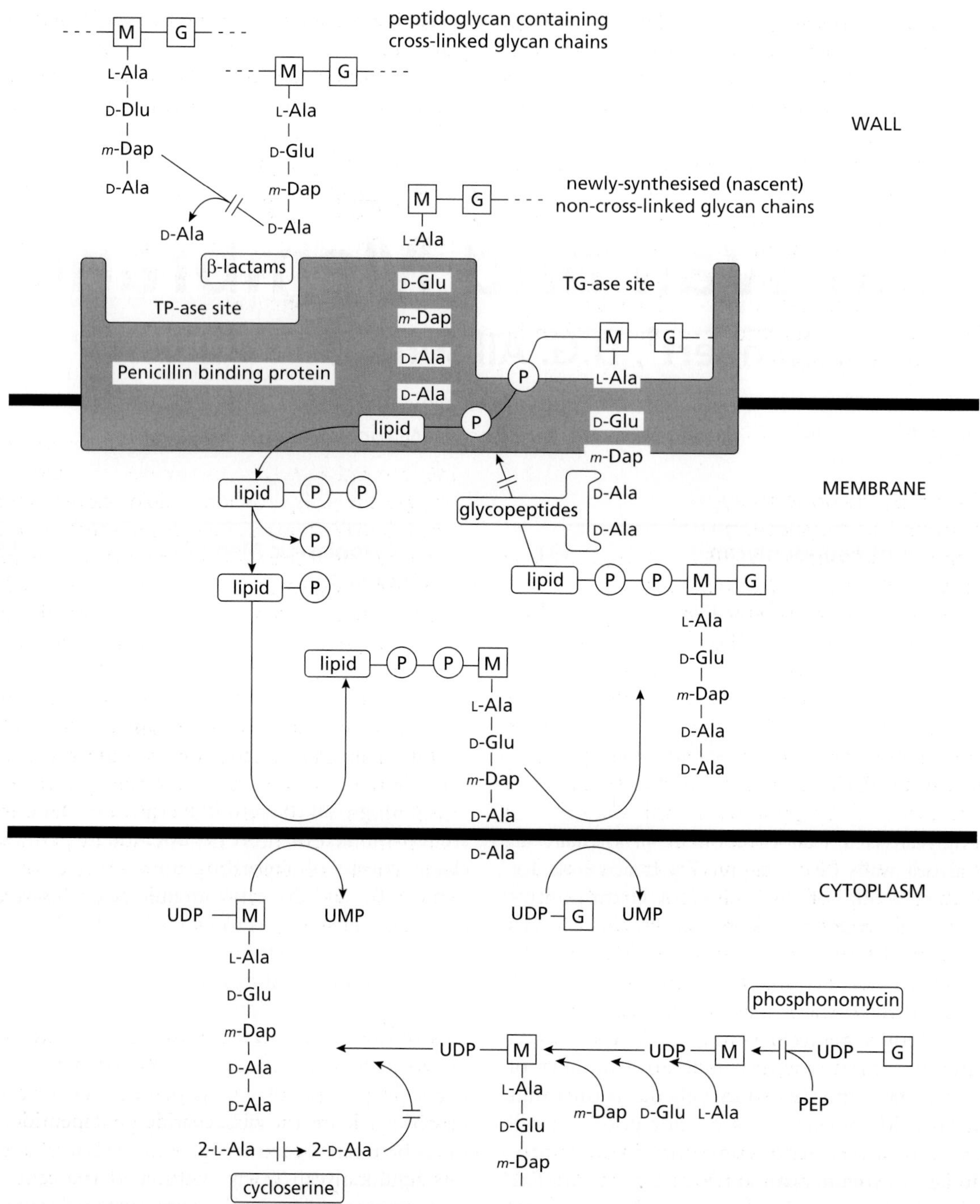

Fig. 1 Pathway for the biosynthesis of peptidoglycan. Three stages are shown: the cytoplasmic stage in which the precursors are synthesised (inhibited by phosphonomycin and cycloserine); the membrane stage in which precursor units are transported to the cell wall by a lipid carrier molecule; the cell-wall stage in which the precursors are first inserted into linear, non-cross-linked glycan strands by transglycosylase (inhibited by glycopeptides) then cross-linked to existing cell wall peptidoglycan by transpeptidase (inhibited by β-lactams). M, *N*-acetylmuramic acid; G, *N*-acetylglucosamine; lipid, undecaprenyl lipid carrier. The penicillin-binding proteins contain both transglycosylase and transpeptidase active sites.

cell wall, interference with any stage in its synthesis and assembly is an effective means of inhibiting cell growth. Some of the most valuable antibiotics act in this way, including the β-lactams (penicillins, cephalosporins, monobactams and carbapenems) and the glycopeptides (vancomycin and teicoplanin).

Peptidoglycan biosynthesis can be divided into three stages (**Fig. 1**). The first step occurs in the cytoplasm, where the precursors are produced. The second stage occurs in the cell membrane where the precursors are progressively coupled together on a lipid carrier molecule, and transported from the cytoplasm to the cell wall. The third stage occurs in the cell wall, where the precursors are integrated into the existing peptidoglycan network. This final stage involves two separate reactions which are each the potential target of antibiotic action. The first involves a transglycosylation reaction in which linear, non-cross-linked glycan chains are formed, and the second involves cross-linking of the peptide substituents on the glycan chains by transpeptidation reactions.

Cytoplasmic Stages of Peptidoglycan Synthesis

Inhibition by Phosphonomycin (fosfomycin) and Cycloserine

The sugar nucleotides, UDP-*N*-acetylglucosamine (UDP-G) and UDP-NAMA pentapeptide (UDP-M-pentapeptide) are the two precursors of peptidoglycan produced in the cytoplasm. The synthesis of the UDP-M-pentapeptide precursor in the cytoplasm takes place in a number of sequential steps. One of these involves the synthesis of UDP-M from UDP-G by addition of a three-carbon pyruvate fragment from phosphoenolpyruvate (PEP) to the 3 position of G. This reaction is catalysed by pyruvate transferase, which is inhibited by phosphonomycin (fosfomycin), a structural analogue of PEP. The mechanism of inhibition involves binding of the antibiotic to a cysteine residue of the enzyme (Skarzynski *et al.*, 1996; De Smet *et al.*, 1999; Samland *et al.*, 1999). Phosphonomycin appears to inhibit the enzyme selectively, because other reactions that involve PEP are unaffected. Phosphonomycin is not widely used for the treatment of infections, however, because relatively high concentrations are required and resistance can rapidly develop.

A series of ligase reactions is involved in the course of the addition of the five amino acid residues to UDP-M. The first three amino acids that are added in turn, are L-alanine, D-glutamic acid and *m*-diaminopimelic acid (Dap). The third amino acid plays a vital role in the later cross-linking stage of peptidoglycan assembly. On its side-chain it has an additional free amino group, which is involved in the cross-linking reaction. In many Gram-negative and Gram-positive organisms this third amino acid is Dap, in staphylococci it is frequently L-lysine. The two final amino acids are added as the dipeptide, D-alanyl-D-alanine, which is produced by two reactions: first two L-alanine molecules are converted to two D-alanines by a specific L-alanine racemase, then the two molecules of D-alanine are coupled by a D-alanine ligase to form the D-alanyl-D-alanine dipeptide. Both the racemase and the ligase are inhibited by the antibiotic D-cycloserine, which bears some structural similarities to D-alanine. Cycloserine binds covalently to the pyridoxal phosphate cofactor of the enzymes, effectively preventing them from forming D-alanyl-D-alanine (Caceres *et al.*, 1997). Cycloserine has some application as a secondary drug used in combination with other drugs for the treatment of tuberculosis.

Precursor Translocation across the Cytoplasmic Membrane

The M-pentapeptide and G units are transferred from the respective UDP nucleotides to undecaprenyl phosphate, a hydrophobic lipid carrier molecule, which initially contains a single phosphate residue and is located exclusively in the cytoplasmic membrane. Another phosphate is coupled to it from UDP-M-pentapeptide to form undecaprenyl pyrophosphoryl-M-pentapeptide, and the remainder of the nucleotide is released as UMP. The G residue of UDP-G is then coupled to the M on the lipid carrier by formation of a glycosidic linkage with release of the remaining UDP. The UDP nucleotides provide the energy to couple the glycan precursors to the lipid carrier. The nucleotides UMP and UDP remain in the cytoplasm and play no further part in the synthetic pathway. The disaccharide pentapeptide is transported across the membrane on the lipid carrier to be inserted into the cell wall at a growing point.

Peptidoglycan Assembly

Inhibition by Glycopeptides and β-Lactams

Glycopeptide and β-lactam antibiotics act at the third and final stage of the peptidoglycan biosynthetic pathway. Here the disaccharide-pentapeptide is first incorporated into the expanding cell wall linked to its lipid carrier. Newly synthesised (nascent) peptidoglycan in the cell wall consists of non-cross-linked linear glycan chains with their pentapeptide substituents on each NAMA residue. This growing (non-cross-linked) glycan-peptide chain is transferred in turn to each molecule of lipid carrier as it delivers its disaccharide-pentapeptide precursor to the cell wall. Each lipid carrier molecule in turn acts to hold the growing linear glycan strand before returning through the membrane to the cytoplasmic face. The lipid carrier itself does not leave the cell membrane, but is recycled to participate in another round of precursor transport. It loses a single phosphate group during its return to the cytoplasmic face of the membrane to collect another disaccharide-pentapeptide from the cytoplasm.

Incorporation of each disaccharide-pentapeptide into the linear glycan in the wall is catalysed by a transglycosylase reaction and this step is blocked by the glycopeptide antibiotics (Barna and Williams, 1984; Billot-Klein *et al.*, 1997). They do not directly interact with the transglycosylases. Rather, they bind very tightly to the terminal D-alanyl-D-alanine on each pentapeptide by formation of five hydrogen bonds (Loll *et al.*, 1999). The presence of the bulky glycopeptides on each D-alanyl-D-alanine residue prevents extension of the linear glycan peptide in the cell wall by denying access of the transglycosylases to the substrate. Vancomycin does not penetrate the cell membrane of bacteria, but is thought to bind to the disaccharide-pentapeptides on the outer face of the cytoplasmic membrane. It has been suggested that two vancomycin molecules form a back-to-back dimer, which bridges between pentapeptides on separate glycan chains, thus preventing further peptidoglycan assembly (Loll *et al.*, 1998). Teicoplanin also binds tightly to the D-alanyl-D-alanine region of the peptidoglycan precursor, but, as a lipoglycopeptide, it may act slightly differently from vancomycin, by locating itself in the outer face of the cytoplasmic membrane and binding the pentapeptide as the precursors are transferred through the membrane. Resistance to the glycopeptides can occur, particularly in *Enterococcus faecium* and *E. faecalis*, which are termed vancomycin-resistant enterococci (VRE). Resistance results from a modification of the terminal D-alanyl-D-alanine of the peptidoglycan precursor to D-alanyl-D-lactate. The lactate is joined to the D-alanine by an ester link to form a depsipeptide in place of the peptide link in D-alanyl-D-alanine. This change does not alter the ability of the transpeptidases to recognise the unit as a substrate, break the ester linkage and proceed to cross-link the C-terminal of the remaining D-alanine to the peptide chain on another glycan strand. It does, however, prevent the glycopeptides from binding tightly, since modification of the peptide bond presumably results in loss of one of the five sites of hydrogen bonding between the peptidoglycan precursor and the glycopeptide. Two major forms of resistance by this mechanism, termed VanA and VanB, occur among VRE (Reynolds, 1998). The multiple genes responsible for VanA and VanB resistance occur in clusters associated with transposable elements that are responsible for transfer of the genes between organisms. The genes involved in resistance in both VanA and VanB isolates include pyruvate dehydrogenases, which convert pyruvate to lactate, D,D-peptidases which cleave existing terminal D-alanines from the peptidoglycan precursors and ligases that add the lactate to the precursor. The VanA and VanB types differ in their regulation of expression, and especially in their response to vancomycin and teicoplanin as inducers. Teicoplanin does not induce expression in VanB isolates, hence VanA isolates are resistant to vancomycin and teicoplanin whereas VanB isolates remain susceptible to teicoplanin. It is of particular concern that glycopeptide resistance can spread to *Staphylococcus aureus*, especially methicillin-resistant strains (MRSA) against which glycopeptides are the sole remaining effective agents.

Cross-linking of Peptidoglycan

Inhibition by Penicillins, Cephalosporins, Carbapenems and Monobactams

Antibiotics of the β-lactam type block the final cross-linking stage of peptidoglycan biosynthesis, which takes place in the cell wall. Here the linear glycan strands are cross-linked to the mature peptidoglycan in the cell wall via their peptide chains by a group of transpeptidases located on the outer face of the cytoplasmic membrane. They first remove the terminal D-alanine residue from each pentapeptide on the linear glycan, which involves breakage of the peptide between the two D-alanine residues on the linear glycan. The energy released is thought to be used in the formation of a new peptide bond between the remaining D-alanine on the glycan chain and an acceptor amino group on existing cross-linked peptidoglycan. In *E. coli* this acceptor is the free amino group on the side-chain of Dap (the third amino acid on each NAMA). In other organisms, such as *Staph. aureus*, it is the free amino group on lysine (replacing Dap) which acts as the acceptor. It should be noted that although there is considerable variation in the composition of the peptide cross-link in different species of bacteria, the essential transpeptidase mechanism is the same. Therefore, virtually all bacteria can be inhibited by interference with this group of enzymes. The β-lactam antibiotics effectively inhibit the transpeptidases by acting as alternative substrates. They mimic the structure of D-alanyl-D-alanine residues and react with the transpeptidases (**Fig. 2**).

Instead of the equivalent peptide bond that joins the alanine residues, the β-lactam bond is broken but the remaining ring system in the β-lactam (a thiazolidine in penicillins) is not released (**Fig. 2**). The transpeptidase remains linked to the hydrolysed antibiotic with a half life of 10–15 minutes. While it is bound to the β-lactam, the transpeptidase cannot participate in further rounds of peptidoglycan

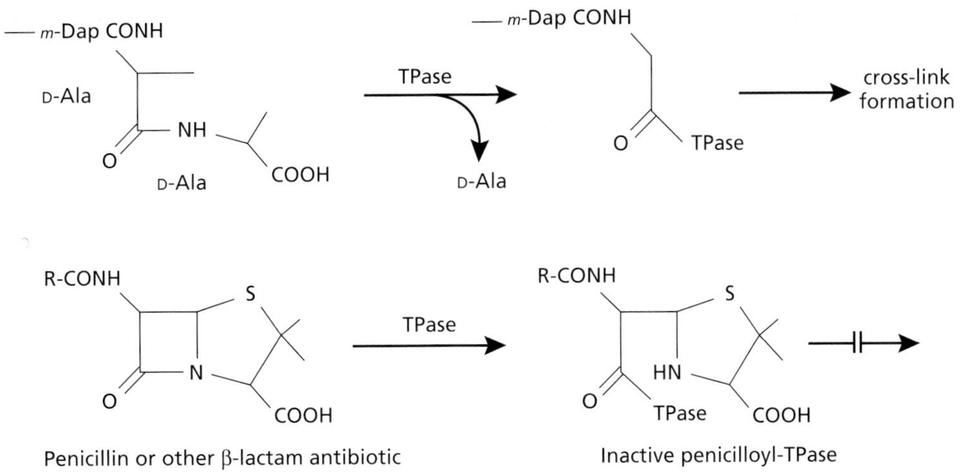

Fig. 2 Interaction between transpeptidase and the D-alanyl-D-alanine region of the non-cross-linked peptidoglycan, resulting in release of the terminal D-alanine and subsequent cross-linking to existing peptidoglycan. The reaction of transpeptidase with penicillin results in the formation of an inactive penicilloyl–enzyme complex.

cross-linking by reaction with its true substrate. All β-lactam antibiotics – penicillins, cephalosporins, carbapenems and monobactams – are thought to act in a similar way through interaction of their β-lactam ring with transpeptidases. There is, however, considerable variation in the morphological effects of different β-lactams on bacterial cells, and this is due to the existence of several types of transpeptidases. The transpeptidase enzymes are usually referred to as penicillin-binding proteins (PBPs), because they can be separated and studied after reaction with ^{14}C-labelled penicillin. This step is necessary because very few copies of each enzyme are present in each cell. PBPs are usually separated by electrophoresis according to their size and are numbered PBP1, PBP2 etc., starting from the species of highest molecular weight (Spratt, 1994). In Gram-negative bacteria the high-molecular-weight transpeptidases have a dual function in the final stages of peptidoglycan synthesis and appear also to possess transglycosylase activity (Ghuysen, 1991; 1994). The transglycosylase and transpeptidase activities are located in separate regions of the protein structures (Terrak *et al.*, 1999). Furthermore, the different transpeptidases have specialised functions in the cell, all cross-link peptidoglycan but some are involved with maintenance of cell integrity, some regulate cell shape, and others produce new cross-wall between elongating cells and secure chromosome segregation before cell division (Bramhill, 1997). Recognition of the existence of multiple transpeptidase targets and their relative sensitivity to different β-lactams helps to explain the different morphological effects observed on treated bacteria. For example, benzyl penicillin (penicillin G), ampicillin and cephaloridine are particularly effective in causing rapid lysis of Gram-negative bacteria such as *E. coli*. These antibiotics act primarily on PBP 1B, the major transpeptidase of the organism. Other β-lactams have little activity against this PBP; mecillinam, for example, binds preferentially to PBP2 and produces a pronounced change in the cells from a rod shape to an oval form. Many of the cephalosporins, such as cephalexin, cefotaxime and ceftazidime bind to PBP3, resulting in the formation of elongated, filamentous cells. The lower-molecular-weight PBPs (4, 5 and 6) do not possess transpeptidase activity. These are carboxypeptidases which remove the terminal D-alanine from the pentapeptides on the linear glycans in the cell wall but do not catalyse the cross-linkage. Their role in the cells is to regulate the degree of cross-linking by denying the D-alanyl-D-alanine substrate to the transpeptidases, but they are not essential for cell growth. Up to 90% of the antibiotic that reacts with the cells may be consumed in inhibiting the carboxypeptidases, without lethal consequences for the cells. PBP4 also possesses endopeptidase activity which hydrolyses peptide cross-links in peptidoglycan, thus reversing the cross-linking function of the transpeptidases. The product of endopeptidase action is a tetrapeptide chain on the NAMA which cannot be re-cross-linked by transpeptidase, because of the absence of the terminal D-alanyl-D-alanine substrate. This endopeptidase activity may be involved in re-structuring of the peptidoglycan during cell growth.

Gram-positive bacteria also have multiple transpeptidases, but fewer than Gram-negative organisms.

Changes in shape are less evident than with Gram-negative rod-shaped organisms. Cell death follows lysis of the cells mediated by the action of endogenous autolytic enzymes (autolysins) present in the cell wall which are activated as a consequence of β-lactam action. Autolytic enzymes able to hydrolyse peptidoglycan are present in most bacterial walls, and are needed to re-shape the wall during growth and to aid cell separation during division. Their activity is regulated by binding to wall components, such as the wall and membrane teichoic acids. When peptidoglycan assembly is disrupted by β-lactam action, some of the teichoic acids are released from the cells, which are then susceptible to attack by their own autolysins.

Expression of β-lactamase enzymes is the most important mechanism by which organisms become resistant to β-lactams. The majority of the enzymes have a serine residue at their active site and bear structural and mechanistic similarities to the carboxypeptidases from which they are presumed to have evolved. Many different types of serine-based enzymes have been identified and classified according to their sequence and biochemical properties, including substrate profiles and susceptibility to inhibitors. A number of successful inhibitors, including clavulanic acid, sulbactam and tazobactam have been developed for use in combination with susceptible β-lactams, to protect them from inactivation by the β-lactamases. Metallo-β-lactamases are a separate group of enzymes which pose a particular threat to the carbapenems. These possess a zinc atom at their active site and are not inactivated by the inhibitors of the serine-based enzymes. In spite of the deployment of inhibitors and the design of new β-lactam structures intrinsically resistant to β-lactamases, organisms continue to evolve new forms of β-lactamases that confer resistance. Other mechanisms by which resistance to β-lactams can arise also exist. Methicillin-resistant strains of *Staph. aureus* (MRSA) display an altered pattern of PBPs in which an alternative PBP called 2a or 2′ is produced. This PBP, the product of the *mecA* gene, is a transpeptidase with a very low susceptibility to inhibition by methicillin or other β-lactam antibiotics. MRSA strains can continue to cross-link peptidoglycan in the presence of β-lactams and are, therefore, resistant to their action. Methicillin resistance levels are controlled by a number of chromosomal genes present in normally sensitive strains. The *fem* genes (factors essential for methicillin resistance) include genes that are thought to influence the chemical nature of the peptide substituents on the NAMA units of peptidoglycan. In *Staph. aureus* the peptidoglycan is cross-linked by an pentaglycine interpeptide bridge, formation of which may be influenced by the *femA* and *femB* genes. *FemC* may influence insertion of isoglutamic acid into the NAMA-peptide (glutamate is the second amino acid substituent in the NAMA-pentapeptide). Thus, a combination of modified peptidoglycan and expression of PBP2′ is responsible for the overall level of resistance expressed by MRSA. Penicillin-resistant strains of *Streptococcus pneumoniae* are increasingly encountered worldwide. These naturally competent organisms become resistant partly by acquisition of PBP genes from other Gram-positive bacteria. The genes have become re-assembled in *Strep. pneumoniae* by recombination to form new PBPs which are resistant to inhibition by penicillin, but retain their transpeptidase activity and still function in peptidoglycan cross-linking. Other changes in the structure of the peptidoglycan are also responsible for the emerging penicillin resistance in *Strep. pneumoniae*, particularly alterations in the peptide substituents on their glycan strands. The free amino group of the third amino acid (lysine) in the peptide chain on each NAMA becomes extended by an additional stem dipeptide (usually alanyl-alanine or seryl-alanine). This modification may assist the strains with altered PBPs to cross-link their peptidoglycan.

Mycolic Acid and Arabinogalactan Synthesis in Mycobacteria

The cell walls of mycobacteria contain an arabinogalactan polysaccharide in addition to the peptidoglycan, plus a variety of high-molecular-weight lipids including the mycolic acids. The lipid-rich nature of the mycobacterial wall is responsible for its characteristic acid-fastness and serves as a penetration barrier to many antibiotics. Isoniazid and ethambutol have long been known as specific antimycobacterial agents, but their mechanisms of action have only recently been established. Isoniazid interferes with mycolic acid synthesis by inhibiting a dehydrogenase that inserts a double bond into the extending fatty acid chain at the 24-carbon stage (Khoo *et al.*, 1999; Alland *et al.*, 2000; Lei *et al.*, 2000; Marrkchi *et al.*, 2000). The active component is a free-radical species produced from isoniazid within the mycobacteria by a catalase-peroxidase, KatG. Ethambutol is thought to block assembly of the arabinogalactan polysaccharide by inhibition of an arabinotransferase enzyme (Alcaide *et al.*, 1997; Lety *et al.*, 1997; Sreevatsan *et al.*, 1997; Chatterjee and Khoo, 1998). Cells treated with ethambutol accumulate the isoprenoid intermediate decaprenylarabinose, which supplies arabinose units for assembly in the arabinogalactan polymer.

The Cytoplasmic Membrane

Selective Disruption of Cytoplasmic Membranes

The integrity of the cytoplasmic membrane is vital for normal cell function. Bacterial membranes do not contain sterols and, in this respect, differ from membranes of fungi and mammalian cells. Ergosterol is the predominant sterol component of fungal membranes, whereas mammalian cells contain cholesterol. Gram-negative bacteria contain an additional outer membrane structure which provides a protective penetration barrier to potentially harmful substances, including many antibiotics. The outer membrane has an unusual asymmetric structure in which phospholipids occupy the inner face, and the lipopolysaccharide (LPS) occupies the outer face. The outer membrane is attached to the peptidoglycan by proteins and lipoproteins. The stability of all membranes is maintained by a combination of non-covalent interactions between the constituents involving ionic, hydrophobic and hydrogen bonding. The balance of these interactions can be disturbed by the intrusion of membrane-active agents, which destroy the integrity of the membrane, thereby causing leakage of cytoplasmic contents or impairment of metabolic functions associated with the membrane. Most membrane-active agents which function in this way, such as the alcohols, quaternary ammonium compounds and bisbiguanides, have very poor selectivity. They cannot be used systemically because of their damaging effects on mammalian cells; instead they are used as skin antiseptics, disinfectants and preservatives, although a few agents can be used therapeutically: for instance the polymyxins, which act principally on the outer membrane of Gram-negative bacteria, and the antifungal polyenes, imidazoles and allylamines.

Colistin

Colistin E is a polymyxin used for the treatment of serious Gram-negative bacterial infections, particularly those caused by *Pseudomonas aeruginosa*. It consists of a cyclic peptide containing positively charged groups linked by a tripeptide to a hydrophobic branched chain fatty acid. Colistin is also available as the sulfomethate derivative, which is hydrolysed in the body to the active colistin base. Colistin binds tightly to negatively-charged phosphate groups on LPS in the outer membrane of Gram-negative bacteria (Pristovsek and Kidric, 1999). The outer leaflet of the membrane structure is distorted, segments of it are released and the permeability barrier destroyed. Colistin can then penetrate to the cytoplasmic membrane where it binds to phospholipids, to disrupt membrane integrity and cause lethal leakage of cytoplasmic components. The detergent-like properties are a key feature of this membrane-damaging action, which is similar to that of quaternary ammonium compounds. The high affinity of colistin for LPS is an advantage in treatment of *Ps. aeruginosa* lung infections where neutralisation of the endotoxic action of LPS released from the organisms may help to control inflammation.

References

Alcaide F, Pfyffer GE, Telenti A (1997) Role of *embB* in natural and acquired resistance to ethambutol in mycobacteria. *Antimicrob. Agents Chemother.* 41: 2270–2273.

Alland D, Steyn AJ, Weisbrod T *et al.* (2000) Characterization of the *Mycobacterium tuberculosis iniBAC* promoter, a promoter that responds to cell wall biosynthesis inhibition. *J. Bacteriol.* 182: 1802–1811.

Barna JCJ, Williams DH (1984) Structure and mode of action of glycopeptide antibiotics of the vancomycin group. *Ann. Rev. Microbiol.* 38: 339–357.

Billot-Klein D, Shlaes D, Bryant D *et al.* (1997) Presence of UDP-*N*-acetylmuramyl-hexapeptides and -heptapeptides in enterococci and staphylococci after treatment with ramoplanin, tunicamycin, or vancomycin. *J. Bacteriol.* 179: 4684–4688.

Bramhill D (1997) Bacterial cell division. *Annu. Rev. Cell. Dev. Biol.* 13: 395–424.

Caceres NE, Harris NB, Wellehan JF *et al.* (1997) Overexpression of the D-alanine racemase gene confers resistance to D-cycloserine in *Mycobacterium smegmatis. J. Bacteriol.* 179: 5046–5055.

Chatterjee D, Khoo KH (1998) Mycobacterial lipoarabinomannan: an extraordinary lipoheteroglycan with profound physiological effects. *Glycobiology* 8: 113–120.

De Smet KA, Kempsell KE, Gallagher A *et al.* (1999) Alteration of a single amino acid residue reverses fosfomycin resistance of recombinant MurA from *Mycobacterium tuberculosis. Microbiology* 145: 3177–3184.

Ghuysen JM (1991) Serine beta-lactamases and penicillin-binding proteins. *Ann. Rev. Microbiol.* 45: 37–67.

Ghuysen JM (1994) Molecular structures of penicillin-binding proteins and β-lactamases. *Trends Microbiol.* 2: 372–380.

Khoo KH, Jarboe E, Barker A *et al.* (1999) Altered expression profile of the surface glycopeptidolipids in drug-resistant clinical isolates of *Mycobacterium avium* complex. *J. Biol. Chem.* 274: 9778–9785.

Lei B, Wei CJ, Tu SC (2000) Action mechanism of antitubercular isoniazid, activation by *Mycobacterium tuberculosis* KatG, isolation, and characterization of *inha* inhibitor. *J. Biol. Chem.* 275: 2520–2526.

Lety MA, Nair S, Berche P, Escuyer V (1997) A single point mutation in the *embB* gene is responsible for resistance to ethambutol in *Mycobacterium smegmatis*. *Antimicrob. Agents Chemother.* 41: 2629–2633.

Loll PJ, Kaplan J, Selinsky BS, Axelsen PH (1999) Vancomycin binding to low-affinity ligands: delineating a minimum set of interactions necessary for high-affinity binding. *J. Med. Chem.* 42: 4714–4719.

Loll PJ, Miller R, Weeks CM, Axelsen PH (1998) A ligand-mediated dimerization mode for vancomycin. *Chem. Biol.* 5: 293–298.

Marrakchi H, Laneelle G, Quemard A (2000) InhA, a target of the antituberculous drug isoniazid, is involved in a mycobacterial fatty acid elongation system, FAS-II. *Microbiology* 146: 289–296.

Pristovsek P, Kidric J (1999) Solution structure of polymyxins B and E and effect of binding to lipopolysaccharide: an NMR and molecular modeling study. *J. Med. Chem.* 42: 4604–4613.

Reynolds PE (1998) Control of peptidoglycan synthesis in vancomycin-resistant enterococci: D,D-peptidases and D,D-carboxypeptidases. *Cell. Mol. Life. Sci.* 54:325–331.

Samland AK, Amrhein N, Macheroux P (1999) Lysine 22 in UDP-*N*-acetylglucosamine enolpyruvyl transferase from *Enterobacter cloacae* is crucial for enzymatic activity and the formation of covalent adducts with the substrate phosphoenolpyruvate and the antibiotic fosfomycin. *Biochemistry* 38: 13162–13169.

Skarzynski T, Mistry A, Wonacott A *et al.* (1996) Structure of UDP-*N*-acetylglucosamine enolpyruvyl transferase, an enzyme essential for the synthesis of bacterial peptidoglycan, complexed with substrate UDP-*N*-acetylglucosamine and the drug fosfomycin. *Structure* 4: 1465–1474.

Spratt BG (1994) Resistance to antibiotics mediated by target alterations. *Science* 264: 388–393.

Sreevatsan S, Stockbauer KE, Pan X *et al.* (1997) Ethambutol resistance in *Mycobacterium tuberculosis*: critical role of *embB* mutations. *Antimicrob. Agents Chemother.* 41: 1677–1681.

Terrak M, Ghosh TK, van Heijenoort J *et al.* (1999) The catalytic, glycosyl transferase and acyl transferase modules of the cell wall peptidoglycan-polymerizing penicillin-binding protein 1β of *Escherichia coli*. *Mol. Microbiol.* 34: 350–364.

28

Antibiotics That Act on Nucleic Acids and Protein Biosynthesis

P. Gilbert,[1] D.G. Allison[1] and P.A. Lambert[2]

[1] *University of Manchester, Manchester, UK*
[2] *Aston University, Birmingham, UK*

Nucleic Acid as a Target for Selective Toxicity

Living cellular systems store the genetic information associated with their growth, division and survival as a sequence of bases within their DNA. The essence of the cellular life process is the translation of these base sequences into structural and functional elements of the cell. Thus, the DNA must be capable not only of accurate self-replication and segregation into daughter cells, but also of accurate transcription and eventual translation into protein. Structural proteins provide some stability to the cells. Most proteins, however, function as catalysts, either in the biosynthesis of non-protein elements (anabolism), in catabolic events leading to the capture and utilisation of chemical of energy (catabolism) or in the transport and movement of substrates, solutes and waste products into and out of the cell (osmoregulation and active transport). Double-stranded DNA is not only replicated for cell division but may also be transcribed into single-stranded RNA. This RNA may simply provide templates for the biosynthesis of protein (mRNA) or it may provide the translational and regulatory machinery both as ribosomal RNA and as aminoacyl-tRNA. Additionally, some of the transcription and translation products, together with the products associated with protein expression, act as activators and suppressors that in turn modulate transcription and the nature of the expressed phenotype. The precise mechanisms associated with nucleic acid replication, transcription and translation are very similar in all living systems whether bacterial or eukaryotic, plant or animal (see Chapter 14). Although they therefore represent prime targets for biocidal action, selective toxicity at such targets must exploit fine differences in these divergently evolved cellular processes (Table 1).

Precursor Synthesis

A common element of all living cells is the use of folic acid, in its reduced form (tetrahydrofolate), as a carrier of carbon fragments involved in the biosynthesis of the nucleic acid bases. Of primary importance is the folate-dependent enzyme thymidylate synthetase. Mammalian systems acquire their folate from their

 doi:10.1006/bkmm.2001.0028

Table 1 Antibiotics targeted at replication, transcription and translation of DNA

Target	Antibiotic	Mechanism	Selectivity
DNA replication			
Precursor biosynthesis	Sulfonamides	Inhibits folate synthesis	Target unique
	Trimethoprim	Inhibits dihydrofolate reductase	Target unique
	Pyrimethamine	Inhibits dihydrofolate reductase	Target unique
	Trimetrexate	Inhibits dihydrofolate reductase	Target unique
DNA architecture	Quinolones	Inhibit DNA gyrase	Target specificity
	Metronidazole	DNA strand breakage	Requires anaerobiosis
	Nitrofurantoin	DNA strand breakage	Requires anaerobiosis
Transcription	Rifamycin	Inhibits RNA polymerase	Target specificity
Protein synthesis	Aminoglycosides	Distort 30S ribosome subunit	40S subunit immune
	Tetracyclines	Block 30S ribosomal subunit	Access to target
	Chloramphenicol	Inhibits peptidyl transferase	Target specificity
	Macrolides, azalides	Block translocation	Target specificity
	Fusidic acid	Inhibits elongation factor	Access to target
	Mupirocin	Inhibit isoleucyl-tRNA synthesis	Target specificity

diet, but prokaryotes and protozoa must synthesise it for themselves. The enzymes associated with folate biosynthesis thus represent a unique biochemical target in many pathogenic species.

Chromosome Organisation

Although the mechanisms associated with chromosome replication and function are broadly similar in prokaryotic and eukaryotic cells, some important differences provide an opportunity for the selective targeting of toxic agents. Eukaryotic DNA is folded and packaged in conjunction with histone elements, whereas bacterial DNA is histone-free and highly coiled. Positive supercoiling – twisting in the same rotational direction as the double helix – renders the bacterial DNA more compact. Negative supercoiling – twisting in a direction opposite to the rotational direction of the double helix – makes the DNA less compact and amenable, by local strand separation, to replication and transcription. Since the bacterial chromosome is circular, changes in coiling in one region will cause stress in the entire structure and lead to the opposite form of coiling elsewhere in the molecule. The DNA topoisomerases control these topological changes in DNA architecture. Since the chromosome of eukaryotes is linear and associated with histones, the topoisomerases that control the circular histone-free bacterial chromosome provide unique biochemical targets for antibiotic action. *Escherichia coli* has four topoisomerase enzymes, which cut one or both DNA strands and impart or relieve torsion before resealing the DNA. Type I topoisomerases cut one strand of the DNA, whereas type II topoisomerases cut both strands. Topoisomerases I and III of *E. coli* are both type I enzymes, whereas topoisomerases II and IV are type II enzymes. Topoisomerase II is also known as DNA gyrase, the cellular target for the quinolone antibiotics.

DNA Replication

The separated DNA strands are kept apart during replication by a specialised protein (Albert's protein) and, with the separated strands as templates, a series of enzymes produce new strands of DNA. An RNA polymerase then forms short primers of RNA on each strand at specific initiator sites, and DNA polymerase III synthesises and joins short DNA strands onto the RNA primers. DNA polymerase I, which possesses nucleotidase activity, then removes the primers and replaces them with DNA strands. Finally a DNA ligase joins the DNA strands together to produce two daughter chromosomes. The entire process is closely surveyed and regulated by proof-reading stages to ensure that each nucleotide is incorporated according to the sequence specified in the template. So far, no therapeutic antimicrobials are available that specifically target the DNA polymerases.

Transcription

During transcription a single strand of RNA is produced from an appropriate coding region within an unfolded segment of DNA. The RNA product may be mRNA subsequently used as the template for protein synthesis, or it may be a structural or functional

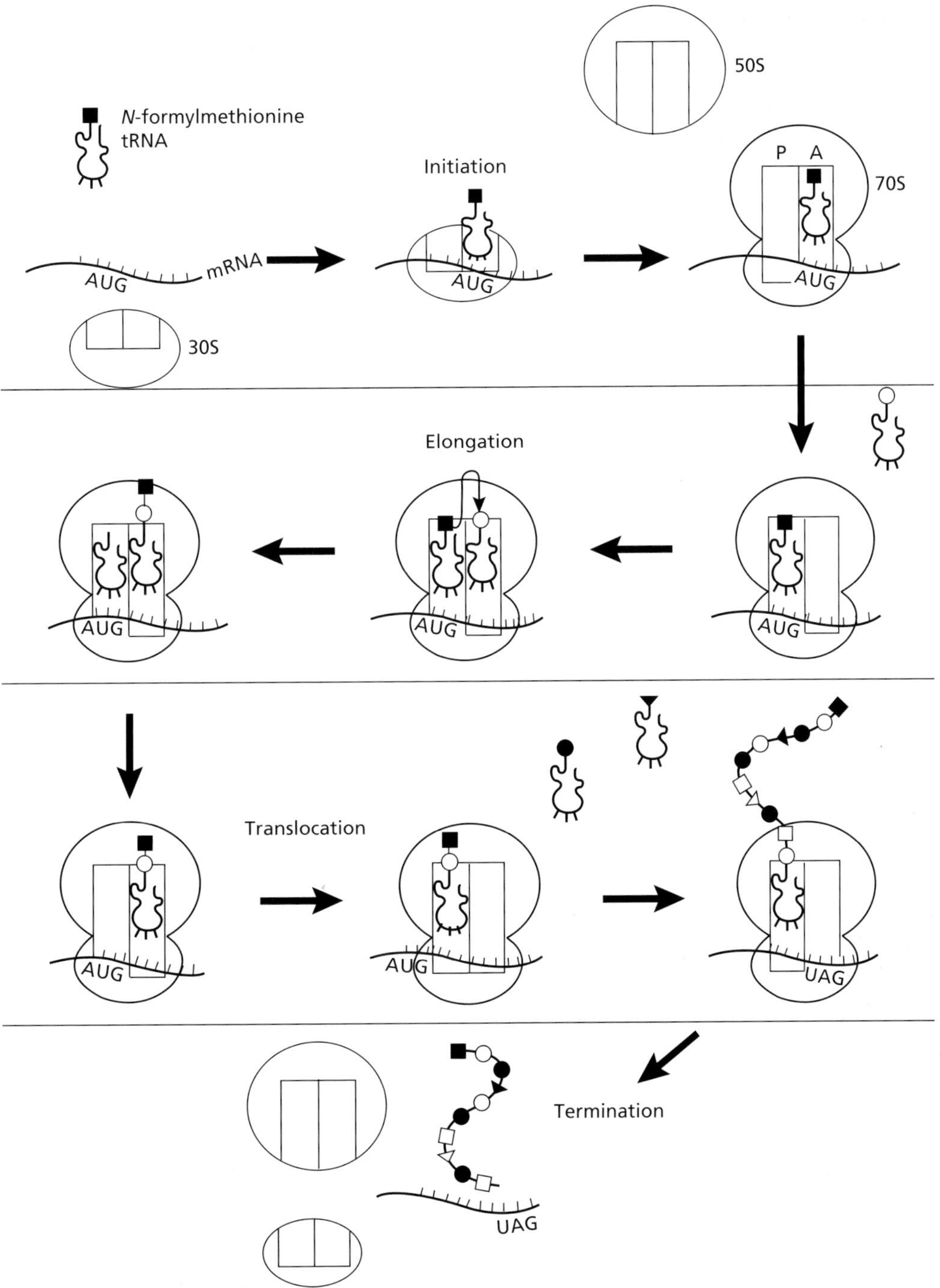

Fig. 1 Outline of the main events in protein synthesis: initiation, elongation, translocation and termination. AUG is an initiation codon; it codes for *N*-formylmethionine and initiates the formation of the peptide chain on the 70S ribosome. UAG is a termination codon, which does not code for an amino acid and brings about the termination of peptide chain synthesis.

element of the translational process (ribosomal RNA or tRNA). The key enzyme in this process is RNA polymerase, which is a complex of four proteins (two α, one β and one β' subunit) which make the core enzyme. A small protein σ-factor joins the core enzyme and binds to the promotor region of the DNA preceding the gene that is to be transcribed. The σ-factor is responsible for recognition of the initiation signal for transcription; the core enzyme possesses the activity to sequester and join the appropriate nucleotides. Mammalian cells possess a similar RNA polymerase complex but sufficient differences in structure permit the selective targeting of inhibitors such as rifampicin.

Translation into Protein

The translation of mRNA into catalytic and structural elements of the cell involves not only the translation of a base sequence encoded within the messenger into a peptide sequence, but also its post-translational folding, modification and assembly into functional elements. The process of protein synthesis, which involves the ribosome, mRNA, a series of aminoacyl-tRNAs and accessory proteins involved in initiation, elongation and termination is outlined in **Fig. 1**. Although the essential elements of protein synthesis are the same in prokaryotes and eukaryotes, the differences are sufficient to allow the targeting of numerous groups of antibacterial agent to this process (**Table 1**).

Folate Antagonists

As we have seen above, folate biosynthesis is essential to the growth and survival of bacteria, and the reduced form, tetrahydrofolate (THF), acts as a carrier for the carbon fragments used in the biosynthesis of bases that form the coding sequence of the nucleic acids. Additionally, the synthesis of thymidine is folate-dependent and produces dTMP by the transfer of a methyl group from tetrahydrofolate to dUMP. In this and other folate-dependent reactions, THF is converted to dihydrofolate (DHF) but it must be converted back to the reduced form before it can participate in further rounds of synthesis. Dihydrofolate reductase (DHFR), with the cofactor nucleotide $NADPH_2$, is responsible for DHF reduction to THF. Bacteria, protozoa, yeasts and mammalian cells all contain DHFR because it is associated with cycling of the folate rather than its synthesis, but sufficient differences in the enzyme

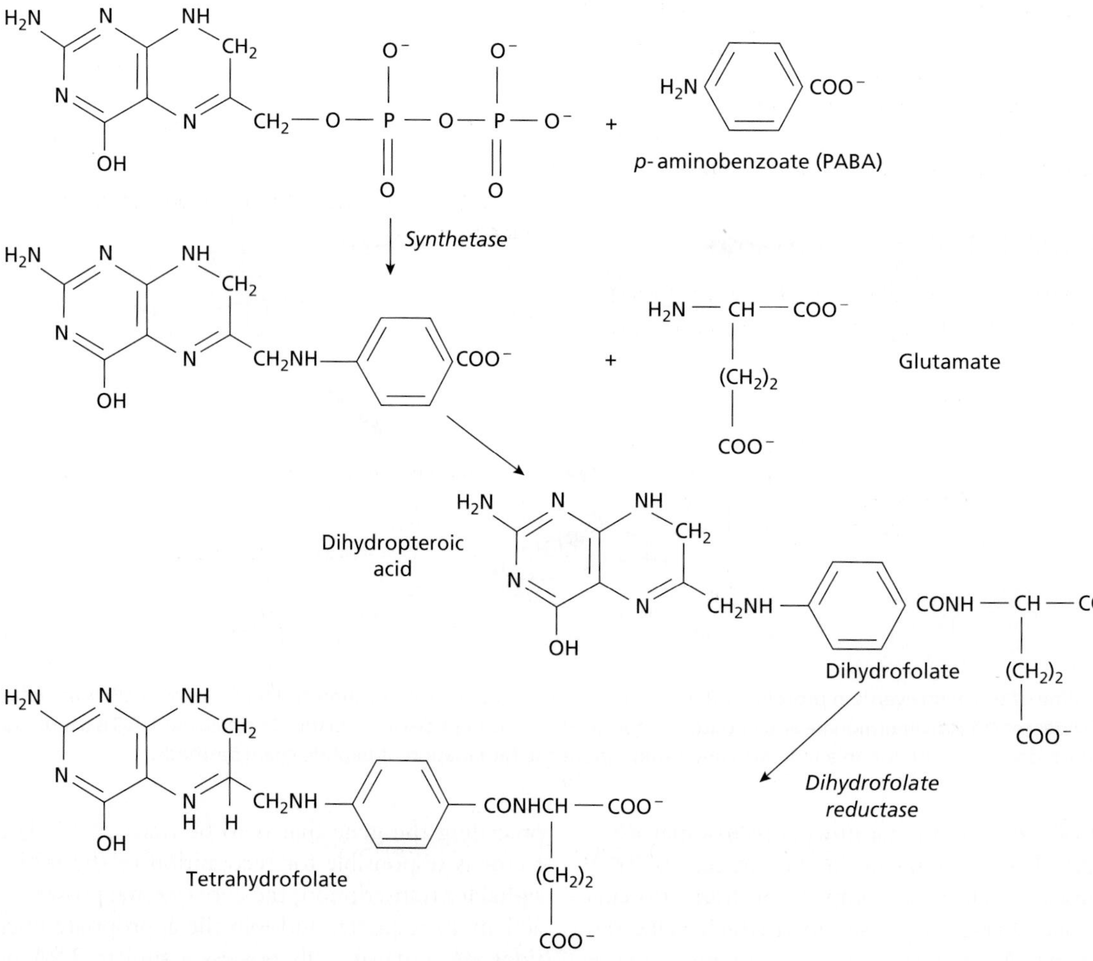

Fig. 2 Final steps in the biosynthesis of tetrahydrofolate by bacteria.

structure for the selective allow targeting of inhibitors to the bacterial and protozoan enzymes. Since the biosynthesis of folate is unique to the pathogens it represents another obvious target for chemotherapeutic agents. A series of reactions leads first to the synthesis of dihydropteroic acid from one molecule of pteridine and *p*-amino benzoic acid (PABA). Glutamic acid is then added to form DHF which is then reduced by DHFR to THF. The bacterial route to THF is illustrated in Fig. 2. The drugs used to target folate biosynthesis and folate action include agonists of DHFR (trimethoprim) and dihydropteroic acid synthetase (sulphonamides). Combinations of such agonists, such as co-trimoxazole, are a powerful double blockade of folate biosynthesis and utilisation.

Sulfonamides

Sulfonamides are structural analogues of PABA (Fig. 3) and competitively inhibit the incorporation of PABA into dihydropteroic acid. There is some evidence that they are synthesised into 'false' folate analogues that inhibit further metabolism. Since the inhibition is competitive, it can easily be reversed by the accumulation of PABA, or by the presence of thymine, adenine, guanine and methionine. Fortunately these bases are not usually present at the sites of infections for which treatment with sulphonamides is used.

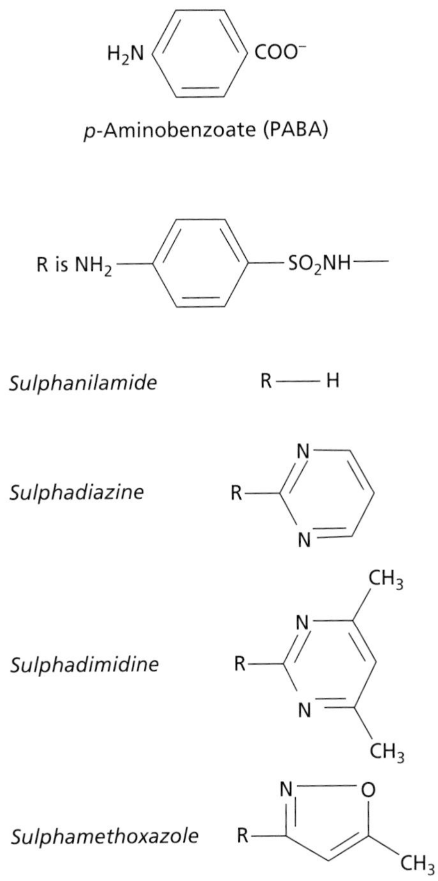

Dihydrofolate Reductase Inhibitors

Trimethoprim is a selective inhibitor of bacterial DHFR. The bacterial enzyme is several thousand times

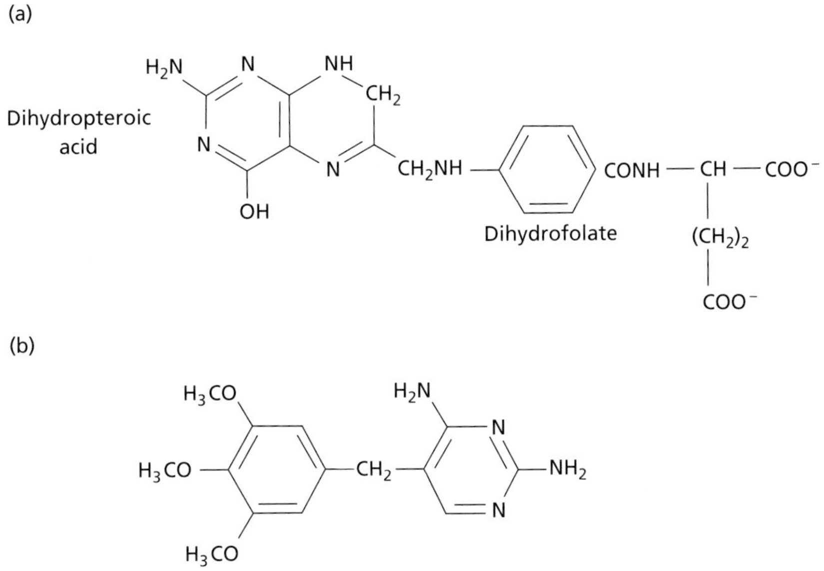

H_2N COO^-

p-Aminobenzoate (PABA)

R is NH_2 — SO_2NH—

Sulphanilamide R—H

Sulphadiazine R—

Sulphadimidine R— CH_3 CH_3

Sulphamethoxazole R— CH_3

Fig. 3 Chemical structures of *p*-aminobenzoate and some sulphonamides.

(a)

Dihydropteroic acid

H_2N N NH CH_2 N N OH CH_2NH CONH — CH — COO^- $(CH_2)_2$ COO^-

Dihydrofolate

(b)

H_3CO H_3CO H_3CO CH_2 H_2N N N NH_2

Fig. 4 Chemical structures of (a) dihydropteroic acid dihydrofolate and (b) trimethoprim.

more sensitive than its mammalian counterpart. Likewise, pyrimethamine is a selective inhibitor of the plasmodial DHFR. Both agents are structural analogues of the dihydropteroic acid portion of DHF (Fig. 4). Both these inhibitors bind avidly to the active site on DHFR usually occupied by the DHF substrate. A pattern of strong hydrogen bonding between amino acid residues and water molecules that line the site reinforces the binding, which cannot be reversed by the accumulation of the natural substrate. These inhibitors therefore act in a non-competitive fashion. Proguanil, another DHFR inhibitor, is used for the treatment of plasmodial infection. It is a pro-drug that is metabolised in the liver to its active form (cytoguanil, Fig. 4). Methotrexate is also a potent DHFR inhibitor, but it has structural analogy to the whole of the DHF molecule, including the glutamate residue. It therefore has little or no selectivity towards the bacterial or protozoan form of the enzyme, but it can be used for the treatment of some types of cancer.

Sulfonamide Combinations

Combinations of a sulphonamide with DHFR inhibitors, such as trimethoprim, cause a double blockade of the synthesis and cycling of folates in the bacterial metabolism. The action of such combinations is far less susceptible to reversion through substrate accumulation and point mutation than is the action of the individual agents. Such combinations, with suitable matching pharmacokinetic parameters, are characterised by the antibacterial combination of sulphamethoxazole with trimethoprim (co-trimoxazole) and the antimalarial combination of sulphadoxine with pyrimethamine (fansidar).

Antibiotics Active against DNA Architecture

Quinolones

The quinolones selectively inhibit the DNA gyrase (topoisomerase I). Work on quinolone antibiotics goes back to the late 1960s when nalidixic acid was introduced into clinical practice. It had only moderate activity against Gram-negative bacteria and no activity against Gram-positive organisms and pseudomonads. Its low systemic availability and rapid renal excretion made it useful only for the treatment of urinary tract infections. In spite of this poor beginning, nalidixic acid was the origin of more than 5000 quinolone agents with significant antibacterial activity. Those with significant activity can be related to any of nine basic chemical ring structures (Fig. 5).

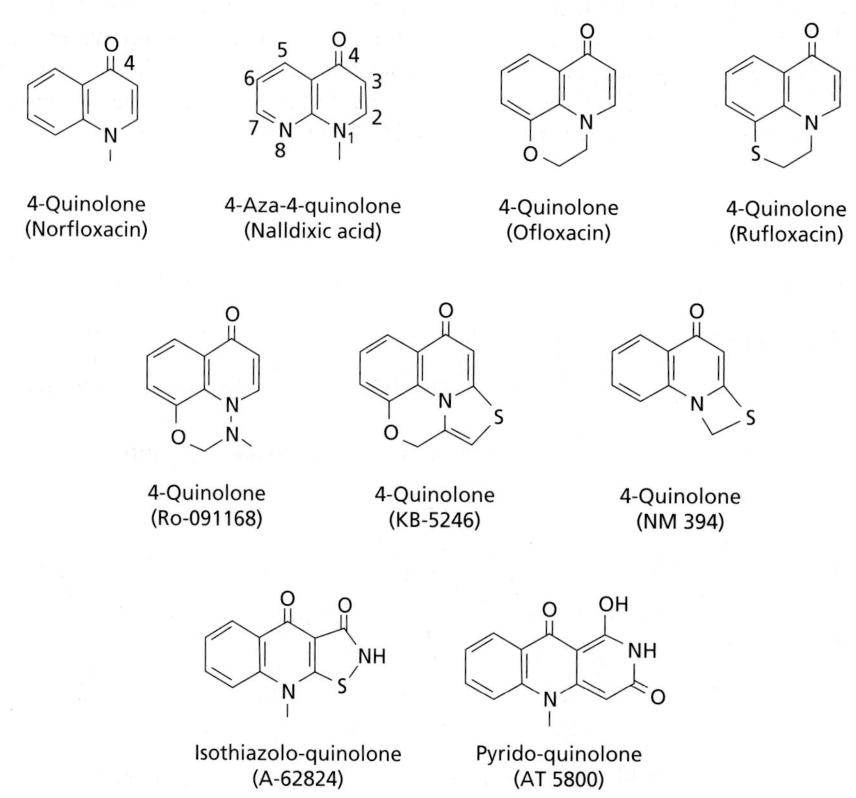

Fig. 5 Chemical structures of some quinolones.

During the early 1980s fluorinated quinolone antibacterials were discovered in the form of norfloxacin, pefloxacin and ciprofloxacin. These have greatly improved activity against Gram-negative bacteria and good activity against Gram-positive organisms. Improved activity was subsequently reported for some second-generation fluoroquinolones including, enoxacin, ofloxacin, fleroxacin, lomefloxacin, tosufloxacin and sparfloxacin (Fig. 6). All of the quinolone antibacterial agents act in a similar fashion. The target DNA gyrase, a tetramer of two A subunits and two B subunits, is capable of catalysing various topological changes in the conformation of bacterial DNA. These include both the introduction of negative supercoiling and the removal of positive supercoiling. Both processes allow the DNA structure to relax and un-knot so that the gyrase can remove linked structures (decatenation). These processes allow the DNA to separate after replication and segregate into the progeny. The gyrase enzyme binds to the DNA at the crossing point of two separate double-stranded regions of DNA (a supercoiled region). The A subunits produce a staggered cut (four-base separation) in both strands of one DNA molecule through which the other strand is passed. The DNA is then rejoined by the action of a DNA ligase. Energy for the cleavage is derived through the B subunits which hydrolyse ATP. Action of the gyrase alone will, therefore, fragment the chromosome according to the degree of supercoiling. Consequently, inhibition of gyrase activity is strongly bacteriostatic by preventing DNA segregation. The newer quinolones appear also to act on topoisomerase IV, which is particularly important in the segregation processes in staphylococci, and this may account for their enhanced activity against these organisms. Killing of bacterial cells by the fluoroquinolones is an anomaly currently thought to be due to the introduction of repair-resistant strand breakage by the partially inhibited gyrase. Indeed, the action of the quinolones is itself inhibited by the presence of inhibitors of protein biosynthesis (chloramphenicol) and of transcription (rifamycin), and at higher concentration they are themselves mild protein biosynthesis inhibitors.

Nitroimidazoles and Nitrofurans

These agents cause direct DNA strand breakage, rather than the indirect breakages introduced through an inhibition of DNA topoisomerase. These molecules are active only under anaerobic conditions and are therefore preferentially active against anaerobic bacteria. Host cell DNA resistance to these substances is conferred by the maintenance of an aerobic state. Under anaerobic conditions the nitrate group of the drug molecule is converted to a nitronate radical. These are chemically highly reactive and, unlike other radicals produced during the normal functioning of a cell – the breakdown products of hydrogen peroxide, superoxide and hydroxyl radicals – they are not detoxified by natural protective enzymes. Nitronate radicals react directly with the DNA and break the strands. Drugs that act in this fashion include metronidazole and nitrofurantoin.

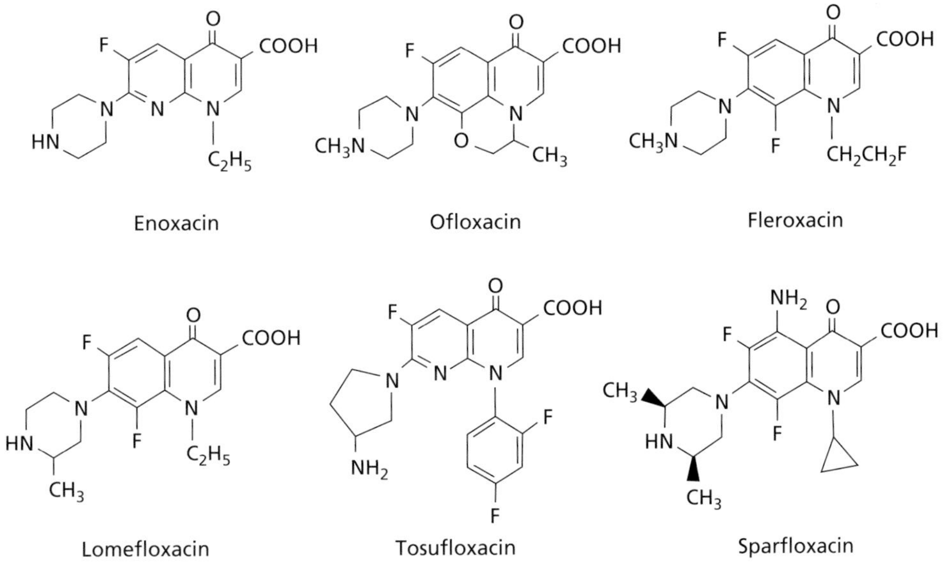

Enoxacin Ofloxacin Fleroxacin

Lomefloxacin Tosufloxacin Sparfloxacin

Fig. 6 Chemical structures of some fluorinated second-generation quinolones.

Inhibitors of DNA Transcription

Rifamycin (rifampicin) acts against the β subunit of RNA polymerase by blocking the initiation process, and each molecule of rifampicin neutralises one RNA polymerase molecule. After initiation has commenced, rifampicin is excluded from the enzyme which must become dissociated from the DNA in order for it to be affected. The action of rifampicin is, therefore, to prevent the initiation of new rounds of RNA transcription. It has been suggested than the structure of rifampicin has a marked similarity to that of two adenosine nucleotides in RNA and that this structural similarity forms the basis of the inhibition of the β subunit. A major problem with rifampicin is that resistance to it may relatively easily be conferred by changes of amino-acid sequence in one small region of the β subunit, which does not affect polymerase activity. Similar differences between the bacterial and mammalian RNA polymerase enzymes account for the specificity of the antibiotic.

Inhibitors of Translation in Protein Biosynthesis

Inhibitors of the translation of mRNA into protein represent a large and diverse group of therapeutically valuable antibiotics, including the aminoglycosides (streptomycin, kanamycin, gentamicin, etc.), the tetracyclines, chloramphenicol, macrolides (erythromycin), azolides (azithromycin), the lincosamides (lincomycin, clindamycin), fusidic acid and mupirocin. A detailed account of protein synthesis is to be found in Chapter 15.

Bacterial ribosomes are smaller than their eukaryotic counterparts and consist of one 30S and one 50S subunit rather than the 40S and 60S subunits of the mammalian protein synthesis apparatus. The 30S subunit consists of a single strand of 16S ribosomal RNA and more than 20 different proteins that are bound to it. The larger 50S subunit comprises two strands of ribosomal RNA (23S and 5S) and more than 30 different proteins. Each of the group of agents acts at a distinct stage of translation and protein biosynthesis (see **Fig. 1**) and exploits fundamental differences in either the structure and function of bacterial and mammalian ribosomes or in the access of such agents to the various cell types.

Aminocyclitol (Aminoglycoside) Antibiotics

Streptomycin is probably the most notable of all the aminoglycoside antibiotics, and the one on which our understanding of mechanism for the entire aminoglycoside (aminocyclitol) group is based. Other aminoglycoside antibiotics include neomycin, gentamicin and kanamycin; the aminocyclitols include spectinomycin. A major effect of the aminoglycosides is interference with the initiation stages of protein biosynthesis, that is the assembly of the ribosomal subunits onto mRNA. During assembly of the initiation complex, *N*-formylmethionyl-tRNA (fmet-tRNA) binds to the ribosome-binding site on the untranslated 5′ end of the mRNA together with the 30S subunit. Three accessory proteins, termed initiation factors (IF_{1-3}), and a single molecule of guanosine triphosphate (GTP), position the fmet-tRNA on the AUG initiation codon on the messenger. IF_1 and IF_2 are released, GTP is hydrolysed to GDP and released together with IF_2, as the 50S ribosomal subunit is sequestered to the initiation site. The fmet-tRNA occupies the peptidyl site (P site), leaving the adjacent acceptor site (A site) free to interact with a suitably coded aminoacyl-tRNA. Streptomycin and its related antibiotics bind tightly to the protein component of the 30S subunit that normally receives IF_3, and initiation of protein synthesis is thus inhibited. Rounds of protein synthesis that have already commenced can also be affected by streptomycin. Binding of the agent to the IF_3 receptor distorts the A site, making it less discerning in terms of its reception of aminoacyl-tRNA. Thus, existing rounds of protein synthesis will continue but the translational process will be error-prone. The fact that protein synthesis is error-prone in the presence of streptomycin is the reason for its lethality to the cell; simple inhibition of protein synthesis would merely be bacteriostatic. It appears that the synthesis of large amounts of toxic, non-functional proteins is very detrimental to the cell life. The misreading of the mRNA sequence is not random; rather the cell misreads U as A or C, and C is read as A or U. Damage may be initiated when the cell attempts to export the non-functional proteins.

In order for aminoglycosides to inhibit protein synthesis, they must first gain access to the target cell cytosol. This is a three-stage process. The first stage is a non-specific interaction of the cationic drug with the anionic groupings of lipopolysaccharide or teichoic and teichuronic acids on the cell surface. This destabilises the outer membrane of Gram-negative bacteria, but it is not lethal to the cells. Under some circumstances, however, it can promote the access of other antibiotics across the cell envelope of such organisms, and this accounts for the use of streptomycin in certain combination therapies. The second and third stages of uptake are energy-dependent and require the cell to be active in aerobic respiration. It is these energy-dependent stages that account for the lack of effectiveness of the aminoglycosides towards anaerobic infections.

Tetracyclines

This group of antibiotics is actively transported into susceptible bacterial cells, probably as an ion complex with magnesium. This enables the drug to be concentrated as much as 50-fold in the bacterial cytosol. Since mammalian cells do not have this uptake mechanism, they are far less susceptible to the inhibitory effects of the tetracyclines. In a similar fashion, resistance to tetracyclines is generally mediated through the acquisition of efflux pumps that actively remove the agent before protein synthesis is inhibited. The inhibitory action of the tetracyclines is not specifically directed to the bacterial ribosome, rather it binds to the 3′ end of the 16S RNA associated with the 30S subunit and also to a number of accessory proteins and magnesium ions. The net effect is to block sterically the binding of aminoacyl-tRNA to the A site and halt protein synthesis. In this respect the tetracyclines, unlike the aminoglycosides, are bacteriostatic rather than bactericidal.

Chloramphenicol

Chloramphenicol selectively inhibits protein synthesis in bacterial cells by binding to the 50S ribosomal subunit in the region of the A site involving the 23S ribosomal RNA. Normal binding of the aminoacyl-tRNA is affected, so that it becomes inaccessible to the peptidyl transferase needed to form a new peptide bond between the nascent peptide at the P site and the aminoacyl-tRNA associated with the A site. Experiments *in vitro* suggest that the inhibition is brought about by a direct interaction of the chloramphenicol with the aminoacyl-tRNA causing distortion adjacent to the position where the amino acid is carried. Although chloramphenicol has a broad spectrum of activity and is active against a number of potentially lethal pathogens, such as *Salmonella typhi*, it has toxicity for certain mammalian cells unrelated to its action as a protein synthesis inhibitor. This toxicity affects rapidly metabolising mammalian cells such as liver and particularly bone marrow stem cells, which may lead to potentially life-threatening bone marrow aplasia. The systemic use of chloramphenicol is therefore limited to serious infections.

Macrolide and Azalide Antibiotics

This group of protein synthesis inhibitors, the most notable of which is erythromycin, act on protein elongation, rather than initiation. They have a broad spectrum of action, and affect a wide range of Gram-positive and Gram-negative bacteria. As with chloramphenicol, these drugs bind to the 23S ribosomal RNA associated with the 50S ribosomal subunit. Changes in methylation of 23S ribosomal RNA can confer resistance to both macrolide and azalide antibiotics but this is associated with slightly different regions of the ribosomal RNA molecule. Binding of chloramphenicol to the 50S ribosome can also inhibit binding of macrolides, and vice versa. This implies a common but different target for each agent. Unlike chloramphenicol, the macrolides block translocation of the ribosomal complex along the mRNA. After the peptidyl transferase has formed a peptide bond between the nascent peptide bound to the P site and the aminoacyl-tRNA at the A site, the whole ribosome must relocate to the next codon for translation to proceed. Energy for this process is derived by hydrolysis of GTP to GDP mediated through the accessory elongation factor (Ef-G). By blocking the translocation incomplete peptides are lost from the ribosome–mRNA complex, the ribosomal subunits dissociate and initiation must recommence. It is assumed that the azalides have a mode of action similar to erythromycin, but they have better intracellular penetration and are less readily metabolised by the host with a resulting improved serum half-life.

Lincomycin and Clindamycin

These agents have binding sites for the 23S ribosomal RNA of the 50S subunit close to those of chloramphenicol and the macrolides. Their action, however, is to block the elongation of protein biosynthesis by a direct inhibition of the peptidyltransferase enzyme.

Fusidic Acid

This steroidal antibiotic acts directly on the elongation factor Ef-G and not on the ribosome. Ef-G, an accessory protein, hydrolyses GTP to GDP in order to provide energy for the translocation of the ribosome. A second elongation factor (Ef-Tu) promotes binding of the aminoacyl-tRNA to the ribosomal A site, also with hydrolysis of GTP. The two elongation factors have overlapping binding sites on the ribosome. Fusidic acid binds to the Ef-G : GDP complex to the ribosome after GTP hydrolysis, sterically blocking the next stage of Ef-Tu : GTP binding. As with the tetracyclines, fusidic acid is selectively toxic only because it is concentrated in bacterial cells by an active transport process. Equivalent elongation factors in mammalian cell systems are susceptible in cell-free culture.

Mupirocin

Mupirocin targets a group of enzymes that couple amino acids to their appropriate tRNA for delivery to the ribosomal A site. One particular enzyme, involved in synthesising isoleucyl-tRNA, is inhibited by mupirocin. This is brought about by a structural similarity between the mupirocin and isoleucine. Protein synthesis is halted when an isoleucine codon is encountered at the A site of the mRNA.

Further Reading

Coulson CJ (1994) *Molecular Mechanisms of Drug Action.* London: Taylor and Francis.

Franklin JJ, Snow GA (1989) *Biochemistry of Antimicrobial Action*, 4th edn. London: Chapman & Hall.

Greenwood D (1995) *Antimicrobial Chemotherapy*, 3rd edn. Oxford: Oxford University Press.

Hooper DC, Wolfson JS (1991) *Quinolone Antimicrobial Agents.* Washington, DC: American Society for Microbiology Press.

Russell AD, Chopra I (1996) *Understanding Antibacterial Action and Resistance*, 2nd edn. New York: Ellis Horwood.

Williams RAD, Lambert PA, Singleton P (1996) *Antimicrobial Drug Action.* Oxford: Bios.

29

Design of Antibacterial Agents

Michael H. Block[1] and Wright W. Nichols[2]

[1]AstraZeneca, Alderley Park, Macclesfield, UK
[2]AstraZeneca R&D, Boston, MA, USA

'Design'

The term 'design' provokes the idea of a conceived new object of desired function, based on an intuitive understanding of the properties of the materials from which the object is to be made. Thus one can imagine designing clothes, a building, a car, tableware. Clearly the object should perform its intended function and also, where appropriate, it should be safe. The Design Council of the United Kingdom (http://www.design-council.org.uk) would add that a well-conceived design should also be aesthetically pleasing. However in the invention of a candidate drug, its beauty lies more in its fitness for its task rather than in any appeal to other aesthetic senses. Inevitably the intuitive understanding of the properties of the chosen materials will be incomplete, so that the designed object is likely initially to be less than satisfactory; and trial-and-error changes to the materials or to other properties of the object, such as shape, size, colour, etc., will be undertaken in order to perfect it. Designing a drug is no exception to this need for trial and error. Again, the sense of 'perfecting' the object generally refers to its appeal to the human senses as well as to its function; but this is not the case for drugs.

When considering the design of antibacterial agents, one's first thoughts might be of chemists using molecular modelling to define compounds that, because of the spatial configuration of the chemical groups, will bind to a specific biochemical target and elicit the desired response. This has been strikingly exemplified in the design of HIV protease inhibitors for the treatment of AIDS (Wlodawer and Erickson, 1993; Kempf, 1994; Vacca, 1994; Romines and Chrusciel, 1995; De Lucca *et al.*, 1997). However, this represents just one aspect of the design of a chemical for medicinal use. Not only must a compound be able to exert its effect at the enzyme or receptor level, it must also have the correct macroscopic properties to reach the biochemical target in the patient at a high enough concentration, and for a long enough period of time. Furthermore, it must be free of unacceptable toxicity. Designing a molecule whose chemical and physical properties will produce the desired profile is a very significant challenge. Indeed, the published literature is probably misleading because, on the whole, failures remain unreported. In the case study described in this chapter, we hope to demonstrate that structure-based drug design can nevertheless have an impact in the field of antibacterial drug research.

There is no doubt that the goal of designing molecules that fulfil their expectations has been made more attainable by developments in structural biology and molecular modelling, which have provided more

 doi:10.1006/bkmm.2001.0029

detailed information about biochemical target sites, and tools with which to try and exploit that information. Moreover, there are working empirical models of how macroscopic properties, such as molecular mass, lipophilicity, amphiphilicity and pK_a affect pharmacokinetics. Our understanding of the behaviour of biochemical targets within tissues and of the environment in which the compound is expected to display its designed properties remains far from perfect, however, and this is one reason why there is still the unavoidable element of trial and error in the design of drugs, including antibacterial agents.

In the case of drug design, 'trial' and detection of 'error' consist of laboratory tests that allow one to infer whether a particular compound approaches the desired drug more closely than another compound does. In other words, assays that determine whether or not a compound behaves as intended remain a vital part of the drug-design process (**Fig. 1**). Such assays might be applied to every compound or might be specific and intended to test a hypothesis concerning the unexpected behaviour of a particular compound. They provide the necessary information for the next round of synthesis of a modified compound, and are of course themselves the outcome of a different design process.

It is clear, then, that the design of drugs requires several steps, each of which is essential. A crucial part of the design process, and indeed the first step, must be to choose the medical 'target drug'; which can be viewed as the design brief. An antibacterial example might be:

> an orally-active broad-spectrum bactericidal or bacteristatic compound, that does not select spontaneously resistant variants with any great frequency, that is not compromised by any known resistance mechanisms, that can be given up to four times daily, and that can be used to treat infection at any body site.

The second step is the construction of the set of biological assays, which may or may not be based on a specific biochemical target, that guide the subsequent iterative cycle of conceived compound/chemical synthesis/evaluation/next conceived compound (**Fig. 1**). We do not discuss these assays *per se*, but mention some principles in the choice of biochemical target and the importance of monitoring the distribution of compounds in animals. The third step is the design of compounds having microscopic and macroscopic properties that will result in the desired activity. As well as discussing the principles of such chemical design, we use as an example the design of inhibitors of the B subunit of bacterial DNA gyrase.

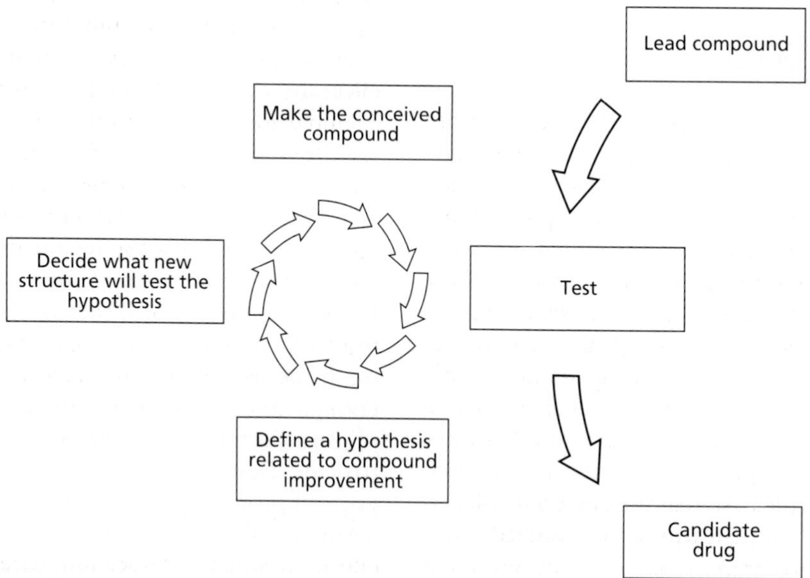

Fig. 1 The 'make-test' cycle. This is the trial-and-error process referred to in the text. The tests in this process can be 'efficacy' tests, such as enzyme-inhibition potency or antibacterial potency, or 'non-efficacy' tests, such as solubility, protein binding, pharmacokinetic profiling in animals, etc. The cycle continues until a compound produces the test results desired of the target drug. That compound is then a candidate for clinical trials. The make-test cycle is part of traditional medicinal chemistry. The 'make' step can now consist of the synthesis of chemical libraries (e.g. of several hundreds of compounds) starting with a synthetically tractable core structure derived from, or related to, the lead compound. By testing all these compounds, many potential hypotheses are tested simultaneously, so that the eventual number of make-test cycles needed to produce the candidate drug is reduced as compared with the number needed when the compounds are made singly.

One further point is worth making in this introductory section. 'Design' is often coupled with the word 'rational', implying that the alternative approach of screening for antibacterial activity among large numbers of compounds, such as natural products or combinatorial libraries, is somehow irrational. This is a misconception that we wish to allay, and so we advocate the term 'structure-based design'. Indeed, the search for antibacterial compounds among natural products and their semi-synthetic derivatives has been a decidedly rational, albeit empirical, process that has produced many compounds of great utility: one need only consider the β lactams (Rolinson, 1985), the aminoglycosides (Phillips and Shannon, 1997) and vancomycin (Felmingham, 1997) to be convinced of this. Moreover, although much of that work has not depended on biochemical-target-based design as such, it has always required the construction of biological screening assays that have identified compounds of further interest, and often the design of derivative compounds with the required macroscopic physical properties. In other words, any procedure with a good chance of discovering new antibacterial agents that are useful in human medicine is rational: whether it be based on three-dimensional structure, on random screening of chemical libraries or on the synthesis of analogues of known agents.

The 'Lead' Compound or Series of Compounds

The pharmaceutical industry, like many other specialist fields, uses concepts that can be extremely useful but that are not always defined precisely. One such is the idea of a 'lead' compound, or 'lead series' of compounds. We use this concept below, so we define here what we mean. Regardless of the initial approach to drug discovery, be it random screening or three-dimensional-structure, one can only start the cycle of chemical synthesis/biological assay/decision about the structure of the next compound/chemical synthesis (**Fig. 1**), when one has a starting compound. This starting compound is generally one with certain promising attributes giving rise to the expectation that making analogues of it will be likely to result eventually in the discovery of a related compound with the properties desired of the candidate drug. Such a starting compound is termed a *lead compound*, usually abbreviated to 'lead'. Often, analogues of this first promising compound already exist in chemical collections, and the first action on identifying the putative lead is to test these analogues in the biological assays. From this exercise one might find that certain of the chemical groups on the putative lead compound appear to be responsible for the desirable properties, and others for undesirable properties. This link between the structural features of a series of molecules and the activity they display is called the structure–activity relationship (SAR). The group of related compounds that have several desirable properties in common is termed a *lead series*.

We have purposely couched these definitions in general terms, because individuals differ on what constitutes a 'good' lead or lead series: i.e. one that is more likely than another possible lead series to generate a candidate drug, with perhaps fewer cycles of synthesis and biological testing of analogues.

Drug Profile

Two Broad Types of Antibacterial Agent

In this chapter, we restrict ourselves to compounds that act directly on the bacteria: we exclude compounds that act on the human host, such as vaccines or other immunostimulants. What, then, are the desirable properties of a medically useful antibacterial agent? Ignoring for the moment the attributes of absorption, distribution, metabolism and excretion which are determined by human biology, and focusing on the antibacterial properties *per se*, one can attempt to inhibit the growth of bacteria infecting the host in two broad ways. The first way would be to design a classical antibacterial agent that inhibits the growth of bacteria, or even kills them, wherever they are: in the patient, in the test tube or on agar. The alternative is to try to compromise the bacterium's ability to grow in the host: for example, to interfere with its ability to evade the host's immune response. This chapter considers only the design of the former type of antibacterial agent, because this is the type that has been clinically successful over many years. Designing the second type of antibacterial agent is the subject of much debate (alluded to briefly by Knowles, 1997), but so far there are no examples of the clinical development of a compound that acts as an 'anti-virulence' agent. The concept is thus, as yet, untested.

The Design Brief

Let us therefore consider the design brief as follows. We want a compound that inhibits the growth of particular bacteria, both *in vitro* and when they infect a patient. Let us also stipulate that when the compound is taken orally, it should distribute to the tissues and remain there at an effective level long enough for four-times-daily dosing to suppress the infecting bacteria. We shall assume that suppression of bacterial growth will be adequate to achieve a 'cure'. For simplicity, we are not specifying whether the compound should be 'bacteriostatic' (often abbreviated to 'static') or 'bactericidal' (often abbreviated to 'cidal'). Indeed it is not completely understood how these *in vitro*-measured parameters change the clinical outcome of patients treated with those classes of compound (Amsterdam, 1991).

We limit ourselves to biochemical-target-based discovery of antibacterial drugs. As discussed above, it is possible to screen natural products and other libraries for compounds with the antibacterial and chemical properties desired of a lead compound. It is also possible to modify that lead in order to optimise the empirically testable properties of antibacterial potency and spectrum, as witnessed by the evolution of β lactam compounds over the years (Rolinson, 1986). However, that approach exercises only limited elements of the design process and it would specifically exclude the design of molecules that inhibit the activity of a biochemical target and penetrate the bacterial cell's envelope layers to reach that target.

Choice and Validation of the Biochemical Target

'Essentiality'

The immediately apparent set of potential protein targets is those whose genes are 'essential'. That is to say, if the gene for that protein is deleted, then the bacterium cannot grow on typical susceptibility-testing media, such as Mueller–Hinton (Amsterdam, 1991). Often, this is detected as the observation that a mutation within that gene can be identified that causes a conditional phenotype; most commonly temperature-sensitivity for growth. In the case study below we discuss the essential protein target, the enzyme DNA gyrase (Luttinger, 1995).

Ideally one would like to show that the gene product is not only essential for growth *in vitro*, but that it is also essential for growth in an animal infection model. This is not often experimentally demonstrable. Techniques are, however, available that show some genes to have the different property of being essential *in vivo*, but not *in vitro* (Hensel *et al.*, 1995). General methods that demonstrate a bacterial gene product to be essential for growth in an animal, e.g. by using an experimentally controllable promoter to switch expression on and off at will in the host animal, are not available. The question of whether or not the activity of a particular protein is essential for growth in an animal therefore cannot be answered where a model inhibitory compound does not exist and where it is not known whether metabolites from the host can overcome the need for the activity of that protein. For example, the synthesis of *p*-aminobenzoate cannot be rendered inessential by acquisition of that compound from the host (Hoiseth and Stocker, 1981), whereas most amino acids can be so acquired (Bacon *et al.*, 1950), rendering the enzymes responsible for their synthesis 'inessential' for bacteria to be able to grow in a mammalian host. In other words, an enzyme on one of the amino acid biosynthetic pathways is unlikely to be chosen as a biochemical target for the design of antibacterial agents (unless it also happened to be necessary for the production of some other metabolite that could not be acquired from the host).

For some potential biochemical targets, model compounds exist that are not good lead compounds in themselves, but by their inhibition of bacterial growth through inhibiting the activity of that target, demonstrate it to be essential. Sometimes such compounds can also show that the activity of the biochemical target is necessary for bacteria to survive in animal infection models. One example is the diazaborine series of compounds that inhibit the enoyl-acyl-carrier-protein reductase of *Escherichia coli* (Baldock *et al.*, 1998).

One can also envisage compounds that exert their inhibition of growth by binding not to a bacterial protein, but to some other biological molecule. Such non-enzyme targets might be DNA or RNA: the most familiar example is vancomycin and related glycopeptides. Agents of this class are understood to bind directly to the D-Ala-D-Ala moiety of the immediate precursor of peptidoglycan transglycosylation (Williams and Waltho, 1988). In other words, the biochemical target is not an enzyme.

If a biochemical target is chosen whose essentiality is unknown, then it is generally prudent to check, by gene knockout for instance, that indeed it is essential. This is not a trivial task and is outside the scope of this chapter. Suffice it to say that there is currently much exploratory work in academia and in the pharmaceutical industry on the development of easy reliable technology to allow this to be assessed for any gene in any bacterium of possible interest.

Genomics and 'Essentiality'

The advent of the base sequences of complete bacterial genomes (e.g. Fleischmann *et al.*, 1995; Blattner *et al.*, 1997) makes it feasible to submit every gene within a bacterium whose genome sequence is known to a test of whether that gene is essential, and thus to a test of whether it should be included in the set of possible biochemical targets for antibacterial drug design

(Chu *et al.*, 1996; Knowles, 1997; Allsop, 1998; Moir *et al.*, 1999).

Selective Toxicity

In order to design compounds that inhibit bacterial growth, but do not have mechanism-dependent toxicity, the biochemical target should ideally be absent from humans. Conventionally, the absence of a biochemical target from humans has been inferred by comparing the metabolic pathway in which that biochemical target participates with the metabolic pathways found in humans. A well-studied and essential bacterial metabolic pathway that is absent from humans is that of peptidoglycan biosynthesis. Hence, vancomycin derivatives that bind to peptidoglycan precursors fit the criterion of an expectation that no toxicity should result from the mechanism of action. This is one reason why many past endeavours at drug design have concentrated on this pathway (e.g. Duncan *et al.*, 1990). However, other such pathways exist too: for example, lipopolysaccharide biosynthesis in Gram-negative bacteria (Hammond *et al.*, 1987; Onishi *et al.*, 1996), and lipoteichoic acid biosynthesis in Gram-positive bacteria (Canepari *et al.*, 1990).

Genomics and Selective Toxicity

As mentioned above, the basis of microbiological knowledge is undergoing a revolution owing to the acquisition of complete base sequences of the genomes of bacteria and fungi. Human biology is undergoing a similar revolution in its knowledge base with the acquisition of a large proportion of the base sequences of transcribed DNA. Thus, more and more, one can answer the question of whether a putative bacterial target protein is closely homologous to a protein in humans by using bioinformatics to compare the inferred amino acid sequence of that bacterial protein with the inferred sequences of all known human proteins (Moir *et al.*, 1999). Bacterial proteins that show little homology with human proteins are biochemical targets that should avoid mechanism-based toxicity if a potent inhibitor were to be designed. Bioinformatics cannot, however, be used to infer whether putative biochemical targets that are not protein, RNA or DNA (e.g. peptidoglycan precursors) are absent from humans.

The Desired Antibacterial Spectrum

A concept used in the practice of antibacterial chemotherapy is that of the 'spectrum' of the antibacterial agent. This simply means the collection of bacterial species that are susceptible to a particular compound. Many compounds, such as the 2-deoxystreptamine-containing aminoglycosides gentamicin and tobramycin, can be used to treat infections caused by a 'broad spectrum' of both Gram-positive and Gram-negative bacteria (Phillips and Shannon, 1997). Some compounds have a narrower spectrum, however, such as the oxazolidinones, which are effective only against Gram-positive bacteria (Ford *et al.*, 1997). Isoniazid is an example of a drug with an extremely narrow spectrum: its only use is in the treatment of tuberculosis (Inderlied, 1991).

Ideally, when constructing the design brief one would define the spectrum of bacteria against which the conceived drug should be active. Say one wanted a compound to be active against bacterially caused sexually transmitted disease, then this would define the desired spectrum of the eventual compound (it would have to inhibit the growth of *Neisseria gonorrhoea*, *Chlamydia trachomatis* and *Treponema pallidum*, for example). This means that the biochemical target chosen for the design of one's drug would need to be present in all these bacteria at least.

Genomics and Antibacterial Spectrum of the Eventual Compound

If the complete gene sequences were known for all bacteria that cause infectious disease, it would be possible to answer by bioinformatics the question of whether any putative protein target for antibacterial drug-discovery was present in all the bacteria of interest. Using the examples mentioned above, an inhibitor of lipopolysaccharide biosynthesis would not be expected to affect the growth of Gram-positive bacteria; and similarly, inhibition of lipoteichoic acid biosynthesis would not be expected to affect the growth of Gram-negative bacteria. On the other hand, a potent inhibitor of one of the enzymes of peptidoglycan bio-synthesis might be expected to have broad-spectrum activity against Gram-negative and Gram-positive bacteria (but not against bacteria that do not synthesise a peptidoglycan cell wall, such as mycoplasma).

Spectrum Dissociated from Taxonomic Sub-Grouping?

In principle, complete genome sequences could allow one to choose a biochemical target present in – say – all bacteria that cause urinary tract infections, or all those that cause intracellular infections, or as in the above example all those that cause sexually transmitted diseases, etc. In other words, one can imagine matching a drug-design programme with a clinical syndrome, rather than with a taxonomic sub-group of bacteria (e.g. just Gram-positive bacteria). This is not yet feasible for all syndromes at the time of writing, but might not be far off given the rate of base-sequencing and analysis of bacterial genomes (Koonin *et al.*, 1996; Bernal *et al.*, 2001).

Penetration to Periplasmic and Cytoplasmic Biochemical Target Sites

A question that arises in the choice of which biochemical target to pursue is that of its location within the bacterial cell. Some targets, such as DNA gyrase discussed in the case study below, are found in the cytoplasm. Others, such as the targets of vancomycin or the β lactams, are in the periplasm of Gram-negative bacteria and thus a compound designed to inhibit them would have to cross the outer membrane to exert its antibacterial effect. Gram-positive bacteria, on the other hand, do not possess an outer membrane, so the vancomycin and β lactam targets appear not to be behind any diffusion barrier. The fact that there are compounds that inhibit bacterial growth by binding to biochemical targets in the cytoplasm, and in the periplasm of Gram-positive bacteria demonstrates that those compartments are accessible to antibacterial agents. However, a compound that only needs to penetrate the surface peptidoglycan layer of a Gram-positive bacterium to reach its target has fewer permeability barriers to overcome than a compound that must diffuse to a target site within, say, the cytoplasm of a Gram-negative bacterium.

The 'permeability barriers' of Gram-negative bacteria are two lipid bilayer membranes and also various outflow pumps that appear to accept solutes located within the cytoplasmic membrane, at either its periplasmic or its cytoplasmic side, and transport them to the aqueous phase outside the outer membrane (Nikaido, 1998). As far as we understand, there is only one permeability barrier in Gram-positive bacteria, that of the cytoplasmic membrane. However, again, this membrane is also the site of outflow pumps that can reduce the steady-state cytoplasmic concentration of particular antibacterial agents (Paulsen *et al.*, 1996). Nikaido and Thanassi (1993) have provided an excellent analysis of the penetration of antibacterial agents to their target sites within the bacterial cell.

Design of Compounds that Must Cross the Cytoplasmic Membrane

What must be considered when attempting to design compounds that are affected minimally by permeability barriers and outflow pumps? A variety of compounds of different mechanisms of action appear to diffuse across the lipid bilayer cytoplasmic membrane at a rate fast enough to reach a near-steady-state concentration in the cytoplasm high enough to inhibit growth. Examples are chloramphenicol, tetracyclines and the fluoroquinolones. As pointed out by Nikaido and Thanassi (1993), there are features of these compounds that promote their penetration across the cytoplasmic membrane. The physical property of lipophilicity is one such feature. Another important property appears to be the presence of dissociable groups whose pK_a values are near enough to neutrality that there is enough non-electrically-charged compound for it to enter the lipid bilayer reasonably rapidly (i.e. for it to lose its water of solvation without a prohibitively large input of energy). The section on DNA gyrase below gives some examples of the design of compounds intended to cross the cytoplasmic membrane of Gram-positive bacteria to reach the biochemical target site in the cytoplasm.

Aminoglycosides appear to be unique in crossing the cytoplasmic membrane through a voltage-gated channel (Miller, 1995). The properties that a compound must have in order to access this channel are unknown. Indeed, an increase in the aminoglycoside uptake rate occurs a few minutes after the exposure of *E. coli* or *Staphylococcus aureus* to the compound, and this increase in uptake rate depends on the biochemical action of the aminoglycoside within the cell (Taber *et al.*, 1987; Miller, 1995). Thus, because of the special nature of how the aminoglycosides act, it is not practical to attempt to emulate their penetration in compounds designed to inhibit a new cytoplasmic biochemical target.

The macrolide antibiotics, e.g. erythromycin, act at bacterial ribosomes, which are in the cytoplasm. How these compounds cross the cytoplasmic membrane has not been studied very extensively, but data are consistent with its being by transmembrane diffusion, like the fluoroquinolones, rather than by a protein-determined channel or transporter, like the aminoglycosides (Bauernfeind and Georgopapadakou, 1994). Reviews on the mechanism of action of macrolides tend not to address how the compounds enter the bacterial cytoplasm (e.g. Mazzei *et al.*, 1993); it is an important question, however, because they are large, water-soluble molecules that would not intuitively be expected to cross lipid bilayer membranes particularly rapidly.

Design of Compounds that Must Cross the Gram-negative Bacterial Outer Membrane

The one feature that is needed in a compound to ensure that it enters the periplasm at a reasonable rate is a relative molecular mass (M_r) lower than about 500; larger compounds do enter, though more slowly, and the transmembrane diffusion rate can be influenced by the carriage of electric charge (Nikaido and Vaara, 1985). For example, vancomycin is a narrow-spectrum compound that is active only against Gram-positive bacteria because it is too large to diffuse through the water-filled channels that cross the outer membranes of Gram-negative bacteria.

Another preferred design feature, if the biochemical target is in the periplasm, is that the compound should be reasonably soluble in water (i.e. that it should be hydrophilic). Examples are the β lactams (Nikaido and Vaara, 1985). Hydrophobic compounds also cross the outer membrane; for example, rifampicin is active at low external concentrations against many Gram-negative bacteria. Nevertheless, some hydrophobic compounds, particularly ones that are amphiphilic in having hydrophobic regions and hydrophilic regions, do not appear to accumulate to very high steady-state levels in Gram-negative bacteria in any of the compartments that are internal to the outer membrane. The reason for this is the presence of outflow pumps of very broad specificity which can be very active in clinical isolates of *Pseudomonas aeruginosa*, for instance (Nikaido, 1998).

Biochemical Targets for which One Would Not Need to Design Properties that Promote Penetration in a Candidate Drug

To re-state: clinically useful compounds do exist that act within the bacterial cytoplasm, e.g. aminoglycosides, fluoroquinolones and macrolides. Nevertheless, when choosing between potential biochemical targets against which to mount an antibacterial drug design programme one would, other things being equal, choose an externally facing biochemical target or one located in the periplasm, to minimise the design complexity.

Genomics and the Location of the Biochemical Target

A full gene and protein index to bacteria of interest, that described whether each mature protein was located externally to the permeability barriers of the cytoplasmic membrane and/or the outer membrane, would allow one to choose only externally facing protein targets, or periplasmic ones. As was the case for the bioinformatic analysis relevant to 'spectrum', enough information is not yet available to do this at the time of writing (mid-1998). It is however still some way off.

Pharmacokinetics and Bioavailability

Solubility, Distribution and Free Concentrations of Compounds

When attempting to design a compound that binds to and inhibits the function of a bacterial biochemical target, two traditional aspects of medicinal chemistry must be borne in mind. The first is that the compound must be sufficiently soluble in aqueous systems that it can be suitably formulated for parenteral or oral administration, whichever is desired. A related consideration is that the compound must not be so hydrophobic that its free concentration in a human is vanishingly low because of binding to extracellular proteins and fatty tissues. This is an important counterbalance to one aspect of designing a compound that enters a Gram-positive bacterium very well. Hydrophobic (lipophilic) compounds can generally be very potent inhibitors of the growth of Gram-positive bacteria *in vitro*. The more hydrophobic a compound is, however, the less likely it is either to remain in solution in the human or to stay at a very high concentration because of binding to serum proteins and fatty regions in the tissues. This proved to be a problem in the design of compounds active against bacterial DNA gyrase (see below).

Oral Bioavailability

When the aim is to design a drug that will eventually be administered orally, a second traditional medicinal chemistry feature must be considered to ensure that a substantial fraction of the compound is absorbed into the circulation. It is not possible to predict how much of, and how rapidly, a compound of a particular structure will be absorbed when given orally. However, some trends can be described, based on physical properties of the compound, such as M_r, lipophilicity and its number of hydrogen-bond acceptors and donors (Lipinski *et al.*, 1997). These must thus be borne in mind when designing orally active drugs, regardless of the therapeutic area for which they are intended. We cannot review the topic fully here, but for this and other fundamental aspects of medicinal chemistry the reader is referred to the excellent introductory text by Taylor and Kennewell (1993).

Chemical Design Approaches

Natural Products and Chemical Libraries

As mentioned earlier, the field of antibacterial agents has traditionally been dominated by natural products and their semisynthetic and synthetic analogues. Examples are the β lactams, macrolides, tetracyclines, glycopeptides and aminoglycosides. Less frequently, classes of compounds have been derived from synthetic chemicals that were prepared initially for an entirely different purpose; the quinolones would probably fall into this category. An immensely rich diversity of chemical structure exists among natural products as a result of evolution. One might thus argue that natural products are still the obvious place to look for lead compounds. On the other hand, developments in chemical technology now allow the rapid preparation of huge 'libraries' of synthetic compounds that offer the possibility of identifying a starting point for a drug discovery program by random screening. This approach will undoubtedly play a major role in the antibacterial field in the future.

Structure-based Design

Developments in computational chemistry and structural biology over the last three decades have provided an alternative approach to identifying and optimising novel pharmacologically active compounds, namely 'structure-based design'. In principle the availability of three-dimensional structural information on target enzymes or receptors offers the possibility of designing new chemicals that will interact with that target and produce the desired effect. X-ray crystallography and, more recently, nuclear magnetic resonance (NMR) techniques have improved enormously and there has been an explosion in the number of biological macromolecules whose three-dimensional structures are known. With these structures chemists use computational tools to help design small molecules intended to bind to specific targets. In doing this, they attempt to derive binding energy through specific non-covalent interactions with the target, for example through ionic interactions or hydrogen bonds. A key component in the binding of small molecules with proteins and receptors is often the hydrophobic interactions formed between lipophilic groups on the ligand and hydrophobic pockets in the target. Binding energy can also be gained 'entropically' by displacing fixed water molecules from the target, thus allowing them to regain random motion.

Having conceived and synthesised a compound with some activity, albeit perhaps weak, one would hope to be able to optimise that activity. This could be by an empirical approach, but would preferably be based on further structural information derived from the biochemical target with bound inhibitor. If the latter can be achieved, then an iterative structure/design/synthesis cycle may be possible. This structure-based design approach seems obvious and highly attractive, yet in practice there have been few successes in the design of hitherto unprecedented compounds in any area of pharmaceutical chemistry. Two examples can be cited, however. The first is a fine piece of work undertaken by the Agouron group to design and optimise thymidylate synthase inhibitors using protein crystal structures and computational molecular modelling (Appelt *et al.*, 1991). The second example is in the area of HIV protease inhibitors, where the research carried out by a number of groups arguably represents a *tour de force* of the techniques currently available and stands as the finest example of what can be achieved in structure-based drug design (Wlodawer and Erickson, 1993; Kempf, 1994; Vacca, 1994; Romines and Chrusciel, 1995; De Lucca *et al.*, 1997).

To reiterate: successes in structure-based design have been limited, particularly so in the field of antibacterial research. Protein crystal structures of dihydrofolate reductases have been used post-rationally to explain the selectivity of the antibacterial agent trimethoprin for the bacterial enzyme, as opposed to that of the antileukaemia agent methotrexate for the mammalian enzyme. Structural information from these comparisons has also led to the design of more-potent inhibitors of bacterial dihydrofolate reductases, but not to more-potent antibacterial agents (Kuyper *et al.*, 1985). Herein lies the crux of the problem for antibacterial research. Even when structure-based design can aid the identification of enzyme inhibitors of increased potency, it still cannot help in the challenge of designing compounds that attain high steady-state concentrations inside bacterial cells.

In recent years structural information has become available for a whole range of potential antibacterial target sites. These include DNA gyrase, the essential bacterial enzyme involved in supercoiling DNA (Wigley *et al.*, 1991; Lewis *et al.*, 1994, 1996; Tsai *et al.*, 1997); D-alanine:D-alanine ligase, an essential enzyme required for the synthesis of peptidoglycan (Fan *et al.*, 1994); enoyl-acyl-carrier-protein reductase, a key enzyme in the biosynthesis of fatty acids and the target of the diazaborines and triclosan (Baldock *et al.*, 1996, 1998; Levy *et al.*, 1999); and InhA, the corresponding enzyme from *Mycobacterium tuberculosis* (Dessen *et al.*, 1995; Rozwarski *et al.*, 1998). High-resolution structural information now exists for much of a bacterial ribosome (Cech, 2000). Given the continuing developments in the area of structure-based design it would be surprising if successful applications of these approaches to the field of antibacterial research were not reported in the near future.

Case Study: The Design of Antibacterial Agents that Inhibit DNA Gyrase at the B Subunit

DNA gyrase is the bacterial enzyme responsible for the process of negatively supercoiling DNA (for a critical review of gyrase structure and function see Reece and Maxwell, 1991). It is an essential enzyme (as defined above under 'essentiality') that is present in both Gram-negative and Gram-positive bacteria but that has no direct counterpart in mammals. It consequently represents an attractive biochemical target for the design of antibacterial agents. The work that has been done in trying to design such compounds illustrates some of the potential of molecular design techniques in the field of antibacterial agents, along with some of the pitfalls.

The enzyme has two discrete subunits (A and B; also called proteins GyrA and GyrB) and it functions as an A_2B_2 tetramer in which the A subunits are responsible for the binding, breaking and re-joining of the DNA, and the B subunits couple the process to the hydrolysis of ATP.

Antibacterial Agents that Inhibit DNA Gyrase

Two classes of clinically effective antibiotics act at DNA gyrase. These are the quinolones and the coumarins, typified respectively by ciprofloxacin **1** and the natural product novobiocin **2** (Reece and Maxwell, 1991). A third class of compounds currently under investigation includes the natural product cyclothialidine **3** (Nakada *et al.*, 1993, 1994; Goetschi *et al.*, 1994; Kamiyama *et al.*, 1994; Watanabe *et al.*, 1994), the closely-related GR 122222X **4** (Oram *et al.*, 1996), and a variety of synthetic analogues (Arisawa *et al.*, 1992; Geiwiz *et al.*, 1995).

The quinolones (e.g. **1**) bind to the A subunits forming a stable ternary complex between DNA, enzyme and drug, which is responsible for the

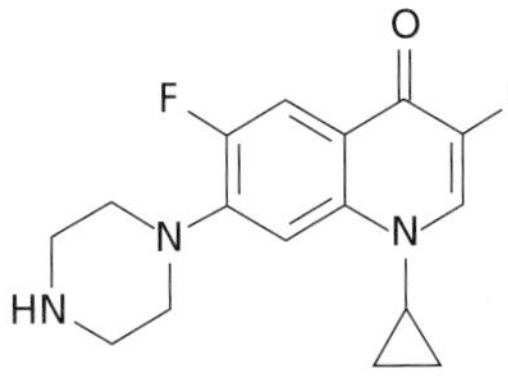

F
CO$_2$H
O
N
N
HN

1
Ciprofloxacin

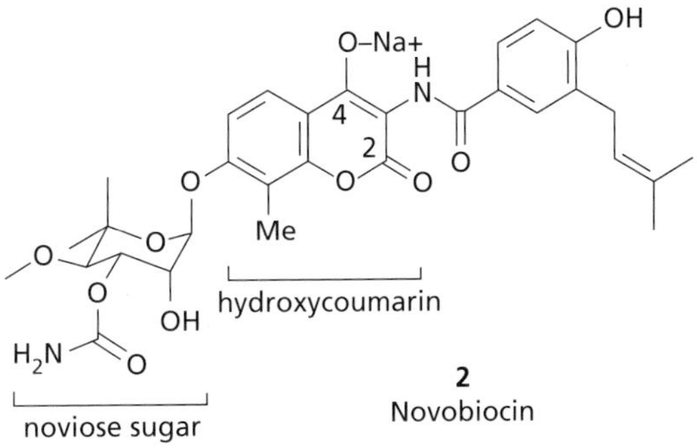

O–Na+
H
N
OH
4
2
O
O
O
O
Me
O
O
O
OH
H$_2$N
O
hydroxycoumarin
noviose sugar

2
Novobiocin

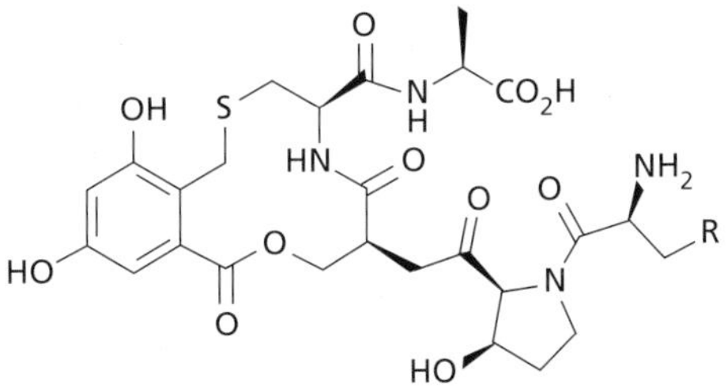

O
N
H
CO$_2$H
OH
S
HN
O
NH$_2$
O
O
R
HO
O
N
O
HO

3 R=OH; Cyclothialidine
4 R=H; GR122222X

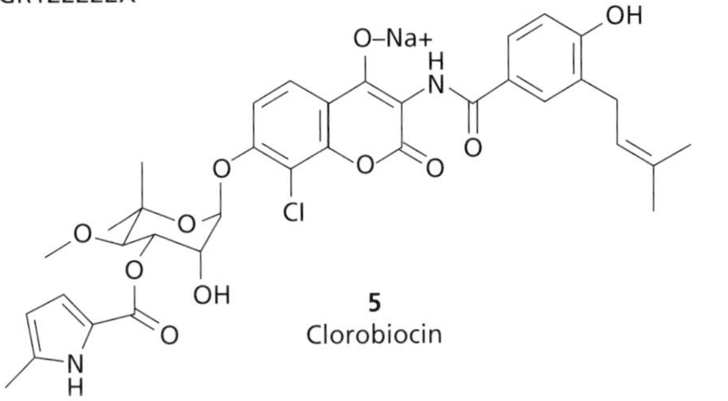

O–Na+
H
N
OH
O
O
O
O
Cl
O
O
O
OH
O
N
H

5
Clorobiocin

subsequent cell death. These compounds have found widespread use against infections caused by both Gram-negative and Gram-positive bacteria. They are not without limitations, however; they are contraindicated for paediatric use because of impairment of cartilage development in young animals although use in children is increasing (Alghasham and Nahata, 2000). They are also known in some cases to potentiate the side effects of non-steroidal anti-inflammatory agents on the central nervous system, and they have been known to be phototoxic. The greatest shortcoming of these agents in clinical use is the development of resistance (Domagala, 1994).

The coumarins, such as novobiocin **2**, bind to the B subunit of DNA gyrase causing inhibition of the ATPase activity. Novobiocin **2** itself has found limited clinical use, mostly against infections caused by Gram-positive bacteria, but suffers from problems of toxicity and rapidly developing resistance (O'Grady, 1997). The clinical development of coumarins of higher potency, such as coumermycin A1 and clorobiocin **5**, has been limited owing to their physical properties, such as poor solubility (Berger *et al.*, 1965; Kawaguchi *et al.*, 1965; Ninet *et al.*, 1972). Cyclothialidine **3** and GR122222X **4** also inhibit DNA gyrase by competitive binding to the B subunit.

Design Based on Previously Known Compounds

Considerable effort has already been aimed at developing an improved antibacterial agent by modifying the coumarin natural products (for example: Keil *et al.*, 1968; Berger and Batcho, 1978; Ryan, 1979; Ueda *et al.*, 1994). The chemistry was complicated, and in trying to establish SAR there was always a concern that weak levels of activity might simply be due to contamination with small amounts of the highly potent parent compound. The structural information available for novobiocin **2** (Boles and Taylor, 1975) and more recently for a complex between novobiocin and a 24-kDa protein fragment derived from the B subunit of *E. coli* DNA gyrase (Lewis *et al.*, 1994, 1996) has, however, provided a basis for designing novel, wholly synthetic compounds that mimic novobiocin.

In one study (Bell *et al.*, 1997) the crystal structure of novobiocin itself (**Fig. 2**) was the starting point. A two-dimensional ^{1}H NMR ('proton NMR') experiment confirmed that the key conformational features seen in the solid state were also evident in solution. This information was used in conjunction with some understanding of the SAR within the coumarins to design compounds that mimicked features of both the

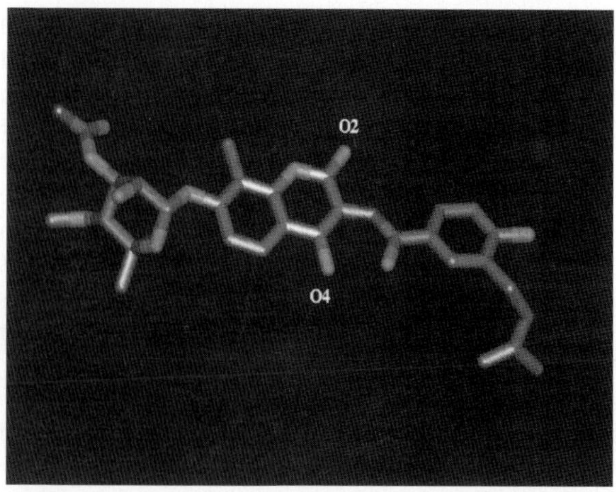

Fig. 2 The crystal structure of novobiocin. The colour coding is as follows: green, carbon; blue, nitrogen; red, oxygen. Hydrogen atoms are not shown explicitly: they are to be inferred. The positions 02 and 04 refer to the ring numbering as shown on the hydroxycoumarin ring of structure 2 (See also Colour Plate 7).

aromatic portion and the sugar which were viewed as essential for activity. In particular, in several bacteria, resistance to novobiocin **2** can result from a variety of point mutations that alter a key arginine residue (equivalent to Arg136 of the *E. coli* enzyme) (Holmes and Dyall-Smith, 1991; Contreras and Maxwell, 1992; Stieger *et al.*, 1996). Therefore, the interaction between novobiocin **2** and Arg136 is important, and it was assumed that the 4-oxy group on the coumarin (as its anion) was the most likely group responsible for this interaction (shown as the sodium salt in the structure of **2** above). Compounds such as **6** were designed to contain a cyclohexyl mimic of the noviose sugar that would carry similar functionality (for example the pendant hydroxyl and carbamoyl groups) to the natural product and place an aromatic ring in the same vicinity as the hydroxycoumarin, and this ring would bear a hydrogen bond acceptor (the oxygens of the carboxylate group) in a similar position to the 4-oxy group of the coumarin (**Fig. 2**). We hoped that by making the structure rigid in an appropriate conformation, the binding would be enhanced by reducing the 'entropic' energy required to constrain a flexible structure.

Further Structural Information

During the course of this research, further highly relevant structural information became available, namely, the X-ray crystal structure of a complex between novobiocin **2** and a 24-kDa protein fragment derived from *E. coli* GyrB (**Fig. 3**) (Lewis *et al.*, 1994, 1996). The conformation of novobiocin **2** in the

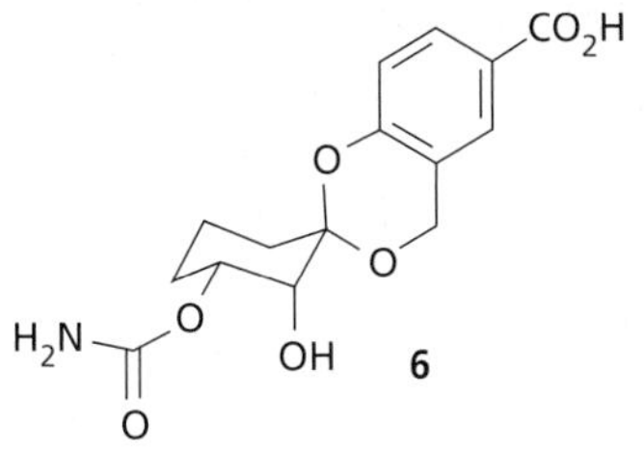

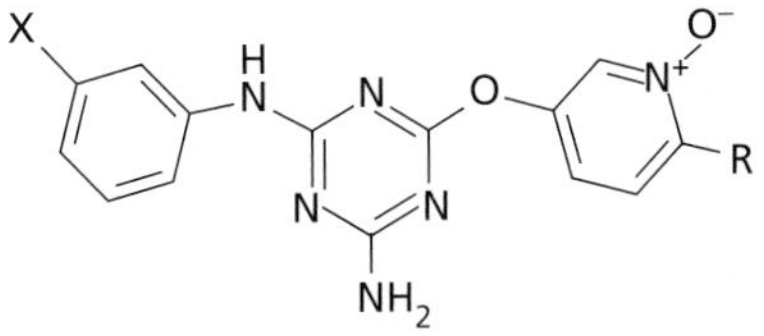

11 X = 3-F, R = H
12 X = 3-Cl, R = Me
13 X = 3,5-di-Cl, R = Me

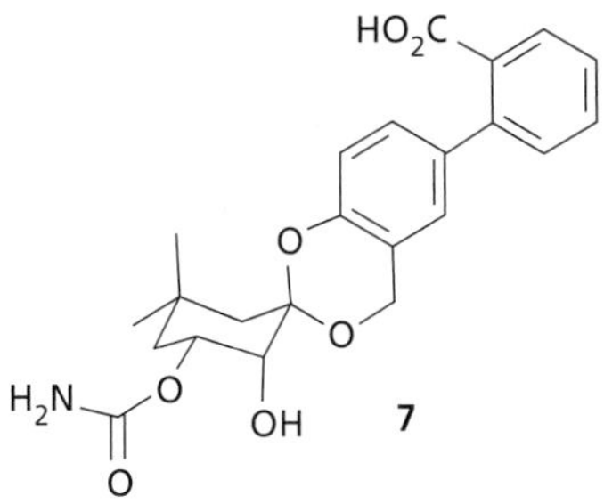

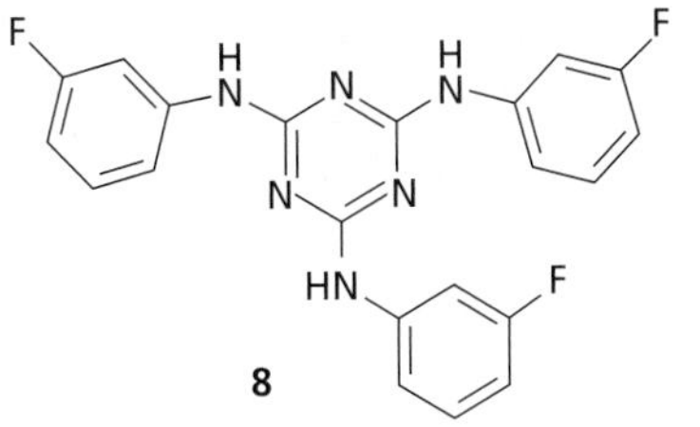

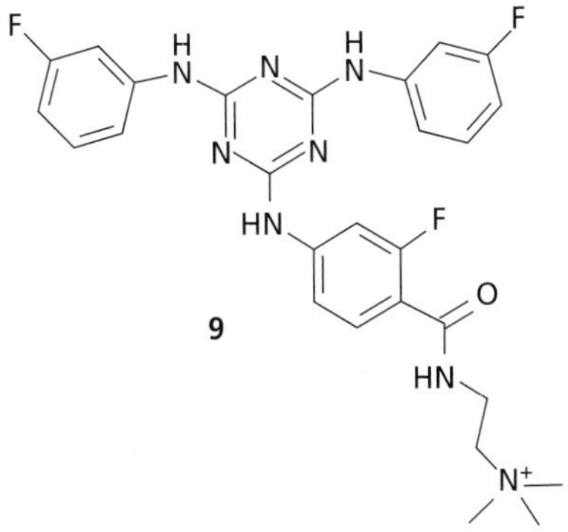

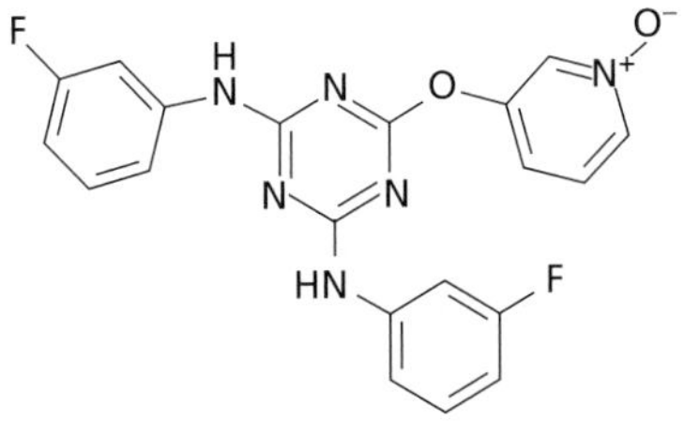

protein–ligand complex was also solved in solution by NMR, and this agreed with that observed in the solid state (Tsai *et al.*, 1997). Not only did the crystal structure provide information on the conformation of novobiocin **2** in a 'biologically relevant' environment, but it also clarified the nature of the interactions of novobiocin **2** with those parts of the B subunit contained within the protein fragment.

On the whole, comparison of the two structures shows the conformations to be quite similar (**Figs 2** and **3**), but the terminal phenolic group of the bound novobiocin appears to be bent back on itself and there is a slightly increased twist of the coumarin ring relative to the sugar. The protein crystal structure revealed, however, that it is the lactone 2-carbonyl (as its oxyanion, which can be pictured by re-arrangement of the 2–4 double-bond system of **2**), rather than the 4-oxy group, that binds to Arg136. This information implied that the carboxy group of **6** would not be in the correct position to pick up the key interactions with Arg136. This led to the design of structures such as 7 in which the natural twist imparted by the biphenyl group put the terminal aromatic ring back into co-planarity with the original coumarin and placed the carboxyl group in a very favourable position to pick up the hydrogen bonding with Arg136 (**Fig. 4**). Compound 7 and a variety of analogues were prepared, and their (non-protein-bound) X-ray crystal structures gave experimental support to the modelling predictions and structural design ideas. None of the compounds showed any significant inhibition of the enzyme, however, nor of bacterial growth.

Although several possible explanations were proposed for the above findings, it is still very disappointing that compounds which were believed to have the correct conformation and to carry much of the chemical structure thought to be important for binding, and which modelled so well into the protein, showed no activity. This remains a salutary lesson that structure-based drug design is not easy.

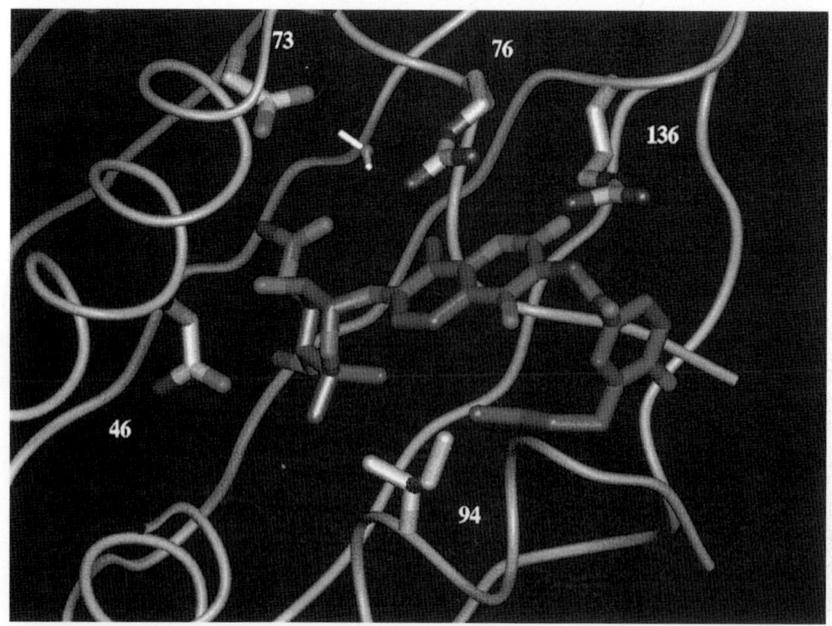

Fig. 3 The crystal structure of the complex of novobiocin and the 24-kDa fragment of the B-subunit of DNA gyrase from *E. coli*. The colour coding of novobiocin is the same as in Fig. 2. Again hydrogen atoms are not shown on that molecule. The pink line represents the shape of the peptide backbone. The numbers refer to amino acid positions in the primary sequence of intact GyrB. The side-chains of the numbered amino acids are shown (Asn46, Asp73, Arg76, Ile94 and Arg136), with colour coding as for novobiocin, except that carbon atoms are pale green. Between Asp73 and Arg76 there is a bound water molecule, whose hydrogen atoms are shown in white. (See also Colour Plate 8).

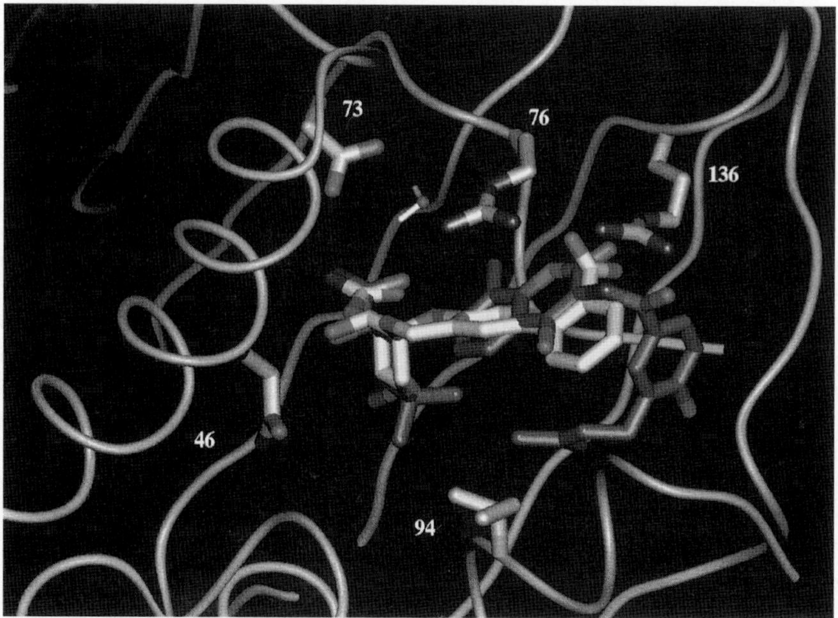

Fig. 4 The crystal structure of the complex of compound **7** and the 24-kDa fragment of the B subunit of DNA gyrase from *E. coli*, compared with novobiocin when it occupies the same site. The conventions are as in Fig. 3, except that the carbon atoms of compound **7** are coloured yellow (See also Colour Plate 9).

A New Lead Series of Compounds

At about the same time as the work described above, a series of *tris*-anilino-triazines, including **8**, were identified by reporter-gene-based random screening as potent inhibitors of DNA gyrase (IC_{50} of **8** = 0.75 μmol/L) that bound to the B subunit. The compounds also inhibited the growth of Gram-positive

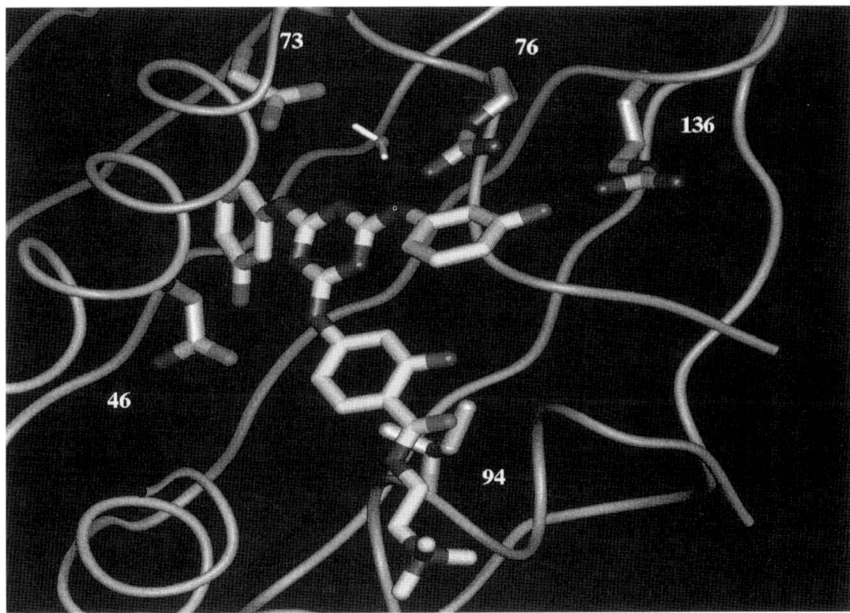

Fig. 5 The crystal structure of the complex of the soluble triazine **9** and the 24-kDa fragment of the B subunit of DNA gyrase from *E. coli*. The conventions are as in Fig. 3, except that the carbon atoms of compound **9** are coloured yellow and the fluorine atoms green (See also Colour Plate 10).

bacteria (the MIC of **8** was 1 mg/L against *Staph. aureus* NCTC6571, the 'Oxford' strain), but there was little activity against Gram-negative bacteria. The compounds were generally poorly soluble, but the synthesis of the deliberately soluble analogue **9**, and its co-crystallisation with the 24-kDa fragment of GyrB resulted in the generation of the crystal structure of the complex, thus providing specific information on the mode of binding (Fig. 5).

There were, however, concerns that the antibacterial activity of compounds like **8** and **9** might not be primarily due to inhibition of DNA gyrase but could be a consequence of other, possibly non-specific, effects. This fear was confirmed by the demonstration that these compounds inhibited oxygen uptake by *Staph. aureus* NCTC6571; which implied that their inhibition of bacterial growth could potentially be by non-DNA-gyrase-mediated mechanisms.

The combination of a novel series of enzyme inhibitors and structural information on their mode of binding undoubtedly represented a reasonable starting point for a medicinal chemistry programme. Three key questions needed to be addressed in the early phase of the project, however:

1. How could the confusing (and potentially toxic) inhibition of oxygen uptake be eliminated?
2. How could DNA-gyrase-mediated inhibition of bacterial growth be achieved?
3. How could we demonstrate that the inhibition of growth was a consequence of the desired mechanism?

The approach that was followed illustrates several elements of the design process (see Table 1 for properties of the compounds concerned).

Design of Novel Compounds Based on the New Lead Series

We speculated that the inhibition of oxygen uptake by the *tris*-anilinotriazines such as **8** and **9** might be potentiated by their high lipophilicity, and thus that hydrophilic compounds might not show this effect. Furthermore, it was felt that lowering lipophilicity in conjunction with enhanced enzyme activity should result in more rapid penetration of drug to the cytoplasm (rather than just dissolution of the compound in the hydrophobic region of the cytoplasmic membrane). This would be expected to lead to more potent inhibition of DNA gyrase and hence greater specificity in the inhibition of bacterial growth. We planned to improve the enzyme inhibitory activity through forming non-covalent bonds with Arg136 (numbered according to the *E. coli* enzyme). This in turn would allow us to identify DNA-gyrase-mediated inhibition of bacterial growth by observing the difference in MIC between *Staph. aureus* strain 601122, which is resistant to novobiocin owing to a mutation at that

Table 1 The first series of 'designed' compounds; and their properties in tests used to evaluate them as antibacterial agents acting by inhibition of the B subunit of DNA gyrase

Compound	log P^a	$C\log P^a$	IC_{50} for inhibition of DNA gyrase (μmol/L)	MIC (mg/L)[b]		IC_{50} for inhibition of oxygen uptake (mg/L)[c]
				novo S	novo R	
Novobiocin	nd	4.1	0.23	0.06	8	> 261
8	nd	7.6	0.75	1	1	1.7
9	nd	–	2.4	8	8	10
10	nd	4.2	0.4	> 128	> 128	nd
11	nd	1.2	16	> 128	> 128	nd
12	1.9	2.3	1.5	16	> 128	> 160
13	nd	3.0	0.1	4	16	> 261

Novobiocin is included for comparison. See the text for the design rationale.
nd, not determined; –, data not available

[a] log *P* is the octanol/water partition co-efficient; it is a guide to the ability of the compound to partition into membranes. To a very crude approximation, compounds that are suitable for susceptibility-testing and for dosing to humans usually possess a log *P* in the region of 1–3. There are exceptions, however. *C* log *P* is simply a calculated estimate of the log *P*.

[b] Values of MIC (minimum inhibitory concentration) are given for the inhibition of the growth of two strains of *Staph. aureus*: a novobiocin-susceptible strain (strain NCTC6571, the 'Oxford' strain) and a novobiocin-resistant strain derived from it (designated Zeneca strain 601122).

[c] Oxygen uptake by *Staph. aureus* NCTC6571 was measured polarographically at 37°C in liquid Mueller–Hinton medium.

arginine residue (Arg143 in *Staph. aureus*), and the wild-type parent strain, *Staph. aureus* NCTC6571. Consequently compounds **10–12,** a series of substituted triazines, were prepared that incorporated an appropriately-positioned hydrogen bond acceptor group that would non-covalently bond with Arg136, while lowering lipophilicity and also maintaining the other known non-covalent bonds with GyrB.

Compounds such as **10** were designed on the basis of the crystal structure in conjunction with the knowledge that pyridine *N*-oxide is an excellent hydrogen-bond acceptor. In particular the structural information implied that one, and only one, of the aniline NH groups could be replaced by oxygen and that this linker would point towards the Arg136-binding site. Results for **10** were encouraging in that they showed increased potency in inhibiting the enzyme, despite a substantially lower lipophilicity, implying that a new interaction had occurred. The results were also consistent with the idea that the 'non-specific' antibacterial activity could be reduced by lowering the lipophilicity, because **10** no longer inhibited bacterial growth at 128 mg/L (Table 1). However, it was disappointing that inhibition of DNA gyrase at sub-micromolar concentrations did not translate to antibacterial activity.

It was suspected that compounds such as **10** were still too lipophilic. One of the aniline groups was therefore replaced by an amino group which gave compounds like **11** with even lower lipophilicities, and reasonable potencies in inhibiting the enzyme. For synthetic purposes a methyl group was incorporated into the pyridine ring and on the basis of SAR from *bis*-anilino amino triazines (structures not shown), the *m*-fluoro group was changed to *m*-chloro. These two changes led to compound **12**, which inhibited DNA gyrase activity with comparable potency (50% inhibition at 1.5 μmol/L) to compound **8**. However, compound **12** possessed a log *P* of 1.9, indicating a lipophilicity some 100 000 times lower than that of compound **8** (*C*log *P* 7.6) (see Table 1 for definitions). Furthermore, **12** showed moderate inhibition of the growth of *Staph. aureus* that depended on whether the strain was mutated at Arg143 to novobiocin-resistance (for strain NCTC6571, MIC = 16 mg/L; and for strain 122, MIC > 128 mg/L). In common with novobiocin **1** and other antibacterial compounds in clinical use, **12** also did not inhibit oxygen uptake by *Staph. aureus* at 10 times the growth-inhibiting concentration, which was in direct contrast to compounds such as **8** and **9**. The addition of a second *m*-chloro group to the aniline gave **13**, which displayed a further increased potency in both the inhibition of DNA gyrase (50% inhibition at 0.1 μmol/L) and the inhibition of bacterial growth (MIC against *Staph. aureus* NCTC6571 = 4 mg/L).

When *Staph. aureus* NCTC6571 was cultured in the presence of an inhibitory concentration of **12**, it was possible to isolate resistant mutants. We cloned and sequenced the *gyrB* genes from two of these, designated strains R1 and R2. Using numbering based on the *E. coli* amino acid sequence, isoleucine 94 had changed to valine in strain R1 and to threonine in strain R2. The generation of these mutants confirmed that the antibacterial mode of action of **12** was indeed the inhibition of DNA gyrase at the B subunit.

The outcome of this work demonstrates that three-dimensional structural information can have an impact in the field of antibacterial research. Progress was made from an initial enzyme inhibitor that possessed the complicating property of possibly inhibiting bacterial growth by another mechanism, to a compound whose mechanism of antibacterial action was inhibition of the target enzyme. Compound **13** was only poorly soluble, however. A 'nano-suspension' of it injected subcutaneously into mice at 80 mg/kg failed to protect them against an intraperitoneal challenge with *Staph. aureus.* Moreover, when dosed subcutaneously at 10 mg/kg, it could not be detected in the blood at a level of 50 ng/mL or greater. This poor solubility meant that it could not be considered as a good lead for further conventional medicinal chemistry (i.e. to tailor the properties of absorption, distribution, metabolism and excretion).

An Alternative Design Strategy

An obvious approach to making use of the available structural information was to design compounds which combined fragments that were thought to be important for the binding of different inhibitors. Inspection of the structures of novobiocin and the triazine compound **9** bound to the 24-kDa GyrB fragment (Figs 3 and 5) made it clear that these two inhibitors had significant areas of overlap. We thus hypothesised that a combination of fragments of the two compounds might be highly active. Compound **14** was designed on this premise and was expected to be a very potent gyrase inhibitor. Disappointingly, it turned out to be only weakly active as a gyrase inhibitor and did not inhibit bacterial growth at 128 mg/L (Table 2). The reasons for the poor activity were not clear from molecular modelling, but the degree of enzyme inhibition by compounds **15** and **16** (isomers of **13**) led us to suggest that a methyl group *ortho* to the oxygen linkage detracted from the tightness of binding to the enzyme, hence reducing potency. In other words, we proposed that the methyl group on the coumarin ring of **14** was responsible for its poor activity. In consequence we designed and synthesised

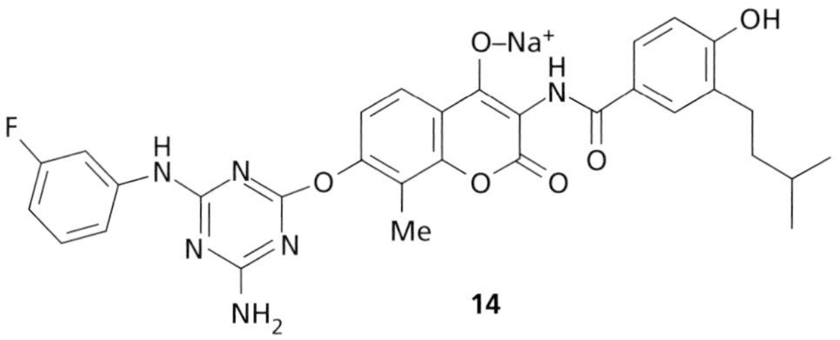

14

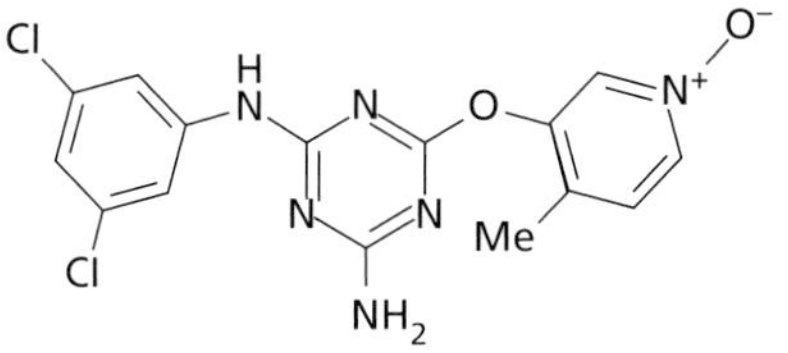

15

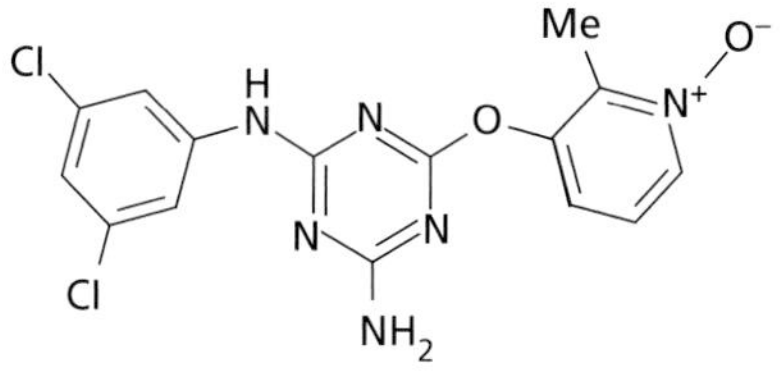

16

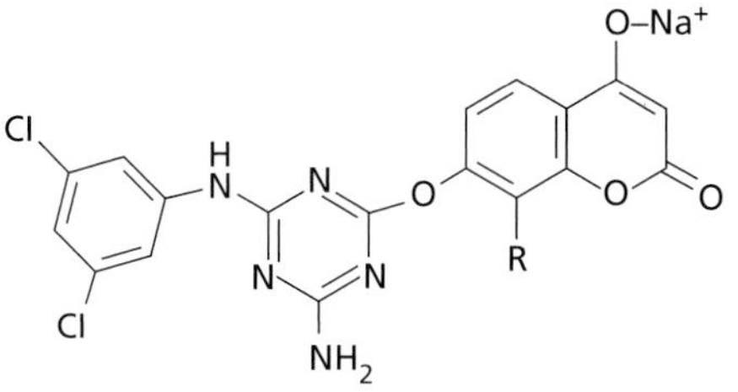

17 R = H
18 R = Me

Table 2 The second series of 'designed' compounds, and their properties in tests used to evaluate them as antibacterial agents that act by inhibition of the B-subunit of DNA gyrase. See text for the design rationale

Compound	log P[a]	C log P[a]	IC_{50} for inhibition of DNA gyrase (μmol/L)	MIC (mg/L)[b] novo S	MIC (mg/L)[b] novo R	IC_{50} for inhibition of oxygen uptake (mg/L)[c]
14	nd	8.0[d]	170	> 128	> 128	nd
15	nd	3.0	> 525	> 128	> 128	nd
16	nd	3.0	6	> 128	> 128	nd
17	nd	5.8[d]	0.07	16	> 128	213
18	nd	6.3[d]	16	> 128	> 128	nd

nd, not determined.

[a] For an explanation of log P and C log P, see the footnote to Table 1.

[b] Data are given for novobiocin-susceptible and -resistant strains of *Staph. aureus* (see foonote to Table 1).

[c] Oxygen uptake by *Staph. aureus* NCTC6571 was measured polarographically (see the foonote to Table 1).

[d] These figures are very high and may be unrealistic. In any event because hydroxycoumarins are fairly acidic (pK_a novobiocin ~4.3) the partition co-efficient for these compounds including the ionised species (termed 'log D') will be up to 3 log units lower.

compound **17**, which combined features of the anilino triazine class and the coumarins, but lacked the methyl group on the coumarin. The design process in this case did prove to be successful. When this work was done (1996), **17** was one of the most potent DNA gyrase inhibitors that had been reported (50% inhibition at 0.07 μmol/L in our standard assay), although this potency was not reflected in the antibacterial activity (MIC against *Staph. aureus* NCTC6571 = 16 mg/L). We do not know the reason for the weak antibacterial activity of such a potent DNA gyrase enzyme inhibitor: we can only speculate that at external concentrations below 16 mg/L the compound does not penetrate into the cytoplasm of *Staph. aureus* to reach a high enough steady-state concentration at the DNA gyrase target to inhibit growth. If this explanation is correct, one can assume that the steady-state cytoplasmic concentration will be a balance between inward transport and diffusion and outward pumping processes. It is not clear, however, which of these is likely to be more important in limiting access of compound **17** to its target. With reference to the biochemical target, we tested the above '*ortho*-methyl' hypothesis by preparing compound **18**. It was about 100 times less potent in inhibiting DNA gyrase than was **17**, which was consistent with the hypothesis being correct.

Lessons from Designing Inhibitors of DNA Gyrase as Antibacterial Agents

In summary, the work on gyrase B demonstrates that structure-based design can play a major role in generating novel and potent enzyme inhibitors and antibacterial agents. The results from this type of approach are not entirely predictable, however, and encouraging modelling results are by no means a guarantee of success.

Acknowledgements

Many of the ideas in this chapter were discussed with others, and indeed conceived by others, at ZENECA Pharmaceuticals (now AstraZeneca). We thus owe a debt to many people, including Keith Barrett-Bee, Peter Barth, Mike Betts, Mark Egerton, Neil Hales, Ruth Illingworth, Richard Pauptit, Phil Poyser, Dave Roberts, Dave Timms, Alex Tucker, Simon Weston, Geoff Wilson, and Alan Wookey.

References

Alghasham AA, Nahata MC (2000) Clinical use of fluoroquinolones in children. *Ann. Pharmacother.* 34: 347–359.

Allsop AE (1998) Bacterial genome sequencing and drug discovery. *Curr. Opin. Biotechnol.* 9: 637–642.

Amsterdam D (1991) Susceptibility testing of antimicrobials in liquid media. In: Lorian V (ed.) *Antibiotics in Laboratory Medicine*, 3rd edn. Baltimore: Williams & Wilkins, pp. 53–105.

Appelt K, Bacquet RJ, Bartlett CA *et al.* (1991) Design of enzyme inhibitors using iterative protein crystallographic analysis. *J. Med. Chem.* 34: 1925–1934.

Arisawa M, Goetshi E, Kamiyama T *et al.* (1992) Novel DNA gyrase inhibitors, process for their manufacture and pharmaceutical preparations containing same. World Patent WO 9218490 A1.

Bacon GA, Burrows TW, Yates M (1950) The effects of biochemical mutation on the virulence of *Bacterium typhosum*: the virulence of mutants. *Br. J. Exp. Pathol.* 31: 714–724.

Baldock C, Rafferty JB, Sedelnikova SE *et al.* (1996) A mechanism of drug action revealed by structural studies of enoyl reductase. *Science* 274: 2107–2110.

Baldock C, de Boer G-J, Rafferty JB, Stuitje AR, Rice DW (1998) The mechanism of action of the diazaborines. *Biochem. Pharmacol.* 55: 1541–1549.

Bauernfeind A, Georgopapadakou NH (1995) Clinical significance of antibacterial transport. In: Georgopapadakou NH (ed.) *Drug Transport in Antimicrobial and Anticancer Chemotherapy.* New York: Marcel Dekker, pp. 1–19.

Bell W, Block MH, Cook C, Grant JA, Timms D (1997) Design, synthesis and evaluation of a novel series of spiroketals based on the antibacterial gyrase inhibitor novobiocin. *J. Chem. Soc. Perkin* 1: 2789–2801.

Berger J, Schochter AJ, Batcho AD *et al.* (1965) Production and isolation and synthesis of the coumermycins (sugordomycins), a new Streptomyces antibiotic complex. *Antimicrob. Agents Chemother.* 5: 778–785.

Berger J, Batcho AD (1978) Coumerin-glycoside antibiotics. *J. Chromatogr. Library* 15: 101–156.

Bernal MO, Ear U, Kirpides N (2001) Genomes on-line database (GOLD): a monitor of genome projects worldwide. *Nucl. Acids Res.* 29: 126–127.

Blattner FR, Plunkett G, Bloch CA *et al.* (1997) The complete genome sequence of *Escherichia coli* K-12. *Science* 277: 1453–1462.

Boles MO, Taylor DJ (1975) The crystal structure of novobiocin. *Acta Crystallogr.* B31: 1400–1406.

Canepari P, Boaretti M, Del Mar Lleo M, Satta G (1990) Lipoteichoic acid as a new target for activity of antibiotics: mode of action of daptomycin, LY-146032. *Antimicrob. Agents Chemother.* 34: 1220–1226.

Cech TR (2000) The ribosome is a ribozyme. *Science* 289: 878–879.

Chu DTW, Plattner JJ, Katz L (1996) New directions in antibacterial research. *J. Med. Chem.* 39: 3853–3874.

Contreras A, Maxwell A (1992) gyrB mutations which confer coumarin resistance also affect DNA supercoiling and ATP hydrolysis by *Escherichia coli* DNA gyrase. *Mol. Microbiol.* 6: 1617–1624.

De Lucca GV, Erickson-Viitanen S, Lam PYS (1997) Cyclic HIV protease inhibitors capable of displacing the active site structural water molecule. *Drug Discov. Today* 2: 6–18.

Dessen A, Quemard A, Blanchard JS, Jacobs WR, Sacchettini JC (1995) Crystal structure and function of the isoniazid target of *Mycobacterium tuberculosis. Science* 267: 1638–1641.

Domagala JM (1994) Structure–activity and structure–side-effect relationships for the quinolone antibacterials. *J. Antimicrob. Chemother.* 33: 685–706.

Duncan K, van Heijenoort J, Walsh CT (1990) Purification and characterisation of the D-alanyl-D-alanine adding enzyme from *Escherichia coli. Biochemistry* 29: 2379–2386.

Fan C, Moews PC, Walsh CT, Knox JR (1994) Vancomycin resistance: structure of D-alanine:D-alanine ligase at 2.3 Å resolution. *Science* 266: 439–443.

Felmingham D (1997) Glycopeptides. In: O'Grady F, Lambert HP, Finch RG, Greenwood D (eds) *Antibiotics and Chemotherapy*, 7th edn. Edinburgh: Churchill Livingstone, pp. 363–368.

Fleischmann RD, Adams MD, White O *et al.* (1995) Whole-genome random sequencing and assembly of *Haemophilus influenzae* Rd. *Science* 269: 496–512.

Ford CW, Hamel JC, Stapert D *et al.* (1997) Oxazolidinones: new antibacterial agents. *Trends Microbiol.* 5: 196–200.

Geiwiz J, Goetschi E, Hebeisen P, Link H, Luebbers T (1995) Mono- or bicyclic DNA gyrase inhibitors. European patent EP 0675122 A2.

Goetschi E, Angehrn P, Gmünder H *et al.* (1994) Cyclothialidine and its congeners: a new class of DNA gyrase inhibitor. *Pharmacol. Therapeut.* 60: 367–380.

Hammond SM, Claesson A, Jansson AM *et al.* (1987) A new class of synthetic antibacterials acting on lipopolysaccharide biosynthesis. *Nature* 327: 730–732.

Hensel M, Shea JE, Gleeson C, Jones MD, Dalton E, Holden DW (1995) Simultaneous identification of bacterial virulence genes by negative selection. *Science* 269: 400–403.

Hoiseth SK, Stocker BAD (1981) Aromatic-dependent *Salmonella typhimurium* are non-virulent and effective as live vaccines. *Nature (London)* 291: 238–239.

Holmes ML, Dyall-Smith ML (1991) Mutations in DNA gyrase result in Novobiocin resistance in halophilic archaebacteria. *J. Bacteriol.* 173: 642–648.

Inderlied CB (1991) Antimycobacterial agents: *in vitro* susceptibility testing, spectrums of activity, mechanisms of action and resistance, and assays for activity in biological fluids. In: Lorian V (ed.) *Antibiotics in Laboratory Medicine*, 3rd edn. Baltimore: Williams & Wilkins, pp. 134–197.

Kamiyama T, Shimma N, Ohtsuka T *et al.* (1994) Cyclothialidine, a novel DNA gyrase inhibitor 2. Isolation, characterisation and structure elucidation. *J. Antibiotics* 47: 37–45.

Kater MM, Koningstein GM, Nijkamp HJJ, Stuitje AR (1994) The use of a hybrid genetic system to study the functional relationship between prokaryotic and plant multi-enzyme fatty acid synthetase complexes. *Plant Mol. Biol.* 25: 771–790.

Kawaguchi H, Tsukiura H, Okanishi M *et al.* (1965) Coumermycin, a new antibiotic. I. Production, isolation and characterisation of coumermycin A1. *J. Antibiot. (Tokyo)* 18: 1–10.

Keil JG, Hooper IR, Cron MJ *et al.* (1968) Semisynthetic coumermycins. *J. Antibiot. (Tokyo)* 21: 551–566.

Kempf DL (1994) Design of symmetry based, peptidomimetic inhibitors of human immunodeficiency virus protease. *Meth. Enzymol.* 241: 334–354.

Knowles DJC (1997) New strategies for antibacterial drug design. *Trends Microbiol.* 5: 379–383.

Koonin EV, Mushegian AR, Rudd KE (1996) Sequencing and analysis of bacterial genomes. *Curr. Biol.* 6: 404–416.

Kuyper LF, Roth B, Baccanari DP *et al.* (1985) Receptor based design of dihydrofolate reductase inhibitors: comparison of crystallographically determined enzyme binding with enzyme affinity in a series of carboxy-substituted trimethoprim derivatives. *J. Med. Chem.* 28: 303–311.

Levy CE, Roujeinikova A, Sedelnikova S *et al.* (1999) Molecular basis of triclosan activity. *Nature.* 398: 383–384.

Lewis RJ, Singh OMP, Smith CV *et al.* (1994) Crystallisation of inhibitor complexes of an N-terminal 24-kDa fragment of the DNA gyrase B protein. *J. Mol. Biol.* 241: 128–130.

Lewis RJ, Singh OMP, Smith CV *et al.* (1996) The nature of inhibition of DNA gyrase by the coumarins and the cyclothialidines revealed by X-ray crystallography. *EMBO J.* 15: 1412–1420.

Lipinski CA, Lombardo F, Dominy BW, Feeney PJ (1997) Experimental and computational approaches to estimate solubility and permeability in drug discovery and development settings. *Adv. Drug Deliv. Rev.* 23: 3–25.

Luttinger A (1995) The twisted 'life' of DNA in the cell: bacterial topoisomerases. *Mol. Microbiol.* 15: 601–606.

Mazzei T, Mini E, Novelli A, Periti P (1993) Chemistry and mode of action of macrolides. *J. Antimicrob. Chemother.* 31, Suppl C, 1–9.

Miller MH (1995) Aminoglycoside transport. In: Georgopapadakou NH (ed.) *Drug Transport in Antimicrobial and Anticancer Chemotherapy.* New York: Marcel Dekker, pp. 175–219.

Moir DT, Shaw KJ, Hare RS, Vovis GF (1999) Genomics and antimicrobial drug discovery. *Antimicrob. Agents Chemother.* 43: 439–446.

Nakada N, Shimada H, Hirata T *et al.* (1993) Biological characterisation of cyclothialidine, a new DNA gyrase inhibitor. *Antimicrob. Agents Chemother.* 37: 2656–2661.

Nakada N, Gmunder H, Hirata T, Arisawa M (1994) Mechanism of inhibition of DNA gyrase by cyclothialidine, a novel DNA gyrase inhibitor. *Antimicrob. Agents Chemother.* 38: 1966–1973.

Nikaido H (1998) Antibiotic resistance caused by Gram-negative multidrug efflux pumps. *Clin. Infect. Dise.* 27 (Supplement 1), S32–S41.

Nikaido H, Thanassi DG (1993) Penetration of lipophilic agents with multiple protonation sites into bacterial cells: tetracyclines and fluoroquinolones as examples. *Antimicrob. Agents Chemother.* 37: 1393–1399.

Nikaido H, Vaara M (1985) Molecular basis of bacterial outer membrane permeability. *Microbiol. Rev.* 49: 1–32.

Ninet L, Benazet F, Chapentie Y *et al.* (1972) Clorobiocin (18.631R.P.) a new chlorinated antibiotic produced by several streptomyces species. *C. R. Acad. Sci. Paris Ser. C* 275: 455.

O'Grady F (1997) Coumarins. In: O'Grady F, Lambert HP, Finch RG, Greenwood D (eds) *Antibiotic and Chemotherapy,* 7th edn. Edinburgh: Churchill Livingstone, pp. 333–335.

Onishi HR, Pelak BA, Gerckens LS *et al.* (1996) Antibacterial agents that inhibit lipid A biosynthesis. *Science* 274: 980–982.

Oram M, Dosanjh B, Gormley NA *et al.* (1996) Mode of action of GR 122222X, a novel inhibitor of bacterial DNA gyrase. *Antimicrob. Agents Chemother.* 40: 473–476.

Paulsen IT, Brown MH, Skurray RA (1996) Proton-dependent multidrug efflux systems. *Microbiol. Rev.* 60: 575–608.

Phillips I, Shannon KP (1997) Aminoglycosides and aminocyclitols. In: O'Grady F, Lambert HP, Finch RG, Greenwood D (eds) *Antibiotics and Chemotherapy,* 7th edn. Edinburgh: Churchill Livingstone, pp. 164–201.

Reece RJ, Maxwell A (1991) DNA gyrase: structure and function. *Crit. Rev. Biochem. Mol. Biol.* 26: 335–375.

Rolinson GN (1986) β-Lactam antibiotics. *J. Antimicrob. Chemother.* 17: 5–36.

Romines KR, Chrusciel RA (1995) 4-Hydroxypyrones and related templates as nonpeptidic protease inhibitors. *Curr. Med. Chem.* 2: 825–838.

Rozwarski DA, Grant GA, Barton DHR, Jacobs WR, Sacchettini JC (1998) Modification of the NADH of the isoniazid target (InhA) from *Mycobacterium tuberculosis. Science* 279: 98–102.

Ryan MJ (1979) Novobiocin and coumermycin A1. In: Hahn FE (ed.) *Antibiotics,* Vol. V. Berlin: Springer Verlag, pp. 214–234.

Stieger M, Angehrn P, Wohlgensinger B, Gmünder H (1996) GyrB mutations in *Staphylococcus aureus* strains resistant to cyclothialidine, coumermycin and novobiocin. *Antimicrob. Agents Chemother.* 40: 1060–1062.

Taber HW, Mueller JP, Miller PF, Arrow AS (1987) Bacterial uptake of aminoglycoside antibiotics. *Microbiol. Rev.* 51: 439–457.

Taylor JB, Kennewell PD (1993) *Modern Medicinal Chemistry.* Chichester: Ellis Horwood.

Tsai FTF, Singh OMP, Skarzynski T *et al.* (1997) The high resolution crystal structure of a 24-kDa gyrase B fragment from *Escherichia coli* complexed with one of the most potent coumarin inhibitors, clorobiocin. *Proteins Struct. Funct. Genet.* 28: 41–52.

Ueda Y, Chuang JM, Fung-Tome J, Partyka RA (1994) Synthesis and *in-vitro* antibacterial activity of new semi-synthetic noviosyl coumarin antibiotics: chemical modification at the C-3 ester. *Biorg. Med. Chem. Lett.* 4: 1623–1628.

Vacca JP (1994) Design of tight binding human immunodeficiency virus type I protease inhibitors. *Meth. Enzymol.* 241: 311–334.

Watanabe J, Nakada N, Sawaira S *et al.* (1994) Cyclothialidine, a novel DNA gyrase inhibitor 1. Screening, taxonomy, fermentation and biological activity. *J. Antibiotics (Tokyo)* 47: 32–36.

Wigley DB, Davies GJ, Dodson EJ, Maxwell A, Dodson G (1991) Crystal structure of the N terminal fragment of the DNA gyrase B protein. *Nature* 351: 624–629.

Williams DH, Waltho JP (1988) Molecular basis of the activity of antibiotics of the vancomycin group. *Biochem. Pharmacol.* 17: 133–141.

Wlodawer A, Erickson JW (1993) Structure-based inhibitors of HIV-1 protease. *Ann. Rev. Biochem.* 62: 543–585.

PART 5

MECHANISMS OF PATHOGENESIS AND PREVENTION

6

30

Adhesion and Colonisation

S.N. Abraham[1], N. Sharon[2] and I. Ofek[3]

[1]Duke University Medical Center, Durham, North Carolina, USA
[2]Weizmann Institute of Science, Rehovot, Israel
[3]Tel-Aviv University, Israel

An overwhelming number of infectious diseases are initiated by bacterial colonisation of the mucosal surfaces of the genitourinary, gastrointestinal or respiratory tracts. Since mucosal colonisation by bacteria is preceded by bacterial binding to epithelial cells, or to mucins that coat mucosal cells, intensive studies have focused on the cellular and molecular aspects that underlie the adhesion of bacteria. These have revealed that many bacteria express adhesins on their surfaces, frequently in the form of specialised organelles, that seek and bind to cognate receptors on the surface of mucosal cells. The specific binding interaction between the bacterial adhesins and host receptors allows the bacteria to attach firmly to particular sites on the mucosal surfaces and thereby resist dislocation by the hydrokinetic forces that typically act on these surfaces. Furthermore, such bacteria have better access to nutrients and resist damaging effects of antimicrobial agents in the surrounding milieu (Zafriri *et al.*, 1987).

Although adhesion of bacteria to mucosal surfaces is an important determinant of mucosal colonisation, especially in determining its site and density, it is becoming increasingly clear that this is only part of the story. Several critical post-adhesion events are necessary for the bacteria to establish themselves successfully on the mucosal surfaces and to initiate infection. These events, triggered by the adhesion of the bacteria to their complementary receptor on the mucosal cells, include up-regulation of the expression of virulence factors by the bacteria and induction of physiological changes on the mucosal surface. The latter include proliferation of epithelial cells, increased mucus secretion, endocytosis of adherent bacteria and release of pro- and anti-inflammatory mediators by mucosal and sub-mucosal cells. This chapter deals with the adherence of uropathogenic *Escherichia coli* to the urinary tract. Selected post-adhesion events, and how these influence bacterial infection, are discussed. Finally, we consider how this information provides a basis for the development of anti-adhesion strategies that can block and possibly even reverse bacterial colonisation of mucosal surfaces and the harmful effects that follow adhesion. Whenever possible, we focus on the adhesion of uropathogenic *E. coli* in order to capture the emerging concepts of the role of adhesion in the infectious process.

Adhesins and Receptors

The adhesins and their cognate receptors of many bacterial pathogens have been identified (Karlsson *et al.*, 1994; Ofek and Doyle, 1994). There are three main types of adhesin-receptor interactions (Table 1). One type shared by the majority of bacterial pathogens involves lectin-carbohydrate recognition. Many pathogens express surface lectins that serve as adhesins

 doi:10.1006/bkmm.2001.0030

Table 1 Molecular interactions with host cells in bacterial adhesion

Bacterial adhesin	Host cell receptor	Example		Reference
		Adhesin	Receptor	
Lectin–carbohydrate interactions				
Lectin	Glycoprotein or glycolipid	P fimbriae	Globotetraosyl on uroepithelial cells	Karlsson (1995)
Glycoconjugate	Lectin	Type 1 fimbriae	Uroplakin on bladder cells	Wu *et al.* (1996)
		Klebsiella capsule	Mannose receptor on macrophage	Ofek (1995)
Pasteurella LPS			Undefined on porcine tracheal cell	
Protein–protein interactions				
Fibronectin protein	Fibronectin-binding	*Strep. pyogenes* F protein	Fibronectin on respiratory cells	Hanski *et al.* (1996)
Hydrophobin–protein interactions				
Glycolipid	Lipid receptors?	Lipoteichoic acid	Lipid-binding region of fibronectin on epithelial cells	Hasty and Sokurenko (2000)
Lipid-binding protein	Membrane lipids	*Campylobacter* surface protein	Phospholipids and sphingolipids of animal cells	Szymanski and Armstrong (1996) Sylvester *et al.* (1996)

that bind the bacteria to carbohydrate moieties of cell surface glycoconjugates, e.g. glycoproteins or glycolipids (Table 2). In some cases lectins on animal cells (e.g. macrophages) bind bacteria by complementary polysaccharides of either the capsule or the outer-membrane lipopolysaccharides (Ofek, 1995). The second type involves protein–protein interaction on the surfaces of the host cells and the bacteria. The third type involves hydrophobic interactions between a hydrophobin on the bacterial surface and a complementary hydrophobic moiety of protein on the animal cell (Ofek and Doyle, 1994).

Table 2 Some bacterial surface lectins and their cognate receptors on animal cells

Organism	Target tissue	Receptor specificity	Form of receptor
A. naeslundi	Oral cavity	GalNAc(β1–31–3)Gal	Glycoprotein
E. coli			
Fimbrial adhesins			
Type 1	Urinary tract	Man(α1–3)[Man(α1–6)] Man	Glycoprotein
P	Urinary tract	Gal(α1–4)Gal	Glycolipid
F1C	Urinary tract	GalNAc(β1–4)Gal	Glycolipid[a]
S	Neural	NeuAc(α2–3)Gal(β1–3)GalNAc	Glycolipid
CFA/1	Intestinal	NeuAc(α2–8)	Glycoprotein
K99	Intestinal	NeuGc(α2–3)Gal(β1–4)Glc	Glycolipid
H. pylori	Stomach	NeuAc(α2–3)Gal(β1–4)Glc(NAc)	Glycolipid
		Fuc(α12)Gal(β1–3)[Fuca(1–4)]Gal	Glycoprotein
K. pneumonia	Respiratory	GlcNAcβ3Gal	Glycoprotein
N. gonorrhoea	Genital	Gal(β1–4)Glc	Glycolipid
		NeuAc(α2–3)Gal(β1–4)Glc(NAc)	Glycoprotein
S. typhimurium	Intestinal	OligoMan	Glycoprotein

Based on Ofek and Doyle (1994); Karlsson (1995); Sharon and Lis (1996).
[a] Khan *et al.* (2000).

Multiple Adhesins

Adhesins are presented on bacterial surface as a complex molecular structure (Ofek and Doyle, 1994). The most common structure is that of the fimbrial (pili) adhesins (Sauer *et al.*, 2000). The fimbriae (or pili) are protein appendages of varying lengths and diameters that consist of several hundreds of major subunits tipped or interspersed with minor subunits, one of which carries the adhesive function. In at least in one case, however, two minor proteins associated with same type of fimbriae function as adhesins with distinct receptor specificity. Thus, for example, two subunits on *E. coli* P fimbriae have been implicated in adhesion. The first is PapG, which mediates the characteristic Galα1–4Gal binding properties of P fimbriae; the other is PapE, which is responsible for promoting bacterial binding to fibronectin (Westerlund *et al.*, 1989, 1991; Tennent *et al.*, 1990). Interestingly, the PapG and PapE subunits are both located at the distal ends of the peritrichous fimbrial filaments and make up the unique fibrillum tip structure.

Adhesins as Lectins: Inter- and Intra-species Variation in Fine Sugar Specificity

Many bacteria produce lectins that bind to carbohydrate moieties of glycoproteins and glycolipids on the surface of the host cells (Table 2) (Ofek and Doyle, 1994). The best-studied lectin-producing bacteria are those that colonise animal tissues, but environmental organisms also produce surface lectins. The role of surface lectins in mediating adhesion of bacteria to animal tissue is well established, but under some circumstances an intracellular lectin may also participate in the adhesion process. For example, periplasmic lectins released from *Pseudomonas aeruginosa* during lysis can form bridges between the surviving *Pseudomonas* and target tissue (Garber *et al.*, 1992). Virtually all animal tissue colonisers produce one or more lectins/adhesins, but bacteria that colonise inanimate and animate substrata in the environment also use lectins to bind. For example, *Vibrio furnissi* binds to a glycocalyx-containing inanimate substratum (Yu *et al.*, 1987), and *Vibrio shiloi* colonises coral by binding via a galactose-specific lectin (Rosenberg *et al.*, 1999).

Bacterial lectins are generally classified according to the structure of the carbohydrates they recognise, and their specificity is usually defined as the simplest carbohydrate structure – usually a monosaccharide – that best inhibits the lectin-mediated adhesion. This is referred to as the *primary sugar specificity* of the lectin (Table 2). Differences in the binding of different oligosaccharides are often observed among lectins that posses the same primary sugar specificity. This is referred to as the *fine sugar specificity*. The determination of the fine sugar specificity of a bacterial lectin is usually based on two different types of assay. In one assay, the concentration of various derivatives of the primary sugar inhibitor that gives 50% inhibition of adhesion (IC_{50}) is established for the various isolates. In the other assay, the magnitude of adhesion to various glycoproteins that contain the primary sugar in different contexts is measured. In this chapter we focus on a particular class of bacterial lectins in order to give a better illustration of the potential biological effects of the diversification of the fine sugar specificity found in a given family of lectins. This class of lectins is expressed on the bacterial surface in the form of fimbriae, and they mediate the adhesion of many species of enterobacteria to animal tissue, including the urinary tract.

Enterobacterial fimbriae share a number of common features (Ofek and Doyle, 1994). They are all encoded by complex operons and have genes for regulatory proteins, a periplasmic chaperone, an outer-membrane assembly platform and major as well as minor structural subunits. In most cases, the lectin subunits are minor components of the fimbrial superstructure. These lectin subunits are complexed with other minor and/or major subunits, which comprise the fimbrial shaft and are situated at the fimbrial tip to be optimally positioned to mediate interaction with host cells. The lectin subunit was thought to contain two separate domains, one for saccharide binding in the N-terminal region and one for polymerisation with shaft subunits in the C-terminal region (Hultgren *et al.*, 1989; Klemm and Krogfeld, 1994), and recent crystallographic and site mutation studies have confirmed this prediction (Choudhury *et al.*, 1999; Schembri *et al.*, 2000).

The primary sugar specificities of a number of enterobacterial fimbrial adhesins have been characterised (review by Ofek and Doyle, 1994). Structure-function studies of the fine sugar specificity of enterobacterial lectins have been carried out in detail for two types of fimbriae, type 1 fimbriae of several enterobacteria and P fimbriae of *E. coli* (Table 3). Only type 1 fimbrial lectins are considered here. Type 1 fimbriae share a common primary sugar specificity, mannose, but vary in their fine sugar specificities. There is diversity between different species (inter-species) as well as between different isolates of the same species (intra-species) (Table 3). Two molecular mechanisms have been suggested to account for this diversity. One of these involves the association between the FimH adhesin and one or more subunits of the fimbrial shaft (i.e. FimA, FimF and/or FimG). This mechanism

Table 3 Types of diversity in the fine sugar specificity of bacterial lectins that serve as adhesins

Species or strain	Primary sugar specificity	Fine sugar specificity	Relative adhesion to cells
Inter-species diversity due to the fimbrial shaft presenting the lectin			
E. coli (T1F)	Man	pNPαMan ≫ Man	Uro ≫ Entero
K. pneumoniae (T1F)	Man	pNPαMan > Man	Uro > Entero
S. typhimurium (T1F)	Man	Man > pNPα Man	Entero > Uro
E. coli ($Fim^{E/S}$/$Fshaft^{S}$)	Man	Man > pNPα Man	Entero > Uro
Intra-species diversity due to variations in the gene that encodes the lectin			
E. coli			
UPEC	Man	$Man_1 = Man_3$	Buccal ≫ Uro
Faecal	Man	$Man_3 \gg Man_1$	Buccal = Uro[b]
UPEC (PF-G1)	Galα4Gal	GbO3	
UPEC (PF-G2)	Galα4Gal	GbO3	
UPEC (PF-G3)	Galα4Gal	GbO3	

Buccal, epithelial cells scraped from the buccal mucosa of healthy volunteers; Entero, mouse enterocyte cell line SI-H10, from ATCC; $Fim^{E/S}$/$Fshaft^{S}$, a hybrid FimH lectin composed of the N-terminal region of *E. coli* FimH and the C-terminal region of *Salmonella* Fim H expressed in the host *E. coli* strain on shaft fimbriae of *S. typhimurium* (adapted from Thankavel *et al*., 1999; Abraham *et al*. unpublished); G1, G2, G3, PapG adhesin subclasses; Gal, galactose; GalNAc, N-acetylgalactosamine; GbO3, Galα4Galβ4GlcβCer; GbO4, GalNAcα2Galα4Galβ4GlcβCer; GbO5, GalNAcα2GalNacα2Galα4Galβ4GlcβCer (from Stromberg *et al*., 1991); Glc, glucose. Man, mannose; Man_3, Manα3Manα6Man; PF, P fimbriae; pNPα-Man, paranitrophenylα mannoside; T1F, type 1 fimbriae; Uro, Uroepithelial mouse cell lines MM45T.BL, from ATCC (for inter-species differences) and J82 human cell line (for intra-species differences); Uro[b], J82 human uroepithelial cell line, from ATCC (adapted from Sokurenko *et al*., 1997); UPEC, uropathogenic *E. coli*.
Data adapted from Ofek and Sharon (1990); Hasty and Sokurenko (2000).

accounts for the inter-species diversities observed (Madison *et al*., 1994; Thankavel *et al*., 1999). Another mechanism involves allelic variations of the *fimH* gene of *E. coli* which accounts for intra-species diversity in binding among type 1 fimbriate *E. coli* (Sokurenko *et al*., 1997, 1999).

Influence of Fimbrial Shaft on Fine Sugar Specificity

The unique way of presenting adhesin molecules as multi-subunit complexes in association with other proteins may affect the specificity of the adhesins and the tissue tropism of the adhesin-bearing bacteria. For example, inter-species differences in the fine specificity of FimH, the mannose-binding subunit of type 1 fimbriae, are influenced by the fimbrial shaft (Madison *et al*., 1994).

If different aromatic mannosides are used to inhibit binding interactions mediated type 1 by fimbriae, distinct differences in the IC_{50} are observed between *E. coli*, *Klebsiella pneumoniae* and *Salmonella typhimurium* (Firon *et al*., 1983; Ofek, 1995). Evidence that the fimbrial subunits of the shaft that presents the FimH adhesin at the fimbrial tip (i.e. FimA, FimF and/or FimG) influence the fine sugar specificity of this lectin was first described for the type 1 fimbriae of *K. pneumoniae* and *E. coli* (Madison *et al*., 1994). When the genes that encode the fimbrial shaft and the fimbrial adhesin were exchanged between these species, the fimbrial shaft influenced the fine sugar specificity of fimbriae from each species. For example, the relative inhibitory activity of aromatic mannosides was 3–4-fold greater than that of aliphatic mannosides (e.g. Methyl αMan) for a recombinant strain expressing *E. coli* shaft and *K. pneumoniae* FimH than it was for a recombinant strain expressing *K. pneumoniae* shaft and FimH. Conversely, the relative inhibitory activity of a recombinant strain expressing the *K. pneumoniae* shaft and *E. coli* FimH was 5–6 times less than that for a recombinant strain expressing *E. coli* shaft and FimH. This is consistent with the high degree of identity in the C-terminal regions of *E. coli* and *K. pneumoniae* FimH that associate with the shaft (GenBank accession number L19338). In contrast to the high degree of homology between *E. coli* FimH and *K. pneumoniae* FimH, there is limited sequence homology between these proteins and the FimH of *S. typhimurium* (Thankavel *et al*., 1999). To determine whether the marked difference in the fine sugar specificity between *S. typhimurium* FimH ($FimH^{S}$) and *E. coli* FimH ($FimH^{E}$) in affinity for aromatic mannosides (Table 3) is due to the primary structure of FimH or to the FimH-presenting shaft, a chimeric FimH was constructed (Thankavel *et al*., 1999). This fusion protein consisted of the N-terminus of *E. coli* FimH ($FimH^{E}$), which contains the carbohydrate-binding domain, and the C-terminus of *S. typhimurium* FimH ($FimH^S$), which contains the region that associates with the fimbrial shaft. When this $FimH^{E/S}$ fusion was expressed in a recombinant strain and presented on the *S. typhimurium* type 1 fimbrial shaft, the recombinant strain bound aromatic mannosides with an affinity similar to that of $FimH^s$ (Abraham *et al.*, unpublished). These data taken together support the concept that the affinity of FimH of enterobacteria for aromatic mannosides is significantly affected by the fimbrial shaft proteins.

Influence of Allelic Variation of Fimbrial Adhesin on Fine Sugar Specificity

Determination of the sequencing of a large number of the FimH lectins derived from different *E. coli* strains revealed that there are naturally occurring phenotypic variants of the protein (Sokurenko *et al.*, 1994, 1995). These alleles are 99% conserved among the *E. coli* strains and vary by as little as one amino acid substitution in the N-terminal region of the FimH molecule in most of the strains studied. Since, as discussed above, this region contains the mannose-binding site, it was reasoned that at least some of these alleles may exhibit differences in their fine sugar specificity. In fact, all *E. coli fimH* alleles so far studied encode subunits that mediate high levels of binding to trimannose structures ($Man_3 = Man\alpha 3Man\alpha 6Man$), but binding to mannose (Man_1) residues among the FimH alleles can differ up to 15-fold (Table 3; Sokurenko *et al.*, 1997).

Since manno-oligosaccharides are naturally presented as glycoproteins on animal cell surfaces, a battery of glycoproteins was tested for their ability to serve as effective ligands for type 1 fimbriae-mediated adhesion (Sokurenko *et al.*, 1997). *E. coli* that carry the Fim H alleles classified as Man_1 binders exhibited a distinctly different pattern of adhesion from the panel of glycoproteins than did bacteria carrying the Man_3 *fimH* allele. For example, Man_1-binding allelic variants bound well to mannan and RNAse B, whereas Man_3-binding allelic variants bound well only to RNAse B. The binding of each of these allelic variants to the various ligands was inhibited by the simple sugar mannose, suggesting that the differences in adhesion patterns reflected differences in the fine sugar specificity.

Bacterial Glycoconjugates as Adhesins

Mammalian macrophages express lectins that recognise complementary carbohydrate structures on bacterial surfaces and mediate non-opsonic phagocytosis of bacteria. Although phagocytosis, termed lectinophagocytosis, of a number of bacterial species involves macrophage lectins, the identity of the surface glycoconjugates that mediate binding to the macrophage lectin has been defined for only a few bacteria. The mannose receptor of macrophages recognises *K. pneumoniae* capsules that contain Manα2/3Man or Rhaα2/3Rha sequences and the arabinomanan on the surface of *Mycobacterium tuberculosis* (Athamna *et al.*, 1991; Schlesinger *et al.*, 1994). A comprehensive reviews on the role of the macrophage lectin-bacterial polysaccharide interaction in the infectious process has been published by Ofek (1995). It has been suggested that the lipo-oligosaccharide/lipopolysaccharide (LOS/LPS) of the outer membrane of Gram-negative bacteria mediates adhesion to non-professional phagocytes, including mucosal cells, and to mucus constituents (Nassif and So, 1995; Jacques, 1996). The evidence for this effect is not conclusive but is based on the following observations: (1) binding to epithelial cells of mutants that lack the O-side chain of LPS is lower than that of the parent strains and isolated LPS acts as inhibitor of the binding; (2) isolated *Vibrio mimicus* LPS haemagglutinates of rabbit erythrocytes (Alam *et al.*, 1996); (3) the heptose-3-deoxy-D-manno -2-octulosonic acid disaccharide present in the inner core region of LPS is recognised by a lectin-like molecule on the plasma membrane of rat hepatocytes (Parent, 1990) and (4) binding to and internalisation by corneal epithelial cells of *P. aeruginosa* requires the presence of intact inner-core LPS with a terminal glucose residue (Zaidi *et al.*, 1996). In a few cases aggregation of bacteria may be mediated by an interaction between a lectin on one bacterial cell and the LOS on the surface of another cell (Blake *et al.*, 1995). The animal lectin galectin-3 recognises bacterial LPS of Gram-negative bacteria (Mey *et al.*, 1996). In no case, however, has a definitive proof or identification been presented of a lectin on mucosal cells that is involved in binding of complementary carbohydrates of pathogenic bacteria.

Tissue Tropism and Receptor Specificity of the Bacterial Adhesins

As a first approach to investigating the contribution of the observed differences in the fine sugar specificity to the infectivity of the bacteria, studies on type 1 fimbriae-mediated adhesion to various animal cells were carried out. The interspecies differences between *E. coli*, *K. pneumoniae* and *S. typhimurium* reflected their relative-binding affinity to mouse uroepithelial and enterocyte cell lines (Table 3). *E. coli* binds several-fold better to uroepithelial cells than to enterocytes, but the opposite is true for *S. typhimurium*. The binding pattern of *K. pneumoniae* is similar to that of *E. coli*, but the differences in binding to uroepithelial cells and enterocytes are much less pronounced. Recombinant strains were used to test the effect of the fimbrial shaft on this binding. When *E. coli* FimH was expressed on the *S. typhimurium* fimbrial shaft, the resulting recombinant strain bound to cells in a *S. typhimurium*-like pattern. Furthermore, when *fimH* genes of the three enterobacterial species were fused with a maltose-binding protein (MBP) gene, blots of the soluble fusion proteins bound intact uroepithelial and enterocyte cells equally well (Abraham *et al.*, unpublished). All of these cell-binding activities were inhibited by similar concentrations of mannose, suggesting that the magnitude of differences reflected differences in the fine sugar specificity. Collectively, these data suggest that inter-species differences in type 1 fimbria-mediated binding to various animal cell types may be influenced by the fimbrial shaft on which the FimH is borne.

Intra-species differences in adhesion to mannose-containing glycoprotein were reflected in the binding of wild-type and recombinant strains to various cell lines. For example, Man_1-binding FimH variants bound to uroepithelial cell lines to a much greater extent than trimannose-only-binding FimH variants (Table 3). The two FimH variants, however, bound to buccal epithelial cells in an equivalent manner (Sokurenko *et al.*, 1998). As in all other cases, binding was equally inhibited by mannose. The differences in fine sugar specificities of FimH alleles and in binding to animal cells resulted in marked differences in bladder colonisation by *E. coli*. The Man_1-binding FimH variants bound 15-fold more effectively to uroepithelial cell lines and also colonised mouse bladders in 15-fold higher numbers (Sokurenko *et al.*, 1998).

Tissue tropism may be influenced by factors other than the fine sugar specificity of the adhesins. For example, the affinity of adhesins for cells is influenced by the physical state of the adhesin structure, as in the case of *Neisseria meningitidis* which adheres better to tissue culture cells when its fimbriae or pili are in the form of bundles than when they are not (Marceau *et al.*, 1995). Moreover, intra-strain variants of the meningococci exhibit post-translational glycosylation of their pilus with Galβ4Galα3(2,4-diacetamido-2,4,6-trideoxyhexose) which is *O*-linked to serine residue 63 (Virji *et al.*, 1996). It is not clear how such glycosylation of the pilus adhesin affects the receptor specificity of the meningococci. Even variations in the location of the adhesin genes may affect the receptor specificity, as shown for the MrkD fimbrial adhesin encoded by the *mrk* gene cluster of both the plasmid and chromosome of *K. pneumoniae* (Hornick *et al.*, 1995); the chromosomal genes encode for adhesin with a receptor specificity distinct from that encoded by plasmid *mrk* genes.

Receptors may also undergo subtle changes and thus greatly affect susceptibility to infection. For example, K99-bearing *E. coli* binds to glycolipids that contain *N*-glycolylneuraminic acid, which is found on intestinal cells of newborn piglets, but is replaced by glycolipids containing *N*-acetylneuraminic acid in adult pigs (Yuyama *et al.*, 1993). The two sugar substitutes differ in only a single hydroxyl group, which is present in the acyl substituent on the 4-NH group of the piglet glycolipid but absent from the glycolipids of adult pigs and of humans. This subtle age-related change in the glycolipids may explain why *E. coli* K99 can cause diarrhoea in piglets, but not in adult pigs or in humans.

Adhesion-dependent Events

Although the contribution of fimbria-mediated bacterial adherence to colonisation of mucosal surfaces has been well recognised for many years (Ofek *et al.*, 1995), its involvement in subsequent steps of the infectious process was until recently largely unclear. Several activities intimately related to the infectious process have emerged that are so-called adhesion-dependent or are immediate consequences of adhesion (reviewed by Finlay and Cossart, 1997). Apparently, the specific interaction between bacterial fimbriae and their complementary receptors on mucosal cells elicits a variety of distinct responses in the host cells and in the bacteria that can markedly affect the course of the infectious process. These findings point to intriguing and novel functions for bacterial fimbriae on mucosal surfaces and in deeper tissues. The biological consequences of bacterial adhesion for various animal cell types has been best studied in uropathogenic *E. coli*, particularly in nephritogenic strains.

The Adhesins of Nephritogenic *E. coli*

An overwhelming number of bladder and upper urinary tract infections (UTI) are caused by the *E. coli* and enteric bacteria (Hagberg *et al.*, 1981). There may be several contributing factors, but the expression of organelles of adhesion or fimbriae by these bacteria appears to be critical. The urinary tract is a relatively inhospitable site for bacterial colonisation. A potential pathogen must withstand the dilution effects of the frequent exfoliation of epithelial cells as well as the flushing action associated with the voiding of urine. Thus, the adhesive property of a bacterium appears to be an essential determinant for colonisation of the urinary tract. Over 80% of uropathogenic bacteria have the capacity to adhere to uroepithelial cells, and the most commonly expressed organelles of adhesion on these bacteria are type P fimbriae and type 1 fimbriae (Hagberg *et al.*, 1981; Gander and Thomas, 1987). Indeed, these isolates express either or both fimbrial types (Gander and Thomas, 1987). Interestingly, the expression of the P and type 1 fimbriae on multi-fimbriate *E. coli* is differentially regulated, and the two fimbrial types are sometimes expressed by separate cells. This makes the infecting bacterial population heterogeneous with regard to fimbriation. The P fimbriae are characterised by their ability to bind Galα(1–4)Gal-residues, whereas type 1 fimbriae bind specifically to mannose-containing residues. Although these fimbriae are morphologically indistinguishable by standard electron microscopy and are heteropolymeric structures with a minor subunit serving as the adhesin moiety, there are a number of distinguishing structural features. The adhesin moiety of P fimbriae, PapG, is exclusively expressed on the distal tips of the fimbriae (Lindberg *et al.*, 1987), whereas the adhesin moiety of type 1 fimbriae, FimH, is preferentially expressed on the distal tips with a certain proportion localised intermittently along the fimbrial length (Ofek *et al.*, 1995). Both fimbriae are composite structures that consist of a long rigid rod and a distal thin fibrillar structure, but the fibrillum structure of P fimbriae appears markedly longer and more flexible than that of type 1 fimbriae (Kuehn *et al.*, 1992; Jones *et al.*, 1995).

When these fimbrial adhesins of uropathogenic *E. coli* couple with their cognate receptor on the uroepithelium, the bacteria adhere to the uroepithelium. In the case of type 1 fimbriae, the adhesins may sometimes dock with Tamm–Horsfall protein (Kuriyama *et al.*, 1986), a highly mannosylated compound secreted by the kidneys, which adsorbs on to the surface of the uroepithelium before the adhesins eventually make contact with the epithelium. The complementary epithelial cell receptors for P fimbriae are globosides that contain Galα(1–4)Gal receptor epitopes (Leffler *et al.*, 1981). For type 1 fimbriae these are the mannose-rich uroplakins found to coat 70–80% of the apical surface of the bladder (Wu *et al.*, 1986). Although the binding affinity between individual fimbria and their cognate receptor may not be high, cumulatively the binding interactions mediated by several fimbriae and their cognate receptors on host cells results in a powerful interaction that evokes specific responses not only in the host cell but also in the bacteria.

Induction of Bacterial Virulence Genes

As already mentioned, the urinary tract is relatively inhospitable to bacterial colonisation. In addition to resisting the constant hydrokinetic forces, a potential pathogen must multiply sufficiently rapidly to compensate for the diluting effects of urine. Thus, the capacity to grow in urine appears essential for bacterial survival. Urine is a complex body fluid that contains a variety of excreted products, but it is growth-limiting for bacteria, in part because it contains minimal amounts of free iron. The intrinsic iron acquisition machinery of uropathogenic *E. coli* is activated on the binding of the PapG fimbrial adhesin to its Galα(1–4)Gal-containing globoseries receptor (Zhang and Normark, 1996). When P-fimbriate bacteria bind to their immobilised receptor, transcriptional activation of a sensor-regulator protein, AirS, is detected. The *airS* gene product is a sensor protein located in the cytoplasmic membrane, which belongs to the two-component family of signal transduction factors (Zhang and Normark, 1996). Although the precise mechanism is as yet undetermined, AirS is believed to regulate the bacterial iron acquisition system, which comprises high-affinity iron-binding proteins or siderophores and iron-regulated membrane proteins (IRMPs), which serve as receptors for the iron-siderophore complexes and facilitate the translocation of iron into the bacterium. When the *airS* gene is knocked out in uropathogenic *E. coli*, the resulting mutant loses its capacity to generate siderophores and IRMPs, and consequently loses its capacity for growth in urine. Evidence that binding of PapG to its host-cell receptor is the critical signal necessary for transcriptional activation of *airS* comes from the observation that no *airS* gene transcription can be detected in an isogenic PapG$^-$ mutant when exposed to the Galα(1–4)Gal-containing globoseries receptor (Zhang and Normark, 1996). It appears that uropathogenic bacteria possess the capacity to 'sense' the urinary tract environment by means of PapG and elicit an appropriate response to facilitate colonisation of the body site. These findings point to an intriguing new function for bacterial P fimbriae – that of a sensory organelle. The strategic location of PapG at the distal tips of the peritrichous fimbriae probably facilitates this suggested role. This finding is one of an increasing number showing that bacterial pathogens are intrinsically capable of responding to important cues from host cells after interactions between complementary cell-surface molecules (Cotter and Miller, 1996; Finlay and Falkow, 1997). In addition, these observations provide a molecular basis for previous findings that various bacteria obtain a growth advantage after binding to host cells, as demonstrated for type 1 fimbriate *E. coli* and *N. gonorrheae*, which have shorter lag periods when adhering to tissue culture cells (Bessen and Gotschlich, 1986; Zafriri *et al.*, 1987).

Induction of Cytokine Responses from Mucosal Cells

In addition to evoking responses amongst adherent bacteria, the specific binding of bacterial adhesins to their receptors on the mucosal cells also elicits a range of cellular responses. For example, the specific binding of the P-fimbrial adhesin to its complementary Galα(1–4)Gal-containing receptor epitopes on uroepithelial cells elicits the release of several immunoregulatory cytokines, including interleukin (IL)-1α, IL-1β, IL-6 and IL-8 (Svanborg *et al.*, 1996; Hedlund *et al.*, 1996). Some intriguing clues have emerged to the underlying intracellular molecular links between fimbria-mediated attachment to the cell surface and cellular cytokine production. The specific attachment of PapG adhesin of P fimbriae to the Galα(1–4)Gal oligosaccarides on globoseries glycolipids, localised on the outer leaflet of the lipid bilayer of the epithelial cell membrane, triggers the intracellular release of ceramides (Hedlund *et al.*, 1996). These PapG-elicited ceramides may be derived by the generation of endogenous sphingomyelinases from the globoseries receptor itself or from neighbouring sphingomyelin molecules (Hedlund *et al.*, 1996, 1999; Svanborg *et al.*, 1999). Ceramide is known to be a critical second messenger in signal transduction processes in the cell and can activate the serine/threonine family of protein kinases and phosphatases with the serine/threonine specificity, eventually leading to cytokine production. Interestingly, this bacterial adhesin-mediated mechanism of signalling is not unlike pathways known to be utilised by immunoregulatory cytokines such as tumour necrosis factor α (TNFα) and IL-1 when evoking cellular responses (Svanborg *et al.*, 1999). Thus, in this instance, the bacterial adhesin appears functionally to mimic the host immunoregulatory molecules. Although the type 1 fimbriae of uropathogenic *E. coli* also stimulates a cytokine response from uroepithelial cells, the array of cytokines released is different from those elicited by P fimbriae (Connell *et al.*, 1996b; Svanborg *et al.*, 1999). The transmembrane signalling pathway of type 1 fimbriae appears to involve phosphoinositide 3-kinase (PI 3-kinase) and host protein tyrosine phosphorylation (Martinez *et al.*, 2000). Phosphorylation of focal adhesin kinase (FAK) is observed as well as the formation of complexes between FAK and PI 3-kinase and between α-actinin and vinculin (Martinez *et al.*, 2000).

Induction of Cytokine Responses from Inflammatory Cells

The capacity of bacterial adhesins to elicit cytokine response is not confined to mucosal cells. Indeed, type 1 fimbriae of uropathogenic *E. coli* are capable of binding to and eliciting immunoregulatory products from a wide range of inflammatory cells, including macrophages, neutrophils, mast cells and B and T lymphocytes *in vitro* (Connell *et al.*, 1996a, b). The association of type 1 fimbriate enteric bacteria with mast cells of mouse origin is shown in **Fig. 1**. Evidence that these interactions may occur *in vivo* with significant physiological effects comes from intraperitoneal injection into mice of *E. coli* expressing type 1 fimbriae, which results in the generation of extracellular TNFα in the peritoneal fluid (Malaviya *et al.*, 1996). Evidence for the role of FimH as the fimbrial adhesin is that intraperitoneal challenge with an FimH$^-$ isogenic mutant shows only a limited TNFα response (Malaviya *et al.*, 1996). We know that the cellular source of TNFα in the mouse peritoneum is the mast cells, because mice genetically deficient in mast cells show a limited TNFα response after intraperitoneal injection of type 1 fimbriae. Notably, the TNFα response of mast cells is accompanied by a large influx of neutrophils into the peritoneum, which is consistent with TNFα as a potent neutrophil chemoattractant (Malaviya *et al.*, 1996). Thus, one of the immediate outcomes of type 1 fimbriae-mediated activation of mast cells is recruitment of neutrophils to sites of bacterial challenge. Since mast cells are found mainly in mucosal surfaces, including the urinary tract, the interaction of *E. coli* type 1 fimbriae with mast cells

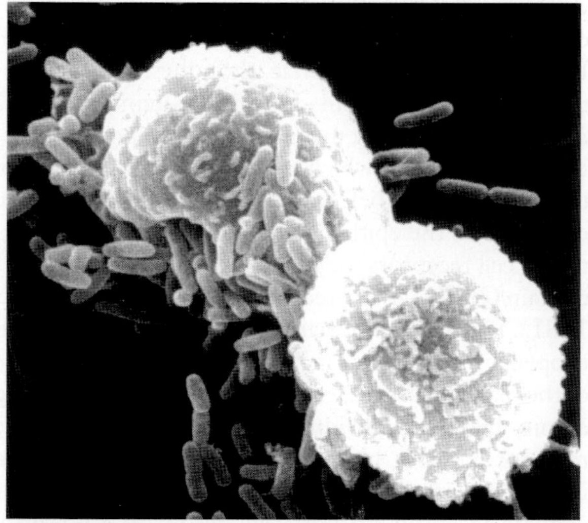

Fig. 1 Scanning electron micrograph of mast cells with adherent type 1 fimbriate bacteria.

close to the uroepithelium may contribute to the influx of neutrophils and their translocation through the epithelial barrier into the lumen of the urinary tract. Indeed, the presence of neutrophils in the urine is an early and often diagnostic indication of UTI. The excessive transepithelial migration of neutrophils during infections may predispose this barrier to increased bacterial penetration as has been suggested for other mucosal surfaces (Finlay and Cossart, 1997; Finlay and Falkow, 1997) and raises the possibility that the host immune response may be co-opted by pathogenic bacteria to facilitate their virulence.

Impact of Cytokine Responses Elicited by Bacteria

It is generally difficult to evaluate the physiological effects of some of the adhesin-elicited cytokines at sites of bacterial infection because their effects are numerous and complex (Henderson *et al.*, 1996; Abraham and Malaviya, 1997). Some of the various responses evoked in the urinary tract after the adherence of uropathogenic *E. coli* to the urinary mucosa are illustrated in Fig. 2. In general, these cytokines serve to initiate and regulate the innate and specific immune responses of the host to an infectious agent. These responses include increased mucus secretion, proliferation of epithelial cells and the recruitment and activation of a variety of phagocytic cells, all of which could potentially effect the early elimination of the pathogen (Henderson *et al.*, 1996; Abraham and Malaviya, 1997). Not all these responses have their intended effects, however, indeed, some of the adhesin-triggered secreted products of host cells may have severe pathophysiological effects on the surrounding tissue, particularly when released in excess or at inopportune times (Abraham and Malaviya, 1997). Although direct evidence for this is still lacking, considerable circumstantial evidence supports the notion that many proteases, oxygen radicals and cytotoxic cytokines are secreted by inflammatory cells after activation by type 1 fimbriate *E. coli* (Tewari *et al.*, 1993; Malaviya *et al.*, 1994, 1996), and that these are detrimental to the host and foster bacterial pathogenesis. For example, the neutrophil elastases, oxygen radicals and other cytotoxic agents produced in the kidney by neutrophils after their interaction with *E. coli* type 1 fimbriae are a major contributor to renal scarring (Steadman *et al.*, 1988; Topley *et al.*, 1989). In several instances, the critical determinant of

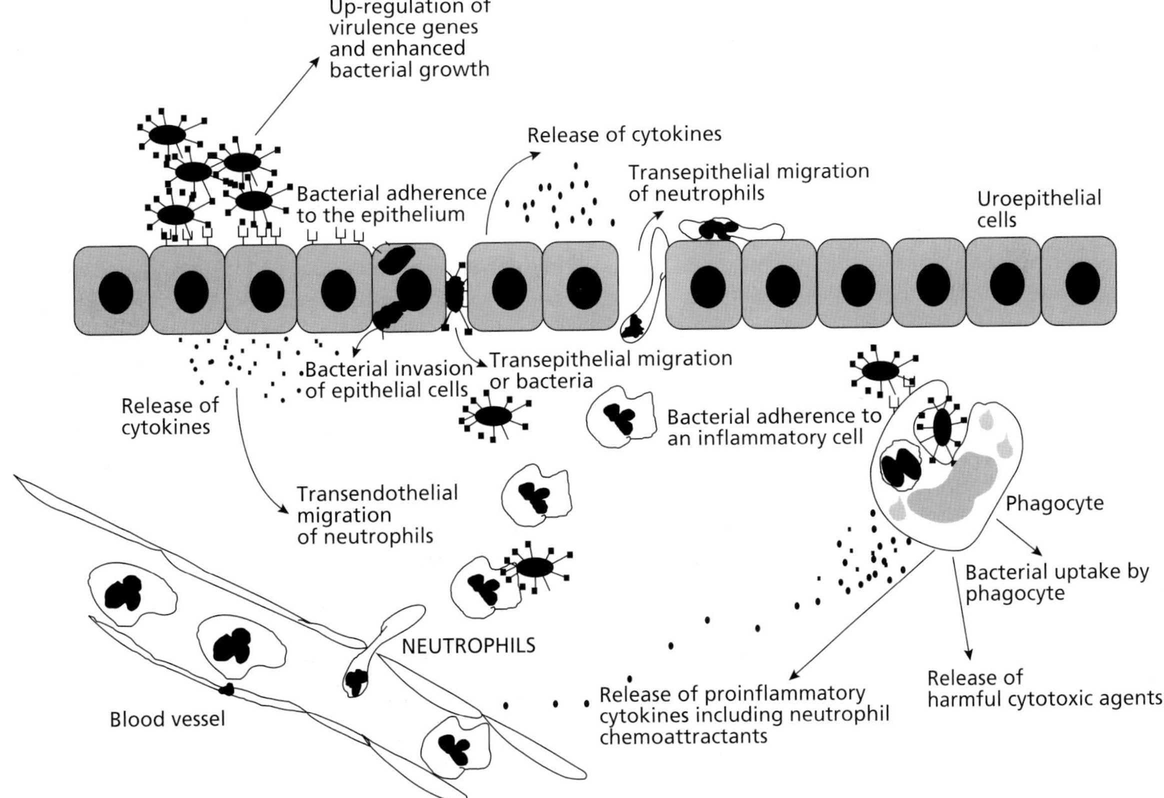

Fig. 2 Bacterial adherence to uroepithelial cells and immune cells and various bacterial and host responses.

whether an inflammatory response favours the host or pathogen depends on other factors including the host immune status and the intrinsic virulence of the pathogen. Since certain bacteria have the capacity of 'quorum sensing', the number of bacteria at the site of infection is another potentially critical factor (Passador *et al.*, 1993). That is, the organisms can 'measure' their population density at a given site and, on reaching a critical density, they can co-ordinately switch on the expression of a battery of new virulence factors, presumably to mount an effective challenge to the host immune response.

Induction of Bacterial Phagocytosis

In addition to eliciting a variety of pharmacologically active mediators from various host cells, bacterial adhesins also elicit uptake of bacteria by host phagocytic cells under serum-free conditions. The process has been termed lectinophagocytosis, by analogy with opsonophagocytosis, in which recognition between the micro-organisms and the phagocytic cells is mediated by serum opsonins, mainly IgG and the C3b and C3bi fragments of the C3 component of complement (Ofek *et al.*, 1995; Ofek and Sharon, 1988). The best-characterised system of lectinophagocytosis is that of bacteria carrying mannose-specific type 1 fimbrial lectins. It is something of a paradox that a bacterial adhesin that promotes bacterial colonisation and infection also promotes ingestion by phagocytic cells. New evidence has, however, emerged to suggest that the type 1 fimbria-elicited bacterial phagocytosis by macrophages is actually advantageous to the bacterial population (Baorto *et al.*, 1997). Survival assays *in vitro* have revealed that, unlike *E. coli* phagocytosed by opsonin-mediated processes, *E. coli* internalised via type 1 fimbriae survive much of the intracellular killing process of macrophages. It has been suggested that by associating with CD48, a glycosylphosphoinisitol-linked moiety on the surface of macrophages, the bacteria gain access to a lipid trafficking pathway that bypasses the normal phagocytic killing mechanisms of the macrophages (Baorto *et al.*, 1997). The precise mechanism by which type 1 fimbriate *E. coli* are internalised by phagocytes has been elucidated. It appears that CD48 is present in distinct glycolipid and cholesterol-rich micro-domains on the plasmalemma surface of mast cells and macrophages termed caveolae (Shin *et al.*, 2000). Upon ligation of these receptors by bacterial FimH, these caveolar structures enlarge and mediate the encapsulation of the bacteria within intracellular compartments the membrane of which consists of caveolar material (Shin *et al.*, 2000). Caveolae are rich in signalling molecules and are mostly known for mediating transmembrane signals in the cell (Anderson, 1998). This finding is notable not only because it provides a basis for how fimbrial adhesins activate phagocytes, but also because it reveals a role for cellular caveolae in endocytic functions of phagocytic cells (Shin *et al.*, 2000). This provides a molecular basis for earlier observations which showed that, compared with bacteria ingested via opsonophagocytosis, bacteria subjected to lectinophagocytosis are markedly less sensitive to killing by phagocytes (reviewed by Ofek *et al.*, 1995). It is noteworthy that lectinophagocytosis comes into play only at sites with poor opsonising capacity, such as the urinary mucosa, and this may help to explain why type 1 fimbriate *E. coli* is the single most frequent pathogen at this site (Hagberg *et al.*, 1981).

Internalisation by Non-phagocytic Cells

Remarkably, the FimH adhesin of type 1 fimbriae has been shown to mediate entry of bacteria into human uroepithelial cells, which are usually regarded as non-phagocytic. Moreover, inert beads coated with FimH are internalised by these cells (Martinez *et al.*, 2000). It is conceivable that this behaviour is restricted to uroepithelial cells, since if enterocytes were also to exhibit this activity the many enteric type 1 fimbriate bacteria of the normal gut flora would also penetrate the mucosal barrier. In any case, uroepithelial phagocytosis supports the earlier findings *in vivo* demonstrating the invasion of the mouse uroepithelium by type 1 fimbriate *E. coli*, which are regarded by many as classical extracellular bacteria (Mulvey *et al.*, 1998). Taken together, these observations suggest that the binding of type 1 fimbriate bacteria to their complementary receptors on immune and epithelial cells triggers not only exocytic responses (cytokine secretion) by the host cell but also endocytic responses by which adherent bacteria are internalised. This capacity of bacteria to survive in host cells allows the pathogens to avoid the bactericidal actions of host immune cells as well as antibiotics that are administered, and so they may contribute to the recurrence or persistence of UTI.

This notion seems not to be limited to uroepithelial cells but also occurs with other bacteria including Gram-positive bacteria. For example, it has been suggested that in the case of *Streptococcus pyogenes* internalisation into epithelial cells may lead to asymptomatic carriage (Neeman *et al.*, 1998). The large bowel serves as the main reservoir for *K. pneumoniae* where the organisms rarely cause symptomatic infection but have the capacity preferentially to invade intestinal cells (Sahly *et al.*, 2001). These findings suggest that bacterial pathogens have adopted various strategies to penetrate and survive within the epithelial cells of mucosal surfaces in order to evade deleterious agents and host immune defences and at the same time to colonise for extended periods of time.

Anti-adhesion-based Therapy and Prevention of Bacterial Infection

The influence of bacterial adhesion on the infectious process as discussed above has stimulated numerous studies to explore the possibility that inhibition of this process will result in the prevention or termination of the infection. Thus, anti-adhesion therapy involves the use of agents that prevent the attachment of the bacteria to host tissues or displace bacteria from the tissues at the early stages of infection. It is expected, therefore, that anti-adhesion therapy would be targeted on mucosal surfaces and will eventually lead to the eradication of the bulk of the population of pathogens without killing them. Experiments in animals have shown that it is possible to prevent infections by blocking the adhesion of the pathogen to target tissue. These findings have provided an impetus for the development of anti-adhesion drugs for the prevention and therapy of microbial infections in humans (reviewed by Ofek *et al.*, 1996). New classes of drugs such as these are necessary, because of the increased prevalence of pathogens resistant to conventional antibiotics. From a medical point of view, anti-adhesion intervention may be regarded mild, gentle and safe as compared with current chemotherapeutic approaches. In the following, a summary of the various approaches for the development of anti-adhesion drugs is presented.

Anti-adhesion-based Vaccines

Since bacterial adhesins were identified and their importance in the infectious process was established, efforts have been made to produce pure protein vaccines to induce anti-adhesion antibodies (Ofek and Doyle, 1994). Special efforts have been made to evoke such antibodies in mucosal secretions in order to prevent colonisation and subsequent invasion. A number of studies in experimental infections have reported the efficacy of adhesin-based vaccines in preventing bacterial infections. Perhaps the most convincing results are those recently obtained with urinary pathogens as the vaccine target.

In view of the important role played by type 1 fimbriae in initiating and facilitating infection, attempts have been made to test the efficacy of vaccines that consist only of the fimbrial FimH moiety or its functionally relevant fragments to protect against experimentally induced UTI in mice (Langermann *et al.*, 1997; Thankavel *et al.*, 1997). These studies reported remarkable success in affording protection in actively and passively immunised mice. Much of the protection was ascribed to the ability of fimbrial adhesin-specific antibodies to be filtered through the kidneys and block bacterial binding to the epithelial surface of the urinary tract (Thankavel *et al.*, 1997). It is also conceivable that these antibodies modulate the interaction between *E. coli* and the many phagocytes recruited to the site of infection. It was also reported that in a primate model of *E. coli* bladder infection, a uropathogenic strain expressing P fimbriae with its PapG adhesin, generated a protective immune response that lasted for at least 150 days against challenge from the same strain (Soderhall *et al.*, 1997). These findings make the development of an efficacious fimbrial adhesin vaccine against UTI a real possibility.

Sugars as Receptor Analogues to Block Adhesion

Soluble carbohydrates recognised by the bacterial lectins block adhesion of the bacteria to a variety of animal cells *in vitro*. Such saccharides can also detach bacteria that are bound to cells by their surface lectins (Ofek *et al.*, 1996). The ability of sugars to protect animals against experimental infection by lectin-carrying bacteria was first reported two decades ago, in a model of UTI in mice (Aronson *et al.*, 1979). Co-administration of methyl α-mannoside with type 1 fimbriate *E. coli* into the urinary bladder of the mice reduced the rate of UTI by two-thirds, whereas methyl α-glucoside, which is not inhibitory to the fimbriae, was without effect. The protective effect of anti-adhesive sugars was subsequently demonstrated in a variety of studies with different pathogenic bacteria and animals (Table 4). For instance, colostrum-deprived newborn calves given a lethal dose of *E. coli* K99 could be cured by drinking water containing glycopeptides prepared from cow plasma non-immunoglobulin glycoproteins (Mouricout *et al.*, 1990). The treatment decreased by 100-fold the number of bacteria that adhered to the intestinal epithelial cells of the calves. Although the glycopeptides used were not analysed in detail, it is very likely that they contained the NeuGc(α2–3)Gal-R sequence recognised by the bacteria. Derivatives of galabiose that inhibit the adhesion of P fimbriate *E. coli* to animal cells *in vitro* prevented UTIs by the bacteria in mice and monkeys. In a clinical trial in humans, patients with otitis externa

Table 4 Inhibitors of carbohydrate-specific adhesion that prevent bacterial infection *in vivo* (see text for references)

Bacterial pathogen	Animal and site	Inhibitor
E. coli type 1	Mouse UT	MeαMan
	Mouse GIT	Mannose
	Mouse UT	Globotetrose
E. coli P	Mouse UT	Globotetrose
	Monkey UT	Galα(1α4)GalβOMe
E. coli K99	Calf GIT	Glycopeptides
H. pylori	Piglet GIT	Sia3′LacNAc
	Monkey GIT	Sia3′Lac
K. pneumoniae type 1	Rat UT	MeαMan
Shigella flexneri type 1	Guinea-pig eye	Mannose
Strep. pneumoniae	Mouse lung	*N*-acetyglucosamine
	Rabbit lungs	Lacto-*N*-neotetraose

GIT, gastrointestinal tract; UT, urinary tract.

caused by *Ps. aeruginosa* were treated with a solution of galactose, mannose and *N*-acetylneuraminic acid (Beuth *et al.*, 1996). The results were comparable with those obtained with conventional antibiotic treatment. In rabbits and infant rats, experimental pneumonia caused by *Strep. pneumoniae* was markedly reduced by intranasal (Idanpaan-Heikkila *et al.*, 1997) or intratracheal administration of free oligosaccharides or neoglycoproteins (Karlsson, 1995).

Very recently, results of the treatment of 12 *H. pylori*-positive rhesus monkeys with the sodium salt of Sia3′Lac alone or in combination with either one of the commonly used gastric coating anti-ulcer drugs, bismuth subsalicylate, or the proton-pump inhibitor omeprazole, have been reported (Mysore *et al.*, 1999). Of the six monkeys given the trisaccharide only, two were cured permanently, and a third was transiently cleared, while three of the animals remained persistently colonised. The other six animals received the trisaccharide with the anti-ulcer drugs in different combinations. Transient decreases in colony counts were observed in three of these animals, but gastritis was suppressed only in the two animals that became persistently *H. pylori*-negative. No side effects were observed in any of the animals given Sia3′Lac. The authors concluded that, 'anti-adhesive therapy is safe; it can cure or decrease *H. pylori* colonisation in some rhesus monkeys, but the addition of a proton pump inhibitor or bismuth subsalicylate does not increase cure rate'.

Anti-adhesive sugars, in addition to blocking the attachment of the bacteria to tissues, may act similarly on the binding of toxins such as those of *Shigella dysenteriae* type 1, or of the homologous toxins of *E. coli*, called verotoxins, that are Gal(α1–4)Gal-specific. Also in this system polyvalent ligands, when appropriately designed, are markedly more inhibitory than monovalent ligands; the latter are active at the millimolar level only, but the former are active at the nanomolar level (Kitov *et al.*, 2000).

Dietary Inhibitors of Bacterial Adhesion

Since several of the inhibitory oligosaccharides are common constituents of human milk, these substances play a protective role against different infectious agents in breast-fed babies (Ashkenazi, 1994; Coppa *et al.*, 1999). This is the basis of attempts made by a biotechnology company to develop oligosaccharides for use in nutritional and other non-prescription products (Glaser, 1997). Breast-feeding has long been associated with a reduced incidence of infantile infections, particularly gastroenteritis (Ashkenazi, 1994). Recent work has shown that this is due in part to the oligosaccharides, glycoproteins and glycolipids of which the level is much higher in human milk than in cows' milk (Ashkenazi, 1996). These glycoconjugates may act as receptor analogues of human pathogens, thereby interfering with binding of enteric pathogens or their toxins to the intestine.

Cranberry juice has been studied for some time for its bacterial anti-adhesive properties and has been found to inhibit adhesion of *E. coli in vitro* (Zafriri *et al.*, 1987; Ofek *et al.*, 1991, 1996). Convincing evidence suggests that a high-molecular-weight component in cranberry juice is responsible for the beneficiary effect of cranberry consumption on UTIs because of its potent anti-adhesion activity toward uropathogens (Burger *et al.*, 2001). Moreover, daily consumption of cranberry juice reduces UTIs caused by *E. coli* in humans (Avorn *et al.*, 1994). The high-molecular-weight constituent isolated from cranberry acts as an inhibitor not only of uropathogenic *E. coli* but also of co-aggregation of certain oral bacteria (Weiss *et al.*, 1998). The cranberry material also inhibits the adhesion of *H. pylori* to human gastric mucus. Proanthocyanidin extracts of cranberries inhibit the adhesion of P-fimbriate *E. coli* to uroepithelial cells (Howell *et al.*, 1998), but the relationship of this compound to the high-molecular-weight constituent inhibitor is not clear.

Conclusion

The binding of the bacterial adhesins to host receptors is a very dynamic process which signals many responses by the bacteria from host cells. In the urinary tract, adherent *E. coli* appear capable of inducing the expression of bacterial genes that would favour their growth in the urine and elicit a range of immune responses from host cells, aspects of which may be co-opted by the bacteria to favour their survival (Fig. 2). Thus, fimbrial adhesins may be viewed as direct mediators of bidirectional signalling between bacteria and host cells. Future studies in this area should prove highly informative for the design of novel strategies to modulate these signalling pathways and to curb bacterial infections and their harmful sequelae. For example, the great potential of inhibitors of adhesion to mucosal surfaces in the prevention and, perhaps, treatment of bacterial infections raises hopes for the development of anti-adhesive drugs for human use. The development of anti-adhesion therapy targeted at the microbial lectins has been hampered by the great difficulty in large-scale synthesis of the required inhibitory saccharides. An alternative is the use of glycomimetics, compounds that structurally mimic the inhibitory carbohydrates, and which may be more readily obtainable. Eventually, a cocktail of inhibitors or a polyvalent inhibitor will have to be used, since many infectious agents express multiple specificities. Pathogens employ multiple molecular mechanisms of adhesion to initiate infection and effective anti-adhesion drugs will probably have to consist of a cocktail of inhibitors. The biotechnology required to achieve this goal is promising.

Acknowledgements

The work by the authors described in this chapter was supported by NIH grants AI-35678 and DK-50814.

References

Abraham SN, Malaviya R (1997) Mast cells in infection and immunity. *Infect. Immun.* 65: 3501–3508.

Alam M, Miyoshi SI, Tomochika KI, Shinoda S (1996) Purification and characterization of novel hemagglutinins from *Vibrio mimicus*: a 39-kilodalton major outer membrane protein and lipopolysaccharide. *Infect. Immun.* 64: 4035–4041.

Anderson RG (1998) The caveolae membrane system. *Annu. Rev. Biochem.* 67: 199–225.

Aronson M, Medalia O, Schori L *et al.* (1979) Prevention of colonization of the urinary tract of mice with *Escherichia coli* by blocking of bacterial adherence with methyl alpha-D-mannopyranoside. *J. Infect. Dis.* 139: 329–332.

Ashkenazi S (1994) A review of the effect of human milk fractions on the adherence of diarrheogenic *Escherichia coli* to the gut in an animal model. *Isr. J. Med. Sci.* 30: 335–338.

Ashkenazi S (1996) Role of human milk constituents in blocking the adherence of enteric pathogens. *Adv. Exp. Med. Biol.* 408: 187–192.

Athamna A, Ofek I, Keisari Y *et al.* (1991) Lectinophagocytosis of encapsulated *Klebsiella pneumoniae* mediated by surface lectins of guinea pig alveolar macrophages and human monocyte- derived macrophages. *Infect. Immun.* 59: 1673–1682.

Avorn J, Monane M, Gurwitz JH *et al.* (1994) Reduction of bacteriuria and pyuria after ingestion of cranberry juice [see comments]. *JAMA* 271: 751–754.

Baorto DM, Gao Z, Malaviya R *et al.* (1997) Survival of FimH-expressing enterobacteria in macrophages relies on glycolipid traffic. *Nature* 389: 636–639.

Bessen D, Gotschlich EC (1986) Interactions of gonococci with HeLa cells: attachment, detachment, replication, penetration, and the role of protein II. *Infect. Immun.* 54: 154–160.

Beuth J, Stoffel B, Pulverer G (1996) Inhibition of bacterial adhesion and infections by lectin blocking. *Adv. Exp. Med. Biol.* 408: 51–56.

Blake MS, Blake CM, Apicella MA, Mandrell RE (1995) Gonococcal opacity: lectin-like interactions between Opa proteins and lipooligosaccharide. *Infect. Immun.* 63: 1434–1439.

Burger O, Ofek I, Tabak M *et al.* (2000) A high molecular weight constituent of cranberry juice inhibits *Helicobacter pylori* adhesion to human gastric mucus. *FEMS Med. Microbiol. Immunol.* 29: 295–301.

Choudhury D, Thompson A, Stojanoff V *et al.* (1999) X-ray structure of the FimC-FimH chaperone-adhesin complex from uropathogenic *Escherichia coli* [see comments]. *Science* 285: 1061–1066.

Connell H, Agace W, Hedlund M *et al.* (1996b) Fimbriae-mediated adherence induces mucosal inflammation and bacterial clearance. Consequences for anti-adhesion therapy. *Adv. Exp. Med. Biol.* 408: 73–80.

Connell I, Agace W, Klemm P *et al.* (1996a) Type 1 fimbrial expression enhances *Escherichia coli* virulence for the urinary tract. *Proc. Natl Acad. Sci. USA* 93: 9827–9832.

Coppa GV, Pierani P, Zampini L *et al.* (1999) Oligosaccharides in human milk during different phases of lactation. *Acta Paediatr. Suppl.* 88: 89–94.

Cotter PA, Miller JF (1996) Triggering bacterial virulence [comment]. *Science* 273: 1183–1184.

Finlay BB, Cossart P (1997) Exploitation of mammalian host cell functions by bacterial pathogens [published erratum appears in *Science* 1997 278(5337):373] [see also comments]. *Science* 276: 718–725.

Finlay BB, Falkow S (1997) Common themes in microbial pathogenicity revisited. *Microbiol. Mol. Biol. Rev.* 61: 136–169.

Firon N, Ofek I, Sharon N (1983) Carbohydrate specificity of the surface lectins of *Escherichia coli*, *Klebsiella pneumoniae*, and *Salmonella typhimurium*. *Carbohydr. Res.* 120: 235–249.

Gander RM, Thomas VL (1987) Distribution of type 1 and P pili on uropathogenic *Escherichia coli* O6. *Infect. Immun.* 55: 293–297.

Garber N, Guempel U, Belz A, Gilboa-Garber N, Doyle RJ (1992) On the specificity of the D-galactose-binding lectin (PA-I) of *Pseudomonas aeruginosa* and its strong binding to hydrophobic derivatives of D-galactose and thiogalactose. *Biochim. Biophys. Acta* 1116: 331–333.

Glaser V (1997) Noise technologies sets the stage for large-scale testing of its anti-ulcer drug. *Genetic Eng. News* 26: 45.

Hagberg L, Jodal U, Korhonen TK *et al.* (1981) Adhesion, hemagglutination, and virulence of *Escherichia coli* causing urinary tract infections. *Infect. Immun.* 31: 564–570.

Hanski E, Jaffe J, Ozeri V (1996) Proteins F1 and F2 of *Streptococcus pyogenes. Adv. Exp. Med. Biol.* 408: 141–150.

Hasty DL, Sokuernko EV (2000) FimH, an adaptable adhesin In: Doyle RJ (ed.) *Glycomicrobiology.* New York: Kluwer Academic/Plenum, pp. 481–515.

Hedlund M, Svensson M, Nilsson A, Duan RD, Svanborg C (1996) Role of the ceramide-signaling pathway in cytokine responses to P-fimbriated *Escherichia coli. J. Exp. Med.* 183: 1037–1044.

Hedlund M, Wachtler C, Johansson E *et al.* (1999) P fimbriae-dependent, lipopolysaccharide-independent activation of epithelial cytokine responses. *Mol. Microbiol.* 33: 693–703.

Henderson B, Poole S, Wilson M (1996) Bacterial modulins: a novel class of virulence factors which cause host tissue pathology by inducing cytokine synthesis. *Microbiol. Rev.* 60: 316–341.

Hornick DB, Thommandru J, Smits W, Clegg S (1995) Adherence properties of an mrkD-negative mutant of *Klebsiella pneumoniae. Infect. Immun.* 63: 2026–2032.

Howell AB, Vorsa N, Der Marderosian A, Foo LY (1998) Inhibition of the adherence of P-fimbriated *Escherichia coli* to uroepithelial-cell surfaces by proanthocyanidin extracts from cranberries [letter] [published erratum appears in *N. Engl. J. Med.* 1998; 339(19): 1408]. *N. Engl. J. Med.* 339: 1085–1086.

Hultgren SJ, Lindberg F, Magnusson G *et al.* (1989) The PapG adhesin of uropathogenic *Escherichia coli* contains separate regions for receptor binding and for the incorporation into the pilus. *Proc. Natl Acad. Sci. USA* 86: 4357–4361.

Idanpaan-Heikkila I, Simon PM, Zopf D *et al.* (1997) Oligosaccharides interfere with the establishment and progression of experimental pneumococcal pneumonia. *J. Infect. Dis.* 176: 704–712.

Jacques M (1996) Role of lipo-oligosaccharides and lipopolysaccharides in bacterial adherence [see comments]. *Trends Microbiol.* 4: 408–409.

Jones CH, Pinkner JS, Roth R *et al.* (1995) FimH adhesin of type 1 pili is assembled into a fibrillar tip structure in the Enterobacteriaceae. *Proc. Natl Acad. Sci. USA* 92: 2081–2085.

Karlsson KA (1995) Microbial recognition of target-cell glycoconjugates. *Curr. Opin. Struct. Biol.* 5: 622–635.

Karlsson KA, Abul Milh M, Andersson C, Angstrom J (1994) Carbohydrate attachment sites for microbes on animal cells: Aspects on the possible use of analogs to treat infections. In: Bock K, Clausen H (ed.) *Complex Carbohydrates in Drug Research.* Copenhagen: Munksgaard, pp. 1–11.

Khan AS, Kniep B, Oelschlaeger TA *et al.* (2000) Receptor structure for F1C fimbriae of uropathogenic *Escherichia coli. Infect. Immun.* 68: 3541–3547.

Kitov PI, Sadowska JM, Mulvey G *et al.* (2000) Shiga-like toxins are neutralized by tailored multivalent carbohydrate ligands. *Nature* 403: 669–672.

Klemm P, Krogfeld KA (1994) Type 1 fimbriae of *Escherichia coli.* In: Klemm P (ed.) *Fimbriae, Adhesion, Genetics, Biogenesis and Vaccines.* Boca Raton, FL: CRC Press, pp. 9–26.

Kuehn MJ, Heuser J, Normark S, Hultgren SJ (1992) P pili in uropathogenic *E. coli* are composite fibres with distinct fibrillar adhesive tips. *Nature* 356: 252–255.

Kuriyama SM, Silverblatt FJ (1986) Effect of Tamm-Horsfall urinary glycoprotein on phagocytosis and killing of type I-fimbriated *Escherichia coli. Infect. Immun.* 51: 193–198.

Langermann S, Palaszynski S, Barnhart M *et al.* (1997) Prevention of mucosal *Escherichia coli* infection by FimH-adhesin-based systemic vaccination [see comments]. *Science* 276: 607–611.

Leffler H, Svanborg-Eden C (1981) Glycolipid receptors for uropathogenic *Escherichia coli* on human erythrocytes and uroepithelial cells. *Infect. Immun.* 34: 920–929.

Lindberg F, Lund B, Johansson L, Normark S (1987) Localization of the receptor-binding protein adhesin at the tip of the bacterial pilus. *Nature* 328: 84–87.

Madison B, Ofek I, Clegg S, Abraham SN (1994) Type 1 fimbrial shafts of *Escherichia coli* and *Klebsiella pneumoniae* influence sugar-binding specificities of their FimH adhesins. *Infect. Immun.* 62: 843–848.

Malaviya R, Ross E, Jakschik BA, Abraham SN (1994) Mast cell degranulation induced by type 1 fimbriated *Escherichia coli* in mice. *J. Clin. Invest.* 93: 1645–1653.

Malaviya R, Ikeda T, Ross E, Abraham SN (1996) Mast cell modulation of neutrophil influx and bacterial clearance at sites of infection through TNF-alpha [see comments]. *Nature* 381: 77–80.

Marceau M, Beretti JL, Nassif X (1995) High adhesiveness of encapsulated *Neisseria meningitidis* to epithelial cells is associated with the formation of bundles of pili. *Mol. Microbiol.* 17: 855–863.

Martinez JJ, Mulvey MA, Schilling JD, Pinkner JS, Hultgren SJ (2000) Type 1 pilus-mediated bacterial invasion of bladder epithelial cells. *EMBO J.* 19: 2803–2812.

Mey A, Leffler H, Hmama Z, Normier G, Revillard JP (1996) The animal lectin galactin-3 interacts with bacterial lipopolysaccharides via two independent sites. *J. Immunol.* 156: 1572–1577.

Mouricout M, Petit JM, Carias JR, Julien R (1990) Glycoprotein glycans that inhibit adhesion of *Escherichia coli* mediated by K99 fimbriae: treatment of experimental colibacillosis. *Infect. Immun.* 58: 98–106.

Mulvey MA, Lopez-Boado YS, Wilson CL *et al.* (1998) Induction and evasion of host defenses by type 1-piliated uropathogenic *Escherichia coli* [published erratum appears in *Science* 1999 283(5403):795]. *Science* 282: 1494–1497.

Mysore JV, Wigginton T, Simon PM *et al.* (1999) Treatment of *Helicobacter pylori* infection in rhesus monkeys using a novel antiadhesion compound. *Gastroenterology* 117: 1316–1325.

Nassif X, So M (1995) Interaction of pathogenic neisseriae with nonphagocytic cells. *Clin. Microbiol. Rev.* 8: 376–388.

Neeman R, Keller N, Barzilai A, Korenman Z, Sela S (1998) Prevalence of internalisation-associated gene, prtF1, among persisting group-A streptococcus strains isolated from asymptomatic carriers [see comments]. *Lancet* 352: 1974–1977.

Ofek I (1995) Enzyme-linked immunosorbent-based adhesion assays. *Meth. Enzymol.* 253: 528–536.

Ofek I, Doyle R (1994) *Bacterial Adhesion to Cells and Tissues.* London: Chapman & Hall.

Ofek I, Sharon N (1988) Lectinophagocytosis: a molecular mechanism of recognition between cell surface sugars and lectins in the phagocytosis of bacteria. *Infect. Immun.* 56: 539–547.

Ofek I, Sharon N (1990) Adhesins as lectins: Specificity and role in infection. In: Jann K, Jann B (eds) *Bacterial Capsules and Adhesin: Facts and Principles.* Current Topics in Microbiology and Immunology vol 151. Berlin: Springer, pp. 91–113.

Ofek I, Goldhar J, Zafriri D *et al.* (1991) Anti-*Escherichia coli* adhesin activity of cranberry and blueberry juices [letter]. *N. Engl. J. Med.* 324: 1599.

Ofek I, Goldhar J, Keisari Y, Sharon N (1995) Nonopsonic phagocytosis of microorganisms. *Annu. Rev. Microbiol.* 49: 239–276.

Ofek I, Kahane I, Sharon N (1996) Toward anti-adhesion therapy for microbial diseases. *Trends Microbiol.* 4: 297–299.

Parent JB (1990) Membrane receptors on rat hepatocytes for the inner core region of bacterial lipopolysaccharides. *J. Biol. Chem.* 265: 3455–3461.

Passador L, Cook JM, Gambello MJ, Rust L, Iglewski BH (1993) Expression of *Pseudomonas aeruginosa* virulence genes requires cell-to-cell communication. *Science* 260: 1127–1130.

Rosenberg E, Ben-Haim Y, Toren A *et al.* (1999) Effects of temperature on bacterial bleaching of corals. In: Rosenberg E (ed.) *Microbial Ecology and Infectious Diseases.* Washington, DC: ASM Press, pp. 242–254.

Sahly HRP, Oelschlaeger TA, Greiwe M *et al.* (2001) Capsule impedes adhesion and invasion of epithelial cells by *Klebsiella pneumonia. Infect. Immun.* 68: 6744–6749.

Sauer FG, Mulvey MA, Schilling JD, Martinez JJ, Hultgren SJ (2000) Bacterial pili: molecular mechanisms of pathogenesis. *Curr. Opin. Microbiol.* 3: 65–72.

Schembri MA, Sokurenko EV, Klemm P (2000) Functional flexibility of the FimH adhesin: insights from a random mutant library. *Infect. Immun.* 68: 2638–2646.

Schlesinger LS, Hull SR, Kaufman TM (1994) Binding of the terminal mannosyl units of lipoarabinomannan from a virulent strain of *Mycobacterium tuberculosis* to human macrophages. *J. Immunol.* 152: 4070–4079.

Sharon N, Lis H (1996) Microbial lectins and their receptors. In: Montreuil J, Vliegenthart JFG, Schachter H (eds) *Glycoproteins.* Amsterdam: Elsevier, pp. 475–506.

Shin JS, Gao Z, Abraham SN (2000) Involvement of cellular caveolae in bacterial entry into mast cells [see comments]. *Science* 289: 785–788.

Soderhall M, Normark S, Ishikawa K *et al.* (1997) Induction of protective immunity after *Escherichia coli* bladder infection in primates. Dependence of the globoside-specific P-fimbrial tip adhesin and its cognate receptor. *J. Clin. Invest.* 100: 364–372.

Sokurenko EV, Courtney HS, Ohman DE, Klemm P, Hasty DL (1994) FimH family of type 1 fimbrial adhesins: functional heterogeneity due to minor sequence variations among fimH genes. *J. Bacteriol.* 176: 748–755.

Sokurenko EV, McMackin VA, Hasty DL (1995) Bacterial adhesion measured by growth of adherent organisms. *Meth. Enzymol.* 253: 519–528.

Sokurenko EV, Chesnokova V, Doyle RJ, Hasty DL (1997) Diversity of the *Escherichia coli* type 1 fimbrial lectin. Differential binding to mannosides and uroepithelial cells. *J. Biol. Chem.* 272: 17880–17886.

Sokurenko EV, Chesnokova V, Dykhuizen DE *et al.* (1998) Pathogenic adaptation of *Escherichia coli* by natural variation of the FimH adhesin. *Proc. Natl Acad. Sci. USA* 95: 8922–8926.

Sokurenko EV, Hasty DL, Dykhuizen DE (1999) Pathoadaptive mutations: gene loss and variation in bacterial pathogens. *Trends Microbiol.* 7: 191–195.

Steadman R, Topley N, Jenner DE, Davies M, Williams JD (1988) Type 1 fimbriate *Escherichia coli* stimulates a unique pattern of degranulation by human polymorphonuclear leukocytes. *Infect. Immun.* 56: 815–822.

Stromberg N, Nyholm PG, Pascher I, Normark S (1991) Saccharide orientation at the cell surface affects glycolipid receptor function. *Proc. Natl Acad. Sci. USA* 88: 9340–9344.

Svanborg C, Hedlund M, Connell H *et al.* (1996) Bacterial adherence and mucosal cytokine responses. Receptors and transmembrane signaling. *Ann N Y Acad. Sci.* 797: 177–190.

Svanborg C, Godaly G, Hedlund M (1999) Cytokine responses during mucosal infections: role in disease pathogenesis and host defence. *Curr. Opin. Microbiol.* 2: 99–105.

Sylvester FA, Philpott D, Gold B, Lastovica A, Forstner JF (1996) Adherence to lipids and intestinal mucin by a recently recognized human pathogen, Campylobacter upsaliensis. *Infect. Immun.* 64: 4060–4066.

Szymanski CM, Armstrong GD (1996) Interactions between *Campylobacter jejuni* and lipids. *Infect. Immun.* 64: 3467–3474.

Tennent JM, Hultgren S, Macklund B-I *et al.* (1990) Genetics of adhesin expression in *Escherichia coli*. San Diego: Academic Press.

Tewari R, MacGregor JI, Ikeda T *et al.* (1993) Neutrophil activation by nascent FimH subunits of type 1 fimbriae purified from the periplasm of *Escherichia coli. J. Biol. Chem.* 268: 3009–3015.

Thankavel K, Madison B, Ikeda T *et al.* (1997) Localization of a domain in the FimH adhesin of *Escherichia coli* type 1 fimbriae capable of receptor recognition and use of a domain-specific antibody to confer protection against experimental urinary tract infection. *J. Clin. Invest.* 100: 1123–1136.

Thankavel K, Shah AH, Cohen MS *et al.* (1999) 3rd, Abraham SN. Molecular basis for the enterocyte tropism exhibited by Salmonella typhimurium type 1 fimbriae. *J. Biol. Chem.* 274: 5797–5809.

Topley N, Steadman R, Mackenzie R, Knowlden JM, Williams JD (1989) Type 1 fimbriate strains of *Escherichia coli* initiate renal parenchymal scarring. *Kidney Int.* 36: 609–616.

Virji M, Stimson E, Makepeace K *et al.* (1996) Posttranslational modifications of meningococcal pili. Identification of a common trisaccharide substitution on variant pilins of strain C311. *Ann. N Y Acad. Sci.* 797: 53–64.

Weiss EI, Lev-Dor R, Kashamn Y *et al.* (1998) Inhibiting interspecies coaggregation of plaque bacteria with a cranberry juice constituent [published erratam appear in *J. Am. Dent. Assoc.* 1999 130(1): 36 and 130(3): 332. *J. Am. Dent. Assoc.* 129: 1719–1723.

Westerlund B, Kuusela P, Vartio T, van Die I , Korhonen TK (1989) A novel lectin-independent interaction of P fimbriae of *Escherichia coli* with immobilized fibronectin. *FEBS Lett.* 243: 199–204.

Westerlund B, van Die I, Kramer C *et al.* (1991) Multifunctional nature of P fimbriae of uropathogenic *Escherichia coli*: mutations in fsoF and fsoF influence fimbrial binding to renal tubuli and immobilized fibronectin. *Mol. Microbiol.* 5: 2965–2975.

Wu XR, Sun TT, Medina JJ (1996) *In vitro* binding of type 1-fimbriated *Escherichia coli* to uroplakins Ia and Ib: relation to urinary tract infections. *Proc. Natl Acad. Sci. USA* 93: 9630–9635.

Yu C, Lee AM, Roseman S (1987) The sugar-specific adhesion/deadhesion apparatus of the marine bacterium *Vibrio furnissii* is a sensorium that continuously monitors nutrient levels in the environment. *Biochem. Biophys. Res. Commun.* 149: 86–92.

Yuyama Y, Yoshimatsu K, Ono E, Saito M, Naiki M (1993) Postnatal change of pig intestinal ganglioside bound by *Escherichia coli* with K99 fimbriae. *J. Biochem. (Tokyo)* 113: 488–492.

Zafriri D, Oron Y, Eisenstein BI, Ofek I (1987) Growth advantage and enhanced toxicity of *Escherichia coli* adherent to tissue culture cells due to restricted diffusion of products secreted by the cells. *J. Clin. Invest.* 79: 1210–1216.

Zaidi TS, Fleiszig SM, Preston MJ, Goldberg JB, Pier GB (1996) Lipopolysaccharide outer core is a ligand for corneal cell binding and ingestion of *Pseudomonas aeruginosa. Invest. Ophthalmol. Vis. Sci.* 37: 976–986.

Zhang JP, Normark S (1996) Induction of gene expression in *Escherichia coli* after pilus-mediated adherence [see comments]. *Science* 273: 1234–1236.

31 Invasion

Chihiro Sasakawa

University of Tokyo, Tokyo, Japan

Some pathogenic bacteria have an ability to direct their own internalisation into non-phagocytic cells such as epithelial cells. This bacterial capacity is often termed invasiveness and such pathogens termed invasive bacteria. Bacterial internalisation into epithelium results in either colonisation within or translocation across the mucosal barrier and, in some cases, the pathogen becomes sequestered within an infected organ or gains further access to deeper or systemic tissues by way of the blood or lymphatic vessels. Thus, the ability to invade non-phagocytic cells is a prominent feature of bacterial pathogenesis.

Entry into non-phagocytic cells, such as epithelial and endothelial cells or stromal cells, obviously benefits the pathogen. Host cell cytoplasm offers a safe environment (a 'niche') for bacterial multiplication, where the pathogen has no competitor for growth, and the pathogen may evade the host immune system or avoid killing by antibiotics. Although the supply of nutrients in intracellular environments is limited, invasive bacteria exclusively possess various sensing systems which respond to change in the environment and develop import systems to obtain necessary nutrients from the host.

Bacteria use various mechanisms to enter host cells, and on this basis they are categorised into two major classes: those in which a microbial ligand interacts with a host cell receptor, and those in which entry is mediated by the delivery into the host cells of bacterial proteins, often called effectors. These trigger membrane ruffling and macro-pinocytosis. The first type of invasion is represented by *Yersinia pseudotuberculosis* and is mediated by a 'zipper-like' mechanism. In this case, the bacterial internalisation event limits its own uptake by the host cells. The second class of invasion is represented by *Salmonella typhimurium* and allows uptake of other particles together with the invasive pathogen. In spite of these differences, invasive bacteria can variously remodel the host cell surface, for example by stimulating Rho GTPases, protein tyrosine phosphorylation or lipid metabolism, which lead to actin or tubulin dynamism in eukaryotic cells. Although exactly how invasive pathogens act is still largely unclear, the current *in vitro* approach of 'cellular microbiology' and studies with appropriate animal models *in vivo* has yielded considerable insight into the important roles of bacterial and host cellular functions. In parallel with this, the marked accumulation of microbial genomic information is making it possible to identify potent virulence factors. A comprehensive genomic analysis of various pathogens has indicated that bacteria often share common genetic elements or strategies for the establishment of infection. This chapter will focus on how invasive pathogens such as *S. typhimurium*, *Shigella flexneri*, *Campylobacter jejuni*, *Y. pseudotuberculosis*, *Listeria monocytogenes*, *Neisseria gonorrhoeae* or *Streptococcus pyogenes* direct their own internalisation into epithelial cells.

Salmonella

Salmonella are enteric pathogens that infect a variety of hosts, including humans, and cause diseases that range

 doi:10.1006/bkmm.2001.0031

from localised gastroenteritis (salmonellosis) to more serious systemic, host-specific illnesses such as typhoid fever. The outcome of *Salmonella* infection varies, depending on the bacterial serotype as well as on the infected host. In general, host-adapted pathogenic *Salmonella* species cause systemic enteric fevers, while non-host-adapted *Salmonella* strains cause gastroenteritis as the result of localised infection of the small intestine. Typically, *S. typhimurium* has a murine reservoir where it causes a typhoid fever-like disease, but it can also infect a variety of other animals including humans to cause non-specific, usually short-lived, superficial gastroenteritis (Jones and Falkow, 1996).

When *S. typhimurium* is introduced into the stomachs of mice, the bacteria invade the mucosa of the small intestine and display a tropism for the M cells of Peyer's patches (Carter and Collins, 1974; Hohmann *et al.*, 1978; Clark *et al.*, 1994; Jones *et al.*, 1994). Entry of *S. typhimurium* into M cells is accompanied by the formation of membrane ruffle in a manner similar to that seen in the invasion of epithelial cells in tissue culture (Francis *et al.*, 1992, 1993; Ginocchio *et al.*, 1992), and M cell internalisation of *S. typhimurium* causes a destruction of the cytoskeletal architecture (Jones *et al.*, 1994; Jones and Falkow, 1996). After translocation, *Salmonella* invades the resident macrophages beneath the M cells, where the infected macrophages are killed by *Salmonella*-induced apoptosis (Monack *et al.*, 1996; Chen *et al.*, 1996a). Although, at the initial stage of infection, adjacent enterocytes appear to be spared, after a few hours significant invasion of enterocytes can occur (Jones and Falkow, 1996). After intestinal penetration, *S. typhimurium* causes a transient bacteraemia and disseminates to systemic sites of infection, most importantly the liver and spleen, where the bacteria multiply. It is at present unclear, however, whether non-host-adapted *Salmonella* serotypes that cause salmonellosis in humans actually initiate infection by the M cell route, the apical membrane of enterocytes, or the tight junction between epithelial cells.

Mode of Invasion

Invasion of cultured polarised epithelial cells, such as MDCK and HEp-2 cells, by *S. typhimurium* shows that the pathogen can efficiently enter the host cell from the apical surface, where membrane ruffling and subsequent macro-pinocytosis are evoked on the epithelial cell surface to engulf the contact bacteria (**Fig. 1**). The bacteria are surrounded by a membrane vacuole, multiply in the vacuole, and are then transcytosed toward the outside of the host cell. Unlike *Yersinia*, *Salmonella* invasion involves the triggering of host cellular signalling events that lead to the membrane ruffling and macro-pinocytosis around the site of bacterial entry. Through this mode of bacterial entry, other non-invasive bacteria can be entrapped and can co-infect together with the pathogen (Francis *et al.*, 1993). Such cellular responses are triggered upon delivery of a subset of *Salmonella*

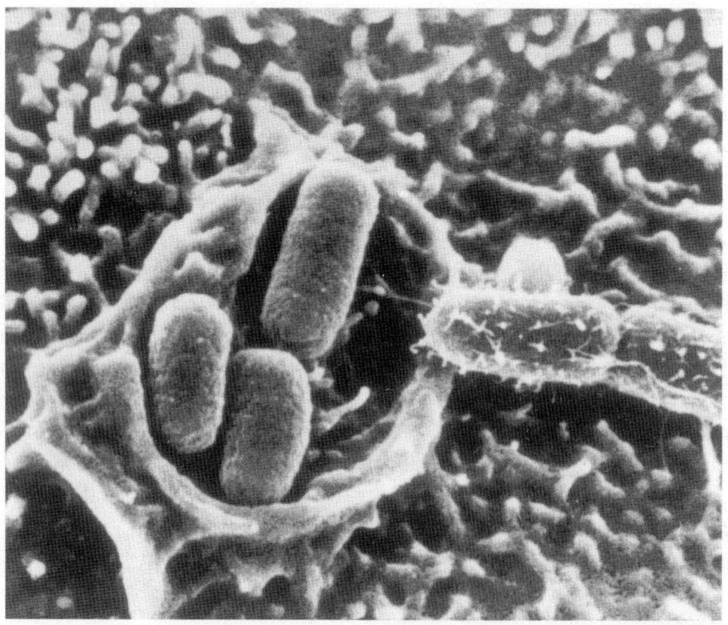

Fig. 1 Membrane ruffling evoked on cultured MDCH cells by *S. typhimurium*. The picture shows bacteria associated with membrane ruffles (Ginocchio *et al.*, 1994). The photo was kindly provided by J.E. Galán with permission by Cell Press.

effector proteins through the type III secretion machinery after contact with the host cell (see below). *Salmonella*, once taken up by the host cells, remains within a membrane vacuole which is surrounded by large (5–10 μm) dense structures composed of various cytoskeletal components including F-actin, α-actinin, ezrin, talin and tropmyosin (Garcia-del Portillo *et al.*, 1994; Garcia-del Portillo and Finlay, 1995).

Host Cell Signalling

The initial intracellular signalling event after contact of *S. typhimurium* with host cells is a rapid increase in the level of inositol triphosphate. This results from the activation of host cell phospholipase C (PLC) and Ca^{2+} flux, where activated PLC appears to induce an increase in the intracellular level of Ca^{2+} ion (Pace *et al.*, 1993). In Ca^{2+} flux triggered by *Salmonella* infection, this response ensues within a few minutes of the internalisation, reaching a maximum level of over 1 μM of intracellular Ca^{2+} ion within 30 minutes after infection (Pace *et al.*, 1993). Since chelators of intracellular Ca^{2+} can block *Salmonella* invasion, and a non-invasive *S. typhimurium* mutant fails to cause a rise in Ca^{2+} ion and rearrangement of actin cytoskeletons, Ca^{2+} ion is thought to serve as the messenger that transmits the invasion signal toward downstream effector molecules to initiate uptake of bacteria.

Members of the Rho subfamily of GTPases, RhoA, Rac1 and Cdc42, regulate various actin-dependent processes in mammalian cells, such as formation of focal adhesions, lamellipodia (membrane ruffling or macro-pinocytosis) and filopodia. In particular Cdc42, and to a lesser extent Rac1, play crucial roles in *Salmonella*-stimulated actin cytoskeletal rearrangements (Chen *et al.*, 1996b). Galán and his co-workers showed that expression of a dominant interfering mutant of Cdc42 prevents *S. typhimurium*-induced cytoskeletal rearrangement and bacterial internalisation into host cells, while cells that express a constitutively active form of Cdc42, allow internalisation of a non-invasive *S. typhimurium* mutant (Chen *et al.*, 1996b). *S. typhimurium* secretes an activator of Rho GTPases that can induce membrane ruffling and nuclear responses in host cells (Hardt *et al.*, 1998). The activator of *S. typhimurium* is identical to the SopE protein of *S. dublin*, which has been cloned by reverse genetics (Wood *et al.*, 1996; Hardt *et al.*, 1998), and was later shown to be one of the *S. typhimurium* effector proteins delivered into the host cells by the type III secretion machinery (Hardt *et al.*, 1998). Microinjection of purified SopE protein into Cos-1 cells provokes lamellipodia and ruffles in as little as 5 minutes. SopE-mediated membrane ruffling and nuclear responses, such as JNK activation, occur in a Cdc42- and Rac1-dependent manner. In mammalian cells, over-expression of a dominant interfering Cdc42 or Rac1 completely abrogates SopE-induced cytoskeletal rearrangements. Purified SopE protein can stimulate GDP/GTP nucleotide exchange in Cdc42 and Rac1 *in vitro*, suggesting that SopE mimics eukaryotic GEF, the stimulator for exchange of GDP for GTP, but SopE does not share significant amino acid similarity with known GEF proteins (Hardt *et al.*, 1998; Rudolph *et al.*, 1999).

Secreted Effector Proteins and Type III Secretion System

Salmonella typhimurium secretes at least 12 proteins via the type III secretion system and delivers them into the host cell. The secreted proteins include SipA, SipB, SipC, SipD, InvJ (SpaN), SicA, IacP, SicP, SptP, AvrA, SopB (SigD) and SopE (Darwin and Miller, 1999). SipA, SipB, SipC, SipD, InvJ, SicA, IacP, SicP, SptP and AvrA are encoded by the *Salmonella* pathogenicity island 1 (SPI-1) genes, while SopB and SopE are encoded by genes located outside SPI-1. The roles of the secretion proteins in *Salmonella* invasion, including the pathogenicity, are still partly unclear but some, such as SipA, SipB, SipC, SptP, SopB and SopE, serve as effectors to modulate the mammalian cell surface driven by the rearrangements of actin cytoskeletons. SipA, which shares amino acid identity with IpaA of *Shigella*, can bind F-actin, leading to a decrease in the actin critical concentration and repression of F-actin depolymerisation *in vitro* (Zhou *et al.*, 1999). The interaction of SipA with F-actin seems to contribute to stabilisation of the architecture of prominent lamellipodium and membrane ruffling induced by *Salmonella* invasion (Zhou *et al.*, 1999). SipB and SipC share amino acid similarity with IpaB, and IpaC of *Shigella* is required for invasion. SipB and SipC, like IpaB and IpaC, are involved in delivering the other effector proteins, and they serve as a molecular bridge between the bacterium and host cell. When SipB is delivered from *S. typhimurium* into macrophages, it induces apoptosis (Monack *et al.*, 1996; Chen *et al.*, 1996a). SipC acts *in vitro* to direct actin nucleation and bundling. Upon microinjection of SipC into HeLa cells, actin condensation and cytoskeletal rearrangements are induced (Hayward and Koronakis, 1999). The N-terminal region of SptP shares sequence similarity with ExoS of *Pseudomonas aeruginosa* and YopE of *Yersinia* spp. (Kaniga *et al.*, 1996). A recent study has indicated that SptP encodes GAP (a family of GTPase-activating proteins) activity and can preferentially bind to Rac1, and that SptP is involved in the down-regulation of the cellular responses required to rebuild the actin cytoskeletons altered by *Salmonella* internalisation (Fu and Galán, 1999). SopB (also called SigD) appears to be involved in invasion, since a *sigD* mutation affects *S. typhimurium* invasion of CHO and HEp-2 cells (Galyov *et al.*, 1997; Hong and Miller, 1998). Interestingly, SopB has sequence homology with mammalian inositol polyphosphate 4-phosphatase, and recombinant SopB protein prepared from *S. dublin* shows inositol phosphate phosphatase activity *in vitro* (Norris *et al.*, 1998).

The supramolecular structure of the type III secretion machinery of *S. typhimurium*, including its membrane location has recently been elucidated (Kubori *et al.*, 1998). The type III secretion machinery is located in the *Salmonella* envelope and spans the inner and outer membrane. The purified type III complex is a hollow structure approximately 120 nm long with basal and needle components. In the type III complex observed in an osmotically shocked *S. typhimurium*, the needle part protrudes outside the outer membrane, while the basal part, which resembles a flagellar basal body, is embedded in the envelope. The type III complex contains at least three proteins, InvG, PrgH and PrgK. Although in basic features the type III complex resembles a flagellar basal body, the InvG, PrgH and PrgK proteins share only limited sequence similarity with it (Galán and Collmer, 1999). The precise mechanisms that underlie the type III export system, including the delivery of effectors into mammalian cells, are still poorly understood. However, effector proteins including the needle components appear to be exported through the hollow conduit of the machinery in the same way as has been proposed for the flagellin export system. It is postulated that the tip of the type III needle, secreted SipB and SipC, is inserted into the host mammalian cell membrane to form a membrane pore through which *Salmonella* effector proteins, including SipB and SipC, are injected into the target cell.

Regulation of Invasion Genes

SPI-1 genes are required for *Salmonella* invasion, and SPI-1 possesses over 30 genes that encode type III secretion machinery, effector proteins, chaperones and regulatory proteins. The expression of SPI-1 genes is affected by various environmental conditions. For example, *Salmonella* invasion peaks its expression at 37°C *in vitro*, and invasiveness is enhanced by growing bacteria to stationary phase in high osmotic medium under anaerobic conditions (Darwin and Miller, 1999). SPI-1 expression is governed by a multiple regulatory systems involving PhoP/PhoQ, HilA, SirA, SprA and InvF proteins (Darwin and Miller, 1999; Eichelberg *et al.*, 1999; Eichelberg and Galàn, 1999). The *hilA*, *invF* and *sprA* genes are located in SPI-1. The PhoP/PhoQ two-component regulatory system affects *hilA* expression, while HilA requires activation of *invF* expression. HilA can bind to the *invF* promoter region, since HilA possesses a putative DNA-binding domain that shares amino acid similarity with the DNA-binding domain of OmpR and ToxR. InvF and SprA belong to the AraC family and can also bind to DNA. Recent studies have indicated that *invF* expression is governed by a multiple activation system that involves HilA, SprA and SirA, in which *hilA* expression is positively controlled by SprA and SirA (Darwin and Miller, 1999; Eichelberg and Galàn, 1999; Eichelberg *et al.*, 1999). The disruption of the *invF* gene causes a marked decline in the invasiveness of *Salmonella*, since InvF directly regulates downstream gene expression of *sipC*, *sptP* and *sopB*, the *Salmonella* effector proteins (Darwin and Miller, 1999; Eichelberg and Galán, 1999). The disruption of the *sprA* gene little affects *Salmonella* invasion, but its over-expression can markedly enhance the invasiveness (Eichelberg *et al.*, 1999). Although the roles of each of the regulatory proteins in the expression of the SPI-1 genes are still to be determined, the existence of differential regulatory systems in *S. typhimurium* may be important in producing and delivering the appropriate effectors at the appropriate time, to the appropriate target, and in the appropriate order during the infection process (Eichelberg and Galàn, 1999).

Shigella

Shigella are highly adapted human pathogens which cause bacillary dysentery, a disease that provokes a severe bloody and mucous diarrhoea. Bacillary dysentery is also caused by enteroinvasive *Escherichia coli* (EIEC) which closely resembles *Shigella*. *Shigella* and EIEC possess a 200–230-kb plasmid on which the major virulence functions are encoded. When *Shigella* reaches the colon, bacteria translocate through the epithelial barrier by way of the M cells that overly the solitary lymphoid nodules (Wassef *et al.*, 1989; Sansonetti *et al.*, 1991, 1996) (Fig. 2). Once they reach the underlying M cells, *Shigella* infect the resident macrophages and induce apoptosis (Zychlinsky *et al.*, 1992). The infected macrophages release massive amounts of IL-1β, thus triggering a strong inflammatory response (Zychlinsky *et al.*, 1994, 1996), which eventually leads to an increase in the

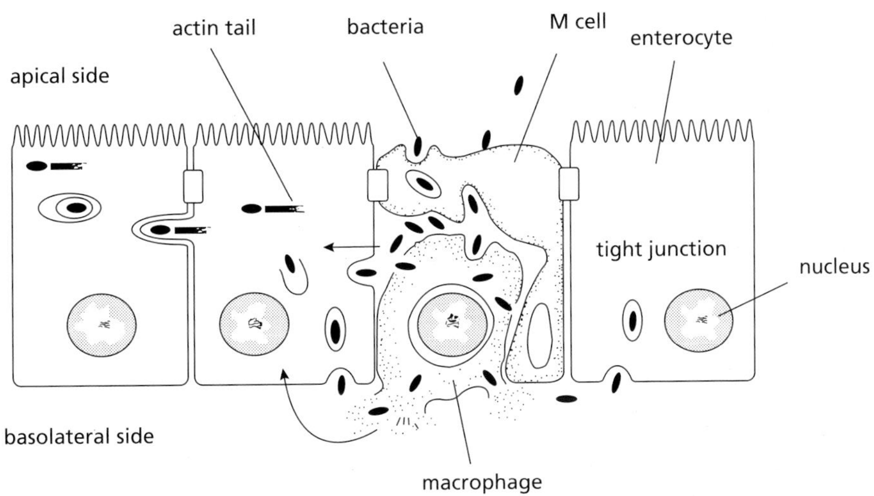

Fig. 2 A simplified model for the infection of colonic epithelial cells by *Shigella*. The diagram shows the M cell entry pathway of *Shigella* into enterocytes proposed by Sansonetti and his co-workers.

permeability of the epithelial barrier to *Shigella* and migration of polymorphonuclear leucocytes (PMN) (Perdomo *et al.*, 1994).

Meanwhile, the bacteria released from the killed macrophages enter into enterocytes from the basolateral surface by inducing membrane ruffling and macro-pinocytosis. Once the bacterium is surrounded by a membrane vacuole, it immediately disrupts the vacuole membrane to escape into the cytoplasm (Sansonetti *et al.*, 1986). Within the cytoplasm, *Shigella* can multiply and move by inducing actin polymerisation at one pole of the bacterium, allowing intracellular bacterium to spread within the cytoplasm as well as into the adjacent epithelial cells (Makino *et al.*, 1986; Bernardini *et al.*, 1989). The bacteria do not normally disseminate systemically, but are confined within the epithelial barriers. In response to bacterial invasion, colonic epithelial cells express a large array of pro-inflammatory cytokines, especially IL-8, thus further promoting local inflammation and attracting more PMNs. Thus, the predominant feature of the pathogenicity of *Shigella* is the ability to invade epithelial cells, including dissemination into adjacent epithelial cells and to infect macrophages.

Basolateral Entry

Invasion of non-differentiated epithelial cells by *Shigella* takes place through the edges of peripheral cells, where the transferrin receptor, a basolateral marker protein, is preferentially expressed (Mounier *et al.*, 1992). However, infection of fibroblasts by *Shigella* tends to be from areas where focal adhesion plaques have formed. *Shigella* infection of polarised Caco-2 cells reveals that the bacteria do not penetrate the apical surface, instead, they enter from the basolateral surface (Mounier *et al.*, 1992). This mode of entry is consistent with events during infection of rabbit ligated ileal loops *in vivo*, where bacteria move to the basolateral surface via the M cells (Sansonetti *et al.*, 1996) (Fig. 2). *Shigella*, therefore, appears to have an affinity for the basolateral surface of the polarised enterocytes for its entry. *Shigella* can induce a highly dynamic re-arrangement of the actin cytoskeleton during the entry that leads to a macro-pinocytotic event. Shortly after contact with epithelial cells, the organisms induce focal adhesion-like actin dense patches beneath the contact point followed by membrane protrusions and extension. The invading *Shigella* finally becomes enclosed by a large membrane vacuole, but the pathogen can immediately disrupt the membrane vacuole and escape into the host cell cytoplasm.

Invasion Genes and their Regulation

Genetic studies have revealed that *Shigella* invasion of epithelial cells requires the numerous genes encoded by the 31-kb pathogenicity island on the 230-kb plasmid (Sasakawa *et al.*, 1988). The pathogenicity island of *Sh. flexneri* contains 28 invasion-associated genes arranged in several transcribed regions. One of these contains the *ipa* genes whose products, such as IpaB and IpaC, are required for bacterial-cell interaction, including induction of macro-pinocytosis from host cells, while the *mxi* and *spa* regions are mostly involved in the formation of type III secretion machinery. The other loci contain genes that encode a molecular chaperone (IpgC) required for IpaB and IpaC expression in *Shigella* (Ménard *et al.*, 1994a, 1994b), and a regulator, *virB*, required for the expression of invasion genes at 37°C (Adler *et al.*, 1989).

Shigella encounter an increase in growth temperature to 37°C upon infection of the human host. This change in environment results in the expression of the virulence genes in the 31 kb pathogenicity island invasion gene, which are expressed at 37°C (Adler *et al.*, 1989). VirF, a member of AraC family, binds the promoter region of *virB*, thus activating *virB* transcription, which, in turn, activates several operons encoding *ipa*, *mxi*, *spa*, *ics* and *ipg* genes. The activation of *virB* transcription by VirF is highly sensitive to changes in DNA super-helicity around the *virB* promoter region and is antagonised at 30°C by H-NS (Tobe *et al.*, 1993, 1995). Thus, the cellular level of VirF at 37°C is critical for the expression of the invasiveness. Indeed, the expression of *virF* does not seem to be constitutive at 30°C, rather it is induced at 37°C (Colona *et al.*, 1995; Falconi *et al.*, 1998).

Secretion of Effector Proteins and Host Cellular Responses

Shigella induces macro-pinocytosis by host cells through two host cell signalling pathways. One involves signalling linked to the interaction of secreted Ipa proteins, such as IpaB and IpaC, with putative host surface receptors. The second involves signalling evoked by bacterial effectors, such as IpaA, IpaB, IpaC, IpgD and VirA (Tran van Nhieu and Sansonetti, 1999). The roles of Ipa proteins during invasion are complicated, because these proteins have multiple functions. For example, IpaA, IpaB and IpaC act as the secreted effector proteins in target host cells to modify actin dynamism, while IpaB and IpaD act as the molecular plug for regulating the type III secretion system (Ménard *et al.*, 1994b, 1996; Tran van Nhieu and Sansonetti, 1999). IpaB and IpaC, like *Salmonella* SipB and SipC, serve as a membrane pore as part of the type III machinery in the host cell plasma membrane, allowing the translocation of secreted effector proteins into the host cells (Blocker *et al.*, 1999).

IpaB and IpaC proteins secreted into the culture supernatant form high molecular matrix-like structures that seem to promote *Shigella* invasion by interacting with the putative host receptors, such as $\alpha5\beta1$ integrin and CD44, thus mediating outside-in signalling to induce rearrangement of actin cytoskeletons. $\alpha5\beta1$ integrin localises at focal adhesions and the cytoplasmic domain of the $\beta1$ integrin acts as a cytoskeleton linker via the focal adhesion proteins, while the cytoplasmic moiety of CD44 also acts as a cytoskeleton linker via the association with ERM (ezrin-radixin-moesin) proteins. Ezrin is recruited at the periphery of the extended membrane ruffles at the point of *Shigella* entry (Skoudy *et al.*, 1999), while at the site of bacterial entry, $\alpha5\beta1$ integrin appears to be accumulated together with F-actin, vinculin, talin, α-actin and tyrosine-phosphorylated proteins, major components of focal adhesions (Watarai *et al.*, 1996, 1997). This interaction would not, however, be sufficient to elicit the macro-pinocytotic events required for bacterial uptake by epithelial cells. Although latex beads coated with IpaB–IpaC complex can be taken up by host cells, they are unable to stimulate membrane ruffles as large as those induced by *Shigella* (Ménard *et al.*, 1996).

Further large-scale rearrangement of actin dynamism would require a second signalling event within the host cytoplasm by the delivery of IpaA and IpaC and perhaps other effectors, such as VirA and IpgD. IpaA binds vinculin, a component of focal adhesions (Tran van Nhieu *et al.*, 1997). The membrane ruffling induced by an *ipaA* mutant of *Sh. flexneri* and the invasive efficiency are reduced compared with that evoked by the wild type. Subsequently it was shown that the IpaA–vinculin complex with F-actin promotes depolymerisation of actin filaments, which is thought to be required for modification of the *Shigella*-induced membrane protrusions (Bourdet-Sicard *et al.*, 1999). IpaC may modulate actin dynamism, since filopodium and lamellipodium formation is induced when purified IpaC protein is added to permeabilised Swiss 3T3 cells or an *ipaC* clone is transfected into HeLa cells (Tran van Nhieu *et al.*, 1999). IpaC-induced membrane protrusions can be inhibited by a dominant interfering form of Cdc42, while a dominant interfering form of Rac causes inhibition of lamellipodium formation, suggesting that IpaC induces activation of Cdc42 which in turn, causes activation of Rac.

Other studies have also indicated that Rho GTPases, such as RhoA, Rac1 and Cdc42, control the entry of *Sh. flexneri* into epithelial cells, since the dominant interfering form of any of these GTPases causes a significant reduction in efficiency of *Shigella* invasion of these mutants host cells (Mounier *et al.*, 1999). *Shigella* also requires activation of the focal adhesion kinase (FAK) and Src tyrosine kinase to induce cytoskeletal rearrangements during entry (Watarai *et al.*, 1997; Duménil *et al.*, 1998). Over-expression of a dominant interfering form of $pp60^{c\text{-}Src}$ leads to inhibition of *Shigella*-induced cytoskeletal rearrangements and decreases phosphorylation of cortactin (Duménil *et al.*, 1998), an Src substrate recruited at the site of *Shigella* entry (Dehio *et al.*, 1995). Since focal adhesion formation depends on Rho and Src activation, an early cellular event triggered by *Shigella* contact may be associated with such a cellular function (Adam *et al.*, 1996; Ménard *et al.*, 1996; Watarai *et al.*, 1997).

Campylobacter

Campylobacter is a spiral, Gram-negative rod with polar flagella. *Campylobacter jejuni* and *C. coli* are major causes of human diarrhoeal diseases and are often contracted from their natural avian hosts by consumption of undercooked poultry or contaminated drinking water. The clinical symptoms range widely from a protracted watery diarrhoea to bloody diarrhoea with fever, abdominal cramps and the presence of leucocytes in the faeces (Ketley, 1997; Wassenaar and Blaser, 1999). In addition, a recent study has suggested several *C. jejuni* serotypes as causative factors of post-diarrhoeal Guillain-Barré syndrome, an acute paralysing neuromuscular disorder (Nachamkin *et al.*, 1998).

Mode of Invasion

Studies with models of infection *in vitro* and *in vivo* have indicated that *C. jejuni* and *C. coli* are highly invasive for epithelial cells and that this process plays an important role in *Campylobacter*-mediated diseases (Ketley, 1997; Wooldridge and Ketley, 1997). In experimental infection of infant macaque monkeys, as well as several other animal species, and in intestinal biopsy specimens from *C. jejuni*-positive patients with colitis, the pathogen has been shown to be taken up by enterocytes, where bacteria are found both within membrane vacuoles and free in the host cell cytoplasm (Blaser *et al.*, 1979, 1983; Duffy *et al.*, 1980; Black *et al.*, 1988).

Another possible route for translocation across the mucosal barrier is through M cells in the intestine. *Campylobacter* was shown to interact with and translocate through M cells in infected rabbit intestine (Walker *et al.*, 1988). Invasion of enterocytes by *Campylobacter* is accompanied by a cytopathic effect that is believed to be the primary mechanism of damage to the colonic mucosa, leading to severe inflammation and bloody diarrhoea (Russell *et al.*, 1993). The invasive capacity of *Campylobacter* has been shown in a variety of epithelial cell lines, although the invasion efficiency and dependency on the host cytoskeletons vary according to bacterial strain, mammalian cell line and conditions of infection. In particular, a dependency on host cell cytoskeletons has been reported. Some workers indicated that infection depends on microfilaments (MFs) (Duffy *et al.*, 1980; Konkel and Joens, 1989; de Melo *et al.*, 1989), while others consider that it depends on microtubules (MTs) (Oelschlaeger *et al.*, 1993; Hu and Kopecko, 1999). Yet other workers consider that it depends on neither MFs of MTs (Russell and Blake, 1994).

Although no consensus on the host cell factors has yet been arrived at, some earlier reports suggested that actin polymerisation is important, since treatment of HEp-2 cells with cytochalasin B caused a significant decrease in invasiveness (Konkel and Joens, 1989; Konkel *et al.*, 1992). Consistent with this, F-actin aggregation may be induced at the sites of *Campylobacter* attachment to HEp-2 cells (Konkel and Joens, 1989; de Melo *et al.*, 1989), while in the invasion of INT407 cells, an invagination of the epithelial cell membrane through formation of filopodia can be evoked (Konkel *et al.*, 1992), suggesting that bacterial entry involves a highly dynamic rearrangement of actin-related cytoskeletons. Furthermore, experiments with various inhibitors of tyrosine phosphorylation, activity of heterodimeric G proteins or phosphatidylinositol 3-kinase (PI3-kinase) have indicated a reduction of invasiveness of *C. jejuni* (Wooldridge *et al.*, 1996). Although no functional assay has yet been undertaken, these results strongly suggest that host cellular signals involved in modulating actin dynamism leading to formation of filopodia and lamellipodia would be required for uptake of *C. jejuni*.

Other studies, however, have indicated that *Campylobacter* invasion takes place independently of actin polymerisation, rather the uptake by INT407 or Caco-2 cells is strongly dependent on MTs. In fact, pre-treatment of mammalian cells with MT-depolymerising agents such as colchicine, vinblastine or vincristine greatly reduces *C. jejuni* invasion (Oelschlaeger *et al.*, 1993; Hu and Kopecko, 1999). In the latter study, the treatment of epithelial cells with cytochalasin D, a specific inhibitor of actin polymerisation, actually stimulated uptake of *Campylobacter* by epithelial cells. Examination by immunofluorescence microscopy of *Campylobacter* for association with MTs and MFs indicated that bacteria can initially bind to the tips of host cell membranous extensions that contain MTs. They then align in parallel with MTs during entry, co-localise specifically with MTs and dynine, a motor protein associated with MTs, and subsequently move toward the nuclear membrane of the host cells (Hu and Kopecko, 1999). Furthermore, orthovanadate, an inhibitor of dynine activity, inhibits *C. jejuni* entry, suggesting that a molecular motor associated with MTs is involved in entry and vacuole trafficking during the internalisation of *C. jejuni* (Hu and Kopecko, 1999). Irrespective of the reason for these controversial results, the consensus is that *C. jejuni* invasion requires highly dynamic rearrangement of the host cell cytoskeletons.

Proteins Involved in Invasion

Although *Campylobacter* factors involved in invasion are still poorly understood, at least two bacterial components, flagellum and CiaB, have been genetically defined as required for invasion. The invasive efficiency of *C. jejuni* is affected by the expression of flagellin, including flagellin-mediated bacterial motility. A *flaA* mutant of *C. jejuni* showed a decrease in invasive capacity as compared with the isogenic wild type strain, even through the mutant could adhere fully to epithelial cells. Furthermore, a non-motile mutant with intact, but non-functional flagella, also showed reduced invasiveness, implying that the functional flagellum is required for internalisation into host cells (Wassenarr *et al.*, 1991; Grant *et al.*, 1993; Yao *et al.*, 1994). Expression of flagella by *Campylobacter* species shows antigenic and phase variation, although not all *Campylobacter* isolates show both types of variation. Studies with rabbits and human volunteers have shown that flagellated *C. jejuni* became predominant in the intestine during infection, supporting the importance of flagellin in the pathogenesis. *C. jejuni* like *Salmonella*, can traverse monolayers of polarised colonic epithelial cells by being surrounded by membrane vacuoles (Everest *et al.*, 1992; Konkel *et al.*, 1992; Grant *et al.*, 1993). The ability of *C. jejuni* and *C. coli* to translocate across the epithelial cell barrier is considered to contribute to pathogenesis, since the translocation permits pathogens to gain further access to submucosal tissue, thus leading to the tissue damage and inflammation, often represented by *Campylobacter*-associated enteritis. In experimental infection of polarised Caco-2 cells with non-flagellated or non-motile mutants of *C. jejuni*, either the motility or the product of the *flaA* gene were shown to be essential factors in translocation of polarised epithelial cell monolayers (Grant *et al.*, 1993).

CiaB has been identified as one of the secretion proteins required for invasion of INT407 cells (Konkel *et al.*, 1999). Interestingly, the 73-kDa CiaB protein shares significant amino acid similarity with type III secreted proteins such as *Salmonella* SipB (45%), *Shigella* IpaB (41%) and *Yersinia* YopB (45%). Indeed, like SipB, IpaB and YopB proteins, CiaB can be secreted from *C. jejuni* into the culture medium and also delivered into INT407 cells through contact. Culture of *C. jejuni* with INT407 cells or in INT407-conditioned medium can also stimulate secretion of at least eight proteins, including CiaB, ranging from 12.8 to 108 kDa in size, into the culture medium. Since *C. jejuni ciaB*-null mutants are unable to support the secretion of all eight proteins, CiaB is presumed to serve as a part of the secretion system itself. Although, based on the DNA sequence surrounding the *ciaB* gene, *ciaB* does not seem to be present within the putative pathogenicity island, the properties of CiaB and the amino acid similarity to other type III secreted proteins may imply that *C. jejuni* also possesses a protein export system similar to that of type III secretion systems (Konkel *et al.*, 1999).

Yersinia

The genus *Yersinia*, which comprises three species, is pathogenic for rodents and humans. *Yersinia pestis* causes bubonic plague, *Yersinia pseudotuberculosis* causes mesenteric adenitis and septicaemia, and *Yersinia enterocolitica* causes acute enteritis and lymphadenitis. *Y. pestis* is contracted from flea bites or by inhalation of contaminated aerosol from human cases, while the other two are food-borne enteric pathogens. *Y. pseudotuberculosis* and *Y. enterocolitica* both have an affinity for Peyer's patches and translocate the intestinal mucosa through M cells. In a mouse infection model, efficient translocation of *Y. enterocolitica* and *Y. pseudotuberculosis* into the Peyer's patches requires the bacterial outer membrane protein invasin (Pepe *et al.*, 1995; Marra and Isberg, 1997). Once *Yersinia* reaches the underlying tissue, they enter the mesenteric lymph nodes where the three *Yersinia* spp. exhibit a common capacity to evade the bactericidal activity of phagocytes by the inhibition of phagocytosis, cytokine release, oxidative burst and induction of apoptosis, thus allowing bacteria to survive and multiply in lymphatic tissues (Cornelis *et al.*, 1998).

The capacity of *Yersinia* to evade phagocytic cells is mostly encoded by the 70-kb plasmid present in *Yersinia* spp. Although the precise mechanisms underlying evasion of phagocytosis are still to be determined, it is evident that the co-operative action of virulence-associated proteins, named Yops (*Yersinia* outer membrane proteins), is involved (Cornelis *et al.*, 1998). Yops shared by the three *Yersinia* spp. are encoded by the 70-kb plasmid. Upon bacterial contact with the target host cells, Yops are directly delivered through the type III machinery from the attached bacterium into the target host cell cytoplasm, where they act as effectors to disrupt variously host functions. The major *Yersinia* effectors include YopH (a tyrosine phosphatase), YopE (cytotoxic factor), YopT (RhoA modification), YopP (apoptosis inducer), YpkA (a threonine/serine kinase) and YopM (a protein with homology to thrombin-binding protein) (Cornelis *et al.*, 1998). The properties of Yop effectors together with electron microscopic examination of mice infected with *Y. pseudotuberculosis* (Simonet *et al.*, 1990) have indicated that *Yersinia* are extracellular pathogens, and that their basic strategy for survival is to avoid the non-specific immune systems (Cornelis *et al.*, 1998).

Invasin-mediated Internalisation

The first critical step in *Y. enterocolitica* and *Y. pseudotuberculosis* infection is attachment to the intestinal mucosa including Peyer's patch M cells, followed by invasion through the M cells into underlying tissue. In a tissue culture cell infection model, attachment to and invasion of epithelial cells by enteropathogenic *Yersinia* is a tightly coupled event mediated by invasin (see below). Internalisation of *Yersinia* into cultured epithelial cells is also mediated by other outer membrane proteins, Ail and YadA, albeit to a lesser extent than by invasin (Corneli *et al.*, 1998). Isberg and Falkow (1985) identified the invasin gene, *inv*, on the chromosome of *Y. pseudotuberculosis*. The introduction of the cloned *inv* gene confers an efficient invasive capacity upon non-invasive *E. coli* strain HB101. Similarly, when the *ail* or *yadA* genes were introduced into non-invasive *E. coli* strains, the bacteria also became invasive (Miller and Falkow, 1988; Miller *et al.*, 1989). Invasin protein specifically binds five different members of the $\beta 1$ integrin family of cell adhesion molecules (Isberg and Leong, 1990), while YadA binds $\alpha 5\beta 1$ integrin (Cornelis *et al.*, 1998). The receptor for Ail remains unknown.

The *Y. pseudotuberculosis* invasin is a 986-amino acid outer membrane protein. Homologous invasin proteins have been found in *Y. enterocolitica* (Miller and Falkow, 1988) and are very similar to a family of proteins, the intimins, involved in intimate attachment to host cells of enteropathogenic *E. coli* (EPEC), enterohaemorrhagic *E. coli* (EHEC), *Hafnia alvei* and *Citrobacter rodentium* (Jerse *et al.*, 1990; Schauer and Falkow, 1993; Frankel *et al.*, 1994). Analysis of truncated derivatives of invasin shows that a region corresponding to the extreme C-terminal 192 amino acids of invasin is sufficient to promote integrin-dependent bacterial attachment to and invasion of host cells (Leong *et al.*, 1990, 1995; Rankin *et al.*, 1992). One of the striking features of the invasin C-terminal region is the presence of an intramolecular disulphide bond between residues 907 and 982 that is essential for integrin recognition (Leong and Isberg, 1993). Another critical characteristic of invasin is its strong binding capacity for its integrin receptors such as $\alpha 5\beta 1$, which is approximately 100-fold greater ($K_d = 5 \times 10^{-9}$) than that of fibronectin, the natural ligand for the integrin receptor (Leong *et al.*, 1990; Isberg, 1991). Electron microscopic observation indicates that binding of invasin to integrin at a number of points along the bacterium and host cell interface provokes the extension of tight-fitting pseudopodia around individual bacteria, each of which is internalised into a membrane-bound vacuole (Isberg, 1989; Tran van Nhieu *et al.*, 1996). Thus, a 'zipper-like' internalisation process may take place as a consequence of the high receptor and ligand affinity at the interface of the bacterium/host cell contact at which additional $\beta 1$ integrin seems to be recruited for sustaining high receptor density (Isberg and Tran van Nhieu, 1994).

Interestingly, a recent study has indicated that invasin efficiently interacts with the integrin formed as a result of dimer formation through invasin-to-invasin association (Dersch and Isberg, 1999). The monomeric derivative encompassing the C-terminal 197 amino acids is less efficient in promoting the entry of latex beads coated with the C-terminal portion into HEp-2 cells, as compared with beads coated with the dimeric form. A region responsible for the dimerisation of invasin is located N-terminal to the cell adhesion domain of invasin. Interaction of invasin dimers with multiple integrins may somehow promote clustering and activation of integrin, thus leading to stimulation of downstream signal transduction pathways required for bacterial internalisation (Dersch and Isberg, 1999).

Role of Invasin in *Yersinia* Entry into M Cells

One role of invasin in *Yersinia* infection appears to be in the translocation step across the intestinal mucosa through M cells. Several studies have shown that although *Y. enterocolitica inv* mutants kill mice with the same time course and same LD_{50} as the wild type after oral inoculation, the *inv* mutant did not colonise Peyer's patch M cells as efficiently as the wild type (Pepe and Miller, 1993; Pepe *et al.*, 1995; Marra and Isberg, 1997). Importantly, $\beta 1$ integrin is expressed in the apical membranes of M cells but not enterocytes (Clark *et al.*, 1998). Indeed, M cell targeting by *Yersinia in vivo* is mediated primarily by the interaction of invasin with M cell surface $\beta 1$ integrin. Other studies have indicated that YadA, which is also able to interact with $\alpha 5\beta 1$ integrin, can also mediate the interaction with M cell in an invasin-independent manner, since the *inv* mutant of *Y. pseudotuberculosis* retained the capacity, albeit a very reduced one, to invade intestinal M cells (Marra and Isberg, 1997). Thus, enteropathogenic *Yersinia* appear to have invasin-dependent and invasin-independent pathways for translocation across the Peyer's patch in the intestinal epithelium.

Host Cell Signalling

The clustering of integrins at the interface of the bacterium/host cell contact and the resulting activation of host protein tyrosine kinases seem to be required for invasin-mediated uptake by epithelial cells. Bacteria that encode invasin efficiently, bind staurosporin-treated cells, but the attached bacteria remain extracellularly localised (Rosenshine *et al.*, 1992). The activity of focal adhesion kinase (FAK) is involved in the invasin-mediated uptake, since over-expression of a dominant interfering form of FAK reduces the efficiency of invasin-mediated bacterial entry. The auto-phosphorylation site of FAK, a site for c-Src kinase binding, appears to be important for bacterial internalisation. Host cells over-expressing dominant interfering variants of Src kinase show reduced bacterial uptake, indicating the involvement of a Src family kinase in invasion-mediated uptake (Alrutz and Isberg, 1998). The importance of protein phosphorylation of FAK and $p130^{CAS}$ in promoting invasin-mediated uptake is also supported by data revealing the specific role of YopH in *Yersinia* blockage of bacterial uptake by mammalian cells (Fällman *et al.*, 1997). The YopH protein, a tyrosine phosphatase acting as the specific inhibitor for FAK and $p130^{CAS}$ (Persson *et al.*, 1997), can antagonise the ability of invasin-promoted bacterial uptake, since bacteria capable of injecting YopH into the target host cells can bind host cells by the invasin but fail to enter it (Persson *et al.*, 1999). In agreement with this, a point mutation that destroys the tyrosine phosphatase activity of YopH can eliminate this antagonism (Bliska *et al.*, 1992). Upon infection of HeLa cells by *Y. pseudotuberculosis*, YopH delivered through the type III machinery into host cells exhibits affinity for focal complexes, at which YopH specifically blocks the activity of FAK and $p130^{CAS}$ (Black and Bliska, 1997; Persson *et al.*, 1999). Thus, these studies clearly demonstrate that linkage of the cytoplasmic domain of integrin with actin cytoskeletons via the scaffold proteins of the focal complexes plays a critical role in invasin-mediated *Yersinia* internalisation into the target epithelial cells such as M cells (Fällman *et al.*, 1997).

Listeria monocytogenes

Listeria monocytogenes is a Gram-positive, facultative intracellular opportunist pathogen for humans and a wide range of animal species. Especially in infants, the elderly and in immunocompromised individuals, *L. monocytogenes* often causes serious disease (listeriosis). The major route of bacterial entry in infection in listeriosis is thought to be through the intestinal barrier, in particular, by translocation through the M cells of Peyer's patches (Marco *et al.*, 1992; Pron *et al.*, 1998). After crossing the intestinal barrier, *L. monocytogenes* translocates to the mesenteric lymph nodes and subsequently disseminates via lymphatics and blood to the spleen and liver. If the pathogens are not cleared from the primary target organs by the immune cells, they are transported by blood to secondary target organs such as the central nervous system, placenta or fetus by breaching the blood-brain barrier or transplacental barrier. Thus, a prominent feature of the pathogenesis of *L. monocytogenes* is its ability to invade a wide variety of mammalian cell types.

Infection experiments *in vitro* with enterocyte-like Caco-2 cells have indicated that uptake of *L. monocytogenes* begins in a manner similar to that of *Yersinia pseudotuberculosis*, a process closely related to the 'zipper-like' mechanism. At the site of bacterial contact, the intimately apposed membrane undulates, and a local invagination evoked by the bacterium results in the engulfment without the appearance of membrane ruffling (Mengaud *et al.*, 1996). Once inside the cell, the bacterium immediately lyses the vacuolar membrane and escapes into the host cytoplasm (Gaillard *et al.*, 1987; Bielecki *et al.*, 1990). Like *Shigella*, in the cytoplasm, *Listeria* can multiply and move, and some of the motile bacteria can spread into adjacent epithelial cells, which is another prominent feature of the pathogenesis. The intracellular and intercellular movement of *L. monocytogenes* is based on its ability to direct actin polymerisation at one pole of the bacterium (Tilney and Portnoy, 1989; Domann *et al.*, 1992; Kocks *et al.*, 1992). A similar actin-based motility of pathogens in mammalian cells is shown in *Shigella*, *Rickettsia* and vaccinia virus.

InlA-mediated Invasion

Cossart and colleagues have identified two major *L. monocytogenes* genes directly involved in invasion of non-phagocytic cells (Gaillard *et al.*, 1991; Dramsi *et al.*, 1995, 1997). These two genes, *inlA* and *inlB*, encode surface proteins called internalin A (InlA) and internalin B (InlB), which are associated with the cell wall. InlA and InlB share considerable homology, with 15 and 7 leucine-rich repeats (LRR), respectively (Gaillard *et al.*, 1991). There is evidence to suggest that InlA and InlB mediate distinctive invasion pathways in different cell types. InlA promotes entry into enterocyte-like cells (Gaillard *et al.*, 1991), while InlB mediates entry into a variety of cells including hepatocytes, fibroblasts, and endothelial and epithelial cells (Dramsi *et al.*, 1995; Lignau *et al.*, 1995; Gaillard *et al.*, 1996; Gregory *et al.*, 1996; Ireton *et al.*, 1996; Cossart and Lecuit, 1998).

InlA is an 800-amino-acid surface-exposed protein composed of several distinct domains including the signal sequence, LRR region, inter-repeat (IR) region, B-repeat region and hydrophobic C-terminal end. InlA itself confers invasiveness on other non-invasive bacteria such as *L. innocua* and *Enterococcus faecalis*, and InlA-coated latex beads can also be internalised by mammalian cells (Lecuit *et al.*, 1997). The N-terminal portion of InlA encompassing LRR and IR regions is sufficient to promote bacterial entry (Lecuit *et al.*, 1997). The specific host cell receptor for InlA was identified as E-cadherin (Mengaud *et al.*, 1996), a transmembrane glycoprotein that mediates Ca^{2+}-dependent cell-to-cell adhesion (Takeichi, 1990). E-cadherin consists of a cytoplasmic domain, which interacts with cytoskeletal components such as catenins, a transmembrane region and an extracellular domain required for cell-to-cell adhesion. E-cadherin has also been shown to be expressed in many types of cells other than enterocytes; it is present on hepatocytes and dendritic cells, on the polarised epithelial cells in the choroid plexus, at the sites of cell-to-cell contact of intracerebral microvascular endothelial cells, and on chorionic villi of the placenta (Lecuit *et al.*, 1999). These cells are known as potential target cells in *Listeria* infection, but experimental evidence for the involvement in InlA-dependent *Listeria* infection is not yet available.

Importantly, InlA–E-cadherin interaction is vital, not only for the efficient invasion of mammalian cells by *L. monocytogenes*, but also for determining the host specificity for the infection. Recently, the amino acid residue in E-cadherin critical for recognition by InlA has been identified within the first 94 amino acids. Mammalian cells that express E-cadherin (hEcad), L-CAM (the chicken hEcad homologue) or guinea-pig E-cadherin permit efficient entry of *L. monocytogenes*, while mammalian cells that express mouse E-cadherin (mEcad) or rat E-cadherin were not (Lecuit *et al.*, 1999). Analyses of members of the E-cadherin family for their ability to interact with InlA as the invasion receptor have clearly shown that proline residue 16 of E-cadherin is crucial for conferring on *L. monocytogenes* its host specificity, since some of the E-cadherin members (human, guinea-pig and chicken) permissive for interaction with InlA possess exclusively a proline residue at position 16, while non-permissive E-cadherin members (mouse and rat) possess glutamic acid at position 16 (Lecuit *et al.*, 1999). Indeed, when a series of E-cadherin chimaeras created by exchanging hEcad and mEcad domains was stably expressed in transfected fibroblasts. The specificity was shown to reside within the first 94 amino acids of hEcad. Substitution of proline residue 16 with glutamic acid in hEcad, or the glutamic acid in mEcad with proline residue, confirms the critical role of the proline residue in promoting InlA-mediated *L. monocytogenes* internalisation. Thus the inability of mEcad to mediate *Listeria* invasion *in vitro* provides a molecular explanation for the fact that a role has not been found for InlA *in vivo* in a mouse model. Indeed, wild-type *L. monocytogenes* and its *inlA* mutant can translocate across the intestinal barrier and reach the liver and spleen with the same low efficiency, and their LD_{50} is identical (Lecuit *et al.*, 1999).

Although the precise mechanism that underlies InlA-mediated *Listeria* invasion is still unknown, the notion proposed by Lecuit *et al.* (1999) is important as the first report that a single amino acid difference in a bacterial invasive protein is critical for determining host tropism.

InlB-mediated Invasion

The *inlB* gene, located downstream from *inlA*, encodes a 630-amino acid protein (Gaillard *et al.*, 1991). Although the exact role of InlB in listeriosis remains to be determined, it seems to play a role in the hepatic phase of the infection in a mouse model (Dramsi *et al.*, 1995; Gaillard *et al.*, 1996). InlB is loosely associated with the *Listeria* surface by a 232-amino acid C-terminal region, and thus is released to some extent into the culture supernatant (Dramsi *et al.*, 1997; Braun *et al.*, 1997). When InlB is added externally, it can associate with the bacterial surface, and confer invasiveness to an *inlB* mutant or non-invasive species, such as *L. innocua* and *Staphylococcus carnosus*, indicating that InlB itself has the ability to mediate internalisation (Braun *et al.*, 1997, 1998; Müller *et al.*, 1998). Although the mammalian receptor for InlB has yet to be defined, as in the case of the entry of *L. monocytogenes*, uptake of InlB-coated beads by Vero cells requires the host protein tyrosine phosphorylation, PI3-kinase activity and cytoskeletal rearrangement. This suggests that InlB has a potency to act as an effector to modulate host cellular signalling involved in uptake of bacteria (Ireton *et al.*, 1999).

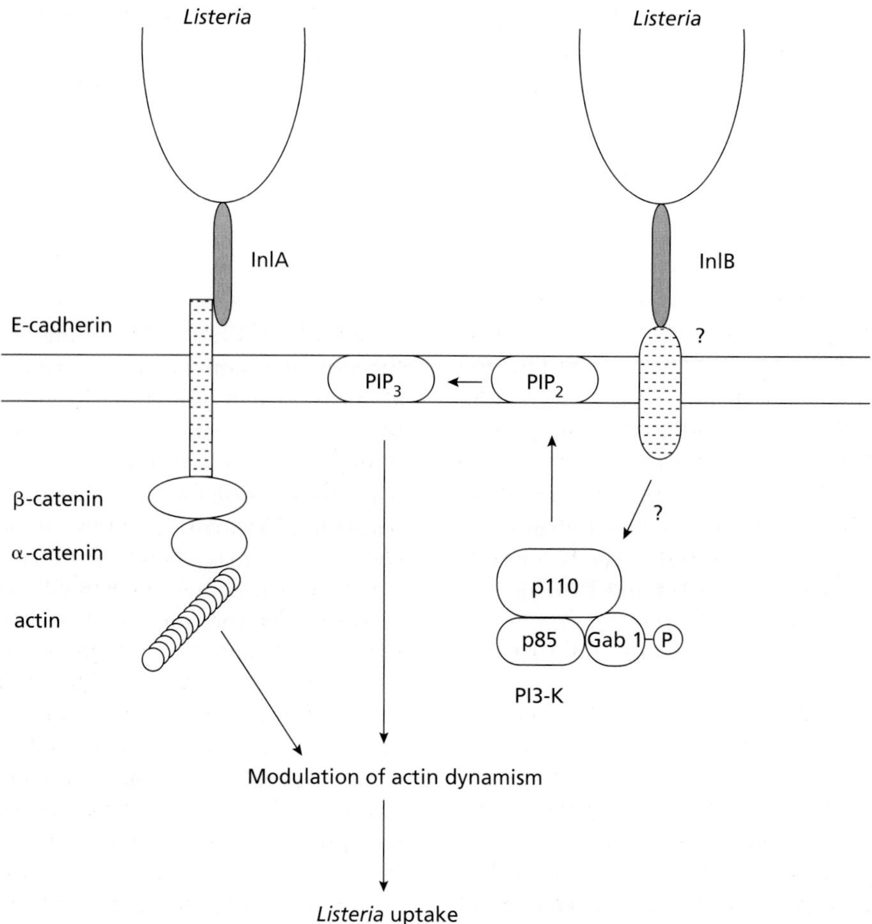

Fig. 3 Simplified models for the InlA- or InlB-dependent pathways of *L. monocytogenes* invasion of mammalian cells. See details in the text and Cossart and Lecuit (1998).

Cellular Signalling

Entry of *L. monocytogenes* requires tyrosine phosphorylation and other signal transduction events that are likely to occur downstream of engagement of the InlA receptor (E-cadherin) or the unknown InlB receptor (**Fig. 3**). Although the cellular signal transduction cascade stimulated by the interaction between InaA and E-cadherin has not yet been elucidated, the cellular events required for InlB-mediated invasion are much clearer (Ireton *et al.*, 1996, 1999; Cossart and Lecuit, 1998). *L. monocytogenes* invasion elicits rapid increases in cellular amounts of $PI(3,4)P_2$ and $PI(3,4,5)P_3$ in Vero cells. The cellular responses are not evoked by the *inlB* mutant, suggesting that InlB is involved in the activation of PI3-kinase. Activation of p85–p110 proteins, heterodimeric PI3-kinase subunits, is required for uptake of *L. monocytogenes*, since invasion can be blocked by genetic or pharmacological inhibition of p85–p110. Bacterial stimulation of PI3-kinase is accompanied by interaction of p85 with one or more tyrosine phosphorylated proteins (Ireton *et al.*, 1999). Purified InlB can elicit large-scale changes in host actin cytoskeletons. These include membrane ruffling in Vero cells, which depends on PI3-kinase activity, in which InlB-treated Vero cells cause rapid and transient increases in the phosphorylation of Gab1, Cbl and Shc, the mammalian adapter proteins known to be associated with p85 of PI3-kinase. Since a *inlB Listeria* mutant, defective for entry into Vero cells, is also defective in promoting phosphorylation of Gab1, but not Cbl and Shc, it is assumed that Gab1 plays an important role in signalling during InlB-mediated invasion (Ireton *et al.*, 1999).

The downstream host effector protein(s) directed by PI3-kinase, which mediates uptake of *Listeria*, is still unknown. InlB can elicit membrane ruffling by Vero cells, but ruffle formation is not associated with *L. monocytogenes* invasion. Although the reason for the different cellular responses to InlB-mediated bacterial invasion and addition of purified InlB protein remains elusive, the local ligand density of InlB may affect the extent of the actin dynamism in mammalian cells.

Neisseria gonorrhoeae

Neisseria gonorrhoeae and *Neisseria meningitidis* primarily colonise mucosal surfaces of humans and, in some cases, they cause serious diseases. *N. gonorrhoeae* colonises the mucosa and submucosal stroma of the conjunctiva, throat, urogenital tract and rectum, often leading to the sexually transmitted disease gonorrhoea. *N. meningitidis* colonises the nasopharyngeal epithelium and, rarely, disseminates by way of the circulation and crosses the blood-brain barrier to cause meningitis. Although the primary routes of infection for *N. gonorrhoeae* and *N. meningitidis* are different, their binding specificities for human tissues are similar.

An early essential step in the infection of epithelial surface by *N. gonorrhoeae* is attachment by the hair-like type IV pili. Recent studies have indicated that the PilC protein, the tip component of the pili, mediates binding to CD46 (membrane co-factor protein) and stimulates Ca^{2+} flux within the target cells, causing activation of the cellular signalling cascades (Kallstrom *et al.*, 1997, 1998). Other studies have indicated that attachment of gonococci can trigger activation of PAK (stress response kinases) and the downstream signal transduction cascade in host cells, which in turn leads to activation of AP-1 (basic leucine zipper transcription factor activator protein 1), thus promoting the production of pro-inflammatory cytokines (Naumann *et al.*, 1997, 1998).

The attachment, including the induced signalling, however, is not sufficient to mediate the internalisation of the bacteria into the epithelial cells. Shortly after initial attachment, gonococci become more tightly attached to the epithelial cell surface. This intimate binding and uptake are mediated by the outer membrane phase-variable opacity-associated (Opa) proteins. *N. gonorrhoeae* possesses up to 11 *opa* genes scattered on the chromosome. The expression of each Opa protein undergoes a high-frequency phase variation. This involves reversible sequence variation in the number of pentanucleotide-coding repeat units in the leader sequence of *opa* structural genes, resulting in a frameshift in the *opa* gene. Thus a single strain can easily generate heterogeneous gonococcal populations that express none, one, or multiple Opa variants at the same time (Bhat, 1991; Kupsch *et al.*, 1993). Recent studies have also shown that the 11 Opa proteins possess distinctive repertoires for targeting their cellular receptors, and that the repertoires play important roles in determining the bacterial tropism for the target host epithelia, endothelia and phagocytic cells (van Putten and Duensing, 1997; Nassif *et al.*, 1999; Naumann *et al.*, 1999).

Opa-mediated Invasion and the Receptor Tropism

Opa proteins can temporarily be classified into two groups according to their receptor tropism. The first group is Opa50 (also called Opa30), which possesses multiple binding ability for several host receptors including heparan sulphate proteoglycan (HSPG) of the syndecan family and $\alpha v\beta 5$ and $\alpha 5\beta 1$ integrins (**Fig. 4**). The receptor tropism for Opa50 also varies, depending on the epithelial cell types. For example, $\alpha v\beta 5$ integrin-dependent gonococcal internalisation is displayed in the infection of HeLa and CHO cells, while $\alpha 5\beta 1$ integrin-dependent internalisation is represented in the infection of HEp-2 cells. HSPG and the integrins are, however, predominantly expressed on the basolateral surface of polarised epithelial cells. Thus, it is assumed that the primary reason for gonococci expressing Opa50 in the infection is not associated with apical entry into the mucosal epithelium, rather that it is required when the mucosal barrier is disrupted, thus being exposed to the basolateral surface of the polarised epithelial cells, or at a later stage of infection after translocation of epithelial barriers (Dehio *et al.*, 1998; Naumann *et al.*, 1999). In support of this view, an additional interaction between Opa50 and a serum component such as vitronectin (Vn) or fibronectin (Fn) is shown to promote gonococcal invasion of epithelial cells. In this process, Vn and Fn are thought to serve as the molecular bridge to connect between Opa50 and the specific receptors, $\alpha v\beta 5$ or $\alpha 5\beta 1$ integrins, respectively, thus mediating a tight binding which includes clustering of integrins that trigger gonococcal uptake (Duensing and van Putten, 1997; Gomez-Duarte *et al.*, 1997). In Opa50-mediated invasion, heparinase, RGD-containing peptides or anti-$\alpha 5\beta 1$ integrin antibodies inhibit gonococcal uptake by epithelial cells. Furthermore, when Opa50-expressing gonococci infect endothelial cells, the bacteria also display selective binding to the integrins and are capable of invading the endothelial cells, supporting the above notion that Opa50 is needed at the infection stage underlying epithelial barriers (Duensing and van Putten, 1997; Gomez-Duarte *et al.*, 1997; Dehio *et al.*, 1998; van Putten *et al.*, 1998a).

The second class of Opa proteins, comprising the rest of the 10 Opa variants, can also mediate efficient invasion of epithelial cells as well as

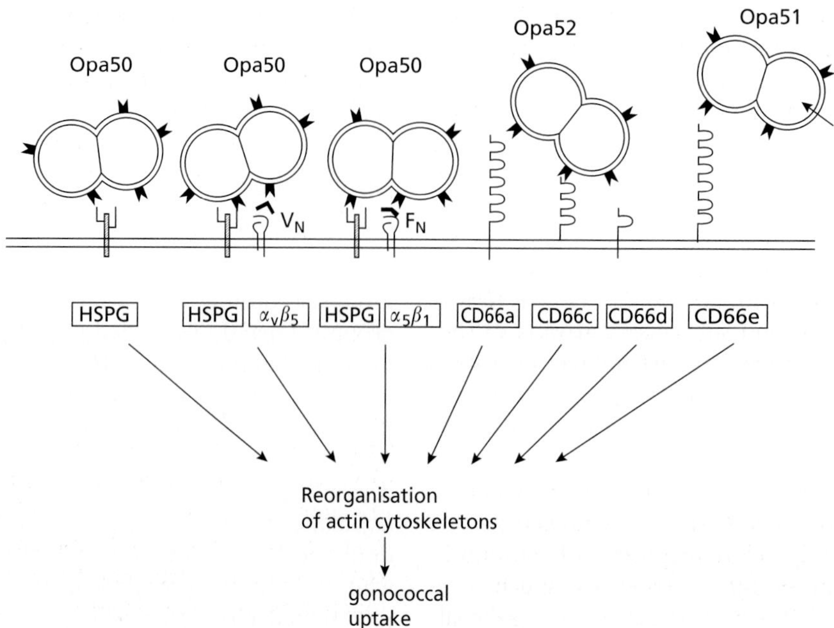

Fig. 4 Opa-mediated interaction with host receptors involved in gonococcal invasion.

Table 1 Gonococcal Opa proteins and cellular receptors

Opa protein	Cellular receptor
50	HSPG, Fn, Vn
51	CD66e
52	CD66a, CD66c, CD66d, CD66e
53	CD66a
54	CD66a, CD66e
55	CD66e
56	CD66e
57	CD66a, CD66c, CD66d, CD66e
58	CD66a, CD66c, CD66d, CD66e
59	CD66a, CD66e
60	CD66a, CD66c, CD66d, CD66e

non-opsonophagocytosis by PMNs (Fig. 4). This class of Opa proteins can mediate internalisation through interaction with the CD66 family, a member of the immunoglobulin superfamily, whose members are differentially expressed on various human tissues, including epithelial cells and phagocytes (Virji *et al.*, 1996). Members of the CD66 family involved in gonococcal internalisation are CD66a (biliary glycoprotein, BGP), CD66c (non-specific cross-reacting antigen, NCA), CD66d (carcinoembryonic antigens (CEA) gene family member 1; CGN1) and CD66e (CEA), but not CD66b. The CD66 members are also involved in *N. meningitidis* invasion (Virji, 1996). The Opa protein repertoire for the CD66 family in *N. gonorrhoeae* strain SM11 is shown in Table 1. Although the precise mechanisms underlying CD66-mediated uptake pathways are still partly unclear, since CD66 family proteins are expressed on apical surface of polarised epithelial cells, these 10 Opa proteins are postulated as playing roles in gonococcal invasion and subsequent transcytosis from the apical surface (Virji *et al.*, 1996; Dehio *et al.*, 1998).

Porin-mediated Invasion

Porin is the major outer membrane protein. It consists of a hydrophilic pore that mediates the passage of small nutrients and waste products across the outer membrane of Gram-negative bacteria. Gonococci produce two porins, PorA and PorB. The PorB protein has been implicated in mediating bacterial internalisation into epithelial cells and phagocytes (van Putten *et al.*, 1998b). PorB-mediated invasion seems to occur by two distinct means, Opa-independent and Opa50-dependent pathways (Bauer *et al.*, 1999). In infection of Chang human conjunctiva cells, when phosphate is depleted from the medium, PorB can mediate uptake of gonococci independently of Opa50 (van Putten *et al.*, 1998b). In Opa50-dependent invasion, however, PorB is thought to serve as a co-stimulator with Opa50 (Bauer *et al.*, 1999).

Gonococcal porin is a GTP-binding protein displaying various biological activities for mammalian cells (Nassif *et al.*, 1999; Naumann *et al.*, 1999). Purified PorB can selectively interfere with the signalling of PMNs required for degranulation (Haines *et al.*, 1988). Externally added PorB can be translocated into eukaryotic cell membrane, in which PorB porin forms a voltage-gated channel, causing a transient change in membrane potential, and modulates host cell signalling events (Ulmer *et al.*, 1992). A recent study has reported that porin can stimulate a transient Ca^{2+} flux in epithelial cells and phagocytes (Müller *et al.*, 1999). Although the mechanisms of PorB-stimulated gonococcal internalisation are still largely unknown, a strong transient Ca^{2+} flux induced by PorB is thought to activate PKC, thus promoting a local rearrangement of actin cytoskeletons (Bauer *et al.*, 1999).

Cellular Signalling

Since gonococcal entry cannot be accompanied by internalisation of other adherent bacteria, gonococcal invasion is a highly selective process, representing invasion mediated through a 'zipper-like' mechanism (Grassmé *et al.*, 1996; van Putten and Duensing, 1997). Nevertheless, cellular responses to gonococcal invasion seem to be dynamic, involving local rearrangements of actin cytoskeletons, accumulation of protein phosphotyrosine, microvillus extension and pedestal formation around the site of bacterial contact (Grassmé *et al.*, 1996; Merz and So, 1997; Giardina *et al.*, 1998).

As mentioned above, Opa50-mediated gonococcal invasion of epithelial cells occurs by several distinct mechanisms, depending on the target epithelial cell lines. Invasion of Chang cells or ME-180 human cervical carcinoma cells depends on a cellular signal cascade required by activation of PC-PLC (phosphatidylcholine-specific phopholipase C) and acidic sphingomyelinase (ASM), which generate the second messengers diacylglycerol and cerimide (Grassmé *et al.*, 1997). Indeed, genetic or pharmacological blockage of ASM or PC-PLC activity results in inhibition of gonococcal internalisation into epithelial cells. In Opa50-mediated gonococcal invasion, however, nothing is known about downstream effector molecules directly mediating the gonococcal uptake. The Opa50-mediated invasion of other epithelial cells, such as HeLa and CHO cells, is stimulated by addition of Vn, while invasion into HEp-2 cells can be promoted by addition of Fn. Activation of PKC is involved in HSPG-dependent gonococcal invasion, since PKC inhibitors can block Vn-mediated invasion. Although the signal transduction pathways induced in HeLa, Caco-2 or HEp-2 cells are still unknown, it is assumed that Opa50-mediated invasion requires stimulation of a downstream signal transduction cascade of the receptor integrin (Dehio *et al.*, 1998; van Putten *et al.*, 1998a).

Members of the CD66 family are expressed differently by epithelial and endothelial cells and PMNs. Although the Opa-mediated binding to CD66a, CD66c or CD66d is able to activate an Src-like tyrosine kinase- and Rac1-dependent signalling pathway (Hauck *et al.*, 1998), neither the downstream signal cascades nor the host effector molecules directly involved in gonococcal uptake are known.

Streptococcus pyogenes

Streptococcus pyogenes is an important human pathogen that causes a wide variety of infections. These range from trivial skin and throat infections to severe, life-threatening illnesses such as septicaemia, necrotising fasciitis, toxic shock syndrome and the post-streptococcal sequelae, rheumatic fever and acute glomerulonephritis. In *Strep. pyogenes*-associated infections, a first critical step is bacterial adherence leading to colonisation of mucosal or skin surfaces. The persistence of bacteria in host tissues and the development of invasive diseases are thought to require an enhanced capacity to invade host cells and evade host defense mechanisms. Indeed, traditionally *Strep. pyogenes* had been viewed as an extracellular pathogen, however, recent studies have established that *Strep. pyogenes* can efficiently invade various human epithelial cells, and this suggests that invasive ability is an important contributing factor in more severe, invasive infections (La Penta *et al.*, 1994; Greco *et al.*, 1995; Molinari and Chhatwal, 1998, 1999). In streptococcal invasion, the two major surface proteins M protein and protein F1 (protein F or SfbI) mediate bacterial entry into various types of epithelial cells.

M protein-mediated Invasion

M protein is an important virulence protein expressed on the surface of *Strep. pyogenes* and plays multiple roles in streptococcal infection, including resistance to phagocytosis, adherence to epidermal keratinocytes, microcolony formation and invasion of epithelial cells. M protein is highly variable in antigenicity, and there are more than 80 different serotypes. In spite of serotype differences, all M proteins are α-helical coiled-coil molecules able to interact with various plasma proteins. Some M serotypes, such as M1, M3, M6 and M12, mediate efficient internalisation of streptococci into epithelial cells (Molinari and Chhatwal, 1999). Cleary and co-workers initially reported that *Strep. pyogenes* of serotypes M1 and M12 can efficiently invade human lung carcinoma A549 cells and persisting in the cultured cells (La Penta *et al.*, 1994). Their recent study has revealed that a streptococcal strain expressing M1 protein, but not protein F1, or M1 protein-coated latex beads can efficiently enter HeLa cells. Scanning electron microscopy of HeLa cells infected by the streptococci or incubated with latex beads coated with M1 protein indicates that at the entry site the plasma membrane engulfs the streptococcal chain or the latex beads (Dombek *et al.*, 1999). In the initial stage of entry, host cell microvilli and pseudopodia are often extended toward and over the streptococci. Scanning electron micrographs, however, indicate that the uptake occurs by the close juxtaposition of the epithelial membrane around the streptococci or the beads. Thus, although M1-mediated streptococcal invasion elicits highly dynamic rearrangement of actin cytoskeletons at the interface of the streptococci/host cell contact, the internalisation process itself is thought to be mediated by a 'zipper-like' mechanism (Dombek *et al.*, 1999).

Different serotypes of M protein appear to interact with different host surface receptors, although the relationship between a given M serotype and the host receptor has not yet been conclusively established. Streptococcal invasion of HEp-2 cells mediated by M1 (and M3) is strongly (~50-fold) stimulated by addition of mammalian serum, extracellular matrix proteins such as fibronectin (Fn), laminin (Lm) or RGD-containing peptides, but the extent varies between the proteins (Cue and Cleary, 1997; Cue *et al.*, 1998; Berkower *et al.*, 1999). Addition of an antibody directed against Fn or integrin $\beta1$ inhibits Fn-stimulated invasion by M1 serotype streptococcal strain, but does not inhibit Lm- or RGD peptide-stimulated invasion. While streptococcal invasion of epithelial cells stimulated by RGD peptides occurs regardless of whether the background is $M1^+$ or $M1^-$, suggesting that Fn and Lm are distinct as M1-dependent invasion agonists, RGD peptide stimulates M1-independent internalisation (Cue *et al.*, 1998).

Berkower *et al.* (1999) constructed a series of isogenic M protein- and protein F1-deficient *Strep. pyogenes* strains that expressed none, M3, M6 or M18, and investigated their ability to adhere to and invade two types of epithelial cells, HEp-2 and HaCaT cells. The M3-expressing strain adhered to both cells, but its invasion of HEp-2 cell was serum-stimulated. The M6-expressing strain adhered to and invaded HEp-2 and HaCaT cells, while the M18-expressing strain was able to adhere to both cells but was unable to invade. Although neither of the precise invasion pathways has as yet been defined, since the streptococcal strains used in the studies lacked protein F, the M proteins appear to be able to mediate multiple invasion pathways.

Since M protein is highly variable in serotype and different M serotype proteins can mediate streptococcal entry into various epithelial cells (Molinari and Chhatwal, 1999), differential expression of individual M serotype on *Strep. pyogenes* is thought to confer cell tropism on bacteria for targeting host cells or tissues by their recognition of distinct cellular receptors.

Protein F1-mediated Invasion

Protein F1 or SfbI (an allelic variant of protein F1) is a fibronectin-binding protein of *Strep. pyogenes*, which plays a central role at the initial stage in the streptococcal adherence to the host epithelial surface of the pharynx or skin tissue (Hanski and Caparon, 1992; Hanski *et al.*, 1992; Okada *et al.*, 1994). The deduced sequence of protein F1 (and SfbI) reveals that it represents a highly homologous family that shares a number of features in common with other fibronectin-binding proteins of *Streptococcus dysgalactiae* and *Staphylococcus aureus*. For example, a characteristic common feature is the presence of tandem repeats of 32–50 amino acids located adjacent to the conserved C-terminal cell attachment domain, which is sufficient to mediate binding to Fn and host cells (Ozeri *et al.*, 1996; Okada *et al.*, 1997). Protein F1 or the Fn-binding portion can mediate efficient entry of *Strep. pyogenes* into epithelial cells (Molinari *et al.*, 1997; Jadoun *et al.*, 1998; Okada *et al.*, 1998; Ozeri *et al.*, 1998). Latex beads coated with protein F1 or the Fn-binding segment can be taken up efficiently by epithelial cells. (Molinari *et al.*, 1997; Ozeri *et al.*, 1998). In addition, the surface density of protein F1 is a critical factor that affects the efficiency of uptake by epithelial cells (Okada *et al.*, 1998).

Ozeri *et al.* (1998) used a defined *Strep. pyogenes* mutant expressing protein F1 but not M protein and showed that entry of *Strep. pyogenes* into various epithelial cells is mediated by different integrin receptors. In HeLa cells, protein F1-mediated bacterial entry is promoted by interaction with $\alpha5\beta1$ integrin, in which invasion can be stimulated by adding Fn. In a mouse embryonic cell line, GD25, a $\beta1$-null mutant, the bacteria can still invade the cell, but the process depends on the presence of other integrins such as $\alpha v\beta3$ integrin. In both $\alpha5\beta1$- and $\alpha v\beta3$-dependent streptococcal entry into mammalian cells, the ligand occupancy of integrin receptor by protein F1 via Fn or vitronectin appears to be involved in triggering host cell signal transduction pathways that lead to uptake of bacteria (Ozeri *et al.*, 1998).

A similar invasion strategy has recently been shown by *Staph. aureus* FnBPs, the Fn-binding proteins, in which Fn also serves as a molecular bridge between FnBPs and host $\alpha5\beta1$ integrin (Sinha *et al.*, 1999).

Other Components Associated with Invasion

Streptococcal pyogenic exotoxin A, SpeA, has been shown to affect invasion efficiency. An M1 serotype *Streptococcus*, M1inv$^+$, invades human lung epithelial cells more efficiently than the isogenic *speA*$^-$ mutant (La Penta *et al.*, 1994). Streptococcal pyogenic toxin B, SpeB, also affects streptococcal invasiveness, but the effect of SpeB expression on invasion is controversial (Burns *et al.*, 1998; Tsai *et al.*, 1998). Interestingly, one of the SpeB variants, mSpeB2, contains an RGD sequence, a triplet motif commonly recognised by integrins, and the mSpeB2 variant expressed in streptococci can mediate binding to transfected mammalian cells expressing $\alpha v\beta3$ integrin (Stockbauer *et al.*, 1999). Although the precise roles in streptococcal invasion await further investigation, the potential role of RGD-containing SpeB variants in mediating adhesion may contribute to streptococcal internalisation.

Cellular Signalling

As indicated by the functional linkage between protein F1, including some serotype M proteins and host integrin via the extracellular matrix, a host cellular signal cascade downstream activation of integrin at least takes part in promoting streptococcal internalisation (Ozeri *et al.*, 1998). As mentioned above, streptococcal invasion induces various cellular responses such as the appearance of microvilli or lamellipodia or accumulation of protein tyrosine phosphorylation, but some signal cascades other than the integrin-mediated signalling may also be involved in the uptake (La Penta *et al.*, 1994; Molinari *et al.*, 1997; Pancholi and Fischetti, 1997; Okada *et al.*, 1998; Ozeri *et al.*, 1998; Dombek *et al.*, 1999). Besides, although no direct role in streptococcal invasion has yet been revealed, streptococcal surface dehydrogenase (SDH), which has multiple binding activity to a variety of mammalian proteins, may participate in inducing signal transduction events at an early stage of infection (Pancholi and Fishetti, 1992, 1997). Whole bacteria and purified SDH can induce tyrosine phosphorylation of several eukaryotic proteins, including 30-kDa and 32-kDa host membrane proteins that interact directly with SDH. Although much about the cellular signalling required for *Strep. pyogenes* invasion is still poorly understood, *Strep. pyogenes* invasion events may involve a variety of cellular signalling cascades, depending on streptococcal surface proteins, the host cellular receptors and the type of targeting host cells and tissues.

References

Adam T, Giry M, Boquet P, Sansonetti PJ (1996) Rho-dependent membrane folding causes *Shigella* entry into epithelial cells. *EMBO J.* 15: 3315–3321.

Adler B, Sasakawa C, Okada N, Makino S, Yoshikawa M (1989) A dual transcriptional activation system for the 230 kb plasmid genes coding for virulence-associated antigens of *Shigella flexneri. Mol. Microbiol.* 3: 627–635.

Alrutz MA, Isberg RR (1998) Involvement of focal adhesion kinase in invasin-mediated uptake. *Proc. Natl Acad. Sci. USA* 95: 13658–13668.

Bauer FJ, Rudel T, Stein M, Meyer TF (1999) Mutagenesis of the *Neisseria gonorrhoeae* porin reduces invasion in epithelial cells and enhances phagocyte responsiveness. *Mol. Microbiol.* 31: 903–913.

Bielecki J, Youngman P, Connelly P, Portnoy DA (1990) *Bacillus subtilis* expressing a haemolysin gene from *Listeria monocytogenes* can be grown in mammalian cells. *Nature* 345: 175–176.

Berkower C, Ravins M, Moses AE, Hanski E (1999) Expression of different group A streptococcal M proteins in an isogenic background demonstrates diversity in adherence to and invasion of eukaryotic cells. *Mol. Microbiol.* 31: 1463–1475.

Bernardini ML, Mounier J, d'Hauteville H, Coquis-Rondon M, Sansonetti PJ (1989) Identification of *icsA*, a plasmid locus of *Shigella flexneri* that governs bacterial intra- and intercellular spreading through interaction with F-actin. *Proc. Natl Acad. Sci. USA* 86: 3867–3871.

Bhat KS, Gibbs CP, Barrera O *et al.* (1991) The repertories of opacity proteins displayed by *Neisseria gonorrhoeae* MS11 outer surface are encoded by a family of 11 complete genes. *Mol. Microbiol.* 5: 1889–1901.

Black DS, Bliska JB (1997) Identification of p130^{Cas} as a substrate of *Yersinia* YopH (Yop51), a bacterial protein tyrosine phosphatase that translocates into mammalian cells and targets focal adhesions. *EMBO J.* 16: 2730–2741.

Black RE, Levine MM, Clements ML, Hughes TP, Blaser MJ (1988) Experimental *Campylobacter jejuni* infection in humans. *J. Infect. Dis.* 157: 472–479.

Blaser MJ, Berkowitz FM, LaForce FM, Cravens J, Reller LB, Wang WI (1979) *Campylobacter* enteritis: clinical and epidemiological features. *Ann. Intern. Med.* 9: 179–185.

Blaser MJ, Wells JG, Feldman RA *et al.* (1983) *Campylobacter enteritis* in the United States. A multicenter study. *Ann. Intern. Med.* 98: 360–365.

Bliska JB, Clemens JC, Dixson JE, Falkow S (1992) The *Yersinia* tyrosine phosphatase: specificity of a bacterial virulence determinant for phosphoproteins in the J774.1 macrophage. *J. Exp. Med.* 176: 1625–1630.

Blocker A, Gounon P, Larquet E *et al.* (1999) The tripartite type III secretion of *Shigella flexneri* inserts IpaB and IpaC into host membranes. *J. Cell Biol.* 147: 683–693.

Bourdet-Sicard R, Rüdiger M, Jockush BM, Gounon P, Sansonetti PJ, Tran van Nhieu G (1999) Binding of the *Shigella* protein IpaA to vinculin induces F-actin depolymerization. *EMBO J.* 18: 5853–5862.

Braun L, Dramsi S, Dehoux P, Bierne H, Lindahl G, Cossart P (1997) InlB: an invasion protein of *Listeria monocytogenes* with a novel type of surface association. *Mol. Microbiol.* 25: 175–176.

Braun L, Ohayon H, Cossart P (1998) The InlB protein of *Listeria monocytogenes* is sufficient to promote entry into mammalian cells. *Mol. Microbiol.* 27: 1077–1087.

Burns EH Jr, Lukomski S, Rurangirwa J, Podbielski A, Musser JM (1998) Genetic inactivation of the extracellular cysteine protease enhances *in vitro* internalization of group A streptococci by human epithelial and endothelial cells. *Microb. Pathog.* 24: 333–339.

Carter PB, Collins FM (1974) The route of enteric infection in normal mice. *J. Exp. Med.* 139: pp. 1189–1203.

Chen LM, Kaniga K, Galán JE (1996a) *Salmonella* spp. are cytotoxic for cultured macrophages. *Mol. Microbiol.* 21: 1101–1115.

Chen LM, Hobbie S, Galán JE (1996b) Requirement of Cdc42 for *Salmonella*-induced cytoskeletal and nuclear responses. *Science* 274: 2115–2118.

Clark MA, Jepson MA, Simmons NL, Hirst BH (1994) Preferential interaction of *Salmonella typhimurium* with mouse Peyer's patch M cells. *Res. Microbiol.* 145: 543–552.

Clark MA, Hirst BH, Jepson MA (1998) M-cell surface $\beta 1$ integrin expression and invasin-mediated targeting of *Yersinia pseudotuberculosis* to mouse Peyer's patch M cells. *Infect. Immun.* 66: 1237–1243.

Colonna B, Casalino M, Fradiani PA *et al.* (1995) H-NS regulation of virulence gene expression in enteroinvasive *Escherichia coli* harboring the virulence plasmid integrated into the host chromosome. *J. Bacteriol.* 177: 4703–4712.

Cornelis GR, Boland A, Boyd AP *et al.* (1998) The virulence plasmid of *Yersinia*, anti host genome. *Microbiol. Mol. Biol. Rev.* 62: 1315–1352.

Cossart P, Lecuit M (1998) Interaction of *Listeria monocytogenes* with mammalian cells during entry and actin-based movement: bacterial factors, cellular ligands and signaling. *EMBO J.* 17: 3797–3806.

Cue DR, Cleary PP (1997) High-frequency invasion of epithelial cells by *Streptococcus pyogenes* can be activated by fibrinogen and peptides containing the sequence RGD. *Infect. Immun.* 65: 2759–2764.

Cue D, Dombek PE, Lam H, Cleary PP (1998) Serotype M1 *Streptococcus pyogenes* encodes multiple pathways for entry into epithelial cells. *Infect. Immun.* 66: 4593–4601.

Darwin KH, Miller VL (1999) Molecular basis of the interaction of *Salmonella* with the intestinal mucosa. *Clin. Microbiol. Rev.* 12: 405–428.

Dehio C, Prévost MC, Sansonetti PJ (1995) Invasion of epithelial cells by *Shigella flexneri* induces tyrosine phosphorylation of cortactin by a pp60$^{c\text{-}src}$-mediated signalling pathway. *EMBO J.* 14: 2471–2482.

Dehio C, Gray-Owen SD, Meyer TF (1998) The role of neisserial Opa proteins in interactions with host cells. *Trends Microbiol.* 6: 489–495.

de Melo MA, Gabbioni G, Péchére JC (1989) Cellular events and intracellular survival of *Campylobacter jejuni* during infection of HEp-2 cells. *Infect. Immun.* 57: 2214–2222.

Dersch P, Isberg RR (1999) A region of the *Yersinia pseudotuberculosis* invasin protein enhances integrin-mediated uptake into mammalian cells and promotes self-association. *EMBO J.* 18: 1199–1213.

Domann E, Wehland J, Rohde M *et al.* (1992) A novel bacterial gene in *Listeria monocytogenes* required for host cell microfilament interaction with homology to the proline-rich region of vinculin. *EMBO J.* 11: 1981–1990.

Dombek PE, Cue D, Sedgewick J *et al.* (1999) High-frequency intracellular invasion of epithelial cells by serotype M1 group A streptococci: M1 protein-mediated invasion and cytoskeletal rearrangements. *Mol. Microbiol.* 31: 859–870.

Dramsi S, Biswas I, Maguin E, Braun L, Mastroeni P, Cossart P (1995) Entry of *L. monocytogenes* into hepatocytes expression of InlB, a surface protein of the internalin multigene family. *Mol. Microbiol.* 16: 251–261.

Dramsi S, Dohoux P, Lebrun M, Goossens P, Cossart P (1997) Identification of four new members of the internalin multigene family in *Listeria monocytogenes* strain EGD. *Infect. Immun.* 65: 1615–1625.

Duffy MC, Benson JB, Rubin SJ (1980) Mucosal invasion in *Campylobacter* enteritis. *Am. J. Clin. Pathol.* 73: 706–708.

Duménil G, Olivo JC, Pellegrini S, Fellous M, Sansonetti PJ, Tran van Nhieu G (1998) Interferon α inhibits a Src-mediated pathway necessary for *Shigella*-induced cytoskeletal rearrangements in epithelial cells. *J. Cell Biol.* 143: 1003–1012.

Duensing TD, van Putten JPM (1997) Vitronectin mediates internalization of *Neisseria gonorrhoeae* by Chinese hamster ovary cells. *Infect. Immun.* 65: 965–970.

Eichelberg K, Galán JE (1999) Differential regulation of *Salmonella typhimurium* Type III secretion proteins by pathogenicity island 1 (SPI-1)-encoded trasncriptional activators InvF and HilA. *Infect. Immun.* 67: 4099–4105.

Eichelberg K, Hardt WD, Galán JE (1999) Characterization of SprA, an AraC-like transcriptional regulator encoded within the *Salmonella typhimurium* pathogenicity island 1. *Mol. Microbiol.* 33: 139–152.

Everest PH, Goossens H, Butzler JP *et al.* (1992) Differentiated Caco-2 cells as a model for enteric invasion by *Campylobacter jejuni* and *Campylobacter coli. J. Med. Microbiol.* 37: 319–325.

Falconi M, Colonna B, Prosseda G, Micheli G, Gualerzi CO (1998) Thermoregulation of *Shigella* and *Escherichia coli* EIEC pathogenicity. A temperature-dependent structural transition of DNA modulates accessibility of *virF* promoter to transcriptional repressor H-NS. *EMBO J.* 17: 7033–7043.

Fällman M, Persson C, Wolf-Watz H (1997) *Yersinia* proteins that target host cell signaling pathways. *J. Clin. Invest.* 99: 1153–1157.

Francis M, Starnbach MN, Falkow S (1992) Morphological and cytoskeletal change in epithelial cells occur immediately upon interaction with *Salmonella typhimurium* grown under low-oxygen conditions. *Mol. Microbiol.* 6: 3077–3087.

Francis CL, Ryan TA, Jones BD, Smith SJ, Falkow S (1993) Ruffles induced by *Salmonella* and other stimuli direct macropinocytosis of bacteria. *Nature* 364: 639–642.

Frankel G, Candy DCA, Everest P, Dougan G (1994) Characterization of the C-terminal domains of intimin-like proteins of enteropathogenic and enterohemorrhagic *Escherichia coli*, *Citrobacter freundii* and *Hafnia alvei*. *Infect. Immun.* 62: 1835–1842.

Fu Y, Galán JE (1999) A *Salmonella* protein antagonizes Rac-1 and Cdc42 to mediate host-cell recovery after bacterial invasion. *Nature* 401: 293–297.

Gaillard JL, Berche P, Mounier J, Richard S, Sansonetti PJ (1987) *In vitro* model of penetration and intracellular growth of *L. monocytogenes* in the human enterocyte-like cell line Caco-2. *Infect. Immun.* 55: 2822–2829.

Gaillard JL, Berche P, Frehel C, Gouin E, Cossart P (1991) Entry of *L. monocytogenes* into cells is mediated by internalin a repeat protein reminiscent of surface antigens from gram-positive cocci. *Cell* 65: 1127–1141.

Gaillard JL, Jaubert F, Berche P (1996) The *inlAB* locus mediates the entry of *Listeria monocytogenes* into hepatocytes in vivo. *J. Exp. Med.* 183: 359–369.

Galán JE, Collmer A (1999) Type III secretion machines: bacterial devices for protein delivery into host cells. *Science* 284: 1322–1328.

Galyov EE, Wood MW, Rosqvust R *et al.* (1997) A secreted effector protein of *Salmonella dublin* is translocated into eucaryotic cells and mediates inflammation and fluid secretion in infected ileal mucosa. *Mol. Microbiol.* 25: 903–912.

Garcia-del Porillo F, Finlay BB (1995) The varied lifestyles of intracellular pathogens within eukaryotic vacuolar compartments. *Trends Microbiol.* 3: 373–380.

Garcia-del Portillo F, Pucciarelli MG, Jefferies WA, Finlay BB (1994) *Salmonella typhimurium* induces selective aggregation and internalization of host cell surface proteins during invasion of epithelial cells. *J. Cell Sci.* 107: 2005–2020.

Giardina PC, Williams R, Lubaroff D, Apicella MA (1998) *Neisseria gonorrhoeae* induces focal polymerization of actin in primary human urethral epithelium. *Infect. Immun.* 66: 3416–3419.

Ginocchio CC, Pace J, Galán (1992) Identification and molecular characterization of a *Salmonella typhimurium* gene involved in triggering the internalization of salmonellae into cultured epithelial cells. *Proc. Natl Acad. Sci. USA* 89: 5976–5980.

Ginocchio CC, Olmsted SB, Wells CL, Galán (1994) Contact with epithelial cells induces the formation of surface appendages on *Salmonella typhimurium*. *Cell* 76: 717–724.

Gomez-Duarte OG, Dehio M, Guzman CA, Chhatwal GS, Dehio C, Meyer TF (1997) Binding of vitronectin to Opa-expressing *Neisseria gonorrhoeae* mediates invasion of HeLa cells. *Infect. Immun.* 65: 3857–3866.

Grant CCR, Konkel ME, Cieplak W Jr, Tompkins LS (1993) Role of flagella in adherence, internalization, and translocation of *Campylobacter jejuni* in nonpolarized and polarized epithelial cells cultures. *Infect. Immun.* 61: 1764–1771.

Grassmé HU, Ireland RM, van Putten JPM (1996) Gonococcal opacity protein promotes bacterial entry-associated rearrangements of the epithelial cell actin cytoskeleton. *Infect. Immun.* 64: 1621–1630.

Grassmé HU, Gulbins E, Brenner B *et al.* (1997) Acidic sphingomyelinase mediates entry of *N. gonorrhoeae* into nonphagocytic cells. *Cell* 91: 605–615.

Greco R, DeMartio L, Donnarumma G, Conte MP, Seganti L, Valenti P (1995) Invasion of cultured human cells by *Streptococcus pyogenes*. *Res. Microbiol.* 146: 551–560.

Gregory SH, Sagnimeni AJ, Wing EJ (1996) Expression of the *inlAB* operon by *Listeria monocytogenes* is not required for entry into hepatic cells *in vivo*. *Infect. Immun.* 64: 3983–3986.

Haines KA, Yeh L, Blake MS, Cristello P, Korchak H, Weissmann G (1988) Protein I, a translocatable ion channel from *Neisseria gonorrhoeae*, selectively inhibits exocytosis from human neutrophils without inhibiting O_2-generation. *J. Biol. Chem.* 263: 945–951.

Hanski E, Caparon MG (1992) Protein F, a fibronectin-binding protein, is an adhesion of the group A streptococcus-*Streptococcus pyogenes*. *Proc. Natl Acad. Sci. USA* 89: 6172–6176.

Hanski E, Horwitz PA, Caparon M (1992) Expression of protein F, the fibronectin-binding protein of *Streptococcus pyogenes* JRS4, in heterologous streptococcal and enterococcal strains promotes their adherence to respiratory epithelial cells. *Infect. Immun.* 60: 5119–5125.

Hardt W-D, Chen LM, Schuebel KE, Bustelo XR, Galán JE (1998) *S. typhimurium* encodes an activator of rho GTPases that induces membrane ruffling and nuclear responses in host cells. *Cell* 93: 815–826.

Hauck CR, Meyer TF, Lang F, Gulbins E (1998) CD66-mediated phagocytosis of Opa_{52} *Neisseria gonorrhoeae* requires a Src-like tyrosine kinase- and Rac1-dependent signalling pathway. *EMBO J.* 17: 443–454.

Hayward RD, Koronakis V (1999) Direct nucleation and bundling of actin by the SipC protein of invasive *Salmonella*. *EMBO J.* 18: 4926–4934.

Hohmann AW, Schmidt G, Rowley D (1978) Intestinal colonization and virulence of *Salmonella* in mice. *Infect. Immun.* 22: 763–770.

Hong KH, Miller VL (1998) Identification of a novel *Salmonella* invasion locus homologous to *Shigella ipg DE*. *J. Bacteriol.* 180: 1793–1802.

Hu L, Kopecko DJ (1999) *Campylobacter jejuni* 81–176 associates with microtubules and dynein invasion of human intestinal cells. *Infect. Immun.* 67: 4171–4182.

Ireton K, Payrastre B, Chap H *et al.* (1996) A role for phosphoinositide 3-kinase in bacterial invasion. *Science* 274: 780–782.

Ireton K, Payrastre B, Cossart P (1999) The *Listeria monocytogenes* protein InlB is an agonist of mammalian phosphoinositide 3-kinase. *J. Biol. Chem.* 274: 17025–17032.

Isberg RR (1989) Mammalian cell adhesion functions and cellular penetration of enteropathogenic *Yersinia* species. *Mol. Microbiol.* 3: 1449–1453.

Isberg RR (1991) Discrimination between intracellular uptake and surface adhesion of bacterial pathogens. *Science* 252: 934–938.

Isberg RR, Falkow S (1985) A single genetic locus encoded by *Yersinia pseudotuberculosis* permits invasion of cultured animal cells by *Escherichia coli* K12. *Nature* 317: 262–264.

Isberg RR, Leong JM (1990) Multiple β1 chain integrins are receptors for invasin, a protein that promoted bacterial penetration into mammalian cells. *Cell* 60: 861–871.

Isberg RR, Tran van Nhieu G (1994) Two mammalian cell internalization strategies used by pathogenic bacteria. *Annu. Rev. Genet.* 27: 395–422.

Jadoun J, Ozeri V, Burstein E, Skutelsky E, Hanski E, Sela S (1998) Protein F1 is required for efficient entry of *Streptococcus pyogenes* into epithelial cells. *J. Infect. Dis.* 178: 147–158.

Jerse AE, Yu J, Tall BD, Kaper JB (1990) A genetic locus of enteropathogenic *Escherichia coli* necessary for the production of attaching and effacing lesions on tissue culture cells. *Proc. Natl Acad. Sci. USA* 87: 7839–7843.

Jones BD, Falkow S (1996) Salmonellosis: host immune responses and bacterial virulence determinants. *Annu. Rev. Immunol.* 14: 533–561.

Jones BD, Ghori N, Falkow S (1994) *Salmonella typhimurium* initiates murine infection by penetrating and destroying the specialized epithelial M cells of the Peyer's patches. *J. Exp. Med.* 180: 15–23.

Kallstrom H, Liszewski MK, Atkinson JP, Jonsson AB (1997) Membrane cofactor protein (MCP or CD46) is a cellular pilus receptor for pathogenic *Neisseria*. *Mol. Microbiol.* 25: 639–647.

Kallstrom H, Islam S, Berggren PO, Jonsson AB (1998) Cell signalling by the type IV pili of pathogenic *Neisseria*. *J. Biol. Chem.* 273: 21777–21782.

Kaniga K, Uralil J, Bliska JB, Galán JE (1996) A secreted protein tyrosine phosphatase with modulator effector domains in the bacterial pathogen *Salmonella typhimurium*. *Mol. Microbiol.* 21: 633–641.

Ketley JM (1997) Pathogenesis of enteric infection by *Campylobacter*. *Microbiology* 143: 5–21.

Kocks C, Gouin E, Tabouret M, Berche P, Ohayon H, Cossart P (1992) *Listeria monocytogenes*-induced actin assembly requires the *actA* gene product, a surface protein. *Cell* 68: 521–531.

Konkel ME, Joens LA (1989) Adhesion to and invasion of HEp-2 cells by *Campylobacter* spp. *Infect. Immun.* 57: 2984–2990.

Konkel ME, Hayes SF, Joens LA, Cieplak W Jr (1992) Characteristics of the internalization and intracellular survival of *Campylobacter jejuni* in human epithelial cells cultures. *Microb. Pathog.* 13: 357–370.

Konkel ME, Kim BJ, Rivera-Amill V, Garvis SG (1999) Bacterial secreted proteins are required for the internalization of *Campylobacter jejuni* into cultured mammalian cells. *Mol. Microbiol.* 32: 691–701.

Kubori T, Matsushima Y, Nakamura D *et al.* (1998) Supramolecular structure of the *Salmonella typhimurium* type III protein secretion system. *Science* 280: 602–605.

Kupsch E.-M, Knepper B, Kuroki T, Heuer I, Meyer TF (1993) Variable opacity (Opa) outer membrane proteins account for the cell tropism displayed by *Neisseria gonorrhoeae* for human leukocytes and epithelial cells. *EMBO J.* 12: 641–650.

La Penta D, Rubens DC, Chi E, Cleary PP (1994) Group A streptococci efficiently invade human respiratory epithelial cells. *Proc. Natl Acad. Sci. USA* 91: 12115–12119.

Lecuit M, Ohayon H, Braun L, Mengaud J, Cossart P (1997) Internalization of *Listeria monocytogenes* with an intact leucine-rich repeat region is sufficient to promote internalization. *Infect. Immun.* 65: 5309–5313.

Lecuit M, Dramsi S, Gottardi C, Fedor-Chaiken M, Gumbiner B, Cossart P (1999) A single amino acid in E-cadherin responsible for host specificity towards the human pathogen *Listeria monocytogenes*. *EMBO J.* 18: 3956–3963.

Leong JM, Isberg RR (1993) The formation of a disulfide bond is required for efficient integrin binding by invasin. *J. Biol. Chem.* 268: 20524–20532.

Leong JM, Fournier R, Isberg RR (1990) Identification of the integrin binding domain of the *Yersinia pseudotuberculosis* invasion protein. *EMBO J.* 9: 1979–1989.

Leong JM, Morrissey PE, Isberg RR (1995) An aspartate residue of the *Yersinia pseudotuberculosis* invasion protein that is critical for integrin binding. *EMBO J.* 14: 422–431.

Lignau A, Domann E, Hudel M *et al.* (1995) Expression of the *Listeria monocytogenes* EGD *inlA* and *inlB* genes, whose products mediate bacterial entry into tissue culture cell lines, by PrfA-dependent and -independent mechanisms. *Infect. Immun.* 63: 3896–3903.

Makino S, Sasakawa C, Kamata K, Kurata T, Yoshikawa M (1986) A genetic determinant required for continuous reinfection of adjacent cells on a large plasmid in *S. flexneri* 2a. *Cell* 46: 551–555.

Marco AJ, Prats N, Ramos JA *et al.* (1992) A microbiological, histopathological and immunohistological study of the intragastric inoculation of *Listeria monocytogenes* in mice. *J. Comp. Pathol.* 107: 1–9.

Marra A, Isberg RR (1997) Invasin-dependent and invasin-independent pathways for translocation of *Yersinia pseudotuberculosis* across the Peyer's patch intestinal epithelium. *Infect. Immun.* 65: 3412–3421.

Ménard R, Sansonetti PJ, Parsot C (1994a) The secretion of the *Shigella flexneri* Ipa invasins is induced by the epithelial cells and controlled by IpaB and IpaD. *EMBO J.* 13: 5293–5302.

Ménard R, Sansonetti PJ, Parsot C, Vasselon T (1994b) Extracellular association and cytoplasmic partitioning of the IpaB and IpaC invasins of *Shigella flexneri*. *Cell* 79: 515–525.

Ménard R, Prévost MC, Gounon P, Sansonetti PJ, Dehio C (1996) The secreted Ipa complex of *Shigella flexneri* promotes entry into mammalian cells. *Proc. Natl Acad. Sci. USA* 93: 1254–1258.

Mengaud J, Ohayon H, Gounon P, Mége RM, Cossart P (1996) E-cadherin is the receptor for internalin, a surface protein required for entry of *Listeria monocytogenes* into epithelial cells. *Cell* 84: 923–932.

Merz AJ, So M (1997) Attachment of piliated, Opa^- and Opc^- gonococci and meningococci to epithelial cells elicits cortical actin rearrangements and clustering of tyrosine-phosphorylated proteins. *Infect. Immun.* 65: 4341–4349.

Miller VL, Falkow S (1988) Evidence for two genetic loci in *Yersinia enterocolitica* that can promote invasion of epithelial cells. *Infect. Immun.* 56: 1242–1248.

Miller VL, Farmer III JJ, Hill WE, Falkow S (1989) The *ail* locus is found uniquely in *Yersinia enterocolitica* serotypes commonly associated with disease. *Infect. Immun.* 57: 121–131.

Molinari G, Chhatwal GS (1998) Invasion and survival of *Streptococcus pyogenes* in eukaryotic cells correlates with the source of the clinical isolates. *J. Infect. Dis.* 177: 1600–1607.

Molinari G, Chhatwal GS (1999) Streptococcal invasion. *Curr. Opin. Microbiol.* 2: 56–61.

Molinari G, Talay SR, Valentin-Weigand P, Rohde M, Chhatwal GS (1997) The fibronectin-binding protein of *Streptococcus pyogenes*, SfbI, is involved in the internalization of group A streptococci by epithelial cells. *Infect. Immun.* 65: 1357–1363.

Monack DM, Raupach B, Hromockyj AE, Falkow S (1996) *Salmonella typhimurium* invasion induces apoptosis in infected macrophages. *Proc. Natl Acad. Sci. USA* 93: 9833–9838.

Mounier J, Vasselon T, Hellio R, Lesourd M, Sansonetti PJ (1992) *Shigella flexneri* enters human colonic Caco-2 epithelial cells through the basolateral pole. *Infect. Immun.* 60: 237–248.

Mounier J, Laurent V, Hall A *et al.* (1999) Rho family GTPases control entry of *Shigella flexneri* into epithelial cells but not intracellular motility. *J. Cell Sci.* 112: 2069–2080.

Müller S, Hain T, Pashalidis P *et al.* (1998) Purification of the *inlB* gene product of *Listeria monocytogenes* and demonstration of its biological activity. *Infect. Immun.* 66: 3128–3133.

Müller N, Gunther D, Naumann M, Meyer TF, Rudel T (1999) Neisserial porin (PorB) causes rapid calcium influx in target cells and induces apoptosis by the activation of cystein proteases. *EMBO J.* 18: 339–352.

Nachamkin I, Mishu Allos B, Ho T (1998) *Campylobacter* species and Guillain-Barré syndrome. *Clin. Microbiol. Rev.* 11: 555–567.

Nassif X, Pujol C, Morand P, Eugéne E (1999) Interactions of pathogenic *Neisseria* with host cells. Is it possible to assemble the puzzle? *Mol. Microbiol.* 32: 1124–1132.

Naumann M, Wessler S, Bartsch C, Wieland B, Meyer TF (1997) *Neisseria gonorrhoeae* epithelial cell infection leads to the activation of the transcription factors nuclear factor kappaB and activator protein 1 and the induction of inflammatory cytokine. *J. Exp. Med.* 186: 247–258.

Naumann M, Rudel T, Wieland B, Bartsch C, Meyer TF (1998) Coordinate activation of activator protein 1 and inflammatory cytokines in response to *Neisseria gonorrhoeae* epithelial cell contact involves stress response kinases. *J. Exp. Med.* 188: 1277–1286.

Naumann M, Rudel T, Meyer TF (1999) Host cell interactions and signalling with *Neisseria gonorrhoeae*. *Curr. Opin. Microbiol.* 2: 62–70.

Norris FA, Wilson MP, Wallis TS, Galyov EE, Majerus PW (1998) SopB, a protein required for virulence of *Salmonella dublin*, is an inositol phosphate phosphatase. *Proc. Natl Acad. Sci. USA* 95: 14057–14059.

Oelschlaeger TA, Guerry P, Kopecko DJ (1993) Unusual microtubule-dependent endocytosis mechanisms triggered by *Campylobacter jejuni* and *Citrobacter freundii*. *Proc. Natl Acad. Sci. USA* 99: 6884–6888.

Okada N, Pentland AP, Falk P, Caparon MG (1994) M protein and protein F act as important determinants of cell-specific tropism of *Streptococcus pyogenes* in skin tissue. *J. Clin. Invest.* 94: 965–977.

Okada N, Watarai M, Ozeri V, Hanski E, Caparon M, Sasakawa C (1997) A matrix form of fibronectin mediates enhanced binding of *Streptococcus pyogenes* to host tissue. *J. Biol. Chem.* 43: 26978–26984.

Okada N, Tatsuno I, Hanski E, Caparon M, Sasakawa C (1998) *Streptococcus pyogenes* protein F promotes invasion of HeLa cells. *Microbiology* 144: 3079–3086.

Ozeri V, Tovi A, Burstein I *et al.* (1996) A two domain mechanisms for group A streptococcal adherence through protein F to the extracellular matrix. *EMBO J.* 15: 989–998.

Ozeri V, Rosenshine I, Mosher DF, Fässler R, Hanski E (1998) Roles of integrins and fibronectin in the entry of *Streptococcus pyogenes* into cells via protein F1. *Mol. Microbiol.* 30: 625–637.

Pace J, Hayman MJ, Galán JE (1993) Signal transduction and invasion of epithelial cells by *S. typhimurium*. *Cell* 72: 505–514.

Pancholi V, Fischetti VA (1992) A major surface protein on group A streptococci is a glyceraldehyde-3-phosphate-dehydrogenase with multiple binding activity. *J. Exp. Med.* 176: 415–426.

Pancholi V, Fischetti VA (1997) Regulation of the phosphorylation of human pharyngeal cell proteins by group A streptococcal surface dehydrogenase: signal transduction between streptococci and pharyngeal cell. *J. Exp. Med.* 186: 1633–1643.

Pepe JC, Miller VL (1993) *Yersinia enterocolitica* invasin: a primary role in the initiation of infection. *Proc. Natl Acad. Sci. USA* 90: 6473–6477.

Pepe JC, Wachtel MR, Wager E, Miller VL (1995) Pathogenesis of defined invasion mutants of *Yersinia enterocolitica* in a BALB/c mouse model of infection. *Infect. Immun.* 63: 4837–4848.

Perdomo JJ, Gounon P, Sansonetti PJ (1994) Polymorphonuclear leukocyte transmigration promotes invasion of colonic epithelial monolayer by *Shigella flexneri*. *J. Clin. Invest.* 93: 633–643.

Persson C, Carballeira N, Wolf-Watz H, Fällman M (1997) YopH inhibits uptake of *Yersinia*, tyrosine phosphorylation of p130^{Cas} and FAK, the associated accumulation of these proteins in peripheral focal adhesions. *EMBO J.* 16: 2307–2318.

Persson C, Nordfelth R, Persson K, Forsberg Å, Wolf-Watz H, Fällman M (1999) Localization of the *Yersinia* PTPase to focal complexes is an important virulence mechanism. *Mol. Microbiol.* 33: 828–838.

Pron B, Boumaila C, Jaubert F *et al.* (1998) Comprehensive study of the intestinal stage of listeriosis in a rat ligated ileal loop system. *Infect. Immun.* 66: 747–755.

Rankin S, Isberg RR, Leong JM (1992) The integrin-binding domain of invasin is sufficient to allow bacterial entry into mammalian cell. *Infect. Immun.* 60: 3909–3912.

Rosenshine I, Duronio V, Finlay BB (1992) Tyrosine protein kinase inhibitors block invasin-promoted bacterial uptake by epithelial cells. *Infect. Immun.* 60: 2211–2217.

Rudolph MG, Weise C, Mirold S *et al.* (1999) Biochemical analysis of SopE from *Salmonella typhimurium*, a highly efficient guanosine nucleotide exchange factor for RhoGTPases. *J. Biol. Chem.* 274: 30501–30509.

Russell RG, Blake DC (1994) Cell association and invasion of Caco-2 cells by *Campylobacter jejuni*. *Infect. Immun.* 62: 3773–3779.

Russell RG, O'Donnoghue M, Blake DC Jr, Zulty J, DeTolla LJ (1993) Early colonic damage and invasion of *Campylobacter jejuni* in experimentally challenged *Macaca mulatta*. *J. Infect. Dis.* 168: 210–215.

Sansonetti PJ, Ryter A, Clerc P, Maurelli AT, Mounier J (1986) Multiplication of *Shigella flexneri* within HeLa cells: lysis of the phagocytic vacuole and plasmid-mediated contact hemolysis. *Infect. Immun.* 51: 461–469.

Sansonetti PJ, Arondel J, Fontaine A, d'Hauteville H, Bernardini L (1991) *ompB* (osmo-regulation) and *icsA* (cell to cell spreading) mutants of *Shigella flexneri*: vaccine candidates and probes to study the pathogenesis of shigellosis. *Vaccine* 9: 416–422.

Sansonetti PJ, Arondel J, Cantey JR, Prévost MG, Huerre M (1996) Infection of rabbit Peyer's patches by *Shigella flexneri* effect of adhesion or invasive bacterial phenotypes on follicle-associated epithelium. *Infect. Immun.* 64: 2752–2764.

Sasakawa C, Kamata K, Sakai T *et al.* (1988) Virulence-associated genetic regions comprising 31 kilobase of the 230-kilobase plasmid in *Shigella flexneri* 2a. *J. Bacteriol.* 170: 2480–2484.

Schauer DB, Falkow S (1993) The *eae* gene of *Citrobacter frundii* biotype 4280 is necessary for colonialization in transmissible murine colonic hyperplasia. *Infect. Immun.* 61: 4654–4661.

Simonet M, Richard S, Berche P (1990) Electron microscopic evidence for *in vivo* extracellular localization of *Yersinia pseudotuberculois* harboring the pYV plasmid. *Infect. Immun.* 58: 841–845.

Sinha B, François PP, Nüsse O *et al.* (1999) Fibronectin-binding protein acts as *Staphylococcus aureus* invasin via fibronectin bridging to integrin $\alpha 5\beta 1$. *Cell. Microbiol.* 1: 101–117.

Skoudy A, Tran van Nhieu GT, Mantis N *et al.* (1999) A functional role for ezrin during *Shigella flexneri* entry into epithelial cells. *J. Cell Sci.* 112: 2059–2068.

Stockbauer KE, Magoun L, Liu M *et al.* (1999) A natural variant of the cysteine protease virulence factor of group A streptococcus with an arginine-glycine-aspartic acid (RGD) motif preferentially binds human integrins $\alpha v\beta 3$ and $\alpha IIb\beta 3$. *Proc. Natl Acad. Sci. USA* 96: 242–247.

Takeichi M (1990) Cadherins: a molecular family important in selective cell-cell adhesion. *Annu. Rev. Biochem.* 59: 237–252.

Tilney LG, Portnoy DA (1989) Actin filaments and the growth, movement, spread of the intracellular bacterial parasite *Listeria monocytogenes*. *J. Cell Biol.* 109: 1597–1608.

Tobe T, Yoshikawa M, Mizuno T, Sasakawa C (1993) Transcriptional control of the invasion regulatory gene *virB* of *Shigella flexneri*: activation by VirF and repression by H-NS. *J. Bacteriol.* 175: 6142–6149.

Tobe T, Yoshikawa M, Sasakawa C (1995) Thermoregulation of *virB* transcription in *Shigella flexneri* by sensing of changes in local DNA superhelicity. *J. Bacteriol.* 177: 1094–1097.

Tran van Nhieu G, Sansonetti PJ (1999) Mechanism of *Shigella* entry into and promotes bacterial entry. *Curr. Opin. Microbiol.* 22: 327–338.

Tran van Nhieu G, Krukonis ES, Reszka AA, Horwitz AF, Isberg RR (1996) Mutations in the cytoplasmic domain of the $\beta 1$ integrin chain indicate a role for endocytosis factors in bacterial internalization. *J. Biol. Chem.* 271: 7665–7672.

Tran van Nhieu G, Benz-Ze'ev A, Sansonetti PJ (1997) Modulation of bacterial entry into epithelial cells by interaction between vinculin and the *Shigella* IpaA invasin. *EMBO J.* 16: 2717–2729.

Tran van Nhieu G, Caron E, Hall A, Sansonetti PJ (1999) IpaC induces actin polymerization and filopodia formation during *Shigella* entry into epithelial cells. *EMBO J.* 18: 3249–3262.

Tsai P-J, Kuo CF, Lin KY *et al.* (1998) Effect of group A streptococcal cystein protease on invasion of epithelial cells. *Infect. Immun.* 66: 1460–1466.

Ulmer JB, Burke CJ, Shi C, Friedman A, Donnelly JJ, Liu MA (1992) Pore formation and mitogenicity in blood cells by the class 2 protein of *Neisseria meningitidis*. *J. Biol. Chem.* 267: 19266–19271.

van Putten JMP, Duensing D (1997) Infection of mucosal epithelial cells by *Neisseria gonorrhoeae*. *Rev. Med. Microbiol.* 8: 51–59.

van Putten JMP, Duensing TD, Cole RL (1998a) Entry of OpaA(+) gonococci into HEp-2 cells requires concerted action of glycosaminoglycans, fibronectin and integrin receptors. *Mol. Microbiol.* 29: 369–379.

van Putten JMP, Duensing TD, Carlson J (1998b) Gonococcal invasion of epithelial cells driven by P. IA, a bacterial ion channel with GTP binding properties. *J. Exp. Med.* 188: 941–952.

Virji M (1996) Microbial utilization of human signalling molecules. *Microbiology* 142: 3319–3336.

Virji M, Makepeace K, Ferguson DJP, Watt SM (1996) Carcinoembryonic antigens (CD66) on epithelial cells and neutrophils are receptors for Opa proteins of pathogenic nesseriae. *Mol. Microbiol.* 22: 941–950.

Walker RI, Schmauder-Chock EA, Parker JL, Burr D (1988) Selective association and transport of *Campylobacter jejuni* through M cells of rabbit Peyer's patches. *Can. J. Microbiol.* 34: 1142–1147.

Wassef JS, Keren DF, Mailloux JL (1989) Role of M cells in initial antigen uptake and in ulcer formation in the rabbit intestinal loop model of shigellosis. *Infect. Immun.* 57: 858–863.

Wassenaar TM, Blaser MJ (1999) Pathophysiology of *Campylobacter jejuni* infection of humans. *Microb. Infect.* 1: 1023–1033.

Wassenaar TM, Bleumink-Pluym NM, van der Zeijst BAM (1991) Inactivation of *Campylobacter jejuni* flagellin genes by homologous recombination demonstrates that *flaA* but not *flaB* is required for invasion. *EMBO J.* 10: 2055–2061.

Watarai M, Funato S, Sasakawa C (1996) Interaction of Ipa proteins of *Shigella flexneri* with $\alpha5\beta1$ integrin promotes entry of the bacteria into mammalian cells. *J. Exp. Med.* 183: 991–999.

Watarai M, Kamata Y, Kozaki S, Sasakawa C (1997) Rho, a small GTP-binding protein, is essential for *Shigella* invasion of epithelial cells. *J. Exp. Med.* 185: 281–292.

Wood MW, Rosqvist R, Mullan PB, Edwards MH, Galyov EE (1996) SopE, a secreted protein of *Salmonella dublin* is translocated into the target eukaryotic cell via a *sip*-dependent mechanism and promotes bacterial entry. *Mol. Microbiol.* 22: 327–338.

Wooldridge KG, Ketley JM (1997) *Campylobacter*-host cell interactions. *Trends Microbiol.* 5: 96–102.

Wooldridge KG, Williams PH, Ketley JM (1996) Host signal transduction and endocytosis of *Campylobacter jejuni*. *Microb. Pathog.* 21: 299–305.

Zhou D, Mooseker M, Galán JE (1999) Role of the *S. typhimurium* actin-binding protein SipA in bacterial internalization. *Science* 283: 2092–2095.

Zychlinsky A, Prévost MC, Sansonetti PJ (1992) *Shigella flexneri* induces apoptosis in infected macrophages. *Nature* 358: 167–169.

Zychlinsky A, Fitting C, Cavaillon JM, Sansonetti PJ (1994) Interleukin 1 is released by murine macrophages during apoptosis induced by *Shigella flexneri*. *J. Clin. Invest.* 94: 1328–1332.

Zychlinsky A, Thirumalai K, Arondel J, Cantey JR, Aliprantis AO, Sansonetti PJ (1996) In vivo apoptosis in apoptosis in *Shigella flexneri* infection. *Infect. Immun.* 64: 5357–5365.

Yao RJ, Burr DH, Doig P, Trust TJ, Nlu H, Guerry P (1994) Isolation of motile and nonmotile insertional mutants of *Campylobacter jejuni*: the role of motility in adherence and invasion of eukaryotic cells. *Mol. Microbiol.* 14: 883–893.

32

Microbial Growth and Nutrition in the Host

H. Chart[1] and E. Griffiths[2]

[1] *PHLS, London, UK*
[2] *WHO, Geneva, Switzerland*

The ability of bacterial pathogens to evade host defences, such as the bactericidal action of plasma, or to resist phagocytosis, and to survive *in vivo*, can be seen as playing an important part in establishing an infection. This is, however, but one component of the complex interactions between pathogen and host (Table 1). The ability of bacteria to multiply successfully under the environmental conditions found in host tissues is also vital for the production of disease. Bacteria may establish an infection either by increasing in number in selected tissues or by increasing in number and disseminating throughout the host. The rate of bacterial multiplication under the conditions found in host tissues, or within host cells, is an important parameter in determining the outcome of infection, and it is here too that pathogens generally must produce the full complement of virulence determinants required for pathogenicity. Indeed, the growth of bacteria *in vivo* can affect their ability to resist some host defences.

Factors that might be expected to influence bacterial growth and multiplication *in vivo* include temperature, pH, oxygen levels, and the availability of both organic and inorganic nutrients.

Bacteria experience different environmental conditions depending on whether they are in, for example, water or soil, or in an eukaryotic host as a commensal or as a pathogen. Irrespective of the environment, however, bacteria must acquire all of the various inorganic and organic nutrients necessary for cellular metabolism if they are to grow and multiply, otherwise cell division will be suspended. This chapter will consider the general aspects of microbial growth and nutrition *in vivo*, but will focus on selected inorganic ions. Several of these are essential for numerous bacterial functions, but their availability depends on the prevailing environmental conditions.

In studying the mechanisms involved in the uptake of specific nutrients and ions it is important to realise that bacteria have evolved to survive and grow in a range of environments, usually those they are likely to encounter in their natural life cycles. In the case of pathogenic bacteria this will obviously include the environment of a particular host. Bacteria that live on a host surface, such as the skin, on the gut mucosa or in the nasopharynx, face very different ecological conditions, which will be quite different again from those encountered by bacteria living within host cells or in

 doi:10.1006/bkmm.2001.0032

Table 1 Crucial elements of microbial pathogenicity

Colonisation/invasion of host /tissue and host specificity
Survival on mucosal surfaces, in blood, body fluids, and intracellularly
Growth and multiplication, extracellularly/intracellularly *in vivo*
Inhibition/evasion of host defence mechanisms
Production of virulence determinants/toxins/ damage to host

Table 2 Inorganic nutrients that affect the expression of toxins and other potential virulence determinants

Nutrient	Toxins and virulence determinants
Magnesium	PhoP/Q-regulated virulence genes: *bvgA*/*bvgS* (*Bordetella pertussis*)
Calcium	*Yersinia* outer membrane proteins (Yops)
Sodium	Toxic shock syndrome toxin 1
	Haemolysin
	Surface protein A (*S. aureus*)
	Hyphal formation in *C. albicans*
Molybdenum	Molybdate transport/nitrate reductase expression important for anaerobic respiration (*E. coli*)
Zinc	Elastase production (*P. aeruginosa*)
Iron	High-affinity iron uptake
	Diphtheria toxin production
	Shiga toxin
	Pseudomonas exotoxin A
	Verocytoxin
	Enteroinvasive *E. coli* toxin
	Haemolysin
	Pilin production

the bloodstream. Different ecological niches will be characterised by different ionic conditions. However, because of the complexity of host–pathogen interactions, in particular in the context of changing cellular and host tissue environments encountered by invading pathogens during the course of infection, it is a very difficult task to identify the important parameters. A key fact to remember when studying the uptake of ions by pathogenic bacteria is that they have not evolved to grow on the laboratory media that have frequently been used in an attempt to obtain a better understanding of bacterial behaviour *in vivo*. The results of such studies often misrepresent the real situation in host tissues. With a better understanding of bacterial biochemistry and of an appreciation of media composition, however, the potential ionic concentrations experienced *in vivo* can be mimicked *in vitro*, and in this way important information obtained. In addition, newer genomic and molecular genetic technologies are being applied to this problem, and many exciting developments can be expected in the near future.

This chapter describes how pathogenic bacteria acquire essential ions in the course of the colonisation and/or invasion of mammalian hosts, and how differences in the concentration of some of the ions in host tissues have been exploited by pathogens as an environmental signalling device to regulate bacterial characteristics important for pathogenicity (**Table 2**).

Magnesium

Magnesium is the most abundant cation in bacterial cells (Hantke, 1997) with cytoplasmic concentrations of approximately 100 nM (Véscovi *et al.*, 1996). Magnesium ions are vital co-factors in numerous enzymatic reactions and are known to be essential for the stabilisation of ribosomes, the Gram-negative bacterial outer membrane and other cellular structures (Leive, 1965; Haque and Russell, 1976; Altura, 1992). For certain pathogenic bacteria, the concentration of Mg^{2+} ions also acts as an environmental signal that controls the expression of important virulence determinants (Smith and Maguire, 1998). The role of magnesium ions in bacterial biochemistry has frequently been explored using the active binding of divalent cations with ethylenediamine tetraacetic acid (EDTA) and observing the effect of replacing magnesium ions. For example, the incubation of *Pseudomonas aeruginosa* with EDTA causes cells to become osmotically fragile and the addition of magnesium ions re-establishes membrane stability (Haque and Russell, 1976; Ayres *et al.*, 1998). For certain strains of enteroaggregative *Escherichia coli* (EAggEC), adhesion involves the expression of certain outer membrane-associated proteins (Chart *et al.*, 1994, 1995). Although it is known that in strains belonging to serotypes O44:H18 and O126:H27 the expression of these proteins requires magnesium ions (Chart *et al.*, 1997), the actual function of Mg^{2+} in their expression is not known.

Acquisition of Mg^{2+}

Bacteria acquire magnesium ions by at least three distinct transport systems: the constitutive CorA system (Smith and Maguire, 1998; Kehres *et al.*, 1998; Smith *et al.*, 1998a), and the two inducible MgtA and MgtB transport systems (Smith and Maguire, 1998). *Salmonella typhimurium* has all three of these systems (Snavely *et al.*, 1991a; Moncrief and Maguire, 1998; Tao *et al.*, 1998), while magnesium uptake systems based on CorA and MgtA operate in strains of *E. coli*

(Hantke, 1997). The CorA magnesium transport system was first shown in *E. coli* and subsequently demonstrated in *S. typhimurium* (Snavely *et al.*, 1991b). Magnesium ions enter the periplasmic space after passage through porins, and are transported across the inner membrane by the CorA transport system. This involves a constitutive 37-kDa membrane protein encoded by the single *corA* gene. The CorA protein lacks homology to any known protein and is capable by itself of mediating Mg^{2+} influx (D.L. Smith *et al.*, 1993). At high levels of magnesium, CorA may also be involved with the efflux of magnesium from the bacterial cell (Smith *et al.*, 1998b).

In contrast, MgtA and MgtB are inducible P-type ATPases (Snavely *et al.*, 1991b; Tao *et al.*, 1995), and belong to a large family of homologous proteins responsible for membrane transport of cations (Pedersen and Carafoli, 1987). However, MgtA and MgtB proteins themselves have little similarity to other prokaryotic P-type ATPases, but much greater similarity to the eukaryotic P-type ATPases, especially the muscle sarcoplasmic reticulum Ca^{2+}-ATPase (Snavely *et al.*, 1991a; R.L. Smith *et al.*, 1993). MgtA is encoded by a single gene but MgtB is the product of a two-gene operon, *mgtCB*. MgtC is a 23-kDa protein, but its function is unknown. Its absence has little or no effect on the transport properties of MgtB (Snavely *et al.*, 1991a, 1991b; Tao *et al.*, 1995). Both the *mgtA* and *mgtCB* loci are highly repressed in bacteria growing in normal laboratory media, but their transcription is increased several thousand-fold when the organisms are grown in media limiting in Mg^{2+} (Snavely *et al.*, 1989, 1991a, 1991b; Tao *et al.*, 1995).

PhoPQ and Mg^{2+} Signalling

In *S. typhimurium*, the transcription of *mgtA* and *mgtB*, and many other genes, is under the control of the two-component regulatory system called PhoPQ, which is known to be an important virulence determinant in this organism (Groisman *et al.*, 1989; Gunn and Miller, 1996; Guo *et al.*, 1997; S.L. Miller *et al.*, 1989). Binding of Mg^{2+} to the PhoQ membrane sensor protein (kinase) mediates the regulation of multiple genes (Garcia-Vescori *et al.*, 1996; Soncini *et al.*, 1996). Under normal extracellular Mg^{2+} concentrations, Mg^{2+} binds to PhoQ and represses transcription of genes regulated by PhoP. When the extracellular Mg^{2+} concentration falls, transcription is no longer repressed and MgtA, MgtB and MgtC, as well as other proteins, are selectively expressed. It has been suggested that sulphate rather than Mg^{2+}, or perhaps in addition to Mg^{2+}, may play a role in gene expression (Gross and Rappuoli, 1989; Scarlato and Rappuoli, 1991). Nevertheless, in *S. typhimurium* the ability of Mg^{2+} to repress transcription is independent of the counteranions chloride or sulphate, and sulphate alone had no effect on transcription (Tao *et al.*, 1998). The data suggest, however, that although PhoPQ mediates much of the transcriptional regulation of *mgtA* and *mgtCB*, regulation of *mgtA* is more complex and has additional levels of regulation independent of PhoPQ (Tao *et al.*, 1998; Smith *et al.*, 1998a). The data are also consistent with the idea that regulation of both *mgtA* and *mgtCB* is sensitive to both Mg^{2+} and pH (Bearson *et al.*, 1998; Smith *et al.*, 1998a).

Transcription of *mgtCB*, and to a lesser extent *mgtA*, are markedly increased when *S. typhimurium* invades cells (Garcia-del Portillo *et al.*, 1992; Smith *et al.*, 1998b). This link between the virulence determinant PhoPQ, Mg^{2+} and invasion of cells suggests that increased transcription of the *mgtA* and *mgtCB* genes is needed because of limiting Mg^{2+} levels inside the cell vacuole within which the invading pathogen initially resides. The hypothesis is that transcription of *mgtA* and/or *mgtCB* is induced on invasion in order to scavenge the Mg^{2+} needed for replication in the eukaryotic cell. The vacuole formed by *S. typhimurium* on invasion is limiting in iron and Mg^{2+}, and initially it also has an acid pH (Finlay and Cossart, 1997; Apulche-Aranda *et al.*, 1992; Garcia-del Portillo *et al.*, 1992; Rathman *et al.*, 1996; Ernst *et al.*, 1999). Other data do not, however, entirely support this idea (see below).

Role of MgtA, MgtB and MgtC

Although *mgtA*, *mgtB* and *mgtC* are markedly induced on invasion of a eukaryotic cell by *S. typhimurium*, none of these genes is necessary either for invasion *per se*, or for short-term survival in the eukaryotic cell, or extra scavenging of Mg^{2+}. It appears that the bacteria do not require additional Mg^{2+} transport capacity after invasion, since the constitutive CorA system remains functional at the acid pH of the vacuole (Smith *et al.*, 1998a). Indeed, the capacity of the CorA transporter in mediating Mg^{2+} uptake is greater at pH 5.2 than at pH 7.4. The capacity of CorA is quite large, at least as large as that of the induced MgtA and MgtB transporters.

The presence of a functional MgtA has no significant effect on short- or long-term survival of *S. typhimurium* in eukaryotic cells, which suggests that it is not associated with virulence (Blanc-Potard and Groisman, 1997). However, MgtB is necessary for long-term survival in the cell, and it is suggested that this protein may have functions additional to Mg^{2+} transport. Likewise, MgtC is essential for *S. typhimurium* virulence, since it is required for long-term (18 hours) survival in eukaryotic cells. *S. typhimurium* that carries an insertional mutation in the *mgtC* region is markedly attenuated (Blanc-Potard and Groisman, 1997). Thus, while Mg^{2+} and various Mg^{2+} transporter systems are important for virulence, their actual functions in the eukaryotic cell after bacterial invasion remain unclear (**Table 2**). It may be that the function of MgtC is completely unrelated to the transport or use of Mg^{2+}, and that it is merely associated with Mg^{2+} by the fact that the level of Mg^{2+} controls its expression. This situation would be similar to that seen with a number of other bacterial virulence determinants, such as diphtheria toxin, the expression of which is regulated by the availability of iron, but which themselves have no role in iron metabolism (see below).

In *Bordetella pertussis* there is a two-component regulator called *bvgA/bvgS*, which is analogous to the PhoPQ system and which also responds to Mg^{2+} (**Table 2**). The *bvg* regulon regulates its own expression and that of other genes, including filamentous haemagglutinin (FHA), pertussis toxin and adenylate cyclase. The BvgS protein senses different stimuli, including temperature and magnesium sulphate. Thus, *in vitro*, $MgSO_4$ represses the expression of virulence genes in *B. pertussis* (Scarlato and Rappuoli, 1991), but the factors that activate the BvgA/BvgS system *in vivo* have not been characterised.

Calcium

Calcium ions play a key role in several bacterial processes, including protein phosphorylation, in the stability of the bacterial cell wall and in the adhesion of bacteria to eukaryotic cells. In common with magnesium, the role of calcium in bacterial biochemistry has been studied extensively by removing it from organisms by treating them with EDTA and observing the effect of replacing Ca^{2+} ions. Early studies with this approach showed that removal of calcium ions increased cell wall permeability (Leive, 1965; Schnaitman and Klena, 1993), making the cell envelope of bacteria such as *E. coli* fragile and more susceptible to physical disruption by sonication (H. Chart and E. Griffiths, unpublished data). In Gram-negative bacteria, calcium ions stabilise the cell envelope by forming an association between lipopolysaccharide (LPS) molecules and outer membrane proteins (Nikaido and Vaara, 1985) by binding to phosphate residues on LPS molecules (Schnaitman and Klena, 1993).

For other bacteria, such as *S. typhimurium*, calcium enhances invasion of eukaryotic cells (Niesel and Peterson, 1987) and for certain strains of EAggEC, calcium ions are essential for the expression of outer membrane-associated proteins that mediate the adhesion of these bacteria to HEp-2 cells (Spencer *et al.*, 1998). The growth of these organisms in media containing agents that chelate divalent cations results in the repression of protein expression and an inability to adhere to HEp-2 cells (Spencer *et al.*, 1998). It has not been determined whether the calcium ions are required for protein synthesis as such or for the assembly of protein subunits on the surface of the bacterial cell.

In the case of some bacteria, such as some strains of *E. coli*, levels of intracellular calcium are approximately 0.1 μM and some 1000-fold less than those typically found in the surrounding environment (Norris *et al.*, 1996). Many bacteria have transmembrane calcium transporter systems that exclude calcium ions. Since human blood contains approximately 2–3 mM calcium, it would seem that bacteria could readily acquire calcium ions in the course of a systemic infection. In contrast, the levels of calcium within human cells is considerably lower and bacteria such as *Streptococcus* (Trombe, 1993; Trombe *et al.*, 1994) and *Yersinia* have specialised calcium uptake mechanisms.

Members of the genus *Yersinia*, including *Y. pestis*, *Y. enterocolitica* and *Y. pseudotuberculosis*, use calcium ions as a switch for the expression of plasmid-encoded *Yersinia* outer membrane proteins (Yops) (Boland and Havaux, 1998; Monack *et al.*, 1998), which are secreted by a type III secretion system (Nilles *et al.*, 1997; Williams and Straley, 1998) when they are grown at 37°C in media deprived of calcium ions (Cornelis, 1998). It has been suggested that the low level of calcium in eukaryotic cells forms a switch for the expression of the Yop proteins (Payne and Straley, 1999), which seems to have a number of roles in virulence, such as interfering with phagocytosis (Finlay and Cossart, 1997). For *Y. pestis* a low-Ca^{2+} response stimulus is responsible for the temperature- and Ca^{2+}-regulated expression and secretion of the Yops (Payne and Straley, 1998) (**Table 2**). Whether calcium ions, or some other stimulus such as contact with phagocytes, or both, are responsible for induction of Yops *in vivo* is not clear (Straley *et al.*, 1993). Experimental virulence in *Streptococcus pneumoniae* also appears to be modulated by the kinetics of Ca^{2+} transport (Trombe, 1999).

Phosphorus

Inorganic phosphorus (P_i) is essential for many metabolic processes, including biochemical reactions involved with the production of energy-rich compounds. Furthermore, P_i forms an integral part of membrane lipids, phosphoproteins, LPS and nucleic acids. Bacteria such as *E. coli* may contain as much as 15 mg of P_i per g dry cell weight. Two mechanisms are involved in the uptake of P_i, an inducible high-affinity system (Pst: phosphate-specific transport) regulated by the phosphate (Pho) regulon and a low-affinity constitutive system (Pit: phosphate inorganic transport; Surin *et al.*, 1985). The expression of the Pho regulon is inhibited by excess P_i, and under phosphate starvation the Pst P_i uptake system becomes activated. An outer membrane protein PhoE is expressed, as are elements encoded by the Pst system. The *pstA* and *pstC* genes encode membrane channel-forming proteins, and a 34-kDa PstS protein forms a periplasmic receptor for P_i and the P_i uptake process is fuelled by the ATPase, PstB.

The availability of P_i to bacteria multiplying within a given host depends on the host tissues involved. Human serum contains 5–7 mg100 cm^{-3} of inorganic phosphate, which suggests that in this location P_i would be readily available.

Sodium

In common with other ions, bacteria such as *E. coli* maintain intracellular levels of sodium lower than the surrounding medium and possess specific antiportation systems to pump extraneous sodium ions out of the cell. The ability of bacteria to acquire sodium ions from the environment depends on the location that is colonised. For putative bacterial pathogens, however, the levels of sodium within, for example, human serum are approximately 130–150 mM, suggesting that sodium ions would not be a limiting factor for bacterial multiplication.

Recent evidence suggests, however, that subtle differences in sodium ion concentration may in some cases play a role in regulating the expression of virulence determinants. The effect of NaCl on virulence-determinant gene expression was examined in *Staphylococcus aureus* by Chan and Foster (1998), who found that high levels of Na^+ (1 M) lead to a dramatic decrease in the expression of toxic shock syndrome toxin 1 (encoded by the *tst* gene), α-haemolysin (*hla* gene) and surface protein A (*spa* gene) (Table 2).

The relevance of such a high sodium concentration to the situation *in vivo* in a mammalian host is not altogether clear. It is also not clear in the production by *Staph. aureus* of menstrual toxic shock syndrome, which is associated with the use of high-absorbency tampons (Kreiswirth, 1989). A number of divalent metal ions and other factors, such as osmolarity and quorum-sensing may be involved.

Nevertheless, a possible role for sodium ions is in influencing the growth and dimorphic potential of the human opportunistic pathogenic yeast *Candida albicans* (Northrop *et al.*, 1997). The ability of the yeast form of *C. albicans* to switch to the 'hyphal' form may allow penetrative infection of internal organs, the transition between different morphological forms being triggered by a number of environmental signals including pH, temperature and nutritional status (Gow, 1997). In spite of a preponderance of sodium ions and its balancing chloride anion in the microenvironment encountered by *C. albicans in vivo*, the data suggest that subtle differences in concentrations, combined with other parameters such as pH, may play an important part in influencing dimorphic transition in this pathogenic yeast (Table 2). The Na^+ and Cl^- concentrations of defined media used in studies of *C. albicans* vary in their resemblance to those of tissue fluids. For example, the Na^+, K^+ and Cl^- concentrations of Lee's medium (86 mM, 28 mM, 86 mM, respectively; Lee *et al.*, 1975) are similar to those reported for vaginal fluid (61 mM, 23 mM, 62 mM; Levin and Wagner, 1977), but differ from the general composition of tissue fluid (132 mM, 4 mM, 102 mM; Levin and Wagner, 1977). Media such as Yeast Nitrogen Base have little physiological relevance in terms of electrolyte composition (Na^+, K^+, Cl^-: 0.1 mM, 0.5 mM, 0.09 mM, respectively; Northrop *et al.*, 1997).

The results of the study by Northrop *et al.* (1997) show that increasing the Na^+ concentration to 200 mM leads to inhibition of germ tube formation but suggest that this is not the only controlling factor. The precise site of Na^+ action is unclear. Recently, however, a Na^+/H^+ antiporter gene, *CNH1*, has been isolated and characterised in *C. albicans* (Soong *et al.*, 2000). This gene has a role in Na^+ and H^+ transport, salt tolerance and morphogenesis. Further studies can be expected to throw new light on its role in pathogenesis.

Potassium

Potassium is a co-factor for certain enzymes, but it also plays a major role in osmoregulation, and bacteria maintain an intracellular potassium ion concentration of between 0.1 and 0.5 M. *E. coli* has three known mechanisms for the acquisition of potassium, one a high-affinity inducible uptake system (potassium-dependent growth: Kdg) and two low-affinity constitutive systems (transport of potassium: Trk; and potassium uptake: Kup). The Kdg high-affinity uptake system involves five genes, *kdgA–E*. The gene product of *kdgB* is a membrane ATPase and that of *KgpD* is a cell membrane sensor for osmotic pressure, also involved in the regulation of the Kdp uptake system. The low-affinity Trk uptake system involves the *trkA*, *trkE*, *trkG* and *trkH* genes and the Kup system; both are constitutive.

The bacterial intracellular concentration of potassium is some 20-fold higher than, for example, that in human serum, which contains some 4–5 mM potassium. The amounts of potassium in the host would seem quite high and the acquisition of potassium by extra-intestinal bacteria would not appear to be a major problem for pathogenic organisms. However, nothing is known about its role in environmental signalling for invading pathogens.

Ammonium

Ammonium ions are the preferred source of nitrogen for many micro-organisms and are generally assimilated into organic constituents by two pathways. The first is into glutamate by a reaction catalysed by L-glutamate dehydrogenase and the second is by a cycle of reactions catalysed by glutamine and glutamate synthases. *E. coli* grows poorly when deprived of ammonium ions, and it has an active transport mechanism to concentrate ammonium ions (Soupene *et al.*, 1998). A major component in the ammonium ion transport is the ammonium/methylammonium transport B (AmtB) protein, which has been described as a cytoplasmic protein with transmembrane spanning segments (Soupene *et al.*, 1998).

The ability of enteric bacteria to acquire ammonium ions during infection is poorly understood but must depend on the concentration of these ions in the intestine, and this probably depends on dietary intake. In contrast, extra-intestinal bacteria seem to rely on the presence of ammonium ions in host tissues, which in human blood is 10–50 μM (this varies between arterial and venous blood). Clearly, bacteria multiplying in host tissues require a source of nitrogen, but it is not known whether extra-intestinal bacteria use host ammonium ions.

Molybdenum

Although molybdenum is required in only trace amounts, it plays a critical role in several metabolic pathways, primarily those involved in energy production. Under anaerobic growth conditions, *E. coli* produces several molybdenum-containing enzymes, almost all of which serve as terminal enzymes in anaerobic respiratory pathways (Gennis and Stewart, 1996). One, the molybdenum-containing formate dehydrogenase (FHL), is unique in not being a component of anaerobic respiration. The FHL complex, which catalyses cleavage of formate to H_2 and CO_2, contains a formate dehydrogenase isoenzyme (FDH-H; *fdhF*), a hydrogenase isoenzyme (HYD 3; *hyc*), and intermediate electron carriers, which are also encoded by the *hyc* operon (Böck and Sawers, 1996). FDH-H contains selenium and molybdenum, respectively in the form of selenocysteine and molybdopterin guanine dinucleotide (Zinoni *et al.*, 1986; Boyington *et al.*, 1997).

Molybdate is transported in bacteria by a high-affinity transport system involving three proteins encoded by the *modA*, *modB* and *modC* genes (Grunden and Shanmugam, 1997). The protein encoded by *modA* is a 27.4-kDa periplasmic protein, which actively binds molybdate ions and other ions such as tungstate. The ModB protein is a 24-kDa protein thought to form transmembrane channels for molybdate ions to gain access to the cell cytoplasm (Grunden and Shanmugam, 1997). The molybdate uptake system is fuelled by the 39-kDa ModC energy-transducing protein and regulated by the repressor protein ModE.

Molybdenum is involved in the transcriptional regulation of molybdo-enzyme synthesis as well as regulation of molybdate transport components (*modABCD*) (Self *et al.*, 1999). ModE-molybdate is a repressor of the *modABCD* operon and is a secondary transcriptional activator for the *hyc* and *nar* operons, the latter encoding nitrate reductase. For a shift to nitrate respiration during anaerobic growth, *E. coli* requires the presence of nitrate, the substrate for the molybdo-enzyme nitrate reductase, the presence of intracellular molybdate and the absence of O_2 (Gunsalus, 1992; Gennis and Stewart, 1996) (Table 2).

The part played in infection by these anaerobic systems and by molybdenum is unknown. There seems to be little doubt, however, that tissue hypoxia plays an important role in the development of sepsis and organ failure (Ward and Bullen, 1999; Griffiths and Bullen, 1999; Dobson and Bullen, 1999), and many bacteria have considerable reducing powers as a result of natural metabolic activities. For example, *E. coli* growing in glucose broth can reduce the Eh to −380 mV in an hour. The Eh and pH in tissues infected with various organisms has been little investigated, but low oxygen tension and low pH have been recorded in drainage fluid from septic peritonitis where several samples had a zero pO_2 (Simmen and Blaser, 1993). The critical role of Eh and pH, and their effect on the availability of iron in sepsis is considered by Bullen *et al.* (2000). However little appears to be known about the availability of molybdenum in host tissues.

Nickel

Certain bacteria require low levels of nickel, principally as co-factors for metallo-enzymes. For example, *Helicobacter pylori* expresses a urease which hydrolyses urea to ammonia and carbon dioxide, whereupon the ammonia is used to neutralise gastric acid to reduce the environmental pH. This urease requires nickel as a co-factor (Collins and D'Orazio, 1993; Bauerfeind *et al.*, 1996; McGee *et al.*, 1999). Otherwise, little is known about its role in relation to infective processes.

Zinc

Many bacteria have metallo-proteases that require zinc ions for enzymatic activity (Hase and Finkelstein, 1993), and these include certain forms of phospholipase C (Titball, 1993). A phosphatidylcholine phospholipase C expressed by *Listeria monocytogenes* is a 29-kDa zinc-dependent protein, which is important for virulence in a mouse model (Raveneau *et al.*, 1992; Poyart *et al.*, 1993). The action of neurotoxins, such as tetanus and botulinum toxins, also involves endopeptidases with a zinc co-factor (Tonello *et al.*, 1996). Elastase, a toxic factor produced by *Pseudomonas aeruginosa*, is another zinc metallo-protease (Kharazmi, 1989; Saulnier *et al.*, 1989). The 32-kDa elastase degrades elastin and collagen and thereby provides a route by which *Ps. aeruginosa* can invade the host (Morihara and Tsuzuki, 1975). In addition, elastase can degrade a variety of other host proteins, including IgA and IgG, thereby decreasing the ability of the host to combat infection.

Zinc, along with iron, has also been implicated in the regulation of the translation of the genes encoding elastase in *Ps. aeruginosa* (Brumlik and Storey, 1992). By means of a *lasB-lacZ* translational fusion to examine expression of the elastase operon (*lasB*), it has been shown that increasing the zinc content of the medium has a positive influence on *lasB-lacZ* expression. The results suggest that zinc regulates the expression of elastase at the translational level. Iron concentration also affected the expression of *lasB*.

The specific mechanisms for zinc accumulation and transport in bacteria are poorly understood. However, a periplasmic zinc-binding protein has been identified in *Haemophilus influenzae* (Lu *et al.*, 1997). Whether bacteria utilise zinc-specific outer membrane binding proteins or transmembrane pores to acquire zinc ions is unknown.

Iron

One of the best-understood properties of the environment encountered by pathogens in host tissues and its effects on bacterial characteristics and growth, concerns the availability of iron. In contrast to the ions discussed above, the situation with iron appears rather special. Although an abundance of iron is present in the body fluids of healthy humans and animals, the amount readily available to invading bacteria is normally extremely small. Most body iron is to be found intracellularly in ferritin, haemosiderin or haem. Extracellular iron in fluids such as plasma and mucosal secretions is bound to the high-affinity specific iron-binding glycoproteins transferrin and lactoferrin. A related protein, ovotransferrin, is found in avian egg white. These proteins have association constants for iron of about 10^{36} and so bind iron extremely tightly. They are also only partially saturated with iron and it has been calculated that the amount of free iron in equilibrium with an iron-binding protein at neutral pH is of the order of 10^{-18} M, which is far too low to support bacterial growth. The reason why such extreme measures are taken to keep 'free' iron so extremely low is that under physiological conditions ferric iron would otherwise oxidise, hydrolyse and polymerise into insoluble forms. In addition, loosely bound iron is highly toxic under aerobic conditions (Griffiths, 1999). Thus, the very advantages that make iron such an excellent catalyst and an essential component of numerous metabolic processes, also make it a potentially hazardous metal, since reactions that involve oxygen lead to the formation of damaging free radicals. Iron-binding proteins bind iron tightly so that it is unable to catalyse free-radical cascades (Evans *et al.*, 1999).

During infections, iron availability is further limited by a phenomenon called hypoferraemia of infection, which is a well-recognised component of the acute-phase response to infection. Hypoferraemia of infection involves the lowering still further of the total amount of iron bound to serum transferrin. This response can be induced by endotoxin (Baker and Wilson, 1965) and mediated by release of various cytokines, especially interleukin-1 (Westmacott *et al.*, 1986; Gordenk *et al.*, 1988; Uchida *et al.*, 1991). Also, lactoferrin levels increase in the bloodstream and the protein accumulates at sites of inflammation.

Unlike transferrin, the iron-binding ability of which diminishes below pH 6, lactoferrin retains its iron-binding properties in the more acidic conditions that prevail at such sites (Morgan, 1981; Crichton and Ward, 1992).

Iron is an essential element for the growth of almost all living cells (Griffiths, 1999). It plays an important part in a wide range of cellular functions, including transport and storage of oxygen and electron transport processes. In the important process of DNA replication, iron-containing ribonucleotide reductases catalyse the conversion of ribonucleotides to deoxyribonucleotides. All known bacterial pathogens require iron to multiply and have developed different strategies to deal with the acquisition of sufficient of the metal for multiplication *in vivo*. In some cases the role of the iron uptake system is very clear and its loss results in the conversion of a virulent pathogen into a non-pathogenic organism. For example, highly virulent strains of *Vibrio anguillarum*, which is responsible for a devastating septicaemic disease of fish, carry a plasmid-encoded iron uptake system that allows the organism to grow rapidly under iron-restricted conditions. With loss of the plasmid, *V. anguillarum* loses both its ability to grow under such conditions and its virulence (Crosa, 1984). In other cases, the contribution of a particular iron-uptake system to virulence is less obvious since alternative mechanisms may operate. Part of the difficulty in assessing the role of any single iron-uptake mechanism in natural infection may also be related to the type of animal model used. Our understanding of the way the host normally restricts the availability of essential ferric ions and of the way invading pathogens respond to iron restriction, has increased enormously in recent years and there is now a considerable literature on the relationships between iron and pathogenicity (Bullen and Griffiths, 1999).

Table 3 Microbial strategies for acquiring iron bound by transferrin, lactoferrin and ovotransferrin

- (a) Proteolytic cleavage of the iron-binding glycoprotein, disruption of the iron-binding site and release of iron
- (b) Reduction of ferric to ferrous iron with consequent release of the ferrous iron from the iron-binding protein
- (c) Production of iron-chelating compounds (siderophores) which remove iron from the ferric-glycoprotein complex and transport it back to the bacterial cell via outer membrane siderophore receptors
- (d) Direct interaction between bacterial cell surface transferrin or lactoferrin receptors and the appropriate glycoprotein in a manner analogous to the interaction between transferrin and reticulocytes

Strategies for Acquiring Iron *in vivo*

In spite of the virtual absence of freely available iron in normal body fluids, pathogenic bacteria can clearly multiply successfully *in vivo* to establish infection. They must, therefore, be able to adapt to the severely iron-restricted environment usually found *in vivo* and produce mechanisms to assimilate protein-bound iron, or to acquire it from haem. It is not known whether ferritin can act as an iron source. Invading pathogens could also move rapidly into local environments where iron may be more readily available, such as inside cells. Little is known about the availability of intracellular iron, but it seems that inside at least some cells, iron is more readily available (Lawlor *et al.*, 1987; Byrd and Horwitz, 1989). The nature of this iron is unknown but in view of the toxicity of the loosely bound metal, it is likely to represent only a very small portion of cellular iron.

Four possible mechanisms can be used by bacteria to assimilate iron bound to the high-affinity iron-binding glycoproteins (**Table 3**). There is little evidence that pathogens in general degrade transferrin or lactoferrin to obtain sufficient iron to multiply (mechanism a, **Table 3**), nor that mechanisms to reduce Fe^{3+} to Fe^{2+} play much of a role *in vivo*. Most bacterial pathogens use siderophore-mediated iron-uptake systems or can acquire iron directly from host transferrin or lactoferrin by a receptor-mediated process. Some pathogens possess both mechanisms.

The co-ordinated response of many organisms to low iron levels involves the synthesis of iron chelators (siderophores), which take iron from the iron-binding proteins. Membrane receptors and enzymes are then involved in the uptake and release of iron from the ferric-siderophores. The strict requirement for a ferric-siderophore receptor for siderophore-mediated iron uptake is shown by the absence of transport activity in strains that lack such proteins (Grewel *et al.*, 1982; Carmel *et al.*, 1990). Examples of pathogenic bacteria that have been reported to produce siderophores are listed in **Table 4**. Siderophores and siderophore-receptors are expressed in host tissues during infection (Griffiths and Williams, 1999). Some pathogens are able to produce two or more siderophores or to obtain iron by means of a variety of exogenous siderophores produced by other micro-organisms. It is not, however, clear whether this characteristic plays a role in natural infections, or whether it may be a tactic to facilitate iron scavenging under a wide range of environmental conditions outside the host.

Table 4 Some bacterial pathogens that produce siderophore-mediated iron-uptake systems

Aeromonas spp.
Bordetella pertussis
Corynebacterium diphtheriae
Escherichia coli
Klebsiella pneumoniae
Mycobacteria
Pasteurella multocida
Pseudomonas spp.
Salmonella spp.
Shigella spp.
Staphylococcus aureus
Vibrio spp.

Table 5 Some bacterial pathogens that express transferrin/lactoferrin receptors during iron-restricted growth

Transferrin	Lactoferrin
Bordetella pertussis	*Bordetella pertussis*
Haemophilus influenzae	*Staphylococcus aureus*
Neisseria meningitidis	*Neisseria meningitidis*
Pasteurella haemolytica	*Neisseria gonorrhoeae*
Staphylococcus spp.	*Treponema pallidum*

Table 6 Some pathogenic organisms able to utilise haem as a source of iron

Candida albicans
Escherichia coli
Haemophilus spp.
Legionella pneumophila
Neisseria spp.
Yersinia spp.
Pasteurella spp.
Porphyromonas gingivalis
Staphylococcus aureus
Shigella flexneri
Vibrio spp.
Streptococcus pneumoniae

Whilst siderophore-based iron uptake mechanisms may be the most intensively investigated systems by which bacteria sequester iron from the host, they may not be the most effective means for acquiring iron *in vivo*. An increasing number of human and animal pathogens have been identified as exploiting siderophore-independent receptor-mediated mechanisms. These depend on direct contact between the bacterial cell surface and the iron-binding glycoprotein, and involve bacterial cell surface receptors that are highly specific for the transferrin or lactoferrin of the pathogen's natural host. Such mechanisms are used effectively by *Neisseria* spp., *Haemophilus*, *Actinobacter*, *Pasteurella* and *Staphylococcus*. In these organisms, transferrin- and lactoferrin-binding proteins have been identified and characterised and there is evidence for their expression *in vivo* during infection (Griffiths and Williams, 1999). Much progress has also been made in understanding the molecular basis of receptor recognition of transferrin/lactoferrin and in the identification of regions on the iron-binding proteins recognised by the bacterial receptors. Some features of these systems resemble those for iron assimilation by eukaryotic cells (Evans *et al.*, 1999). Unlike the eukaryotic system, however, where the transferrin molecule is internalised by the cell, it is not internalised in the bacterial system. Uptake of iron from ferrated transferrin bound to the bacterial receptor requires the protein-bound iron first to be released and then internalised. Examples of bacterial pathogens that express transferrin and/or lactoferrin receptors are given in Table 5.

In addition to obtaining iron from the host iron-binding proteins, many bacteria can also obtain iron from cell-free haemoglobin or haem (Lee, 1995) (Table 6). Normally, however, only a trace of haem is free in serum, and this is bound to haemopexin or serum albumin. Likewise, cell-free haemoglobin is rapidly bound by haptoglobin. Bacterial receptors have been identified for haemoglobin, haemoglobin–haptoglobin, haem and haem–haemopexin complexes (Lee, 1992; Frangipane *et al.*, 1994; Jih *et al.*, 1996; Ochsner *et al.*, 2000). These several iron-acquisition systems may contribute to virulence, since the ability of pathogenic micro-organisms to scavenge for iron and/or haem from the host is fundamental for survival and for production of disease. In some instances, such as that of *Helicobacter pylori*, the organism possesses multiple haem-acquisition systems (Worst *et al.*, 1999), which again makes sense because *in vivo* the pathogen encounters different environments that may vary not only in the availability of iron and haem but also in iron and haem composition.

Regulation by Iron

The restricted availability of iron *in vivo* not only presents microbial pathogens with the problem of acquiring sufficient for multiplication, but it also constitutes a major environmental signal which regulates the expression of virulence genes unrelated to iron metabolism, such as those encoding toxins (Griffiths and Chart, 1999; Crosa, 1999). Table 2 lists a number of toxins and other virulence determinants whose expression is regulated by the availability of iron. Some of the genes for high-affinity iron-uptake systems and other virulence determinants are chromosomal, whereas others are carried by plasmids or phages.

Much progress has been made in understanding the molecular mechanism by which iron co-ordinately regulates the expression of a large number of unlinked genes. Some are regulated by a direct and fairly simple on–off negative regulatory system controlled by iron. In other cases regulation is more complex and involves additional positive regulatory elements responsive to other signals. Genes controlled by negative regulation are automatically transcribed unless they are switched off by the regulator protein (repressor). In many organisms, expression of iron-controlled genes is negatively regulated at the transcriptional level by the global repressor protein Fur, which uses ferrous (Fe^{2+}) iron as the co-repressor (Bagg and Neilands, 1987a, 1987b; O'Halloran, 1993). Binding sites for the Fe^{2+}-Fur-repressor have been identified in the promoters of a large number of iron-controlled negatively regulated genes, including the *fur* gene itself (Bagg and Neilands, 1987a, 1987b). These are the 'iron-boxes', the DNA of which consists of a highly AT-rich palindrome. During iron restriction, the intracellular iron concentration falls and less Fe^{2+} is available for binding to Fur. Fur in its metal-free form binds poorly, or not at all to the iron boxes and this leads to de-repression of an iron-regulated gene, or of a number of genes in an iron-regulated regulon, and subsequently to the synthesis of high-affinity iron uptake systems as well as the production of toxins.

For genes under positive control by iron, expression occurs when an active regulator protein, the activator, is present, and is involved with the initiation of transcription of mRNA from DNA. It is the expression of the activator that is negatively iron-controlled by Fur. Once switched on, the positive regulators are sensitive to additional regulatory signals and initiate transcription only when other conditions, in addition to iron-restriction are appropriate.

A regulatory system analogous to that dependent on Fur operates in a number of Gram-positive pathogens like *Corynebacterium diphtheriae*. The phage-encoded diphtheria toxin and the siderophore-mediated iron uptake system are both co-ordinately regulated by the chromosomal iron repressor DtxR, which is analogous to the Fur protein (Tao and Murphy, 1992; Tao *et al.*, 1992). DtxR-binding sites have been found in iron-regulated genes of a number of Gram-positive bacteria and a DtxR homologue (IdeR) has been identified in *Mycobacterium tuberculosis* and several other mycobacteria (Schmitt *et al.*, 1995; De Voss *et al.*, 1999).

In addition to Fur-controlled regulation, iron-controlled global regulatory systems based on the under-modification of several transfer RNAs (tRNAs) also appear to operate in *E. coli* and *S. typhimurium*. Iron-dependent modification of adenosine 37, located adjacent to the anti-codon of certain tRNAs, induces strong pleiotropic effects on cell physiology, in some instances by effects on translational efficiency of the tRNA, and in others by altering codon context and in yet others by modulating spontaneous mutation frequencies, an event which may have a role in adapting the bacterial cell to environmental stress (Buck and Griffiths, 1982; Bouadloun *et al.*, 1986; Connolly and Winkler, 1989, 1991; Persson, 1993).

Co-ordinated Control of Bacterial Virulence by the Availability of Metal Ions and Other Environmental Signals

The ability of invading pathogens to multiply successfully under conditions found in host tissues is an essential factor in all infections. Bacterial pathogens *in vivo* must produce the full complement of virulence determinants needed for pathogenicity. In spite of the fact that much was known about the ability of bacteria to alter their metabolism in response to environmental changes in the laboratory, in the past most investigations into bacterial virulence were carried out with organisms grown *in vitro* under conditions that did not necessarily reflect the microbial environment in host tissues.

Smith (1980) and Griffiths (1983) argued that the lack of knowledge about the changes that may occur in pathogens as they adapt to and multiply in the environment of the host, and about the effects of these changes on the ability of the organisms to proliferate *in vivo* was a major gap in the understanding of the mechanisms of microbial pathogenicity. Over the past 10–15 years overwhelming evidence has accumulated to show that many of the crucial virulence determinants of pathogenic bacteria are indeed environmentally controlled. The recognition that the restricted availability of iron in mammalian tissues constitutes not only a problem for pathogens in acquiring sufficient for multiplication but also acts as an environmental signal for the bacterium, by regulating the expression of other virulence genes unrelated to iron metabolism, has been a major step forward in our understanding of microbial pathogenicity. A similar situation exists for Mg^{2+} and Ca^{2+}, where again a low level of an essential ion is used as an environmental signal to regulate not only the expression of high-affinity ion-uptake mechanisms, but also unrelated virulence determinants, such as PhoP/Q. The levels of essential ions in different body niches are, therefore, utilised by invading pathogens as a means of monitoring their location within the host, or in other environments, and to ensure that virulence determinants are switched on only when necessary.

There are, of course, a number of other environmental signals, in addition to those based on the levels of ions, which regulate bacterial characteristics important for pathogenicity, such as temperature, pH, levels of O_2, osmolarity and the availability of other nutrients (Table 7). Indeed, special nutritional conditions that occur only in certain tissues may be expected to lead to the localisation of bacteria, or at least to their rapid multiplication, at specific body sites. This has been shown for certain strains of the genus *Brucella* for which the carbohydrate erythritol is a preferred metabolite and growth stimulant. Erythritol is present in significant amounts in ungulate placental tissue and promotes an intense placentitis in *Brucella abortus* infections that may lead to abortion (Smith *et al.*, 1962; Keppie *et al.*, 1965). In many other animals brucellosis is a relatively mild and chronic infection and the bacteria do not have a marked affinity for any particular tissue. *Brucella abortus* strain B19 is a

Table 7 Environmental signals that regulate bacterial virulence determinants

Signal	Reference
Temperature	Maurelli and Sansonetti, 1988; Mekalanos, 1992; Konkle and Tilly, 2000
Osmolarity	Herrington *et al.*, 1988; Bernadini *et al.*, 1990; Mekalanos, 1992
pH	J.F. Miller *et al.*, 1989
O_2	Buck and Ames, 1984; Tardat and Touti, 1993
Calcium	This chapter
Magnesium	This chapter
Sodium	This chapter
Molybdenum	This chapter
Zinc	This chapter
Iron	This chapter
Specific nutrients, e.g. erythritol	Smith *et al.*, 1962; Keppie *et al.*, 1965; Sangari *et al.*, 2000

spontaneously attenuated mutant strain which lacks D-erythrulose-1-phosphate dehydrogenase, a critical enzyme in the erythritol catabolic pathway, and this strain is widely used to vaccinate cattle. Erythritol actually inhibits, rather than promotes the growth of the B19 strain (Sperry and Robertson, 1975). This defect causes both the accumulation of the toxic intermediate D-erythrulose-1-phosphate and ATP depletion, leading to the inhibition of bacterial growth.

Although this has been known for many years, the exact role of erythritol metabolism as a virulence factor in *Brucella* spp. has remained obscure and only recently has work begun to clarify the situation. To date, results have shown that the genes for erythritol catabolism are organised as an inducible operon in *B. abortus* (Sangari *et al.*, 2000). These workers did not address the growth-promoting effect of erythritol, but they suggested that a link may exist between erythritol and the aromatic biosynthetic pathway. Erythrose-4-phosphate, the precursor for the biosynthesis of aromatic compounds via the aromatic biosynthetic pathway, which includes phenylalanine, tryptophan and tyrosine as well as catechols, may easily be obtained from the intermediates of erythritol catabolism. They suggested that erythritol may be crucial to obtaining these physiologically important aromatic molecules in strains of *Brucella* spp.

Pathogenicity depends generally on a variety of bacterial properties rather than on any one single trait, and responses to a number of environmental signals are often important. Thus, the expression of some virulence factors is controlled not by one signal alone, but by a combination of two or more environmental signals acting in concert. This concept of control mediated by a combination of two or more signals appears in several environmentally regulated systems. Sometimes the resulting changes are dramatic and leave no doubt that the cell phenotype has been drastically altered. For example, in *Vibrio parahaemolyticus* iron limitation combined with an interference with polar flagellar rotation, induces swarmer cell differentiation, by which the organism becomes covered with numerous lateral flagella that allow it to spread and colonise surfaces (McCarter and Silverman, 1989). Growth of the same organism in liquid media results in a swimmer cell phenotype.

In most instances, however, environmentally controlled changes are subtle, but nonetheless have far-reaching consequences for the capacity of the organism to cause disease. Thus, Shiga-like toxin production in *E. coli* is regulated by a combination of temperature and iron-restriction (Chart *et al.*, 1987; Weinstein *et al.*, 1988), as is the mini-pilin of *Aeromonas hydrophila* (Ho *et al.*, 1990). The 'mini-pilin' of *A. hydrophila* is produced during growth under iron-restriction at 22°C but not at 37°C. It is suggested that the pathogenic and ecological roles of this pilus are related to the existence of the bacterium as a free-living organism in aquatic environments and its ability to cause infection in both cold-blooded vertebrates and in the human intestine. Similarly, as already mentioned, several virulence factors in *Yersinia* spp. are co-ordinately controlled by temperature and the environmental concentration of Ca^{2+} (J.F. Miller *et al.*, 1989; Maurelli, 1989).

Conclusion

Studies over the past 20 years or so have begun to unravel the complex interactions of pathogens with their hosts, especially with the ionic environment in the host. In certain studies, the role of essential ions has been examined by adding reagents known to bind the ion of interest. Certain chelating agents may, however, bind more than one ion and caution should be exercised when assessing the results of chelator-based experiments. For example, a study designed to examine the role of divalent cations on the expression of membrane-associated proteins by certain strains of EAggEC (Chart *et al.*, 1997), used EDTA to bind calcium and magnesium ions. It was noted that growth of these particular strains of EAggEC in media containing EDTA resulted in the expression of high-affinity iron-uptake proteins. This showed that EDTA can also bind ferric ions and that certain chelators may bind essential ions in addition to those of interest. Norris *et al.* (1996) described how the traditionally accepted calcium ion chelator ethylene-bis(oxy-ethylenenitrilo)-tetraacetic acid (EGTA) binds manganese, zinc and iron in preference to calcium ions. Similarly, early studies of the role of iron in the expression of bacterial toxins used chelators such as Chelex-100. Ion binding by Chelex-100 is, however, hierarchical and ions other than iron are bound in preference. Furthermore, in the case of iron, the use of iron-depleted media produced by chelation leads to a growth environment quite different from the natural situation *in vivo*, in which sufficient iron is usually present, but not readily available to the invading pathogen (Stevenson and Griffiths, 1994). In iron-depleted media, organisms may become iron starved, a condition quite different from that encountered *in vivo*.

In other cases, the use of low-molecular-weight iron chelators may be problematic if the chelator enters the bacterial cell. For example, α,α'-dipyridyl can enter bacterial cells and lead to a loss of viability (Chart *et al.*, 1986). Caution is therefore necessary when the results of *in vitro* experiments are interpreted. In the case of iron, qualitative and quantitative differences have been observed in the iron-regulated proteins expressed by some bacteria, depending on the chelator used (West and Sparling, 1985; Chart *et al.*, 1986). Media that contain transferrin or ovotransferrin seem to be the best choice to mimic conditions *in vivo*, but other systems might be considered for routine production of particular iron-regulated proteins or siderophores (Stevenson and Griffiths, 1994).

The limitations of *in vitro* studies are well illustrated by the expression of *Vibrio cholerae* virulence genes during infection. Because of methodological limitations, studies of the regulation of *V. cholerae* virulence gene expression have mostly been carried out *in vitro*. Environmental stimuli may include temperature, pH and certain amino acids (Miller and Mekalanos, 1984; Schuhmaker and Klose, 1999). However, the laboratory conditions required for maximum expression of cholera toxin (CT) and toxin-co-regulated pilus (TCP) in the classical biotype (30°C, pH 6.5 and low osmolarity), seem dissimilar to those expected in the small intestine, and repressing conditions *in vitro* (37°C, pH 8.5 and high osmolarity) are similar to conditions *in vivo* (Chakrabarti *et al.*, 1999). Further studies, with *in vivo* expression technology, have shown that CT and TCP are expressed sequentially during infection, and that full CT expression occurs only after, and depends on, colonisation (Lee *et al.*, 1999). It is also possible that a quorum-dependent signal is involved in the process. Quorum sensing is a process by which cell–cell communications are mediated by the synthesis, secretion and detection of small extracellular signal molecules (Surette and Bassler, 1998).

With the development of sophisticated genome-based technologies, including *in vivo* expression systems to probe host environments, significant new insights into the complexities of host–pathogen interactions are now being gained (Strauss and Falkow, 1997; Harb and Kwaik, 1999; Manabe *et al.*, 1999; Salama and Falkow, 1999; Marcus *et al.*, 2000). Many of the subtle host micro-environments that affect the expression of bacterial virulence determinants are those controlled by essential ionic nutrients. By responding to such environmental signals, pathogenic micro-organisms can be primed ready for adherence, invasion or multiplication, as the case may be, in the appropriate host environment. This involves not only the switching on of certain genes, but also, in some instances, the down-regulation of others. The use of regulatory systems sensitive to two or more environmental signals allows sensitive environmental monitoring.

References

Altura BM (1992) Basic biochemistry and physiology of magnesium: a brief review. *Magnesium Trace Elements* 10: 167–171.

Apulche-Aranda CM, Swanson JA, Loomis WP, Miller SI (1992) *Salmonella typhimurium* activates virulence gene transcription with acidified macrophage phagosomes. *Proc. Natl Acad. Sci. USA* 89: 10079–10083.

Ayres HM, Furr JR, Russell AD (1998) Effect of divalent cations on permeabilizer-induced lysozyme lysis of *Pseudomonas aeruginosa. Lett. Appl. Microbiol.* 27: 372–374.

Bagg A, Neilands JB (1987a) Molecular mechanism of regulation of siderophore mediated iron assimilation. *Microbiol. Rev.* 51: 509–518.

Bagg A, Neilands JB (1987b) Ferric uptake regulation protein acts as a repressor employing iron II as a co-factor to bind to the operator of an iron transport operon in *Escherichia coli. Biochemistry* 26: 5471–5477.

Baker PJ, Wilson JB (1965) Hypoferremia in mice and its application to the assay of endotoxin. *J. Bacteriol.* 90: 903–910.

Bauerfeind P, Garner RM, Mobley LT (1996) Allelic exchange mutagenesis of *nixA* in *Helicobacter pylori* results in reduced nickel transport and urease activity. *Infect. Immun.* 64: 2877–2880.

Bearson B, Wilson L, Foster JW (1998) A low pH-inducible PhoPQ-dependent acid tolerance response protects *Salmonella typhimurium* against inorganic acid stress. *J. Bacteriol.* 180: 2409–2417.

Bernadini ML, Fontaine A, Sansonetti PJ (1990) The two-component regulatory system OmpR-EnvZ controls the virulence of *Shigella flexneri. J. Bacteriol.* 172: 6274–6281.

Blanc-Potard AB, Groisman EA (1997) The *Salmonella typhimurium selC* locus contains a pathogenicity island mediating intramacrophage survival. *EMBO J.* 16: 5376–5385.

Böck A, Sawers G (1996) Fermentation. In: Neidhardt FC *et al.* (eds) *Escherichia coli and Salmonella: Cellular and Molecular Biology*, 2nd edn. Washington, DC: American Society for Microbiology, pp. 262–282.

Boland A, Havaux S, Cornelis GR (1998) Heterogeneity of the yersinia YopM protein. *Microb. Pathogen.* 25: 343–348.

Bouadloun F, Srichaiyo T, Isaksson LA, Bjork GR (1986) Influence of modification next to the anticodon in tRNA on codon context, sensitivity of translational suppression and accuracy. *J. Bacteriol.* 166: 1022–1027.

Boyington JC, Gladyshev VN, Kangulov SV, Stadtman TC, Sun PD (1997) Crystal structure of formate dehydrogenase H: catalysis involving Mo, molybdopterin, selenocysteine and an Fe_4S_4 cluster. *Science* 275: 1305–1308.

Brumlik MJ, Storey DG (1992) Zinc and iron regulate translation of the gene encoding *Pseudomonas aeruginosa* elastase. *Mol. Microbiol.* 6: 337–344.

Buck M, Ames BN (1984) A modified nucleotide in tRNA as a possible regulator of anaerobiosis: synthesis of cis-2- methylthio-ribosylzeatin in the tRNA of Salmonella. *Cell* 36: 523–531.

Buck M, Griffiths E (1982) Iron-mediated methylthiolation of tRNA as a regulator of operon expression in *Escherichia coli. Nucleic Acids Res.* 10: 2609–2624.

Bullen JJ, Griffiths E (1999) *Iron and Infection*, 2nd edn. Chichester: John Wiley & Sons.

Bullen JJ Griffiths E, Rogers H, Ward G (2000) Sepsis, the critical role of iron. *Microbes Infect.* 2: 409–415.

Byrd TF, Horwitz MA (1989) Interferon gamma-activated human monocytes down regulate transferrin receptors and inhibit the intracellular multiplication of *Legionella pneumophila* by limiting the availability of iron. *J. Clin. Invest.* 83: 11457–11465.

Carmel G, Hellstress D, Hennrig D, Coulton JW (1990) Insertion mutagenesis of the gene encoding the ferrichrome-iron receptor of *Escherichia coli* K12. *J. Bacteriol.* 172: 1861–1869.

Chakrabarti S, Sengupta N, Chowdhury N (1999) Role of DnaK in *in vitro* and *in vivo* expression of virulence factors of *Vibrio cholerae. Infect. Immun.* 67: 1025–1033.

Chan PF, Foster SJ (1998) The role of environmental factors in the regulation of virulence-determinant expression in *Staphylococcus aureus 8325–8334. Microbiology* 144: 2469–2479.

Chart H, Buck M, Stevenson P, Griffiths E (1986) Iron regulated outer membrane proteins of *Escherichia coli*: variation in the expression due to the chelator used to restrict the availability of iron. *J. Gen. Microbiol.* 132: 1373–1378.

Chart H, Scotland SM, Rowe B (1987) Production of Vero cytotoxin by strains of *Escherichia coli* as related to the availability of iron. *FEMS Microbiol. Lett.* 48: 385–390.

Chart H, Smith HR, Rowe B (1994) Antigenic cross-reactions between fimbriae expressed by *Salmonella enteritidis S. dublin* and an 18 kDa outer membrane associated protein expressed by *Escherichia coli* O126:H27. *FEMS Microbiol. Lett.* 121: 19–24.

Chart H, Smith HR, Rowe B (1995) Enteroaggregative strains of *Escherichia coli* belonging to serotypes O126:H27 and O44:H18 express antigenically similar 18 kDa outer membrane-associated proteins. *FEMS Microbiol. Lett.* 132: 17–22.

Chart H, Spencer J, Smith HR, Rowe B (1997) Magnesium ions are required for HEp-2 cell adhesion by enteroaggregative strains of *Escherichia coli* O126:H27 and O44:H18. *FEMS Microbiol. Lett.* 148: 49–52.

Collins CM, D'Orazio SE (1993) Bacterial ureases: structure, regulation of expression and role in pathogenesis. *Mol. Microbiol.* 9: 907–913.

Connolly DM, Winkler ME (1989) Genetic and physiological relationships among the *mia* A gene, 2-methylthio-N6- (2 isopentenyl-adenosine) tRNA modification and spontaneous mutagenesis in *Escherichia coli* K12. *J. Bacteriol.* 171: 3233–3246.

Connolly DM, Winkler ME (1991) Structure of *Escherichia coli* K-12 *miaA* and characterization of the mutator phenotype caused by *miaA* insertion mutations. *J. Bacteriol.* 173: 1711–1721.

Cornelis GR (1998) The *Yersinia* deadly kiss. *J. Bacteriol.* 180: 5495–5504.

Crichton RR, Ward RJ (1992) Structure and molecular biology of iron-binding proteins and the regulation of 'free' iron pools. In: Lauffer RB (ed.) *Iron and Human Disease.* Boca Raton, Florida: CRC Press, pp. 23–75.

Crosa JH (1984) The relationship of plasmid mediated iron transport and bacterial virulence. *Annu. Rev. Microbiol.* 38: 69–89.

Crosa JH (1999) Molecular genetics of iron transport as a component of bacterial virulence. In: Bullen JJ, Griffiths E (eds) *Iron and Infection*, 2nd edn. Chichester: John Wiley & Sons, pp. 255–288.

De Voss JJ, Rutter K, Schroeder BG, Barry III CE (1999) Iron acquisition and metabolism by *Mycobacteria. J. Bacteriol.* 181: 4443–4451.

Dobson A, Bullen JJ (1999) Measurement and control of Eh. In: Bullen JJ, Griffiths E (eds) *Iron and Infection*, 2nd edn. Chichester: John Wiley & Sons, pp. 505–508.

Ernst RK, Guina T, Miller SI (1999) How intracellular bacteria survive: surface modifications that promote resistance to host innate immune responses. *J. Infect. Dis.* 179: S326–330.

Evans RW, Crawley JB, Joannou CL, Sharma ND (1999) Iron proteins. In: Bullen JJ, Griffiths E (eds) *Iron and Infection*, 2nd edn. Chichester: John Wiley & Sons, pp. 27–86.

Finlay BB, Cossart P (1997) Exploitation of mammalian host cell functions by bacterial pathogens. *Science* 276: 718–725.

Frangipane ME, Morton DW, Wooten JA, Pozsgay JM, Stull TL (1994) Binding of human hemoglobin by *Haemophilus influenze. FEMS Microbiol. Lett.* 188: 243–248.

Garcia-del Portillo F, Foster JW, Maguire ME, Finlay BB (1992) Characterization of the micro-environment of *Salmonella typhimurium*-containing vacuoles within MDCK epithelial cells. *Mol. Microbiol.* 6: 3289–3297.

Garcia-Vescovi E, Soncini FC, Groisman EA (1996) Mg^{2+} as an extracellular signal: environmental regulation of *Salmonella* virulence. *Cell* 84: 165–174.

Gennis RB, Stewart V (1996) Respiration. In: Neidhardt FC *et al.* (eds) *Escherichia coli and Salmonella: Cellular and Molecular Biology*, 2nd edn. Washington, DC: American Society for Microbiology, pp. 217–261.

Gordenk VR, Printhviraj P, Doliner T, Brittenham GM (1988) Interleukin-1 administration in mice produces hypoferremia despite neutropenia. *J. Clin. Invest.* 82: 1934–1938.

Gow NAR (1997) Germ tube growth of *Candida albicans. Curr. Topics Trop. Med. Mycol.* 8: 43–55.

Grewel KK, Warner PJ, Williams PH (1982) An inducible outer membrane protein involved in aerobactin mediated iron transport by ColV strains of *Escherichia coli. FEMS Microbiol. Lett.* 140: 27–30.

Griffiths E (1983) Adaptation and multiplication of bacteria in host tissues. *Phil. Trans. R. Soc. Lond. B* 303: 85–96.

Griffiths E (1999) Iron in biological systems. In: Bullen JJ, Griffiths E (eds) *Iron and Infection*, 2nd edn. Chichester: John Wiley & Sons, pp. 1–26.

Griffiths E, Bullen JJ (1999) Iron and infection: future prospects. In: Bullen JJ, Griffiths E (eds) *Iron and Infection*, 2nd edn. Chichester: John Wiley & Sons, pp. 451–481.

Griffiths E, Chart H (1999) Iron as a regulatory signal. In: Bullen JJ, Griffiths E (eds) *Iron and Infection*, 2nd edn. Chichester: John Wiley & Sons, pp. 213–253.

Griffiths E, Williams P (1999) The iron-uptake systems of pathogenic bacteria, fungi and protozoa. In: Bullen JJ, Griffiths E (eds) *Iron and Infection*, 2nd edn. Chichester: John Wiley & Sons, pp. 87–212.

Groisman EA, Chiao E, Lipps CJ, Heffron F (1989) *Salmonella typhimurium pho*P virulence gene is a transcriptional regulator. *Proc. Natl Acad. Sci. USA* 86: 7077–7081.

Gross R, Rappuoli R (1989) Pertussis toxin promoter sequences involved in modulation. *J. Bacteriol.* 171: 4026–4030.

Grunden AM, Shanmugam KT (1997) Molybdate transport and regulation in bacteria. *Arch. Microbiol.* 168: 345–354.

Gunn JS, Miller SI (1996) PhoP-PhoQ activates transcription of pmrAB, encoding a two component regulatory system involved in *Salmonella typhimurium* antimicrobial peptide resistance. *J. Bacteriol.* 178: 6857–6864.

Gunsalus RP (1992) Control of electron flow in *Escherichia coli*: co-ordinate transcription of respiratory pathway genes. *J. Bacteriol.* 174: 7069–7074.

Guo L, Lim KB, Gunn JS, Bainbridge B, Darveau RP, Hackett M, Miller SI (1997) Regulation of lipid A modifications by *Salmonella typhimurium* virulence genes *pho* P-*pho* Q. *Science* 276: 250–253.

Hantke K (1997) Ferrous iron uptake by a magnesium transport system is toxic for *Escherichia coli* and *Salmonella typhimurium. J. Bacteriol.* 179: 6201–6204.

Haque H, Russell AD (1976) Cell envelopes of gram-negative bacteria: composition, response to chelating agents and susceptibility of whole cells to antibacterial agents. *J. Appl. Bacteriol.* 40: 89–99.

Harb OS, Kwaik YA (1999) Probing the microenvironment of intracellular bacterial pathogens. *Microbes Immun.* 1: 445–453.

Hase CC, Finkelstein RA (1993) Bacterial extracellular zinc-containing metalloproteases. *Microbiol. Rev.* 57: 823–837.

Herrington DA, Hall RH, Losonsky G, Mekalanos JJ, Taylor RK, Levine MM (1988) Toxin, toxin-coregulated pili and the toxR regulon are essential for *Vibrio cholerae* pathogenesis in humans. *J. Exp. Med.* 168, 1487–1492.

Ho ASY, Mietzner TA, Smith AJ, Schoolnik GK (1990) The pili of *Aeromonas hydrophila*: identification of an environmentally-regulated mini pilin. *J. Exp. Med.* 172: 759–806.

Jih H, Ren Z, Pozsgay JM, Elkins C, Whitby PW, Morton DJ, Stull TL (1996) Cloning of a DNA fragment encoding a heme repressible hemoglobin-binding outer membrane protein from *Haemophilus influenzae*. *Infect. Immun.* 64: 3134–3141.

Kehres DG, Lawyer CH, Maguire ME (1998) The CorA transporter gene family. *Microb. Comp. Genomics* 3: 151–169.

Keppie J, Williams AE, Witt K, Smith H (1965) The role of erythritol in the tissue localisation of the brucellae. *Br. J. Exp. Pathol.* 46: 104–108.

Kharazmi A (1989) Interactions of *Pseudomonas aeruginosa* proteases with cells of the immune system. *Antibiot. Chemother.* 42: 42–49.

Konkel ME, Tilly K (2000) Temperature-regulated expression of bacterial virulence genes. *Microbes Infection* 2: 157–166.

Kreiswirth BN (1989) Genetics and expression of toxic shock syndrome toxin-1: an overview. *Rev. Infect. Dis.* 11: S97–S100.

Lawlor KM, Daskaleras PA, Robinson RE, Payne SM (1987) Virulence of iron-transport mutants of *Shigella flexneri* and utilization of host iron compounds. *Infect. Immun.* 55: 594–595.

Lee BC (1992) Isolation of an outer membrane haemin-binding protein from *Haemophilus influenzae* type b. *Infect. Immun.* 60: 810–816.

Lee BC (1995) Quelling the red menace: haem capture by bacteria. *Mol. Microbiol.* 18: 383–390.

Lee KL, Buckley HR, Campbell CC (1975) An amino acid liquid synthetic medium for the development of mycelial and yeast forms of *Candida albicans*. *Sabouraudia* 13: 148–153.

Lee SL, Hava DL, Waldor MK, Camilli A (1999) Regulation and temporal expression patterns of *Vibrio cholerae* virulence genes during infection. *Cell* 99: 625–634.

Leive L (1965) Release of lipopolysaccharide by EDTA treatment of *E. coli*. *Biochem. Biophys. Res. Commun.* 21: 290–296.

Levin RJ, Wagner G (1977) Human vaginal fluid–ionic composition and modification by sexual arousal. *J. Physiol. Lond.* 266: 62–63.

Lu D, Boyd B, Lingwood CA (1997) Identification of the key protein for zinc uptake in *Hemophilus influenzae*. *J. Biol. Chem.* 272: 29033–29038.

McCarter L, Silverman M (1989) Iron regulation of swarmer cell differentiation of *Vibrio parahaemolyticus*. *J. Bacteriol.* 171: 731–736.

McGee DJ, May CA, Garner RM, Himpsl JM, Mobley HL (1999) Isolation of *Helicobacter pylori* genes that modulate urease activity. *J. Bacteriol.* 181: 2477–2484.

Manabe YC, Savioloa BJ, Sun L, Murphy JR, Bishai WR (1999) Attenuation of virulence in *Mycobacterium tuberculosis*. Expressing a constitutively active iron repressor. *Proc. Natl Acad. Sci. USA* 94: 12844–12848.

Marcus SL, Brumell JH, Pfeifer CG, Finlay BB (2000) Salmonella pathogenicity islands: big virulence in small packages. *Microbes Infection* 2: 615–619.

Maurelli AT (1989) Temperature regulation of virulence genes in pathogenic bacteria: a general strategy for human pathogens? *Microb. Pathogenicity* 7: 1–10.

Maurelli AT, Sansonetti PJ (1988) Identification of a chromosomal gene controlling temperature-regulated expression of *Shigella* virulence. *Proc. Natl Acad. Sci. USA* 85: 2820–2824.

Mekalanos JJ (1992) Environmental signals controlling the expression of virulence determinants in bacteria. *J. Bacteriol.* 174: 1–7.

Miller JF, Mekalanos JJ, Falkow S (1989) Co-ordinate regulation and sensory transduction in the control of bacterial virulence. *Science* 243: 916–922.

Miller SL, Kukral AM, Mekalanos JJ (1989) A two component regulatory system (phoP/phoQ) controls *Salmonella typhimurium* virulence. *Proc. Natl Acad. Sci. USA* 86: 5054–5058.

Miller VL, Mekalanos JJ (1984) Synthesis of cholera toxin is positively regulated at the transcriptional level by *tox* R. *Proc. Natl Acad. Sci. USA* 81: 3471–3475.

Monack DM, Mecsas J, Bouley D, Falkow S (1998) Yersinia-induced apoptosis in vivo aids in the establishment of a systemic infection of mice. *J. Exp. Med.* 188: 2127–2137.

Moncrief MB, Maguire ME (1998) Magnesium and the role of MgtC in growth of *Salmonella typhimurium*. *Infect. Immun.* 66: 3802–3809.

Morgan EH (1981) Transferrin, biochemistry, physiology and clinical significance. *Mol. Aspects Med.* 4: 1–123.

Morihara K, Tsuzuki H (1975) *Pseudomonas aeruginosa* elastase: affinity chromatography and some properties as a metallo-neutral proteinase. *Agric. Biol. Chem.* 39: 1123–1128.

Niesel DW, Peterson JW (1987) Calcium enhances *Salmonella typhimurium* invasion of HeLa cells. *FEMS Microbiol. Lett.* 41: 299–304.

Nikaido H, Vaara M (1985) Molecular basis for bacteria outer membrane permeability. *Microbiol. Rev.* 49: 1–32.

Nilles ML, Williams AW, Skrzypek E, Straley SC (1997) *Yersinia pestis* LcrV forms a stable complex with LcrG and may have a secretion-related regulatory role in the low-Ca^{2+} response. *J. Bacteriol.* 179: 1307–1316.

Norris V, Grant S, Freestone P, Canvin J, Sheikh FN, Toth I, Trinei M, Modha K, Norman RI (1996) Calcium signalling in bacteria. *J. Bacteriol.* 178: 3677–3682.

Northrop FD, Ljubojevic S, Davies JM (1997) Influence of Na^+ and anions on the dimorphic transition of *Candida albicans*. *Microbiology* 143: 3757–3765.

O'Halloran TV (1993) Transition metals in control of gene expression. *Science* 261: 715–725.

Ochsner UA, Johnson Z, Vasil ML (2000) Genetics and regulation of two distinct haem uptake systems, phu and has, in *Pseudomonas aeruginosa*. *Microbiology* 146: 185–198.

Payne PL, Straley SC (1998) YscO of *Yersinia pestis* is a mobile core component of the Yop secretion system. *J. Bacteriol.* 180: 3882–3890.

Payne PL, Straley SC (1999) YscP of *Yersinia pestis* is a secreted component of the Yop secretion system. *J. Bacteriol.* 181: 2852–2862.

Pederson PL, Carafoli E (1987) Ion motive ATPases I: ubiquity, properties and significance to cell functions. *Trends Biochem. Sci.* 12: 146–150.

Persson BC (1993) Modification of tRNA as a regulatory device. *Mol. Microbiol.* 8: 1011–1016.

Poyart C, Abachin E, Razafimanantsoa I, Berche P. (1993) The zinc metalloprotease of *Listeria monocytogenes* is required for maturation of phosphatidylcholine phospholipase C: direct evidence obtained by gene complementation. *Infect. Immun.* 61: 1576–1580.

Rathman M, Sjaastad MD, Falkow S (1996) Acidification of phagosomes containing *Salmonella typhimurium* in murine macrophages. *Infect. Immun.* 64: 2765–2773.

Raveneau J, Geoffroy C, Beretti JL, Gaillard JL, Alouf JE, Berche P (1992) Reduced virulence of a *Listeria monocytogenes* phospholipase-deficient mutant obtained by transposon insertion into the zinc metalloprotease gene. *Infect. Immun.* 60: 916–921.

Salama NR, Falkow S (1999) Genomic clues for defining bacterial pathogenicity. *Microbes Infection* 1: 615–619.

Sangari FJ, Agüero J, Garcia-Lobo JM (2000) The genes for erythritol catabolism are organized as an inducible operon in *Brucella arbortus. Microbiology* 146: 487–495.

Saulnier JM, Curtil FM, Duclos MC, Wallach JM (1989) Elastolytic activity of *Pseudomonas aeruginosa* elastase. *Biochem. Biophys. Acta* 995: 285–290.

Scarlato V, Rappuoli R (1991) Differential response of the *bvg* virulence regulon of *Bordetella pertussis* to Mg SO_4 modulation. *J. Bacteriol.* 173: 7401–7404.

Schmitt MP, Predich M, Doukhan L, Smith I, Holmes RK (1995) Characterization of an iron-dependent regulatory protein (IdeR) of *Mycobacterium tuberculosis* as a functional homolog of the diphtheria toxin repressor (DtxR) from *Corynebacterium diphtheriae. Infect. Immun.* 63: 4284–4289.

Schnaitman CA, Klena JD (1993) Genetics of lipo-polysaccharide biosynthesis in enteric bacteria. *Microbiol. Rev.* 57: 655–682.

Schuhmacher DA, Klose KE (1999) Environmental signals modulate Tox T-dependent virulence factor expression in *Vibrio cholerae. J. Bacteriol.* 181: 1508–1514.

Self WT, Grunden AM, Hasona A, Shanmugam KT (1999) Transcriptional regulation of molybdoenzyme synthesis in *Escherichia coli* in response to molybdenum: ModE-molybdate, a repressor of the *mod* ABCD (molybdate transport)operon is a secondary transcriptional activator for the *hyc* and *nar* operons. *Microbiology* 145: 41–45.

Simmen HP, Blaser J (1993) Analysis of pH and pO_2 in abscesses, peritoneal fluid, and drainage fluid in the presence or absence of bacterial infection during and after abdominal surgery. *Am. J. Surg.* 166: 24–27.

Smith DL, Tao T, Maguire ME (1993) Membrane topology of a P-type ATPase: the MgtB Mg^{2+} transport protein of *Salmonella typhimurium. J. Biol. Chem.* 268: 22469–22479.

Smith H (1980) The effect of environmental conditions *in vivo* and *in vitro* on the determinants of microbial pathogenicity. In: Smith H, Skehel JJ, Turner HJ (eds) *The Molecular Basis of Microbial Pathogenicity.* Dahlem Konferenzen, Weinheim, Germany: Verlag-Chemie, pp. 159–172.

Smith H, Williams AE, Pearce JH *et al.* (1962) Foetal erythritol: a cause of the localization of *Brucella arbortus* in bovine contagious abortion. *Nature (Lond.)* 193: 47–49.

Smith RL, Maguire ME (1998) Microbial magnesium transport: unusual transporters searching for identity. *Mol. Microbiol.* 28: 217–226.

Smith RL, Banks JL, Snavely MD, Maguire ME (1993) Sequence and topology of the CorA magnesium transport systems of *Salmonella typhimurium* and *Escherichia coli.* Identification of a new class of transport proteins. *J. Biol. Chem.* 268: 14071–14080.

Smith RL, Kaczmarek MT, Kucharski LM, Maguire ME (1998a) Magnesium transport in *Salmonella typhimurium*: regulation of mgtA and mgtCB during invasion of epithelial and macrophage cells. *Microbiology* 144: 1835–1843.

Smith RL, Gottlieb E, Kucharski LM, Maguire ME (1998b) Functional similarity between archaeal and bacterial CorA magnesium transporters. *J. Bacteriol.* 180: 2788–2795.

Snavely MD, Florer JB, Miler CG, Maguire ME (1989) Magnesium transport in *Salmonella typhimurium*: 28 Mg^{2+} transport by the Cor A, MgtA and MgtB systems. *J. Bacteriol.* 171: 4761–4766.

Snavely MD, Gravina SA, Cheung TT, Miller CG, Maguire ME (1991a) Magnesium transport in *Salmonella typhimurium*. Regulation of *mgt*A and *mgt*B expression. *J. Biol. Chem.* 266: 824–829.

Snavely MD, Miller CG, Maguire ME (1991b) The mgtB Mg^{2+} transport locus of *Salmonella typhimurium* encodes a P-type ATPase. *J. Biol. Chem.* 266: 815–823.

Soncini FC, Garcia-Vescovi E, Solomon F, Groisman EA (1996) Molecular basis of the magnesium deprivation response in *Salmonella typhimurium*: identification of PhoP-regulated genes. *J. Bacteriol.* 178: 5092–5099.

Soong T-W, Yong T-F, Ramanan N, Wang Y (2000) The *Candida albicans* antiporter gene CNH1 has a role in Na^+ and H^+ transport, salt tolerance and morphogenesis. *Microbiology* 146: 1035–1044.

Soupene E, He L, Yan D, Kustu S (1998) Ammonia acquisition in enteric bacteria: Physiological role of the ammonium/methylammonium transport B (amt B) protein. *Proc. Natl Acad. Sci. USA* 95: 7030–7034.

Spencer J, Chart H, Smith HR, Rowe B (1998) Expression of membrane-associated proteins by strains of enteroaggregative *Escherichia coli. FEMS Microbiol. Lett.* 161: 325–330.

Sperry JF, Robertson DC (1975) Inhibition of growth by erythritol catabolism in *Brucella arbortus*. *J. Bacteriol.* 124: 391–397.

Stevenson P, Griffiths E (1994) Environmental regulation of bacterial characteristics. In: Chart H (ed.) *Methods in Practical Laboratory Bacteriology*. Boca Raton, London, New York: CRC Press, pp. 57–72.

Straley SC, Plano GV, Skrzpek E, Hadix PL, Fields KA (1993) Regulation of Ca^{2+} in the Yersinia low-Ca^{2+} response. *Mol. Microbiol.* 8: 1005–1010.

Strauss A, Falkow S (1997) Microbial pathogenesis: genomics and beyond. *Science* 276: 707–712.

Surette MG, Bassler BL (1998) Quorum sensing in *Escherichia coli* and *Salmonella typhimurium*. *Proc. Natl Acad. Sci. USA* 95: 7046–7050.

Surin BP, Rosenburg H, Cox GB (1985) Phosphate-specific transport system of *Escherichia coli*: nucleotide sequence and gene-polypeptide relationships. *J. Bacteriol.* 161: 189–198.

Tardat B, Touati D (1993) Iron and oxygen regulation of *Escherichia coli* MnSOD expression: competition between the global regulators Fur and ArcA for binding to DNA. *Mol. Microbiol.* 9: 53–63.

Tao T, Murphy JR (1992) Binding off the metalloregulatory protein DtxR to the diphtheria tox operator requires a divalent heavy metal ion and protects the palindromic sequence from DNAase I digestion. *J. Biol. Chem.* 267: 21761–21764.

Tao T, Boyd J, Murphy JR (1992) Specific binding of the tox regulatory element DtxR to the tox operator requires divalent heavy metal ions and a 9-base-pair interrupted palindromic sequence. *Proc. Natl Acad. Sci. USA* 89: 5897–5901.

Tao T, Snavely MD, Gura S, Maguire ME (1995) Magnesium transport in *Salmonella typhimurium*: the *mgt* A locus encodes a P-type ATPase regulated similarly to *mgt* B. *J. Bacteriol.* 177: 2663–2672.

Tao T, Grulich PF, Kucharski LM, Smith RL, Maguire ME (1998) Magnesium transport in *Salmonella typhimurium*: biphasic magnesium and time dependence of the transcription of the mgtA and mgtCB loci. *Microbiology* 144: 655–664.

Titball RW (1993) Bacterial phospholipases C. *Microbiol. Rev.* 57: 347–366.

Tonello F, Morante S, Rossetto O, Schiavo G, Montecucco C (1996) Tetanus and botulism neurotoxins: a novel group of zinc-endopeptidases. *Adv. Exp. Med. Biol.* 389: 251–260.

Trombe M-C (1993) Characterization of a calcium porter of *Streptococcus pneumonia* involved in calcium regulation and competence. *J. Gen. Microbiol.* 139: 433–439.

Trombe M-C (1999) Calcium signalling in *Streptococcus pneumoniae*: implication of the kinetics by calcium transport. *Microb. Drug Resist.* 5: 247–252.

Trombe M-C, Rieux V, Baile F (1994) Mutations which alter the calcium transport alter the regulation competence in *Streptococcus pneumoniae*. *J. Bacteriol.* 176: 1992–1996.

Uchida T, Yamagiwa A, Nakamura K (1991) The effect of interleukin-1 on iron metabolism in rats. *Eur. J. Haematol.* 46: 1–5.

Véscovi EG, Soncini FC, Groisman EA (1996) Mg^{2+} as an extracellular signal: Environmental regulation of *Salmonella* virulence. *Cell* 84: 165–174.

Ward CG, Bullen JJ (1999) Clinical and physiological aspects. In: Bullen JJ, Griffiths E (eds) *Iron and Infection*, 2nd edn. Chichester: John Wiley & Sons, pp. 369–450.

Weinstein DL, Holmes RK, O'Brien AD (1988) Effects of iron and temperature on Shiga-like toxin production by *Escherichia coli*. *Infect. Immun.* 56: 106–111.

West SEH, Sparling PF (1985) Response of *Neisseria gonorrhoeae* to iron limitation: alterations in expression of membrane proteins without apparent siderophore production. *Infect. Immun.* 47: 388–394.

Williams AW, Straley SC (1998) YopD of *Yersinia pestis* plays a role in negative regulation of the low-calcium response in addition to its role in translocation of Yops. *J. Bacteriol.* 180: 350–358.

Worst DJ, Maaskant J, Vandenbrouche-Grauls CMJE, Kusters JG (1999) Multiple haem-utilization loci in *Helicobacter pylori*. *Microbiology* 145: 681–688.

Westmacott D, Hawkes JE, Hill RP, Clarke LE, Bloxham DP (1986) Comparison of the effects of murine and human interleukin-1 *in vitro* and *in vivo*. *Lymphokine Res.* 5 (Supplement): 87–91.

Zinoni F, Birkman A, Stadtman TC, Böck A (1986) Nucleotide sequence and expression of selenocysteine-containing polypeptide of formate dehydrogenase (formate-hydrogenlysase-linked) from *Escherichia coli*. *Proc. Natl Acad. Sci. USA* 83: 4650–4654.

33

The Generation of Inflammatory Responses

Didier Heumann, Michel Pierre Glauser and Thierry Calandra

CHUV, Lausanne, Switzerland

Inflammation is an essential host response to the invasion of tissues by pathogens, and plays a profound role in health and disease. The presence of bacteria in mammalian tissues elicits an inflammatory response, which involves a complex series of reactions, including: (1) increased blood vessel permeability and exudation of fluids, and (2) migration of leucocytes into the inflammatory focus, which results in phagocytosis and killing of the micro-organisms. The response is very rapid and is initially mediated by innate immune reactions, which do not usually require an adaptive response.

It must be emphasised that inflammatory processes are critical for host defence, helping to clear the infection. The inflammatory response is under tight control. In most instances it leads to the elimination of the micro-organisms without producing clinically detectable signs of inflammation. In severe infections, however, inflammatory reactions may contribute to pathological processes that culminate in multiple organ failure, shock and death. The reasons for the different courses of these response patterns are poorly understood, but it is likely that in most circumstances the host response rather than the micro-organism *per se* determines these differences. It appears that patients with an underlying disease are more likely to develop a potentially adverse response to bacteria.

Most forms of acute and chronic inflammation involve the recruitment of humoral and cellular components of the immune system. Micro-organisms, recognised as foreign by the host, react with humoral components (complement, immunoglobulins) or with inflammatory cells (neutrophils, macrophages, T cells, NK cells), resulting in activation of an amplification system that initiates production of a large array of mediators, including specialised cytokines and chemokines. These mediators alter blood flow and vascular permeability, and increase the adherence of circulating professional phagocytes to endothelial cells, thus promoting migration of leucocytes into the tissue.

The affected individual responds with an inflammatory process not only to micro-organisms but also to other antigens, such as allergens or immune complexes. This chapter will focus on the current understanding of the processes involved in the production of inflammation by bacteria. While knowledge is still fragmentary, it seems that a very complex interplay between cells and mediators is the hallmark of the inflammatory response. In addition to cells, many systems are activated, including the coagulation, the contact, and the complement systems, culminating either in the resolution of the inflammatory process or in overt circulatory failure, reflected by multiple organ failure or septic shock.

 doi:10.1006/bkmm.2001.0033

One of the critical roles of innate immunity is to recognise the microbial invaders. Janeway proposed that pattern-recognition receptors present on immune cells bind exclusively to conserved structures such as lipopolysaccharide (LPS), which are not normally found on the surface of host cells. This is the basic concept underlying innate immunity. Given the impressive recent development in our understanding of the mechanisms of innate immunity, attention will be focused on molecules such as LBP, CD14 and Toll-like receptors, which play a central role in innate immunity.

Microbial Components Responsible for Inflammatory Responses

Infections are common, and innumerable micro-organisms can infect humans. These micro-organisms induce asymptomatic, localised or overt infection, depending on the invasiveness and virulence of the bacteria and on the host response. On the one hand, bacteria have developed mechanisms that recognise receptors on host cells that promote bacterial adhesion and colonisation. On the other hand, host cells have developed mechanisms to recognise bacterial cell surface components that promote and enhance phagocytosis and killing of the micro-organisms. Much of our knowledge of the interactions that occur during an inflammatory episode is derived from studies of host responses to lipopolysaccharide (LPS). During the last decade, however, several other bacterial components have been shown to induce inflammation (Table 1). For example, in addition to LPS, exotoxins and lipoarabinomannan (LAM) are powerful inducers of inflammatory responses and there is evidence that an additional wide range of bacterial components and secreted products also have the capacity to stimulate host immune cell activation and cytokine synthesis. Some of these components are common to all bacteria, while others are specific for individual species. It is likely that inflammation results from the combined action of these bacterial compounds.

The objective of this chapter is to focus on the bacterial components that underlie the production of the inflammatory response.

Inflammatory Components of Gram-negative Bacteria

In Gram-negative bacteria, the outer membrane surrounding the bacteria is composed of phospholipids, lipopolysaccharides, enzymes and other proteins, including lipoproteins. Peptidoglycan is a less compact structure in Gram-negative bacteria than in Gram-positive bacteria, and is usually present as a monolayer.

Table 1 Bacterial components known to induce inflammation

Micro-organisms	Inflammatory components
Gram-negative bacteria	Lipopolysaccharide
	Peptidoglycan
	Porins
	Lipoproteins
	Lipopeptides
	Lipid A-associated proteins
	Pili
Gram-positive bacteria	Exotoxins
	Peptidoglycan
	Lipoteichoic and teichoic acids
Mycobacteria	Lipoarabinomannan
	Phosphoinositolmannoside
	Lipomannan
	Mycolylarabinogalactan-peptidoglycan
Fungi	Mannoproteins
	β-Glucan

Lipid A is the active structural component of all LPS. The fact that lipid A is a highly conserved molecule among Gram-negative bacteria implies that it is recognised by a unique system on target cells. LPS is one of the most powerful microbial products in terms of induction of the inflammatory response. When Gram-negative bacteria multiply or die, LPS is released either in a free form, or as a complex of LPS and bacterial surface proteins. A large body of evidence implicates LPS in the inflammatory response (reviewed by Rietschel *et al.*, 1994). The concept that LPS is the major virulence factor of Gram-negative bacteria is based on evidence that purified LPS or chemically synthesised lipid A is able to reproduce in humans or animals manifestations similar to those evoked by whole bacteria. This observation does not, however, rule out the possibility that molecules other than LPS can also manifest inflammatory properties. Indeed, most, if not all, components of Gram-negative bacteria activate host immune cells. Peptidoglycan is far less abundant in Gram-negative than in Gram-positive organisms, but its composition is very similar in the two types of bacteria. Evidence has been presented that peptidoglycan from Gram-negative bacteria elicits inflammatory reactions, since glycopeptides isolated from Gram-negative peptidoglycans increase permeability of the blood–brain barrier (Spellerberg *et al.*, 1995).

The outer cell membrane of Gram-negative bacteria also contains a large variety of proteins that play an active role in cell wall structure and function. For example, porins allow hydrophilic solutes to traverse the outer membrane. The channel-forming porins are trimers that span the outer bacterial cell membrane. They stimulate the production of the potent inflammatory mediator platelet-activating factor (PAF) by neutrophils (Tufano *et al.*, 1992), and the production of pro- and anti-inflammatory cytokines (Galdiero *et al.*, 1993).

Other outer cell membrane proteins (lipoproteins and lipopeptides) are present in large quantities in Gram-negative bacteria. Many of these molecules are covalently linked to the peptidoglycan monolayer and possess immunological activity. For example, purified outer membrane lipoproteins of *Borrelia burgdorferi* possess B cell mitogenic activity and cytokine stimulatory properties (Tai *et al.*, 1994).

The possible interactions between these outer membrane components and LPS remain to be determined. The role of the so-called lipid A-associated proteins, also known as endotoxin-associated proteins (EP) or lipid A-associated proteins (LAP) is also still uncertain. These structures consist of a complex of four or five major proteins linked to glucosamine and fatty acids, but they do not contain keto-deoxyoctulonate (KDO) of endotoxin. EPs are powerful mitogens for C3H/HeJ mouse cells, which do not respond to LPS (Sultzer and Goodman, 1976), and they are as effective as LPS in stimulating cytokine production by LPS-sensitive cells (Mangan *et al.*, 1992).

Pili are one of the means by which bacteria attach to host cells. They thus serve as virulence factors by promoting the colonisation process. Microbial attachment to host cells is characterised by specific lectin-like receptors. Attaching bacteria may activate the receptor-bearing cells to produce cytokines. For example, P-fimbriated uropathogenic *Escherichia coli* can induce cytokine production by urinary tract epithelial cells (Hedlund *et al.*, 1996).

Inflammatory Components of Gram-positive Bacteria

Exotoxins of Gram-positive bacteria exhibit powerful inflammatory activities, but only a minority of Gram-positive organisms produce such toxins. By analogy with Gram-negative bacteria, it has been thought that other components of Gram-positive bacteria may also induce inflammation. Attention has recently focused on Gram-positive cell wall fragments as potential triggers for the activation of host cells, in spite of the difficulty in characterising these components chemically.

Exotoxins are small, single-chain proteins secreted mainly by Gram-positive organisms (Table 2). Exotoxins of *Staphylococcus aureus* and of streptococci are well characterised. Other Gram-positive bacteria produce similar toxins, but most studies have concentrated on staphylococcal and streptococcal toxins (reviewed by Schlievert *et al.*, 1995). These exotoxins are also known as microbial superantigens because of their ability to stimulate subsets of T cells through an interaction with the Vβ-chain of the T-cell receptor. The exotoxins share a significant structural homology and also a number of biological properties such as pyrogenicity, cytokine-inducing capacity and enhancement of the susceptibility of experimental shock to lethal shock by LPS (reviewed by Heeg *et al.*, 1996).

The cell wall of Gram-positive bacteria is composed of multiple layers of peptidoglycan, to which teichoic acids are attached. Lipoteichoic acids are anchored in the plasma membrane and extend through the peptidoglycan layer to the outer surface. Peptidoglycan is a poorly soluble material and is very heterogeneous in terms of molecular mass and structure. For these reasons, the basic functional unit capable of activating immune cells has not yet been identified. Attention has focused on synthetic muramyl dipeptide (MDP), the smallest chemical unit of peptidoglycans. MDP and its analogues stimulate cytokine production by monocytic cells. However, synthetic MDP is not a natural product of cellular degradation of peptidoglycan, so that results obtained with this compound may not have much physiological relevance. On the other hand, there is growing evidence that whole cell walls or fragments of the cell wall possess stimulatory activity for cells of the monocyte lineage (reviewed by Verhoef and Mattsson, 1995). In this connection, sugars and sugarless stem-branched peptides induce monocytes to release cytokines (Majcherczyk *et al.*, 1999).

Teichoic acids (TA) are found in all Gram-positive bacteria, but are absent from Gram-negative bacteria. They are covalently linked to peptidoglycan through muramic acid and the phosphate group of the ribitol or glycerol phosphate. Lipoteichoic acids (LTA) are very similar in structure to teichoic acids. They are not

Table 2 Major exotoxins of Gram-positive bacteria

Bacterial species	Exotoxin
Staphylococci	Toxic shock syndrome toxin (TSST) Enterotoxins (A, B, C, D, E, G) Exfoliative toxins (A, B)
Streptococci	Streptococcal pyrogenic exotoxins (SpeA, SpeB, SpeC, SpeF)

linked to peptidoglycan but are anchored in the plasma membrane by a glycolipid. As with LPS of Gram-negative bacteria, Gram-positive bacteria spontaneously release TAs and LTAs. Antibiotic therapy may also trigger release of TAs and LTAs. The role played by TA in the generation of inflammation is not known and the role of LTA is controversial. Marked differences in their capacity to stimulate cytokine production by monocytes have been reported. LTA of *Staph. aureus* and of *Strep. pneumoniae*, for example, is inactive, whereas LTA of enterococcal species is stimulatory, though to a lesser extent than LPS (Bhakdi *et al.*, 1991; Keller *et al.*, 1992). However, Standiford *et al.* (1994) found that LTA of *Staph. aureus* induces monocyte activation. The difficulty in purifying LTA may account for this variability. The observation that synthetic LTA analogues are biologically inactive also hinders definition of the structure–function relationship of these molecules (Takada *et al.*, 1995).

Inflammatory Components of Mycobacteria

Mycobacteria are a group of eubacteria that are Gram-positive. Most of our knowledge of the structure of these bacteria was established for *Mycobacterium leprae* and *Mycobacterium tuberculosis*, two major pathogens in humans. Mycobacteria have a more complicated structure than other Gram-positive cells such as staphylococci or streptococci. The cell wall is composed of a complex structure called mycolylarabinogalactan (mAG), which is covalently linked to peptidoglycan through a phosphodiester bond. In addition, there are several other membrane lipids that are not covalently linked to mAG. Among these glycolipids are lipo-oligosaccharides (LOSs), glycolipids such as phenolic glycolipids (PGLs), and lipoarabinomannan (LAM) attached to the plasma membrane by a phosphatidylinositol link. Phosphatidylinositol mannosides (PIMs) are similarly attached to the plasma membrane.

LAM and the related structural analogues lipomannan and phosphatidylinositol mannosides are major constituents of mycobacterial cell walls. These molecules are often considered to be the equivalents of LPS of Gram-negative bacteria. LAM consists of highly branched carbohydrate polymers of mannose and arabinose terminated by a glycosylphosphatidylinositol (GPI) anchor. Lipomannan is in essence an arabinose-free version of LAM. Phosphatidylinositol mannosides are polymers of mannose linked to phosphatidylinositol. All these compounds are likely to display cytokine-inducing activity, as shown for example by the production of IL-8 by alveolar macrophages (Zhang *et al.*, 1995).

Two major classes of LAM have been isolated from mycobacteria. One form, ManLAM, which is expressed mainly by virulent strains, is characterised by the presence of numerous mannose residues on the arabinan termini (hence its name). In contrast, some avirulent mycobacterial strains contain LAM with mannose-free arabinan termini, designated as AraLAM. The latter induces strong cytokine production by monocytes *in vitro*, while ManLAM is a weak inducer (Roach *et al.*, 1993). This marked biological difference in activity has been attributed to the heavy mannose capping of ManLAM, and may partially explain the virulence and the persistence of some mycobacterial infections, because of the lack of an adequate cytokine stimulation.

Mycolylarabinogalactan (mAG) is a peptidoglycan linked to arabinogalactan and fatty acids, which shares with LAM some similarities of the arabinan motifs. The observation that the mAG of *Mycobacterium leprae* induces cytokine synthesis by monocytes (Dockrell *et al.*, 1996) suggests that the cytokine-inducing capacity of LAM and of mAG can be attributed to very similar chemical structures. Mycobacteria are endowed with specific antigenic glycolipids that form the basis of the serodiagnosis of mycobacterial infections. The phenolic glycolipids of *Mycobacterium leprae* and the oligosaccharides of *Mycobacterium kansasii* are good examples. In contrast to LAM or to mAG, the potential of these glycolipids to induce cytokine synthesis has not so far been reported.

Other Inflammatory Microbial Products

Inflammation can also be induced by other microbial products. Some of these are common to many microbes, such as capsules or DNA, while others are specific for defined strains, such as the pore-forming toxins.

Many micro-organisms produce a polysaccharide or polypeptide capsule. When this material becomes sufficiently plentiful it is referred to as slime. Capsules are generally high-molecular-weight polysaccharide polymers. These capsules are important virulence factors, because of their antiphagocytic properties, which enable the pathogenic bacteria, in the absence of anticapsular antibodies, to evade the host defence mechanisms. In contrast, other capsules enhance the bactericidal activity of phagocytic cells (Cross, 1990). The glucuronoxylmannan (GXM) capsules of *Cryptococcus neoformans*, have cytokine-inducing activity (Chaka *et al.*, 1997a). Although the precise chemical structure responsible for this activity has not been defined for most strains, capsules comprising β1,4-linked polyuronic acids bind to CD14, the LPS receptor (Espevik *et al.*, 1993).

Many pathogens produce and secrete water-soluble pore-forming toxins, such as the α-toxin of *Staph. aureus*, the streptolysin O of *Strep. pyogenes* and the haemolysin of *E. coli*, which have properties that make them ideal contributors to the pathogenesis of severe inflammation. They bind to as yet undefined receptors on target cells and insert into the membrane to create transmembrane pores. Less attention has been paid to their non-lytic activities on host cells, but it seems that the spectrum of cellular effects is as wide as that elicited by endotoxins, including cell death and tissue necrosis. Pore-forming toxins stimulate cells to release cytokines (Hackett and Stevens, 1992; Houldsworth *et al.*, 1994). Thus, *E. coli* haemolysin activates granulocytes at much lower than haemolytic concentrations (Grimminger *et al.*, 1991). In addition, streptolysin O and *E. coli* haemolysin have the capacity to induce massive shedding of the LPS receptor, CD14 (Walew *et al.*, 1996), indicating that they can interfere with the host inflammatory response.

Unmethylated cytosine–guanine (CpG) dinucleotides are encountered more frequently in the genomes of bacteria and viruses than of vertebrates. CpG groups have an impact on immune cell function by an as yet unidentified mechanism. After the original demonstration that CpG groups of bacterial DNA activate B cells (Krieg *et al.*, 1995), it was reported that similar DNA motifs can induce TNF release and mediate shock in mice (Sparwasser *et al.*, 1999).

Recognition of Microbial Components by Macrophages

When bacterial components are shed into the infected individual, they activate host monocytes/macrophages, neutrophils and endothelial cells, resulting in the development of an inflammatory response, which may be local at the site of infection or in the bloodstream.

Cell Receptors for Bacterial Products

Although many microbial components can activate host cells, knowledge of the interactions between microbial products and host cell receptors is fragmentary. Many of the microbial components, when present in high concentrations, can interact directly with membranes because of their glycolipid nature, but it is more likely that they bind to and activate specific receptors on responsive cells. Knowledge about the recognition of bacterial components by specific receptors is derived mostly from studies of the interactions between LPS or exotoxins and host cells.

Receptors for Gram-negative Bacterial Products

Initially, the binding of lipopolysaccharide to host cells, which results in cell activation, was believed to be non-specific and due to the intercalation of LPS in target cell membranes (Rietschel *et al.*, 1994). However, it is now known that LPS is recognised by specific mechanisms (Schumann *et al.*, 1990; Wright *et al.*, 1990). The various aspects of recognition of LPS by the innate immune system, by LPS-binding protein (LBP), and membrane-bound and soluble CD14 have been reviewed (Ulevitch and Tobias, 1994, 1999; Tobias *et al.*, 1999).

LPS-binding Protein, a Lipid-transfer Protein Involved in LPS-mediated Activation of Macrophages LPS released from Gram-negative bacteria is present as aggregates, because of the amphiphilic structure of the molecule. From these aggregates, LPS is transformed into monomers by the action of LPS-binding protein (LBP) (**Fig. 1**). Some of the LPS is then transferred to high-density lipoprotein (HDL), which results in neutralisation of the LPS in the bloodstream, but some of the LPS–LBP complexes bind to CD14 on the surface of myelomonocytic cells, or to soluble CD14 (sCD14). This leads to activation of myelomonocytic cells and endothelial cells, respectively.

LBP is constitutively synthesised by hepatocytes as a 55 kDa polypeptide and is released into the plasma after glycosylation. LBP is an acute-phase protein, the levels of which are increased after exposure to LPS and to cytokines, such as TNFα, IL-1 and IL-6. LBP is a lipid transfer molecule, catalysing movement of phospholipids, and in particular LPS monomers, from LPS aggregates to phospholipids or to CD14. This process results either in cell activation or LPS neutralisation. Thus, the rate of each of these processes will determine the response of the host to LPS. Kinetic studies have shown that LPS–LBP complexes first bind to CD14, and are then transferred to HDL, so that the cells of the immune system first encounter LPS and become activated, before LPS is neutralised to prevent over-stimulation of the immune system.

CD14, a Membrane-bound Molecule Involved in LPS-mediated Activation of Monocytes CD14 may serve two different functions: (1) as a functional receptor for LPS; (2) as a scavenger for LPS. These possibilities are based on the observation that certain anti-CD14 monoclonal antibodies (mAbs) suppress LPS-induced activation but do not interfere with internalisation, while other anti-CD14 mAbs suppress LPS internalisation without affecting stimulation. This suggests that, while LPS signal transduction and

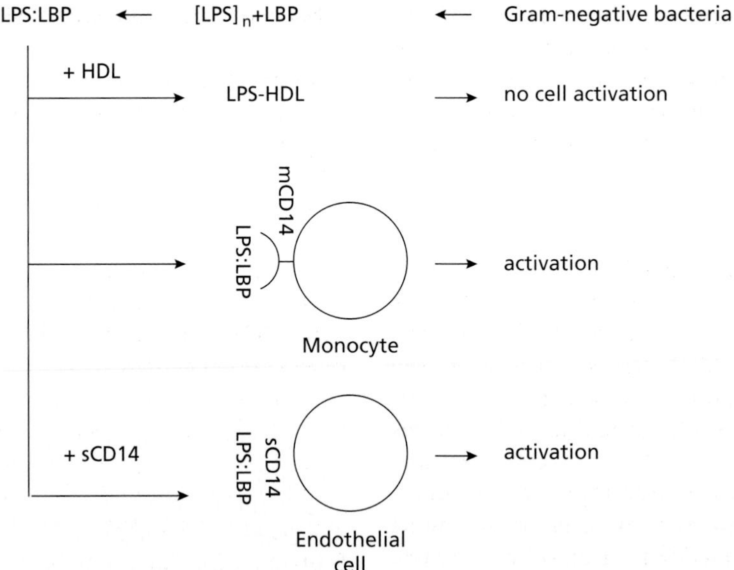

Fig. 1 The activation of monocytes and endothelial cells by lipopolysaccharides (LPS). HDL, high-density lipoprotein; LBP, LPS-binding protein.

LPS clearance utilise both LBP and membrane-bound CD14 (mCD14), these two pathways separate after LPS binds to mCD14. The role of CD14 as a receptor for LPS was confirmed by showing that transfection of CD14-negative cells with CD14 greatly enhances sensitivity to LPS.

The mCD14, which is a 53-kDa glycoprotein present at the surface of the myelomonocytic cells, is embedded in the plasma membrane by a glycerophosphate inositol (GPI) anchor. This does not allow signal transduction, but the mechanism by which CD14 confers LPS responsiveness has been partially clarified by the definition of the critical role played by Toll-like molecules in the process (see below).

CD14 and LBP: Key Components of Innate Immune Responses Against Gram-negative Infections As a result of its essential role in presenting LPS to CD14 and in the ensuing cytokine response, the acute-phase LBP enhances LPS-induced lethality, as has been shown in mice with a disrupted LBP gene (Jack *et al.*, 1997) and in wild-type mice treated with anti-LBP antibodies (Le Roy *et al.*, 1999). Thus, mice with a disrupted CD14 gene are resistant to LPS-induced toxicity (Haziot *et al.*, 1996), while antibodies to CD14 protect rabbits and primates from death after exposure to endotoxin (Leturcq *et al.*, 1996; Schimke *et al.*, 1998).

While LBP and CD14 are critical and detrimental components of the pathophysiological events resulting from endotoxaemia, their role in bacterial infections is still unclear. It has been proposed that LBP and CD14 are required for the initiation of host defence against Gram-negative bacteria, alerting the host to the presence of minute amounts of LPS. In favour of this hypothesis, *LBP−/−* mice, while resistant to LPS, are highly susceptible to infection by *Salmonella typhimurium* (Jack *et al.*, 1997). Rabbits treated with a neutralising anti-CD14 mAb have a 50-fold increase in intestinal mucosal invasion by *Shigella* and display more severe injury than controls (Wenneras *et al.*, 2000). Our unpublished data confirm an essential role for LBP and CD14 in innate immunity. Indeed, while antibody-mediated blockade of LBP or CD14 is protective in a lethal model of endotoxaemia, in a model of sepsis induced by *Klebsiella pneumoniae* it is associated with impairment of normal bacterial clearance and with early death.

Other Receptors for LPS-induced Activation Molecules other than CD14, with the capacity to recognise LPS, have been identified at the surface of the monocyte/macrophage. Several of these molecules, including the scavenger receptor and the integrin CD11/CD18 receptor, mediate uptake of LPS, a process that may lead to detoxification of LPS. However, they contribute only marginally to LPS-induced activation of cells. CD18 may be involved at high concentrations of LPS, as indicated by studies in Chinese hamster ovary (CHO) cells transfected with CD11c/CD18 (Ingalls and Golenbock, 1995). No evidence has been reported for the involvement of the scavenger receptor in LPS-induced activation of cells.

Several other membrane proteins of 18, 25, 38, 40 and 80 kDa, expressed on various cells, have been implicated in LPS binding (reviewed by Fenton and Golenbock, 1998). The biological significance of these proteins in LPS-mediated activation remains to be elucidated.

Thus, CD14 appears to be the predominant LPS receptor, at least on monocytes. In Gram-negative bacteraemia, CD14 is probably the main LPS receptor, in agreement with the observation that CD14-deficient knockout mice are hyporesponsive to intravenous LPS injections. Much less is known about the role of CD14 in the responses of tissue macrophages to LPS, since macrophages differ from blood monocytes in the relative expression of CD14 and other receptors. Thus, in tissue infections, macrophages may encounter the LPS of Gram-negative bacteria under situations different from those described for blood monocytes. Other mechanisms of LPS-induced activation probably also exist and remain to be defined. While most studies demonstrating the important contribution of plasma LBP and membrane-bound CD14 in LPS-induced activation have been performed with circulating monocytes or cell lines, other reports have suggested that differentiated macrophages may be activated by LPS without requiring serum LBP or CD14 (Ingalls and Golenbock, 1995; Jungi *et al.*, 1997; Perera *et al.*, 1997). Whether LBP- and CD14-independent pathways play an important role in the response of tissues to LPS remains to be investigated. Studies of mice with deletions of the CD14 or the LBP gene, however, clearly indicate that these two molecules play an essential role in cytokine production and lethality after exposure to low doses of LPS.

The Role of Soluble CD14 CD14 also exists as a soluble form (sCD14), by shedding of membrane-bound CD14 or by the production of GPI-free CD14 molecules. It can interact directly with LPS, a process that is catalysed by the presence of LBP. The sCD14/LPS complex can then bind to CD14-negative cells, such as endothelial cells and some epithelial cells, to induce cytokine production. Many epithelial cells do not, however, respond to LPS even in the presence of sCD14 (Hedlund *et al.*, 1996). The contribution of pathways involving sCD14 to activation of endothelial and epithelial cells in pathophysiology has not been clearly defined, but these cells may be activated indirectly by a pathway involving cytokines produced by CD14-positive monocytes (Pugin *et al.*, 1995).

Besides mediating LPS-induced activation of CD14-negative cells, sCD14 can also antagonise LPS–LBP-induced activation of CD14-positive cells, presumably by competing with mCD14 for LPS binding. However, this process requires sCD14 concentrations far higher than those present in blood (Haziot *et al.*, 1994), which casts doubt on the inhibitory role of sCD14 on the response of CD14 cells to LPS under physiological conditions. It has also been demonstrated, in a mouse model of bacterial meningitis, that increased levels of sCD14 derived from invading neutrophils augment TNF and IL-6 levels in the cerebrospinal fluid (Cauwels *et al.*, 1999). It seems that sCD14 may exert different effects on different tissues and in the circulation.

In addition, like LBP, sCD14 facilitates the transfer of LPS to HDL, resulting in the neutralisation of LPS. This suggests that LBP and sCD14 function as shuttle molecules for LPS, mediating both activating and suppressive effects via mCD14 and HDL, respectively. Conversely, locally in the tissues, as for example in the brain, sCD14 may serve a pro-inflammatory function.

Receptors for Gram-positive Bacteria

Much less information is available about the receptors for microbial components of Gram-positive bacteria. While cell receptors for exotoxins are clearly defined, knowledge of the identity of receptors for bacterial cell wall fragments is still patchy. There is an imperative need to define these receptors since, in the past few years, Gram-positive cocci have emerged as a frequent cause of infection.

Receptors for Exotoxins Certain staphylococcal and streptococcal exotoxins are superantigens characterised by their ability both to bind class II major histocompatibility complex (MHC) molecules and to induce activation of T cells in a Vβ-selective manner. Although a modest direct cytokine-inducing effect on monocytes has been observed, a requirement for T cells appears to be critical for adequate cytokine stimulation by superantigens. Isolated monocytes produced either few or no cytokines on stimulation with toxic shock syndrome toxin 1 (TSST-1) or staphylococcal enterotoxins (SE), whereas mixtures of monocytes and circulating lymphocytes produced copious cytokines (See *et al.*, 1992). For a majority of exotoxins, cytokine production is also strictly dependent on binding to class II molecules. MHC class II-deficient mice do not respond to SEB, SEA or TSST-1, superantigens that bind exclusively to class II molecules, whereas they can be stimulated by SEC, a superantigen that does not bind to class II molecules, but to ICAM-1 (Avery *et al.*, 1994).

From the point of view of mechanisms, exotoxins first bind to MHC class II molecules on monocytes or accessory cells. The superantigen–MHC complex then interacts with the variable region of the β chain of the T-cell receptor. A recognition event occurs between particular subsets of the Vβ chain and a given superantigen, so that all T cells bearing a defined Vβ chain will be stimulated. Consequently, a trimolecular complex of T-cell receptor, superantigen and MHC is formed which causes cross-linking and immobilisation of the T-cell receptor and activation of the respective T cell. A dramatic proliferation of T cells causes a significant release of lymphokines and cytokines, including IL-1, TNFα, TNFβ, IL-2 and IFNγ, by T cells and monocytes.

Receptors for Peptidoglycan, Lipoteichoic Acids, and the Role of CD14 Peptidoglycan is a major component of Gram-positive bacteria. Soluble peptidoglycan and LPS exert similar effects *in vivo*, such as fever, acute-phase responses, inflammation and activation of macrophages. Antibodies that block the action of CD14 on monocytes, peptidoglycan and LPS activate monocytes by a similar pathway (Heumann *et al.*, 1994; Pugin *et al.*, 1994; Mattsson *et al.*, 1994).

It is not clear which components of the huge peptidoglycan structure are responsible for immune cell activation. Most studies have been with large and poorly soluble molecules. These compounds may not be ideal for solving structure–activity relationships. On a weight basis, peptidoglycan of Gram-positive bacteria is 100- to 1000-fold less active than LPS in its ability to stimulate TNF production by monocytes (Majcherczyk *et al.*, 1999). Thus, peptidoglycan is either poorly active or only part of it is required for cell activation. To test this hypothesis, we recently digested peptidoglycan from *Strep. pneumoniae* with enzymes and separated the digested components by chromatography. Among these compounds, structures enriched in sugars and sugar-free branched peptides both demonstrated cytokine-inducing activity (Majcherczyk *et al.*, 1999). Thus, sugars and branched peptides of Gram-positive bacteria may trigger monocyte activation. The precise mechanisms of recognition of these compounds by CD14 remain to be determined. In Gram-positive bacteria, not only peptidoglycan, but also LTA induces cytokine production in monocytes by a CD14-dependent pathway (Cleveland *et al.*, 1996).

By analogy with LPS, peptidoglycan-induced activation of monocytes is optimal in the presence of serum/plasma (Heumann *et al.*, 1994; Mattsson *et al.*, 1994). However, the factors that augment the monocyte response in serum/plasma remain to be defined. Peptidoglycan is made up of various components that must be processed in order to activate cells, but information about this processing is not available. It seems possible that either plasma factors or enzymes present in the bacteria may fragment the peptidoglycan to smaller pieces with cytokine-inducing activity. Autolysins and hydrolases that hydrolyse bonds in the glycan strand or bonds between the glycan strand and the peptide units may contribute to this processing. Indeed, small derivatives of the peptidoglycan, such as a naturally produced tetrapeptide containing *N*-acetylglucosaminyl and muramic acid (GMTetra) have cytokine-inducing activity (Dokter *et al.*, 1994).

CD14 as a Pattern Recognition Receptor for Bacterial Components in Innate Immunity CD14 was originally defined as a specific receptor for LPS. It now appears that common cellular recognition pathways may be involved in the responses to a variety of pathogens (**Table 3**). In support of this hypothesis, the LPS of Gram-negative bacteria, the peptidoglycan and LTA of Gram-positive bacteria,

Table 3 Bacterial components shown to bind to myelomonocytes

Ligands	Origin	Enhancement by serum/plasma (factors involved)
Lipopolysaccharide	Gram-negative bacteria	Yes (LBP, sCD14)
Peptidoglycan	Gram-positive bacteria	Yes (?)
Lipoteichoic acid (LTA)	Gram-positive bacteria	?
Lipoarabinomannan (LAM)	*Candida*	Yes (LBP, others?)
Rhamnose-glucose polymers	*Streptococcus*	Yes
Polyuronic acids	Bacteria	Yes (LBP)
Acylpolygalactosyl	*K. pneumoniae*	No
Chitin	yeast	Yes (LBP)
Lipoproteins	*B. burgdorferi*, *T. pallidum*	?
Amphiphilic molecules	*Staph. aureus*	Yes (sCD14)
WI-1 protein (120 kDa)	*Blastomyces dermatidis*	?

and the LAM of *M. tuberculosis* activate a monocytic cell line by CD14-dependent mechanisms (Zhang *et al.*, 1993). In addition, β1,4-linked polyuronic acid polymers from different origins and chitins from bacteria or yeasts are able to stimulate human monocytes in a CD14-dependent manner (Espevik *et al.*, 1993; Otterlei *et al.*, 1994; Jahr *et al.*, 1997). A similar CD14-dependency has also been identified for rhamnose glucose polymers isolated from streptococci (Soell *et al.*, 1995), for an acylpolygalactoside of *Klebsiella pneumoniae* (Hmama *et al.*, 1994), and for as yet unidentified amphiphilic molecules of *Staph. aureus* (Kusunoki and Wright, 1996). In addition to these sugar polymers or glycolipid molecules, a fungal protein has also been shown to bind CD14 (Newman *et al.*, 1995), as well as lipoproteins of *Treponema pallidum* and *Borrelia burgdorferi* (Sellati *et al.*, 1998). Very preliminary data indicate that these molecules require serum/plasma factors for optimal binding to CD14. LBP, soluble CD14 and possibly other molecules may be candidates, but the requirement for one or other helping factor may differ among bacterial products (Table 3).

Mechanisms that Limit the Inflammatory Function of Bacterial Components

Macrophages are among the first cells to interact with foreign micro-organisms and their components. Depending on the recognition receptors present at the cell surface, they interact with the bacterial components to produce an inflammatory response. Macrophages also phagocytose and kill micro-organisms, and present antigens to the adaptive system of immunity that includes T and B cells. Indeed, it is likely that a major function of macrophages is the elimination of micro-organisms. While the inflammatory response induced by the bacterial components is essential, the physical elimination of bacteria is also necessary to avoid overwhelming the inflammatory response. Macrophages therefore express a series of receptors involved in these complex responses, such as complement receptors, mannose receptors and scavenger receptors. In addition, soluble molecules with very similar properties may react with bacteria or bacterial components, contributing to the control of the inflammatory response.

Collectins and C1q

Collectins are soluble effector proteins also implicated in innate immunity. This family of proteins comprises mannose-binding protein (MBP), lung surfactant protein D (SP-D) and lung surfactant protein A (SP-A). Collectins derive their name from the fact that they have a collagen and a lectin domain. They recognise a large variety of Gram-negative bacteria, Gram-positive bacteria and yeasts (reviewed by Epstein *et al.*, 1996; Eggleton and Reid, 1999). MBP and SP-A specifically recognise sugars that are widely represented in all classes of micro-organisms, such as carbohydrates with terminal mannose, *N*-acetylglucosamine or fucose residues of microbial cell walls. SP-D recognises maltose and glucose moieties. MBP activates the complement system, unlike SP-A and SP-D.

The complement system can be activated by three pathways: the classical pathway initiated by antibody–antigen complexes, the alternative pathway initiated by structures present on microbial surfaces, and a recently recognised antibody-independent pathway initiated by the binding of MBP to carbohydrates. The mannose-binding lectin pathway also contributes to the deposition of C3b molecules required for the effective opsonisation and killing of bacteria. Surfactants that do not activate complement are thought to agglutinate bacteria, thus facilitating phagocytosis by neutrophils and macrophages. Low levels of MBP predispose to recurrent infections, and SP-A knockout mice are less effective than wild-type mice in clearing bacterial lung infections, indicating the important role of collectins in the control of infections.

The recognition component of the classical component pathway, C1q, is similar to collectins. Like MBP it contributes to triggering the soluble complement cascade, resulting in lysis or clearance of the micro-organisms. However, C1q and collectins may directly interact with specific receptors on a variety of cells, thereby influencing cellular defence responses, including phagocytosis and modulation of cytokine production. C1qRp, a transmembrane receptor that binds C1q, MBP and SP-A, and SPR210, which binds SP-A, enhance phagocytosis of aggregated micro-organisms by as yet ill-defined mechanisms (reviewed by Nicholson-Weller and Kllickstein, 1999; Tenner, 1999). Further work is needed to clarify specific ligand–receptor interactions and the correlation of specific molecules with the induction of function. As more is learnt about soluble and membrane-bound complement and collectin pathways, the exact role of collectins in the development of the inflammatory response will be defined.

The complement system and the collectins have evolved to limit the inflammatory response by eliminating bacteria and bacterial components. It appears generally that collectins and C1q do not trigger the release of pro-inflammatory cytokines but rather down-regulate their secretion. SP-A binds to LPS and reduces the TNF activity of LPS-stimulated macrophages. Some of these compounds may, however, also contribute to the inflammatory process. For example, induction of TNF by human peripheral blood mononuclear cells stimulated with mannoprotein of *Cryptococcus neoformans* requires MBP (Chaka *et al.*, 1997b). Thus, the mechanisms by which the same component induces either pro- or anti-inflammatory responses remain to be clarified.

Mannose and Scavenger Receptors

The mannose receptor is a transmembrane macrophage receptor with a spectrum of recognition similar to that of soluble MBP (reviewed by Stahl and Ezekowitz, 1998). The mannose receptor is able to recognise a wide range of Gram-negative and Gram-positive bacteria, yeasts and mycobacteria. When activated by micro-organisms, it may act as a signal-transducing receptor that triggers a variety of responses, including phagocytosis, and secretion of mediators or cytokines. For example, attachment of *Candida albicans* to the mannose receptor results in the expression of cytokines, such as IL-1, IL-6 or GM-CSF (Yamamoto *et al.*, 1997). How the mannose receptor and the CD14 receptor are involved in specific responses to bacterial components is not known. For example, CD14 is a signalling receptor for the lipoarabinomannan of mycobacteria that lack terminal mannosyl units (AraLAM), whereas terminally mannosylated lipoarabinomannan (MAnLAM) binds to the mannose receptor. Evidence suggests that the mannose receptor and the CD14 receptor must be colligated to induce responses in macrophages (Bernardo *et al.*, 1998).

The scavenger receptors (SR) also have a collagen domain and broad ligand-binding properties to bind lipoproteins, anionic polysaccharides, phospholipids, as well as LPS and many Gram-positive and Gram-negative bacteria, so that one of the principal functions of SR is clearance of micro-organisms and/or microbial products or apoptotic cells (reviewed by Pearson, 1996). Scavenger receptors are defined by their ability to bind modified low-density lipoproteins. Three different SR are implicated in host defences. Macrophages express two SR (SR-AI and SR-AII), and MARCO (macrophage receptor with collagen structure). SR-As do not mediate LPS-induced macrophage activation. They limit LPS-activating effects, probably by competing with CD14, as shown by the fact that mice with disrupted *SR-AI* and *SR-AII* genes produce more cytokines in response to LPS than wild-type mice, and therefore are more susceptible to endotoxic shock (Haworth *et al.*, 1997).

The Global Concept of Innate Immunity

Janeway and colleagues have proposed that pattern-recognition receptors present on immune cells function to bind to conserved structures (such as LPS) of microbial pathogens, and to discriminate between self and non-self (Hoffmann *et al.*, 1999). This proposal is the basic concept of innate immunity and it is a tantalising one. Many soluble and membrane-bound proteins, including CD14, the mannose receptor and the collectins, are members of the innate immune system. The basic theme is that most of the bacterial components share a common receptor, i.e. CD14. The latter recognises various sugar or glycolipid motifs in bacteria, acting as a lectin-like receptor. The precise nature of the structures involved in these interactions remains undefined. LPS, peptidoglycan and mannan all share disaccharide units made up of *N*-acetylglucosamine or muramic acids. This is not, however, an unique characteristic of bacterial components, and alternative motifs do bind to CD14, such as sugar polymers found in LAM or other molecules.

This scheme does not exclude the possibility that interaction with CD14-independent pathways may be the rule for other bacterial components. For instance, Gram-negative bacterial glycosphingolipids activate cytokine production in human mononuclear cells by a CD14-independent pathway (Krziwon *et al.*, 1995). Similarly, fimbriae-associated adhesins of *E. coli* do not bind to CD14, but bind to and activate cells through glycosphingolipids present at the cell surface (Hedlund *et al.*, 1996).

Toll-like Receptors as Signal-transducing Molecules for Bacterial Compounds

Cell activation of specific genes occurs either directly in response to stimulation by bacterial molecules or indirectly in response to cytokines produced by the inflammatory response. During the past 10 years, understanding of the mechanisms that underlie cell signalling has progressed markedly, yet we are still far from a complete picture of these complex systems. Their very complexity poses the question of their control and regulation. These pathways, which are mostly arranged as modular systems, are not isolated from each other and 'cross-talk' takes place. Such a network, linking distinct signalling systems, offers the cell a sophisticated means of sensing a particular external signal. A detailed discussion of all the components of the various cascades is beyond the scope of this chapter. The focus will be on some aspects of this complex signal-transduction system.

Macrophages, which are the major target of LPS, are also the target of most of the cytokines which, in turn, also possess their own signalling mechanisms. This presents a major difficulty in understanding the mechanisms of LPS-induced signal transduction. In other words, the events secondary to the stimulation of macrophages by LPS result not only from the primary binding to an LPS receptor, but also from the co-ordinated triggering of a very wide range of parallel signalling pathways. LPS is such a powerful stimulant that it activates all currently defined signalling pathways. **Figure 2** summarises the main pathways implicated in LPS signalling (for details see Han *et al.*, 1994; Reinmann *et al.*, 1994; Raingeaud *et al.*, 1995; Herrera-Velit and Reiner, 1996). It is not known how and to what extent these signals are interconnected, and the basis of the specificity of the signal induced by LPS and by other exogenous stimuli is unclear.

Upstream of the cytoplasmic activation cascades, the recently identified Toll-like receptors (TLRs), and in particular TLR4, are likely candidates for the transmission of the LPS signal from membrane-bound CD14 to the cytoplasm (reviewed by Anderson, 2000; Beutler, 2000). The role of TLR4 in innate immunity was revealed by studies performed in *Drosophila*, where Toll molecules were first identified as cell receptors involved in embryogenesis and in the defence against micro-organisms (Hoffmann *et al.*, 1999). All Toll-like molecules have an extracellular leucine-rich domain and a conserved intracellular domain, the 'Toll domain'. Nine human receptors structurally related to the *Drosophila* Toll have so far been identified. Evidence for the participation of TLR4 was confirmed by the observation that LPS-resistant mutant mice (C3H/HeJ) have a point mutation in the Toll domain of TLR4 and that another LPS-resistant mouse strain (C57BL/10ScCr) has a total deletion of the TLR4 gene (Poltorak *et al.*, 1999). While confirming these observations, it was noticed that mice deficient in the TLR4 gene do not respond to LPS (Takeuchi *et al.*, 1999). It was further demonstrated, by transfection experiments, that the introduction of TLR4 into LPS-unresponsive human embryonic kidney cells (HEK) restored LPS responses (Chow *et al.*, 1999), an effect that was strikingly amplified by co-transfection with MD2, an adapter molecule that links LPS and TLR4 (Shimazu *et al.*, 1999).

The role of other TLRs for LPS signal transduction, in particular TLR2, is controversial (Chow *et al.*, 1999). Mice made deficient in the TLR2 gene respond to LPS to the same extent as wild-type mice (Takeuchi *et al.*, 1999), ruling out a significant contribution of TLR2 to the LPS responses of mice. It seems possible, however, that different mammalian species may use different combinations of TLRs. Indeed, while human cells respond to lipid A but not to enzymatically de-acylated or tetra-acyl lipid A, mouse cells respond to both stimuli. It has recently been shown, by genetic complementation, that macrophages from LPS-unresponsive C3H/HeJ mice respond to both lipid A and tetra-acyl lipid A after the introduction of the mouse TLR4 gene, while they respond only to lipid A when the human TLR4 gene is added (Poltorak *et al.*, 2000). This suggests that the two genes, human and mouse, interpret the chemical structure of LPS differently. In addition, since the adapter molecule MyD88 is a dimer (Burns *et al.*, 1998), dimers of TLRs, such as TLR4/TLR4 or TLR4 associated with some other TLR, should be recognisable.

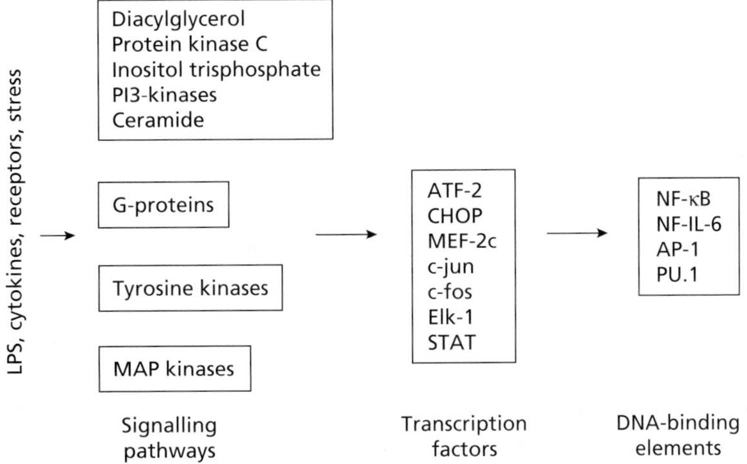

Fig. 2 The main pathways implicated in signalling by lipopolysaccharides.

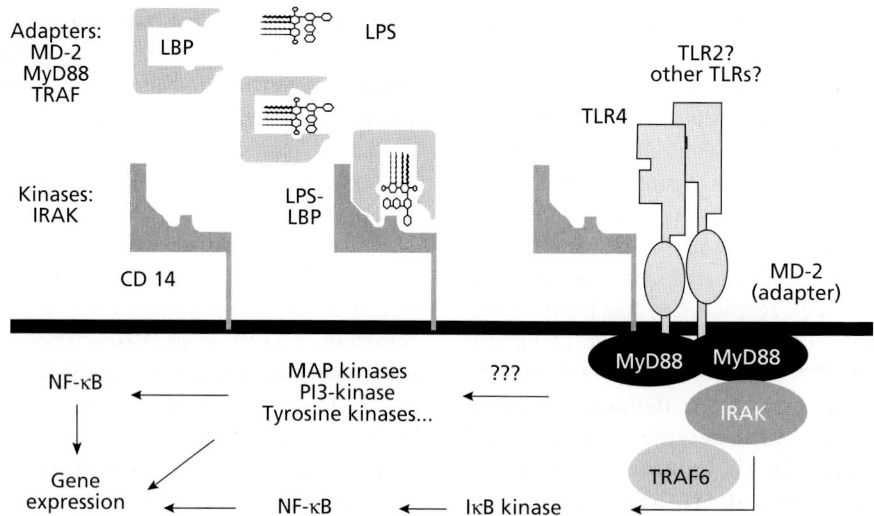

Fig. 3 Hypothetical scheme to show current views of the lipopolysaccharide receptor and pathways to gene expression.

Table 4 Microbial products recognised by Toll-like receptor 2

Bacterial compounds	Amount
Staph. aureus	10^8 CFU/mL
Strep. pneumoniae	10^7 CFU/mL
Soluble peptidoglycan (sPGN)	10 nM
Staph. aureus	6×10^6 CFU/mL
B. subtilis	9×10^9 CFU/mL
Lipoteichoic acid (LTA), sPGN	10 μg/mL
E. coli lipoprotein	200 pM
Synthetic lipopeptide	200 pM
B. burgdorferi lipoprotein	500 ng/mL
B. burgdorferi lipoprotein	300 ng/mL
T. pallidum lipoprotein	10 μg/mL
M. tuberculosis	10 CFU/cell
Lipoarabinomannan (LAM)	10 μg/mL
Arabino-galactan	10 μg/mL
Mycoplasma lipopeptide-2	10 pg/mL

A hypothesis of the current view of the LPS receptor is presented in Fig. 3. Signalling by TLR4 occurs by sequential recruitment of the adapter molecules MyD88 and IRAK (Muzio *et al.*, 1998), so that LPS signalling seems to operate in a manner that is very similar to IL-1 signalling. The latter is initiated by the formation of a complex composed of IL-1, IL-1 receptor I, and IL-1 receptor-associated accessory protein (IL-1R-AcP). The adapter MyD88, which is a cytoplasmic Toll-like molecule, binds to this complex by Toll/Toll interactions, followed by the recruitment to the complex of the IL-1 receptor-associated kinase (IRAK). Subsequently, IRAK dissociates from the receptor and interacts with TRAF6, which results in cell activation through NF-κB or JNK/SAP kinases, two systems involved in the activation of numerous inflammation-associated genes. The key role of the *MyD88* gene has been confirmed in these two pathways, since the disruption of *MyD88* results in loss of IL-1-mediated function (Adachi *et al.*, 1998) or unresponsiveness to LPS (Kawai *et al.*, 1999).

A number of questions remain unsolved with regard to LPS signalling: (1) Does TLR4 interact with LPS alone or with the complex composed of LPS, LBP, CD14 and potentially other molecules? (2) Do other TLRs form heteromeric complexes with TLR4? (3) Are there other members in the MyD88 family? and (4) How do other signalling pathways, including MAP kinases, tyrosine kinases, PI3-kinase, G proteins (Figs 2 and 3) connect with the upstream signalling system?

With regard to the action of Gram-positive components, there is a general consensus that TLR2 plays a key role in cell signalling. All major bacterial components of Gram-positive bacteria, including peptidoglycan, various lipopeptides and lipoarabinomannan, have been found, by a transfection-based approach (HEK cells transfected with TLR2 or TLR4), to signal cells exclusively through TLR2, but not through TLR4 (Table 4) (reviewed by Lien *et al.*, 1999). This was confirmed with mice deficient in the *TLR2* or the *TLR4* gene. The effect of Gram-positive lipoteichoic acid (LTA) is not clear, since LTA signals through TLR2 by the transfection-base approach (Schwandner *et al.*, 1999) or through TLR4 in gene-deficient mice (Takeuchi *et al.*, 1999). Although evidence for TLR2 involvement is supported by studies with cells from TLR2-deficient animals, caution seems appropriate with regard to the role of TLR2 for signalling by Gram-positive components. First, as is shown in Table 4, in contrast to LPS, very high doses of Gram-positive components are necessary to trigger a signal. Second, all these compounds are chemically distinct, and the exact and minimal structure required to activate cells is unknown, so that there is some doubt as to whether they use the same receptor. These questions necessitate identification of the subcomponents that comprise the active part of the Gram-positive organisms.

The discovery that common cellular pathways are shared between very different stimuli, from *Drosophila* to humans, opens a new exciting topic in innate immunity that may lead to the identification of strategies for the treatment of bacterial infections. It is anticipated that gene inactivation, as well as careful identification in terms of chemical structure of the various toxic bacterial products, will uncover the microbial ligands that trigger responses from each member of the Toll family. A recent report seems to indicate that mutations in the extracellular domain of TLR4 may be associated with LPS hyporesponsiveness in humans. This points to the potential for new therapeutic approaches to modulation of the LPS signalling system (Arbour *et al.*, 2000).

Mediators of the Inflammatory Response

This chapter reviews the mediators that are most probably involved in the pathophysiological process that underlies inflammation. Special emphasis is devoted to sepsis, characterised by profound alterations of the inflammatory response. If many mediators have a pro-inflammatory function, it seems likely that there are also counter-regulatory mechanisms that oppose the effects of pro-inflammatory mediators. These inhibitors reduce the severity of inflammation, so that the net biological response to pro- and anti-inflammatory mediators determines the outcome of the inflammatory process. In most episodes of infection, the two countervailing processes are well matched, but the balance can be changed by overproduction of pro-inflammatory mediators and by failure of the anti-inflammatory pathway to respond adequately. Septic shock develops under these circumstances, which may be due to the pre-existing health of the patient or to genetic factors.

A distinction must be made between the local effects of inflammatory mediators and the consequences of their systemic effects. For the effective elimination of the invading organisms, inflammation is necessary. It is important to recognise that the same mediators are produced in mild and severe infections. If there is systemic inflammation, however, large amounts of these mediators may induce hypotension, shock and death. The amount of mediators produced and their spill-over into the circulation distinguishes between these two manifestations of infections.

Basically, the inflammatory process results in complex interactions between bacterial components and cells of the immune system (monocytes, neutrophils, platelets and endothelial cells). In addition, bacterial components can directly affect humoral factors (**Fig. 4**). For the sake of simplicity, two pathways can be defined: (1) a direct pathway (or humoral pathway) by which bacterial components interfere directly with the complement and the coagulation proteins and

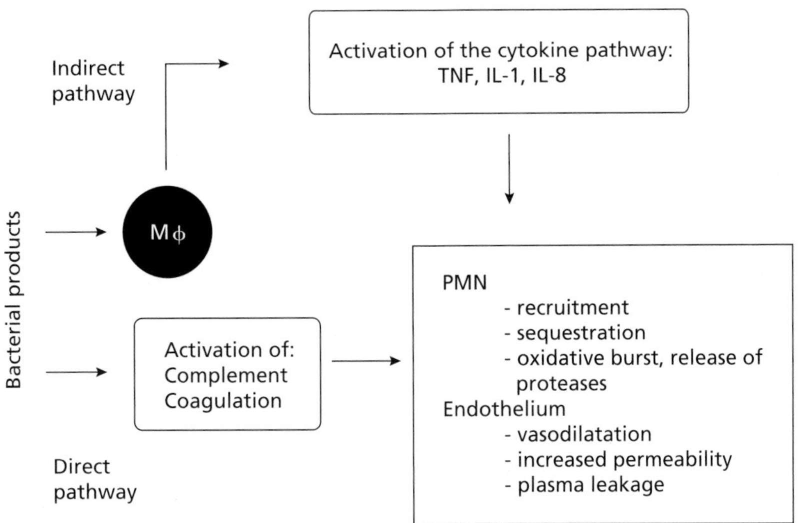

Fig. 4 Pathways of polymorphonuclear leucocyte and endothelial cell activation. Direct activation is via the complement and coagulation cascades. Indirect activation is by macrophage activation and cytokine production.

(2) an indirect pathway (or cellular pathway) that leads to production of cytokines mainly by macrophages. This indirect pathway amplifies dramatically the effects observed as a result of activation of the direct pathway. The targets of these two pathways are, (1) neutrophils, which become hyperactive, thus inducing adherence to vessels and degranulation, and (2) endothelial cells, which become permeable, inducing vasodilatation and plasma leakage.

Direct Activation

Bacterial components can directly activate humoral factors present in the bloodstream, including especially two homeostatic systems: the complement cascade and the coagulation cascade.

Activation of Complement

The complement system is essential for defence against bacterial infections, as shown by the observation that animals deficient in complement C3 and C4 are highly susceptible to endotoxaemia and bacterial infections (Quezado *et al.*, 1994; Wessels *et al.*, 1995; Fischer *et al.*, 1997). Excessive activation of complement may, however, lead to the generation of molecules that mediate inflammation and tissue injury.

Evidence for the key role of complement in the elimination of bacteria is so extensive that it will not be presented here. Briefly, complement acts by the direct lysis of bacteria, by helping recruitment of phagocytic cells, and by facilitating phagocytosis of bacteria by opsonisation. After interaction with bacteria, complement is activated and peptides are generated (C3e, C5a, C5a des Arg), which increase the number of circulating neutrophils, promote adherence to vascular endothelium, and attract neutrophils to sites of microbial invasion. Fragments of C3 (C3b, C3bi) promote recognition of particles. The complement-dependent killing of serum-sensitive organisms then occurs by insertion of the C5–C9 complex into the bacterial membrane. In bacteria resistant to complement-mediated lysis, C3 fragments deposited on bacterial surfaces can facilitate ingestion and killing by phagocytic cells, by binding of opsonised bacteria to the complement receptors present on professional phagocytes.

On the other hand, complement is an important element in the inflammatory response. Complement activation in tissues, caused by micro-organisms, results in the production of a series of activation peptides – the anaphylatoxins (C3a, C4a, C5a) – that have powerful biological activities (**Table 5**). They increase vascular permeability and cause histamine release from mast cells. Among the complement factors, C5a has been singled out as crucial to the part played by the complement system in inflammation. C5a has many important biological activities that appear to be mediated by its interaction with specific receptors on a variety of cells. The binding of C5a to neutrophils stimulates an oxidative burst, degranulation and enhanced adherence of these cells to one another and to endothelium (reviewed by Hugli and Muller-Eberhard, 1978).Complement-stimulated neutrophils damage endothelium by releasing toxic oxygen molecules and proteases. The lungs appear to represent the favoured target for this C5a-mediated reaction, and the stasis of neutrophils in capillaries is an early event in the adult respiratory distress syndrome (ARDS). As an anaphylatoxin, C5a can bind to mast cells, basophils and platelets, triggering the discharge of molecules present in granules, including histamine, serotonin, prostaglandins, leukotrienes and other lipids (Hugli and Muller-Eberhard, 1978). The result is vascular dilation and damage. C5a is also a potent stimulant of cytokine secretion by macrophages (Okusawa *et al.*, 1988a), and these cytokines, in turn, can activate neutrophils, endothelial cells and platelets.

Table 5 Biological activities of anaphylatoxins

Target cells	Major responses
Neutrophils	Migration Increased oxidative metabolism Increased adherence to endothelium Secretion of proteases Aggregation Increased expression of complement receptors
Macrophages	Migration Increased oxidative metabolism Increased adherence to endothelium Secretion of proteases Aggregation Increased expression of complement receptors Cytokine production
Endothelial cells	Increased permeability
Smooth muscle cells	Contraction
Mast cells	Release of histamine, serotonin

Complement activation occurs in many infections, especially in bacteraemia with shock. Circulating levels of C5a have been associated with ARDS and meningoccaemia. C5a plays a toxic role in models of endotoxaemia in mice and in models of acute lung inflammatory injury in rats (Barton and Warren, 1993), as well as in peritonitis (Czermak *et al.*, 1999).

Activation of Coagulation

Activation of coagulation leading to either localised or disseminated intravascular clotting, depending on the type of infection, localised or generalised, is a hallmark of inflammation. In local infections, inflammation is manifest by pain and swelling, and a sense of heat of the injured part. Redness and increased temperature are due to the fact that the blood supply to the affected part increases, as a result of vasodilatation and increased capillary permeability. This effect is necessary to recruit neutrophils, antibodies and complement in order to eliminate the micro-organisms from the infected tissue. When localised, coagulation of the affected area limits the process of recruitment. In severe infections, after generalised spill-over of bacterial components, disseminated intravascular coagulation may occur. The current view is that the coagulation system is activated by the extrinsic rather than by the intrinsic pathway (Fig. 5).

One of the earliest events is activation of factor XII (Hageman factor) by bacterial products, and this in turn activates the complement and the coagulation cascades, which illustrates the association of the functions and disorders in two major systems of homeostasis. The contact system (intrinsic clotting pathway) consists of four primary proteins: Hageman factor (HF, XII), pre-kallikrein (PK), high-molecular-weight kininogen (HMKW) and coagulation factor XI. Secondary reactions with plasminogen, factor VII and complement proteins extend the physiologic effects of the contact system. Activation of factor XII by microbial components initiates the kinin pathway. Limited proteolysis of factor XII is necessary for activation. In turn, PK is activated after limited proteolytic digestion by activated factor XII. Tissue kallikrein and plasma kallikrein cleave high-molecular-weight kallikrein (HMWK) at two internal bonds to release bradykinin, which is one of the most potent vasodilators, resulting in increased permeability of the vascular endothelium. In addition, bradykinin also contributes, like C5a, to neutrophil margination. It appears that on activation of factor XII, the kinin pathway rather than the coagulation cascade is initiated. Indeed, experiments in baboons aimed at blocking the contact system with anti-factor XII antibodies showed a reduction of the hypotensive response, but did not show any effect on the coagulation response (Pixley *et al.*, 1993). There is evidence that the contact system is activated locally in tissues during infection. Although injection of low doses of LPS did not result in systemic activation of the contact system, there were signs of activation in ARDS patients, as illustrated by the presence of reduced levels of factor XII and pre-kallikrein (Hesselvik *et al.*, 1989).

Coagulation is initiated *in vivo* by the extrinsic pathway. The initial step is the synthesis of tissue factor expressed on endothelial cells and monocytes. Resting endothelium does not promote thrombin formation, but bacterial components or cytokines (especially TNFα and IL-1) may up-regulate tissue factor on circulating monocytes or in the lumen of the endothelium (Bevilacqua *et al.*, 1986). The main natural inhibitors of activated pro-coagulant factors (reviewed by Sala *et al.*, 1993) include, anti-thrombin III, which inhibits thrombin, protein C/thrombomodulin, which is essential for the inactivation of activated

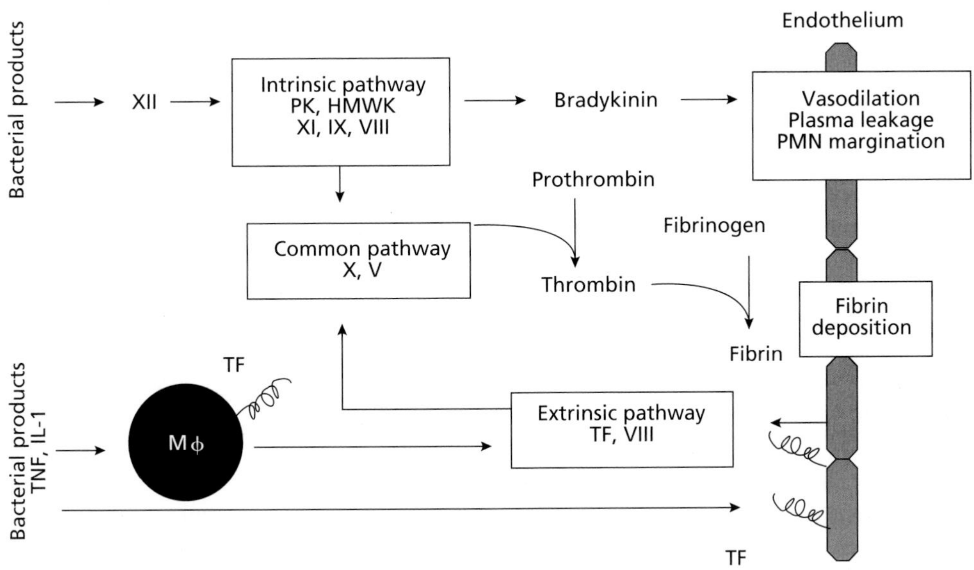

Fig. 5 Routes of activation of the coagulation cascade.

coagulation factors Va and VIIIa, the tissue factor pathway inhibitor that counteracts the effects of tissue factor, and the fibrinolytic system that dissolves clotted factors.

There is evidence for an imbalance between coagulation and anti-coagulation in severe infections. Injection of tissue factor pathway inhibitor reduces mortality and prevents coagulopathy in baboons infected with *E. coli*, which supports the role of altered coagulation in the pathogenesis of septic shock (Creasey *et al.*, 1993). In severe infections and in shock, coagulation disorders and abnormalities of fibrinolysis are common. Thrombin–antithrombin complexes are sensitive markers of thrombin formation and high levels of such complexes are common in septic patients. To counter-regulate coagulation, an important function of the vascular endothelium is modulation of the fibrinolytic pathway. Endothelial cells first synthesise tissue plasminogen activator (t-PA) followed by its inhibitor, plasminogen activator inhibitor (PAI). Injection of LPS into healthy volunteers causes an initial increase in t-PA release, followed by PAI release. These changes are associated with an initial increase, followed by a subsequent reduction in fibrin formation (Suffredini *et al.*, 1989). In severe infections and shock, however, fibrinolysis is inhibited by the release of PAI-1 (Pralong *et al.*, 1989), so that high levels of PAI-1 may contribute to the development of multi-organ failure.

Another counter-regulatory coagulation mechanism involves inhibition of thrombin formation by protein C. The latter is activated by thrombomodulin after binding thrombin at the endothelial surface. Decreased protein C synthesis, as well as a reduction in protein C activation by thrombomodulin (which is down-regulated by pro-inflammatory cytokines) may account for the coagulopathy observed in severe meningococcaemia (Fijnvandraat *et al.*, 1995). Experimentally, it has also been reported that activated protein C may prevent the coagulopathic and lethal effects of *E. coli* infusion in baboons (Taylor *et al.*, 1987) or attenuate the LPS-induced pulmonary vascular injury by inhibiting activated neutrophils in rats (Murakami *et al.*, 1996). Interestingly, preliminary data (unpublished) in a recent clinical trial of treatment with activated protein C in patients with severe sepsis and shock, seems to have been beneficial.

In summary, severe infections and shock alter the normal anticoagulant function of the vascular endothelium. Endothelial surfaces may become strongly pro-coagulant, causing local fibrin deposition and interfering with tissue perfusion. In addition, the kinin pathway contributes to overt plasma leakage and vasodilatation. While it is likely that these mechanisms are integral parts of the events that occur locally in an infected tissue, their systemic generalisation to the capillaries and small vessels in severe infections and shock contribute to multi-organ failure.

Indirect Activation

Bacterial components bind to, activate and recruit host cells, which include especially monocytes/macrophages, as well as neutrophils and endothelial cells. This response, which is synonymous with cytokine production, may take place locally or in the bloodstream and represents the basis of the indirect pathway of activation by bacterial components.

Cytokines are powerful communication molecules capable of regulating a wide spectrum of biological activities. They share some characteristics: (1) pleiotropism indicated by the ability of a given cytokine to react with different targets, (2) redundancy, as reflected by the ability of different cytokines to perform identical functions, (3) a propensity to induce each other, resulting in complex cascades of activation and/or inactivation. Furthermore, cytokines exert autocrine and paracrine functions that may differ markedly. Cytokines are classified into families based on their main biological activities. When LPS or some other bacterial product is infused into animals or human volunteers *in vivo*, or when monocytes are stimulated *in vitro* with these agents, a wide array of cytokines is produced (**Fig. 6**). While a typical pro-inflammatory response occurs, anti-inflammatory cytokines or factors are also secreted at the same time to prevent the effects associated with an excess of pro-inflammatory mediators. In this chapter, the biological activities of the major cytokines involved in the generation of the inflammatory response will be reviewed, and their biochemical characteristics and function will be described.

Pro-inflammatory Cytokine Response

Bacterial products trigger many of the pathophysiological features of severe inflammation and sepsis by inducing the release of many pro-inflammatory cytokines, including TNFα, IL-1 and macrophage migration inhibitory factor (MIF). Another family of cytokines, including IL-12, IL-18 and IFNγ also contributes to the pro-inflammatory response, together with lipids, nitric oxide and oxygen species.

Tumour Necrosis Factor α and Interleukin 1

TNFα belongs to a family of ligands that activate a corresponding family of structurally related receptors, the TNF/nerve growth factor family, the p55 TNF receptor, the p75 TNF receptor, the

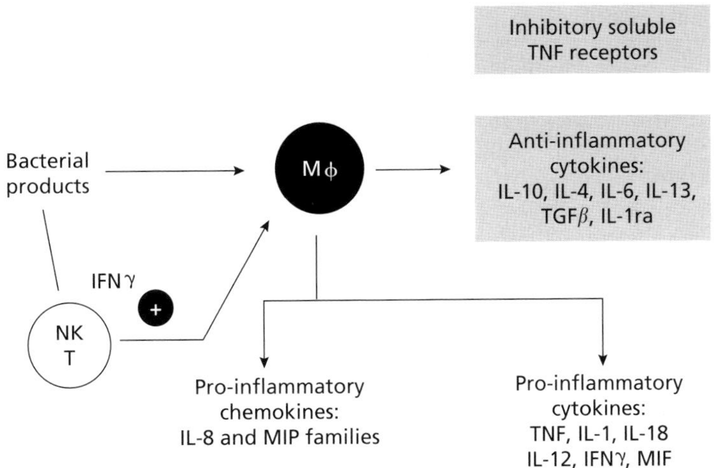

Fig. 6 Induction of cytokine production by monocytes.

lymphotoxin-β receptor (LTβR) (reviewed by Wallach *et al.*, 1999). Major members of the family are TNFα, lymphotoxin-α, lymphotoxin-β and Fas ligand. All members except lymphotoxin-α (which is entirely secreted) and TNFα (which is predominantly secreted), are transmembrane proteins that act by cell-to-cell contact. These receptors are essential in a wide range of innate and adaptive immune responses, including cell proliferation and activation or apoptosis of the cells of the immune system.

Among these factors, TNFα is predominant because it is highly toxic in animals and humans. It is synthesised as a precursor protein of 26 kDa, which is processed to a secreted 17-kDa mature form by enzymatic cleavage. TNFα engages two receptors, the p55 and the p75 TNF receptors. Although most cell lines and tissues co-express both receptor types, cellular responses to TNFα are dominated by the interaction with the p55 TNF receptor, the low-affinity p75 receptor merely contributing to the response by favouring ligand-passing activity to the p55 TNF receptor (Tartaglia *et al.*, 1993).

IL-1 exists as two forms, IL-1α and IL-1β, and is actively produced by monocytes on stimulation with a large variety of bacterial products. IL-1α is a 17-kDa protein which is secreted in mice but not in humans. In humans, IL-1α is synthesised as a 31-kDa precursor protein and either remains in the cytoplasm or is expressed at the cell surface. Cytoplasmic IL-1 may be an autocrine growth factor. Membrane IL-1α is biologically active and can be neutralised by anti-IL-1α antibodies. IL-1β is a secreted cytokine originating from cytoplasmic pro-IL-1β by enzymatic cleavage (ICE, caspase-1). Two primary receptors, IL-1RI and IL-1RII, and one receptor accessory protein (IL-1R-AcP) have been identified. IL-1RI transduces the signal, whereas IL-1RII does not transduce the signal, but acts as a sink for IL-1β. IL-1RI is an 80-kDa glycoprotein found on virtually all cells. A comprehensive review of the biological activity of IL-1 has been published by Dinarello (1998).

During inflammation, injury or infection, TNFα and IL-1, the most important proximal mediators of the inflammatory response, are produced in copious amounts primarily by activated monocytes and macrophages. TNFα shares a considerable range of activities, mostly pro-inflammatory activities, with IL-1 (Table 6). TNFα and IL-1 act synergistically in nearly every *in vitro* and *in vivo* model of local or systemic infection. When both cytokines are blocked by appropriate antibodies and inhibitors, the severity of inflammation is reduced. Infusion of TNFα or IL-1 into humans or experimental animals induces a rapid fall in blood pressure by mechanisms that involve the generation of low-molecular-weight mediators, cyclooxygenase components, PAF and nitric oxide (NO) (Beasley *et al.*, 1991). The only major difference between the two cytokines is that animals lacking IL-1β are not resistant to LPS-induced shock, in contrast to mice lacking either the TNFα gene or the p55TNF receptor gene, suggesting a more toxic contribution of TNFα to endotoxaemia.

Macrophage Migration Inhibitory Factor During the past 5 years, macrophage migration inhibitory factor (MIF) has been shown to be a pivotal cytokine of the innate immune system and to play an important role in the control of inflammatory responses. First described as a T-cell cytokine, MIF was then identified as a cytokine released by pituitary cells, monocytes and macrophages, in response to infection and stress (Bernhagen *et al.*, 1993). It is pro-inflammatory and induces the expression of pro-inflammatory mediators by macrophages and also

Table 6 Biological effects of IL-1 and TNFα

Biological activity	IL-1	TNFα
Activation of T cells	+	+
Activation of B cells	+	+
Induction of acute phase proteins	+	+
Pyrogenicity	+	+
Activation of monocytes	+	+
Adhesion to endothelial cells	+	+
Activation of osteoclasts	+	+
Degradation of articular cartilage	+	+
Angiogenesis	+	+
Activation of synovial cells	+	+
Induction of CSF	+	+
Radioprotection	+	+
Tumour cytotoxicity	+	+
Adjuvancity	+	+
Sickness behaviour in mice	+	+
Resistance to infection	+	+
Induction of IL-1, TNFα, IL-6 and IL-8	+	+
Induction of PGE2 and cyclo-oxygenase	+	+
Cachexia and lipolysis	+	+
Generation of Shwartzman reaction	+	+
Neutrophilia	+	+
Induction of shock	+	+
Release of corticotrophin	+	+
Insulin production	+	−
Stimulation of dendritic cells	+	−
Stimulation of haematopoiesis	+	−
Maturation of haematopoietic cells	−	+
Activation of neutrophils	−	+
Tumour necrosis	−	+

Data originating from Durum and Oppenheim (1989) and Dinarello (1991).

activates T cells, thus promoting inflammatory and immune responses (Bernhagen *et al.*, 1993) The critical role of MIF in the immune system has been further emphasised by the observation that MIF is induced by glucocorticoids and has the ability to over-ride the anti-inflammatory and immunosuppressive effects of glucocorticoids on macrophages and T cells (Calandra *et al.*, 1995, 2000).

Interleukin 12, Interleukin 18 and Interferon γ IL-12, IL-18 and IFNγ are functionally closely related molecules. IL-18 and IL-12 are two cytokines with potent IFNγ-inducing activities. IFNγ is mainly produced by NK cells and activated T cells.

IL-12 is a heterodimeric cytokine comprised of covalently linked p35 and p40 subunits (reviewed by Biron and Gazzinelli, 1995; Lamont and Adorini, 1996). The IL-12 receptor is composed of three chains and belongs to the gp130-like cytokine receptor super-family. IL-12 is produced by macrophages, dendritic cells and B lymphocytes in response to bacterial products, intracellular parasites, or by the ligation of cell surface CD40 in co-operation with T cells. IL-12 plays a critical role in the early inflammatory response by augmenting the production of IFNγ from NK cells and T cells, by stimulating the proliferation of activated T cells and NK cells, and supporting the generation of cytolytic lymphocytes. IFNγ enhances the release of IL-12 by macrophages, thereby inducing positive feedback interactions that are important for phagocytosis and T-cell differentiation.

IL-18 was originally identified as IFNγ-inducing factor, a molecule induced in the serum of mice injected with LPS (reviewed by Akira, 2000). IL-18 is expressed in a wide range of cells including macrophages, T cells and B cells in response to bacterial products. Like IL-1β, IL-18 is synthesised as a precursor and processed to mature and bioactive IL-18 by ICE (caspase-1). In collaboration with IL-12, IL-18 participates in IFNγ production by T cells and NK cells. Although IL-18 does not induce IFNγ production by naïve T cells, because of their lack of IL-18 receptor expression, IL-12 and IL-18 synergistically induce IFNγ production. This synergy is mediated by the induction of IL-18 receptors on naïve T cells by IL-12.

Activation of T lymphocytes, NK cells, and also monocytes by LPS, superantigens, and many other bacterial products induces the release of IFNγ. This is a 17-kDa protein that binds to a unique receptor (reviewed by Billiau, 1988; Belardelli, 1995). Specifically, IFNγ induces Fc receptors and human leucocyte antigen complex (HLA-DR) antigens, to potentiate the release of hydrogen peroxide in macrophages, and to be the most potent macrophage-activating factor, also enhancing the expression of cellular receptors for TNFα, and acting strongly and synergistically with TNFα. IFNγ is necessary for the activation of macrophages to a microbicidal state.

The expression of protective immunity to intracellular pathogens is largely mediated by activated macrophages and T cells. A number of intracellular pathogens, of which *Listeria monocytogenes* and mycobacteria are prototypes, induce T-cell and NK-cell responses characterised by the generation of IL-12, IL-18 and IFNγ. In this context, both IFNγ and TNFα play a determinant role. Indeed, disruption of the TNFα gene, the TNFp55 TNF receptor gene, the IFNγ gene or of the IFNγ receptor gene renders mice highly susceptible to bacterial infections with intracellular pathogens (see below under animal models). In humans, deficient production of IL-12 or the lack of functional IL-12 receptors is associated with severe mycobacterial and *Salmonella* infections, confirming the essential role of these two cytokines in the control of intracellular infections (De Jong *et al.*, 1998; Haraguchi *et al.*, 1998). Not only intracellular bacteria but also bacterial components may stimulate the production of IL-12 and IFNγ from macrophages and NK cells, respectively. Since there is strong synergy between IFNγ and TNF (Billiau, 1988), the net result is amplification of the pro-inflammatory response.

Phospholipid Components Inflammation is partially related to TNFα- or IL-1-mediated stimulation of phospholipase A2 that results in release of platelet-activating factor (PAF) and arachidonic acid (reviewed by Pruzanski and Vadas, 1991; Bone, 1992) (**Figure 7**). TNFα, IL-1 and PAF are capable of promoting their own release and the release of other mediators in an autocrine manner. This in turns creates a feedback loop which activates macrophages to produce more cytokines. PAF is the most potent lipid mediator yet discovered (reviewed by Snyder, 1990) and stimulates a wide range of biological responses ranging from aggregation of platelets and neutrophils to cytokine and chemokine production. Its effect is thus essentially pro-inflammatory, and many studies with PAF antagonists have demonstrated that PAF contributes largely to the haemodynamic changes that occur in endotoxaemia models and in multi-organ failure.

Arachidonic acid is the rate-limiting step in the production of thromboxane A2, prostaglandins and leukotrienes. Thromboxane A2 and prostaglandins are released via the cyclo-oxygenase pathway, and leukotrienes by the lipo-oxygenase pathway. These phospholipids have powerful effects on tissue and inflammatory cells. Thromboxane A2 and leukotrienes

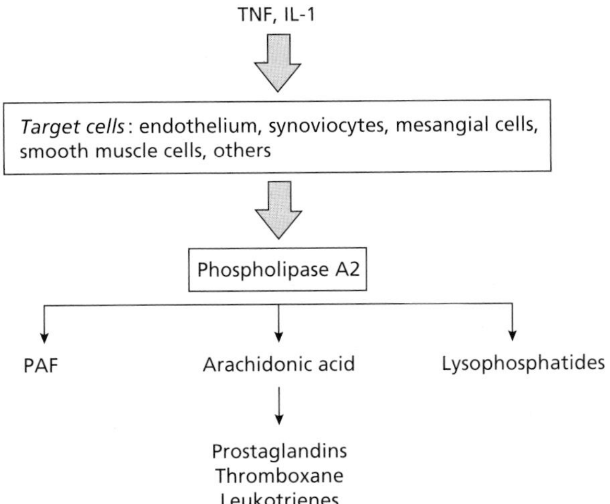

Fig. 7 The effect of tumour necrosis factor α and interleukin-1 on target cells.

are pro-inflammatory, causing platelet aggregation, neutrophil accumulation and increased vascular permeability. Prostaglandins, PGE2 and prostacyclin PGI2, are vasodilators and increase blood flow, and therefore exert beneficial effects.

Nitric Oxide and Oxygen Species The generation of nitric oxide (NO) is also a key factor in inflammation, a process that is under the control of three distinct isoforms of the enzyme nitric oxide synthase (Nathan, 1992; Nathan and Xie, 1994). Two of the enzymes, the endothelial cNOS and the neuronal cNOS, are constitutive. NO derived from these two isoforms acts as a physiological regulator, by relaxing vascular smooth muscle or by functioning as a neurotransmitter. In marked contrast, the third enzyme is inducible and not constitutive. The pro-inflammatory cytokines (TNFα, IL-1, IFNγ) and bacterial components induce transcription and translation of the inducible form of the nitric oxide synthase (iNOS), mainly by macrophages and hepatocytes. Expression of iNOS is optimal when macrophages stimulated by bacterial components have been first primed with IFNγ (Bogdan and Nathan, 1993).

The discovery of NO as a mediator of TNF-induced hypotension (Kilbourn *et al.*, 1990) was a major breakthrough in understanding the mechanisms that underlie hypotension in severe inflammation. Treatment with inhibitors of NO synthase inhibits or reverses the hypotension caused by LPS or TNF in animal models and in septic patients, but without improvement in survival. In contrast, in several experimental models, treatment with NO synthase inhibitors worsens the outcome (Evans *et al.*, 1994a).

NO is also essential for the killing of intracellular micro-organisms, and shares this function with oxygen radicals. Superoxide (O_2^-) is produced by activated phagocytes and a variety of epithelial and endothelial cells under conditions that induce NO production. Superoxide is toxic when produced in increased amounts by cells stimulated by pro-inflammatory molecules. It is even more toxic when combined with NO, forming a highly potent oxidising agent, peroxynitrite ($ONOO^-$), which is beneficial for controlling intracellular infections but may cause profound tissue damage when released in excess.

Chemokines and Adhesion Molecules A fundamental property of pro-inflammatory cytokines is their ability to induce a variety of genes that affect the vascular endothelium and other cells in the local tissue environment. TNFα and IL-1 may up-regulate or down-regulate other members of the cytokine family or up-regulate adhesion molecules on the endothelium. Leucocyte migration takes place at sites of local injection of bacterial components. This migration, which represents the hallmark of inflammation, is the result of two separate sets of events, (1) induction of chemokines that attract leucocytes to the sites of infection, and (2) up-regulation of the endothelial adhesion molecules, a necessary step which permits leucocyte margination and diapedesis.

A number of chemotactic proteins and their receptors orchestrate the directed migration of leucocytes to inflammatory sites. Some of these chemo-attractants, such as formyl-methionyl-leucyl-phenylalanine (fMLP), originate directly from the membrane of bacteria, while others, such as C5a, leukotrienes and PAF are derived from the host (reviewed by Ben-Baruch *et al.*, 1995). A super-family of closely related proteins which attract leucocytes to sites of injury has been identified, and designated 'chemokines' (for chemo-attractant cytokines) (reviewed by Ward and Westwick, 1998). This super-family, which comprises more than 40 members, is divided into four subfamilies according to the position of four cysteine residues in highly conserved positions (Table 7).

The two main subfamilies are the α and the β subfamilies. The α subfamily is designated CXC, because of the presence of an amino acid X between the pair of cysteine residues. Interleukin 8 is the prototype of this family, and is involved in neutrophil recruitment in a number of human inflammatory conditions. Murine macrophage inflammatory protein-2 (MIP-2) is a murine CXC chemokine that is considered functionally analogous to human IL-8.

The β subfamily is designated CC, because of the absence of an amino acid between the pair of cysteine residues. Principal members of the β-chemokine family include monocyte chemo-attractant protein-1 (MCP-1), MIP-1α and MIP-1β. While the α subfamily members act as attractants of polymorphonuclear leucocytes (PMN), monocytes and T cells, the β subfamily members primarily chemo-attract monocytes and T cells.

In addition to the α and β subfamilies, lymphotactin, which attracts lymphocytes, but not monocytes, defines the C (or γ) subfamily, so-named because it has only a single cysteine residue near the N-terminus. A recently identified chemokine (fractalkine) defines the CX_3C (or δ) subfamily with three intervening amino acids between the N-terminal cysteine residues.

The specific effects of chemokines mediated by seven transmembrane-spanning G protein-coupled receptors have been reviewed by Ward and Westwick, 1998. Although over 40 chemokines have been identified to date, fewer chemokine receptors have been identified, with overlapping characteristics (Table 7). This implies that each chemokine may recruit multiple types of cell, even if they express different types of receptors. Conversely, each cell may respond to multiple chemokines.

The expression of chemokines can be induced by a variety of cytokines, TNFα and IL-1 being the most potent. Chemokines may be also released by direct contact between monocytes and endothelium. All chemokines promote adherence of leucocytes to endothelium. They also elicit neutrophil and monocyte degranulation and respiratory bursts. Chemokines are generated by infection, ischaemia and trauma, and in all conditions in which TNFα and IL-1 are detected (Baggiolini *et al.*, 1995). Lung diseases with high neutrophil influx are of particular interest. Evidence

Table 7 Chemokine family members

Family	Structure	Main members	Chemokine receptors
CXC (α)	CXC...C...C	Interleukin 8 (IL-8)	CXCR1, CXCR2
		Growth-related oncogene (Groα/β)	CXCR2
		Neutrophil-activating protein (Nap-2)	CXCR2
		Interferon γ-inducible protein (IP-10)	
		Monocyte induced by interferon γ (Mig)	
		Macrophage inflammatory protein (MIP-2, mouse)	CXCR3
CC (β)	CC...C...C	Monocyte chemotactic protein (MCP-1,2,3,4)	CCR1, CCR2, CCR3
		Regulated on activation, T cell expressed and secreted (RANTES)	CCR1, CCR3, CCR5
		Eotaxin	CCR3, CCR10
		Macrophage inflammatory protein (MIP-1α, MIP-1β)	CCR1, CCR5, CCR8, CCR4, CCR10
		Macrophage-derived chemokine (MDC)	CCR4
		Thymus and activation-regulated chemokine (TARC)	CCR4
		Liver and activation-regulated chemokine (LARC)	Unknown
C (γ)	C...C	Lymphotactin	Unknown
CX_3C (δ)	CXXXC...C...C	Fractalkine	CX3CR1

for the role of chemokines arose from mice with a disrupted IL-8 receptor gene (Cacalano *et al.*, 1994), in which neutrophils exhibited markedly reduced migration.

Migration of leucocytes into tissues is regulated by specific combinations of adhesion receptors and chemo-attractants. Selectins initiate leucocyte attachment to vascular endothelium by mediating leucocyte rolling along inflamed endothelium, while integrins and molecules of the Ig superfamily allow firm adhesion and diapedesis. Pro-inflammatory cytokines are potent inducers of up-regulation of adhesion receptors on the surface of leucocytes and endothelial cells. The growth of knowledge about the interactions between leucocytes and endothelial cells in the last 10 years has been spectacular. Several comprehensive reviews have been published, in particular Kansas (1996), Konstantopoulos and McIntire (1996), Etzioni (1996) and Hynes (1992). The events starting from the contact phase between leucocytes and endothelium and ending in transmigration or diapedesis of leucocytes through the endothelium will be briefly summarised (Fig. 8, Table 8).

On activation of the endothelium, specific adhesion molecules allow binding of neutrophils, causing them to roll along the endothelium. Initial contact and

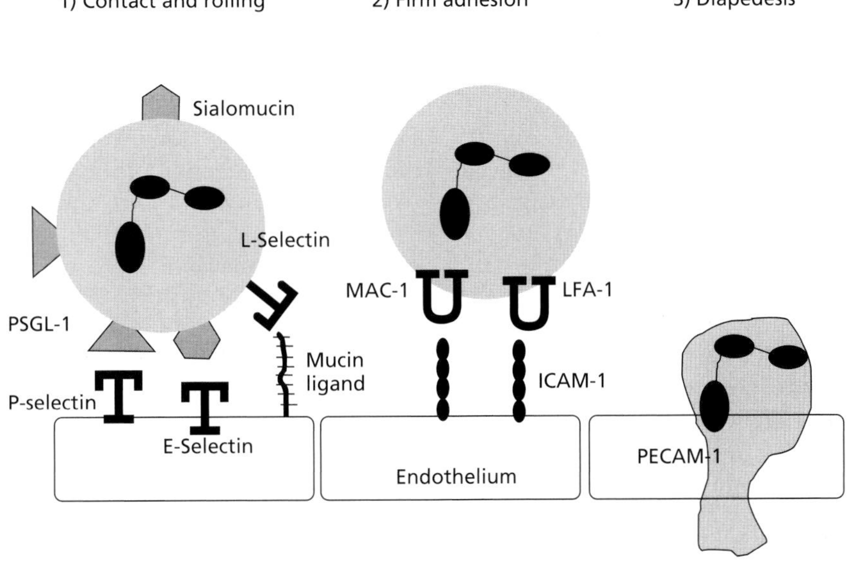

Fig. 8 Events in the interaction between leucocytes and endothelial cells.

Table 8 Adhesion molecules in leucocyte–endothelial interaction

	Localization	Ligand	Induced by:	Main function for leucocytes
Selectins				
E-selectin	E	Sialyl, fucosyl residues	TNFα, IL-1	Rolling
P-selectin	E	Sialyl, fucosyl residues	Histamine, thrombin	Rolling
L-selectin	L, N	Sialyl, fucosyl residues	Constitutional	Rolling
Integrins				
Mac-1 (CD11b/CD18)	N, M, NK	ICAM-1	TNFα	Adherence
LFA-1 (CD11a/CD18)	All leucocytes	ICAM-1,2,3	TNFα, IL-1, IL-8, C5a	Adherence
CD11c/CD18	N, M, NK	?	?	Adherence
VLA-4	L, NK	VCAM-1	Constitutional	Adherence
Ig superfamily				
ICAM-1	L, E, N, M	LFA-1, Mac-1	TNFα, IL-1	Firm adhesion
ICAM-2	E	LFA-1	Constitutional	Firm adhesion
VCAM-1	E	VLA-4	TNFα, IL-1	Adhesion
PECAM-1	E, N	–	Constitutional	Diapedesis

L, lymphocytes; E, endothelial cells; M, macrophages; N, neutrophils; NK, natural killer cells.

Table 9 Cytokines and soluble cytokine receptors with anti-inflammatory properties

Cytokines	Biological activity
IL-10	Suppression of lymphocyte and monocyte/macrophage functions
IL-4	Reduction of LPS-induced cytokine synthesis by monocytes and macrophages
IL-13	Down-regulation of monocyte and macrophage functions
IL-6	Inhibition of TNFα and IL-1 production
TGFβ	Deactivation of monocytes and macrophages (decreased expression of MHC class II, inhibition of TNFα and IL-1 synthesis)
IL-1ra	Inhibition of IL-1-mediated cell activation by blocking IL-1 binding to its receptor
Soluble receptors	
TNFR p55, p75	Binding of TNF and prevention of ligand–receptor interaction
IL-6R	Prolongation of the half-life of IL-6

rolling is mediated predominantly by selectins, which are carbohydrate-recognising glycoproteins. E- and P-selectins are expressed on activated endothelial cells, and L-selectin is constitutively expressed on leucocytes. All selectins bind to sialylated and fucosylated oligosaccharides, such as sialyl Lewis X. Platelet–endothelial adhesion molecule-1 (PGSL-1), a ligand of protein origin, has been identified only for P-selectin. Expression of E-selectin and P-selectin is inducible, while that of L-selectin is constitutive.

Integrins, molecules expressed at the surface of leucocytes, are activated by bacterial components and cytokines such as TNFα or IL-1. After recognition by selectins and their ligands, the up-regulation of integrins on leucocytes allows firm adhesion of leucocytes to endothelium through endothelial ligands of the Ig super-family, especially ICAM-1 and ICAM-2. The final step is diapedesis of leucocytes through the endothelium, a process that depends on PECAM-1, another member of the Ig super-family.

In summary, the co-ordinated action of chemokines and adhesion molecules allows margination, aggregation and activation of neutrophils in the inflamed tissue. In generalised infections, this process extends to capillaries in various organs, principally in the lungs, kidneys and liver.

Anti-inflammatory Cytokine Response

The mechanisms that limit the effects of the pro-inflammatory response include, (1) anti-inflammatory cytokines, (2) soluble receptors and receptor antagonists and (3) low-molecular-weight products. This chapter will review current knowledge about cytokines with anti-inflammatory properties that are likely to be involved in the pathophysiology of inflammation and infection.

Cytokines are often classified into two main categories depending on whether they exert pro-inflammatory or anti-inflammatory activities. Yet, the pro-inflammatory versus anti-inflammatory dichotomy is an oversimplification, since several cytokines may exert both pro-inflammatory and anti-inflammatory effects, depending on their target cells. Cytokines and soluble cytokine receptors with anti-inflammatory properties are listed in Table 9. All these molecules inhibit the production of the pro-inflammatory cytokines TNFα and IL-1 and, as such, are considered to be anti-inflammatory mediators. Anti-inflammatory cytokines not only inhibit pro-inflammatory cytokines, but also contribute to the inhibition of the production of toxic oxygen and reactive nitrogen intermediates by macrophages. It is important to emphasise that these cytokines may also produce immunostimulatory effects, since they may enhance B-cell responses and activate T cells.

Mosmann *et al.* (1986) reported that clones of murine helper T (TH) cells can be divided into two subsets based on the type of cytokines that they produce. TH1 clones produce IFNγ, IL-2 and other cytokines, but not IL-4, IL-5 and IL-10, whereas TH2 clones produce IL-4, IL-5, IL-10 and other cytokines, but not IFNγ and IL-2. TH1 cells play a central role in cell-mediated immunity (delayed-type hypersensitivity reactions and activation of macrophages), whereas TH2 cells are critical for the development of humoral immunity (antibody production and IgE-mediated immune responses). Each subset also exerts inhibitory functions on its counterpart. IFNγ released by TH1 cells and IL-12 produced by phagocytic cells, inhibit the growth of antigen-stimulated TH2 cells and the production of IL-4 by these cells, while IL-4 and IL-10 released by TH2 cells inhibit cellular responses and cytokine production by TH1 cells. Since these landmark observations, it has become apparent that TH1 and TH2 clones represent two extremes and that most T-cell clones express mixed cytokine patterns.

Interleukin 10

Interleukin 10 (IL-10) was originally identified as a product of murine T helper 2

(TH2) clones that inhibit the production of TH1 cytokines, including IL-2 and IFNγ (reviewed by Opal *et al.*, 1998). IL-10 is also produced by monocytes/macrophages and B cells. It exhibits profound immunosuppressive properties, but also some immunostimulatory effects on B cells, because it up-regulates MHC class II expression and augments the proliferation and differentiation of activated B cells. IL-10 is a potent anti-inflammatory molecule that suppresses cell-mediated immune responses and inhibits macrophage-dependent T-cell proliferation and cytokine production by activated macrophages, neutrophils, TH1 and NK cells. It inhibits most of the cytokines involved in inflammatory and immune responses, including TNFα, IL-1, IL-6, IL-8, IL-12, GM-CSF, G-CSF, MIP-1, MIP-1α, IL-2, IL-3 and IFNγ, and is therefore the prototype of an anti-inflammatory agent (Wang *et al.*, 1994). In addition, IL-10 down-regulates surface expression of TNF receptors and up-regulates shedding of p75 and p55 soluble TNF receptors.

Interleukin 4 Interleukin 4 (IL-4) is an 18-kDa cytokine produced by activated T cells, mast cells and basophils. Like most cytokines, it acts on a variety of cells expressing high-affinity IL-4 type I receptor (IL-4Rα). This is a heterodimer of a 140 kDa IL-4-specific ligand-binding chain (α chain) and a signalling chain (γ chain), also called the common γ chain (γ_c), that is shared with several other cytokines such as IL-2 and IL-7 (Callard *et al.*, 1996). A number of non-T cells express an alternative form of IL-4R, the type II receptor, consisting of IL-4Rα and IL-13Rα and can therefore also transduce the IL-4 signal. IL-4 was originally discovered as a B-cell growth factor. Subsequent studies have shown that it is essential for immunoglobulin class switch, augmenting the production and release of IgG1 and IgE. It also induces the expression of cell surface molecules on B cells, such as the class II MHC molecules. IL-4 is critical for the development of the TH2 subtype of the CD4+ T helper cell and can also influence the development of CD8+ effector cells. Moreover, IL-4 plays an important role in haematopoiesis by stimulating precursor cells, and in allergic responses it exerts stimulatory effects on cells that contribute to inflammation, including macrophages, endothelial cells, stromal cells and fibroblasts (reviewed by Brown and Hural, 1997). Therefore, IL-4 is in many respects an essential component of late-phase inflammatory reactions.

Like IL-10, IL-4 has anti-inflammatory properties. It suppresses the production of several pro-inflammatory molecules, including the monocyte/macrophage cytokines TNFα and IL-1 (Essner *et al.*, 1989). In contrast to IL-10, IL-4 suppresses the release of p75 TNF receptors.

Interleukin 13 Interleukin 13 (IL-13) is a recently described IL-4-like cytokine, produced mainly by activated T cells, that acts on B cells and on monocytes, but not on T cells (Zurawski and de Vries, 1994). Human IL-13 is primarily expressed by TH2 cells, but is also released by TH1 and TH0 clones. It shares many biological activities with IL-4, but the IL-13-mediated effects are more restricted and IL-13 is generally less potent than IL-4. This is partly due to the fact that IL-4 and IL-13 interact and compete for the high-affinity IL-4 type I receptors (IL-4Rα).

The structure of the IL-13 receptor is complex. IL-13, but not IL-4, binds to the IL-13Rα1 and IL13Rα2 chains. Different combinations of the α-chains (IL-13Rα1, IL-13Rα2), the IL-4Rα and the IL2Rγ_c chains have been detected on various cells (Murata *et al.*, 1997). However, the exact composition of the receptor complexes engaging each individual cytokine has not been completely elucidated. IL-13 exerts growth-promoting effects on B cells and induces immunoglobulin class switching (IgG4 and IgE). Like IL-4, IL-13 enhances B-cell expression of MHC class II. It directly stimulates IFNγ production by NK cells, but also exerts synergistic effects with IL-4. In contrast to IL-4, human IL-13 does not activate human T cells and does not inhibit T-cell proliferation induced by IL-4.

IL-13 is a very potent modulator of monocyte/macrophage functions, affecting cell morphology, expression of surface molecules, and production of cytokines. Interestingly, IL-13 exhibits dual effects on monocytes, both stimulatory and inhibitory. Like IL-10 and IL-4, IL-13 inhibits the production of pro-inflammatory cytokines and chemokines by LPS-stimulated monocytes and macrophages. Of note, IL-4 and IL-13 inhibitory activities occur even in the presence of anti-IL-10 antibodies, indicating that their effects are not mediated by IL-10 (Shen and Stavnezer, 1998). IL-13 also potentiates the production of IL-6 after LPS exposure of mice, while at the same time inhibiting that of TNFα (Di Santo *et al.*, 1997).

Transforming Growth Factor β The transforming growth factor β (TGFβ) family of cytokines is composed of five isoforms, of which three, TGFβ1, TGFβ2 and TGFβ3, are expressed in mammals. Macrophages, T and B lymphocytes, endothelial cells, platelets, fibroblasts, astrocytes, osteoblasts and osteoclasts are important cellular sources of TGFβ. In contrast to most cytokines, TGFβ is expressed constitutively by many cells and tissues and is up-regulated on stimulation. It regulates the growth and differentiation of several cell types (TGFβ2 and TGFβ3) and plays an important role in inflammation, immune responses and tissue repair (TGFβ1) (Letterio and Roberts, 1997).

TGFβ has pro-inflammatory and immunosuppressive activities, which promote inflammatory and immune responses at the tissue level, while acting as a systemic suppressor of the immune system. It is released in response to tissue injury, and inflammation stimulates chemotaxis of leucocytes (neutrophils, monocytes, lymphocytes), mast cells and fibroblasts and induces the production of pro-inflammatory mediators (TNFβ, IL-1 and IL-6). TGFβ was one of the first cytokines reported to deactivate monocytes and macrophages, and evidence supporting this was clearly established by targeted disruption of the TGFβ1 gene. Mice deficient in TGFβ1 develop multiple foci of inflammation, characterised by extensive leucocyte infiltration, and over-express molecules such as MHC class I and II and the cytokines TNFβ, MIP-1α and IFNγ, causing the death of the animals (Shull *et al.*, 1992). TGFβ suppresses the proliferation and differentiation of T cells and B cells and inhibits the synthesis of cytokines (IL-2, IFNγ, IL-3, GM-CSF, TNFα) induced by mitogens. Like IL-10, IL-4 and IL-13, TGFβ up-regulates production of interleukin-1 receptor antagonist (IL-1ra) by LPS-stimulated monocytes and down-regulates the expression of the IL-1 receptor type I (Vannier *et al.*, 1992; Cassatella *et al.*, 1994).

Interleukin 1 Receptor Antagonist Interleukin 1 receptor antagonist (IL-1ra) is a further member of the IL-1 gene family. IL-1α and IL-1β are agonists, whereas IL-1ra is a naturally occurring, specific receptor antagonist. IL-1ra exists in soluble IL-1ra (sIL-1ra) and intracellular IL-1ra (icIL-1raI) forms, and icL-1raII, that is either soluble or cytoplasmic, constitutive or inducible (reviewed by Dinarello, 1998). Like IL-1β, IL-1ra is induced by viruses, bacteria, yeasts, soluble microbial components and the pro-inflammatory cytokines IL-1 and TNFα (Vannier *et al.*, 1992; Cassatella *et al.*, 1994). IL-1ra competes with IL-1 for binding to IL-1 receptors, resulting in a strong inhibition of IL-1-mediated cell stimulation. The formation of a heterodimer of IL-1 receptor 1 and IL-1 accessory protein (IL-1R-AcP) is believed to play a key role in cell activation by IL-1. It is postulated that IL-1ra binds to the IL-1 type I receptor, but cannot bind to IL-1R-AcP, thus providing the basis for the inhibitory effect of IL-1ra. IL-1ra does not trigger a signal and is devoid of any agonist activity, even when injected intravenously into humans at concentrations 10^6-fold greater than those of IL-1α or IL-1β. It is beyond the scope of this chapter to review all the existing data regarding the biological activities of IL-1ra in healthy volunteers and in models of inflammation, infection, autoimmune and malignant diseases. The reader is referred to an excellent review by Dinarello (1998).

Interleukin 6 Interleukin 6 (IL-6) is a secreted single glycoprotein of about 26 kDa (reviewed by Barton, 1997). Its receptor is composed of two peptide chains, the ligand-binding chain (gp80), and the signal-transducing peptide (gp130). The IL-6 receptor is found on many cells, such as lymphocytes, monocytes, cardiac cells, endothelial cells and hepatocytes. It plays a major role in the differentiation of cells of the myeloid lineage, and is necessary, together with IL-1, to induce T-cell proliferation and to differentiate B cells and promote immunoglobulin secretion. IL-6 is the predominant cytokine in the induction of acute-phase proteins from hepatocytes.

IL-6 has long been viewed as a pro-inflammatory cytokine. This assumption is based primarily on the observation that it is secreted in copious amounts together with TNFα and IL-1 after LPS challenge or during acute infections and that these three cytokines are involved in the initiation of the acute phase response. Unlike TNFα and IL-1, however, IL-6 does not up-regulate major pro-inflammatory mediators, such as NO or prostaglandins, nor does it induce the synthesis of adhesion molecules, such as ICAM-1. While low doses of TNFα and IL-1 are toxic, fairly high doses of IL-6 can be tolerated without inducing shock. Thus, IL-6 should be regarded as an anti-inflammatory rather than a pro-inflammatory molecule. Indeed, it inhibits TNFα and IL-1 synthesis *in vitro* and *in vivo*.

The acute-phase response is mediated mainly by IL-6, while TNFα and IL-1 contribute less. Although the pro-inflammatory and the anti-inflammatory nature of these acute-phase proteins has not been clearly defined, IL-6 may be responsible for the synthesis of proteins with anti-inflammatory potential, such as inhibitors of proteases.

Soluble Receptors TNFα is an important component of the cytokine network that is activated during sepsis. The biological activity of TNFα is exerted through two TNF receptors present on target cells, TNF receptor I (TNFRI) (or TNFR p55) and TNFRII (or TNFR p75). These receptors are shed from the membrane on cell activation and circulate as soluble molecules. Soluble TNFR p55 and TNFR p75 are naturally occurring inhibitors of TNF activity that serve two functions (Spinas *et al.*, 1992; VanZee *et al.*, 1992). First, shedding of the TNF receptors in response to inflammatory stimuli results in a transient desensitisation of the cell, thereby providing a mechanism to inhibit TNF activity. Second, soluble TNF receptors also bind TNF and may thus inhibit TNFα bioactivity by preventing its binding to the membrane-bound receptor. Conversely, these soluble receptors may exert a detrimental effect, especially at low concentrations, by prolonging TNF activity, allowing the complex to dissociate later in the course of the disease (Aderka *et al.*, 1992). This occurs with the soluble p75 TNF receptor (Mohler *et al.*, 1993; Evans *et al.*, 1994b; Moller *et al.*, 1996). Current understanding of the *in vivo* interactions between TNF and the soluble TNF receptors is, however, far from complete.

In addition to inhibiting the effects of their ligands, soluble receptors can also facilitate ligand-mediated signalling. This mechanism is common among the IL-6 family of soluble receptors, for which both agonistic and antagonistic effects have been observed (Heaney and Golde, 1996). Similarly to TNFα, complexes of IL-6 and soluble IL-6 receptor markedly prolong the plasma half-life of IL-6 (Peters *et al.*, 1996).

Low-molecular-weight Products Cyclo-oxygenase inhibitors such as ibuprofen reduce inflammation and sequelae such as fever and myalgia caused by bacterial components. In addition, these drugs also reduce neutrophil adherence and responses to chemotactic factors in animals challenged with TNFα or IL-1 (Okusawa *et al.*, 1988b).

Whereas prostaglandins and leukotrienes in general increase the pro-inflammatory response, their effects on the production of pro-inflammatory cytokines is not clear. In humans injected with LPS and pre-treated with cyclo-oxygenase inhibitors, the circulating levels of TNFα are higher than in controls not receiving the drug. This is consistent with the observation that prostaglandins suppress TNFα (Spinas *et al.*, 1991). The addition of cyclo-oxygenase products does not affect, or marginally reduces, the IL-1 response (Dinarello, 1998). This may indicate that prostaglandins play a dual role, since many changes induced by IL-1 and TNFα are mediated by prostaglandins. They contribute to the pro-inflammatory response, but at the same time they exert a negative feed-back influence on the production of a pro-inflammatory response, particularly that produced by TNFα.

Corticosteroids are very potent inhibitors of the pro-inflammatory response. They inhibit the transcription of nearly all cytokines, but do not suppress IL-1ra production. Since pro-inflammatory cytokines induce endogenous corticosteroid production, it is likely that this represents an intrinsic anti-inflammatory negative feed-back mechanism. This may be modulated by MIF, which acts as a counter-regulator to the inhibitory effects of corticosteroids (Calandra, 1995).

Animal Models of Inflammation Caused by Bacterial Components

The concept that severe inflammation and multiple organ failure are related to an uncontrolled systemic pro-inflammatory state is derived mainly from models of LPS or cytokine infusions in animals. This is based on the discovery that injection of LPS or recombinant pro-inflammatory cytokines can experimentally reproduce a syndrome indistinguishable from the response to severe infections. Although LPS undoubtedly plays an important role in the pathophysiology of Gram-negative infections, its administration does not completely mimic the complex host–pathogen interactions during sepsis and septic shock.

Blockade of Pro-Inflammatory Cytokines TNFα, IL-1 and MIF

Most attempts to block the pro-inflammatory effects of TNFα, IL-1 and MIF have been performed in mice treated with anti-TNFα antibodies, TNFR constructs, IL-1 receptor antagonist, and anti-MIF antibodies (Table 10). Confirmatory experiments have sometimes been performed in higher species, including baboons, to allow pre-clinical evaluation of drugs. Mice deficient in the TNFα gene, the TNF receptor gene or the MIF gene have also been studied.

Endotoxaemia and Systemic Injection of Bacteria

TNFα appears to be the most toxic of the pro-inflammatory cytokines, so that the pioneering studies of Beutler *et al.* (1985) have opened a new line of research, by demonstrating that anti-cytokine therapies may be a useful approach to the treatment of endotoxaemia and Gram-negative infections. Since these studies, many attempts have been made with anti-TNFα antibodies to confirm the role played by TNFα in models based on the parenteral injection of LPS or of Gram-negative bacteria. In a few studies, Gram-positive bacteria were also evaluated. In most cases animals were protected against an intraperitoneal

Table 10 Role of TNF, IL-1 and MIF in animal models of endotoxaemia and sepsis

Model	Species	Blocking agent	Effect
Endotoxaemia			
LPS (low dose, galactosamine)	Mouse, rhesus	Anti-TNF	Beneficial
LPS (low dose, galactosamine)	Mouse	TNF-deficient	Beneficial
LPS (low dose, galactosamine)	Mouse	TNFR-IgG1/3, p55	Beneficial
LPS (low dose, galactosamine)	Mouse	TNFR p55-deficient	Beneficial
LPS (high dose)	Mouse, rabbit, rat, baboon	Anti-TNF	Beneficial
LPS (high dose)	Rat	TNFR-IgG1, p55	Beneficial
LPS (high dose)	Mouse	TNFR p55-deficient	No effect
LPS (very high dose)	Mouse	Anti-TNF	No effect
LPS (high dose)	Rabbit, mouse, baboon	IL-1ra	Beneficial
LPS (high dose)	Mouse, rat	Anti-MIF	Beneficial
LPS (high dose)	Mouse	MIF-deficient	Beneficial
Gram-positive cell walls, enterotoxins			
Peptidoglycan (*Strep. pneumoniae*)	Mouse	Anti-TNF	Beneficial
SEB (*Staph. aureus*)	Mouse	Anti-TNF	Beneficial
TSST-1 (*Staph. aureus*)	Mouse	Anti-TNF	Beneficial
SEB (*Staph. aureus*)	Mouse	MIF-deficient	Beneficial
TSST-1 (*Staph. aureus*)	Mouse	Anti-MIF	Beneficial
Systemic injections of bacteria			
E. coli	Baboon, mouse, pig	Anti-TNF	Beneficial
Staph. aureus	Baboon	Anti-TNF	Beneficial
E. coli	Mouse, baboon	TNFR-IgG1/3, p55	Beneficial
Strep. pyogenes	Mouse	Anti-TNF	No effect
K. pneumoniae	Mouse	Anti-TNF	No effect
Staph. aureus	Mouse	TNF-deficient	Harmful
E. coli	Rabbit, baboon	IL-1ra	Beneficial
Staph. epidermidis	Rabbit	IL-1ra	Beneficial
Local injections of bacteria or focal infections			
CLP (caecal ligation puncture)	Mouse	Anti-TNF	Harmful
Ps. aeruginosa (oral)	Rat	Anti-TNF	Beneficial
E. coli (peritonitis)	Rat, mouse	Anti-TNF	No effect
Strep. pneumoniae (pneumonia)	Mouse	Anti-TNF	Harmful
Ps. aeruginosa (pneumonia)	Mouse	Anti-TNF	Harmful
N. meningitidis (peritonitis)	Rat	Anti-TNF	Beneficial
Group B streptococcus (peritonitis)	Rat	Anti-TNF	No effect
Strep. pneumoniae (peritonitis)	Mouse	TNFR-IgG3, p55	No effect
K. pneumoniae (pneumonia)	Mouse	TNR-IgG1, p55	Harmful
Strep. pneumoniae (peritonitis)	Mouse	TNFR p55-deficient	Harmful
E. coli (peritonitis, high inoculum)	Mouse	TNFR p55-deficient	No effect
E. coli (peritonitis, low inoculum)	Mouse	TNFR p55-deficient	Harmful
K. pneumoniae sc	Rat	IL-1ra	Harmful
CLP	Mouse	Anti-MIF	Beneficial
E. coli (peritonitis)	Mouse	Anti-MIF	Beneficial
Intracellular and fungal infections			
L. monocytogenes, C. albicans	Mouse	Anti-TNF	Harmful
L. monocytogenes	Mouse	TNFR p55-deficient	Harmful
L. monocytogenes, C. albicans, C. neoformans	Mouse	TNF-deficient	Harmful
C. trachomatis, S. typhimurium	Mouse	Anti-TNF	Harmful

Studies presented in the tables are not referenced, but selected studies are quoted in the text to illustrate the main observations.

or intravenous challenge with LPS, provided that the anti-TNFα antibodies were given before or during the infusion of LPS. Protection was observed in models of low-dose LPS challenge (galactosamine model) (Zanetti *et al.*, 1992), and in models involving injection of high doses of LPS (usually 2–20 mg/kg) (e.g. Mathison *et al.*, 1988; Jin *et al.*, 1994). At very high doses of LPS (> 50 mg/kg), however, protection with anti-TNFα antibodies did not occur (Eskandari *et al.*, 1992). Protection was lost if antibody administration was delayed until after the LPS challenge.

Administration of anti-TNFα antibody after intravenous challenge with *Escherichia coli*, was protective in various animal species (e.g. Silva *et al.*, 1990). Treatment with anti-TNFα antibodies did not protect mice against the lethal effect of challenge with Gram-negative species other than *E. coli*, such as *Klebsiella* (Silva *et al.*, 1990). Hinshaw *et al.* (1992) observed protection after intravenous challenge with *Staph. aureus*, but Martin *et al.* (1993) did not find any protection against *Strep. pyogenes*. Overall, administration of anti-TNFα antibody after parenteral challenge with live bacteria is uniformly protective. Early treatment generally gives better protection, while delay of treatment is less effective than treatment given before bacterial challenge.

Another approach to TNFα blockade is based on the use of soluble TNF receptors. Fusion proteins have been developed in which each of the two TNFα receptors (p55, p75) is linked with hinge regions of the human IgG. This dimeric construct extends the half-life of the molecule *in vivo* and increases the affinity for TNFα. Four variants have been constructed with IgG1 or IgG3, sTNFR-IgG1 p75, sTNFR-IgG3 p75, sTNFR-IgG1 p55, and sTNFR-IgG3 p55, and investigated in animals. All but one of the published experiments were conducted in mice, mostly with parenteral LPS injections. Protection was very effective when the sTNFR p55 construct was given prophylactically (e.g. Lesslauer *et al.*, 1991). Similarly, the sTNFR p55 construct appeared to be effective in a model of intravenous bacterial challenge (Evans *et al.*, 1994b). No effect was generally observed with the sTNFR p75 construct (Peschon *et al.*, 1998). Taken together, these experiments show a promising effect in limiting TNFα toxicity for the sTNFR p55 constructs, which was greater than TNFα antibody. In some models, successful treatment with sTNFR p55 could be delayed for several hours, but this was not true for all models. In contrast, the potential use of the sTNFR p75 constructs appears to be limited.

An alternative to TNFα blockade involves blocking the actions of IL-1 by means of the IL-1 receptor antagonist (IL-1ra), and this was very promising as an experimental tool in animal models. IL-1ra improves various haemodynamic parameters and survival in models of parenteral LPS or bacterial challenges (e.g. Fischer *et al.*, 1992).

MIF, which strongly promotes inflammatory responses, has emerged as a pivotal cytokine of the innate immune system. Immuno-neutralisation of MIF in mice confers protection against lethal endotoxaemia (Bernhagen *et al.*, 1993) and reduces acute lung injury in rats (Makita *et al.*, 1998).

Experiments in mice deficient in the genes for TNFα, the TNF receptor p55 or MIF generally confirm the harmful role played by excess TNFα production in models of parenteral injections with LPS or bacteria.

Gram-positive Cell Walls and Exotoxins

While LPS constitutes the main pathogenic factor of Gram-negative bacteria, cell wall components and secreted exotoxins initiate the inflammatory cascade induced by Gram-positive organisms. Heat-killed Gram-positive bacteria and purified cell walls of *Strep. mitis* cause death by a TNF-dependent pathway in galactosamine-sensitised mice (Le Roy *et al.*, 1996). A subcomponent of *Staph. aureus* peptidoglycan acts in synergy with lipoteichoic acid as a cause of shock in mice (Kengatharan *et al.*, 1998). Staphylococcal enterotoxins (SEB, TSST-1) also cause death in galactosamine-sensitised mice by TNF- and MIF-dependent pathways (Miethke *et al.*, 1993; Calandra *et al.*, 1998).

Local Injection of Bacteria or Focal Infections

While blockade of pro-inflammatory TNFα or IL-1 increase survival after parenteral injections of LPS or Gram-negative bacteria, these treatments have deleterious effects in models of local infections caused by these organisms. These adverse effects are observed in various murine and rat models of peritonitis and pneumonia, including the caecal ligature/puncture peritonitis model (CLP), when TNFα is blocked with anti-TNF antibodies or with TNFR p55 constructs (e.g. Echtenacher *et al.*, 1990; Laichalk *et al.*, 1996; van der Poll *et al.*, 1997). A beneficial effect of TNFα blockade has been observed in a model of *Neisseria meningitis* peritonitis (Nassif *et al.*, 1992), and in a model of oral *Pseudomonas aeruginosa* administration (Opal *et al.*, 1991). While a single injection of IL-1ra reduces the lethality of *Klebsiella pneumoniae* sepsis in new-born rabbits, a second injection 24 hours after challenge enhanced the lethal effect (Mancilla *et al.*, 1993).

In many of these models mice are inoculated with large numbers of bacteria. When mice are infected with low inocula of bacteria (LD_{10-20}), generation of a pro-inflammatory response is a prerequisite for the activation of professional phagocytes to phagocytose and/or to kill bacteria. Animals with a disrupted TNFα gene or a p55 mutant TNF receptor succumb to very low bacterial doses (Heumann *et al.*, 1996). It is of note that mice with bacterial peritonitis and protected with anti-MIF do not have an impairment of their natural defences against challenge with a small number of *E. coli* (LD_{10}) (Calandra *et al.*, 2000).

Intracellular Infections

The essential role of TNFα in the control of intracellular infections has been well demonstrated, since mice deficient in TNFα or TNF receptor p55, or in which TNFα is blocked, show increased lethality or toxicity.

These results were obtained by preventing normal macrophage activation, which would lead to bacterial killing and/or phagocytosis, in mice infected with a broad range of micro-organisms, including intracellular bacteria or fungi (*Listeria monocytogenes*, *Salmonella typhimurium*, *Chlamydia trachomatis*, *Candida albicans*, *Cryptococcus neoformans*) (e.g. Rothe *et al.*, 1993; Netea *et al.*, 1999).

Blockade of Pro-inflammatory Cytokines IL-18, IL-12 and IFNγ

All the reported experiments to block the pro-inflammatory cytokines IFNγ, IL-12, and IL-18 were carried out in mice treated with the relevant anti-cytokine antibodies (**Table 11**). Similarly, mice deficient in the corresponding cytokine genes were evaluated.

Endotoxaemia and Systemic Injection of Bacteria

It has frequently been reported that IFNγ exerts strong synergy with TNFα in LPS-induced responses *in vivo* and *in vitro*. The mechanism that underlies this synergy is ill-defined, but probably involves the control of the TNFα gene by molecules induced by IFNγ, because addition of recombinant IFNγ increases serum TNFα levels and the mortality of mice challenged with

Table 11 Role of IFNγ, IL-12 and IL-18 in animal models

Model	Species	Blocking agent	Effect
Endotoxaemia			
LPS (low dose, galactosamine)	Mouse	Anti-IFNγ	No effect
LPS (low dose, galactosamine)	Mouse	IFNγR-deficient	Beneficial
LPS (high dose)	Mouse	Anti-IFNγ	Beneficial
LPS (high dose)	Mouse	IFNγR-deficient	Beneficial
LPS (high dose)	Mouse	Anti-IL-12	Beneficial
LPS (low dose, BCG-primed mice)	Mouse	Anti-IL-12	Beneficial
LPS (high dose)	Mouse	Anti-IL-18	Beneficial
LPS (low dose, *P. acnes*-primed mice)	Mouse	IL-18-deficient	Beneficial
LPS (low dose, *P. acnes*-primed mice)	Mouse	Anti-IL-18	Beneficial
Systemic injections of bacteria			
E. coli	Mouse	Anti-IFNγ	Beneficial
Group B streptococcus	Mouse	Anti-IL-12	Harmful
S. aureus	Mouse	IL-18-deficient	Beneficial
Local injections of bacteria or focal infections			
E. coli (peritonitis)	Mouse	Anti-IFNγ	Beneficial
E. coli (peritonitis)	Mouse	Anti-IL-12	Harmful
CLP	Mouse	Anti-IL-12	Harmful
K. pneumoniae (pneumonia)	Mouse	Anti-IL-12	Harmful
Intracellular and fungal infections			
M. tuberculosis, M. bovis, L. monocytogenes	Mouse	IFNγ-deficient	Harmful
L. monocytogenes, Y. enterocolitica, S. typhimurium	Mouse	Anti-IFNγ	Harmful
C. neoformans, M. tuberculosis	Mouse	IL-12-deficient	Harmful
Y. enterocolitica	Mouse	Anti-IL-12	Harmful
M. tuberculosis, M. bovis	Mouse	IL-18-deficient	Harmful
Y. enterocolitica, S. typhimurium	Mouse	Anti-IL-18	Harmful

LPS (Heinzel, 1990). On the other hand, anti-IFNγ antibodies decrease the production of TNFα and thus protect mice from lethality induced by large doses of LPS (Doherty *et al.*, 1992). The suppression of TNFα is not complete, as is shown by models of endotoxaemia in galactosamine-sensitised mice, which are killed by very low doses of TNFα and are resistant to treatment with anti-IFNγ antibodies (Matthys *et al.*, 1995). However, mice deficient in the IFNγR gene are resistant to LPS (Car *et al.*, 1994).

IL-18 and IL-12 induce IFNγ production by T cells and NK cells. Consequently, anti-IL-12 or anti-IL-18 antibodies protect mice from lethal endotoxaemia by reducing IFNγ and TNFα levels. Similar conclusions were derived from studies of mice deficient in the IL-18 gene or in the IFNγ receptor gene (e.g. Sakao *et al.*, 1999).

The roles of IL-18, IL-12 and IFNγ have been little studied in models involving parenteral injection of bacteria. Blockade of IFNγ is beneficial after the systemic injection of *E. coli* (Silva and Cohen, 1992). Mice with a deletion of the IL-18 gene are protected from a lethal intravenous challenge with *Staph. aureus* (Wei *et al.*, 1999), while blockade of IL-12 is associated with an adverse response after intravenous injection of streptococci (Mancuso *et al.*, 1997).

Local Injection of Bacteria and Focal Infections

Though Kohler *et al.* (1993) suggested that IFNγ may contribute to toxicity in peritonitis induced by *E. coli*, other reports suggest that this family of molecules is necessary to control focal infections after blocking IL-12 in mice with antibodies or by using mice deficient in the IFNγ receptor (e.g. Zisman *et al.*, 1997a; Zantl *et al.*, 1998).

Intracellular Infections

The essential role of IFNγ, IL-12 and IL-18, as well as of TNFα in the control of intracellular infections has repeatedly been confirmed. Studies aimed at blocking this family of molecules, treatment of mice with neutralising antibodies and use of mice deficient in IFNγ, IL-12 or IL-18 have all shown a detrimental effect in infections with intracellular bacteria or fungi (e.g. Nakane *et al.*, 1989; Bohn and Autenrieth, 1996; Sugawara *et al.*, 1999).

Anti-inflammatory Cytokines

All experiments but one have been performed in mice, which received either recombinant IL-10, IL-13, IL-6 or IL-4, or in which these cytokines were blocked with antibodies (**Table 12**). Mice deficient for these genes have also been studied.

Endotoxaemia and Systemic Injection of Bacteria

Anti-inflammatory cytokines, such as IL-10, IL-13 or IL-4, all protect mice from LPS, either in the galactosamine model or in models with higher doses of LPS (e.g. Gérard *et al.*, 1993; Baumhofer *et al.*, 1998). This reflects the cytokine-induced suppression of LPS-evoked TNFα production, since this is a critical determinant of death in these models.

Treatment with neutralising anti-IL-6 antibodies has yielded variable and conflicting results. Anti-IL-6 neutralising antibodies have a paradoxical effect *in vivo*, supporting rather than inhibiting IL-6 (May *et al.*, 1993). Understanding of the role of IL-6 therefore depends on experiments in mice deficient in the IL-6 gene or in mice treated with recombinant IL-6. IL-6 is less effective than IL-10 or IL-13 in reducing TNFα levels in models of LPS injection. However, IL-6 partially reduces TNFα production and thus LPS-induced toxicity in mice (Xing *et al.*, 1998).

The role of this family of molecules in the context of systemic injections of bacteria has not been investigated.

Exotoxins

A protective anti-inflammatory effect of IL-6 and IL-10 has been observed in models of staphylococcal enterotoxin B injections (e.g. Barton *et al.*, 1996; Hasko *et al.*, 1998).

Local Injection of Bacteria and Focal Infections

While anti-inflammatory cytokines markedly improve survival when tested in endotoxin models, their contribution in models of focal infections is less clear. Very different results, ranging from protection to worsening of the infection, have been observed, depending on the cytokine evaluated, the bacteria used for challenge and the type of infection. For example, treatment with IL-10 is helpful in *Pseudomonas aeruginosa* pneumonia (Sawa *et al.*, 1997), but it is detrimental in *Strep. pneumoniae* pneumonia (van der Poll *et al.*, 1996). In peritonitis, treatment with IL-10 is without effect (Remick *et al.*, 1998), while antibody-mediated blockade of IL-10 increases mortality (van der Poll *et al.*, 1995). Similarly, the results were unpredictable with IL-4 or IL-13 in various models of focal infection. In contrast, although IL-6 appears to contribute only marginally to endotoxaemia, it has repeatedly been found to play a beneficial role in the resolution of some bacterial infections (e.g. van der Poll *et al.*, 1998).

Intracellular Infections

Since pro-inflammatory cytokines (TNFα, IFNγ, IL-12 and IL-18) are key molecules in the control and

Table 12 Role of IL-10, IL-4, IL-13 and IL-6 in animal models

Model	Species	Cytokine	Effect
Endotoxaemia			
LPS (high dose)	Mouse	IL-10	Beneficial
LPS (high dose)	Mouse	Anti-IL-10	Harmful
LPS (high dose)	Mouse	IL-6	No effect
LPS (high dose)	Mouse	IL-6-deficient	Harmful
LPS (low dose, galactosamine)	Mouse	IL-4	Beneficial
LPS (low dose, galactosamine)	Mouse	IL-13	Beneficial
LPS (high dose)	Mouse	IL-13	Beneficial
Enterotoxins			
SEB (*S. aureus*)	Mouse	IL-10	Beneficial
SEB (*S. aureus*)	Mouse	IL-10-deficient	Harmful
SEB (*S. aureus*)	Mouse	IL-6	Beneficial
Local injections of bacteria or focal infections			
CLP	Mouse	IL-10	No effect
CLP	Mouse	Anti-IL-10	Harmful
E. coli, *B. fragilis* (peritonitis)	Rat	IL-10	No effect
Group B streptococci (sc)	Mouse	IL-10	Beneficial
Strep. pneumoniae (pneumonia)	Mouse	IL-10	Harmful
Ps. aeruginosa (pneumonia)	Mouse	IL-10	Beneficial
K. pneumoniae (pneumonia)	Mouse	Anti-IL-10	Beneficial
E. coli	Mouse	IL-6-deficient	Harmful
S. pneumoniae (pneumonia)	Mouse	IL-6-deficient	Harmful
Group B streptococci (sc)	Mouse	IL-6	Beneficial
B. burgdorferi (sc)	Mouse	IL-4	Beneficial
Ps. aeruginosa (pneumonia)	Mouse	IL-4	Beneficial
Staph. aureus (arthritis)	Mouse	IL-4-deficient	Beneficial
CLP	Mouse	Anti-IL-13	Harmful
Intracellular and fungal infections			
L. monocytogenes	Mouse	IL-10-deficient	Beneficial
L. monocytogenes	Mouse (pup)	Anti-IL-10	Beneficial
L. monocytogenes	Mouse (adult)	Anti-IL-10	No effect
C. albicans	Mouse	IL-10	Harmful
L. monocytogenes	Mouse	IL-6	Beneficial
L. monocytogenes, *M. tuberculosis*	Mouse	IL-6-deficient	Harmful
C. albicans	Mouse	IL-4	Harmful
C. albicans	Mouse	IL-4-deficient	No effect
L. monocytogenes	Mouse	Anti-IL-4	Beneficial
L. monocytogenes	Mouse	Anti-IL-4	No effect
S. typhimurium	Mouse	IL-4-deficient	Beneficial
L. monocytogenes	Mouse	IL-13	Beneficial

eradication of intracellular and fungal infections, it seems possible that anti-inflammatory cytokines can exacerbate such infections. Indeed, this has been observed in studies of IL-10 (e.g. Tonnetti *et al.*, 1995; Dai *et al.*, 1997). Other observations have been of IL-13 and IL-4. For example, IL-13 has an anti-listerial effect, unlike IL-4 (Tonnetti *et al.*, 1995; Flesch *et al.*, 1997). IL-6 is also beneficial rather than detrimental in intracellular infections (e.g. Ladel *et al.*, 1997).

Chemokines

Most studies of chemokines have been carried out in mice on MIP-2, the equivalent of human IL-8, and a chemo-attractant for neutrophils, and MCP-1, a chemo-attractant for monocytes (**Table 13**). The contribution of chemokine receptors has been little studied.

Endotoxaemia

Chemokines recruit activated neutrophils that are known to release a large array of toxic molecules, and

Table 13 Role of chemokines in animal models

Model	Species	Cytokine	Effect
Endotoxaemia			
LPS (lung)	Rabbit	Anti-IL-8	Beneficial
LPS (lung)	Rat	Anti-MIP-2	Beneficial
LPS (lung)	Rat	Anti-MIP-1α	Beneficial
LPS (ip)	Mouse	Anti-MCP-1	Harmful
LPS (ip)	Mouse	CCR4-deficient	Beneficial
Local injections of bacteria or focal infections			
CLP	Mouse	Anti-MIP-2	Beneficial
H. influenzae (meningitis)	Mouse	Anti-MIP-2	Beneficial
H. influenzae (meningitis)	Mouse	Anti-MIP-1a	Beneficial
K. pneumoniae (peritonitis)	Mouse	Anti-MIP-2	Harmful
CLP	Mouse	Anti-MCP-1	Harmful
Intracellular and fungal infections			
Aspergillus fumigatus	Mouse	Anti-CXCR2	Harmful
Mycobacterium tuberculosis	Mouse	MCP-1-deficient	No effect

more especially oxygen radicals. In endotoxaemia models, blockade of chemokines such as IL-8, MIP-1α and MIP-2 by antibodies prevents recruitment of neutrophils and thus decreases the harmful effect of the endotoxaemia and confers some protection, especially on the lungs of experimental animals (Broaddus *et al.*, 1994; Shanley *et al.*, 1995; Schmal *et al.*, 1996). In contrast, the monocyte chemo-attractant protein MCP-1 is necessary for resistance to endotoxaemia (Zisman *et al.*, 1997b). Mice deficient in CCR4, the receptor for the chemokines thymus and activation-regulated chemokine (TARC), macrophage-derived chemokine complex (MDC), MIP-1α, are resistant to LPS injection, possibly because of a decrease in the number of monocytes (Chvatchko *et al.*, 2000).

Local Injection of Bacteria and Focal Infections

In models of peritonitis and meningitis, MIP-2 and MIP-1α contribute adversely to the inflammatory response and overall mortality, possibly by increasing recruitment of activated neutrophils (Walley *et al.*, 1997; Diab *et al.*, 1999). On the other hand, MIP-2 is essential for clearing lung infections (Greenberger *et al.*, 1996). In peritonitis, MCP-1 is also helpful, perhaps by increasing leucocyte recruitment (Matsukawa *et al.*, 1999).

Intracellular Infections

The contribution of chemokines to the control of intracellular and fungal infections has been little investigated. In mice, CXCR2 is the major receptor of CXC chemokines, and immuno-neutralisation of CXCR2 reduces neutrophil influx, and thus it adversely influences the outcome of fungal infections (Mehrad *et al.*, 1999).

Inferences from Animal Models

The dichotomous role of cytokines in the activation of the host response is evident. In endotoxaemia, an excess of pro-inflammatory cytokines is certainly detrimental. While LPS is important in the pathophysiology of sepsis induced by Gram-negative bacteria, its actions do not replicate the complexity of host–pathogen infections. Indeed, pro-inflammatory cytokines have both beneficial and deleterious effects on the host in the presence of an invasive microbial infection. The net effect of a given cytokine depends on many factors, including the type of bacteria, the size of the inoculum, and the site of infection.

Parenteral models of challenge with bacteria should be viewed as models of intoxination rather than true models of infections. In models of parenteral challenge with live bacteria, the inoculum used in most studies has been 10^9 to 10^{12} cfu/kg.

Parenteral challenge with LPS, as well as the large bacterial numbers, both induce extremely high concentrations of cytokines, usually 20–150 ng/ml of TNFα. Such high concentrations of TNFα are probably toxic for the host, as is suggested by the fact that prophylactic TNFα blockade protects from death. Models of tissue infection in which bacterial inocula are smaller (less than 10^6 cfu) are more likely to reflect true infections, because bacteria must multiply to invade the tissues and eventually the bloodstream. The time-course of cytokine production is, therefore, different from parenteral challenge. In mice challenged intraperitoneally with *E. coli*, TNFα levels are at least 10 times lower than after parenteral challenge (usually less than 1 ng/ml), and in contrast to parenteral challenge models, they are sustained during the whole observation period (Zanetti *et al.*, 1992). Indeed, cytokine profiles in these models of tissue infection are reminiscent of the profiles observed in patients with shock. In view of the fact that blockade of TNFα and IL-1 has failed in models of focal infection or even worsened the prognosis, these results can be interpreted as indicating that pro-inflammatory cytokines do not play a toxic role in these conditions, but rather help to fight infection. The majority of experiments reported in **Table 10** support the view that TNFα blockade is detrimental in focal infections. Furthermore, our observations that blockade of innate immune responses by blocking CD14, LBP and TNFα prevents natural bacterial clearance, thus leading to accelerated death, is also in favour of this hypothesis. It is likely, therefore, that in infections caused by intracellular and by extracellular pathogens, the pro-inflammatory response is necessary to control infection, and that it is detrimental to block the pro-inflammatory response under conditions in which the host is challenged with small bacterial inocula.

In models of focal infections, such as peritonitis, pneumonia or meningitis, the bacterial species used to induce the infection and the size of the inoculum (i.e. the severity of the infection) are critical factors to be taken into account. For a given pathogen and a given experimental model it is often not possible to predict whether an anti-inflammatory strategy will be beneficial or detrimental. For example, meningitis and peritonitis are very different diseases, and the part played by inflammation, compared with the part played by bacterial invasion, may be different. Furthermore, inhibitory cytokines do not always exert comparable effects in a given model of infection. Indeed, evaluation of the role of inhibitory cytokines is confusing, since experiments in which exogenous recombinant cytokines have been administered or experiments in which the same cytokines have been blocked with antibodies often give different results. These observations illustrate the difficulty of modulating the immune response in experimental sepsis. The fact that a molecule suppresses LPS-induced TNFα or IL-1 production by monocytes *in vitro* does not mean that this anti-inflammatory effect is the main function of that mediator. Indeed, cytokines such as IL-4, IL-6, IL-13 and TGFβ also exhibit critical pro-inflammatory and immuno-stimulatory functions within the immune system.

In conclusion, animal studies serve to emphasise the need for caution in the interpretation of a defined role for a given cytokine, in a given experimental model, especially because the majority of experiments have been performed in mice. Rodents, and mice in particular, are very resistant to the effects of LPS and it is not yet clear whether other mammalian species, humans in particular, share with rodents the same pattern of cytokine expression during severe inflammation.

Cytokines in Sepsis

Despite apparently encouraging pre-clinical studies in animals, more than 20 clinical trials aimed at blocking the pro-inflammatory response, targeting mainly TNFα and IL-1, have failed to protect patients with sepsis (reviewed by Bone, 1996). Undoubtedly, a severe, pro-inflammatory reaction contributes to sepsis. A massive compensatory anti-inflammatory reaction, however, also appears to be important. Indeed, pro-inflammatory cytokines and anti-inflammatory mediators have both been detected in blood and body fluids during all inflammatory states, including sepsis. The data are so numerous that they cannot all be reviewed here. Several problems related to the timing and processing of samples and to the cytokine assays complicate the interpretation. The problems are further compounded by the fact that it is not completely understood how mediator networks are regulated, nor the role of individual cytokines in the pathogenesis of sepsis and septic shock. For convenience, cytokine levels are usually measured in the bloodstream. They may, however, play a more important autocrine or paracrine roles at the primary site of infection or tissue injury, but little information on that topic is available. Mostly there is a correlation between the amount of circulating mediators and the severity of the disease. In general, blood levels have been higher in non-survivors than in survivors. For example, in fulminating meningococcal septic shock, mediators are released in larger amounts than in the course of sepsis due to other bacteria (e.g. Calandra *et al.*, 1990; Casey *et al.*, 1994).

Since it is generally believed that sepsis arises from failure to control the pro-inflammatory response, the outstanding problem remains why anti-inflammatory mechanisms are inadequate in sepsis. It appears that levels of soluble TNFR p55 range between 5 and 20 ng/ml and levels of soluble TNFR p75 between 20 and 100 ng/ml (e.g. van Deuren *et al.*, 1995). Levels of IL-10 are especially elevated in patients with meningococcaemia (10–20 ng/ml, and sometimes up to 100 ng/ml), but they do not usually exceed 1 ng/ml in patients with septic shock caused by other organisms (Derkx *et al.*, 1995). Levels of IL-1ra do not exceed 20 ng/ml (van Deuren *et al.*, 1995). IL-4 and

IL-13 have rarely been detected in the circulation of septic patients and, when present, occur in small amounts, usually below 1 ng/ml (van Deuren *et al.*, 1995; Kasai *et al.*, 1997). Each of IL-10, IL-4, IL-13, and TGFβ decreases, or may even completely suppress, the synthesis of pro-inflammatory cytokines *in vitro*. Overall, studies performed in septic patients suggest that the anti-inflammatory mediators circulate at concentrations in the range (although in lesser amounts) of those shown to inhibit cytokine production by mononuclear cells activated by LPS. It is therefore possible that they modulate, at least partially, the magnitude of the pro-inflammatory response *in vivo*. In whole blood assays, the presence of anti-inflammatory agents may inhibit pro-inflammatory events triggered by LPS. Brandtzaeg *et al.* (1996) observed that circulating anti-inflammatory cytokines exert profoundly inhibitory effects. The IL-10 in the plasma of patients with meningococcal sepsis was found to suppress the cytokine response of LPS-stimulated monocytes of healthy donors. Plasma levels of IL-10 may be as high as 40 ng/ml in patients with meningococcal septic shock, but they usually do not exceed 1 ng/ml in other septic conditions and may therefore only partly inhibit the production of pro-inflammatory molecules. Therefore, this observation may not be relevant when septic shock is caused by other microorganisms. Similar considerations also apply for the other anti-inflammatory molecules, such as TGFβ, IL-1 receptor antagonist or the soluble forms of the TNF receptors.

The pathogenesis of septic shock is very complex and it is unclear whether the results obtained *in vitro* with isolated cells stimulated with bacterial toxins can be extrapolated to the situation in a patient with septic shock caused by live bacteria. Studies *in vitro* have shown that the cytokines IL-1ra, IL-10, IL-4, IL-13, TGFβ, IL-6 and the soluble TNF receptors exert suppressive effects on lymphocytes, monocytes and macrophages, by inhibiting or down-modulating a wide range of pro-inflammatory mediators – hence their designation as anti-inflammatory mediators.

Moreover, these inhibitory functions represent only one facet of the complex biological activities of these molecules, which also have critical immuno-stimulatory functions in the immune system. Whether patients with septic shock die because the anti-inflammatory response is insufficient to control a superabundant inflammatory response, or because it has induced a state of relative immunosuppression that impairs host defence mechanisms, is difficult to establish. Indeed, except for IL-1ra, TNFR p55, TNFR p75 and possibly IL-10, it is difficult to determine whether the anti-inflammatory properties of these mediators exceed the pro-inflammatory effects that are also probably activated during sepsis. Further research is necessary to improve our understanding of the very complex pathophysiology of sepsis and septic shock.

Conclusion

Like many other areas of medical research, knowledge of inflammation has undergone extensive growth in the past few years. Although the present chapter has tried to summarise the overall organisation of the immune and anti-infective system, it is clear that the system is highly complex and includes a wide range of effector functions, while its activities are subject to regulatory processes that are only partially understood. Extensive studies are needed to define more clearly the role played by the various bacterial components in the initiation of the inflammatory response. A further challenge involves the need for the precise determination of their receptors and of the signal transduction mechanisms.

An additional problem, which is still not well understood, is the need to discriminate between all activation signals that contribute to inflammation, locally or systemically, depending on the type of infection. In addition, it remains necessary further to define the pathophysiology of severe inflammation and sepsis. The failure of clinical trials aimed at blocking the pro-inflammatory response in spite of apparently encouraging data *in vitro* and in animal studies requires further attention. Given the enormous interest in this topic, it is to be hoped that understanding of the basic science and its clinical evaluation will be reflected by better therapy for severe inflammation and sepsis.

References

Adachi O, Kawai T, Takeda K *et al.* (1998) Targeted disruption of the *MyD88* gene results in loss of IL-1 and IL-18-mediated function. *Immunity* 9: 143–150.

Aderka D, Engelmann H, Maor YBC, Wallach D (1992) Stabilization of the bioactivity of tumor necrosis factor by its soluble receptors. *J. Exp. Med.* 175: 323–329.

Akira S (2000) The role of IL-18 in innate immunity. *Curr. Opin. Immunol.* 12: 59–63.

Anderson KV (2000) Toll signaling pathways in the innate immune response. *Curr. Opin. Immunol.* 12: 13–19.

Arbour NC, Lorenz E, Schutte BC *et al.* (2000) *TLR4* mutations are associated with endotoxin hyporesponsiveness in humans. *Nature Med.* 25: 187–191.

Avery AC, Markowitz JS, Grusby MJ, Glimcher LH, Cantor H (1994) Activation of T cells by superantigen in class II-negative mice. *J. Immunol.* 153: 4853–4861.

Baggiolini M, Loetscher P, Moser B (1995) Interleukin-8 and the chemokine family. *Int. J. Immunopharmac.* 17: 103–108.

Barton BE (1997) IL-6: insights into novel biological activities. *Clin. Immunol. Immunopathol.* 85: 16–20.

Barton BE, Shortall J, Jackson JV (1996) Interleukins 6 and 11 protect mice from mortality in a staphylococcal enterotoxin-induced toxic shock model. *Infect. Immun.* 64: 714–718.

Barton PA, Warren JS (1993) Complement component C5 modulates the systemic tumor necrosis factor response in murine endotoxin shock. *Infect. Immun.* 61: 1474–1481.

Baumhofer JM, Beinhauer BG, Wang JE *et al.* (1998) Gene transfer with IL-4 and IL-13 improves survival in lethal endotoxemia in the mouse and ameliorates peritoneal macrophages immune competence. *Eur. J. Immunol.* 28: 610–615.

Beasley D, Schwartz JH, Brenner BM (1991) Interleukin 1 induces prolonged L-arginine-dependent cyclic guanosine monophosphate and nitrite production in rat vascular smooth muscle cells. *J. Clin. Invest.* 87: 602–608.

Belardelli F (1995) Role of interferons and other cytokines in the regulation of the immune response. *APMIS* 103: 161–179.

Ben-Baruch A, Michiel DF, Oppenheim JJ (1995) Signals and receptors involved in recruitment of inflammatory cells. *J. Biol. Chem.* 270: 11703–11706.

Bernardo J, Billingslea AM, Blumenthal RL, Seetoo KF, Simons ER, Fenton MJ (1998) Differential responses of human mononuclear phagocytes to mycobacterial lipoarabinimannans: role of CD14 and the mannose receptor. *Infect. Immun.* 66: 28–35.

Bernhagen J, Calandra T, Mitchell RA *et al.* (1993) Macrophage migration inhibitory factor (MIF) is a pituitary-derived cytokine that potentiates lethal endotoxaemia. *Nature* 365: 756–759.

Beutler B (2000) Tlr4: central component of the sole mammalian LPS sensor. *Curr. Opin. Immunol.* 12: 20–26.

Beutler B, Milsark IW, Cerami A (1985) Passive immunization against cachectin/tumor necrosis factor protects mice from lethal effect of endotoxin. *Science* 229: 869–871.

Bevilacqua MP, Pober JS, Majeau GR, Fiers W, Cotran RS, Gimbrone MA (1986) Recombinant tumor necrosis factor induces procoagulant activity in cultured vascular endothelium: characterization and comparison with the actions of interleukin 1. *Proc. Natl Acad. Sci. USA* 83: 4533–4537.

Bhakdi S, Klonisch T, Nuber P, Fischer W (1991) Stimulation of monokine production by lipoteichoic acids. *Infect. Immun.* 59: 4614–4620.

Billiau A (1988) Gamma-interferon: the match that light the fire? *Immunol. Today* 9: 37–40.

Biron CA, Gazzinelli RT (1995) Effects of IL-12 on immune responses to microbial infections: a key mediator in regulating disease outcome. *Curr. Opin. Immunol.* 7: 485–496.

Bogdan C, Nathan C (1993) Modulation of macrophage production by transforming growth factor β, interleukin-4, and interleukin-10. *Ann. NY Acad. Sci.* 685: 713–739.

Bohn E, Autenrieth IB (1996) IL-12 is essential for resistance against *Yersinia enterocolitica* by triggering IFNγ production in NK cells and CD4+ T cells. *J. Immunol.* 156: 1458–1468.

Bone RC (1992) Phospholipids and their inhibitors: a critical evaluation of their role in the treatment of sepsis. *Crit. Care Med.* 20: 884–890.

Bone RC (1996) Why sepsis trials fail. *JAMA* 276: 565–566.

Brandtzaeg P, Osnes L, Ovstebo R, Joo GB, Westwik AB, Kierulf P (1996) Net inflammatory capacity of human septic shock plasma evaluated by a monocyte-based target cell assay: identification of interleukin-10 as a major functional deactivator of human monocytes. *J. Exp. Med.* 184: 51–60.

Broaddus VC, Boylan AM, Hoeffel JM *et al.* (1994) Neutralization of IL-8 inhibits neutrophil influx in a rabbit model of endotoxin-induced pleurisy. *J. Immunol.* 152: 2960–2967.

Brown MA, Hural J (1997) Functions of Il-4 and control of its expression. *Crit. Rev. Immunol.* 17: 1–32.

Burns K, Martinon F, Esslinger C *et al.* (1998) MyD88, an adapter protein involved in interleukin-1 signaling. *J. Biol. Chem.* 273: 12203–12209.

Cacalano G, Lee J, Kikly K *et al.* (1994) Neutrophila and B cell expansion in mice that lack the murine IL-8 receptor homolog. *Science* 265: 682–684.

Calandra T, Baumgartner JD, Grau GE *et al.* and the Swiss–Dutch J5 study group. (1990) Prognostic values of tumor necrosis factor/cachectin, interleukin-1, α-interferon and γ-interferon in the serum of patients with septic shock. *J. Infect. Dis.* 161: 982–987.

Calandra T, Bernhagen J, Metz CN *et al.* (1995) MIF is a glucocorticoid-induced modulator of cytokine production. *Nature* 377: 68–71.

Calandra T, Spiegel LA, Metz CN, Bucala R (1998) Macrophage migration inhibitory factor is a critical mediator of the activation of innate immune cells by exotoxins of Gram-positive bacteria. *Proc. Natl Acad. Sci. USA* 95: 11383–11388.

Calandra T, Echtenacher B, Le Roy D *et al.* (2000) Protection from septic shock by neutralization of macrophage migration inhibitory factor. *Nature Med.* 6: 164–170.

Callard RE, Matthews DJ, Hibbert L (1996) IL-4 and IL-13 receptors: are they one and the same? *Immunol. Today* 17: 108–110.

Car BD, Eng VM, Schnyder B *et al.* (1994) Interferon γ receptor deficient mice are resistant to endotoxin shock. *J. Exp. Med.* 179: 1437–1444.

Casey LC, Balk RA, Bone RC (1994) Plasma cytokine and endotoxin levels correlate with survival in patients with sepsis syndrome. *Ann. Intern. Med.* 119: 771–778.

Cassatella MA, Meda L, Gasperini S, Calzetti F, Bonora S (1994) Interleukin 10 (IL-10) upregulates IL-1 receptor antagonist production from lipopolysaccharide-stimulated human polymorphonuclear leukocytes by delaying mRNA degradation. *J. Exp. Med.* 179: 1695–1699.

Cauwels A, Frei K, Sansano S *et al.* (1999) The origin and function of soluble CD14 in experimental bacterial meningitis. *J. Immunol.* 162: 4762–4772.

Chaka W, Verheul AFM, Vaishnav VV *et al.* (1997a) *Cryptococcus neoformans* and cryptococcal glucuronoxylomannan, galactoxylomannan, and mannoprotein induce different levels of tumor necrosis factor alpha in human peripheral blood mononuclear cells. *Infect. Immun.* 65: 272–278.

Chaka W, Verheul AFM, Vaishnav VV *et al.* (1997b) Induction of TNF-α in human peripheral blood mononuclear cells by the mannoprotein of *Cryptococcus neoformans* involves human mannose binding protein. *J. Immunol.* 159: 2979–2985.

Chow JC, Young DW, Golenbock DT, Christ WJ, Gusovsky F (1999) Toll-like receptor-4 mediates lipopolysaccharide-induced signal transduction. *J. Biol. Chem.* 274: 19689–19692.

Chvatchko Y, Hoogewerf A, Meyer A *et al.* (2000) A key role for CC chemokine receptor 4 in lipopolysaccharide-induced endotoxic shock. *J. Exp. Med.* 191: 1755–1763.

Cleveland MG, Gorham JD, Murphy TL, Tuomanen E, Murphy KM (1996) Lipoteichoic acid preparations of Gram-positive bacteria induce interleukin-12 through a CD14-dependent pathway. *Infect. Immun.* 64: 1906–1912.

Creasey AA, Chang ACK, Feigen L, Wün TC, Taylor FB, Hinshaw LB (1993) Tissue factor pathway inhibitor reduces mortality from *Escherichia coli* septic shock. *J. Clin. Invest.* 91: 2850–2860.

Cross AS (1990) The biological significance of bacterial encapsulation. *Curr. Top. Microbiol. Immunol.* 150: 87–95.

Czermak BJ, Sarma V, Pierson CL *et al.* (1999) Protective effects of C5a blockade in sepsis. *Nature Med.* 5: 788–792.

Dai W, Köhler G, Brombacher F (1997) Both innate and acquired immunity to *Listeria monocytogenes* infections are increased in IL-10-deficient mice. *J. Immunol.* 158: 2259–2267.

De Jong R, Altare F, Haagen IA *et al.* (1998) Severe mycobacterial and *Salmonella* infections in interleukin-12 receptor-deficient patients. *Science* 280: 1435–1438.

Derkx B, Marchant A, Goldman M, Bijlmer R, van Deventer S (1995) High levels of interleukin-10 during the initial phase of fulminant meningococcal septic shock. *J. Infect. Dis.* 171: 229–232.

Di Santo E, Meazza C, Sironi M *et al.* (1997) IL-13 inhibits TNF production but potentiates that of IL-6 in vivo and ex vivo in mice. *J. Immunol.* 159: 379–382.

Diab A, Abdalla H, Li HL *et al.* (1999) Neutralization of macrophage protein 2 (MIP-2) and MIP-1α attenuates neutrophil recruitment in the central nervous system during experimental bacterial meningitis. *Infect. Immun.* 67: 2590–2601.

Dinarello CA (1991) The proinflammatory cytokines interleukin-1 and tumor necrosis factor and treatment of the septic shock syndrome. *J. Infect. Dis.* 163: 1177–1184.

Dinarello CA (1998) Interleukin 1, interleukin-1 receptors and interleukin-1 receptor antagonist. *Intern. Rev. Immunol.* 16: 457–499.

Dockrell HM, Young SK, Britton K *et al.* (1996) Induction of Th 1 cytokine responses by mycobacterial antigens in leprosy. *Infect. Immun.* 64: 4385–4389.

Doherty GM, Lange JR, Langstein HN, Alexander HR, Buresh CM, Norton JA (1992) Evidence for IFNγ as a mediator of the lethality of endotoxin and tumor necrosis factor-α. *J. Immunol.* 149: 1666–1670.

Dokter WHA, Dijkstra AJ, Koopmans SB *et al.* (1994) G(Anh)M Tetra, a natural bacterial cell wall breakdown product, induces interleukin-1β and interleukin-6 expression in human monocytes. *J. Biol. Chem.* 269: 4201–4206.

Durum SK, Oppenheim JJ (1989) Macrophage-derived mediators: interleukin-1, tumor necrosis factor, interleukin-6, interferon and related cytokines. In: Paul WEW (ed.) *Fundamental Immunology*, pp. 639–661. New York: Raven Press.

Echtenacher B, Falk W, Männel DN, Krammer PH (1990) Requirement of endogenous tumor necrosis factor/cachectin for recovery from experimental peritonitis. *J. Immunol.* 145: 3762–3766.

Eggleton P, Reid KBM (1999) Lung surfactant proteins involved in innate immunity. *Curr. Opin. Immunol.* 11: 28–33.

Epstein J, Eichbaum Q, Sheriff S, Ezekowitz RAB (1996) The collections in innate immunity. *Curr. Opin. Immunol.* 8: 29–35.

Eskandari MK, Bolgos G, Miller C, Nguyen DT, DeForge LE, Remick DG (1992) Anti-tumor necrosis factor antibody therapy fails to prevent lethality after cecal ligation and puncture or endotoxemia. *J. Immunol.* 148: 2724–2730.

Espevik T, Otterlei M, Skjak-Braek G, Ryan L, Wright SD, Sundan A (1993) The involvement of CD14 in stimulation of cytokine production by uronic acid polymers. *Eur. J. Immunol.* 23: 255–261.

Essner R, Rhoades K, McBride WH, Morton DL, Economou JS (1989) IL-4 down-regulates IL-1 and TNF gene expression in human monocytes. *J. Immunol.* 3857–3861.

Etzioni A (1996) Adhesion molecules – their roles in health and disease. *Pediatr. Res.* 39: 191–198.

Evans T, Carpenter A, Silva A, Cohen J (1994a) Inhibition of nitric oxide synthase in experimental Gram-negative sepsis. *J. Infect. Dis.* 169: 343–349.

Evans TJ, Moyes D, Carpenter A, Martin R, Loetscher H, Lesslauer W, Cohen J (1994b) Protective effect of 55- but not 75-kD soluble tumor necrosis factor receptor-immunoglobulin G fusion proteins in an animal model of Gram-negative sepsis. *J. Exp. Med.* 180: 2173–2179.

Fenton MJ, Golenblock DT (1998) LPS-binding proteins and receptors. *J. Leuk. Biol.* 64: 25–32.

Fijnvandraat K, Derkx HHF, Peters M, Bijlmer R, Sturk A, van Deventer SJH, Ten Cate JW (1995) Coagulation activation and tissue necrosis in meningococcal septic shock: severely reduced protein C predicts a high mortality. *Haemost. Thromb.* 73: 15–20.

Fischer E, Marano MA, Van Zee KJ *et al.* (1992) Interleukin-1 receptor blockade improves survival and hemodynamic performance in *Escherichia coli* septic shock, but fails to alter host responses to sublethal endotoxemia. *J. Clin. Invest.* 89: 1551–1557.

Fischer MB, Prodeus AP, Nicholson-Weller A *et al.* (1997) Increased susceptibility to endotoxin shock in complement C3- and C4-deficient mice is corrected by C1 inhibitor replacement. *J. Immunol.* 159: 976–982.

Flesch IE, Wandersee A, Kaufmann SH (1997) Effects of IL-13 on murine listeriosis. *Intern. Immunol.* 9: 467–474.

Galdiero F, Cipollaro de l'Ero G, Benedetto N, Galdiero M, Tufano MA (1993) Release of cytokines induced by *Salmonella typhimurium* porins. *Infect. Immun.* 61: 155–161.

Gérard C, Bruyns C, Marchant A *et al.* (1993) Interleukin 10 reduces the release of tumor necrosis factor and prevents lethality in experimental endotoxemia. *J. Exp. Med.* 177: 547–550.

Greenberger MJ, Strieter RM, Kunkel SL *et al.* (1996) Neutralization of macrophage inflammatory protein-2 attenuates neutrophil recruitment and bacterial clearance in murine *Klebsiella* pneumonia. *J. Infect. Dis.* 173: 159–165.

Grimminger F, Sibelius U, Bhakdi S, Suttorp N, Seeger W (1991) *Escherichia coli* hemolysin is a potent inductor of phosphoinositide hydrolysis and related metabolic responses in human neutrophils. *J. Clin. Invest.* 88: 1531–1539.

Hackett SP, Stevens DL (1992) Streptococcal toxic shock syndrome: synthesis of tumor necrosis factor and interleukin-1 by monocytes stimulated with pyrogenic exotoxin A and streptolysin O. *J. Infect. Dis.* 165: 879–885.

Han J, Lee JD, Bibbs L, Ulevitch RJ (1994) A MAP kinase targeted by endotoxin and hyperosmolarity in mammalian cells. *Science* 265: 808–811.

Haraguchi S, Day NK, Nelson RP, Emmanuel P, Duplantier JE, Christodoulou CS, Good RA (1998) Interleukin 12 deficiency associated with recurrent infections. *Proc. Natl Acad. Sci. USA* 95: 13125–13129.

Hasko G, Virag L, Egnaczyk G, Salzman AL, Szabo C (1998) The crucial role of IL-10 in the suppression of the immunological response in mice exposed to staphylococcal enterotoxin B. *Eur. J. Immunol.* 28: 1417–1425.

Haworth R, Platt N, Keshav S *et al.* (1997) The macrophage scavenger receptor type A is expressed by activated macrophages and protects the host against lethal endotoxic shock. *J. Exp. Med.* 186: 1431–1439.

Haziot A, Rong GW, Bazil V, Silver J, Goyert SM (1994) Recombinant soluble CD14 inhibits LPS-induced tumor necrosis factor-α production by cells in whole blood. *J. Immunol.* 152: 5868–5876.

Haziot A, Ferrero E, Köntgen F *et al.* (1996) Resistance to endotoxin shock and reduced dissemination of Gram-negative bacteria in CD14-deficient mice. *Immunity* 4: 407–414.

Heaney ML, Golde DW (1996) Soluble cytokine receptors. *Blood* 87: 847–857.

Hedlund M, Svensson M, Nilsson A, Duan RD, Svanborg C (1996) Role of ceramide-signaling pathway in cytokine responses to P-fimbriated *Escherichia coli. J. Exp. Med.* 183: 1037–1044.

Heeg K, Miethke T, Wagner H (1996) Superantigen-mediated lethal shock: the functional state of ligand-reactive T cells. *Curr. Top. Microbiol. Immunol.* 216: 83–100.

Heinzel FP (1990) The role of IFNγ in the pathology of experimental endotoxemia. *J. Immunol.* 145: 2920–2924.

Herrera-Velit P, Reiner NE (1996) Bacterial lipopolysaccharide induces the association and coordinate activation of p53/56lyn and phosphatidylinositol 3-kinase in human monocytes. *J. Immunol.* 156: 1157–1165.

Hesselvik JF, Blombäck M, Brodin B, Maller R (1989) Coagulation, fibrinolysis, and kallikrein systems in sepsis: relation to outcome. *Crit. Care Med.* 17: 724–733.

Heumann D, Barras C, de Jonge B, Severin A, Glauser MP, Tomasz A (1994) Gram positive cell walls stimulate the synthesis of TNF alpha and IL-6 by human monocytes. *Infect. Immun.* 62: 2715–2721.

Heumann D, Le Roy D, Zanetti G *et al.* (1996) Contribution of TNF/TNF receptor and of Fas ligand to toxicity in murine models of endotoxemia and bacterial peritonitis. *J. Inflam.* 47: 173–179.

Hinshaw LB, Emerson TE, Taylor FB *et al.* (1992) Lethal *Staphylococcus aureus*-induced shock in primates: prevention of death with anti-TNF antibody. *J. Trauma* 33: 568–573.

Hmama Z, Mey A, Normier G, Binz H, Revillard JP (1994) CD14 and CD11b mediate serum-independent binding to human monocytes of an acylpolygalactoside isolated from *Klebsiella pneumoniae. Infect. Immun.* 62: 1520–1527.

Hoffmann JA, Kafatos FC, Janeway CA, Ezekowitz RAB (1999) Phylogenetic perspectives in innate immunity. *Science* 284: 1313–1318.

Houldsworth S, Andrew PW, Mitchell TJ (1994) Pneumolysin stimulates production of tumor necrosis factor alpha and interleukin-1β by human mononuclear phagocytes. *Infect. Immun.* 62: 1501–1503.

Hugli TE, Muller-Eberhard HJ (1978) Anaphylatoxins: C3a and C5a. *Adv. Immunol.* 26: 1–53.

Hynes RO (1992) Integrins: versatility, modulation, and signaling in cell adhesion. *Cell* 69: 11–25.

Ingalls RR, Golenbock DT (1995) CD11c/CD18, a transmembrane signaling receptor for lipopolysaccharide. *J. Exp. Med.* 181: 1473–1479.

Jack RS, Fan XL, Bernheiden M *et al.* (1997) Lipopolysaccharide-binding protein is required to combat a Gram-negative bacterial infection. *Nature* 389: 742–745.

Jahr TG, Ryan L, Sundan A, Lichenstein HS, Skjak-Braek G, Espevik T (1997) Induction of tumor necrosis factor production from monocytes stimulated with mannuronic acid polymers and involvement of lipopolysaccharide-binding protein, CD14, and bactericidal/permeability-increasing factor. *Infect. Immun.* 65: 89–94.

Jin H, Yang R, Marsters SA *et al.* (1994) Protection against rat endotoxic shock by p55 tumor necrosis factor (TNF) receptor immunoadhesin: comparison with anti-TNF monoclonal antibody. *J. Infect. Dis.* 170: 1323–1326.

Jungi TW, Sager H, Adler H, Brcic M, Pfister H (1997) Serum factors, cell membrane CD14, and β_2 integrins are not required for activation of bovine macrophages by lipopolysaccharide. *Infect. Immun.* 65: 3577–3584.

Kansas GS (1996) Selectins and their ligands: current concepts and controversies. *Blood* 88: 3259–3287.

Kasai T, Inada K, Takakuwa T *et al.* (1997) Anti-inflammatory cytokine levels in patients with septic shock. *Res. Commun. Mol. Pathol. Pharmacol.* 98: 34–42.

Kawai T, Adachi O, Ogawa T, Takeda K, Akira S (1999) Unresponsiveness of MyD88-deficient mice to endotoxin. *Immunity* 11: 115–122.

Keller R, Fischer W, Keist R, Bassetti S (1992) Macrophage response to bacteria: induction of marked secretory and cellular activities by lipoteichoic acids. *Infect. Immun.* 60: 3664–3672.

Kengatharan KM, de Kimpe SJ, Robson C, Foster SJ, Thiemermann C (1998) Mechanisms of Gram-positive shock: identification of peptidoglycan and lipoteichoic acid moieties essential in the induction of nitric oxide synthase, shock and multiple organ failure. *J. Exp. Med.* 188: 305–315.

Kilbourn RG, Gross SS, Jubran A, Griffith OW, Levi R, Lodato RF (1990) N^G-methyl-arginine inhibits tumor necrosis factor-induced hypotension: implications for the involvement of nitric oxide. *Proc. Natl Acad. Sci. USA* 87: 3629–3632.

Kohler J, Heumann D, Garotta G *et al.* (1993) Interferon-γ involvement in the severity of Gram negative infections in mice. *J. Immunol.* 151: 916–921.

Konstantopoulos K, McIntire LV (1996) Perspective series: cell adhesion in vascular biology. Effects of fluid dynamic forces on vascular cell adhesion. *J. Clin. Invest.* 98: 2661–2665.

Krieg AM, Yi AK, Matson S *et al.* (1995) CpG motifs in bacterial DNA trigger direct B-cell activation. *Nature* 374: 546–549.

Krziwon C, Zähringer U, Kawahara K *et al.* (1995) Glycosphingolipids from *Sphingomonas paucimobilis* induce monokine production in human mononuclear cells. *Infect. Immun.* 63: 2899–2905.

Kusunoki T, Wright SD (1996) Chemical characteristics of *Staphylococcus aureus* molecules that have CD14-dependent stimulating activity. *J. Immunol.* 157: 5112–5117.

Ladel CH, Blum C, Dreher A, Reifenberg K, Kopf M, Kaufmann SHE (1997) Lethal tuberculosis in interleukin-6-deficient mutant mice. *Infect. Immun.* 65: 4843–4849.

Laichalk LL, Kunkel SL, Strieter RM, Danforth JM, Bailie MB, Standiford TJ (1996) Tumor necrosis factor mediates lung antibacterial host defense in murine *Klebsiella* pneumonia. *Infect. Immun.* 64: 5211–5218.

Lamont AG, Adorini L (1996) IL-12: a key cytokine in immune regulation. *Immunol. Today* 17: 214–217.

Le Roy D, Morand P, Lengacher S *et al.* (1996) *Streptococcus mitis* cell walls and lipopolysaccharide induce lethality in D-galactosamine-sensitized mice by a tumor necrosis factor-dependent pathway. *Infect. Immun.* 64: 1846–1849.

Le Roy D, Di Padova F, Tees R *et al.* (1999) Monoclonal antibodies to murine lipopolysaccharide (LPS)-binding protein (LBP) protect mice from lethal endotoxemia by blocking either the binding of LPS to LBP or the presentation of LPS/LBP complexes to CD14. *J. Immunol.* 162: 7454–7460.

Lesslauer W, Tabuchi H, Gentz R *et al.* (1991) Recombinant soluble tumor necrosis factor receptor proteins protect mice from lipopolysaccharide-induced lethality. *Eur. J. Immunol.* 21: 2883–2886.

Letterio JJ, Roberts AB (1997) TGFβ: a critical modulator of immune cell function. *Clin. Immunol. Immunopathol.* 84: 244–250.

Leturcq D, Moriarty AM, Talbott G, Winn RK, Martin TR, Ulevitch RJ (1996) Antibodies against CD14 protect primates from endotoxin-induced shock. *J. Clin. Invest.* 98: 1533–1538.

Lien E, Sellati TJ, Yoshimura A *et al.* (1999) Toll-like receptor 2 functions as a pattern recognition receptor for diverse bacterial products. *J. Biol. Chem.* 274: 33419–33425.

Majcherczyk PA, Langen H, Heumann D, Fountoulakis M, Glauser MP, Moreillon P (1999) Digestion of *Streptococcus pneumoniae* cell walls in its major peptidoglycan hydrolase releases branched stem peptides carrying proinflammatory activity. *J. Biol. Chem.* 274: 12537–12543.

Makita H, Nishimura M, Miyamoto K *et al.* (1998) Effect of anti-macrophage migration inhibitory factor antibody on lipopolysaccharide-induced pulmonary neutrophil accumulation. *Am. J. Respir. Crit. Care Med.* 158: 573–579.

Mancilla J, Garcia P, Dinarello CA (1993) The interleukin-1 receptor antagonist can either reduce or enhance the lethality of *Klebsiella pneumoniae* sepsis in newborn rats. *Infect. Immun.* 61: 926–932.

Mancuso G, Cusumano V, Genovese F, Gambuzza M, Beninati C, Teti G (1997) Role of interleukin 12 in experimental neonatal sepsis caused by Group B streptococci. *Infect. Immun.* 65: 3731–3735.

Mangan DF, Wahl SM, Sultzer BM, Mergenhagen SE (1992) Stimulation of human monocytes by endotoxin-associated protein: inhibition of programmed cell death (apoptosis) and potential significance in adjuvancity. *Infect. Immun.* 60: 1684–1686.

Martin RA, Silva AT, Cohen J (1993) Effect of anti-TNF-alpha treatment in an antibiotic treated murine model of shock due to *Streptococcus pyogenes*. *FEMS Microbiol. Lett.* 110: 175–178.

Mathison JC, Wolfson E, Ulevitch RJ (1988) Participation of tumor necrosis factor in the mediation of bacterial lipopolysaccharide-induced injury in rabbits. *J. Clin. Invest.* 81: 1925–1937.

Matsukawa A, Hogaboam CM, Lukacs NW, Lincoln PM, Strieter RM, Kunkel SL (1999) Endogenous monocyte chemoattractant protein-1 (MCP-1) protects mice in a model of acute septic peritonitis: cross-talk between MCP-1 and leukotriene B_4. *J. Immunol.* 163: 6148–6154.

Matthys P, Mitera T, Heremans H, Van Damme J, Billiau A (1995) Anti-gamma interferon and anti-interleukin 6 antibodies affect staphylococcal enterotoxin B-induced weight loss, hypoglycemia, and cytokine release in D-galactosamine-sensitized and unsensitized mice. *Infect. Immun.* 63: 1158–1164.

Mattsson E, Rollof J, Verhoef J, van Dick H, Fleer A (1994) Serum-induced potentiation of tumor necrosis factor alpha production by human monocytes in response to staphylococcal peptidoglycan: involvement of different serum factors. *Infect. Immun.* 62: 3837–3843.

May LT, Neta R, Moldawer LL, Kenney JS, Patel K, Sehgal PB (1993) Antibodies chaperone circulating IL-6. Paradoxical effects of anti-IL-6 neutralizing antibodies *in vivo*. *J. Immunol.* 151: 3225–3236.

Mehrad BM, Strieter RM, Moore TA, Tsai WC, Lira SA, Standiford TJ (1999) CXC chemokine receptor-2 ligands are necessary components of neutrophil-mediated host defense in invasive pulmonary aspergillosis. *J. Immunol.* 163: 6086–6094.

Miethke T, Duschek K, Wahl C, Heeg K, Wagner H (1993) Pathogenesis of the toxic shock syndrome: T cell mediated lethal shock caused by the superantigen TSST-1. *Eur. J. Immunol.* 23: 1494–1500.

Mohler KM, Torrance DS, Smith CA *et al.* (1993) Soluble tumor necrosis factor (TNF) receptors are effective therapeutic agents in lethal endotoxemia and function simultaneously as both TNF carriers and TNF antagonists. *J. Immunol.* 151: 1548–1561.

Moller B, Ellermann-Eriksen S, Storgaard M, Obel N, Bendtzen K, Petersen CM (1996) Soluble tumor necrosis factor (TNF) receptors conserve TNF bioactivity in meningitis patient spinal fluid. *J. Infect. Dis.* 174: 557–563.

Mosmann TR, Cherwinski H, Bond MW, Giedlin MA, Coffman RL (1986) Two types of murine helper T-cell clone. I. Definition according to profiles of lymphokine activities and secreted proteins. *J. Immunol.* 136: 2348–2357.

Murakami K, Okajima K, Uchiba M *et al.* (1996) Activated protein C attenuates endotoxin-induced pulmonary vascular injury by inhibiting activated leukocytes in rats. *Blood* 87: 642–647.

Murata T, Obiri NI, Debinski W, Puri RK (1997) Structure of the IL-13 receptor: analysis of subunit composition in cancer and immune cells. *Biochem. Biophys. Res. Commun.* 238: 90–94.

Muzio M, Natoli G, Saccani S, Levrero M, Mantovani A (1998) The human Toll signaling pathway: divergence of nuclear factor κB and JNK/SAPK activation upstream of tumor necrosis factor receptor-associated factor 6 (TRAF6). *J. Exp. Med.* 187: 2097–2101.

Nakane A, Minagawa T, Kohanawa M *et al.* (1989) Interactions between endogenous gamma interferon and tumor necrosis factor in host resistance against primary and secondary *Listeria monocytogenes* infections. *Infect. Immun.* 57: 3331–3337.

Nassif X, Mathison JC, Wolfson E, Koziol JA, Ulevitch RJ, So M (1992) Tumour necrosis factor alpha antibody protects against lethal meningococcaemia. *Mol. Microbiol.* 6: 591–597.

Nathan C (1992) Nitric oxide as a secretory product of mammalian cells. *FASEB J.* 6: 3051–3064.

Nathan C, Xie Q (1994) Nitric oxide synthases: roles, tolls, and controls. *Cell* 78: 915–918.

Netea MG, van Tits LJH, Curfs JHA *et al.* (1999) Increased susceptibility of TNF-α lymphotoxin-β double knockout mice to systemic candidiasis through impaired recruitment of neutrophils and phagocytosis of *Candida albicans*. *J. Immunol.* 163: 1498–1505.

Newman SL, Chaturvedi S, Klein BS (1995) The Wl-1 antigen of *Blastomyces dermatidis* yeasts mediates binding to human macrophage CD11b/CD18 (CR3) and CD14. *J. Immunol.* 154: 753–761.

Nicholson-Weller A, Kllickstein LB (1999) C1q-binding proteins and C1q receptors. *Curr. Opin. Immunol.* 11: 42–46.

Okusawa S, Gelfand JA, Ikejima T, Connolly RJ, Dinarello CA (1988a) Interleukin 1 induces a shock-like state in rabbits. Synergism with tumor necrosis factor and the effect of cyclooxygenase inhibition. *J. Clin. Invest.* 81: 1162–1172.

Okusawa S, Yancey KB, Van der Meer JWM *et al.* (1988b) C5a stimulates secretion of tumor necrosis factor from human mononuclear cells in vitro. Comparison with secretion of interleukin 1β and interleukin 1α. *J. Exp. Med.* 168: 443–448.

Opal SM, Cross AS, Sadoff JC *et al.* (1991) Efficacy of antilipopolysaccharide and anti-tumor necrosis factor monoclonal antibodies in a neutropenic rat model of *Pseudomonas sepsis*. *J. Clin. Invest.* 88: 885–890.

Opal SM, Wherry JC, Grint P (1998) Interleukin-10: potential benefits and possible risks in clinical infectious diseases. *Clin. Infect. Dis.* 27: 1497–1507.

Otterlei M, Varum KM, Ryan L, Espevik T (1994) Characterization of binding and TNFα inducing ability of chitosans on monocytes: the involvement of CD14. *Vaccine* 12: 825–832.

Pearson AM (1996) Scavenger receptors in innate immunity. *Curr. Opin. Immunol.* 8: 20–28.

Perera PY, Vogel SN, Detore GR, Haziot A, Goyert SM (1997) CD14-dependent and CD14-independent signaling pathways in murine macrophages from normal and CD14 knockout mice stimulated with lipopolysaccharide or taxol. *J. Immunol.* 158: 4422–4429.

Peschon JJ, Torrance DS, Stocking KL *et al.* (1998) TNF receptor-deficient mice reveal divergent roles for p55 and p75 in several models of inflammation. *J. Immunol.* 160: 943–952.

Peters M, Jacobs S, Ehlers M *et al.* (1996) The function of the soluble interleukin-6 (IL-6) receptor in vivo: sensitization of human soluble IL-6 receptor transgenic mice towards IL-6 and prolongation of the plasma half-life of IL-6. *J. Exp. Med.* 183: 1399–1406.

Pixley RA, De la Cadena R, Page JD *et al.* (1993) The contact system contributes to hypotension but not disseminated intravascular coagulation in lethal bacteremia. In vivo use of a monoclonal anti-factor XII antibody to block contact activation in baboons. *J. Clin. Invest.* 91: 61–68.

Poltorak A, He X, Smirnova I *et al.* (1999) Defective LPS signaling in C3H/HeJ and C57BL/10ScCr mice; mutations in *Tlr4* gene. *Science* 282: 2085–2088.

Poltorak A, Ricciardi-Castagnoli P, Citterio S, Beutler B (2000) Physical contact between lipopolysaccharide and Toll-like receptor 4 by genetic complementation. *Proc. Natl Acad. Sci. USA* 97: 2163–2167.

Pralong G, Calandra T, Glauser MP *et al.* (1989) Plasminogen activator inhibitor 1: a new prognostic marker in septic shock. *Thromb. Haemostas.* 61: 459–462.

Pruzanski W, Vadas P (1991) Phospholipase A_2 – a mediator between proximal and distal effectors of inflammation. *Immunol. Today* 12: 143–146.

Pugin J, Heumann D, Tomasz A *et al.* (1994) CD14 is a pattern recognition receptor. *Immunity* 1: 509–516.

Pugin J, Ulevitch RJ, Tobias PS (1995) Tumor necrosis factor-α and interleukin-1β mediate endothelial cell activation in blood at low endotoxin concentrations. *J. Inflam.* 45: 49–55.

Quezado ZMN, Hoffman WD, Winkelstein JA *et al.* (1994) The third component of complement protects against *Escherichia coli* endotoxin-induced shock and multiple organ failure. *J. Exp. Med.* 179: 569–578.

Raingeaud J, Gupta S, Rogers JS *et al.* (1995) Pro-inflammatory cytokines and environmental stress cause p38 mitogen-activated protein kinase activation by dual phosphorylation on tyrosine and threonine. *J. Biol. Chem.* 270: 7420–7426.

Reinmann T, Büscher D, Hipskind RA, Krautwald S, Lohmann-Matthes ML, Baccarini M (1994) Lipopolysaccharide induces activation of the Raf-1/MAP kinase pathway. *J. Immunol.* 153: 5740–5749.

Remick DG, Garg SJ, Newcomb DE, Wollenberg G, Huie TK, Bolgos GL (1998) Endogenous IL-10 fails to decrease the mortality of morbidity of sepsis. *Crit. Care Med.* 26: 895–904.

Rietschel ET, Kirikae T, Schade FU *et al.* (1994) Bacterial endotoxin: molecular relationships of structure to activity and function. *FASEB J.* 8: 217–225.

Roach TIA, Barton CH, Chatterje D, Blackwell JM (1993) Macrophage activation: lipoarabinomannan from avirulent and virulent strains of *Mycobacterium tuberculosis* differentially induces the early c-fos, KC, JE, and tumor necrosis factor-α. *J. Immunol.* 150: 1886–1896.

Rothe J, Lesslauer W, Lötscher H *et al.* (1993) Mice lacking the tumour necrosis factor receptor 1 are resistant to TNF-mediated toxicity but highly susceptible to infection by *Listeria monocytogenes. Nature* 364: 798–802.

Sakao Y, Takeda K, Tsutsui H *et al.* (1999) IL-18-deficient mice are resistant to endotoxin-induced liver injury but highly susceptible to endotoxin shock. *Intern. Immunol.* 11: 471–480.

Sala N, Fontcuberta J, Rutllant MLL (1993) New concepts of coagulation inhibitors. *Intensive Care Med.* 19: S3–S7.

Sawa T, Corry DB, Gropper MA, Ohara M, Kurahashi K, Wiener-Kronisch JP (1997) IL-10 improves lung injury and survival in *Pseudomonas aeruginosa* pneumonia. *J. Immunol.* 159: 2858–2866.

Schimke J, Mathison J, Morgiewicz J, Ulevitch RJ (1998) Anti-CD14 mAb treatment provides therapeutic benefit after exposure to endotoxin. *Proc. Natl Acad. Sci. USA* 95: 13875–13880.

Schlievert PM, Bohach GA, Ohlendorf DH *et al.* (1995) Molecular structure of Staphylococcus and Streptococcus superantigens. *J. Clin. Immunol.* 15: 4S–10S.

Schmal H, Shanley TP, Jones ML, Friedl HP, Ward PA (1996) Role for macrophage inflammatory protein-1 in lipopolysaccharide-induced lung injury in rats. *J. Immunol.* 156: 1963–1972.

Schumann RR, Leong SR, Flaggs GW *et al.* (1990) Structure and function of lipopolysaccharide binding protein. *Science* 249: 1429–1431.

Schwandner R, Dziarski R, Wesche H, Rothe M, Kirschning CJ (1999) Peptidoglycan- and lipoteichoic acid-induced cell activation is mediated by Toll-like receptor 2. *J. Biol. Chem.* 274: 17406–17409.

See RH, Kum WWS, Chang AH, Goh SH, Chow AW (1992) Induction of tumor necrosis factor and interleukin-1 by purified staphylococcal toxic shock syndrome toxin 1 requires the presence of both monocytes and T lymphocytes. *Infect. Immun.* 60: 2612–2618.

Sellati TJ, Bouis DA, Kitchens RL *et al.* (1998) *Treponema pallidum* and *Borrelia burgdorferi* lipoproteins and synthetic lipopeptides activate monocytic cells via a CD14-dependent pathway distinct from that used by liopopolysaccharide. *J. Immunol.* 160: 5455–5464.

Shanley TP, Schmal H, Friedl HP, Jones ML, Ward PA (1995) Role of macrophage inflammatory protein-1α (MIP-1α) in acute lung injury in rats. *J. Immunol.* 154: 4793–4802.

Shen CH, Stavnezer J (1998) Interaction of Stat6 and NF-κB: direct association and synergistic activation of interleukin-4 induced transcription. *Mol. Cell Biol.* 18: 3395–3404.

Shimazu R, Akashi S, Ogata H, Nagai YFK, Miyake K, Imoto M (1999) MD-2, a molecule that confers lipopolysaccharide responsiveness on Toll-like receptor 4. *J. Exp. Med.* 189: 1777–1782.

Shull MM, Ormsby I, Kier AB *et al.* (1992) Targeted disruption of the mouse transforming growth factor-β1 gene results in multifocal inflammatory disease. *Nature* 359: 693–699.

Silva AT, Cohen J (1992) Role of interferon-γ in experimental Gram-negative sepsis. *J. Infect. Dis.* 166: 331–335.

Silva AT, Bayston KF, Cohen J (1990) Prophylactic and therapeutic effects of a monoclonal antibody to tumor necrosis factor-α in experimental gram-negative shock. *J. Infect. Dis.* 162: 421–427.

Snyder F (1990) Platelet-activating factor and related acetylated lipids as potent biologically active cellular mediators. *Am. J. Physiol.* 259: C697–C708.

Soell M, Lett E, Holveck F, Schöller M, Wachsmann D, Klein JP (1995) Activation of human monocytes by streptococcal rhamnose glucose polymers is mediated by CD14 antigen, and mannan binding protein inhibits TNFα release. *J. Immunol.* 154: 851–860.

Sparwasser T, Miethke T, Lipford G *et al.* (1999) Bacterial DNA causes septic shock. *Nature* 386: 336–337.

Spellerberg B, Prasad S, Cabellos C, Burroughs M, Cahill P, Tuomanen E (1995) Penetration of the blood–brain barrier: enhancement of drug delivery and imaging by bacterial glycopeptides. *J. Exp. Med.* 182: 1037–1044.

Spinas GA, Bloesch D, Keller U, Zimmerli W, Cammisuli S (1991) Pretreatment with ibuprofen augments circulating tumor necrosis factor-α, interleukin-6, and elastase during acute endotoxemia. *J. Infect. Dis.* 163: 89–95.

Spinas GA, Keller U, Brockhaus M (1992) Release of soluble receptors for tumor necrosis factor (TNF) in relation to circulating TNF during experimental endotoxinemia. *J. Clin. Invest.* 90: 533–536.

Stahl PD, Ezekowitz RAB (1998) The mannose receptor is a pattern recognition involved in host defense. *Curr. Opin. Immunol.* 10: 50–55.

Standiford TJ, Arenberg DA, Danforth JM, Kunkel SL, VanOttern GM, Strieter RM (1994) Lipoteichoic acid induces secretion of interleukin-8 from human blood monocytes: a cellular and molecular analysis. *Infect. Immun.* 62: 119–125.

Suffredini AF, Harpel PC, Parillo JE (1989) Promotion and subsequent inhibition of plasminigen activation after administration of intravenous endotoxin to normal subjects. *New Engl. J. Med.* 320: 1165–1172.

Sugawara I, Yamada H, Kaneko H, Mizuno S, Takeda K, Akira S (1999) Role of interleukin-18 (IL-18) in mycobacterial infection in IL-18-gene-disrupted mice. *Infect. Immun.* 67: 2585–2589.

Sultzer BM, Goodman GW (1976) Endotoxin protein: a B-cell mitogen and polyclonal activator of C3H/HeJ lymphocytes. *J. Exp. Med.* 144: 821–827.

Tai KF, Ma Y, Weis JJ (1994) Normal human B lymphocytes and mononuclear cells respond to the mitogenic and cytokine-stimulatory activities of *Borrelia burgdorferi* and its lipoprotein OspA. *Infect. Immun.* 62: 520–528.

Takada H, Kawabata Y, Arakaki R *et al.* (1995) Molecular and structural requirements of a lipoteichoic acid from *Enterococcus hirae* ATCC 9790 for cytokine-inducing, antitumor, and antigenic activities. *Infect. Immun.* 63: 57–65.

Takeuchi O, Hoshino K, Kawai T *et al.* (1999) Differential roles of TLR2 and TLR4 in recognition of Gram-negative and Gram-positive bacterial cell wall components. *Immunity* 11: 443–451.

Tartaglia LA, Pennica D, Goeddel DV (1993) Ligand passing: the 75-kDa tumor necrosis factor (TNF) receptor recruits TNF for signaling by the 55-kDa TNF receptor. *J. Biol. Chem.* 268: 18542–18548.

Taylor FB, Chang A, Esmon T, D'angelo A, Vigano-D'Angelo S, Blick KE (1987) Protein C prevents the coagulopathic and lethal effects of Escherichia coli infusion in the baboon. *J. Clin. Invest.* 79: 918–925.

Tenner AJ (1999) Membrane receptors for soluble defense collagens. *Curr. Opin. Immunol.* 11: 34–41.

Tobias PS, Tapping RI, Gegner JA (1999) Endotoxin interactions with lipopolysaccharide-responsive cells. *Clin. Infect. Dis.* 28: 476–481.

Tonnetti L, Spaccapelo R, Cenci E *et al.* (1995) Interleukin-4 and -10 exacerbate candidiasis in mice. *Eur. J. Immunol.* 25: 1559–1565.

Tufano MA, Tetta C, Biancone L *et al.* (1992) *Salmonella typhimurium* porins stimulate platelet-activating factor by human polymorphonuclear neutrophils. *J. Immunol.* 149: 1023–1030.

Ulevitch RJ, Tobias PS (1994) Recognition of endotoxin by cells leading to transmembrane signaling. *Curr. Opin. Immunol.* 6: 125–130.

Ulevitch RJ, Tobias PS (1999) Recognition of Gram-negative bacteria and endotoxin by the innate immune system. *Curr. Opin. Immunol.* 11: 19–22.

van der Poll T, Marchant A, Buurman WA *et al.* (1995) Endogenous IL-10 protects mice from death during septic peritonitis. *J. Immunol.* 155: 5397–5401.

van der Poll T, Marchant A, Keogh CV, Goldman M, Lowry SF (1996) Interleukin-10 impairs host defense in murine pneumococcal pneumonia. *J. Infect. Dis.* 174: 994–1000.

van der Poll T, Keogh CV, Buurman WA, Lowry SF (1997) Passive immunization against tumor necrosis factor-α impairs host defense during pneumococcal pneumonia in mice. *Am. J. Respir. Crit. Care Med.* 155: 603–608.

van der Poll T, Keogh CV, Guirao X, Buurman WA, Kopf M, Lowry SF (1998) Interleukin-6 gene-deficient mice show impaired defense against pneumococcal pneumonia. *J. Infect. Dis.* 176: 439–444.

van Deuren M, van der Ven-Jongekrijk J, Bartelink AKM, van Dalen R, Sauerwein RW, Van der Meer JWM (1995) Correlation between proinflammatory cytokines and antiinflammatory mediators and the severity of disease in meningococcal infections. *J. Infect. Dis.* 172: 433–439.

Vannier E, Miller MC, Dinarello CA (1992) Coordinated antiinflammatory effects of interleukin 4: interleukin 4 suppresses interleukin 1 production but up-regulates gene expression and synthesis of interleukin 1 receptor antagonist. *Proc. Natl Acad. Sci. USA* 89: 4076–4080.

VanZee KJ, Kohno T, Fischer E, Rock CS, Moldawer LL, Lowry SF (1992) Tumor necrosis factor soluble receptors circulate during experimental and clinical inflammation and can protect against excessive tumor necrosis factor α in vitro and in vivo. *Proc. Natl Acad. Sci. USA* 89: 4845–4849.

Verhoef J, Mattsson E (1995) The role of cytokines in Gram-positive bacterial shock. *Trends Microbiol.* 3: 136–140.

Walew I, Vollmer P, Palmer M, Bhakdi S, Rose-Jone S (1996) Pore-forming toxins trigger shedding of receptors for interleukin 6 and lipopolysaccharide. *Proc. Natl Acad. Sci. USA* 93: 7882–7887.

Wallach D, Varfolomeev EE, Malinin NL, Goltsev YV, Kovalenko AV, Boldin MP (1999) Tumor necrosis factor receptor and Fas signaling mechanisms. *Annu. Rev. Immunol.* 17: 331–367.

Walley KR, Lukacs NW, Standiford TJ, Strieter RM, Kunkel SL (1997) Elevated levels of macrophage inflammatory protein 2 in severe murine peritonitis increase neutrophil recruitment and mortality. *Infect. Immun.* 65: 3847–3851.

Wang P, Wu P, Siegel MI, Egan RW, Billah MM (1994) IL-10 inhibits transcription of cytokine genes in human peripheral blood mononuclear cells. *J. Immunol.* 153: 811–816.

Ward SG, Westwick J (1998) Chemokines: understanding their role in T-lymphocyte biology. *Biochem. J.* 333: 457–470.

Wei XQ, Leung BP, Niedbala W *et al.* (1999) Altered immune responses and susceptibility to *Leishmania major* and *Staphylococcus aureus* infection in IL-18-deficient mice. *J. Immunol.* 163: 2821–2828.

Wenneras C, Ave P, Huerre M *et al.* (2000) Blockade of CD14 increases *Shigella*-mediated invasion and tissue destruction. *J. Immunol.* 164: 3214–3221.

Wessels MR, Butko P, Ma M, Warren HB, Lage AL (1995) Studies of group B streptococcal infection in mice deficient in complement component C3 or C4 demonstrate an essential role for complement in both innate and acquired immunity. *Proc. Natl Acad. Sci. USA* 92: 11490–11494.

Wright SD, Ramos RA, Tobias PS, Ulevitch RJ, Mathison JC (1990) CD14, a receptor for complexes of lipopolysaccharide (LPS) and LPS binding protein. *Science* 249: 1431–1433.

Xing Z, Gauldie J, Cox G *et al.* (1998) IL-6 is an antiinflammatory cytokine required for controlling local or systemic acute inflammatory responses. *J. Clin. Invest.* 101: 311–320.

Yamamoto Y, Klein TW, Friedman H (1997) Involvement of mannose receptor in cytokine interleukin-1β (IL-1β), IL-6, and granulocyte–macrophage-colony stimulating factor responses, but not chemokine macrophage inflammatory protein 1β (MIP-1β), and KC response, caused by attachment of *Candida albicans* to macrophages. *Infect. Immun.* 65: 1077–1082.

Zanetti G, Heumann D, Gérain J *et al.* (1992) Cytokine production after intravenous or peritoneal Gram negative bacterial challenge in mice. Comparative protective efficacy of antibodies to TNFα and to LPS. *J. Immunol.* 148: 1890–1897.

Zantl N, Uebe A, Neumann B *et al.* (1998) Essential role of gamma interferon in survival of colon ascendens stent peritonitis, a novel murine model of abdominal sepsis. *Infect. Immun.* 66: 2300–2309.

Zhang Y, Doerfler M, Lee TC, Guillemin B, Rom WM (1993) Mechanisms of stimulation of interleukin-1β and tumor necrosis factor-α by *Mycobacterium tuberculosis* components. *J. Clin. Invest.* 91: 2076–2083.

Zhang Y, Broser M, Cohen H *et al.* (1995) Enhanced interleukin-8 release and gene expression in macrophages after exposure to *Mycobacterium tuberculosis* and its components. *J. Clin. Invest.* 95: 586–592.

Zisman DA, Kunkel SL, Strieter RM *et al.* (1997a) Anti-interleukin-12 therapy protects mice in lethal endotoxemia but impairs bacterial clearance in murine *Escherichia coli* peritoneal sepsis. *Shock* 8: 349–356.

Zisman DA, Kunkel SL, Strieter RM *et al.* (1997b). MCP-1 protects mice in lethal endotoxemia. *J. Clin. Invest.* 99: 2832–2836.

Zurawski G, de Vries JE (1994) Interleukin 13, and interleukin-4-like cytokine that act on monocytes and B cells, but not on T cells. *Immunol. Today* 15: 19–26.

34

Strategies for Survival of Extracellular Pathogens

S. Bhakdi

Institute of Medical Microbiology and Hygiene, University of Mainz, Germany

Few pathogenic bacteria can be regarded as strictly extracellular or intracellular parasites. Most are able to exist in either form, and so discussions of extracellular versus intracellular survival strategies cannot but contain overlaps. *Listeria monocytogenes*, for example, is generally regarded as the prototype of an intracellular bacterium, but extracellular infections also frequently occur. Mechanisms that usually operate to promote intracellular survival, such as toxin production, may then also become relevant in the context of extracellular survival. Similarly, *Staphylococcus aureus* is usually considered a classical extracellular pathogen, but it is now apparent that this bacterium often persists intracellularly. Hence, differentiation between extra- and intracellular survival strategies is necessarily artificial and, in most cases, it is not strictly possible. The following discussion approaches the difficult task of highlighting mechanisms that possibly tend to tip the balance toward extracellular infections. Two topics are excluded: adhesive mechanisms responsible for the primary interaction of microbes with host cells and tissues (see Chapter 30), and the problems of nutrient acquisition by invading bacteria (see Chapter 32).

The space available here would scarcely be able to accommodate a simple listing of microbial factors that may contribute to extracellular survival. A reductive approach must therefore be taken, in which certain themes are highlighted at the expense of others. I will endeavour to outline principles and discuss prototypes, leaving details to the chapters on the specific organisms.

Strategies for extracellular survival can often be equated with pathogenicity mechanisms, and counteractive measures can therefore form the basis for prophylaxis or therapy of the respective infections. These aspects will be alluded to throughout the chapter, since they highlight how basic research into microbial virulence mechanisms can lead to advances in clinical medicine.

For didactic reasons, it appears useful to define two major categories of survival strategies. The first involves mechanisms that afford protection to the micro-organism, which can be grouped into three main topics: hiding, camouflage, and immune evasion and escape. The second set of strategies comprises the active attack mechanisms of micro-organisms towards the host. These are also dealt with under three main headings: toxins, toxic injection, and enzymatic tissue degradation.

Protection Strategies

Most bacteria that cause extracellular infections cannot afford to be engulfed by professional phagocytes, since they are usually highly susceptible to the microbicidal mechanisms of these cells. A recurring theme underlying most strategies for extracellular survival, therefore, relates to how microbes avoid phagocytosis. Hiding and camouflage are two frequently encountered strategies. Additionally or alternatively, pathogens endeavour to evade and escape phagocytes by

 doi:10.1006/bkmm.2001.0034

employing mechanisms to counteract the opsonising effects of antibody and complement.

Hiding

Bacteria may seek refuge at the single-cell level: the classical and most important mechanism is the production of a polysaccharide capsule. Alternatively, they may collectively seek shelter in a protected niche within the infected host. These two basic mechanisms are, of course, not mutually exclusive. However, diseases caused by bacteria that more heavily exploit one or the other strategy tend to follow different patterns. Thus, micro-organisms that enjoy the protective properties of capsules have a tendency to invade and disseminate, whereas bacteria that seek out survival niches tend to become sequestrated and to cause more localised infections.

Hiding Within a Capsule

Capsules (see Chapter 4) may be envisaged as porous, three-dimensional cages that permit the diffusion of macromolecules to the underlying bacterial wall. In a simplistic sense, they can be viewed as shields that render bacteria invisible to phagocytes, since the sentinel cells lack receptors for capsular polysaccharides (**Fig. 1**). In contrast to several components of the bacterial cell wall, also, capsules do not primarily activate complement, and so opsonins do not become deposited on the bacterial surface. Although complement components may diffuse to and become activated by structures located beneath the capsule, phagocytic receptors cannot probe the depth of the polysaccharide shields and are therefore incapable of binding the opsonins. Since most invasive pathogens can withstand an attack by the terminal complement sequence, they can afford to permit the entire complement cascade to be activated. In effect, the invading micro-organisms enjoy the status of invisible and invincible offenders that cannot efficiently be cleared from the bloodstream and tissues. This detrimental situation can be rectified when antibodies directed against a capsular polysaccharide are generated that eventually lay down opsonins on the bacterial surface.

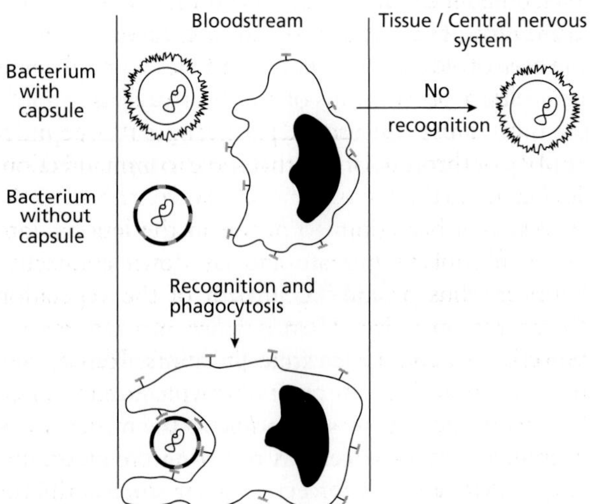

Fig. 1 Hiding within a capsule. Encapsulated pathogens are not recognised by phagocytes and can therefore disseminate by the haematogenous route to reach peripheral organs.

Encapsulated human pathogens often have the capacity to cross the blood–brain barrier. The reasons are poorly understood, but they may quite simply be related to the prolonged sojourn of the bacteria in the circulation that results from ineffective clearance. It is hardly a coincidence that the major causative agents of acute bacterial meningitis are encapsulated, with *Neisseria meningitidis*, *Streptococcus pneumoniae*, *Haemophilus influenzae* B, group B streptococci and *Escherichia coli* K1 leading the list. That the capsules allow these micro-organisms to survive and to disseminate in the bloodstream is undoubted and has, of course, formed the basis for the development of vaccines (Adams *et al.*, 1993; De Velasco *et al.*, 1995). This straightforward rationale has repeatedly been successful; antibodies bound to the capsule activate complement, so that opsonins that can be recognised by phagocytic receptors are deposited on the surface. Thus, bacteria will be ingested and killed by blood neutrophils and tissue macrophages. Use of *H. influenzae* B vaccine exploits this mechanism and has led to a dramatic fall in invasive *H. influenzae* infections. Similarly, multivalent *S. pneumoniae* capsule vaccine affords significant, albeit less complete protection, as does the capsule vaccine against *N. meningitidis* C. The essential role of antibodies in protection against systemic infections with encapsulated pathogens is highlighted by the paradigm of *E. coli* K1 and *N. meningitidis* B. The capsule polysaccharide structures of these organisms are identical to mammalian glycoconjugates, and are therefore not immunogenic. As a consequence, endeavours to produce vaccines have failed, and immunoprophylaxis against these agents may never become possible.

Hiding in a Niche

Sequestration in a protective niche is a widely employed strategy for extracellular survival. Two following examples illustrate how diverse the underlying mechanisms may be.

Hiding in Biofilms: Slime Production

Classical examples are infections caused by *Staphylococcus epidermidis*, and chronic lung infections caused by *Pseudomonas aeruginosa*. In both cases, the micro-organisms produce copious amounts of extracellular polysaccharides that shield them from attack by host immune effector mechanisms (Costerton *et al.*, 1995). *Staph. epidermidis*, the major commensal of the intact epidermis, did not play a prominent role in medicine before plastic medical devices were invented. The bacterium possesses a unique protein adhesin that mediates its avid-binding to plastic. Oddly, after the bacteria adhere to surfaces genes are turned on that regulate the synthesis of polysaccharides, which rapidly accumulate to form a mass of protective slime and simultaneously promote adherence of bacteria to one another. The discovery of these mechanisms has been a telling lesson in microbiology, which illustrates the fact that pathogenicity mechanisms need not have evolved through selection or necessity. The capacity of *Staph. epidermidis* to colonise plastic materials and survive in self-produced biofilms has catapulted this bacterium to the top of the list of agents responsible for hospital-associated infections.

An analogous situation exists for *Ps. aeruginosa*. Primarily an environmental organism, in contrast to *Staph. epidermidis*, it produces an array of toxic factors that render it an aggressive, invasive pathogen which typically causes acute, purulent infections. Yet, *Ps. aeruginosa* keeps one survival strategy in reserve for situations in which chronicity of infection is necessary: the prime example is respiratory tract infection in cystic fibrosis, in which mass production of alginate occurs. Like *Staph. epidermidis*, the bacterium turns on genes that lead to mass production of a polymer which provides protection against attack by the host.

Apart from the common strategy of hiding in a mass of polysaccharide slime, *Staph. epidermidis* and *Ps. aeruginosa* represent two extremes. *Staph. epidermidis* has few weapons with which it can inflict damage to host tissues, so that infections are relatively benign and tend to remain localised to the primary sites of colonisation. In contrast, *Ps. aeruginosa* is armed with an array of destructive factors which it uses to mount an attack on the host from its sheltered niche. The protective mass of polysaccharides also provides a diffusion barrier against antimicrobial agents. Hence, chronic *Ps. aeruginosa* infections inexorably lead to organ destruction, as is tragically illustrated by pulmonary lesions in patients with cystic fibrosis and its associated high mortality.

Sequestration

Bacteria embedded in a sequestrated mass of cell debris are protected against host defence mechanisms (Fig. 2). *Staph. aureus* capitalises on this principle by secreting coagulase, which binds to and activates prothrombin, to cause local fibrin formation. Within the 'battle-wall', cytotoxins and degradative microbial enzymes destroy host tissues and cells, and an abscess is thus created. Corroboration that such a mechanism underlies the tendency of *Staph. aureus* infections to remain localised and manifest themselves as abscesses has been obtained from animal experiments with coagulase-deficient mutants. Such strains have a reduced capacity to cause abscesses, and this is accompanied by reduced virulence. Sequestration of bacteria within abscesses also shelters them from antibiotics given systemically. Hence, despite all advances in antimicrobial therapy, the surgeon's classic motto 'ubi pus – ibi evacuo' ([where there is pus, drain it] still holds in the era of modern medicine.

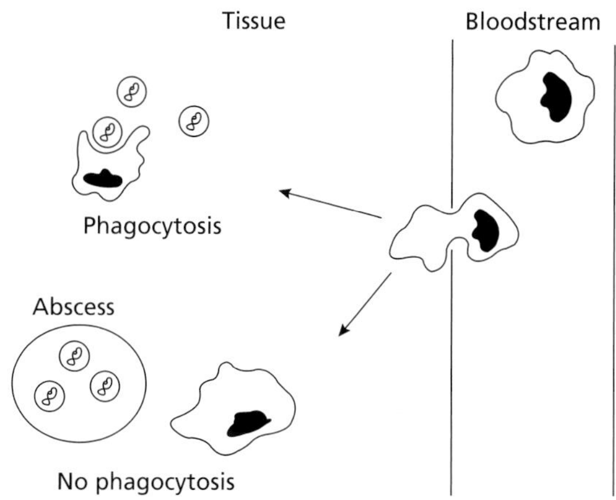

Fig. 2 Sequestration. Micro-organisms sequestered within abscesses are sheltered from host immune effectors.

Another sequestration strategy is encountered in bacterial endocarditis. After they have adhered to the endothelium of heart valves, bacteria form foci for the generation of microthrombi, and so become enmeshed in a mass of blood cells, platelets and fibrin. The platelets may play a protective role by the secretion of antimicrobial substances termed platelet microbial peptides (PMP), or thrombocidins, that serve to limit infection. Resistance to the action of PMPs may contribute to the success of a bacterium to persist in the lesions, into which fibroblasts migrate and lay down connective matrices, thus promoting growth of the vegetation to macroscopic size. Host defence mechanisms are powerless to counteract these processes. Hence, bacterial endocarditis, although characteristically caused by agents such as *Strep. viridans* that are devoid of prominent attack mechanisms, takes an inexorable course that can be halted only by vigorous antibiotic therapy. The vegetations represent a unique form of sequestered bacterial colonies. Once their growth has exceeded a critical mass, surgical removal remains the last life-saving therapeutic measure.

Small-Colony Variants

A possibly relevant strategy for extracellular survival is the formation of small-colony variants (SCV). Investigations in this area have only recently commenced, and so the true significance of the phenomenon cannot yet be gauged. The findings are, however, suggestive (Proctor and Peters, 1998). During infections, auxotrophic mutants of pathogens such as *Staph. aureus* and *Ps. aeruginosa* with an impaired electron transport may exist in a state of quasi-hibernation. They have been termed SCV because of the atypical, small colonies they form on nutrient agar. SCVs appear to increase in numbers as infections become chronic, and they are believed to be selected during prolonged antibiotic therapy. They appear to have novel mechanisms for antibiotic resistance that are related to their impaired electron transport. In addition to increased antibiotic resistance, SCVs also appear to have an enhanced capacity for intracellular persistence. Hence, selection of SCVs may represent a survival strategy that is common to both intra- and extracellular pathogens.

Camouflage

This term is used to describe situations in which the bacteria cannot be recognised by host defence cells. Active camouflage occurs when the bacterium itself produces surface molecules that mimic host structures. Passive camouflage is encountered when bacterial surfaces become coated with host macromolecules.

Active Camouflage: Molecular Mimicry

Microbial surface structures can sometimes resemble host molecules to an extent that renders them unrecognisable. The capsule of group B meningococci represents a classical paradigm; the capsular polysaccharide consists of β1,4-linked neuraminic acid polymers that are identical to certain mammalian conjugates (neuronanal cell-adhesion molecules, N-CAM).

Passive Camouflage: Coating with Host Macromolecules

Fibrinogen *Staph. aureus* clumping factor (CF) is the classic example of a fibrinogen-binding microbial surface protein. It is expressed on most strains of *Staph. aureus* and is absent from coagulase-negative staphylococci. Detection of CF is widely used as a rapid test for presumptive identification of *Staph. aureus*, which reflects rapid and avid interaction with fibrinogen. It is easy to envisage that fibrinogen-coated bacteria enjoy camouflage and thus more easily escape phagocyte surveillance. Indeed, CF-deficient mutants display reduced virulence in animal models, and so the contention that CF is an important determinant underlying extracellular survival is well justified.

Many other microbes bind fibrinogen, but the biological significance and relevance of this for the pathogenesis of disease is less clear. In part, this is due to the difficulty of producing mutants with an isolated deficiency in fibrinogen-binding capacity. A case in point is streptococcal M-protein, perhaps the most versatile camouflage element in the realm of microbial pathogenicity factors. M-protein can bind so many host molecules that defining the significance of a single interaction currently poses an unresolved problem. For example, M-protein also counteracts opsonisation by suppressing complement activation. Hence, it is difficult to determine whether any one property of M-protein is predominantly responsible for its antiphagocytic function.

Immunoglobulins Staphylococcal protein A binds the Fc-domain of IgG, and the 'inverse' orientation of the bound antibodies counteracts recognition and uptake of the bacteria by host phagocytes (**Fig. 3**). This can be regarded as camouflage in the sense that micro-organisms coat themselves with a host molecule which becomes oriented in such a way as to make it indistinguishable from 'self'. It is sometimes argued that expression of protein A by isolates of *Staph. aureus* varies so that its role in extracellular survival may be questioned. An equally reasonable argument, however, is that *Staph. aureus* possesses so many virulence factors that it can afford to vary the expression of any single one. In fact, mutants of protein-A-deficient *Staph. aureus* display reduced virulence in animal models compared with their isogenic strains, and so it is reasonable to retain protein A on the list of pathogenicity factors that capitalise on the camouflage principle.

Fibronectin and Related Proteins Studies on the binding of fibronectin, vitronectin and related proteins currently enjoy popularity, which is understandable because these proteins play key roles in mediating cell–cell and cell–matrix interactions. Numerous bacteria bind these proteins, and it has often been inferred that binding plays a significant role in promoting colonisation. Furthermore, the possibility of camouflage immediately presents itself but it must be noted that stringent evidence for a biological meaning of these interactions is in general lacking. Increased attention to interaction affinities, molecular identification of the binding sites, the creation of defined deletion mutants and their use in valid animal

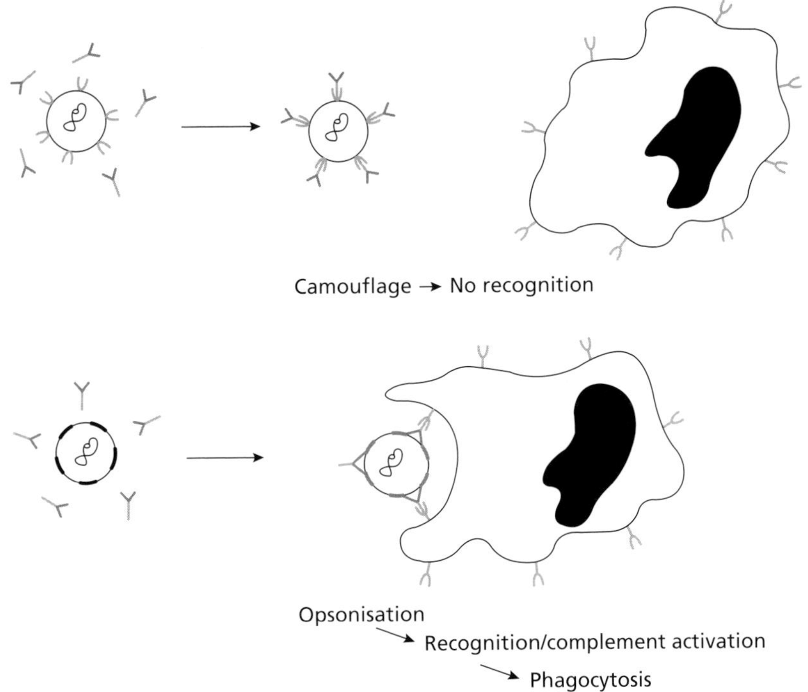

Fig. 3 Protein A. Inverse orientation of host antibodies counteracts recognition and uptake of the bacteria by phagocytes.

models should pave the way to a better appreciation of the true significance of these interactions.

Polysulphated glycosaminoglycans A recently discovered and potentially important camouflage mechanism is the binding to bacteria of polysulphated glycosaminoglycans. It appears that a surprisingly heterogeneous range of micro-organisms, including *Staph. aureus*, *Strep. pyogenes*, pathogenic *Neisseria* and *Helicobacter pylori* express such binding sites. In turn, the glycosaminoglycans become sites for binding of many mammalian heparin-binding proteins, such as fibronectin/vitronectin and many cytokines (e.g. chemokines and fibroblast growth factor). An intriguing possibility is that absorption of the cytokines leads to their local elimination and, thus, to the creation within tissues of survival niches for the micro-organisms.

Immune Evasion

Further to concealment strategies, pathogenic bacteria have evolved mechanisms actively to undermine humoral defence mechanisms. These may be divided into three major categories: immune inhibition and inactivation, immune resistance, and immune decoy and escape.

Immune Inhibition and Inactivation

These terms allude to mechanisms for inactivating or inhibiting antibodies and complement. By counteracting the opsonic mechanisms of the humoral immune system, micro-organisms are able to escape phagocytosis.

Ig-proteases

Proteases produced by pathogenic *Neisseria* and *Haemophilus influenzae* cleave and destroy IgA molecules. It is believed that this process enables the bacteria to remove the obstructive coat of host immunoglobulins, and to thus expose the epitopes required for binding to epithelial cells. The biological relevance of Ig-proteases requires further exploration.

C5a-peptidase

The C5a-peptidase of group A streptococci has the capacity to cleave and inactivate C5. The destruction of the chemokine *in vitro* has marked antiphagocytic effects. Yet, proof that C5a-peptidase truly represents a virulence factor is subject to difficulties, because infection experiments in animal models have not yielded concordant results. This is an example of a situation where attempts to define the relevance of a putative virulence factor can encounter problems even when isogenic bacterial strains and defined animal models are available.

Secreted inhibitor of Complement

This secreted protein of *Strep. pyogenes* (SIC) has the singular capacity to become incorporated into and to inactivate complement C5b-9 complexes (Fig. 4).

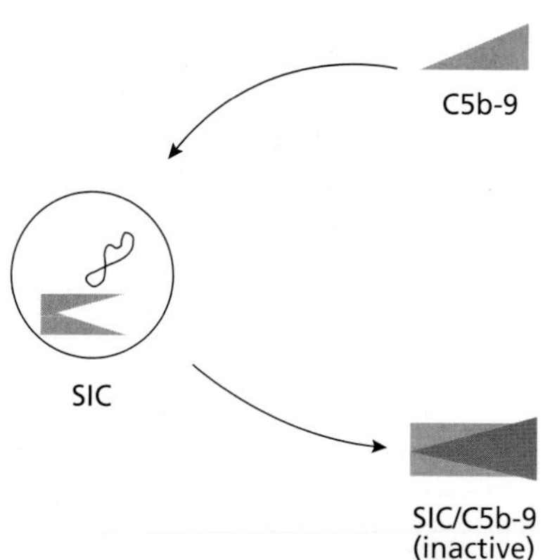

Fig. 4 Secreted inhibitor of complement (SIC). The released inhibitor complexes to and inactivates C5b-9.

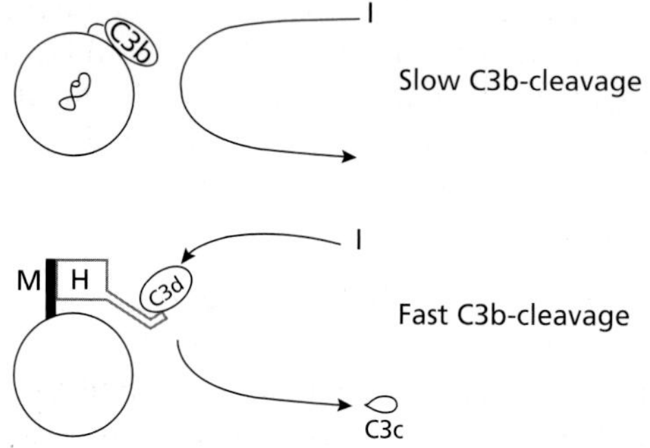

Fig. 5 Factor H binding to bacteria. Bound factor H accelerates the cleavage of C3b by I.

It appears that rapid selection of SIC variants occurs during epidemic waves of *Strep. pyogenes* infections. The level of polymorphism far exceeds that of other genes in these bacteria, indicating that natural selection contributes to this variation. The conclusion can be drawn from those findings that SIC plays a dominant role in enabling bacteria to survive within the host.

One may wonder *a priori* why streptococci should produce such a complement inactivator, which targets a molecule (C5b-9) that cannot in the first place kill Gram-positive organisms. In this context, it is of interest that clusterin and vitronectin, the physiological inactivators of C5b-9, are multifunctional mammalian proteins whose prime relevance probably does not lie in the inactivation of C5b-9. By analogy, it is possible that SIC may have other, still undiscovered activities that underlie its role as a virulence factor.

Factor H Binding

Streptococcal M-protein, an integral component of the cell wall, is the best-studied microbial molecule endowed with the capacity to bind factor H; it possesses a domain that binds factor H. The latter binds C3b, and accelerates its destruction by I, the C3b-cleaving enzyme, and in this way a major opsonin is eliminated (**Fig. 5**). This mechanism is thought to be of prime importance for the antiphagocytic capacity of the M-protein (Zipfel *et al.*, 1999).

Immune Resistance

Mechanisms for the extracellular extermination of bacteria are killing by the terminal complement membrane attack complexes (C5b-9), and by secreted lysosomal components, in particular lysozyme and defensins. In addition, current research suggests roles for lactoferrin, platelet granule components and secreted phospholipase A_2 (sPLA_2). Immune resistance denotes the capacity of micro-organisms to withstand attack by these humoral components.

Resistance to Complement: Serum Resistance

The cytotoxic effector molecule of complement is the terminal C5b-9 complex, which assembles on and inserts into target lipid bilayers to form water-filled pores. Complex assembly is initiated when C5a is cleaved by particle-bound C5a convertases, which are formed by the assembly of earlier complement components. After generation of C5, C6–C9 components spontaneously assemble into the C5b-9 complex, which exposes amphipathic amino acid sequences that insert into the membrane. The half-life of these complexes is very short, and they are rapidly inactivated by the binding of serum inactivators. The major principle that underlies microbial resistance to C5b-9 is the increase in distance between the site of C5b generation and the threatened target membrane. All Gram-positive organisms enjoy this protection because of their extended three-dimensional peptidoglycan wall. This natural resistance to complement killing is often not adequately appreciated; it is probably one important reason why Gram-positive organisms dominate as causes of extracellular infections.

In spite of decades of research, the mechanism of complement killing of Gram-negative bacteria is still not fully understood. C5b-9 complexes appear to generate pores that traverse the entire bacterial wall – i.e. outer membrane, peptidoglycan layer, and inner membrane – but how this happens remains a mystery. In any event, the critical site that bacteria need to protect is their outer membrane. A major strategy employed to this end is the production of lipopolysaccharide (LPS) molecules with elongated side-chains. By analogy to the situation with Gram-positive organisms, complement activation then takes place at sites remote enough from the outer membrane to ensure that C5b-9 inactivation occurs before the molecules reach their target (Rautemaa and Meri, 1999) (**Fig. 6**).

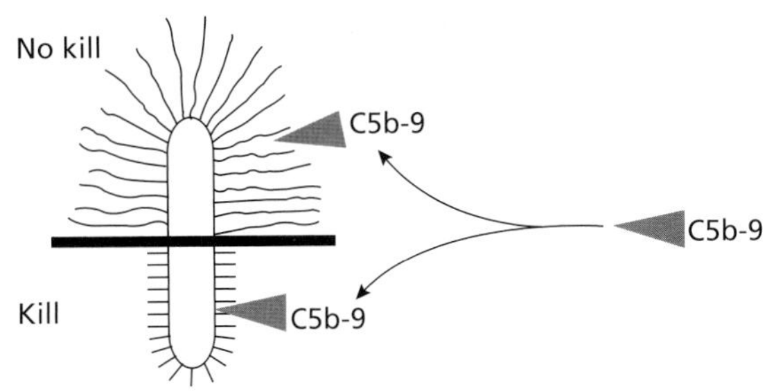

Fig. 6 Serum resistance. C5b-9 cannot exert its bactericidal effect if generated at a distance from the target lipid membrane.

The significance of serum resistance is established. The majority of Gram-negative commensals that colonise the colon are serum-susceptible. Therefore, when the mucosal barrier is compromised and bacteria enter the bloodstream, most are rapidly killed. Thus, patients with appendicitis or diverticulitis characteristically have high fever but their blood cultures are sterile or contain only few viable bacteria. When bacterial dissemination occurs, only the serum-resistant isolates can be recovered from extra-intestinal sites.

Bacterial meningitis typically evolves by haematogeneous dissemination of the micro-organisms, which accordingly must be serum-resistant. *H. influenzae* seems to be able rapidly to convert from a serum-sensitive to a serum-resistant phenotype. In the case of *N. meningitidis*, it appears that protective antibodies directed against outer-membrane components target and focus complement attack to critical sites that are close enough to the outer membrane to cause its perforation. In the absence of such antibodies, complement activation is ineffective because C5b-9 deposition occurs at sites that are not vulnerable. Thus, antibodies against meningococci may provide protection through two pathways: antibodies against the capsule promote opsonisation and phagocytosis, whereas antibodies against certain outer membrane proteins directly mediate complement killing.

Lysozyme Resistance

Lysozyme is a ubiquitous antimicrobial agent. In addition to representing an important microbicidal component of phagocyte lysosomes, it is also present in blood and body fluids. Whenever serum bactericidal activity against a Gram-positive organism is observed, the action of lysozyme is likely to be the underlying cause.

Lysozyme disrupts peptidoglycan (murein) by cleaving the $\beta 1-4$ glycosidic bond between *N*-acetylmuramic acid and *N*-acetylglucosamine. Staphylococcal peptidoglycan is markedly resistant to degradation, partially because of the exceptionally high degree of peptide cross-linking. This may be yet another reason for the dominance of these bacteria as causes of extracellular infections.

The action of lysozyme on Gram-negative organisms is more complicated, because the enzyme cannot pass the intact outer membrane barrier to reach its target. This can occur, however, when complement C5b-9 pores perforate the outer membrane. Lysozyme then acts synergistically with complement to promote killing of Gram-negative organisms. Mechanisms that may render these bacteria resistant to lysozyme action have not been described.

Resistance to Other Microbicidal Mechanisms

Lactoferrin In addition to its iron-binding function, lactoferrin appears to possess domains that are directly bactericidal. The spectrum of lactoferrin-susceptible microbes is reportedly broad, and lactoferrin is thus widely assumed to represent the major antimicrobial component of human milk. Resistance mechanisms of micro-organisms have not been characterised.

Defensins and Platelet Microbicidal Peptides Cationic antimicrobial peptides are widely present in cell lysosomes and in neutrophil and platelet granules. Some bacteria display a natural resistance to PMPs; in animal models of infectious endocarditis, such strains are more virulent. This finding has obvious implications; platelet activation is a common event in extracellular infections, and it is intriguing to speculate that bacteria enmeshed in clots may be exposed to toxic concentrations of PMPs. Questions relating to the relevance of resistance against these molecules have yet to be addressed in a clinical setting.

Investigations into possible mechanisms that may render microbes resistant against leukocidal defensins are already forthcoming. Thus, addition of positive charges, e.g. to phosphates of LPS, confers resistance to Gram-negative organisms against the action of cationic defensins. This and related phenomena may emerge as a novel survival strategy of both intra- and extracellular pathogens.

Secreted Phospholipase A_2

Another host component with potential antimicrobial activity is the secreted phospholipase A_2 ($sPLA_2$), which can be released in quantity from many cells, in particular from macrophages. $sPLA_2$ can directly kill Gram-positive organisms *in vitro*, apparently by cleaving phospholipids in the outer leaflet of the bacterial plasma membrane. Gram-negative organisms are not primarily susceptible but, in analogy with lysozyme, $sPLA_2$ appears to act in synergy with complement. Thus, complement-mediated damage appears to render the outer membrane permeable for $sPLA_2$, which can then attack the inner membrane. Time will tell whether these processes are truly relevant in the context of host-defence against extracellular bacteria, and whether microbes have evolved strategies to thwart attack by this enzyme.

Immune Decoy and Escape

Here, we allude to active mechanisms by which bacteria evade immune effectors. They are directed against the opsonic action of antibodies and complement, which aid the invaders to persist unidentified in the midst of phagocytic sentinels in blood and tissues.

Antigenic Variation

The classical paradigm is the variation of surface antigens in *Borrelia recurrentis*, which explains the clinical presentation of relapsing fever. Serotype specificity is determined by variable lipoprotein antigens in the spirochaete outer membrane (see Chapter 98). Variants appear spontaneously and flourish in the host until specific antibodies are produced (**Fig. 7**). Then fever subsides, to return after 5–10 days when new variants have multiplied (Haake, 2000).

Borrelia burgdorferi also expresses lipoprotein antigens. Not quite unexpectedly, but none the less disturbingly, antigenic variation has also been observed in this prevalent bacterium (see Chapter 97). There is an immediate possibility that, under selective pressure, antigenic variation may evolve to become a relevant mechanism for this pathogen to survive immune attack. The mechanism of antigenic variation in *B. burgdorferi* is similar to that operating in *N. gonorrhoeae*, which leads to the generation of pilus variants. The latter phenomenon contributes to the fact that acquired immunity against gonorrhoea is poor, and it also compromises efforts to produce an efficacious vaccine. Whether the ospA vaccine currently being tested for the prevention of Lyme disease will be successful remains to be established.

Shedding of Cell-Wall Components

The shedding of cell-wall components is widespread and its possible significance in the context of immune decoy deserves mention. Release of LPS from Gram-negative bacteria, or of teichoic acids from Gram-positive organisms, might serve to bind antibody and complement at sites remote from the bacteria. LPS is spontaneously released in quantity from *N. meningitidis*, and may be shed to various degrees from other Gram-negative organisms. By analogy, Gram-positive pathogens may shed antibody-binding and complement-consuming cell-wall constituents, such as peptidoglycan and teichoic acids. Release of immune decoy molecules in tissues may direct the action of complement away from the bacteria and thus help the micro-organisms to evade the immune system (**Fig. 8**).

Immune Paralysis

This term has been used to describe a state of depression of the immune system that is characteristically observed in systemic inflammatory reaction syndromes and septic shock. Immune paralysis appears to result from dominance of inhibitory mediators such as prostaglandin E_2 and certain cytokines, in particular transforming growth factor (TGF)-β and interleukin (IL)-10, whose production follows after the first wave of pro-inflammatory cytokines. Characteristically, MHC II molecules on antigen-presenting cells then become down-regulated and a systemically subdued immune response ensues.

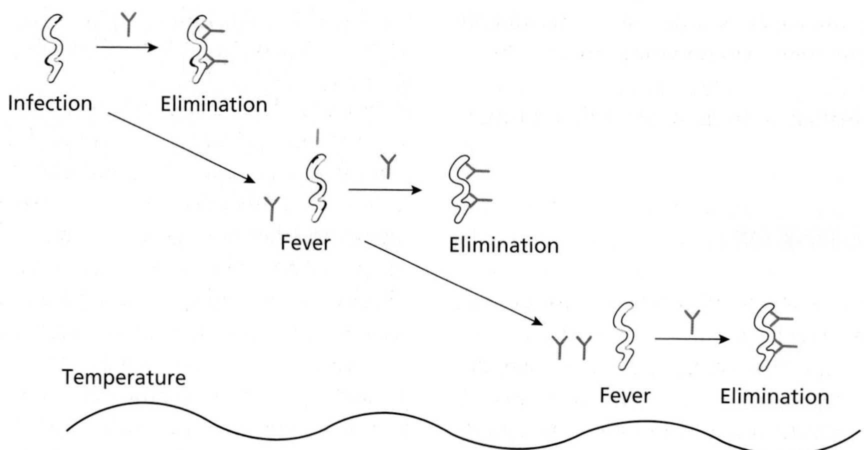

Fig. 7 Antigenic variation. Variants escape elimination by antibodies/phagocytes.

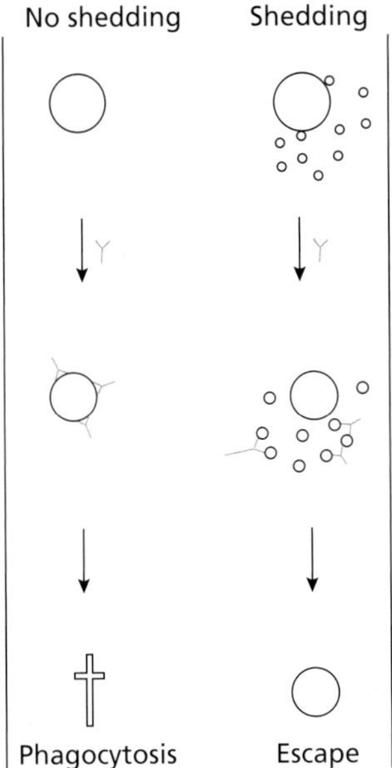

Fig. 8 Shedding of cell-wall components. Shed-wall components may serve as a decoy for host antibodies, thus helping the bacteria to escape phagocytosis.

It is possible that certain pathogens actively exploit immune paralysis as a survival strategy. *Yersinia pestis*, once one of the most dreaded of all pathogens, is armed with an array of virulence factors that enable it rapidly to multiply and disseminate both extra- and intracellularly in the host. *Y. pestis* secretes a soluble protein called LcrV (V-antigen, a product of the low calcium response stimulon). This product appears to be essential for extracellular dissemination, because antibodies against the protein confer resistance against experimental plague. In animals, the isolated V-antigen prevents production of tumour necrosis factor α (TNFα) and interferon γ (IFNγ), possibly via overinduction of IL-10. It is intriguing to postulate that systemic immune paralysis ensuing as a consequence of such cytokine imbalance is a decisive determinant that underlies bacterial dissemination in the host. Whether other pathogens utilise similar strategies to paralyse the mammalian immune system calls for future investigation.

Attack Strategies

Attack is the best strategy of defence. Virtually all extracellular pathogens employ mechanisms to damage host cells and destroy tissues, thus enhancing their chances of survival. To bring some system into the discussion of these multi-faceted microbial weapons, the topic will be treated under three headings – toxins, toxic injection and enzymes – but overlaps in the discussion are unavoidable. For example, certain toxins are enzymes, and their classification as toxins is based on an arbitrary definition, i.e. that cells are the targets for toxins. On the other hand, endotoxins and superantigens derange physiological functions, and do not primarily inflict damage on host cells, as might be demanded of genuine toxins. 'Toxin injection' is used to denote the direct delivery of macromolecules from bacteria into host cells. Many of these molecules are actually toxins with modes of action similar to those of their secreted counterparts.

Microbial Toxins

These are actively secreted or surface-associated microbial products that are traditionally classified as exotoxins or endotoxins. The molecular structure and mechanisms of action of endotoxin and major microbial exotoxins are discussed elsewhere in this book. Here an attempt is made to provide a general overview and to discuss their relevance to disease pathogenesis. It will, therefore, sometimes not be evident whether a toxin truly contributes to the extracellular survival of the producing micro-organism. A discussion directed exclusively to survival strategies would, for example, not include the neurotoxins. On the other hand, the latter are of prime medical importance, and unravelling their mechanisms of action has represented major advances in cell biology. The following discussion is broadened in scope to cover the topic of microbial pathogenesis in general.

Endotoxin

The term endotoxin is synonymous with bacterial LPS and, more specifically, with the lipid A component of this remarkable macromolecule (Rietschel *et al.*, 1996) (see also Chapter 6). LPS is an integral constituents of the outer membrane of all Gram-negative bacteria, and the structure of lipid A is highly conserved. Gram-negative organisms display a variable and little-studied tendency spontaneously to shed outer-membrane blebs rich in LPS. Endotoxin is released naturally when bacteria are killed by humoral immune effectors (complement) and by antibiotics. Furthermore, endotoxin embedded in intact bacterial walls also appears to be biologically active. Thus, endotoxin effects can be elicited both locally by contact of bacteria with host cells, and systemically by liberated LPS. It is possible that the immune system, in particular the mucosa-associated lymphoid system, is continuously primed by small quantities of endotoxin. This may even serve a positive, physiological function by keeping the immune system in a state of low-grade activation. The detrimental effects of endotoxin, which can result in local and systemic organ dysfunction, derive from exaggerated physiological responses. In a sense, these effects may thus be regarded as a special type of immunopathological reaction.

It is seldom appreciated that the biological activity of endotoxins from different bacterial species can vary considerably as the result of differences in chemical structure. The number and length of the fatty acids and the extent of phosphorylation in the lipid A moiety represent decisive elements that apparently affect the molecular conformation and, concomitantly, the biological activity of LPS. The specific activities of LPS from *Bacteroides*, *Legionella* and *Chlamydia* are orders of magnitude lower than the activity of LPS from Enterobacteriaceae and *Neisseria*, as judged by their capacity to promote the release of inflammatory cytokines from human macrophages. By contrast, the reactivity of different LPS in the *Limulus* coagulation test, the standard procedure for their quantification in body fluids, is quite uniform. Consequently, it is virtually impossible to differentiate between levels of 'active' and 'poorly active' circulating endotoxins, and this impedes the interpretation of most clinical studies. Investigations in patients with meningococcal sepsis have been the most fruitful, and have established that the severity of symptoms and the clinical prognosis correlate with levels of endotoxin and inflammatory cytokines. Paired with the fact that meningococci lack other attack mechanisms, the *N. meningitidis* model has become the classical paradigm that illustrates the relevance of endotoxin in human disease.

Structure and Mechanism of Action Enterobacterial LPS molecules have three structural domains: the lipid A moiety, consisting of a disaccharide unit, with six fatty acids attached, the core oligosaccharide and the O-specific side-chain (Fig. 9).

Although the structure of lipid A is highly conserved, variations occur amongst bacterial species that markedly influence biological activity. Lipid A that contains C14 fatty acids displays the highest endotoxic activity. Lipid A from *Chlamydia* and *Legionella* have very long fatty-acyl chains (C22–C32) and are much

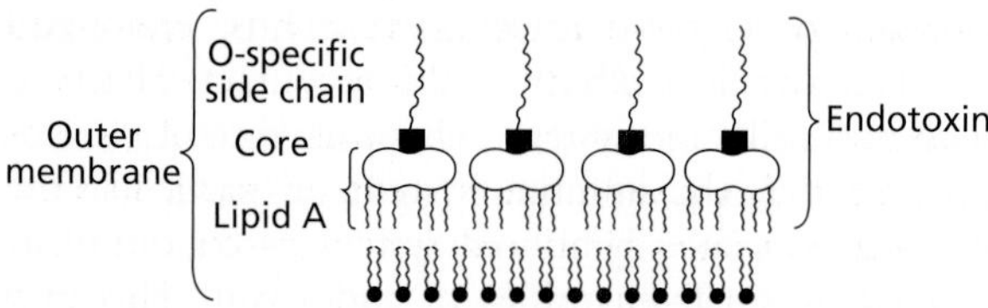

Fig. 9 Endotoxin structure. Endotoxin (LPS), composed of lipid A, core polysaccharide and the O-specific side-chain, is localised in the outer leaflet of the outer bacterial membrane.

less active. Lipid A from *Bacteroides* spp. lacks one phosphate group and is also less active.

The inner core region contains two 2-keto-3-deoxy-octulosonic acid (KDO) residues and several (usually 3) heptose residues. The structure of this region is highly conserved. It is linked to the more variable outer core region that comprises a number (usually three) of hexose residues.

Linked to the outer core region is the O-specific chain, a polymer of repeating oligosaccharide units that can vary greatly in composition and length. The O-specific chain is not an essential part of LPS; it has no toxic potential and bacteria with short O-specific chains are phagocytosed and killed by complement. Long O-specific side-chains are therefore important determinants that influence the capacity of pathogens to survive extracellularly.

Lipid A is an amphiphile that exists in two physical states, the monomer and the aggregate. It is not known which of these dominates *in vivo*, and which mediates toxicity but a large number of host macromolecules have the capacity to bind LPS. These include high-density lipoprotein (HDL), low-density lipoprotein (LDL), and several cationic proteins, of which the bactericidal permeability-increasing protein (BPI) and cationic antibacterial peptide (CAP) lead the list. Binding of LPS to these proteins is accompanied by its neutralisation. In contrast, one protein that binds LPS enhances its endotoxic activity, and this LPS-binding protein (LBP) is considered to be a major mediator of endotoxin action in mammalian hosts. LBP appears to capture and deliver LPS to cell-surface receptors, in particular to CD14 (Fig. 10). The latter can bind LPS, but since it is an inositol phosphate-anchored protein, it cannot itself deliver signals to cells. There is compelling evidence that signal transduction involves Toll-like receptor 4 (TLR4). Thus, in experimental knockout models, deletion of TLR4 completely abolishes LPS stimulation. Moreover, in humans and mice naturally occurring TLR4-defects exist that result in abrogated responsiveness to LPS. In humans, this is associated with increased susceptibility to Gram-negative infections, which illustrates how important the physiological cytokine response to endotoxin is in the context of host defence. TLRs, which were first discovered in insects, are emerging as key recognition molecules of the innate immune system. Whereas TLR4 is essential for LPS-recognition, by analogy TLR2 appears to 'see' peptidoglycan and in particular, bacterial lipoproteins (Janeway and Medzhitov, 1999; Beutler, 2000).

The ecto-domain of TLR4 exhibits marked interspecies variation, which contrasts with the much more conserved cytoplasmic domain. Intriguingly, this is

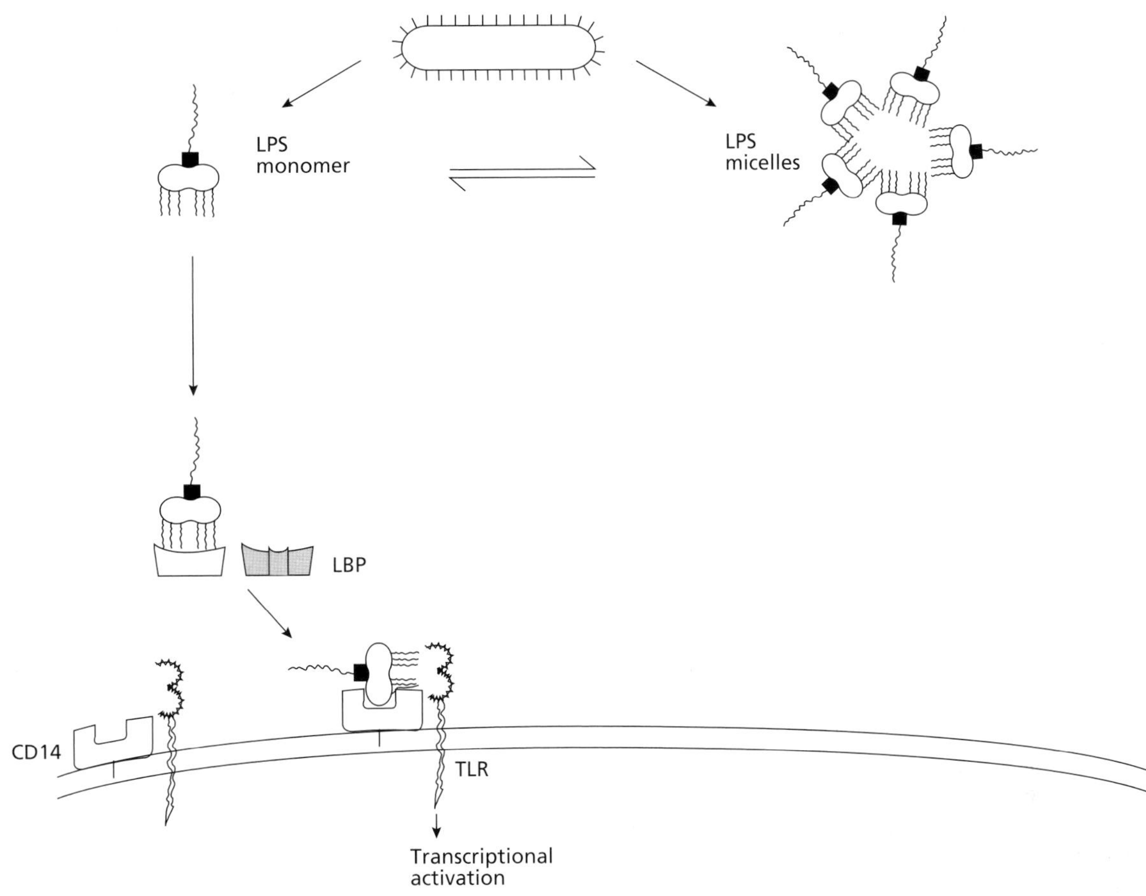

Fig. 10 A model for LPS signalling. LPS is released by bacteria. The monomer is bound by LBP and then delivered to CD14. Co-interaction with CD14-associated TLR4 results in transcriptional activation with production of inflammatory cytokines.

mirrored by varying responsiveness to structural variants of lipid A. For example, tetraacyl-lipid A stimulates murine but not human cells, and cross-complementation studies show that this is due solely to the respective TLR4. It is possible that TLRs directly contact the stimulatory microbial molecules, with CD14 playing a major role as a deliverer of LPS. CD14 is expressed mainly on macrophages, so that the primary action of LPS is probably directed to these cells. However, activated macrophages cleave and shed soluble CD14 molecules (sCD14), which retain their capacity to bind LPS. There is evidence to indicate that sCD14/LPS complexes can subsequently activate bystander cells, presumably by interacting with TLRs. The situation appears analogous to '*trans*-signalling' by IL-6/sIL-6R complexes, which bind to their signal-transducing element (gpl30) on bystander cells. *Trans*-signalling by CD14/LPS would satisfactorily account for the disseminated effects of LPS observed in systemic inflammatory reactions and septic shock.

Signalling via TLRs leads to a multitude of cellular events including induction of synthesis of pro-inflammatory cytokines (IL-1, IL-6, IL-8, IL-12, TNFα), and in up-regulation of adhesion molecules on endothelial cells and on white blood cells. De-regulation of these processes is responsible for the detrimental effects of endotoxin. Seen very schematically, endothelial dysfunction promotes adherence and activation of granulocytes and platelets so that thrombotic foci are created. The adherent granulocytes are thought to invoke endothelial and tissue damage through their release of reactive oxygen metabolites and granule constituents. Disseminated formation of microthrombi follows. Pathological shunt flow leads to ventilation-perfusion mismatch in the lung and to derangement of oxygen supply in other organs. Unchecked consumption of coagulation factors leads to coagulopathy that manifests first as petechial and later as massive bleeding. Modern intensive-care medicine has not been able to improve the outcome of endotoxin shock significantly, and mortality in such patients even now is in the range of 70–80%.

Endotoxin as a Future Target for Therapy As already mentioned, LPS is not a toxin in the strict sense, since its primary effects are not detrimental to cells. LPS-induced production of cytokines is *per se* probably important for the correct functioning of the immune system. In fact, suppression of TNFα is harmful and even augments bacterial infections in animals. The toxic properties of LPS thus probably do not serve any primary purpose in promoting extracellular survival, and bacterial dissemination that occurs, for example, during meningococcal sepsis is probably an epiphenomenon. Although systemic organ failure caused by endotoxin is the consequence of cytokine overload, therapeutic attempts directed at immune suppression, as with single cytokine antagonists and antibodies, will probably fail because of the intricate nature of the cytokine network. Rather, the logical approach to the development of effective therapeutic strategies is the neutralisation and reduction of endotoxin load (Rietschel *et al.*, 1996). What, therefore, are the prospects?

Lipid A carries the toxic activity of LPS. When separated from other LPS components, lipid A is immunogenic. Unfortunately, however, anti-lipid A-antibodies do not efficiently bind to their targets in intact bacteria or in released outer membrane vesicles, and hence they cannot protect against endotoxin *in vivo*. Accordingly, efforts to use such antibodies for the therapy of sepsis and septic shock have failed.

The core region with its highly conserved structure is a much better candidate target, because it is accessible to antibodies both in intact bacteria and in shed LPS molecules. Antibodies raised against the inner core of one *E. coli* strain have been found to react with all other *E. coli* strains and also to cross-react with *Salmonella*. Intriguingly, binding of endotoxin by this antibody results in neutralisation, and the application of the antibody was found to be highly protective in animal models of bacterial sepsis.

Certain cationic molecules can bind to and inactivate LPS. BPI is one candidate molecule that is at present undergoing clinical trials in patients with meningococcal sepsis. Another promising approach is the application of polymyxin B coupled to dextran carrier molecules. Carrier-bound polymyxin B retains its capacity to bind and neutralise endotoxin, while toxic properties are largely abrogated. Loss of cellular cytotoxicity may be due to the inability of carrier-bound polymyxin to insert and oligomerise in cell membranes. Dextran-coupled polymyxin also has a markedly prolonged half-life, and so current hopes that a novel agent may become available for therapy of endotoxin-mediated systemic disease appear justified.

Endotoxin Antagonists LPS molecules with altered acyl chains can be rendered inactive with retention of their ability to bind and saturate cellular LPS receptors (CD14 and TLR4). It is theoretically possible that such LPS analogues may prove to be of therapeutic value if applied at timely stages in the course of systemic inflammatory reactions.

In summary, endotoxins may be regarded as accidental weapons of extracellular pathogens that indirectly inflict a multitude of deleterious events on the host. Research in this area involves all fields in biology and medicine, from chemistry to molecular cell biology and immunology. Despite spectacular successes, innovative approaches need to be taken in order to develop effective therapeutic strategies in the future.

Protein Toxins

Most extracellular pathogens produce at least one important protein toxin, and research on these molecules has become a major topic in molecular and cell biology. Hundreds of bacterial protein toxins have been described and characterised, often at the molecular and even atomic level (**Fig. 11**). These toxins are grouped into three major categories: toxins with intracellular targets, toxins that damage the cytoplasmic membrane and toxins that cause dysfunction of the immune system (superantigens).

Toxins with Intracellular Targets All the toxins that have intracellular targets have been identified as enzymes. These molecules must be delivered to the cytosol of target cells in order to exert their toxic action. The first pressing question relates to how such delivery occurs (Olsnes *et al.*, 2000). Primary interaction with susceptible cells is determined by the presence and number of toxin receptors, which are often glycosylated surface components, i.e. glycolipids and glycoproteins. Bound toxins are usually endocytosed, but thereafter the sites from which they enter the cytosol varies. For example, diphtheria toxin and anthrax toxin translocate from acidic late endosomes, whereas shiga toxin undergoes retrograde transport to the Golgi and the endoplasmic reticulum, to be delivered from these compartments.

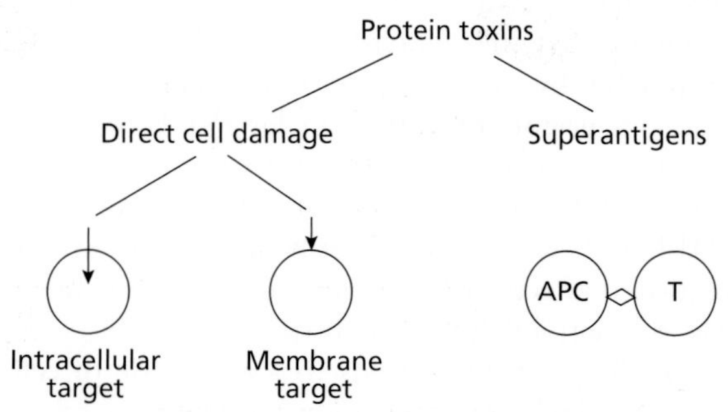

Fig. 11 The action of protein toxins.

The mechanism of toxin translocation from late endosomes is understood best and displays analogies to the mechanism of infection by many enveloped viruses. The acid pH in the endosomes provokes conformational changes leading to exposure of domains within the toxin molecules that can spontaneously insert into lipid bilayers. This leads to the creation of translocation channels through which the catalytic domains are threaded. Diphtheria toxin is composed of two chains linked by a disulphide bond. The B-chain harbours the translocating, pore-forming sequence, which comprises an amphipathic α helix (Fig. 12). It is not known whether the A-chain is threaded through a channel composed of B-chain oligomers, or whether the pore is lined by the α helix on one side and by lipid on the other. Anthrax toxin represents another translocation prototype. The cell-binding and translocating element is designated the protective antigen. This component binds to cell receptors and heptamerises on the membrane surface. The heptamers then bind the catalytic entity, termed the lethal factor. The complex comprising lethal factor and protective antigen is endocytosed as a unit. In the acidic endosomes, one stretch of amino acids inserts from each protective antigen monomer into the membrane, leading to creation of a water-filled heptameric pore. Lethal factor is then threaded through this channel into the cytosol.

The mechanisms of translocation of most other intracellular toxins have not been delineated in similar detail. Toxin translocation from the endoplasmic reticulum may operate by reversal of the physiological 'Sec' import machinery, similar to what has been described for the escape of viral proteins from the endoplasmic reticulum back to the cytosol.

Having reached the cytosol, the catalytic domains of bacterial toxins can provoke a plethora of effects depending on their molecular targets, which are usually proteins. Two predominant types of modification can be differentiated: proteolytic cleavage, or covalent modification with a cytosolic molecule.

The two types of covalent modification identified to date are ADP-ribosylation and glycosylation. ADP-ribosylating toxins catalyse two closely linked reactions. First, they cleave NAD to generate ADP-ribose; then they covalently couple the ADP-ribose to specific protein targets. This causes functional impairment of the ribosylated domains. Different toxins ADP-ribosylate different proteins, and this accounts for the widely varying consequences of their action (Masignani *et al.*, 2000).

Glycosylating enzymes utilise cellular UDP-monosaccharides as substrates and covalently couple the sugars to specific residues on small GTPases, causing their inactivation (Just *et al.*, 2000). Loss of G protein function leads to diverse detrimental cellular events.

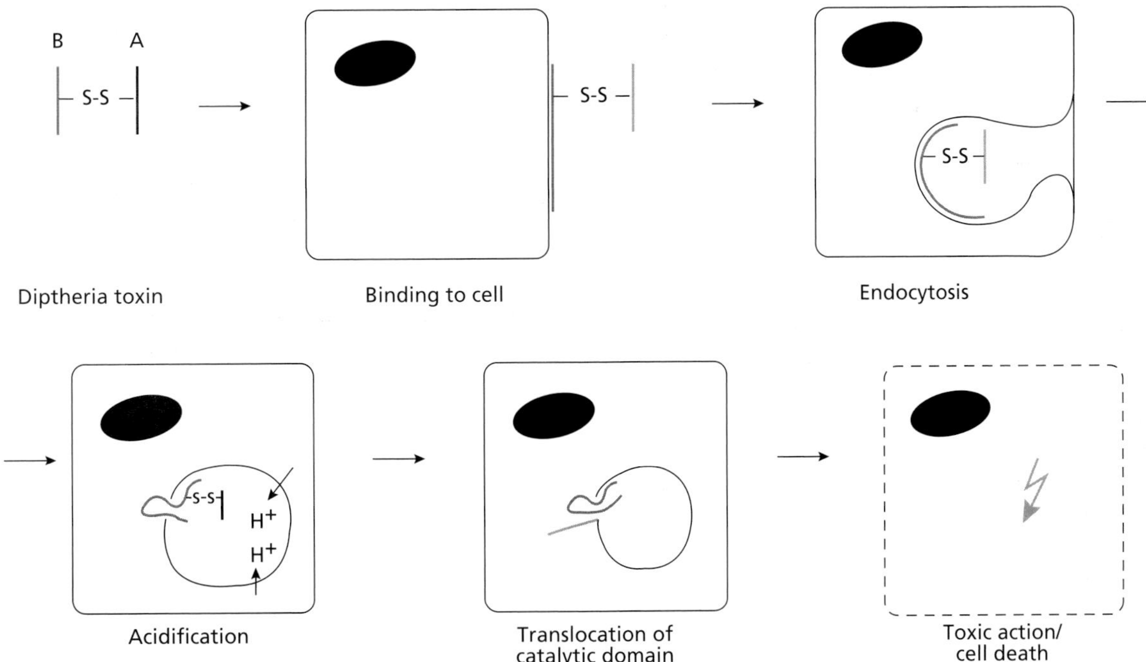

Fig. 12 Mechanism of translocation of diphtheria toxin.

Toxins that Inhibit Protein Synthesis Prototypes of these toxins are diphtheria toxin and the structurally −1 related exotoxin A of *Ps. aeruginosa*. Both toxins ADP-ribosylate elongation factor-2 (EF-2), which results in cessation of protein synthesis. Very few diphtheria toxin molecules are sufficient to kill a cell. This explains the dramatic clinical course taken by diphtheria. Production of exotoxin A by *Ps. aeruginosa* undoubtedly contributes to the virulence of this bacterium, and it is one reason why this species plays a more prominent role in medicine than other members of the genus. It should be re-called that diphtheria toxin is the only virulence factor of *Corynebacterium diphtheriae*. Since the bacterium does not disseminate, neutralisation of the toxin suffices as the single measure to prevent and cure the disease. In contrast, *Ps. aeruginosa* possesses an array of virulence factors and has a tendency to disseminate. Accordingly, identification of protective antigens and production of effective vaccines have encountered great difficulties. *Ps. aeruginosa* continues to be the major cause of pneumonia in cystic fibrosis patients, and is a prominent pathogen in wound- and hospital-acquired infections.

Toxins that Deregulate Adenylate Cyclase Two different paths can converge to cause hyperactivity of adenylate cyclase. One pathway involves locking the stimulatory G protein α subunit in its active state. The other is more complicated and functions by repression of an inhibitory G protein. The first pathway is utilised by enteric pathogens and targets the intestinal epithelium. The second pathway underlies the action of pertussis toxin in the respiratory tract. Oddly, the consequences are entirely different; enhanced adenylate cyclase activity in intestinal epithelial cells causes diarrhoea, whereas pertussis toxin is centrally involved in the pathogenesis of whooping cough.

Cholera toxin and the structurally related *E. coli* heat-labile toxin represent the classic prototypes of ADP-ribosyltransferases that cause watery diarrhoea (Sears and Kaper, 1996; Patton *et al*., 2000). After their adherence to the epithelial cell surface and multiplication of the pathogens, the secreted enterotoxins bind to gangliosides and translocate their catalytic domains to the cytosol. The bacteria do not invade the tissues and thus there is no inflammatory reaction. The catalytic domains of the enterotoxins ADP-ribosylate the α subunit of heterotrimeric, stimulatory G proteins, which are responsible for activating adenylate cyclase. ADP-ribosylation inactivates the GTPase domain, so the α subunit is locked in its active state. As a consequence, adenylate cyclase activity cannot be halted, and massive accumulation of AMP occurs. Elevated cAMP in turn renders intestinal epithelial cells incapable of absorbing chloride ions and this leads to a massive efflux of salt and water into the intestinal lumen.

This is a rather indirect strategy for extracellular survival, because the offending micro-organisms are never directly threatened by the host defence system. From a teleological view, after their multiplication, the bacteria harness host physiology to effect their ejection into the environment, where new hosts can be found. Contaminated drinking-water is the major route responsible for rapid spread of cholera. There has been progress in vaccine development, and a live oral vaccine is currently in use. The vaccine strain produces a truncated form of cholera toxin that lacks the catalytic domain. IgA antibodies produced in response to vaccination inhibit binding of cholera toxin to intestinal cells. Full protection cannot be expected, however, since the toxin secreted by cell-adherent bacteria is likely to be less accessible to the antibodies and, as with neurotoxins, toxin molecules that have been delivered to the cytosol cannot be neutralised. The same difficulty is faced by research in the *E. coli* enterotoxin field, and enterotoxic *E. coli* continues to cause millions of diarrhoeal episodes worldwide each year. Although this is a self-limiting disease, dehydration can be a major cause of death in developing countries.

Although whooping cough is an ancient disease of major medical importance, the intricacies of its pathogenesis have not been fully unravelled. A number of virulence factors act in concert, with pertussis toxin representing one important component. The catalytic domain of this toxin ADP-ribosylates the α subunit of an inhibitory G protein and a suppression of G_1 results in hyperactivity of adenylate cyclase. Deregulation of this enzyme in respiratory tract epithelia, together with the effects of other toxins elaborated by *B. pertussis*, lead to the remarkable clinical manifestation of whooping cough.

Microbial Adenylate Cyclases Stringent control of adenylate cyclase activity is crucial for normal cell function, and hence microbial toxins that target adenylate cyclase are very dangerous. The most direct way to disrupt the adenylate-cyclase-dependent regulatory network is by 'smuggling' extraneous enzyme molecules into target cells. This strategy is realised by several important pathogens. The adenylate cyclase cytotoxin of *B. pertussis* is a remarkable molecule that actually harbours two entities: one is the enzyme adenylate cyclase, the other is a pore-forming toxin (PFT) belonging to the RTX family (see below) that forms a channel through which the adenylate cyclase is delivered.

B. anthracis also produces an adenylate cyclase, termed oedema factor. By analogy with lethal factor, this uses the protective antigen entry system to gain access to the cell cytosol. Enhanced cAMP causes egress of cellular fluid to the interstitium and massive tissue oedema that is characteristic of cutaneous anthrax follows.

Proteolytic Toxins Major representatives of toxins with proteolytic activity are the neurotoxins of *Clostrodium tetani* and *Clostridium botulinum*, which are zinc metalloproteinases that specifically cleave and destroy components of the neuronal secretion machinery. As a consequence, vesicles that contain neurotransmitters cannot dock and fuse with the plasma membrane, and secretion is inhibited. Neurotoxins are long-lived enzymes and this accounts for the duration of clinical symptoms (Schiavo *et al.*, 1995).

As with all protein toxins, preventive immunological measures rest on the capacity of antibodies to inhibit the binding of the toxins to cells. Antibodies directed against tetanus toxin capture the toxin during its haematogeneous dissemination, and therefore efficiently prevent tetanus. Although antibodies against botulinum toxin are normally given to patients with suspected or established botulism, their value has never been proved, since manifestation of clinical symptoms show that the toxin has already gained access to the cells, and is thus no longer susceptible to the neutralisation by antibodies.

A recently discovered protease is the lethal factor of *B. anthracis*. This enzyme cleaves mitogen-activated protein kinase kinases (MAPKK), disrupting their interaction with the MAPKase extracellular signal-regulated kinases, ERK1/ERK2. As a consequence, ERK1/ERK2 are free to become phosphorylated and to translocate to the nucleus. As-yet-undelineated pathways then lead to cytokine production in macrophages, production of oxygen radicals, apoptotic cell death and the release of IL-1β. Protection against anthrax can be achieved by immunisation with the protective antigen. The respective antibodies prevent binding of the translocating machinery, so that lethal factor cannot gain entry into the cells.

Toxins that Disrupt the Cytoskeleton ADP-ribosylating toxins can directly and indirectly cause breakdown of the actin cytoskeleton, which results in rounding of cells and loss of adherence (Fig. 13). *Cl. botulinum* C2 toxin is composed of two subunits. Subunit C2-II carries the binding domain, and subunit C2-I is enzymatically active. It is reminiscent of the anthrax protective antigen, in that membrane-bound C2-II subunits oligomerise, bind C2-I and are endocytosed. Once delivered to the cytosol, C2-I ADP-ribosylates actin monomers, rendering them unable to polymerise. This directly causes disruption of the cytoskeleton.

The C3-toxin of *Cl. botulinum* acts indirectly. Its catalytic domain ADP-ribosylates rhoA, the small G protein that is essential for actin polymerisation. Ribosylated rhoA is inactive and cells deprived of this important molecule display an array of functional defects, including a disrupted cytoskeleton (Bouquet *et al.*, 1998).

Glycosylating Toxins The prototype of this toxin family is the cytotoxin of *Cl. difficile*, the major cause of antibiotic-associated colitis. This and other members of the family are referred to as 'large clostridial toxins'. Each is composed of a single polypeptide chain with a molecular mass of $\geq$300 kDa that contains a binding, a translocating and a catalytic domain. How *Cl. difficile* cytotoxin, or any other large clostridial toxin, reaches the cytosol remains a mystery. In the cell, the toxin covalently attaches one glucose molecule to a serine residue in the catalytically active pocket of rhoA. The consequence is analogous to that of ADP-ribosylation by *Cl. botulinum* C3-toxin. Amongst other effects, loss of ρ function causes the rounding-up of the cells and detachment from their substratum (Fig. 13).

Ribonucleases The prominent member in this category is the cytotoxin of *Shigella*, known as shiga toxin. It is endowed with the capacity selectively to cleave ribosomal RNA, thus destroying ribosome function and the ensuing cessation of protein synthesis causes cell death. Related toxins are produced by invasive *E. coli* strains. They have been termed shiga-like toxins or verotoxin, and are probably important factors that contribute to the pathogenicity and invasive capacity of these bacteria (O'Brien *et al.*, 1992).

The list of biochemical reactions catalysed by intracellular toxins will never be complete. The nature of a modification can be so subtle that it may only be detected by a directed search with the use of high-resolution analytical techniques. An example is the deamidation of 'rho' catalysed by *E. coli* lethal distending toxin. This reaction, which inactivates rho and causes cell death, reduces the molecular mass of the G protein by only 1 Da!

Membrane-damaging Toxins The plasma membrane of animal cells is naturally the most exposed target for invading microbes, and it is unsurprising that most pathogens exploit this route of attack. Membrane damage can be inflicted by two major

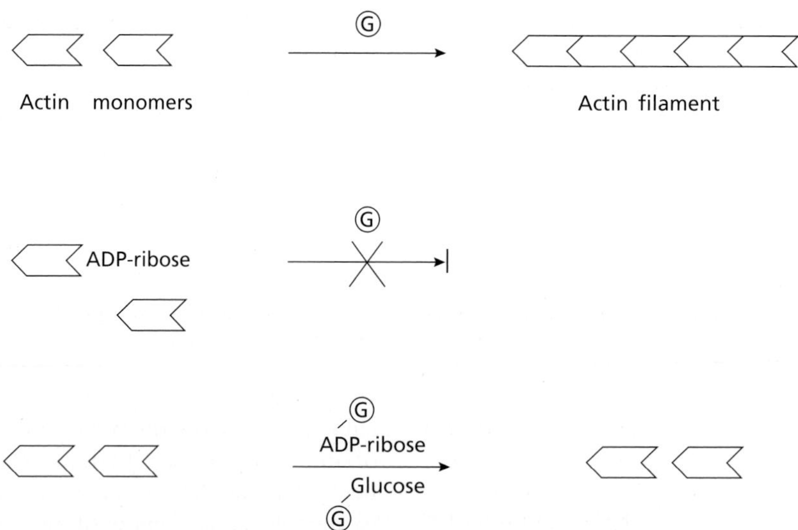

Fig. 13 Pathways to the disruption of the actin cytoskeleton. Monomeric GTPases (G) are required for actin filament formation. Actin filament formation can be inhibited by ADP ribosylation of monomeric actin, or by inactivation of the G protein via ADP ribosylation or glucosylation.

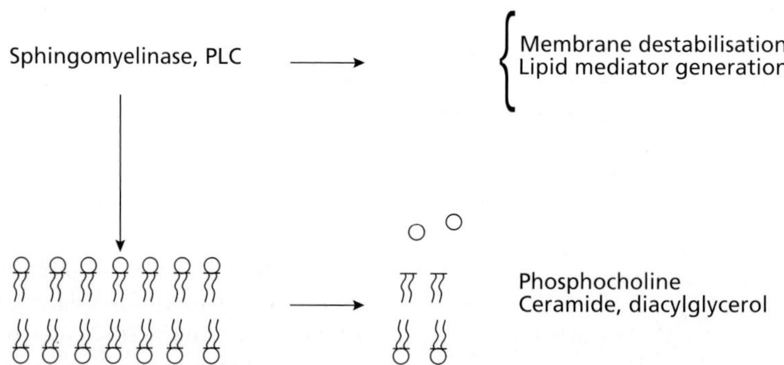

Fig. 14 Action of sphingomyelinase and phospholipase C on membranes. Removal of phosphorylcholine groups leads to membrane destabilisation and the generation of lipid mediators (ceramide, diacylglycerol).

mechanisms. The first is the enzymatic degradation of lipid membrane components. The second is the physical disruption of membrane bilayer integrity, primarily by the formation of transmembrane pores.

Membrane-damaging Enzymes Phospholipase C (PLC) is the major representative of this group. The enzyme removes hydrophilic head groups from phospholipid molecules that are exposed on the outer-membrane leaflet (**Fig. 14**). This destabilises the bilayer structure and is directly cytotoxic. PLC of *Clostridium perfringens*, which is synonymous with *Cl. perfringens* α toxin, has been intensively studied. It represents a major virulence factor that contributes to the pathogenesis of gas gangrene. For a long time, a major puzzle related to why this phospholipase possesses much higher haemolytic activity than other PLC molecules produced by many other bacteria. Further, only the PLC of *Cl. perfringens* has potent lethal activity. Enigmatically, other phospholipases were found to possess essentially the same catalytic domain and exhibit comparable specific activities when assays were conducted with isolated phospholipids. The riddle was solved when it was found that lethality and haemolytic activity of *Cl. perfringens* PLC were abrogated when the C-terminus was removed, despite the fact that the catalytic domain was still contained in the truncated α toxin. It emerged that the C-terminus, which is unique to *Cl. perfringens* α toxin, is required for the enzyme to interact efficiently with cells. This explains why *Cl. perfringens* PLC assumes such a prominent position as a virulence factor in this category of microbial enzymes (Titball, 1998).

An important emerging topic relates to the roles of lipid metabolites as signalling and messenger molecules. Diacylglycerol, a degradation product of PLC, is a prime example. This lipid plays a central role in regulating the activity of several cellular enzymes, in particular of protein kinase C (PKC). Hence, one consequence of PLC action is the activation of PKC, which subsequently triggers a wide spectrum of further cellular events.

Sphingomyelinases that are normally present in mammalian cells influence diverse processes through the generation of ceramide, an important second messenger, from sphingomyelin. Since sphingomyelin is abundantly present in the outer leaflet of the plasma membrane, it is also readily accessible to attack by secreted microbial enzymes. Sphingomyelinases are produced by many important extracellular pathogens including *Staph. aureus*. Their possible contribution to disease pathogenesis has received little attention, but they may one day be recognised as playing more important roles than is currently appreciated.

Pore-forming Toxins PFTs are produced by bacteria as water-soluble proteins. They bind by interaction with cell-surface receptors or by non-specific absorption to lipid bilayers. After binding, amino acid sequences insert into the target membrane to create water-filled transmembrane pores (Fig. 15). This process is often preceded by oligomerisation of toxin monomers in the membrane with the formation of high-molecular-weight entities in the membrane. Oligomerisation appears to provide the energy required to induce conformational changes that drive the membrane-inserting polypeptide domain into the lipid bilayer (Bhakdi *et al.*, 1996).

The size of transmembrane pores can vary greatly and is a significant factor determining the spectrum

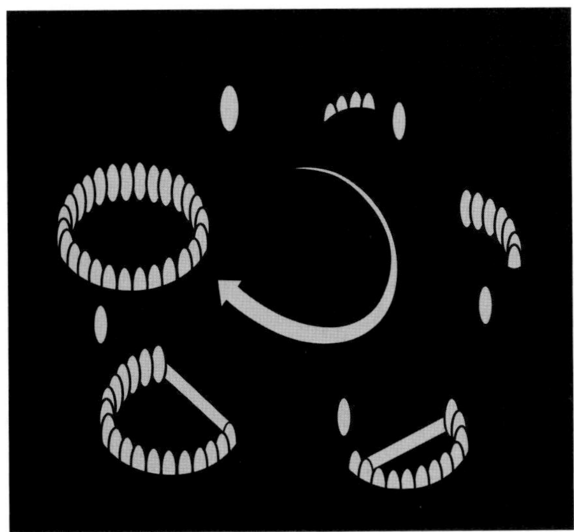

Fig. 15 Membrane damage by PFTs. These toxins often form oligomeric structures that represent transmembrane water-filled channels.

of reactions that will be provoked in the attacked cells.

Staphylococcal toxin displays structural and functional similarities to the pore-forming leukocidal toxins that are also produced by *Staph. aureus*. The α toxin is secreted as a water-soluble monomer (M_r 34 000). The toxin binds efficiently and with high-affinity to artificial phospholipid bilayers derived from proteins. A long-standing puzzle has been that high-affinity binding sites similar to those exposed on pure phospholipid liposomes are present on certain cells (e.g. rabbit erythrocytes, human platelets and endothelial cells) but not on others (e.g. human erythrocytes). Possibly it is not the presence of specific-binding sites, but rather the concealment of the phospholipids, that determines the susceptibility of cells to the toxin.

Bound toxin molecules diffuse in the membrane plane and oligomerise to form ring-structured heptamers. Heptamerisation is accompanied by insertion of a single amino acid sequence emanating from each protomer into the bilayer. The size of the α toxin pores produced in mammalian target cells is of the order of 1 nm.

Cholesterol-binding Toxins Streptolysin-O and perfringolysin are prototypes of this family, to which at least 14 other cytotoxins of Gram-positive organisms belong. A highly conserved C-terminal domain is responsible for the binding to cholesterol molecules in the bilayer. It is not understood why efficient binding requires that membranes must have a high content of cholesterol (over 20 molar % of lipid). After binding, a rate-limiting nucleation process follows that probably involves the formation of dimers, which act as crystallisation points for successive attachment of further monomers. Toxin complexes known to date can reach 30–35 nm in diameter.

RTX Toxins *E. coli* haemolysin is the prototype of this toxin family, which encompasses at least 14 members of homologous proteins elaborated by Gram-negative organisms. *E. coli* haemolysin is a single-chain polypeptide (M_r 107 000), which requires acylation with fatty acids at two lysine residues to obtain pore-forming activity. The mechanisms underlying pore formation by *E. coli* haemolysin remain largely unknown. It is not yet clear whether pore formation involves assembly of oligomers. *E. coli* haemolysin pores have a diameter of 1.5–2 nm and are thus permissive for calcium and nucleotides, but not for proteins.

Consequences of Toxin Attack

Cell death is the most obvious and inevitable consequence of transmembrane pore formation, if the lesion is not removed or repaired. Cell death can have immediate detrimental consequences. Tissue necrosis generates niches for bacterial survival and multiplication. Death of phagocytes and lymphocytes fosters microbial persistence and invasion. Death of endothelial cells can lead to major perturbations in the microcirculation. For example, endothelial cell destruction in the pulmonary vasculature will cause rapid haemorrhagic oedema formation. Damage to renal epithelial cells evoked, for example, by *E. coli* haemolysin may be responsible for renal dysfunction during acute pyelonephritis. The cytotoxic action of pneumolysin is probably partly responsible for lung tissue destruction during pneumococcal pneumonia. The action of this and related toxins on epithelial cells lining the cochlear round window (*fenestra cochleae*) membrane may be one cause of sensory neural hearing loss in acute otitis media. There is also suggestive evidence that the cytotoxic action of pore-formation on epithelial cells of the gastrointestinal tract contributes to the pathogenesis of diarrhoea. Overall, it should be note that, although most PFTs have been designated as haemolysins, their significance *in vivo* is certainly due in the main to their actions on cells other than erythrocytes.

Secondary reactions mounted by cells under attack by a PFT are of prime importance, and they can often explain the long-range effects of these agents (Bhakdi *et al.*, 1996). These reactions may be grouped into three broad categories:

- Pores that allow passage of potassium and sodium ions but restrict passage of calcium ions and larger molecules elicit unexpected effects. Triggering of programmed cell death has been observed in T lymphocytes and processing and transport of IL-1β has been observed in monocytes. The latter phenomenon was due to enhanced cleavage of pro-IL-1 by the interleukin-converting enzyme caspase-1, and since related proteases are also involved in the apoptosis pathway, the above findings may be linked to one another. There is evidence to indicate that K^+ depletion stimulates the activity of calcium-independent phospholipase A_2, and that the latter then accelerates the process of IL-1β maturation. Calcium-dependent reactions occur when calcium-permissive pores are generated. Well-studied phenomena include secretion, activation of phospholipases with the generation of lipid mediators, contraction of cytoskeletal elements and the stimulation of constitutive NO synthetase. By stimulating secretion of pro-coagulatory substances from platelets, α toxin exerts potent pro-coagulatory effects. *E. coli* haemolysin provokes the secretion of granular constituents from human granulocytes and the uncontrolled liberation of elastase and other enzymes may be pathophysiologically relevant. PFT induce the generation of biologically active lipid mediators in various cell targets, and this may contribute to local and systemic inflammatory responses. PFT also provoke cytoskeletal dysfunction in endothelial cells. Cellular contraction leads to the formation of intercellular gaps that permit rapid leakage of macro molecules across confluent monolayers. This may be a significant event underlying development of pulmonary oedema during severe pneumonia.
- Certain reactions not recognisably associated with ion flux. Short-circuiting of G protein-dependent signalling pathways is a property characteristic of *E. coli* haemolysin and probably other RTX toxins. For unknown reasons, the initial interaction of *E. coli* haemolysin with the target membrane is linked to triggering of G protein-dependent pathways, which results in reactions typically elicited by chemokines.
- Proteolytic shedding of membrane proteins. Many membrane protein-associated molecules can be released from cells by a proteolytic cleavage process called shedding. The liberated molecules often retain their ligand-binding capacity and cause an array of secondary reactions including those derived from trans-signalling. This term denotes the transfer of sensitivity towards a ligand onto cells that primarily lack their respective receptor. Trans-signalling can occur when the soluble receptor binds to a bystander cell. PFTs induce rapid and massive shedding such as CD14 and IL-6R. Release of receptors for IL-6 and LPS in biologically active form is a potentially significant mechanism that contributes to the systemic inflammatory response during infections with toxin-producing organisms.

PFTs can provoke acute organ dysfunction and are lethal in experimental animals. In these respects, they are no less potent than endotoxins. Indeed, in some experimental models, such as the isolated perfused and ventilated lung, they are more toxic than their LPS counterparts. Detailed investigations have been performed on the isolated lung model with α toxin and *E. coli* haemolysin. Each of these toxins rapidly evokes profound pathophysiological alterations in the pulmonary microvasculature, and sub-nanomolar concentrations lead to the development of irreversible pulmonary oedema. The underlying mechanisms are complex but encompass direct toxic action on endothelial cells and the production of eicosanoids that in turn promote pulmonary arterial hypertension. These experiments have revealed that pore-forming exotoxins have the capacity to provoke acute pulmonary failure, such as is typically found in patients presenting with severe respiratory distress syndrome during bacterial infections.

The major PFTs are also lethal when applied in experimental animals. In the case of α toxin and *E. coli* haemolysin, it is of interest that animals succumb with symptoms that may be explained from the observations *in vitro*. Thus, α toxin provokes acute thrombocytopenia and pulmonary oedema in rabbits and monkeys. Rabbits challenged with *E. coli* haemolysin present with acute leukopenia and haemorrhagic pulmonary oedema. Thus, the action of pore-formers *in vivo* can rapidly lead to multi-organ failure and death.

Synergies Between Pore-forming and Other Toxins Synergies exist between endotoxin and PFTs. For example, priming of pulmonary cells in isolated lungs with endotoxin potentiates the vascular abnormality in response to α toxin and *E. coli* haemolysin. Such synergy and the resulting vascular abnormalities are probably relevant to the pathogenesis of organ failure in systemic infections. Similar synergies are may operate between superantigens and PFTs.

PFTs as Vaccine Candidates PFTs are usually immunogenic, and most toxins fall within a limited range of related protein families with antigenic cross-reactivity. Development of vaccines against the major toxins is therefore feasible. Most toxins have been cloned and it is easy to generate inactive mutants. In the case of staphylococcal α toxin, the replacement of a single amino acid residue generates an immunogenic but completely non-toxic protein that elicits high antibody titres in rabbits. High-titre human antibodies against α toxin have also been produced with the use of a conventional toxoid, and a monoclonal human antibody against α toxin is also available. All these antibodies are able to neutralise the detrimental effects of α toxin *in vitro* and in animal experiments. Analogous protective efficacy has been reported for antibodies against *E. coli* haemolysin and pneumolysin. Apart from the possibility of passive immunisation in patients with severe infections, the possibility of vaccinating patients who are to undergo high-risk operations with a combination of potentially relevant vaccines may be considered. Such a prophylactic regimen would be cheap and safe, and might provide partial protection for patients undergoing elective surgery.

Superantigens From a conceptual viewpoint, these protein toxins may be regarded as counterparts of endotoxins. They induce cytokine responses that may primarily be of benefit to the host, but become detrimental when over-production occurs. Endotoxins are limited to Gram-negative bacteria, but most microbial producers of superantigens so far identified are Gram-positive. Common to superantigens is the possession of two binding domains, one for the Vβ subunit of the T cell receptor, and one for the constant region of MHC molecules. Thus, superantigens are able to cross-link antigen-presenting cells with CD4 cells in an antigen-independent manner. As a result, polyclonal T cell activation occurs, which leads to clonal proliferation and massive cytokine production (**Fig. 16**). Depending on the Vβ-specificity of the superantigen, polyclonal expansion occurs in 2–5% of the peripheral T cells. After days, this is followed by apoptosis and immune paralysis. When these processes occur excessively, systemic deregulation of the cytokine network ensues and leads to disturbances in the micro-circulation and to multi-organ dysfunction in a manner analogous to septic shock induced by endotoxin (Marrack and Kappler, 1990; Rago and Schlievert, 1998).

Medically important superantigens are produced by *Staph. aureus* and group A streptococci. The *Staph. aureus* superantigens comprise the structurally related staphylococcal enterotoxins (SE) A–E, and the more remote toxic shock syndrome toxin (TSST-1). The three-dimensional structure of SEB has been solved and the binding domains for the T cell receptor and MHC II have been identified. The streptococcal superantigens are synonymous with erythrogenic toxins (ET); ETA and ETC are related and their three-dimensional structure resembles remarkably that of staphylococcal superantigens.

Recognition of toxic shock syndrome heralded research into the medical aspects of superantigen action. It should be recalled that *Staph. aureus* strains

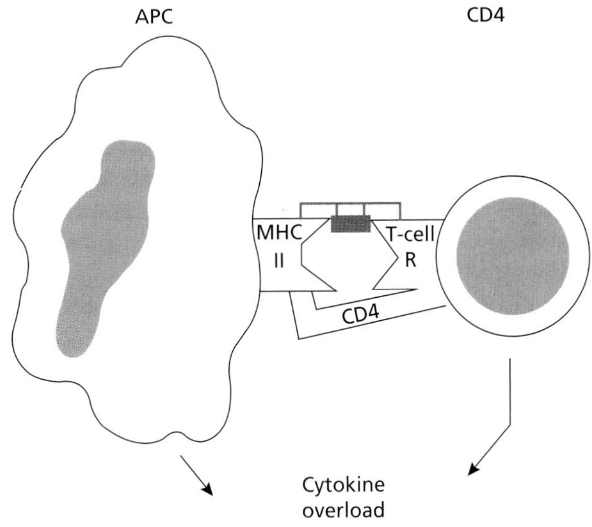

Fig. 16 Mechanism of action of superantigens.

isolated from patients suffering from toxic shock syndrome were initially discovered to produce the toxin. After it was realised that superantigens are produced by approximately 50% of clinical *Staph. aureus* isolates, it became clear that superantigen-mediated cytokine over-induction probably also contributes to the pathogenesis of many severe *Staph. aureus* infections. Similarly, the streptococcal superantigens have received much attention as important causes of shock syndromes encountered during severe invasive streptococcal infections. The streptococcal superantigens are undoubtedly responsible for the characteristic clinical manifestations of scarlet fever. It is useful to recollect that antibodies directed against a streptococcal superantigen have long been known to confer protection against a second attack of type-specific scarlet fever. The capacity of antibodies to neutralise superantigens may be an important explanation for the widely varying clinical course of severe infections with agents that produce these toxins. Presumably, patients with protective antibodies will be less prone to develop septic shock than those who lack the antibodies. Superantigens are currently being considered as candidates for multi-component staphylococcal and streptococcal vaccines.

In addition to their systemic effects, which clearly depend on immune stimulation, bacterial superantigens exert strong emetic effects and produce diarrhoea after oral ingestion. The staphylococcal enterotoxins are consequently responsible for the characteristic symptoms of acute *Staph. aureus* food poisoning. The pathogenesis of vomiting and diarrhoea has remained enigmatic. It is logical to envisage a link with deregulated cytokine responses in the mucosa-associated lymphoid tract, but clear evidence for such a connection has been difficult to obtain. Because of their structural stability, staphylococcal enterotoxins withstand denaturation despite heating at 100°C for several minutes, and they thus belong to the small select group of heat-stable exotoxins. By analogy with diarrhoea-producing toxins of Gram-negative organisms, the local host response to Gram-positive enterotoxins may initially serve to purge the gastrointestinal tract of noxious agents and potentially dangerous micro-organisms.

Superantigens certainly play dominant roles in the pathogenesis of Gram-positive shock, but it should be realised that their action is flanked by other equally important processes that are not mediated by superantigens but, rather, by tissue-destroying cytotoxins and enzymes. It should also be emphasised that the diverse pathogenetic factors produced by invading bacteria often act in synergy with each other to debilitate the host. This holds not only for potentiating interactions between products of one bacterial species, e.g. superantigens and cytotoxins of staphylococci and streptococci, but also for synergies between toxins of Gram-positive and Gram-negative bacteria. For example, superantigens potently augment the action of LPS. This may not only play a role in the context of mixed infections. It is probable that small quantities of endotoxin continuously leak from the gastrointestinal tract to the mucosa-associated lymphoid tract. This spontaneous translocation of endotoxin is enhanced in trauma patients and in those suffering from severe infections. If this is accompanied by production of microbial superantigens, cytotoxins, or other sensitisers to LPS, the consequences may far surpass those primarily expected from the presence of any one agent alone.

Superantigen-induced, abrupt expansion of the relevant T-cell clones is later followed by apoptosis and clonal anergy. As a result, a condition of immune paralysis can ensue that bears resemblance to that observed during endotoxin-mediated sepsis.

Stimulatory Bacterial DNA Sequences There has been a surge of interest in a remarkable phenomenon first described in 1995 that relates to the immune-activating capacity of specific bacterial DNA sequences. These are rich in GcP motifs and display pseudo-superantigenic activity (Krieg *et al.*, 2000). Superantigenic DNA induces cytokine production in isolated mouse macrophages. Injection of superantigenic DNA into mice produces the same symptoms as those observed after injection of LPS or superantigen. Surprisingly, human macrophages are not responsive and the true role of superantigenic DNA in human medicine is certainly far from clear. A promising lead is that human dendritic cells appear to respond to specific bacterial oligonucleotides. The phenomenon is intriguing, and ongoing studies are devoted to elucidating a possible role not only in the context of acute bacterial infections, but also in the pathogenesis of enigmatic diseases such as rheumatoid arthritis, for which noxious microbial agents or products have long been suspected to be responsible, but have never been identified. Questions relating to the mechanisms of cytokine induction are entirely open, as are basic problems regarding the generation, distribution, half-life and cellular uptake mechanisms of the stimulatory DNA sequences.

A potentially useful spin-off from this research has been the introduction of stimulatory DNA sequences as vaccine adjuvants, since immune responses are markedly augmented in their presence. It must be noted, however, that this effect has only so far been demonstrated in rodents.

Toxic Injection

The term 'toxic injection' is used here to describe the direct introduction of proteins by Gram-negative bacteria into target eukaryotic cells. This is rendered possible by contact-dependent assembly of a remarkable bacterial export machinery, termed the 'type III secretion system'. The injected proteins are, essentially, toxins – hence the term toxic injection – and some of these produce cell pathological effects similar to conventional exotoxins (Lee and Schneewind, 1999; Galan and Colmer, 1999).

Toxic injection by the type III secretion machinery was discovered by researchers in the *Yersinia* and *Shigella* fields during their investigations into the molecular mechanisms of bacterial invasion into cells. Hence, it has largely been a domain of research into mechanisms of intracellular microbial pathogenesis. However, important extracellular pathogens, including *Ps. aeruginosa*, also utilise this machinery to attack the host. The translocating element is assembled when bacteria contact eukaryotic cells, and it consists of a number of outer-membrane proteins that create a pore to traverse the target plasma membrane and the bacterial wall. The mechanisms that underlie pore assembly are currently under investigation. In the case of *Yersinia*, the translocating pore has been isolated and initial ultrastructural analysis indicates a hollow cylindrical structure akin to oligomerising PFTs. After pore assembly, diverse bacterial products can be injected to the target cell cytosol; the critical molecular signals that determine export are not understood in detail. The microbial proteins can subsequently invoke a multitude of effects, several of which are triggered by phosphorylation and dephosphorylation events. For example, *Y. enterocolitica* and *Salmonella typhimurium* inject tyrosine phosphatases that ultimately cause perturbation of the actin cytoskeleton. Yop H, the phosphatase of *Yersinia*, targets p130^{cas} and paxillin, two proteins that are involved in connecting integrins to the actin cytoskeleton in epithelial cells. It is probable that pathological actin re-arrangements that result from the action of Yop H contribute to epithelial damage during intestinal *Yersinia* infections. By analogy, the tyrosine phosphatase SptP of *S. typhimurium* and its analogues in *S. typhi* and *S. paratyphi* probably exert similar effects, but their substrates *in vivo* have not yet been identified.

Oddly, the *Salmonella* SptP protein possesses a second *in vitro* activity, it is able to assume the role of a GTPase-activating protein (GAP) with specificity for two monomeric G proteins, Rac and Cdc 42, which have key functions in regulating actin dynamics. The relevance of this finding to disease pathogenesis remains to be clarified.

Salmonella can also secrete inositol phosphatases with broad substrate-specificities *in vitro*, which may be responsible for many pathological effects including disturbances in ionic homeostasis and production of diarrhoea.

Other injected toxins have phosphorylating activity, but their targets and the consequences of their action are less well known. Kinases injected by *Y. enterocolitica* (YopO) and *Y. pestis* (YpKA) are autophosphorylating enzymes with homology to eukaryotic serine/threonine kinases. Deletion of YpKA attenuates virulence, but the underlying mechanisms remain unclear.

The link to extracellular pathogens was made with the discovery that *Ps. aeruginosa* uses the type III secretion system to inject an important toxin, exotoxin S, into targets. Exotoxin S appears to be a bifunctional toxin. The N-terminal domain causes rho-dependent cell rounding and actin depolymerisation, whereas the C-terminal half of the molecule has cytotoxic ADP-ribosyltransferase activity. One target for this enzyme is *ras*, but the functional consequences of ADP-ribosylation have yet to be precisely defined in this case. As is to be expected, other toxic components of *Pseudomonas* have also been found to utilise the transport system. These include ExoY, an adenylate cyclase, and ExoU, a cytotoxin with an unknown mechanism of action. *Ps. aeruginosa* induces type-III-secretion-dependent apoptosis of macrophages and epithelial cells which is probably due to the injection of several, as yet unidentified, molecular entities. That toxic injection is important for the extracellular survival of *Ps. aeruginosa* is supported by the finding that deletion of the type III secretion system diminishes virulence in diverse animal models of infection (Fig. 17).

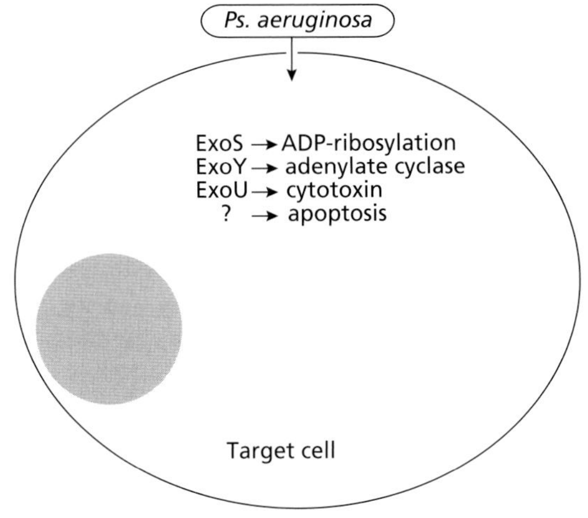

Fig. 17 Toxic injection. The type III secretion pathways used by Gram-negative organisms to inject toxin proteins into the target cells. *Ps. aeruginosa* makes abundant use of this principle.

The importance of toxic injection in the pathogenesis of other Gram-negative infections is being investigated, and many suggestive leads are already available. For example, exploitation of the type III secretion system by enteropathogenic *E. coli* has multiple consequences *in vitro*. As in other cases, the contact pore itself is potentially cytotoxic. An intriguing finding has been that an injected component of these bacteria, termed Tir, is first modified in the host cytosol and then becomes inserted into the membrane to assume the role of a bacterial receptor. Thus, enteropathogenic *E. coli* appear to affect the transfer of their own receptor into mammalian membranes via the type III secretion machinery. Other injected products of these bacteria mediate and derange various signalling events. Among others, tyrosine phosphorylation of host proteins induces cytoskeletal rearrangement that may mediate bacterial uptake, but which may also underlie the epithelial effacing capacity of these enteric pathogens.

Enzymes

Proteases assume the dominant role in this attack strategy. It is useful to distinguish between situations in which a protease remains bound to the bacterial cell surface and those in which enzymes are liberated into the surrounding tissue. Another aspect concerns the origin of the protease. Although many bacteria produce their own enzymes, others have involved mechanisms to harness host proteases to their advantage. Some examples follow.

Cell-bound Proteases The strategy of expressing protease activity at the cell surface has been most intensively studied in group A streptococci. This activity may derive from the presence of both a streptococcal enzyme and of absorbed host proteases (predominantly plasminogen), and likely contributes to the capacity of group A streptococci to invade and persist in tissues.

Streptococcal Cysteine Proteinase Streptococcal cysteine proteinase (SCP) is secreted as an inactive precursor, of which a fraction remains bound to the cell surface and is proteolytically converted to its active form. SCP is endowed with many activities that are potentially relevant to disease pathogenesis. These include the destruction of C5a, a major chemokine-like molecule, cleavage of fibronectin and cleavage of cell-bound immunoglobulins. SCP also cleaves streptococcal M-protein and it has been proposed that this may promote detachment of the bacteria from host cells and thus enhance their spread in tissues. SCP is identical to streptococcal pyrogenic toxin B, but the mechanisms that underlie its systemic action are poorly understood.

Plasminogen Plasminogen, the major proteinase precursor in blood, is normally activated by the action of endogenous enzymes (plasminogen activators). *Strep. pyogenes* has evolved mechanisms that converge to recruit plasmin activity to the cell surface. A key step is the initial binding of plasminogen to the bacteria, which can occur through high-affinity reactions of the protein with M-related proteins or with surface-expressed streptococcal α enolase (**Fig. 18**). Plasmin activity can subsequently be generated both by the action of physiological plasminogen activators and by streptokinase, which is secreted by the micro-organisms themselves. Streptokinase in turn can be concentrated on the bacterial surface by a fibrinogen bridge; the bacterial enzyme binds to fibrinogen which is captured by M-protein. These events together lead to efficient generation of plasmin activity at the bacterial surface (Lottenberg *et al.*, 1994).

Plasminogen-binding activities have also been detected on *B. burgdorferi* and *Y. pestis* and may contribute to the tissue-invasive capacity of these pathogens.

Secreted Proteases Most extracellular pathogens secrete one or more proteases. Because of their abundance and wide variety, examples cannot be cited for a major role of any single protease in the pathogenesis of disease. The capacity of these enzymes to promote pathological changes can readily be demonstrated *in vitro*, however, and secreted proteases may at present be underestimated virulence factors. The following are examples for detrimental effects that may play a significant role in the pathogenesis of extracellular infections (Maeda and Yamamoto, 1996).

Tissue Destruction Cytotoxins kill cells, but microbial proteases destroy the extracellular matrices and foster bacterial invasion. The elastase elaborated by *Ps. aeruginosa* is important in the pathogenesis of keratitis and it is probably also relevant in infections of the lung and skin. Clostridial proteases are probably relevant to the pathogenesis of gas gangrene. *Porphyromonas gingivalis* proteases may be involved in the initiation and progression of gingivitis and periodontitis.

Mediator Generation The activation of the kinin-generating cascade can be induced by a large number of microbial proteinases, including those produced by such important extracellular pathogens as *Staph. aureus*, *Strep. pyogenes*, *Ps. aeruginosa*, and Clostridia. Bradykinin is generated via a cascade of proteolytic activation steps, commencing with cleavage of

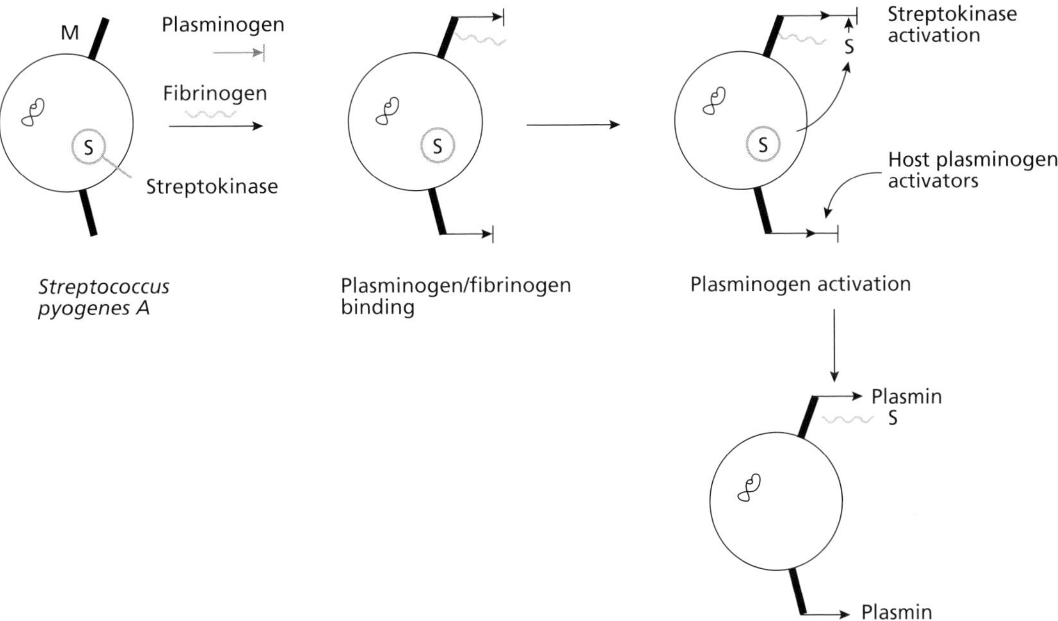

Fig. 18 Recruitment of plasminogen/plasmin to bacterial surfaces. Plasminogen and fibrinogen bind to M-protein, and streptokinase can bind to fibrinogen. Surface-bound plasminogen can subsequently be activated by streptokinase or by host plasminogen activators. Surface-associated plasmin supports bacterial invasion in tissues.

Hageman factor (factor XII) and continuing by activation of pre-kallikrein and high-molecular-weight kininogen. Microbial proteases may intervene positively along the entire pathway. Bradykinin causes pain, vasodilatation and enhanced vascular permeability. Massive generation of bradykinin during several infections is possibly also a contributing factor to the development of septic shock.

Coagulation Dysfunction A straightforward path leading to coagulation is the activation of Hageman factor with subsequent triggering of the intrinsic coagulation cascade. A more indirect pro-coagulant mechanism is the destruction of antithrombin III by bacterial enzymes. Thrombin may subsequently be dissolved, for example by plasminogen, to allow local infiltration of the micro-organisms in destroyed tissues.

Complement Activation Systemic complement activation often occurs during severe bacterial infections. It is widely believed that microbial cellular components, in particular endotoxin, are mainly responsible. The concentrations of endotoxin present both locally and systemically would hardly suffice to provoke massive complement consumption, however, and microbial proteases may play a less appreciated, but significant contributing role. Cleavage of Hageman factor may provide a direct link to the classical pathway of complement activation. Furthermore, certain microbial proteinases, for example the elastase of *Serratia*, destroy the C1-inhibitor, which is essential to control Cl activity. Complement activation by microbial proteases may in the future emerge as a hitherto underestimated pathogenic mechanism.

Activation of Host Proteases In addition to plasminogen, many other precursor proteases, such as metalloproteases secreted by macrophages, can be activated by microbial enzymes. Also of potential significance is the inactivation of endogenous protease inhibitors (e.g. α_1 anti-trypsin) by microbial proteases, which would lead to enhanced activity of many proteases of both host and microbial origin.

Shedding and *Trans*-signalling by Host Surface Molecules Removal of host surface molecules by microbial proteases may directly compromise cell function. Furthermore, secondary disturbances may ensue via *trans*-signalling mechanisms if the released molecules retain their ligand-binding activity. This can happen, for example, after release of IL-6 receptor and CD14 by microbial metalloproteinases. The soluble receptors subsequently bind their ligands and have the capacity to activate bystander cells. This may be important in promoting long-range effects of both IL-6 and LPS in the host organism.

Many bacterial exotoxins require proteolytic processing to acquire biological activation, and microbial proteinases can sometimes fulfill this function. This would be another indirect way in which microbial proteinases can cause damage to the host.

Conclusion

This chapter ends with reflections on some key mechanisms that come into play during the evolution of two classic extracellular infections, wound infections caused by *Staph. aureus* and bronchopneumonia caused by *Ps. aeruginosa* in cystic fibrosis patients.

Many pathogenic mechanisms act in concert to promote bacterial multiplication and extracellular survival of *Staph. aureus* during wound infections. Fibrinogen bound to staphylococcal clumping factor provides camouflage. Interaction of protein A with immunoglobulins helps the bacteria to avoid phagocytosis. Secreted cytotoxins damage host cells and enlarge the survival niche for the bacteria. Cell-wall components activate complement and phagocytes, but the ensuing pro-inflammatory reactions may not lead to bacterial elimination if staphylococcal cytolysins and leukotoxins overcome host defence cells. The scenario may be worsened as superantigens come into play and derange inflammatory reactions, such that physiological responses are deviated to augment pathology. Coagulase promotes fibrin generation, thus contributing to a demarcation of the infected site. As cells die and microbial enzymes locally digest host cells and tissues, the abscesses form that are so characteristic of *Staph. aureus* infections (Fig. 19).

Bronchopneumonia in cystic fibrosis patients is typically first caused by *Staph. aureus*. Subsequently, *Ps. aeruginosa* takes the lead and replaces *Staph. aureus* as the major infectious agent. Copious production of alginate generates a protective shield around the micro-organisms. At the same time, host cells are hit by an array of toxins. Macrophages can receive lethal injections of exotoxin S as they come into contact with the bacteria. Exotoxin A can kill cells at a distance and endotoxin can exaggerate and derange the inflammatory response. Should the situation turn in favour of the host, for example through the action of antibiotics, both *Staph. aureus* and *Ps. aeruginosa* may transform to SCV that persist in the lesions, biding their time until conditions for re-activation return. Mixed infections with *Staph. aureus* and *Ps. aeruginosa* probably represent situations in which synergy of virulence factors is

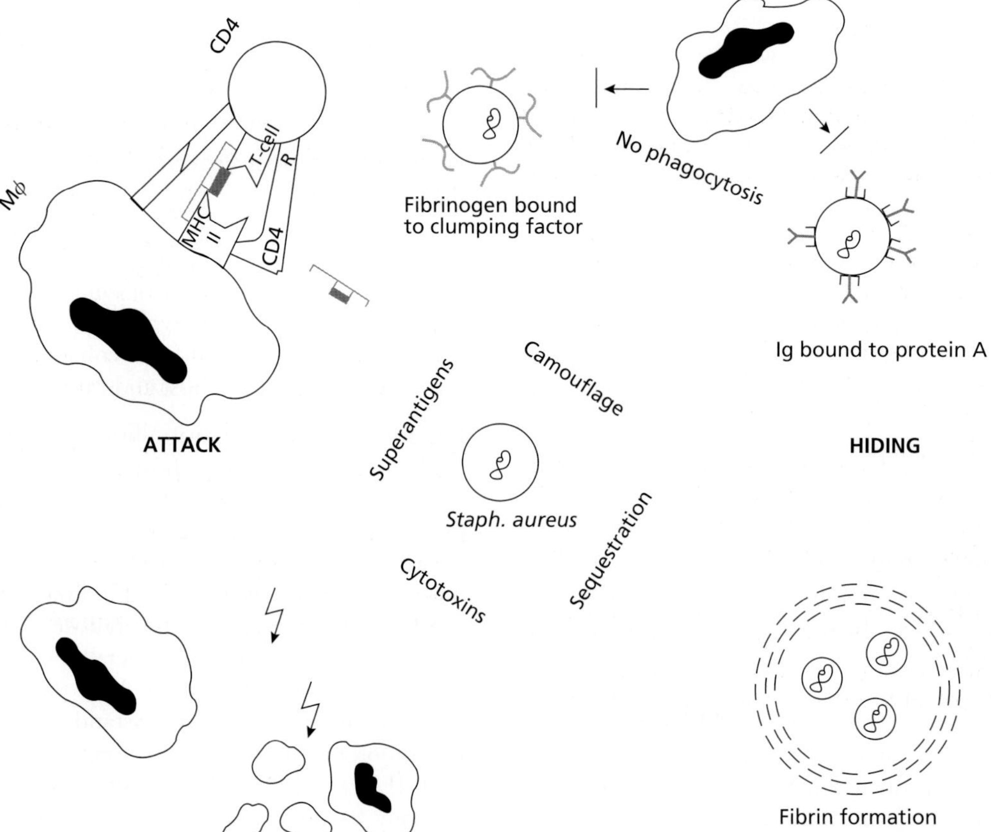

Fig. 19 Pathogenetic principles underlying *Staph. aureus* wound infection.

relevant, the LPS of *Ps. aeruginosa* acting in concert with PFTs and with superantigens of *Staph. aureus.* As toxic agents accumulate, processes expand to reach sites that are remote from infectious foci. Ultimately, the detrimental effects culminate to produce the clinical presentation of global pulmonary failure.

Acknowledgement

The author is grateful to Katharina Strach who prepared sketches for the illustrations.

References

Adams G, Deaver KA, Cochi SL *et al.* (1993) Decline of childhood *Haemophilus influenzae* type b (Hib) disease in the Hib vaccine era. *JAMA* 269: 221–226.

Beutler B (2000) Endotoxin, toll-like receptor 4, and the afferent limb of innate immunity. *Curr. Opin. Microbiol.* 3: 23–28.

Bhakdi S, Bayley H, Valeva A *et al.* (1996) Staphylococcal alpha-toxin, Streptolysin-O and *Escherichia coli* hemolysin: prototypes of pore-forming bacterial cytolysins. *Arch. Microb.* 165: 73–79.

Bouquet P, Munro P, Fiorentini C, Just I (1998) Toxins from anaerobic bacteria: Specificity and molecular mechanisms of action. *Curr. Opin. Microbiol.* 1: 66–74.

Costerston JW, Lewandowski Z, Caldwell DE *et al.* (1995) Microbial films. *Annu. Rev. Microbiol.* 49: 711–745.

De Velasco A, Verheul AF, Verhoef J, Snippe H (1995) *Streptococcus pneumoniae*: virulence factors, pathogenesis, and vaccines. *Microbiol. Rev.* 59: 591–603.

Galan JE, Colmer A (1999) Type III secretion machines: bacterial devices for protein delivery into host cells. *Science* 284: 1322–1328.

Haake DA (2000) Spirochaetal lipoproteins and pathogenesis. *Microbiology* 146: 149–1504.

Janeway CA Jr, Medzhitov R (1999) Innate immunity: Lipoproteins take their toll on the host. *Curr. Biol.* 9: R879–R882.

Just I, Hofmann F, Aktories K (2000) Molecular mechanisms of action of the large clostridial cytotoxins. In: Aktories K, Just I (eds) *Bacterial Protein Toxins; Handbook of Experimental Pharmacology*. Berlin: Springer-Verlag, pp. 307–331.

Krieg AM, Hartmann G, Yi AK (2000) Mechanism of action of CPG-DNA. *Curr. Top. Microbiol. Immunol.* 247: 1–22.

Lee VT, Schneewind O (1999) Type III secretion machines and the pathogenesis of enteric infections caused by *Yersinia* and *Salmonella* spp. *Immunol. Rev.* 168: 241–255.

Lottenberg R, Minning-Wenz D, Boyel MDP (1994) Capturing host plasmin(ogen): a common mechanism for invasive pathogens? *Trends. Microbiol.* 2: 20–24.

Maeda H, Yamamoto T (1996) Pathogenic mechanisms induced by microbial proteases in microbial infections. *Biol. Chem.* 377: 217–226.

Marrack P, Kappler J (1990) The staphylococcal enterotoxins and their relatives. *Science* 248: 705.

Masignani V, Pizza M, Rappuoli R (2000) Common features of ADP-ribosyltransferases. In: Aktories K, Just I (eds) *Bacterial Protein Toxins; Handbook of Experimental Pharmacology*. Berlin: Springer-Verlag, pp. 21–44.

O'Brien AD, Tesh VL, Donohue-Rolfe A *et al.* (1992) Shiga toxin: Biochemistry, genetics, mode of action, and role in pathogenesis. *Curr. Top. Microbiol. Immunol.* 180: 65–94.

Olsnes S, Wesche J, Fames PO (2000) Uptake of protein toxins acting inside cells. In: Aktories K, Just I (eds) *Bacterial Protein Toxins; Handbook of Experimental Pharmacology*. Berlin: Springer-Verlag, pp. 1–19.

Patton WA, Vitale N, Moss J, Vaughn M (2000) Mechanism of cholera toxin action: ADP-ribosylation and effectors in intracellular vesicular trafficking events. In: Aktories K, Just I (eds) *Bacterial Protein Toxins; Handbook of Experimental Pharmacology*. Berlin: Springer-Verlag, pp. 133–165.

Proctor RA, Peters G (1998) Small colony variants in staphylococcal infections; diagnostic and therapeutic implications. *Clin. Infect. Dis.* 27: 419–422.

Rago JV, Schlievert PM (1998) Mechanisms of pathogenesis of staphylococcal and streptococcal superantigens. *Curr. Top. Microbiol. Immunol.* 225(8): 1–97.

Rautemaa R, Men S (1999) Complement-resistance mechanism of bacteria. *Microbes Infect.* 1: 785–794.

Rietschel ET, Kirikae T, Schade FU *et al.* (1994) Bacterial endotoxin: molecular relationships if structure to activity and function. *FASEB J.* 8: 217–225.

Schiavo G, Rosetto O, Tonello F, Montecucco C (1995) Intracellular targets and metalloprotease activity of tetanus and botulism neurotoxin. *Curr. Top. Microbiol. Immunol.* 195: 257–274.

Sears CL, Kaper JB (1996) Enteric bacterial toxins: Mechanisms of action and linkage to intestinal secretion. *Microbiol. Rev.* 60: 167–215.

Titball RW (1998) Bacterial phospholipases. *Soc. Appl. Bacteriol. Symp. Ser.* 27: 1278–137S.

Zipfel PF, Hellwege J, Friese MA *et al.* (1999) a complement regulator affects vital body functions. *Mol. Immunol.* 36: 241–248.

35

Strategies for Survival of Intracellular Pathogens

Kathleen A. McDonough

Wadsworth Center, Albany, NY, USA

Professional phagocytes, including monocytes, macrophages and polymorphonuclear leucocytes (PMNs), are a critical part of the innate immunity system that provides the first-line defence against invading microbes. Whereas PMNs and monocytes (immature macrophages), circulate in the bloodstream, macrophages are found primarily in the tissues and lymphatics. All of these phagocytes have microbicidal capabilities, but macrophages also function as antigen-presenting cells. In their role as antigen-presenting cells, macrophages provide an important link between the innate and induced immune response to specific microbes.

The success of many pathogens depends upon their ability to either withstand or avoid the microbicidal capabilities of phagocytes. Some bacteria use anti-phagocytic mechanisms to deal with this problem, but others have evolved to survive within these cells. This chapter will discuss the strategies used by the intracellular bacterial pathogens that fall into the 'survivor' group.

The focus of this chapter is limited to bacterium–macrophage interactions because macrophages are the phagocytes most often exploited by bacterial pathogens, and have been most studied by those interested in microbial pathogenesis. As relatively long-lived cells, macrophages provide a reasonably secure niche, making survival and replication an efficient way of persisting within the host in a privileged environment.

Macrophage Functions

Particle Uptake Mechanisms

Phagocytic cells ingest particles by a variety of mechanisms, including pinocytosis, macropinocytosis, receptor-mediated endocytosis, and phagocytosis (Sandvig and van Deurs, 1994; Swanson and Watts, 1995; Aderem and Underhill, 1999). Pinocytosis and macropinocytosis are relatively non-specific uptake mechanisms associated with cell ruffling that result in the clathrin-independent uptake of plasma membrane along with liquids and solutes. Large particles, including some pathogenic micro-organisms, may be ingested by macropinocytosis (see below). Receptor-mediated endocytosis is a clathrin-dependent form of pinocytosis whereby macromolecules and small particles, such as viruses, are taken up into endosomes. In contrast, phagocytosis is the process by which particles larger than about 0.5 μm, including most bacterial pathogens, are internalised into membrane-bound vacuoles (phagosomes). Phagocytosis is often, but not always, receptor-mediated, requires localised actin polymerisation, and is clathrin-independent.

Phagocytic Receptors

A variety of receptors have been implicated in bacterial entry into macrophages, including the Fc (FcR),

 doi:10.1006/bkmm.2001.0035

complement (CR), mannose (MR), and scavenger receptors (SR) (Unkeless and Wright, 1988; Mosser, 1994; Greenberg, 1995; Aderem and Underhill, 1999). Microbes that are opsonised with IgG are taken up primarily through one of several classes of Fcγ family of receptors found on macrophages (FcγRI, FcγRII, or FcγRIII in humans, and FcγRI or FcγRIII in mice). Entry through the FcγRs is a Ca^{2+}-independent process that requires no additional stimulatory signals, triggers a respiratory burst, and is considered a major microbicidal clearance pathway. Depending on the activation state of the macrophage, this respiratory burst includes oxygen or oxygen and nitrogen radicals that are bactericidal for many micro-organisms (see below).

Unlike with FcγR-mediated phagocytosis, entry through the complement receptors CR1, CR3 or CR4 requires co-stimulation with a mitogen or lipopolysaccharide (LPS) and does not lead to a respiratory burst. Binding to these receptors is mediated by opsonisation of the microbe with fragments of the third component of complement, C3b or iC3b of C3. CR3 can also mediate the uptake of non-opsonised bacteria through a lectin-binding site that binds independently of the C3 binding site. Uptake by the lectin-binding site may lead to differential trafficking of the ingested particle (Ehlers and Daffe, 1998). CR3 and CR4 both belong to the integrin family of receptors and contain a β2 chain (Springer, 1990). Other integrin receptors, including those for fibronectin and vitronectin, are also used by bacteria to gain access to non-professional phagocytes such as epithelial cells (Greenberg, 1995; Isberg and Van Nhieu, 1995; Rabinovitch, 1995). Many of these receptors are also found on macrophages, but their role in bacterial uptake by these professional phagocytes has not been established.

The MR contains a number of lectin-like carbohydrate-binding domains that recognise both mannose and fucose on many bacterial pathogens (Stahl and Ezekowitz, 1998; Aderem and Underhill, 1999). Like FcγR-mediated phagocytosis, MR-mediated uptake is a pro-inflammatory process that stimulates a respiratory burst.

Understanding of the role of SRs in microbial clearance is increasing. This diverse group of receptors also participates in clearance of apoptotic cells (Aderem and Underhill, 1999). At least three class A SRs, SR-AI, SR-AII and MARCO (macrophage receptor with collagenous structure), bind intact bacteria, leading to their uptake and clearance (Pearson, 1996; Palecanda *et al.*, 1999; van der Laan *et al.*, 1999). This list will undoubtedly grow as more of these receptors are identified.

Phagocytic Mechanisms: Zipper Versus Trigger Models

Uptake of particles by macrophages can occur by either a zipper or trigger mechanism (Swanson and Baer, 1995). The zipper mechanism requires continuous contact between host-cell receptors and the entire surface of an uptake substrate to complete the process of internalisation. In contrast, according to the trigger model, only an initiating signal, or trigger, is needed for a particle to be completely engulfed. Classical phagocytosis, as here defined, occurs by the zipper mechanism and is the predominant means of bacterial engulfment by macrophages. There is, however, convincing evidence that some pathogens enter macrophages and epithelial cells by macropinocytosis, which occurs by a trigger rather than zipper-like mechanism (Fig. 1) (Galan *et al.*, 1992; Francis *et al.*, 1993; Alpuche-Aranda *et al.*, 1994).

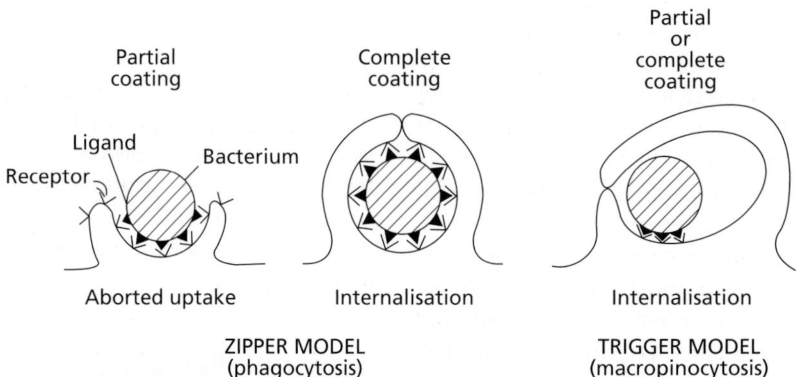

Fig. 1 Schematic of the outcomes predicted by the zipper and trigger models of particle uptake, depending upon whether the substrate particles are fully or partially coated with ligands that engage phagocytic receptors on the cell.

The zipper uptake mechanism is a highly specific and localised event that requires successive engagement of ligands on the surface of the particle with receptors on the phagocyte (Griffin and Silverstein, 1974). This process has been best studied with Fc receptors, where pseudopodia advance around the surface of an IgG-coated particle to effect engulfment. The requirement for sequential receptor–ligand engagement predicted by the zipper model was proven by elegant studies in which particles fully opsonised with antibody were taken up, but partially coated particles were not (Griffin *et al.*, 1976). The advance of pseudopodia terminated when either the receptors or the ligands became unavailable for continuous engagement on the partially coated particles.

Particle uptake via complement receptor CR3 also occurs by a variation of the zipper mechanism, which differs from Fc-mediated uptake in several important ways (Kaplan, 1977; Swanson and Baer, 1995; Aderem and Underhill, 1999). Although both pathways require actin polymerisation, Fc-mediated uptake is not affected by the microtubule inhibitors nocodazole and colchicine, whereas CR-mediated uptake requires microtubules and is mechanistically very similar to cell spreading. Morphologically, particles engulfed via CR3 appear to sink into the cell and the pseudopodia are less prominent than with the Fc receptors. In contrast, with Fc-mediated uptake, the actin-based pseudopodia extend out from the cell surface to surround the particle, eventually forming a phagosome with a tightly apposed membrane. CR-mediated uptake also results in a looser association of the phagosome membrane with the engulfed particle than is observed with Fc-mediated phagocytosis.

Macropinocytosis is an alternative phagocytic mechanism used by several pathogens, and occurs by what has been referred to as the trigger model (Griffin *et al.*, 1975; Swanson and Baer, 1995; Swanson and Watts, 1995). Unlike the highly specific zipper mechanism associated with the Fc and complement receptors, the trigger model of phagocytosis does not require continuous receptor–ligand interaction for particle uptake. Macropinocytosis is mechanistically more like cell ruffling than cell spreading and is enhanced by stimulation of growth-factor receptors. Cell ruffling during macropinocytosis occurs in a patch of membrane larger than that associated with zipper phagocytosis, and adherent particles are taken up when they are trapped in the 'ruffles' or folds of cell membrane. Macropinocytosis results in large spacious phagosomes and is susceptible to 'bystander' uptake, which occurs when nearby non-invasive bacteria or inert latex beads are engulfed with the invasive bacteria that triggered the uptake process.

Phagosome Maturation and Fusion with Lysosomes

Maturation is a Gradual Process

Once formed, phagosomes mature by a series of fusion and fission events that culminate in fusion of the phagosome with lysosomes to form a phagolysosome (Desjardins *et al.*, 1994b; Claus *et al.*, 1998; Jahraus *et al.*, 1998). Phagolysosomes are acidic compartments that contain a vast array of hydrolytic enzymes and few nutrients, an environment that is extremely hostile to most microbes. Phagosome maturation is a gradual process that includes a regulated series of fission and fusion reactions. Surface receptors are removed from the phagosome by a membrane budding process. Meanwhile, the developing phagosome fuses with a variety of endocytic organelles to acquire the properties associated with mature phagolysosomes. Vesicles encountered along the endocytic pathway are generally classified into three broad temporal classes with specific characteristics, including early endosomes, late endosomes, and lysosomes.

Vesicle fusion depends on microtubules and is mediated by Rab, SNAP (soluble NSF (*N*-ethylmaleimide-sensitive factor) attachment protein), and SNARE (SNAP receptor) proteins, whereas vesicle budding requires COPs (coat proteins) and ARFs (ADP ribosylation factors) (Rothman, 1994; Stamnes *et al.*, 1995). The precise roles of each of these proteins in vesicle fusion has still to be determined, but it is thought that the SNAREs are required for vesicle docking, while the Rabs provide specificity. Phagosomal trafficking also requires annexins, a family of proteins that bind phospholipids in a calcium-dependent manner and mediate interactions between the phagosome and the cytoskeleton. Several annexins (I, II, III and IV) are associated with phagosomes at all stages, but annexin IV is primarily associated with the more mature vacuoles in macrophages (Desjardins *et al.*, 1994a; Diakonova *et al.*, 1997).

Kinetics of Phagosome Maturation

The rate of phagosome maturation is affected by the size and the surface properties of the engulfed particles (Oh and Swanson, 1996; de Chastellier and Thilo, 1997). Phagolysosome (P-L) fusion of vacuoles containing opsonised sheep red blood cells is complete in less than 30 minutes. In contrast, phagosomes loaded with hydrophobic latex beads may take hours to fuse with late endosomes and lysosomes, or they may not fuse at all. These hydrophobic particles are associated with close-fitting phagosome membranes, and it has been suggested that intimate interaction of the particle surface with the vacuole membrane interferes with fusion events (de Chastellier and Thilo, 1997). In addition, many pathogens modify the fusion behaviour of their phagosomes, as will be discussed below.

Acquisition of markers occurs gradually, and many markers are present only transiently, being removed from the phagosome as new markers are introduced. Endocytic membrane and content markers are not acquired by phagosomes with the same kinetics, and fluid-phase markers are generally transferred ahead of membrane markers (Desjardins *et al.*, 1994b; Desjardins *et al.*, 1997; Claus *et al.*, 1998). For example, the fluid-phase marker horseradish peroxidase (HRP) transfers from late endosomes to latex bead-containing phagosomes within 60–90 minutes, whereas acquisition of the late endosome/lysosome LAMP membrane proteins by the same phagosomes requires more than 12 hours.

One of the earliest lysosome hydrolases acquired by phagosomes is cathepsin H, which is active at pH 6.5 and is associated with early endosomes (Claus *et al.*, 1998). Cathepsin S accumulates in phagosomes when they fuse with late endosomes, but only small amounts of other hydrolytic enzymes, including cathepsin B, β glucosidase, and a variety of other acid hydrolases, are found in phagosomes prior to fusion with lysosomes.

Membrane markers are also acquired sequentially as phagosomes mature, with the small GTPase Rab5 being one of the earliest to appear (Desjardins *et al.*, 1994b; Pizon *et al.*, 1994). Rab5, which is associated primarily with early endosomes, mediates homotypic fusion between early endosomes (Gorvel *et al.*, 1991). Rab7, which is thought to play a role in late endosome fusion events, accumulates on the phagosome membrane as Rab5 disappears (Desjardins *et al.*, 1994b; Feng *et al.*, 1995). Phagosomes also acquire the mannose-6-phosphate receptor (M-6-PR) and the lysosome-associated membrane proteins LAMP1 and LAMP2 at the late endosome stage of maturation (Griffiths *et al.*, 1988; Kornfeld and Mellman, 1989). Rab7 and M-6-PR are then removed and the LAMPs reach their maximum concentration as the phagosome becomes a phagolysosome upon fusion with lysosomes (Rabinowitz *et al.*, 1992; Desjardins *et al.*, 1994b; Desjardins, 1995). Vacuolar ATPase is also acquired during phagosome maturation, and contributes to the pH drop that occurs between the early endosome (pH ~6.5) and lysosome (pH 4.5–5.0) stages.

Lysosomes may be somewhat heterogeneous, a possibility that has important implications for pathogens that modulate their contact with lysosome components. In addition to the previously identified tubular lysosomes, which have late endosome markers with many lysosome properties, terminal (or secondary) lysosomes can be separated into two functional groups with respect to their pH requirements for fusion with late endosome vesicles (Swanson *et al.*, 1987; Claus *et al.*, 1998). The specific functions and trafficking patterns associated with each of these different sub-populations remain to be determined, but their presence suggests that phagolysosomes may vary in their microbicidal potential.

The 'Kiss-and-Run' Model of Maturation

The dynamic nature of maturing phagosomes and the differential kinetics of content versus membrane marker transfer suggest that phagosome maturation occurs by a series of transient contacts between endocytic organelles and phagosomes as opposed to one or more complete fusion events. This gradual model for phagosome maturation is called the 'kiss-and-run' hypothesis (Desjardins, 1995; Desjardins *et al.*, 1997). The kiss-and-run model proposes that endosome organelles engage in multiple incomplete fusion interactions that allow intermittent transfer of vacuolar contents through a narrow bridge formed transiently between two vacuoles ('kiss'). Rapid separation of the vacuoles ('run') would restrict intermixing of the vacuoles, allowing each to preserve its own integrity.

In addition to content transfer, incomplete fusion events allow for gradual acquisition of membrane markers that may be required for complete fusion with vacuoles at a later stage in the pathway. This model also explains the observed size dependence of the kinetics of vacuolar content transfer, as smaller markers would be predicted to transfer through the small transient pore more rapidly than larger particles.

Respiratory Burst

Oxidative Response

Phagocytosis of bacteria by Fc and mannose receptors triggers a respiratory burst, which is microbicidal to many micro-organisms, and generally occurs as the particle is being engulfed. The respiratory burst is associated with a large increase in non-mitochondrial oxygen consumption, oxidation of glucose through the hexosemonophosphate shunt, and production of highly reactive oxygen species such a superoxide (O_2^-), hydrogen peroxide (H_2O_2), and other oxygen radicals (Borregaard, 1988; Robinson, 1998).

The primary enzyme responsible for respiratory burst activity is NADPH oxidase, which is required for production of superoxide. Superoxide then gives rise to the other reactive and/or toxic intermediates, including hydrogen peroxide and hydroxyl radicals. Hydrogen peroxide is formed in a spontaneous or superoxide dismutase-catalysed reaction between superoxide and hydrogen ions. Toxic hydroxyl radicals are formed in a Haber–Weiss reaction, in which superoxide reduces hydrogen peroxide in a reaction that is catalysed by iron chelates (Haber and Weiss, 1934).

Neutrophils and some monocytes also contain the haem-enzyme myeloperoxidase, which reacts hydrogen peroxide with chloride to form highly toxic hypochlorous acid. Macrophages do not produce myeloperoxidase, but can acquire it when it is released by nearby polymorphonuclear cells during an inflammatory response (Heifets *et al.*, 1980).

NADPH oxidase is located in phagosome and plasma membranes (Bokoch, 1995; Robinson, 1998). It is not active in resting cells, because the plasma membrane form of the oxidase in these cells is only partially assembled. The assembly process has been best studied in neutrophils, where it requires mobilisation of the proteins p47*phox* (phagocyte oxidase, which should be distinguished from mitochondrial respiration) and p67*phox* from the cytoplasm to the membrane-bound cytochrome b complex before the addition of Rac-GTP (Bokoch, 1995). Binding of a G protein, Rac1 in guinea-pig macrophages and Rac2 in human neutrophils, is also required for full oxidase enzyme activity. The identity of this NADPH oxidase system, and its regulation, remain the subject of intense investigation.

Reactive Nitrogen Intermediates

Reactive nitrogen intermediates are an additional antimicrobial defence of activated macrophages (Fang, 1997; MacMicking *et al.*, 1997a). Many microbes that resist killing by reactive oxygen intermediates cannot survive in the presence of reactive nitrogen intermediates, the best studied of which is nitric oxide (NO). NO production is also associated with maintaining a variety of intracellular pathogens in their latent state, including *Mycobacterium tuberculosis* (MacMicking *et al.*, 1997b), *Toxoplasma gondii* (Adams *et al.*, 1990), and *Leishmania* spp. (Stenger *et al.*, 1996). In each case, inhibition of NO production results in re-activation of acute disease in an experimental model, but the mechanism by which NO protects is unclear. In addition to its antimicrobial effects, NO influences an enormous range of diverse physiological functions that affect the host response to infection, including vasodilatation and host cell cytotoxicity (Nathan, 1992).

The inducible nitric oxide synthase (iNOS, NOS2) of murine macrophages is stimulated by inflammatory cytokines such as IFNγ, TNFα, IL-1 and IL-2, as well as by microbial products, including LPS and lipoteichoic acid (LTA). iNOS catalyses the oxidation of L-arginine at a guanidino nitrogen to yield nitric oxide (NO) and L-citrulline. Like superoxide, NO is highly reactive, and readily becomes oxidised to form a number of other toxic nitrogen intermediates, including peroxynitrite ($OONO^-$), nitrogen dioxide (NO_2), dinitrogen trioxide (N_2O_3), and dinitrogen tetroxide (N_2O_4) (Fang, 1997; Robinson, 1998).

Oxygen and Nitrogen Intermediates Interact

Although they are regulated differently, there is a great deal of interaction between the reactive oxygen species produced by NADPH oxidase and the nitrogen intermediates resulting from iNOS. This synergy provides an important expansion of the macrophage antimicrobial arsenal, because the susceptibility of different microbes to individual oxidative molecules varies greatly. For example, interaction of NO with O_2^- produces peroxynitrite, which is more toxic than NO for *S. typhimurium* (De Groote *et al.*, 1995) and *E. coli* (Pacelli *et al.*, 1995). Peroxynitrite can also be formed by the interaction of hydrogen peroxide with nitrous acid (HNO_2). Likewise, NO enhances the production of hydrogen peroxide from superoxide by the Haber–Weiss reaction, because NO reduces Fe^{3+} to Fe^{2+}, which catalyses this reaction.

Many of the specific antibacterial mechanisms of these reactive nitrogen intermediates remain unclear. Probable targets for which there is experimental support include, DNA damage, inactivation of critical enzymes, especially including those with reactive thiol groups, haem or iron sulphur groups, and membrane damage, which results from lipid peroxidation (Fang, 1997; MacMicking *et al.*, 1997a). In addition, inactivation of haem-containing host enzymes involved in iron acquisition can deplete iron stores on which microbes depend.

NO Production in Human versus Murine Macrophages

NO-mediated antimicrobial effects are well established in murine macrophages, but have been difficult to demonstrate in human cells *in vitro*. The controversy surrounding this problem centres on whether NO plays an important role in human infection or whether human macrophages use only alternative NO-independent mechanisms to kill invading microbes.

NO-independent antimicrobial mechanisms have been reported in human cells, and it is clear that NO production is not induced in human cells by the same factors that stimulate production of large amounts of NO in murine cells (Fang, 1997). None the less, iNOS mRNA and protein are present in human macrophages, and low levels of L-arginine-dependent NO production have been found in response to an assortment of stimuli (De Maria *et al.*, 1994; Bukrinsky *et al.*, 1995).

Additional evidence that NO may be an important antimicrobial defence in humans comes from clinical studies. Macrophages from patients with active pulmonary tuberculosis have high levels of NOS2 protein and increased NO production (Kumar *et al.*, 1995; Nicholson *et al.*, 1996). Similar results have been found in children with malaria and patients with a variety of inflammatory conditions, including acute respiratory distress syndrome, rheumatoid arthritis and glomerulonephritis (Condino-Neto *et al.*, 1993; St Clair *et al.*, 1996; Anstey *et al.*, 1999). Overall, it is likely that NO synthesis plays an essential role in controlling human infection, albeit with unique features and activation signals different from those in mice.

Intracellular Bacterial Survival Themes: An Overview

In spite of the considerable antimicrobial capacity of macrophages, a surprising number of pathogenic bacteria have developed ways to survive and multiply within these host cells. Numerous strategies have been identified, and undoubtedly more are yet to be discovered. While it is absolutely essential for some bacteria to avoid the host microbicidal mechanisms, others thrive in some or all of these hostile environments. For example, *Coxiella burnetti* multiplies in phagolysosomes and requires acidification for replication, as do *Francisella* and *Brucella* spp. *Salmonella* requires vacuole acidification to induce expression of intracellular survival genes, and *Listeria* depends on acidification to escape from its vacuole. Although each organism uses its own unique overall approach to intracellular survival, some strategies are shared by several organisms, albeit with variations and in different combinations (Fig. 2).

Some bacteria facilitate development of their preferred niche within macrophages by modulating their route of entry into host cells and avoiding conventional phagocytosis. *Ehrlichia*, *Chlamydia* and *Francisella* enter cells by receptor-mediated endocytosis, whereas *Listeria*, *Salmonella* and *Shigella* stimulate macropinocytosis to facilitate their uptake.

Vacuole escape can also be a mechanism for avoiding bactericidal mechanisms, and/or a means by which bacteria acquire access to nutrients and signalling pathways for long-term survival and replication.

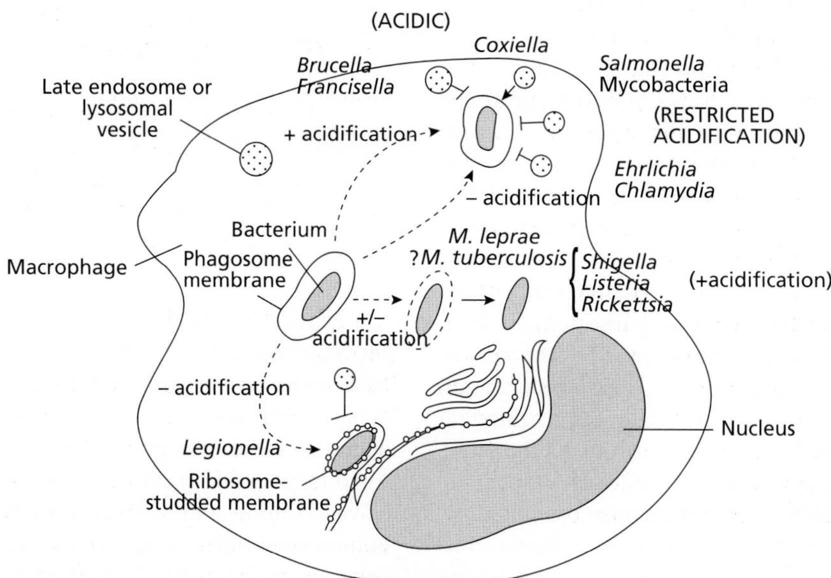

Fig. 2 Overview of bacterial trafficking patterns in macrophages. The wide range of possible interactions with the macrophage endo-lysosomal pathway used by the intracellular bacteria, as discussed in this chapter, have been divided into three groups. The largest group includes bacteria that remain within vacuoles that retain at least some interaction with the endo-lysosomal pathway. Nearly all of these pathogens in some fashion modulate their vacuoles. A second group includes bacteria that escape from their vacuoles to reside free in the cytoplasm. *Legionella* is shown separately, because it diverts from the endo-lysosomal pathway within minutes of uptake. Instead, it traffics to a perinuclear position, where it associates with rough endoplasmic reticulum. Some bacteria, such as Mycobacteria and *Salmonella*, are shown more than once, or in ambiguous positions, because they are capable of several interactions within macrophages.

Shigella, *Listeria*, *Rickettsia* and *M. leprae* escape fully from their phagosomes by lysing the vacuolar membrane. *M. tuberculosis* appears to permeabilise vacuolar membranes, but often stops short of complete membrane lysis. Most other bacteria remain within limiting vacuolar membranes, but many modify the compartments in which they survive. This often requires communication with the signalling machinery in the host-cell cytoplasm. Several organisms have acquired type 3 (*Salmonella*, *Shigella*) or type 4 (*Legionella*, *Brucella*) secretion systems that allow them to penetrate the restrictive barrier of the host phagosome and/or cytoplasmic membranes, allowing them to control events in the host cell during or after their uptake.

Nutrient acquisition is another major concern for these intravacuolar bacteria, because in this regard the phagosome is thought to be a very restrictive environment. Many pathogens have acquired special strategies to address this problem. *Coxiella* and *Ehrlichia* increase their access to the essential nutrient iron by up-regulating the expression of transferrin receptors in the cell. Likewise, to meet their energy needs, *Rickettsia* and *Chlamydia* express specialised ATP/ADP translocases that allow them to tap into the host-cell ATP supply.

Bacterial survival strategies can also include sophisticated interactions with the host signalling pathways. *Ehrlichia* interferes with the Jak/STAT signalling pathway to prevent IFNγ activation of its host cell. *Shigella* and *Salmonella* induce apoptosis in their macrophage hosts by activating the caspase1 cell-death protein. In contrast, *Rickettsia* activates NFκB expression to prevent apoptosis of its host. *Listeria* activates NFκB expression too, but it is unknown whether it does this to block apoptosis.

Overall, the interactions between bacterial pathogens and their host cells are complex, and in many cases only just beginning to be understood. The first part of this chapter reviewed some of the basic macrophage functions; the remainder describes the specific survival strategies developed by a variety of intracellular bacterial pathogens to deal with the challenges they confront by choosing to parasitise macrophages.

Bacterial Survival Strategies

Brucella

Brucella spp. are facultative intracellular parasites capable of surviving within both professional and non-professional phagocytes (see Chapter 94). Survival within macrophages correlates with the ability of *Brucella* spp. to cause disease, and is considered an essential virulence determinant of this pathogen.

Role of Lipopolysaccharide

LPS has been proposed as a virulence determinant for some *Brucella* spp. because, in animal models of infection, smooth isolates of *B. abortus*, *B. melitensis* and *B. suis* are more virulent than their rough counterparts (Baldwin and Winter, 1994). Survival in macrophages also correlates with the smoothness of these strains, as determined by the length of the O-polysaccharide side-chain. This is not, however, a universal phenomenon in *Brucella* spp., because rough isolates of *B. ovis* and *B. canis* are more virulent than their smooth counterparts.

Brucella LPS is unusual in that it is much less host-reactive compared with enterobacterial LPS (Moriyon *et al.*, 1987; Rasool *et al.*, 1992). It is relatively non-toxic for endotoxin-sensitive mice, rabbits and chicken embryos, has low macrophage toxicity, and is not highly pyrogenic. It is also a poor inducer of IFN and TNF, although it does stimulate IL-12 production. *Brucella* spp. may also counter LPS induction of TNFα by releasing a protease-sensitive inhibitor that specifically blocks TNFα expression in activated macrophages (Caron *et al.*, 1996).

Transposon knockout of LPS in *B. abortus* has shown that it is not essential for survival in macrophages (Price *et al.*, 1990a), but it is necessary for full virulence in mice (Godfroid *et al.*, 1998). *Brucella* LPS is not degraded in macrophages and is eventually recycled from lysosomes to the plasma membrane, where it can remain for months (Forestier *et al.*, 1999). The role of this LPS in the host macrophage plasma membrane is unknown.

Uptake and Survival

Opsonisation is necessary for the efficient uptake of *Brucella* by macrophages *in vitro* (Baldwin and Winter, 1994). Survival of bacteria in unactivated macrophages is not affected by the route of entry, but opsonisation may lead to more efficient bacterial killing in IFNγ-activated macrophages. The route of entry into macrophages *in vivo* in the absence of antibody is not known. However, the fact that *Brucella* enter a variety of non-professional phagocytes suggests that bacterially mediated uptake mechanisms exist and that they may be important *in vivo*.

Although the number of bacilli in murine and human macrophages initially decreases by 30–50%, the surviving bacilli begin to grow logarithmically approximately 10 hours post-infection. This initial killing in unactivated macrophages is probably due primarily to superoxide and hydrogen peroxide, because it is specifically inhibited by addition of catalase and superoxide dismutase (Jiang *et al.*, 1993). Macrophages from *Brucella*-resistant cattle also show higher levels of oxidative bursts than those of susceptible cattle (Price *et al.*, 1990b).

Intracellular Localisation

Brucella spp. remain within an acidified membrane-bound vacuole during macrophage infection, but in other non-professional phagocytes they are located in the lumen of the rough endoplasmic reticulum (Baldwin and Winter, 1994; Pizarro-Cerda *et al.*, 1998a). These bacterial phagosomes rapidly acidify to pH 4.0–4.5, and this pH drop is required for survival and replication in macrophages (Porte *et al.*, 1999). Neutralisation of these phagosomes with bafilomycin A, ammonium chloride or monensin within an hour of infection prevents bacterial growth and results in bacterial clearance by 48 hours post-infection. Vacuole neutralisation has no effect if delayed until 7 hours post-infection, indicating that acidification is required only during the early phase of bacterial establishment within the macrophage. Viable bacteria are not required for the acidification of the vacuoles, but viable bacteria may promote acidification to a small degree. Vacuoles that contain live or gentamicin-killed *Brucella* acidify with similar kinetics, but the lowest pH levels achieved (4.5–4.7) with dead organisms are slightly higher than those achieved with live bacteria.

In spite of acidification, phagosomes that contain virulent *B. abortus* or *B. suis* do not fuse with lysosomes in HeLa cells or macrophages (Frenchick *et al.*, 1985; Pizarro-Cerda *et al.*, 1998b). This failure is an active process that requires viable bacilli and may be mediated by a water-soluble component that has not yet been fully characterised. This inhibitor is not produced by the attenuated *B. abortus* strain 19, which fails to inhibit phagolysosome (P-L) fusion and survives poorly in macrophages.

Molecular and Genetic Components

A two-component regulatory system required for the inhibition of P-L fusion in HeLa cells has been identified in *B. abortus* (Sola-Landa *et al.*, 1998). This BvrR–BvrS two-component system is also required for efficient uptake, survival and replication of *B. abortus* in HeLa cells and murine macrophages. Further characterisation of the BvrR–BvrS regulon, which is related to ChvI–ExoS in *Rhizobium meliloti* and to ChvI–ChvG in *Agrobacterium tumefaciens*, is likely to provide a great deal of information about the interaction of *Brucella* spp. with their host cells.

A putative type IV secretion system is also important to intracellular *B. suis* (O'Callaghan *et al.*, 1999). This secretion system is similar to the VirB DNA secretion apparatus in *A. tumefaciens* and the Ptl secretion apparatus in *Bordetella pertussis*, which directs the secretion of pertussis toxin into mammalian cells (Christie, 1997). Similar secretion systems are required for virulence and pathogen–host cell interactions in *Helicobacter pylori* and *Legionella pneumophila* (Roy *et al.*, 1998; Covacci *et al.*, 1999).

The genes that encode this putative VirB secretion apparatus in *Brucella* are located within a 12-kb region of the smaller chromosome (O'Callaghan *et al.*, 1999). The *B. suis virB* locus was originally identified by transposon insertion mutations that rendered the bacteria unable to grow in HeLa cells. Several of these genes are also necessary for survival and replication in THP1 cells and primary human monocyte-derived macrophages. The role of VirB in *Brucella* is unclear. It is not required for secretion of the protease sensitive TNFα inhibitor within macrophages. An alternative possibility that has not been tested is that the VirB secretion apparatus releases effector molecules into or through the phagosome membrane to influence the trafficking of this vacuole, as does *Legionella* (Roy *et al.*, 1998; Segal *et al.*, 1998).

Interaction with Activated Macrophages

IFNγ activation of macrophages makes them brucellacidal, whereas a number of other cytokines, including GM-CSF, TNF, IL-4 and IL-1 have no effect (Baldwin *et al.*, 1993; Liautard *et al.*, 1996). Unlike many other organisms, such as *Francisella*, for which iron supplementation facilitates survival, iron loading of IFNγ-treated macrophages increases their anti-*Brucella* activity (Baldwin *et al.*, 1993; Leonard *et al.*, 1997). It has been suggested that the excess iron facilitates the production of toxic oxygen and nitrogen species by the Haber–Weiss reaction. In PMNs, virulent *B. abortus* inhibits the myeloperoxidase–hydrogen peroxide–halide system and the oxidative burst (Canning *et al.*, 1986). cAMP and cGMP have been implicated in inhibition of the myeloperoxidase–hydrogen peroxide–halide system by preventing degranulation of PMNs, but the mechanism for inhibition of the oxidative burst, which might also function in macrophages, is unknown.

Rickettsia

Role of ATP/ADP Translocase

Rickettsia are obligate intracellular pathogens that survive within host endothelial cells and macrophages (Winkler and Turco, 1994) (see Chapter 102). The inability of *Rickettsia* spp. to generate their own ATP means that they must satisfy their energy requirements by obtaining ATP from their host cells. This is achieved by an ATP/ADP translocase (Tlc) that resides within the bacterial cell membrane and exchanges ADP for the ATP from the host (Alexeyev and Winkler, 1999). *Chlamydia psittaci*, another obligate intracellular pathogen (see below), contains a similar transporter. The ATP/ADP transport proteins of *Rickettsia* and *Chlamydia* are similar to transporters of plant plastids, but distinct from mitochondrial transporters, which serve an opposite function in that they move ATP out of the mitochondrion into the cytoplasm (Winkler and Neuhaus, 1999).

Intracellular Localisation

After engulfment by macrophages or endothelial cells, *Rickettsia* escape from the phagosome into the host cytoplasm (Winkler and Turco, 1994). They then recruit and polymerise host-cell actin to facilitate movement in the cytoplasm by a mechanism distinct from that observed with *Listeria* and *Shigella* (Gouin *et al.*, 1999; Heinzen *et al.*, 1999a). Actin-based motility is less apparent with the typhus group *Rickettsia*; *R. prowazeckii* does not exhibit actin-based motility and *R. typhi* does so very inefficiently. The importance of actin-based motility for rickettsial virulence and survival in macrophages has not been explored.

Uptake and Survival

One of two major immunodominant surface-exposed proteins, rOmpA, has been implicated in the binding of spotted fever group *Rickettsia* to non-professional phagocytes (Li and Walker, 1998). The role of rOmpA in macrophage phagocytosis and phagosome escape has not been investigated. The typhus group *Rickettsia* do not express rOmpA or any identifiable homologue, and these organisms are likely to have additional adhesins.

Most *Rickettsia* spp. survive and kill their host macrophages when ingested in the presence of normal serum (Winkler and Turco, 1994). The means by which *Rickettsia* kill their host cells has not been established, but apoptosis is not generally observed. This may be because *R. rickettsii* prevents apoptosis of infected endothelial cells in culture by activating NF-κB (Clifton *et al.*, 1998; Sahni *et al.*, 1998). Apoptosis occurs within four hours of infections when NFκB activation is blocked, which suggests that, to allow sufficient time for bacterial replication and spread prior to necrotic cell death, *Rickettsia* actively prevent apoptosis in some host cells. This anti-apoptotic effect has not been investigated in macrophages.

Opsonisation with specific antibodies present in immune serum leads to macrophage-mediated killing of the bacterium by an unknown mechanism (Winkler and Turco, 1994). This killing of *Rickettsia* by unactivated macrophages does not involve a respiratory burst, but NO is probably important in activated macrophages (see below). *R. prowazeckii* does not induce a respiratory burst on entry, and IgG-coated *Rickettsia* are killed by a mutant macrophage cell line that is unable to produce a respiratory burst.

The presence of complement in immune serum can override the bactericidal effects usually associated with immune serum and instead results in bacterial survival and macrophage death. The simplest interpretation of these data is that complement-mediated entry allows the bacteria to bypass a lethal respiratory burst that would be triggered during entry via the FcRγ receptor. Other data suggest, however, that trafficking of the bacteria through the Fc receptor does not fully explain the observed killing of antibody-opsonised bacteria. *Rickettsia* opsonised with F(ab′)2 were found in phagolysosomes and were killed by murine macrophages to the same extent as antibody-opsonised bacteria. This shows that the bacterial killing associated with immune sera may be independent of Fc-mediated entry into the macrophages. In addition, increased bacterial killing is observed only when the *Rickettsia* are pre-coated with antibodies, but not when bacteria are ingested under conditions in which the same anti-*Rickettsia* antibodies are pre-bound to the surface of the macrophages rather than the bacteria (Beaman and Wisseman, 1976). In this case of cytophilic antibody binding, bacterial phagocytosis is enhanced but bacterial killing is not.

These experiments indicate that antibody-mediated bacterial uptake alone is not always sufficient for the killing of *Rickettsia*. It is possible that the use of different Fc receptors leads to the various outcomes, or that the different distributions of the bound antibodies in each case affects receptor activation. Alternatively, antibody opsonisation may prevent an active moiety on the surface of the bacterium from carrying out its prescribed function. These antibody-mediated effects may involve either a specific inactivation or steric hindrance that precludes a necessary interaction between the engulfed bacterium and the phagocytic membrane.

Interaction with Activated Macrophages

Activation of macrophages with IFNγ and LPS results in increased macrophage survival and greatly reduced infection of RAW macrophages or L929 fibroblasts by *Rickettsia* (Turco *et al.*, 1998). Cell entry by *Rickettsia* requires viable bacteria, and the decreased number of *Rickettsia* taken up by activated murine macrophages is thought to be due to NO-mediated killing of *Rickettsia* in the extracellular space before their uptake. *Rickettsia* are exquisitely sensitive to killing by NO, and the decreased *Rickettsia* infection levels are reversed by the NO scavenger oxyhaemoglobin. Nearly 20-fold more oxyhaemoglobin is required to ameliorate the anti-*Rickettsia* activity of activated macrophages than of L929 cells. This is consistent with the higher levels of nitrate present in murine macrophages as compared with fibroblasts. Cytokine-treated human cells also restrict rickettsial growth, but it is unclear whether in these cells the anti-*Rickettsia* activity is NO-dependent.

Coxiella

Coxiella burnetti is an obligate intracellular pathogen classified in the family Rickettsiaceae (see Chapter 80), but its interaction with host macrophages is very different from that of other *Rickettsia* species. *C. burnetti* exists in either of two morphologically distinct developmental forms, the small-cell variant (SCV) and the large-cell variant (LCV) (Heinzen *et al.*, 1999b). SCVs are metabolically dormant and optimised for long-term extracellular survival under harsh conditions. In contrast, the LCVs are metabolically active, and fragile outside of the host-cell environment. The transition between SCV and LCV occurs in host cells, including macrophages. The developmental cycle of *Coxiella* in cells is similar to that of *Chlamydia trachomatis* (see below), another obligate intracellular pathogen.

Entry and Survival

Entry into macrophages and a variety of non-professional phagocytes occurs by a parasite-directed microfilament-dependent process, but the receptor–ligand interactions have not been defined (Baca *et al.*, 1993). Energy expenditure on the part of the bacterium is not required to stimulate uptake, since viable and non-viable bacteria enter into these cells with the same efficiency. *Coxiella* usually enter host cells as SCVs, but the LCV form is equally infectious *in vitro*.

Intracellular Localisation

Coxiella burnetti resides in a membrane-bound parasitophorous vacuole (PV) that has many characteristics of secondary lysosomes. PVs are highly fusigenic, and individual phagosomes coalesce to form large multibacillary vacuoles (Hechemy *et al.*, 1993). Although ingestion of *Coxiella* fails to stimulate a significant respiratory burst, the bacterial phagosomes appear to mature normally along the endosome pathway. The PVs acidify to a pH of about 4.8 and acquire late endosome and lysosome markers including 5′ nucleotidase, acid phosphatase, cathepsin D and LAMPS 1 and 2 (Maurin *et al.*, 1992b; Heinzen *et al.*, 1996; Heinzen *et al.*, 1999b).

Acidification is required for *Coxiella* replication, and is thought to be one of the signals within the PV that allows conversion of SCVs to the metabolically active LCV form (Hackstadt and Williams, 1981; Maurin *et al.*, 1992a). Nonetheless, a typical PV contains a continuum of both SCV and LCV morphological forms, and each form maintains its own characteristic protein expression profile. It is puzzling how the two forms co-exist and maintain different regulatory responses to the same environment. Breakthroughs in the genetic transformation of *Coxiella* are likely to facilitate research into this question (Suhan *et al.*, 1996; Lukacova *et al.*, 1999).

Accumulation of Transferrin Receptors

Pathogen-mediated modification of the PVs also occurs, as shown by an increased expression of transferrin receptors (TfRs) in J774A.1 macrophages infected with viable, but not aldehyde- or heat-killed *C. burnetti* (Howe and Mallavia, 1999). The metabolic activity of *C. burnetti* is sensitive to iron concentration, and the increased TfR expression is thought to be a mechanism for providing the bacterium with sufficient iron in the PV. A similar intracellular survival strategy is used by *Ehrlichia* spp. (see below and Chapter 102). Iron overload interferes with IFNγ activation of *Ehrlichia*-infected macrophages (Barnewell and Rikihisa, 1994), so it is possible that TfR induction also serves an immunomodulatory function that benefits *Coxiella* in its host macrophages.

Chlamydia

Like *Coxiella*, *Chlamydia* are obligate intracellular pathogens that undergo a biphasic developmental cycle that includes the infectious elemental body (EB) and vegetative reticulate body (RB) (Moulder, 1991)(see Chapter 87). EB are compact, metabolically inert, and have a heavily cross-linked rigid cell wall that makes them impervious to a variety of environmental stresses. EB represent the extracellular transmissible form of this pathogen that can bind to and trigger uptake by host cells. In contrast, the RB form is non-infectious and unable to survive outside of host cells. RBs are the metabolically active form of this pathogen and replicate within host cells by binary fission.

The infectious life cycle of *Chlamydia* begins when EBs enter a new host cell to reside within a specialised membrane-bound vacuole that has traditionally been called an inclusion. Inclusions undergo high levels of homotypic fusion, and individual vacuoles coalesce into a single large inclusion within each cell. The EBs differentiate into metabolically active RBs, which proceed to replicate within the inclusion. After several rounds of replication, the RBs begin to convert back to the infectious EB form. The infected cell eventually undergoes lysis and releases the parasites to infect neighbouring cells.

Uptake and Survival

Survival of different *Chlamydia* spp. in macrophages varies. As a result, most macrophage studies have been done with the *C. trachomatis* LCG biovar and avian *C. psittaci* strains, which are more capable of intramacrophage survival and replication than the less invasive *C. trachomatis* oculogenital and mammalian *C. psittaci* strains.

Uptake of *Chlamydia* into macrophages is cytochalasin B-independent (Gregory *et al.*, 1979; Nakajo *et al.*, 1990). For unknown reasons, high levels of uptake (> 10 bacteria/macrophage) causes immediate macrophage toxicity (Kuo, 1978). The plasma membrane receptors that trigger uptake of *Chlamydia* have not been identified, but binding to cultured cells is inhibited by trypsin, heat treatment and heparin (La Verda and Byrne, 1994; Sinai and Joiner, 1997). Several bacterial ligands have been implicated, including a major outer membrane protein, (MOMP), an EB-specific cell surface protein, OMP2, and a sulphated oligosaccharide-modified adhesin that binds to heparin sulphate receptors. A Mip (macrophage infectivity potentiator) homologue has also been identified in *Chlamydia* (Lundemose *et al.*, 1991), but it is not known whether this protein causes the same enhancement of macrophage infection and virulence in *Chlamydia* as it does in *Legionella* (Cianciotto *et al.*, 1990).

Once internalised, the phagosome does not fuse with lysosomes and resists acidification (La Verda and Byrne, 1994; Heinzen *et al.*, 1996; Sinai and Joiner, 1997). The pH of inclusions remains > 6.0, which is consistent with the absence of vacuolar ATPase in the inclusion membrane. Lack of P-L fusion has been clearly demonstrated, and it has been shown that *Chlamydia* inclusions do not acquire soluble lysosome markers, such as lysosome acid phosphatase, cathepsin D, thorotrast, ferritin, or the fixed lysosome membrane marker glycoproteins LAMP1 and LAMP2. Isolated EB envelopes are sufficient to inhibit P-L fusion, but heating EBs before infection results in P-L fusion and bacterial killing. Ultraviolet irradiation does not destroy the ability to inhibit P-L fusion for up to 12 hours, which suggests that a heat-labile membrane component, rather than active bacterial metabolism, is required for this behaviour. Bacterial protein synthesis is, however, required immediately after bacterial uptake for *Chlamydia* survival and continued inhibition of P-L fusion beyond 12 hours post-infection (Scidmore *et al.*, 1996).

It is debated whether antibody opsonisation leads to P-L fusion and killing of *Chlamydia*. Early studies reported that *Chlamydia* opsonised with specific antibody were trafficked to phagolysosomes and killed (Wyrick *et al.*, 1978; Hammerschlag *et al.*, 1985). More recently it was found that antibody-opsonised bacteria ingested by HeLa or CHO cells that express FcR avoided P-L fusion and trafficked similarly to non-opsonised bacteria (Scidmore *et al.*, 1996). Polyclonal antisera were used in the early studies, while monoclonal antibodies were used in the latter. This or the difference in cell types used may be responsible for the disparate results.

Intracellular Localisation

Chlamydia inclusions also appear to remain segregated from the endocytic and transferrin recycling pathways (Heinzen *et al.*, 1996). They fail to acquire exogenous fluid-phase endocytic markers, such as lucifer yellow, fluorescein dextran, or HRP; transferrin-HRP does not traffic to these inclusions, and neither transferrin nor transferrin receptors can be detected by immunofluorescent staining. It has been suggested that inclusions have selective access to the exocytic pathway for Golgi-derived glycosphingolipids, but not glycoproteins (Hackstadt *et al.*, 1996). C6-NBD-ceramide is a vital stain for Golgi that is converted to sphingomyelin. When C6-NBD-ceramide is added to *Chlamydia*-infected cells, a large amount of the label is transported to inclusions and is incorporated into the bacterial cell wall. This transport is inhibited by Brefeldin A, which blocks transport from the Golgi. C6-NBD-sphingomyelin in the inclusion is derived from the Golgi rather than from the plasma membrane, but Golgi glycoproteins did not accumulate in the inclusion with the sphingomyelin. This indicates that lipid delivery to the inclusion is a specific and selective phenomenon that occurs directionally from the Golgi. It also suggests that glycoprotein and lipid trafficking from the Golgi are distinct, a possibility not previously considered.

Molecular and Genetic Determinants

Chlamydia are strict energy parasites, and must acquire ATP and nucleotides from the host cell. *C. psittaci*, but not *C. trachomatis*, inclusions are surrounded by mitochondria, which may be a source of ATP. Two nucleotide transporters, Npt1ct and Npt2ct, have been identified and characterised in *C. trachomatis* (Tjaden *et al.*, 1999), and these are similar to the *Rickettsia* ATP/ADP Tlc transporter (Alexeyev and Winkler, 1999). Npt1ct catalyses the exchange of ADP for ATP, presumably by providing the bacteria with a means of acquiring ATP from their hosts. In contrast, Npt2ct is highly specific for GTP and other nucleotide triphosphates, and is energised by a proton motive force. The Npt2ct transporter probably functions in nutrient acquisition.

An unresolved problem is how these nucleotides are transported across the inclusion membrane. Unlike *Rickettsia*, which are free in the cytoplasm, *Chlamydia* are restricted from the host-cell cytoplasm by a vacuolar membrane. Projections from the bacterial surface have been identified that may function as transport channels, but these would serve only bacilli proximal to the inclusion membrane. Non-specific pores have not been identified, and low-molecular-weight molecules from the cytoplasm do not diffuse into the inclusion (Heinzen and Hackstadt, 1997). Nonetheless, it is possible that highly selective pores or specific transporters exist within the inclusion membrane to provide the pathogen with essential nutrients from the host.

The protein composition of the inclusion membrane has not been well characterised, but several *Chlamydia*-specific proteins have been identified. IncA is a serine/threonine phosphoprotein on the cytoplasmic face of the *C. trachomatis*, but not of *C. psittaci* inclusion membranes (Rockey *et al.*, 1997). This protein is present as a moiety of higher molecular weight in infected cells than in RBs and it has been suggested that the size increase is due to its phosphorylation by host-cell kinases. Several additional *Chlamydia*-specific proteins identified with antisera against inclusion membranes remain to be characterised (Rockey and Rosquist, 1994). The presence of these proteins in the inclusion membrane suggest, nonetheless, that *Chlamydia* have extensively remodelled their parasitophorous vacuoles to facilitate selective interaction with their host cells.

Interaction with Activated Macrophages

Chlamydia are unable to replicate in cytokine-activated macrophages, but they survive in these cells *in vitro* for several days before they die (Byrne *et al.*, 1989). IFNγ is essential and sufficient for this effect, but additional cytokines may act synergistically to broaden the anti-*Chlamydia* activity. Several lines of evidence suggest that *Chlamydia*-stasis results from an oxygen-independent mechanism (Byrne and Faubion, 1983; Murray *et al.*, 1983). At low multiplicity of infection (< 10 bacteria per cell), bacterial growth is restricted in spite of the absence of a respiratory burst. In addition, a macrophage cell line unable to produce superoxide, and oxygen-competent cells, restrict growth of *Chlamydia*, and various oxygen radical scavengers fail to reverse the bacteriostatic effects.

Tryptophan starvation has been proposed as the most likely explanation for the anti-*Chlamydia* effects observed in IFNγ-activated human macrophages (La Verda and Byrne, 1994). IFNγ depletes the intracellular tryptophan pools in human macrophages by the action of indoleamine 2,3-dioxygenase (IDO). Since *Chlamydia* are auxotrophic for tryptophan, they depend on their host cell for this essential nutrient. Tryptophan reverses bacteriostasis in a dose-dependent manner, and increasing tryptophan is required to suppress the effects of higher doses of IFNγ. Bacterial growth is rescued only if tryptophan is provided during the first three days after macrophage infection. IFNγ does not induce IDO in murine cells, so that killing of *Chlamydia* by these cells must occur by a different mechanism (Zhong and de la Maza, 1988).

Ehrlichia

Ehrlichia (see Chapter 102) are small obligate intracellular pathogens classified within the family Rickettsiaceae, but like *Coxiella* their behaviour within phagocytes differs considerably from that of other Rickettsia (see Chapter 102). Many *Ehrlichia* spp., including the human pathogen *E. chaffeensi*, which causes human monocytic ehrlichiosis (HME), reside within monocytes and macrophages. A related group of *Ehrlichia* spp. infects granulocytes rather than monocytes and causes human granulocytic ehrlichiosis (HGE) (Bakken *et al.*, 1994).

Uptake and Survival

The means by which *Ehrlichia* enter monocytes/macrophages is poorly understood, but it is predominantly actin-independent and in the case of *E. risticcii* requires tyrosine phosphorylation of host proteins (Messick and Rikihisa, 1993; Rikihisa *et al.*, 1994; Zhang and Rikihisa, 1997). Binding of *E. chaffeensis* to host macrophages also induces high levels of protein kinase A (PKA) activity, which may play a key role in *Ehrlichia* survival in activated macrophages (Lee and Rikihisa, 1998) (see below). Several *Ehrlichia* outer-membrane proteins bind to monocytes and may play a role in uptake (Rikihisa, 1999). One of these is encoded in a multi-gene family that contains as many as 17 gene copies. Such a high degree of redundancy in a small genome probably means that this gene product plays an important role during the *Ehrlichia* infection cycle.

Intracellular Localisation

Ehrlichia reside within large membrane-bound vacuoles termed 'morulae' (**Fig. 3**). These slightly acidic inclusions fail to fuse with lysosomes; instead they retain features of early endosomes (Barnewall *et al.*, 1997). They are negative for the fixed lysosome membrane markers LAMP1 and CD63, but positive for the early endosome markers Rab5 and early endosome antigen 1. Interestingly, co-infection of HL-60 cells with *E. chaffeensis* and the agent of HGE results in separate inclusions for each bacterial type (Mott *et al.*, 1999). This suggests that these related organisms modify their vacuoles in distinct ways, but both avoid contact with lysosomes. Inclusions that contain the agent of HGE remain on the endocytic pathway in that they continue to acquire endocytosed colloidal gold particles (Webster *et al.*, 1998). Like *Ehrlichia*, HGE-bacterial inclusions fail to acquire the lysosome membrane glycoprotein LAMP1, but they stain with antibodies to the mannose-6-phosphate receptor, a late endosome marker. In contrast to *E. chaffeensis*, HGE-bacterial inclusions do not stain for the transferrin receptor, suggesting that these organisms use different mechanisms for iron acquisition.

Mature *E. chaffeensis* morulae label strongly with antibodies against TfR. TfRs begin to accumulate in the inclusions shortly after infection and continue to do so until nearly all of the cell supply of this molecule co-localises to the inclusion. *E. chaffeensis* infection also induces cellular TfR mRNA expression, by a process that requires *de novo* protein synthesis within internalised bacteria (Barnewall *et al.*, 1997). Since growth of *E. chaffeensis* in macrophages is inhibited by desferoxamine, a cell permeant iron chelator, *Ehrlichia* may obtain its iron directly from iron-bound transferrin complexes (Barnewell and Rikihisa, 1994). This suggests that these bacteria depend on intracellular iron pools, but lack siderophores capable of competing with desferoxamine.

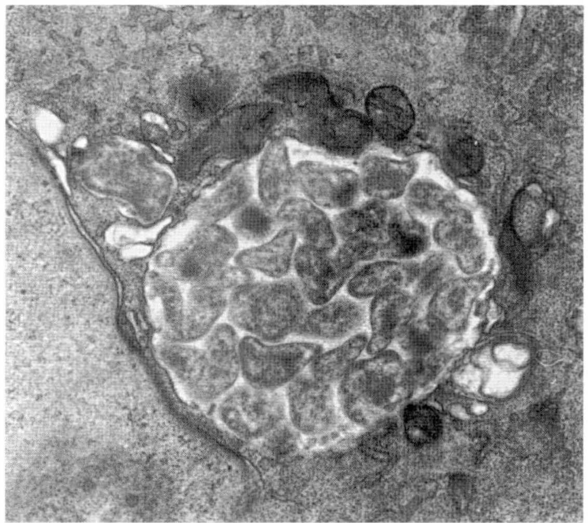

Fig. 3 Electron micrograph showing polymorphic *Ehrlichia chaffeensis* in an inclusion termed a 'morula'. The macrophage nucleus is at the bottom left of the figure, and several mitochondria appear closely associated with the membrane of the morula. Courtesy of S.R. Bawa.

Interaction with Activated Macrophages

Scavengers of reactive oxygen intermediates are unable to suppress the killing of *E. chaffeensis* by IFNγ-treated human macrophages (Barnewell and Rikihisa, 1994), indicating that the killing mechanism is oxygen-independent. NO generation appears to be the primary anti-*Ehrlichia* mechanism in murine but not human macrophages, although iron starvation may play a role in the anti-*Ehrlichia* activity of both host species (Rikihisa, 1999). Transferrin-dependent cytoplasmic iron stores are depleted by down-regulation of TfR in IFNγ-treated cells, which reduces the available iron supply for *Ehrlichia*. In human cells, iron-saturated, but not iron-free, transferrin can reverse IFNγ-induced inhibition of *E. chaffeensis*. In murine cells, NO binds to the active sites of various iron sulphur enzymes, inhibiting functions that *Ehrlichia* may need for iron acquisition. In addition, reactive nitrogen intermediates may have direct anti-*Ehrlichia* effects.

Interestingly, IFNγ is effective at restricting *Ehrlichia* growth only if macrophages are activated before *Ehrlichia* infection (Barnewell and Rikihisa, 1994). This is most probably due to the ability of *Ehrlichia* to up-regulate TfR mRNA expression as a counter to the inhibitory effects of IFNγ on TfR production. The cellular PKA activation that occurs when *Ehrlichia* bind to host cells blocks IFNγ-mediated activation of host macrophages by preventing tyrosine phosphorylation of Jak (Janus kinase) and STAT (signal transducers and activators of transcription), so abrogating the effectiveness of this key IFNγ cell-activation pathway (Briscoe *et al.*, 1996; Lee and Rikihisa, 1998). PKA activation is caused by binding of an *Ehrlichia* protein to an unknown cellular receptor. This phenotype does not require bacterial internalisation, and is unaffected by periodate treatment, suggesting that it is not a lectin-type interaction (Rikihisa, 1999). Identification of the bacterial ligand and host-cell receptor will provide new clues to how *Ehrlichia* manipulates the signalling pathways of its host cell.

Francisella

Entry and Survival

Francisella tularensis (see Chapter 101) is nearly always found within macrophages *in vivo*, and its uptake by these cells is not inhibited by cytochalasin B (Fortier *et al.*, 1994). As with *Chlamydia* and *Ehrlichia*, this actin-independence suggests that *F. tularensis* entry into macrophage occurs by a process more similar to receptor-mediated endocytosis than phagocytosis, as defined above.

Francisella also fail to trigger a respiratory burst on entry into macrophages and reside in membrane-bound vacuoles (Anthony *et al.*, 1991; Fortier *et al.*, 1992a). These vacuoles do not appear to fuse with lysosomes, as measured by their failure to acquire thorium particles from within pre-labelled lysosomes. Nonetheless, these phagosomes must acidify to allow bacterial replication, and bacterial growth can be restricted with any of a variety of agents that raise the vacuolar pH. This low-pH requirement probably reflects the strict need of this bacterium to acquire iron for growth (Chamberlain, 1965). Internalised transferrin receptors must be acidified to release bound iron into the vacuole, and it is likely that *F. tularensis* depends on this iron source for survival.

Surprisingly, *F. tularensis* grows faster (2–4 hours) within inflammatory macrophages than in differentiated tissue macrophages (6–8 hours) (Fortier *et al.*, 1994), which suggests that inflammation may be advantageous for the bacterium, at least in the short term. *F. tularensis* is susceptible to oxidative killing by PMNs, so that entry into inflammatory macrophages may be an effective strategy to avoid these cells.

Molecular and Genetic Determinants

The genetic response of *F. tularensis* to the intracellular macrophage environment has been examined by metabolic radiolabelling with $[S^{35}]$methionine and two-dimensional gel electrophoresis (Golovliov *et al.*, 1997). The results suggest that changes in overall protein expression profiles of *Francisella* while in macrophages may be less extensive than that of other facultative intracellular bacteria. As compared with tissue culture media, only four bacterial proteins are induced during growth in macrophages. One of these is a novel 23-kDa polypeptide of unknown function that is also highly produced in the presence of hydrogen peroxide.

Two other proteins of unknown function, MglA and MglB, (macrophage growth loci) are essential for intra-macrophage survival of *Francisella* (Baron and Nano, 1998). At the deduced amino acid level they are similar to SspA and SspB of *E. coli* and *Haemophilus* spp. In *E. coli*, SspA and SspB are thought to be genetic regulatory proteins that control gene expression during nutrient stress conditions (Williams *et al.*, 1994).

Interaction with Activated Macrophages

IFNγ-activated murine macrophages are efficient killers of *Francisella* (Anthony *et al.*, 1992; Fortier *et al.*, 1992b; Fortier *et al.*, 1994). In murine macrophages, killing is mediated by NO, but can be reversed in the presence of excess exogenous free iron. L-arginine-derived NO reacts with free iron to form iron–nitrosyl complexes in activated macrophages, thereby restricting the availability of iron to the bacteria. IFNγ-activated human monocytes are also bactericidal for *Francisella*, but the pathway may involve an NO-independent iron-depletion mechanism.

Legionella

Outside a human host, *Legionella* often survives within amoebae and ciliated protozoa, which are thought to provide an environment similar to macrophages. Previous residence in amoebae affects subsequent interactions with host macrophages and may enhance the ability of *L. pneumophila* to establish infection in their human hosts (Cirillo *et al.*, 1999).

Uptake and Survival

L. pneumophila enters macrophages by a conventional or an unusual form of coiling phagocytosis and begins to replicate within 2–4 hours post-infection. The host cell ruptures approximately 20 hours later, and the bacteria move on to infect other cells. The bacteria must avoid phagolysosome fusion to survive within macrophages, and probably also in their free-living amoebae (Sinai and Joiner, 1997; Roy, 1999). They do this by directing the trafficking of their phago somes (replicative vacuoles) away from the default endolysosomal pathway. The decision to avoid phagolysosome fusion is made at the time of uptake or very soon thereafter, because mutant bacteria begin to accumulate markers of the default endosome/late endosome fusion within 5 minutes of infection (Roy *et al.*, 1998b; Wiater *et al.*, 1998). This trafficking is independent of the route of entry into the cell, because opsonisation and complement-mediated uptake result in similar outcomes. Activated macrophages are, however, able to override the agenda of the pathogen and kill the organism by a NO-dependent mechanism (Skerrett and Martin, 1996).

Intracellular Localisation

Vacuoles that contain virulent *L. pneumophila* fail to acidify and sequentially associate with smooth vesicles, mitochondria, and rough ER. Eventually they end up in a perinuclear membrane-bound replicative vacuole that fails to acidify and is surrounded with rough ER. Electron microscopically, these replicative vacuoles look like ribosome-studded organelles. The reason for this intimate association with host ribosomes is not clear, but it is possible that they provide the replicating bacteria with nutrients.

Molecular and Genetic Determinants

Two classes of genes are required for the ultimate success of *L. pneumophila* in macrophages (Roy, 1999). These include the establishment genes that direct formation of the replicative vacuole, and the maintenance genes that are required for bacterial multiplication within the replicative vacuole. Most of the genes that have been described fall into the establishment group, and mutations in these genes have drastic effects on bacterial survival in macrophages. The *dot* (defective organelle trafficking) and *icm* (intracellular multiplication) genes are required to establish the replicative vacuole. In contrast, *thy*A and *mip* (macrophage infectivity potentiator) are examples of genes in the latter maintenance group (Cianciotto *et al.*, 1990; Swanson and Isberg, 1996). Mutations in either gene allow formation of a normal replicative phagosome, but the bacteria fail to multiply efficiently within this compartment.

The *dot/icm* loci include 16–18 genes encoded in two large regions of the chromosome (Berger *et al.*, 1994; Segal and Shuman, 1997; Segal *et al.*, 1998). Some of these genes are required for a large pore-forming apparatus that has homology to broad host–range bacterial conjugation systems such as *Vir*B in *A. tumefaciens* (Christie, 1997; Kirby *et al.*, 1998). Although this complex can transfer broad host–range plasmids, such as RSF1010 (Vogel *et al.*, 1998), its primary function is probably to transport effector proteins that direct the trafficking of the replicative vacuole (Roy, 1999; Zuckman *et al.*, 1999). These effector proteins may be inserted into the vacuole membrane as it forms to redirect its trafficking. Alternatively, such effector molecules may be secreted into the host-cell cytoplasm to interfere with the normal cell response to the incoming bacterium (Fig. 4). Similar type IV secretory systems are required for virulence-associated protein secretion in *H. pylori* and *B. pertussis* (Covacci *et al.*, 1999). Homologous genes have been identified in *R. prowazekii* and *B. suis*, but their roles in virulence have not been established.

The DotA secretory apparatus must be present in the outer membrane of *L. pneumophila* before macrophage entry, because cytochalasin D and bacterial binding studies suggest that pore formation occurs in nascent vacuoles as bacteria are being engulfed (Roy *et al.*, 1998; Segal *et al.*, 1998). The effector protein(s) secreted by the DotA apparatus have not been identified, but the product of *icm*W has been proposed as a chaperone for the effector protein(s) (Zuckman *et al.*, 1999). An *icm*W mutant forms pores but does not inhibit phagolysosome fusion. This indicates that IcmW is essential for interruption of the host signalling response to bacterial entry, but does not play a structural role in assembly of the DotA secretory apparatus. However, the restriction of IcmW to the cytoplasm of intracellular bacteria, and its similarity to chaperones of the *Yersinia* type III secretion system suggest that it plays a supportive, rather than a direct, role in the effector functions of the DotA secretory apparatus.

The rapid cytotoxicity that occurs when more than five bacteria are present per cell is also consistent with DotA-mediated pore formation in the plasma membrane of host cells at the time of bacterial uptake (Segal

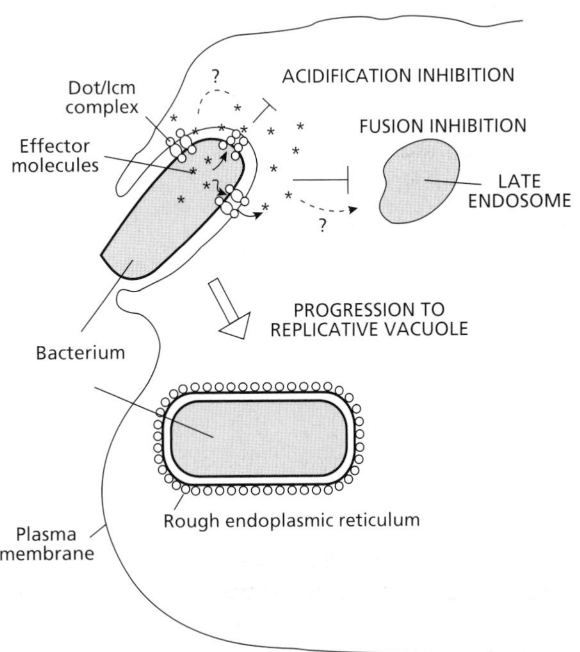

Fig. 4 Model showing a possible role for the Dot/Icm type IV secretion apparatus in the trafficking of *Legionella* on entry into macrophages. The Dot/Icm complex is shown inserted across the bacterial cell wall and into the macrophage membrane to create a pore through which effector molecules can be secreted into the macrophage cytoplasm or membranes to mediate trafficking events. The bacterium is shown in the process of uptake, but the timing of Dot/Icm-mediated secretion remains unclear and may not occur until after uptake.

et al., 1998). This rapid cell killing is not caused by metalloprotease and legiolysin. Cell death is primarily necrotic, and depends on the presence of the (conjugation) DotA pore-forming assembly. Lower multiplicities of infection (MOIs), for example one bacterium per cell, lead to productive infection rather than immediate cell killing, and are probably physiologically more relevant. Nonetheless, the rapid cell-killing phenotype is a useful laboratory correlate of virulence and has helped researchers to identify a number of virulence-associated genes.

Mycobacteria

Members of the genus *Mycobacterium* (see Chapters 81 and 83 [Collins, Snewin *et al.*]) are classified as tuberculous (*M. tuberculosis* or Mtb complex) or non-tuberculous (NTM) based on the diseases they cause. The Mtb complex includes the agents *M. tuberculosis*, *M. bovis*, *M. africanum* and *M. microti*. *M. bovis* BCG is an attenuated derivative of *M. bovis* that is used as a vaccine in some parts of the world. *M. avium* and *M. intracellulare* form the *M. avium* complex (MAC) (see Chapter 83), a special class of NTM. *M. leprae* is an obligate intracellular pathogen and the only *Mycobacterium* species that cannot be cultured *in vitro* (see Chapter 82). Little is known about how *M. leprae* interacts with its host cells, and this section will focus primarily on the Mtb and MAC organisms, which have received much recent attention.

Uptake and Survival

Macrophages are the primary residence of Mtb complex and MAC bacteria in their hosts, but *in vitro* they enter a variety of non-professional phagocytes. Unopsonised or serum-opsonised mycobacteria enter macrophages by conventional phagocytosis by use of a wide range of receptors, including the complement receptors CR3, CR1 and CR4, mannose receptor (MR), class A scavenger receptors (SR) (Ehlers and Daffe, 1998; Ernst, 1998). CR3 and CR4, which are both $\beta2$ integrin receptors, mediate uptake of iC3b-opsonised mycobacteria and unopsonised bacteria. Unopsonised bacteria bind directly to a lectin site distinct from the C3bi binding site on CR3.

The importance of the uptake route to subsequent events within macrophages is not clear. Normal or immune serum alters the trafficking of Mtb after entry into macrophages, but bacterial viability is not affected (Armstrong and Hart, 1975; McDonough *et al.*, 2000). Zimmerli *et al.* (1996) found no difference in bacterial survival when either the CRs or MR were blocked. Taken together, this supports a model in which bacterial survival is independent of entry route. Nonetheless, other studies suggest that the bacterial entry route may be more important to escape the microbicidal mechanisms of activated macrophages. Mtb and *M. avium* enhance their ability to enter non-professional and professional phagocytes by non-opsonic routes after a period of residence in macrophages (McDonough and Kress, 1995; Bermudez *et al.*, 1997; McDonough *et al.*, 2000). Mtb traffics differently and is more cytotoxic to host cells, and *M. avium* resists TNFγ-mediated growth restriction, after pre-growth in other macrophages. Further work is necessary to determine whether entry route or other bacteria factors cause these effects.

Intracellular Localisation

Trafficking Armstrong and Hart (1971) provided the first evidence that Mtb inhibits phagolysosome fusion in macrophages. Since then, it has been confirmed that Mtb, *M. avium* and *M. marinum* can redirect the trafficking of the vacuoles in which they reside (Barker *et al.*, 1997; Deretic *et al.*, 1997). These vacuoles fail to acquire most late endosome/lysosome markers, including CD63, M-6-PR, Rab7, cathepsin D and tracer molecules BSA-gold that has been chased to secondary lysosomes. They do, however, accumulate the late endosome/lysosome markers LAMPs 1 and 2.

Vacuoles that contain mycobacteria retain early endosome markers such as Rab5 and TfR, and remain accessible to endosome tracers (Clemens and Horwitz, 1996; Russell *et al.*, 1996; Deretic *et al.*, 1997). Mtb and *M. avium* phagosomes also fail fully to acidify, and in the case of *M. avium* this is probably due to a lack of a vacuole ATPase in their membranes (Crowle *et al.*, 1991; Sturgill-Koszycki *et al.*, 1994). One interpretation of these data is that the vacuoles are frozen at an early endosome stage of the endolysosomal pathway. Retention of a host membrane protein, TACO, in Mtb vacuoles has been implicated as a mechanism to inhibit phagosome maturation, because TACO must usually be released from vesicles before fusion with lysosomes can occur (Ferrari *et al.*, 1999). The lack of vacuole acidification, which is required for some late endosome and lysosome fusion events, may also contribute to this inhibition of phagolysosomal fusion (Aniento *et al.*, 1996).

Nonetheless, in spite of its clear ability to do so, Mtb does not always inhibit P-L fusion on entry into macrophages, and its survival does not require that it do so. Armstrong and Hart (1975) later showed that Mtb-mediated inhibition of P-L fusion can be reversed

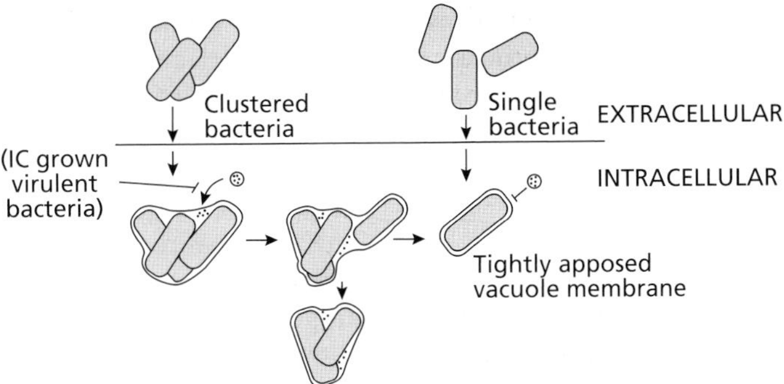

Fig. 5 Model showing the different trafficking of *Mycobacterium* engulfed by macrophages singly, or in clusters of multiple bacilli. With the exception of virulent bacteria that have been pre-grown in macrophages, vacuoles that contain clusters of mycobacteria acquire late endosomal/lysosomal content markers more readily than do vacuoles that contain single bacilli.

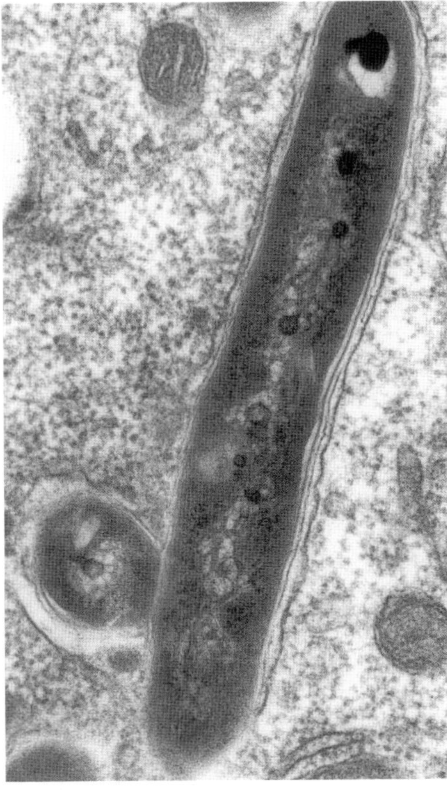

Fig. 6 Electron micrograph showing longitudinal and cross-sectional views of *M. tuberculosis* bacilli in a tightly apposed vacuole membrane. Courtesy of Y. Kress.

if bacteria are pre-opsonised with normal or immune serum at the time of entry, and that the resulting fusion had no negative effects on bacterial multiplication. Others have also confirmed that Mtb and *M. avium* can survive and replicate in phagosomes that have acquired lysosome characteristics, and in one case P-L fusion correlated with enhanced Mtb replication (Brown *et al.*, 1969; McDonough *et al.*, 1993; Gomes *et al.*, 1999).

Single Bacteria Traffic Differently from Clumped Bacteria Bacterial clumping also leads to higher levels of phagolysosome fusion (Clemens and Horwitz, 1995; McDonough *et al.*, 2000), and this probably explains some of the discrepancies in the recent literature. Whereas vesicles that contain single tubercle bacilli rarely acquire lysosome markers, vacuoles that contain tubercle bacilli taken up as part of a cluster often fuse with lysosomes, but the extent of this fusion may partially be restricted (**Fig. 5**) (Clemens and Horwitz, 1995; de Chastellier *et al.*, 1999; McDonough *et al.*, 2000). These observations may be important, because macrophages encounter mycobacteria singly and in larger clusters during natural infection, because of the enormous propensity of these organisms to aggregate and form long cord-like structures.

The low-fusion phenotype associated with single mycobacteria can be reproduced with hydrophobic latex beads enclosed in tightly apposed vacuole membranes, but not with hydrophilic beads that are surrounded by much looser membranes. This led to the suggestion that tight apposition of a vacuole membrane to the surface of an engulfed particle can inhibit vesicle maturation (de Chastellier *et al.*, 1995; de Chastellier and Thilo, 1997). This possibility is consistent with the observation that intracellular mycobacteria are often enclosed in tightly apposed membranes (**Fig. 6**), but the irregular shape of bacterial clusters prevents portions of vacuolar membrane from adhering tightly to a bacterial surface (**Fig. 5**) (McDonough *et al.*, 1993; de Chastellier *et al.*, 1995; Moreira *et al.*, 1997).

Nonetheless, additional factors are also likely to be important in Mtb phagosome trafficking. Low fusion with clumps of virulent Mtb, but not avirulent BCG bacteria, have been observed after bacterial pre-growth in other macrophages (McDonough *et al.*, 2000). This behaviour is serum-dependent, and fusion inhibition occurs only when re-infection occurs in the absence of serum (Armstrong and Hart, 1971, 1975). It is not clear whether the mechanism by which clumps of virulent bacteria avoid fusion is the same as that used by single bacilli (Fig. 6).

Taken together, these studies suggest that mycobacterial trafficking in macrophages is complex, and may vary at different stages of infection. Fusion inhibition is likely to be essential for survival in partially activated macrophages (Schaible *et al.*, 1998; Via *et al.*, 1998), whereas contact with lysosome components may be tolerated, or even exploited, in unactivated macrophages. Exposure to late endosome/lysosome components in unactivated macrophages may provide environmental stimulatory signals that control gene expression in intracellular bacteria. These changes in gene expression may enhance the survival of the bacteria when they encounter activated macrophages on release from their original host macrophages as the infection progresses.

Vacuole Escape The intracellular localisation of Mtb with respect to the phagosome membrane remains controversial. Immunological data are most consistent with a model in which virulent Mtb permeabilise or destroy their vacuolar membranes to gain access to the host-cell cytoplasm, as is observed with *M. leprae* (Kaufmann, 1993; Britton *et al.*, 1994). Although several studies have found virulent Mtb bacteria free in the host-cell cytoplasm (Myrvik *et al.*, 1984; McDonough *et al.*, 1993; Florczyk *et al.*, 2001), others have reported only intra-vacuolar bacilli (Xu *et al.*, 1994; Clemens and Horwitz, 1996; Moreira *et al.*, 1997). A recent study showed that BCG bacteria are surrounded by permeable vacuolar membranes that allow entry of fluorescent-tagged dextrans as large as 70 kDa from the cytoplasm (Teitelbaum *et al.*, 1999). In contrast, there is no evidence that *M. avium* disrupts its phagosome membrane.

Both CD4+ and CD8+ T cells are necessary to protect against Mtb, and it has been shown that a class I MHC-mediated immune response is required to control virulent Mtb infection, but not BCG or *M. avium* infection in mice (Orme and Collins, 1984; Flynn *et al.*, 1992; Bermudez and Petrofsky, 1999). Class I antigen presentation is generally restricted to cytoplasmic peptides transported into the lumen of the endoplasmic reticulum by the transporter protein, TAP, for loading on to MHC molecules, while class II peptides are derived from phagolysosmes. Class I antigen presentation of Mtb and, to a more limited extent, BCG peptides is TAP-dependent, so that TAP knockout mice cannot control Mtb infection (Mazzaccaro *et al.*, 1996; Behar *et al.*, 1999).

We have re-examined the intracellular localisation of Mtb and BCG in three-dimensional reconstructions by high-voltage electron microscopy (Florczyk *et al.*, 2001). BCG was always in intact phagosome membranes, while Mtb vacuoles remained intact 4 hours after infection. However, nearly 30% of Mtb pre-grown in other macrophages were associated with damaged or absent vacuolar membranes by 4 hours post-infection, indicating a regulated phenotype. The same conditions were associated with the highest levels of Mtb-mediated necrosis of lung epithelial cells and macrophages (McDonough and Kress, 1995).

Taken together, these data suggest that Mtb has a tightly regulated membranolytic potential, induction of which may result in a continuum of phenotypes ranging from permeabilised vacuolar membranes to cell necrosis (Fig. 7) (Myrvik *et al.*, 1984; McDonough and Kress, 1995; Teitelbaum *et al.*, 1999). In contrast, conditions that restrict induction of this phenotype would result in intact membranes (Xu *et al.*, 1994; Clemens and Horwitz, 1995; Moreira *et al.*, 1997).

Molecular and Genetic Determinants

Like many other facultative intracellular bacteria, Mtb modifies its gene expression while in macrophages. This has been demonstrated with metabolic labelling with ^{35}S-methionine and two-dimensional gel electrophoresis in which significant differences in protein expression were detected between intracellular and extracellular bacteria (Lee and Horwitz, 1995; McDonough *et al.*, 2000). Overall, about 15% of Mtb

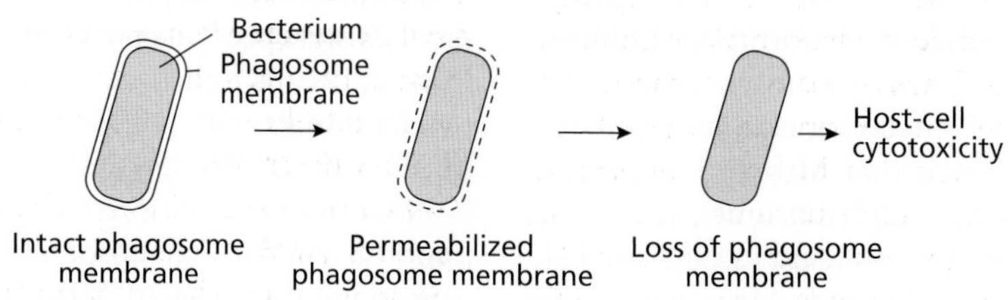

Fig. 7 Schematic showing the effects of increasing membrane lysis caused by a bacterium in a phagosome. Such activity may cause a continuum of phenotypes, ranging limited permeability of the vacuolar membrane to necrosis of the host cell.

proteins are expressed at higher levels, and about 10% are reduced in expression when bacteria are in macrophages, compared with control bacteria growing in tissue culture media. Several of these proteins have been identified as heat-shock and iron-responsive proteins.

Additional intracellularly(IC)-induced genes have been identified in Mtb, *M. avium*, and *M. marinum* by promoter : reporter gene fusion and RNA-based assays, but few of these have been characterised (Dellagostin *et al.*, 1995; Barker *et al.*, 1998; Yuan *et al.*, 1998). With the exception of the *acr* promoter, which is highly induced in macrophages, the change in transcriptional expression of regulated genes in mycobacteria is often only several-fold, and very rarely exceeds 10-fold. Several additional genes likely to be induced in macrophages have been implicated in intramacrophage survival because of their ability to enhance the survival rates of *M. smegmatis* in THP1 macrophages, or protection of non-mycobacterial species from nitrosative and oxidative stress (Ruan *et al.*, 1999; Miller and Shinnick, 2000).

The α-crystallin homologue, encoded by *acr/hspX*, is a heat-shock protein with a chaperone function, and is one of the better studied IC-induced Mtb genes (Chang *et al.*, 1996; Yuan *et al.*, 1998). Although Acr is required for intramacrophage growth, its role remains unknown. Acr expression is also induced by NO, low oxygen conditions, stationary phase, and is essential for transition of Mtb bacteria to a non-replicating state, which is thought to resemble the condition of latent bacteria in granulomas (Yuan *et al.*, 1996; Garbe *et al.*, 1999; Hu and Coates, 1999). Regulation of *acr* expression is complex, and occurs at both the transcriptional and post-transcriptional levels (Hu and Coates, 1999; Manabe *et al.*, 1999). Elucidation of the role of this protein in Mtb intramacrophage survival may lead to a number of additional contributing factors.

Interaction with Activated Macrophages

Macrophages are a key defence against mycobacteria, but they are unable to kill these organisms unless appropriately activated. The requirements for activation vary with the host species. Murine macrophages kill Mtb by a NO-dependent mechanism when activated by IFNγ and LPS, LAM (lipoarabinomannan) or TNFα (Chan *et al.*, 1992). It is more difficult to stimulate mycobactericidal activity in human monocytes/macrophages *in vitro*, though it undoubtedly occurs *in vivo*. Protection from oxidative damage is provided by a combination of catalase-peroxidase and alkyl hydroperoxide enzymes, as well as the mycobacterial cell wall LAM, which scavenges oxygen free radicals (Chan *et al.*, 1991; Manca *et al.*, 1999). Additional genes may provide some protection from reactive nitrogen and oxygen intermediates (Ehrt *et al.*, 1997; Hu *et al.*, 1999; Ruan *et al.*, 1999).

Listeria

Uptake and Survival

Listeria monocytogenes (see Chapter 69) survives in macrophages by escaping from the phagosome to reside and replicate in the host-cell cytoplasm. Activation of NFκB, MAP kinase(s) ERK-1 and/or ERK-2 occurs in J774 cells during *L. monocytogenes* uptake, suggesting that the bacteria are signalling uptake pathways different from those for latex beads (Kugler *et al.*, 1997). Like *Salmonella typhimurium*, the bacteria are internalised into spacious macrophage phagosomes by a macropinocytosis-like process (Beauregard *et al.*, 1997). Tight phagosomes are rarely observed in macrophages immediately after *L. monocytogenes* uptake, but the large phagosomes shrink with time. This process differs from internalin-mediated entry of *L. monocytogenes* into non-professional phagocytes (Mengaud *et al.*, 1996). In addition, phagocytosis of *L. monocytogenes* by different macrophage populations varies, and the outcome of infection may be affected by the route of infection (Drevets *et al.*, 1992).

Intracellular Localisation

Escape from the phagolysosome requires the bacterial haemolysin (listeriolysin O, LLO), which is encoded by the *hly* gene and is active only at acid pH (Portnoy *et al.*, 1992; Beauregard *et al.*, 1997). Intact vacuoles begin to acidify shortly after bacterial uptake; and on reaching a pH of approximately 6.0, LLO first perforates and then lyses the phagosome membrane. The initial perforation results in a rapid rise in pH, as the membrane potential equilibrates. This transient effect is followed by release of the vacuole contents. Membrane perforation and lysis does not occur with LLO mutants or when vacuole acidification is inhibited with bafilomycin or NH_4Cl.

In addition to LLO, *Listeria* require PlcA, a phosphatidylinositol-specific phospholipase C, for efficient vacuole escape (Portnoy *et al.*, 1992) (**Fig. 8**). After escape, the bacteria move within the host-cell cytosol on comet-like tails of polymerised actin, similar to the process described for *Shigella* (Tilney and Portnoy, 1989; Dabiri *et al.*, 1990). This actin-based movement requires ActA to mediate actin assembly. Bacterial spread to other cells occurs by extension of bacteria on long actin tails into adjacent cells. This requires ActA-based movement, along with LLO and PlcB, a second

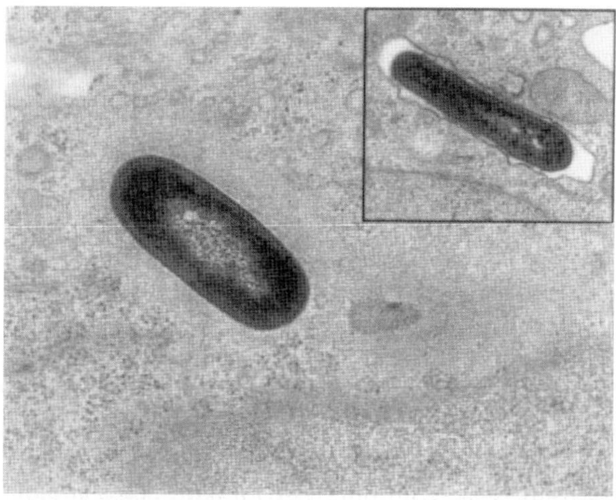

Fig. 8 Electron micrograph of *Listeria monocytogenes*. The main image is a wild-type bacterium free in the macrophage cytoplasm after lysing out of its vacuole. Note the halo of actin that surrounds this bacterium, and the actin tail that is beginning to form (bottom right). The insert shows an LLO$^-$ mutant that remains within a phagosome. Bacteria for these images were provided by D. Portnoy; photo is courtesy of M. Florczyk.

phospholipase C with lecithinase activity. PlcB is required for escape from the double-membrane-bound vacuole that surrounds the bacterium after entry into the newly infected neighbouring cell.

Molecular and Genetic Determinants

The genes that encode these virulence determinants, *hly*, *plc*A, *act*A and *plc*B, are located close together on the chromosome in a gene cluster that includes *prf*A (Leimeister-Wachter *et al.*, 1990). PrfA protein transcriptionally regulates *hly*, *plc*A, *act*A and *plc*B (Chakraborty *et al.*, 1992). Transcription of *hly*, *plc*A and *plc*B is activated predominantly in the phagosome, whereas actA expression occurs mostly in the cytosol (Bubert *et al.*, 1999). This expression pattern is consistent with where the protein products are needed during infection, and lead to the suggestion that PlcB can substitute for PlcA in vacuole escape (Bubert *et al.*, 1999). Interestingly, *prf*A and *plc*A are poorly expressed in epithelial cells as compared with macrophages, suggesting that the bacteria encounter macrophage-specific environmental signals and challenges within these phagocytes.

PrfA is a member of the Crp/Fnr family of transcriptional regulatory proteins (Lampidis *et al.*, 1994). It possesses a leucine zipper-like structural motif at its C terminal end that is required for the binding of PrfA to its target sequence. The target sequence is called a PrfA box, and consists of a 14- to 16-bp DNA sequence with dyad symmetry located 40–42 bp upstream of the transcriptional start site of each PrfA-regulated gene or operon (Freitag *et al.*, 1993). PrfA-mediated gene regulation is complex and incompletely understood. Recent evidence suggests that PrfA may exist in a functionally inactive form that requires activation by protein factor Paf (PrfA activating factor) to carry out its regulatory function.

Interaction with Activated Macrophages

Macrophages activated with IFNγ and TNFα kill *L. monocytogenes* primarily by a NO-dependent mechanism (Leenen *et al.*, 1994), but oxidative intermediates may play a role in cells activated before bacterial entry (Ohya *et al.*, 1998). The mechanism by which NO kills *Listeria* is not known, but the low intracellular iron concentrations that result from down-regulation of transferrin receptors in activated macrophages may play a role (Alford *et al.*, 1991).

Shigella

Molecular Determinants of Uptake and Survival

Survival in macrophages is central to the success of *Shigella* in its human host, but they pass through host macrophages only briefly during the course of their infectious cycle (see Chapter 61). *Shigella* gains access from the lumen of the intestine to resident macrophages in Peyer's patches by passing through M cells (Sansonetti and Phalipon, 1999; Sansonetti *et al.*, 1999). After uptake by resident macrophages, *Shigella* escape from their phagosomes, and then kills their host cells by inducing apoptosis. On release from the dying macrophages the bacteria infect neighbouring epithelial cells from their basolateral surface.

Only invasive *Shigella* escape from the macrophage phagosome, and they must reach the cytoplasm to induce apoptosis. At least 30 gene products encoded on a 31-kbp fragment of the 220-kbp plasmid present in virulent *Shigella* may be required for invasion of non-professional phagocytes and subsequent events in phagocytes and epithelial cells (Maurelli *et al.*, 1985). Most of these gene products are associated with a type III secretion system, termed a 'secreton', encoded by the *mxi* and *spa* operons (Parsot and Sansonetti, 1996). This complex secretory apparatus forms a flagellar structure that injects bacterial effector molecules directly into the host-cell membranes or cytoplasm. This secretory apparatus is similar to that found in *Salmonella typhimurium*, and it assembles on the bacterial surface on contact with host cells.

Invasion of epithelial cells requires IpaA, IpaB, IpaC and IpaD, and occurs by triggering a cell ruffling process that leads to macropinocytosis (Swanson and

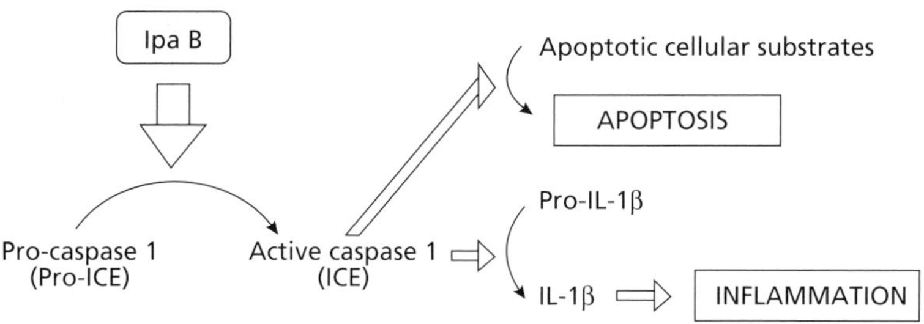

Fig. 9 Schematic of the pathway by which IpaB causes both apoptosis and inflammation after infection of macrophages. See text for additional details.

Watts, 1995; Sansonetti *et al.*, 1999). These invasion proteins are secreted via the Mxi/Spa secreton. Much less is known about entry into macrophages, but phagocytic clearance of *Shigella* may be mediated by the CD14 receptor. CD14 is a pattern recognition receptor that engulfs bacteria opsonised with LPS binding protein (LBP) (Mosser, 1994). Neutralisation of the CD14 receptor leads to overgrowth of *Shigella* in mucosal tissues (Sansonetti, 1999).

IpaB and IpaC are also necessary for vacuole lysis in macrophages and epithelial cells (Barzu *et al.*, 1997; de Geyter *et al.*, 1997; Sansonetti *et al.*, 1999). IpaB is thought to be the main effector, but IpaC may also contribute. Purified IpaC lyses lipid vesicles in a pH-dependent manner, and IpaC mutants are contact haemolysin-negative. Intracellular movement through the cytoplasm and lateral spread through the epithelium are mediated by IcsA, which forms a ternary complex with host proteins N-WASP (neural Wiskott–Aldrich syndrome protein) and Arp2/3 (actin related proteins) to nucleate actin polymerisation in a polar fashion (Egile *et al.*, 1999). This results in the formation of long actin tails, similar to those of *Listeria*, that propel bacteria through the host-cell cytoplasm and into neighbouring cells (Gouin *et al.*, 1999). IcsB lyses endocytosed bacterial protrusions from their double-membrane vacuoles after IcsA propels the bacteria to adjacent cells (Allaoui *et al.*, 1992). The need for IcsA and IcsB in macrophages is less clear.

Apoptosis

Apoptosis is a natural cell-death programme triggered by a number of pathways, only one of which is inflammatory (Steller, 1995; Hilbi *et al.*, 1997b; Zychlinsky and Sansonetti, 1997). This inflammatory pathway requires the activity of the cysteine protease caspase 1, (formerly ICE, for interleukin-1b converting enzyme), which is a homologue of the cell-death protein Ced3 in *Caenorhabditis elegans.*

IpaB was identified as the apoptotic factor of *Shigella* by using isogenic mutants in a background that contained *E. coli* haemolysin (Zychlinsky *et al.*, 1994). Its role was confirmed by demonstrating that microinjection of purified IpaB protein is sufficient for the apoptosis of peritoneal macrophages (Chen *et al.*, 1996b). IpaB induces apoptosis by binding directly to caspase 1, which goes on to mediate apoptosis (**Fig. 9**). Addition of the YVAD-CHO tetrapeptide, which inhibits cysteine proteases of the ICE family, protects macrophages from *Shigella*-induced apoptosis. Interestingly, *Shigella* do not cause apoptosis in PMNs or epithelial cells, but the reasons are not clear.

In addition to causing apoptosis, IpaB-activated caspase 1 triggers inflammation by cleaving inactive Pro-IL-1β in the cytoplasm to form active IL-1β, which is released from the dying macrophages (Hilbi *et al.*, 1997a; Thirumalai *et al.*, 1997). IL-1β is an inflammatory cytokine that recruits PMNs to the site of infection. The crucial role of IL-1 in *Shigella* pathogenesis was proved with IL-1ra (IL-1 receptor antagonist) experiments that blocked inflammation and disease (Sansonetti *et al.*, 1995).

Interaction with Activated Macrophages

Shigella are killed by IFNγ-activated macrophages *in vitro*, and in an experimental model of infection the LD_{50} is reduced by five orders of magnitude in IFNγ-deficient mice. (Way *et al.*, 1998). These data indicate that IFNγ is critical for innate resistance to *Shigella* infection, but the mechanism by which IFNγ mediates killing of *Shigella* in macrophages is not clear.

Salmonella

Salmonella enterica typhi (hereafter *S. typhi*) (see Chapter 66), is a strict human pathogen but *S. typhimurium* causes a typhoid-like systemic disease in mice. Some of these differences in host specificity may result from differential survival efficiency in human versus murine macrophages (Schwan and Kopecko, 1997), but *S. typhimurium* is often used in murine models of infection to gain insight into the more invasive *S. typhi*-like illness in humans.

Uptake and Survival

Salmonella enters macrophages and non-professional phagocytes by macropinocytosis rather than zipper-like phagocytosis, and survives in spacious phagosomes (Alpuche-Aranda *et al.*, 1994; Swanson and Watts, 1995). The cell ruffling required for bacterial uptake is triggered through a CDC42 G- protein-dependent mechanism (Francis *et al.*, 1993; Chen *et al.*, 1996a). Trafficking of the vacuole in macrophages remains controversial (Oh *et al.*, 1996; Rathman *et al.*, 1997). Some workers have found that *S. typhimurium*-containing vacuoles fuse completely with lysosomes, whereas others observed various degrees of fusion inhibition (Buchmeier and Heffron, 1991). Nonetheless, survival of *Salmonella* in macrophages is reduced if acidification of the vacuoles are prevented, and low pH is required for induction of genes associated with intracellular survival and replication (Alpuche-Aranda *et al.*, 1992; Rathman *et al.*, 1996).

Molecular and Genetic Determinants

Role of Pathogenicity Islands Five large pathogenicity islands, designated SPI-1 to SPI-5, have been identified in *S. typhi* (Mecsas and Strauss, 1996; Groisman and Ochman, 2000). Genes associated with the bacterium–macrophage interaction have been identified on all of these islands except SPI-5, which is required for enteropathogenicity. SPI-1 encodes type III secretion system (TTSS) genes required for the invasion of epithelial and M cells, and macrophage apoptosis (Mecsas and Strauss, 1996; Galan and Collmer, 1999; Hersh *et al.*, 1999). SPI-2 encodes a variety of genes needed for intramacrophage survival and systemic infections, including a second TTSS and its presumed effector proteins (Cirillo *et al.*, 1998; Hensel *et al.*, 1998). This latter is expressed only when the bacteria are in host cells. These SPI-2 TTSS genes are regulated by SsrA/SsrB rather than PhoP/PhoQ and their products may be involved in modulating the *Salmonella* vacuole. MgtC is of unknown function and is encoded on SPI-3 that is required for long-term intramacrophage survival and virulence in a mouse model (Blanc-Potard *et al.*, 1999). SPI-4 contains a macrophage survival locus, but has otherwise not been well characterised (Groisman and Ochman, 2000).

Macrophage Survival Genes Many additional genes have been identified that are important for *Salmonella* survival and growth in macrophages, but many of the mechanisms by which these gene products act remain unclear. It may also be important to differentiate between bacterial growth versus survival in macrophages, which can be difficult to separate in simple survival (bacterial recovery) experiments. Growth of a small population can mask killing of another population, or vice versa, and bacterial death can counteract the increases in population and lead to underestimates of true bacterial growth. These two parameters were examined individually by using a purine auxotroph (*pur*D) unable to grow in macrophages. It was shown that J774 cells and peritoneal exudate macrophages kill similar numbers of *Salmonella*, but only J774 cells were permissive for bacterial growth (Buchmeier and Libby, 1997).

The PhoQ/PhoP two-component signal transduction system regulates a large number of genes required for *Salmonella* survival in macrophages, including those required for LPS modification, resistance to antimicrobial peptides and nutrient scavenging (Groisman and Heffron, 1995). Other genes needed by *Salmonella* bacteria in macrophages include *rec*A and *rec*BC, whose products are required for DNA repair, and *sodA*, which is presumably also required for protection from oxidative stress (Buchmeier *et al.*, 1993; Tsolis *et al.*, 1995). Several other catalase-associated proteins, KatE, KatG and OxyR, protect *Salmonella* from oxidative compounds *in vitro*, but not in phagocytes (Papp-Szabo *et al.*, 1994; Buchmeier *et al.*, 1995). This is consistent with studies *in vivo* which show that neither *ahp*C nor *oxy*R are required for virulence in a mouse model (Taylor *et al.*, 1998).

SlyA is a transcriptional regulator required for adaptation to intracellular environments (Buchmeier *et al.*, 1997). SlyA regulates expression of a variety of proteins during either the stationery phase or intramacrophage growth, and *sly*A mutants are unable to survive in macrophages. The alternative sigma factor, sigma E, is also essential for survival in macrophages and virulence in a mouse model, and *sig*E mutants have heightened sensitivity to hydrogen peroxide, superoxide and antimicrobial peptides *in vitro* (Humphreys *et al.*, 1999). One of the genes regulated by *sig*E is *htr*A, which is required for intramacrophage survival and virulence. The effects of a *sig*E knockout are, however, far more dramatic than a *htr*A knockout, suggesting that additional genes in the sigma E regulon are critical for macrophage survival and virulence in *Salmonella*.

Magnesium Transporters Three magnesium transporters, CorA, MgtA and MgtB, have been identified in *Salmonella* (Smith *et al.*, 1998). MgtA and MgtB are regulated by the virulence-associated PhoP/PhoQ system, and activated at least 100-fold after invasion of macrophages and epithelial cells. The CorA transport system is ubiquitous in prokaryotes, and it is thought to be the primary Mg^{2+} influx pathway for most eubacteria and archaebacteria. In contrast, MgtA is present in most enteric bacterial species, but MgtB has been found in only three of 10 enteric species, and it is not present in *E. coli*.

The roles of MgtA and MgtB in *Salmonella* infection remain unclear. MgtB is expressed from a two-gene operon on SPI-3 that also encodes MgtC, a protein of unknown function essential for virulence in a mouse model and required for long-term survival in macrophages (Blanc-Potard and Groisman, 1997). Neither MgtA nor MgtB is, however, required for invasion of eukaryotic cells, survival within macrophages, or virulence in a mouse model of infection. Nonetheless, the regulation of these genes by the PhoP/PhoQ system is intriguing and it has been proposed that MgtA and MgtB transport something other than Mg^{2+} in the macrophage phagosome. It is possible that these proteins play an important, but redundant, role in infection that has yet to be identified.

Apoptosis

Like *Shigella*, *Salmonella* induces apoptosis of its host macrophage; but unlike *Shigella*, *Salmonella* delays killing of its host until after significant levels of bacterial replication have taken place. Apoptosis is mediated by SipB, an analogue of IpaB in *Shigella*, and SipB can complement for IpaA in HeLa cell invasion by *S. flexneri* and phagosome escape in macrophages (Hermant *et al.*, 1995; Hersh *et al.*, 1999).

Like IpaB, SipB triggers cell death by binding directly to caspase-1 which in turn causes apoptosis. It is not clear how SipB gains access to the host-cell cytoplasm, as *Salmonella* do not escape from their phagosomes in the same manner as *Shigella*. In addition, the role of apoptosis in *Salmonella* infection is not known, but it may be required to establish infection efficiently after oral inoculation.

Conclusion

Despite the tremendous progress in understanding the complex interactions of intracellular bacteria with their macrophage hosts, in most cases only the outlines of the bacterium–host relationships are visible. Coming years are likely to bring an unprecedented increase in the rate of discovery in this area, as the emerging fields of genomics and proteomics, and the application of global genetic approaches to identify specific classes of environmentally regulated genes, reach their full potential. The only certainty is that the journey ahead promises to be as exciting as it will be challenging.

References

Adams LB, Hibbs JB, Taintor RR, Krahenbuhl JL (1990) Microbiostatic effect of murine-activated macrophages for *Toxoplasma gondii*: role for synthesis of inorganic nitrogen oxides from L-arginine. *J. Immunol.* 144: 2725–2729.

Aderem A, Underhill DM (1999) Mechanisms of phagocytosis in macrophages. *Annu. Rev. Immunol.* 17: 593–623.

Alexeyev MF, Winkler HH (1999) Membrane topology of the *Rickettsia prowazekii* ATP/ADP translocase revealed by novel dual pho-lac reporters. *J. Mol. Biol.* 285: 1503–1513.

Alford CE, King TE, Campbell A (1991) Role of transferrin, transferrin receptors, and iron in macrophage listericidal activity. *J. Exp. Med.* 174: 459–466.

Allaoui A, Mournier J, Prevost MC, Sansonetti J, Parsot C (1992) *icsB*: a *Shigella flexneri* virulence gene necessary for the lysis of protrusions during intercellular spread. *Mol. Microbiol.* 6: 1605–1616.

Alpuche-Aranda CM, Swanson JA, Loomis W, Miller SI (1992) *Salmonella typhimurium* activates virulence gene transcription within acidified macrophage phagosomes. *Proc. Natl Acad. Sci. USA* 89: 10079–10083.

Alpuche-Aranda CM, Racoosin EL, Swanson JA, Miller SI (1994) *Salmonella* stimulate macrophage macropinocytosis and persist within spacious phagosomes. *J. Exp. Med.* 179: 601–608.

Aniento F, Gu F, Parton RG, Gruenberg J (1996) An endosomal beta-COP is involved in the pH-dependent formation of transport vesicles destined for late endosomes. *J. Cell Biol.* 133: 29–41.

Anstey NM, Weinberg JB, Wang Z, Mwaikambo ED, Duffy E, Granger DL (1999) Effects of age and parasitemia on nitric oxide production/leucocyte nitric oxide synthase type 2 expression in asymptomatic, malaria-exposed children. *Am. J. Trop. Med. Hygiene* 61: 253–258.

Anthony LD, Burke RD, Nano FE (1991) Growth of *Francisella* sp. in rodent macrophages. *Infect. Immun.* 59: 3291–3296.

Anthony LSD, Morrissey J, Nano FE (1992) Growth inhibition of *Francisella tularensis* live vaccine strain by IFN-gamma-activated macrophages is mediated by reactive nitrogen intermediates derived from L-arginine metabolism. *J. Immunol.* 148: 1829–1834.

Armstrong JA, Hart D (1971) Response of cultured macrophages to *Mycobacterium tuberculosis*, with observations on fusion of lysosomes with phagosomes. *J. Exp. Med.* 134: 713–740.

Armstrong JA, Hart D (1975) Phagosome-lysosome interactions in cultured macrophages infected with virulent tubercle bacilli: reversal of the usual nonfusion pattern and observations on bacterial survival. *J. Exp. Med.* 142: 1–16.

Baca OG, Klassen DA, Aragon AS (1993) Entry of *Coxiella burnetii* into host cells. *Acta Virol.* 37: 143–155.

Bakken JS, Dumler JS, Chen SM, Eckman MR, Van Etta LL, Walker DH (1994) Human granulocytic ehrlichiosis in the upper midwest United States: a new species emerging? *JAMA* 272: 212–218.

Baldwin CL, Winter AJ (1994) Macrophages and *Brucella*. In: Zwilling BS, Eisenstein TK (eds) *Macrophage–Pathogen Interactions*. New York: Marcel Dekker, pp. 363–380.

Baldwin CL, Jiang X, Fernandes DM (1993) Macrophage control of *Brucella abortus*: influence of cytokines and iron. *Trends Microbiol.* 1: 99–104.

Barker L, George KM, Falkow S, Small L (1997) Differential trafficking of live and dead *Mycobacterium marinum* organisms in macrophages. *Infect. Immun.* 65: 1497–1504.

Barker L, Brooks DM, Small L (1998) The identification of *Mycobacterium marinum* genes differentially expressed in macrophage phagosomes using promoter fusions to green fluorescent protein. *Mol. Microbiol.* 29: 1167–1177.

Barnewell RE, Rikihisa Y (1994) Abrogation of gamma interferon-induced inhibition of *Ehrlichia chaffeensis* infection in human monocytes with iron transferrin. *Infect. Immun.* 62: 4804–4810.

Barnewall RE, Rikihisa Y, EH Lee (1997) *Ehrlichia chaffeensis* inclusions are early endosomes which selectively accumulate transferrin receptor. *Infect. Immun.* 65: 1455–1461.

Baron GS, Nano FE (1998) MglA and MglB are required for the intramacrophage growth of *Francisella novicida*. *Mol. Microbiol.* 29: 247–259.

Barzu S, Benjelloun-Touimi Z, Phalipon A, Sansonetti J, Parsot C (1997) Functional analysis of the *Shigella flexneri* IpaC invasion by insertional mutagenesis. *Infect. Immun.* 65: 1599–1605.

Beaman L, Wisseman CL (1976) Mechanisms of immunity in typhus infections: differential opsonizing and neutralizing action of human typhus *Rickettsia*-specific cytophilic antibodies in cultures of human macrophages. *Infect. Immun.* 14: 1071–1076.

Beauregard KE, Lee KD, Collier RJ, Swanson JA (1997) pH-dependent perforation of macrophage phagosomes by listeriolysin O from *Listeria monocytogenes*. *J. Exp. Med.* 186: 1159–1163.

Behar SM, Dascher CC, Grusby MJ, Wang C-R, Brenner MB (1999) Susceptibility of mice deficient in CD1D or TAP1 to infection with *Mycobacterium tuberculosis*. *J. Exp. Med.* 189: 1973–1980.

Berger KH, Merriam JJ, Isberg RR (1994) Altered intracellular targeting properties associated with mutations in the *Legionella pneumophila* dotA gene. *Mol. Microbiol.* 14: 809–822.

Bermudez LE, Petrofsky M (1999) Host defense against *Mycobacterium avium* does not have an absolute requirement for major histocompatibility complex Class I-restricted T cells. *Infect. Immun.* 67: 3108–3111.

Bermudez LE, Parker A, Goodman JR (1997) Growth within macrophages increases the efficiency of *Mycobacterium avium* in invading other macrophages by a complement receptor-independent pathway. *Infect. Immun.* 65: 1916–1925.

Bermudez LE, Goodman J, Petrofsky M (1999) Role of complement receptors in uptake of *Mycobacterium avium* by macrophages *in vivo*: evidence from studies using CD18-deficient mice. *Infect. Immun.* 67: 4912–4916.

Bermudez LE, Goodman J, Petrofsky M (1999) Role of complement receptors in uptake of *Mycobacterium avium* by macrophages *in vivo*: evidence from studies using CD18-deficient mice. *Infect. Immun.* 67: 4912–4916.

Blanc-Potard A-B, Groisman EA (1997) The *Salmonella* selC locus contains a pathogenicity island mediating intramacrophage survival. *EMBO J.* 16: 5376–5385.

Blanc-Potard AB, Solomon F, Kayser J, Groisman EA (1999) The SPI-3 pathogenicity island of *Salmonella enterica*. *J. Bacteriol.* 181: 998–1004.

Bokoch GM (1995) Regulation of the phagocyte burst by small GTP-binding proteins. *Trends Cell Biol.* 5: 109–113.

Borregaard N (1988) The respiratory burst: an overview. In: Sbarra AJ, Strau RR (eds) *The Respiratory Burst and its Physiological Significance*. New York: Plenum Press, pp. 1–31.

Briscoe J, Kohlhuber F, Muller M (1996) JAKs and STATs branch out. *Trends Cell Biol.* 6: 336–340.

Britton WJ, Roche W, Winter N (1994) Mechanisms of persistence of mycobacteria. *Trends Microbiol.* 2: 284–288.

Brown CA, Draper P, Hart PD (1969) Mycobacteria and lysosomes: a paradox. *Nature* 221: 658–660.

Bubert A, Sololovic Z, Chun S-K, Papatheodourou L, Simm A, Goebel W (1999) Differential expression of *Listeria monocytogenes* virulence genes in mammalian host cells. *Mol. Gen. Genet.* 261: 323–336.

Buchmeier NA, Heffron F (1991) Inhibition of macrophage phagosome-lysosome fusion by *Salmonella typhimurium*. *Infect. Immun.* 59: 2232–2238.

Buchmeier NA, Libby SJ (1997) Dynamics of growth and death within a *Salmonella typhimurium* population during infection of macrophages. *Can. J. Microbiol.* 43: 29–34.

Buchmeier NA, Lipps CJ, So MYH, Heffron F (1993) Recombination-deficient mutants of *Salmonella typhimurium* are avirulent and sensitive to oxidative burst of macrophages. *Mol. Microbiol.* 7: 933–936.

Buchmeier NA, Libby SJ, Xu Y *et al.* (1995) DNA repair is more important than catalase for *Salmonella virulence*. *J. Clin. Invest.* 95: 1047–1053.

Buchmeier NA, Bossie S, Chen C-Y, Fang FC, Guiney DG, Libby SJ (1997) SlyA, a transcriptional regulator of *Salmonella typhimurium*, is required for resistance to oxidative stress and is expressed in the intracellular environment of macrophages. *Infect. Immun.* 65: 3725–3730.

Bukrinsky MI, Nottet HS, Schmidtmayerova H *et al.* (1995) Regulation of nitric oxide synthase activity in human immunodeficiency virus type 1 (HIV-1)-infected monocytes: implications for HIV-associated neurological disease. *J. Exp. Med.* 181: 735–745.

Byrne GI, Faubion CL (1983) Inhibition of *Chlamydia psittaci* in oxidatively active thioglycolate-elicited macrophages: distinction between lymphokine-mediated oxygen-dependent oxygen-independent macrophage activation. *Infect. Immun.* 40: 464–471.

Byrne GI, Carlin JM, Merkert T, Arter DL (1989) Long-term effects of gamma interferon on *Chlamydia*-infected host cells: microbicidal activity follows microbistasis. *Infect. Immun.* 57: 1318–1320.

Canning C, Roth JA, Deyoe BL (1986) Release of 5′-guanosine monophosphate and adenine by *B. abortus* and their role in the intracellular survival of the bacteria. *J. Infect. Dis.* 154: 464–470.

Caron E, Gross A, Liautard J, Dornand J (1996) *Brucella* species release a specific, protease-sensitive, inhibitor of TNF-alpha expression, active on human macrophage-like cells. *J. Immunol.* 156: 2885–2893.

Chakraborty T, Leimeister-Wachter M, Domann E *et al.* (1992) Coordinate regulation of virulence genes in *Listeria monocytogenes* requires the product of the prfA gene. *J. Bacteriol.* 174: 568–574.

Chamberlain RE (1965) Evaluation of live tularemia vaccine prepared in a chemically defined medium. *Appl. Microbiol.* 13: 232–235.

Chan J, Fan XD, Hunter SW, Brennan J, Bloom BR (1991) Lipoarabinomannan, a possible virulence factor involved in persistence of *Mycobacterium tuberculosis* within macrophages. *Infect. Immun.* 59: 1755–1761.

Chan J, Xing Y, Magliozzo RS, Bloom BR (1992) Killing of virulent *Mycobacterium tuberculosis* by reactive nitrogen intermediates produced by activated murine macrophages. *J. Exp. Med.* 175: 1111–1122.

Chang Z, Primm T, Jakana J *et al.* (1996) *Mycobacterium tuberculosis* 16-kDa antigen (Hsp 163) functions as an oligomeric structure *in vitro* to suppress thermal aggregation. *J. Biol. Chem.* 271: 7218–7223.

Chen LM, Hobbie S, Galan JE (1996a) Requirement of CDC42 for *Salmonella*-induced cytoskeletal and nuclear responses. *Science* 274: 2115–2118.

Chen Y, Smith MR, Thirumalai K, Zychlinsky A (1996b) A bacterial invasin induces macrophage apoptosis by binding directly to ICE. *EMBO J.* 15: 3853–3860.

Christie J (1997) *Agrobacterium tumefaciens* T-complex transport apparatus: a paradigm for a new family of multifunctional transporters in eubacteria. *J. Bacteriol.* 179: 3085–3094.

Cianciotto N, Eisenstein BI, Mody CH, Engleberg NC (1990) A mutation in the mip gene results in an attenuation of *Legionella pneumophila* virulence. *J. Infect. Dis.* 162: 121–126.

Cirillo DM, Valdivia RH, Monack DM, Falkow S (1998) Macrophage-dependent induction of the *Salmonella* pathogenicity island 2 type III secretion system and its role in intracellular survival. *Mol. Microbiol.* 30: 175–188.

Cirillo JD, Cirillo SLG, Yan L, Bermudez LE, Falkow S, Tompkins LS (1999) Intracellular growth in *Acanthamoeba castellanii* affects monocyte entry mechanisms and enhances virulence of *Legionella pneumophila*. *Infect. Immun.* 67: 4427–4434.

Claus V, Jahraus A, Tjelle T *et al.* (1998) Lysosomal enzyme trafficking between phagosomes, endosomes, and lysosomes in J774 macrophages: enrichment of cathepsin H in early endosomes. *J. Biol. Chem.* 273: 9842–9851.

Clemens DL, Horwitz MA (1995) Characterization of the *Mycobacterium tuberculosis* phagosome and evidence that phagosomal maturation is inhibited. *J. Exp. Med.* 181: 257–270.

Clemens DL, Horwitz MA (1996) The *Mycobacterium tuberculosis* phagosome interacts with early endosomes and is accessible to exogenously administered transferrin. *J. Exp. Med.* 184: 1349–1355.

Clifton DR, Goss RA, Sahni SK *et al.* (1998) NF-κB-dependent inhibition of apoptosis is essential for host cell survival during *Rickettsia rickettsii* infection. *Proc. Natl Acad. Sci. USA* 95: 4646–4651.

Condino-Neto A, Muscara MN, Grumach AS, Carneiro-Sampaio MM, De Nucci G (1993) Neutrophils and mononuclear cells from patients with chronic granulomatous disease release nitric oxide. *Br. J. Clin. Pharmacol.* 35: 485–490.

Covacci A, Telford JL, Del Giudice G, Parsonnet J, Rappuoli R (1999) *Helicobacter pylori* virulence and genetic geography. *Science* 284: 1328–1333.

Crowle AJ, Dahl R, Ross E, May MH (1991) Evidence that vesicles containing living, virulent *Mycobacterium tuberculosis* or *Mycobacterium avium* in cultured human macrophages are not acidic. *Infect. Immun.* 59: 1823–1831.

Dabiri GA, Sanger JM, Portnoy DA, Southwick FS (1990) *Listeria monocytogenes* moves rapidly through the host cell cytoplasm by inducing directional actin assembly. *Proc. Natl Acad. Sci. USA* 87: 6068–6072.

de Chastellier C, Thilo L (1997) Phagosome maturation and fusion with lysosomes in relation to surface property and size of the phagocytic particle. *Eur. J. Cell Biol.* 74: 49–62.

de Chastellier C, Lang T, Thilo L (1995) Phagocytic processing of the macrophage endoparasite, *Mycobacterium avium*, in comparison to phagosomes which contain *Bacillus subtilis* or latex beads. *Eur. J. Cell Biol.* 68: 167–182.

de Chastellier C, Thibon M, Rabinovitch M (1999) Construction of chimeric phagosomes that shelter *Mycobacterium avium* and *Coxiella burnetii* (phase II) in doubly infected mouse macrophages: an ultrastructural study. *Eur. J. Cell Biol.* 78: 580–592.

de Geyter C, Vogt B, Benjelloun-Touimi Z *et al.* (1997) Purification of IpaC, a protein involved in entry of *S. flexneri* into epithelial cells, and characterization of its interaction with lipid membranes. *FEBS Lett.* 400: 149–154.

De Groote MA, Granger D, Xu Y, Campbell G, Prince R, Fang FC (1995) Genetic and redox determinants of nitric oxide cytotoxicity in a *Salmonella typhimurium* model. *Proc. Natl Acad. Sci. USA* 92: 6399–6403.

De Maria R, Cifone MG, Trotta R *et al.* (1994) Triggering of human monocyte activation through CD69, a member of the natural killer cell gene complex family of signal transducing receptors. *J. Exp. Med.* 180: 1999–2004.

Dellagostin OA, Esposito G, Eales LJ, Dale JW, McFadden J (1995) Activity of mycobacterial promoters during intracellular and extracellular growth. *Microbiology* 141: 1785–1792.

Deretic V, Via LE, Fratti RA, Deretic D (1997) Mycobacterial phagosome maturation, rab proteins, and intracellular trafficking. *Electrophoresis* 18: 2542–2547.

Desjardins M (1995) Biogenesis of phagolysosomes: the 'kiss-and-run' hypothesis. *Trends Cell Biol.* 5: 183–186.

Desjardins M, Celis JE, van Meer G *et al.* (1994a) Molecular characterization of phagosomes. *J. Biol. Chem.* 269: 32194–32200.

Desjardins M, Huber LA, Parton RG, Griffiths G (1994b) Biogenesis of phagolysosomes proceeds through a sequential series of interactions with the endocytic apparatus. *J. Cell Biol.* 124: 677–688.

Desjardins M, Nzala NN, Corsini R, Rondeau C (1997) Maturation of phagosomes is accompanied by changes in their fusion properties and size-selective acquisition of solute materials from endosomes. *J. Cell Sci.* 110: 2303–2314.

Diakonova M, Gerke V, Ernst J, Liautard J, van der Vusse G, Griffiths G (1997) Localization of five annexins in J774 macrophages and on isolated phagosomes. *J. Cell Sci.* 110: 1199–1213.

Drevets DA, Canono B, Campbell A (1992) Listericidal and nonlistericidal mouse macrophages differ in complement receptor type 3-mediated phagocytosis of *L. monocytogenes* and in preventing escape of the bacteria into the cytoplasm. *J. Leucocyte Biol.* 52: 70–79.

Egile C, Loisel T, Laurent V *et al.* (1999) Activation of the CDC42 effector N-WASP by the *Shigella flexneri* IcsA protein promotes actin nucleation by Arp2/3 complex and bacterial actin-based motility. *J. Cell Biol.* 146: 1319–1332.

Ehlers MR, Daffe M (1998) Interactions between *Mycobacterium tuberculosis* and host cells: are mycobacterial sugars the key? *Trends Microbiol.* 6: 328–335.

Ehrt S, Shiloh MU, Ruan J *et al.* (1997) A novel antioxidant gene from *Mycobacterium tuberculosis. J. Exp. Med.* 186: 1885–1896.

Ernst JD (1998) Macrophage receptors for *Mycobacterium tuberculosis. Infect. Immun.* 66: 1277–1281.

Fang FC (1997) Perspectives series. Host/pathogen interactions: mechanisms of nitric oxide-related antimicrobial activity. *J. Clin. Invest.* 99: 2818–2825.

Feng Y, Press B, Wandinger-Ness A (1995) Rab 7: an important regulator of late endocytic membrane traffic. *J. Cell Biol.* 131: 1435–1452.

Ferrari G, Langen H, Naito M, Pieters J (1999) A coat protein on phagosomes involved in the intracellular survival of mycobacteria. *Cell* 97: 435–447.

Florczyk MA, Turner JN, McDonough KA (2001) High voltage electron microscopic analysis of intracellular localization of *Mycobacterium tuberculosis* and *Mycobacterium bovis* BCG within macrophages (in preparation).

Flynn JL, Goldstein MM, Triebold KJ, Koller B, Bloom BR (1992) Major histocompatibility complex class I-restricted T cells are required for resistance to *Mycobacterium tuberculosis* infection. *Proc. Natl Acad. Sci. USA* 89: 12013–12017.

Forestier C, Moreno E, Pizarro-Cerda J, Gorvel J (1999) Lysosomal accumulation and recycling of lipopolysaccharide to the cell surface of murine macrophages, an *in vitro* and *in vivo* study. *J. Immunol.* 162: 6784–6791.

Fortier AH, Nacy CA, Meltzer MS (1992a) Growth of *Francisella tularensis* Live Vaccine Strain (LVS) in macrophages requires a low pH compartment. *J. Leucocyte Biol.* S3: 28–36.

Fortier AH, Polsinelli T, Green SJ, Nacy CA (1992b) Activation of macrophages for destruction of *Francisella tularensis*: identification of cytokines, effector cells, and effector molecules *Infect. Immun.* 60: 817–825.

Fortier AH, Green SJ, Polsinelli T *et al.* (1994) Life and death of an intracellular pathogen: *Francisella tularensis* and the macrophage. In: Zwilling BS, Eisenstein TK (eds) *Macrophage–Pathogen Interactions.* New York: Marcel Dekker, pp. 349–361.

Francis CL, Ryan TA, Jones BD, Smith SJ, Falkow S (1993) Ruffles induced by *Salmonella* and other stimuli direct macropinocytosis of bacteria. *Nature* 364: 639–642.

Freitag NE, Rong L, Portnoy DA (1993) Regulation of the prfA transcriptional activator of *Listeria monocytogenes*: multiple promoter elements contribute to intracellular growth and cell-to-cell spread. *Infect. Immun.* 61: 2537–2544.

Frenchick J, Markham RJF, Cochrane AH (1985) Inhibition of phagosome–lysosome fusion in macrophages by soluble extracts of virulent *Brucella abortus. Am. J. Vet. Res.* 46: 332–335.

Galan JE, Collmer A (1999) Type III secretion machines: bacterial devices for protein delivery into host cells. *Science* 284: 1322–1328.

Galan JE, Pace J, Hayman MJ (1992) Involvement of the epidermal growth factor receptor in the invasion of cultured mammalian cells by *Salmonella typhimurium. Nature* 357: 588–589.

Garbe TR, Hibler NS, Deretic V (1999) Response to reactive nitrogen intermediates in *Mycobacterium tuberculosis*: induction of the 16-kilodalton a-crystallin homolog by exposure to nitric oxide donors. *Infect. Immun.* 67: 460–465.

Godfroid F, Taminiau B, Danese I *et al.* (1998) Identification of the perosamine synthetase gene of *Brucella melitensis* 16M and involvement of lipopolysaccharide O side chain in *Brucella* survival in mice and in macrophages. *Infect. Immun.* 66: 5485–5493.

Golovliov I, Ericsson M, Sandstrom G, Tarnvik A, Sjostedt A (1997) Identification of proteins of *Francisella tularensis* induced during growth in macrophages and cloning of the gene encoding a prominently induced 23-kilodalton protein. *Infect. Immun.* 65: 2183–2189.

Gomes MS, Paul S, Moreira AL, Appleberg R, Rabinovitch M, Kaplan G (1999) Survival of *Mycobacterium avium* and

Mycobacterium tuberculosis in acidified vacuoles of murine macrophages. *Infect. Immun.* 67: 3199–3206.

Gorvel JP, Chavrier P, Zerial M, Gruenberg J (1991) Rab5 controls early endosome fusion *in vitro*. *Cell* 64: 915–925.

Gouin E, Gantelet H, Egile C *et al.* (1999) A comparative study of the actin-based motilities of the pathogenic bacteria *Listeria monocytogenes*, *Shigella flexneri* and *Rickettsia conorii*. *J. Cell Sci.* 112: 1697–1708.

Greenberg S (1995) Signal transduction of phagocytosis. *Trends Cell Biol.* 5: 93–99.

Gregory WW, Byrne GI, Gardner M, Moulder JW (1979) Cytochalasin B does not inhibit ingestion of *Chlamydia psittaci* by mouse fibroblasts (L-cells) and mouse peritoneal macrophages. *Infect. Immun.* 25, 463–466.

Griffin FM, Silverstein SC (1974) Segmental response of the macrophage plasma membrane to a phagocytic stimulus. *J. Exp. Med.* 139: 323–336.

Griffin FM, Griffin JA, Leider JE, Silverstein SC (1975) Studies on the mechanism of phagocytosis: I. Requirements for circumferential attachment of particle-bound ligands to specific receptors on the macrophage plasma membrane. *J. Exp. Med.* 142: 1263–1282.

Griffin FM, Griffin JA, Silverstein SC (1976) Studies on the mechanism of phagocytosis: II. The interaction of macrophages with anti-immunoglobulin IgG-coated bone marrow-derived lymphocytes. *J. Exp. Med.* 144: 788–809.

Griffiths G, Hoflack B, Simons K, Mellman I, Kornfeld S (1988) The mannose 6-phosphate receptor and the biogenesis of lysosomes. *Cell* 52: 329–341.

Groisman EA, Heffron F (1995) Regulation of *Salmonella* virulence by two-component regulatory systems. In: Hoch JA, Silhavy TJ (eds) *Two-Component Signal Transduction*. Washington, DC: ASM Press, pp. 319–332.

Groisman EA, Ochman H (2000) The path to *Salmonella*. *ASM News* 66: 21–27.

Haber F, Weiss J (1934) The catalytic decomposition of hydrogen peroxide by iron salts. *Proc. R. Soc. Lond. Ser. A* 147: 332–351.

Hackstadt T, Williams JC (1981) Biochemical stratagem for obligate parasitism of eukaryotic cells by *Coxiella burnetii*. *Proc. Natl Acad. Sci. USA* 78: 3240–3244.

Hackstadt T, Rockey DD, Heinzen RA, Scidmore MA (1996) *Chlamydia trachomatis* interrupts an exocytic pathway to acquire endogenously synthesized sphingomyelin in transit from the Golgi apparatus to the plasma membrane. *EMBO J.* 15: 964–977.

Hammerschlag M, Suntharalingam K, Fikrig S (1985) The effect of *Chlamydia trachomatis* on luminol-dependent chemiluminescence of human polymorphonuclear leucocytes: requirements for opsonization. *J. Infect. Dis.* 151: 1045–1051.

Hechemy KE, McKee M, Marko M, Samsonoff WA, Roman M, Baca O (1993) Three-dimensional reconstruction of *Coxiella burnetii*-infected L929 cells by high-voltage electron microscopy. *Infect. Immun.* 61: 4485–4488.

Heifets L, Imai K, Goren MB (1980) Expression of peroxidase-dependent iodination by macrophages ingesting neutrophil debris. *J. Reticuloend. Soc.* 28: 391–404.

Heinzen RA, Hackstadt T (1997) The *Chlamydia trachomatis* parasitophorous vacuolar membrane is not passively permeable to low-molecular-weight compounds. *Infect. Immun.* 65: 1088–1094.

Heinzen RA, Scidmore MA, Rockey DD, Hackstadt T (1996) Differential interaction with endocytic and exocytic pathways distinguish parasitophorous vacuoles of *Coxiella burnetii* and *Chlamydia trachomatis*. *Infect. Immun.* 64: 796–809.

Heinzen RA, Grieshaber SS, Van Kirk LS, Devin CJ (1999a) Dynamics of actin-based movement by *Rickettsia rickettsii* in vero cells. *Infect. Immun.* 67: 4201–4207.

Heinzen RA, Hackstadt T, Samuel JE (1999b) Developmental biology of *Coxiella burnettii*. *Trends Microbiol.* 7: 149–154.

Hensel M, Shea JE, Waterman SR *et al.* (1998) Genes encoding putative effector proteins of the type III secretion system of *Salmonella* pathogenicity island 2 are required for bacterial virulence and proliferation in macrophages. *Mol. Microbiol.* 30: 163–174.

Hermant D, Menard R, Arricau N, Parsot C, Popoff MY (1995) Functional conservation of the *Salmonella* and *Shigella* effectors of entry into epithelial cells. *Mol. Microbiol.* 17: 781–789.

Hersh D, Monack DM, Smith MR, Ghori N, Falkow S, Zychlinsky A (1999) The *Salmonella* invasin SipB induces macrophage apoptosis by binding to caspase-1. *Proc. Natl Acad. Sci. USA* 96: 2396–2401.

Hilbi H, Chen Y, Thirumalai K, Zychlinsky A (1997a) The interleukin-1 beta-converting enzyme, caspase 1, is activated during *Shigella flexneri*-induced apoptosis in human monocyte-derived macrophages. *Infect. Immun.* 65: 5165–5170.

Hilbi H, Zychlinsky A, Sansonetti J (1997b) Macrophage apoptosis in microbial infections. *Parasitology* 115: S79–S87.

Howe D, Mallavia L (1999) *Coxiella burnetii* infection increases transferrin receptors on J774A 1 cells. *Infect. Immun.* 67: 3236–3241.

Hu Y, Coates ARM (1999) Transcription of the stationary-phase associated hspX gene of *Mycobacterium tuberculosis* is inversely related to the synthesis of the 16-kilodalton protein. *J. Bacteriol.* 181: 1380–1387.

Hu Y, Butcher D, Mangan JA, Rajandream M-A, Coates ARM (1999) Regulation of hmp gene transcription in *Mycobacterium tuberculosis*: effects of oxygen limitation and nitrosative and oxidative stress. *J. Bacteriol.* 181: 3486–3493.

Humphreys S, Stevenson A, Bacon A, Weinhardt AB, Roberts M (1999) The alternative sigma factor, sigmaE, is critically important for the virulence of *Salmonella typhimurium*. *Infect. Immun.* 67: 1560–1568.

Isberg RR, Van Nhieu GT (1995) The mechanism of phagocytic uptake promoted by invasin–integrin interaction. *Trends Cell Biol.* 5: 120–124.

Jahraus A, Tjelle TE, Berg T *et al.* (1998) *In vitro* fusion of phagosomes with different endocytic organelles from J774 macrophages. *J. Biol. Chem.* 273: 30379–30390.

Jiang X, Leonard B, Benson R, Baldwin CL (1993) Macrophage control of *Brucella abortus*: role of reactive oxygen intermediates and nitric oxide. *Cell. Immun.* 151: 309–319.

Kaplan G (1977) Differences in the mode of phagocytosis with Fc and C3 receptors in macrophages. *Scand. J. Immunol.* 6: 797–807.

Kaufmann SHE (1993) Immunity to intracellular bacteria. *Ann. Rev. Immunol.* 11: 129–163.

Kirby JE, Vogel J, Andrews HL, Isberg RR (1998) Evidence for pore-forming ability by *Legionella pneumophila*. *Mol. Microbiol.* 27: 323–336.

Kornfeld S, Mellman I (1989) The biogenesis of lysosomes. *Ann. Rev. Cell Biol.* 5: 483–525.

Kugler S, Schuller S, Goebel W (1997) Involvement of MAP-kinases and -phosphatases in uptake and intracellular replication of *Listeria monocytogenes* in J774 macrophage cells. *FEMS Microbiol. Lett.* 157: 131–136.

Kumar V, Jindal SK, Ganguly NK (1995) Release of reactive oxygen and nitrogen intermediates from monocytes of patients with pulmonary tuberculosis. *Scand. J. Clin. Lab. Invest.* 55: 163–169.

Kuo CC (1978) Immediate cytotoxicity of *Chlamydia trachomatis* for mouse peritoneal macrophages. *Infect. Immun.* 20: 613–618.

La Verda D, Byrne GI (1994) Interactions between macrophages and Chlamydiae. In: Zwilling BS, Eisenstein TK (eds) *Macrophage–Pathogen Interactions.* New York: Marcel Dekker, pp. 381–399.

Lampidis R, Gross R, Sokolovic A, Goebel W, Kreft J (1994) The virulence regulator protein of *Listeria ivanovii* is highly homologous to PrfA from *Listeria monocytogenes* and both belong to the Crp-Fnr family of transcription regulators. *Mol. Microbiol.* 13: 141–151.

Lee BY, Horwitz MA (1995) Identification of macrophage and stress-induced proteins of *Mycobacterium tuberculosis*. *J. Clin. Invest.* 96: 245–249.

Lee EH, Rikihisa Y (1998) Protein kinase A-mediated inhibition of gamma interferon-induced tyrosine phosphorylation of Janus kinases and latant cytoplasmic transcription factors in human monocytes by *Ehrlichia chaffeensis*. *Infect. Immun.* 66: 2514–2520.

Leenen J, Canono B, Drevets A, Voerman JS, Campbell A (1994) TNF-α and IFN-γ stimulate a macrophage precursor cell line to kill *Listeria monocytogenes* in a nitric oxide-dependent manner. *J. Immunol.* 153: 5141–5147.

Leimeister-Wachter M, Haffner C, Domann E, Goebel W, Chakraborty T (1990) Identification of a gene that positively regulates expression of listeriolysin, the major virulence factor of *Listeria monocytogenes*. *Proc. Natl Acad. Sci. USA* 87: 8336–8340.

Leonard BA, Lopez-Goni I, Baldwin CL (1997) *Brucella abortus* siderophore 2,3-dihydroxybenzoic acid protects Brucellae from killing by macrophages. *Vet. Res.* 28: 87–92.

Li H, Walker DH (1998) rOmpA is a critical protein for the adhesion of *Rickettsia rickettsii* to host cells. *Microb. Pathogen.* 24: 289–298.

Liautard J, Gross A, Dornand J, Kohler S (1996) Interactions between professional phagocytes and *Brucella* sp [review]. *Microbiologia* 12: 197–206.

Lukacova M, Valkova D, Quevedo Diaz M, Perecko D, Barak I (1999) Green fluorescent protein as a detection marker for *Coxiella burnetii* transformation. *FEMS Microbiol. Lett.* 175: 255–260.

Lundemose AG, Birkelund S, Fey SJ, Larson M, Christiansen G (1991) *Chlamydia trachomatis* contains a protein similar to the *Legionella pneumophila* mip gene product. *Mol. Microbiol.* 5: 109–115.

MacMicking JD, Xie QW, Nathan C (1997a) Nitric oxide and macrophage function. *Ann. Rev. Immunol.* 15: 323–350.

MacMicking JD, North RJ, LaCourse R, Mudgett JS, Shah SK, Nathan CF (1997b) Identification of nitric oxide synthase as a protective locus against tuberculosis. *Proc. Natl Acad. Sci. USA* 94: 5243–5248.

Manabe YC, Chen JM, Ko CG *et al.* (1999) Conditional sigma factor expression, using the inducible acetamidase promoter, reveals that the *Mycobacterium tuberculosis* sigF gene modulates expression of the 16-kilodalton a-crystallin homologue. *J. Bacteriol.* 181: 7629–7633.

Manca C, Paul S, Barry CE, Freedman VH, Kaplan G (1999) *Mycobacterium tuberculosis* catalase and peroxidase activities and resistance to oxidative killing in human monocytes *in vitro*. *Infect. Immun.* 67: 74–79.

Maurelli AT, Baudry B, d'Hauteville H, Hale TL, Sansonetti PJ (1985) Cloning of plasmid DNA sequences involved in invasion of HeLa cells by *Shigella flexneri*. *Infect. Immun.* 49: 164–171.

Maurin M, Benoliel AM, Bongrand P, Raoult D (1992a) Phagolysosomal alkalinization and the bactericidal effect of antibiotics: the *Coxiella burnetii* paradigm. *J. Infect. Dis.* 166: 1097–1102.

Maurin M, Benoliel AM, Bongrand P, Raoult D (1992b) Phagolysosomes of *Coxiella burnetii*-infected cell lines maintain an acidic pH during persistent infection. *Infect. Immun.* 60: 5013–5016.

Mazzaccaro RJ, Gedde M, Jensen ER *et al.* (1996) Major histocompatibility class I presentation of soluble antigen facilitated by *Mycobacterium tuberculosis* infection. *Proc. Natl Acad. Sci. USA* 93: 11786–11791.

McDonough KA, Kress Y (1995) Cytotoxicity for lung epithelial cells is a virulence-associated phenotype of *Mycobacterium tuberculosis*. *Infect. Immun.* 63: 4802–4811.

McDonough KA, Kress Y, Bloom BR (1993) Pathogenesis of tuberculosis: interaction of *Mycobacterium tuberculosis* with macrophages. *Infect. Immun.* 61: 2763–2773.

McDonough KA, Florczyk MA, Kress Y (2000) Intracellular passage within macrophages affects the trafficking of virulent tubercle bacilli upon reinfection of other macrophages in a serum-dependent manner. *Tuber. Lung Dis.* 80: 259–271.

Mecsas JJ, Strauss EJ (1996) Molecular mechanisms of bacterial virulence: type III secretion and pathogenicity islands. *Emerg. Infect. Dis.* 2: 270–288.

Mengaud J, Ohayon H, Gounon P, Mege RM, Cossart P (1996) E-cadherin is the receptor for internalin, a surface protein required for entry of *L. monocytogenes* into epithelial cells. *Cell* 84: 923–932.

Messick JB, Rikihisa Y (1993) Characterization of *Ehrlichia risticii* binding, internalization, and proliferation in host cells by flow cytometry. *Infect. Immun.* 61: 3803–3810.

Miller BH, Shinnick TM (2000) Evaluation of *Mycobacterium tuberculosis* genes involved in resistance to killing by human macrophages. *Infect. Immun.* 68: 387–390.

Moreira AL, Wang J, Tsenova-Berkova L, Hellmann W, Freedman VH, Kaplan G (1997) Sequestration of *Mycobacterium tuberculosis* in tight vacuoles *in vivo* in lung macrophages of mice infected by the respiratory route. *Infect. Immun.* 65: 305–308.

Moriyon I, Gamazo C, Diaz R (1987) Properties of the outer membrane of *Brucella. Ann. Inst. Pasteur Microbiol.* 138: 89–91.

Mosser DM (1994) Receptors on phagocytic cells involved in microbial recognition. In: Zwilling BS, Eisenstein TK (eds) *Macrophage–Pathogen Interactions.* New York: Marcel Dekker, pp. 99–110.

Mott J, Barnewall R, Rikihisa Y (1999) Human granulocytic ehrlichiosis agent and *Ehrlichia chaffeensis* reside in different cytoplasmic compartments in HL-60 cells. *Infect. Immun.* 67: 1368–1398.

Moulder JW (1991) Interaction of Chlamydiae and host cells *in vitro. Microbiol. Rev.* 55: 143–190.

Murray HW, Byrne GI, Rothermel CD, Cartelli DM (1983) Lymphokine enhances oxygen-independent activity against intracellular pathogens. *J. Exp. Med.* 158: 234–239.

Myrvik QN, Leake ES, Wright MJ (1984) Disruption of phagosomal membranes of normal alveolar macrophages by the H37Rv strain of *Mycobacterium tuberculosis*: a correlate of virulence. *Am. Rev. Respir. Dis.* 129: 322–328.

Nakajo MN, Roblin M, Hammerschlag MR *et al.* (1990) Chlamydicidal activity of human alveolar macrophages. *Infect. Immun.* 58: 3640–3644.

Nathan C (1992) Nitric oxide as a secretory product of mammalian cells. *FASEB J.* 6: 3051–3064.

Nhieu GT, Sansonetti J (1999) Mechanism of *Shigella* entry into epithelial cells. *Curr. Opin. Microbiol.* 2: 51–55.

Nicholson S, Bonecini-Almeida MDG, Lapa-Silva JR *et al.* (1996) Inducible nitric oxide synthase in pulmonary alveolar macrophages from patients with tuberculosis. *J. Exp. Med.* 183: 2293–2302.

O'Callaghan D, Cazevieille C, Allardet-Servent A *et al.* (1999) A homologue of the *Agrobacterium tumefaciens* VirB and *Bordetella pertussis* Ptl type IV secretion systems is essential for intracellular survival of *Brucella suis. Mol. Microbiol.* 33: 1210–1220.

Oh YK, Swanson JA (1996) Different fates of phagocytosed particles after delivery into macrophage lysosomes. *J. Cell Biol.* 132: 585–593.

Oh YK, Alpuche-Aranda C, Berthiaume E, Jinks T, Miller SI, Swanson JA (1996) Rapid and complete fusion of macrophage lysosomes with phagosomes containing *Salmonella typhimurium. Infect. Immun.* 64: 3877–3883.

Ohya S, Tanabe Y, Makino M *et al.* (1998) The contributions of reactive oxygen intermediates and reactive nitrogen intermediates to listericidal mechanisms differ in macrophages activated pre- and postinfection. *Infect. Immun.* 66: 4043–4049.

Orme I, Collins F (1984) Adoptive protection of the *Mycobacterium tuberculosis*-infected lung. *Cell Immunol.* 84: 113–120.

Pacelli R, Wink DA, Cook JA *et al.* (1995) Nitric oxide potentiates hydrogen peroxide-induced killing of *Escherichia coli. J. Exp. Med.* 182: 1469–1479.

Palecanda A, Paulauskis J, Al-Mutairi E *et al.* (1999) Role of the scavenger receptor MARCO in alveolar macrophage binding of unopsonized environmental particles. *J. Exp. Med.* 189: 1497–1506.

Papp-Szabo E, Firtel M, Josephy D (1994) Comparison of the sensitivities of *Salmonella typhimurium oxy*R and *kat*G mutants to killing by human neutrophils. *Infect. Immun.* 62: 2662–2668.

Parsot C, Sansonetti J (1996) Invasion and the pathogenesis of *Shigella* infections. *Curr. Top. Microbiol. Immunol.* 209: 25–42.

Pearson AM (1996) Scavenger receptors in innate immunity. *Curr. Opin. Immunol.* 8: 20–28.

Pizarro-Cerda J, Meresse S, Parton RG *et al.* (1998a) *Brucella abortus* transits through the autophagic pathway and replicates in the endoplasmic reticulum of nonprofessional phagocytes. *Infect. Immun.* 66: 5711–5724.

Pizarro-Cerda J, Moreno E, Sanguedolce V, Mege JL, Gorvel J (1998b) Virulent *Brucella abortus* prevents lysosome fusion and is distributed within autophagosome-like compartments. *Infect. Immun.* 66: 2387–2392.

Pizon V, Desjardins M, Bucci C, Parton RG, Zerial M (1994) Association of Rap1a and Rap1b proteins with late endocytic/phagocytic compartments and Rap2a with the Golgi complex. *J. Cell Sci.* 107: 1661–1670.

Plum G, Brenden M, Clark-Curtiss JE, Pulverer G (1997) Cloning, sequencing, and expression of the mig gene of *Mycobacterium avium*, which codes for a secreted macrophage-induced protein. *Infect. Immun.* 65: 4548–4557.

Porte F, Liautard J, Kohler S (1999) Early acidification of phagosomes containing *Brucella suis* is essential for intracellular survival in murine macrophages. *Infect. Immun.* 67: 4041–4047.

Portnoy DA, Chakraborty T, Goebel W *et al.* (1992) Molecular determinants of *Listeria monocytogenes* pathogenesis. *Infect. Immun.* 60: 1263–1267.

Price RE, Templeton JW, Adams LG (1990a) Survival of smooth, rough and transposon mutant strains of *Brucella abortus* in bovine mammary macrophages. *Vet. Immunol. Immunopathol.* 26: 353–365.

Price RE, Templeton JW, Smith R, Adams LG (1990b) Ability of mononuclear phagocytes from cattle naturally resistant or susceptible to brucellosis to control *in vitro* intracellular survival of *Brucella abortus. Infect. Immun.* 58: 879–886.

Rabinovitch M (1995) Professional and non-professional phagocytes: an introduction. *Trends Cell Biol.* 5: 85–87.

Rabinowitz S, Horstmann H, Gordon S, Griffiths G (1992) Immunocytochemical characterization of the endocytic and phagolysosomal compartments in peritoneal macrophages. *J. Cell Biol.* 116: 95–112.

Rasool O, Freer E, Moreno E, Jarstrand C (1992) Effect of *Brucella abortus* lipopolysaccharide on oxidative metabolism and lysozyme release by human neutrophils. *Infect. Immun.* 60: 1699–1702.

Rathman M, Sjaastad MD, Falkow S (1996) Acidification of phagosomes containing *Salmonella typhimurium* in murine macrophages. *Infect. Immun.* 64: 2765–2773.

Rathman M, Barker L, Falkow S (1997) The unique trafficking pattern of *Salmonella typhimurium*-containing phagosomes in murine macrophages is independent of the mechanism of bacterial entry. *Infect. Immun.* 65: 1475–1485.

Rikihisa Y (1999) Clinical and biological aspects of infection caused by *Ehrlichia chaffeensis. Microb. Infect.* 1: 367–376.

Rikihisa Y, Zhang Y, Park J (1994) Inhibition of infection of macrophages with *Ehrlichia risticii* by cytochalasins, monodansylcadaverine, and taxol. *Infect. Immun.* 62: 5126–5132.

Robinson J (1998) Oxygen and nitrogen reactive metabolites and phagocytic cells. In: Robinson J, Babcock GF (eds) *Phagocyte Function: A Guide for Research and Clinical Evaluation.* New York: Wiley–Liss, pp. 217–254.

Rockey DD, Rosquist JL (1994) Protein antigens of *Chlamydia psittaci* present in infected cells but not detected in the infectious elementary body. *Infect. Immun.* 62: 106–112.

Rockey DD, Grosenbach D, Hruby DE, Peacock MG, Heinzen RA, Hackstadt T (1997) *Chlamydia psittaci* IncA is phosphorylated by the host cell and is exposed on the cytoplasmic face of the developing inclusion. *Mol. Microbiol.* 24: 217–228.

Rothman JE (1994) Mechanisms of intracellular protein transport. *Nature* 372: 55–63.

Roy CR (1999) Trafficking of the *Legionella pneumophila* phagosome. *ASM News* 65: 416–421.

Roy CR, Berger KH, Isberg RR (1998) *Legionella pneumophila* DotA protein is required for early phagosome trafficking decisions that occur within minutes of bacterial uptake. *Mol. Microbiol.* 28: 663–674.

Ruan J, St John G, Ehrt S, Riley L, Nathan C (1999) *nox*R3, a novel gene from *Mycobacterium tuberculosis*, protects *Salmonella typhimurium* from nitrosative and oxidative stress. *Infect. Immun.* 67: 3276–3283.

Russell DG, Dant J, Sturgill-Koszycki S (1996) *Mycobacterium avium*- and *Mycobacterium tuberculosis*-containing vacuoles are dynamic, fusion-competent vesicles that are accessible to glycosphingolipids from the host cell plasmalemma. *J. Immunol.* 156: 4764–4773.

Sahni SK, Van Antwerp DJ, Eremeeva ME, Silverman DJ, Marder VJ, Sporn LA (1998) Proteasome-independent activation of nuclear factor kappaB in cytoplasmic extracts from human endothelial cells by *Rickettsia rickettsii. Infect. Immun.* 66: 1827–1833.

Sandvig K, van Deurs B (1994) Endocytosis without clathrin. *Trends Cell Biol.* 4: 275–277.

Sansonetti J (1999) *Shigella* plays dangerous games. *ASM News* 65: 611–617.

Sansonetti J, Phalipon A (1999) M cells as ports of entry for enteroinvasive pathogens: mechanisms of interaction, consequences for the disease process. *Semin. Immunol.* 11: 193–203.

Sansonetti J, Arondel J, Cavaillon J-M, Huerre M (1995) Role of IL-1 in the pathogenesis of experimental shigellosis. *J. Clin. Invest.* 96: 884–892.

Sansonetti J, Tran Van Nhieu G, Egile C (1999) Rupture of the intestinal epithelial barrier and mucosal invasion by *Shigella flexneri. Clin. Infect. Dis.* 28: 466–475.

Schaible UE, Sturgill-Koszycki S, Schlesinger H, Russell DG (1998) Cytokine activation leads to acidification and increases maturation of *Mycobacterium avium*-containing phagosomes in murine macrophages. *J. Immunol.* 160: 1290–1296.

Schwan WR, Kopecko DJ (1997) Serovar specific differences in *Salmonella* survival within macrophage cells. *Adv. Exp. Med. Biol.* 412: 277–278.

Scidmore MA, Rockey DD, Fischer ER, Heinzen RA, Hackstadt T (1996) Vesicular interactions of the *Chlamydia trachomatis* inclusion are determined by Chlamydial early protein synthesis rather than route of entry. *Infect. Immun.* 64: 5366–5372.

Segal G, Shuman HA (1997) Characterization of a new region required for macrophage killing by *Legionella pneumophila. Infect. Immun.* 65: 5057–5066.

Segal G, Purcell M, Shuman HA (1998) Host cell killing and bacterial conjugation require overlapping sets of genes within a 22-kb region of the *Legionella pneumophila* genome. *Proc. Natl Acad. Sci. USA* 95: 1669–1674.

Sinai A, Joiner KA (1997) Safe haven: the cell biology of nonfusogenic pathogen vacuoles. *Ann. Rev. Microbiol.* 51: 415–462.

Skerrett SJ, Martin TR (1996) Roles for tumor necrosis factor a and nitric oxide in resistance of rat alveolar macrophages to *Legionella pneumophila. Infect. Immun.* 64: 3236–3243.

Smith RL, Kaczmarek MT, Kucharski LM, Maguire ME (1998) Magnesium transport in *Salmonella typhimurium*: regulation of mgtA and mgtCB during invasion of epithelial and macrophage cells. *Microbiology* 144: 1835–1843.

Sola-Landa A, Pizarro-Cerda J, Grillo MJ *et al.* (1998) A two-component regulatory system playing a critical role in plant pathogens and endosymbionts is present in *Brucella abortus* and controls cell invasion and virulence. *Mol. Microbiol.* 29: 125–138.

Springer TA (1990) Integrin receptors. *Nature* 346: 425–434.

St Clair EW, Wilkinson WE, Lang T *et al.* (1996) Increased expression of blood mononuclear cell nitric oxide synthase type 2 in rheumatoid arthritis patients. *J. Exp. Med.* 184: 1173–1178.

Stahl D, Ezekowitz RA (1998) The mannose receptor is a pattern recognition receptor involved in host defense. *Curr. Opin. Immunol.* 10: 50–55.

Stamnes MA, Craighead MW, Hoe MH *et al.* (1995) An integral membrane component of coatomer-coated transport vesicles defines a family of proteins involved in budding. *Proc. Natl Acad. Sci. USA* 92: 8011–8015.

Steller H (1995) Mechanisms and genes of cellular suicide. *Science* 267: 1445–1449.

Stenger S, Donhauser N, Thuring H, Rollinghoff M, Bogdan C (1996) Reactivation of latent leishmaniasis by inhibition of inducible nitric oxide synthase. *J. Exp. Med.* 183: 1501–1514.

Sturgill-Koszycki S, Schlesinger H *et al.* (1994) Lack of acidification in *Mycobacterium* phagosomes produced by exclusion of the vesicular proton-ATPase. *Science* 263: 678–681.

Suhan ML, Chen SY, Thompson HA (1996) Transformation of *Coxiella burnetii* to ampicillin resistance. *J. Bacteriol.* 178: 2701–2708.

Swanson JA, Baer SC (1995) Phagocytosis by zippers and triggers. *Trends Cell Biol.* 5: 89–93.

Swanson JA, Watts C (1995) Macropinocytosis. *Trends Cell Biol.* 5: 424–428.

Swanson J, Bushnell A, Silverstein SC (1987) Tubular lysosome morphology and distribution within macrophages depend on the integrity of cytoplasmic microtubules. *Proc. Natl Acad. Sci. USA* 84: 1921–1925.

Swanson MS, Isberg RR (1996) Analysis of the intracellular fate of *Legionella pneumophila* mutants. *Ann. NY Acad. Sci.* 797: 8–18.

Taylor D, Inchley CJ, Gallagher M (1998) The *Salmonella typhimurium* AhpC polypeptide is not essential for virulence in BALB/c mice but is recognized as an antigen during infection. *Infect. Immun.* 66: 3208–3217.

Teitelbaum R, Cammer M, Maitland ML, Freitag NE, Condeelis J, Bloom BR (1999) Mycobacterial infection of macrophages results in membrane-permeable phagosomes. *Proc. Natl Acad. Sci. USA* 96: 15190–15195.

Thirumalai K, Kim KS, Zychlinsky A (1997) IpaB, a *Shigella flexneri* invasin, colocalizes with interleukin-1 beta-converting enzyme in the cytoplasm of macrophages. *Infect. Immun.* 65: 787–793.

Tilney LG, Portnoy DA (1989) Actin filaments and the growth, movement, and spread of the intracellular parasite, *Listeria monocytogenes*. *J. Cell Biol.* 109: 1597–1608.

Tjaden J, Winkler HH, Schwoppe C, van der Laan M, Mohlmann T, Neuhaus HE (1999) Two nucleotide transport proteins in *Chlamydia trachomatis*, one for net nucleoside triphosphate uptake and the other for transport of energy. *J. Bacteriol.* 181: 1196–1202.

Tsolis RM, Baumler AJ, Heffron F (1995) Role of *Salmonella typhimurium* Mn-superoxide dismutase (SodA) in protection against early killing by J774 macrophages. *Infect. Immun.* 63: 1739–1744.

Turco J, Liu H, Gottlieb SF, Winkler HH (1998) Nitric oxide-mediated inhibition of the ability of *Rickettsia prowazekii* to infect mouse fibroblasts and mouse macrophagelike cells. *Infect. Immun.* 66: 558–566.

Unkeless JC, Wright SD (1988) Phagocytic cells: Fcγ and complement receptors. In: Gallin JI, Goldstein IM, Snyderman R (eds) *Inflammation: Basic Principles and Clinical Correlates*, New York: Raven Press, pp. 343–362.

van der Laan LJ, Dopp EA, Haworth R *et al.* (1999) Regulation and functional involvement of macrophage scavenger receptor MARCO in clearance of bacteria *in vivo*. *J. Immunol.* 162: 939–947.

Via LE, Fratti RA, McFalone M, Pagan-Ramos E, Deretic D, Deretic V (1998) Effects of cytokines on mycobacterial phagosome maturation. *J. Cell Sci.* 111: 897–905.

Vogel J, Andrews HL, Wong SK, Isberg RR (1998) Conjugative transfer by the virulence system of *Legionella pneumophila*. *Science* 279: 873–876.

Way SS, Borczuk AC, Dominitz R, Goldberg MB (1998) An essential role for gamma interferon in innate resistance to *Shigella flexneri* infection. *Infect. Immun.* 66: 1342–1348.

Webster JW *et al.* (1998) The agent of human granulocytic ehrlichiosis resides in an endosomal compartment. *J. Clin. Invest.* 101: 1932–1941.

Wiater LA, Dunn K, Maxfield FR, Shuman HA (1998) Early events in phagosome establishment are required for intracellular survival of *Legionella pneumophila*. *Infect. Immun.* 66: 4450–4460.

Williams MD, Ouyang TX, Flickinger MC (1994) Starvation-induced expression of SspA and SspB: the effects of a null mutation in sspA on *E. coli* protein synthesis and survival during growth and prolonged starvation. *Mol. Microbiol.* 11: 1029–1043.

Winkler HH, Neuhaus HE (1999) Non-mitochondrial ATP transport. *Trends Biochem. Sci.* 24: 64–68.

Winkler HH, Turco J (1994) Rickettsiae and macrophages. In: Zwilling BS, Eisenstein TK (eds) *Macrophage–Pathogen Interactions*. New York: Marcel Dekker, pp. 401–414.

Wyrick B, Brownridge EA, Ivins BE (1978) Interaction of *Chlamydia psittaci* with mouse peritoneal macrophages. *Infect. Immun.* 19 pp. 1061–1067.

Xu S, Cooper A, Sturgill-Koszycki S *et al.* (1994) Intracellular trafficking in *Mycobacterium tuberculosis*- and *Mycobacterium avium*-infected macrophages. *J. Immunol.* 153: 2568–2578.

Yuan Y, Crane DD, Barry CE (1996) Stationary phase-associated protein expression in *Mycobacterium tuberculosis*: function of the mycobacterial a-crystallin homolog. *J. Bacteriol.* 178: 4484–4492.

Yuan Y, Crane DD, Simpson RM *et al.* (1998) The 16-kda a-crystallin (*acr*) protein of *Mycobacterium tuberculosis* is required for growth in macrophages. *Proc. Natl Acad. Sci. USA* 95: 9578–9583.

Zhang Y, Rikihisa Y (1997) Tyrosine phosphorylation is required for Ehrlichial internalization and replication. *Infect. Immun.* 65: 2959–2964.

Zhong G, de la Maza LM (1988) Activation of mouse peritoneal macrophages *in vitro* or *in vivo* by recombinant murine gamma interferon inhibits the growth of *Chlamydia trachomatis* serovar L1. *Infect. Immun.* 56: 3322–3325.

Zimmerli S, Edwards S, Ernst JD (1996) Selective receptor blockade during phagocytosis does not alter the survival and growth of *Mycobacterium tuberculosis* in human macrophages. *Am. J. Respir. Cell Mol. Biol.* 15: 760–770.

Zuckman D, Hung J, Roy CR (1999) Pore-forming activity is not sufficient for *L. pneumophila* phagosome trafficking and intracellular growth. *Mol. Microbiol.* 32: 990–1001.

Zychlinsky A, Kenny B, Menard R, Prevost MC, Holland IB, Sansonetti J (1994) IpaB mediates macrophage apoptosis induced by *Shigella flexneri*. *Mol. Microbiol.* 11: 619–627.

Zychlinsky A, Sansonetti J (1997) Apoptosis as a proinflammatory event: what can we learn from bacteria-induced cell death? *Trends Microbiol.* 201: 201–204.

36

Interactions with the Compromised Host

Alan S. Cross[1] and Steven M. Opal[2]

[1] *University of Maryland School of Medicine, Baltimore, Maryland, USA*
[2] *Brown University School of Medicine, Providence, Rhode Island, USA*

The consequences of an encounter between a bacterium and human host is largely determined by the virulence determinants of the micro-organism and the defences mounted by the host. Some bacteria can cause disease in normal individuals, usually by virtue of the elaboration of an extracellular toxin (e.g. *Vibrio cholerae*, *Corynebacterium diphtheriae*) or through mechanisms that can subvert normal host defence (e.g. *Shigella dysenteriae*, *Salmonella*). A far larger number of bacteria, however, are ubiquitous in the environment, yet cause disease only when normal host defences are compromised. These opportunistic pathogens take advantage of specific defects in host defences that are congenital, that develop concomitantly with a disease process or that are induced iatrogenically. In these situations, opportunistic pathogens usually handled efficiently by the host can now cause potentially lethal infection.

In protecting against microbial invasion the host immune system has two different tasks. First, it must defend against a universe of potential pathogens, many of which it might not have encountered and, second, it must efficiently mobilise its components toward off frequently encountered pathogens. Natural or innate immunity must hold the pathogens at bay to provide sufficient time for the specific or adaptive division of the immune system to develop more effective defences against highly specific antigens on pathogens that it can quickly recall during subsequent encounters (Table 1). In recent years significant advances have been made in knowledge of the cell biology of T and B lymphocyte cell function and in the molecular changes that accompany the development of specific immunity. In contrast, understanding of the innate immune system has developed more slowly. New cellular components (e.g. $\gamma\delta$ T cells, NK 1.1 T cells) or new pattern recognition systems (e.g. CD1, CD14 or scavenger receptors) of the innate immune system are being identified, and new functions are being elucidated for previously identified cells (e.g. dendritic and natural killer cells). These developments ensure that current understanding of the immune system will undergo considerable refinement in the coming years.

While extensive evidence underscores the interrelatedness of diverse components of the immune system, it has been convenient to divide the immune system into humoral and cellular components. The types of infections likely to be present during defects in cellular and humoral immunity are quite predictable and undoubtedly reflect the relative importance of these arms of the host immune system against specific pathogens. In this chapter we will focus on how defects in individual components of the immune system increase the susceptibility to bacterial and mycobacterial pathogens.

 doi:10.1006/bkmm.2001.0036

Table 1 Correlation of specific immune defects with likely infection

	Role	Conditions	Expected organisms
Innate immune system			
Phagocytes			
Neutrophils	Uptake and killing of microbes	Acute leukaemia, chemotherapy	Streptococci, staphylococci, *Enterobacteriaceae*, *Candida* spp., *Aspergillus* spp.
Macrophage/monocytes	Uptake and killing of intracellular pathogens	Steroid therapy, NRAMP1 deficiency, IFNR deficiency	*Salmonella*, *Listeria*, mycobacterial disease
NK cells	Kills virus-infected cells; mediates ADCC; early source of IFNγ	Unknown	Poorly defined
$\gamma\delta$ T cells	Early source of IFNγ, 'pattern' recognition	Unknown	Poorly defined
Humoral factors			
Mannose-binding protein	Opsonisation and/or lysis	Congenital defects	Increased susceptibility to bacterial infection
Alternate complement pathway			
Late complement components			Neisserial infection
Adaptive immune system			
B lymphocyte	Differentiates to antibody-producing cell; antigen presentation	Lymphoma, multiple myeloma, chronic lymphocytic leukaemia, congenital immunodeficiencies	Encapsulated bacteria, echovirus
T lymphocyte (CD4/CD8)	CD4:antigen-induced increase of IL-2 and expansion of clone CD8: antigen-specific cytotoxic cells	Hodgkin's disease, HIV	Intracellular bacteria (listeria, nocardia, mycobacteria); cryptococci, toxoplasma, *Pneumocystis*, cytomegalovirus, varicella zoster virus

Humoral Immunity

Neutrophils

Neutrophils and other phagocytic cells of the innate immune response system represent the first line of defence when the integument (skin or mucosa) is breached by pathogenic bacteria. Bacterial infections occur more frequently in the presence of quantitative and qualitative abnormalities of neutrophils.

Chronic Granulomatous Disease (CGD)

This is a family of inherited defects in the ability of the host phagocyte to form NADPH oxidase, the rate-limiting enzyme in the bactericidal oxidative response (for review, see Ezekowitz, 1992). The clinical syndrome is characterised by recurrent, life-threatening infections of the upper airway, bone and visceral organs caused by bacteria and fungi, with granuloma formation due to the recruitment of mononuclear phagocytes. Serious infections occur at the rate of approximately one every 12 months, but with antibiotic prophylaxis these are reduced to an incidence of approximately one in every 40 months.

The NADPH oxidase is required for the transfer of an electron to molecular oxygen, which results in the formation of superoxide anion, hydrogen peroxide and hypochlorous acid (HOCl, or bleach). Upon neutrophil stimulation, a functioning oxidase is formed in a highly regulated interaction between phagocytic oxidase (phox) proteins. Activation begins with the translocation of the cytosolic components of the oxidase, 47-kDa ($p47^{phox}$) and 67-kDa ($p67^{phox}$) proteins, and a low-molecular-weight guanosine triphosphatase, *rac2*, to cytochrome b_{558}. Cytochrome b_{558} is a heterodimeric transmembrane flavoprotein formed by the tight association of the 91-kDa glycoprotein ($gp91^{phox}$) and $p22^{phox}$ in the phagocyte membrane.

These cytosolic proteins bind to the cytoplasmic domain of the flavocytochrome b_{558}, probably through SH3-p47^{phox} regions of one protein and proline-rich regions of another protein. The role of the membrane p22^{phox} component is unclear, but it is required for either the stability and/or synthesis of gp91^{phox} and oxidase activity. In the absence of a functioning oxidative respiratory burst, these patients are susceptible to bacterial infections, often with strains that are catalase-positive (e.g. *Staphylococcus aureus*, Gram-negative bacilli).

While there are many phagocyte oxidase protein components of the membrane oxidase, four separate genotypes produce essentially the same phenotype that is present in the population with a frequency of one case for every 250 000 people. An X-linked deficiency occurs in the gene encoding the gp91^{phox} component of flavocytochrome b_{558} (71%), and autosomal deficiencies occur in the p47^{phox} (23%), and p67^{phox} (6%) proteins. Many distinct types of defects have been defined at each of the genes encoding these proteins. A mouse model of CGD, formed by a knockout of p47^{phox}, is susceptible to similar kinds of infections as are seen in humans (Jackson *et al.*, 1995). Because of the well-characterised genetic defect in CGD, this is one of the first diseases for which gene therapy has been attempted (Malech and Nauseef, 1997). In addition, CGD patients successfully respond to chronic interferon γ (IFNγ) therapy, but the basis for their response has not been identified (Gallin *et al.*, 1991).

Leukocyte Adhesion Deficiency

While CGD patients exhibit an inability to kill bacteria by oxidative mechanisms, patients with structural and/or functional defects in the expression of ligands critical to neutrophil interaction with endothelial cells during diapedesis are also susceptible to lethal bacterial infection. A deficiency in the surface expression of either β2 integrins (leucocyte adhesion deficiency-1; LAD-1) or sialyl Lewis X ligands (LAD-2) impairs the ability of neutrophils to emigrate from the bloodstream to an extravascular inflammatory focus (Springer *et al.*, 1984; Etzioni *et al.*, 1992).

Neutrophil Diapedesis

During tissue inflammation, mediators such as tumour necrosis factor α (TNFα), IL-1 and IL-8 induce the expression of cell adhesion molecules in adjacent venular endothelial cells. Circulating neutrophils interact with counter-structures on these activated endothelial surfaces, initially through a weak interaction mediated by constitutively expressed selectins (CD62L) on the neutrophil surface. This results in low-affinity adherence of neutrophils to the underlying endothelium which enables them to resist the shear force of the blood flow and to initiate slow rolling along the activated endothelial surface ('rolling adhesion'). If a rolling neutrophil becomes activated, for example by endothelial cell-derived IL-8 or platelet-activating factor, a tight adhesion forms between neutrophil and endothelial cells which is a necessary prelude to the emigration of neutrophils to the underlying inflamed extravascular compartment ('diapedesis').

Two classes of inherited defects have been described, LAD types 1 and 2, which are characterised by a similar clinical syndrome: recurrent bacterial infection without pus formation, marked increase in circulating white cell numbers, recurrent or progressive necrotic soft tissue infection, severe and early-onset gingivitis, impaired wound healing and delayed umbilical cord severance. In LAD-1 there is an inherited defect in a gene on chromosome 21q22.3 that encodes CD18, the common β chain of the β2 integrin family. This defect occurs with an estimated frequency of 1 per million, with most patients dying in their first decade. Five distinct abnormalities have been identified in the β chain: two classes affect the gross structure of the β subunit and result in either an abnormally large or small β subunit. Another mutation results in a lack of detectable mRNA or protein precursor. A fourth class of mutation is characterised by the synthesis of small amounts of β subunit precursor and low levels of mRNA, while a fifth class of mutation results in the synthesis of an apparently normal β subunit, but the subunit is unable to associate with the α subunit.

Tight adhesion of activated neutrophils to endothelial tissue is mediated through the binding by the β2 integrin, CD11b/CD18, a member of a family of heterodimeric proteins composed of different α chains (CD11a, b and c) coupled to a common β protein subunit (CD18). The CD11/CD18 on the surface of the activated neutrophil binds to a ligand in the immunoglobulin superfamily, primarily to ICAM-1, on the activated endothelial cell. This integrin, also known as complement receptor type 3 (CR3) or Mac1, after the monoclonal antibody that recognises the epitope, is critical not only for binding of neutrophils to endothelial cells but it also recognises complement factor iC3b that opsonises bacteria and thereby brings bacteria in close apposition to the phagocyte (Arnaout *et al.*, 1983). Consequently, in the absence of β2 integrin, neutrophils are less able to emigrate from the circulation to sites of infection, and are less efficient in the engulfment of bacterial targets. Patients born with this deficiency usually succumb to overwhelming infection.

In the more recently described LAD-2, patients have normal CD11/CD18 integrins, but an inability to synthesise fucosylated carbohydrate molecules. These structures include sialyl Lewis X and related molecules that are necessary for the initial step in diapedesis, rolling adherence of non-activated neutrophils along the activated endothelium (i.e. the carbohydrate targets of selectins are not synthesised). These patients have the same susceptibility to recurrent infections as those with LAD-1 but also have neurological defects. They also have a Bombay (hh) erythrocyte phenotype, because of the glycosylation defect. Through these experiments of nature, the importance to neutrophil function of selectin–carbohydrate-initiated rolling adherence and subsequent activation-induced $\beta 2$ integrin–ICAM-1 (protein–protein) interactions have been demonstrated. While prophylactic antibiotics may delay the onset of infection, patients with LAD-1 may require allogeneic bone marrow transplantation for optimal treatment.

More recently, a 10-year-old with recurrent bacterial infections and impaired pus formation was found to have normal neutrophil function, including normal levels of CD18, L-selectin (CD62L) and sialyl Lewis X (CD15). The endothelial cells of patient did not express E-selectin, although E-selectin mRNA was present in endothelial cells and significantly elevated levels of soluble E-selectin were present in the serum (DeLisser *et al.*, 1999). This is the first potentially inherited dysfunction of an endothelial cell adhesion molecule involved in diapedesis to be identified.

Myeloperoxidase Deficiency

Myeloperoxidase (MPO) is a protein found in the primary or azurophilic granules of neutrophils which, in the presence of chloride, catalyses the formation of oxidase-induced hydrogen peroxide to hypochlorous acid (HOCl). Complete MPO deficiency occurs in approximately 1 in 4000 individuals, but there is no evidence of increased susceptibility to infection. Human MPO has been cloned and the genetic defects associated with MPO deficiency have provided an insight into the post-translational processing of this protein (Malech *et al.*, 1997).

Specific Granule Deficiency

Inheritance of this rare, autosomal recessive condition renders the individual susceptible to recurrent bacterial infections, primarily in the bronchopulmonary system and skin. Neutrophils from these patients are defective in both degranulation and chemotaxis. While the gene responsible for this condition has not been cloned, evidence suggests that there may be a disorder of haematopoietic differentiation. The gene encoding lactoferrin, a secondary granular protein, is expressed in non-neutrophil secretory cells but not in myeloid cells of these patients. A defect in a myeloid-specific regulatory element may account for this discrepancy.

Chediak–Higashi Syndrome (CHS)

Patients with this rare, autosomal recessive condition are at increased risk of infection, have a mild bleeding tendency and, in an accelerated phase, can develop a life-threatening, lymphoma-like syndrome. During their second decade, such patients develop a progressive peripheral neuropathy. While the neutrophils of CHS patients are characterised by giant, perinuclear vesicles, defective chemotaxis and degranulation, the defect in granule formation is not limited to neutrophils. The gene responsible for CHS, *LYST*, is homologous with the *lyst* defect in beige mice known to have defects in NK cell function. The gene in mice and humans encodes proteins with a C-terminal prenylation motif with many potential phosphorylation sites. The proteins have structural homology with a vesicular sorting protein in yeast that may play a role in the stabilisation of microtubules (Barbosa *et al.*, 1996).

Neutropenia

Decreases in the number of circulating neutrophils, such as may occur following chemotherapy, are associated with a highly predictable susceptibility to infection. There is little change in the incidence of bacterial infection with decreases in absolute neutrophil count to a level of 1000 cells/mm^3, but at neutrophil counts below 1000 there is an increasing risk of bacterial infection which increases further at counts $\leq$100 cells/mm^3 (Bodey *et al.*, 1996). At an absolute neutrophil count below 100/mm^3 there is an 80% likelihood of bacterial infection. Gram-negative bacterial pathogens, particularly *Pseudomonas aeruginosa*, have historically been the most important pathogens, in part, because of the fulminant nature of infection, but now infections caused by Gram-positive bacteria, especially coagulase-negative staphylococci and viridans streptococci have become more common (Klastersky, 1994). Neutropenia is also associated with disseminated fungal infection caused by *Candida* and *Aspergillus* spp., thereby establishing the importance of neutrophils in host defences against these fungi.

Discovery of colony-stimulating factors, which have since been cloned and expressed as recombinant proteins, has made it possible to shorten the duration of neutropenia. With the ability to increase neutrophil counts to normal levels, there is a shorter time to bone marrow recovery and a decreased risk of infection, particularly with aspergillosis and candidiasis.

Cytokine-related Defects

The central importance of cytokines to the innate immune response to bacterial pathogens is illustrated by recent detailed studies of children with recurrent mycobacterial infection. Patients who were not infected with HIV and not deficient in CD4 lymphocytes were found to be susceptible to disseminated infection with non-tuberculous mycobacteria. Six apparently healthy Maltese children, three of whom were related, were diagnosed with disseminated atypical mycobacterial disease (Levin *et al.*, 1995). Investigation of macrophages *in vitro* isolated from these children and their parents revealed an impaired generation of TNFα in response to LPS and IFNγ, and reduced IFNγ production by T cells after exposure to mycobacterial and recall antigens. Subsequent studies identified a family in Tunisia with disseminated BCG infection after immunisation and an additional child with disseminated *Mycobacterium smegmatis* infection. These infections were associated with mutations in the IFNγ R1 receptor gene (Jouanguy *et al.*, 1997). It is of great interest that no other opportunistic infections were diagnosed in these children, suggesting that IFNγ is essential to host responses to mycobacteria. Defects in the IL-12 ligand (Altare *et al.*, 1998b) or the IL-12 receptor (Altare *et al.*, 1998a) also led to increased susceptibility to mycobacterial infections in affected patients.

More recently, *bcg*, a gene previously associated with resistance to infection in mice with mycobacteria, *Leishmania* and *Salmonella*, was found to encode for a transmembrane protein named natural resistance-associated macrophage protein-1 (Nramp-1) (Bellamy *et al.*, 1998). A human homologue of this gene was found on chromosome 2, and alleles of *Nramp1* were associated with increased susceptibility to tuberculosis in a West African population (Bloom and Small, 1998). While the DNA sequence of this protein is similar to those of nitrate or iron transporters, its function or role in susceptibility to infection remains unknown.

The TNFα response to endotoxin, measured after recovery from meningococcal sepsis by the production of TNFα in whole blood samples *ex vivo*, was correlated with disease severity and clinical outcome (Westendorp *et al.*, 1995). Interestingly, blood obtained from first-degree relatives of these patients had altered cytokine responses when similarly stimulated (Westendorp *et al.*, 1997). Among patients with septic shock, the presence of the *TNF2* gene, a TNFα promoter polymorphism, was associated with septic shock susceptibility and mortality (Mira *et al.*, 1999). Similarly, polymorphisms of the IL-1 genes have been associated with outcome after development of meningococcal infection (Read *et al.*, 1999). Moreover, a DNA polymorphism in the intron 2 position of the human IL-1 receptor antagonist (IL-1Ra) gene exists and this polymorphism determines the amount of IL-1Ra (an anti-inflammatory molecule that inhibits IL-1 activity) released during infection. The IL-1Ra A2 allele is associated with high levels of expression of IL-1Ra and an increased risk of sepsis (Fang *et al.*, 1999). Thus, there may be innate, familial cytokine responses to infections that affect both the clinical presentation and severity of disease. A 15-year-old patient with recurrent bacterial infections was found to be refractory to the effects of endotoxin both *in vitro* and *in vivo*, after intravenous challenge, and to IL-1 *in vitro*. The defect was most probably due to a defect very early in the signal transduction pathway (Kuhns *et al.*, 1997).

Complement Deficiency

Patients with inherited homozygous or acquired complement deficiencies are susceptible to bacterial infections and the development of autoimmune diseases. Homozygous or acquired C3 deficiency renders patients highly susceptible to life-threatening, pyogenic infections with encapsulated bacteria. Deficiency in complement regulatory proteins, factor I or factor H, results in a functional C3 deficiency due to an inability to control the consumption of C3 by C3 convertases.

The absence of complement components C5–C8 have all been associated with impaired neutrophil function and continued infection. Unlike patients with C3 deficiencies, these patients appear otherwise healthy and become infected beyond childhood, usually with *Neisseria* spp., which require terminal complement components for their killing. Multiple family pedigrees examined after recurrent infections with *Neisseria* spp. lacked the ability to assemble the membrane attack complex from terminal complement components by which the bacteria are lysed (Merino *et al.*, 1983). While diminished amounts of complement protein are detected in the serum of patients heterozygous for those terminal complement component alleles, a marked reduction in circulating terminal complement components is found in the serum of patients homozygous for this disease. Patients deficient in alternative complement pathway components, properdin and factor D, are also susceptible to *Neisseria* infections. If patients with complement deficiency are not promptly treated with antibiotics, they can die with overwhelming *Neisseria* infection, such as meningococcal meningitis.

Evasion of Complement-mediated Killing by Gonococci

The interaction of gonococci with complement-dependent host defences is an excellent demonstration of how pathogenic bacteria appropriate host defence mechanisms that enable them to evade host defences and cause infection. Gonococci usually are of a phenotype with lipo-oligosaccharide (LOS) in their outer membrane, which allows binding of complement components to the surface of the bacteria. This complement binding to the bacteria may facilitate uptake of the bacteria by white blood cells or even their lysis (see above). Gonococci that can evade these host defences are able to use the sialyltransferase in human neutrophils to add sialic acid from host serum onto the bacterial LOS. Complement is unable to bind to surfaces that contain sialic acid. Most mammalian cells have sialic acid on their surface which enables them to avoid routine complement-mediated autoimmune disease. Thus, by appropriating host enzymes and sialic acid, bacteria are able to assume the appearance of a 'self' antigen, avoid immune detection and thereby facilitate their replication and dissemination throughout the body. Other bacteria have encapsulated themselves in sialic acid-containing capsules (*E. coli* K1, group B meningococci, group B streptococcus, type III) thereby escaping detection by the host immune system (Parsons *et al.*, 1988). Sialylated surfaces are unable to activate the alternative complement pathway and additionally may mask underlying antigens (Reutter *et al.*, 1982).

Mannose-binding Protein

Mannose-binding protein (MBP), an acute phase protein that belongs to a group of calcium-dependent lectins called collectins, is now recognised as an important component of the innate immune system (Turner, 1996). This lectin binds to mannose-terminated glycoproteins, as well as to other sugars, such as *N*-acetylglucosamine and fucose. In the circulation, MBP, with its three chain subunits, looks like a bouquet of tulips with stems formed by collagen domains and the flower heads formed by the mannose-binding domains. The C1q subcomponent of C1, which is structurally similar to the collectins, lacks sugar-binding (lectin) activity. In binding to collectin receptors expressed on various cell types, MBP can function as an opsonin for microbes. Moreover, by directly cleaving C3, MBP activates the classical complement pathway in the absence of C1q, perhaps through the activity of its own protease, mannose-binding protein-associated serine protease (MASP). It is the only serum lectin with such activity.

MBP is functionally analogous to IgM, IgG and C1q and therefore has been called the 'universal antibody' or 'ante-antibody'. While the avidity of MBP binding is weak, the multiplicity of binding sites results in significant functional avidity towards a microbial surface, much like IgM. Most of its sugar targets are not normally exposed on mammalian surfaces at high densities, and therefore MBP prefers microbial structures to 'self'. While MBP binds to a wide range of Gram-positive and Gram-negative organisms, its binding ability is impeded by the presence of bacterial capsule. However, when bound to mannose-rich O-polysaccharide of *Salmonella montevideo*, the organism attaches to phagocytes and is taken up and killed.

MBP deficiency was initially associated with an opsonic defect in 10 children with persistent infections and no other known immunological defect (Sumiya and Summerfield, 1996; Summerfield *et al.*, 1997). While MBP deficiency is present in the 30% of the population who are heterozygotes for one variant allele, in a smaller subset this condition is associated with recurrent, life-threatening infections, particularly in children. Low levels of MBP were traced to three point mutations in exon 1 of the MBP gene. Mutations in two of the codons alters the ability of MBP to form stable, functional multimers. Subjects homozygous for any of the three mutations usually lack detectable MBP, while heterozygotes have reduced levels.

Immunoglobulin Deficiency

Gammaglobulin plays a role in host defences by facilitating the immune clearance of bacteria. Circulating natural antibodies in concert with complement form a first line of defence against invading pathogens. In addition, infecting bacteria elicit the formation of new antibodies with higher affinities and perhaps different specificities than natural antibody. Both natural and acquired antibodies facilitate the clearance of bacteria by functioning as opsonins. Bacterial surface-bound antibodies are recognised by Fc receptors of phagocytes. IgM antibodies, which are primarily located within the vascular space, promote bactericidal killing of target bacteria, often in conjunction with complement. In contrast, IgA isotype antibodies are an integral part of the mucosal immune system found in the gastrointestinal and respiratory tracts.

Low or, less frequently, absent levels of immunoglobulin are associated with increased susceptibility to bacterial infection. In 1951 Bruton recognised a patient with a congenital absence of circulating immunoglobulin. This young boy had recurrent life-threatening infections due to an X-linked deficiency in immunoglobulin synthesis. Severe combined immunodeficiency diseases are a group of immunologic disorders characterised by a profound defect in T cell and B cell function. These disorders are caused by a variety of specific immune defects including (1) dysfunctional cytokine receptor signalling (e.g. Jak3 deficiency), (2) enzyme deficiencies (e.g. adenosine deaminase deficiency), (3) defects in T cell-specific receptors or signalling molecules (TCR/CD3 deficiency) and (4) defective MHC expression (Leonard, 1996). Antibody replacement, usually with immunoglobulin for intravenous use (IVIg), is routinely given to patients lacking adequate levels of immunoglobulin on the basis of a congenital deficiency. Such treatment results in a marked decrease in the frequency of infections.

In clinical practice acquired hypogammaglobulinaemia is more commonly encountered. This is seen most frequently in premature infants, but also occurs when acquired diseases affect the normal function of both B and T lymphocytes (e.g. AIDS, see below). It is also increasingly recognised in older populations in the absence of diagnosed disease. In these situations it would seem reasonable to supplement existing low levels of circulating antibodies with additional exogenous antibody for the prevention of infection. It has been difficult to show, however, that the infusion of IVIg leads to significant improvement in the treatment of infection.

The current methods used to prepare immunoglobulin from plasma fractions results primarily in IgG, but very little IgM and IgA isotypes, though one method of preparation currently available only in Europe does enrich for IgM. While most current methods lack appreciable amounts of IgA, there is considerable evidence that adequate amounts of IgG can prevent infection, even with mucosal pathogens, especially viral pathogens, like polio. It is hypothesised that when there is exposure to a small microbial inoculum, the IgG present on the mucosal surfaces is sufficient to permit host clearance of the pathogen. In contrast, larger inocula of pathogens may require IgA for clearance from these sites (Robbins *et al.*, 1995).

IgA deficiency is the most common immune deficiency in the general population, and is present in about 5% of the general population. To date, it is believed that such patients are not at an increased risk of bacterial infection, unless the deficiency is accompanied by an IgG subclass deficiency. The compensatory mechanism by which these patients clear potential pathogens is not well known, but local IgG may compensate and contribute to the mucosal immune response. The IgG response to polysaccharides is predominantly of the IgG2 subclass. Consequently, individuals with selective deficiency of IgG2 are unable to generate antibodies to bacterial capsular polysaccharides. Since immunisation with protein–polysaccharide conjugate vaccines induces antibodies of the IgG1 and some IgG2 antibodies in these patients, the unresponsiveness of those with IgG2 deficiency is attributable to a defect in immunoregulation among T and B lymphocytes and/or antigen-presenting cells (Insel and Anderson, 1986). Patients with extreme elevations in serum IgE levels may be subject to skin abscesses, pneumonia with formation of pneumatoceles, and recurrent eczematoid rashes (Grimbacher *et al.*, 1999). The specific primary immunologic defect in the hyper-IgE syndrome has not been defined.

Defective Cell-mediated Immunity

HIV Infection

Since the beginning of the AIDS epidemic, the crucial role of the T helper cell in acquired host defence against microbial pathogens has become abundantly clear. The clinical manifestations of AIDS, particularly the dramatic increase in opportunistic infections, are primarily the result of progressive loss of cellular immunity as the population of CD4+ T lymphocytes declines. It soon became apparent that the specific deletion of this subgroup of T cells is essentially incompatible with life. As such, AIDS has become the paradigm for the study of loss of cell-mediated immune function acquired in the post-natal period (Fauci, 1993; Fauci *et al.*, 1996).

Bacterial Infections in Patients with AIDS

While *Mycobacterium avium-intracellulare* (MAI) infections remain the most common opportunistic bacterial infection in AIDS (Gordin *et al.*, 1997), a wide variety of common, encapsulated, extracellular pyogenic bacterial pathogens and previously unrecognised bacteria also may cause serious infections in HIV-infected patients (Caiaffa *et al.*, 1993; Berger *et al.*, 1994; Hirschstick *et al.*, 1995; Brettle, 1997). The remarkable susceptibility of these patients to bacteraemic *Strep. pneumoniae* and *H. influenzae* infections and difficulties with *Staphylococcus aureus* infections was not appreciated at the beginning of the AIDS era (Janoff *et al.*, 1992; Rodriguez-Barradas *et al.*, 1992; Hirschstick *et al.*, 1995; Brettle, 1997). Unusual infections by obscure organisms such as

Table 2 Common bacterial pathogens found in HIV-infected patients

Frequently associated pathogens	Less common pathogens	Possible associations with HIV
Mycobacterium avium complex	*Campylobacter* spp.	*Pseudomonas aeruginosa*
Mycobacterium tuberculosis	*Helicobacter* spp.	*Legionella pneumophila*
Streptococcus pneumoniae	*Nocardia* spp.	*Aeromonas hydrophila*
Salmonella spp.	*Listeria monocytogenes*	*Treponema pallidum*
Staphylococcus aureus	*Rhodococcus equi*	*Haemophilus ducreyi*
Haemophilus influenzae	*Bartonella quintana*	Other mycobacterial species
	Bartonella henselae	*Chromobacterium violaceum*
	Shigella spp.	*Moraxella catarrhalis*
	Clostridium difficile	*Ehrlichia chaffeensis*

Table 3 Immune defects associated with bacterial infections in HIV infection

Cell line	Quantitative defects	Qualitative immune defects	Associated bacterial infections
CD4+ T cells	Severe losses (++++)	Loss of cytokine production, IL-2, IFNγ, lymphotoxin-α	*Mycobacteria*, *Listeria*, *Salmonella*, *Nocardia* spp. infections
CD8+ T cells	Moderate losses (++)	Loss of CTL responses, poor intracellular killing	*Mycobacteria*, *Salmonella* spp. infections
B cells	Minimal losses (+)	Poor cytokine synthesis, poor response to neo-antigens	*Streptococcus*, *Haemophilus* spp. infections
NK cells	Minimal losses(+)	Poor cytokine production, cytotoxic activity	*Mycobacteria* spp. infections
Neutrophils	Variable losses (++)	Variable defects in chemotaxis, phagocytosis, bactericidal activity	*Staphylococcus*, *Pseudomonas*, *Haemophilus* spp. infections
Macrophages	Minimal losses (+)	Decreased antigen presentation, MHC class 2 expression, IL-12 and other cytokine production, microbial killing	*Mycobacteria*, *Listeria*, *Salmonella* spp. infections

Rhodococcus spp. (Sasal *et al.*, 1992) and previously unrecognised bacterial infections, including bacillary angiomatosis (Schwartzman, 1992), may complicate HIV infection. A review of the bacterial pathogens seen with increased frequency in patients with HIV infection is listed in Table 2.

The remarkable capacity of HIV to disrupt the host immune response has been the subject of many reviews (Fauci, 1993; Pantaleo and Fauci, 1995; Fauci *et al.*, 1996; Brettle, 1997) and this topic will be summarised only briefly in this chapter. The principal immunological defects associated with HIV infection that predispose to bacterial infections are outlined in Table 3.

Mechanism

The increased susceptibility of HIV-infected patients to bacterial infections, including those previously considered to be associated primarily with defects of humoral immunity, demonstrates the critical role of CD4 cells in orchestrating host defences. HIV specifically attaches to and replicates within CD4+ immune effector cells. Upon integration into the genome of these cells, HIV infection may alter their function. During the productive phase of HIV infection, the virus may deplete the body of CD4− cells either by cytolysis or by the action of cytotoxic T lymphocyte (CTL) responses directed against viral-specific antigens expressed on their cell surface (Fauci, 1993; Pantaleo and Fauci, 1995).

The loss of CD4 cells has a profoundly devastating effect on host immune defences. HIV-infected patients are rendered susceptible to a broad array of bacterial, viral, parasitic and fungal opportunistic pathogens which would otherwise be innocuous for a normal host. CD4 lymphocytes are principally responsible for the coordination of the cell-mediated immune response within the host. CD4 cells normally synthesise a large number of pro-inflammatory cytokines such as interleukin-2, IFNγ and lymphotoxin-α in response to microbial infection (Fauci *et al.*, 1996). The loss of these lymphokines interferes with cell-to-cell communication between components of the functional cell-mediated immune response. This carefully regulated and highly integrated cellular immune system rapidly becomes ineffective as the CD4 T cell populations are depleted.

Similarly, the progressive loss of MHC class I-restricted cytotoxic T cell responses by CD8+ T lymphocytes occurs throughout the course of HIV infection. This loss of cell-mediated effector function further compromises the host response to invasive bacterial pathogens, particularly those that exist in the intracellular location (Fauci, 1993; Pantaleo and Fauci, 1995) (Table 2).

Several explanations account for quantitative reductions in CD8 cells and NK cells in AIDS. Co-infection with human herpesvirus type 6 (HHV-6) may contribute to the immunosuppression of AIDS (Lusso *et al.*, 1991). It has been observed *in vitro* that HHV-6 infection in CD8 cells and NK cells increases the synthesis and surface expression of CD4 antigens on the cell surface of these cells, and renders them susceptible to direct infection from HIV via CD4-mediated uptake. These results demonstrate that CD4 may be re-acquired on the cell surface of CD8 cells even after thymic conditioning and selection has occurred. Up-regulation of CD4 by the ubiquitous presence of HHV-6 infection may explain the qualitative loss of CD8 cells and NK cells in the late phases of HIV infection (Fauci, 1993; Levy, 1997). It is also proposed that gp120 of HIV, in the absence of CD4, may form a complex with the cell membrane of CD8 cells within lymphoid organs where close cell-to-cell contact occurs. These CD8 cells may then undergo apoptosis from surface contact with HIV-derived conventional antigens and superantigens (Fauci, 1993).

Since B lymphocyte populations responsible for antibody formation are also affected by HIV infection, patients with HIV have a strikingly increased incidence of bacteraemia with pyogenic organisms such as *Streptococcus pneumoniae*, group B streptococci and *Haemophilus influenzae* (Berger *et al.*, 1994; Schuchat *et al.*, 1994; Tumbarello *et al.*, 1995; Brettle, 1997). The rate of pneumococcal bacteraemia in HIV-infected persons in San Francisco is 100-fold higher than in the historical control group. Most of the patients with pneumococcal bacteraemia had HIV infection but not fully advanced AIDS. HIV-associated *Strep. pneumoniae* infection is manifest by a higher incidence of bacteraemia (up to 60%). Recurrent bacterial pneumonia is so striking that it is now recognised as sufficient for a diagnosis of AIDS in an HIV-infected individual (Centers for Disease Control, 1992). Bacteraemia with pyogenic encapsulated bacterial pathogens is a particular problem in paediatric AIDS (Bernstein *et al.*, 1985).

The ability of memory B cells to recall antigens is generally preserved during the course of HIV infection (Fauci, 1993), but ability to respond to neo-antigens is impaired and may predispose to systemic infection (Rodriguez-Barradas *et al.*, 1992). This is a particular problem in paediatric AIDS, where infants and children have not had an immunological experience with common encapsulated bacterial pathogens (Lusso *et al.*, 1991). Recurrent bacterial infections with bacteraemia, meningitis and multi-lobar pneumonia is a major problem in paediatric AIDS. Long-term prophylactic IVIg is successful in reducing the incidence of serious bacterial infections in children with symptomatic HIV infection (NIH *et al.*, 1991). The data to support intravenous immunoglobulin therapy in adult patients is less clear.

Pneumococcal vaccine is recommended for all HIV-infected patients. Vaccine responsiveness is preserved at least in the early phases of HIV infection (Janoff *et al.*, 1992; Brettle, 1997). Trimethoprim-sulphamethoxazole prophylaxis routinely given for *Pneumocystis carinii* pneumonia also provides protection against pneumococcal infection. The same is true with macrolide agents used for prophylaxis for *Mycobacterium avium* infections. Highly active antiretroviral therapy (HAART) is changing the outlook for patients living with AIDS and may alleviate much of the infectious morbidity associated with bacterial pathogens.

Antigen-presenting cells of the monocyte/macrophage cell line are susceptible to infection with HIV, because these cells express cell surface CD4. While HIV infection occurs, replication and cytolysis of macrophages generally does not occur in response to HIV infection. Nonetheless, HIV-infected macrophages may have functional defects in association with HIV infection. HIV infection up-regulates synthesis of IL-6 and TNFα by infected mononuclear cells. This may lead to B cell activation and, along with IL-4, contribute to the hypergammaglobulinaemia observed in the early phases of HIV infection (Fauci, 1993). Recent evidence suggests that the B-cell stimulatory cytokine IL-15 also participates in the elevated immunoglobulin levels in HIV infection (Kacani *et al.*, 1997).

Deficient IL-12 synthesis by monocyte/macrophage cells may result in the failure to maintain an active TH1 response that is characteristically observed in later phases of HIV infection (Pantaleo and Fauci, 1995; Harrison and Levitz, 1997). Diminished IL-12 synthesis may be the principal explanation for the increased frequency of opportunistic intracellular pathogens such as *Mycobacterium avium* infection in AIDS (Gordin *et al.*, 1997).

Various defects of macrophage function, including abnormal chemotaxis, have been described. This may be related to down-regulation of chemotactic receptors in HIV-infected mononuclear cells (Fauci, 1993).

The opposite is true in HIV-infected B cells where IL-8 receptor surface expression is up-regulated by HIV infection (Jinquan *et al.*, 1997). Complement receptor and complement-mediated opsonophagocytosis may be diminished in HIV-infected mononuclear cells.

A deficiency of IFNγ production may contribute to decreased IL-12 synthesis in HIV-infected mononuclear cells (Trinchieri, 1997). Antigen processing and presentation is impaired as MHC class II antigen expression is reduced on macrophages in AIDS patients. HIV itself may directly interfere with antigen handling, since gp120 may already occupy the CD4 antigen and disrupt T cell–macrophage interactions (Fauci, 1993). Additionally, MHC class II expression is regulated by IFNγ production from CD4 T cells and NK cells. Since IL-12 is deficient in HIV-infected mononuclear cells, the co-stimulatory effect of IFNγ is lost and hence MHC class II expression is diminished (Trinchieri, 1997).

A variety of neutrophil defects have also been described in HIV infection (Ellis *et al.*, 1988; Murphy *et al.*, 1988; Corazzo *et al.*, 1991; Pitrak *et al.*, 1993; Vecchiarelli *et al.*, 1995). While these defects are minor compared with T-cell defects in HIV, defective granulocyte function combined with granulocytopenia may have important clinical consequences in some HIV-infected patients. Bone marrow suppression is a frequent in HIV infection. This may relate to susceptibility of CD34+ bone marrow precursor cells to HIV infection (Folks *et al.*, 1988). CD34+ progenitor cells include both uncommitted stem cells and committed myeloid progenitor cells. Granulocytopenia may also occur in relationship to drug-induced cytopenia from numerous medications given to HIV-infected patients (trimethoprim-sulphamethoxazole, ganciclovir, reverse transcriptase inhibitors). Diffuse bone marrow replacement with lymphomas, granulomatous diseases or systemic infections with parvovirus B19 or herpesvirus infections (CMV, HHV-6) may also contribute to myelosuppression and granulocytopenia. Even with adequate neutrophil production there may be an impaired recruitment of these cells to inflammatory sites during late stages of disease, because of a reduced IL-8 production upon exposure to C3a and C5a. Fcγ receptor function was preserved throughout the stages of HIV infection (Monari *et al.*, 1999).

Neutrophil dysfunction is manifest by an increased incidence of infections from *Staph. aureus* and other encapsulated bacterial organisms which complicates vascular access in these patients. Central intravenous catheter infections are commonplace in HIV and greatly complicate the ability to provide parenteral nutrition and other long-term intravenous therapies to HIV-infected patients (Brettle, 1997).

Bacterial Infections Associated with Trauma

The effects of multiple trauma on the immune system have only recently been recognised and are currently the focus of extensive clinical and experimental investigation (Ertel *et al.*, 1990; Faist *et al.*, 1992). Early observations found that multi-trauma patients with skin test anergy had an increased susceptibility to infectious complications and an excess mortality rate (MacLean *et al.*, 1975). It was subsequently shown that a marked depression in both cellular and humoral immunity may occur in multi-trauma patients (Nohr *et al.*, 1984; Ertel *et al.*, 1990; Faist *et al.*, 1992).

Current evidence indicates that stress associated with multiple trauma induces a switch from a TH1 response to a TH2 response (Hernandez-Pando and Rook, 1994; Brinkmann and Kristofic, 1995). After an initial period of intense mononuclear cell activation, multi-trauma patients may develop a period of suppressed immune activity, which is now referred to as 'immunoparalysis' or the immune refractory period. During this time, measurable levels of pro-inflammatory mediators such as IL-1, TNFα and IL-2 are diminished and blood levels of the anti-inflammatory cytokines IL-6, IL-10 and IL-4 are frequently elevated (Ertel *et al.*, 1990; Hernandez-Pando and Rook, 1994).

The endogenous stress hormone response during trauma results in increased levels of intracellular cyclic AMP (cAMP) in response to catecholamine and prostaglandin synthesis. Increased cAMP levels inhibit the synthesis of the inflammatory cytokines and promote the synthesis of anti-inflammatory cytokines (Endres *et al.*, 1991). Stress-induced glucocorticoid production results in the synthesis of IκB, which inhibits the nuclear translocation of the transcriptional activating protein, NF-κB (Ramirez *et al.*, 1996). The anti-inflammatory cytokines IL-4, IL-6, IL-10 and IL-13 are simultaneously up-regulated. These cytokines stimulate the synthesis of endogenous cytokine inhibitors, IL-1 receptor antagonist and soluble TNFα receptors, and attenuate the production of IFNγ along with that of TNFα and IL-1 (Vannier *et al.*, 1992; Chernoff *et al.*, 1995). The end result is an immunological state characterised by a diminished capacity to eliminate invasive, microbial pathogens (Ertel *et al.*, 1990; Faist *et al.*, 1992).

The effects of trauma are further complicated by the frequent presence of extensive haemorrhage and the need for multiple blood transfusions. Extensive haemorrhage may lead to prolonged hypotension with splanchnic vasoconstriction and ischaemia of the gastrointestinal mucosa. Ischaemia–reperfusion injury may adversely affect epithelial cell function with increased translocation of gut bacteria and bacterial endotoxin to the blood (Wang *et al.*, 1992). This may result in an initial, transient pro-inflammatory cytokine excess (Ayala *et al.*, 1990), followed by a prolonged period of down-regulation of cytokine synthesis (Faist *et al.*, 1992; Ayala *et al.*, 1994).

Multiple transfusions in the process of resuscitation for blood loss may also impair the host immune response. The delivery of multiple allo-antigens on the cell surface of transfused platelets and white blood cells, and the transfusion of immune competent cells from donated blood may adversely affect cell-mediated immune function in the transfusion recipient (Grafter *et al.*, 1996).

The need to implant foreign materials such as orthopaedic devices and vascular grafts is a frequent occurrence in multi-trauma patients. Bio-prosthetic materials interfere with neutrophil function and restrict clearance of microbial contamination with foreign bodies (Kurz *et al.*, 1996). Neutrophils become activated upon attachment to bio-prosthetic materials and this leads to degranulation and inefficient phagocytosis and killing when a microbial pathogen is present in the vicinity of the prosthetic device.

Trauma patients may also suffer from hypothermia in response to environmental stress in the trauma setting. In a recent double-blind, placebo-controlled trial, it was shown that surgical patients who have a normal temperature at the time of surgery have a statistically significant lower incidence of wound infections than those who were hypothermic at the time of the surgery (Kaplan *et al.*, 1996). This may be related to poor wound healing and diminished neutrophil function in response to hypothermia.

For numerous reasons, multi-trauma patients suffer a prolonged period of hypo-functioning, cell-mediated immune clearance after an initial traumatic insult. This vulnerable period increases susceptibility to secondary infectious complications, sepsis and increased mortality. Therapeutic measures designed to promote the host immune response following multi-trauma with the administration of pro-inflammatory cytokines such as IFNγ and granulocyte colony-stimulating factor may prove to be beneficial in the future management of the trauma patient.

Host Response to Burns

The care of the patient with severe burns has dramatically improved over the last 30 years. Careful clinical investigations have demonstrated the value of early massive fluid resuscitation, initiation of enteral hyper-alimentation and skilled supportive care for smoke inhalation and acute lung injury. Bacterial infections remain a leading cause of death, but mortality from bacterial infections has been dramatically reduced by aggressive burn eschar excision, quantitative bacterial cultures of burn wounds, and improvements in antimicrobial therapy. In the 1950s, burn wounds of approximately 30% total body surface area burn were associated with a 50% mortality rate. In the 1990s this has at least doubled, and 50% mortality in many burn centres is now observed only in patients with over 70% total body surface burn wounds (Deitch, 1990).

Success with early resuscitation and burn wound management in thermally injured patients has been remarkably successful in the early phases of their care. However, many patients now develop multi-organ dysfunction after a period of initial stabilisation. The aetiology of this remains unclear but probably has an immune pathogenesis similar to that observed after massive trauma (Huang and Yang, 1992; Bone, 1996). After an initial period of intense cytokine release, a period of immune suppression exists in which monocyte/macrophage cytokine levels are diminished (Hansbrough *et al.*, 1983). Recently, it has been demonstrated that T cell reactivity shifts to a TH2-type cytokine response after severe burns (Zedar *et al.*, 1999). This cytokine shift appears to be mediated primarily by CD8-positive T lymphocyte production of IL-4.

While levels of interleukin1 and TNFα are elevated in the early period after burn injury (Marano *et al.*, 1990), a persistent deficiency of circulating IL-1β levels may also occur and correlates with a poor outcome (Cannon *et al.*, 1992). It has been suggested that an intact pro-inflammatory cytokine response with IL-1 is an essential mediator in the host defence against infection in burn patients. In experimental studies, mice given IL-1 after a burn wound infection with *Ps. aeruginosa* had an increased survival when compared with animals that did not receive IL-1 (Silver *et al.*, 1990). IL-6 levels are elevated in burn patients and this portends an unfavourable outcome (Schluter *et al.*, 1991).

Patients with increased IL-6 levels that continue to increase over time have a less favourable prognosis than those with a short duration of increased IL-6 production (Schluter *et al.*, 1991). This is compatible with the hypothesis that after the stress of severe burns, there is initial cytokine excess, perhaps associated with tissue destruction, followed by a prolonged period of relative immune suppression (Hansbrough *et al.*, 1983;

Huang and Yang, 1992; Bone, 1996). This leads to an increased risk of burn wound infection and sepsis.

A variety of defects in neutrophil function have also been described in thermally injured patients (Deitch, 1990). These include poor chemotactic activity (Grogan, 1976), reduced complement-mediated opsonic activity in burn wound tissues (Deitch *et al.*, 1987), and excessive and diffuse margination and degranulation of neutrophils in the severely burned patient (Yurt and Pruitt, 1985). It is apparent that skilled and aggressive management of burn wounds, adequate resuscitation, and the provision of nutritional components which maintain immune reactivity are essential for optimal care of thermally injured patients (Deitch, 1990). Whether recent efforts with immunostimulatory agents such as G-CSF and IFNγ, immunonutrition, such as diet supplements, or growth hormone will ultimately be shown to be beneficial remains to be demonstrated in large clinical trials.

The Host Response Following Splenectomy

The spleen is a remarkable organ which plays a critical role in the systemic immune response to microbial pathogens. Under normal circumstances, the spleen represents no more than 1% of total body weight yet contains up to 25% of the body's entire lymphocyte population (Brown, 1992). The spleen behaves as a lymphatic organ for the systemic blood supply with surveillance and microbial clearance functions. In post-natal life it primarily functions to remove particulate antigens from the bloodstream, such as damaged red cells and encapsulated microbial pathogens (Lynch and Kapila, 1996). The micro-circulation of the spleen is organised to allow efficient removal of new microbial antigens which have not been fully opsonised (Hosea *et al.*, 1981).

A variety of immune defects have been noted after splenectomy. These include poor IgM responses to neo-antigens, low concentrations of complement components, tuftsin deficiency (a low-molecular-weight opsonic peptide synthesised in the spleen), and poor antigen processing and presentation (Styrt, 1990; Brigden, 1992). While these defects may contribute to increased susceptibility to infection, the inability to clear incompletely opsonised antigens appears to be the principal cause for life-threatening infections after splenectomy (Hosea *et al.*, 1981). A syndrome known as overwhelming post-splenectomy infection (OPSI) is a well-known complication of surgical splenectomy or functional asplenia. This is particularly true in immunocompromised patients or in immunologically inexperienced patients in the first decade of life. Splenectomy should be delayed if at all possible until after early childhood (Waghorn, 1993). This provides an opportunity for the child to develop the full complement of immune responses toward common bacterial pathogens in the environment. Patients with congenital asplenia or those undergoing splenectomy at a young age have a particularly high risk of infection from encapsulated bacterial pathogens such as *Strep. pneumoniae*, *H. influenzae* and *N. meningitidis*. This risk extends to adults after splenectomy, but is less common in older patient populations (Styrt, 1990; Brigden, 1992).

Infections from a variety of other bacteria may occur after splenectomy, such as *Capnocytophaga* spp. bacteraemia after animal bites. High-grade bacteraemia with detection of bacteria in the peripheral bloodstream by the direct Gram stain is characteristic of post-splenectomy infection syndromes. In the absence of a spleen, intravascular replication of bacterial pathogens may occur to levels that may exceed 10^5 cfu/mL. Such patients may rapidly develop septic shock with a fatal outcome if the syndrome is not immediately recognised and appropriately treated.

The problem of OPSI can be obviated to some degree by penicillin prophylaxis for children with asplenia (Cohn and Schiffman, 1987), timing of splenectomy and provision of available immunisations against common encapsulated bacterial organisms. The pneumococcal vaccine should be offered to every patient, if possible before splenectomy. The *H. influenzae* type B conjugate vaccine and probably the meningococcal vaccine should be given as well (Cohn and Schiffman, 1987). This is particularly true for children and immunocompromised adults at greatest risk for OPSI. Post-splenectomy patients will always be at some increased risk for overwhelming infection from encapsulated bacteria and parasitic infections. They should be advised to seek immediate medical attention if a febrile illness develops after splenectomy. Provision of empirical high-dose oral antibiotics for post-splenectomy patients with sudden, unexplained, febrile illnesses may be able to abort OPSI, if immediate medical care is unavailable (Cullingford *et al.*, 1991).

Conclusions

Patterns of infection by specific organisms with various disease states were recognised long before our current understanding of the immune system. Now, with a more detailed understanding of the role of phagocytes and lymphocyte subsets in the pathogenesis of infections and the mechanisms by which their interactions are orchestrated by cytokines and other inflammatory mediators, it has been possible to correlate these infections with qualitative or quantitative defects in specific immune function, such as lack of CD4 T cells in HIV infection with *Pn. carinii* infection and lack of terminal complement components with *Neisseria* infection. Some of these conditions are, or ultimately might be, corrected by the replacement of the defective or missing immune component, as is the case in patients with agammaglobulinaemia treated with gammaglobulin. As our understanding of deficits in immune function progresses to the molecular level, as in the case of neutrophil adhesins in LAD and with defective IFNγ receptor in mycobacterial infection, therapy for these conditions may require gene therapy. Undoubtedly our ability to correlate specific infections with defined immune defects will be further refined as understanding of the immune system continues to evolve.

References

Altare F, Durandy A, Lamas D *et al.* (1998a) Impairment of mycobacterial immunity in human interleukin-12 receptor deficiency. *Science* 280: 1432–1435.

Altare F, Lammas D, Revy P *et al.* (1998b) Inherited interleukin 12 deficiency in a child with bacillus Calmette-Guerin and *Salmonella enteritidis* disseminated infection. *J. Clin. Invest.* 102: 2035–2040.

Arnaout MA, Todd RF III, Dana N, Melamed J, Schlossman SF, Colten HR (1983) Inhibition of phagocytosis of complement C3- or immunoglobulin G-coated particles and of C3bi: binding by monoclonal antibody to a monocyte-granulocyte membrane glycoprotein (Mo1). *J. Clin. Invest.* 72: 171–179.

Ayala A, Perrin MM, Meldrum DR (1990) Hemorrhage-induced increase in serum TNF which is not associated with elevated levels of endotoxin. *Cytokine* 2: 170–174.

Ayala A, Lehman DL, Herdon CD (1994) Mechanism for enhanced susceptibility to sepsis following hemorrhage. *Arch. Surg.* 129: 1172–1178.

Barbosa MDFS, Nguyen QA, Tchernev VT *et al.* (1996) Identification of the homologous beige and Chediak-Higashi syndrome genes. *Nature* 382: 262–265.

Bellamy R, Ruwende C, Corrah T, McAdam KPWJ, Whittle HC, Hill AVS (1998) Variations in the *NRAMP 1* gene and susceptibility to tuberculosis in West Africans. *N. Engl. J. Med.* 338: 640–644.

Berger BJ, Hassain F, Roistacher K (1994) Bacterial infections in HIV-infected patients. *Infect. Dis. Clin. N. Am.* 8: 449–464.

Bernstein LJ, Kreiger BZ, Novick B (1985) Bacterial infection in the acquired immunodeficiency syndrome of children. *Pediatr. Infect. Dis.* 4: 472–479.

Bloom BR, Small PM (1998) The evolving relation between humans and *Mycobacterium tuberculosis*. *N. Engl. J. Med.* 338: 677–678.

Bodey GP, Buckley M, Sathe YS, Freireich EJ (1996) Quantitative relationships between circulating leucocytes and infection in patients with acute leukemia. *Ann. Intern. Med.* 64: 328–340.

Bone RC (1996) Toward a theory regarding the pathogenesis of the systemic inflammatory response syndrome: what we do and do not know about cytokine regulation. *Crit. Care Med.* 24: 163–172.

Brettle RP (1997) Bacterial infections in HIV: the extent and nature of the problem. *Int. J. Sexually Transmitted Dis. AIDS* 8: 5–15.

Brigden ML (1992) Overwhelming post-splenectomy infection: still a problem. *West. J. Med.* 157: 440–443.

Brinkmann V, Kristofic C (1995) Regulation by corticosteroids of Th1 and Th2 cytokine production by human CD4+ effector T cells generated from CD45 RO− and CD45 RO+ subsets. *J. Immunol.* 155: 3322–3328.

Brown AR (1992) Immunological functions of splenic B lymphocytes. *Crit. Rev. Immunol.* 11: 395–403.

Caiaffa WT, Graham NMH, Vlahov D (1993) Bacterial pneumonia in adult populations with HIV infection. *Am. J. Epidemiol.* 138: 909–922.

Cannon JG, Friedberg JS, Gelfand JA (1992) Circulating interleukin-1β and tumor necrosis factor-α concentrations after burn injury in humans. *Crit. Care Med.* 20: 1414–1419.

Centers for Disease Control (1992) Revised classification system for HIV infection and expanded surveillance case definition for AIDS among adolescents and adults. *MMWR Morb. Mortal. Wkly Rep.* 41(RR-17): 1–19.

Chernoff AE, Granowitz EV, Shapiro L (1995) A randomized, controlled trial of interleukin-10 in humans: Inhibition of inflammatory cytokine production and immune responses. *J. Immunol.* 154: 5492–5499.

Cohn DA, Schiffman G (1987) Immunoregulatory role of the spleen in antibody responses to pneumococcal polysaccharide antigens. *Infect. Immun.* 55: 1375–1380.

Corazzo GR, Zoli G, Ginaldi L (1991) Tuftsin deficiency in AIDS. *Lancet* 337: 12–13.

Cullingford GL, Watkins DN, Watts AD, Mallon DF (1991) Severe late post-splenectomy infection. *Br. J. Surg.* 78: 716–721.

Deitch EA (1990) The management of burns. *N. Engl. J. Med.* 323: 1249–1250.

Deitch EA, Bridges RM, Dobke M, McDonald JC (1987) Burn wound sepsis may be promoted by a failure of local antibacterial host defenses. *Ann. Surg.* 206: 340–348.

DeLisser HM, Christofidou-Solomidou M, Sun J, Nakada MT, Sullivan KE (1999) Loss of endothelial surface expression of E-selectin in a patient with recurrent infections. *Blood* 94: 884.

Ellis M, Gupta S, Galant S (1988) Impaired neutrophil function in patients with AIDS or AIDS-related complex: a comprehensive evaluation. *J. Infect. Dis.* 158: 1268–1276.

Endres S, Fuelle H-J, Sinha B (1991) Cyclic nucleotides differentially regulate the synthesis of tumour necrosis factor α and interleukin-1β by human mononuclear cells. *Immunology* 72: 56–62.

Ertel W, Faist E, Nestle C (1990) Kinetics of interleukin-2 and interleukin-6 synthesis following major mechanical trauma. *J. Surg. Res.* 48: 622–628.

Etzioni A, Frydman M, Pollack S (1992) Recurrent severe infections caused by a novel leucocyte adhesion deficiency. *N. Engl. J. Med.* 327: 1789–1792.

Ezekowitz RAB (1992) Chronic granulomatous disease: an update and a paradigm for the use of interferon-γ as adjunct immunotherapy in infectious diseases. *Curr. Top. Microbiol. Immunol.* 181: 283–292.

Faist E, Storck M, Hultner L (1992) Functional analysis of monokine activity through synthesis patterns of proinflammatory cytokines and neopterin in patients in surgical intensive care. *Surgery* 112: 562–572.

Fang XM, Schroder S, Hoeft A, Stuber F (1999) Comparison of two polymorphisms of the interleukin-1 gene family: interleukin-1 receptor antagonist polymorphism contributes to susceptibility to severe sepsis. *Crit. Care Med.* 27: 1330–1334.

Fauci AS (1993) Multifactorial nature of human immunodeficiency virus disease: implications for therapy. *Science* 262: 1011–1018.

Fauci AS, Pantaleo G, Stanley S, Weissman D (1996) Immunopathogenic mechanisms of HIV infection. *Ann. Intern. Med.* 24: 654–667.

Folks TM, Kessler SW, Orenstein JM, Justement JS, Jaffe ES, Fauci AS (1988) Infection and replication of HIV-I in purified progenitor cells of normal human bone marrow. *Science* 242: 919–922.

Gallin JI, Malech HL, Melnick DA and the International Chronic Granulomatous Disease Study Group (1991) A controlled trial of interferon gamma to prevent infections in chronic granulomatous disease. *N. Engl. J. Med.* 324: 509–516.

Gordin FM, Cohn DL, Sullam PM *et al.* (1997) Early manifestations of disseminated *Mycobacterium avium* complex disease: a prospective evaluation. *J. Infect. Dis.* 176: 126–132.

Grafter U, Kalechman Y, Sredni B (1996) Blood transfusion enhances production of T-helper-2 cytokines and transforming growth factor beta in humans. *Clin. Sci.* 91: 519–523.

Grimbacher B, Holland SM, Gallin JI (1999) Hyper-IgE syndrome with recurrent infections – an autosomal dominant multisystem disorder. *N. Engl. J. Med.* 340: 692.

Grogan JB (1976) Suppressed in vitro chemotaxis in burn neutrophils. *J. Trauma* 16: 985–988.

Hansbrough JF, Peterson V, Kortz E *et al.* (1983) Postburn immunosuppression in an animal model: Monocyte dysfunction induced in burned tissue. *Surgery* 93: 415–423.

Harrison TS, Levitz SM (1997) Priming with IFNγ restores deficient IL-12 production by peripheral blood mononuclear cells from HIV-seropositive donors. *J. Immunol.* 158: 459–463.

Hernandez-Pando R, Rook GAW (1994) The role of TNF alpha in T cell-mediated inflammation depends upon the Th1/Th2 cytokine balance. *J. Immunol.* 82: 591–595.

Hirschstick RE, Glassroth J, Jordan MC (1995) Bacterial pneumonia in persons infected with the human immunodeficiency virus. *N. Engl. J. Med.* 333: 845–851.

Hosea SW, Brown EJ, Hamburger MI, Frank MM (1981) Opsonic requirements for intravascular clearance after splenectomy. *N. Engl. J. Med.* 304: 245–250.

Huang YS, Yang ZC (1992) Clinical studies on post burn multiple organ failure: Its aetiological factors and monitoring. *Burns* 18: 26–29.

Insel RA, Anderson PW (1986) Response to oligosaccharide conjugate vaccine against *Hemophilus influenzae* b in two patients with IgG_2 deficiency unresponsive to capsular polysaccharide vaccine. *N. Engl. J. Med.* 315: 499–503.

Jackson SH, Gallin JI, Holland SM (1995) The $p47^{phox}$ mouse knock-out model of chronic granulomatous disease. *J. Exp. Med.* 182: 751–758.

Janoff EN, Breiman RF, Daley CL, Hopewell PC (1992) Pneumococcal disease during HIV infection. *Ann. Intern. Med.* 117: 314–324.

Jinquan T, Moller B, Storgaard M, Mukaida N, Bonde J, Grunnet N (1997) Chemotaxis and IL-8 receptor expression in B cells from normal and HIV-infected subjects. *J. Immunol.* 158: 475–484.

Jouanguy E, Lamhamedi-Cherradi S, Altare F *et al.* (1997) Partial interferonγ receptor 1 deficiency in a child with tuberculoid bacillus Calmette-Guerin infection and a sibling with clinical tuberculosis. *J. Clin. Invest.* 100: 2658–2664.

Kacani L, Steiber H, Dierich MD (1997) Role of IL-15 in HIV-associated hypergammaglobulinemia. *Clin. Exp. Immunol.* 108: 14–18.

Kaplan SS, Basford RE, Jueng MH, Simmons RL (1996) Biomaterial-neutrophil interactions: dysregulation of oxidative functions of fresh neutrophils induced by prior neutrophil-biomaterial interactions. *J. Biomed. Materials Res.* 30: 67–75.

Klastersky J (1994) Empirical therapy for bacterial infections in neutropenic patients. *Supportive Care Cancer* 2: 347–354.

Kuhns DB, Priel DAL, Gallin JI (1997) Endotoxin and IL-1 hyporesponsiveness in a patient with recurrent bacterial infections. *J. Immunol.* 158: 3959.

Kurz A, Sessler DI, Lenhardt R (1996) Perioperative normothermia to reduce the incidence of surgical wound infections and shorten hospital stay. *N. Engl. J. Med.* 334: 1209–1212.

Leonard WJ (1996) Dysfunctional cytokine receptor signaling in severe combined immunodeficiency. *J. Invest. Med.* 44: 304–310.

Levin M, Newport MJ, D'Souza S *et al.* (1995) Familial disseminated atypical mycobacterial infection in childhood: a human mycobacterial susceptibility gene? *Lancet* 345: 79–83.

Levy JA (1997) Three new human herpes viruses (HH6, 7, and 8). *Lancet* 349: 558–562.

Lusso P, De Maria A, Malnati M *et al.* (1991) Induction of CD4 and susceptibility to HIV-I infection in human CD8+ T lymphocytes by human herpes virus 6. *Nature* 349: 533–535.

Lynch AM, Kapila R (1996) Overwhelming post-splenectomy infection. *Infect. Dis. Clin. N. Am.* 10: 693–708.

MacLean LD, Meakins JL, Taguchi K (1975) Host resistance in sepsis and trauma. *Ann. Surg.* 192: 207–211.

Malech HL, Nauseef WM (1997). Primary inherited defects in neutrophil function: etiology and treatment. *Semin. Hematol.* 34: 279–290.

Malech HL, Bauer TR Jr, Hickstein DD (1997) Prospects for gene therapy of neutrophil defects. *Semin. Hematol.* 34: 355–361.

Marano MA, Fong Y, Moldawer LL (1990) Serum cachectin/tumor necrosis factor in critically ill patients with burns correlates with infection and mortality. *Surg. Gynecol. Obstet.* 170: 32–38.

Merino J, Rodriguez-Valverde V, Lamelas JA, Riestra JL, Casanueva B (1983) Prevalence of deficits of complement components in patients with recurrent meningococcal infections. *J. Infect. Dis.* 148: 331.

Mira J-P, Cariou A, Grall F *et al.* (1999) Association of *TNF2*, a TNFα promoter polymorphism, with septic shock susceptibility and mortality. *JAMA* 282: 561.

Monari C, Casadevall A, Pietrella D, Bistoni F, Vecchiarelli A (1999) Neutrophils from patients with advanced human immunodeficiency virus infection have impaired complement receptor function and preserved Fcγ receptor function. *J. Infect. Dis.* 180: 1542–1549.

Murphy PM, Lain C, Fauci AS, Gallin JI (1988) Impairment of neutrophil bactericidal capacity in patients with AIDS. *J. Infect. Dis.* 158: 627–629.

NIH and Human Development Intravenous Immunoglobulin Study Trial Group (1991) Efficacy of intravenous immunoglobulin for the prevention of bacterial infections in children with symptomatic HIV infection. *N. Engl. J. Med.* 325: 73–79.

Nohr CW, Christou NV, Rode H (1984) *In vivo* and *in vitro* humoral immunity in surgical patients. *Ann. Surg.* 200: 373–380.

Pantaleo G, Fauci AS (1995) New concepts in the immunopathogenesis of HIV infection. *Annu. Rev. Immunol.* 13: 487–512.

Parsons NJ, Patel PV, Tan EL (1988) Cytidine 5′-monophospho-*N*-acetyl neuraminic acid and a low molecular weight factor from human blood cells induce lipopolysaccharide alteration in gonococci when conferring resistance to killing by human serum. *Microb. Pathogen.* 5: 303–309.

Pitrak DL, Bak PM, DeMarais P (1993) Depressed neutrophil superoxide production in HIV infection. *J. Infect. Dis.* 167: 1406–1410.

Ramirez F, Fowell DJ, Puklavec M, Simmonds S, Mason D (1996) Glucocorticoids promote a Th2 cytokine response by CD4+ T cells *in vitro*. *J. Immunol.* 156: 2406–2416.

Read RC, Camp NJ, Borrow R *et al.* (1999) An interleukin-1 genotype determines outcome in meningococcal disease. *39th Interscience Conference on Antimicrobial Agents and Chemotherapy (ICAAC), San Francisco, CA, 1999* (Abstract 496). Washington DC: American Society for Thermobiology.

Reutter W, Kottgen E, Bauer C, Gerok W (1982) Biological significance of sialic acids. In: Schauer R (ed.) *Sialic Acids: Chemistry, Metabolism and Function*. Vienna, New York: Springer-Verlag, pp. 265–305.

Robbins JB, Schneerson R, Szu SC (1995) Perspective hypothesis: Serum IgG antibody is sufficient to confer protection against infectious diseases by inactivating the inoculum. *J. Infect. Dis.* 171: 1387–1398.

Rodriguez-Barradas MC, Musher DM, Lahart C (1992) Antibody to capsular polysaccharides of *Streptococcus pneumoniae* after immunization of HIV-infected subjects with 23 valent pneumococcal vaccine. *J. Infect. Dis.* 165: 553–556.

Sasal M, Roig J, Cervantes M (1992) Good response to antibiotic treatment of lung infection due to *Rhodococcus equi* in a patient infected with human immunodeficiency virus. *Clin. Infect. Dis.* 15: 747.

Schluter B, Konig B, Bergmann U (1991) Interleukin-6 – A potent mediator lethal sepsis after major thermal trauma: evidence for increase in IL-6 production in peripheral blood mononuclear cells. *J. Trauma* 33: 1663–1670.

Schuchat A, Bloome CV, Hightower A (1994) Use of surveillance of invasive pneumococcal disease to estimate the size of the immunosuppressed HIV-infected population. *JAMA* 265: 3275–3280.

Schwartzman WA (1992) Infections due to *Rochalimaea*: the expanding clinical spectrum. *Clin. Infect. Dis.* 15: 893–900.

Silver GM, Gamelli RL, O'Reilly M (1990) Effect of interleukin 1α on survival in a murine model of burn wound sepsis. *Arch. Surg.* 125: 922–925.

Springer TA, Thompson WS, Miller J, Schmalsteig FC, Anderson DC (1984) Inherited deficiency of the Mac-1, LFA-1, p150.95 glycoprotein family and its molecular basis. *J. Exp. Med.* 160: 1901–1918.

Styrt B (1990) Infection associated with asplenia: risks, mechanisms, and prevention. *Am. J. Med.* 88: 35–42.

Sumiya M, Summerfield JA (1996) Mannose-binding protein, genetic variants and the risk of infection. *Q. J. Med.* 89: 723–726.

Summerfield JA, Sumiya M, Levin M, Turner MW (1997) Association of mutations in mannose binding protein gene with childhood infection in consecutive hospital series. *BMJ* 314: 1229–1232.

Trinchieri G (1997) Function and clinical use of interleukin-12. *Curr. Opin. Hematol.* 4: 59–66

Tumbarello M, Tacconelli E, Caponera S (1995) The impact of bacteraemia in HIV infection. Nine years experience in a large Italian university hospital. *J. Infect.* 31: 123–131.

Turner MW (1996) Mannose-binding lectin: the pluripotent molecule of the innate immune system. *Immunol. Today* 17: 532–540.

Vannier E, Miller LC, Dinarello CA (1992) Coordinated anti-inflammatory effect of IL-4: IL-4 down regulates IL-1 synthesis but up regulates IL-1ra synthesis. *Proc. Natl Acad. Sci. USA* 89: 4076–4080.

Vecchiarelli A, Monari C, Baldelli F *et al.* (1995) Beneficial effect of recombinant human granulocyte colony-stimulating factor on fungicidal activity of polymorphonuclear leucocytes from patients with AIDS. *J. Infect. Dis.* 171: 1448–1457.

Waghorn DJ (1993) Prevention of postsplenectomy sepsis. *Lancet* 341: 248–249.

Wang P, Ba ZF, Morrison MH (1992) Mechanism of the beneficial effect of pentoxifylline on hepatocellular function after trauma hemorrhage and resuscitation. *Surgery* 112: 451–458.

Westendorp RGJ, Langermans JAM, de Bel CE *et al.* (1995) Release of tumor necrosis factor: an innate host characteristic that may contribute to the outcome of meningococcal disease. *J. Infect. Dis.* 171: 1057.

Westendorp RGJ, Langermans JAM, Huizinga TWJ *et al.* (1997) Genetic influence on cytokine production and fatal meningococcal disease. *Lancet* 349: 170.

Yurt RW, Pruitt BA (1985) Decreased wound neutrophils and indiscriminate margination in the pathogenesis of wound infection. *Surgery* 95: 191–198.

Zedar S, Bone RC, Baue AE, Donnersmarck GHV, Faist E (1999) T cell reactivity and its predictive role in immunosuppression after burns. *Crit. Care Med.* 27: 66–72.

37 Vaccine Development

Rino Rappuoli[1] and Margaret Liu[2]

[1]*IRIS, Chiron S.p.A., Siena, Italy*
[2]*Bill and Melinda Gates Foundation, Seattle, WA, USA*

The concept that those who survive an infectious disease do not develop the same disease again is as old as history itself. The historian Thucydides, describing the Peloponnesian War, reports that during the plague of Athens in 430 BCE it was a common practice to employ those who had recovered from the disease to take care of the sick, because 'the same man was never attacked twice' (Silverstein, 1989). This observation has been made by many writers since then, among them Procopius (CE 541), Fracastoro (1483–1553) and the novelist Alessandro Manzoni. Describing the terrible plague that killed half of the population of Milan in 1630 Manzoni recounted that 'those that had recovered were quite a privileged class' because they could walk without fear anywhere, while the rest of the population were hiding themselves, fearing the constant risk of infection and death (Manzoni, 1984).

Today it is popular knowledge that this is due to our immune system, which during infection recognises as non-self the antigens of the infecting agent, and rearranges the genetic repertoire of B and T cells to code for molecules that specifically recognise them. This generates specific B and T cell clones that combat the invading agent by neutralising it with antibodies, or by killing the cells that have been infected by it. After infection, the specific B and T cells are retained as memory cells capable of recognising and proliferating if they meet the same agent again. Vaccination takes advantage of this property of the immune system and induces it to produce B and T cells able to combat the infectious agent in the absence of a dangerous infection. To do so, killed or attenuated micro-organisms – either whole, or parts of them – are presented to the immune system, usually by injection.

The first report of the induction of immunity by the deliberate infection of healthy individuals comes from Asia in CE 590, but it is likely that the practice is much older. In those early procedures, infected material from a mild smallpox lesion, sometimes dried with little cotton, was transferred to healthy people to render them resistant to subsequent exposures to the same disease (variolation). During the Middle Ages, the practice spread to Persia, India, Egypt, Libya and Turkey. The practice of variolation spread widely in Europe after the reports to the Royal Society of London, in 1713 by Emanuele Timoni, the family physician of the British ambassador to Constantinople, and in 1716 by Jacob Pylarini, the Venetian consul in Smyrna. A few years later, the wife of the British ambassador in Turkey, Lady Mary Wortley Montagu (1689–1762), who had seen the 'miracles' of variolation in that country, promoted the practice so efficiently that the sons of the British Royal family were vaccinated in 1722. Nevertheless, the practice was still a desperate undertaking, because up to 4% of those exposed to variolation developed a severe form of the disease and died; this was, however, much lower than the 20–30% fatality rate from natural smallpox. The solution finally came in 1796 when Edward Jenner, an English physician, noticed during his practice in the countryside that farmers exposed to infected materials from cows with cowpox lesions did not develop the disease but acquired immunity to smallpox. Jenner decided to use the less dangerous material derived from the bovine vaccinia lesions to vaccinate a boy, James Phipps, and showed that he was immune to a subsequent challenge with smallpox. The scientific approach to vaccination came only a century

 doi:10.1006/bkmm.2001.0037

Table 1 Landmarks in vaccine development

Disease (infectious agent)	Vaccine type	Year	Consequence on human health
Smallpox	Live attenuated	1796	Eradication of disease and micro-organism
Rabies	Live attenuated	1885	Elimination of post-exposure mortality
	Killed virus	1907	
Salmonella typhi	Killed bacteria	1912	Protection from disease
	Live attenuated (T2IA)	1991	
	Purified capsular polysaccharide	1991	
Tuberculosis	Live attenuated bacterium (BCG)	1921	Tuberculosis is still a major problem
Diphtheria	Toxoid	1939	Protection from disease and micro-organism elimination
Yellow fever	Live attenuated virus (strain 17D)	1940	Protection from disease
Tetanus	Toxoid	1949	Protection from disease
Pertussis	Killed bacterial cells	1953	Protection from severe disease and mortality associated with it
	Acellular (Japan)	1981	
	Recombinant	1993	
	Acellular (USA)	1996	
Poliovirus	Killed virus (Salk)	1954	Protection from disease and micro-organism elimination
	Live attenuated (Sabin)	1961	
Influenza	Killed virus	1958	Decreased incidence and attenuated disease in vaccinated people
	Detergent-split virus	1968	
	Purified virus subunit	1976	
	Adjuvanted vaccine	1997	
	Live attenuated	1998	
Cholera	Killed bacteria (injectable)	1960	Protection from disease
	Killed bacterium$^+$ CTB (oral)	1994	
	Live attenuated, recombinant (CVD 103 HgrR)	1995	
Measles	Live attenuated virus	1963	Protection from disease for several years after vaccination
Mumps	Live attenuated virus	1966	Protection from disease for several years after vaccination
Rubella	Live attenuated virus	1969	Protection from disease for several years after vaccination
Meningococcus	Purified capsular polysaccharide	1969	Protection from disease in vaccinated adolescents and adults
	(tetravalent A, C, Y, W135)	1984	
	Conjugated C	2000	
Pneumococcus	Purified capsular polysaccharide	1977 (14-valent) 1983 (23-valent)	Protection from disease in vaccinated adults and elderly people
	Conjugated	2000	
Hepatitis B	Plasma-derived	1981	Protection from infections, chronic hepatitis B and liver cancer
	Recombinant	1986	

Table 1 (Continued)

Disease (infectious agent)	Vaccine type	Year	Consequence on human health
Haemophilus influenzae	Polysaccharide Conjugated (toddlers) Conjugated (infants)	1985 1987 1990	Protection from disease and micro-organism elimination
Hepatitis A	Killed virus	1994	Protection from disease
Varicella	Live attenuated virus	1995	Protection from disease
Lyme disease	Recombinant	1998	Protection from disease
Rotavirus	Live attenuated	1998	Protection from disease*

*The vaccine was subsequently withdrawn because of a few cases of intussusception.

later, when Louis Pasteur introduced the concept that infectious diseases were caused by micro-organisms and discovered that these could be attenuated by growing them under adverse conditions. Using this empirical approach he developed the first live-attenuated bacterial and viral vaccines – the chicken cholera bacillus and rabies, respectively (Silverstein, 1989; Ada, 1993). Large-scale vaccination came only after the discovery by Glenny and Hopkins (1923) and Ramon (1924) of safe and reproducible ways to inactivate toxins and pathogens by the use of formaldehyde, and of the stable attenuation of pathogens by serial passage *in vitro*. For most of the last century, until the very recent biotechnological revolution, these simple, basic methods have been the only means used for vaccine development.

The progress made during this period has been remarkable. As shown in Table 1, many infectious diseases that caused millions of deaths at the beginning of the twentieth century have now been completely eradicated, as a consequence of simple and cheap methods such as vaccination. Because of their success, vaccines are now considered to be the most potent and cost-effective means of combating infectious diseases. The role of vaccination will increase in future, because the new technologies allow the molecular design of vaccines targeted to any infectious agent, making it possible to eradicate most of the remaining infectious diseases. Furthermore, new methods for the delivery of vaccines by the oral or mucosal routes, or by direct injection of nucleic acid that encodes the vaccine molecule, will expand the use and the target of vaccines (WHO, 1997). Indeed, although existing vaccines are at present used only for prevention, their use to cure chronic diseases, tumors and allergies is beginning (Rappuoli *et al.*, 1997).

Existing Bacterial Vaccines

Toxoids

Some bacterial diseases, such as diphtheria and tetanus, can be fully reproduced by toxins that are released into the extracellular medium by the bacteria. Diphtheria toxin is a 58-kDa protein molecule that can be divided into two functionally different moieties (Collier, 1982). The A fragment is an enzyme that causes cell death by ADP-ribosylating the elongation factor 2, which once modified is unable to carry out protein synthesis. The B fragment has a C-terminal moiety capable of binding the heparin-binding EGF-like growth factor precursor (Naglich *et al.*, 1992), and a central hydrophobic portion that can insert into membranes and facilitate the translocation of the toxic A fragment across the cell membrane (Choe *et al.*, 1992; Domenighini *et al.*, 1995; Rappuoli, 1997b).

Tetanus toxin is a 150.5-kDa protein molecule that can also be divided into two moieties. One is the C-terminal region (fragment C) which binds to receptors at the peripheral portion of nerve cells and is moved by retrograde transport to the body of the spinal cord neurons. The other is the N-terminal region, a zinc-metalloprotease that cleaves synaptobrevin, an integral membrane protein of presynaptic vesicles, thus inhibiting vesicle fusion with cell membranes and neurotransmitter release (Schiavo *et al.*, 1992; Montecucco and Schiavo, 1993; Rappuoli, 1997b).

When a disease is caused by a toxin, it is sufficient to induce toxin-neutralising antibodies to induce protection from disease. The existing vaccines against diphtheria and tetanus were developed in the 1920s and contain partially purified toxins extracted from the culture supernatant of the bacteria and treated with formaldehyde to inactivate them (toxoids) (Glenny and Hopkins, 1923; Ramon, 1924). These vaccines are routinely used for infant immunisation and administered as a diphtheria, tetanus and pertussis (whooping cough) combination vaccine. Booster vaccinations with diphtheria and tetanus toxoids are periodically given to adults to maintain immunity throughout life.

It is now possible to manufacture vaccines against diphtheria and tetanus that are more pure than in the past, and some manufacturers purify the toxins before chemical detoxification (toxoiding). Molecular vaccines are also now possible; recombinant fragment C (Makoff *et al.*, 1989; Norton *et al.*, 1997) has been shown to be sufficient to induce a protective immune response against tetanus, and a non-toxic mutant of diphtheria toxin (CRM 197) can be used against diphtheria (Giannini *et al.*, 1984). The conventional diphtheria and tetanus toxoids are, however, incorporated in all recently developed combination vaccines because of their long record of safety and success (Rappuoli, 1997b). Chemical detoxification of bacterial toxins has been extremely useful in developing effective vaccines, and the method has been used very recently to detoxify pertussis toxin that has been included in many acellular vaccines (Edwards *et al.*, 1995).

More effective technologies have now become available to eliminate toxicity by the manipulation of toxin-encoding genes. The first example of this came at Harvard in the 1970s from the laboratory of J.A.M. Pappenheimer, where by random mutagenesis T. Uchida modified the gene that encodes diphtheria toxin and isolated a strain that produced a non-toxic mutant of the toxin (Uchida *et al.*, 1971). This protein, named cross-reacting material (CRM 197), was subsequently shown to differ from the wild-type in only one amino acid (Gly52→Glu52) (Giannini *et al.*, 1984). It is now widely used in vaccination as a carrier for conjugate vaccines against *Haemophilus influenzae* and *Neisseria meningitidis* (Costantino *et al.*, 1992; Hall *et al.*, 1993). A rational approach has recently been introduced to eliminate the toxicity of pertussis toxin. The gene was cloned (Nicosia *et al.*, 1986) and the amino acids involved in the enzymatic activity responsible for the toxicity were identified (Pizza *et al.*, 1988). Then the two amino acids important for catalysis were changed by site-directed mutagenesis (**Fig. 1a**) (Pizza *et al.*, 1989). The resulting molecule, which was devoid of any toxicity, was tested in clinical trials and had an immunogenicity approximately 10 times higher than the wild-type molecule detoxified by conventional formaldehyde treatment (**Fig. 1b**) (Rappuoli, 1997a). In large-scale trials, the vaccine containing this molecule gave excellent protection from disease (Greco *et al.*, 1996b), which appeared earlier and lasted longer than that induced by conventional vaccines (Greco *et al.*, 1996a; Rappuoli, 1997a). The greater immunogenicity was obtained with lower doses of protein (Hewlett, 1996), suggesting that in future this should be the way to produce new toxin-based vaccines. A similar approach was used later genetically to detoxify toxins produced by *Vibrio cholerae* (CT) and enterotoxigenic *Escherichia coli* (Pizza *et al.*, 1994). The crystal structure of wild-type heat-labile toxin (LT) (Sixma *et al.*, 1991) and of a non-toxic mutant (LTK63) (Van den Akker *et al.*, 1997), which is a candidate for development of future

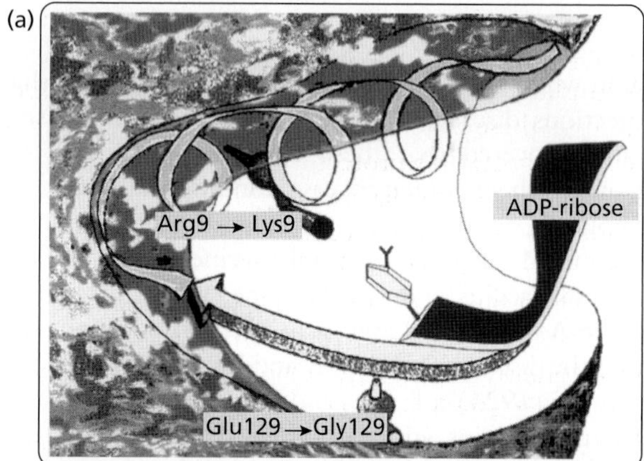

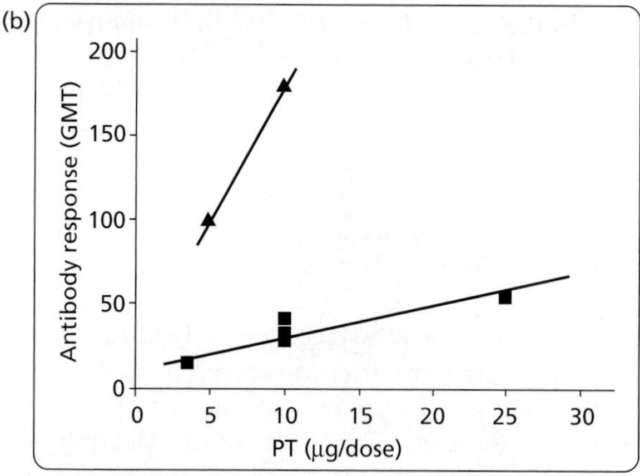

Fig. 1 (a) Cavity containing the active site of pertussis toxin. The amino acids Arg9 and Glu129 are located within the cavity essential for the enzymatic activity. By mutating them to Lys9 and Gly129, respectively, a mutant molecule (PT9K/129G) was obtained that is non-toxic and more immunogenic than the chemically detoxified toxoids (see Fig. 1b). This mutant molecule is the main component of the recombinant vaccine against pertussis (Pizza *et al.*, 1989; Greco *et al.*, 1996b; Rappuoli 1997). (b) Immunogenicity of the genetically-detoxified pertussis toxin (PT9K/129G, described in Fig. 1a; triangles), compared with conventional chemically modified toxoids (squares).

vaccines against diarrhoea and also promising mucosal adjuvant (Giuliani *et al.*, 1998) is shown in **Fig. 2**. The arrow in the figure shows the substitution of a serine with lysine that fills a cavity in the catalytic site, which then becomes inactive; as shown by its X-ray structure, the rest of the molecule is indistinguishable from wild-type.

Whole-cell Vaccines

Since killed, injectable whole-cell vaccines are easy to develop, they were the first to be introduced against most bacteria, including pertussis (Manclark and Cowell, 1984), cholera (Finkelstein, 1975) and Salmonella (Ashcroft *et al.*, 1964; Yugoslav Typhoid Committee, 1964; Hejfec *et al.*, 1966). These vaccines usually consisted of bacteria killed with formaldehyde or another treatment. Although they are still licensed, most of these vaccines are no longer used, usually because of their low protective effects, as in the case of cholera and Salmonella, and their high reactogenicity. Among the whole-cell vaccines, pertussis vaccine is the one that is still widely used. In recent clinical trials it has been shown to be more than 90% effective, and the high efficacy is believed to be sufficient to compensate for its high reactogenicity (Olin *et al.*, 1997; Plotkin and Cadoz 1997; Simondon *et al.*, 1997). It is predicted, therefore, that combination vaccines containing whole-cell pertussis vaccine will be used for many years in the future in most developing countries and some developed countries, such as the UK. Some whole-cell pertussis vaccines have, however, also been shown to be quite ineffective (36%), so that caution is necessary when they are used (Greco *et al.*, 1996b; Gustafsson *et al.*, 1996).

An oral vaccine against cholera that contains 10^{11} bacteria per dose and 1 mg of the B subunit of cholera toxin has been developed (Holmgren *et al.*, 1987, 1989). This vaccine has been shown to be effective in a large-scale clinical trial in an endemic area, and represents the first example of a whole-cell vaccine delivered by a mucosal route (Clemens *et al.*, 1988; Black *et al.*, 1990; Sanchez *et al.*, 1993; Vanloon *et al.*, 1996).

Live-attenuated Vaccines

Non-virulent derivatives of pathogens can be used to mount a harmless and limited infection that induces a protective immune response. Existing bacterial vaccines of this type are the bacille Calmette–Guérin (BCG), the Ty21a against typhoid fever, and the CVD–HgR recombinant vaccine against cholera. BCG is a strain of *Mycobacterium tuberculosis* introduced for human use in 1921, after attenuation by 213 passages for 13 years (Calmette, 1927). This vaccine is widely used to immunise against tuberculosis, especially in developing countries. The efficacy of this vaccine is difficult to establish, because in clinical trials the efficacy varies from 0% to 80% (Collins, 1984). Although this vaccine is still in use, there is a strong motivation to develop new vaccines against tuberculosis to replace the existing BCG.

(a)
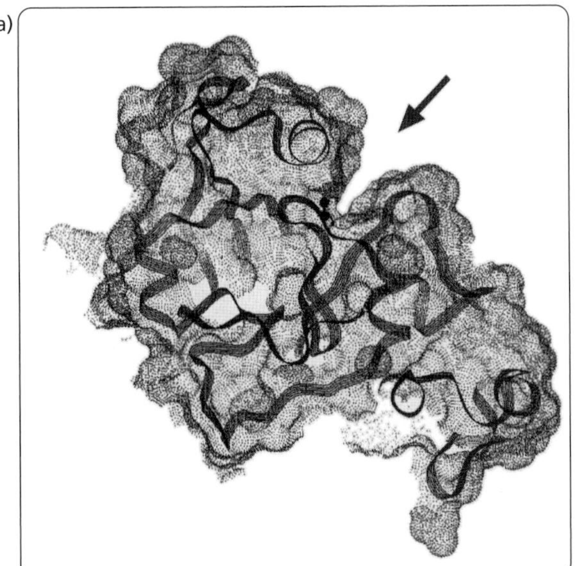

(b)
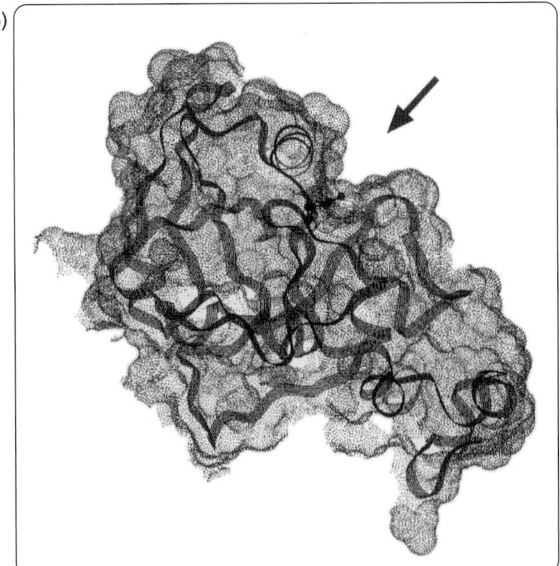

Fig. 2 X-ray structure of wild-type LT (a) and the non-toxic derivative LTK63 (b). The arrow indicates the catalytic cavity that in wild-type LT is empty and available to bind NAD and the substrate. After mutagenesis of Ser63 to Lys, the cavity is filled by the long side-chain of lysine and is no more available for the enzymatic reaction. The rest of the structure of LT and LTK63 is identical.

The Ty21a strain of *Salmonella typhi* was obtained in the 1970s by random mutagenesis of the bacterial genome with nitrosoguanidine (Germanier and Furer, 1975). Although the bacterium is *galE*-negative, it is not clear which are the mutations responsible for the attenuation of virulence, because the genome was mutagenised in multiple sites. The vaccine is given orally and it gave a protection of approximately 65% in different clinical trials (Germanier, 1984; Levine *et al.*, 1987, 1989; Black *et al.*, 1990).

The CVD–HgR strain of cholera is the first recombinant live-attenuated bacterial vaccine introduced for human use. It was constructed by rational deletion from the genome of the gene encoding the A subunit of cholera toxin and, to mark the vaccine strain, the introduction of the mercury resistance gene (*HgR*) into the haemolysin gene (Gottuzzo *et al.*, 1983; Levine *et al.*, 1988; Cryz *et al.*, 1995). The strain does not produce the cholera toxin responsible for the severe watery diarrhoea typical of cholera, and after oral delivery it induces an asymptomatic infection of the gut that results in a long-lasting immunity.

Molecular tools are now available for rational design of live-attenuated derivatives of any bacterial pathogen. Pioneering work has been done on *S. typhi*, in which it has been shown that inactivation of the metabolic genes that code for the synthesis of aromatic compounds (*aro* mutants) makes the bacteria unable to grow *in vivo* where such compounds are not available (Hoiseth and Stocker, 1981; Stocker *et al.*, 1983; Dougan *et al.*, 1988). Similarly, regulatory genes such as *cya–crp* (Curtiss and Kelly, 1987) and *phoP–phoQ* (Galan and Curtiss, 1989), genes coding for stress proteins (*htrA*) (Chatfield *et al.*, 1992), and nucleic acid metabolism (*purA*) and recombination, such as *recA*, are necessary for survival *in vivo*, and can therefore be used for the attenuation of pathogens.

The ultimate dream is an attenuated bacterium that has been engineered to produce protective antigens *in vivo* derived from many different pathogens, so that it is possible to deliver many different vaccines by a single immunisation. Attempts in this direction include strains of Salmonella (Chatfield and Dougan, 1997), BCG (Stover *et al.*, 1993; Fennelly *et al.*, 1997), *Shigella* (Noriega *et al.*, 1996). A strain of *V. cholerae* has been engineered to vaccinate against diphtheria, tetanus and pertussis, and against hepatitis B, malaria, *Shigella*, *E. coli*, etc. (Butterton *et al.*, 1993, 1997; Chen *et al.*, 1998).

Polysaccharide and Conjugate Vaccines

Many pathogenic bacteria, including *N. meningitidis*, *Streptococcus pneumoniae* and *H. influenzae* type b, produce a polysaccharide capsule composed of many repeating units of a simple sequence of one or more sugars (see Chapter 4). Vaccines that contain purified, high-molecular-weight polysaccharides and induce protective immunity in adults were introduced in the 1970s. A tetravalent vaccine of this kind against serotypes A, C, Y and W135 is available (Gotschlich *et al.*, 1969a, b; Zollinger, 1997). Also available is a similar vaccine that contains 23 different serotypes of polysaccharides from *Strep. pneumoniae* (Butler *et al.*, 1993). These vaccines induce T-cell-independent immunity, because polysaccharides are unable to bind the T-cell receptor and therefore stimulate only B cells. T-cell-independent vaccines induce an IgM-mediated antibody response that cannot be boosted by subsequent immunisation and it is effective only after 2 years of age. In infants polysaccharide vaccines usually do not work at all. To overcome the limitation of the T-cell independence of this type of vaccine, bacterial polysaccharides or oligosaccharides containing many repeating units have been chemically coupled to protein antigens that are T-cell dependent. The resulting semi-synthetic conjugate vaccines are T-cell-dependent and induce an IgG-based immune response with memory, so that immunity can be boosted by subsequent infection and immunisation. Furthermore, the conjugate vaccines have the great advantage of being effective in infants under the age of 2 years. The development of conjugate vaccines has allowed the introduction of mass vaccination of infants against *H. influenzae*, which has been one of the most successful vaccinations introduced during the last decade (Booy *et al.*, 1994; Barbour *et al.*, 1995; Madore, 1996). In 5 years this vaccination has virtually eliminated infant meningitis due to *H. influenzae* from the USA and all countries where the vaccination has been introduced (WHO, 1998). Clinical trials are at present in progress with conjugate vaccines against meningococcus A and C (Costantino *et al.*, 1992; Twumasi *et al.*, 1995; Lieberman *et al.*, 1996) and against the most common serotypes of pneumococcus. These vaccines will be introduced within a few years and are expected to eradicate most causes of bacterial meningitis. A polysaccharide vaccine based on the Vi antigen has also been developed against *S. typhi* and conjugate vaccines against *Salmonella* (Szu *et al.*, 1994) and *Shigella* are being tested (Polotsky *et al.*, 1994).

Acellular Vaccines

The term ‘acellular vaccine’ for vaccines composed of one or more purified components of bacteria was introduced to differentiate them from conventional whole-cell vaccines.

In the case of pertussis, they may contain a detoxified form of pertussis toxin either alone or combined with other surface-exposed molecules of the bacterium, such as filamentous haemagglutinin (FHA), pertactin (PRN) and the agglutinogens (AGG). Acellular pertussis vaccines were developed to reduce the side-effects of the whole-cell vaccines and in a number of clinical trials they have been shown to be very safe and protective (Greco *et al.*, 1996b; Gustafsson *et al.*, 1996; Olin *et al.*, 1997; Rappuoli, 1997a). Another effective is that against Lyme disease, which is based on the recombinant outer-membrane lipoprotein OspA (Edelman, 1997; Probert *et al.*, 1997).

The Future of Bacterial Vaccines

Designing a Bacterial Vaccine in the Golden Era of Genomics

Depending on the complexity of their genomes, pathogenic bacteria may produce between 1500 and 6000 proteins, and one or more major polysaccharides. Identification of the protective antigens is therefore quite a complex problem that often requires a combination of biochemical, genetic and immunological studies. Vaccine targets are usually those antigens that are surface-exposed or secreted (Fig. 3). Among these, the most popular are bacterial toxins, which prevent disease by inducing toxin-neutralising antibodies, and capsular polysaccharides that induce bactericidal and opsonising antibodies. Other surface-exposed vaccine candidates include, amongst others, adhesins, outer-membrane proteins, fimbriae, polysaccharide O antigens, lipo-oligosaccharides, flagella, surface-exposed receptors and transporters. These, in addition to inducing opsonising bactericidal antibodies, may lead to the production of antibodies that prevent bacterial adhesion and colonisation of host tissues. At present cytoplasmic antigens are not considered to be primary vaccine candidates, but ribosomal preparations have sometimes been used in the past. So far, an empirical approach has been to identify those molecules that are present in pathogenic bacteria but are absent from non-virulent variants of the same species. Examples of these are bacterial toxins and capsular polysaccharides that are the basis for many successful vaccines. Also, in the case of *B. pertussis*, antigens included in acellular vaccines (PT, FHA, PRN and fimbriae) were selected from among those proteins produced only by virulent (phase I) bacteria. Unfortunately, these approaches, although successful in many cases, have so far been limited because conventional techniques could not identify more than one or two dozens of target protein or polysaccharide antigens (usually the most abundant ones), and when these were not good enough to make a vaccine, no further rational approach was available to solve the problem. This was often the case when the most abundant surface molecules showed antigenic variation, so that conventional approaches are impractical. Examples of the latter are the fimbriae and outer-membrane proteins of gonococci and meningococci, streptococcal M antigen and *Strep. pneumoniae* polysaccharide capsules.

The availability of the genomic sequence of many pathogenic bacteria (Fleischman *et al.*, 1995; Smith *et al.*, 1995; Strauss and Falkow, 1997; Tomb *et al.*, 1997) will increase the possibility of developing vaccines against these organisms. Indeed, with a knowledge of all the proteins encoded by the genome it should, at least in theory, be possible to develop a vaccine against any bacterial pathogen. Such genomic information can be used for three different purposes:

- Identification of all possible protein antigens. After computer analysis, all proteins that are predicted to have a surface location and to be exported can be tested as vaccines in experimental models, either with the genes used as DNA vaccines or with expressed recombinant proteins. This approach may be facilitated by reducing the number of target proteins in several ways. For instance, high-resolution two-dimensional gel electrophoresis can be used to identify proteins that are present only in culture supernatants or in outer-membrane preparations, or that are not expressed by non-virulent variants of pathogenic bacteria. The identified proteins can be N-terminally sequenced directly from the two-dimensional gel, or identified by mass spectrometry of partially digested spots. The corresponding genes can then be used to express recombinant proteins for use in experimental vaccine models. An example of two-dimensional gels showing the proteins that are differentially expressed in more virulent bacteria is shown in Fig. 3. The combination of computer analysis and the experimental approach outlined above should theoretically allow the identification of all possible vaccine candidates.
- The biosynthetic pathways for polysaccharides and lipo-oligosaccharides can be dissected and manipulated to construct strains that produce modified new antigens which express only common, non-variable portions of these structures that can be used for vaccination. An approach of this type is being used to identify a possible vaccine candidate against the meningococcus (Moxon, 1997).

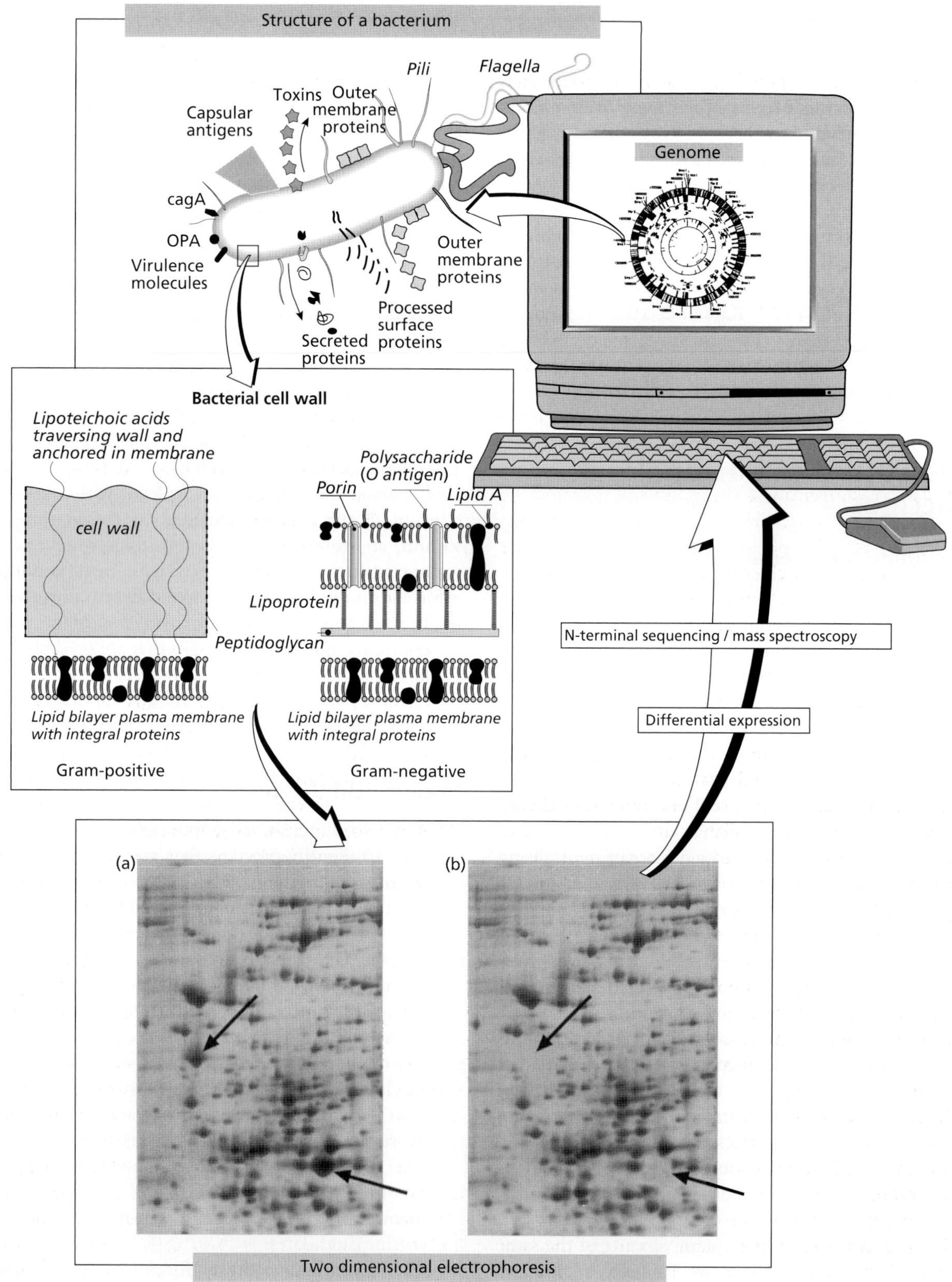

Fig. 3 Modern approach to bacterial vaccine development. 1. Targets for vaccines are proteins, polysaccharides and other molecules located on the surface of bacteria. 2. The organisation of the bacterial surface differs greatly in Gram-positive and Gram-negative bacteria. 3. Two-dimensional electrophoresis can identify more than a thousand of the expressed proteins in one bacterium. By comparing virulent bacteria (a) with non-virulent bacteria from the same species (b) one can identify the proteins that are unique to virulent bacteria and therefore likely vaccine candidates. These proteins can be identified directly from two-dimensional gels by N-terminal sequencing or mass spectrometry. 4. Availability of the genome sequence provides information about the primary structure of all potential antigens.

- Construction of novel and safer live-attenuated strains. Knowledge of all genes involved in regulatory and biosynthetic pathways, will allow the construction of strains that are capable of replicating well *in vitro* but which are defective *in vivo*, and can be used as carriers for recombinant antigens.

From Injectable to Mucosal Vaccines

With few exceptions, most vaccines are at present delivered by intramuscular injection, but this is a primitive way of delivering drugs. In the 1960s this was also true for conventional drugs, and up to 80% of the important drugs were delivered by this route. Today, after a relentless effort by the pharmaceutical industry, only 15% of conventional drugs are still delivered by injection: the rest are delivered by more convenient routes such as oral, mucosal or transdermal. In the case of vaccines, the research effort necessary to move away from delivery by injection has not yet been applied, mainly because the low profit margins offered by the sale of vaccines does not allow the necessary investment in research. After the introduction of 'biotech vaccines', the increased profit margins should allow more private investment in this sector and the effort is now beginning. As more research in this field is carried out, it is expected that within the next few decades a more convenient oral and mucosal delivery of vaccines will become available.

Development of vaccines that can be delivered by the mucosal route will not only provide a more convenient way of delivery, as has happened with other drugs, but will stimulate an immune response at mucosal sites, which are not usually involved by systemic vaccines. Mucosal surfaces are exposed to the environment and are usually the first barrier encountered by pathogens during infection. The mucosal surface is enormous; the human body has a total mucosal surface of 400 m^2, an area equivalent to that of a tennis court. In order to resist pathogens at their portal of entry, in addition to fighting them once they are within the body (systemic immunity), the immune system has developed a special arm localised in mucosal tissues. This system utilises different types of antibodies (IgA instead of IgG) and different T cells that home specifically to the mucosal tissues (Czerkinsky and Holmgren, 1995; O'Hagan, 1997).

In addition to pharmaceutical techniques for drug presentation, the development of mucosal vaccines requires the use of special adjuvants. Indeed, mucosal tissues are constantly exposed to millions of non-pathogenic substances introduced by eating, drinking, breathing, contact, etc., and these tissues have learned not to worry about them. Therefore, in order to be recognised as pathogenic and so to alert the immune system, substances must be presented in a distinct way with adjuvants. These help the host to recognise that they are associated with a pathogen and allow them to reach special mucosal cells (M cells), that sample the surroundings and present these substances to the immune system. It is not yet clear what renders molecules special in order to be recognised as dangerous or pathogenic at mucosal level, but it is known that some molecules have the peculiar property of stimulating the mucosal immune system. These are toxic molecules such as cholera toxin (CT) and the LT of *E. coli*. When they are delivered orally or intranasally, they provoke a powerful mucosal immune response against themselves, and also against molecules delivered together with them. We asked ourselves whether it is possible to render these molecules non-toxic without affecting the ability of the immune system to recognise them as dangerous material. We have, indeed, found that mutants of CT and LT, genetically detoxified by the same approach as described for *B. pertussis* toxin (Fig. 2), although they were no longer toxic, retained the property to stimulate the mucosal immune system (Douce *et al.*, 1995, 1997a, b; Giuliani *et al.*, 1998).

Nucleic Acid Vaccines

DNA vaccines are bacterial plasmids that encode a protein antigen and utilise a promoter that has a high degree of activity in mammalian cells rather than in bacteria. After injection into the recipient host, the plasmid expresses the protein antigen directly in transduced cells, and immune responses against the antigen are subsequently induced. Thus, DNA vaccines deliver a gene rather than an antigen or pathogen, while utilising the same types of plasmids employed for the production of recombinant proteins. Although certain elements such as the promoter differ, the vaccinated host acts as the bioreactor for production of the protein.

It was somewhat unexpected that DNA plasmids can be directly delivered into cells *in vivo*, since the plasmids obviously lack the viral structures employed for attachment or entry into cells. Moreover, the plasmids can be delivered by simple injection in saline without any requirement for delivery agents such as cationic lipids, which are utilised for transfection *in vitro*. If the DNA is precipitated on to gold beads they can then be propelled into tissue by a biolistic device or so-called 'gene gun'.

The ability of DNA vaccines to generate protective immune responses was initially demonstrated in a model of viral infection and disease (influenza) (Ulmer *et al.*, 1993). The immunogenicity and protective efficacy of DNA vaccines have now, however, been shown for a variety of pre-clinical systems, including a number of viral, bacterial, parasitic and tumour models (for review see Donnelly *et al.*, 1997). Antibody and cellular immune responses have both been observed, and the species in which the vaccines have been shown to have efficacy have ranged from mice to non-human primates. Early-phase human clinical studies are underway.

In the initial study to demonstrate the protective efficacy of DNA vaccines, DNA that encodes the internal protein nucleoprotein (NP) of influenza A virus was injected intramuscularly into mice. The animals produced cytotoxic T lymphocytes (CTL) specific for syngeneic cells presenting the appropriate MHC class I-restricted epitope. The CTL were also able to kill influenza-infected syngeneic target cells and NP DNA-immunised mice survived lethal challenge with a sub-type of influenza A virus (H3N2) that was different from the sub-type of the virus (H1N1) from which the NP gene had been cloned. This ability of DNA vaccines to generate the MHC class I-restricted CTL responses has been a significant reason why these vaccines have attracted interest. This is because it is generally accepted that for CTL responses to be generated, antigen must be present in the cytoplasm of professional antigen-presenting cells in order to enter the correct antigen-processing and presentation pathway for peptide fragments of the antigen to associate with the MHC class I molecules to generate the antigen–MHC class I complex which binds to the T-cell receptor of CD8+ T cells. Since exogenous proteins are usually taken up into the endolysosomal pathway of cells rather than entering the cytoplasm, a variety of approaches have been taken to try to introduce the antigen directly into the cytoplasm or to have it produced there by the use of vector systems, such as bacteria or viruses that express heterologous proteins. These CTL responses are thought to be particularly important for controlling or eliminating viral infections, parasites and tumours. Cellular responses are, however, also important for certain bacterial diseases, notably those caused by intracellular bacteria. Indeed, the mycobacterium BCG (Lagranderie *et al.*, 1997) and *Listeria monocytogenes* (Dietrich *et al.*, 1998) have been utilised as vector systems to introduce the genes for antigens from other pathogens in efforts to induce such cell-mediated immunity.

The generation of cellular responses is thought to be an important component of a number of vaccines. This is not only because of the ability of CTL to kill infected cells, and thereby contain the spread of infection, but also because in certain pathogens the surface proteins that are the targets of antibody responses are highly variable, making it difficult to devise vaccines with a sufficiently broad efficacy to deal with the wide variety of existing or newly evolving strains. In addition, with certain pathogens, such as *M. tuberculosis*, other components of the cellular immune response, such as the production of certain cytokines, are thought to be important (Orme *et al.*, 1993).

DNA vaccines have also been shown to be capable of inducing protective antibody responses. For example, when DNA that encodes influenza haemagglutinin (HA) is injected into mice, ferrets and non-human primates, functional antibodies are generated (Donnelly *et al.*, 1995). These antibodies prevent the agglutination of red blood cells induced by influenza virus; they are the so-called haemagglutination-inhibiting antibodies that are considered the surrogates of virus neutralising antibodies. In mice or ferrets that can be challenged with virus, the DNA vaccine protects the animals from subsequent challenge with the homologous virus. In the case of influenza HA and for other proteins, such as the major capsid protein L1 of papilloma virus, the viral proteins were produced with the necessary oligomerisation for successful expression and formation of the conformational epitopes necessary to induce virus-neutralising antibodies.

M. tuberculosis is one of the most studied bacterial pathogens from the standpoint of DNA vaccines. Efforts to make new vaccines against tuberculosis have focused on proteins that are secreted by the mycobacterium. These proteins are the ones that would have access to the cytoplasm of the infected cell, and thus be able to enter the antigen processing pathway associated with sensitising antigen-presenting cells for generation of MHC class I-restricted CTL. DNA constructs that encode secreted proteins, such as the antigen 85 complex (Ag85) (Huygen *et al.*, 1996), hsp70, a heat-shock protein (Tascon *et al.*, 1996), and a 38-kDa lipoprotein (Zhu *et al.*, 1997) have been used to generate protective immunity from subsequent challenge with BCG or *M. tuberculosis.* In the case of the Ag85 constructs, DNA plasmids encoding Ag85A–C were made, as were variations of the Ag85A construct that utilised either the native signal sequence, a heterologous signal sequence (the mammalian tissue plasminogen activator leader sequence, tPA), or no signal sequence. CTL could be obtained from *in vitro* re-stimulated splenocytes of mice immunised with all of the constructs, but the levels of bulk CTL were greatest with the constructs with the tPA promoter.

BCG culture filtrates contain secreted proteins, many of which are thought to be similar or identical to the proteins secreted by *M. tuberculosis.* It is most significant that when splenocytes from mice immunised with the Ag85A DNA constructs were restimulated with such filtrates or purified native Ag85, the splenocytes from immunised but not those from naive mice secreted significant quantities of interleukin 2 (IL-2), interferon γ, tumour necrosis factor (TNFα) and granulocyte macrophage colony-stimulating factor (GMCSF), but little IL-4 or IL-10. Thus, the phenotype of the T cell help generated is consistent with the Th-1 type of help associated with the desired form of cellular immunity thought to be important for a tuberculosis vaccine. The cytokine secretion was antigen-specific in that splenocytes from immunised animals secreted the cytokines only when cultured with BCG culture filtrate.

Possibly the most interesting observation, which has significant implications for the use of DNA vaccines for bacterial vaccines, is the post-translational modifications of the bacterial proteins when expressed by the mammalian host, rather than when made by mycobacterial replication as occurs during infection. Examination of the sequence of Ag85C revealed that it has seven potential N-linked glycosylation sites, whereas Ag85B has three and Ag85A has one. Indeed, in glycosidase digestion experiments, N-linked glycosylation of these sites occurs when Ag85A, B and C are expressed in mammalian cells (Montgomery *et al.*, unpublished observations). Thus, whereas the mammalian host cell expression of viral or tumour proteins provides a potential advantage for these antigens overheterologous recombinant production of these proteins (e.g., in bacterial or insect expression systems), the expression of bacterial proteins by mammalian cells after DNA immunisation may result in proteins with glycosylation patterns or with other post-translational modifications different from those that result from bacterial synthesis of the protein. Although in this case the sequences of the antigens could readily be altered to eliminate the glycosylation sites, the variety of post-translational modifications that could potentially differ is, clearly, a problem to be considered when DNA vaccines for bacterial proteins are used.

Huygen *et al.*, (1996) and Denis *et al.* (1998) compared the repertoire of epitopes for T-cell help that is recognised when a DNA vaccine is used with that in *M. tuberculosis* infection. This was done by immunising either with a DNA vaccine that encodes Ag85A or live *M. tuberculosis*, and then using a peptide library of the entire protein to determine which epitopes resulted in the generation of T-cell proliferative responses. Immunisation with the DNA version of the protein generated T helper cells that responded to a broader variety of peptides than did immunisation with the corresponding protein. Whether this observation can be generalised to other mycobacterial antigens or whether it is unique to this particular protein is unknown. It is, however, interesting in that, given the haplotypic diversity of the human population, a vaccine with the ability to induce a broad array of responses for T-cell epitopes may be more effective than one with a limited range of epitopic responses.

DNA vaccines are being used as a tool to evaluate different proteins from a pathogen as potential vaccine components. There are certainly a number of benefits in engendering CTL responses to DNA rather than to recombinant proteins. These include ease of making the DNA plasmids as against purification or the recombinant expression of the protein followed by the purification of several individual proteins. In addition, the obvious difficulties in the routine use of exogenous proteins to induce CTL responses contrasts with the general ability of DNA constructs to generate MHC class I-restricted responses and T-helper responses. Combined with the burgeoning gene sequence information available for the entire genomes of certain bacterial pathogens (*Helicobacter pylori*, *M. tuberculosis* and *H. influenzae*), it would be possible to use DNA vaccines to screen for protective immune responses in animal challenge models, but it is important to remember that proteins expressed in mammals may differ from the native versions.

An additional approach is to utilise expression libraries constructed in DNA vaccine plasmids to search for protective antigens or epitopes (Barry *et al.*, 1995). In this technique, the genome of the pathogen is fragmented into random libraries which are inserted into expression plasmids. Animals are then immunised with pools of these plasmids, which contain thousands of different constructs, most of which encode fragments of proteins. Any pools shown to confer protection in an animal challenge model could then be simplified to determine which antigen or antigens conferred protection. Although libraries so far have been shown to be protective in various disease models, such as *Mycoplasma pneumoniae*, a library has not yet been successfully decoded. While the use of gene fragments may have some advantages, if expression of a protein fragment is greater than that of the entire protein, the size limitation of the expressed proteins ($\sim$100–150 amino acids) may mean that conformational epitopes are not expressed.

Thus, DNA vaccines offer not only a potential new type of vaccine, but also a useful tool for antigen screening. Although they may have particular value in diseases for which a cellular immune response is critical, the ability to induce antibody responses may also be important for bacterial vaccines, provided that the mammalian expression of the proteins results in a protein with the correct structure.

References

Ada GL (1993) Vaccines. In: Paul WE (ed.) *Fundamental Immunology*, 3rd ed. New York : Raven Press, pp. 1309–1352.

Ashcroft MT, Morrison-Ritchie J, Nicholson CC (1964) Controlled field trial in British Guyana schoolchildren of heat-killed-phenolized and acetone-killed lyophilized typhoid vaccines. *Am. J. Hyg.* 79: 196–206.

Barbour ML, Mayon-White RT, Coles C, Crook DWM, Moxon ER (1995) The impact of conjugate vaccine on carriage of *Haemophilus influenzae* type b. *J. Infect. Dis.* 171: 93–98.

Barry MA, Lai WC, Johnston SA (1995) Protection against mycoplasma infection using expression-library immunization. *Nature* 377: 632–635.

Black RE, Levine MM, Ferreccio C *et al.* (1990) Efficacy of one or two doses of Ty2la *Salmonella typhi* vaccine in enteric-coated capsules in a controlled field trial. Chilean Typhoid Committee. *Vaccine* 8(1): 81–84.

Booy R, Hodgson S, Carpenter L *et al.* (1994) Efficacy of *Haemophilus influenzae* type b conjugate vaccine PRP-T. *Lancet* 344: 362–366.

Butler JC, Breiman RF, Campbell JF *et al.* (1993) Pneumococcal polysaccharide vaccine efficacy: an evaluation of current recommendations. *JAMA* 270: 1826–1831.

Butterton JR, Boyko SA, Calderwood SB (1993) Use of the *Vibrio cholerae* irgA gene as a locus for insertion and expression of heterologous antigens in cholera vaccine strains. *Vaccine* 11: 1327–1335.

Butterton JR, Calderwood SB (1997) Attenuated *Vibrio cholerae* as a live vector for expression of foreign antigens. In: Cobon GS (ed.) *New Generation Vaccines*, 2nd ed. New York: Marcel Dekker, pp. 379–385.

Calmette A (1927) *La vaccination preventive contre la tuberculose par le 'BCG'.* Paris: Masson.

Chatfield SN, Strahan K, Pickard D *et al.* (1992) Evaluation of *Salmonella typhimurium* strains harbouring defined mutations in htrA and aroA in the murine salmonellosis model. *Microb. Pathog.* 12: 145–151.

Chatfield SN, Dougan G (1997) Attenuated *Salmonella* as a live vector for expression of foreign antigens. Part I: Expressing bacterial antigens. In: Levine MM, Woodrow GC, Kaper JB, Cobon GS (ed.) *New Generation Vaccines*, 2nd ed. New York: Marcel Dekker, pp. 331–341.

Chen I, Finn TM, Yanqing L *et al.* (1998) A recombinant live-attenuated strain of *Vibrio cholerae* induces immunity against tetanus toxin and *Bordetella pertussis* tracheal colonization factor. *Infect. Immun.* 66(4): 1648–1653.

Choe S, Bennett MJ, Fujii G *et al.* (1992) The crystal structure of diphtheria toxin. *Nature* 357: 216–222.

Clemens JD, Harris JR, Sack DA *et al.* (1988) Field trial of oral cholera vaccines in Bangladesh: Results of one year follow-up. *J Infect Dis* 158: 60–69.

Clemens JD, Sack DA, Harris JR *et al.* (1990) Field trial of oral cholera vaccines in Bangladesh: results from three-year follow-up. *Lancet* 335: 270–273.

Collier RJ (1982) Structure and activity of diphtheria toxin. In: Hayashi D, Ueda K (ed.) *ADP-ribosylation Reactions.* New York: Academic Press, pp. 575–592.

Collins FM (1984) Tuberculosis. In: Germanier R (ed.) *Bacterial Vaccines.* New York: Marcel Dekker, pp. 373–418.

Costantino P, Viti S, Podda A *et al.* (1992) Development and phase I clinical testing of a conjugate vaccine against meningococcus A and C. *Vaccine* 10: 691–698.

Cryz SJ, Kaper J, Tacket C, Nataro J, Levine MM (1995) Vibrio cholerae CVD103-HgR live oral attenuated vaccine: Construction, safety, immunogenicity, excretion and non-target effects. In: Brown F (ed.) *Non-Target Effects of Live Vaccines* . Karger, Basel, 84. pp. 237–244.

Curtiss R III, Kelly SM (1987) *Salmonella typhimurium* deletion mutants lacking adenylate cyclase and cyclic AMP receptor protein are avirulent and immunogenic. *Infect. Immun.* 55: 3035–3043.

Czerkinsky C, Holmgren J (1995) Mucosally-induced immune responses: from vaccine development to specific immunotherapy. In: Ferguson A, Auricchio S, Troncone G (eds) *Mucosal Immunity and the Gut Epithelium.* Interactions in Health and Disease 4. Basel: Karger, pp. 161–173.

Denis O, Tanghe A, Paifliet K *et al.* (1998) Vaccination with plasmid DNA encoding Ag85A stimulates a broader CD4+ and CD8+ T cell epitopic repertoire than infection with *M. tuberculosis* H37Rv. *Infect. Immun.* 66: 1527–1533.

Dietrich G, Bubert A, Gentschev I *et al.* (1998) Delivery of antigen-encoding plasmid DNA into the cytosol of macrophages by attenuated suicide *Listeria monocytogenes. Nat. Biotech.* 16(1): 181–185.

Domenighini M, Pizza M, Rappuoli R (1995) Bacterial ADP-Ribosyltransferases. In: Moss J, Iglewski B, Vaughan M, Tu AT (ed.) *Bacterial Toxins and Virulence Factors in Disease.* New York: Marcel Dekker, pp. 59–80.

Donnelly JJ, Friedman A, Martinez D *et al.* (1995) Preclinical efficacy of a prototype DNA vaccine: enhanced protection against antigenic drift in influenza virus. *Nat. Med.* 1(6): 583–587.

Donnelly JJ, Ulmer JB, Shiver JW, Liu M (1997) DNA vaccines. *Annu Rev. Immunol.* 15: 617–648.

Douce G, Turcotte C, Cropley I *et al.* (1995) Mutants of *Escherichia coli* heat-labile toxin lacking ADP-ribosyltransferase activity act as nontoxic, mucosal adjuvants. *Proc. Natl Acad. Sci. USA* 92: 1644–1648.

Douce G, Dougan G, Pizza M, Rappuoli R (1997a) Mutant pertussis and *Escherichia coli* heat-labile toxins as adjuvants for enhancing local and systemic immune responses to coadministered, nonliving antigens. In: Levine MM, Woodrow GC, Kaper JB, Cobon GS (ed.)

New Generation Vaccines, 2nd ed. New York: Marcel Dekker, pp. 253–262.

Douce G, Fontana M, Pizza M, Rappuoli R, Dougan G (1997b) Intranasal immunogenicity and adjuvanticity of site-directed mutant derivatives of cholera toxin. *Infect. Immun.* 65: 2821–2828.

Dougan G, Chatfield S, Pickard D *et al.* (1988) Construction and characterisation of vaccine strains of *Salmonella* harbouring mutations in two different aro genes. *J. Infect. Dis.* 158: 1329–1335.

Edelman R (1997) Vaccines against lyme disease. In: Levine MM, Woodrow GC, Kaper JB, Cobon GS (eds) *New Generation Vaccines*, 2nd ed. New York: Marcel Dekker, pp. 897–912.

Edwards KM, Meade BD, Decker MD *et al.* (1995) Comparison of thirteen acellular pertussis vaccines: overview and serologic response. *Pediatrics* 96(3Pt2): 548–557.

Fennelly GJ, Jacobs WR Jr, Bloom BR (1997) BCG as a recombinant vaccine vector. In: Levine MM, Woodrow GC, Kaper JB, Cobon GS (eds) *New Generation Vaccines*, 2nd ed. New York: Marcel Dekker, pp. 363–377.

Finkelstein RA (1975) Immunology of cholera. *Curr. Top. Microbiol. Immunol.* 69: 137–196.

Fleischmann RD, Adams MD, White O *et al.* (1995) Whole-genome random sequencing and assembly of *Haemophilus influenzae* rd. *Science* 269: 496–512.

Galan JE, Curtiss RI (1989) Virulence and vaccine potential of phoP mutants of *Salmonella typhimurium*. *Microb. Pathog.* 6: 433–443.

Germanier R, Furer E (1975) Isolation and characterization of galE mutant and Ty21a of *Salmonella typhi*: a candidate strain for a live oral typhoid vaccine. *J. Infect. Dis.* 141: 553.

Germanier R (1984) Typhoid fever. In: Germanier R (ed.) *Bacterial Vaccines*. London: Academic Press, pp. 137–165.

Giannini G, Rappuoli R, Ratti G (1984) The amino-acid sequence of two non-toxic mutants of diphtheria toxin: *CRM45* and CRM1 97. *Nucleic Acids Res.* 12: 4063–4069.

Giuliani MM, Giudice G Del, Giannelli V *et al.* (1998) Mucosal adjuvanticity and immunogenicity of LTR72, a novel mutant of *Escherichia coli* heat-labile enterotoxin with partial knock-out of ADP-ribosyltransferase activity. *J. Exp. Med.* 187(7): 1123–1132.

Glenny AT, Hopkins BE (1923) Diphtheria toxoid as an immunizing agent. *Br. J. Exp. Pathol.* 4: 283–288.

Gotschlich EC, Goldschneider I, Artenstein MS (1969a) Human immunity to the meningococcus: IV. Immunogenicity of group A and group C meningococcal polysaccharides in human volunteers. *J. Exp. Med.* 129: 1367–1384.

Gotschlich EC, Liu TY, Artenstein MS (1969b) Human immunity to the meningococcus: III. Group A, group B, and group C meningococcal polysaccharides. *J. Exp. Med.* 129: 1349–1365.

Gotuzzo E, Butron B, Seas C *et al.* (1993) Safety, immunogenicity, and excretion pattern of single-dose live oral cholera vaccine CVD 103-HgR in Peruvian adults of high and low socioeconomic levels. *Infect. Immun.* 61: 3994–3997.

Greco D, Salmaso S, Mastrantonio P *et al.* (1996a) A difference in relative efficacy of two DTaP vaccines in continued blinded observation of children following a clinical trial (abstract). *Pediatr. Res.* 39: 173A.

Greco D, Salmaso S, Mastrantonio P *et al.* (1996b) A controlled trial of two acellular vaccines and one whole-cell vaccine against pertussis. *N. Engl. J. Med.* 334: 341–348

Gustafsson L, Hallander HO, Olin P, Reizenstein E, Storsaeter J (1996) A controlled trial of a two-component acellular, a five-component acellular, and a whole-cell pertussis vaccine. *N. Engl. J. Med.* 334(55): 349–355.

Hall CB, Granoff DM, Gromisch DS *et al.* (1993) *Haemophilus influenzae* type-b conjugate vaccines. Recommendations for immunization with recently and previously licensed vaccines. Committee on Infectious Diseases. *Pediatrics* 92: 480–488.

Hejfec LB, Salmin LV, Leytman MZ *et al.* (1966) A controlled field trial and laboratory study of five typhoid vaccines in the USSR. *Bull. WHO* 34: 321–339.

Hewlett EL (1996) Acellular pertussis trial. *Pediatrics* 98: 800.

Hoiseth SK, Stocker BA (1981) Aromatic-dependent *Salmonella typhimurium* are non-virulent and effective as live vaccines. *Nature* 291: 238–239.

Holmgren J, Svennerholm AM, Clemens J *et al.* (1987) An oral B subunit-whole cell vaccine against cholera: from concept to successful field trial. *Adv. Exp. Med. Biol.* 216B: 1649–1660.

Holmgren J, Clemens J, Sack DA, Sanchez J, Svennerholm AM (1989) Oral immunization against cholera. *Curr. Top. Microbiol. Immunol.* 146: 197–204.

Huygen K, Content J, Denis O *et al.* (1996) Immunogenicity and protective efficacy of a tuberculosis DNA vaccine. *Nat. Med.* 2(8): 893–898.

Langrandene M, Balazuc AM, Gicquel B, Gheorghiu M (1997) Oral immunization with recombinant *Mycobacterium bovis* BCG simian immunodeficiency virus ned induces local and systemic cytotoxic T-lymphocyte responses in mice. *J. Virol.* 71: 2303–2309.

Levine MM, Ferreccio C, Black RE, Gennanier R (1987) Large-scale field trial of Ty21a live oral typhoid vaccine in enteric-coated capsule formulation. *Lancet* 1: 1049–1052.

Levine MM, Herrington D, Losonsky G *et al.* (1988) Safety, immunogenicity and efficacy of recombinant live oral cholera vaccines, CVD 103 and CVD 103-HgR. *Lancet* 2: 467–470.

Levine MM, Ferreccio C, Black RE, Tacket CO, Germanier R (1989) Progress in vaccines against typhoid fever. *Rev. Infect. Dis.* 11 (Suppl 3): S552–S567.

Lieberman JM, Chiu SS, Wong VK *et al.* (1996) Safety and immunogenicity of a serogroups A/C *Neisseria meningitidis* oligosaccharide-protein conjugate vaccine in young children: A randomized controlled trial. *JAMA* 275: 1499–1503.

Madore DV (1996) Impact of immunization on *Haemophilus influenzae* type b disease. *Infect. Agents Dis.* 5: 8–20.

Makoff AJ, Ballantine SP, Smallwood AE, Fairweather NF (1989) Expression of tetanus toxin fragment C in *E. coli*: its purification and potential use as a vaccine. *Biotechnology* 7: 1043–1046.

Manclark CR, Cowell JL (1984) Pertussis. In: Germanier R (ed.) *Bacterial Vaccines*. Orlando, FL: Academic Press, pp. 69–106.

Manzoni A (1984) The Betrothed. *Translated by Bruce Penman*. London: Penguin, p. 720.

Montecucco C, Schiavo G (1993) Tetanus and botulism neurotoxins – a new group of zinc proteases. *Trends Biochem. Sci.* 18: 324–327.

Moxon ER (1997) Applications of molecular microbiology to vaccinology. *Lancet* 350: 1240–1244.

Naglich JG, Metherall JE, Russell DW, Eidels L (1992) Expression cloning of a diphtheria toxin receptor: identity with a heparin-binding EGF-like growth factor precursor. *Cell* 69: 1051–1061.

Nicosia A, Perugini M, Franzini C *et al.* (1986) Cloning and sequencing of the pertussis toxin genes: operon structure and gene duplication. *Proc. Natl Acad. Sci. USA* 83: 4631–4635.

Noriega FR, Losonsky G, Lauderbaugh C *et al.* (1996) Engineered Delta guaB-A Delta virG *Shigella flexneri* 2a strain CVD 1205: Construction, safely, immunogenicity, and potential efficacy as a mucosal vaccine. *Infect. Immun.* 64: 3055–3061.

Norton PM, Wells JM, Brown HWG, Macpherson AM, Lepage RWF (1997) Protection against tetanus toxin in mice nasally immunized with recombinant *Lactococcus lactis* expressing tetanus toxin fragment C. *Vaccine* 15: 616–619.

O'Hagan D (1997) Prospects for the development of new and improved vaccines through the use of microencapsulation technology. In: Levine MM, Woodrow GC, Kaper JB, Cobon GS (eds) *New Generation Vaccines*, 2nd ed. New York: Marcel Dekker, pp. 215–228.

Olin P and others for the Ad Hoc Group for the Study of Pertussis Vaccines (1997) Randomised controlled trial of two-component, three-component, and five-component acellular pertussis vaccines compared with whole-cell pertussis vaccine. *Lancet* 350(9091): 1569–1577.

Orme IM, Andersen P, Boom WH (1993) T cell response to *Mycobacterium tuberculosis*. *J. Infect. Dis.* 167: 1481–1497.

Pizza M, Bartoloni A, Prugnola A, Silvestri S, Rappuoli R (1988) Subunit S1 of pertussis toxin: mapping of the regions essential for ADP-ribosyltransferase activity. *Proc. Natl Acad. Sci. USA* 85: 7521–7525.

Pizza M, Covacci A, Bartoloni A *et al.* (1989) Mutants of pertussis toxin suitable for vaccine development. *Science* 246: 497–500.

Pizza MG, Domenighini M, Hol W *et al.* (1994) Probing the structure-activity relationships of K *colt* LT-A by site-directed mutagenesis. *Mol. Microbiol.* 14(1): 51–60.

Plotkin SA, Cadoz M (1997) The acellular pertussis vaccine trials: an interpretation. *Pediatr. Infect. Dis. J.* 16(5): 508–517.

Polotsky VY, Robbins JB, Bryla D, Schneerson R (1994) Comparison of conjugates composed of lipopolysaccharide from *Shigella flexneri* type 2A Detoxified by two methods and bound to tetanus toxoid. *Infect. Immun.* 62: 210–214.

Probert WS, Crawford M, Cadiz RB, LeFebvre RB (1997) Immunization with outer surface protein (Osp) A, but not OspC, provides cross-protection of mice challenged with North American isolates of *Borrelia burgdorferi*. *J. Infect. Dis.* 175(2): 400–405.

Ramon G (1924) Sur la toxine et sur l'anatoxine diphtheriques. *Ann. Inst. Pasteur* 38: 1–10.

Rappuoli R (1997a) Rational design of vaccines. The recombinant pertussis vaccine induces early and long-lasting protection. *Nat. Med.* 3(4): 374–376.

Rappuoli R (1997b) New and improved vaccines against diphtheria and tetanus. In: Levine MM, Woodrow GC, Kaper JB, Cobon GS (eds) *New Generation Vaccines*, 2nd ed. New York: Marcel Dekker, pp. 4.

Rappuoli R, Abrignani S, Grandi G (1997) Les vaccins. *Pour la Science* (edition française de *Scientific American*) 241: 90–96.

Smith HO, Tomb JF, Dougherty BA, Fleischmann RD, Venter JC (1995) Frequency and distribution of DNA uptake signal sequences in the *Haemophilus influenzae* Rd genome. *Science* 269: 538–540.

Sanchez JL, Trofa AF, Taylor DN *et al.* (1993) Safety and immunogenicity of the oral, whole cell/recombinant B-subunit cholera vaccine in North American volunteers. *J. Infect. Dis.* 167: 1446–1449.

Schiavo G, Benfenati F, Poulain B *et al.* (1992) Tetanus and botulinum-B neurotoxins block neurotransmitter release by proteolytic cleavage of synaptobrevin [see comments]. *Nature* 359: 832–835.

Silverstein AM (1989) *A History of Immunology*. San Diego, Academic Press.

Simondon F, Preziosi M-P, Yam A *et al.* (1997) A randomized double-blind trial comparing a two-component acellular to a whole-cell pertussis vaccine in Senegal. *Vaccine* 15(15): 1606–1612.

Sixma TK, Pronk SE, Kalk KH *et al.* (1991) Crystal structure of a cholera toxin-related heat-labile enterotoxin from *E. coli*. (1991). *Nature* 351: 371–377.

Stocker BA, Hoiseth SK, Smith BP (1983) Aromatic-dependent '*Salmonella* sp.' as live vaccine in mice and calves. *Dev. Biol. Stand.* 53: 47–54.

Stover CK, Bansal GP, Hanson MS *et al.* (1993) Protective immunity elicited by recombinant bacille Calmette-Guerin (BCG) expressing outer surface protein A (OspA) lipoprotein: a candidate lyme disease vaccine. *J. Exp. Med.* 178: 197–209.

Strauss EJ, Falkow S (1997) Microbial pathogenesis: genomics and beyond. *Science* 276(5313): 707–712.

Szu SC, Taylor DN, Trofa AC *et al.* (1994) Laboratory and preliminary clinical characterization of Vi capsular polysaccharide-protein conjugate vaccines. *Infect. Immun.* 62: 4440–4444.

Tascon RE, Colston MJ, Ragno S *et al.* (1996) Vaccination against tuberculosis by DNA injection. *Nat. Med.* 2(8): 888–892.

Tomb JF, White O, Kerlavage AR *et al.* (1997) The complete genome sequence of the gastric pathogen *Helicobacter pylori*. *Nature* 388: 539–547.

Twumasi PA, Kurnah S, Leach A *et al.* (1995) A trial of a group A plus group C meningococcal polysaccharide-protein conjugate vaccine in African infants. *J. Infect. Dis.* 171: 632–638.

Uchida T, Gill DM, Pappenheimer AMJ (1971) Mutation in the structural gene for diphtheria toxin, carried by the temperate phage beta. *Nature* 233: 8–11.

Ulmer JB, Donnelly JJ, Parker SE *et al.* (1993) Heterologous protection against influenza by injection of DNA encoding a viral protein. *Science* 259: 1745–1749.

Van den Akker F, Pizza M, Rappuoli R, Hol WGJ (1997) Crystal structure of a non-toxic mutant of heat-labile enterotoxin which is a potent mucosal adjuvant. *Pro. Sci.* 6: 2650–2654.

Vanloon FPL, Clemens JD, Chakraborty J *et al.* (1996) Field trial of inactivated oral cholera vaccines in Bangladesh: Results from 5 years of follow-up. *Vaccine* 14: 162–166.

World Health Organization (1997) The *CVI* Strategic Plan. *Managing Opportunity and Change: A Vision Of Vaccination for the 21st Century*. Geneva: CVI, pp. 1–92.

World Health Organization (1998) The WHO position paper on *Haemophilus influenzae* type b conjugate vaccines. *Wkly Epidemiol. Rec.* 73: 64–68.

Yugoslav Typhoid Committee (1964) A controlled field trial of the effectiveness of acetone-dried and inactivated and heat-phenol-inactivated typhoid vaccines in Yugoslavia. *Bull. WHO* 30: 623–630.

Zhu X, Venkataprasad N, Thangaraj HS *et al.* (1997) Functions and specificity of T cells following nucleic acid vaccination of mice against *Mycobacterium tuberculosis* infection. J. Immunol. 158: 5921–5926.

Zollinger WD (1997) New and improved vaccines against meningococcal disease. In: Levine MM, Woodrow GC, Kaper JB, Cobon GS (eds) *New Generation Vaccines*, 2nd ed. New York: Marcel Dekker, pp. 469–488.

Index

Note: page numbers in *italics* refer to figures and tables. *Salmonella* organisms are entered under old and new classification, according to how they are described in respective chapters.

B

C

G

I

L

M

N

O

P

S

T

W

X

Y

Z